Physical Constants

Name	Symbol	SI Units	cgs Units
Avogadro's number	N	6.022137×10^{23}/mol	6.022137×10^{23}/mol
Boltzmann constant	k	1.38066×10^{-23} J/K	1.38066×10^{-16} erg/K
Curie	Ci	3.7×10^{10} d/s	3.7×10^{10} d/s
Electron charge	e	1.602177×10^{-19} coulomb†	4.80321×10^{-10} esu
Faraday constant	$\mathcal{F}$	96485 J/V · mol	9.6485×10^{11} erg/V · mol
Gas constant*	R	8.31451 J/K · mol	8.31451×10^{7} erg/K · mol
Gravity acceleration	g	9.80665 m/s^2	980.665 cm/s^2
Light speed (vacuum)	c	2.99792×10^{8} m/s	2.99792×10^{10} cm/s
Planck's constant	h	6.626075×10^{-34} J · s	6.626075×10^{-27} erg · s

*Other values of R: 1.9872 cal/K · mol = 0.082 liter · atm/K · mol

†1 coulomb = 1 J/V

Conversion Factors

Energy: 1 Joule = 10^7 ergs = 0.239 cal
1 cal = 4.184 Joule

Length: 1 nm = 10 Å = $1 \times ^{-7}$ cm

Mass: 1 kg = 1000 g = 2.2 lb
1 lb = 453.6 g

Pressure: 1 atm = 760 torr = 14.696 psi
1 torr = 1 mm Hg

Temperature: K = °C + 273
C = (5/9)(°F − 32)

Volume: 1 liter = 1×10^{-3} m^3 = 1000 cm^3

Useful Equations

The Henderson–Hasselbalch Equation
$pH = pK_a + \log([A^-]/[HA])$

The Michaelis–Menten Equation
$v = V_{max}[S]/(K_m + [S])$

Temperature Dependence of the Equilibrium Constant
$\Delta H° = -R d(\ln K_{eq})/d(1/T)$

Free Energy Change under Non-Standard-State Conditions
$\Delta G = \Delta G° + RT \ln([C][D]/[A][B])$

Free Energy Change and Standard Reduction Potential
$\Delta G°' = -n\mathcal{F}\Delta\mathcal{E}_o'$

Reduction Potentials in a Redox Reaction
$\Delta\mathcal{E}_o' = \mathcal{E}_o'(\text{acceptor}) - \mathcal{E}_o'(\text{donor})$

The Proton-Motive Force
$\Delta p = \Delta\Psi - (2.3RT/\mathcal{F})\Delta pH$

Passive Diffusion of a Charged Species
$\Delta G = G_2 - G_1 = RT \ln(C_2/C_1) + Z\mathcal{F}\Delta\Psi$

Biochemistry

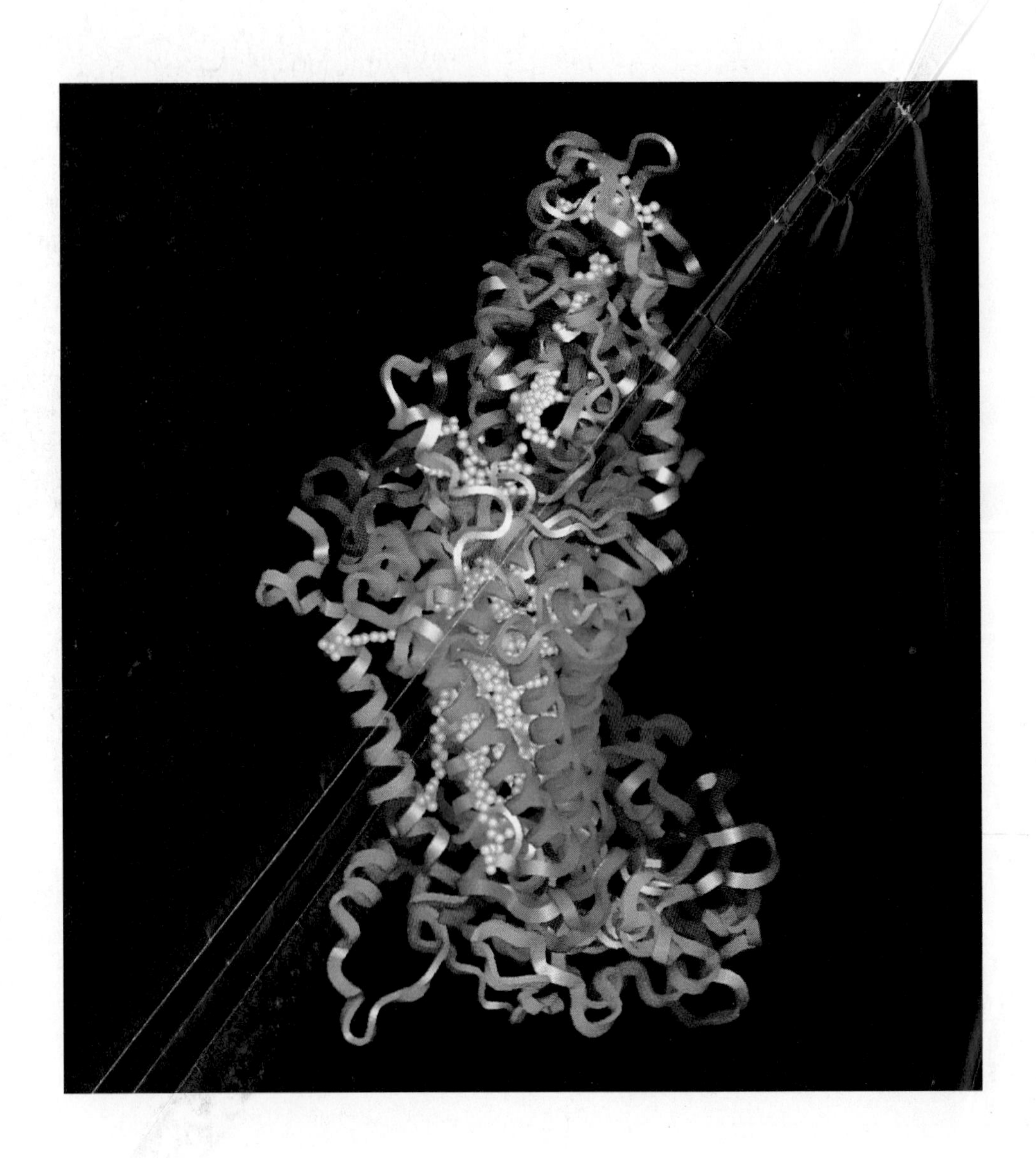

Biochemistry

Reginald H. Garrett

Charles M. Grisham

University of Virginia

SAUNDERS COLLEGE PUBLISHING

HARCOURT BRACE COLLEGE PUBLISHERS

Fort Worth • Philadelphia • San Diego • New York • Orlando • Austin

San Antonio • Toronto • Montreal • London • Sydney • Tokyo

Text Typeface: Baskerville
Compositor: York Graphic Services, Inc.
Acquisitions Editor: John J. Vondeling
Developmental Editor: Sandra Kiselica
Managing Editor: Carol Field
Project Editors: Becca Gruliow, Sarah Fitz-Hugh
Copy Editor: Zanae Rodrigo
Manager of Art and Design: Carol Bleistine
Art Director: Carol Bleistine, Anne Muldrow
Art Assistant: Sue Kinney
Text Designer: Rebecca Lemna
Cover Designer: Lawrence R. Didona
Text Artwork: J/B Woolsey Associates
Layout Artwork: Claudia Durrell
Director of EDP: Tim Frelick
Production Manager: Charlene Squibb
Marketing Manager: Marjorie Waldron
Field Product Manager: Laura Coaty

Cover Credit: J/B Woolsey Associates

Printed in the United States of America

5 6 7 8 9 0 1 2 3 032 10 9 8 7 6 5 4 3

ISBN 0-03-009758-4

Library of Congress Catalog Card Number: 93-087782

We dedicate this book to every student of biochemistry and to the professors who taught us biochemistry, especially

Alvin Nason	*Kenneth R. Schug*
William D. McElroy	*Ronald E. Barnett*
Maurice J. Bessman	*Albert S. Mildvan*
Ludwig Brand	*Rufus Lumry*

Preface

Scientific understanding of the molecular nature of life is growing at an astounding rate. Significantly, society is the prime beneficiary of this increased understanding. Cures for diseases, better public health, remediations for environmental pollution, and the development of cheaper and safer natural products are just a few practical results of this knowledge.

In addition, this expansion of information furthers what Thomas Jefferson called "*the illimitable freedom of the human mind.*" Scientists can use the tools of biochemistry and molecular biology to explore all aspects of an organism—from basic questions about its chemical composition, through inquiries into the complexities of its metabolism, its differentiation and development, to analysis of its evolution and even its behavior. Biochemistry is a science whose boundaries now encompass all aspects of biology, from molecules to cells, to organisms, and to ecology.

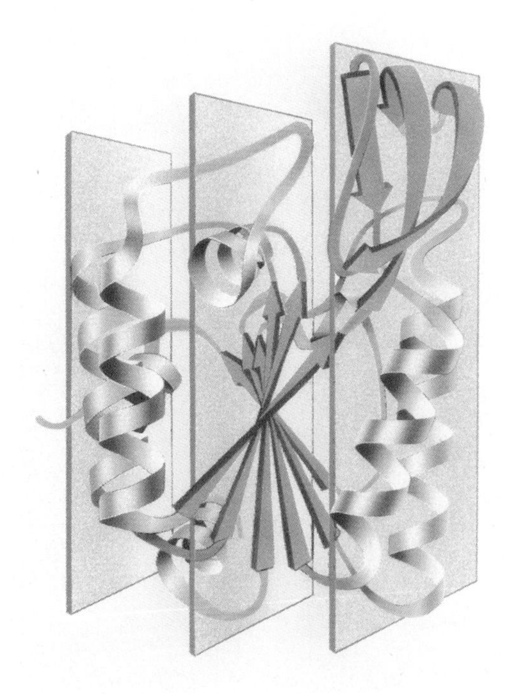

As biochemistry increases in prominence among the natural sciences, its inclusion in undergraduate and beginning-graduate curricula in biology and chemistry becomes imperative. And the challenge to authors and instructors is a formidable one: how to provide a comprehensive description of biochemistry in an introductory course or textbook. Fortunately, the increased scope of knowledge allows scientists to make generalizations connecting the biochemical properties of living systems with the character of their constituent molecules. As a consequence, these generalizations, validated by repetitive examples, emerge in time as principles of biochemistry, principles that are useful in discerning and describing new relationships between diverse biomolecular functions and in predicting the mechanisms underlying newly discovered biomolecular phenomena.

We are both biochemists—one of us is in a biology department, and the other is in a chemistry department. Undoubtedly, our approaches to biochemistry are influenced by the academic perspectives of our respective disciplines. We believe, however, that our collaboration on this textbook represents a melding of our perspectives that will provide new dimensions of appreciation and understanding for all students.

FEATURES AND ORGANIZATION

The organizational approach we have taken in this textbook is traditional in that it builds from the simple to the complex. Part I, *Molecular Components of Cells,* begins with a presentation of the structure and chemistry of biomolecules. This part of the text creates a continuum between biochemistry and the course in organic chemistry that should be a prerequisite to the use of this text. Chapters 4 and 5 are devoted to a fresh and comprehensive description of the molecular anatomy of proteins, illustrated by a gallery of ribbon diagrams and space-filling computer graphics that capture the diversity of these structurally most sophisticated of all the biological macromolecules. Chapters 6 and 7 describe the chemistries and strategies employed in analysis of the sequential order of monomeric units in nucleic acids and the arrangements of these units into higher ordered structures, such as the A, B, and Z forms of the DNA double helix. A chapter on recombinant DNA technology (Chapter 8) is provided early in the text to acquaint students with the basic concepts and applications of this methodology. This technology is one of the newest tools available to biochemical research, and its enormous success in illuminating the relationships between genetic information, biomolecular structure, and biomolecular function is unparalleled. Part I concludes with chapters on the structure of lipids and carbohydrates (Chapters 9 and 10).

Some instructors might prefer to present lipids and carbohydrates prior to a discussion of nucleotides and nucleic acids (Chapters 6 and 7), and Chapters 6–10 have been written so that this alternative sequence of presentation will go smoothly. Highlights of Chapters 9 and 10 include the structural chemistry of lipid-anchored membrane proteins and the complex proteoglycan structures that adorn cell surfaces. The chapters of Part I, like all the chapters in this text, feature text boxes of two kinds: *A Deeper Look* and *Critical Developments in Biochemistry*. These text boxes delve a little more deeply into selected topics or experimental observations, providing the student greater insight into the logic and historical context of significant biochemical advances. (A list of these boxes follows the Table of Contents).

Part II, *Enzymes and Energetics,* begins with a quantitative but easily grasped discussion of the kinetics of enzyme-catalyzed reactions. Also described are the exciting new discoveries of catalytic roles for certain RNA molecules (so-called *ribozymes*) and the purposeful design of antibodies with catalytic properties (dubbed *abzymes*). Chapter 12 reviews the mechanisms by which enzymes—and hence metabolic processes—are regulated. Highlights of this chapter include the detailed description of hemoglobin and its allosteric properties and a close scrutiny of structure-function relationships in *Escherichia coli* aspartate transcarbamoylase. The atomic structure of this paradigm of allosteric enzymes has been solved by X-ray crystallography, and the details of its structure provide many insights into the catalytic and regulatory properties of proteins.

Chapter 13 presents the basic physicochemical principles of transition-state stabilization that underlie the catalytic power of enzymes. It also examines a gallery of enzyme mechanisms, including the well-studied serine proteases and aspartic proteases. This latter class includes the clinically important HIV-1 protease. (HIV is the causative agent in AIDS.) Chapter 14 collects in one chapter a discussion of the principal coenzymes of metabolism. These substances are typically organic molecules whose structure is based on a vitamin. This chapter might be viewed as optional in a short course in biochemistry, or it might serve simply as a source of reference supporting lectures on metabolism. Text boxes in this chapter highlight the colorful history of the discovery of vitamins. Chapter 15 presents fundamental thermodynamic relationships useful to understanding the energetics of cellular metabolism. The thermodynamics of protein folding is illustrated as an apt example for the relative contributions of enthalpy and entropy changes to free energy changes in biological systems. This chapter is a self-contained unit, suitable for presentation early in a course if the instructor so chooses. Chapter 16 deals with the particular chemical features of ATP and other biomolecules that make these substances effective as cellular energy carriers.

Part III, *Metabolism and Its Regulation,* encompassing Chapters 17 through 27, describes the metabolic pathways that orchestrate the synthetic and degradative chemistry of life. Particular emphasis is placed on the chemical logic of intermediary metabolism. Chapter 17 points out the fundamental similarities in metabolism that unite all organisms and gives a survey of the underlying principles of metabolism. Fundamental aspects of catabolic metabolism are described in Chapters 18, 19, and 20. The contemporary view that energy coupling in mitochondria is probably nonstoichiometric, yielding a P/O ratio of 2.5 rather than the traditionally held value of 3.0, is a pertinent feature of Chapter 20. Chapter 21 expands the discussion of carbohydrate metabolism beyond glycolysis and stresses the interconnections and interrelationships between the different pathways of carbohydrate conversion.

The photosynthetic processes that provide the energy and fundamental carbohydrate synthesis upon which all life depends are described in Chapter 22. Focal points of this chapter include the newly emerging descriptions of the molecular structure of photosynthetic reaction centers and the mechanism and regulation of ribulose bisphosphate carboxylase, the key CO_2-fixing enzyme. A two-chapter discussion of lipid metabolism (Chapters 23 and 24) follows; the latter chapter details the novel properties of acetyl-CoA carboxylase, which catalyzes the principal regulated step in fatty acid biosynthesis, and synthesis of cholesterol and the biologically active lipids (eicosanoids and steroid hormones). Chapter 25 is unique among textbook chapters in defining the essentially unidirectional nature of metabolic pathways and the stoichiometric role of ATP in driving vital processes that are thermodynamically unfavorable. This chapter also addresses the organ level of metabolic specificity in mammals.

Part III concludes with an examination of the degradation and biosynthesis of amino acids and nucleotides. Chapter 26 gives greater emphasis than other textbook chapters to inorganic nitrogen metabolism and the biological acquisition of inorganic nitrogen from the inanimate environment. Chapter 27 profiles the biochemical basis of genetic defects in purine and pyrimidine metabolism and the potential for pharmaceutical intervention in these pathways to control unwanted cell proliferation, such as occurs in cancer or microbial infections.

Part IV, *Genetic Information,* addresses the essential role of information storage and transmission in organisms. The historical documentation of DNA molecules as the repositories of inheritable information is presented in Chapter 28, along with the latest discoveries unraveling the molecular mechanisms underlying genetic recombination. Chapter 29 treats the biochemistry involved in maintenance and replication of genetic information for transmission to daughter cells, accenting exciting new information on the complex enzymatic choreography of DNA replication. Chapter 30 then characterizes the expression of this information through synthesis of RNA and protein molecules. Highlights of this chapter include accounts of the molecular structure and mechanism of RNA polymerase and the DNA-binding transcription factors that modulate its activity.

Chapter 31 recounts the biochemical approaches that "cracked" the genetic code and describes the molecular events that underlie the "second" genetic code—how aminoacyl-tRNA synthetases uniquely recognize their specific tRNA acceptors. Chapter 32 presents the structure and function of ribosomes as the agents of protein synthesis, with emphasis on the interesting new realization that 23S rRNA is the peptidyl transferase enzyme responsible for peptide bond formation. This chapter also discusses the new appreciation of heat-shock proteins as important participants in protein folding.

Heredity is typified by high-fidelity replication of information in the short term and changes in genetic content in the long term (evolution). A chapter not previously found in standard biochemistry texts, Chapter 33 presents an account of evolution as a process that is fundamentally molecular in terms of changes within the hereditary material and biochemical in the expression or function of the information of heredity. Because the sequence of nucleotides in DNA can easily be determined and analyzed to reveal the similarities and differences between homologous sequences in related organisms (or genes), the field of molecular evolution is generating many new insights illuminating the comparative biochemistry of cells, organisms, species, and populations.

Part V is published as a separate paperback volume, *Molecular Aspects of Cell Biology*. Part V describes the biochemistry that underlies a number of complex biological structures and systems. It begins with a discussion of the spontaneous assembly of large biomolecular aggregates. Chapter 34 examines the thermodynamic basis for spontaneous self-assembly and then describes several typical self-assembling systems, including microtubules and viruses. It concludes with a detailed examination of the structure and assembly of HIV, the AIDS virus. (Other classic self-assembling systems such as ribosomes are described in detail elsewhere in the text.) Chapter 35 explains biological transport processes, and Chapter 36 discusses the structure and function of muscle. Chapter 35 is notable for its coverage of recent developments related to several novel transport systems (including the multidrug transporter and the osteoclast proton pumps of bone) and amphipathic, ion-translocating peptides. Part V concludes with chapters on hormonal signaling (Chapter 37) and neurotransmission (Chapter 38). These chapters provide a fresh and up-to-date perspective on the rapidly changing fields of cellular signaling and signal transduction, with coverage of the superfamilies of membrane receptors, oncogenes, tumor suppressor genes, sensory transduction, and the biochemistry of neurological disorders.

Acknowledgments

We are indebted to our colleagues David Jemiolo (Department of Biology, Vassar College) and Barton K. Hawkins (Department of Chemistry and Biochemistry, University of South Carolina), who carefully reviewed the entire manuscript at several stages, and to all the many other outstanding and invaluable expert reviewers of the manuscript.

We wish to warmly and gratefully acknowledge many other people who assisted and encouraged us in this endeavor. These include the outstanding staff at Saunders College Publishing, including our publisher, John Vondeling, who recruited us to this monumental task, and also Sandi Kiselica, Sarah Fitz-Hugh, Becca Gruliow, Carol Bleistine, Anne Muldrow, Carol Field, Tim Frelick, and Pauline Mula. The beautiful illustrations that grace this textbook are a testament to the creative and tasteful work of John and Bette Woolsey, Patrick Lane, and the entire staff of J/B Woolsey Associates. We are grateful to our many colleagues who provided original art and graphic images for this work, particularly Professor Jane Richardson of Duke University, who provided numerous original line drawings of the protein ribbon structures, and Mindy Whaley, who prepared the molecular graphic displays. We owe a very special thank-you to our devoted wives, Catherine Leigh Touchton and Rosemary Jurbala Grisham, to our children, Jeffrey, Randal, and Robert Garrett, and David, Emily, and Andrew Grisham, and finally to Cassie the cat, who stayed close to the project throughout its many stages. With the publication of this book we celebrate and commemorate the lives of our parents, William W. Garrett and Lelia B. Bosley, and Ernest M. Grisham and Mary Charlotte Markell Grisham.

Reginald H. Garrett — **Advance Mills, VA**

Charles M. Grisham — **Ivy, VA**

February 1994

Support Package

Molecular Aspects of Cell Biology is a paperback supplement to BIOCHEMISTRY. It provides an up-to-date perspective on topics in cell biology, including the rapidly changing fields of cellular signaling and signal transduction. Coverage includes the superfamilies of membrane receptors, oncogenes, tumor suppressor genes, sensory transduction, and the biochemistry of neurological disorders.

The **Student Study Guide** by David Jemiolo (Vassar College) includes summaries of the chapters, detailed solutions to all end-of-chapter problems, a guide to the key points of each chapter, important definitions, and illustrations of major metabolic pathways.

The **Test Bank** by William Scovell (Bowling Green State University) includes over 900 multiple-choice questions for professors to use as tests, quizzes, and homework assignments. The bank of questions is also available in computerized form for IBM-compatible and Macintosh computers.

A set of **Overhead Transparency Acetates** includes 175 of the pedagogically most important figures in the text.

Monthly updates of **UVa Images,** a collection of macromolecular structures with explanatory text is available on Internet by accessing the computer facility at the University of Virginia. Users have the ability to manipulate the images in any dimension.

Saunders Chemistry Videodisc Version 3 Multimedia Package includes still images from the text, as well as hundreds from other Saunders chemistry texts. The disc can be operated via a computer, a bar code reader, or a hand-controlled keypad. It also features molecular simulations and demonstrations.

Lecture Active™ Software enables instructors to customize their lectures with the Videodisc. Available for both IBM and Macintosh computers.

List of Reviewers

Hugh A. Akers
Lamar University

Christian B. Anfinsen
The Johns Hopkins University

Dean R. Appling
University of Texas, Austin

Daniel E. Atkinson
University of California, Los Angeles

John R. Baker
University of Alabama, Birmingham

Ronald H. Bauerle
University of Virginia

Shelby L. Berger
National Institutes of Health

Maurice J. Bessman
The Johns Hopkins University

Lutz Birnbaumer
Baylor College of Medicine

Richard P. Boyce
University of Florida, School of Medicine

Paul D. Boyer
University of California, Los Angeles

Rodney F. Boyer
Hope College

Clive Bradbeer
University of Virginia, School of Medicine

Michael Brown
Emory School of Medicine

Larry G. Butler
Purdue University

James W. Campbell
Rice University

Mary Campbell
Mount Holyoke College

Scott W. Champney
East Tennessee State University, College of Medicine

Jeffrey A. Cohlberg
California State University, Long Beach

Vincent Davisson
Purdue University

Lenny Dawidowicz
National Institutes of Health

David J. DeRosier
Brandeis University

John T. Edsall
Harvard University

Don W. Fawcett
University of Montana

Jay W. Fox
University of Virginia, School of Medicine

Perry A. Frey
University of Wisconsin, Madison

Eric Gouaux
Massachusetts Institute of Technology

Charles C. Griffin
Miami University of Ohio

Hans M. Gunderson
Northern Arizona University

Martyn Gunn
Texas A&M University

Barbara A. Hamkalo
University of California, Irvine

Gordon G. Hammes
University of California, Santa Barbara

J. Norman Hansen
University of Maryland

Barton K. Hawkins
University of South Carolina

Winston Hide
Baylor College of Medicine

Peter C. Hinkle
Cornell University

Joel Hockensmith
University of Virginia, School of Medicine

David Jemiolo
Vassar College

Larry L. Jackson
Montana State University

Robert J. Kadner
University of Virginia, School of Medicine

Peter J. Kennelly
Virginia Polytechnic Institute

Thomas Kodadek
University of Texas, Austin

Kenneth N. Kreuzer
Duke University Medical Center

Robert Kuchta
University of Colorado, Boulder

Sidney Kushner
University of Georgia

James C. Lee
University of Texas Medical Branch, Galveston

Y. C. Lee
The Johns Hopkins University

Weng-Hsiung Li
University of Texas at Houston

Robert Lindquist
San Francisco State University

Ariel Lowey
Haverford College

Ponzy Lu
University of Pennsylvania

Dawn S. Luthe
Mississippi State University

Celia Marshak
San Diego State University

Alan G. Marshall
Ohio State University

Richard E. McCarty
The Johns Hopkins University

Mark McNamee
University of California, Davis

Richard Mintel
University of Illinois, College of Medicine at Urbana

Henry M. Miziorko
Medical College of Wisconsin, Milwaukee

Thomas Nowak
University of Notre Dame

Neil Osheroff
Vanderbilt University, School of Medicine

Alan T. Phillips
Pennsylvania State University

Simon J. Pilkis
State University of New York, Health Science Center

Paul D. Ray
University of North Dakota, School of Medicine

Saul Roseman
The Johns Hopkins University

Robert W. Roxby
University of Maine

Paul R. Schimmell
Massachusetts Institute of Technology

Jessup M. Shively
Clemson University

William M. Scovell
Bowling Green State University

Lewis M. Siegel
Duke University, School of Medicine

Ram P. Singhal
Wichita State University

Gerald Smith
Fred Hutchinson Cancer Research Center, Seattle

Andrew P. Somlyo
University of Virginia

Paul A. Srere
VA Medical Center, Dallas

Alfred Stracher
State University of New York, Health Science Center

Michael P. Timko
University of Virginia

David C. Turner
State University of New York, Health Science Center

Dennis E. Vance
University of Alberta

Richard L. Weiss
University of California, Los Angeles

Peter J. Wejksnora
University of Wisconsin, Milwaukee

Beulah M. Woodfin
University of New Mexico

About the Authors

Reginald H. Garrett was educated in the Baltimore city public schools and at the Johns Hopkins University, where he received his Ph.D. in biology in 1968. Since that time, he has been at the University of Virginia, where he is currently professor of biology. He is the author of numerous papers and review articles on biochemical, genetic, and molecular biological aspects of inorganic nitrogen metabolism. Since 1964, his research interests have centered on the pathway of nitrate assimilation in filamentous fungi. His investigations have contributed substantially to our understanding of the enzymology, genetics, and regulation of this major pathway of biological nitrogen acquisition. His research has been supported by grants from the National Institutes of Health, the National Science Foundation, and private industry. He is a former Fulbright Scholar and has been a Visiting Scholar at the University of Cambridge on two sabbatical occasions. He has taught biochemistry at the University of Virginia for 26 years. He is a member of the American Society for Biochemistry and Molecular Biology.

Charles M. Grisham was born and raised in Minneapolis, Minnesota, and educated at Benilde High School. He received his B.S. in chemistry from the Illinois Institute of Technology in 1969 and his Ph.D. in chemistry from the University of Minnesota in 1973. Following a postdoctoral appointment at the Institute for Cancer Research in Philadelphia, he joined the faculty of the University of Virginia, where he is professor of chemistry. He has authored numerous papers and review articles on active transport of sodium, potassium, and calcium in mammalian systems, on protein kinase C, and on the applications of NMR and EPR spectroscopy to the study of biological systems. His work has been supported by the National Institutes of Health, the National Science Foundation, the Muscular Dystrophy Association of America, the Research Corporation, and the American Chemical Society. He is a Research Career Development Awardee of the National Institutes of Health, and in 1983 and 1984 he was a Visiting Scientist at the Aarhus University Institute of Physiology, Aarhus, Denmark. He has taught biochemistry and physical chemistry at the University of Virginia for 20 years. He is a member of the American Society for Biochemistry and Molecular Biology.

Contents in Brief

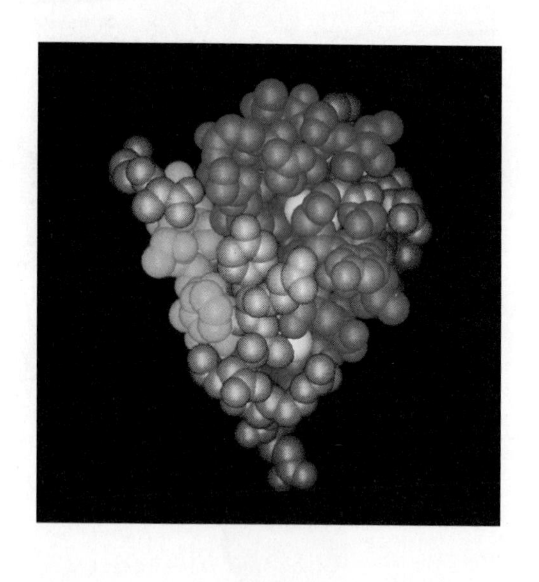

Part I: Molecular Components of Cells 1

Chapter 1: Chemistry Is the Logic of Biological Phenomena 2
Chapter 2: Water, pH, and Ionic Equilibria 32
Chapter 3: Amino Acids 55
Chapter 4: Proteins: Their Biological Functions and Primary Structure 81
Chapter 5: Proteins: Secondary, Tertiary, and Quaternary Structure 136
Chapter 6: Nucleotides and Nucleic Acids 180
Chapter 7: Structure of Nucleic Acids 208
Chapter 8: Recombinant DNA: Cloning and Creation of Chimeric Genes 247
Chapter 9: Lipids and Membranes 276
Chapter 10: Carbohydrates and Cell Surfaces 310

Part II: Enzymes and Energetics 351

Chapter 11: Enzyme Kinetics 352
Chapter 12: Enzyme Specificity and Allosteric Regulation 387
Chapter 13: Mechanisms of Enzyme Action 424
Chapter 14: Coenzymes and Vitamins 462
Chapter 15: Thermodynamics of Biological Systems 511
Chapter 16: ATP and Energy-Rich Compounds 523

Part III: Metabolism and Its Regulation 543

Chapter 17: Metabolism—An Overview 544
Chapter 18: Glycolysis 569
Chapter 19: The Tricarboxylic Acid Cycle 598
Chapter 20: Electron Transport and Oxidative Phosphorylation 627

Chapter 21: Gluconeogenesis, Glycogen Metabolism, and the Pentose Phosphate Pathway 660
Chapter 22: Photosynthesis 698
Chapter 23: Fatty Acid Catabolism 731
Chapter 24: Lipid Biosynthesis 757
Chapter 25: Metabolic Integration and the Unidirectionality of Pathways 803
Chapter 26: Nitrogen Acquisition and Amino Acid Metabolism 826
Chapter 27: The Synthesis and Degradation of Nucleotides 871

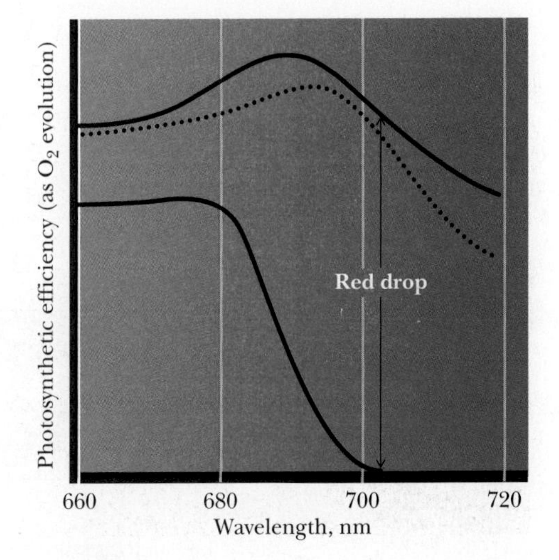

P*art* IV: Genetic Information 899

Chapter 28: DNA: Genetic Information, Recombination, and Mutation 900
Chapter 29: DNA Replication and Repair 936
Chapter 30: Transcription and the Regulation of Gene Expression 963
Chapter 31: The Genetic Code 1017
Chapter 32: Protein Synthesis and Degradation 1040
Chapter 33: Molecular Evolution 1075

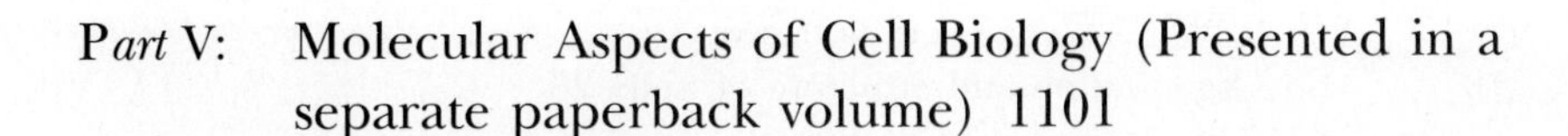

P*art* V: Molecular Aspects of Cell Biology (Presented in a separate paperback volume) 1101

Chapter 34: Self-Assembling Macromolecular Complexes 1102
Chapter 35: Membrane Transport 1125
Chapter 36: Muscle Contraction 1156
Chapter 37: The Molecular Basis of Hormone Action 1180
Chapter 38: Excitable Membranes, Neurotransmission, and Sensory Systems 1218

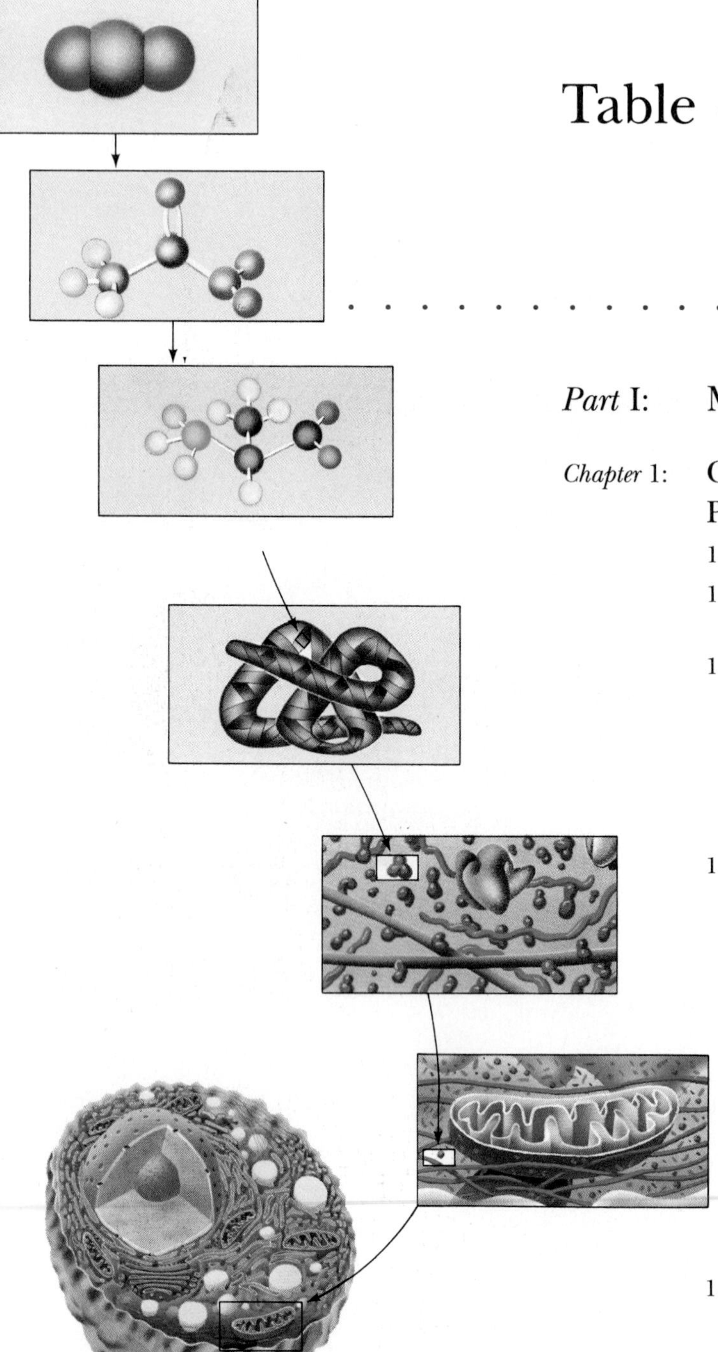

Table of Contents

Part I: Molecular Components of Cells 1

Chapter 1: Chemistry Is the Logic of Biological Phenomena 2

1.1 Distinctive Properties of Living Systems 3

1.2 Biomolecules: The Molecules of Life 5
- *Biomolecules Are Carbon Compounds* 6

1.3 A Biomolecular Hierarchy: Simple Molecules Are the Units of Complex Structures 8
- *Metabolites and Macromolecules* 9
- *Organelles* 9
- *Membranes* 11
- *The Unit of Life Is the Cell* 11

1.4 Properties of Biomolecules Reflect Their Fitness to the Living Condition 11
- *Biological Macromolecules and Their Building Blocks Have Directionality* 12
- *Biological Macromolecules Are Informational* 13
- *Biomolecules Have Characteristic Three-Dimensional Architecture* 13
- *Weak Forces Determine Biological Structure and Biomolecular Interactions* 13
- *Structural Complementarity Determines Biomolecular Interactions* 16
- *Biomolecular Recognition Is Mediated by Weak Chemical Forces* 16
- *Weak Forces Restrict Organisms to a Narrow Range of Environmental Conditions* 18
- *Enzymes* 18

1.5 Organization and Structure of Cells 20
- *Early Evolution of Cells* 20
- *Structural Organization of Prokaryotic Cells* 22
- *Structural Organization of Eukaryotic Cells* 24

1.6 Viruses Are Supramolecular Assemblies Acting as Cell Parasites 26

Chapter 2: Water, pH, and Ionic Equilibria 32

2.1 Properties of Water 33
- *Unusual Properties* 33
- *Structure of Water* 33

Structure of Ice 34
Molecular Interactions in Liquid Water 35
Solvent Properties 35
Ionization of Water 40

2.2 pH 42
Dissociation of Strong Electrolytes 43
Dissociation of Weak Electrolytes 44
Henderson–Hasselbalch Equation 45
Titration Curves 46
Phosphoric Acid Has Three Dissociable H^+ 47

2.3 Buffers 48
Phosphate System 49
Histidine System 50
The Bicarbonate Buffer System of Blood Plasma 50
"Good" Buffers 52

2.4 Water's Unique Role in the Fitness of the Environment 52

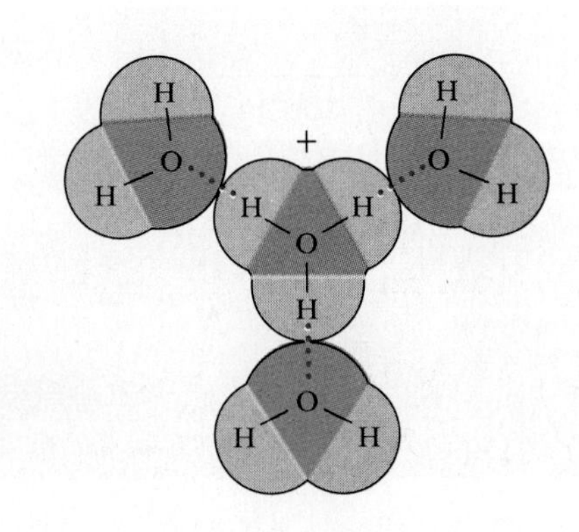

Chapter 3: Amino Acids 55

3.1 Amino Acids: Building Blocks of Proteins 56
Structure of a Typical Amino Acid 56
Amino Acids Can Join via Peptide Bonds 56
Common Amino Acids 56
Uncommon Amino Acids 60
Amino Acids Not Found in Proteins 61

3.2 Acid-Base Chemistry of Amino Acids 61
Amino Acids Are Weak, Polyprotic Acids 61
Ionization of Side Chains 65

3.3 Reactions of Amino Acids 66
Carboxyl and Amino Group Reactions 66
The Ninhydrin Reaction 67
Specific Reactions of Amino Acid Side Chains 67

3.4 Optical Activity and Stereochemistry of Amino Acids 70
Amino Acids Are Chiral Molecules 70
Nomenclature for Chiral Molecules 71

3.5 Spectroscopic Properties of Amino Acids 72
Ultraviolet Spectra 72
Nuclear Magnetic Resonance Spectra 73

3.6 Separation and Analysis of Amino Acid Mixtures 74
Chromatographic Methods 74
Ion Exchange Chromatography 75

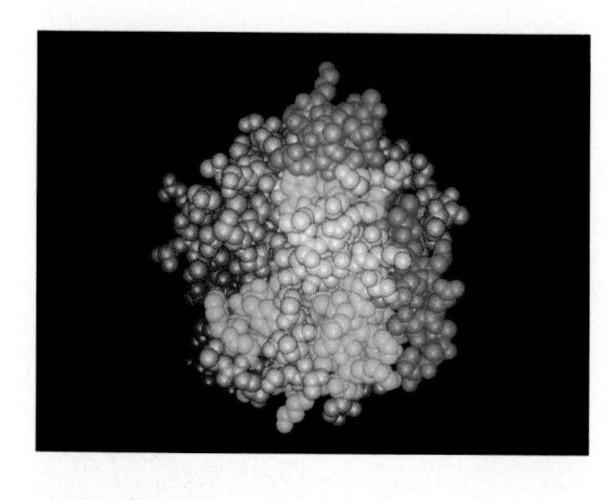

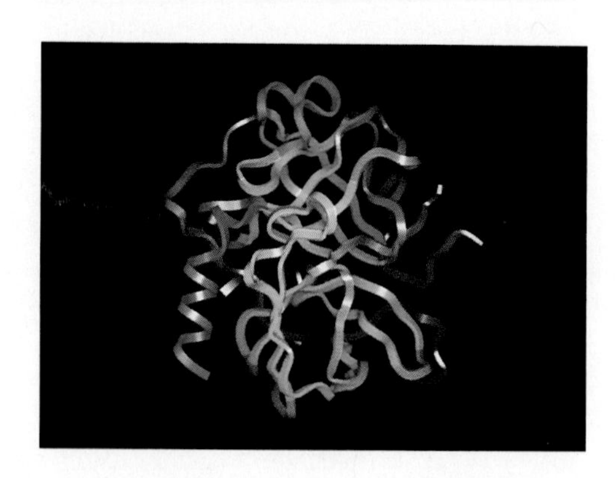

Chapter 4: **Proteins: Their Biological Functions and Primary Structure 81**

4.1 Proteins Are Linear Polymers of Amino Acids 82
The Peptide Bond Has Partial Double Bond Character 82
The Polypeptide Backbone Is Relatively Polar 84
Peptide Classification 84
Proteins Are Composed of One or More Polypeptide Chains 84
Acid Hydrolysis of Proteins 86
Amino Acid Analysis of Proteins 87
The Sequence of Amino Acids in Proteins 87

4.2 Architecture of Protein Molecules 89
The Levels of Protein Structure 89
Protein Conformation 93

4.3 The Many Biological Functions of Proteins 93
Enzymes 94
Regulatory Proteins 96
Transport Proteins 97
Storage Proteins 98
Contractile and Motile Proteins 98
Structural Proteins 98
Protective and Exploitive Proteins 98
Exotic Proteins 99

4.4 Some Proteins Have Chemical Groups Other Than Amino Acids 99

4.5 Reactions of Peptides and Proteins 101

4.6 The Purification of Protein Mixtures 102
Separation Methods 102
A Typical Protein Purification Scheme 104

4.7 The Primary Structure of a Protein: Determining the Amino Acid Sequence 105
Protein Sequencing Strategy 105
Step 1. Separation of Polypeptide Chains 105
Step 2. Cleavage of Disulfide Bridges 106
Step 3. Analysis of Amino Acid Composition 106
Step 4. Identification of the N- and C-Terminal Residues 106
Steps 5 and 6. Fragmentation of the Polypeptide Chain 111
Step 7. Reconstruction of the Overall Amino Acid Sequence 116
Step 8. Location of Disulfide Bridges 116
Sequence Databases 117

4.8 The Nature of Amino Acid Sequences 117
Homologous Proteins from Different Organisms Have Homologous Amino Acid Sequences 119
Related Proteins Share a Common Evolutionary Origin 120
Disparate Proteins May Share a Common Ancestry 122
Mutant Proteins 124

4.9 Synthesis of Polypeptides in the Laboratory 125
Synthetic Strategies 125
Solid Phase Peptide Synthesis 125

Appendix: Protein Techniques 131

Chapter 5: Proteins: Secondary, Tertiary, and Quaternary Structure 136

5.1 Forces Influencing Protein Structure 137
Hydrogen Bonds 137
Hydrophobic Interactions 137
Electrostatic Interactions 137
van der Waals Interactions 138

5.2 Role of the Amino Acid Sequence in Protein Structure 139

5.3 Secondary Structure in Proteins 140
Consequences of the Amide Plane 140
The Alpha-Helix 141
Other Helical Structures 144
The Beta-Pleated Sheet 145
The Beta-Turn 147
The Beta-Bulge 148

5.4 Protein Folding and Tertiary Structure 148
Fibrous Proteins 149
Globular Proteins 156
Classification of Globular Proteins 161
Molecular Chaperones: Proteins That Help Fold Globular Proteins 167
Protein Modules: Nature's Modular Strategy for Protein Design 167
Prediction of Secondary and Tertiary Structure in Proteins 170

5.5 Subunit Interactions and Quaternary Structure 172
Symmetry of Quaternary Structure 173
Forces Driving Quaternary Association 174
Modes and Models for Quaternary Structures 176
Open Quaternary Structures and Polymerization 177
Structural and Functional Advantages of Quaternary Association 177

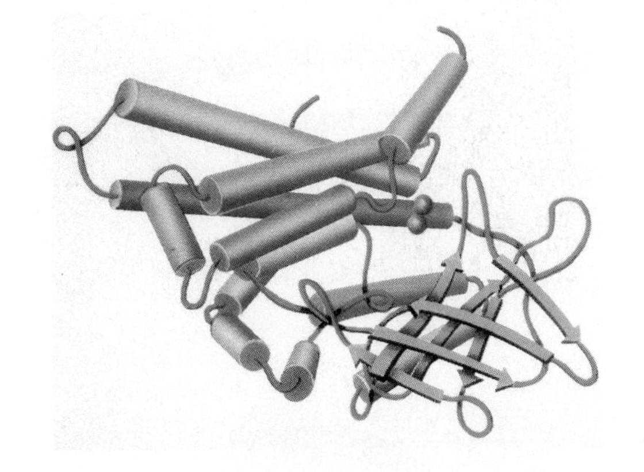

Chapter 6: Nucleotides and Nucleic Acids 180

6.1 Nitrogenous Bases 181
Common Pyrimidines and Purines 181
Properties of Pyrimidines and Purines 182

6.2 The Pentoses of Nucleotides and Nucleic Acids 183

6.3 Nucleosides Are Formed by Joining a Nitrogenous Base to a Sugar 184
Nucleoside Conformation 184
Nucleosides Are More Water-Soluble Than Free Bases 185

6.4 Nucleotides Are Nucleoside Phosphates 185
Cyclic Nucleotides 186
Nucleoside Diphosphates and Triphosphates 186
NDPs and NTPs Are Polyprotic Acids 187
Nucleoside 5'-Triphosphates Are Carriers of Chemical Energy 187
The Bases of Nucleotides Serve as "Recognition Units" 187

6.5 Nucleic Acids Are Polynucleotides 188
Shorthand Notations for Polynucleotide Structures 188
Base Sequence 190

6.6 Classes of Nucleic Acids 190
DNA 191
RNA 194
Significance of Chemical Differences Between DNA and RNA 197

6.7 Hydrolysis of Nucleic Acids 199
Hydrolysis by Acid or Base 199

Enzymatic Hydrolysis 200
Nuclease Specificity 200
RNases 201
DNases 202
Restriction Enzymes 202
Type II Restriction Endonucleases 203
Restriction Mapping 206

Chapter 7: Structure of Nucleic Acids 208

7.1 The Primary Structure of Nucleic Acids 209
Sequencing Nucleic Acids 209
Chain Termination or Dideoxy Method 209
Base-Specific Chemical Cleavage Method 212
Automated DNA Sequencing 216

7.2 The ABZ's of DNA Secondary Structure 216
Structural Equivalence of Watson–Crick Base Pairs 218
The DNA Double Helix Is a Stable Structure 219
Conformational Variation in Double-Helical Structures 220
Alternative Form of Right-Handed DNA 221
Z-DNA: A Left-Handed Double Helix 223
The Double Helix in Solution 225

7.3 Supercoils and Cruciforms: Tertiary Structure in DNA 227
Supercoils 227
Cruciforms 228

7.4 Denaturation and Renaturation of DNA 230
Thermal Denaturation and Hyperchromic Shift 230
pH Extremes or Strong H-Bonding Solutes Also Denature DNA Duplexes 230
DNA Renaturation 231
Renaturation Rate and DNA Sequence Complexity-c_0t Curves 231
Nucleic Acid Hybridization 232
Buoyant Density of DNA 233

7.5 Chromosome Structure 233
Nucleosomes 233
Organization of Chromatin and Chromosomes 234

7.6 Chemical Synthesis of Nucleic Acids 236
Phosphoramidite Chemistry 237
Chemically Synthesized Genes 237

7.7 Secondary and Tertiary Structure of RNA 238
Transfer RNA 238
Ribosomal RNA 241

Appendix: Isopycnic Centrifugation and Buoyant Density of DNA 245

Chapter 8: Recombinant DNA: Cloning and Creation of Chimeric Genes 247

8.1 Cloning 248
Plasmids 248
Bacteriophage λ as a Cloning Vector 255
Vectors From Filamentous Phage 255
Shuttle Vectors 255
Artificial Chromosomes 257

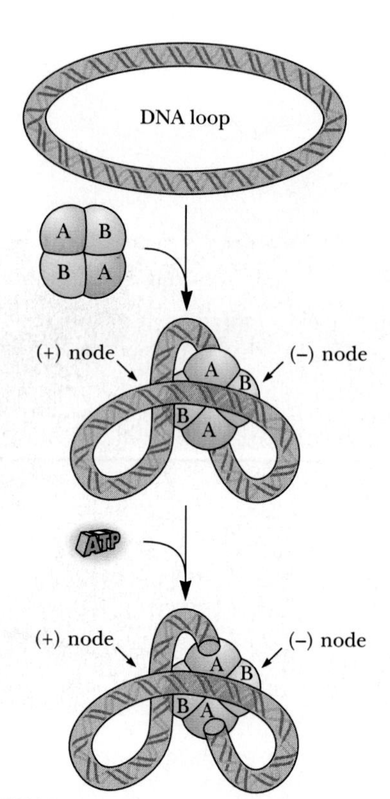

DNA is cut and a conformational change allows the DNA to pass through. Gyrase religates the DNA and then releases it.

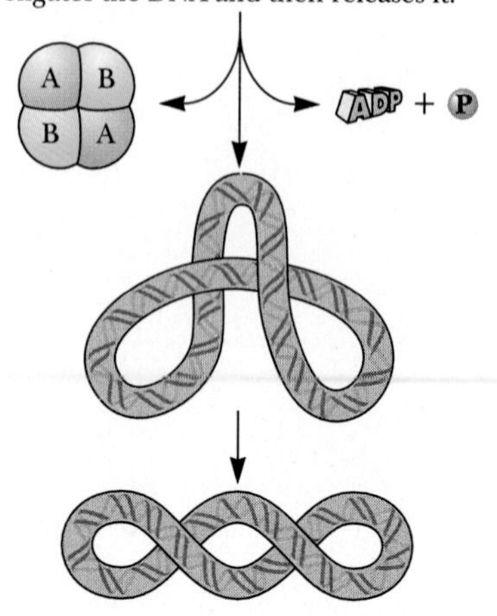

8.2 DNA Libraries 257
Genomic Libraries 257
Screening Libraries 258
Probes for Southern Hybridization 258
Chromosome Walking 262
cDNA Libraries 262
Expression Vectors 265
Reporter Gene Constructs 269

8.3 Polymerase Chain Reaction (PCR) 271
In Vitro *Mutagenesis* 273

8.4 Recombinant DNA Technology: An Exciting Scientific Frontier 273

Chapter 9: **Lipids and Membranes 276**

9.1 Classes of Lipids 277
Fatty Acids 277
Triacylglycerols 279
Glycerophospholipids 281
Sphingolipids 286
Waxes 287
Terpenes 288
Steroids 289

9.2 Membranes 291
Spontaneously Formed Lipid Structures 293
Fluid Mosaic Model 295
Membranes Are Asymmetric Structures 296
Membrane Phase Transitions 301

9.3 Structure of Membrane Proteins 303
Integral Membrane Proteins 303
Lipid-Anchored Membrane Proteins 306

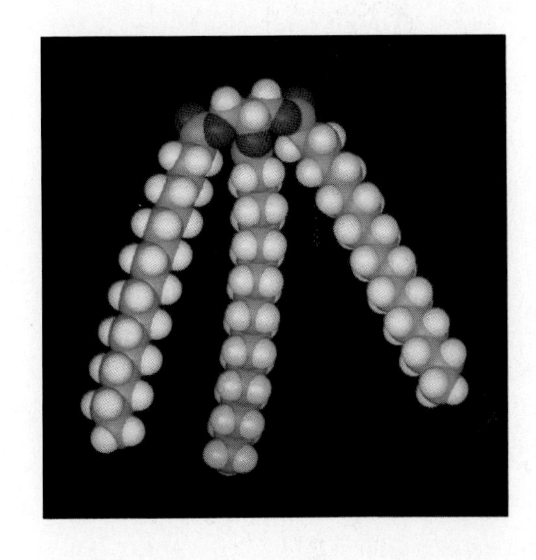

Chapter 10: **Carbohydrates and Cell Surfaces 310**

10.1 Carbohydrate Nomenclature 311

10.2 Monosaccharides 311
Classification 311
Stereochemistry 311
Cyclic Structures and Anomeric Forms 314
Haworth Projections 315
Derivatives of Monosaccharides 318

10.3 Oligosaccharides 322
Disaccharides 322
Higher Oligosaccharides 324

10.4 Polysaccharides 326
Structure and Nomenclature 326
Polysaccharide Functions 327
Storage Polysaccharides 327
Structural Polysaccharides 330
Bacterial Cell Walls 335
Cell Surface Polysaccharides 339

10.5 Glycoproteins 340
Antifreeze Glycoproteins 342
N-Linked Oligosaccharides 343
Oligosaccharide Cleavage as a Timing Device for Protein Degradation 343

10.6 Proteoglycans 345
Functions of Proteoglycans 346
Proteoglycans May Modulate Cell Growth Processes 347
Proteoglycans Make Cartilage Flexible and Resilient 348

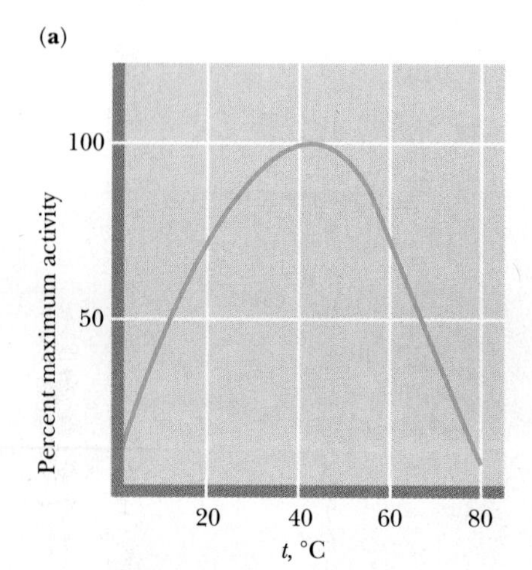

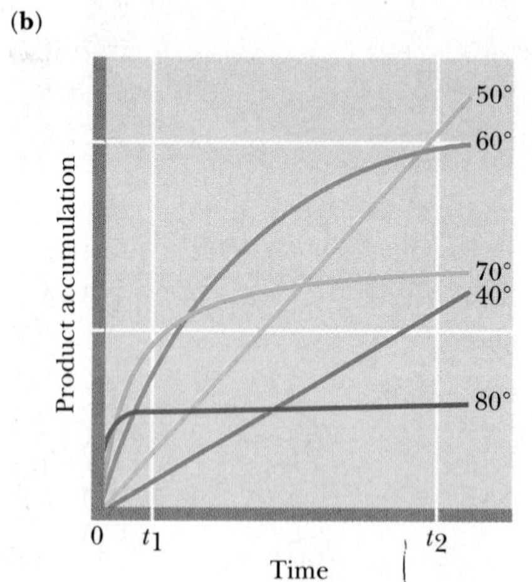

Part II: Enzymes and Energetics 351

Chapter 11: Enzyme Kinetics 352

11.1 Enzymes—Catalytic Power, Specificity, and Regulation 353
Catalytic Power 353
Specificity 353
Regulation 354
Enzyme Nomenclature 354
Coenzymes 356

11.2 Introduction to Enzyme Kinetics 356
Review of Chemical Kinetics 357
Bimolecular Reactions 358
Pseudo First-Order Reactions 358
Free Energy of Activation and the Action of Catalysts 359
Decreasing $\Delta G^{\ddagger}$ *Increases Reaction Rate* 360

11.3 Kinetics of Enzyme-Catalyzed Reactions 360
The Michaelis–Menten Equation 361
Steady-State Assumption 361
Initial Velocity Assumption 361
The Michaelis Constant, K_m 362
When $[S] = K_m$, $v = V_{max}/2$ 363
Relationships Between V_{max}, K_m, and Reaction Order 363
Enzyme Units 364
Turnover Number 364
k_{cat}/K_m 365
Linear Plots Can Be Derived from the Michaelis–Menten Equation 366
The Advantage of Linear Plots 368
Departures from Linearity: A Hint of Regulation? 368
Effect of pH on Enzymatic Activity 368
Effect of Temperature on Enzymatic Activity 369

11.4 Enzyme Inhibition 369
Reversible versus Irreversible Inhibition 369
Reversible Inhibition 370
Irreversible Inhibition 373

11.5 Kinetics of Enzyme-Catalyzed Reactions Involving Two or More Substrates 374
Random, Single-Displacement Reactions 376
Ordered, Single-Displacement Reactions 377
Double-Displacement (Ping-Pong) Reactions 379
Diagnosis of Bisubstrate Mechanisms 379
Multisubstrate Reactions 381

11.6 RNA and Antibody Molecules as Enzymes: Ribozymes and Abzymes 381
Catalytic RNA Molecules: Ribozymes 381
Catalytic Antibodies: Abzymes 384

Chapter 12: Enzyme Specificity and Allosteric Regulation 387

12.1 Specificity Is the Result of Molecular Recognition 387
The "Lock and Key" Hypothesis 388
The "Induced Fit" Hypothesis 388
"Induced Fit" and the Transition-State Intermediate 388
Specificity and Reactivity 388

12.2 Controls Over Enzymatic Activity—General Considerations 389
Zymogens 391
Isozymes 392
Modulator Proteins 392

12.3 The Allosteric Regulation of Enzymatic Activity 394
General Properties of Regulatory Enzymes 394

12.4 Hemoglobin and Myoglobin 394
The Comparative Biochemistry of Myoglobin and Hemoglobin 395
Myoglobin 396
The Mb Polypeptide Cradles the Heme Group 396
O_2 Binding to Mb 396
O_2 Binding Alters Mb Conformation 397
The Physiological Significance of Cooperative Binding of O_2 by Hemoglobin 398
The Structure of the Hemoglobin Molecule 398
Oxygenation Markedly Alters the Quaternary Structure of Hemoglobin 399
Movement of the Heme Iron by Less Than 0.04 nm Induces the Conformational Change in Hemoglobin 400
Oxy and Deoxy Hemoglobin Represent Two Different Conformational States 400
A Model for the Allosteric Behavior of Hemoglobin 402
H^+ Promotes Dissociation of Oxygen from Hemoglobin 402
CO_2 Also Promotes the Dissociation of O_2 from Hemoglobin 403
2,3-Bisphosphoglycerate Is an Important Allosteric Effector for Hemoglobin 404
The Physiological Significance of BPG Binding 405
Fetal Hb Has a Higher Affinity for O_2 Because It Has a Lower Affinity for BPG 405
Sickle-Cell Anemia 406
Sickle-Cell Anemia Is a Molecular Disease 407

12.5 *Escherichia coli* Aspartate Transcarbamoylase—An Allosteric Enzyme 407
ATCase Shows Cooperativity in Substrate Binding 408
CTP Is a Feedback Inhibitor of ATCase 408
ATP Is an Allosteric Activator of ATCase 408
*Desensitization of ATCase with p*HMB 409
The Subunit Organization of ATCase 410

12.6 Allosteric Models 410
The Symmetry Model of Monod, Wyman, and Changeux 410
Heterotropic Effectors 412
Positive Effectors 412
Negative Effectors 412
K *Systems and* V *Systems* 413
K *Systems and* V *Systems Fill Different Biological Roles* 414
The Sequential Allosteric Model of Koshland, Nemethy, and Filmer 414
The Allosteric Mechanism of E. coli *Aspartate Transcarbamoylase* 415

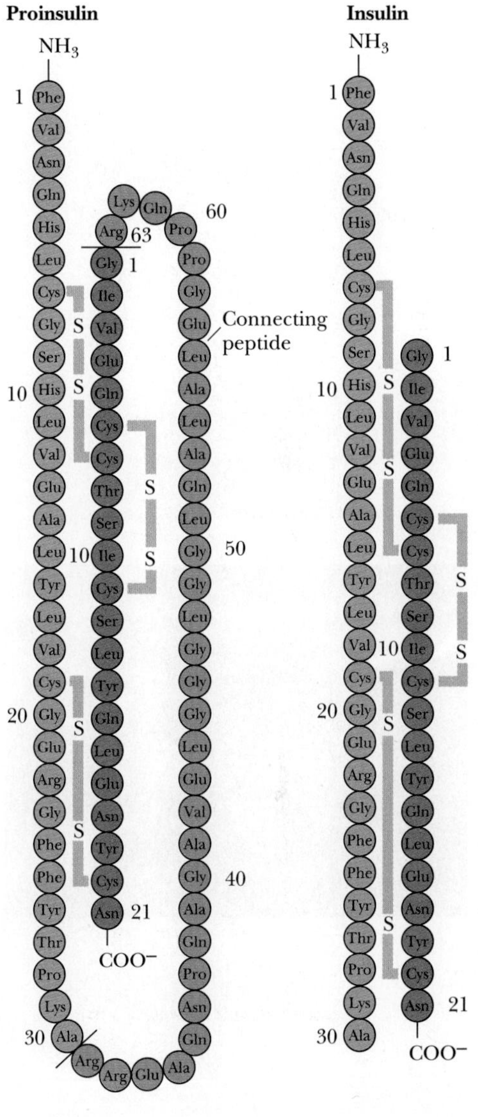

The Quaternary Structure of ATCase 416
Function of Active-Site Residues in ATCase 417
Structural Changes in ATCase in the T to R Transition 418
The Allosteric Mechanism 418

Appendix: The Oxygen-Binding Curves of Myoglobin and Hemoglobin 420

Chapter 13: Mechanisms of Enzyme Action 424

13.1 The Basic Principle—Stabilization of the Transition State 425
13.2 Enzymes Provide Enormous Rate Accelerations 426
13.3 The Binding Energy of ES Is Crucial to Catalysis 427
13.4 Entropy Loss and Destabilization of the ES Complex 428
13.5 Transition-State Analogs Bind Very Tightly to the Active Site 430
13.6 Covalent Catalysis 431
13.7 General Acid-Base Catalysis 433
13.8 Metal Ion Catalysis 434
13.9 Proximity 434
13.10 Typical Enzyme Mechanisms 436
13.11 Serine Proteases 436
The Digestive Serine Proteases 436
The Chymotrypsin Mechanism in Detail: Kinetics 438
The Serine Protease Mechanism in Detail: Events at the Active Site 440
13.12 The Aspartic Proteases 440
The Mechanism of Action of Aspartic Proteases 443
The AIDS Virus HIV-1 Protease Is an Aspartic Protease 444
13.13 Lysozyme 446
Model Studies Reveal a Strain-Induced Destabilization of a Bound Substrate on Lysozyme 448
The Lysozyme Mechanism Involves General Acid-Base Catalysis 451
13.14 Carboxypeptidase A 451
13.15 Liver Alcohol Dehydrogenase 455
The LAD Active Site Includes Structural and Catalytic Zinc Ions 456
The LAD Mechanism—Proton Abstraction Followed by Hydride Transfer 458

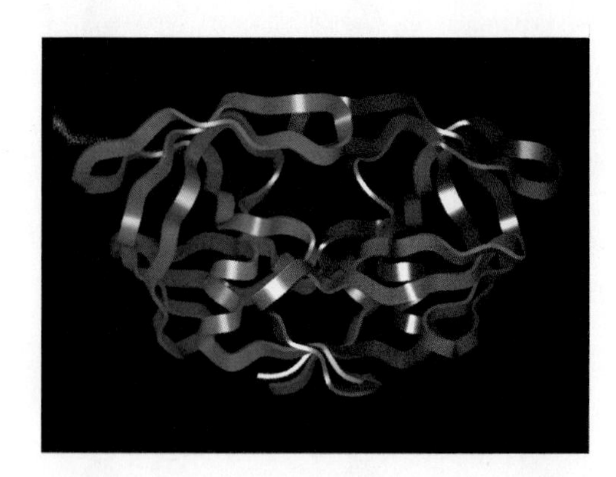

Chapter 14: Coenzymes and Vitamins 462

14.1 Vitamin B_1: Thiamine and Thiamine Pyrophosphate 463
14.2 Vitamins Containing Adenine Nucleotides 466
14.3 Nicotinic Acid and the Nicotinamide Coenzymes 468
14.4 Riboflavin and the Flavin Coenzymes 474
14.5 Pantothenic Acid and Coenzyme A 478
14.6 Vitamin B_6: Pyridoxine and Pyridoxal Phosphate 480
14.7 Vitamin B_{12}: Cyanocobalamin 485
B_{12}-Catalyzed Intramolecular Rearrangements 488
14.8 Vitamin C: Ascorbic Acid 489
Ascorbic Acid in the Brain 490
Ascorbic Acid Mobilizes Iron and Prevents Anemia 491

Ascorbic Acid Ameliorates Allergic Responses 491
Ascorbic Acid Can Stimulate the Immune System 491
14.9 Biotin 492
14.10 Lipoic Acid 495
14.11 Folic Acid 498
14.12 The Vitamin A Group 502
14.13 The Vitamin D Group 505
14.14 Vitamin E: Tocopherol 507
14.15 Vitamin K: Naphthoquinone 508

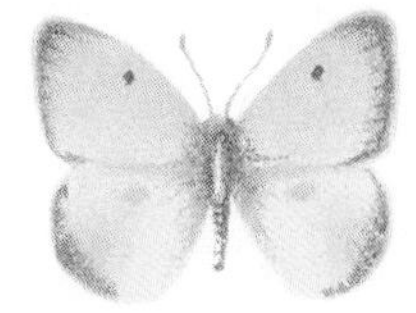

Chapter 15: Thermodynamics of Biological Systems 511
15.1 Basic Thermodynamic Concepts 512
15.2 The First Law: Heat, Work, and Other Forms of Energy 512
15.3 Enthalpy: A More Useful Function for Biological Systems 513
15.4 The Second Law and Entropy: An Orderly Way of Thinking About Disorder 514
15.5 The Third Law: Why Is "Absolute Zero" So Important? 516
15.6 Free Energy: A Hypothetical but Useful Device 516
The Standard-State Free Energy Change 516
15.7 The Physical Significance of Thermodynamic Properties 518
15.8 The Effect of pH on Standard-State Free Energies 519
15.9 The Important Effect of Concentration on Net Free Energy Changes 520
15.10 Irreversible Thermodynamics—Life in the Nonequilibrium Lane 520
15.11 The Importance of Coupled Processes in Living Things 521

Chapter 16: ATP and Energy-Rich Compounds 523
16.1 The High-Energy Biomolecules 524
ATP Is an Intermediate Energy-Shuttle Molecule 528
Group Transfer Potential 528
16.2 Classes of High-Energy Compounds 529
Phosphoric Acid Anhydrides 529
A Comparison of the Free Energy of Hydrolysis of ATP, ADP, and AMP 532
Phosphoric–Carboxylic Anhydrides 532
Enol Phosphates 533
Guanidinium Phosphates 534
Cyclic Nucleotides 535
Amino Acid Esters of tRNA Molecules 535
Thiol Esters 536
Pyridine Nucleotides 536
Other Compounds 538
16.3 Complex Equilibria Involved in ATP Hydrolysis 539
The Multiple Ionization States of ATP and the pH Dependence of $\Delta G^{\circ\prime}$ 539
The Effect of Metal Ions on the Free Energy of Hydrolysis of ATP 540
16.4 The Effect of Concentration on the Free Energy of Hydrolysis of ATP 540
16.5 The Daily Human Requirement for ATP 541

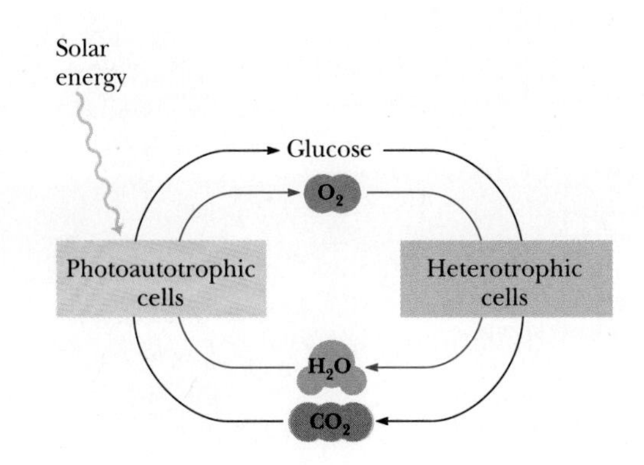

Part III: Metabolism and Its Regulation 543

Chapter 17: Metabolism—An Overview 543

17.1 Virtually All Organisms Have the Same Basic Set of Metabolic Pathways 546
- *Metabolic Diversity* 546
- *The Role of O_2 in Metabolism* 548
- *The Flow of Energy in the Biosphere and the Carbon and Oxygen Cycles Are Intimately Related* 549
- *The Nitrogen Cycle* 549

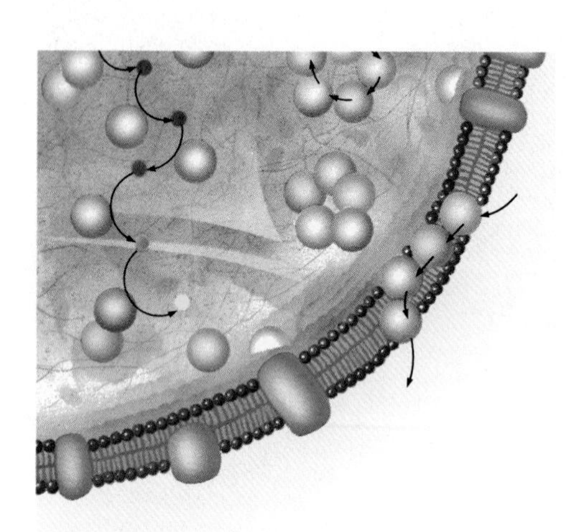

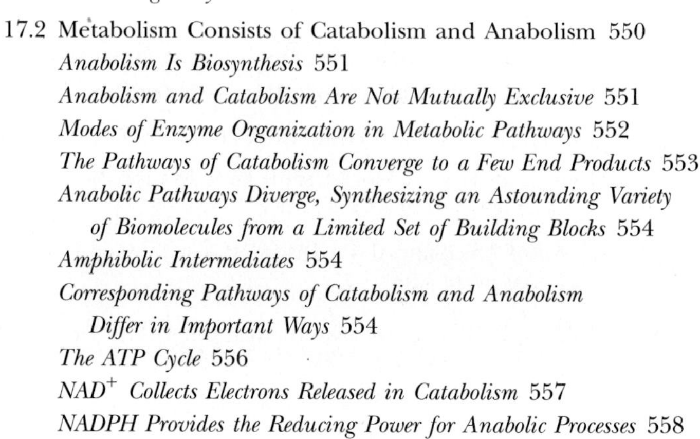

17.2 Metabolism Consists of Catabolism and Anabolism 550
- *Anabolism Is Biosynthesis* 551
- *Anabolism and Catabolism Are Not Mutually Exclusive* 551
- *Modes of Enzyme Organization in Metabolic Pathways* 552
- *The Pathways of Catabolism Converge to a Few End Products* 553
- *Anabolic Pathways Diverge, Synthesizing an Astounding Variety of Biomolecules from a Limited Set of Building Blocks* 554
- *Amphibolic Intermediates* 554
- *Corresponding Pathways of Catabolism and Anabolism Differ in Important Ways* 554
- *The ATP Cycle* 556
- *NAD^+ Collects Electrons Released in Catabolism* 557
- *NADPH Provides the Reducing Power for Anabolic Processes* 558

17.3 Intermediary Metabolism Is a Tightly Regulated, Integrated Process 559
- *Organisms Have Carbon Reserves but Not Nitrogen Reserves* 559
- *Energy Metabolism Quickly Adjusts to Demands* 559
- *Metabolic Regulation Is Achieved by Regulating Enzyme Activity* 559

17.4 Experimental Methods To Reveal Metabolic Pathways 562
- *Mutations Create Specific Metabolic Blocks* 563
- *Isotopic Tracers as Metabolic Probes* 564
- *NMR as a Metabolic Probe* 565
- *Metabolic Pathways Are Compartmentalized Within Cells* 567

Chapter 18: Glycolysis 569

18.1 Overview of Glycolysis 570
- *Rates and Regulation of Glycolytic Reactions Vary Among Species* 570

18.2 The Importance of Coupled Reactions in Glycolysis 570

18.3 The First Phase of Glycolysis 573
- *Reaction 1: Phosphorylation of Glucose by Hexokinase or Glucokinase—The First Priming Reaction* 573
- *Reaction 2: Phosphoglucoisomerase Catalyzes the Isomerization of Glucose-6-Phosphate* 576
- *Reaction 3: Phosphofructokinase—The Second Priming Reaction* 577
- *Reaction 4: Cleavage of Fructose-1,6-bisP by Fructose Bisphosphate Aldolase* 580
- *Reaction 5: Triose Phosphate Isomerase* 580

18.4 The Second Phase of Glycolysis 582
- *Reaction 6: Glyceraldehyde-3-Phosphate Dehydrogenase* 584
- *Reaction 7: Phosphoglycerate Kinase* 585
- *Reaction 8: Phosphoglycerate Mutase* 586
- *Reaction 9: Enolase* 588
- *Reaction 10: Pyruvate Kinase* 589

18.5 The Metabolic Fates of NADH and Pyruvate—The Products of Glycolysis 590

18.6 Anaerobic Pathways for Pyruvate 591

Lactate Accumulates Under Anaerobic Conditions in Animal Tissues 591

18.7 The Energetic Elegance of Glycolysis 592

18.8 Utilization of Other Substrates in Glycolysis 593

The Entry of Mannose into Glycolysis 594

The Special Case of Galactose 595

Lactose Intolerance 596

Glycerol Can Also Enter Glycolysis 596

Chapter 19: The Tricarboxylic Acid Cycle 598

19.1 Hans Krebs and the Discovery of the TCA Cycle 600

19.2 The TCA Cycle—A Brief Summary 601

The Chemical Logic of the TCA Cycle 601

19.3 The Bridging Step: Oxidative Decarboxylation of Pyruvate 603

19.4 Entry into the Cycle: The Citrate Synthase Reaction 603

The Structure of Citrate Synthase 604

Regulation of Citrate Synthase 605

19.5 The Isomerization of Citrate by Aconitase 605

Aconitase Utilizes an Iron-Sulfur Cluster 606

Fluoroacetate Blocks the TCA Cycle 606

19.6 Isocitrate Dehydrogenase—The First Oxidation in the Cycle 607

Isocitrate Dehydrogenase Links the TCA Cycle and Electron Transport 608

19.7 α-Ketoglutarate Dehydrogenase—A Second Decarboxylation 609

19.8 Succinyl-CoA Synthetase—A Substrate-Level Phosphorylation 609

The Mechanism of Succinyl-CoA Synthetase 610

The First Five Steps of the TCA Cycle Produce NADH, CO_2, GTP (ATP), and Succinate 610

19.9 Succinate Dehydrogenase—An Oxidation Involving FAD 610

19.10 Fumarase Catalyzes *Trans*-Hydration of Fumarate 611

19.11 Malate Dehydrogenase—Completing the Cycle 612

19.12 A Summary of the Cycle 613

The Fate of the Carbon Atoms of Acetyl-CoA in the TCA Cycle 613

19.13 The TCA Cycle Provides Intermediates for Biosynthetic Pathways 616

19.14 The Anaplerotic, or "Filling Up," Reactions 618

19.15 Regulation of the TCA Cycle 619

Regulation of Pyruvate Dehydrogenase 619

Regulation of Isocitrate Dehydrogenase 622

19.16 The Glyoxylate Cycle of Plants and Bacteria 622

The Glyoxylate Cycle Operates in Specialized Organelles 623

Isocitrate Lyase Short-Circuits the TCA Cycle by Producing Glyoxylate and Succinate 624

The Glyoxylate Cycle Helps Plants Grow in the Dark 624

Glyoxysomes Must Borrow Three Reactions from Mitochondria 625

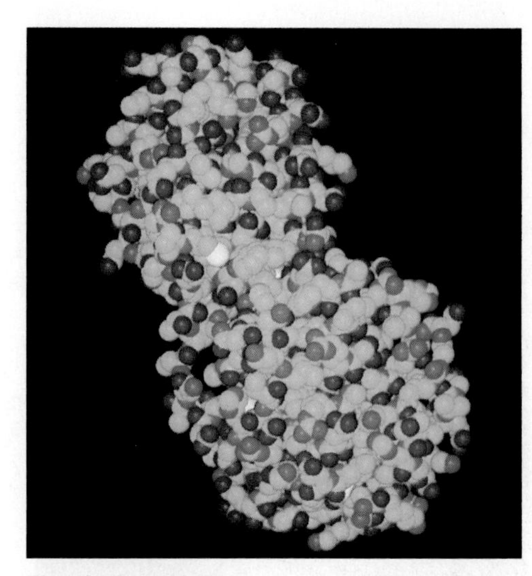

Phosphofructokinase

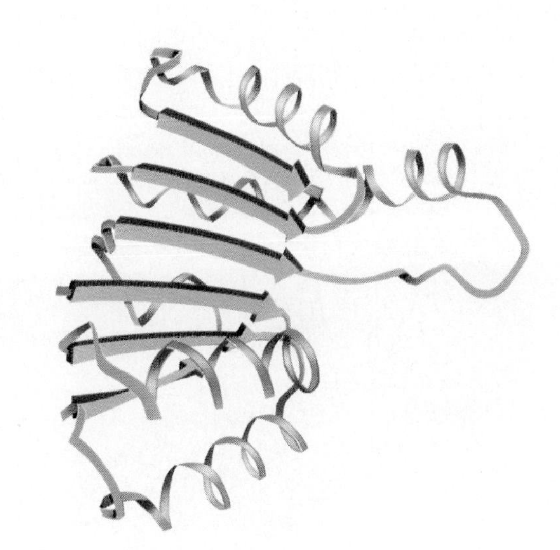

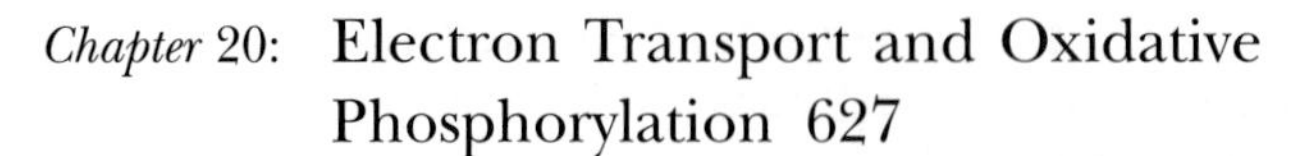

Chapter 20: Electron Transport and Oxidative Phosphorylation 627

20.1 Electron Transport and Oxidative Phosphorylation Are Membrane-Associated Processes 628
The Mitochondrial Matrix Contains the Enzymes of the TCA Cycle 629

20.2 Reduction Potentials—An Accounting Device for Free Energy Changes in Redox Reactions 629
Measurement of Standard Reduction Potentials 630
The Significance of $\mathcal{E}_o'$ 631
Coupled Redox Reactions 632
The Dependence of the Reduction Potential on Concentration 632

20.3 The Electron Transport Chain—An Overview 633
The Electron Transport Chain Can Be Isolated in Four Complexes 634

20.4 Complex I: NADH-Coenzyme Q Reductase 635
Complex I Transports Protons from the Matrix to the Cytosol 637

20.5 Complex II: Succinate-Coenzyme Q Reductase 637

20.6 Complex III: Coenzyme Q-Cytochrome *c* Reductase 638
Complex III Drives Proton Transport 640
The Q Cycle Is an Unbalanced Proton Pump 641
Cytochrome c *Is a Mobile Electron Carrier* 641

20.7 Complex IV: Cytochrome *c* Oxidase 641
Electron Transfer in Complex IV Involves Two Hemes and Two Copper Sites 642
Complex IV Also Transports Protons Across the Inner Mitochondrial Membrane 644
Independence of the Four Carrier Complexes 644
A Dynamic Model of Electron Transport 644
The $H^+/2e^-$ *Ratio for Electron Transport Is Uncertain* 645

20.8 The Thermodynamic View of Chemiosmotic Coupling 647

20.9 ATP Synthase 647
ATP Synthase Consists of Two Complexes—F_1 *and* F_0 648
Racker and Stoeckenius Confirmed the Mitchell Model in a Reconstitution Experiment 648
ATP Hydrolysis by ATP Synthase Is a One-Step S_N2 *Displacement* 649
Boyer's ^{18}O *Exchange Experiment Identified the Energy-Requiring Step* 649

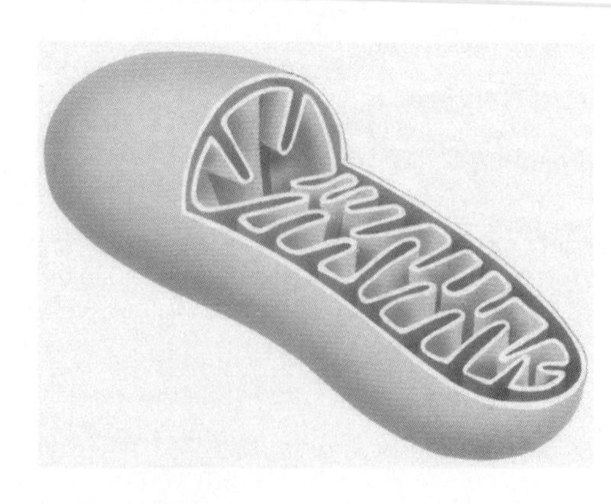

20.10 Inhibitors of Oxidative Phosphorylation 651
Inhibitors of Complexes I, II, and III Block Electron Transport 651
Cyanide, Azide, and Carbon Monoxide Inhibit Complex IV 651
Oligomycin and DCCD Are ATP Synthase Inhibitors 652

20.11 Uncouplers Disrupt the Coupling of Electron Transport and ATP Synthase 652
Endogenous Uncouplers Enable Organisms To Generate Heat 653

20.12 Shuttle Systems Feed the Electrons of Cytosolic NADH into Electron Transport 653
The Glycerophosphate Shuttle Ensures Efficient Use of Cytosolic NADH 653
The Malate-Aspartate Shuttle Is Reversible 655

20.13 ATP Exits the Mitochondria via an ATP-ADP Translocase 655
Outward Movement of ATP Is Favored Over Outward ADP Movement 655

20.14 What is the P/O Ratio for Mitochondrial Electron Transport and Oxidative Phosphorylation? 656
The Net Yield of ATP from Glucose Oxidation Depends on the Shuttle Used 657
3.5 Billion Years of Evolution Have Resulted in a System That Is 54% Efficient 658

Chapter 21: Gluconeogenesis, Glycogen Metabolism, and the Pentose Phosphate Pathway 660

21.1 Gluconeogenesis 661
The Substrates of Gluconeogenesis 661
Nearly All Gluconeogenesis Occurs in the Liver and Kidneys in Animals 661
Gluconeogenesis Is Not Merely the Reverse of Glycolysis 661
Gluconeogenesis—Something Borrowed, Something New 662
The Unique Reactions of Gluconeogenesis 663

21.2 Regulation of Gluconeogenesis 667
Gluconeogenesis Is Regulated by Allosteric and Substrate-Level Control Mechanisms 668

21.3 Glycogen Catabolism 671
Dietary Glycogen and Starch Breakdown 671
Metabolism of Tissue Glycogen 672
The Glycogen Phosphorylase Reaction 672
The Structure of Glycogen Phosphorylase 673
The Phosphorylase Reaction Mechanism 674

21.4 Glycogen Synthesis 674
Glucose Units Are Activated for Transfer by Formation of Sugar Nucleotides 675
UDP-Glucose Synthesis Is Driven by Pyrophosphate Hydrolysis 676
Glycogen Synthase Catalyzes Formation of $\alpha(1\rightarrow 4)$ *Glycosidic Bonds in Glycogen* 677
Glycogen Branching Occurs by Transfer of Terminal Chain Segments 678

21.5 Control of Glycogen Metabolism 678
Glycogen Metabolism Is Highly Regulated 678
Regulation of Glycogen Phosphorylase and Glycogen Synthase by Allosteric Effectors 678
Regulation of Glycogen Phosphorylase and Glycogen Synthase by Covalent Modification 679
Enzyme Cascades Regulate Glycogen Phosphorylase and Glycogen Synthase 682
Hormones Regulate Glycogen Synthesis and Degradation 683

21.6 The Pentose Phosphate Pathway 687
An Overview of the Pathway 687
The Oxidative Steps of the Pentose Phosphate Pathway 689
The Nonoxidative Steps of the Pentose Phosphate Pathway 690
Utilization of Glucose-6-P Depends on the Cell's Need for ATP, NADPH, and Ribose-5-P 693

Chapter 22: Photosynthesis 698

22.1 General Aspects of Photosynthesis 699
Photosynthesis Occurs in Membranes 699

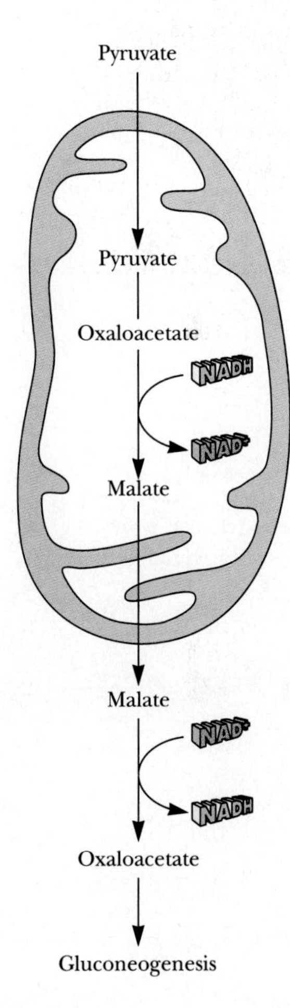

Photosynthesis Consists of Both Light and Dark Reactions 701
Water is the Ultimate e$^-$ *Donor for Photosynthetic NADP$^+$ Reduction* 701

22.2 Photosynthesis Depends on the Photoreactivity of Chlorophyll 702
Chlorophylls and Accessory Light-Harvesting Pigments 703
Photosynthetic Units Have Many Chlorophyll Molecules But Only a Single Reaction Center 705

22.3 Eukaryotic Phototrophs Possess Two Distinct Photosystems 705
Chlorophyll Exists in Plant Membranes in Association with Proteins 706
The Roles of PSI and PSII 707

22.4 The Z Scheme of Photosynthesis 707
O_2 Evolution Requires the Accumulation of Four Oxidizing Equivalents in PSII 710
Light-Driven Electron Flow from H_2O Through PSII 710
Electron Transfer Within the Cytochrome b_6*/ Cytochrome* f *Complex* 710
Electron Transfer from the Cytochrome b_6*/ Cytochrome* f *Complex to PSI* 711
The Initial Events in Photosynthesis Are Very Rapid Electron-Transfer Reactions 711

22.5 The Molecular Architecture of Photosynthetic Reaction Centers 711
Structure of the R. viridis *Photosynthetic Reaction Center* 712
Photosynthetic Electron Transfer in the R. viridis *Reaction Center* 712
Eukaryotic Reaction Centers: The Molecular Architecture of PSII 713
The Molecular Architecture of PSI 714
PSI and PSII Show a Nonuniform Distribution in the Thylakoid Membrane 715

22.6 The Quantum Yield of Photosynthesis 715
Photosynthetic Energy Requirements for Hexose Synthesis 716

22.7 Light-Driven ATP Synthesis—Photophosphorylation 716
The Mechanism of Photophosphorylation Is Chemiosmotic 716
Chloroplast CF_1CF_0 ATP Synthase Resembles Mitochondrial F_1F_0 ATP Synthase 718
Cyclic and Noncyclic Photophosphorylation 719
Cyclic Photophosphorylation 719

22.8 Carbon Dioxide Fixation 720
Ribulose-1,5-Bisphosphate Is the CO_2 Acceptor in CO_2 Fixation 720
The Ribulose-1,5-Bisphosphate Carboxylase Reaction 721

22.9 The Ribulose Bisphosphate Oxygenase Reaction: Photorespiration 721

22.10 The Calvin-Benson Cycle 723
The Enzymes of the Calvin Cycle 723
Balancing the Calvin Cycle Reactions To Account for Net Hexose Synthesis 725

22.11 Regulation of Carbon Dioxide Fixation 726
Light-Induced pH Changes in Chloroplast Compartments 726
Light-Induced Generation of Reducing Power 726
Light-Induced Mg^{2+} Efflux from Thylakoid Vesicles 726

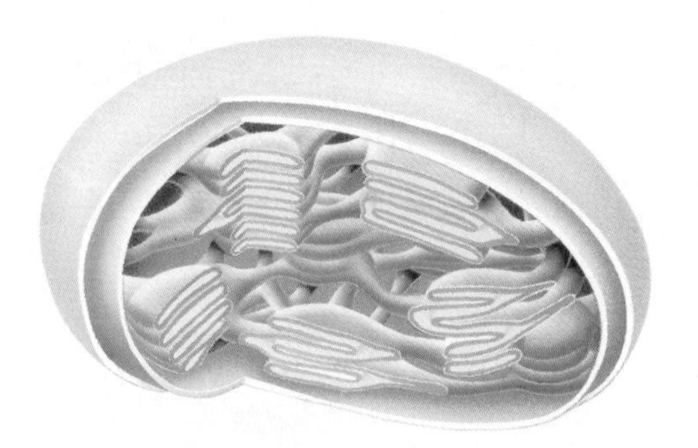

22.12 The C-4 Pathway of CO_2 Fixation 727
Intracellular Transport of Each CO_2 via a C-4 Intermediate Costs 2 ATP 727

22.13 Crassulacean Acid Metabolism 728

Chapter 23: Fatty Acid Catabolism 731

23.1 Mobilization of Fats from Dietary Intake and Adipose Tissue 732
Modern Diets Are Often High in Fat 732
Triacylglycerols Are a Major Form of Stored Energy in Animals 732
Hormones Signal the Release of Fatty Acids from Adipose Tissue 732
Degradation of Dietary Fatty Acids Occurs Primarily in the Duodenum 733

23.2 β-Oxidation of Fatty Acids 735
Franz Knoop and the Discovery of β-Oxidation 735
Coenzyme A Activates Fatty Acids for Degradation 737
Carnitine Carries Fatty Acyl Groups Across the Inner Mitochondrial Membrane 738
β-Oxidation Involves a Repeated Sequence of Four Reactions 739
Repetition of the β-Oxidation Cycle Yields a Succession of Acetate Units 745
Complete β-Oxidation of One Palmitic Acid Yields 106 Molecules of ATP 745
Migratory Birds Travel Long Distances on Energy from Fatty Acid Oxidation 746
Fatty Acid Oxidation Is an Important Source of Metabolic Water for Some Animals 746

23.3 β-Oxidation of Odd-Carbon Fatty Acids 747
β-Oxidation of Odd-Carbon Fatty Acids Yields Propionyl-CoA 747
A B_{12}-Catalyzed Rearrangement Yields Succinyl-CoA from L-Methylmalonyl-CoA 748
Net Oxidation of Succinyl-CoA Requires Conversion to Acetyl-CoA 749

23.4 β-Oxidation of Unsaturated Fatty Acids 750
An Isomerase and a Reductase Facilitate the β-Oxidation of Unsaturated Fatty Acids 750
Degradation of Polyunsaturated Fatty Acids Requires 2,4-Dienoyl-CoA Reductase 750

23.5 Other Aspects of Fatty Acid Oxidation 750
Peroxisomal β-Oxidation Requires FAD-Dependent Acyl-CoA Oxidase 750
Branched-Chain Fatty Acids and α-Oxidation 753
Refsum's Disease Is a Result of Defects in α-Oxidation 753
ω-Oxidation of Fatty Acids Yields Small Amounts of Dicarboxylic Acids 753

23.6 Ketone Bodies 753
Ketone Bodies Are a Significant Source of Fuel and Energy for Certain Tissues 753
Ketone Bodies and Diabetes Mellitus 755

Chapter 24: Lipid Biosynthesis 757

24.1 The Fatty Acid Biosynthesis and Degradation Pathways Are Different 757
Formation of Malonyl-CoA Activates Acetate Units for Fatty Acid Synthesis 758

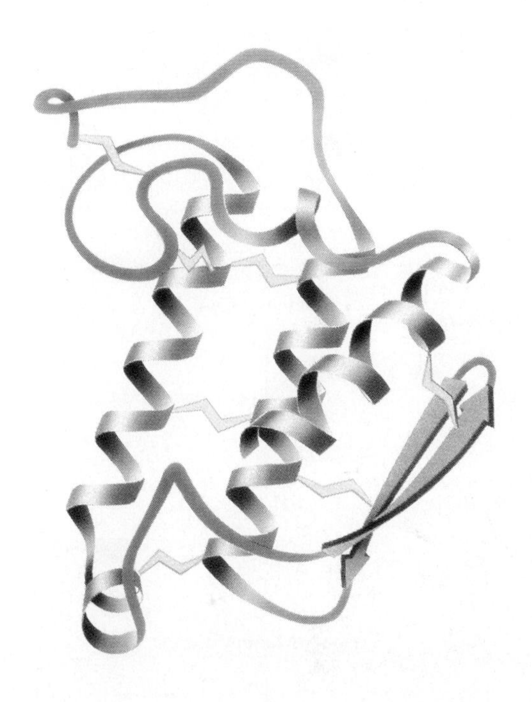

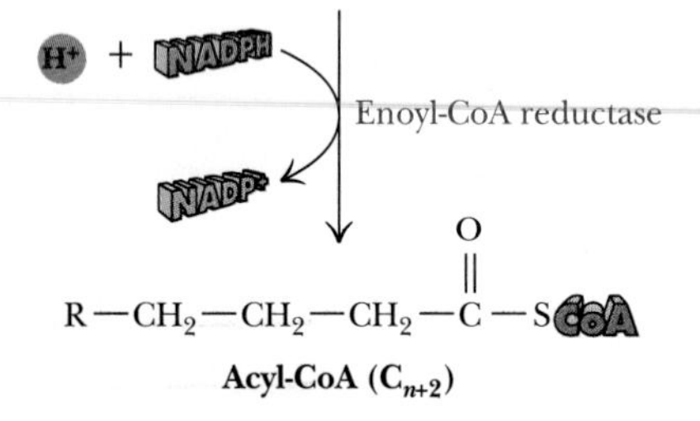

Fatty Acid Biosynthesis Depends on the Reductive Power of NADPH 758
Providing Cytosolic Acetyl-CoA and Reducing Power for Fatty Acid Synthesis 758
Acetate Units Are Committed to Fatty Acid Synthesis by Formation of Malonyl-CoA 760
Acetyl-CoA Carboxylase Is Biotin-Dependent and Displays Ping-Pong Kinetics 760
Acetyl-CoA Carboxylase in Animals Is a Multifunctional Protein 761
Phosphorylation of ACC Modulates Activation by Citrate and Inhibition by Palmitoyl-CoA 761
Acyl Carrier Proteins Carry the Intermediates in Fatty Acid Synthesis 761
Fatty Acid Synthesis in Bacteria and Plants 762
Decarboxylation Drives the Condensation of Acetyl-CoA and Malonyl-CoA 763
Reduction of the β-Carbonyl Group Follows a Now-Familiar Route 765
Fatty Acid Synthesis in Eukaryotes Occurs on a Multienzyme Complex 765
The Mechanism of Fatty Acid Synthase 767
Further Processing of C_{16} Fatty Acids 768
Unsaturation Reactions Occur in Eukaryotes in the Middle of an Aliphatic Chain 768
The Unsaturation Reaction May Be Followed by Chain Elongation 769
Biosynthesis of Polyunsaturated Fatty Acids 769
Arachidonic Acid Is Synthesized from Linoleic Acid by Mammals 770
Regulatory Control of Fatty Acid Metabolism—An Interplay of Allosteric Modifiers and Phosphorylation–Dephosphorylation Cycles 771
Hormonal Signals Regulate ACC and Fatty Acid Biosynthesis 772

24.2 Biosynthesis of Complex Lipids 773
Glycerolipid Biosynthesis 773
Eukaryotes Synthesize Glycerolipids from CDP-Diacylglycerol or Diacylglycerol 773
Phosphatidylethanolamine Is Synthesized from Diacylglycerol and CDP-Ethanolamine 776
Exchange of Ethanolamine for Serine Converts Phosphatidylethanolamine to Phosphatidylserine 776
Eukaryotes Synthesize Other Phospholipids via CDP-Diacylglycerol 777
Dihydroxyacetone Phosphate Is a Precursor to the Plasmalogens 777
Platelet Activating Factor 777
Sphingolipid Biosynthesis 777
Ceramide Is the Precursor for Other Sphingolipids and Cerebrosides 780

24.3 Eicosanoid Biosynthesis and Function 780
Eicosanoids Are Local Hormones 784
Prostaglandins Are Formed from Arachidonate by Oxidation and Cyclization 784
A Variety of Stimuli Trigger Arachidonate Release and Eicosanoid Synthesis 784
"Take Two Aspirin and ..." Inhibit Your Prostaglandin Synthesis 785

24.4 Cholesterol Biosynthesis 785
Mevalonate Is Synthesized from Acetyl-CoA via HMG-CoA Synthase 786

A Thiolase Brainteaser 789
Squalene Is Synthesized from Mevalonate 789
Conversion of Lanosterol to Cholesterol Requires 20 Additional Steps 789

24.5 Transport of Many Lipids Occurs via Lipoprotein Complexes 792
The Structure and Synthesis of the Lipoproteins 793
Lipoproteins in Circulation Are Progressively Degraded by Lipoprotein Lipase 795
Structure of the LDL Receptor 797
Defects in Lipoprotein Metabolism Can Lead to Elevated Serum Cholesterol 797

24.6 Biosynthesis of Bile Acids 798

24.7 Synthesis and Metabolism of Steroid Hormones 799
Pregnenolone and Progesterone Are the Precursors of All Other Steroid Hormones 799
Steroid Hormones Modulate Transcription in the Nucleus 799
Cortisol and Other Corticosteroids Regulate a Variety of Body Processes 801
Anabolic Steroids Have Been Used Illegally To Enhance Athletic Performance 801

Chapter 25: Metabolic Integration and the Unidirectionality of Pathways 803

25.1 A Systems Analysis of Metabolism 804
Only a Few Intermediates Interconnect the Major Metabolic Sequences 805
ATP and NADPH Couple Anabolism and Catabolism 805
Phototrophs Have an Additional Metabolic System—The Photochemical System 805

25.2 Metabolic Stoichiometry and ATP Coupling 806
The Significance of 38 ATP/Glucose in Cellular Respiration 807
The Significance of Large K_{eq} 808
The ATP Equivalent 808
The ATP Value of NADH and $FADH_2$ 809
The ATP Value of NADPH 809
The Nature and Magnitude of the ATP Equivalent 809
ATP and the Solvent Capacity of the Cell 811
Substrate Cycles Revisited 812

25.3 Unidirectionality 814
Cellular Respiration Versus CO_2 *Fixation—A Vivid Illustration of the Role of ATP Stoichiometry* 815
ATP Has Two Metabolic Roles 815
Energy Storage in the Adenylate System 816
Adenylate Kinase Interconverts ATP, ADP, and AMP 816
Energy Charge 816
The Response of Enzymes to Energy Charge 817
Phosphorylation Potential 818

25.4 Metabolism in a Multicellular Organism 819
Organ Specializations 819
Ethanol Metabolism Alters the $NAD^+/NADH$ *Ratio* 824

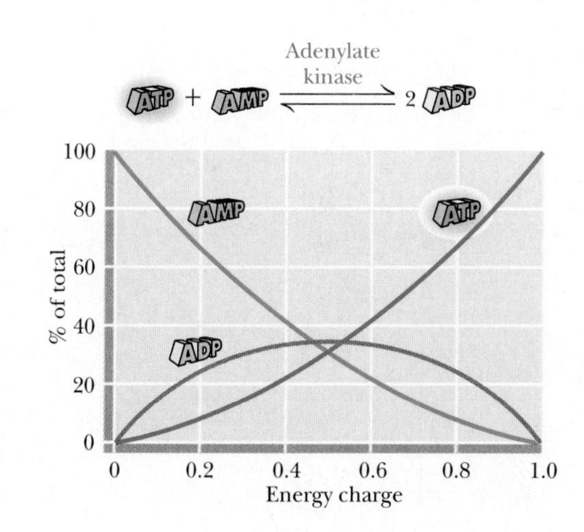

Uric acid

Excreted by primates, birds, reptiles, insects

2 H_2O + O_2 → CO_2 + H_2O_2 (Urate oxidase)

Allantoin

Excreted by other mammals

H_2O (Allantoinase)

Allantoic acid

Excreted by teleost fish

H_2O (Allantoicase)

Glyoxylic acid

2 Urea

Excreted by cartilaginous fish and amphibia

2 H_2O → 2 CO_2 (Urease)

4 NH_3

Excreted by marine invertebrates

Chapter 26: Nitrogen Acquisition and Amino Acid Metabolism 826

26.1 The Two Major Pathways of Biological N Acquisition 826
An Overview of the Biochemistry of N Acquisition 827

26.2 The Fate of Ammonium 832
Pathways of Ammonium Assimilation 833

26.3 *Escherichia coli* Glutamine Synthetase: A Case Study in Enzyme Regulation 835
AT:P_{IIA} Adenylylates GS; AT:P_{IID} Deadenylylates GS 838

26.4 Amino Acid Biosynthesis 839
Transamination 840
The Pathways of Amino Acid Biosynthesis 840
The α-Ketoglutarate Family of Amino Acids 841
The Urea Cycle 842
The Aspartate Family of Amino Acids 847
The Pyruvate Family of Amino Acids 850
The 3-Phosphoglycerate Family of Amino Acids 853
Biosynthesis of the Aromatic Amino Acids 855

26.5 Metabolic Degradation of Amino Acids 863
The 20 Common Amino Acids Are Degraded by 20 Different Pathways That Converge to Just 7 Metabolic Intermediates 865
Hereditary Defects in Phe Catabolism: Alcaptonuria and Phenylketonuria 869
Nitrogen Excretion 870

Chapter 27: The Synthesis and Degradation of Nucleotides 871

27.1 Nucleotide Biosynthesis 872

27.2 The Biosynthesis of Purines 872
IMP Biosynthesis: Inosinic Acid Is the Immediate Precursor to GMP and AMP 872
AMP and GMP Are Synthesized from IMP 876
Regulation of the Purine Biosynthetic Pathway 877
ATP-Dependent Kinases Form Nucleoside Diphosphates and Triphosphates from the Nucleoside Monophosphates 878

27.3 Purine Salvage 879
Lesch-Nyhan Syndrome: HGPRT Deficiency Leads to Severe Clinical Disorder 879

27.4 Purine Degradation 881
The Major Pathways of Purine Catabolism Lead to Uric Acid 881
Severe Combined Immunodeficiency Syndrome: A Lack of Adenosine Deaminase Is One Cause of This Inherited Disease 881
The Purine Nucleoside Cycle: An Anaplerotic Pathway in Skeletal Muscle 882
Xanthine Oxidase 884
Gout: An Excess of Uric Acid 884
The Fate of Uric Acid 884

27.5 The Biosynthesis of Pyrimidines 886
Pyrimidine Biosynthesis in Mammals Is Another Example of "Metabolic Channeling" 888
Synthesis of the Prominent Ribonucleotides UTP and CTP 888
Regulation of Pyrimidine Biosynthesis 889

27.6 Pyrimidine Degradation 889

27.7 Deoxyribonucleotide Biosynthesis 890

E. coli *Ribonucleotide Reductase* 890
The Reducing Power for Ribonucleotide Reductase 892
Regulation of Ribonucleotide Reductase Specificity and Activity 894
27.8 Synthesis of Thymine Nucleotides 895

Part IV: Genetic Information 899

Chapter 28: DNA: Genetic Information, Recombination, and Mutation 900
28.1 Genetic Information: The One-Gene, One-Enzyme Hypothesis 901
28.2 The Discovery That DNA Carries Genetic Information 901
The Transforming Principle Is DNA 903
DNA Is the Hereditary Molecule of Bacteriophage 904
The Quantity and Composition of DNA per Cell Is Constant 905
Each Species' DNA Has a Characteristic Base Composition 906
28.3 Genetic Information in Bacteria: Its Organization, Transfer, and Rearrangement 906
Identifying Bacterial Mutants 908
Mapping the Structure of Bacterial Chromosomes 908
Sexual Conjugation in Bacteria 910
High Frequency of Recombination 910
Transduction 913
Transformation and Transfection 914
28.4 The Molecular Mechanism of Recombination 914
General Recombination 915
The Holliday Model 916
The Enzymology of General Recombination 918
The RecBCD Enzyme Complex 918
RecBCD Cleaves ssDNA at Chi *Sites* 918
The RecA Protein 918
Resolution of the Holliday Junction Yields "Patch" and "Splice" Heteroduplexes 921
Transposons 921
28.5 The Immunoglobulin Genes: Generating Protein Diversity Using Genetic Recombination 923
The Immune Response 923
The Immunoglobulin G Molecule 923
The Organization of Immunoglobulin Genes 924
DNA Rearrangements Assemble an L-Chain Gene by Combining Three Separate Genes 925
DNA Rearrangements Assemble an H-Chain Gene by Combining Four Separate Genes 926
Mechanism of V-J and V-D-J Joining in Light- and Heavy-Chain Gene Assembly 927
Imprecise Joining 928
Antibody Diversity 928
28.6 The Molecular Nature of Mutation 928
Point Mutations 928
Mutations Induced by Base Analogs 929
Chemical Mutagens 930
Insertions and Deletions 932
28.7 RNA as Genetic Material 932
28.8 Transgenic Animals 933

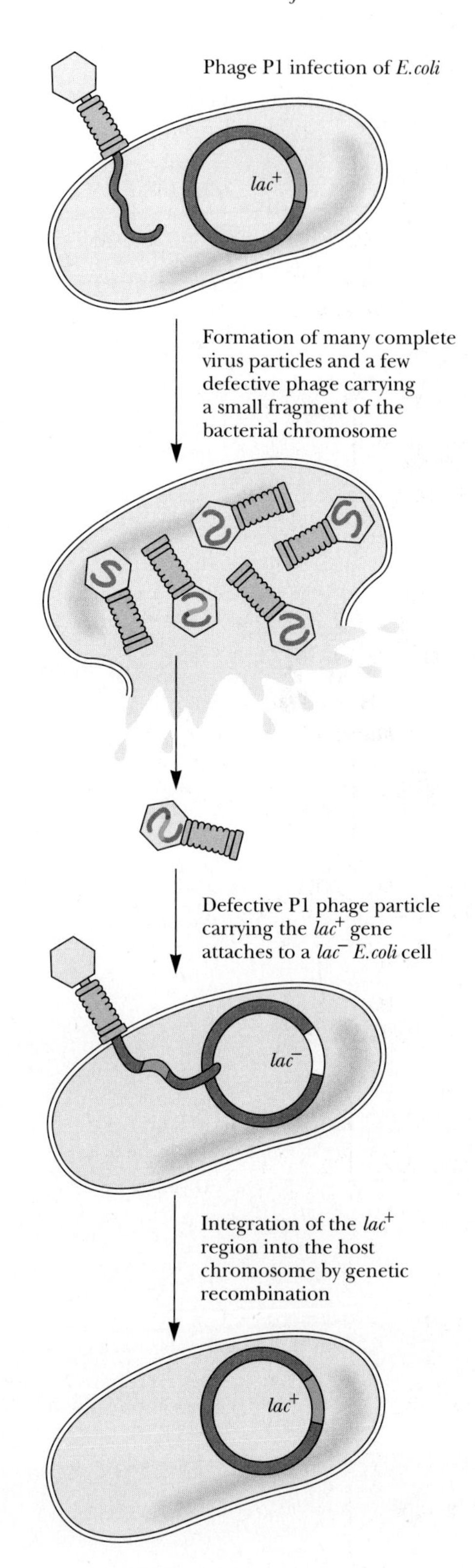
Phage P1 infection of *E.coli*
lac^+
Formation of many complete virus particles and a few defective phage carrying a small fragment of the bacterial chromosome
Defective P1 phage particle carrying the lac^+ gene attaches to a lac^- *E.coli* cell
lac^-
Integration of the lac^+ region into the host chromosome by genetic recombination
lac^+

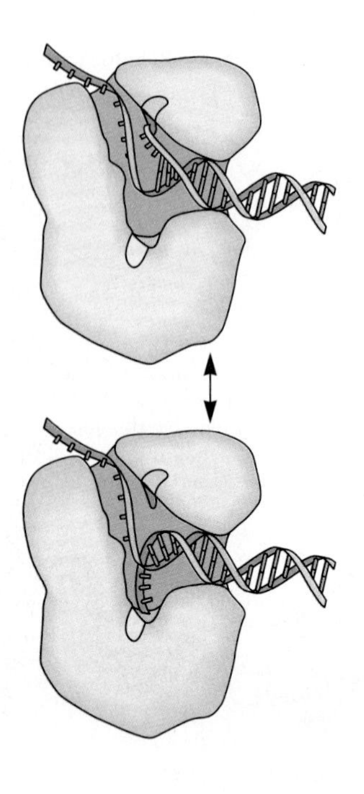

Chapter 29: DNA Replication and Repair 936

29.1 DNA Replication Is Semiconservative 937

29.2 The Enzymology of DNA Replication 940
E. coli *DNA Polymerase I* 940
E. coli *DNA Polymerase I Is Its Own Proofreader and Editor* 941
DNA Polymerase I Has Three Active Sites on Its Single Polypeptide Chain 942
E. coli *DNA Polymerase III* 943
"Core" DNA Polymerase III 943
DNA Polymerase III Holoenzyme 944

29.3 General Features of DNA Replication 944
Replication Is Bidirectional 944
Unwinding the DNA Helix 944
Replication Is Semidiscontinuous 945
The Lagging Strand Is Formed from Okazaki Fragments 946
The Chemistry of DNA Synthesis Favors Semidiscontinuous Replication 947
Synthesis of DNA Is Primed by RNA 948
E. coli *DNA Polymerase I Excises the RNA Primer* 948
DNA Ligase 949
General Features of a Replication Fork 949

29.4 The Mechanism of DNA Replication in *E. coli* 950
Initiation 950
Elongation 951
Termination 954

29.5 Eukaryotic DNA Replication 955
Eukaryotic DNA Polymerases 955
SV40 DNA Replication 956
T Antigen and the Initiation of SV40 Replication 956
T Antigen Function Is Regulated by Protein Phosphorylation 957

29.6 Reverse Transcriptase: An RNA-Directed DNA Polymerase 957
The Enzymatic Activities of Reverse Transcriptases 957

29.7 DNA Repair 958
Molecular Mechanisms of DNA Repair 959
The SOS Response 960

Chapter 30: Transcription and the Regulation of Gene Expression 963

30.1 Transcription in Prokaryotes 964
The Structure and Function of E. coli *RNA Polymerase* 964
The Steps of Transcription in Prokaryotes 965

30.2 Transcription in Eukaryotes 970
The Structure and Function of RNA Polymerase II 972
Transcription Initiation by RNA Polymerase II 973

30.3 The Regulation of Transcription in Prokaryotes 974
Transcription of Operons Is Controlled by Induction and Repression 975
lac: *The Paradigm of Operons* 976
The lac *Operator* 978
Interactions of lac *Repressor with DNA* 979
Positive Control of the lac *Operon by CAP* 980
Positive Versus Negative Control 981

The araBAD *Operon: Positive and Negative Control by* AraC 982
The trp *Operon: Attenuation as a Mechanism To Regulate Gene Expression* 984
Attenuation 985
Transcription Is Regulated by a Diversity of Mechanisms 985

30.4 Transcription Regulation in Eukaryotes 988
Eukaryotic Promoters, Enhancers, and Response Elements 989
DNA Looping 991

30.5 Structural Motifs in DNA-Binding Regulatory Proteins 992
α-Helices Fit Snugly into the Major Groove of B-DNA 992
Proteins with the Helix-Turn-Helix Motif 992
Proteins with Zn-Finger Motifs 996
Proteins with the Leucine Zipper Motif 999
The Zipper Motif: Intersubunit Interaction of Leucine Side Chains 999

30.6 Post-Transcriptional Processing of mRNA in Eukaryotic Cells 1001
Eukaryotic Genes Are Split Genes 1002
The Organization of Split Genes 1002
Post-Transcriptional Processing of Messenger RNA Precursors 1003
3′-Polyadenylation of Eukaryotic mRNAs 1003
Nuclear Pre-mRNA Splicing 1005
The Splicing Reaction: Lariat Formation 1006
Splicing Depends on snRNPs 1006
snRNPs Form the Spliceosome 1007
Alternative RNA Splicing 1009
The Fast Skeletal Muscle Troponin T Gene—An Example of Alternative Splicing 1011

Appendix: DNA:Protein Interactions 1014

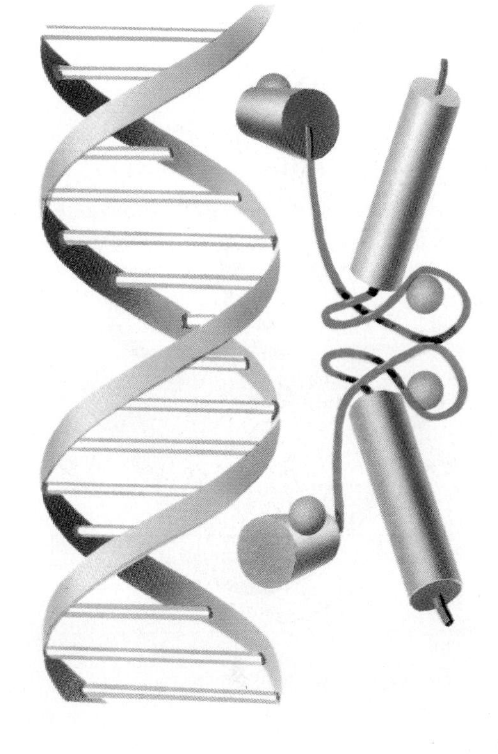

Chapter 31: The Genetic Code 1017

31.1 The Collinearity of Gene Structure and Protein Structure 1018

31.2 Elucidating the Genetic Code 1019
The General Nature of the Genetic Code 1020
Elucidating the Genetic Code Through Biochemistry 1020
Synthetic mRNAs from Nucleotide Copolymers 1020
Alternating Nucleotide Copolymers as mRNAs 1021
Repeating Nucleotide Copolymers as mRNAs 1022
Trinucleotides Bound to Ribosomes Promote the Binding of Specific Aminoacyl-tRNAs 1022

31.3 The Nature of the Genetic Code 1023

31.4 Amino Acid Activation for Protein Synthesis: Aminoacyl-tRNA Synthetases 1027
The Aminoacyl-tRNA Synthetase Reaction 1027
Selective tRNA Recognition by Aminoacyl-tRNA Synthetases 1029
tRNA Recognition Sites in E. coli *Glutaminyl-tRNA^{Gln} Synthetase* 1030
The Identity Elements for Some tRNAs Reside in the Anticodon 1030
Five Different Bases in Yeast tRNA^{Phe} Serve as Its Identity Elements 1030
Twelve Nucleotides in Common Define the tRNA^{Ser} Family 1031
A Single G:U Base Pair Defines tRNA^{Ala}s 1031

31.5 Codon-Anticodon Pairing, Third-Base Degeneracy, and the Wobble Hypothesis 1033
The Wobble Hypothesis 1033
The Purpose of Wobble 1035

31.6 Codon Usage 1035

31.7 Nonsense Suppression 1038

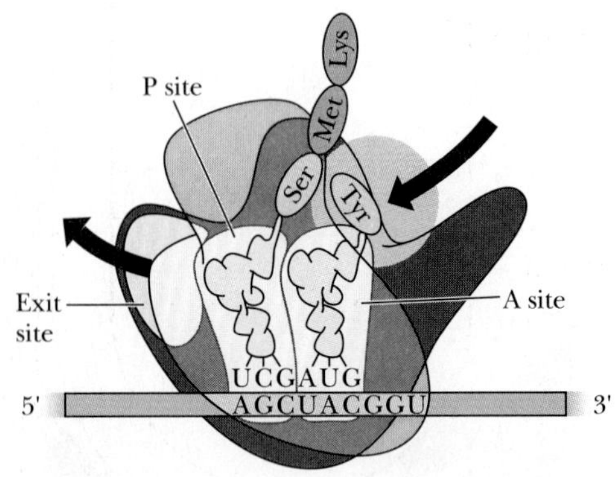
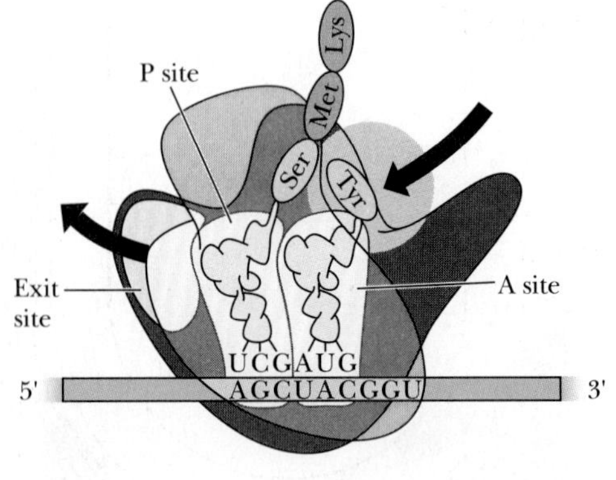

Chapter 32: Protein Synthesis and Degradation 1040

32.1 Ribosome Structure and Assembly 1040
The Composition of Prokaryotic Ribosomes 1041
Ribosomal Proteins 1041
rRNAs 1042
Self-Assembly of Ribosomes 1042
Ribosome Architecture 1043
Eukaryotic Ribosomes 1044
Conserved Bases in rRNA Are Clustered in Single-Stranded Regions 1045

32.2 The Mechanics of Protein Synthesis 1045
Peptide Chain Initiation in Prokaryotes 1047
Peptide Chain Elongation 1051
The Elongation Cycle 1051
23S rRNA Is the Peptidyl Transferase Enzyme 1053
GTP Fuels the Conformational Changes That Drive Ribosomal Functions 1055
Peptide Chain Termination 1055
The Ribosome Life Cycle 1057
Polyribosomes Are the Active Structures of Protein Synthesis 1057
The Relationship Between Transcription and Translation in Prokaryotes 1058

32.3 Protein Synthesis in Eukaryotic Cells 1058
Peptide Chain Initiation in Eukaryotes 1058
Regulation of Eukaryotic Peptide Chain Initiation 1060
Peptide Chain Elongation in Eukaryotes 1061
Eukaryotic Peptide Chain Termination 1061

32.4 Inhibitors of Protein Synthesis 1063
Streptomycin 1063
Puromycin 1064
Diphtheria Toxin 1064
Ricin 1064

32.5 Protein Folding 1066

32.6 Post-Translational Processing of Proteins 1066
Proteolytic Cleavage of the Polypeptide Chain 1066
Protein Translocation 1068
Prokaryotic Protein Translocation 1068
Eukaryotic Protein Sorting and Translocation 1069

32.7 Protein Degradation 1071
The Ubiquitin Pathway for Protein Degradation in Eukaryotes 1071

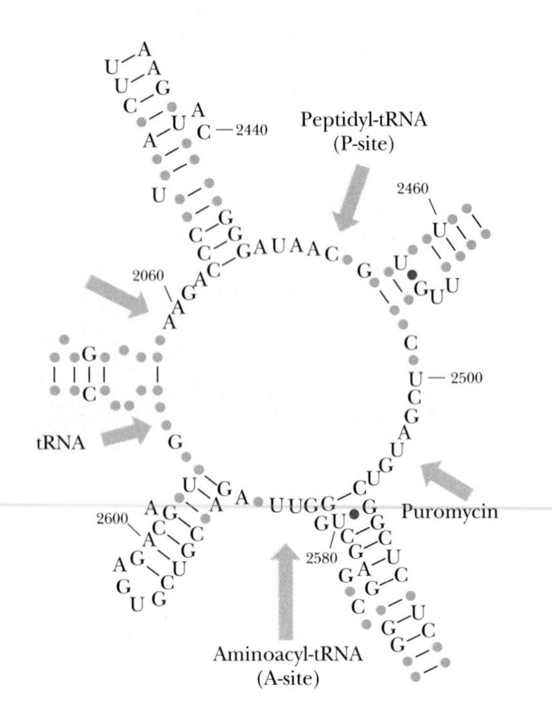

Chapter 33: Molecular Evolution 1075

33.1 An Overview of Evolution 1076

33.2 Evolutionary Change in Nucleotide Sequences 1076
Number of Nucleotide Substitutions Between Two Sequences 1077
Coding Regions Versus Noncoding Regions 1078

Sequence Alignments 1078
Rates of Nucleotide Substitutions 1080
33.3 Molecular Clocks 1081
Small Subunit rRNA 1084
33.4 Nonrandom Codon Usage 1084
33.5 Evolution by Gene Duplication and Exon Shuffling 1084
Exons and the Evolution of Proteins 1085
Correlations of Exons with Protein Structure 1086
Mosaic Proteins 1088
Is Exon Shuffling a Major Mechanism in Eukaryotic Evolution? 1090
33.6 Gene Sharing 1091
33.7 Proton-Translocating ATPases: Clues to Early Evolution 1093
Eukaryotic Cells Have Three Classes of H^+-ATPases 1094
H^+-ATPase Genes 1096
Archaebacterial H^+-ATPases 1096

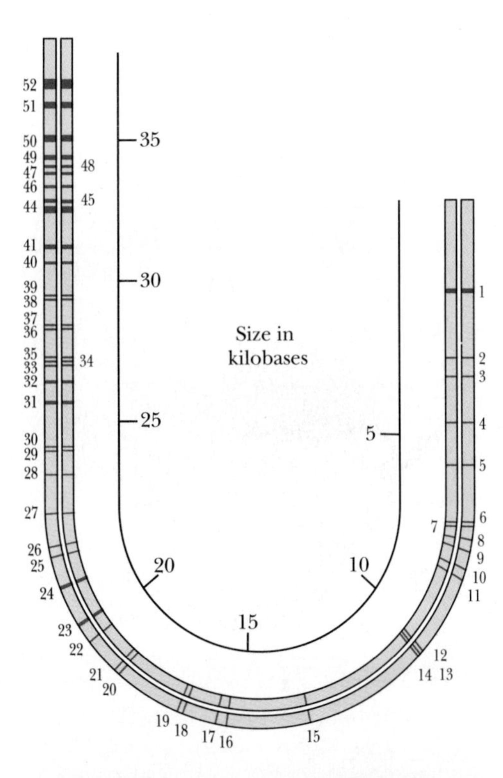

The following chapters are being published
separately in a paperback edition entitled
MOLECULAR ASPECTS OF CELL BIOLOGY

Part V: Molecular Aspects of Cell Biology 1101

Chapter 34: Self-Assembling Macromolecular Complexes 1102
34.1 Basic Principles of Spontaneous Self-Assembly 1103
Large Biological Structures Formed from Repeating Elements Are Highly Symmetric 1104
Self-Assembling Systems Seek the Lowest Possible Energy State 1104
Equilibrium and Free Energy Changes for Self-Assembly Processes 1104
Many Self-Assembly Processes are Entropy-Driven 1104
34.2 Tubulin, Microtubules, and Related Structures 1105
Microtubules Are Constituents of the Cytoskeleton 1106
Microtubules Are the Fundamental Structural Units of Cilia and Flagella 1106
The Mechanism of Ciliary Motion 1107
Microtubules Also Mediate Intercellular Motion of Organelles and Vesicles 1107
Dyneins Move Organelles in a Plus-to-Minus Direction, Kinesins in a Minus-to-Plus Direction 1109
34.3 Virus Structure and Assembly 1110
34.4 Tobacco Mosaic Virus 1110
Two-Layer Disks and Proto-helices Are Nucleation Points for Formation of Viral Particles 1110
Two-Layer Disks and Proto-Helices Recognize Specific Segments of TMV RNA 1111
The Assembly of TMV 1112
34.5 Spherical and Icosahedral Viruses 1113
A Simple Icosahedral Virus—Satellite Tobacco Necrosis Virus 1114
Tomato Bushy Stunt Virus—A Quasi-Equivalent Icosahedral Structure 1115
A Model for Icosahedral Virus Assembly 1116
Rhinovirus—An Icosahedral Virus That Causes the Common Cold 1119

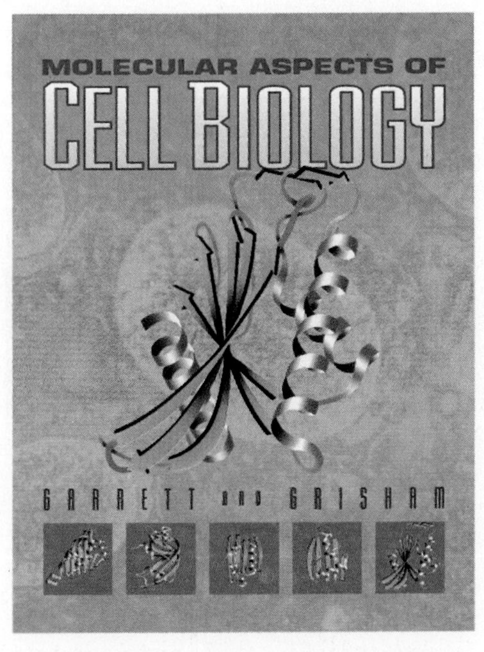

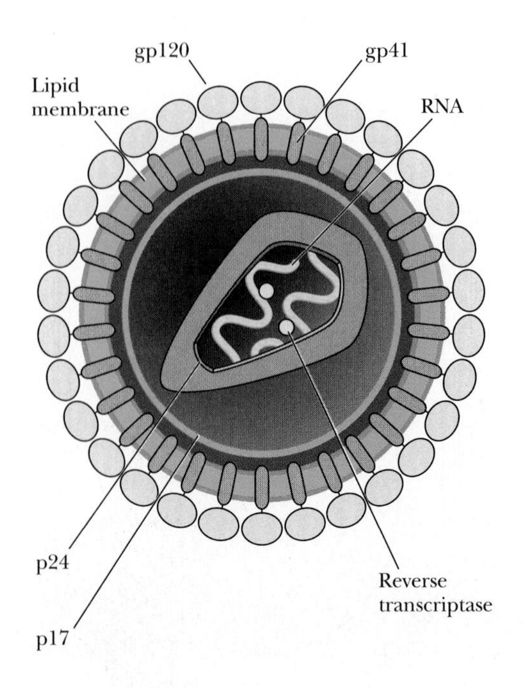

34.6 Membrane-Coated Viruses 1119
AIDS and the Human Immunodeficiency Virus 1121
The Structure of HIV 1121
The Mechanism of HIV Infection 1121
The Mechanism of HIV Assembly 1122

Chapter 35: **Membrane Transport 1125**

35.1 Passive Diffusion 1126
Passive Diffusion of a Charged Species 1127

35.2 Facilitated Diffusion 1127
Glucose Transport in Erythrocytes Occurs by Facilitated Diffusion 1128
The Anion Transporter of Erythrocytes Also Operates by Facilitated Diffusion 1130

35.3 Active Transport Systems 1131
All Active Transport Systems Are Energy-Coupling Devices 1131

35.4 Transport Processes Driven by ATP 1132
Monovalent Cation Transport: Na^+,K^+*-ATPase* 1132
Na^+,K^+*-ATPase Is Inhibited by Cardiac Glycosides* 1134
Calcium Transport: Ca^{2+}*-ATPase* 1134
The Gastric H^+,K^+*-ATPase* 1137
Bone Remodeling by Osteoclast Proton Pumps 1137
ATPases That Transport Peptides and Drugs 1138

35.5 Transport Processes Driven by Light 1140
A Model for Light-Driven Proton Transport 1140
Light-Driven Chloride Transport in H. halobium 1140

35.6 Transport Processes Driven by Ion Gradients 1141
Amino Acid and Sugar Transport 1141
Lactose Permease Actively Transports Lactose into E. coli 1142

35.7 Group Translocation 1143

35.8 Specialized Membrane Pores 1145
Porins in Gram-Negative Bacterial Membranes 1145
Gap Junctions in Mammalian Cell Membranes 1148

35.9 Ionophore Antibiotics 1149
Valinomycin Is a Mobile Carrier Ionophore 1150
Gramicidin Is a Channel-Forming Ionophore 1152
Amphipathic Helices Form Transmembrane Ion Channels 1152

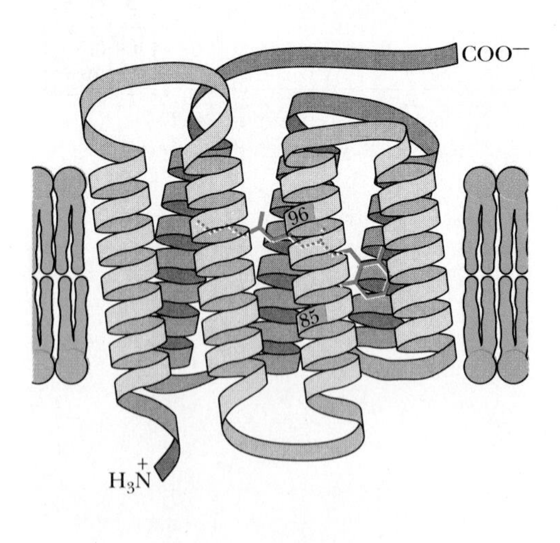

Chapter 36: **Muscle Contraction 1156**

36.1 The Morphology of Muscle 1157
Structural Features of Skeletal Muscle 1157

36.2 The Molecular Structure of Skeletal Muscle 1159
The Composition and Structure of Thin Filaments 1159
The Composition and Structure of Thick Filaments 1160
Proteolysis of Myosin Produces Meromyosin Fragments 1162
Repeating Structural Elements Are the Secret of Myosin's Coiled Coils 1163
The Associated Proteins of Striated Muscle 1164

36.3 The Mechanism of Muscle Contraction 1168
The Sliding Filament Model 1168
Albert Szent-Györgyi's Discovery of the Effects of Actin on Myosin 1169
The Coupling Mechanism: ATP Hydrolysis Drives Conformation Changes in the Myosin Heads 1171

36.4 Control of the Contraction–Relaxation Cycle by Calcium Channels and Pumps 1171

36.5 Regulation of Contraction by Ca^{2+} 1174

The Interaction of Ca^{2+} with Troponin C 1175

36.6 The Structure of Cardiac and Smooth Muscle 1175

The Structure of Smooth Muscle Myocytes 1176

36.7 The Mechanism of Smooth Muscle Contraction 1176

Chapter 37: The Molecular Basis of Hormone Action 1180

37.1 Classes of Hormones

37.2 Signal-Transducing Receptors Transmit the Hormonal Message

37.3 Intracellular Second Messengers

Cyclic AMP and the Second Messenger Model

Synthesis and Degradation of Cyclic AMP

37.4 GTP-Binding Proteins: The Hormonal Missing Link

G Proteins

Stimulatory and Inhibitory G Protein Effects

G_s is the Site of Action of Cholera Toxin

Pertussis Toxin Causes ADP-Ribosylation of G_i

Fluoride Ion Stimulation of Adenylyl Cyclase

ras *and the Small GTP-Binding Proteins*

37.5 The 7-TMS Receptors

37.6 Specific Phospholipases Release Second Messengers

Inositol Phospholipid Breakdown Yields Inositol-1,4,5-trisphosphate and Diacylglycerol

Activation of Phospholipase C Is Mediated by G Proteins or by Tyrosine Kinases

The Metabolism of Inositol-Derived Second Messengers

Second Messengers Are Derived from Phosphatidylcholine Breakdown

Sphingomyelin and Glycosphingolipids Also Generate Second Messengers

37.7 Calcium as a Second Messenger

Calcium-Induced Calcium Release

Calcium Oscillations

Intracellular Calcium-Binding Proteins

Calmodulin Target Proteins Possess a Basic Amphiphilic Helix

37.8 Protein Kinase C Transduces the Signals of Two Second Messengers

Cellular Target Proteins Are Dephosphorylated by Phosphoprotein Phosphatases

37.9 The Single-TMS Receptors

Non-Receptor Tyrosine Kinases

Receptor Tyrosine Kinases

Receptor Tyrosine Kinases Are Membrane-Associated Allosteric Enzymes

Receptor Tyrosine Kinases Phosphorylate a Variety of Cellular Target Proteins

The Polypeptide Hormones

The Processing of Gastrin

Protein-Tyrosine Phosphatases

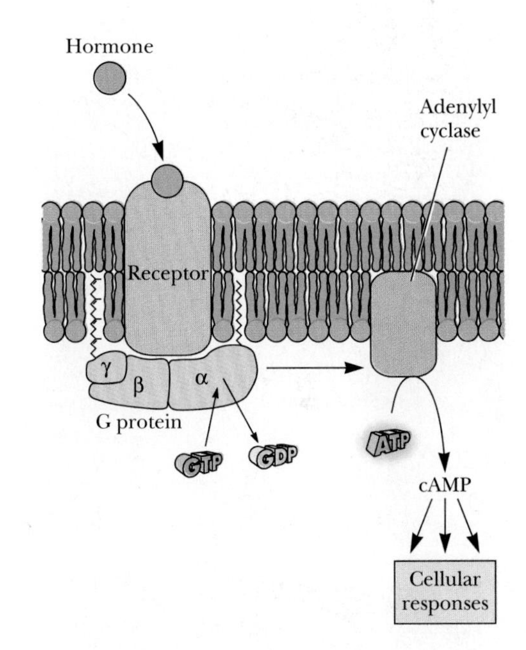

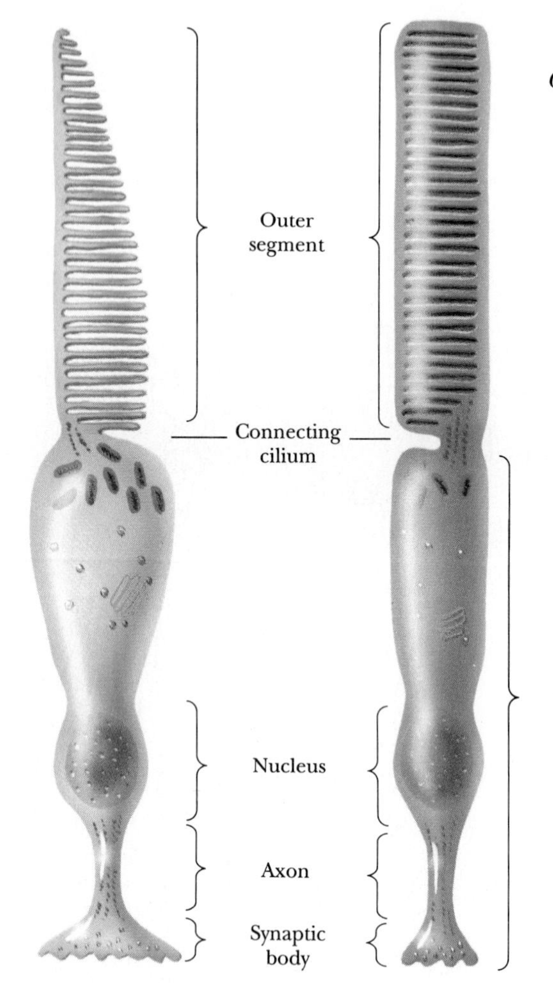

Membrane-Bound Guanylyl Cyclases Are Single-TMS Receptors
Soluble Guanylyl Cyclases Are Receptors for Nitric Oxide

37.10 Steroid Hormones
Receptor Proteins Carry Steroids to the Nucleus
Extracellular Effects of the Steroid Hormones

Chapter 38: **Excitable Membranes, Neurotransmission, and Sensory Systems 1218**

38.1 The Cells of Nervous Systems

38.2 Ion Gradients: Source of Electrical Potentials in Neurons

38.3 The Action Potential
The Action Potential Is Mediated by the Flow of Na^+ and K^+ Ions

38.4 The Voltage-Gated Sodium and Potassium Channels

38.5 Cell–Cell Communication at the Synapse
The Cholinergic Synapses
Acetylcholine Release Is Quantized
Two Classes of Acetylcholine Receptor
The Nicotinic Acetylcholine Receptor Is a Ligand-Gated Ion Channel
Acetylcholinesterase Degrades Acetylcholine in the Synaptic Cleft
Muscarinic Receptor Function Is Mediated by G Proteins

38.6 Other Neurotransmitters and Synaptic Junctions
Glutamate and Aspartate: Excitatory Amino Acid Neurotransmitters
γ-Aminobutyric Acid and Glycine: Inhibitory Neurotransmitters
The Catecholamine Neurotransmitters
The Peptide Neurotransmitters

38.7 Sensory Transduction
Vision
Olfaction
Hearing

List of Boxes

Chapter 2: Water, pH, and Ionic Equilibria 32

A Deeper Look

How the Bicarbonate Buffer System Works 51

Chapter 3: Amino Acids 55

Critical Developments in Biochemistry

Discovery of Optically Active Molecules and Determination of Absolute Configuration 70

Rules for Description of Chiral Centers in the R,S System 73

Chapter 4: Proteins: Their Biological Functions and Primary Structure 81

A Deeper Look

The Virtually Limitless Number of Different Amino Acid Sequences 90

Estimation of Protein Concentrations in Solutions of Biological Origin 103

Chapter 5: Proteins: Secondary, Tertiary, and Quaternary Structure 136

Critical Developments in Biochemistry

In Bed with a Cold, Pauling Stumbles onto the α-Helix and a Nobel Prize 145

Thermodynamics of the Folding Process in Globular Proteins 166

A Deeper Look

Charlotte's Web Revisited: Helix-Sheet Composites in Spider Dragline Silk 152

Faster-Acting Insulin: Genetic Engineering Solves a Quaternary Structure Problem 178

Chapter 6: Nucleotides and Nucleic Acids 180

A Deeper Look

Adenosine: A Nucleoside with Physiological Activity 187

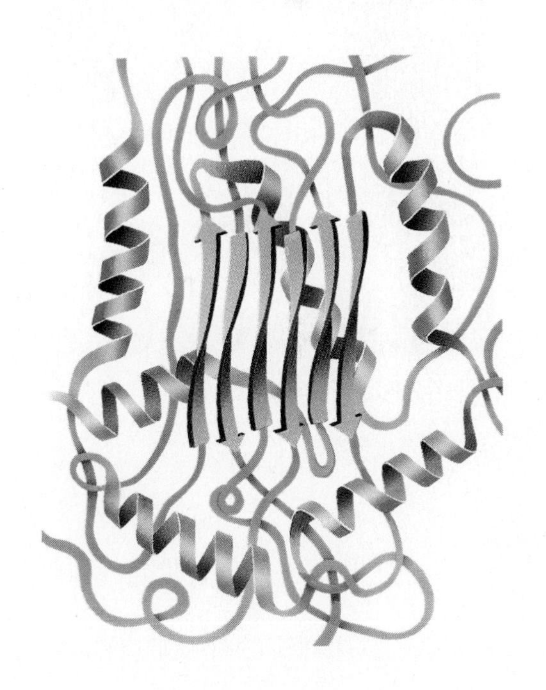

Chapter 8: Recombinant DNA: Cloning and Creation of Chimeric Genes 247

Critical Developments in Biochemistry

Identifying Specific DNA Sequences by Southern Blotting (Southern Hybridization) 260

The Human Genome Project 264

Gene Therapy 273

Chapter 9: Lipids and Membranes 276

A Deeper Look

Fatty Acids in Food: Saturated Versus Unsaturated 279

Polar Bears Use Triacylglycerols To Survive Long Periods of Fasting 281

Prochirality 282

Glycerophospholipid Degradation: One of the Effects of Snake Venoms 284

Platelet Activating Factor: A Potent Glyceroether Mediator 285

Moby Dick and Spermaceti: A Valuable Wax from Whale Oil 288

Chapter 11: Enzyme Kinetics 352

A Deeper Look

The Equations of Competitive Inhibition 371

Chapter 12: Enzyme Specificity and Allosteric Regulation 387

A Deeper Look

The Physiological Significance of Hb:O_2 Interactions 399

Changes in the Heme Iron upon O_2-Binding 401

Chapter 13: Mechanisms of Enzyme Action 424

A Deeper Look

Relating Rate Acceleration to Free Energies of Activation 426

Critical Developments in Biochemistry

Transition-State Stabilization in the Serine Proteases 440

The pH Dependence of Aspartic Proteases and HIV-1 Protease 446

Chapter 14: Coenzymes and Vitamins 462

A Deeper Look

Thiamine and Beriberi 464

Niacin and Pellagra 473

Riboflavin and Old Yellow Enzyme 476

Fritz Lipmann and Coenzyme A 480

Vitamin B_6 483

Vitamin B_{12} and Pernicious Anemia 487

Ascorbic Acid and Scurvy 491

Biotin 495

Lipoic Acid 496

Folic Acid, Pterins, and Insect Wings 500

Mechanism of Methotrexate Inhibition 502

β-Carotene and Vision 504

Vitamin D and Rickets 507

Vitamin E 508

Vitamin K and Blood Clotting 509

Chapter 15: Thermodynamics of Biological Systems 511

A Deeper Look

Entropy, Information, and the Importance of "Negentropy" 515

Chapter 17: Metabolism—An Overview 544

A Deeper Look

Calcium Carbonate—A Biological Sink for CO_2 551

Chapter 18: Glycolysis 569

A Deeper Look

The Chemical Evidence for the Schiff Base Intermediate in Class I Aldolase 582

Chapter 20: Electron Transport and Oxidative Phosphorylation 627

Critical Developments in Biochemistry

Oxidative Phosphorylation—The Clash of Ideas and Energetic Personalities 646

Chapter 21: Gluconeogenesis, Glycogen Metabolism, and the Pentose Phosphate Pathway 660

A Deeper Look

Carbohydrate Utilization in Exercise 684

Chapter 22: Photosynthesis 698

Critical Developments in Biochemistry

Experiments with Isolated Chloroplasts Provided the First Direct Evidence for the Chemiosmotic Hypothesis 717

Chapter 23: Fatty Acid Catabolism 731

A Deeper Look

The Akee Tree 742

Chapter 24: Lipid Biosynthesis 757

A Deeper Look

Choosing the Best Organism for the Experiment 763

The Discovery of Prostaglandins 785

Critical Developments in Biochemistry

The Long Search for the Route of Cholesterol Biosynthesis 788

Lovastatin Lowers Serum Cholesterol Levels 792

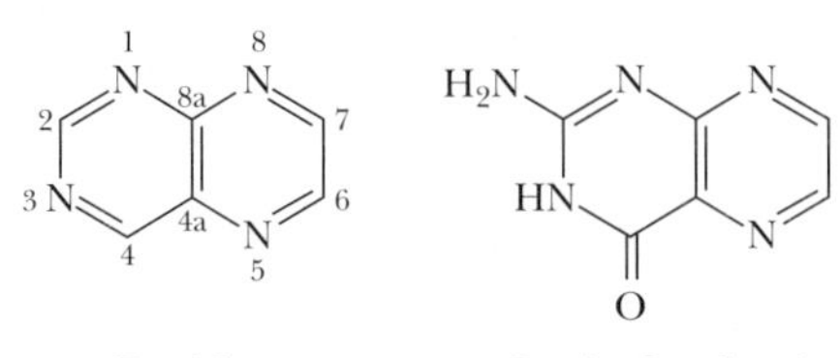

Pteridine **Pterin: 2-amino-4-oxopteridine**

Xanthopterin (yellow) **Leucopterin (white)**

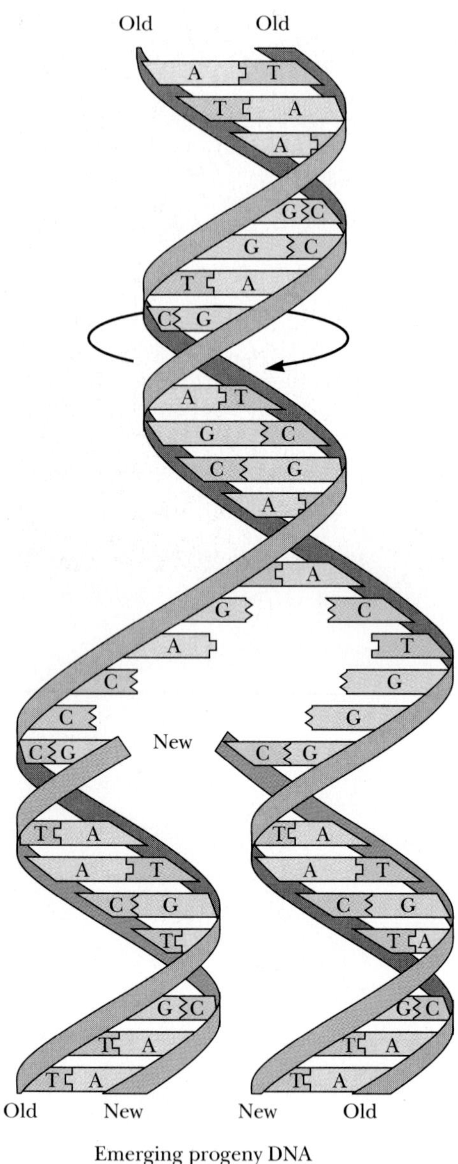

Emerging progeny DNA

Chapter 25: Metabolic Integration and the Unidirectionality of Pathways 803

A Deeper Look

ATP Changes the K_{eq} for a Process by a Factor of 10^8 810

Chapter 26: Nitrogen Acquisition and Amino Acid Metabolism 826

A Deeper Look

The Urea Cycle as Both an Ammonium and a Bicarbonate Disposal Mechanism 845

Amino Acid Biosynthesis Inhibitors as Herbicides 863

Histidine—A Clue to Understanding Early Evolution? 864

Chapter 27: The Synthesis and Degradation of Nucleotides 871

Critical Developments in Biochemistry

Enzyme Inhibition by Fluoro Compounds 897

Chapter 29: DNA Replication and Repair 936

A Deeper Look

DNA Repair and Aging 961

Chapter 30: Transcription and the Regulation of Gene Expression 963

A Deeper Look

Conventions Employed in Expressing the Sequences of Nucleic Acids and Proteins 965

Quantitative Evaluation of *lac* repressor:DNA Interactions 979

Chapter 34: Self-Assembling Macromolecular Complexes 1102

Critical Developments in Biochemistry

Microtubule Polymerization Inhibitors as Therapeutic Agents 1109

Michael Rossman's Canyon Hypothesis 1120

A Deeper Look

Out of Africa? The Origin and Discovery of HIV 1122

Chapter 35: Membrane Transport 1125

A Deeper Look

Cardiac Glycosides: Potent Drugs from Ancient Times 1135

Chapter 36: Muscle Contraction 1156

Critical Developments in Biochemistry

The Molecular Defect in Duchenne Muscular Dystrophy Involves an Actin-Anchoring Protein 1166

A Deeper Look

Viscous Solutions Reflect Long-Range Molecular Interactions 1169

Smooth Muscle Effectors Are Useful Drugs 1178

Chapter 37: The Molecular Basis of Hormone Action 1180

A Deeper Look

Cancer, Oncogenes, and Tumor Suppressor Genes

PI Metabolism and the Pharmacology of Li^{+}

Okadaic Acid: A Marine Toxin and Tumor Promoter

Nitric Oxide, Nitroglycerin, and Alfred Nobel

The Acrosome Reaction

Chapter 38: Excitable Membranes, Neurotransmission, and Sensory Systems 1218

A Deeper Look

The Actual Transmembrane Potential Difference

Tetrodotoxin and Other Na^{+} Channel Toxins

Potassium Channel Toxins

The Biochemistry of Neurological Disorders

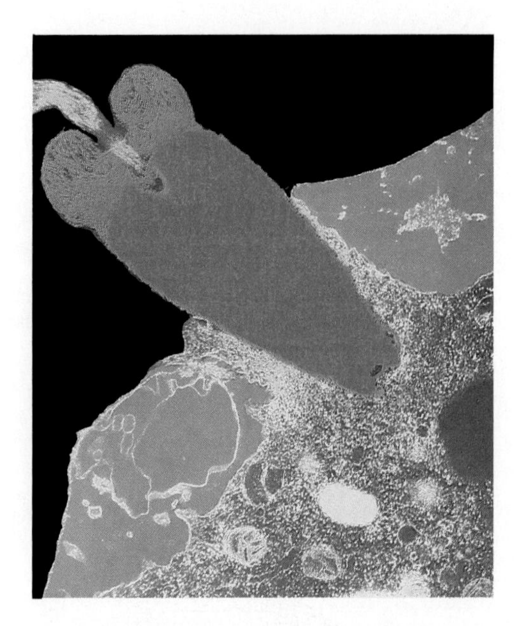

Synopsis of Icon and Color Use in Illustrations

The following symbols and colors are used in this text to help in illustrating structures, reactions, and biochemical principles

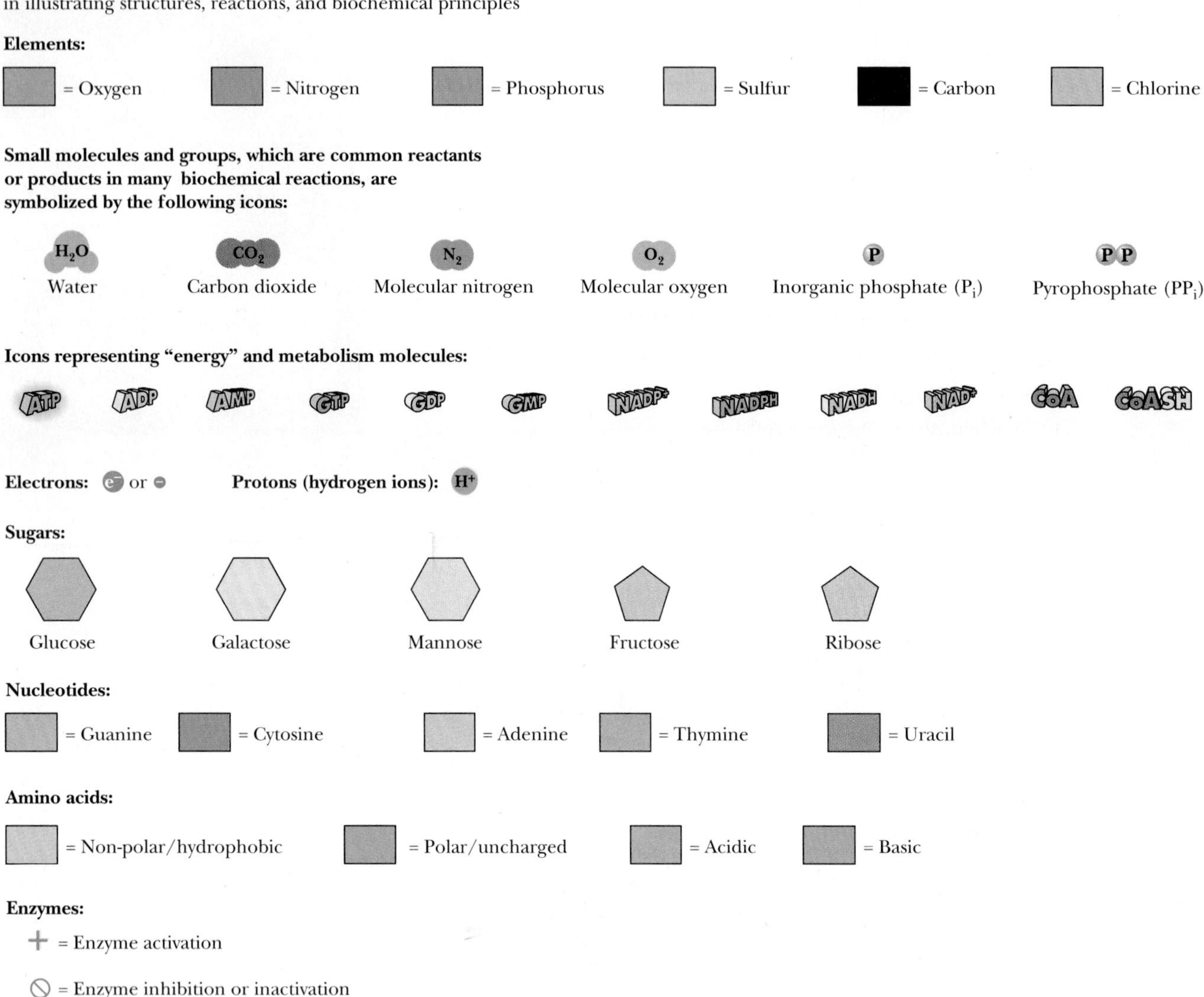

In reactions, blocks of color over parts of molecular structures are used so that discrete parts of the reaction can be easily followed from one intermediate to another and it is easy to see where the reactants originate and how the products are produced.

Some examples:

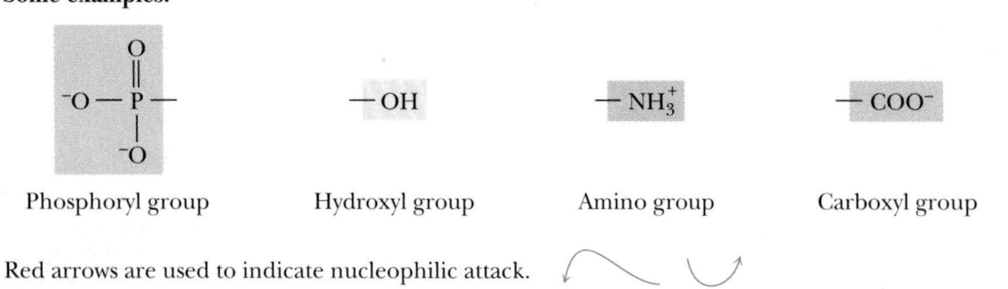

Red arrows are used to indicate nucleophilic attack.

These colors are internally consistent within reactions and are generally consistent within the scope of a chapter or treatment of a particular topic.

Part I

Molecular Components of Cells

Phosphorylase (Domain 2)

"Living things are composed of lifeless molecules."

A. Lehninger, *Biochemistry,* Worth Publishing (1970)

Chapter 1

Chemistry Is the Logic of Biological Phenomena

Organisms exhibit great diversity in form and function, yet share a common biochemistry.

Outline

1.1 Distinctive Properties of Living Systems

1.2 Biomolecules: The Molecules of Life

1.3 A Biomolecular Hierarchy: Simple Molecules Are the Units for Building Complex Structures

1.4 Properties of Biomolecules Reflect Their Fitness to the Living Condition

1.5 Organization and Structure of Cells

1.6 Viruses Are Supramolecular Assemblies Acting as Cell Parasites

Molecules are lifeless. Yet, in appropriate complexity and number, molecules compose living things. These living systems are distinct from the inanimate world because they have certain extraordinary properties. They can grow, move, perform the incredible chemistry of metabolism, respond to stimuli from the environment, and most significantly, replicate themselves with exceptional fidelity. The complex structure and behavior of living organisms veils the basic truth that their molecular constitution can be described and understood. The chemistry of the living cell resembles the chemistry of organic reactions. Indeed, cellular constituents or **biomolecules** must conform to the chemical and physical principles that govern all matter. Despite the spectacular diversity of life, the intricacy of biological structures, and the complexity of vital mechanisms, life functions are ultimately interpretable in chemical terms. *Chemistry is the logic of biological phenomena.*

logic: a system of reasoning, using principles of valid inference

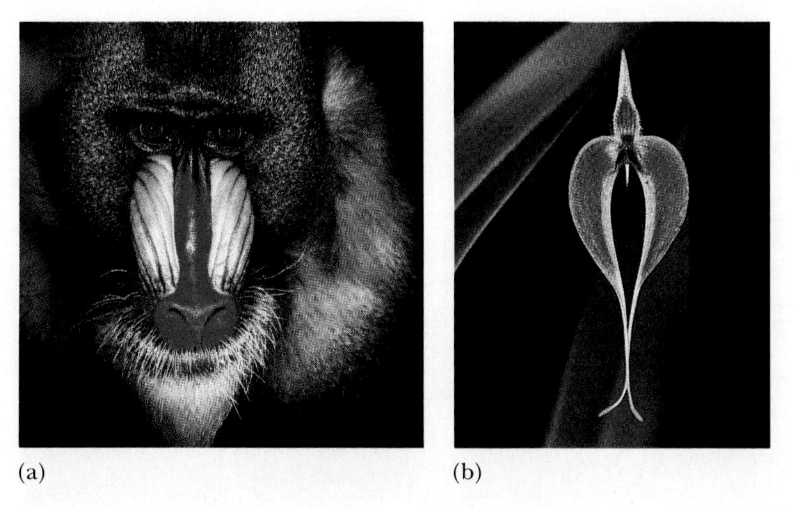

(a) (b)

***Figure* 1.1** (a) Mandrill (*Mandrillus sphinx*), a baboon native to West Africa. (b) Tropical orchid (*Bulbophyllum blumei*), New Guinea.

1.1 Distinctive Properties of Living Systems

First, the most obvious quality of **living organisms** is that they are *complicated and highly organized* (Figure 1.1). For example, organisms large enough to be seen with the naked eye are composed of many **cells,** typically of many types. In turn, these cells possess subcellular structures or **organelles,** which are complex assemblies of very large polymeric molecules or **macromolecules.** These macromolecules themselves show an exquisite degree of organization in their intricate three-dimensional architecture, even though they are composed of simple sets of chemical building blocks, such as sugars and amino acids. Indeed, the complex three-dimensional structure or **conformation** characteristic of macromolecules is a consequence of interactions between the monomeric units, according to their individual chemical properties.

Second, *biological structures serve functional purposes.* That is, biological structures have a role in terms of the organism's existence. From parts of organisms, such as limbs and organs, down to the chemical agents of metabolism, such as enzymes and metabolic intermediates, a purpose can be given for each component. Indeed, it is this functional characteristic of biological structures that separates the science of biology from studies of the inanimate world such as chemistry, physics, and geology. In biology, it is always appropriate to ask why observed structures, organizations, or patterns exist, that is, to ask what functional role they serve within the organism. It is nonsense to pose "why" questions about the properties of inanimate matter.

Third, *living systems are actively engaged in energy transformations.* The maintenance of the highly organized structure and activity of living systems is dependent upon their ability to extract energy from the environment. The ultimate source of energy is the sun. Solar energy flows from *photosynthetic organisms* (those organisms able to capture light energy by the process of photosynthesis) through food chains to herbivores and ultimately to carnivorous predators at the apex of the food pyramid (Figure 1.2). The biosphere is thus a system through which energy flows. Organisms capture some of this energy, be it from photosynthesis or the metabolism of food, by forming special energized biomolecules, of which **ATP** and **NADPH** are the two most prominent examples (Figure 1.3). ATP and NADPH are energized biomolecules because they represent chemically useful forms of stored energy. When these molecules react with other molecules in the cell, the energy released can be used to drive thermodynamically unfavorable processes. That is, ATP, NADPH, and related compounds are the power sources that drive the energy-

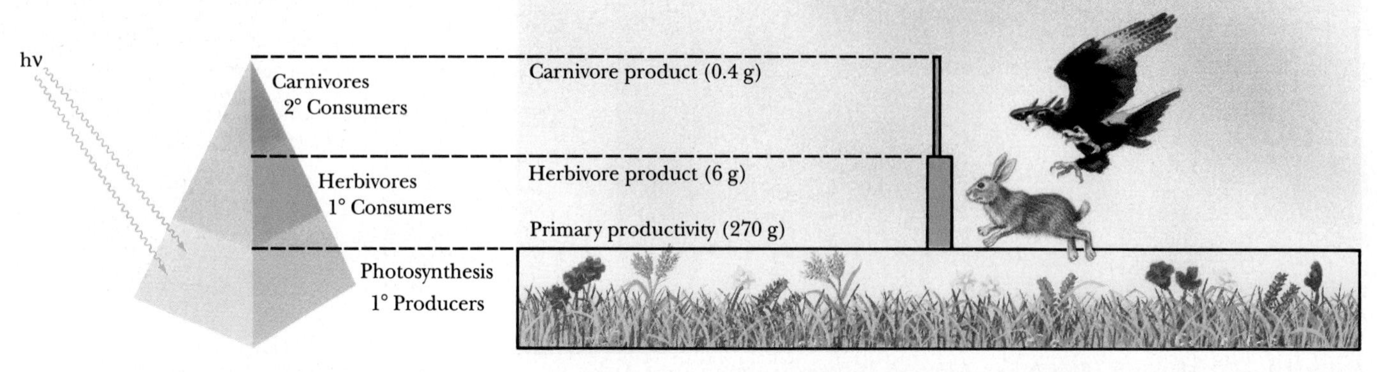

***Figure* 1.2** The food pyramid. Photosynthetic organisms at the base capture light energy. Herbivores and carnivores derive their energy ultimately from these primary producers.

requiring activities of the cell, including biosynthesis, movement, osmotic work against concentration gradients, and in special instances, light emission (bioluminescence). Only upon death does an organism reach equilibrium with its inanimate environment. *The living state is characterized by the flow of energy through the organism.* At the expense of this energy flow, the organism can maintain its intricate order and activity far removed from equilibrium with its surroundings, yet exist in a state of apparent constancy over time. This state of apparent constancy, or so-called **steady state,** is actually a very dynamic condition: energy and material are consumed by the organism and used to maintain its harmonious stability and order. In contrast, inanimate matter, as exemplified by the universe in totality, is moving to a condition of increasing disorder, or in thermodynamic terms, maximum entropy.

Fourth, *living systems have a remarkable capacity for self-replication.* Generation after generation, organisms reproduce virtually identical copies of themselves. This self-replication can proceed by a variety of mechanisms, ranging from simple division in bacteria to sexual reproduction in plants and animals, but in every case, it is characterized by an astounding degree of fidelity (Figure 1.4). Indeed, if the accuracy of self-replication were significantly greater, the evolution of organisms would be hampered because evolution depends

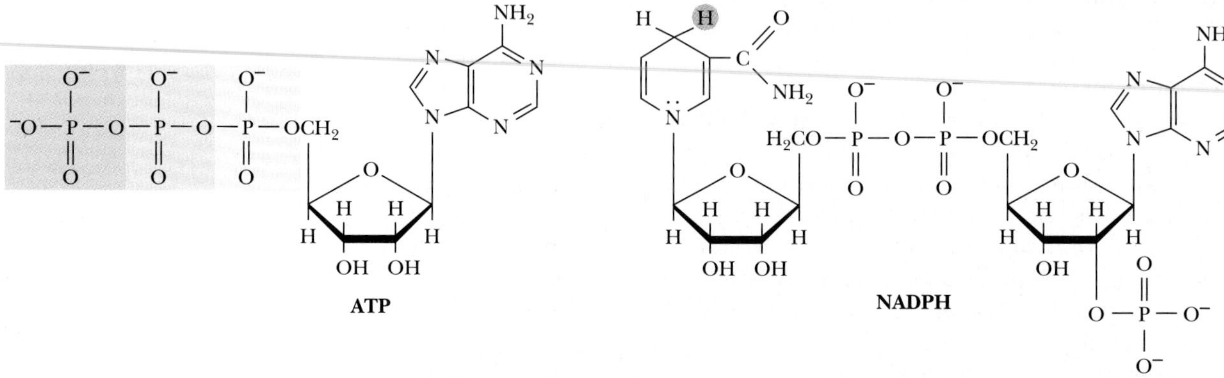

ATP

NADPH

***Figure* 1.3** ATP and NADPH, two biochemically important energy-rich compounds.

Figure **1.4** Organisms resemble their parents. (a) Reg Garrett with sons Robert and Jeffrey. (b) Orangutan with infant. (c) The Grishams: Rosemary, Emily, David, Andrew, and Charles.

upon natural selection operating on individual organisms that vary slightly in their adaptation to the environment. The fidelity of self-replication resides ultimately in the chemical nature of the genetic material. This substance consists of polymeric chains of deoxyribonucleic acid, or **DNA,** which are structurally complementary to one another (Figure 1.5). These molecules can generate new copies of themselves in a rigorously executed polymerization process that ensures a faithful reproduction of the original DNA strands. In contrast, the molecules of the inanimate world lack this capacity, with the trivial exception of crystal growth. A crude mechanism of replication, or specification of unique chemical structure according to some blueprint, must have existed at life's origin. This primordial system no doubt shared the property of **structural complementarity** (see later section) with the highly evolved patterns of replication prevailing today.

complementary: completing, making whole or perfect by combining or filling a deficiency

1.2 Biomolecules: The Molecules of Life

The elemental composition of living matter differs markedly from the relative abundance of elements in the earth's crust (Table 1.1). Hydrogen, oxygen, carbon, and nitrogen constitute more than 99% of the atoms in the human body, with most of the H and O occurring as H_2O. Oxygen, silicon, aluminum, and iron are the most abundant atoms in the earth's crust, with hydrogen, carbon, and nitrogen being relatively rare (less than 0.2% each). Nitrogen as dinitrogen (N_2) is the predominant gas in the atmosphere, and carbon dioxide (CO_2) is present at a level of 0.05%, a small but critical amount.

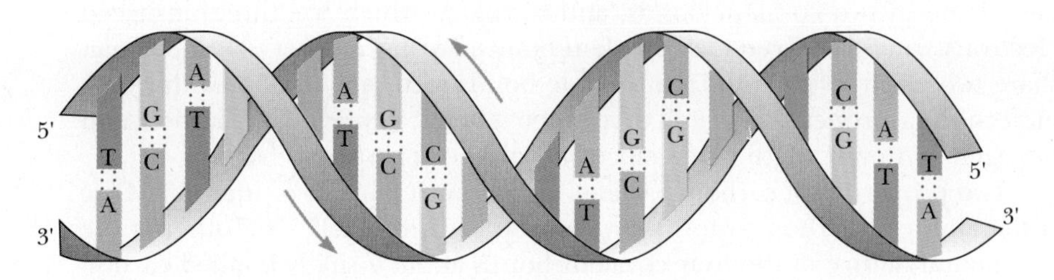

Figure **1.5** The DNA double helix. Two complementary polynucleotide chains running in opposite directions can pair through hydrogen bonding between their nitrogenous bases. Their complementary nucleotide sequences give rise to structural complementarity.

***Table* 1.1**

Composition of the Earth's Crust, Seawater, and the Human Body*

Earth's Crust		**Seawater**		**Human Body**†	
Element	%	Compound	m*M*	Element	%
O	47	Cl^-	548	H	63
Si	28	Na^+	470	O	25.5
Al	7.9	Mg^{2+}	54	C	9.5
Fe	4.5	SO_4^{2-}	28	N	1.4
Ca	3.5	Ca^{2+}	10	Ca	0.31
Na	2.5	K^+	10	P	0.22
K	2.5	HCO_3^-	2.3	Cl	0.08
Mg	2.2	NO_3^-	0.01	K	0.06
Ti	0.46	HPO_4^{2-}	<0.001	S	0.05
H	0.22			Na	0.03
C	0.19			Mg	0.01

*Figures for the earth's crust and the human body are presented as percentages of the total number of atoms; seawater data are millimoles per liter. Figures for the earth's crust do *not* include water, while figures for the human body do.

†Trace elements found in the human body serving essential biological functions include Mn, Fe, Co, Cu, Zn, Mo, I, Ni, and Se.

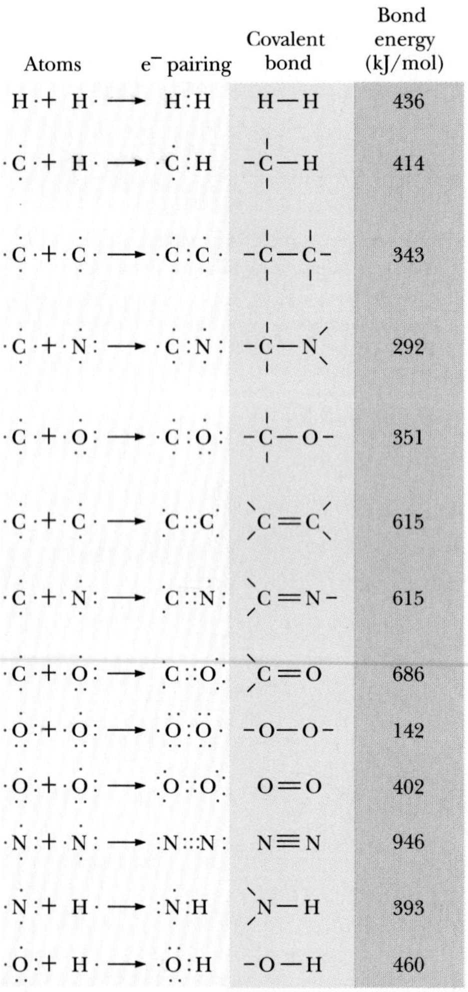

Atoms	e^- pairing	Covalent bond	Bond energy (kJ/mol)
H· + H· →	H:H	H—H	436
·C· + H· →	·C:H	—C—H	414
·C· + ·C· →	·C:C·	—C—C—	343
·C· + ·N: →	·C:N:	—C—N<	292
·C· + ·O: →	·C:O:	—C—O—	351
·C· + ·C· →	C::C	>C=C<	615
·C· + ·N: →	C::N:	>C=N—	615
·C· + ·O: →	C::O:	>C=O	686
·O: + ·O: →	·O:O·	—O—O—	142
·O: + ·O: →	:O::O:	O=O	402
·N: + ·N: →	:N:::N:	N≡N	946
·N: + H· →	:N:H	>N—H	393
·O: + H· →	·O:H	—O—H	460

***Figure* 1.6** Covalent bond formation by e^- pair sharing.

Oxygen is also abundant in the atmosphere and in the oceans. What property unites H, O, C, and N and renders these atoms so appropriate to the chemistry of life? It is their ability to form covalent bonds by electron-pair sharing. Furthermore, H, C, N, and O are among the lightest elements of the periodic table capable of forming such bonds (Figure 1.6). Since the strength of covalent bonds is inversely proportional to the atomic weights of the atoms involved, H, C, N, and O form the strongest covalent bonds. Two other covalent bond-forming elements, phosphorus (as phosphate ($—PO_4^{2-}$) derivatives) and sulfur, also play important roles in biomolecules.

Biomolecules Are Carbon Compounds

All biomolecules contain carbon. The prevalence of C is due to its unparalleled versatility in forming stable covalent bonds by electron-pair sharing. Carbon can form as many as four such bonds by sharing each of the four electrons in its outer shell with electrons contributed by other atoms. Atoms commonly found in covalent linkage to C are C itself, H, O, and N. Hydrogen can form one such bond by contributing its single electron to formation of an electron pair. Oxygen, with two unpaired electrons in its outer shell, can participate in two covalent bonds, and nitrogen, which has three unshared electrons, can form three such covalent bonds. Furthermore, C, N, and O can share two electron pairs to form double bonds with one another within biomolecules, a property that enhances their chemical versatility. Carbon and nitrogen can even share three electron pairs to form triple bonds.

Two properties of carbon covalent bonds merit particular attention. One is the ability of carbon to form covalent bonds with itself. The other is the tetrahedral nature of the four covalent bonds about a singly bonded carbon atom. Together these properties hold the potential for an incredible variety of linear, branched, and cyclic compounds of C whose diversity is further multiplied by the possibilities for including N, O, and H atoms in these compounds (Figure 1.7). We can therefore envision the ability of C to generate complex

LINEAR ALIPHATIC:

Stearic acid

$HOOC-(CH_2)_{16}-CH_3$

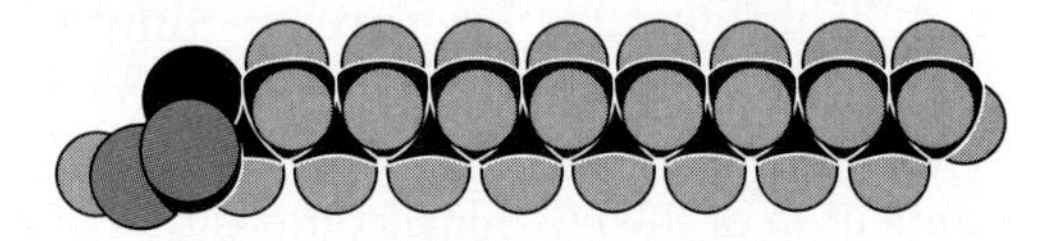

CYCLIC:

Cholesterol

BRANCHED:

β-carotene

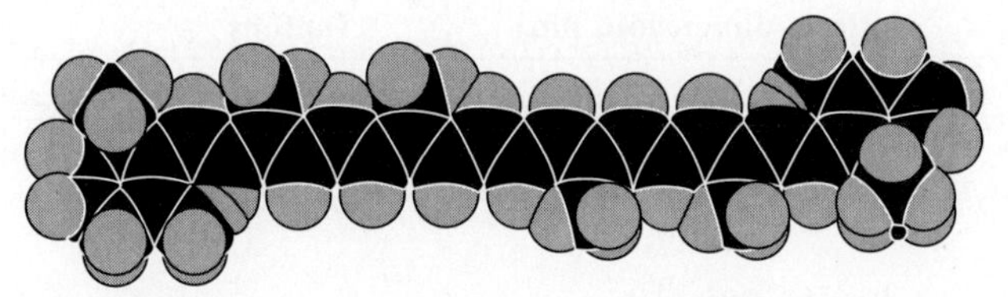

PLANAR:

Chlorophyll *a*

***Figure* 1.7** Examples of the versatility of C—C bonds in building complex structures: linear aliphatic, cyclic, branched, and planar.

structures in three dimensions. These structures, by virtue of appropriately included N, O, and H atoms, can display unique chemistries suitable to the living state. Thus, we may ask, is there any pattern or underlying organization that brings order to this astounding potentiality?

1.3 A Biomolecular Hierarchy: Simple Molecules Are the Units for Building Complex Structures

Examination of the chemical composition of cells reveals a dazzling variety of organic compounds covering a wide range of molecular dimensions (Table 1.2). As this complexity is sorted out and biomolecules are classified according to the similarity of their sizes and chemical properties, an organizational pattern emerges. The molecular constituents of living matter do not reflect randomly the infinite possibilities for combining carbon, hydrogen, oxygen, and nitrogen atoms. Instead, only a limited set of the many possibilities is found, and these collections share certain properties essential to the establishment and maintenance of the living state. We will first characterize the hier-

Table **1.2**

Biomolecular Dimensions

The dimensions of mass* and length for biomolecules are given typically in daltons and nanometers,† respectively. One dalton (D) is the mass of one hydrogen atom, 1.67×10^{-24} g. One nanometer (nm) is 10^{-9} m, or 10 Å (angstroms).

Biomolecule	Length (long dimension, nm)	Mass	
		Daltons	Picograms
Water	0.3	18	
Alanine	0.5	89	
Glucose	0.7	180	
Phospholipid	3.5	750	
Ribonuclease (a small protein)	4	12,600	
Immunoglobulin G (IgG)	14	150,000	
Myosin (a large muscle protein)	160	470,000	
Ribosome (bacteria)	18	2,520,000	
Bacteriophage ϕx174 (a very small bacterial virus)	25	4,700,000	
Pyruvate dehydrogenase complex (a multienzyme complex)	60	7,000,000	
Tobacco mosaic virus (a plant virus)	300	40,000,000	6.68×10^{-5}
Mitochondrion (liver)	1,500		1.5
Escherichia coli cell	2,000		2
Chloroplast (spinach leaf)	8,000		60
Liver cell	20,000		8,000

*Molecular mass is expressed in units of daltons (D) or kilodaltons (kD) in this book; alternatively, the dimensionless term *molecular weight*, symbolized by M_r and defined as the ratio of the mass of a molecule to 1 dalton of mass, is used.

†Prefixes used for powers of 10 are

10^{6}	mega M	10^{-3}	milli m
10^{3}	kilo k	10^{-6}	micro μ
10^{-1}	deci d	10^{-9}	nano n
10^{-2}	centi c	10^{-12}	pico p
		10^{-15}	femto f

archical relationships of biomolecules and then briefly consider what properties they possess that have rendered them so appropriate for the condition of life.

Metabolites and Macromolecules

The major precursors for the formation of biomolecules are water, carbon dioxide, and three inorganic nitrogen compounds—ammonium (NH_4^+), nitrate (NO_3^-), and dinitrogen (N_2). Metabolic processes assimilate and transform these inorganic precursors through ever more complex levels of biomolecular order (Figure 1.8). In the first step, precursors are converted to **metabolites,** simple organic compounds that are intermediates in cellular energy transformation and in the biosynthesis of various sets of **building blocks:** amino acids, sugars, nucleotides, fatty acids, and glycerol. By covalent linkage of these building blocks, the **macromolecules** are constructed: proteins, polysaccharides, polynucleotides (DNA and RNA), and lipids. (Strictly speaking, lipids contain relatively few building blocks and are therefore not really "macromolecules"; however, lipids are important contributors to higher levels of complexity.) Interactions among macromolecules lead to the next level of structural organization, **supramolecular complexes.** Here, various members of one or more of the classes of macromolecules come together to form specific assemblies serving important subcellular functions. Examples of these supramolecular assemblies are multifunctional enzyme complexes, ribosomes, chromosomes, and cytoskeletal elements. For example, a eukaryotic ribosome contains four different RNA molecules and at least 70 unique proteins. These supramolecular assemblies are an interesting contrast to their components because their structural integrity is maintained by noncovalent forces, not by covalent bonds. Hydrogen bonds, ionic attractions, van der Waals forces, and hydrophobic interactions between macromolecules maintain these assemblies in a highly ordered functional state. Although noncovalent forces are weak (less than 40 kJ/mol), they are numerous in these assemblies and thus can collectively maintain the essential architecture of the supramolecular complex under conditions of temperature, pH, and ionic strength that are consistent with cell life.

Organelles

The next higher rung in the hierarchical ladder is occupied by the **organelles,** entities of considerable dimensions compared to the cell itself. Organelles are found only in **eukaryotic cells,** that is, the cells of "higher" organisms (eukaryotic cells are described in Section 1.5). Several kinds, such as mitochondria and chloroplasts, are believed to have arisen from bacteria that gained entry to the eukaryotic cytoplasm. Organelles share two attributes: they are cellular inclusions, usually membrane bounded, and are dedicated to important cellular tasks. Organelles include the nucleus, mitochondria, chloroplasts, endoplasmic reticulum, Golgi apparatus, and vacuoles as well as other relatively small cellular inclusions, such as peroxisomes, lysosomes, and chromoplasts. The **nucleus** is the repository of genetic information as contained within the linear sequences of nucleotides in the DNA of chromosomes. **Mitochondria** are the "power plants" of cells by virtue of their ability to carry out the energy-releasing aerobic metabolism of carbohydrates and fatty acids with the concomitant capture of energy in metabolically useful forms such as ATP. **Chloroplasts** endow cells with the ability to carry out photosynthesis. They are the biological agents for harvesting light energy and transforming it into metabolically useful chemical forms.

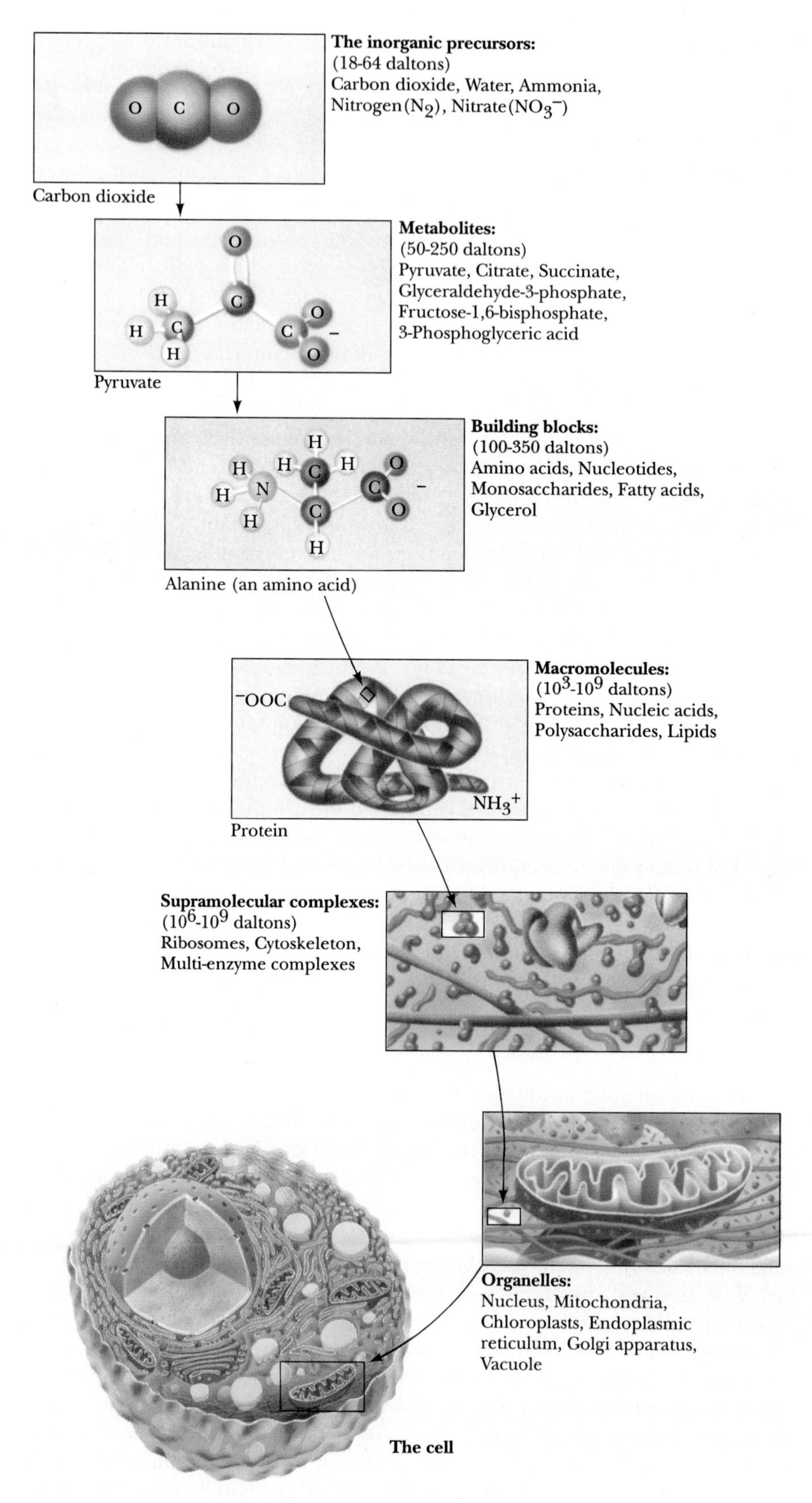

***Figure* 1.8** Molecular organization in the cell is a hierarchy.

Membranes

Membranes define the boundaries of cells and organelles. As such, they are not easily classified as supramolecular assemblies or organelles, although they share the properties of both. Membranes resemble supramolecular complexes in their construction since they are complexes of proteins and lipids maintained by noncovalent forces. **Hydrophobic interactions** are particularly important in maintaining membrane structure. These interactions reflect the tendency of nonpolar molecules to come together as they are excluded by the polar solvent, water. The spontaneous assembly of membranes in the aqueous environment where life arose and exists is the natural result of their hydrophobic ("water-fearing") character. Hydrophobic interactions are the creative means of membrane formation and the driving force that presumably established the boundary of the first cell. Membranes are the frontiers of cells that demarcate "self" from "nonself" for the living system. The membranes of organelles, such as nuclei, mitochondria, and chloroplasts, differ from one another, with each having a unique protein and lipid composition suited to the organelle's function. Furthermore, the creation of discrete volumes or **compartments** within cells (compartmentation) is not only an inevitable consequence of the presence of membranes but usually an essential condition for proper organellar function.

The Unit of Life Is the Cell

The **cell** is characterized as the unit of life, the smallest entity capable of displaying the attributes associated uniquely with the living state: growth, metabolism, stimulus response, and replication. In the previous discussions, we explicitly narrowed the infinity of chemical complexity potentially available to organic life, and we previewed an organizational arrangement, moving from simple to complex, that provides interesting insights into the functional and structural plan of the cell. Nevertheless, we find no obvious explanation within these features for the living attributes of the cell. Can we find other themes represented within biomolecules that are explicitly chemical yet anticipate or illuminate the living condition?

1.4 Properties of Biomolecules Reflect Their Fitness to the Living Condition

If we consider what attributes of biomolecules render them so fit as components of growing, replicating systems, several biologically relevant themes of structure and organization emerge. Furthermore, as we study biochemistry, we will see that these themes serve as principles of biochemistry. Prominent among them is *the necessity for information and energy in the maintenance of the living state.* This principle leads us to anticipate that some biomolecules must have the capacity to contain the information or "recipe" of life. Other biomolecules must have the capacity to translate this information so that the blueprint is transformed into the functional, organized structures essential to life. There must also exist an orderly mechanism for abstracting energy from the environment to drive the exquisitely dynamic, intricate processes that collectively constitute life. What properties of biomolecules endow them with the potential for such remarkable qualities?

Biological Macromolecules and Their Building Blocks Have a "Sense" or Directionality

The macromolecules of cells are built of units—amino acids in proteins, nucleotides in nucleic acids, and carbohydrates in polysaccharides—that have **structural polarity.** That is, these molecules are not symmetrical, and so they can be thought of as having a "head" and a "tail." Polymerization of these units to form macromolecules occurs by head-to-tail linear connections. Because of this, the polymer will also have a head and a tail, and hence, the macromolecule has a "sense" or direction to its structure (Figure 1.9).

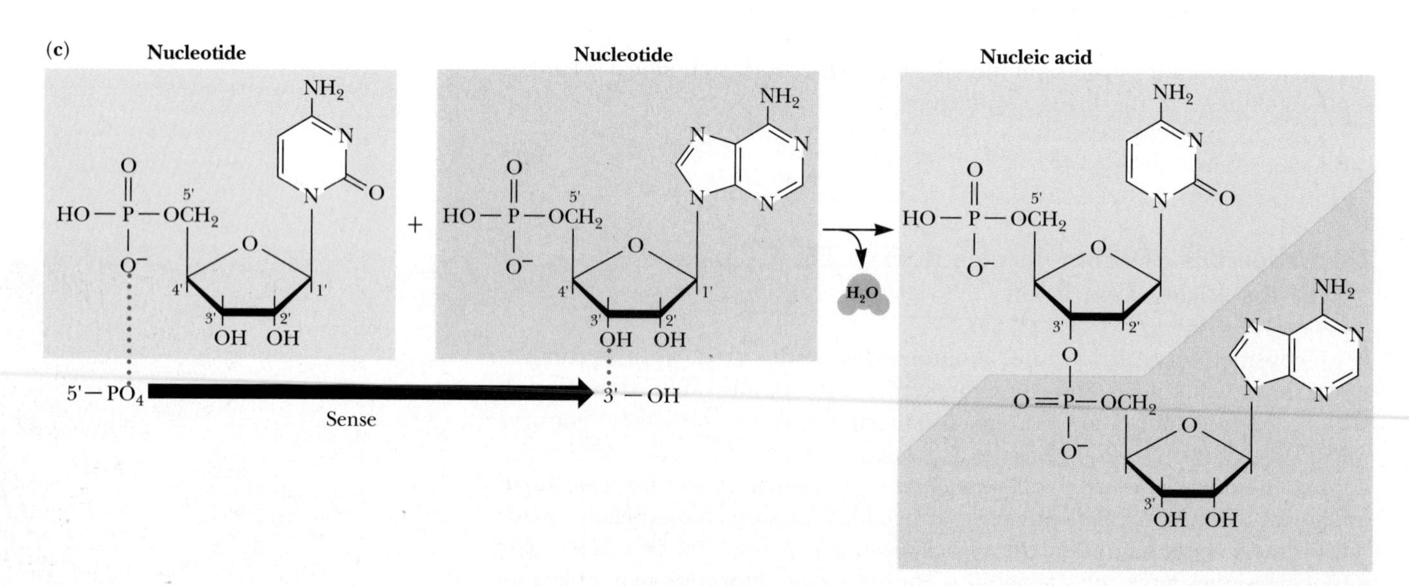

Figure **1.9** (a) Amino acids build proteins by connecting the α-carboxyl C atom of one amino acid to the α-amino N atom of the next amino acid in line. (b) Polysaccharides are built by combining the C-1 of one sugar to the C-4 O of the next sugar in the polymer. (c) Nucleic acids are polymers of nucleotides linked by bonds between the 3′-OH of the ribose ring of one nucleotide to the 5′-PO_4 of its neighboring nucleotide. All three of these polymerization processes involve bond formations accompanied by the elimination of water.

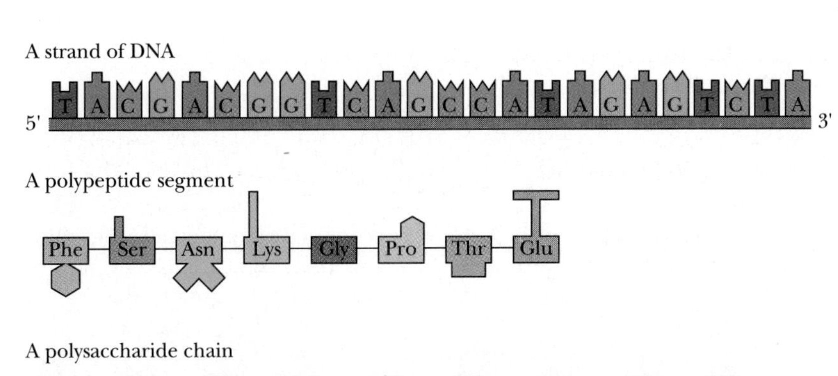

***Figure* 1.10** The sequence of monomeric units in a biological polymer has the potential to contain information if the diversity and order of the units are not overly simple or repetitive. Nucleic acids and proteins are information-rich molecules; polysaccharides are not.

Biological Macromolecules Are Informational

Because biological macromolecules have a sense to their structure, their component building blocks, when read along the length of the molecule, have the capacity to specify information in the same manner that the letters of the alphabet can form words when arranged in a linear sequence (Figure 1.10). Not all biological macromolecules are rich in information. Polysaccharides are often composed of the same sugar unit repeated over and over, as in cellulose or starch, which are homopolymers of many glucose units. Proteins and polynucleotides, in contrast, are typically composed of building blocks arranged in no obvious repetitive way; that is, their sequences are unique, akin to the letters and punctuation that form this descriptive sentence. In these unique sequences lies meaning. To discern the meaning, however, requires some mechanism for recognition.

Biomolecules Have Characteristic Three-Dimensional Architecture

The structure of any molecule is a unique and specific aspect of its identity. Molecular structure reaches its pinnacle in the intricate complexity of biological macromolecules, particularly the proteins. Although proteins are linear sequences of covalently linked amino acids, the course of the protein chain can turn, fold, and coil in the three dimensions of space to establish a specific, highly ordered architecture that is an identifying characteristic of the given protein molecule (Figure 1.11).

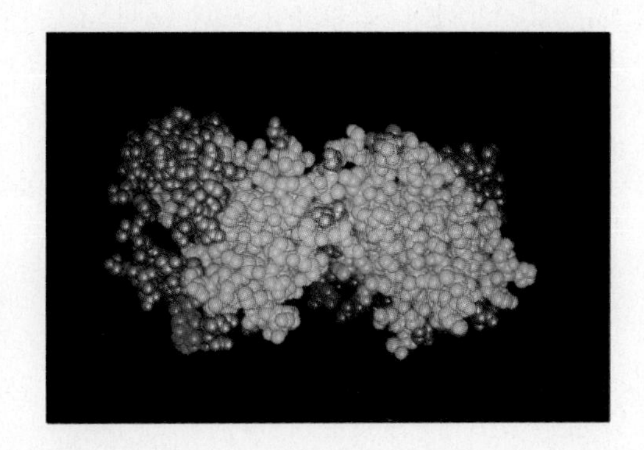

***Figure* 1.11** Three-dimensional representation of part of a protein molecule, the antigen-binding domain of immunoglobulin G (IgG). Immunoglobulin G is a major type of circulating antibody.

Weak Forces Maintain Biological Structure and Determine Biomolecular Interactions

Covalent bonds hold atoms together so that molecules are formed. **Weak chemical forces,** which include hydrogen bonds, van der Waals forces, ionic bonds, and hydrophobic interactions, are intramolecular or intermolecular attractions between atoms. None of these forces, which typically range from 4 to 30 kJ/mol, are strong enough to bind free atoms together (Table 1.3). The average kinetic energy of molecules at 25°C is 2.5 kJ/mol, so the energy of weak forces is only several times greater than the dissociating tendency due to thermal motion of molecules. Thus, these weak forces create bonds that are constantly forming and breaking at physiological temperature, unless by cumulative number they impart stability to the structures generated by their

Table **1.3**

Weak Chemical Forces and Their Relative Strengths and Distances

Force	Strength (kJ/mol)	Distance (nm)	Description
Van der Waals interactions	0.4–4.0	0.2	Strength depends on the relative size of the atoms or molecules and the distance between them. The size factor determines the area of contact between two molecules: the greater the area, the stronger the interaction. Attractive force is inversely proportional to the sixth power of the distance, *r*, separating two atoms or molecules: $F \approx 1/r^6$.
Hydrogen bonds	12–30	0.3	Relative strength is proportional to the polarity of the H bond donor and H bond acceptor. More polar atoms form stronger H bonds.
Ionic bonds	20	0.25	Strength also depends on the relative polarity of the interacting charged species. Some ionic bonds are also H bonds: $—NH_3^+ \cdots {}^-OOC—$
Hydrophobic interactions	<40	—	Force is a complex phenomenon determined by the degree to which the structure of water is disordered as discrete hydrophobic molecules or molecular regions coalesce.

collective action. These weak forces merit further discussion because their attributes profoundly influence the nature of the biological structures they build.

Van der Waals Attractive Forces

Van der Waals forces are the result of induced electrical interactions between closely approaching atoms or molecules as their negatively charged electron clouds fluctuate instantaneously in time. These fluctuations allow attractions to occur between the positively charged nuclei and the electron density of the incoming atom. The limit of approach of two atoms is determined by the sum of their van der Waals radii (Table 1.4), and the strength of van der Waals forces varies inversely with the sixth power of the distance between atoms. Therefore, van der Waals attractions operate only over a limited interatomic distance and are an effective bonding interaction at physiological temperatures only when a number of atoms in a molecule can interact with several atoms in a neighboring molecule. For this to occur, the atoms on interacting molecules must pack together neatly. That is, their molecular surfaces must possess a degree of structural complementarity (Figure 1.12).

Hydrogen Bonds

Hydrogen bonds form between a hydrogen atom covalently bonded to an electronegative atom (such as oxygen or nitrogen) and a second electronegative atom that serves as the hydrogen bond acceptor. Several important biological examples are given in Figure 1.13. Hydrogen bonds, at a strength of 12 to 30 kJ/mol, are stronger than van der Waals forces and have an additional property: H bonds tend to be highly directional, forming straight bonds between donor, hydrogen, and acceptor atoms. Hydrogen bonds are also more specific than van der Waals interactions because they require the presence of complementary hydrogen donor and acceptor groups.

Table **1.4**
Radii of the Common Atoms of Biomolecules

Atom	Van der Waals radius, nm	Covalent radius, nm	Atom represented to scale
H	0.1	0.037	
C	0.17	0.077	
N	0.15	0.070	
O	0.14	0.066	
P	0.19	0.096	
S	0.185	0.104	
Half-thickness of an aromatic ring	0.17	—	

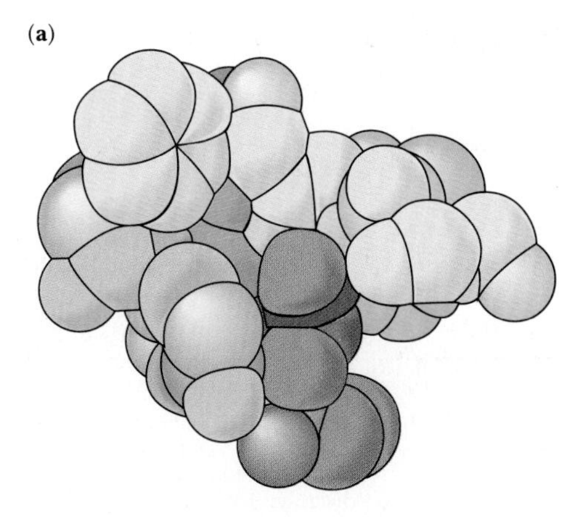

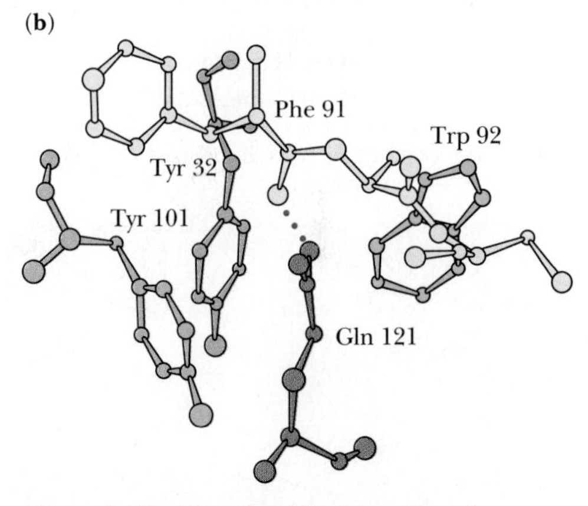

***Figure* 1.12** Van der Waals packing is enhanced in molecules that are structurally complementary. Gln^{121} represents a surface protuberance on the protein lysozyme. This protuberance fits nicely within a pocket (formed by Tyr^{101}, Tyr^{32}, Phe^{91}, and Trp^{92}) in the antigen-binding domain of an antibody raised against lysozyme. (See also Figure 1.15.) (a) A space-filling representation. (b) A ball-and-stick model. *(From Science 233:751 (1986), figure 5.)*

Ionic Bonds

Ionic bonds are the result of attractive forces between oppositely charged polar functions, such as negative carboxyl groups and positive amino groups (Figure 1.14). These electrostatic forces average about 20 kJ/mol in aqueous solutions. Because the electrical charge is typically radially distributed, these bonds can be considered to lack the directionality of hydrogen bonds or the precise fit of van der Waals interactions. Nevertheless, since the opposite charges are restricted to sterically defined positions, ionic bonds can impart a high degree of structural specificity.

Hydrophobic Interactions

Hydrophobic interactions are due to the strong tendency of water to exclude nonpolar groups or molecules (see Chapter 2). Hydrophobic interactions arise not so much because of any intrinsic affinity of nonpolar substances for one another (although van der Waals forces do promote the weak bonding of nonpolar substances), but because water molecules prefer the stronger interactions that they share with one another. Since the strongest chemical interaction that is possible between two molecules actually determines what occurs, the preferential hydrogen bonding interactions between polar water molecules excludes nonpolar groups. It is this exclusion that drives the tendency of nonpolar substances to cluster in aqueous solution. Thus, nonpolar regions of biological macromolecules are often buried in the molecule's interior to exclude them from the aqueous milieu. The formation of oil droplets as hydro-

milieu: the environment or surroundings; from the French *mi* meaning "middle" and *lieu* meaning "place" phobic nonpolar lipid molecules coalesce upon addition to water is an approximation of this phenomenon. These tendencies have important consequences in the creation and maintenance of the macromolecular structures and supramolecular assemblies of living cells.

Structural Complementarity Determines Biomolecular Interactions

ligand: something that binds; a molecule that is bound to another molecule; from the Latin *ligare* meaning "to bind"

Structural complementarity is the means of recognition in biomolecular interactions. The complicated and highly organized patterns of life are dependent upon the ability of biomolecules to recognize and interact with one another in very specific ways to accomplish the ends of metabolism, growth, replication, and other vital processes. The interaction of one molecule with another, a protein with a metabolite, for example, can be most precisely achieved if the structure of one is complementary to the structure of the other, as in two connecting pieces of a puzzle or, in the more popular analogy for macromolecules and their **ligands,** a lock and its key (Figure 1.15). This principle of **structural complementarity** is the very essence of biomolecular recognition. *Structural complementarity is the significant clue to understanding the functional properties of biological systems.* Biological systems from the macromolecular level to the cellular level operate via specific molecular recognition mechanisms based on structural complementarity: a protein recognizes its specific metabolite, a strand of DNA recognizes its complementary strand, sperm recognize an egg. All these interactions involve structural complementarity between molecules.

Bonded atoms	Approximate bond length*
O—H---O	0.27 nm
O—H---O⁻	0.26 nm
O—H---N	0.29 nm
N—H---O	0.30 nm
N⁺—H---O	0.29 nm
N—H---N	0.31 nm

*Lengths given are distances from the atom covalently linked to the H to the atom H-bonded to the hydrogen:

O—H---O
|←0.27 nm→|

Functional groups which are important H bond donors and acceptors:

Donors | Acceptors

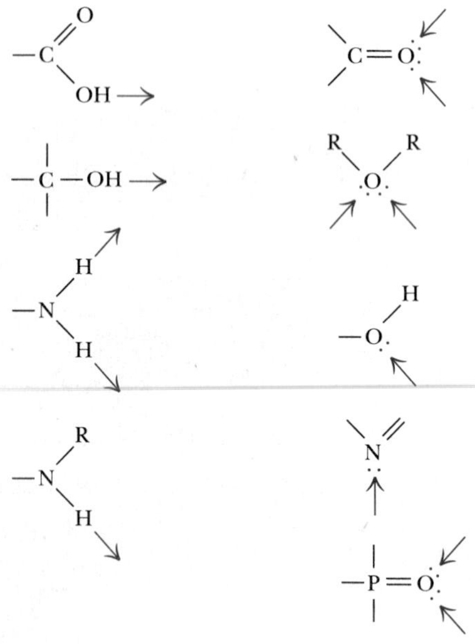

Figure 1.13 Biologically important H bonds and functional groups that are important H bond donors and acceptors.

Biomolecular Recognition Is Mediated by Weak Chemical Forces

The biomolecular recognition events that occur through structural complementarity are mediated by the weak chemical forces previously discussed. It is important to realize that, since these interactions are sufficiently weak, they are readily reversible under physiological conditions. Consequently, biomolecular interactions tend to be transient; rigid, static lattices of biomolecules that might paralyze cellular activities are not formed. Instead, there is a dynamic interplay between metabolites and macromolecules, hormones and receptors, and all the other participants in the panoply of life processes that operate by the principle of structural complementarity. This interplay is initiated upon specific recognition between complementary molecules and ultimately culminates in unique physiological activities. *Biological function is achieved through mechanisms based on structural complementarity and weak chemical interactions.*

This principle of structural complementarity extends to higher interactions essential to the establishment of the living condition. For example, the formation of supramolecular complexes occurs because of recognition and interaction between their various macromolecular components as governed by the weak forces formed between them. If a sufficient number of weak bonds can be formed, as in macromolecules complementary in structure to one another, larger structures will assemble spontaneously. The tendency for nonpolar molecules and parts of molecules to come together through hydrophobic interactions also promotes the formation of supramolecular assemblies. Very complex subcellular structures are actually spontaneously formed in an assembly process that is driven by weak forces accumulated through structural complementarity.

Magnesium ATP

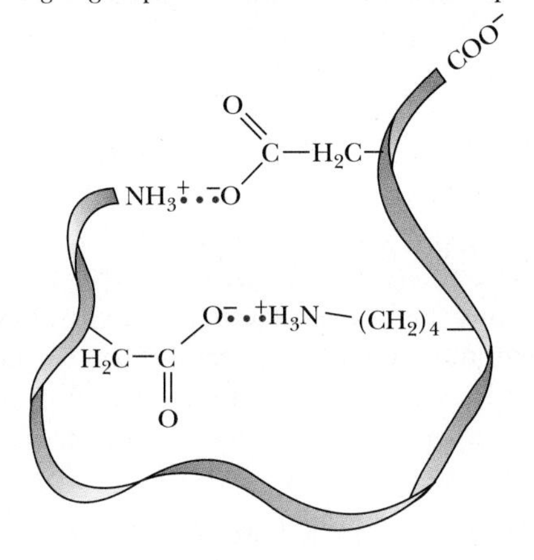

Intramolecular ionic bonds between oppositely charged groups on amino acid residues in a protein

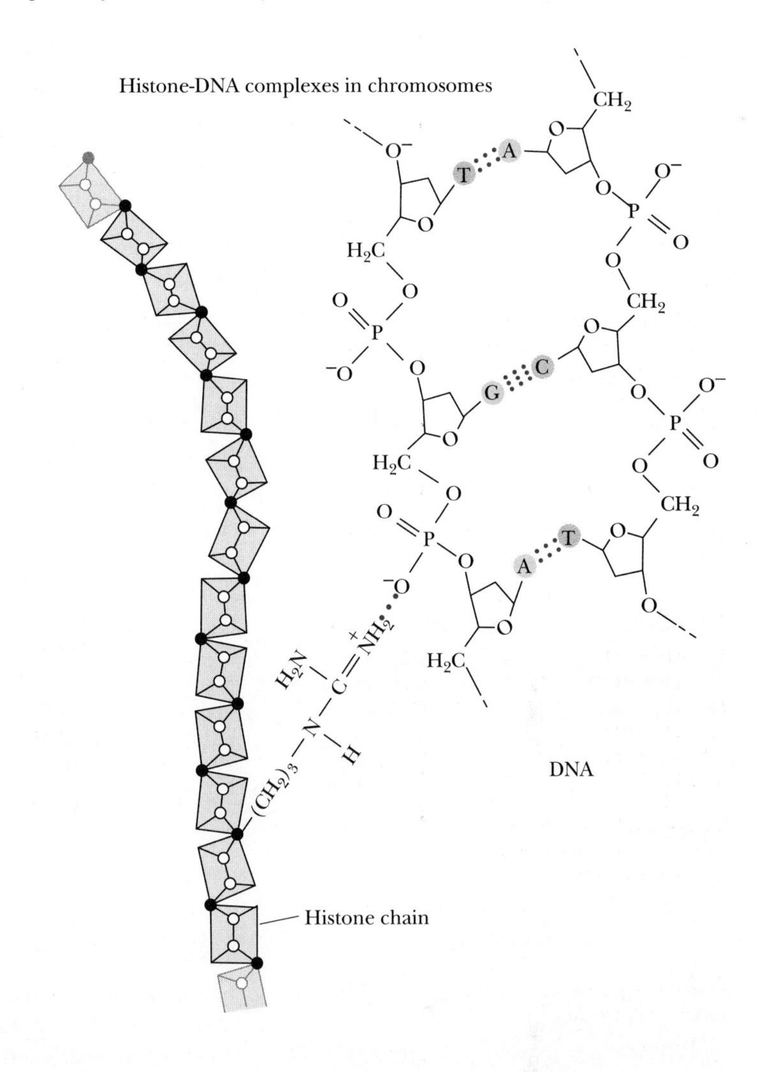

***Figure* 1.14** Ionic bonds in biological molecules.

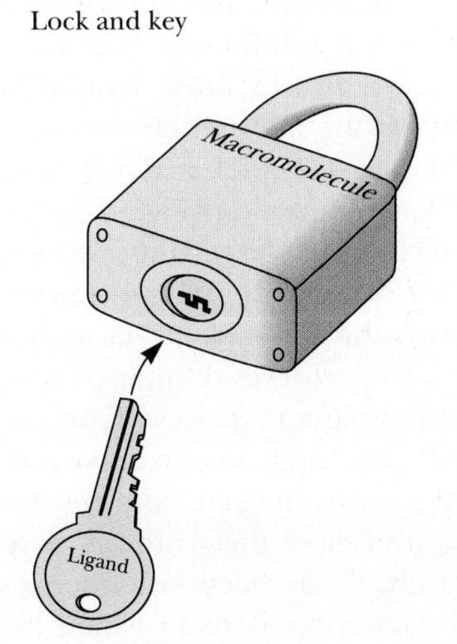

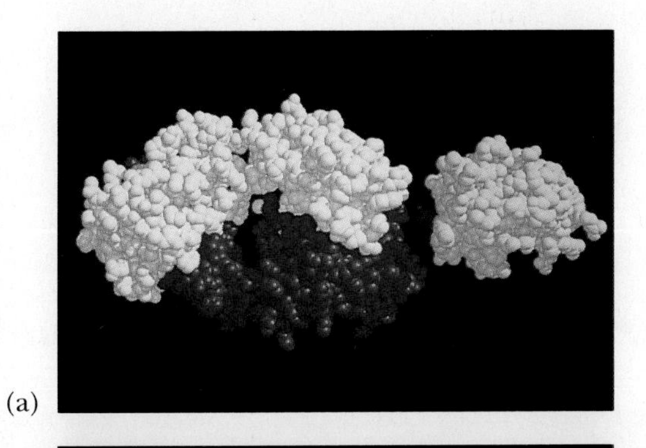

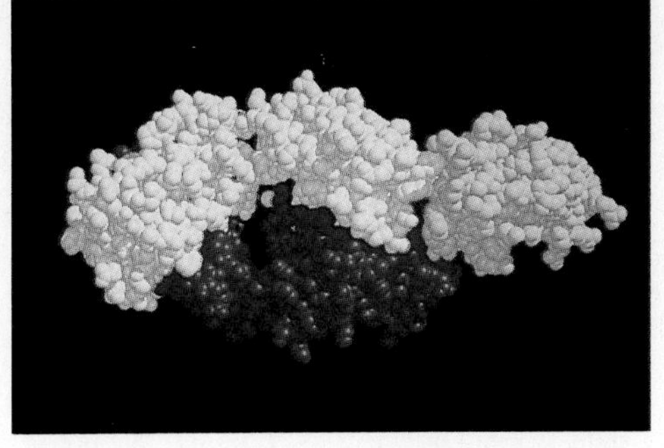

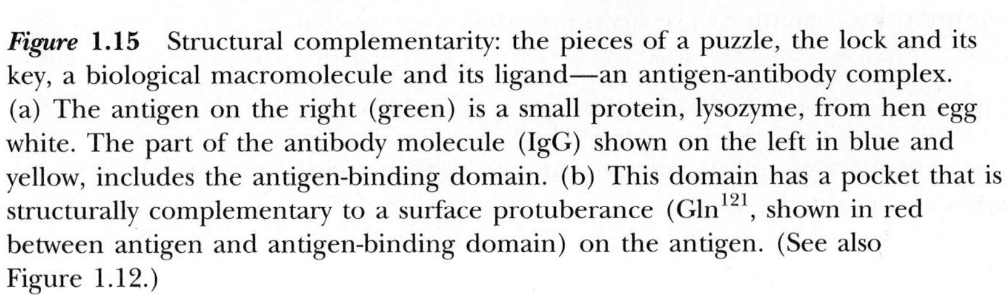

***Figure* 1.15** Structural complementarity: the pieces of a puzzle, the lock and its key, a biological macromolecule and its ligand—an antigen-antibody complex. (a) The antigen on the right (green) is a small protein, lysozyme, from hen egg white. The part of the antibody molecule (IgG) shown on the left in blue and yellow, includes the antigen-binding domain. (b) This domain has a pocket that is structurally complementary to a surface protuberance (Gln^{121}, shown in red between antigen and antigen-binding domain) on the antigen. (See also Figure 1.12.)

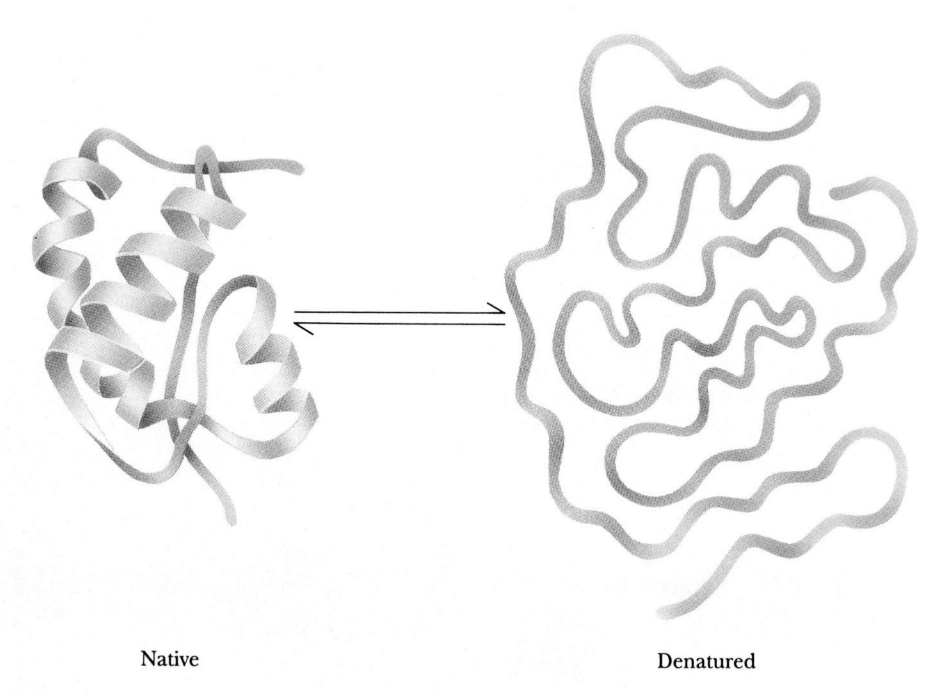

***Figure* 1.16** Denaturation and renaturation of the intricate structure of a protein.

Weak Forces Restrict Organisms to a Narrow Range of Environmental Conditions

The central role of weak forces in biomolecular interactions restricts living systems to a narrow range of environmental conditions. Energy transformations in cells represent another instance in which the range of chemical possibilities is restricted to conform with the unique requirements of life. Biological macromolecules are functionally active only within a narrow range of environmental conditions, such as temperature, ionic strength, and relative acidity. Extremes of these conditions disrupt the weak forces essential to maintaining the intricate structure of macromolecules. The loss of structural order in these complex macromolecules, so-called **denaturation,** is accompanied by loss of function (Figure 1.16). As a consequence, cells cannot tolerate reactions in which large amounts of energy are released. Nor can they generate a large energy burst to drive energy-requiring processes. Instead, such transformations take place via sequential series of chemical reactions whose overall effect achieves dramatic energy changes, even though any given reaction in the series proceeds with only modest input or release of energy (Figure 1.17). These sequences of reactions are organized to provide for the release of useful energy to the cell from the breakdown of food or to take such energy and use it to drive the synthesis of biomolecules essential to the living state. Collectively, these reaction sequences constitute cellular **metabolism:** the ordered reaction pathways by which cellular chemistry proceeds and biological energy transformations are accomplished.

Enzymes

The sensitivity of cellular constituents to environmental extremes places another constraint on the reactions of metabolism. The rate at which cellular reactions proceed is a very important factor in maintenance of the living state. However, the common ways chemists accelerate reactions are not available to

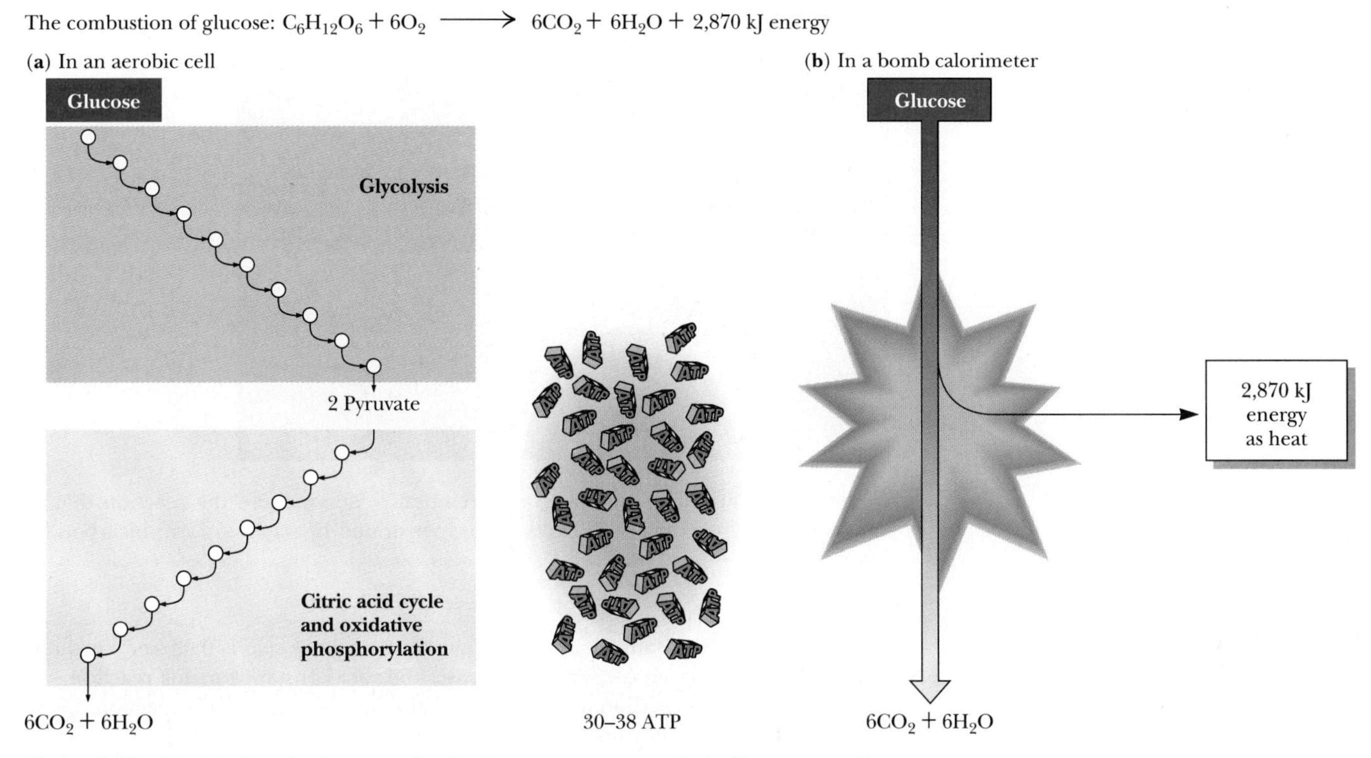

***Figure* 1.17** Metabolism is the organized release or capture of small amounts of energy in processes whose overall change in energy is large. (a) For example, the combustion of glucose by cells is a major pathway of energy production, with the energy captured appearing as 30 to 38 equivalents of ATP, the principal energy-rich chemical of cells. The ten reactions of glycolysis, the nine reactions of the citric acid cycle, and the successive linked reactions of oxidative phosphorylation release the energy of glucose in a stepwise fashion and the small "packets" of energy appear in ATP. (b) Combustion of glucose in a bomb calorimeter results in an uncontrolled, explosive release of energy in its least useful form, heat.

cells; the temperature cannot be increased, acid or base cannot be added, the pressure cannot be raised, and concentrations cannot be markedly changed. Instead, proteinaceous catalysts mediate cellular reactions by accelerating their rates many orders of magnitude and by ensuring the selectivity or specificity of the substances undergoing reaction. These catalysts are called **enzymes,** and virtually every metabolic reaction is served by an enzyme whose sole biological purpose is to catalyze its specific reaction (Figure 1.18).

Metabolic Regulation Is Achieved by Controlling the Activity of Enzymes

Thousands of reactions mediated by an equal number of enzymes are occurring at any given instant within the cell. Since metabolism has many branch points, cycles, and interconnections, as a glance at a metabolic pathway map will reveal (Figure 1.19), the need for metabolic regulation is obvious. All of these reactions, many of which are at apparent cross-purposes in the cell, must be fine-tuned and integrated so that metabolism and, in turn, life proceed harmoniously. This metabolic regulation is achieved through controls on enzyme activity so that the rates of cellular reactions are appropriate to cellular requirements.

Despite the organized pattern of metabolism and the thousands of enzymes required, cellular reactions nevertheless conform to the same thermodynamic principles that govern any chemical reaction. Enzymes have no influence over energy changes (the thermodynamic component) in their

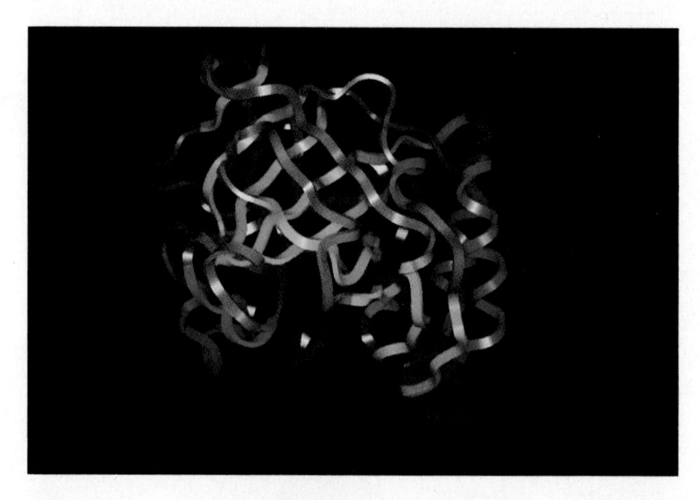

Figure **1.18** Carbonic anhydrase, a representative enzyme and the reaction that it catalyzes. Dissolved carbon dioxide is slowly hydrated by water to form bicarbonate ion and H^+:

$$CO_2 + H_2O \rightleftharpoons HCO_3^- + H^+$$

At 20°C, the rate constant for this uncatalyzed reaction, k_{uncat}, is 0.03/sec. In the presence of the enzyme carbonic anhydrase, the rate constant for this reaction, k_{cat}, is 10^6/sec. Thus carbonic anhydrase accelerates the rate of this reaction 3.3×10^7 times. Carbonic anhydrase is a 29-kD protein.

reactions. They only influence reaction rate. Thus, cells are systems that take in food, release waste, and carry out complex degradative and biosynthetic reactions essential to their survival while operating under conditions of essentially constant temperature and pressure and maintaining a constant internal environment **(homeostasis)** with no outwardly apparent changes. *Cells are open thermodynamic systems exchanging matter and energy with their environment and functioning as highly regulated isothermal chemical engines.*

1.5 Organization and Structure of Cells

All living cells fall into one of two broad categories—**prokaryotic** or **eukaryotic.** The distinction is based on whether or not the cell has a nucleus. Prokaryotes are single-celled organisms that lack nuclei and other organelles; the word is derived from *pro* meaning "prior to" and *karyote* meaning "nucleus." In conventional biological classification schemes, prokaryotes are grouped together as members of the kingdom Monera, represented by bacteria and cyanobacteria (formerly called blue-green algae). The other four living kingdoms are all eukaryotes—the single-celled Protists, such as amoebae, and all multicellular life forms, including the Fungi, Plant, and Animal kingdoms. Eukaryotic cells have true nuclei and other organelles such as mitochondria, with the prefix *eu* meaning "true."

Early Evolution of Cells

Until recently, most biologists accepted the idea that eukaryotes evolved from the simpler prokaryotes in some linear progression from simple to complex over the course of geological time. Contemporary evidence favors the view that present-day organisms are better grouped into three classes or lineages: eukaryotes and two prokaryotic groups, the **eubacteria** and the **archaebacteria** (more recently designated as **archaea**), no one of which appears to be the progenitor of any other. All are believed to have evolved approximately 3.5

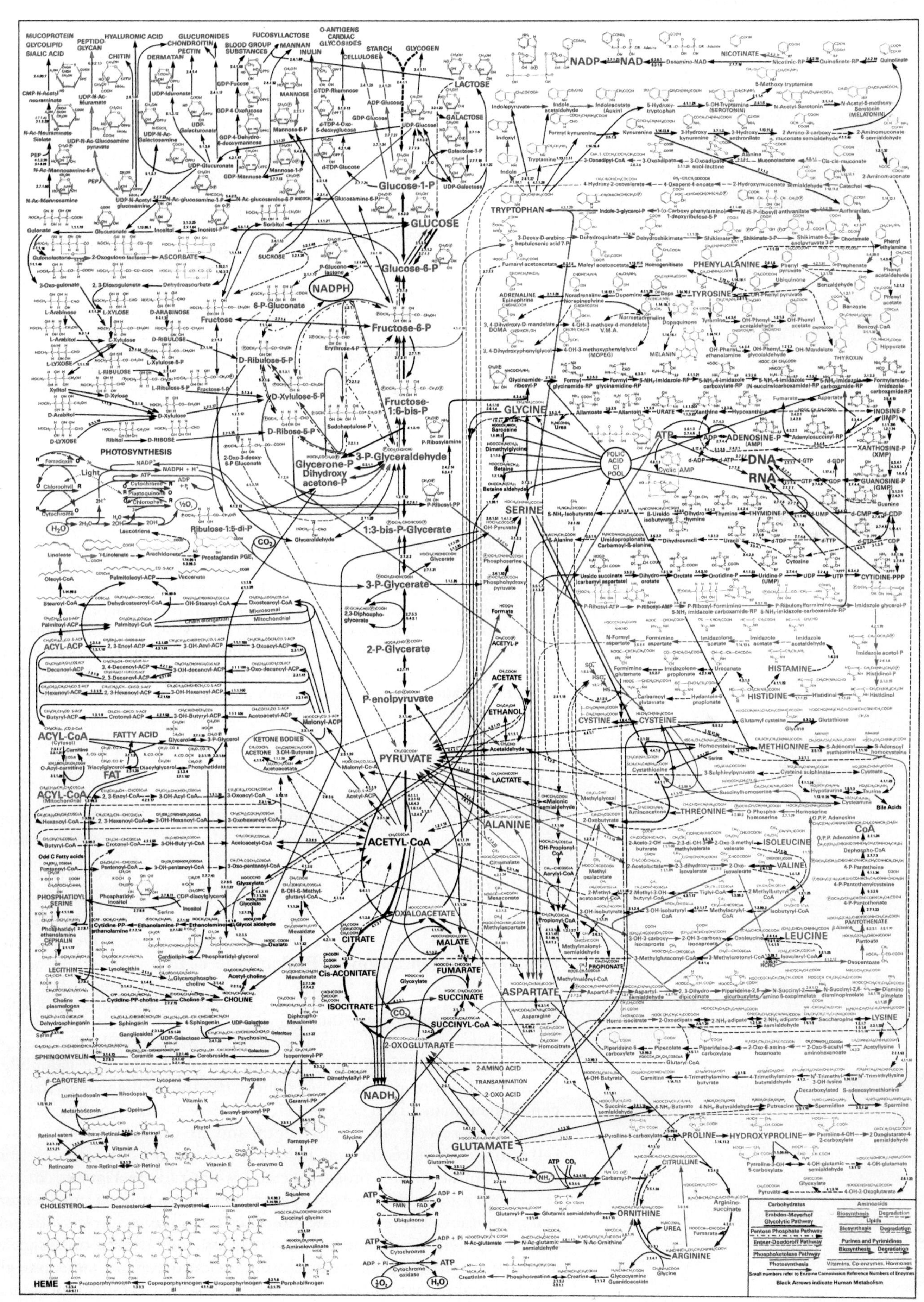

Figure **1.19** Reproduction of a metabolic map.

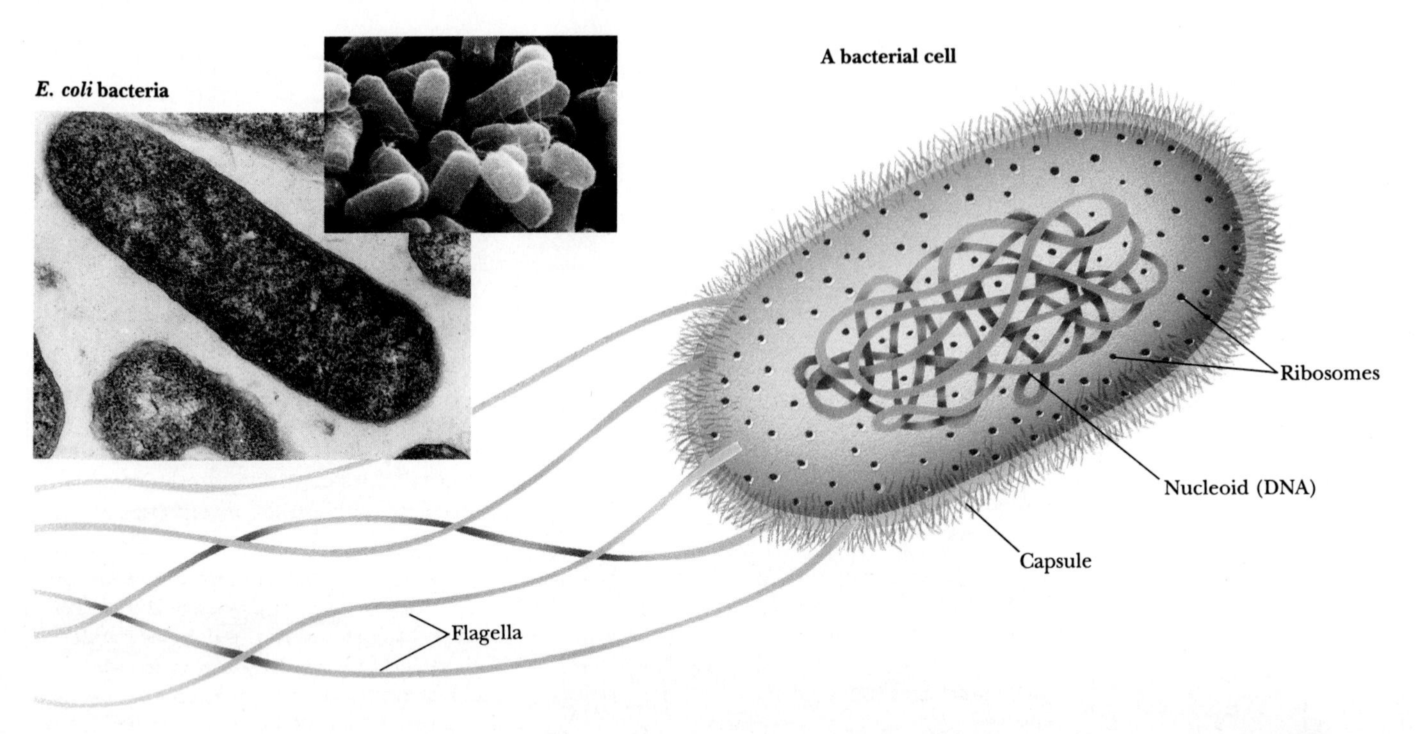

Figure **1.20** This bacterium is *Escherichia coli,* a member of the coliform group of bacteria that colonize the intestinal tract of humans. *E. coli* have rather simple nutritional requirements. They grow and multiply quite well if provided with a simple carbohydrate source of energy (such as glucose), ammonium ions as a source of nitrogen, and a few mineral salts. The simple nutrition of this "lower" organism means that its biosynthetic capacities must be quite advanced. When growing at 37°C on a rich organic medium, *E. coli* cells divide every 20 minutes. Subcellular features include the cell wall, plasma membrane, nuclear region, ribosomes, storage granules, and cytosol (Table 1.5).

billion years ago from a common ancestral form called the **progenote.** It is now understood that eukaryotic cells are, in reality, composite cells derived from various prokaryotic contributions. Thus, the dichotomy between prokaryotic cells and eukaryotic cells, though useful, is an artificial distinction.

Structural Organization of Prokaryotic Cells

Among prokaryotes (the simplest cells), eubacteria are the most numerous and widely spread group; the archaea are found only in unusual environments where other cells cannot survive. Archaea include the **thermoacidophiles** (heat- and acid-loving bacteria) of hot springs, the **halophiles** (salt-loving bacteria) of salt lakes and ponds, and the **methanogens** (bacteria that generate methane from CO_2 and H_2). Prokaryotes are typically very small, on the order of several microns in length, and are usually surrounded by a rigid **cell wall** that protects the cell and gives it its shape. The characteristic structural organization of a prokaryotic cell is depicted in Figure 1.20.

Prokaryotic cells have only a single membrane, the **plasma membrane** or **cell membrane.** Since they have no other membranes, prokaryotic cells contain no nucleus or organelles. Nevertheless, they possess a distinct **nuclear area** where a single circular chromosome is localized, and some have an internal membranous structure called a **mesosome** that is derived from and con-

Table 1.5
Major Features of Prokaryotic Cells

Structure	Molecular Composition	Function
Cell wall	Peptidoglycan: a rigid framework of polysaccharide cross-linked by short peptide chains. Some bacteria possess a lipopolysaccharide- and protein-rich outer membrane.	Mechanical support, shape, and protection against swelling in hypotonic media. The cell wall is a porous nonselective barrier allowing most small molecules to pass.
Cell membrane	The cell membrane is composed of about 45% lipid and 55% protein. The lipids form a bilayer that is a continuous nonpolar hydrophobic phase in which the proteins are embedded.	The cell membrane is a highly selective permeability barrier that controls the entry of most substances into the cell. Important enzymes in the generation of cellular energy are located in the membrane.
Nuclear area or nucleoid	The genetic material is a single tightly coiled DNA molecule 2 nm in diameter but over 1 mm in length (molecular mass of *E. coli* DNA is 3×10^9 daltons; 4.72×10^6 nucleotide pairs).	DNA is the blueprint of the cell, the repository of the cell's genetic information. During cell division, each strand of the double-stranded DNA molecule is replicated to yield two double-helical daughter molecules. Messenger RNA (mRNA) is transcribed from DNA to direct the synthesis of cellular proteins.
Ribosomes	Bacterial cells contain about 15,000 ribosomes. Each is composed of a small (30S) subunit and a large (50S) subunit. The mass of a single ribosome is 2.3×10^6 daltons. It consists of 65% RNA and 35% protein.	Ribosomes are the sites of protein synthesis. The mRNA binds to ribosomes, and the mRNA nucleotide sequence specifies the protein that is synthesized.
Storage granules	Bacteria contain granules that represent storage forms of polymerized metabolites such as sugars or β-hydroxybutyric acid.	When needed as metabolic fuel, the monomeric units of the polymer are liberated and degraded by energy-yielding pathways in the cell.
Cytosol	Despite its amorphous appearance, the cytosol is now recognized to be an organized gelatinous compartment that is 20% protein by weight and rich in the organic molecules that are the intermediates in metabolism.	The cytosol is the site of intermediary metabolism, the interconnecting sets of chemical reactions by which cells generate energy and form the precursors necessary for biosynthesis of macromolecules essential to cell growth and function.

tinuous with the cell membrane. Reactions of cellular respiration are localized on these membranes. In photosynthetic prokaryotes such as the **cyanobacteria,** flat, sheetlike membranous structures called **lamellae** are formed from cell membrane infoldings. These lamellae are the sites of photosynthetic activity, but in prokaryotes, they are not contained within **plastids,** the organelles of photosynthesis found in higher plant cells. Prokaryotic cells also lack a cytoskeleton; the cell wall maintains their structure. Some bacteria have **flagella,** single, long filaments used for motility. Prokaryotes largely reproduce by asexual division, although sexual exchanges can occur. Table 1.5 lists the major features of prokaryotic cells.

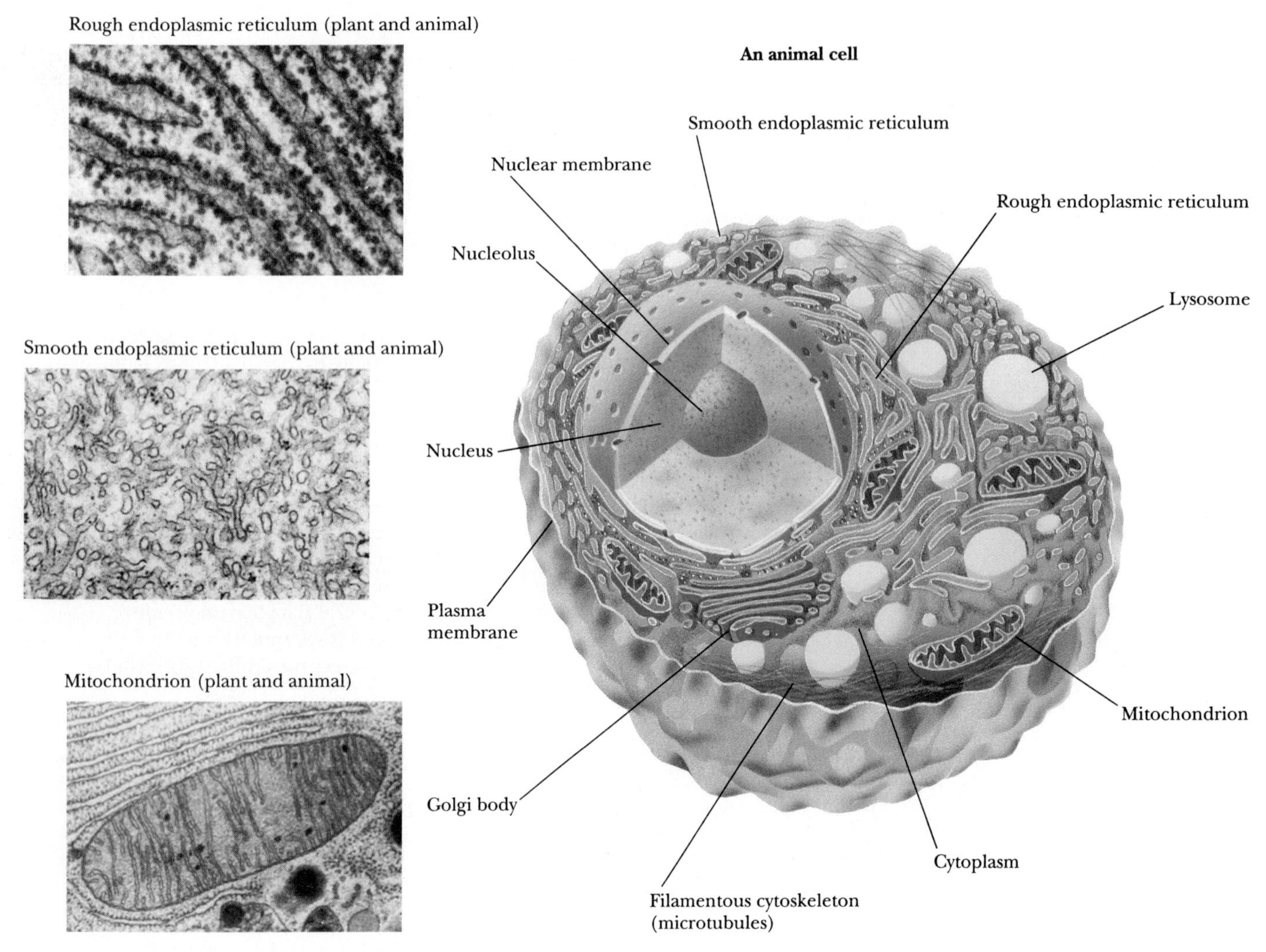

***Figure* 1.21** This figure diagrams a rat liver cell, a typical higher animal cell in which the characteristic features of animal cells are evident, such as a nucleus, nucleolus, mitochondria, Golgi bodies, lysosomes, and endoplasmic reticulum (ER). Microtubules and the network of filaments constituting the cytoskeleton are also depicted.

Structural Organization of Eukaryotic Cells

In comparison to prokaryotic cells, eukaryotic cells are much greater in size, typically having cell volumes 10^3 to 10^4 times larger. Also, they are much more complex. These two features require that eukaryotic cells partition their diverse metabolic processes into organized compartments, with each compartment dedicated to a few particular functions. A system of internal membranes accomplishes this partitioning. A typical animal cell is shown in Figure 1.21; a typical plant cell in Figure 1.22.

Eukaryotic cells possess a discrete, membrane-bounded **nucleus,** the repository of the cell's genetic material, which is distributed among a few or many **chromosomes.** During cell division, equivalent copies of this genetic material must be passed to both daughter cells through duplication and orderly partitioning of the chromosomes by the process known as **mitosis.** Like

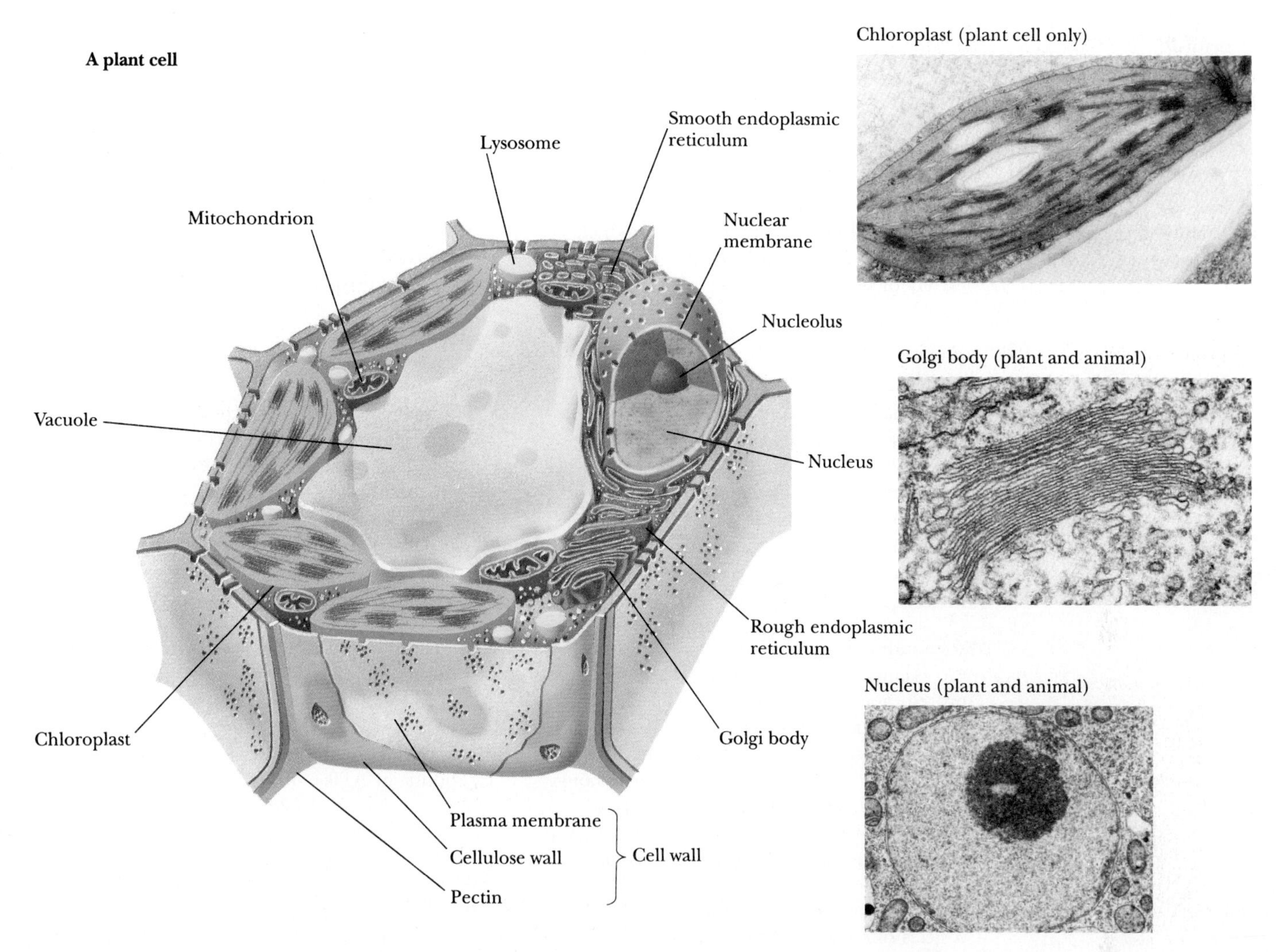

***Figure* 1.22** This figure diagrams a cell in the leaf of a higher plant. The cell wall, membrane, nucleus, chloroplasts, mitochondria, vacuole, ER, and other characteristic features are shown.

prokaryotic cells, eukaryotic cells are surrounded by a plasma membrane. Unlike prokaryotic cells, eukaryotic cells are rich in internal membranes that are differentiated into specialized structures such as the **endoplasmic reticulum (ER)** and the **Golgi apparatus.** Membranes also surround certain organelles (**mitochondria** and **chloroplasts,** for example) and various vesicles, including **vacuoles, lysosomes,** and **peroxisomes.** The common purpose of these membranous partitionings is the creation of cellular compartments that have specific, organized metabolic functions, such as the mitochondrion's role as the principal site of cellular energy production. Eukaryotic cells also have a **cytoskeleton** composed of arrays of filaments that give the cell its shape and its capacity to move. Some eukaryotic cells also have long projections on their surface—cilia or flagella—which provide propulsion. Tables 1.6 and 1.7 list the major features of a typical animal cell and a higher plant cell, respectively.

Table **1.6**

Major Features of a Typical Animal Cell

Structure	Molecular Composition	Function
Extracellular matrix	The surfaces of animal cells are covered with a flexible and sticky layer of complex carbohydrates, proteins, and lipids.	This complex coating is cell-specific, serves in cell–cell recognition and communication, creates cell adhesion, and provides a protective outer layer.
Cell membrane (plasma membrane)	Roughly 50:50 lipid:protein as a 5-nm-thick continuous sheet of lipid bilayer in which a variety of proteins are embedded.	The plasma membrane is a selectively permeable outer boundary of the cell, containing specific systems—pumps, channels, transporters—for the exchange of nutrients and other materials with the environment. Important enzymes are also located here.
Nucleus	The nucleus is separated from the cytosol by a double membrane, the nuclear envelope. The DNA is complexed with basic proteins (histones) to form chromatin fibers, the material from which chromosomes are made. A distinct RNA-rich region, the nucleolus, is the site of ribosome assembly.	The nucleus is the repository of genetic information encoded in DNA and organized into chromosomes. During mitosis, the chromosomes are replicated and transmitted to the daughter cells. The genetic information of DNA is transcribed into RNA in the nucleus and passes into the cytosol where it is translated into protein by ribosomes.
Mitochondria	Mitochondria are organelles surrounded by two membranes that differ markedly in their protein and lipid composition. The inner membrane and its interior volume, the matrix, contain many important enzymes of energy metabolism. Mitochondria are about the size of bacteria, $\approx 1\ \mu m$. Cells contain hundreds of mitochondria, which collectively occupy about one-fifth of the cell volume.	Mitochondria are the power plants of eukaryotic cells where carbohydrates, fats, and amino acids are oxidized to CO_2 and H_2O. The energy released is trapped as high-energy phosphate bonds in ATP.
Golgi apparatus	A system of flattened membrane-bounded vesicles often stacked into a complex. Numerous small vesicles are found peripheral to the Golgi and contain secretory material packaged by the Golgi.	Involved in the packaging and processing of macromolecules for secretion and for delivery to other cellular compartments.

1.6 Viruses Are Supramolecular Assemblies Acting as Cell Parasites

Viruses are supramolecular complexes of nucleic acid, either DNA or RNA, encapsulated in a protein coat and, in some instances, surrounded by a membrane envelope (Figure 1.23). The bits of nucleic acid in viruses are, in reality, mobile elements of genetic information. The protein coat serves to protect the nucleic acid and allows it to gain entry to the cells that are its specific hosts. Viruses unique for all types of cells are known. Viruses infecting bacteria are called **bacteriophages** (''bacteria eaters''); different viruses infect animal cells and plant cells. Once the nucleic acid of a virus gains access to its specific host, it typically takes over the metabolic machinery of the host cell, diverting it to the production of virus particles. The host metabolic functions are subjugated to the synthesis of viral nucleic acid and proteins. Mature virus particles arise by encapsulating the nucleic acid within a protein coat called the **capsid.** (Assembly of infectious virus particles from component nucleic

(text continues on page 29)

Table **1.6**
Continued

Structure	Molecular Composition	Function
Endoplasmic reticulum (ER) and ribosomes	Flattened sacs, tubes, and sheets of internal membrane extending throughout the cytoplasm of the cell and enclosing a large interconnecting series of volumes called *cisternae.* The ER membrane is continuous with the outer membrane of the nuclear envelope. Portions of the sheetlike areas of the ER are studded with ribosomes, giving rise to *rough ER.* Eukaryotic ribosomes are larger than prokaryotic ribosomes.	The endoplasmic reticulum is a labyrinthine organelle where both membrane proteins and lipids are synthesized. Proteins made by the ribosomes of the rough ER pass through the outer ER membrane into the cisternae and can be transported via the Golgi to the periphery of the cell. Other ribosomes unassociated with the ER carry on protein synthesis in the cytosol.
Lysosomes	Lysosomes are vesicles 0.2–0.5 μm in diameter, bounded by a single membrane. They contain hydrolytic enzymes such as proteases and nucleases which, if set free, could degrade essential cell constituents. They are formed by budding from the Golgi apparatus.	Lysosomes function in intracellular digestion of materials entering the cell via phagocytosis or pinocytosis. They also function in the controlled degradation of cellular components.
Peroxisomes	Like lysosomes, peroxisomes are 0.2–0.5 μm single-membrane-bounded vesicles. They contain a variety of oxidative enzymes that use molecular oxygen and generate peroxides. They are formed by budding from the smooth ER.	Peroxisomes act to oxidize certain nutrients, such as amino acids. In doing so, they form potentially toxic hydrogen peroxide, H_2O_2, and then decompose it to H_2O and O_2 by way of the peroxide-cleaving enzyme catalase.
Cytoskeleton	The cytoskeleton is composed of a network of protein filaments: actin filaments (or microfilaments), 7 nm in diameter; intermediate filaments, 8–10 nm; and microtubules, 25 nm. These filaments interact in establishing the structure and functions of the cytoskeleton. This interacting network of protein filaments gives structure and organization to the cytoplasm.	The cytoskeleton determines the shape of the cell and gives it its ability to move. It also mediates the internal movements that occur in the cytoplasm, such as the migration of organelles and mitotic movements of chromosomes. The propulsion instruments of cells—cilia and flagella—are constructed of microtubules.

(a)

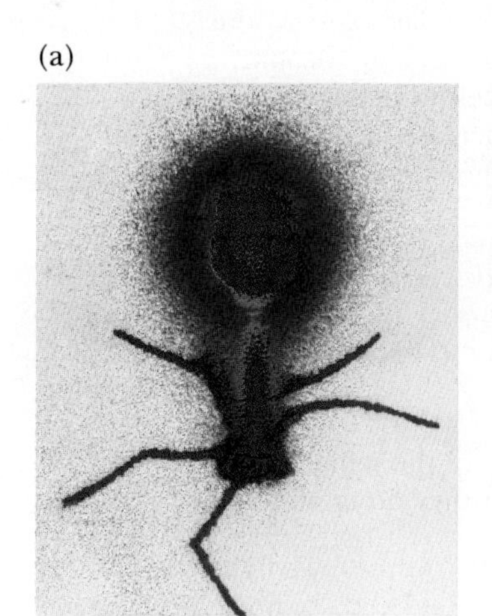

(b)

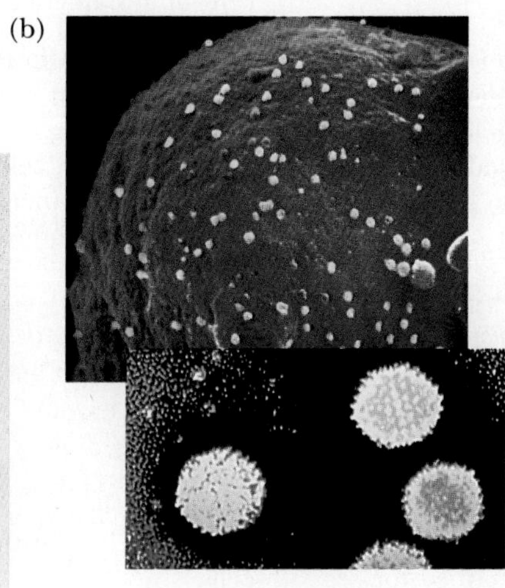

(c)

Figure **1.23** Viruses are genetic elements enclosed in a protein coat. Viruses are not free-living and can only reproduce within cells. Viruses show an almost absolute specificity for their particular host cells, infecting and multiplying only within those cells. Viruses are known for virtually every kind of cell. Shown here are examples of (a) a bacterial virus, bacteriophage T_4; (b) an animal virus, adenovirus (inset at greater magnification); and (c) a plant virus, tobacco mosaic virus.

***Table* 1.7**

Major Features of a Higher Plant Cell: A Photosynthetic Leaf Cell

Structure	Molecular Composition	Function
Cell wall	Cellulose fibers embedded in a polysaccharide/protein matrix; it is thick (>0.1 μm), rigid, and porous to small molecules.	Protection against osmotic or mechanical rupture. The walls of neighboring cells interact in cementing the cells together to form the plant. Channels for fluid circulation and for cell–cell communication pass through the walls. The structural material confers form and strength on plant tissue.
Cell membrane	Plant cell membranes are similar in overall structure and organization to animal cell membranes, but differ in lipid and protein composition.	The plasma membrane of plant cells is selectively permeable, containing transport systems for the uptake of essential nutrients and inorganic ions. A number of important enzymes are localized here.
Nucleus	The nucleus, nucleolus, and nuclear envelope of plant cells are like those of animal cells.	Chromosomal organization, DNA replication, transcription, ribosome synthesis, and mitosis in plant cells are grossly similar to the analogous features in animals.
Chloroplasts	Plant cells contain a unique family of organelles, the plastids, of which the chloroplast is the prominent example. Chloroplasts have a double membrane envelope, an inner volume called the **stroma,** and an internal membrane system rich in thylakoid membranes, which enclose a third compartment, the thylakoid **lumen.** Chloroplasts are significantly larger than mitochondria. Other plastids are found in specialized structures such as fruits, flower petals, and roots and have specialized roles.	Chloroplasts are the site of photosynthesis, the reactions by which light energy is converted to metabolically useful chemical energy in the form of ATP. These reactions occur on the thylakoid membranes. The formation of carbohydrate from CO_2 takes place in the stroma. Oxygen is evolved during photosynthesis. Chloroplasts are the primary source of energy in the light.
Mitochondria	Plant cell mitochondria resemble the mitochondria of other eukaryotes in form and function.	Plant mitochondria are the main source of energy generation in photosynthetic cells in the dark and in nonphotosynthetic cells under all conditions.
Vacuole	The vacuole is usually the most obvious compartment in plant cells. It is a very large vesicle enclosed by a single membrane called the **tonoplast.** Vacuoles tend to be smaller in young cells, but in mature cells, they may occupy more than 50% of the cell's volume. Vacuoles occupy the center of the cell, with the cytoplasm being located peripherally around it. They resemble the lysosomes of animal cells.	Vacuoles function in transport and storage of nutrients and cellular waste products. By accumulating water, the vacuole allows the plant cell to grow dramatically in size with no increase in cytoplasmic volume.
Golgi apparatus, endoplasmic reticulum, ribosomes, lysosomes, peroxisomes, and cytoskeleton	Plant cells also contain all of these characteristic eukaryotic organelles, essentially in the form described for animal cells.	These organelles serve the same purposes in plant cells that they do in animal cells.

acid and protein subunits is discussed in detail in Chapter 34.) Viruses are thus supramolecular assemblies that act as parasites of cells (Figure 1.24).

Usually, viruses cause the lysis of the cells they infect. It is their cytolytic properties that are the basis of viral disease. In certain circumstances, the viral genetic elements may integrate into the host chromosome and become quiescent. Such a state is termed **lysogeny.** Typically, damage to the host cell acti-

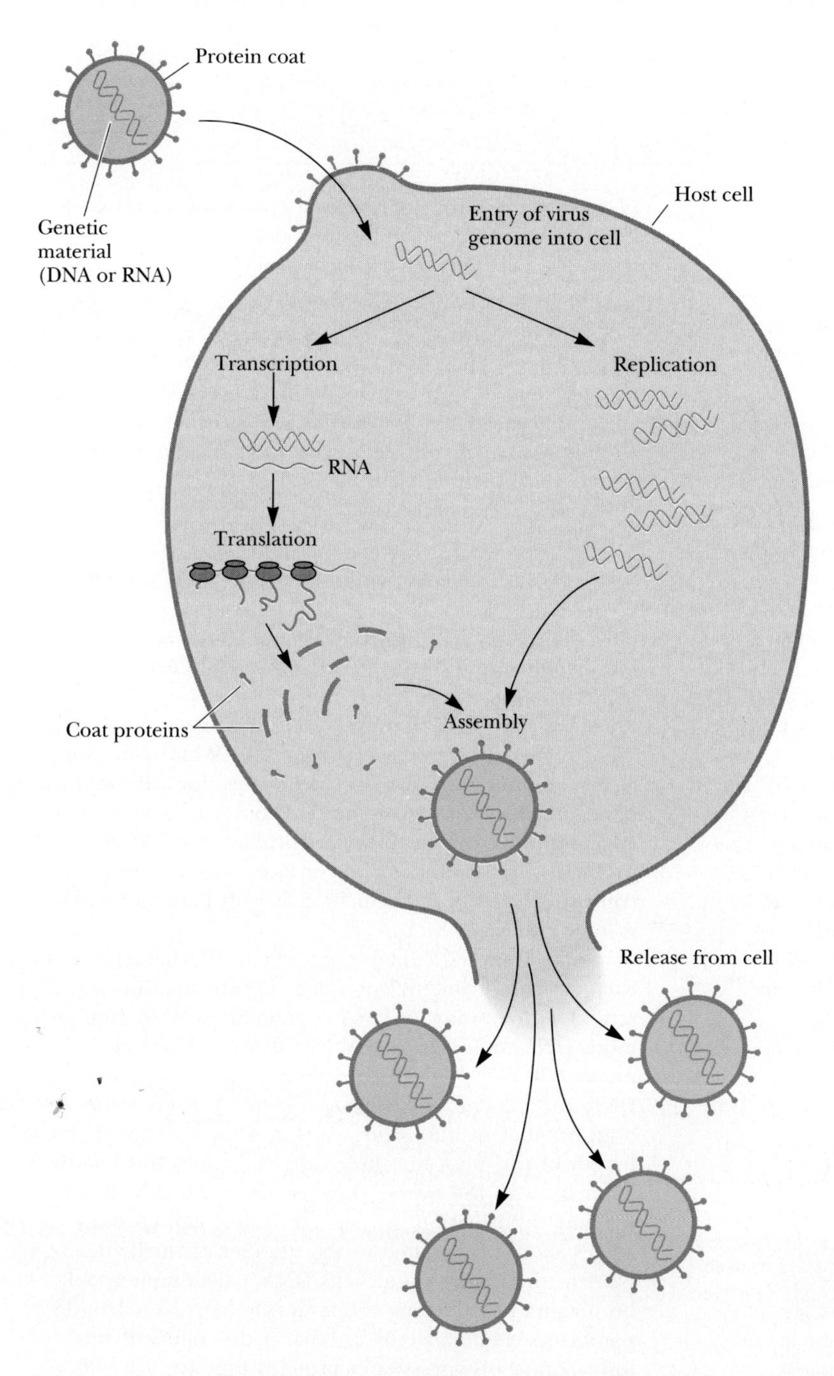

***Figure* 1.24** The virus life cycle. Viruses are mobile bits of genetic information encapsulated in a protein coat. The genetic material can be either DNA or RNA. Once this genetic material gains entry to its host cell, it takes over the host machinery for macromolecular synthesis and subverts it to the synthesis of viral-specific nucleic acids and proteins. These virus components are then assembled into mature virus particles that are released from the cell. Often, this parasitic cycle of virus infection leads to cell death and disease.

vates the replicative capacities of the quiescent viral nucleic acid, leading to viral propagation and release. Viruses are implicated in transforming cells into a cancerous state, that is, in converting their hosts to an unregulated state of cell division and proliferation. Since all viruses are heavily dependent on their host for the production of viral progeny, viruses must have arisen after cells were established in the course of evolution. Presumably, the first viruses were fragments of nucleic acid that developed the ability to replicate independently of the chromosome and then acquired the necessary genes enabling protection, autonomy, and transfer between cells.

Problems

1. The nutritional requirements of *Escherichia coli* cells are far simpler than those of humans, yet the macromolecules found in bacteria are about as complex as those of animals. Since bacteria can make all their essential biomolecules while subsisting on a simpler diet, do you think bacteria may have more biosynthetic capacity and hence more metabolic complexity than animals? Organize your thoughts on this question, pro and con, into a rational argument.
2. Without consulting chapter figures, sketch the characteristic prokaryotic and eukaryotic cell types and label their pertinent organelle and membrane systems.
3. *Escherichia coli* cells are about 2 μm (microns) long and 0.8 μm in diameter.
   a. How many *E. coli* cells laid end to end would fit across the diameter of a pin head? (Assume a pinhead diameter of 0.5 mm.)
   b. What is the volume of an *E. coli* cell? (Assume it is a cylinder, with the volume of a cylinder given by $V = \pi r^2 h$, where $\pi = 3.14$.)
   c. What is the surface area of an *E. coli* cell? What is the surface-to-volume ratio of an *E. coli* cell?
   d. Glucose, a major energy-yielding nutrient, is present in bacterial cells at a concentration of about 1 m*M*. How many glucose molecules are contained in a typical *E. coli* cell? (Recall that Avogadro's number = 6.023×10^{23}.)
   e. A number of regulatory proteins are present in *E. coli* at only one or two molecules per cell. If we assume that an *E. coli* cell contains just one molecule of a particular protein, what is the molar concentration of this protein in the cell?
   f. An *E. coli* cell contains about 15,000 ribosomes, which carry out protein synthesis. Assuming ribosomes are spherical and have a diameter of 20 nm (nanometers), what fraction of the *E. coli* cell volume is occupied by ribosomes?
   g. The *E. coli* chromosome is a single DNA molecule whose mass is about 3×10^9 daltons. This macromolecule is actually a linear array of nucleotide pairs. The average molecular weight of a nucleotide pair is 660 and each pair imparts 0.34 nm to the length of the DNA molecule. What is the total length of the *E. coli* chromosome? How does this length compare with the overall dimensions of an *E. coli* cell? How many nucleotide pairs does this DNA contain? The average *E. coli* protein is a linear chain of 360 amino acids. If three nucleotide pairs in a gene encode one amino acid in a protein, how many different proteins can the *E. coli* chromosome encode? (The answer to this question is a reasonable approximation of the maximum number of different kinds of proteins that can be expected in bacteria.)
4. Assume that mitochondria are cylinders 1.5 μm in length and 0.6 μm in diameter.
   a. What is the volume of a single mitochondrion?
   b. Oxaloacetate is an intermediate in the citric acid cycle, an important metabolic pathway localized in the mitochondria of eukaryotic cells. The concentration of oxaloacetate in mitochondria is about 0.03 μ*M*. How many molecules of oxaloacetate are in a single mitochondrion?
5. Assume that liver cells are cuboidal in shape, 20 μm on a side.
   a. How many liver cells laid end to end would fit across the diameter of a pin head? (Assume a pinhead diameter of 0.5 mm.)
   b. What is the volume of a liver cell? (Assume it is a cube.)
   c. What is the surface area of a liver cell? What is the surface-to-volume ratio of a liver cell? How does this compare to the surface-to-volume ratio of an *E. coli* cell (compare this answer to that of problem 3c)? What problems must cells with low surface-to-volume ratios confront that do not occur in cells with high surface-to-volume ratios?
   d. A human liver cell contains two sets of 23 chromosomes, each set being roughly equivalent in information content. The total mass of DNA contained in these 46 enormous DNA molecules is 4×10^{12} daltons. Since each nucleotide pair contributes 660 daltons to the mass of DNA and 0.34 nm to the length of DNA, what is the total number of nucleotide pairs and the complete length of the DNA in a liver cell? How does this length compare with the overall dimensions of a liver cell? The maximal information in each set of liver cell chromosomes should be related to the number of nucleotide pairs in the chromosome set's DNA. This number can be obtained by dividing the total number of nucleotide pairs calculated above by 2. What is this value? If this information is expressed in proteins that average 400 amino acids in length and three nucleotide pairs encode one amino acid in a protein, how many different kinds of proteins might a liver cell be able to produce? (In reality, liver cells express at most about 30,000 different proteins. Thus, a large discrepancy exists between the theoretical information content of DNA in liver cells and the amount of information actually expressed.)

6. Biomolecules interact with one another through molecular surfaces that are structurally complementary. How can various proteins interact with molecules as different as simple ions, hydrophobic lipids, polar but uncharged carbohydrates, and even nucleic acids?
7. What structural features allow biological polymers to be informational macromolecules? Is it possible for polysaccharides to be informational macromolecules?
8. Why is it important that weak forces, not strong forces, mediate biomolecular recognition?
9. Why does the central role of weak forces in biomolecular interactions restrict living systems to a narrow range of environmental conditions?
10. Describe what is meant by the phrase "cells are steady-state systems."

Further Reading

Alberts, B., Bray, D., Lewis, J., et al., 1989. *Molecular Biology of the Cell,* 2nd ed. New York: Garland Press.

Goodsell, D. S., 1991. Inside a living cell. *Trends in Biochemical Sciences* **16**:203–206.

Lloyd, C., ed., 1986. Cell organization. *Trends in Biochemical Sciences* **11**:437–485.

Loewy, A. G., Siekevitz, P., Menninger, J. R., Gallant, J. A. N., 1991. *Cell Structure and Function.* Philadelphia: Saunders College Publishing.

Solomon, E. P., Berg, L. R., Martin, D. W., and Villee, C., 1993. *Biology,* 3rd ed. Philadelphia: Saunders College Publishing.

Wald, G., 1964. The origins of life. *Proceedings of the National Academy of Science, U.S.A.* **52**:595–611.

Watson, J. D., Hopkins, N. H., Roberts, J. W., et al., 1987. *Molecular Biology of the Gene,* 4th ed. Menlo Park, CA: Benjamin/Cummings Publishing Co.

"Water is the origin of all things."

Thales (c. 634–c. 546 B.C.)

Chapter 2

Water, pH, and Ionic Equilibria

All life depends on water; all organisms are aqueous chemical systems.

Outline

2.1 Properties of Water

2.2 pH

2.3 Buffers

2.4 Water's Unique Role in the Fitness of the Environment

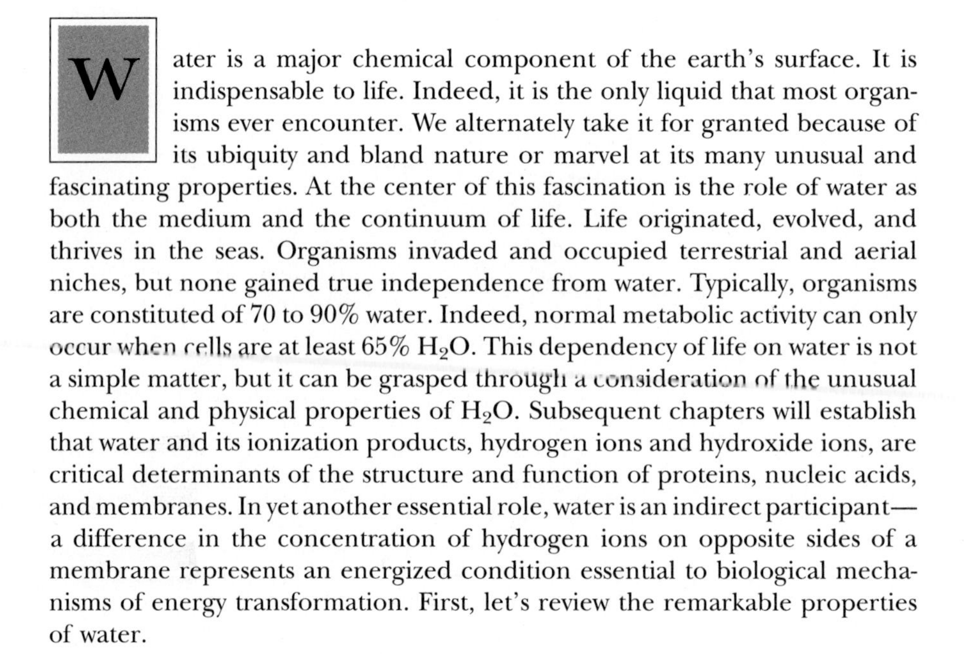

Water is a major chemical component of the earth's surface. It is indispensable to life. Indeed, it is the only liquid that most organisms ever encounter. We alternately take it for granted because of its ubiquity and bland nature or marvel at its many unusual and fascinating properties. At the center of this fascination is the role of water as both the medium and the continuum of life. Life originated, evolved, and thrives in the seas. Organisms invaded and occupied terrestrial and aerial niches, but none gained true independence from water. Typically, organisms are constituted of 70 to 90% water. Indeed, normal metabolic activity can only occur when cells are at least 65% H_2O. This dependency of life on water is not a simple matter, but it can be grasped through a consideration of the unusual chemical and physical properties of H_2O. Subsequent chapters will establish that water and its ionization products, hydrogen ions and hydroxide ions, are critical determinants of the structure and function of proteins, nucleic acids, and membranes. In yet another essential role, water is an indirect participant—a difference in the concentration of hydrogen ions on opposite sides of a membrane represents an energized condition essential to biological mechanisms of energy transformation. First, let's review the remarkable properties of water.

2.1 Properties of Water

Unusual Properties

In comparison with chemical compounds of similar atomic organization and molecular size, water displays rather aberrant properties. For example, compare water, the hydride of oxygen, with hydrides of oxygen's nearest neighbors in the periodic table, namely, ammonia (NH_3) and hydrogen fluoride (HF), or with the hydride of its nearest congener, sulfur (H_2S). Water has a substantially higher boiling point, melting point, heat of vaporization, and surface tension. Indeed, all of these physical properties are anomalously high for a substance of this molecular weight that is neither metallic nor ionic. These properties suggest that intermolecular forces of attraction between H_2O molecules are high. Thus, the internal cohesion of this substance is high. Furthermore, water has an unusually high dielectric constant, its maximum density is found in the liquid (not the solid) state, and it has a negative volume of melting (that is, the solid form, ice, occupies more space than does the liquid form, water). It is truly remarkable that so many eccentric properties should occur together in a single substance. As chemists, we expect to find an explanation for these apparent anomalies in the structure of water. The key to its intermolecular attractions must be intrinsic to its atomic constitution. Indeed, *the fact crucial to understanding water is its unrivaled ability to form hydrogen bonds.*

Structure of Water

The two hydrogen atoms of water are linked covalently to oxygen, each sharing an electron pair, to give a nonlinear arrangement (Figure 2.1). This "bent" structure of the H_2O molecule is of enormous significance to its properties. If H_2O were linear, it would be a nonpolar substance. In the bent configuration, however, the electronegative O atom and the two H atoms form a dipole that renders the molecule distinctly polar. Furthermore, this structure is ideally suited to H bond formation. Water can serve as both an H donor and an H acceptor in H bond formation. The potential to form four H

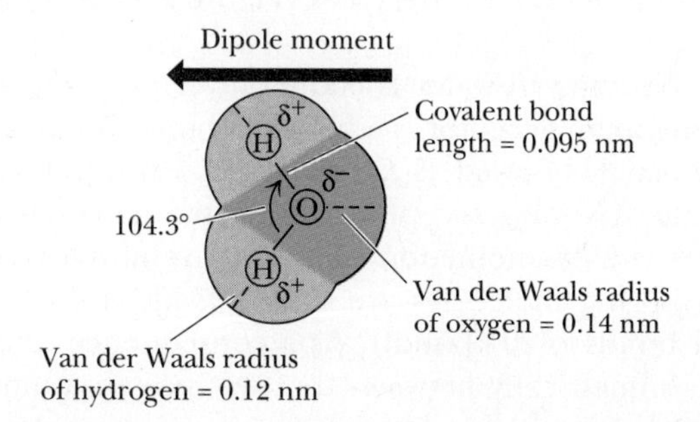

***Figure* 2.1** The structure of water. Two lobes of negative charge formed by the lone-pair electrons of the oxygen atom lie above and below the plane of the diagram. This electron density contributes substantially to the large dipole moment and polarizability of the water molecule. The dipole moment of water corresponds to the O–H bonds having 33% ionic character. Note that the H–O–H angle is 104.3°, *not* 109°, the angular value found in molecules with tetrahedral symmetry, such as CH_4. Many of the important properties of water derive from this angular value, such as the decreased density of its crystalline state, ice.

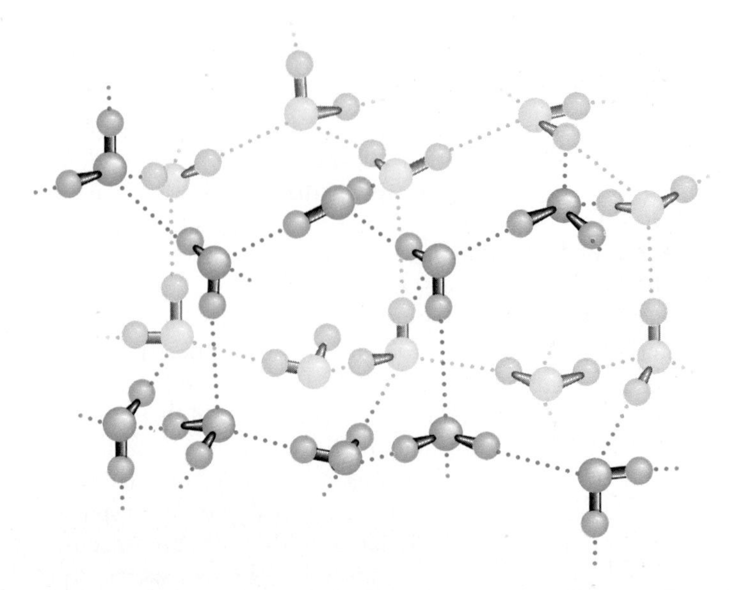

***Figure* 2.2** The structure of normal ice. The hydrogen bonds in ice form a three-dimensional network. The smallest number of H_2O molecules in any closed circuit of H-bonded molecules is six, so that this structure bears the name *hexagonal ice.* Covalent bonds are represented as solid lines, whereas hydrogen bonds are shown as dashed lines. The directional preference of H bonds leads to a rather open lattice structure for crystalline water and, consequently, a low density for the solid state. The distance between neighboring oxygen atoms linked by a hydrogen bond is 0.274 nm. Since the covalent H–O bond is 0.095 nm, the H–O hydrogen bond length in ice is 0.18 nm.

bonds per water molecule is the source of the strong intermolecular attractions that endow this substance with its anomalously high boiling point, melting point, heat of vaporization, and surface tension. In ordinary ice, the common crystalline form of water, each H_2O molecule has four nearest neighbors to which it is hydrogen bonded: each H atom donates an H bond to the O of a neighbor, while the O atom serves as an H bond acceptor from H atoms bound to two different water molecules (Figure 2.2). A local tetrahedral symmetry results.

Hydrogen bonding in water is cooperative. That is, an H-bonded water molecule serving as an acceptor is a better H bond donor than an unbonded molecule (and an H-bonded H_2O molecule serving as an H bond donor becomes a better H bond acceptor). Thus, participation in H bonding by H_2O molecules is a phenomenon of mutual reinforcement. The H bonds between neighboring molecules are weak (23 kJ/mol each) relative to the H–O covalent bonds (420 kJ/mol). As a consequence, the hydrogen atoms are situated asymmetrically between the two oxygen atoms along the O–O axis. There is never any ambiguity about which O atom the H atom is chemically bound to, nor to which O it is H bonded.

Structure of Ice

In ice, the hydrogen bonds form a space-filling, three-dimensional network. These bonds are directional and straight; that is, the H atom lies on a direct line between the two O atoms. This linearity and directionality mean that the resultant H bonds are strong. In addition, the directional preference of the H bonds leads to an open lattice structure. For example, if the water molecules are approximated as rigid spheres centered at the positions of the O atoms in the lattice, then the observed density of ice is actually only 57% of that expected for a tightly packed arrangement of such spheres. The H bonds in ice hold the water molecules apart. Melting involves breaking some of the H bonds that maintain the crystal structure of ice so that the molecules of water (now liquid) can actually pack closer together. Thus, the density of ice is slightly less than the density of water. Ice floats, a property of great importance to aquatic organisms in cold climates.

In liquid water, the rigidity of ice is replaced by fluidity, and the crystalline periodicity of ice gives way to spatial homogeneity. The H_2O molecules in liquid water form a random, H-bonded network with each molecule having an average of 4.4 close neighbors situated within a center-to-center distance of 0.284 nm (2.84 Å). At least half of the hydrogen bonds have nonideal orientations (that is, they are not perfectly straight); consequently, liquid H_2O lacks the regular latticelike structure of ice. The space about an O atom is not defined by the presence of four hydrogens, but can be occupied by other water molecules randomly oriented so that the local environment, over time, is essentially uniform. Nevertheless, the heat of melting for ice is but a small fraction (13%) of the heat of sublimation for ice (the energy needed to go from the solid to the vapor state). This fact indicates that the majority of H bonds between H_2O molecules survive the transition from solid to liquid. At 10°C, 2.3 H bonds per H_2O molecule remain, and the tetrahedral bond order persists even though substantial disorder is now present.

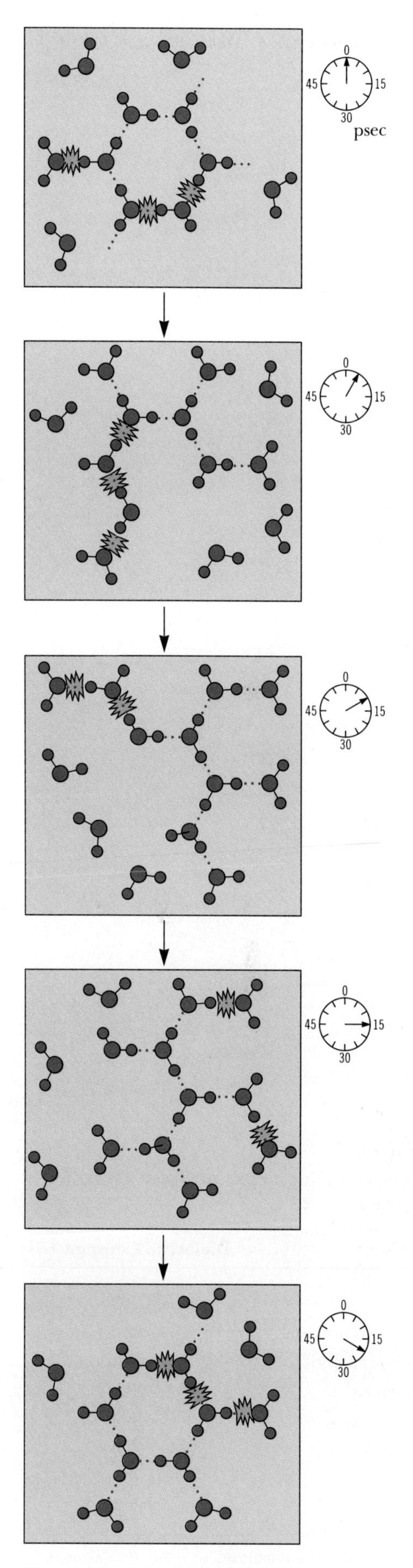

Figure **2.3** The fluid network of H bonds linking water molecules in the liquid state.

Molecular Interactions in Liquid Water

The present interpretation of water structure is that water molecules are connected by uninterrupted H bond paths running in every direction, spanning the whole sample. The picture of water molecules persisting in tetrahedrally arrayed crystalline sets or "icebergs" suspended in a sea of unbonded water molecules is *not* an accurate description of the structure of water. Instead, the participation of each water molecule in an average state of H bonding to its neighbors means that each molecule is connected to every other in a fluid network of H bonds. The average lifetime of an H-bonded connection between two H_2O molecules in water is 9.5 psec (picoseconds, where 1 psec = 10^{-12} sec). Thus, about every 10 psec, the average H_2O molecule moves, reorients, and interacts with new neighbors, as illustrated in Figure 2.3.

In summary, pure liquid water consists of H_2O molecules held in a random, three-dimensional network that has a local preference for tetrahedral geometry but contains a large number of strained or broken hydrogen bonds. The presence of strain creates a kinetic situation in which H_2O molecules can switch H bond allegiances; fluidity ensues.

Solvent Properties

Because of its highly polar nature, water is an excellent solvent for ionic substances such as salts; nonionic but polar substances such as sugars, simple alcohols, and amines; and carbonyl-containing molecules such as aldehydes and ketones. Although the electrostatic attractions between the positive and negative ions in the crystal lattice of a salt are very strong, water readily dissolves salts. For example, sodium chloride is dissolved because dipolar water molecules participate in strong electrostatic interactions with the Na^+ and Cl^- ions, leading to the formation of **hydration shells** surrounding these ions (Figure 2.4). Although hydration shells are stable structures, they are also dynamic. Each water molecule in the inner hydration shell around a Na^+ ion is replaced on average every 2 to 4 nsec (nanoseconds, where 1 nsec = 10^{-9} sec) by another H_2O. Consequently, a water molecule is trapped only

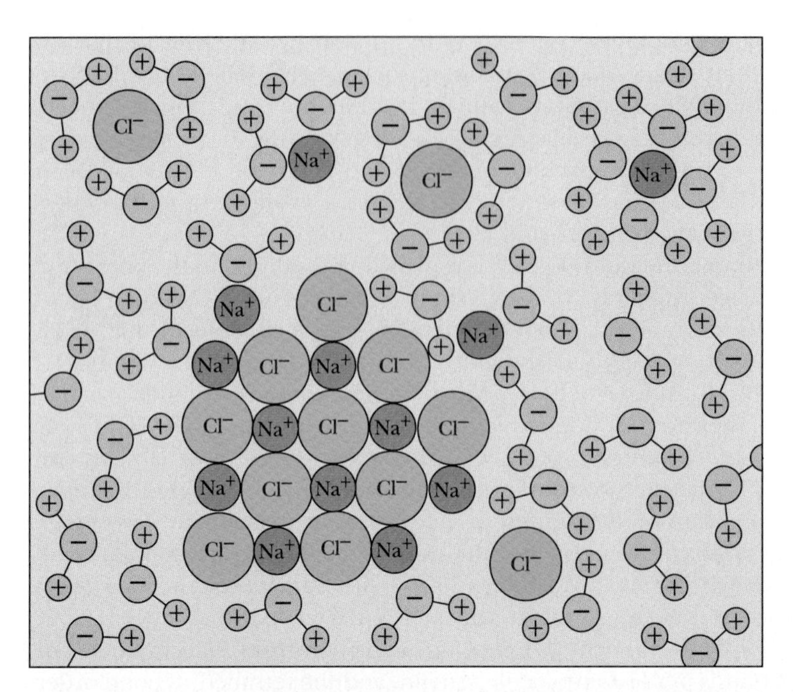

***Figure* 2.4** Hydration shells surrounding ions in solution. Water molecules orient so that the electrical charge on the ion is sequestered by the water dipole. For positive ions (cations), the partially negative oxygen atom of H_2O is toward the ion in solution. Negatively charged ions (anions) attract the partially positive hydrogen atoms of water in creating their hydration shells.

several hundred times longer by the electrostatic force field of an ion than it is by the H-bonded network of water. (Recall that the average lifetime of H bonds between water molecules is about 10 psec.)

Water Has a High Dielectric Constant

The attractions between the water molecules interacting with, or **hydrating,** ions are much greater than the tendency of oppositely charged ions to attract one another. The ability of water to surround ions in dipole interactions and diminish their attraction for one another is a measure of its **dielectric constant,** *D*. Indeed, ionization in solution depends on the dielectric constant of the solvent; otherwise the strongly attracted positive and negative ions would unite to form neutral molecules. The strength of the dielectric constant is related to the force, *F,* experienced between two ions of opposite charge separated by a distance, *r,* as given in the relationship

$$F = e_1 e_2 / Dr^2$$

where e_1 and e_2 are the charges on the two ions. Table 2.1 lists the dielectric constants of some common liquids. Note that the dielectric constant for water is more than twice that of methanol and more than forty times that of hexane.

***Table* 2.1**

Dielectric Constants* of Some Common Solvents at 25°C

Solvent	Dielectric Constant (*D*)
Water	78.5
Methyl alcohol	32.6
Ethyl alcohol	24.3
Acetone	20.7
Acetic acid	6.2
Chloroform	5.0
Benzene	2.3
Hexane	1.9

*The dielectric constant is also referred to as *relative permittivity* by physical chemists.

Water Forms H Bonds with Polar Solutes

In the case of nonionic but polar compounds such as sugars, the excellent solvent properties of water are attributed to its ability to readily form hydrogen bonds with the polar functional groups on these compounds, such as hydroxyls, amines, and carbonyls. These polar interactions between solvent and solute are stronger than the intermolecular attractions between solute molecules caused by van der Waals forces and weaker hydrogen bonding. Thus, the solute molecules readily dissolve in water.

Hydrophobic Interactions

The behavior of water toward nonpolar solutes is different from the interactions just discussed. Nonpolar solutes (or nonpolar functional groups on biological macromolecules) do not readily H bond to H_2O, and as a result, such compounds tend to be only sparingly soluble in water. The process of dissolving such substances is accompanied by significant reorganization of the water surrounding the solute so that the response of the solvent water to such solutes can be equated to "structure making." Since nonpolar solutes must occupy space, the random H bond network of water must reorganize to accommodate them. At the same time, the water molecules participate in as many H-bonded interactions with one another as the temperature will permit. Consequently, the H-bonded water network rearranges toward formation of a local cagelike **(clathrate)** structure surrounding each solute molecule (Figure 2.5). A major consequence of this rearrangement is that the molecules of H_2O participating in the cage layer have markedly reduced orientational options. Water molecules tend to straddle the nonpolar solute such that two or three tetrahedral directions (H-bonding vectors) are tangential to the space occupied by the inert solute. This "straddling" means that no water H-bonding capacity is lost because no H bond donor or acceptor of the H_2O is directed toward the caged solute. The water molecules forming these clathrates are involved in highly ordered structures. That is, clathrate formation is accompanied by significant ordering of structure or negative entropy.

Under these conditions, nonpolar solute molecules experience a net attraction for one another that is called **hydrophobic interaction.** The basis of this interaction is that when two nonpolar molecules meet, their joint solvation cage involves less overall ordering of the water molecules than in their separate cages and the attraction is an entropy-driven process, that is, a net

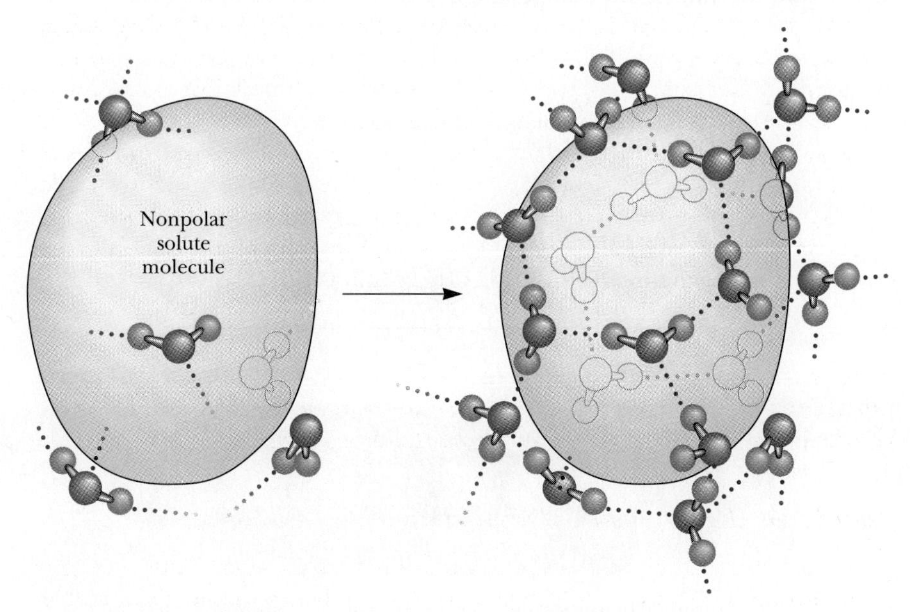

***Figure* 2.5** Formation of a clathrate structure by water molecules surrounding a hydrophobic solute.

decrease in order among the H_2O molecules. To be specific, hydrophobic interactions between nonpolar molecules are maintained not so much by direct interactions between the inert solutes themselves as by the stability achieved when the water cages coalesce and reorganize. Because interactions between nonpolar solute molecules and the water surrounding them are of uncertain stoichiometry and do not share the equality of atom-to-atom participation implicit in chemical bonding, the term *hydrophobic interaction* is more correct than the misleading expression *hydrophobic bond.*

Amphiphilic Molecules

amphiphilic molecules, amphipathic molecules: compounds containing both strongly polar and strongly nonpolar groups

Compounds containing both strongly polar and strongly nonpolar groups are called **amphiphilic molecules** (from the Greek *amphi* meaning "both," and *philos* meaning "loving"), also referred to as **amphipathic molecules** (from the Greek *pathos* meaning "passion, suffering"). Salts of fatty acids are a typical example that has biological relevance. They have a long nonpolar hydrocarbon tail and a strongly polar carboxyl head group, as in the sodium salt of palmitic acid (Figure 2.6). Their behavior in aqueous solution reflects the combination of the contrasting polar and nonpolar nature of these substances. The ionic carboxylate function hydrates readily, whereas the long hydrophobic tail is intrinsically insoluble. Thus, sodium palmitate, a soap, shows little tendency to form a true ionic solution in water. Nevertheless, sodium palmitate and other amphiphilic molecules readily disperse in water because the hydrocarbon tails of these substances are joined together in hydrophobic interactions as their polar carboxylate functions are hydrated in typical hydrophilic fashion. Such clusters of amphipathic molecules are termed **micelles;** Figure 2.7 depicts their structure. Of enormous biological significance is the contrasting solute behavior of the two ends of amphipathic molecules upon introduction into aqueous solutions. The polar ends express their hydrophilicity in ionic interactions with the solvent, whereas their nonpolar counterparts are excluded from the water into a hydrophobic domain constituted from the hydrocarbon tails of many like molecules. It is exactly this behavior that accounts for the formation of membranes, the structures that define the limits and compartments of cells (see Chapter 9).

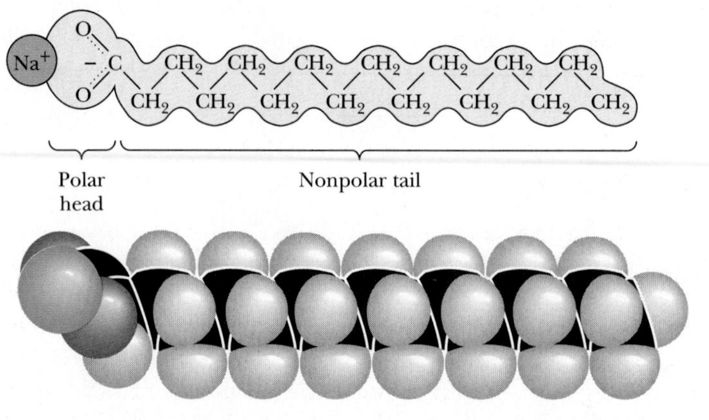

***Figure* 2.6** An amphiphilic molecule: sodium palmitate. Amphiphilic molecules are frequently symbolized by a ball and zig-zag line structure, where the ball represents the hydrophilic polar head and the zig-zag represents the nonpolar hydrophobic hydrocarbon tail.

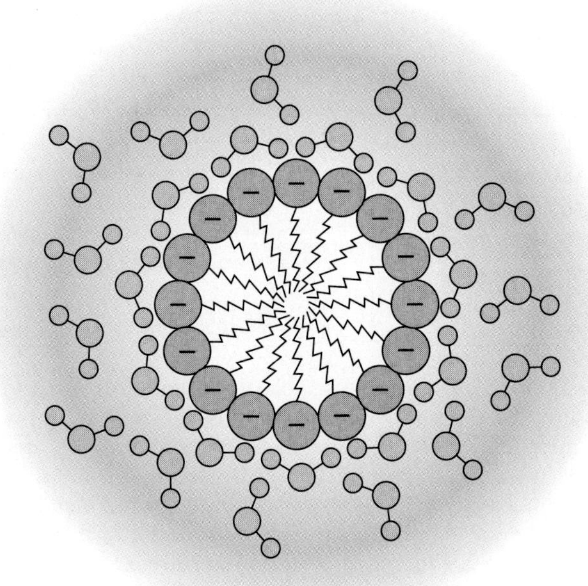

Figure **2.7** Micelle formation by amphiphilic molecules in aqueous solution. Negatively charged carboxylate head groups orient to the micelle surface and interact with the polar H_2O molecules via H bonding. The nonpolar hydrocarbon tails cluster in the interior of the spherical micelle, driven by hydrophobic exclusion from the solvent and the formation of favorable van der Waals interactions. Because of their negatively charged surfaces, neighboring micelles repel one another and thereby maintain a relative stability in solution.

Influence of Solutes on Water Properties

The presence of dissolved substances disturbs the structure of liquid water so that its properties change. The dynamic hydrogen-bonding pattern of water must now accommodate the intruding substance. The net effect is that solutes, regardless of whether they are polar or nonpolar, fix nearby water molecules in a more ordered array. Ions, by the establishment of hydration shells through interactions with the water dipoles, create local order. Hydrophobic effects, for various reasons, make structures within water. To put it another way, by limiting the orientations that neighboring water molecules can assume, solutes give order to the solvent and diminish the dynamic interplay among H_2O molecules that occurs in pure water.

Colligative Properties

This influence of the solute on water is reflected in a set of characteristic changes in behavior that are termed **colligative properties,** or properties related by a common principle. These alterations in solvent properties are related in that they all depend only on the number of solute particles per unit volume of solvent and not on the chemical nature of the solute. These effects include freezing point depression, boiling point elevation, vapor pressure lowering, and osmotic pressure effects. For example, 1 mol of an ideal solute dissolved in 1000 g of water (a 1 *m*, or molal, solution) at 1 atm pressure depresses the freezing point by 1.86°C, raises the boiling point by 0.543°C, lowers the vapor pressure in a temperature-dependent manner, and yields a

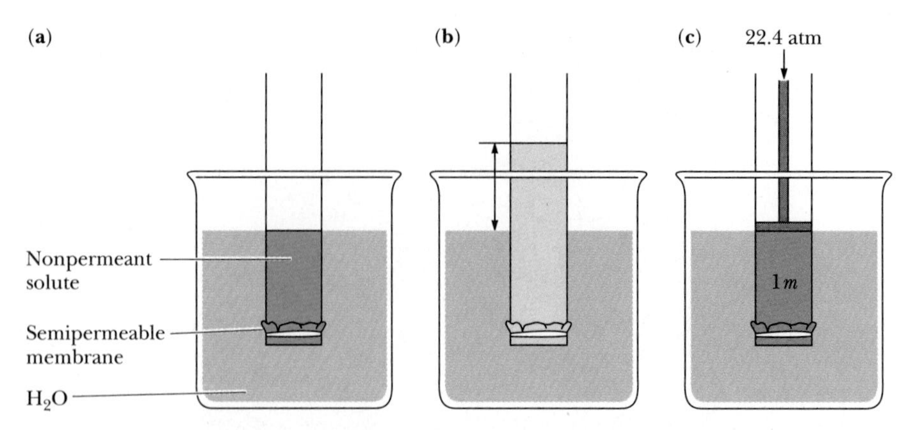

***Figure* 2.8** The osmotic pressure of a 1 molal (*m*) solution is equal to 22.4 atmospheres of pressure. (a) If a nonpermeant solute is separated from pure water by a semipermeable membrane through which H_2O passes freely, (b) water molecules enter the solution (osmosis) and the height of the solution column in the tube rises. The pressure necessary to push water back through the membrane at a rate exactly equaled by the water influx is the osmotic pressure of the solution. (c) For a 1 *m* solution, this force is equal to 22.4 atm of pressure. Osmotic pressure is directly proportional to the concentration of the nonpermeant solute.

solution whose osmotic pressure relative to pure water is 22.4 atm. In effect, by imposing local order on the water molecules, solutes make it more difficult for water to assume its crystalline lattice (freeze) or escape into the atmosphere (boil or vaporize). Furthermore, when a solution (such as the 1 *m* solution discussed here) is separated from a volume of pure water by a semipermeable membrane, the solution will draw water molecules across this barrier. The water molecules are moving from a region of higher effective concentration (pure H_2O) to a region of lower effective concentration (the solution). This movement of water into the solution dilutes the effects of the solute that is present. The osmotic force exerted by each mole of solute is so strong that it requires the imposition of 22.4 atm of pressure to be negated (Figure 2.8).

Osmotic pressure from high concentrations of dissolved solutes is a serious problem for cells. Bacterial and plant cells have strong, rigid cell walls to contain these pressures. In contrast, animal cells are bathed in extracellular fluids of comparable osmolarity, so no net osmotic gradient exists. Also, to minimize the osmotic pressure created by the contents of their cytosol, cells tend to store substances such as amino acids and sugars in polymeric form. For example, a molecule of glycogen or starch containing 1000 glucose units exerts only 1/1000 the osmotic pressure that 1000 free glucose molecules would.

Ionization of Water

Water shows a small but finite tendency to form ions. This tendency is demonstrated by the electrical conductivity of pure water, a property that clearly establishes the presence of charged species (ions). Water ionizes because the larger, strongly electronegative oxygen atom strips the electron from one of its hydrogen atoms, leaving the proton to dissociate (Figure 2.9):

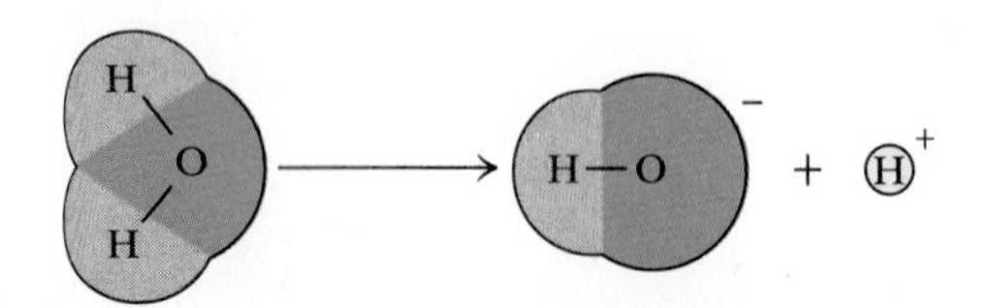

***Figure* 2.9** The ionization of water.

$$H{-}O{-}H \longrightarrow H^+ + OH^-$$

Two ions are thus formed: protons or **hydrogen ions,** H^+, and **hydroxyl ions,** OH^-. Free protons are immediately hydrated to form **hydronium ions,** H_3O^+:

$$H^+ + H_2O \longrightarrow H_3O^+$$

Indeed, since most hydrogen atoms in liquid water are hydrogen bonded to a neighboring water molecule, this protonic hydration is an instantaneous process and the ion products of water are H_3O^+ and OH^-:

$$\text{H}_2\text{O}\cdots\text{H—OH} \longrightarrow \text{H}_2\text{O—H}^+ + \text{OH}^-$$

The amount of H_3O^+ or OH^- in 1 L (liter) of pure water at 25°C is 1×10^{-7} mol; the concentrations are equal since the dissociation is stoichiometric.

While it is important to keep in mind that the hydronium ion, or hydrated hydrogen ion, represents the true state in solution, the convention is to speak of hydrogen ion concentrations in aqueous solution, even though "naked" protons are virtually nonexistent. Indeed, H_3O^+ itself attracts a hydration shell by H bonding to adjacent water molecules to form an $H_9O_4{}^+$ species (Figure 2.10) and even more highly hydrated forms. Similarly, the hydroxyl ion, like all other highly charged species, is also hydrated.

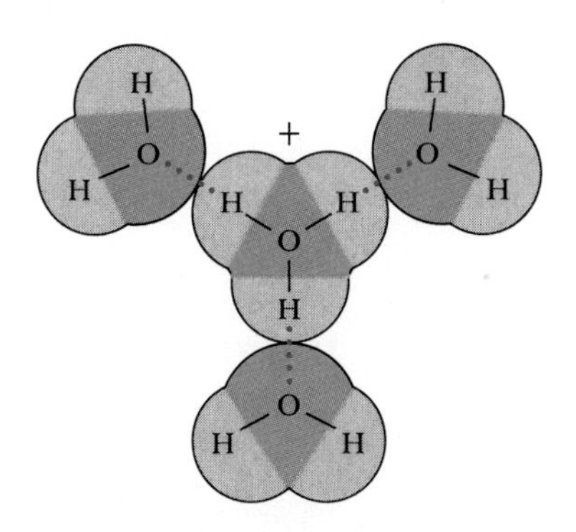

***Figure* 2.10** The hydration of H_3O^+. Solid lines denote covalent bonds; dashed lines represent the H bonds formed between the hydronium ion and its waters of hydration.

Proton Jumping

Because of the high degree of hydrogen bonding in water, H^+ ions show an apparent rate of migration in an electrical field that is vastly greater than other univalent cations in aqueous solution, such as Na^+ and K^+. In effect, the net transfer of a proton from molecule to molecule throughout the H-bonded network accounts for this apparent rapidity of migration (Figure 2.11).

$$\text{H}^+ \rightarrow \text{HO—H}\cdots\text{HO—H}\cdots\text{HO—H}\cdots\text{HO—H}\cdots\text{HO—H}\cdots\text{HO—H}\cdots\text{HO—H}\cdots\text{HO—H}$$

$$\downarrow$$

$$\text{H}_2\text{O} \quad \text{H}_2\text{O} \quad \text{H}_2\text{O} \quad \text{H}_2\text{O} \quad \text{H}_2\text{O} \quad \text{H}_2\text{O} \quad \text{H}_2\text{O} \quad \text{H}_2\text{O} \rightarrow \text{H}^+$$

***Figure* 2.11** Proton jumping via the hydrogen-bonded network of water molecules.

That is, the H-bonded network provides a natural route for rapid H^+ transport. This phenomenon of **proton jumping** thus occurs with little actual movement of the water molecules themselves. Ice has an electrical conductivity close to that of water because such proton jumps also readily occur even when the water molecules are fixed in a crystal lattice. Such conduction of protons via H-bonded molecules has been offered as an explanation for a number of rapid proton transfers of biological significance.

K_w, the Ion Product of Water

The dissociation of water into hydrogen ions and hydroxyl ions occurs to the extent that 10^{-7} mol of H^+ and 10^{-7} mol of OH^- are present at equilibrium in 1 L of water at 25°C.

$$H_2O \longrightarrow H^+ + OH^-$$

The equilibrium constant for this process is

$$K_{eq} = \frac{[H^+][OH^-]}{[H_2O]}$$

where brackets denote concentrations in moles per liter. Since the concentration of H_2O in 1 L of pure water is equal to the number of grams in a liter divided by the gram molecular weight of H_2O, or 1000/18, the molar concentration of H_2O in pure water is 55.5 *M* (molar). The decrease in H_2O concentration as a result of ion formation $[H^+]$, $[OH^-] = 10^{-7}$ *M*) is negligible in comparison and thus its influence on the overall concentration of H_2O can be ignored. Thus,

$$K_{eq} = \frac{(10^{-7})(10^{-7})}{55.5} = 1.8 \times 10^{-16}$$

Since the concentration of H_2O in pure water is essentially constant, a new constant, $\boldsymbol{K_w}$, the **ion product of water,** can be written as

$$K_w = 55.5K_{eq} = 10^{-14} = [H^+][OH^-]$$

The equation has the virtue of revealing the reciprocal relationship between H^+ and OH^- concentrations of aqueous solutions. If a solution is acidic, that is, of significant $[H^+]$, then the ion product of water dictates that the OH^- concentration is correspondingly less. For example, if $[H^+]$ is 10^{-2} *M*, $[OH^-]$ must be 10^{-12} *M* ($K_w = 10^{-14} = [10^{-2}][OH^-]$; $[OH^-] = 10^{-12}$ *M*). Similarly, in an alkaline, or basic, solution in which $[OH^-]$ is great, $[H^+]$ is low.

2.2 pH

To avoid the cumbersome use of negative exponents to express concentrations that range over 14 orders of magnitude, Sørensen, a Danish biochemist, devised the **pH scale** by defining **pH** as *the negative logarithm of the hydrogen ion concentration*[1]*:*

$$pH = -\log_{10} [H^+]$$

Table 2.2 gives the pH scale. Note again the reciprocal relationship between $[H^+]$ and $[OH^-]$. Also because the pH scale is based on negative logarithms, low pH values represent the highest H^+ concentrations (and the lowest OH^- concentrations, as K_w specifies). Note also that

$$pK_w = pH + pOH = 14$$

The pH scale is widely used in biological applications because hydrogen ion concentrations in biological fluids are very low, about 10^{-7} *M* or 0.0000001 *M*, a value more easily represented as pH 7. The pH of blood plasma, for example, is 7.4 or 0.00000004 *M* H^+. Certain disease conditions may lower the plasma pH level to 6.8 or less, a situation that may result in death. At pH 6.8, the H^+ concentration is 0.00000016 *M*, four times greater than at pH 7.4.

At pH 7, $[H^+] = [OH^-]$; that is, there is no excess acidity or basicity. The point of **neutrality** is at pH 7, and solutions having a pH of 7 are said to be at

[1] To be precise in physical chemical terms, the *activities* of the various components, *not* their molar concentrations, should be used in these equations. The activity (a) of a solute component is defined as the product of its molar concentration, c, and an *activity coefficient*, γ: $a = [c]\gamma$. Most biochemical work involves dilute solutions, and the use of activities instead of molar concentrations is usually neglected. However, the concentration of certain solutes may be very high in living cells.

Table 2.2
pH Scale

The hydrogen ion and hydroxyl ion concentrations are given in moles per liter at 25°C.

pH	$[H^+]$		$[OH^-]$	
0	(10^0)	1.0	0.00000000000001	(10^{-14})
1	(10^{-1})	0.1	0.0000000000001	(10^{-13})
2	(10^{-2})	0.01	0.000000000001	(10^{-12})
3	(10^{-3})	0.001	0.00000000001	(10^{-11})
4	(10^{-4})	0.0001	0.0000000001	(10^{-10})
5	(10^{-5})	0.00001	0.000000001	(10^{-9})
6	(10^{-6})	0.000001	0.00000001	(10^{-8})
7	$\mathbf{(10^{-7})}$	**0.0000001**	**0.0000001**	$\mathbf{(10^{-7})}$
8	(10^{-8})	0.00000001	0.000001	(10^{-6})
9	(10^{-9})	0.000000001	0.00001	(10^{-5})
10	(10^{-10})	0.0000000001	0.0001	(10^{-4})
11	(10^{-11})	0.00000000001	0.001	(10^{-3})
12	(10^{-12})	0.000000000001	0.01	(10^{-2})
13	(10^{-13})	0.0000000000001	0.1	(10^{-1})
14	(10^{-14})	0.00000000000001	1.0	(10^0)

neutral pH. The pH values of various fluids of biological origin or relevance are given in Table 2.3. Since the pH scale is a logarithmic scale, two solutions whose pH values differ by one pH unit have a 10-fold difference in $[H^+]$. For example, grapefruit juice at pH 3.2 contains more than 12 times as much H^+ as orange juice at pH 4.3.

Dissociation of Strong Electrolytes

Substances that are almost completely dissociated to form ions in solution are called **strong electrolytes.** The term **electrolyte** describes substances capable of generating ions in solution and thereby causing an increase in the electrical conductivity of the solution. Many salts (such as NaCl and K_2SO_4) fit this category, as do strong acids (such as HCl) and strong bases (such as NaOH). Recall from general chemistry that acids are proton donors and bases are proton acceptors. In effect, the dissociation of a strong acid such as HCl in water can be treated as a proton transfer reaction between the acid HCl and the base H_2O to give the **conjugate acid** H_3O^+ and the **conjugate base** Cl^-:

$$HCl + H_2O \longrightarrow H_3O^+ + Cl^-$$

The equilibrium constant for this reaction is

$$K = \frac{[H_3O^+][Cl^-]}{[H_2O][HCl]}$$

Customarily, since the term $[H_2O]$ is essentially constant in dilute aqueous solutions, it is incorporated into the equilibrium constant K to give a new term, K_a, the acid dissociation constant (where $K_a = K[H_2O]$). Also, the term $[H_3O^+]$ is often replaced by H^+, such that

$$K_a = \frac{[H^+][Cl^-]}{[HCl]}$$

Table 2.3
The pH of Various Common Fluids

Fluid	pH
Household lye	13.6
Bleach	12.6
Household ammonia	11.4
Milk of magnesia	10.3
Baking soda	8.4
Seawater	8.0
Pancreatic fluid	7.8–8.0
Blood plasma	7.4
Intracellular fluids	
Liver	6.9
Muscle	6.1
Saliva	6.6
Urine	5–8
Boric acid	5.0
Beer	4.5
Orange juice	4.3
Grapefruit juice	3.2
Vinegar	2.9
Soft drinks	2.8
Lemon juice	2.3
Gastric juice	1.2–3.0
Battery acid	0.35

For HCl, the value of K_a is exceedingly large because the concentration of HCl in aqueous solution is vanishingly small. Because this is so, the pH of HCl solutions is readily calculated from the amount of HCl used to make the solution:

$$[H^+] \text{ in solution} = [HCl] \text{ added to solution}$$

Thus, a 1 *M* solution of HCl has a pH of 0; a 0.01 *M* HCl solution has a pH of 2. Similarly, a 0.1 *M* NaOH solution has a pH of 13. (Since $[OH^-] = 0.1$ *M*, $[H^+]$ must be 10^{-13} *M*.)

Viewing the dissociation of strong electrolytes another way, we see that the ions formed show little affinity for one another. For example, in HCl in water, Cl^- has very little affinity for H^+:

$$HCl \rightarrow H^+ + Cl^-$$

and in NaOH solutions, Na^+ has little affinity for OH^-. The dissociation of these substances in water is effectively complete.

Dissociation of Weak Electrolytes

Substances with only a slight tendency to dissociate to form ions in solution are called **weak electrolytes.** Acetic acid, CH_3COOH, is a good example:

$$CH_3COOH + H_2O \rightleftharpoons CH_3COO^- + H_3O^+$$

The acid dissociation constant K_a for acetic acid is 1.74×10^{-5}:

$$K_a = \frac{[H^+][CH_3COO^-]}{[CH_3COOH]} = 1.74 \times 10^{-5}$$

The term K_a is also called an **ionization constant** because it states the extent to which a substance forms ions in water. The relatively low value of K_a for acetic acid reveals that the un-ionized form, CH_3COOH, predominates over H^+ and CH_3COO^- in aqueous solutions of acetic acid. Viewed another way, CH_3COO^-, the acetate ion, has a high affinity for H^+.

Example

What is the pH of a 0.1 *M* solution of acetic acid? Or, to restate the question, what is the final pH when 0.1 mol of acetic acid (HAc) is added to water and the volume of the solution is adjusted to equal 1 L?

Answer

The dissociation of HAc in water can be written simply as

$$HAc \rightleftharpoons H^+ + Ac^-$$

where Ac^- represents the acetate ion, CH_3COO^-. In solution, some amount x of HAc dissociates, generating x amount of Ac^- and an equal amount x of H^+. Ionic equilibria characteristically are established very rapidly. At equilibrium, the concentration of HAc + Ac^- must equal 0.1 *M*. So, [HAc] can be represented as $(0.1 - x)$ *M*, and $[Ac^-]$ and $[H^+]$ then both equal x molar. From $1.74 \times 10^{-5} = ([H^+][Ac^-])/[HAc]$, we get $1.74 \times 10^{-5} = x^2/[0.1 - x]$. The solution to quadratic equations of this form ($ax^2 + bx + c = 0$) is $x = (-b \pm \sqrt{b^2 - 4ac})/2a$. However, the calculation of x can be simplified by noting that, since K_a is quite small, $x \ll 0.1$ *M*. Therefore, K_a is essentially equal to $x^2/0.1$. This simplification yields $x^2 = 1.74 \times 10^{-6}$, or $x = 1.32 \times 10^{-3}$ *M* and pH = 2.88.

Henderson–Hasselbalch Equation

Consider the ionization of some weak acid, HA, occurring with an acid dissociation constant, K_a. Then,

$$HA \rightleftharpoons H^+ + A^-$$

and

$$K_a = \frac{[H^+][A^-]}{[HA]}$$

Rearranging this expression in terms of the parameter of interest, $[H^+]$, we have

$$[H^+] = \frac{K_a[HA]}{[A^-]}$$

Taking the logarithm of both sides gives

$$\log [H^+] = \log K_a + \log_{10}\frac{[HA]}{[A^-]}$$

If we change the signs and define $pK_a = -\log K_a$, we have

$$pH = pK_a - \log_{10}\frac{[HA]}{[A^-]}$$

or

$$\mathbf{pH = pK_a + \log_{10}\frac{[A^-]}{[HA]}}$$

This relationship is known as the **Henderson–Hasselbalch equation.** Thus, the pH of a solution can be calculated, provided K_a and the concentrations of the weak acid HA and its conjugate base A^- are known. Note particularly that when $[HA] = [A^-]$, $pH = pK_a$. For example, if equal volumes of 0.1 *M* HAc and 0.1 *M* sodium acetate are mixed, then

$$pH = pK_a = 4.76$$
$$pK_a = -\log K_a = -\log_{10} (1.74 \times 10^{-5}) = 4.76$$

(Sodium acetate, the sodium salt of acetic acid, is a strong electrolyte and dissociates completely in water to yield Na^+ and Ac^-.)

The Henderson–Hasselbalch equation provides a general solution to the quantitative treatment of acid–base equilibria in biological systems. Table 2.4 gives the acid dissociation constants and pK_a values for some weak electrolytes of biochemical interest.

Example

What is the pH when 100 mL of 0.1 *N* NaOH is added to 150 mL of 0.2 *M* HAc if pK_a for acetic acid = 4.76?

Answer

100 mL 0.1 *N* NaOH = 0.01 mol OH^-, which neutralizes 0.01 mol of HAc, giving an equivalent amount of Ac^-:

$$OH^- + HAc \longrightarrow Ac^- + H_2O$$

0.02 mol of the original 0.03 mol of HAc remains essentially undissociated. The final volume is 250 mL.

$$pH = pK_a + \log_{10}\frac{[Ac^-]}{[HAc]} = 4.76 + \log (0.01 \text{ mol})/(0.02 \text{ mol})$$
$$pH = 4.76 - \log_{10} 2 = 4.46$$

Table 2.4

Acid Dissociation Constants and pK_a Values for Some Weak Electrolytes (at 25°C)

Acid	K_a (*M*)	pK_a
HCOOH (formic acid)	1.78×10^{-4}	3.75
CH_3COOH (acetic acid)	1.74×10^{-5}	4.76
CH_3CH_2COOH (propionic acid)	1.35×10^{-5}	4.87
$CH_3CHOHCOOH$ (lactic acid)	1.38×10^{-4}	3.86
$HOOCCH_2CH_2COOH$ (succinic acid) pK_1*	6.16×10^{-5}	4.21
$HOOCCH_2CH_2COO^-$ (succinic acid) pK_2	2.34×10^{-6}	5.63
H_3PO_4 (phosphoric acid) pK_1	7.08×10^{-3}	2.15
$H_2PO_4^-$ (phosphoric acid) pK_2	6.31×10^{-8}	7.20
HPO_4^{2-} (phosphoric acid) pK_3	3.98×10^{-13}	12.40
$C_3N_2H_5^+$ (imidazole)	1.02×10^{-7}	6.99
$C_6O_2N_3H_{11}^+$ (histidine–imidazole group) pK_R†	9.12×10^{-7}	6.04
H_2CO_3 (carbonic acid) pK_1	1.70×10^{-4}	3.77
HCO_3^- (bicarbonate) pK_2	5.75×10^{-11}	10.24
$(HOCH_2)_3CNH_3^+$ (*tris*-hydroxymethyl aminomethane)	8.32×10^{-9}	8.07
NH_4^+ (ammonium)	5.62×10^{-10}	9.25
$CH_3NH_4^+$ (methylammonium)	2.46×10^{-11}	10.62

*These pK values listed as pK_1, pK_2, or pK_3 are in actuality pK_a values for the respective dissociations. This simplification in notation is used throughout this book.

†pK_R refers to the imidazole ionization of histidine.

Data from *CRC Handbook of Biochemistry*, The Chemical Rubber Co., 1968.

If 150 mL of 0.2*M* HAc had merely been diluted with 100 mL of water, this would leave 250 mL of a 0.12*M* HAc solution. The pH would be given by:

$$K_a = \frac{[H^+][Ac^-]}{[HAc]} = \frac{x^2}{0.12\ M} = 1.74 \times 10^{-5}$$
$$x = 1.44 \times 10^{-3} = [H^+]$$
$$pH = 2.84$$

Clearly, the presence of sodium hydroxide has mostly neutralized the acidity of the acetic acid through formation of acetate ion.

Titration Curves

Titration is the analytical method used to determine the amount of acid in a solution. A measured volume of the acid solution is titrated by slowly adding a solution of base, typically NaOH, of known concentration. As incremental amounts of NaOH are added, the pH of the solution is determined and a plot of the pH of the solution versus the amount of OH^- added yields a **titration curve.** The titration curve for acetic acid is shown in Figure 2.12. In considering the progress of this titration, keep in mind two important equilibria:

$$1.\ HAc \rightleftharpoons H^+ + Ac^- \qquad K_a = 1.74 \times 10^{-5}$$
$$2.\ H^+ + OH^- \longrightarrow H_2O \quad K = \frac{[H_2O]}{K_w} = 5.55 \times 10^{15}$$

As the titration begins, mostly HAc is present, plus some H^+ and Ac^- in amounts that can be calculated (see the Example on page 44). Addition of a solution of NaOH allows hydroxide ion to neutralize any H^+ present. Note that reaction (2) as written is strongly favored; its apparent equilibrium constant is greater than 10^{15}! As H^+ is neutralized, more HAc will dissociate to H^+ and Ac^-. As further NaOH is added, the pH gradually increases as Ac^- accumulates at the expense of diminishing HAc and the neutralization of H^+. At the point where half of the HAc has been neutralized, that is, where 0.5 equivalent of OH^- has been added, the concentrations of HAc and Ac^- are equal and pH = pK_a for HAc. Thus, we have an experimental method for determining the pK_a values of weak electrolytes. These pK_a values lie at the midpoint of their respective titration curves. After all of the acid has been neutralized (that is, when one equivalent of base has been added), the pH rises exponentially.

The shapes of the titration curves of weak electrolytes are identical, as Figure 2.13 reveals. Note, however, that the midpoints of the different curves vary in a way that characterizes the particular electrolytes. The pK_a for acetic acid is 4.76, the pK_a for imidazole is 6.99, and that for ammonium is 9.25. These pK_a values are directly related to the dissociation constants of these substances, or viewed the other way, to the relative affinities of the conjugate bases for protons. NH_3 has a high affinity for protons compared to Ac^-; NH_4^+ is a poor acid compared to HAc.

Phosphoric Acid Has Three Dissociable H^+

Figure 2.14 shows the titration curve for phosphoric acid, H_3PO_4. This substance is a *polyprotic acid,* meaning it has more than one dissociable proton. Indeed, it has three, and thus three equivalents of OH^- are required to neutralize it, as Figure 2.14 shows. Note that the three dissociable H^+ are lost in discrete steps, each dissociation showing a characteristic pK_a. Note that pK_1

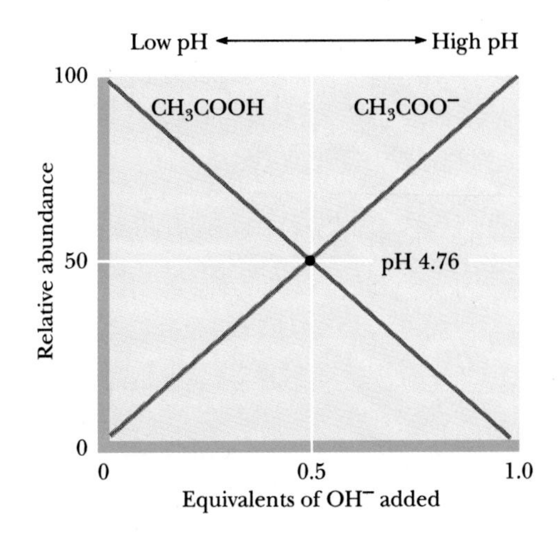

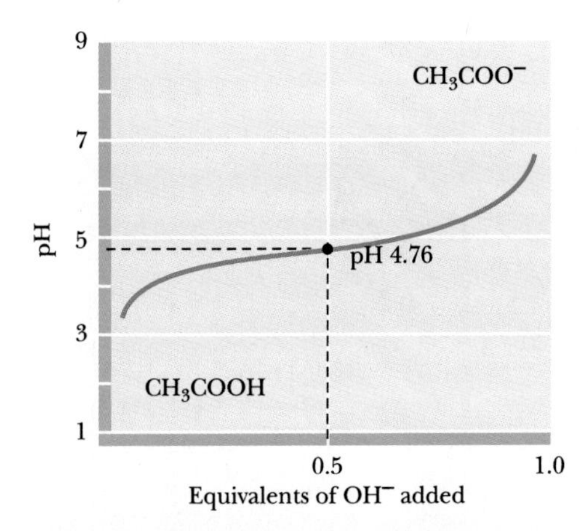

***Figure* 2.12** The titration curve for acetic acid. Note that the titration curve is relatively flat at pH values near the pK_a; in other words, the pH changes relatively little as OH^- is added in this region of the titration curve.

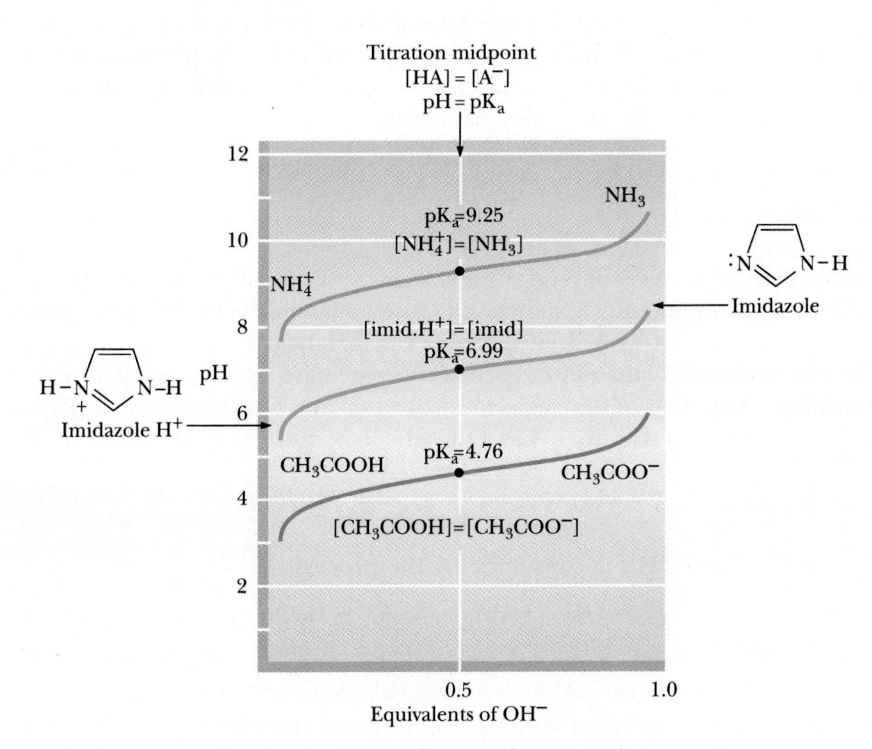

***Figure* 2.13** The titration curves of several weak electrolytes: acetic acid, imidazole, and ammonium. Note that the shape of these different curves is identical. Only their position along the pH scale is displaced, in accordance with their respective affinities for H^+ ions, as reflected in their differing pK_a values.

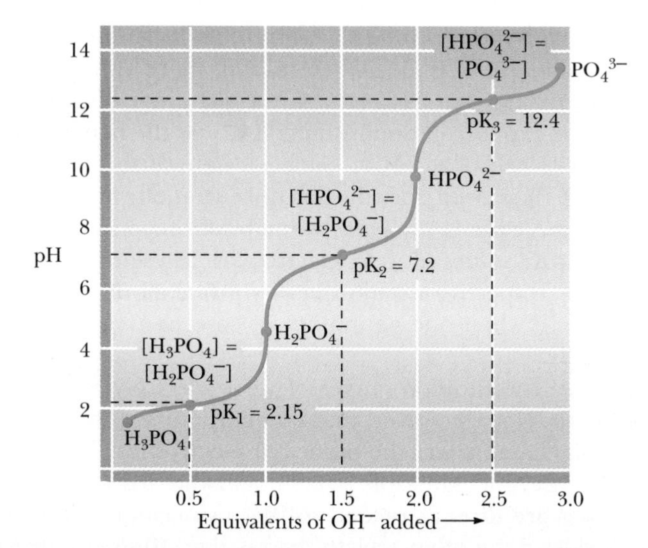

***Figure* 2.14** The titration curve for phosphoric acid. The chemical formulas show the prevailing ionic species present at various pH values. Phosphoric acid (H_3PO_4) has three titratable hydrogens and therefore three midpoints are seen: at pH 2.15 (pK_1), pH 7.20 (pK_2), and pH 12.4 (pK_3).

occurs at pH = 2.15, and the concentrations of the acid H_3PO_4 and the conjugate base $H_2PO_4^-$ are equal. As the next dissociation is approached, $H_2PO_4^-$ is treated as the acid and HPO_4^{2-} is its conjugate base. Their concentrations are equal at pH 7.20, so $pK_2 = 7.20$. (Note that at this point, 1.5 equivalents of OH^- have been added.) As more OH^- is added, the last dissociable hydrogen is titrated, and pK_3 occurs at pH = 12.4, where $[HPO_4^{2-}] = [PO_4^{3-}]$.

A biologically important point is revealed by the basic shape of the titration curves of weak electrolytes: in the region of the pK_a, pH remains relatively unaffected as increments of OH^- (or H^+) are added. The weak acid and its conjugate base are acting as a buffer.

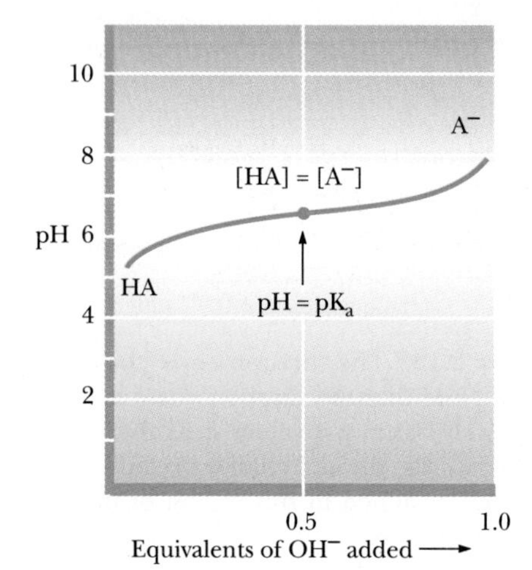

***Figure* 2.15** Buffer systems consist of a weak acid, HA, and its conjugate base, A^-. The pH varies only slightly in the region of the titration curve where [HA] = $[A^-]$. The unshaded box denotes this area of greatest buffering capacity. Buffer action: when HA and A^- are both available in sufficient concentration, the solution can absorb input of either H^+ or OH^-, and pH is maintained essentially constant.

2.3 Buffers

Buffers are solutions that tend to resist changes in their pH as acid or base is added. Typically, a buffer system is composed of a weak acid *and* its conjugate base. A solution of a weak acid that has a pH nearly equal to its pK_a by definition contains an amount of the conjugate base nearly equivalent to the weak acid. Note that in this region, the titration curve is relatively flat (Figure 2.15). Addition of H^+ then has little effect because it is absorbed by the following reaction:

$$H^+ + A^- \longrightarrow HA$$

Similarly, added OH^- is consumed by the process

$$OH^- + HA \longrightarrow A^- + H_2O$$

The pH then remains relatively constant. The components of a buffer system are chosen such that the pK_a of the weak acid is close to the pH of interest. It is at the pK_a that the buffer system shows its greatest buffering capacity. At pH values more than 1 pH unit from the pK_a, buffer systems become ineffective because the concentration of one of the components is too low to absorb the influx of H^+ or OH^-. The molarity of a buffer is defined as the *sum* of the concentrations of the acid and conjugate base forms.

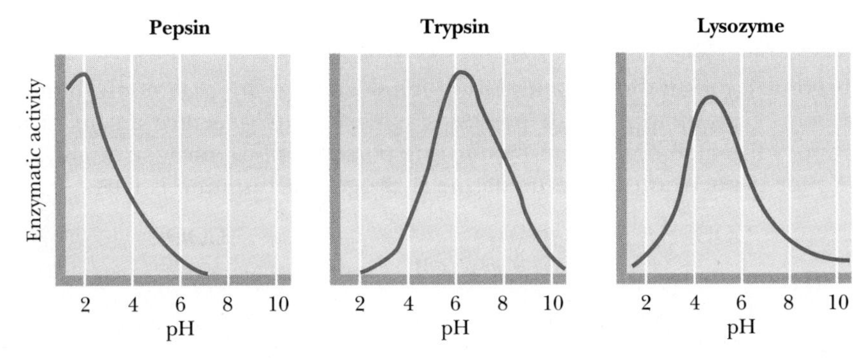

***Figure* 2.16** pH versus enzymatic activity. The activity of enzymes is very sensitive to pH. The pH optimum of an enzyme is one of its most important characteristics. Pepsin is a protein-digesting enzyme active in the gastric fluid. Trypsin is also a proteolytic enzyme, but it acts in the more alkaline milieu of the small intestine. Lysozyme digests the cell walls of bacteria; it is found in tears.

Maintenance of pH is vital to all cells. Cellular processes such as metabolism are dependent on the activities of enzymes, and in turn, enzyme activity is markedly influenced by pH, as the graphs in Figure 2.16 show. Consequently, changes in pH would be very disruptive to metabolism for reasons that will become apparent in later chapters. Organisms have a variety of mechanisms to keep the pH of their intra- and extracellular fluids essentially constant, but the primary protection against harmful pH changes is provided by buffer systems. The buffer systems selected reflect both the need for a pK_a value near pH 7 and the compatibility of the buffer components with the metabolic machinery of cells. Two buffer systems act to maintain intracellular pH essentially constant—the phosphate ($HPO_4^{2-}/H_2PO_4^-$) system and the histidine system. The pH of the extracellular fluid that bathes the cells and tissues of animals is maintained by the bicarbonate/carbonic acid (HCO_3^-/H_2CO_3) system.

Phosphate System

The **phosphate system** serves to buffer the intracellular fluid of cells at physiological pH because pK_2 lies near this pH value. The intracellular pH of most cells is maintained in the range between 6.9 and 7.4. Phosphate is an abundant anion in cells, both in inorganic form and as an important substituent of organic molecules that serve as metabolites or macromolecular precursors. In both organic and inorganic forms, its characteristic pK_2 means that the ionic species present at physiological pH are sufficient to donate or accept hydrogen ions to buffer any changes in pH, as inspection of the titration curve for H_3PO_4 in Figure 2.14 reveals. For example, if the total cellular concentration of phosphate is 20 mM (millimolar) and the pH is 7.4, the distribution of the major phosphate species is given by

$$\mathrm{pH} = \mathrm{p}K_2 + \log_{10}\frac{[\mathrm{HPO_4}^{2-}]}{[\mathrm{H_2PO_4}^-]}$$

$$7.4 = 7.20 + \log_{10}\frac{[\mathrm{HPO_4}^{2-}]}{[\mathrm{H_2PO_4}^-]}$$

$$\frac{[\mathrm{HPO_4}^{2-}]}{[\mathrm{H_2PO_4}^-]} = 1.58$$

Thus, if $[HPO_4^{2-}] + [H_2PO_4^-] = 20$ mM, then

$$[\mathrm{HPO_4}^{2-}] = 12.25\ \mathrm{m}M \quad \text{and} \quad [\mathrm{H_2PO_4}^-] = 7.75\ \mathrm{m}M$$

Histidine System

Histidine is one of the 20 naturally occurring amino acids commonly found in proteins (see Chapter 3). It possesses as part of its structure an imidazole group, a five-membered heterocyclic ring possessing two nitrogen atoms. The pK_a for dissociation of the imidazole hydrogen of histidine is 6.04.

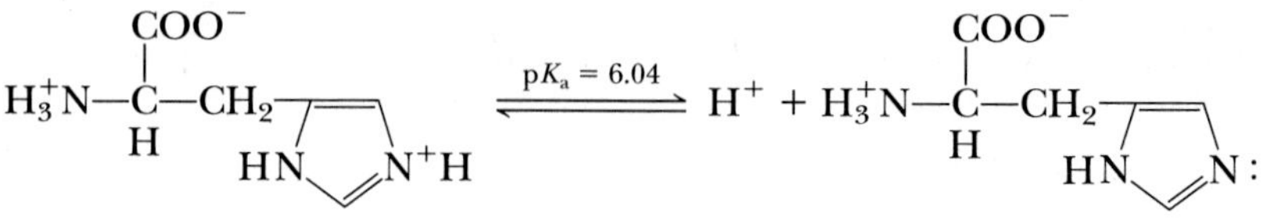

$$H_3^+N{-}CH(COO^-){-}CH_2{-}\text{(imidazole, HN–N}^+\text{H)} \underset{}{\overset{pK_a = 6.04}{\rightleftharpoons}} H^+ + H_3^+N{-}CH(COO^-){-}CH_2{-}\text{(imidazole, HN–N:)}$$

In cells, histidine occurs as the free amino acid, as a constituent of proteins, and as part of dipeptides in combination with other amino acids. Because the concentration of free histidine is low and its imidazole pK_a is more than 1 pH unit removed from prevailing intracellular pH, its role in intracellular buffering is minor. However, protein-bound and dipeptide histidine may be the dominant buffering system in some cells. In combination with other amino acids, as in proteins or dipeptides, the imidazole pK_a may increase substantially. For example, the imidazole pK_a is 7.04 in **anserine,** a dipeptide containing β-alanine and histidine (Figure 2.17). Thus, this pK_a is near physiological pH, and some histidine peptides are well suited for buffering at physiological pH.

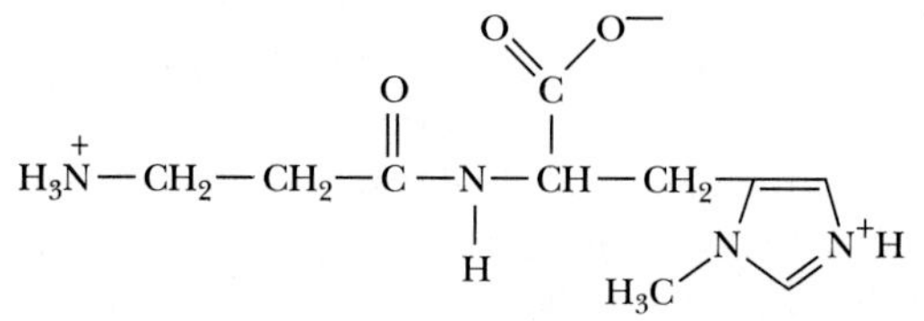

$$H_3\overset{+}{N}{-}CH_2{-}CH_2{-}\overset{O}{\overset{\|}{C}}{-}\underset{H}{\underset{|}{N}}{-}\overset{COO^-}{\overset{|}{C}H}{-}CH_2{-}\text{(3-methylimidazole, } H_3C{-}N\text{, N}^+\text{H)}$$

Figure **2.17** Anserine (*N*-β-alanyl-3-methyl-L-histidine) is an important dipeptide buffer in the maintenance of intracellular pH in some tissues. The structure shown is the predominant ionic species at pH 7. pK_1 (COOH) = 2.64; pK_2 (imidazole–N^+H) = 7.04; pK_3 (NH_3^+) = 9.49.

The Bicarbonate Buffer System of Blood Plasma

The important buffer system of blood plasma is the bicarbonate/carbonic acid couple:

$$H_2CO_3 \rightleftharpoons H^+ + HCO_3^-$$

The relevant pK_a is pK_1 for carbonic acid, which has a value of 3.80 at 37°C, far removed from the normal pH of blood plasma (pH 7.4). At pH 7.4, the concentration of H_2CO_3 is a minuscule fraction of the HCO_3^- concentration, and thus the plasma would appear to be poorly protected against an influx of OH^- ions.

$$pH = 7.4 = 3.8 + \log_{10}\frac{[HCO_3^-]}{[H_2CO_3]}$$

$$\frac{[HCO_3^-]}{[H_2CO_3]} = 3981$$

For example, if $[HCO_3^-] = 4$ mM, then $[H_2CO_3]$ is barely 1 μM (10^{-6} M), and an equivalent amount of OH^- would swamp the buffer system, causing a dangerous rise in the plasma pH. How, then, can this bicarbonate system function effectively? The bicarbonate buffer system works well because the critical concentration of H_2CO_3 is maintained relatively constant through equilibrium with dissolved CO_2 produced in the tissues and available as a gaseous CO_2 reservoir in the lungs.[2]

"Good" Buffers

Not many common substances have pK_a values in the range from 6 to 8. Consequently, biochemists conducting *in vitro* experiments were limited in their choice of buffers effective at or near physiological pH. In 1966, N. E.

A Deeper Look

How the Bicarbonate Buffer System Works

Gaseous carbon dioxide from the lungs and tissues is dissolved in the blood plasma, symbolized as CO_2(d), and hydrated to form H_2CO_3:

$$CO_2(g) \rightleftharpoons CO_2(d)$$
$$CO_2(d) + H_2O \rightleftharpoons H_2CO_3$$
$$H_2CO_3 \rightleftharpoons H^+ + HCO_3^-$$

Thus, the concentration of H_2CO_3 is itself buffered by the available pools of CO_2. The hydration of CO_2 is actually mediated by an enzyme, *carbonic anhydrase,* which facilitates the equilibrium by rapidly catalyzing the reaction

$$H_2O + CO_2(d) \rightleftharpoons H_2CO_3$$

Under the conditions of temperature and ionic strength prevailing in mammalian body fluids, the equilibrium for this reaction lies far to the left, such that about 500 CO_2 molecules are present in solution for every molecule of H_2CO_3. Since dissolved CO_2 and H_2CO_3 are in equilibrium, the proper expression for H_2CO_3 availability is $[CO_2(d)] + [H_2CO_3]$, the so-called total carbonic acid pool, consisting primarily of CO_2(d). The overall equilibrium for the bicarbonate buffer system then is

$$CO_2(d) + H_2O \overset{K_h}{\rightleftharpoons} H_2CO_3$$
$$H_2CO_3 \overset{K_a}{\rightleftharpoons} H^+ + HCO_3^-$$

An expression for the ionization of H_2CO_3 under such conditions (that is, in the presence of dissolved CO_2) can be obtained from K_h, the equilibrium constant for the hydration of CO_2, and from K_a, the first acid dissociation constant for H_2CO_3:

$$K_h = \frac{[H_2CO_3]}{[CO_2(d)]}$$

Thus,

$$[H_2CO_3] = K_h[CO_2(d)]$$

Putting this value for $[H_2CO_3]$ into the expression for the first dissociation of H_2CO_3 gives

$$K_a = \frac{[H^+][HCO_3^-]}{[H_2CO_3]} = \frac{[H^+][HCO_3^-]}{K_h[CO_2(d)]}$$

Therefore, the overall equilibrium constant for the ionization of H_2CO_3 in equilibrium with CO_2(d) is given by

$$K_aK_h = \frac{[H^+][HCO_3^-]}{[CO_2(d)]}$$

and K_aK_h, the product of two constants, can be defined as a new equilibrium constant, $\boldsymbol{K_{overall}}$. The value of K_h is 0.0031 at 37°C and K_a, the ionization constant for H_2CO_3, is $10^{-3.8}$ = 0.000158. Therefore,

$$K_{overall} = (0.000158)(0.0031) = 4.91 \times 10^{-7}$$
$$pK_{overall} = 6.31$$

which yields the following Henderson–Hasselbalch relationship:

$$pH = pK_{overall} + \log_{10}\frac{[HCO_3^-]}{[CO_2(d)]}$$

Although the prevailing blood pH of 7.4 is more than 1 pH unit away from p$K_{overall}$, the bicarbonate system is still an effective buffer. At a pH of more than 1 pH unit below blood pH, the concentration of the acid component of the buffer is less than 10% of the conjugate base component. One might imagine that this buffer component could be overwhelmed by relatively small amounts of alkali, with consequent disastrous rises in blood pH. However, the acid component is the total carbonic acid pool, that is, $[CO_2(d)] + [H_2CO_3]$, which is stabilized by its equilibrium with CO_2(g). The gaseous CO_2 buffers any losses from the total carbonic acid pool by entering solution as CO_2(d), and blood pH is effectively maintained. Thus, the bicarbonate buffer system is an *open system.* The natural presence of CO_2 gas at a partial pressure of 40 mm Hg in the alveoli of the lungs and the equilibrium

$$CO_2(g) \rightleftharpoons CO_2(d)$$

keeps the concentration of CO_2(d) (the principal component of the total carbonic acid pool in blood plasma) in the neighborhood of 1.2 m*M*. Plasma $[HCO_3^-]$ is slightly less than 15 m*M* under such conditions.

[2]Well-fed human adults exhale about 1 kg of CO_2 daily. Imagine the excretory problem if CO_2 were not a volatile gas!

Good devised a set of synthetic buffers to remedy this problem, and over the years the list has expanded so that a ''good'' selection is available over the pH range from 6 to 10 (Figure 2.18).

2.4 Water's Unique Role in the Fitness of the Environment

The remarkable properties of water render it particularly suitable to its unique role in living processes and the environment, and its presence in abundance favors the existence of life. Let's examine water's physical and chemical properties to see the extent to which they provide conditions that are advantageous to organisms.

As a *solvent* water is powerful yet innocuous. No other chemically inert solvent compares with water for the substances it can dissolve. Also, it is very important to life that water is a ''poor'' solvent for nonpolar substances. Thus, through hydrophobic interactions, lipids coalesce, membranes form, boundaries are created delimiting compartments, and the cellular nature of life is established. Because of its very high dielectric constant, water is a medium for ionization. Ions enrich the living environment in that they enhance the variety of chemical species and introduce an important class of chemical reactions. They provide electrical properties to solutions and therefore to organisms. Aqueous solutions are the prime source of ions.

The *thermal properties* of water are especially relevant to its environmental fitness. It has great power as a buffer resisting thermal (temperature) change. Its heat capacity, or specific heat (4.1840 J/g°C), is remarkably high; it is ten times greater than iron, five times greater than quartz or salt, and twice as great as hexane. Its heat of fusion is 335 J/g. Thus, at 0°C, it takes a loss of 335 J to change the state of 1 g of H_2O from liquid to solid. Its heat of vaporization, 2.24 kJ/g, is exceptionally high. These thermal properties mean that it takes substantial changes in heat content to alter the temperature and especially the *state* of water. Water's thermal properties allow it to buffer the climate through such processes as condensation, evaporation, melting, and freezing. Furthermore, these properties allow effective temperature regulation in living organisms. For example, heat generated within an organism as a result of metabolism can be efficiently eliminated by evaporation or conduction. The thermal conductivity of water is very high in comparison with other liquids. The anomalous expansion of water as it cools to temperatures near its freezing point is a unique attribute of great significance to its natural fitness. As water cools, H bonding increases because the thermal motions of the molecules are lessened. Hydrogen bonding tends to separate the water molecules (Figure 2.2), and thus the density of water decreases. These changes in density mean that, at temperatures below 4°C, cool water rises and, most importantly, ice freezes on the surfaces of bodies of water, forming an insulating layer protecting the liquid water underneath.

Water has the highest *surface tension* (75 dyn/cm) of all common liquids (except mercury). Together, surface tension and density determine how high a liquid will rise in a capillary system. Capillary movement of water plays a prominent role in the life of plants. Lastly, consider *osmosis,* the bulk movement of water in the direction from a dilute aqueous solution to a more concentrated one across a semipermeable boundary. Such bulk movements determine the shape and form of living things.

Water is truly a crucial determinant of the fitness of the environment. In a very real sense, organisms are aqueous systems in a watery world.

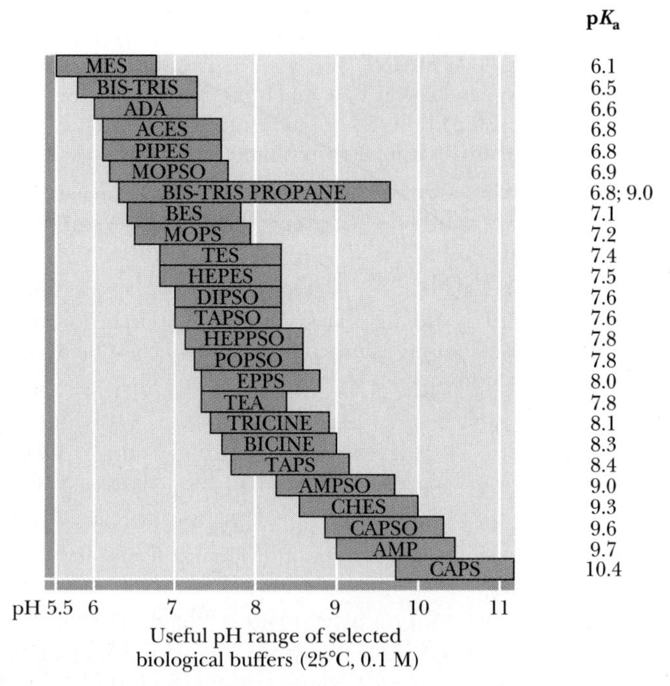

***Figure* 2.18** Biological buffers. These buffers cover the pH range of most biochemical interest.

ACES: 2-[(2-amino-2-oxoethyl)-amino]ethanesulfonic acid, pK_a = 6.8
ADA: *N*-[2-acetamido]-2-iminodiacetic acid, pK_a = 6.6
AMP: 2-amino-2-methyl-1-propanol, pK_a = 9.8
AMPSO: 3-[(1,1-dimethyl-2-hydroxyethyl)amino]-2-hydroxypropanesulfonic acid, pK_a = 9.0
BES: *N,N-bis*[2-hydroxyethyl]-2-aminoethanesulfonic acid, pK_a = 7.1
BICINE: *N,N-bis*[2-hydroxyethyl]glycine, pK_a = 8.3
BIS-TRIS: *bis*[2-hydroxyethyl]iminotris[hydroxymethyl]methane, pK_a = 6.5
BIS-TRIS PROPANE: 1,2-*bis*[*tris*(hydroxymethyl)] methylaminopropane, pK_{a1} = 6.8, pK_{a2} = 9.0
CAPS: 3-[cyclohexylamino]-1-propanesulfonic acid, pK_a = 10.4
CAPSO: 3-[cyclohexylamino]-2-hydroxy-1-propanesulfonic acid, pK_a = 9.6
CHES: 2-[*N*-cyclohexylamino]ethanesulfonic acid, pK_a = 9.3
DIPSO: 3-[*N,N*-bis(2-hydroxyethyl)amino]-2-hydroxypropanesulfonic acid, pK_a = 7.6
EPPS: *N*-[2-hydroxyethyl]piperazine-*N'*-[3-propanesulfonic acid], pK_a = 8.0
HEPES: *N*-[2-hydroxyethyl]piperazine-*N'*-[2-ethanesulfonic acid], pK_a = 7.5
HEPPSO: *N*-[2-hydroxyethyl]piperazine-*N'*-[2-hydroxypropanesulfonic acid], pK_a = 7.8
MES: 2-[*N*-morpholino]ethanesulfonic acid, pK_a = 6.1
MOPS: 3-[*N*-morpholino]propanesulfonic acid, pK_a = 7.2
MOPSO: 3-[*N*-morpholino]-2-hydroxypropanesulfonic acid, pK_a = 6.9
PIPES: piperazine-*N,N'-bis*[2-ethanesulfonic acid], pK_a = 6.8
POPSO: piperazine-*N,N'-bis*[2-hydroxypropanesulfonic acid], pK_a = 7.8
TAPS: *N-tris*[hydroxymethyl]methyl-3-aminopropanesulfonic acid, pK_a = 8.4
TAPSO: 3-[*N-tris*(hydroxymethyl)methylamino]-2-hydroxypropanesulfonic acid, pK_a = 7.6
TEA: triethanolamine, pK_a = 7.8
TES: *N-tris*[hydroxymethyl]methyl-2-aminoethanesulfonic acid, pK_a = 7.4
TRICINE: *N-tris*[hydroxymethyl]methylglycine, pK_a = 8.1

Problems

1. Calculate the pH of the following.
   a. 5×10^{-4} *M* HCl
   b. 7×10^{-5} *M* NaOH
   c. 2×10^{-6} *M* HCl
   d. 3×10^{-2} *M* KOH
   e. 4×10^{-5} *M* HCl
   f. 6×10^{-9} *M* HCl

2. Calculate the following from the pH values given in Table 2.3.
   a. $[H^+]$ in vinegar
   b. $[H^+]$ in saliva
   c. $[H^+]$ in household ammonia
   d. $[OH^-]$ in milk of magnesia
   e. $[OH^-]$ in beer
   f. $[H^+]$ inside a liver cell
3. The pH of a 0.02 *M* solution of an acid was measured at 4.6.
   a. What is the $[H^+]$ in this solution?
   b. Calculate the acid dissociation constant K_a and pK_a for this acid.
4. The K_a for formic acid is 1.78×10^{-4} *M*.
   a. What is the pH of a 0.1 *M* solution of formic acid?
   b. 150 mL of 0.1 *M* NaOH is added to 200 mL of 0.1 *M* formic acid, and water is added to give a final volume of 1 L. What is the pH of the final solution?
5. Given 0.1 *M* solutions of acetic acid and sodium acetate, describe the preparation of 1 L of 0.1 *M* acetate buffer at a pH of 5.4.
6. If the internal pH of a muscle cell is 6.8, what is the $[HPO_4^{2-}]/[H_2PO_4^-]$ ratio in this cell?
7. Given 0.1 *M* solutions of Na_3PO_4 and H_3PO_4, describe the preparation of 1 L of a phosphate buffer at a pH of 7.5. What are the molar concentrations of the ions in the final buffer solution, including Na^+ and H^+?
8. BICINE is a compound containing a tertiary amino group whose relevant pK_a is 8.3 (Figure 2.18). Given 1 L of 0.05 *M* BICINE with its tertiary amino group in the unprotonated form, how much 0.1 *N* HCl must be added to have a BICINE buffer solution of pH 7.5? What is the molarity of BICINE in the final buffer?
9. What are the approximate fractional concentrations of the following phosphate species at pH values of 0, 2, 4, 6, 8, 10, and 12?
   a. H_3PO_4 b. $H_2PO_4^-$ c. HPO_4^{2-} d. PO_4^{3-}
10. Citric acid, a tricarboxylic acid important in intermediary metabolism, can be symbolized as H_3A. Its dissociation reactions are

$$H_3A \rightleftharpoons H^+ + H_2A^- \qquad pK_1 = 3.13$$
$$H_2A^- \rightleftharpoons H^+ + HA^{2-} \qquad pK_2 = 4.76$$
$$HA^{2-} \rightleftharpoons H^+ + A^{3-} \qquad pK_3 = 6.40$$

    If the *total* concentration of the acid *and* its anion forms is 0.02 *M*, what are the individual concentrations of H_3A, H_2A^-, HA^{2-}, and A^{3-} at pH 5.2?
11. a. If 50 mL of 0.01 *M* HCl is added to 100 mL of 0.05 *M* phosphate buffer at pH 7.2, what is the resultant pH? What are the concentrations of $H_2PO_4^-$ and HPO_4^{2-} in the final solution?
    b. If 50 mL of 0.01 *M* NaOH is added to 100 mL of 0.05 *M* phosphate buffer at pH 7.2, what is the resultant pH? What are the concentrations of $H_2PO_4^-$ and HPO_4^{2-} in this final solution?

Further Reading

Cooper, T. G., 1977. *The Tools of Biochemistry,* Chap. 1. New York: John Wiley & Sons.

Darvey, I. G., and Ralston, G. B., 1993. Titration curves—misshapen or mislabeled? *Trends in Biochemical Sciences* **18**:69–71.

Edsall, J. T., and Wyman, J., 1958. Carbon dioxide and carbonic acid, in *Biophysical Chemistry,* Vol. 1, Chap. 10. New York: Academic Press.

Franks, F., ed., 1982. *The Biophysics of Water.* New York: John Wiley & Sons.

Henderson, L. J., 1913. *The Fitness of the Environment.* New York: Macmillan Co. (Republished 1970. Gloucester, MA: P. Smith).

Hille, B., 1992. *Ionic Channels of Excitable Membranes,* 2nd ed., Chap. 10. Sunderland, MA: Sinauer Associates.

Masoro, E. J., and Siegel, P. D., 1971. *Acid–Base Regulation: Its Physiology and Pathophysiology.* Philadelphia: W. B. Saunders Co.

Segel, I. H., 1976. *Biochemical Calculations,* 2nd ed., Chap. 1. New York: John Wiley & Sons.

Stillinger, F. H., 1980. Water revisited. *Science* **209**:451–457.

"To hold, as 'twere, the mirror up to nature."

William Shakespeare, *Hamlet*

Chapter 3

Amino Acids

All objects have mirror images. Like many biomolecules, amino acids exist in mirror-image forms (stereoisomers) that are not superimposable. Only the L-isomers of amino acids commonly occur in nature.

Outline

3.1 Amino Acids: Building Blocks of Proteins

3.2 Acid–Base Chemistry of Amino Acids

3.3 Reactions of Amino Acids

3.4 Optical Activity and Stereochemistry of Amino Acids

3.5 Spectroscopic Properties of Amino Acids

3.6 Separation and Analysis of Amino Acid Mixtures

Proteins are the indispensable agents of biological function, and **amino acids** are the building blocks of proteins. The stunning diversity of the thousands of proteins found in nature arises from the intrinsic properties of only 20 commonly occurring amino acids. These features include (1) the capacity to polymerize, (2) novel acid–base properties, (3) varied structure and chemical functionality in the amino acid side chains, and (4) chirality. This chapter describes each of these properties, laying a foundation for discussions of protein structure (Chapters 4 and 5), enzyme function (Chapters 11–14), and many other subjects in later chapters.

3.1 Amino Acids: Building Blocks of Proteins

Structure of a Typical Amino Acid

The structure of a single typical amino acid is shown in Figure 3.1. Central to this structure is the tetrahedral alpha (α) carbon (C_α), which is covalently linked to both the amino group and the carboxyl group. Also bonded to this α-carbon is a hydrogen and a variable side chain. It is the side chain, the so-called R group, that gives each amino acid its identity. The detailed acid–base properties of amino acids are discussed in the following sections. It is sufficient for now to realize that, in neutral solution (pH 7), the carboxyl group exists as $—COO^-$ and the amino group as $—NH_3^+$. Since the resulting amino acid contains one positive and one negative charge, it is a neutral molecule called a **zwitterion.** Amino acids are also *chiral* molecules. With four different groups attached to it, the α-carbon is said to be *asymmetric.* The two possible configurations for the α-carbon constitute nonidentical mirror image isomers or *enantiomers.* Details of amino acid stereochemistry are discussed in Section 3.4.

Amino Acids Can Join via Peptide Bonds

The crucial feature of amino acids that allows them to polymerize to form peptides and proteins is the existence of their two identifying chemical groups: the amino ($—NH_3^+$) and carboxyl ($—COO^-$) groups, as shown in Figure 3.2. The amino and carboxyl groups of amino acids can react in a head-to-tail fashion, eliminating a water molecule and forming a covalent amide linkage, which, in the case of peptides and proteins, is typically referred to as a **peptide bond.** The equilibrium for this reaction in aqueous solution favors peptide bond hydrolysis. For this reason, biological systems as well as peptide chemists in the laboratory must carry out peptide bond formation in an indirect manner or with energy input.

Iteration of the reaction shown in Figure 3.2 produces **polypeptides** and **proteins.** The remarkable properties of proteins, which we shall discover and come to appreciate in later chapters, all depend in one way or another on the unique properties and chemical diversity of the 20 common amino acids found in proteins.

Common Amino Acids

The structures and abbreviations for the 20 amino acids commonly found in proteins are shown in Figure 3.3. All the amino acids except proline have both free α-amino and free α-carboxyl groups (Figure 3.1). There are several ways to classify the common amino acids. The most useful of these classifications is based on the polarity of the side chains. Thus, the structures shown in Figure 3.3 are grouped into the following categories: (1) nonpolar or hydrophobic

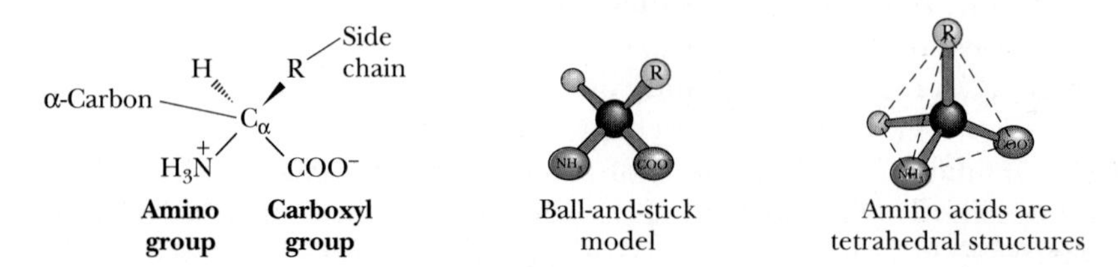

***Figure* 3.1** Anatomy of an amino acid. Except for proline and its derivatives, all of the amino acids commonly found in proteins possess this type of structure.

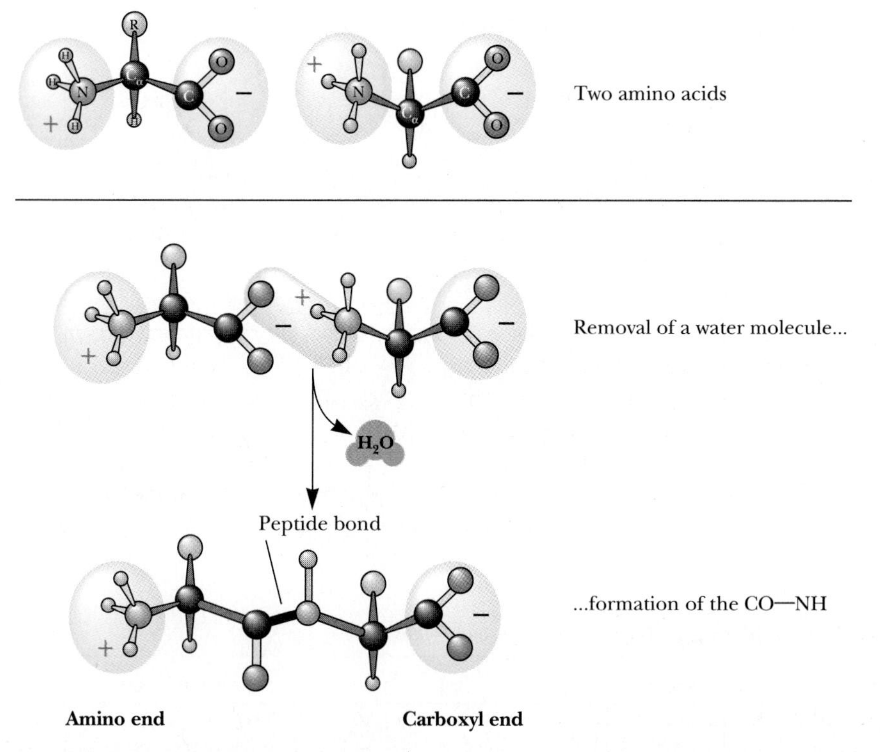

***Figure* 3.2** The α-COOH and α-NH_3^+ groups of two amino acids can react with the resulting loss of a water molecule to form a covalent amide bond.
(Figure courtesy of Irving Geis.)

amino acids, (2) neutral (uncharged) but polar amino acids, (3) acidic amino acids (which have a net negative charge at pH 7.0), and (4) basic amino acids (which have a net positive charge at neutral pH). In later chapters, the importance of this classification system for predicting protein properties will become clear. Also shown in Figure 3.3 are the three-letter and one-letter codes used to represent the amino acids. These codes are useful when displaying and comparing primary sequences of proteins in shorthand form. (Note that several of the one-letter abbreviations are phonetic in origin: arginine = "Rginine" = R, phenylalanine = "Fenylalanine" = F, aspartic acid = "asparDic" = D.)

Nonpolar Amino Acids

The nonpolar amino acids (Figure 3.3a) include all those with alkyl chain R groups (alanine, valine, leucine, and isoleucine), as well as proline (with its unusual cyclic structure), methionine (one of the two sulfur-containing amino acids), and two aromatic amino acids, phenylalanine and tryptophan. Tryptophan is sometimes considered a borderline member of this group since it can interact favorably with water via the N–H moiety of the indole ring. Proline, strictly speaking, is not an amino acid but rather an α-imino acid.

Polar, Uncharged Amino Acids

The polar, uncharged amino acids (Figure 3.3b) except for glycine contain R groups that can form hydrogen bonds with water. Thus, these amino acids are usually more soluble in water than the nonpolar amino acids. Several exceptions should be noted. Tyrosine displays the lowest solubility in water of the 20 common amino acids (0.453 g/L at 25°C). Also, proline is very soluble in

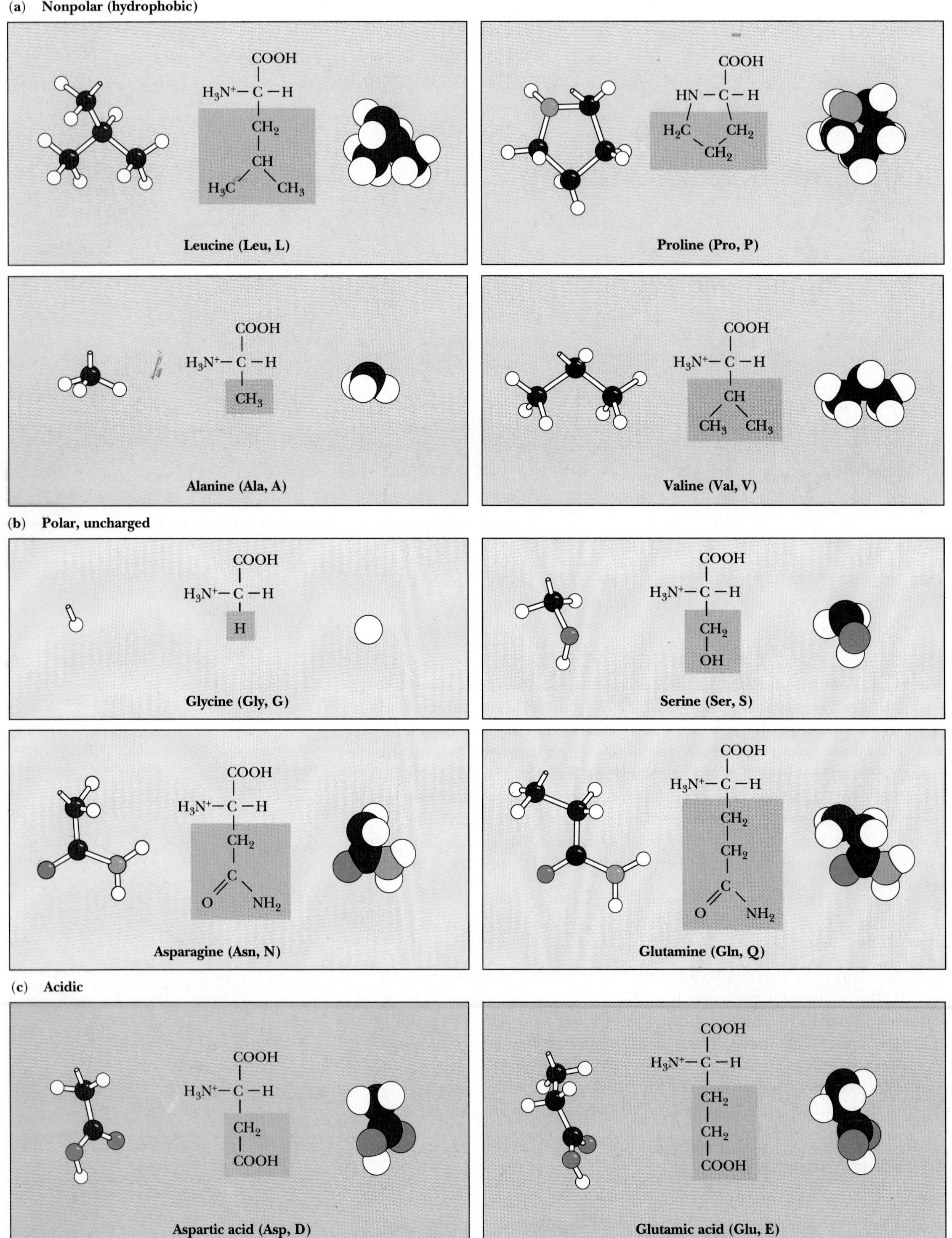

Figure 3.3 The 20 amino acids that are the building blocks of most proteins can be classified as (a) nonpolar (hydrophobic), (b) polar, neutral, (c) acidic, or (d) basic. Also shown are the one-letter and three-letter codes used to denote amino acids. For each amino acid, the ball-and-stick (left) and space-filling (right) models show only the side chain.

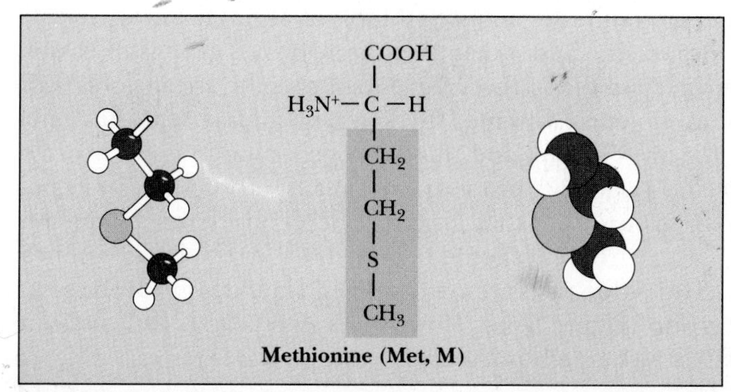

Methionine (Met, M)

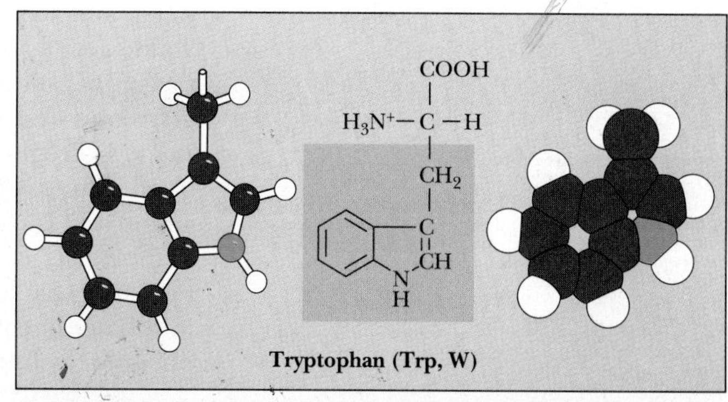

Tryptophan (Trp, W)

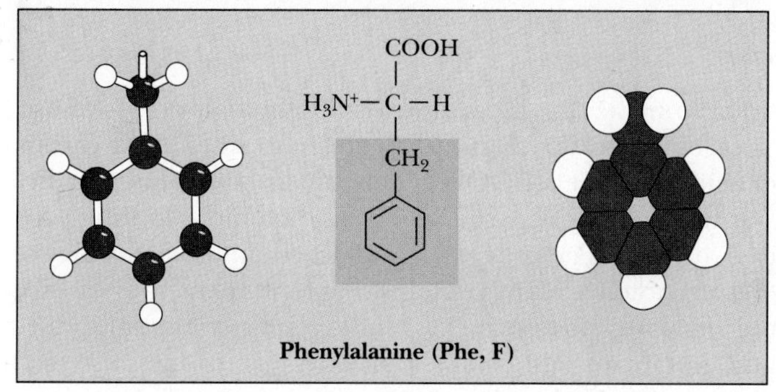

Phenylalanine (Phe, F)

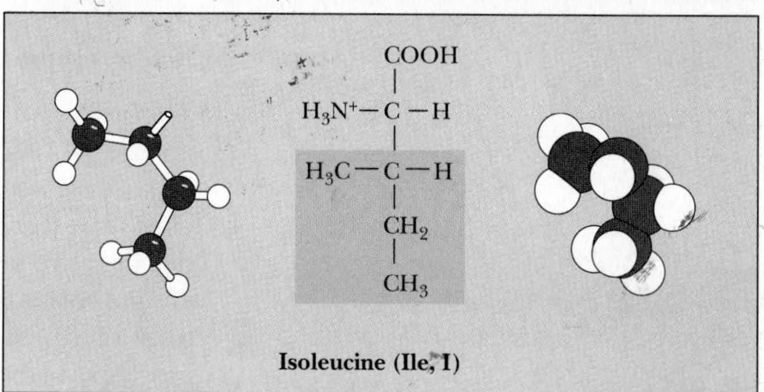

Isoleucine (Ile, I)

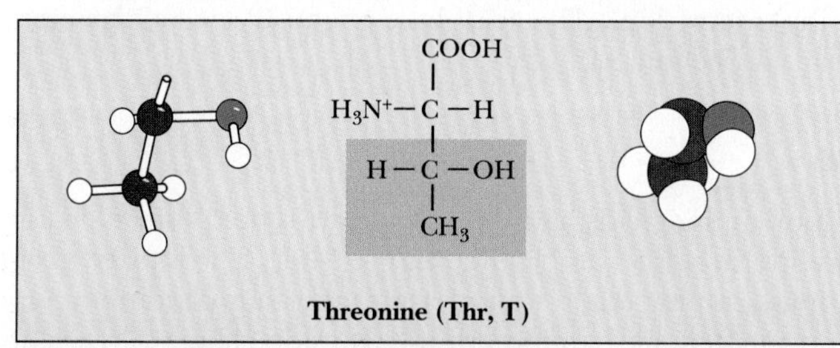

Threonine (Thr, T)

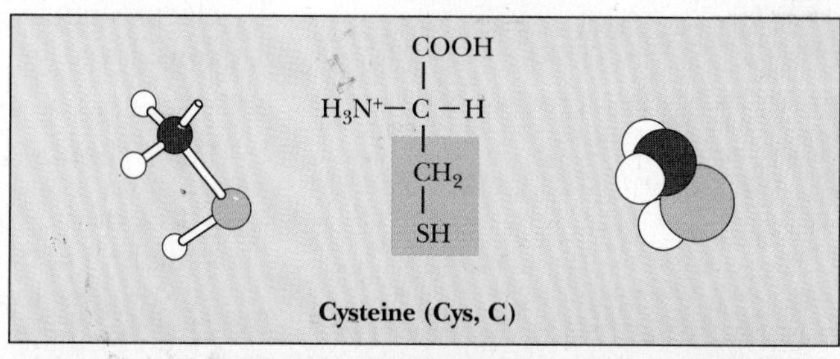

Cysteine (Cys, C)

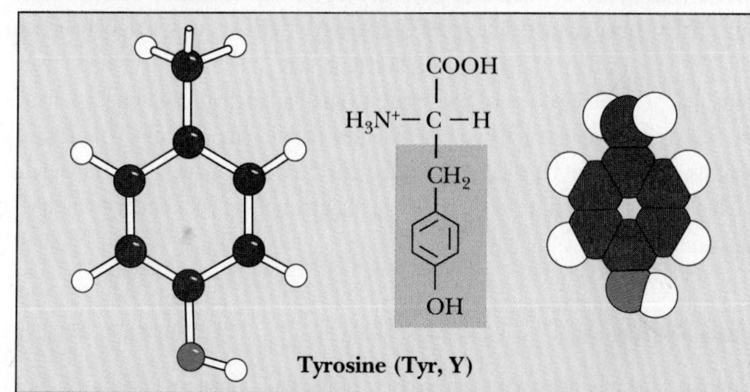

Tyrosine (Tyr, Y)

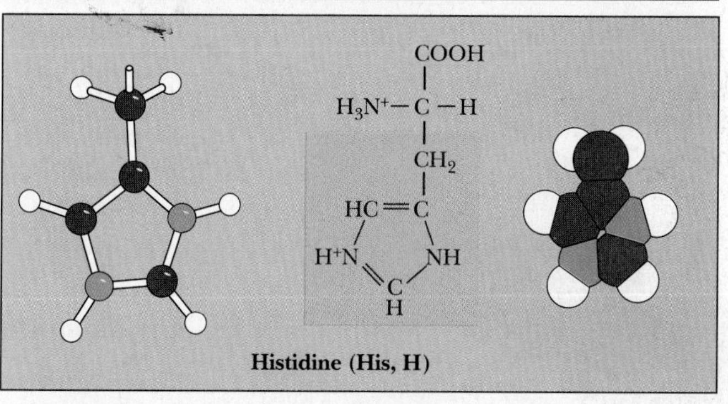

Histidine (His, H)

(d) Basic

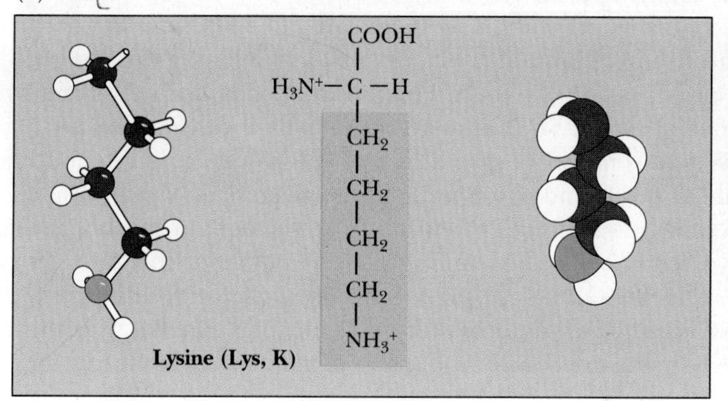

Lysine (Lys, K)

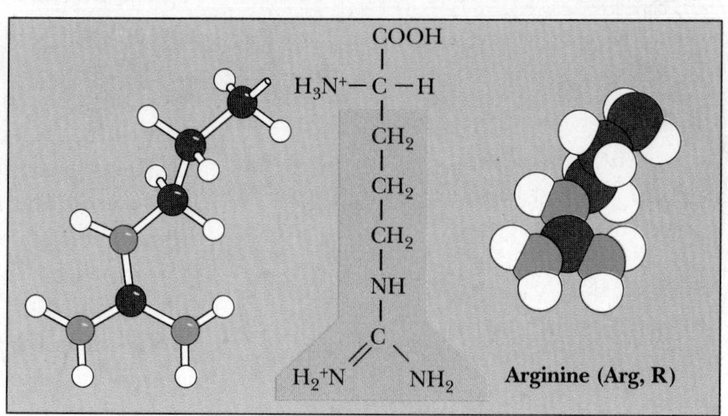

Arginine (Arg, R)

water, and alanine and valine are about as soluble as arginine and serine. The amide groups of asparagine and glutamine; the hydroxyl groups of tyrosine, threonine, and serine; and the sulfhydryl group of cysteine are all good hydrogen bond–forming moieties. Glycine, the simplest amino acid, has only a single hydrogen for an R group, and this hydrogen is not a good hydrogen bond former. Glycine's solubility properties are mainly influenced by its polar amino and carboxyl groups, and thus glycine is best considered as a member of the polar, uncharged group. It should be noted that tyrosine has significant nonpolar characteristics due to its aromatic ring and could arguably be placed in the nonpolar group (Figure 3.3a). However, with a pK_a of 10.1, tyrosine's phenolic hydroxyl is a charged, polar entity at high pH.

Acidic Amino Acids

There are two acidic amino acids—aspartic acid and glutamic acid—whose R groups contain a carboxyl group (Figure 3.3c). These side chain carboxyl groups are weaker acids than the α-COOH group, but are sufficiently acidic to exist as $—COO^-$ at neutral pH. Aspartic acid and glutamic acid thus have a net negative charge at pH 7. These negatively charged amino acids play several important roles in proteins. Many proteins that bind metal ions for structural or functional purposes possess metal binding sites containing one or more aspartate and glutamate side chains. Carboxyl groups may also act as nucleophiles in certain enzyme reactions and may participate in a variety of electrostatic bonding interactions. The acid–base chemistry of such groups is considered in detail in Section 3.2.

Basic Amino Acids

Three of the common amino acids have side chains with net positive charges at neutral pH: histidine, arginine, and lysine (Figure 3.3d). The ionized group of histidine is an imidazolium, while that of arginine is a guanidinium, and lysine contains a protonated alkyl amino group. The side chains of the latter two amino acids are fully protonated at pH 7, but histidine, with a side chain pK_a of 6.0 is only 10% protonated at pH 7. With a pK_a near neutrality, histidine side chains play important roles as proton donors and acceptors in many enzyme reactions. Histidine-containing peptides are important biological buffers, as was discussed in Chapter 2. Arginine and lysine side chains, which are protonated under physiological conditions, participate in electrostatic interactions in proteins.

Uncommon Amino Acids

There are several amino acids that occur only rarely in proteins (Figure 3.4). These include **hydroxylysine** and **hydroxyproline,** which are found mainly in the collagen and gelatin proteins, and **thyroxine,** an iodinated amino acid that is only found in thyroglobulin, a protein produced by the thyroid gland. Certain muscle proteins contain methylated amino acids, including **methylhistidine, ε-*N*-methyllysine,** and **ε-*N,N,N*-trimethyllysine** (Figure 3.4). **γ-Carboxyglutamic acid** is found in several proteins involved in blood clotting, and **pyroglutamic acid** is found in a unique light-driven proton-pumping protein called bacteriorhodopsin, which is discussed elsewhere in this book. Certain proteins involved in cell growth and regulation are reversibly phosphorylated on the −OH groups of serine, threonine, and tyrosine residues. **Aminoadipic acid** is found in proteins isolated from corn. Finally, ***N*-methylarginine** and ***N*-acetyllysine** are found in histone proteins associated with chromosomes.

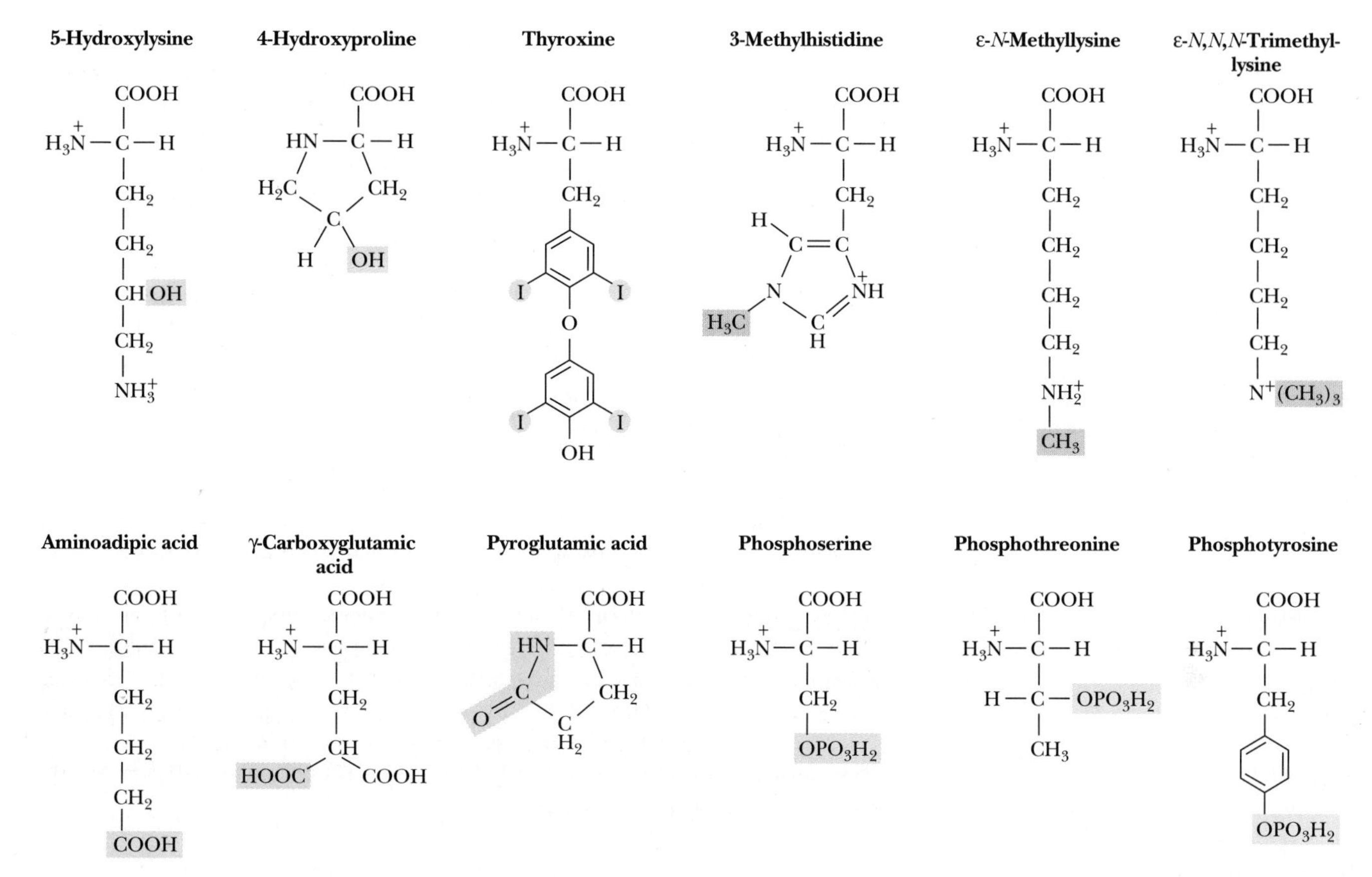

***Figure* 3.4** The structures of several amino acids that are less common but nevertheless found in certain proteins. Hydroxylysine and hydroxyproline are found in connective-tissue proteins, pyroglutamic acid is found in bacteriorhodopsin (a protein in *Halobacterium halobium*), and aminoadipic acid is found in proteins isolated from corn.

Amino Acids Not Found in Proteins

Certain amino acids and their derivatives, while not found in proteins, nonetheless are biochemically important. A few of the more notable examples are shown in Figure 3.5. **γ-Aminobutyric acid,** or **GABA,** is produced by the decarboxylation of glutamic acid and is a potent neurotransmitter. **Histamine,** which is synthesized by decarboxylation of histidine, and **serotonin,** which is derived from tryptophan, similarly function as neurotransmitters and regulators. **β-Alanine** is found in nature in the peptides carnosine and anserine and is a component of pantothenic acid (a vitamin), which comprises part of coenzyme A. **Epinephrine** (also known as **adrenaline**), derived from tyrosine, is an important hormone. **Penicillamine** is a constituent of the penicillin antibiotics. **Ornithine, betaine, homocysteine,** and **homoserine** are important metabolic intermediates. **Citrulline** is the immediate precursor of arginine.

3.2 Acid–Base Chemistry of Amino Acids

Amino Acids Are Weak Polyprotic Acids

From a chemical point of view, the common amino acids are all weak polyprotic acids. The ionizable groups are not strongly dissociating ones, and the degree of dissociation thus depends on the pH of the medium. All the amino acids contain at least two dissociable hydrogens.

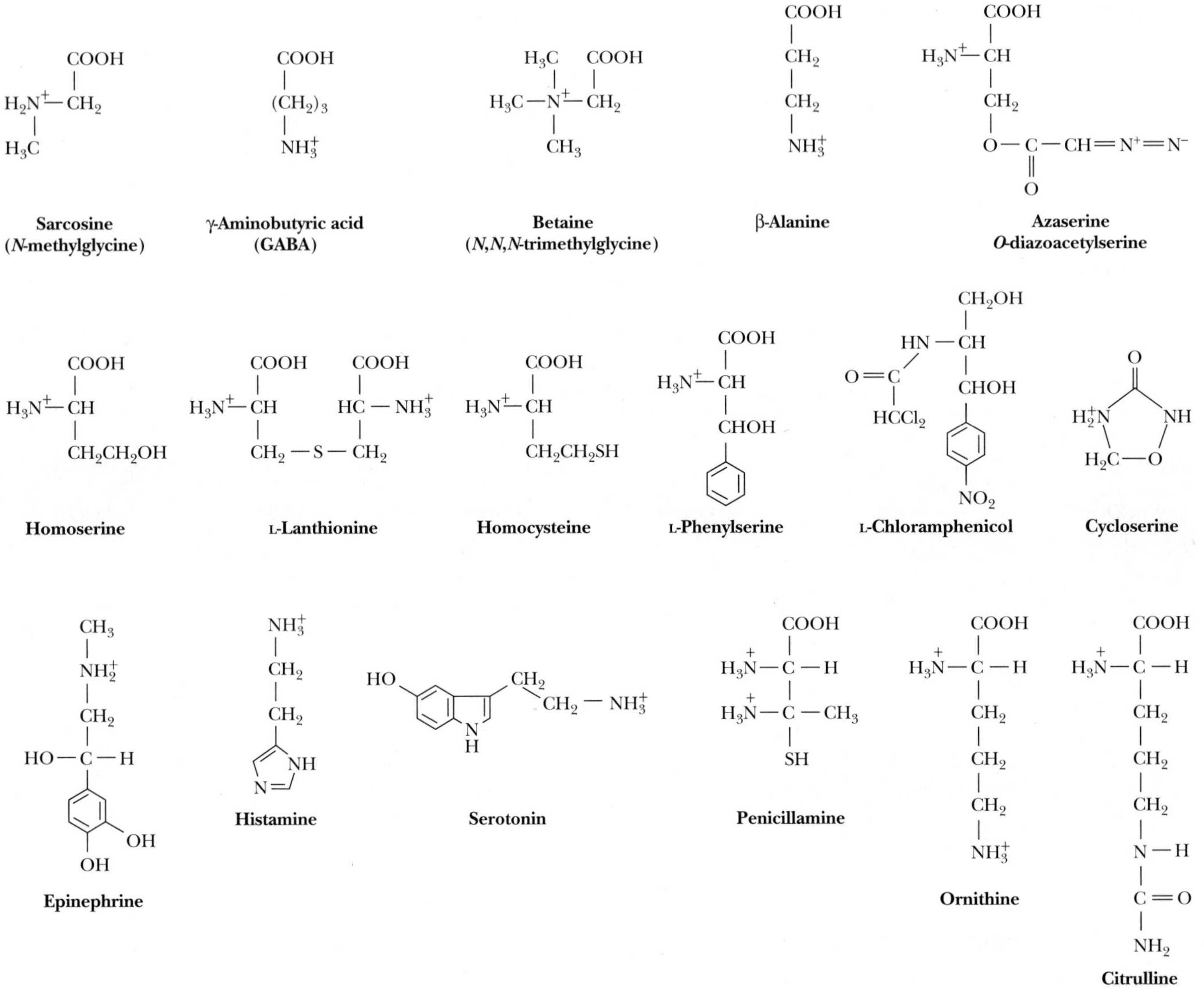

Sarcosine (*N*-methylglycine)

γ-Aminobutyric acid (GABA)

Betaine (*N,N,N*-trimethylglycine)

β-Alanine

Azaserine *O*-diazoacetylserine

Homoserine

L-Lanthionine

Homocysteine

L-Phenylserine

L-Chloramphenicol

Cycloserine

Epinephrine

Histamine

Serotonin

Penicillamine

Ornithine

Citrulline

***Figure* 3.5** The structures of some amino acids that are not normally found in proteins but that perform other important biological functions. Epinephrine, histamine, and serotonin, while not amino acids, are derived from and closely related to amino acids.

Consider the acid–base behavior of glycine, the simplest amino acid. At low pH, both the amino and carboxyl groups are protonated and the molecule has a net positive charge. If the counterion in solution is a chloride ion, this form is referred to as **glycine hydrochloride.** If the pH is increased, the carboxyl group is the first to dissociate, yielding a neutral zwitterionic species (Figure 3.6). Further increase in pH eventually results in dissociation of the amino group to yield the negatively charged **glycinate.** If we denote these three forms as Gly^+, Gly^0, and Gly^-, we can write the first dissociation of Gly^+ as

$$Gly^+ + H_2O \rightleftharpoons Gly^0 + H_3O^+$$

and the dissociation constant K_1 as

$$K_1 = \frac{[Gly^0][H_3O^+]}{[Gly^+]}$$

Values for K_1 for the common amino acids are typically 0.4 to 1.0×10^{-2} *M*, so that typical values of pK_1 center around values of 2.0 to 2.4 (see Table 3.1). In a similar manner, we can write the second dissociation reaction as

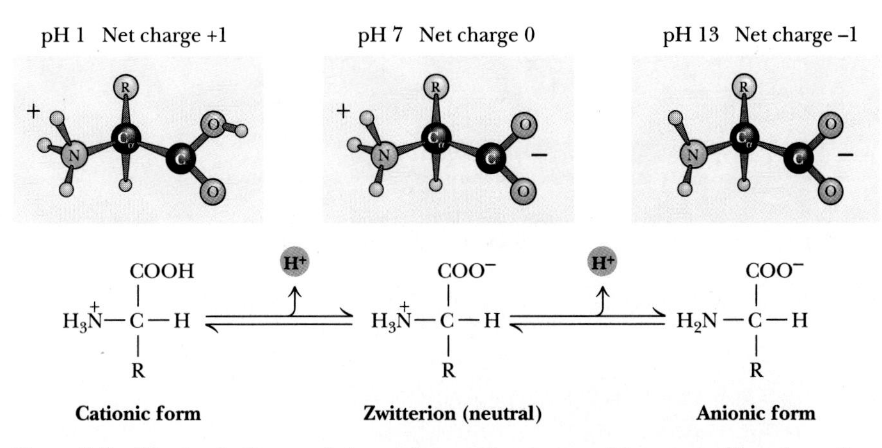

***Figure* 3.6** The ionic forms of the amino acids, shown without consideration of any ionizations on the side chain. The cationic form is the low pH form, and the titration of the cationic species with base will yield the zwitterion and finally the anionic form.

(Figure courtesy of Irving Geis.)

$$\mathrm{Gly}^0 + \mathrm{H_2O} \rightleftharpoons \mathrm{Gly}^- + \mathrm{H_3O}^+$$

and the dissociation constant K_2 as

$$K_2 = \frac{[\mathrm{Gly}^-][\mathrm{H_3O}^+]}{[\mathrm{Gly}^0]}$$

Typical values for pK_2 are in the range of 9.0 to 9.8. At physiological pH, the α-carboxyl group of a simple amino acid (with no ionizable side chains) is completely dissociated, whereas the α-amino group has not really begun its dissociation. The titration curve for such an amino acid is shown in Figure 3.7.

***Table* 3.1**
pK_a Values of Common Amino Acids

Amino Acid	α-COOH pK_a	α-NH_3^+ pK_a	R group pK_a
Alanine	2.4	9.7	
Arginine	2.2	9.0	12.5
Asparagine	2.0	8.8	
Aspartic acid	2.1	9.8	3.9
Cysteine	1.7	10.8	8.3
Glutamic acid	2.2	9.7	4.3
Glutamine	2.2	9.1	
Glycine	2.3	9.6	
Histidine	1.8	9.2	6.0
Isoleucine	2.4	9.7	
Leucine	2.4	9.6	
Lysine	2.2	9.0	10.5
Methionine	2.3	9.2	
Phenylalanine	1.8	9.1	
Proline	2.1	10.6	
Serine	2.2	9.2	~13
Threonine	2.6	10.4	~13
Tryptophan	2.4	9.4	
Tyrosine	2.2	9.1	10.1
Valine	2.3	9.6	

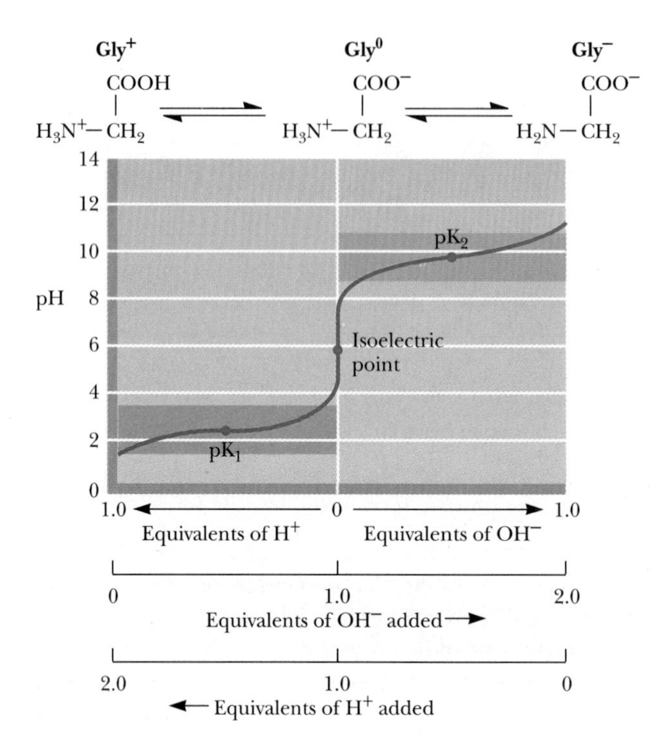

***Figure* 3.7** Titration of glycine, a simple amino acid.

Example

What is the pH of a glycine solution in which the α-NH_3^+ group is one-third dissociated?

Solution

The appropriate Henderson–Hasselbalch equation is

$$pH = pK_a + \log_{10}\frac{[Gly^-]}{[Gly^0]}$$

If the α-amino group is one-third dissociated, there is one part Gly^- for every two parts Gly^0. The important pK_a is the pK_a for the amino group. The glycine α-amino group has a pK_a of 9.6. The result is

$$pH = 9.6 + \log_{10}(1/2)$$
$$pH = 9.3$$

Note that the dissociation constants of both the α-carboxyl and α-amino groups are affected by the presence of the other group. The adjacent α-amino group makes the α-COOH group more acidic (that is, it lowers the pK_a) so that it gives up a proton more readily than simple alkyl carboxylic acids. Thus, the pK_1 of 2.0 to 2.1 for α-carboxyl groups of amino acids is substantially lower than that of acetic acid ($pK_a = 4.76$), for example. What is the chemical basis for the low pK_a of the α-COOH group of amino acids? The α-NH_3^+ (ammonium) group is strongly electron-withdrawing, and the positive charge of the amino group exerts a strong field effect and stabilizes the carboxylate anion. (The effect of the α-COOH group on the pK_a of the α-NH_3^+ group is the basis for Problem 4 at the end of this chapter.)

Ionization of Side Chains

As we have seen, the side chains of several of the amino acids also contain dissociable groups. Thus, aspartic and glutamic acids contain an additional carboxyl function, while lysine possesses an aliphatic amino function. Histidine contains an ionizable imidazolium proton, while arginine carries a guanidinium function. Typical values for the pK_a values of these groups are shown in Table 3.1. The β-carboxyl group of aspartic acid and the γ-carboxyl side chain of glutamic acid exhibit pK_a values intermediate to the α-COOH on the one hand and typical of aliphatic carboxyl groups on the other hand. In a similar fashion, the ϵ-amino group of lysine exhibits a pK_a that is higher than the α-amino group but similar to that for a typical aliphatic amino group. These intermediate values for side-chain pK_a values reflect the slightly diminished effect of the α-carbon dissociable groups that lie several carbons removed from the side-chain functional groups. Figure 3.8 shows typical titration curves for glutamic acid and lysine, along with the ionic species that predominate at various points in the titration. The only other side-chain groups that exhibit any significant degree of dissociation are the *para*-OH group of tyrosine and the —SH group of cysteine. The pK_a of the cysteine sulfhydryl is 8.32, so that it is approximately 8 to 10% dissociated at pH 7.0. The tyrosine *para*-OH group is a very weakly acidic group, with a pK_a of about 10.1. This group is essentially fully protonated and uncharged at pH 7.0.

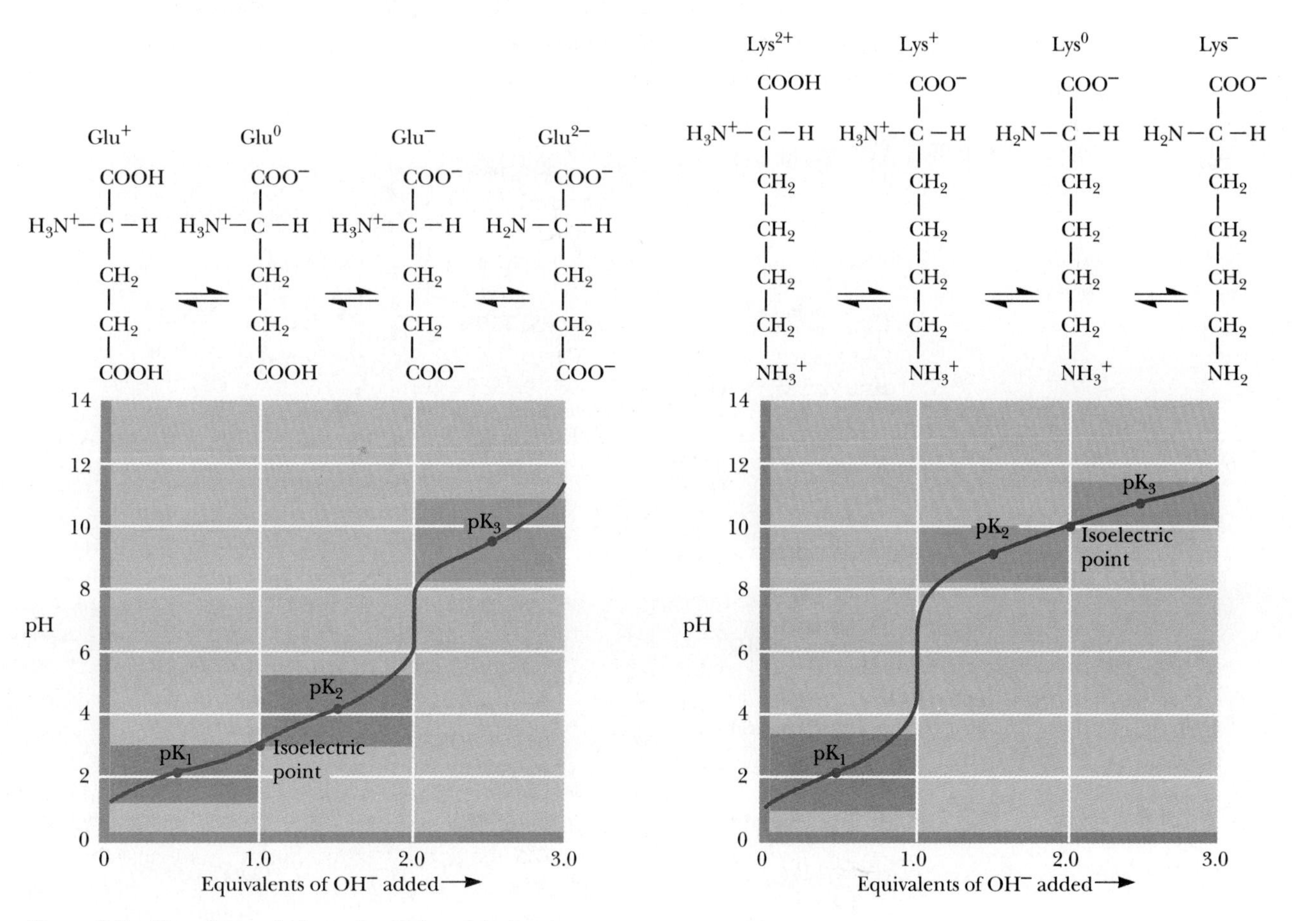

***Figure* 3.8** Titrations of glutamic acid and lysine.

3.3 Reactions of Amino Acids

Carboxyl and Amino Group Reactions

The α-carboxyl and α-amino groups of all amino acids exhibit similar chemical reactivity. The side chains, however, exhibit specific chemical reactivities, depending on the nature of the functional groups. Whereas all of these reactivities are important in the study and analysis of isolated amino acids, it is the characteristic behavior of the side chain that governs the reactivity of amino acids incorporated into proteins. There are three reasons to consider these reactivities. Proteins can be chemically modified in very specific ways by taking advantage of the chemical reactivity of certain amino acid side chains. The detection and quantification of amino acids and proteins often depend on reactions that are specific to one or more amino acids and that result in color, radioactivity, or some other quantity that can be easily measured. Finally and most importantly, the biological functions of proteins depend on the behavior and reactivity of specific R groups.

The carboxyl groups of amino acids undergo all the simple reactions common to this functional group. Reaction with ammonia and primary amines yields unsubstituted and substituted amides, respectively (Figure 3.9a,b). Esters and acid chlorides are also readily formed. Esterification pro-

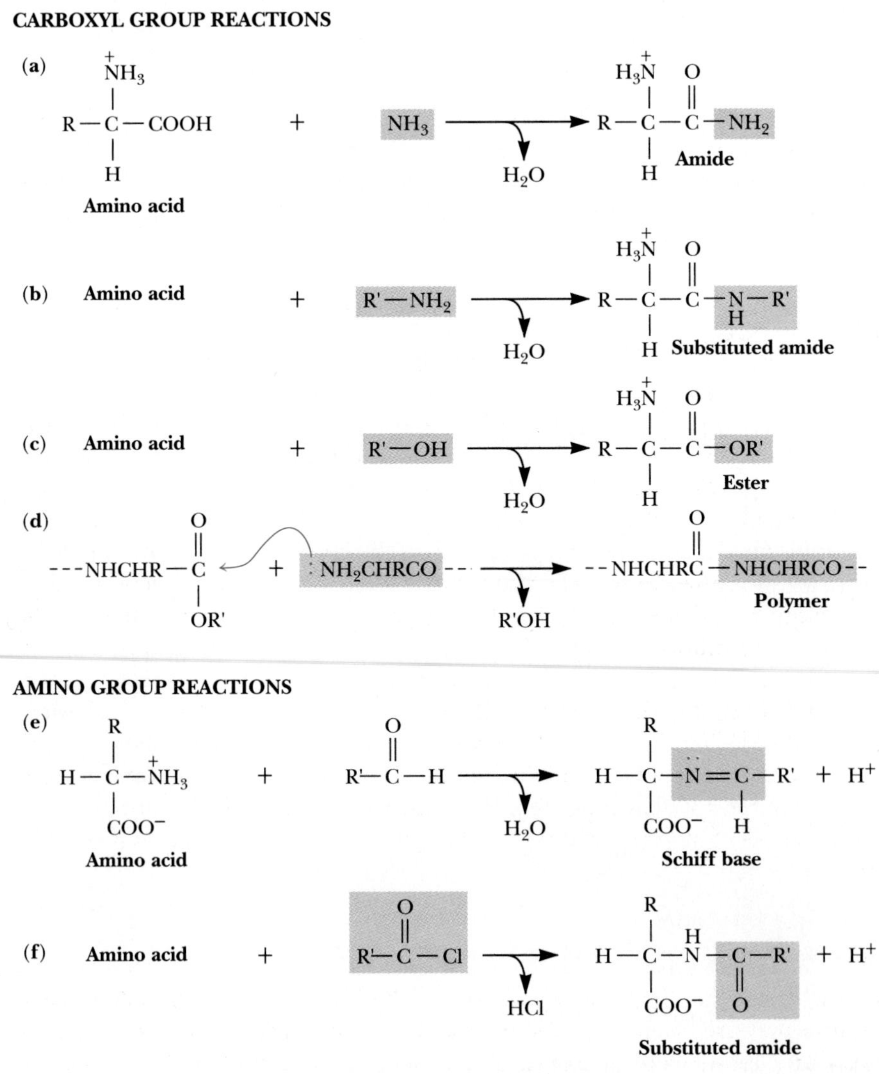

CARBOXYL GROUP REACTIONS

(a) $R{-}C(\overset{+}{N}H_3)(H){-}COOH + NH_3 \xrightarrow{-H_2O} R{-}C(H_3\overset{+}{N})(H){-}C(=O){-}NH_2$ (Amino acid → Amide)

(b) Amino acid + $R'{-}NH_2 \xrightarrow{-H_2O} R{-}C(H_3\overset{+}{N})(H){-}C(=O){-}NH{-}R'$ (Substituted amide)

(c) Amino acid + $R'{-}OH \xrightarrow{-H_2O} R{-}C(H_3\overset{+}{N})(H){-}C(=O){-}OR'$ (Ester)

(d) $---NHCHR{-}C(=O){-}OR' + :NH_2CHRCO-- \xrightarrow{-R'OH} --NHCHRC(=O){-}NHCHRCO--$ (Polymer)

AMINO GROUP REACTIONS

(e) $H{-}C(R)(COO^-){-}\overset{+}{N}H_3 + R'{-}C(=O){-}H \xrightarrow{-H_2O} H{-}C(R)(COO^-){-}\ddot{N}{=}C(H){-}R' + H^+$ (Amino acid → Schiff base)

(f) Amino acid + $R'{-}C(=O){-}Cl \xrightarrow{-HCl} H{-}C(R)(COO^-){-}N(H){-}C(=O){-}R' + H^+$ (Substituted amide)

***Figure* 3.9** Typical reactions of the common amino acids (see text for details).

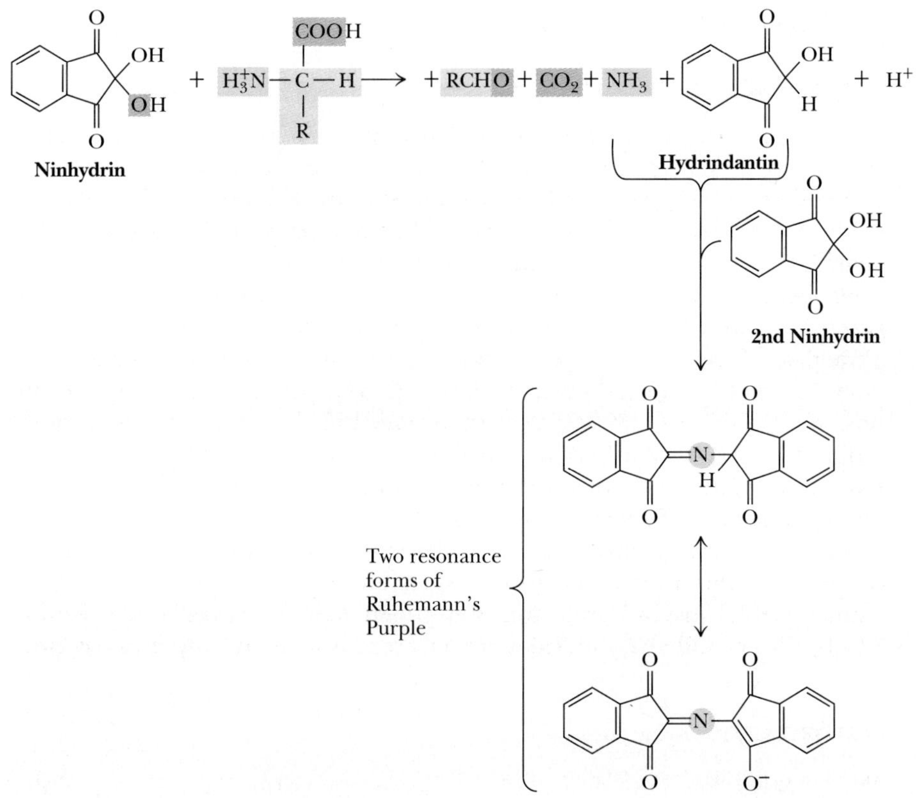

Figure **3.10** The pathway of the ninhydrin reaction, which produces a colored product called "Ruhemann's Purple" that absorbs light at 570 nm. Note that the reaction involves and consumes two molecules of ninhydrin.

ceeds in the presence of the appropriate alcohol and a strong acid (Figure 3.9c). Polymerization can occur by repetition of the reaction shown in Figure 3.9d. Free amino groups may react with aldehydes to form Schiff bases (Figure 3.9e) and can be acylated with acid anhydrides and acid halides (Figure 3.9f).

The Ninhydrin Reaction

Amino acids can be readily detected and quantified by reaction with ninhydrin. As shown in Figure 3.10, *ninhydrin,* or triketohydrindene hydrate, is a strong oxidizing agent and causes the oxidative deamination of the α-amino function. The products of the reaction are the resulting aldehyde, ammonia, carbon dioxide, and hydrindantin, a reduced derivative of ninhydrin. The ammonia produced in this way can react with the hydrindantin and another molecule of ninhydrin to yield a purple product (Ruhemann's Purple) which can be quantified spectrophotometrically at 570 nm. The appearance of CO_2 can also be monitored. Indeed, CO_2 evolution is diagnostic of the presence of an α-amino acid. α-Imino acids, such as proline and hydroxyproline, give bright yellow ninhydrin products with absorption maxima at 440 nm, allowing these to be distinguished from the α-amino acids. Since amino acids are one of the components of human skin secretions, the ninhydrin reaction was once used extensively by law enforcement and forensic personnel for fingerprint detection. More sensitive fluorescent reagents are now used routinely for this purpose.

Specific Reactions of Amino Acid Side Chains

A number of reactions of amino acids have become important in recent years because they are essential to the degradation, sequencing, and chemical synthesis of peptides and proteins. These reactions are discussed in detail in Chapter 4.

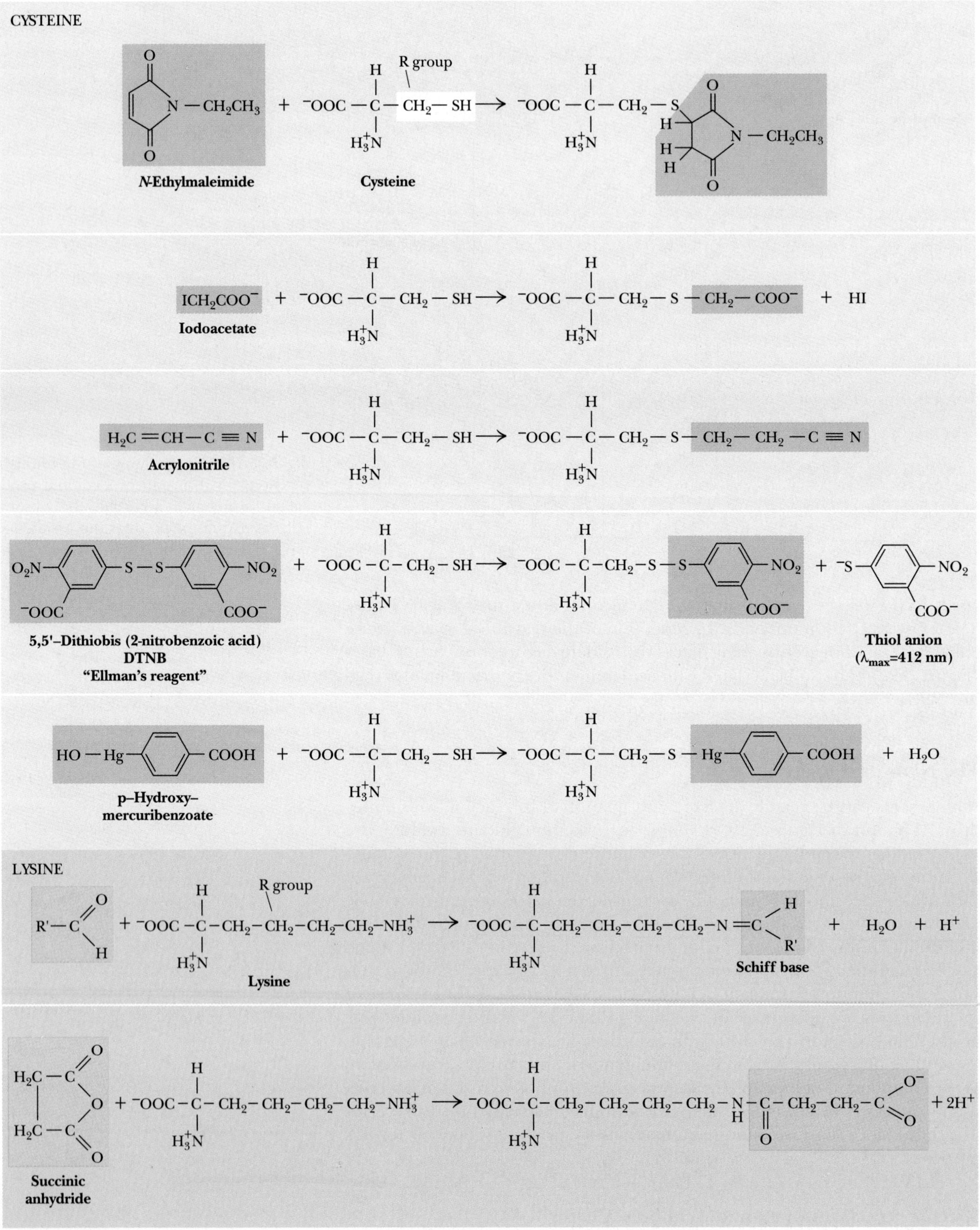

CYSTEINE

N-Ethylmaleimide + **Cysteine** (R group: CH_2-SH) →

$$^{-}OOC-CH(H_3^{+}N)-CH_2-SH \longrightarrow\ ^{-}OOC-CH(H_3^{+}N)-CH_2-S-(\text{succinimide ring})N-CH_2CH_3$$

$$ICH_2COO^{-} + {}^{-}OOC-CH(H_3^{+}N)-CH_2-SH \longrightarrow {}^{-}OOC-CH(H_3^{+}N)-CH_2-S-CH_2-COO^{-} + HI$$

Iodoacetate

$$H_2C=CH-C\equiv N + {}^{-}OOC-CH(H_3^{+}N)-CH_2-SH \longrightarrow {}^{-}OOC-CH(H_3^{+}N)-CH_2-S-CH_2-CH_2-C\equiv N$$

Acrylonitrile

$$O_2N-C_6H_3(COO^{-})-S-S-C_6H_3(COO^{-})-NO_2 + {}^{-}OOC-CH(H_3^{+}N)-CH_2-SH \longrightarrow {}^{-}OOC-CH(H_3^{+}N)-CH_2-S-S-C_6H_3(COO^{-})-NO_2 + {}^{-}S-C_6H_3(COO^{-})-NO_2$$

5,5'-Dithiobis (2-nitrobenzoic acid)
DTNB
"Ellman's reagent"

Thiol anion
(λ_{max}=412 nm)

$$HO-Hg-C_6H_4-COOH + {}^{-}OOC-CH(H_3^{+}N)-CH_2-SH \longrightarrow {}^{-}OOC-CH(H_3^{+}N)-CH_2-S-Hg-C_6H_4-COOH + H_2O$$

p-Hydroxy-mercuribenzoate

LYSINE

$$R'-CHO + {}^{-}OOC-CH(H_3^{+}N)-CH_2-CH_2-CH_2-CH_2-NH_3^{+} \longrightarrow {}^{-}OOC-CH(H_3^{+}N)-CH_2-CH_2-CH_2-CH_2-N=CH-R' + H_2O + H^{+}$$

Lysine (R group) — **Schiff base**

$$(\text{succinic anhydride}) + {}^{-}OOC-CH(H_3^{+}N)-CH_2-CH_2-CH_2-CH_2-NH_3^{+} \longrightarrow {}^{-}OOC-CH(H_3^{+}N)-CH_2-CH_2-CH_2-CH_2-NH-C(=O)-CH_2-CH_2-COO^{-} + 2H^{+}$$

Succinic anhydride

***Figure* 3.11** Reactions of amino acid side-chain functional groups.

TYROSINE

R group

$N\text{-}C(=O)\text{-}CH_3$ + $^{-}OOC-C(H)(H_3\overset{+}{N})-CH_2-C_6H_4-OH \longrightarrow {}^{-}OOC-C(H)(H_3\overset{+}{N})-CH_2-C_6H_4-O-C(=O)-CH_3$ + N NH

Acetyl imidazole

Tyrosine

TRYPTOPHAN

R group

O NBr O + $^{-}OOC-C(H)(H_3\overset{+}{N})-CH_2-$ (indole, N H) $\longrightarrow {}^{-}OOC-C(H)(H_3\overset{+}{N})-CH_2-$ (oxindole, O, N H)

***N*-Bromosuccinimide**

Tryptophan

ARGININE

R group

O =O + $^{-}OOC-C(H)(H_3\overset{+}{N})-CH_2-CH_2-CH_2-N(H)-C(=NH_2^{+})-NH_2 \longrightarrow {}^{-}OOC-C(H)(H_3\overset{+}{N})-CH_2-CH_2-CH_2-N=C$ (N, N, O)

Cyclohexane–1,2–dione

Arginine

SERINE

R group

H_3C CH_3 HC $O=P-O-CH(CH_3)_2$ F + $^{-}OOC-C(H)(H_3\overset{+}{N})-CH_2-OH \longrightarrow {}^{-}OOC-C(H)(H_3\overset{+}{N})-CH_2-O-P(=O)-CH(CH_3)_2$ O C H H_3C CH_3

Diisopropyl fluorophosphate

Serine

In recent years, biochemists have developed an arsenal of reactions that are relatively specific to the side chains of particular amino acids. These reactions can be used to identify functional amino acids at the active sites of enzymes or to label proteins with appropriate reagents for further study. Cysteine residues in proteins, for example, react with one another to form disulfide species and also react with a number of reagents, including maleimides (typically *N*-ethylmaleimide), as shown in Figure 3.11. Cysteines also react

Critical Developments in Biochemistry

Discovery of Optically Active Molecules and Determination of Absolute Configuration

The optical activity of quartz and certain other materials was first discovered by Jean-Baptiste Biot in 1815 in France, and in 1848 a young chemist in Paris named Louis Pasteur made a related and remarkable discovery. Pasteur noticed that preparations of optically inactive sodium ammonium tartrate contained two visibly different kinds of crystals that were mirror images of each other. Pasteur carefully separated the two types of crystals, dissolved them each in water, and found that each solution was *optically active*. Even more intriguing, the specific rotations of these two solutions were equal in magnitude and of opposite sign. Because these differences in optical rotation were apparent properties of the dissolved molecules, Pasteur eventually proposed that the molecules themselves were mirror images of each other, just like their respective crystals. Based on this and other related evidence, in 1847 van't Hoff and LeBel proposed the tetrahedral arrangement of valence bonds to carbon.

In 1888, Emil Fischer decided that it should be possible to determine the *relative* configuration of (+)-glucose, a six-carbon sugar with four asymmetric centers (see figure).

CHO
H—C—OH
HO—C—H
H—C—OH
H—C—OH
CH_2OH

The absolute configuration of (+)-glucose.

Since each of the four C could be either of two configurations, glucose conceivably could exist in any one of 16 possible isomeric structures. It took three years to complete the solution of an elaborate chemical and logical puzzle. By 1891, Fischer had reduced his puzzle to a choice between two enantiomeric structures. (Methods for determining *absolute* configuration were not yet available, so Fischer made a simple guess, selecting the structure shown in the figure.) For this remarkable feat, Fischer received the Nobel Prize in chemistry in 1902. Sadly, Fischer, a brilliant but troubled chemist, later committed suicide.

The absolute choice between Fischer's two enantiomeric possibilities would not be made for a long time. In 1951, J. M. Bijvoet in Utrecht, The Netherlands, used a new X-ray diffraction technique to determine the absolute configuration of (among other things) the sodium rubidium salt of (+)-tartaric acid. Since the tartaric acid configuration could be related to that of glyceraldehyde, and since sugar and amino acid configurations could all be related to glyceraldehyde, it became possible to determine the absolute configuration of sugars and the common amino acids. The absolute configuration of tartaric acid determined by Bijvoet turned out to be the configuration that, up to then, had only been assumed. This meant that Emil Fischer's arbitrary guess 60 years earlier had been correct.

It was M. A. Rosanoff, a chemist and instructor at New York University, who first proposed (in 1906) that the isomers of glyceraldehyde be the standards for denoting the stereochemistry of sugars and other molecules. Later, when experiments showed that the configuration of (+)-glyceraldehyde was related to (+)-glucose, (+)-glyceraldehyde was given the designation D. Emil Fischer rejected the **Rosanoff convention,** but it was universally accepted. Ironically, this nomenclature system is often mistakenly referred to as the **Fischer convention.**

effectively with iodoacetic acid to yield *S*-carboxymethyl cysteine derivatives. There are numerous other reactions involving specialized reagents specific for particular side chain functional groups. Figure 3.11 presents a representative list of these reagents and the products that result. It is important to realize that few if any of these reactions are truly specific for one functional group, and consequently, care must be exercised in their use.

Perspective drawing

Fischer projections

***Figure* 3.12** Enantiomeric molecules based on a chiral carbon atom. Enantiomers are nonsuperimposable mirror images of each other.

3.4 Optical Activity and Stereochemistry of Amino Acids

Amino Acids Are Chiral Molecules

Except for glycine, all of the amino acids isolated from proteins have four different groups attached to the α-carbon atom. In such a case, the α-carbon is said to be **asymmetric** or **chiral** (from the Greek *cheir,* meaning "hand"),

and the two possible configurations for the α-carbon constitute nonsuperimposable mirror image isomers, or **enantiomers** (Figure 3.12). Enantiomeric molecules display a special property called **optical activity**—the ability to rotate the plane of polarization of plane-polarized light. Clockwise rotation of incident light is referred to as **dextrorotatory** behavior, and counterclockwise rotation is called **levorotatory** behavior. The magnitude and direction of the optical rotation depend on the nature of the amino acid side chain. The temperature, the wavelength of the light used in the measurement, the ionization state of the amino acid, and therefore the pH of the solution, can also affect optical rotation behavior. As shown in Table 3.2, some protein-derived amino acids at a given pH are dextrorotatory and others are levorotatory, even though all of them are of the L configuration. The direction of optical rotation can be specified in the name by using a (+) for dextrorotatory compounds and a (−) for levorotatory compounds, as in L(+)-leucine.

***Table* 3.2**
Specific Rotations for Some Amino Acids

Amino Acid	Specific Rotation $[\alpha]_D^{25}$, degrees
L-Alanine	+1.8
L-Arginine	+12.5
L-Aspartic acid	+5.0
L-Glutamic acid	+12.0
L-Histidine	−38.5
L-Isoleucine	+12.4
L-Leucine	−11.0
L-Lysine	+13.5
L-Methionine	−10.0
L-Phenylalanine	−34.5
L-Proline	−86.2
L-Serine	−7.5
L-Threonine	−28.5
L-Tryptophan	−33.7
L-Valine	+5.6

Nomenclature for Chiral Molecules

The discoveries of optical activity and enantiomeric structures (see the box, page 70) made it important to develop suitable nomenclature for chiral molecules. Two systems are in common use today: the so-called D,L system and the (*R,S*) system.

In the **D,L system** of nomenclature, the (+) and (−) isomers of glyceraldehyde are denoted as **D-glyceraldehyde** and **L-glyceraldehyde,** respectively (Figure 3.13). Absolute configurations of all other carbon-based molecules are referenced to D- and L-glyceraldehyde. When sufficient care is taken to avoid racemization of the amino acids during hydrolysis of proteins, it is found that all of the amino acids derived from natural proteins are of the L configuration. Amino acids of the D configuration are nonetheless found in nature, especially as components of certain peptide antibiotics, such as valinomycin, gramicidin, and actinomycin D, and in the cell walls of certain microorganisms.

In spite of its widespread acceptance, there are problems with the D,L system of nomenclature. For example, this system can be ambiguous for molecules with two or more chiral centers. To address such problems, the ***(R,S)* system** of nomenclature for chiral molecules was proposed in 1956 by Robert Cahn, Sir Christopher Ingold, and Vladimir Prelog. In this more versatile system, priorities are assigned to each of the groups attached to a chiral center on the basis of atomic number, atoms with higher atomic numbers having higher priorities (see the box, page 73).

The newer *(R,S)* system of nomenclature is superior to the older D,L system in one important way. The configuration of molecules with more than one chiral center can be more easily, completely, and unambiguously described with *(R,S)* notation. There are several amino acids, including isoleucine, threonine, hydroxyproline, and hydroxylysine, which have two chiral centers. In the *(R,S)* system, L-threonine is (2*S*,3*R*)-threonine. A chemical compound with n chiral centers can exist in 2^n isomeric structures, and the four amino acids just listed can thus each take on four different isomeric configurations. This amounts to two pairs of enantiomers. Isomers that differ in configuration at only one of the asymmetric centers are non–mirror image isomers or **diastereomers.** The four stereoisomers of isoleucine are shown in Figure 3.14. The isomer obtained from digests of natural proteins is arbitrarily designated L-isoleucine. In the *(R,S)* system, L-isoleucine is (2*S*,3*S*)-isoleucine. Its diastereomer is referred to as L-allo-isoleucine. The D-enantiomeric pair of isomers is named in a similar manner.

CHO
HO — C — H
CH_2OH
L-Glyceraldehyde

CHO
H — C — OH
CH_2OH
D-Glyceraldehyde

COOH
$H_3\overset{+}{N}$ — C — H
CH_2OH
L-Serine

COOH
H — C — $\overset{+}{N}H_3$
CH_2OH
D-Serine

***Figure* 3.13** The configuration of the common L-amino acids can be related to the configuration of L(−)-glyceraldehyde as shown. These drawings are known as Fischer projections. The horizontal lines of the Fischer projections are meant to indicate bonds coming out of the page from the central carbon, while vertical lines represent bonds extending behind the page from the central carbon atom.

COOH	COOH	COOH	COOH
$H_3\overset{+}{N}$—C—H	H—C—$\overset{+}{N}H_3$	$H_3\overset{+}{N}$—C—H	H—C—$\overset{+}{N}H_3$
H_3C—C—H	H—C—CH_3	H—C—CH_3	H_3C—C—H
C_2H_5	C_2H_5	C_2H_5	C_2H_5
L-Isoleucine	**D-Isoleucine**	**L-Alloisoleucine**	**D-Alloisoleucine**
(2*S*,3*S*)-Isoleucine	**(2*R*,3*R*)-Isoleucine**	**(2*S*,3*R*)-Isoleucine**	**(2*R*,3*S*)-Isoleucine**

COOH	COOH	COOH	COOH
$H_3\overset{+}{N}$—C—H	H—C—$\overset{+}{N}H_3$	$H_3\overset{+}{N}$—C—H	H—C—$\overset{+}{N}H_3$
H—C—OH	HO—C—H	HO—C—H	H—C—OH
CH_3	CH_3	CH_3	CH_3
L-Threonine	**D-Threonine**	**L-Allothreonine**	**D-Allothreonine**

***Figure* 3.14** The stereoisomers of isoleucine and threonine. The structures at the far left are the naturally occurring isomers.

3.5 Spectroscopic Properties of Amino Acids

One of the most important and exciting advances in modern biochemistry has been the application of **spectroscopic methods,** which measure the absorption and emission of energy of different frequencies by molecules and atoms. Spectroscopic studies of proteins, nucleic acids, and other biomolecules are providing many new insights into the structure and dynamic processes in these molecules.

Ultraviolet Spectra

Many details of the structure and chemistry of the amino acids have been elucidated or at least confirmed by spectroscopic measurements. None of the amino acids absorb light in the visible region of the electromagnetic spec-

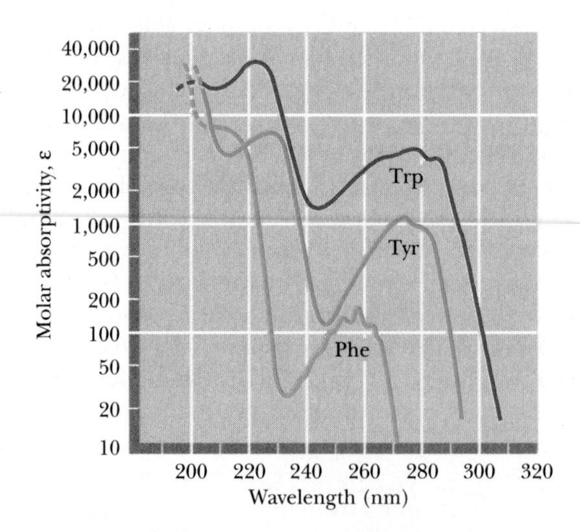

***Figure* 3.15** The ultraviolet absorption spectra of the aromatic amino acids at pH 6.

*(From Wetlaufer, D. B., 1962. Ultraviolet spectra of proteins and amino acids. Advances in Protein Chemistry **17**:303–390.)*

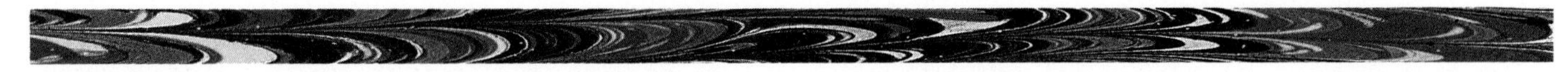

Critical Developments in Biochemistry

Rules for Description of Chiral Centers in the (R,S) System

Naming a chiral center in the *(R,S)* system is accomplished by viewing the molecule from the chiral center to the atom with the lowest priority. If the other three atoms facing the viewer then decrease in priority in a clockwise direction, the center is said to have the *(R)* configuration (where *R* is from the Latin *rectus* meaning "right"). If the three atoms in question decrease in priority in a counterclockwise fashion, the chiral center is of the *(S)* configuration (where *S* is from the Latin *sinistrus* meaning "left"). If two of the atoms coordinated to a chiral center are identical, the atoms bound to these two are considered for priorities. For such purposes, the priorities of certain functional groups found in amino acids and related molecules are in the following order:

$$SH > OH > NH_2 > COOH > CHO > CH_2OH > CH_3$$

From this, it is clear that D-glyceraldehyde is *(R)*-glyceraldehyde, and L-alanine is *(S)*-alanine (see figure). Interestingly, the α-carbon configuration of all the L-amino acids *except for cysteine* is *(S)*. Cysteine, by virtue of its thiol group, is in fact *(R)*-cysteine.

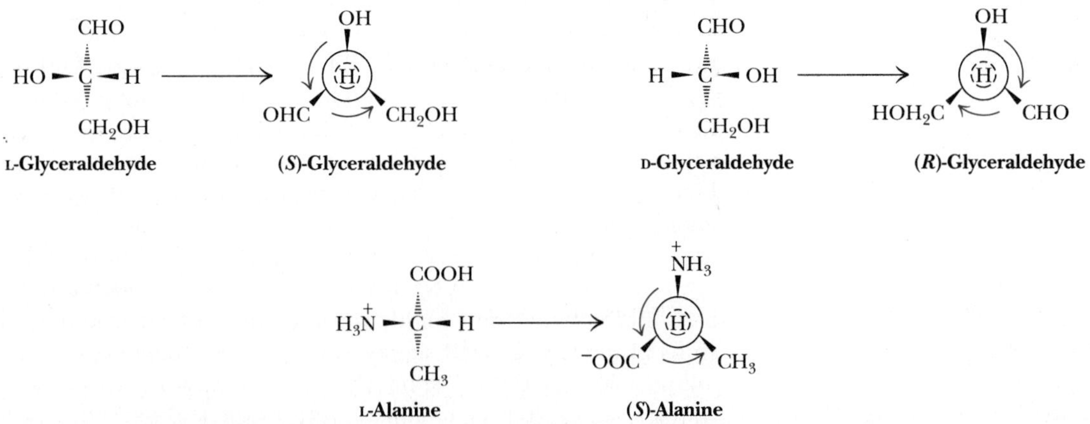

L-Glyceraldehyde → **(*S*)-Glyceraldehyde**

D-Glyceraldehyde → **(*R*)-Glyceraldehyde**

L-Alanine → **(*S*)-Alanine**

The assignment of *(R)* and *(S)* notation for glyceraldehyde and L-alanine.

trum. Several of the amino acids, however, do absorb **ultraviolet** radiation, and all absorb in the **infrared** region. The absorption of energy by electrons as they rise to higher electronic states occurs in the ultraviolet/visible region of the energy spectrum. Only the aromatic amino acids phenylalanine, tyrosine, and tryptophan exhibit significant ultraviolet absorption above 250 nm, as shown in Figure 3.15. These strong absorptions are the basis for spectroscopic determinations of protein concentration, as will be seen in Chapter 4. The aromatic amino acids also exhibit relatively weak fluorescence, and it has recently been shown that tryptophan can exhibit *phosphorescence*—a relatively long-lived emission of light. These fluorescence and phosphorescence properties are especially useful in the study of protein structure and dynamics (see Chapter 5).

Nuclear Magnetic Resonance Spectra

The development in the 1950s of **nuclear magnetic resonance** (NMR), a spectroscopic technique that involves the absorption of radio frequency energy by certain nuclei in the presence of a magnetic field, played an important part in

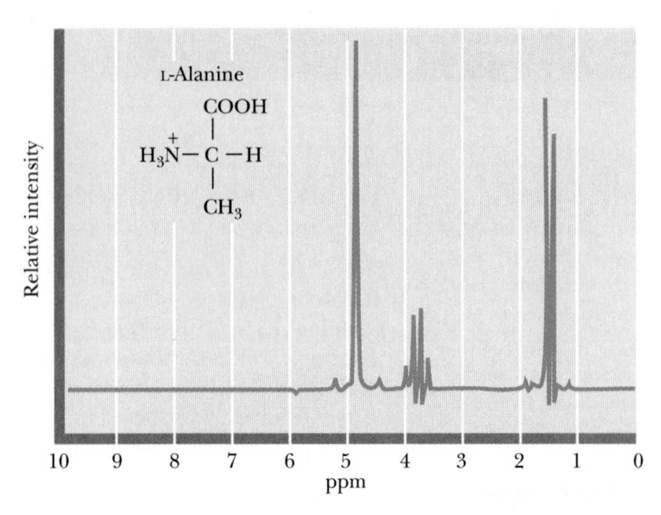

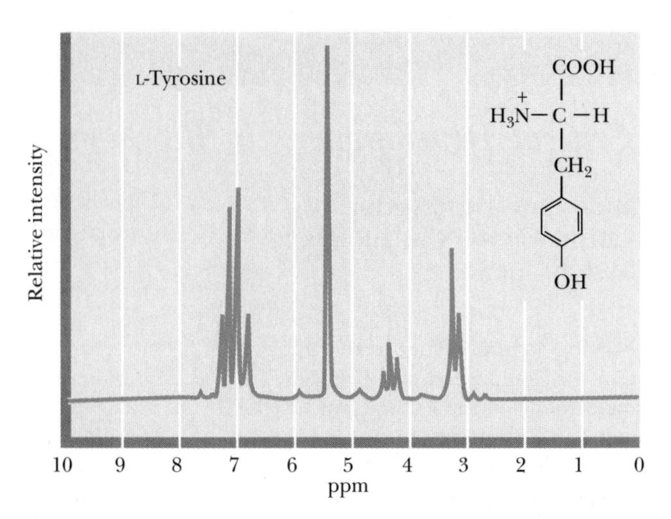

***Figure* 3.16** Proton NMR spectra of several amino acids. Zero on the chemical shift scale is defined by the resonance of tetramethylsilane (TMS).
(Adapted from Aldrich Library of NMR Spectra.)

the chemical characterization of amino acids and proteins. Several important principles rapidly emerged from these studies. First, the **chemical shift**[1] of amino acid protons depends on their particular chemical environment and thus on the state of ionization of the amino acid. Second, the change in electron density during a titration is transmitted throughout the carbon chain in the aliphatic amino acids and the aliphatic portions of aromatic amino acids, as evidenced by changes in the chemical shifts of relevant protons. Finally, the magnitude of the **coupling constants** between protons on adjacent carbons depends in some cases on the ionization state of the amino acid. This apparently reflects differences in the preferred conformations in different ionization states. Proton NMR spectra of two amino acids are shown in Figure 3.16. Because they are highly sensitive to their environment, the chemical shifts of individual NMR signals can detect the pH-dependent ionizations of amino acids. Figure 3.17 shows the ^{13}C chemical shifts occurring in a titration of lysine. Note that the chemical shifts of the carboxyl C, C_α, and C_β carbons of lysine are sensitive to dissociation of the nearby α-COOH and α-NH_3^+ protons (with pK values of about 2 and 9, respectively), whereas the C_δ and C_ϵ carbons are sensitive to dissociation of the ϵ-NH_3^+ group. Such measurements have been very useful for studies of the ionization behavior of amino acid residues in proteins. More sophisticated NMR measurements at very high magnetic fields are also used to determine the three-dimensional structures of peptides and even small proteins.

3.6 Separation and Analysis of Amino Acid Mixtures

Chromatographic Methods

The purification and analysis of individual amino acids from complex mixtures was once a very difficult process. Today, however, the biochemist has a wide variety of methods available for the separation and analysis of amino acids, or for that matter, any of the other biological molecules and macromol-

[1]The chemical shift for any NMR signal is the difference in resonant frequency between the observed signal and a suitable reference signal. If two nuclei are magnetically coupled, the NMR signals of these nuclei will be split, and the separation between such split signals, known as the coupling constant, is likewise dependent on the structural relationship between the two nuclei.

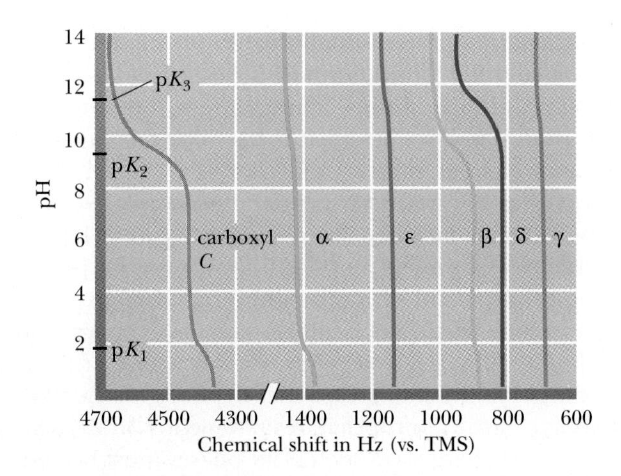

Figure* 3.17** A plot of chemical shifts versus pH for the carbons of lysine. Changes in chemical shift are most pronounced for atoms near the titrating group. Note the correspondence between the pK_a values and the particular chemical shift changes. All chemical shifts are defined relative to tetramethylsilane (TMS). *(From Suprenant, H., et al., 1980. Journal of Magnetic Resonance* ***40*:231–243.)*

ecules we shall encounter. All of these methods take advantage of the relative differences in the physical and chemical characteristics of amino acids, particularly ionization behavior and solubility characteristics. The methods important for amino acids include separations based on **partition** properties (the tendency to associate with one solvent or phase over another) and separations based on **electrical charge.** In all of the partition methods discussed here, the molecules of interest are allowed (or forced) to flow through a medium consisting of two phases—solid–liquid, liquid–liquid, or gas–liquid. In all of these methods, the molecules must show a preference for associating with one or the other phase. In this manner, the molecules partition, or distribute themselves, between the two phases in a manner based on their particular properties. The ratio of the concentrations of the amino acid (or other species) in the two phases is designated the *partition coefficient.*

In 1903, a separation technique based on repeated partitioning between phases was developed by M. Tswett for the separation of plant pigments (carotenes and chlorophylls). Tswett poured solutions of the pigments through columns of finely divided alumina and other solid media, allowing the pigments to partition between the liquid solvent and the solid support. Due to the colorful nature of the pigments thus separated, Tswett called his technique **chromatography.** This term is now applied to a wide variety of separation methods, regardless of whether the products are colored or not. The success of all chromatography techniques depends on the repeated microscopic partitioning of a solute mixture between the available phases. The more frequently this partitioning can be made to occur within a given time span or over a given volume, the more efficient is the resulting separation. Chromatographic methods have advanced rapidly in recent years, due in part to the development of sophisticated new solid-phase materials. Methods important for amino acid separations include ion exchange chromatography, gas chromatography (GC), and high-performance liquid chromatography (HPLC).

Ion Exchange Chromatography

The separation of amino acids and other solutes is often achieved by means of **ion exchange chromatography,** in which the molecule of interest is *exchanged* for another ion onto and off of a charged solid support. In a typical procedure, solutes in a liquid phase, usually water, are passed through columns filled with a porous solid phase, usually a bed of synthetic resin particles, containing charged groups. Resins containing positive charges attract negatively charged solutes and are referred to as *anion exchangers*. Solid supports possessing negative charges attract positively charged species and are referred to as *cation exchangers*. Several typical cation and anion exchange resins with different types of charged groups are shown in Figure 3.18. The strength of the acidity or basicity of these groups and their number per unit volume of resin determine the type and strength of binding of an exchanger. Fully ionized acidic groups such as sulfonic acids result in an exchanger with a negative charge which binds cations very strongly. Weakly acidic or basic groups yield resins whose charge (and binding capacity) depends on the pH of the eluting solvent. The choice of the appropriate resin depends on the strength of binding desired. The bare charges on such solid phases must be counterbalanced by oppositely charged ions in solution ("counterions"). Washing a cation exchange resin, such as Dowex-50, which has strongly acidic phenyl-SO_3^- groups, with a NaCl solution results in the formation of the so-called sodium form of the resin (see Figure 3.19). When the mixture whose separation is desired is added to the column, the positively charged solute molecules displace the Na^+ ions and bind to the resin. A gradient of an appropriate salt is then applied to the column, and the solute molecules are competitively (and sequentially) displaced (eluted) from the column by the rising concentration of cations in the gradient, in an order that is inversely proportional to their

(a) Cation Exchange Media	Structure
Strongly acidic, polystyrene resin (Dowex–50)	$-C_6H_4-SO_2-O^-$
Weakly acidic, carboxymethyl (CM) cellulose	$-O-CH_2-C(=O)O^-$
Weakly acidic, chelating, polystyrene resin (Chelex–100)	$-C_6H_4-CH_2-N(CH_2C(=O)-O^-)_2$

(b) Anion Exchange Media	Structure
Strongly basic, polystyrene resin (Dowex–1)	$-C_6H_4-CH_2-N^+(CH_3)_3$
Weakly basic, diethylaminoethyl (DEAE) cellulose	$-OCH_2CH_2-N^+H(CH_2CH_3)_2$

***Figure* 3.18** Cation (a) and anion (b) exchange resins commonly used for biochemical separations.

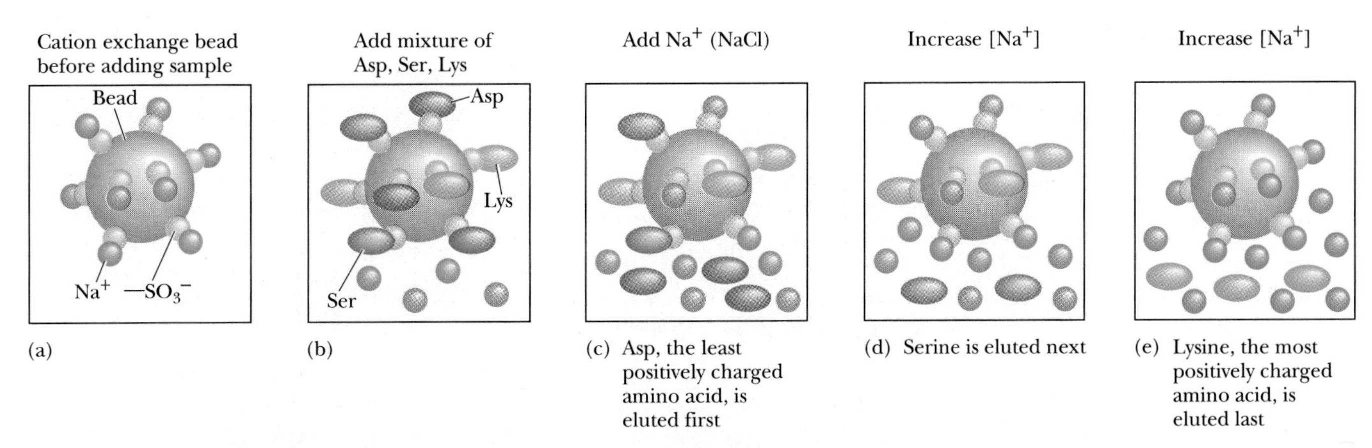
Cation exchange bead before adding sample
Bead
Na^+ $—SO_3^-$
(a)
Add mixture of Asp, Ser, Lys
Asp
Lys
Ser
(b)
Add Na^+ (NaCl)
(c) Asp, the least positively charged amino acid, is eluted first
Increase $[Na^+]$
(d) Serine is eluted next
Increase $[Na^+]$
(e) Lysine, the most positively charged amino acid, is eluted last

***Figure* 3.19** Operation of a cation exchange column, separating a mixture of Asp, Ser, and Lys. (a) The cation exchange resin in the beginning, Na^+ form. (b) A mixture of Asp, Ser, and Lys is added to the column containing the resin. (c) A gradient of the eluting salt (e.g., NaCl) is added to the column. Asp, the least positively charged amino acid, is eluted first. (d) As the salt concentration increases, Ser is eluted. (e) As the salt concentration is increased further, Lys, the most positively charged of the three amino acids, is eluted last.

affinities for the column. The separation of a mixture of amino acids on such a column is shown in Figures 3.19 and 3.20. Figure 3.21, taken from a now-classic 1958 paper by Stanford Moore, Darrel Spackman, and William Stein, shows a typical separation of the common amino acids. The events occurring in this separation are essentially those depicted in Figures 3.19 and 3.20. The amino acids are applied to the column at low pH (3.25), under which condi-

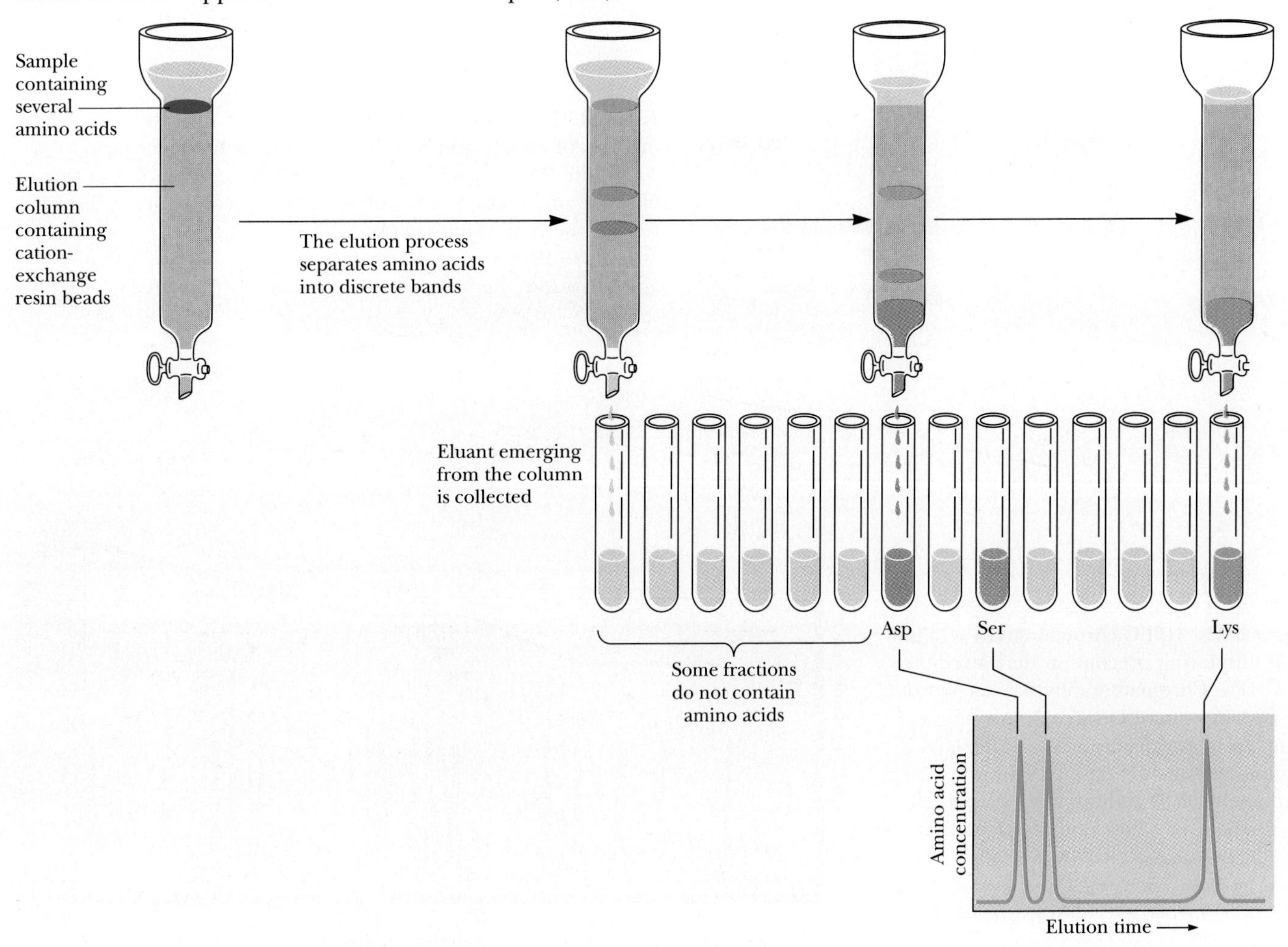
Sample containing several amino acids
Elution column containing cation-exchange resin beads
The elution process separates amino acids into discrete bands
Eluant emerging from the column is collected
Asp
Ser
Lys
Some fractions do not contain amino acids
Amino acid concentration
Elution time ⟶

***Figure* 3.20** The separation of amino acids on a cation exchange column.

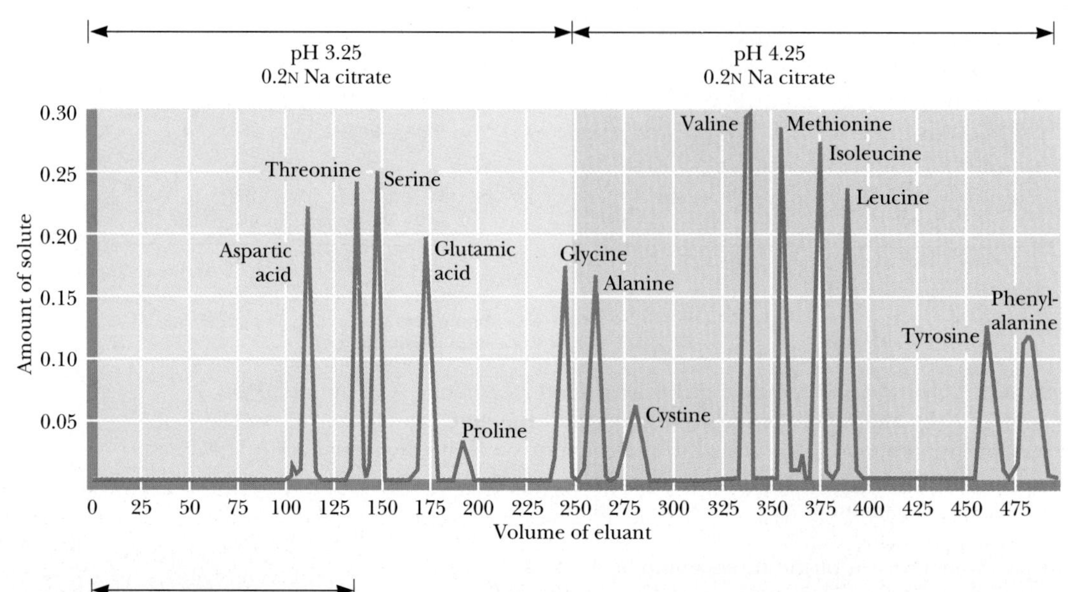

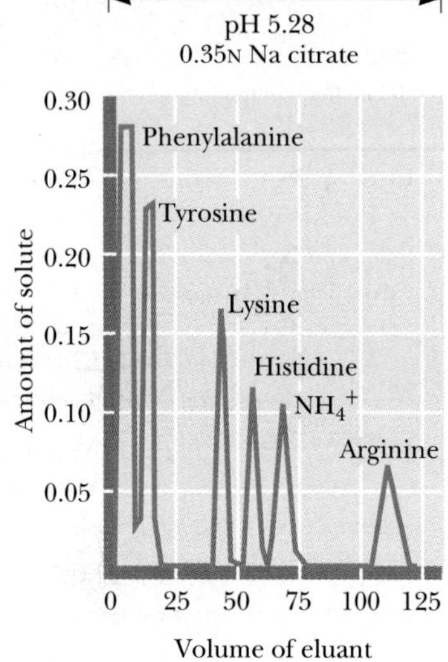

***Figure* 3.21** Chromatographic fractionation of a synthetic mixture of amino acids on ion exchange columns using Amberlite IR-120, a sulfonated polystyrene resin similar to Dowex-50. A second column with different buffer conditions is used to resolve the basic amino acids.
(Adapted from Moore, S., Spackman, D., and Stein, W., 1958. Chromatography of amino acids on sulfonated polystyrene resins. Analytical Chemistry 30:1185–1190.)

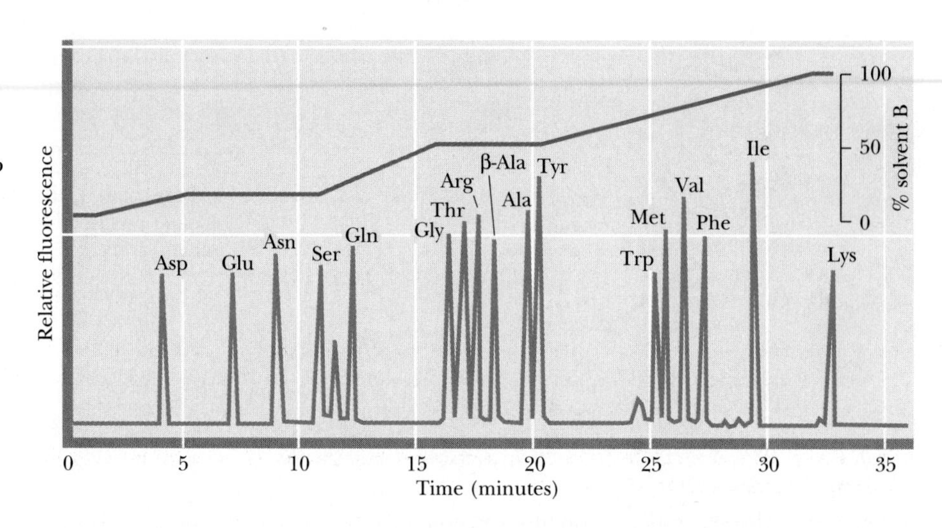

***Figure* 3.22** HPLC chromatogram of amino acids employing precolumn derivatization with OPA. Chromatography was carried out on an Ultrasphere ODS column using a complex tetrahydrofuran:methanol:0.05 *M* sodium acetate (pH 5.9) 1:19:80 to methanol:0.05 *M* sodium acetate (pH 5.9) 4:1 gradient at a flow rate of 1.7 mL/min.
(Adapted from Jones, B. N., Pääbo, S., and Stein, S., 1981. Amino acid analysis and enzymic sequence determination of peptides by an improved o-phthaldialdehyde precolumn labeling procedure. Journal of Liquid Chromatography 4:565–586.)

tions the acidic amino acids (aspartate and glutamate, among others) are weakly bound and the basic amino acids, such as arginine and lysine, are tightly bound. Sodium citrate solutions, at two different concentrations and three different values of pH, are used to elute the amino acids gradually from the column.

A typical HPLC chromatogram using precolumn derivatization of amino acids with *o*-phthaldialdehyde (OPA) is shown in Figure 3.22. HPLC has rapidly become the chromatographic technique of choice for most modern biochemists. The very high resolution, excellent sensitivity, and high speed of this technique usually outweigh the disadvantage of relatively low capacity.

Problems

1. Without consulting chapter figures, draw Fischer projection formulas for glycine, aspartate, leucine, isoleucine, methionine, and threonine.
2. Without reference to the text, give the one-letter and three-letter abbreviations for asparagine, arginine, cysteine, lysine, proline, tyrosine, and tryptophan.
3. Write equations for the ionic dissociations of alanine, glutamate, histidine, lysine, and phenylalanine.
4. How is the pK_a of the α-NH_3^+ group affected by the presence on an amino acid of the α-COOH?
5. Draw an appropriate titration curve for aspartic acid, labeling the axes and indicating the equivalence points and the pK_a values.
6. Calculate the concentrations of all ionic species in a 0.25 *M* solution of histidine at pH 2, pH 6.4, and pH 9.3.
7. Calculate the pH at which the γ-carboxyl group of glutamic acid is two-thirds dissociated.
8. Calculate the pH at which the ϵ-amino group of lysine is 20% dissociated.
9. Calculate the pH of a 0.3 *M* solution of (a) leucine hydrochloride, (b) sodium leucinate, and (c) isoelectric leucine.
10. Quantitative measurements of optical activity are usually expressed in terms of the specific rotation, $[\alpha]_D^{25}$, defined as

$$[\alpha]_D^{25} = \frac{\text{Measured rotation in degrees} \times 100}{(\text{Optical path in dm}) \times (\text{conc. in g/mL})}$$

For any measurement of optical rotation, the wavelength of the light used and the temperature must both be specified. In this case, D refers to the "D line" of sodium at 589 nm and 25 refers to a measurement temperature of 25°C. Calculate the concentration of a solution of L-arginine that rotates the incident light by 0.35° in an optical path length of 1 dm (decimeter).

11. Absolute configurations of the amino acids are referenced to D- and L-glyceraldehyde on the basis of chemical transformations that can convert the molecule of interest to either of these reference isomeric structures. In such reactions, the stereochemical consequences for the asymmetric centers must be understood for each reaction step. Propose a sequence of reactions that would demonstrate that L(−)-serine is stereochemically related to L(−)-glyceraldehyde.
12. Describe the stereochemical aspects of the structure of cystine, the structure that is a disulfide-linked pair of cysteines.
13. Draw a simple mechanism for the reaction of a cysteine sulfhydryl group with iodoacetamide.
14. Describe the expected elution pattern for a mixture of aspartate, histidine, isoleucine, valine, and arginine on a column of Dowex-50.
15. Assign *(R,S)* nomenclature to the threonine isomers of Figure 3.14.

Further Reading

Barker, R., 1971. *Organic Chemistry of Biological Compounds,* Chap. 4. Englewood Cliffs, NJ: Prentice-Hall.

Barrett, G. C., ed., 1985. *Chemistry and Biochemistry of the Amino Acids.* New York: Chapman and Hall.

Bovey, F. A., and Tiers, G. V. D., 1959. Proton N.S.R. spectroscopy. V. Studies of amino acids and peptides in trifluoroacetic acid. *Journal of the American Chemical Society* **81:**2870–2878.

Cahn, R. S., 1964. An introduction to the sequence rule. *Journal of Chemical Education* **41:**116–125.

Greenstein, J. P., and Winitz, M., 1961. *Chemistry of the Amino Acids.* New York: John Wiley & Sons.

Heiser, T., 1990. Amino acid chromatography: the "best" technique for student labs. *Journal of Chemical Education* **67:**964–966.

Herod, D. W., and Menzel, E. R., 1982. Laser detection of latent fingerprints: ninhydrin. *Journal of Forensic Science* **27:**200–204.

Iizuka, E., and Yang, J. T., 1964. Optical rotatory dispersion of L-amino acids in acid solution. *Biochemistry* **3:**1519–1524.

Kauffman, G. B., and Priebe, P. M., 1990. The Emil Fischer–William Ramsey Friendship. *Journal of Chemical Education* **67:**93–101.

Mabbott, G., 1990. Qualitative amino acid analysis of small peptides by GC/MS. *Journal of Chemical Education* **67:**441–445.

Meister, A., 1965. *Biochemistry of the Amino Acids,* 2nd ed., Vol. 1. New York: Academic Press.

Moore, S., Spackman, D., and Stein, W. H., 1958. Chromatography of amino acids on sulfonated polystyrene resins. *Analytical Chemistry* **30:**1185–1190.

Roberts, G. C. K., and Jardetzky, O., 1970. Nuclear magnetic resonance spectroscopy of amino acids, peptides and proteins. *Advances in Protein Chemistry* **24:**447–545.

Segel, I. H., 1976. *Biochemical Calculations,* 2nd ed. New York: John Wiley & Sons.

Suprenant, H. L., Sarneski, J. E., Key, R. R., Byrd, J. T., and Reilley, C. N., 1980. Carbon-13 NMR studies of amino acids: chemical shifts, protonation shifts, microscopic protonation behavior. *Journal of Magnetic Resonance* **40:**231–243.

". . . by small and simple things are great things brought to pass."
Alma 37.6, *The Book of Mormon*

Chapter 4

Proteins: Their Biological Functions and Primary Structure

Outline

4.1 Proteins Are Linear Polymers of Amino Acids
4.2 Architecture of Protein Molecules
4.3 The Many Biological Functions of Proteins
4.4 Some Proteins Have Chemical Groups Other Than Amino Acids
4.5 Reactions of Peptides and Proteins
4.6 Purification of Protein Mixtures
4.7 The Primary Structure of a Protein: Determining the Amino Acid Sequence
4.8 Nature of Amino Acid Sequences
4.9 Synthesis of Polypeptides in the Laboratory

Proteins are sequences of amino acids joined by peptide bonds, much as this painting is a sequence of images linked by time.

Proteins are the most abundant class of biomolecules, constituting more than 50% of the dry weight of cells. This abundance reflects the ubiquitous role of proteins in virtually all aspects of cell structure and function. The incredibly diverse repertoire of cellular activity is possible only because of the versatility inherent in proteins, each of which is specifically tailored to its biological role. The pattern by which each is tailored resides within the genetic information of cells, encoded in a specific sequence of nucleotide bases in DNA. Each such segment of encoded information defines a gene, and expression of the gene leads to synthesis of the specific protein encoded by it, endowing the cell with the functional attributes unique to that particular protein. Proteins are the agents of biological function; they are also the expressions of genetic information.

4.1 Proteins Are Linear Polymers of Amino Acids

Chemically, proteins are unbranched polymers of amino acids linked head to tail, from carboxyl group to amino group, through formation of covalent **peptide bonds,** a type of amide linkage (Figure 4.1).

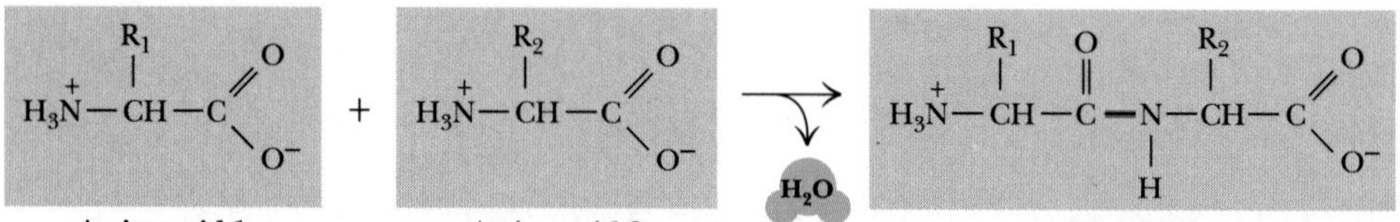

$$H_3\overset{+}{N}-\underset{|}{\overset{R_1}{CH}}-COO^- \;+\; H_3\overset{+}{N}-\overset{R_2}{CH}-COO^- \;\longrightarrow\; H_3\overset{+}{N}-\overset{R_1}{CH}-\overset{O}{\overset{\|}{C}}-\underset{H}{N}-\overset{R_2}{CH}-COO^- + H_2O$$

Amino acid 1 **Amino acid 2** **Dipeptide**

***Figure* 4.1** Peptide formation is the creation of an amide bond between the carboxyl group of one amino acid and the amino group of another amino acid. R_1 and R_2 represent the R groups of two different amino acids.

Peptide bond formation results in the release of H_2O. The peptide "backbone" of a protein consists of the repeated sequence —N—C_α—C—, where the N represents the amide nitrogen, the C_α is the α-carbon atom of an amino acid in the polymer chain, and the final C is the carbonyl carbon of the amino acid, which in turn is linked to the amide N of the next amino acid down the line. The geometry of the peptide backbone is shown in Figure 4.2. Note that the carbonyl oxygen and the amide hydrogen are *trans* to each other in this figure. This conformation is favored energetically since it results in less steric hindrance between nonbonded atoms in neighboring amino acids. Since the α-carbon atom of the amino acid is an asymmetric center (in all amino acids except glycine), the polypeptide chain is inherently asymmetric. Only L-amino acids are found in proteins.

The Peptide Bond Has Partial Double Bond Character

The peptide linkage is usually portrayed by a single bond between the carbonyl carbon and the amide nitrogen (Figure 4.3a). Therefore, in principle, rotation may occur about any covalent bond in the polypeptide backbone

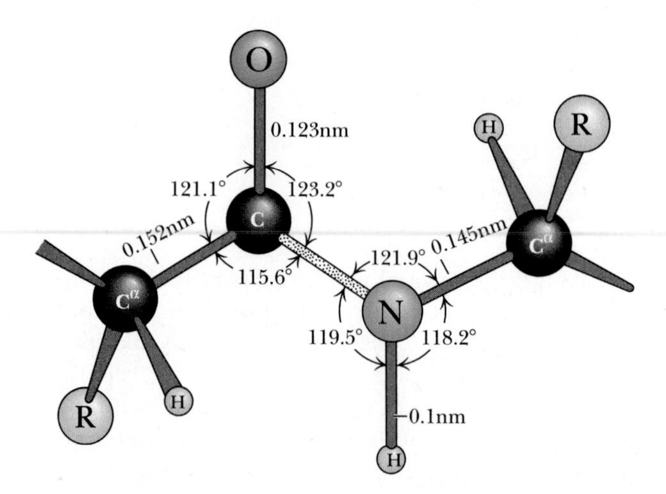

***Figure* 4.2** The peptide bond is shown in its usual *trans* conformation of carbonyl O and amide H. The C_α atoms are the α-carbons of two adjacent amino acids joined in peptide linkage. The dimensions and angles are the average values observed by crystallographic analysis of amino acids and small peptides. The peptide bond is stippled.

(Adapted from Ramachandran, G. N., et al., 1974. Biochimica Biophysica Acta 359:298–302.)

Table 4.1
Size of Protein Molecules*

Protein	M_r	Number of Residues per Chain	Subunit Organization
Insulin (bovine)	5,733	21 (A) 30 (B)	$\alpha\beta$
Cytochrome *c* (equine)	12,500	104	α_1
Ribonuclease A (bovine pancreas)	12,640	124	α_1
Lysozyme (egg white)	13,930	129	α_1
Myoglobin (horse)	16,980	153	α_1
Chymotrypsin (bovine pancreas)	22,600	13 (α) 132 (β) 97 (γ)	$\alpha\beta\gamma$
Hemoglobin (human)	64,500	141 (α) 146 (β)	$\alpha_2\beta_2$
Serum albumin (human)	68,500	550	α_1
Hexokinase (yeast)	96,000	200	α_4
γ-Globulin (horse)	149,900	214 (α) 446 (β)	$\alpha_2\beta_2$
Glutamate dehydrogenase (liver)	332,694	500	α_6
Myosin (rabbit)	470,000	1800 (heavy, *h*) 190 (α) 149 (α') 160 (β)	$h_2\alpha_1\alpha'_2\beta_2$
Ribulose bisphosphate carboxylase (spinach)	560,000	475 (α) 123 (β)	$\alpha_8\beta_8$
Glutamine synthetase (*E. coli*)	600,000	468	α_{12}

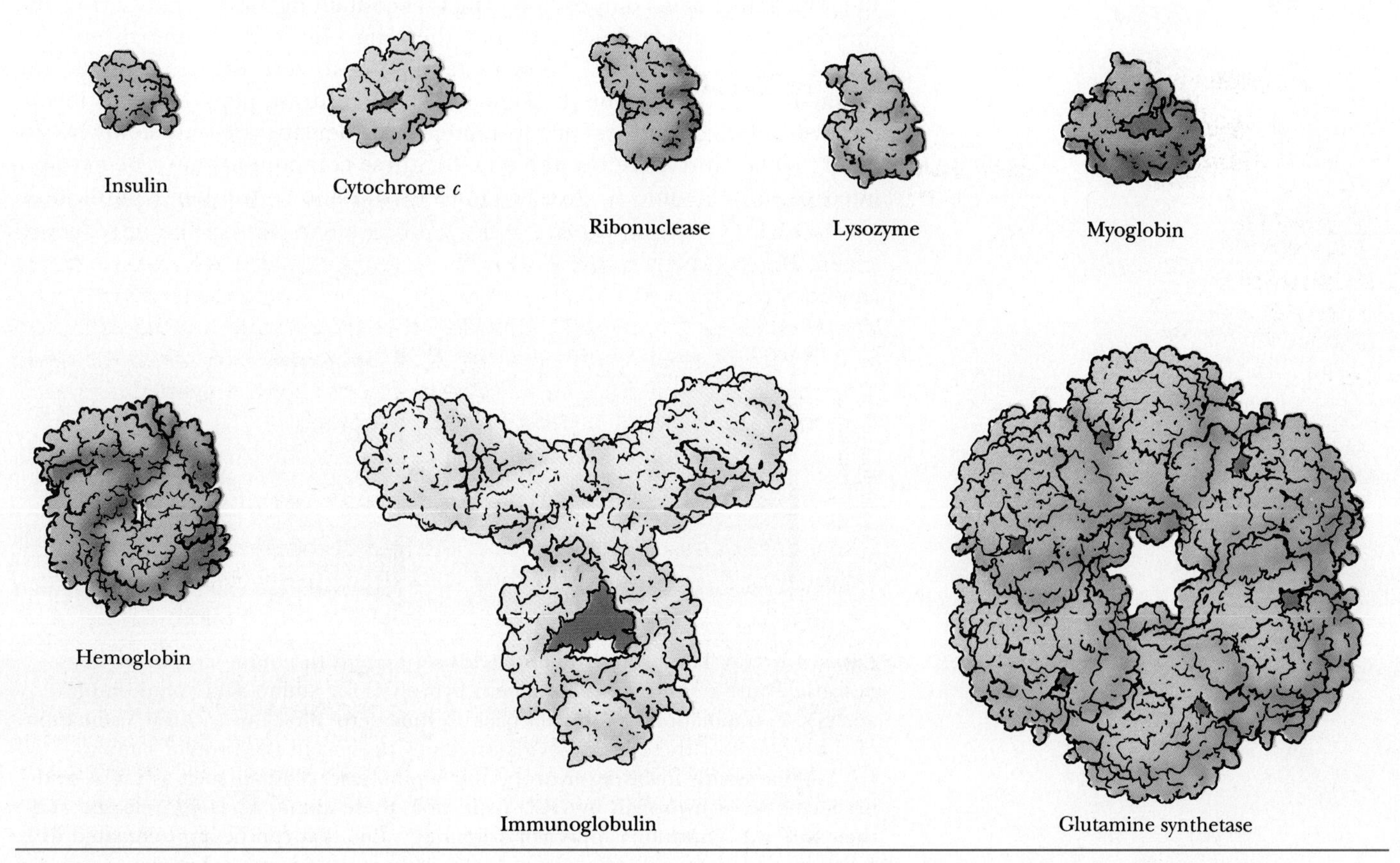

*Illustrations of selected proteins listed in Table 4.1 are drawn to constant scale. (*Adapted from Goodsell and Olson, 1993. Trends in Biochemical Sciences* ***18:65–68.***)

The polypeptide chains of proteins range in length from about 100 amino acids to as many as 1800, the number found in each of the two polypeptide chains of myosin, the contractile protein of muscle. The polypeptide chains of myosin are among the longest known. The average molecular weight of polypeptide chains in eukaryotic cells is about 31,700, corresponding to about 270 amino acid residues. Table 4.1 is a representative list of proteins according to size. The molecular weights (M_r) of proteins can be estimated by a number of physicochemical methods such as polyacrylamide gel electrophoresis or ultracentrifugation (see Chapter Appendix). Precise determinations of protein molecular masses are best obtained by simple calculations based on knowledge of their amino acid sequence. No simple generalizations correlate the size of proteins with their functions. For instance, the same function may be fulfilled in different cells by proteins of different molecular weight. The *Escherichia coli* enzyme responsible for glutamine synthesis (a protein known as *glutamine synthetase*) has a molecular weight of 600,000, whereas the analogous enzyme in brain tissue has a molecular weight of just 380,000.

Acid Hydrolysis of Proteins

Peptide bonds of proteins are hydrolyzed by either strong acid or strong base. Because acid hydrolysis proceeds without racemization and with less destruction of certain amino acids (Ser, Thr, Arg, and Cys) than alkaline treatment, it is the method of choice in analysis of the amino acid composition of proteins and polypeptides. Typically, samples of a protein are hydrolyzed with 6 *N* HCl at 110°C for 24, 48, and 72 hr in sealed glass vials. Tryptophan is destroyed by acid and must be estimated by other means to determine its contribution to the total amino acid composition. The OH-containing amino acids serine and threonine are slowly destroyed, but the data obtained for the three time points (24, 48, and 72 hr) allow extrapolation to zero time to estimate the original Ser and Thr content (Figure 4.5). In contrast, peptide bonds involving hydrophobic residues such as valine and isoleucine are only slowly hydrolyzed in acid, and therefore the data for these components must be extrapolated to infinite time to give accurate quantitation. Another complication

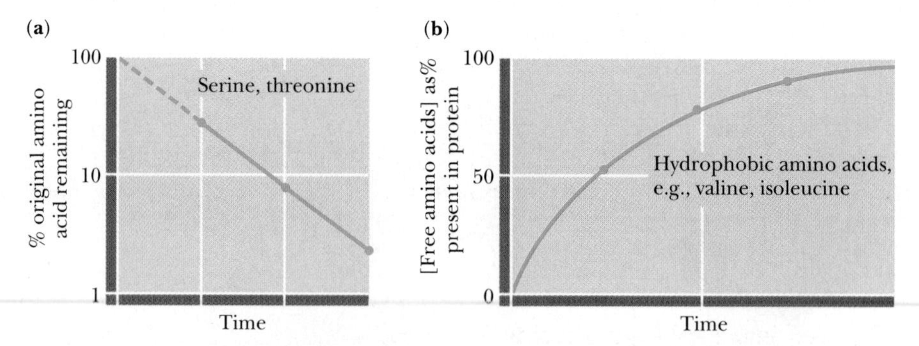

***Figure* 4.5** (a) The hydroxy amino acids serine and threonine are slowly destroyed during the course of protein hydrolysis for amino acid composition analysis. Extrapolation of the data back to time zero allows an accurate estimation of the amount of these amino acids originally present in the protein sample. (b) Peptide bonds involving hydrophobic amino acid residues such as valine and isoleucine resist hydrolysis by HCl. With time, these amino acids are released and their free concentrations approach a limiting value that can be approximated with reliability.

arises because the β- and γ-amide linkages in asparagine (Asn) and glutamine (Gln) are acid labile. The amide nitrogen is released as free ammonia, and all of the Asn and Gln residues of the protein become aspartic acid (Asp) and glutamic acid (Glu), respectively. The amount of ammonia released during acid hydrolysis gives an estimate of the total number of Asn and Gln residues in the original protein, but not the amounts of either. Accordingly, the concentrations of Asp and Glu determined in amino acid analysis are expressed as Asx and Glx, respectively. Since the relative contributions of [Asn + Asp] or [Gln + Glu] cannot be derived from the data, this information must be obtained by alternative means.

Amino Acid Analysis of Proteins

The complex amino acid mixture in the hydrolysate obtained after digestion of a protein for one to three days in 6 *N* HCl is resolved by ion exchange chromatography, which separates the component amino acids. They can then be individually quantified by spectrophotometric assay (see Chapter 3). The chromatographic separation and analysis is completely automated in instruments called **amino acid analyzers,** which can perform the determination on as little as 1 nmol of protein in less than 1 hr.

Table 4.2 gives the amino acid composition of several selected proteins: ribonuclease A, alcohol dehydrogenase, myoglobin, histone H3, and collagen. Each of the 20 naturally occurring amino acids is usually represented at least once in a polypeptide chain. However, some small proteins may not have a representative of every amino acid. Note that ribonuclease (12.6 kD, 124 amino acid residues) does not contain any tryptophan. Amino acids virtually never occur in equimolar ratios in proteins, indicating that proteins are not composed of repeating arrays of amino acids. There are a few exceptions to this rule. Collagen, for example, contains large proportions of glycine and proline, and much of its structure is composed of (Gly-*x*-Pro) repeating units, where *x* is any amino acid. Other proteins show unusual abundances of various amino acids. For example, histones are rich in positively charged amino acids such as arginine and lysine. Histones are a class of proteins found associated with the anionic phosphate groups of eukaryotic DNA.

Amino acid analysis itself does not directly give the number of residues of each amino acid in a polypeptide, but it does give amounts from which the percentages or ratios of the various amino acids can be obtained (Table 4.2). If the molecular weight *and* the exact amount of the protein analyzed are known (or the number of amino acid residues per molecule is known), the molar ratios of amino acids in the protein can be calculated. Amino acid analysis provides no information on the order or sequence of amino acid residues in the polypeptide chain. Since the polypeptide chain is unbranched, it has only two ends, an amino-terminal or **N-terminal end** and a carboxyl-terminal or **C-terminal end.**

The Sequence of Amino Acids in Proteins

The unique characteristic of each protein is the distinctive sequence of amino acid residues in its polypeptide chain(s). Indeed, it is the **amino acid sequence** of proteins that is encoded by the nucleotide sequence of DNA. This amino acid sequence, then, is a form of genetic information. By convention, the amino acid sequence is read from the N-terminal end of the polypeptide chain through to the C-terminal end. As an example, every molecule of ribonuclease A from bovine pancreas has the same amino acid sequence, begin-

***Table* 4.2**
Amino Acid Composition of Some Selected Proteins

Values expressed are percent representation of each amino acid.

	Proteins*				
Amino Acid	RNase	ADH	Mb	Histone H3	Collagen
Ala	6.9	7.5	9.8	13.3	11.7
Arg	3.7	3.2	1.7	13.3	4.9
Asn	7.6	2.1	2.0	0.7	1.0
Asp	4.1	4.5	5.0	3.0	3.0
Cys	6.7	3.7	0	1.5	0
Gln	6.5	2.1	3.5	5.9	2.6
Glu	4.2	5.6	8.7	5.2	4.5
Gly	3.7	10.2	9.0	5.2	32.7
His	3.7	1.9	7.0	1.5	0.3
Ile	3.1	6.4	5.1	5.2	0.8
Leu	1.7	6.7	11.6	8.9	2.1
Lys	7.7	8.0	13.0	9.6	3.6
Met	3.7	2.4	1.5	1.5	0.7
Phe	2.4	4.8	4.6	3.0	1.2
Pro	4.5	5.3	2.5	4.4	22.5
Ser	12.2	7.0	3.9	3.7	3.8
Thr	6.7	6.4	3.5	7.4	1.5
Trp	0	0.5	1.3	0	0
Tyr	4.0	1.1	1.3	2.2	0.5
Val	7.1	10.4	4.8	4.4	1.7
Acidic	8.4	10.2	13.7	8.1	7.5
Basic	15.0	13.1	21.8	24.4	8.8
Aromatic	6.4	6.4	7.2	5.2	1.7
Hydrophobic	18.0	30.7	27.6	23.0	6.5

*Proteins are as follows:

RNase: Bovine ribonuclease A, an enzyme; 124 amino acid residues. Note that RNase lacks tryptophan.

ADH: Horse liver alcohol dehydrogenase, an enzyme; dimer of identical 374 amino acid polypeptide chains. The amino acid composition of ADH is reasonably representative of the norm for water-soluble proteins.

Mb: Sperm whale myoglobin, an oxygen-binding protein; 153 amino acid residues. Note that Mb lacks cysteine.

Histone H3: Histones are DNA-binding proteins found in chromosomes; 135 amino acid residues. Note the very basic nature of this protein due to its abundance of Arg and Lys residues. It also lacks tryptophan.

Collagen: Collagen is an extracellular structural protein; 1052 amino acid residues. Collagen has an unusual amino acid composition; it is about one-third glycine and is rich in proline. Note that it also lacks Cys and Trp and is deficient in aromatic amino acid residues in general.

ning with N-terminal lysine at position 1 and ending with C-terminal valine at position 124 (Figure 4.6). Given the possibility of any of the 20 amino acids at each position, the number of unique amino acid sequences is astronomically large. The astounding sequence variation possible within polypeptide chains provides a key insight into the incredible functional diversity of protein molecules in biological systems, which is discussed shortly.

Figure **4.6** Bovine pancreatic ribonuclease A contains 124 amino acid residues, none of which are tryptophan. Four intrachain disulfide bridges (S—S) form cross-links in this polypeptide between Cys_{26} and Cys_{84}, Cys_{40} and Cys_{95}, Cys_{58} and Cys_{110}, and Cys_{65} and Cys_{72}. These disulfides are depicted by yellow bars. Such disulfide bridges form when two cysteine SH-groups react, releasing two H^+ and two e^-: Cys-SH + HS-Cys → Cys-S-S-Cys + 2 H^+ + 2 e^-.

4.2 Architecture of Protein Molecules

Protein Shape

As a first approximation, proteins can be assigned to one of two great classes on the basis of shape: fibrous or globular (Figure 4.7). **Fibrous proteins** tend to have relatively simple, regular linear structures. These proteins often serve structural roles in cells. Typically, they are insoluble in water or in dilute salt solutions. In contrast, **globular proteins** are roughly spherical in shape. The polypeptide chain is compactly folded so that hydrophobic amino acid side chains are in the interior of the molecule and the hydrophilic side chains are on the outside exposed to the solvent, water. Consequently, globular proteins are usually very soluble in aqueous solutions. Most soluble proteins of the cell, such as the cytosolic enzymes, are globular in shape.

The Levels of Protein Structure

The architecture of protein molecules is quite complex. Nevertheless, this complexity can be resolved by defining various levels of structural organization.

Primary Structure

The amino acid sequence is the **primary (1°) structure** of a protein, such as that shown in Figure 4.6, for example.

A Deeper Look

The Virtually Limitless Number of Different Amino Acid Sequences

For a chain of n residues, there are 20^n possible sequence arrangements. To portray this, consider the number of tripeptides possible if there were only three different amino acids, A, B, and C (tripeptide = 3 = n; $n^3 = 3^3 = 27$):

AAA	BBB	CCC
AAB	BBA	CCA
AAC	BBC	CCB
ABA	BAB	CBC
ACA	BCB	CAC
ABC	BAA	CBA
ACB	BCC	CAB
ABB	BAC	CBB
ACC	BCA	CAA

For a polypeptide chain of 100 residues in length, a rather modest size, the number of possible sequences is 20^{100}, or since $20 = 10^{1.3}$, 10^{130} unique possibilities. These numbers are more than astronomical! Since an average protein molecule of 100 residues would have a mass of 13,800 daltons (average molecular mass of an amino acid residue = 138), 10^{130} such molecules would have a mass of 1.38×10^{134} daltons. The mass of the observable universe is estimated to be 10^{80} proton masses (about 10^{80} daltons). Thus, the universe lacks enough material to make just one molecule of each possible polypeptide sequence for a protein only 100 residues in length.

Secondary Structure

Through hydrogen bonding interactions between adjacent amino acid residues (discussed in detail in Chapter 5), the polypeptide chain can arrange itself into characteristic patterns, such as helical or pleated segments. These segments constitute structural conformities, so-called **regular structures,** that extend along one dimension, like the coils of a spring. Such architectural features of a protein are designated as **secondary (2°) structures** (Figure 4.8). Secondary structures are just one of the higher levels of structure that represent the three-dimensional arrangement of the polypeptide in space.

(a)

Collagen, a fibrous protein

(b)

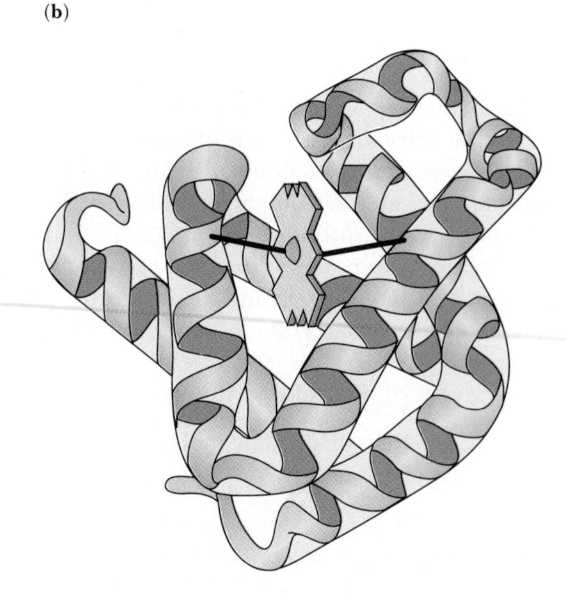

Myoglobin, a globular protein

***Figure* 4.7** (a) Proteins having structural roles in cells are typically fibrous and often water insoluble. Collagen is a good example. Collagen is composed of three polypeptide chains that intertwine. (b) Proteins serving metabolic functions can be characterized as compactly folded globular molecules, such as myoglobin. The folding pattern puts hydrophilic amino acid side chains on the outside and buries hydrophobic side chains in the interior, making the protein highly water soluble.

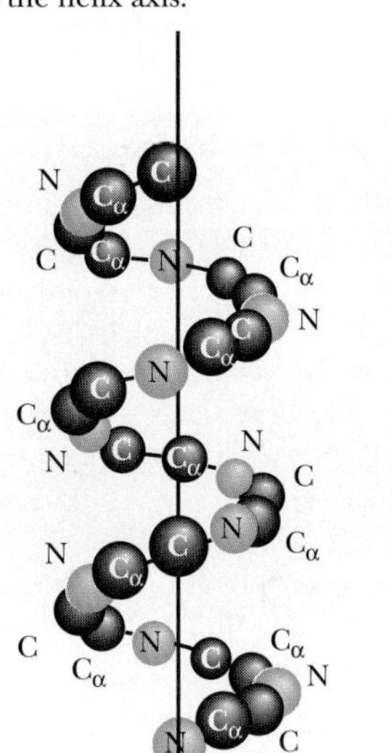

α-Helix
only the N–Cα–C backbone
is represented. The vertical line
is the helix axis.

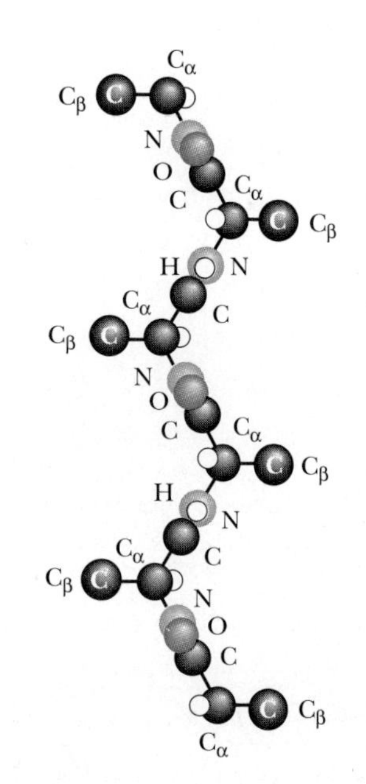

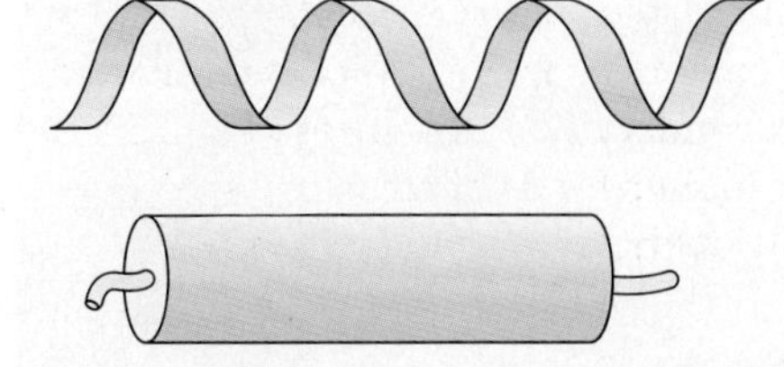

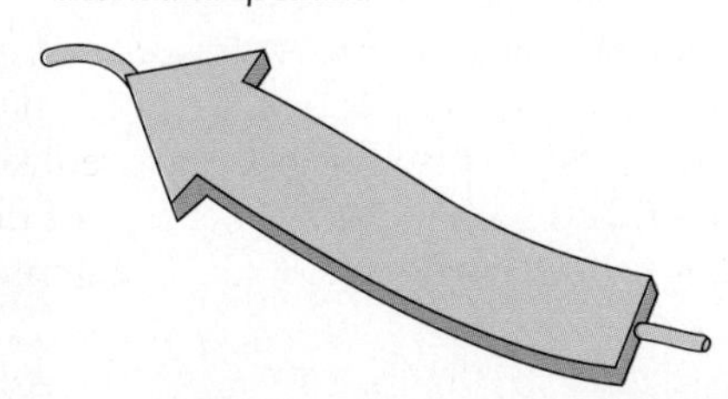

***Figure* 4.8** Two structural motifs that arrange the primary structure of proteins into a higher level of organization predominate in proteins: the α-helix and the β-pleated strand. Atomic representations of these secondary structures are shown here, along with the symbols used by structural chemists to represent them: the flat, helical ribbon for the α-helix and the flat, wide arrow for β-structures. Both of these structures owe their stability to the formation of hydrogen bonds between N—H and O═C functions along the polypeptide backbone (see Chapter 5).

Tertiary Structure

When the polypeptide chains of protein molecules bend and fold in order to assume a more compact three-dimensional shape, a **tertiary (3°) level of structure** is generated (Figure 4.9). It is by virtue of their tertiary structure that proteins adopt a globular shape, especially those proteins commonly existing in solution in the aqueous compartments of cells. A globular conformation gives the lowest surface-to-volume ratio, shielding much of the protein from interaction with the solvent.

Quaternary Structure

Many proteins consist of two or more interacting polypeptide chains of characteristic tertiary structure, each of which is commonly referred to as a **subunit** of the protein. Subunit organization in proteins constitutes another level in the hierarchy of protein structure, defined as the protein's **quaternary (4°) structure** (Figure 4.10). Questions of quaternary structure address the various kinds of subunits within a protein molecule, the number of each, and the ways in which they interact with one another.

(**a**)

H_2N–CGVPAIQPVL$_{10}$SGL[SR]IVNGE$_{20}$EAVPGSWPWQ$_{30}$VSLQDKTGFH$_{40}$FCGGSLINEN$_{50}$WVVTAAHCGV$_{60}$TTSDVVVAGE$_{70}$FDQGSSSEKI$_{80}$QKLKIA
KVFK$_{90}$NSKYNSLTIN$_{100}$NDITLLKLST$_{110}$AASFSQTVSA$_{120}$VCLPSASDDF$_{130}$AAGTTCVTTG$_{140}$WGLTRY[TN]AN$_{150}$TPDRLQQASL$_{160}$PLLSNTNCK
K$_{170}$YWGTKIKDAM$_{180}$ICAGASGVSS$_{190}$CMGDSGGPLV$_{200}$CKKNGAWTLV$_{210}$GIVSWGSSTC$_{220}$STSTPGVYAR$_{230}$VTALVNWVQQ$_{240}$TLAAN–COOH

(**b**) Chymotrypsin tertiary structure

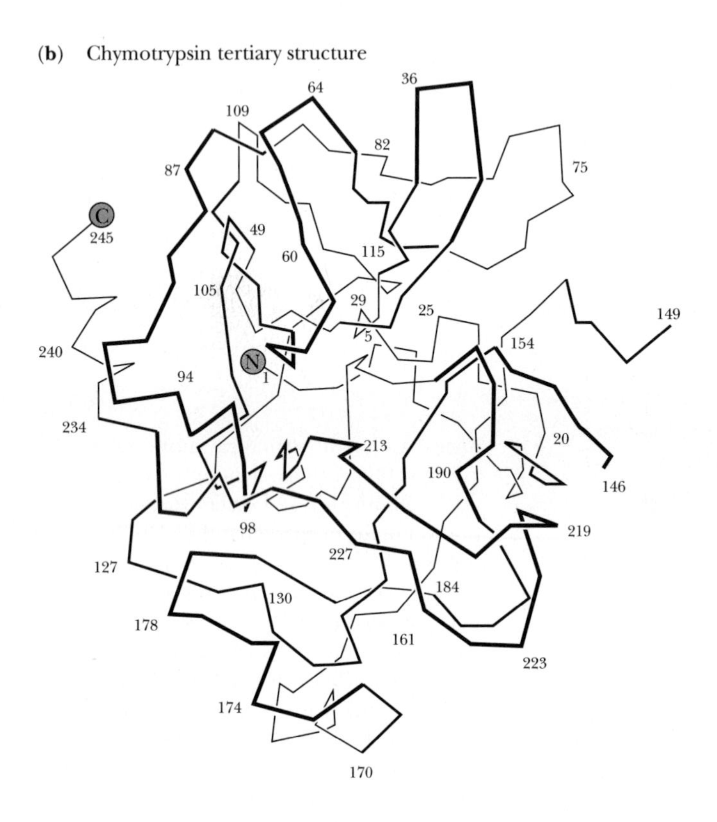

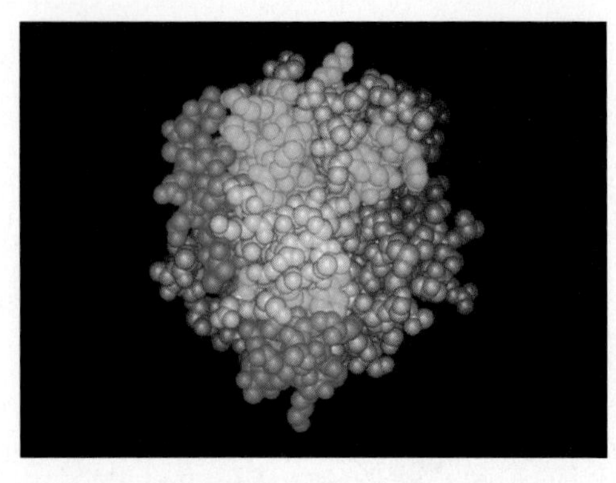

Chymotrypsin space-filling model

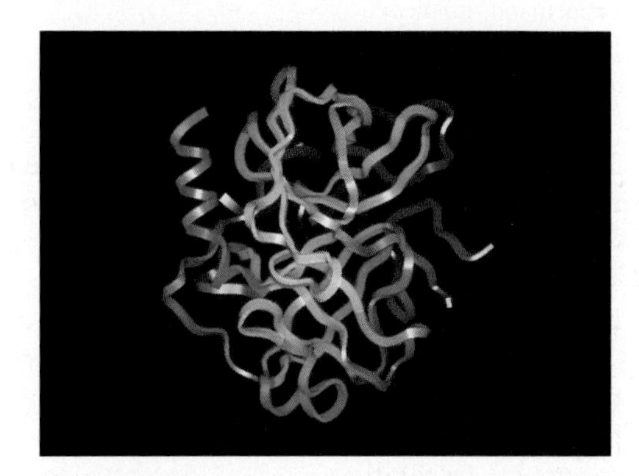

Chymotrypsin ribbon

***Figure* 4.9** Folding of the polypeptide chain into a compact, roughly spherical conformation creates the tertiary level of protein structure. (a) The primary structure and (b) a representation of the tertiary structure of chymotrypsin, a proteolytic enzyme, are shown here. The tertiary representation on the left shows the course of the chymotrypsin folding pattern by successive numbering of the amino acids in its sequence. (Residues 14 and 15 and 147 and 148 are missing because these residues are proteolytically removed when chymotrypsin is formed from its larger precursor, chymotrypsinogen.) The ribbon diagram depicts the three-dimensional track of the polypeptide in space.

Whereas the primary structure of a protein is determined by the covalently linked amino acid residues in the polypeptide backbone, secondary and higher orders of structure are determined principally by noncovalent forces such as hydrogen bonds, ionic bonds, and van der Waals and hydrophobic interactions. It is important to emphasize that *all the information necessary for a protein molecule to achieve its intricate architecture is contained within its 1° structure,* that is, within the amino acid sequence of its polypeptide chain(s). Chapter 5

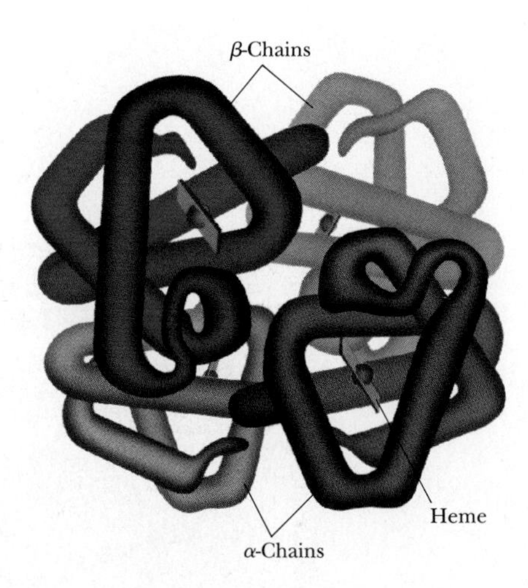

Figure **4.10** Hemoglobin, which consists of two α and two β polypeptide chains, is an example of the quaternary level of protein structure. In this drawing, the β-chains are the two uppermost polypeptides and the two α-chains are the lower half of the molecule. The two closest chains (darkest colored) are the β_2-chain (upper left) and the α_1-chain (lower right). The heme groups of the four globin chains are represented by rectangles with spheres (the heme iron atom). Note the symmetry of this macromolecular arrangement.

presents a detailed discussion of the 2°, 3°, and 4° structure of protein molecules.

Protein Conformation

The overall three-dimensional architecture of a protein is generally referred to as its **conformation.** This term is not to be confused with **configuration,** which denotes the geometric possibilities for a particular set of atoms (Figure 4.11). In going from one configuration to another, covalent bonds must be broken and rearranged. In contrast, the conformational possibilities of a molecule are achieved without breaking any covalent bonds. In proteins, rotations about each of the single bonds along the peptide backbone have the potential to alter the course of the polypeptide chain in three-dimensional space. These rotational possibilities create many possible orientations for the protein chain, referred to as its *conformational possibilities.* Of the great number of theoretical conformations a given protein might adopt, only a very few are favored energetically under physiological conditions. At this time, little is understood regarding the rules that direct the folding of protein chains into energetically favorable conformations.

Later we return to an analysis of the 1° structure of proteins and the methodology used in determining the amino acid sequence of polypeptide chains, but let's first consider the extraordinary variety and functional diversity of these most interesting macromolecules.

4.3 The Many Biological Functions of Proteins

Proteins are the agents of biological function. Virtually every cellular activity is dependent on one or more particular proteins. Thus, a convenient way to classify the enormous number of proteins is by the biological roles they fill. Table 4.3 summarizes the classification of proteins by function and gives examples of representative members of each class.

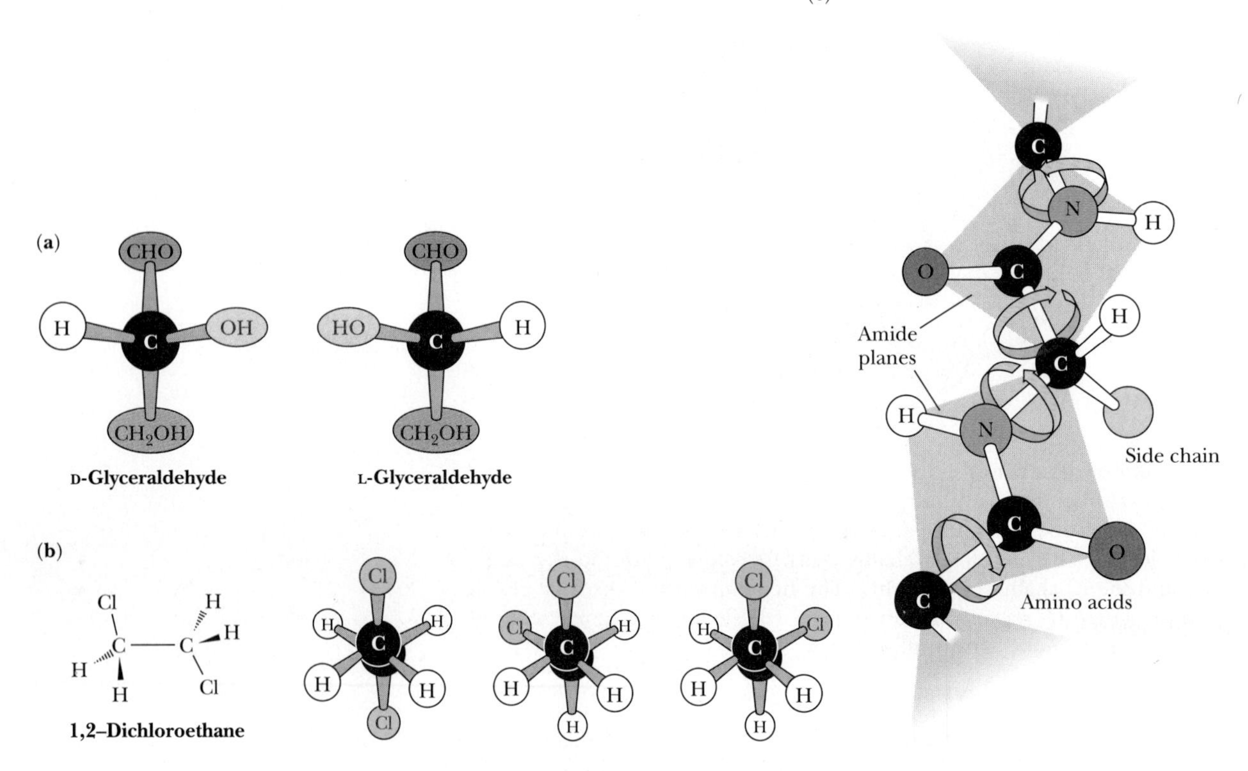

Figure **4.11** Configuration and conformation are *not* synonymous. (a) Rearrangements between configurational alternatives of a molecule can only be achieved by breaking and remaking bonds, as in the transformation between the D- and L-configurations of glyceraldehyde. No possible rotational reorientation of bonds linking the atoms of D-glyceraldehyde will yield geometric identity with L-glyceraldehyde, even though they are mirror images of each other. (b) The intrinsic free rotation about single covalent bonds creates a great variety of three-dimensional conformations, even for relatively simple molecules. Consider 1,2-dichloroethane. Viewed end-on in a Newman projection, three principal rotational orientations or conformations predominate. Steric repulsion between eclipsed and partially eclipsed conformations keeps the possibilities at a reasonable number. (c) Imagine the conformational possibilities for a protein, where two out of every three bonds along its backbone are freely rotating single bonds.

Enzymes

By far the largest class of proteins is enzymes. More than 2000 different enzymes are listed in *Enzyme Nomenclature,* the standard reference volume on enzyme classification. **Enzymes** are catalysts that accelerate the rates of biological reactions. Each enzyme is very specific in its function and acts only in a particular metabolic reaction. Virtually every step in metabolism is catalyzed by an enzyme. The catalytic power of enzymes far exceeds that of synthetic catalysts. Enzymes can enhance reaction rates in cells as much as 10^{20} times the uncatalyzed rate. Enzymes are systematically classified according to the nature of the reaction that they catalyze, such as the transfer of a phosphate group *(phosphotransferase)* or an oxidation–reduction *(oxidoreductase)*. The formal names of enzymes come from the particular reaction within the class that they catalyze, as in ATP:D-fructose-6-phosphate 1-phosphotransferase and alcohol:NAD^+ oxidoreductase. Often, enzymes have common names in addition to their formal names. ATP:D-fructose-6-phosphate 1-phosphotransferase is more commonly known as *phosphofructokinase* (*kinase* is a common name

***Table* 4.3**
Biological Functions of Proteins and Some Representative Examples

Functional Class	Examples
Enzymes	Ribonuclease
	Trypsin
	Phosphofructokinase
	Alcohol dehydrogenase
	Catalase
	"Malic" enzyme
Regulatory proteins	Insulin
	Somatotropin
	Thyrotropin
	lac repressor
	trp repressor
	Catabolite activator protein
Transport proteins	Hemoglobin
	Serum albumin
	Glucose transporter
Storage proteins	Ovalbumin
	Casein
	Zein
	Phaseolin
	Ferritin
Contractile and motile proteins	Actin
	Myosin
	Tubulin
	Dynein
	Kinesin
Structural proteins	α-Keratins
	Collagen
	Elastin
	Fibroin
	Proteoglycans
Protective proteins	Immunoglobulins
	Thrombin
	Fibrinogen
	Snake and bee venom proteins
	Diphtheria toxin
	Ricin
Exotic proteins	Antifreeze proteins
	Monellin
	Resilin
	Glue proteins

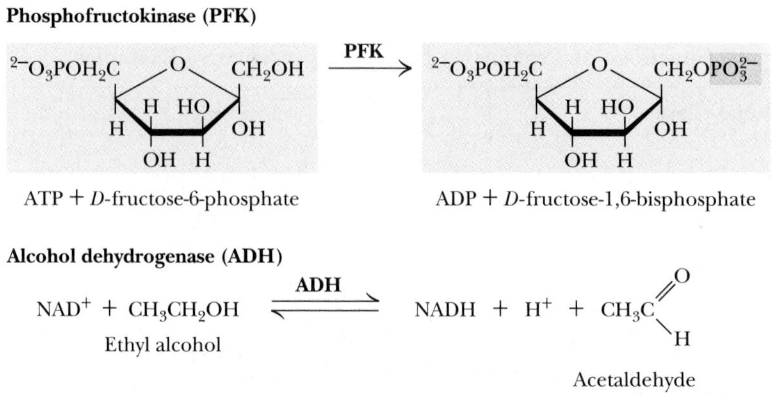

Phosphofructokinase (PFK)

ATP + *D*-fructose-6-phosphate $\xrightarrow{PFK}$ ADP + *D*-fructose-1,6-bisphosphate

Alcohol dehydrogenase (ADH)

$NAD^+ + CH_3CH_2OH \underset{}{\overset{ADH}{\rightleftharpoons}} NADH + H^+ + CH_3CHO$

Ethyl alcohol — Acetaldehyde

Figure **4.12** Enzymes are classified according to the specific biological reaction that they catalyze. Cells contain thousands of different enzymes. Two common examples drawn from carbohydrate metabolism are phosphofructokinase (PFK), or more precisely, ATP:D-fructose-6-phosphate 1-phosphotransferase, and alcohol dehydrogenase (ADH), or alcohol:NAD^+ oxidoreductase, which catalyze the reactions shown here.

given to ATP-dependent phosphotransferases). Similarly, alcohol:NAD^+ oxidoreductase is casually referred to as *alcohol dehydrogenase.* The reactions catalyzed by these two enzymes are shown in Figure 4.12. Other enzymes are known by trivial names that have historical roots, such as *catalase* (systematic name, hydrogen-peroxide:hydrogen-peroxide oxidoreductase), and sometimes these trivial names have descriptive connotations as well, as in *malic enzyme* (systematic name, L-malate:$NADP^+$ oxidoreductase).

Regulatory Proteins

A number of proteins do not perform any obvious chemical transformation, but nevertheless can regulate the ability of other proteins to carry out their physiological functions. Such proteins are referred to as **regulatory proteins.** A well-known example is *insulin,* the hormone regulating glucose metabolism in animals. Insulin is a relatively small protein (5.7 kD) and consists of two polypeptide chains held together by disulfide cross-bridges. Other hormones that are also proteins include the pituitary *somatotropin* (21 kD)and *thyrotropin* (28 kD), which stimulates the thyroid gland. Another group of regulatory proteins is involved in the regulation of gene expression. These proteins characteristically act by binding to DNA sequences that are adjacent to coding regions of genes, either activating or inhibiting the transcription of genetic information into RNA. Examples include **repressors,** which, because they block transcription, are considered negative control elements. Two prokaryotic representatives are *lac repressor* (37 kD), which controls expression of the enzyme system responsible for the metabolism of lactose (milk sugar), and *trp repressor,* which prevents synthesis of the enzymes involved in tryptophan biosynthesis. Positively acting control elements are also known. For example, the *E. coli catabolite gene activator protein* **(CAP)** (44 kD), under appropriate metabolic conditions, can bind to specific sites along the *E. coli* chromosome and increase the rate of transcription of adjacent genes. Transcription in eukaryotic cells is also governed by DNA-binding regulatory proteins; examples include steroid receptors and proteins with characteristic structural motifs, such as leucine zipper and zinc finger proteins (Chapter 30).

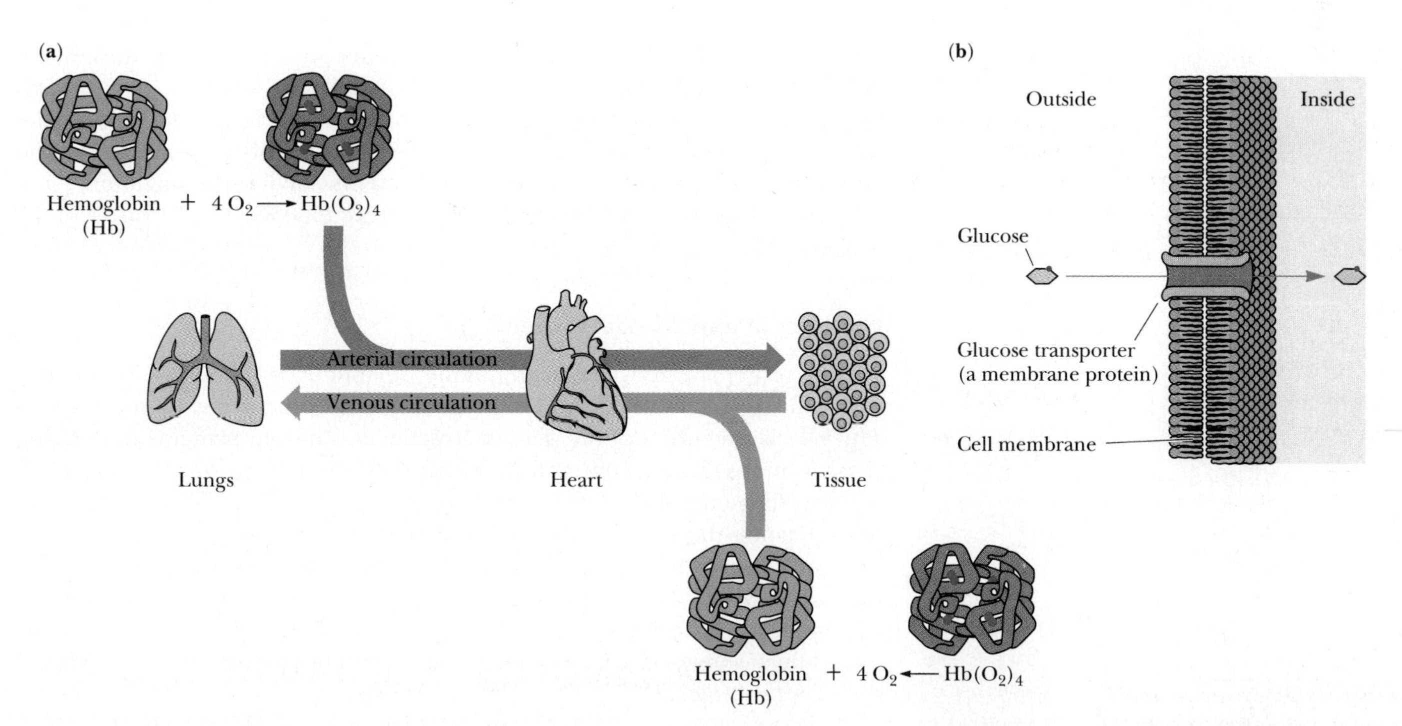

***Figure* 4.13** Two basic types of biological transport are (a) transport within or between different cells or tissues and (b) transport into or out of cells. Proteins function in both of these phenomena. For example, the protein hemoglobin transports oxygen from the lungs to actively respiring tissues. Transport proteins of the other type are localized in cellular membranes where they function in the uptake of specific nutrients, such as glucose (shown here) and amino acids, or the export of metabolites and waste products.

Transport Proteins

A third class of proteins are the **transport proteins.** These proteins function to transport specific substances from one place to another. One type of transport is exemplified by the transport of oxygen from the lungs to the tissues by *hemoglobin* (Figure 4.13) or by the transport of fatty acids from adipose tissue to various organs by the blood protein *serum albumin.* A very different type is the transport of metabolites across permeability barriers such as cell membranes, as mediated by specific membrane proteins. These *membrane transport proteins* take up metabolite molecules on one side of a membrane, transport them across the membrane, and release them on the other side. Examples include the transport proteins responsible for the uptake of essential nutrients into the cell, such as glucose or amino acids (Figure 4.13). All naturally occurring membrane transport proteins studied thus far form channels in the membrane through which the transported substances are passed.

Storage Proteins

Proteins whose biological function is to provide a reservoir of an essential nutrient are called **storage proteins.** Since proteins are amino acid polymers and since nitrogen is commonly a limiting nutrient for growth, organisms have exploited proteins as a means to provide sufficient nitrogen in times of need. For example, *ovalbumin,* the protein of egg white, provides the developing bird embryo with a source of nitrogen during its isolation within the egg. *Casein* is the most abundant protein of milk and thus the major nitrogen source for mammalian infants. The seeds of higher plants often contain as much as 60% storage protein to make the germinating seed nitrogen-sufficient during this crucial period of plant development. In corn (*Zea mays* or *maize*), a family of low molecular weight proteins in the kernel called *zeins* serve this purpose; peas (the seeds of *Phaseolus vulgaris*) contain a storage protein called *phaseolin.* The use of proteins as a reservoir of nitrogen is more efficient than storing an equivalent amount of amino acids. Not only is the osmotic pressure minimized, but the solvent capacity of the cell is taxed less in solvating one molecule of an amino acid polymer than in dissolving, for example, 100 molecules of free amino acids. Proteins can also serve to store nutrients other than the more obvious elements composing amino acids (N, C, H, O, and S). As an example, *ferritin* is a protein found in animal tissues that binds iron, retaining this essential metal so that it will be available for the synthesis of important iron-containing proteins such as hemoglobin. One molecule of ferritin (460 kD) binds as many as 4500 atoms of iron (35% by weight).

Contractile and Motile Proteins

Certain proteins endow cells with unique capabilities for movement. Cell division, muscle contraction, and cell motility represent some of the ways in which cells execute motion. The **contractile** and **motile proteins** underlying these motions share a common property: they are filamentous or polymerize to form filaments. Examples include *actin* and *myosin,* the filamentous proteins forming the contractile systems of cells, and *tubulin,* the major component of microtubules (the filaments involved in the mitotic spindle of cell division as well as in flagella and cilia). Another class of proteins involved in movement includes *dynein* and *kinesin,* so-called **motor proteins** that propel intracellular movement of vesicles, granules, and organelles along established cytoskeletal "tracks."

Structural Proteins

An apparently passive but very important role of proteins is their function in creating and maintaining biological structures. **Structural proteins** provide strength and protection to cells and tissues. Monomeric units of structural proteins typically polymerize to generate long fibers (as in hair) or protective sheets of fibrous arrays, as in cowhide (leather). *α-Keratins* are insoluble fibrous proteins making up hair, horns, and fingernails. *Collagen,* another insoluble fibrous protein, is found in bone, connective tissue, tendons, cartilage, and hide, where it forms inelastic fibrils of great strength. One-third of the total protein in a vertebrate animal is collagen. A structural protein having elastic properties is, appropriately, *elastin,* an important component of ligaments. Because of the way elastin monomers are cross-linked in forming polymers, elastin can stretch in two dimensions. Certain insects make a structurally useful protein, *fibroin* (a β-keratin), the major constituent of cocoons (silk) and spider webs. An important protective barrier for animal cells is the *ground substance,* an extracellular matrix containing *proteoglycans,* covalent protein–polysaccharide complexes that cushion and lubricate.

Protective and Exploitive Proteins

In contrast to the passive protective nature of some structural proteins, another group can be more aptly classified as **protective** or **exploitive proteins** because of their biologically active role in cell defense, protection, or exploitation. Prominent among the protective proteins are the *immunoglobulins* or *antibodies* produced by the lymphocytes of vertebrates. Antibodies have the remarkable ability to "ignore" molecules that are an intrinsic part of the host organism, yet they can specifically recognize and neutralize "foreign" molecules resulting from the invasion of the organism by bacteria, viruses, or other infectious agents. Another group of protective proteins are the blood-clotting proteins, *thrombin* and *fibrinogen,* which prevent the loss of blood when the circulatory system is damaged. In addition, various proteins serve defensive or exploitive roles for organisms, including the lytic and neurotoxic proteins of snake and bee venoms and toxic plant proteins, such as *ricin,* whose apparent purpose is to thwart predation by herbivores. Another class of exploitive proteins includes the toxins produced by bacteria, such as diphtheria toxin or cholera toxin.

Exotic Proteins

Some proteins display rather exotic functions that do not quite fit the previous classifications. Arctic and Antarctic fishes have *antifreeze proteins* to protect their blood against freezing in the below-zero temperatures of high-latitude seas. *Monellin,* a protein found in an African plant, has a very sweet taste and is being considered as an artificial sweetener for human consumption. *Resilin,* a protein having exceptional elastic properties, is found in the hinges of insect wings. Certain marine organisms such as mussels secrete *glue proteins,* allowing them to attach firmly to hard surfaces. It is worth repeating that the great diversity of function in proteins, as reflected in this survey, is attained using just 20 amino acids.

4.4 Some Proteins Have Chemical Groups Other Than Amino Acids

Many proteins consist of only amino acids and contain no other chemical groups. The enzyme ribonuclease and the contractile protein actin are two such examples. Such proteins are called **simple proteins.** However, many other proteins contain various chemical constituents as an integral part of their structure. These proteins are termed **conjugated proteins** (Table 4.4). If the nonprotein part is crucial to the protein's function, it is referred to as a **prosthetic group.** If the nonprotein moiety is not covalently linked to the protein, it can usually be removed by denaturing the protein structure. However, if the conjugate is covalently joined to the protein, it may be necessary to carry out acid hydrolysis of the protein into its component amino acids in order to release it. Conjugated proteins are typically classified according to the chemical nature of their nonamino acid component; a representative selection of them is given here and in Table 4.4. (Note that comparisons of Tables 4.3 and 4.4 reveal two distinctly different ways of considering the nature of proteins—function versus chemistry.)

Glycoproteins. Glycoproteins are proteins that contain carbohydrate. Proteins destined for an extracellular location are characteristically glycoproteins. For example, fibronectin and proteoglycans are important components of the extracellular matrix that surrounds the cells of most tissues in multicellular organisms. Immunoglobulin G molecules are the principal antibody species found circulating free in the blood plasma. Many membrane proteins are glycosylated on their extracellular segments.

Lipoproteins. Blood plasma lipoproteins are prominent examples of the class of proteins conjugated with lipid. The plasma lipoproteins function primarily in the transport of lipids to sites of active membrane synthesis. Serum levels of *low density lipoproteins (LDLs)* are often used as a clinical index of susceptibility to vascular disease.

Nucleoproteins. Nucleoprotein conjugates have many roles in the storage and transmission of genetic information. Ribosomes are the sites of protein synthesis. Virus particles and even chromosomes are protein–nucleic acid complexes.

***Table* 4.4**
Representative Conjugated Proteins

Class	Prosthetic Group	Percent by Weight (approx.)
Glycoproteins contain carbohydrate		
Fibronectin		
γ-Globulin		
Proteoglycan		
Lipoproteins contain lipid		
Blood plasma lipoproteins:		
High density lipoprotein (HDL) (α-lipoprotein)	Triacylglycerols, phospholipids, cholesterol	75
Low density lipoprotein (LDL) (β-lipoprotein)	Triacylglycerols, phospholipids, cholesterol	67
Nucleoprotein complexes contain nucleic acid		
Ribosomes	RNA	50–60
Tobacco mosaic virus	RNA	5
Adenovirus	DNA	
HIV-1 (AIDS virus)	RNA	
Phosphoproteins contain phosphate		
Casein	Phosphate groups	
Glycogen phosphorylase *a*	Phosphate groups	
Metalloproteins contain metal atoms		
Ferritin	Iron	35
Alcohol dehydrogenase	Zinc	
Cytochrome oxidase	Copper and iron	
Nitrogenase	Molybdenum and iron	
Pyruvate carboxylase	Manganese	
Hemoproteins contain heme		
Hemoglobin		
Cytochrome *c*		
Catalase		
Nitrate reductase		
Ammonium oxidase		
Flavoproteins contain flavin		
Succinate dehydrogenase	FAD	
NADH dehydrogenase	FMN	
Dihydroorotate dehydrogenase	FAD and FMN	
Sulfite reductase	FAD and FMN	

Phosphoproteins. These proteins have phosphate groups esterified to the hydroxyls of serine, threonine, or tyrosine residues. Casein, the major protein of milk, contains many phosphates and serves to bring essential phosphorus to the growing infant. Many key steps in metabolism are regulated between states of activity or inactivity, depending on the presence or absence of phosphate groups on proteins, as we shall see in Chapter 12. Glycogen phosphorylase *a* is one well-studied example.

Metalloproteins. Metalloproteins are either metal storage forms, as in the case of ferritin, or enzymes in which the metal atom participates in a catalytically important manner. We will encounter many examples throughout this book of the vital metabolic functions served by metalloenzymes.

Hemoproteins. These proteins are actually a subclass of metalloproteins because their prosthetic group is **heme,** the name given to iron protoporphyrin IX (Figure 4.14). Since heme-containing proteins enjoy so many prominent biological functions, they are considered a class by themselves.

Flavoproteins. *Flavin* is an essential substance for the activity of a number of important oxidoreductases. We discuss the chemistry of flavin and its derivatives, FMN and FAD, in Chapter 14 on vitamins and coenzymes.

4.5 Reactions of Peptides and Proteins

The chemical properties of peptides and proteins are most easily considered in terms of the chemistry of their component functional groups. That is, they possess reactive amino and carboxyl termini and they display reactions characteristic of the chemistry of the R groups of their complement of amino acids.

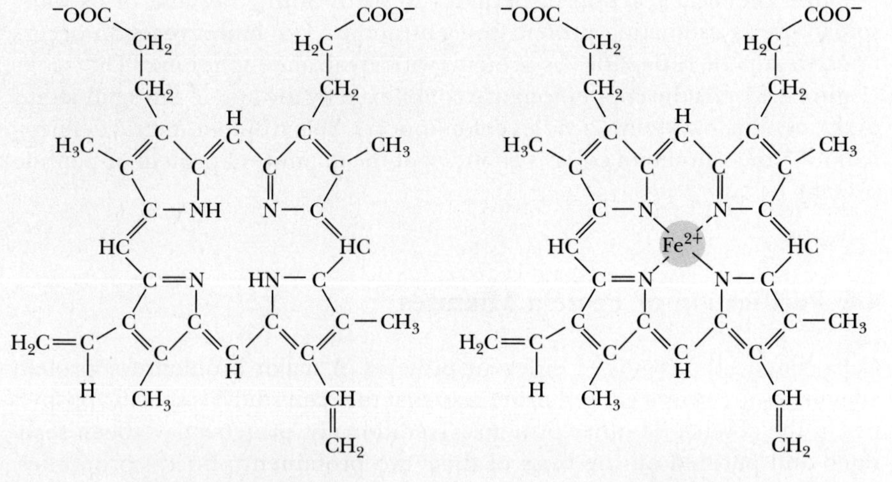

Protoporphyrin IX

Heme
(Fe-protoporphyrin IX)

***Figure* 4.14** Heme consists of protoporphyrin IX and an iron atom. Protoporphyrin, a highly conjugated system of double bonds, is composed of four 5-membered heterocyclic rings (pyrroles) fused together to form a tetrapyrrole macrocycle. The specific isomeric arrangement of methyl, vinyl, and propionate side chains shown is protoporphyrin IX. Coordination of an atom of ferrous iron (Fe^{2+}) by the four pyrrole nitrogen atoms yields heme.

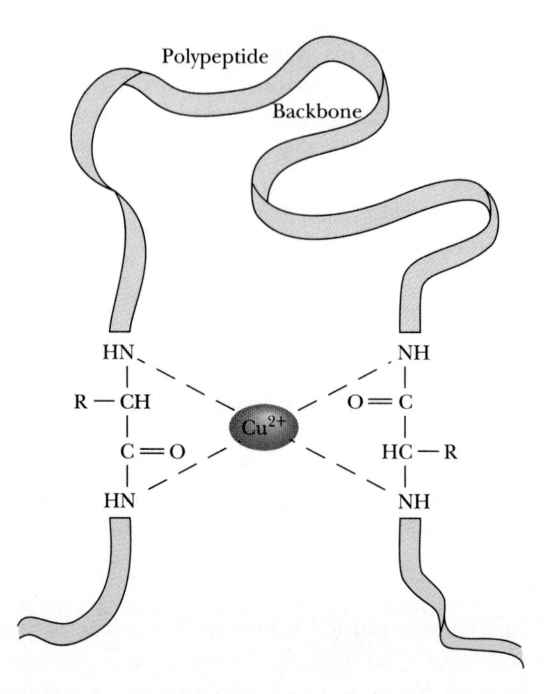

***Figure* 4.15** Treatment of proteins with a dilute alkaline solution of Cu^{2+} ions results in the formation of a complex between the copper and amide N atoms of the polypeptide backbone. The dilute copper solution is pale blue, but its complex with the protein has a strong reddish violet color that can be spectrophotometrically quantified because of its absorption of 540 nm light. The reaction is named for biuret, an amide complex formed on heating urea.

These reactions are familiar to us from Chapter 3 and from the study of organic chemistry and need not be repeated here.

One chemical reaction of peptides is worth noting because of its widespread use in estimating protein concentrations. The **biuret reaction** occurs upon treatment of peptides or proteins with an alkaline solution of Cu^{2+} ions (Figure 4.15). As the copper ions are complexed by the amide nitrogen atoms of the peptide backbone, a violet color appears. Spectrophotometric determination of the amount of color is an index of the quantity of protein or peptide present.

4.6 Purification of Protein Mixtures

Cells contain thousands of different proteins. A major problem for protein chemists is to purify a chosen protein so that they can study its specific properties in the absence of other proteins. Traditionally, proteins have been separated and purified on the basis of their two prominent physical properties: size and electrical charge.

Separation Methods

Separation methods based on size include size exclusion chromatography, ultrafiltration, and ultracentrifugation (see Chapter Appendix). The ionic properties of peptides and proteins are determined principally by their complement of amino acid side chains. Furthermore, the ionization of these groups is pH-dependent.

A Deeper Look

Estimation of Protein Concentrations in Solutions of Biological Origin

Biochemists are often interested in knowing the protein concentration in various preparations of biological origin. Such quantitative analysis is not straightforward. Cell extracts are complex mixtures that typically contain protein molecules of many different molecular weights, so the results of protein estimations cannot be expressed on a molar basis. Also, aside from the rather unreactive repeating peptide backbone, there is little common chemical identity among the many proteins found in cells that might be readily exploited for exact chemical analysis. Most of their chemical properties vary with their amino acid composition, for example, nitrogen or sulfur content or the presence of aromatic, hydroxyl, or other functional groups.

Lowry Procedure

A method that has been the standard of choice for many years is the **Lowry procedure.** This method combines the copper-dependent biuret reaction (see Figure 4.15) with the use of *Folin–Ciocalteau reagent,* a combination of phosphomolybdic and phosphotungstic acid complexes that react with Cu^+. Cu^+ is generated from Cu^{2+} by readily oxidizable protein components, such as cysteine or the phenols and indoles of tyrosine and tryptophan. While the precise chemistry of the Lowry method remains uncertain, the Cu^{1+} reaction with the Folin reagent gives intensely colored products measurable spectrophotometrically. The sensitivity of this method (about 10 μg/mL) is about 100 times better than the biuret reaction alone, but its obvious limitation is the variable amount of Cys, Tyr, and Trp in different proteins. Typically, a well-studied protein such as bovine serum albumin is assayed, and its reaction in the Lowry procedure as a function of its concentration allows the construction of a standard curve. The degree of Lowry reaction of an unknown protein mixture run simultaneously can be compared with the standard curve to obtain an estimate of the unknown's protein content. (Since no information is usually available regarding the Cys, Tyr, and Trp content of the unknown sample, the assay is empirical.) The results are expressed as milligrams of protein per milliliter of solution. Despite widespread use, the procedure is troublesome because many simple substances (sugars, detergents, and certain buffers) interfere with its accuracy.

BCA Method

Recently, a reagent that reacts more efficiently with Cu^+ than Folin–Ciocalteau reagent has been developed for protein assays. *Bicinchoninic acid (BCA)* forms a purple complex with Cu^+ in alkaline solution, enhancing the response of the biuret reaction with protein (see illustration below). The BCA technique is simpler, more tolerant of interfering substances, and slightly more sensitive than the Lowry procedure.

Assays Based on Dye Binding

Several other protocols for protein estimation enjoy prevalent usage in biochemical laboratories. The **Bradford assay** is a rapid and reliable technique that uses a dye called *Coomassie Brilliant Blue G-250,* which undergoes a change in its color upon noncovalent binding to proteins. The binding is quantitative and less sensitive to variations in the protein's amino acid composition. The color change is easily measured by a spectrophotometer. The Bradford procedure easily detects protein concentrations as low as 5 μg/mL. A similar, very sensitive method capable of quantifying nanogram amounts of protein is based on the shift in color of colloidal gold upon binding to proteins.

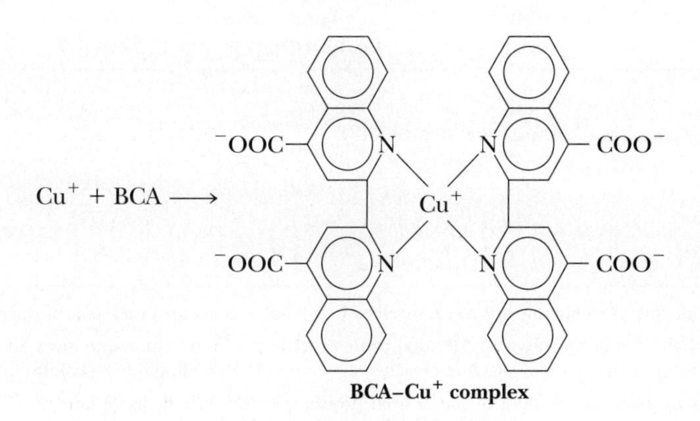

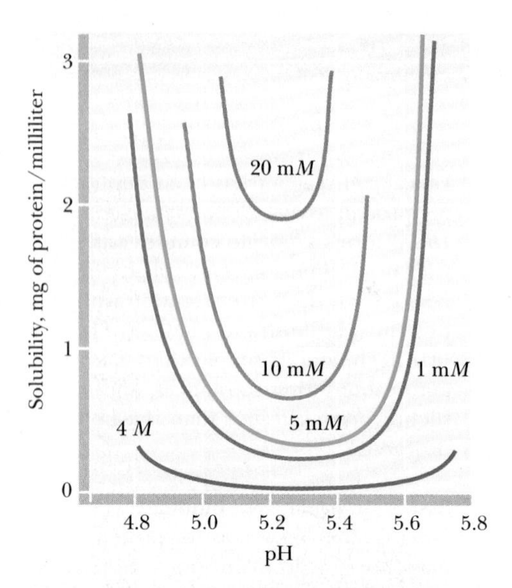

Figure **4.16** The solubility of most globular proteins is markedly influenced by pH and ionic strength. This figure shows the solubility of a typical protein as a function of pH and various salt concentrations.

A variety of procedures exploit electrical charge as a means of discriminating between proteins, including ion exchange chromatography (see Chapter 3), electrophoresis (see Chapter Appendix), and solubility. Proteins tend to be least soluble at their **isoelectric point,** the pH value where the sum of their positive and negative electrical charges is zero. At this pH, electrostatic repulsion between protein molecules is minimal and they are more likely to coalesce and precipitate out of solution. Ionic strength also profoundly influences protein solubility. Most globular proteins tend to become increasingly soluble as the ionic strength is raised. This phenomenon, the salting-in of proteins, is attributed to the diminishment by the salt ions of electrostatic attractions between the protein molecules. Such electrostatic interactions would otherwise lead to precipitation. However, as the salt concentration reaches high levels (greater than 1 *M*), the effect reverses and the protein is salted-out of solution. At this point, the numerous salt ions begin to compete with the protein for waters of solvation, and, as they win out, the protein becomes insoluble. The solubility properties of a typical protein are shown in Figure 4.16.

Although the side chains of most nonpolar amino acids in proteins are usually buried in the interior of the protein away from contact with the aqueous solvent, a portion of them is exposed at the protein's surface, giving it a partially hydrophobic character. *Hydrophobic interaction chromatography* is a protein purification technique that exploits this hydrophobicity (see Chapter Appendix).

A Typical Protein Purification Scheme

Most purification procedures for a particular protein are developed in an empirical manner, the overriding principle being purification of the protein to a homogeneous state with acceptable yield. Table 4.5 presents a summary of a purification scheme for a selected protein. Note that the **specific activity** of the protein (the enzyme xanthine dehydrogenase) in the immuno-affinity purified fraction (fraction 5) has been increased 152/0.108, or 1407 times the specific activity in the crude extract (fraction 1). Thus, xanthine dehydrogenase in fraction 5 versus fraction 1 is enriched more than 1400-fold by the purification procedure.

Table **4.5**

Example of a Protein Purification Scheme: Purification of the Enzyme Xanthine Dehydrogenase from a Fungus

Fraction	Volume (mL)	Total Protein (mg)	Total Activity*	Specific Activity†	Percent Recovery‡
1. Crude extract	3,800	22,800	2,460	0.108	100
2. Salt precipitate	165	2,800	1,190	0.425	48
3. Ion exchange chromatography	65	100	720	7.2	29
4. Molecular sieve chromatography	40	14.5	555	38.3	23
5. Immunoaffinity chromatography§	6	1.8	275	152	11

*The relative enzymatic activity of each fraction in catalyzing the xanthine dehydrogenase reaction is cited as arbitrarily defined units.

†The specific activity is the total activity of the fraction divided by the total protein in the fraction. This value gives an indication of the increase in purity attained during the course of the purification as the samples become enriched for xanthine dehydrogenase protein.

‡The percent recovery of total activity is a measure of the yield of the desired product, xanthine dehydrogenase.

§The last step in the procedure is an affinity method in which antibodies specific for xanthine dehydrogenase are covalently coupled to a chromatography matrix and packed into a glass tube to make a chromatographic column through which fraction 4 is passed. The enzyme is bound by this immunoaffinity matrix while other proteins pass freely out. The enzyme is then recovered by passing a strong salt solution through the column, which dissociates the enzyme–antibody complex.

(*Adapted from Lyon, E. S., and Garrett, R. H., 1978. Journal of Biological Chemistry.* ***253****:2604–2614.*)

4.7 The Primary Structure of a Protein: Determining the Amino Acid Sequence

In 1953, Frederick Sanger of Cambridge University in England reported the amino acid sequences of the two polypeptide chains composing the protein insulin (Figure 4.17). Not only was this a remarkable achievement in analytical chemistry but it helped to demystify speculation about the chemical nature of proteins. Sanger's results clearly established that all of the molecules of a given protein have a fixed amino acid composition, a defined amino acid sequence, and therefore an invariant molecular weight. In short, proteins are well defined chemically. Today, the amino acid sequences of almost 20,000 different proteins are known. While many sequences have been determined from application of the principles first established by Sanger, most are now deduced from knowledge of the nucleotide sequence of the gene that encodes the protein.

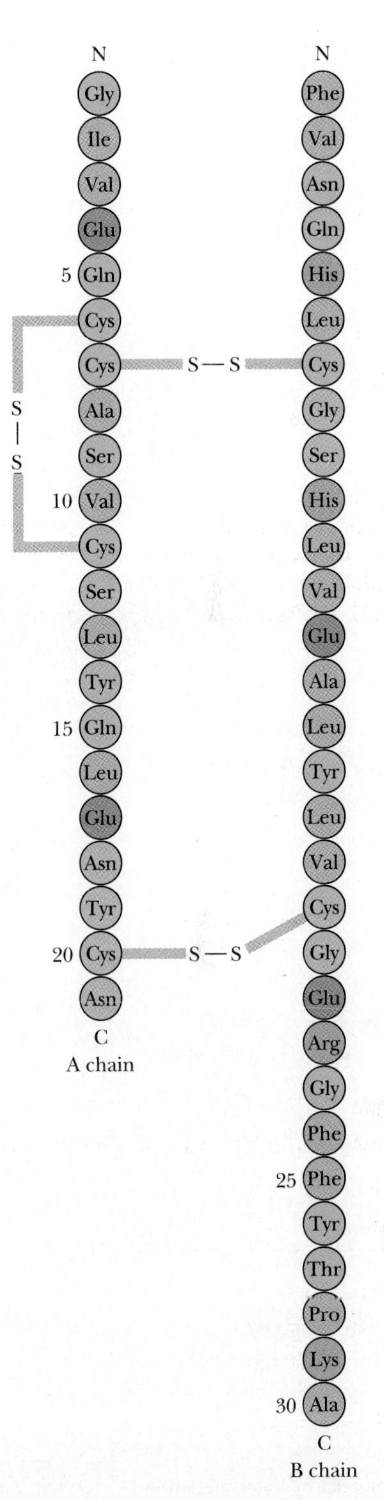

***Figure* 4.17** The hormone insulin consists of two polypeptide chains, A and B, held together by two disulfide cross-bridges (S—S). The A chain has 21 amino acid residues and an intrachain disulfide; the B polypeptide contains 30 amino acids. The sequence shown is for bovine insulin.

Protein Sequencing Strategy

The usual strategy for determining the amino acid sequence of a protein involves eight basic steps:

1. If the protein contains more than one polypeptide chain, the chains are separated and purified.
2. Intrachain S—S (disulfide) cross-bridges between cysteine residues in the polypeptide chain are cleaved. (If these disulfides are interchain linkages, then step 2 precedes step 1.)
3. The amino acid composition of each polypeptide chain is determined.
4. The N-terminal and C-terminal residues are identified.
5. Each polypeptide chain is cleaved into smaller fragments, and the amino acid composition and sequence of each fragment is determined.
6. Step 5 is repeated, using a different cleavage procedure to generate a different and therefore overlapping set of peptide fragments.
7. The overall amino acid sequence of the protein is reconstructed from the sequences in overlapping fragments.
8. The positions of S—S cross-bridges formed between cysteine residues are located.

Each of these steps is discussed in greater detail in the following sections.

Step 1. Separation of Polypeptide Chains

If the protein of interest is a **heteromultimer** (composed of more than one type of polypeptide chain), then the protein must be dissociated and its component polypeptide subunits must be separated from one another and sequenced individually. Subunit associations in multimeric proteins are typically maintained solely by noncovalent forces, and therefore most multimeric proteins can usually be dissociated by exposure to pH extremes, 8 *M* urea, 6 *M* guanidinium hydrochloride, or high salt concentrations. (All of these treatments disrupt polar interactions such as hydrogen bonds both within the protein molecule and between the protein and the aqueous solvent.) Once dissociated, the individual polypeptides can be isolated from one another on the basis of differences in size and/or charge. Occasionally, heteromultimers are linked together by interchain S—S bridges. In such instances, these cross-links must be cleaved prior to dissociation and isolation of the individual chains. The methods described under step 2 are applicable for this purpose.

Step 2. Cleavage of Disulfide Bridges

A number of methods exist for cleaving disulfides (Figure 4.18). An important consideration is to carry out these cleavages so that the original or even new S—S links do not form.

A. Performic Acid Oxidation

Oxidation of a disulfide by performic acid results in the formation of two equivalents of cysteic acid. Because these cysteic acid side chains are ionized SO_3^- groups, electrostatic repulsion (as well as altered chemistry) prevents S—S recombination.

B. Reduction and Alkylation of the Disulfide

Sulfhydryl compounds such as 2-mercaptoethanol readily reduce S—S bridges to regenerate two cysteine—SH side chains. However, these SH groups will recombine to re-form either the original disulfide link or, if other free Cys—SHs are available, new disulfide links. To prevent this, S—S reduction must be followed by treatment with alkylating agents such as iodoacetate or acrylonitrile, which modify the SH groups and block disulfide bridge formation (Figure 4.18).

Step 3. Analysis of Amino Acid Composition

The standard protocol for analysis of the amino acid composition of proteins was discussed in Section 4.1. Results of such analyses allow the researcher to anticipate which methods of polypeptide fragmentation might be useful for the protein.

Step 4. Identification of the N- and C-Terminal Residues

End-group analysis reveals several things. First, it identifies the N- and C-terminal residues in the polypeptide chain. Second, it can be a clue to the number of ends in the protein. That is, if the protein consists of two or more different polypeptide chains, then more than one end group may be discovered, alerting the investigator to the presence of multiple polypeptides.

A. N-Terminal Analysis

The amino acid residing at the N-terminal end of a protein can be identified in a number of ways, some of which are destructive to the protein sample. One method, Edman degradation, has become the procedure of choice because it allows the sequential identification of a series of residues beginning at the N-terminus.

I. Sanger's Reagent. Sanger's reagent is **dinitrofluorobenzene (DNFB).** Figure 4.19 shows the reaction of DNFB with the free amino terminus of a polypeptide. HF is liberated as the amino group is tagged with the **dinitrophenyl (DNP)** function. Subsequent acid hydrolysis of the polypeptide releases its component residues as free α-amino acids, with the exception of the amino terminal residue, which is a **DNP-amino acid.** The nonpolar DNP group ren-

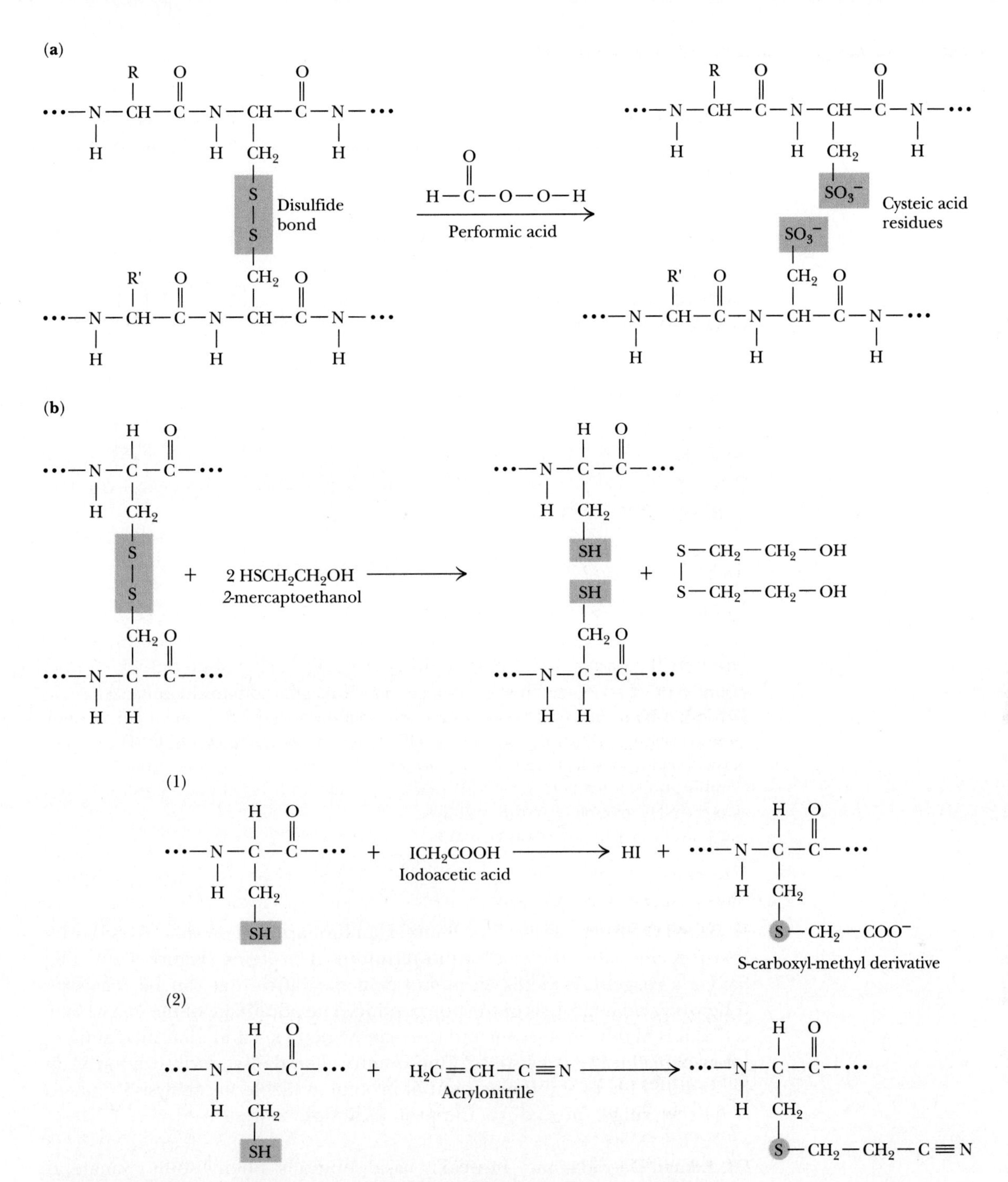

***Figure* 4.18** Methods for cleavage of disulfide bonds in proteins. (a) Oxidative cleavage by reaction with performic acid. (b) Reductive cleavage with sulfhydryl compounds. Disulfide bridges can be broken by reduction of the S—S link with sulfhydryl agents such as 2-mercaptoethanol or dithiothreitol. Because reaction between the newly reduced —SH groups to reestablish disulfide bonds is a likelihood, S—S reduction must be followed by —SH modification: (1) alkylation with iodoacetate (ICH_2COOH), or (2) modification with acrylonitrile ($H_2C{=}CHCN$).

ders this DNP-amino acid ether-extractable from the acid hydrolysate. Comparison of the chromatographic behavior of the yellow DNP-amino acid isolated from the ether extract with known DNP-amino acid derivatives allows identification of the N-terminal residue in the protein. Unfortunately, the protein sample is destroyed during the acid hydrolysis.

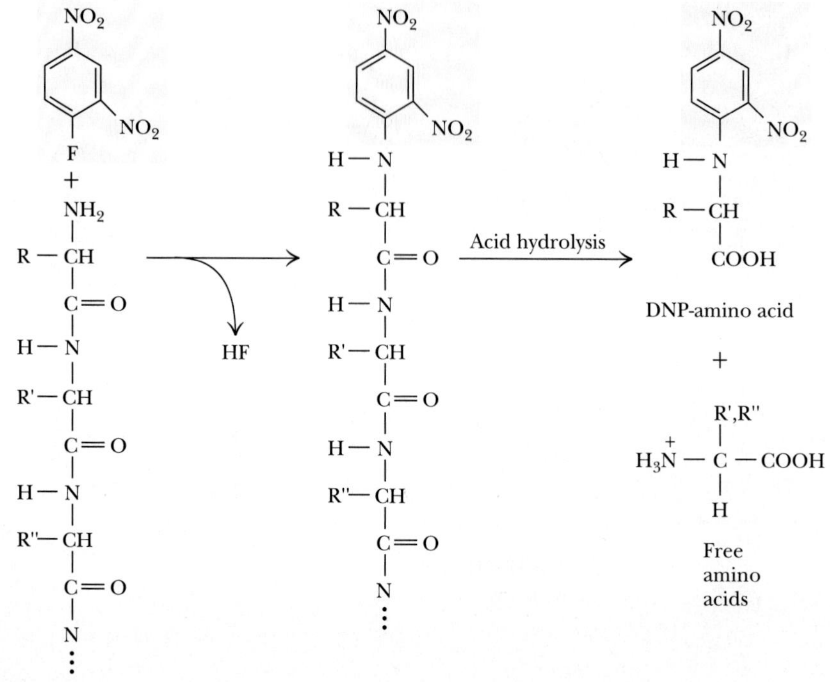

***Figure* 4.19** N-Terminal analysis using Sanger's reagent, dinitrofluorobenzene (DNFB). DNFB will react with the free amino-terminus of a peptide chain to yield a dinitrophenyl (DNP) derivative and HF as a by-product. Since the DNP derivatives are intensely yellow, they are easily observed on chromatograms. Complications arise because DNFB reacts generally with free amino groups, such as the ϵ-NH_2 groups of lysine residues.

II. Dansyl Chloride. Dansyl (or dimethylaminonaphthalenesulfonyl) chloride also reacts to label the free amino terminus of proteins (Figure 4.20). Like Sanger's reagent, it yields an amino acid derivative that can be recovered following acid hydrolysis of the polypeptide. The advantage of the dansyl label is that it is highly fluorescent and thus can be detected and identified at much lower concentrations by spectrofluorometry. The dansyl derivatization is the procedure of choice if the amount of protein available for analysis is limited. As in the Sanger procedure, the protein sample is destroyed.

III. Edman Degradation. In weakly basic solutions, phenylisothiocyanate, or **Edman's reagent** (phenyl—N=C=S), will combine with the free amino terminus of a protein (Figure 4.21), which can be excised from the end of the polypeptide chain and recovered as a phenylthiohydantoin (PTH) derivative. This PTH derivative can be identified by chromatographic methods. Importantly, in this procedure, the rest of the polypeptide chain remains intact and can be subjected to further rounds of Edman degradation to identify successive amino acid residues in the chain. Often, the carboxyl terminus of the polypeptide under analysis is coupled to an insoluble matrix, allowing the polypeptide to be easily recovered by filtration following each round of Edman reaction. Thus, Edman reaction not only identifies the N-terminus of proteins but can also reveal additional information regarding sequence. Instruments automated to carry out the reaction cycles of the procedure, so-called Edman sequenators, can yield the sequence of the first 30 to 60 amino acid residues in a polypeptide under appropriate conditions.

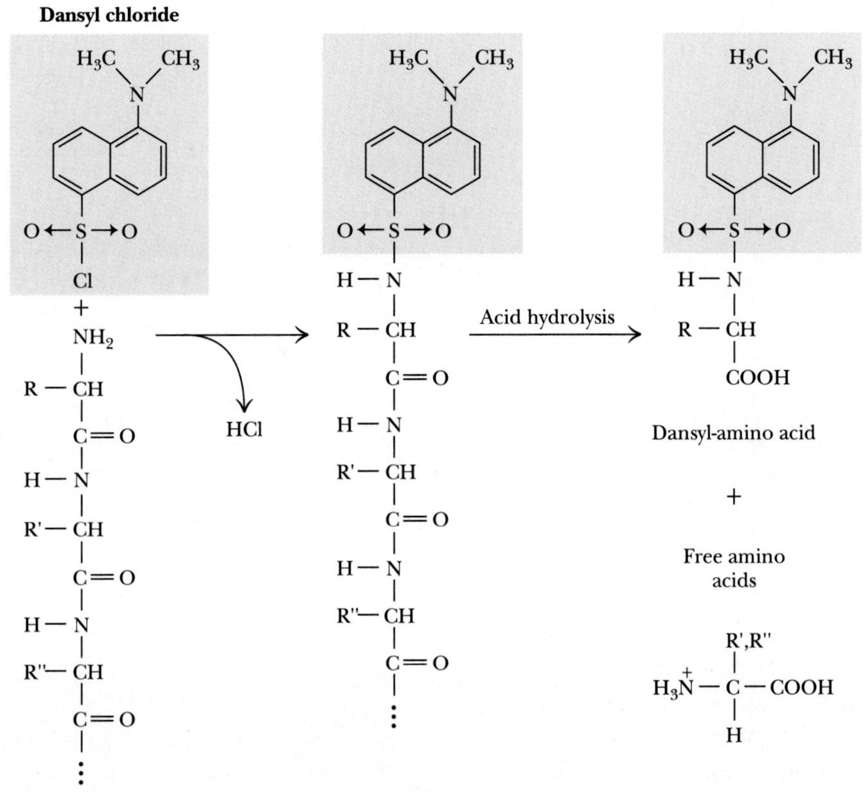

Dansyl chloride

HCl

Acid hydrolysis

Dansyl-amino acid

+

Free amino acids

Figure **4.20** N-Terminal analysis using dimethylaminonaphthalenesulfonyl (dansyl) chloride. Dansyl chloride will react with free amino groups, such as the N-termini of peptide chains, to form dansylated derivatives. Because of the intense fluorescence of the dansyl moiety under ultraviolet illumination, the product is easily visualized and therefore dansyl–amino acid derivatives can be recognized on chromatograms of peptide hydrolysates and identified by comparison to a chromatographic separation of standard dansyl derivatives of known amino acids.

IV. Leucine Aminopeptidase. An enzymatic method of N-terminal analysis uses leucine aminopeptidase, an enzyme from hog kidney, to hydrolyze the peptide bond linking the N-terminal amino acid to the polypeptide chain. It works best when nonpolar residues such as Leu are N-terminal. Leucine aminopeptidase will not attack peptide bonds in which the N is contributed by proline.

B. C-Terminal Analysis

Fewer methods are available for the identification of the C-terminal residue of polypeptides. Two chemical procedures and an enzymatic approach are in common use.

I. Hydrazinolysis. In what is called the *Akabori reaction,* polypeptides are treated with **hydrazine**[2] ($H_2N—NH_2$) at 100°C. Reaction with hydrazine lyses all the peptide linkages, resulting in the liberation of every amino acid as a hydrazide derivative, with the exception of the C-terminal residue, which ap-

[2]Hydrazine is also used as rocket fuel.

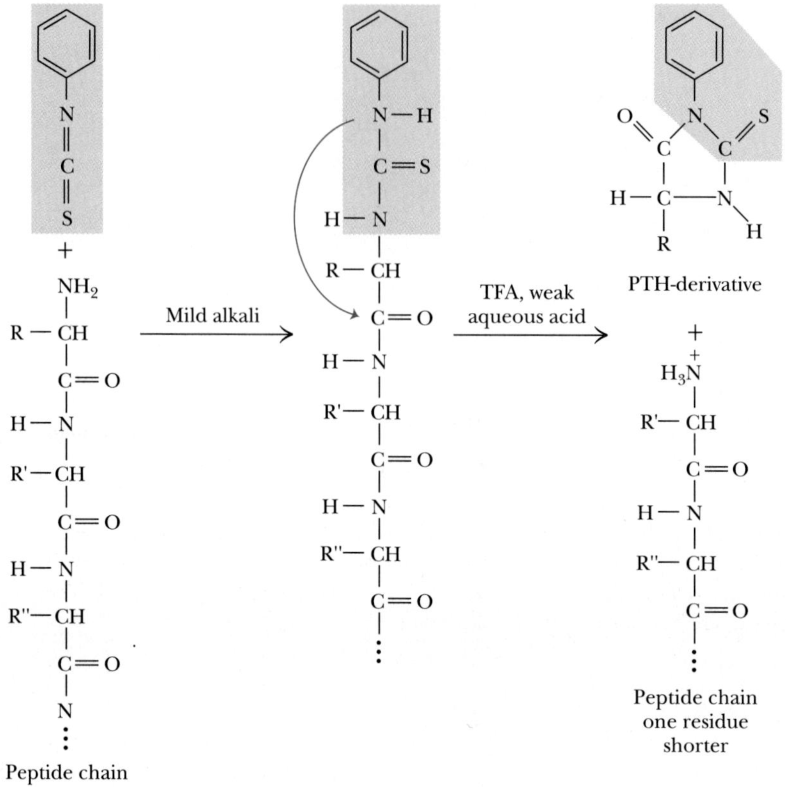

Peptide chain

Mild alkali

TFA, weak aqueous acid

PTH-derivative

Peptide chain one residue shorter

Figure **4.21** N-Terminal analysis using Edman's reagent, phenylisothiocyanate. Phenylisothiocyanate will combine with the N-terminus of a peptide under mildly alkaline conditions to form a phenylthiocarbamoyl substitution. Upon treatment with TFA (trifluoroacetic acid), this cyclizes to release the N-terminal amino acid residue, but the other peptide bonds are not hydrolyzed. Organic extraction and treatment with aqueous acid yields the amino acid as a phenyl thiohydantoin (PTH)–derivative.

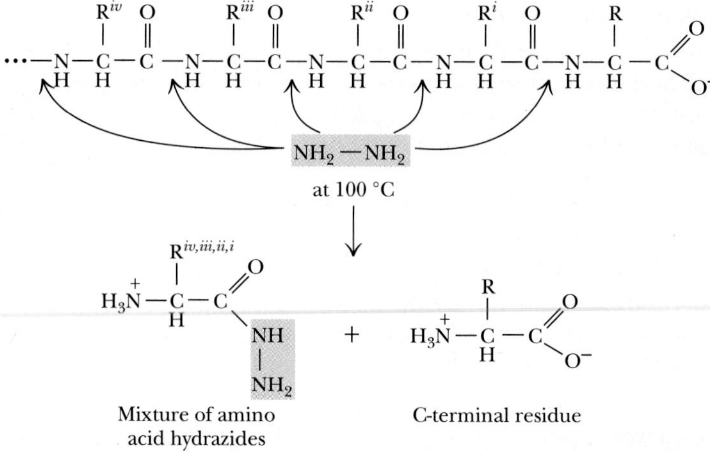

NH_2-NH_2

at 100 °C

Mixture of amino acid hydrazides

C-terminal residue

Figure **4.22** C-Terminal analysis by hydrazinolysis. All peptide linkages in a polypeptide can be cleaved by hydrazine (NH_2-NH_2) treatment at 100°C, with the component amino acids appearing as amino acyl hydrazide derivatives. The sole exception is the C-terminal amino acid, whose carboxyl group does not participate in a peptide bond. Identification of this free amino acid reveals the C-terminus. Hydrazinolysis is known as the *Akabori reaction* in honor of its discoverer, S. Akabori.

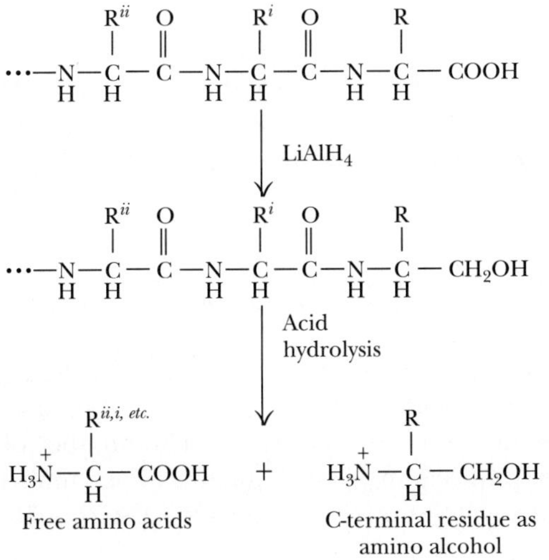

$$\cdots\text{NH}-\text{CH}(\text{R}^{ii})-\text{CO}-\text{NH}-\text{CH}(\text{R}^{i})-\text{CO}-\text{NH}-\text{CH}(\text{R})-\text{COOH}$$

$$\downarrow \text{LiAlH}_4$$

$$\cdots\text{NH}-\text{CH}(\text{R}^{ii})-\text{CO}-\text{NH}-\text{CH}(\text{R}^{i})-\text{CO}-\text{NH}-\text{CH}(\text{R})-\text{CH}_2\text{OH}$$

$$\downarrow \text{Acid hydrolysis}$$

$$\underset{\text{Free amino acids}}{\overset{+}{\text{H}_3\text{N}}-\text{CH}(\text{R}^{ii,i,\,etc.})-\text{COOH}} \;+\; \underset{\text{C-terminal residue as amino alcohol}}{\overset{+}{\text{H}_3\text{N}}-\text{CH}(\text{R})-\text{CH}_2\text{OH}}$$

***Figure* 4.23** C-Terminal analysis with lithium aluminum hydride ($LiAlH_4$), a powerful reducing agent capable of converting a carboxyl group to an alcohol. Reduction of peptides yields a C-terminal alcohol, which, upon acid hydrolysis of the peptide, is released as an amino alcohol.

pears as a free amino acid (Figure 4.22). This free amino acid is the sole residue whose carboxyl group was not joined in an amide bond. Because the free amino acid has different properties from the amino acid hydrazides, it can be isolated and identified. Despite complicating side reactions, this is the most satisfactory chemical method presently available.

II. Reduction of C-Terminus by $LiAlH_4$. The free α-COOH group at the end of a peptide chain can be reduced to an alcohol by the strong reducing agent **lithium aluminum hydride** ($LiAlH_4$) (Figure 4.23). Subsequent acid hydrolysis of the protein yields a mixture of free amino acids and a single free amino alcohol representing the C-terminal residue.

III. Enzymatic Analysis with Carboxypeptidases. Carboxypeptidases are enzymes that cleave amino acid residues from the C-termini of polypeptides in a successive fashion. Four carboxypeptidases are in general use: A, B, C, and Y. *Carboxypeptidase A* (from bovine pancreas) works well in hydrolyzing the C-terminal peptide bond of all residues except proline, arginine, and lysine. The analogous enzyme from hog pancreas, *carboxypeptidase B,* is effective only when Arg or Lys are the C-terminal residues. Thus, a mixture of carboxypeptidases A and B will liberate any C-terminal amino acid except proline. *Carboxypeptidase C* from citrus leaves and *carboxypeptidase Y* from yeast act on any C-terminal residue. Since the nature of the amino acid residue at the end often determines the rate at which it is cleaved and since these enzymes remove residues successively, care must be taken in interpreting results.

Steps 5 and 6. Fragmentation of the Polypeptide Chain

The aim at this step is to produce fragments amenable to sequence analysis. The cleavage methods employed are usually enzymatic, but proteins can also be fragmented by specific or nonspecific chemical means (such as partial acid hydrolysis). Proteolytic enzymes offer an advantage in that they may hydrolyze only specific peptide bonds, and this specificity immediately gives information about the peptide products. As a first approximation, fragments produced upon cleavage should be small enough to yield their sequences through end-group analysis and Edman degradation, yet not so small that an overabundance of products must be resolved before analysis. However, the determination of total sequences for proteins predates the Edman procedure, and alternative approaches obviously exist.

A. Trypsin

The digestive enzyme *trypsin* is the most commonly used reagent for specific proteolysis. Trypsin is specific in hydrolyzing only peptide bonds in which the carbonyl function is contributed by an arginine or a lysine residue. That is, trypsin cleaves on the C-side of Arg or Lys, generating a set of peptide fragments having Arg or Lys at their C-termini. The number of smaller peptides resulting from trypsin action is equal to the total number of Arg and Lys residues in the protein *plus* one—the protein's C-terminal peptide fragment (Figure 4.24).

B. Chymotrypsin

Chymotrypsin shows a strong preference for hydrolyzing peptide bonds formed by the carboxyl groups of the aromatic amino acids, phenylalanine, tyrosine, and tryptophan. However, over time chymotrypsin will also hydrolyze amide

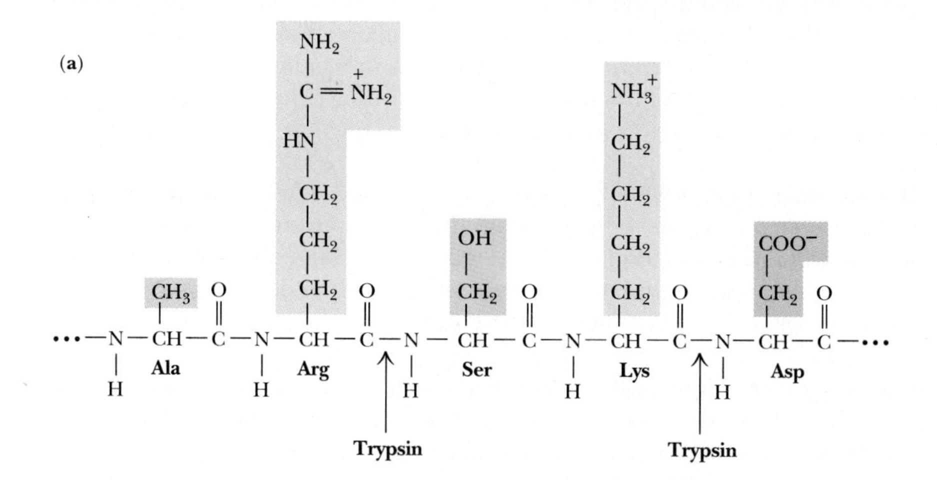

(b)

N—Asp—Ala—Gly—Arg—His—Cys—Lys—Trp—Lys—Ser—Glu—Asn—Leu—Ile—Arg—Thr—Tyr—C

Trypsin

Asp—Ala—Gly—Arg

His—Cys—Lys

Trp—Lys

Ser—Glu—Asn—Leu—Ile—Arg

Thr—Tyr

***Figure* 4.24** Trypsin is a proteolytic enzyme, or *protease,* that specifically cleaves only those peptide bonds in which arginine or lysine contribute the carbonyl function. The products of the reaction are a mixture of peptide fragments with C-terminal Arg or Lys residues *and* a single peptide derived from the polypeptide's C-terminal end.

bonds involving amino acids other than Phe, Tyr, or Trp. Peptide bonds having leucine-donated carboxyls become particularly susceptible. Thus, the specificity of chymotrypsin is only relative. Since chymotrypsin produces a very different set of products than trypsin, treatment of samples of a protein with these two enzymes will generate fragments whose sequences overlap. Resolution of the order of amino acid residues in the fragments yields the amino acid sequence in the original protein.

C. Clostripain

Clostripain is a protease with a specificity range even narrower than trypsin's. Like trypsin, it prefers positively charged amino acids, but it attacks the carboxyl peptide linkage of arginine residues much more avidly than lysine residues. This greater specificity also means that the products of clostripain action are larger.

D. Staphylococcal Protease

Specific enzymatic cleavage of peptides at the acidic residues, aspartic and glutamic acid, can be achieved using *staphylococcal protease* to generate a set of fragments with C-terminal Asp or Glu, if the reaction is run in phosphate buffer. Substitution of acetate or bicarbonate buffer for phosphate renders the enzyme even more specific, causing it to act only upon peptide linkages involving the carboxyl group of glutamic acid.

E. Relatively Nonspecific Endopeptidases

In addition to the enzymes listed previously, a number of other *endopeptidases* (proteases that cleave peptide bonds within the interior of a polypeptide chain) are occasionally used in sequence investigations, even though they do not display a high degree of specificity. These enzymes are handy for digesting large tryptic or chymotryptic fragments. *Pepsin, papain, subtilisin, thermolysin,* and *elastase* are some examples. Papain is the active ingredient in meat tenderizer and in soft contact lens cleaner as well as in some laundry detergents. The abundance of papain in papaya, and a similar protease (bromelain) in pineapple, causes the hydrolysis of gelatin and prevents the preparation of Jell-O® containing either of these fresh fruits. Cooking these fruits thermally denatures their proteolytic enzymes so that they can be used in gelatin desserts.

F. Cyanogen Bromide

Several highly specific chemical methods of proteolysis are available, the most widely used being *cyanogen bromide (CNBr)* cleavage. CNBr acts upon methionine residues (Figure 4.25). The nucleophilic sulfur atom of Met reacts with CNBr, yielding a sulfonium ion that undergoes a rapid intramolecular rearrangement to form a cyclic iminolactone. Water readily hydrolyzes this iminolactone, cleaving the polypeptide and generating peptide fragments having C-terminal homoserine lactone residues at the former Met positions.

Other Methods of Fragmentation

A number of other chemical methods give specific fragmentation of polypeptides, including cleavage at asparagine–glycine bonds by hydroxylamine (NH_2OH) at pH 9 and selective hydrolysis at aspartyl–prolyl bonds under mildly acidic conditions. Table 4.6 summarizes the various procedures described here for polypeptide cleavage. These methods are only a partial list of the arsenal of reactions available to protein chemists. Cleavage products gen-

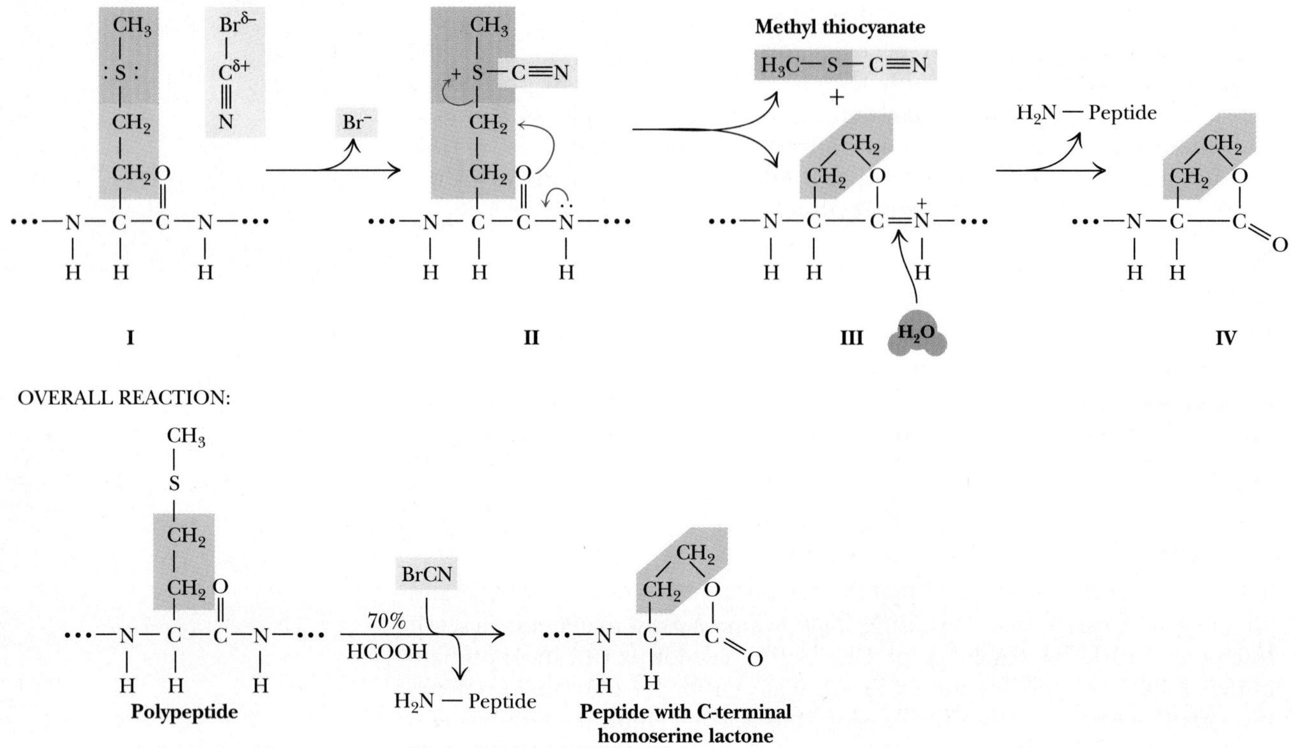

Methyl thiocyanate

$H_3C—S—C≡N$

H_2N — Peptide

I II III IV

OVERALL REACTION:

BrCN

70% HCOOH

H_2N — Peptide

Polypeptide

Peptide with C-terminal homoserine lactone

Figure **4.25** Cyanogen bromide (CNBr) is a highly selective reagent for cleavage of peptides only at methionine residues. (I) The reaction occurs in 70% formic acid via nucleophilic attack of the Met S atom on the —C≡N carbon atom, with displacement of Br^-. (II) The cyano intermediate undergoes nucleophilic attack by the Met carbonyl oxygen atom on the R group, (III) resulting in formation of the cyclic derivative, which is unstable in aqueous solution. (IV) Hydrolysis ensues, producing cleavage of the Met peptide bond and release of peptide fragments with C-terminal homoserine lactone residues where Met residues once were. One peptide will not have a C-terminal homoserine lactone: the original C-terminal end of the polypeptide.

erated by these procedures must be isolated and individually analyzed with respect to amino acid composition, end-group identity, and amino acid sequence to accumulate the information necessary to reconstruct the protein's total amino acid sequence. In the past, sequence was often deduced from exhaustive study of the amino acid composition and end-group analysis of small, overlapping peptides. Peptide sequencing today is most commonly done by Edman degradation of relatively large peptides.

Sequence Determination by Mass Spectrometric Methods

Mass spectrometry is rapidly coming into vogue for the sequence determination of peptides of 15 or fewer residues. This technique has great advantages of sensitivity and accuracy, but conventional instruments require that the polar functions of peptide backbones and side chains be modified to render the peptide sufficiently volatile for evaporation in the vacuum of the mass spectrometer. Amino groups can be acetylated with acetic anhydride, and then carboxyl, hydroxyl, and —NH groups are permethylated with methyl iodide (Figure 4.26) to accomplish this purpose. The modified peptide is introduced into the spectrometer and fragmented in an electron beam to produce a complex set of cationic species, which are then identified solely by

***Table* 4.6**
Specificity of Representative Polypeptide Cleavage Procedures Used in Sequence Analysis

Method	Peptide Bond on Carboxyl (C) or Amino (N) Side of Susceptible Residue	Susceptible Residue(s)
Proteolytic enzymes		
Trypsin	C	Arg or Lys
Chymotrypsin	C	Phe, Trp, or Tyr; Leu
Clostripain	C	Arg
Staphylococcal protease	C	Asp or Glu
Chemical methods		
Cyanogen bromide	C	Met
NH_2OH	Asn-Gly bonds	
pH 2.5, 40°C	Asp-Pro bonds	

their mass to charge ratio. An enormous advantage of mass spectrometry is that the peptides need not be pure. Peptide mixtures can be fractionally distilled within the spectrometer by gradually increasing the temperature. The signals in the mass spectrum attributable to a given peptide rise and fall together, so that complex spectral patterns can be resolved according to the volatility of each component.

Recent technical advances have even obviated chemical modification of the peptide. In *fast atom bombardment mass spectrometry (FAB-MS)* and in *liquid scattered ion mass spectrometry (Liquid SIMS),* the sample need not be volatile. Furthermore, tandem arrangement of two or more mass spectrometers *(tandem mass spectrometry)* allows one instrument to be used to separate peptides, a second to fragment them by ion bombardment, and a third to analyze the masses of the fragments. In such devices, oligopeptides and proteins up to 13 kD in molecular weight can be analyzed directly, and only 1 μg of the protein is consumed in the analysis.

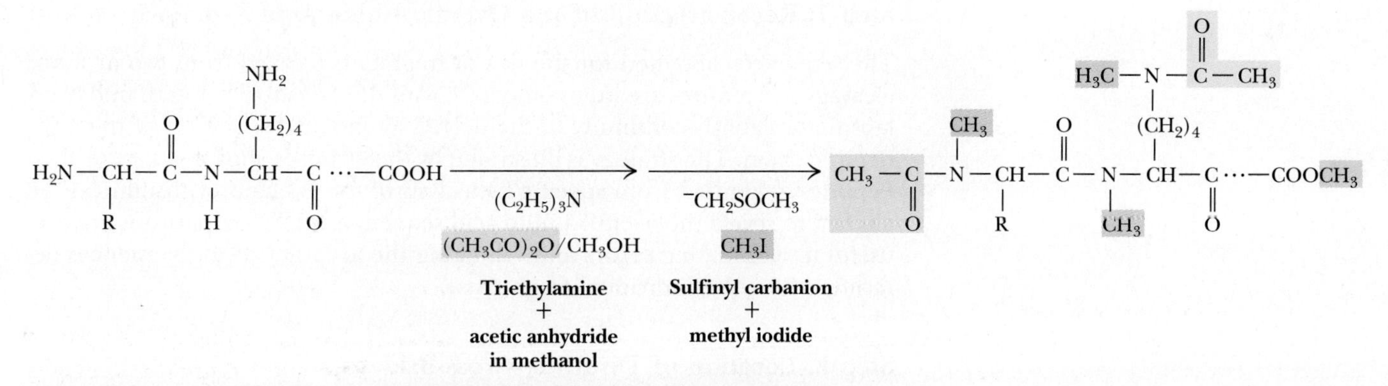

***Figure* 4.26** Certain modification reactions are used to render peptides volatile for sequencing by mass spectrometry. First, the peptide is treated with triethylamine, $(C_2H_5)_3N$:, to give alkaline conditions. Acetic anhydride in methanol, $(CH_3CO)_2O/CH_3OH$, is then added to acetylate all amino groups. Next, the peptide is permethylated with the strong base methyl sulfinyl carbanion ($^-CH_2SOCH_3$) (which abstracts H atoms) and methyl iodide (CH_3I) (to replace the H atoms with methyl groups). If the peptide contains Arg residues, additional modifications are necessary.

(Adapted from Morris, H. R., 1980. Nature 286:447–452.)

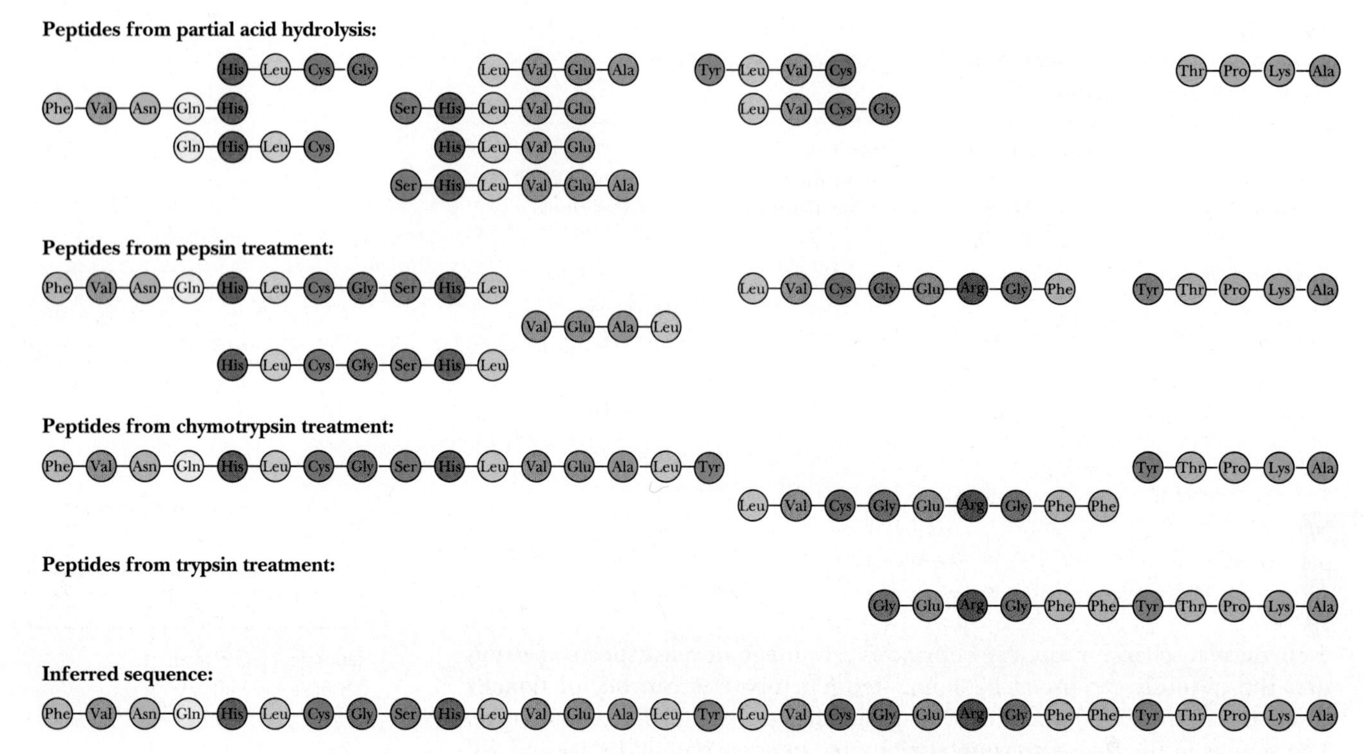

***Figure* 4.27** The amino acid sequence of a polypeptide is elucidated by establishing overlaps within fragments produced by different cleavage procedures. The results are taken from sequence studies on the B chain of insulin. The inferred sequence of the B chain is shown at the bottom. These original results contain exceptions to some of the specificities taken as fact in sequence determinations. For example, the tryptic peptide has an internal Arg residue, which should have been cleaved by trypsin. Similarly, aromatic residues are found within the chymotryptic peptides. A possible reason for these exceptions may be incomplete hydrolysis. Nevertheless, these exceptions should alert the reader to be more critical of the certainty of experimental expectations.
(Adapted from Thompson, E.O.P., 1955. Scientific American 192:36-41.)

Step 7. Reconstruction of the Overall Amino Acid Sequence

The sequences obtained for the sets of fragments derived from two or more cleavage procedures are now compared, with the objective being to find overlaps that establish continuity of the overall amino acid sequence of the polypeptide chain. The strategy is illustrated by the example shown in Figure 4.27. Peptides generated from specific hydrolysis of the B chain of insulin can be aligned to reveal the overall amino acid sequence. Such comparisons are also useful in eliminating errors and validating the accuracy of the sequences determined for the individual fragments.

Step 8. Location of Disulfide Cross-Bridges

Strictly speaking, the disulfide bonds formed between cysteine residues in a protein are not a part of its primary structure. Nevertheless, information about their location can be obtained by procedures used in sequencing, provided the disulfides are not broken prior to cleaving the polypeptide chain. Since these covalent bonds are stable under most conditions used in the cleavage of polypeptides, intact disulfides will link the peptide fragments containing their specific cysteinyl residues and thus these linked fragments can be isolated and identified within the protein digest.

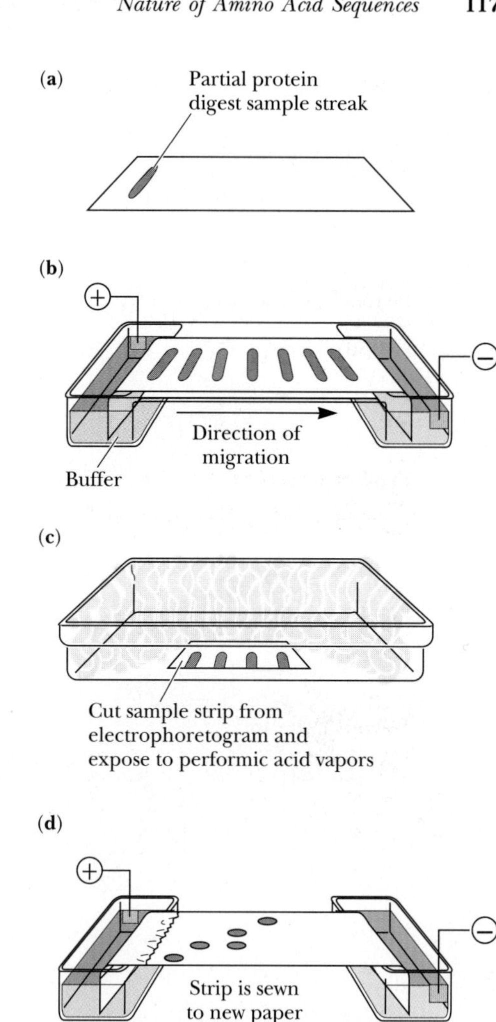

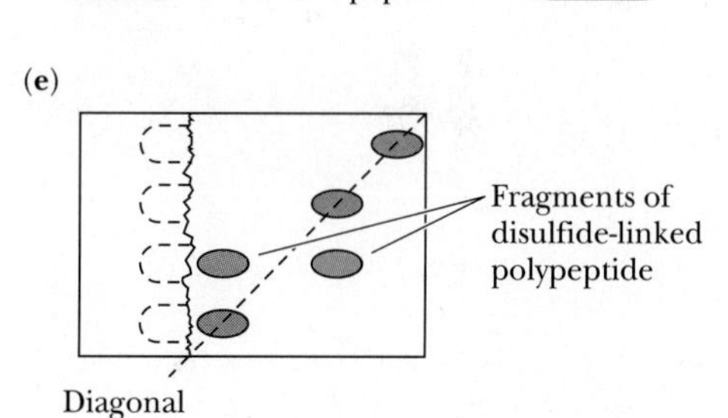

***Figure* 4.28** Disulfide bridges typically are cleaved prior to determining the primary structure of a polypeptide. Consequently, the positions of disulfide links are not obvious from the primary sequence data. To determine their location, a sample of the polypeptide with intact S—S bonds can be fragmented and the sites of any disulfides can be elucidated from fragments that remain linked. *Diagonal electrophoresis* is a technique for identifying such fragments. (a) A protein digest in which any disulfide bonds remain intact and link their respective Cys-containing peptides is streaked along the edge of a filter paper and (b) subjected to electrophoresis. (c) A strip cut from the edge of the paper is then exposed to performic acid fumes to oxidize any disulfide bridges. (d) Then the paper strip is attached to a new filter paper so that a second electrophoresis can be run in a direction perpendicular to the first. (e) Peptides devoid of disulfides experience no mobility change, and thus their pattern of migration defines a diagonal. Peptides that had disulfides migrate off this diagonal and can be easily identified, isolated, and sequenced to reveal the location of cysteic acid residues formerly involved in disulfide bridges.

An effective way to isolate these fragments is through **diagonal electrophoresis** (the basic technique of *electrophoresis* is described in the Chapter Appendix). In diagonal electrophoresis, a protein is cleaved by some method (trypsin digestion, for example) that yields a mixture of peptide fragments with any disulfides still intact. The mixture is then streaked along the edge of a piece of filter paper (Figure 4.28a) and subjected to electrophoresis (Figure 4.28b). Next, a sample strip containing the pattern of separated peptides is cut from one end of the paper and exposed to performic acid fumes to cleave any disulfide bonds (Figure 4.28c). Then the strip is joined to the edge of a second filter paper, and electrophoresis is conducted in a direction perpendicular to the first run (Figure 4.28d). Peptides lacking disulfides will have the same electrophoretic mobility they had before and thus will align along the diagonal of the paper. Peptides that were originally linked by disulfides now migrate as distinct species and are obvious by their location off the diagonal (Figure 4.28e). These cysteic acid–containing peptides are then isolated from the paper and sequenced. From this information, the positions of the disulfides in the protein can be stipulated.

Sequence Databases

A database of protein sequences collected by protein chemists can be found in the *Atlas of Protein Sequence and Structure.* As of 1979, over 1100 completed protein sequences were compiled in this atlas. Today, however, most protein sequence information is derived from translating the nucleotide sequences of genes into codons and, thus, amino acid sequences (see Chapter 7). Sequencing the order of nucleotides in cloned genes is a more rapid, efficient, and informative process than determining the amino acid sequences of proteins. A number of electronic databases containing continuously updated sequence information are readily accessible by personal computer. Prominent among these are PIR (Protein Identification Resource Protein Sequence Database), GenBank (Genetic Sequence Data Bank), and EMBL (European Molecular Biology Laboratory Data Library).

4.8 Nature of Amino Acid Sequences

With a knowledge of the methodology in hand, let's review the results of amino acid composition and sequence studies on proteins. Table 4.7 lists the relative frequencies of the amino acids in various proteins. It is very unusual

Table **4.7**

Frequency of Occurrence of Amino Acid Residues in Proteins

Amino Acid		M_r*	Occurrence in Proteins (%)†
Alanine	Ala A	71.1	9.0
Arginine	Arg R	156.2	4.7
Asparagine	Asn N	114.1	4.4
Aspartic acid	Asp D	115.1	5.5
Cysteine	Cys C	103.1	2.8
Glutamine	Gln Q	128.1	3.9
Glutamic acid	Glu E	129.1	6.2
Glycine	Gly G	57.1	7.5
Histidine	His H	137.2	2.1
Isoleucine	Ile I	113.2	4.6
Leucine	Leu L	113.2	7.5
Lysine	Lys K	128.2	7.0
Methionine	Met M	131.2	1.7
Phenylalanine	Phe F	147.2	3.5
Proline	Pro P	97.1	4.6
Serine	Ser S	87.1	7.1
Threonine	Thr T	101.1	6.0
Tryptophan	Trp W	186.2	1.1
Tyrosine	Tyr Y	163.2	3.5
Valine	Val V	99.1	6.9

*Molecular weight of amino acid *minus* that of water.

†Frequency of occurrence of each amino acid residue in the polypeptide chains of 207 unrelated proteins of known sequence. Values from Klapper, M. H., 1977. *Biochemical and Biophysical Research Communications* **78:**1018–1024.

for a globular protein to have an amino acid composition that deviates substantially from these values. Apparently, these abundances reflect a distribution of amino acid polarities that is optimal for protein stability in an aqueous milieu. Membrane proteins have relatively more hydrophobic and fewer ionic amino acids, a condition consistent with their location. Fibrous proteins may show compositions that are atypical with respect to these norms, indicating an underlying relationship between the composition and the structure of these proteins.

This similarity in composition is *not* due to a nonrandom distribution of possible amino acid neighbors within proteins. The likelihood of finding an amino acid next to any other amino acid is proportional to the relative frequencies of the two amino acids in proteins. Thus, dipeptide sequences within proteins are essentially random. However, the overall sequence of a protein is emphatically nonrandom. Proteins have unique amino acid sequences, and it is this uniqueness of sequence that ultimately gives each protein its own particular personality. Because the number of possible amino acid sequences in a protein is astronomically large, the probability that two proteins will, by chance, have similar amino acid sequences is negligible. Consequently, sequence similarities between proteins imply evolutionary relatedness.

Homologous Proteins from Different Organisms Have Homologous Amino Acid Sequences

Proteins sharing a significant degree of sequence similarity are said to be **homologous.** Proteins that perform the same function in different organisms are also referred to as homologous. For example, the oxygen transport protein, hemoglobin, serves a similar role and has a similar structure in all vertebrates. The study of the amino acid sequences of homologous proteins from different organisms provides very strong evidence for their evolutionary origin within a common ancestor. Homologous proteins characteristically have polypeptide chains that are nearly identical in length, and their sequences share identity in direct correlation to the relatedness of the species from which they are derived.

Cytochrome c

The electron transport protein, cytochrome *c,* found in the mitochondria of all eukaryotic organisms, provides the best-studied example of homology. The polypeptide chain of cytochrome *c* from most species contains slightly more than 100 amino acids and has a molecular weight of about 12.5 kD. Amino acid sequencing of cytochrome *c* from more than 40 different species has revealed that there are 28 positions in the polypeptide chain where the same amino acid residues are always found (Figure 4.29). These **invariant residues** apparently serve roles crucial to the biological function of this protein, and thus substitutions of other amino acids at these positions cannot be tolerated.

Furthermore, as shown in Figure 4.30, the number of amino acid differences between two cytochrome *c* sequences is proportional to the phylogenetic difference between the species from which they are derived. The cytochrome *c* in humans and in chimpanzees is identical; human and another mammalian (sheep) cytochrome *c* differ at 10 residues. The human cytochrome *c* sequence has 14 variant residues from a reptile sequence (rattlesnake), 18 from a fish (carp), 29 from a mollusc (snail), 31 from an insect (moth), and more than 40 from yeast or higher plants (cauliflower).

The Phylogenetic Tree for Cytochrome c

Figure 4.31 displays a **phylogenetic tree** (a diagram illustrating the evolutionary relationships among a group of organisms) constructed from the sequences of cytochrome *c.* The tips of the branches are occupied by contemporary species whose sequences have been determined. The tree has been deduced by computer analysis of these sequences to find the minimum number of mutational changes connecting the branches. Other computer methods can be used to infer potential ancestral sequences represented by *nodes,* or branch points, in the tree. Such analysis ultimately suggests a primordial cyto-

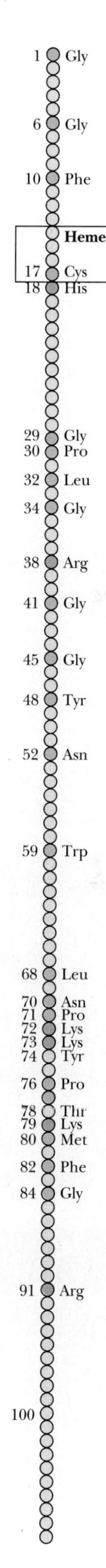

***Figure* 4.29** Cytochrome *c* is a small protein consisting of a single polypeptide chain of 104 residues in terrestrial vertebrates, 103 or 104 in fishes, 107 in insects, 107 to 109 in fungi and yeasts, and 111 or 112 in green plants. Analysis of the sequence of cytochrome *c* from more than 40 different species reveals that 28 residues are invariant. These invariant residues are scattered irregularly along the polypeptide chain, except for a cluster between residues 70 and 80. All cytochrome *c* polypeptide chains have a cysteine residue at position 17, and all but one have another Cys at position 14. These Cys residues serve to link the heme prosthetic group of cytochrome *c* to the protein, a role explaining their invariable presence.

	Chimpanzee	Sheep	Rattlesnake	Carp	Snail	Moth	Yeast	Cauliflower	Parsnip
Human	0	10	14	18	29	31	44	44	43
Chimpanzee		10	14	18	29	31	44	44	43
Sheep			20	11	24	27	44	46	46
Rattlesnake				26	28	33	47	45	43
Carp					26	26	44	47	46
Garden snail						28	48	51	50
Tobacco hornworm moth							44	44	41
Baker's yeast (iso-1)								47	47
Cauliflower									13

***Figure* 4.30** The number of amino acid differences among the cytochrome *c* sequences of various organisms can be compared. The numbers bear a direct relationship to the degree of relatedness between the organisms. Each of these species has a cytochrome *c* of at least 104 residues, so any given pair of species have more than half their residues in common.
(Adapted from Creighton, 1983. Proteins: Structure and Molecular Properties. W. H. Freeman and Co.)

chrome *c* sequence lying at the base of the tree. Evolutionary trees constructed in this manner, that is, solely on the basis of amino acid differences occurring in the primary sequence of one selected protein, show remarkable fidelity to phylogenetic relationships derived from more classic approaches and have given rise to the field of *molecular evolution* (see Chapter 33).

Related Proteins Share a Common Evolutionary Origin

Amino acid sequence analysis reveals that proteins with related functions often show a high degree of sequence similarity. Such findings suggest a common ancestry for these proteins.

Oxygen-Binding Heme Proteins

The oxygen-binding heme protein of muscle, **myoglobin,** consists of a single polypeptide chain of 153 residues. **Hemoglobin,** the oxygen transport protein of erythrocytes, is a tetramer composed of two **α-chains** (141 residues each) and two **β-chains** (146 residues each). These globin polypeptides—myoglobin, α-globin, and β-globin—share a strong degree of sequence homology (Figure 4.32). Myoglobin and the α-globin chain show 38 amino acid identities, whereas α-globin and β-globin have 64 residues in common. The relatedness suggests an evolutionary sequence of events in which chance mutations led to amino acid substitutions and divergence in primary structure. The ancestral myoglobin gene diverged first, after duplication of a primordial globin gene had given rise to its progenitor and an ancestral hemoglobin gene (Figure 4.33). Subsequently, the ancestral hemoglobin gene duplicated to generate the progenitors of the present-day α-globin and β-globin genes. The ability to bind O_2 via a heme prosthetic group is retained by all three of these polypeptides.

Serine Proteases

While the globins provide an example of gene duplication giving rise to a set of proteins in which the biological function has been highly conserved, other sets of proteins united by strong sequence homology show more divergent

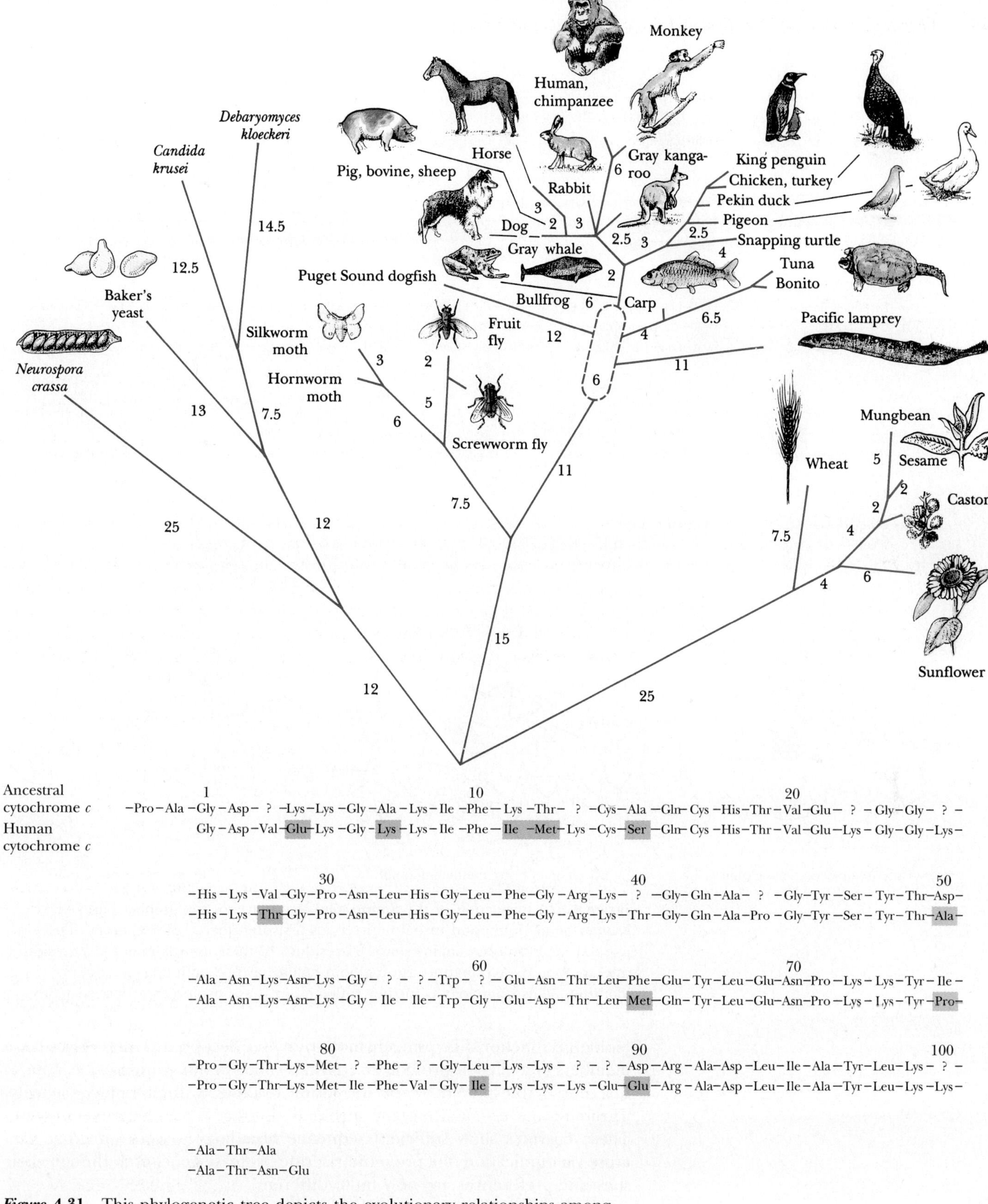

***Figure* 4.31** This phylogenetic tree depicts the evolutionary relationships among organisms as determined by the similarity of their cytochrome *c* amino acid sequences. The numbers along the branches give the amino acid changes between a species and a hypothetical progenitor. Note that extant species are located only at the tips of branches. Below, the sequence of human cytochrome *c* is compared with an inferred ancestral sequence represented by the base of the tree. Uncertainties are denoted by question marks.

(Adapted from Creighton, 1983. Proteins: Structure and Molecular Properties. W. H. Freeman and Co.)

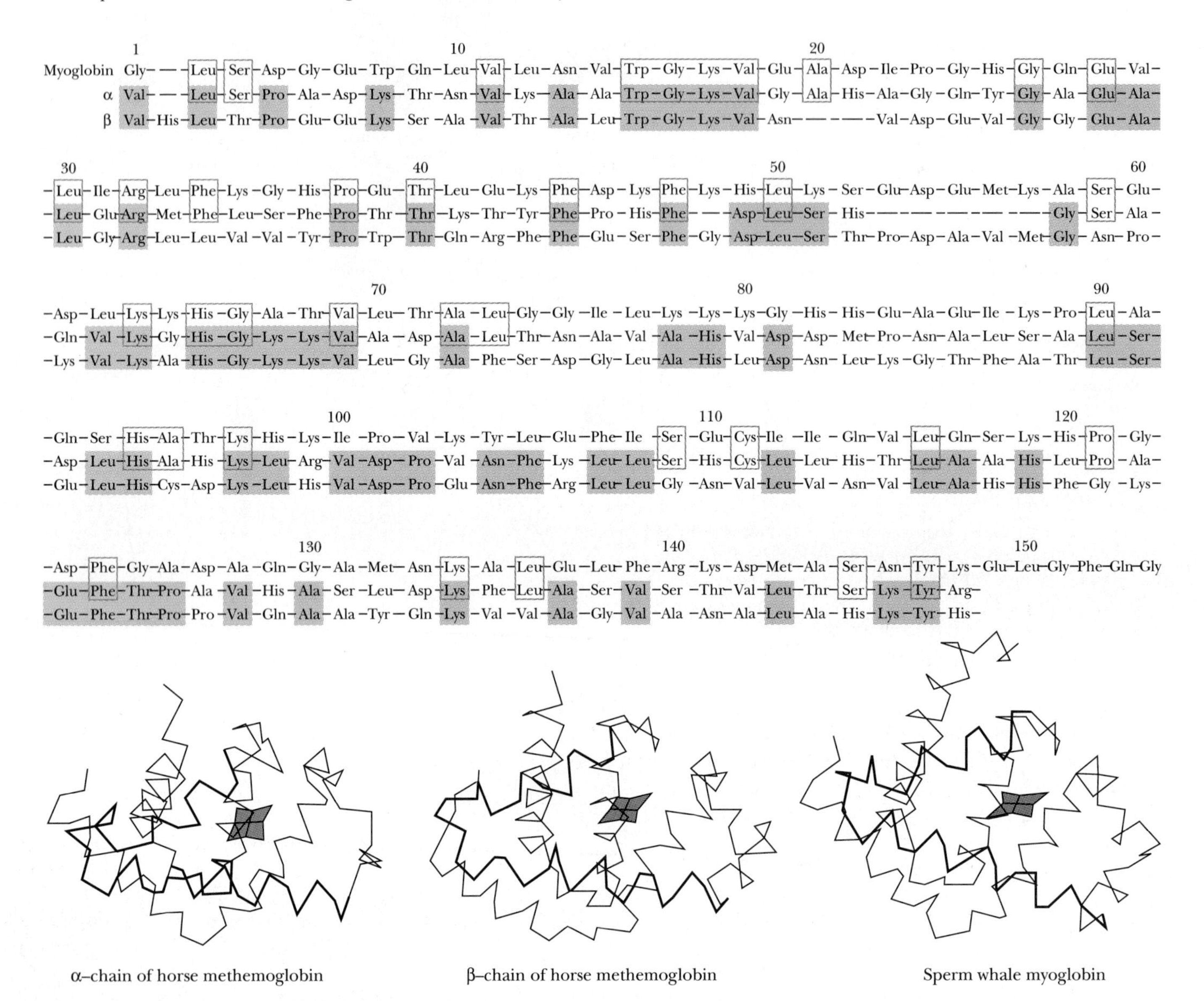

***Figure* 4.32** Inspection of the amino acid sequences of the globin chains of human hemoglobin and myoglobin reveals a strong degree of homology. The α-globin and β-globin chains share 64 residues of their approximately 140 residues in common. Myoglobin and the α-globin chain have 38 amino acid sequence identities. This homology is further reflected in these proteins' tertiary structure.

biological functions. Trypsin, chymotrypsin (see Section 4.7), and elastase are members of a class of proteolytic enzymes called **serine proteases** because of the central role played by specific serine residues in their catalytic activity. **Thrombin,** an essential enzyme in blood clotting, is also a serine protease. These enzymes show sufficient sequence homology to conclude that they arose via duplication of a progenitor serine protease gene, even though their substrate preferences are now quite different.

Disparate Proteins May Share a Common Ancestry

A more remarkable example of evolutionary relatedness is inferred from sequence homology between hen egg white **lysozyme** and human milk **α-lactalbumin,** proteins of rather disparate biological activity and origin. Lysozyme (129 residues) and α-lactalbumin (123 residues) are identical at 48 positions. Lysozyme hydrolyzes the polysaccharide wall of bacterial cells, whereas

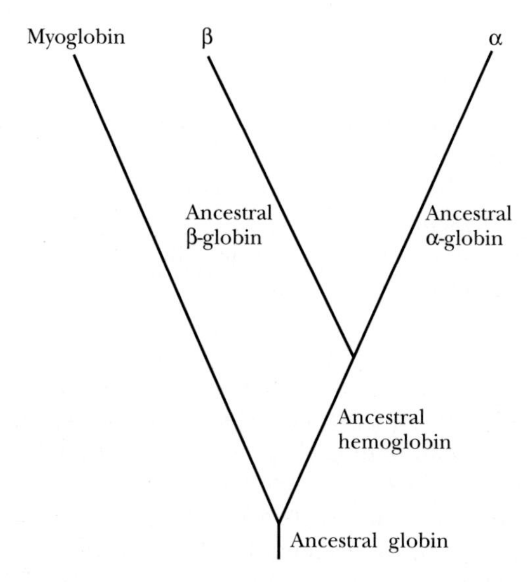

***Figure* 4.33** This evolutionary tree is inferred from the homology between the amino acid sequences of the α-globin, β-globin, and myoglobin chains. Duplication of an ancestral globin gene allowed the divergence of the myoglobin and ancestral hemoglobin genes. Another gene duplication event subsequently gave rise to ancestral α and β forms, as indicated. Gene duplication is an important evolutionary force in creating diversity.

α-lactalbumin regulates milk sugar (lactose) synthesis in the mammary gland. Although both proteins act in reactions involving carbohydrates, their functions show little similarity otherwise. Nevertheless, their tertiary structures are strikingly similar (Figure 4.34). It is conceivable that many proteins are related in this way, but time and the course of evolutionary change erased most evidence of their common ancestry.

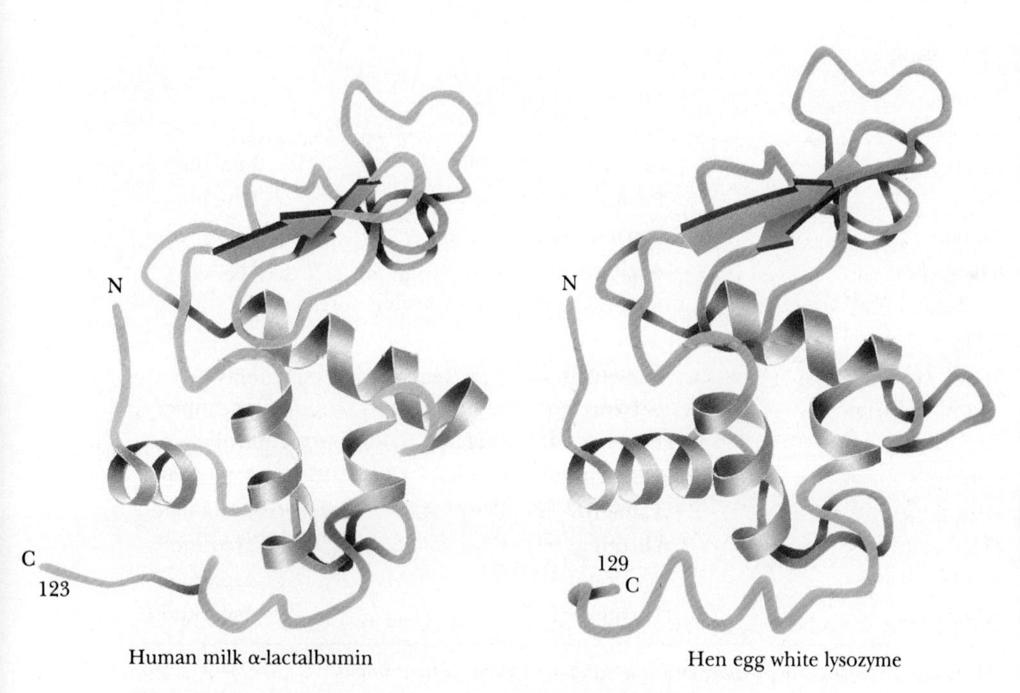

***Figure* 4.34** The tertiary structures of hen egg white lysozyme and human α-lactalbumin are very similar.

(Adapted from Acharya, et al., 1990. Journal of Protein Chemistry 9:549–563; and Acharya, et al., 1991. Journal of Molecular Biology 221:571–581.

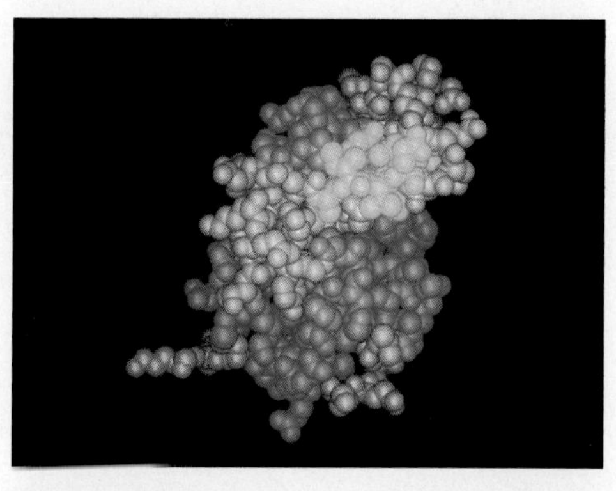

α-Lactalbumin

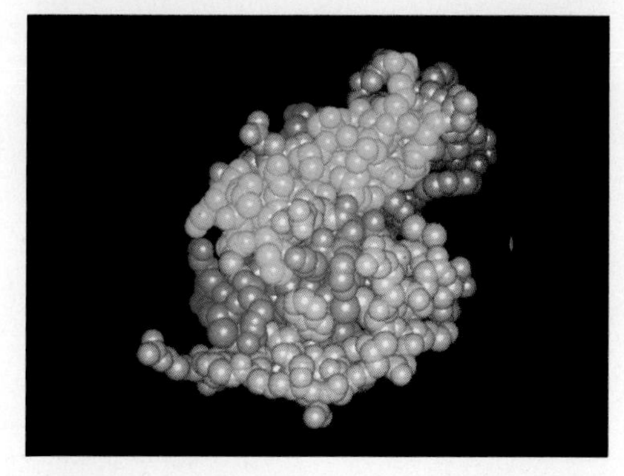

Lysozyme

Mutant Proteins

Given a large population of individuals, a considerable number of sequence variants can be found for a protein. These variants are a consequence of **mutations** in a gene (base substitutions in DNA) that have arisen naturally within the population. Gene mutations lead to mutant forms of the protein in which the amino acid sequence is altered at one or more positions. Many of these mutant forms will be "neutral" in that the functional properties of the protein will be unaffected by the amino acid substitution. Others may be nonfunctional (if loss of function is not lethal to the individual), and yet others may display a range of aberrations between these two extremes. The severity of the effects on function will depend on the nature of the amino acid substitution and its role in the protein. These conclusions are exemplified by the more than 300 human hemoglobin variants that have been discovered to date. Some of these are listed in Table 4.8.

A variety of effects on the hemoglobin molecule are seen in these mutants, including alterations in oxygen affinity, heme affinity, stability, solubility, and subunit interactions between the α-globin and β-globin polypeptide chains. Some variants show no apparent changes, while others, such as HbS, sickle-cell hemoglobin (see Chapter 12), result in serious illness. This diversity of response substantiates the conclusion that various amino acid residues play widely different roles in the function of a protein.

***Table* 4.8**
Some Pathological Sequence Variants of Human Hemoglobin

Abnormal Hemoglobin*	Normal Residue and Position	Substitution
Alpha chain		
Torino	Phenylalanine 43	Valine
M_{Boston}	Histidine 58	Tyrosine
Chesapeake	Arginine 92	Leucine
$G_{Georgia}$	Proline 95	Leucine
Tarrant	Aspartate 126	Asparagine
Suresnes	Arginine 141	Histidine
Beta chain		
S	Glutamate 6	Valine
Riverdale-Bronx	Glycine 24	Arginine
Genova	Leucine 28	Proline
Zurich	Histidine 63	Arginine
$M_{Milwaukee}$	Valine 67	Glutamate
$M_{Hyde\ Park}$	Histidine 92	Tyrosine
Yoshizuka	Asparagine 108	Aspartate
Hiroshima	Histidine 146	Aspartate

*Hemoglobin variants are often given the geographical name of their origin.

Adapted from Dickerson, R. E., and Geis, I. 1983. *Hemoglobin: Structure, Function, Evolution and Pathology.* Menlo Park, CA: Benjamin-Cummings Publishing Co.

4.9 Synthesis of Polypeptides in the Laboratory

The chemical synthesis of peptides and polypeptides of defined sequence has the potential of being a very useful tool for a variety of purposes: (1) verification of structures determined for naturally occurring peptides, (2) preparation of analogs of known polypeptides for studies on structure–function relationships, (3) synthesis of clinically important, pharmacologically active peptides, and (4) synthesis of antigenic peptides for antibody production. While formation of peptide bonds to link amino acids together is not a chemically complex process, making a specific peptide presents a challenging problem because the diverse functional groups found as side chains on amino acids are also reactive under the conditions used to form peptide bonds. Furthermore, if correct sequences are to be synthesized, the α-COOH group of residue x must be linked to the α-NH_2 group of neighboring residue y by a method that precludes reaction of the amino group of x with the carboxyl group of y.

Synthetic Strategies

Ingenious synthetic strategies are required to circumvent these formidable technical problems. In essence, the functional groups to be excluded from reaction must be blocked while the desired coupling reactions proceed. Also, the blocking groups must be removable later under conditions in which the peptide bonds formed are stable. Since these limitations mean that addition of each amino acid requires several steps, all reactions must proceed with high yield if peptide recoveries are to be acceptable. Peptide formation between amino and carboxyl groups is not spontaneous under normal conditions (see Chapter 3), and therefore one or the other of these groups must be activated to facilitate the reaction. Despite these difficulties, biologically active peptides and polypeptides have been synthesized in the laboratory. Milestones include the pioneering synthesis of the nonapeptide posterior pituitary hormones oxytocin and vasopressin by du Vigneaud in 1953, and in later years, the blood pressure–regulating hormone bradykinin (9 residues), melanocyte-stimulating hormone (24 residues), adrenocorticotropin (39 residues), insulin (21 A-chain and 30 B-chain residues), and ribonuclease *A* (124 residues).

Solid Phase Peptide Synthesis

Bruce Merrifield and his collaborators have found a clever solution to the problem of recovering intermediate products in the course of a synthesis. The carboxyl-terminal residues of synthesized peptide chains were covalently anchored to an insoluble resin particle large enough to be removed from reaction mixtures simply by filtration. After each new residue was added successively at the free amino-terminus, the elongated product was recovered by filtration and readied for the next synthetic step. Because the growing peptide chain was coupled to an insoluble resin bead, the method is called **solid phase synthesis.**

The procedure is detailed in Figure 4.35. The carboxyl-terminal amino acid residue is shown linked to the solid resin bead by its carboxyl group but with its —NH_2 free. The incoming amino acid has its amino group blocked by reaction with *tertiary*-butyloxycarbonyl chloride (*t*BocCl) or with *t*-butyloxycarbonyl azide ($t\text{BocN}_3$). In the presence of the carboxyl-activating agent, dicyclohexylcarbodiimide, the —COOH of the incoming residue is activated to react with the free amino function of the attached C-terminal amino acid to

(a)

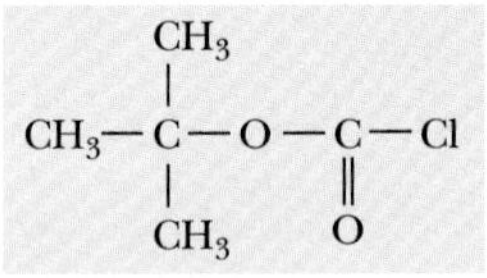

$$CH_3-\underset{CH_3}{\overset{CH_3}{\underset{|}{\overset{|}{C}}}}-O-\underset{O}{\underset{\|}{C}}-Cl$$

BocCl

(b)

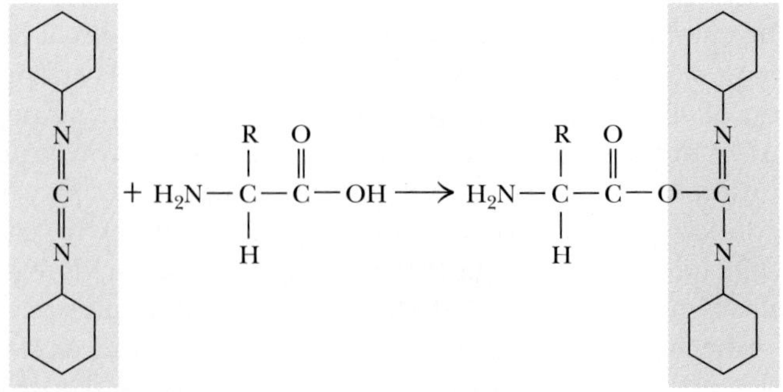

$$\text{Dicyclohexyl-N=C=N-cyclohexyl} + H_2N-\underset{H}{\underset{|}{\overset{R}{\overset{|}{C}}}}-\overset{O}{\overset{\|}{C}}-OH \longrightarrow H_2N-\underset{H}{\underset{|}{\overset{R}{\overset{|}{C}}}}-\overset{O}{\overset{\|}{C}}-O-C(=N\text{-cyclohexyl})-NH\text{-cyclohexyl}$$

Dicyclohexyl-carbodiimide **Activated amino acid**

Figure **4.35** Solid phase synthesis of a peptide. (a) Tertiary butyloxycarbonyl chloride (*t*BocCl) is an excellent reagent for blocking amino groups of amino acids during organic synthesis. (b) Dicyclohexylcarbodiimide is a powerful agent for activating carboxyl groups to condense with amino groups to form peptide bonds. (c) By coupling the carboxyl group of the carboxyl-terminal amino acid to an insoluble resin particle, the growing peptide chain is easily recovered after cyclic additions of amino acids simply by filtering or centrifuging the reaction mixture.

form an amino-blocked dipeptide covalently coupled to the resin bead. Acidification decomposes the *t*Boc blocking agent into the gaseous products, isobutylene and carbon dioxide, freeing the dipeptide's N-terminus for another round of amino acid addition. This cyclic process has been automated and computer controlled so that the reactions take place in a small cup with reagents being pumped in and removed as programmed. The 124-residue-long bovine pancreatic ribonuclease A sequence was synthesized, using this technology, with 18% yield in Merrifield's laboratory. Although catalytically active, the final product was not pure, but was contaminated by fragments and products lacking the complete sequence. While the technology has limitations, the prospects of specific peptide synthesis by chemical means is an appealing frontier.

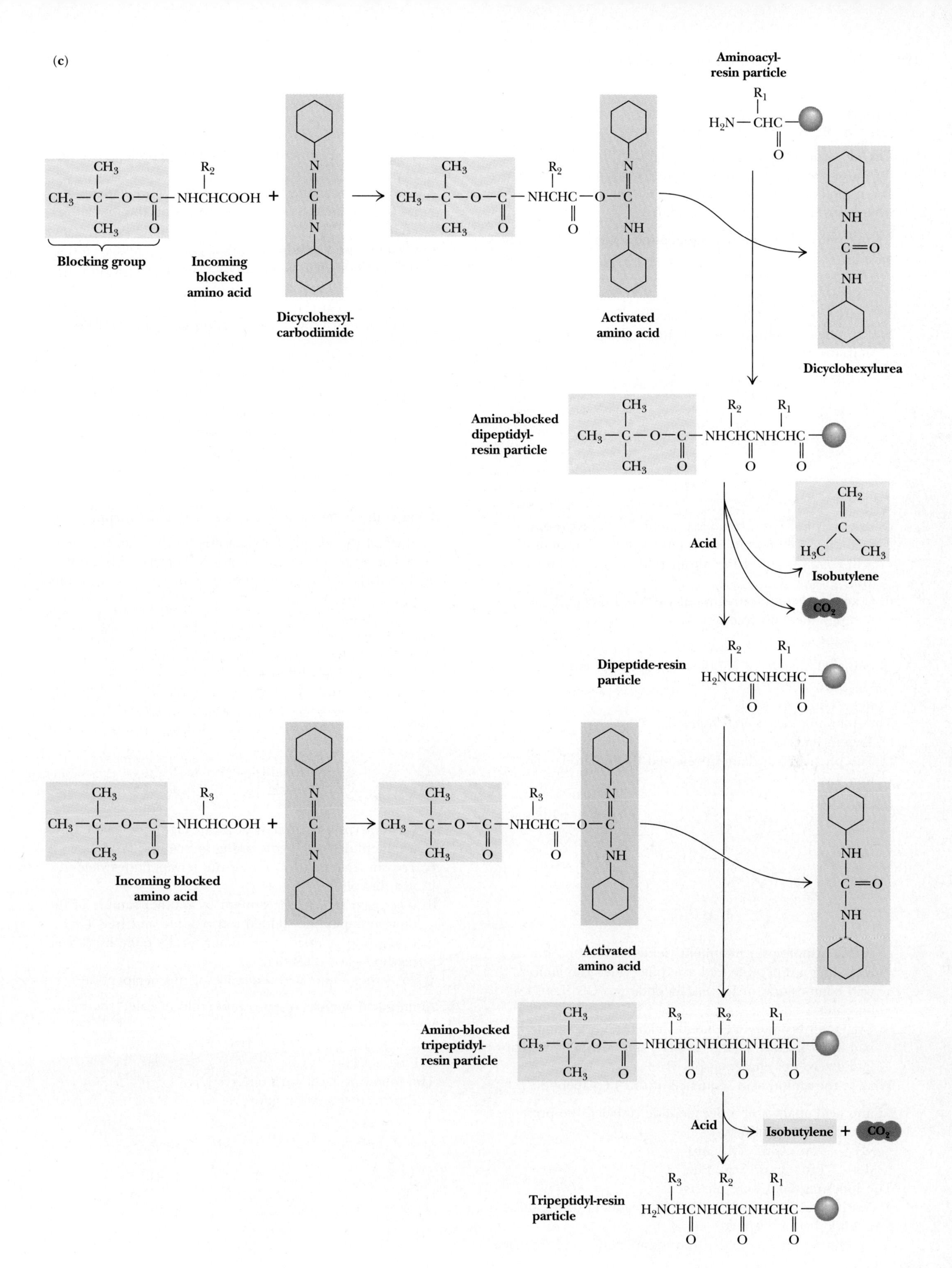
(c)
Aminoacyl-resin particle
Blocking group
Incoming blocked amino acid
Dicyclohexyl-carbodiimide
Activated amino acid
Dicyclohexylurea
Amino-blocked dipeptidyl-resin particle
Acid
Isobutylene
CO2
Dipeptide-resin particle
Incoming blocked amino acid
Activated amino acid
Amino-blocked tripeptidyl-resin particle
Acid
Isobutylene + CO2
Tripeptidyl-resin particle

Problems

1. The element selenium (atomic weight 78.96) constitutes 0.34% of the weight of the enzyme glutathione peroxidase. If the molecular weight of glutathione peroxidase is 80,000, what is its likely quaternary structure?

2. Amino acid analysis of an oligopeptide seven residues long gave

   Asp Leu Lys Met Phe Tyr

   The following facts were observed:

   a. Trypsin treatment had no apparent effect.

   b. The phenylthiohydantoin released by Edman degradation was

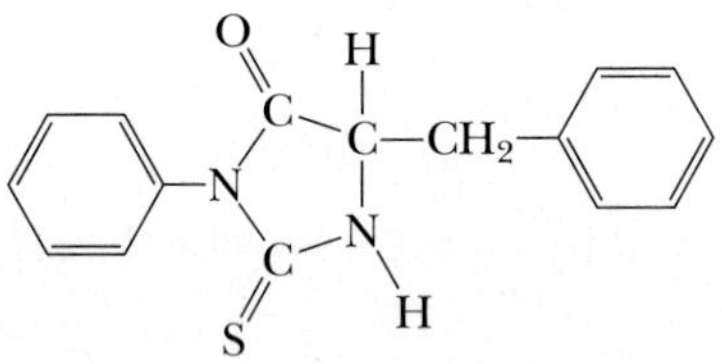

   O H
   C—C—CH_2—(phenyl)
   (phenyl)—N
   C—N
   S H

   c. Brief chymotrypsin treatment yielded several products, including a dipeptide and a tetrapeptide. The amino acid composition of the tetrapeptide was Leu, Lys, and Met.

   d. Cyanogen bromide treatment yielded a dipeptide, a tetrapeptide, and free Lys.

   What is the amino acid sequence of this heptapeptide?

3. Amino acid analysis of another heptapeptide gave

   Asp Glu Leu Lys

   Met Tyr Trp NH_4^+

   The following facts were observed:

   a. Trypsin had no effect.

   b. The phenylthiohydantoin released by Edman degradation was

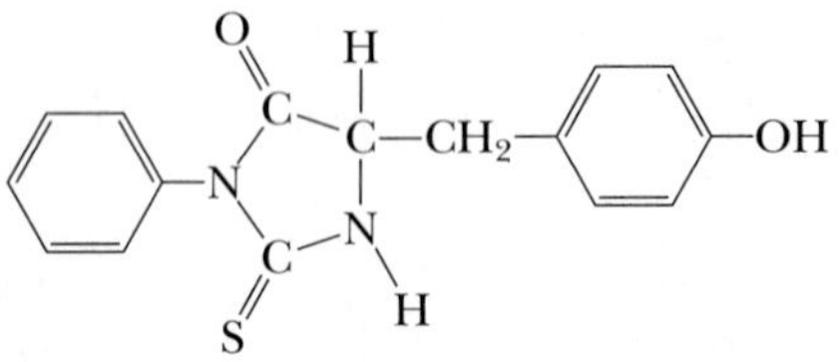

   O H
   C—C—CH_2—(phenyl)—OH
   (phenyl)—N
   C—N
   S H

   c. Brief chymotrypsin treatment yielded several products, including a dipeptide and a tetrapeptide. The amino acid composition of the tetrapeptide was Glx, Leu, Lys, and Met.

   d. Cyanogen bromide treatment yielded a tetrapeptide that had a net positive charge at pH 7 and a tripeptide that had a zero net charge at pH 7.

   What is the amino acid sequence of this heptapeptide?

4. Amino acid analysis of a decapeptide revealed the presence of the following products:

   NH_4^+ Asp Glu Tyr Arg

   Met Pro Lys Ser Phe

   The following facts were observed:

   a. Neither carboxypeptidase A or B treatment of the decapeptide had any effect.

   b. Trypsin treatment yielded two tetrapeptides and free Lys.

   c. Clostripain treatment yielded a tetrapeptide and a hexapeptide.

   d. Cyanogen bromide treatment yielded an octapeptide and a dipeptide of sequence NP (using the one-letter codes).

   e. Chymotrypsin treatment yielded two tripeptides and a tetrapeptide. The N-terminal chymotryptic peptide had a net charge of −1 at neutral pH and a net charge of −3 at pH 12.

   f. One cycle of Edman degradation gave the PTH derivative

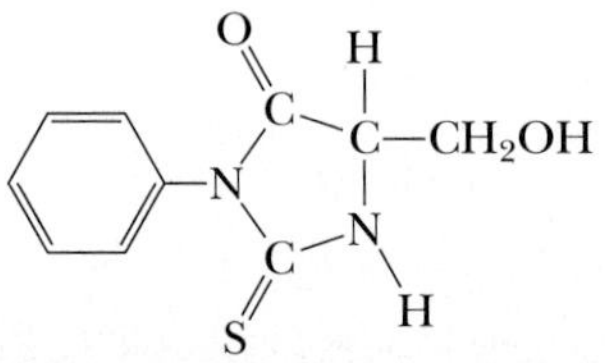

   O H
   C—C—CH_2OH
   (phenyl)—N
   C—N
   S H

   What is the amino acid sequence of this decapeptide?

5. Analysis of the blood of a catatonic football fan revealed large concentrations of a psychotoxic octapeptide. Amino acid analysis of this octapeptide gave the following results:

   2 Ala 1 Arg 1 Asp 1 Met

   2 Tyr 1 Val 1 NH_4^+

   The following facts were observed:

   a. Partial acid hydrolysis of the octapeptide yielded a dipeptide of the structure

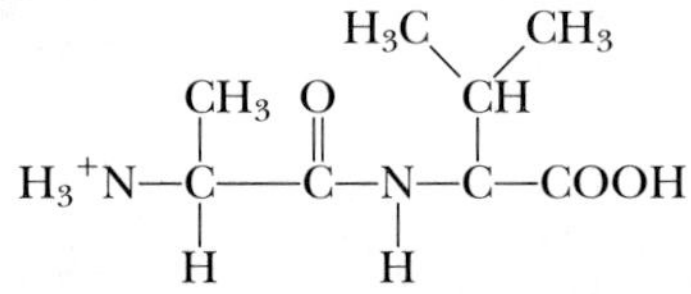

   H_3C CH_3
   CH_3 O CH
   $H_3{}^+N$—C—C—N—C—COOH
   H H

   b. Chymotrypsin treatment of the octapeptide yielded two tetrapeptides, each containing an alanine residue.

   c. Trypsin treatment of one of the tetrapeptides yielded two dipeptides.

   d. Cyanogen bromide treatment of another sample of the same tetrapeptide yielded a tripeptide and free Tyr.

   e. End-group analysis of the other tetrapeptide by Sanger's method gave DNP-Asp.

   What is the amino acid sequence of this octapeptide?

6. Amino acid analysis of an octapeptide revealed the following composition:

   2 Arg 1 Gly 1 Met 1 Trp

   1 Tyr 1 Phe 1 Lys

   The following facts were observed:

   a. Edman degradation gave

   O H
   C—C—H
   (phenyl)—N
   C—N
   S H

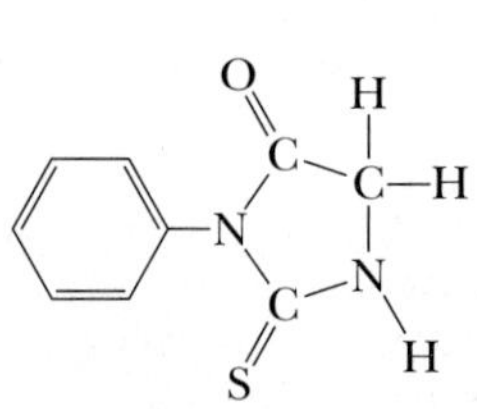

b. CNBr treatment yielded a pentapeptide and a tripeptide containing phenylalanine.
c. Chymotrypsin treatment yielded a tetrapeptide containing a C-terminal indole amino acid and two dipeptides.
d. Trypsin treatment yielded a tetrapeptide, a dipeptide, and free Lys and Phe.
e. Clostripain yielded a pentapeptide, a dipeptide, and free Phe.

What is the amino acid sequence of this octapeptide?

7. Amino acid analysis of an octapeptide gave the following results:

   1 Ala 1 Arg 1 Asp 1 Gly
   3 Ile 1 Val 1 NH_4^+

   The following facts were observed:
   a. Trypsin treatment yielded a pentapeptide and a tripeptide.
   b. Lithium aluminum hydride treatment and subsequent acid hydrolysis yielded 2-aminopropanol.
   c. Partial acid hydrolysis of the tryptic pentapeptide yielded, among other products, two dipeptides, each of which contained C-terminal isoleucine. One of these dipeptides migrated as an anionic species upon electrophoresis at neutral pH.
   d. The tryptic tripeptide was degraded in an Edman sequenator, yielding first **A**, then **B**:

   A.

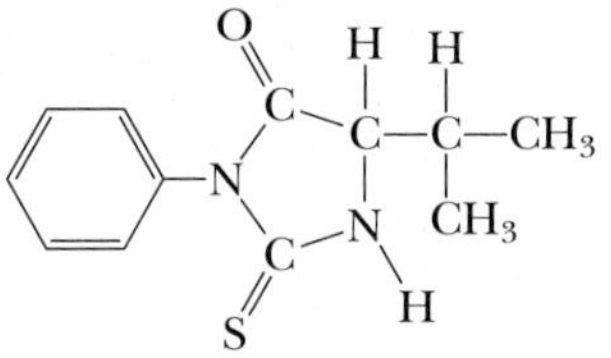

   B.

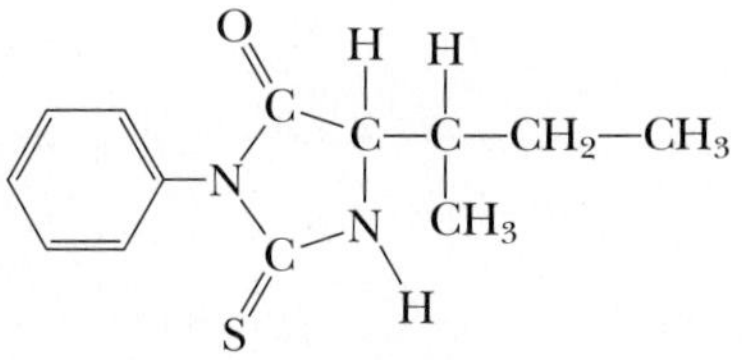

   What is an amino acid sequence of the octapeptide? Four sequences are possible, but only one suits the authors. Why?

8. An octapeptide consisting of 2 Gly, 1 Lys, 1 Met, 1 Pro, 1 Arg, 1 Trp, and 1 Tyr was subjected to sequence studies. The following was found:
   a. Edman degradation yielded

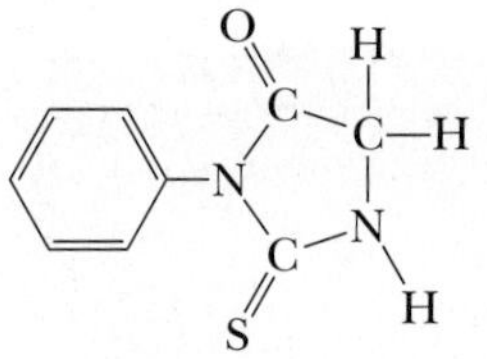

   b. Upon treatment with carboxypeptidases A, B, and C, only carboxypeptidase C had any effect.
   c. Trypsin treatment gave two tripeptides and a dipeptide.
   d. Chymotrypsin treatment gave two tripeptides and a dipeptide. Acid hydrolysis of the dipeptide yielded only Gly.
   e. Cyanogen bromide treatment yielded two tetrapeptides.
   f. Clostripain treatment gave a pentapeptide and a tripeptide.

   What is the amino acid sequence of this octapeptide?

9. Amino acid sequence of an oligopeptide containing nine residues revealed the presence of the following amino acids:

   Arg Cys Gly Leu
   Met Pro Tyr Val

   The following was found:
   a. Carboxypeptidase A treatment yielded no free amino acid.
   b. Edman analysis of the intact oligopeptide released

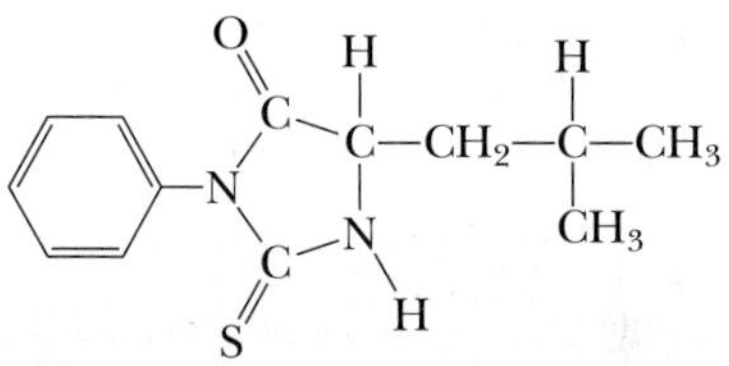

   c. Neither trypsin nor chymotrypsin treatment of the nonapeptide released smaller fragments. However, combined trypsin and chymotrypsin treatment liberated free Arg.
   d. CNBr treatment of the eight-residue fragment left after combined trypsin and chymotrypsin action yielded a six-residue fragment containing Cys, Gly, Pro, Tyr, and Val; and a dipeptide.
   e. Treatment of the six-residue fragment with β-mercaptoethanol yielded two tripeptides. Brief Edman analysis of the tripeptide mixture yielded only PTH-Cys. (The sequence of each tripeptide, as read from the N-terminal end, is alphabetical if the one-letter designation for amino acids is used.)

   What is the amino acid sequence of this octapeptide?

10. Amino acid analysis of an octapeptide yielded the following results:

    Lys Thr Ser Met
    Arg Trp Tyr Glu

    The following was found:
    a. Treatment of the octapeptide with Sanger's reagent followed by acid hydrolysis yielded

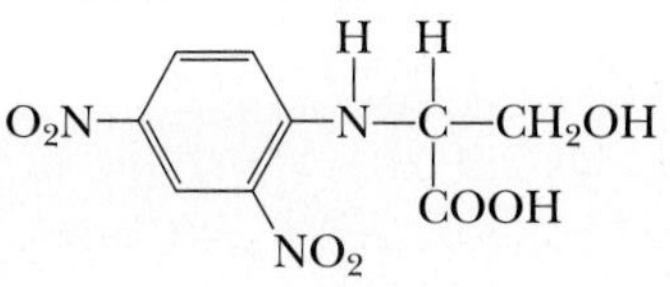

    b. Chymotrypsin treatment yielded a pentapeptide, a dipeptide containing Tyr, and free Glu.
    c. Trypsin treatment gave two tripeptides and a dipeptide.
    d. Cyanogen bromide treatment gave two tetrapeptides, one of which gave the following upon treatment with Edman's reagent:

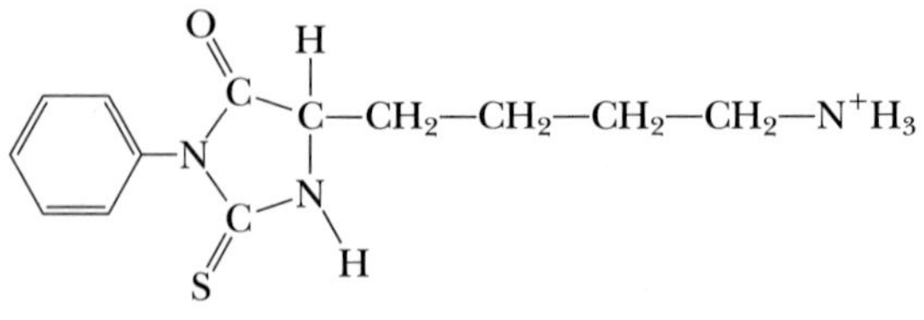

O H

C—C—CH_2—CH_2—CH_2—CH_2—N^+H_3

—N

C—N

S H

What is the amino acid sequence of this octapeptide?

11. Describe the synthesis of the dipeptide Lys-Ala by Merrifield's solid phase chemical method of peptide synthesis. What pitfalls might be encountered if you attempted to add a leucine residue to Lys-Ala to make a tripeptide?

Further Reading

Creighton, T. E., 1983. *Proteins: Structure and Molecular Properties.* San Francisco: W. H. Freeman and Co., 515 pp.

Dayhoff, M. O., 1972–1978. *The Atlas of Protein Sequence and Structure,* Vols. 1–5. Washington, D.C.: National Medical Research Foundation.

Deutscher, M. P., ed., 1990. *Guide to Protein Purification.* Vol. 182, *Methods in Enzymology.* San Diego: Academic Press, 894 pp.

Goodsell, D. S., and Olson, A. J., 1993. Soluble proteins: size, shape and function. *Trends in Biochemical Sciences* **18:**65–68.

Heijne, G. von, 1987. *Sequence Analysis in Molecular Biology: Treasure Trove or Trivial Pursuit?* San Diego: Academic Press.

Hill, R. L., 1965. Hydrolysis of proteins. *Advances in Protein Chemistry* **20:**37–107.

Hirs, C. H. W., ed., 1967. *Enzyme Structure.* Vol. XI, *Methods in Enzymology.* New York: Academic Press, 987 pp.

Hirs, C. H. W., and Timasheff, S. E., eds., 1977–1986. *Enzyme Structure,* Parts E–L. New York: Academic Press.

Hunt, D. F., et al., 1987. Tandem quadrupole Fourier transform mass spectrometry of oligopeptides and small proteins. *Proceedings of the National Academy of Sciences, U.S.A.* **84:**620–623.

Merrifield, B., 1986. Solid phase synthesis. *Science* **232:**341–347.

Appendix to Chapter 4

Protein Techniques[3]

Size Exclusion Chromatography

Size exclusion chromatography is also known as *gel filtration chromatography* or *molecular sieve chromatography*. In this method, fine, porous beads are packed into a chromatography column. The beads are composed of dextran polymers *(Sephadex)*, agarose *(Sepharose)*, or polyacrylamide *(Sephacryl* or *BioGel P)*. The pore sizes of these beads approximate the dimensions of macromolecules. The total bed volume of the packed chromatography column, V_t, is equal to the volume outside the porous beads (V_o) plus the volume inside the beads (V_i) plus the volume actually occupied by the bead material (V_g): $V_t = V_o + V_i + V_g$. (V_g is typically less than 1% of V_t and can be conveniently ignored in most applications.)

As a solution of molecules is passed through the column, the molecules passively distribute between V_o and V_i, depending on their ability to enter the pores (that is, their size). If a molecule is too large to enter at all, it is totally excluded from V_i and emerges first from the column at an elution volume, V_e, equal to V_o (Figure A4.1). If a particular molecule can enter the pores in the gel, its distribution is given by the *distribution coefficient,* K_D:

$$K_D = (V_e - V_o)/V_i$$

where V_e is the molecule's characteristic elution volume (Figure A4.1). The chromatographic run is complete when a volume of solvent equal to V_t has passed through the column.

Dialysis and Ultrafiltration

If a solution of protein is separated from a bathing solution by a semipermeable membrane, small molecules and ions can pass through the semipermeable membrane to equilibrate between the protein solution and the bathing solution, called the *dialysis bath* or *dialysate* (Figure A4.2). This method is useful for removing small molecules from macromolecular solutions or for altering the composition of the protein-containing solution.

Ultrafiltration is an improvement on the dialysis principle. Filters having pore sizes over the range of biomolecular dimensions are used to filter solutions to select for molecules in a particular size range. Because the pore sizes in these filters are microscopic, high pressures are often required to force the solution through the filter. This technique is useful for concentrating dilute solutions of macromolecules. The concentrated protein can then be diluted into the solution of choice.

[3]Although this appendix is entitled *Protein Techniques,* these methods are also applicable to other macromolecules such as nucleic acids.

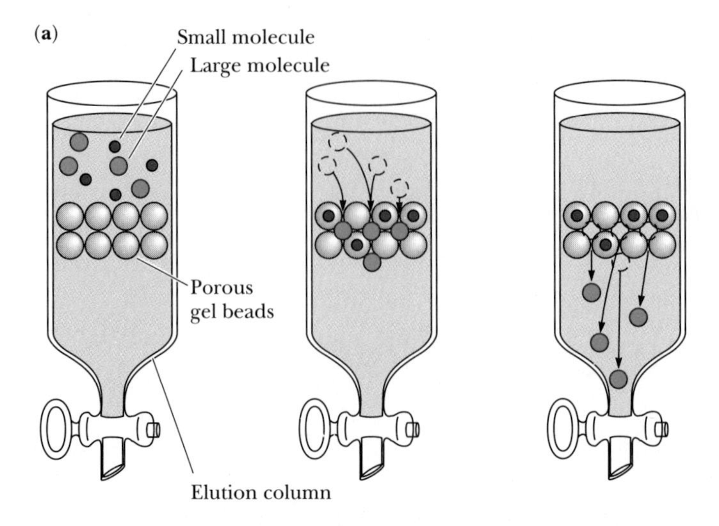

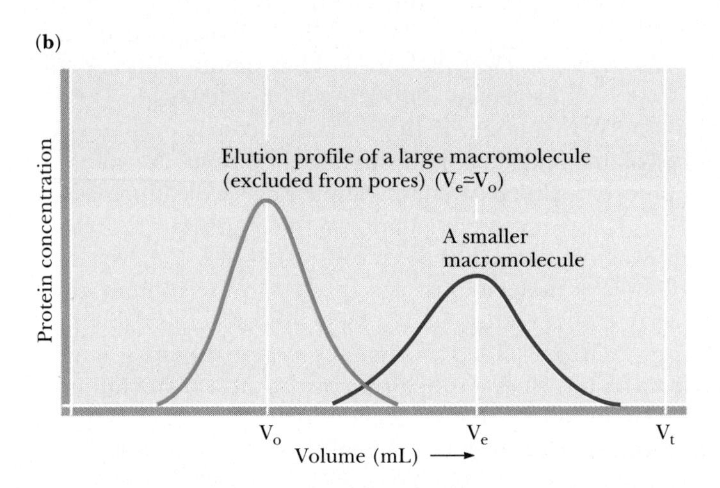

***Figure* 4A.1** A gel filtration chromatography column. Larger molecules are excluded from the gel beads and emerge from the column sooner than smaller molecules, whose migration is retarded because they can enter the beads.

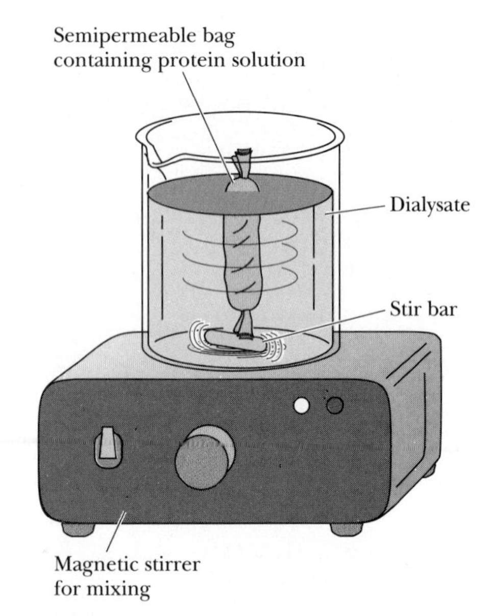

***Figure* 4A.2** A dialysis experiment. The solution of macromolecules to be dialyzed is placed in a semipermeable membrane bag, and the bag is immersed in a bathing solution. A magnetic stirrer gently mixes the solution to facilitate equilibrium of diffusible solutes between the dialysate and the solution contained in the bag.

Electrophoresis

Electrophoretic techniques are based on the movement of ions in an electrical field. An ion of charge q experiences a force F given by $F = Eq/d$, where E is the voltage (or *electrical potential*) and d is the distance between the electrodes. In a vacuum, F would cause the molecule to accelerate. In solution, the molecule experiences *frictional drag*, F_f, due to the solvent:

$$F_f = 6\pi r\eta v$$

where r is the radius of the charged molecule, η is the viscosity of the solution, and v is the velocity at which the charged molecule is moving. So, the velocity of the charged molecule is proportional to its charge q and the voltage E, but inversely proportional to the viscosity of the medium η and d, the distance between the electrodes.

Generally, electrophoresis is carried out *not* in free solution but in a porous support matrix such as polyacrylamide or agarose, which retards the movement of molecules according to their dimensions relative to the size of the pores in the matrix.

SDS-Polyacrylamide Gel Electrophoresis (SDS-PAGE)

SDS is sodium dodecylsulfate (sodium lauryl sulfate) (Figure A4.3). The hydrophobic tail of dodecylsulfate interacts strongly with polypeptide chains.

$$Na^+ \ {}^-O-\overset{\overset{\displaystyle O}{\|}}{\underset{\underset{\displaystyle O^-\ Na^+}{|}}{S}}-O-CH_2-CH_2-CH_2-CH_2-CH_2-CH_2-CH_2-CH_2-CH_2-CH_2-CH_2-CH_3$$

***Figure* 4A.3** The structure of sodium dodecylsulfate (SDS).

The number of SDS molecules bound by a polypeptide is proportional to the length (number of amino acid residues) of the polypeptide. Each dodecylsulfate contributes two negative charges. Collectively, these charges overwhelm any intrinsic charge that the protein might have. SDS is also a detergent that disrupts protein folding (protein 3° structure). SDS-PAGE is usually run in the presence of sulfhydryl-reducing agents such as β-mercaptoethanol so that any disulfide links between polypeptide chains are broken. The electrophoretic mobility of proteins upon SDS-PAGE is inversely proportional to the logarithm of the protein's molecular weight (Figure A4.4). SDS-PAGE is often used to determine the molecular weight of a protein.

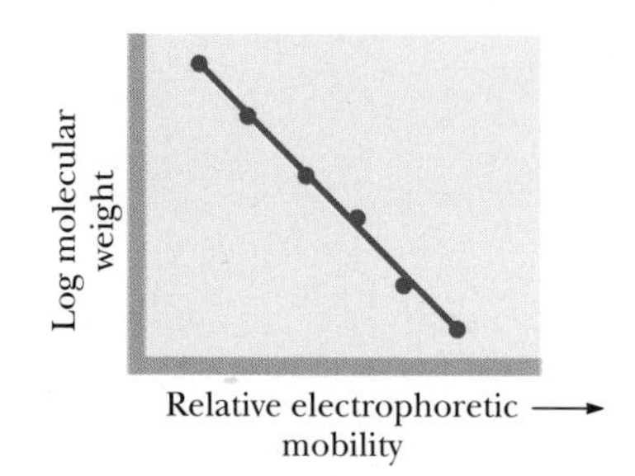

***Figure* 4A.4** A plot of the relative electrophoretic mobility of proteins in SDS-PAGE versus the log of the molecular weights of the individual proteins.

Isoelectric Focusing

Isoelectric focusing is an electrophoretic technique for separating proteins according to their *isoelectric points* (pIs). A solution of *ampholytes* (amphoteric electrolytes) is first electrophoresed through a gel, usually contained in a small tube. The migration of these substances in an electric field establishes a pH gradient in the tube. Then a protein mixture is applied to the gel and electrophoresis is resumed. As the protein molecules move down the gel, they experience the pH gradient and migrate to a position corresponding to their respective pIs. At its pI, a protein has no net charge and thus moves no farther.

amphoteric: having the characteristics of both an acid and a base

Two-Dimensional Gel Electrophoresis

This separation technique uses isoelectric focusing in one dimension and SDS-PAGE in the second dimension to resolve protein mixtures. The proteins in a mixture are first separated according to pI by isoelectric focusing in a polyacrylamide gel in a tube. The gel is then removed and laid along the top of an SDS-PAGE slab, and the proteins are electrophoresed into the SDS-polyacrylamide gel, where they are separated according to size (Figure A4.5). The gel slab can then be stained to reveal the locations of the individual proteins. Using this powerful technique, researchers have the potential to visualize and construct catalogs of virtually *all* the proteins present in particular cell types.

Affinity Chromatography

Recently, *affinity purification* strategies for proteins have been developed to exploit the biological function of the target protein. In most instances, proteins carry out their biological activity through binding or complex formation

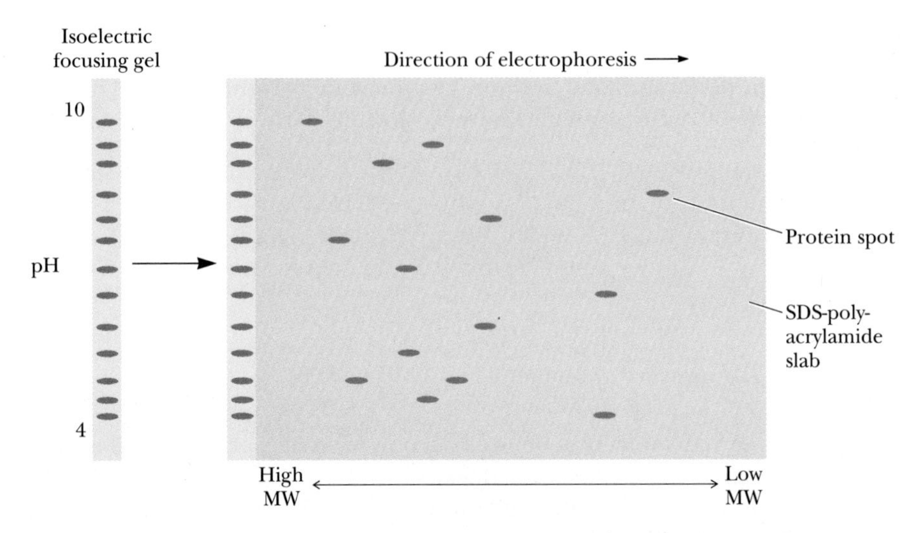

***Figure* 4A.5** A two-dimensional electrophoresis separation. A mixture of macromolecules is first separated according to charge by isoelectric focusing in a tube gel. The gel containing separated molecules is then placed next to an SDS-PAGE slab, and the molecules are electrophoresed into the SDS-PAGE gel where they are separated according to size.

with specific small biomolecules, or *ligands,* as in the case of an enzyme binding its substrate. If this small molecule can be immobilized through covalent attachment to an insoluble matrix, such as a chromatographic medium like cellulose or polyacrylamide, then the protein of interest, in displaying affinity for its ligand, will become bound and immobilized itself. It can then be removed from contaminating proteins in the mixture by simple means such as filtration and washing the matrix. Finally, the protein is dissociated or eluted from the matrix by the addition of high concentrations of the free ligand in solution. Figure A4.6 depicts the protocol for such an *affinity chromatography* scheme. Since this method of purification relies on the biological specificity of the protein of interest, it is a very efficient procedure and proteins can be purified several thousand-fold in a single step.

Hydrophobic Interaction Chromatography (HIC)

Hydrophobic interaction chromatography (HIC) exploits the hydrophobic nature of proteins in purifying them. Proteins are passed over a chromatographic column packed with a support matrix to which hydrophobic groups are covalently linked. *Phenyl Sepharose®*, an agarose support matrix to which phenyl groups are affixed, is a prime example of such material. In the presence of high salt concentrations, proteins bind to the phenyl groups by virtue of hydrophobic interactions. Proteins in a mixture can be differentially eluted from the phenyl groups by lowering the salt concentration or by adding solvents such as polyethylene glycol to the elution fluid.

High Performance Liquid Chromatography

The principles exploited in *high performance* (or high pressure) *liquid chromatography (HPLC)* are the same as used in the common chromatographic methods such as ion exchange chromatography or size exclusion chromatography. Very high resolution separations can be achieved in HPLC.

Ultracentrifugation

Centrifugation methods separate macromolecules on the basis of their characteristic densities. Particles will tend to "fall" through a solution if the density of the solution is less than the density of the particle. The velocity of the particle through the medium will be proportional to the difference in density between the particle and the solution. The tendency of any particle to move through a solution under centrifugal force is given by the *sedimentation coefficient, S:*

$$S = (\rho_p - \rho_m) V/f$$

where ρ_p is the density of the particle or macromolecule, ρ_m is the density of the medium or solution, V is the volume of the particle, and f is the frictional coefficient, given by

$$f = F_f/v$$

where v is the velocity of the particle and F_f is the frictional drag. Nonspherical molecules have larger frictional coefficients and thus smaller sedimentation coefficients. The smaller the particle and the more its shape deviates from spherical, the more difficult it is to sediment that particle in a centrifuge.

Centrifugation can be used either as a preparative technique for separating and purifying macromolecules and cellular components or as an analytical technique to characterize the hydrodynamic properties of macromolecules such as proteins and nucleic acids.

A protein interacts with a metabolite. The metabolite is thus a ligand which binds specifically to this protein

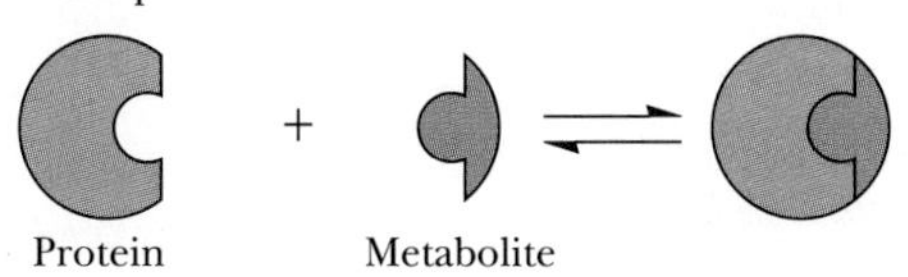

The metabolite can be immobilized by covalently coupling it to an insoluble matrix such as an agarose polymer. Cell extracts containing many individual proteins may be passed through the matrix.

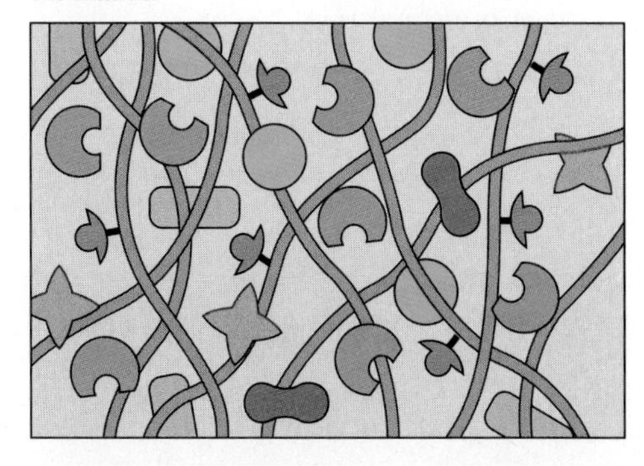

Specific protein binds to ligand. All other unbound material is washed out of the matrix.

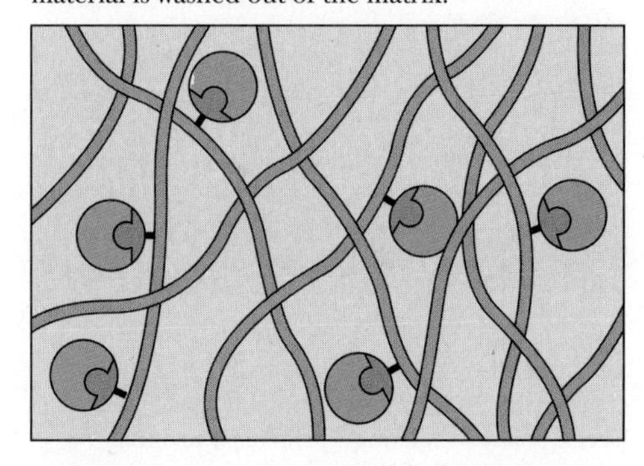

Adding an excess of free metabolite that will compete for the bound protein dissociates the protein from the chromatographic matrix. The protein passes out of the column complexed with free metabolite.

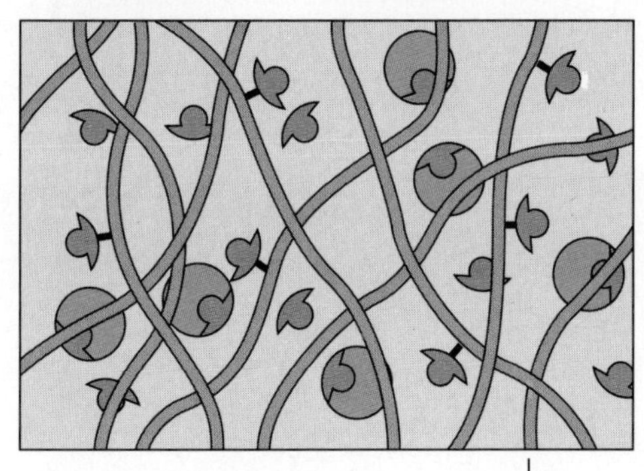

Purifications of proteins as much as 1000-fold or more are routinely achieved in a single affinity chromatographic step like this.

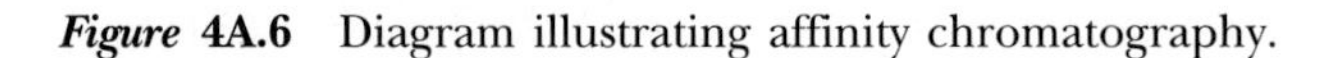

***Figure* 4A.6** Diagram illustrating affinity chromatography.

Growing in size and complexity
Living things, masses of atoms, DNA, protein
Dancing a pattern ever more intricate.

Out of the cradle onto the dry land
Here it is standing
Atoms with consciousness
Matter with curiosity.

Stands at the sea
Wonders at wondering
I

A universe of atoms
An atom in the universe.

Richard P. Feynman (1918–1988)

From *"The Value of Science"* in Edward Hutchings, Jr. (ed.), *Frontiers of Science: A Survey*. New York: Basic Books, 1958.

Like the Greek sea god Proteus, who could assume different forms, proteins act through changes in conformation. Proteins (from the Greek *proteios,* meaning "primary") are the primary agents of biological function.

Chapter 5

Proteins: Secondary, Tertiary, and Quaternary Structure

Outline

5.1 Forces Influencing Protein Structure
5.2 Role of the Amino Acid Sequence in Protein Structure
5.3 Secondary Structure in Proteins
5.4 Protein Folding and Tertiary Structure
5.5 Subunit Interactions and Quaternary Structure

Nearly all biological processes involve the specialized functions of one or more protein molecules. Proteins function to produce other proteins, control all aspects of cellular metabolism, regulate the movement of various molecular and ionic species across membranes, convert and store cellular energy, and carry out many other activities. Essentially all of the information required to initiate, conduct, and regulate each of these functions must be contained in the structure of the protein itself. The previous chapter described the details of primary protein structure. However, proteins do not normally exist as fully extended polypeptide chains but rather as compact, folded structures, and the function of a given protein is rarely if ever dependent only on the amino acid sequence. Instead, the ability of a particular protein to carry out its function in nature is normally determined by its overall three-dimensional shape or *conformation.* This native, folded structure of the protein is dictated by several factors: (1) interactions with solvent molecules (normally water), (2) the pH and ionic composition of the solvent, and most important, (3) the sequence of the protein. The first two of these effects are intuitively reasonable, but the third, the role of the primary sequence, may not be. In ways that are just now beginning to be understood, the primary structure facilitates the development of short-range interactions among adjacent parts of the sequence and also long-range interactions among distant parts of the sequence. Although the resulting overall structure of the complete protein molecule may at first look like a disorganized and random arrangement, it is in nearly all cases a delicate and sophisticated balance of numerous forces that combine to determine the protein's unique conformation. This chapter considers the details of protein structure and the forces that maintain these structures.

5.1 Forces Influencing Protein Structure

Several different kinds of noncovalent interactions are of vital importance in protein structure. Hydrogen bonds, hydrophobic interactions, electrostatic bonds, and van der Waals forces are all noncovalent in nature, yet are extremely important influences on protein conformations. The stabilization free energies afforded by each of these interactions may be highly dependent on the local environment within the protein, but certain generalizations can still be made.

Hydrogen Bonds

Hydrogen bonds are generally made wherever possible within a given protein structure. In most protein structures that have been examined to date, component atoms of the peptide backbone tend to form hydrogen bonds with one another. Furthermore, side chains capable of forming H bonds are usually located on the protein surface and form such bonds primarily with the water solvent. While each hydrogen bond may contribute an average of only about 12 kJ/mol in stabilization energy for the protein structure, the number of H bonds formed in the typical protein is very large. For example, in α-helices, the C=O and N—H groups of every residue participate in H bonds. The importance of H bonds in protein structure cannot be overstated.

Hydrophobic Interactions

Hydrophobic "bonds," or more accurately, *interactions*, form because nonpolar side chains of amino acids and other nonpolar solutes prefer to cluster in a nonpolar environment rather than to intercalate in a polar solvent such as water. The forming of hydrophobic bonds minimizes the interaction of nonpolar residues with water and is therefore highly favorable. Such clustering is entropically driven. In proteins, nonpolar residues are often oriented toward the interior of the protein structure, and it is unusual to find significant numbers of nonpolar residues at the protein surface.

intercalate: to insert between

Electrostatic Interactions

"Ionic" bonds arise either as electrostatic attractions between opposite charges or repulsions between like charges. Chapter 3 discussed the ionization behavior of amino acids. Amino acid side chains can carry positive charges, as in the case of lysine, arginine, and histidine, or negative charges, as in aspartate and glutamate. In addition, the NH_2-terminal and COOH-terminal residues of a protein or peptide chain usually exist in ionized states and carry positive or negative charges, respectively. All of these may experi-

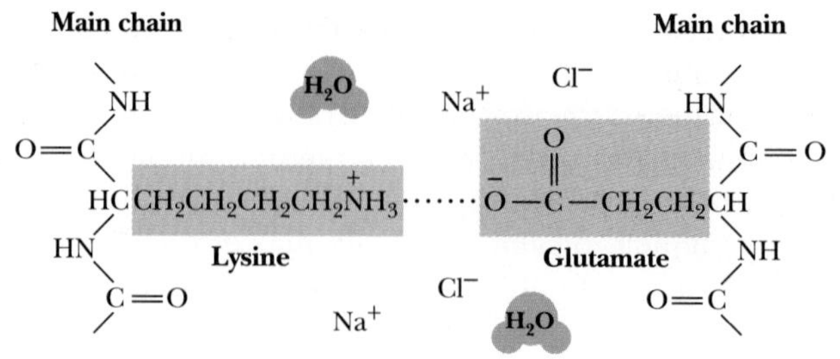

***Figure* 5.1** An electrostatic interaction between the ε-amino group of a lysine and the γ-carboxyl group of a glutamate residue.

ence electrostatic interactions in a protein structure. Charged residues are normally located on the protein surface, where they may interact optimally with the water solvent. It is energetically unfavorable for an ionized residue to be located in the hydrophobic core of the protein. Electrostatic interactions between charged groups on a protein surface are often complicated by the presence of salts in the solution. For example, the ability of a positively charged lysine to attract a nearby negative glutamate may be weakened by dissolved NaCl (Figure 5.1). The Na^+ and Cl^- ions are highly mobile, compact units of charge, compared to the amino acid side chains, and thus compete effectively for charged sites on the protein. In this manner, electrostatic interactions among amino acid residues on protein surfaces may be damped out by high concentrations of salts. Nevertheless, these interactions are important for protein stability.

The strength of electrostatic interactions is highly dependent on the nature of the interacting species and the distance (r) between them. Electrostatic interactions may involve **ions** (species possessing discrete charges), **permanent dipoles** (with a permanent separation of positive and negative charge), and **induced dipoles** (with a temporary separation of positive and negative charge induced by the environment). Between two ions, the energy falls off as $1/r$. The interaction energy between permanent dipoles falls off as $1/r^6$, whereas the energy between an ion and an induced dipole falls off as $1/r^4$.

Van der Waals Interactions

Both attractive forces and repulsive forces are included in van der Waals interactions. The attractive forces are due to instantaneous induced dipoles that arise because of fluctuations in the electron charge distributions of nearby nonbonded atoms. This amounts to a dipole–induced dipole interaction, and the interaction energy falls off as $1/r^6$. At best, these interactions are weak ones and individually contribute 0.4 to 1.2 kJ/mol of stabilization energy. As with hydrogen bonds, however, there are typically many such interactions in a protein, and the sum of these interactions can be substantial. For example, model studies of heats of sublimation show that each methylene group in a crystalline hydrocarbon accounts for 8 kJ, and each C—H group in a benzene crystal contributes 7 kJ of van der Waals energy per mole. Calculations indicate that the attractive van der Waals energy between the enzyme lysozyme (see Chapter 13) and a model sugar substrate, which it binds, is about 60 kJ/mol.

Repulsive van der Waals forces are different in nature. When two atoms approach each other so closely that their electron clouds interpenetrate, a strong and characteristic repulsion occurs. This repulsion follows an inverse twelfth power dependence on $r(1/r^{12})$, as shown in Figure 5.2. Clearly, between the repulsive and attractive domains, there lies a low point in the potential curve. The corresponding distance is called the **van der Waals contact distance,** which is the interatomic distance that results if only van der Waals forces hold two atoms together. The equation describing the total potential energy (U) is

$$U = \frac{B}{r^{12}} - \frac{A}{r^{6}}$$

where A and B are empirical constants that determine the balance of the repulsive and attractive forces and thus the van der Waals contact distance. Particular atoms have characteristic van der Waals radii (which are additive; see Chapter 1). Thus, the optimal distance of contact between two atoms can be found by adding their two van der Waals radii (see Table 1.4).

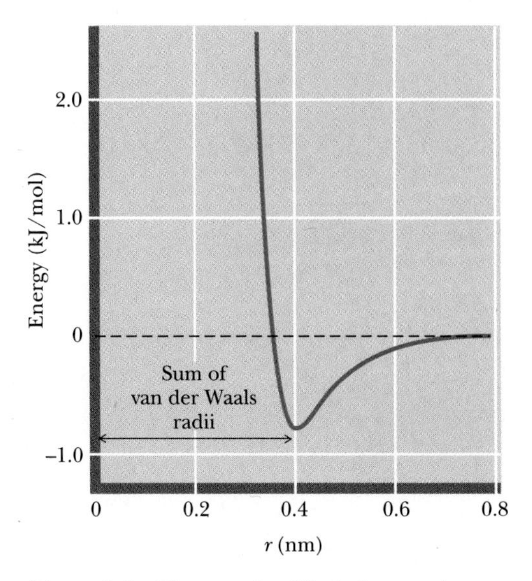

***Figure* 5.2** The van der Waals interaction energy profile as a function of the distance, r, between the centers of two atoms. The energy was calculated using the empirical equation $U = B/r^{12} - A/r^{6}$. (Values for the parameters $B = 11.5 \times 10^{-6}$ kJnm^{12}/mol and $A = 5.96 \times 10^{-3}$ kJnm^{6}/mol for the interaction between two carbon atoms are from Levitt, M., 1974, *Journal of Molecular Biology* **82**:393–420.)

5.2 Role of the Amino Acid Sequence in Protein Structure

It can be inferred from the first sections of this chapter that many different forces work together in a delicate balance to determine the overall three-dimensional structure of a protein. These forces operate both within the protein structure itself and also between the protein and the water solvent. How, then, does nature dictate the manner of protein folding to generate the three-dimensional structure that optimizes and balances these many forces? *All of the information necessary for folding the peptide chain into its "native" structure is contained in the primary amino acid sequence of the peptide.* This principle was first appreciated by C. B. Anfinsen and F. White, whose work in the early 1960s dealt with the chemical denaturation and subsequent renaturation of bovine pancreatic ribonuclease. Ribonuclease was first denatured with urea and mercaptoethanol, a treatment that cleaved the four covalent disulfide cross-bridges (S—S) in the protein. Subsequent air oxidation permitted random formation of disulfide cross-bridges, most of which were incorrect. Thus, the air-oxidized material showed little enzymatic activity. However, treatment of these inactive preparations with small amounts of mercaptoethanol allowed a reshuffling of the disulfide bonds and permitted formation of significant amounts of active native enzyme. It may be difficult to imagine "uncooking" a hard-boiled egg, but experiments such as these indicate that ovalbumin, the principal egg white protein, could also in theory be renatured and returned to its native state by a suitable sequence of steps. In such experiments, the only road map for the protein, that is, the only "instructions" it has, are those directed by its primary structure, the linear sequence of its amino acid residues.

Just how proteins recognize and interpret the information that is stored in the polypeptide sequence is not well understood yet. It may be assumed that certain loci along the peptide chain act as nucleation points, which initiate folding processes that eventually lead to the correct structures. Regardless of how this process operates, it must take the protein correctly to the final native structure, without getting trapped in a local energy-minimum state which, though stable, may be different from the native state itself. A long-range goal of many researchers in the protein structure field is the prediction of three-dimensional conformations from the amino acid sequence. As the details of secondary and tertiary structure are described in this chapter, the complexity and immensity of such a prediction will be more fully appreciated. This gap is perhaps the greatest uncharted frontier remaining in molecular biology.

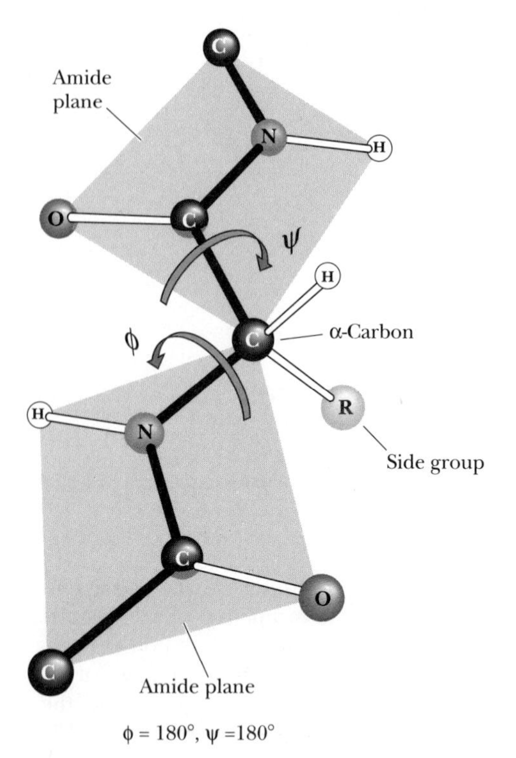

***Figure* 5.3** The amide or peptide bond planes are joined by the tetrahedral bonds of the α-carbon. The rotation parameters are ϕ and ψ. The conformation shown corresponds to $\phi = 180°$ and $\psi = 180°$. Note that positive values of ϕ and ψ correspond to clockwise rotation as viewed from C_α. Starting from 0°, a rotation of 180° in the clockwise direction (+180°) is equivalent to a rotation of 180° in the counterclockwise direction (−180°).

5.3 Secondary Structure in Proteins

Any discussion of protein folding and structure must begin with the *peptide bond,* the fundamental structural unit in all proteins. As we saw in Chapter 4, the resonance structures experienced by a peptide bond constrain the oxygen, carbon, nitrogen, and hydrogen atoms of the peptide group, as well as the adjacent α-carbons, to all lie in a plane. The resonance stabilization energy of this planar structure is approximately 88 kJ/mol, and substantial energy is required to twist the structure about the C—N bond. A twist of θ degrees involves a twist energy of 88 $\sin^2 \theta$ kJ/mol.

Consequences of the Amide Plane

The planarity of the peptide bond means that there are only two degrees of freedom per residue for the peptide chain. Rotation is allowed about the bond linking the α-carbon and the carbon of the peptide bond and also about the bond linking the nitrogen of the peptide bond and the adjacent α-carbon. As shown in Figure 5.3, each α-carbon is the joining point for two planes defined by peptide bonds. The angle about the C_α—N bond is denoted by the Greek letter ϕ (phi) and that about the C_α—C_o is denoted by ψ (psi). For either of these bond angles, a value of 0° corresponds to an orientation with the amide plane bisecting the H—C_α—R (side-chain) plane and a *cis* configuration of the main chain around the rotating bond in question (Figure 5.4). In any case, the entire path of the peptide backbone in a protein is known if the ϕ and ψ rotation angles are all specified. Some values of ϕ and ψ are not allowed due to steric interference between nonbonded atoms. As shown in Figure 5.4, values of $\phi = 180°$ and $\psi = 0°$ are not allowed because of the forbidden overlap of the N—H hydrogens. Similarly, $\phi = 0°$ and $\psi = 180°$ is forbidden because of unfavorable overlap between the carbonyl oxygens.

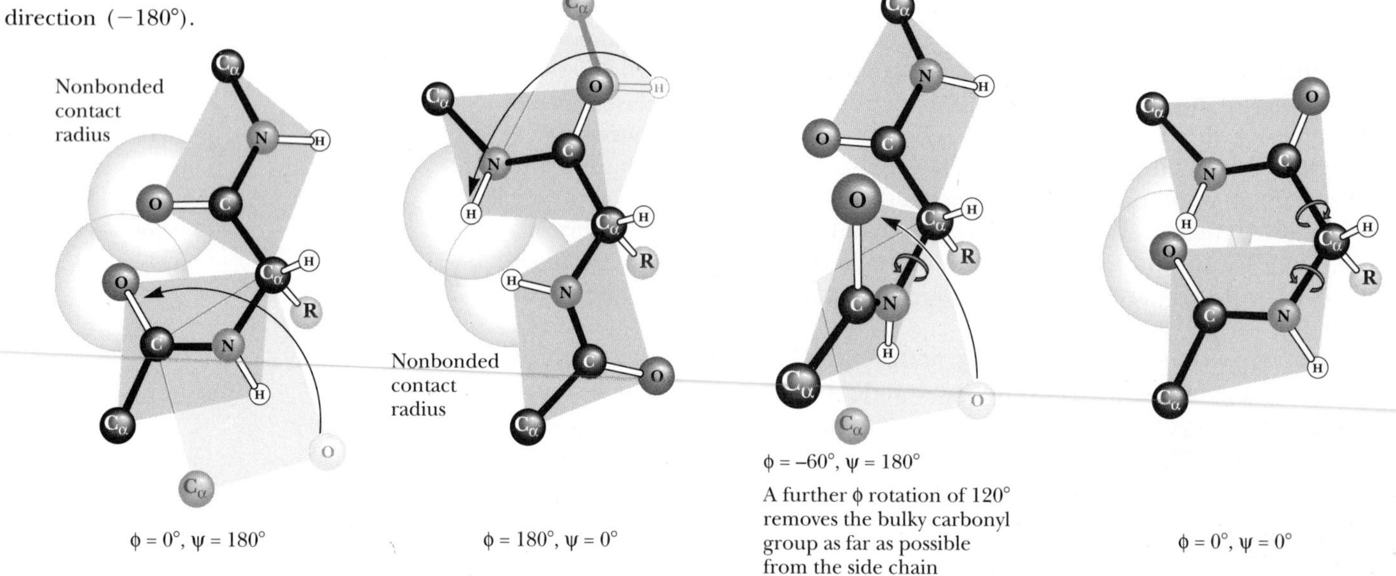

***Figure* 5.4** Many of the possible conformations about an α-carbon between two peptide planes are forbidden because of steric crowding. Several noteworthy examples are shown here.

Note: The formal IUPAC-IUB Commission on Biochemical Nomenclature convention for the definition of the torsion angles ϕ and ψ in a polypeptide chain (*Biochemistry* **9**:3471–3479, 1970) is different from that used here, where the C_α atom serves as the point of reference for both rotations, but the result is the same.

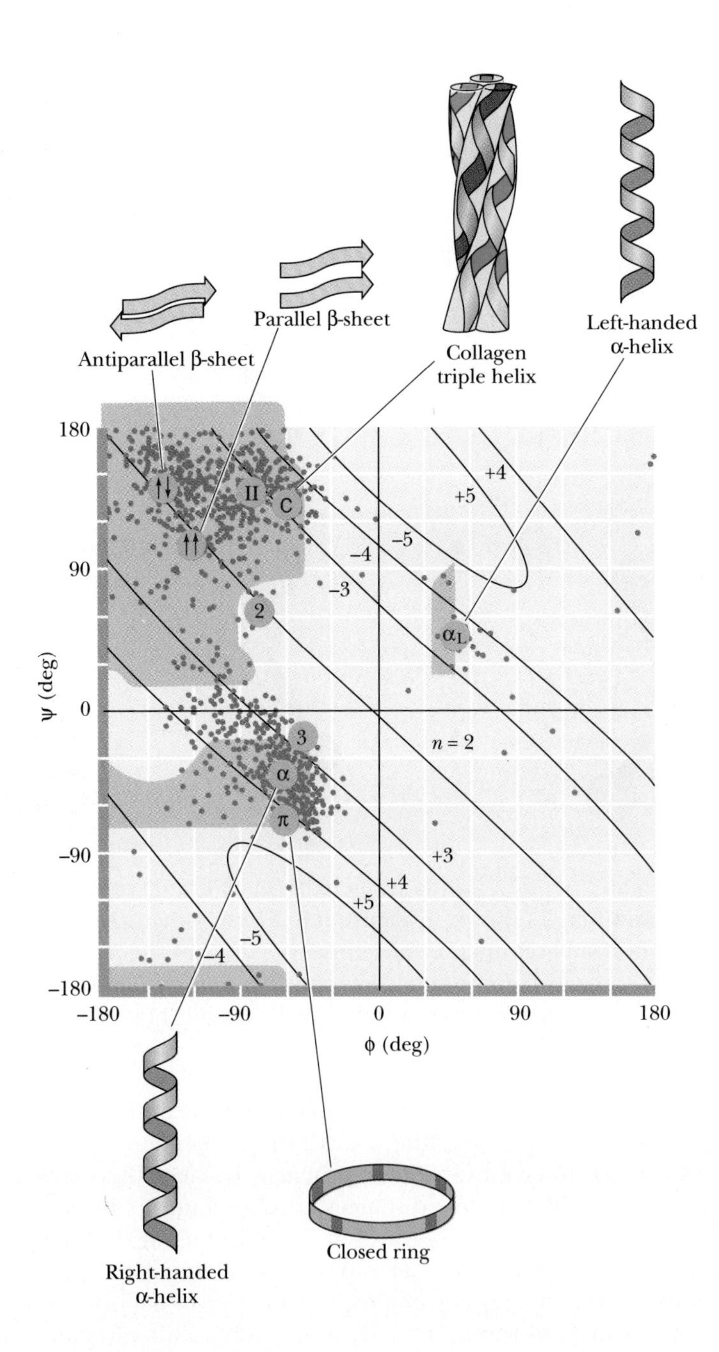

***Figure* 5.5** A Ramachandran diagram showing the sterically reasonable values of the angles ϕ and ψ. The shaded regions indicate particularly favorable values of these angles. Dots in purple indicate actual angles measured for 1000 residues (excluding glycine, for which a wider range of angles is permitted) in eight proteins.

(After Richardson, J. S., 1981, Advances in Protein Chemistry **34**:167–339.)

G. N. Ramachandran and his co-workers in Madras, India, first showed that it was convenient to plot ϕ values against ψ values to show the distribution of allowed values in a protein or in a family of proteins. A typical **Ramachandran plot** is shown in Figure 5.5. Note the clustering of ϕ and ψ values in a few regions of the plot. Most combinations of ϕ and ψ are sterically forbidden, and the corresponding regions of the Ramachandran plot are sparsely populated. The combinations that are sterically allowed represent the subclasses of structure described in the remainder of this section.

The Alpha-Helix

The discussion of hydrogen bonding in Section 5.1 pointed out that the carbonyl oxygen and amide hydrogen of the peptide bond could participate in H bonds either with water molecules in the solvent or with other H-bonding groups in the peptide chain. In nearly all proteins, the carbonyl oxygens and

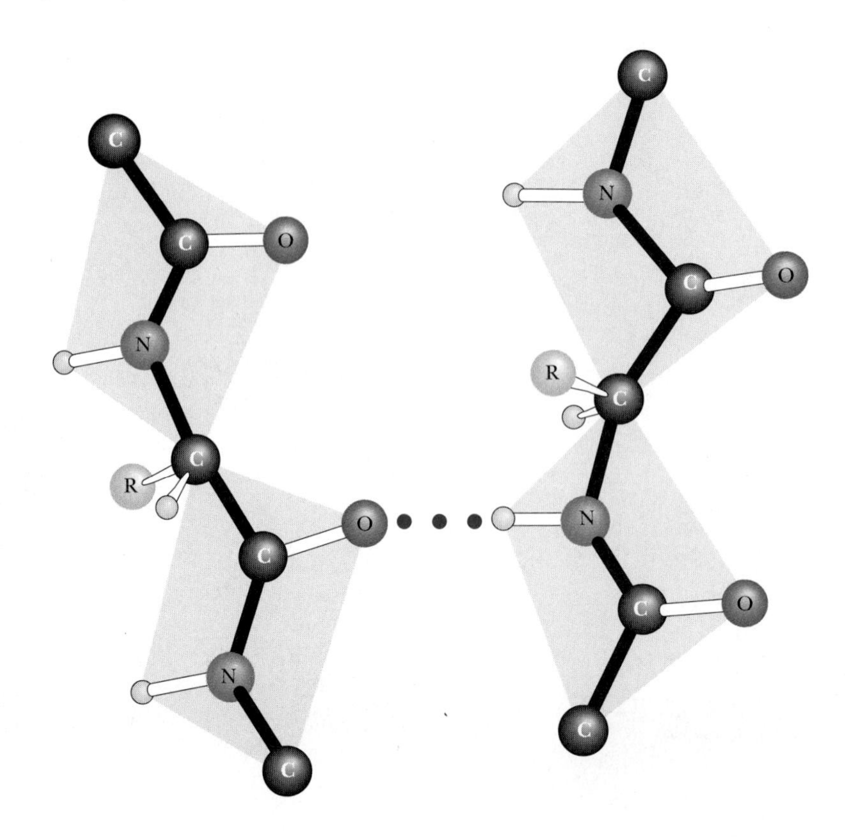

***Figure* 5.6** A hydrogen bond between the amide proton and carbonyl oxygen of adjacent peptide groups.

the amide protons of many peptide bonds participate in H bonds that link one peptide group to another, as shown in Figure 5.6. These structures tend to form in cooperative fashion and involve substantial portions of the peptide chain. Structures resulting from these interactions constitute **secondary structure** for proteins (see Chapter 4). When a number of hydrogen bonds form between portions of the peptide chain in this manner, two basic types of structures can result: *α-helices* and *β*-pleated *sheets.*

Evidence for helical structures in proteins was first obtained in the 1930s in studies of fibrous proteins. However, there was little agreement at that time, about the exact structure of these helices, primarily because there was also lack of agreement about interatomic distances and bond angles in peptides. In 1951, Linus Pauling, Robert Corey, and their colleagues at the California Institute of Technology summarized a large volume of crystallographic data in a set of dimensions for polypeptide chains. (A summary of data similar to what they reported is shown in Figure 4.2.) With these data in hand, Pauling, Corey, and their colleagues proposed a new model for a helical structure in proteins, which they called the **α-helix.** The report from Caltech was of particular interest to Max Perutz in Cambridge, England, a crystallographer who was also interested in protein structure. By taking into account a critical but previously ignored feature of the X-ray data, Perutz realized that the α-helix existed in keratin, a protein from hair, and also in several other proteins. Since then, the α-helix has proved to be a fundamentally important peptide structure. Several representations of the α-helix are shown in Figure 5.7. One turn of the helix represents 3.6 amino acid residues. (A single turn of the α-helix involves 13 atoms from the O to the H of the H bond. For this reason, the α-helix is sometimes referred to as the 3.6_{13} helix.) This is in fact the feature that most confused crystallographers before the Pauling and Corey α-helix model. Crystallographers were so accustomed to finding twofold, threefold, sixfold, and similar integral axes in simpler molecules that the notion of a nonintegral number of units per turn was never taken seriously before Pauling and Corey's work.

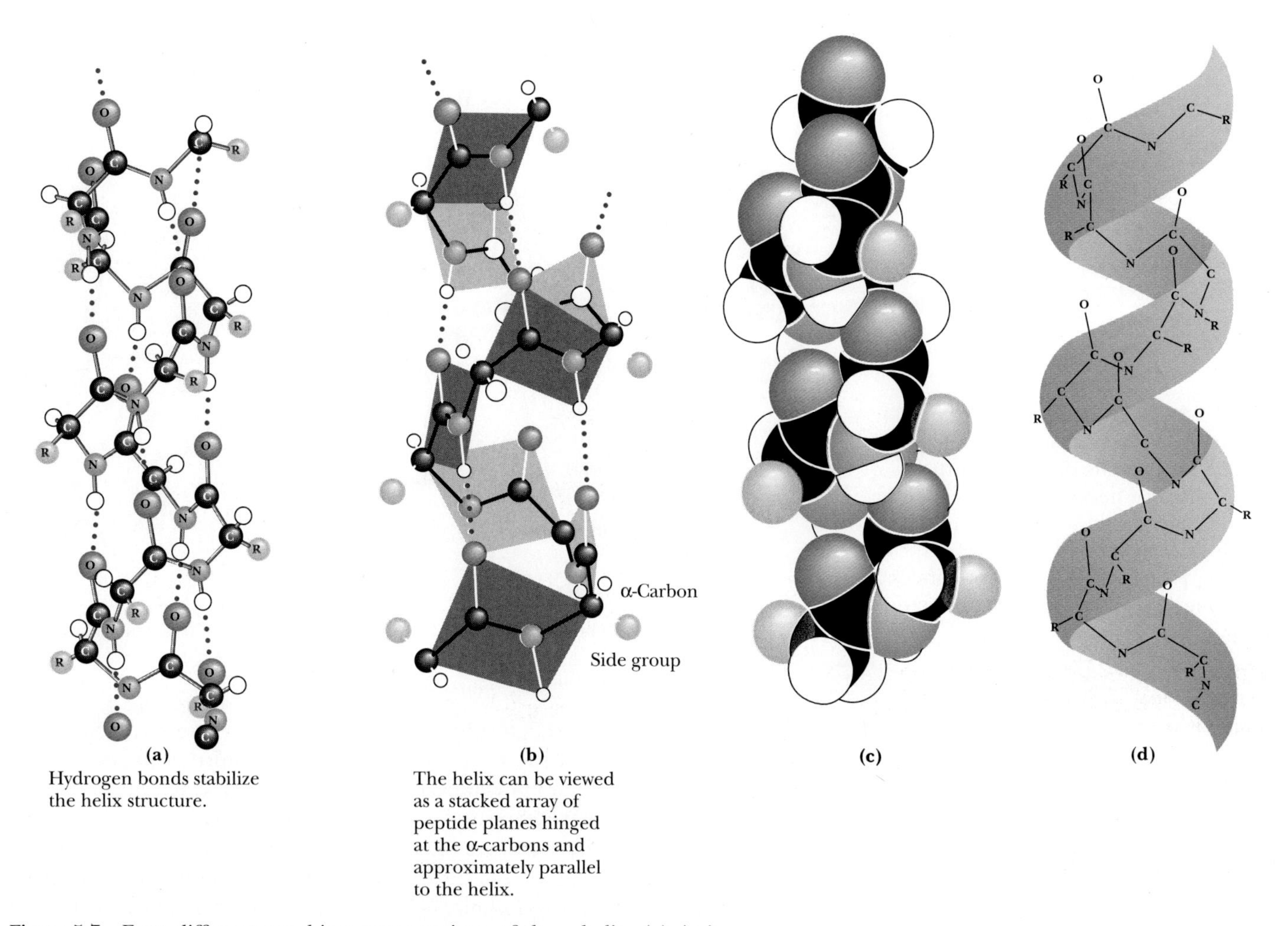

Hydrogen bonds stabilize the helix structure.

The helix can be viewed as a stacked array of peptide planes hinged at the α-carbons and approximately parallel to the helix.

***Figure* 5.7** Four different graphic representations of the α-helix. (a) As it originally appeared in Pauling's 1960 *The Nature of the Chemical Bond.* (b) Showing the arrangement of peptide planes in the helix. (c) A space-filling computer graphic presentation. (d) A "ribbon structure" with an inlaid stick figure, showing how the ribbon indicates the path of the polypeptide backbone.

Each amino acid residue extends 1.5 Å (0.15 nm) along the helix axis. With 3.6 residues per turn, this amounts to 3.6 × 1.5 Å or 5.4 Å (0.54 nm) of travel along the helix axis per turn. This is referred to as the translation distance or the **pitch** of the helix. If one ignores side chains, the helix is about 6 Å in diameter. The side-chains, extending outward from the core structure of the helix, are removed from steric interference with the polypeptide backbone. As can be seen in Figure 5.7, each peptide carbonyl is hydrogen bonded to the peptide N—H group four residues farther up the chain. Note that all of the H bonds lie parallel to the helix axis and that all of the carbonyl groups are pointing in one direction along the helix axis while the N—H groups are pointing in the opposite direction. Recall that the entire path of the peptide backbone can be known if the ϕ and ψ twist angles are specified for each residue. The α-helix is formed if the values of ϕ are approximately $-60°$ and the values of ψ are in the range of -45 to $-50°$. Figure 5.8 shows the structures of several proteins that contain α-helical segments. The number of residues involved in a given α-helix varies from helix to helix and from protein to protein. On average, there are about 10 residues per helix. Myoglobin, one of the first proteins in which α-helices were observed, has eight stretches of α-helix that form a box to contain the heme prosthetic group. The structures of the α and β subunits of hemoglobin are strikingly similar, with only a few

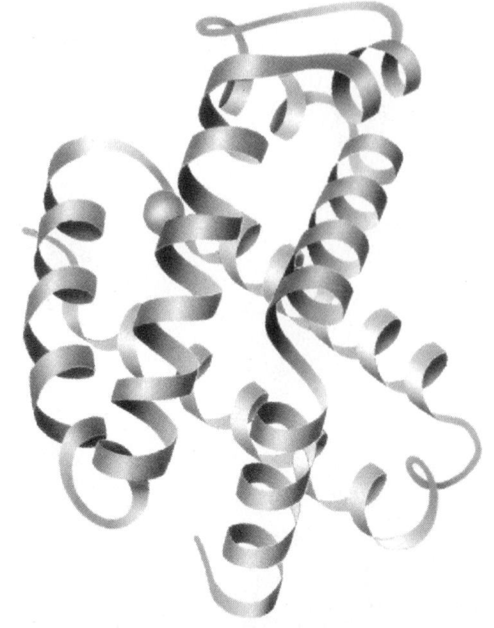

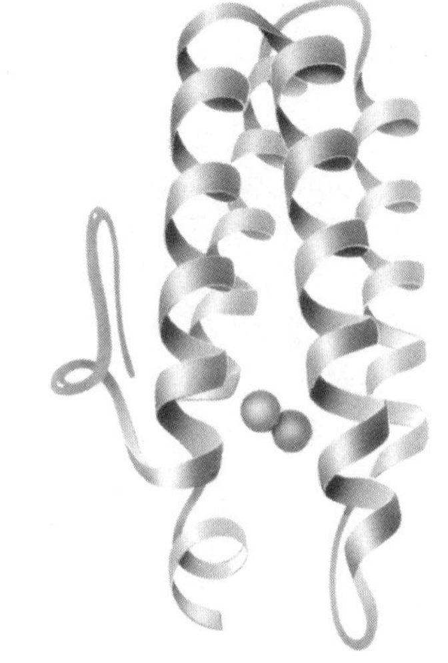

Figure **5.8** The three-dimensional structures of two proteins that contain substantial amounts of α-helix in their structures. The helices are represented by the regularly coiled sections of the ribbon drawings. Myohemerythrin is the oxygen-carrying protein in certain invertebrates, including *Sipunculids,* a phylum of marine worm.

differences at the C- and N-termini and on the surfaces of the structure that contact or interact with the other subunits of this multisubunit protein.

Careful studies of the **polyamino acids,** polymers in which all the amino acids are identical, have shown that certain amino acids tend to occur in α-helices, whereas others are less likely to be found in them. Polyleucine and polyalanine, for example, readily form α-helical structures. In contrast, polyaspartic acid and polyglutamic acid, which are highly negatively charged at pH 7.0, form only random structures because of strong charge repulsion between the R groups along the peptide chain. At pH 1.5 to 2.5, however, where the side chains are protonated and thus uncharged, these latter species spontaneously form α-helical structures. In similar fashion, polylysine is a random coil at pH values below about 11, where repulsion of positive charges prevents helix formation. At pH 12, where polylysine is a neutral peptide chain, it readily forms an α-helix.

Table **5.1**

Helix-Forming and Helix-Breaking Behavior of the Amino Acids

Amino Acid		Helix Behavior*	
A	Ala	H	(I)
C	Cys	Variable	
D	Asp	Variable	
E	Glu	H	
F	Phe	H	
G	Gly	I	(B)
H	His	H	(I)
I	Ile	H	(C)
K	Lys	Variable	
L	Leu	H	
M	Met	H	
N	Asn	C	(I)
P	Pro	B	
Q	Gln	H	(I)
R	Arg	H	(I)
S	Ser	C	(B)
T	Thr	Variable	
V	Val	Variable	
W	Trp	H	(C)
Y	Tyr	H	(C)

*H = helix former; I = indifferent; B = helix breaker; C = random coil; ( ) = secondary tendency.

The tendencies of the amino acids to stabilize or destabilize α-helices are different in typical proteins than in polyamino acids. The occurrence of the common amino acids in helices is summarized in Table 5.1. Notably, proline (and hydroxyproline) act as helix breakers due to their unique structure, which fixes the value of the C_α—N—C bond angle. Helices can be formed from either D- or L-amino acids, but a given helix must be composed entirely of amino acids of one configuration. α-Helices cannot be formed from a mixed copolymer of D- and L-amino acids. An α-helix composed of D-amino acids is left-handed.

Other Helical Structures

There are several other far less common types of helices found in proteins. The most common of these is the 3_{10} helix, which contains 3.0 residues per turn (with 10 atoms in the ring formed by making the hydrogen bond three residues up the chain). It normally extends over shorter stretches of sequence than the α-helix. Other helical structures include the 2_7 ribbon and the π-helix, which has 4.4 residues and 16 atoms per turn and is thus called the 4.4_{16} helix.

Critical Developments in Biochemistry

In Bed with a Cold, Pauling Stumbles onto the α-Helix and a Nobel Prize[1]

As high technology continues to transform the modern biochemical laboratory, it is interesting to reflect on Linus Pauling's discovery of the α-helix. It involved only a piece of paper, a pencil, scissors, and a sick Linus Pauling, who had tired of reading detective novels. The story is told in the excellent book *The Eighth Day of Creation* by Horace Freeland Judson:

> From the spring of 1948 through the spring of 1951 . . . rivalry sputtered and blazed between Pauling's lab and (Sir Lawrence) Bragg's—over protein. The prize was to propose and verify in nature a general three-dimensional structure for the polypeptide chain. Pauling was working up from the simpler structures of components. In January 1948, he went to Oxford as a visiting professor for two terms, to lecture on the chemical bond and on molecular structure and biological specificity. "In Oxford, it was April, I believe, I caught cold. I went to bed, and read detective stories for a day, and got bored, and thought why don't I have a crack at that problem of alpha keratin." Confined, and still fingering the polypeptide chain in his mind, Pauling called for paper, pencil, and straightedge and attempted to reduce the problem to an almost Euclidean purity. "I took a sheet of paper—I still have this sheet of paper—and drew, rather roughly, the way that I thought a polypeptide chain would look if it were spread out into a plane." The repetitious herringbone of the chain he could stretch across the paper as simply as this—
>
> —putting in lengths and bond angles from memory. . . . He knew that the peptide bond, at the carbon-to-nitrogen link, was always rigid:
>
> And this meant that the chain could turn corners only at the alpha carbons. . . . "I creased the paper in parallel creases through the alpha carbon atoms, so that I could bend it and make the bonds to the alpha carbons, along the chain, have tetrahedral value. And then I looked to see if I could form hydrogen bonds from one part of the chain to the next." He saw that if he folded the strip like a chain of paper dolls into a helix, and if he got the pitch of the screw right, hydrogen bonds could be shown to form, N—H· ·O—C, three or four knuckles apart along the backbone, holding the helix in shape. After several tries, changing the angle of the parallel creases in order to adjust the pitch of the helix, he found one where the hydrogen bonds would drop into place, connecting the turns, as straight lines of the right length. He had a model.

[1] The discovery of the α-helix structure was only one of many achievements that led to Pauling's Nobel Prize in chemistry in 1954. The official citation for the prize was "for his research into the nature of the chemical bond and its application to the elucidation of the structure of complex substances."

The Beta-Pleated Sheet

In addition to helices, another type of structure is commonly observed in proteins that forms because of local, cooperative formation of hydrogen bonds. This is the pleated sheet, or β-structure, often called the **β-pleated sheet.** This structure was also first postulated by Pauling and Corey in 1951 and has now been observed in many natural proteins. A β-pleated sheet can be visualized by laying thin, pleated strips of paper side by side to make a "pleated sheet" of paper (Figure 5.9). Each strip of paper can then be pictured as a single peptide strand in which the peptide backbone makes a zigzag pattern along the strip, with the α-carbons lying at the folds of the pleats. The pleated sheet can exist in both parallel and antiparallel forms. In the

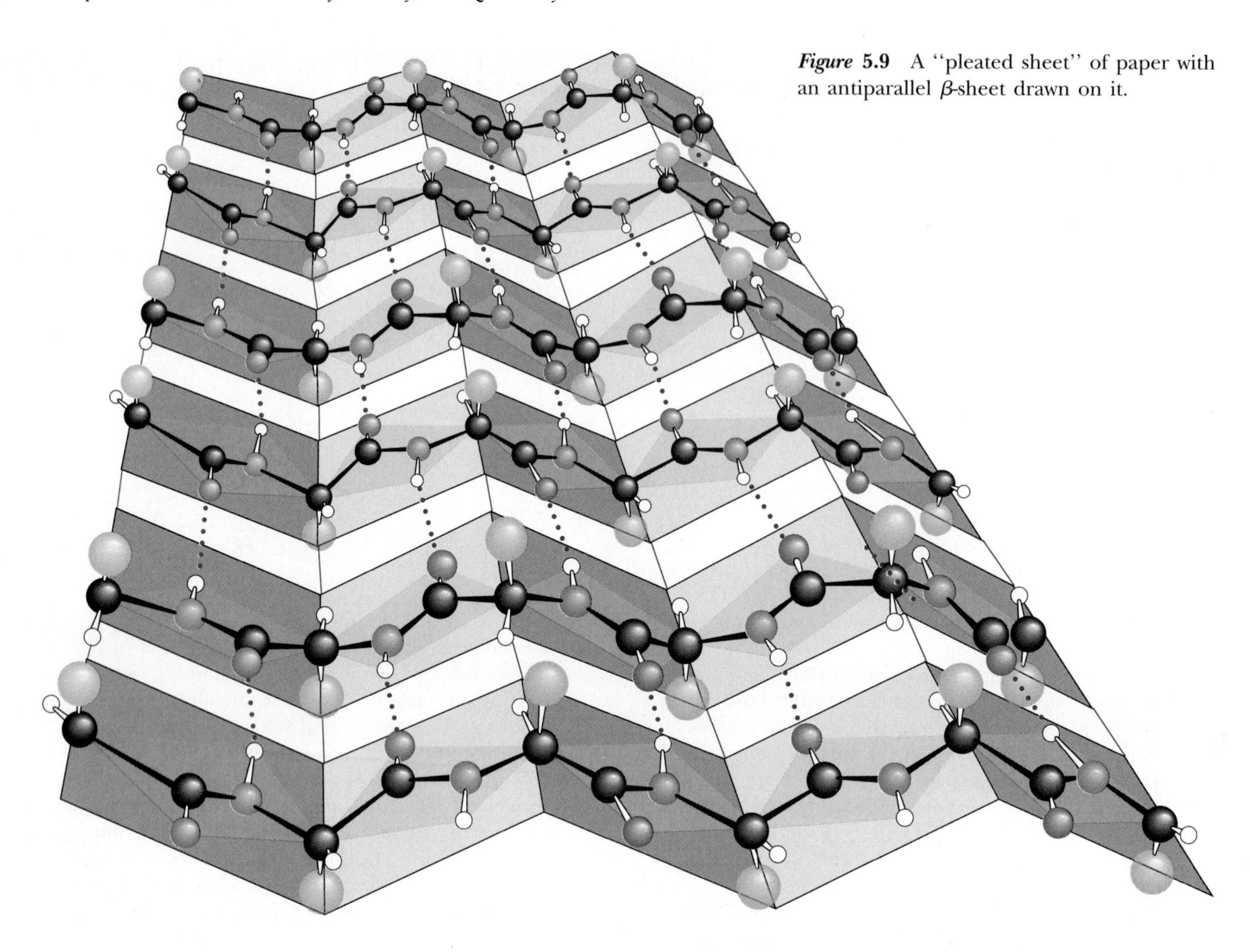

***Figure* 5.9** A "pleated sheet" of paper with an antiparallel β-sheet drawn on it.

parallel β-pleated sheet, adjacent chains run in the same direction (N → C or C → N). In the **antiparallel β-pleated sheet,** adjacent strands run in opposite directions.

Each single strand of the β-sheet structure can be pictured as a twofold helix, that is, a helix with two residues per turn. The arrangement of successive amide planes has a pleated appearance due to the tetrahedral nature of the C_α atom. It is important to note that the hydrogen bonds in this structure are essentially *inter*strand rather than *intra*strand. The peptide backbone in the β-sheet is in its most extended conformation (sometimes called the **ϵ-conformation**). At the same time, the optimum formation of H bonds in the parallel pleated sheet results in a slightly less extended configuration than in the antiparallel sheet. The H bonds thus formed in the parallel β-sheet are bent significantly. The distance between residues is 0.347 nm for the antiparallel pleated sheet, but only 0.325 nm for the parallel pleated sheet. Figure 5.10 shows examples of both parallel and antiparallel β-pleated sheets. Note that the side chains in the pleated sheet are oriented perpendicular or normal to the plane of the sheet, extending out from the plane on alternating sides.

Parallel β-sheets tend to be more regular than antiparallel β-sheets. The range of ϕ and ψ angles for the peptide bonds in parallel sheets is much smaller than that for antiparallel sheets. Parallel sheets are typically large structures; those composed of less than five strands are rare. Antiparallel sheets, however, may consist of as few as two strands. Parallel sheets characteristically distribute hydrophobic side chains on both sides of the sheet, while antiparallel sheets are usually arranged with all their hydrophobic residues on

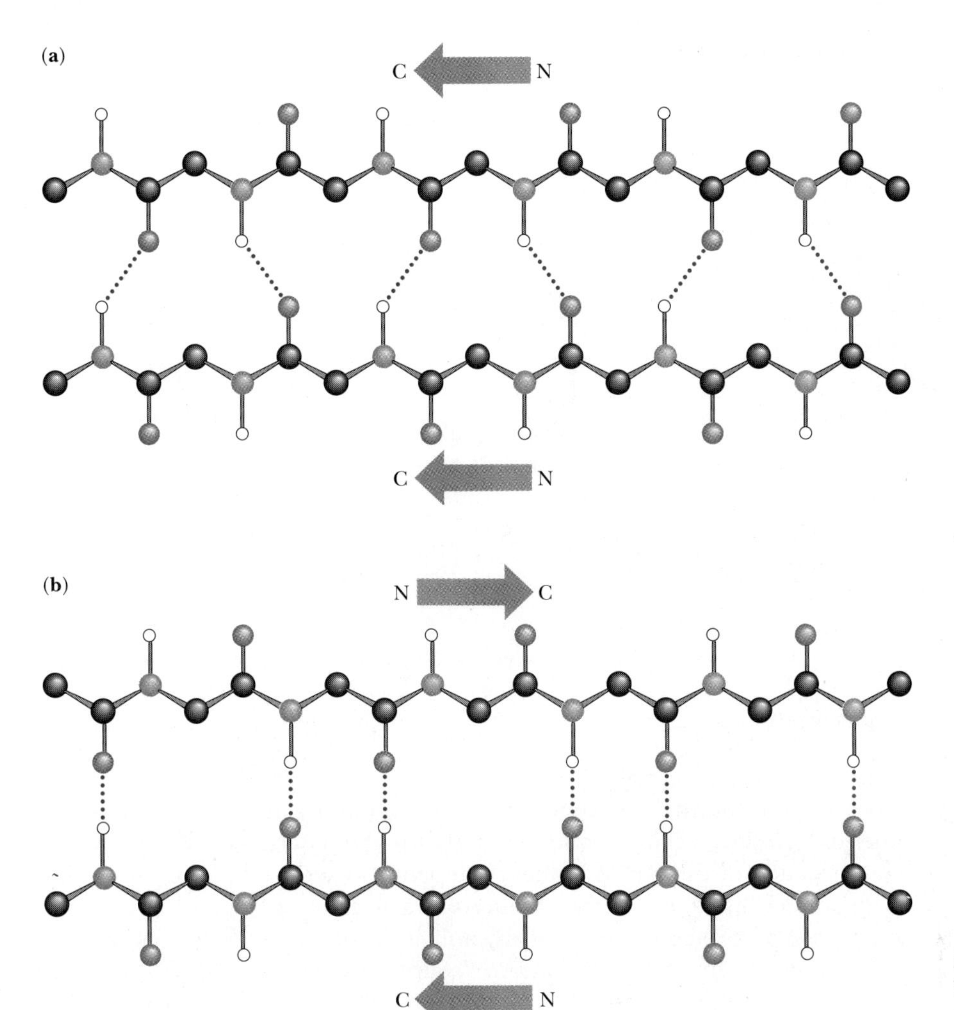

Figure **5.10** The arrangement of hydrogen bonds in (a) parallel and (b) antiparallel β-pleated sheets.

one side of the sheet. This requires an alternation of hydrophilic and hydrophobic residues in the primary structure of peptides involved in antiparallel β-sheets, since alternate side chains project to the same side of the sheet (see Figure 5.9).

Antiparallel pleated sheets are the fundamental structure found in silk, with the polypeptide chains forming the sheets running parallel to the silk fibers. The silk fibers thus formed have properties consistent with those of the β-sheets that form them. They are quite flexible, but cannot be stretched or extended to any appreciable degree. Antiparallel structures are also observed in many other proteins, including immunoglobulin G, superoxide dismutase from bovine erythrocytes, and concanavalin A. Many proteins, including carbonic anhydrase, egg lysozyme, and glyceraldehyde phosphate dehydrogenase, possess both α-helices and β-pleated sheet structures within a single polypeptide chain.

The Beta-Turn

Most proteins are globular structures. The polypeptide chain must therefore possess the capacity to bend, turn, and reorient itself to produce the required compact, globular structures. A simple structure observed in many proteins is the **β-turn** (also known as the *tight turn* or *β-bend*), in which the peptide chain forms a tight loop with the carbonyl oxygen of one residue hydrogen-bonded with the amide proton of the residue three positions down the chain. This H bond makes the β-turn a relatively stable structure. As shown in Figure 5.11, the β-turn allows the protein to reverse the direction of its peptide chain. This

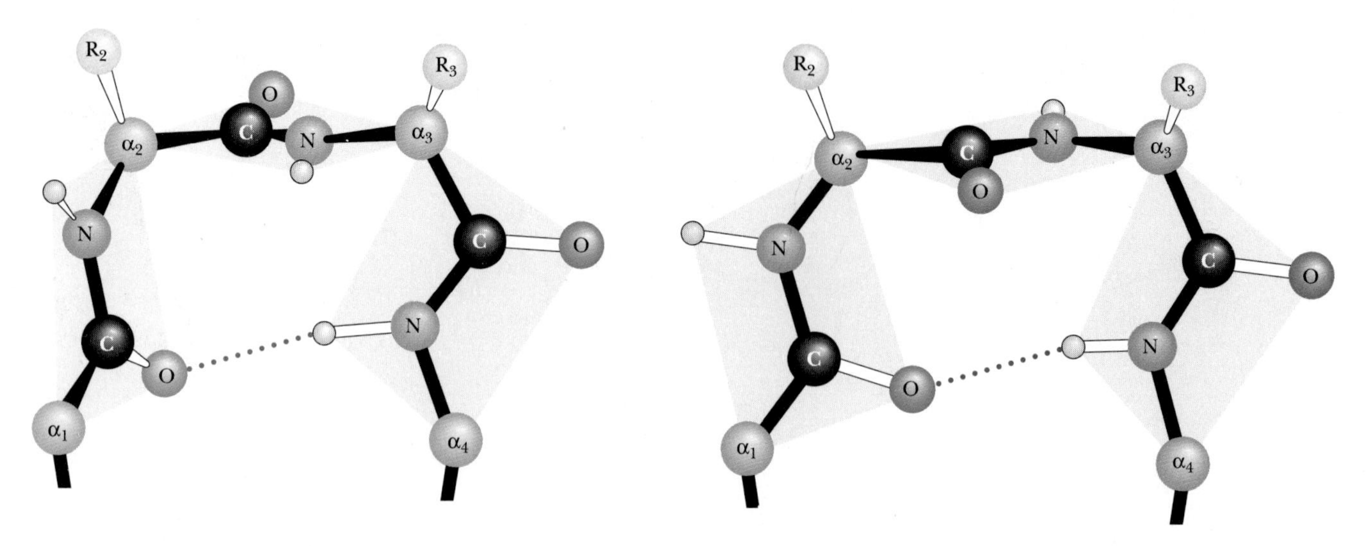

***Figure* 5.11** The structures of two kinds of β-turns (also called tight turns or β-bends)
(Figure © Irving Geis.)

figure shows the two major types of β-turns, but a number of less common types are also found in protein structures. Certain amino acids, such as proline and glycine, occur frequently in β-turn sequences, and the particular conformation of the β-turn sequence depends to some extent on the amino acids composing it. Due to the absence of a side chain, glycine is sterically the most adaptable of the amino acids, and it accommodates conveniently to other steric constraints in the β-turn. Proline, however, has a cyclic structure and a fixed ϕ angle, so, to some extent, it forces the formation of a β-turn, and in many cases this facilitates the turning of a polypeptide chain upon itself. Such bends promote formation of antiparallel β-pleated sheets.

The Beta-Bulge

One final secondary structure, the **β-bulge,** is a small piece of nonrepetitive structure that can occur by itself, but most often occurs as an irregularity in antiparallel β-structures. A β-bulge occurs between two normal β-structure hydrogen bonds and comprises two residues on one strand and one residue on the opposite strand. Figure 5.12 illustrates typical β-bulges. The extra residue on the longer side, which causes additional backbone length, is accommodated partially by creating a bulge in the longer strand and partially by forcing a slight bend in the β-sheet. Bulges thus cause changes in the direction of the polypeptide chain, but to a lesser degree than tight turns do. Over 100 examples of β-bulges are known in protein structures.

The secondary structures we have described here are all found commonly in proteins in nature. In fact, it is hard to find proteins that do not contain one or more of these structures. The energetic (mostly H bond) stabilization afforded by α-helices, β-pleated sheets, and β-turns is important to proteins, and they seize the opportunity to form such structures wherever possible.

5.4 Protein Folding and Tertiary Structure

The folding of a single polypeptide chain in three-dimensional space is referred to as its **tertiary structure.** As discussed in Section 5.2, all of the information needed to fold the protein into its native tertiary structure is contained within the primary structure of the peptide chain itself. With this in mind, it was disappointing to the biochemists of the 1950s when the early

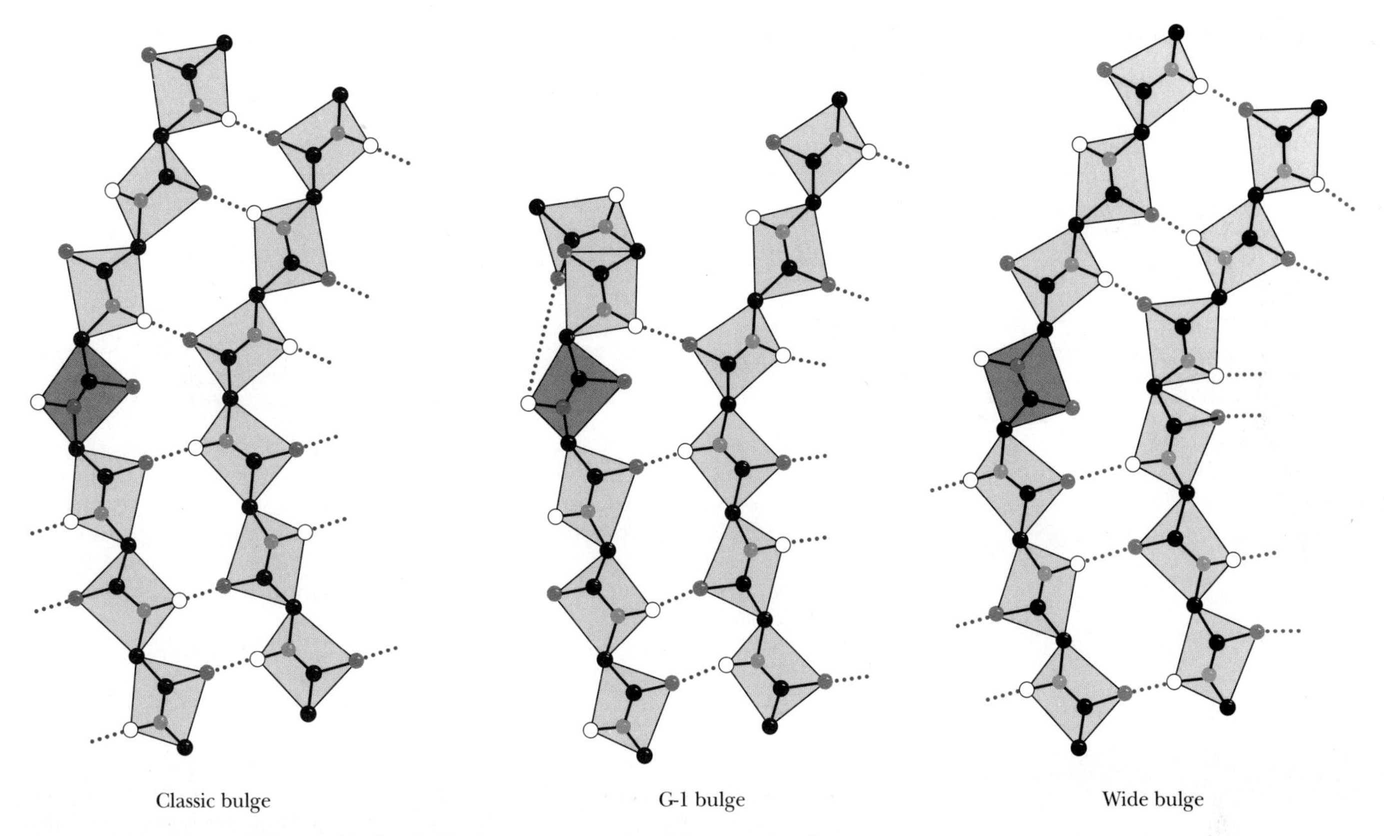

***Figure* 5.12** Three different kinds of β-bulge structures involving a pair of adjacent polypeptide chains.
(Adapted from Richardson, J. S., 1981. Advances in Protein Chemistry 34:167–339.)

protein structures did not reveal the governing principles in any particular detail. It soon became apparent that the proteins knew how they were supposed to fold into tertiary shapes, even if the biochemists did not. Vigorous work in many laboratories has slowly brought important principles to light.

First, secondary structures—helices and sheets—will form whenever possible as a consequence of the formation of large numbers of hydrogen bonds. Second, α-helices and β-sheets often associate and pack close together in the protein. No protein is stable as a single-layer structure, for reasons that will become apparent later. There are a few common methods for such packing to occur. Third, since the peptide segments between secondary structures in the protein tend to be short and direct, the peptide does not execute complicated twists and knots as it moves from one region of a secondary structure to another. A consequence of these three principles is that protein chains are usually folded so that the secondary structures are arranged in one of a few common patterns. For this reason, there are families of proteins that have similar tertiary structure, with little apparent evolutionary or functional relationship among them. Finally, proteins generally fold so as to form the most stable structures possible. The stability of most proteins arises from (1) the formation of large numbers of intramolecular hydrogen bonds and (2) the reduction in the surface area accessible to solvent that occurs upon folding.

Fibrous Proteins

In Chapter 4, we saw that proteins can be grouped into two large classes based on their three-dimensional structure: *fibrous proteins* and *globular proteins.* Fibrous proteins contain polypeptide chains organized approximately parallel along a single axis, producing long fibers or large sheets. Such proteins tend to be mechanically strong and resistant to solubilization in water and dilute salt solutions. Fibrous proteins often play a structural role in nature (see Chapter 4).

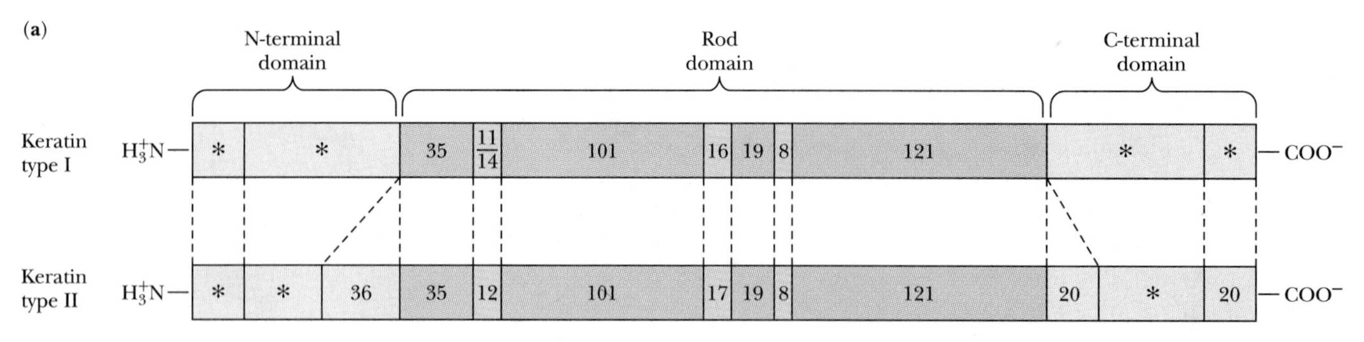

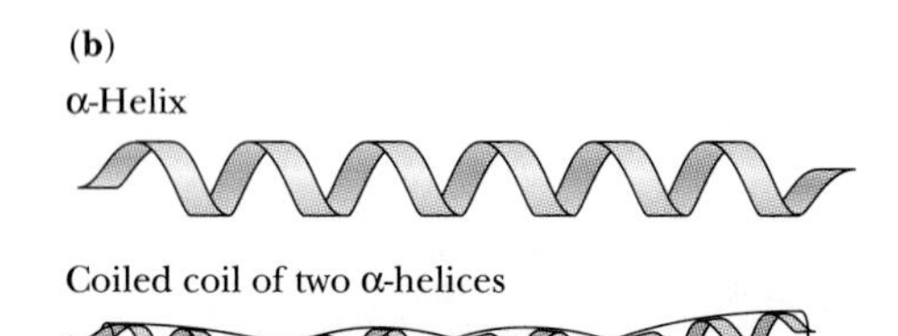

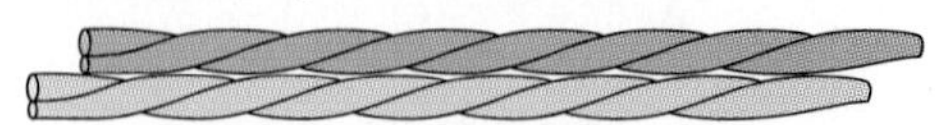

***Figure* 5.13** (a) Both type I and type II α-keratin molecules have sequences consisting of long, central rod domains with terminal cap domains. The numbers of amino acid residues in each domain are indicated. Asterisks denote domains of variable length. (b) The rod domains form coiled coils consisting of intertwined right-handed α-helices. These coiled coils then wind around each other in a left-handed twist. Keratin filaments consist of twisted protofibrils (each a bundle of four coiled coils).

(Adapted from Steinert, P., and Parry, D., 1985. Annual Review of Cell Biology 1:41–65 and Cohlberg, J., 1993. Trends in Biochemical Sciences 18:360–362.)

α-Keratin

As their name suggests, the structure of the ***α*-keratins** is dominated by α-helical segments of polypeptide. The amino acid sequence of α-keratin subunits is composed of central α-helix-rich rod domains about 311 to 314 residues in length, flanked by nonhelical N- and C-terminal domains of varying size and composition (Figure 5.13a). The structure of the central rod domain of a typical α-keratin is shown in Figure 5.13b. It consists of four helical strands arranged as twisted pairs of two-stranded **coiled coils.** X-ray diffraction patterns show that these structures resemble α-helices, but with a pitch of 0.51 nm rather than the expected 0.54 nm. This is consistent with a tilt of the helix relative to the long axis of the fiber, as in the two-stranded "rope" in Figure 5.13.

The primary structure of the central rod segments of α-keratin consists of quasi-repeating seven-residue segments of the form $(a\text{-}b\text{-}c\text{-}d\text{-}e\text{-}f\text{-}g)_n$. These units are not true repeats, but residues *a* and *d* are usually nonpolar amino acids. In α-helices, with 3.6 residues per turn, these nonpolar residues are arranged in an inclined row or stripe that twists around the helix axis. These nonpolar residues would make the helix highly unstable if they were exposed to solvent, but the association of hydrophobic strips on two coiled coils to form the two-stranded rope effectively buries the hydrophobic residues and forms a highly stable structure (Figure 5.13). The helices clearly sacrifice some stability in assuming this twisted conformation, but they gain stabilization energy from the packing of side chains between the helices. In addition to these interactions, covalent disulfide bonds form between cysteine residues of adjacent fibers, making the overall structure rigid, inextensible, and insoluble—important properties for structures such as claws, fingernails, hair, and horns in animals. How and where these disulfides form determines the amount of curling in hair and wool fibers. When a hairstylist creates a permanent wave (simply called a "permanent") in a hair salon, disulfides in the hair are first reduced and cleaved, then reorganized and reoxidized to change the degree of curl or wave. In contrast, a "set" that is created by wetting the hair, setting it with curlers, and then drying it represents merely a rearrangement of the hydrogen bonds between helices and between fibers. (On humid or rainy days, the hydrogen bonds in curled hair may rearrange, and the hair will become "frizzy.")

Fibroin and β-Keratin: β-Sheet Proteins

The **fibroin** proteins found in silk fibers represent another type of fibrous protein. These are composed of stacked antiparallel β-sheets, as shown in Figure 5.14. In the polypeptide sequence of silk proteins, there are large stretches in which every other residue is a glycine. As previously mentioned, the residues of a β-sheet extend alternately above and below the plane of the sheet. As a result, the glycines all end up on one side of the sheet and the other residues (mainly alanines and serines) compose the opposite surface of the sheet. Pairs of β-sheets can then pack snugly together (glycine surface to glycine surface or alanine–serine surface to alanine–serine surface). The β-keratins found in bird feathers are also made up of stacked β-sheets.

Collagen: A Triple Helix

Collagen is a rigid, inextensible fibrous protein that is a principal constituent of connective tissue in animals, including tendons, cartilage, bones, teeth, skin, and blood vessels. The high tensile strength of collagen fibers in these structures makes possible the various animal activities such as running and jumping that put severe stresses on joints and skeleton. Broken bones and tendon and cartilage injuries to knees, elbows, and other joints involve tears or hyperextensions of the collagen matrix in these tissues.

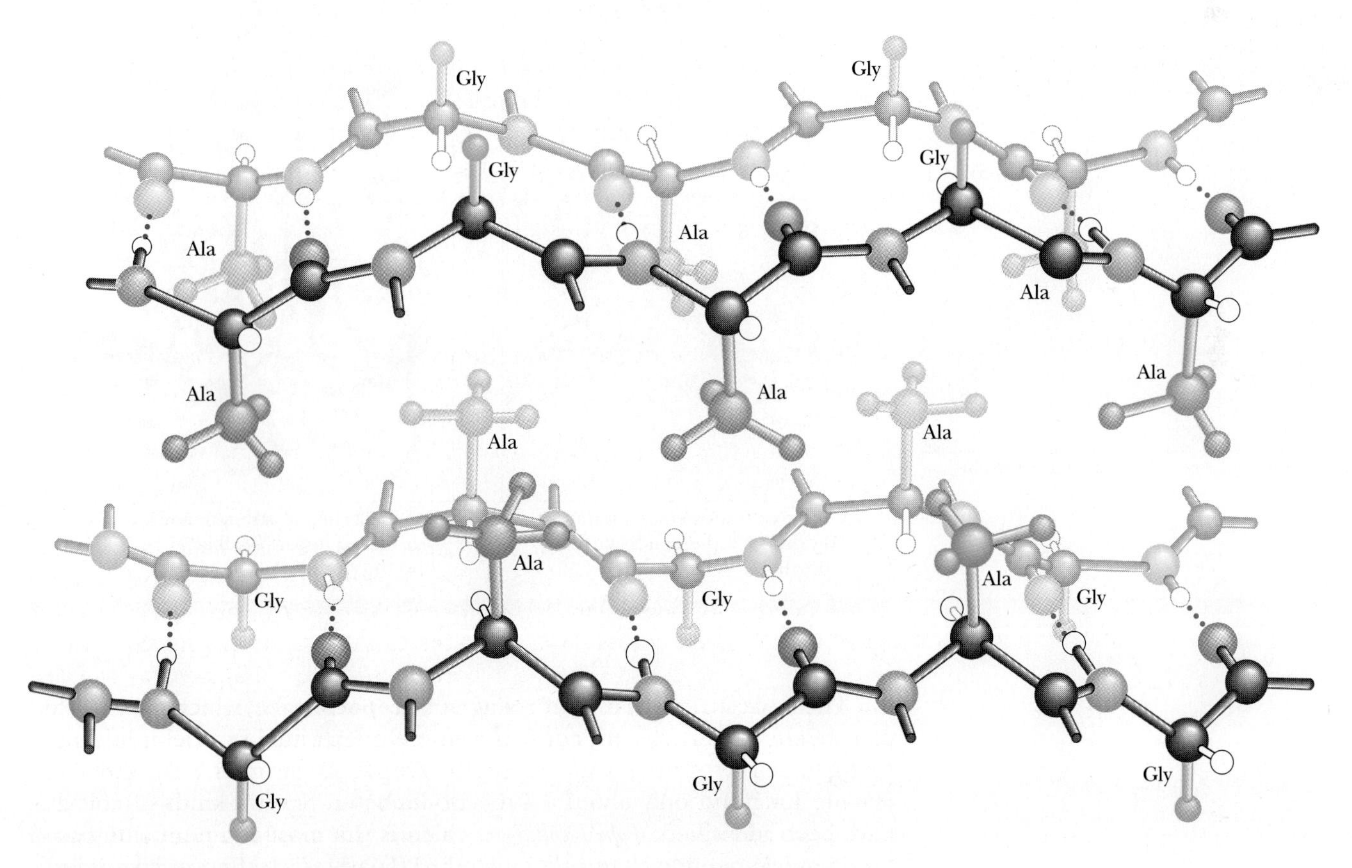

***Figure* 5.14** Silk fibroin consists of a unique stacked array of β-sheets. The primary structure of fibroin molecules consists of long stretches of alternating glycine and alanine or serine residues. When the sheets stack, the more bulky alanine and serine residues on one side of a sheet interdigitate with similar residues on an adjoining sheet. Glycine hydrogens on the alternating faces interdigitate in a similar manner, but with a smaller intersheet spacing.

A Deeper Look

Charlotte's Web Revisited: Helix–Sheet Composites in Spider Dragline Silk

E. B. White's endearing story *Charlotte's Web* centers around the web-spinning feats of Charlotte the spider. Though the intricate designs of spiderwebs are eye- (and fly-) catching, it might be argued that the composition of web silk itself is even more remarkable. Spider silk is synthesized in special glands in the spider's abdomen. The silk strands produced by these glands are both strong and elastic. *Dragline silk* (that from which the spider hangs) has a tensile strength of 200,000 psi (pounds per square inch)—stronger than steel and similar to Kevlar, the synthetic material used in bulletproof vests! This same silk fiber is also flexible enough to withstand strong winds and other natural stresses.

This combination of strength and flexibility derives from the *composite nature* of spider silk. As keratin protein is extruded from the spider's glands, it endures shearing forces that break the H bonds stabilizing keratin α-helices. These regions then form microcrystalline arrays of β-sheets. These microcrystals are surrounded by the keratin strands, which adopt a highly disordered state composed of α-helices and random coil structures.

The β-sheet microcrystals contribute strength, and the disordered array of helix and coil make the silk strand flexible. The resulting silk strand resembles modern human-engineered composite materials. Certain tennis racquets, for example, consist of fiberglass polymers impregnated with microcrystalline graphite. The fiberglass provides flexibility, and the graphite crystals contribute strength. Modern high technology, for all its sophistication, is merely imitating nature—and Charlotte's web—after all.

(a) Spider web

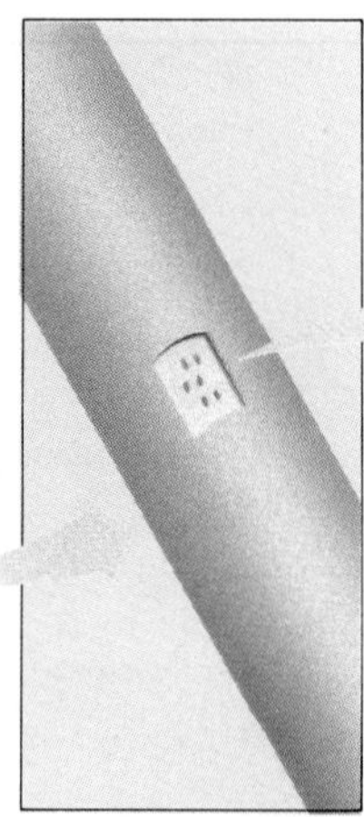

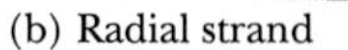

(b) Radial strand

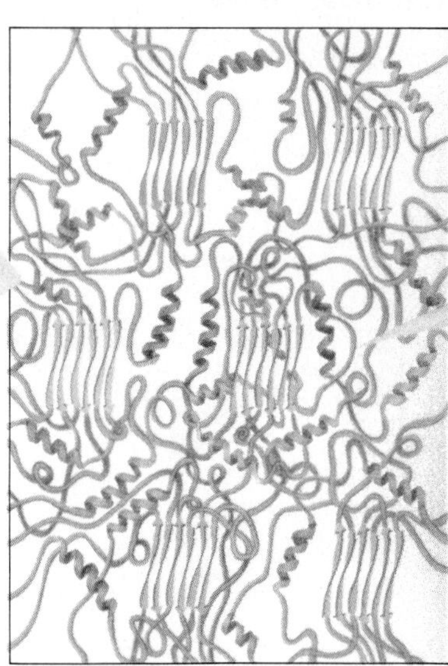

(c) Ordered β-sheets surrounded by disordered α-helices and β-bends.

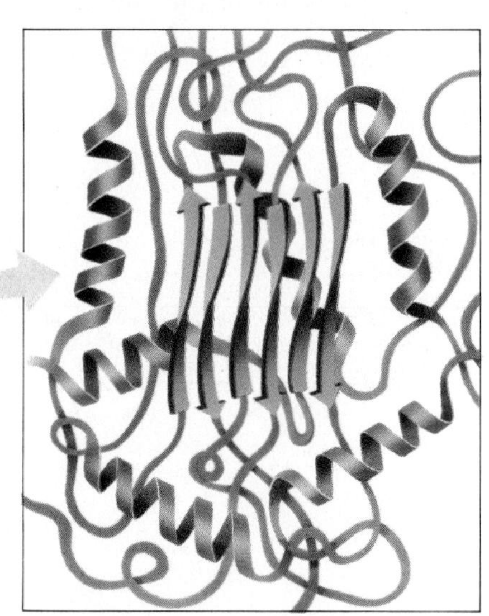

(d) β-sheets impart strength and α-helices impart flexibility to the strand.

The basic structural unit of collagen is **tropocollagen,** which has a molecular weight of 285,000 and consists of three intertwined polypeptide chains, each about 1000 amino acids in length. Tropocollagen molecules are about 300 nm long and only about 1.4 nm in diameter. Several kinds of collagen have been identified. *Type I collagen,* which is the most common, consists of two identical peptide chains designated $\alpha 1(\mathrm{I})$ and one different chain designated $\alpha 2(\mathrm{I})$. Type I collagen predominates in bones, tendons, and skin. *Type II collagen,* found in cartilage, and *type III collagen,* found in blood vessels, consist of three identical polypeptide chains.

Collagen has an amino acid composition that is unique and is crucial to its three-dimensional structure and its characteristic physical properties. Nearly one residue out of three is a glycine, and the proline content is also unusually high. Three unusual modified amino acids are also found in colla-

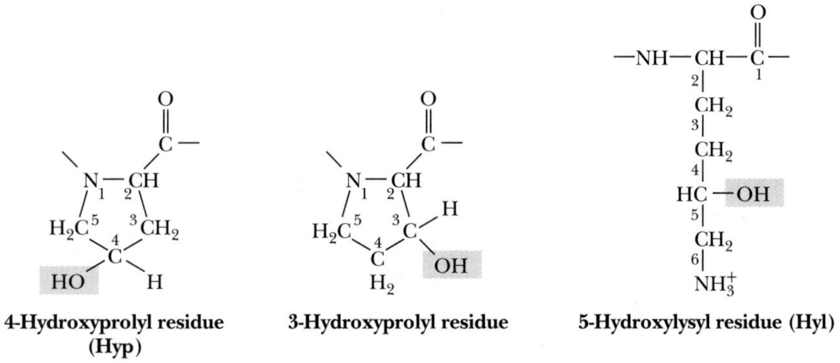

4-Hydroxyprolyl residue (Hyp) **3-Hydroxyprolyl residue** **5-Hydroxylysyl residue (Hyl)**

***Figure* 5.15** The hydroxylated residues typically found in collagen.

gen: 4-hydroxyproline (Hyp), 3-hydroxyproline, and 5-hydroxylysine (Hyl) (Figure 5.15). Proline and Hyp together compose up to 30% of the residues of collagen. Interestingly, these three amino acids are formed from normal proline and lysine *after* the collagen polypeptides are synthesized. The modifications are affected by two enzymes: *prolyl hydroxylase* and *lysyl hydroxylase.* The prolyl hydroxylase reaction (Figure 5.16) requires molecular oxygen, α-ketoglutarate, and ascorbic acid (vitamin C) and is activated by Fe^{2+}. The hydroxylation of lysine is similar. These processes are referred to as **posttranslational modifications** since they occur after genetic information from DNA has been *translated* into newly formed protein.

Because of their high content of glycine, proline, and hydroxyproline, collagen fibers are incapable of forming traditional structures such as α-helices and β-sheets. Instead, collagen polypeptides intertwine to form a unique **triple helix,** with each of the three strands arranged in a helical fash-

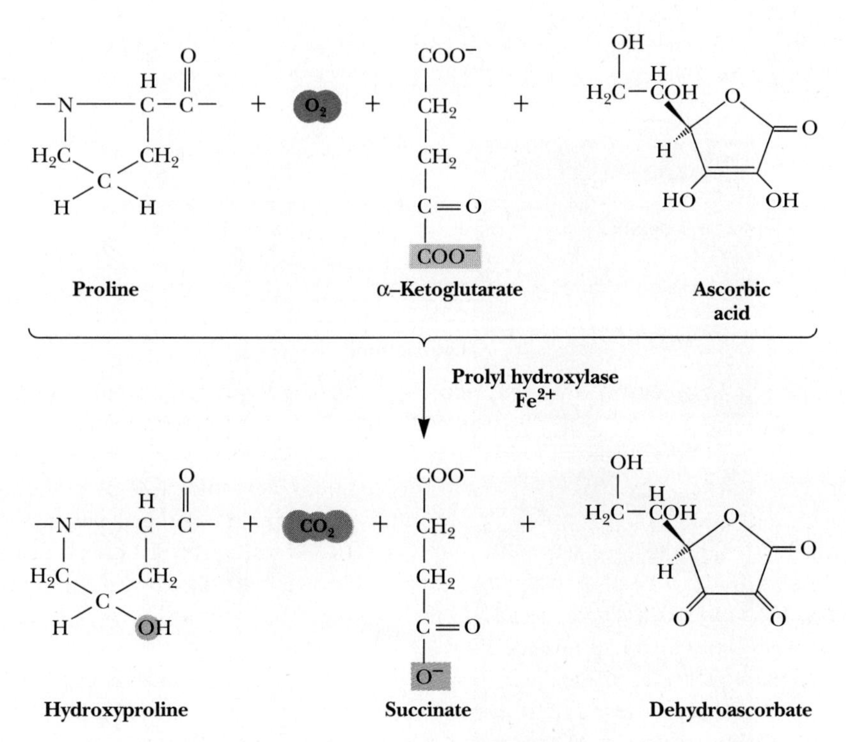

***Figure* 5.16** Hydroxylation of proline residues is catalyzed by prolyl hydroxylase. The reaction requires α-ketoglutarate and ascorbic acid (vitamin C).

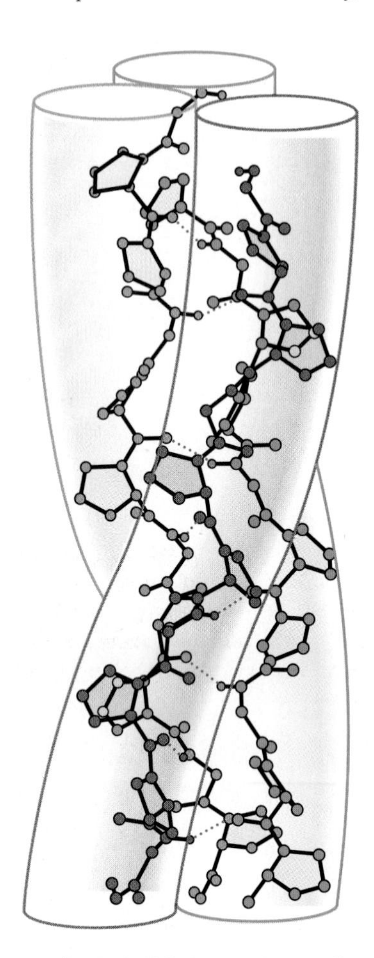

***Figure* 5.17** Poly(Gly-Pro-Pro), a collagen-like right-handed triple helix composed of 3 left-handed helical chains.

(Adapted from Miller, M. H., and Scheraga, H. A., 1976, Calculation of the structures of collagen models. Role of interchain interactions in determining the triple-helical coiled-coil conformation. I. Poly(glycyl-prolyl-prolyl). Journal of Polymer Science Symposium No. 54:171–200.)

ion (Figure 5.17). Compared to the α-helix, the collagen helix is much more extended, with a rise per residue along the triple helix axis of 2.9 Å, compared to 1.5 Å for the α-helix. There are about 3.3 residues per turn of each of these helices. *The triple helix is a structure that forms to accommodate the unique composition and sequence of collagen.* Long stretches of the polypeptide sequence are repeats of a Gly-x-y motif, where x is frequently Pro and y is frequently Pro or Hyp. In the triple helix, every third residue faces or contacts the crowded center of the structure. This area is so crowded that only Gly will fit, and thus every third residue must be a Gly (as observed). Moreover, the triple helix is a *staggered* structure, such that Gly residues from the three strands stack along the center of the triple helix and the Gly from one strand lies adjacent to an x residue from the second strand and to a y from the third. This allows the N—H of each Gly residue to hydrogen bond with the C=O of the adjacent x residue. The triple helix structure is further stabilized and strengthened by the formation of interchain H bonds involving hydroxyproline.

Collagen types I, II, and III form strong, organized **fibrils,** consisting of staggered arrays of tropocollagen molecules (Figure 5.18). The periodic arrangement of triple helices in a head-to-tail fashion results in banded patterns in electron micrographs. The banding pattern typically has a periodicity (repeat distance) of 68 nm. Since collagen triple helices are 300 nm long, this means that there are 40-nm gaps between adjacent collagen molecules in a row along the long axis of the fibrils and that the pattern repeats every five rows (5 × 68 nm = 340 nm). The 40-nm gaps are referred to as *hole regions,* and they are important in at least two ways. First, sugars are found covalently attached to 5-hydroxylysine residues in the hole regions of collagen (Figure 5.19). The occurrence of carbohydrate in the hole region has led to the proposal that it plays a role in organizing fibril assembly. Second, the hole regions may play a role in bone formation. Bone consists of microcrystals of **hydroxyapatite,** $Ca_5(PO_4)_3OH$, imbedded in a matrix of collagen fibrils. When new bone tissue forms, the formation of new hydroxyapatite crystals occurs at intervals of 68 nm. The hole regions of collagen fibrils may be the sites of nucleation for the mineralization of bone.

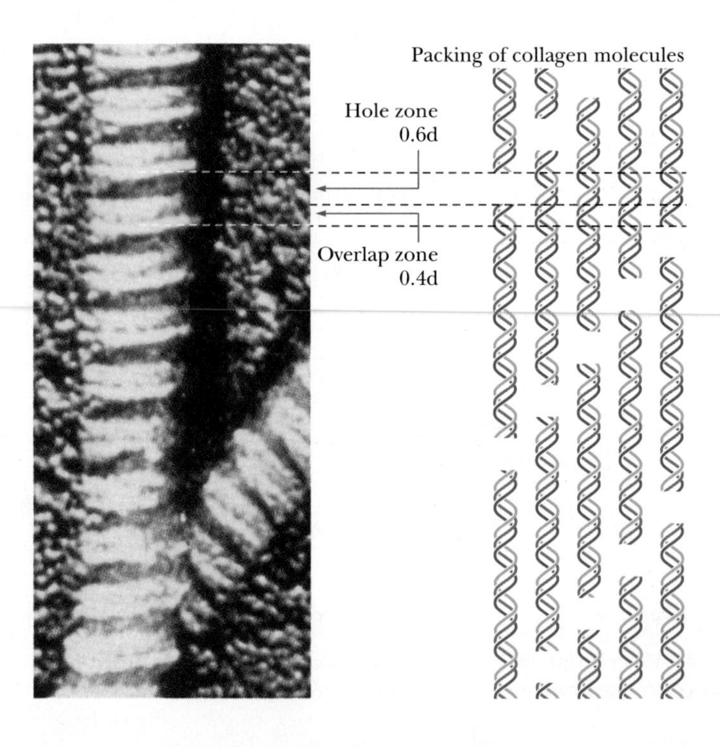

***Figure* 5.18** In the electron microscope, collagen fibers exhibit alternating light and dark bands. The dark bands correspond to the 40-nm gaps or "holes" between pairs of aligned collagen triple helices. The repeat distance, *d*, for the light and dark banded pattern is 68 nm. The collagen molecule is 300 nm long, which corresponds to 4.41*d*. The molecular repeat pattern of five staggered collagen molecules corresponds to 5*d*.

The collagen fibrils are further strengthened and stabilized by the formation of both *intramolecular* (within a tropocollagen molecule) and *intermolecular* (between tropocollagen molecules in the fibril) cross-links. Intramolecular cross-links are formed between lysine residues in the (nonhelical) N-terminal region of tropocollagen in a unique pair of reactions shown in Figure 5.20. The enzyme *lysyl oxidase* catalyzes the formation of aldehyde groups at the lysine side chains in a copper-dependent reaction. The aldehyde groups of two such side chains then link covalently in a spontaneous nonenzymatic *aldol condensation.* The intermolecular cross-linking of tropocollagens involves the formation of a unique **hydroxypyridinium** structure from one lysine and two hydroxylysine residues (Figure 5.21). These cross-links form between the N-terminal region of one tropocollagen and the C-terminal region of an adjacent tropocollagen in the fibril.

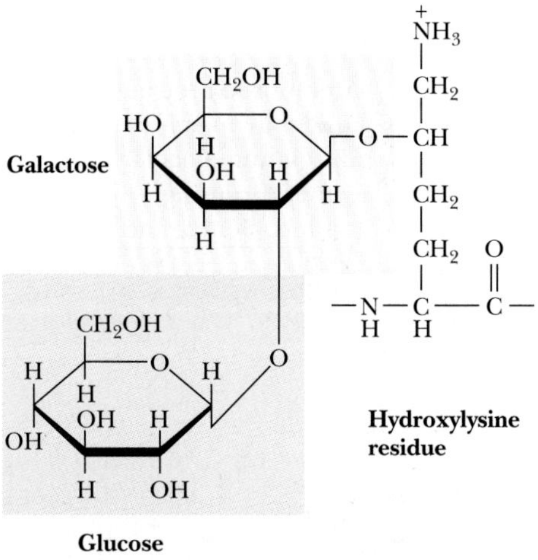

Figure **5.19** A disaccharide of galactose and glucose is covalently linked to the 5-hydroxyl group of hydroxylysines in collagen by the combined action of the enzymes galactosyl transferase and glucosyl transferase.

Collagen-Related Diseases

Collagen provides an ideal case study of the molecular basis of physiology and disease. For example, the nature and extent of collagen cross-linking depends on the age and function of the tissue. Collagen from young animals is predominantly un-cross-linked and can be extracted in soluble form, whereas collagen from older animals is highly cross-linked and thus insoluble. The loss of flexibility of joints with aging is probably due in part to increased cross-linking of collagen.

Several serious and debilitating diseases involving collagen abnormalities are known. **Lathyrism** occurs in animals due to the regular consumption of

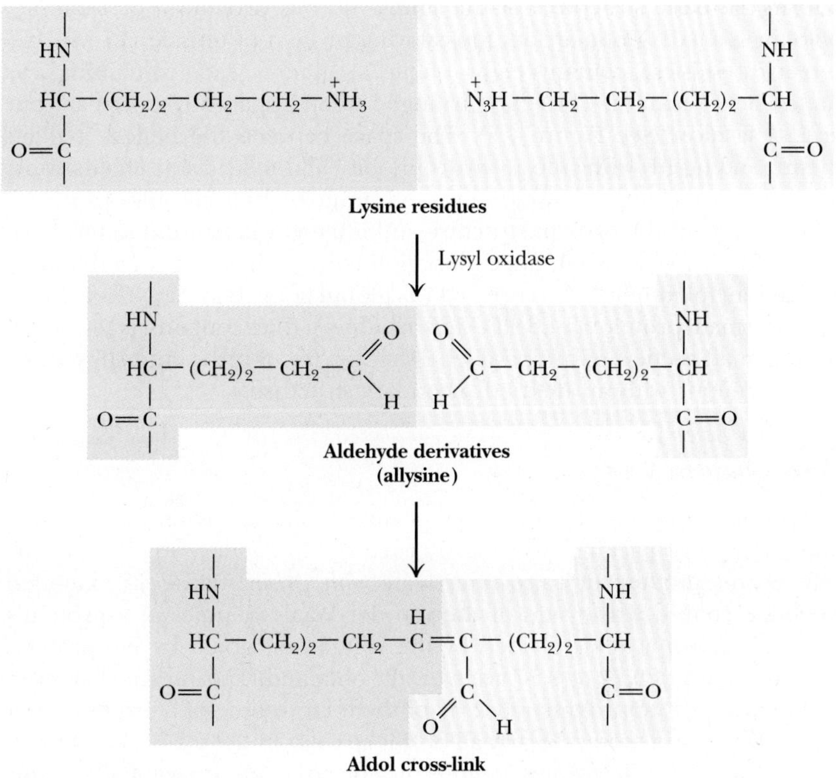

Figure **5.20** Collagen fibers are stabilized and strengthened by Lys—Lys cross-links. Aldehyde moieties formed by lysyl oxidase react in a spontaneous nonenzymatic aldol reaction.

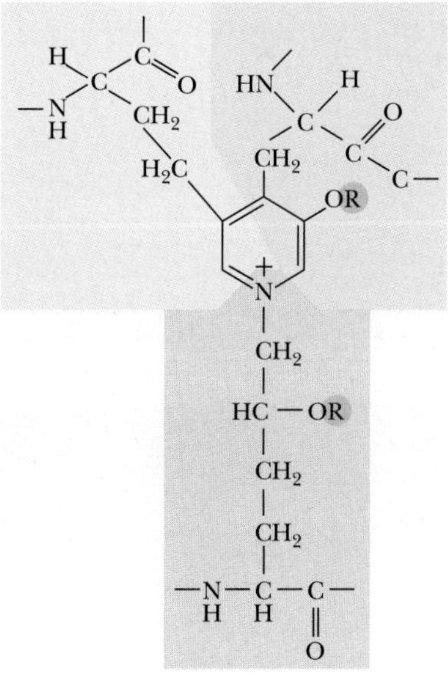

Figure **5.21** The hydroxypyridinium structure formed by the cross-linking of a Lys and two hydroxy Lys residues.

$N \equiv C-CH_2-CH_2-\overset{+}{N}H_3$

***Figure* 5.22** β-Aminopropionitrile (present in sweet peas) covalently inactivates lysyl oxidase, preventing intramolecular cross-linking of collagen and causing abnormalities in joints, bones, and blood vessels.

seeds of *Lathyrus odoratus*, the sweet pea, and involves weakening and abnormalities in blood vessels, joints, and bones. These conditions are caused by **β-aminopropionitrile** (Figure 5.22), which covalently inactivates lysyl oxidase and leads to greatly reduced intramolecular cross-linking of collagen in affected animals (or humans).

Scurvy results from a dietary vitamin C deficiency and involves the inability to form collagen fibrils properly. This is the result of reduced activity of prolyl hydroxylase, which is vitamin C–dependent, as previously noted. Scurvy leads to lesions in the skin and blood vessels, and, in its advanced stages, it can lead to grotesque disfiguration and eventual death. Although rare in the modern world, it was a disease well known to sea-faring explorers in earlier times who did not appreciate the importance of fresh fruits and vegetables in the diet (see the box in Chapter 14).

A number of rare genetic diseases involve collagen abnormalities, including *Marfan's syndrome* and the *Ehlers–Danlos syndromes*, which result in hyperextensible joints and skin. The formation of *atherosclerotic plaques*, which cause arterial blockages in advanced stages, is due in part to the abnormal formation of collagenous structures in blood vessels.

Globular Proteins

Fibrous proteins, while interesting for their structural properties, represent only a small percentage of the proteins found in nature. **Globular proteins,** so named for their approximately spherical shape, are far more numerous.

Helices and Sheets in Globular Proteins

Globular proteins exist in an enormous variety of three-dimensional structures, but nearly all contain substantial amounts of the α-helices and β-sheets that form the basic structures of the simple fibrous proteins. For example, myoglobin, a small, globular, oxygen-carrying protein of muscle (17 kD, 153 amino acid residues), contains eight α-helical segments, each containing 7 to 26 amino acid residues. These are arranged in an apparently irregular (but invariant) fashion (see Figure 4.7). The space between the helices is filled efficiently and tightly with (mostly hydrophobic) amino acid side chains. Most of the polar side chains in myoglobin (and in most other globular proteins) face the outside of the protein structure and interact with solvent water. Myoglobin's structure is unusual, since most globular proteins contain a relatively small amount of α-helix. A more typical globular protein (Figure 5.23) is *lysozyme,* a small protein (14.6 kD, 129 residues) that contains a few short helices, a small section of antiparallel β-sheet, a few β-turns, and substantial peptide segments without defined secondary structure.

Packing Considerations

The secondary and tertiary structures of myoglobin and lysozyme illustrate the importance of packing in tertiary structures. Secondary structures pack closely to one another and also intercalate with (insert between) extended polypeptide chains. If the sum of the van der Waals volumes of a protein's constituent amino acids is divided by the volume occupied by the protein, packing densities of 0.72 to 0.77 are typically obtained. This means that, even with close packing, approximately 25% of the total volume of a protein is not occupied by protein atoms. Nearly all of this space is in the form of very small cavities. Cavities the size of water molecules or larger do occasionally occur, but they make up only a small fraction of the total protein volume. It is likely that such cavities provide flexibility for proteins and facilitate conformation changes and a wide range of protein dynamics (discussed later).

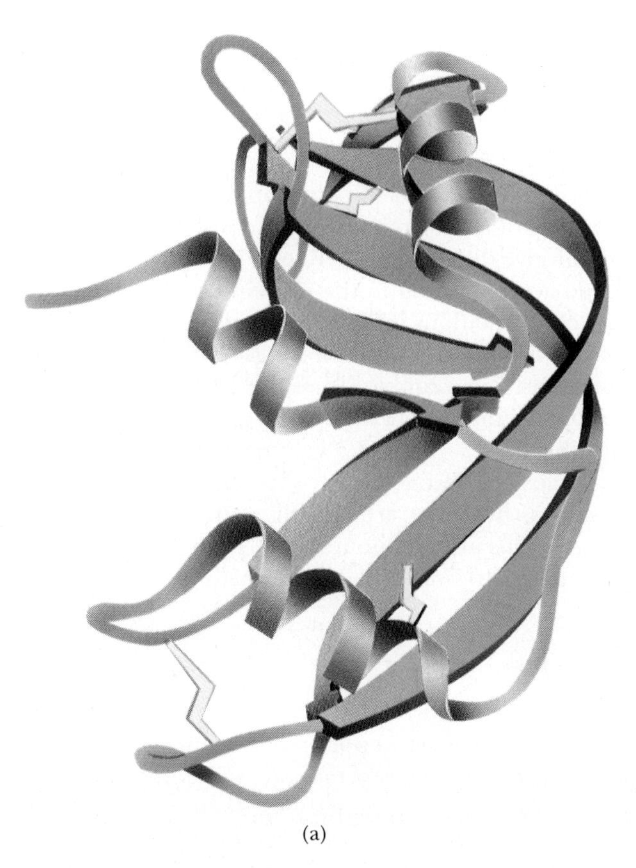

(a)

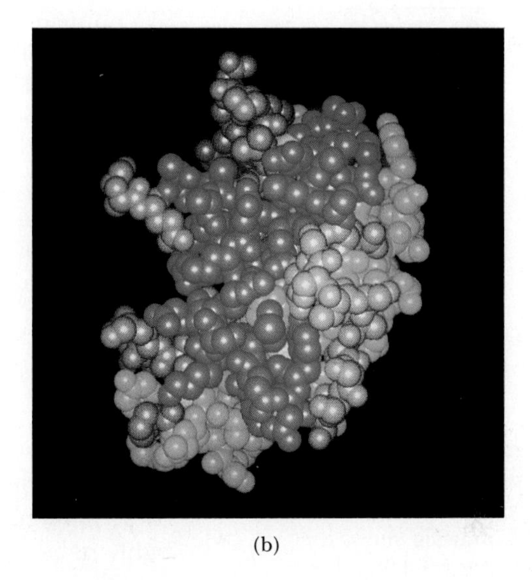

(b)

Figure **5.23** The three-dimensional structure of lysozyme, showing the α-helices as ribbons.

Ordered, Nonrepetitive Structures

In any protein structure, the segments of the polypeptide chain that cannot be classified as defined secondary structures, such as helices or sheets, have been traditionally referred to as *coil* or *random coil.* Both these terms are misleading. Most of these segments are neither coiled nor random, in any sense of the words. These structures are every bit as highly organized and stable as the defined secondary structures. They are just more variable and difficult to describe. These so-called coil structures are strongly influenced by side-chain interactions. Few of these interactions are well understood, but a number of interesting cases have been described. In his early studies of myoglobin structure, John Kendrew found that the —OH group of threonine or serine often hydrogen bonds to a backbone NH at the beginning of an α-helix. The same stabilization of an α-helix by a serine is observed in the three-dimensional structure of pancreatic trypsin inhibitor (Figure 5.24). Also in this same structure, an asparagine residue adjacent to a β-strand is found to form H bonds that stabilize the β-structure.

Nonrepetitive but well-defined structures of this type form many important features of enzyme active sites. In some cases, a particular arrangement of "coil" structure providing a specific type of functional site recurs in several functionally related proteins. The peptide loop that binds iron–sulfur clusters in both ferredoxin and high potential iron protein is one example. Another is the central loop portion of the *E-F hand structure* that binds a calcium ion in several calcium-binding proteins, including calmodulin, carp parvalbumin, troponin C, and the intestinal calcium-binding protein. This loop, shown in Figure 5.25, connects two short α-helices. The calcium ion nestles into the pocket formed by this structure.

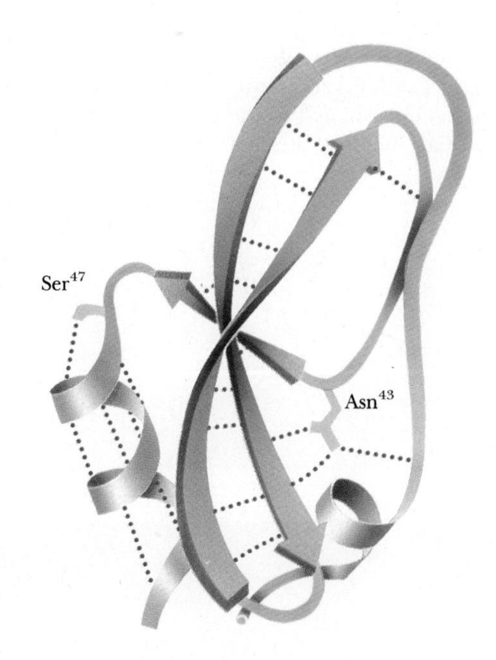

Pancreatic trypsin inhibitor

Figure **5.24** The three-dimensional structure of bovine pancreatic trypsin inhibitor. Note the stabilization of the α-helix by a hydrogen bond to Ser^{47} and the stabilization of the β-sheet by Asn^{43}.

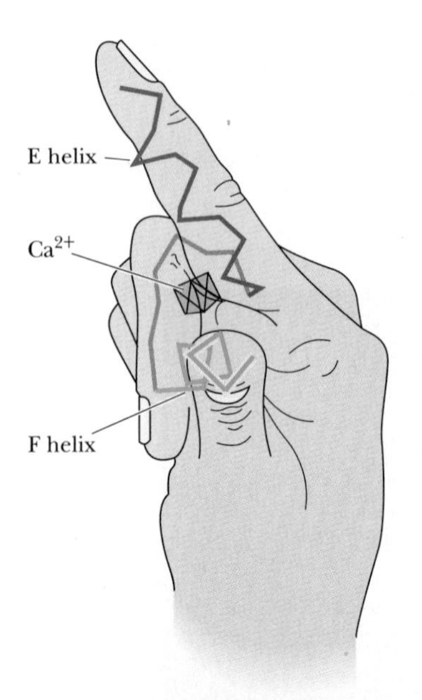

***Figure* 5.25** A representation of the so-called E-F hand structure, which forms calcium-binding sites in a variety of proteins. The stick drawing shows the peptide backbone of the E-F hand motif. The "E" helix extends along the index finger, a loop traces the approximate arrangement of the curled middle finger, and the "F" helix extends outward along the thumb. A calcium ion (Ca^{2+}) snuggles into the pocket created by the two helices and the loop. Kretsinger and co-workers originally assigned letters alphabetically to the helices in parvalbumin, a protein from carp. The E-F hand derives its name from the letters assigned to the helices at one of the Ca^{2+}-binding sites.

Flexible, Disordered Segments

In addition to nonrepetitive but well-defined structures, which exist in all proteins, genuinely disordered segments of polypeptide sequence also occur. These sequences either do not show up in electron density maps from X-ray crystallographic studies or give diffuse or ill-defined electron densities. These segments either undergo actual motion in the protein crystals themselves or take on many alternate conformations in different molecules within the protein crystal. Such behavior is quite common for long, charged side chains on the surface of many proteins. For example, 16 of the 19 lysine side chains in myoglobin have uncertain orientations beyond the δ-carbon, and five of these are disordered beyond the β-carbon. Similarly, a majority of the lysine residues are disordered in trypsin, rubredoxin, ribonuclease, and several other proteins. Arginine residues, however, are usually well ordered in protein structures. For the four proteins just mentioned, 70% of the arginine residues are highly ordered, compared to only 26% of the lysines.

Motion in Globular Proteins

Although we have distinguished between well-ordered and disordered segments of the polypeptide chain, it is important to realize that even well-ordered side chains in a protein undergo motion, sometimes quite rapid. These motions should be viewed as momentary oscillations about a single, highly stable conformation. *Proteins are thus best viewed as dynamic structures.* The allowed motions may be motions of individual atoms, groups of atoms, or even whole sections of the protein. Furthermore, they may arise from either thermal energy or specific, triggered conformational changes in the protein. **Atomic fluctuations** such as vibrations typically are random, very fast, and usually occur over small distances (less than 0.5 Å), as shown in Table 5.2. These motions arise from the kinetic energy within the protein and are a function of temperature. These very fast motions can be modeled by molecular dynamics calculations and studied by X-ray diffraction.

A class of slower motions, which may extend over larger distances, is **collective motions.** These are movements of groups of atoms covalently linked in such a way that the group moves as a unit. Such groups range in size from a few atoms to hundreds of atoms. Whole structural domains within a protein may be involved, as in the case of the flexible antigen-binding domains of immunoglobulins, which move as relatively rigid units to selectively bind separate antigen molecules. Such motions are of two types—(1) those that occur quickly but infrequently, such as tyrosine ring flips, and (2) those that occur slowly, such as *cis-trans* isomerizations of prolines. These collective motions also arise from thermal energies in the protein and operate on a time scale of 10^{-12} to 10^{-3} sec. These motions can be studied by nuclear magnetic resonance (NMR) and fluorescence spectroscopy.

Conformational changes involve motions of groups of atoms (individual side chains, for example) or even whole sections of proteins. These motions occur on a time scale of 10^{-9} to 10^{3} sec, and the distances covered can be as large as 1 nm. These motions may occur in response to specific stimuli or arise from specific interactions within the protein, such as hydrogen bonding, electrostatic interactions, and ligand binding. More will be said about conformational changes when enzyme catalysis and regulation are discussed (see Chapters 12 and 13).

Forces Driving the Folding of Globular Proteins

As already pointed out, the driving force for protein folding and the resulting formation of a tertiary structure is the formation of the most stable structure

Table **5.2**
Motion and Fluctuations in Proteins

Type of Motion	Spatial Displacement (Å)	Characteristic Time (sec)	Source of Energy
Atomic vibrations	0.01–1	10^{-15}–10^{-11}	Kinetic energy
Collective motions	0.01–5 or more	10^{-12}–10^{-3}	Kinetic energy
1. Fast: Tyr ring flips; methyl group rotations			
2. Slow: hinge bending between domains			
Triggered conformation changes	0.5–10 or more	10^{-9}–10^{3}	Interactions with triggering agent

Adapted from Petsko and Ringe (1984).

possible. There are two forces at work here. The peptide chain must both (1) satisfy the constraints inherent in its own structure and (2) fold so as to "bury" the hydrophobic side chains, minimizing their contact with solvent. The polypeptide itself does not usually form simple straight chains. Even in chain segments where helices and sheets are not formed, an extended peptide chain, being composed of L-amino acids, has a tendency to twist slightly in a right-handed direction. As shown in Figure 5.26, this tendency is apparently the basis for the formation of a variety of tertiary structures having a right-handed sense. Principal among these are the right-handed twists in arrays of β-sheets and right-handed cross-overs in parallel β-sheet arrays. Right-handed twisted β-sheets are found at the center of a number of proteins and provide an extended, highly stable structural core. Phosphoglycerate mutase, adenylate kinase, and carbonic anhydrase, among others, exist as smoothly twisted planes or saddle-shaped structures. Triose phosphate isomerase, soybean trypsin inhibitor, and domain 1 of pyruvate kinase contain right-handed twisted cylinders or barrel structures at their cores.

Connections between β-strands are of two types—hairpins and cross-overs. **Hairpins,** as shown in Figure 5.26, connect adjacent antiparallel

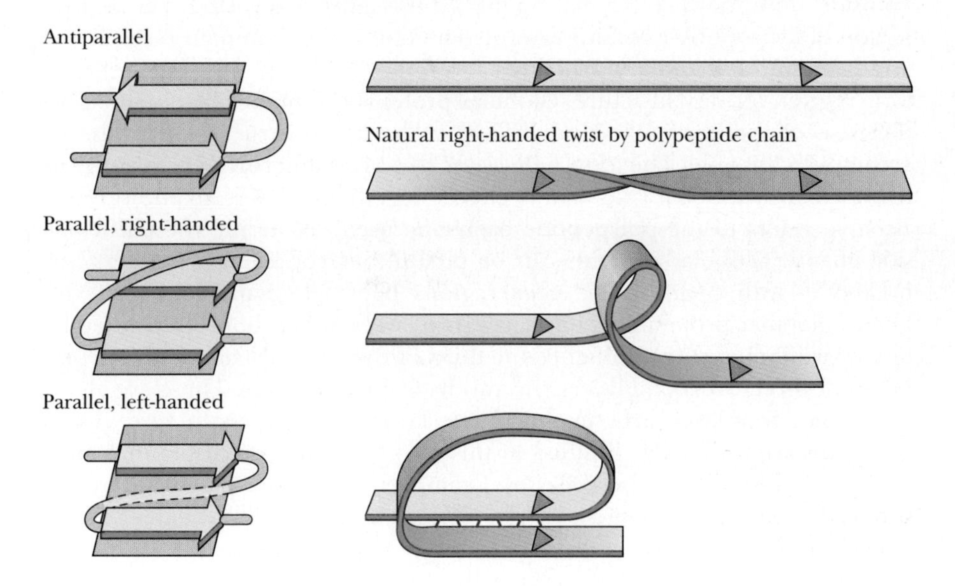

***Figure* 5.26** The natural right-handed twist exhibited by polypeptide chains, and the variety of structures that arise from this twist.

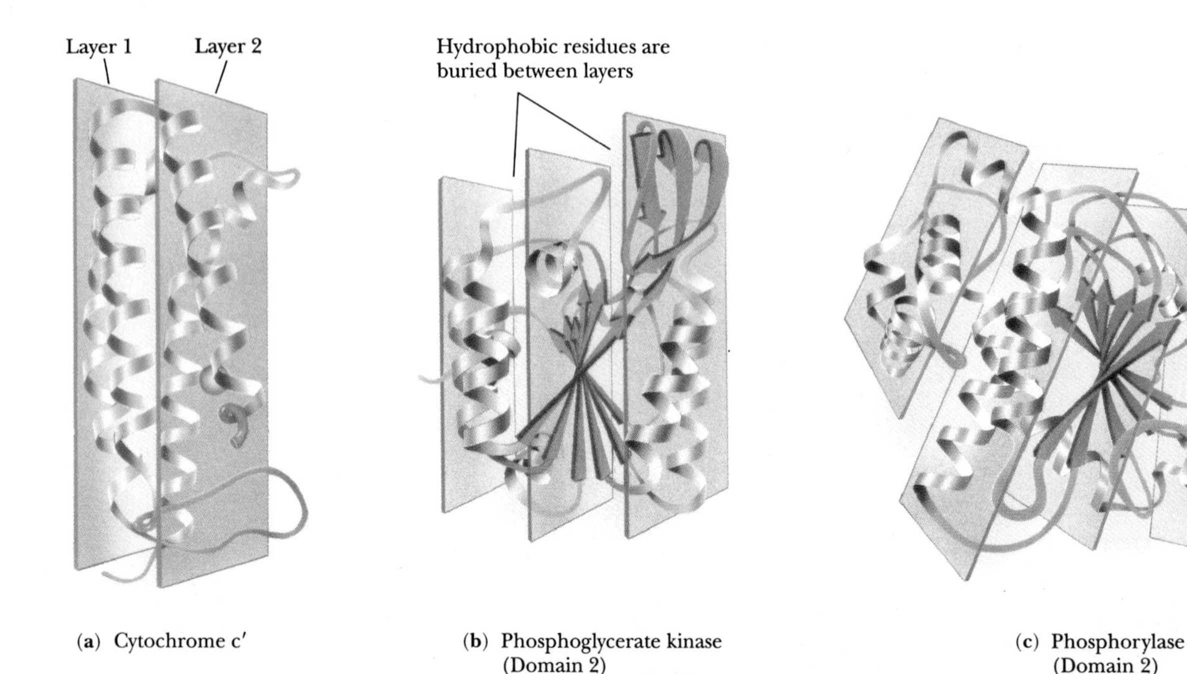

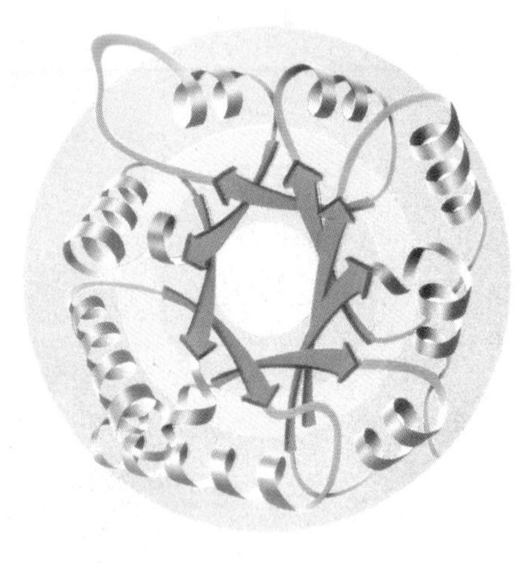

***Figure* 5.27** Examples of protein domains with different numbers of layers of backbone structure. (a) Cytochrome *c*′ with two layers of α-helix. (b) Domain 2 of phosphoglycerate kinase, composed of a β-sheet layer between two layers of helix, three layers overall. (c) An unusual five-layer structure, domain 2 of glycogen phosphorylase, a β-sheet layer sandwiched between four layers of α-helix. (d) The concentric "layers" of β-sheet (inside) and α-helix (outside) in triose phosphate isomerase. Hydrophobic residues are buried between these concentric layers in the same manner as in the planar layers of the other proteins. The hydrophobic layers are shaded yellow.

β-strands. **Cross-overs** are necessary to connect adjacent (or nearly adjacent) parallel β-strands. Nearly all cross-over structures are right-handed. Only in subtilisin and glucose phosphate isomerase have isolated left-handed cross-overs been identified. In many cross-over structures, the cross-over connection itself contains an α-helical segment. This is referred to as a **βαβ-loop.** As shown in Figure 5.26, the strong tendency in nature to form right-handed cross-overs, the wide occurrence of α-helices in the cross-over connection, and the right-handed twists of β-sheets can all be understood as arising from the tendency of an extended polypeptide chain of L-amino acids to adopt a right-handed twist structure. This is a chiral effect. Proteins composed of D-amino acids would tend to adopt left-handed twist structures.

The second driving force that affects the folding of polypeptide chains is the need to bury the hydrophobic residues of the chain, protecting them from solvent water. From a topological viewpoint, then, all globular proteins must have an "inside" where the hydrophobic core can be arranged and an "outside" toward which the hydrophilic groups must be directed. The sequestration of hydrophobic residues away from water is the dominant force in the arrangement of secondary structures and nonrepetitive peptide segments to form a given tertiary structure. Globular proteins can mainly be classified on the basis of the particular kind of core or backbone structure they use to accomplish this goal. The term *hydrophobic core,* as used here, refers to a region in which hydrophobic side chains cluster together, away from the solvent. *Backbone* refers to the polypeptide backbone itself, excluding the particular side chains. Globular proteins can be pictured as consisting of "layers" of backbone, with hydrophobic core regions between them. Over half the known globular protein structures have two layers of backbone (separated by one hydrophobic core). Roughly one-third of the known structures are composed of three backbone layers and two hydrophobic cores. There are also a few known four-layer structures and one known five-layer structure. A few structures are not easily classified in this way, but it is remarkable that most proteins fit into one of these classes. Examples of each are presented in Figure 5.27.

Classification of Globular Proteins

In addition to classification based on layer structure, proteins can be grouped according to the type and arrangement of secondary structure. There are four such broad groups: antiparallel α-helix, parallel or mixed β-sheet, antiparallel β-sheet, and the small metal- and disulfide-rich proteins.

It is important to note that the similarities of tertiary structure within these groups do not necessarily reflect similar or even related functions. Instead, **functional homology** usually depends on structural similarities on a smaller and more intimate scale.

Antiparallel α-Helix Proteins

Antiparallel α-helix proteins are structures heavily dominated by α-helices. The simplest way to pack helices is in an antiparallel manner, and most of the proteins in this class consist of bundles of antiparallel helices. Many of these exhibit a slight (15°) left-handed twist of the helix bundle. Figure 5.28 shows a representative sample of antiparallel α-helix proteins. Many of these are regular, uniform structures, but in a few cases (uteroglobin, for example) one of the helices is tilted away from the bundle. Tobacco mosaic virus protein has small, highly twisted antiparallel β-sheets on one end of the helix bundle, with two additional helices on the other side of the sheet. Notice in Figure 5.28 that most of the antiparallel α-helix proteins are made up of four-helix bundles.

The so-called globin proteins are an important group of α-helical proteins. These include hemoglobins and myoglobins from many species. The globin structure can be viewed as two layers of helices, with one of these layers perpendicular to the other and the polypeptide chain moving back and forth between the layers.

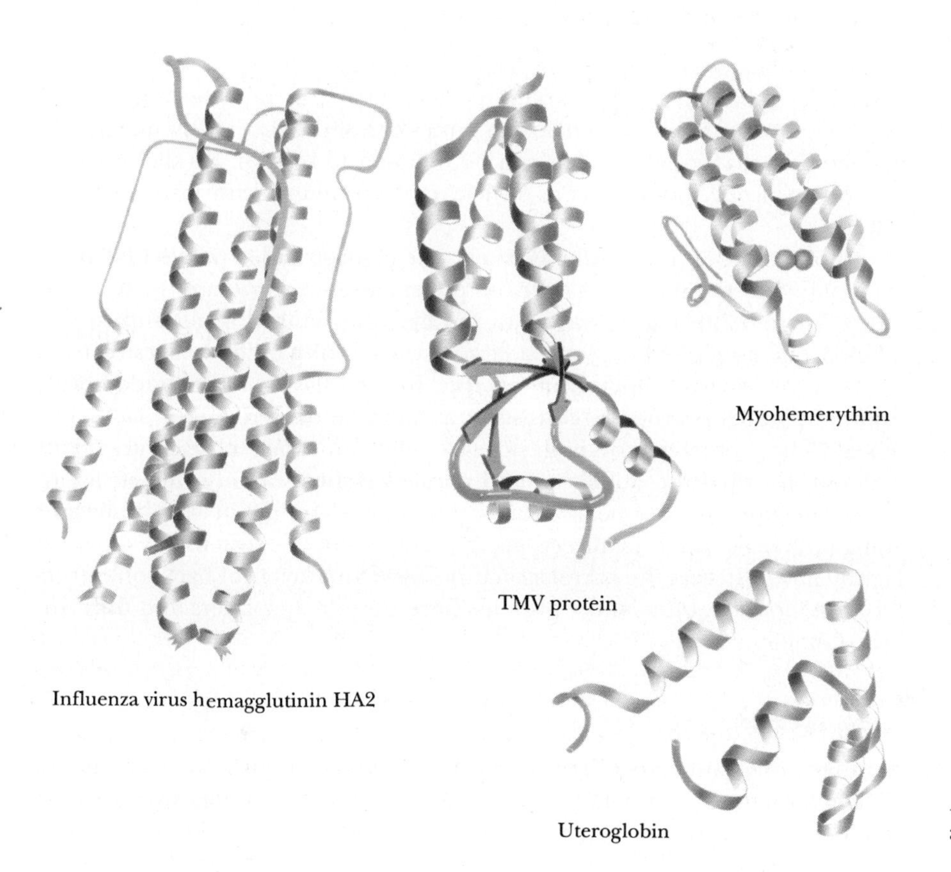

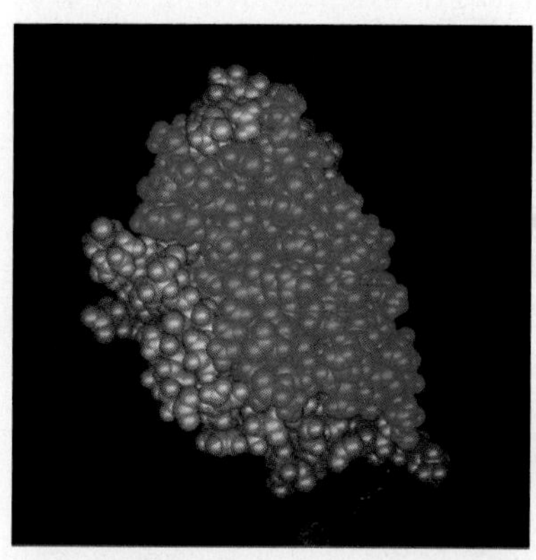
Myohemerythrin

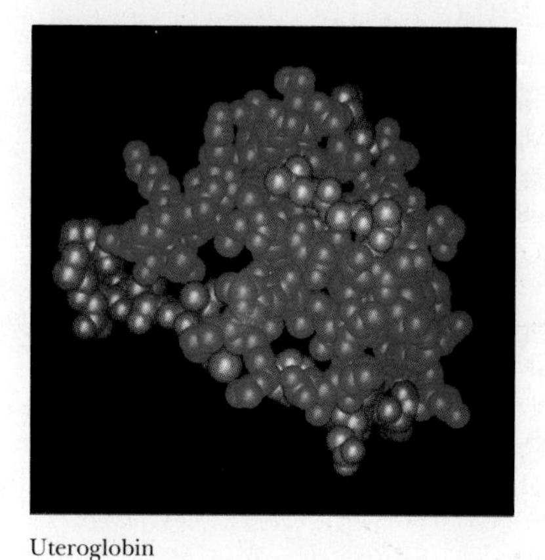
Uteroglobin

***Figure* 5.28** Several examples of antiparallel α-proteins.

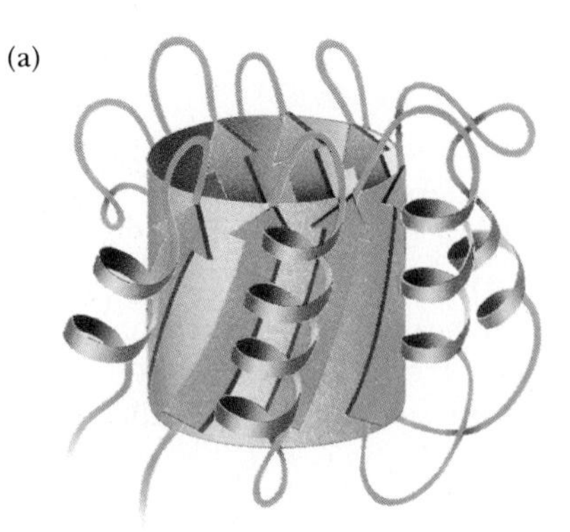

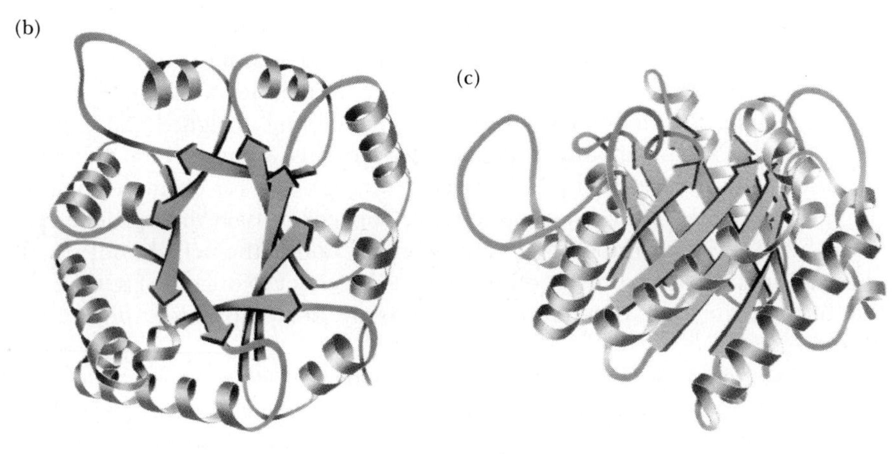

Figure **5.29** Parallel β-array proteins—the eight-stranded β-barrels of triose phosphate isomerase (a, side view, and b, top view) and (c) pyruvate kinase.

Parallel or Mixed β-Sheet Proteins

The second major class of protein structures contains structures based around **parallel** or **mixed β-sheets.** Parallel β-sheet arrays, as previously discussed, distribute hydrophobic side chains on both sides of the sheet. This means that neither side of parallel β-sheets can be exposed to solvent. Parallel β-sheets are thus typically found as core structures in proteins, with little access to solvent.

Another important parallel β-array is the eight-stranded **parallel β-barrel,** exemplified in the structures of triose phosphate isomerase and pyruvate kinase (Figure 5.29). Each β-strand in the barrel is flanked by an antiparallel α-helix. The α-helices thus form a larger cylinder of parallel helices concentric with the β-barrel. Both cylinders thus formed have a right-handed twist. Another parallel β-structure consists of an internal twisted wall of parallel or mixed β-sheet protected on both sides by helices or other substructures. This structure is called the **doubly wound parallel β-sheet** because the structure can be imagined to have been wound by strands beginning in the middle and going outward in opposite directions. The essence of this structure is shown in Figure 5.30. Whereas the barrel structures have four layers of backbone structure, the doubly wound sheet proteins have three major layers and thus two hydrophobic core regions.

Antiparallel β-Sheet Proteins

Another important class of tertiary protein conformations is the **antiparallel β-sheet** structures. Antiparallel β-sheets, which usually arrange hydrophobic residues on just one side of the sheet, can exist with one side exposed to solvent. The minimal structure for an antiparallel β-sheet protein is thus a

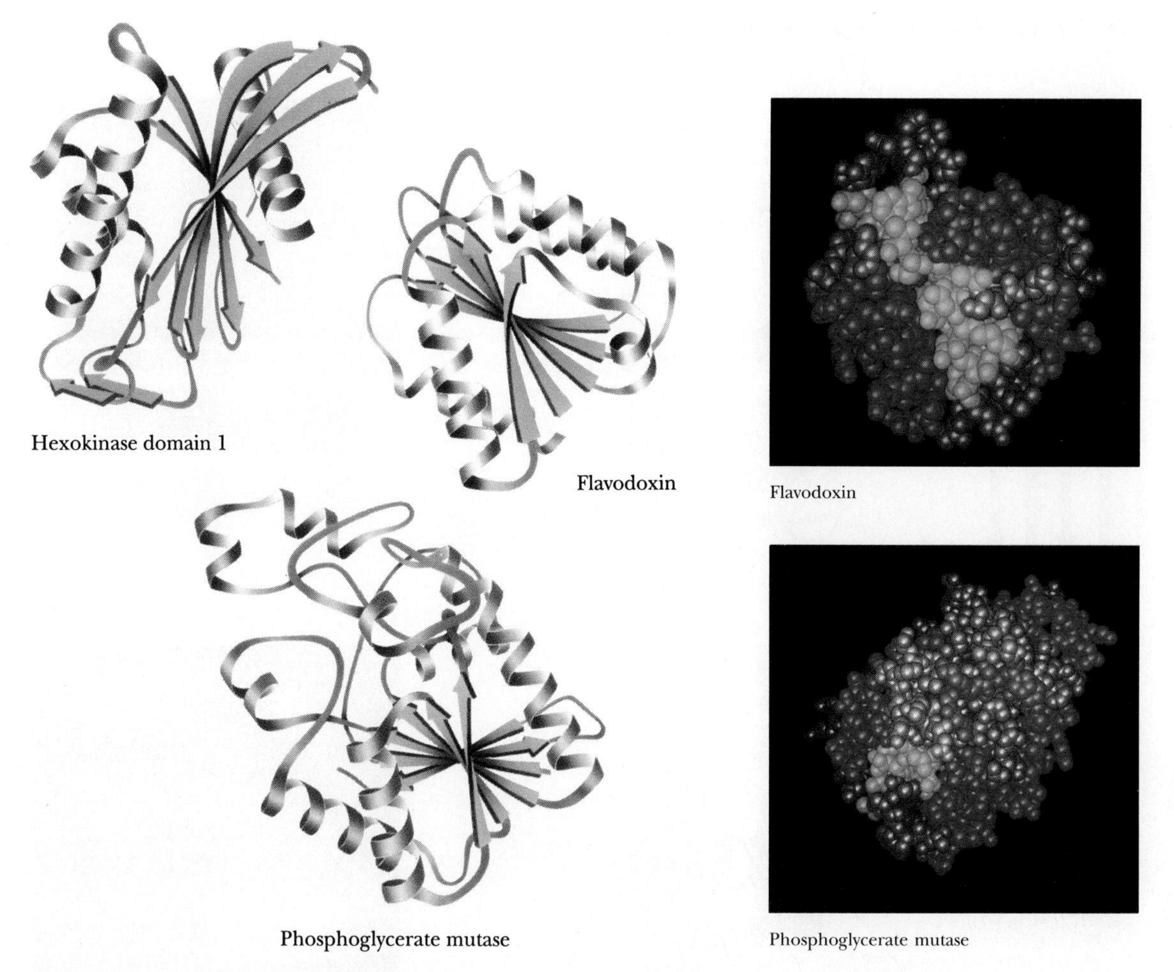

***Figure* 5.30** Several typical doubly wound parallel β-sheet proteins.

two-layered structure, with hydrophobic faces of the two sheets juxtaposed and the opposite faces exposed to solvent. Such domains consist of β-sheets arranged in a cylinder or barrel shape. These structures are usually less symmetric than the singly wound parallel barrels and are not as efficiently hydrogen bonded, but they occur much more frequently in nature. Barrel structures tend to be either all parallel or all antiparallel and usually consist of even numbers of β-strands. Good examples of antiparallel structures include soybean trypsin inhibitor, rubredoxin, and domain 2 of papain (Figure 5.31). Topology diagrams of antiparallel β-sheet barrels reveal that many of them

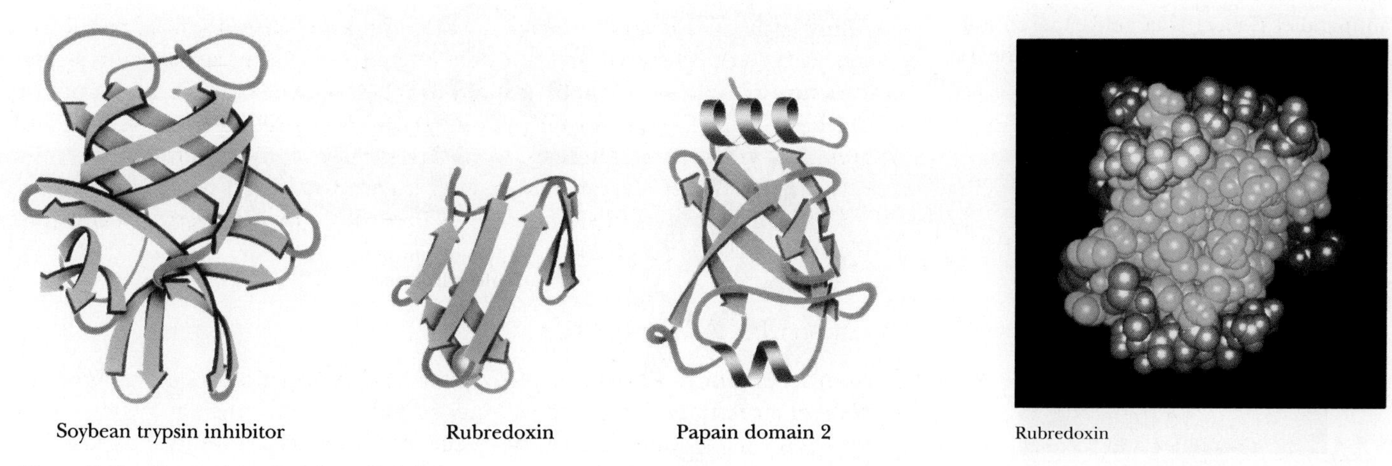

***Figure* 5.31** Examples of antiparallel β-sheet structures in proteins.

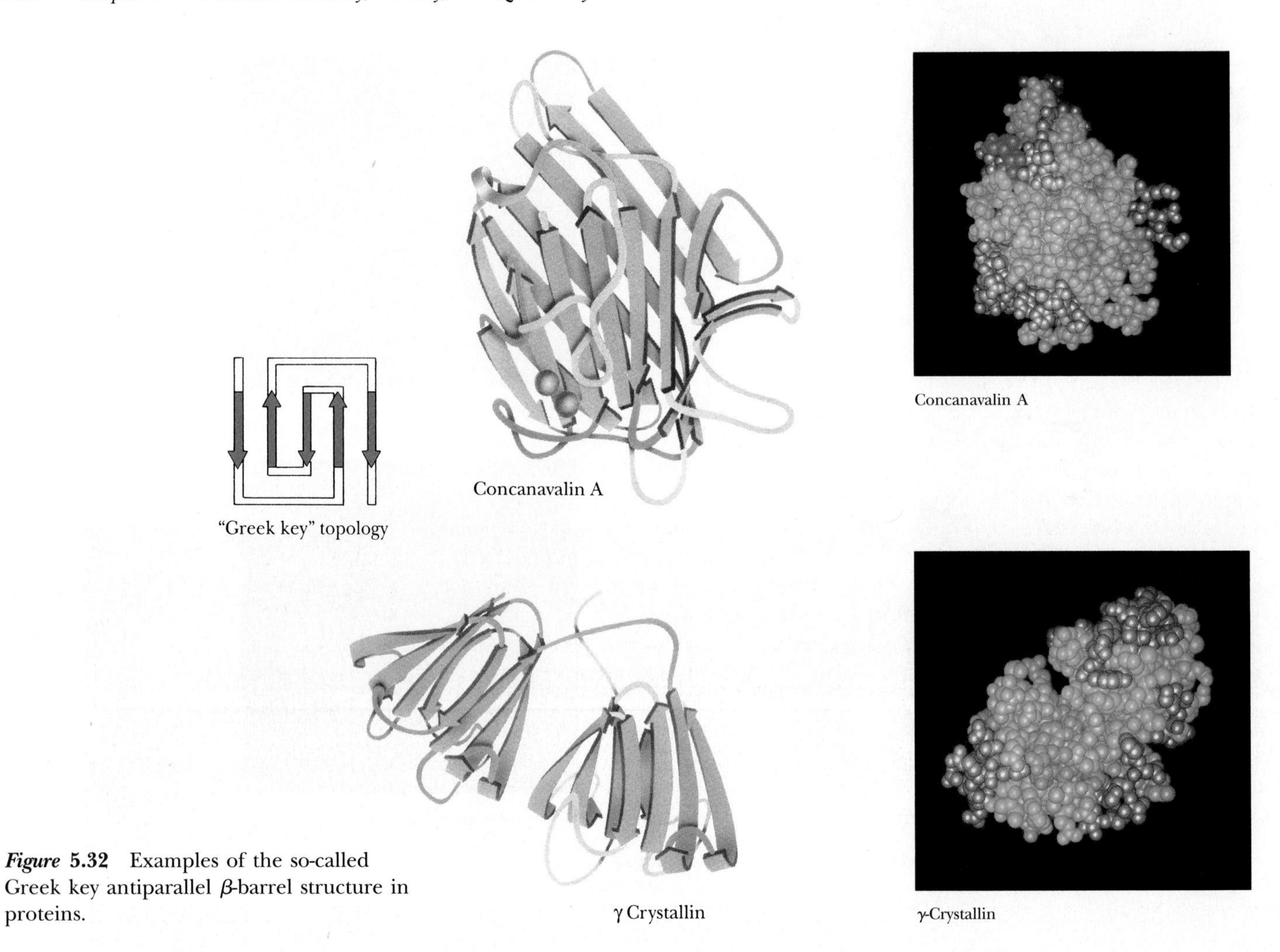

***Figure* 5.32** Examples of the so-called Greek key antiparallel β-barrel structure in proteins.

***Figure* 5.33** Sheet structures formed from antiparallel arrangements of β-strands. (a) *Streptomyces* subtilisin inhibitor, (b) glutathione reductase domain 3, and (c) the second domain of glyceraldehyde-3-phosphate dehydrogenase represent minimal antiparallel β-sheet domain structures. In each of these cases, an antiparallel β-sheet is largely exposed to solvent on one face and covered by helices and random coils on the other face.

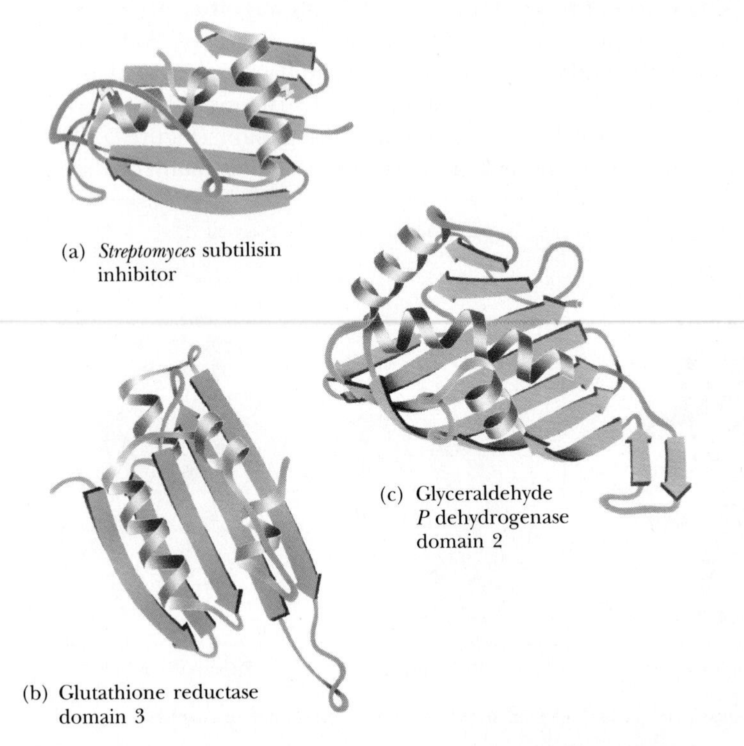

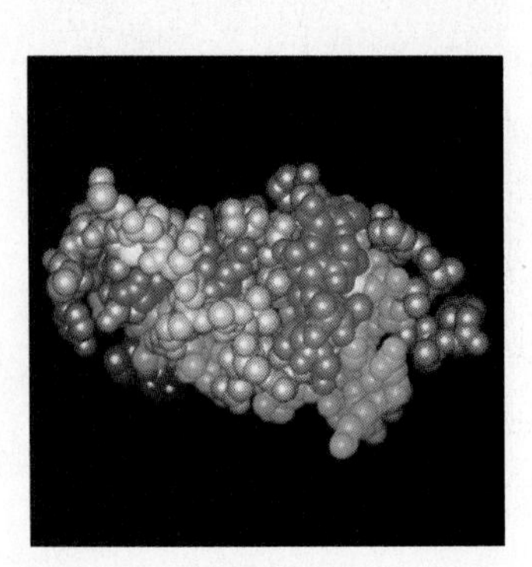

arrange the polypeptide sequence in an interlocking pattern reminiscent of patterns found on ancient Greek vases (Figure 5.32) and are thus referred to as a *Greek key topology*. Several of these, including concanavalin A and γ-crystallin, contain an extra swirl in the Greek key pattern (see Figure 5.32). Antiparallel arrangements of β-strands can also form sheets as well as barrels. Glyceraldehyde-3-phosphate dehydrogenase, *Streptomyces* subtilisin inhibitor, and glutathione reductase are examples of single-sheet, double-layered topology (Figure 5.33).

Metal- and Disulfide-Rich Proteins

Other than the structural classes just described and a few miscellaneous structures that do not fit nicely into these categories, there is only one other major class of protein tertiary structures—the small metal-rich and disulfide-rich structures. These proteins or fragments of proteins are usually small (<100 residues), and their conformations are heavily influenced by their high content of either liganded metals or disulfide bonds. The structures of disulfide-rich proteins are unstable if their disulfide bonds are broken. Figure 5.34 shows several representative disulfide-rich proteins, including insulin, phospholipase A_2, and crambin (from the seeds of *Crambe abyssinica*), as well as several metal-rich proteins, including ferredoxin and high potential iron protein (HiPIP). The structures of some of these proteins bear a striking resemblance to structural classes that have already been discussed. For example,

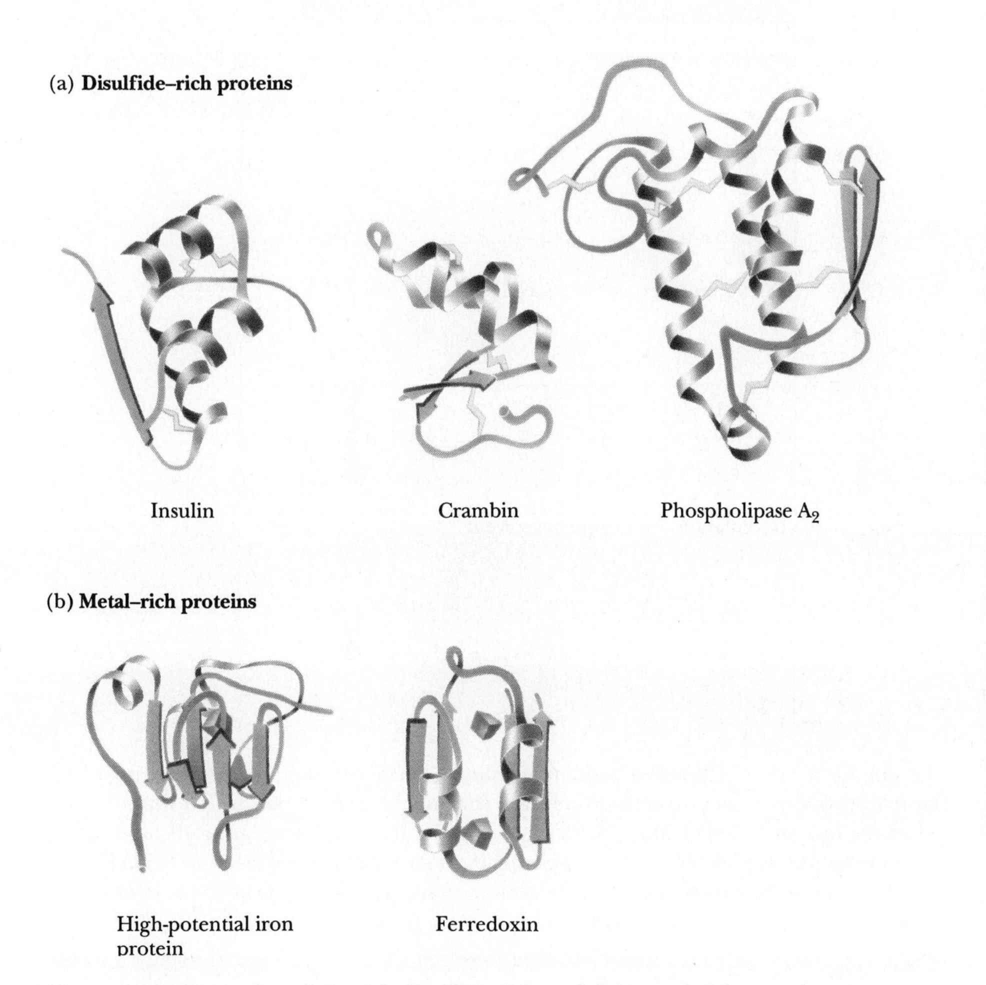

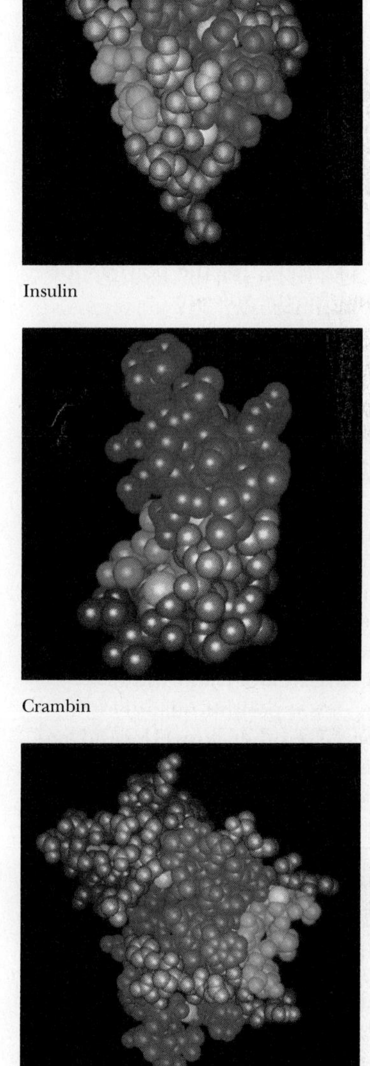

***Figure* 5.34** Examples of the (a) disulfide-rich and (b) metal-rich proteins.

Critical Developments in Biochemistry

Thermodynamics of the Folding Process in Globular Proteins

Section 5.1 considered the noncovalent binding energies that stabilize a protein structure. However, the folding of a protein depends ultimately on the difference in Gibbs free energy (ΔG) between the folded (F) and unfolded (U) states at some temperature T:

$$\Delta G = G_F - G_U = \Delta H - T\Delta S \quad (5.1)$$
$$= (H_F - H_U) - T(S_F - S_U)$$

In the unfolded state, the peptide chain and its R groups interact with solvent water, and any measurement of the free energy change upon folding must consider contributions to the enthalpy change (ΔH) and the entropy change (ΔS) both for the polypeptide chain and for the solvent:

$$\Delta G_{total} = \Delta H_{chain} + \Delta H_{solvent} - T\Delta S_{chain} - T\Delta S_{solvent} \quad (5.2)$$

If each of the four terms on the right side of this equation is understood, the thermodynamic basis for protein folding should be clear. A summary of the signs and magnitudes of these quantities for a typical protein is shown in the accompanying figure. The folded protein is a highly ordered structure compared to the unfolded state, so ΔS_{chain} is a negative number and thus $-T\Delta S_{chain}$ is a positive quantity in the equation. The other terms depend on the nature of the particular ensemble of R groups. The nature of ΔH_{chain} depends on both residue–residue interactions and residue–solvent interactions. Nonpolar groups in the folded protein interact mainly with one another via weak van der Waals forces. Interactions between nonpolar groups and water in the unfolded state are stronger since the polar water molecules induce dipoles in the nonpolar groups, producing a significant electrostatic interaction. As a result, ΔH_{chain} is positive for nonpolar groups and favors the unfolded state. $\Delta H_{solvent}$ for nonpolar groups, however, is negative and favors the folded state. This is because folding allows many water molecules to interact (favorably) with one another rather than (less favorably) with the nonpolar side chains. The magnitude of ΔH_{chain} is smaller than that of $\Delta H_{solvent}$, but both these terms are small and usually do not dominate the folding process. However, $\Delta S_{solvent}$ for nonpolar groups is large and positive and strongly favors the folded state. This is because nonpolar groups force order upon the water solvent in the unfolded state.

For polar side chains, ΔH_{chain} is positive and $\Delta H_{solvent}$ is negative. Since solvent molecules are ordered to some extent around polar groups, $\Delta S_{solvent}$ is small and positive. As shown in the figure, ΔG_{total} for the polar groups of a protein is near zero. Comparison of all the terms considered here makes it clear that *the single largest contribution to the stability of a folded protein is $\Delta S_{solvent}$ for the nonpolar residues.*

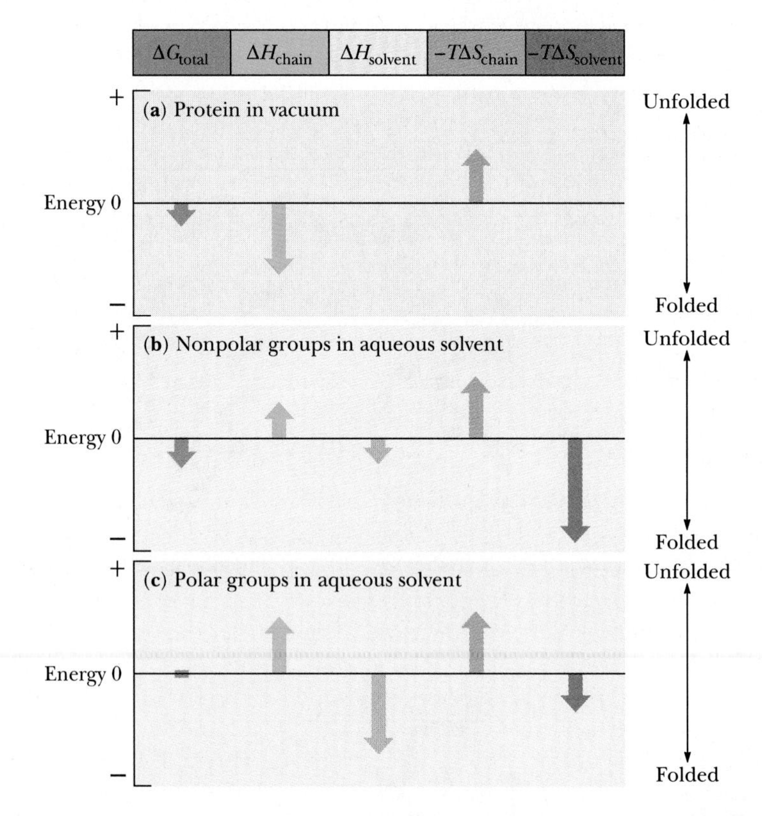

The energetic contributions to protein stability. Thermodynamic factors favoring the native, folded structure of a protein are shown with arrows pointing down, below the energy 0 (zero) line. Factors favoring the unfolded state are indicated by arrows pointing up. *The largest factor favoring the native, folded structure is the $-T\Delta S$ (entropy) term for the solvent (water) in the accommodation of nonpolar protein side chains.* *(Adapted from Shulz, G., and Schirmer, R., 1979, Principles of Protein Structure.* Berlin: Springer-Verlag, p. 37.)

phospholipase A_2 is a distorted α-helix cluster, whereas HiPIP is a distorted β-barrel structure. Others among this class (such as insulin and crambin), however, are not easily likened to any of the standard structure classes.

Molecular Chaperones: Proteins That Help Fold Globular Proteins

The landmark experiments by Christian Anfinsen on the refolding of ribonuclease clearly showed that the refolding of a pure, denatured protein *in vitro* is a spontaneous process. As noted above, this refolding is driven by the small Gibbs free energy difference between the unfolded and folded states. It has also been generally assumed that all the information necessary for the correct folding of a polypeptide chain is contained in the primary structure and requires no additional molecular factors. However, the folding of proteins in the cell is a different matter. The highly concentrated protein matrix in the cell may adversely affect the folding process by causing aggregation of unfolded or partially folded proteins. Also, it may be necessary to accelerate slow steps in the folding process or to suppress or reverse incorrect or premature folding. Recent studies have uncovered a family of proteins, known as **molecular chaperones,** that appear to be essential for the correct folding of certain polypeptide chains *in vivo,* for their assembly into oligomers, and for preventing inappropriate liaisons with other proteins during their synthesis, folding, and transport. Many of these proteins were first identified as **heat shock proteins,** which are induced in cells by elevated temperature or other stress. The most thoroughly studied proteins are **hsp70,** a 70-kD heat shock protein, and the so-called **chaperonins,** also known as **cpn60s** or **hsp60s,** a class of 60-kD heat shock proteins. Another well-characterized chaperonin is **GroEL,** an *E. coli* protein that has been shown to affect the folding of several proteins.

The way in which molecular chaperones interact with polypeptides during the folding process is not well understood. What *is* clear is that chaperones bind effectively to partially folded structures. These folding intermediates are less compact than the native folded protein. They contain large amounts of secondary and even some tertiary structure, but they undergo relatively large conformational fluctuations. It is possible that chaperone proteins recognize exposed helices or other secondary structure elements on their target proteins. This initial interaction may then allow the chaperone to guide or regulate the subsequent events of folding (Figure 5.35).

Protein Modules: Nature's Modular Strategy for Protein Design

Now that many thousands of proteins have been sequenced (more than 20,000 sequences are known), it has become obvious that certain protein sequences that give rise to distinct structural domains are used over and over again in modular fashion. These **protein modules** may occur in a wide variety of proteins, often being used for different purposes, or they may be used repeatedly in the same protein. Figure 5.36 shows the tertiary structures of five protein modules, and Figure 5.37 presents several proteins that contain versions of these modules. These modules typically contain about 40 to 100 amino acids and often adopt a stable tertiary structure when isolated from their parent protein. One of the best-known examples of a protein module is the **immunoglobulin module,** which has been found not only in immunoglobulins but also in a wide variety of cell surface proteins, including cell adhesion molecules and growth factor receptors, and even in *twitchin,* an intracellular protein found in muscle. It is likely that more protein modules will be identified. (The role of protein modules in evolution is discussed in Chapter 33.)

(text continues on page 170)

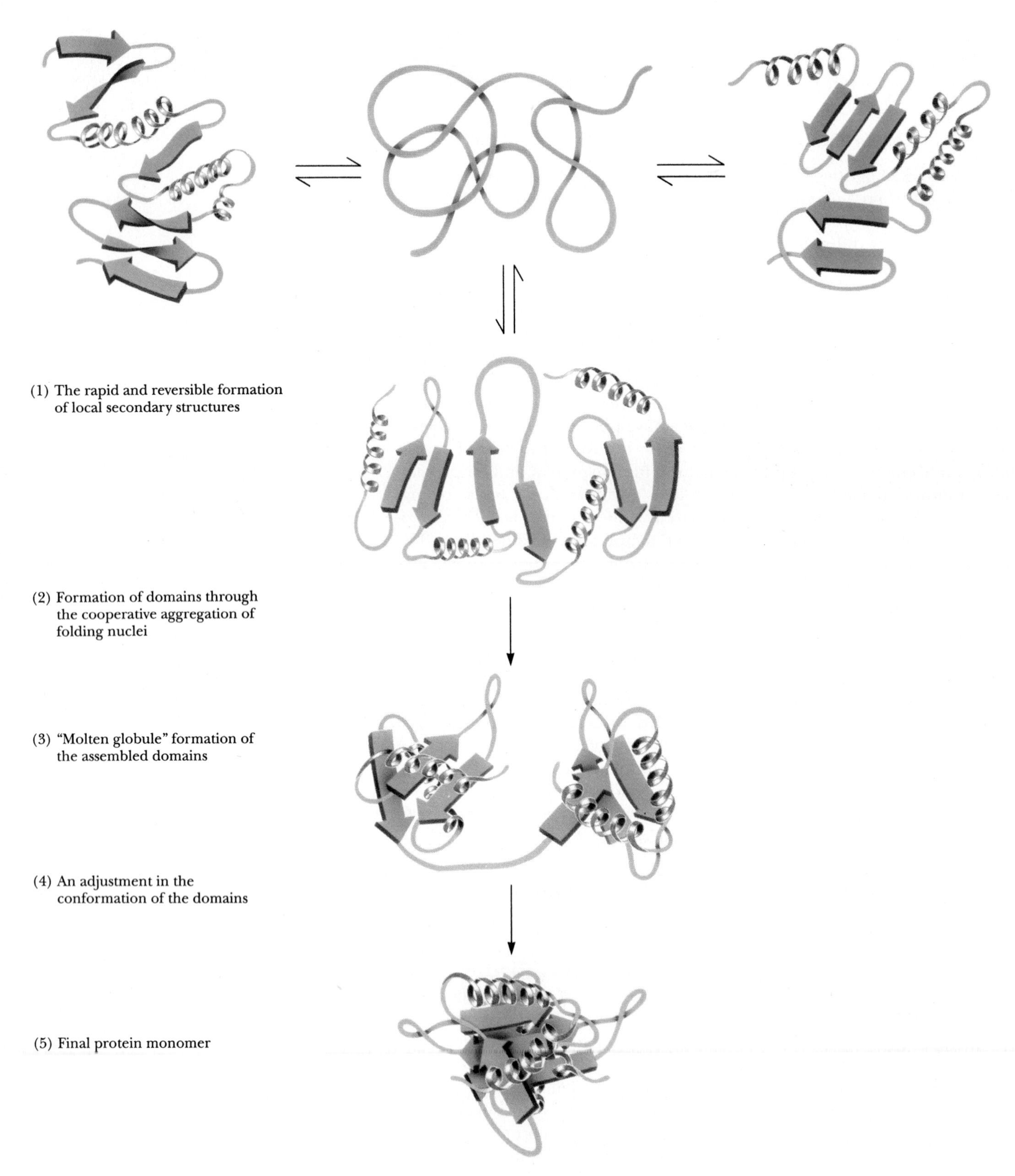
(1) The rapid and reversible formation of local secondary structures
(2) Formation of domains through the cooperative aggregation of folding nuclei
(3) "Molten globule" formation of the assembled domains
(4) An adjustment in the conformation of the domains
(5) Final protein monomer

***Figure* 5.35** A model for the steps involved in the folding of globular proteins. Chaperone proteins may assist in the initiation of the folding of domains.

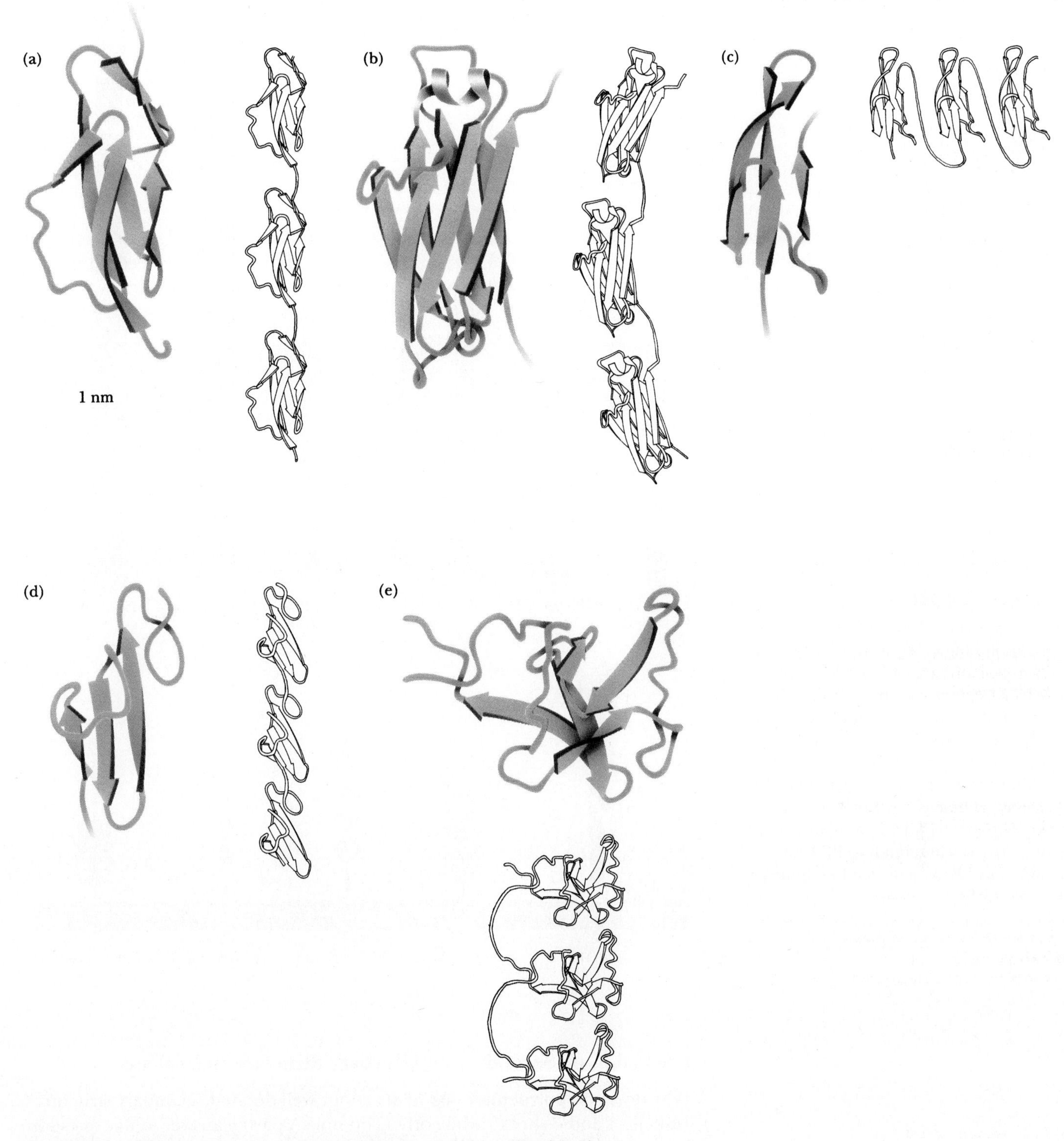

***Figure* 5.36** Ribbon structures of several protein modules utilized in the construction of complex multimodule proteins. (a) The complement control protein module. (b) The immunoglobulin module. (c) The fibronectin type I module. (d) The growth factor module. (e) The kringle module.

(Adapted from Baron, M., Norman, D., and Campbell, I., 1991, Protein modules. Trends in Biochemical Sciences 16:13–17.)

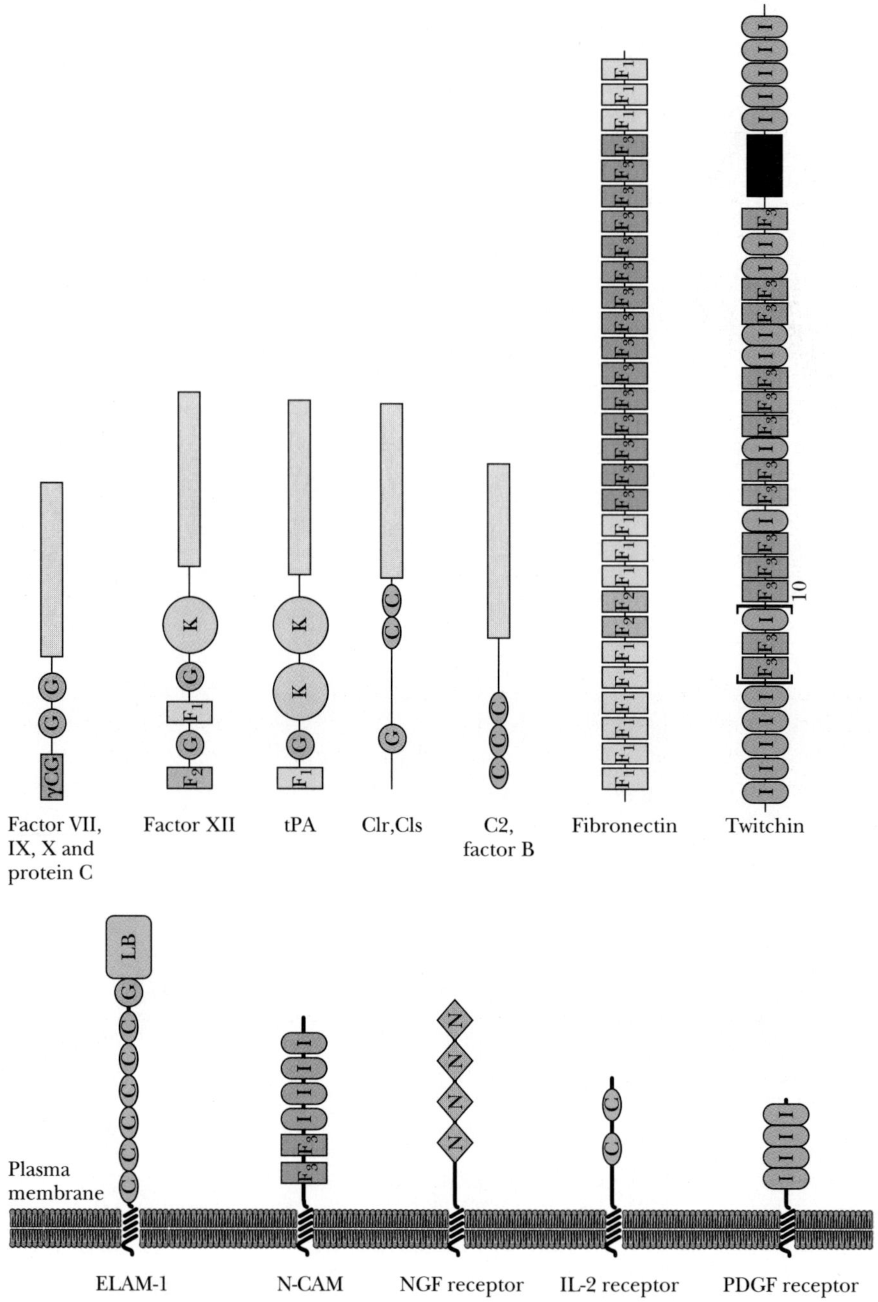

***Figure* 5.37** A sampling of proteins that consist of mosaics of individual protein modules. The modules shown include: γCG, a module containing γ-carboxyglutamate residues; G, an epidermal growth-factor–like module; K, the "kringle" domain, named for a Danish pastry; C, which is found in complement proteins; F1, F2, and F3, first found in fibronectin; I, the immunoglobulin superfamily domain; N, found in some growth factor receptors; E, a module homologous to the calcium-binding EF hand domain; and LB, a lectin module found in some cell surface proteins.

(Adapted from Baron, M., Norman, D., and Campbell, I., 1991, Protein modules. Trends in Biochemical Sciences 16:13–17.)

Prediction of Secondary and Tertiary Structure in Proteins

Why do some polypeptide segments adopt well-defined secondary structures (α-helices and β-sheets) while others do not? A partial answer to this question has come from surveys of the frequency with which various residues appear in helices and sheets. As shown in Figure 5.38, some residues, such as alanine, glutamate, and methionine, occur much more frequently in α-helices than others do. In contrast, glycine and proline are the least likely residues to be found in an α-helix. Likewise, certain residues, including valine, isoleucine, and the aromatic amino acids, are more likely to be found in β-sheets than other residues, and aspartate, glutamate, and proline are much less likely to be found in β-sheets.

Such observations have led to many efforts to predict the occurrence of secondary structure in proteins from knowledge of the peptide sequence. Such **predictive algorithms** consider the composition of short segments of a

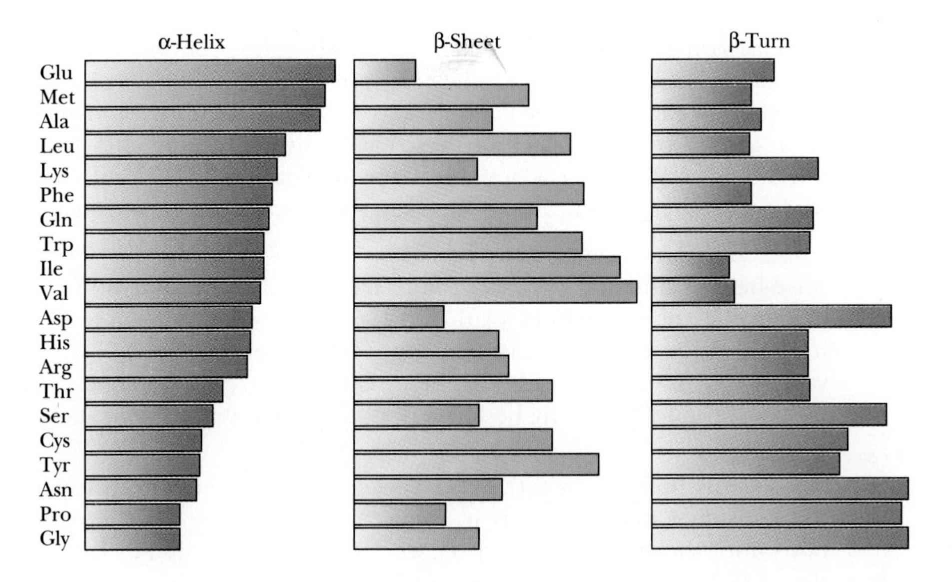

***Figure* 5.38** Relative frequencies of occurrence of amino acid residues in α-helices, β-sheets, and β-turns in proteins of known structure.

(Adapted from Bell, J. E., and Bell, E. T., 1988, Proteins and Enzymes. Englewood Cliffs, NJ: Prentice-Hall.)

polypeptide. If these segments are rich in residues that are found frequently in helices or sheets, then that segment is judged likely to adopt the corresponding secondary structure. The predictive algorithm designed by Peter Chou and Gerald Fasman in 1974 used data like that in Figure 5.38 to classify the 20 amino acids for their α-helix-forming and β-sheet-forming **propensities, P_α and P_β** (Table 5.3). These residues are classified as strong helix form-

***Table* 5.3**

Chou–Fasman Helix and Sheet Propensities (P_α and P_β) of the Amino Acids

Amino Acid	P_α	Helix Classification	P_β	Sheet Classification
A Ala	1.42	H_α	0.83	i_β
C Cys	0.70	i_α	1.19	h_β
D Asp	1.01	I_α	0.54	B_β
E Glu	1.51	H_α	0.37	B_β
F Phe	1.13	h_α	1.38	h_β
G Gly	0.57	B_α	0.75	b_β
H His	1.00	I_α	0.87	h_β
I Ile	1.08	h_α	1.60	H_β
K Lys	1.16	h_α	0.74	b_β
L Leu	1.21	H_α	1.30	h_β
M Met	1.45	H_α	1.05	h_β
N Asn	0.67	b_α	0.89	i_β
P Pro	0.57	B_α	0.55	B_β
Q Gln	1.11	h_α	1.10	h_β
R Arg	0.98	i_α	0.93	i_β
S Ser	0.77	i_α	0.75	b_β
T Thr	0.83	i_α	1.19	h_β
V Val	1.06	h_α	1.70	H_β
W Trp	1.08	h_α	1.37	h_β
Y Tyr	0.69	b_α	1.47	H_β

Source: Chou, P. Y., and Fasman, G. D., 1978. *Annual Review of Biochemistry.* **47:**258.

ers (H_α), helix formers (h_α), weak helix formers (I_α), indifferent helix formers (i_α), helix breakers (b_α), and strong helix breakers (B_α). Similar classes were established by Chou and Fasman for β-sheet-forming ability. By studying the patterns of occurrence of each of these classes in helices and sheets of proteins with known structures, Chou and Fasman formulated three rules to predict the occurrence of helices and sheets in sequences of unknown structure:

1. An α-helix is initiated by a six-residue segment that contains four or more H_α or h_α residues (with two I_α residues counting as one h_α). The helix will extend in both directions until the average value of P_α for a four-residue segment is less than 1.0. If present, proline residues may only be located at the amino-terminus of the helix.
2. A β-sheet is initiated by a five-residue segment that contains at least three residues rated as H_β or h_β. Sheets may extend in both directions until the average value of P_β for a four-residue segment is less than 1.0.
3. When H_α, h_α and H_β, h_β residues occur together, the peptide is predicted to be helical if the average of P_α is greater than P_β. If P_β is greater than P_α, a β-sheet is predicted.

Such algorithms are only modestly successful in predicting the occurrence of helices and sheets in proteins.

5.5 Subunit Interactions and Quaternary Structure

Many proteins exist in nature as **oligomers,** complexes composed of (often symmetric) noncovalent assemblies of two or more monomer subunits. In fact, subunit association is a common feature of macromolecular organization in biology. Most intracellular enzymes are oligomeric and may be composed either of a single type of monomer subunit (*homomultimers*) or of several different kinds of subunits (*heteromultimers*). The simplest case is a protein composed of identical subunits. Liver alcohol dehydrogenase, shown in Figure 5.39, is such a protein. More complicated proteins may have several different subunits in one, two, or more copies. Hemoglobin, for example, contains two each of two different subunits and is referred to as an $\alpha_2\beta_2$-complex. An interesting counterpoint to these relatively simple cases is made by the proteins that form polymeric structures. Tubulin is an αβ-dimeric protein, which polymerizes to form microtubules of the formula $(\alpha\beta)_n$. The way in which separate folded monomeric protein subunits associate to form the oligomeric protein constitutes the **quaternary structure** of that protein. Table 5.4 lists several proteins and their subunit compositions (see also Table 4.1). Clearly, proteins with two to four subunits dominate the list, but many cases of higher numbers exist.

***Table* 5.4**
Aggregation Symmetries of Globular Proteins

Protein	Number of Subunits
Alcohol dehydrogenase	2
Immunoglobulin	4
Malate dehydrogenase	2
Superoxide dismutase	2
Triose phosphate isomerase	2
Glycogen phosphorylase	2
Alkaline phosphatase	2
6-Phosphogluconate dehydrogenase	2
Wheat germ agglutinin	2
Glucose phosphate isomerase	2
Tyrosyl-tRNA synthetase	2
Glutathione reductase	2
Aldolase	3
Bacteriochlorophyll protein	3
TMV protein disc	17
Concanavalin A	4
Glyceraldehyde-3-phosphate dehydrogenase	4
Lactate dehydrogenase	4
Prealbumin	4
Pyruvate kinase	4
Phosphoglycerate mutase	4
Hemoglobin	2 + 2
Insulin	6
Aspartate transcarbamoylase	6 + 6
Glutamine synthetase	12
Apoferritin	24
Coat of tomato bushy stunt virus	180

The subunits of an oligomeric protein typically fold into apparently independent globular conformations and then interact with other subunits. The particular surfaces at which protein subunits interact are similar in nature to the interiors of the individual subunits. These interfaces are closely packed and involve both polar and hydrophobic interactions. Interacting surfaces must therefore possess complementary arrangements of polar and hydrophobic groups.

Oligomeric associations of protein subunits can be divided into those between **identical** subunits and those between **nonidentical** subunits. Interactions among identical subunits can be further distinguished as either **isologous** or **heterologous.** In isologous interactions, the interacting surfaces are identical, and the resulting structure is necessarily dimeric and closed,

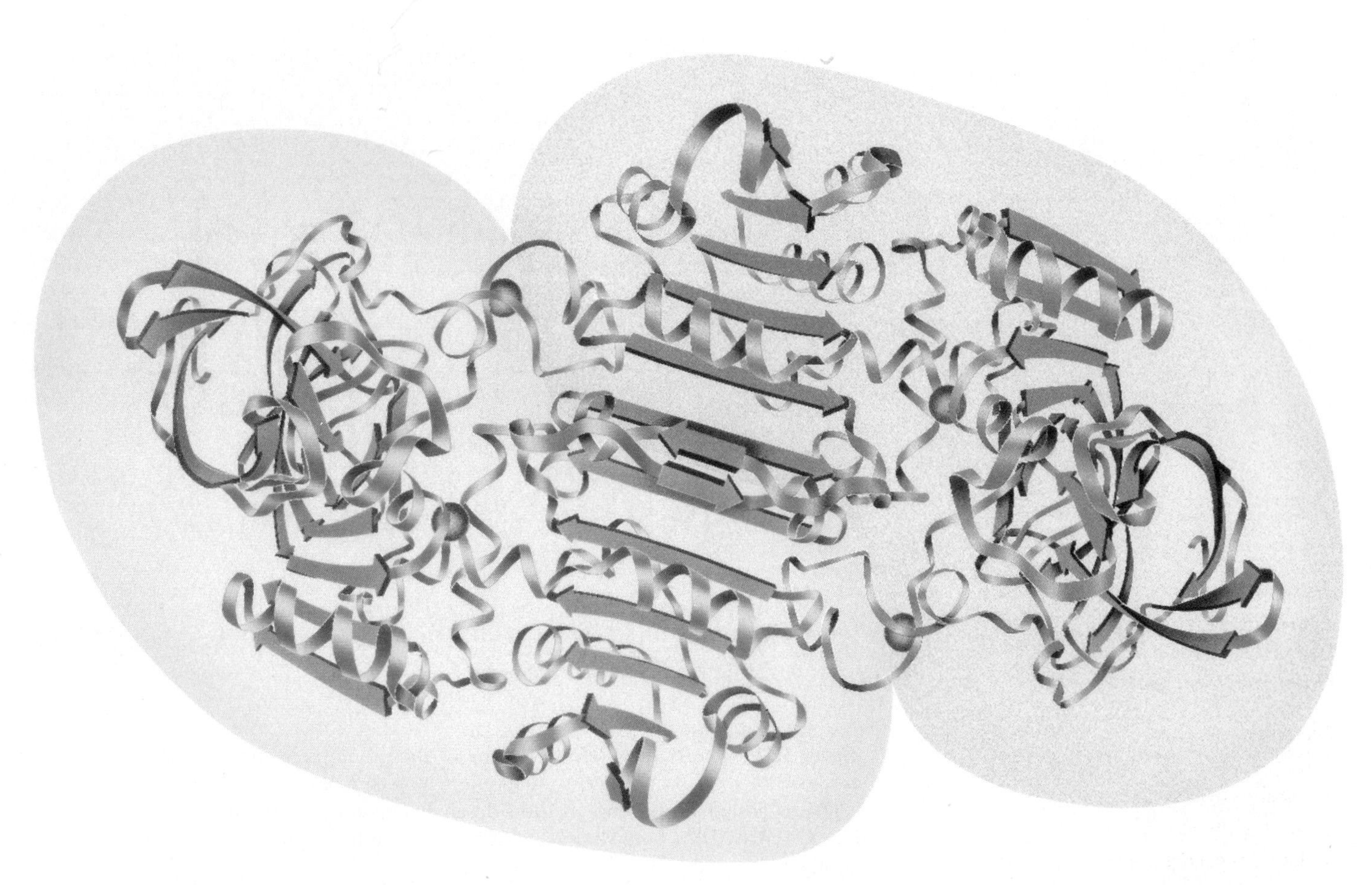

Figure 5.39 The quaternary structure of liver alcohol dehydrogenase. Within each subunit there is a six-stranded parallel sheet. Between the two subunits is a two-stranded antiparallel sheet. The point in the center is a C_2 symmetry axis.

with a twofold axis of symmetry (Figure 5.40). If any additional interactions occur to form a trimer or tetramer, these must use different interfaces on the protein's surface. Many proteins, including concanavalin and prealbumin, form tetramers by means of two sets of isologous interactions, one of which is shown in Figure 5.41. Such structures possess three different twofold axes of symmetry. In contrast, heterologous associations among subunits involve nonidentical interfaces. These surfaces must be complementary, but they are generally not symmetric. As shown in Figure 5.40, heterologous interactions are necessarily open-ended. This can give rise either to a closed cyclic structure, if geometric constraints exist, or to large polymeric assemblies. The closed cyclic structures are far more common and include the trimers of aspartate transcarbamoylase catalytic subunits and the tetramers of neuraminidase and hemerythrin.

Symmetry of Quaternary Structures

One useful way to consider quaternary interactions in proteins involves the symmetry of these interactions. Globular protein subunits are always asymmetric objects. All of the polypeptide's α-carbons are asymmetric, and the polypeptide nearly always folds to form a low-symmetry structure. (The long helical arrays formed by some synthetic polypeptides are an exception.) Thus, protein subunits do not have mirror reflection planes, points, or axes of inversion. The only symmetry operation possible for protein subunits is a rotation. The most common symmetries observed for multisubunit proteins are cyclic symmetry and dihedral symmetry. In **cyclic symmetry,** the subunits are arranged around a single rotation axis, as shown in Figure 5.42. If there are two subunits, the axis is referred to as a *twofold rotation axis.* Rotating the quaternary structure 180° about this axis gives a structure identical to the original

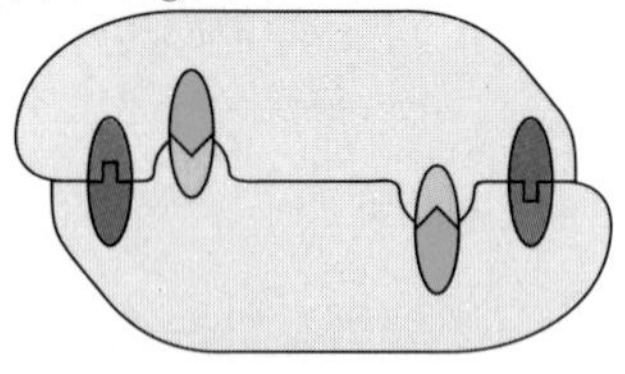

(b) Heterologous association

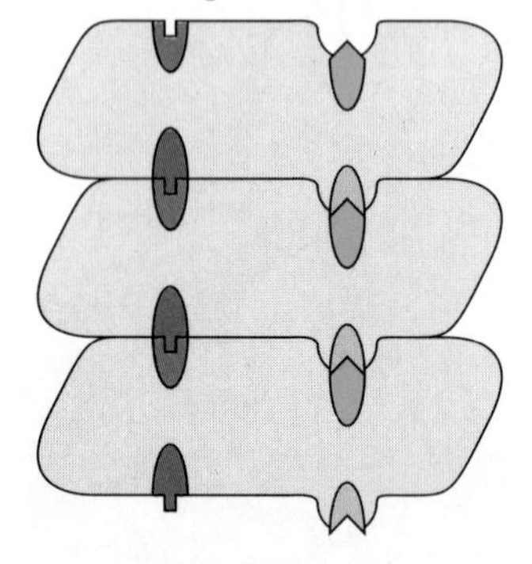

(c) Heterologous tetramer

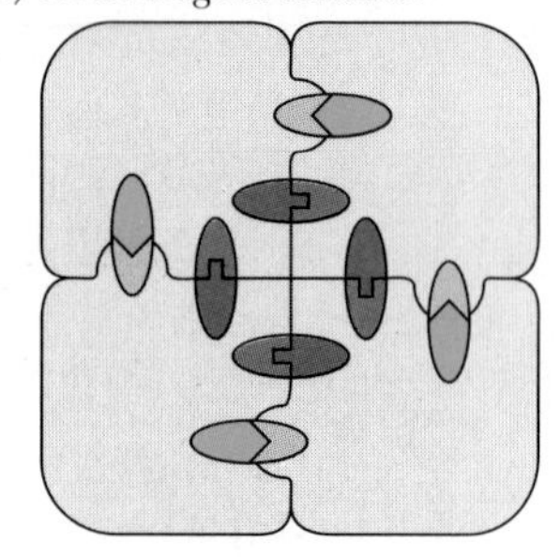

(d) Isologous tetramer

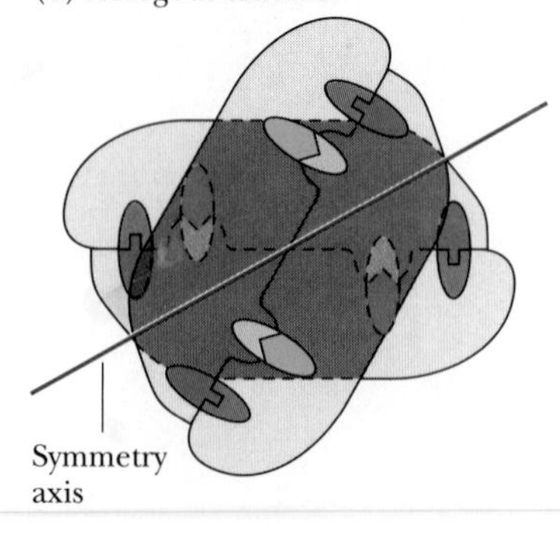

Figure **5.40** Isologous and heterologous associations between protein subunits. (a) An isologous interaction between two subunits with a twofold axis of symmetry perpendicular to the plane of the page. (b) A heterologous interaction that could lead to the formation of a long polymer. (c) A heterologous interaction leading to a closed structure—a tetramer. (d) A tetramer formed by two sets of isologous interactions.

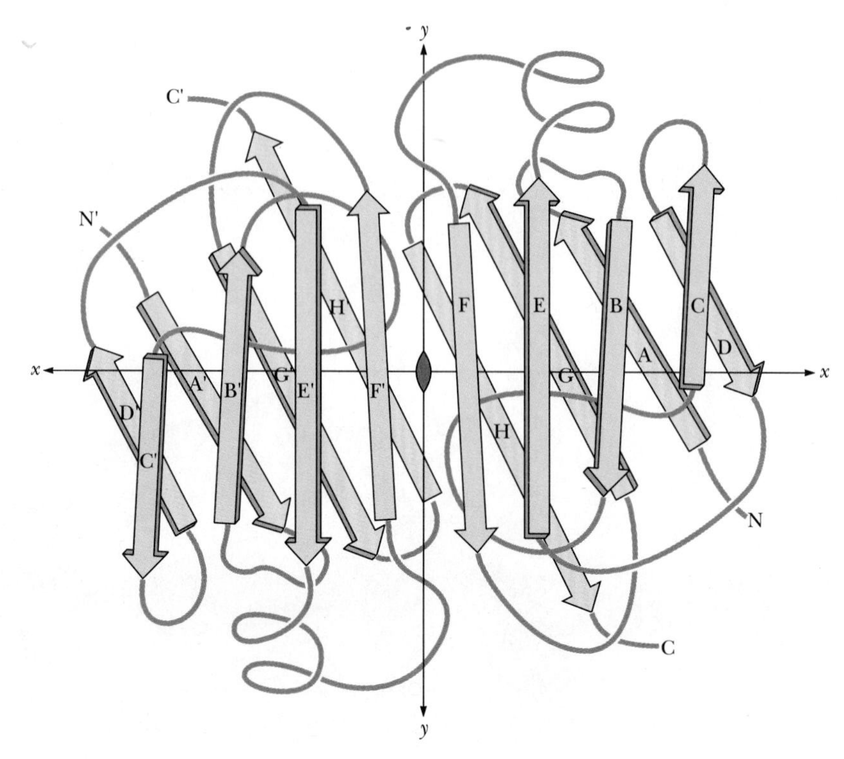

Figure **5.41** The polypeptide backbone of the prealbumin dimer. The monomers associate in a manner that continues the β-sheets. A tetramer is formed by isologous interactions between the side chains extending outward from sheet D′A′G′H′HGAD in both dimers, which pack together nearly at right angles to one another.

one. With three subunits arranged about a threefold rotation axis, a rotation of 120° about that axis gives an identical structure. **Dihedral symmetry** occurs when a structure possesses at least one twofold rotation axis perpendicular to another *n*-fold rotation axis. This type of subunit arrangement (Figure 5.42) occurs in concanavalin A (where $n = 2$) and in insulin (where $n = 3$). Higher symmetry groups, including the tetrahedral, octahedral, and icosahedral symmetries, are much less common among multisubunit proteins, partially because of the large number of asymmetric subunits required to assemble truly symmetric tetrahedra and other high symmetry groups. For example, a truly symmetric tetrahedral protein structure would require 12 identical monomers arranged in triangles, as shown in Figure 5.42. Simple four-subunit tetrahedra of protein monomers, which actually possess dihedral symmetry, are more common in biological systems.

Forces Driving Quaternary Association

The forces that stabilize quaternary structure have been evaluated for a few proteins. Typical dissociation constants for simple two-subunit associations range from 10^{-8} to 10^{-16} *M*. These values correspond to free energies of association of about 50 to 100 kJ/mol at 37°C. Dimerization of subunits is accompanied by both favorable and unfavorable energy changes. The favorable interactions include van der Waals interactions, hydrogen bonds, ionic bonds, and hydrophobic interactions. However, considerable entropy loss occurs when subunits interact. When two subunits move as one, three translational degrees of freedom are lost for one subunit since it is constrained to move with the other one. In addition, many peptide residues at the subunit

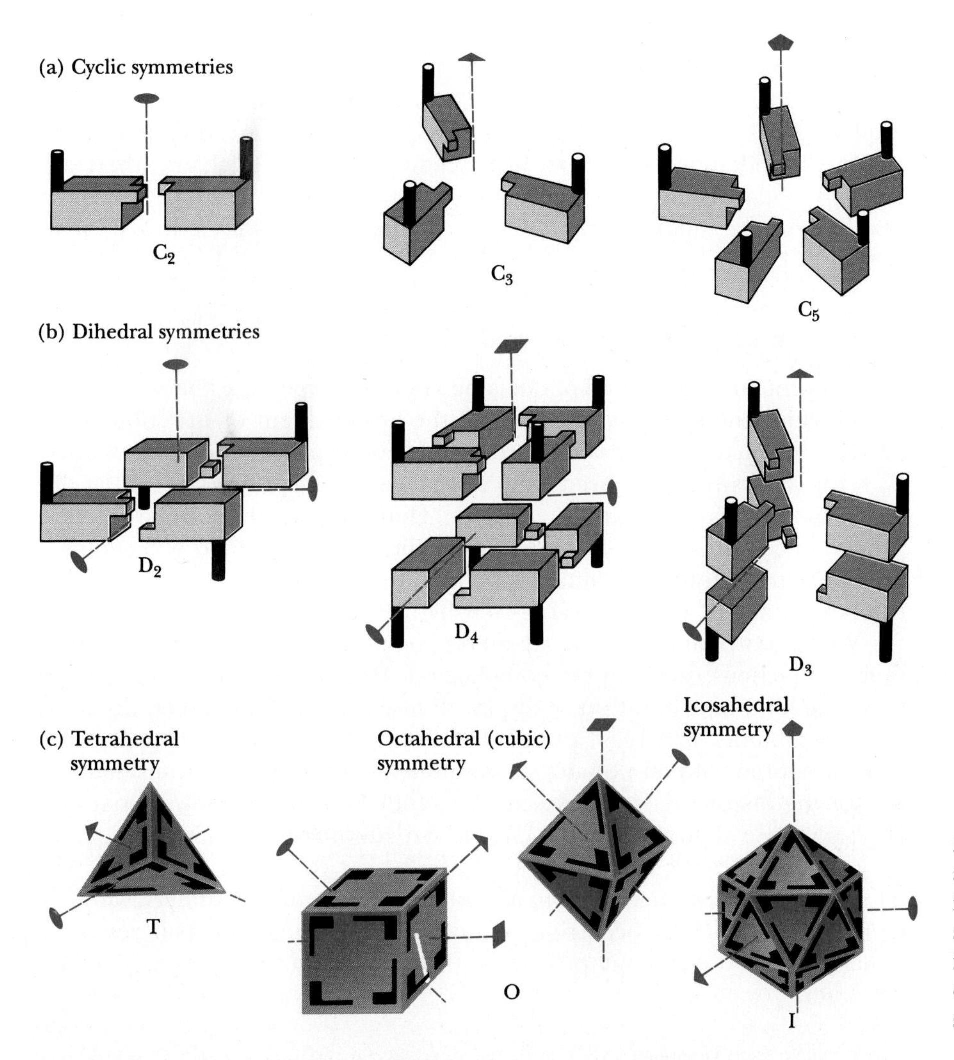

***Figure* 5.42** Several possible symmetric arrays of identical protein subunits, including (a) cyclic symmetry, (b) dihedral symmetry, and (c) cubic symmetry, including examples of tetrahedral (T), octahedral (O), and icosahedral (I) symmetry.

interface, which were previously free to move on the protein surface, now have their movements restricted by the subunit association. This unfavorable energy of association is in the range of 80 to 120 kJ/mol for temperatures of 25 to 37°C. Thus, to achieve stability, the dimerization of two subunits must involve approximately 130 to 220 kJ/mol of favorable interactions.[2] Van der Waals interactions at protein interfaces are numerous, often running to several hundred for a typical monomer–monomer association. This would account for about 150 to 200 kJ/mol of favorable free energy of association. However, when solvent is removed from the protein surface to form the subunit–subunit contacts, nearly as many van der Waals associations are lost as are made. One subunit is simply trading water molecules for peptide residues in the other subunit. As a result, the energy of subunit association due to van der Waals interactions actually contributes little to the stability of the dimer. Hydrophobic interactions, however, are generally very favorable. For many proteins, the subunit association process effectively buries as much as 20 nm^2 of surface area that had previously been exposed to solvent, resulting in as much as 100 to 200 kJ/mol of favorable hydrophobic interactions. Together with whatever polar interactions occur at the protein–protein interface, this is sufficient to account for the observed stabilization that occurs when two protein subunits associate.

[2]For example, 130 kJ/mol of favorable interaction minus 80 kJ/mol of unfavorable interaction equals a net free energy of association of 50 kJ/mol.

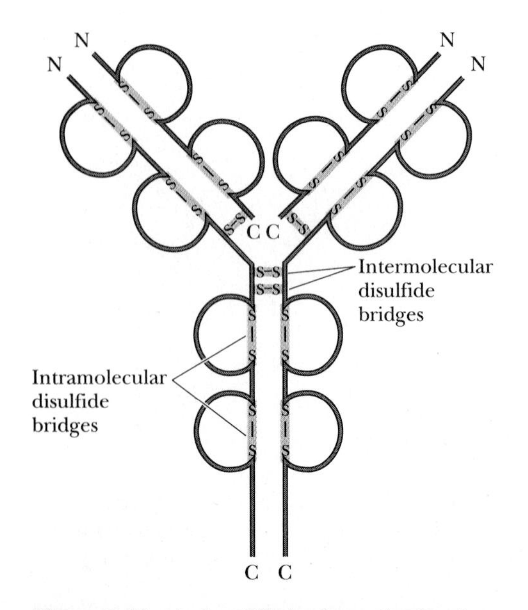

***Figure* 5.43** A simplified schematic drawing of an immunoglobulin molecule showing the intramolecular and intermolecular disulfide bridges. (A space-filling model of the same molecule is shown in Figure 1.11.)

An additional and important factor contributing to the stability of subunit associations for some proteins is the formation of disulfide bonds between different subunits. All antibodies are $\alpha_2\beta_2$-tetramers composed of two heavy chains (53–75 kD) and two relatively light chains (23 kD). In addition to *intrasubunit* disulfide bonds (four per heavy chain, two per light chain), there are two *intersubunit* disulfide bridges holding the two heavy chains together and a disulfide bridge linking each of the two light chains to a heavy chain (Figure 5.43).

Modes and Models for Quaternary Structures

When a protein is composed of only one kind of polypeptide chain, the manner in which the subunits interact and the arrangement of the subunits to produce the quaternary structure are usually simple matters. Sometimes, however, the same protein derived from several different species can exhibit different modes of quaternary interactions. Hemerythrin, the oxygen-carrying protein in certain species of marine invertebrates, is composed of a compact arrangement of four antiparallel α-helices. It is capable of forming dimers, trimers, tetramers, octamers, and even higher aggregates (Figure 5.44).

When two or more distinct peptide chains are involved, the nature of their interactions can be quite complicated. Multimeric proteins with more than one kind of subunit often display different affinities between different pairs of subunits. Whereas strongly denaturing solvents may dissociate the protein entirely into monomers, more subtle denaturing conditions may dissociate the oligomeric structure in a carefully controlled stepwise manner. Hemoglobin is a good example. Strong denaturants dissociate hemoglobin into α- and β-monomers. Using mild denaturing conditions, however, it is possible to dissociate hemoglobin almost completely into $\alpha\beta$-dimers, with few or no free monomers occurring. In this sense, hemoglobin behaves functionally like a two-subunit protein, with each "subunit" composed of an $\alpha\beta$-dimer.

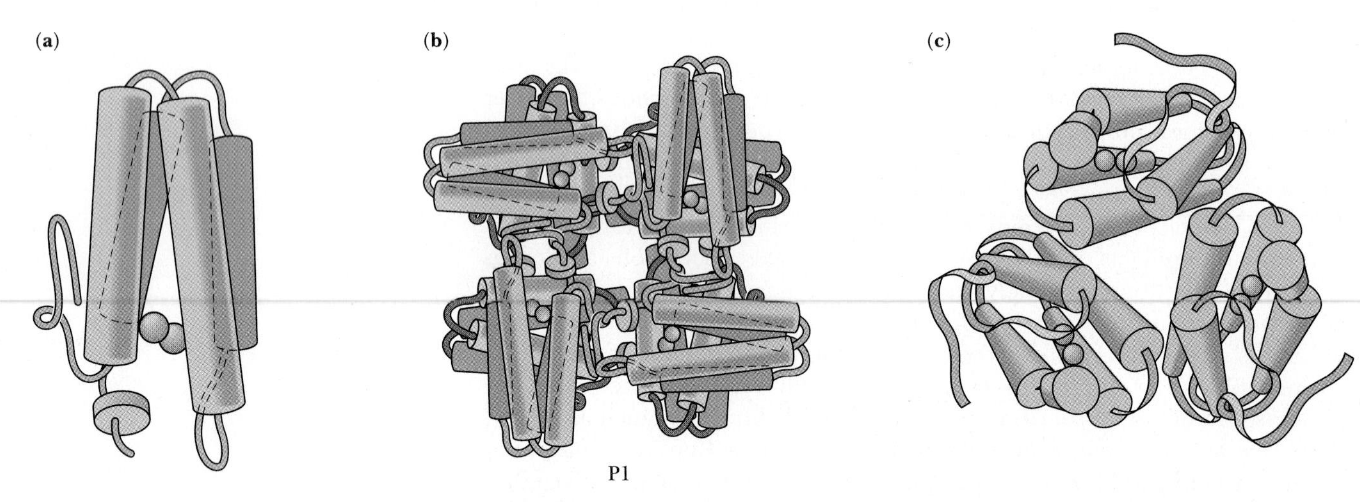

***Figure* 5.44** The oligomeric states of hemerythrin in various marine worms. (a) The hemerythrin in *Thermiste zostericola* crystallized as a monomer; (b) the octameric hemerythrin crystallized from *Phascolopsis gouldii*; (c) the trimeric hemerythrin crystallized from *Siphonosoma* collected in mangrove swamps in Fiji.

Open Quaternary Structures and Polymerization

All of the quaternary structures we have considered to this point have been **closed** structures, with a limited capacity to associate. Many proteins in nature associate to form **open** heterologous structures, which can polymerize more or less indefinitely, creating structures that are both esthetically attractive and functionally important to the cells or tissue in which they exist. One such protein is **tubulin,** the $\alpha\beta$-dimeric protein that polymerizes into long, tubular structures, which are the structural basis of cilia, flagella, and the cytoskeletal matrix. The *microtubule* thus formed (Figure 5.45) may be viewed as consisting of 13 parallel filaments arising from end-to-end aggregation of the tubulin dimers. The polymerization of the coat protein of tobacco mosaic virus (discussed in Chapter 34) is another example of extended quaternary structures. Human immunodeficiency virus, HIV, the causative agent of AIDS (also discussed in Chapter 34), is enveloped by a spherical shell composed of hundreds of coat protein subunits, a large-scale quaternary association.

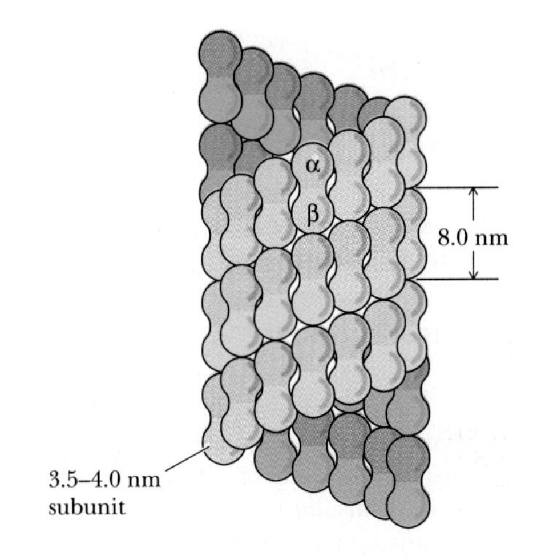

***Figure* 5.45** The structure of a typical microtubule, showing the arrangement of the α- and β-monomers of the tubulin dimer.

Structural and Functional Advantages of Quaternary Association

There are several important reasons for protein subunits to associate in oligomeric structures.

Stability

One general benefit of subunit association is a favorable reduction of the protein's surface-to-volume ratio. The surface-to-volume ratio becomes smaller as the radius of any particle or object becomes larger. (This is because surface area is a function of the radius squared and volume is a function of the radius cubed.) Since interactions within the protein usually tend to stabilize the protein energetically, and since the interaction of the protein surface with solvent water is often energetically unfavorable, it is usually the case that decreased surface-to-volume ratios result in more stable proteins. Subunit association may also serve to shield hydrophobic residues from solvent water. Subunits that recognize either themselves or other subunits avoid any errors arising in genetic translation by binding mutant forms of the subunits less tightly.

Genetic Economy and Efficiency

Oligomeric association of protein monomers is genetically economical for an organism. Less DNA is required to code for a monomer that assembles into a homomultimer than for a large polypeptide of the same molecular mass. Another way to look at this is to realize that virtually all of the information that determines oligomer assembly and subunit–subunit interaction is contained in the genetic material needed to code for the monomer. For example, HIV protease, an enzyme that is a dimer of identical subunits, performs a catalytic function similar to homologous cellular enzymes that are single polypeptide chains of twice the molecular mass (Chapter 13).

Bringing Catalytic Sites Together

Many enzymes (see Chapters 11 to 14) derive at least some of their catalytic power from oligomeric associations of monomer subunits. This can happen

A Deeper Look

Faster-Acting Insulin: Genetic Engineering Solves a Quaternary Structure Problem

Insulin is a peptide hormone, secreted by the pancreas, that regulates glucose metabolism in the body. Insufficient production of insulin or failure of insulin to stimulate target sites in liver, muscle, and adipose tissue leads to the serious metabolic disorder known as *diabetes mellitus.* Diabetes afflicts millions of people worldwide. Diabetic individuals typically exhibit high levels of glucose in the blood, but insulin injection therapy allows diabetic individuals to maintain normal levels of blood glucose.

Insulin is composed of two peptide chains covalently linked by disulfide bonds (Figures 4.17 and 5.34). This "monomer" of insulin is the active form that binds to receptors in target cells. However, in solution, insulin spontaneously forms dimers, which themselves aggregate to form hexamers. The surface of the insulin molecule that self-associates to form hexamers is also the surface that binds to insulin receptors in target cells. Thus, hexamers of insulin are inactive.

Insulin released from the pancreas is monomeric and acts rapidly at target tissues. However, when insulin is administered (by injection) to a diabetic patient, the insulin hexamers dissociate slowly, and the patient's blood glucose levels typically drop slowly (over a period of several hours).

In 1988, G. Dodson showed that insulin could be genetically engineered to prefer the monomeric (active) state. Dodson and his colleagues used recombinant DNA technology (discussed in Chapter 8) to produce insulin with an aspartate residue replacing a proline at the contact interface between adjacent subunits. The negative charge on the Asp side chain creates electrostatic repulsion between subunits and increases the dissociation constant for the hexamer ⇄ monomer equilibrium. Injection of this mutant insulin into test animals produced more rapid decreases in blood glucose than did ordinary insulin. The Danish pharmaceutical company Novo is conducting clinical trials of the mutant insulin, which may eventually replace ordinary insulin in treatment of diabetes.

in several ways. The monomer may not constitute a complete enzyme active site. Formation of the oligomer may bring all the necessary catalytic groups together to form an active enzyme. For example, the active sites of bacterial glutamine synthetase are formed from pairs of adjacent subunits. The dissociated monomers are inactive.

Oligomeric enzymes may also carry out different but related reactions on different subunits. Thus, tryptophan synthase is a tetramer consisting of pairs of different subunits, $\alpha_2\beta_2$. Purified α-subunits catalyze the following reaction:

$$\text{Indoleglycerol phosphate} \rightleftharpoons \text{Indole} + \text{glyceraldehyde-3-phosphate}$$

and the β-subunits catalyze this reaction:

$$\text{Indole} + \text{L-serine} \rightleftharpoons \text{L-Tryptophan}$$

Indole, the product of the α-reaction and the reactant for the β-reaction, is passed directly from the α-subunit to the β-subunit and cannot be detected as a free intermediate.

Cooperativity

There is another, more important reason for monomer subunits to associate into oligomeric complexes. Most oligomeric enzymes regulate catalytic activity by means of subunit interactions, which may give rise to cooperative phenomena. Multisubunit proteins typically possess multiple binding sites for a given ligand. If the binding of ligand at one site changes the affinity of the protein for ligand at the other binding sites, the binding is said to be **cooperative.** Increases in affinity at subsequent sites represent positive cooperativity, whereas decreases in affinity correspond to negative cooperativity. The points of contact between protein subunits provide a mechanism for communication between the subunits. This in turn provides a way in which the binding of ligand to one subunit can influence the binding behavior at the other subunits. Such cooperative behavior, discussed in greater depth in Chapter 12, is the underlying mechanism for regulation of many biological processes.

Problems

1. The central rod domain of a keratin protein is approximately 312 residues in length. What is the length (in Å) of the keratin rod domain? If this same peptide segment were a true α-helix, how long would it be? If the same segment were a β-sheet, what would its length be?
2. A teenager can grow 4 in. in a year during a "growth spurt." Assuming that the increase in height is due to vertical growth of collagen fibers (in bone), calculate the number of collagen helix turns synthesized per minute.
3. Discuss the potential contributions to hydrophobic and van der Waals interactions and ionic and hydrogen bonds for the side chains of Asp, Leu, Tyr, and His in a protein.
4. Figure 5.38 shows that Pro is the amino acid least commonly found in α-helices but most commonly found in β-turns. Discuss the reasons for this behavior.
5. For flavodoxin in Figure 5.30, identify the right-handed cross-overs and the left-handed cross-overs in the parallel β-sheet.
6. Choose any three regions in the Ramachandran plot and discuss the likelihood of observing that combination of ϕ and ψ in a peptide or protein. Defend your answer using suitable molecular models of a peptide.
7. A new protein of unknown structure has been purified. Gel filtration chromatography reveals that the native protein has a molecular weight of 240,000. Chromatography in the presence of 6 *M* guanidine hydrochloride yields only a peak for a protein of M_r 60,000. Chromatography in the presence of 6 *M* guanidine hydrochloride and 10 m*M* β-mercaptoethanol yields peaks for proteins of M_r 34,000 and 26,000. Explain what can be determined about the structure of this protein from these data.
8. Two polypeptides, A and B, have similar tertiary structures, but A normally exists as a monomer, whereas B exists as a tetramer, B_4. What differences might be expected in the amino acid composition of A versus B?
9. The hemagglutinin protein in influenza virus contains a remarkably long α-helix, with 53 residues.
   a. How long is this α-helix (in nm)?
   b. How many turns does this helix have?
   c. Each residue in an α-helix is involved in two H bonds. How many H bonds are present in this helix?

Further Reading

Abeles, R., Frey, P., and Jencks, W., 1992. *Biochemistry.* Boston: Jones and Bartlett.

Cantor, C. R., and Schimmel, P. R., 1980. *Biophysical Chemistry, Part I: The Conformation of Biological Macromolecules.* New York: W. H. Freeman and Co.

Chothia, C., 1984. Principles that determine the structure of proteins. *Annual Review of Biochemistry* **53:**537–572.

Creighton, T. E., 1983. *Proteins: Structure and Molecular Properties.* New York: W. H. Freeman and Co.

Dickerson, R. E., and Geis, I., 1969. *The Structure and Action of Proteins.* New York: Harper and Row.

Englander, S. W., and Mayne, L., 1992. Protein folding studied using hydrogen exchange labeling and two-dimensional NMR. *Annual Review of Biophysics and Biomolecular Structure* **21**:243–265.

Hardie, D. G., and Coggins, J. R., eds., 1986. *Multidomain Proteins: Structure and Evolution.* New York: Elsevier.

Judson, H. F., 1979. *The Eighth Day of Creation.* New York: Simon and Schuster.

Klotz, I. M., Langerman, N. R., and Darnell, D. W., 1970. Quaternary structure of proteins. *Annual Review of Biochemistry* **39:**25–62.

Petsko, G. A., and Ringe, D., 1984. Fluctuations in protein structure from X-ray diffraction. *Annual Review of Biophysics and Bioengineering* **13:**331–371.

Richardson, J. S., 1981. The anatomy and taxonomy of protein structure. *Advances in Protein Chemistry* **34:**167–339.

Rossman, M. G., and Argos, P., 1981. Protein folding. *Annual Review of Biochemistry* **50:**497–532.

Salemme, F. R., 1983. Structural properties of protein β-sheets. *Progress in Biophysics and Molecular Biology* **42:**95–133.

Torchia, D. A., 1984. Solid state NMR studies of protein internal dynamics. *Annual Review of Biophysics and Bioengineering* **13:**125–144.

Wagner, G., Hyberts, S., and Havel, T., 1992. NMR structure determination in solution: A critique and comparison with X-ray crystallography. *Annual Review of Biophysics and Biomolecular Structure* **21**:167–242.

"We have discovered the secret of life!"

Proclamation by Francis H. C. Crick to patrons of The Eagle, a pub in Cambridge, England (1953)

Chapter 6

Nucleotides and Nucleic Acids

James Watson and Francis Crick point out features of their model for the structure of DNA.

Outline

6.1 Nitrogenous Bases

6.2 The Pentoses of Nucleotides and Nucleic Acids

6.3 Nucleosides Are Formed by Joining a Nitrogenous Base to a Sugar

6.4 Nucleotides Are Nucleoside Phosphates

6.5 Nucleic Acids Are Polynucleotides

6.6 Classes of Nucleic Acids

6.7 Hydrolysis of Nucleic Acids

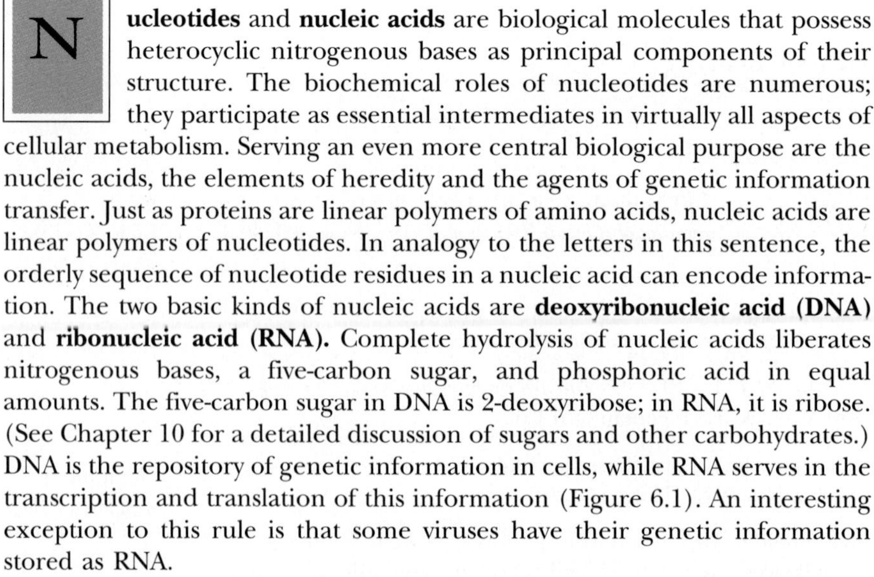

Nucleotides and **nucleic acids** are biological molecules that possess heterocyclic nitrogenous bases as principal components of their structure. The biochemical roles of nucleotides are numerous; they participate as essential intermediates in virtually all aspects of cellular metabolism. Serving an even more central biological purpose are the nucleic acids, the elements of heredity and the agents of genetic information transfer. Just as proteins are linear polymers of amino acids, nucleic acids are linear polymers of nucleotides. In analogy to the letters in this sentence, the orderly sequence of nucleotide residues in a nucleic acid can encode information. The two basic kinds of nucleic acids are **deoxyribonucleic acid (DNA)** and **ribonucleic acid (RNA).** Complete hydrolysis of nucleic acids liberates nitrogenous bases, a five-carbon sugar, and phosphoric acid in equal amounts. The five-carbon sugar in DNA is 2-deoxyribose; in RNA, it is ribose. (See Chapter 10 for a detailed discussion of sugars and other carbohydrates.) DNA is the repository of genetic information in cells, while RNA serves in the transcription and translation of this information (Figure 6.1). An interesting exception to this rule is that some viruses have their genetic information stored as RNA.

This chapter describes the chemistry of nucleotides and the major classes of nucleic acids. Chapter 7 presents methods for determination of nucleic acid primary structure (nucleic acid sequencing) and describes the higher orders of nucleic acid structure. Chapter 8 introduces the *molecular biology of recombinant DNA:* the construction and characterization of novel DNA molecules assembled by combining segments from other DNA molecules.

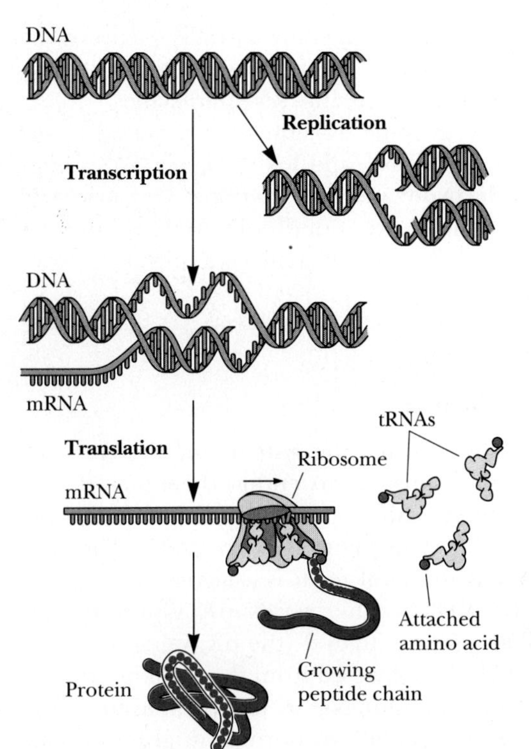

Replication
DNA replication yields two DNA molecules identical to the original one, ensuring transmission of genetic information to daughter cells with exceptional fidelity.

Transcription
The sequence of bases on DNA are recorded as complementary bases on a single-stranded mRNA molecule.

Translation
Three-base codons on the mRNA corresponding to specific amino acids direct the sequence of building a protein. These codons are recognized by tRNAs (transfer RNAs) carrying the appropriate amino acids. Ribosomes are the "machinery" for protein synthesis.

Figure **6.1** The fundamental process of information transfer in cells. Information encoded in the nucleotide sequence of DNA is transcribed through synthesis of an RNA molecule whose sequence is dictated by the DNA sequence. As the sequence of this RNA is read (as groups of three consecutive nucleotides) by the protein synthesis machinery, it is translated into the sequence of amino acids in a protein. This information transfer system is encapsulated in the dogma:
DNA → RNA → protein.

6.1 Nitrogenous Bases

The bases of nucleotides and nucleic acids are derivatives of either **pyrimidine** or **purine.** Pyrimidines are six-membered heterocyclic aromatic rings containing two nitrogen atoms (Figure 6.2a). The atoms are numbered in a clockwise fashion, as shown in the figure. The purine ring structure is represented by the combination of a pyrimidine ring with a five-membered imidazole ring to yield a fused ring system (Figure 6.2b). The nine atoms in this system are numbered according to the convention shown.

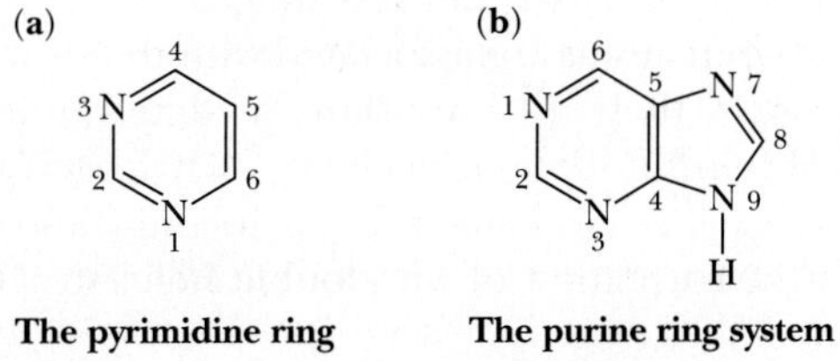

(a) **The pyrimidine ring**

(b) **The purine ring system**

Figure **6.2** (a) The pyrimidine ring system; by convention, atoms are numbered as indicated. (b) The purine ring system, atoms numbered as shown.

The pyrimidine ring system is planar, while the purine system deviates somewhat from planarity in having a slight pucker between its imidazole and pyrimidine portions. Both are relatively insoluble in water, as might be expected from their pronounced aromatic character.

Cytosine (2-oxy-4-amino pyrimidine)

Uracil (2-oxy-4-oxy pyrimidine)

Thymine (2-oxy-4-oxy 5-methyl pyrimidine)

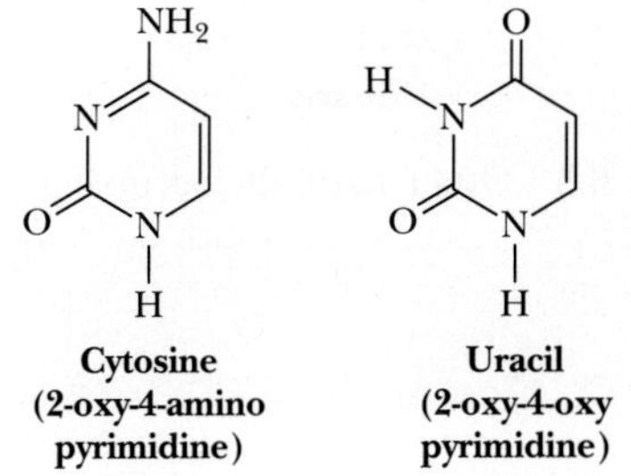

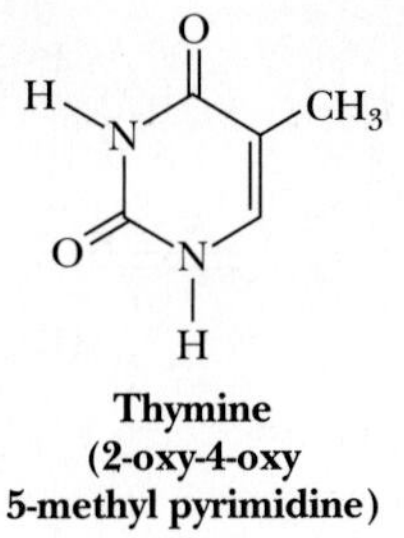

Figure **6.3** The common pyrimidine bases—cytosine, uracil, and thymine—in the tautomeric forms predominant at pH 7.

Common Pyrimidines and Purines

The common naturally occurring pyrimidines are **cytosine, uracil,** and **thymine** (5-methyluracil) (Figure 6.3). Cytosine and thymine are the pyrimidines typically found in DNA, whereas cytosine and uracil are common in RNA. To

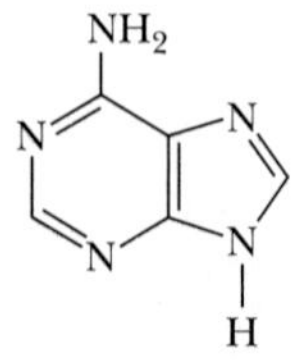

Adenine
(6-amino purine)

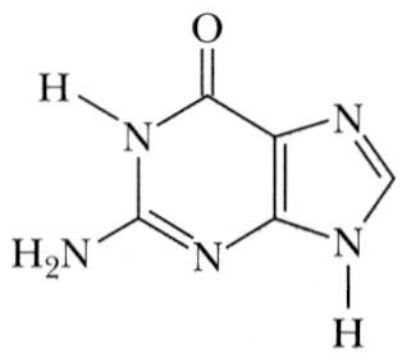

Guanine
(2-amino-6-oxy purine)

Figure **6.4** The common purine bases—adenine and guanine—in the tautomeric forms predominant at pH 7.

view this generality another way, the uracil component of DNA occurs as the 5-methyl variety, thymine. Various pyrimidine derivatives, such as dihydrouracil, are present as minor constituents in certain RNA molecules.

Adenine (6-amino purine) and **guanine** (2-amino-6-oxy purine), the two common purines, are found in both DNA and RNA (Figure 6.4). Other naturally occurring purines include **hypoxanthine, xanthine,** and **uric acid** (Figure 6.5). Hypoxanthine and xanthine are found only rarely as constituents of nucleic acids. Uric acid, the most oxidized state for a purine, is never found in nucleic acids.

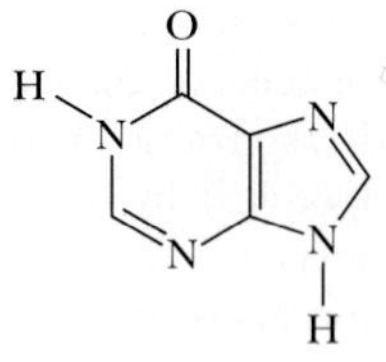

Hypoxanthine

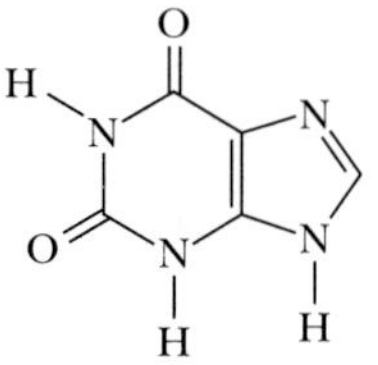

Xanthine

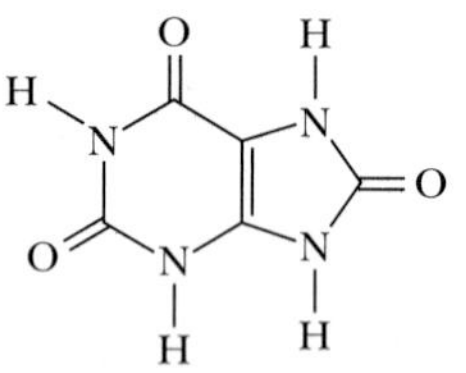

Uric acid

Figure **6.5** Other naturally occurring purines—hypoxanthine, xanthine, and uric acid.

Properties of Pyrimidines and Purines

The aromaticity of the pyrimidine and purine ring systems and the electron-rich nature of their —OH and $—NH_2$ substituents endow them with the capacity to undergo **keto–enol tautomeric shifts.** That is, pyrimidines and purines exist as tautomeric pairs, as shown in Figure 6.6 for uracil. The keto tautomer is called a **lactam,** whereas the enol form is a **lactim.** The lactam form vastly predominates at neutral pH. In other words, pK_a values for ring nitrogen atoms 1 and 3 in uracil are greater than 8 (the pK_a value for N-3 is 9.5) (Table 6.1). In contrast, as might be expected from the form of cytosine that predominates at pH 7, the pK_a value for N-3 in this pyrimidine is 4.5. Similarly, tautomeric forms can be represented for purines, as given for guanine in Figure 6.7.

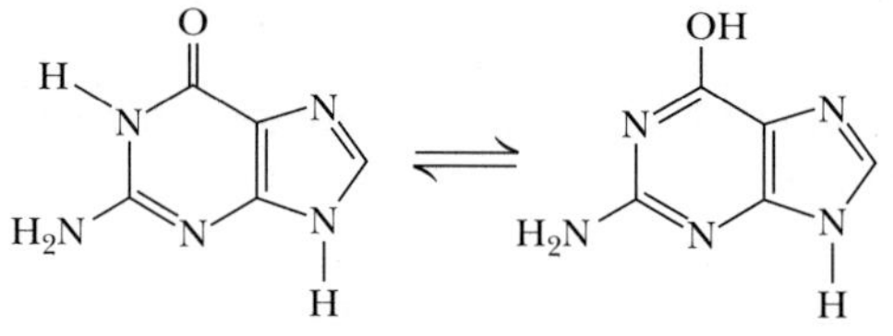

Keto form **Enol form**

Figure **6.7** The tautomerism of the purine, guanine.

Here, the pK_a value is 9.4 for N-1 and less than 5 for N-3. These pK_a values specify whether hydrogen atoms are associated with the various ring nitrogens at neutral pH. As such, they are important in determining whether these nitrogens serve as H bond donors or acceptors. Hydrogen bonding between purine and pyrimidine bases is fundamental to the biological functions of nucleic acids, as in the formation of the double helix structure of DNA (see

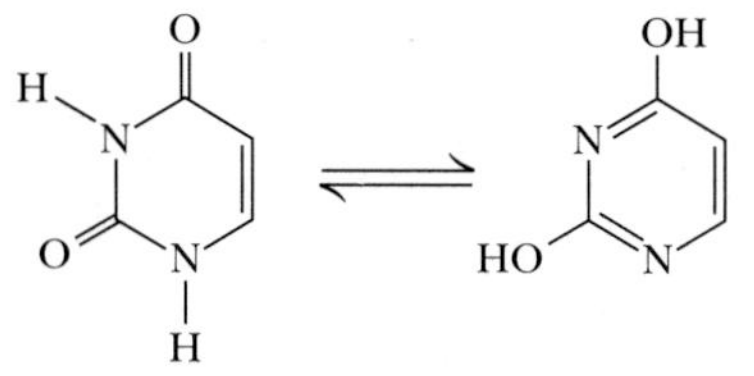

Lactam **Lactim**

Figure **6.6** The keto/enol tautomerism of uracil.

Table **6.1**
Proton Dissociation Constants (pK_a Values) for Nucleotides

Nucleotide	pK_a Base-N	pK_1 Phosphate	pK_2 Phosphate
5′-AMP	3.8 (N-1)	0.9	6.1
5′-GMP	9.4 (N-1)	0.7	6.1
	2.4 (N-7)		
5′-CMP	4.5 (N-3)	0.8	6.3
5′-UMP	9.5 (N-3)	1.0	6.4

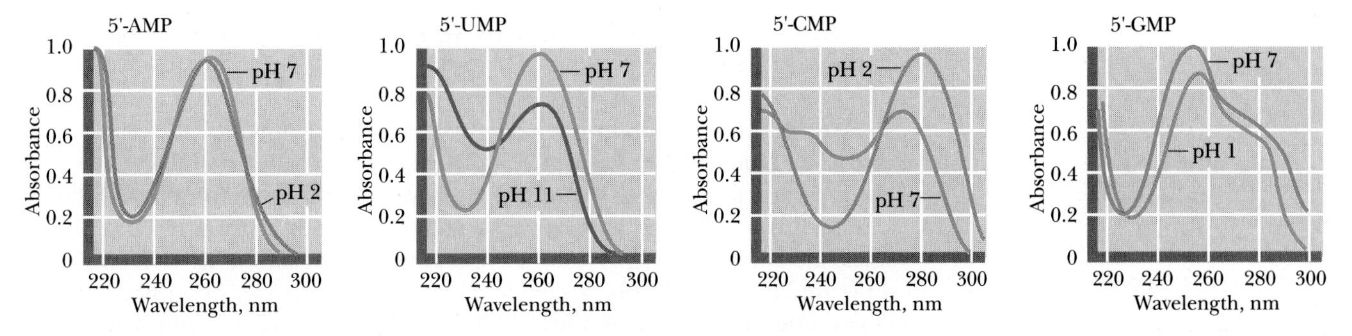

***Figure* 6.8** The UV absorption spectra of the common ribonucleotides.

Section 6.6). The important functional groups participating in H-bond formation are the amino groups of cytosine, adenine, and guanine; the ring nitrogens at position 3 of pyrimidines and position 1 of purines; and the strongly electronegative oxygen atoms attached at position 4 of uracil and thymine, position 2 of cytosine, and position 6 of guanine (see Figure 6.21).

Another property of pyrimidines and purines is their strong absorbance of ultraviolet (UV) light, which is also a consequence of the aromaticity of their heterocyclic ring structures. Figure 6.8 shows characteristic absorption spectra of several of the common bases of nucleic acids—adenine, uracil, cytosine, and guanine—in their nucleotide forms: AMP, UMP, CMP, and GMP (see Section 6.4). This property is particularly useful in quantitative and qualitative analysis of nucleotides and nucleic acids.

6.2 The Pentoses of Nucleotides and Nucleic Acids

Five-carbon sugars are called **pentoses** (see Chapter 10). RNA contains the pentose D-ribose, while 2-deoxy-D-ribose is found in DNA. In both instances, the pentose is in the five-membered ring form known as *furanose:* D-ribofuranose for RNA and 2-deoxy-D-ribofuranose for DNA (Figure 6.9). When these ribofuranoses are found in nucleotides, their atoms are numbered as 1′, 2′, 3′, and so on to distinguish them from the ring atoms of the nitrogenous bases. As we shall see, the seemingly minor difference of a hydroxyl group at the 2′-position has far-reaching effects on the secondary structures available to RNA and DNA, as well as their relative susceptibilities to chemical and enzymatic hydrolysis.

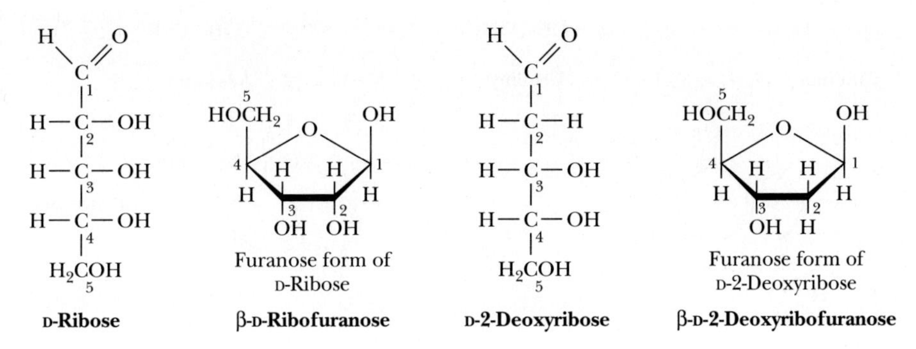

***Figure* 6.9** Furanose structures—ribose and deoxyribose.

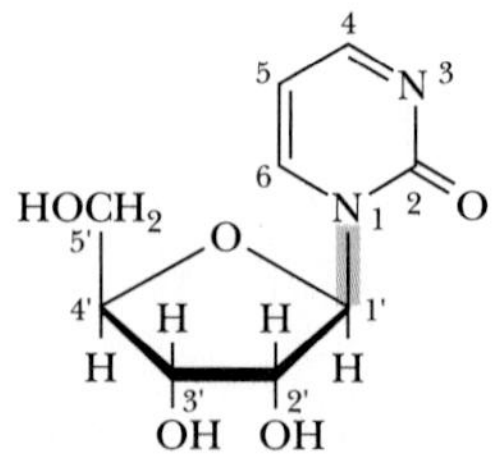

β-N_1-glycosidic bond in pyrimidine ribonucleosides

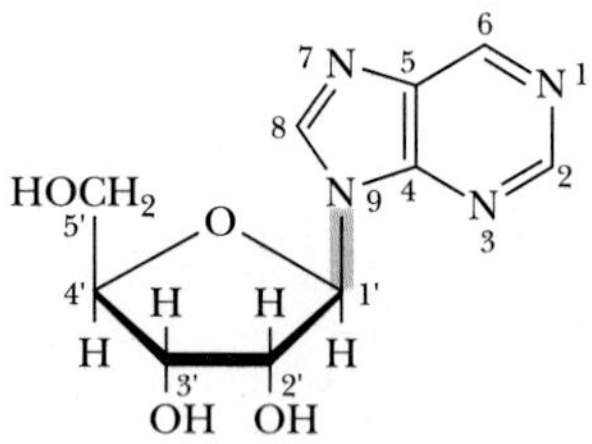

β-N_9-glycosidic bond in purine ribonucleosides

Figure **6.10** Glycosidic bonds link nitrogenous bases and sugars to form nucleosides.

6.3 Nucleosides Are Formed by Joining a Nitrogenous Base to a Sugar

Nucleosides are compounds formed when a base is linked to a sugar via a **glycosidic bond** (Figure 6.10). Glycosidic bonds by definition involve the carbonyl carbon atom of the sugar, which in cyclic structures is joined to the ring O atom (see Chapter 10). Such carbon atoms are called **anomeric.** In nucleosides, the bond is an *N*-glycoside since it connects the anomeric C-1′ to N-1 of a pyrimidine or to N-9 of a purine. Glycosidic bonds can be either α or β, depending on their orientation relative to the anomeric C atom. Glycosidic bonds in nucleosides and nucleotides are always of the β-configuration, as represented in Figure 6.10. Nucleosides are named by adding the ending *-idine* to the root name of a pyrimidine or *-osine* to the root name of a purine. The common nucleosides are thus **cytidine, uridine, thymidine, adenosine,** and **guanosine.** The structures shown in Figure 6.11 are *ribonucleosides. Deoxyribonucleosides,* in contrast, lack a 2′-OH group on the pentose. The nucleoside formed by hypoxanthine and ribose is **inosine.**

Nucleoside Conformation

In nucleosides, rotation of the base about the glycosidic bond is sterically hindered, principally by the hydrogen atom on the C-2′ carbon of the furanose. (This hindrance is most easily seen and appreciated by manipulating accurate molecular models of these structures.) Consequently, nucleosides and nucleotides (see next section) exist in either of two conformations, designated *syn* and *anti* (Figure 6.12). For pyrimidines in the syn conformation, the oxygen substituent at position C-2 lies immediately above the furanose ring; in the anti conformation, this steric interference is avoided. Consequently, pyrimidine nucleosides favor the anti conformation. Purine nucleosides can adopt either the syn or anti conformation. In either conformation, the roughly planar furanose and base rings are not coplanar but lie at approximately right angles to one another.

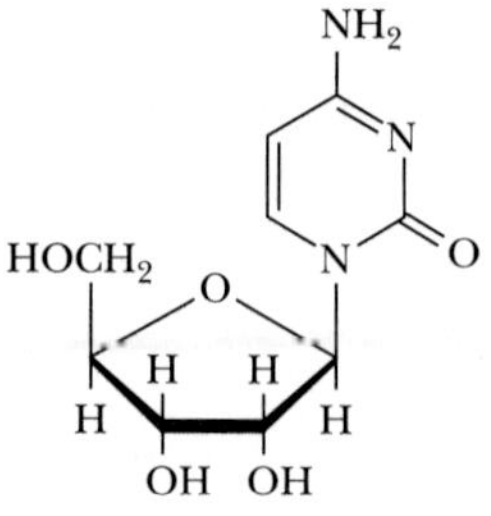

Cytidine 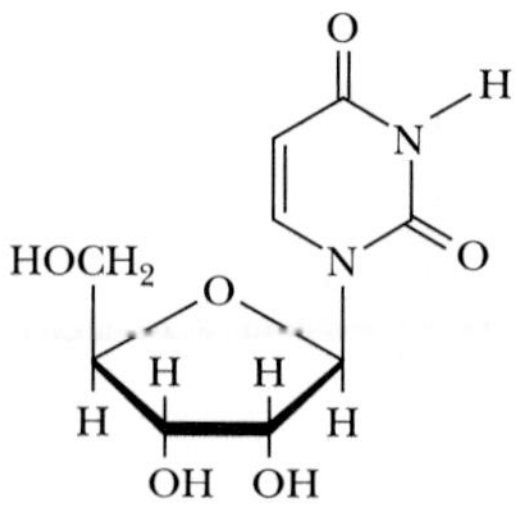 **Uridine** 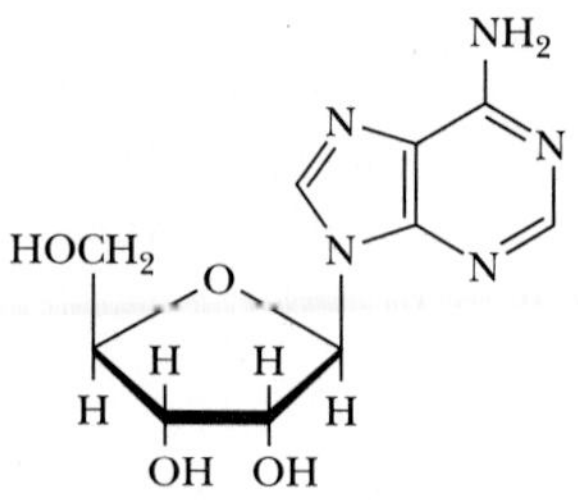 **Adenosine**

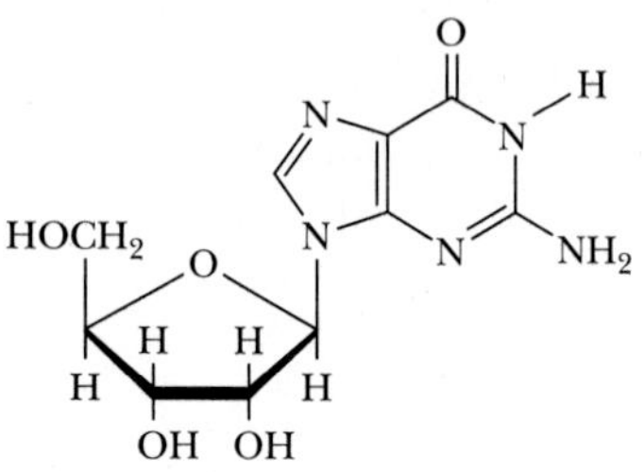

Guanosine 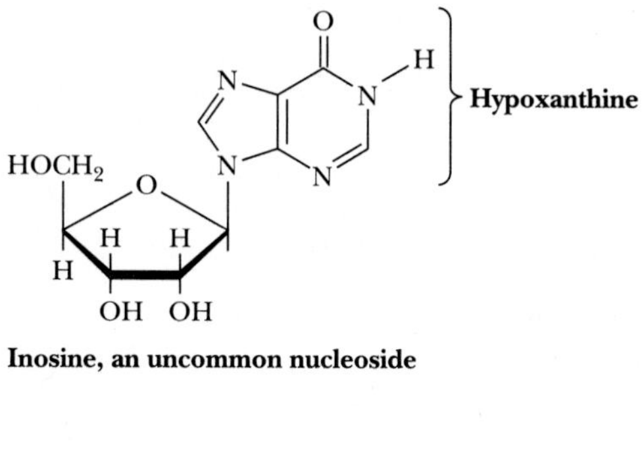 **Inosine, an uncommon nucleoside** **Hypoxanthine**

Figure **6.11** The common ribonucleosides—cytidine, uridine, adenosine, and guanosine. Also, inosine drawn in anti conformation.

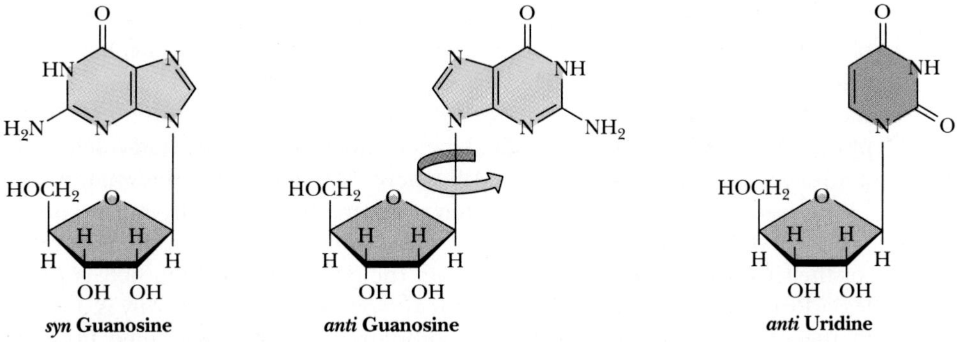

***syn* Guanosine** ***anti* Guanosine** ***anti* Uridine**

***Figure* 6.12** Rotation around the glycosidic bond is sterically hindered; syn versus anti conformations in nucleosides are shown.

Nucleosides Are More Water-Soluble Than Free Bases

Nucleosides are much more water-soluble than the free bases because of the hydrophilicity of the sugar moiety. Like glycosides (see Chapter 10), nucleosides are relatively stable in alkali. Pyrimidine nucleosides are also resistant to acid hydrolysis, but purine nucleosides are easily hydrolyzed in acid to yield the free base and pentose.

6.4 Nucleotides Are Nucleoside Phosphates

A **nucleotide** results when phosphoric acid is esterified to a sugar —OH group of a nucleoside. The nucleoside ribose ring has three —OH groups available for esterification, at C-2′, C-3′, and C-5′ (although 2′-deoxyribose has only two). The vast majority of monomeric nucleotides in the cell are **ribonucleotides** having 5′-phosphate groups. Figure 6.13 shows the structures of the

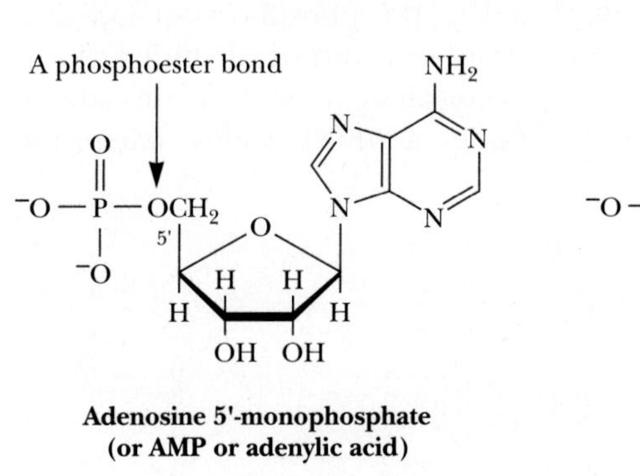

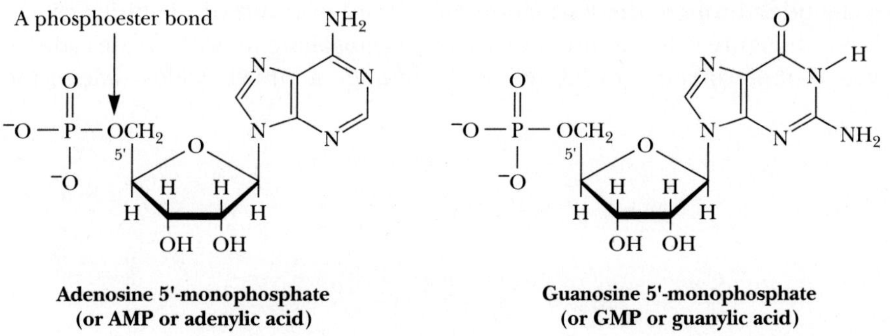

Adenosine 5'-monophosphate (or AMP or adenylic acid)

Guanosine 5'-monophosphate (or GMP or guanylic acid)

Uridine 5'-monophosphate (or UMP or uridylic acid)

Cytidine 5'-monophosphate (or CMP or cytidylic acid)

A nucleoside 3'-monophosphate 3'-AMP

***Figure* 6.13** Structures of the four common ribonucleotides—AMP, GMP, CMP, and UMP—together with their two sets of full names, for example, adenosine 5′-monophosphate and adenylic acid.

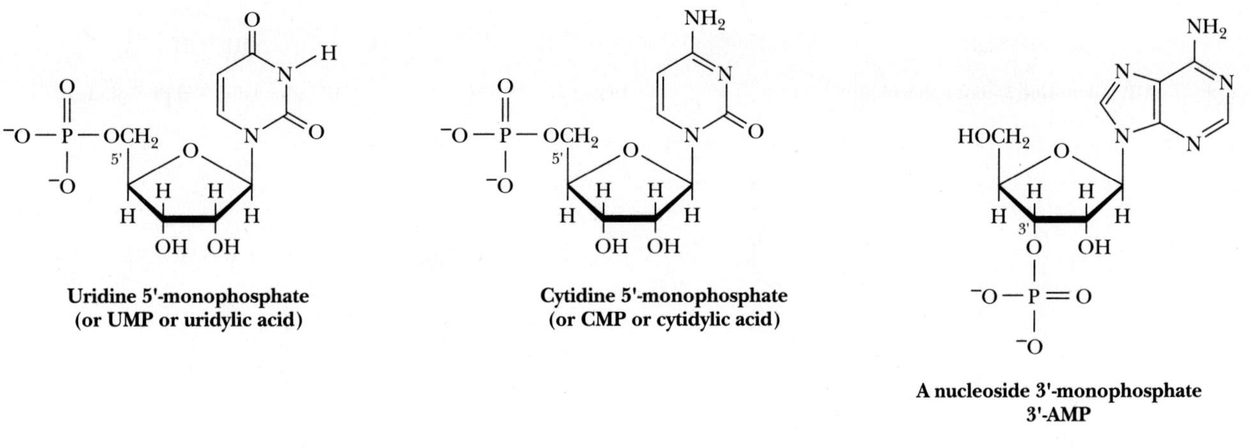

common four *ribonucleotides,* whose formal names are **adenosine 5′-monophosphate, guanosine 5′-monophosphate, cytidine 5′-monophosphate,** and **uridine 5′-monophosphate.** These compounds are more often referred to by their abbreviations: **5′-AMP, 5′-GMP, 5′-CMP,** and **5′-UMP,** or even more simply as **AMP, GMP, CMP,** and **UMP.** Nucleoside 3′-phosphates and nucleoside 2′-phosphates (3′-NMP and 2′-NMP, where N is a generic designation for "nucleoside") do not occur naturally, but are biochemically important as products of polynucleotide or nucleic acid hydrolysis. Because the pK_a value for the first dissociation of a proton from the phosphoric acid moiety is 1.0 or less (Table 6.1), the nucleotides have acidic properties. This acidity is implicit in the other names by which these substances are known—**adenylic acid, guanylic acid, cytidylic acid,** and **uridylic acid.** The pK_a value for the second dissociation, pK_2, is about 6.0, so at neutral pH or above, the net charge on a nucleoside monophosphate is −2. Nucleic acids, which are polymers of nucleoside monophosphates, derive their name from the acidity of these phosphate groups.

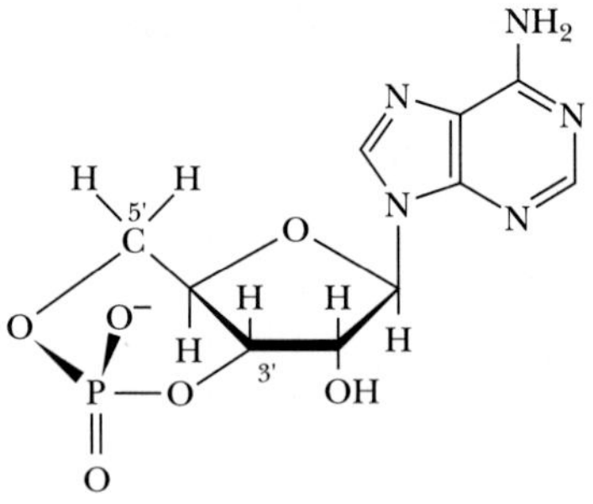

3',5'-Cyclic AMP

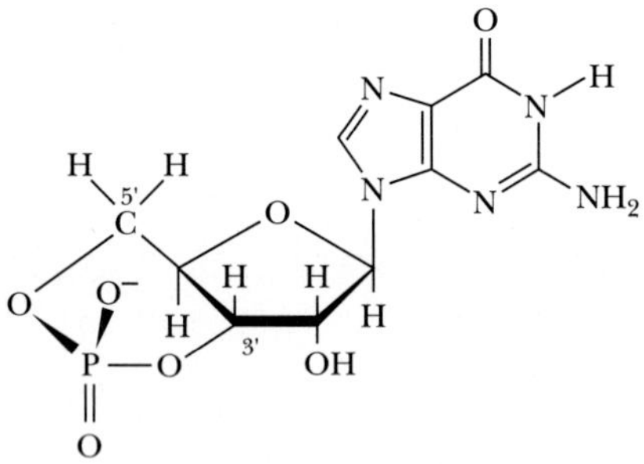

3',5'-Cyclic GMP

Figure **6.14** Structures of the cyclic nucleotides cAMP and cGMP.

Cyclic Nucleotides

Nucleoside monophosphates in which the phosphoric acid is esterified to *two* of the available ribose hydroxyl groups (Figure 6.14) are found in all cells. Forming two such ester linkages with one phosphate results in a cyclic structure. **3′,5′-cyclic AMP,** often abbreviated **cAMP,** and its guanine analog **3′,5′-cyclic GMP,** or **cGMP,** are important regulators of cellular metabolism (see Part III: Metabolism and Its Regulation).

Nucleoside Diphosphates and Triphosphates

Additional phosphate groups can be linked to the phosphoryl group of a nucleotide through the formation of phosphoric anhydride linkages, as shown in Figure 6.15. Addition of a second phosphate to AMP creates **adenosine 5′-diphosphate,** or **ADP,** and adding a third yields **adenosine**

A phosphoric anhydride

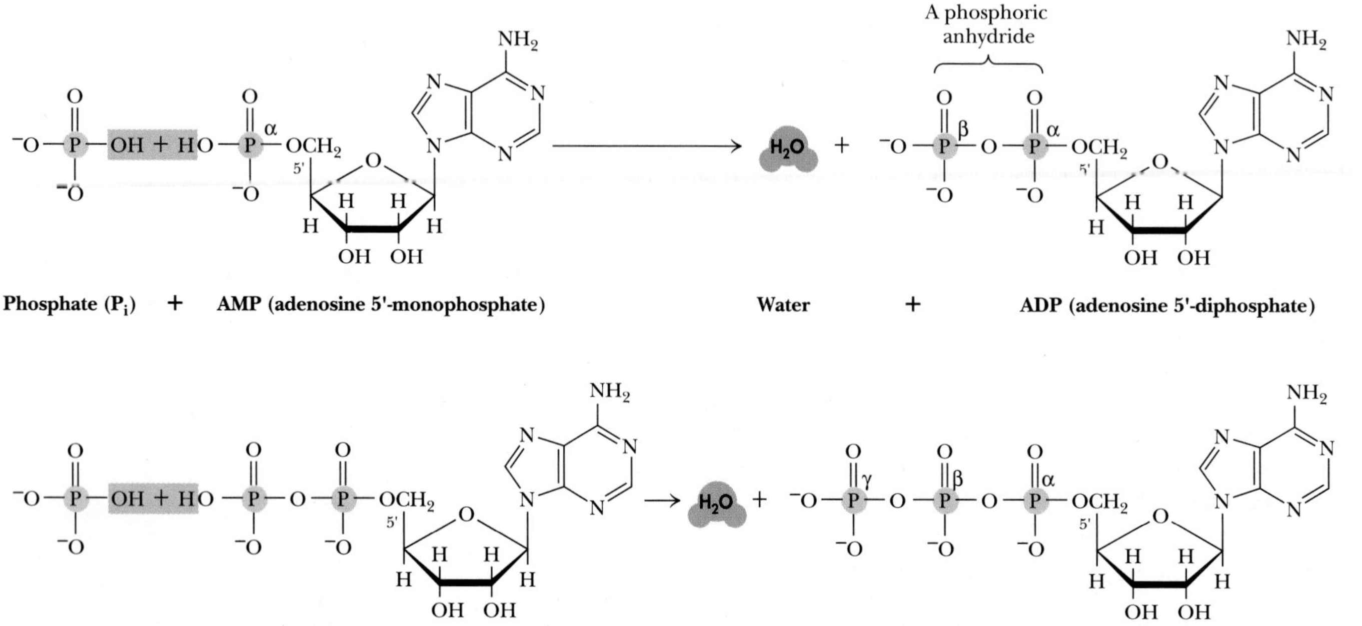

Phosphate (P_i) + AMP (adenosine 5'-monophosphate) → Water + ADP (adenosine 5'-diphosphate)

Phosphate + ADP → ATP (adenosine 5'-triphosphate)

Figure **6.15** Formation of ADP and ATP by the successive addition of phosphate groups via phosphoric anhydride linkages. Note the removal of equivalents of H_2O.

A Deeper Look

Adenosine: A Nucleoside with Physiological Activity

For the most part, nucleosides have no biological role other than to serve as precursors to nucleotides. Adenosine is an exception. In mammals, adenosine also functions as an **autocoid,** or "local hormone." This nucleoside circulates in the bloodstream, acting locally on specific cells to influence such diverse physiological phenomena as blood vessel dilation, smooth muscle contraction, neuronal discharge, neurotransmitter release, and metabolism of fat.

5′-triphosphate, or **ATP.** The respective phosphate groups are designated by the Greek letters α, β, and γ, starting with the α-phosphate as the one linked directly to the pentose. The abbreviations **GTP, CTP,** and **UTP** represent the other corresponding nucleoside 5′-triphosphates. Like the nucleoside 5′-monophosphates, the nucleoside 5′-diphosphates and 5′-triphosphates all occur in the free state in the cell, as do their deoxyribonucleoside phosphate counterparts, represented as dAMP, dADP, and dATP; dGMP, dGDP, and dGTP; dCMP, dCDP, and dCTP; dUMP, dUDP, and dUTP; and dTMP, dTDP, and dTTP.

NDPs and NTPs Are Polyprotic Acids

Nucleoside 5′-diphosphates (NDPs) and **nucleoside 5′-triphosphates (NTPs)** are relatively strong *polyprotic acids,* in that they dissociate three and four protons, respectively, from their phosphoric acid groups. The resulting phosphate anions on NDPs and NTPs form stable complexes with divalent cations such as Mg^{2+} and Ca^{2+}. Since Mg^{2+} is present at high concentrations (5 to 10 mM) intracellularly, NDPs and NTPs occur primarily as Mg^{2+} complexes in the cell. The phosphoric anhydride linkages in NDPs and NTPs are readily hydrolyzed by acid, liberating inorganic phosphate (often symbolized as P_i) and the corresponding NMP. A diagnostic test for NDPs and NTPs is quantitative liberation of P_i upon treatment with 1 N HCl at 100°C for 7 min.

Nucleoside 5′-Triphosphates Are Carriers of Chemical Energy

Nucleoside 5′-triphosphates are indispensable agents in metabolism because the phosphoric anhydride bonds they possess are a prime source of chemical energy to do biological work. ATP has been termed the energy currency of the cell (see Chapter 16). GTP is the major energy source for protein synthesis (see Chapter 32), CTP is an essential metabolite in phospholipid synthesis (see Chapter 24), and UTP forms activated intermediates with sugars that go on to serve as substrates in the biosynthesis of complex carbohydrates and polysaccharides (see Chapter 21). The evolution of metabolism has led to the dedication of one of these four NTPs to each of the major branches of metabolism. To complete the picture, the four NTPs and their dNTP counterparts are the substrates for the synthesis of the remaining great class of biomolecules—the nucleic acids.

The Bases of Nucleotides Serve as "Recognition Units"

Virtually all of the biochemical reactions of nucleotides involve either *phosphate* or *pyrophosphate group transfer:* the release of a phosphoryl group from an

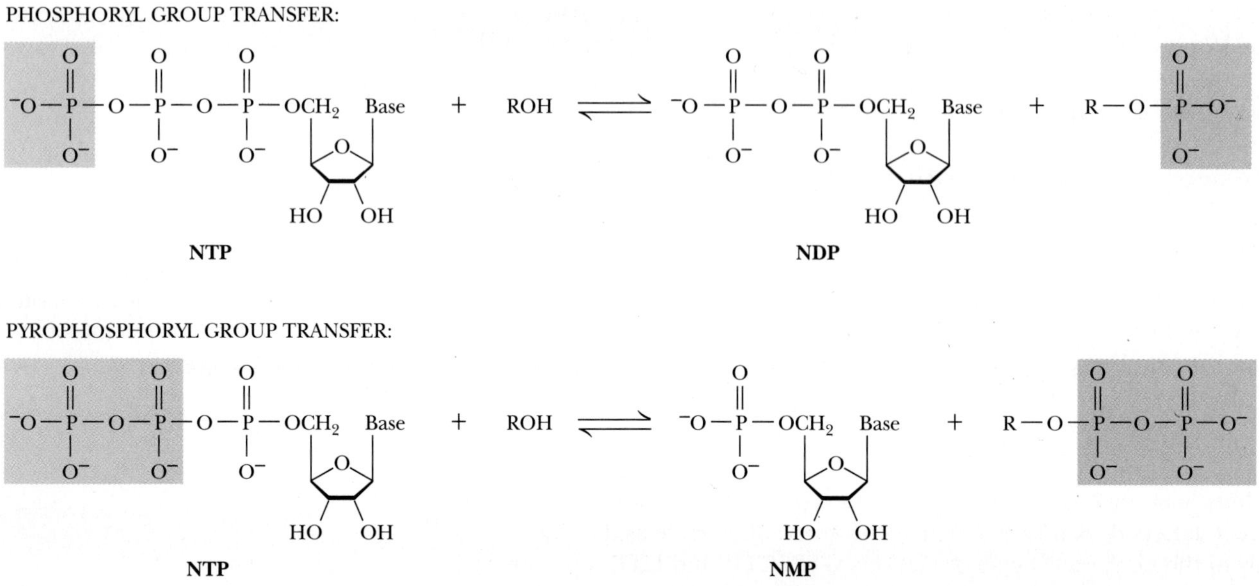

PHOSPHORYL GROUP TRANSFER:

NTP + ROH ⇌ NDP + R—O—PO$_3^{2-}$

PYROPHOSPHORYL GROUP TRANSFER:

NTP + ROH ⇌ NMP + R—O—P$_2$O$_6^{3-}$

***Figure* 6.16** Phosphoryl and pyrophosphoryl group transfer, the major biochemical reactions of nucleotides.

NTP to give an NDP, the release of a pyrophosphoryl group to give an NMP unit, or the acceptance of a phosphoryl group by an NMP or an NDP to give an NDP or an NTP (Figure 6.16). Interestingly, the pentose and the base are *not* directly involved in this chemistry. However, a "division of labor" directs ATP to serve as the primary nucleotide in central pathways of energy metabolism, while GTP, for example, is used to drive protein synthesis. Thus, the various nucleotides are channeled in appropriate metabolic directions through specific recognition of the base of the nucleotide. That is, the bases of nucleotides serve solely as *recognition units,* aloof from the covalent bond chemistry that goes on. This role as recognition units extends to nucleotide polymers, the nucleic acids, where the bases serve as the recognition units for the code of genetic information.

6.5 Nucleic Acids Are Polynucleotides

Nucleic acids are linear polymers of nucleotides linked 3′ to 5′ by **phosphodiester bridges** (Figure 6.17). They are formed as 5′-nucleoside monophosphates are successively added to the 3′-OH group of the preceding nucleotide, a process that gives the polymer a directional sense. Polymers of ribonucleotides are named **ribonucleic acid,** or **RNA.** Deoxyribonucleotide polymers are called **deoxyribonucleic acid,** or **DNA.** Since C-1′ and C-4′ in deoxyribonucleotides are involved in furanose ring formation and since there is no 2′-OH, only the 3′- and 5′-hydroxyl groups are available for internucleotide phosphodiester bonds. In the case of DNA, a polynucleotide chain may contain hundreds of millions of nucleotide units. Any structural representation of such molecules would be cumbersome at best, even for a short oligonucleotide stretch.

Shorthand Notations for Polynucleotide Structures

Several conventions have been adopted to convey the sense of polynucleotide structures. A repetitious uniformity exists in the covalent backbone of polynu-

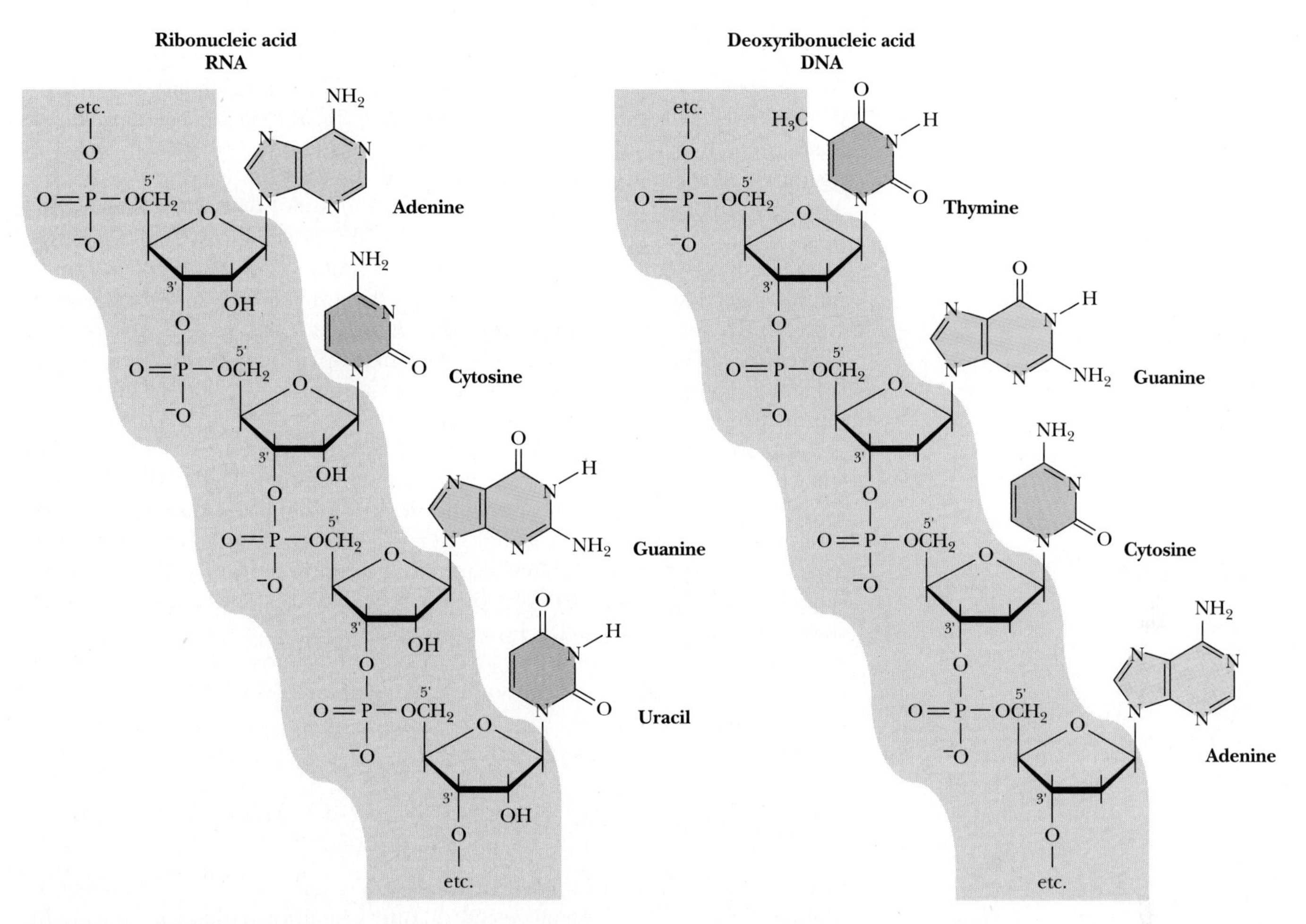

***Figure* 6.17** 3′-5′ phosphodiester bridges link nucleotides together to form polynucleotide chains.

cleotides, in which the chain can be visualized as running from 5′ to 3′ along the atoms of one furanose and thence across the phosphodiester bridge to the furanose of the next nucleotide in line. Thus, this backbone can be portrayed by the symbol of a vertical line representing the furanose and a slash representing the phosphodiester link, as shown in Figure 6.18. The diagonal slash runs from the middle of a furanose line to the bottom of an adjacent one to indicate the 3′- (middle) to 5′- (bottom) carbons of neighboring furanoses joined by the phosphodiester bridge. The base attached to each furanose is indicated above it by a one-letter designation: A, C, G, or U (or T). The convention in all notations of nucleic acid structure is to read the polynucleotide chain from the 5′-end of the polymer to the 3′-end. Note that this reading direction actually passes through each phosphodiester from 3′ to 5′.

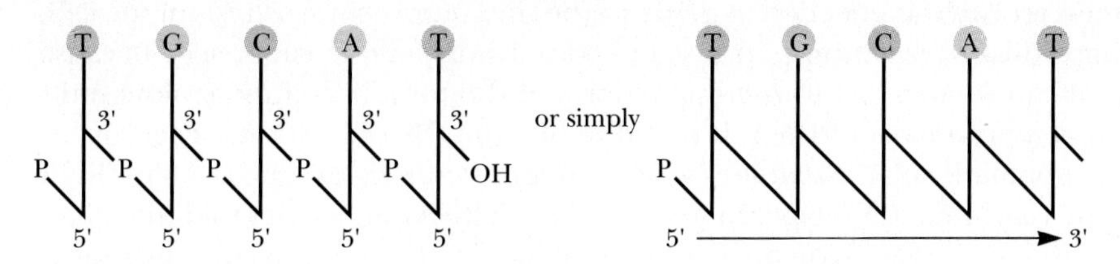

***Figure* 6.18** Furanoses are represented by lines; phosphodiesters are represented by diagonal slashes in this shorthand notation for nucleic acid structures.

Base Sequence

The only significant variation that commonly occurs in the chemical structure of nucleic acids is the nature of the base at each nucleotide position. These bases are not part of the sugar–phosphate backbone but instead serve as distinctive side chains, like the R groups attached to a polypeptide backbone. They give the polymer its unique identity. A simple notation of these structures is merely to list the order of bases in the polynucleotide using single capital letters—A, G, C, and U (or T). Occasionally, a lowercase "p" is written between each successive base to indicate the phosphodiester bridge, as in GpApCpGpUpA. A "p" preceding the sequence indicates that the nucleic acid carries a PO_4 on its 5′-end, as in pGpApCpGpUpA; a "p" terminating the sequence connotes the presence of a phosphate on the 3′-OH end, as in GpApCpGpUpAp.

A more common method of representing nucleotide sequences is to omit the "p" and write only the order of bases, such as GACGUA. This notation assumes the presence of the phosphodiesters joining adjacent nucleotides. The presence of 3′- or 5′-phosphate termini, however, must still be specified, as in GACGUAp for a 3′-PO_4 terminus. To distinguish between RNA and DNA sequences, DNA sequences are typically preceded by a lowercase "d" to denote deoxy, as in d-GACGTA. From a simple string of letters such as this, any biochemistry student should be able to draw the unique chemical structure for a pentanucleotide, even though it may contain over 200 atoms.

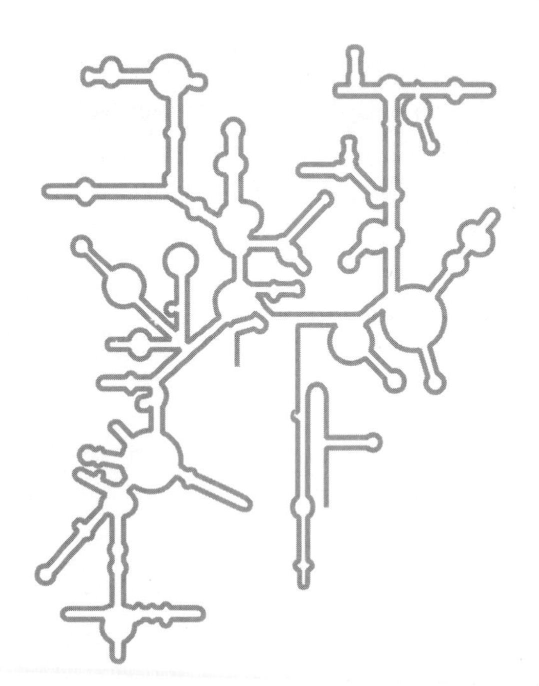

Ribosomal RNA.

6.6 Classes of Nucleic Acids

The two major classes of nucleic acids are DNA and RNA. DNA has only one biological role, but it is the more central one. The information to make all the functional macromolecules of the cell (even DNA itself) is preserved in DNA and accessed through transcription of the information into RNA copies. Coincident with its singular purpose, there is only a single DNA molecule (or "chromosome") in simple life forms such as viruses or bacteria. Such DNA molecules must be quite large in order to embrace enough information for making the macromolecules necessary to maintain a living cell. The *Escherichia coli* chromosome has a molecular mass of 2.9×10^9 daltons and contains 9.5 million nucleotides. Eukaryotic cells have many chromosomes, and DNA is found principally in two copies in the diploid chromosomes of the nucleus, but it also occurs in mitochondria and in chloroplasts, where it encodes a restricted set of proteins and RNAs unique to these organelles.

In contrast, RNA occurs in multiple copies and various forms (Table 6.2). Cells contain up to eight times as much RNA as DNA. RNA has a number of important biological functions, and on this basis, RNA molecules are categorized into several major types: **messenger RNA, ribosomal RNA,** and **transfer RNA.** Eukaryotic cells contain an additional type, **small nuclear RNA (snRNA)** (see Chapter 30). Messenger RNA (**mRNA**) serves to carry the information or "message" that is encoded in genes to the sites of protein synthesis in the cell, where this information is translated into a polypeptide sequence. Because mRNA molecules are transcribed copies of the protein-coding genetic units that comprise most of DNA, it is said to be "the DNA-like RNA." Ribosomes, the supramolecular assemblies where protein synthesis occurs, are 65% RNA of the ribosomal RNA type. Ribosomal RNA (**rRNA**) molecules fold into characteristic secondary structures as a consequence of intramolecular hydrogen bond interactions. There are three major species of rRNA in ribosomes, and they are generally referred to according to their sedimentation coefficients

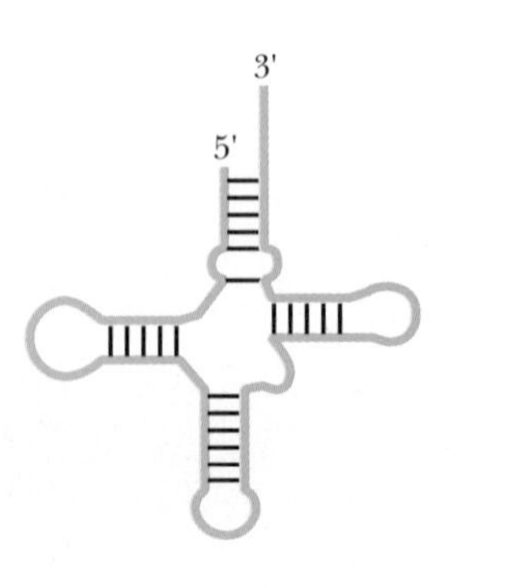

tRNA.

Table **6.2**
Various Kinds of RNA Found in an *E. coli* Cell

Type	Sedimentation Coefficient	Molecular Weight	Number of Nucleotide Residues	Percentage of Total Cell RNA
mRNA	6–25	25,000–1,000,000	75–3,000	~2
tRNA	~4	23,000–30,000	73–94	16
rRNA	5	35,000	120	
	16	550,000	1542	82
	23	1,100,000	2904	

(see the Appendix to Chapter 4), which are a rough measure of their relative size (Table 6.2). Transfer RNA (**tRNA**) serves as a carrier of amino acid residues for protein synthesis. tRNA molecules also fold into a characteristic secondary structure. The amino acid is attached as an aminoacyl ester to the 3′-terminus of the tRNA. Aminoacyl-tRNAs are the substrates for protein biosynthesis. These are the smallest RNAs, falling in the size range of 23 to 30 kD.

With these basic definitions in mind, let's now briefly consider the chemical and structural nature of DNA and the various RNAs. Chapter 7 elaborates on methods to determine the primary structure of nucleic acids by sequencing methods and discusses the secondary and tertiary structures of DNA and RNA. Part IV, Genetic Information, is devoted to a detailed treatment of the dynamic role of nucleic acids in the molecular biology of the cell.

DNA

The DNA isolated from different cells and viruses characteristically consists of two polynucleotide strands wound together to form a long, slender, helical molecule, the **DNA double helix.** The strands run in opposite directions; that is, they are *antiparallel* and are held together in the double helical structure through *interchain hydrogen bonds* (Figure 6.19). These H bonds pair the bases of nucleotides in one chain to complementary bases in the other, a phenomenon called **base pairing.**

Chargaff's Rules

A clue to the chemical basis of base pairing in DNA came from the analysis of the base composition of various DNAs by Erwin Chargaff in the late 1940s. His data showed that the four bases commonly found in DNA (A, C, G, and T) do not occur in equimolar amounts and that the relative amounts of each vary from species to species (Table 6.3). Nevertheless, Chargaff noted that certain pairs of bases, namely, adenine and thymine, and guanine and cytosine, are always found in a 1:1 ratio and that the number of pyrimidine residues always equals the number of purine residues. These findings are known as *Chargaff's rules:* **[A] = [T]; [C] = [G]; [pyrimidines] = [purines].**

Watson and Crick's Double Helix

James Watson and Francis Crick, working in the Cavendish Laboratory at Cambridge University in 1953, took advantage of Chargaff's results and observations from X-ray diffraction studies on the structure of DNA to conclude

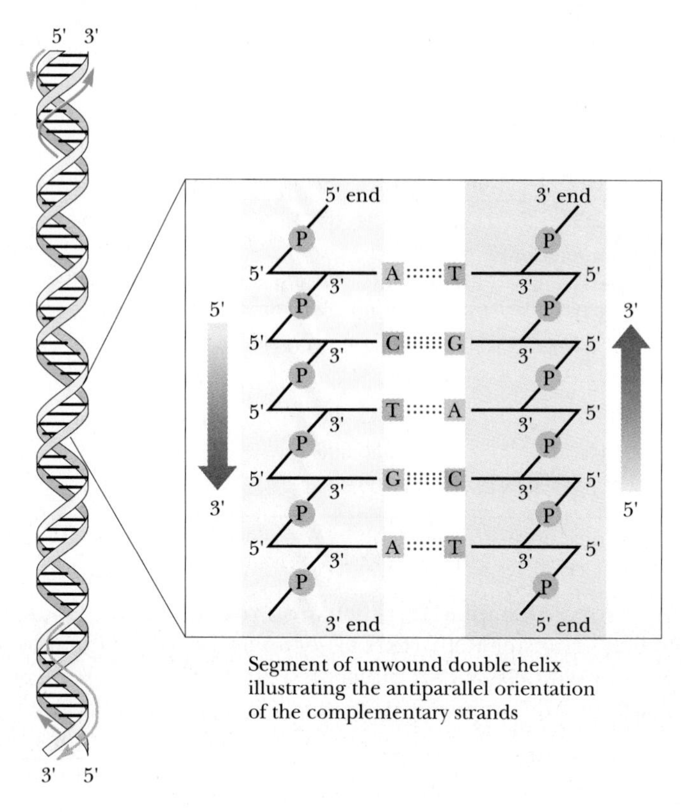

***Figure* 6.19** The antiparallel nature of the DNA double helix.

that DNA was a *complementary double helix.* Two strands of deoxyribonucleic acid (sometimes referred to as the *Watson strand* and the *Crick strand*) are held together by hydrogen bonds formed between unique base pairs, always consisting of a purine in one strand and a pyrimidine in the other. Base pairing is very specific: if the purine is adenine, the pyrimidine must be thymine. Similarly, guanine pairs only with cytosine (Figure 6.20). Thus, if an A occurs in one strand of the helix, T must occupy the complementary position in the opposing strand. Likewise, a G in one dictates a C in the other.

***Table* 6.3**

Molar Ratios Leading to the Formulation of Chargaff's Rules

Source	Adenine to Guanine	Thymine to Cytosine	Adenine to Thymine	Guanine to Cytosine	Purines to Pyrimidines
Ox	1.29	1.43	1.04	1.00	1.1
Human	1.56	1.75	1.00	1.00	1.0
Hen	1.45	1.29	1.06	0.91	0.99
Salmon	1.43	1.43	1.02	1.02	1.02
Wheat	1.22	1.18	1.00	0.97	0.99
Yeast	1.67	1.92	1.03	1.20	1.0
Hemophilus influenzae	1.74	1.54	1.07	0.91	1.0
E. coli K-12	1.05	0.95	1.09	0.99	1.0
Avian tubercle bacillus	0.4	0.4	1.09	1.08	1.1
Serratia marcescens	0.7	0.7	0.95	0.86	0.9
Bacillus schatz	0.7	0.6	1.12	0.89	1.0

Source: After Chargaff, E., 1951. Federation Proceedings **10**:654–659.

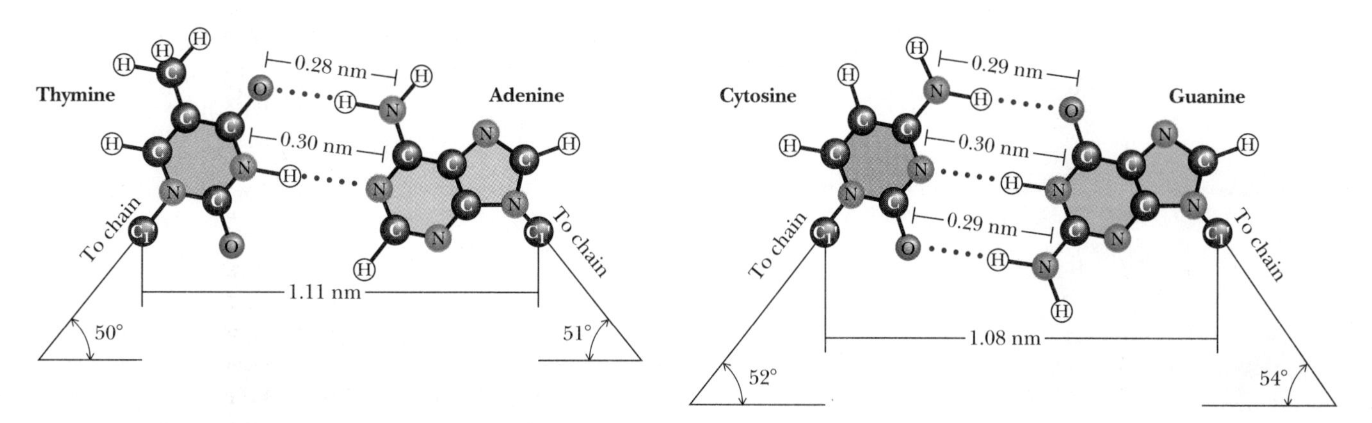

Figure **6.20** The Watson–Crick base pairs A:T and G:C.

Such a molecule not only conforms to Chargaff's rules but also has a profound property relating to heredity: *The sequence of bases in one strand has a complementary relationship to the sequence of bases in the other strand.* That is, the information contained in the sequence of one strand is conserved in the sequence of the other. Therefore, separation of the two strands and faithful replication of each, through a process in which base pairing specifies the sequence in the newly synthesized strand, leads to two progeny molecules identical in every respect to the parental double helix (Figure 6.21). Elucidation of the double helical structure of DNA represented one of the most significant events in the history of science. This discovery more than any other marked the beginning of molecular biology. Indeed, upon solving the structure of DNA, Crick proclaimed in The Eagle, a pub just across from the Cavendish lab, "We have discovered the secret of life!"

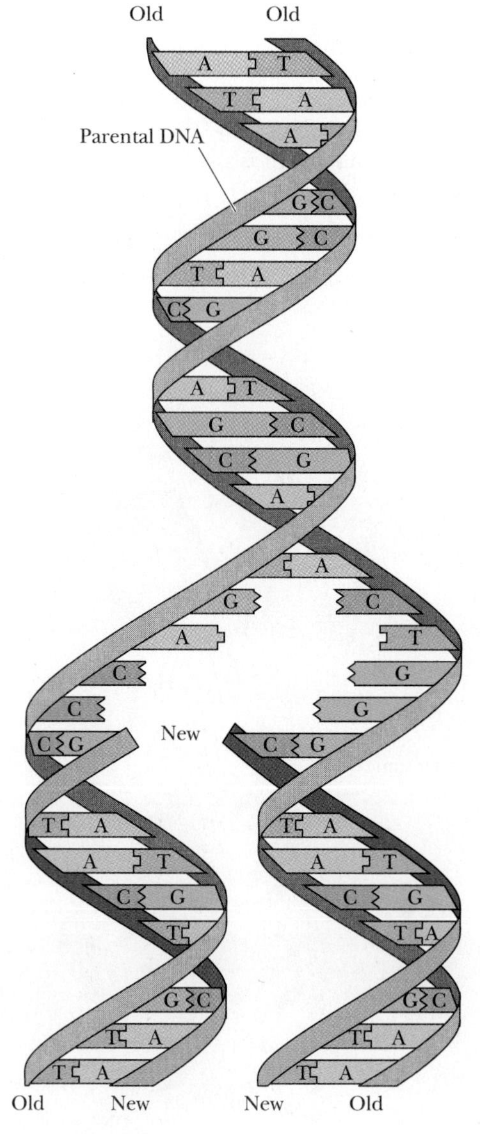

Figure **6.21** Semiconservative replication of DNA gives identical progeny molecules since base pairing is the mechanism determining the nucleotide sequence synthesized within each of the new strands during replication.

Size of DNA Molecules

Because of the double helical nature of DNA molecules, their size can be represented in terms of the numbers of nucleotide base pairs they contain. For example, the *E. coli* chromosome consists of 4.72×10^6 base pairs (abbreviated bp) or 4.72×10^3 kilobase pairs (kbp). DNA is a threadlike molecule. The diameter of the DNA double helix is only 2 nm, but the length of the DNA molecule forming the *E. coli* chromosome is over 1.6×10^6 nm (1.6 mm). Since the long dimension of an *E. coli* cell is only 2000 nm (0.002 mm), its chromosome must be highly folded. Because of their long, threadlike nature, DNA molecules are easily sheared into shorter fragments during isolation procedures, and it is difficult to obtain intact chromosomes even from the simple cells of prokaryotes.

DNA in the Form of Chromosomes

DNA occurs in various forms in different cells. The single chromosome of prokaryotic cells (Figure 6.22) is typically a circular DNA molecule. Little protein is associated with prokaryotic chromosomes. In contrast, the DNA molecules of eukaryotic cells, each of which defines a chromosome, are linear and richly adorned with proteins. A class of arginine- and lysine-rich basic proteins called **histones** interact ionically with the anionic phosphate groups in the DNA backbone to form **nucleosomes,** structures in which the DNA double helix is wound around a protein "core" composed of pairs of four different histone polypeptides (Figure 6.23; see also Section 7.5 in Chapter 7). Chromosomes also contain a varying mixture of other proteins, so-called **non-**

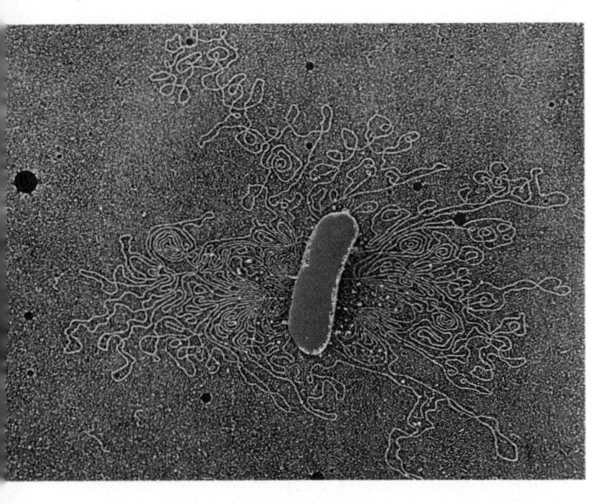

Figure **6.22** If the cell walls of bacteria such as *Escherichia coli* are partially digested and the cells are then osmotically shocked by dilution with water, the contents of the cells are extruded to the exterior. In electron micrographs, the most obvious extruded component is the bacterial chromosome, shown here surrounding the cell.

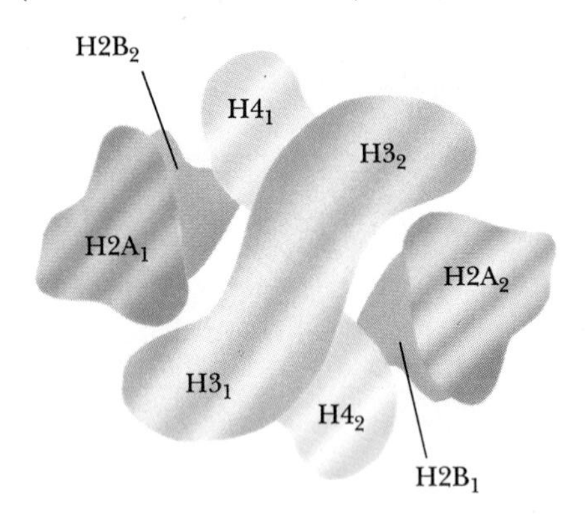

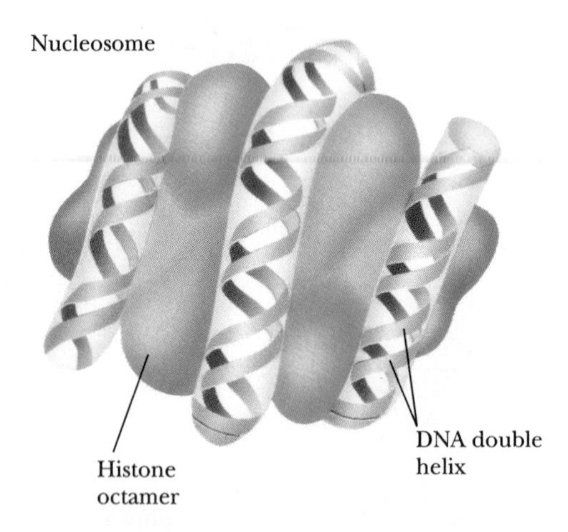

Figure **6.23** A diagrammatic representation of the histone octamer. Nucleosomes consist of two turns of DNA supercoiled about a histone "core" octamer.

histone chromosomal proteins, many of which are involved in regulating which genes in DNA are transcribed at any given moment. The amount of DNA in a diploid mammalian cell is typically more than 1000 times that found in an *E. coli* cell. Some higher plant cells contain more than 50,000 times as much.

RNA

Messenger RNA

Messenger RNA is synthesized during **transcription,** an enzymatic process in which an RNA copy is made of the sequence of bases along one strand of DNA. This mRNA then directs the synthesis of a polypeptide chain as the information that is contained within its nucleotide sequence is translated into an amino acid sequence by the protein-synthesizing machinery of the ribosomes. rRNA and tRNA molecules are also synthesized by transcription of DNA sequences, but unlike mRNA molecules, these RNAs are not subsequently translated to form proteins. Only the genetic units of DNA sequence that encode proteins are transcribed into mRNA molecules. In prokaryotes, a single mRNA may contain the information for the synthesis of several polypeptide chains within its nucleotide sequence (Figure 6.24). In contrast, eukaryotic mRNAs encode only one polypeptide, but are more complex in that they are synthesized in the nucleus in the form of much larger precursor molecules called **heterogeneous nuclear RNA,** or **hnRNA.** hnRNA molecules contain stretches of nucleotide sequence that have no protein-coding capacity. These noncoding regions are called **intervening sequences** or **introns** because they intervene between coding regions, which are called **exons.** Introns interrupt the continuity of the information specifying protein amino acid sequence and must be spliced out before the message can be translated. In addition, eukaryotic hnRNA and mRNA molecules have a run of 100 to 200 adenylic acid residues attached at their 3′-ends, so-called **poly(A) tails.** This polyadenylation occurs after transcription has been completed and is believed to contribute to mRNA stability. The properties of messenger RNA molecules as they move through transcription and translation in prokaryotic versus eukaryotic cells are summarized in Figure 6.24.

Ribosomal RNA

Ribosomes, the agents of protein synthesis, are composed of two subunits of different sizes that dissociate from each other if the Mg^{2+} concentration is below 10^{-3} *M*. Each subunit is a supramolecular assembly of proteins and RNA and has a total mass of 10^6 daltons or more. Because of this, ribosomal subunit sizes are usually expressed in terms of their ultracentrifugal **sedimentation coefficients,** or **S** values (see Appendix to Chapter 4). The *E. coli* ribosomal subunits have sedimentation coefficients of 30S (the small subunit) and 50S (the large subunit). Eukaryotic ribosomes are somewhat larger than prokaryotic ribosomes, consisting of 40S and 60S subunits. Ribosomes are about 65% RNA and 35% protein. The properties of ribosomes and their rRNAs are summarized in Figure 6.25. The 30S subunit of *E. coli* contains a single RNA chain of 1542 nucleotides. This small subunit rRNA itself has a sedimentation coefficient of 16S. The large *E. coli* subunit has two rRNA molecules, a 23S (2904 nucleotides) and a 5S (120 nucleotides). The ribosomes of a typical eukaryote, the rat, have rRNA molecules of 18S (1874 nucleotides) and 28S (4718 bases), 5.8S (160 bases), and 5S (120 bases). The 18S rRNA is in the 40S subunit and the latter three are all part of the 60S subunit.

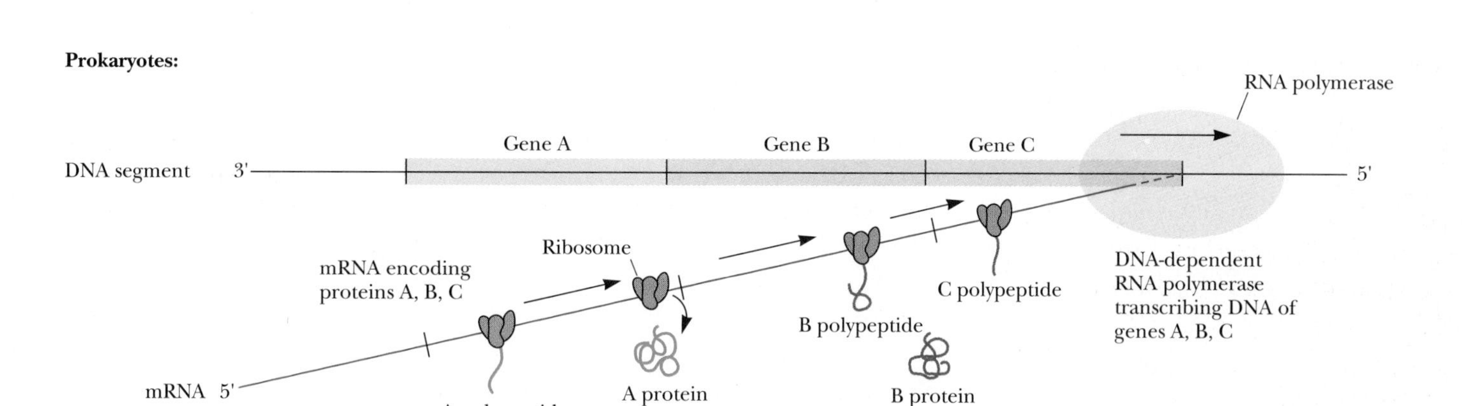

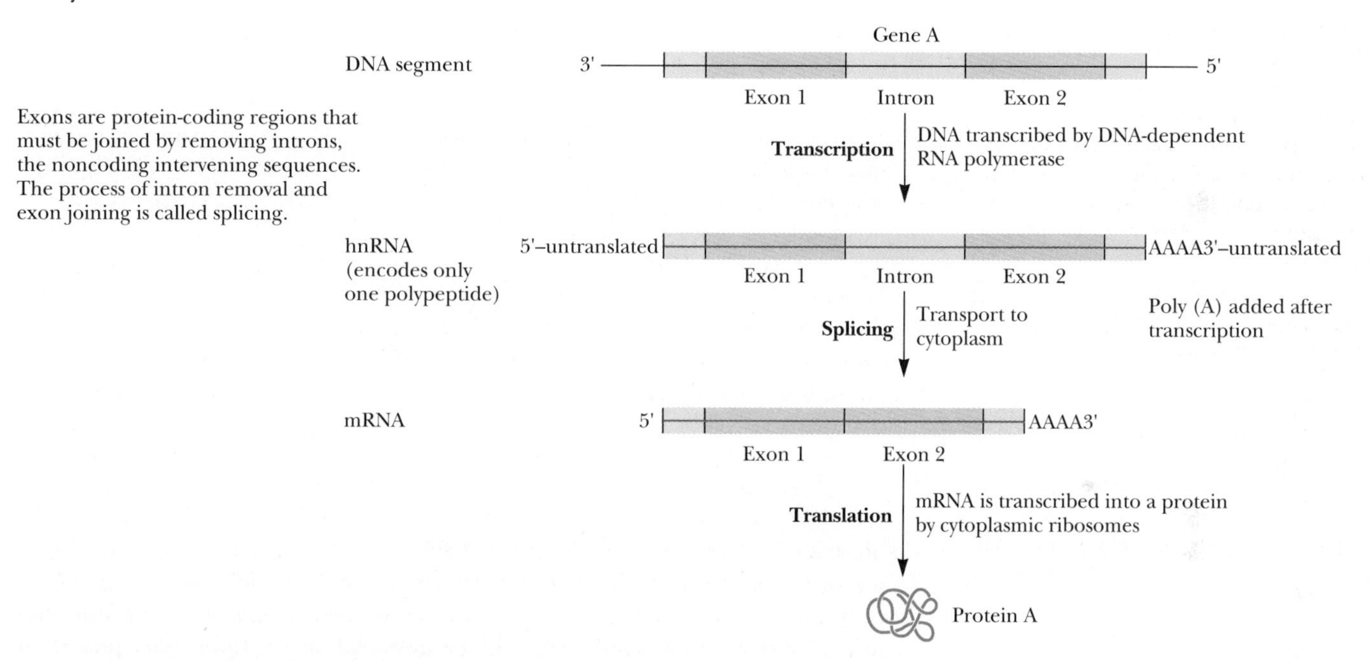

***Figure* 6.24** The properties of mRNA molecules in prokaryotic versus eukaryotic cells during transcription and translation.

Ribosomal RNAs characteristically contain a number of specially modified nucleotides, including **pseudouridine** residues, **ribothymidylic acid,** and **methylated bases** (Figure 6.26). Ribosomal RNAs have a high degree of secondary structure (see Chapter 7), which, in addition to other functions, serves as a scaffold for the many ribosomal proteins that assemble to form the complete ribosome structure. The central role of ribosomes in the biosynthesis of proteins is treated in detail in Chapter 32. Here we briefly note the significant point that genetic information in the nucleotide sequence of an mRNA is translated into the amino acid sequence of a polypeptide chain by ribosomes.

Transfer RNA

Transfer RNA molecules are relatively small polynucleotides containing 73 to 94 residues, a substantial number of which are methylated or otherwise unusually modified. tRNA derives its name from its role as the carrier of amino acids during the process of protein synthesis. Each of the 20 amino acids of proteins has at least one unique tRNA species dedicated to chauffeuring its insertion into growing polypeptide chains, and some amino acids are served

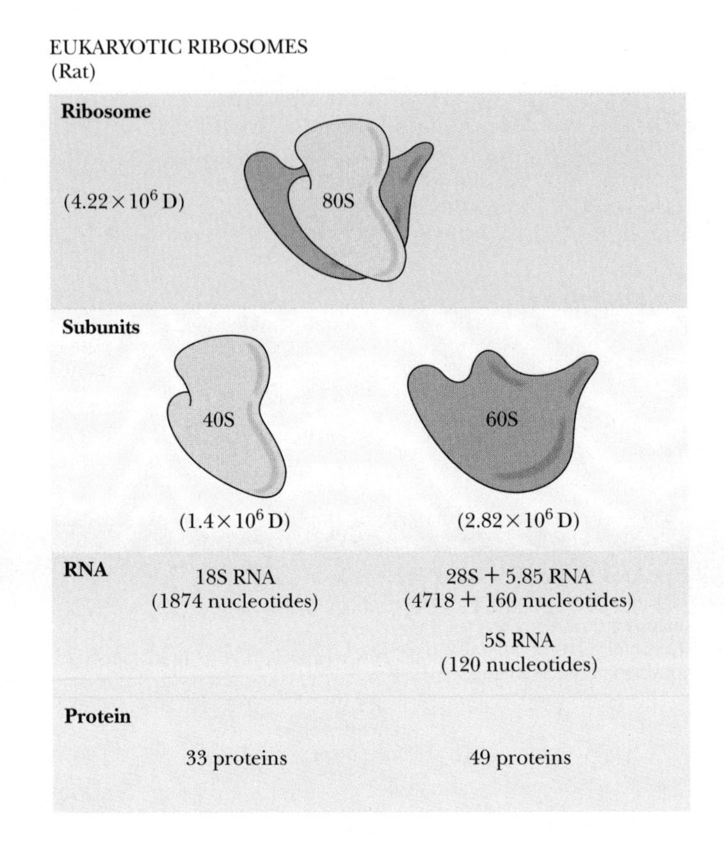

***Figure* 6.25** The organization and composition of prokaryotic and eukaryotic ribosomes.

by several tRNAs. For example, five different tRNAs act in the transfer of leucine into proteins. In eukaryotes, there are even discrete sets of tRNA molecules for each site of protein synthesis—the cytoplasm, the mitochondrion, and, in plant cells, the chloroplast. All tRNA molecules possess a 3′-terminal nucleotide sequence that reads **-CCA,** and the amino acid is carried to the ribosome attached as an acyl ester to the free 3′-OH of the terminal A residue. These **aminoacyl-tRNAs** are the substrates of protein synthesis, the amino acid being transferred to the carboxyl end of a growing polypeptide. The peptide bond-forming reaction is a catalytic process intrinsic to ribosomes.

Small Nuclear RNAs

Small nuclear RNAs, or **snRNAs,** are a recently recognized class of RNA molecules found only in eukaryotic cells, principally in the nucleus. They are neither tRNA nor small rRNA molecules, although they are similar in size to these species. They contain from 100 to about 200 nucleotides, some of which, like tRNA and rRNA, are methylated or otherwise modified. No snRNA exists as naked RNA. Instead, snRNA is found in stable complexes with specific proteins forming **small nuclear ribonucleoprotein particles,** or **snRNPs,** which are about 10S in size. Their occurrence only in eukaryotes, their location in the nucleus, and their relative abundance (1 to 10% of the number of ribosomes) are significant clues to their biological purpose: snRNPs are important in the processing of eukaryotic gene transcripts (hnRNA) into mature messenger RNA for export from the nucleus to the cytoplasm.

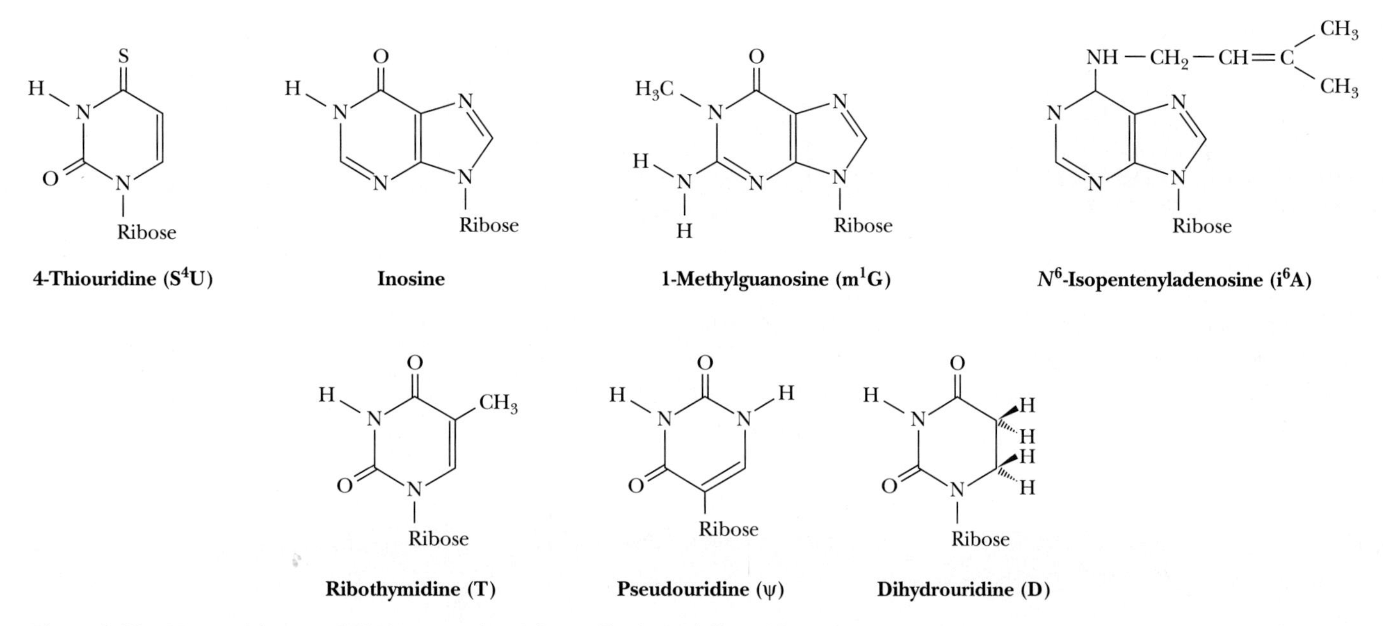

Figure **6.26** Unusual bases of RNA—pseudouridine, ribothymidylic acid, and various methylated bases.

Significance of Chemical Differences Between DNA and RNA

Two fundamental chemical differences distinguish DNA from RNA. (1) DNA contains 2-deoxyribose instead of ribose, and (2) DNA contains thymine instead of uracil. What are the consequences of these differences and do they hold any significance in common? An argument can be made that these differences are the result of factors that render DNA a more stable polymeric form than RNA, with obvious significance for the roles these macromolecules have assumed in heredity.

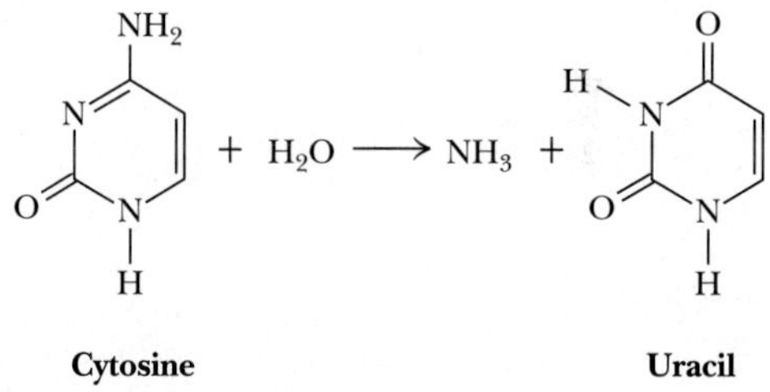

Consider first why DNA contains thymine instead of uracil. The key observation is that *cytosine deaminates to form uracil* at a finite rate *in vivo* (Figure 6.27). Since C pairs with G in the opposing strand, whereas U would pair with A, conversion of a C to a U could potentially result in a heritable change in nucleotide sequence, that is, a *mutation*. To prevent this reaction from leading to changes in nucleotide sequence, a cellular repair mechanism "proofreads" DNA, and when a U arising from C deamination is encountered, it is treated as inappropriate and is replaced by a C. If DNA normally contained U rather than T, this repair system could not readily distinguish U formed by C deamination from U correctly paired with A. However, the U in DNA is "5-methyl-U" or, as it is conventionally known, thymine (Figure 6.28). That is, the 5-methyl group on T labels it to say "this U belongs; do not replace it."

The ribose 2′-OH group of RNA is absent in DNA. Consequently, the ubiquitous 3′-O of polynucleotide backbones lacks a vicinal hydroxyl neighbor in DNA. This difference leads to a greater resistance of DNA to hydrolysis, examined in detail in the following section. To view it another way, RNA is less stable than DNA because its vicinal 2′-OH group makes the 3′-phosphodiester bond susceptible to nucleophilic cleavage (Figure 6.29). For just this reason, it is selectively advantageous for the heritable form of genetic information to be DNA rather than RNA.

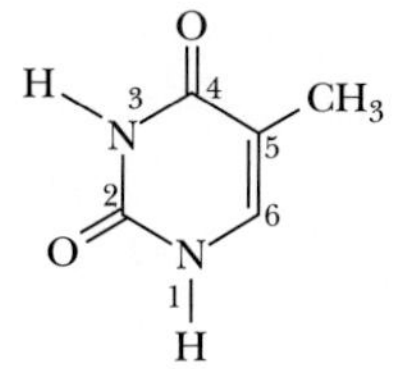

Figure **6.27** Deamination of cytosine forms uracil.

Figure **6.28** The 5-methyl group on thymine labels it as a special kind of uracil.

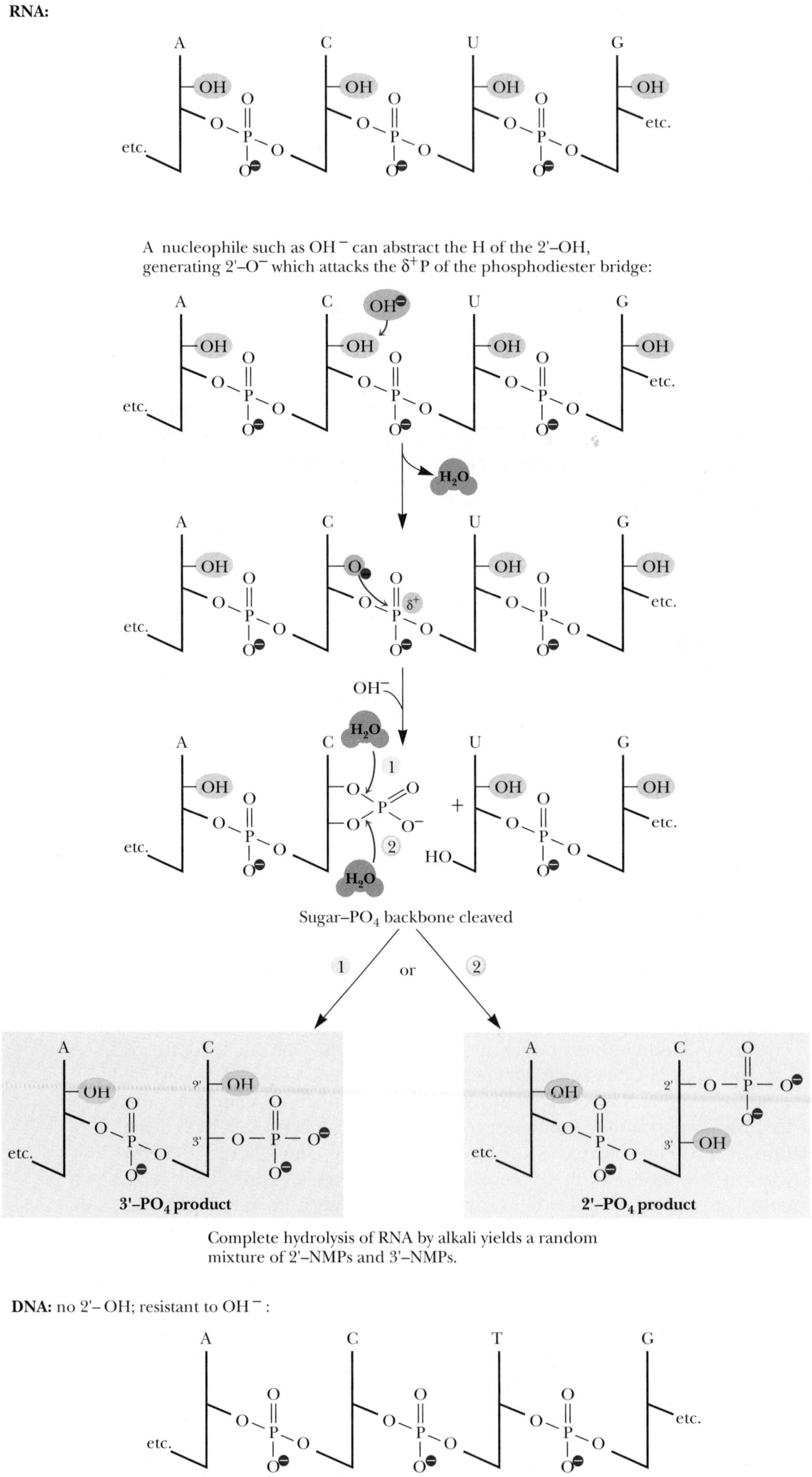

***Figure* 6.29** The vicinal —OH groups of RNA are susceptible to nucleophilic attack leading to hydrolysis of the phosphodiester bond and fracture of the polynucleotide chain; DNA lacks a 2′-OH vicinal to its 3′-O-phosphodiester backbone. Alkaline hydrolysis of RNA results in the formation of a mixture of 2′- and 3′-nucleoside monophosphates.

6.7 Hydrolysis of Nucleic Acids

The vast majority of reactions leading to nucleic acid hydrolysis break bonds in the polynucleotide backbone. Such reactions are important because they can be used to manipulate these molecules. For example, hydrolysis of polynucleotides may generate fragments that are more amenable to sequence determination.

Hydrolysis by Acid or Base

RNA is relatively resistant to the effects of dilute acid, but gentle treatment of DNA with 1 m*M* HCl leads to selective removal of its purine bases by cleavage of the purine glycoside bonds. The glycosidic bonds between pyrimidine bases and 2′-deoxyribose are not affected, and, in this case, the polynucleotide's sugar–phosphate backbone remains intact. The purine-free polynucleotide product is called **apurinic acid.**

DNA is not susceptible to alkaline hydrolysis. On the other hand, RNA is alkali labile and is readily hydrolyzed by dilute sodium hydroxide. Cleavage is random in RNA, and the ultimate products are a mixture of nucleoside 2′- and 3′-monophosphates. These products provide a clue to the reaction mechanism (Figure 6.29). Abstraction of the 2′-OH hydrogen by hydroxyl anion leaves a 2′-O$^-$ that carries out a nucleophilic attack on the δ^+ phosphorus atom of the phosphate moiety, resulting in cleavage of the 5′-phosphodiester bond and formation of a cyclic 2′,3′-phosphate. This cyclic 2′,3′-phosphodiester is unstable and decomposes randomly to either a 2′- or 3′-phosphate ester. DNA has no 2′-OH, therefore DNA is alkali stable.

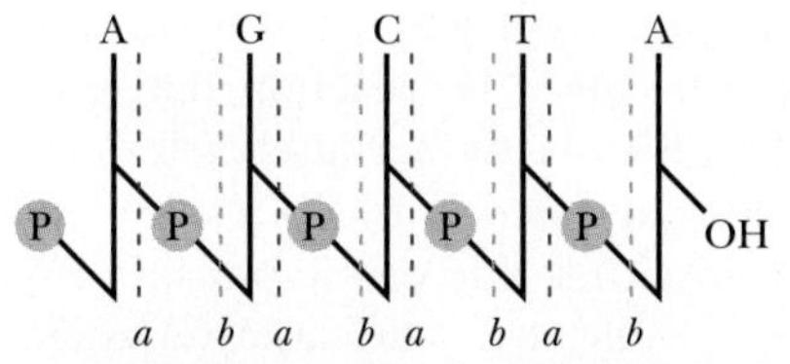

Convention: The 3'-side of each phosphodiester is termed *a*; the 5 '-side is termed *b*.

Hydrolysis of the *a* bond yields 5'–PO_4 products:

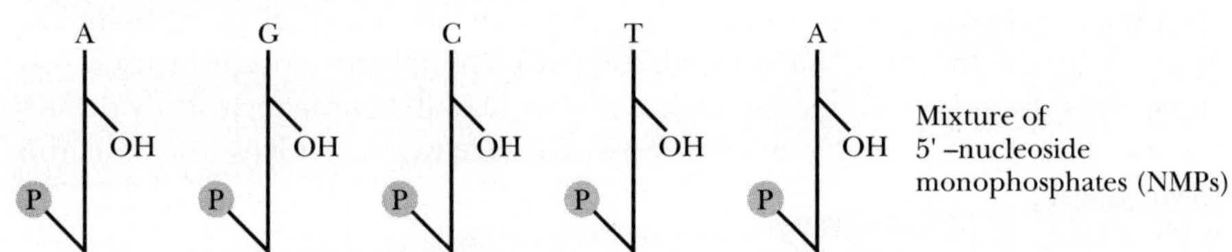

Hydrolysis of the *b* bond yields 3'–PO_4 products:

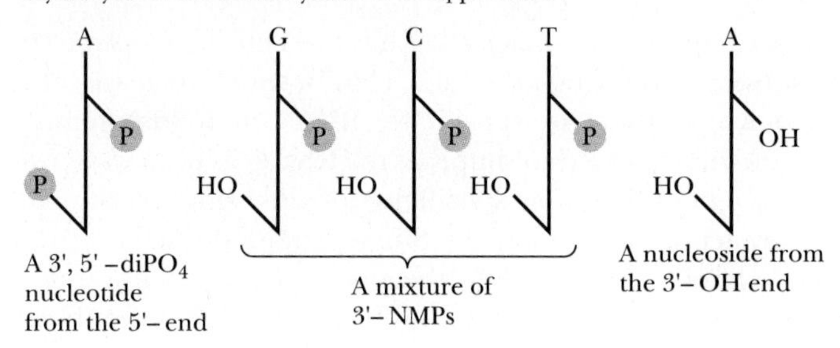

***Figure* 6.30** Cleavage in polynucleotide chains: *a* cleavage yields 5′-phosphate products, whereas *b* cleavage gives 3′ phosphate products.

Table 6.4
Specificity of Various Nucleases

Enzyme	DNA, RNA, or Both	*a* or *b*	Specificity
Exonucleases			
Snake venom phosphodiesterase	Both	*a*	Starts at 3′-end, 5′-NMP products
Spleen phosphodiesterase	Both	*b*	Starts at 5′-end, 3′-NMP products
Endonucleases			
RNase A (pancreas)	RNA	*b*	Where 3′-PO_4 is to pyrimidine; oligos with pyrimidine 3′-PO_4 ends
Bacillus subtilis RNase	RNA	*b*	Where 3′-PO_4 is to purine; oligos with purine 3′-PO_4 ends
RNase T_1	RNA	*b*	Where 3′-PO_4 is to guanine
RNase T_2	RNA	*b*	Where 3′-PO_4 is to adenine
DNase I (pancreas)	DNA	*a*	Preferably between Py and Pu; nicks dsDNA, creating 3′-OH ends
DNase II (spleen, thymus, *Staphylococcus aureus*)	DNA	*b*	Oligo products
Nuclease S1	Both	*a*	Cleaves single-stranded but not double-stranded nucleic acids

Enzymatic Hydrolysis

Enzymes that hydrolyze nucleic acids are called **nucleases.** Virtually all cells contain various nucleases that serve important housekeeping roles in the normal course of nucleic acid metabolism. Organs that provide digestive fluids, such as the pancreas, are rich in nucleases and secrete substantial amounts to hydrolyze ingested nucleic acids. Fungi are often good sources of nucleases. As a class, nucleases are **phosphodiesterases** because the reaction that they catalyze is the cleavage of phosphodiester bonds by H_2O. Because each internal phosphate in a polynucleotide backbone is involved in two phosphodiester linkages, cleavage can potentially occur on either side of the phosphorus (Figure 6.30). Convention labels the 3′-side as *a* and the 5′-side as *b.* Cleavage on the *a* side leaves the phosphate attached to the 5′-position of the adjacent nucleotide, while *b*-side hydrolysis yields 3′-phosphate products. Enzymes or reactions that hydrolyze nucleic acids are characterized as acting at either *a* or *b.* A second convention denotes whether the nucleic acid chain was cleaved at some internal location, *endo,* or whether a terminal nucleotide residue was hydrolytically removed, *exo.* Note that exo *a* cleavage characteristically occurs at the 3′-end of the polymer, whereas exo *b* cleavage involves attack at the 5′-terminus.

Nuclease Specificity

Like most enzymes (see Chapter 11), nucleases exhibit selectivity or *specificity* for the nature of the substance on which they act. That is, some nucleases act only on DNA **(DNases),** while others are specific for RNA (the **RNases**). Still others are nonspecific and are referred to simply as **nucleases,** as in *nuclease S1* (see Table 6.4). Nucleases may also show specificity for only single-stranded nucleic acids or may only act on double helices. Single-stranded nucleic acids are abbreviated by an *ss* prefix, as in ssRNA; the prefix *ds* denotes double-stranded. Nucleases may also display a decided preference for acting only at

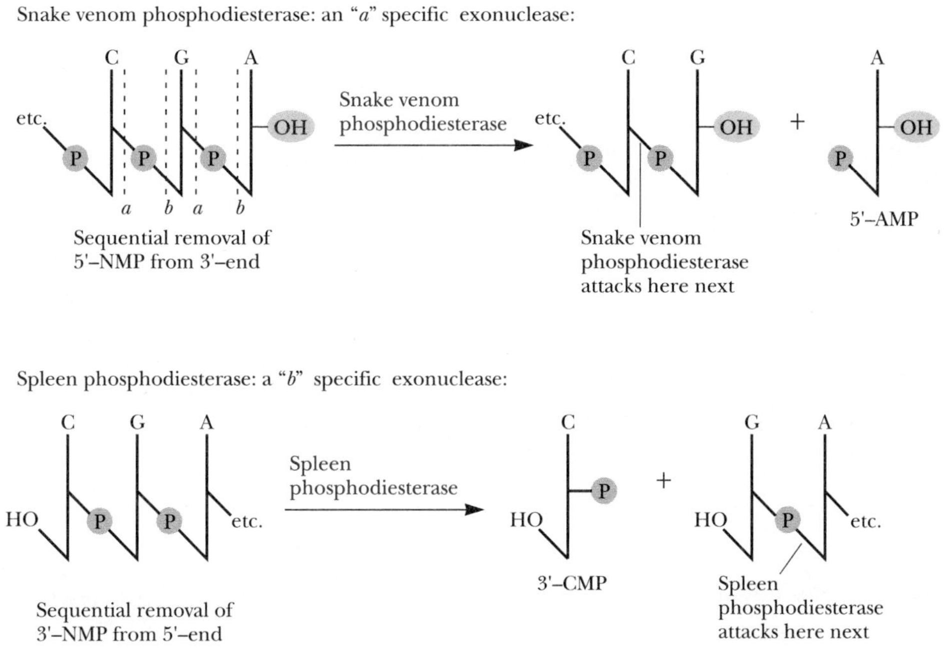

Snake venom phosphodiesterase: an "*a*" specific exonuclease:

C G A
etc. P P P OH
a b a b
Sequential removal of
5'–NMP from 3'–end
Snake venom
phosphodiesterase
etc. P P OH + P OH
C G A
Snake venom
phosphodiesterase
attacks here next
5'–AMP

Spleen phosphodiesterase: a "*b*" specific exonuclease:

C G A
HO P P etc.
Sequential removal of
3'–NMP from 5'–end
Spleen
phosphodiesterase
C
HO P + HO P etc.
G A
3'–CMP
Spleen
phosphodiesterase
attacks here next

***Figure* 6.31** Snake venom phosphodiesterase and spleen phosphodiesterase are exonucleases that degrade polynucleotides from opposite ends.

certain bases in a polynucleotide, or, as we shall see for *restriction endonucleases,* some nucleases will act only at a particular nucleotide sequence four to eight nucleotides in length. Table 6.4 lists the various permutations in specificity displayed by these nucleases and gives prominent examples of each. To the molecular biologist, nucleases are the surgical tools for the dissection and manipulation of nucleic acids in the laboratory.

Exonucleases degrade nucleic acids by sequentially removing nucleotides from their ends. Two in common use are *snake venom phosphodiesterase* and *bovine spleen phosphodiesterase* (Figure 6.31). Since they act on either DNA or RNA, they are referred to by the generic name *phosphodiesterase.* These two enzymes have complementary specificities. Snake venom phosphodiesterase acts by *a* cleavage and starts at the free 3′-OH end of a polynucleotide chain, liberating nucleoside 5′-monophosphates. In contrast, the bovine spleen enzyme starts at the 5′-end of a nucleic acid, cleaving *b* and releasing 3′-NMPs.

RNases

A number of RNases have proven useful in the analysis of RNA. The best studied and most widely used ribonuclease is **RNase A** from bovine pancreas. This enzyme is an endonuclease that catalyzes *b* cleavage where the 3′-PO_4 bond is to a pyrimidine (Figure 6.32). The final product of RNase A action is a mixture of oligoribonucleotides having pyrimidine 3′-phosphate ends. RNase A is a remarkably stable enzyme and can withstand heating to 100°C without loss of activity. *Bacillus subtilis* contains an RNase whose specificity is the analog of pancreatic RNase. It is a *b*-type endonuclease that acts where purines occur on the 3′-side of the susceptible bond. Its products are oligonucleotides with 3′-terminal purine 3′-PO_4 residues. RNases of even greater specificity toward phosphodiester bonds involving purine nucleotides are the fungal *RNases T1* and *T2.* RNase T1 acts endo *b,* where G residues are 3′ to the phosphodiester bridge; RNase T2 acts similarly, except A residues must occupy the 3′-position. Such enzymes can be used in determining the nucleotide sequence of relatively short ribonucleotides (see problems at the end of this chapter).

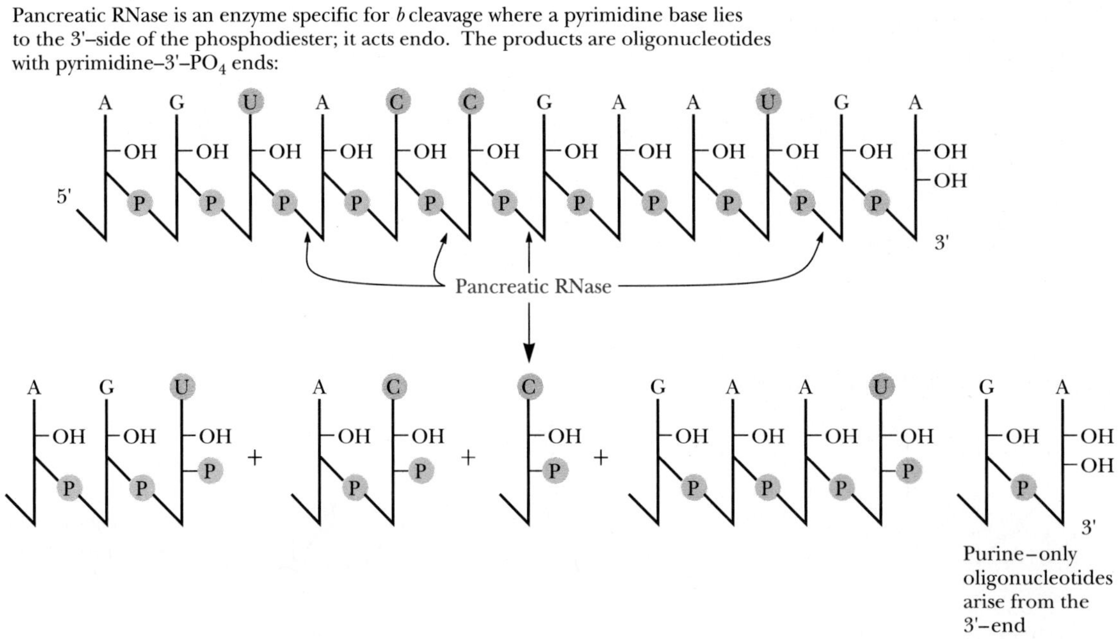

Pancreatic RNase is an enzyme specific for *b* cleavage where a pyrimidine base lies to the 3'–side of the phosphodiester; it acts endo. The products are oligonucleotides with pyrimidine–3'–PO_4 ends:

Pancreatic RNase

Purine–only oligonucleotides arise from the 3'–end

***Figure* 6.32** The specificity of RNA hydrolysis by bovine pancreatic RNase, which cleaves *b* at 3′-pyrimidines, yielding oligonucleotides with pyrimidine 3′-PO_4 ends.

RNases appear to be ubiquitous in the environment, to the extent that their presence often makes it difficult to isolate RNA intact. mRNA is particularly labile to the effects of these endogenous RNases, and biochemists must often resort to the use of specially prepared "RNase-free" glassware and solutions to obtain good-quality RNA. Since RNase is difficult to inactivate, its ubiquity poses a serious nuisance.

DNases

In general, DNases hydrolyze either single-stranded (**ss**) or double-stranded (**ds**) deoxyribonucleotide chains. Two of opposing specificity are *DNase I* of bovine pancreas and *DNase II,* a type of DNase that can be isolated from a variety of sources, including spleen or thymus tissue or the bacterium *Staphylococcus aureus.* DNase I is an *a*-acting endonuclease that shows some preference for cleaving between pyrimidines and purines. It can be used at low concentration to introduce "nicks" in dsDNA by creating 3′-OH ends at random internal positions. Nicked dsDNA is an important precursor to the *in vitro* synthesis of radioactively labeled DNA molecules for nucleic acid sequencing (see Chapter 7) and for various analytical applications in molecular biological research (see Chapter 8). DNase II is a *b* endonuclease. It yields a mixture of oligodeoxyribonucleotides with 3′-PO_4 ends as products.

Restriction Enzymes

Restriction endonucleases are enzymes, isolated chiefly from bacteria, that have the ability to cleave double-stranded DNA. The term *restriction* comes from the capacity of prokaryotes to defend against or "restrict" the possibility of takeover by foreign DNA that might gain entry into their cells. Prokaryotes degrade foreign DNA by using their unique restriction enzymes to chop it into relatively large but noninfective fragments. Restriction enzymes are classified into three types, I, II, or III. Types I and III require ATP to hydrolyze DNA and can also catalyze chemical modification of DNA through addition of methyl groups to specific bases. Type I restriction endonucleases cleave DNA randomly, while type III recognize specific nucleotide sequences within dsDNA and cut the DNA at or near these sites.

Type II Restriction Endonucleases

Type II restriction enzymes have received widespread application in the cloning and sequencing of DNA molecules. Their hydrolytic activity is not ATP-dependent, and they do not modify DNA by methylation or other means. Most importantly, they cut DNA within or near particular nucleotide sequences that they specifically recognize. These recognition sequences are typically four or six nucleotides in length and have a twofold axis of symmetry. For example, *E. coli* has a restriction enzyme, *Eco*RI, that recognizes the following hexanucleotide sequence:

```
5′------G-A-A-T-T-C------3′
        · · · · · ·
        · · · · · ·
3′------C-T-T-A-A-G------5′
```

Note the twofold symmetry: the sequence read $5' \rightarrow 3'$ is the same in both strands.

When *Eco*RI encounters this sequence in dsDNA, it causes a staggered, double-stranded break by hydrolyzing each chain between the G and A residues:

```
5′------G A-A-T-T-C------3′
        · · · · · ·
        · · · · · ·
3′------C-T-T-A-A G------5′
```

Staggered cleavage results in fragments with protruding 5′-ends:

```
5′------G                 5′A-A-T-T-C------3′
        ·                             ·
        ·                             ·
3′------C-T-T-A-A5′                   G------5′
```

Because the protruding termini of *Eco*RI fragments have complementary base sequences, they can form base pairs with one another:

```
------G A-A-T-T-C------
      · · · · · ·
      · · · · · ·
------C-T-T-A-A G------
```

Therefore, DNA restriction fragments having such "sticky" ends can be joined together to create new combinations of DNA sequence. If the fragments are derived from DNA molecules of different origin, novel recombinant forms of DNA are created.

*Eco*RI leaves staggered 5′-termini. Other restriction enzymes, such as *Pst*I, which recognizes the sequence 5′-CTGCAG-3′ and cleaves between A and G, produce cohesive staggered 3′-ends. Still others, such as *Bal*I, act at the center of the twofold symmetry axis of their recognition site and generate blunt ends that are noncohesive. *Bal*I recognizes 5′-TGGCCA-3′ and cuts between G and C.

Table 6.5 lists many of the commonly used restriction endonucleases and their unique recognition sites. Since these sites all have twofold symmetry, only the sequence on one strand needs to be designated. Close to 1000 restriction enzymes have been characterized. They are named by italicized three-letter codes, the first, a capital letter denoting the genus of the organism of

***Table* 6.5**

Restriction Endonucleases

With only one exception (*Nci*I), all type II restriction endonucleases generate fragments with 5′-PO_4 and 3′-OH ends.

Enzyme	Common Isoschizomers	Recognition Sequence	Compatible Cohesive Ends
*Alu*I		AG↓CT	Blunt
*Aos*I		TGC↓GCA	Blunt
*Apy*I	*Atu*I, *Eco*RII	CC↓$\binom{A}{T}$GG	
*Asu*I		G↓GNCC	
*Asu*II		TT↓CGAA	*Cla*I, *Hpa*II, *Taq*I
*Ava*I		G↓PyCGPuG	*Sal*I, *Xho*I, *Xma*I
*Ava*II		G↓G$\binom{A}{T}$CC	*Sau*96I
*Avr*II		C↓CTAGG	
*Bal*I		TGG↓CCA	Blunt
*Bam*HI		G↓GATCC	*Bcl*I, *Bgl*II, *Mbo*I, *Sau3*A, *Xho*II
*Bcl*I		T↓GATCA	*Bam*HI, *Bgl*II, *Mbo*I, *Sau*3A, *Xho*II
*Bgl*II		A↓GATCT	*Bam*HI, *Bcl*I, *Mbo*I, *Sau*3A, *Xho*II
*Bst*EII		G↓GTNACC	
*Bst*NI		CC↓$\binom{A}{T}$GG	
*Bst*XI		CCANNNNN↓NTGG	
*Cla*I		AT↓CGAT	*Acc*I, *Acy*I, *Asy*II, *Hpa*II, *Taq*I
*Dde*I		C↓TNAG	
*Eco*RI		G↓AATTC	
*Eco*RII	*Atu*I, *Apy*I	↓CC$\binom{A}{T}$GG	
*Fnu*4HI		GC↓NGC	
*Fnu*DII	*Tha*I	CG↓CG	Blunt
*Hae*I		$\binom{A}{T}$GG↓CC$\binom{T}{A}$	Blunt
*Hae*II		PuGCGC↓Py	
*Hae*III		GG↓CC	Blunt
*Hha*I	*Cfo*I	GCG↓C	
*Hinc*II		GTPy↓PuAC	Blunt
*Hind*II		GTPy↓PuAC	Blunt
*Hind*III		A↓AGCTT	
*Hinf*I		G↓ANTC	
*Hpa*I		GTT↓AAC	Blunt
*Hpa*II		C↓CGG	*Acc*I, *Acy*I, *Asu*II, *Cla*I, *Taq*I
*Kpn*I		GGTAC↓C	*Bam*HI, *Bcl*I, *Bgl*II, *Xho*II
*Mbo*I	*Sau*3A	↓GATC	
*Msp*I		C↓CGG	
*Mst*I		TGC↓GCA	Blunt
*Not*I		GC↓GGCCGC	
*Pst*I		CTGCA↓G	
*Pru*II		CAG↓CTG	Blunt
*Rsa*I		GT↓AC	Blunt
*Sac*I	*Sst*I	GAGCT↓C	
*Sac*II		CCGC↓GG	

continued

***Table* 6.5 (*continued*)**

Enzyme	Common Isoschizomers	Recognition Sequence	Compatible Cohesive Ends
*Sal*I		G↓TCGAC	*Ava*I, *Xho*I
*Sau*3A		↓GATC	*Bam*HI, *Bcl*I, *Bgl*II, *Mbo*I, *Xho*II
*Sau*96I		G↓GNCC	
*Sfi*I		GGCCNNNN↓NGGCC	
*Sma*I	*Xma*I	CCC↓GGG	Blunt
*Sph*I		GCATG↓C	
*Sst*I	*Sac*I	GAGCT↓C	
*Sst*II		CCGC↓GG	
*Taq*I		T↓CGA	*Acc*I, *Acy*I, *Asu*II, *Cla*I, *Hpa*II
*Tha*I	*Fnu*DII	CG↓CG	Blunt
*Xba*I		T↓CTAGA	
*Xho*I		C↓TCGAG	*Ava*I, *Sal*I
*Xho*II		($^{A}_{G}$)↓GATC($^{T}_{C}$)	*Bam*HI, *Bcl*I, *Bgl*II, *Mbo*I, *Sau*3A
*Xma*I	*Sma*I	C↓CCGGG	*Ava*I
*Xma*III		C↓GGCCG	
*Xor*II		CGATC↓G	

origin, while the next two letters are an abbreviation of the particular species. Since prokaryotes often contain more than one restriction enzyme, the various representatives are assigned letter and number codes as they are identified. Thus, *Eco*RI is the initial restriction endonuclease isolated from *E. coli* strain R.

Isoschizomers. Different restriction enzymes sometimes recognize and cleave within identical target sequences. For example, *Mbo*I and *Sau*3A recognize the same tetranucleotide run: 5′-GATC-3′. Both cleave the DNA strands at the same position, namely, on the 5′-side of the G. Such enzymes are called **isoschizomers,** meaning that they cut at the same site. The enzyme *Bam*HI is an isoschizomer of *Mbo*I and *Sau*3A except that it has greater specificity because it acts only at hexanucleotide sequences reading GGATCC. *Bam*HI cuts between the two G's, leaving cohesive 5′-ends that can match up with *Mbo*I or *Sau*3A fragments.

schizo: from the Greek *schizein,* to split

Restriction Fragment Size. Assuming random distribution and equimolar proportions for the four nucleotides in DNA, a particular tetranucleotide sequence should occur once every 4^4 nucleotides, or every 256 bases. Therefore, the fragments generated by a restriction enzyme that acts at a four-nucleotide sequence should average about 250 bp in length. "Six-cutters," enzymes such as *Eco*RI or *Bam*HI, will find their unique hexanucleotide sequences on the average once in every 4096 (4^6) bp of length. Since the genetic code is a triplet code with three bases of DNA specifying one amino acid in a polypeptide sequence, and since polypeptides typically contain at most 1000 amino acid residues, the fragments generated by six-cutters are approximately the size of prokaryotic genes. This property makes these enzymes useful in the construction and cloning of genetically useful recombinant DNA molecules.

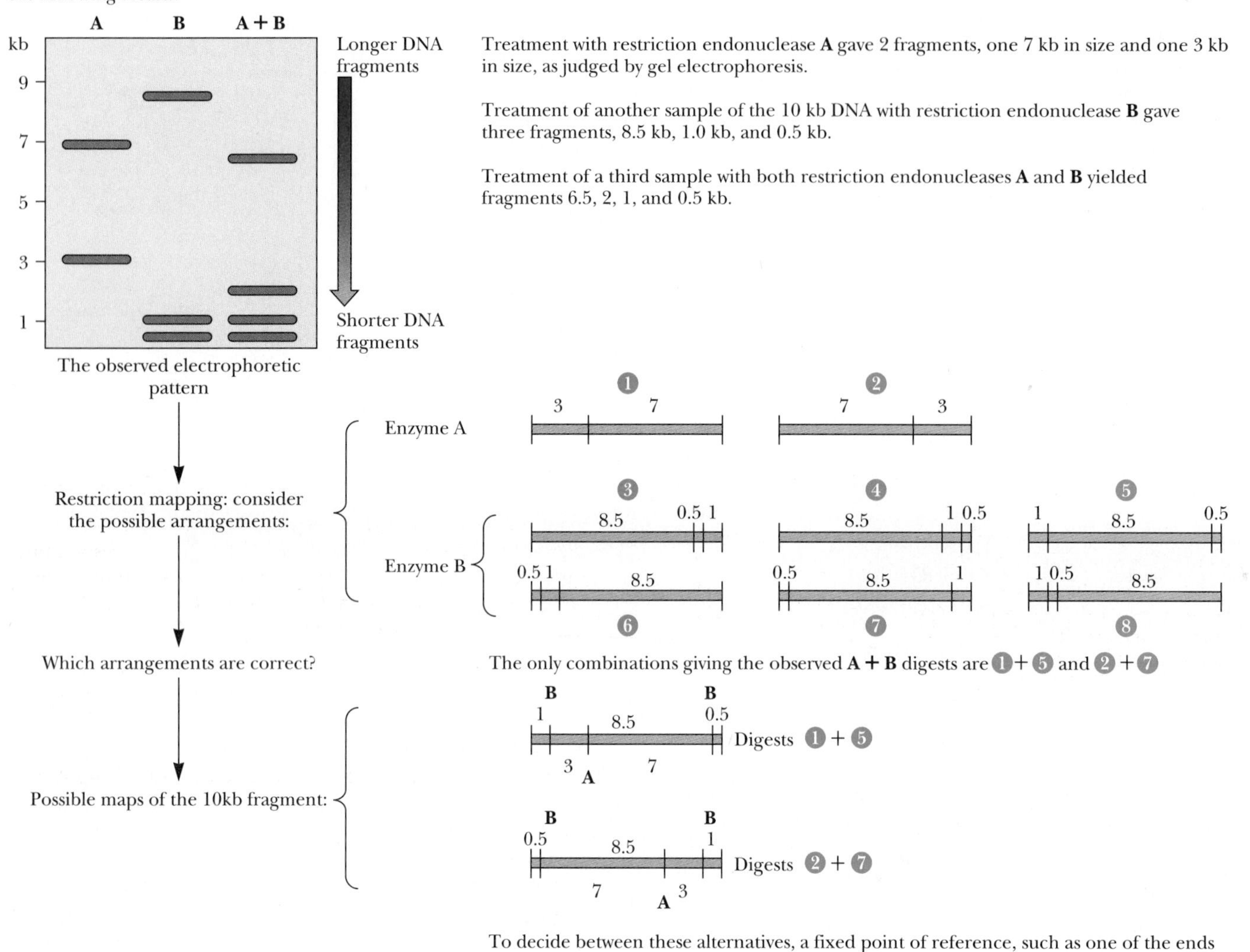

Treatment of a linear 10kb DNA molecule with endonucleases gave the following results:
A
B
A + B
kb
9
7
5
3
1
Longer DNA fragments
Shorter DNA fragments
The observed electrophoretic pattern
Treatment with restriction endonuclease A gave 2 fragments, one 7 kb in size and one 3 kb in size, as judged by gel electrophoresis.
Treatment of another sample of the 10 kb DNA with restriction endonuclease B gave three fragments, 8.5 kb, 1.0 kb, and 0.5 kb.
Treatment of a third sample with both restriction endonucleases A and B yielded fragments 6.5, 2, 1, and 0.5 kb.
Restriction mapping: consider the possible arrangements:
Enzyme A
Enzyme B
Which arrangements are correct?
The only combinations giving the observed A + B digests are 1 + 5 and 2 + 7
Possible maps of the 10kb fragment:
Digests 1 + 5
Digests 2 + 7
To decide between these alternatives, a fixed point of reference, such as one of the ends of the fragment, must be identified or labeled. The task increases in complexity as DNA size, number of restriction sites, and/or number of restriction enzymes used increases.

Figure **6.33** Restriction mapping of a DNA molecule as derived from an analysis of the electrophoretic pattern obtained for different restriction endonuclease digests.

For the isolation of even larger nucleotide sequences, such as those of genes encoding large polypeptides (or those of eukaryotic genes that are disrupted by large introns), partial or limited digestion of DNA by restriction enzymes can be employed. Recently, however, restriction endonucleases that cut only at specific nucleotide sequences 8 or even 13 nucleotides in length have become available, such as *Not*I (which recognizes GCGGCCGC and cleaves after the first C) and *Sfi*I (which is specific for the sequence arrangement GGCCNNNNNGGCC and cuts the DNA between the fourth and fifth variable nucleotide, N).

Restriction Mapping

The application of these sequence-specific nucleases to problems in molecular biology is considered in detail in Chapter 8, but one prevalent application merits discussion here. Since restriction endonucleases cut dsDNA at unique sites to generate large fragments, they provide a means for mapping DNA molecules that are many kilobase pairs in length. Restriction digestion of a DNA molecule is in many ways analogous to proteolytic digestion of a protein by an enzyme such as trypsin (see Chapter 4): the restriction endonuclease acts only at its specific sites so that a discrete set of nucleic acid fragments is generated. This action is analogous to trypsin cleaving only at Arg and Lys residues to yield a particular set of tryptic peptides from a given protein. The restriction fragments represent a unique collection of different-sized DNA pieces. Fortunately, this complex mixture can be resolved by *electrophoresis* (see the Appendix to Chapter 4). Electrophoresis of DNA molecules on gels of restricted pore size (as formed in agarose or polyacrylamide media) separates them according to size, the largest being retarded in their migration through the gel pores while the smallest move relatively unhindered. Figure 6.33 shows a hypothetical electrophoretogram obtained for a DNA molecule treated with two different restriction nucleases, alone and in combination. Just as cleavage of a protein with different proteases to generate overlapping fragments allows an ordering of the peptides, restriction fragments can be ordered or "mapped" according to their sizes, as deduced from the patterns depicted in Figure 6.33.

Problems

1. Draw the chemical structure of pACG.
2. Chargaff's results (Table 6.3) yielded a molar ratio of 1.56 for A to G in human DNA, 1.75 for T to C, 1.00 for A to T, and 1.00 for G to C. Given these values, what are the mole fractions of A, C, G, and T in human DNA?
3. Adhering to the convention of writing nucleotide sequences in the $5' \rightarrow 3'$ direction, what is the nucleotide sequence of the DNA strand that is complementary to d-ATCGCAACTGTCACTA?
4. Messenger RNAs are synthesized by RNA polymerases that read along a DNA template strand in the $3' \rightarrow 5'$ direction, polymerizing ribonucleotides in the $5' \rightarrow 3'$ direction (see Figure 6.24). Give the nucleotide sequence ($5' \rightarrow 3'$) of the DNA template strand from which the following mRNA segment was transcribed: 5′-UAGUGACAGUUGCGAU-3′.
5. The DNA strand that is complementary to the template strand copied by RNA polymerase during transcription has a nucleotide sequence identical to that of the RNA being synthesized (except T residues are found in the DNA strand at sites where U residues occur in the RNA). This DNA strand is the so-called sense strand; the template strand is an antisense strand. A promising strategy to thwart the deleterious effects of genes activated in disease states (such as cancer) is to generate antisense RNAs in affected cells. These antisense RNAs would form double-stranded hybrids with mRNAs transcribed from the activated genes and prevent their translation into protein. Suppose transcription of a cancer-activated gene yielded an mRNA whose sequence included the segment 5′-UACGGUCUAAGCUGA. What is the corresponding nucleotide sequence ($5' \rightarrow 3'$) of the template strand in a DNA duplex that might be introduced into these cells so that an antisense RNA could be transcribed from it?
6. The following analyses were performed to deduce a possible nucleotide sequence for an oligonucleotide of base composition $A_2C_4G_2U$:
   a. RNase A treatment: The products liberated were (in molar equivalents) 2 Cp, a dinucleotide containing A and U, a dinucleotide containing G and C, and a trinucleotide containing A, C, and G.
   b. RNase T_1: The products liberated were C, pGp, and a heptanucleotide containing 3C, 2A, 1U, and 1G.
   c. RNase T_2: The products were a dinucleotide containing A and G, a trinucleotide containing A and C, and a tetranucleotide containing C, G, and U.
   d. Limited digestion with snake venom phosphodiesterase yielded some pC.

   Give a sequence.
7. A 10-kb DNA fragment digested with restriction endonuclease *Eco*RI yielded fragments 4 kb and 6 kb in size. When digested with *Bam*HI, fragments 1, 3.5, and 5.5 kb were generated. Concomitant digestion with both *Eco*RI and *Bam*HI yielded fragments 0.5, 1, 3, and 5.5 kb in size. Give a possible restriction map for the original fragment.

Further Reading

Adams, R. L. P., Knowler, J. T., and Leader, D. P., 1986. *The Biochemistry of the Nucleic Acids,* 10th ed. New York: Chapman and Hall (Methuen and Co., distrib.).

Gray, M. W., and Cedergren, R., eds., 1993. The New Age of RNA. *The FASEB Journal* **7**:4–239. A collection of articles emphasizing the new appreciation for RNA in protein synthesis, in evolution, and as a catalyst.

Judson, H. F., 1979. *The Eighth Day of Creation.* New York: Simon and Schuster.

Maniatis, T., Frisch, E. F., and Sambrook, J., 1989. *Molecular Cloning: A Laboratory Manual,* 2nd ed. Cold Spring Harbor, NY: Cold Spring Harbor Laboratory.

Watson, J. D., Hopkins, N. H., Roberts, J. W., Steitz, J. A., and Weiner, A. M., 1987. *The Molecular Biology of the Gene,* Vol. I, *General Principles,* 4th ed. Menlo Park, CA: Benjamin/Cummings.

"The Structure of DNA: 'A melody for the eye of the intellect, with not a note wasted.'"

Horace Freeland Judson, *The Eighth Day of Creation*

Chapter 7

Structure of Nucleic Acids

"Scherzo in D & A" (detail) by David E. Rodale (1955–1985)

Outline

7.1 The Primary Structure of Nucleic Acids

7.2 The ABZ's of DNA Secondary Structure

7.3 Supercoils and Cruciforms: Tertiary Structure in DNA

7.4 Denaturation and Renaturation of DNA

7.5 Chromosome Structure

7.6 Chemical Synthesis of Nucleic Acids

7.7 Secondary and Tertiary Structure of RNA

Chapter 6 presented the structure and chemistry of nucleotides and how these units are joined via phosphodiester bonds to form nucleic acids, the biological polymers for information storage and transmission. In this chapter, we investigate biochemical methods that reveal this information by determining the sequential order of nucleotides in a polynucleotide, the so-called **primary structure** of nucleic acids. Then, the higher orders of structure in the nucleic acids, the secondary and tertiary levels, are considered. Although the focus here is primarily on the structural and chemical properties of these macromolecules, it is fruitful to keep in mind the biological roles of these remarkable substances. Nucleic acids are the embodiment of genetic information (see Part IV). We can anticipate that cellular mechanisms for accessing this information, as well as reproducing it with high fidelity, will be illuminated by knowledge of the chemical and structural qualities of these polymers.

7.1 The Primary Structure of Nucleic Acids

As recently as 1975, determining the primary structure of nucleic acids (the nucleotide sequence) was thought to present a theoretically more formidable problem than amino acid sequencing of proteins, simply because nucleic acids contain only four unique monomeric units while proteins have twenty. With only four, there are *apparently* fewer specific sites for selective cleavage, distinctive sequences are more difficult to recognize, and the likelihood of ambiguity is greater. A further difficulty is imposed by the much greater number of monomeric units in most polynucleotides as compared to polypeptides. Two important breakthroughs have reversed this outlook so that now sequencing nucleic acids is substantially easier than sequencing polypeptides. One is the discovery of *restriction endonucleases* that cleave DNA at specific oligonucleotide sites, generating unique fragments of manageable size (see Chapter 6). The second is the power of *polyacrylamide gel electrophoresis* separation methods to resolve nucleic acid fragments that differ from one another in length by only a single nucleotide.

Sequencing Nucleic Acids

Two basic protocols for nucleic acid sequencing are in widespread use: the **chain termination** or **dideoxy method** of F. Sanger and the **base-specific chemical cleavage method** developed by A. M. Maxam and W. Gilbert. Since both methods are carried out on nanogram amounts of DNA, very sensitive analytical techniques are used to detect the DNA chains following electrophoretic separation on polyacrylamide gels. Typically, the DNA molecules are labeled with radioactive ^{32}P,[1] and following electrophoresis, the pattern of their separation is visualized by **autoradiography.** A piece of X-ray film is placed over the gel and the radioactive disintegrations emanating from ^{32}P decay create a pattern on the film that is an accurate image of the resolved oligonucleotides. Recently, sensitive biochemical and chemiluminescent methods have begun to supersede the use of radioisotopes as tracers in these experiments.

Chain Termination or Dideoxy Method

To appreciate the rationale of the chain termination or dideoxy method, we first must briefly examine the biochemistry of DNA replication. DNA is a double-helical molecule. In the course of its replication, the sequence of nucleotides in one strand is copied in a complementary fashion to form a new second strand by the enzyme **DNA polymerase.** Each original strand of the double helix serves as **template** for the biosynthesis that yields two daughter DNA duplexes from the parental double helix (Figure 7.1). DNA polymerase will carry out this reaction *in vitro* in the presence of the four deoxynucleotide monomers and will copy single-stranded DNA, provided a double-stranded region of DNA is artificially generated by adding a **primer.** This primer is merely an oligonucleotide capable of forming a short stretch of dsDNA by base pairing with the ssDNA (Figure 7.2). The primer must have a free 3′-OH end from which the new polynucleotide chain can grow as the first residue is added in the initial step of the polymerization process. DNA polymerases synthesize new strands by adding successive nucleotides in the $5' \rightarrow 3'$ direction.

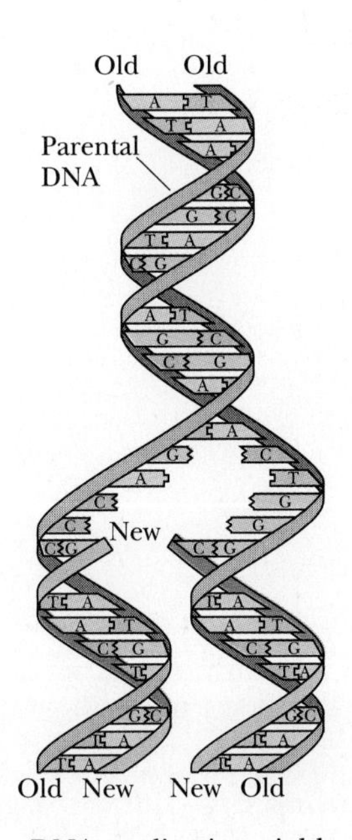

Figure **7.1** DNA replication yields two daughter DNA duplexes identical to the parental DNA molecule. Each original strand of the double helix serves as a template, and the sequence of nucleotides in each of these strands is copied to form a new complementary strand by the enzyme DNA polymerase. By this process, biosynthesis yields two daughter DNA duplexes from the parental double helix.

[1] Because its longer half-life and lower energy make it safer to handle, ^{35}S is replacing ^{32}P as the radioactive tracer of choice in sequencing by the Sanger method. ^{35}S-α-labeled deoxynucleotide analogs provide the source for incorporating radioactivity into DNA.

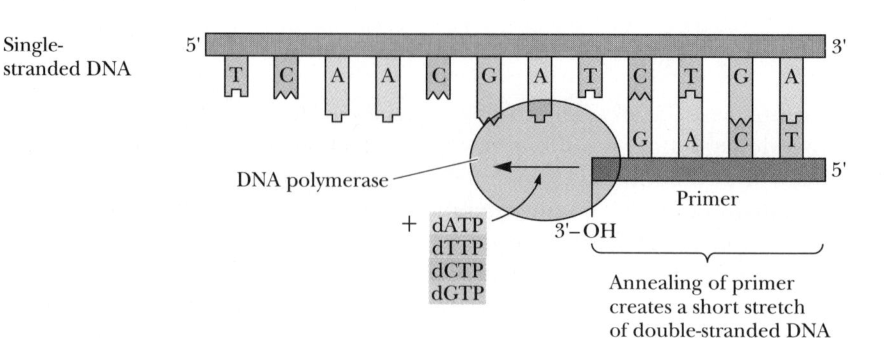

***Figure* 7.2** The *E. coli* enzyme DNA polymerase I will copy ssDNA *in vitro* in the presence of the four deoxynucleotide monomers, provided a double-stranded region of DNA is artificially generated by adding a primer, an oligonucleotide capable of forming a short stretch of dsDNA by base pairing with the ssDNA. The primer must have a free 3′-OH end from which the new polynucleotide chain can grow as the first residue is added in the initial step of the polymerization process.

Chain Termination Protocol

In the chain termination method of DNA sequencing, a DNA fragment of unknown sequence serves as template in a polymerization reaction using some type of DNA polymerase, usually *Sequenase 2®*, a genetically engineered version of bacteriophage T7 DNA polymerase that lacks all traces of exonuclease activity that might otherwise degrade the DNA. The primer requirement is met by an appropriate oligonucleotide (this method is also known as the **primed synthesis method** for this reason). Four parallel reactions are run; all four contain the four deoxynucleoside triphosphates dATP, dGTP, dCTP, and dTTP, which are the substrates for DNA polymerase (Figure 7.3 on the facing page). In each of the four reactions, a different 2′,3′-**di**deoxynucleotide is included, and it is these dideoxynucleotides that give the method its name.

Because dideoxynucleotides lack 3′-OH groups, these nucleotides cannot serve as acceptors for 5′-nucleotide addition in the polymerization reaction, and thus the chain is terminated where they become incorporated. The concentrations of the four deoxynucleotides and the single dideoxynucleotide in each reaction mixture are adjusted so that the dideoxynucleotide is incorporated infrequently. Therefore, base-specific premature chain termination is only a random, occasional event, and a population of new strands of varying length is synthesized. Four reactions are run, one for each dideoxynucleotide, so that terminations, although random, can occur everywhere in the sequence. In each mixture, each newly synthesized strand has a dideoxynucleotide at its 3′-end, and its presence at that position demonstrates that a base of that particular kind was specified by the template. A radioactively labeled dNTP is included in each reaction mixture to provide a tracer for the products of the polymerization process.

Reading Dideoxy Sequencing Gels

The sequencing products are visualized by autoradiography (or similar means) following their separation according to size by polyacrylamide gel electrophoresis (Figure 7.3). Since the smallest fragments migrate fastest upon electrophoresis and since fragments differing by only single nucleotides in length are readily resolved, the autoradiogram of the gel can be read from bottom to top, noting which lane has the next largest band at each step. Thus, the gel in Figure 7.3 is read ATCGTTGA (5′ → 3′). Because of the way DNA polymerase acts, this observed sequence is complementary to the corresponding unknown template sequence. Knowing this, the template sequence now can be written TCAACGAT (5′ → 3′). **Arabinose** derivatives of the nucleotides have also been used as chain terminators. Arabinose is the C-2 epimer of ribose (Figure 7.4), and DNA polymerase will incorporate such nucleotides into chains but will not add subsequent nucleotides to them.

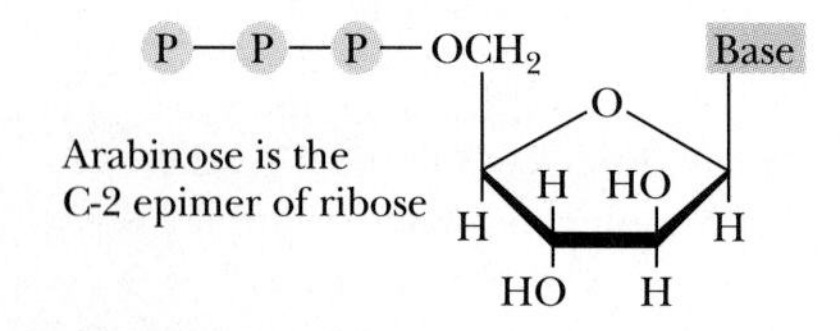

***Figure* 7.4** Structure of an arabinose derivative of a nucleotide. Arabinoside derivatives can also be used in the Sanger chain termination protocol.

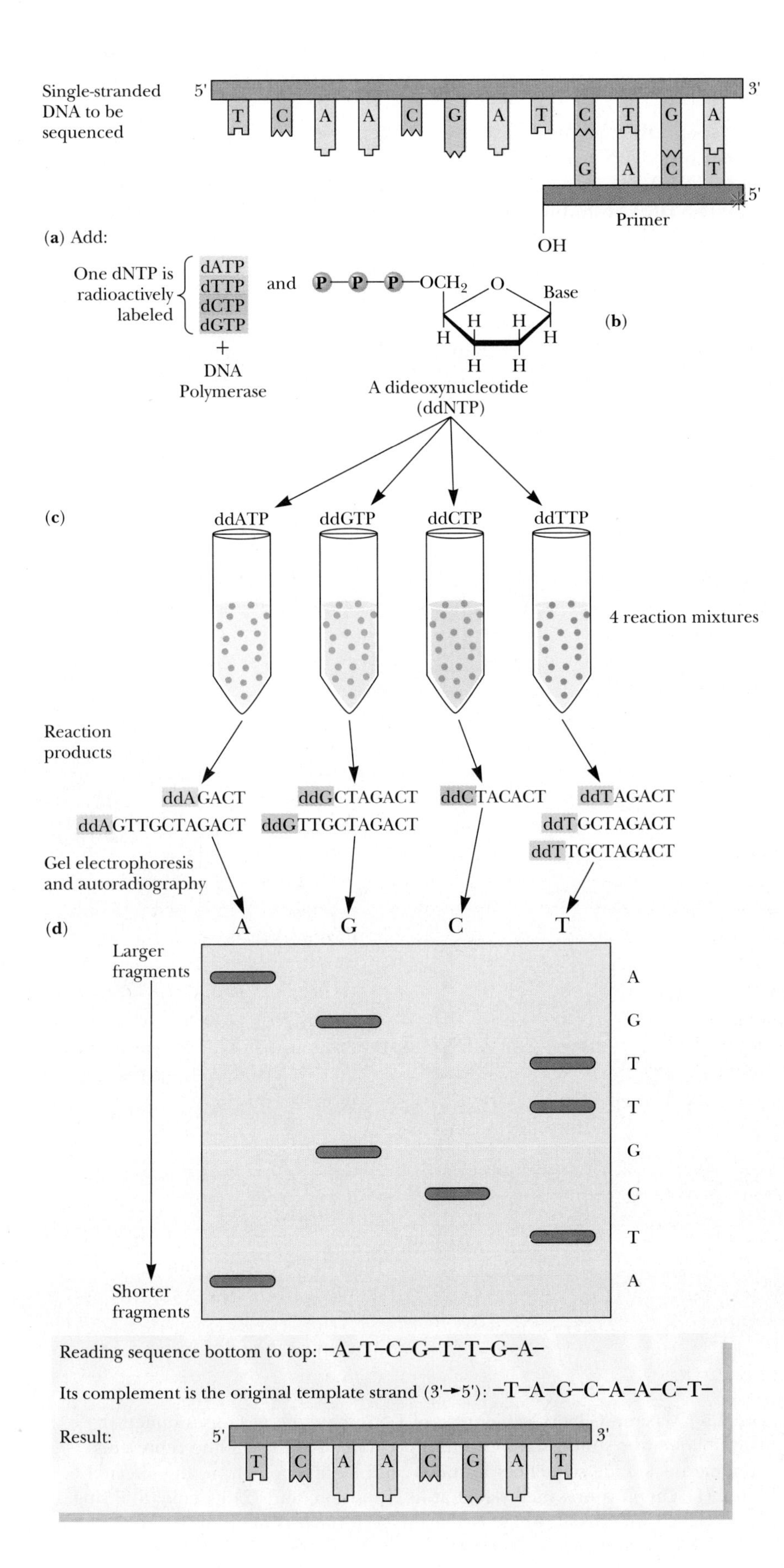

***Figure* 7.3** The chain termination or dideoxy method of DNA sequencing. (a) DNA polymerase reaction. (b) Structure of dideoxynucleotide. (c) Four reaction mixtures with nucleoside triphosphates plus one dideoxynucleoside triphosphate. (d) Electrophoretogram. Note that the nucleotide sequence as read from the bottom to the top of the gel is the order of nucleotide addition carried out by DNA polymerase.

Base-Specific Chemical Cleavage Method

The base-specific chemical cleavage (or Maxam–Gilbert) method starts with a single-stranded DNA that is labeled at one end with radioactive ^{32}P. (Double-stranded DNA can be used if only one strand is labeled at only one of its ends.) The DNA strand is then randomly cleaved by reactions that specifically fragment its sugar–phosphate backbone only where certain bases have been chemically removed. There is no unique reaction for each of the four bases. However, there is a reaction specific to G only and a purine-specific reaction that removes A or G. Thus, the difference in these two reactions is a specific indication of where A occurs. Similarly, there is a cleavage reaction specific for the pyrimidines (C+T), which, if run in the presence of 1 or 2 M NaCl,

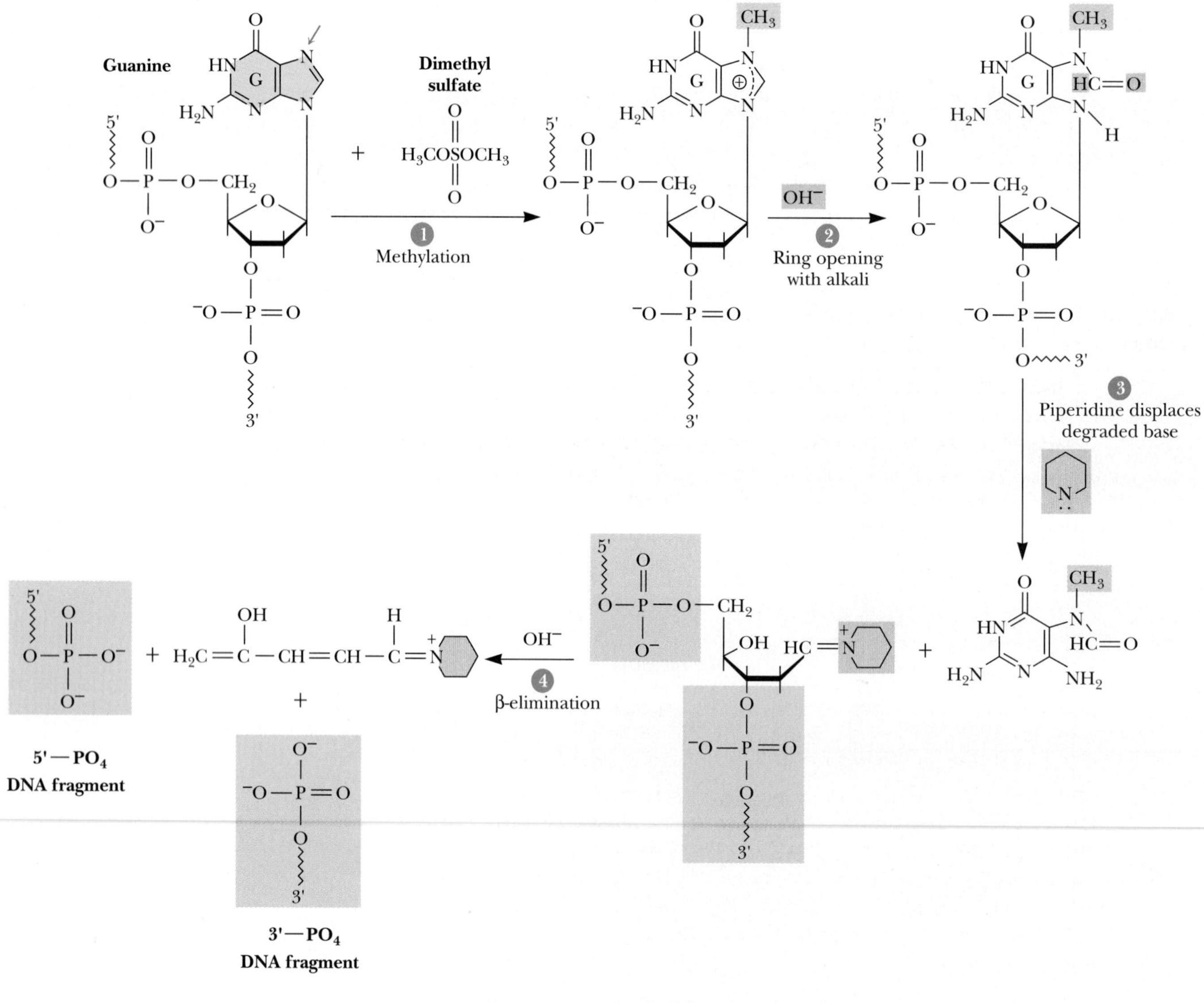

Guanine
Dimethyl sulfate
$H_3COSOCH_3$
1 Methylation
OH^-
2 Ring opening with alkali
3 Piperidine displaces degraded base
OH^-
4 β-elimination
$5'-PO_4$ DNA fragment
$3'-PO_4$ DNA fragment

Figure **7.5** Maxam–Gilbert sequencing of DNA: cleavage at G uses dimethyl sulfate, followed by strand scission with piperidine. The zigzag line represents continuing nucleotide sequences in the 5′- and 3′-directions from the affected G residue. **(1)** The G is first methylated at its 7-position, and **(2)** its imidazole ring is then opened with alkali, which is actually performed concurrently with **(3)** piperidine treatment. **(4)** Piperidine displaces the degraded base and triggers strand scission by *β*-elimination to yield $5'$-PO_4 and $3'$-PO_4 oligonucleotide fragments.

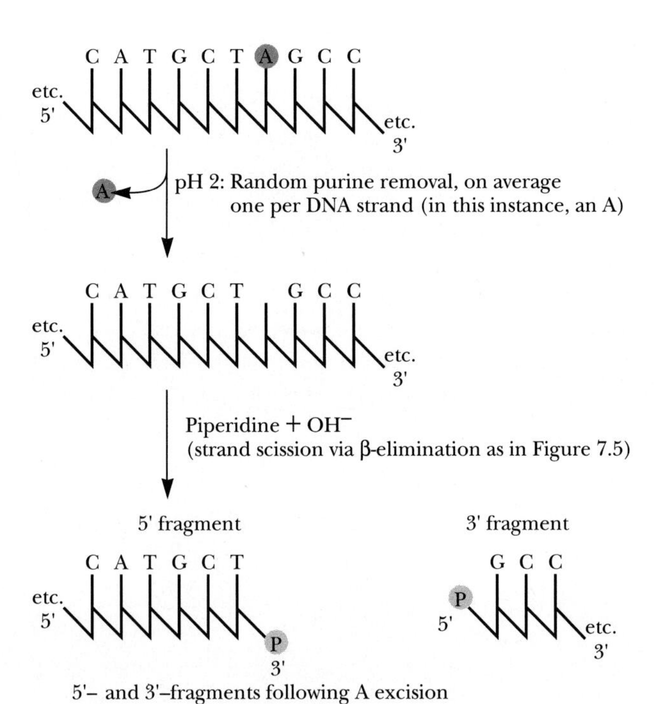

***Figure* 7.6** Maxam–Gilbert sequencing of DNA: cleavage at A and G. Mild acid treatment of DNA randomly depurinates it because the N-9 glycosidic bonds linking purines to the sugar–phosphate backbone are susceptible to acid hydrolysis (at pH 2). In the presence of piperidine and NaOH, the depurinated deoxyribose residues undergo strand scission by β-elimination.

becomes uniquely specific for C. Differences in these two are thus attributable to the presence of T in the nucleotide sequence.

Two **purine-dependent** reactions are used in this Maxam–Gilbert type sequencing.

scission: the act of cutting

1. *Cleavage at G using dimethyl sulfate, followed by strand scission with piperidine* (Figure 7.5). Dimethyl sulfate reacts with guanine to methylate it in the 7-position. This substitution leads to instability of the N-9 glycosidic bond, so that in the presence of OH^- and the secondary amine **piperidine,** the purine ring is released and degraded. A β-elimination reaction facilitated by piperidine then causes the excision of the naked deoxyribose moiety from the sugar–phosphate backbone, with consequent scission of the DNA strand to yield 5′- and 3′-fragments.
2. *Cleavage at A and G* (Figure 7.6). Mild acid treatment of DNA leads to random loss of purines, referred to as *depurination.* That is, the N-9 glycosidic bonds linking purines to the sugar–phosphate backbone are susceptible to acid hydrolysis. If such treatment is carried out at pH 2 in the presence of piperidine and NaOH is then added, the depurinated deoxyribose residues undergo strand scission by β-elimination as in reaction 1. The DNA fragments remaining are 5′-oligonucleotides and 3′-oligonucleotides that were previously linked through a purine nucleotide that was chemically destroyed.

The two remaining reactions of Maxam–Gilbert sequencing are **pyrimidine dependent:**

3. *Hydrolysis of pyrimidine rings by hydrazine* (Figure 7.7). Hydrazine ($H_2N—NH_2$) attacks across the C-4 and C-6 atoms of pyrimidines to open the ring. This degradation subsequently leads to modification of the deoxyribose, rendering it susceptible to β-elimination by piperidine in the presence of hydroxide ion. As in reactions 1 and 2, 5′- and 3′-fragments are produced.

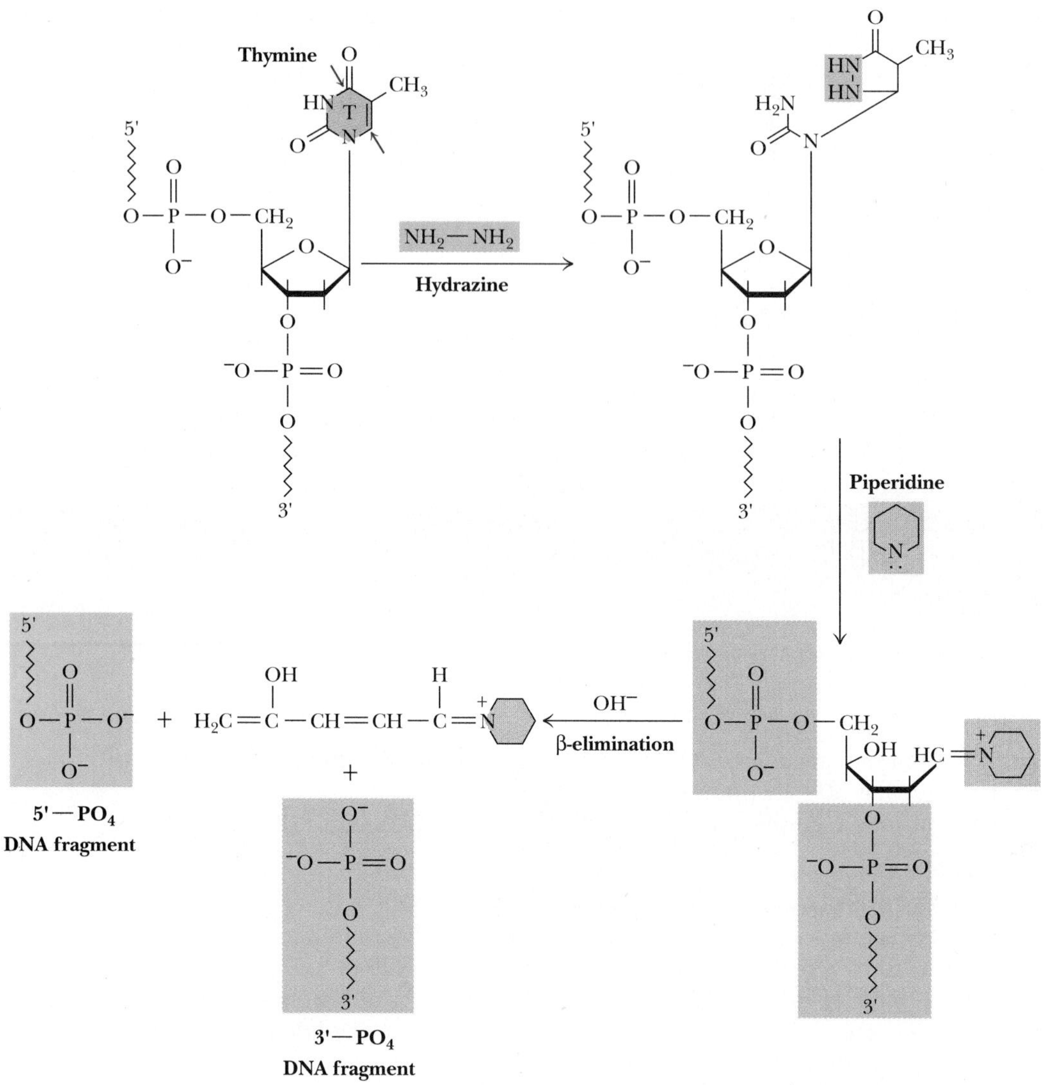

Figure **7.7** Maxam–Gilbert sequencing of DNA: hydrolysis of pyrimidine rings by hydrazine ($H_2N—NH_2$). Hydrazine attacks across the C-4 and C-6 atoms of pyrimidines to open the ring. This degradation leaves the naked sugar–phosphate backbone susceptible to *β*-elimination by piperidine in the presence of hydroxide ion. Shown here is the excision of a T residue. As in Figures 7.5 and 7.6, 5′- and 3′-fragments are produced.

4. *Specific hydrazinolysis at cytosine residues.* The presence of high salt concentrations protects T from reaction with hydrazine. In the presence of 2 *M* NaCl, the reaction shown in Figure 7.7 occurs only at C.

Note that the key to Maxam–Gilbert sequencing is to modify a base chemically so that it is removed from its sugar. Then piperidine excises the sugar from its 5′- and 3′-links in a *β*-elimination reaction. The conditions of chemical cleavage described in reactions 1 through 4 are generally adjusted so that, on average, only a single scission occurs per DNA molecule. However, since a very large number of DNA molecules exist in each reaction mixture, the products are a random collection of different-sized fragments wherein the occurrence of any base is represented by its unique pair of 5′- and 3′-cleavage products. These products form a complete set, the members of which differ in

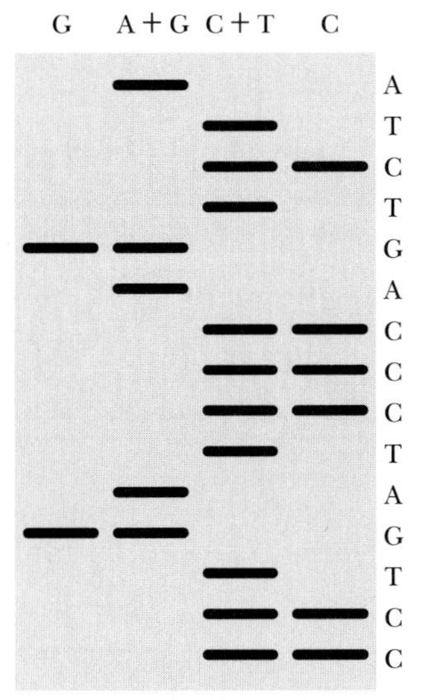

Figure **7.8** Autoradiogram of a hypothetical electrophoretic pattern obtained for four reaction mixtures performed as described in Figures 7.5 through 7.7 and run in the four lanes G, A+G, C+T, and C, respectively. Reading this pattern from the bottom up yields the sequence CCTGATCCCAGTCTA. The correct 5′ → 3′ order is determined by knowing which end of the ssDNA was ^{32}P-labeled. If the 5′-end was ^{32}P-labeled, only the 5′-fragments will be evident on the autoradiogram; the 3′-ends will be invisible. Similarly, if the 3′-end was originally labeled, only the 3′-fragments light up the autoradiogram. Assuming that the 5′-end was labeled, the sequence would be CCTGATCCCAGTCTA. If it were the 3′-end, the sequence read in the 5′ → 3′ convention would be ATCTGACCCTAGTCC. (Indication of T as the end-labeled nucleotide is arbitrary.)

length by only one nucleotide, and they can be resolved by gel electrophoresis into a "ladder," which can be visualized by autoradiography of the gel if the DNA fragments are radioactively labeled (Figure 7.8).

An interesting feature of the Maxam–Gilbert sequencing procedure is that the base that is "read" in the ladder is actually not present in the oligonucleotide that identifies it. Thus, an unidentified base bears the label at the end of the smallest fragment; this unidentified base is the one that preceded the first identified base. For example, an oligonucleotide of either

$$^{32}\text{P-5}'\text{-}\mathbf{(A,C,G,T)CCTGATCCCAGTCTA}\text{-3}'$$

or

$$\text{5}'\text{-}\mathbf{ATCTGACCCTAGTCC(A,C,G,T)}\text{-3}'\text{-}^{32}\text{P}$$

would yield the same pattern in the autoradiogram.

In principle, the Maxam–Gilbert method can provide the total sequence of a dsDNA molecule just by determining the purine positions on one strand (using reactions 1 and 2) and then the purines on the complementary strand. Complementary base-pairing rules then reveal the pyrimidines along each strand, T complementary to where A is, C complementary to where G occurs. (The analogous approach of using reactions 3 and 4 to find the pyrimidine locations on each strand would also provide sufficient information to write the total sequence.)

With current technology, it is possible to read the order of as many as 400 bases from the autoradiogram of a sequencing gel (Figure 7.9). The actual chemical or enzymatic reactions, electrophoresis, and autoradiography are now routine, and a skilled technician can sequence about 1 kbp per week using these manual techniques. The major effort in DNA sequencing is in the isolation and preparation of fragments of interest, such as cloned genes.

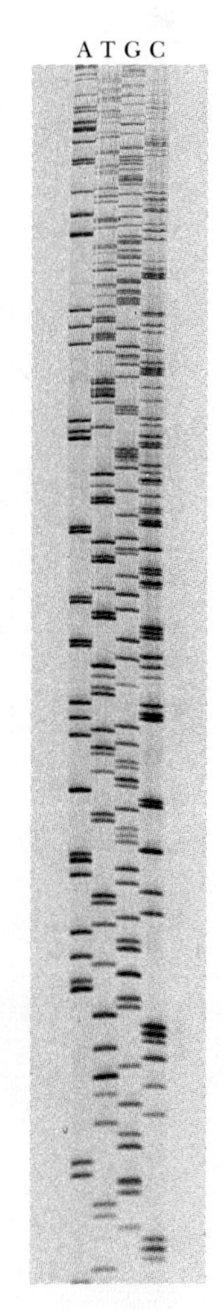

Figure **7.9** A photograph of the autoradiogram from an actual sequencing gel. A portion of the DNA sequence of *nit-6*, the *Neurospora* gene encoding the enzyme nitrite reductase.

Automated DNA Sequencing

In recent years, automated DNA sequencing machines capable of identifying about 10^4 bases per day have become commercially available. One clever innovation has been the use of fluorescent dyes of different colors to uniquely

Figure **7.10** Schematic diagram of the methodology used in fluorescent labeling and automated sequencing of DNA. Four reactions are set up, one for each base, and the primer in each is end-labeled with one of four different fluorescent dyes; the dyes serve to color-code the base-specific sequencing protocol (a unique dye is used in each dideoxynucleotide reaction). The four reaction mixtures are then combined and run in one lane. Thus, each lane in the gel represents a different sequencing experiment. As the differently sized fragments pass down the gel, a laser beam excites the dye in the scan area. The emitted energy passes through a rotating color filter and is detected by a fluorometer. The color of the emitted light identifies the final base in the fragment.

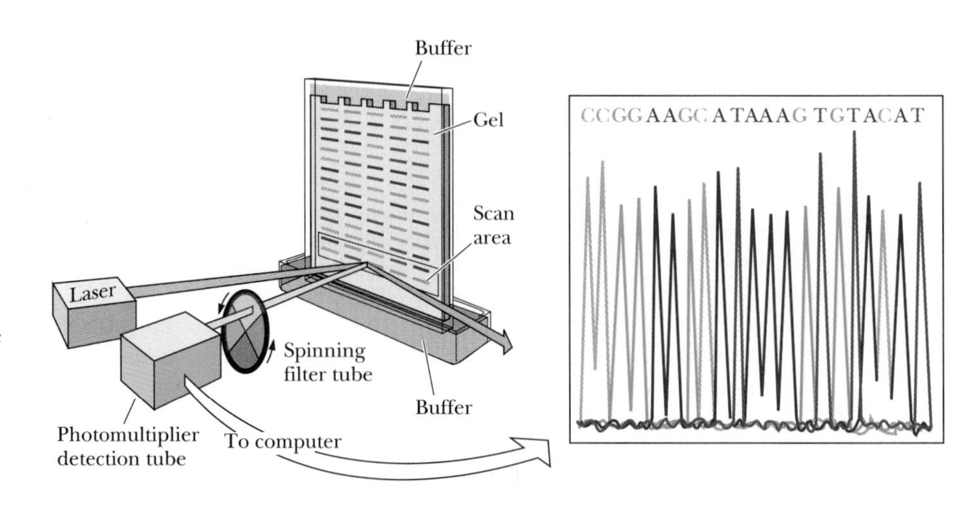

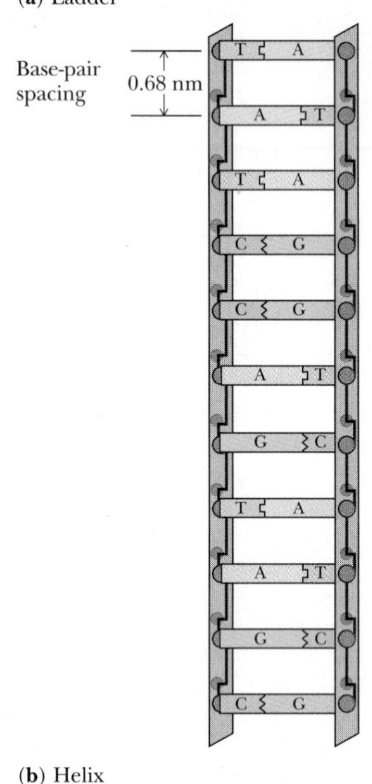

label the primer DNA introduced into the four sequencing reactions; for example, red for the A reaction, blue for T, green for G, and yellow for C. Then, all four reaction mixtures can be combined and run together on one electrophoretic gel slab. As the oligonucleotides are separated and pass to the bottom of the gel, each is illuminated by a low power argon laser beam that causes the dye attached to the primer to fluoresce. The color of the fluorescence is detected automatically, revealing the identity of the primer, and hence the base, immediately (Figure 7.10). The development of such automation has opened the possibility for sequencing the entire human genome, some 2.9 billion bp. Even so, if 100 automated machines operating at peak efficiency were dedicated to the task, it would still take at least 8 years to complete!

7.2 The ABZ's of DNA Secondary Structure

Double-stranded DNA molecules assume one of three secondary structures, termed A, B, and Z. Fundamentally, double-stranded DNA is a regular two-chain structure with hydrogen bonds formed between opposing bases on the two chains (see Chapter 6). Such H bonding is possible only when the two chains are antiparallel. The polar sugar–phosphate backbones of the two chains are on the outside. The bases are stacked on the inside of the structure; these heterocyclic bases, as a consequence of their π-electron clouds, are hydrophobic on their flat sides. One purely hypothetical conformational possibility for a two-stranded arrangement would be a ladderlike structure (Figure 7.11) in which the base pairs are fixed at 0.68 nm apart by the nature of the ladder. Since H_2O molecules would be accessible to the spaces between the hydrophobic surfaces of the bases, this conformation is energetically unfavorable. This ladderlike structure converts to a helix when given a simple right-handed twist. This brings the base pair rungs of the ladder closer together, stacking them 0.34 nm apart. Since this helix repeats itself approximately every 10 bp, its **pitch** is 3.4 nm. This is the major conformation of DNA in solution and is called **B-DNA.**

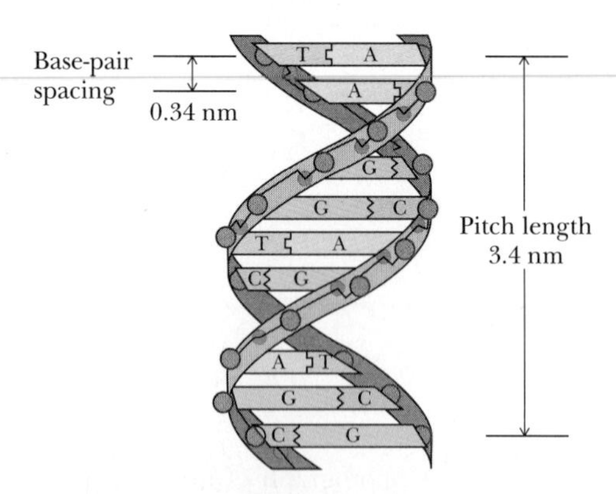

Figure **7.11** (a) Double-stranded DNA as an imaginary ladderlike structure. (b) A simple right-handed twist converts the ladder to a helix.

Structural Equivalence of Watson–Crick Base Pairs

The base pairing in DNA is very specific: the purine adenine pairs with the pyrimidine thymine; the purine guanine pairs with the pyrimidine cytosine. Because exceptions to this exclusive pairing of A only with T and G only with C are rare, these pairs are taken as the standard or accepted law, and the A:T and G:C base pairs are often referred to as **canonical.** As Watson recognized from testing various combinations of bases using structurally accurate models, the A:T pair and the G:C pair form spatially equivalent units (Figure 7.12). The backbone-to-backbone distance of an A:T pair is 1.11 nm, virtually identical to the 1.08 nm chain separation in G:C base pairs. Sharing this insight with

canon: rule, standard, or law

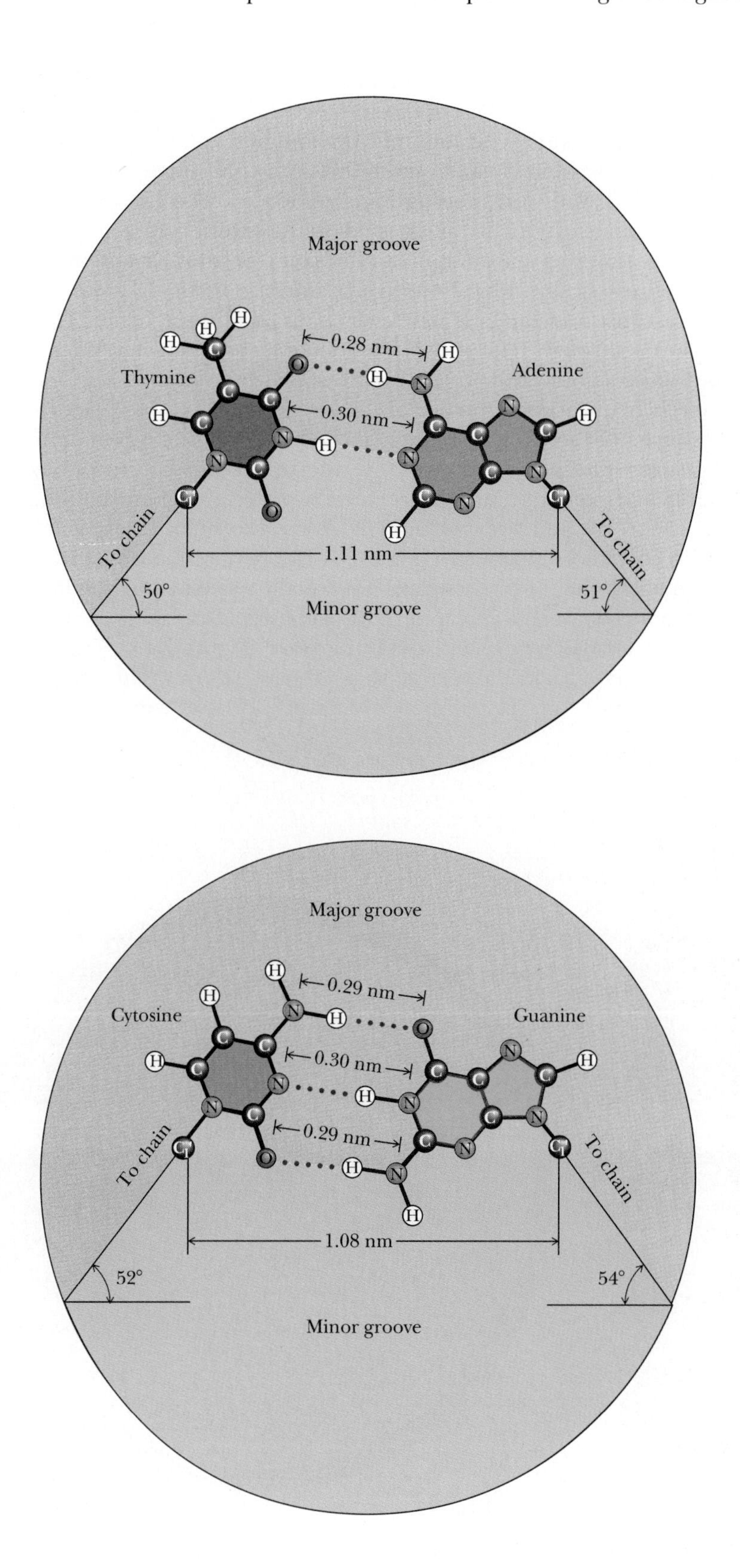

***Figure* 7.12** Watson–Crick A:T and G:C base pairs. All H bonds in both base pairs are straight, with each H atom pointing directly at its acceptor N or O atom. Straight H bonds are the strongest. The mandatory binding of larger purines with smaller pyrimidines leads to base pairs that have virtually identical dimensions, allowing the two sugar–phosphate backbones to adopt identical helical conformations.

Crick led them to a crucial realization: units of such similarity could serve as spatially invariant substructures to build a polymer whose exterior dimensions would be uniform along its length, regardless of the sequence of bases.

The DNA Double Helix Is a Stable Structure

A number of factors account for the stability of the double-helical structure of DNA. First, both internal and external hydrogen bonds stabilize the double helix. The two strands of DNA are held together by H bonds that form between the complementary purines and pyrimidines, two in an A:T pair and three in a G:C pair, while polar atoms in the sugar–phosphate backbone form external H bonds with surrounding water molecules. Second, the negatively charged phosphate groups are all situated on the exterior surface of the helix in such a way that they have minimal effect on one another and are free to interact electrostatically with cations in solution such as Mg^{2+}. Third, the core of the helix consists of the base pairs, which, in addition to being H-bonded, stack together through hydrophobic interactions and van der Waals forces that contribute significantly to the overall stabilizing energy.

A stereochemical consequence of the way A:T and G:C base pairs form is that the sugars of the respective nucleotides have opposite orientations, and thus the sugar–phosphate backbones of the two chains run in opposite or "antiparallel" directions. Furthermore, the two glycosidic bonds holding the bases in each base pair are not directly across the helix from each other, defining a common diameter (Figure 7.13). Consequently, the sugar–phosphate backbones of the helix are not equally spaced along the helix axis and the grooves between them are not the same size. Instead, the intertwined chains create a **major groove** and a **minor groove** (Figure 7.13). The edges of the base pairs have a specific relationship to these grooves. The "top" edges of the base pairs ("top" as defined in Figure 7.12) are exposed along the

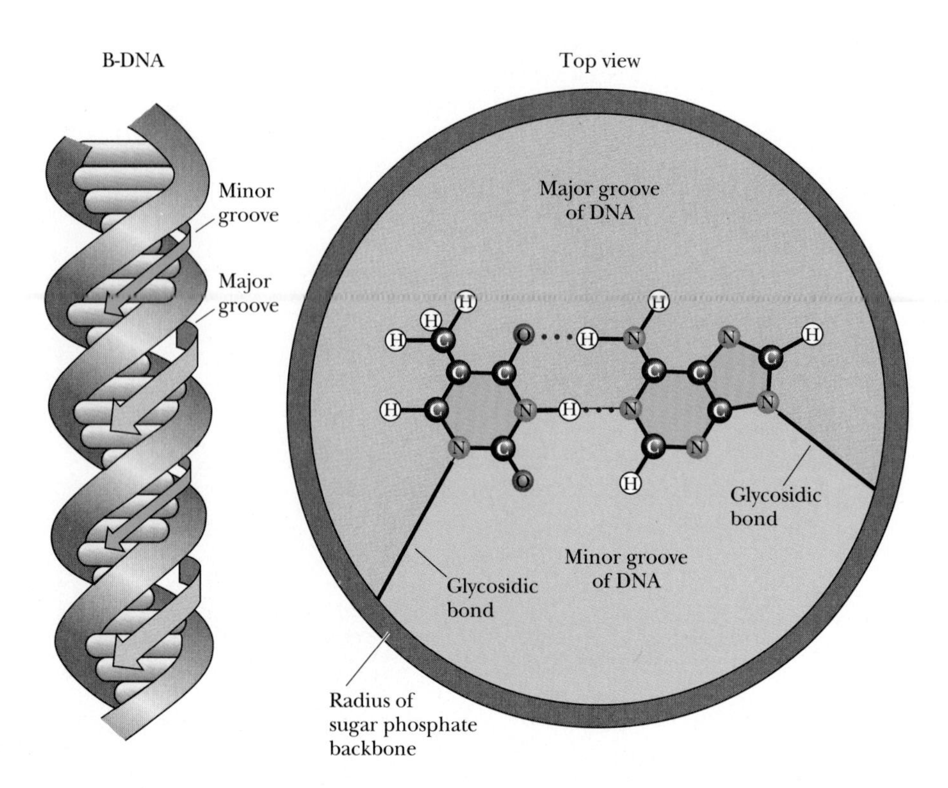

***Figure* 7.13** The bases in a base pair are not directly across the helix axis from one another along some diameter but rather slightly displaced. This displacement, and the relative orientation of the glycosidic bonds linking the bases to the sugar–phosphate backbone, leads to differently sized grooves in the cylindrical column created by the double helix, the major groove and the minor groove, each coursing along its length.

interior surface or "floor" of the major groove; the base pair edges nearest to the glycosidic bond form the interior surface of the minor groove. Some proteins that bind to DNA can actually recognize specific nucleotide sequences by "reading" the pattern of H-bonding possibilities presented by the edges of the bases in these grooves. Such DNA–protein interactions provide one step toward understanding how cells regulate the expression of genetic information encoded in DNA (see Chapter 30).

Conformational Variation in Double-Helical Structures

In solution, DNA ordinarily assumes the structure we have been discussing: B-DNA. However, nucleic acids also occur naturally in other double-helical forms. The base-pairing arrangement remains the same, but the sugar–phosphate groupings that constitute the backbone are inherently flexible and can adopt different conformations. Each deoxyribose–PO_4 segment of the backbone has six degrees of freedom (Figure 7.14a) as a consequence of the six successive single bonds that compose the covalent links of the chain. The furanose rings of the pentoses are not planar but instead are puckered in variant conformations, four of which are depicted in Figure 7.14b. In B-DNA, the 2′-endo conformation is the favored form. A seventh degree of freedom per nucleotide unit arises owing to free rotation about the C1′–N glycosidic

(a) The six degrees of freedom in the sugar–PO_4 backbone:

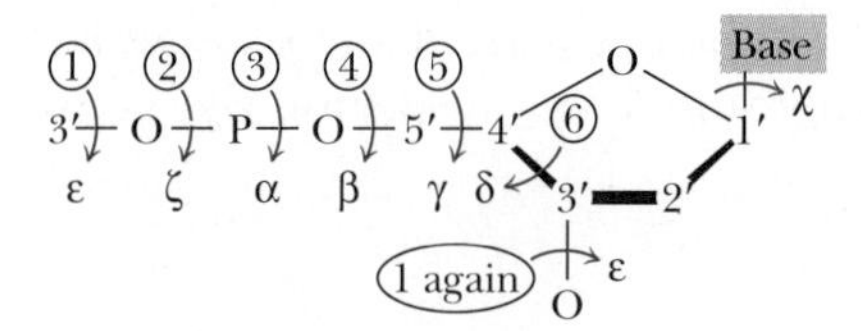

Rotation about bonds 1, 2, 3, 4, 5, and 6 correspond to 6 degrees of freedom designated α, β, γ, δ, ε, and ζ as indicated.

(b) Four puckered conformations of furanose rings:

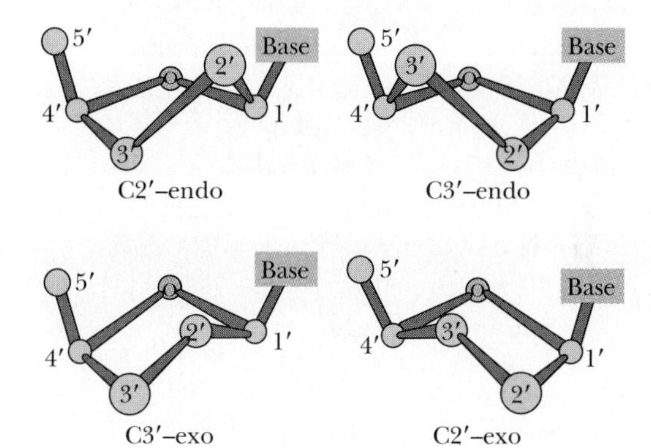

(c) Free rotation about C1′–N glycosidic bond (7th degree of freedom):

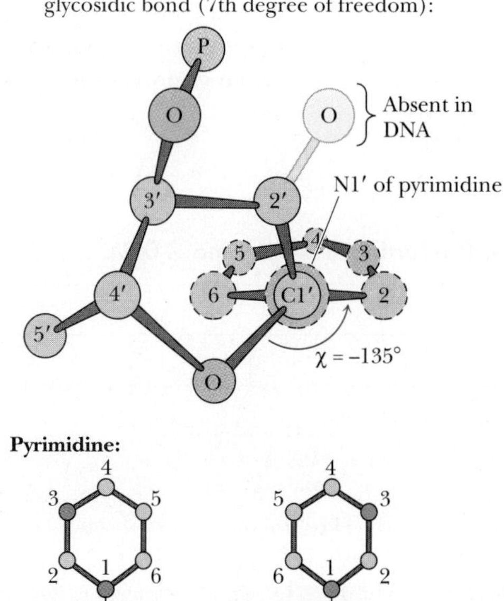

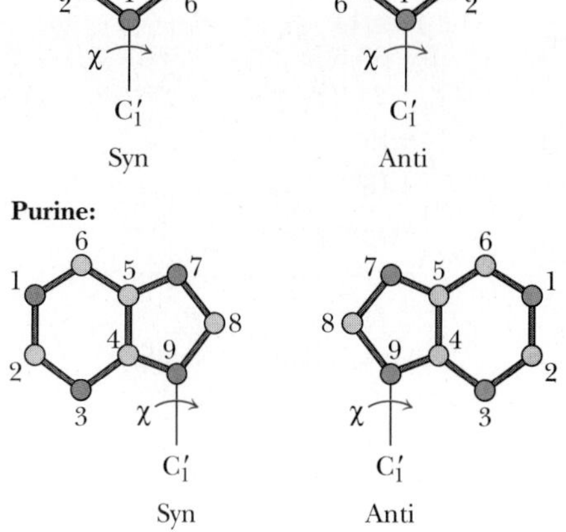

Figure **7.14** (a) The six degrees of freedom in the deoxyribose–PO_4 units of the polynucleotide chain. (b) Four puckered conformations of the furanose rings. (c) Free rotation about the C1′–N glycosidic bond.

bond. This freedom allows the plane of the base to rotate relative to the course of the backbone chain (Figure 7.14c). Alternative double-helical structures are the end result of these conformational variations.

Alternative Form of Right-Handed DNA

An alternative form of the right-handed double helix is **A-DNA.** The A-form usually occurs only when relatively little water is available to hydrate the double helix. A-DNA molecules differ in a number of ways from B-DNA. The pitch, or distance required to complete one helical turn, is different. In B-DNA, it is 3.4 nm, whereas in A-DNA, it is 2.46 nm. One turn in A-DNA requires 11 bp to complete. Depending on local sequence, 10 to 10.6 bp define one helical turn in B-form DNA. The deoxyribose conformation has changed from 2′-endo to 3′-endo, and in A-DNA, the base pairs are no longer nearly perpendicular to the helix axis but instead are tilted 19° with respect to this axis. Successive base pairs occur every 0.23 nm along the axis, as opposed to 0.332 nm in B-DNA. The B-form of DNA is thus longer and thinner than the short, squat A-form, which has its base pairs displaced around, rather than centered on, the helix axis. Figure 7.15 shows the relevant structural characteristics of the A- and B-forms of DNA. (Z-DNA, another form of DNA to be discussed shortly, is also depicted in Figure 7.15.) A comparison of the structural properties of A-, B-, and Z-DNA is summarized in Table 7.1.

Although relatively dehydrated DNA fibers can be shown to adopt the A-conformation under physiological conditions, it is unclear whether DNA ever assumes this form *in vivo.* However, double-helical DNA:RNA hybrids probably have an A-like configuration. The 2′-OH in RNA sterically prevents double-helical regions of RNA chains from adopting the B-form helical arrangement. Importantly, double-stranded regions in RNA chains assume an A-like conformation, with their bases strongly tilted with respect to the helix axis.

***Table* 7.1**
Comparison of the Structural Properties of A-, B-, and Z-DNA

	Double Helix Type		
	A	B	Z
Overall proportions	Short and broad	Longer and thinner	Elongated and slim
Rise per base pair	2.3 Å	3.32 Å ± 0.19 Å	3.8 Å
Helix packing diameter	25.5 Å	23.7 Å	18.4 Å
Helix rotation sense	Right-handed	Right-handed	Left-handed
Base pairs per helix repeat	1	1	2
Base pairs per turn of helix	~11	~10	12
Mean rotation per base pair	33.6°	35.9° ± 4.2°	−60°/2
Pitch per turn of helix	24.6 Å	33.2 Å	45.6 Å
Base-pair tilt from the perpendicular	+19°	−1.2° ± 4.1°	−9°
Base-pair mean propeller twist	+18°	+16° ± 7°	~0°
Helix axis location	Major groove	Through base pairs	Minor groove
Major groove proportions	Extremely narrow but very deep	Wide and with intermediate depth	Flattened out on helix surface
Minor groove proportions	Very broad but shallow	Narrow and with intermediate depth	Extremely narrow but very deep
Glycosyl bond conformation	anti	anti	anti at C, syn at G

Adapted from Dickerson, R. L., et al., 1982. *Cold Spring Harbor Symposium on Quantitative Biology* **47**:14.

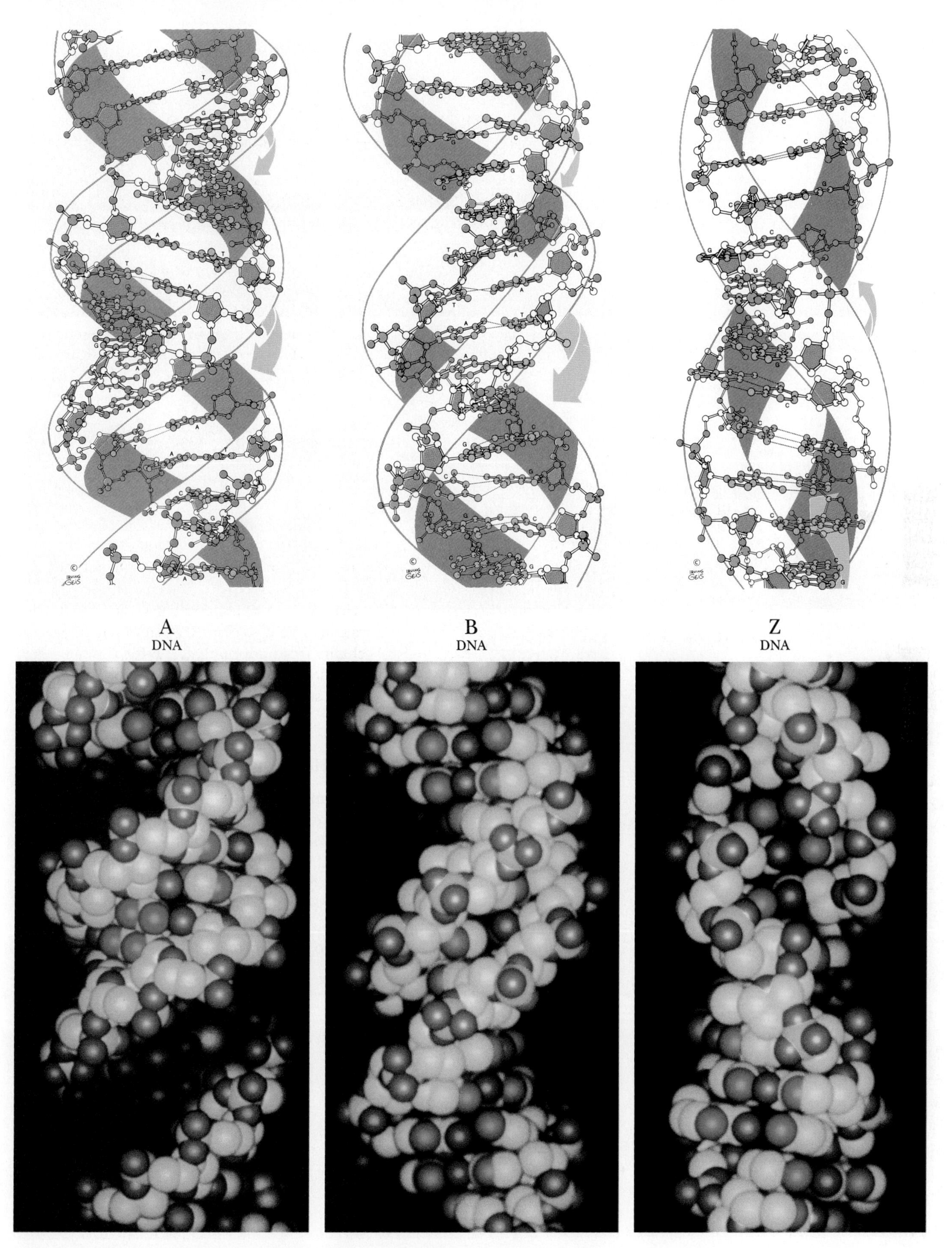

***Figure* 7.15** Comparison of the A-, B-, and Z-forms of the DNA double helix. The distance required to complete one helical twin is shorter in A-DNA than it is in B-DNA. The alternating pyrimidine–purine sequence of Z-DNA is the key to the ''left-handedness'' of this helix.

(continued on next page)

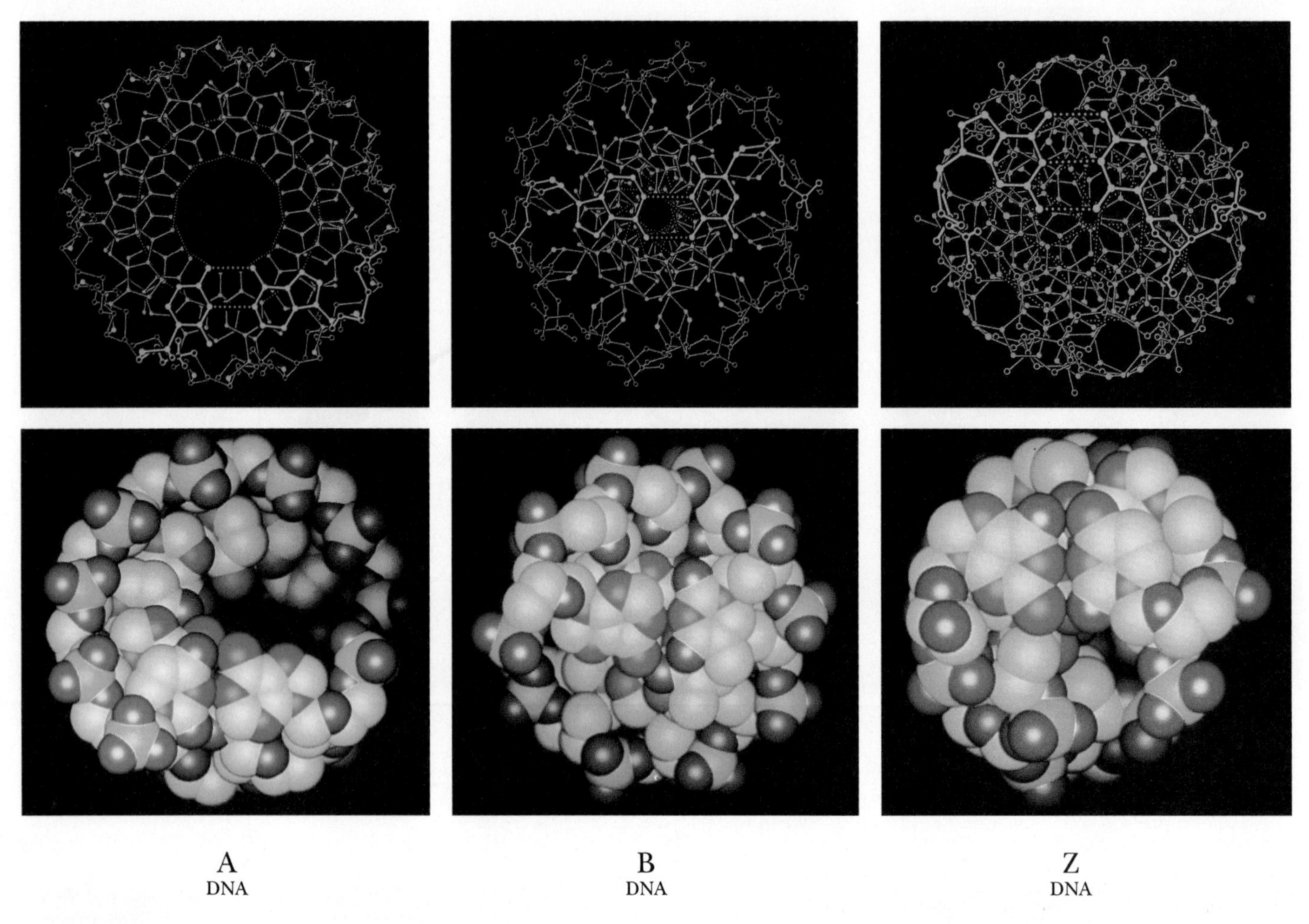

***Figure* 7.15** (*continued from previous page*)

Z-DNA: A Left-Handed Double Helix

Z-DNA was first recognized by Alexander Rich and his colleagues at MIT in X-ray analysis of the synthetic deoxynucleotide dCpGpCpGpCpG, which crystallized into an antiparallel double helix of unexpected conformation. The alternating pyrimidine–purine (Py–Pu) sequence of this oligonucleotide is the key to its unusual properties. The *N*-glycosyl bonds of G residues in this alternating copolymer are rotated 180° with respect to their conformation in B-DNA, so that now the purine ring is in the syn rather than the anti conformation (Figure 7.16). The deoxyribose conformation adjusts to this change

***Figure* 7.16** Comparison of the deoxyguanosine conformation in B- and Z-DNA. In B-DNA, the C1′–N-9 glycosyl bond is always in the anti position (left). In contrast, in the left-handed Z-DNA structure, this bond rotates (as shown) to adopt the syn conformation. The deoxyribose ring also changes conformation from C2′ endo to C3′ endo.

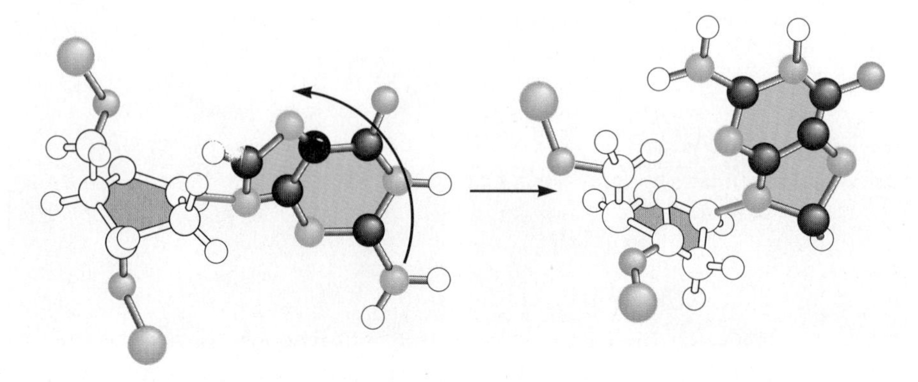

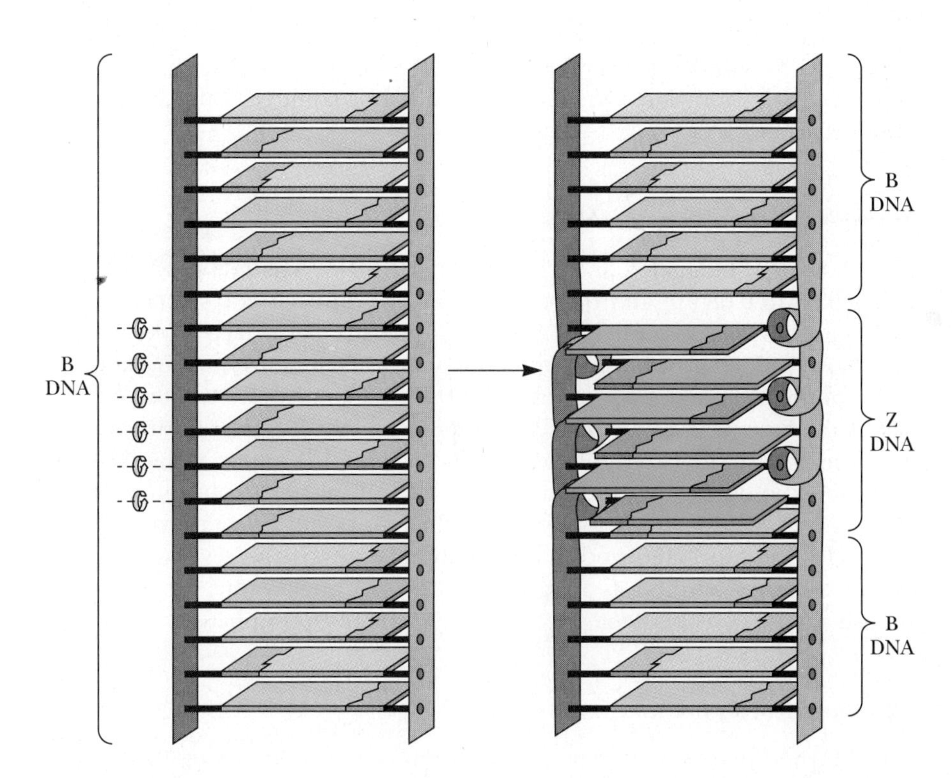

***Figure* 7.17** The change in topological relationships of base pairs from B- to Z-DNA. A six-base-pair segment of B-DNA is converted to Z-DNA through rotation of the base pairs, as indicated by the curved arrows. The purine rings (green) of the deoxyguanosine nucleosides rotate via an anti to syn change in the conformation of the guanine–deoxyribose glycosidic bond; the pyrimidine rings (blue) are rotated by flipping the entire deoxycytidine nucleoside (base *and* deoxyribose). As a consequence of these conformational changes, the base pairs in the Z-DNA region no longer share π,π stacking interactions with adjacent B-DNA regions.

by becoming C3′-endo. The C residues remain in the anti form. The alternating *anti*-Py/*syn*-Pu arrangement favors transition into a left-handed helix. (In right-handed DNA, the bases are all in the anti conformation with respect to the deoxyribose ring.) Because the G ring is "flipped," the C ring must also flip to maintain normal Watson–Crick base pairing. Since this is not achieved by rotation of the base relative to the furanose, the whole C nucleoside (base and sugar) must flip 180° (Figure 7.17). The deoxycytidine furanose remains C2′-endo.

Structural Characteristics of Z-DNA

It is topologically possible for the G to go syn and the C nucleoside to undergo rotation by 180° without breaking and re-forming the G:C hydrogen bonds. In other words, the B to Z structural transition can take place without disruption of the bonding relationships among the atoms involved. Because alternate nucleotides assume different conformations, the repeating unit on a given strand in the Z-helix is the dinucleotide. That is, for any number of bases, n, along one strand, $n - 1$ dinucleotides must be considered. For example, a GpCpGpC subset of sequence along one strand is comprised of *three* successive dinucleotide units: GpC, CpG, and GpC. (In B-DNA, the nucleotide conformations are essentially uniform and the repeating unit is the mononucleotide.) It follows that the CpG sequence is distinct conformationally from the GpC sequence along the alternating copolymer chains in the Z-double helix. The conformational alterations going from B to Z realign the sugar–phosphate backbone along a zigzag course that has a left-handed orientation (Figure 7.15), thus the designation *Z-DNA*. Keep in mind that the sugar–phosphate grouping is the important structural unit in the backbone of nucleic acids. Note that in any GpCpGp subset, the sugar–phosphates of GpC form the horizontal "zig" while the CpG backbone segment forms the vertical "zag." The mean rotation angle circumscribed around the helix axis is −15°

for a CpG step and $-45°$ for a GpC step (giving $-60°$ for the dinucleotide repeat). The minus sign denotes a left-handed or counterclockwise rotation about the helix axis.

Overall Appearance of Z-DNA

Z-DNA is more elongated and slimmer than B-DNA. The rise per base pair, ΔZ, is 0.36 to 0.38 nm and there are 12 bp (six dinucleotide segments) per turn, giving this helix a pitch of 4.56 nm. The external surface of the Z-helix has a single, deep groove running between the sugar–phosphate backbones, corresponding to the minor groove of B-DNA. The narrowness of this groove places the base pairs to the outside of the helix. This feature, and the anti to syn transition in the G residues, causes the purine ring atoms N-7 and C-8 to become exposed to the exterior (Figure 7.15). The sugar–phosphate backbone is not as extended in Z-DNA as in B-DNA, as can be seen from Figure 7.16. The Z-conformation occurs in solution in alternating Py–Pu DNA copolymers, provided there is a high concentration of Na^+ ions present to diminish interactions between the negatively charged phosphates.

Cytosine Methylation and Z-DNA

The Z-form can arise in sequences that are not strictly alternating Py–Pu. For example, the hexanucleotide ^{m5}CGAT^{m5}CG, a Py-Pu-Pu-Py-Py-Pu sequence containing two 5-methylcytosines (^{m5}C), crystallizes as Z-DNA. Indeed, the *in vivo* methylation of C at the 5-position is believed to favor a B to Z switch since, in B-DNA, these hydrophobic methyl groups would protrude into the aqueous environment of the major groove and destabilize its structure. In Z-DNA, the same methyl groups can form a stabilizing hydrophobic patch. It is likely that the Z-conformation naturally occurs in specific regions of cellular DNA, which otherwise is predominantly in the B-form. Furthermore, since methylation is implicated in gene regulation, the occurrence of Z-DNA may affect the expression of genetic information (see Part IV, Genetic Information).

The Double Helix in Solution

B-DNA in solution is not a rigid, linear rod but instead behaves as a dynamic, flexible molecule responding to localized thermal fluctuations that temporarily distort and deform its structure over short regions. Base and backbone ensembles of atoms undergo elastic motions on a time scale of nanoseconds. To some extent, these effects represent changes in rotational angles of the bonds comprising the polynucleotide backbone. These changes are also influenced by sequence-dependent variations in base-pair stacking. The consequence is that the helix bends gently. When these variations are summed over the great length of a DNA molecule, the net result of these bending motions is that, at any given time, the double helix assumes a roughly spherical shape, as might be expected for a long, semi-rigid rod undergoing apparently random coiling. It is also worth noting that, on close scrutiny, the surface of the double helix is *not* that of a totally featureless, smooth, regular "barber pole" structure. Different base sequences impart their own special signatures to the molecule by subtle influences on such factors as the groove width, the angle between the helix axis and base planes, and the mechanical rigidity. Certain regulatory proteins bind to specific DNA sequences and participate in activating or suppressing expression of the information encoded therein. These proteins bind at unique sites by virtue of their ability to recognize novel structural characteristics imposed on the DNA by the local nucleotide sequence.

Intercalating Agents Distort the Double Helix

intercalate: to insert between others

Aromatic macrocycles, flat hydrophobic molecules composed of fused, heterocyclic rings, such as **ethidium bromide, acridine orange,** and **actinomycin D** (Figure 7.18), can insert between the stacked base pairs of DNA. The bases are forced apart to accommodate these so-called **intercalating agents,** causing an unwinding of the helix to a more ladderlike structure. The deoxyribose–phosphate backbone is almost fully extended as successive base pairs are displaced 0.7 nm from one another, and the rotational angle about the helix axis between adjacent base pairs is reduced from 36° to 10°.

Dynamic Nature of the DNA Double Helix in Solution

Intercalating substances insert with ease into the double helix, indicating that the van der Waals bonds they form with the bases sandwiching them are more favorable than similar bonds between the bases themselves. Furthermore, the fact that these agents slip in suggests that the double helix must temporarily unwind and present gaps for these agents to occupy. That is, the DNA double helix in solution must be represented by a set of metastable alternatives to the standard B-conformation. These alternatives constitute a flickering repertoire of dynamic structures.

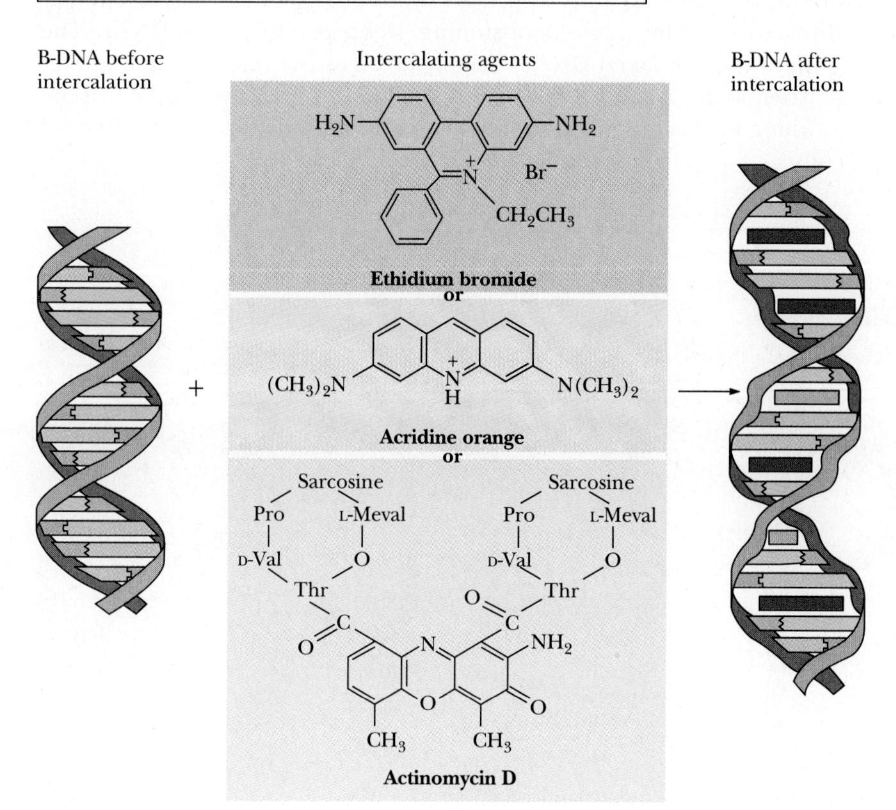

Figure **7.18** The structures of ethidium bromide, acridine orange, and actinomycin D, three intercalating agents, and their effects on DNA structure.

7.3 Supercoils and Cruciforms: Tertiary Structure in DNA

The conformations of DNA discussed thus far are variations sharing a common secondary structural theme, the double helix, in which the DNA is assumed to be in a regular, linear form. DNA can also adopt regular structures of higher complexity in several ways. For example, many DNA molecules are circular. Most, if not all, bacterial chromosomes are covalently closed, circular DNA duplexes, as are almost all plasmid DNAs. **Plasmids** are naturally occurring, self-replicating, circular, extrachromosomal DNA molecules found in bacteria; plasmids carry genes specifying novel metabolic capacities advantageous to the host bacterium. Various animal virus DNAs are circular as well.

Supercoils

In duplex DNA, the two strands are wound about each other once every 10 bp, that is, once every turn of the helix. Double-stranded circular DNA (or linear DNA duplexes whose ends are not free to rotate), can form **supercoils** if the strands are underwound (*negatively supercoiled*) or overwound (*positively supercoiled*) (Figure 7.19). DNA supercoiling is analogous to twisting or untwisting a multistranded rope so that it is torsionally stressed. Negative supercoiling introduces a torsional stress that favors unwinding of the right-handed B-DNA double helix, while positive supercoiling overwinds such a helix. Both forms of supercoiling compact the DNA so that it sediments faster upon ultracentrifugation or migrates more rapidly in an electrophoretic gel in comparison to **relaxed DNA** (DNA that is not supercoiled).

Linking Number

The basic parameter characterizing supercoiled DNA is the **linking number** (L). This is the number of times the two strands are intertwined, and provided both strands remain covalently intact, L cannot change. In a relaxed circular DNA duplex of 400 bp, L is 40 (assuming 10 bp per turn in B-DNA). The linking number for relaxed DNA is usually taken as the reference parameter and is written as L_0. L can be equated to the **twist** (T) and **writhe** (W) of the duplex, where twist is the number of helical turns and writhe is the number of supercoils:

$$L = T + W$$

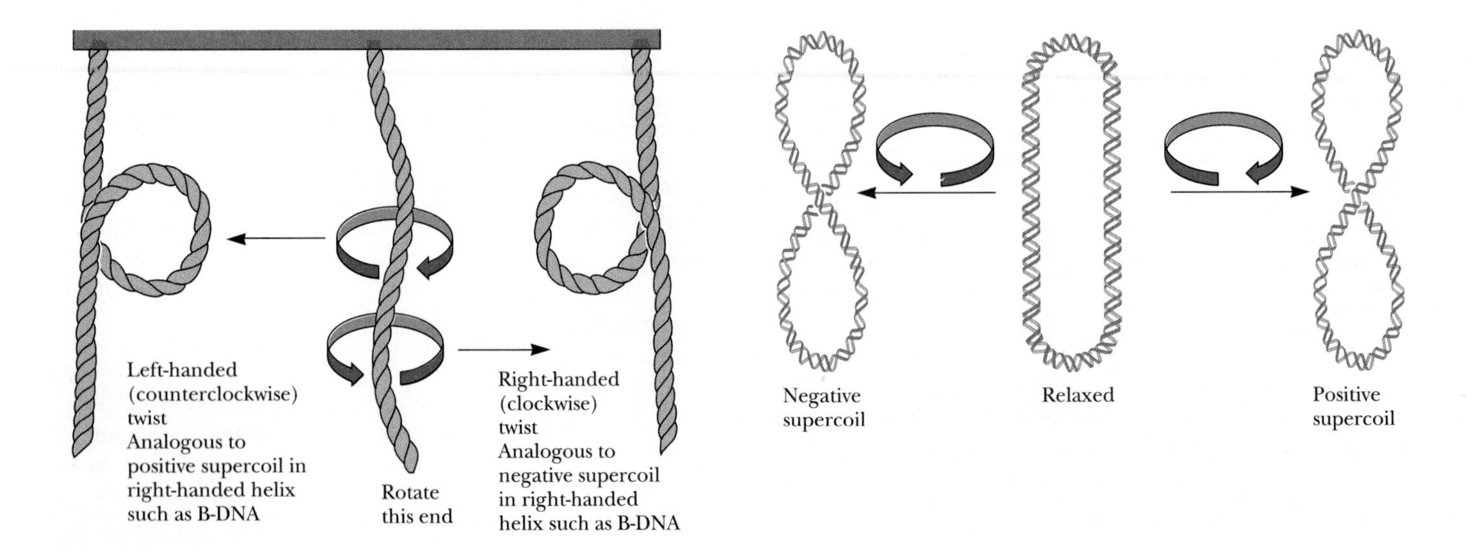

***Figure* 7.19** Supercoiled DNA topology. The DNA double helix can be approximated as a two-stranded, right-handed coiled rope. If one end of the rope is rotated counterclockwise, the strands begin to separate (negative supercoiling). If the rope is twisted clockwise (in a right-handed fashion), the rope becomes overwound (positive supercoiling). Get a piece of right-handed multistrand rope, and carry out these operations to convince yourself.

Thus, in a 400-bp closed, circular DNA duplex that is relaxed, $W = 0$. Since there are 40 helical turns, $T = L = 40$. This linking number can only be changed by breaking one or both strands of the DNA, winding them tighter or looser, and rejoining the ends. Enzymes capable of carrying out such reactions are called **topoisomerases** because they change the topological state of DNA. Topoisomerases are important players in DNA replication (see Chapter 29).

DNA Gyrase

The bacterial enzyme **DNA gyrase** is a topoisomerase that introduces negative supercoils into DNA in the manner shown in Figure 7.20. Suppose DNA gyrase puts four negative supercoils into the 400-bp circular duplex, then $W = -4$, T remains the same, and $L = 36$ (Figure 7.21). In actuality, the negative supercoils cause a torsional stress on the molecule so that T tends to decrease, that is, the helix becomes a bit unwound so that base pairs are separated. The extreme would be that T would decrease by 4 and the supercoiling would be removed ($T = 36$, $L = 36$, and $W = 0$). Usually the real situation is a compromise in which the negative value of W is reduced, T decreases slightly, and these changes are distributed over the length of the circular duplex so that no localized unwinding of the helix ensues. While the parameters T and W are conceptually useful, neither can be measured experimentally at the present time.

Superhelix Density

The difference between the linking number of a DNA and the linking number of its relaxed form is ΔL: $\Delta L = (L - L_0)$. In our example with four negative supercoils, $\Delta L = -4$. The **superhelix density** or **specific linking difference** is defined as $\Delta L/L_0$ and is sometimes termed *sigma*, σ. For our example, $\sigma = -4/40$, or -0.1. As a ratio, σ is a measure of supercoiling that is independent of length. Its sign reflects whether the supercoiling tends to unwind (*negative* σ) or overwind (*positive* σ) the helix. In other words, the superhelix density states the number of supercoils per 10 bp, which also is the same as the number of supercoils per B-DNA repeat. Circular DNA isolated from natural sources is always found in the underwound, negatively supercoiled state. Negative supercoiling is also an important factor in the stabilization of Z-DNA.

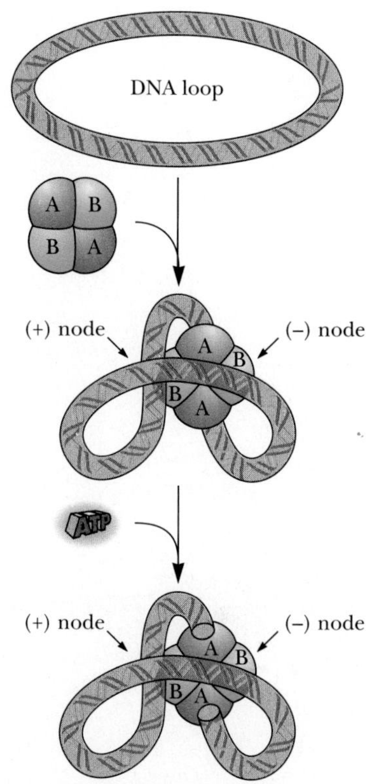

DNA is cut and a conformational change allows the DNA to pass through. Gyrase religates the DNA and then releases it.

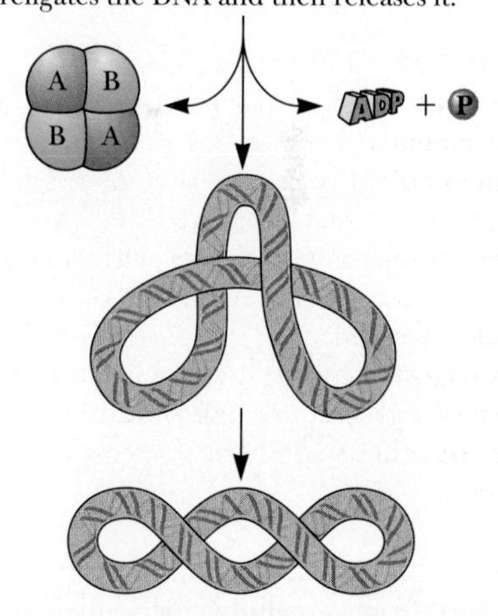

***Figure* 7.20** A simple model for the action of bacterial DNA gyrase (topoisomerase II). The *A*-subunits cut the DNA duplex and then hold onto the cut ends. Conformational changes occur in the enzyme that allow a continuous region of the DNA duplex to pass between the cut ends and into an internal cavity of the protein. The cut ends are then re-ligated, and the intact DNA duplex is released from the enzyme. The released intact circular DNA now contains two negative supercoils as a consequence of DNA gyrase action.

Cruciforms

Palindromes are words, phrases, or sentences that are the same when read backward or forward, such as "radar," "sex at noon taxes," "Madam, I'm Adam," and "a man, a plan, a canal, Panama." DNA sequences that are **inverted repeats,** or palindromes, have the potential to form a tertiary structure known as a **cruciform** (literally meaning "cross-shaped") if the normal interstrand base pairing is replaced by intrastrand pairing (Figure 7.22). In effect, each DNA strand folds back on itself in a hairpin structure to align the palindrome in base-pairing register. Such cruciforms are never as stable as normal DNA duplexes because an unpaired segment must exist in the loop region. However, negative supercoiling causes a localized disruption of hydrogen bonding between base pairs in DNA and may promote formation of cruciform loops. Cruciform structures have a two-fold rotational symmetry about their centers and potentially create distinctive recognition sites for specific DNA-binding proteins.

Figure **7.21** A 400-bp circular DNA molecule in different topological states: (a) relaxed, (b) negative supercoils distributed over the entire length, and (c) negative supercoils creating a localized single-stranded region. Negative supercoiling has the potential to cause localized unwinding of the DNA double helix so that single-stranded regions (or bubbles) are created.

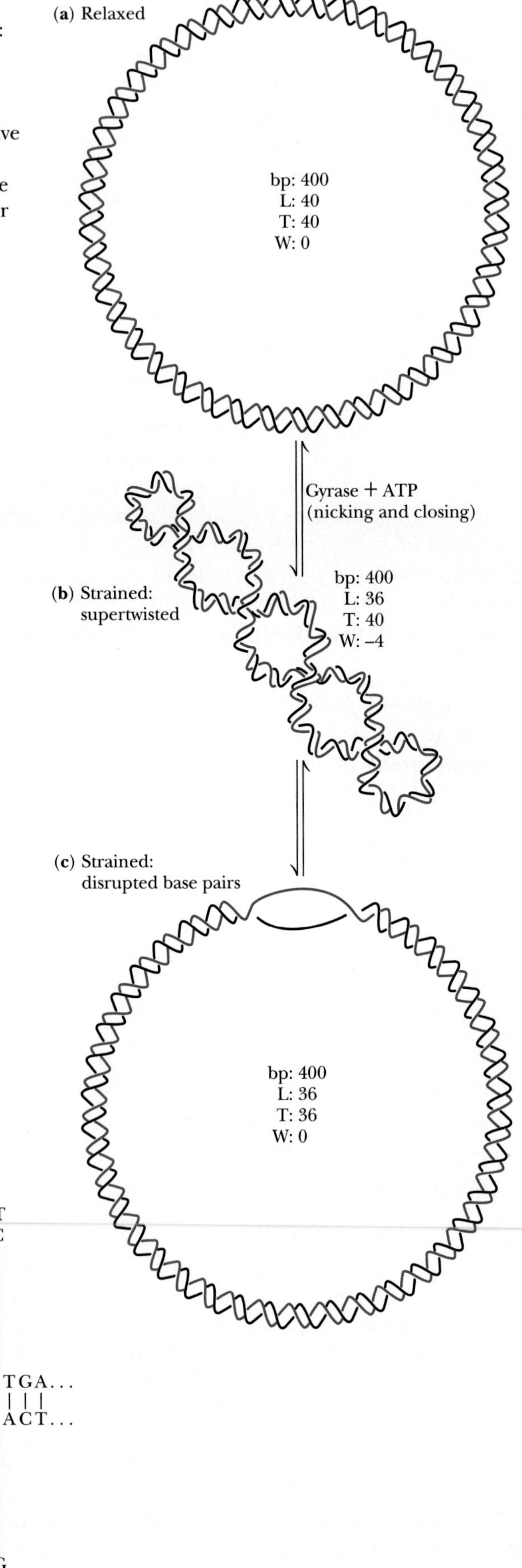

Figure **7.22** The formation of a cruciform structure from a palindromic sequence within DNA. The self-complementary inverted repeats can rearrange to form hydrogen-bonded cruciform loops.

```
...CATGAACGTCCTATTGTCGGACGTTCTGA...
   ||||||||||||||||||||||||||||||
...GTACTTGCAGGATAACAGCCTGCAAGACT...
```

```
        T T G
      A       T
      T       C
        C—G
        C—G
        T—A
        G—C
        C—G
        A—T
        A—T
        G—C
...CAT         TGA...
   |||         |||
...GTA         ACT...
        C—G
        T—A
        T—A
        G—C
        C—G
        A—T
        G—C
        G—C
      A       G
      T       A
        A A C
```

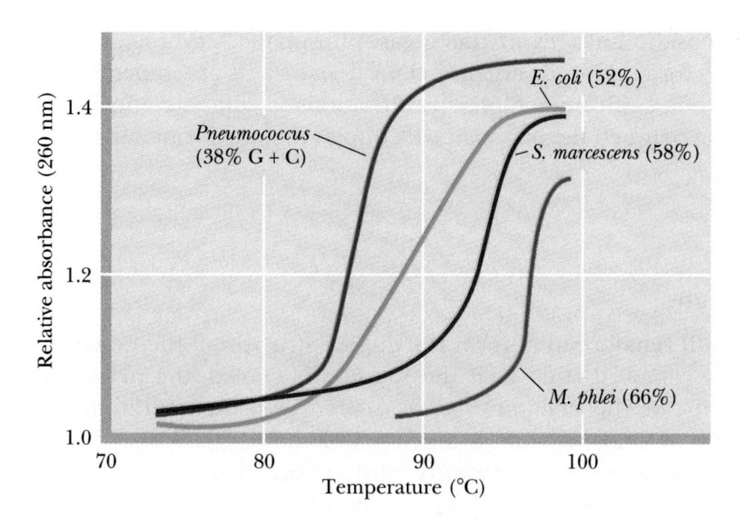

***Figure* 7.23** Heat denaturation of DNA from various sources, so-called melting curves. The midpoint of the melting curve is defined as the melting temperature, T_m.
(From Marmur, J., 1959. Nature 183:1427–1429.)

7.4 Denaturation and Renaturation of DNA

Thermal Denaturation and Hyperchromic Shift

When duplex DNA molecules are subjected to conditions of pH, temperature, or ionic strength that disrupt hydrogen bonds, the strands are no longer held together. That is, the double helix is **denatured** and the strands separate as individual random coils. If temperature is the denaturing agent, the double helix is said to *melt.* The course of this dissociation can be followed spectrophotometrically because the relative absorbance of the DNA solution at 260 nm increases as much as 40% as the bases unstack. This absorbance increase, or **hyperchromic shift,** is due to the fact that the aromatic bases in DNA interact via their π electron clouds when stacked together in the double helix. Since the UV absorbance of the bases is a consequence of π electron transitions, and since the potential for these transitions is diminished when the bases stack, the bases in duplex DNA absorb less 260-nm radiation than expected for their numbers. Unstacking alleviates this suppression of UV absorbance. The rise in absorbance coincides with strand separation, and the midpoint of the absorbance increase is termed the **melting temperature,** T_m (Figure 7.23). DNAs differ in their T_m values because they differ in relative G+C content. The higher the G+C content of a DNA, the higher its melting temperature because G:C pairs are held by three H bonds whereas A:T pairs have only two. The dependence of T_m on the G+C content is depicted in Figure 7.24. Also note that T_m is dependent on the ionic strength of the solution; the lower the ionic strength, the lower the melting temperature. At 0.2 M Na^+, $T_m = 69.3 + 0.41(\%\ G{+}C)$. Ions suppress the electrostatic repulsion between the negatively charged phosphate groups in the complementary strands of the helix, thereby stabilizing it. (DNA in pure water melts even at room temperature.) At high concentrations of ions, T_m is raised and the transition between helix and coil is sharp.

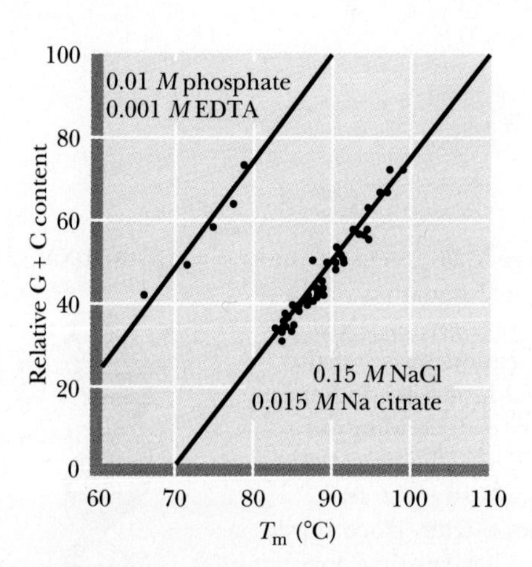

***Figure* 7.24** The dependence of melting temperature on relative (G+C) content in DNA. Note that T_m increases if ionic strength is raised at constant pH (pH 7); 0.01 M phosphate + 0.001 M EDTA versus 0.15 M NaCl/0.015 M Na citrate. In 0.15 M NaCl/0.015 M Na citrate, duplex DNA consisting of 100% A:T pairs melts at less than 70°C, while DNA of 100% G:C has a T_m greater than 110°C.
(From Marmur, J., and Doty, P., 1962. Journal of Molecular Biology 5:120.)

pH Extremes or Strong H-Bonding Solutes Also Denature DNA Duplexes

At pH values greater than 11.5, extensive deprotonation of the bases occurs, destroying their hydrogen bonding potential and denaturing the DNA duplex. Similarly, extensive protonation of the bases below pH 2.3 disrupts base pairing. Alkali is the preferred denaturant because, unlike acid, it does not hydrolyze the glycosidic linkages in the sugar–phosphate backbone. Small solutes that readily form H bonds are also DNA denaturants at temperatures below T_m if present in sufficiently high concentrations to compete effectively with the H bonding between the base pairs. Examples include formamide and urea.

DNA Renaturation

Denatured DNA will **renature** to re-form the duplex structure if the denaturing conditions are removed (that is, if the solution is cooled, the pH is returned to neutrality, or the denaturants are diluted out). Renaturation requires reassociation of the DNA strands into a double helix, a process termed **reannealing.** For this to occur, the strands must realign themselves so that their complementary bases are once again in register and the helix can be zippered up (Figure 7.25). Renaturation is dependent both on DNA concentration and time. Many of the realignments are imperfect, and thus the strands must dissociate again to allow for proper pairings to be formed. The process occurs more quickly if the temperature is warm enough to promote diffusion of the large DNA molecules but not so warm as to cause melting.

Renaturation Rate and DNA Sequence Complexity—c_0t Curves

The renaturation rate of DNA is an excellent indicator of the sequence complexity of DNA. For example, bacteriophage T_4 DNA contains about 2×10^5 nucleotide pairs, whereas *Escherichia coli* DNA possesses 4.72×10^6. *E. coli* DNA is considerably more complex in that it encodes more information. Expressed another way, for any given amount of DNA (in grams), the sequences represented in an *E. coli* sample are more heterogeneous, that is, more dissimilar from one another, than those in an equal weight of phage T_4 DNA. Therefore, it will take the *E. coli* DNA strands longer to find their complementary partners and reanneal. This situation can be analyzed quantitatively.

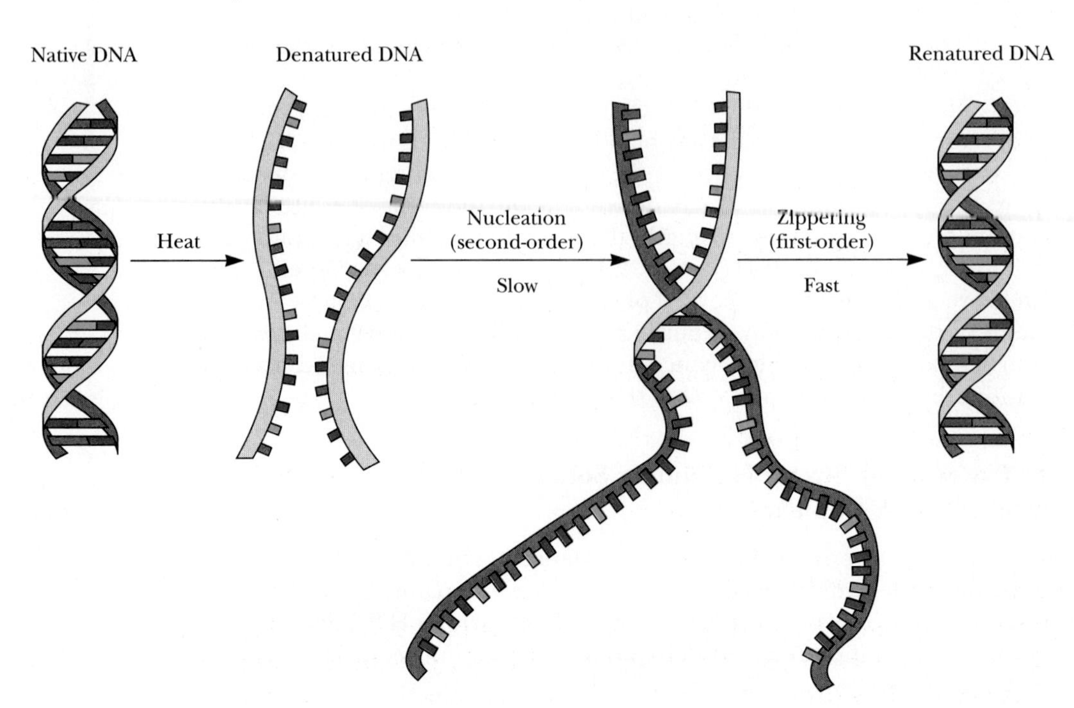

***Figure* 7.25** Steps in the thermal denaturation and renaturation of DNA. The nucleation phase of the reaction is a second-order process depending on sequence alignment of the two strands. This process takes place slowly since it takes time for complementary sequences to encounter one another in solution and then align themselves in register. Once the sequences are aligned, the strands zipper-up quickly.

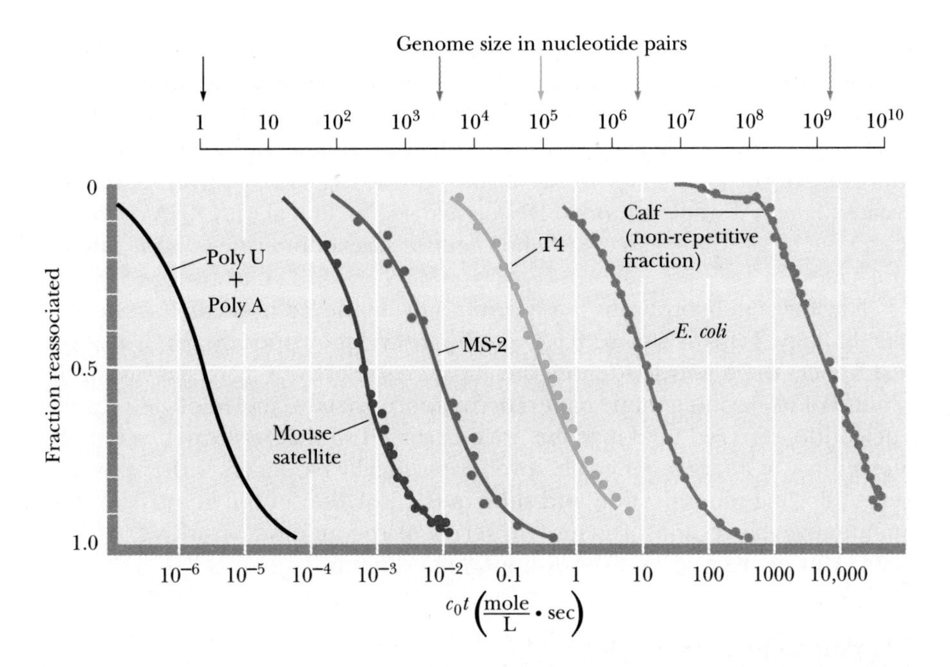

***Figure* 7.26** These c_0t curves show the rates of reassociation of denatured DNA from various sources and illustrate how the rate of reassociation is inversely proportional to genome complexity. The DNA sources are as follows: poly A + poly U, a synthetic DNA duplex of poly A and poly U polynucleotide chains; mouse satellite DNA, a fraction of mouse DNA in which the same sequence is repeated many thousands of times; MS-2 dsRNA, the double-stranded form of RNA found during replication of MS-2, a simple bacteriophage; T4 DNA, the DNA of a more complex bacteriophage; *E. coli* DNA, bacterial DNA; calf DNA (nonrepetitive fraction), mammalian DNA (calf) from which the highly repetitive DNA fraction (satellite DNA) has been removed. Arrows indicate the genome size (in bp) of the various DNAs.

(From Britten, R. J., and Kohne, D. E., 1968. Science 161:529–540.)

If c is the concentration of single-stranded DNA at time t, then the second-order rate equation for two complementary strands coming together is given by the rate of disappearance of c:

$$-dc/dt = k_2c^2$$

where k_2 is the second-order rate constant. Starting with a concentration, c_0, of completely denatured DNA at $t = 0$, the amount of single-stranded DNA remaining at some time t is

$$c/c_0 = 1/(1 + k_2c_0t)$$

where the units of c are mol of nucleotide per L and t is in seconds. The time for half of the DNA to renature (when $c/c_0 = 0.5$) is defined as $t = t_{1/2}$. Then,

$$0.5 = 1/(1 + k_2c_0t_{1/2}) \quad \text{and thus} \quad 1 + k_2c_0t_{1/2} = 2$$

yielding

$$c_0t_{1/2} = 1/k_2$$

A graph of the fraction of single-stranded DNA reannealed (c/c_0) as a function of c_0t on a semilogarithmic plot is referred to as a $\boldsymbol{c_0t}$ (pronounced "cot") **curve** (Figure 7.26). The rate of reassociation can be followed spectrophotometrically by the UV absorbance decrease as duplex DNA is formed. Note that relatively more complex DNAs take longer to renature, as reflected by their greater $c_0t_{1/2}$ values. Poly A and poly U (Figure 7.26) are minimally complex in sequence and anneal rapidly to form a double-stranded A:U polynucleotide. *Mouse satellite DNA* is a highly repetitive subfraction of mouse DNA. Its lack of sequence heterogeneity is seen in its low $c_0t_{1/2}$ value. MS-2 is a small bacteriophage whose genetic material is RNA. Calf thymus DNA is the mammalian representative in Figure 7.26.

Nucleic Acid Hybridization

If DNA from two different species are mixed, denatured, and allowed to cool slowly so that reannealing can occur, artificial **hybrid duplexes** may form, provided the DNA from one species is similar in nucleotide sequence to the

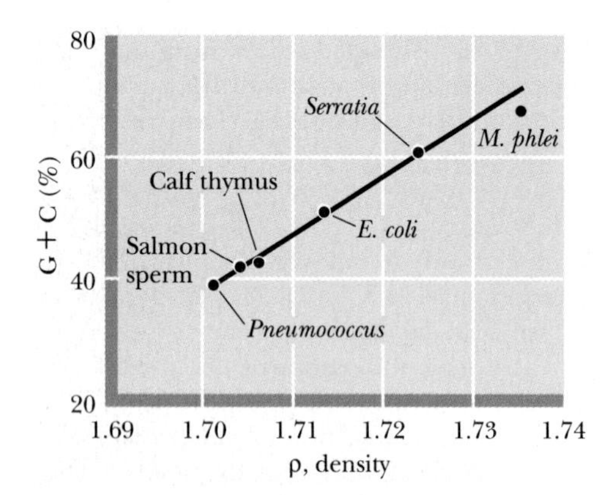

***Figure* 7.27** The relationship of the densities (in g/mL) of DNA from various sources and their G:C content.

(From Doty, P., 1961. Harvey Lectures 55:103.)

DNA of the other. The degree of hybridization is a measure of the sequence similarity or *relatedness* between the two species. Depending on the conditions of the experiment, about 25% of the DNA from a human will form hybrids with mouse DNA, implying that some of the nucleotide sequences (genes) in humans are very similar to those in mice. Mixed RNA:DNA hybrids can be created *in vitro* if single-stranded DNA is allowed to anneal with RNA copies of itself, such as those formed when genes are transcribed into mRNA molecules.

Nucleic acid hybridization is a commonly employed procedure in molecular biology. First, it can reveal evolutionary relationships. Second, it gives researchers the power to identify specific genes selectively against a vast background of irrelevant genetic material. An appropriately labeled oligo- or polynucleotide, referred to as a **probe,** is constructed so that its sequence is complementary to a target gene. The probe specifically base pairs with the target gene, allowing identification and subsequent isolation of the gene. Also, the quantitative expression of genes (in terms of the amount of mRNA synthesized) can be assayed by hybridization experiments.

Buoyant Density of DNA

Not only the melting temperature of DNA but also its density in solution is dependent on relative G:C content. G:C-rich DNA has a significantly higher density than A:T-rich DNA. Furthermore, a linear relationship exists between the buoyant densities of DNA from different sources and their G:C content (Figure 7.27). The density of a DNA, ρ (in g/mL), as a function of its G:C content is given by the equation $\rho = 1.660 + 0.098(GC)$, where (GC) is the mole fraction of (G+C) in the DNA. Because of its relatively high density, DNA can be purified from cellular material by a form of density gradient centrifugation known as *isopycnic centrifugation* (see Appendix to this chapter).

7.5 Chromosome Structure

A typical human cell is 20 μm in diameter. Its genetic material consists of 23 pairs of dsDNA molecules in the form of **chromosomes,** the average length of which is 3×10^9 bp/23 or 1.3×10^8 nucleotide pairs. At 0.34 nm/bp in B-DNA, this represents a DNA molecule 5 cm long. Together, these 46 dsDNA molecules amount to more than 2 m of DNA that must be packaged into a nucleus perhaps 5 μm in diameter! Clearly, the DNA must be condensed by a factor of more than 10^5. This remarkable task is accomplished by neatly wrapping the DNA around protein spools called **nucleosomes** and then packing the nucleosomes to form a helical filament that is arranged in loops associated with the **nuclear matrix,** a skeleton or scaffold of proteins providing a structural framework within the nucleus.

Nucleosomes

The DNA in a eukaryotic cell nucleus during the interphase between cell divisions exists as a nucleoprotein complex called **chromatin.** The proteins of chromatin fall into two classes: **histones** and **nonhistone chromosomal proteins.** Histones are abundant structural proteins, whereas the nonhistone class is represented only by a few copies each of many diverse proteins involved in genetic regulation. The histones are relatively small, positively charged arginine- or lysine-rich proteins that interact via ionic bonds with the negatively charged phosphate groups on the polynucleotide backbone. Five distinct his-

Table 7.2
Properties of Histones

Histone	Ratio of Lysine to Arginine	M_r	Copies per Nucleosome
H1	59/3	21.2	1 (not in bead)
H2A	13/13	14.1	2 (in bead)
H2B	20/8	13.9	2 (in bead)
H3	13/17	15.1	2 (in bead)
H4	11/14	11.4	2 (in bead)

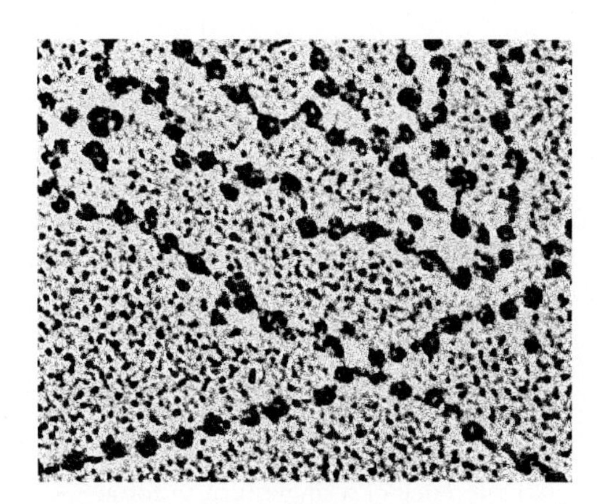

Figure 7.28 Electron micrograph of *Drosophila melanogaster* chromatin after swelling reveals the presence of nucleosomes as "beads on a string."
(Electron micrograph courtesy of Oscar L. Miller, Jr. of the University of Virginia.)

tones are known: **H1, H2A, H2B, H3,** and **H4** (Table 7.2). Pairs of histones H2A, H2B, H3, and H4 aggregate to form an octameric core structure, the **nucleosome,** around which the DNA helix is wound (Figure 6.23).

If chromatin is swelled suddenly in water and prepared for viewing in the electron microscope, the nucleosomes are evident as "beads on a string," dsDNA being the string (Figure 7.28). The structure of the histone octamer core of the nucleosome has been determined by X-ray crystallography (Figure 7.29). About two full turns of DNA wind around this octamer to form a left-handed superhelix, following the surface landmarks of the histone octamer, itself a left-handed protein superhelix. Two full turns require about 168 base pairs, and the distance between the ends of the two full turns is 7.5 nm. Histone H1, a three-domain protein, serves to seal the ends of the turns to the nucleosome core. An additional 40 to 60 base pairs of duplex DNA link consecutive nucleosomes, which are stabilized through interactions between their respective H1 components to create the 30-nm filament (see following section).

solenoid: a coil wound in the form of a helix

Organization of Chromatin and Chromosomes

A higher order of chromatin structure is created when the nucleosomes, in their characteristic beads-on-a-string motif, are wound in the fashion of a *solenoid* having six nucleosomes per turn (Figure 7.30). The resulting 30-nm filament contains about 1200 bp in each of its solenoid turns. This 30-nm filament then forms long DNA loops of variable length, each containing on

Figure 7.29 Four orthogonal views of the histone octamer as determined by X-ray crystallography: (a) front view; (b) top view; and (c) disk view; that is, as viewed down the long axis of the chromatin fiber. In the (c) perspective, the DNA duplex would wrap around the octamer, with the axis of the DNA supercoil perpendicular to the plane of the picture. (d) The nucleosome wrapped with DNA.
(Photographs courtesy of Evangelos N. Moudrianakis of the Johns Hopkins University.)

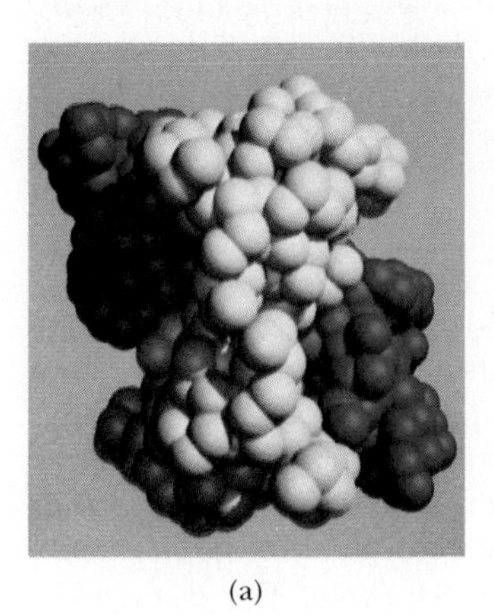

(a)

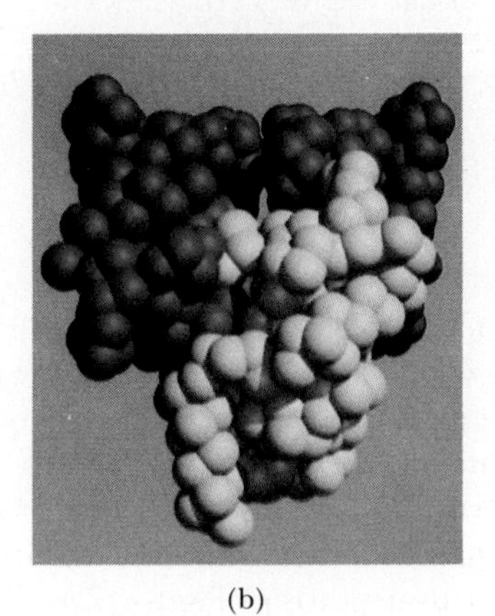

(b)

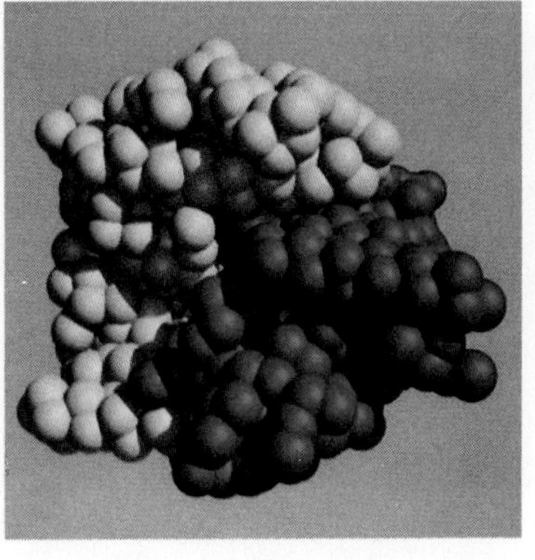

(c)

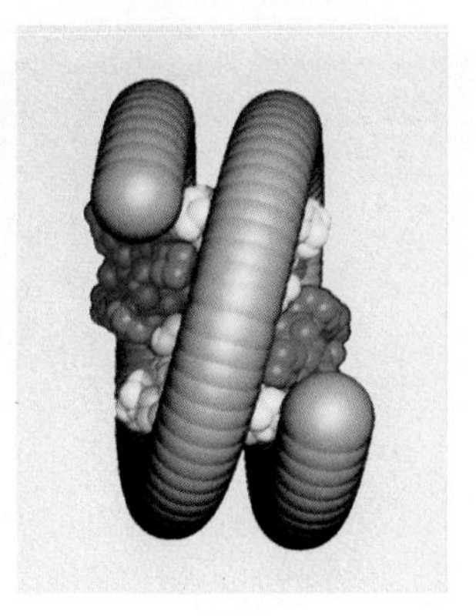

(d)

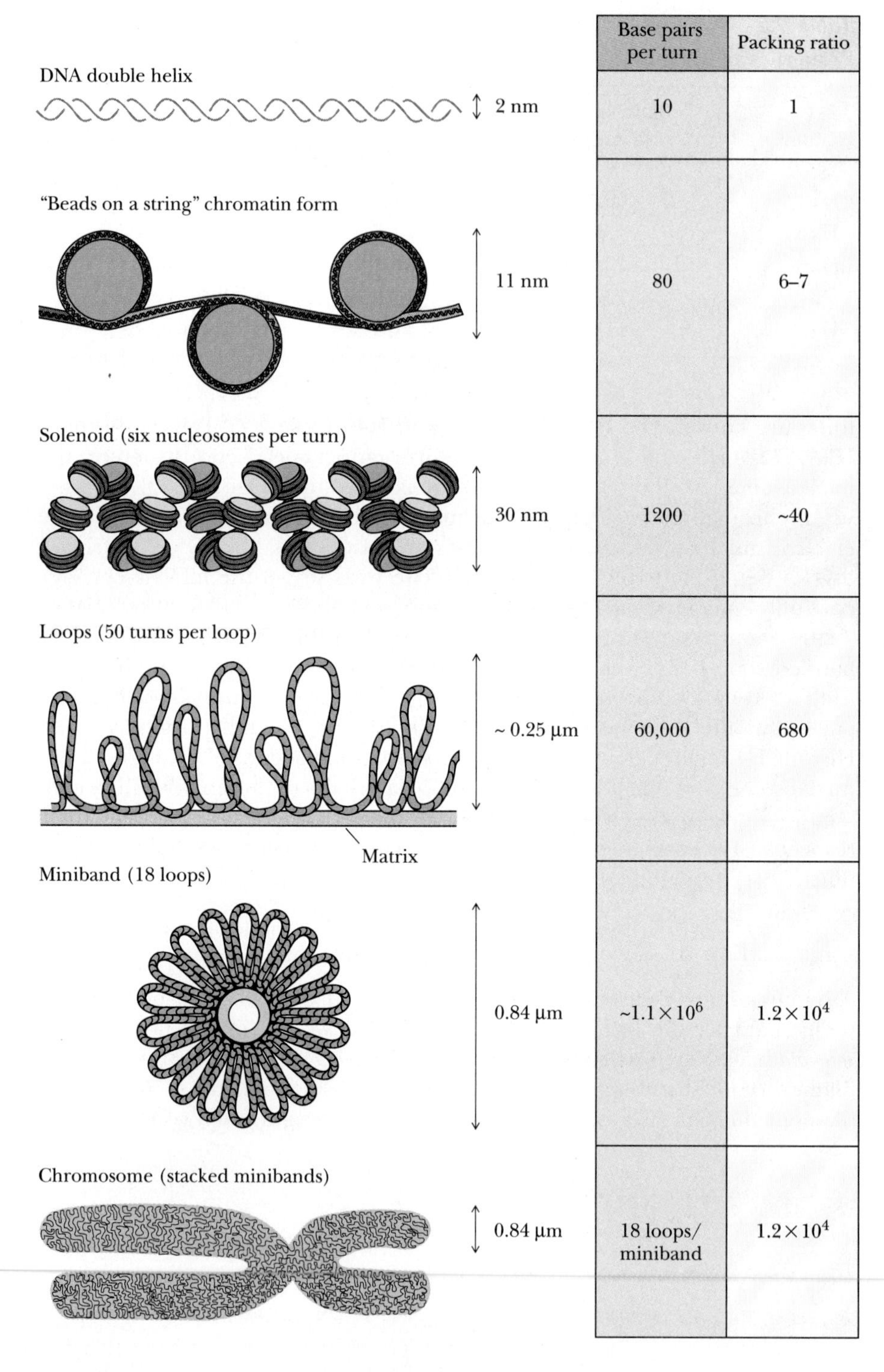

Base pairs per turn	Packing ratio
10	1
80	6–7
1200	~40
60,000	680
~1.1×10^6	1.2×10^4
18 loops/ miniband	1.2×10^4

***Figure* 7.30** A model for chromosome structure, human chromosome 4. The 2-nm DNA helix is wound twice around histone octamers to form 10-nm nucleosomes, each of which contains 160 bp (80 per turn). These nucleosomes are then wound in solenoid fashion with 6 nucleosomes per turn to form a 30-nm filament. In this model, the 30-nm filament forms long DNA loops, each containing about 60,000 bp, which are attached at their base to the nuclear matrix. Eighteen of these loops are then wound radially around the circumference of a single turn to form a miniband unit of a chromosome. Approximately 10^6 of these minibands occur in each chromatid of human chromosome 4 at mitosis.

average between 60,000 and 150,000 bp. Electron microscopic analysis of human chromosome 4 suggests that 18 such loops are then arranged radially about the circumference of a single turn to form a **miniband unit** of the chromosome. According to this model, approximately 10^6 of these minibands are arranged along a central axis in each of the chromatids of human chromosome 4 that form at mitosis (Figure 7.30).

7.6 Chemical Synthesis of Nucleic Acids

Laboratory synthesis of oligonucleotide chains of defined sequence presents some of the same problems encountered in chemical synthesis of polypeptides (see Chapter 4). First, functional groups on the monomeric units (in this case, bases) are reactive under conditions of polymerization and therefore must be protected by blocking agents. Second, to generate the desired sequence, a phosphodiester bridge must be formed between the 3′-O of one nucleotide (B) and the 5′-O of the preceding one (A) in a way that precludes the unwanted bridging of the 3′-O of A with the 5′-O of B. Finally, recoveries at each step must be high so that overall yields in the multistep process are acceptable. As in peptide synthesis (see Chapter 4), *solid phase methods* are used to overcome some of these problems. Commercially available automated instruments, called **DNA synthesizers** or "gene machines," are capable of carrying out the synthesis of oligonucleotides of 150 bases or more.

Phosphoramidite Chemistry

Phosphoramidite chemistry is currently the accepted method of oligonucleotide synthesis. The general strategy involves the sequential addition of nucleotide units as *nucleoside phosphoramidite* derivatives to a nucleoside covalently attached to the insoluble resin. Excess reagents, starting materials, and side products are removed after each step by filtration. After the desired oligonucleotide has been formed, it is freed of all blocking groups, hydrolyzed from the resin, and purified by gel electrophoresis. The four-step cycle is shown in Figure 7.31. Chemical synthesis takes place in the 3′ → 5′ direction (the reverse of the biological polymerization direction).

Chemically Synthesized Genes

Table 7.3 lists some of the genes that have been chemically synthesized. Since protein-coding genes are characteristically much larger than the 150-bp practical limit on oligonucleotide synthesis, their synthesis involves joining a series of oligonucleotides to assemble the overall sequence. A prime example is the gene for rhodopsin (Figure 7.32). This gene is 1057 base pairs long and encodes the 348 amino-acid photoreceptor protein of the vertebrate retina.

***Table* 7.3**
Some Chemically Synthesized Genes

Gene	Size (bp)	Gene	Size (bp)
tRNA	126	Tissue plasminogen activator	1610
α-Interferon	542	c-Ha-ras	576
Secretin	81	RNase T1	324
γ-Interferon	453	Cytochrome b_5	330
Rhodopsin	1057	Bovine intestinal Ca-binding protein	298
Proenkephalin	77	Hirudin	226
Connective tissue activating peptide III	280	RNase A	375
Lysozyme	385		

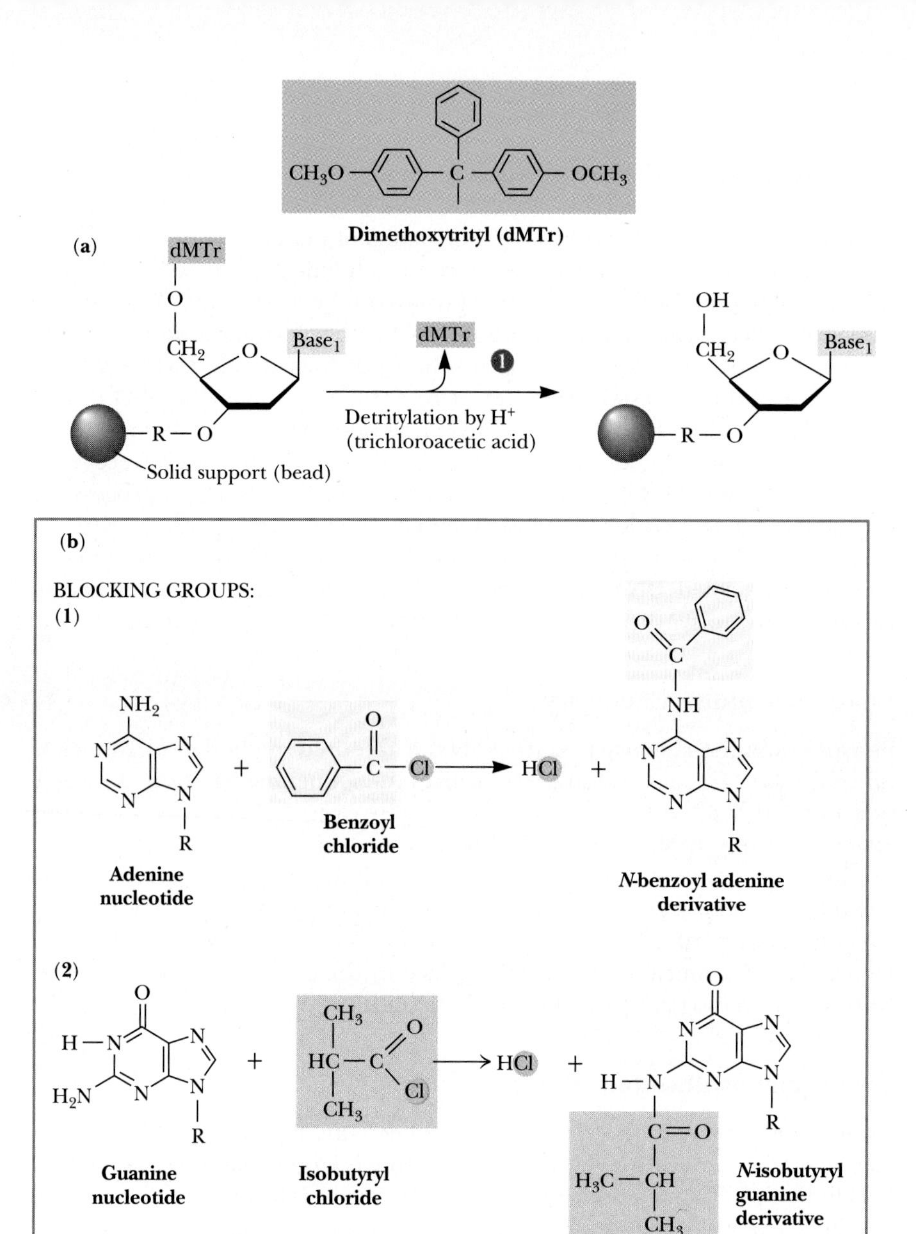

Figure **7.31** Solid phase oligonucleotide synthesis. The four-step cycle starts with the first base in nucleoside form (N-1) attached by its 3′-OH group to an insoluble, inert resin or matrix, typically either controlled pore glass (CPG) or silica beads. Its 5′-OH is blocked with a dimethoxytrityl (DMTr) group (a). If the base has reactive —NH_2 functions, as in A, G, or C, then *N*-benzoyl or *N*-isobutyryl derivatives are used to prevent their reaction (b). In step 1, the DMTr protecting group is removed by trichloroacetic acid treatment. Step 2 is the coupling step: the second base (N-2) is added in the form of a nucleoside phosphoramidite derivative whose 5′-OH bears a DMTr blocking group so it cannot polymerize with itself (c). The presence of a weak acid, such as tetrazole, activates the phosphoramidite, and it rapidly reacts with the free 5′-OH of N-1, forming a dinucleotide linked by a phosphite group. Chemical synthesis thus takes place in the 3′ → 5′ direction. Unreacted free 5′-OHs of N-1 (usually only 2–6% of the total) are blocked from further participation in the polymerization process by acetylation with acetic anhydride in step 3, referred to as *capping*. The phosphite linkage between N-1 and N-2 is highly reactive and, in step 4, it is oxidized by aqueous iodine (I_2) to form the desired more stable phosphate group. This completes the cycle. Subsequent cycles add successive residues to the resin-immobilized chain. When the chain is complete, it is cleaved from the support with NH_4OH, which also removes the *N*-benzoyl and *N*-isobutyryl protecting groups from the amino functions on the A, G, and C residues.

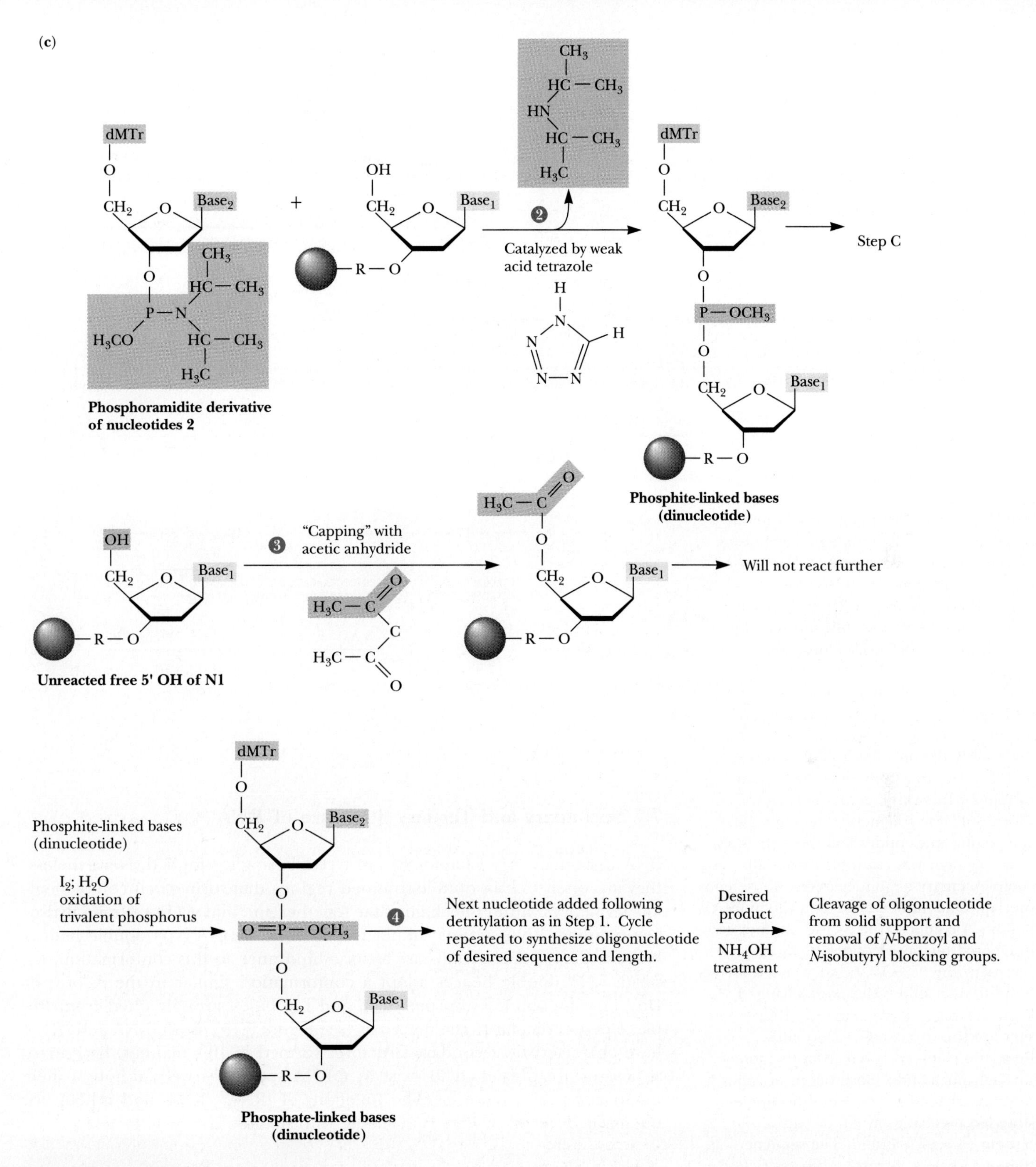

Figure 7.31 (*continued*)

The complete synthesis of genes is becoming increasingly easy and rapid, largely because of greater efficiency in the synthesis of defined oligonucleotides. Theoretically, no gene is beyond the scope of these methods, opening the door to an incredibly exciting range of possibilities for investigating structure-function relationships in the organization and expression of hereditary material.

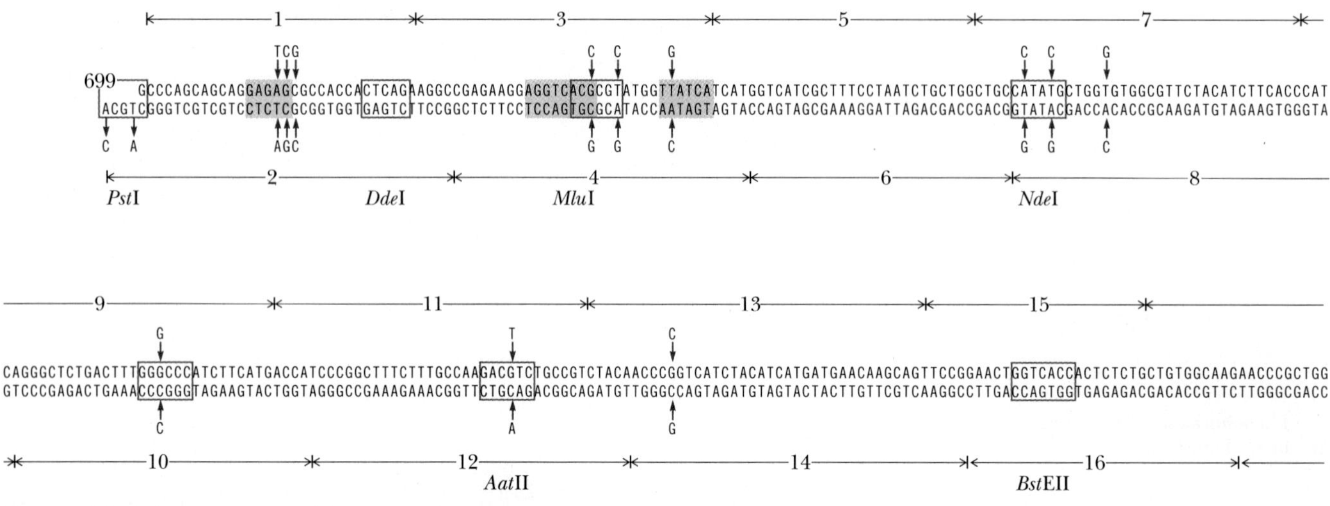

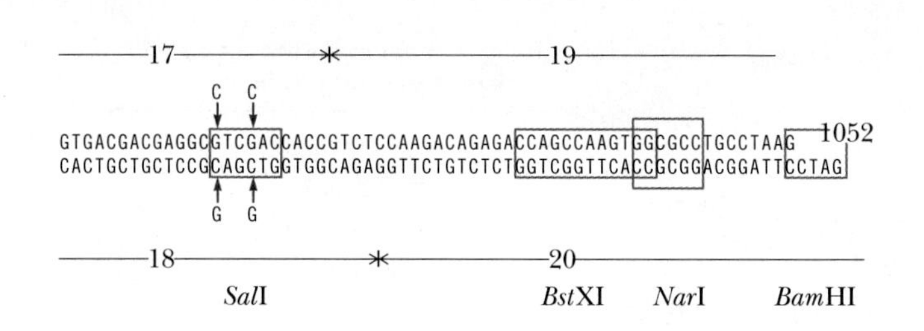

***Figure* 7.32** The strategy used in the total synthesis of the gene for bovine rhodopsin. Total synthesis of this gene was achieved by joining 72 synthetic oligonucleotides, 36 representing one strand and 36 the complementary strand. These oligonucleotides are overlapping. Once synthesized, the various oligonucleotides, each 15 to 40 nucleotides long, were assembled by annealing and enzymatic ligation into three large fragments, representing nucleotides −5 to 338 (−5 meaning 5 nucleotides before the start of the coding region), 335 to 702, and 699 to 1052. The total gene was then created by joining these fragments. This figure shows only one fragment (Fragment PB, comprising nucleotides 699 through 1052). This fragment was assembled from 20 complementary oligonucleotides whose ends overlap. Odd-numbered oligonucleotides (1, 3, 5, . . .) compose the $5' \rightarrow 3'$ strand; even-numbered oligonucleotides (2, 4, 6, . . .) represent the $3' \rightarrow 5'$ strand. Vertical arrows indicate nucleotides that were changed from the native gene sequence. Restriction sites are shown boxed in blue lines. Restriction sites removed from the gene through nucleotide substitutions are shown as yellow shaded boxes. Note the single-stranded overhangs at either end of the bottom ($3' \rightarrow 5'$) strand. The sequences at these overhangs correspond to restriction endonuclease sites (*Pst*I and *Bam*H1), which will facilitate subsequent manipulation of the fragment in gene assembly and cloning.

7.7 Secondary and Tertiary Structure of RNA

RNA molecules (see Chapter 6) are typically single stranded. Nevertheless, they are often rich in double-stranded regions that form when complementary sequences within the chain come together and join via **intrastrand hydrogen bonding.** RNA strands cannot fold to form B-DNA type double helices because their 2′-OH groups are a steric hindrance to this conformation. Instead, RNA double helices adopt a conformation similar to the A-form of DNA, having about 11 bp per turn, and the bases strongly tilted from the plane perpendicular to the helix axis (see Figure 7.15). Both tRNA and rRNA have characteristic secondary structures formed in this manner. Secondary structures are presumed to exist in mRNA species as well, although their nature is as yet unknown. (The functions of tRNA, rRNA, and mRNA are discussed in detail in Part IV: Genetic Information.)

Transfer RNA

tRNA molecules contain from 73 to 94 nucleotides in a single chain; a majority of the bases are hydrogen bonded to one another. Figure 7.33 shows the structure that typifies tRNAs. *Hairpin turns* bring complementary stretches of bases in the chain into contact so that double-helical regions form. Because of the arrangement of the complementary stretches along the chain, the overall pattern of H bonding can be represented as a *cloverleaf.* Each cloverleaf consists of four H-bonded segments—three loops and the stem where the 3′- and 5′-ends of the molecule meet. These four segments are designated the ***acceptor stem,*** the ***D loop,*** the ***anticodon loop,*** and the ***TψC loop.***

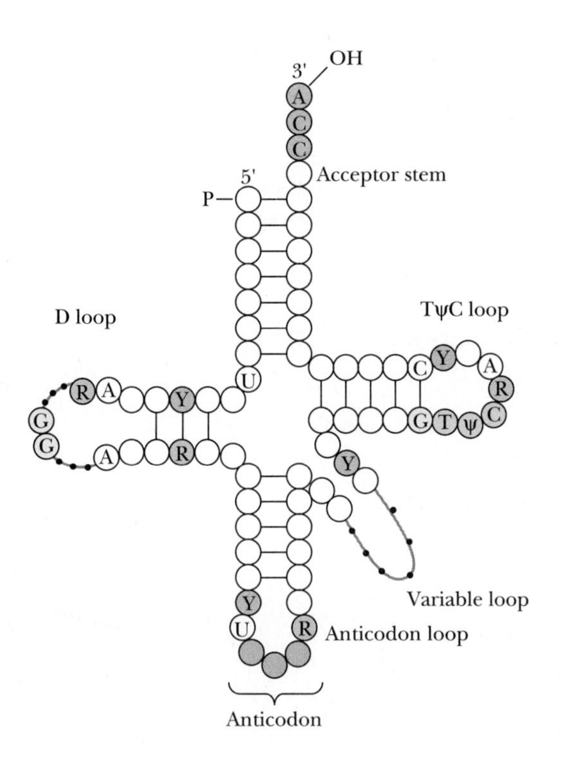

Invariant G in D loop
Invariant pyrimidine, Y
Invariant TψC
Invariant purine, R
Anticodon
CCA 3' end

***Figure* 7.33** A general diagram for the structure of tRNA. The positions of invariant bases as well as bases that seldom vary are shown in color. The numbering system is based on yeast $tRNA^{Phe}$. R = purine and Y = pyrimidine. Dotted lines denote sites in the D loop and variable loop regions where varying numbers of nucleotides are found in different tRNAs.

tRNA Secondary Structure

The *acceptor stem* is where the amino acid is linked to form the aminoacyl-tRNA derivative, which serves as the amino acid-donating species in protein synthesis; this is the physiological role of tRNA. The amino acid adds to the 3′-OH of the 3′-terminal A nucleotide (Figure 7.34). The 3′-end of tRNA is invariantly CCA-3′-OH. This CCA sequence plus a fourth nucleotide extends beyond the double-helical portion of the acceptor stem. The *D loop* is so named because this tRNA loop often contains dihydrouridine, or D, residues. In addition to dihydrouridine, tRNAs characteristically contain a number of unusual bases, including inosine, thiouridine, pseudouridine, and hypermethylated purines (see Figure 6.26). The *anticodon loop* consists of a double-helical segment and seven unpaired bases, three of which are the **anticodon.** (The anticodon is the three-nucleotide unit that recognizes and base pairs with a particular mRNA **codon,** a complementary three-base unit in mRNA which is the genetic information that specifies an amino acid.) Reading $3' \rightarrow 5'$, the anticodon is invariably preceded by a purine (often an alkylated one) and followed by a U. Anticodon base pairing to the codon on mRNA allows a particular tRNA species to deliver its amino acid to the protein-synthesizing apparatus. It represents the key event in translating the information in the nucleic acid sequence so that the appropriate amino acid is inserted at the right place in the amino acid sequence of the protein being synthesized. Next along the tRNA sequence in the $5' \rightarrow 3'$ direction comes a loop that varies from tRNA to tRNA in the number of residues that it has, the so-called **extra** or **variable loop.** The last loop in the tRNA, reading $5' \rightarrow 3'$, is the **T*ψ*C loop,** which contains seven unpaired bases among which is virtually always the sequence T*ψ*C, where *ψ* is the symbol for **pseudouridine.** Ribosomes bind tRNAs through recognition of this T*ψ*C loop. Almost all of the invariant residues common to tRNAs lie within the non-hydrogen-bonded regions of the cloverleaf structure (Figure 7.33). Figure 7.35 depicts the complete nucleotide sequence and cloverleaf structure of yeast alanine tRNA.

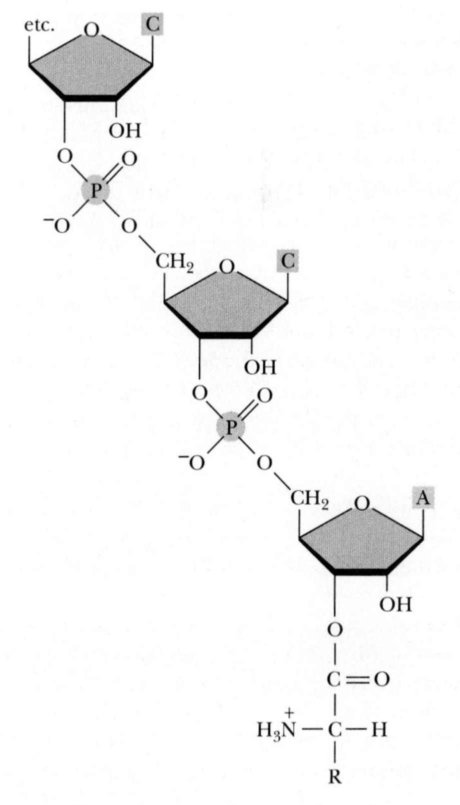

***Figure* 7.34** Amino acids are linked to the 3′-OH end of tRNA molecules by an ester bond formed between the carboxyl group of the amino acid and either the 2′- or 3′-OH of the terminal ribose of the tRNA. (Here, the amino acid is covalently linked to the 3′-O.)

Figure **7.35** The complete nucleotide sequence and cloverleaf structure of yeast alanine tRNA.

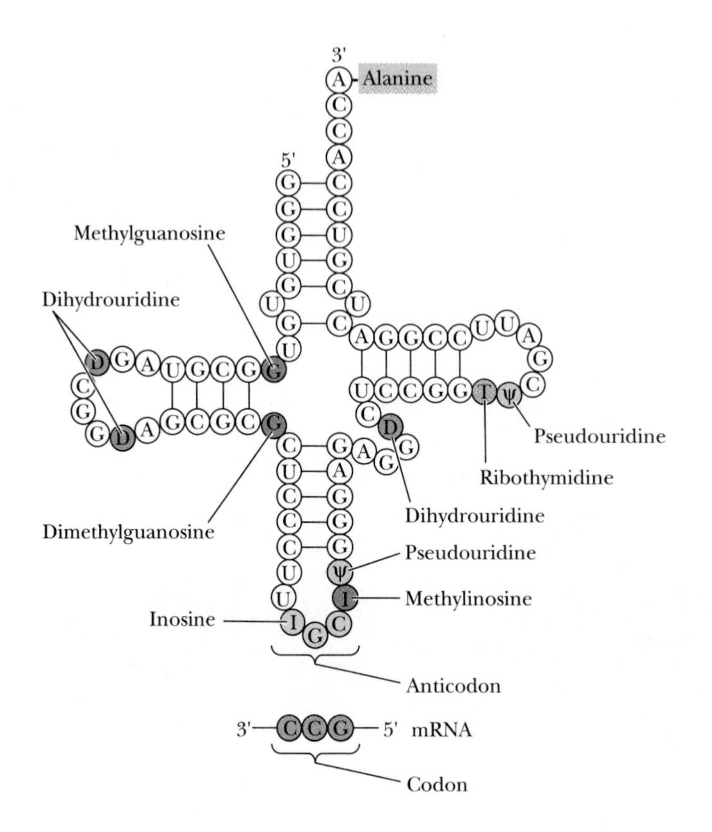

Figure **7.36** Tertiary interactions in yeast phenylalanine tRNA. The molecule is presented in the conventional cloverleaf secondary structure generated by intrastrand hydrogen bonding. Solid lines connect bases that are H-bonded when this cloverleaf pattern is folded into the characteristic tRNA tertiary structure (see also Figure 7.37).

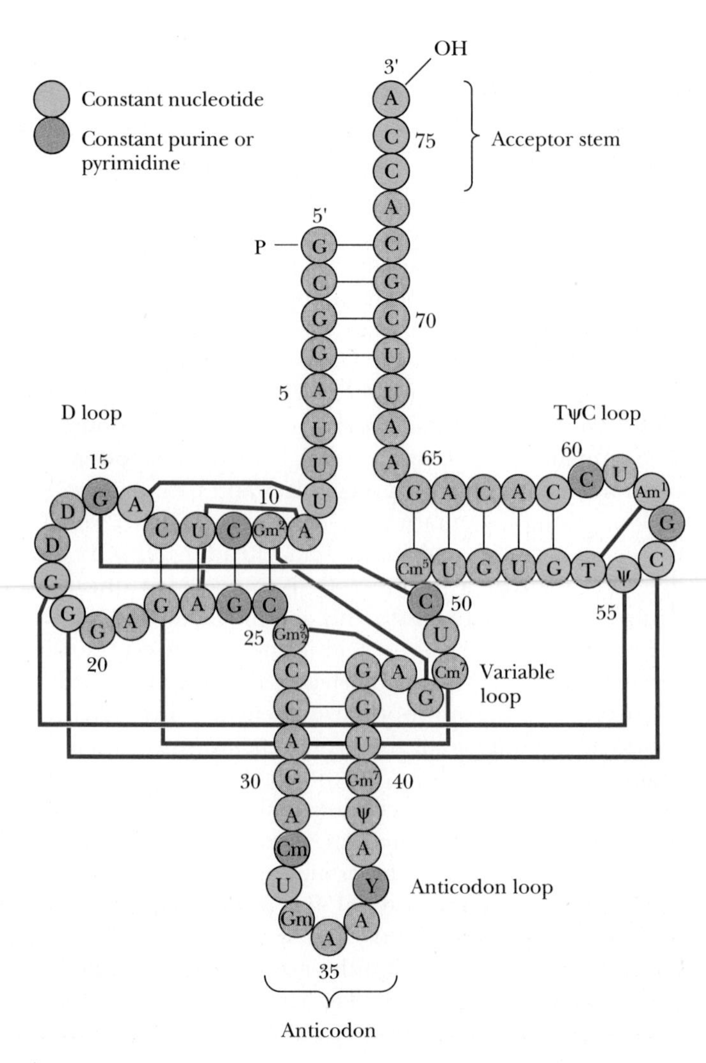

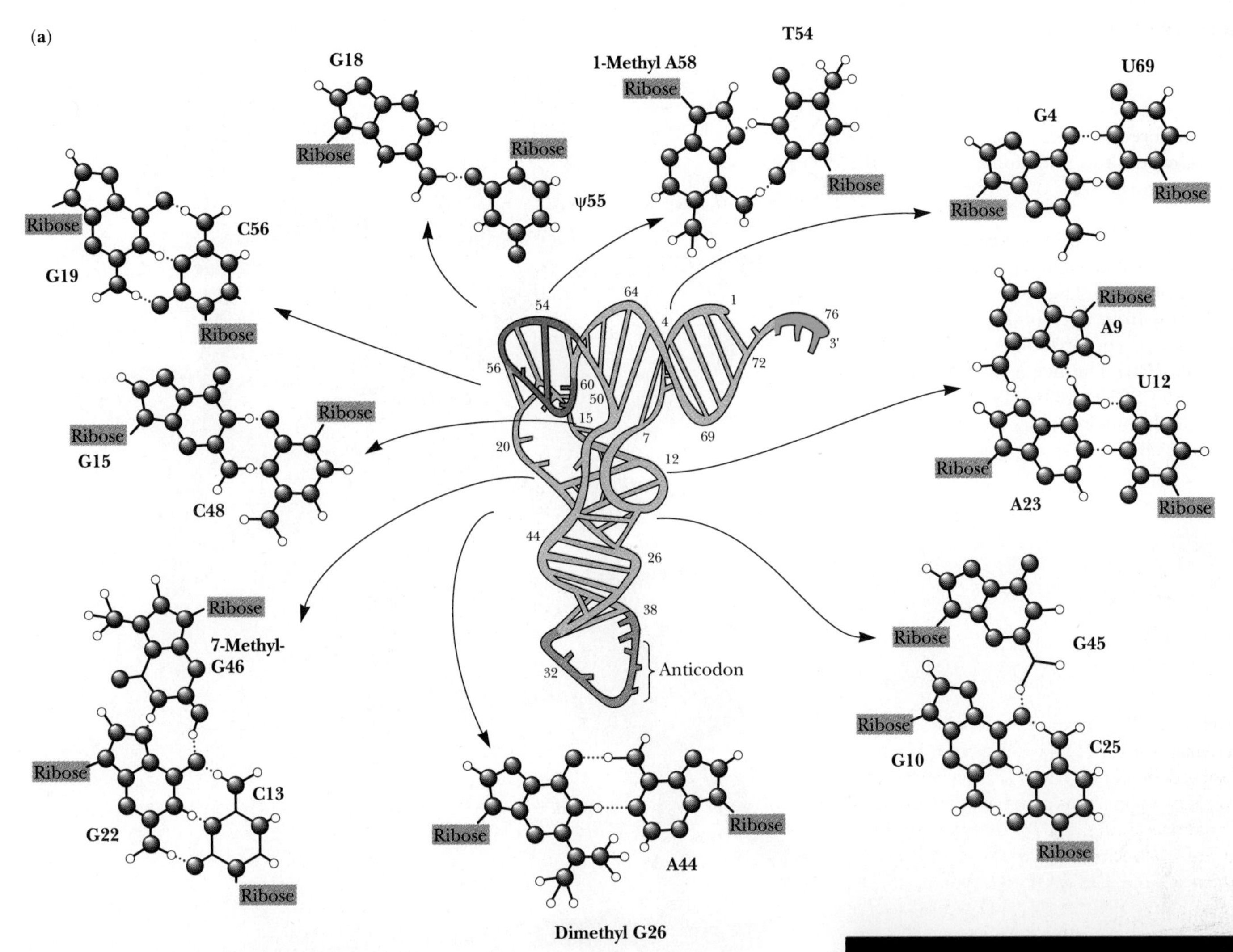

(b)

Figure **7.37** (a) The three-dimensional structure of yeast phenylalanine tRNA as deduced from X-ray diffraction studies of its crystals. The tertiary folding is illustrated in the center of the diagram with the ribose–phosphate backbone presented as a continuous ribbon; H bonds are indicated by crossbars. Unpaired bases are shown as short, unconnected rods. The anticodon loop is at the bottom, and the -CCA 3′-OH acceptor end is at the top right. The various types of noncanonical H-bonding interactions observed between bases surround the central molecule. Three of these lower structures show examples of unusual H-bonded interactions involving three bases; these interactions aid in establishing tRNA tertiary structure. (b) A space-filling model of the molecule.

(After Kim, S. H., in Schimmel, P., Söll, D., and Abelson, J. N., eds., 1979. Transfer RNA: Structure, Properties and Recognition. Cold Spring Harbor Laboratory, New York.)

tRNA Tertiary Structure

Tertiary structure in tRNA arises from hydrogen-bonding interactions between bases in the D loop with bases in the variable and TψC loops, as shown for yeast phenylalanine tRNA in Figure 7.36. Note that these H bonds involve the invariant nucleotides of tRNAs, thus emphasizing the importance of the tertiary structure they create to the function of tRNAs in general. These H bonds fold the D and TψC arms together and bend the cloverleaf into the stable L-shaped tertiary form (Figure 7.37). Many of these H bonds involve base pairs that are not canonical A:T or G:C pairings (Figure 7.37). The

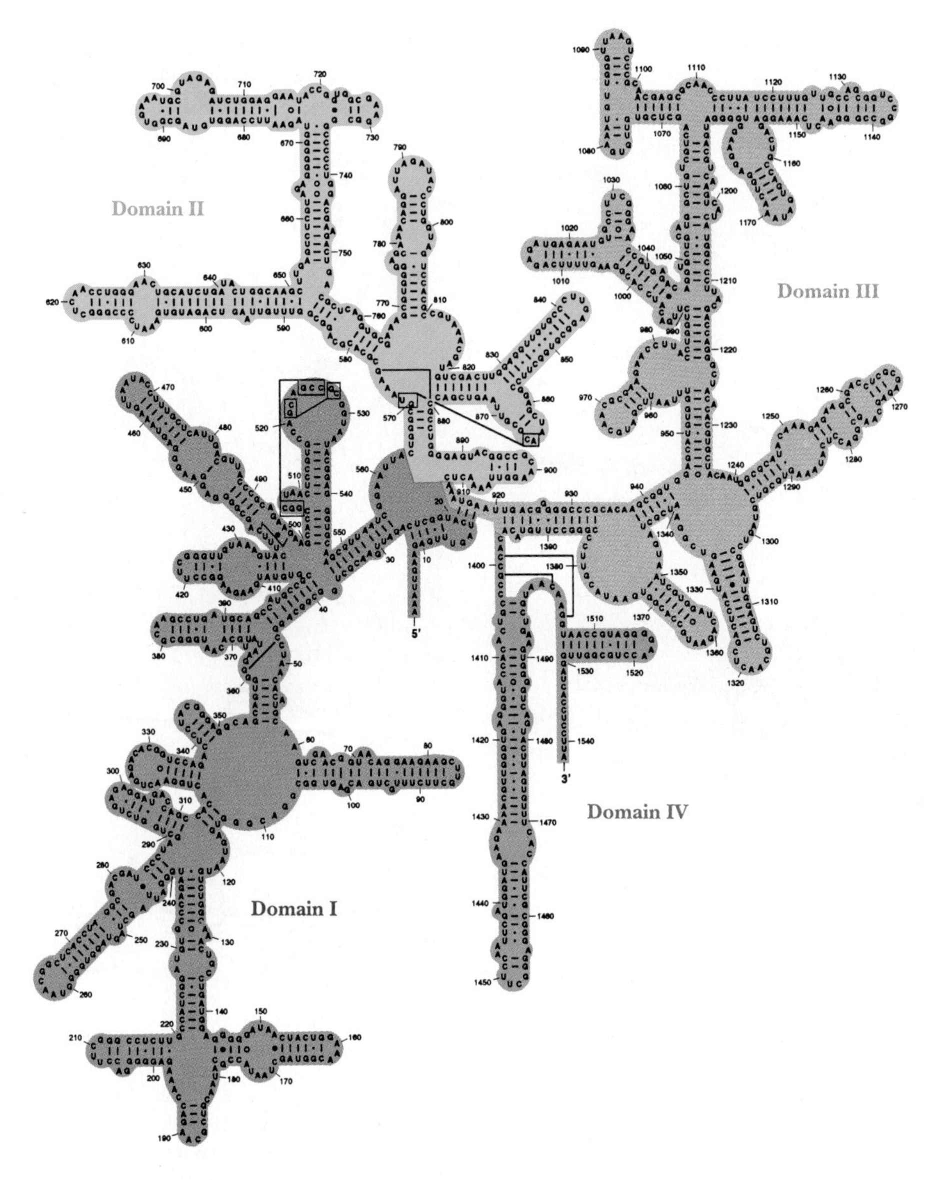

***Figure* 7.38** The proposed secondary structure for *E. coli* 16S rRNA, based on comparative sequence analysis in which the folding pattern is assumed to be conserved across different species. The molecule can be subdivided into four domains—I, II, III, and IV—on the basis of contiguous stretches of the chain that are closed by long-range base-pairing interactions. (I) The 5′-domain, which includes nucleotides 1 through 563. (II) The central domain, which runs from nucleotide 564 to 912. Two domains comprise the 3′-end of the molecule. (III) The major one comprises nucleotides 923 to 1391. (IV) The 3′-terminal domain covers residues 1392 to 1542.

amino acid acceptor stem is at one end of the L, separated by 7 nm or so from the anticodon at the opposite end of the L. The D and TψC loops form the corner of the L. In the L-conformation, the bases are oriented to maximize hydrophobic stacking interactions between their flat faces. Such stacking is a second major factor contributing to L-form stabilization.

Ribosomal RNA

rRNA Secondary Structure

Ribosomes, the protein-synthesizing machinery of cells, are composed of two **subunits,** called **small** and **large,** and ribosomal RNAs are integral components of these subunits (see Table 6.2). A large degree of *intrastrand sequence complementarity* is found in all rRNA strands, and all assume a highly folded pattern that allows base pairing between these complementary segments. Figure 7.38 shows the secondary structure assigned to the *E. coli* 16S rRNA. This structure is based on alignment of the nucleotide sequence into H-bonding segments. The reliability of these alignments is then tested through a comparative analysis of whether identical secondary structures can be predicted from primary sequences of 16S-like rRNAs from other species. If so, then such structures are apparently conserved. The approach is based on the thesis that, since ribosomal RNA species (regardless of source) serve common roles in protein synthesis, it may be anticipated that they share structural features. The structure is marvelously rich in short, helical segments separated and punctuated by single-stranded loops.

Comparison of rRNAs from Various Species

If a phylogenetic comparison is made of the 16S-like rRNAs from an archaebacterium (*Halobacterium volcanii*), a eubacterium (*E. coli*), and a eukaryote (the yeast *Saccharomyces cerevisiae*), a striking similarity in secondary structure emerges (Figure 7.39). Remarkably, these secondary structures are similar despite a low degree of similarity in the nucleotide sequences of these rRNAs. Apparently, evolution is acting at the level of rRNA secondary structure, not rRNA nucleotide sequence. Similar conserved folding patterns are seen for the 23S-like and 5S-like rRNAs that reside in the large ribosomal subunits of various species. An insightful conclusion may be drawn regarding the persistence of such strong secondary structure conservation despite the millennia that have passed since these organisms diverged: *all ribosomes are constructed to a common design and all function in a similar manner.*

rRNA Tertiary Structure

Despite the unity in secondary structural patterns, little is known about the three-dimensional, or tertiary, structure of rRNAs. Even less is known about the quaternary interactions that occur when ribosomal proteins combine with rRNAs and when the ensuing ribonucleoprotein complexes, the small and large subunits, come together to form the complete ribosome. Furthermore, assignments of functional roles to rRNA molecules are still tentative and approximate. (We return to these topics in Chapter 32.)

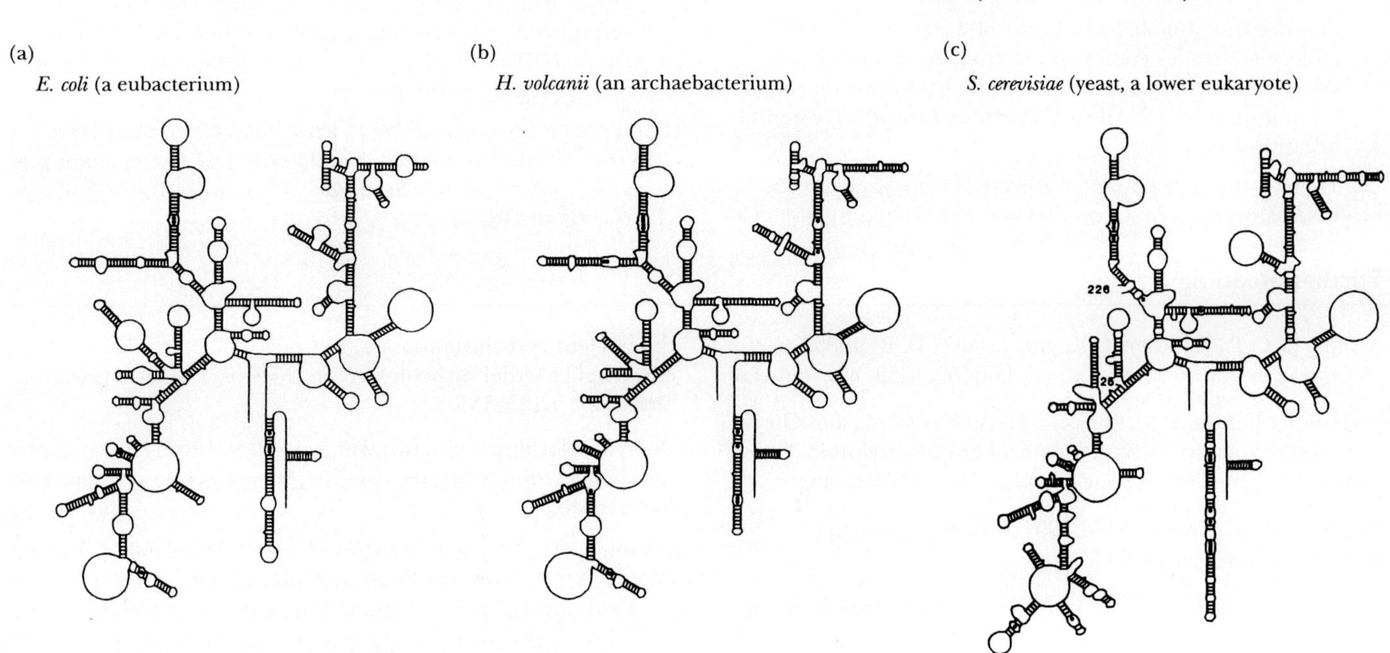

***Figure* 7.39** Phylogenetic comparison of secondary structures of 16S-like rRNAs from (a) a eubacterium (*E. coli*), (b) an archaebacterium (*H. volcanii*), and (c) a eukaryote (*S. cerevisiae*, a yeast).

Problems

1. The oligonucleotide d-ATGCCTGACT was subjected to sequencing by (a) Sanger's dideoxy method and (b) Maxam and Gilbert's chemical cleavage method, and the products were analyzed by electrophoresis on a polyacrylamide gel. Draw diagrams of the gel banding patterns obtained for (a) and (b).

2. The result of sequence determination of an oligonucleotide as performed by the Sanger dideoxy chain termination method is displayed at right.

   What is the sequence of the original oligonucleotide? A second sample of the oligonucleotide was 3′-end labeled with ^{32}P and then subjected to the Maxam–Gilbert chemical cleavage sequencing protocol. Draw a diagram depicting the pattern seen on the autoradiogram of the Maxam–Gilbert sequencing gel.

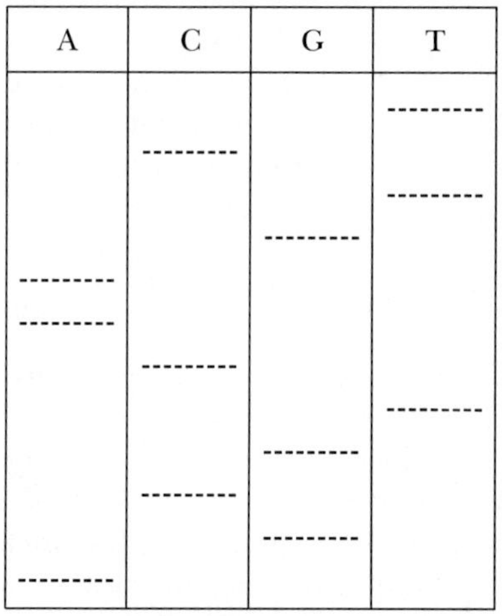

A	C	G	T

---------			

3. X-ray diffraction studies indicate the existence of a novel double-stranded DNA helical conformation in which ΔZ (the rise per base pair) = 0.32 nm and P (the pitch) = 3.36 nm. What are the other parameters of this novel helix: (a) the number of base pairs per turn, (b) $\Delta\phi$ (the mean rotation per base pair), and (c) c (the true repeat)?

4. A 41.5-nm-long duplex DNA molecule in the B-conformation upon dehydration adopts the A-conformation. How long is it now? What is its approximate number of base pairs?

5. If 80% of the base pairs in a duplex DNA molecule (12.5 kbp) are in the B-conformation and 20% are in the Z-conformation, what is the length of the molecule?

6. A "relaxed," circular, double-stranded DNA molecule (1600 bp) is in a solution where conditions favor 10 bp per turn. What is the value of L_0 for this DNA molecule? Suppose DNA gyrase introduces 12 negative supercoils into this molecule. What are the values of L, W, and T now? What is the superhelical density, σ?

7. Suppose one double-helical turn of a superhelical DNA molecule changes conformation from B-form to Z-form. What are the changes in L, W, and T? Why do you suppose the transition of DNA from B-form to Z-form is favored by negative supercoiling?

8. There is one nucleosome for every 200 bp of eukaryotic DNA. How many nucleosomes are in a diploid human cell? Nucleosomes can be approximated as disks 11 nm in diameter and 6 nm long. If all the DNA molecules in a diploid human cell are in the B-conformation, what is the sum of their lengths? If this DNA is now arrayed on nucleosomes in the "beads-on-a-string" motif, what is its approximate total length?

9. The characteristic secondary structures of tRNA and rRNA molecules are achieved through intrastrand hydrogen bonding. Even for the small tRNAs, remote regions of the primary sequence interact via H bonding when the molecule adopts the cloverleaf pattern. Using Figure 7.33 as a guide, draw the primary structure of a tRNA and label the positions of its various self-complementary regions.

10. Using the data in Table 6.3, arrange the DNAs from the following sources in order of increasing T_m: human, salmon, wheat, yeast, *E. coli.*

11. The DNAs from mice and rats have (G+C) contents of 44% and 40%, respectively. Calculate the T_ms for these DNAs in 0.2 M NaCl. If samples of these DNAs were inadvertently mixed, how might they be repurified from one another? Describe the procedure and the results (hint: see the Appendix to this chapter).

12. Calculate the density (ρ) of avian tubercle bacillus DNA from the data presented in Table 6.3 and the equation $\rho = 1.660 + 0.098(GC)$, where (GC) is the mole fraction of (G+C) in DNA.

Further Reading

Adams, R. L. P., Knowler, J. T., and Leader, D. P., 1986. *The Biochemistry of the Nucleic Acids,* 10th ed. London: Chapman and Hall.

Ferretti, L., Karnik, S. S., Khorana, H. G., Nassal, M., and Oprian, D. D., 1986. Total synthesis of a gene for bovine rhodopsin. *Proceedings of the National Academy of Sciences, U.S.A.* **83:**599–603.

Kornberg, A., and Baker, T. A., 1991. *DNA Replication,* 2nd ed. New York: W. H. Freeman and Co.

Noller, H. F., 1984. Structure of ribosomal RNA. *Annual Review of Biochemistry* **53:**119–162.

Pienta, K. J., and Coffey, D. S., 1984. A structural analysis of the role of the nuclear matrix and DNA loops in the organization of the nucleus and chromosomes. In Cook, P. R., and Laskey, R. A., eds., Higher Order Structure in the Nucleus. *Journal of Cell Science Supplement* **1:**123–135.

Rich, A., Nordheim, A., and Wang, A. H.-J., 1984. The chemistry and biology of left-handed Z-DNA. *Annual Review of Biochemistry* **53:**791–846.

Watson, J. D., Hopkins, N. H., Roberts, J. W., Steitz, J. A., and Weiner, A. M., 1987. *The Molecular Biology of the Gene,* Vol. I, *General Principles,* 4th ed. Menlo Park, CA: Benjamin/Cummings.

Watson, J. D., ed., 1983. Structures of DNA. *Cold Spring Harbor Symposia on Quantitative Biology,* Volume XLVII. Cold Spring Harbor Laboratory, New York.

Appendix to Chapter 7

Isopycnic Centrifugation and Buoyant Density of DNA

Density gradient ultracentrifugation is a variant of the basic technique of ultracentrifugation (discussed in the Appendix to Chapter 4). Density gradient centrifugation can be used to isolate DNA. The densities of DNAs are about the same as concentrated solutions of cesium chloride, CsCl (1.6 to 1.8 g/mL). Centrifugation of CsCl solutions at very high rotational speeds, where the centrifugal force becomes 10^5 times stronger than the force of gravity, causes the formation of a density gradient within the solution. This gradient is the result of a balance that is established between the sedimentation of the salt ions toward the bottom of the tube and their diffusion upward toward regions of lower concentration. If DNA is present in the centrifuged CsCl solution, it will move to a position of equilibrium in the gradient equivalent to its buoyant density (Figure A7.1). For this reason, this technique is also called **isopycnic centrifugation.**

isopycnic: same density

Cesium chloride centrifugation is an excellent means of removing RNA and proteins in the purification of DNA. The density of DNA is typically slightly greater than 1.7 g/cm^3, while the density of RNA is more than 1.8 g/cm^3. Proteins have densities less than 1.3 g/cm^3. In CsCl solutions of appropriate density, the DNA bands near the center of the tube, RNA pellets to the bottom, and the proteins float near the top. Single-stranded DNA is denser than double-helical DNA. The irregular structure of randomly coiled ssDNA allows the atoms to pack together through van der Waals interactions. These interactions compact the molecule into a smaller volume than that occupied by a hydrogen-bonded double helix.

The net movement of solute particles in an ultracentrifuge is the result of two processes: diffusion (from regions of higher concentration to regions of lower concentration) and sedimentation due to centrifugal force (in the direction away from the axis of rotation). In general, diffusion rates for molecules are inversely proportional to their molecular weight—larger molecules diffuse more slowly than smaller ones. On the other hand, sedimentation rates increase with increasing molecular weight. A macromolecular species that has reached its position of equilibrium in isopycnic centrifugation has formed a concentrated band of material.

Essentially three effects are influencing the movement of the molecules in creating this concentration zone: (1) diffusion away to regions of lower concentration; (2) sedimentation of molecules situated at positions of slightly lower solution density in the density gradient; and (3) flotation (buoyancy or "reverse sedimentation") of molecules that have reached positions of slightly greater solution density in the gradient. The consequence of the physics of these effects is that, at equilibrium, *the width of the concentration band established by the macromolecular species is inversely proportional to the square root of its molecular weight.* That is, a population of large molecules will form a concentration band that is narrower than the band formed by a population of small molecules. For example, the band width formed by dsDNA will be less than the band width formed by the same DNA when dissociated into ssDNA.

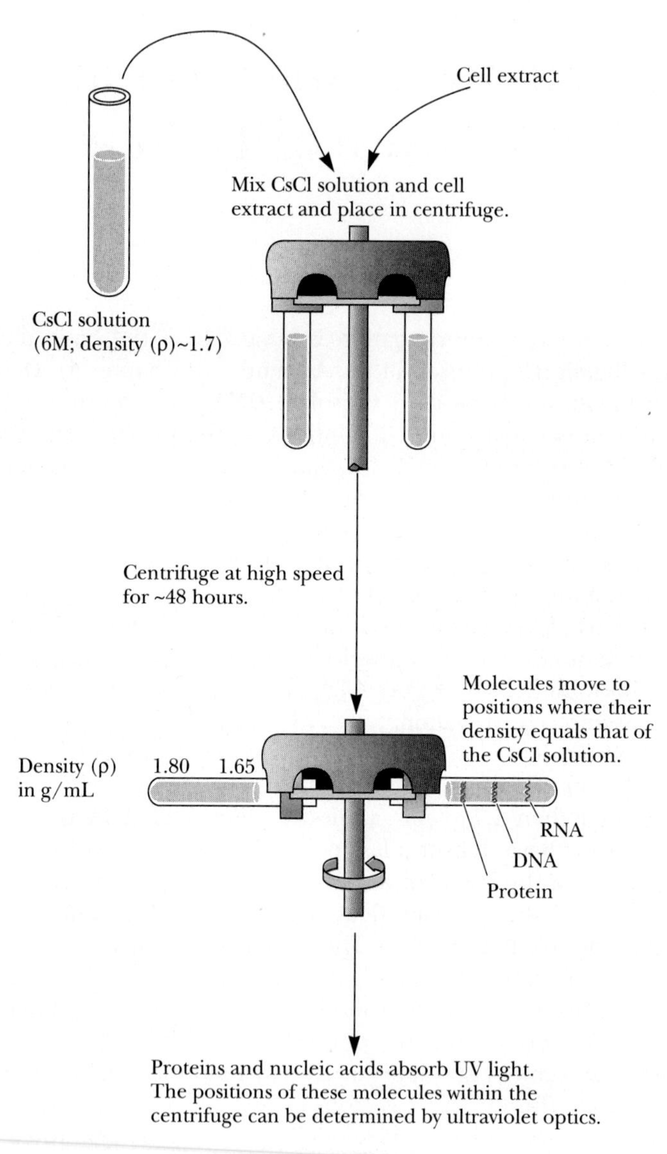

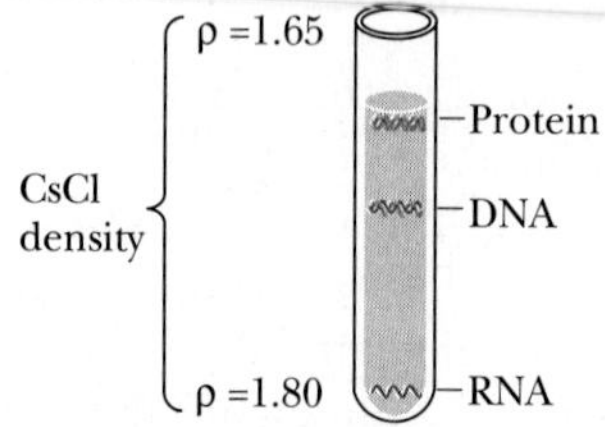

***Figure* A7.1** Density gradient centrifugation is a common method of separating macromolecules, particularly nucleic acids, in solution. A cell extract is mixed with a solution of CsCl to a final density of about 1.7 g/cm^3 and centrifuged at high speed (40,000 rpm, giving relative centrifugal forces of about 200,000 *g*). The biological macromolecules in the extract will move to equilibrium positions in the CsCl gradient that reflect their buoyant densities.

Chapter 8

". . . how many vain chimeras have you created? . . . Go and take your place with the seekers after gold."

Leonardo da Vinci, *The Notebooks* (1508–1518), Volume II, Chapter 25

Recombinant DNA: Cloning and Creation of Chimeric Genes

Outline

8.1 Cloning

8.2 DNA Libraries

8.3 Polymerase Chain Reaction (PCR)

8.4 Recombinant DNA Technology: An Exciting Scientific Frontier

Chimeric animals like the minotaur existed only in the imagination of the ancients. But the ability to create chimeric DNA molecules is a very real technology that has opened up a whole new field of scientific investigation.

In the early 1970s, emerging technologies for the laboratory manipulation of nucleic acids led to the construction of DNA molecules composed of nucleotide sequences taken from different sources. The products of these innovations, **recombinant DNA molecules,**[1] opened exciting new avenues of investigation in molecular biology and genetics, and a new field was born—**recombinant DNA technology. Genetic engineering** is the application of this technology to the study of genes. These advances were made possible by methods for **amplification** of any particular DNA segment, regardless of source, within bacterial host cells. Or, in the language of recombinant DNA technology, the **cloning** of almost any chosen DNA sequence became feasible.

amplification: the production of multiple copies

[1]The advent of molecular biology, like that of most scientific disciplines, has generated a jargon all its own. Learning new fields often requires gaining familiarity with a new vocabulary. We will soon see that many words—*vector, amplification,* and *insert* are but a few examples—have been bent into new meanings to describe the marvels of this new biology.

8.1 Cloning

In classical biology, a *clone* is a population of identical organisms derived from a single parental organism. For example, the members of a colony of bacterial cells that arise from a single cell on a petri plate are clones. Molecular biology has borrowed the term to mean a collection of molecules or cells all identical to an original molecule or cell. So, if the original cell on the petri plate harbored a recombinant DNA molecule in the form of a plasmid, the plasmids within the millions of cells in a bacterial colony represent a clone of the original DNA molecule, and these molecules can be isolated and studied. Furthermore, if the cloned DNA molecule is a gene (or part of a gene), that is, it encodes a functional product, a new avenue to isolating and studying this product has opened. Recombinant DNA methodology offers exciting new vistas in biochemistry.

Plasmids

Plasmids are naturally occurring, circular, extrachromosomal DNA molecules (see Chapter 7). Natural strains of the common colon bacterium *Escherichia coli* isolated from various sources harbor diverse plasmids. Often these plasmids carry genes specifying novel metabolic activities that are advantageous to the host bacterium. These activities range from catabolism of unusual organic substances to metabolic functions that endow the host cells with resistance to antibiotics, heavy metals, or bacteriophages. Plasmids that are able to perpetuate themselves in *E. coli,* the established workhorse of bacterial geneticists and molecular biologists, have become the darlings of recombinant DNA technology. Since restriction endonuclease digestion of plasmids can generate fragments with overlapping or ''sticky'' ends, artificial plasmids can be constructed by ligating different fragments together. Such artificial plasmids were among the earliest recombinant DNA molecules. As long as they still possess a site signaling where DNA replication can begin (a so-called origin of replication or *ori* sequence), these recombinant molecules can be autonomously replicated, and hence propagated, in suitable bacterial host cells.

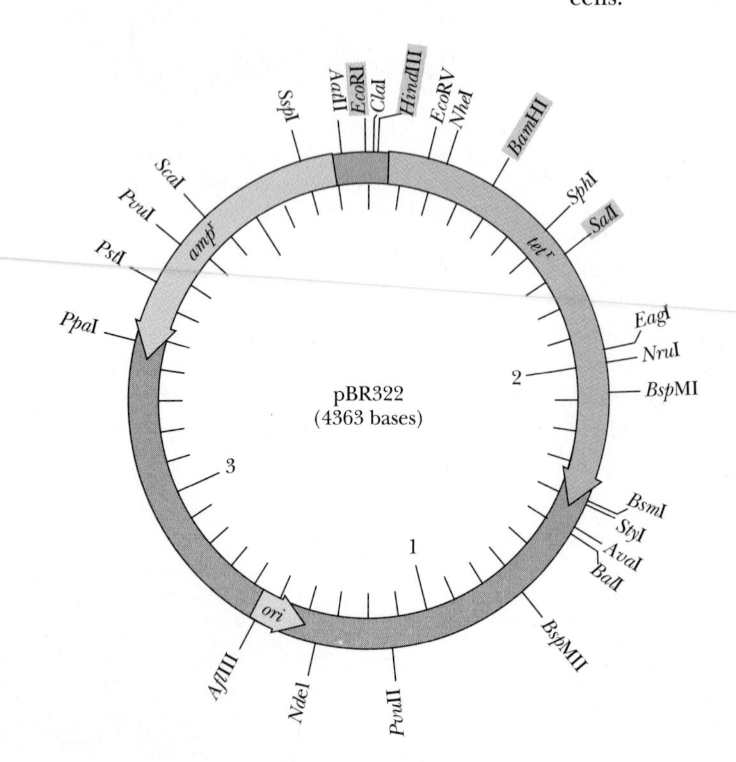

Figure **8.1** One of the first widely used cloning vectors, the plasmid *pBR322.* This 4363-bp plasmid contains an origin of replication (*ori*) and genes encoding resistance to the drugs ampicillin (amp^r) and tetracycline (tet^r). The locations of restriction endonuclease cleavage sites are indicated.

Plasmids as Cloning Vectors

The idea arose that "foreign" DNA sequences could be inserted into artificial plasmids and that these foreign sequences would be carried into *E. coli* and propagated as part of the plasmid. That is, these plasmids could serve as **cloning vectors** to carry genes. (The word *vector* is used here in the sense of "a vehicle or carrier.") Plasmids useful as cloning vectors possess three common features: **a replicator, a selectable marker,** and **a cloning site** (Figure 8.1). A **replicator** is an origin of replication, or *ori*. The **selectable marker** is typically a gene conferring resistance to an antibiotic. Only those cells containing the cloning vector will grow in the presence of the antibiotic. Therefore, growth on antibiotic-containing media "selects for" plasmid-containing cells. Typically, the *cloning site* is a sequence of nucleotides representing one or more restriction endonuclease cleavage sites. Cloning sites are located where the insertion of foreign DNA will neither disrupt the plasmid's ability to replicate nor inactivate essential markers.

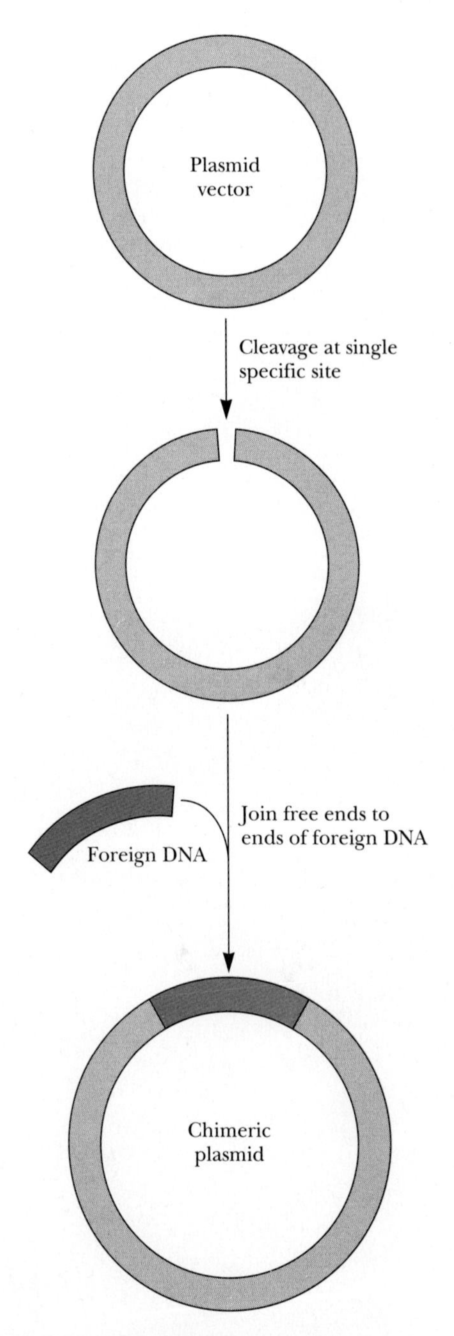

***Figure* 8.2** Foreign DNA sequences can be inserted into plasmid vectors by opening the circular plasmid with a restriction endonuclease. The ends of the linearized plasmid DNA are then joined with the ends of a foreign sequence, reclosing the circle to create a chimeric plasmid.

Virtually Any DNA Sequence Can Be Cloned

Nuclease cleavage at a restriction site opens, or *linearizes,* the circular plasmid so that a foreign DNA fragment can be inserted. The ends of this linearized plasmid are joined to the ends of the fragment so that the circle is closed again, creating a recombinant plasmid (Figure 8.2). **Recombinant plasmids** are hybrid DNA molecules consisting of plasmid DNA sequences plus inserted DNA elements (called *inserts*). Such hybrid molecules are also called **chimeric constructs** or **chimeric plasmids.** (The term *chimera* is borrowed from mythology and refers to a beast composed of the body parts of several creatures, usually a lion's head, goat's body, and a serpent's tail.) The presence of foreign DNA sequences does not adversely affect replication of the plasmid, so chimeric plasmids can be propagated in bacteria just like the original plasmid. Bacteria often harbor several hundred copies of common cloning vectors per cell. Hence, large amounts of a cloned DNA sequence can be recovered from bacterial cultures. The enormous power of recombinant DNA technology stems in part from the fact that *virtually any DNA sequence can be selectively cloned and amplified in this manner.* DNA sequences that are intractable to cloning include inverted repeats, origins of replication, centromeres, and telomeres. The only practical limitation is the size of the foreign DNA segment: most plasmids with inserts larger than about 10 kbp are not replicated efficiently.

Bacterial cells may harbor one or many copies of a particular plasmid, depending on the nature of the plasmid replicator. That is, plasmids are classified as *high copy number* or *low copy number.* The copy number of most plasmids is high (200 or so), but some are lower.

Construction of Chimeric Plasmids

Creation of chimeric plasmids requires joining the ends of the foreign DNA insert to the ends of a linearized plasmid (Figure 8.2). This ligation is facilitated if the ends of the plasmid and the insert have complementary, single-stranded overhangs. Then these ends can base-pair with one another, annealing the two molecules together. One way to generate such ends is to cleave the DNA with restriction enzymes that make staggered cuts; many such restriction endonucleases are available (see Table 6.5). For example, if the sequence to be inserted is an *Eco*RI fragment and the plasmid is cut with *Eco*RI, the single-stranded sticky ends of the two DNAs can anneal (Figure 8.3). The interrup-

ligation: the act of joining

(text continues on page 262)

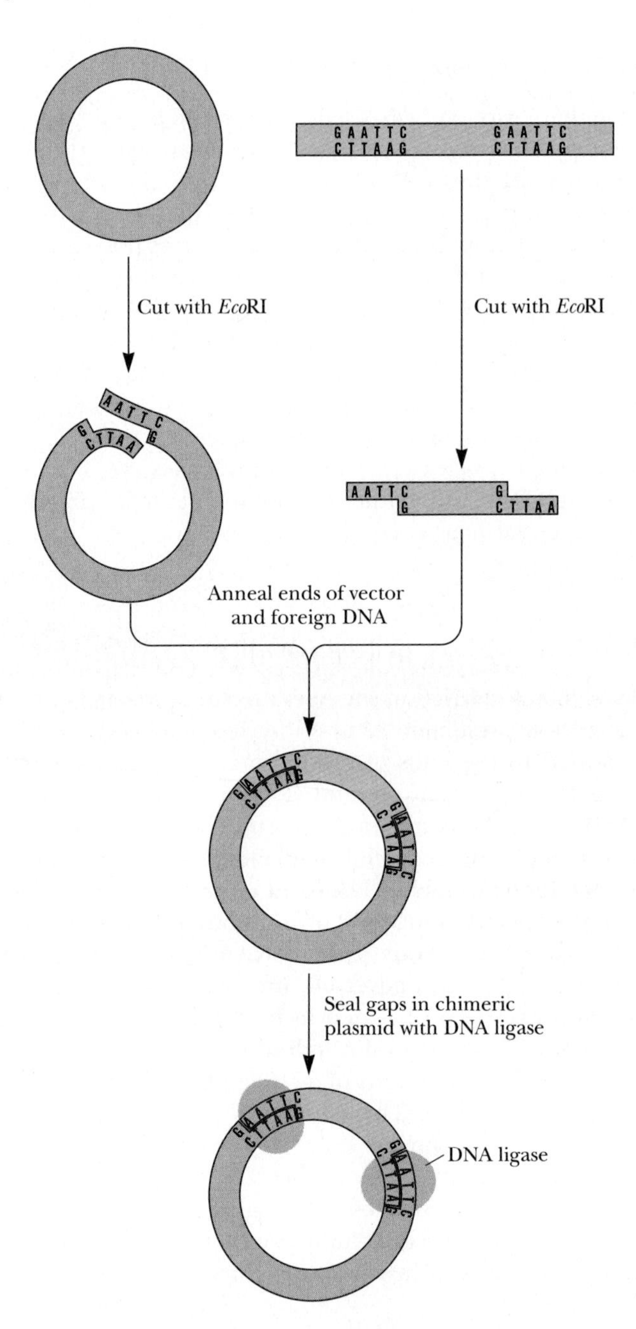

***Figure* 8.3** Restriction endonuclease *Eco*RI cleaves double-stranded DNA. The recognition site for *Eco*RI is the hexameric sequence GAATTC:

5′···NpNpNpNp**GpApApTpTpC**pNpNpNpNp···3′
3′···NPNpNpNp**CpTpTpApApG**pNpNpNpNp···5′

Cleavage occurs at the G residue on each strand so that the DNA is cut in a staggered fashion, leaving 5′-overhanging single-stranded ends (sticky ends):

5′···NpNpNpNp**G** **pApApTpTpC**pNpNpNpNp···3′
3′···NpNpNpNp**CpTpTpApAp** **G**pNpNpNpNp···5′

An *Eco*RI restriction fragment of foreign DNA can be inserted into a plasmid having an *Eco*RI cloning site by (a) cutting the plasmid at this site with *Eco*RI, annealing the linearized plasmid with the *Eco*RI foreign DNA fragment, and (b) sealing the nicks with DNA ligase.

tions in the sugar–phosphate backbone of DNA can then be sealed with DNA ligase to yield a covalently closed, circular chimeric plasmid. DNA ligase is an enzyme that covalently links adjacent 3′-OH and 5′-PO_4 groups. An inconvenience of this strategy is that *any* pair of *Eco*RI sticky ends can anneal with each other. So, plasmid molecules can reanneal with themselves, as can the foreign DNA restriction fragments. These DNAs can be eliminated by selection schemes designed to identify only those bacteria containing chimeric plasmids.

Blunt-end ligation is an alternative method for joining different DNAs. This method depends on the ability of **phage T4 DNA ligase** to covalently join the ends of any two DNA molecules (even those lacking 3′- or 5′-overhangs) (Figure 8.4). Some restriction endonucleases cut DNA so that blunt ends are formed (see Table 6.5). Since there is no control over which pair of DNAs are blunt-end ligated by T4 DNA ligase, strategies to identify the desired products must be applied.

A great number of variations on these basic themes have emerged. For example, short synthetic DNA duplexes whose nucleotide sequence consists of little more than a restriction site can be blunt-end ligated onto any DNA. These short DNAs are known as **linkers.** Cleavage of the ligated DNA with the restriction enzyme then leaves tailor-made sticky ends useful in cloning reactions (Figure 8.5). Similarly, many vectors contain a **polylinker** cloning site, a short region of DNA sequence bearing numerous restriction sites.

Promoters and Directional Cloning

Note that the strategies discussed thus far create hybrids in which the orientation of the DNA insert within the chimera is random. Sometimes it is desirable to insert the DNA in a particular orientation. For example, an experimenter might wish to insert a particular DNA (a gene) in a vector so that its

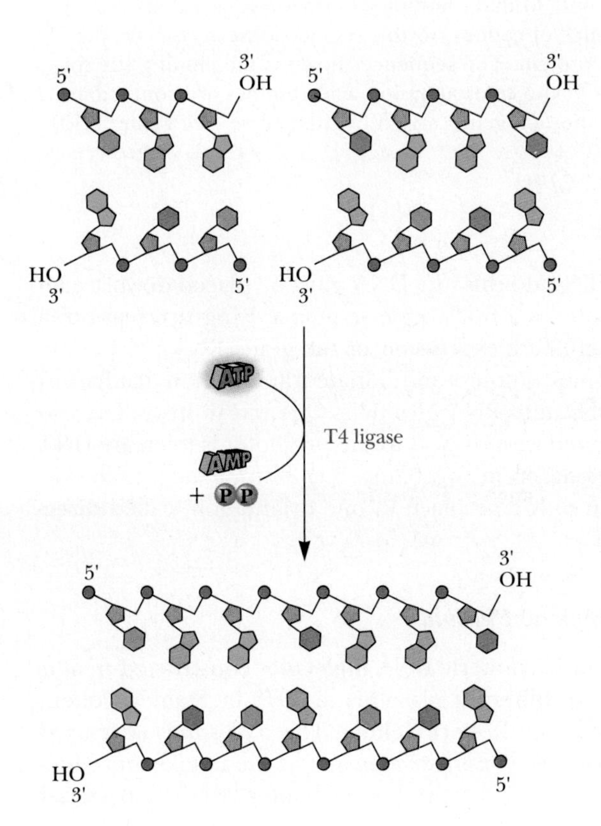

***Figure* 8.4** Blunt-end ligation using phage T4 DNA ligase, which catalyzes the ATP-dependent ligation of DNA molecules. AMP and PP_i are by-products.

(a)

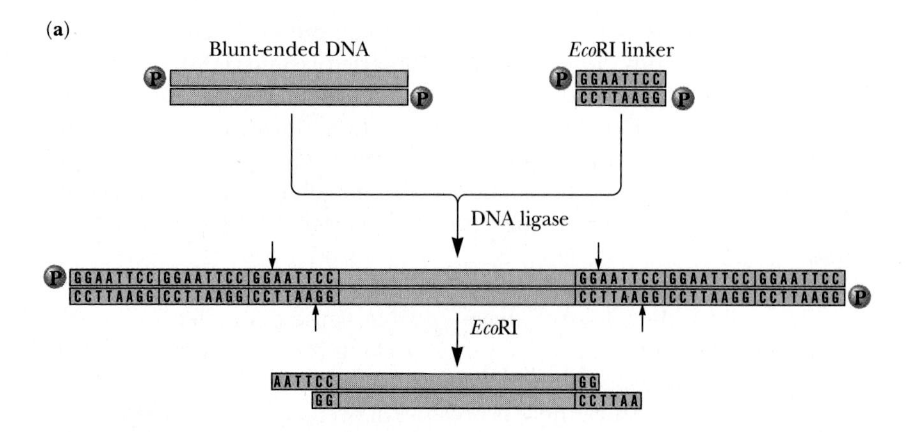

(b) A vector cloning site containing multiple restriction sites, a so-called *polylinker.*

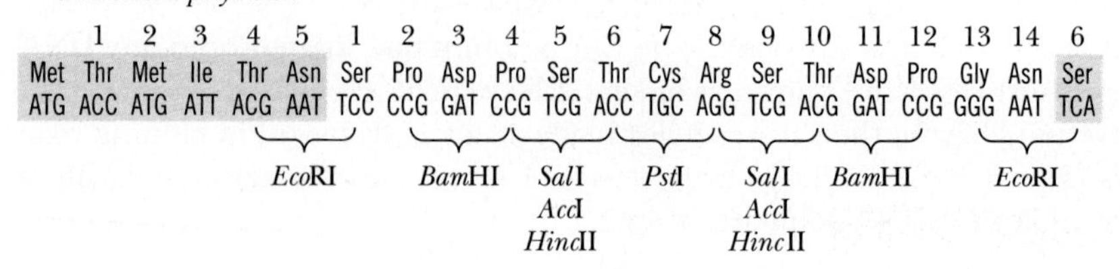

***Figure* 8.5** (a) The use of linkers to create tailor-made ends on cloning fragments. Synthetic oligonucleotide duplexes whose sequences represent *Eco*RI restriction sites are blunt-end ligated to a DNA molecule using T4 DNA ligase. Note that the ligation reaction can add multiple linkers on each end of the blunt-ended DNA. *Eco*RI digestion removes all but the terminal one, leaving the desired 5′-overhangs. (b) Cloning vectors often have polylinkers consisting of a multiple array of restriction sites at their cloning sites, so restriction fragments generated by a variety of endonucleases can be incorporated into the vector. Note that the polylinker is engineered not only to have multiple restriction sites but also to have an uninterrupted sequence of codons, so this region of the vector has the potential for translation into protein. The sequence shown is the cloning site for the vectors M13mp7 and pUC7; the colored amino acid residues are contiguous with the coding sequence of the *lacZ* gene carried by this vector (see Figure 8.19). *(a, Adapted from Figure 3.16.3; b, adapted from Figure 1.14.2, Ausubel, F. M., et al., 1987, Current Protocols in Molecular Biology. New York: John Wiley and Sons.)*

gene product is synthesized. To do this, the DNA must be placed downstream from a **promoter.** A promoter is a nucleotide sequence lying upstream of a gene that is involved in regulating expression of the gene. RNA polymerase molecules bind specifically at promoters and initiate transcription of adjacent genes, copying template DNA into RNA products. One way to insert DNA so that it will be properly oriented with respect to the promoter is to create DNA molecules whose ends have different overhangs. Ligation of such molecules into the plasmid vector can only take place in one orientation, called **directional cloning** (Figure 8.6).

Biologically Functional Chimeric Plasmids

The first biologically functional chimeric DNA molecules constructed *in vitro* were assembled from parts of different plasmids in 1973 by Stanley Cohen, Annie Chang, Herbert Boyer, and Robert Helling. These plasmids were used to **transform** recipient *E. coli* cells (*transformation* means the uptake and replication of exogenous DNA by a recipient cell; see Chapter 28). The bacterial

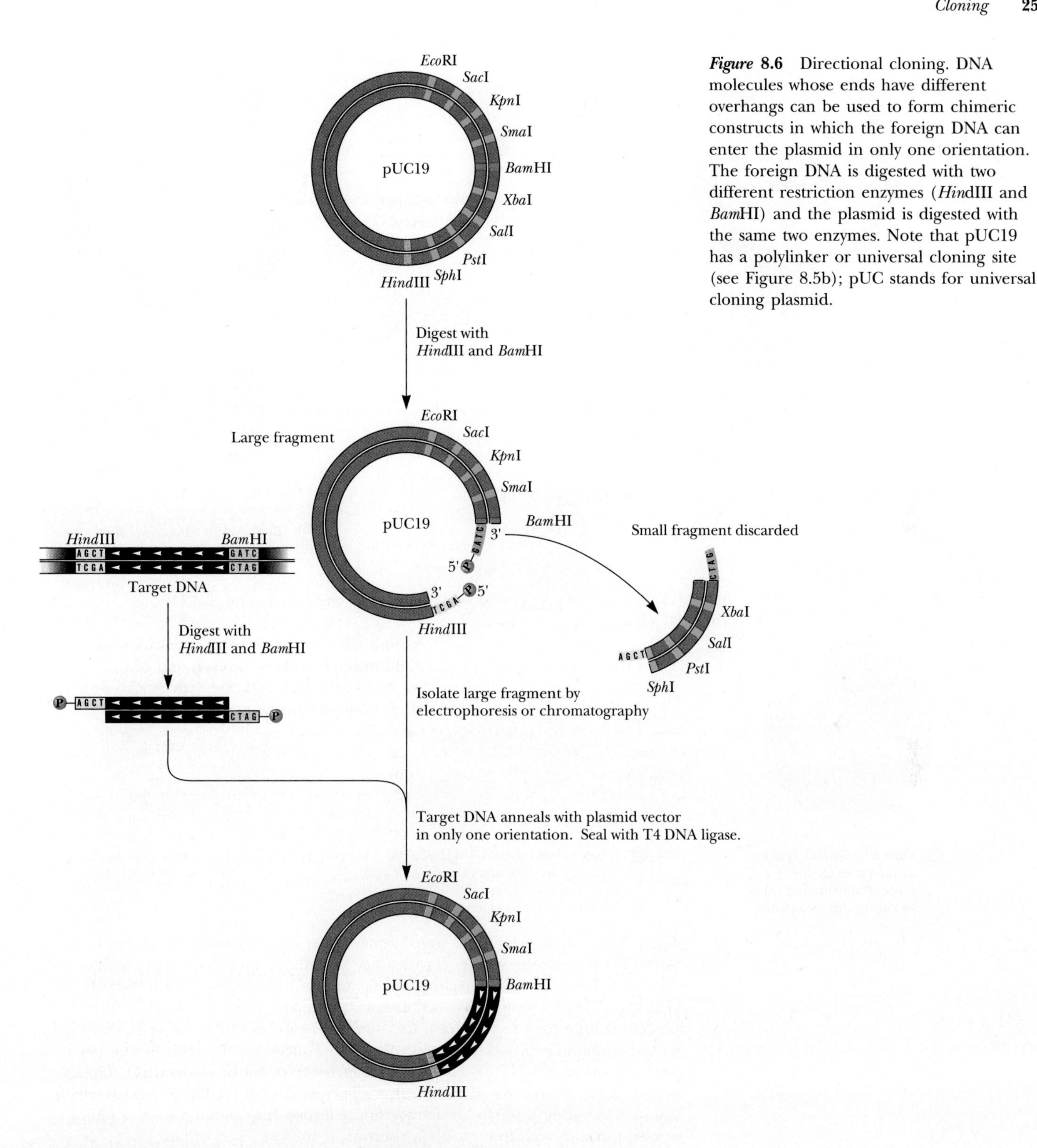

***Figure* 8.6** Directional cloning. DNA molecules whose ends have different overhangs can be used to form chimeric constructs in which the foreign DNA can enter the plasmid in only one orientation. The foreign DNA is digested with two different restriction enzymes (*Hin*dIII and *Bam*HI) and the plasmid is digested with the same two enzymes. Note that pUC19 has a polylinker or universal cloning site (see Figure 8.5b); pUC stands for universal cloning plasmid.

cells were rendered somewhat permeable to DNA by Ca^{2+} treatment and a brief 42°C heat shock. Although less than 0.1% of the Ca^{2+}-treated bacteria became competent for transformation following such treatment, transformed bacteria could be selected by their resistance to certain antibiotics (Figure 8.7). Consequently, the chimeric plasmids must have been biologically functional in at least two aspects: they replicated stably within their hosts and they expressed the drug resistance markers they carried.

In general, plasmids used as cloning vectors are engineered to be small, 2.5 kbp to about 10 kbp in size, so that the size of the insert DNA can be

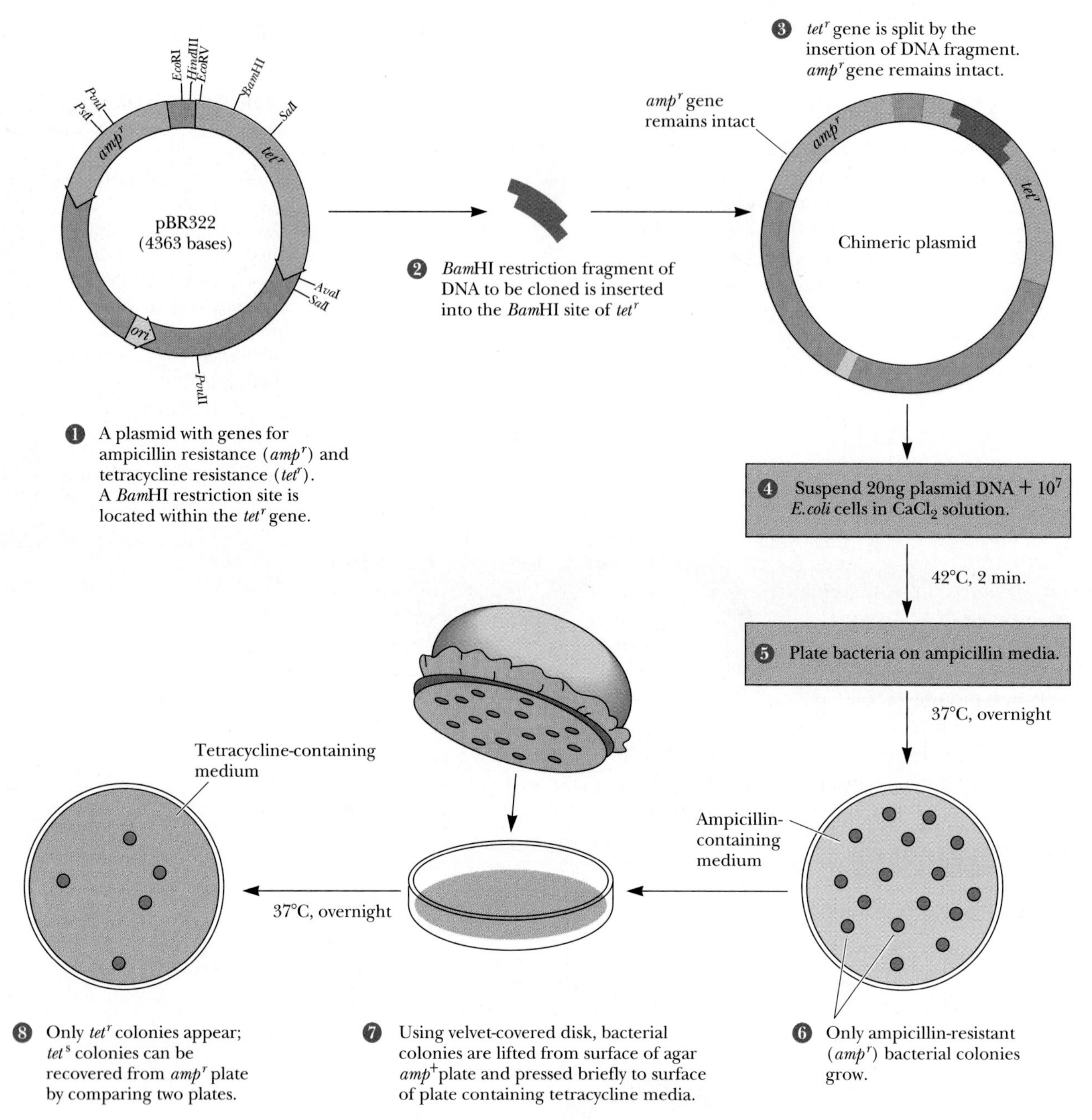

EcoRI
HindIII
EcoRV
BamHI
PvuI
PstI
SalI
amp^r
tet^r
pBR322
(4363 bases)
AvaI
SalI
ori
PvuII
1 A plasmid with genes for ampicillin resistance (amp^r) and tetracycline resistance (tet^r). A BamHI restriction site is located within the tet^r gene.
2 BamHI restriction fragment of DNA to be cloned is inserted into the BamHI site of tet^r
3 tet^r gene is split by the insertion of DNA fragment. amp^r gene remains intact.
amp^r gene remains intact
Chimeric plasmid
4 Suspend 20ng plasmid DNA + 10^7 E.coli cells in CaCl2 solution.
42°C, 2 min.
5 Plate bacteria on ampicillin media.
37°C, overnight
Ampicillin-containing medium
Tetracycline-containing medium
37°C, overnight
6 Only ampicillin-resistant (amp^r) bacterial colonies grow.
7 Using velvet-covered disk, bacterial colonies are lifted from surface of agar amp+plate and pressed briefly to surface of plate containing tetracycline media.
8 Only tet^r colonies appear; tet^s colonies can be recovered from amp^r plate by comparing two plates.

***Figure* 8.7** A typical bacterial transformation experiment. Here the plasmid pBR322 is the cloning vector. (1) Cleavage of pBR322 with restriction enzyme *Bam*H1, followed by (2) annealing and ligation of inserts generated by *Bam*H1 cleavage of some foreign DNA, (3) creates a chimeric plasmid. (4) The chimeric plasmid is then used to transform Ca^{2+}-treated heat-shocked *E. coli* cells, and the bacterial sample is plated on a petri plate. (5) Following incubation of the petri plate overnight at 37°C, (6) colonies of *amp*^{r} bacteria will be evident. (7) Replica plating these bacteria on plates of tetracycline-containing media (8) reveals which colonies are *tet*^{r} and which are tetracycline sensitive (*tet*^{s}). Only the *tet*^{s} colonies possess plasmids with foreign DNA inserts.

maximized. These plasmids have only a single origin of replication, so the time necessary for complete replication depends on the size of the plasmid. Under selective pressure in a growing culture of bacteria, overly large plasmids are prone to delete any nonessential ''genes,'' such as any foreign inserts. Such deletion would confound the aims of most cloning strategies. The useful upper limit on cloned inserts in plasmids is about 10 kbp. Many eukaryotic genes exceed this size.

Bacteriophage λ as a Cloning Vector

The genome of bacteriophage λ (lambda) (Figure 8.8) is a 48.5-kbp linear DNA molecule that is packaged into the phage head. The middle one-third of this genome is not essential to phage infection, so λ phage DNA has been engineered so that foreign DNA molecules up to 16 kbp can be inserted into this region for cloning purposes. *In vitro* packaging systems are then used to package the chimeric DNA into phage heads which, when assembled with phage tails, form infective phage particles. Bacteria infected with these recombinant phage produce large numbers of phage progeny before they lyse, and large amounts of recombinant DNA can be easily purified from the lysate.

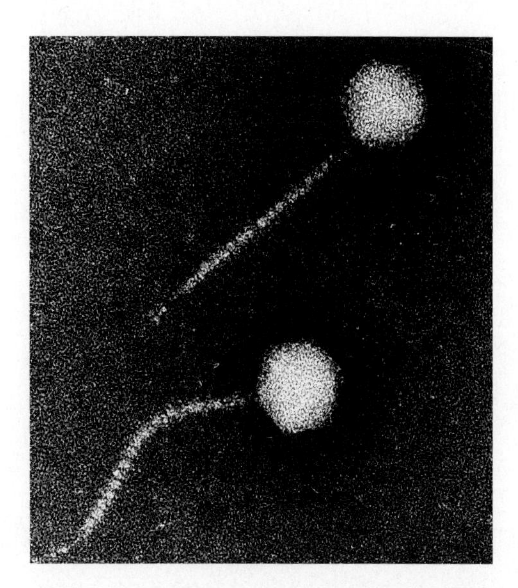

***Figure* 8.8** Electron micrograph of bacteriophage λ.

Cosmids

The DNA incorporated into phage heads by bacteriophage λ packaging systems must meet only limited specifications. It must possess a 14-bp sequence known as *cos* (which stands for *co*hesive end *s*ite) at each of its ends, and these *cos* sequences must be separated by no fewer than 36 kbp and no more than 51 kbp of DNA. Essentially any DNA satisfying these minimal requirements will be packaged and assembled into an infective phage particle. Other cloning features such as an *ori,* selectable markers, and a polylinker are joined to the *cos* sequence so that the cloned DNA can be propagated and selected in host cells. These features have been achieved by placing *cos* sequences on either side of cloning sites in plasmids to create **cosmid vectors** that are capable of carrying DNA inserts about 40 kbp in size (Figure 8.9). Since cosmids lack essential phage genes, they reproduce in host bacteria as plasmids.

Vectors from Filamentous Phage

Filamentous phage particles are single-stranded circular DNA molecules about 6.4 kbp long, packaged into long, thin tubes constructed from thousands of protein monomers. A set of vectors derived from the filamentous phage **M13** resemble plasmids in having one or more cloning sites, a selectable marker, and an origin of replication. Foreign DNA inserted into the cloning site can be recovered in either single-stranded or double-stranded forms since M13-derived vectors replicating in host bacteria give rise to both double-stranded circles as well as single-stranded circular copies of one of the original DNA strands. Bacteria transformed with M13-derived recombinant DNA continue to grow slowly, continually secreting phage particles at the rate of about 200 phage per cell per generation. These particles are an excellent source of single-stranded DNA for DNA sequencing or for *in vitro* mutagenesis (see Section 8.3).

Shuttle Vectors

Shuttle vectors are plasmids capable of propagating and transferring ("shuttling") genes between two different organisms, one of which is typically a prokaryote (*E. coli*) and the other a eukaryote (yeast). Shuttle vectors must have unique origins of replication for each cell type as well as different markers for selection of transformed host cells harboring the vector (Figure 8.10). Shuttle vectors have the advantage that eukaryotic genes can be cloned in bacterial hosts, yet the expression of these genes can be analyzed in appropriate eukaryotic backgrounds.

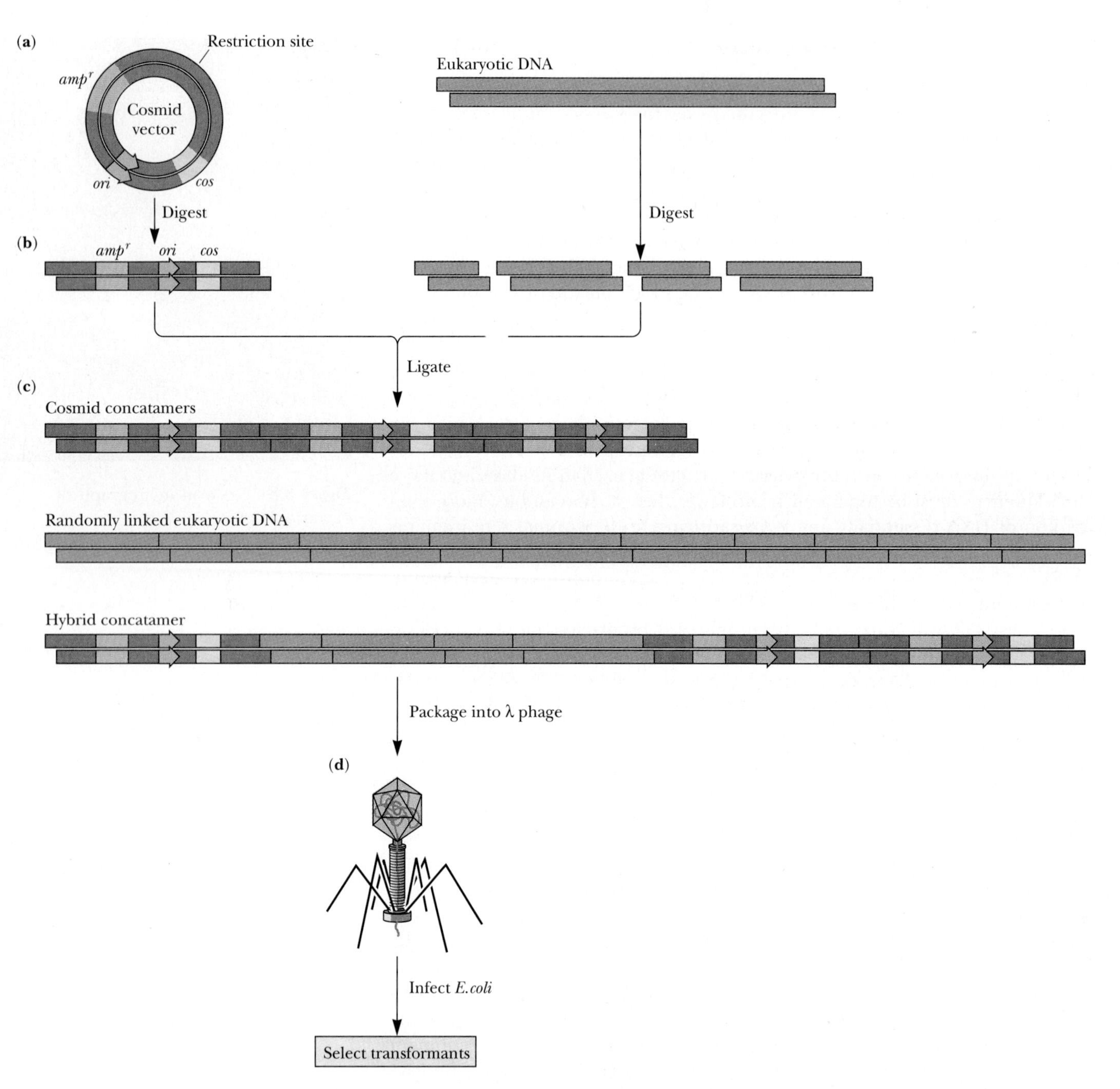

***Figure* 8.9** Cosmid vectors for cloning large DNA fragments. (a) Cosmid vectors are plasmids that carry a selectable marker such as *amp*^{r}, an origin of replication (*ori*), a polylinker suitable for insertion of foreign DNA, and (b) a *cos* sequence. Both the plasmid and the foreign DNA to be cloned are cut with a restriction enzyme and the two DNAs are then ligated together. (c) The ligation reaction leads to the formation of hybrid concatamers, molecules in which plasmid sequences and foreign DNAs are linked in series in no particular order. The bacteriophage λ packaging extract contains the restriction enzyme that recognizes *cos* sequences and cleaves at these sites. (d) DNA molecules of the proper size (36 to 51 kbp) are packaged into phage heads, forming infective phage particles. (e) The *cos* sequence is

↓
5′-TACG**GGGCGGCGACCT**CGCG-3′
3′-ATGC**CCCGCCGCTGGA**GCGC-5′
↑

Endonuclease cleavage at the sites indicated by arrows leaves 12-bp cohesive ends.
(a-d, Adapted from Figure 1.10.7, Ausubel, et al., eds., 1987. Current Protocols in Molecular Biology. New York: John Wiley and Sons; e, from Figure 4, Murialdo, 1991, Annual Review of Biochemistry 60:136.)

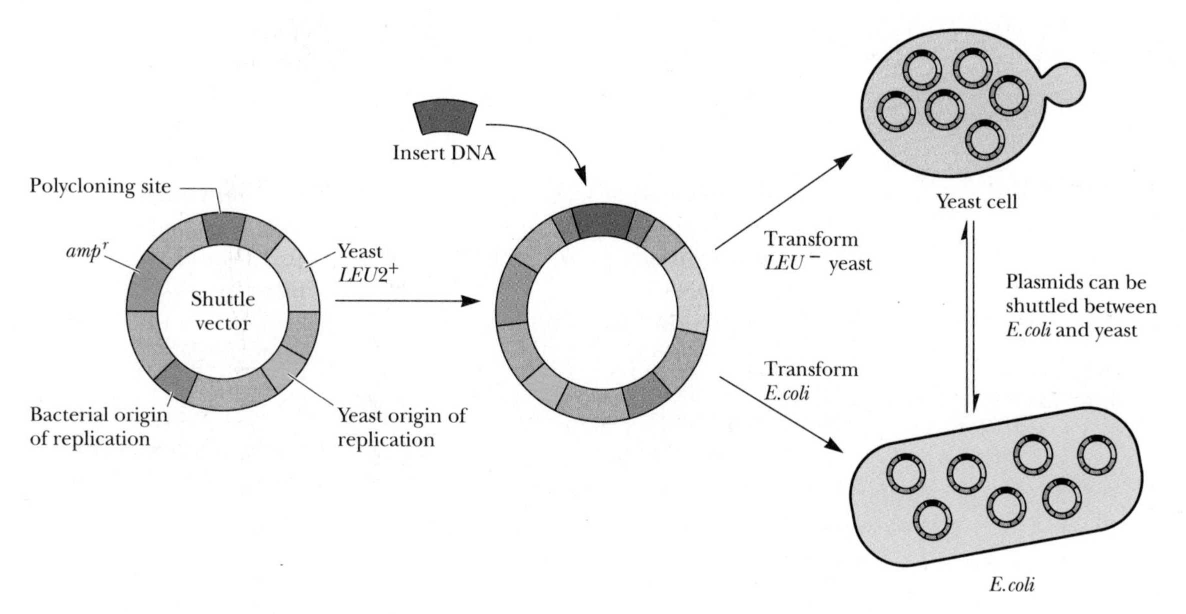

***Figure* 8.10** A typical shuttle vector. This vector has both yeast and bacterial origins of replication, amp^r (ampicillin resistance gene for selection in *E. coli*) and $LEU2^+$, a gene in the yeast pathway for leucine biosynthesis. The recipient yeast cells are $LEU2^-$ (defective in this gene) and thus require leucine for growth. $LEU2^-$ yeast cells transformed with this shuttle vector can be selected on medium lacking any leucine supplement.

(Adapted from Figure 19-5, Watson, J. D., et al., 1987. The Molecular Biology of the Gene. Menlo Park, CA: Benjamin-Cummings.)

Artificial Chromosomes

DNA molecules several hundred kilobase pairs in length have been successfully propagated in yeast by creating **yeast artificial chromosomes** or **YACs.** For these large DNAs to be replicated in the yeast cell, YAC constructs must include not only an origin of replication (known in yeast terminology as an *autonomously replicating sequence* or *ARS*) but also a centromere and telomeres. Recall that centromeres provide the site for attachment of the chromosome to the spindle during mitosis and meiosis. Telomeres are nucleotide sequences defining the ends of chromosomes. Telomeres are essential for proper replication of the chromosome.

8.2 DNA Libraries

A DNA library is a set of cloned fragments that collectively represent the genes of a particular organism. Particular genes can be isolated from DNA libraries, much as books can be obtained from conventional libraries. The secret is knowing where and how to look.

Genomic Libraries

Any particular gene constitutes only a small part of an organism's genome. For example, if the organism is a mammal whose entire genome encompasses some 10^6 kbp and the gene is 10 kbp, then the gene represents only 0.001% of the total nuclear DNA. It is impractical to attempt to recover such rare sequences directly from isolated nuclear DNA because of the overwhelming amount of extraneous DNA sequences. Instead, a **genomic library** is prepared by isolating total DNA from the organism, digesting it into fragments of suitable size, and cloning the fragments into an appropriate vector. This approach is called *shotgun cloning* since the strategy has no way of targeting a particular gene but instead seeks to clone all the genes of the organism at one time. The intent is that at least one recombinant clone will contain at least part of the gene of interest. Usually, the isolated DNA is only partially digested by the chosen restriction endonuclease so that not every restriction site is cleaved in every DNA molecule. Then, even if the gene of interest contains a susceptible restriction site, some intact genes might still be found in the digest. Genomic libraries have been prepared from hundreds of different species.

Many clones must be created to be confident that the genomic library contains the gene of interest. The probability, P, that some number of clones, N, contains a particular fragment representing a fraction, f, of the genome is

$$P = 1 - (1 - f)^N$$

Thus,

$$N = \ln (1 - P)/\ln (1 - f)$$

For example, if the library consists of 10-kbp fragments of the *E. coli* genome (4720 kbp total), over 2000 individual clones must be screened to have a 99% probability ($P = 0.99$) of finding a particular fragment. (Since $f = 10/4720 = 0.0021$ and $P = 0.99$, $N = 2193$). For a 99% probability of finding a particular sequence within the 3×10^6 kbp human genome, N would equal almost 1.4 million if the cloned fragments averaged 10 kbp in size. The need for cloning vectors capable of carrying very large DNA inserts becomes obvious from these numbers.

Screening Libraries

A common method of screening plasmid-based genomic libraries is to carry out a **colony hybridization experiment.** The protocol is similar for phage-based libraries except that bacteriophage plaques, not bacterial colonies, are screened. In a typical experiment, host bacteria containing either a plasmid-based or bacteriophage-based library are plated out on a petri dish and allowed to grow overnight to form colonies (or in the case of phage libraries, plaques) (Figure 8.11). A replica of the bacterial colonies (or plaques) is then obtained by overlaying the plate with a nitrocellulose disc. The disc is removed, treated with alkali to dissociate bound DNA duplexes into single-stranded DNA, dried, and placed in a sealed bag with labeled probe (see the box on Southern blotting). If the probe DNA is duplex DNA, it must be denatured by heating at 70°C. The probe and target DNA complementary sequences must be in a single-stranded form if they are to hybridize with one another. Any DNA sequences compiementary to probe DNA will be revealed by autoradiography of the nitrocellulose disc. Bacterial colonies (phage plaques) containing clones bearing target DNA are identified on the film and can be recovered from the master plate.

Probes for Southern Hybridization

Clearly, specific probes are essential reagents if the goal is to identify a particular gene against a background of innumerable DNA sequences. Usually, the probes that are used to screen libraries are nucleotide sequences that are complementary to some part of the target gene. To make useful probes re-

(Text continues on page 262)

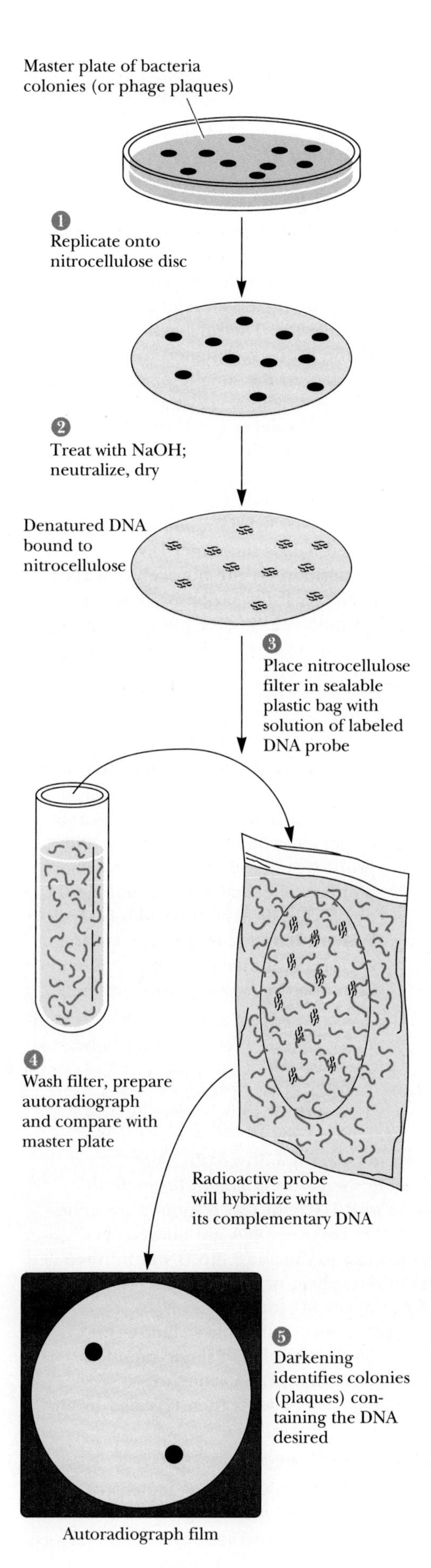
Master plate of bacteria colonies (or phage plaques)
1 Replicate onto nitrocellulose disc
2 Treat with NaOH; neutralize, dry
Denatured DNA bound to nitrocellulose
3 Place nitrocellulose filter in sealable plastic bag with solution of labeled DNA probe
4 Wash filter, prepare autoradiograph and compare with master plate
Radioactive probe will hybridize with its complementary DNA
5 Darkening identifies colonies (plaques) containing the DNA desired
Autoradiograph film

Figure **8.11** Screening a genomic library by colony hybridization (or plaque hybridization). Host bacteria transformed with a plasmid-based genomic library or infected with bacteriophage-based genomic library are plated on a petri plate and incubated overnight to allow bacterial colonies (or phage plaques) to form. A replica of the bacterial colonies (or plaques) is then obtained by overlaying the plate with a nitrocellulose disc (1). Nitrocellulose strongly binds nucleic acids; single-stranded nucleic acids are bound more tightly than double-stranded nucleic acids. Once the nitrocellulose disc has taken up an impression of the bacterial colonies (or plaques), it is removed and the petri plate is set aside and saved. The disc is treated with 2 *M* NaOH, neutralized, and dried (2). NaOH both lyses any bacteria (or phage particles) and dissociates the DNA strands. When the disc is dried, the DNA strands become immobilized on the filter. The dried disc is placed in a sealable plastic bag and a solution containing heat-denatured (single-stranded) labeled probe is added (3). The bag is incubated to allow annealing of the probe DNA to any target DNA sequences that might be present on the nitrocellulose. The filter is then washed, dried, and placed on a piece of X-ray film to obtain an autoradiogram (4). The position of any spots on the X-ray film reveals where the labeled probe has hybridized with target DNA (5). The location of these spots can be used to recover the genomic clone from the bacteria (or plaques) on the original petri plate.

Critical Developments in Biochemistry

Identifying Specific DNA Sequences by Southern Blotting (Southern Hybridization)

Any given DNA fragment is unique solely by virtue of its specific nucleotide sequence. The only practical way to find one particular DNA segment among a vast population of different DNA fragments (such as you might find in genomic DNA preparations) is to exploit its sequence specificity to identify it. In 1975, E. M. Southern invented a technique capable of doing just that.

Electrophoresis

Southern first fractionated a population of DNA fragments according to size by gel electrophoresis (see step 2 in figure). The electrophoretic mobility of a nucleic acid is inversely proportional to its molecular mass. Polyacrylamide gels are suitable for separation of nucleic acids of 25 to 2000 bp. Agarose gels are better if the DNA fragments range up to 10 times this size. Most preparations of genomic DNA show a broad spectrum of sizes, from less than 1 kbp to more than 20 kbp. Typically, no discrete-size fragments are evident following electrophoresis, just a "smear" of DNA throughout the gel, from the electrophoretic origin to the front of migration.

Blotting

Once the fragments have been separated by electrophoresis (step 3), the gel is soaked in a solution of NaOH. Alkali denatures duplex DNA, converting it to single-stranded DNA. After the pH of the gel is adjusted to neutrality with buffer, a sheet of nitrocellulose soaked in a concentrated salt solution is then placed over the gel (c), and salt solution is drawn through the gel in a direction perpendicular to the direction of electrophoresis (step 4). The salt solution is pulled through the gel in one of three ways: capillary action (*blotting*), suction (*vacuum blotting*), or electrophoresis (*electroblotting*). The movement of salt solution through the gel carries the DNA to the nitrocellulose sheet. Nitrocellulose binds single-stranded DNA molecules very tightly, effectively immobilizing them in place on the sheet.[2] Note that the distribution pattern of the electrophoretically separated DNA is maintained when the single-stranded DNA molecules bind to the nitrocellulose sheet (step 5 in figure). Next, the nitrocellulose is dried by baking in a vacuum oven;[3] baking tightly fixes the single-stranded DNAs to the nitrocellulose. Next, in the *prehybridization step,* the nitrocellulose sheet is incubated with a solution containing protein (serum albumin, for example) and/or a detergent such as sodium dodecyl sulfate. The protein and detergent molecules saturate any remaining binding sites for DNA on the nitrocellulose. Thus, no more DNA can bind nonspecifically to the nitrocellulose sheet.

Hybridization

To detect a particular DNA within the electrophoretic smear of countless DNA fragments, the prehybridized nitrocellulose sheet is incubated in a sealed plastic bag with a solution of specific probe molecules (step 6 in figure). A **probe** is usually a single-stranded DNA of defined sequence that is distinctively labeled, either with a radioactive isotope (such as ^{32}P) or some other easily detectable tag. The nucleotide sequence of the probe is designed to be complementary to the sought-for or *target* DNA fragment. The single-stranded probe DNA will **anneal** with the single-stranded target DNA bound to the nitrocellulose through specific base pairing to form a DNA duplex. This annealing, or **hybridization** as it is usually called, labels the target DNA, revealing its position on the nitrocellulose. For example, if the probe is ^{32}P-labeled, its location can be detected by autoradiographic exposure of a piece of X-ray film laid over the nitrocellulose sheet (step 7 in figure).

Southern's procedure has been extended to the identification of specific RNA and protein molecules. In a play on Southern's name, the identification of particular RNAs following separation by gel electrophoresis, blotting, and probe hybridization is called **Northern blotting.** The analogous technique for identifying protein molecules is termed **Western blotting.** In Western blotting, the probe of choice is usually an antibody specific for the target protein.

The Southern blotting technique involves the transfer of electrophoretically separated DNA fragments to a nitrocellulose sheet and subsequent detection of specific DNA sequences. A preparation of DNA fragments [typically a restriction digest, (1)] is separated according to size by gel electrophoresis (2). The separation pattern can be visualized by soaking the gel in ethidium bromide to stain the DNA and then illuminating the gel with UV light (3). Ethidium bromide molecules intercalated between the hydrophobic bases of DNA are fluorescent under UV light. The gel is soaked in strong alkali to denature the DNA and then neutralized in buffer. Next, the gel is placed on a sheet of nitrocellulose, and concentrated salt solution is passed through the gel (4) to carry the DNA fragments out of the gel where they are bound tightly to the nitrocellulose (5). Incubation of the nitrocellulose sheet with a solution of labeled single-stranded probe DNA (6) allows the probe to hybridize with target DNA sequences complementary to it. The location of these target sequences is then revealed by an appropriate means of detection, such as autoradiography (7).

[2]The underlying cause of DNA binding to nitrocellulose is not clear, but probably involves a combination of hydrogen bonding, hydrophobic interactions, and salt bridges.

[3]Vacuum drying is essential because nitrocellulose reacts violently with O_2 if heated.

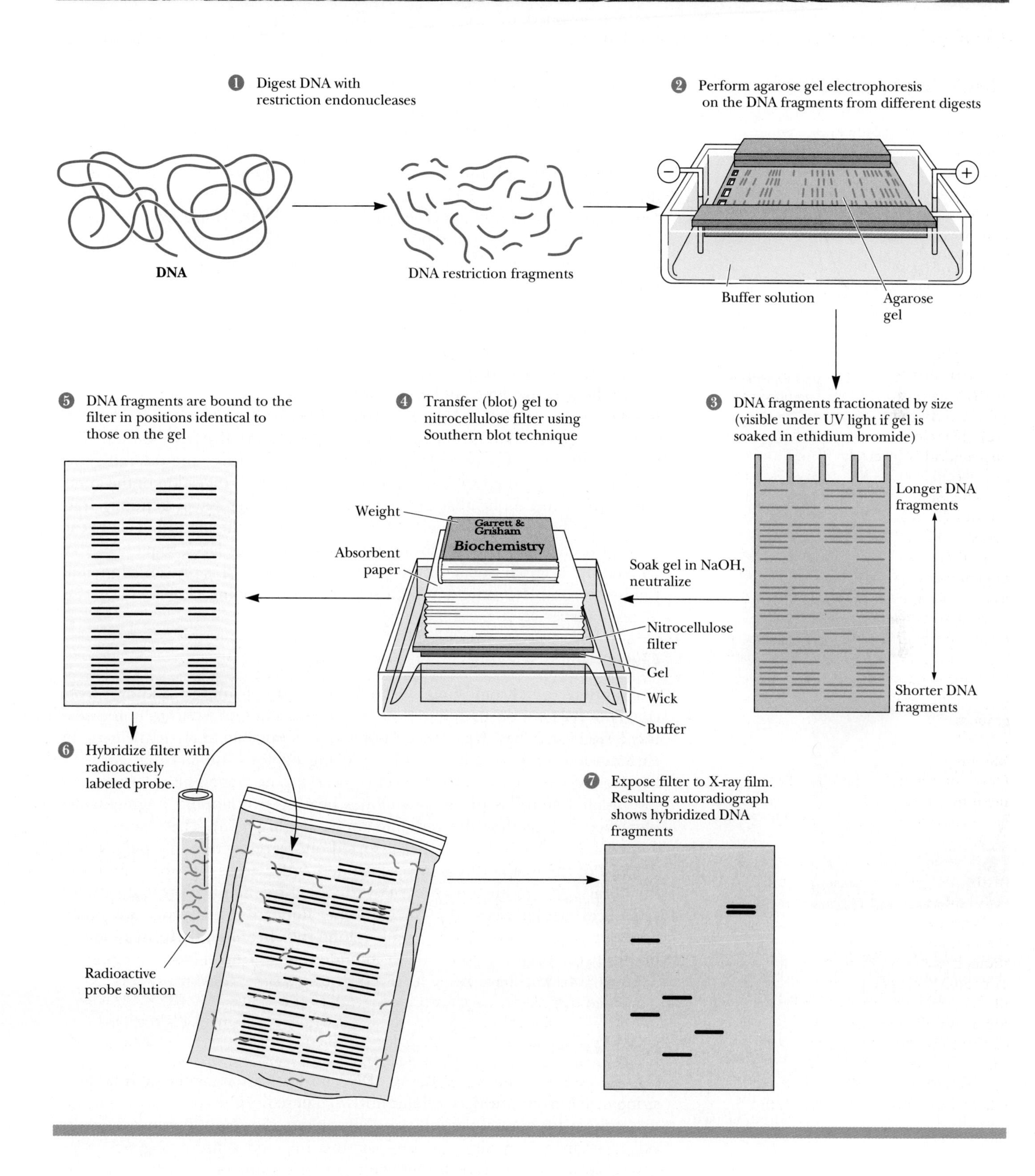
1 Digest DNA with restriction endonucleases
2 Perform agarose gel electrophoresis on the DNA fragments from different digests
DNA
DNA restriction fragments
Buffer solution
Agarose gel
5 DNA fragments are bound to the filter in positions identical to those on the gel
4 Transfer (blot) gel to nitrocellulose filter using Southern blot technique
3 DNA fragments fractionated by size (visible under UV light if gel is soaked in ethidium bromide)
Weight
Garrett & Grisham
Biochemistry
Absorbent paper
Soak gel in NaOH, neutralize
Nitrocellulose filter
Gel
Wick
Buffer
Longer DNA fragments
Shorter DNA fragments
6 Hybridize filter with radioactively labeled probe.
Radioactive probe solution
7 Expose filter to X-ray film. Resulting autoradiograph shows hybridized DNA fragments

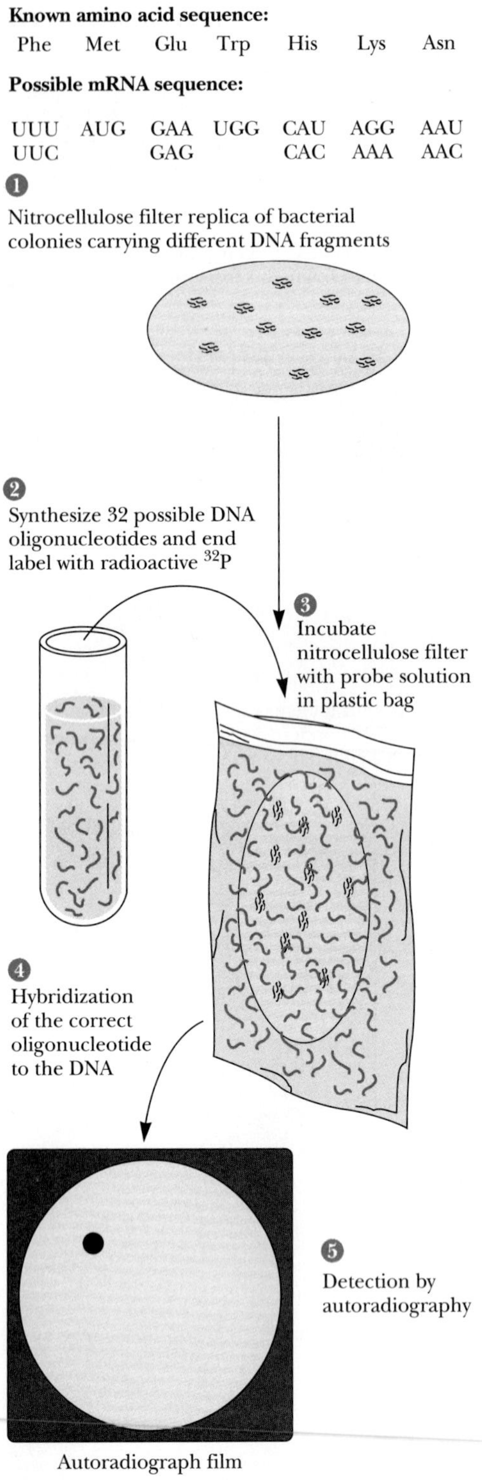

Known amino acid sequence:
Phe Met Glu Trp His Lys Asn

Possible mRNA sequence:

UUU	AUG	GAA	UGG	CAU	AGG	AAU
UUC		GAG		CAC	AAA	AAC

1 Nitrocellulose filter replica of bacterial colonies carrying different DNA fragments

2 Synthesize 32 possible DNA oligonucleotides and end label with radioactive ^{32}P

3 Incubate nitrocellulose filter with probe solution in plastic bag

4 Hybridization of the correct oligonucleotide to the DNA

5 Detection by autoradiography

Autoradiograph film

***Figure* 8.12** Cloning genes using oligonucleotide probes designed from a known amino acid sequence. A radioactively labeled set of DNA (degenerate) oligonucleotides representing all possible mRNA coding sequences are synthesized. (In this case, there are 2^5, or 32.) The complete mixture is used to probe the genomic library by colony hybridization (see Figure 8.11).

(Adapted from Figure 19-18, Watson, J. D., et al., 1987. Molecular Biology of the Gene. Menlo Park, CA: Benjamin-Cummings.)

quires some information about the gene's nucleotide sequence. Sometimes such information is available. Alternatively, if the amino acid sequence of the protein encoded by the gene is known, it is possible to work backward through the genetic code to the DNA sequence (Figure 8.12). Since the genetic code is *degenerate* (that is, several codons may specify the same amino acid; see Chapter 31), probes designed by this approach are usually **degenerate oligonucleotides** about 17 to 50 residues long (such oligonucleotides are so-called 17- to 50-mers). The oligonucleotides are synthesized so that different bases are incorporated at sites where degeneracies occur in the codons. The final preparation thus consists of a mixture of equal-length oligonucleotides whose sequences vary to accommodate the degeneracies. Presumably, one oligonucleotide sequence in the mixture will hybridize with the target gene. These oligonucleotide probes are at least 17-mers, because shorter degenerate oligonucleotides might hybridize with sequences unrelated to the target sequence.

A piece of DNA from the corresponding gene in a related organism can also be used as a probe in screening a library for a particular gene. Such probes are termed **heterologous probes** because they are not derived from the homologous (same) organism.

Problems arise if a complete eukaryotic gene is the cloning target; eukaryotic genes can be tens or even hundreds of kilobase pairs in size. Genes this size are fragmented in most cloning procedures. Thus, the DNA identified by the probe may represent a clone that carries only part of the desired gene. However, most cloning strategies are based on a partial digestion of the genomic DNA, a technique that generates an overlapping set of genomic fragments. This being so, DNA segments from the ends of the identified clone can now be used to probe the library for clones carrying DNA sequences that flanked the original isolate in the genome. Repeating this process ultimately yields the complete gene among a subset of overlapping clones.

Chromosome Walking

Quite often, the chromosomal location of a gene of interest is known from conventional genetic mapping. If some clone that lies on a chromosome near the desired gene has already been isolated, it is possible to clone the gene by **chromosome walking.** Chromosome walking exploits the fact that genomic libraries represent an enormous set of overlapping fragments. Starting with the isolated clone as probe, any clones bearing overlapping fragments are identified by screening the genomic library (Figure 8.13). The relative order and arrangement of these clones is sorted out by restriction mapping. Then, the end of the mapped region is subcloned and used as a probe to rescreen the library, generating a new set of overlapping fragments representing DNA sequences farther along the chromosome. Repeating this procedure many times allows an investigator to "walk" along the chromosome from an identified site (the original clone) to an intended gene. Chromosomal regions as large as 1000 kbp have been surveyed in this way.

cDNA Libraries

cDNAs are DNA copies of mRNAs. cDNA libraries are constructed by synthesizing cDNA from purified cellular mRNA. These libraries present an alternative strategy for gene isolation, especially eukaryotic genes. Because different cell types in eukaryotic organisms express selected subsets of genes, RNA preparations from cells or tissues where genes of interest are selectively transcribed will be enriched for the desired mRNAs. Since most eukaryotic

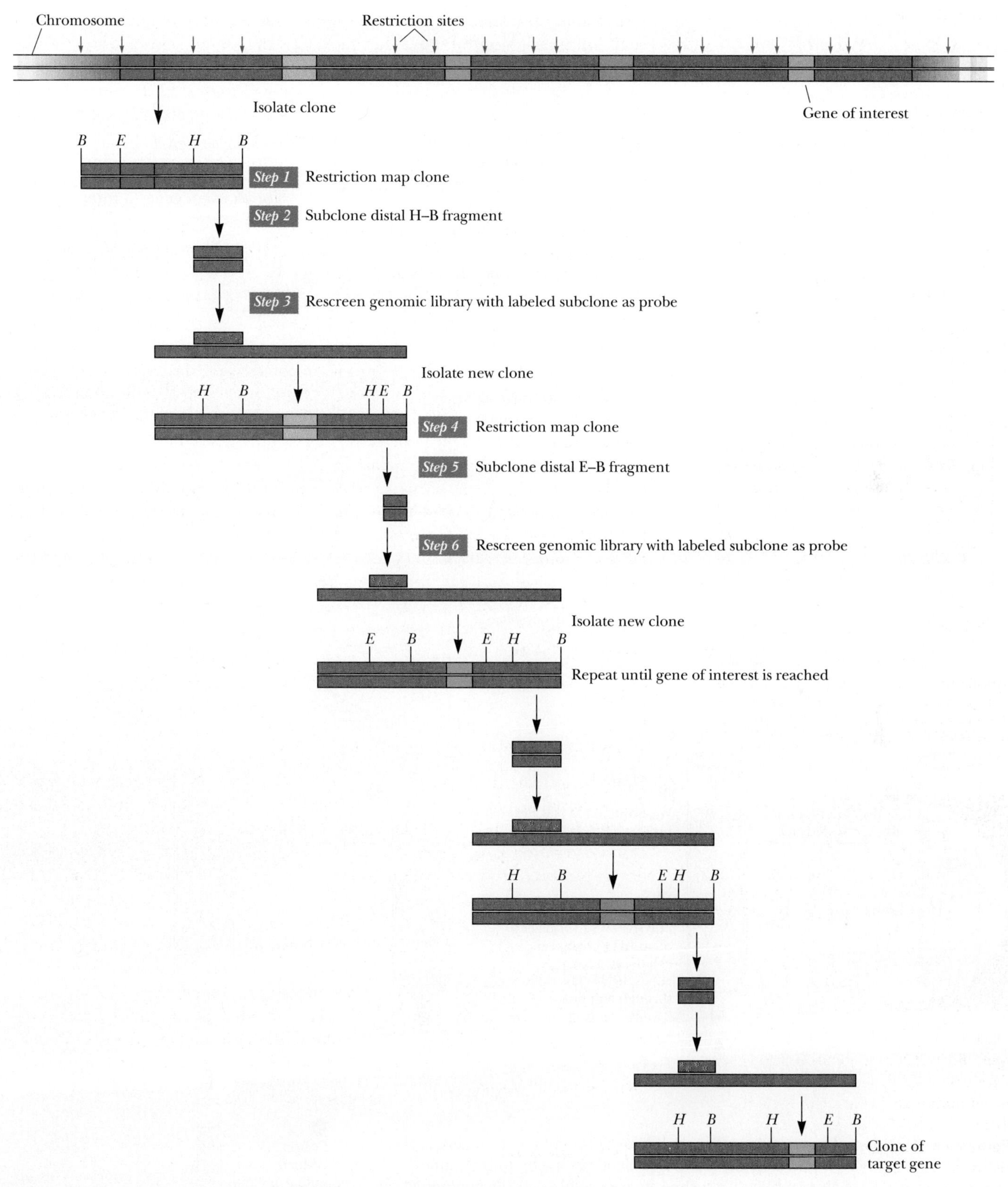

***Figure* 8.13** Chromosome walking. A cloned restriction fragment is used to screen a genomic library and identify all overlapping fragments (fragments derived from sequences neighboring the original isolate). *B, E,* and *H* identify the restriction sites on the clones for *Bam*H1, *Eco*R1, and *Hin*dIII, respectively. New clones are ordered by restriction mapping, and the distal end of the mapped region is subcloned and used as a probe to rescreen the library. The process is repeated until the desired gene has been reached.

The Human Genome Project

The Human Genome Project is a collaborative international, government- and private-sponsored effort to map and sequence the entire human genome, some 3 billion base pairs distributed among the two sex chromosomes (**X** and **Y**) and 22 **autosomes** (chromosomes that are not sex chromosomes). The first goal is to identify and map at least 3000 genetic **markers** (genes or other recognizable loci on the DNA), evenly distributed throughout the chromosomes. At the same time, determination of the entire nucleotide sequence of the human genome is under way, starting from defined positions. An ancillary part of the project is sequencing the genomes of other species (such as yeast, *Drosophila melanogaster* [the fruit fly], mice, and *Arabidopsis thaliana* [a plant]) to reveal comparative aspects of genetic and sequence organization.

A number of human diseases have been traced to genetic defects, whose positions within the human genome have been identified. Among these are:

- *cystic fibrosis* gene
- *Duchenne muscular dystrophy* gene[4] (at 2.4 megabases, the largest known gene in any organism)
- *Huntington's disease* gene
- *neurofibromatosis* gene
- *neuroblastoma* gene (a form of brain cancer)
- *amyotrophic lateral sclerosis* gene (Lou Gehrig's disease)
- *fragile X-linked mental retardation* gene[4]

as well as genes associated with the development of diabetes, breast cancer, colon cancer, and affective disorders such as *schizophrenia* and *bipolar affective disorder* (manic depression).

[4]X-chromosome linked gene. As of 1992, over 100 disease-related genes had been mapped to this chromosome.

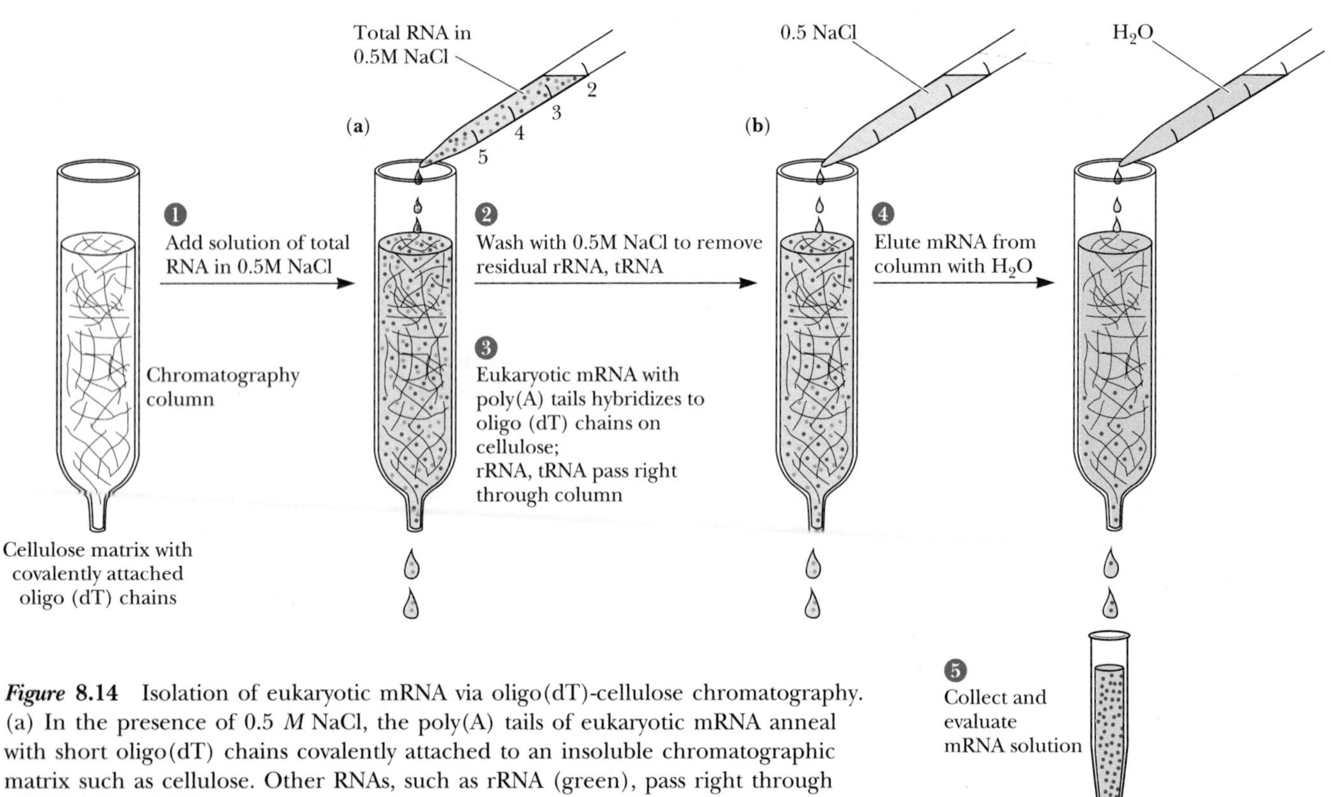

***Figure* 8.14** Isolation of eukaryotic mRNA via oligo(dT)-cellulose chromatography. (a) In the presence of 0.5 *M* NaCl, the poly(A) tails of eukaryotic mRNA anneal with short oligo(dT) chains covalently attached to an insoluble chromatographic matrix such as cellulose. Other RNAs, such as rRNA (green), pass right through the chromatography column. (b) The column is washed with more 0.5 *M* NaCl to remove residual contaminants. (c) Then the poly(A) mRNA is recovered by washing the column with water because the base pairs formed between the poly(A) tails of the mRNA and the oligo(dT) chains are unstable in solutions of low ionic strength.

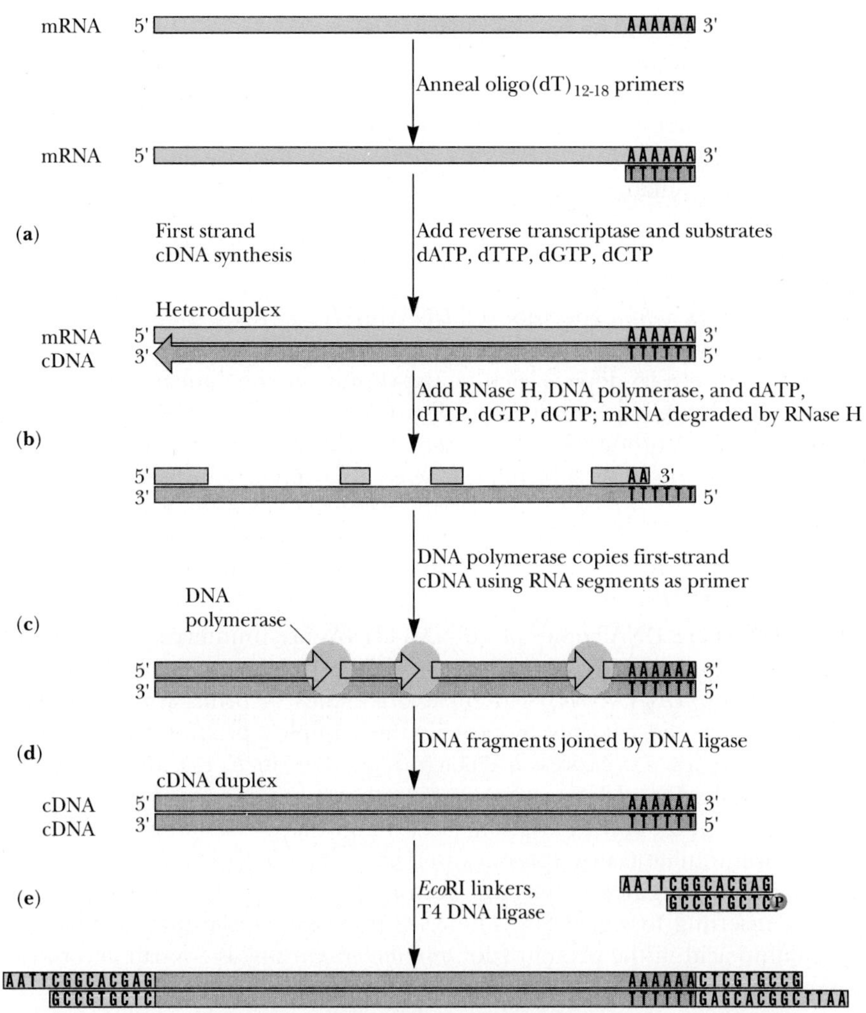

Figure **8.15** Reverse transcriptase-driven synthesis of cDNA from oligo(dT) primers annealed to the poly(A) tails of purified eukaryotic mRNA. (a) Oligo(dT) chains serve as primers for synthesis of a DNA copy of the mRNA by reverse transcriptase. Following completion of first-strand cDNA synthesis by reverse transcriptase, RNase H and DNA polymerase are added (b). RNase H specifically digests RNA strands in DNA:RNA hybrid duplexes. DNA polymerase copies the first-strand cDNA, using as primers the residual RNA segments after RNase H has created nicks and gaps (c). DNA polymerase has a $5' \rightarrow 3'$ exonuclease activity that removes the residual RNA as it fills in with DNA. The nicks remaining in the second-strand DNA are sealed by DNA ligase (d), yielding duplex cDNA. *Eco*RI adapters with 5′-overhangs are then ligated onto the cDNA duplexes (e) using phage T4 DNA ligase to create *Eco*RI-ended cDNA for insertion into a cloning vector.

mRNAs carry 3′-poly(A) tails, mRNA can be selectively isolated from preparations of total cellular RNA by oligo(dT)-cellulose chromatography (Figure 8.14). DNA copies of the purified mRNAs are synthesized by first annealing short oligo(dT) chains to the poly(A) tails. These oligo(dT) chains serve as primers for reverse transcriptase-driven synthesis of DNA (Figure 8.15). (Random oligonucleotides can also be used as primers, with the advantages being less dependency on poly(A) tracts and increased likelihood of creating clones representing the 5′-ends of mRNAs.) **Reverse transcriptase** is an enzyme that synthesizes a DNA strand, copying RNA as the template. DNA polymerase is then used to copy the DNA strand and form a double-stranded (duplex DNA) molecule. Linkers are then added to the DNA duplexes rendered from the mRNA templates, and the cDNA is cloned into a suitable vector. Once a cDNA derived from a particular gene has been identified, the cDNA becomes an effective probe for screening genomic libraries for isolation of the gene itself.

Expression Vectors

Expression vectors are engineered so that any cloned insert can be transcribed into RNA, and in many instances, even translated into protein. cDNA expression libraries can be constructed in specially designed vectors derived

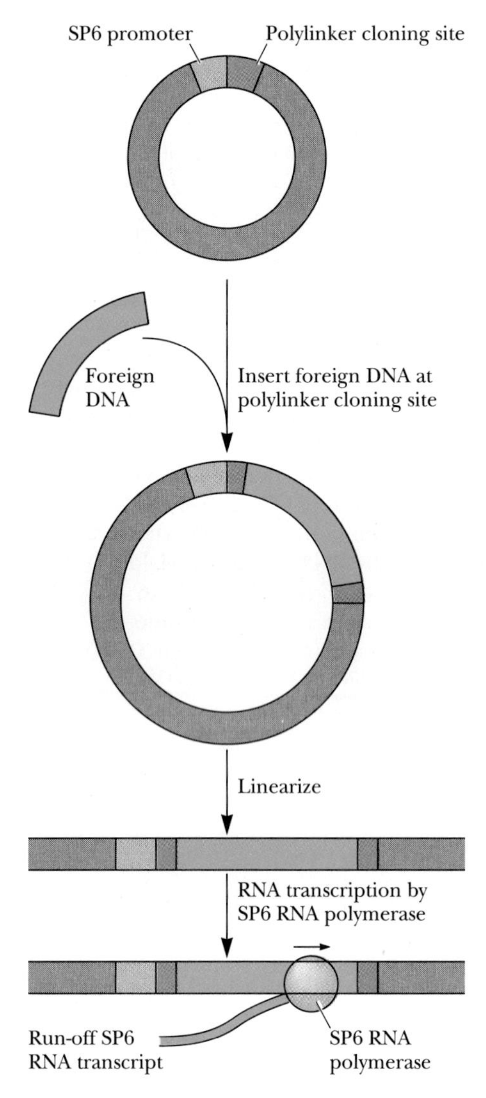

***Figure* 8.16** Expression vectors carrying the promoter recognized by the RNA polymerase of bacteriophage SP6 are useful for making RNA transcripts *in vitro.* SP6 RNA polymerase works efficiently *in vitro* and recognizes its specific promoter with high specificity. These vectors typically have a polylinker adjacent to the SP6 promoter. Successive rounds of transcription initiated by SP6 RNA polymerase at its promoter lead to the production of multiple RNA copies of any DNA inserted at the polylinker. Before transcription is initiated, the circular expression vector is linearized by a single cleavage at or near the end of the insert so that transcription terminates at a fixed point.

from either plasmids or bacteriophage λ. Proteins encoded by the various cDNA clones within such expression libraries can be synthesized in the host cells, and if suitable assays are available to identify a particular protein uniquely, its corresponding cDNA clone can be identified and isolated. Expression vectors designed for RNA expression or for protein expression, or for both, are available.

RNA Expression

A vector for *in vitro* expression of DNA inserts as RNA transcripts can be constructed by putting a highly efficient promoter adjacent to a versatile cloning site. Figure 8.16 depicts such an expression vector. Linearized recombinant vector DNA is transcribed *in vitro* using SP6 RNA polymerase. Large amounts of RNA product can be obtained in this manner; if radioactive ribonucleotides are used as substrates, labeled RNA molecules useful as probes are made.

Protein Expression

Since cDNAs are DNA copies of mRNAs, cDNAs are uninterrupted copies of the exons of expressed genes. Because cDNAs lack introns, it is feasible to express these cDNA versions of eukaryotic genes in prokaryotic hosts that otherwise lack the capacity to process the complex primary transcripts of eukaryotic genes. To express a eukaryotic protein in *E. coli,* the eukaryotic cDNA must be cloned in an *expression vector* that contains regulatory signals for both transcription and translation. Accordingly, a *promoter* where RNA polymerase will initiate transcription as well as a *ribosome binding site* to facilitate translation are engineered into the vector just upstream from the restriction site for inserting foreign DNA. The AUG initiation codon that specifies the first amino acid in the protein (the *translation start site*) is contributed by the insert (Figure 8.17).

Strong promoters have been constructed that drive the synthesis of foreign proteins to levels equal to 30% or more of total *E. coli* cellular protein. An example is the hybrid promoter, p_{tac}, which was created by fusing part of the promoter for the *E. coli* genes encoding the enzymes of lactose metabolism (the *lac* promoter) with part of the promoter for the genes encoding the enzymes of tryptophan biosynthesis (the *trp* promoter) (Figure 8.18). In cells carrying p_{tac} expression vectors, the p_{tac} promoter is not induced to drive transcription of the foreign insert until the cells are exposed to *inducers* that lead to its activation. Analogs of lactose (a β-galactoside) such as *isopropyl-β-thiogalactoside,* or **IPTG,** are excellent inducers of p_{tac}. Thus, expression of the foreign protein is easily controlled. (See Chapter 30 for detailed discussions of inducible gene expression.) The bacterial production of valuable eukaryotic proteins represents one of the most important uses of recombinant DNA technology. For example, human insulin for the clinical treatment of diabetes is now produced in bacteria.

Analogous systems for expression of foreign genes in eukaryotic cells include vectors carrying promoter elements derived from mammalian viruses, such as *simian virus 40 (SV40),* the *Epstein–Barr virus,* and the human *cytomegalovirus (CMV).* A system gaining widespread use for high level expression of foreign genes uses insect cells infected with the *baculovirus* expression vector. **Baculoviruses** infect *lepidopteran* insects (butterflies and moths). In engineered baculovirus vectors, the foreign gene is cloned downstream of the promoter for **polyhedrin,** a major viral-encoded structural protein, and the recombinant vector is incorporated into insect cells grown in culture. Expression from the polyhedrin promoter can lead to accumulation of the foreign gene product to levels as high as 500 mg/L.

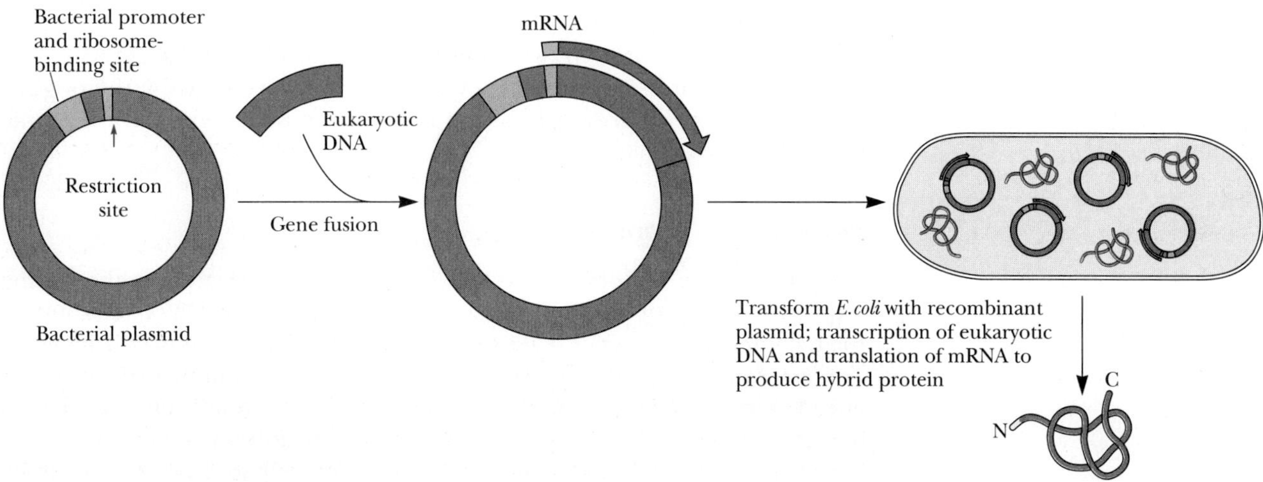

***Figure* 8.17** A typical expression-cloning vector. Eukaryotic coding sequences are inserted at the restriction site just downstream from a promoter region where RNA polymerase binds and initiates transcription. Transcription proceeds through a region encoding a bacterial ribosome-binding site and into the cloned insert. The presence of the bacterial ribosome-binding site in the RNA transcript ensures that the RNA can be translated into protein by the ribosomes of the host bacteria.

Screening cDNA Expression Libraries with Antibodies

Antibodies that specifically cross-react with a particular protein of interest are often available. If so, these antibodies can be used to screen a cDNA expression library to identify and isolate cDNA clones encoding the protein. The cDNA library is introduced into host bacteria, which are plated out and grown overnight, as in the colony hybridization scheme previously described. Nitrocellulose filters are placed on the plates to obtain a replica of the bacterial colonies. The filter is then incubated under conditions that induce protein synthesis from the cloned cDNA inserts, and the cells are treated to release the synthesized protein. The synthesized protein binds tightly to the filter,

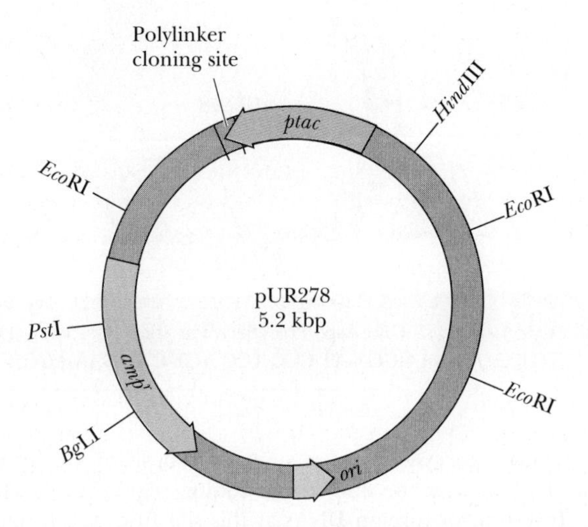

***Figure* 8.18** A p_{tac} protein expression vector contains the hybrid promoter p_{tac} derived from fusion of the *lac* and *trp* promoters. Expression from p_{tac} is more than ten times greater than expression from either the *lac* or *trp* promoter alone. Isopropyl-β-D-thiogalactoside, or IPTG, induces expression from p_{tac} as well as *lac*.

which can then be incubated with the specific antibody. Binding of the antibody to its unique protein product reveals the position of any cDNA clones expressing the protein, and these clones can be recovered from the original plate. Like other libraries, expression libraries can be screened with oligonucleotide probes, too.

Fusion Protein Expression

Some expression vectors carry cDNA inserts cloned directly into the coding sequence of a vector-borne protein-coding gene (Figure 8.19). Translation of the recombinant sequence leads to synthesis of a *hybrid protein* or *fusion protein.* The N-terminal region of the fused protein represents amino acid sequences encoded in the vector, while the remainder of the protein is encoded by the foreign insert. Keep in mind that the triplet codon sequence within the cloned insert must be in phase with codons contributed by the vector sequences to make the right protein. The N-terminal protein sequence contributed by the vector can be chosen to suit purposes. Furthermore, adding an N-terminal signal sequence that targets the hybrid protein for secretion from the cell simplifies recovery of the fusion protein. A variety of gene fusion systems have been developed to facilitate isolation of a specific protein encoded by a cloned insert. The isolation procedures are based on affinity chromatography purification of the fusion protein through exploitation of the unique ligand-binding properties of the vector-encoded protein (Table 8.1).

β-Galactosidase and Blue or White Selection

One version of these fusion protein expression vectors places the cloning site at the end of the coding region of the protein *β-galactosidase,* so that among other things, the fusion protein is attached to *β*-galactosidase and can be

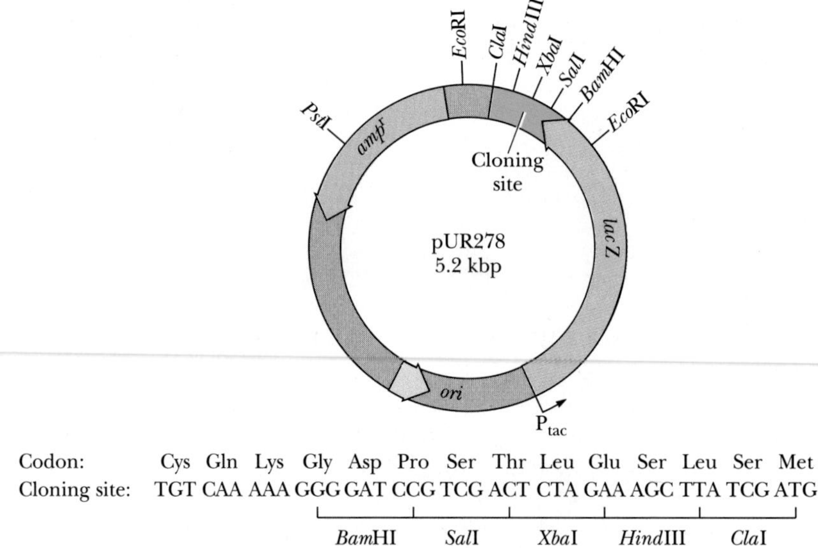

***Figure* 8.19** A typical expression vector for the synthesis of a hybrid protein. The cloning site is located at the end of the coding region for the protein *β*-galactosidase. Insertion of foreign DNAs at this site fuses the foreign sequence to the *β*-galactosidase coding region (the *lacZ* gene). IPTG induces the transcription of the *lacZ* gene from its promoter p_{lac}, causing expression of the fusion protein.

(Adapted from Figure 1.5.4, Ausubel, F. M., et al., 1987. Current Protocols in Molecular Biology. New York: John Wiley and Sons.)

***Table* 8.1**

Gene Fusion Systems for Isolation of Cloned Fusion Proteins

Gene Product	Origin	Molecular Mass (kD)	Secreted?*	Affinity Ligand
β-Galactosidase	*E. coli*	116	No	*p*-Aminophenyl-β-D-thiogalactoside (APTG)
Protein A	*S. aureus*	31	Yes	Immunoglobulin G (IgG)
Chloramphenicol acetyltransferase (CAT)	*E. coli*	24	Yes	Chloramphenicol
Streptavidin	*Streptomyces*	13	Yes	Biotin
Glutathione-S-transferase (GST)	*E. coli*	26	No	Glutathione
Maltose-binding protein (MBP)	*E. coli*	40	Yes	Starch

*This indicates whether combined secretion–fusion gene systems have led to secretion of the protein product from the cells, which simplifies its isolation and purification.

(Adapted from Uhlen, M., and Moks, T., 1990. Gene fusions for purpose of expression: an introduction. Methods in Enzymology ***185****:129–143.)*

recovered by purifying the β-galactosidase activity. Alternatively, placing the cloning site within the β-galactosidase coding region means that cloned inserts disrupt the β-galactosidase amino acid sequence, inactivating its enzymatic activity. This property has been exploited in developing a visual screening protocol that distinguishes those clones in the library that bear inserts from those that lack them.

Cells that have been transformed with a plasmid-based β-galactosidase expression cDNA library (or infected with a similar library constructed in a bacteriophage λ–based β-galactosidase fusion vector) are plated on media containing *5-bromo-4-chloro-3-indolyl-β-D-galactopyranoside,* or **X-gal** (Figure 8.20). X-gal is a *chromogenic substrate,* a colorless substance that upon enzymatic reaction yields a colored product. Following induction with IPTG, bacterial colonies (or plaques) harboring vectors in which the β-galactosidase gene is intact (those vectors lacking inserts) will express an active β-galactosidase that cleaves X-gal, liberating the 5-bromo-4-chloro-indoxyl, which dimerizes to form an indigo blue product. These blue colonies (or plaques) represent clones that lack inserts. The β-galactosidase gene is inactivated in clones with inserts, so those colonies (or plaques) that remain "white" (actually, colorless) are recombinant clones.

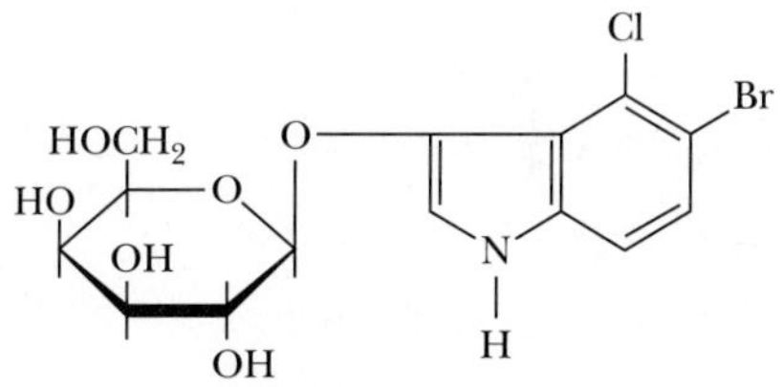

***Figure* 8.20** The structure of 5-bromo-4-chloro-3-indolyl-β-D-galactopyranoside, or X-gal.

Reporter Gene Constructs

Potential regulatory regions of genes (such as promoters) can be investigated by placing these regulatory sequences into plasmids upstream of a gene, called a **reporter gene,** whose expression is easy to measure. Such chimeric plasmids are then introduced into cells of choice (including eukaryotic cells) to assess the potential function of the nucleotide sequence in regulation since expression of the reporter gene serves as a report on the effectiveness of the regulatory element. A number of different genes have been used as reporter genes, including the *lacZ* gene, the *CAT* gene (which encodes **c**hloramphenicol **a**cetyl**t**ransferase) (Figure 8.21), and the firefly gene encoding *luciferase* (Figure 8.22). The action of the enzyme luciferase on its substrate *luciferin* is the biochemical basis of *bioluminescence* (the emission of light by biological systems). Since exceedingly low levels of light emission can be detected electronically using photomultiplier tubes, the luciferase–luciferin assay provides an exquisitely sensitive reporter system.

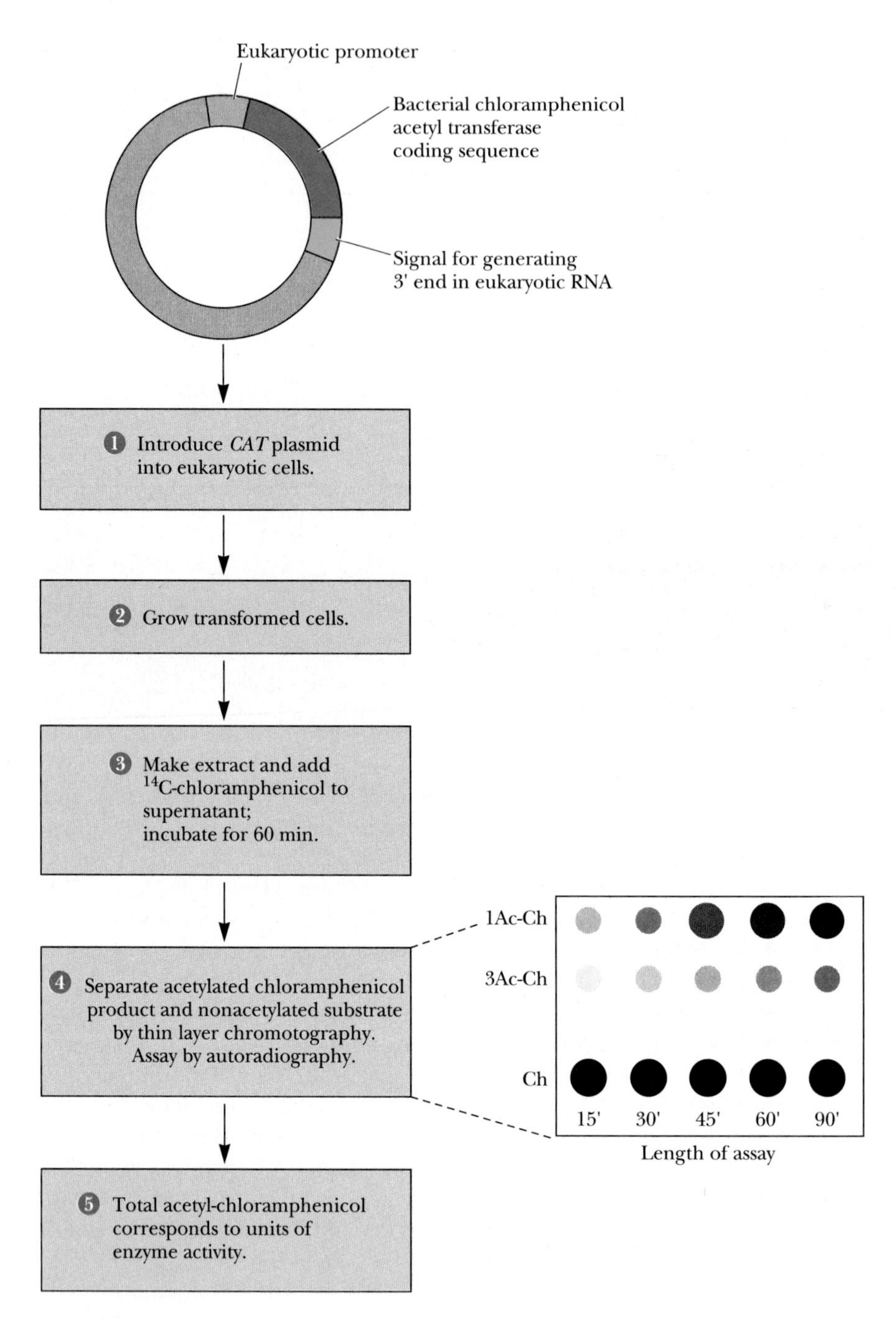
Eukaryotic promoter
Bacterial chloramphenicol acetyl transferase coding sequence
Signal for generating 3' end in eukaryotic RNA
1 Introduce CAT plasmid into eukaryotic cells.
2 Grow transformed cells.
3 Make extract and add 14C-chloramphenicol to supernatant; incubate for 60 min.
4 Separate acetylated chloramphenicol product and nonacetylated substrate by thin layer chromotography. Assay by autoradiography.
5 Total acetyl-chloramphenicol corresponds to units of enzyme activity.
1Ac-Ch
3Ac-Ch
Ch
15'
30'
45'
60'
90'
Length of assay

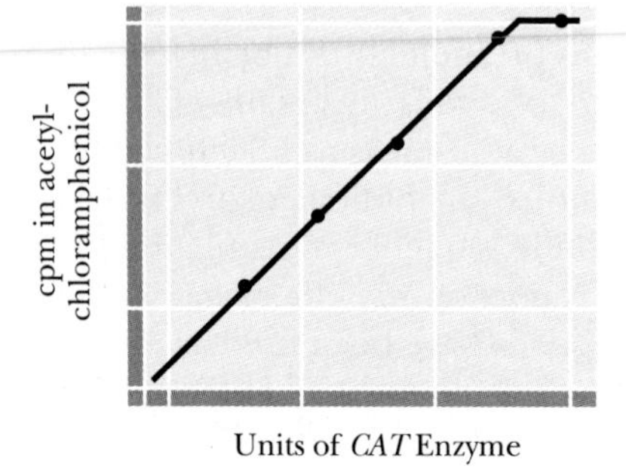
cpm in acetyl-chloramphenicol
Units of CAT Enzyme

***Figure* 8.21** The use of reporter genes, in this case, *CAT,* to assay the activity of gene regulatory sequences such as promoters. A eukaryotic promoter to be tested for its potential to drive gene expression is cloned upstream of the *CAT* gene, and the recombinant plasmid is introduced into cells. After the cells have grown, they are collected and extracts are prepared and assayed for the presence of chloramphenicol acetyltransferase activity by determining the amount of 1-acetylchloramphenicol and 3-acetylchloramphenicol formed. This *CAT* activity is proportional to the amount of chloramphenicol acetyltransferase in the cell extract and serves as an index of the eukaryotic promoter's efficiency in expressing the downstream *CAT* gene.
(Adapted from Figure 23-15, Lewin, Genes IV. Cambridge, MA: Cell Press.)

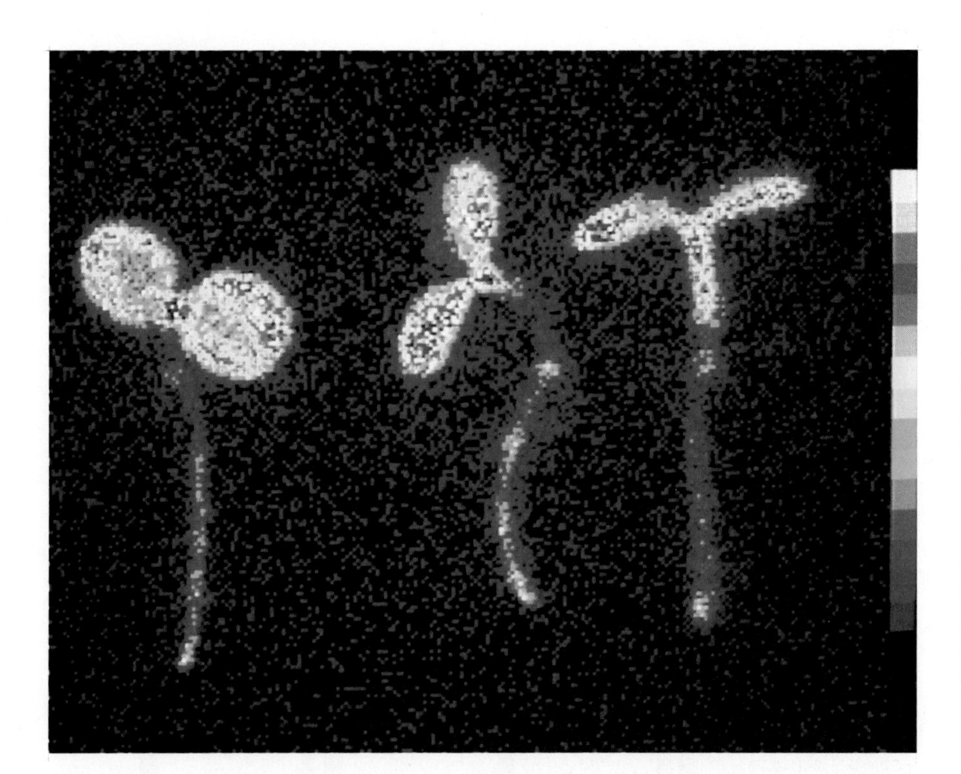

Figure **8.22** Bioluminescence is the consequence of luciferase action on luciferin. The firefly luciferase gene (*Luc*) is used here as a reporter gene. *Luc* was placed downstream of the promoter for the plant *cab2* gene. The *cab2* gene encodes a chlorophyll *a/b* binding protein, an essential component of the light-harvesting system of plants. The recombinant DNA construct was then transformed into plants, and seeds of plants carrying the construct were collected and germinated. Seedlings were then sprayed with firefly luciferin, the substrate for the enzyme luciferase, *and the intensity of bioluminescent light emission was digitally transformed into color (with white-to-red denoting the greatest light emission, as indicated by the spectral bar scale on the right).* The figure shows that light emission is confined *principally* to the cotyledons of the seedling, indicating that expression of *cab* genes occurs specifically within photosynthetic tissue, such as these primordial leaves. *(Courtesy of Steve A. Kay, University of Virginia.)*

8.3 Polymerase Chain Reaction (PCR)

Polymerase chain reaction or **PCR** is a technique for dramatically amplifying the amount of a specific DNA segment. A preparation of denatured DNA containing the segment of interest serves as template for DNA polymerase, and two specific oligonucleotides serve as primers for DNA synthesis. These primers, designed to be complementary to the two 3′-ends of the specific DNA segment to be amplified, are added in excess amounts of 1000 times or greater (Figure 8.23). They prime the DNA polymerase-catalyzed synthesis of the two complementary strands of the desired segment, effectively doubling its concentration in the solution. Then the DNA is heated to dissociate the DNA duplexes and then cooled so that primers bind to both the newly formed and the old strands. Another cycle of DNA synthesis ensues. The reaction is limited by the efficiency of DNA polymerase to segments 10 kbp or less in size. The protocol has been automated through the invention of **thermal cyclers** that alternately heat the reaction mixture to 95°C to dissociate the DNA, followed by cooling, annealing of primers, and another round of DNA synthesis. The isolation of heat-stable DNA polymerases from thermophilic bacteria (such as the *Taq* DNA polymerase from *Thermus aquaticus*) has made it unnecessary to add fresh enzyme for each round of synthesis. Since the amount of target DNA theoretically doubles each round, 25 rounds would increase its concentration about 33 million times. In practice, the increase is actually more like a million times, which is more than ample for gene isolation. Thus, starting with a tiny amount of total genomic DNA, a particular sequence can be produced in quantity in less than 4 hr.

PCR amplification is an effective cloning strategy if sequence information for the design of appropriate primers is available. Because DNA from a single cell can be used as template, the technique has enormous potential for the clinical diagnosis of infectious diseases and genetic abnormalities. With PCR techniques, DNA from a single hair or sperm can be analyzed to identify particular individuals in criminal cases without ambiguity.

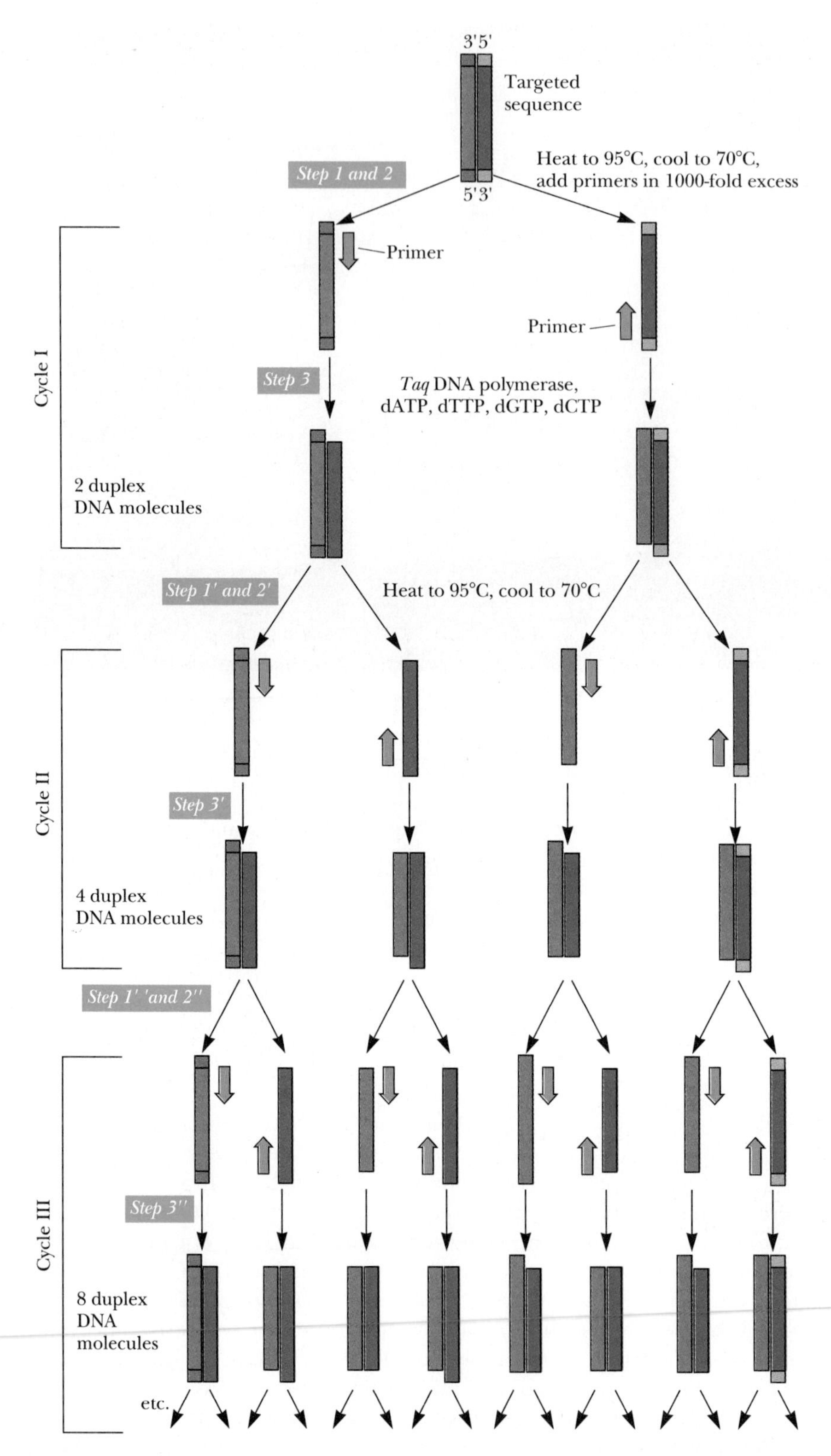

***Figure* 8.23** Polymerase chain reaction (PCR). Oligonucleotides complementary to a given DNA sequence prime the synthesis of only that sequence. Heat-stable *Taq* DNA polymerase survives many cycles of heating. Theoretically, the amount of the specific primed sequence is doubled in each cycle.

Gene Therapy

Gene therapy seeks to repair the damage caused by a genetic deficiency through introduction of a functional version of the defective gene. To achieve this end, a cloned variant of the gene must be incorporated into the organism in such a manner that it is expressed only at the proper time *and* only in appropriate cell types. At this time, these conditions impose serious technical and clinical difficulties. Gene therapies that have received approval from the National Institutes of Health for trials in human patients include gene constructs designed to cure an inherited immune deficiency, neuroblastoma, and cystic fibrosis. The gene defective in cystic fibrosis codes for a membrane protein that pumps Cl^- out of cells. If this Cl^- pump is defective, Cl^- ions remain in cells, which then take up water from the surrounding mucus by osmosis. The mucus thickens and accumulates in various organs, including the lungs, where its presence favors infections such as pneumonia. Left untreated, children with cystic fibrosis seldom survive past the age of five.

In Vitro Mutagenesis

The advent of recombinant DNA technology has made it possible to clone genes, manipulate them *in vitro,* and express them in a variety of cell types under various conditions. The function of any protein is ultimately dependent on its amino acid sequence, which in turn can be traced to the nucleotide sequence of its gene. The introduction of purposeful changes in the nucleotide sequence of a cloned gene represents an ideal way to make specific structural changes in a protein to assess the effects of these changes on the protein's function. Such changes constitute *mutations* of the gene introduced *in vitro.* ***In vitro* mutagenesis** makes it possible to alter the nucleotide sequence of a cloned gene systematically, as opposed to the chance occurrence of mutations in natural genes.

One efficient technique for *in vitro* mutagenesis is **PCR-based mutagenesis.** A mutant primer is added to a PCR reaction in which the gene (or segment of a gene) is undergoing amplification. The *mutant primer* is a primer whose sequence has been specifically altered to introduce a directed change at a particular place in the nucleotide sequence of the gene being amplified. Mutant versions of the gene can then be cloned and expressed to determine any effects of the mutation on the function of the gene product. PCR itself is prone to introduce mutations because *Taq* DNA polymerase is prone to make base-substitution errors, though at low frequency. Frameshift mutations due to insertion or deletion of bases also arise during PCR.

8.4 Recombinant DNA Technology: An Exciting Scientific Frontier

The strategies and methodologies described in this chapter are but an overview of the repertoire of experimental approaches that have been devised by molecular biologists in order to manipulate DNA and the information inherent in it. The enormous success of recombinant DNA technology means that the molecular biologist's task in searching genomes for genes is now akin to that of a lexicographer compiling a dictionary, a dictionary in which the "letters," i.e., the nucleotide sequences, spell out not words, but genes and what they mean. Molecular biologists have no index or alphabetic arrangement to serve as a guide through the vast volume of information in a genome; nevertheless, this information and its organization are rapidly being disclosed by the imaginative efforts and diligence of these scientists and their growing arsenal of analytical schemes.

Recombinant DNA technology now verges on the ability to engineer at will the genetic constitution of organisms for desired ends. The commercial production of therapeutic biomolecules in microbial cultures is already established (for example, the production of human insulin in quantity in *Escherichia coli* cells). Agricultural crops with desired attributes, such as enhanced resistance to herbicides, are in cultivation. The rat growth hormone gene has been cloned and transferred into mouse embryos, creating *transgenic mice* that at adulthood are twice normal size (see Chapter 28). Already, transgenic versions of domestic animals such as pigs, sheep, and even fish have been developed for human benefit. Perhaps most important, in a number of instances, clinical trials have been approved for **gene replacement therapy** (or, more simply, *gene therapy*) to correct particular human genetic disorders.

Problems

1. A DNA fragment isolated from an *Eco*RI digest of genomic DNA was combined with a plasmid vector linearized by *Eco*RI digestion so sticky ends could anneal. Phage T4 DNA ligase was then added to the mixture. List all possible products of the ligation reaction.

2. The nucleotide sequence of a polylinker in a particular plasmid vector is

–GAATTCCCGGGGATCCTCTAGAGTCGACCTGCAGGCATGC–

This polylinker contains restriction sites for *Bam*HI, *Eco*RI, *Pst1*, *Sal*I, *Sma*I, *Sph*I, and *Xba*I. Indicate the location of each restriction site in this sequence. (See Table 6.5 of restriction enzymes for their cleavage sites.)

3. A vector has a polylinker containing restriction sites in the following order: *Hin*dIII, *Sac*I, *Xho*I, *Bgl*II, *Xba*I, and *Cla*I.
   a. Give a possible nucleotide sequence for the polylinker.
   b. The vector is digested with *Hin*dIII and *Cla*I. A DNA segment contains a *Hin*dIII restriction site fragment 650 bases upstream from a *Cla*I site. This DNA fragment is digested with *Hin*dIII and *Cla*I, and the resulting *Hin*dIII–*Cla*I fragment is directionally cloned into the *Hin*dIII–*Cla*I–digested vector. Give the nucleotide sequence at each end of the vector and the insert and show that the insert can be cloned into the vector in only one orientation.

4. Yeast (*Saccharomyces cerevisiae*) has a genome size of 1.35×10^7 bp. If a genomic library of yeast DNA was constructed in a bacteriophage λ vector capable of carrying 16 kbp inserts, how many individual clones would have to be screened to have a 99% probability of finding a particular fragment?

5. The South American lungfish has a genome size of 1.02×10^{11} bp. If a genomic library of lungfish DNA was constructed in a cosmid vector capable of carrying inserts averaging 45 kbp in size, how many individual clones would have to be screened to have a 99% probability of finding a particular DNA fragment?

6. Given the following short DNA duplex of sequence (5′ → 3′)

ATGCCGTAGTCGATCATTACGATAGCATAGCACAGGGATCA-
CACATGCACACACATGACATAGGACAGATAGCAT

what oligonucleotide primers (17-mers) would be required for PCR amplification of this duplex?

7. Figure 8.5b shows a polylinker that falls within the β-galactosidase coding region of the *lacZ* gene. This polylinker serves as a cloning site in a fusion protein expression vector where the cloned insert will be expressed as a β-galactosidase fusion protein. Assume the vector polylinker was cleaved with *Bam*HI and then ligated with an insert whose sequence reads

GATCCATTTATCCACCGGAGAGCTGGTATCCCCAAAAGACGGCC . . .

What is the amino acid sequence of the fusion protein? Where is the junction between β-galactosidase and the sequence encoded by the insert? (Consult the genetic code table on the inside front cover to decipher the amino acid sequence.)

8. The amino acid sequence across a region of interest in a protein is

Asn-Ser-Gly-Met-His-Pro-Gly-Lys-Leu-Ala-Ser-Trp-Phe-Val-Gly-Asn-Ser-

The nucleotide sequence encoding this region begins and ends with an *Eco*RI site, making it easy to clone out the sequence and amplify it by the polymerase chain reaction (PCR). Give the nucleotide sequence of this region. Suppose you wished to change the middle Ser residue to a Cys to study the effects of this change on the protein's activity. What would be the sequence of the mutant oligonucleotide you would use for PCR amplification?

Further Reading

Ausubel, F. M., Brent, R., Kingston, et al., eds., 1987. *Current Protocols in Molecular Biology.* New York: John Wiley and Sons. A popular cloning manual.

Berger, S. L., and Kimmel, A. R., eds., 1987. *Guide to Molecular Cloning Techniques. Methods in Enzymology,* Volume 152. New York: Academic Press.

Bolivar, F., Rodriguez, R. L., Greene, P. J., et al., 1977. Construction and characterization of new cloning vehicles. II. A multipurpose cloning system. *Gene* **2**:95–113. A paper describing one of the early plasmid cloning systems.

Cohen, S. N., Chang, A. C. Y., Boyer, H. W., and Helling, R. B., 1973. Construction of biologically functional bacterial plasmids in vitro. *Proceedings of the National Academy of Sciences, U.S.A.* **70**:3240–3244. The classic paper on the construction of chimeric plasmids.

Goeddel, D. V., ed., 1990. *Gene Expression Technology. Methods in Enzymology,* Volume 185. New York: Academic Press.

Grunstein, M., and Hogness, D. S., 1975. Colony hybridization: a specific method for the isolation of cloned DNAs that contain a specific gene. *Proceedings of the National Academy of Sciences, U.S.A.* **72**:3961–3965. Article describing the colony hybridization technique for specific gene isolation.

Guyer, M., 1992. A comprehensive genetic linkage map for the human genome. *Science* **258**:67–86. M. Guyer is the corresponding author for the American/French NIH/CEPH (National Institutes of Health/Centre Etude du Polymorphisme Humain) Collaborative Mapping Group. This article is part of the October 2, 1992, issue of *Science* (Volume 258, Number 1), which presents an update of progress on the Human Genome Project.

Jackson, D. A., Symons, R. H., and Berg, P., 1972. Biochemical method for inserting new genetic information into DNA of simian virus 40: circular SV40 DNA molecules containing lambda phage genes and the galactose operon of E. coli. *Proceedings of the National Academy of Sciences, U.S.A.* **69**:2904–2909.

Luckow, V. A., and Summers, M. D., 1988. Trends in the development of baculovirus expression vectors. *Biotechnology* **6**:47–55.

Maniatis, T., Hardison, R. C., Lacy, E., et al., 1978. The isolation of structural genes from libraries of eucaryotic DNA. *Cell* **15**:687–701.

Morgan, R. A., and Anderson, W. F., 1993. Human gene therapy. *Annual Review of Biochemistry* **62**:191–217.

Murialdo, H., 1991. Bacteriophage lambda DNA maturation and packaging. *Annual Review of Biochemistry* **60**:125–153. Review of the biochemistry of packaging DNA into λ phage heads.

Saiki, R. K., Gelfand, D. H., Stoeffel, B., et al., 1988. Primer-directed amplification of DNA with a thermostable DNA polymerase. *Science* **239**:487–491. Discussion of the polymerase chain reaction procedure.

Sambrook, J., Fritsch, E. F., and Maniatis, T., 1989. *Molecular Cloning,* 2nd ed. Long Island: Cold Spring Harbor Laboratory Press. (This cloning manual is familiarly referred to as "Maniatis.")

Southern, E. M., 1975. Detection of specific sequences among DNA fragments separated by gel electrophoresis. *Journal of Molecular Biology* **98**:503–517. The classic paper on the identification of specific DNA sequences through hybridization with unique probes.

Timmer, W. C., and Villalobos, J. M., 1993. The polymerase chain reaction. *The Journal of Chemical Education* **70**:273–280.

Young, R. A., and Davis, R. W., 1983. Efficient isolation of genes using antibody probes. *Proceedings of the National Academy of Sciences, U.S.A.* **80**:1194–1198. Using antibodies to screen protein expression libraries to isolate the structural gene for a specific protein.

"A feast of fat things, a feast of wines on the lees."

Isaiah 25:6

Chapter 9

Lipids and Membranes

"The mighty whales which swim in a sea of water, and have a sea of oil swimming in them."

Herman Melville. "Extracts." Moby-Dick. *New York: Penguin Books, 1972.*

Outline

9.1 Classes of Lipids

9.2 Membranes

9.3 Structure of Membrane Proteins

Membranes serve a number of essential cellular functions. They constitute the boundaries of cells and intracellular organelles, and they provide a surface where many important biological reactions and processes occur. Membranes have proteins that mediate and regulate the transport of metabolites, macromolecules, and ions. Hormones and many other biological signal molecules and regulatory agents exert their effects via interactions with membranes. Photosynthesis, electron transport, oxidative phosphorylation, muscle contraction, and electrical activity all depend on membranes and membrane proteins.

Biological membranes are uniquely organized arrays of lipids and proteins (either of which may be modified with carbohydrate groups). **Lipids** are a class of biological molecules defined by low solubility in water and high solubility in nonpolar solvents. As molecules that are largely hydrocarbon in nature, lipids represent highly reduced forms of carbon and, upon oxidation in metabolism, yield large amounts of energy. Lipids are thus the molecules of choice for metabolic energy storage.

The lipids found in biological systems are **amphipathic,** which means they possess both polar and nonpolar groups. The hydrophobic nature of lipid molecules allows membranes to act as effective barriers to more polar molecules. The polar moieties of amphipathic lipids typically lie at the surface of membranes, where they interact with water. Proteins interact with the lipids of membranes in a variety of ways. Some proteins associate with membranes via electrostatic interactions with polar groups on the membrane surface, whereas other proteins are embedded to various extents in the hydrophobic core of the membrane. Other proteins are *anchored* to membranes via covalently bound lipid molecules that associate strongly with the hydrophobic membrane core.

This chapter discusses the composition, structure, and dynamic processes of biological membranes. We begin with a discussion of the lipid molecules found in living things.

9.1 Classes of Lipids

Fatty Acids

A **fatty acid** is composed of a long hydrocarbon chain (''tail'') and a terminal carboxyl group (or ''head''). The carboxyl group is normally ionized under physiological conditions. Fatty acids occur in large amounts in biological systems, but rarely in the free, uncomplexed state. They typically are esterified to glycerol or other backbone structures. Most of the fatty acids found in nature have an even number of carbon atoms (usually 14 to 24). Certain marine organisms, however, contain substantial amounts of fatty acids with odd numbers of carbon atoms. Fatty acids are either **saturated** (all carbon–carbon bonds are single bonds) or **unsaturated** (with one or more double bonds in the hydrocarbon chain). If a fatty acid has a single double bond, it is said to be **monounsaturated,** and if it has more than one, **polyunsaturated.** Fatty acids can be named or described in at least three ways, as listed in Table 9.1. For example, a fatty acid composed of an 18-carbon chain with no double bonds can be called by its systematic name (**octadecanoic acid**), its common name (stearic acid), or its shorthand notation, in which the number of carbons is followed by a colon and the number of double bonds in the molecule (18:0 for stearic acid). The structures of several fatty acids are given in Figure 9.1. **Stearic acid** (18:0) and **palmitic acid** (16:0) are the most common saturated fatty acids in nature.

Free rotation around each of the carbon–carbon bonds makes saturated fatty acids extremely flexible molecules. Owing to steric constraints, however, the fully extended conformation (Figure 9.1) is the most stable for saturated fatty acids. Nonetheless, the degree of stabilization is slight, and (as will be seen) saturated fatty acid chains adopt a variety of conformations.

Unsaturated fatty acids are slightly more abundant in nature than saturated fatty acids, especially in higher plants. The most common unsaturated

***Table* 9.1**
Common Biological Fatty Acids

Number of Carbons	Common Name	Systematic Name	Symbol	Structure
Saturated fatty acids				
12	Lauric acid	Dodecanoic acid	12:0	$CH_3(CH_2)_{10}COOH$
14	Myristic acid	Tetradecanoic acid	14:0	$CH_3(CH_2)_{12}COOH$
16	Palmitic acid	Hexadecanoic acid	16:0	$CH_3(CH_2)_{14}COOH$
18	Stearic acid	Octadecanoic acid	18:0	$CH_3(CH_2)_{16}COOH$
20	Arachidic acid	Eicosanoic acid	20:0	$CH_3(CH_2)_{18}COOH$
22	Behenic acid	Docosanoic acid	22:0	$CH_3(CH_2)_{20}COOH$
24	Lignoceric acid	Tetracosanoic acid	24:0	$CH_3(CH_2)_{22}COOH$
Unsaturated fatty acids (all double bonds are cis)				
16	Palmitoleic acid	9-Hexadecenoic acid	16:1	$CH_3(CH_2)_5CH{=}CH(CH_2)_7COOH$
18	Oleic acid	9-Octadecenoic acid	18:1	$CH_3(CH_2)_7CH{=}CH(CH_2)_7COOH$
18	Linoleic acid	9,12-Octadecadienoic acid	18:2	$CH_3(CH_2)_4(CH{=}CHCH_2)_2(CH_2)_6COOH$
18	α-Linolenic acid	9,12,15-Octadecatrienoic acid	18:3	$CH_3CH_2(CH{=}CHCH_2)_3(CH_2)_6COOH$
18	γ-Linolenic acid	6,9,12-Octadecatrienoic acid	18:3	$CH_3(CH_2)_4(CH{=}CHCH_2)_3(CH_2)_3COOH$
20	Arachidonic acid	5,8,11,14-Eicosatetraenoic acid	20:4	$CH_3(CH_2)_4(CH{=}CHCH_2)_4(CH_2)_2COOH$
24	Nervonic acid	15-Tetracosenoic acid	24:1	$CH_3(CH_2)_7CH{=}CH(CH_2)_{13}COOH$

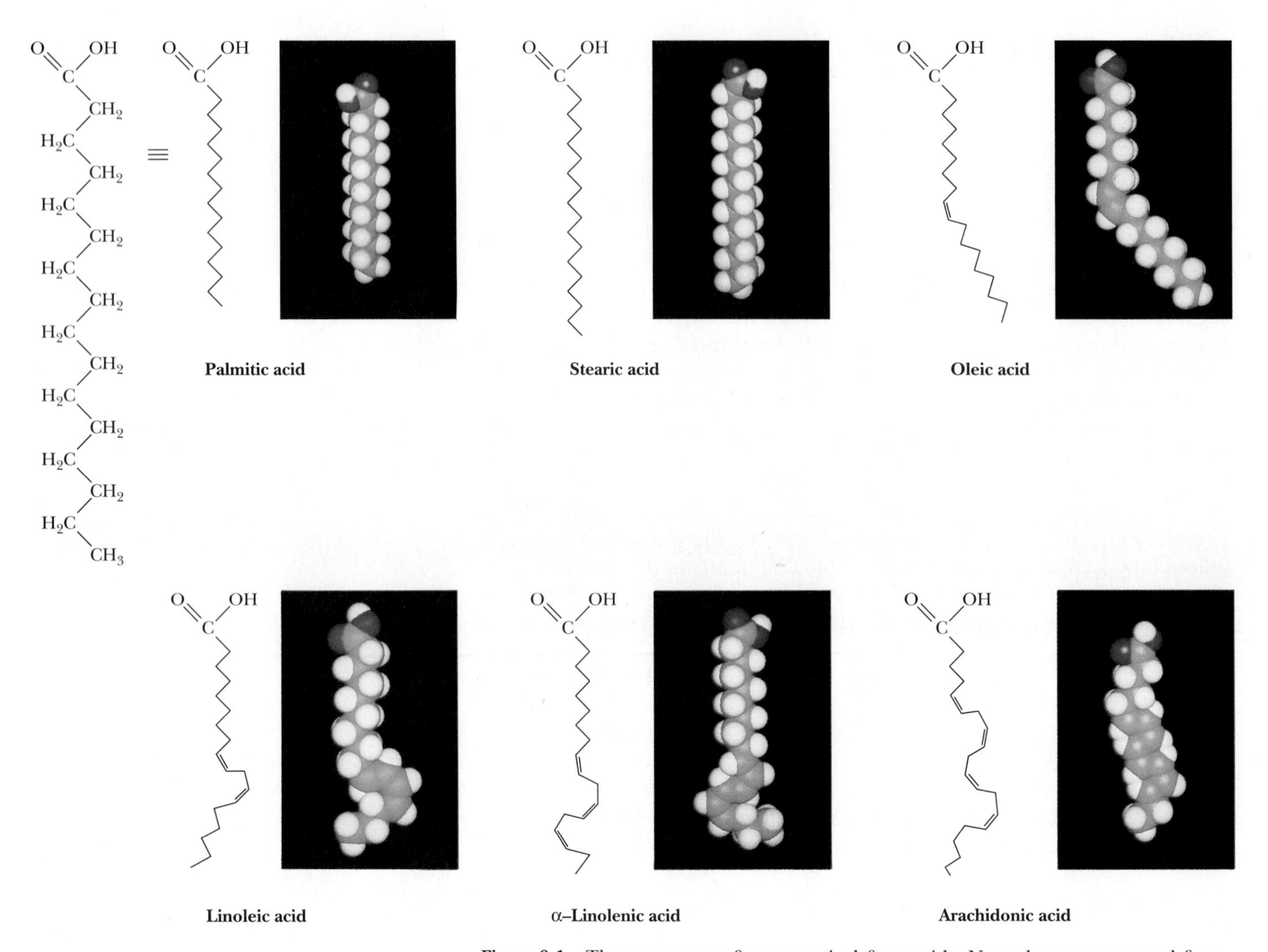

Figure **9.1** The structures of some typical fatty acids. Note that most natural fatty acids contain an even number of carbon atoms and that the double bonds are nearly always cis and rarely conjugated.

fatty acid is **oleic acid,** or 18:1(9), with the number in parentheses indicating that the double bond is between carbons 9 and 10. The number of double bonds in an unsaturated fatty acid varies typically from one to four, but, in the fatty acids found in most bacteria, this number rarely exceeds one.

The double bonds found in fatty acids are nearly always in the cis configuration. As shown in Figure 9.1, this causes a bend or "kink" in the fatty acid chain. This bend has very important consequences for the structure of biological membranes. Saturated fatty acid chains can pack closely together to form ordered, rigid arrays under certain conditions, but unsaturated fatty acids prevent such close packing and produce flexible, fluid aggregates.

Some fatty acids are not synthesized by mammals and yet are necessary for normal growth and life. These *essential fatty acids* include **linoleic** and **γ-linolenic acids.** These must be obtained by mammals in their diet (specifically from plant sources). **Arachidonic acid,** which is not found in plants, can only be synthesized by mammals from linoleic acid. At least one function of the essential fatty acids is to serve as a precursor for the synthesis of **eicosanoids,** such as *prostaglandins,* a class of compounds that exert hormone-like effects in many physiological processes (discussed in Chapter 24).

A Deeper Look

Fatty Acids in Food: Saturated Versus Unsaturated

Fats consumed in the modern human diet vary widely in their fatty acid compositions. The table below provides a brief summary. The incidence of cardiovascular disease is correlated with diets high in saturated fatty acids. By contrast, a diet that is relatively higher in unsaturated fatty acids (especially polyunsaturated fatty acids) may reduce the risk of heart attacks and strokes. Corn oil, abundant in the United States and high in (polyunsaturated) linoleic acid, is an attractive dietary choice.

Margarine made from corn, safflower, or sunflower oils is much lower in saturated fatty acids than is butter, which is made from milk fat. However, margarine may present its own health risks. Its fatty acids contain *trans*-double bonds (introduced by the hydrogenation process), which may also contribute to cardiovascular disease. (Margarine was invented by a French chemist, H. Mège Mouriès, who won a prize from Napoleon III in 1869 for developing a substitute for butter.)

Although vegetable oils usually contain a higher proportion of unsaturated fatty acids than do animal oils and fats, several plant oils are actually high in saturated fats. Palm oil is low in polyunsaturated fatty acids and particularly high in (saturated) palmitic acid (whence the name *palmitic*). Coconut oil is particularly high in lauric and myristic acids (both saturated) and contains very few unsaturated fatty acids.

Fatty Acid Compositions of Some Dietary Lipids*

Source	Lauric and Myristic	Palmitic	Stearic	Oleic	Linoleic
Beef	5	24–32	20–25	37–43	2–3
Milk		25	12	33	3
Coconut	74	10	2	7	—
Corn		8–12	3–4	19–49	34–62
Olive		9	2	84	4
Palm		39	4	40	8
Safflower		6	3	13	78
Soybean		9	6	20	52
Sunflower		6	1	21	66

Data from *Merck Index,* 10th ed. Rahway, NJ: Merck and Co., and Wilson et al., 1967, *Principles of Nutrition,* 2nd ed. New York: Wiley.

*Values are percentages of total fatty acids.

In addition to unsaturated fatty acids, several other modified fatty acids are found in nature. Microorganisms, for example, often contain branched-chain fatty acids, such as **tuberculostearic acid** (Figure 9.2). When these fatty acids are incorporated in membranes, the methyl group constitutes a local structural perturbation in a manner similar to the double bonds in unsaturated fatty acids (see later). Some bacteria also synthesize fatty acids containing cyclic structures such as cyclopropane, cyclopropene, and even cyclopentane rings.

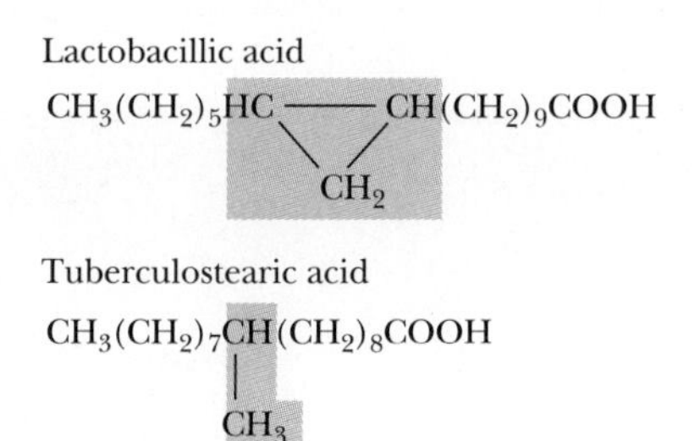

***Figure* 9.2** Structures of two unusual fatty acids: lactobacillic acid, a fatty acid containing a cyclopropane ring, and tuberculostearic acid, a branched-chain fatty acid.

Triacylglycerols

A significant number of the fatty acids in plants and animals exist in the form of **triacylglycerols** (also called **triglycerides**). Triacylglycerols are a major energy reserve and the principal neutral derivatives of glycerol found in animals. These molecules consist of a glycerol esterified with three fatty acids (Figure

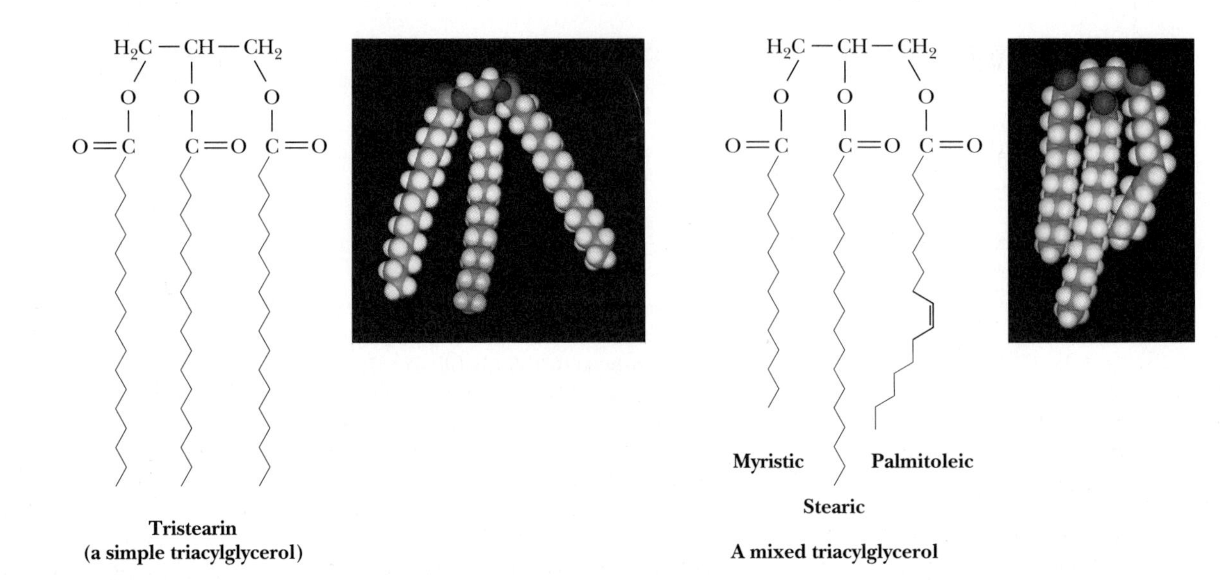

***Figure* 9.3** Triacylglycerols are formed from glycerol and fatty acids.

9.3). If all three fatty acid groups are the same, the molecule is called a simple triacylglycerol. Examples include **tristearoylglycerol** (common name *tristearin*) and **trioleoylglycerol** (*triolein*). Mixed triacylglycerols contain two or three different fatty acids. Triacylglycerols in animals are found primarily in the adipose tissue (body fat), which serves as a depot or storage site for lipids. Monoacylglycerols and diacylglycerols also exist, but are far less common than the triacylglycerols. Most natural plant and animal fat is composed of mixtures of simple and mixed triacylglycerols.

Acylglycerols can be hydrolyzed by heating with acid or base or by treatment with lipases. Hydrolysis with alkali is called **saponification** and yields salts of free fatty acids and glycerol. This is how **soap** (a metal salt of an acid derived from fat) was made by our ancestors. One method used potassium hydroxide (*potash*) leached from wood ashes to hydrolyze animal fat (mostly triacylglycerols). (The tendency of such soaps to be precipitated by Mg^{2+} and Ca^{2+} ions in hard water makes them less useful than modern detergents.) When the fatty acids esterified at the first and third carbons of glycerol are different, the second carbon is asymmetric. The various acylglycerols are normally soluble in benzene, chloroform, ether, and hot ethanol. Although triacylglycerols are insoluble in water, mono- and diacylglycerols readily form organized structures in water (discussed later), owing to the polarity of their free hydroxyl groups.

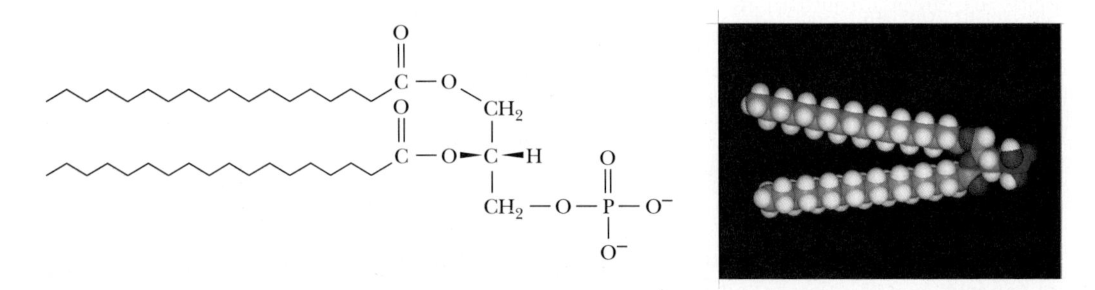

***Figure* 9.4** Phosphatidic acid, the parent compound for glycerophospholipids.

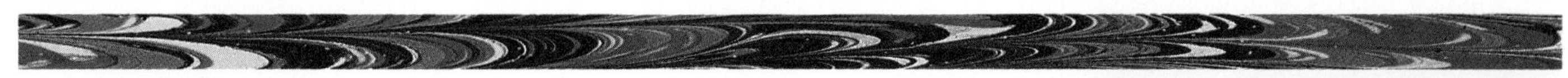

A Deeper Look

Polar Bears Use Triacylglycerols to Survive Long Periods of Fasting

The polar bear is magnificently adapted to thrive in its harsh Arctic environment. Research by Malcolm Ramsey (at the University of Saskatchewan in Canada) and others has shown that polar bears eat only during a few weeks out of the year and then fast for periods of eight months or more, consuming no food or water during that time. Eating mainly in the winter, the adult polar bear feeds almost exclusively on seal blubber (largely composed of triacylglycerols), thus building up its own triacylglycerol reserves. Through the Arctic summer, the polar bear maintains normal physical activity, roaming over long distances, but relies entirely on its body fat for sustenance, burning as much as 1 to 1.5 kg of fat per day. It neither urinates nor defecates for extended periods. All the water needed to sustain life is provided from the metabolism of triacylglycerides (because oxidation of fatty acids yields carbon dioxide and water).

Ironically, the word *Arctic* comes from the ancient Greeks, who understood that the northernmost part of the earth lay under the stars of the constellation Ursa Major, the Great Bear. Although unaware of the polar bear, they called this region *Arktikós,* which means "the country of the great bear."

Triacylglycerols are rich in highly reduced carbons and thus yield large amounts of energy in the oxidative reactions of metabolism. Complete oxidation of 1 g of triacylglycerols yields about 38 kJ of energy, whereas proteins and carbohydrates yield only about 17 kJ/g. Also, their hydrophobic nature allows them to aggregate in highly anhydrous forms, whereas polysaccharides and proteins are highly hydrated. For these reasons, triacylglycerols are the molecules of choice for energy storage in animals. Body fat (mainly triacylglycerols) also provides good insulation. Whales and Arctic mammals rely on body fat for both insulation and energy reserves.

Glycerophospholipids

A 1,2-diacylglycerol that has a phosphate group esterified at carbon atom 3 of the glycerol backbone is a **glycerophospholipid,** also known as a *phosphoglyceride* or a *glycerol phosphatide* (Figure 9.4). These lipids form one of the largest classes of natural lipids and one of the most important. They are essential components of cell membranes and are found in small concentrations in other parts of the cell. It should be noted that all glycerophospholipids are members of the broader class of lipids known as **phospholipids.**

The numbering and nomenclature of glycerophospholipids present a dilemma in that the number 2 carbon of the glycerol backbone of a phospholipid is asymmetric. It is possible to name these molecules either as D- or L-isomers. Thus, glycerol phosphate itself can be referred to either as D-glycerol-1-phosphate or as L-glycerol-3-phosphate (Figure 9.5). Instead of naming the glycerol phosphatides in this way, biochemists have adopted the *stereospecific numbering* or *sn-* system. In this system, the *pro-S* position of a prochiral atom is denoted as the *1-position,* the prochiral atom as the *2-position,* and so on. When this scheme is used, the prefix *sn-* precedes the molecule name (glycerol phosphate in this case) and distinguishes this nomenclature from other approaches. In this way, the glycerol phosphate in natural phosphoglycerides is named *sn*-glycerol-3-phosphate.

pro-*S* position ⟶ CH_2OH ; HO—C—H ; pro-*R* position ⟶ $CH_2OPO_3^{2-}$ ≡ $CH_2OPO_3^{2-}$; H—C—OH ; CH_2OH

L-Glycerol-3-phosphate ≡ D-Glycerol-1-phosphate

sn-Glycerol-3-phosphate

***Figure* 9.5** The absolute configuration of *sn*-glycerol-3-phosphate. The pro-(*R*) and pro-(*S*) positions of the parent glycerol are also indicated.

The Most Common Phospholipids

Phosphatidic acid, the parent compound for the glycerol-based phospholipids (Figure 9.4), consists of *sn*-glycerol-3-phosphate, with fatty acids esterified at the 1- and 2-positions. Phosphatidic acid is found in small amounts in most natural systems and is an important intermediate in the biosynthesis of the more common glycerophospholipids (Figure 9.6). In these compounds, a variety of polar groups are esterified to the phosphoric acid moiety of the molecule. The phosphate, together with such esterified entities, is referred to as a "head" group. Phosphatides with choline or ethanolamine are referred to as **phosphatidylcholine** (known commonly as **lecithin**) or **phosphatidylethanolamine,** respectively. These phosphatides are two of the most common constituents of biological membranes. Other common *head groups* found in phosphatides include glycerol, serine, and inositol (Figure 9.6). Another kind of glycerol phosphatide found in many tissues is **diphosphatidylglycerol.** First observed in heart tissue, it is also called **cardiolipin.** In cardiolipin, a phosphatidylglycerol is esterified through the C-1 hydroxyl group of the glycerol moiety of the head group to the phosphoryl group of another phosphatidic acid molecule.

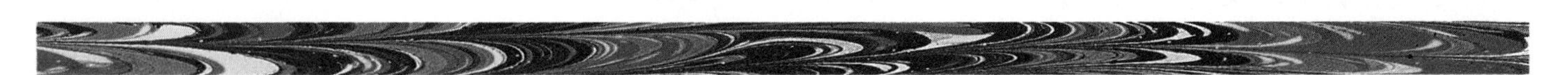

A Deeper Look

Prochirality

If a tetrahedral center in a molecule has two identical substituents, it is referred to as **prochiral** since, if either of the like substituents was converted to a different group, the tetrahedral center would then be chiral. Consider glycerol:

The central carbon of glycerol is prochiral since replacing either of the $—CH_2OH$ groups would make the central carbon chiral. Nomenclature for prochiral centers is based on the (*R,S*) system (in Chapter 3). To name the otherwise identical substituents of a prochiral center, imagine increasing slightly the priority of one of them (by substituting a deuterium for a hydrogen, for example):

The resulting molecule has an (*S*)-configuration about the (now chiral) central carbon atom. The group that contains the deuterium is thus referred to as the *pro-S* group. As a useful exercise, you should confirm that labeling the other CH_2OH group with a deuterium produces the (*R*)-configuration at the central carbon, so that this latter CH_2OH group is the *pro-R* substituent.

HOH_2C, CH_2OH, C, H, OH

Glycerol

D, $^3HOH_2C$, ^1CHOH, ^2C, H, OH

1-d, 2(S)-Glycerol
(S-configuration at C-2)

Phosphatidylcholine

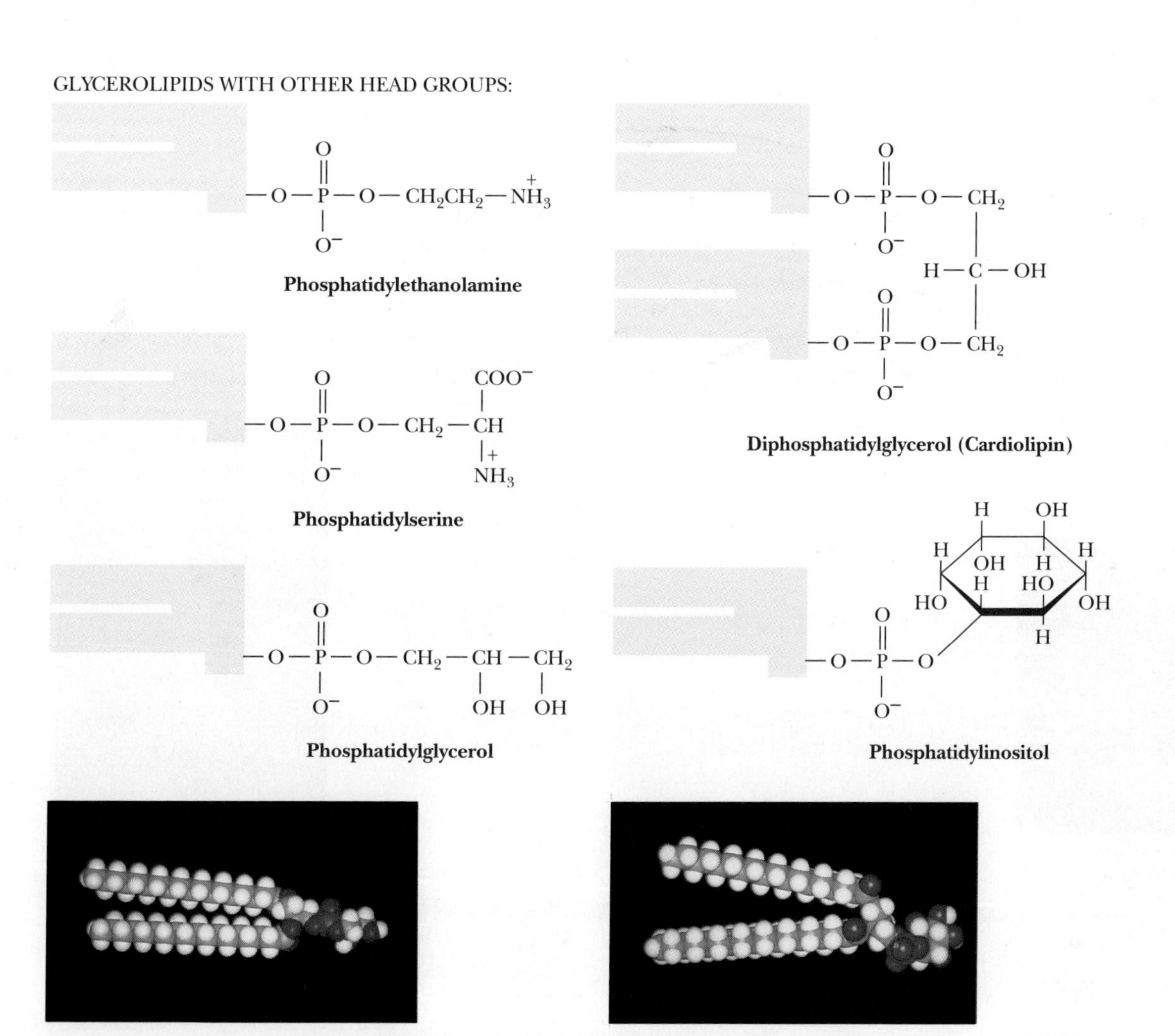

***Figure* 9.6** Structures of several glycerophospholipids and space-filling models of phosphatidylcholine, phosphatidylglycerol, and phosphatidylinositol.

Phosphatides exist in many different varieties, depending on the fatty acids esterified to the glycerol group. As we shall see, the nature of the fatty acids can greatly affect the chemical and physical properties of the phosphatides and the membranes that contain them. In most cases, glycerol phosphatides have a saturated fatty acid at position 1 and an unsaturated fatty acid at position 2 of the glycerol. Thus, **1-stearoyl-2-oleoyl-phosphatidylcholine** (Figure 9.7) is a common constituent in natural membranes, but **1-linoleoyl-2-palmitoylphosphatidylcholine** is not.

A Deeper Look

Glycerophospholipid Degradation: One of the Effects of Snake Venoms

The venoms of poisonous snakes contain (among other things) a class of enzymes known as **phospholipases,** enzymes that cause the breakdown of phospholipids. For example, the venoms of the eastern diamondback rattlesnake (*Crotalus adamanteus*) and the Indian cobra (*Naja naja*) both contain phospholipase A_2, which catalyzes the hydrolysis of fatty acids at the C-2 position of glycerophospholipids.

The phospholipid breakdown product of this reaction, *lysolecithin,* acts as a detergent and dissolves the membranes of red blood cells, causing them to rupture. Indian cobras kill several thousand people each year.

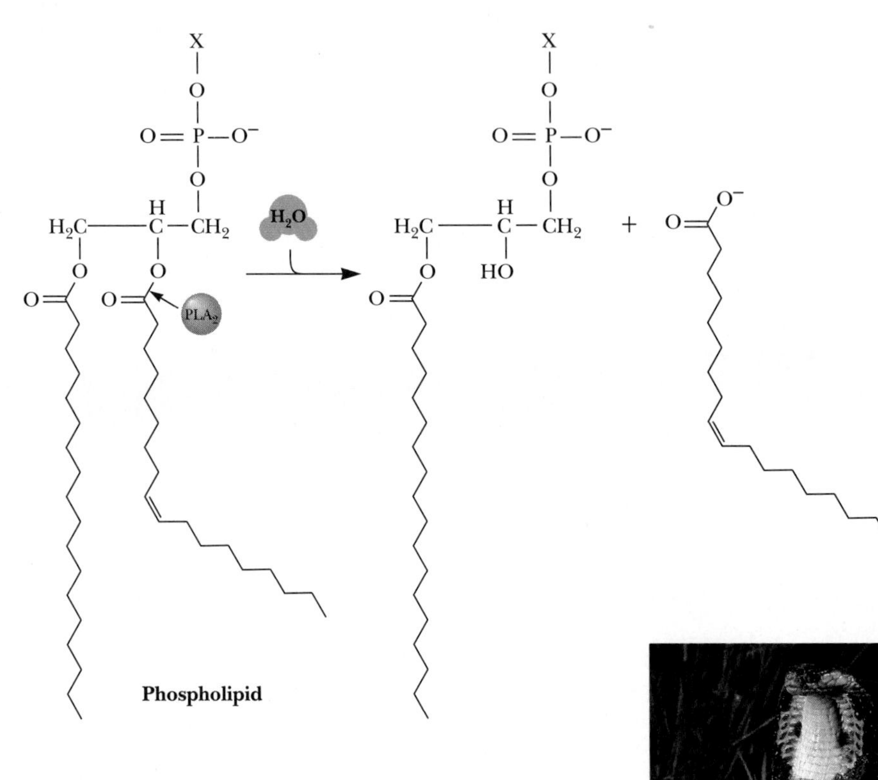

Eastern diamondback rattlesnake

Indian cobra

Figure **9.7** A space-filling model of 1-stearoyl-2-oleoyl-phosphatidylcholine.

Both structural and functional strategies govern the natural design of the many different kinds of glycerophospholipid head groups and fatty acids. The structural roles of these different glycerophospholipid classes are described later in this chapter. Certain phospholipids, including phosphatidylinositol and phosphatidylcholine, participate in complex cellular signaling events. These roles, appreciated only in recent years, are described in Chapter 37.

Ether Glycerophospholipids

Ether glycerophospholipids possess an ether linkage instead of an acyl group at the C-1 position of glycerol (Figure 9.8). One of the most versatile biochemical signal molecules found in mammals is **platelet activating factor** or **PAF,** a

A Deeper Look

Platelet Activating Factor: A Potent Glyceroether Mediator

Platelet activating factor (PAF) was first identified by its ability (at low levels) to cause platelet aggregation and dilation of blood vessels, but it is now known to be a potent mediator in inflammation, allergic responses, and shock. PAF effects are observed at tissue concentrations as low as 1×10^{-12} *M*. PAF causes a dramatic inflammation of air passages and induces asthma-like symptoms in laboratory animals. **Toxic-shock syndrome** occurs when fragments of destroyed bacteria act as toxins and induce the synthesis of PAF. This results in a drop in blood pressure and a reduced volume of blood pumped by the heart, which leads to shock and, in severe cases, death.

Beneficial effects have also been attributed to PAF. In reproduction, PAF secreted by the fertilized egg is instrumental in the implantation of the egg in the uterine wall. PAF is produced in significant quantities in the lungs of the fetus late in pregnancy and may stimulate the production of fetal lung surfactant, a protein–lipid complex that prevents collapse of the lungs in a newborn infant.

unique ether glycerophospholipid (Figure 9.9). The alkyl group at C-1 of PAF is typically a 16-carbon chain, but the acyl group at C-2 is a 2-carbon acetate unit. By virtue of this acetate group, PAF is much more water-soluble than other lipids, allowing PAF to function as a soluble messenger in signal transduction.

Plasmalogens are ether glycerophospholipids in which the alkyl moiety is cis-α,β-unsaturated (Figure 9.10). Common plasmalogen head groups include choline, ethanolamine, and serine. These lipids are referred to as phosphatidal choline, phosphatidal ethanolamine, and phosphatidal serine.

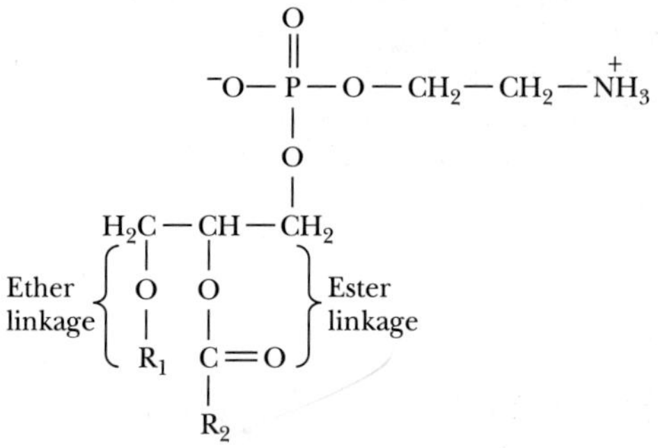

Ether linkage
Ester linkage

Figure **9.8** A 1-alkyl 2-acyl-phosphatidylethanolamine (an ether glycerophospholipid).

Platelet activating factor

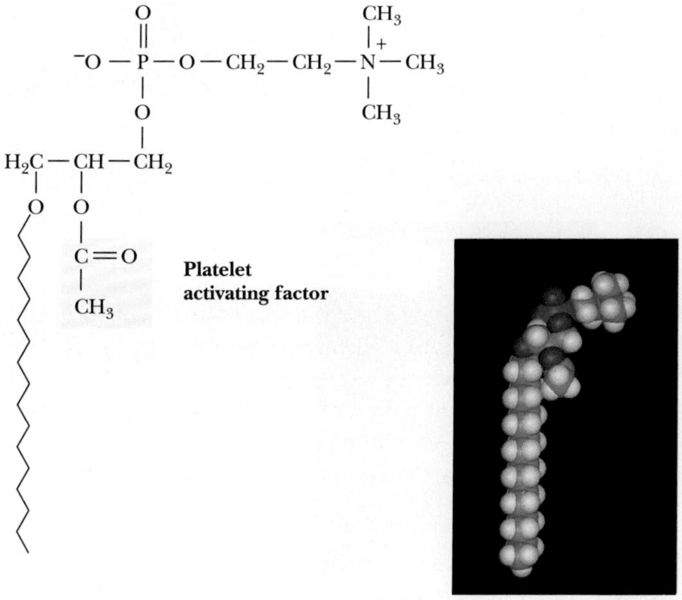

Figure **9.9** The structure of 1-alkyl 2-acetyl-phosphatidylcholine, also known as platelet activating factor or PAF.

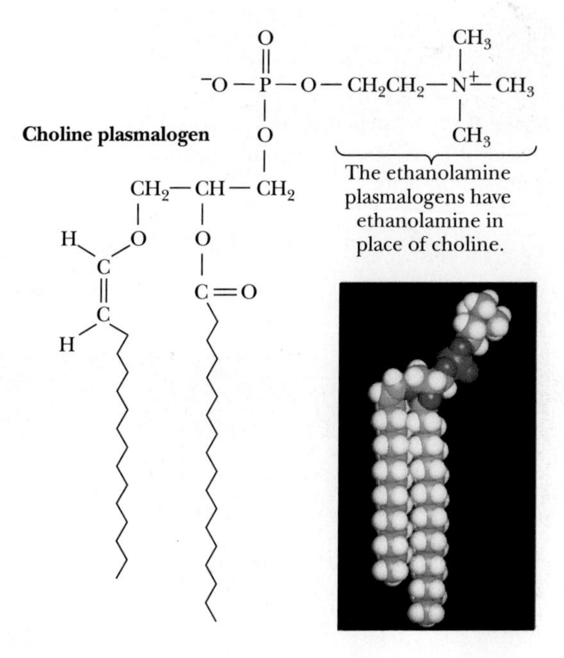

Figure **9.10** The structure and a space-filling model of a choline plasmalogen.

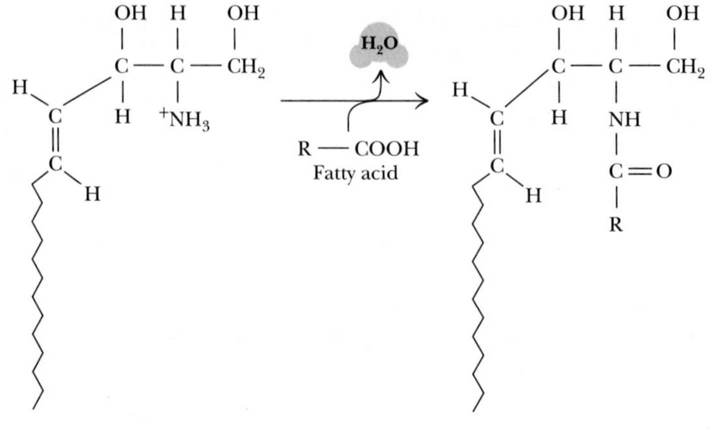

***Figure* 9.11** Formation of an amide linkage between a fatty acid and sphingosine produces a ceramide.

Sphingolipids

Sphingolipids represent another class of lipids found frequently in biological membranes. An 18-carbon amino alcohol, **sphingosine** (Figure 9.11), forms the backbone of these lipids rather than glycerol. Typically, a fatty acid is joined to a sphingosine via an amide linkage to form a **ceramide. Sphingomyelins** represent a phosphorus-containing subclass of sphingolipids and are especially important in the nervous tissue of higher animals. A **sphingomyelin** is formed by the esterification of a phosphorylcholine or a phosphorylethanolamine to the 1-hydroxy group of a ceramide (Figure 9.12).

There is another class of ceramide-based lipids which, like the sphingomyelins, are important components of muscle and nerve membranes in animals. These are the **glycosphingolipids,** and they consist of a ceramide with one or more sugar residues in a β-glycosidic linkage at the 1-hydroxyl moiety. The neutral glycosphingolipids contain only neutral (uncharged) sugar residues. When a single glucose or galactose is bound in this manner, the molecule is a

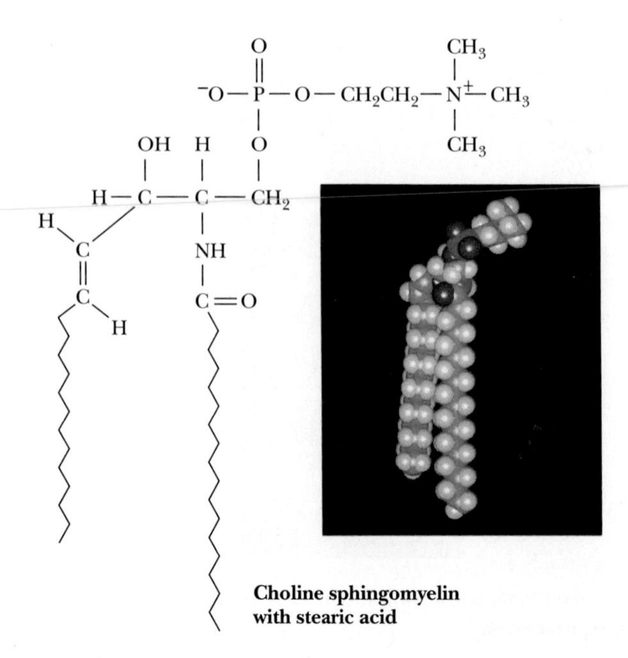

***Figure* 9.12** A structure and a space-filling model of a choline sphingomyelin formed from stearic acid.

cerebroside (Figure 9.13). Another class of lipids is formed when a sulfate is esterified at the 3-position of the galactose to make a **sulfatide. Gangliosides** (Figure 9.14) are more complex glycosphingolipids that consist of a ceramide backbone with three or more sugars esterified, one of these being a **sialic acid** such as ***N*-acetylneuraminic acid.** These latter compounds are referred to as *acidic glycosphingolipids,* and they have a net negative charge at neutral pH.

The glycosphingolipids have a number of important cellular functions, despite the fact that they are present only in small amounts in most membranes. Glycosphingolipids at cell surfaces appear to determine, at least in part, certain elements of tissue and organ specificity. Cell–cell recognition and tissue immunity appear to depend upon specific glycosphingolipids. Gangliosides are present in nerve endings and appear to be important in nerve impulse transmission. A number of genetically transmitted diseases involve the accumulation of specific glycosphingolipids due to an absence of the enzymes needed for their degradation. Such is the case for ganglioside G_{M2} in the brains of *Tay-Sachs disease* victims (see Chapter 24), a rare but fatal disease characterized by a red spot on the retina, gradual blindness, and loss of weight, especially in infants and children.

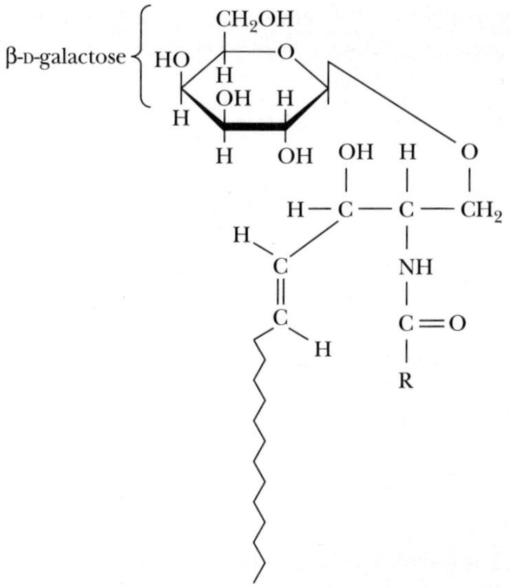

β-D-galactose

A cerebroside

***Figure* 9.13** The structure of a cerebroside. Note the sphingosine backbone.

Waxes

Waxes are esters of long-chain alcohols with long-chain fatty acids. The resulting molecule can be viewed (in analogy to the glycerolipids) as having a weakly polar head group (the ester moiety itself) and a long, nonpolar tail (the hydrocarbon chains) (Figure 9.15). Fatty acids found in waxes are usually saturated. The alcohols found in waxes may be saturated or unsaturated and may include sterols, such as cholesterol (see later section). Waxes are highly insoluble due to the weakly polar nature of the ester group. As a result, this class of molecules confers water-repellant character to animal skin, to the

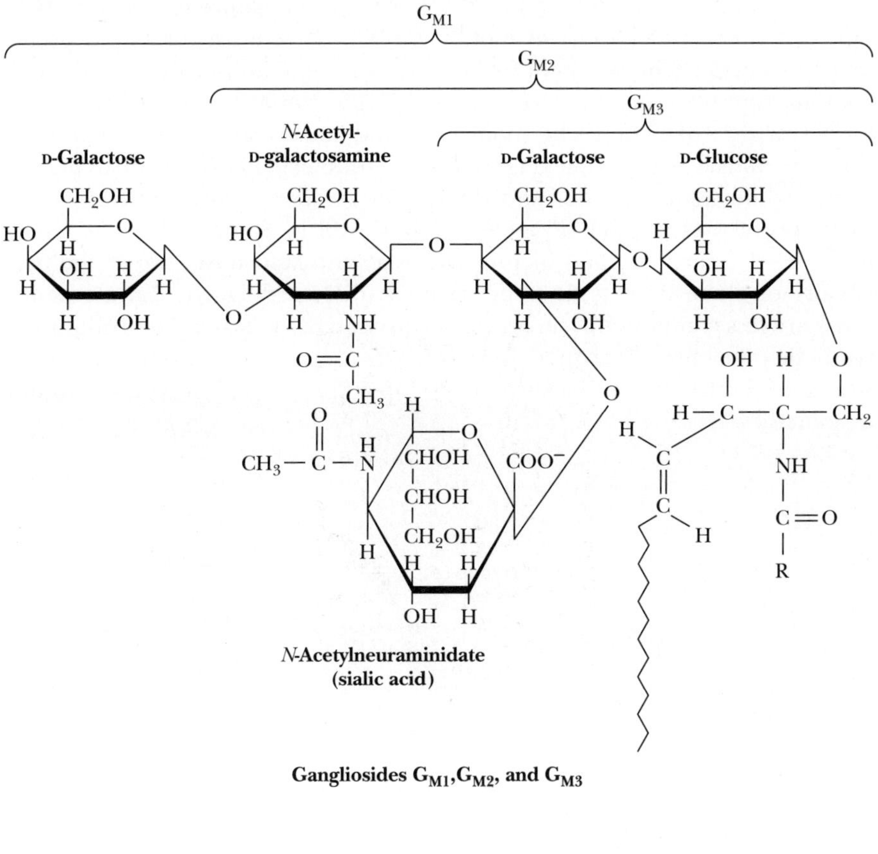

G_{M1} G_{M2} G_{M3}

D-Galactose ***N*-Acetyl-D-galactosamine** **D-Galactose** **D-Glucose**

***N*-Acetylneuraminidate (sialic acid)**

Gangliosides G_{M1}, G_{M2}, and G_{M3}

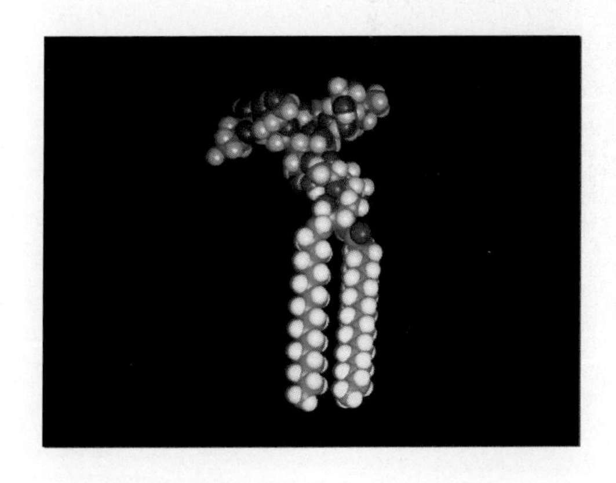

***Figure* 9.14** The structures of several important gangliosides. Also shown is a space-filling model of ganglioside G_{M1}.

A Deeper Look

Moby Dick and Spermaceti: A Valuable Wax from Whale Oil

When oil from the head of the sperm whale is cooled, **spermaceti,** a translucent wax with a white, pearly luster, crystallizes from the mixture. Spermaceti, which makes up 11% of whale oil, is composed mainly of the wax **cetyl palmitate:**

$$CH_3(CH_2)_{14}{-}COO{-}(CH_2)_{15}CH_3$$

as well as smaller amounts of cetyl alcohol:

$$HO{-}(CH_2)_{15}CH_3$$

Spermaceti and cetyl palmitate have been widely used in the making of cosmetics, fragrant soaps, and candles.

In the literary classic *Moby-Dick,* Herman Melville describes Ishmael's impressions of spermaceti, when he muses that the waxes "discharged all their opulence, like fully ripe grapes their wine; as I snuffed that uncontaminated aroma—literally and truly, like the smell of spring violets."*

*Melville, H., *Moby-Dick,* Octopus Books, London, 1984, p. 205. (Adapted from *Chemistry in Moby Dick,* Waddell, T. G., and Sanderlin, R. R. (1986), Journal of Chemical Education **63:**1019–1020.)

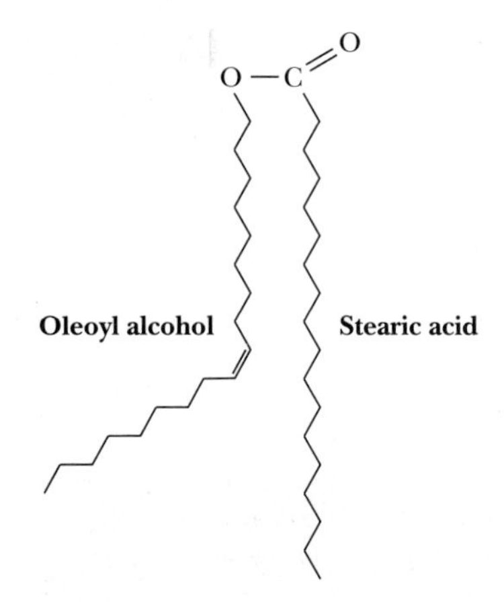

***Figure* 9.15** An example of a wax. Oleoyl alcohol is esterified to stearic acid in this case.

leaves of certain plants, and to bird feathers. The glossy surface of a polished apple results from a wax coating. **Carnauba wax,** obtained from the fronds of a species of palm tree in Brazil, is a particularly hard wax used for high gloss finishes, such as in automobile wax, boat wax, floor wax, and shoe polish. **Lanolin,** a component of wool wax, is used as a base for pharmaceutical and cosmetic products because it is rapidly assimilated by human skin.

Terpenes

The **terpenes** are a class of lipids formed from combinations of two or more molecules of 2-methyl-1,3-butadiene, better known as **isoprene** (a five-carbon unit that is abbreviated C_5). A **monoterpene** (C_{10}) consists of two isoprene units, a **sesquiterpene** (C_{15}) consists of three isoprene units, a **diterpene** (C_{20}) has four isoprene units, and so on. Isoprene units can be linked in terpenes to form straight chain or cyclic molecules, and the usual method of linking isoprene units is head to tail (Figure 9.16). Monoterpenes occur in all higher plants, while sesquiterpenes and diterpenes are less widely known. Several examples of these classes of terpenes are shown in Figure 9.17. The **triterpenes** are C_{30} terpenes and include **squalene** and **lanosterol,** two of the precursors of cholesterol and other steroids (discussed later). **Tetraterpenes** (C_{40}) are less common but include the carotenoids, a class of colorful photosynthetic pigments. β-Carotene is the precursor of vitamin A, while lycopene, similar to β-carotene but lacking the cyclopentene rings, is a pigment found in tomatoes.

***Figure* 9.16** The structure of isoprene (2-methyl-1,3-butadiene) and the structure of head-to-tail and tail-to-tail linkages. Isoprene itself is formed by distillation of natural rubber, a linear head-to-tail polymer of isoprene units.

H, CH_2, C, C, H_2C, CH_3 ≡

Isoprene

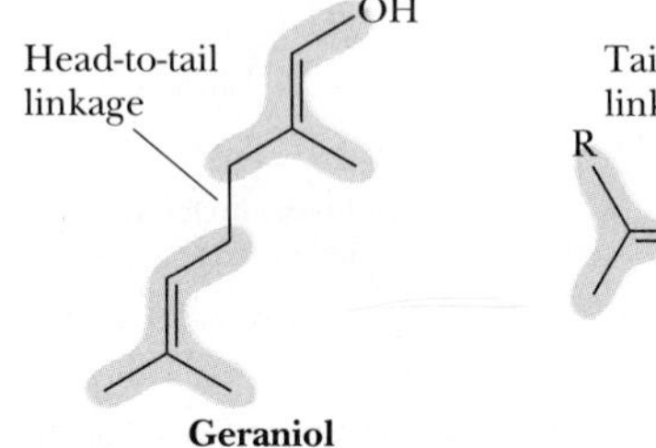

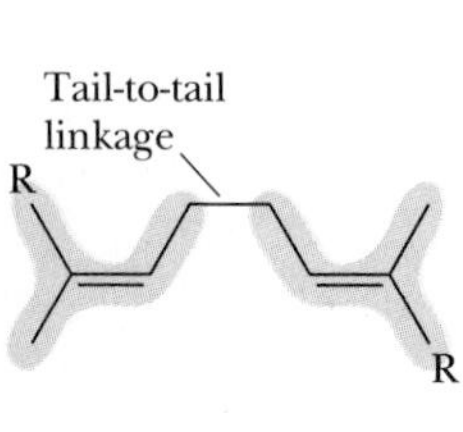

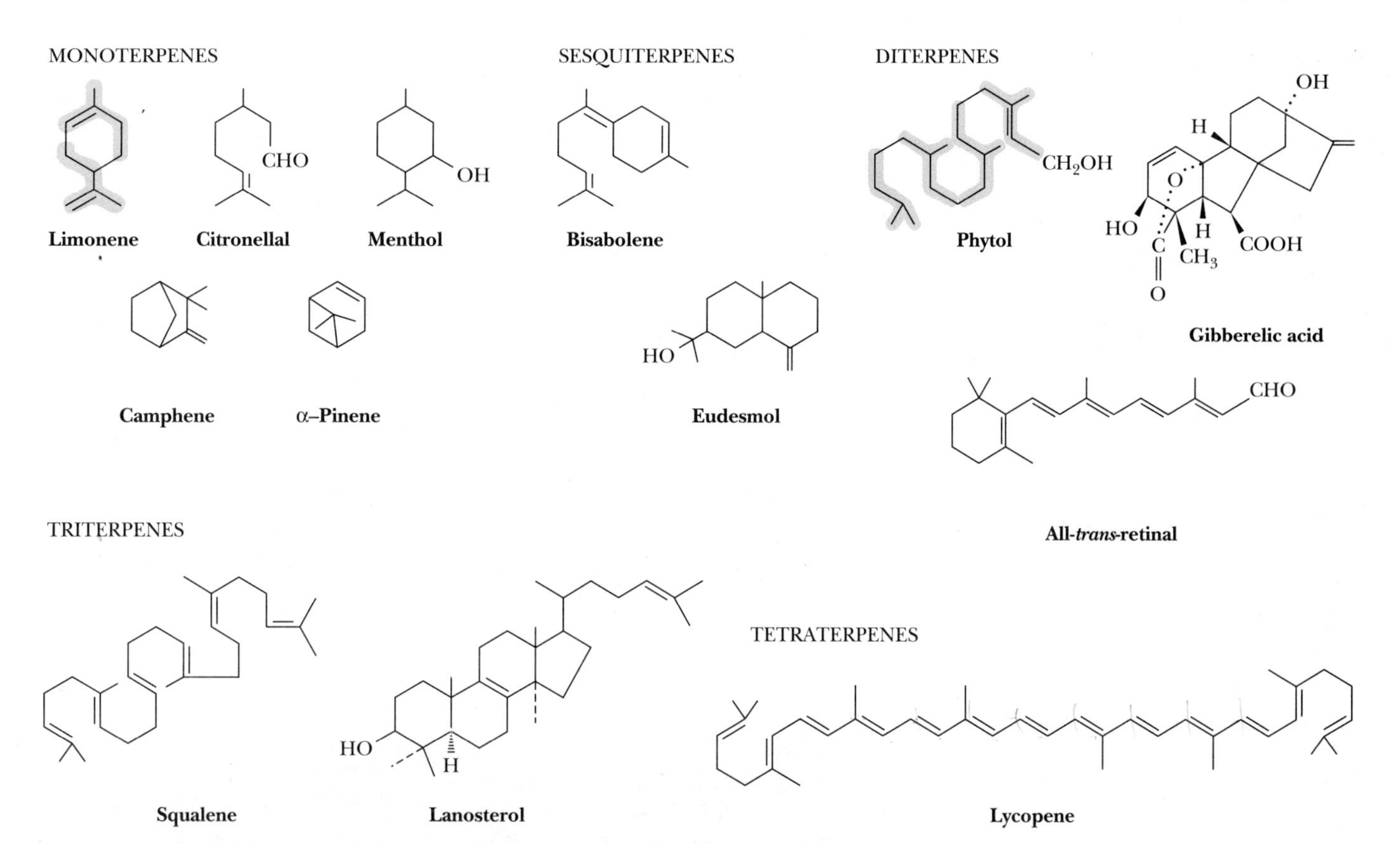

Figure **9.17** Many monoterpenes are readily recognized by their characteristic flavors or odors (limonene in lemons; citronellal in roses, geraniums, and some perfumes; pinene in turpentine; and menthol from peppermint, used in cough drops and nasal inhalers). The diterpenes, which are C_{20} terpenes, include retinal (the essential light-absorbing pigment in rhodopsin, the photoreceptor protein of the eye), phytol (a constituent of chlorophyll), and the gibberellins (potent plant hormones). The triterpene lanosterol is a constituent of wool fat. Lycopene is a carotenoid found in ripe fruit, especially tomatoes.

Long-chain polyisoprenoid molecules with a terminal alcohol moiety are called **polyprenols.** The **dolichols,** one class of polyprenols (Figure 9.18), consist of 16 to 22 isoprene units and, in the form of dolichyl phosphates, function to carry carbohydrate units in the biosynthesis of glycoproteins in animals. Polyprenyl groups serve to *anchor* certain proteins to biological membranes (discussed later).

Steroids

Cholesterol

A large and important class of terpene-based lipids is the **steroids.** This molecular family, whose members effect an amazing array of cellular functions, is based on a common structural motif of three six-membered rings and one five-membered ring all fused together. **Cholesterol** (Figure 9.19) is the most common steroid in animals and the precursor for all other animal steroids. The numbering system for cholesterol applies to all such molecules. Many steroids contain methyl groups at positions 10 and 13 and an 8- to 10-carbon alkyl side chain at position 17. The polyprenyl nature of this compound is particularly evident in the side chain. Many steroids contain an oxygen at C-3, either a hydroxyl group in sterols or a carbonyl group in other steroids. Note

$H{-}[CH_2{-}C(CH_3){=}CH{-}CH_2]_{13-23}{-}CH_2{-}CH(CH_3){-}CH_2{-}CH_2{-}O{-}P(=O)(O^-){-}O^-$

Dolichol phosphate

CH_3O, CH_3O, CH_3, O, O, $[CH_2CH{=}C(CH_3)CH_2]_{10}{-}H$

Coenzyme Q (Ubiquinone, UQ)

CH_3, H_3C, HO, CH_3, O, CH_3, $[H_3C\ H]_3$, H

Vitamin E (α-tocopherol)

O, O, 3

Vitamin K_1 (phylloquinone)

$H_3C{-}C(CH_3){=}CH{-}CH_2{-}[CH_2{-}C(CH_3){=}CH{-}CH_2]_9{-}CH_2{-}C(CH_3){=}CH{-}CH_2OH$

Undecaprenyl alcohol (bactoprenol)

O, O, 8

Vitamin K_2 (menaquinone)

Figure **9.18** Dolichol phosphate is an initiation point for the synthesis of carbohydrate polymers in animals. The analogous alcohol in bacterial systems, *undecaprenol,* also known as *bactoprenol,* consists of 11 isoprene units. Undecaprenyl phosphate delivers sugars from the cytoplasm for the synthesis of cell wall components such as peptidoglycans, lipopolysaccharides, and glycoproteins. Polyprenyl compounds also serve as the side chains of vitamin K, the ubiquinones, plastoquinones, and tocopherols (such as vitamin E).

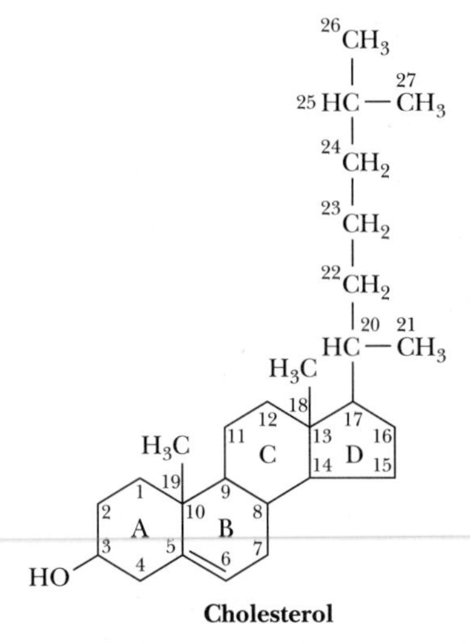

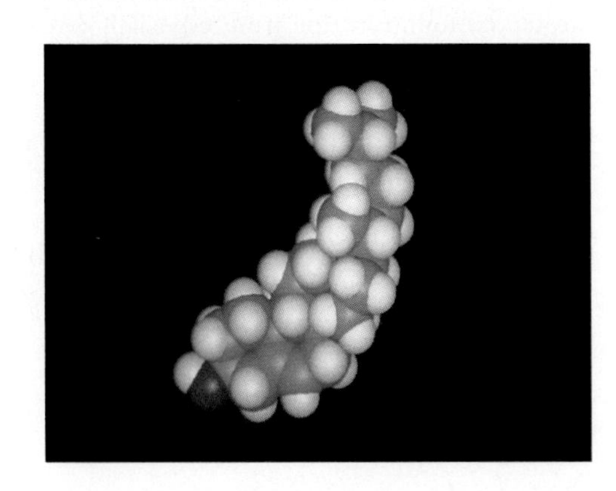

Figure **9.19** The structure of cholesterol, shown with steroid ring designations and carbon numbering.

also that the carbons at positions 10 and 13 and the alkyl group at position 17 are nearly always oriented on the same side of the steroid nucleus, the *β*-orientation. Alkyl groups that extend from the other side of the steroid backbone are in an *α*-orientation.

Cholesterol is a principal component of animal cell plasma membranes, and much smaller amounts of cholesterol are found in the membranes of intracellular organelles. The relatively rigid fused ring system of cholesterol

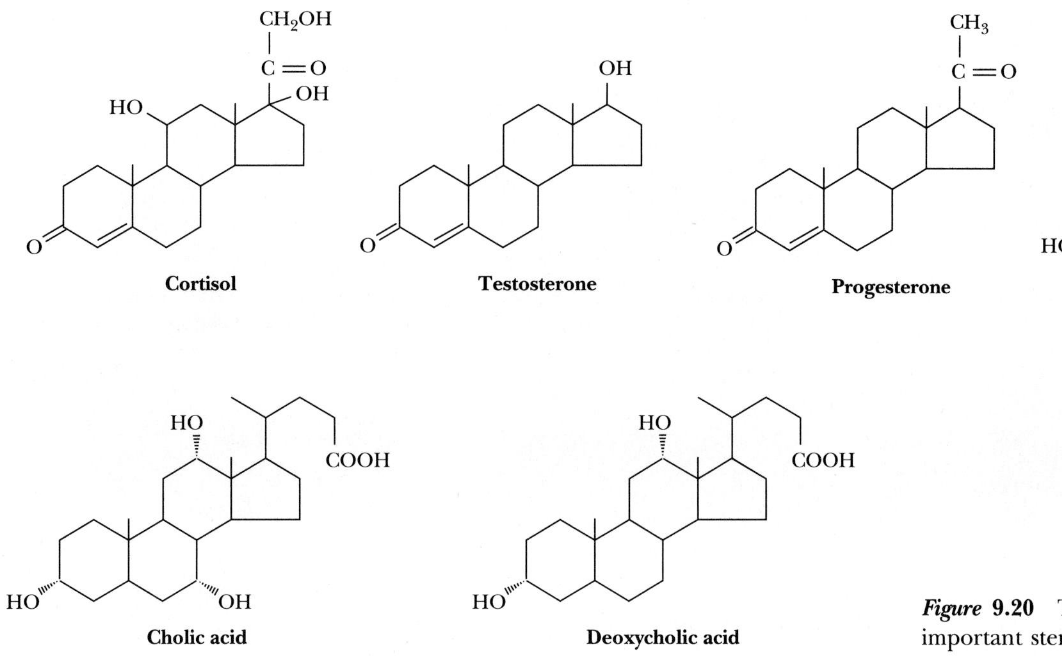

***Figure* 9.20** The structures of several important sterols derived from cholesterol.

and the weakly polar alcohol group at the C-3 position have important consequences for the properties of plasma membranes. Cholesterol is also a component of *lipoprotein complexes* in the blood, and it is one of the constituents of *plaques* that form on arterial walls in *atherosclerosis.*

Steroid Hormones

Steroids derived from cholesterol in animals include five families of hormones: the androgens, estrogens, progestins, glucocorticoids and mineralocorticoids, and bile acids (Figure 9.20). **Androgens** such as **testosterone** and **estrogens** such as **estradiol** mediate the development of sexual characteristics and sexual function in animals. The **progestins** such as **progesterone** participate in control of the menstrual cycle and pregnancy. **Glucocorticoids** (**cortisol,** for example) participate in the control of carbohydrate, protein, and lipid metabolism, whereas the **mineralocorticoids** regulate salt (Na^+, K^+, and Cl^-) balances in tissues. The **bile acids** (including **cholic** and **deoxycholic acid**) are detergent molecules secreted in bile from the gallbladder that assist in the absorption of dietary lipids in the intestine.

9.2 Membranes

Cells make use of many different types of membranes. All cells have a cytoplasmic membrane, or *plasma membrane,* that functions (in part) to separate the cytoplasm from the surroundings. In the early days of biochemistry, the plasma membrane was not accorded many functions other than this one of partition. We now know that the plasma membrane is also responsible for (1) the exclusion of certain toxic ions and molecules from the cell, (2) the accumulation of cell nutrients, and (3) energy transduction. It functions in (4) cell locomotion, (5) reproduction, (6) signal transduction processes, and (7) interactions with molecules or other cells in the vicinity.

Even the plasma membranes of prokaryotic cells (bacteria) are complex (Figure 9.21). With no intracellular organelles to divide and organize the

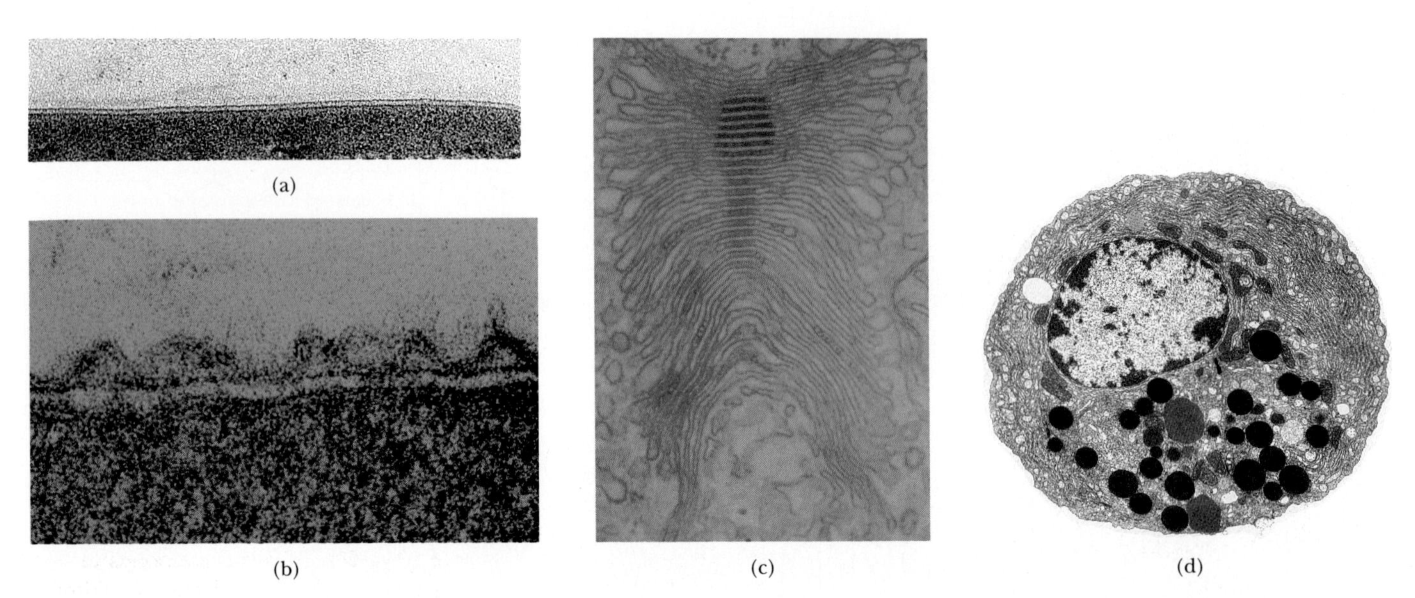

Figure **9.21** Electron micrographs of several different membrane structures: (a) *Menoidium,* a protozoan; (b) Gram-negative envelope of *Aquaspirillum serpens;* (c) Golgi apparatus; (d) pancreatic acinar cell.

work, bacteria must carry out all processes either at the plasma membrane or in the cytoplasm itself. Eukaryotic cells, however, contain numerous intracellular organelles that perform specialized tasks. Nucleic acid biosynthesis is handled in the nucleus; mitochondria are the site of electron transport, oxidative phosphorylation, fatty acid oxidation, and the tricarboxylic acid cycle; and secretion of proteins and other substances is handled by the endoplasmic reticulum and the Golgi apparatus. This partitioning of labor is not the only contribution of the membranes in these cells. Many of the processes occurring in these organelles (or in the prokaryotic cell) actively involve membranes. Thus, some of the enzymes involved in nucleic acid metabolism are membrane-associated. The electron transfer chain and its associated system for ATP synthesis are embedded in the mitochondrial membrane. Many enzymes responsible for aspects of lipid biosynthesis are located in the endoplasmic reticulum membrane.

Figure **9.22** Several spontaneously formed lipid structures.

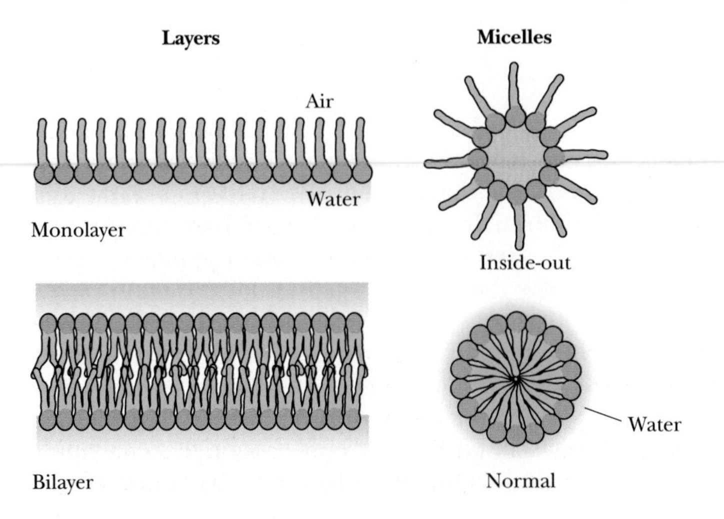

Structure	M_r	CMC	Micelle M_r
Triton X-100 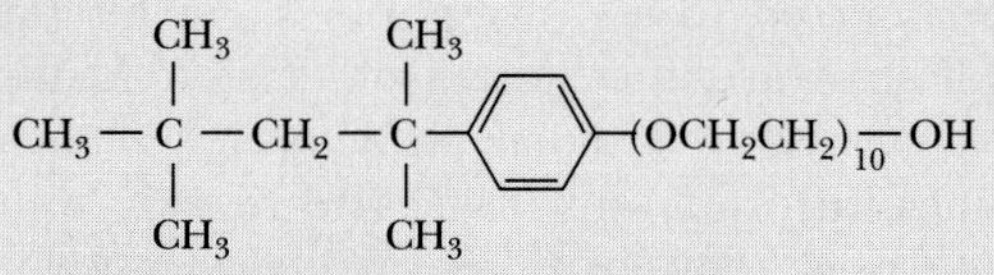 $CH_3-C(CH_3)_2-CH_2-C(CH_3)_2-C_6H_4-(OCH_2CH_2)_{10}-OH$	625	0.24 mM	90–95,000
Octyl glucoside 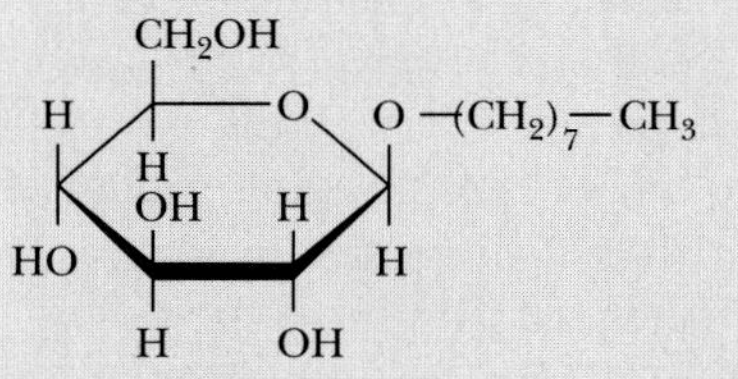 (glucose ring with CH_2OH, H, OH, HO substituents) $O-(CH_2)_7-CH_3$	292	25 mM	
$C_{12}E_8$ (Dodecyl octaoxyethylene ether) $C_{12}H_{25}-(OCH_2CH_2)_8-OH$	538	0.071 mM	

Figure **9.23** The structures of some common detergents and their physical properties. Micelles formed by detergents can be quite large. Triton X-100, for example, typically forms micelles with a total molecular mass of 90 to 95 kD. This corresponds to approximately 150 molecules of Triton X-100 per micelle.

Spontaneously Formed Lipid Structures

Monolayers and Micelles

Amphipathic lipids spontaneously form a variety of structures when added to aqueous solution. All these structures form in ways that minimize contact between the hydrophobic lipid chains and the aqueous milieu. For example, when small amounts of a fatty acid are added to an aqueous solution, a monolayer is formed at the air–water interface, with the polar head groups in contact with the water surface and the hydrophobic tails in contact with the air (Figure 9.22). Few lipid molecules are found as monomers in solution.

Further addition of fatty acid eventually results in the formation of micelles. **Micelles** formed from an amphipathic lipid in water position the hydrophobic tails in the center of the lipid aggregation with the polar head groups facing outward. Amphipathic molecules that form micelles are characterized by a unique **critical micelle concentration,** or **CMC.** Below the CMC, individual lipid molecules predominate. Nearly all the lipid added above the CMC, however, spontaneously forms micelles. Micelles are the preferred form of aggregation in water for detergents and soaps. Some typical CMC values are listed in Figure 9.23.

Lipid Bilayers

Lipid bilayers consist of back-to-back arrangements of monolayers (Figure 9.22). Phospholipids prefer to form bilayer structures in aqueous solution because their pairs of fatty acyl chains do not pack well in the interior of a micelle. Phospholipid bilayers form rapidly and spontaneously when phospholipids are added to water, and they are stable structures in aqueous solution. As opposed to micelles, which are small, self-limiting structures of a few hundred molecules, bilayers may form spontaneously over large areas (10^8 nm^2 or more). Since exposure of the edges of the bilayer to solvent is highly unfavorable, extensive bilayers normally wrap around themselves and form closed vesicles (Figure 9.24). The nature and integrity of these vesicle structures are very much dependent on the lipid composition. Phospholipids can form either *unilamellar vesicles* (with a single lipid bilayer) known as *lipo-*

(a) **Bilayer**

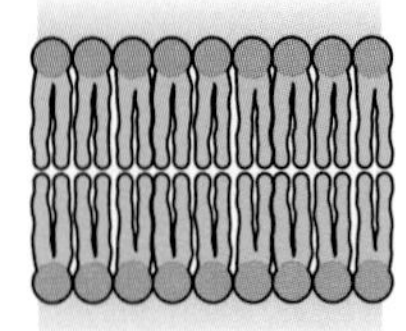

(b) **Unilamellar vesicle**

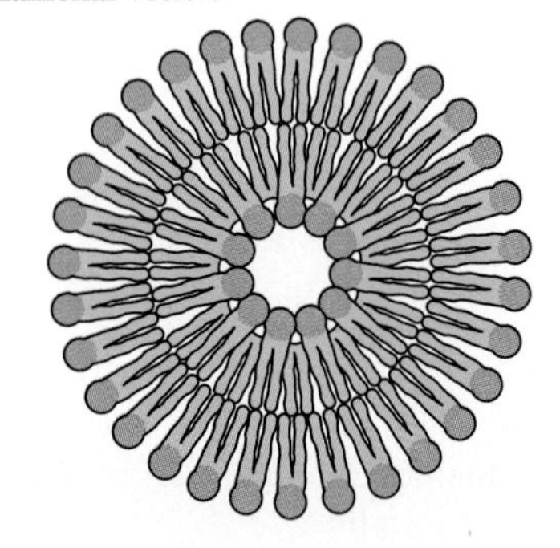

(c) **Multilamellar vesicle**

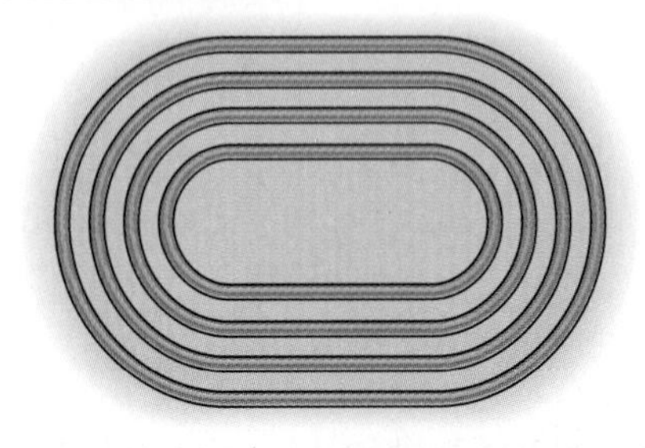

(d)

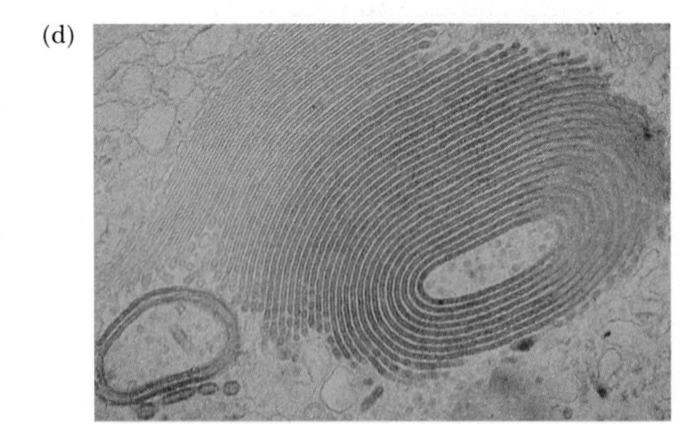

***Figure* 9.24** Drawings of (a) a bilayer, (b) a unilamellar vesicle, (c) a multilamellar vesicle, and (d) an electron micrograph of a multilamellar Golgi structure (×94,000).

somes, or *multilamellar vesicles.* These latter structures are reminiscent of the layered structure of onions. Multilamellar vesicles were discovered by Sir Alex Bangham and are sometimes referred to as "Bangosomes" in his honor.

Liposomes are highly stable structures that can be subjected to manipulations such as gel filtration chromatography and dialysis. With such methods, it is possible to prepare liposomes with different inside and outside solution compositions. Liposomes can be used as drug and enzyme delivery systems in therapeutic applications. For example, liposomes can be used to introduce contrast agents into the body for diagnostic imaging procedures, including *computerized tomography (CT)* and *magnetic resonance imaging (MRI)* (Figure 9.25). Liposomes can fuse with cells, mixing their contents with the intracellular medium. If methods can be developed to target liposomes to selected cell populations, it may be possible to deliver drugs, therapeutic enzymes, and contrast agents to particular kinds of cells (such as cancer cells).

That vesicles and liposomes form at all is a consequence of the amphipathic nature of the phospholipid molecule. Ionic interactions between the polar head groups and water are maximized, whereas hydrophobic interactions (see Chapter 2) facilitate the association of hydrocarbon chains in the interior of the bilayer. The formation of vesicles results in a favorable increase in the entropy of the solution, since the water molecules are not required to order themselves around the lipid chains. It is important to consider for a moment the physical properties of the bilayer membrane, which is the basis of vesicles and also of natural membranes. Bilayers have a polar surface and a nonpolar core. This hydrophobic core provides a substantial barrier to ions and other polar entities. The rates of movement of such species across membranes are thus quite slow. However, this same core also provides a favorable environment for nonpolar molecules and hydrophobic proteins. We will encounter numerous cases of hydrophobic molecules that interact with membranes and regulate biological functions in some way by binding to or embedding themselves in membranes.

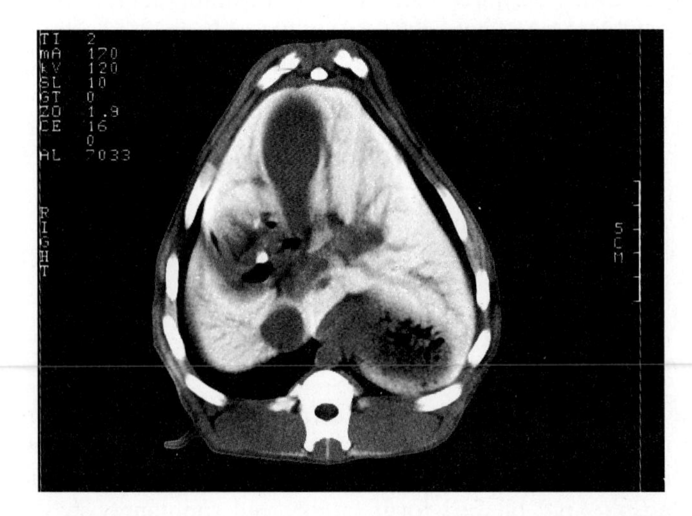

***Figure* 9.25** A computerized tomography (CT) image of the upper abdomen of a dog, following administration of liposome-encapsulated iodine, a *contrast agent* that improves the light/dark contrast of objects in the image. The spine is the bright white object at the bottom and the other bright objects on the periphery are ribs. The liver (white) occupies most of the abdominal space. The gallbladder (bulbous object at the center top) and blood vessels appear dark in the image. The liposomal iodine contrast agent has been taken up by Kuppfer cells, which are distributed throughout the liver, except in tumors. The dark object in the lower right is a large tumor. None of these anatomical features would be visible in a CT image in the absence of the liposomal iodine contrast agent.

Fluid Mosaic Model

In 1972, S. J. Singer and G. L. Nicolson proposed the **fluid mosaic model** for membrane structure, which suggested that membranes are dynamic structures composed of proteins and phospholipids. In this model, the phospholipid bilayer is a *fluid* matrix, in essence, a two-dimensional solvent for proteins. Both lipids and proteins are capable of rotational and lateral movement.

Singer and Nicolson also pointed out that proteins can be associated with the surface of this bilayer or embedded in the bilayer to varying degrees (Figure 9.26). They defined two classes of membrane proteins. The first, called **peripheral proteins** (or **extrinsic proteins**), includes those that do not penetrate the bilayer to any significant degree and are associated with the membrane by virtue of ionic interactions and hydrogen bonds between the membrane surface and the surface of the protein. Peripheral proteins can be dissociated from the membrane by treatment with salt solutions or by changes in pH (treatments that disrupt hydrogen bonds and ionic interactions). **Integral proteins** (or **intrinsic proteins**), in contrast, possess hydrophobic surfaces that can readily penetrate the lipid bilayer itself, as well as surfaces that prefer contact with the aqueous medium. These proteins can either insert into the membrane or extend all the way across the membrane and expose themselves to the aqueous solvent on both sides. Singer and Nicolson also suggested that a portion of the bilayer lipid interacts in specific ways with integral membrane proteins and that these interactions might be important for the function of certain membrane proteins. Because of these intimate associations with membrane lipid, integral proteins can only be removed from the membrane by agents capable of breaking up the hydrophobic interactions within the lipid bilayer itself (such as detergents and organic solvents). The fluid mosaic model has become the paradigm for modern studies of membrane structure and function.

Membrane Bilayer Thickness

The Singer–Nicolson model suggested a value of approximately 5 nm for membrane thickness, the same thickness as a lipid bilayer itself. Low angle X-ray diffraction studies in the early 1970s showed that many natural membranes were approximately 5 nm in thickness and that the interiors of these

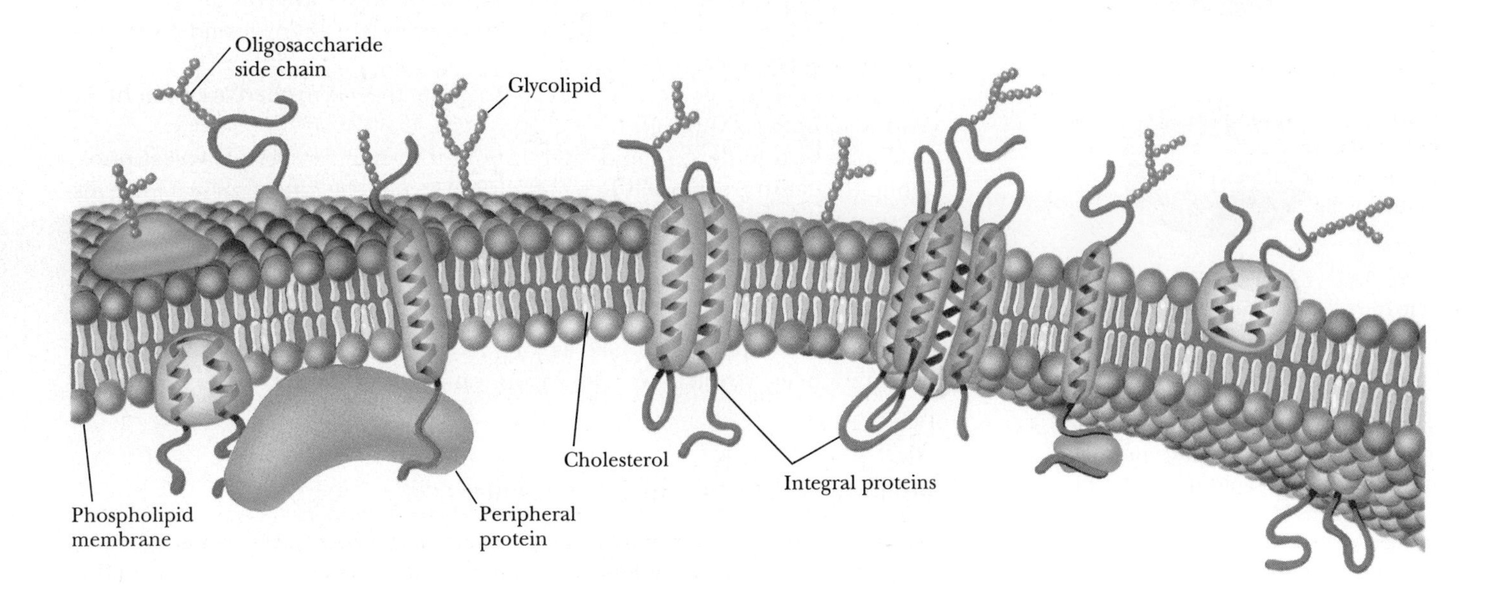

***Figure* 9.26** The fluid mosaic model of membrane structure proposed by S. J. Singer and G. L. Nicolson. In this model, the lipids and proteins are assumed to be mobile, so that they can move rapidly and laterally in the plane of the membrane. Transverse motion may also occur, but it is much slower.

membranes were low in electron density. This is consistent with the arrangement of bilayers having the hydrocarbon tails (low in electron density) in the interior of the membrane. The outside edges of these same membranes were shown to be of high electron density, which is consistent with the arrangement of the polar lipid head groups on the outside surfaces of the membrane.

Hydrocarbon Chain Orientation in the Bilayer

An important aspect of membrane structure is the orientation or ordering of lipid molecules in the bilayer. In the bilayers sketched in Figures 9.22 and 9.24, the long axes of the lipid molecules are portrayed as being perpendicular (or normal) to the plane of the bilayer. In fact, the hydrocarbon tails of phospholipids may tilt and bend and adopt a variety of orientations. Typically, the portions of a lipid chain near the membrane surface lie most nearly perpendicular to the membrane plane, and lipid chain ordering decreases toward the end of the chain (toward the middle of the bilayer).

Membrane Bilayer Mobility

The idea that lipids and proteins could move rapidly in biological membranes was a relatively new one when the fluid mosaic model was proposed. Many of the experiments designed to test this hypothesis involved the use of specially designed probe molecules. The first experiment demonstrating protein lateral movement in the membrane was described by L. Frye and M. Edidin in 1970. In this experiment, human cells and mouse cells were allowed to fuse together. Frye and Edidin used fluorescent antibodies to determine whether integral membrane proteins from the two cell types could move and intermingle in the newly formed, fused cells. The antibodies specific for human cells were labeled with rhodamine, a red fluorescent marker, and the antibodies specific for mouse cells were labeled with fluorescein, a green fluorescent marker. When both types of antibodies were added to newly fused cells, the binding pattern indicated that integral membrane proteins from the two cell types had moved laterally and were dispersed throughout the surface of the fused cell (Figure 9.27). This clearly demonstrated that integral membrane proteins possess significant lateral mobility.

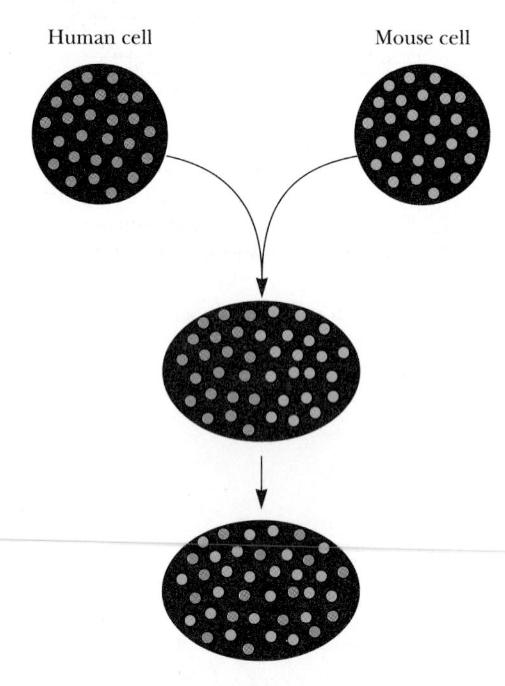

***Figure* 9.27** The Frye–Edidin experiment. Human cells with membrane antigens for red fluorescent antibodies were mixed and fused with mouse cells having membrane antigens for green fluorescent antibodies. Treatment of the resulting composite cells with red- and green-fluorescent-labeled antibodies revealed a rapid mixing of the membrane antigens in the composite membrane. This experiment demonstrated the lateral mobility of membrane proteins.

Just how fast can proteins move in a biological membrane? Many membrane proteins can move laterally across a membrane at a rate of a few microns per minute. On the other hand, some integral membrane proteins are much more restricted in their lateral movement, with diffusion rates of about 10 nm/sec or even slower. These latter proteins are often found to be anchored to the *cytoskeleton* (Chapters 34 and 36), a complex latticelike structure that maintains the cell's shape and assists in the controlled movement of various substances through the cell.

Lipids also undergo rapid lateral motion in membranes. A typical phospholipid can diffuse laterally in a membrane at a linear rate of several microns per second. At that rate, a phospholipid could travel from one end of a bacterial cell to the other in less than a second or traverse a typical animal cell in a few minutes. On the other hand, *transverse* movement of lipids (or proteins) from one face of the bilayer to the other is much slower (and much less likely). For example, it can take as long as several days for half the phospholipids in a bilayer vesicle to "flip" from one side of the bilayer to the other.

Membranes Are Asymmetric Structures

Biological membranes are **asymmetric** structures. There are several kinds of asymmetry to consider. Both the lipids and the proteins of membranes exhibit lateral and transverse asymmetries. **Lateral asymmetry** arises when lipids or proteins of particular types cluster in the plane of the membrane.

Lipids Exhibit Lateral Membrane Asymmetry

Lipids in model systems are often found in asymmetric clusters (see Figure 9.28). Such behavior is referred to as a **phase separation,** which arises either spontaneously or as the result of some extraneous influence. Phase separations can be induced in model membranes by divalent cations, which interact with negatively charged moieties on the surface of the bilayer. For example, Ca^{2+} induces phase separations in membranes formed from phosphatidylserine (PS) and phosphatidylethanolamine (PE) or from PS, PE, and phosphatidylcholine. Ca^{2+} added to these membranes forms complexes with the negatively charged serine carboxyls, causing the PS to cluster and separate from the other lipids. Such metal-induced lipid phase separations have been shown to regulate the activity of membrane-bound enzymes.

There are other ways in which the lateral organization (and asymmetry) of lipids in biological membranes can be altered. For example, cholesterol can intercalate between the phospholipid fatty acid chains, its polar hydroxyl group associated with the polar head groups. In this manner, patches of cholesterol and phospholipids can form in an otherwise homogeneous sea of pure phospholipid. This lateral asymmetry can in turn affect the function of membrane proteins and enzymes. The lateral distribution of lipids in a membrane can also be affected by proteins in the membrane. Certain integral membrane proteins prefer associations with specific lipids. Proteins may select unsaturated lipid chains over saturated chains or may prefer a specific head group over others.

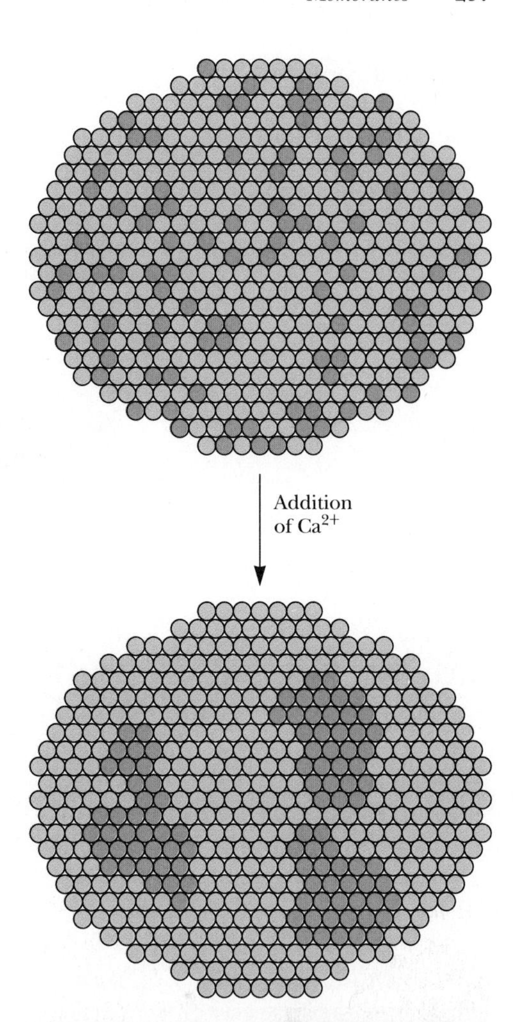

Figure **9.28** An illustration of the concept of lateral phase separations in a membrane. Phase separations of phosphatidylserine (green circles) can be induced by divalent cations such as Ca^{2+}.

Membrane Proteins Are Visible in Electron Micrographs

Membrane proteins in many cases are randomly distributed through the plane of the membrane. This was one of the corollaries of the fluid mosaic model of Singer and Nicholson and has been experimentally verified using the **freeze fracture** and **freeze etch** techniques of electron microscopy. Daniel Branton showed in 1966 that the two monolayers or leaflets of the bilayer membrane could be separated from each other and viewed separately in the electron microscope. To do this, membranes or whole cells are frozen rapidly with liquid nitrogen (so that the sample temperature approaches 77 K) and placed in a vacuum of $<1 \times 10^{-6}$ torr. The sample is then warmed carefully to 158 to 163 K and sliced or "fractured" with a sharp, cold knife called a *microtome.* When this is done, some of the membranes in the sample fall in the path of the fracture caused by the knife and are cleaved so that the two monolayers are separated (Figure 9.29). The sample is immediately coated with a thin layer of heavy metal atoms (often platinum) and then viewed in the electron microscope. The inner surfaces of the monolayers thus separated typically reveal a randomly distributed array of "bumps" known to microscopists as **intramembrane particles,** or **IMPs.** These particles, 5 to 9 nm in diameter, are the integral membrane proteins that have remained with one or the other monolayer during the fracturing process (Figure 9.30). (In some cases, the proteins may also be fractured, but most appear to survive the cleaving process.)

Freeze fracture electron micrographs of pure phospholipid bilayers and myelin membranes, which contain very little protein, reveal smooth surfaces with no IMPs. Treatment of natural membranes with proteases prior to freeze fracture likewise yields no evidence of IMPs. These observations are consistent with the notion that IMPs represent integral membrane proteins. A modification of the freeze fracture technique involves **freeze etching** to reveal the

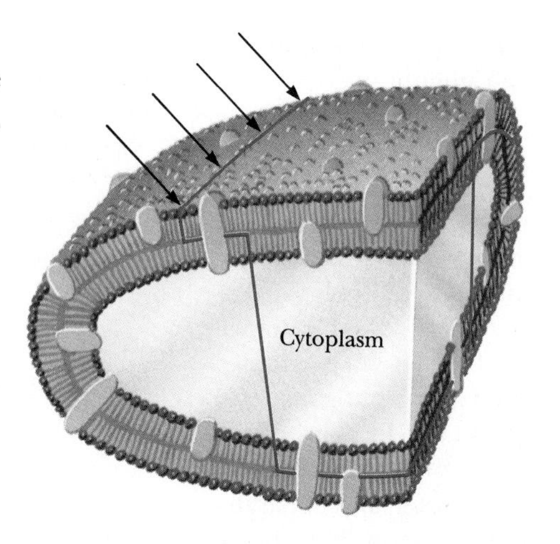

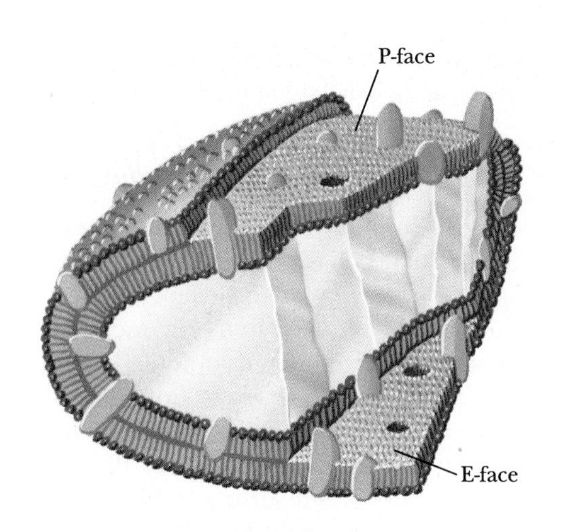

***Figure* 9.29** A diagram illustrating the freeze fracture technique used in electron microscopy of membranes and membrane components.

extracellular surface of the outer monolayer. As shown in Figure 9.31, freeze fracture can be followed immediately by warming slightly to 173 K. At this temperature in the vacuum, water and other volatile molecules on the membrane surface are sublimed, revealing the extracellular surface of the outer monolayer. A freeze fracture, freeze etch electron micrograph can thus reveal simultaneously the extracellular surface and both inner surfaces of a cell membrane. Most such micrographs show that integral membrane proteins are randomly distributed in the membrane, with no apparent long range order.

Electron Microscopy Reveals Protein Lateral Asymmetry

Membrane proteins can also be distributed in nonrandom ways across the surface of a membrane. This can occur for several reasons. Some proteins must interact intimately with certain other proteins, forming multisubunit complexes that perform specific functions in the membrane. A few integral membrane proteins are known to *self-associate* in the membrane, forming large multimeric clusters. **Bacteriorhodopsin,** a light-driven proton pump protein, forms such clusters, known as "purple patches," in the membranes of *Halobacterium halobium.* The bacteriorhodopsin protein in these purple patches forms highly ordered, two-dimensional crystals.

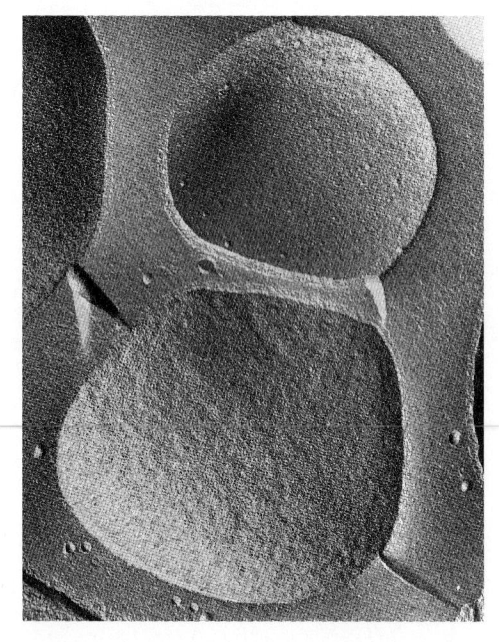

***Figure* 9.30** Freeze fracture cleavage of bilayer membranes exposes intramembrane particles (IMPs) 5 to 9 nm in diameter. These are integral membrane proteins that have remained associated with one or the other of the membrane monolayers during the process of freeze fracture.

Transverse Membrane Asymmetry

Membrane asymmetries in the **transverse** direction (from one side of the membrane to the other) can be anticipated when one considers that many properties of a membrane depend upon its two-sided nature. Properties that are a consequence of membrane "sidedness" include membrane transport, which is driven in one direction only, the effects of hormones at the outsides of cells, and the immunological reactions that occur between cells (necessarily involving only the outside surfaces of the cells). One would surmise that the proteins involved in these and other interactions must be arranged asymmetrically in the membrane.

Protein Transverse Asymmetry

Protein transverse asymmetries have been characterized using chemical, enzymatic, and immunological labeling methods. Working with **glycophorin,** the major glycoprotein in the erythrocyte membrane (discussed in Section 9.3),

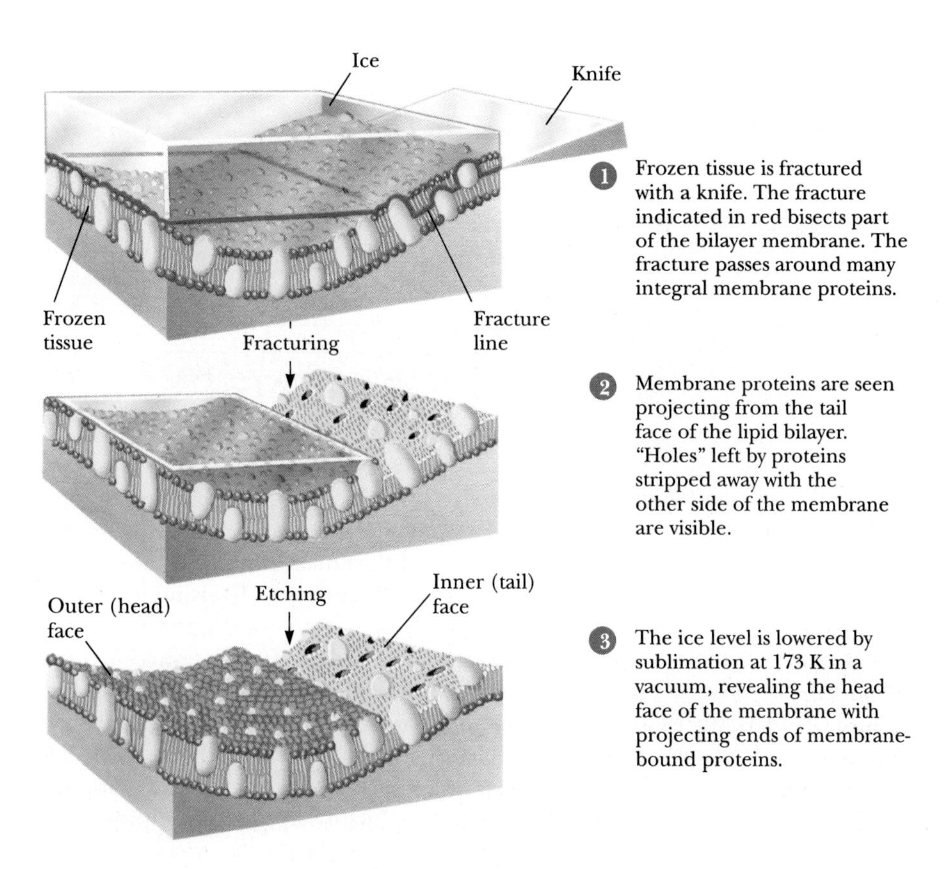

Ice
Knife
Frozen tissue
Fracturing
Fracture line
Outer (head) face
Etching
Inner (tail) face
1 Frozen tissue is fractured with a knife. The fracture indicated in red bisects part of the bilayer membrane. The fracture passes around many integral membrane proteins.
2 Membrane proteins are seen projecting from the tail face of the lipid bilayer. "Holes" left by proteins stripped away with the other side of the membrane are visible.
3 The ice level is lowered by sublimation at 173 K in a vacuum, revealing the head face of the membrane with projecting ends of membrane-bound proteins.

***Figure* 9.31** A freeze etching procedure (warming the sample slightly to allow sublimation of water and other volatile components) following freeze fracture can reveal the extracellular surface of the outer monolayer.

Mark Bretscher was the first to demonstrate the asymmetric arrangement of an integral membrane protein. Treatment of whole erythrocytes with trypsin released the carbohydrate groups of glycophorin (in the form of several small glycopeptides). Since trypsin is much too large to penetrate the erythrocyte membrane, the N-terminus of glycophorin, which contains the carbohydrate moieties, must be exposed to the outside surface of the membrane. Bretscher showed that [^{35}S]-formylmethionylsulfone methyl phosphate could label the C-terminus of glycophorin with ^{35}S in erythrocyte membrane fragments but not in intact erythrocytes. This clearly demonstrated that the C-terminus of glycophorin is uniformly exposed to the interior surface of the erythrocyte membrane. Since that time, many integral membrane proteins have been shown to be oriented uniformly in their respective membranes.

Lipid Transverse Asymmetry

Phospholipids are also distributed asymmetrically across many membranes. It was found that, in the erythrocyte, phosphatidylcholine (PC) comprises about 30% of the total phospholipid in the membrane. Of this amount, 76% is found in the outer monolayer and 24% is found in the inner monolayer. Since this early observation, the lipids of many membranes have been found to be asymmetrically distributed between the inner and outer monolayers. Figure 9.32 shows the asymmetric distribution of phospholipids observed in the human erythrocyte membrane. Asymmetric lipid distributions are important to cells in several ways. The carbohydrate groups of glycolipids (and of glycoproteins) always face the outside surface of plasma membranes where they participate in cell recognition phenomena. Asymmetric lipid distributions may also be important to various integral membrane proteins, which

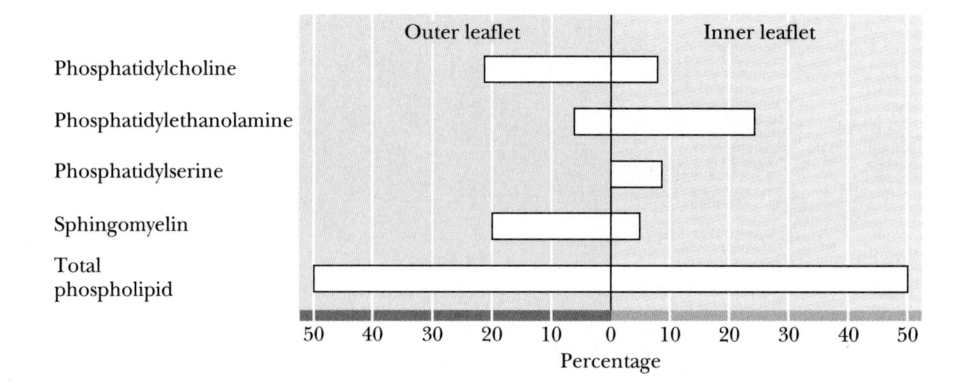

***Figure* 9.32** Phospholipids are arranged asymmetrically in most membranes, including the human erythrocyte membrane, as shown here. Values are mole percentages.
(After Rothman and Lenard, 1977, Science 194:1744.)

may prefer particular lipid classes in the inner and outer monolayers. The total charge on the inner and outer surfaces of a membrane depends on the distribution of lipids. The resulting charge differences affect the membrane potential, which in turn is known to modulate the activity of certain ion channels and other membrane proteins.

How are transverse lipid asymmetries created and maintained in cell membranes? From a thermodynamic perspective, these asymmetries could only occur by virtue of asymmetric syntheses of the bilayer itself or by energy-dependent asymmetric transport mechanisms. Without at least one of these, lipids of all kinds would eventually distribute equally between the two monolayers of a membrane. In eukaryotic cells, phospholipids, glycolipids, and cholesterol are synthesized by enzymes located in (or on the surface of) the endoplasmic reticulum (ER) and the Golgi system (discussed in Chapter 24). Most if not all of these biosynthetic processes are asymmetrically arranged across the membranes of the ER and Golgi. There is also a separate and continuous flow of phospholipids, glycolipids, and cholesterol from the ER and Golgi to other membranes in the cell, including the plasma membrane. This flow is mediated by specific **lipid transfer proteins.** Most cells appear to contain such proteins.

Flippases: Proteins Which Flip Lipids Across the Membrane

Proteins that can "flip" phospholipids from one side of a bilayer to the other have also been identified in several tissues (Figure 9.33). Called **flippases,** these proteins reduce the half-time for phospholipid movement across a membrane from 10 days or more to a few minutes or less. Some of these systems may operate passively, with no required input of energy, but passive transport alone cannot establish or maintain asymmetric transverse lipid dis-

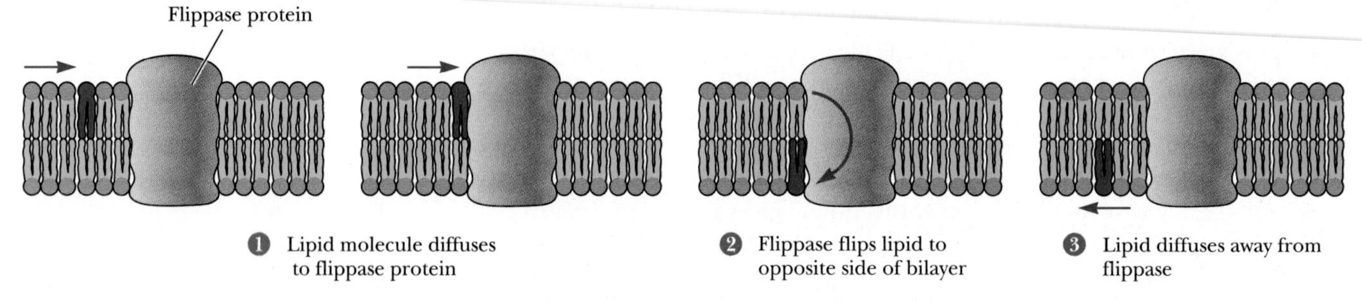

***Figure* 9.33** Phospholipids can be "flipped" across a bilayer membrane by the action of flippase proteins. When, by normal diffusion through the bilayer, the lipid encounters a flippase, it can be moved quickly to the other face of the bilayer.

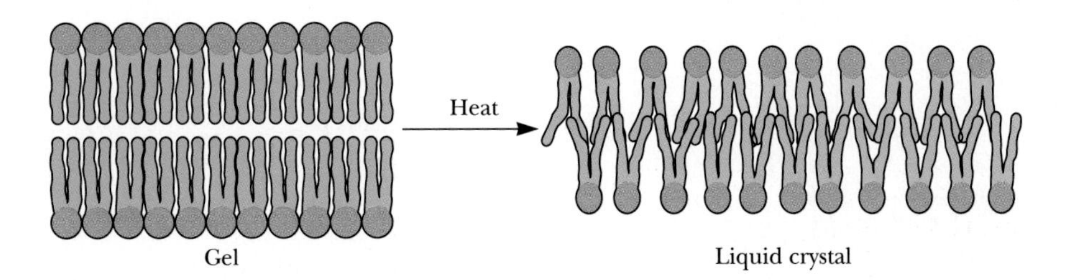

***Figure* 9.34** An illustration of the gel-to-liquid crystalline phase transition. which occurs when a membrane is warmed through the transition temperature, T_m. Notice that the surface area must increase and the thickness must decrease as the membrane goes through a phase transition. The mobility of the lipid chains increases dramatically.

tributions. However, rapid phospholipid movement from one monolayer to the other occurs in an *ATP-dependent* manner in erythrocytes. Energy-dependent lipid flippase activity may be responsible for the creation and maintenance of transverse lipid asymmetries.

Membrane Phase Transitions

Lipids in bilayers undergo radical changes in physical state over characteristic narrow temperature ranges. These changes are in fact true **phase transitions,** and the temperatures at which these changes take place are referred to as **transition temperatures** or **melting temperatures** (T_m). These phase transitions involve substantial changes in the organization and motion of the fatty acyl chains within the bilayer. The bilayer below the phase transition exists in a closely packed gel state, with the fatty acyl chains relatively immobilized in a tightly packed array (Figure 9.34). In this state, the anti configuration is adopted by all the carbon–carbon bonds in the lipid chains. This leaves the lipid chains in their fully extended conformation. As a result, the surface area per lipid is minimal and the bilayer thickness is maximal. Above the transition temperature, a liquid crystalline state exists in which the mobility of fatty acyl chains is intermediate between solid and liquid alkane. In this more fluid, liquid crystalline state, the carbon–carbon bonds of the lipid chains more readily adopt gauche configurations (Figure 9.35). As a result, the surface area per lipid increases and the bilayer thickness decreases by 10 to 15%.

The sharpness of the transition in pure lipid preparations shows that the phase change is a cooperative behavior. This is to say that the behavior of one or a few molecules affects the behavior of many other molecules in the vicinity. The sharpness of the transition then reflects the number of molecules that are acting in concert. Sharp transitions involve large numbers of molecules all "melting" together.

Phase transitions have been characterized in a number of different pure and mixed lipid systems. Table 9.2 shows a comparison of the transition temperatures observed for several different phosphatidylcholines with different fatty acyl chain compositions. General characteristics of bilayer phase transitions include the following:

1. The transitions are always endothermic; heat is absorbed as the temperature increases through the transition (Figure 9.35).
2. Particular phospholipids display characteristic transition temperatures (T_m). As shown in Table 9.2, T_m increases with chain length, decreases with unsaturation, and depends on the nature of the polar head group.
3. For pure phospholipid bilayers, the transition occurs over a narrow temperature range. The phase transition for dimyristoyl lecithin has a peak width of about 0.2°C.
4. Native biological membranes also display characteristic phase transi-

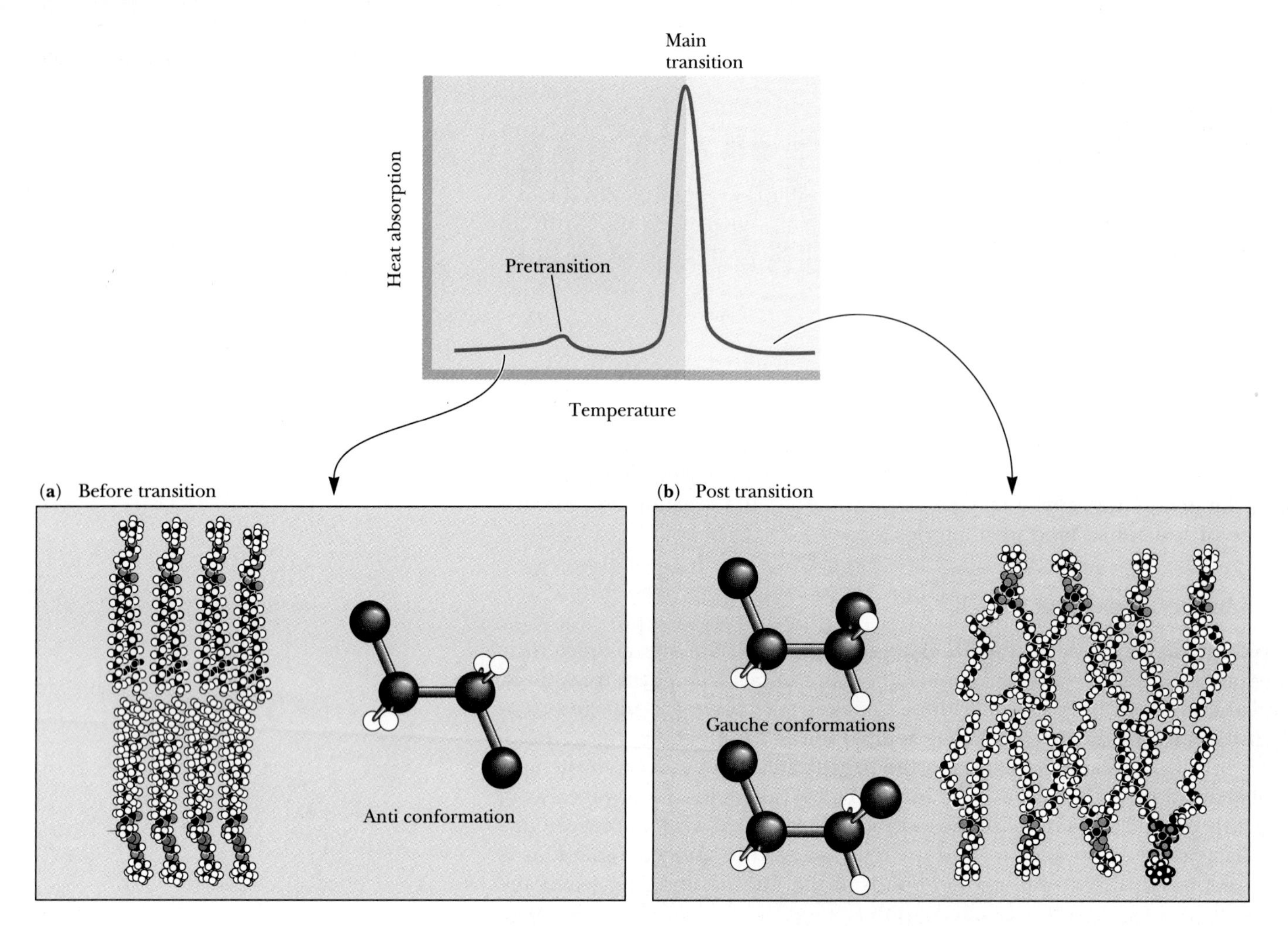

Figure **9.35** Membrane lipid phase transitions can be detected and characterized by measuring the rate of absorption of heat by a membrane sample in a calorimeter (see Chapter 15 for a detailed discussion of calorimetry). Pure, homogeneous bilayers (containing only a single lipid component) give sharp calorimetric peaks. Egg PC contains a variety of fatty acid chains and thus yields a broad calorimetric peak. Below the phase transition, lipid chains primarily adopt the anti conformation. Above the phase transition, lipid chains have absorbed a substantial amount of heat. This is reflected in the adoption of higher-energy conformations, including the gauche configurations shown.

Table **9.2**

Phase Transition Temperatures for Phospholipids in Water

Phospholipid	Transition Temperature (T_m), °C
Dipalmitoyl phosphatidic acid (Di 16:0 PA)	67
Dipalmitoyl phosphatidylethanolamine (Di 16:0 PE)	63.8
Dipalmitoyl phosphatidylcholine (Di 16:0 PC)	41.4
Dipalmitoyl phosphatidylglycerol (Di 16:0 PG)	41.0
Dilauroyl phosphatidylcholine (Di 14:0 PC)	23.6
Distearoyl phosphatidylcholine (Di 18:0 PC)	58
Dioleoyl phosphatidylcholine (Di 18:1 PC)	−22
1-Stearoyl-2-oleoyl-phosphatidylcholine (1-18:0, 2-18:1 PC)	3
Egg phosphatidylcholine (Egg PC)	−15

Adapted from Jain, M., and Wagner, R. C., 1980. *Introduction to Biological Membranes.* New York: John Wiley and Sons; Martonosi, A., ed., 1982. *Membranes and Transport,* Vol. 1. New York: Plenum Press.

tions, but these are broad and strongly dependent on the lipid and protein composition of the membrane.

5. With certain lipid bilayers, a change of physical state referred to as a *pretransition* occurs 5° to 15°C below the phase transition itself. These pretransitions involve a tilting of the hydrocarbon chains.
6. A volume change is usually associated with phase transitions in lipid bilayers.
7. Bilayer phase transitions are sensitive to the presence of solutes that interact with lipids, including multivalent cations, lipid-soluble agents, peptides, and proteins.

9.3 Structure of Membrane Proteins

The lipid bilayer constitutes the fundamental structural unit of all biological membranes. Proteins, in contrast, carry out essentially all of the active functions of membranes, including transport activities, receptor functions, and other related processes. As suggested by Singer and Nicolson, most membrane proteins can be classified as peripheral or integral. The **peripheral proteins** are globular proteins that interact with the membrane mainly through electrostatic and hydrogen-bonding interactions with integral proteins. Although peripheral proteins are not discussed further here, many proteins of this class will be described in the context of other discussions throughout this textbook. **Integral proteins** are those that are strongly associated with the lipid bilayer, with a portion of the protein embedded in, or extending all the way across, the lipid bilayer. Another class of proteins not anticipated by Singer and Nicolson, the **lipid-anchored proteins,** are important in a variety of functions in different cells and tissues. These proteins associate with membranes by means of a variety of covalently linked lipid anchors.

Integral Membrane Proteins

Despite the diversity of integral membrane proteins, most fall into two general classes. One of these includes proteins attached or anchored to the membrane by only a small hydrophobic segment, such that most of the protein extends out into the water solvent on one or both sides of the membrane. The other class includes those proteins that are more globular in shape and more totally embedded in the membrane, exposing only a small surface to the water solvent outside the membrane.

A Protein with a Single Transmembrane Segment

In the case of the proteins that are anchored by a small hydrophobic polypeptide segment, that segment often takes the form of a single α-helix. One of the best examples of a membrane protein with such an α-helical structure is **glycophorin.** Most of glycophorin's mass is oriented on the outside surface of the cell, exposed to the aqueous milieu (Figure 9.36). A variety of hydrophilic oligosaccharide units are attached to this extracellular domain. These oligosaccharide groups constitute the ABO and MN blood group antigenic specificities of the red cell. This extracellular portion of the protein also serves as the receptor for the influenza virus. Glycophorin has a total molecular weight of about 31,000 and is approximately 40% protein and 60% carbohydrate. The glycophorin primary sequence consists of a segment of 19 hydrophobic amino acid residues with a short hydrophilic sequence on one end and a longer hydrophilic sequence on the other end. The 19-residue sequence is just the right length to span the cell membrane if it is coiled in the shape of an

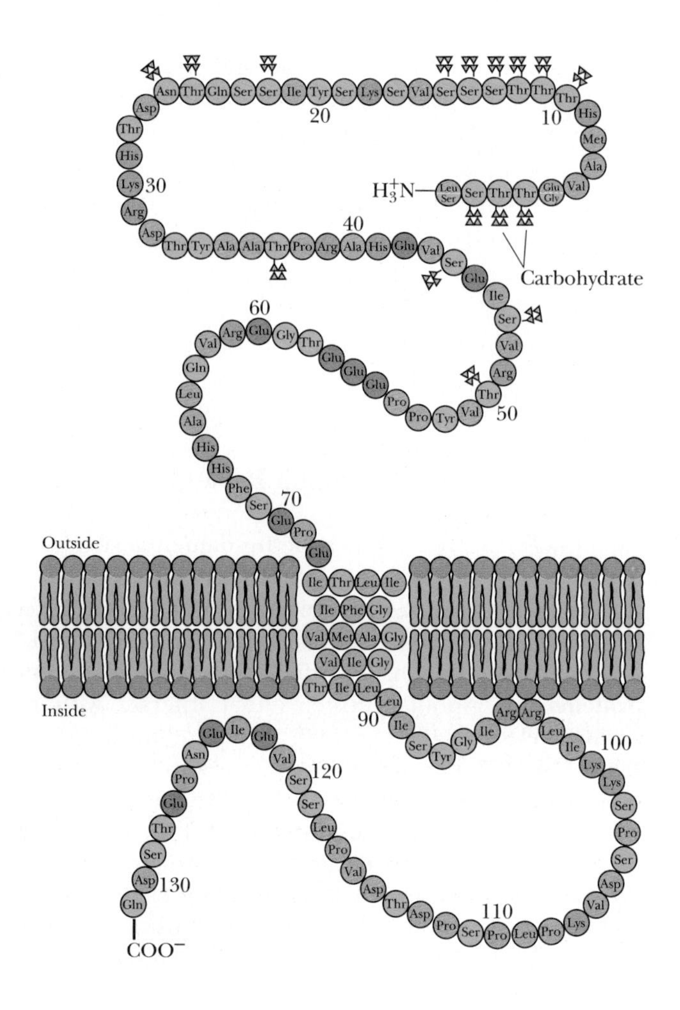

***Figure* 9.36** Glycophorin A spans the membrane of the human erythrocyte via a single α-helical transmembrane segment. The C-terminus of the peptide, whose sequence is shown here, faces the cytosol of the erythrocyte; the N-terminal domain is extracellular. Points of attachment of carbohydrate groups are indicated.

α-helix. The large hydrophilic sequence includes the amino terminal residue of the polypeptide chain.

Numerous other membrane proteins are also attached to the membrane by means of a single hydrophobic α-helix, with hydrophilic segments extending either into the cytoplasm or the extracellular space. These proteins often function as receptors for extracellular molecules or as recognition sites that allow the immune system to recognize and distinguish the cells of the host organism from invading foreign cells or viruses. The proteins that represent the *major transplantation antigens H2* in mice and *human leukocyte associated (HLA) proteins* in humans are members of this class. Other such proteins include the *surface immunoglobulin receptors* on B lymphocytes and the *spike proteins* of many membrane viruses. The function of many of these proteins depends primarily on their extracellular domain, and thus the segment facing the intracellular surface is often a shorter one.

Bacteriorhodopsin: A 7-Transmembrane Segment Protein

Membrane proteins that take on a more globular shape, instead of the rodlike structure previously described, are often involved with transport activities and other functions requiring a substantial portion of the peptide to be embedded in the membrane. In these proteins, the primary sequence may consist of numerous hydrophobic α-helical segments joined by hinge regions so that the protein winds in a zig-zag pattern back and forth across the membrane. A well-characterized example of such a protein is **bacteriorhodopsin,** which clusters in purple patches in the membrane of the bacterium *Halobacterium halobium.* The name *Halobacterium* refers to the fact that this bacterium thrives in

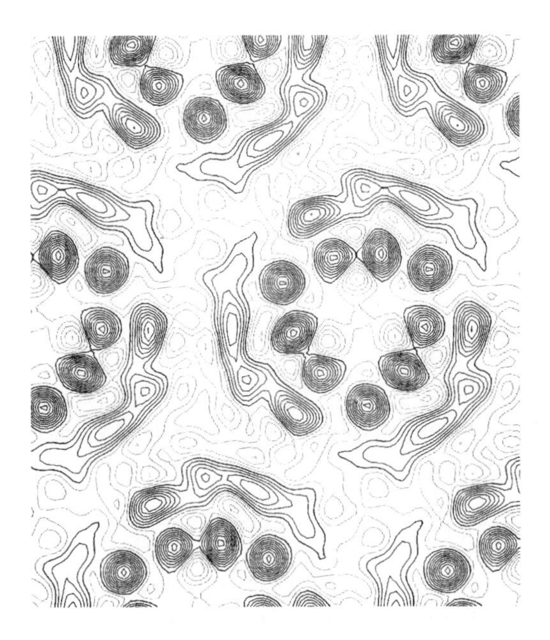

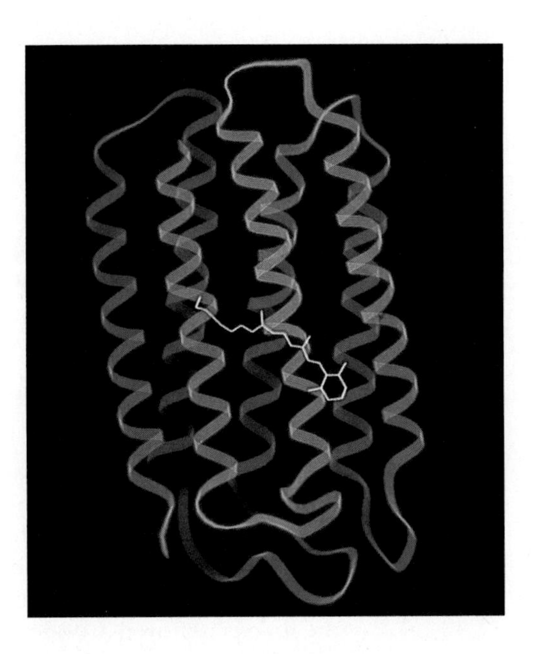

***Figure* 9.37** An electron density profile illustrating the three centers of threefold symmetry in arrays of bacteriorhodopsin in the purple membrane of *Halobacterium halobium,* together with a computer-generated model showing the seven α-helical transmembrane segments in bacteriorhodopsin.

(Electron density map from *Stoecknius, W., 1980. Purple membrane of halobacteria: a new light-energy converter. Accounts of Chemical Research 13:337–344.* Model on right from *Henderson, R., 1990. Model for the structure of bacteriorhodopsin based on high-resolution electron cryo-microscopy. Journal of Molecular Biology 213:899–929.*)

solutions having high concentrations of sodium chloride, such as the salt beds of San Francisco Bay. *Halobacterium* carries out a light-driven proton transport by means of bacteriorhodopsin, named in reference to its spectral similarities to rhodopsin in the rod outer segments of the mammalian retina. When this organism is deprived of oxygen for oxidative metabolism, it switches to the capture of energy from sunlight, using this energy to pump protons out of the cell. The proton gradient generated by such light-driven proton pumping represents potential energy, which is exploited elsewhere in the membrane to synthesize ATP.

Bacteriorhodopsin clusters in hexagonal arrays (Figure 9.37) in the purple membrane patches of *Halobacterium,* and it was this orderly, repeating arrangement of proteins in the membrane that enabled Nigel Unwin and Richard Henderson in 1975 to determine the bacteriorhodopsin structure. The polypeptide chain crosses the membrane seven times, in seven α-helical segments, with very little of the protein exposed to the aqueous milieu (Figure 9.38). The bacteriorhodopsin structure has become a model of globular

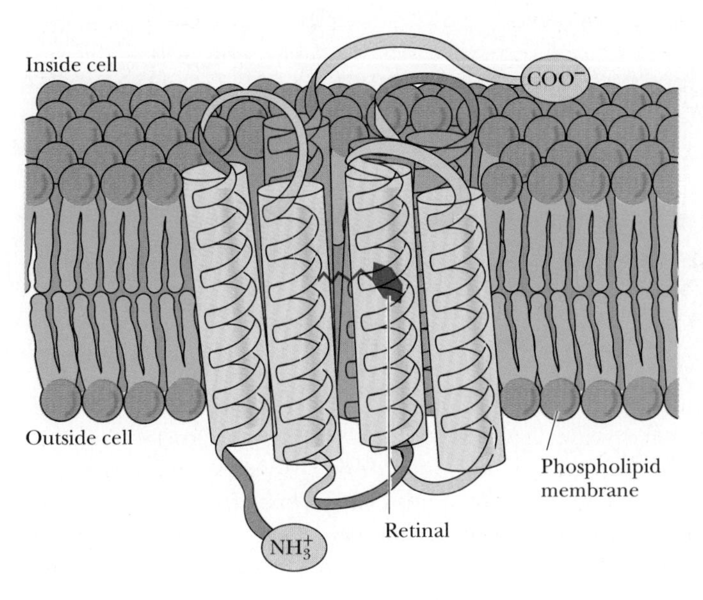

***Figure* 9.38** A model for the structure of bacteriorhodopsin in the purple membrane of *Halobacterium halobium.* The N-terminal segment is outside the cell and the C-terminal segment is intracellular. The retinal chromophore is buried deep within the structure and lies approximately parallel to the plane of the membrane.

membrane protein structure. The primary sequences of many other integral membrane proteins contain numerous hydrophobic sequences which, like those of bacteriorhodopsin, could form α-helical transmembrane segments. For example, the primary sequence of the sodium–potassium transport ATPase contains eight hydrophobic segments of length sufficient to span the plasma membrane. By analogy with bacteriorhodopsin, one would expect that these segments form a globular hydrophobic core that anchors the ATPase in the membrane. The helical segments may also account for the transport properties of the enzyme itself.

Lipid-Anchored Membrane Proteins

Certain proteins are found to be covalently linked to lipid molecules. For many of these proteins, covalent attachment of lipid is required for association with a membrane. The lipid moieties can insert into the membrane bilayer, effectively **anchoring** their linked proteins to the membrane. Some proteins with covalently linked lipid normally behave as soluble proteins; others are integral membrane proteins and remain membrane-associated even when the lipid is removed. Covalently bound lipid in these latter proteins can play a role distinct from membrane anchoring. In many cases, attachment to the membrane via the lipid anchor serves to modulate the activity of the protein.

Four different types of lipid anchoring motifs have been found to date. These are **amide-linked myristoyl** anchors, **thioester-linked fatty acyl** anchors, **thioether-linked prenyl** anchors, and **amide-linked glycosyl phosphatidylinositol** anchors. Each of these anchoring motifs is used by a variety of membrane proteins, but each nonetheless exhibits a characteristic pattern of structural requirements.

Amide-Linked Myristoyl Anchors

Myristic acid may be linked via an amide bond to the α-amino group of the N-terminal glycine residue of selected proteins (Figure 9.39). The reaction is referred to as ***N*-myristoylation** and is catalyzed by *myristoyl–CoA:protein* N*-myristoyltransferase,* known simply as **NMT.** *N*-Myristoyl-anchored proteins include the catalytic subunit of *cAMP-dependent protein kinase,* the *pp60^{src} tyrosine*

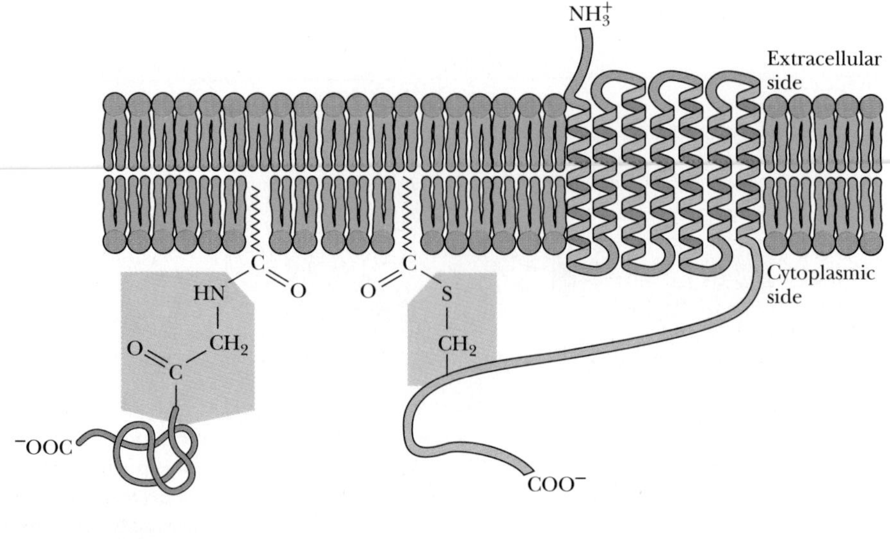

***Figure* 9.39** Certain proteins are anchored to biological membranes by lipid anchors. Particularly common are the *N*-myristoyl- and *S*-palmitoyl-anchoring motifs shown here. *N*-Myristoylation always occurs at an N-terminal glycine residue, whereas thioester linkages occur at cysteine residues within the polypeptide chain. G-protein-coupled receptors, with seven transmembrane segments, may contain one (and sometimes *two*) palmitoyl anchors in thioester linkage to cysteine residues in the C-terminal segment of the protein.

kinase, the phosphatase known as *calcineurin B,* the α-subunit of *G proteins* (involved in GTP-dependent transmembrane signaling events), and the *gag proteins* of certain retroviruses, including the HIV-I virus that causes AIDS.

Thioester-Linked Fatty Acyl Anchors

A variety of cellular and viral proteins contain fatty acids covalently bound via ester linkages to the side chains of cysteine and sometimes to serine or threonine residues within a polypeptide chain (Figure 9.39). This type of fatty acyl chain linkage has a broader fatty acid specificity than *N*-myristoylation. Myristate, palmitate, stearate, and oleate can all be esterified in this way, with the C_{16} and C_{18} chain lengths being most commonly found. Proteins anchored to membranes via fatty acyl thioesters include *G-protein-coupled receptors,* the *surface glycoproteins* of several viruses, and the *transferrin receptor* protein.

Thioether-Linked Prenyl Anchors

As noted in Section 9.1, polyprenyl (or simply prenyl) groups are long-chain polyisoprenoid groups derived from isoprene units. Prenylation of proteins destined for membrane anchoring can involve either **farnesyl** or **geranylgeranyl** groups (Figure 9.40). The addition of a prenyl group typically occurs at the cysteine residue of a carboxy-terminal CAAX sequence of the target protein, where C is cysteine, A is an aliphatic residue, and X can be any amino acid. As shown in Figure 9.40, the result is a thioether-linked farnesyl or geranylgeranyl group. Once the prenylation reaction has occurred, a specific protease cleaves the three carboxy-terminal residues, and the carboxyl group of the now terminal Cys is methylated to produce an ester. All of these modifications appear to be important for subsequent activity of the prenyl-anchored protein. Proteins anchored to membranes via prenyl groups include *yeast mating factors,* the *p21^{ras} protein* (the protein product of the *ras* oncogene; see Chapter 37), and the *nuclear lamins,* structural components of the lamina of the inner nuclear membrane.

Glycosyl Phosphatidylinositol Anchors

Glycosyl phosphatidylinositol, or **GPI,** groups are structurally more elaborate membrane anchors than fatty acyl or prenyl groups. GPI groups modify the carboxy-terminal amino acid of a target protein via an ethanolamine residue linked to an oligosaccharide, which is linked in turn to the inositol moiety of

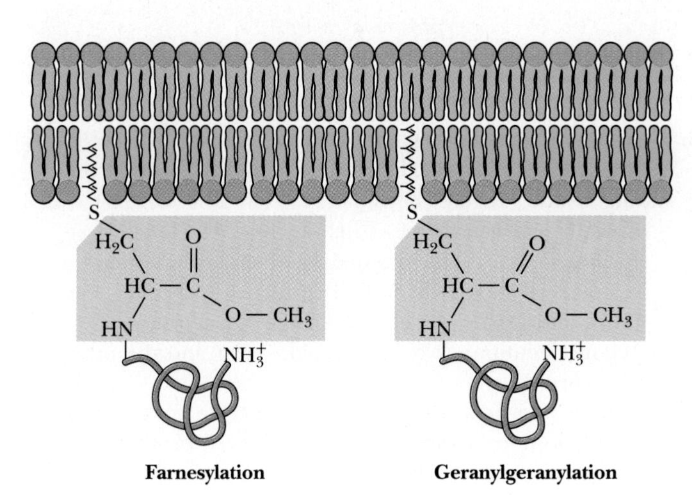

***Figure* 9.40** Proteins containing the C-terminal sequence CAAX can undergo prenylation reactions that place thioether-linked farnesyl or geranylgeranyl groups at the cysteine side chain. Prenylation is accompanied by removal of the AAX peptide and methylation of the carboxyl group of the cysteine residue, which has become the C-terminal residue.

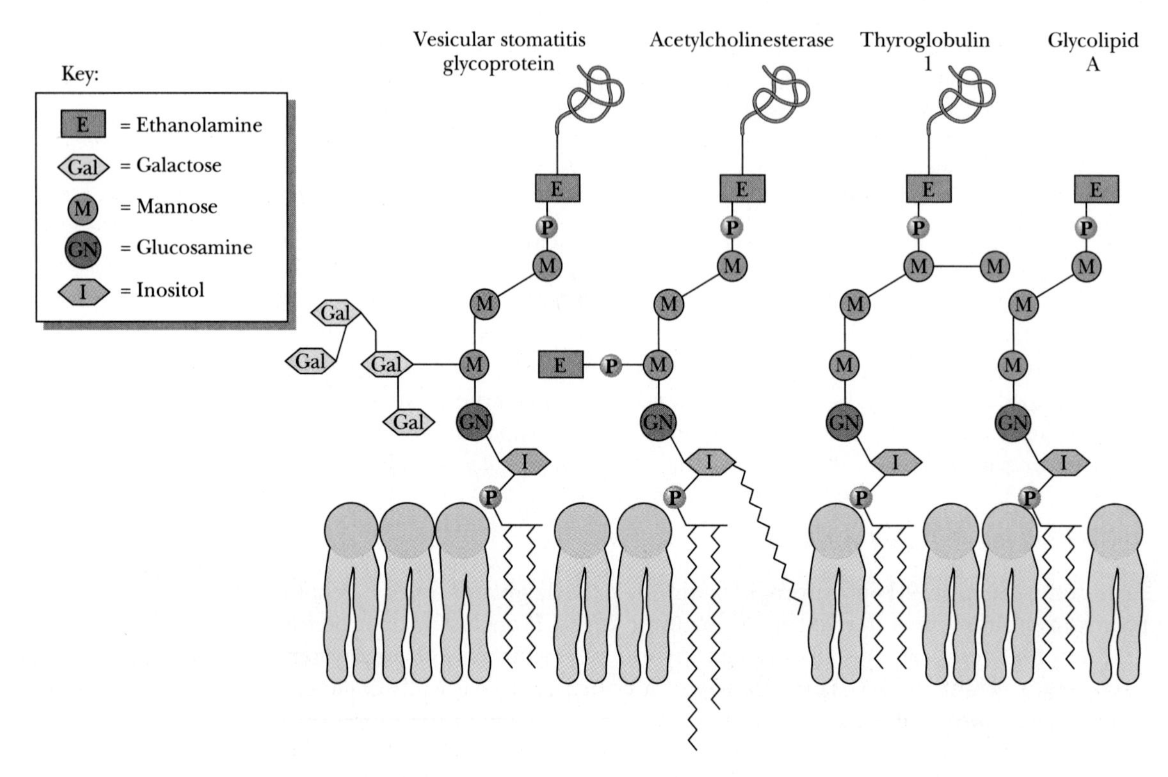

***Figure* 9.41** The glycosyl phosphatidylinositol (GPI) moiety is an elaborate lipid-anchoring group. Note the core of three mannose residues and a glucosamine. Additional modifications may include fatty acids at the inositol and glycerol —OH groups.

a phosphatidylinositol (Figure 9.41). The oligosaccharide typically consists of a conserved tetrasaccharide core of three mannose residues and a glucosamine, which can be altered by modifications of the mannose residues or addition of galactosyl side chains of various sizes, extra phosphoethanolamines, or additional *N*-acetylgalactose or mannosyl residues (Figure 9.41). The inositol moiety can also be modified by an additional fatty acid, and a variety of fatty acyl groups are found linked to the glycerol group. GPI groups anchor a wide variety of *surface antigens, adhesion molecules,* and *cell surface hydrolases* to plasma membranes in various eukaryotic organisms. GPI anchors have not yet been observed in prokaryotic organisms or plants.

Problems

1. Draw the structures of (a) all the possible triacylglycerols that can be formed from glycerol with stearic and arachidonic acid and (b) all the phosphatidylserine isomers that can be formed from palmitic and linolenic acids. Which of the PS isomers are not likely to be found in biological membranes?

2. The purple patches of the *Halobacterium halobium* membrane, which contain the protein bacteriorhodopsin, are approximately 75% protein and 25% lipid. If the protein molecular weight is 26,000 and an average phospholipid has a molecular weight of 800, calculate the phospholipid to protein mole ratio.

3. Sucrose gradients for separation of membrane proteins must be able to separate proteins and protein–lipid complexes having a wide range of densities, typically 1.00 to 1.35 g/mL.
   a. Consult reference books (such as the *CRC Handbook of Biochemistry*) and plot the density of sucrose solutions versus percent sucrose by weight (g sucrose per 100 g solution), and versus percent by volume (g sucrose per 100 mL solution). Why is one plot linear and the other plot curved?
   b. What would be a suitable range of sucrose concentrations for separation of three membrane-derived protein–lipid complexes with densities of 1.03, 1.07, and 1.08 g/mL?

4. Phospholipid lateral motion in membranes is characterized by a diffusion coefficient of about 1×10^{-8} cm^2/sec. The distance traveled in two dimensions (across the membrane) in a given time is $r = (4Dt)^{1/2}$, where r is the distance traveled in centimeters, D is the diffusion coefficient, and t is the time during which diffusion occurs. Calculate the distance traveled by a phospholipid across a bilayer in 10 msec (milliseconds).

5. Protein lateral motion is much slower than that of lipids because proteins are larger than lipids. Also, some membrane proteins can diffuse freely through the membrane, whereas others are bound or anchored to other protein structures in the membrane. The diffusion constant for the membrane protein fibronectin is approximately 0.7×10^{-12} cm^2/sec, whereas that for rhodopsin is about 3×10^{-9} cm^2/sec.
   a. Calculate the distance traversed by each of these proteins in 10 msec.
   b. What could you surmise about the interactions of these proteins with other membrane components?

6. Discuss the effects on the lipid phase transition of pure dimyristoyl phosphatidylcholine vesicles of added (a) divalent cations, (b) cholesterol, (c) distearoyl phosphatidylserine, (d) dioleoyl phosphatidylcholine, and (e) integral membrane proteins.

Further Reading

Bennett, V., 1985. The membrane skeleton of human erythrocytes and its implications for more complex cells. *Annual Review of Biochemistry* **54**:273–304.

Branton, D., 1985. Fracture faces of frozen membranes. *Proceedings of the National Academy of Sciences* **55**:1048–1056.

Bretscher, M., 1985. The molecules of the cell membrane. *Scientific American* **253**:100–108.

Dawidowicz, E. A., 1987. Dynamics of membrane lipid metabolism and turnover. *Annual Review of Biochemistry* **56**:43–61.

Doering, T. L., Masterson, W. J., Hart, G. W., and Englund, P. T., 1990. Biosynthesis of glycosyl phosphatidylinositol membrane anchors. *Journal of Biological Chemistry* **265**:611–614.

Fasman, G. D., and Gilbert, W. A., 1990. The prediction of transmembrane protein sequences and their conformation: an evaluation. *Trends in Biochemical Sciences* **15**:89–92.

Frye, C. D., and Edidin, M., 1970. The rapid intermixing of cell surface antigens after formation of mouse-human heterokaryons. *Journal of Cell Science* **7**:319–335.

Glomset, J. A., Gelb, M. H., and Farnsworth, C. C., 1990. Prenyl proteins in eukaryotic cells: a new type of membrane anchor. *Trends in Biochemical Sciences* **15**:139–142.

Gordon, J. I., Duronio, R. J., Rudnick, D. A., Adams, S. P., and Gokel, G. W., 1991. Protein N-myristoylation. *Journal of Biological Chemistry* **266**:8647–8650.

Jain, M. K., 1988. *Introduction to Biological Membranes,* 2nd ed. New York: John Wiley & Sons.

Jennings, M. L., 1989. Topography of membrane proteins. *Annual Review of Biochemistry* **58**:999–1027.

Marchesi, V. T., 1985. Stabilizing infrastructure of cell membranes. *Annual Review of Cell Biology* **1**:531–561.

Marchesi, V. T., 1984. Structure and function of the erythrocyte membrane skeleton. *Progress in Clinical Biology Research* **159**:1–12.

Op den Kamp, J. A. F., 1979. Lipid asymmetry in membranes. *Annual Review of Biochemistry* **48**:47–71.

Robertson, R. N., 1983. *The Lively Membranes.* Cambridge: Cambridge University Press.

Seelig, J., and Seelig, A., 1981. Lipid conformation in model membranes and biological membranes. *Quarterly Review of Biophysics,* **13**:19–61.

Sefton, B., and Buss, J. E., 1987. The covalent modification of eukaryotic proteins with lipid. *Journal of Cell Biology* **104**:1449–1453.

Singer, S. J., and Nicolson, G. L., 1972. The fluid mosaic model of the structure of cell membranes. *Science* **175**:720–731.

Singer, S. J., and Yaffe, M. P., 1990. Embedded or not? Hydrophobic sequences and membranes. *Trends in Biochemical Sciences* **15**:369–373.

Tanford, C., 1980. *The Hydrophobic Effect: Formation of Micelles and Biological Membranes,* 2nd ed. New York: Wiley-Interscience.

Towler, D. A., Gordon, J. I., Adams, S. P., and Glaser, L., 1988. The biology and enzymology of eukaryotic protein acylation. *Annual Review of Biochemistry* **57**:69–99.

Unwin, N., and Henderson, R., 1984. The structure of proteins in biological membranes. *Scientific American* **250**:78–94.

Wirtz, K. W. A., 1991. Phospholipid transfer proteins. *Annual Review of Biochemistry* **60**:73–99.

"Sugar in the gourd and honey in the horn,
I never was so happy since the hour I was born."
Turkey in the Straw, stanza 6 (classic American folk tune)

Chapter 10

Carbohydrates and Cell Surfaces

"The Discovery of Honey"
—Piero di Cosimo (1462)

Outline

10.1 Carbohydrate Nomenclature
10.2 Monosaccharides
10.3 Oligosaccharides
10.4 Polysaccharides
10.5 Glycoproteins
10.6 Proteoglycans

Carbohydrates are the single most abundant class of organic molecules found in nature. The name *carbohydrate* arises from the basic molecular formula $(CH_2O)_n$, which can be rewritten $(C \cdot H_2O)_n$ to show that these substances are hydrates of carbon, where $n = 3$ or more. Carbohydrates constitute a versatile class of molecules. Energy from the sun captured by green plants, algae, and some bacteria during photosynthesis (see Chapter 22) is stored in the form of carbohydrates. In turn, carbohydrates are the metabolic precursors of virtually all other biomolecules. Breakdown of carbohydrates provides the energy that sustains animal life. In addition, carbohydrates are covalently linked with a variety of other molecules. Carbohydrates linked to lipid molecules, or **glycolipids,** are common components of biological membranes. Proteins that have covalently linked carbohydrates are called **glycoproteins.** These two classes of biomolecules, together called **glycoconjugates,** are important components of cell walls and extracellular structures in plants, animals, and bacteria. In addition to the structural roles such molecules play, they also serve in a variety of processes involving *recognition* between cell types or recognition of cellular structures by other molecules. Recognition events are important in normal cell growth, fertilization, transformation of cells, and other processes.

All of these functions are made possible by the characteristic chemical features of carbohydrates: (1) the existence of at least one and often two or more asymmetric centers, (2) the ability to exist either in linear or ring structures, (3) the capacity to form polymeric structures via *glycosidic* bonds, and (4) the potential to form multiple hydrogen bonds with water or other molecules in their environment.

10.1 Carbohydrate Nomenclature

Carbohydrates are generally classified into three groups: **monosaccharides** (and their derivatives), **oligosaccharides,** and **polysaccharides.** The monosaccharides are also called **simple sugars** and have the formula $(CH_2O)_n$. Monosaccharides cannot be broken down into smaller sugars under mild conditions. Oligosaccharides derive their name from the Greek word *oligo,* meaning "few" and consist of from two to ten simple sugar molecules. Disaccharides are common in nature, and trisaccharides also occur frequently. Four- to six-sugar-unit oligosaccharides are usually bound covalently to other molecules, including glycoproteins. As their name suggests, polysaccharides are polymers of the simple sugars and their derivatives. They may be either linear or branched polymers and may contain hundreds or even thousands of monosaccharide units. Their molecular weights range up to 1 million or more.

10.2 Monosaccharides

Classification

Monosaccharides consist typically of three to seven carbon atoms and are described either as **aldoses** or **ketoses,** depending on whether the molecule contains an aldehyde function or a ketone group. The simplest aldose is glyceraldehyde, and the simplest ketose is dihydroxyacetone (Figure 10.1). These two simple sugars are termed **trioses** because they each contain three carbon atoms. The structures and names of a family of aldoses and ketoses with three, four, five, and six carbons are shown in Figures 10.2 and 10.3. *Hexoses* are the most abundant sugars in nature. Nevertheless, sugars from all these classes are important in metabolism.

Monosaccharides, either aldoses or ketoses, are often given more detailed generic names to describe both the important functional groups and the total number of carbon atoms. Thus, one can refer to *aldotetroses* and *ketotetroses, aldopentoses* and *ketopentoses, aldohexoses* and *ketohexoses,* and so on. Sometimes the ketone-containing monosaccharides are named simply by inserting the letters -ul- into the simple generic terms, such as *tetruloses, pentuloses, hexuloses, heptuloses,* and so on. The simplest monosaccharides are water-soluble, and most taste sweet.

Stereochemistry

Aldoses with at least three carbons and ketoses with at least four carbons contain **chiral centers** (Chapter 3). The nomenclature for such molecules must specify the **configuration** about each asymmetric center, and drawings of these molecules must be based on a system that clearly specifies these configurations. As noted in Chapter 3, the **Fischer projection** system is used almost

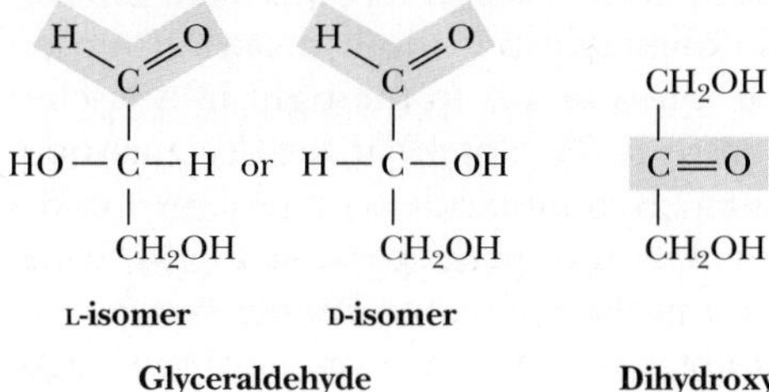

CH_2OH

C=O

CH_2OH

Dihydroxyacetone

***Figure* 10.1** Structure of a simple aldose (glyceraldehyde) and a simple ketose (dihydroxyacetone).

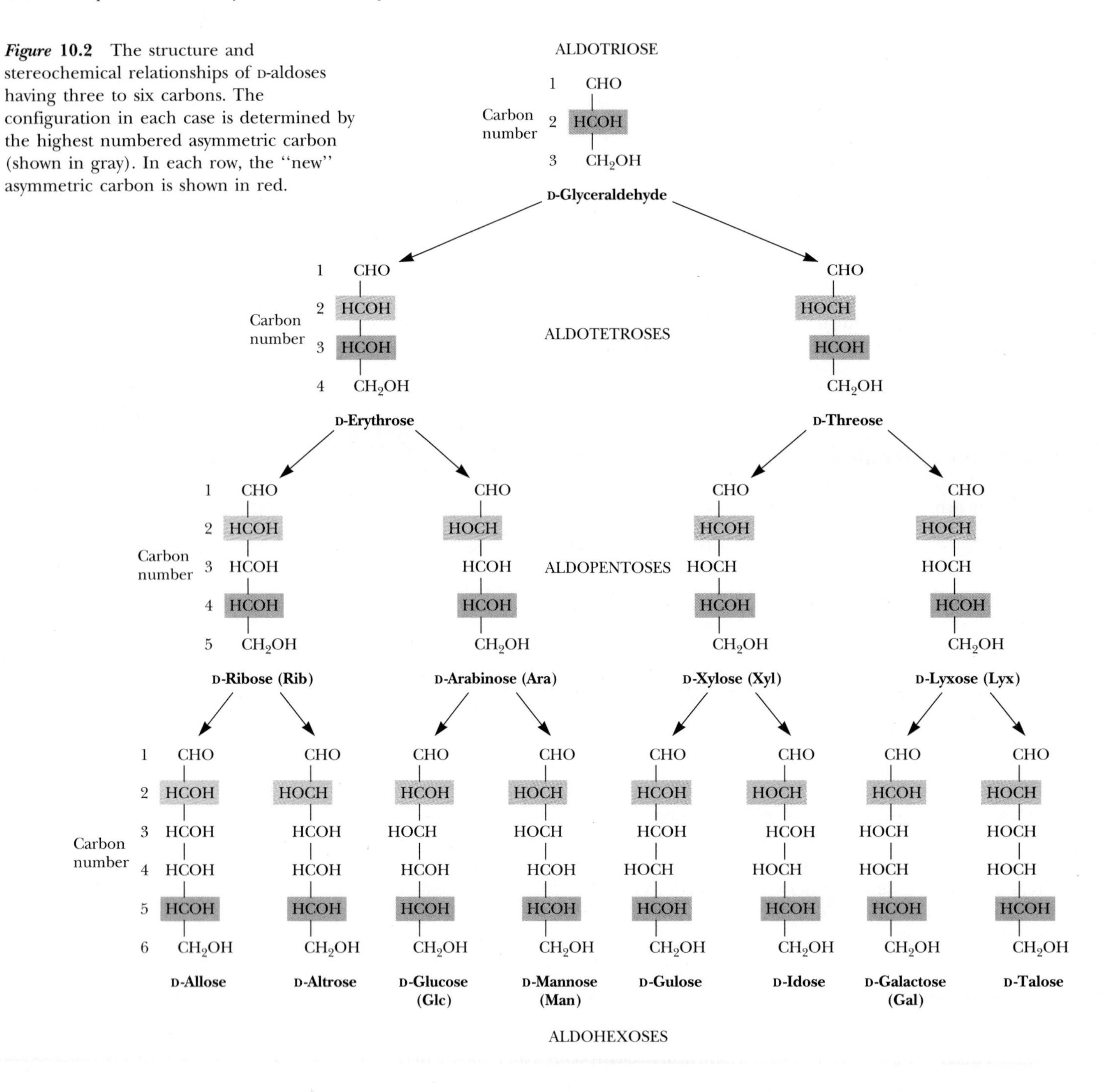

***Figure* 10.2** The structure and stereochemical relationships of D-aldoses having three to six carbons. The configuration in each case is determined by the highest numbered asymmetric carbon (shown in gray). In each row, the "new" asymmetric carbon is shown in red.

universally for this purpose today. The structures shown in Figures 10.2 and 10.3 are Fischer projections. For monosaccharides with two or more asymmetric carbons, the prefixes D or L refer to the configuration of the highest numbered asymmetric carbon (the asymmetric carbon farthest from the carbonyl carbon). A monosaccharide is designated D if the hydroxyl group on the highest numbered asymmetric carbon is drawn to the right in a Fischer projection, as in D-glyceraldehyde (Figure 10.1). Note that the designation D or L merely relates the configuration of a given molecule to that of glyceraldehyde and does *not* specify the sign of rotation of plane-polarized light. If the sign of optical rotation is to be specified in the name, the Fischer convention of D or L designations may be used along with a + (plus) or − (minus) sign. Thus, D-glucose (Figure 10.2) may also be called D(+)-glucose since it is dex-

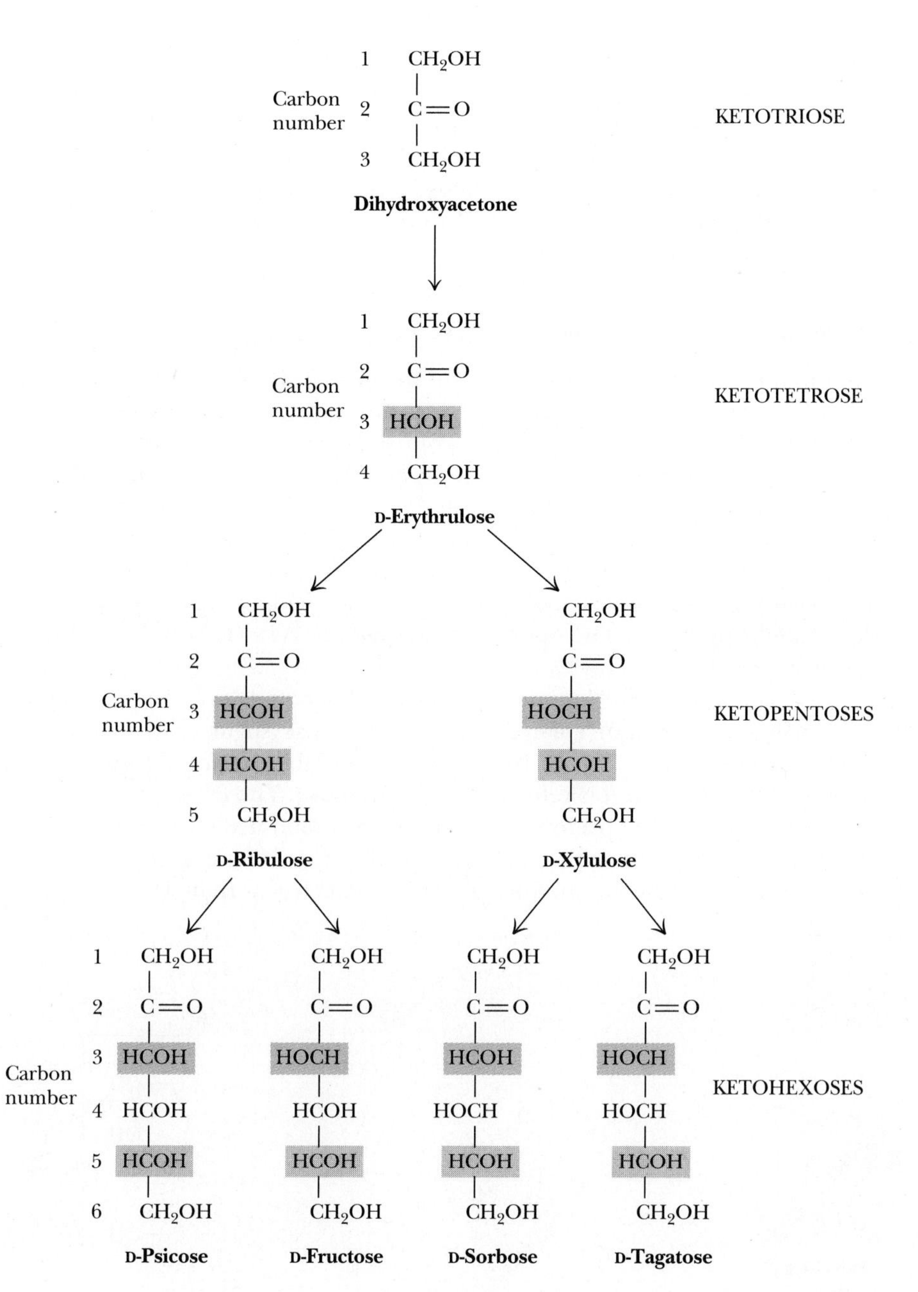

Figure **10.3** The structure and stereochemical relationships of D-ketoses having three to six carbons. The configuration in each case is determined by the highest numbered asymmetric carbon (shown in gray). In each row, the "new" asymmetric carbon is shown in red.

trorotatory, whereas D-fructose (Figure 10.3), which is levorotatory, can also be named D(−)-fructose.

All of the structures shown in Figures 10.2 and 10.3 are D-configurations, and the D- forms of monosaccharides predominate in nature, just as L-amino acids do. These preferences, established in apparently random choices early in evolution, persist uniformly in nature because of the stereospecificity of the enzymes that synthesize and metabolize these small molecules. L-Monosaccharides do exist in nature, serving a few relatively specialized roles. L-Galactose is a constituent of certain polysaccharides, and L-arabinose is a constituent of bacterial cell walls.

According to convention, the D- and L-forms of a monosaccharide are *mirror images* of each other, as shown in Figure 10.4 for fructose. Stereoisomers that are mirror images of each other are called **enantiomers,** or sometimes *enantiomeric pairs.* For molecules that possess two or more chiral centers, more than two stereoisomers can exist. Pairs of isomers that have opposite configu-

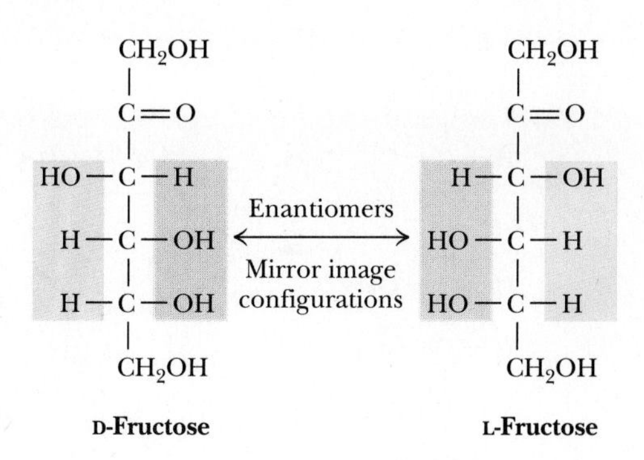

Figure **10.4** D-Fructose and L-fructose, an enantiomeric pair. Note that changing the configuration only at C-5 would change D-fructose to L-sorbose.

rations at one or more of the chiral centers but that are not mirror images of each other are called **diastereomers** or *diastereomeric pairs.* Any two structures in a given row in Figures 10.2 and 10.3 are diastereomeric pairs. Two sugars that differ in configuration at *only one* chiral center are described as **epimers.** For example, D-mannose and D-talose are epimers and D-glucose and D-mannose are epimers, whereas D-glucose and D-talose are *not* epimers but merely diastereomers.

Cyclic Structures and Anomeric Forms

Although Fischer projections are useful for presenting the structures of particular monosaccharides and their stereoisomers, they ignore one of the most interesting facets of sugar structure—*the ability to form cyclic structures with formation of an additional asymmetric center.* Alcohols react readily with aldehydes to form **hemiacetals** (Figure 10.5). The British carbohydrate chemist Sir Norman Haworth showed that the linear form of glucose (and other aldohexoses) could undergo a similar *intramolecular* reaction to form a *cyclic hemiacetal.* The resulting six-membered, oxygen-containing ring is similar to *pyran* and is designated a **pyranose.** The reaction is catalyzed by acid (H^+) or base (OH^-) and is readily reversible.

In a similar manner, ketones can react with alcohols to form **hemiketals.** The analogous intramolecular reaction of a ketose sugar such as fructose yields a *cyclic hemiketal* (Figure 10.6). The five-membered ring thus formed is reminiscent of *furan* and is referred to as a **furanose.** The cyclic pyranose and furanose forms are the preferred structures for monosaccharides in aqueous solution. At equilibrium, the linear aldehyde or ketone structure is only a minor component of the mixture (generally much less than 1%).

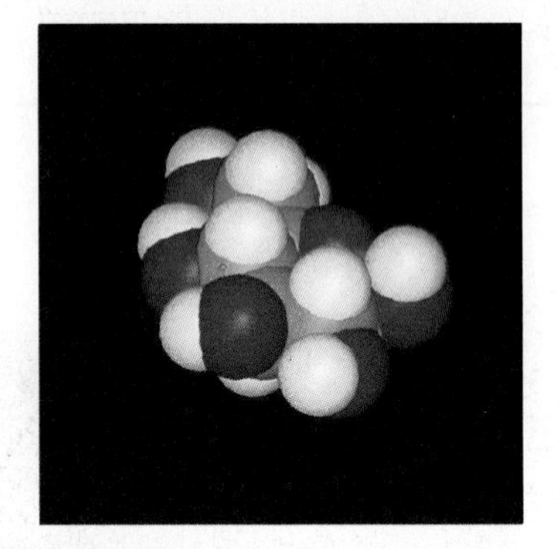

β-D-Glucopyranose

Alcohol **Aldehyde** **Hemiacetal**

D-Glucose **Pyran** **Cyclization**

α-D-Glucopyranose **β-D-Glucopyranose**

HAWORTH PROJECTION FORMULAS

α-D-Glucopyranose **β-D-Glucopyranose**

FISCHER PROJECTION FORMULAS

***Figure* 10.5**

Figure **10.6**

When hemiacetals and hemiketals are formed, the carbon atom that carried the carbonyl function becomes an asymmetric carbon atom. Isomers of monosaccharides that differ only in their configuration about that carbon atom are called **anomers,** designated as α or β, as shown in Figure 10.5, and the carbonyl carbon is thus called the **anomeric carbon.** When the hydroxyl group at the anomeric carbon is on the *same side* of a Fischer projection as the oxygen atom at the highest numbered asymmetric carbon, the configuration at the anomeric carbon is α, as in α-D-glucose. When the anomeric hydroxyl is on the *opposite side* of the Fischer projection, the configuration is β, as in β-D-glucopyranose (Figure 10.5).

The addition of this asymmetric center upon hemiacetal and hemiketal formation alters the optical rotation properties of monosaccharides, and the original assignment of the α and β notations arose from studies of these properties. Early carbohydrate chemists frequently observed that the optical rotation of glucose (and other sugar) solutions could change with time, a process called **mutarotation.** This indicated that a structural change was occurring. It was eventually found that α-D-glucose has a specific optical rotation, $[\alpha]_D^{20}$, of 112.2°, and that β-D-glucose has a specific optical rotation of 18.7°. Mutarotation involves interconversion of α and β forms of the monosaccharide with intermediate formation of the linear aldehyde or ketone, as shown in Figures 10.5 and 10.6.

Haworth Projections

Another of Haworth's lasting contributions to the field of carbohydrate chemistry was his proposal to represent pyranose and furanose structures as hexagonal and pentagonal rings lying perpendicular to the plane of the paper, with thickened lines indicating the side of the ring closest to the reader. Such

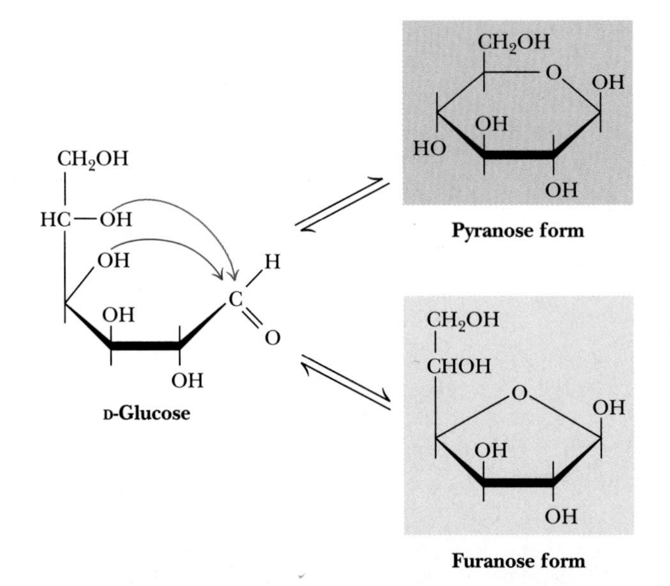

***Figure* 10.7** D-Glucose can cyclize in two ways, forming either furanose or pyranose structures.

Haworth projections, which are now widely used to represent saccharide structures (Figures 10.5 and 10.6), show substituent groups extending either above or below the ring. Substituents drawn to the left in a Fischer projection are drawn above the ring in the corresponding Haworth projection. Substituents drawn to the right in a Fischer projection are below the ring in a Haworth projection. Exceptions to these rules occur in the formation of furanose forms of pentoses and the formation of furanose or pyranose forms of hexoses. In these cases, the structure must be redrawn with a rotation about the carbon whose hydroxyl group is involved in the formation of the cyclic form (Figures 10.7 and 10.8) in order to orient the appropriate hydroxyl group for ring formation. This is merely for illustrative purposes and involves no change in configuration of the saccharide molecule.

The rules previously mentioned for assignment of α- and β-configurations can be readily applied to Haworth projection formulas. For the D-sugars, the anomeric hydroxyl group is below the ring in the α-anomer and above the ring in the β-anomer. For L-sugars, the opposite relationship holds.

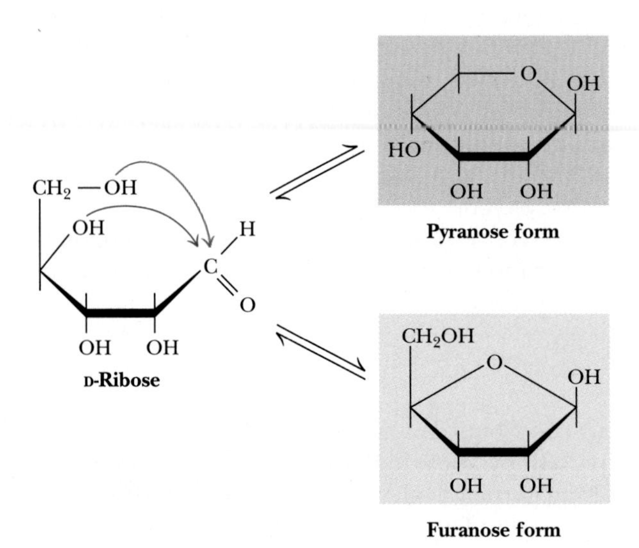

***Figure* 10.8** D-Ribose and other five-carbon saccharides can form either furanose or pyranose structures.

As Figures 10.7 and 10.8 imply, in most monosaccharides there are two or more hydroxyl groups which can react with an aldehyde or ketone at the other end of the molecule to form a hemiacetal or hemiketal. Consider the possibilities for glucose, as shown in Figure 10.7. If the C-4 hydroxyl group reacts with the aldehyde of glucose, a five-membered ring is formed, whereas if the C-5 hydroxyl reacts, a six-membered ring is formed. The C-6 hydroxyl does not react effectively since a seven-membered ring is too strained to form a stable hemiacetal. The same is true for the C-2 and C-3 hydroxyls, and thus five- and six-membered rings are by far the most likely to be formed from six-membered monosaccharides. D-Ribose, with five carbons, readily forms either five-membered rings (α- or β-D-ribofuranose) or six-membered rings (α- or β-D-ribopyranose) (Figure 10.8). In general, aldoses and ketoses with five or more carbons can form *either* furanose or pyranose rings, and the more stable form depends on structural factors. The nature of the substituent groups on the carbonyl and hydroxyl groups and the configuration about the asymmetric carbon will determine whether a given monosaccharide prefers the pyranose or furanose structure. In general, the pyranose form is favored over the furanose ring for aldohexose sugars, although, as we shall see, furanose structures are more stable for ketohexoses.

Although Haworth projections are convenient for display of monosaccharide structures, they do not accurately portray the conformations of pyranose and furanose rings. Given C—C—C tetrahedral bond angles of 109° and C—O—C angles of 111°, neither pyranose nor furanose rings can adopt true planar structures. Instead, they take on puckered conformations, and in the case of pyranose rings, the two favored structures are the **chair conformation** and the **boat conformation,** shown in Figure 10.9. Note that the ring substituents in these structures can be **equatorial,** which means approximately coplanar with the ring, or **axial,** that is, parallel to an axis drawn through the ring as shown. Two general rules dictate the conformation to be adopted by a given saccharide unit. First, bulky substituent groups on such rings are more stable when they occupy equatorial positions rather than axial positions, and second, chair conformations are slightly more stable than boat conformations. For a typical pyranose, such as β-D-glucose, there are two possible chair conformations (Figure 10.9). Of all the D-aldohexoses, β-D-glucose is the only one

(b)

Figure **10.9** (a) Chair and boat conformations of a pyranose sugar. (b) Two possible chair conformations of β-D-glucose.

that can adopt a conformation with all its bulky groups in an equatorial position. With this advantage of stability, it may come as no surprise that β-D-glucose is the most widely occurring organic group in nature and the central hexose in carbohydrate metabolism.

Derivatives of Monosaccharides

A variety of chemical and enzymatic reactions produce **derivatives** of the simple sugars. These modifications produce a diverse array of saccharide derivatives. Some of the most common derivations are discussed here.

Sugar Acids

Sugars with free anomeric carbon atoms are reasonably good reducing agents and will reduce hydrogen peroxide, ferricyanide, certain metals (Cu^{2+} and Ag^{+}), and other oxidizing agents. Such reactions convert the sugar to a **sugar**

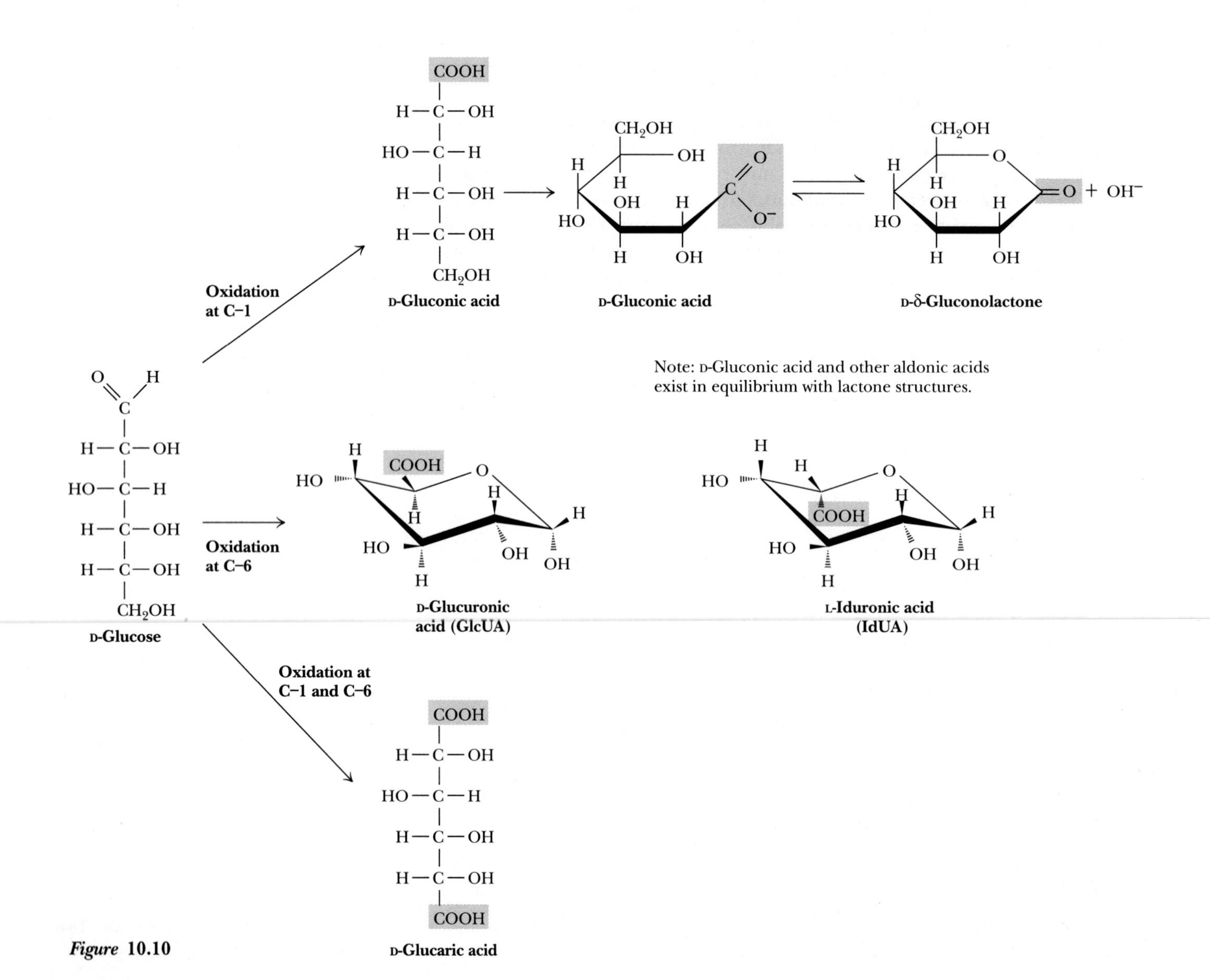

Figure **10.10**

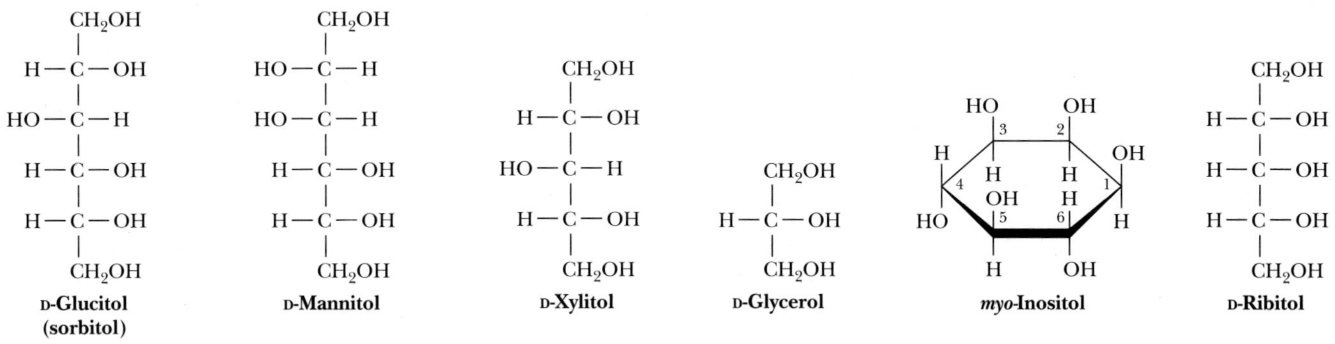

D-Glucitol (sorbitol) **D-Mannitol** **D-Xylitol** **D-Glycerol** ***myo*-Inositol** **D-Ribitol**

Figure **10.11** Structures of some sugar alcohols.

acid. For example, addition of alkaline $CuSO_4$ (called *Fehling's solution*) to an aldose sugar produces a red cuprous oxide (Cu_2O) precipitate:

$$\underset{\text{Aldehyde}}{\mathrm{R\overset{O}{\overset{\|}{C}}{-}H}} + 2\,Cu^{2+} + 5\,OH^- \longrightarrow \underset{\text{Carboxylate}}{\mathrm{R\overset{O}{\overset{\|}{C}}{-}O^-}} + Cu_2O\downarrow + 3\,H_2O$$

and converts the aldose to an **aldonic acid,** such as **gluconic acid** (Figure 10.10). Formation of a precipitate of red Cu_2O constitutes a positive test for an aldehyde. Carbohydrates that can reduce oxidizing agents in this way are referred to as **reducing sugars.** By quantifying the amount of oxidizing agent reduced by a sugar solution, one can accurately determine the concentration of the sugar. *Diabetes mellitus* is a condition that causes high levels of glucose in urine and blood, and frequent analysis of reducing sugars in diabetic patients is an important part of the diagnosis and treatment of this disease. Over-the-counter kits for the easy and rapid determination of reducing sugars have made this procedure a simple one for diabetics.

Monosaccharides can be oxidized enzymatically at C-6, yielding **uronic acids,** such as **D-glucuronic** and **L-iduronic** acids (Figure 10.10). L-Iduronic acid is similar to D-glucuronic acid, except for having an opposite configuration at C-5. Oxidation at both C-1 and C-6 produces **aldaric acids,** such as **D-glucaric acid.**

Sugar Alcohols

Sugar alcohols, another class of sugar derivative, can be prepared by the mild reduction (with $NaBH_4$ or similar agents) of the carbonyl groups of aldoses and ketoses. Sugar alcohols, or **alditols,** are designated by the addition of *-itol* to the name of the parent sugar (Figure 10.11). The alditols are linear molecules that cannot cyclize in the manner of aldoses. Nonetheless, alditols are characteristically sweet tasting, and **sorbitol, mannitol,** and **xylitol** are widely used to sweeten sugarless gum and mints. Sorbitol buildup in the eyes of diabetics is implicated in cataract formation. **Glycerol** and ***myo*-inositol,** a cyclic alcohol, are components of lipids (see Chapter 9). There are nine different stereoisomers of inositol; the one shown in Figure 10.12 was first isolated from heart muscle and thus has the prefix *myo-* for muscle. **Ribitol** is a constituent of flavin coenzymes (see Chapter 13).

Deoxy Sugars

The **deoxy sugars** are monosaccharides with one or more hydroxyl groups replaced by hydrogens. 2-Deoxy-D-ribose (Figure 10.12), whose systematic

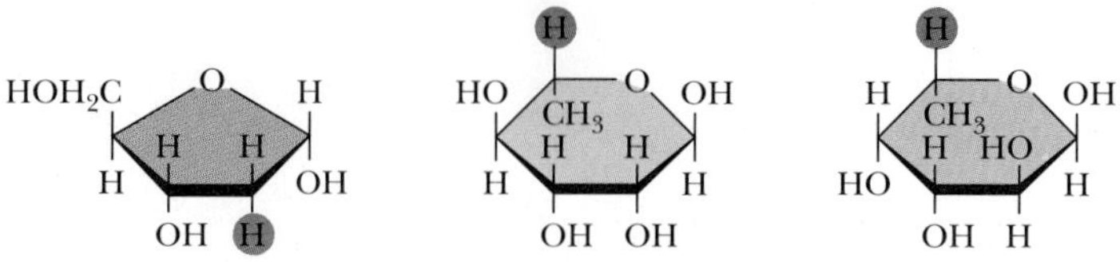

2-Deoxy-α-D-Ribose　α-L-Rhamnose (Rha)　α-L-Fucose (Fuc)

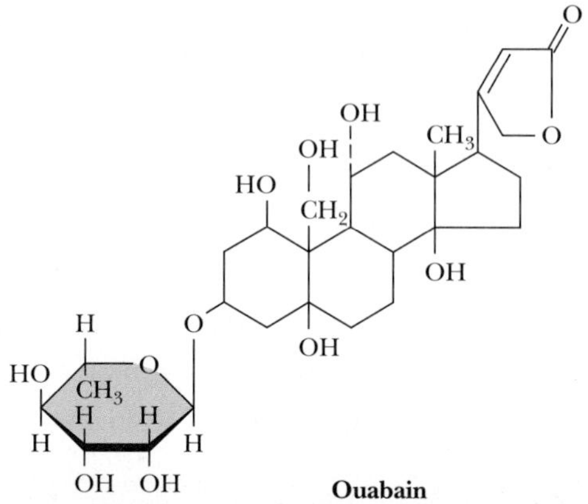

Ouabain

Figure **10.12** Several deoxy sugars and ouabain, which contains α-L-rhamnose (Rha). Hydrogen atoms highlighted in red are "deoxy" positions.

name is 2-deoxy-D-erythropentose, is a constituent of DNA in all living things (see Chapter 6). Deoxy sugars also occur frequently in glycoproteins and polysaccharides. L-Fucose and L-rhamnose, both 6-deoxy sugars, are components of some cell walls, and rhamnose is a component of **ouabain,** a highly toxic *cardiac glycoside* found in the bark and root of the ouabaio tree and used by the East African Somalis as an arrow poison. The sugar moiety is not the toxic part of the molecule (see in Chapter 35).

Sugar Esters

Phosphate esters of glucose, fructose, and other monosaccharides are important metabolic intermediates, and the ribose moiety of nucleotides such as ATP and GTP is phosphorylated at the 5′-position (Figure 10.13).

Amino Sugars

Amino sugars, including **D-glucosamine** and **D-galactosamine** (Figure 10.14), contain an amino group (instead of a hydroxyl group) at the C-2 position. They are found in many oligo- and polysaccharides, including *chitin,* a polysaccharide in the exoskeletons of crustaceans and insects.

Figure **10.13** Several sugar esters important in metabolism.

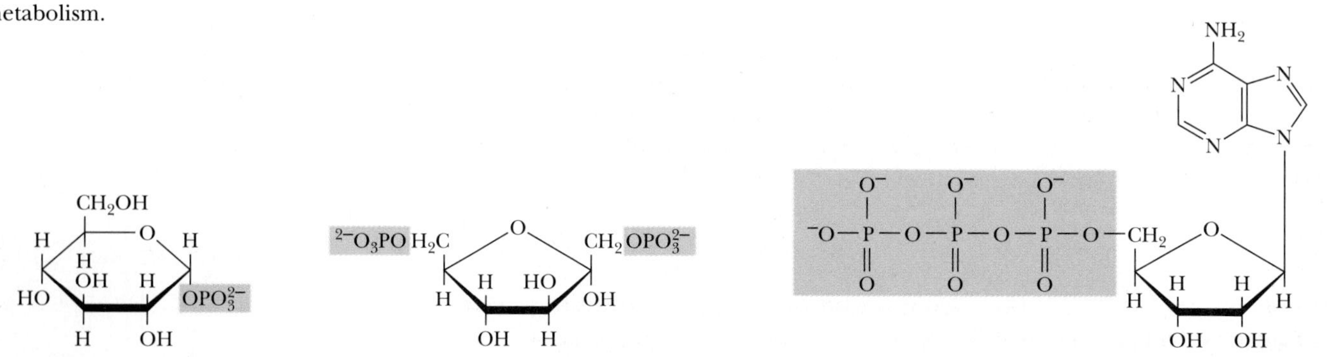

α-D-Glucose-1-phosphate　α-D-Fructose-1,6-bisphosphate　Adenosine-5′-triphosphate

Muramic acid and **neuraminic acid,** which are components of the polysaccharides of cell membranes of higher organisms and also bacterial cell walls, are glucosamines linked to three-carbon acids at the C-1 or C-3 positions. In muramic acid (thus named as an *amine* isolated from bacterial cell wall polysaccharides; *murus* is Latin for "wall"), the hydroxyl group of a lactic acid moiety makes an ether linkage to the C-3 of glucosamine. Neuraminic acid (an *amine* isolated from *neural* tissue) forms a C—C bond between the C-1 of *N*-acetylmannosamine and the C-3 of pyruvic acid (Figure 10.15). The *N*-acetyl and *N*-glycolyl derivatives of neuraminic acid are collectively known as **sialic acids** and are distributed widely in bacteria and animal systems.

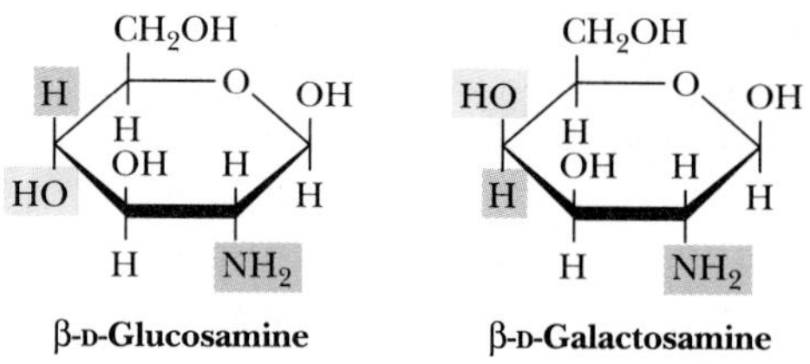

β-D-Glucosamine β-D-Galactosamine

Figure **10.14** Structures of D-glucosamine and D-galactosamine.

Acetals, Ketals, and Glycosides

Hemiacetals and hemiketals can react with alcohols in the presence of acid to form **acetals** and **ketals,** as shown in Figure 10.16. This reaction is another example of a *dehydration synthesis* and is similar in this respect to the reactions undergone by amino acids and nucleotides. The pyranose and furanose forms of monosaccharides react with alcohols in this way to form **glycosides** with retention of the α- or β-configuration at the C-1 carbon. The new bond be-

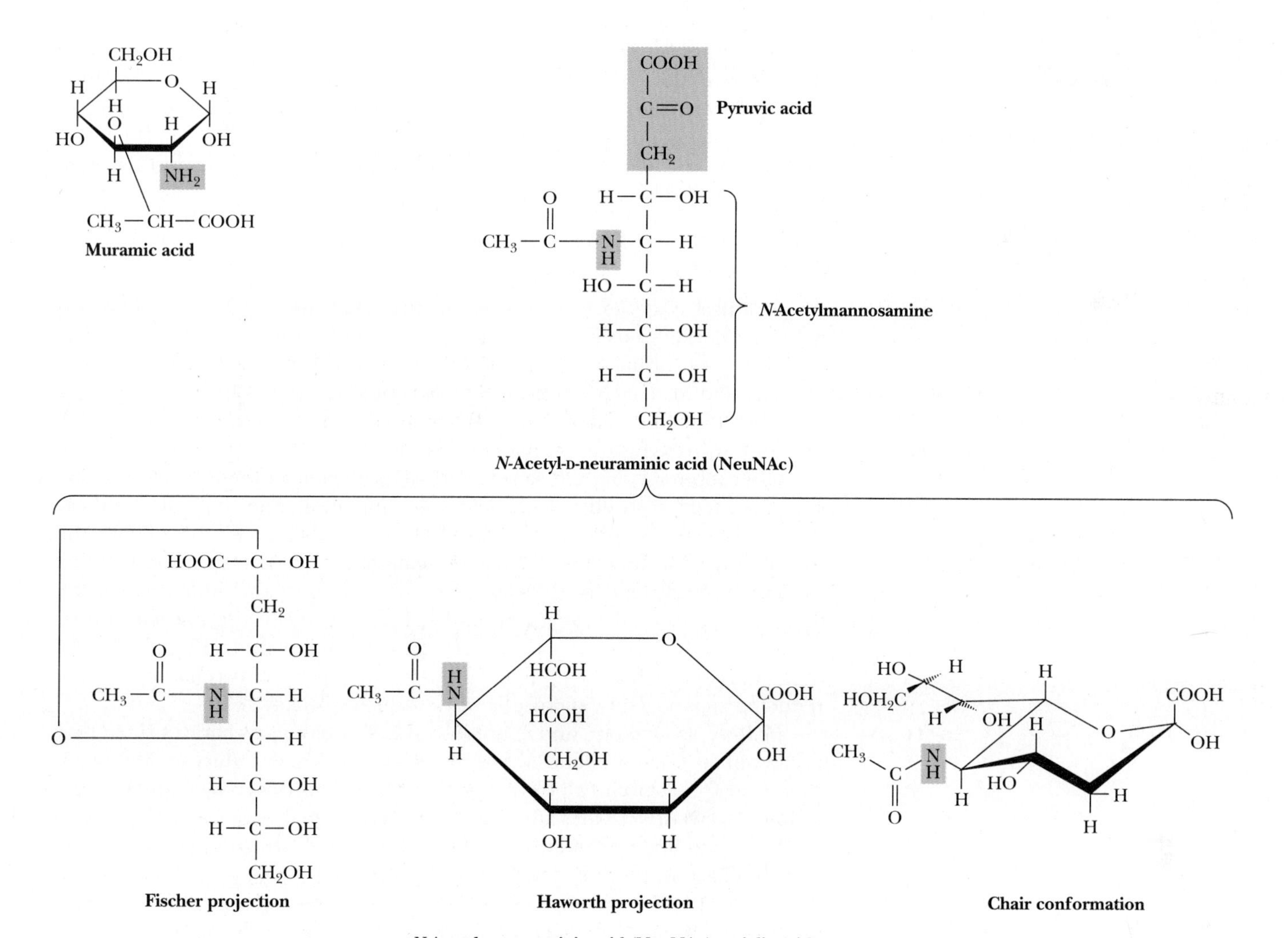

Figure **10.15** Structures of muramic acid and neuraminic acid and several depictions of sialic acid.

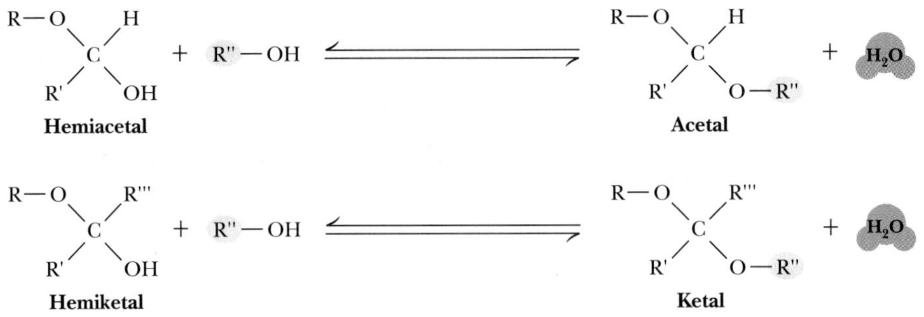

Figure **10.16** Acetals and ketals can be formed from hemiacetals and hemiketals, respectively.

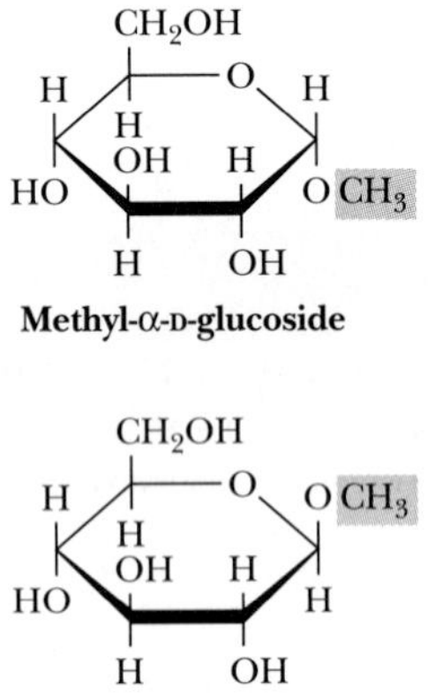

Hemiacetal + R''—OH ⇌ Acetal + H_2O

Hemiketal + R''—OH ⇌ Ketal + H_2O

Methyl-α-D-glucoside

Methyl-β-D-glucoside

Figure **10.17** The anomeric forms of methyl-D-glucoside.

tween the anomeric carbon atom and the oxygen atom of the alcohol is called a **glycosidic bond.** Glycosides are named according to the parent monosaccharide. For example, *methyl-β-D-glucoside* (Figure 10.17) can be considered a derivative of β-D-glucose.

10.3 Oligosaccharides

Given the relative complexity of oligosaccharides and polysaccharides in higher organisms, it is perhaps surprising that these molecules are formed from relatively few different monosaccharide units. (In this respect, the oligo- and polysaccharides are similar to proteins; both form complicated structures based on a small number of different building blocks.) Monosaccharide units include the hexoses glucose, fructose, mannose, and galactose and the pentoses ribose and xylose.

Disaccharides

The simplest oligosaccharides are the **disaccharides,** which consist of two monosaccharide units linked by a glycosidic bond. As in proteins and nucleic acids, each individual unit in an oligosaccharide is termed a *residue.* The disaccharides shown in Figure 10.18 are all commonly found in nature, with sucrose, maltose, and lactose being the most common. Each is a mixed acetal, with one hydroxyl group provided intramolecularly and one hydroxyl from the other monosaccharide. Except for sucrose, each of these structures possesses one free unsubstituted anomeric carbon atom, and thus each of these disaccharides is a reducing sugar. The end of the molecule containing the free anomeric carbon is called the **reducing end,** and the other end is called the **nonreducing end.** In the case of sucrose, both of the anomeric carbon atoms are substituted, that is, neither has a free —OH group. The substituted anomeric carbons cannot be converted to the aldehyde configuration and thus cannot participate in the oxidation–reduction reactions characteristic of reducing sugars. Thus, sucrose is *not* a reducing sugar.

Maltose, isomaltose, and cellobiose are all **homodisaccharides** since they each contain only one kind of monosaccharide, namely, glucose. **Maltose** is produced from starch (a polymer of α-D-glucose produced by plants) by the action of amylase enzymes and is a component of malt, a substance obtained by allowing grain (particularly barley) to soften in water and germinate. The enzyme **diastase,** produced during the germination process, catalyzes the hydrolysis of starch to maltose. Maltose is used in beverages (malted milk, for example), and since it is fermented readily by yeast, it is important in the brewing of beer. In both maltose and cellobiose, the glucose units are **1→4 linked,** meaning that the C-1 of one glucose is linked by a glycosidic

Free anomeric carbon (reducing end)

Lactose (galactose-β-1,4-glucose)

Maltose (glucose-α-1,4-glucose)

Sucrose (glucose-α-1,2-fructose)

Cellobiose (glucose-β-1,4-glucose)

Isomaltose (glucose-α-1,6-glucose)

Simple sugars
Glucose
Galactose
Fructose

Figure **10.18** The structures of several important disaccharides. Note that the notation —HOH means that the configuration can be either α or β. If the —OH group comes up out of the ring, the configuration is termed β. The configuration is α if the —OH group comes down out of the ring as shown. Also note that sucrose has no free anomeric carbon atoms.

bond to the C-4 oxygen of the other glucose. The only difference between them is in the configuration at the glycosidic bond. Maltose exists in the α-configuration, whereas cellobiose is β. **Isomaltose** is obtained in the hydrolysis of some polysaccharides (such as dextran), and **cellobiose** is obtained from the acid hydrolysis of cellulose. Isomaltose also consists of two glucose units in a glycosidic bond, but in this case, C-1 of one glucose is linked to C-6 of the other, and the configuration is α.

The complete structures of these disaccharides can be specified in shorthand notation by using abbreviations for each monosaccharide, α or β to denote configuration, and appropriate numbers to indicate the nature of the linkage. Thus, cellobiose is Glcβ1-4Glc, whereas isomaltose is Glcα1-6Glc. Often the glycosidic linkage is written with an arrow so that cellobiose and isomaltose would be Glcβ1→4Glc and Glcα1→6Glc, respectively. Since the linkage carbon on the first sugar is always C-1, a newer trend is to drop the 1- or 1→ and describe these simply as Glcβ4Glc and Glcα6Glc, respectively. More complete names can also be used, however, so that maltose would be O-α-D-glucopyranosyl-(1→4)-D-glucopyranose. Cellobiose, because of its β-glycosidic linkage, is formally O-β-D-glucopyranosyl-(1→4)-D-glucopyranose.

β-D-lactose (O-β-D-Galactopyranosyl-(1→4)-D-glucopyranose) (Figure 10.18) is the principal carbohydrate in milk and is of critical nutritional importance to mammals in the early stages of their lives. It is formed from D-galactose and D-glucose via a β(1→4) link, and because it has a free anomeric carbon, it is capable of mutarotation and is a reducing sugar. It is an interesting quirk of nature that lactose cannot be absorbed directly into the bloodstream. It must first be broken down into galactose and glucose by **lactase,** an intestinal enzyme that exists in young, nursing mammals but is not produced in significant quantities in the mature mammal. Most humans, with the exception of certain groups in Africa and northern Europe, produce only low levels of lactase. For most individuals, this is not a problem, but some cannot tolerate lactose and experience intestinal pain and diarrhea upon consumption of milk.

Sucrose, in contrast, is a disaccharide of almost universal appeal and tolerance. Produced by many higher plants and commonly known as *table sugar,* it is one of the products of photosynthesis and is composed of fructose and glucose. Sucrose has a specific optical rotation, $[\alpha]_D^{20}$, of +66.5°, but an equimolar mixture of its component monosaccharides has a net negative rotation ($[\alpha]_D^{20}$ of glucose is +52.5° and of fructose is −92°). Sucrose is hydrolyzed by the enzyme **invertase,** so named for the change of optical rotation accompa-

Sucrose

Melezitose (a constituent of honey)

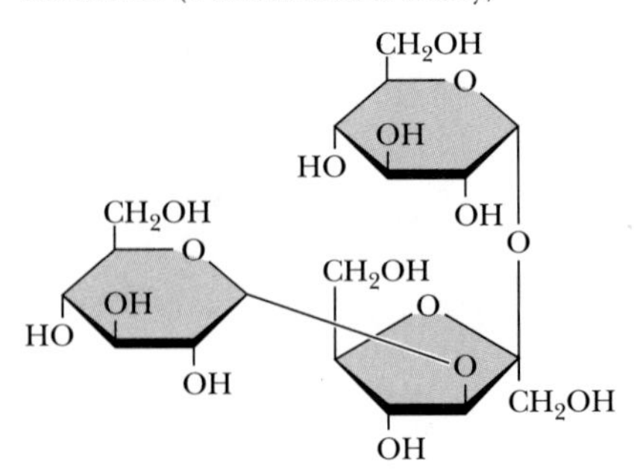

Amygdalin (occurs in seeds of *Rosaceae*, glycoside of bitter almonds, in kernels of cherries, peaches, apricots)

Laetrile (claimed to be an anticancer agent, but there is no rigorous scientific evidence for this)

Stachyose (a constituent of many plants: white jasmine, yellow lupine, soybeans, lentils, etc.; causes flatulence since humans cannot digest it)

Cycloheptaamylose (a breakdown product of starch; useful in chromatographic separations)

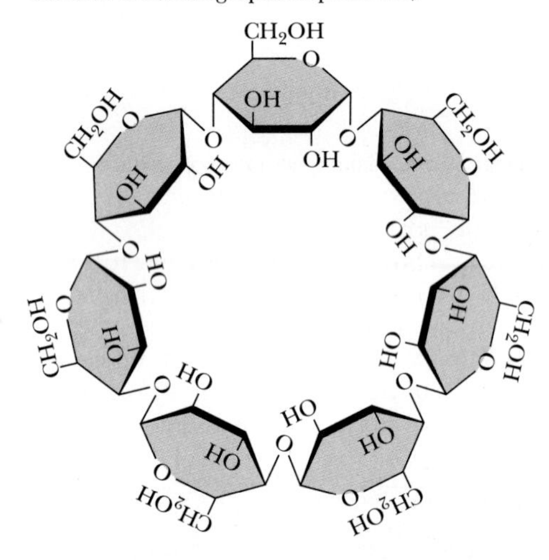

Dextrantriose (a constituent of saké and honeydew)

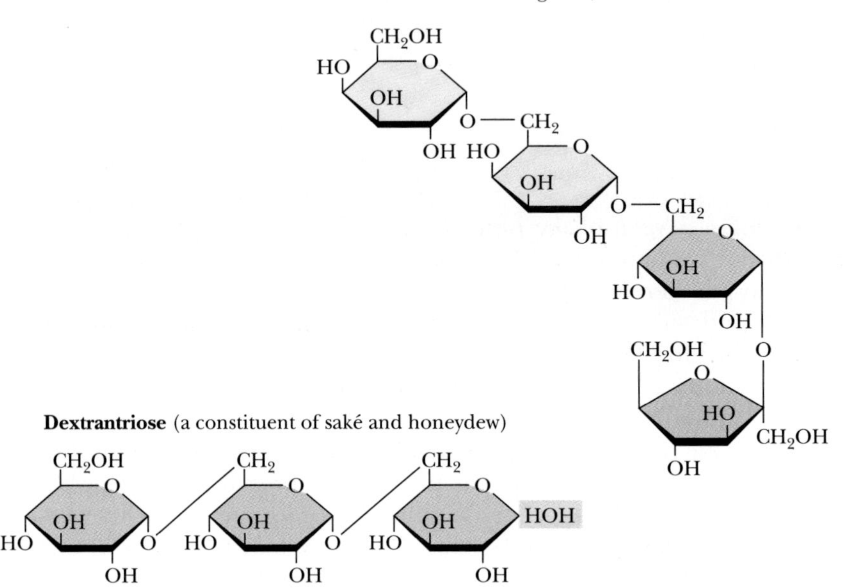

Figure **10.19** The structures of some interesting oligosaccharides.

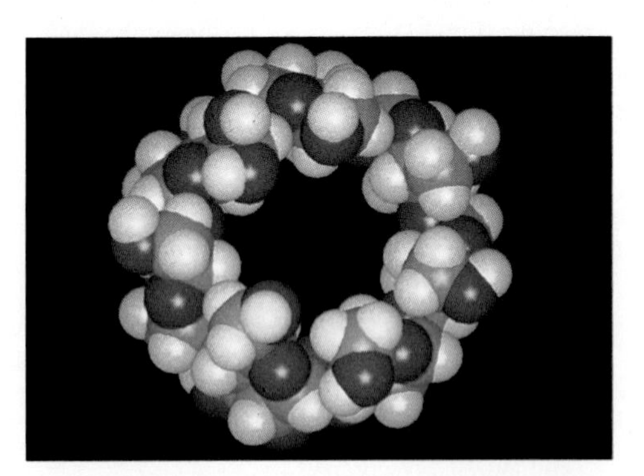

Cycloheptaamylose

Cycloheptaamylose (side view)

nying this reaction. Sucrose is also easily hydrolyzed by dilute acid, apparently because the fructose in sucrose is in the relatively unstable furanose form. Although sucrose and maltose are important to the human diet, they are not taken up directly in the body. In a manner similar to lactose, they are first hydrolyzed by **sucrase** and **maltase,** respectively, in the human intestine.

Higher Oligosaccharides

In addition to the simple disaccharides, many other oligosaccharides are found in both prokaryotic and eukaryotic organisms, either as naturally occurring substances or as hydrolysis products of natural materials. Figure 10.19 lists a number of simple oligosaccharides, along with descriptions of their origins and interesting features. Several are constituents of the sweet nectars or saps exuded or extracted from plants and trees. One particularly interesting and useful group of oligosaccharides are the **cycloamyloses.** These oligosaccharides are cyclic structures, and in solution they form molecular "pockets" of various diameters. These pockets are surrounded by the chiral carbons of the saccharides themselves and are able to form stereospecific inclusion complexes with chiral molecules that can fit into the pockets. Thus, mixtures of stereoisomers of small organic molecules can be separated into pure isomers on columns of **cycloheptaamylose,** for example.

Bleomycin A$_2$ (an antitumor agent used clinically against specific tumors)

Aburamycin C (an antibiotic and antitumor agent)

Sulfurmycin B (active against Gram-positive bacteria, mycobacteria, and tumors)

Streptomycin (a broad spectrum antibiotic)

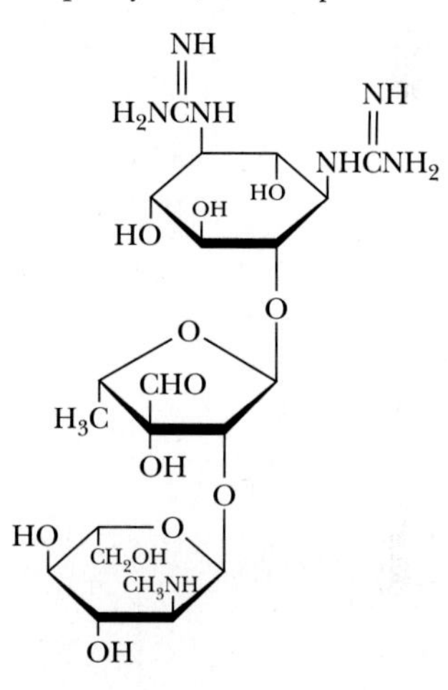

Figure **10.20** Some antibiotics are oligosaccharides or contain oligosaccharide groups.

Stachyose is typical of the oligosaccharide components found in substantial quantities in beans, peas, bran, and whole grains. These oligosaccharides are not digested by stomach enzymes, but *are* metabolized readily by bacteria in the intestines. This is the source of the flatulence that often accompanies the consumption of such foods. Commercial products are now available that assist in the digestion of the gas-producing components of these foods. These products contain an enzyme that hydrolyzes the culprit oligosaccharides in the stomach before they become available to intestinal microorganisms.

Another notable glycoside is **amygdalin,** which occurs in bitter almonds and in the kernels or pits of cherries, peaches, and apricots. Hydrolysis of this substance and subsequent oxidation yields **laetrile,** which has been claimed by some to have anticancer properties. There is no scientific evidence for these claims, and the U.S. Food and Drug Administration has never approved laetrile for use in the United States.

Oligosaccharides also occur widely as components (via glycosidic bonds) of *antibiotics* derived from various sources. Figure 10.20 shows the structures of a few representative carbohydrate-containing antibiotics. Some of these antibiotics also show antitumor activity. One of the most important of this type is **bleomycin A$_2$,** which is used clinically against certain tumors.

10.4 Polysaccharides

Structure and Nomenclature

By far the majority of carbohydrate material in nature occurs in the form of polysaccharides. By our definition, polysaccharides include not only those substances composed only of glycosidically linked sugar residues but also molecules that contain polymeric saccharide structures linked via covalent bonds to amino acids, peptides, proteins, lipids, and other structures.

Polysaccharides, also called **glycans,** consist of monosaccharides and their derivatives. If a polysaccharide contains only one kind of monosaccharide molecule, it is a **homopolysaccharide,** or **homoglycan,** whereas those containing more than one kind of monosaccharide are **heteropolysaccharides.** The most common constituent of polysaccharides is D-glucose, but D-fructose, D-galactose, L-galactose, D-mannose, L-arabinose, and D-xylose are also common. Common monosaccharide derivatives in polysaccharides include the amino sugars (D-glucosamine and D-galactosamine), their derivatives (*N*-acetylneuraminic acid and *N*-acetylmuramic acid), and simple sugar acids (glucuronic and iduronic acids). Homopolysaccharides are often named for the sugar unit they contain, so that glucose homopolysaccharides are called **glucans,** while mannose homopolysaccharides are **mannans.** Other homopolysaccharide names are just as obvious: *galacturonans, arabinans,* and so on. Homopolysaccharides of uniform linkage type are often named by including notation to denote ring size and linkage type. Thus, cellulose is a *(1→4)-β-D-glucopyranan.* Polysaccharides differ not only in the nature of their component monosaccharides but also in the length of their chains and in the amount of chain branching that occurs. Although a given sugar residue has only one anomeric carbon and thus can form only one glycosidic linkage with hydroxyl groups on other molecules, each sugar residue carries several hydroxyls, one or more of which may be an acceptor of glycosyl substituents (Figure 10.21). This ability to form branched structures distinguishes polysaccharides from proteins and nucleic acids, which occur only as linear polymers.

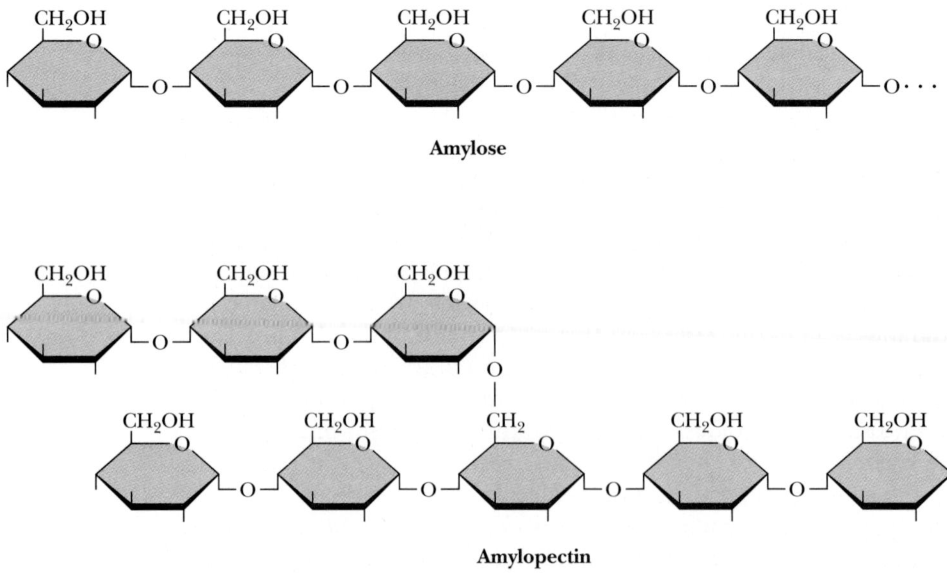

Amylose

Amylopectin

***Figure* 10.21** Amylose and amylopectin are the two forms of starch. Note that the linear linkages are α(1→4), but the branches in amylopectin are α(1→6). Branches in polysaccharides can involve any of the hydroxyl groups on the monosaccharide components. Amylopectin is a highly branched structure, with branches occurring every 12 to 30 residues.

Polysaccharide Functions

The functions of many individual polysaccharides cannot be assigned uniquely, and some of their functions may not yet be appreciated. Traditionally, biochemistry textbooks have listed the functions of polysaccharides, as storage materials, structural components, or protective substances. Thus, *starch, glycogen,* and other storage polysaccharides, as readily metabolizable food, provide energy reserves for cells. *Chitin* and *cellulose* provide strong support for the skeletons of arthropods and green plants, respectively. Mucopolysaccharides, such as the *hyaluronic acids,* form protective coats on animal cells. In each of these cases, the relevant polysaccharide is either a homopolymer or a polymer of small repeating units. Recent research indicates, however, that oligosaccharides and polysaccharides with more unique structures may also be involved in much more sophisticated tasks in cells, including a variety of cellular recognition and intercellular communication events, as discussed later.

Storage Polysaccharides

Storage polysaccharides are an important carbohydrate form in plants and animals. It seems likely that organisms store carbohydrates in the form of polysaccharides rather than as monosaccharides to lower the osmotic pressure of the sugar reserves. Since osmotic pressures depend only on *numbers of molecules,* the osmotic pressure is greatly reduced by formation of a few polysaccharide molecules out of thousands (or even millions) of monosaccharide units.

Starch

By far the most common storage polysaccharide in plants is **starch,** which exists in two forms: **α-amylose** and **amylopectin,** the structures of which are shown in Figure 10.21. Most forms of starch in nature are 10 to 30% α-amylose and 70 to 90% amylopectin. Typical cornstarch produced in the United States is about 25% α-amylose and 75% amylopectin. α-Amylose is composed of linear chains of D-glucose in α(1→4) linkages. The chains are of varying length, having molecular weights from several thousand to half a million. As can be seen from the structure in Figure 10.21, the chain has a reducing end and a nonreducing end. Although poorly soluble in water, α-amylose forms micelles in which the polysaccharide chain adopts a helical conformation (Figure 10.22). Iodine reacts with α-amylose to give a characteristic blue color, which arises from the insertion of iodine into the middle of the hydrophobic amylose helix.

In contrast to α-amylose, amylopectin, the other component of typical starches, is a highly branched chain of glucose units (Figure 10.21). Branches occur in these chains every 12 to 30 residues. The average chain length is between 24 and 30 residues, and molecular weights of amylopectin molecules can range up to 100 million. The linear linkages in amylopectin are α(1→4), whereas the branch linkages are α(1→6). As is the case for α-amylose, amylopectin forms micellar suspensions in water; iodine reacts with such suspensions to produce a red-violet color.

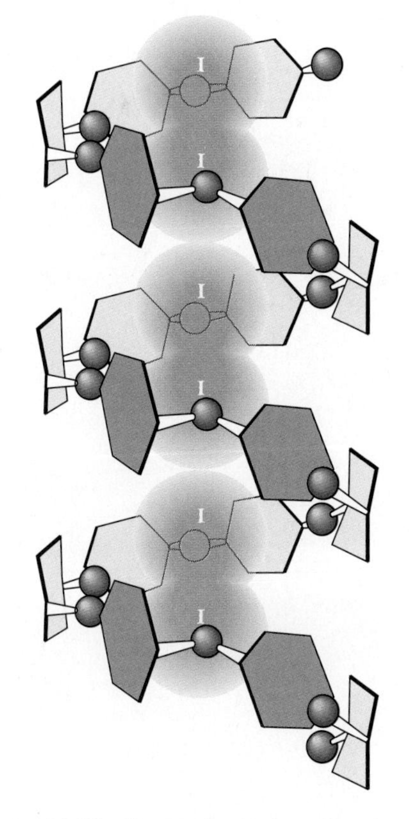

***Figure* 10.22** Suspensions of amylose in water adopt a helical conformation. Iodine (I_2) can insert into the middle of the amylose helix to give a blue color that is characteristic and diagnostic for starch.

Starch is stored in plant cells in the form of granules in the stroma of plastids (plant cell organelles) of two types: **chloroplasts,** in which photosynthesis takes place, and **amyloplasts,** plastids that are specialized starch accumulation bodies. When starch is to be mobilized and used by the plant that

***Figure* 10.23** The starch phosphorylase reaction cleaves glucose residues from amylose, producing α-D-glucose-1-phosphate.

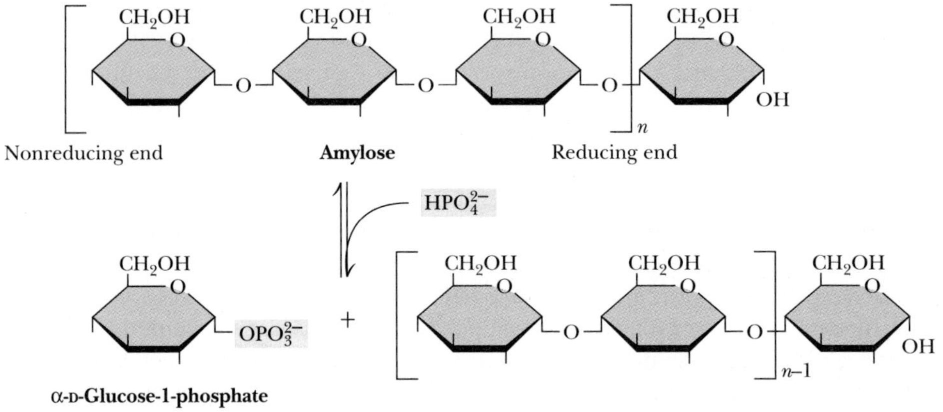

stored it, it must be broken down into its component monosaccharides. Starch is split into its monosaccharide elements by stepwise phosphorolytic cleavage of glucose units, a reaction catalyzed by **starch phosphorylase** (Figure 10.23). This is formally an $\alpha(1\rightarrow4)$-glucan phosphorylase reaction, and at each step, the products are one molecule of glucose-1-phosphate and a starch molecule with one fewer glucose units. In α-amylose, this process continues all along the chain until the end is reached. However, the $\alpha(1\rightarrow6)$ branch points of amylopectin are not susceptible to cleavage by phosphorylase, and thorough digestion of amylopectin by phosphorylase leaves a *limit dextrin,* which must be attacked by an $\alpha(1\rightarrow6)$-glucosidase to cleave the 1→6 branch points and allow complete hydrolysis of the remaining 1→4 linkages. Glucose-1-phosphate units are thus delivered to the plant cell, suitable for further processing in glycolytic pathways (see Chapter 18).

In animals, digestion and use of plant starches begins in the mouth with **salivary α-amylase** ($\alpha(1\rightarrow4)$-glucan 4-glucanohydrolase), the major enzyme secreted by the salivary glands. Although the capability of making and secreting salivary α-amylases is widespread in the animal world, some animals (such as cats, dogs, birds, and horses) do not secrete them. Salivary α-amylase is an **endoamylase** which splits $\alpha(1\rightarrow4)$ glycosidic linkages only within the chain. Raw starch is not very susceptible to salivary endoamylase. However, when suspensions of starch granules are heated, the granules swell, taking up water and causing the polymers to become more accessible to enzymes. Thus, cooked starch is more digestible. In the stomach, salivary α-amylase is inactivated by the lower pH, but pancreatic secretions also contain α-amylase. **β-Amylase,** an enzyme absent in animals but prevalent in plants and microorganisms, cleaves disaccharide (maltose) units from the termini of starch chains and is an **exoamylase.** Neither α-amylase nor β-amylase, however, can cleave the $\alpha(1\rightarrow6)$ branch points of amylopectin, and once again, $\alpha(1\rightarrow6)$-glucosidase is required to cleave at the branch points and allow complete hydrolysis of starch amylopectin.

Glycogen

The major form of storage polysaccharide in animals is **glycogen.** Glycogen is found mainly in the liver (where it may amount to as much as 10% of liver mass) and skeletal muscle (where it accounts for 1 to 2% of muscle mass). Liver glycogen consists of granules containing highly branched molecules, with $\alpha(1\rightarrow6)$ branches occurring every 8 to 12 glucose units. Like amylopectin, glycogen yields a red-violet color with iodine. Glycogen can be hydrolyzed by both α- and β-amylases, yielding glucose and maltose, respectively, as products and can also be hydrolyzed by **glycogen phosphorylase,** an enzyme present in liver and muscle tissue, to release glucose-1-phosphate.

Dextran

Another important family of storage polysaccharides are the **dextrans,** which are $\alpha(1\rightarrow6)$-linked polysaccharides of D-glucose with branched chains found in yeast and bacteria (Figure 10.24). Since the main polymer chain is $\alpha(1\rightarrow6)$ linked, the repeating unit is *isomaltose,* Glc$\alpha(1\rightarrow6)$-Glc. The branch points may be $1\rightarrow2$, $1\rightarrow3$, or $1\rightarrow4$ in various species. The degree of branching and the average chain length between branches depends on the species and strain of the organism. Bacteria growing on the surfaces of teeth produce extracellular accumulations of dextrans, an important component of *dental plaque.* Bacterial dextrans are frequently used in research laboratories as the support medium for column chromatography of macromolecules. Dextran chains cross-linked with epichlorohydrin yield the structure shown in Figure 10.25. These preparations (known by various trade names such as Sephadex and Bio-gel) are extremely hydrophilic and swell to form highly hydrated gels in water. Depending on the degree of cross-linking and the size of the gel particle, these materials form gels containing from 50 to 98% water. Dextran can also be cross-linked with other agents, forming gels with slightly different properties.

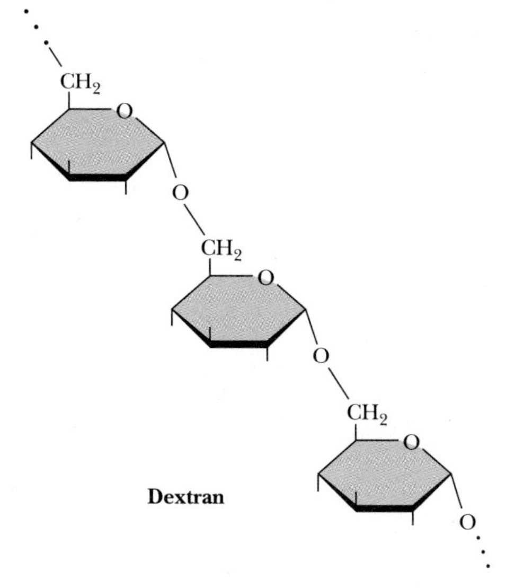

***Figure* 10.24** Dextran is a branched polymer of D-glucose units. The main chain linkage is $\alpha(1\rightarrow6)$, but $1\rightarrow2$, $1\rightarrow3$, or $1\rightarrow4$ branches can occur.

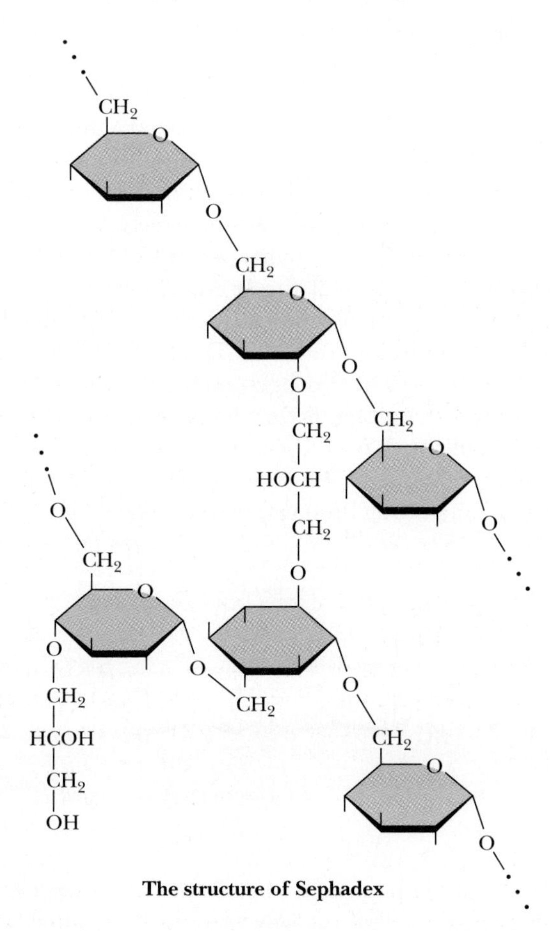

***Figure* 10.25** Sephadex gels are formed from dextran chains cross-linked with epichlorohydrin. The degree of cross-linking determines the chromatographic properties of Sephadex gels. Sephacryl gels are formed by cross-linking of dextran polymers with *N,N′*-methylene bisacrylamide.

Structural Polysaccharides

Cellulose

The **structural polysaccharides** have properties that are dramatically different from those of the storage polysaccharides, even though the compositions of these two classes are similar. The structural polysaccharide **cellulose** is the most abundant natural polymer found in the world. Found in the cell walls of nearly all plants, cellulose is one of the principal components providing physical structure and strength. The wood and bark of trees is an insoluble, highly organized structure formed from cellulose and also from *lignin* (see Figure 26.33). It is awe-inspiring to look at a large tree and realize the amount of weight supported by polymeric structures derived from sugars and organic alcohols. Cellulose also has its delicate side, however. *Cotton,* whose woven fibers make some of our most comfortable clothing fabrics, is almost pure cellulose. Derivatives of cellulose have found wide use in our society. **Cellulose acetates** are produced by the action of acetic anhydride on cellulose in the presence of sulfuric acid and can be spun into a variety of fabrics with particular properties. Referred to simply as *acetates,* they have a silky appearance, a luxuriously soft feel, and a deep luster and are used in dresses, lingerie, linings, and blouses.

Cellulose is a linear homopolymer of D-glucose units, just as in α-amylose. The structural difference, which completely alters the properties of the polymer, is that in cellulose the glucose units are linked by $\beta(1\rightarrow4)$-glycosidic bonds, whereas in α-amylose the linkage is $\alpha(1\rightarrow4)$. The conformational difference between these two structures is shown in Figure 10.26. The $\alpha(1\rightarrow4)$-linkage sites of amylose are naturally bent, conferring a gradual turn to the polymer chain, which results in the helical conformation already described (see Figure 10.22). The most stable conformation about the $\beta(1\rightarrow4)$ linkage involves alternating 180° flips of the glucose units along the chain so that the chain adopts a fully extended conformation, referred to as an **extended ribbon.** Juxtaposition of several such chains permits efficient interchain hydrogen bonding, the basis of much of the strength of cellulose.

The structure of one form of cellulose, determined by X-ray and electron diffraction data, is shown in Figure 10.27. The flattened sheets of the chains lie side by side and are joined by hydrogen bonds. These sheets are laid on top of one another in a way that staggers the chains, just as bricks are staggered to give strength and stability to a wall. Cellulose is extremely resistant to hydrolysis, whether by acid or by the digestive tract amylases described earlier. As a result, most animals (including humans) cannot digest cellulose to any signifi-

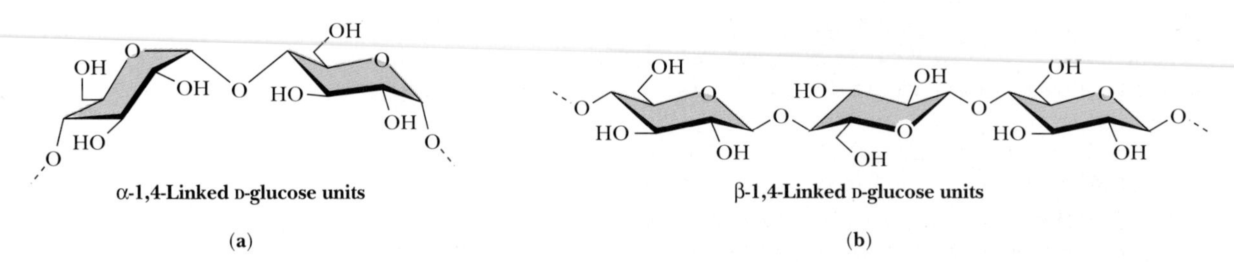

Figure **10.26** (a) Amylose, composed exclusively of the relatively bent $\alpha(1\rightarrow4)$ linkages, prefers to adopt a helical conformation, whereas (b) cellulose, with $\beta(1\rightarrow4)$-glycosidic linkages, can adopt a fully extended conformation with alternating 180° flips of the glucose units. The hydrogen bonding inherent in such extended structures is responsible for the great strength of tree trunks and other cellulose-based materials.

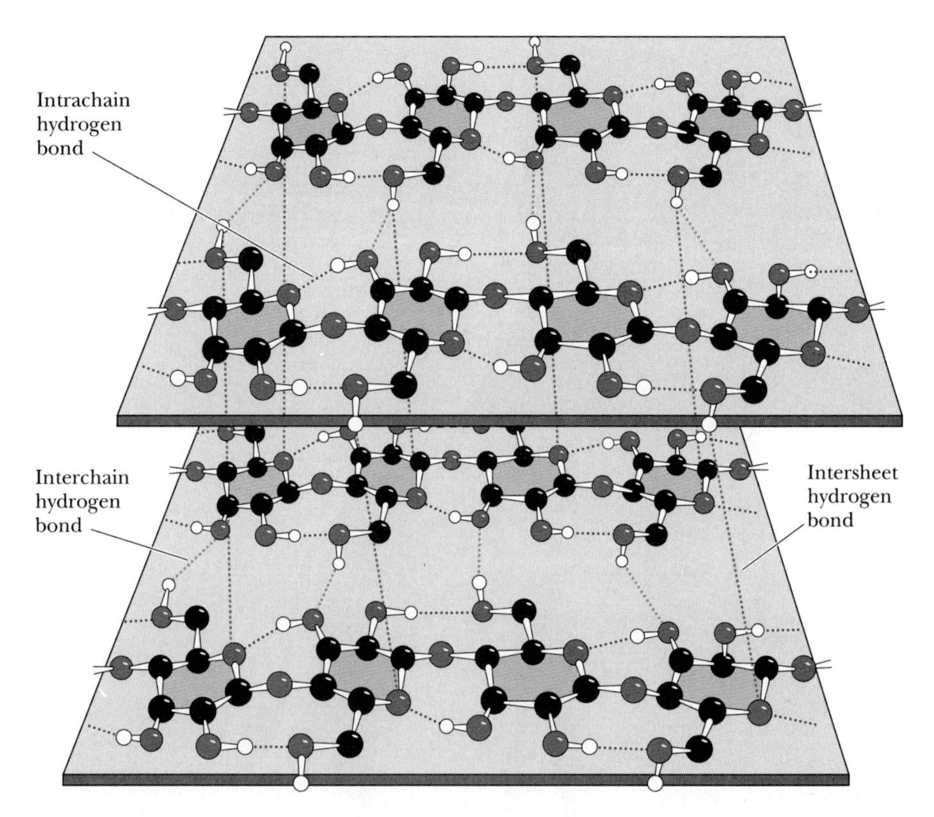

***Figure* 10.27** The structure of cellulose, showing the hydrogen bonds (blue) between the sheets, which strengthen the structure. Intrachain hydrogen bonds are in red and interchain hydrogen bonds are in green.

cant degree. Ruminant animals, such as cattle, deer, giraffes, and camels, are an exception because bacteria that live in the rumen (Figure 10.28) secrete the enzyme **cellulase,** a β-glucosidase effective in the hydrolysis of cellulose. The resulting glucose is then metabolized in a fermentation process to the benefit of the host animal. Termites and shipworms (*Teredo navalis*) similarly digest cellulose because their digestive tracts also contain bacteria that secrete cellulase.

Chitin

A polysaccharide that is similar to cellulose, both in its biological function and its primary, secondary, and tertiary structure, is **chitin.** Chitin is present in the cell walls of fungi and is the fundamental material in the exoskeletons of crustaceans, insects, and spiders. The structure of chitin, an extended ribbon, is identical to cellulose, except that the —OH group on each C-2 is replaced by $—NHCOCH_3$, so that the repeating units are *N-acetyl-D-glucosamines* in $\beta(1\rightarrow4)$ linkage. Like cellulose (Figure 10.27), the chains of chitin form extended ribbons (Figure 10.29) and pack side by side in a crystalline, strongly hydrogen-bonded form. One significant difference between cellulose and chitin is whether the chains are arranged in **parallel** (all the reducing ends together at one end of a packed bundle and all the nonreducing ends together at the other end) or **antiparallel** (each sheet of chains having the chains arranged oppositely from the sheets above and below). Natural cellulose seems to only occur in parallel arrangements. Chitin, however, can occur in three forms, sometimes all in the same organism. *α-Chitin* is an all-parallel arrangement of the chains, whereas *β-chitin* is an antiparallel arrangement. In

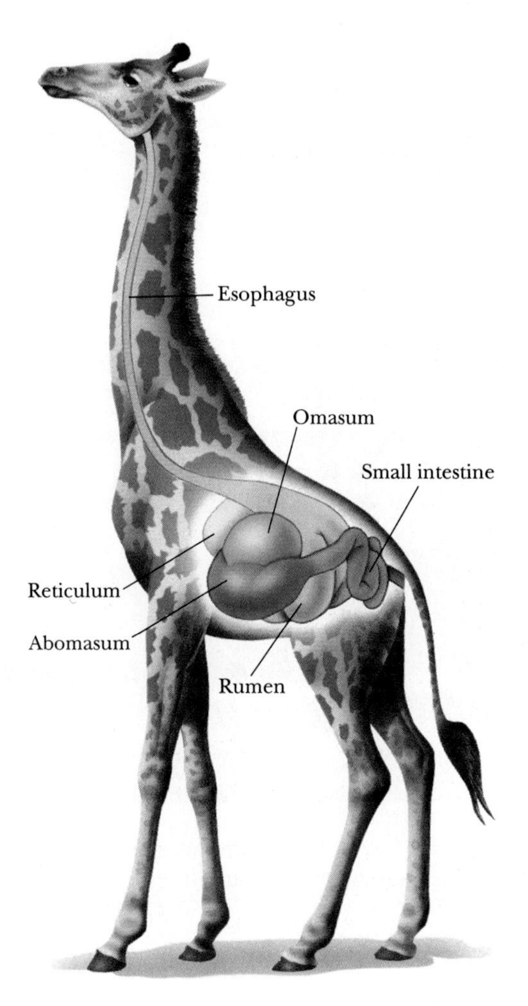

***Figure* 10.28** Giraffes, cattle, deer, and camels are ruminant animals that are able to metabolize cellulose, thanks to bacterial cellulase in the rumen, a large first compartment in the stomach of a ruminant.

δ-chitin, the structure is thought to involve pairs of parallel sheets separated by single antiparallel sheets.

Chitin is the earth's second most abundant carbohydrate polymer (after cellulose), and its ready availability and abundance offer opportunities for industrial and commercial applications. Chitin-based coatings can extend the shelf life of fruits, and a chitin derivative that binds to iron atoms in meat has been found to slow the reactions that cause rancidity and flavor loss. Without such a coating, the iron in meats activates oxygen from the air, forming reactive free radicals that attack and oxidize polyunsaturated lipids, causing most of the flavor loss associated with rancidity. Chitin-based coatings coordinate the iron atoms, preventing their interaction with oxygen.

Alginates

A family of novel extended ribbon structures that bind metal ions, particularly calcium, in their structure are the **alginate** polysaccharides of *marine brown algae (Phaeophyceae).* These include **poly(β-D-mannuronate)** and **poly(α-L-guluronate),** which are (1→4) linked chains formed from *β-D-mannuronic acid* and *α-L-guluronic acid,* respectively. Both of these homopolymers are found

Cellulose

Chitin

N-Acetylglucosamine units

Mannan

Mannose units

Poly (D-Mannuronate)

Poly (L-Guluronate)

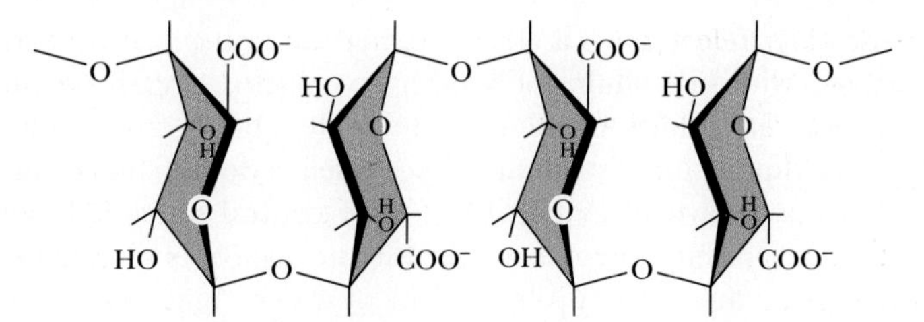

***Figure* 10.29** Like cellulose, chitin, mannan, and poly(D-mannuronate) form extended ribbons and pack together efficiently, taking advantage of multiple hydrogen bonds.

together in most marine alginates, although to widely differing extents, and mixed chains containing both monomer units are also found. As shown in Figure 10.29, the conformation of poly(β-D-mannuronate) is similar to that of cellulose. In the solid state, the free form of the polymer exists in cellulose-like form. However, complexes of the polymer with cations (such as lithium, sodium, potassium, and calcium) adopt a threefold helix structure, presumably to accommodate the bound cations. For poly(α-L-guluronate) (Figure 10.29), the axial–axial configuration of the glycosidic linkage leads to a distinctly buckled ribbon with limited flexibility. Cooperative interactions between such buckled ribbons can only be strong if the interstices are filled effectively with water molecules or metal ions. Figure 10.30 shows a molecular model of a Ca^{2+}-induced dimer of poly(α-L-guluronate).

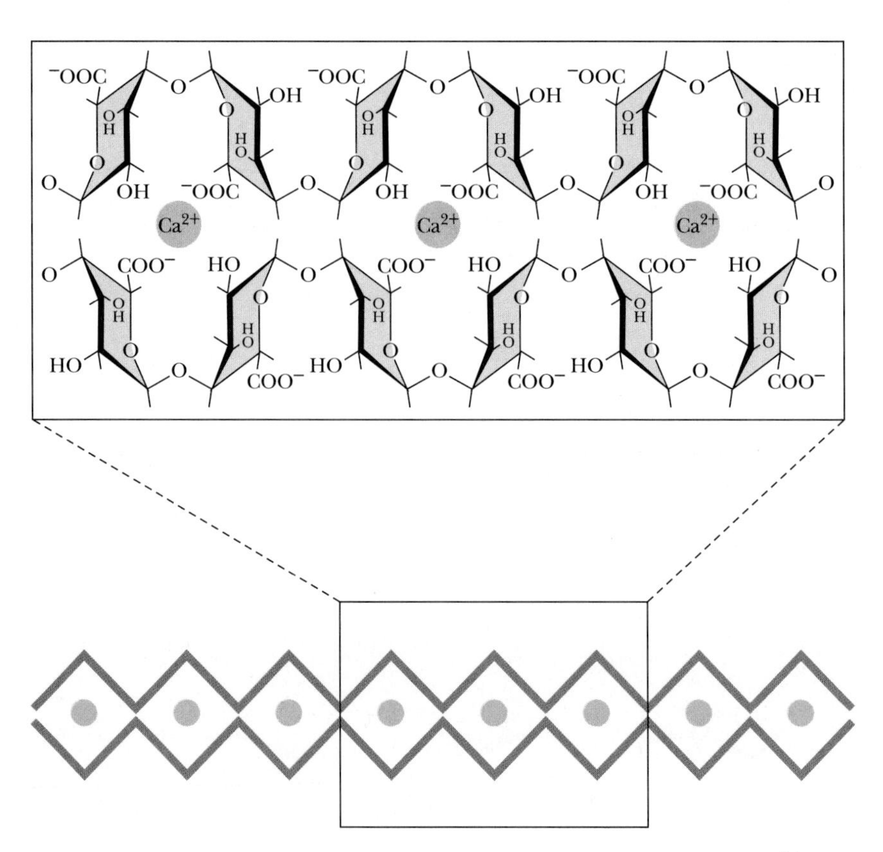

Figure **10.30** Poly(α-L-guluronate) strands dimerize in the presence of Ca^{2+}, forming a structure known as an "egg carton."

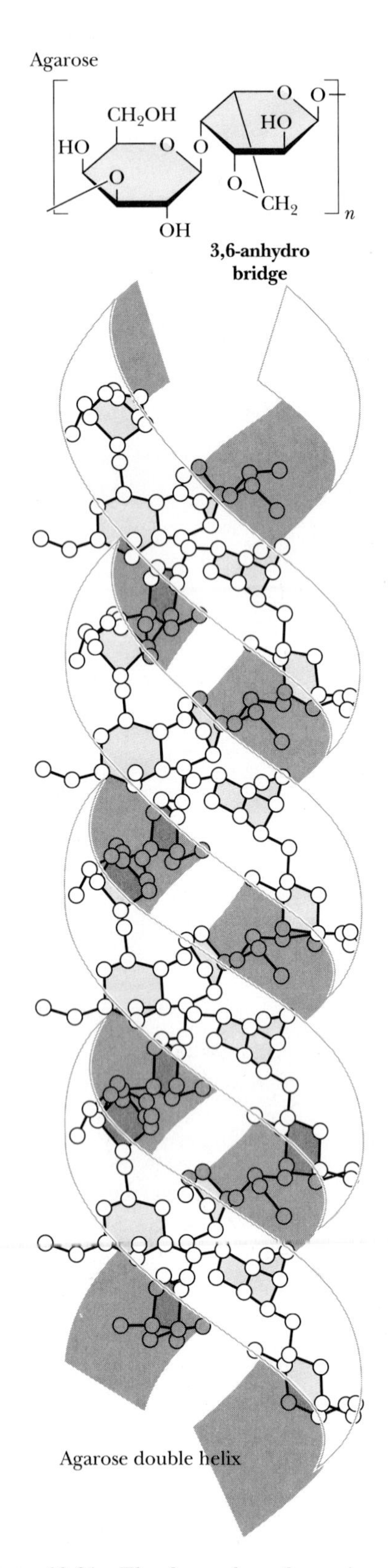

Figure **10.31** The favored conformation of agarose in water is a double helix with a threefold screw axis.

Agarose

An important polysaccharide mixture isolated from marine red algae (*Rhodophyceae*) is **agar,** which consists of two components, **agarose** and **agaropectin.** Agarose (Figure 10.31) is a chain of alternating D-galactose and 3,6-anhydro-L-galactose, with side chains of 6-methyl-D-galactose. Agaropectin is similar, but contains in addition sulfate ester side chains and D-glucuronic acid. The three-dimensional structure of agarose is a double helix with a threefold screw axis, as shown in Figure 10.31. The central cavity is large enough to accommodate water molecules. Agarose and agaropectin readily form gels containing large amounts (up to 99.5%) of water. Agarose can be processed to remove most of the charged groups, yielding a material (trade name Sepharose) useful for purification of macromolecules in gel exclusion chromatography. Pairs of chains form double helices which subsequently aggregate in bundles to form a stable gel, as shown in Figure 10.32.

Glycosaminoglycans

A class of polysaccharides known as **glycosaminoglycans** is involved in a variety of extracellular (and sometimes intracellular) functions. Glycosaminoglycans consist of linear chains of repeating disaccharides in which one of the monosaccharide units is an amino sugar and one (or both) of the monosaccharide units contains at least one negatively charged sulfate or carboxylate group. The repeating disaccharide structures found commonly in glycosaminoglycans are shown in Figure 10.33. **Heparin,** with the highest net negative charge of the disaccharides shown, is a natural anticoagulant substance. It binds strongly to *antithrombin III* (a protein involved in terminating the clotting process) and inhibits blood clotting. **Hyaluronate** molecules may consist of as many as 25,000 disaccharide units, with molecular weights of up to 10^7. Hyaluronates are important components of the vitreous humor in the eye and of

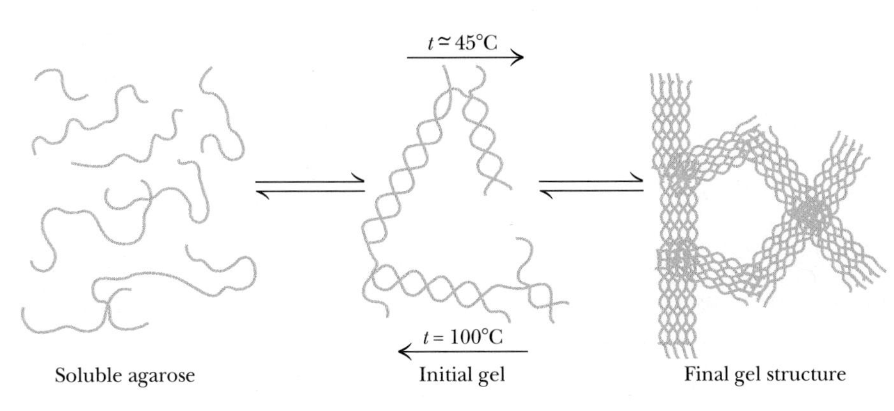

***Figure* 10.32** The ability of agarose to assemble in complex bundles to form gels in aqueous solution makes it useful in numerous chromatographic procedures, including gel exclusion chromatography and electrophoresis. Cells grown in culture can be embedded in stable agarose gel "threads" so that their metabolic and physiological properties can be studied.

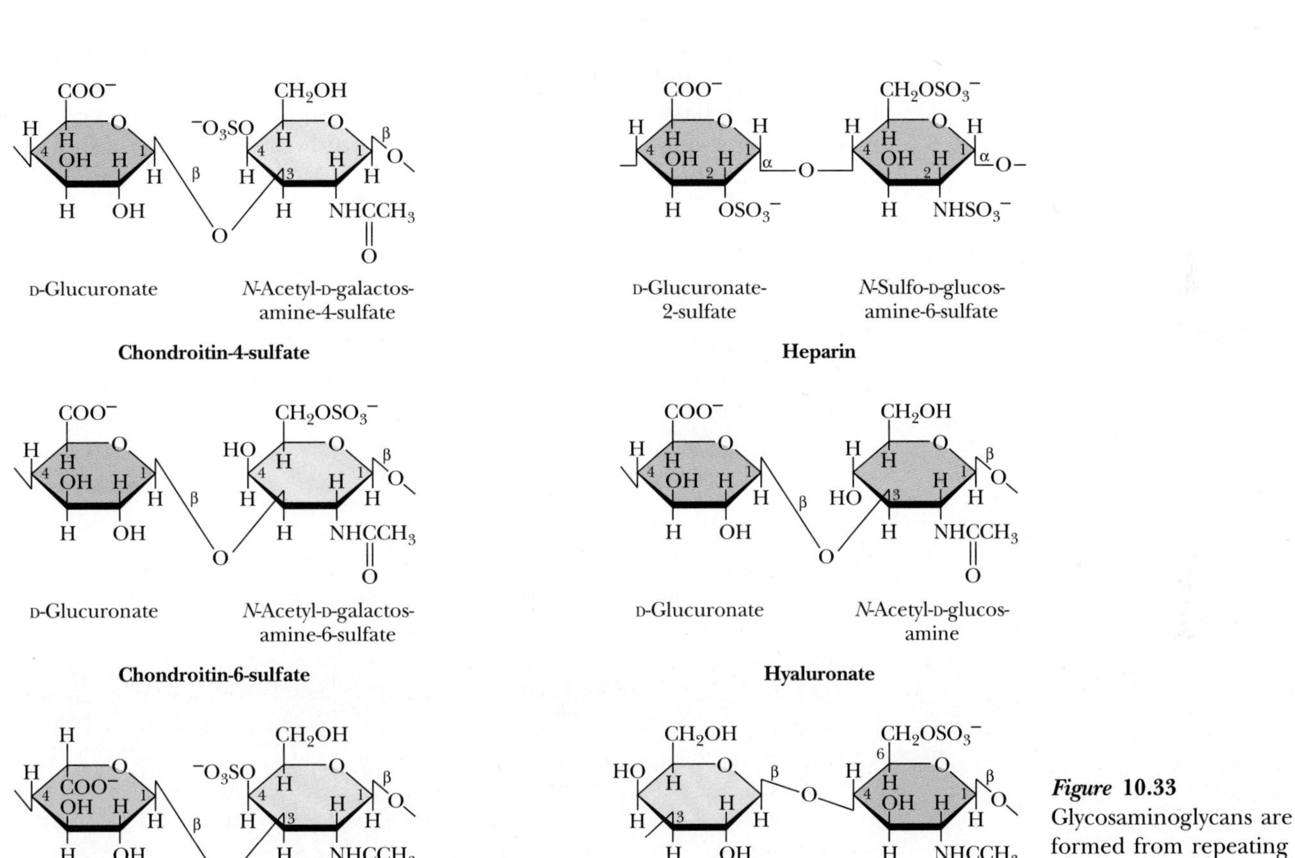

***Figure* 10.33** Glycosaminoglycans are formed from repeating disaccharide arrays. Glycosaminoglycans are components of the proteoglycans.

synovial fluid, the lubricant fluid of joints in the body. The **chondroitins** and **keratan sulfate** are found in tendons, cartilage, and other connective tissue, whereas **dermatan sulfate,** as its name implies, is a component of the extracellular matrix of skin. Glycosaminoglycans are fundamental constituents of *proteoglycans* (discussed later).

Bacterial Cell Walls

Some of nature's most interesting polysaccharide structures are found in *bacterial cell walls.* Given the strength and rigidity provided by polysaccharide structures, it is not surprising that bacteria use such structures to provide protection for their cellular contents. Bacteria normally exhibit high internal osmotic pressures and frequently encounter variable, often hypotonic exterior conditions. The rigid cell walls synthesized by bacteria maintain cell shape and size and prevent swelling or shrinkage that would inevitably accompany variations in solution osmotic strength.

Peptidoglycan

Bacteria are conveniently classified as either **Gram-positive** or **Gram-negative** depending on their response to the so-called *Gram stain* (see Chapter 9). Despite substantial differences in the various structures surrounding these two types of cell, nearly all bacterial cell walls have a strong, protective peptide–polysaccharide layer called **peptidoglycan.** Gram-positive bacteria have a thick (approximately 25 nm) cell wall consisting of multiple layers of peptidoglycan. This thick cell wall surrounds the bacterial plasma membrane. Gram-negative bacteria, in contrast, have a much thinner (2 to 3 nm) cell wall consisting of a single layer of peptidoglycan sandwiched between the inner and outer lipid bilayer membranes. In either case, peptidoglycan, sometimes

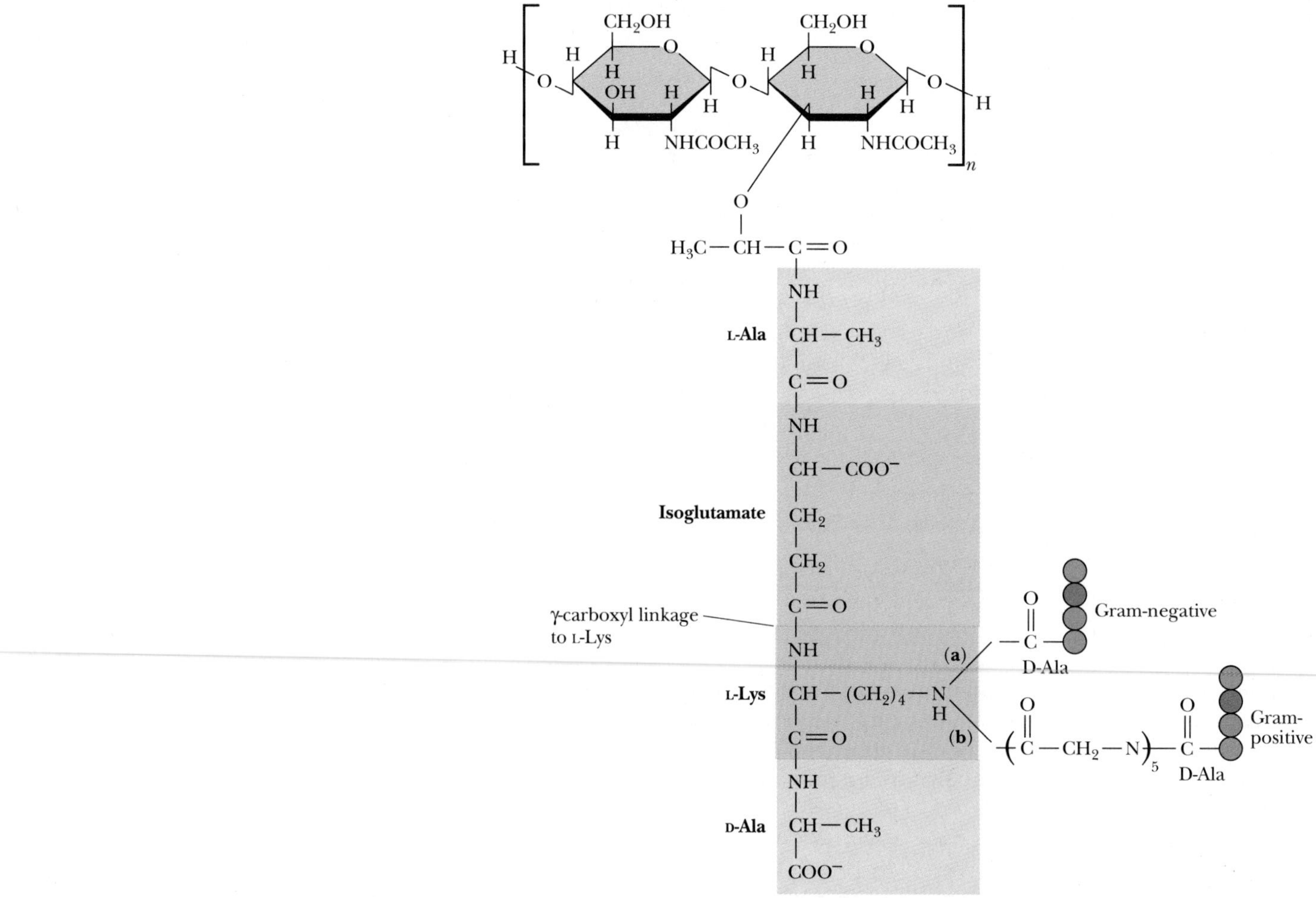

***Figure* 10.34** The structure of peptidoglycan. The tetrapeptides linking adjacent backbone chains contain an unusual γ-carboxyl linkage.

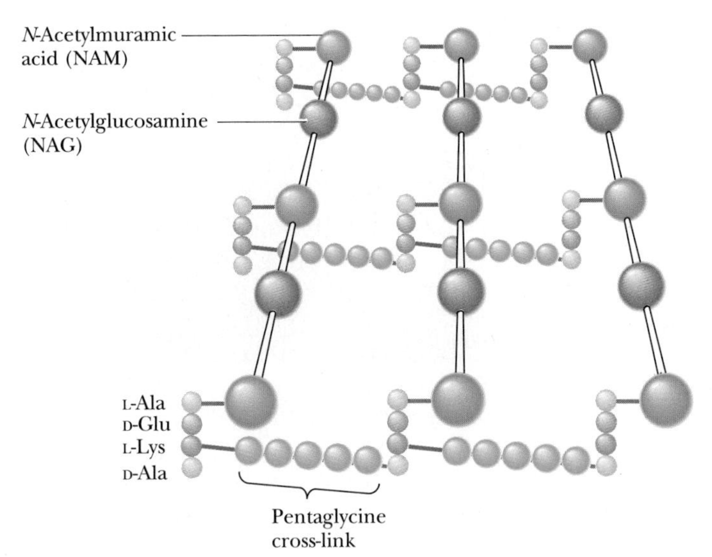

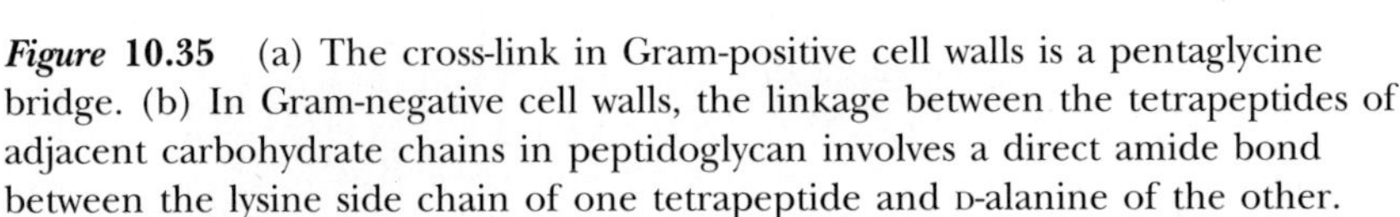

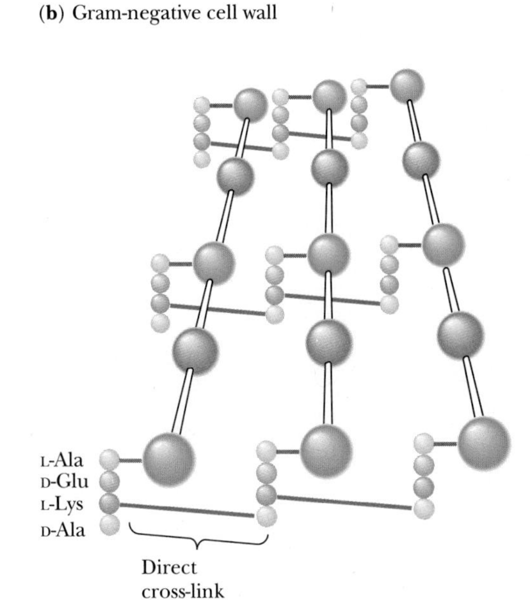

***Figure* 10.35** (a) The cross-link in Gram-positive cell walls is a pentaglycine bridge. (b) In Gram-negative cell walls, the linkage between the tetrapeptides of adjacent carbohydrate chains in peptidoglycan involves a direct amide bond between the lysine side chain of one tetrapeptide and D-alanine of the other.

called **murein** (from the Latin *murus* for "wall"), is a continuous cross-linked structure—in essence, a single molecule—built around the cell. The structure is shown in Figure 10.34. The backbone is a $\beta(1\rightarrow4)$ linked polymer of alternating *N*-acetylglucosamine and *N*-acetylmuramic acid units. This part of the structure is similar to chitin, but it is joined to a tetrapeptide, usually L-Ala·D-Glu·L-Lys·D-Ala, in which the L-lysine is linked to the γ-COOH of D-glutamate. The peptide is linked to the *N*-acetylmuramic acid units via its D-lactate moiety. The ϵ-amino group of lysine in this peptide is linked to the —COOH of D-alanine of an adjacent tetrapeptide. In Gram-negative cell walls, the lysine ϵ-amino group forms a *direct amide bond* with this D-alanine carboxyl (Figure 10.35). In Gram-positive cell walls, a **pentaglycine chain** bridges the lysine ϵ-amino group and the D-Ala carboxyl group.

Cell Walls of Gram-Negative Bacteria

In Gram-negative bacteria, the peptidoglycan wall is the rigid framework around which is built an elaborate membrane structure (Figure 10.36). The peptidoglycan layer encloses the *periplasmic space* and is attached to the outer membrane via a group of **hydrophobic proteins.** These proteins, each having 57 amino acid residues, are attached through amide linkages from the side chains of C-terminal lysines of the proteins to diaminopimelic acid groups on the peptidoglycan. This linkage to the hydrophobic protein replaces one of the D-alanine residues in about 10% of the peptides of the peptidoglycan. On the other end of the hydrophobic protein, the N-terminal residue, a serine, makes a covalent bond to a lipid that is part of the outer membrane.

As shown in Figure 10.36, the outer membrane of Gram-negative bacteria is coated with a highly complex **lipopolysaccharide,** which consists of a lipid group (anchored in the outer membrane) joined to a polysaccharide made up of long chains with many different and characteristic repeating structures

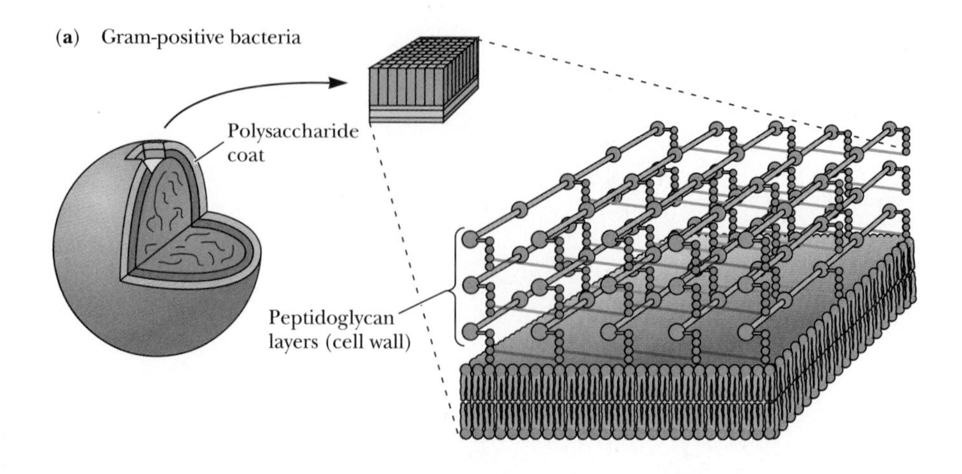

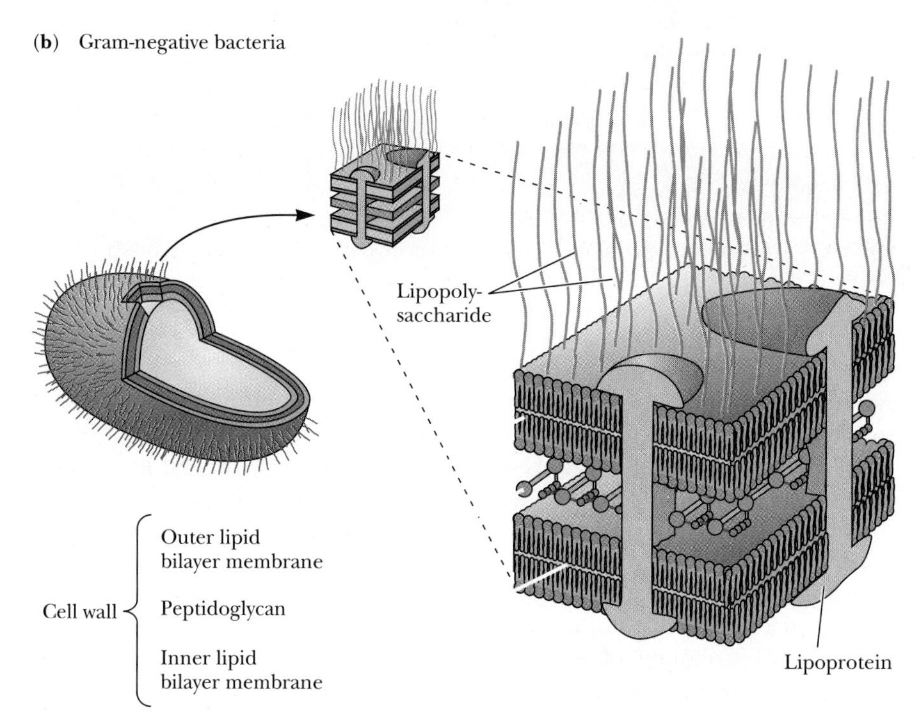

***Figure* 10.36** The structures of the cell wall and membrane(s) in Gram-positive and Gram-negative bacteria. The Gram-positive cell wall is thicker than that in Gram-negative bacteria, compensating for the absence of a second (outer) bilayer membrane.

(Figure 10.37). These many different unique units determine the antigenicity of the bacteria; that is, animal immune systems recognize them as foreign substances and raise antibodies against them. As a group, these **antigenic determinants** are called the **O antigens,** and there are thousands of different ones. The *Salmonella* bacteria alone have well over a thousand known O antigens that have been organized into 17 different groups. The great variation in these O antigen structures apparently plays a role in the recognition of one type of cell by another and in evasion of the host immune system.

Cell Walls of Gram-Positive Bacteria

In Gram-positive bacteria, the cell exterior is less complex than for Gram-negative cells. Having no outer membrane, Gram-positive cells compensate with a thicker wall. Covalently attached to the peptidoglycan layer are **teichoic acids,** which often account for 50% of the dry weight of the cell wall (Figure 10.38). The teichoic acids are polymers of *ribitol phosphate* or *glycerol phosphate* linked by phosphodiester bonds. In these heteropolysaccharides, the free hydroxyl groups of the ribitol or glycerol are often substituted by glycosidically linked monosaccharides (often glucose or *N*-acetylglucosamine) or disaccharides. D-Alanine is sometimes found in ester linkage to the saccharides. Teichoic acids are not confined to the cell wall itself, and they may be present in the inner membranes of these bacteria. Many teichoic acids are antigenic, and they also serve as the receptors for bacteriophages in some cases.

Cell Surface Polysaccharides

Compared to bacterial cells, which are identical within a given cell type (except for O antigen variations), animal cells display a wondrous diversity of structure, constitution, and function. Although each animal cell contains, in its genetic material, the instructions to replicate the entire organism, each differentiated animal cell carefully controls its composition and behavior within the organism. A great part of each cell's uniqueness begins at the cell surface. This surface uniqueness is critical to each animal cell since cells spend their entire life span in intimate contact with other cells and must therefore communicate with one another. That cells are able to pass information among themselves is evidenced by numerous experiments. For example, heart *myocytes,* when grown in culture (in glass dishes) establish *synchrony* when they make contact, so that they "beat" or contract in unison. If they are removed from the culture and separated, they lose their synchronous behavior, but if allowed to reestablish cell-to-cell contact, they spontaneously restore their synchronous contractions. Kidney cells grown in culture with liver cells seek out and make contact with other kidney cells and avoid contact with liver cells. Cells grown in culture grow freely until they make contact with one another, at which point growth stops, a phenomenon well known as **contact inhibition.** One important characteristic of cancerous cells is the loss of contact inhibition.

As these and many other related phenomena show, it is clear that molecular structures on one cell are recognizing and responding to molecules on the adjacent cell or to molecules in the **extracellular matrix,** the complex "soup" of connective proteins and other molecules that exists outside of and among cells. Many of these interactions involve *glycoproteins* on the cell surface and *proteoglycans* in the extracellular matrix. The "information" held in these special carbohydrate-containing molecules is not encoded directly in the genes (as with proteins), but is determined instead by expression of the appropriate enzymes that assemble carbohydrate units in a characteristic way on these molecules. Also, by virtue of the several hydroxyl linkages that can be formed with each carbohydrate monomer, these structures can be more information rich than proteins and nucleic acids, which can form only linear polymers. A few of these glycoproteins and their unique properties are described in the following sections.

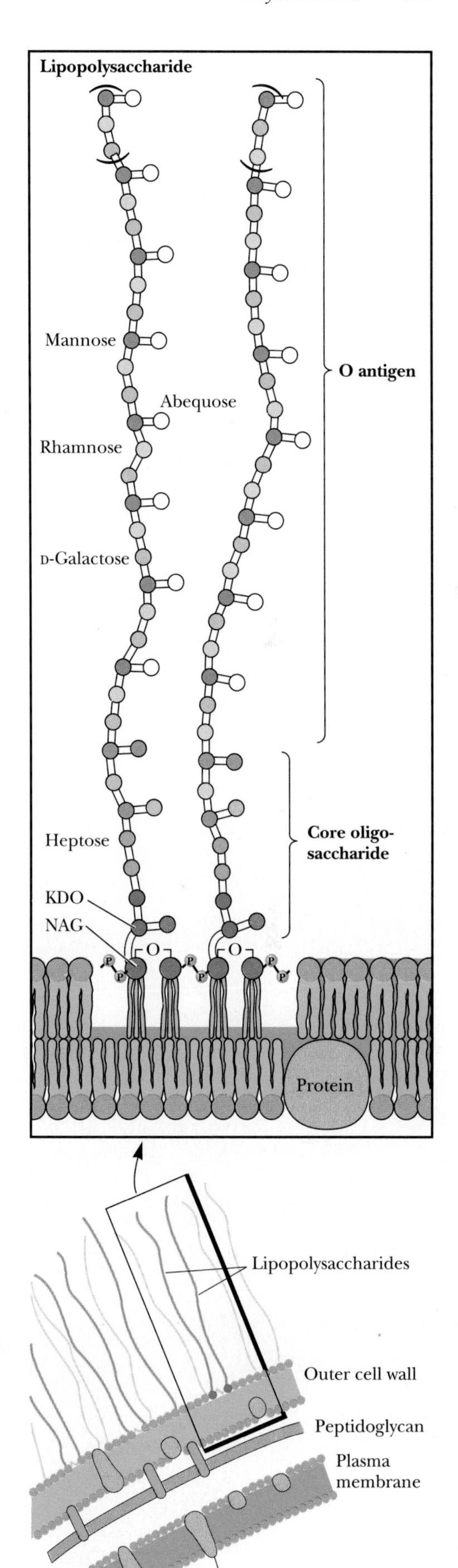

***Figure* 10.37** Lipopolysaccharide (LPS) coats the outer membrane of Gram-negative bacteria. The lipid portion of the LPS is embedded in the outer membrane and is linked to a complex polysaccharide.

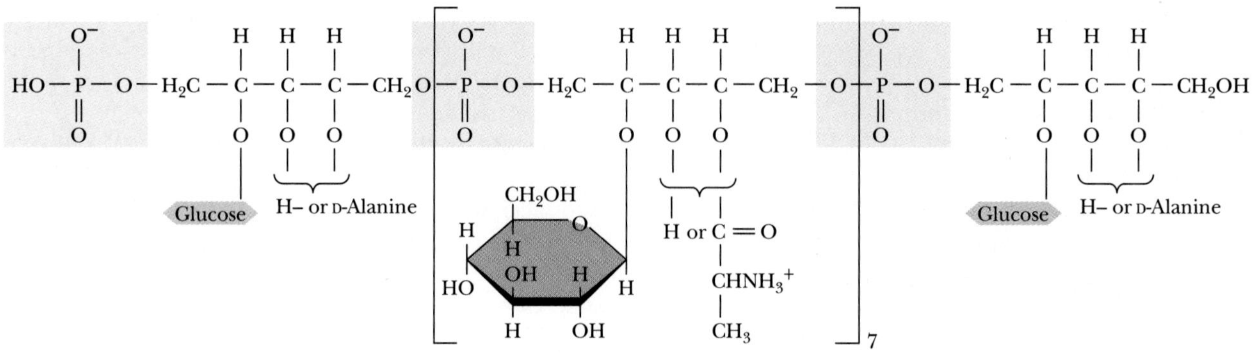

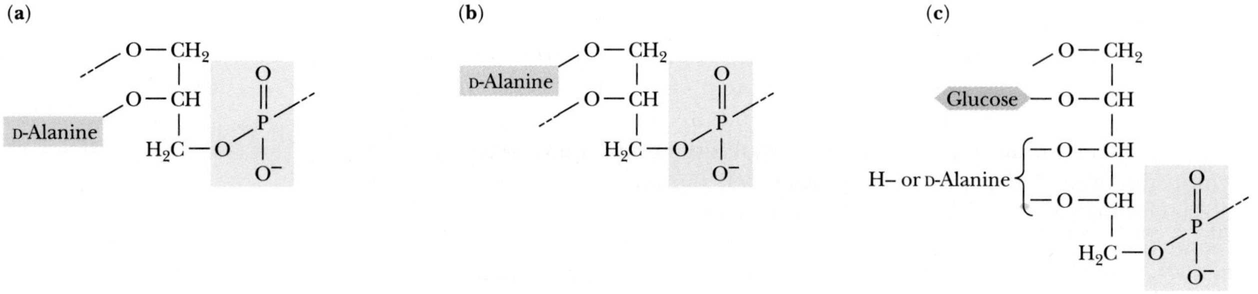

Figure **10.38** Teichoic acids are covalently linked to the peptidoglycan of Gram-positive bacteria. These polymers of (a, b) glycerol phosphate or (c) ribitol phosphate are linked by phosphodiester bonds.

10.5 Glycoproteins

Many proteins found in nature are **glycoproteins** since they contain covalently linked oligo- and polysaccharide groups. The list of known glycoproteins includes structural proteins, enzymes, membrane receptors, transport proteins, and immunoglobulins, among others. In most cases, the precise function of the bound carbohydrate moiety is not understood.

Carbohydrate groups may be linked to polypeptide chains via the hydroxyl groups of serine, threonine, or hydroxylysine residues (in **O-linked saccharides**) (Figure 10.39a) or via the amide nitrogen of an asparagine residue (in **N-linked saccharides**) (Figure 10.39b). The carbohydrate residue linked to the protein in O-linked saccharides is usually an *N*-acetylgalactosamine, but mannose, galactose, and xylose residues linked to protein hydroxyls are also found (Figure 10.39a). Oligosaccharides O-linked to glycophorin (see Chapter 9) involve *N*-acetylgalactosamine linkages and are rich in sialic acid residues (Figure 9.36). N-linked saccharides always have a unique core structure composed of two *N*-acetylglucosamine residues linked to a branched mannose triad (Figure 10.39b, c). Many other sugar units may be linked to each of the mannose residues of this branched core.

O-Linked saccharides are often found in cell surface glycoproteins and in **mucins,** the large glycoproteins that coat and protect mucous membranes in the respiratory and gastrointestinal tracts in the body. Certain viral glycoproteins also contain O-linked sugars. O-Linked saccharides in glycoproteins are often found clustered in richly glycosylated domains of the polypeptide chain. Physical studies on mucins show that they adopt rigid, extended structures so that an individual mucin molecule ($M_r = 10^7$) may extend over a distance of

***Figure* 10.39** The carbohydrate moieties of glycoproteins may be linked to the protein via (a) serine or threonine residues (in the O-linked saccharides) or (b) asparagine residues (in the N-linked saccharides). (c) N-Linked glycoproteins are of three types: high mannose, complex, and hybrid, the latter of which combines structures found in the high mannose and complex saccharides.

(a) O-linked saccharides

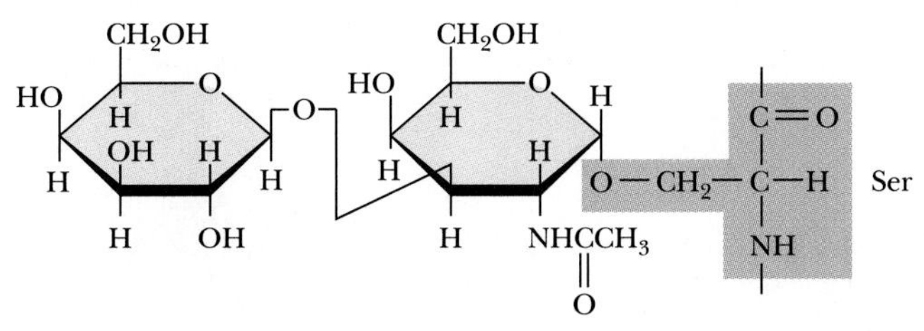

β-Galactosyl–1,3–α-*N*-acetylgalactosyl-serine

$HOCH_2$ O OH CH₃ C=O O−CH−C−H Thr OH NH

α-Xylosyl-threonine

CH_2OH H H OH HO HO O−CH₂−C−H Ser C=O NH

α-Mannosyl-serine

(b) Core oligosaccharides in N-linked glycoproteins

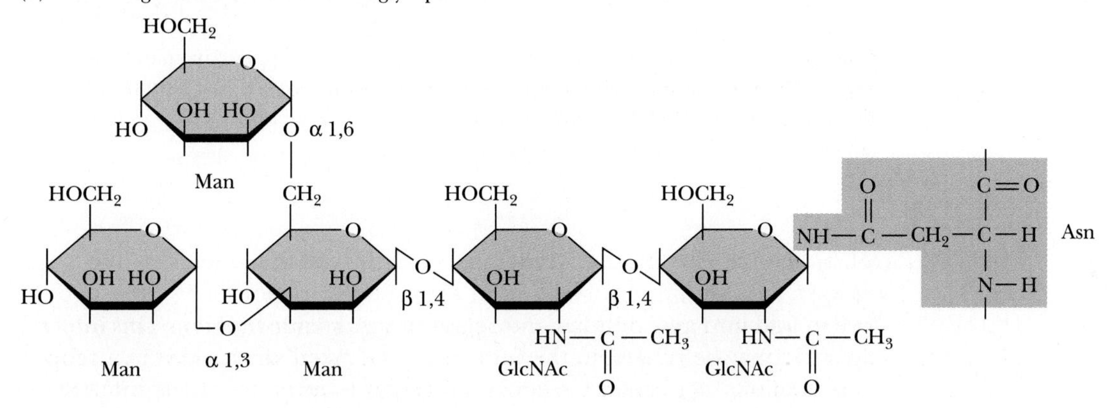

(c) N-linked glycoproteins

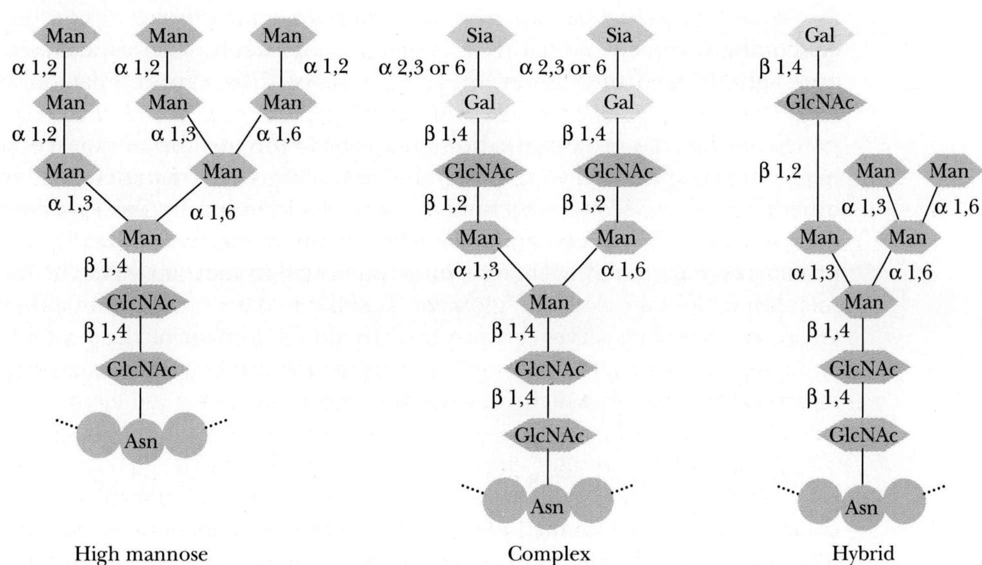

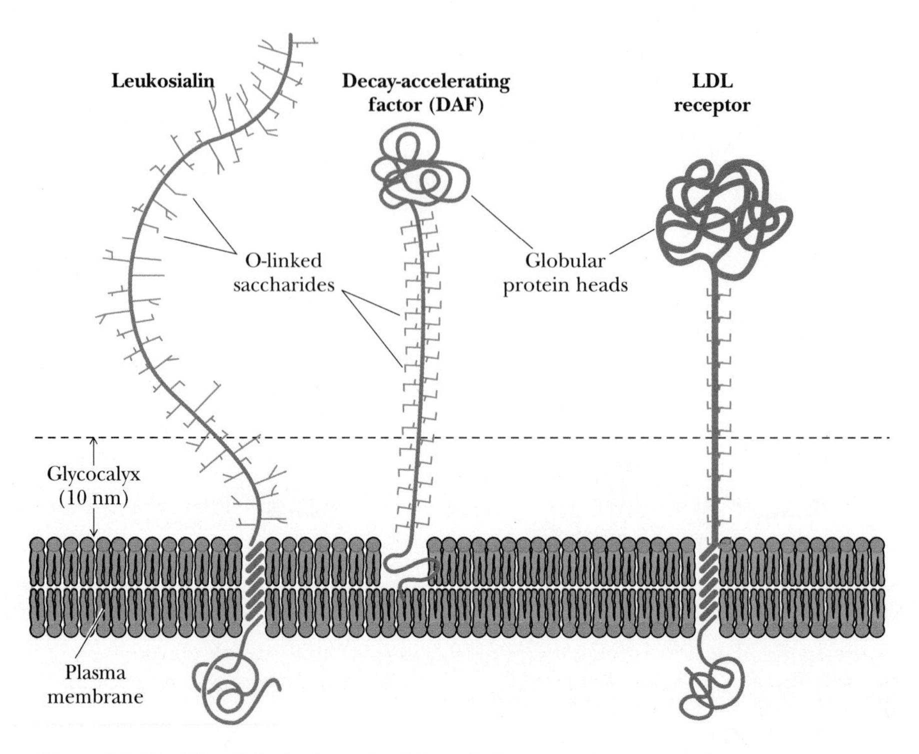

Figure **10.40** The O-linked saccharides of glycoproteins appear in many cases to adopt extended conformations that serve to extend the functional domains of these proteins above the membrane surface.
(Adapted from Jentoft, N., 1990, Trends in Biochemical Sciences 15:291–294.)

150 to 200 nm in solution. Inherent steric interactions between the sugar residues and the protein residues in these cluster regions cause the peptide core to fold into an extended and relatively rigid conformation. This interesting effect may be related to the function of O-linked saccharides in glycoproteins. It allows aggregates of mucin molecules to form extensive, intertwined networks, even at low concentrations. These viscous networks protect the mucosal surface of the respiratory and gastrointestinal tracts from harmful environmental agents.

There appear to be two structural motifs for membrane glycoproteins containing O-linked saccharides. Certain glycoproteins, such as **leukosialin,** are O-glycosylated throughout much or most of their extracellular domain (Figure 10.40). Leukosialin, like mucin, adopts a highly extended conformation, allowing it to project great distances above the membrane surface, perhaps protecting the cell from unwanted interactions with macromolecules or other cells. The second structural motif is exemplified by the **low density lipoprotein (LDL) receptor** and by **decay accelerating factor (DAF).** These proteins contain a highly O-glycosylated stem region that separates the transmembrane domain from the globular, functional extracellular domain. The O-glycosylated stem serves to raise the functional domain of the protein far enough above the membrane surface to make it accessible to the extracellular macromolecules with which it interacts.

Antifreeze Glycoproteins

A unique family of O-linked glycoproteins permits fish to live in the icy sea water of the Arctic and Antarctic regions where water temperature may reach as low as −1.9°C. **Antifreeze glycoproteins (AFGPs)** are found in the blood of

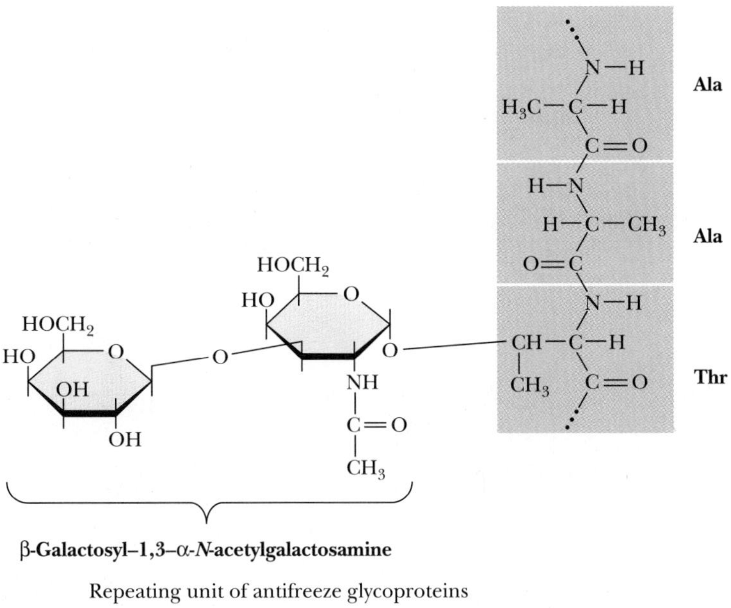

β-Galactosyl–1,3–α-*N*-acetylgalactosamine

Repeating unit of antifreeze glycoproteins

***Figure* 10.41** The structure of the repeating unit of antifreeze glycoproteins, a disaccharide consisting of *β*-galactosyl-(1→3)-*α*-*N*-acetylgalactosamine in glycosidic linkage to a threonine residue.

nearly all Antarctic fish and at least five Arctic fish. These glycoproteins have the peptide structure

$$[\text{Ala-Ala-Thr}]_n\text{-Ala-Ala}$$

where n can be 4, 5, 6, 12, 17, 28, 35, 45, or 50. Each of the threonine residues is glycosylated with the disaccharide *β*-galactosyl-(1→3)-*α*-*N*-acetylgalactosamine (Figure 10.41). This glycoprotein adopts a **flexible rod** conformation with regions of threefold left-handed helix. The evidence suggests that antifreeze glycoproteins may inhibit the formation of ice in the fish by binding specifically to the growth sites of ice crystals, inhibiting further growth of the crystals.

N-Linked Oligosaccharides

N-Linked oligosaccharides are found in many different proteins, including immunoglobulins G and M, ribonuclease B, ovalbumin, and peptide hormones (Figure 10.42). Many different functions are known or suspected for *N*-glycosylation of proteins. Glycosylation can affect the physical and chemical properties of proteins, altering solubility, mass, and electrical charge. Carbohydrate moieties have been shown to stabilize protein conformations and protect proteins against proteolysis. Eukaryotic organisms use posttranslational additions of N-linked oligosaccharides to direct selected proteins to various intracellular organelles.

Oligosaccharide Cleavage as a Timing Device for Protein Degradation

The slow cleavage of monosaccharide residues from N-linked glycoproteins circulating in the blood targets these proteins for degradation by the organism. The liver contains specific receptor proteins that recognize and bind glycoproteins that are ready to be degraded and recycled. Newly synthesized

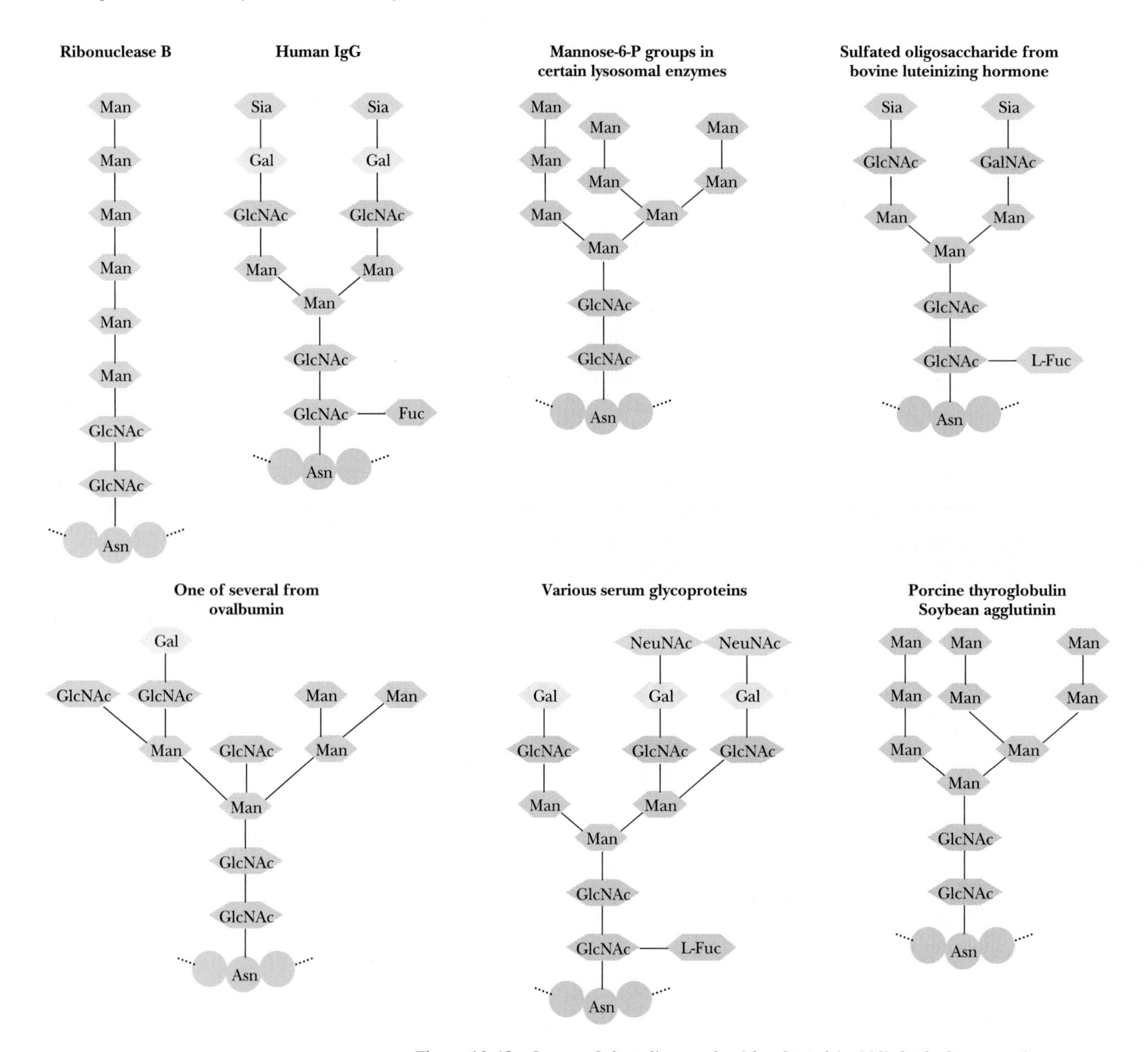

***Figure* 10.42** Some of the oligosaccharides found in N-linked glycoproteins.

serum glycoproteins contain N-linked **triantennary** (three-chain) oligosaccharides having structures similar to those in Figure 10.43, in which sialic acid residues cap galactose residues. As these glycoproteins circulate, enzymes on the blood vessel walls cleave off the sialic acid groups, exposing the galactose residues. In the liver, the **asialoglycoprotein receptor** binds these glycoproteins with exposed galactose residues with very high affinity ($K_D = 10^{-9}$ to 10^{-8} M). The complex of receptor and glycoprotein is then taken into the cell by **endocytosis,** and the glycoprotein is degraded in cellular lysosomes. Highest affinity binding of glycoprotein to the asialoglycoprotein receptor requires three free galactose residues. Oligosaccharides with only one or two exposed galactose residues bind less tightly. This is an elegant way for the body to keep track of how long glycoproteins have been in circulation. Over a period of time, anywhere from a few hours to weeks, the sialic acid groups are

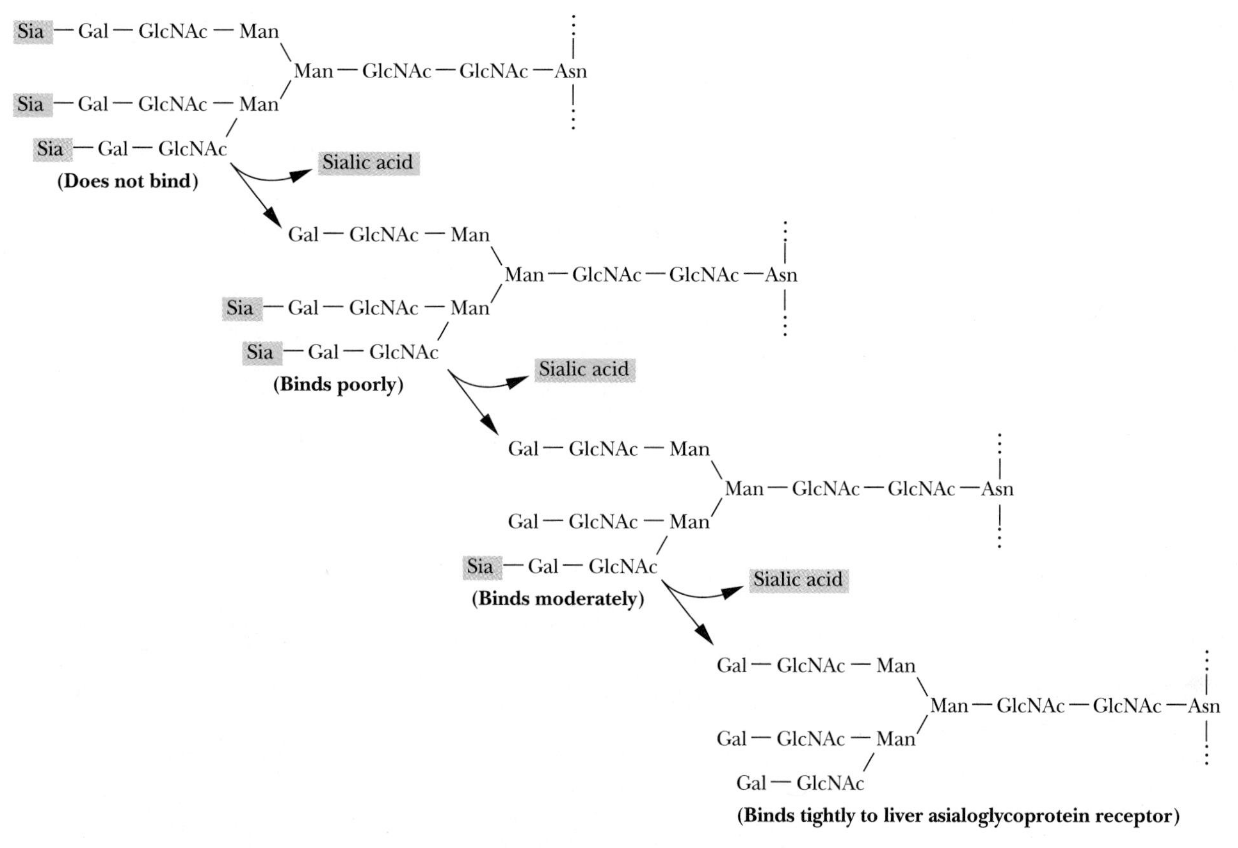

Figure **10.43** Progressive cleavage of sialic acid residues exposes galactose residues. Binding to the asialoglycoprotein receptor in the liver becomes progressively more likely as more Gal residues are exposed.

cleaved one by one. The longer the glycoprotein circulates and the more sialic acid residues are removed, the more galactose residues become exposed so that the glycoprotein is eventually bound to the liver receptor.

10.6 Proteoglycans

Proteoglycans are a family of glycoproteins whose carbohydrate moieties are predominantly **glycosaminoglycans.** The structures of only a few proteoglycans are known (Table 10.1), and even these few display considerable diversity (Figure 10.44). They range in size from **serglycin,** having 104 amino acid residues (10.2 kD) to **versican,** having 2409 residues (265 kD). Each of these proteoglycans contains one or two types of covalently linked glycosaminoglycans (Table 10.1). In the known proteoglycans, the glycosaminoglycan units are O-linked to serine residues of Ser-Gly dipeptide sequences. Serglycin is named for a unique central domain of 49 amino acids composed of alternating serine and glycine residues. The **cartilage matrix proteoglycan** contains 117 Ser-Gly pairs to which chondroitin sulfates attach. *Decorin,* a small proteoglycan secreted by fibroblasts and found in the extracellular matrix of connective tissues, contains only three Ser-Gly pairs, only one of which is normally glycosylated. In addition to glycosaminoglycan units, proteoglycans may also contain other N-linked and O-linked oligosaccharide groups.

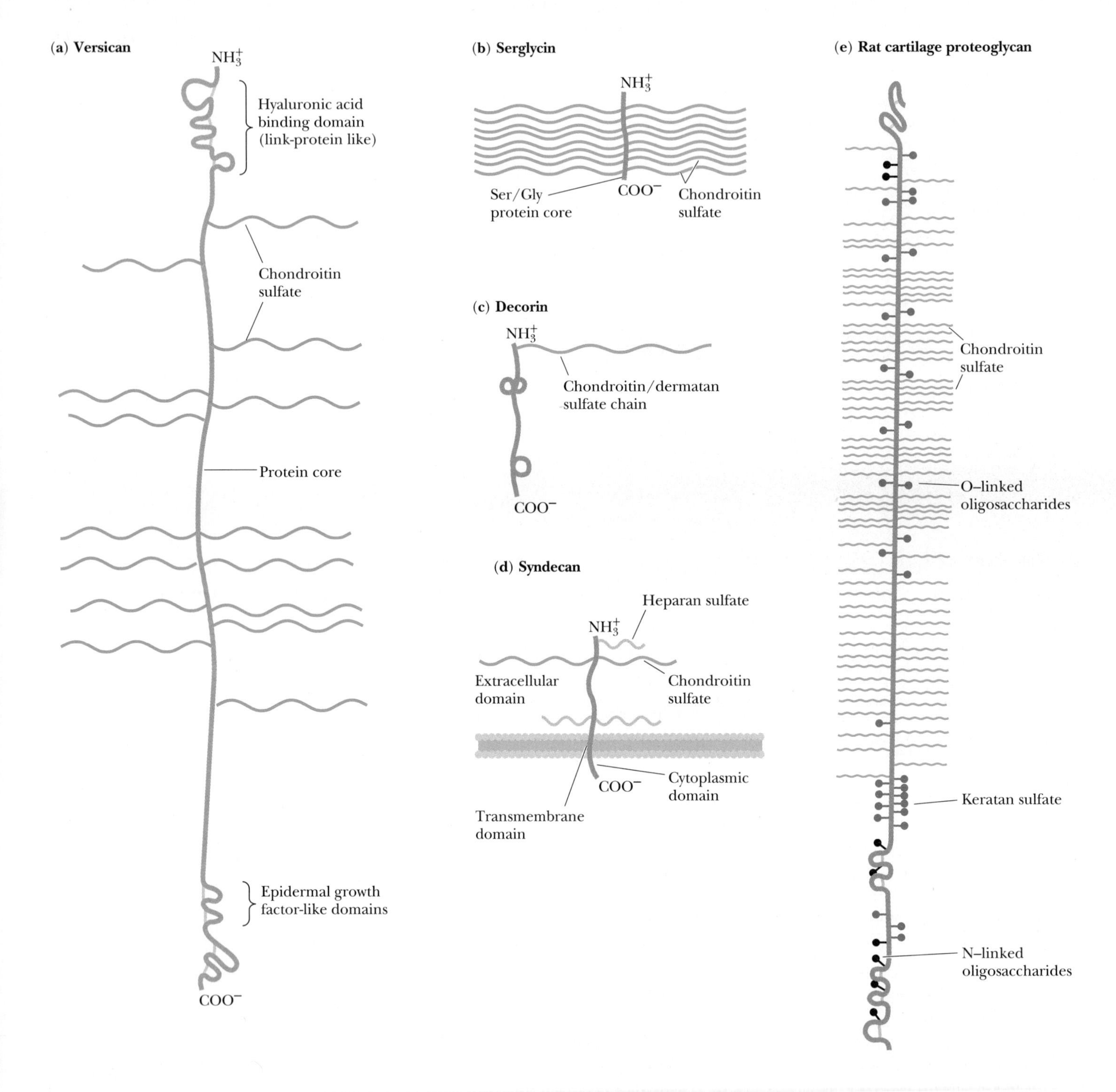

Figure 10.44 The known proteoglycans include a variety of structures. The carbohydrate groups of proteoglycans are predominantly glycosaminoglycans O-linked to serine residues. Proteoglycans include both soluble proteins and integral transmembrane proteins.

Functions of Proteoglycans

Proteoglycans may be *soluble* and located in the extracellular matrix, as is the case for serglycin, versican, and the cartilage matrix proteoglycan, or they may be *integral transmembrane proteins,* such as **syndecan.** Both types of proteoglycan appear to function by interacting with a variety of other molecules through their glycosaminoglycan components and through specific receptor domains in the polypeptide itself. For example, syndecan (from the Greek *syndein* meaning "to bind together") is a transmembrane proteoglycan that associates intracellularly with the actin cytoskeleton (Chapter 34). Outside the cell, it interacts with **fibronectin,** an extracellular protein that binds to several cell surface proteins and to components of the extracellular matrix. The ability of syndecan to participate in multiple interactions with these target molecules allows them to act as a sort of "glue" in the extracellular space, linking com-

Table **10.1**
Some Proteoglycans of Known Sequence

Proteoglycan	Glycosaminoglycan	Protein M_r	Number of Amino Acid Residues
Secreted or extracellular matrix proteoglycans			
Large aggregating cartilage proteoglycans	CS/KS[a]	220,952	2124
Versican	CS/DS	265,048	2409
Decorin	CS/DS	38,000	329
Intracellular granule proteoglycan			
Serglycin (PG19)	CS/DS	10,190	104
Membrane-intercalated proteoglycans			
Syndecan	HS/CS	38,868	311

[a]CS, chondroitin sulfate; DS, dermatan sulfate; HS, heparan sulfate (an analog of heparin); KS, keratan sulfate. These glycosaminoglycans are polymers consisting of the repeating disaccharides: glucuronic acid *N*-acetylgalactosamine (CS), iduronic acid *N*-acetylgalactosamine (DS), iduronic acid *N*-acetylglucosamine (HS and heparin), and galactose *N*-acetylglucosamine (KS). DS, HS, and heparin also contain some disaccharide units in which the uronic acid is glucuronic acid instead of iduronic acid. These glycosaminoglycans and CS are generally bound to the hydroxyl group of a serine residue to give the sequence (disaccharide) *n*GlcUA-Gal-Gal-Xyl-*O* Ser. Keratan sulfate has a different linkage region and can be either O- or N-linked. The sugars in the repeating disaccharide unit are sulfated to various degrees. By comparison, hyaluronic acid is a polymer of glucuronic acid and glucosamine that is not sulfated and does not attach covalently to a protein core.

Adapted from Ruoslahti, E., 1989. *Journal of Biological Chemistry* **264**:13369–13372.

ponents of the extracellular matrix, facilitating the binding of cells to the matrix, and mediating the binding of growth factors and other soluble molecules to the matrix and to cell surfaces (Figure 10.45).

Many of the functions of proteoglycans involve the binding of specific proteins to the glycosaminoglycan groups of the proteoglycan. The glycosaminoglycan binding sites on these specific proteins contain multiple basic amino acid residues. The amino acid sequences BBXB and BBBXXB (where B is a basic amino acid and X is any amino acid) recur repeatedly in these binding domains. Basic amino acids such as lysine and arginine provide charge neutralization for the negative charges of glycosaminoglycan residues, and in many cases, the binding of extracellular matrix proteins to glycosaminoglycans is primarily charge dependent. For example, more highly sulfated glycosaminoglycans bind more tightly to fibronectin. Certain protein–glycosaminoglycan interactions, however, require a specific carbohydrate sequence. A particular pentasaccharide sequence in heparin, for example, binds tightly to antithrombin III (Figure 10.46), accounting for the anticoagulant properties of heparin. Other glycosaminoglycans interact much more weakly.

Proteoglycans May Modulate Cell Growth Processes

Several lines of evidence raise the possibility of modulation or regulation of cell growth processes by proteoglycans. First, heparin and heparan sulfate are known to inhibit cell proliferation in a process involving internalization of the glycosaminoglycan moiety and its migration to the cell nucleus. Second, *fibroblast growth factor* binds tightly to heparin and other glycosaminoglycans, and

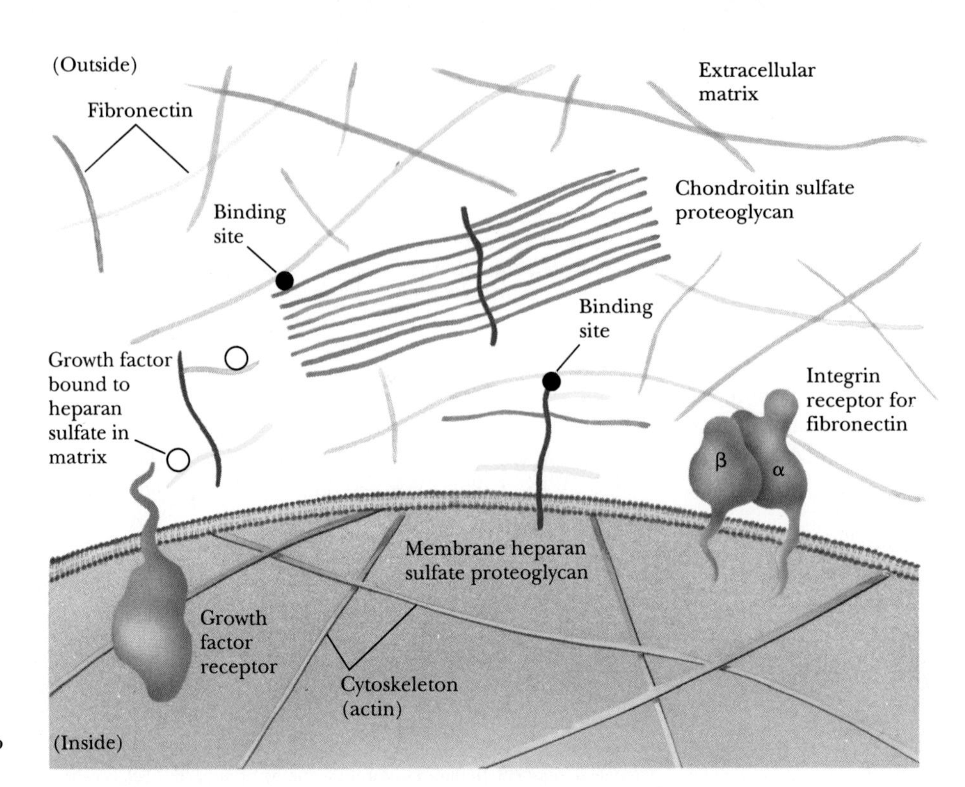

***Figure* 10.45** Proteoglycans serve a variety of functions on the cytoplasmic and extracellular surfaces of the plasma membrane. Many of these functions appear to involve the binding of specific proteins to the glycosaminoglycan groups.

the heparin–growth factor complex protects the growth factor from degradative enzymes, thus enhancing its activity. There is evidence that binding of fibroblast growth factors by proteoglycans and glycosaminoglycans in the extracellular matrix creates a reservoir of growth factors for cells to use. Third, *transforming growth factor β* has been shown to stimulate the synthesis and secretion of proteoglycans in certain cells. Fourth, several proteoglycan core proteins, including versican and *lymphocyte homing receptor,* have domains similar in sequence to *epidermal growth factor* and *complement regulatory factor.* These growth factor domains may interact specifically with growth factor receptors in the cell membrane in processes that are not yet understood.

Proteoglycans Make Cartilage Flexible and Resilient

Cartilage matrix proteoglycan is responsible for the flexibility and resilience of cartilage tissue in the body. In cartilage, long filaments of hyaluronic acid are studded or coated with proteoglycan molecules, as shown in Figure 10.47.

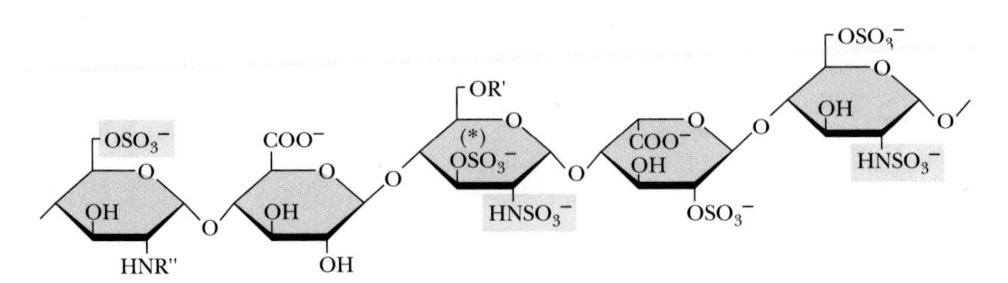

***Figure* 10.46** A portion of the structure of heparin, a carbohydrate having anticoagulant properties. It is used by blood banks to prevent the clotting of blood during donation and storage and also by physicians to prevent the formation of life-threatening blood clots in patients recovering from serious injury or surgery. This sulfated pentasaccharide sequence in heparin binds with high affinity to antithrombin III, accounting for this anticoagulant activity. The 3-O-sulfate marked by an asterisk is essential for high-affinity binding of heparin to antithrombin III.

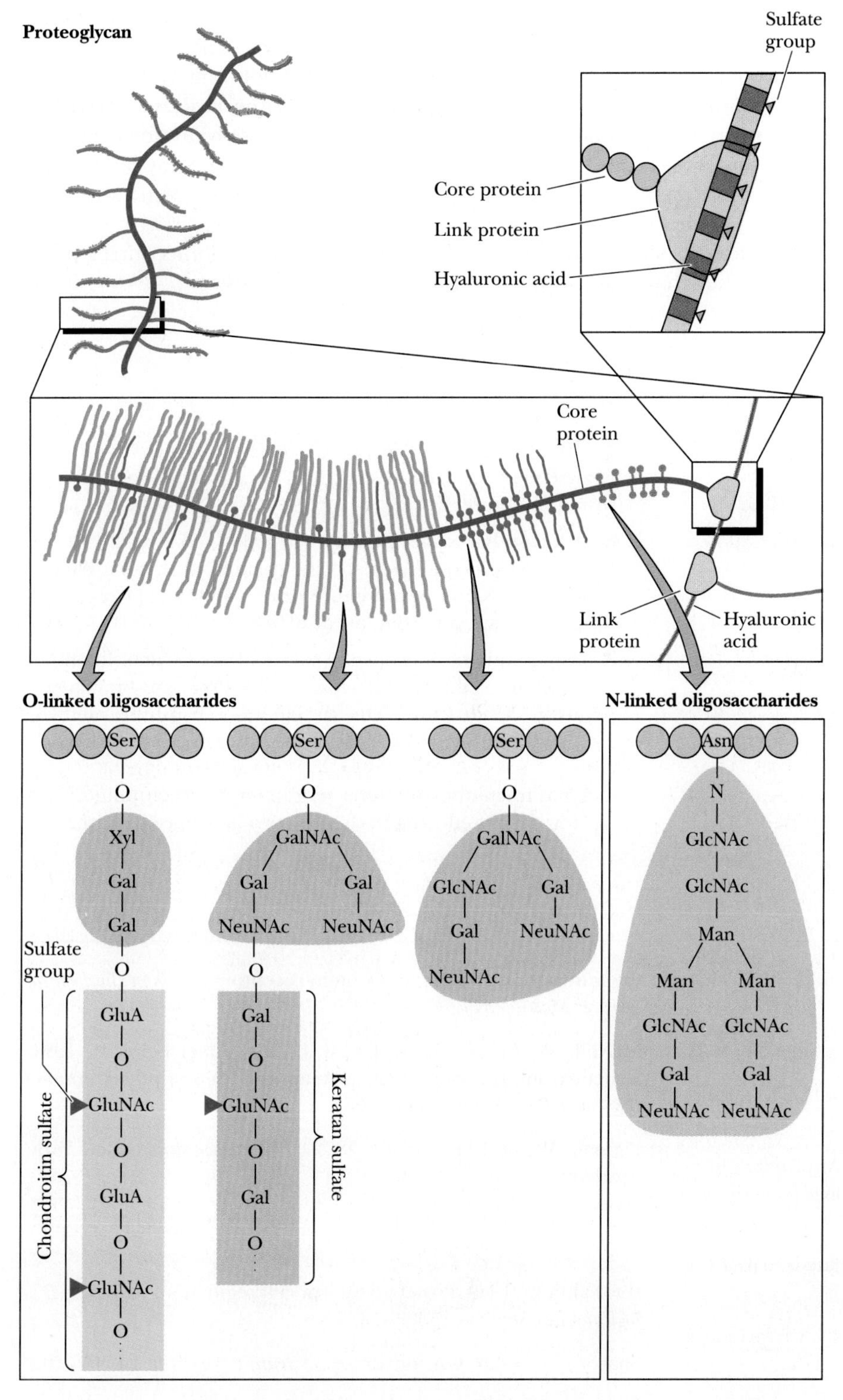

***Figure* 10.47** Hyaluronate (see Figure 10.33) forms the backbone of proteoglycan structures, such as those found in cartilage. The proteoglycan subunits consist of a core protein containing numerous O-linked and N-linked glycosaminoglycans. In cartilage, these highly hydrated proteoglycan structures are enmeshed in a network of collagen fibers. Release (and subsequent reabsorption) of water by these structures during compression accounts for the shock-absorbing qualities of cartilaginous tissue.

The hyaluronate chains can be as long as 4 μm and can coordinate 100 or more proteoglycan units. Cartilage proteoglycan possesses a **hyaluronic acid binding domain** on the NH_2-terminal portion of the polypeptide, which binds to hyaluronate with the assistance of a **link protein.** The proteoglycan–hyaluronate aggregates can have molecular weights of 2 million or more.

The proteoglycan–hyaluronate aggregates are highly hydrated by virtue of strong interactions between water molecules and the polyanionic complex. When cartilage is compressed (such as when joints absorb the impact of walking or running), water is briefly squeezed out of the cartilage tissue and then reabsorbed when the stress is diminished. This reversible hydration gives cartilage its flexible, shock-absorbing qualities and cushions the joints during physical activities that might otherwise injure the involved tissues.

Problems

1. Draw Haworth structures for the two possible isomers of D-altrose (Figure 10.2) and D-psicose (Figure 10.3).
2. Give the systematic name for stachyose (Figure 10.19).
3. Trehalose, a disaccharide produced in fungi, has the following structure:

   a. What is the systematic name for this disaccharide?
   b. Is trehalose a reducing sugar? Explain.
4. Draw Fischer projection structures for L-sorbose (D-sorbose is shown in Figure 10.3).
5. α-D-Glucose has a specific rotation, $[\alpha]_D^{20}$, of +112.2°, whereas β-D-glucose has a specific rotation of +18.7°. What is the composition of a mixture of α-D- and β-D-glucose, which has a specific rotation of 83.0°?
6. A 0.2-g sample of amylopectin was analyzed to determine the fraction of the total glucose residues that are branch points in the structure. The sample was exhaustively methylated and then digested, yielding 50 μmol of 2,3-dimethylglucose.
   a. What fraction of the total residues are branch points?
   b. How many reducing ends does this amylopectin have?

Further Reading

Aspinall, G. O., 1982. *The Polysaccharides,* Vols. 1 and 2. New York: Academic Press.

Collins, P. M., 1987. *Carbohydrates.* London: Chapman and Hall.

Davison, E. A., 1967. *Carbohydrate Chemistry.* New York: Holt, Rinehart and Winston.

Feeney, R. E., Burcham, T. S., and Yeh, Y., 1986. Antifreeze glycoproteins from polar fish blood. *Annual Review of Biophysical Chemistry* **15**:59–78.

Jentoft, N., 1990. Why are proteins O-glycosylated? *Trends in Biochemical Sciences* **15**:291–294.

Kjellen, L., and Lindahl, U., 1991. Proteoglycans: structures and interactions. *Annual Review of Biochemistry* **60**:443–475.

Lennarz, W. J., 1980. *The Biochemistry of Glycoproteins and Proteoglycans.* New York: Plenum Press.

Lodish, H. F., 1991. Recognition of complex oligosaccharides by the multisubunit asialoglycoprotein receptor. *Trends in Biochemical Sciences* **16**:374–377.

McNeil, M., Darvill, A. G., Fry, S. C., and Albersheim, P., 1984. Structure and function of the primary cell walls of plants. *Annual Review of Biochemistry* **53**:625–664.

Pigman, W., and Horton, D. 1972. *The Carbohydrates.* New York: Academic Press.

Rademacher, T. W., Parekh, R. B., and Dwek, R. A., 1988. Glycobiology. *Annual Review of Biochemistry* **57**:785–838.

Ruoslahti, E., 1989. Proteoglycans in cell regulation. *Journal of Biological Chemistry* **264**:13369–13372.

Sharon, N., 1980. Carbohydrates. *Scientific American* **243**:90–102.

Sharon, N., 1984. Glycoproteins. *Trends in Biochemical Sciences* **9**:198–202.

Part II

Enzymes and Energetics

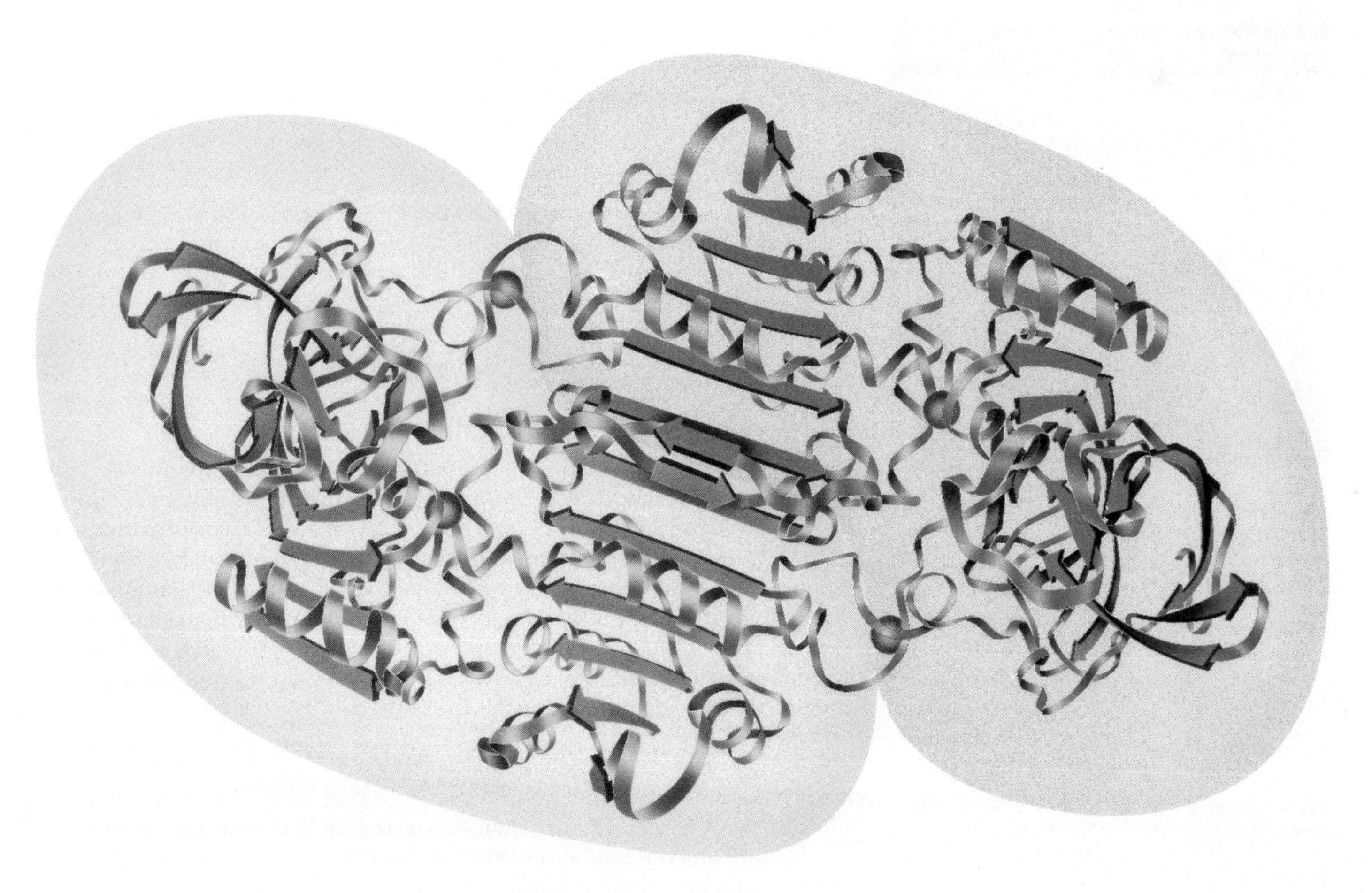

Liver alcohol dehydrogenase

'"Now! Now!" cried the Queen. "Faster! Faster! . . . Now *here,* you see, it takes all the running you can do, to keep in the same place. If you want to get somewhere else, you must run at least twice as fast as that!"'

Lewis Carroll, *Alice's Adventures in Wonderland* (1865)

Chapter 11

Enzyme Kinetics

The Queen of Hearts confronts Alice. Illustration from an original edition of *Alice's Adventures in Wonderland.*

Outline

11.1 Enzymes—Catalytic Power, Specificity, and Regulation

11.2 Introduction to Enzyme Kinetics

11.3 Kinetics of Enzyme-Catalyzed Reactions

11.4 Enzyme Inhibition

11.5 Kinetics of Enzyme-Catalyzed Reactions Involving Two or More Substrates

11.6 RNA and Antibody Molecules as Enzymes: Ribozymes and Abzymes

Living organisms seethe with metabolic activity. Thousands of chemical reactions are proceeding very rapidly at any given instant within all living cells. Virtually all of these transformations are mediated by **enzymes,** proteins specialized to catalyze metabolic reactions. The substances transformed in these reactions are often organic compounds that show little tendency for reaction outside the cell. An excellent example is glucose, a sugar that can be stored indefinitely on the shelf with no deterioration. Most cells quickly oxidize glucose, producing carbon dioxide and water and releasing lots of energy:

$$C_6H_{12}O_6 + 6\ O_2 \longrightarrow 6\ CO_2 + 6\ H_2O + 2870\ \text{kJ of energy}$$

(-2870 kJ/mol is the standard free energy change [$\Delta G°'$] for the oxidation of glucose; see Chapter 15). In chemical terms, 2870 kJ is a large amount of energy, and glucose can be viewed as an energy-rich compound even though at ambient temperature it is not readily reactive with oxygen outside of cells. Stated another way, glucose represents **thermodynamic potentiality:** Its reaction with oxygen is strongly exergonic, but it just doesn't occur under normal conditions. On the other hand, enzymes can catalyze such thermodynamically favorable reactions so that they proceed at extraordinarily rapid rates (Figure 11.1). In glucose oxidation and countless other instances, enzymes endow cells with the remarkable capacity to exert *kinetic control over thermodynamic potentiality.* That is, living systems use enzymes to accelerate and control the rates of vitally important biochemical reactions.

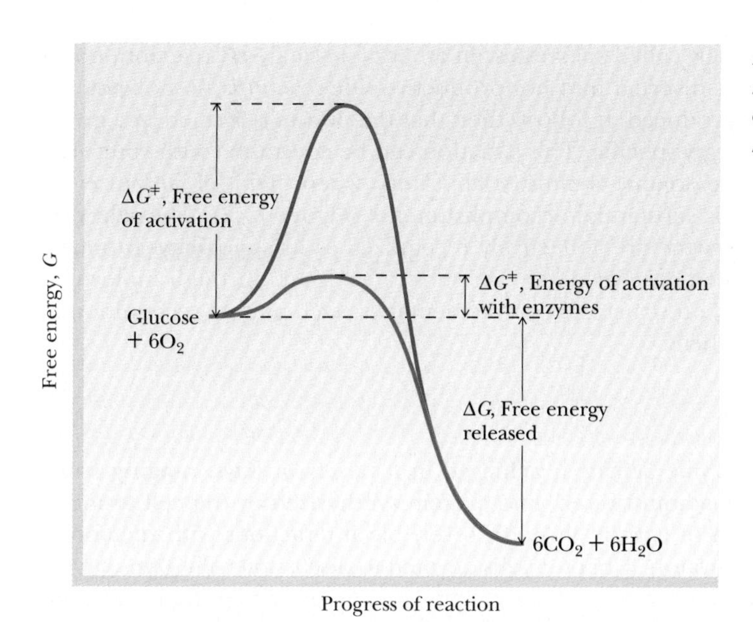

***Figure* 11.1** Reaction profile showing large $\Delta G^{\ddagger}$ for glucose oxidation, free energy change of −2870 kJ/mol; catalysts lower $\Delta G^{\ddagger}$, thereby accelerating rate.

Enzymes Are the Agents of Metabolic Function

Acting in sequential order, enzymes constitute metabolic pathways by which nutrient molecules are degraded, energy is released and converted into metabolically useful forms, and precursors are generated and transformed to create the literally thousands of distinctive biomolecules found in any living cell. Situated at key junctions of metabolic pathways are specialized **regulatory enzymes** capable of sensing the momentary metabolic needs of the cell and adjusting their catalytic rates accordingly. The responses of these enzymes ensure the harmonious integration of the diverse and often divergent metabolic activities of cells so that the living state is promoted and preserved.

11.1 Enzymes—Catalytic Power, Specificity, and Regulation

Enzymes are characterized by three distinctive features: **catalytic power, specificity,** and **regulation.**

Catalytic Power

Enzymes display enormous catalytic power, accelerating reaction rates as much as 10^{16} over uncatalyzed levels, which is far greater than any synthetic catalysts can achieve, and enzymes accomplish these astounding feats in dilute aqueous solutions under mild conditions of temperature and pH. For example, the enzyme jack bean *urease* catalyzes the hydrolysis of urea:

$$H_2N{-}\overset{\overset{\displaystyle O}{\|}}{C}{-}NH_2 + 2\ H_2O + H^+ \longrightarrow 2\ NH_4{}^+ + HCO_3{}^-$$

At 20°C, the rate constant for the enzyme-catalyzed reaction is 3×10^4/sec; the rate constant for the uncatalyzed hydrolysis of urea is 3×10^{-10}/sec. Thus, 10^{14} is the ratio of the catalyzed rate to the uncatalyzed rate of reaction. Such a ratio is defined as the relative **catalytic power** of an enzyme.

Specificity

A given enzyme is very selective, both in the substances with which it interacts and in the reaction that it catalyzes. The substances upon which an enzyme

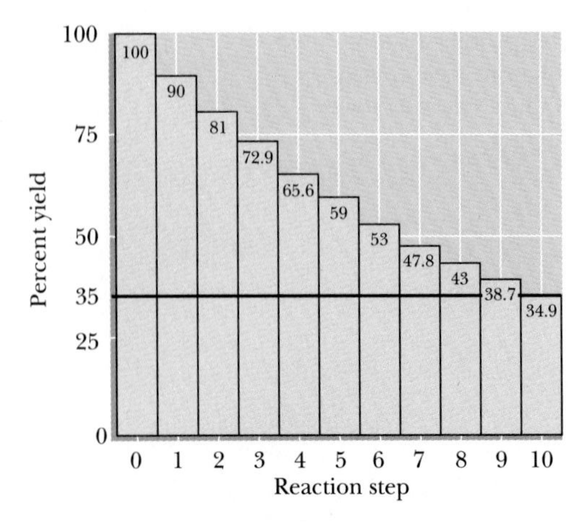

***Figure* 11.2** A 90% yield over 10 steps, for example, in a metabolic pathway, gives an overall yield of 35%. Therefore, yields in biological reactions *must be substantially greater;* otherwise, unwanted by-products would accumulate to unacceptable levels.

acts are traditionally called **substrates.** In an enzyme-catalyzed reaction, none of the substrate is diverted into nonproductive side-reactions, so no wasteful by-products are produced. It follows then that the products formed by a given enzyme are also very specific. This situation can be contrasted with your own experiences in the organic chemistry laboratory, where yields of 50% or even 30% are viewed as substantial accomplishments (Figure 11.2). The selective qualities of an enzyme are collectively recognized as its **specificity.** Intimate interaction between an enzyme and its substrates occurs through molecular recognition based on structural complementarity; such mutual recognition is the basis of specificity.

Regulation

Regulation of enzyme activity is achieved in a variety of ways, ranging from controls over the amount of enzyme protein synthesized by the cell to more rapid modulation of activity through reversible interactions with metabolic inhibitors and activators. Chapter 12 is devoted to discussions of enzyme regulation. Because most enzymes are proteins, we can anticipate that the functional attributes of enzymes can be ascribed to the remarkable versatility found in protein structures.

Enzyme Nomenclature

Traditionally, enzymes often were named by adding the suffix *-ase* to the name of the substrate upon which they acted, as in *urease* for the urea-hydrolyzing enzyme or *phosphatase* for enzymes hydrolyzing phosphoryl groups from organic phosphate compounds. Other enzymes acquired names bearing little resemblance to their activity, such as the peroxide-decomposing enzyme *catalase* or the proteolytic enzymes (*proteases*) of the digestive tract, *trypsin* and *pepsin.* Because of the confusion arising from these trivial designations, an International Commission on Enzymes was established in 1956 to create a systematic basis for enzyme nomenclature. Although common names for many enzymes remain in use, all enzymes now are classified and formally named according to the reaction they catalyze.

Six classes of reactions are recognized (Table 11.1). Within each class are subclasses, and under each subclass are sub-subclasses within which individual enzymes are listed. Classes, subclasses, sub-subclasses, and individual entries are each numbered, so that a series of four numbers serves to specify a particular enzyme. A systematic name, descriptive of the reaction, is also assigned to each entry. To illustrate, consider the enzyme that catalyzes this reaction:

$$\text{ATP} + \text{D-glucose} \longrightarrow \text{ADP} + \text{D-glucose-6-phosphate}$$

A phosphate group is transferred from ATP to the C-6-OH group of glucose, so the enzyme is a *transferase* (Class 2, Table 11.1). Subclass 7 of transferases is *enzymes transferring phosphorus-containing groups,* and sub-subclass 1 covers those *phosphotransferases with an alcohol group as an acceptor.* Entry 2 in this sub-subclass is **ATP:D-glucose-6-phosphotransferase,** and its classification number is **2.7.1.2.** In use, this number is written preceded by the letters **E.C.,** denoting the Enzyme Commission. For example, entry 1 in the same sub-subclass is E.C.2.7.1.1, ATP:D-hexose-6-phosphotransferase, an ATP-dependent enzyme that transfers a phosphate to the 6-OH of hexoses (that is, it is nonspecific regarding its hexose acceptor). These designations can be cumbersome, so in everyday usage, trivial names are employed frequently. The glucose-specific enzyme, E.C.2.7.1.2, is called *glucokinase* and the nonspecific E.C.2.7.1.1 is known as *hexokinase. Kinase* is a trivial term for enzymes that are ATP-dependent phosphotransferases.

***Table* 11.1**
Systematic Classification of Enzymes According to the Enzyme Commission

E.C. Number	Systematic Name and Subclasses
1	*Oxidoreductases* (oxidation–reduction reactions)
1.1	Acting on CH—OH group of donors
1.1.1	With NAD or NADP as acceptor
1.1.3	With O_2 as acceptor
1.2	Acting on the >C=O group of donors
1.2.3	With O_2 as acceptor
1.3	Acting on the CH—CH group of donors
1.3.1	With NAD or NADP as acceptor
2	*Transferases* (transfer of functional groups)
2.1	Transferring C-1 groups
2.1.1	Methyltransferases
2.1.2	Hydroxymethyltransferases and formyltransferases
2.1.3	Carboxyltransferases and carbamoyltransferases
2.2	Transferring aldehydic or ketonic residues
2.3	Acyltransferases
2.4	Glycosyltransferases
2.6	Transferring N-containing groups
2.6.1	Aminotransferases
2.7	Transferring P-containing groups
2.7.1	With an alcohol group as acceptor
3	*Hydrolases* (hydrolysis reactions)
3.1	Cleaving ester linkage
3.1.1	Carboxylic ester hydrolases
3.1.3	Phosphoric monoester hydrolases
3.1.4	Phosphoric diester hydrolases
4	*Lyases* (addition to double bonds)
4.1	C=C lyases
4.1.1	Carboxy lyases
4.1.2	Aldehyde lyases
4.2	C=O lyases
4.2.1	Hydrolases
4.3	C=N lyases
4.3.1	Ammonia-lyases
5	*Isomerases* (isomerization reactions)
5.1	Racemases and epimerases
5.1.3	Acting on carbohydrates
5.2	Cis-trans isomerases
6	*Ligases* (formation of bonds with ATP cleavage)
6.1	Forming C—O bonds
6.1.1	Amino acid–RNA ligases
6.2	Forming C—S bonds
6.3	Forming C—N bonds
6.4	Forming C—C bonds
6.4.1	Carboxylases

***Table* 11.2**

Enzyme Cofactors: Metal Ions and Coenzymes and the Enzymes with Which They Are Associated

Metal Ions and Some Enzymes That Require Them		**Coenzymes Serving as Transient Carriers of Specific Atoms or Functional Groups**		**Representative Enzymes Using Coenzymes**
Metal Ion	Enzyme	Coenzyme	Entity Transferred	
Fe^{2+} or Fe^{3+}	Cytochrome oxidase Catalase Peroxidase	Thiamine pyrophosphate	Aldehydes	Pyruvate dehydrogenase
Cu^{2+}	Cytochrome oxidase	Flavin adenine dinucleotide	Hydrogen atoms	Succinic dehydrogenase
Zn^{2+}	DNA polymerase Carbonic anhydrase Alcohol dehydrogenase	Nicotinamide adenine dinucleotide	Hydride ion (H^-)	Alcohol dehydrogenase
Mg^{2+}	Hexokinase Glucose-6-phosphatase	Coenzyme A	Acyl groups	Acetyl-CoA carboxylase
Mn^{2+}	Arginase	Pyridoxal phosphate	Amino groups	Aspartate aminotransferase
K^+	Pyruvate kinase (also requires Mg^{2+})	5′-Deoxyadenosylcobalamin (vitamin B_{12})	H atoms and alkyl groups	Methylmalonyl-CoA mutase
Ni^{2+}	Urease	Biotin (biocytin)	CO_2	Proprionyl-CoA carboxylase
Mo	Nitrate reductase	Tetrahydrofolate	Other one-carbon groups	Thymidylate synthase
Se	Glutathione peroxidase			

Coenzymes

Many enzymes carry out their catalytic function relying solely on their protein structure. Many others require nonprotein components, called **cofactors,** which may be metal ions or organic molecules referred to as **coenzymes.** Because cofactors are structurally much less complex than proteins, they tend to be stable to heat (incubation in a boiling water bath). Typically, proteins are denatured under such conditions. Many coenzymes are vitamins (Chapter 14) or contain vitamins as part of their structure. Usually coenzymes are actively involved in the catalytic reaction of the enzyme, often serving as intermediate carriers of functional groups in the conversion of substrates to products. In most cases, a coenzyme is firmly associated with its enzyme, perhaps even by covalent bonds, and it is difficult to separate the two. Such tightly bound coenzymes are referred to as **prosthetic groups** of the enzyme. The catalytically active complex of protein and prosthetic group is called the **holoenzyme.** The protein without the prosthetic group is called the **apoenzyme;** it is catalytically inactive. Table 11.2 lists various metal cofactors and organic coenzymes that are essential to certain enzymatic reactions.

11.2 Introduction to Enzyme Kinetics

Kinetics is the branch of science concerned with the rates of chemical reactions. The study of **enzyme kinetics** addresses the biological roles of enzymatic catalysts and how they accomplish their remarkable feats. In enzyme kinetics, we seek to determine the maximum reaction velocity that the enzyme can attain and its binding affinities for substrates and inhibitors. Coupled with studies on the structure and protein chemistry of the enzyme, analysis of the response of enzymatic rate to reaction conditions yields insights regarding the enzyme's mechanism of catalytic action. Such information is essential to an overall understanding of metabolism.

Significantly, this information can be exploited to control and manipulate the course of metabolic events. The science of pharmacology relies on such a strategy. Specific drugs, such as antibiotics, often target a selected metabolic enzyme with the aim of blocking its action in order to overcome infection or to alleviate illness. A detailed knowledge of the enzyme's kinetics is indispensable to rational drug design and successful pharmacological intervention.

Review of Chemical Kinetics

Before beginning a quantitative treatment of enzyme kinetics, we review briefly some basic principles of chemical kinetics. **Chemical kinetics** is the study of the rates of chemical reactions. Consider a reaction of overall stoichiometry

$$\mathrm{A} \longrightarrow \mathrm{P}$$

Although we treat this reaction as a simple, one-step conversion of A to P, it more likely occurs through a sequence of elementary reactions, each of which is a simple molecular process, as in

$$\mathrm{A} \longrightarrow \mathrm{I} \longrightarrow \mathrm{J} \longrightarrow \mathrm{P}$$

where I and J represent intermediates in the reaction. Precise description of all of the elementary reactions in a process is necessary to define the overall reaction mechanism for $\mathrm{A} \rightarrow \mathrm{P}$.

Let us assume that $\mathrm{A} \rightarrow \mathrm{P}$ *is* an elementary reaction and that it is spontaneous and essentially irreversible. Irreversibility is easily assumed if the rate of P conversion to A is very slow *or* the concentration of P (expressed as [P]) is negligible under the conditions chosen. The **velocity,** v, or **rate,** of the reaction $\mathrm{A} \rightarrow \mathrm{P}$ is the amount of P formed or the amount of A consumed per unit time, t. That is,

$$v = \frac{d[\mathrm{P}]}{dt} \quad \text{or} \quad v = \frac{-d[\mathrm{A}]}{dt} \tag{11.1}$$

The mathematical relationship between reaction rate and concentration of reactant(s) is the **rate law.** For this simple case, the rate law is

$$v = \frac{-d[\mathrm{A}]}{dt} = k[\mathrm{A}] \tag{11.2}$$

From this expression, it is obvious that the rate is proportional to the concentration of A, and k is the proportionality constant, or **rate constant.** k has the units of $(\text{time})^{-1}$, usually sec^{-1}. v is a function of [A] to the first power, or, in the terminology of kinetics, v is first-order with respect to A. For an elementary reaction, the **order** for any reactant is given by its exponent in the rate equation. The number of molecules which must simultaneously interact is defined as the **molecularity** of the reaction. Thus, the simple elementary reaction of $\mathrm{A} \rightarrow \mathrm{P}$ is a **first-order reaction.** Figure 11.3 portrays the course of a first-order reaction as a function of time. The rate of decay of a radioactive isotope, like $^{14}\mathrm{C}$ or $^{32}\mathrm{P}$, is a first-order reaction, as is an intramolecular rearrangement, such as $\mathrm{A} \rightarrow \mathrm{P}$. Both are **unimolecular reactions** (the molecularity equals 1).

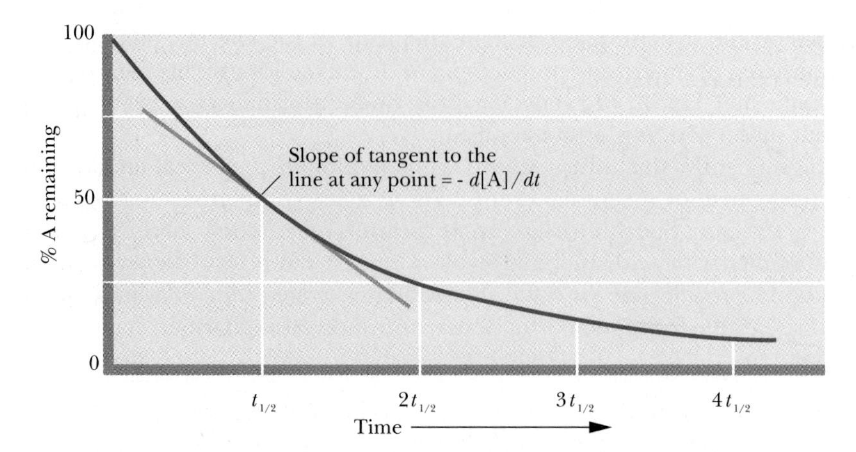

***Figure* 11.3** Plot of the course of a first-order reaction. The half-time, $t_{1/2}$, is the time for one-half of the starting amount of A to disappear.

Bimolecular Reactions

Consider the more complex reaction, where two molecules must react to yield products:

$$A + B \longrightarrow P + Q$$

Assuming this reaction is an elementary reaction, its molecularity is 2; that is, it is a **bimolecular reaction.** The velocity of this reaction can be determined from the rate of disappearance of either A or B, or the rate of appearance of P or Q:

$$v = \frac{-d[A]}{dt} = \frac{-d[B]}{dt} = \frac{d[P]}{dt} = \frac{d[Q]}{dt} \quad (11.3)$$

The rate law is

$$v = k[A][B] \quad (11.4)$$

The rate is proportional to the concentrations of both A and B. Since it is proportional to the product of two concentration terms, the reaction is **second-order** overall, first-order with respect to A and first-order with respect to B. (Were the elementary reaction $2A \rightarrow P + Q$, the rate law would be $v = k[A]^2$, second-order overall and second-order with respect to A.) Second-order rate constants have the units of $(\text{concentration})^{-1}(\text{time})^{-1}$, as in $M^{-1}\ sec^{-1}$.

Pseudo First-Order Reactions

If the situation is such that the starting concentration of B, $[B_0]$, greatly exceeds the starting concentration of A, $[A_0]$, that is, $[B_0] \gg [A_0]$, then $[B_0]$ will remain virtually constant during the course of the elementary reaction. The rate law can be expressed as

$$v = k[A][B_0] = k'[A] \quad (11.5)$$

where $k' = k[B_0]$. k' is a pseudo first-order rate constant.

Such considerations lead to the generalized description for an elementary reaction

$$xA + yB \longrightarrow \text{products}$$

The velocity of such a reaction obeys the general rate law, $v = k[A]^x[B]^y$, a reaction x-order with respect to A, y-order with respect to B, and $(x + y)$-order overall. However, molecularities greater than two are rarely found (and greater than three, never). When the overall stoichiometry of a reaction is greater than two (for example, as in A + B + C → or 2A + B →), the reaction almost always proceeds via uni- or bimolecular elementary steps, and the overall rate obeys a simple first- or second-order rate law.

At this point, it may be useful to remind ourselves of an important caveat that is the first principle of kinetics: *Kinetics cannot prove a hypothetical mechanism.* Kinetic experiments can only rule out various alternative hypotheses because they don't fit the data. However, through thoughtful kinetic studies, a process of elimination of alternative hypotheses leads ever closer to the reality.

Free Energy of Activation and the Action of Catalysts

In a first-order chemical reaction, the conversion of A to P occurs because, at any given instant, a fraction of the A molecules has the energy necessary to achieve a reactive condition known as the **transition state,** an unstable state intermediate between A and P. This unstable hybrid can collapse, or relax, to either form product, P, or re-form reactant, A. This transition state sits at the apex of the energy profile in the energy diagram describing the energetic relationship between A and P (Figure 11.4). The average free energy of A molecules defines the initial state and the average free energy of P molecules is the final state along the reaction coordinate. The rate of any chemical reaction is proportional to the concentration of reactant molecules (A in this case) having this transition-state energy. Obviously, the higher this energy is above the average energy, the smaller the fraction of molecules that will have this energy, and the slower the reaction will proceed. The height of this energy barrier is called the **free energy of activation, $\Delta G^{\ddagger}$.** Specifically, $\Delta G^{\ddagger}$ is the energy required to raise the average energy of one mole of reactant (at a given temperature) to the transition-state energy. The relationship between activation energy and the rate constant of the reaction, k, is given by the **Arrhenius equation:**

$$k = Ae^{-\Delta G^{\ddagger}/RT} \tag{11.6}$$

where A is a constant for a particular reaction (not to be confused with the reactant species, A, that we're discussing). Another way of writing this is $1/k =$

***Figure* 11.4** Energy diagram for a chemical reaction (A → P) and the effects of (a) raising the temperature from T_1 to T_2 or (b) adding a catalyst. Raising the temperature raises the average energy of A molecules, which increases the population of A molecules having energies equal to the activation energy for the reaction, thereby increasing the reaction rate. In contrast, the average free energy of A molecules remains the same in uncatalyzed versus catalyzed reactions (conducted at the same temperature). The effect of the catalyst is to lower the free energy of activation for the reaction.

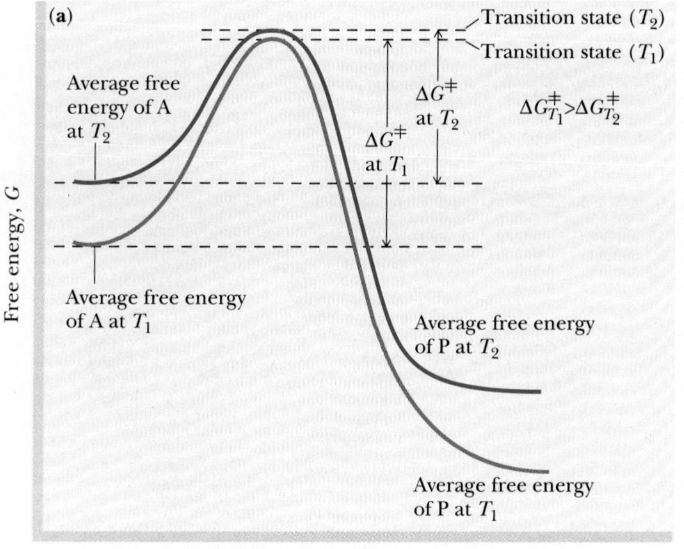

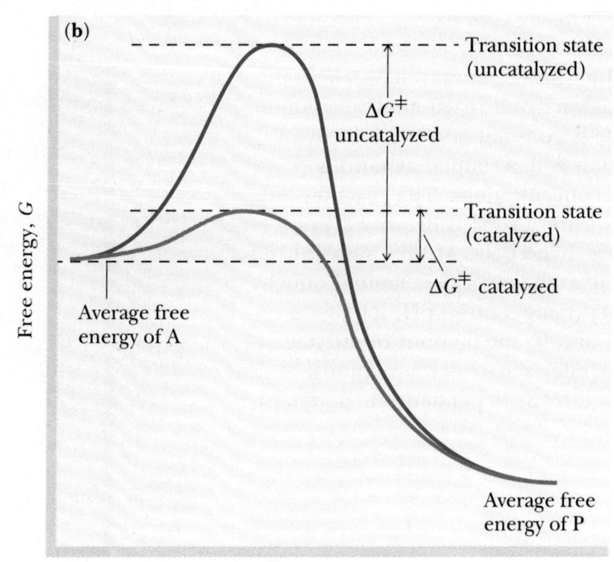

$(1/A)\,e^{\Delta G^{\ddagger}/RT}$, where it can easily be seen that, as the free energy of activation decreases, the reaction rate increases. That is, k is inversely proportional to $e^{\Delta G^{\ddagger}/RT}$. Therefore, if the energy of activation can be decreased, the reaction rate will increase.

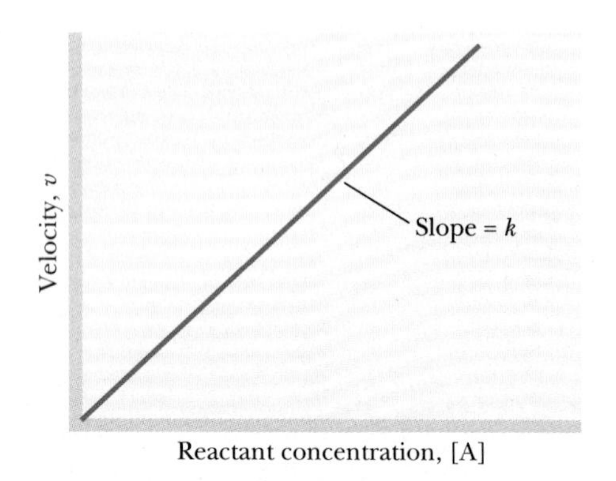

***Figure* 11.5** A plot of v versus [A] for the unimolecular chemical reaction A → P yields a straight line having a slope equal to k.

Decreasing $\Delta G^{\ddagger}$ Increases Reaction Rate

We are familiar with two general ways that rates of chemical reactions may be accelerated. First, the temperature can be raised. This will increase the average energy of reactant molecules, which increases the likelihood that a given molecule can achieve the transition state (Figure 11.4a). The rates of many chemical reactions are doubled by a 10°C rise in temperature. Second, the rates of chemical reactions can also be accelerated by **catalysts,** which work by lowering the energy of activation rather than by raising the average energy of the reactants (Figure 11.4b). Catalysts accomplish this remarkable feat by combining transiently with the reactants in a way that promotes their entry into the reactive, transition-state condition. Two aspects of catalysts are worth noting: (a) They are regenerated after each reaction cycle (A → P), and so can be used over and over again; and (b) catalysts have *no* effect on the overall free energy change in the reaction, the free energy difference between A and P (Figure 11.4b).

11.3 Kinetics of Enzyme-Catalyzed Reactions

Examination of the change in reaction velocity as the reactant concentration is varied is one of the primary measurements in kinetic analysis. Returning to A → P, a plot of the reaction rate as a function of the concentration of A yields a straight line whose slope is k (Figure 11.5). The more A that is available, the greater the rate of the reaction, v. Similar analyses of enzyme-catalyzed reactions involving only a single substrate yield remarkably different results (Figure 11.6). At low concentrations of the substrate S, v is proportional to [S], as expected for a first-order reaction. However, v does not increase proportionally as [S] increases, but instead begins to level off. At high [S], v has become virtually independent of [S], and approaches a maximal limit. The value of v at this limit is written V_{max}. Since rate is no longer dependent on [S] at these high concentrations, the enzyme-catalyzed reaction is now obeying **zero-order kinetics;** that is, the rate is independent of the reactant (substrate) concentra-

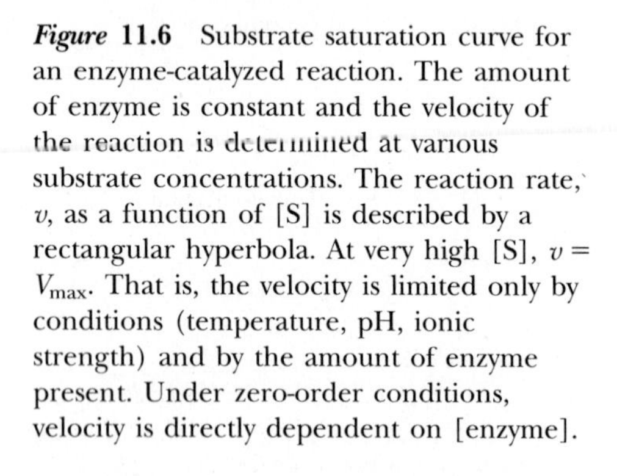

***Figure* 11.6** Substrate saturation curve for an enzyme-catalyzed reaction. The amount of enzyme is constant and the velocity of the reaction is determined at various substrate concentrations. The reaction rate, v, as a function of [S] is described by a rectangular hyperbola. At very high [S], $v = V_{max}$. That is, the velocity is limited only by conditions (temperature, pH, ionic strength) and by the amount of enzyme present. Under zero-order conditions, velocity is directly dependent on [enzyme].

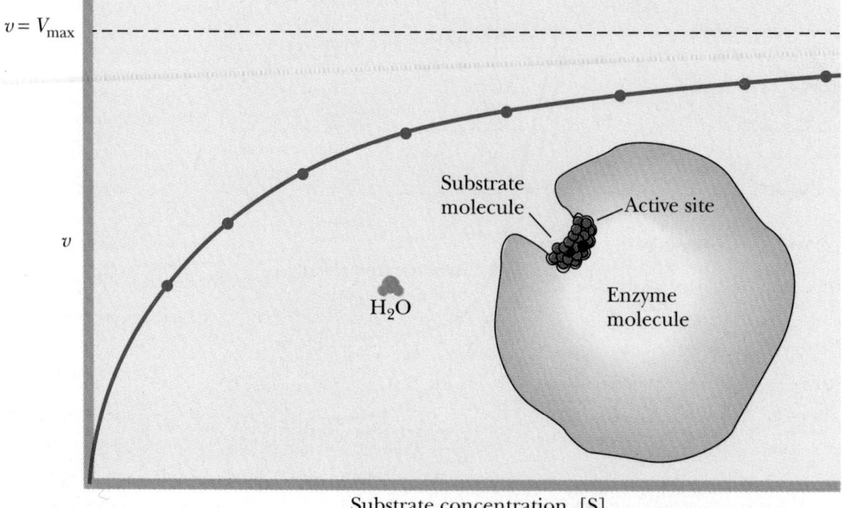

tion. This behavior is a **saturation effect:** When v can increase no further even though [S] is increased, the system is saturated with substrate. Such plots are called **substrate saturation curves.** The physical interpretation is that every enzyme molecule in the reaction mixture has its substrate-binding site occupied by S. Indeed, such curves were the initial clue that an enzyme interacts directly with its substrate by binding it.

The Michaelis–Menten Equation

L. Michaelis and M. L. Menten proposed a general theory of enzyme action in 1913 consistent with observed enzyme kinetics. Their theory was based on the assumption that the enzyme, E, and its substrate, S, associate reversibly to form an enzyme-substrate complex, ES, as developed below.

$$E + S \underset{k_{-1}}{\overset{k_1}{\rightleftharpoons}} ES \tag{11.7}$$

This association/dissociation is assumed to be in rapid equilibrium, and K_S is the *enzyme : substrate dissociation constant.* At equilibrium,

$$k_{-1}[ES] = k_1[E][S] \tag{11.8}$$

and

$$K_S = \frac{[E][S]}{[ES]} = \frac{k_{-1}}{k_1} \tag{11.9}$$

Product, P, is formed in a second step when ES breaks down to yield E + P.

$$E + S \underset{k_{-1}}{\overset{k_1}{\rightleftharpoons}} ES \xrightarrow{k_2} E + P \tag{11.10}$$

E is then free to interact with another molecule of S.

Steady-State Assumption

The interpretations of Michaelis and Menten were refined and extended in 1925 by Briggs and Haldane, by assuming the concentration of the enzyme-substrate complex ES quickly reaches a constant value in such a dynamic system. That is, ES is formed as rapidly from E + S as it disappears by its two possible fates: dissociation to regenerate E + S, and reaction to form E + P. This assumption is termed the **steady-state assumption** and is expressed as

$$\frac{d[ES]}{dt} = 0 \tag{11.11}$$

That is, the change in concentration of ES with time, t, is 0. Figure 11.7 illustrates the time course for formation of the ES complex and establishment of the steady-state condition.

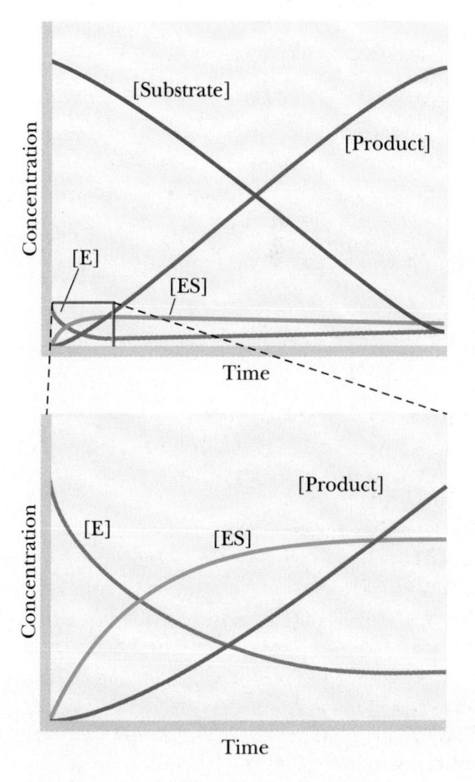

***Figure* 11.7** Time course for the consumption of substrate, the formation of product, and the establishment of a steady-state level of the enzyme-substrate [ES] complex for a typical enzyme obeying the Michaelis–Menten, Briggs–Haldane model for enzyme kinetics. The early stage of the time course is shown in greater magnification in the bottom figure.

Initial Velocity Assumption

One other simplification will be advantageous. Since enzymes accelerate the rate of the reverse reaction as well as the forward reaction, it would be helpful to ignore any back reaction by which E + P might form ES. For example, if we observe only the *initial velocity* for the reaction immediately after E and S are mixed in the absence of *P*, the rate of any back reaction is negligible since its rate will be proportional to [P], and [P] is essentially 0. Given such simplification, we now analyze the system described by Equation (11.10) in order to describe the initial velocity v as a function of [S] and amount of enzyme.

The total amount of enzyme is fixed and is given by the formula

$$\text{Total enzyme, } [E_T] = [E] + [ES] \tag{11.12}$$

where [E] = free enzyme and [ES] = the amount of enzyme in the enzyme-substrate complex. From Equation (11.10), the rate of [ES] formation is

$$v_f = k_1([E_T] - [ES])[S]$$

where

$$[E_T] - [ES] = [E] \tag{11.13}$$

From Equation (11.10), the rate of [ES] disappearance is

$$v_d = k_{-1}[ES] + k_2[ES] = (k_{-1} + k_2)[ES] \tag{11.14}$$

At steady state, $d[ES]/dt = 0$, and therefore, $v_f = v_d$.
So,

$$k_1([E_T] - [ES])[S] = (k_{-1} + k_2)[ES] \tag{11.15}$$

Rearranging gives

$$\frac{([E_T] - [ES])[S]}{[ES]} = \frac{(k_{-1} + k_2)}{k_1} \tag{11.16}$$

The Michaelis Constant, K_m

The ratio of constants $(k_{-1} + k_2)/k_1$ is itself a constant and is defined as the **Michaelis constant, K_m**

$$K_m = \frac{(k_{-1} + k_2)}{k_1} \tag{11.17}$$

Note from (11.16) that K_m is given by the ratio of two concentrations (([E_T] − [S]) and [S]) to one ([ES]), so K_m has the units of *molarity*. From Equation (11.16), we can write

$$\frac{([E_T] - [ES])[S]}{[ES]} = K_m \tag{11.18}$$

which rearranges to

$$[ES] = \frac{[E_T][S]}{K_m + [S]} \tag{11.19}$$

Now, the most important parameter in the kinetics of any reaction is the **rate of product formation.** This rate is given by

$$v = \frac{d[P]}{dt} \tag{11.20}$$

and for this reaction

$$v = k_2[ES] \tag{11.21}$$

Substituting the expression for [ES] from Equation (11.19) into (11.21) gives

$$v = \frac{k_2[E_T][S]}{K_m + [S]} \tag{11.22}$$

The product $k_2[E_T]$ has special meaning. When [S] is high enough to saturate all of the enzyme, the velocity of the reaction, v, is maximal. At saturation, the amount of [ES] complex is equal to the total enzyme concentration, E_T, its maximum possible value. From Equation (11.21), the initial velocity v then equals $k_2[E_T] = V_{max}$. Written symbolically, when $[S] \gg [E_T]$ (and K_m), $[E_T] = [ES]$ and $v = V_{max}$. Therefore,

$$V_{max} = k_2[E_T] \tag{11.23}$$

Substituting this relationship into the expression for v gives the **Michaelis–Menten equation**

$$v = \frac{V_{max}[S]}{K_m + [S]} \tag{11.24}$$

This equation says that the rate of an enzyme-catalyzed reaction, v, at any moment is determined by two constants, K_m and V_{max}, *and* the concentration of substrate at that moment.

When $[S] = K_m$, $v = V_{max}/2$

We can provide an operational definition for the constant K_m by rearranging Equation (11.24) to give

$$K_m = [S]\left(\frac{V_{max}}{v} - 1\right) \tag{11.25}$$

Then, at $v = V_{max}/2$, $K_m = [S]$. That is, K_m is defined by the substrate concentration that gives a velocity equal to one-half the maximal velocity.

Relationships Between V_{max}, K_m, and Reaction Order

The Michaelis–Menten equation (11.24) describes a curve known from analytical geometry as a *rectangular hyperbola.*[1] In such curves, as [S] is increased, v approaches the limiting value, V_{max}, in an asymptotic fashion. V_{max} can be approximated experimentally from a substrate saturation curve (Figure 11.6), and K_m can be derived from $V_{max}/2$, so the two constants of the Michaelis–Menten equation can be obtained from plots of v versus [S]. Note, however, that actual estimation of V_{max}, and consequently K_m is only approximate from such graphs. That is, according to Equation (11.24), in order to get $v = 0.99$ V_{max}, [S] must equal 99 K_m, a concentration that may be difficult to achieve in practice.

From Equation (11.24), when $[S] \gg K_m$, then $v = V_{max}$. That is, v is no longer dependent on [S], so the reaction is obeying zero-order kinetics. Also, when $[S] < K_m$, then $v \approx (V_{max}/K_m)[S]$. That is, the rate, v, will approximate a fit to a first-order rate equation, $v = k'[A]$, where $k' = V_{max}/K_m$.

K_m and V_{max}, once known explicitly, define the rate of the enzyme-catalyzed reaction, *provided:*

1. The reaction involves only one substrate, *or* if the reaction is multisubstrate, the concentration of only one substrate is varied while the concentration of all other substrates is held constant.
2. The reaction $ES \rightarrow E + P$ is irreversible, *or* the experiment is limited to observing only initial velocities where $[P] = 0$.
3. $[S]_0 > [E_T]$ and $[E_T]$ is held constant.

[1] A proof that the Michaelis–Menten equation describes a rectangular hyperbola is given by Naqui, A., 1986, Where are the asymptotes of Michaelis–Menten? *Trends in Biochemical Sciences* 1:64–65.

4. All other variables that might influence the rate of the reaction (temperature, pH, ionic strength, and so on) are constant.

Table 11.3 lists some representative K_m values for the substrates of various enzymes.

***Table* 11.3**
K_m Values for Some Enzymes

Enzyme	Substrate	K_m (mM)
Carbonic anhydrase	CO_2	12
Chymotrypsin	*N*-Benzoyltyrosinamide	2.5
	Acetyl-L-tryptophanamide	5
	N-Formyltyrosinamide	12
	N-Acetyltyrosinamide	32
	Glycyltyrosinamide	122
Hexokinase	Glucose	0.15
	Fructose	1.5
β-Galactosidase	Lactose	4
Glutamate dehydrogenase	NH_4^+	57
	Glutamate	0.12
	α-Ketoglutarate	2
	NAD^+	0.025
	NADH	0.018
Aspartate aminotransferase	Aspartate	0.9
	α-Ketoglutarate	0.1
	Oxaloacetate	0.04
	Glutamate	4
Threonine deaminase	Threonine	5
Arginyl-tRNA synthetase	Arginine	0.003
	$tRNA^{Arg}$	0.0004
	ATP	0.3
Pyruvate carboxylase	HCO_3^-	1.0
	Pyruvate	0.4
	ATP	0.06
Penicillinase	Benzylpenicillin	0.05
Lysozyme	Hexa-*N*-acetylglucosamine	0.006

Enzyme Units

In many situations, the actual molar amount of the enzyme is not known. However, its amount can be expressed in terms of the activity observed. The International Commission on Enzymes established by the International Union of Biochemistry defines **One International Unit** of enzyme as *the amount that catalyzes the formation of one micromole of product in one minute.* (Because enzymes are very sensitive to factors such as pH, temperature, and ionic strength, the conditions of assay must be specified.) Another definition for units of enzyme activity is the **katal.** One katal is *that amount of enzyme catalyzing the conversion of one mole of substrate to product in one second.* Thus, one katal equals 6×10^7 international units.

Turnover Number

The **turnover number** of an enzyme, k_{cat}, is a measure of its maximal catalytic activity. k_{cat} is defined as the number of substrate molecules converted into product per enzyme molecule per unit time when the enzyme is saturated with substrate. The turnover number is also referred to as the **molecular activity** of the enzyme. For the simple Michaelis–Menten situation where there is only one ES complex, V_{max} reveals the turnover number, provided the concentration of enzyme, $[E_T]$, in the reaction mixture is known: at saturating [S], $v = V_{max} = k_2[E_T]$. Thus,

$$k_2 = \frac{V_{max}}{[E_T]} = k_{cat} \quad (11.26)$$

The term k_{cat} represents the kinetic efficiency of the enzyme. Table 11.4 lists turnover numbers for some representative enzymes. Catalase has the highest turnover number known; each molecule of this enzyme can degrade 40 million molecules of H_2O_2 in one second! At the other end of the scale, lysozyme requires 2 seconds to cleave a glycosidic bond in its glycan substrate.

***Table* 11.4**

Values of k_{cat} (Turnover Number) for Some Enzymes

Enzyme	k_{cat} (sec^{-1})
Catalase	40,000,000
Carbonic anhydrase	1,000,000
Acetylcholinesterase	14,000
Penicillinase	2,000
Lactate dehydrogenase	1,000
Chymotrypsin	100
DNA polymerase I	15
Lysozyme	0.5

k_{cat}/K_m

Under physiological conditions, [S] is seldom saturating, and k_{cat} itself is not particularly revealing. To illustrate, the *in vivo* ratio of $[S]/K_m$ usually falls in the range of 0.01 to 1.0, so active sites often are not filled with substrate. Nevertheless, we can derive a meaningful index of the efficiency of Michaelis–Menten-type enzymes under these conditions by employing the following equations. As presented in Equation (11.24), if

$$v = \frac{V_{max}[S]}{K_m + [S]}$$

and $V_{max} = k_{cat}[E_T]$, then

$$v = \frac{k_{cat}[E_T][S]}{K_m + [S]} \quad (11.27)$$

When $[S] \ll K_m$, the concentration of free enzyme, [E], is approximately equal to $[E_T]$, so that

$$v = \left(\frac{k_{cat}}{K_m}\right)[E][S] \quad (11.28)$$

That is, k_{cat}/K_m is an *apparent second-order rate constant* for the reaction of E and S to form a product. This ratio provides an index of the catalytic efficiency of enzymes operating at substrate concentrations substantially below saturation amounts.

An interesting point emerges if we restrict ourselves to the simple case where $k_{cat} = k_2$. Then

$$\frac{k_{cat}}{K_m} = \frac{k_1 k_2}{(k_{-1} + k_2)} \quad (11.29)$$

But k_1 must always be greater than or equal to $k_1k_2/(k_{-1} + k_2)$. That is, the reaction can go no faster than the rate at which E and S come together. Thus, k_1 sets the upper limit for k_{cat}/K_m. In other words, *the catalytic efficiency of an enzyme cannot exceed the diffusion-controlled rate of combination of E and S to form ES.* In H_2O, the rate constant for such diffusion is approximately 10^9/M/sec. Those enzymes that are most efficient in their catalysis have k_{cat}/K_m ratios approaching this value. Their catalytic velocity is limited only by the rate at which they encounter S; enzymes this efficient have achieved so-called *catalytic perfection.* Table 11.5 lists the kinetic parameters of several enzymes in this category. Note that k_{cat} and K_m both show a substantial range of variation in this table, even though their ratio falls around $10^8/M/$sec.

Table **11.5**

Enzymes Whose k_{cat}/K_m Approaches the Diffusion-Controlled Rate of Association with Substrate

Enzyme	Substrate	k_{cat} (sec^{-1})	K_m (M)	k_{cat}/K_m (sec^{-1} M^{-1})
Acetylcholinesterase	Acetylcholine	1.4×10^4	9×10^{-5}	1.6×10^8
Carbonic anhydrase	CO_2	1×10^6	0.012	8.3×10^7
	HCO_3^-	4×10^5	0.026	1.5×10^7
Catalase	H_2O_2	4×10^7	1.1	4×10^7
Crotonase	Crotonyl-CoA	5.7×10^3	2×10^{-5}	2.8×10^8
Fumarase	Fumarate	800	5×10^{-6}	1.6×10^8
	Malate	900	2.5×10^{-5}	3.6×10^7
Triosephosphate isomerase	Glyceraldehyde-3-phosphate*	4.3×10^3	1.8×10^{-5}	2.4×10^8
β-Lactamase	Benzylpenicillin	2.0×10^3	2×10^{-5}	1×10^8

* K_m for glyceraldehyde-3-phosphate is calculated on the basis that only 3.8% of the substrate in solution is unhydrated and therefore reactive with the enzyme.

Adapted from Fersht, A. 1985. *Enzyme Structure and Mechanism,* 2nd ed. New York: W.H. Freeman & Co.

Linear Plots Can Be Derived from the Michaelis–Menten Equation

Because of the hyperbolic shape of v versus [S] plots, V_{max} can only be determined from an extrapolation of the asymptotic approach of v to some limiting value as [S] increases indefinitely (Figure 11.6); and K_m is derived from that value of [S] giving $v = V_{max}/2$. However, several rearrangements of the Michaelis–Menten equation transform it into a straight-line equation.

The **Lineweaver–Burk double-reciprocal plot:**

Taking the reciprocal of both sides of the Michaelis–Menten equation, Equation (11.24), yields the equality

$$\frac{1}{v} = \left(\frac{K_m}{V_{max}}\right)\left(\frac{1}{[S]}\right) + \frac{1}{V_{max}} \tag{11.30}$$

This conforms to $y = mx + b$ (the equation for a straight line), where $y = 1/v$; m, the slope, is K_m/V_{max}; $x = 1/[S]$; and $b = 1/V_{max}$. Plotting $1/v$ versus $1/[S]$ gives a straight line whose x-intercept is $-1/K_m$, whose y-intercept is $1/V_{max}$, and whose slope is K_m/V_{max} (Figure 11.8).

The **Eadie–Hofstee plot:**

Multiplying both sides of Equation (11.24) by $(K_m + [S])$ gives

$$vK_m + v[S] = V_{max}[S] \tag{11.31}$$

or

$$v[S] = -vK_m + V_{max}[S] \tag{11.32}$$

Thus,

$$v = -K_m\left(\frac{v}{[S]}\right) + V_{max} \tag{11.33}$$

Plotting v versus $v/[S]$ as suggested by Eadie and Hofstee yields a straight line of slope $= -K_m$, y-intercept $= V_{max}$, and x-intercept of V_{max}/K_m (Figure 11.9).

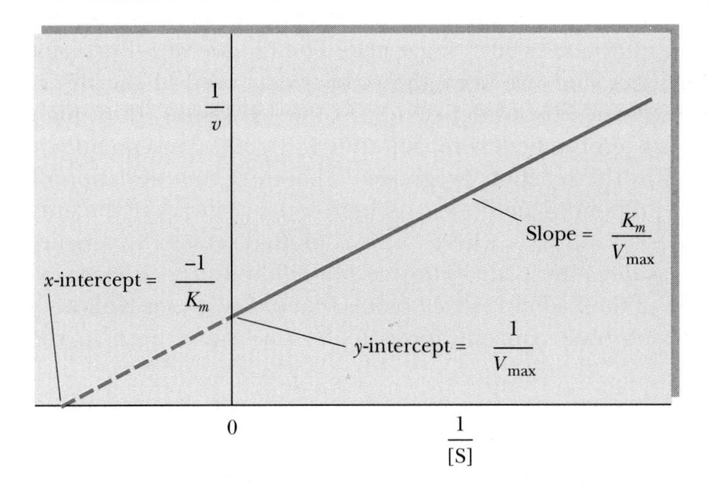

Figure **11.8** The Lineweaver–Burk double reciprocal plot, depicting extrapolations that allow the determination of the *x*- and *y*-intercepts and slope.

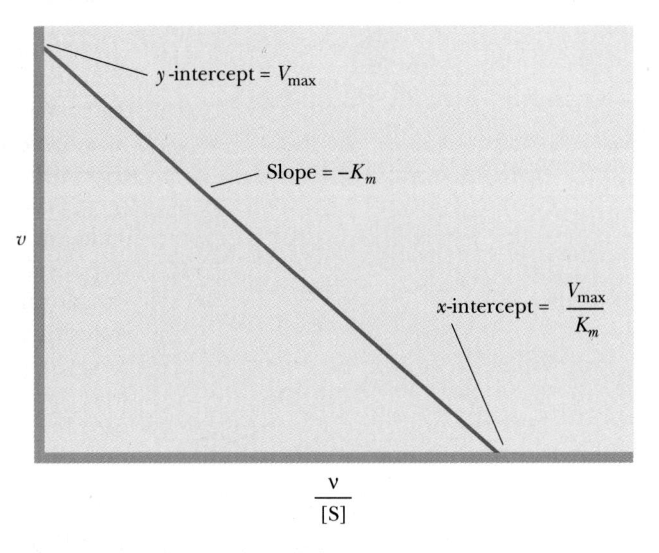

Figure **11.9** An Eadie–Hofstee rearrangement of the Michaelis–Menten equation gives a straight line when v is plotted as a function of $v/[S]$.

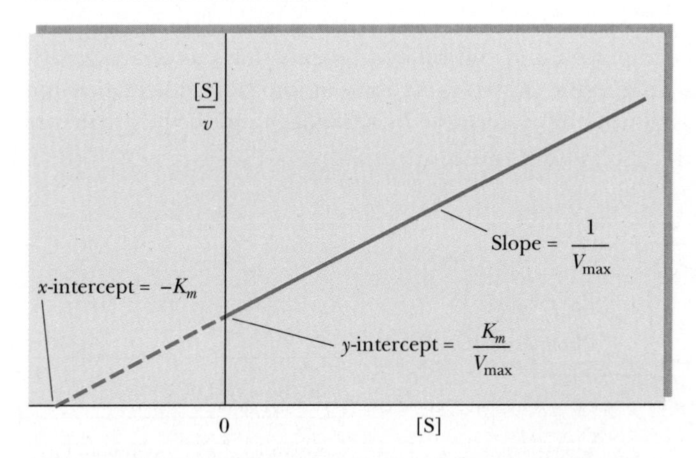

Figure **11.10** A Hanes–Woolf plot of $[S]/v$ versus [S], another straight-line rearrangement of the Michaelis–Menten equation.

The **Hanes–Woolf plot:**
Multiplying both sides of Equation (11.30) by [S] gives

$$\frac{[S]}{v} = [S]\left(\frac{K_m}{V_{max}}\right)\left(\frac{1}{[S]}\right) + \frac{[S]}{V_{max}} = \frac{K_m}{V_{max}} + \frac{[S]}{V_{max}} \tag{11.34}$$

and

$$\frac{[S]}{v} = \left(\frac{1}{V_{max}}\right)[S] + \frac{K_m}{V_{max}} \tag{11.35}$$

Graphing $[S]/v$ versus [S] yields a straight line where the slope is $1/V_{max}$, the *y*-intercept is K_m/V_{max}, and the *x*-intercept is $-K_m$, as shown in Figure 11.10.

The Advantage of Linear Plots

The common advantage of these plots is that they allow both K_m and V_{max} to be accurately estimated by extrapolation of straight lines rather than asymptotes. Why have three alternatives been suggested? The Lineweaver–Burk plot has historical precedence and has been the most widely used of the three. However, it has at least one serious deficiency: At low [S] values (thus, high 1/[S] values), the data tend to be less reliable since [S], and consequently v, changes dramatically as the reaction progresses. Therefore, the deviation of the results at low [S] influences the slope, and hence, the accuracy of the line. In contrast, the Hanes–Woolf plot, where $[S]/v$ is plotted versus [S], ameliorates this problem (as does the Eadie–Hofstee plot, although to a lesser extent). The advantage of the Eadie–Hofstee plot is that the $v/[S]$ term allows a very large range of substrate concentrations to be easily portrayed in the graph.

Departures from Linearity: A Hint of Regulation?

If the kinetics of the reaction disobey the Michaelis–Menten equation, the violation is revealed by a departure from linearity in these straight-line graphs; such departure is best detected in Eadie–Hofstee-type plots. We shall see in the next chapter that such deviations from linearity are characteristic of the kinetics of regulatory enzymes known as **allosteric enzymes.** Such regulatory enzymes are very important in the overall control of metabolic pathways.

Effect of pH on Enzymatic Activity

Enzyme-substrate recognition and the catalytic events that ensue are greatly dependent on pH. An enzyme possesses an array of ionizable side chains and prosthetic groups that not only determine its secondary and tertiary structure

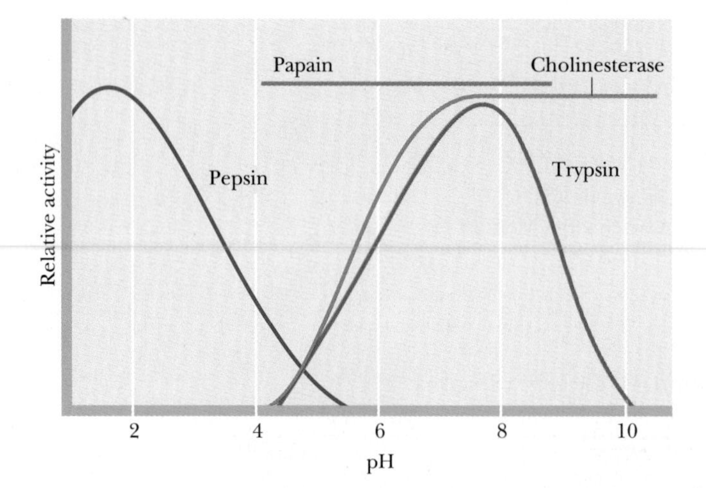

Optimum pH of Some Enzymes

Enzyme	Optimum pH
Pepsin	1.5
Trypsin	7.7
Catalase	7.6
Arginase	9.7
Fumarase	7.8
Ribonuclease	7.8

***Figure* 11.11** The pH activity profiles of four different enzymes. *Trypsin,* an intestinal protease, has a slightly alkaline pH optimum, whereas *pepsin,* a gastric protease, acts in the acidic confines of the stomach and has a pH optimum near 2. *Papain,* a protease found in papaya, is relatively insensitive to pH between 4 and 8. *Cholinesterase* activity is pH-sensitive below pH 7 but not between pH 7 and 10. The cholinesterase pH activity profile suggests an ionizable group with a pK' near 6 is essential to its activity. Might it be a histidine residue within the active site?

but may also be intimately involved in its active site. Further, the substrate itself often has ionizing groups, and one or another of the ionic forms may preferentially interact with the enzyme. Enzymes in general are active only over a limited pH range and most have a particular pH at which their catalytic activity is optimal. These effects of pH may be due to effects on K_m or V_{max} or both. Figure 11.11 illustrates the relative activity of four enzymes as a function of pH. Although the pH optimum of an enzyme often reflects the pH of its normal environment, the optimum may not be precisely the same. This difference suggests that the pH-activity response of an enzyme may be a factor in the intracellular regulation of its activity.

Effect of Temperature on Enzymatic Activity

Like most chemical reactions, the rates of enzyme-catalyzed reactions generally increase with increasing temperature. However, at temperatures above 50° to 60°C, enzymes typically show a decline in activity (Figure 11.12a). Two effects are operating here: (a) the characteristic increase in reaction rate with temperature, and (b) thermal denaturation of protein structure at higher temperatures. Most enzymatic reactions double in rate for every 10°C rise in temperature (that is, $Q_{10} = 2$, where Q_{10} is defined as *the ratio of activities at two temperatures 10° apart*) as long as the enzyme is stable and fully active. Some enzymes, those catalyzing reactions having very high activation energies, show proportionally greater Q_{10} values. The increasing rate with increasing temperature is ultimately offset by the instability of higher orders of protein structure at elevated temperatures, where the enzyme is inactivated. Figure 11.12b illustrates a typical progress curve for an enzymatic reaction at various temperatures. At lower temperatures, the rate is markedly enhanced as the temperature rises. However, at temperatures above 50°C, the enzyme is progressively inactivated over time. Figure 11.12a summarizes the overall situation described by the family of curves in Figure 11.12b. Not all enzymes are quite so thermally labile. For example, the enzymes of thermophilic bacteria (*thermophilic* = "heat-loving") found in geothermal springs retain full activity at temperatures in excess of 85°C.

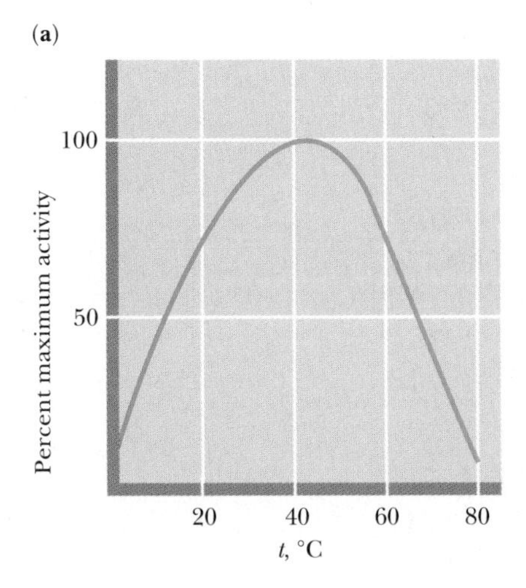

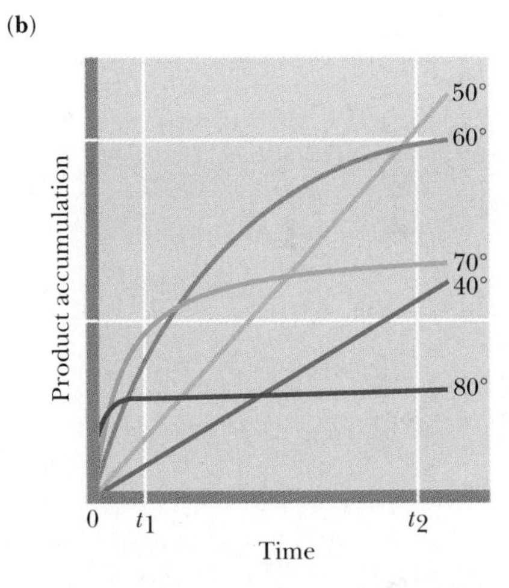

Figure **11.12** The effect of temperature on enzyme activity. (a) The relative activity of an enzymatic reaction as a function of temperature. The decrease in the activity above 50° is due to thermal denaturation. (b) Typical progress curves for an enzymatic reaction at various temperatures. Activity measurements taken between time points t_1 and t_2 give the plot represented in (a). Note the decrease in the amount of product formed as the enzyme declines in activity at higher temperatures.

11.4 Enzyme Inhibition

If the velocity of an enzymatic reaction is decreased, that is, the enzyme is **inhibited** by some compound, the kinetics of the reaction obviously have been perturbed. Systematic perturbations are a basic tool of experimental scientists; much can be learned about the normal workings of any system by inducing changes in it and then observing the effects of the change. Accordingly, the study of enzyme inhibition has contributed significantly to our understanding of enzymes.

Reversible Versus Irreversible Inhibition

Enzyme inhibitors are classified in several ways. The inhibitor may interact either reversibly or irreversibly with the enzyme. **Reversible inhibitors** interact with the enzyme through noncovalent association/dissociation reactions. In contrast, the effects of **irreversible inhibitors** are usually manifested through stable, covalent alterations in the enzyme. The net effect of irreversible inhibition is a decrease in the concentration of active enzyme. The kinetics observed are consistent with this interpretation, as we shall see later.

Reversible Inhibition

Reversible inhibitors fall into two major categories: competitive and noncompetitive (although other more unusual and rare categories are known). **Competitive inhibitors** are characterized by the fact that the substrate and inhibitor compete for the same binding site on the enzyme, the so-called **active site** or **S-binding site.** Thus, increasing the concentration of S favors the likelihood of S binding to the enzyme instead of the inhibitor, I. That is, high [S] can overcome the effects of I. The other major type, noncompetitive inhibition, cannot be overcome by increasing [S]. The two types can be distinguished by the particular patterns obtained when the kinetic data are analyzed in linear plots, such as double-reciprocal (Lineweaver–Burk) plots. A general formulation for common inhibitor interactions in our simple enzyme kinetic model would include

$$\mathrm{E + I \rightleftharpoons EI} \quad \text{and/or} \quad \mathrm{I + ES \rightleftharpoons IES} \qquad (11.36)$$

That is, we consider here reversible combinations of the inhibitor with E and/or ES.

Competitive Inhibition

Consider the following system

$$\mathrm{E + S} \underset{k_{-1}}{\overset{k_1}{\rightleftharpoons}} \mathrm{ES} \xrightarrow{k_2} \mathrm{E + P} \qquad \mathrm{E + I} \underset{k_{-3}}{\overset{k_3}{\rightleftharpoons}} \mathrm{EI} \qquad (11.37)$$

where an inhibitor, I, binds *reversibly* to the enzyme at the same site as S. Therefore, S-binding and I-binding are mutually exclusive, *competitive* processes, and formation of the ternary complex, EIS, where both S and I are bound, is physically impossible. This condition leads us to anticipate that S and I must share a high degree of structural similarity since they bind at the same site on the enzyme. Also notice that, in our model, EI does not react to give rise to E + P; that is, I is not changed by interaction with E. The rate of the product-forming reaction is $v = k_2[\mathrm{ES}]$.

It is revealing to compare the equation for the uninhibited case, Equation (11.24) (the Michaelis–Menten equation) with Equation (11.47) for the rate of the enzymatic reaction in the presence of a fixed concentration of the competitive inhibitor, [I]

$$v = \frac{V_{max}[\mathrm{S}]}{[\mathrm{S}] + K_m}$$

$$v = \frac{V_{max}[\mathrm{S}]}{[\mathrm{S}] + K_m\left(1 + \frac{[\mathrm{I}]}{K_\mathrm{I}}\right)}$$

Table **11.6**

The Effect of Various Types of Inhibitors on the Michaelis–Menten Rate Equation and on Apparent K_m and Apparent V_{max}

Inhibition Type	Rate Equation	Apparent K_m	Apparent V_{max}
None	$v = V_{max}[\mathrm{S}]/(K_m + [\mathrm{S}])$	K_m	V_{max}
Competitive	$v = V_{max}[\mathrm{S}]/([\mathrm{S}] + K_m(1 + [\mathrm{I}]/K_\mathrm{I}))$	$K_m(1 + [\mathrm{I}]/K_\mathrm{I})$	V_{max}
Noncompetitive	$v = (V_{max}[\mathrm{S}]/(1 + [\mathrm{I}]/K_\mathrm{I}))/(K_m + [\mathrm{S}])$	K_m	$V_{max}/(1 + [\mathrm{I}]/K_\mathrm{I})$
Mixed	$v = V_{max}[\mathrm{S}]/((1 + [\mathrm{I}]/K_\mathrm{I})K_m + (1 + [\mathrm{I}]/K_\mathrm{I}')[\mathrm{S}])$	$K_m(1 + [\mathrm{I}]/K_\mathrm{I})/(1 + [\mathrm{I}]/K_\mathrm{I}')$	$V_{max}/(1 + [\mathrm{I}]/K_\mathrm{I}')$

K_I is defined as the enzyme:inhibitor dissociation constant $K_\mathrm{I} = [\mathrm{E}][\mathrm{I}]/[\mathrm{EI}]$; K_I' is defined as the enzyme substrate complex:inhibitor dissociation constant $K_\mathrm{I}' = [\mathrm{ES}][\mathrm{I}]/[\mathrm{ESI}]$

A Deeper Look

The Equations of Competitive Inhibition

Given the relationships between E, S, and I described previously and recalling the steady-state assumption that $d[ES]/dt = 0$, from Equations (11.15) and (11.17) we can write

$$[ES] = \frac{k_1[E][S]}{(k_2 + k_{-1})} = \frac{[E][S]}{K_m} \quad (11.38)$$

Assuming that $E + I \rightleftharpoons EI$ reaches rapid equilibrium, the rate of EI formation, $v_f' = k_3[E][I]$, and the rate of disappearance of EI, $v_d' = k_{-3}[EI]$, are equal. So,

$$k_3[E][I] = k_{-3}[EI] \quad (11.39)$$

Therefore,

$$[EI] = \left(\frac{k_3}{k_{-3}}\right)[E][I] \quad (11.40)$$

If we define K_I as k_{-3}/k_3, an *enzyme-inhibitor dissociation constant,* then

$$[EI] = \frac{[E][I]}{K_I} \quad (11.41)$$

knowing $[E_T] = [E] + [ES] + [EI]$. Then

$$[E_T] = [E] + \frac{[E][S]}{K_m} + \frac{[E][I]}{K_I} \quad (11.42)$$

Solving for [E] gives

$$[E] = \frac{K_I K_m[E_T]}{(K_I K_m + K_I[S] + K_m[I])} \quad (11.43)$$

Since the rate of product formation is given by $v = k_2[ES]$, from Equation (11.38) we have

$$v = \frac{k_2[E][S]}{K_m} \quad (11.44)$$

So,

$$v = \frac{(k_2 K_I[E_T][S])}{(K_I K_m + K_I[S] + K_m[I])} \quad (11.45)$$

Since $V_{max} = k_2[E_T]$,

$$v = \frac{V_{max}[S]}{K_m + [S] + \dfrac{K_m[I]}{K_I}} \quad (11.46)$$

or

$$v = \frac{V_{max}[S]}{[S] + K_m\left(1 + \dfrac{[I]}{K_I}\right)} \quad (11.47)$$

(see also Table 11.6). The K_m term in the denominator in the inhibited case is increased by the factor $(1 + [I]/K_I)$; thus, v is less in the presence of the inhibitor, as expected. Clearly, in the absence of I, the two equations are identical. Figure 11.13 shows a Lineweaver–Burk plot of competitive inhibition. Several features of competitive inhibition are evident. First, at a given [I], v decreases ($1/v$ increases). When [S] becomes infinite, $v = V_{max}$ and is unaffected by I because all of the enzyme is in the ES form. Note that the value of the $-x$-intercept decreases as [I] increases. This $-x$-intercept is often termed the *apparent* K_m (or K_{mapp}) because it is the K_m apparent under these conditions. The diagnostic criterion for competitive inhibition is that V_{max} is unaffected by I; that is, all lines share a common y-intercept. This criterion is also the best experimental indication of binding at the same site by two substances. Competitive inhibitors resemble S structurally.

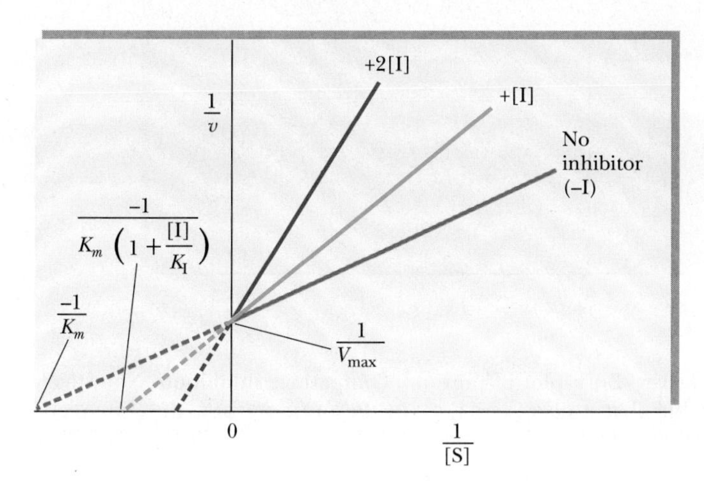

***Figure* 11.13** Lineweaver–Burk plot of competitive inhibition showing lines for no I, [I], and 2[I]. Note that when [S] is infinitely large ($1/[S] = 0$), V_{max} is the same whether I is present or not. In the presence of I, the $-x$-intercept is $-1/K_m(1 + [I]/K_I)$.

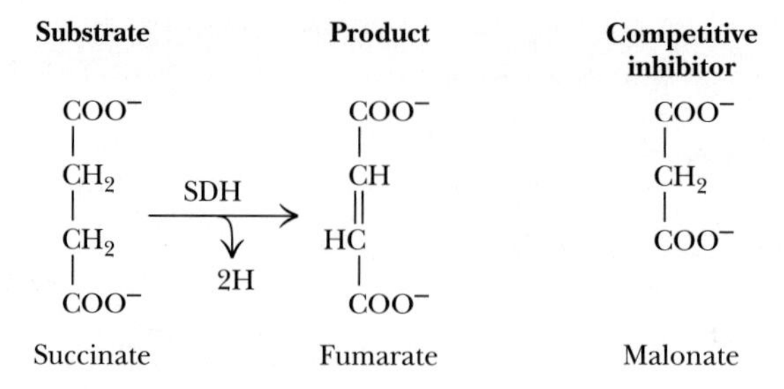

***Figure* 11.14** Structures of succinate, the substrate of succinate dehydrogenase (SDH), and malonate, the competitive inhibitor. Fumarate (the product of SDH action on succinate) is also shown.

Succinate Dehydrogenase—A Classic Example of Competitive Inhibition

The enzyme *succinate dehydrogenase (SDH)* is competitively inhibited by malonate. Figure 11.14 shows the structures of succinate and malonate. The structural similarity between them is obvious and is the basis of malonate's ability to mimic succinate and bind at the active site of SDH. However, unlike succinate, which is oxidized by SDH to form fumarate, malonate cannot lose two hydrogens; consequently, it is unreactive.

Noncompetitive Inhibition

Noncompetitive inhibitors interact with both E and ES (or with S and ES, but this is a rare and specialized case). Obviously then, the inhibitor is not binding to the same site as S, and the inhibition cannot be overcome by raising [S]. There are two types of noncompetitive inhibition: pure and mixed.

Pure Noncompetitive Inhibition

In this situation, the binding of I by E has no effect on the binding of S by E. Consider in the system

$$\mathrm{E} + \mathrm{I} \underset{}{\overset{K_\mathrm{I}}{\rightleftharpoons}} \mathrm{EI} \qquad \mathrm{ES} + \mathrm{I} \underset{}{\overset{K_\mathrm{I}'}{\rightleftharpoons}} \mathrm{IES} \tag{11.48}$$

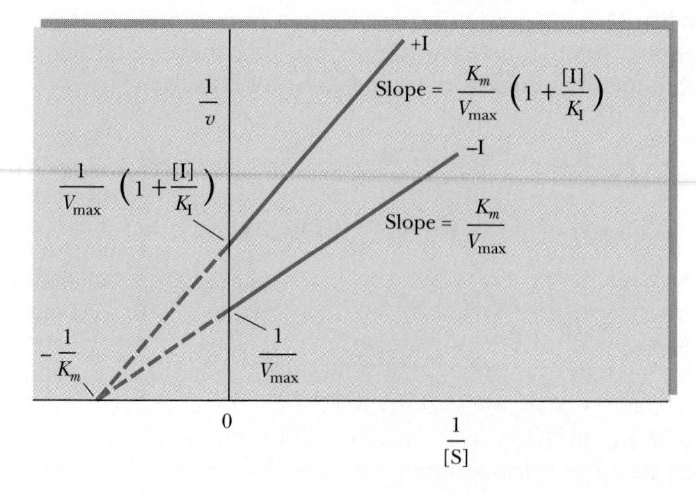

***Figure* 11.15** Lineweaver–Burk plot of pure noncompetitive inhibition. Note that I does not alter K_m, but that it decreases V_{max}. In the presence of I, the *y*-intercept is equal to $(1/V_{max})(1 + [I]/K_I)$.

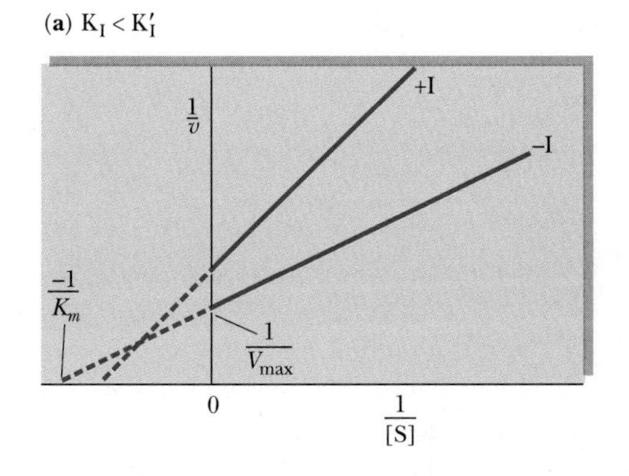

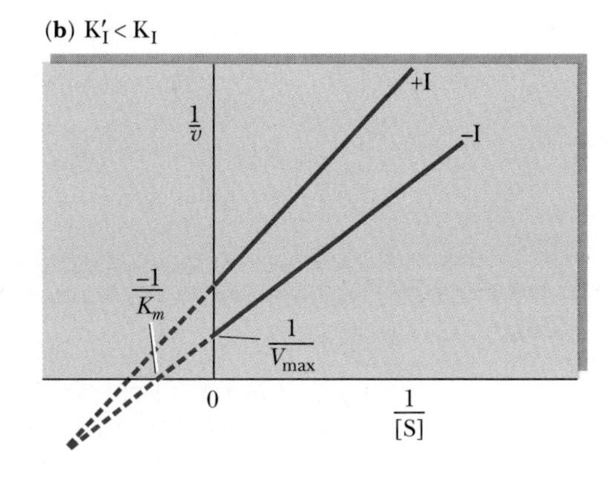

***Figure* 11.16** Lineweaver–Burk plot of mixed noncompetitive inhibition. Note that both intercepts and the slope change in the presence of I: (a) when K_I is less than K_I'; (b) when K_I is greater than K_I'.

Pure noncompetitive inhibition occurs if $K_I = K_I'$. This situation is relatively uncommon; the Lineweaver–Burk plot for such an instance is given in Figure 11.15. Note that K_m is unchanged by I (the *x*-intercept remains the same, with or without I). Note also that V_{max} decreases. A similar pattern is seen if the amount of enzyme in the experiment is decreased. Thus, it is as if I lowered [E].

Mixed Noncompetitive Inhibition

In this situation, the binding of I by E influences the binding of S by E. In this case, K_I and K_I', as defined previously, are not equal. Both K_m and V_{max} are altered by the presence of I, and K_m/V_{max} is not constant (Figure 11.16). This inhibitory pattern is commonly encountered. A reasonable explanation is that the inhibitor is binding at a site distinct from the active site, yet is influencing the binding of S at the active site. Presumably, these effects are transmitted via alterations in the protein's conformation. Table 11.6 includes the rate equations and apparent K_m and V_{max} values for both types of noncompetitive inhibition.

Irreversible Inhibition

If the inhibitor combines irreversibly with the enzyme, for example, by covalent attachment, the kinetic pattern seen is like that of noncompetitive inhibition, since the net effect is a loss of active enzyme. Usually, this type of inhibition can be distinguished from the noncompetitive, reversible inhibition case since the reaction of I with E (and/or ES) is not instantaneous. Instead, there is a *time-dependent decrease in enzymatic activity* as $E + I \rightarrow EI$ proceeds, and the rate of this inactivation can be followed. Also, unlike reversible inhibitions, dilution or dialysis of the enzyme : inhibitor solution does not dissociate the EI complex and restore enzyme activity.

Suicide Substrates—Mechanism-Based Enzyme Inactivators

Suicide substrates are inhibitory substrate analogs designed so that, via normal catalytic action of the enzyme, a very reactive group is generated. This reactive group then forms a covalent bond with a nearby functional group within the active site of the enzyme, thereby causing irreversible inhibition. Suicide substrates, also called *Trojan horse substrates,* are a type of **affinity label.** As substrate analogs, they bind with specificity and high affinity to the enzyme active site; in their reactive form, they become covalently bound to the enzyme. This covalent link effectively labels a particular functional group within the active site, identifying the group as a key player in the enzyme's catalytic cycle.

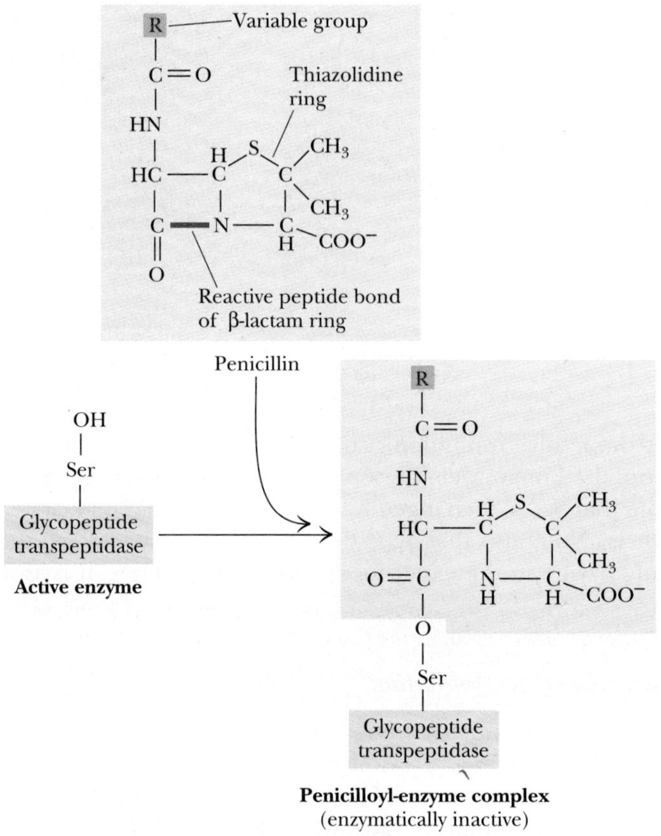

***Figure* 11.17** Penicillin is an irreversible inhibitor of the enzyme *glycoprotein peptidase,* which catalyzes an essential step in bacterial cell wall synthesis. Penicillin consists of a thiazolidine ring fused to a β-lactam ring to which a variable group R is attached. A reactive peptide bond in the β-lactam ring covalently attaches to a serine residue in the active site of the glycopeptide transpeptidase. (The conformation of penicillin around its reactive peptide bond resembles the transition state of the normal glycoprotein peptidase substrate.) The penicilloyl-enzyme complex is catalytically inactive; the bond between the enzyme and penicillin is indefinitely stable, that is, penicillin binding is irreversible.

Penicillin—An Irreversible Inhibitor

Several drugs in current medical use are mechanism-based enzyme inactivators. For example, the antibiotic **penicillin** exerts its effects by covalently reacting with an essential serine residue in the active site of *glycoprotein peptidase,* an enzyme that acts to cross-link the peptidoglycan chains during synthesis of bacterial cell walls (Figure 11.17). Once cell wall synthesis is blocked, the bacterial cells are very susceptible to rupture by osmotic lysis, and bacterial growth is halted.

11.5 Kinetics of Enzyme-Catalyzed Reactions Involving Two or More Substrates

Thus far, we have considered only the simple case of enzymes that act upon a single substrate, S. This situation is not common. Usually, enzymes catalyze reactions in which two (or even more) substrates take part.

Consider the case of an enzyme catalyzing a reaction involving two substrates, A and B, and yielding the products P and Q:

Double-reciprocal form of the rate equation: $\frac{1}{v} = \frac{1}{V_{max}}\left(K_m^A + \frac{K_S^A K_m^B}{[B]}\right)\left(\frac{1}{[A]} + \frac{1}{V_{max}}\left(1+\frac{K_m^B}{[B]}\right)\right)$

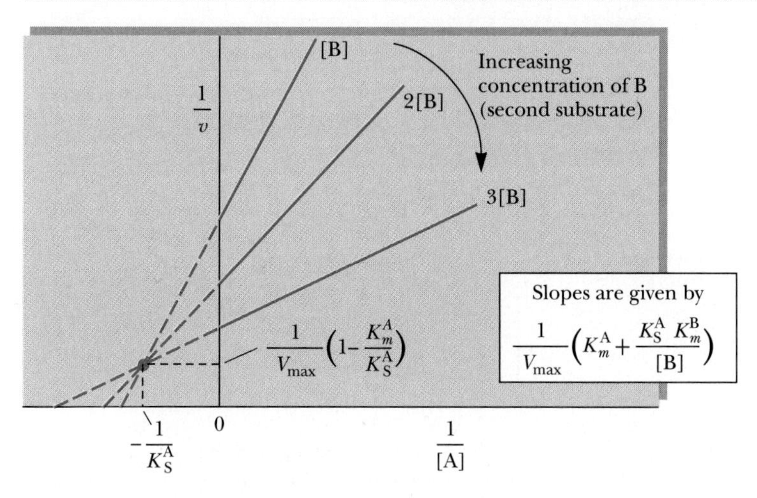

***Figure* 11.18** Single-displacement bisubstrate mechanisms are characterized by Lineweaver–Burk plots in which the lines intersect to the left of the $1/v$ axis when double-reciprocal plots of the rates observed with different fixed concentrations of one substrate (B here) are graphed versus a series of concentrations of A.

$$A + B \xrightarrow{\text{enzyme}} P + Q \tag{11.49}$$

Such a reaction is termed a **bisubstrate reaction.** In general, bisubstrate reactions proceed by one of two possible routes:

1. Both A and B are bound to the enzyme and then reaction occurs to give P + Q:

$$E + A + B \longrightarrow AEB \longrightarrow PEQ \longrightarrow E + P + Q \tag{11.50}$$

   Reactions of this type are defined as **sequential** or **single-displacement reactions.** They can be either of two distinct classes:
   a. **random,** where either A or B may bind to the enzyme first, followed by the other substrate, or
   b. **ordered,** where A, designated the *leading substrate,* must bind to E first before B can be bound.

   Both classes of single-displacement reactions are characterized by lines that intersect to the left of the $1/v$ axis in Lineweaver–Burk double-reciprocal plots (Figure 11.18).

2. One substrate, A, binds to the enzyme and reacts with it to yield a chemically modified form of the enzyme (E′) plus the product, P. The second substrate, B, then reacts with E′, regenerating E and forming the other product, Q.

$$E + A \longrightarrow EA \longrightarrow E'P \longrightarrow E' \longrightarrow E'B \longrightarrow EQ \longrightarrow E + Q \quad (E'P \to P;\ B \to E'B) \tag{11.51}$$

Reactions that fit this model are called **ping-pong** or **double-displacement reactions.** Two distinctive features of this mechanism are the obligatory formation of a modified enzyme intermediate, E′, and the pattern of parallel lines obtained in double-reciprocal plots (Figure 11.19).

Double-reciprocal form of the rate equation: $\frac{1}{v} = \frac{K_m^A}{V_{max}}\left(\frac{1}{[A]}\right) + \left(1 + \frac{K_m^B}{[B]}\right)\left(\frac{1}{V_{max}}\right)$

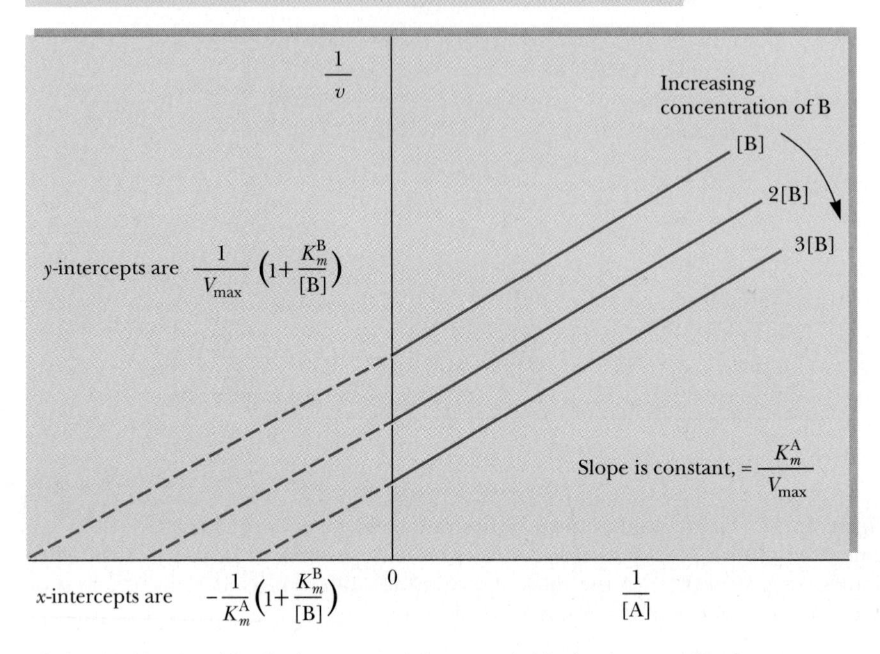

***Figure* 11.19** Double-displacement (ping-pong) bisubstrate mechanisms are characterized by Lineweaver–Burk plots of parallel lines when double-reciprocal plots of the rates observed with different fixed concentrations of the second substrate, B, are graphed versus a series of concentrations of A.

Random, Single-Displacement Reactions

In this type of sequential reaction, all possible binary enzyme:substrate complexes (AE, EB, QE, EP) are formed rapidly and reversibly when the enzyme is added to a reaction mixture containing A, B, P, and Q:

$$\begin{array}{ccccc} A + E \rightleftharpoons AE & \searrow & & \nearrow & QE \rightleftharpoons Q + E \\ & & AEB \rightleftharpoons QEP & & \\ E + B \rightleftharpoons EB & \nearrow & & \searrow & EP \rightleftharpoons E + P \end{array} \qquad (11.52)$$

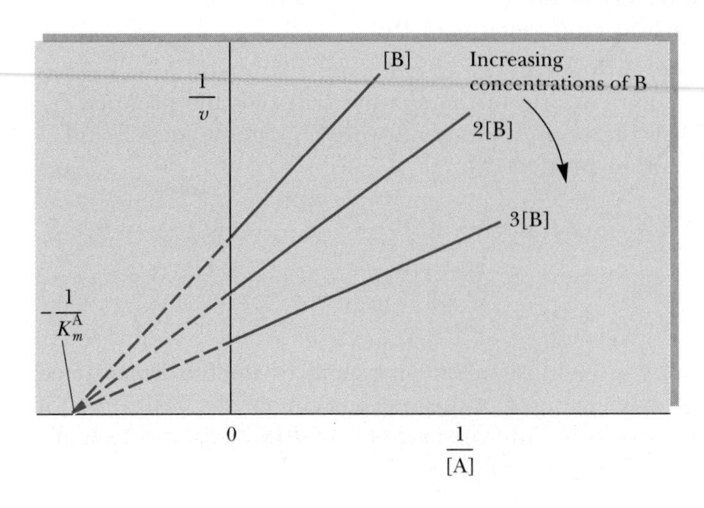

***Figure* 11.20** Random, single-displacement bisubstrate mechanism where A does not affect B binding, and vice versa. Note that the lines intersect at the 1/[A] axis. (If [B] were varied in an experiment with several fixed concentrations of A, the lines would intersect at the 1/[B] axis in a $1/v$ versus 1/[B] plot.)

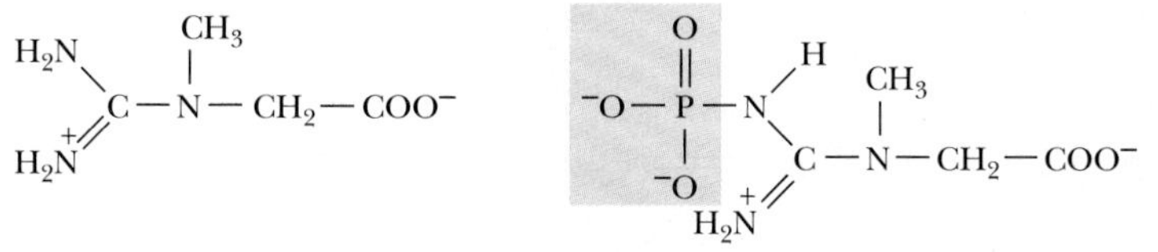

$$H_2N(H_2N^+{=})C{-}N(CH_3){-}CH_2{-}COO^-$$

Creatine

$$^-O{-}P(=O)(O^-){-}NH{-}C(={}^+NH_2){-}N(CH_3){-}CH_2{-}COO^-$$

Creatine–P

***Figure* 11.21** The structures of creatine and creatine phosphate, guanidinium compounds that are important in muscle energy metabolism.

The rate-limiting step is the reaction AEB → QEP. It doesn't matter whether A or B binds first to E, or whether Q or P is released first from QEP. Sometimes, reactions that follow this random order of addition of substrates to E can be distinguished mechanistically from reactions obeying an ordered, single-displacement mechanism, *if* A has *no* influence on the binding constant for B (and vice versa); that is, the mechanism is purely random. Then, the lines in a Lineweaver–Burk plot intersect at the 1/[A] axis (Figure 11.20).

Creatine Kinase Acts by a Random, Single-Displacement Mechanism

An example of a random, single-displacement mechanism is seen in the enzyme *creatine kinase,* a phosphoryl-transfer enzyme that uses ATP as a phosphoryl donor to form creatine phosphate (CrP) from creatine (Cr). Creatine-P is an important reservoir of phosphate-bond energy in muscle cells (Figure 11.21).

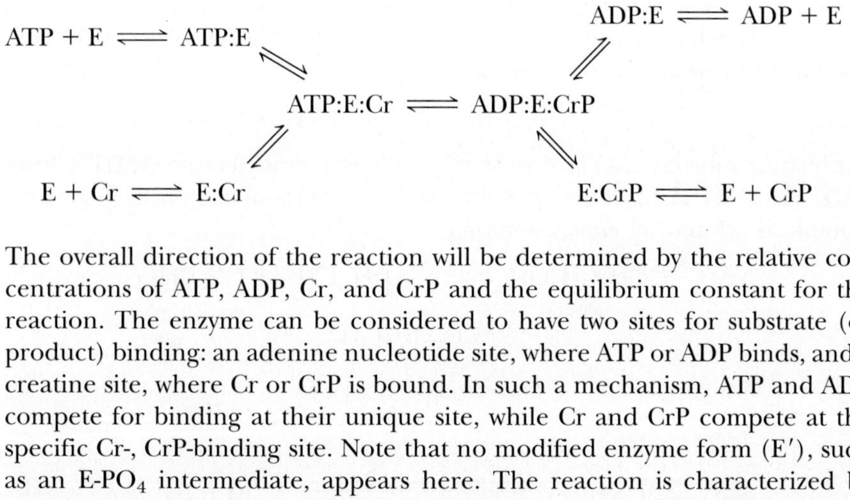

$$\begin{array}{ccccc}
\text{ATP + E} \rightleftharpoons \text{ATP:E} & & & & \text{ADP:E} \rightleftharpoons \text{ADP + E} \\
& \searrow\!\!\nwarrow & & \nearrow\!\!\swarrow & \\
& & \text{ATP:E:Cr} \rightleftharpoons \text{ADP:E:CrP} & & \\
& \nearrow\!\!\swarrow & & \searrow\!\!\nwarrow & \\
\text{E + Cr} \rightleftharpoons \text{E:Cr} & & & & \text{E:CrP} \rightleftharpoons \text{E + CrP}
\end{array}$$

The overall direction of the reaction will be determined by the relative concentrations of ATP, ADP, Cr, and CrP and the equilibrium constant for the reaction. The enzyme can be considered to have two sites for substrate (or product) binding: an adenine nucleotide site, where ATP or ADP binds, and a creatine site, where Cr or CrP is bound. In such a mechanism, ATP and ADP compete for binding at their unique site, while Cr and CrP compete at the specific Cr-, CrP-binding site. Note that no modified enzyme form (E′), such as an E-PO_4 intermediate, appears here. The reaction is characterized by rapid and reversible binary ES complex formation, followed by addition of the remaining substrate, and the rate-determining reaction taking place within the ternary complex.

Ordered, Single-Displacement Reactions

In this case, the **leading substrate,** A (also called the **obligatory** or **compulsory substrate**), must bind first. Then the second substrate, B, binds. Strictly speaking, B cannot bind to free enzyme in the absence of A. Reaction between A and B occurs in the ternary complex, and is usually followed by an ordered release of the products of the reaction, P and Q. In the schemes below, Q is the product of A and is released last. One representation, suggested by W. W. Cleland, follows:

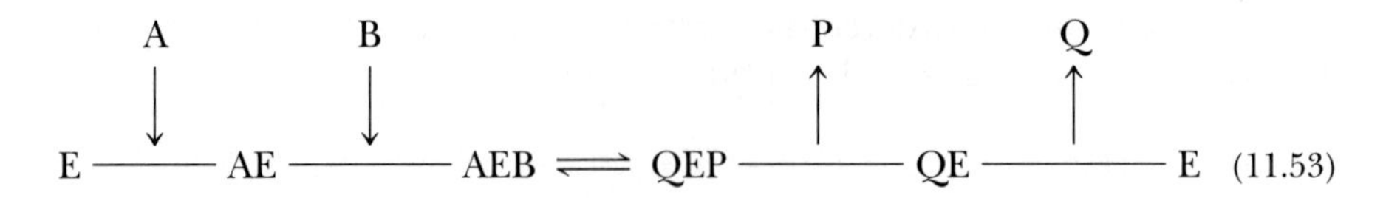

(11.53)

Another way of portraying this mechanism is as follows:

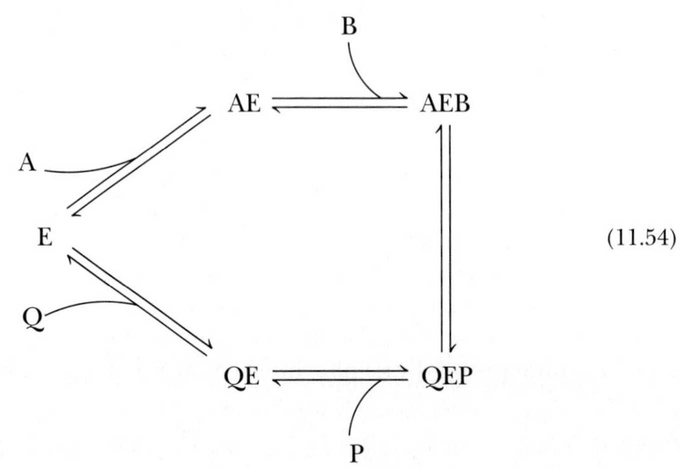

(11.54)

Note that A and Q are competitive for binding to the free enzyme, E, but not A and B (or Q and B).

NAD⁺-Dependent Dehydrogenases Show Ordered Single-Displacement Mechanisms

Nicotinamide adenine dinucleotide (NAD⁺)-dependent dehydrogenases are enzymes that typically behave according to the kinetic pattern just described. A general reaction of these dehydrogenases is

$$NAD^+ + BH_2 \rightleftharpoons NADH + H^+ + B$$

The leading substrate (A) is nicotinamide adenine dinucleotide (NAD^+), and NAD^+ and NADH (product Q) compete for a common site on E. A specific example is offered by *alcohol dehydrogenase (ADH):*

$$\underset{\text{ethanol}}{NAD^+ + CH_3CH_2OH} \rightleftharpoons NADH + H^+ + \underset{\text{acetaldehyde}}{CH_3CHO}$$

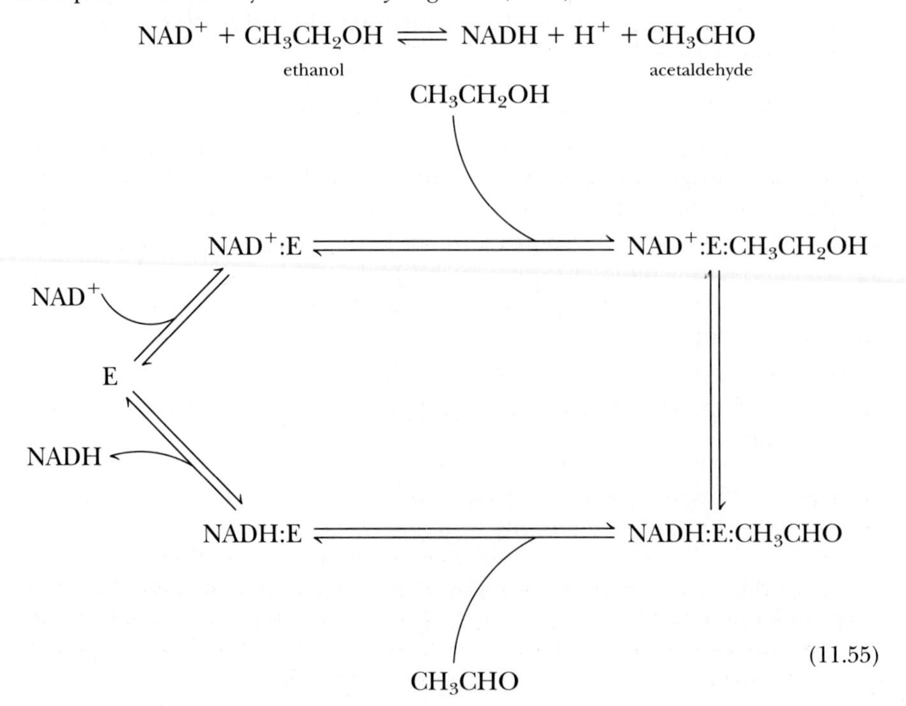

(11.55)

We can verify that this ordered mechanism is not random by demonstrating that no B (ethanol) is bound to E in the absence of A (NAD^+).

Double-Displacement (Ping-Pong) Reactions

Reactions conforming to this kinetic pattern are characterized by the fact that the product of the enzyme's reaction with A (called P in the following schemes) is released *prior* to reaction of the enzyme with the second substrate, B. As a result of this process, the enzyme, E, is converted to a modified form, E′, which then reacts with B to give the second product, Q, and regenerate the unmodified enzyme form, E:

$$E \xrightarrow{\;A\downarrow\;} AE \rightleftharpoons PE' \xrightarrow{\;P\uparrow\;} E' \xrightarrow{\;B\downarrow\;} E'B \rightleftharpoons EQ \xrightarrow{\;Q\uparrow\;} E \qquad (11.56)$$

or

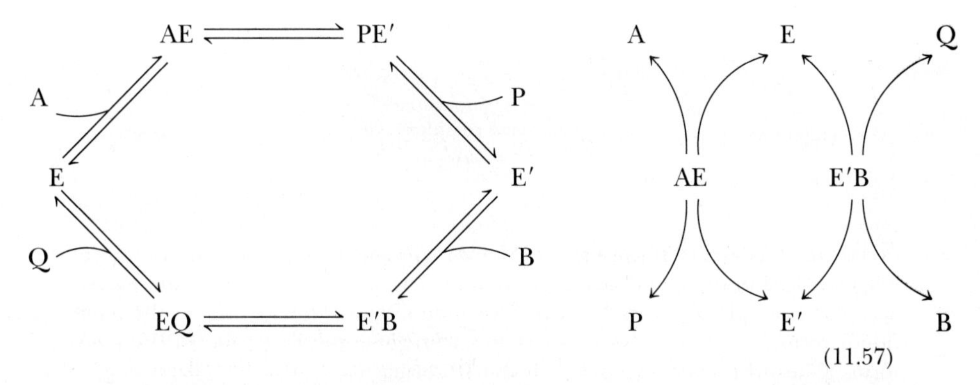

(11.57)

Note that these schemes predict that A and Q compete for the free enzyme form, E, while B and P compete for the modified enzyme form, E′. A and Q do not bind to E′, nor do B and P combine with E.

One class of enzymes that follow a ping-pong-type mechanism are *aminotransferases* (previously known as transaminases). These enzymes catalyze the transfer of an amino group from an amino acid to a keto acid. The products are a new amino acid and the keto acid corresponding to the carbon skeleton of the amino donor:

$$\text{amino acid}_1 + \text{keto acid}_2 \longrightarrow \text{keto acid}_1 + \text{amino acid}_2$$

A specific example would be *glutamate : aspartate aminotransferase.* Figure 11.22 depicts the scheme for this mechanism. Note that glutamate and aspartate are competitive for E, and that oxaloacetate and α-ketoglutarate compete for E′. In glutamate:aspartate aminotransferase, an enzyme-bound coenzyme, *pyridoxal phosphate* (a vitamin B_6 derivative, see Chapter 14), serves as the amino group acceptor/donor in the enzymatic reaction. The unmodified enzyme form, E, has the coenzyme in the aldehydic pyridoxal form, whereas the modified enzyme form, E′, is actually pyridoxamine phosphate (Figure 11.22). Not all enzymes displaying ping-pong-type mechanisms require coenzymes as carriers for the chemical substituent transferred in the reaction.

Diagnosis of Bisubstrate Mechanisms

Kineticists rely on a number of diagnostic tests for the assignment of a reaction mechanism to a specific enzyme. One is the graphic analysis of the kinetic patterns observed. It is usually easy to distinguish between single- and double-displacement reactions in this manner, and examining competitive effects between substrates aids in assigning reactions to random versus or-

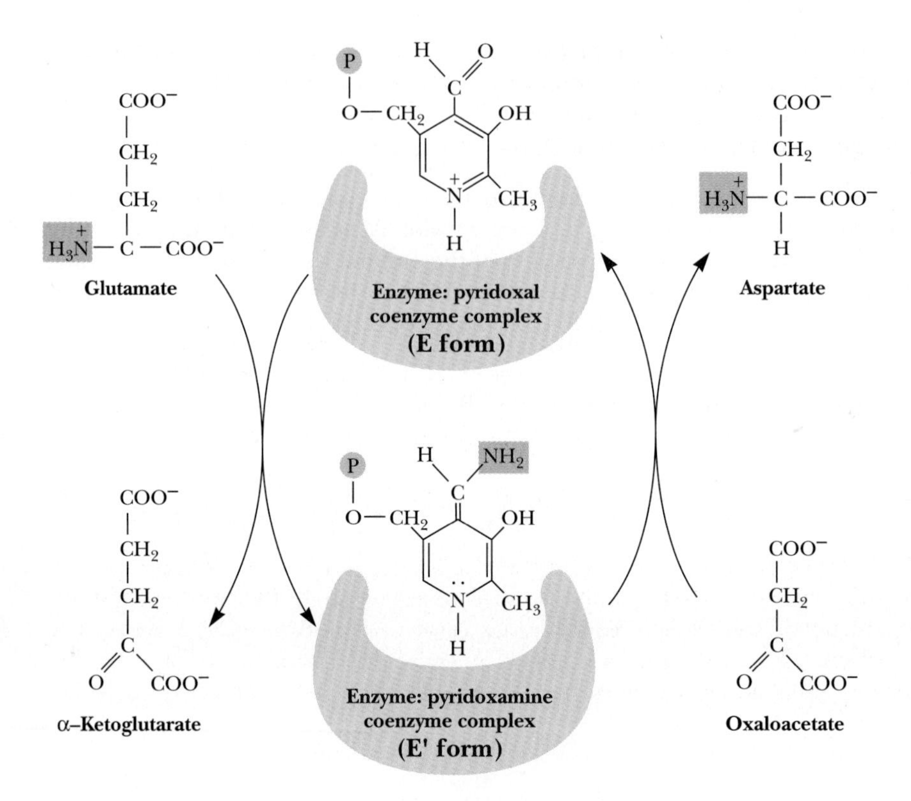

***Figure* 11.22** *Glutamate:aspartate aminotransferase,* an enzyme conforming to a double-displacement bisubstrate mechanism. Glutamate : aspartate aminotransferase is a pyridoxal phosphate-dependent enzyme. The pyridoxal serves as the $-NH_2$ acceptor from glutamate to form pyridoxamine. Pyridoxamine is then the amino donor to oxaloacetate to form aspartate and regenerate the pyridoxal coenzyme form. (The pyridoxamine : enzyme is the E′ form.)

dered patterns of S-binding. A second diagnostic test is to determine whether the enzyme catalyzes an exchange reaction. Consider as an example the two enzymes *sucrose phosphorylase* and *maltose phosphorylase.* Both catalyze the phosphorolysis of a disaccharide and both yield glucose-1-phosphate and a free hexose:

$$\text{sucrose} + P_i \rightleftharpoons \text{glucose-1-phosphate} + \text{fructose}$$
$$\text{maltose} + P_i \rightleftharpoons \text{glucose-1-phosphate} + \text{glucose}$$

Interestingly, in the absence of sucrose and fructose, sucrose phosphorylase will catalyze the exchange of inorganic phosphate, P_i, into glucose-1-phosphate. This reaction can be followed by using $^{32}P_i$ as a radioactive tracer and observing the appearance of ^{32}P into glucose-1-phosphate:

$$^{32}P_i + \text{G-1-P} \rightleftharpoons P_i + \text{G-1-}^{32}\text{P}$$

Maltose phosphorylase cannot carry out a similar reaction. The ^{32}P exchange reaction of sucrose phosphorylase is accounted for by a double-displacement mechanism where E′ = E-glucose:

$$\text{sucrose} + E \rightleftharpoons \text{E-glucose} + \text{fructose}$$
$$\text{E-glucose} + P_i \rightleftharpoons E + \text{glucose-1-phosphate}$$

Thus, in the presence of just $^{32}P_i$ and glucose-1-phosphate, sucrose phosphorylase still catalyzes the second reaction and radioactive P_i is incorporated into glucose-1-phosphate over time.

Maltose phosphorylase proceeds via a single-displacement reaction that necessarily requires the formation of a ternary maltose:E:P_i (or glucose:E:glucose-1-phosphate) complex for any reaction to occur. Exchange reactions are a characteristic of enzymes that obey double-displacement mechanisms at some point in their catalysis.

Multisubstrate Reactions

Thus far, we have considered enzyme-catalyzed reactions involving one or two substrates. How are the kinetics described in those cases where more than two substrates participate in the reaction? An example might be the glycolytic enzyme *glyceraldehyde-3-phosphate dehydrogenase* (Chapter 18):

$$\text{NAD}^+ + \text{glyceraldehyde-3-P} + \text{P}_i \longrightarrow \text{NADH} + \text{H}^+ + \text{1,3-bisphosphoglycerate}$$

Many other multisubstrate examples abound in metabolism. In effect, these situations are managed by realizing that the interaction of the enzyme with its many substrates can be treated as a series of uni- or bisubstrate steps in a multistep reaction pathway. Thus, the complex mechanism of a multisubstrate reaction is resolved into a sequence of steps, each of which obeys the single- and double-displacement patterns just discussed.

11.6 RNA and Antibody Molecules as Enzymes: Ribozymes and Abzymes

The traditional view of enzymes as proteins has been modified by the realization that some RNA molecules act as catalysts. Furthermore, the specificity of an antibody for its particular antigen has been exploited to produce antibody molecules specific for transition-state intermediates such that binding facilitates, and actually catalyzes, a desired reaction.

Catalytic RNA Molecules: Ribozymes

It was long assumed that all enzymes are proteins. However, in recent years, more and more instances of biological catalysis by RNA molecules have been discovered. These catalytic RNAs, or **ribozymes,** satisfy several enzymatic criteria: They are substrate-specific, they enhance the reaction rate, and they emerge from the reaction unchanged. For example, RNase P, an enzyme responsible for the formation of mature tRNA molecules from tRNA precursors, requires an RNA component as well as a protein subunit for its activity in the cell. *In vitro,* the protein alone is incapable of catalyzing the maturation reaction, but the RNA component by itself can carry out the reaction under appropriate conditions. In another case, in the ciliated protozoan *Tetrahymena,* formation of mature ribosomal RNA from a pre-rRNA precursor involves the removal of an internal RNA segment and the joining of the two ends in a process known as **splicing out.** The excision of this intervening internal sequence of RNA and ligation of the ends is, remarkably, catalyzed by the intervening sequence of RNA itself, in the presence of a free molecule of guanosine nucleoside or nucleotide (Figure 11.23). *In vivo,* the intervening sequence RNA probably acts only in splicing itself out; *in vitro,* however, it can act many times, turning over like a true enzyme.

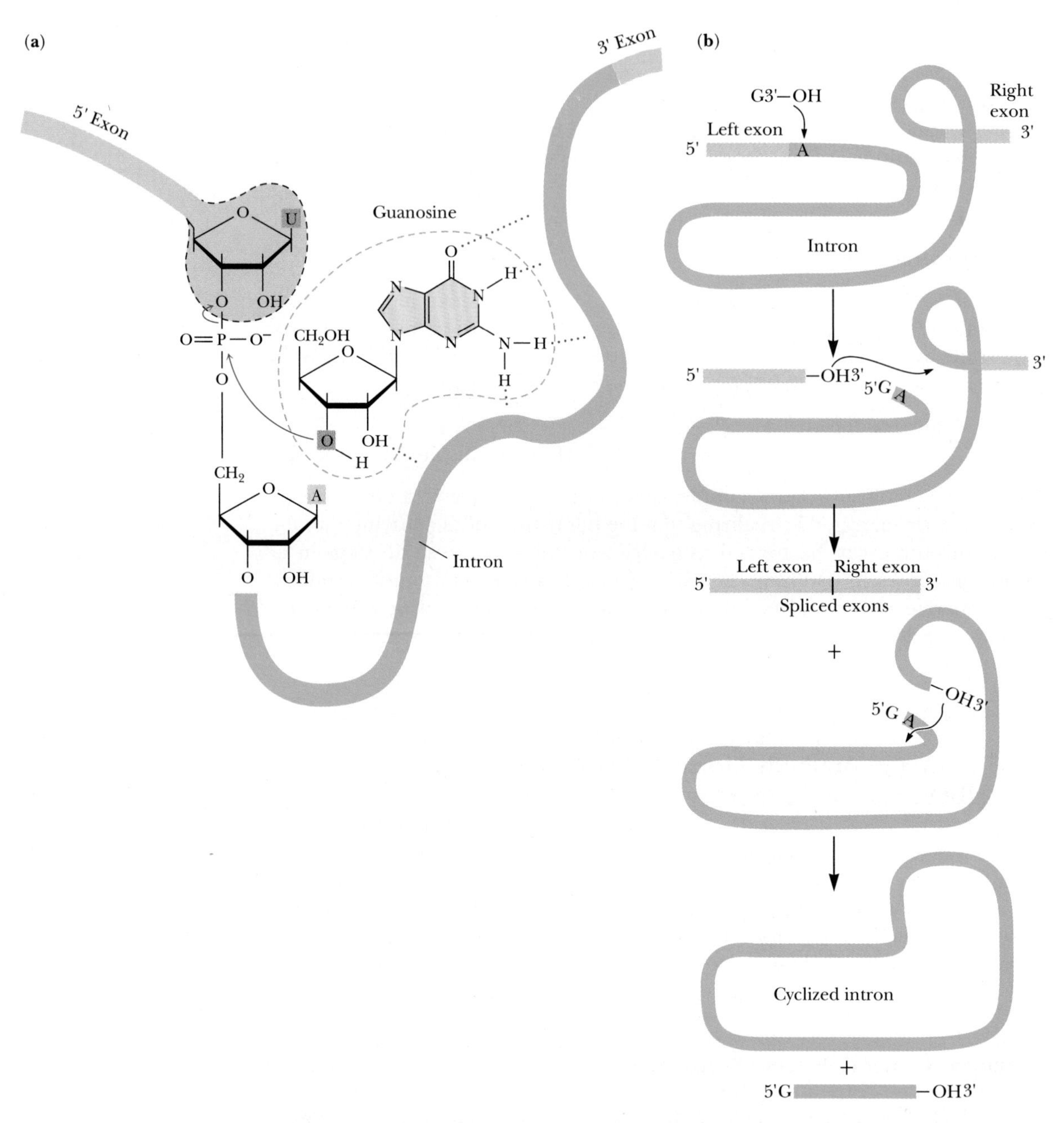

***Figure* 11.23** RNA splicing in *Tetrahymena* rRNA maturation: (a) the guanosine-mediated reaction involved in the autocatalytic excision of the *Tetrahymena* rRNA intron, and (b) the overall splicing process.

Protein-Free 50S Ribosomal Subunits Catalyze Peptide Bond Formation In Vitro

Perhaps the most significant case of catalysis by RNA occurs in protein synthesis. Harry F. Noller and his colleagues have found that the **peptidyl transferase reaction,** which is the reaction of peptide bond formation during protein synthesis (Figure 11.24), can be catalyzed by 50S ribosomal subunits (see Chapter 7) from which virtually all of the protein has been removed. These experiments imply that just the 23S rRNA by itself is capable of catalyzing peptide bond formation. Also, the laboratories of Noller and Thomas Cech (who, along with Sidney Altman, discovered the catalytic properties of RNAs) have demonstrated that the *Tetrahymena* ribozyme can catalyze the hydrolysis

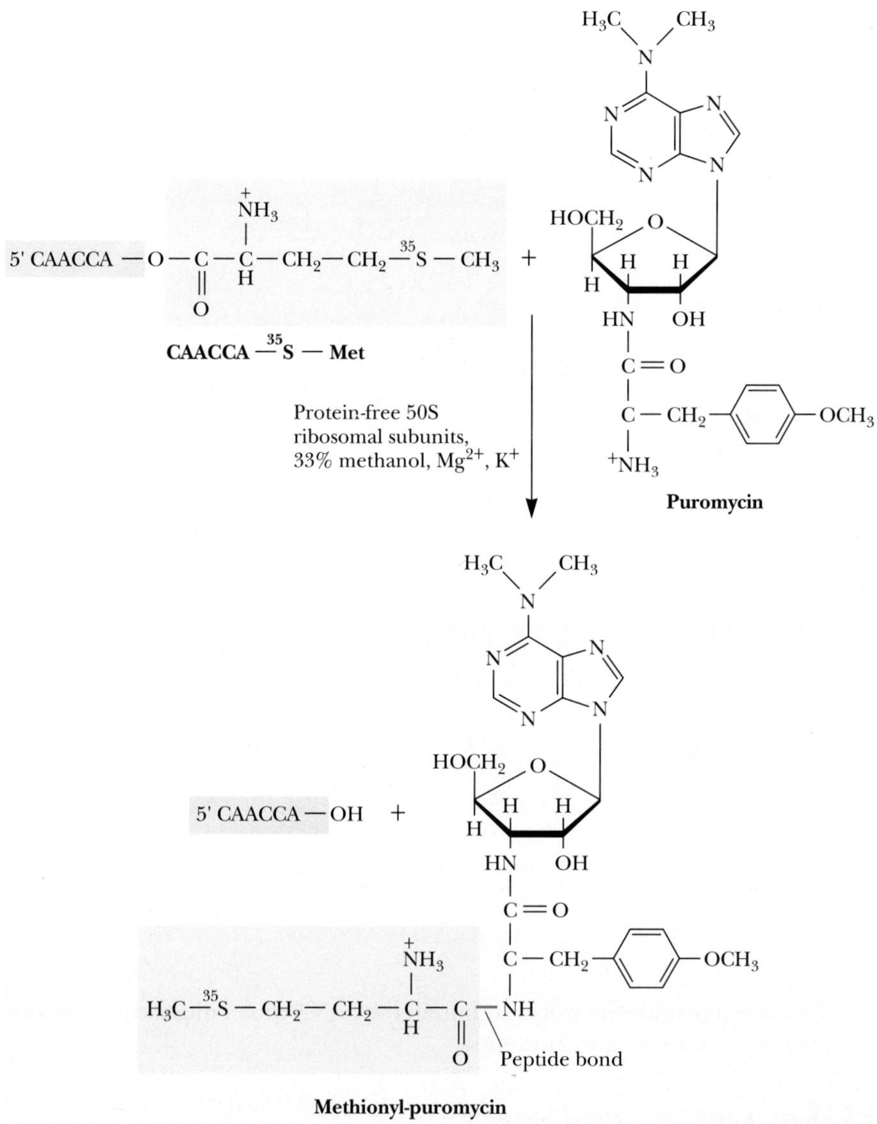

5' CAACCA — O — C(=O) — C(H)(NH_3^+) — CH_2 — CH_2 — ^{35}S — CH_3 + Puromycin

CAACCA — ^{35}S — Met

Protein-free 50S ribosomal subunits, 33% methanol, Mg^{2+}, K^+

Puromycin

5' CAACCA — OH + Methionyl-puromycin

Peptide bond

Methionyl-puromycin

Figure **11.24** Protein-free 50S ribosomal subunits have peptidyl transferase activity. Peptidyl transferase is the name of the enzymatic function that catalyzes peptide bond formation. The presence of this activity in protein-free 50S ribosomal subunits was demonstrated using a model assay for peptide bond formation where an aminoacyl-tRNA analog (a short RNA oligonucleotide of sequence CAACCA carrying ^{35}S-labeled methionine attached at its 3′-OH end) served as the peptidyl donor and puromycin (another aminoacyl-tRNA analog) served as the peptidyl acceptor. Activity was measured by monitoring the formation of ^{35}S-labeled methionyl-puromycin.

of amino acid esters (Figure 11.25). This discovery extends the known repertoire of catalysis by RNA to include reactions involving carbon atoms.

Several features of these "RNA enzymes," or **ribozymes,** lead to the realization that their biological efficiency does not challenge that achieved by proteins. First, RNA enzymes often do not fulfill the criterion of catalysis *in vivo* because they act only once in intramolecular events such as self-splicing. Second, the catalytic rates achieved by RNA enzymes *in vivo* and *in vitro* are significantly enhanced by the participation of protein subunits. Nevertheless, the fact that RNA can catalyze certain reactions is experimental support for

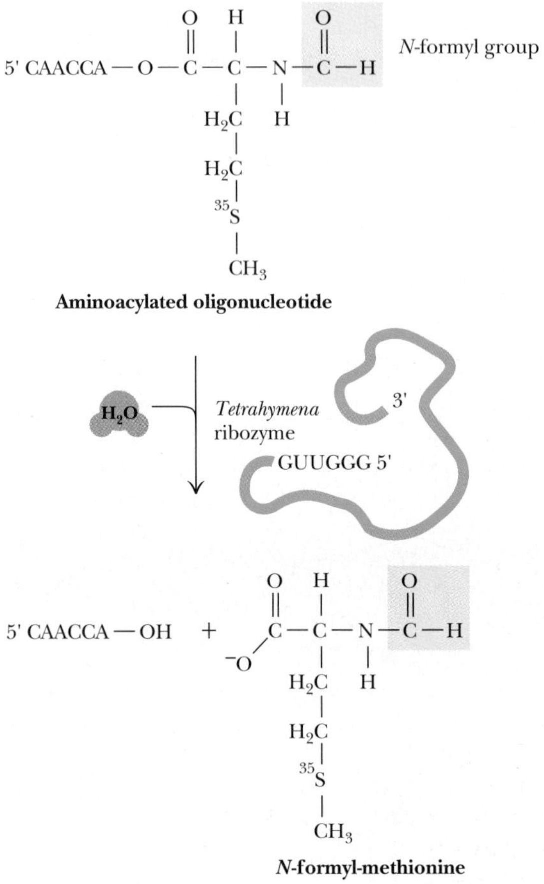

5' CAACCA—O—C(=O)—CH(H$_2$C—H$_2$C—^{35}S—CH$_3$)—NH—C(=O)—H (N-formyl group)

Aminoacylated oligonucleotide

H_2O — *Tetrahymena* ribozyme (3' … GUUGGG 5')

5' CAACCA—OH + $^-$O—C(=O)—CH(H$_2$C—H$_2$C—^{35}S—CH$_3$)—NH—C(=O)—H

***N*-formyl-methionine**

***Figure* 11.25** Aminoacyl esterase activity of *Tetrahymena* ribozyme. The substrate for this reaction is ^{35}S-labeled *N*-formyl-methionine joined in ester linkage to the 3′-OH end of an RNA oligonucleotide (5′-CAACCA). This catalysis establishes that the ribozyme can act on carboxylate esters as well as phosphate esters. Note that the reverse of aminoacyl RNA ester hydrolysis is synthesis of aminoacyl RNA esters (the chemical form in which amino acids are delivered to ribosomes for protein synthesis). This reaction suggests primordial RNAs could catalyze formation of substances analogous to aminoacyl-tRNAs.

the idea that a primordial world dominated by RNA molecules existed before the evolution of DNA and proteins.

Catalytic Antibodies: Abzymes

Antibodies are *immunoglobulins,* which, of course, are proteins. Like other antibodies, **catalytic antibodies,** so-called **abzymes,** are elicited in an organism in response to immunological challenge by a foreign molecule called an **antigen** (see Chapter 28 for discussions on the molecular basis of immunology). In this case, however, the antigen is purposefully engineered to be *an analog of the transition-state intermediate in a reaction.* The rationale is that a protein specific for binding the transition-state intermediate of a reaction will promote entry of the normal reactant into the reactive, transition-state conformation. Thus, a catalytic antibody facilitates, or catalyzes, a reaction by forcing the conformation of its substrate in the direction of its transition state. (A prominent explanation for the remarkable catalytic power of conventional enzymes is their great affinity for the transition-state intermediates in the reactions they catalyze; see Chapter 13.)

One strategy has been to prepare ester analogs by substituting a phosphorus atom for the carbon in the ester group (Figure 11.26). The phospho-compound mimics the natural transition state of ester hydrolysis, and antibodies elicited against these analogs act like enzymes in accelerating the rate of ester hydrolysis as much as 1000-fold. This biotechnology offers the real possibility of creating **"designer enzymes,"** specially tailored enzymes designed to carry out specific catalytic processes.

(a)

Hydroxy ester → Cyclic transition state → δ-Lactone + phenol

(b)

Cyclic phosphonate ester

Figure **11.26** Catalytic antibodies are designed specifically to bind the transition-state intermediate in a chemical reaction. (a) The intramolecular hydrolysis of a hydroxy ester to yield as products a δ-lactone and the alcohol phenol. Note the cyclic transition state. (b) The cyclic phosphonate ester analog of the cyclic transition state. Antibodies raised against this phosphonate ester act as *enzymes:* They are catalysts that markedly accelerate the rate of ester hydrolysis.

Problems

1. What is the v/V_{max} ratio when $[S] = 4K_m$?

2. If $V_{max} = 100\ \mu mol/sec$ and $K_m = 2\ mM$, what is the velocity of the reaction when $[S] = 20\ mM$?

3. For a Michaelis–Menten reaction, $k_1 = 7 \times 10^7/M \cdot sec$, $k_{-1} = 1 \times 10^3/sec$, and $k_2 = 2 \times 10^4/sec$. What are the values of K_S and K_m? Does substrate binding approach equilibrium or does it behave more like a steady-state system?

4. The following kinetic data were obtained for an enzyme in the absence of any inhibitor (1), and in the presence of two different inhibitors (2) and (3) at 5 mM concentration. Assume $[E_T]$ is the same in each experiment.

[S] (mM)	(1) v (μmol/sec)	(2) v (μmol/sec)	(3) v (μmol/sec)
1	12	4.3	5.5
2	20	8	9
4	29	14	13
8	35	21	16
12	40	26	18

   a. Determine V_{max} and K_m for the enzyme.
   b. Determine the type of inhibition and the K_I for each inhibitor.

5. The general rate equation for an ordered, single-displacement reaction where A is the leading substrate is

$$v = \frac{V_{max}[A][B]}{(K_S^A K_m^B + K_m^A[B] + K_m^B[A] + [A][B])}$$

   Write the Lineweaver–Burk (double-reciprocal) equivalent of this equation, and from it calculate algebraic expressions for (a) the slope; (b) the y-intercepts; and (c) the horizontal and vertical coordinates of the point of intersection when $1/v$ is plotted versus $1/[B]$ at various *fixed* concentrations of A.

6. The following graphical patterns obtained from kinetic experiments have several possible interpretations depending on the nature of the experiment and the variables being plotted. Give at least two possibilities for each.

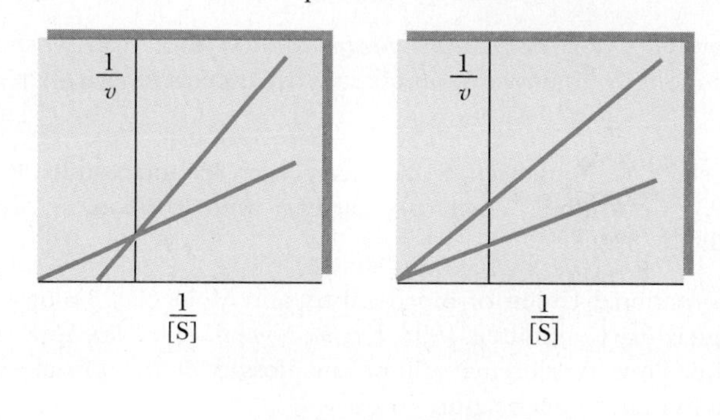

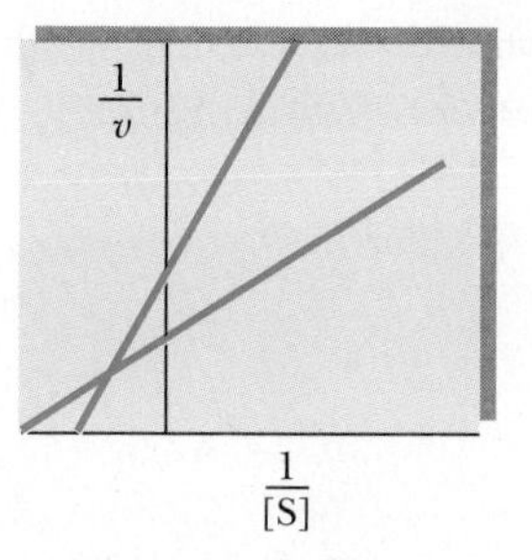

7. Liver alcohol dehydrogenase (ADH) is relatively nonspecific and will oxidize ethanol or other alcohols, including methanol. Methanol oxidation yields formaldehyde, which is quite toxic, causing, among other things, blindness. Mistaking it for the cheap wine he usually prefers, my dog Clancy in-

gested about 50 mL of windshield washer fluid (a solution 50% in methanol). Knowing that methanol would be excreted eventually by Clancy's kidneys if its oxidation could be blocked, and realizing that, in terms of methanol oxidation by ADH, ethanol would act as a competitive inhibitor, I decided to offer Clancy some wine. How much of Clancy's favorite vintage (12% ethanol) must he consume in order to lower the activity of his ADH on methanol to 5% of its normal value if the K_m values of canine ADH for ethanol and methanol are 1 millimolar and 10 millimolar, respectively? (The K_I for ethanol in its role as competitive inhibitor of methanol oxidation by ADH is the same as its K_m.) Both the methanol and ethanol will quickly distribute throughout Clancy's body fluids, which amount to about 15 L. Assume the densities of 50% methanol and the wine are both 0.9 g/mL.

8. From inspection of Figure 11.25, suggest how the *Tetrahymena* ribozyme recognizes its aminoacyl-RNA substrate.

Further Reading

Bell, J. E., and Bell, E. T., 1988. *Proteins and Enzymes.* Englewood Cliffs, NJ: Prentice-Hall. This text describes the structural and functional characteristics of proteins and enzymes.

Boyer, P. D., 1970. *The Enzymes,* vols. 1 and 2. New York: Academic Press. An edition for students seeking advanced knowledge on enzymes.

Cech, T. R., and Bass, B. L., 1986. Biological catalysis by RNA. *Annual Review of Biochemistry* **55:**599–629. A review of the early evidence that RNA can act as an enzyme.

Cech, T. R., et al., 1992. RNA catalysis by a group I ribozyme: Developing a model for transition-state stabilization. *The Journal of Biological Chemistry* **267:**17479–17482.

Dixon, M., et al., 1979. *Enzymes,* 3rd ed. New York: Academic Press. A classic work on enzyme kinetics and the properties of enzymes.

Fersht, A., 1985. *Enzyme Structure and Mechanism,* 2nd ed. Reading, PA: Freeman & Co. A monograph on the structure and action of enzymes.

Gray, C. J. *Enzyme-Catalyzed Reactions.* New York: Van Nostrand Reinhold. A monograph on quantitative aspects of enzyme kinetics.

Hsieh, L. C., Yonkovich, S., Kochersperger, L., and Schultz, P. G., 1993. Controlling chemical reactivity with antibodies. *Science* **260:**337–339.

International Union of Biochemistry and Molecular Biology Nomenclature Committee, 1992. *Enzyme Nomenclature.* New York: Academic Press. A reference volume and glossary on the official classification and nomenclature of enzymes.

Janda, K. D., Shevlin, C. G., and Lerner, R. A., 1993. Antibody catalysis of a disfavored chemical transformation. *Science* **259:**490–493.

Landry, D. W., Zhao, K., Yang, G. X., Glickman, M., and Georgiadis, T. M., 1993. Antibody-catalyzed degradation of cocaine. *Science* **259:**1899–1901.

Napper, A. D., et al., 1987. A stereospecific cyclization catalyzed by an antibody. *Science* **237:**1041–1043.

Noller, H. F., Hoffarth, V., and Zimniak, L., 1992. Unusual resistance of peptidyl transferase to protein extraction procedures. *Science* **256:**1416–1419.

Piccirilli, J. A., et al., 1992. Aminoacyl esterase activity of *Tetrahymena* ribozyme. *Science* **256:**1420–1424.

Segel, I. H., 1976. *Biochemical Calculations,* 2nd ed. New York: John Wiley & Sons. An excellent guide to solving problems in enzyme kinetics.

Silverman, R. B., 1988. *Mechanism-Based Enzyme Inactivation: Chemistry and Enzymology,* vols. 1 and 2. Boca Raton FL: CRC Press.

Smith, W. G., 1992. In vivo kinetics and the reversible Michaelis–Menten model. *Journal of Chemical Education* **12:**981–984.

Watson, J. D., 1987. Evolution of catalytic function. *Cold Spring Harbor Symposium on Quantitative Biology* **52:**1–955. Publications from a symposium on the nature and evolution of catalytic biomolecules (proteins and RNA) prompted by the discovery that RNA could act catalytically.

Zaug, A. J., and Cech, T. R., 1986. The intervening sequence RNA of *Tetrahymena* is an enzyme. *Science* **231:**470–475.

Chapter 12

Enzyme Specificity and Allosteric Regulation

Allostery is the key chemical process that makes possible intracellular and intercellular regulation: ". . . the molecular interactions which ensure the transmission and interpretation of (regulatory) signals rest upon (allosteric) proteins endowed with discriminatory stereospecific recognition properties."

Jacques Monod in *Chance and Necessity*

Outline

12.1 Specificity Is the Result of Molecular Recognition

12.2 Controls Over Enzymatic Activity—General Considerations

12.3 The Allosteric Regulation of Enzyme Activity

12.4 Hemoglobin and Myoglobin

12.5 *Escherichia coli* Aspartate Transcarbamoylase—An Allosteric Enzyme

12.6 Allosteric Models

Sea Scape, a mobile by Alexander Calder (1898–1976), collection of the Whitney Museum of American Art, New York. Metabolic regulation is achieved through an exquisitely balanced interplay among enzymes and small molecules, a process symbolized by the delicate balance of forces in this mobile.

The extraordinary ability of an enzyme to catalyze only one particular reaction is a quality known as **specificity** (Chapter 11). Specificity means an enzyme acts only on a specific substance, its substrate, invariably transforming it into a specific product. That is, an enzyme binds only certain compounds, and then, only a specific reaction ensues. Some enzymes show absolute specificity, catalyzing the transformation of only one specific substrate to yield a unique product. Other enzymes carry out a particular reaction but act on a class of compounds. For example, *hexokinase* (ATP:hexose-6-phosphotransferase) will carry out the ATP-dependent phosphorylation of a number of hexoses at the 6-position, including glucose.

12.1 Specificity Is the Result of Molecular Recognition

An enzyme molecule is typically orders of magnitude larger than its substrate. Its active site comprises only a small portion of the overall enzyme structure. The active site is part of the conformation of the enzyme molecule arranged to create a special pocket or cleft whose three-dimensional structure is complementary to the structure of the substrate. The enzyme and the substrate molecules "recognize" each other through this structural complementarity. The substrate binds to the enzyme through relatively weak forces—H bonds, ionic bonds (salt bridges), and van der Waals interactions between sterically complementary clusters of atoms. Specificity studies on enzymes entail an examination of the rates of the enzymatic reaction obtained with various **structural analogs** of the substrate. By determining which functional and structural groups within the substrate affect binding or catalysis, enzymologists can map the properties of the active site, analyzing questions such as: Can it accommodate sterically bulky groups? Are ionic interactions between E and S important? Are H bonds formed?

The "Lock and Key" Hypothesis

Pioneering enzyme specificity studies at the turn of the century by the great organic chemist Emil Fischer led to the notion of an enzyme resembling a **"lock"** and its particular substrate the **"key."** This analogy captures the essence of the specificity that exists between an enzyme and its substrate, but enzymes are not rigid templates like locks.

The "Induced Fit" Hypothesis

Enzymes are highly flexible, conformationally dynamic molecules, and many of their remarkable properties, including substrate binding and catalysis, are due to their structural pliancy. Realization of the conformational flexibility of proteins led Daniel Koshland to hypothesize that the binding of a substrate (S) by an enzyme is an interactive process. That is, the shape of the enzyme's active site is actually modified upon binding S, in a process of dynamic recognition between enzyme and substrate aptly called **induced fit.** In essence, substrate binding alters the conformation of the protein, so that the protein and the substrate "fit" each other more precisely. The process is truly interactive in that the conformation of the substrate also changes as it adapts to the conformation of the enzyme.

This idea also helps to explain some of the mystery surrounding the enormous catalytic power of enzymes: In enzyme catalysis, precise orientation of catalytic residues comprising the active site is a necessary condition for the reaction to occur; substrate binding induces this precise orientation by the changes it causes in the protein's conformation.

"Induced Fit" and the Transition-State Intermediate

The catalytically active enzyme: substrate complex is an interactive structure in which the enzyme causes the substrate to adopt a form that mimics the transition-state intermediate of the reaction. Thus, a poor substrate would be one that was less effective in directing the formation of an optimally active enzyme: transition-state intermediate conformation. This active conformation of the enzyme molecule is thought to be relatively unstable in the absence of substrate, and free enzyme thus reverts to a conformationally different state.

Specificity and Reactivity

Consider, for example, why hexokinase catalyzes the ATP-dependent phosphorylation of hexoses but not smaller phosphoryl-group acceptors such as glycerol, ethanol, or even water. Surely these smaller compounds are not sterically forbidden from approaching the active site of hexokinase (Figure 12.1). Indeed, water should penetrate the active site easily and serve as a highly effective phosphoryl-group acceptor. Accordingly, hexokinase should display high ATPase activity. It does not. Only the binding of hexoses induces hexokinase to assume its fully active conformation.

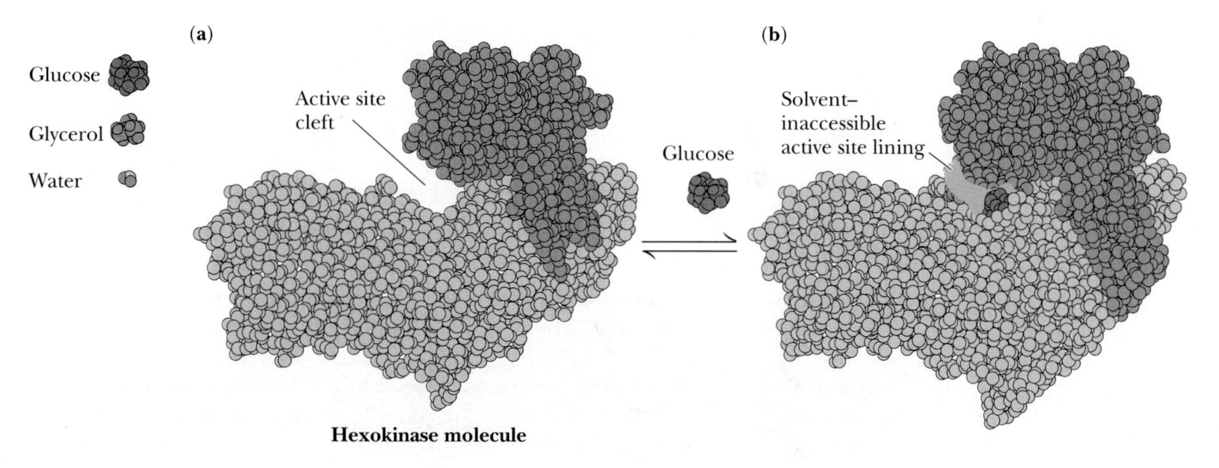

***Figure* 12.1** A drawing, roughly to scale, of H_2O, glycerol, glucose, and an idealized hexokinase molecule. Note the two domains of structure in hexokinase (a), between which the active site is located. Binding of glucose induces a conformational change in hexokinase. The two domains close together, creating the catalytic site (b). The shaded area in (b) represents solvent-inaccessible surface area in the active site cleft that results when the enzyme enters the ES complex.

In Chapter 13, we will explore in greater detail the factors that contribute to the remarkable catalytic power of enzymes and examine specific examples of enzyme reaction mechanisms. Here we focus on another essential feature of enzymes: *the regulation of their activity.*

12.2 Controls Over Enzymatic Activity—General Considerations

The activity displayed by enzymes is subject to a variety of factors, some of which are mere consequences of their role as catalysts, whereas others are essential to the harmony of metabolism.

1. The enzymatic rate "slows down" as product accumulates and equilibrium is approached. The apparent decrease in rate is due to the increasing rate of S formation by the reverse reaction as [P] rises. Once $[P]/[S] = K_{eq}$, no further reaction is apparent. K_{eq} defines thermodynamic equilibrium. Enzymes have no influence on the thermodynamics of a reaction. Also, product inhibition can be a kinetically valid phenomenon: Some enzymes are actually inhibited by the products of their action.
2. The availability of substrates and cofactors will determine the enzymatic reaction rate. In general, enzymes have evolved such that their K_m values approximate the prevailing *in vivo* concentration of their substrates. (It is also true that the concentration of some enzymes in cells is within an order of magnitude or so of the concentrations of their substrates.)
3. There are genetic controls over the amounts of enzyme synthesized (or degraded) by cells. If the gene encoding a particular enzyme protein is turned on or off, changes in the amount of enzyme activity soon follow. **Induction,** which is the activation of enzyme synthesis, and **repression,** which is the shutdown of enzyme synthesis, are important mechanisms for the regulation of metabolism. By controlling the amount of an enzyme that is present at any moment,

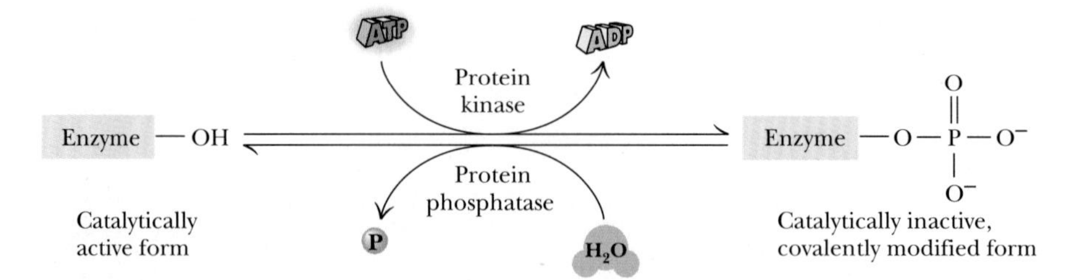

***Figure* 12.2** Enzymes regulated by covalent modification are called **interconvertible enzymes.** The enzymes (*protein kinase* and *protein phosphatase,* in the example shown here) catalyzing the conversion of the interconvertible enzyme between its two forms are called **converter enzymes.** In this example, the free enzyme form is catalytically active, while the phosphoryl-enzyme form represents an inactive state. The —OH on the interconvertible enzyme represents an —OH group on a specific amino acid side chain in the protein (for example, a particular Ser residue), capable of accepting the phosphoryl group.

cells can either activate or terminate various metabolic routes. Genetic controls over enzyme levels have a response time ranging from minutes in rapidly dividing bacteria to hours (or longer) in higher eukaryotes.

4. Enzymes can be regulated by **covalent modification,** the reversible covalent attachment of a chemical group. For example, a fully active enzyme can be modified to an inactive form simply by the covalent attachment of a functional group, such as a phosphoryl moiety (Figure 12.2). Alternatively, some enzymes exist in an inactive state unless specifically modified to the active form through covalent addition of a functional group. Covalent modification reactions are catalyzed by special **modifying enzymes,** or **converter enzymes,** which are themselves subject to metabolic regulation. Although covalent modification represents a stable alteration of the enzyme, a different converter enzyme operates to remove the modification, so that when the conditions that favored modification of the enzyme are no longer present, the process can be reversed, restoring the enzyme to its unmodified state. Many examples of covalent modification at important metabolic junctions will be encountered in our discussions of metabolic pathways. Because covalent modification events are enzyme-catalyzed, they occur very quickly, with response times of seconds or even less for significant changes in metabolic activity. The 1992 Nobel Prize in physiology or medicine was awarded to Edmond Fischer and Edwin Krebs for their pioneering studies of reversible protein phosphorylation as an important means of cellular regulation.
5. Enzymatic activity can also be activated or inhibited through noncovalent interaction of the enzyme with small molecules (metabolites) other than the substrate. This form of control is termed **allosteric regulation,** because the activator or inhibitor binds to the enzyme at a site *other* than (*allo* means "other") the active site. Further, such allosteric regulators, or **effector molecules,** are often quite different sterically from the substrate. Because this form of regulation results simply from reversible binding of regulatory ligands to the enzyme, the cellular response time can be virtually instantaneous.

6. Enzyme regulation is an important matter to cells, and evolution has provided a variety of additional options, including zymogens, isozymes, and modulator proteins.

Zymogens

Most enzymes become fully active as their synthesis is completed and they spontaneously fold into their native, three-dimensional conformations. Some enzymes, however, are synthesized as inactive precursors, called **zymogens** or **proenzymes,** that only acquire full enzymatic activity upon specific proteolytic cleavage of one or several of their peptide bonds. Unlike allosteric regulation or covalent modification, zymogen activation by specific proteolysis is an irreversible process. Activation of enzymes and other physiologically important proteins by specific proteolysis is a strategy frequently exploited by biological systems to switch on processes at the appropriate time and place, as the following examples illustrate.

Insulin. Some protein hormones are synthesized in the form of inactive precursor molecules, from which the active hormone is derived by proteolysis. For instance, **insulin,** an important metabolic regulator, is generated by proteolytic excision of a specific peptide from **proinsulin** (Figure 12.3).

Proteolytic Enzymes of the Digestive Tract. Enzymes of the digestive tract that serve to hydrolyze dietary proteins are synthesized in the stomach and pancreas as zymogens (Table 12.1). Only upon proteolytic activation are these enzymes able to form a catalytically active substrate-binding site. The activation of chymotrypsinogen is an interesting example (Figure 12.4). **Chymotrypsinogen** is a 245-residue polypeptide chain cross-linked by five disulfide bonds. Chymotrypsinogen is converted to an enzymatically active form called π-chymotrypsin when trypsin cleaves the peptide bond joining Arg^{15} and Ile^{16}. The enzymatically active π-chymotrypsin acts upon other π-chymotrypsin molecules, excising two dipeptides, Ser^{14}-Arg^{15} and Thr^{147}-Asn^{148}. The end product of this processing pathway is the mature protease **α-chymotrypsin,** in which the three peptide chains, A (residues 1 through 13), B (residues 16 through 146), and C (residues 149 through 245), remain together because they are linked by two disulfide bonds, one from A to B, and one from B to C. It is interesting to note that the transformation of inactive chymotrypsinogen to active π-chymotrypsin only requires the cleavage of a single specific peptide bond.

Blood Clotting. The formation of blood clots is the result of a series of zymogen activations (Figure 12.5). The amplification achieved by this cascade of

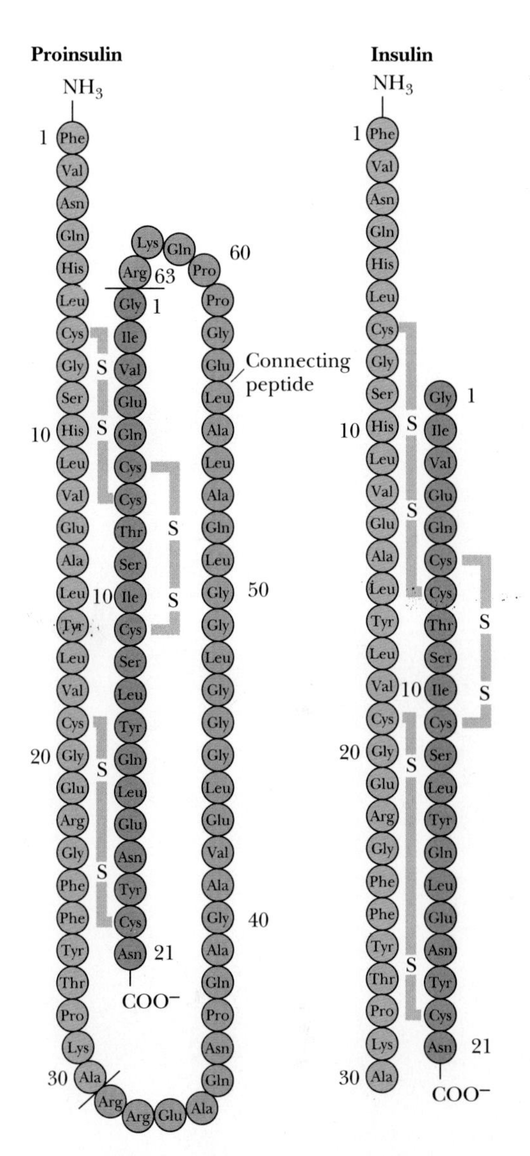

***Figure* 12.3** Proinsulin is an 84-residue precursor to insulin (the sequence shown is for porcine proinsulin). Proteolytic removal of residues 31 through 63 yields insulin. Residues 1 through 30 (the B-chain) remain linked to residues 64 through 84 (the A-chain) by a pair of interchain disulfide bridges.

***Table* 12.1**
Pancreatic and Gastric Zymogens

Origin	Zymogen	Active Protease
Pancreas	Trypsinogen	Trypsin
Pancreas	Chymotrypsinogen	Chymotrypsin
Pancreas	Procarboxypeptidase	Carboxypeptidase
Pancreas	Proelastase	Elastase
Stomach	Pepsinogen	Pepsin

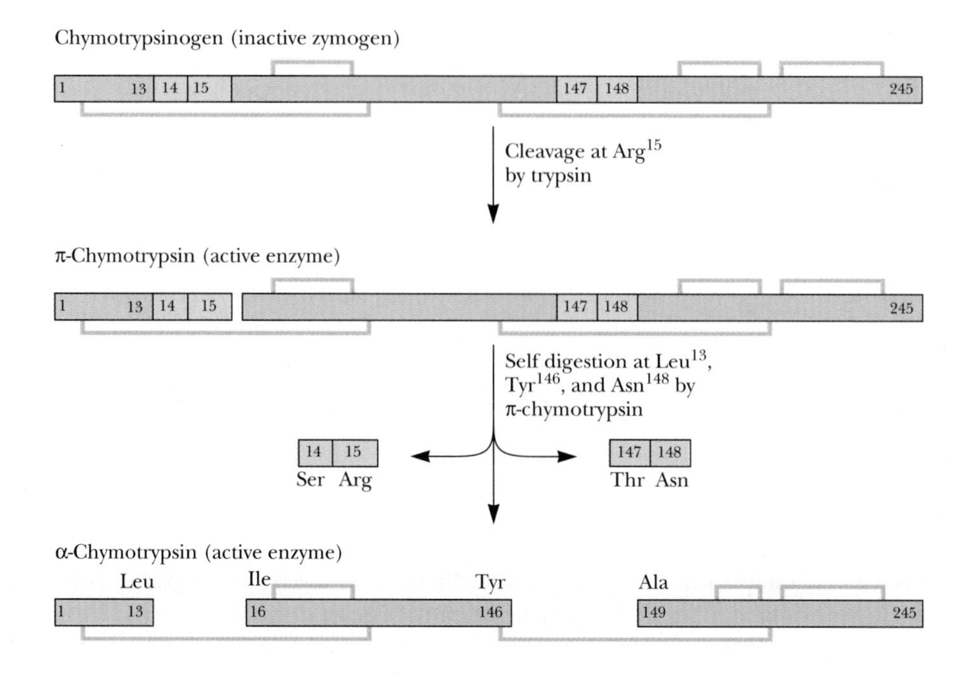

***Figure* 12.4** The proteolytic activation of chymotrypsinogen.

enzymatic activations allows blood clotting to occur rapidly in response to injury. Seven of the clotting factors in their active form are *serine proteases:* **kallikrein, XII$_a$, XI$_a$, IX$_a$, VII$_a$, X$_a$,** and **thrombin.** Two routes to blood clot formation exist. The **intrinsic pathway** is instigated when the blood comes into physical contact with abnormal surfaces caused by injury; the **extrinsic pathway** is initiated by factors released from injured tissues. The pathways merge at Factor X and culminate in clot formation. Thrombin excises peptides rich in negative charge from **fibrinogen,** converting it to **fibrin,** a molecule with a different surface charge distribution. Fibrin then readily aggregates into ordered fibrous arrays that are subsequently stabilized by covalent cross-links. Thrombin specifically cleaves Arg-Gly peptide bonds and is homologous to trypsin, which is also a serine protease (recall that trypsin acts only at Arg and Lys residues).

Isozymes

Other enzymes exist in more than one quaternary form, depending on their relative proportions of structurally equivalent but catalytically distinct polypeptide subunits. A classic example is mammalian *lactate dehydrogenase (LDH),* which exists as five different isozymes, depending on the tetrameric association of two different subunits, A and B: A_4, A_3B, A_2B_2, AB_3, and B_4 (Figure 12.6). The kinetic properties of the various LDH isozymes differ in terms of their relative affinities for the various substrates and their sensitivity to inhibition by product. Different tissues express different isozyme forms, as appropriate to their particular metabolic needs. By regulating the relative amounts of A and B subunits they synthesize, the cells of various tissues control which isozymic forms are likely to assemble, and thus, which kinetic parameters prevail.

Modulator Proteins

Modulator proteins are yet another way that cells mediate metabolic activity. **Modulator proteins** are proteins that bind to enzymes, and by binding, influence the activity of the enzyme. For example, some enzymes, such as *cAMP-*

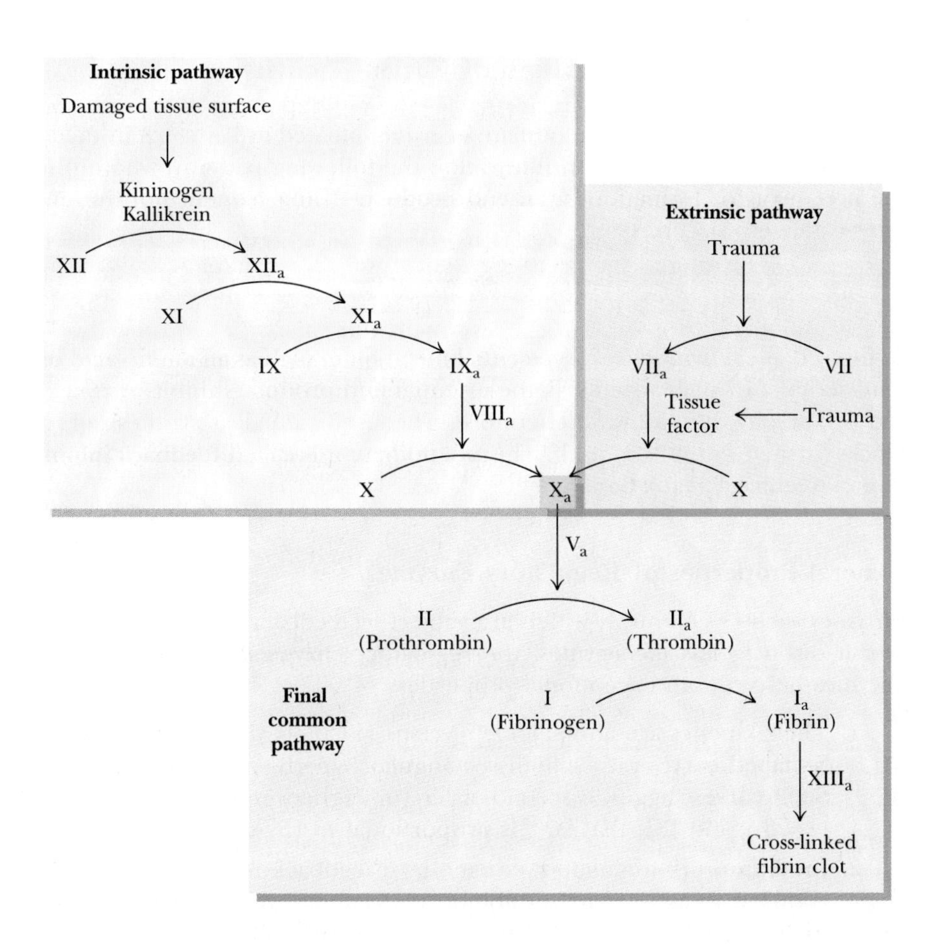

***Figure* 12.5** The cascade of activation steps leading to blood clotting. The intrinsic and extrinsic pathways converge at Factor X, and the final common pathway involves the activation of thrombin and its conversion of fibrinogen into fibrin, which aggregates into ordered filamentous arrays that become cross-linked to form the clot.

dependent protein kinase (Chapter 21), exist as dimers of catalytic subunits and regulatory subunits. These regulatory subunits are *modulator proteins* that suppress the activity of the catalytic subunits. Dissociation of the regulatory subunit (modulator protein) activates the catalytic subunits; reassociation once again suppresses activity. We will meet important representatives of this class as the processes of metabolism unfold in subsequent chapters. For now, let us focus our attention on the fascinating kinetics of allosteric enzymes.

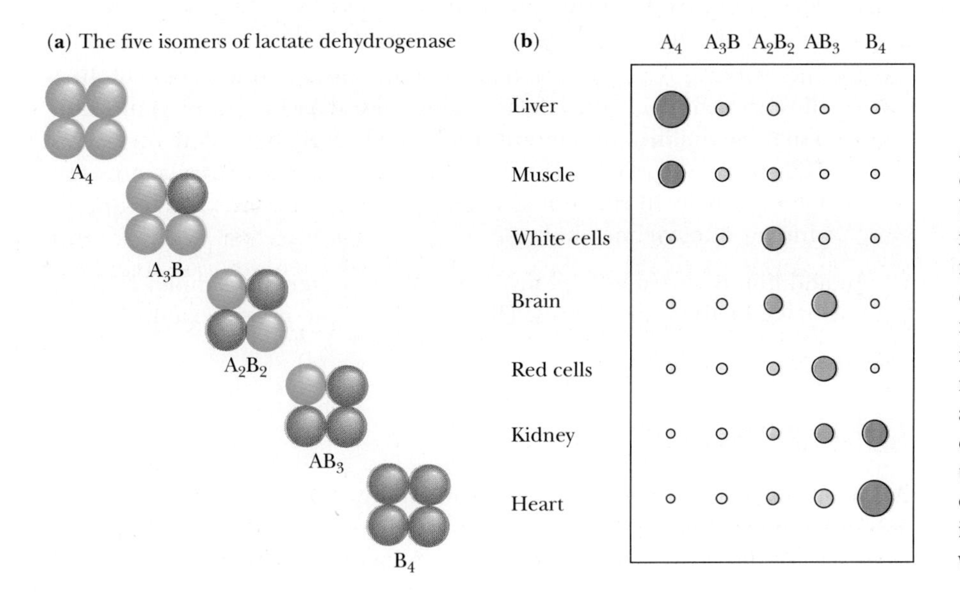

***Figure* 12.6** The isozymes of lactate dehydrogenase (LDH). Active muscle tissue becomes anaerobic and produces pyruvate from glucose via glycolysis. It needs LDH to regenerate NAD^+ from NADH so glycolysis can continue. The lactate produced is released into the blood. The muscle LDH isozyme (A_4) works best in the NAD^+-regenerating direction. Heart tissue is aerobic and uses lactate as a fuel, converting it to pyruvate via LDH and using the pyruvate to fuel the citric acid cycle to obtain energy. The heart LDH isozyme (B_4) is inhibited by excess pyruvate so the fuel won't be wasted.

12.3 The Allosteric Regulation of Enzyme Activity

Allosteric regulation acts to modulate enzymes situated at key steps in metabolic pathways. Consider as an illustration the following pathway, where A is the precursor for formation of an end product, F, in a sequence of five enzyme-catalyzed reactions:

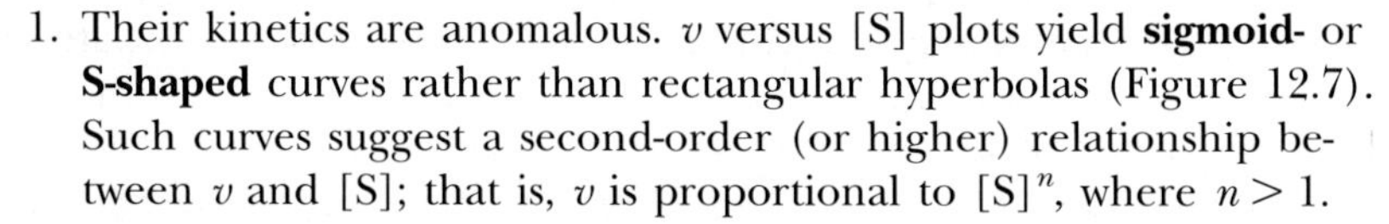

In this scheme, F symbolizes an essential metabolite, such as an amino acid or a nucleotide. In such systems, F, the essential end product, inhibits *enzyme 1,* the *first step* in the pathway. Therefore, when sufficient F is synthesized, it blocks further synthesis of itself. This phenomenon is called **feedback inhibition** or **feedback regulation.**

General Properties of Regulatory Enzymes

Enzymes such as enzyme 1, which are subject to feedback regulation, represent a distinct class of enzymes, the **regulatory enzymes.** As a class, these enzymes have certain exceptional properties:

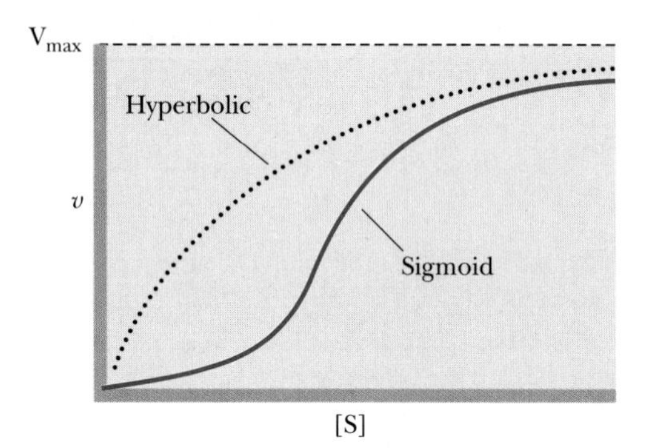

***Figure* 12.7** Sigmoid v versus [S] plot. The dotted line represents the hyperbolic plot characteristic of normal Michaelis–Menten-type enzyme kinetics.

1. Their kinetics are anomalous. v versus [S] plots yield **sigmoid-** or **S-shaped** curves rather than rectangular hyperbolas (Figure 12.7). Such curves suggest a second-order (or higher) relationship between v and [S]; that is, v is proportional to $[S]^n$, where $n > 1$.
2. Inhibition of a regulatory enzyme by a feedback inhibitor does not conform to any normal inhibition pattern, and the feedback inhibitor F bears little structural similarity to A, the substrate for the regulatory enzyme. F apparently acts at a binding site distinct from the substrate-binding site. The term *allosteric* is apt, since F is sterically dissimilar and, moreover, acts at a site other than the site for S. Its effect is called **allosteric inhibition.**
3. Regulatory or allosteric enzymes like enzyme 1 are, in some instances, regulated by activation. That is, whereas some effector molecules such as F exert negative effects on enzyme activity, other effectors show stimulatory, or positive, influences on activity.
4. Allosteric enzymes have an oligomeric organization. They are composed of more than one polypeptide chain (subunit) and have more than one S-binding site per enzyme molecule.
5. The working hypothesis is that, by some means, interaction of an allosteric enzyme with effectors alters the distribution of conformational possibilities or subunit interactions available to the enzyme. That is, the regulatory effects exerted on the enzyme's activity are achieved by conformational changes wrought in the protein upon binding effector metabolites.

In addition to enzymes, various noncatalytic proteins of simpler function exhibit many of these properties. Hemoglobin is the classic example.

12.4 Hemoglobin and Myoglobin

Ancient life forms evolved in the absence of oxygen and were capable only of anaerobic metabolism. As the earth's atmosphere changed over time, so too did living things. Indeed, the biological production of O_2 by photosynthesis

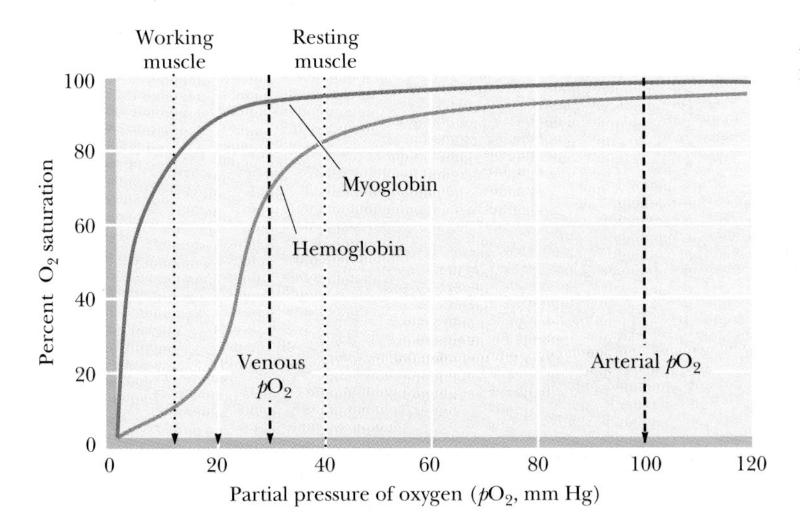

***Figure* 12.8** O_2-binding curves for hemoglobin and myoglobin.

was a major factor in altering the atmosphere! Evolution to an oxygen-based metabolism was highly beneficial. Aerobic metabolism of sugars, for example, yields far more energy than corresponding anaerobic processes. In order that metabolic processes not be limited by the low solubility of O_2 in water, animals evolved two important oxygen-binding proteins: **hemoglobin** in blood and **myoglobin** in muscle. Because hemoglobin and myoglobin are two of the most-studied proteins in nature, they have become paradigms of protein structure and function. Moreover, hemoglobin is a model for protein quaternary structure and allosteric function. The binding of O_2 by hemoglobin and its modulation by protons, CO_2, and 2,3-bisphosphoglycerate depend in exquisite ways on the interactions of polypeptide subunits in a tetrameric protein, revealing much about the functional significance of quaternary associations and allosteric regulation.

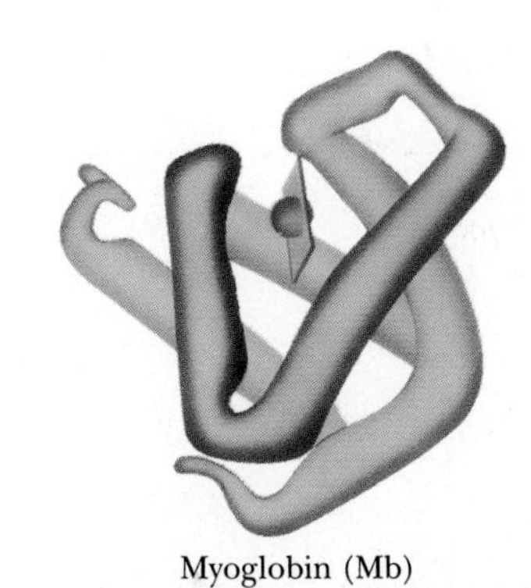

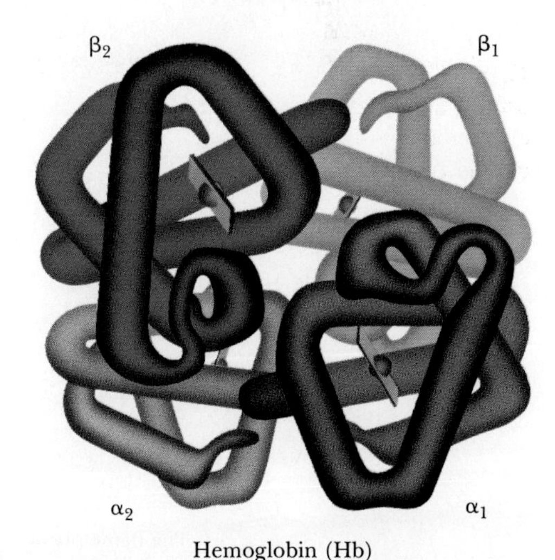

***Figure* 12.9** The myoglobin and hemoglobin molecules. *Myoglobin* (sperm whale): one polypeptide chain of 153 aa residues (mass = 17.2 kD) has one heme (mass = 652 D) and binds one O_2. *Hemoglobin* (human): four polypeptide chains, two of 141 aa residues (α) and two of 146 residues (β); mass = 64.45 kD. Each polypeptide has a heme; the Hb tetramer binds four O_2.

The Comparative Biochemistry of Myoglobin and Hemoglobin

A comparison of the properties of hemoglobin and myoglobin offers insights into allosteric phenomena, even though these proteins are *not* enzymes. Hemoglobin displays sigmoid-shaped O_2-binding curves (Figure 12.8). The unusual shape of these curves was once a great enigma in biochemistry. Such curves closely resemble those seen in allosteric enzyme: substrate saturation graphs (see Figure 12.7). In contrast, myoglobin's interaction with oxygen obeys classical Michaelis–Menten-type substrate saturation behavior.

Before examining myoglobin and hemoglobin in detail, let us first encapsulate the lesson: Myoglobin (symbolized as Mb) is a compact globular protein composed of a single polypeptide chain 153 amino acids in length; its molecular mass is 17.2 kD (Figure 12.9). It contains **heme,** a porphyrin ring system complexing an iron ion, as its prosthetic group (see Figure 4.14). Oxygen binds to Mb via its heme. Hemoglobin (Hb) is also a compact globular protein, but Hb is a tetramer. It consists of four polypeptide chains, each of which is very similar structurally to the myoglobin polypeptide chain, and each bears a heme group. Thus, a hemoglobin molecule can bind four O_2 molecules. In adult human Hb, there are two identical chains of 141 amino acids, the α-chains, and two identical β-chains, each of 146 residues. The human Hb molecule is an $\alpha_2\beta_2$-type tetramer of molecular mass 64.45 kD. The tetrameric nature of Hb is crucial to its biological function: *When a molecule of O_2 binds to a heme in Hb, the heme Fe ion is drawn into the plane of the porphyrin ring. This slight movement sets off a chain of conformational events that are transmitted to*

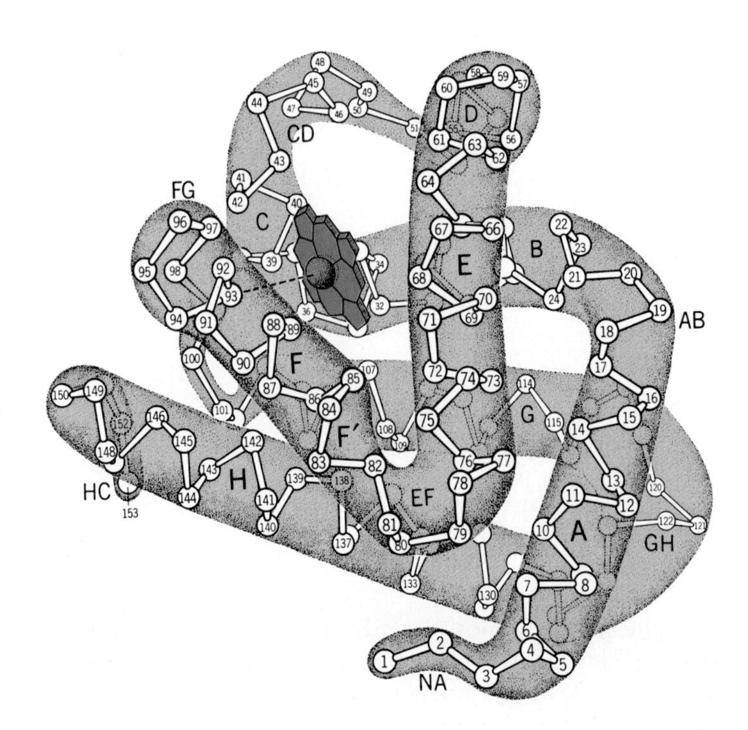

***Figure* 12.10** Detailed structure of the myoglobin molecule. The myoglobin polypeptide chain consists of eight helical segments, designated by the letters A through H, counting from the N-terminus. These helices, ranging in length from 7 to 26 residues, are linked by short, unordered regions that are named for the helices they connect, as in the AB region, or the EF region. The individual amino acids in the polypeptide are indicated according to their position within the various segments, as in His F8, the eighth residue in helix F, or Phe CD1, the first amino acid in the interhelical CD region. Occasionally, amino acids are specified in the conventional way, that is, by the relative position in the chain, as in Gly^{153}. The heme group is cradled within the folded polypeptide chain.

adjacent subunits, dramatically enhancing the affinity of their heme groups for O_2. That is, the binding of O_2 to one heme of Hb makes it easier for the Hb molecule to bind additional equivalents of O_2. This sort of binding phenomenon is termed **cooperativity,** or a **positive cooperative effect.** Hemoglobin is a marvelously constructed molecular machine. Let us dissect its mechanism, beginning with its monomeric counterpart, the myoglobin molecule.

Myoglobin

Myoglobin is the oxygen-storage protein of muscle. The muscles of diving mammals such as seals and whales are especially rich in this protein, which serves as a store for O_2 during the animal's prolonged periods underwater. Myoglobin is abundant in skeletal and cardiac muscle of nondiving animals as well. Myoglobin is the cause of the characteristic red color of muscle.

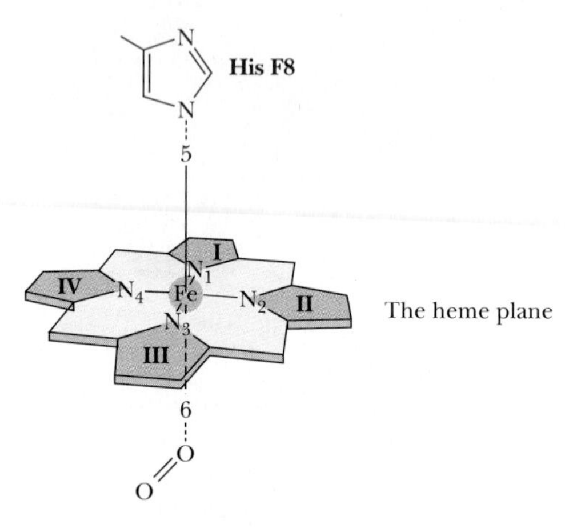

***Figure* 12.11** The six liganding positions of an iron ion. Four ligands lie in the same plane; the remaining two are, respectively, above and below this plane. In myoglobin, His F8 is the fifth ligand; in oxymyoglobin, O_2 becomes the sixth.

The Mb Polypeptide Cradles the Heme Group

The myoglobin polypeptide chain is folded to form a cradle (4.4 × 4.4 × 2.5 nm) that nestles the heme prosthetic group (Figure 12.10). O_2-binding depends on the heme's oxidation state. The iron ion in the heme of myoglobin is in the +2 oxidation state, that is, the *ferrous* form. This is the form that binds O_2. Oxidation of the ferrous form to the +3 *ferric* form yields **metmyoglobin,** which will not bind O_2. It is interesting to note that free heme in solution will readily interact with O_2 also, but the oxygen quickly oxidizes the iron atom to the ferric state. Fe^{3+} : protoporphyrin IX is referred to as **hematin.** Thus, the polypeptide of myoglobin may be viewed as serving three critical functions: It cradles the heme group, it protects the heme iron atom from oxidation, and it provides a pocket into which the O_2 can fit.

O_2-Binding to Mb

Iron ions, whether ferrous or ferric, prefer to interact with six ligands, four of which share a common plane. The fifth and sixth ligands lie above and below

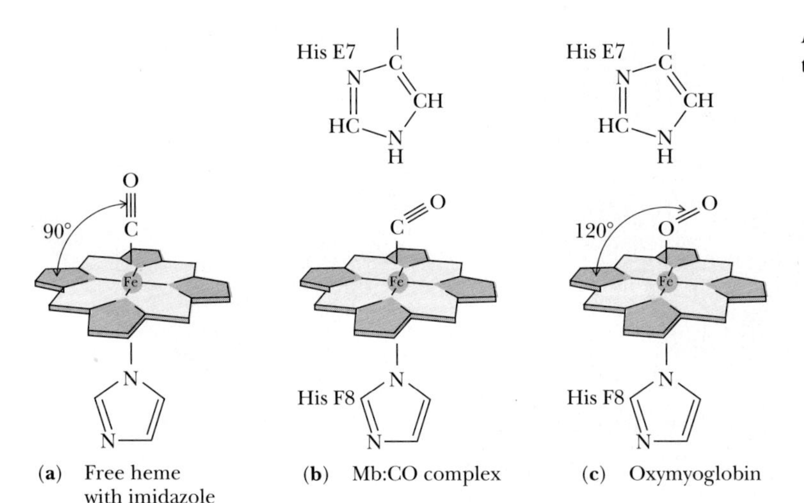

***Figure* 12.12** Oxygen and carbon monoxide binding to the heme group of myoglobin.

this plane (see Figure 12.11). In heme, four of the ligands are provided by the nitrogen atoms of the four pyrroles. A fifth ligand is donated by the imidazole side chain of amino acid residue His F8. When myoglobin binds O_2 to become **oxymyoglobin,** the O_2 molecule adds to the heme iron ion as the sixth ligand (Figure 12.11). O_2 adds end on to the heme iron, but it is not oriented perpendicular to the plane of the heme. Rather, it is tilted about 60° with respect to the perpendicular. In **deoxymyoglobin**, the sixth ligand position is vacant, and in metmyoglobin, a water molecule fills the O_2 site and becomes the sixth ligand for the ferric atom. On the oxygen-binding side of the heme lies another histidine residue, His E7. While its imidazole function lies too far away to interact with the Fe atom, it is close enough to contact the O_2. Therefore, the O_2-binding site is a sterically hindered region. Biologically important properties stem from this hindrance. For example, the affinity of *free heme* in solution for carbon monoxide (CO) is 25,000 times greater than its affinity for O_2. But CO only binds 250 times more tightly than O_2 to the heme of myoglobin, because His E7 forces the CO molecule to tilt away from a preferred perpendicular alignment with the plane of the heme (Figure 12.12). This diminished affinity of myoglobin for CO guards against the possibility that traces of CO produced during metabolism might occupy all of the heme sites, effectively preventing O_2 from binding. Nevertheless, carbon monoxide in the environment is a potent poison, causing death by asphyxiation by precisely this mechanism.

O_2-Binding Alters Mb Conformation

What happens when the heme group of myoglobin binds oxygen? X-ray crystallography has revealed that a crucial change occurs in the position of the iron atom relative to the plane of the heme. In deoxymyoglobin, the ferrous ion has but five ligands, and it lies 0.055 nm above the plane of the heme, in the direction of His F8. The iron:porphyrin complex is therefore dome-shaped. When O_2 binds, the iron atom is pulled back toward the porphyrin plane and is now displaced from it by only 0.026 nm (Figure 12.13). The consequences of this small motion are trivial as far as the biological role of myoglobin is concerned. However, as we shall soon see, this slight movement profoundly affects the properties of hemoglobin. Its action on His F8 is magnified through changes in polypeptide conformation that alter subunit interactions in the Hb tetramer. These changes in subunit relationships are the fundamental cause for the allosteric properties of hemoglobin.

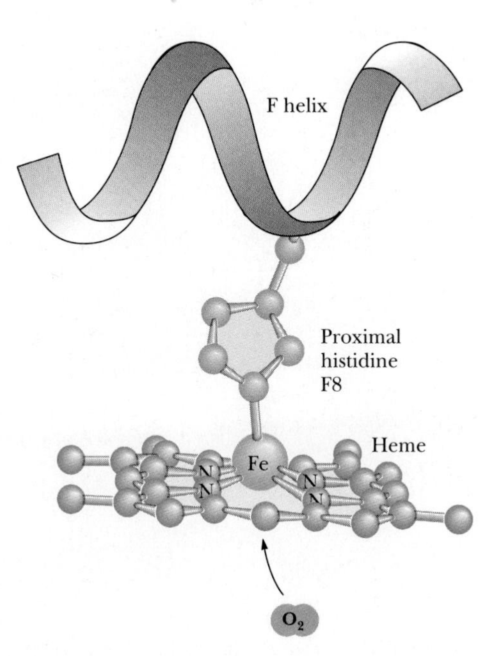

***Figure* 12.13** The displacement of the Fe ion of the heme of deoxymyoglobin from the plane of the porphyrin ring system by the pull of His F8. In oxymyoglobin, the bound O_2 counteracts this effect.

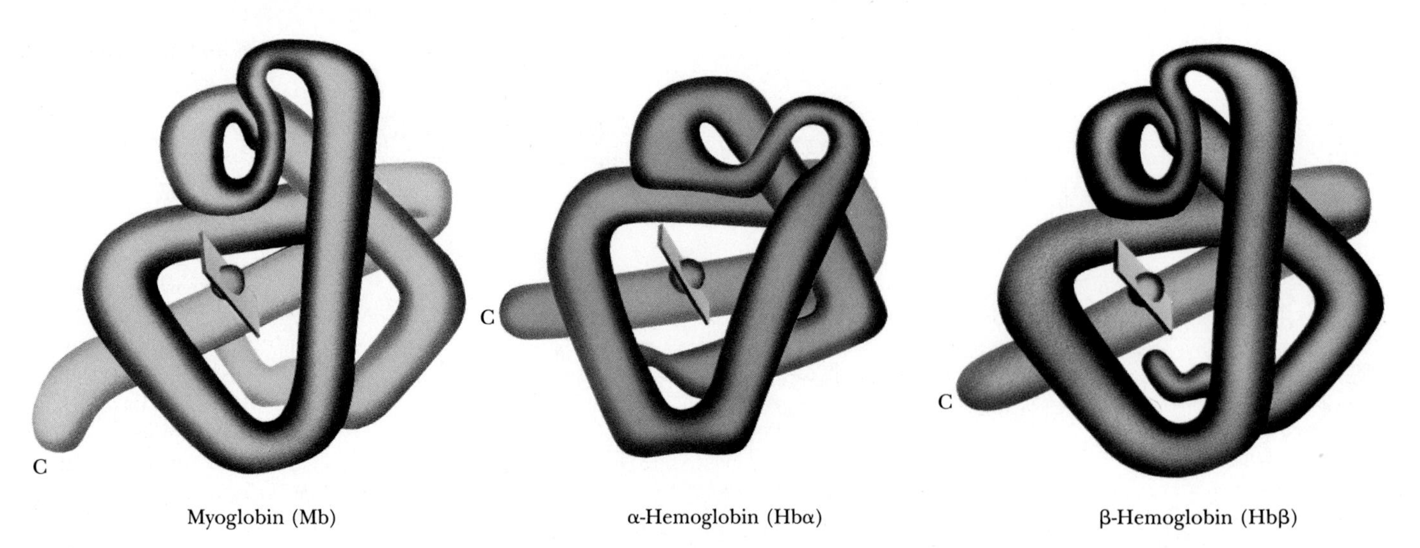

***Figure* 12.14** Conformational drawings of the α- and β-chains of Hb and the myoglobin chain.

The Physiological Significance of Cooperative Binding of Oxygen by Hemoglobin

The oxygen-binding equations for myoglobin and hemoglobin are described in detail in an appendix at the end of this chapter. The relative oxygen affinities of hemoglobin and myoglobin reflect their respective physiological roles (see Figure 12.8). Myoglobin, as an oxygen storage protein, has a greater affinity for O_2 than hemoglobin at all oxygen pressures. Hemoglobin, as the oxygen carrier, becomes saturated with O_2 in the lungs, where the partial pressure of O_2 (pO_2) is about 100 torr.[1] In the capillaries of tissues, pO_2 is typically 40 torr, and oxygen is released from Hb. In muscle, some of it can be bound by myoglobin, to be stored for use in times of severe oxygen deprivation, such as during strenuous exercise.

The Structure of the Hemoglobin Molecule

As noted, hemoglobin is an $\alpha_2\beta_2$ tetramer. Each of the four subunits has a conformation virtually identical to that of myoglobin. Two different types of subunits, α and β, are necessary to achieve cooperative O_2-binding by Hb. The β-chain at 146 amino acid residues is shorter than the myoglobin chain (153 residues), mainly because its final helical segment (the H helix) is shorter. The α-chain (141 residues) also has a shortened H helix and lacks the D helix as well (Figure 12.14). Max Perutz, who has devoted his life to elucidating the atomic structure of Hb, noted very early in his studies that the molecule was highly symmetrical. The actual arrangement of the four subunits with respect to one another is shown in Figure 12.15 for horse methemoglobin. All vertebrate hemoglobins show a three-dimensional structure essentially the same as this. The subunits pack in a tetrahedral array, creating a roughly spherical molecule 6.4 × 5.5 × 5.0 nm. The four heme groups, nestled within the easily recognizable cleft formed between the E and F helices of each polypeptide, are exposed at the surface of the molecule. The heme groups are quite far apart; 2.5 nm separates the closest iron ions, those of hemes α_1 and β_2, and those of hemes α_2 and β_1. The subunit interactions are mostly between dissimilar chains: Each of the α-chains is in contact with both β-chains, but there are few α–α or β–β interactions.

[1] The **torr** is a unit of pressure named for Torricelli, inventor of the barometer. 1 torr corresponds to 1 mm Hg (1/760th of an atmosphere).

A Deeper Look

The Physiological Significance of the Hb : O_2 Interaction

We can determine quantitatively the physiological significance of the sigmoid nature of the hemoglobin oxygen-binding curve, or, in other words, the biological importance of cooperativity. The equation

$$\frac{Y}{(1 - Y)} = \left(\frac{pO_2}{P_{50}}\right)^n$$

describes the relationship between pO_2, the affinity of hemoglobin for O_2 (defined as P_{50}, the partial pressure of O_2 giving half-maximal saturation of Hb with O_2), and the fraction of hemoglobin with O_2 bound, Y, versus the fraction of Hb with no O_2 bound, $(1 - Y)$ (see Appendix Equation [A12.15]). The coefficient n is the *Hill coefficient,* an index of the cooperativity (sigmoidicity) of the hemoglobin oxygen-binding curve (see Appendix for details). Taking pO_2 in the lungs as 100 torr, P_{50} as 26 torr, and n as 2.8, Y, the fractional saturation of the hemoglobin heme groups with O_2, is 0.98. If pO_2 were to fall to 10 torr within the capillaries of an exercising muscle, Y would drop to 0.06. The oxygen delivered under these conditions would be proportional to the difference, $Y_{\text{lungs}} - Y_{\text{muscle}}$, which is 0.92. That is, virtually all the oxygen carried by Hb would be released. Suppose instead that hemoglobin binding of O_2 were not cooperative; in that case, the hemoglobin oxygen-binding curve would be hyperbolic, and $n = 1.0$. Then Y in the lungs would be 0.79 and Y in the capillaries, 0.28, and the difference in Y values would be 0.51. Thus, under these conditions, the cooperativity of oxygen binding by Hb means that 0.92/0.51 or 1.8 times as much O_2 can be delivered.

Oxygenation Markedly Alters the Quaternary Structure of Hemoglobin

Crystals of deoxyhemoglobin shatter when exposed to O_2. Further, X-ray crystallographic analysis reveals that oxy- and deoxyhemoglobin differ markedly in quaternary structure. In particular, specific $\alpha\beta$-subunit interactions change. The $\alpha\beta$ contacts are of two kinds. The $\alpha_1\beta_1$ and $\alpha_2\beta_2$ contacts involve helices B, G, and H and the GH corner. These contacts are extensive and important to subunit packing; they remain unchanged when hemoglobin goes from its deoxy to its oxy form. The $\alpha_1\beta_2$ and $\alpha_2\beta_1$ contacts are called **sliding contacts.** They principally involve helices C and G and the FG corner

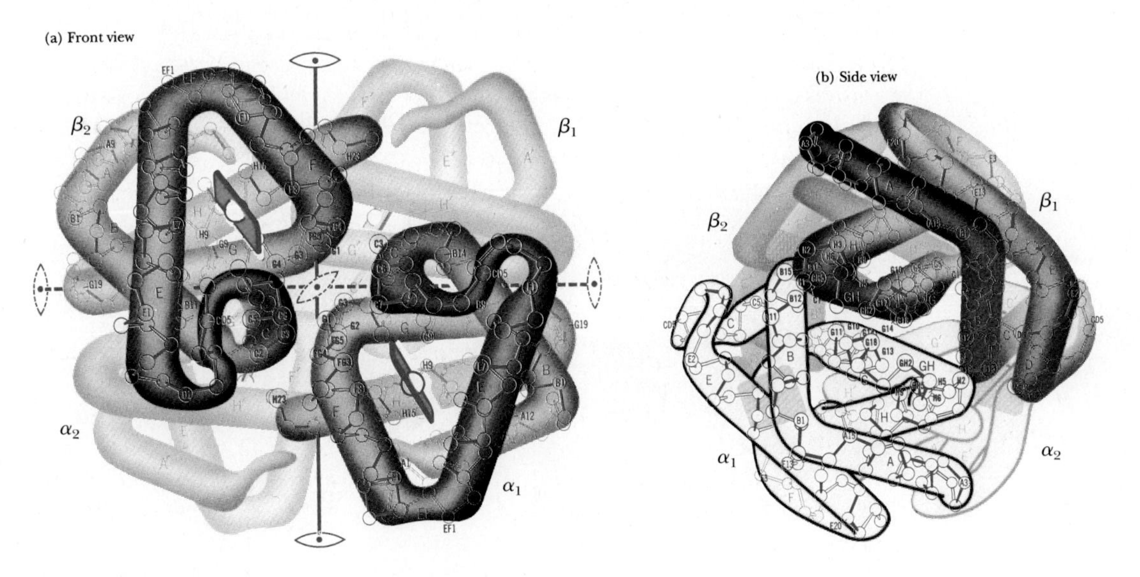

Figure **12.15** The arrangement of subunits in horse methemoglobin, the first hemoglobin whose structure was determined by X-ray diffraction. The iron atoms on metHb are in the oxidized, ferric (Fe^{3+}) state.

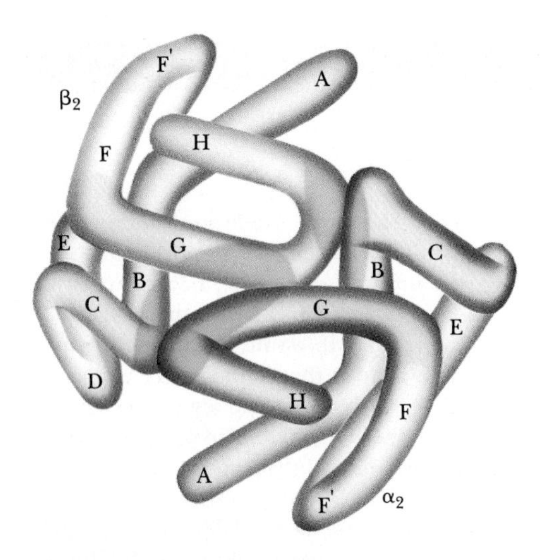

Figure **12.16** Side view of one of the two $\alpha\beta$ dimers in Hb, with packing contacts indicated in blue. The sliding contacts made with the other dimer are shown in yellow. The changes in these sliding contacts are shown in Figure 12.17.

(Figure 12.16). When hemoglobin undergoes a conformational change as a result of ligand binding to the heme, these contacts are altered (Figure 12.17). Hemoglobin, as a conformationally dynamic molecule, consists of two dimeric halves, an $\alpha_1\beta_1$-subunit pair and an $\alpha_2\beta_2$-subunit pair. Each $\alpha\beta$ dimer moves as a rigid body, and the two halves of the molecule slide past each other upon oxygenation of the heme. The two halves rotate some 15° about an imaginary pivot passing through the $\alpha\beta$-subunits; some atoms at the interface between $\alpha\beta$ dimers are relocated by as much as 0.6 nm.

Movement of the Heme Iron by Less Than 0.04 nm Induces the Conformational Change in Hemoglobin

In deoxyhemoglobin, histidine F8 is ligated to the heme iron ion, but steric constraints force the Fe^{2+}:His-N bond to be tilted about 8° from the perpendicular to the plane of the heme. Steric repulsion between histidine F8 and the nitrogen atoms of the porphyrin ring system, combined with electrostatic repulsions between the electrons of Fe^{2+} and the porphyrin π-electrons, forces the iron atom to lie out of the porphyrin plane by about 0.06 nm. Changes in electronic and steric factors upon heme oxygenation allow the Fe^{2+} atom to move about 0.039 nm closer to the plane of the porphyrin, so now it is displaced only 0.021 nm above the plane. It is as if the O_2 were drawing the heme Fe^{2+} into the porphyrin plane (Figure 12.18). This modest displacement of 0.039 nm seems a trivial distance, but its biological consequences are far-reaching. As the iron atom moves, it drags histidine F8 along with it, causing helix F, the EF corner, and the FG corner to follow. These shifts are transmitted to the subunit interfaces, where they trigger conformational readjustments that lead to the rupture of interchain salt links.

The Oxy and Deoxy Forms of Hemoglobin Represent Two Different Conformational States

Hemoglobin resists oxygenation (see Figure 12.8) because the deoxy form is stabilized by specific hydrogen bonds and salt bridges (ion-pair bonds). All of these interactions are broken in oxyhemoglobin, as the molecule stabilizes into a new conformation. A crucial H bond in this transition involves a particular tyrosine residue. Both α- and β-subunits have Tyr as the penultimate

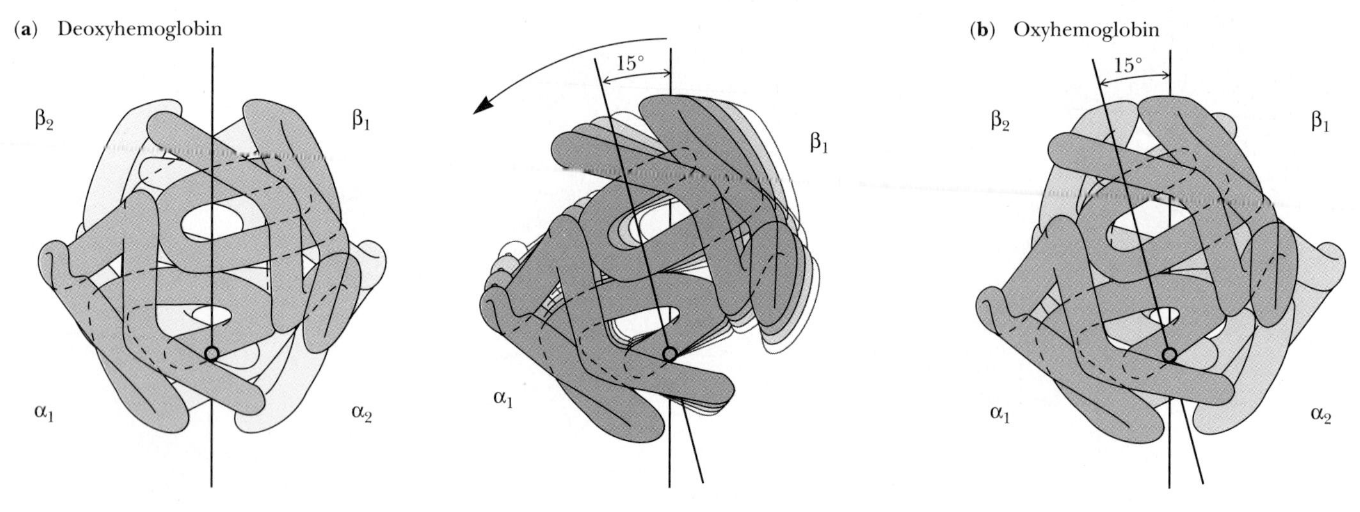

Figure **12.17** Subunit motion in hemoglobin when the molecule goes from the (a) deoxy to the (b) oxy form.

A Deeper Look

Changes in the Heme Iron Upon O_2-Binding

In deoxyhemoglobin, the six *d* electrons of the heme Fe^{2+} exist as four unpaired electrons and one electron pair, and five ligands can be accommodated: the four N-atoms of the porphyrin ring system and histidine F8. In this electronic configuration, the iron atom is paramagnetic and in the **high-spin state.** When the heme binds O_2 as a sixth ligand, these electrons are rearranged into three e^- pairs and the iron changes to the **low-spin state** and is diamagnetic. This change in spin state allows the bond between the Fe^{2+} ion and histidine F8 to become perpendicular to the heme plane and to shorten. In addition, interactions between the porphyrin N-atoms and the iron strengthen. Also, high-spin Fe^{2+} has a greater atomic volume than low-spin Fe^{2+} because its four unpaired e^- occupy four orbitals rather than two when the electrons are paired in low-spin Fe^{2+}. So, low-spin iron is less sterically hindered and able to move nearer to the porphyrin plane.

C-terminal residue (Tyr α140 = Tyr HC2; Tyr β145 = Tyr HC2, respectively[2]). The phenolic —OH groups of these Tyr residues form intrachain H bonds to the peptide C=O function contributed by Val FG5 in deoxyhemoglobin. (Val FG5 is α93 and β98, respectively.) The shift in helix F upon oxygenation leads to rupture of this Tyr HC2:Val FG5 hydrogen bond. Further, eight salt bridges linking the polypeptide chains are broken as hemoglobin goes from the deoxy to the oxy form (Figure 12.19). Six of these salt links are between different subunits. Four of these six involve either carboxyl-terminal or amino-terminal amino acids in the chains; two are between the amino termini and the carboxyl termini of the α-chains, and two join the carboxyl termini of the β-chains to the ϵ-NH_2 groups of the two Lys α140 residues. The other two interchain electrostatic bonds link Arg and Asp residues in the two α-chains. In addition, ionic interactions between Asp β94 and His β146 form an intrachain salt bridge in each β-subunit. In deoxyhemoglobin, with all of these

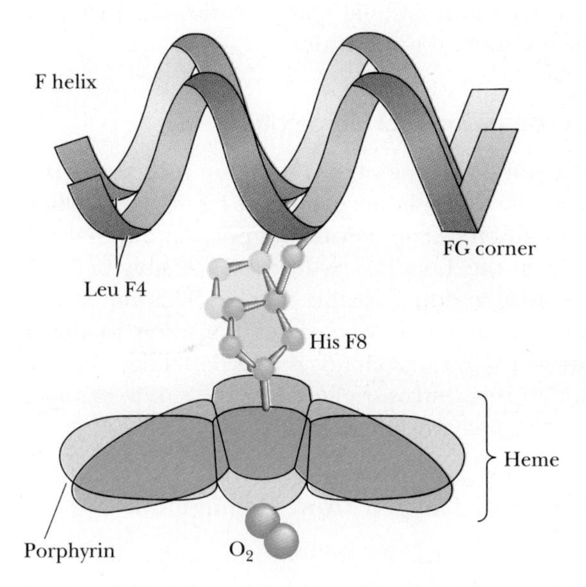

Figure **12.18** Changes in the position of the heme iron atom upon oxygenation lead to conformational changes in the hemoglobin molecule.

[2]C here designates the C-terminus; the H helix is C-terminal in these polypeptides. "C2" symbolizes the next-to-last residue.

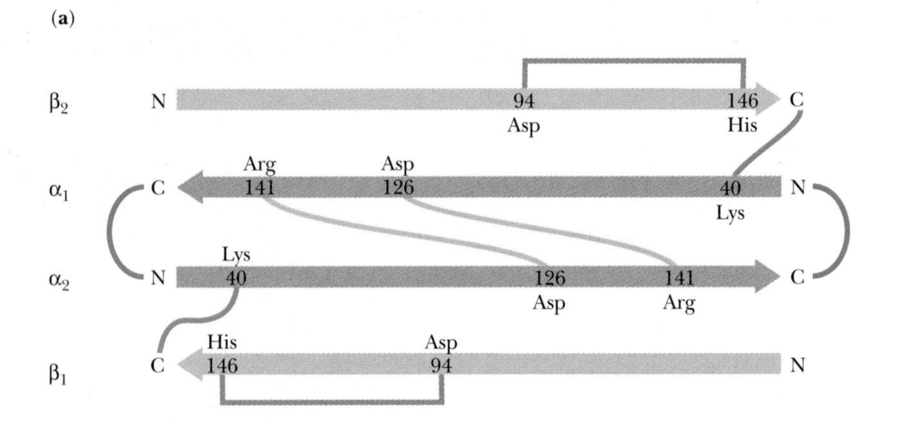

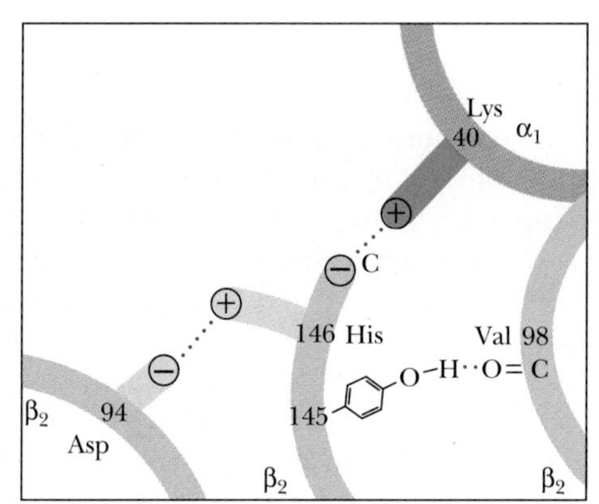

***Figure* 12.19** Salt bridges between different subunits in hemoglobin. These noncovalent, electrostatic interactions are disrupted upon oxygenation. Argα141 and Hisβ146 are the C-termini of the α- and β-polypeptide chains. (a) The various intrachain and interchain salt links formed among the α- and β-chains of deoxyhemoglobin. (b) A focus on those salt bridges and hydrogen bonds involving interactions between N-terminal and C-terminal residues in the α-chains. Note the Cl^- ion, which bridges ionic interactions between the N-terminus of α_2 and the R group of Arg$\alpha_1$141. (c) A focus on the salt bridges and hydrogen bonds in which the residues located at the C-termini of β-chains are involved. All of these links are abolished in the deoxy to oxy transition.

interactions intact, the C-termini of the four subunits are restrained, and this conformational state is termed **T,** the **tense** or **taut form.** In oxyhemoglobin, these C-termini have almost complete freedom of rotation, and the molecule is now in its **R,** or **relaxed, form.** These terms, R and T, are general designations used to describe alternative conformational states for allosteric proteins, the R form having the greater affinity for substrate.

A Model for the Allosteric Behavior of Hemoglobin

A model for the allosteric behavior of hemoglobin is based on recent observations that oxygen is accessible only to the heme groups of the α-chains when hemoglobin is in the T conformational state. Perutz has pointed out that the heme environment of β-chains in the T state is virtually inaccessible because of steric hindrance by amino acid residues in the E helix. This hindrance disappears when the hemoglobin molecule undergoes transition to the R conformational state. Binding of O_2 to the β-chains is thus dependent on the T to R conformational shift, and this shift is triggered by the subtle changes that ensue when O_2 binds to the α-chain heme groups.

H^+ Promotes the Dissociation of Oxygen from Hemoglobin

Protons, carbon dioxide, chloride ions, as well as the metabolite *2,3-bisphosphoglycerate* (or *BPG*) all affect the binding of O_2 by hemoglobin. Their effects have interesting ramifications, which we shall see as we discuss them in turn. Deoxyhemoglobin has a higher affinity for protons than oxyhemoglo-

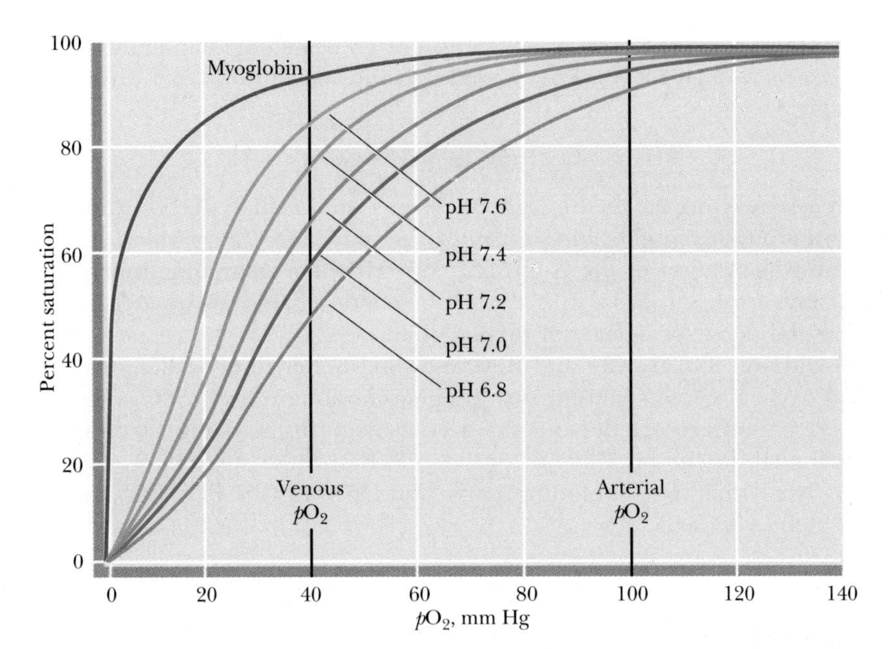

***Figure* 12.20** The oxygen saturation curves for myoglobin and for hemoglobin at five different pH values: 7.6, 7.4, 7.2, 7.0, and 6.8.

bin. Thus, as the pH decreases, dissociation of O_2 from hemoglobin is enhanced. In simple symbolism, ignoring the stoichiometry of O_2 or H^+ involved:

$$HbO_2 + H^+ \rightleftharpoons HbH^+ + O_2$$

Expressed another way, H^+ is an antagonist of oxygen binding by Hb, and the saturation curve of Hb for O_2 is displaced to the right as acidity increases (Figure 12.20). This phenomenon is called the **Bohr effect,** after its discoverer, the Danish physiologist Christian Bohr (the father of Niels Bohr, the atomic physicist). The effect has important physiological significance because actively metabolizing tissues produce acid, promoting O_2 release where it is most needed. About two protons are taken up by deoxyhemoglobin. The N-termini of the two α-chains and the His β146 residues have been implicated as the major players in the Bohr effect. (The pK_a of a free amino terminus in a protein is about 8.0, but the pK_a of a protein histidine imidazole is around 6.5.) Neighboring carboxylate groups of Asp β94 residues help to stabilize the protonated state of the His β146 imidazoles that occur in deoxyhemoglobin, but changes in the conformation of β-chains upon Hb oxygenation move the negative Asp function away, and the dissociation of the imidazole protons is favored.

CO_2 Also Promotes the Dissociation of O_2 from Hemoglobin

Carbon dioxide has an effect on O_2 binding by Hb that is similar to that of H^+, partly because it produces H^+ when it dissolves in the blood:

$$CO_2 + H_2O \xrightleftharpoons{\text{carbonic anhydrase}} \underset{\text{carbonic acid}}{H_2CO_3} \rightleftharpoons H^+ + \underset{\text{bicarbonate}}{HCO_3^-}$$

The enzyme *carbonic anhydrase* promotes the hydration of CO_2. Many of the protons formed upon ionization of carbonic acid are picked up by Hb as O_2 dissociates. The bicarbonate ions are transported with the blood back to the lungs. When Hb becomes oxygenated again in the lungs, H^+ is released and reacts with HCO_3^- to re-form H_2CO_3, from which CO_2 is liberated. The CO_2 is then exhaled as a gas.

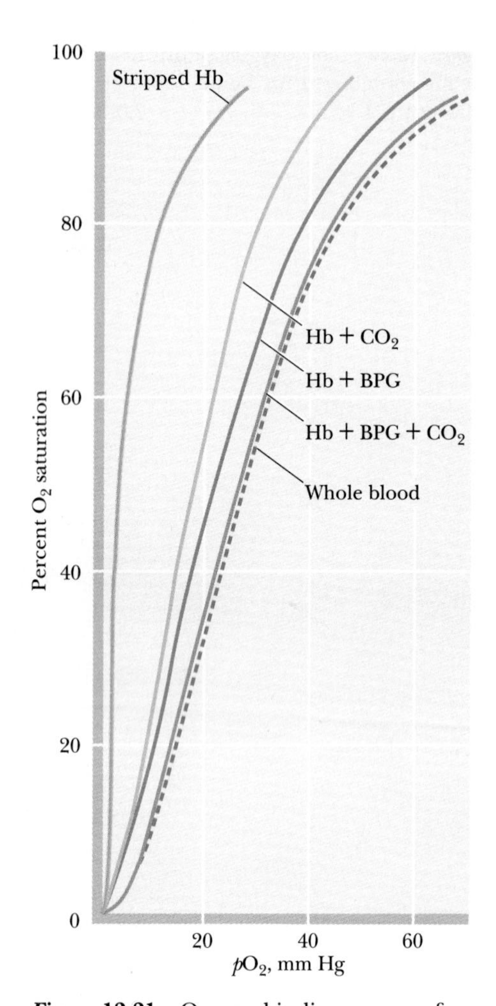

***Figure* 12.21** Oxygen-binding curves of blood and of hemoglobin in the absence and presence of CO_2 and BPG. From left to right: stripped Hb, Hb + CO_2, Hb + BPG, Hb + BPG + CO_2, and whole blood.

In addition, some CO_2 is directly transported by hemoglobin in the form of *carbamate* (—$NHCOO^-$). Free α-amino groups of Hb will react with CO_2 reversibly:

$$R—NH_2 + CO_2 \rightleftharpoons R—NH—COO^- + H^+$$

This reaction is driven to the right in tissues by the high carbon dioxide concentration; the equilibrium shifts the other way in the lungs where [CO_2] is low. Carbamylation of the N-termini converts them to anionic functions, which then form salt links with the cationic side chains of Arg α141 that stabilize the deoxy or T state of hemoglobin.

In addition to CO_2, Cl^- and BPG also bind better to deoxyhemoglobin than to oxyhemoglobin, causing a shift in equilibrium in favor of O_2 release. These various effects are demonstrated by the shift in the oxygen saturation curves for Hb in the presence of one or more of these substances (Figure 12.21). Note that the O_2-binding curve for Hb + BPG + CO_2 fits that of whole blood very well.

2,3-Bisphosphoglycerate Is an Important Allosteric Effector for Hemoglobin

The binding of 2,3-bisphosphoglycerate (BPG) to Hb promotes the release of O_2 (Figure 12.21). Erythrocytes (red blood cells) normally contain about 4.5 mM BPG, a concentration equivalent to that of tetrameric hemoglobin molecules. Interestingly, this equivalence is maintained in the Hb:BPG binding stoichiometry since the tetrameric Hb molecule has but one binding site for BPG. This site is situated within the central cavity formed by the association of the four subunits. The strongly negative BPG molecule (Figure 12.22) is electrostatically bound via interactions with the positively charged functional groups of each Lys β82, His β2, His β143, and the NH_3^+-terminal group of each β-chain. These positively charged residues are arranged to form an electrostatic pocket complementary to the conformation and charge distribution of BPG (Figure 12.23). In effect, BPG cross-links the two β-subunits. The ionic bonds between BPG and the two β-chains aid in stabilizing the conformation of Hb in its deoxy form, thereby favoring the dissociation of oxygen. In oxyhemoglobin, this central cavity is too small for BPG to fit. Or, to put it another way, the conformational changes in the Hb molecule that accompany O_2-binding perturb the BPG-binding site so that BPG can no longer be accommodated. Thus, BPG and O_2 are mutually exclusive allosteric effectors for Hb, even though their binding sites are physically distinct.

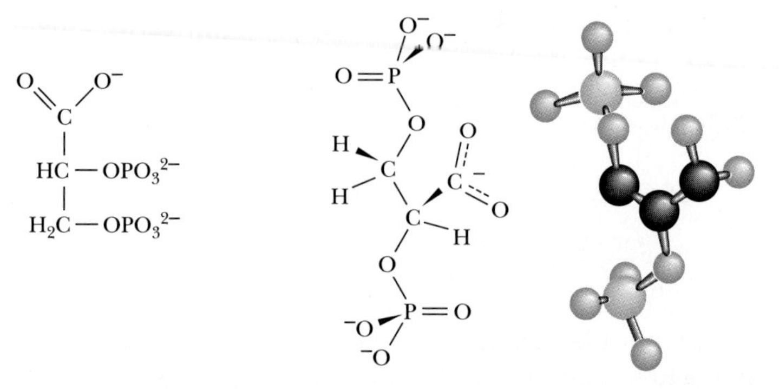

***Figure* 12.22** The structure, in ionic form, of BPG or 2,3-bisphosphoglycerate, an important allosteric effector for hemoglobin.

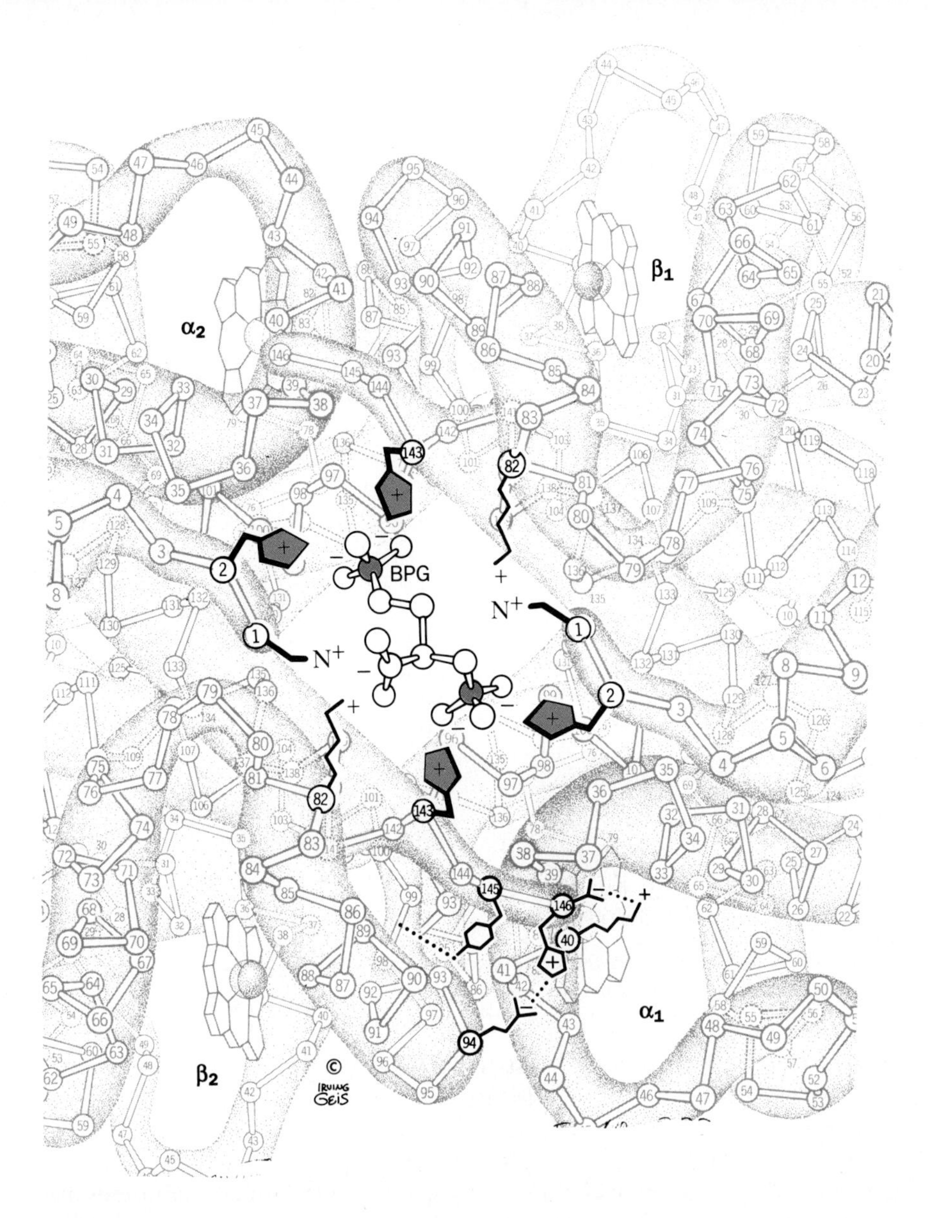

***Figure* 12.23** The ionic binding of BPG to the two β-subunits of Hb.

The Physiological Significance of BPG Binding

The importance of the BPG effect is evident in Figure 12.21: Hemoglobin stripped of BPG is virtually saturated with O_2 at a pO_2 of only 20 torr, and it cannot release its oxygen within tissues, where the pO_2 is typically 40 torr. BPG shifts the oxygen saturation curve of Hb to the right, rendering it an O_2 delivery system eminently suited to the needs of the organism. BPG serves this vital function in humans, most primates, and a number of other mammals. However, the hemoglobins of cattle, sheep, goats, deer, and other animals have an intrinsically lower affinity for O_2, and these Hbs are relatively unaffected by BPG. In fish, whose erythrocytes contain mitochondria, the regulatory role of BPG is filled by ATP or GTP. In reptiles and birds, a different organophosphate serves, namely inositol pentaphosphate (IPP) or inositol hexaphosphate (IHP) (Figure 12.24).

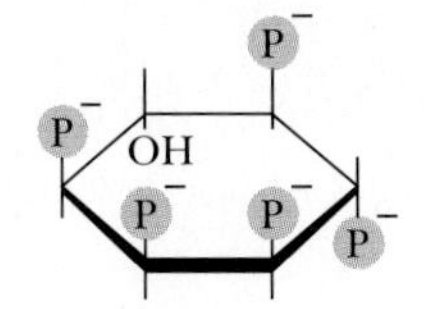

Inositol pentaphosphate (IPP)

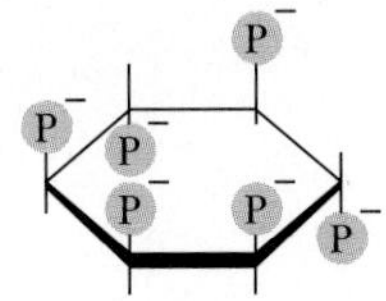

Inositol hexaphosphate (IHP)

***Figure* 12.24** The structures of inositol pentaphosphate and inositol hexaphosphate, the functional analogs of BPG in birds and reptiles.

Fetal Hemoglobin Has a Higher Affinity for O_2 Because It Has a Lower Affinity for BPG

The fetus depends on its mother for an adequate supply of oxygen, but its circulatory system is entirely independent. Gas exchange takes place across the placenta. Ideally then, fetal Hb should be able to absorb O_2 better than maternal Hb so that an effective transfer of oxygen can occur. Fetal Hb differs

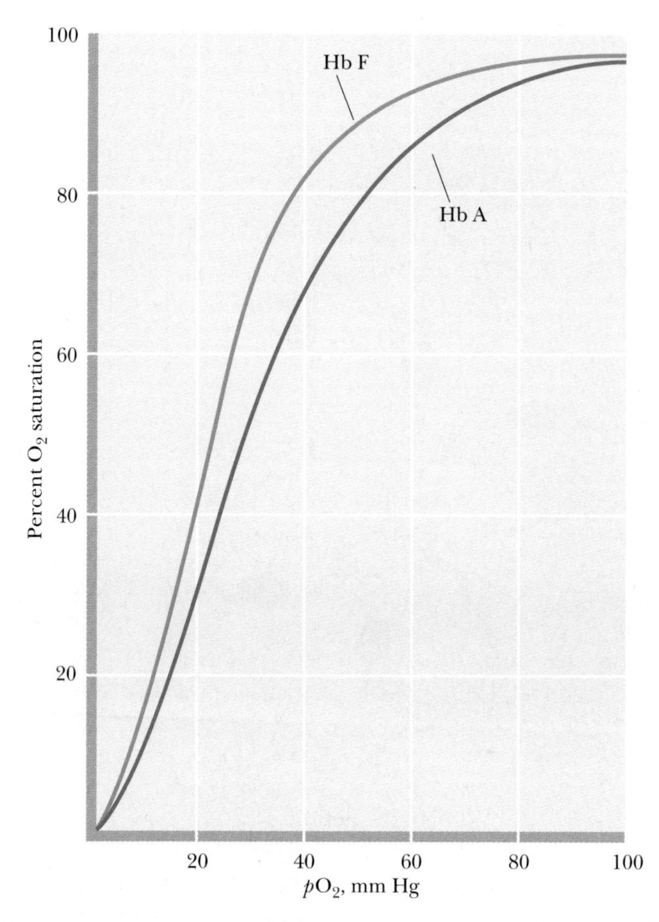

***Figure* 12.25** Comparison of the oxygen saturation curves of Hb A and Hb F under similar conditions of pH and [BPG].

from adult Hb in that the β-chains are replaced by very similar, but not identical, 146-residue subunits called γ-chains (gamma chains). Fetal Hb is thus $\alpha_2\gamma_2$. Recall that BPG functions through its interaction with the β-chains. BPG binds less effectively with the γ-chains of fetal Hb (also called Hb F). (Fetal γ-chains have Ser instead of His at position 143, and thus lack two of the positive charges in the central BPG-binding cavity.) Figure 12.25 compares the relative affinities of adult Hb (also known as Hb A) and Hb F for O_2 under similar conditions of pH and [BPG]. Note that Hb F will bind O_2 at pO_2 values where most of the oxygen has dissociated from Hb A. Much of the difference can be attributed to the diminished capacity of Hb F to bind BPG (compare Figures 12.21 and 12.25); Hb F thus has an intrinsically greater affinity for O_2, and oxygen transfer from mother to fetus is ensured.

Sickle-Cell Anemia

In 1904, a Chicago physician treated a 20-year-old black college student complaining of headache, weakness, and dizziness. The blood of this patient revealed serious anemia—only half the normal number of red cells were present. Many of these cells were abnormally shaped; in fact, instead of the characteristic disc shape, these erythrocytes were elongate and crescentlike in form, a feature that eventually gave name to the disease **sickle-cell anemia.** These sickle cells pass less freely through the capillaries, impairing circulation and causing tissue damage. Further, these cells are more fragile and rupture more easily than normal red cells, leading to anemia.

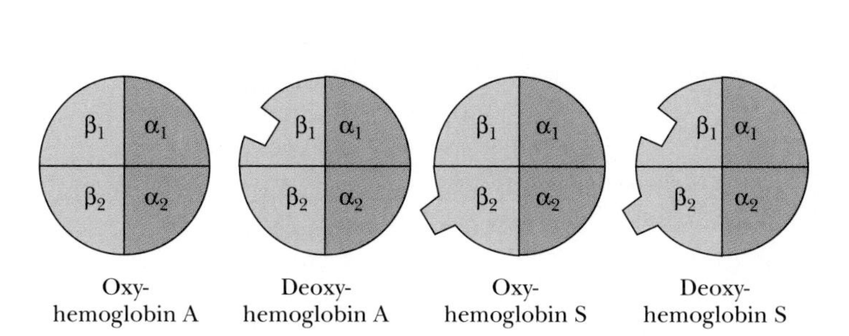

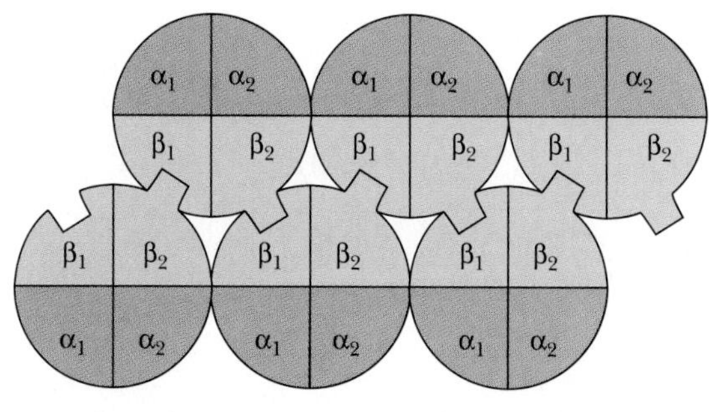

Figure **12.26** The polymerization of Hb S via the interactions between the hydrophobic Val side chains at position $\beta 6$ and the hydrophobic pockets in the EF corners of β-chains in neighboring Hb molecules. The protruding "block" on Oxy S represents the Val hydrophobic protrusion. The complementary hydrophobic pocket in the EF corner of the β-chains is represented by a square-shaped indentation. (This indentation is probably present in Hb A also.) Only the β_2 Val protrusions and the β_1 EF pockets are shown. (The β_1 Val protrusions and the β_2 EF pockets are not involved, although they are present.)

Sickle-Cell Anemia Is a Molecular Disease

A single amino acid substitution in the β-chains of Hb causes sickle-cell anemia. Replacement of the glutamate residue at position 6 in the β-chain by a valine residue marks the only chemical difference between Hb A and sickle-cell hemoglobin, Hb S. The amino acid residues at position $\beta 6$ lie at the surface of the hemoglobin molecule. In Hb A, the ionic R groups of the Glu residues fit this environment. In contrast, the aliphatic side chains of the Val residues in Hb S create hydrophobic protrusions where none existed before. To the detriment of individuals who carry this trait, a hydrophobic pocket forms in the EF corner of each β-chain of Hb when it is in the deoxy state, and this pocket nicely accommodates the Val side chain of a neighboring Hb S molecule (Figure 12.26). This interaction leads to the aggregation of Hb S molecules into long, chainlike polymeric structures. The obvious consequence is that deoxyHb S is less soluble than deoxyHb A. The concentration of hemoglobin in red blood cells is high (about 150 mg/mL), so that even in normal circumstances it is on the verge of crystallization. The formation of insoluble deoxyHb S fibers distorts the red cell into the elongated sickle shape characteristic of the disease.[3]

Although hemoglobin is an informative model system for analysis of allosteric interactions, it is not an enzyme, and the regulation of metabolism is achieved through the modulation of enzyme activity. The paradigm for allosteric enzymes is *Escherichia coli* aspartate transcarbamoylase (aspartate carbamoyltransferase).

12.5 *Escherichia coli* Aspartate Transcarbamoylase—An Allosteric Enzyme

The enzyme *aspartate transcarbamoylase (ATCase)* catalyzes the first step in pyrimidine biosynthesis in *E. coli* (Figure 12.27). This reaction involves the condensation of carbamoyl phosphate and aspartic acid to form *N*-carbamoyl-aspartate. While the reactants participate in a variety of other metabolic pathways, the product, carbamoyl-aspartate, has no other fate than to serve as a precursor to the formation of pyrimidines. For this reason, the aspartate transcarbamoylase reaction is designated as the **committed step** in the pyrimidine biosynthetic pathway. CTP is the end product of this pathway.

[3] In certain regions of Africa, the sickle-cell trait is found in 20% of the people. Why does such a deleterious heritable condition persist in the population? For reasons as yet unknown, individuals with this trait are less susceptible to the most virulent form of malaria. The geographic distribution of malaria and the sickle-cell trait are positively correlated.

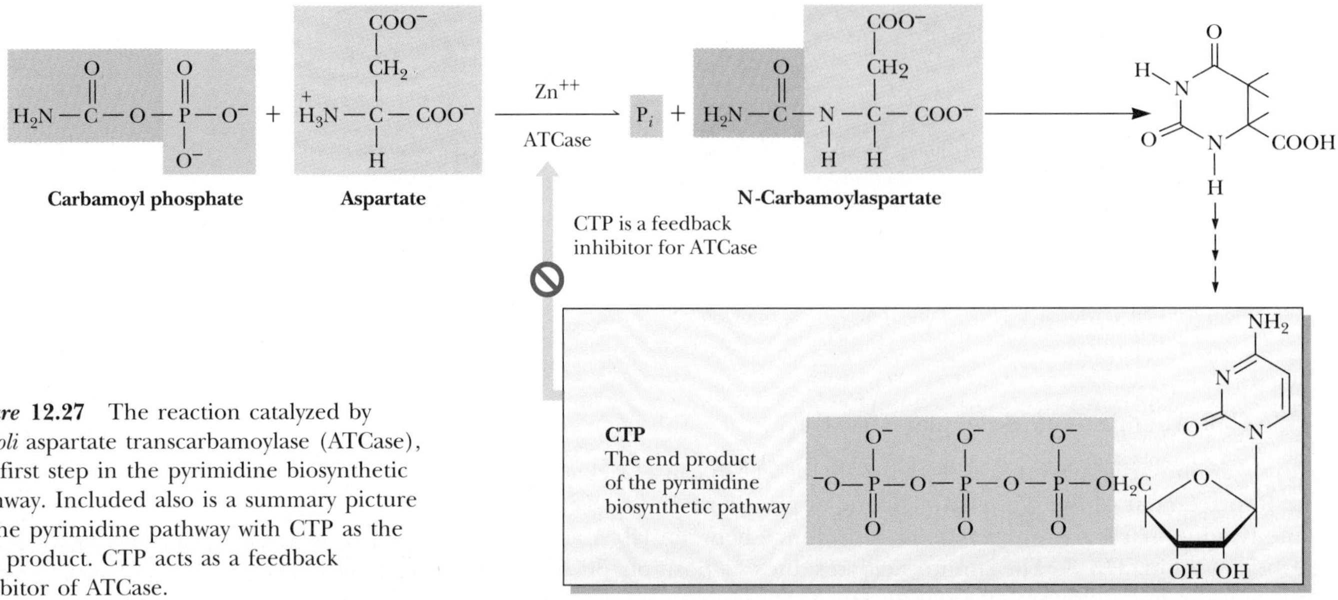

Carbamoyl phosphate + Aspartate $\xrightarrow[\text{ATCase}]{Zn^{++}}$ P_i + N-Carbamoylaspartate

CTP is a feedback inhibitor for ATCase

CTP
The end product of the pyrimidine biosynthetic pathway

Figure **12.27** The reaction catalyzed by *E. coli* aspartate transcarbamoylase (ATCase), the first step in the pyrimidine biosynthetic pathway. Included also is a summary picture of the pyrimidine pathway with CTP as the end product. CTP acts as a feedback inhibitor of ATCase.

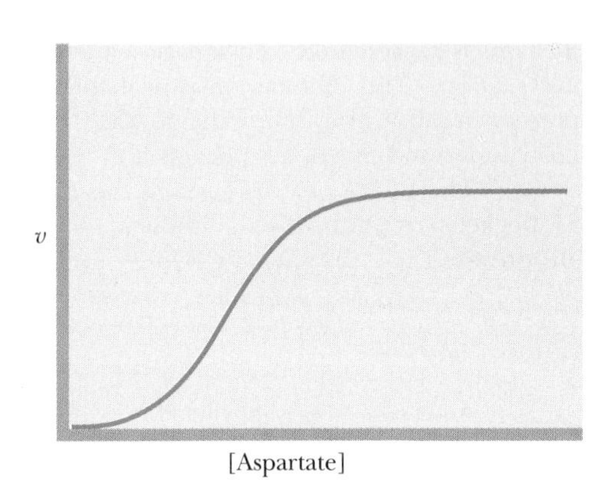

Figure **12.28** The sigmoid response of ATCase activity to the concentration of the substrate aspartate shows a strong positive cooperativity.

ATCase Shows Cooperativity in Substrate Binding

The binding of the substrates *carbamoyl-P* and *aspartate* to ATCase is highly cooperative (Figure 12.28), which allows the enzyme activity to increase markedly over a rather narrow concentration range of substrate availability. In the language of biochemistry, ligands that bind in a cooperative manner, such that the binding of one equivalent enhances the binding of additional equivalents to the same protein molecule, are termed **positive homotropic effectors.** (The prefix *homo* indicates that the ligand influences the binding of itself.) The substrates of ATCase are positive homotropic effectors.

CTP Is a Feedback Inhibitor of ATCase

The end product, CTP, serves as a *feedback inhibitor* of ATCase. Inhibition by CTP of the committed step in the pathway leading to its formation is a very effective way to regulate its biosynthesis. CTP acts by decreasing the affinity of ATCase for its substrates without affecting V_{max} for the reaction (Figure 12.29). CTP is a **negative heterotropic effector;** that is, the binding of CTP has a negative effect on the binding of something other than itself (the substrate in this instance). Note that in Figure 12.29 the substrate saturation curve is displaced to the right in the presence of CTP, and a higher substrate concentration is necessary to achieve half-maximal velocity, $V_{max}/2$. (In sigmoid kinetic responses, the [S] giving $v = V_{max}/2$ is termed $K_{0.5}$, *not* K_m, since the kinetics do not obey the Michaelis–Menten equation.) When concentrations of CTP accumulate to high levels, ATCase is inhibited; when [CTP] is limiting, the activity of ATCase is regulated by the availability of its substrates.

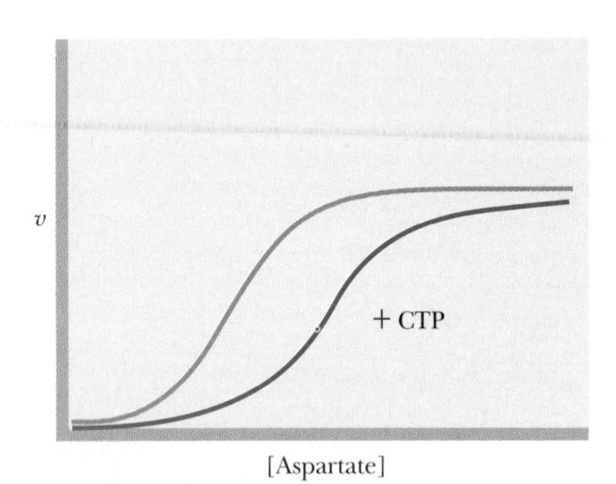

Figure **12.29** CTP is a feedback inhibitor of ATCase which affects the affinity of ATCase for its substrates but does not affect V_{max}.

ATP Is an Allosteric Activator of ATCase

ATP also provides a regulatory signal to ATCase. It binds to the same site as CTP, but it stimulates ATCase rather than inhibiting it (Figure 12.30). ATP acts as a **positive heterotropic effector,** which means it enhances the binding of substrates to ATCase. Sufficient [ATP] indicates that the energy status of the cell is robust and that conditions may be appropriate for cell division. Cell division will require substantial nucleic acid synthesis and a corresponding drain on the available pool of pyrimidine nucleotides. ATP is also the end product of the parallel pathway of purine biosynthesis, and as such, it represents the parallel, or purine counterpart, of CTP. An excess of ATP over CTP signals that the relative amounts of purine versus pyrimidine nucleotides may be unbalanced with respect to the need for them in nucleic acid synthesis, and pyrimidine synthesis should be augmented.

Fluctuation in the levels of ATP and CTP and their competition for binding to the same site on ATCase, with opposite effects, allows these two nucleotides to exert a *rapid and reversible control* over ATCase activity. Such reciprocal regulation ensures that amounts of the pyrimidine nucleotides are commensurate with cellular needs.

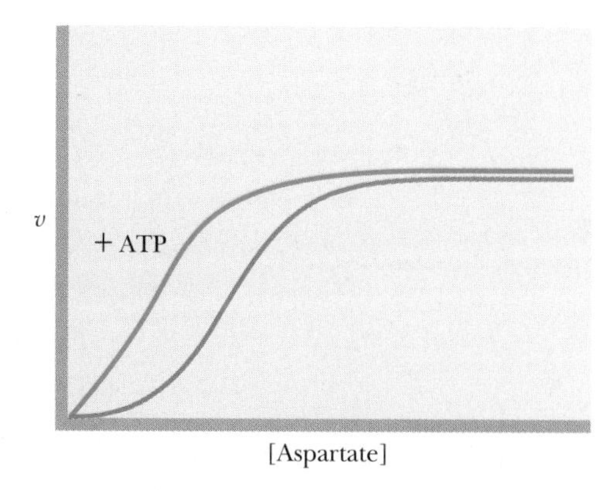

***Figure* 12.30** ATP is a positive heterotropic effector for ATCase. It binds at the same site as CTP; CTP and ATP are competitive. Thus, if CTP is absent, ATP will promote its synthesis. Like CTP, ATP affects the affinity of ATCase for its substrates, but does not affect V_{max}.

Desensitization of ATCase with *p*HMB

Treatment of ATCase with the sulfhydryl-reactive reagent *p*HMB (or *para*-**h**ydroxy**m**ercuri**b**enzoate (Figure 12.31)) abolishes its allosteric properties. CTP and ATP no longer modulate the catalytic activity of the enzyme, and its cooperative substrate-binding behavior is lost, causing it to display the hyperbolic v versus [S] Michaelis–Menten kinetic response characteristic of "normal" enzymes (Figure 12.32). This loss in regulatory properties in an enzyme is referred to as **desensitization.** Note that V_{max} for the enzyme actually increases by 50% upon desensitization. Desensitization by *p*HMB is concomitant with the dissociation of native ATCase holoenzyme into its component subunits.

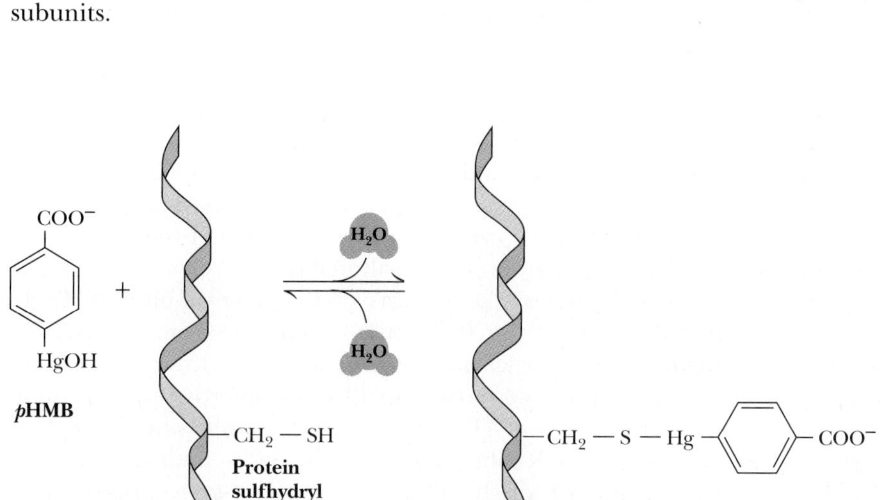

***Figure* 12.31** The structure of *para*-hydroxymercuribenzoate (*p*HMB) and its reaction with —SH groups.

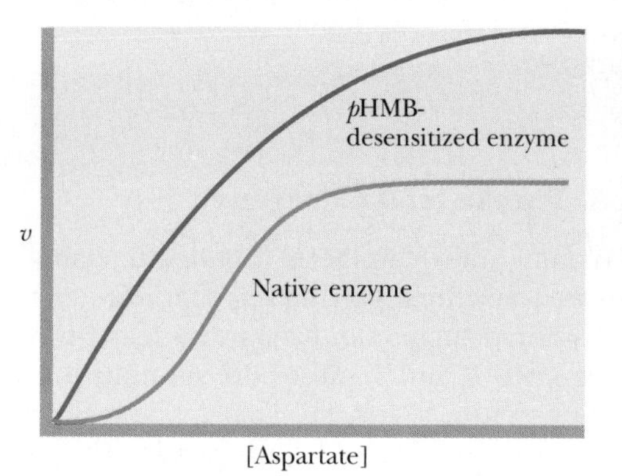

***Figure* 12.32** Treatment of ATCase with *p*HMB abolishes its allosteric properties and causes it to be a "normal" Michaelis–Menten enzyme. The activity of the desensitized ATCase is 50% greater than that of the native ATCase.

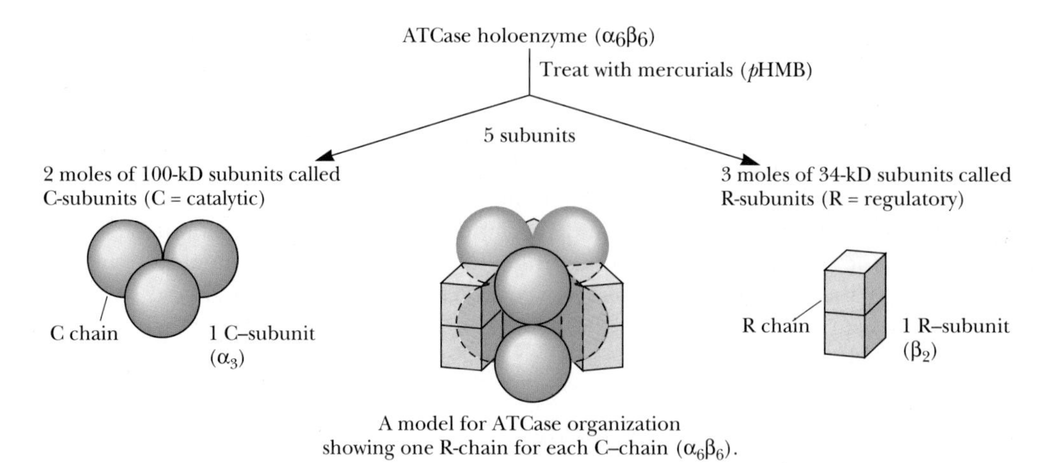

***Figure* 12.33** The subunit organization of ATCase as revealed by *p*HMB dissociation and subsequent analysis.

The Subunit Organization of ATCase

Native ATCase is an $\alpha_6\beta_6$-type heteromultimeric protein, molecular mass 310 kD. Dissociation by *p*HMB yields two equivalents of 100-kD **catalytic** or **C-subunits** and three equivalents of 34-kD **regulatory** or **R-subunits** (Figure 12.33). The C-subunits are enzymatically active, but are not affected by CTP or ATP and do not show cooperative binding of S. Each C-subunit has three binding sites for aspartate and three for carbamoyl-P. Further, each C-subunit is composed of three identical 34-kD polypeptide chains, so-called **C-chains,** which correspond to the α-subunit component of the heteromultimer. Single C-chains are catalytically inactive, indicating that neighboring C-chains must each contribute parts to the formation of ATCase active sites. The R-subunits are catalytically inactive but each can bind two equivalents of the nucleotide effectors CTP or ATP. Each R-subunit is constituted of two polypeptide chains of 17 kD each, called **R-chains,** the β parts of the heteromultimer. Each R-chain contains a Zn^{2+} ion. If the Zn^{2+} ions are removed from native ATCase, it loses its regulatory properties but not its catalytic activity. These results, together with visualization of the enzyme by electron microscopy, lead to a model for the structure of the ATCase $\alpha_6\beta_6$ holoenzyme where the two C-subunits join with the three R-subunits to form the native multimer (Figure 12.33). An important feature of this multimer is its *one-to-one association of catalytic and regulatory chains to form six functional C:R subunits.*

With a basic picture of the subunit organization of this allosteric enzyme in mind, let us now consider models proposed to account for the sigmoid kinetic behavior of allosteric enzymes. We will then return for a closer look at ATCase to see how well this enzyme actually conforms to such models.

12.6 Allosteric Models

The Symmetry Model of Monod, Wyman, and Changeux

In 1965, Jacques Monod, Jeffries Wyman, and Jean-Pierre Changeux elaborated a theoretical model of allosteric transitions postulating that allosteric proteins are oligomers. Further, allosteric proteins can exist in (at least) two different conformational states, designated R and T. All of the subunits in a molecule have the same conformation (either R or T). That is, molecular symmetry is conserved. Specifically, molecules of mixed conformations (having subunits of both R and T states) are not allowed by this model.

(**a**) A dimeric protein can exist in either of two conformational states at equilibrium

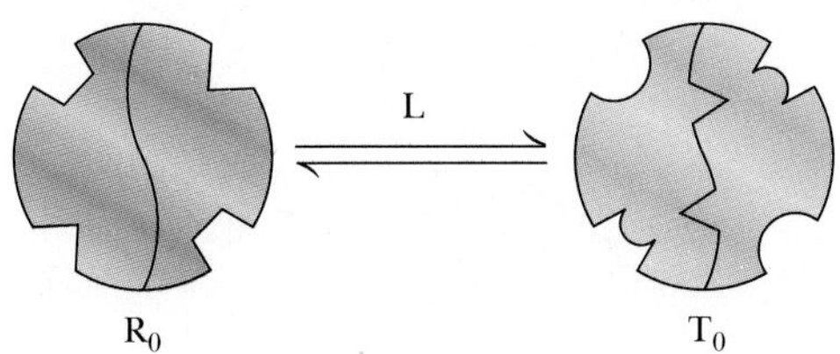

$L = \frac{T_0}{R_0}$ L is large. ($T_0 >> R_0$)

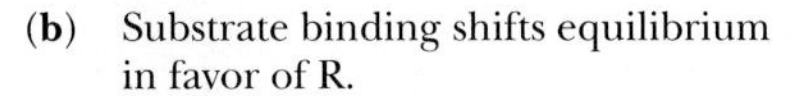

(**b**) Substrate binding shifts equilibrium in favor of R.

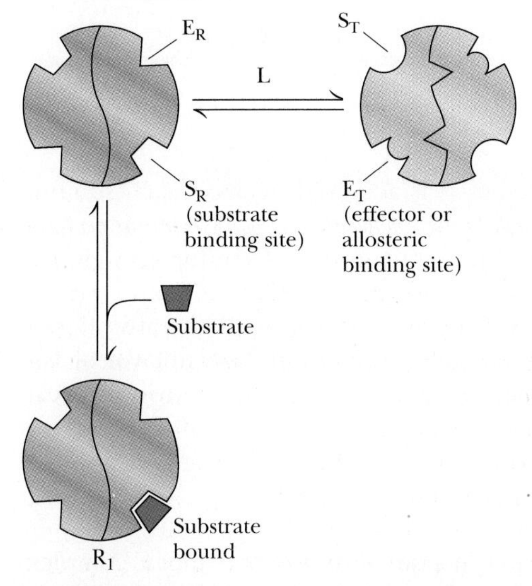

***Figure* 12.34** Monod, Wyman, Changeux (MWC) model for allosteric transitions. Consider a dimeric protein that can exist in either of two conformational states, R or T. Each subunit in the dimer has a binding site for substrate S and an allosteric effector site, E. The subunits are symmetrically related to one another in the protein, and symmetry is conserved regardless of the conformational state of the protein. The different states of the protein, with or without bound ligand, are linked to one another through the various equilibria. Thus, the relative population of protein molecules in the R or T state is a function of these equilibria and the concentration of the various ligands, substrate (S) and effectors (which bind at E_R or E_T). As [S] is increased, the T/R equilibrium shifts in favor of an increased proportion of R-conformers in the total population (that is, more protein molecules in the R conformational state).

In the absence of ligand, the two states of the allosteric protein are in equilibrium:

$$R_0 \rightleftharpoons T_0$$

(Note that the subscript "0" signifies "in the absence of ligand.") The equilibrium constant is termed L: $L = T_0/R_0$. L is assumed to be large; that is, the amount of the protein in the T conformational state is much greater than the amount in the R conformation. Let us suppose that $L = 10^4$.

The affinities of the two states for substrate, S, are characterized by the respective dissociation constants, K_R and K_T. Suppose $K_T \gg K_R$. That is, the affinity of R_0 for S is much greater than the affinity of T_0 for S. Let us choose the extreme where $K_R/K_T = 0$ (that is, K_T is infinitely greater than K_R). In effect, we are picking conditions where S binds only to R. (If K_T is infinite, T does not bind S.)

Given these parameters, consider what happens when S is added to a solution of the allosteric protein at conformational equilibrium (Figure 12.34). Although the relative $[R_0]$ concentration is small, S will bind "only" to R_0, forming R_1. This depletes the concentration of R_0, perturbing the T_0/R_0 equilibrium. To restore equilibrium, molecules in the T_0 conformation undergo a transition to R_0. This shift renders more R_0 available to bind S, yielding R_1, diminishing $[R_0]$, perturbing the T_0/R_0 equilibrium, and so on. Thus, these linked equilibria (Figure 12.34) are such that S-binding by the R_0 state of the allosteric protein perturbs the T_0/R_0 equilibrium with the result that S-binding drives the conformational transition, $T_0 \rightarrow R_0$.

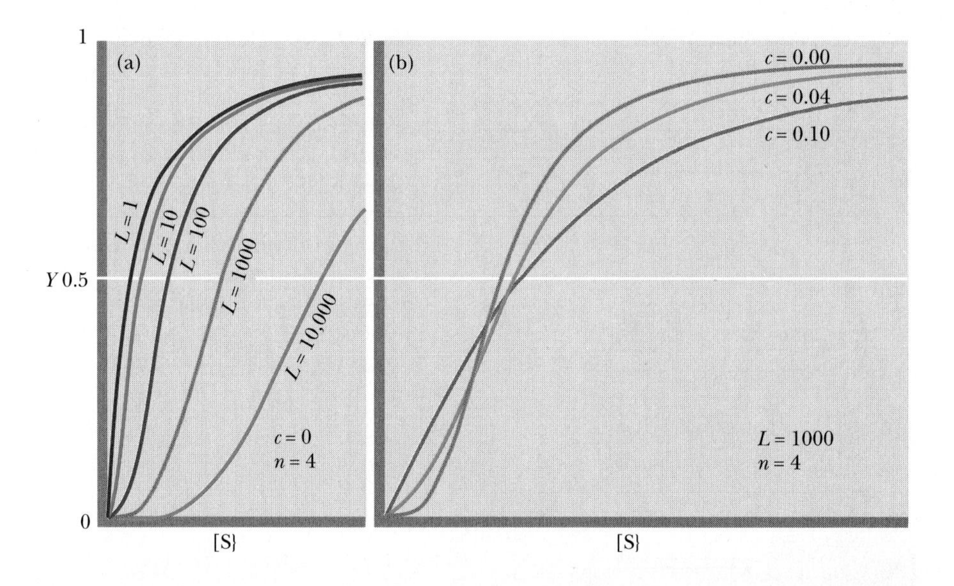

***Figure* 12.35** The Monod, Wyman, Changeux model. Graphs of allosteric effects for a tetramer ($n = 4$) as a function of Y, the saturation function. Y is defined as [ligand-binding sites that are occupied by ligand]/[total ligand-binding sites]. (a) A plot of Y as a function of [S], at various L values. (b) Y as a function of [S], at different c, where $c = K_R/K_T$. (When $c = 0$, K_T is infinite.)

(Adapted from Monod, Wyman, and Changeux, 1965, Journal of Molecular Biology 12:92.)

In just this simple system, *cooperativity* is achieved because each subunit has a binding site for S, and thus, *each protein molecule has more than one binding site for S.* Therefore, the increase in the population of R conformers gives a progressive increase in the number of sites available for S. The extent of cooperativity depends on the relative T_0/R_0 ratio and the relative affinities of R and T for S. If L is large (that is, the equilibrium lies strongly in favor of T_0) and if $K_T \gg K_R$, as in the example we have chosen, cooperativity is great (Figure 12.35).

Heterotropic Effectors

This simple system also provides an explanation for the more complex heterotropic allosteric effects, such as those exerted by CTP and ATP on ATCase. Consider a protein composed of two subunits, each of which has two binding sites: one for the substrate, S, and one to which allosteric effectors bind, *the allosteric site.* Assume that S binds preferentially ("only") to the R conformer; further assume that the *positive heterotropic effector,* A, binds to the allosteric site only when the protein is in the R conformation, and the *negative allosteric effector,* I, binds at the allosteric site only if the protein is in the T conformation. Thus, with respect to binding at the allosteric site, A and I are competitive with each other.

Positive Effectors

If A binds to R_0, forming the new species $R_{1(A)}$, the relative concentration of R_0 is decreased and the T_0/R_0 equilibrium is perturbed (Figure 12.36). As a consequence, a relative $T_0 \rightarrow R_0$ shift occurs in order to restore equilibrium. The net effect is an increase in the number of R conformers in the presence of A, meaning that more binding sites for S are available. For this reason, A leads to a decrease in the cooperativity of the substrate saturation curve, as seen by a shift of this curve to the left (Figure 12.36). Effectively, the presence of A lowers the apparent value of L.

Negative Effectors

The converse situation applies in the presence of I, which binds "only" to T. I-binding will lead to an increase in the population of T conformers, at the expense of R_0 (Figure 12.36). The decline in $[R_0]$ means that it is less likely

A dimeric protein which can exist in either of two states R_0 and T_0. This protein can bind 3 ligands:

1) Substrate (S): A positive homotropic effector that binds only to R at site S
2) Activator (A): A positive heterotropic effector that binds only to R at site E
3) Inhibitor (I): A negative heterotropic effector that binds only to T at site E

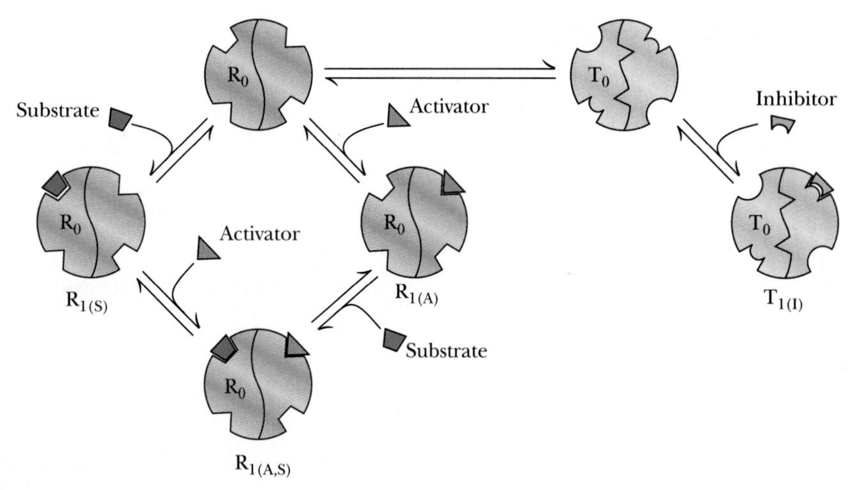

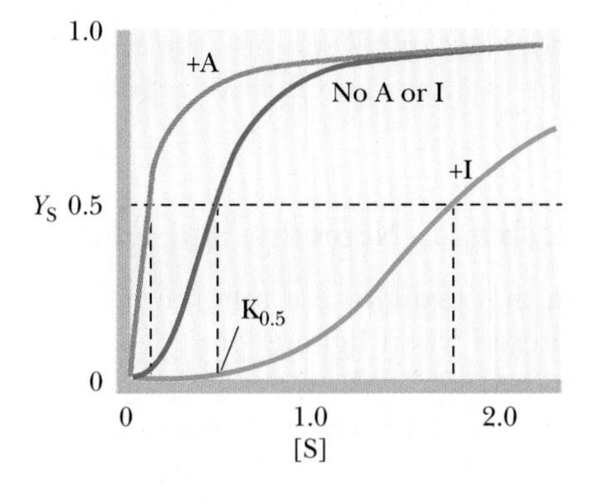

Effects of A:
$A + R_0 \rightarrow R_{1(A)}$
Increase in number of R-conformers shifts $R_0 \rightleftharpoons T_0$ so that $T_0 \rightarrow R_0$

1) More binding sites for S made available
2) Decrease in cooperativity of substrate saturation curve. Effector A lowers the apparent value of L.

Effects of I:
$I + T_0 \rightarrow T_{1(I)}$
Increase in number of T-conformers (decrease in R_0 as $R_0 \rightarrow T_0$ to restore equilibrium).

Thus, I inhibits association of S and A with R by lowering R_0 level. I increases cooperativity of substrate saturation curve. I raises the apparent value of L.

for S (or A) to bind. Consequently, the presence of I increases the cooperativity (that is, the sigmoidicity) of the substrate saturation curve, as evidenced by the shift of this curve to the right (Figure 12.36). The presence of I raises the apparent value of *L*.

***Figure* 12.36** Heterotropic allosteric effects: A and I binding to R and T, respectively. The linked equilibria lead to changes in the relative amounts of R and T and, therefore, shifts in the substrate saturation curve. This behavior, depicted by the graph, defines an allosteric "K" system. The parameters of such a system are: (1) S and A (or I) have different affinities for R and T and (2) A (or I) modifies the apparent $K_{0.5}$ for S by shifting the relative R versus T population.

K Systems and *V* Systems

The allosteric model just presented is called a ***K* system,** since the concentration of substrate giving half-maximal velocity, defined as $K_{0.5}$, changes in response to effectors (Figure 12.36). Note that V_{max} is constant in this system.

An allosteric situation where $K_{0.5}$ is constant but the apparent V_{max} changes in response to effectors is termed a ***V* system.** In a *V* system, all v versus S plots are hyperbolic rather than sigmoid (Figure 12.37). The positive heterotropic effector A activates by raising V_{max}, while I, the negative heterotropic effector, decreases it. Note that neither A nor I affects $K_{0.5}$. This situation arises if R and T have the *same* affinity for the substrate, S, but

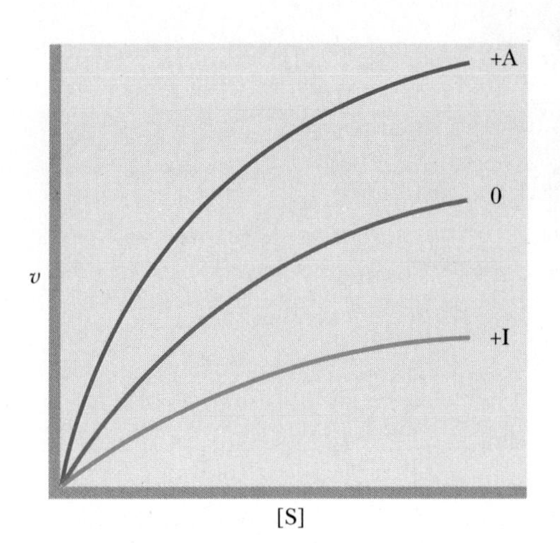

***Figure* 12.37** v versus [S] curves for an allosteric "*V*" system. The *V* system fits the model of Monod, Wyman, and Changeux, given the following conditions: (1) R and T have the *same* affinity for the substrate, S. (2) The effectors A and I have different affinities for R and T, and thus can shift the relative T/R distribution. (That is, A and I change the apparent value of *L*.) Assume, as before, that A binds "only" to the R state and I binds "only" to the T state. (3) R and T differ in their catalytic ability. Assume that R is the enzymatically active form, while T is inactive. Since A perturbs the T/R equilibrium in favor of more R, A increases the apparent V_{max}. I favors transition to the inactive T state.

different affinities for A and I. A and I thus can shift the relative T/R distribution. Acetyl-coenzyme A carboxylase, the enzyme catalyzing the committed step in the fatty acid biosynthetic pathway, behaves as a *V* system in response to its allosteric activator, citric acid (see Chapter 24).

K Systems and *V* Systems Fill Different Biological Roles

The *K* and *V* systems have design features that mean they work best under different physiological situations. "*K* system" enzymes are adapted to conditions where the prevailing substrate concentration is rate-limiting, as when [S] *in vivo* = $K_{0.5}$. On the other hand, when the physiological conditions are such that [S] is usually saturating for the regulatory enzyme of interest, the enzyme conforms to the "*V* system" mode in order to attain an effective regulatory response.

The Sequential Allosteric Model of Koshland, Nemethy, and Filmer

Daniel Koshland has championed the idea that proteins are inherently flexible molecules whose conformations are altered when ligands bind. This notion serves as the fundamental tenet of the "induced-fit hypothesis" discussed earlier. Since this is so, ligand binding can potentially cause conformational changes in the protein. Depending on the nature of these conformational changes, virtually any sort of allosteric interaction is possible. That is, the binding of one ligand could result in conformational transitions that make it easier or harder for other ligands (of the same or different kinds) to bind. In 1966, Koshland and his colleagues proposed an allosteric model in which the ligand-induced conformational changes caused a subsequent transition to a conformational state with altered affinities. Since ligand binding and conformational transitions were distinct steps in a sequential pathway, the Koshland, Nemethy, Filmer (or KNF) model is dubbed the **sequential model** for allosteric transitions. Figure 12.38 depicts the essential features of this model in a hypothetical dimeric protein. Binding of the ligand S induces a conforma-

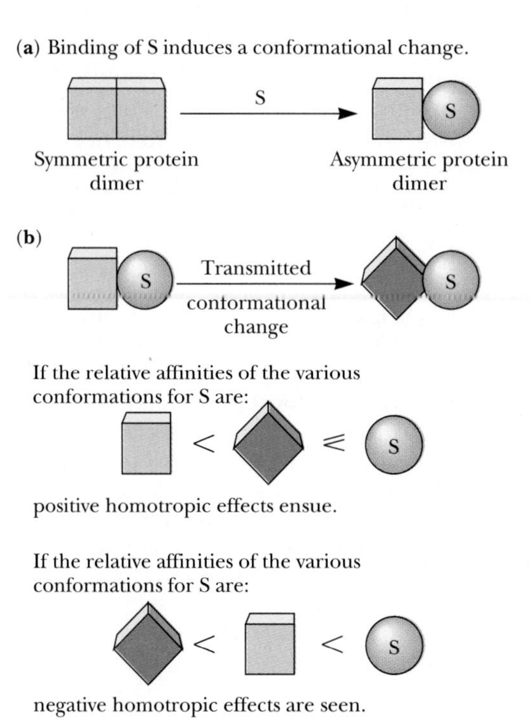
(a) Binding of S induces a conformational change.
S
S
Symmetric protein dimer
Asymmetric protein dimer
(b)
S
Transmitted conformational change
S
If the relative affinities of the various conformations for S are:
S
positive homotropic effects ensue.
If the relative affinities of the various conformations for S are:
S
negative homotropic effects are seen.

***Figure* 12.38** The Koshland, Nemethy, Filmer sequential model for allosteric behavior. (a) S-binding can, by induced fit, cause a conformational change in the protomer to which it binds. (b) If subunit interactions are tightly coupled, binding of S to one protomer may cause the other protomer to assume a conformation having a greater (positive homotropic) or lesser (negative homotropic) affinity for S. That is, the ligand-induced conformational change in one protomer can affect the adjoining protomer. Such effects could be transmitted between neighboring peptide domains by changing alignments of nonbonded amino acid residues.

tional change in the subunit to which it binds. Note that there is no requirement for conservation of symmetry here; the two subunits can assume different conformations (represented as a square and a circle). If the *subunit interactions are tightly coupled,* then binding of S to one subunit could cause the other subunit(s) to assume a conformation having a greater, or a lesser, affinity for S (or some other ligand). The underlying mechanism rests on the fact that the ligand-induced conformational change in one subunit can transmit its effects to neighboring subunits by changing the interactions and alignments of amino acid residues at the interface between subunits. Depending on the relative ligand affinity of the conformation adopted by the neighboring subunit, the overall effect for further ligand binding may be positive, negative, or neutral (Figure 12.38).

With these models for allosteric behavior in mind, let's reexamine ATCase, our paradigm of allosteric enzymes.

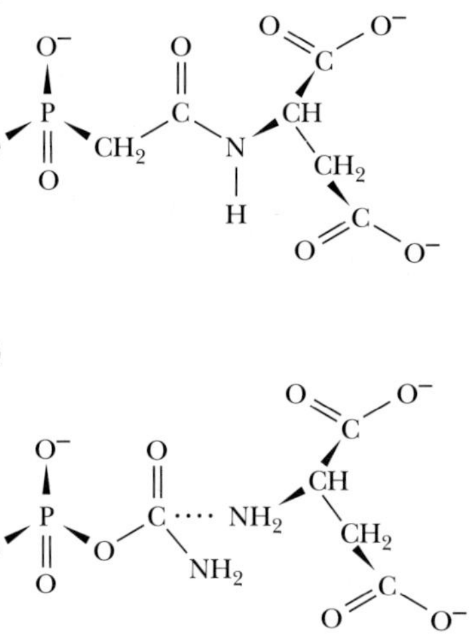

***Figure* 12.39** (a) The structure of *N*-phosphonacetyl-L-aspartate (PALA), an analog of the natural substrates for ATCase, carbamoyl-P and aspartate. (b) PALA strongly resembles the presumed transition-state intermediate for the ATCase reaction. PALA is unreactive because a —CH_2— group replaces the critical phosphoryl O atom. PALA binds very tightly to ATCase; its dissociation constant is about 100 n*M*.

The Allosteric Mechanism of *E. coli* Aspartate Transcarbamoylase

Studies conducted in 1977 by Howard Schachman and colleagues at the Berkeley campus of the University of California demonstrated that, in the presence of CTP, ATCase molecules sedimented about 3% faster in an ultracentrifuge than they did in the presence of ATP. This observation is consistent with two different conformational states for ATCase, corresponding to the R and T states of the MWC model. In the presence of CTP, the enzyme assumed a compact and more rapidly sedimenting T conformation, while in the presence of ATP, the conformation of ATCase expanded into a more open, slower sedimenting form, the R state.

Binding of the substrate analog *N*-phosphonacetyl-L-aspartate (PALA) by ATCase also favors the T→R transition. PALA (Figure 12.39) was designed to mimic the transition-state intermediate for ATCase, and, as such, it binds very tightly to the active sites of ATCase. The Hill coefficient for PALA is 2.0 (Table 12.2), demonstrating that binding occurs cooperatively. ATP and CTP affect PALA binding in expected ways.

William N. Lipscomb and his collaborators at Harvard University have determined the structures for both allosteric forms of ATCase by X-ray crystallographic analysis. The T-form structure was solved for ATCase in the presence of CTP and the R-form in the presence of PALA (Figure 12.40). Comparisons of these structures provide insights into the relation between structure and function in this allosteric enzyme and reveal important clues to the molecular mechanisms of catalysis and cooperativity. The ability to make specific amino acid substitutions at crucial residues in the C-chain sequence by *site-directed mutagenesis* (see Chapter 8) has led to the generation of mutant forms of ATCase for structural analysis. Comparisons of the structure and properties of mutant and wild-type ATCase molecules has aided these investigations immensely.

***Table* 12.2**
Binding of the Bisubstrate Analog PALA to Aspartate Transcarbamoylase

Additions	n_H (Hill Coefficient)	K_D (Dissociation Constant)
None	2.0	110 n*M*
ATP	1.4	65 n*M*
CTP	2.3	270 n*M*

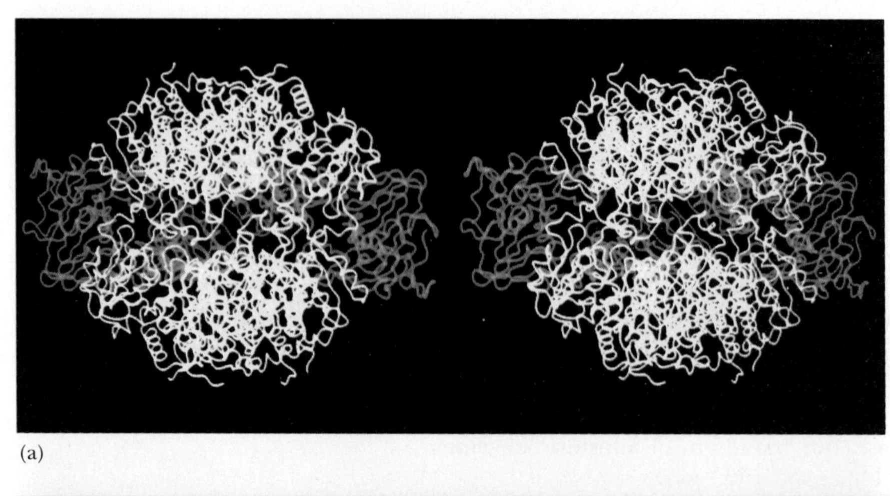

(a)

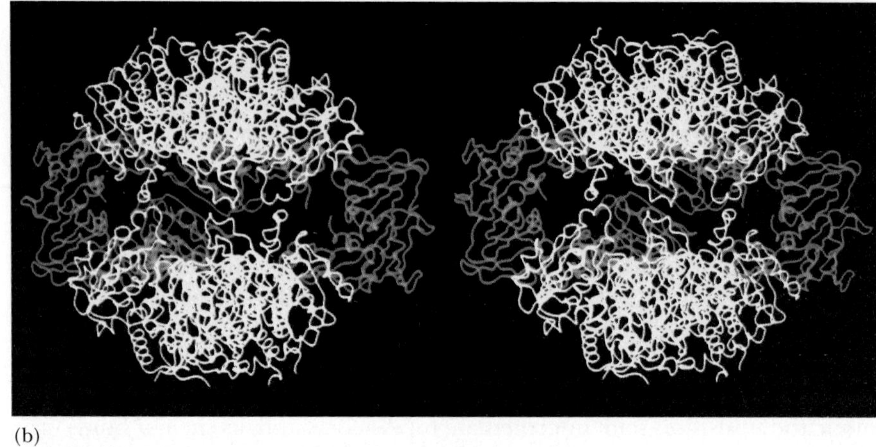

(b)

***Figure* 12.40** Stereo views of the (a) T-state and (b) R-state structures of ATCase obtained by X-ray crystallography. Catalytic chains are shown in yellow, and regulatory chains in red.

(From Kantrowitz, E. R., and Lipscomb, W. N., 1988. Escherichia coli *aspartate transcarbamoylase: The relation between structure and function. Science 241:669–674.)*

The Quaternary Structure of ATCase

Figure 12.41 illustrates the packing of the two catalytic trimers and the three regulatory dimers of ATCase into a highly symmetrical molecule, consisting of six C-chains and six R-chains that form six equivalent catalytic-regulatory functional units. Figure 12.42 presents a stereo view of one catalytic-regulatory (C-chain : R-chain) unit. Each C-chain contains one active site located near the interface between the adjacent C-chains in its catalytic trimer. Amino acid residues involved in substrate binding and catalytic activity are contributed by adjacent pairs of C-chains. The six active sites of ATCase all face a central

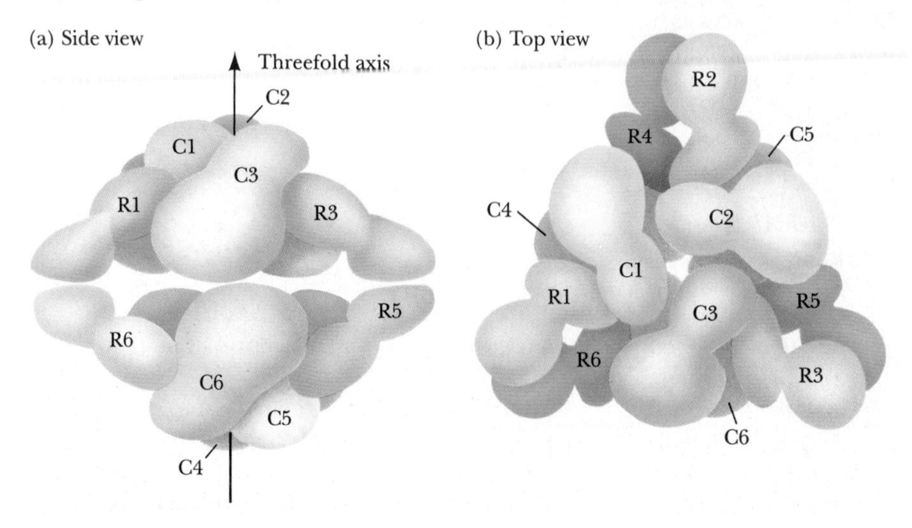

***Figure* 12.41** Schematic diagram of the quaternary structure of aspartate transcarbamoylase.

(From Kantrowitz, E. R., and Lipscomb, W. N., 1988. Escherichia coli *aspartate transcarbamoylase: The relation between structure and function. Science 241:669–674.)*

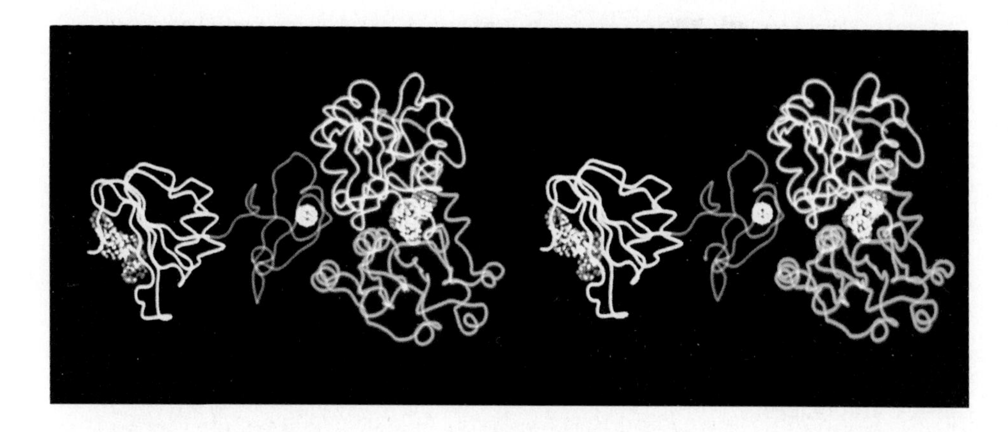

***Figure* 12.42** Stereo view of one catalytic-regulatory unit in the ATCase molecule. (The protein domains are color-coded for easy identification.) Note how far the allosteric site on the R-chain is from the substrate-binding sites on the C-chain. The allosteric (yellow) and Zn^{2+}-binding (red) domains of a regulatory chain are on the left. The aspartate-binding (blue) and carbamoyl phosphate-binding (green) domains of a catalytic chain are on the right. CTP occupies the allosteric site, and PALA is in the active site.

(From Kantrowitz, E. R., and Lipscomb, W. N., 1988. Escherichia coli *aspartate transcarbamoylase: The relation between structure and function. Science 241:669–674.)*

cavity. Via this cavity, substrates outside the molecule are accessible to the active sites in both the T state and the R state.

Function of Active Site Residues in ATCase

The PALA-ligated form of ATCase has revealed much about the binding interactions between the enzyme and its substrates (Figure 12.43). Residues Ser^{52}, Thr^{53}, Arg^{54}, Thr^{55}, Arg^{105}, His^{134}, Arg^{167}, Arg^{229}, Gln^{231}, and Leu^{267} from one C-chain and Ser^{80} and Lys^{84} from the adjacent C-chain interact with PALA. Note that PALA at neutral pH has four negative charges (Figure 12.39), and that the side chains of four Arg and one Lys contribute five positive charges. These electrostatic interactions play a major role in PALA binding. When Lys^{84} was changed to Gln by site-directed mutagenesis, the activity of ATCase was reduced by a factor of 4000. The amino acid at position 83 in the C-chain sequence is also Lys; its replacement by Gln reduces the activity only slightly. Conversion of Arg^{105} to Gln reduced activity a thousandfold.

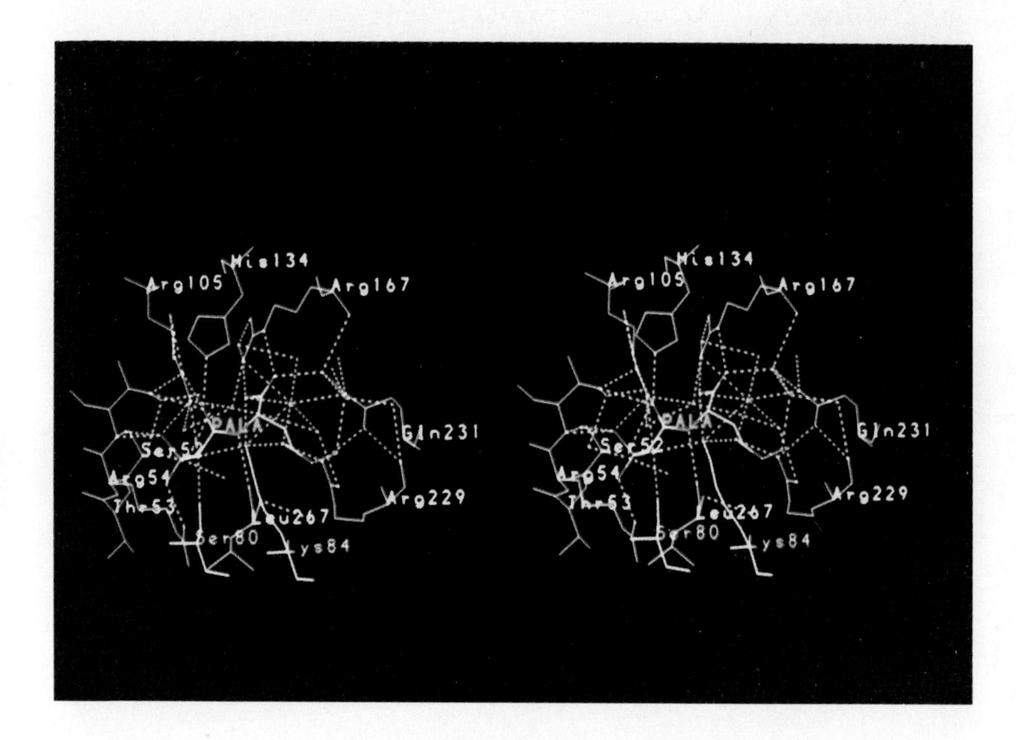

***Figure* 12.43** Stereo view of the PALA-binding site in ATCase. Note that groups interacting in PALA binding are contributed by residues from two adjacent C-chains. PALA is shown in blue. Catalytic side chains contributed by one C-chain are shown in orange; side chains contributed by the other C-chain are shown in yellow.

(From Kantrowitz, E. R., and Lipscomb, W. N., 1988. Escherichia coli *aspartate transcarbamoylase: The relation between structure and function. Science 241:669–674.)*

Structural Changes in ATCase in the T to R Transition

The formation of a complex between a single PALA molecule and ATCase causes the conversion of the five remaining active sites to the R conformation. Therefore, the T→R transition occurs in a concerted fashion. The major change in tertiary structure induced by PALA binding is a substantial shift in the polypeptide stretch spanning residues 225 through 245 of the C-chain (Figure 12.44). This shift closes the active site *and* induces a dramatic change in the quaternary structure. As the T state is converted to the R state, the two C-trimers move apart by 1.2 nm and rotate some 10° relative to each other about the central threefold axis. The R dimers reorient some 15° closer to the perpendicular as the C trimers separate in order to maintain the appropriate C-R-R-C interactions.

The Allosteric Mechanism

We can now suggest a scenario for the allosteric transition in ATCase: Substrate binding causes a tertiary conformational change within the catalytic chain that results in the closure of the two active site-forming domains. This domain closure is more than a simple hinge motion; it is a complex structural rearrangement in which the shift of residues 225 through 245 leads to a major reorientation that is stabilized by new interdomain and intrachain interactions. This tertiary transition cannot occur without conformational changes at the quaternary level. In a concerted transformation from the "closed" T state to the "open" R state, the holoenzyme opens up through expansion along its threefold symmetry axis. Thus, the binding of substrates to only one active site is sufficient to cause all of the five remaining active sites to be converted, in a

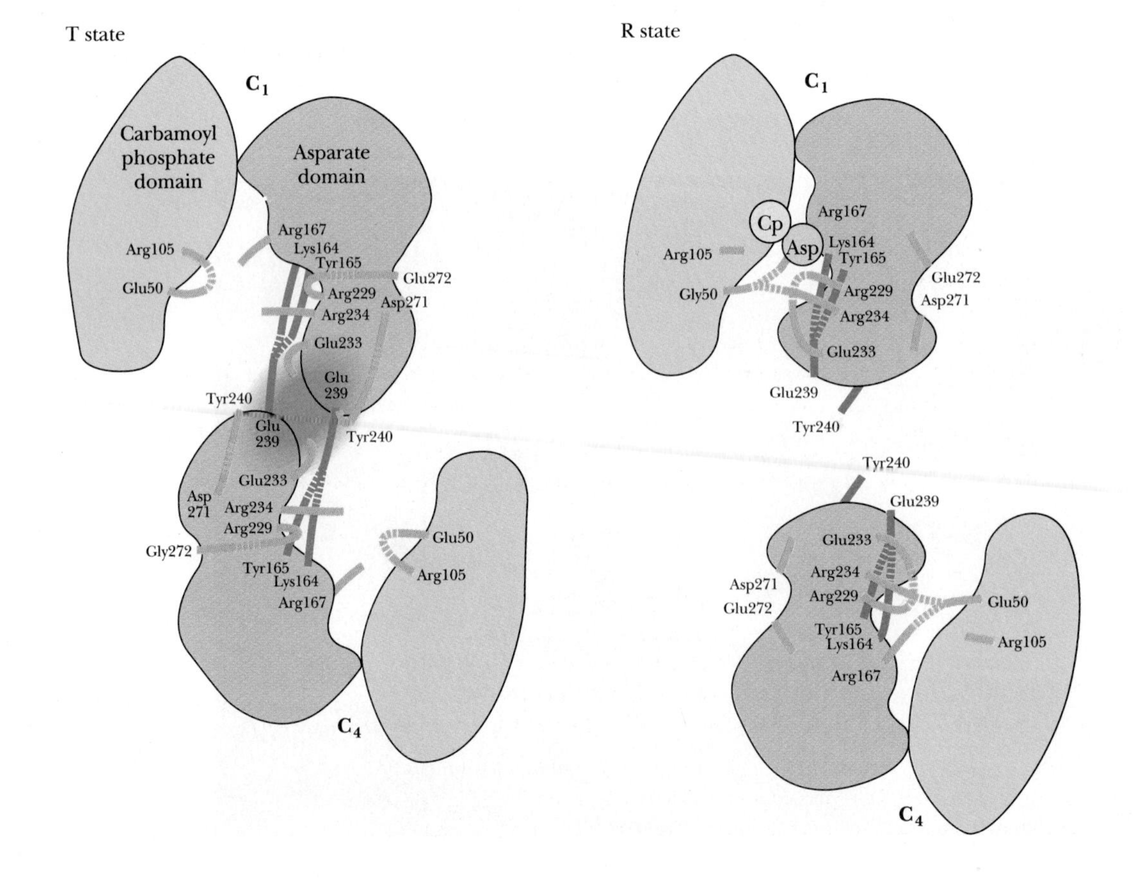

***Figure* 12.44** Tertiary and quaternary changes in the structure of ATCase on substrate binding suggest a mechanism for homotropic cooperativity.

concerted fashion, into the high-affinity, high-activity conformation. Furthermore, crystallographic studies on the role of the heterotropic effectors, CTP and ATP, support the view that the action of these nucleotides is to perturb the T/R equilibrium, as opposed to *directly* altering the nature of the active sites in some way.

Thus, the behavior of ATCase conforms to the postulates of the Monod, Wyman, Changeux model for allosteric transitions. The allosteric behavior of hemoglobin also follows the MWC pattern, although the occurrence of Hb species with only one or two equivalents of O_2 bound suggests that the KNF model is also useful in understanding the mechanics of allostery.

Problems

1. List six general ways in which enzyme activity is controlled.
2. Why do you suppose proteolytic enzymes are often synthesized as inactive zymogens?
3. First draw both Lineweaver–Burk plots and Eadie–Hofstee plots for the following: a Monod–Wyman–Changeux allosteric K enzyme system, showing separate curves for the kinetic response in (1) the absence of any effectors; (2) the presence of allosteric activator A; and (3) the presence of allosteric inhibitor I. Then draw a similar set of curves for a Monod, Wyman, Changeux allosteric V enzyme system.
4. In the Monod, Wyman, Changeux model for allosteric regulation, what values of L and relative affinities of R and T for A will lead activator A to exhibit positive homotropic effects? (That is, under what conditions will the binding of A enhance further A-binding, in the same manner that S-binding shows positive cooperativity?) Similarly, what values of L and relative affinities of R and T for I will lead inhibitor I to exhibit positive homotropic effects? (That is, under what conditions will the binding of I promote further I-binding?)
5. The equation $\frac{Y}{(1-Y)} = \left(\frac{pO_2}{P_{50}}\right)^n$ allows the calculation of Y (the fractional saturation of hemoglobin with O_2), given P_{50} and n (see Box on page 399). Let $P_{50} = 26$ torr and $n = 2.8$. Calculate Y in the lungs where $pO_2 = 100$ torr, and Y in the capillaries where $pO_2 = 40$ torr. What is the efficiency of O_2 delivery under these conditions (expressed as $Y_{lungs} - Y_{capillaries}$)? Repeat the calculations, but for $n = 1$. Compare the values for $Y_{lungs} - Y_{capillaries}$ for $n = 2.8$ versus $Y_{lungs} - Y_{capillaries}$ for $n = 1$ to determine the effect of cooperative O_2-binding on oxygen delivery by hemoglobin.
6. If you move from Key West, Florida (altitude = 10 feet above sea level), to Denver, Colorado (altitude = 5280 feet above sea level), will the 2,3-bisphosphoglycerate (2,3-BPG) concentration in your red blood cells increase, decrease, or stay the same? Explain your answer.
7. If no precautions are taken, blood that has been stored for some time becomes depleted in 2,3-BPG. What will happen if such blood is used in a transfusion?
8. Enzymes have evolved such that their K_m values (or $K_{0.5}$ values) for substrate(s) are roughly equal to the *in vivo* concentration(s) of the substrate(s). Assume that aspartate transcarbamoylase (ATCase) is assayed at [aspartate] $\approx K_{0.5}$ in the absence and presence of ATP or CTP. Estimate from Figures 12.28, 12.29, and 12.30 the relative ATCase activity when (a) neither ATP or CTP is present, (b) ATP is present, and (c) CTP is present.

Further Reading

Creighton, T. E., 1984. *Proteins: Structure and Molecular Properties.* New York: W. H. Freeman and Co. An advanced textbook on the structure and function of proteins.

Dickerson, R. E., and Geis, I., 1983. *Hemoglobin: Structure, Function, Evolution and Pathology.* Menlo Park, CA: Benjamin/Cummings.

Gill, S. J., et al., 1988. New twists on an old story: hemoglobin. *Trends in Biochemical Sciences* **13:**465–467.

Kantrowitz, E. R., and Lipscomb, W. N., 1988. *Escherichia coli* aspartate transcarbamoylase: The relation between structure and function. *Science* **241:**669–674. Description of the structure of ATCase derived from X-ray crystallographic analysis and implications of this structure for the enzyme's function and regulation.

Koshland, D. E., Jr., Nemethy, G., and Filmer, D., 1966. Comparison of experimental binding data and theoretical models in proteins containing subunits. *Biochemistry* **5:**365–385. The KNF model.

Monod, J., Wyman, J., and Changeux, J-P, 1965. On the nature of allosteric transitions: A plausible model. *The Journal of Molecular Biology* **12:**88–118. The classic paper that provided the first theoretical analysis of allosteric regulation.

Schachman, H. K., 1988. Can a simple model account for the allosteric transition of aspartate transcarbamoylase? *The Journal of Biological Chemistry* **263:**18583–18586. Tests of the postulates underlying allosteric models through experiments on ATCase.

Stevens, R. C., Gouaux, J. E., and Lipscomb, W. N., 1990. Structural consequences of effector binding to the T state of aspartate carbamoyltransferase: Crystal structures of the unligated and ATP- and CTP-complexed enzymes at 2.6-Å resolution. *Biochemistry* **29:**7691–7701.

Appendix to Chapter 12

The Oxygen-Binding Curves of Myoglobin and Hemoglobin

Myoglobin

The reversible binding of oxygen to myoglobin,

$$MbO_2 \rightleftharpoons Mb + O_2$$

can be characterized by the equilibrium dissociation constant, K.

$$K = \frac{[Mb][O_2]}{[MbO_2]} \quad \text{(A12.1)}$$

If Y is defined as the **fractional saturation** of myoglobin with O_2, that is, the fraction of myoglobin molecules having an oxygen molecule bound, then

$$Y = \frac{[MbO_2]}{[MbO_2] + [Mb]} \quad \text{(A12.2)}$$

The value of Y ranges from 0 (no myoglobin molecules carry an O_2) to 1.0 (all myoglobin molecules have an O_2 molecule bound). Substituting from Equation (A12.1), $([Mb][O_2])/K$ for $[MbO_2]$ gives

$$Y = \frac{\left(\frac{[Mb][O_2]}{K}\right)}{\left(\frac{[Mb][O_2]}{K} + [Mb]\right)} = \frac{\left(\frac{[O_2]}{K}\right)}{\left(\frac{[O_2]}{K} + 1\right)} = \frac{[O_2]}{[O_2] + K} \quad \text{(A12.3)}$$

and, if the concentration of O_2 is expressed in terms of the partial pressure (in torr) of oxygen gas in equilibrium with the solution of interest, then

$$Y = \frac{pO_2}{pO_2 + K} \quad \text{(A12.4)}$$

(In this form, K has the units of torr.) If we define P_{50} as the partial pressure of O_2 at which 50% of the myoglobin molecules have a molecule of O_2 bound (that is, $Y = 0.5$), then

$$Y = \frac{pO_2}{pO_2 + P_{50}} \quad \text{(A12.5)}$$

(Note from Equation (A12.1) that when $[MbO_2] = [Mb]$, $K = [O_2]$, which is the same as saying when $Y = 0.5$, $K = P_{50}$.)

The relationship defined by Equation (A12.4) plots as a hyperbola. That is, the MbO_2 saturation curve resembles an enzyme:substrate saturation curve. For myoglobin, a partial pressure of 1 torr for pO_2 is sufficient for half-saturation (Figure A12.1).

The ratio of the fractional saturation of myoglobin, Y, to free myoglobin, $1 - Y$, depends on pO_2 and K according to the equation

$$\frac{Y}{1 - Y} = \frac{pO_2}{K} \tag{A12.6}$$

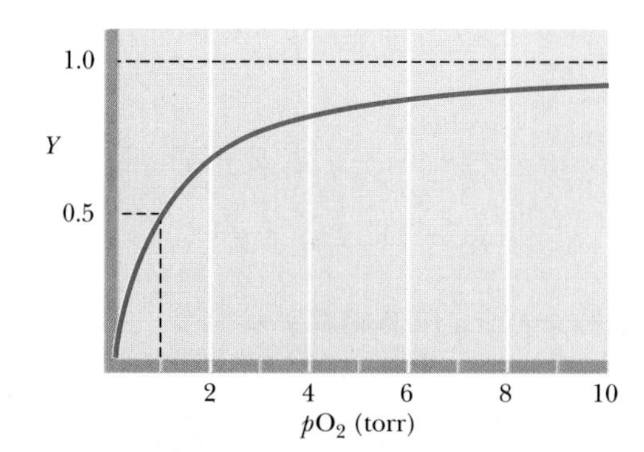

Figure **A12.1** Oxygen saturation curve for myoglobin in the form of Y versus pO_2, showing P_{50} is at a pO_2 of 1 torr (1 mm Hg).

Taking the logarithm yields

$$\log\left(\frac{Y}{1 - Y}\right) = \log pO_2 - \log K \tag{A12.7}$$

A graph of $\log (Y/(1 - Y))$ versus $\log pO_2$ is known as a **Hill plot** (in honor of Archibald Hill, a pioneer in the study of O_2-binding by hemoglobin). A Hill plot for myoglobin (Figure A12.2) gives a straight line. At half-saturation, defined as $Y = 0.5$, $Y/(1 - Y) = 1$, and $\log (Y/(1 - Y)) = 0$. At this value of $\log (Y/(1 - Y))$, the value for $pO_2 = K = P_{50}$. The slope of the Hill plot at $\log (Y/(1 - Y)) = 0$, the midpoint of binding, is known as the **Hill coefficient.** The Hill coefficient for myoglobin is 1.0. A Hill coefficient of 1.0 means that O_2 molecules bind independently of one another to myoglobin, a conclusion entirely logical since each Mb molecule can bind only one O_2.

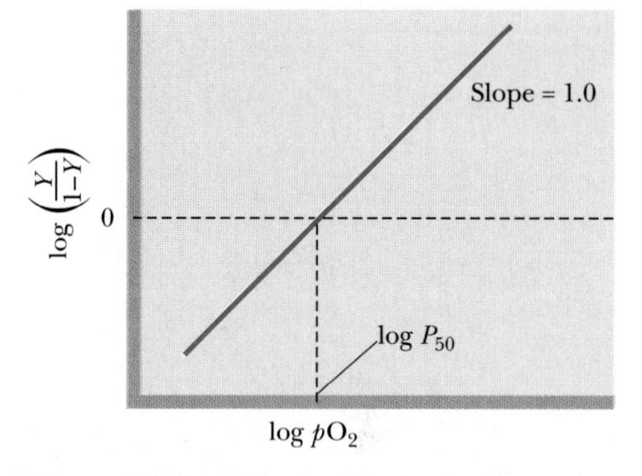

Figure **A12.2** Hill plot for the binding of O_2 to myoglobin. The slope of the line is the **Hill coefficient.** For Mb, the Hill coefficient is 1.0. At $\log(Y/(1 - Y)) = 0$, $\log pO_2 = \log P_{50}$.

Hemoglobin

New properties emerge when four heme-containing polypeptides come together to form a tetramer. The O_2-binding curve of hemoglobin is sigmoid rather than hyperbolic (see Figure 12.8), and Equation (A12.4) does not describe such curves. Of course, each hemoglobin molecule has four hemes and can bind up to four oxygen molecules. Suppose for the moment that O_2-binding to hemoglobin is an "all-or-none" phenomenon, where Hb exists either free of O_2 or with four O_2 molecules bound. This supposition represents the extreme case for cooperative binding of a ligand by a protein with multiple binding sites. In effect, it says that if one ligand binds to the protein molecule, then all other sites are immediately occupied by ligand. Or, to say it another way for the case in hand, suppose that four O_2 molecules bind to Hb simultaneously:

$$Hb + 4\,O_2 \rightleftharpoons Hb(O_2)_4$$

Then the *dissociation* constant, K, would be

$$K = \frac{[Hb][O_2]^4}{[Hb(O_2)_4]} \tag{A12.8}$$

By analogy with Equation (A12.4), the equation for fractional saturation of Hb is given by

$$Y = \frac{[pO_2]^4}{[pO_2]^4 + K} \tag{A12.9}$$

A plot of Y versus pO_2 according to Equation (12.9) is presented in Figure A12.3. This curve has the characteristic sigmoid shape seen for O_2-binding by Hb. Half-saturation is seen at a pO_2 of 26 torr. Note that, when pO_2 is low, the

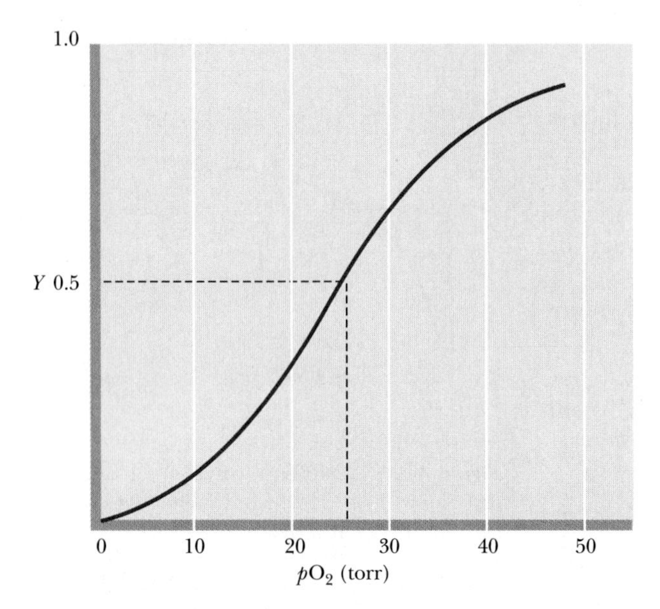

Figure **A12.3** Oxygen saturation curve for Hb in the form of Y versus pO_2, assuming $n = 4$ and $P_{50} = 26$ torr. The graph has the characteristic experimentally observed sigmoid shape.

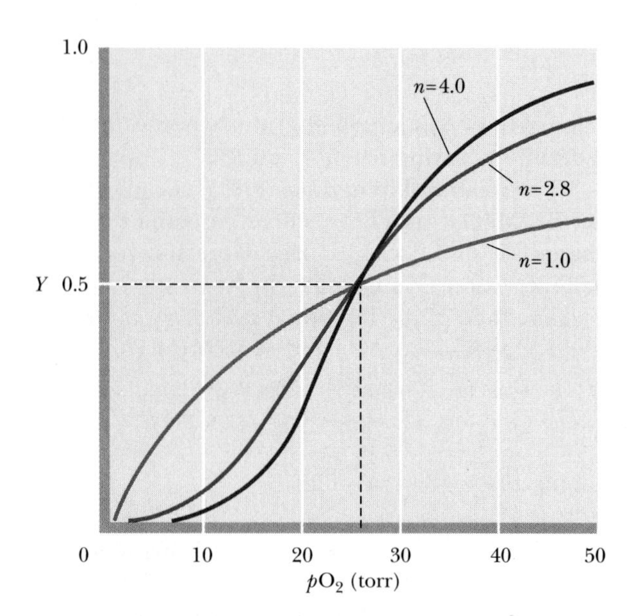

Figure **A12.4** A comparison of the experimentally observed O_2-binding curve for Hb yielding a value for n of 2.8, the hypothetical curve if $n = 4$, and the curve if $n = 1$ (noninteracting O_2-binding sites).

fractional saturation, Y, changes very little as pO_2 increases. The interpretation is that Hb has little affinity for O_2 at these low partial pressures of O_2. However, as pO_2 reaches some threshold value and the first O_2 is bound, Y, the fractional saturation, increases rapidly. Note that the slope of the curve is steepest in the region where $Y = 0.5$. The sigmoid character of this curve is diagnostic of the fact that the binding of O_2 to one site on Hb strongly enhances binding of additional O_2 molecules to the remaining vacant sites on the same Hb, a phenomenon aptly termed **cooperativity.** (If each O_2 bound independently, exerting no influence on the affinity of Hb for more O_2-binding, this plot would be hyperbolic.)

The experimentally observed oxygen-binding curve for Hb does not fit the graph given in Figure A12.3 exactly. If we generalize Equation (A12.9) by replacing the exponent 4 by n, we can write the equation as

$$Y = \frac{[pO_2]^n}{[pO_2]^n + K} \tag{A12.10}$$

Rearranging yields

$$\frac{Y}{1 - Y} = \frac{[pO_2]^n}{K} \tag{A12.11}$$

This equation states that the ratio of oxygenated heme groups (Y) to unoccupied heme $(1 - Y)$ is equal to the nth power of the pO_2 divided by the apparent dissociation constant, K.

Archibald Hill demonstrated in 1913, well before any knowledge about the molecular organization of Hb existed, that the O_2-binding behavior of Hb could be described by Equation (A12.11). If a value of 2.8 is taken for n, Equation (A12.11) fits the experimentally observed O_2-binding curve for Hb very well (Figure A12.4). If the binding of O_2 to Hb were an all-or-none phenomenon, n would equal 4, as discussed above. If the O_2-binding sites on Hb were completely noninteracting, that is, if the binding of one O_2 to Hb had no influence on the binding of additional O_2 molecules to the same Hb, n would equal 1. Figure A12.4 compares these extremes. Obviously, the real situation falls between the extremes of $n = 1$ or 4. The qualitative answer is that O_2-binding by Hb is highly cooperative, and the binding of the first O_2 markedly enhances the binding of subsequent O_2 molecules. However, this binding is not quite an all-or-none phenomenon.

If we take the logarithm of both sides of Equation (A12.11):

$$\log\left(\frac{Y}{1-Y}\right) = n(\log pO_2) - \log K \qquad \text{(A12.12)}$$

this expression is, of course, the generalized form of Equation (A12.7), the *Hill equation,* and a plot of $\log(Y/(1-Y))$ versus $(\log pO_2)$ *approximates* a straight line in the region around $\log(Y/(1-Y)) = 0$. Figure A12.5 represents a *Hill plot* comparing hemoglobin and myoglobin.

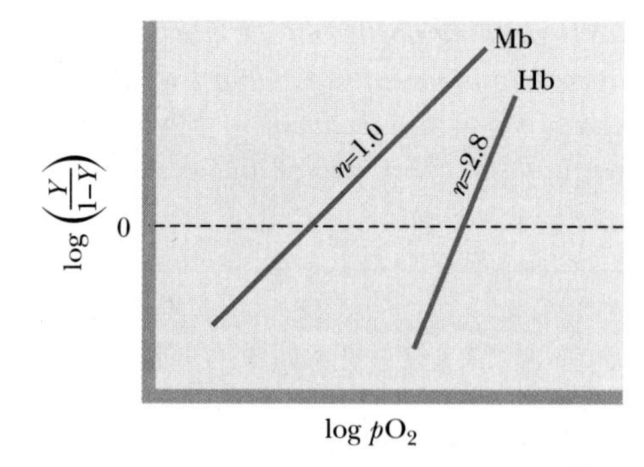

***Figure* A12.5** Hill plot ($\log(Y/(1-Y))$ versus $\log pO_2$) for Mb and Hb, showing that at $\log(Y/(1-Y)) = 0$, that is, $(Y = (1-Y))$, the slope for Mb is 1.0 and for Hb is 2.8. The plot for Hb only approximates a straight line.

Because the binding of oxygen to hemoglobin is cooperative, the Hill plot is actually sigmoid (Figure A12.6). Cooperativity is a manifestation of the fact that the dissociation constant for the *first* O_2 *bound,* K_1, is very different from the dissociation constant for the *last* O_2 *bound,* K_4. The tangent to the lower asymptote of the Hill plot, when extrapolated to the $\log(Y/(1-Y)) = 0$ axis, gives the dissociation constant, K_1, for the binding of the first O_2 by Hb. Note that the value of K_1 is quite large ($>10^2$ torr), indicating a low affinity of Hb for this first O_2 (or conversely, a ready dissociation of the $Hb(O_2)_1$ complex). By a similar process, the tangent to the upper asymptote gives K_4, the dissociation constant for the last O_2 to bind. K_4 has a value of less than 1 torr. The K_1/K_4 ratio exceeds 100, meaning the affinity of Hb for binding the fourth O_2 is over 100 times greater than for binding the first oxygen.

The value P_{50} has been defined above for myoglobin as the pO_2 that gives 50% saturation of the oxygen-binding protein with oxygen. Noting that at 50% saturation, $Y = (1-Y)$, then we have from Equation (A12.12)

$$0 = n(\log pO_2) - \log K = n(\log P_{50}) - \log K \qquad \text{(A12.13)}$$

$$\log K = n(\log P_{50}) \quad \textit{or} \quad K = (P_{50})^n \qquad \text{(A12.14)}$$

That is, the situations for myoglobin and hemoglobin differ; therefore, P_{50} and K cannot be equated for the multiply-binding molecule Hb. The relationship between pO_2 and P_{50} for hemoglobin, by use of Equation (A12.11), becomes

$$\frac{Y}{1-Y} = \left(\frac{pO_2}{P_{50}}\right)^n \qquad \text{(A12.15)}$$

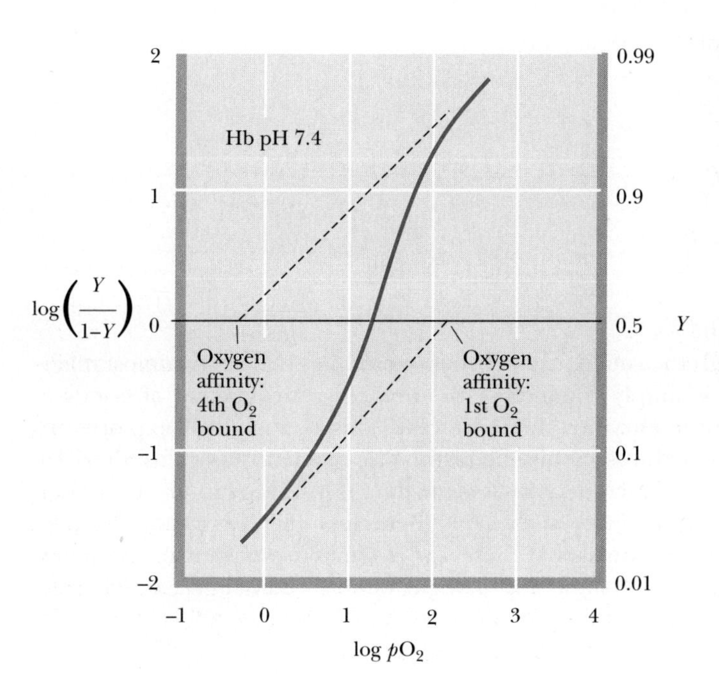

***Figure* A12.6** Hill plot of Hb showing its nonlinear nature and the fact that its asymptotes can be extrapolated to yield the dissociation constants, K_1 and K_4, for the first and fourth oxygens.

*"No single thing abides but all things flow.
Fragment to fragment clings and thus they grow
Until we know them by name.
Then by degrees they change and are no more
the things we know."*

Lucretius (ca. 94 B.C.–50 B.C.)

Chapter 13

Mechanisms of Enzyme Action

Like the workings of an antique clock, the details of enzyme mechanisms are at once complex and simple.

Outline

13.1 The Basic Principle—Stabilization of the Transition State
13.2 Enzymes Provide Enormous Rate Accelerations
13.3 The Binding Energy of ES Is Crucial to Catalysis
13.4 Entropy Loss and Destabilization of the ES Complex
13.5 Transition-State Analogs Bind Very Tightly to the Active Site
13.6 Covalent Catalysis
13.7 General Acid–Base Catalysis
13.8 Metal Ion Catalysis
13.9 Proximity
13.10 Typical Enzyme Mechanisms
13.11 Serine Proteases
13.12 The Aspartic Proteases
13.13 Lysozyme
13.14 Carboxypeptidase A
13.15 Liver Alcohol Dehydrogenase

Although the catalytic properties of enzymes may seem almost magical, it is simply chemistry—the breaking and making of bonds—that gives enzymes their prowess. This chapter will explore the unique features of this chemistry. The mechanisms of hundreds of enzymes have been studied in at least some detail. In this chapter, it will only be possible to examine a few of these. Nonetheless, the chemical principles that influence the mechanisms of these few enzymes are universal, and many other cases will be understandable in light of the knowledge gained from these examples.

13.1 The Basic Principle—Stabilization of the Transition State

In all chemical reactions, the reacting atoms or molecules pass through a state that is intermediate in structure between the reactant(s) and the product(s). Consider the transfer of a proton from a water molecule to a chloride anion:

$$\underset{\text{Reactants}}{\text{H—O—H} + \text{Cl}^-} \rightleftharpoons \underset{\text{Transition state}}{\text{H—O}^{\delta-}\cdots\text{H}\cdots\text{Cl}^{\delta-}} \rightleftharpoons \underset{\text{Products}}{\text{HO}^- \quad + \quad \text{H—Cl}}$$

In the middle structure, the proton undergoing transfer is shared equally by the hydroxyl and chloride anions. This structure represents, as nearly as possible, the transition between the reactants and products, and it is known as the **transition state.**

Chemical reactions in which a substrate (S) is converted to a product (P) can be pictured as involving a transition state (which we will henceforth denote as $X^\ddagger$), a species intermediate in structure between S and P (Figure 13.1). As seen in Chapter 11, the catalytic role of an enzyme is to reduce the energy barrier between substrate and transition state. This is accomplished through the formation of an **enzyme-substrate complex** (ES). This complex is converted to product by passing through a transition state, $EX^\ddagger$ (Figure 13.1). As shown, the energy of $EX^\ddagger$ is clearly lower than $X^\ddagger$. One might be tempted to conclude that this decrease in energy explains the rate acceleration achieved by the enzyme, but there is more to the story.

The energy barrier for the uncatalyzed reaction (Figure 13.1) is of course the difference in energies of the S and $X^\ddagger$ states. Similarly, the energy barrier to be surmounted in the enzyme-catalyzed reaction is the energy difference between ES and $EX^\ddagger$. *Reaction rate acceleration by an enzyme means simply that the energy barrier between ES and $EX^\ddagger$ is less than the energy barrier between S and $X^\ddagger$.* In terms of the free energies of activation, $\Delta G_e^\ddagger < \Delta G_u^\ddagger$.

There are important consequences for this statement. The enzyme must stabilize the transition-state complex, $EX^\ddagger$, more than it stabilizes the substrate complex, ES. Put another way, enzymes are "designed" by nature to bind the transition-state structure more tightly than the substrate (or the product). The dissociation constant for the enzyme-substrate complex is

$$K_S = \frac{[E][S]}{[ES]} \tag{13.1}$$

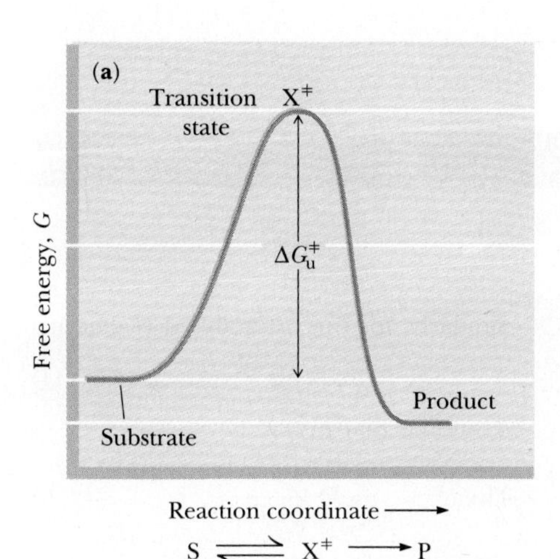

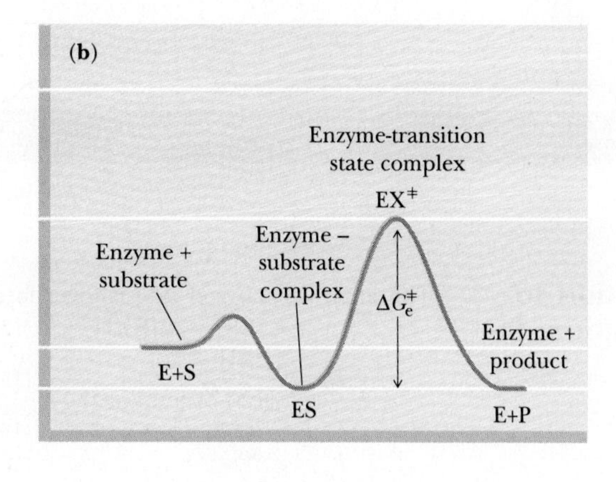

***Figure* 13.1** Enzymes catalyze reactions by lowering the activation energy. Here the free energy of activation for (a) the uncatalyzed reaction, $\Delta G_u^\ddagger$, is larger than that for (b) the enzyme-catalyzed reaction, $\Delta G_e^\ddagger$.

and the corresponding dissociation constant for the transition-state complex is

$$K_T = \frac{[E][X^\ddagger]}{[EX^\ddagger]} \qquad (13.2)$$

Enzyme catalysis requires that $K_T < K_S$. According to **transition-state theory** (see references at end of chapter), the rate constants for the enzyme-catalyzed (k_e) and uncatalyzed (k_u) reactions can be related to K_S and K_T by:

$$k_e/k_u \cong K_S/K_T \qquad (13.3)$$

Thus, the enzymatic rate acceleration is approximately equal to the ratio of the dissociation constants of the enzyme-substrate and enzyme-transition-state complexes.

13.2 Enzymes Provide Enormous Rate Accelerations

Enzymes are powerful catalysts. Enzyme-catalyzed reactions are typically 10^7 to 10^{14} times faster than their uncatalyzed counterparts (Table 13.1). (There is even a report of a rate acceleration of $>10^{16}$ for the alkaline phosphatase-catalyzed hydrolysis of methylphosphate!) From Equation 13.13, one can see that these large rate accelerations correspond to substantial changes in the free energy of activation for the reaction in question. The urease reaction, for example,

$$H_2N-\overset{\overset{\displaystyle O}{\|}}{C}-NH_2 + 2\,H_2O + H^+ \longrightarrow 2\,NH_4^+ + HCO_3^-$$

shows an energy of activation some 84 kJ/mol smaller than the corresponding uncatalyzed reaction. To fully understand any enzyme reaction, it is important to account for the rate acceleration in terms of the structure of the enzyme and its mechanism of action. There are a limited number of catalytic mechanisms or factors that contribute to the remarkable performance of enzymes. These include the following:

1. Entropy loss in ES formation;
2. Destabilization of ES due to strain, desolvation, or electrostatic effects;

A Deeper Look

Relating Rate Acceleration to Free Energies of Activation

Consider the following scheme:

$$\begin{array}{ccccc} S & \underset{}{\overset{K_u^\ddagger}{\rightleftharpoons}} & X^\ddagger & \xrightarrow{k_u'} & P \\ E \updownarrow K_S & & & & \updownarrow E \\ ES & \overset{K_e^\ddagger}{\rightleftharpoons} & EX^\ddagger & \xrightarrow{k_e'} & EP \end{array}$$

The enzyme-catalyzed rate is given by:

$$v = k_e[ES] = k_e'[EX^\ddagger] \qquad (13.4)$$

$$K_e^\ddagger = \frac{[EX^\ddagger]}{[ES]} \qquad (13.5)$$

$$\Delta G_e^\ddagger = -RT \ln K_e^\ddagger \qquad (13.6)$$

$$K_e^\ddagger = e^{-\Delta G_e^\ddagger/RT} \qquad (13.7)$$

$$[EX^\ddagger] = K_e^\ddagger[ES] = e^{-\Delta G_e^\ddagger/RT}[ES] \qquad (13.8)$$

So $k_e[ES] = k_e' e^{-\Delta G_e^\ddagger/RT}[ES]$ (13.9)

or $k_e = k_e' e^{-\Delta G_e^\ddagger/RT}$ (13.10)

Similarly, for the uncatalyzed reaction:

$$k_u = k_u' e^{-\Delta G_u^\ddagger/RT} \qquad (13.11)$$

Assuming that $k_u' \cong k_e'$

Then

$$\frac{k_e}{k_u} = \frac{e^{-\Delta G_e^\ddagger/RT}}{e^{-\Delta G_u^\ddagger/RT}} \qquad (13.12)$$

$$\frac{k_e}{k_u} = e^{(\Delta G_u^\ddagger - \Delta G_e^\ddagger)/RT} \qquad (13.13)$$

Table **13.1**

A Comparison of Enzyme-Catalyzed Reactions and Their Uncatalyzed Counterparts

Reaction	Enzyme	Uncatalyzed Rate, v_u (sec^{-1})	Catalyzed Rate, v_e (sec^{-1})	v_e/v_u
$CH_3—O—PO_3^{2-} + H_2O \longrightarrow CH_3OH + HPO_4^{2-}$	Alkaline phosphatase	1×10^{-15}	14	1.4×10^{16}
$H_2N—C(=O)—NH_2 + 2\ H_2O + H^+ \longrightarrow 2\ NH_4^+ + HCO_3^-$	Urease	3×10^{-10}	3×10^4	1×10^{14}
$R—C(=O)—O—CH_2CH_3 + H_2O \longrightarrow RCOOH + HOCH_2CH_3$	Chymotrypsin	1×10^{-10}	1×10^2	1×10^{12}
Glycogen (n) + $P_i \longrightarrow$ Glycogen $(n-1)$ + Glucose-1-P	glycogen phosphorylase	$<5 \times 10^{-15}$	1.6×10^{-3}	$>3.2 \times 10^{11}$
Glucose + ATP $\longrightarrow$ Glucose-6-P + ADP	Hexokinase	$<1 \times 10^{-13}$	1.3×10^{-3}	$>1.3 \times 10^{10}$
$CH_3CH_2OH + NAD^+ \longrightarrow CH_3C(=O)H + NADH + H^+$	Alcohol dehydrogenase	$<6 \times 10^{-12}$	2.7×10^{-3}	$>4.5 \times 10^8$
$CO_2 + H_2O \longrightarrow HCO_3^- + H^+$	Carbonic anhydrase	10^{-2}	10^5	1×10^7
Creatine + ATP $\longrightarrow$ Cr-P + ADP	Creatine kinase	$<3 \times 10^{-9}$	4×10^{-5}	$>1.33 \times 10^4$

Adapted from Koshland, D., 1956. *Journal of Cellular Comparative Physiology*, Supp. 1, **47**:217.

3. Covalent catalysis;
4. General acid or base catalysis;
5. Metal ion catalysis; and
6. Proximity and orientation.

Any or all of these mechanisms may contribute to the net rate acceleration of an enzyme-catalyzed reaction relative to the uncatalyzed reaction. A thorough understanding of any enzyme would require that the net acceleration be accounted for in terms of contributions from one or (usually) more of these mechanisms. Each of these will be discussed in detail in this chapter, but first it is important to appreciate how the formation of the enzyme-substrate (ES) complex makes all these mechanisms possible.

13.3 The Binding Energy of ES Is Crucial to Catalysis

How is it that $X^‡$ is stabilized more than S at the enzyme active site? To understand this, we must dissect and analyze the formation of the enzyme-substrate complex, ES. There are a number of important contributions to the free energy difference between the uncomplexed enzyme and substrate (E + S) and the ES complex (Figure 13.2). The favorable interactions between the substrate and amino acid residues on the enzyme account for the **intrinsic binding energy, ΔG_b.** The intrinsic binding energy ensures the favorable formation of the enzyme-substrate complex, but, if uncompensated, it makes the activation energy for the enzyme-catalyzed reaction unnecessarily large and wastes some of the catalytic power of the enzyme. Compare the two cases in Figure 13.3. Since the enzymatic reaction rate is determined by the difference in energies between ES and $EX^‡$, the smaller this difference, the faster the enzyme-catalyzed reaction. Tight binding of the substrate deepens the energy well of the ES complex and actually lowers the rate of the reaction.

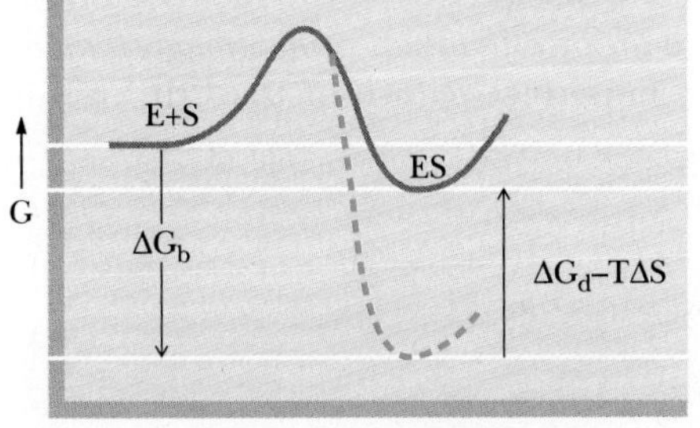

***Figure* 13.2** The intrinsic binding energy of the enzyme-substrate complex (ΔG_b) is compensated to some extent by entropy loss due to the binding of E and S ($T\Delta S$) and by destabilization of ES (ΔG_d) owing to strain, distortion, desolvation, and similar effects. If ΔG_b were not compensated by $T\Delta S$ and ΔG_d, the formation of ES would follow the dashed line.

Figure **13.3** (a) Catalysis does not occur if the ES complex and the transition state for the reaction are stabilized to equal extents. (b) Catalysis *will* occur if the transition state is stabilized to a greater extent than the ES complex. Entropy loss $T\Delta S$ and destabilization of the ES complex ΔG_d ensure that this will be the case.

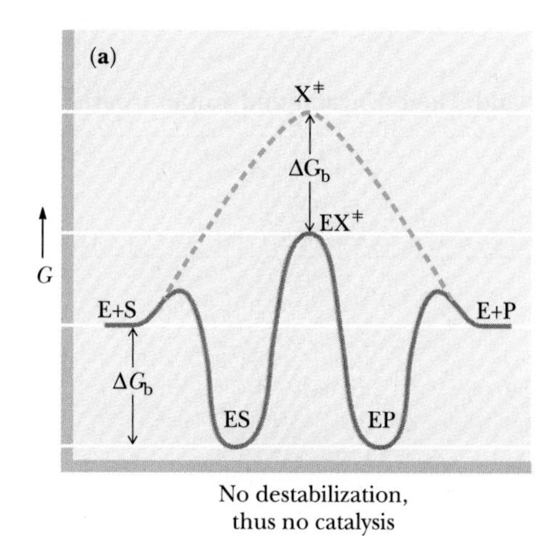

No destabilization, thus no catalysis

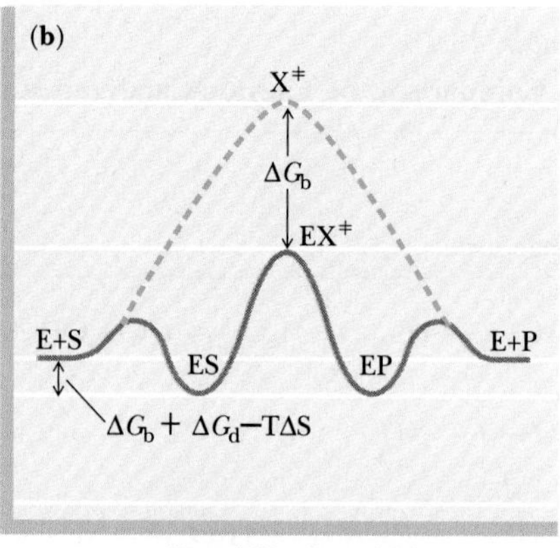

Destabilization of ES facilitates catalysis

13.4 Entropy Loss and Destabilization of the ES Complex

The message of Figure 13.3 is that raising the energy of ES will increase the enzyme-catalyzed reaction rate. This is accomplished in two ways: (a) **loss of entropy** due to the binding of S to E, and (b) **destabilization of ES** by strain, distortion, desolvation, or other similar effects. The entropy loss arises from the fact that the ES complex (Figure 13.4) is a highly organized (low-entropy) entity compared to E + S in solution (a disordered, high-entropy situation). The entry of the substrate into the active site brings all the reacting groups and coordinating residues of the enzyme together with the substrate in just the proper position for reaction, with a net loss of entropy. The substrate and enzyme both possess **translational entropy,** the freedom to move in three dimensions, as well as **rotational entropy,** the freedom to rotate or tumble about any axis through the molecule. Both types of entropy are lost to some extent when two molecules (E and S) interact to form one molecule (the ES complex). Since ΔS is negative for this process, the term $-T\Delta S$ is a positive quantity, and *the intrinsic binding energy of ES is compensated to some extent by the entropy loss that attends the formation of the complex.*

Destabilization of the ES complex can involve **structural strain, desolvation,** or **electrostatic effects.** Destabilization by strain or distortion is usually just a consequence of the fact (noted previously) that *the enzyme is designed to bind the transition state more strongly than the substrate.* When the substrate binds, the imperfect nature of the "fit" results in distortion or strain in the sub-

Substrate

Substrate

Enzyme

Figure **13.4** Formation of the ES complex results in a loss of entropy. Prior to binding, E and S are free to undergo translational and rotational motion. By comparison, the ES complex is a more highly ordered, low-entropy complex.

Substrate (and enzyme) are free to undergo translational motion. A disordered, high-entropy situation

The highly ordered, low-entropy complex

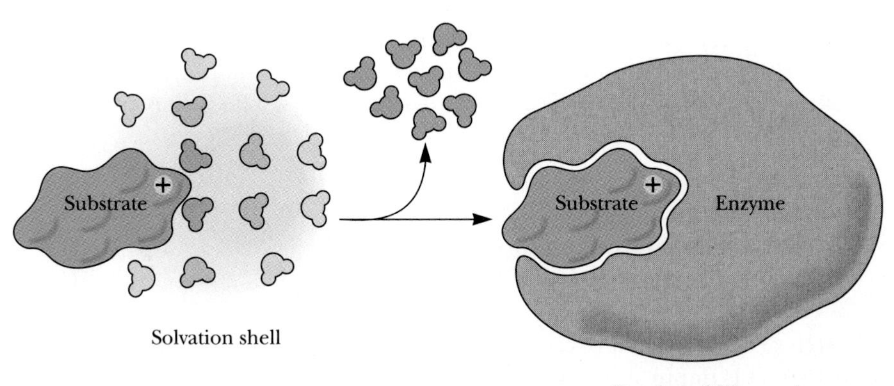

***Figure* 13.5** Substrates typically lose waters of hydration upon formation of the ES complex. Desolvation raises the energy of the ES complex, making it more reactive.

strate, the enzyme, or both. This means that the amino acid residues that make up the active site are oriented to coordinate the transition-state structure perfectly, but will interact with the substrate or product less effectively.

Destabilization may also involve desolvation of charged groups on the substrate upon binding in the active site. Charged groups are highly stabilized in water. For example, the transfer of Na^+ and Cl^- from the gas phase to aqueous solution is characterized by an **enthalpy of solvation, ΔH_{solv},** of -775 kJ/mol. (Energy is given off and the ions become more stable.) When charged groups on a substrate move from water into an enzyme active site (Figure 13.5), they are often desolvated to some extent, becoming less stable and therefore more reactive.

When a substrate enters the active site, charged groups may be forced to interact (unfavorably) with charges of like sign, resulting in **electrostatic destabilization** (Figure 13.6). The reaction pathway acts in part to remove this stress. If the charge on the substrate is diminished or lost in the course of reaction, electrostatic destabilization can result in a rate acceleration.

Whether by strain, desolvation, or electrostatic effects, destabilization raises the energy of the ES complex, and this increase is summed in the term ΔG_d, the free energy of destabilization. As noted in Figure 13.2, the net energy difference between E + S and the ES complex is the sum of the intrinsic binding energy, ΔG_b; the entropy loss on binding, $-T\Delta S$; and the distortion

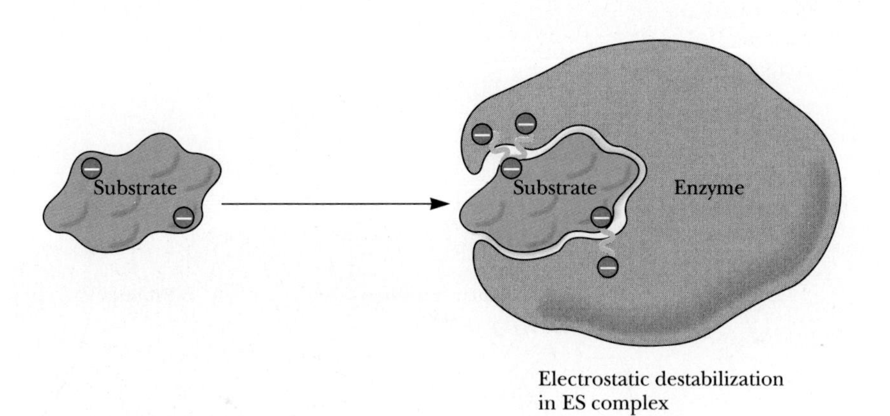

***Figure* 13.6** Electrostatic destabilization of a substrate may arise from juxtaposition of like charges in the active site. If such charge repulsion is relieved in the course of the reaction, electrostatic destabilization can result in a rate increase.

energy, ΔG_d. ES is destabilized (raised in energy) by the amount $\Delta G_d - T\Delta S$. The transition state is subject to no such destabilization, and the difference between the energies of $X^{\ddagger}$ and $EX^{\ddagger}$ is essentially ΔG_b, the full intrinsic binding energy.

13.5 Transition-State Analogs Bind Very Tightly to the Active Site

Though not apparent at first, there are other important implications of Equation 13.3. It is important to consider the magnitudes of K_S and K_T. The ratio k_e/k_u may even exceed 10^{16}, as noted previously. Given a typical ratio of 10^{12} and a typical K_S of $10^{-3}M$, the value of K_T should be $10^{-15}M$! This is the dissociation constant for the transition-state complex from the enzyme, and this very low value corresponds to very tight binding of the transition state by the enzyme.

It is unlikely that such tight binding in an enzyme transition state will ever be measured experimentally, however, since the transition state itself is a "moving target." It exists only for about 10^{-14} to 10^{-13} sec, less than the time required for a bond vibration. The nature of the elusive transition state can be explored, on the other hand, using **transition-state analogs,** stable molecules that are chemically and structurally similar to the transition state. Such molecules should bind more strongly than a substrate and more strongly than competitive inhibitors that bear no significant similarity to the transition state. Hundreds of examples of such behavior have been reported. For example, Robert Abeles studied a series of inhibitors of **proline racemase** (Figure 13.7) and found that *pyrrole-2-carboxylate* bound to the enzyme 160 times more tightly than L-proline, the normal substrate. This analog binds so tightly because it is planar and is similar in structure to the planar transition state for the racemization of proline. Two other examples of transition-state analogs are shown in Figure 13.8. *Phosphoglycolohydroxamate* binds 40,000 times more tightly to yeast aldolase than the substrate, dihydroxyacetone phosphate. Even more remarkable, the 1,6-hydrate of *purine ribonucleoside* has been estimated to bind to adenosine deaminase with a K_i of $3 \times 10^{-13}M$!

It should be noted that transition-state analogs are only approximations of the transition state itself and will never bind as tightly as would be expected for the true transition state. These analogs are, after all, stable molecules and cannot be expected to resemble a true transition state too closely.

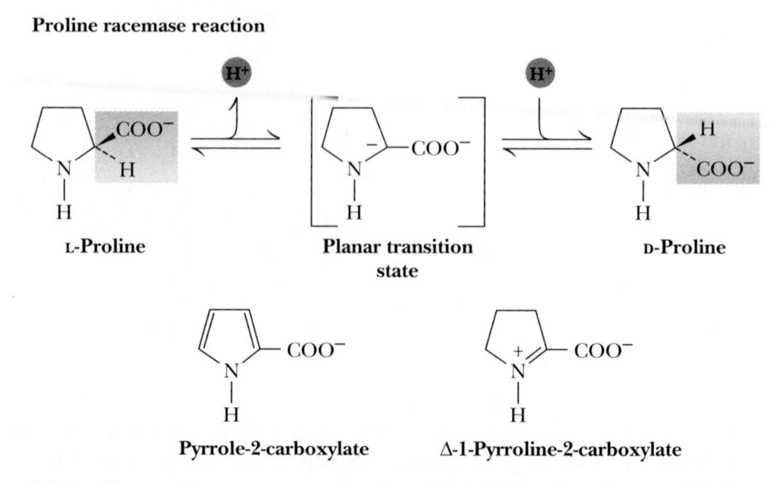

***Figure* 13.7** The proline racemase reaction. Pyrrole-2-carboxylate and Δ-1-pyrroline-2-carboxylate mimic the planar transition state of the reaction.

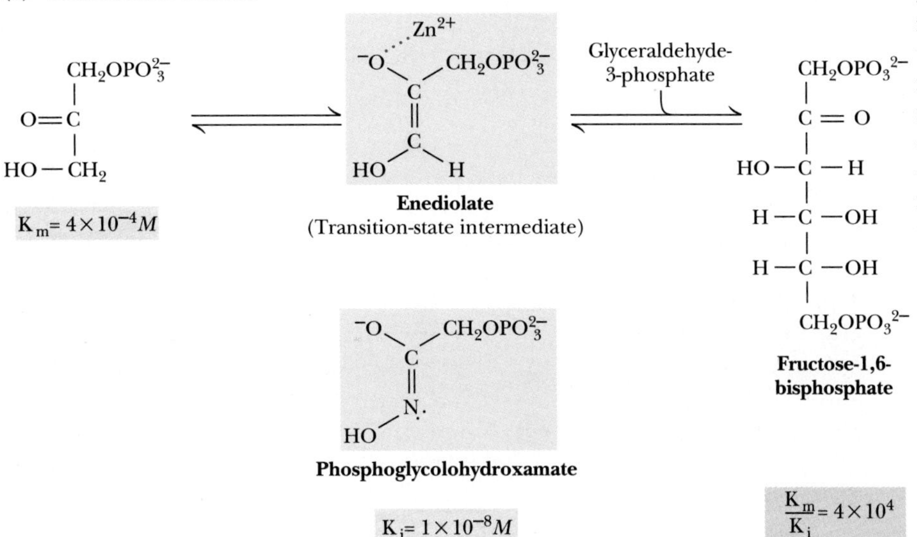

(b) Calf intestinal adenosine deaminase reaction

Adenosine

$K_m = 3 \times 10^{-5} M$

Transition-state intermediate

Inosine

Hydrated form of purine ribonucleoside

$K_i = 3 \times 10^{-13} M$

$\frac{K_m}{K_i} = 1 \times 10^8$

***Figure* 13.8** (a) Phosphoglycolohydroxamate is an analog of the enediolate transition state of the yeast aldolase reaction. (b) Purine riboside, a potent inhibitor of the calf intestinal adenosine deaminase reaction, binds to adenosine deaminase as the 1,6-hydrate. The hydrated form of purine riboside is an analog of the proposed transition state for the reaction.

13.6 Covalent Catalysis

Some enzyme reactions derive much of their rate acceleration from the formation of **covalent bonds** between enzyme and substrate. Consider the reaction:

$$BX + Y \longrightarrow BY + X$$

and an enzymatic version of this reaction involving formation of a **covalent intermediate:**

$$BX + Enz \longrightarrow E{:}B + X + Y \longrightarrow Enz + BY$$

If the enzyme-catalyzed reaction is to be faster than the uncatalyzed case, the acceptor group on the enzyme must be a better attacking group than Y and a better leaving group than X. Note that most enzymes that carry out covalent catalysis have ping-pong kinetic mechanisms.

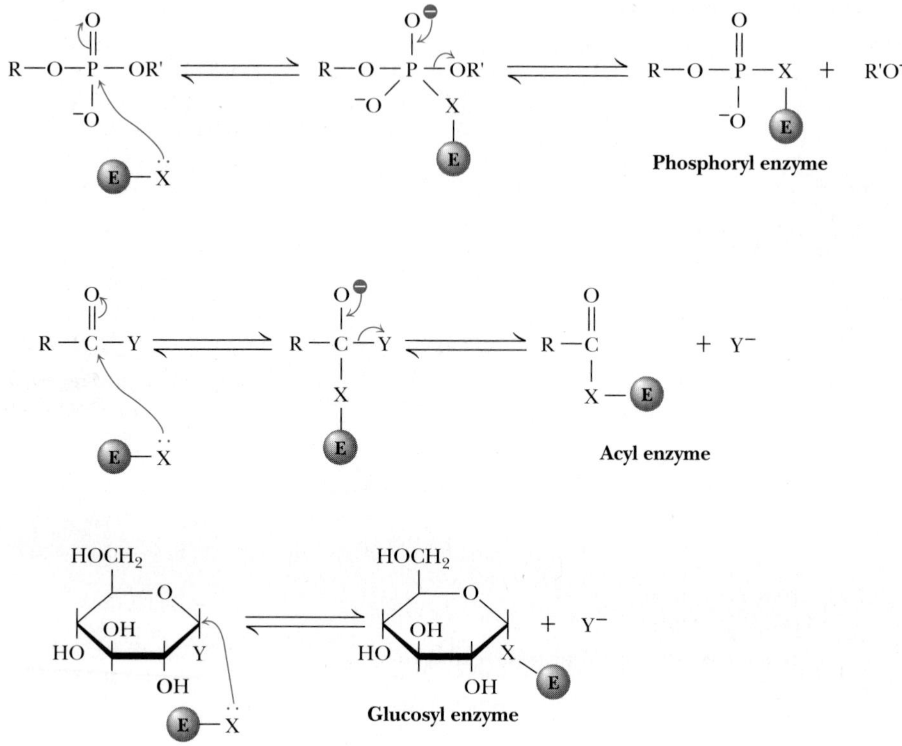

***Figure* 13.9** Examples of covalent bond formation between enzyme and substrate. In each case, a nucleophilic center (X:) on an enzyme attacks an electrophilic center on a substrate.

Phosphoryl enzyme

Acyl enzyme

Glucosyl enzyme

The side chains of amino acids in proteins offer a variety of **nucleophilic** centers for catalysis, including amines, carboxylates, aryl and alkyl hydroxyls, imidazoles, and thiol groups. These groups readily attack electrophilic centers of substrates, forming covalently bonded enzyme-substrate intermediates. Typical electrophilic centers in substrates include phosphoryl groups, acyl groups, and glycosyl groups (Figure 13.9). The covalent intermediates thus formed can be attacked in a subsequent step by a water molecule or a second substrate, giving the desired product. **Covalent electrophilic catalysis** is also observed, but usually involves coenzyme adducts that generate electrophilic centers. Well over 100 enzymes are now known to form covalent intermediates during catalysis. Table 13.2 lists some typical examples, including that of glyceraldehyde-3-phosphate dehydrogenase, which catalyzes the reaction:

$$\text{Glyceraldehyde-3-P} + \text{NAD}^+ + \text{P}_i \longrightarrow \text{1,3-Bisphosphoglycerate} + \text{NADH} + \text{H}^+$$

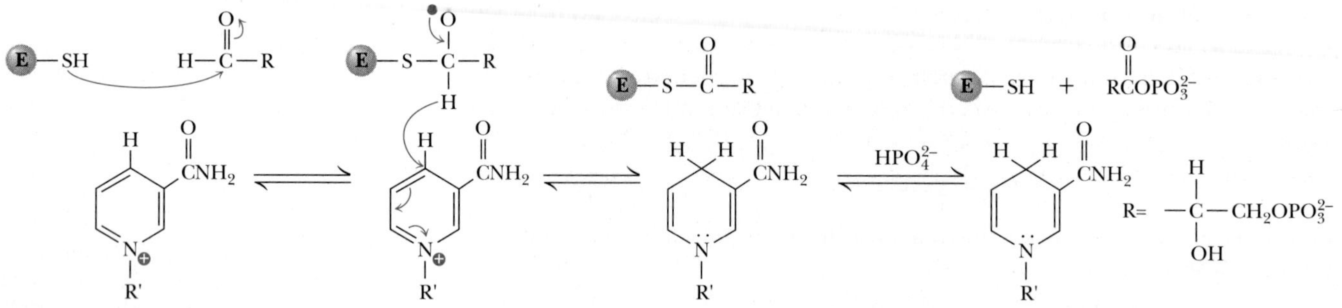

***Figure* 13.10** Formation of a covalent intermediate in the glyceraldehyde-3-phosphate dehydrogenase reaction. Nucleophilic attack by a cysteine —SH group forms a covalent acyl-cysteine intermediate. Following hydride transfer to NAD^+, nucleophilic attack by phosphate yields the product 1,3-bisphosphoglycerate.

Table **13.2**

Enzymes That Form Covalent Intermediates

Enzymes	Reacting Group	Covalent Intermediate
1. Chymotrypsin Elastase Esterases Subtilisin Thrombin Trypsin	$>CH-CH_2-OH$ (Ser)	$>CH-CH_2-O-C(=O)-R$ (Acyl-Ser)
2. Glyceraldehyde-3-phosphate dehydrogenase Papain	$>CH-CH_2-SH$ (Cys)	$>CH-CH_2-S-C(=O)-R$ (Acyl-Cys)
3. Alkaline phosphatase Phosphoglucomutase	$>CH-CH_2-OH$ (Ser)	$>CH-CH_2-O-PO_3^{2-}$ (Phosphoserine)
4. Phosphoglycerate mutase Succinyl-CoA synthetase	$-CH_2$–imidazole (HN, N) (His)	$-CH_2$–imidazole with $O^--P(=O)(O^-)-N$ (Phosphohistidine)
5. Aldolase Decarboxylases Pyridoxal phosphate-dependent enzymes	$R-NH_3^+$ (Amino)	$R-N=C<$ (Schiff base)

As shown in Figure 13.10, this reaction mechanism involves nucleophilic attack by −SH on the substrate glyceraldehyde-3-P to form a covalent *acyl-cysteine* (or *hemithioacetal*) intermediate. Hydride transfer to NAD^+ generates a *thioester* intermediate. Nucleophilic attack by phosphate yields the desired mixed carboxylic-phosphoric anhydride product, 1,3-bisphosphoglycerate. Several examples of covalent catalysis will be discussed in detail in Chapter 14.

13.7 General Acid–Base Catalysis

Nearly all enzyme reactions involve some degree of acid or base catalysis. There are two types of acid–base catalysis: (a) **specific acid–base catalysis,** in which H^+ or OH^- accelerates the reaction, and (b) **general acid–base catalysis,** in which an acid or base other than H^+ or OH^- accelerates the reaction. For ordinary solution reactions, these two cases can be distinguished on the basis of simple experiments. As shown in Figure 13.11, in specific acid or base catalysis, the buffer concentration has no effect. In general acid or base catalysis, however, the buffer may donate or accept a proton in the transition state and thus affect the rate. *By definition, general acid–base catalysis is catalysis in which a proton is transferred in the transition state.* Consider the hydrolysis of *p*-nitrophenylacetate with imidazole acting as a general base (Figure 13.12). Proton transfer apparently stabilizes the transition state here. The water has been made more nucleophilic without generation of a high concentration of OH^- or without the formation of unstable, high-energy species. General acid

(a)

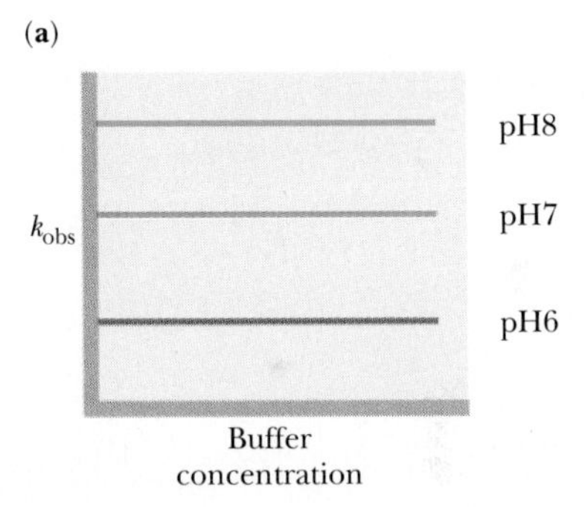

(b)

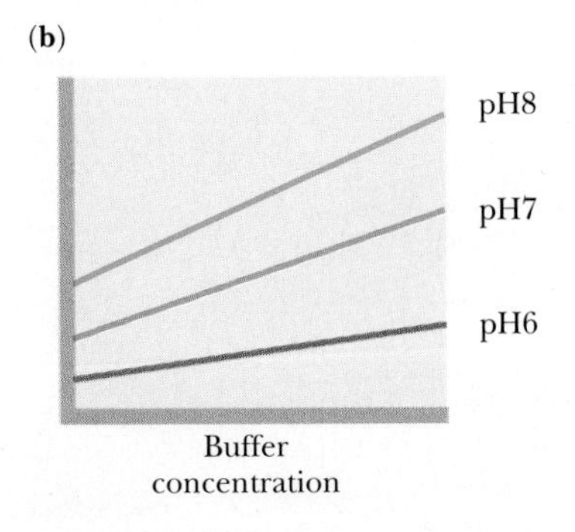

***Figure* 13.11** Specific and general acid–base catalysis of simple reactions in solution may be distinguished by determining the dependence of observed reaction rate constants (k_{obs}) on pH and buffer concentration. (a) In specific acid–base catalysis, H^+ or OH^- concentration affects the reaction rate, k_{obs} is pH-dependent, but buffers (which accept or donate H^+/OH^-) have no effect. (b) In general acid–base catalysis, in which an ionizable buffer may donate or accept a proton in the transition state, k_{obs} is dependent on buffer concentration.

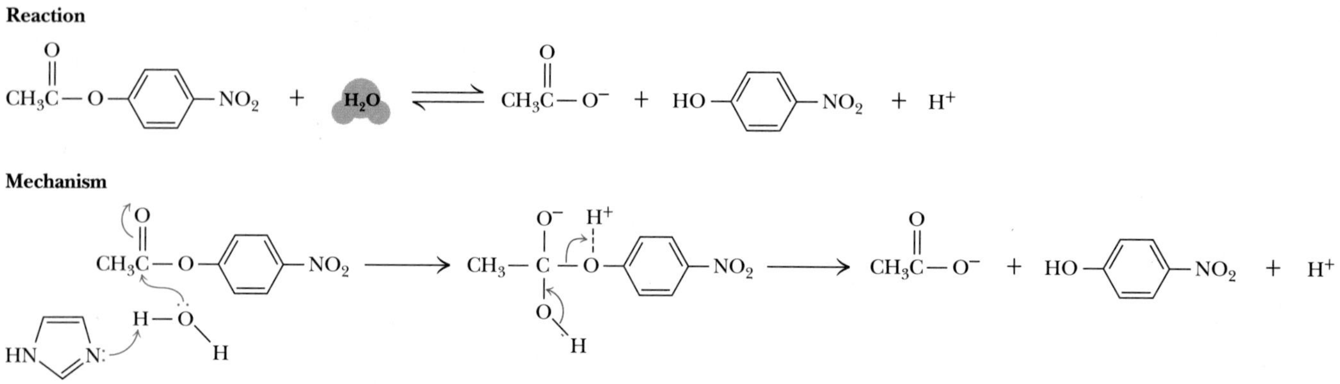

Reaction

$$CH_3C(=O)-O-C_6H_4-NO_2 + H_2O \rightleftharpoons CH_3C(=O)-O^- + HO-C_6H_4-NO_2 + H^+$$

Mechanism

$$CH_3C(=O)-O-C_6H_4-NO_2 \longrightarrow CH_3-C(O^-)(OH)-O-C_6H_4-NO_2 \longrightarrow CH_3C(=O)-O^- + HO-C_6H_4-NO_2 + H^+$$

Figure **13.12** Catalysis of *p*-nitrophenylacetate hydrolysis by imidazole—an example of general base catalysis. Proton transfer to imidazole in the transition state facilitates hydroxyl attack on the substrate carbonyl carbon.

or general base catalysis may increase reaction rates 10- to 100-fold. In an enzyme, ionizable groups on the protein provide the H^+ transferred in the transition state. Clearly, an ionizable group will be most effective as a H^+ transferring agent at or near its pK_a. Since the pK_a of the histidine side chain is near 7, histidine is often the most effective general acid or base. Descriptions of several cases of general acid–base catalysis in typical enzymes follow.

13.8 Metal Ion Catalysis

Many enzymes require metal ions for maximal activity. If the enzyme binds the metal very tightly or requires the metal ion to maintain its stable, native state, it is referred to as a **metalloenzyme.** Enzymes that bind metal ions more weakly, perhaps only during the catalytic cycle, are referred to as **metal activated.** One role for metals in metal-activated enzymes and metalloenzymes is to act as electrophilic catalysts, stabilizing the increased electron density or negative charge that can develop during reactions. Among the enzymes that function in this manner (Figure 13.13) is liver alcohol dehydrogenase, which will be considered in more detail later in this chapter. Another potential function of metal ions is to provide a powerful nucleophile at neutral pH. Coordination to a metal ion can increase the acidity of a nucleophile with an ionizable proton:

$$M^{2+} + \text{NucH} \rightleftharpoons M^{2+}(\text{NucH}) \rightleftharpoons M^{2+}(\text{Nuc}^-) + H^+$$

The reactivity of the coordinated, deprotonated nucleophile is typically intermediate between that of the un-ionized and ionized forms of the nucleophile. Carboxypeptidase (see following discussion) contains an active site Zn^{2+}, which facilitates deprotonation of a water molecule in this manner.

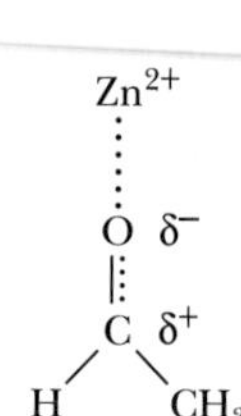

Figure **13.13** In the liver alcohol dehydrogenase reaction, an active-site zinc ion stabilizes negative charge development on the oxygen atom of acetaldehyde, leading to an induced partial positive charge on the C atom, which facilitates hydride transfer to the carbon to form ethanol.

13.9 Proximity

Chemical reactions go faster when the reactants are *in proximity,* that is, near each other. In solution or in the gas phase, this means that increasing the concentrations of reacting molecules, which raises the number of collisions, causes higher rates of reaction. Enzymes, which have specific binding sites for particular reacting molecules, essentially take the reactants out of dilute solu-

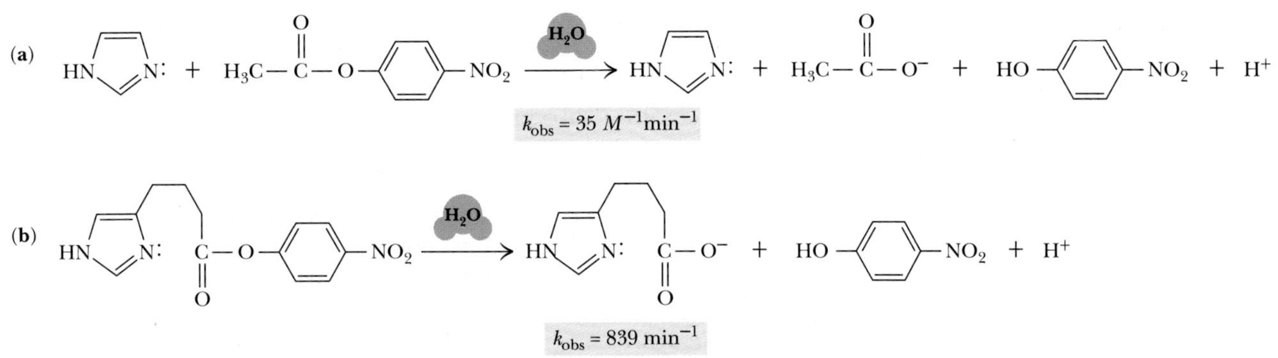

***Figure* 13.14** An example of proximity effects in catalysis. (a) The imidazole-catalyzed hydrolysis of *p*-nitrophenylacetate is slow, but (b) the corresponding intramolecular reaction is 24-fold faster (assuming [imidazole] = 1 M in [a]).

tion and hold them close to each other. This proximity of reactants is said to raise the "effective" concentration over that of the substrates in solution, and leads to an increased reaction rate. In order to measure proximity effects in enzyme reactions, enzymologists have turned to model studies comparing intermolecular reaction rates with corresponding or similar intramolecular reaction rates. A typical case is the imidazole-catalyzed hydrolysis of *p*-nitrophenylacetate (Figure 13.14a). Under certain conditions the rate constant for this bimolecular reaction is 35 $M^{-1}\text{min}^{-1}$. By comparison, the first-order rate constant for the analogous but intramolecular reaction shown in Figure 13.14b is 839 min^{-1}. The ratio of these two rate constants

$$(839\ \text{min}^{-1})/(35\ M^{-1}\text{min}^{-1}) = 23.97\ M$$

has the units of concentration and can be thought of as an effective concentration of imidazole in the intramolecular reaction. Put another way, a concentration of imidazole of 23.9 M would be required in the intermolecular reaction to make it proceed as fast as the intramolecular reaction.

There is more to this story, however. Enzymes not only bring substrates and catalytic groups close together, they orient them in a manner suitable for catalysis as well. Comparison of the rates of reaction of the molecules shown in Figure 13.15 makes it clear that the bulky methyl groups force an orienta-

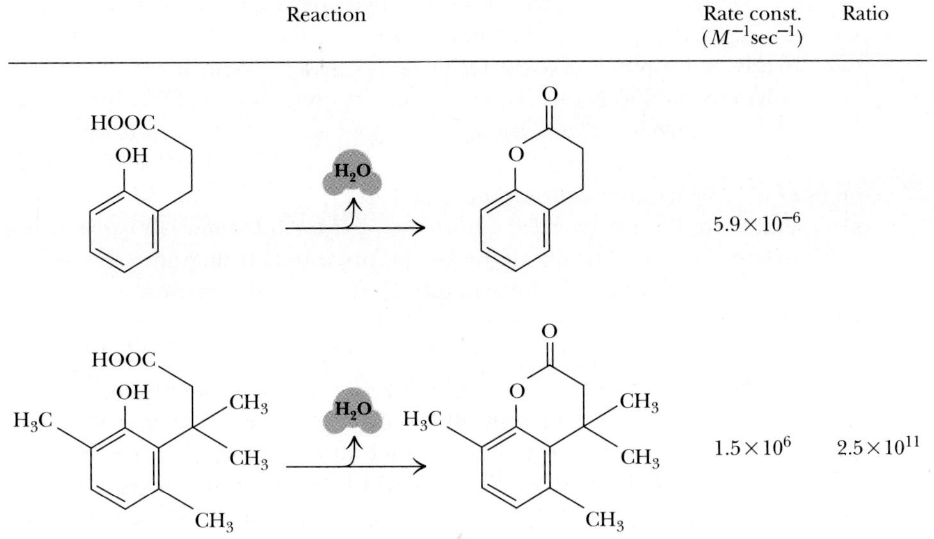

***Figure* 13.15** Orientation effects in intramolecular reactions can be dramatic. Steric crowding by methyl groups provides a rate acceleration of 2.5×10^{11} for the lower reaction compared with the upper reaction. *(Adapted from Milstien, S., and Cohen, L. A., 1972. Stereopopulation control I. Rate enhancements in the lactonization of o-hydroxyhydrocinnamic acid. Journal of the American Chemical Society 94:9158–9165.)*

tion on the alkyl carboxylate and the aromatic hydroxyl groups that makes them approximately 250 billion times more likely to react. Enzymes function similarly by placing catalytically functional groups (from the protein side chains or from another substrate) in the proper position for reaction of a given substrate.

Clearly, proximity and orientation play a role in enzyme catalysis, but there is a problem with each of the above comparisons. In both cases, it is impossible to separate true proximity and orientation effects from the effects of entropy loss when molecules are brought together (described in Section 13.4). The actual rate accelerations afforded by proximity and orientation effects in Figures 13.14 and 13.15, respectively, are much smaller than the values given in these figures. Simple theories based on probability and nearest-neighbor models, for example, predict that proximity effects may actually provide rate increases of only 5- to 10-fold. For any real case of enzymatic catalysis, it is nonetheless important to remember that proximity and orientation effects are significant.

13.10 Typical Enzyme Mechanisms

The balance of this chapter will be devoted to several classic and representative enzyme mechanisms. These particular cases are well understood, because the three-dimensional structures of the enzymes and the bound substrates are known at atomic resolution, and because great efforts have been devoted to kinetic and mechanistic studies. They are important because they represent reaction types that appear again and again in living systems, and because they demonstrate many of the catalytic principles cited above. Enzymes are the catalytic machines that sustain life, and what follows is an intimate look at the inner workings of the machinery.

13.11 Serine Proteases

The **serine proteases** are one of the best-characterized families of enzymes. This family includes *trypsin, chymotrypsin, elastase, thrombin, subtilisin, plasmin, tissue plasminogen activator,* and other related enzymes. The first three of these are digestive enzymes, and are synthesized in the pancreas and secreted into the digestive tract as inactive **proenzymes,** or **zymogens.** Within the digestive tract, the zymogen is converted into the active enzyme form by cleaving off a portion of the peptide chain. Thrombin is a crucial enzyme in the blood-clotting cascade, subtilisin is a bacterial protease, and plasmin breaks down the fibrin polymers of blood clots. Tissue plasminogen activator (TPA) specifically cleaves the proenzyme *plasminogen,* yielding plasmin. Owing to its ability to stimulate breakdown of blood clots, TPA can minimize the harmful consequences of a heart attack, if administered to a patient within 30 minutes of onset. Finally, although not itself a protease, *acetylcholinesterase* is a *serine esterase* and is related mechanistically to the serine proteases. It degrades the neurotransmitter acetylcholine in the synaptic cleft between neurons.

The Digestive Serine Proteases

Trypsin, chymotrypsin, and elastase all carry out the same reaction—the cleavage of a peptide chain—and although their structures and mechanisms are quite similar, they display very different specificities. Trypsin cleaves peptides on the carbonyl side of the basic amino acids, arginine or lysine (see Table

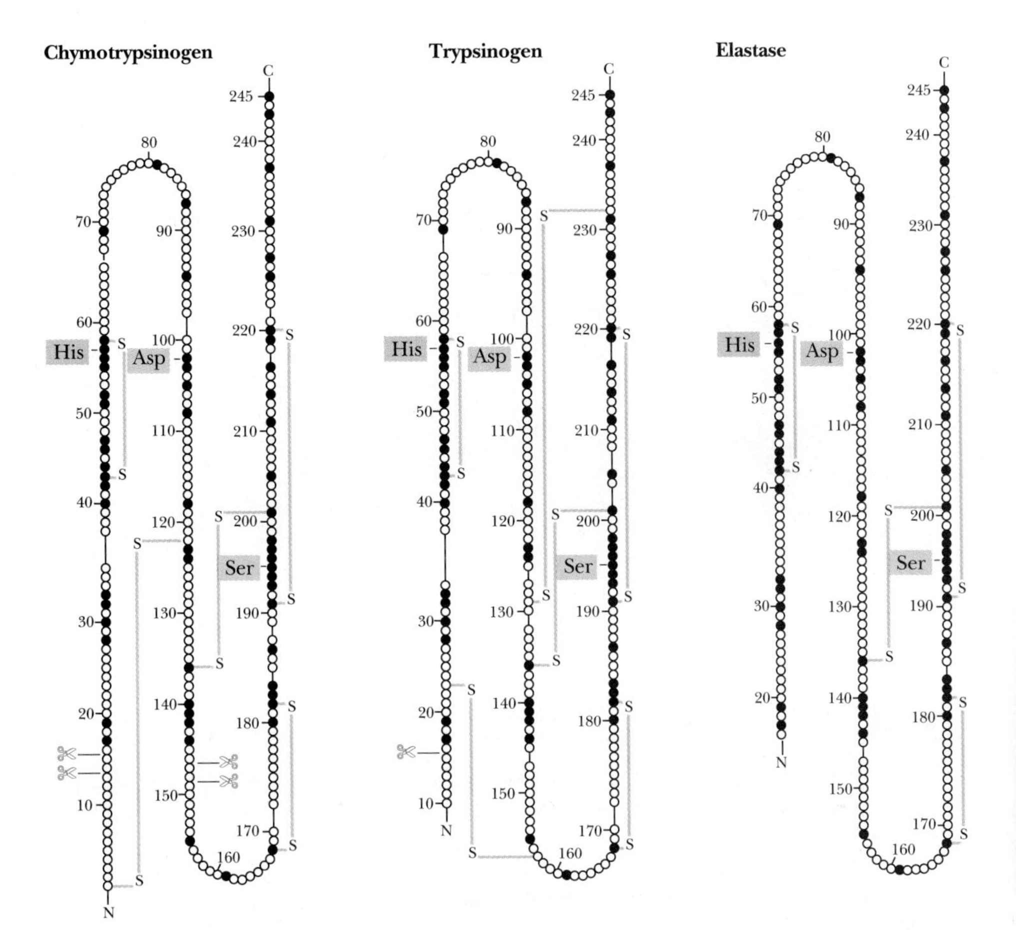

Figure **13.16** Comparison of the amino acid sequences of chymotrypsinogen, trypsinogen, and elastase. Each circle represents one amino acid. Numbering is based on the sequence of chymotrypsinogen. Filled circles indicate residues that are identical in all three proteins. Disulfide bonds are indicated in yellow. The positions of the three catalytically important active-site residues (His^{57}, Asp^{102}, and Ser^{195}) are indicated.

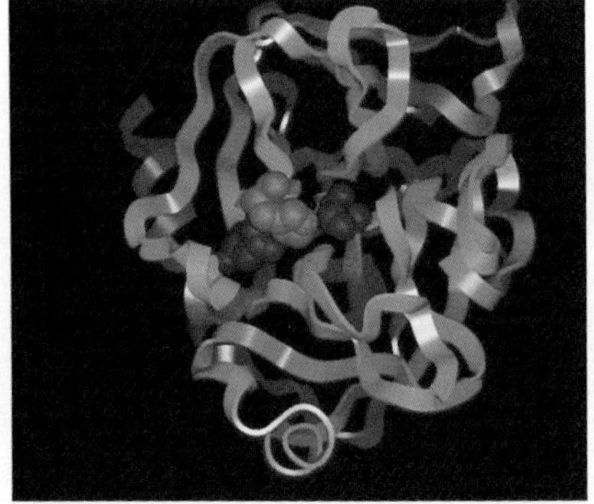

Figure **13.17** Structure of chymotrypsin. The residues of the catalytic triad (His^{57}, Asp^{102}, and Ser^{195}) are highlighted. His^{57} (blue) is flanked on the left by Asp^{102} (red) and on the right by Ser^{195} (red).

4.6, page 115). Chymotrypsin prefers to cleave on the carbonyl side of aromatic residues, such as phenylalanine and tyrosine. Elastase is not as specific as the other two; it mainly cleaves peptides on the carbonyl side of small, neutral residues. These three enzymes all possess molecular weights in the range of 25,000, and all have similar sequences (Figure 13.16) and three-dimensional structures. The structure of chymotrypsin is typical (Figure 13.17). The molecule is ellipsoidal in shape and contains an α-helix at the C-terminal end (residues 230 to 245) and several β-sheet domains. Most of the aromatic and hydrophobic residues are buried in the interior of the protein, and most of the charged or hydrophilic residues are on the surface. Three polar residues—His^{57}, Asp^{102}, and Ser^{195}—form what is known as a **catalytic triad** at the active site (Figure 13.18). These three residues are conserved in trypsin and elastase as well. The active site in this case is actually a depression on the surface of the enzyme, with a small pocket that the enzyme uses to identify the residue for which it is specific (Figure 13.19). Chymotrypsin, for example, has a pocket surrounded by hydrophobic residues and large enough to accommodate an aromatic side chain. The pocket in trypsin has a negative charge (Asp^{189}) at its bottom, facilitating the binding of positively charged arginine and lysine residues. Elastase, on the other hand, has a shallow pocket with bulky threonine and valine residues at the opening. Only small, nonbulky residues can be accommodated in its pocket. The backbone of the peptide substrate is hydrogen bonded in antiparallel fashion to residues 215 to 219 and bent so that the peptide bond to be cleaved is bound close to His^{57} and Ser^{195}.

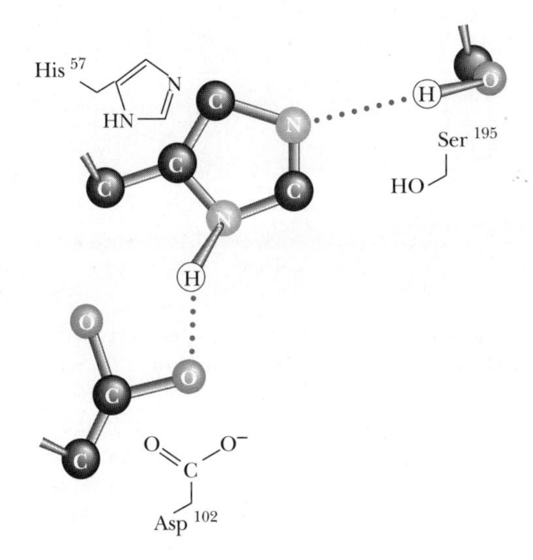

Figure **13.18** The catalytic triad of chymotrypsin.

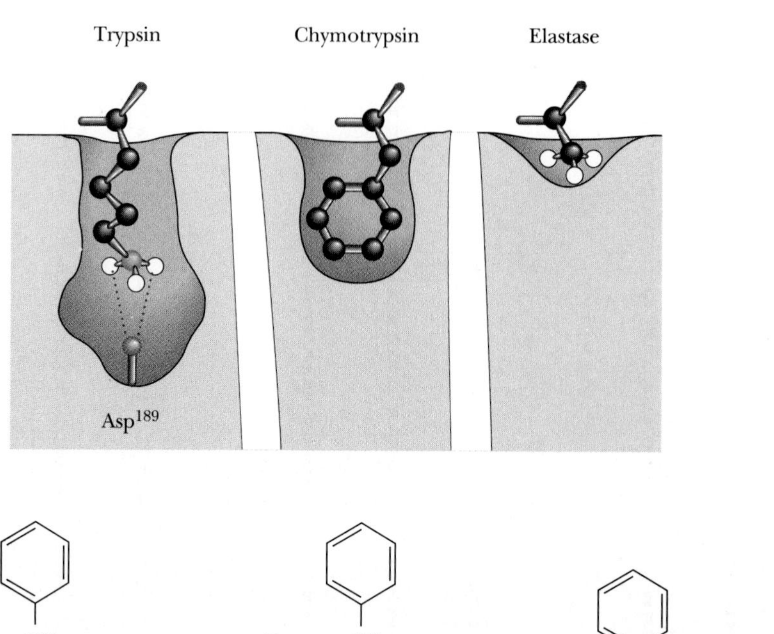

***Figure* 13.19** The substrate-binding pockets of trypsin, chymotrypsin, and elastase.

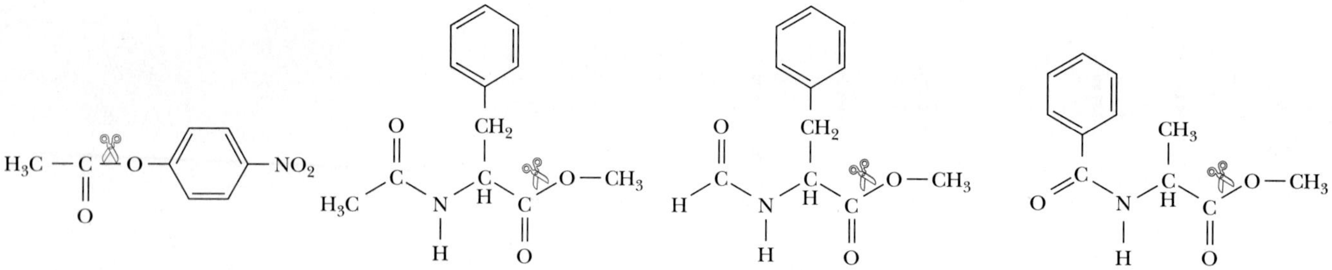

p-Nitrophenylacetate

Acetylphenylalanine methyl ester

Formylphenylalanine methyl ester

Benzoylalanine methyl ester

***Figure* 13.20** Artificial substrates used in studies of the mechanism of chymotrypsin.

The Chymotrypsin Mechanism in Detail: Kinetics

Much of what is known about the chymotrypsin mechanism is based on studies of the hydrolysis of artificial substrates—simple organic esters, such as *p*-nitrophenylacetate, and methyl esters of amino acid analogs, such as formylphenylalanine methyl ester and acetylphenylalanine methyl ester (Figure 13.20). *p*-Nitrophenylacetate is an especially useful model substrate, because the nitrophenolate product is easily observed, owing to its strong absorbance at 400 nm. When large amounts of chymotrypsin are used in kinetic studies with this substrate, a **rapid initial burst** of *p*-nitrophenolate is observed (in an amount approximately equal to the enzyme concentration), followed by a much slower, linear rate of nitrophenolate release (Figure 13.21). Observation of a burst, followed by slower, steady-state product release, is strong evidence for a multistep mechanism, with a fast first step and a slower second step.

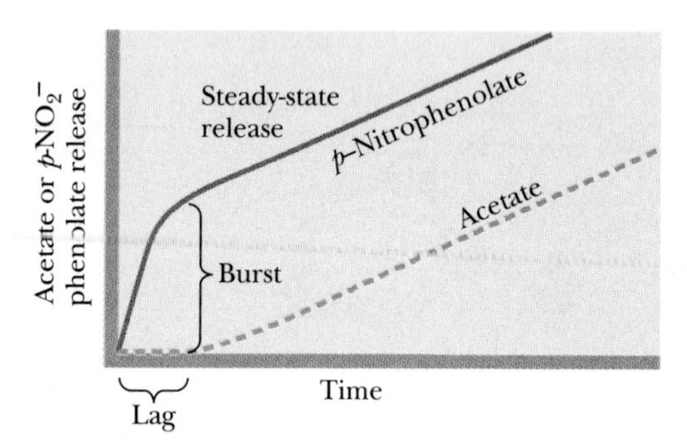

***Figure* 13.21** Burst kinetics observed in the chymotrypsin reaction. A burst of nitrophenolate production is followed by a slower, steady-state release. After an initial lag period, acetate release is also observed. This kinetic pattern is consistent with rapid formation of an acyl-enzyme intermediate (and the burst of nitrophenolate). The slower, steady-state release of products corresponds to rate-limiting breakdown of the acyl-enzyme intermediate.

In the chymotrypsin mechanism, the nitrophenylacetate combines with the enzyme to form an ES complex. This is followed by a rapid second step in which an **acyl-enzyme intermediate** is formed, with the acetyl group covalently bound to the very reactive Ser^{195}. The nitrophenyl moiety is released as nitrophenolate (Figure 13.22), accounting for the burst of nitrophenolate product. Attack of a water molecule on the acyl-enzyme intermediate yields acetate as the second product in a subsequent, slower step. The enzyme is now free to bind another molecule of *p*-nitrophenylacetate, and the *p*-nitrophenolate product produced at this point corresponds to the slower, steady-state formation of product in the upper right portion of Figure 13.21. In this mechanism, the release of acetate is the **rate-limiting step,** and accounts for the observation of **burst kinetics**—the pattern shown in Figure 13.21.

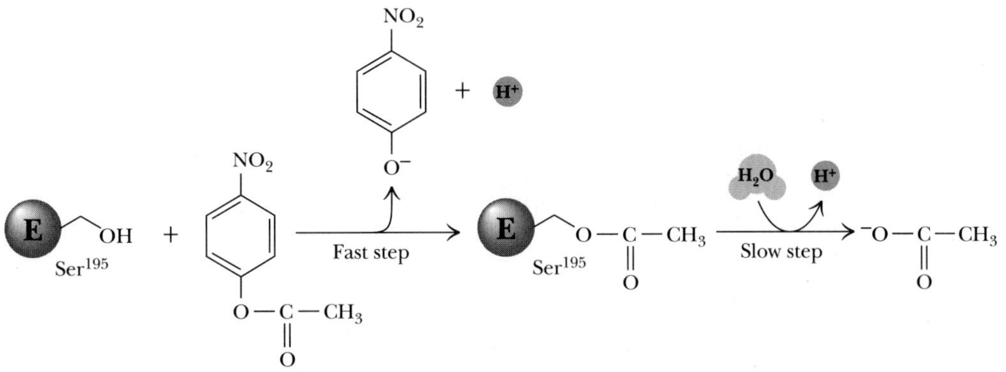

Figure **13.22** Rapid formation of the acyl-enzyme intermediate is followed by slower product release.

Serine proteases like chymotrypsin are susceptible to inhibition by **organic fluorophosphates,** such as *diisopropylfluorophosphate (DIFP,* Figure 13.23). DIFP reacts rapidly with active-site serine residues, such as Ser^{195} of chymotrypsin and the other serine proteases (but not with any of the other serines in these proteins), to form a DIP-enzyme. This covalent enzyme-inhibitor complex is extremely stable, and chymotrypsin is thus permanently inactivated by DIFP.

The Serine Protease Mechanism in Detail: Events at the Active Site

A likely mechanism for peptide hydrolysis is shown in Figure 13.24. As the backbone of the substrate peptide binds adjacent to the catalytic triad, the specific side chain fits into its pocket. Asp^{102} of the catalytic triad positions His^{57} and immobilizes it through a hydrogen bond as shown. In the first step of the reaction, His^{57} acts as a general base to withdraw a proton from Ser^{195}, facilitating nucleophilic attack by Ser^{195} on the carbonyl carbon of the peptide bond to be cleaved. This is probably a *concerted step,* since proton transfer prior to Ser^{195} attack on the acyl carbon would leave a relatively unstable negative charge on the serine oxygen. In the next step, donation of a proton from His^{57} to the peptide's amide nitrogen creates a protonated amine on the covalent, tetrahedral intermediate, facilitating the subsequent bond breaking and dissociation of the amine product. The negative charge on the peptide oxygen is unstable; the tetrahedral intermediate is short-lived and rapidly breaks down to expel the amine product. The acyl-enzyme intermediate that results is reasonably stable; it can even be isolated using substrate analogs for which further reaction cannot occur. With normal peptide substrates, however, subsequent nucleophilic attack at the carbonyl carbon by water generates another transient tetrahedral intermediate (Figure 13.24). His^{57} acts as a general base in this step, accepting a proton from the attacking water molecule. The subsequent collapse of the tetrahedral intermediate is assisted by proton donation from His^{57} to the serine oxygen in a concerted manner. Deprotonation of the carboxyl group and its departure from the active site complete the reaction as shown.

Until recently, the catalytic role of Asp^{102} in trypsin and the other serine proteases had been surmised on the basis of its proximity to His^{57} in struc-

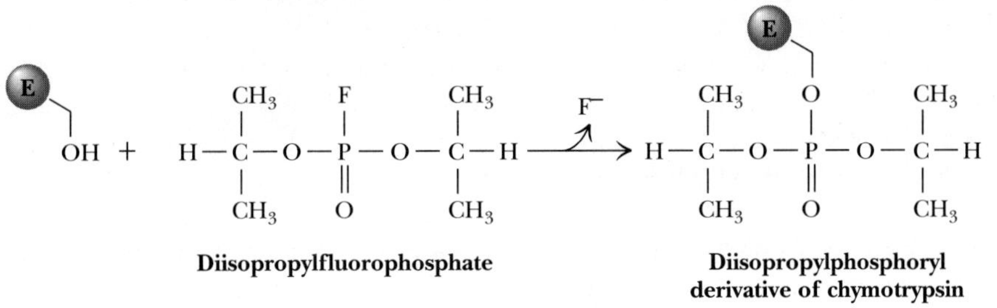

Diisopropylfluorophosphate → **Diisopropylphosphoryl derivative of chymotrypsin**

Figure **13.23** Diisopropylfluorophosphate (DIFP) reacts with active-site serine residues of serine proteases, causing permanent inactivation.

tures obtained from X-ray diffraction studies, but it had never been demonstrated with certainty in physical or chemical studies. As can be seen in Figure 13.17, Asp^{102} is buried at the active site and is normally inaccessible to chemical modifying reagents. In 1987, however, Charles Craik, William Rutter, and their colleagues used site-directed mutagenesis (see Chapter 8) to prepare a mutant trypsin with an asparagine in place of Asp^{102}. This mutant trypsin possessed a hydrolytic activity with ester substrates only 1/10,000 that of native trypsin, demonstrating that Asp^{102} is indeed essential for catalysis and that its ability to immobilize and orient His^{57} is crucial to the function of the catalytic triad.

13.12 The Aspartic Proteases

Mammals, fungi, and higher plants produce a family of proteolytic enzymes known as **aspartic proteases.** These enzymes are active at acidic (or sometimes neutral) pH, and each possesses two aspartic acid residues at the active site. Aspartic proteases carry out a variety of functions (Table 13.3), including

Critical Developments in Biochemistry

Transition-State Stabilization in the Serine Proteases

X-ray crystallographic studies of serine protease complexes with transition-state analogs have shown how chymotrypsin stabilizes the **tetrahedral oxyanion transition states** (structures (c) and (g) in Figure 13.24) of the protease reaction. The amide nitrogens of Ser^{195} and Gly^{193} form an "oxyanion hole" in which the substrate carbonyl oxygen is hydrogen bonded to the amide N-H groups.

Formation of the tetrahedral transition state increases the interaction of the carbonyl oxygen with the amide N-H groups in two ways. Conversion of the carbonyl double bond to the *longer* tetrahedral single bond brings the oxygen atom closer to the amide hydrogens. Also, the hydrogen bonds between the charged oxygen and the amide hydrogens are significantly stronger than the hydrogen bonds with the uncharged carbonyl oxygen.

Transition-state stabilization in chymotrypsin also involves the side chains of the substrate. The side chain of the departing amine product forms stronger interactions with the enzyme upon formation of the tetrahedral intermediate. When the tetrahedral intermediate breaks down (Figure 13.24d and e), steric repulsion between the product amine group and the carbonyl group of the acyl-enzyme intermediate leads to departure of the amine product.

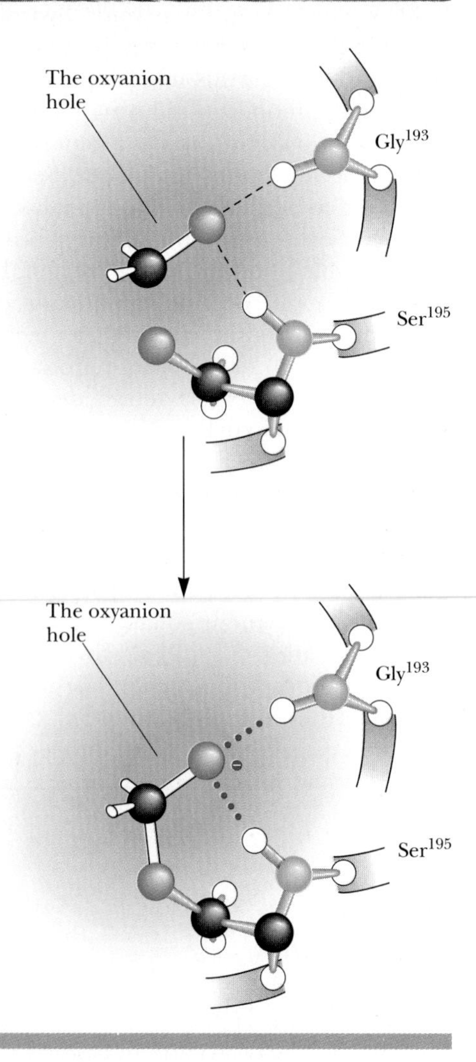

The "oxyanion hole" of chymotrypsin stabilizes the tetrahedral oxyanion transition states of the mechanism in Figure 13.24.

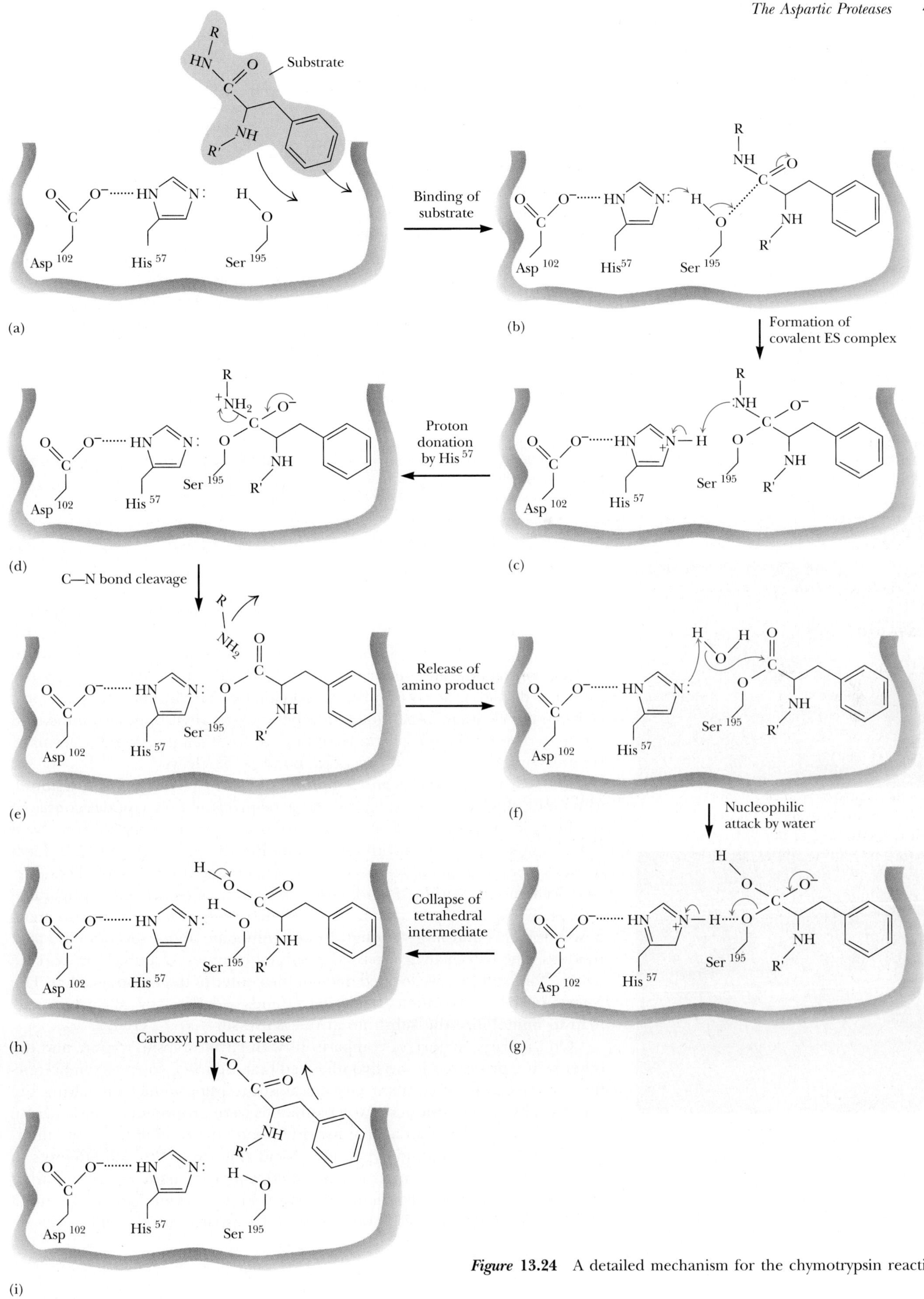

(a) Substrate

Binding of substrate

(b)

Formation of covalent ES complex

(c)

Proton donation by His 57

(d)

C—N bond cleavage

(e)

Release of amino product

(f)

Nucleophilic attack by water

(g)

Collapse of tetrahedral intermediate

(h)

Carboxyl product release

(i)

***Figure* 13.24** A detailed mechanism for the chymotrypsin reaction.

Table **13.3**
Some Representative Aspartic Proteases

Name	Source	Function
Pepsin*	Animal stomach	Digestion of dietary protein
Chymosin†	Animal stomach	Digestion of dietary protein
Cathepsin D	Spleen, liver, and many other animal tissues	Lysosomal digestion of proteins
Renin‡	Kidney	Conversion of angiotensinogen to angiotensin I; regulation of blood pressure
HIV-1 protease§	AIDS virus	Processing of AIDS virus proteins

*The second enzyme to be crystallized (by John Northrup in 1930). Even more than urease before it, pepsin study by Northrup established that enzyme activity came from proteins.

†Also known as rennin, it is the major pepsinlike enzyme in gastric juice of fetal and newborn animals.

‡A drop in blood pressure causes release of renin from the kidneys, which converts more angiotensinogen to angiotensin.

§A dimer of identical monomers, homologous to pepsin.

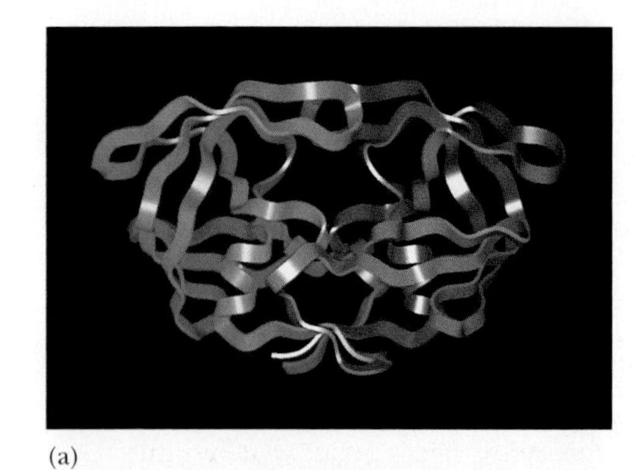

(a)

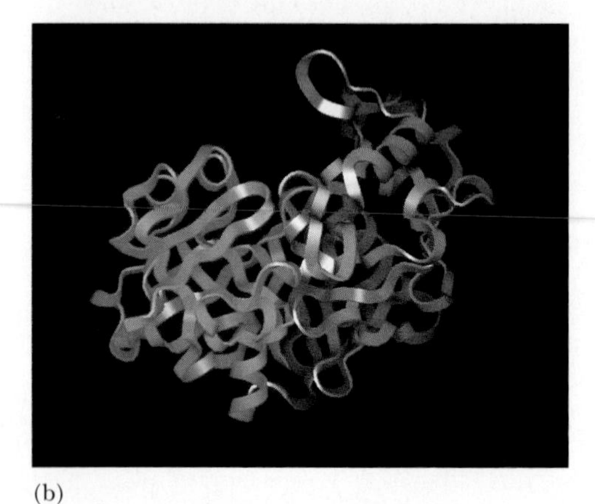

(b)

***Figure* 13.25** Structures of (a) HIV-1 protease, a dimer, and (b) pepsin, a monomer. Pepsin's N-terminal half is shown in red; its C-terminal half is shown in blue.

digestion (*pepsin* and *chymosin*), lysosomal protein degradation (*cathepsin D* and *E*), and regulation of blood pressure (*renin* is an aspartic protease involved in the production of *angiotensin,* a hormone that stimulates smooth muscle contraction and reduces excretion of salts and fluid). The aspartic proteases display a variety of substrate specificities, but normally they are most active in the cleavage of peptide bonds between two hydrophobic amino acid residues. The preferred substrates of pepsin, for example, contain aromatic residues on both sides of the peptide bond to be cleaved.

Most aspartic proteases are comprised of 323 to 340 amino acid residues, with molecular weights near 35,000. Aspartic protease polypeptides consist of two homologous domains that fold to produce a tertiary structure composed of two similar lobes, with approximate twofold symmetry (Figure 13.25). Each of these lobes or domains consists of two β-sheets and two short α-helices. The two domains are bridged and connected by a six-stranded, antiparallel β-sheet. The active site is a deep and extended cleft, formed by the two juxtaposed domains and large enough to accommodate about seven amino acid residues. The two catalytic aspartate residues, residues 32 and 215 in porcine pepsin, for example, are located deep in the center of the active site cleft. The N-terminal domain forms a "flap" that extends over the active site, which may help to immobilize the substrate in the active site.

On the basis, in part, of comparisons with chymotrypsin, trypsin, and the other serine proteases, it was hypothesized that aspartic proteases might function by formation of covalent enzyme-substrate intermediates involving the active-site aspartate residues. Two possibilities were proposed: an acyl-enzyme intermediate involving an acid anhydride bond and an amino-enzyme intermediate involving an amide (peptide) bond (Figure 13.26). All attempts to trap or isolate a covalent intermediate failed, and a mechanism (see following paragraph) favoring noncovalent enzyme-substrate intermediates and general acid–general base catalysis is now favored for aspartic proteases.

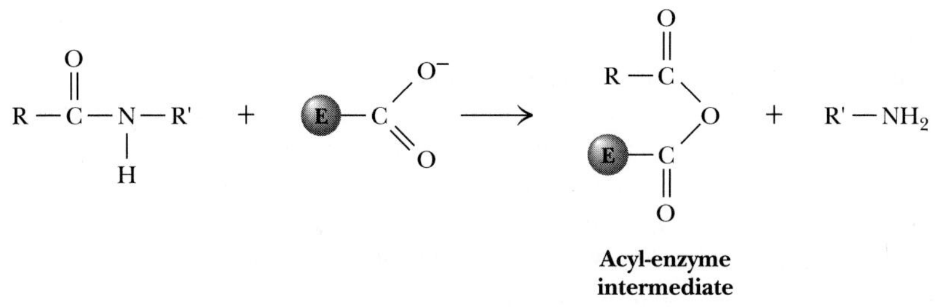

***Figure* 13.26** Acyl-enzyme and amino-enzyme intermediates originally proposed for aspartic proteases were modeled after the acyl-enzyme intermediate of the serine proteases.

The Mechanism of Action of Aspartic Proteases

A crucial datum supporting the general acid–general base model is the pH dependence of protease activity (see Critical Developments in Biochemistry: *The pH Dependence of Aspartic Proteases and HIV-1 Protease,* page 446). Enzymologists hypothesize that the aspartate carboxyl groups function alternately as general acid and general base. This model requires that one of the aspartate carboxyls be protonated and one be deprotonated when substrate binds. X-ray diffraction data on aspartic proteases show that the active-site structure in the vicinity of the two aspartates is highly symmetric. The two aspartates appear to act as a "catalytic diad" (analogous to the catalytic triad of the serine proteases). The diad proton may thus be covalently bound to either of the aspartate groups in the free enzyme or in the enzyme-substrate complex. Thus, in pepsin, for example, Asp^{32} may be deprotonated while Asp^{215} is protonated, or vice versa.

In the most widely accepted mechanism (Figure 13.27), substrate binding is followed by a step in which two concerted proton transfers facilitate nucleophilic attack on the carbonyl carbon of the substrate by water. In the mechanism shown, Asp^{32} acts as a general base, accepting a proton from an active-site water molecule, while Asp^{215} acts a general acid, donating a proton to the oxygen of the peptide carbonyl group. *By virtue of these two proton transfers, nucleophilic attack occurs without explicit formation of hydroxide ion at the active site.*

***Figure* 13.27** A mechanism for the aspartic proteases. In the first step, two concerted proton transfers facilitate nucleophilic attack of water on the substrate carbonyl carbon. In the third step, one aspartate residue (Asp^{32} in pepsin) accepts a proton from one of the hydroxyl groups of the amine dihydrate, and the other aspartate (Asp^{215}) donates a proton to the nitrogen of the departing amine.

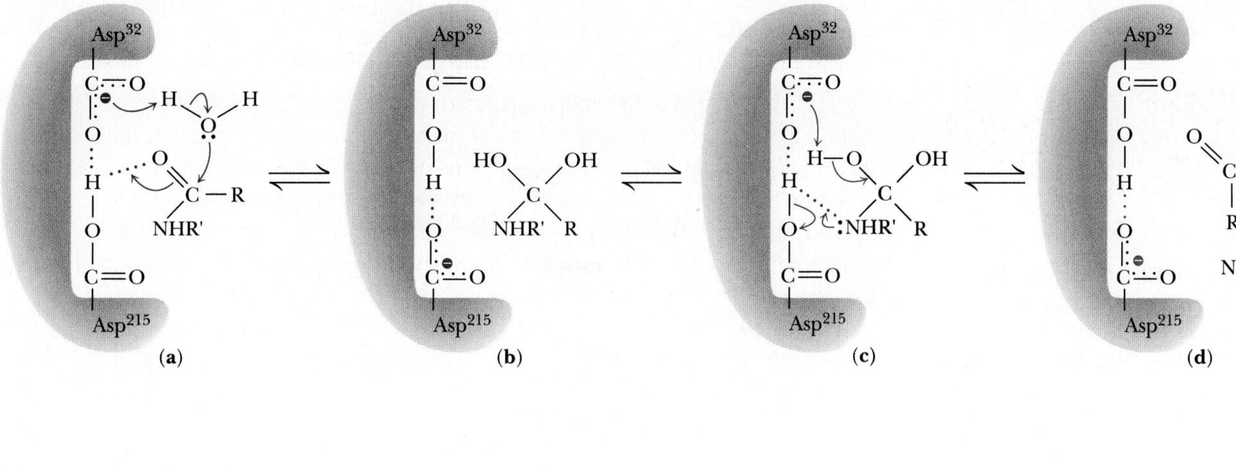

The resulting intermediate is termed an **amide dihydrate.** Note that the protonation states of the two aspartate residues are now opposite to those in the free enzyme (Figure 13.27).

Breakdown of the amide dihydrate occurs by a mechanism similar to its formation. The ionized aspartate carboxyl (Asp^{32} in Figure 13.27) acts as a general base to accept a proton from one of the hydroxyl groups of the amide dihydrate, while the protonated carboxyl of the other aspartate (Asp^{215} in this case) simultaneously acts as a general acid to donate a proton to the nitrogen atom of one of the departing peptide products.

The AIDS Virus HIV-1 Protease Is an Aspartic Protease

Recent research on acquired immune deficiency syndrome (AIDS) and its causative viral agent, the human immunodeficiency virus (HIV-1), has brought a new aspartic protease to light. **HIV-1 protease** cleaves the polyprotein products of the HIV-1 genome (see Chapter 34 for a more thorough discussion of HIV-1), producing several proteins necessary for viral growth and cellular infection. (For this reason, many researchers have explored the possibility that inhibitors of HIV-1 protease could be effective drugs for the treatment of AIDS.) HIV-1 protease cleaves several different peptide linkages in the HIV-1 polyproteins, including those shown in Figure 13.28. For example, the protease cleaves between the Tyr and Pro residues of the sequence Ser-Gln-Asn-Tyr-Pro-Ile-Val, which joins the p17 and p24 HIV-1 proteins.

The HIV-1 protease is a remarkable viral imitation of mammalian aspartic proteases: It is a **dimer of identical subunits** that mimics the two-lobed monomeric structure of pepsin and other aspartic proteases. The HIV-1 protease subunits are 99-residue polypeptides that are homologous with the individual domains of the monomeric proteases. Structures determined by X-ray diffraction studies reveal that the active site of HIV-1 protease is formed at the interface of the homodimer and consists of two aspartate residues, designated Asp^{25} and $Asp^{25'}$, one contributed by each subunit (Figure 13.29). In the homodimer, the active site is covered by two identical "flaps," one from each subunit, in contrast to the monomeric aspartic proteases, which possess only a single, active-site flap.

Enzyme kinetic measurements by Thomas Meek and his collaborators at SmithKline Beecham Pharmaceuticals have shown that the mechanism of

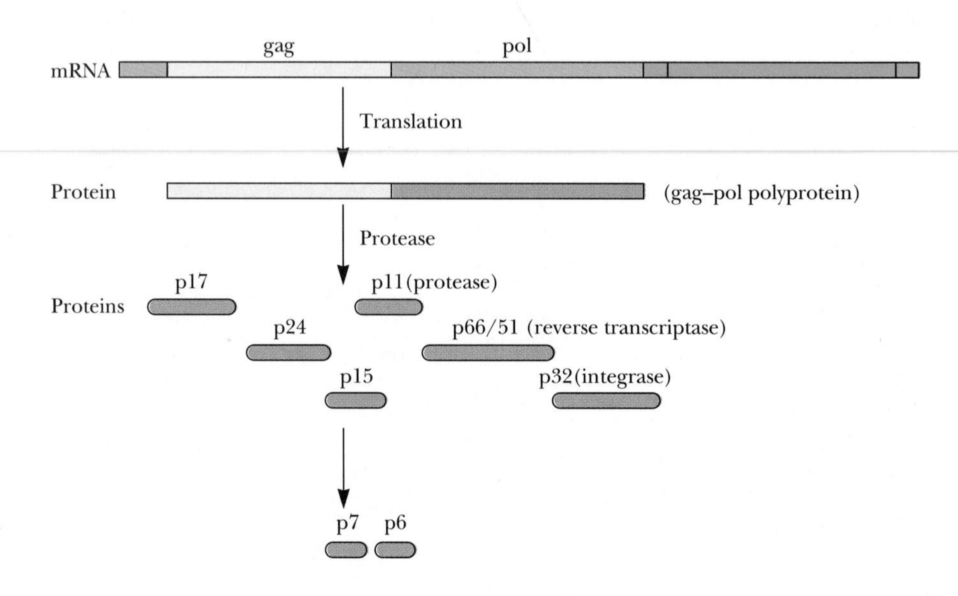

***Figure* 13.28** HIV mRNA provides the genetic information for synthesis of a polyprotein. Proteolytic cleavage of this polyprotein by HIV protease produces the individual proteins required for viral growth and cellular infection.

HIV-1 protease is very similar to those of other aspartic proteases (Figure 13.30). Two concerted proton transfers by the aspartate carboxyls facilitate nucleophilic attack by water on the carbonyl carbon of the peptide substrate. If the protease-substrate complex is incubated with $H_2{}^{18}O$, incorporation of ^{18}O into the peptide carbonyl group can be measured. *Thus, not only is the formation of the amide dihydrate reversible, and the two hydroxyl groups of the dihydrate equivalent, but the protonation states of the active-site carboxyl groups must interchange* (Figure 13.30). The simplest model would involve direct proton exchange across the Asp-Asp diad as shown. The symmetrical nature of the active-site aspartyl groups is consistent with the idea of facile exchange of the proton between the two Asp residues.

The observation that ^{18}O can accumulate in the substrate from $H_2{}^{18}O$ also implies that formation and reversal of the amide dihydrate must be faster than its breakdown to form product. Kinetic studies by Meek and his co-workers are consistent with a model in which breakdown of the dihydrate intermediate is the rate-determining step for the protease reaction. These studies also show that the transition state of this step involves two proton transfers—with one aspartate acting as a general acid to donate a proton to the departing proline, and with the other aspartate acting as a general base to abstract a proton from one of the hydroxyl groups, facilitating collapse of the dihydrate to a carboxyl group. Note that the events of the breakdown to form product are simply the reversal of the steps that formed the intermediate.

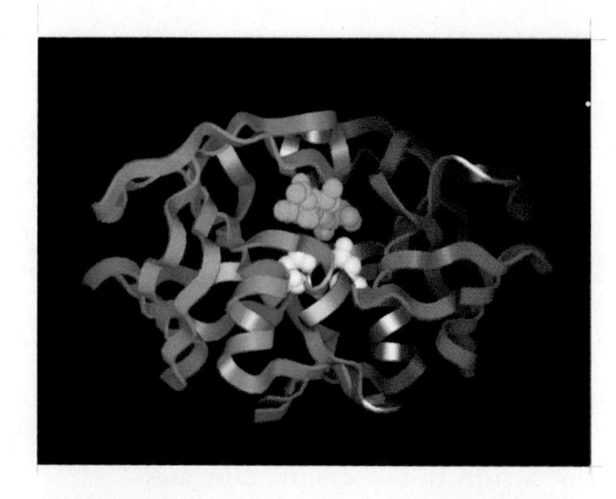

***Figure* 13.29** HIV protease complexed with its inhibitor, acetyl-pepstatin (green). The active-site Asp residues are shown in white.

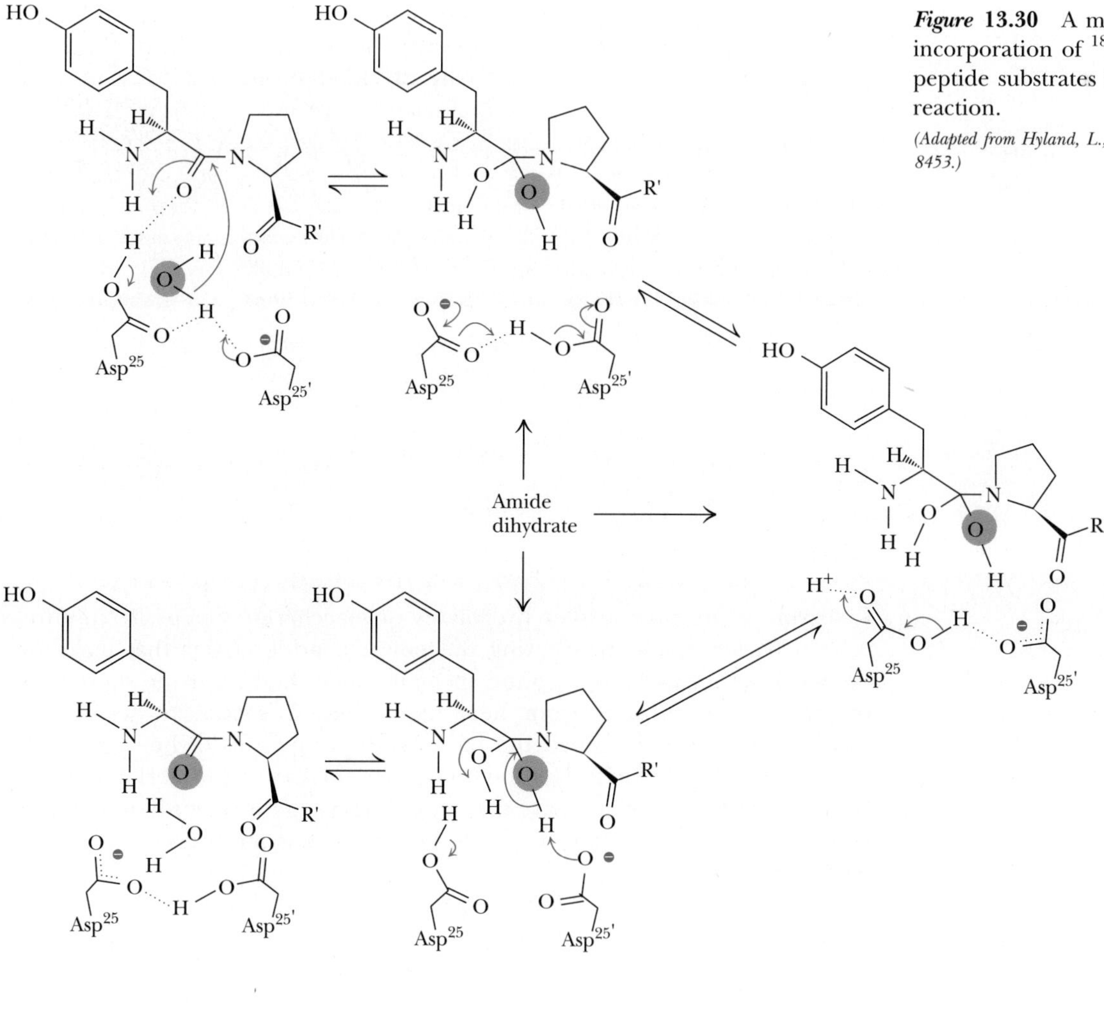

***Figure* 13.30** A mechanism for the incorporation of ^{18}O from $H_2{}^{18}O$ into peptide substrates in the HIV protease reaction.

(Adapted from Hyland, L., et al., 1991. Biochemistry 30:8441–8453.)

Critical Developments in Biochemistry

The pH Dependence of Aspartic Proteases and HIV-1 Protease

The first hint that two active-site carboxyl groups—one protonated and one ionized—might be involved in the catalytic activity of the aspartic proteases came from studies of the pH dependence of enzymatic activity. If an ionizable group in an enzyme active site is essential for activity, a plot of enzyme activity versus pH may look like one of the plots at right.

If activity increases dramatically as pH is increased, catalysis may depend on a deprotonated group that may normally act as a general base, accepting a proton from the substrate or a water molecule, for example (a). Protonation of this group at lower pH prevents it from accepting another proton (from the substrate or water, for example).

On the other hand, if activity decreases sharply as pH is raised, activity may depend on a protonated group, which may act as a general acid, donating a proton to the substrate or a catalytic water molecule (b). At high pH, the proton dissociates and is not available in the catalytic events.

(a)

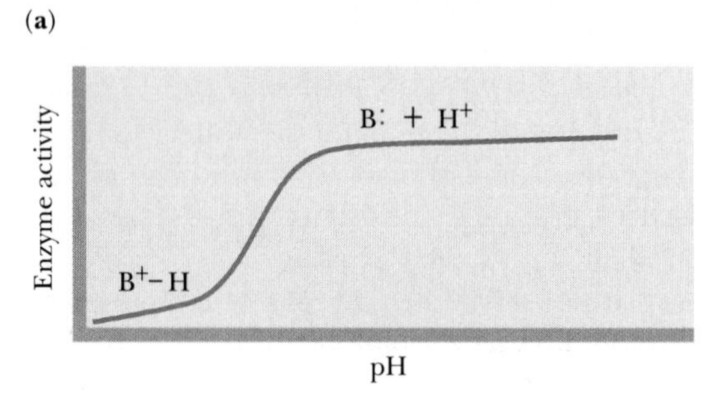

(b)

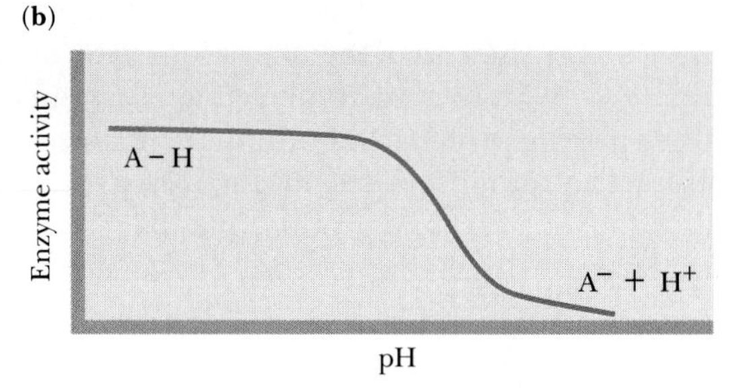

(c)

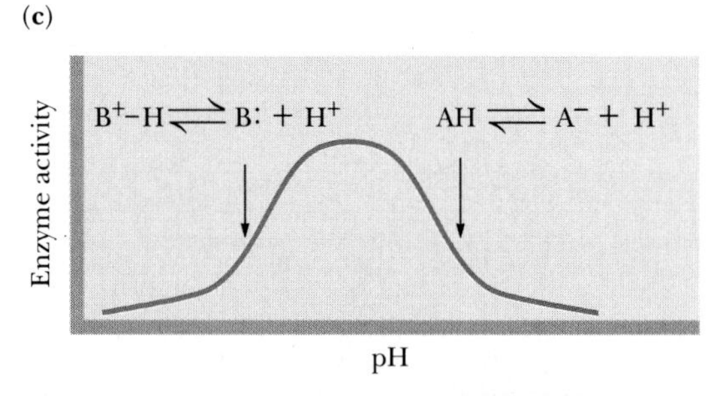

Bell-shaped activity versus pH profiles arise from two separate active-site ionizations. (a) Enzyme activity increases upon deprotonation of B^+-H. (b) Enzyme activity decreases upon deprotonation of A-H. (c) Enzyme activity is maximal in the pH range where one ionizable group is deprotonated (as B:) and the other group is protonated (as AH).

13.13 Lysozyme

Lysozyme is an enzyme that hydrolyzes polysaccharide chains. It ruptures certain bacterial cells by cleaving the polysaccharide chains that make up their cell wall. Lysozyme is found in many body fluids, but the most thoroughly studied form is from hen egg whites. The Russian scientist P. Laschtchenko first described the bacteriolytic properties of hen egg white lysozyme in 1909. In 1922, Alexander Fleming, the London bacteriologist who later discovered penicillin, gave the name *lysozyme* to the agent in mucus and tears that destroyed certain bacteria, because it was an en*zyme* that caused bacterial *lysis.*

As seen in Chapters 9 and 10, bacterial cells are surrounded by a rigid, strong wall of peptidoglycan, a copolymer of two sugar units, *N*-acetylmuramic

The bell-shaped curve in part (c) of the figure combines both of these behaviors, so that activity first increases, then decreases, as pH is increased. This is consistent with the involvement of two ionizable groups—one with a low pK_a that acts as a base above its pK_a, and a second group with a higher pK_a that acts as an acid below its pK_a.

Kinetic studies with pepsin have produced bell-shaped curves for a variety of substrate peptides; see below, (a).

As such data are fitted to calculated curves, dissociation constants can be estimated for the ionizable groups at the active site. In pepsin, the general base exhibits a pK_a of approximately 1.4, whereas the general acid displays a pK_a of about 4.3. Compare these values with the expected pK_a for an aspartate carboxyl group in a protein of 4.2 to 4.6. The value for the aspartyl group acting as a general acid in pepsin is typical, but that for the aspartyl group acting as a general base is unusually low.

The pH dependence of HIV-1 protease has been assessed by measuring the apparent inhibition constant for a synthetic substrate analog (b). The data are consistent with the catalytic involvement of ionizable groups with pK_a values of 3.3 and 5.3. Maximal enzymatic activity occurs in the pH range between these two values. On the basis of the accumulated kinetic and structural data on HIV-1 protease, these pK_a values have been ascribed to the two active-site aspartate carboxyl groups. Note that the value of 3.3 is somewhat low for an aspartate side chain in a protein, whereas the value of 5.3 is somewhat high.

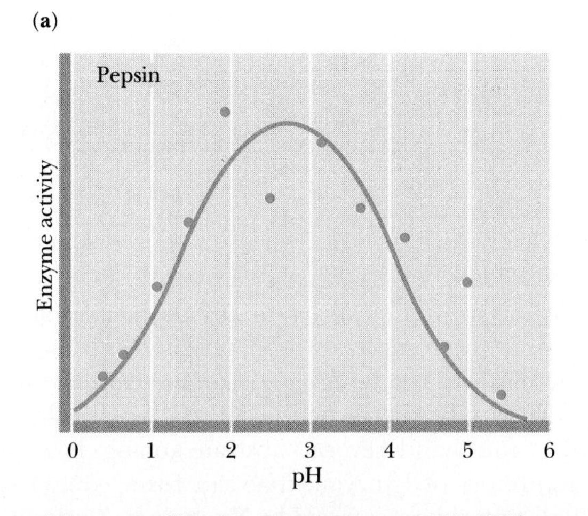

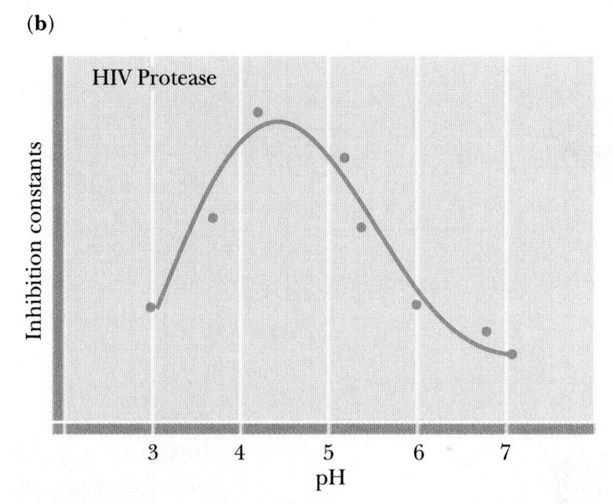

pH-rate profiles for (a) pepsin and (b) HIV protease.
(Adapted from Denburg, J., et al., 1968. Journal of the American Chemical Society 90:479–486, and Hyland, J., 1991. Biochemistry 30:8454–8463.)

acid (NAM) and *N*-acetylglucosamine (NAG). Both of these sugars are *N*-acetylated analogs of glucosamine, and in bacterial cell wall polysaccharides, they are joined in $\beta(1 \rightarrow 4)$ glycosidic linkages (Figure 10.34). Lysozyme hydrolyzes the glycosidic bond between C-1 of NAM and the C-4 of NAG, as shown in Figure 13.31, but does not act on the $\beta(1 \rightarrow 4)$ linkages between NAG and NAM.

Lysozyme is a small globular protein composed of 129 amino acids (14 kD) in a single polypeptide chain. It has eight cysteine residues linked in four disulfide bonds. The structure of this very stable protein was determined by X-ray crystallographic methods in 1965 by David Phillips (Figure 13.32). Although X-ray structures had previously been reported for proteins (hemoglobin and myoglobin), lysozyme was the first enzyme structure to be solved by crystallographic (or any other) methods. Although the location of the

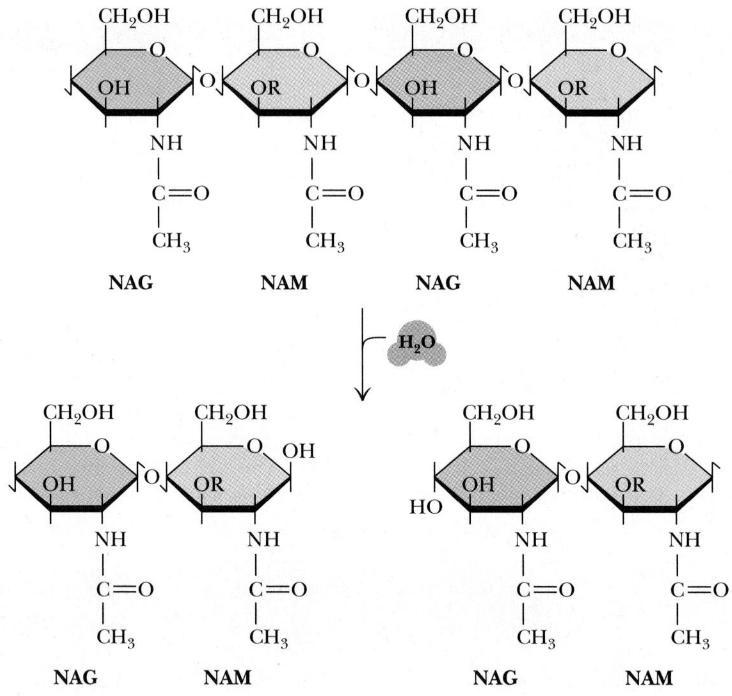

***Figure* 13.31** The lysozyme reaction.

active site was not obvious from the X-ray structure of the protein alone, X-ray studies of lysozyme-inhibitor complexes soon revealed the location and nature of the active site. Since it is an enzyme, lysozyme cannot form stable ES complexes for structural studies, because the substrate is rapidly transformed into products. On the other hand, several substrate analogs have proven to be good competitive inhibitors of lysozyme that can form complexes with the enzyme stable enough to be characterized by X-ray crystallography and other physical techniques. One of the best is a trimer of *N*-acetylglucosamine, $(NAG)_3$ (Figure 13.33), which is hydrolyzed by lysozyme at a rate only 1/60,000 that of the native substrate (Table 13.4). $(NAG)_3$ binds at the enzyme active site by forming five hydrogen bonds with residues located in one-half of a depression or crevice that spans the surface of the enzyme (Figure 13.34). The few hydrophobic residues that exist on the surface of lysozyme are located in this depression, and they may participate in hydrophobic and van der Waals interactions with $(NAG)_3$, as well as the normal substrate. The absence of charged groups on $(NAG)_3$ precludes the involvement of electrostatic interactions with the enzyme. Comparisons of the X-ray structures of the native lysozyme and the lysozyme-$(NAG)_3$ complex reveal that several amino acid residues at the active site move slightly upon inhibitor binding, including Trp^{62}, which moves about 0.75 Å to form a hydrogen bond with a hydroxymethyl group (Figure 13.35).

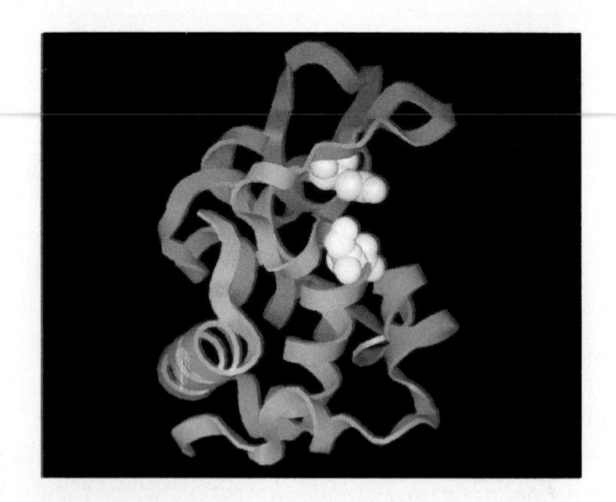

***Figure* 13.32** The structure of lysozyme. Glu^{35} and Asp^{52} are shown in white.

Model Studies Reveal a Strain-Induced Destabilization of a Bound Substrate on Lysozyme

One of the premises of lysozyme models is that the native substrate would occupy the rest of the crevice or depression running across the surface of the enzyme, since there is room to fit three more sugar residues into the crevice, and since the hexamer $(NAG)_6$ is in fact a good substrate for lysozyme (Table 13.4). The model building studies refer to the six sugar residue-binding sub-

NAG NAG NAG

***Figure* 13.33** $(NAG)_3$, a substrate analog, forms stable complexes with lysozyme.

Table **13.4**

Hydrolysis Rate Constants for Model Oligosaccharides with Lysozyme

Oligosaccharide	Rate Constant, $k_{cat}(s^{-1})$
$(NAG\text{-}NAM)_3$	0.5
$(NAG)_6$	0.25
$(NAG)_5$	0.033
$(NAG)_4$	7×10^{-5}
$(NAG)_3$	8×10^{-6}
$(NAG)_2$	2.5×10^{-8}

sites in the crevice with the letters A through F, with A, B, and C representing the part of the crevice occupied by the NAG_3 inhibitor (Figure 13.35). Modeling studies clearly show that NAG residues fit nicely into subsites A, B, C, E, and F of the crevice, but that fitting a residue of the $(NAG)_6$ hexamer into site D requires a substantial distortion of the sugar (out of its preferred chair conformation) to prevent steric crowding and overlap between C-6 and O-6 of the sugar at the D site and Ile^{98} of the enzyme. This distorted sugar residue is adjacent to the glycosidic bond to be cleaved (between sites D and E), and the inference is made that this distortion or strain brings the substrate closer to the transition state for hydrolysis. This is a good example of strain-induced destabilization of an otherwise favorably binding substrate (Section 13.4). Thus, the overall binding interaction of the rest of the sugar substrate would be favorable ($\Delta G < 0$), but distortion of the ring at the D site uses some of this binding energy to raise the substrate closer to the transition state for hydrolysis, *an example of stabilization of a transition state (relative to the simple enzyme-substrate complex)*. As noted in Section 13.4, distortion is one of the molecular mechanisms that can lead to such transition-state stabilization.

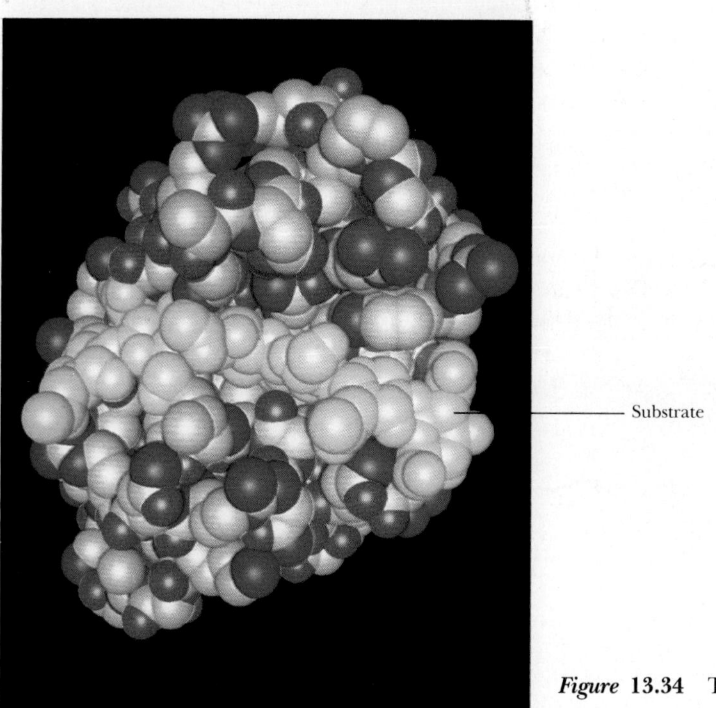

***Figure* 13.34** The lysozyme–enzyme-substrate complex. *(Photo courtesy of John Rupley, University of Arizona)*

***Figure* 13.35** Enzyme-substrate interactions at the six sugar residue-binding subsites of the lysozyme active site.
(Upper figure © Irving Geis.)

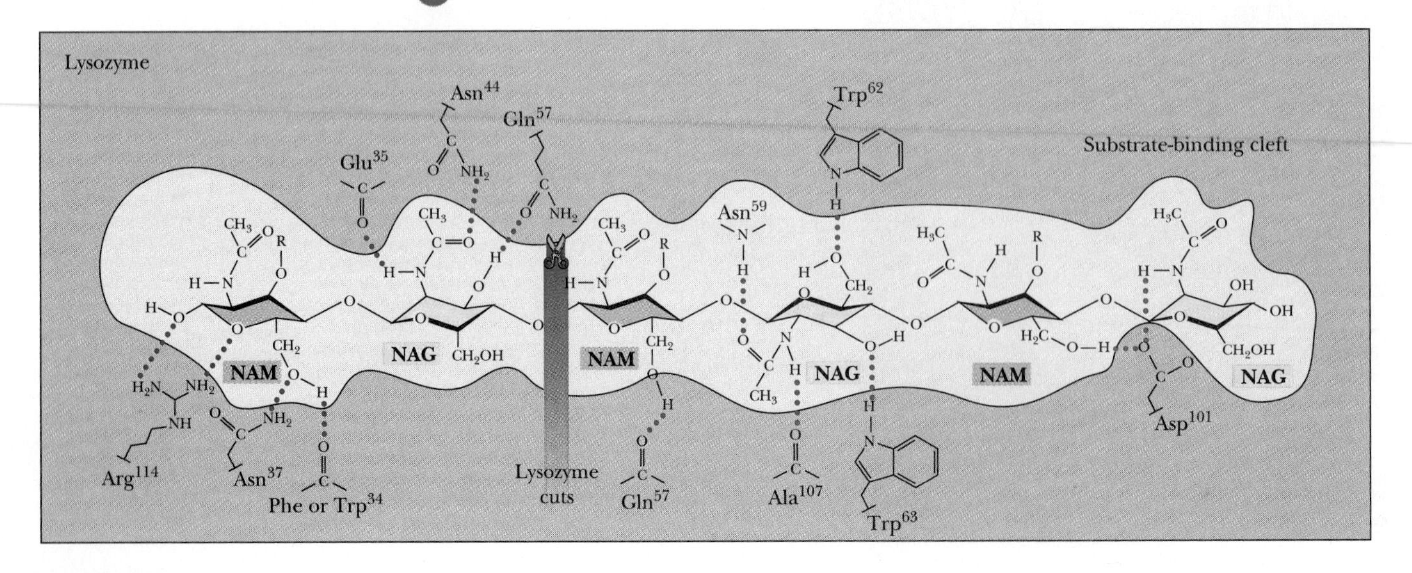

The Lysozyme Mechanism Involves General Acid–Base Catalysis

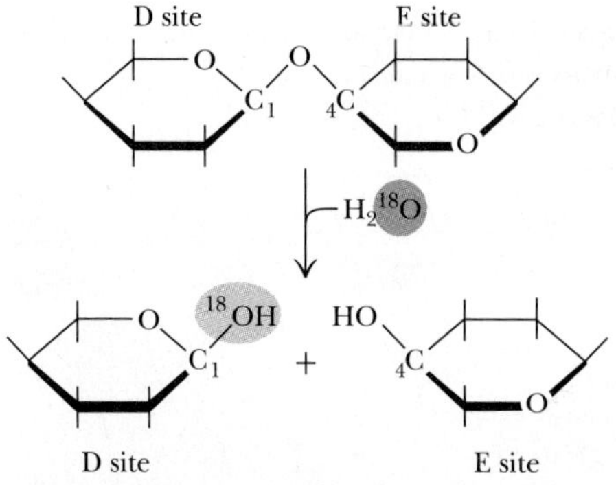

The mechanism of the lysozyme reaction is shown in Figures 13.36 and 13.37. Studies using ^{18}O-enriched water showed that the C_1—O bond is cleaved on the substrate between the D and E sites. Hydrolysis under these conditions incorporates ^{18}O into the C_1 position of the sugar at the D site, not into the oxygen at C_4 at the E site (Figure 13.36). Model building studies place the cleaved bond approximately between protein residues Glu^{35} and Asp^{52}. Glu^{35} is in a nonpolar or hydrophobic region of the protein, whereas Asp^{52} is located in a much more polar environment. Glu^{35} is protonated, but Asp^{52} is ionized (Figure 13.37). Glu^{35} may thus act as a general acid, donating a proton to the oxygen atom of the glycosidic bond and accelerating the reaction. Asp^{52}, on the other hand, probably stabilizes the carbonium ion generated at the D site upon bond cleavage. Formation of the carbonium ion may also be enhanced by the strain on the ring at the D site. Following bond cleavage, the product formed at the E site diffuses away, and the carbonium ion intermediate can then react with H_2O from the solution. Glu^{35} can now act as a general base, accepting a proton from the attacking water. The tetramer of NAG thus formed at sites A through D can now be dissociated from the enzyme.

On the basis of the above, the rate acceleration afforded by lysozyme appears to be due to (a) general acid catalysis by Glu^{35}; (b) distortion of the sugar ring at the D site, which may stabilize the carbonium ion *(and the transition state);* and (c) electrostatic stabilization of the carbonium ion by nearby Asp^{52}. The overall k_{cat} for lysozyme is about 0.5 sec^{-1}, which is quite slow (Table 11.4) compared with that for other enzymes. On the other hand, the destruction of a bacterial cell wall may only require hydrolysis of a few polysaccharide chains. The high osmotic pressure of the cell ensures that cell rupture will follow rapidly. Thus, lysozyme can accomplish cell lysis without a particularly high k_{cat}.

***Figure* 13.36** The C_1—O bond, not the O—C_4 bond, is cleaved in the lysozyme reaction. ^{18}O from $H_2{}^{18}O$ is thus incorporated at the C_1 position.

13.14 Carboxypeptidase A

The crystallographic analysis of lysozyme was reported in 1965. William Lipscomb followed quickly in 1967 with the high resolution crystal structure of a much larger enzyme, **carboxypeptidase A** (34.5 kD). Carboxypeptidase A is an **exopeptidase** secreted along with trypsin, chymotrypsin, and elastase in the pancreatic juice. Carboxypeptidase A is most active on aromatic or bulky ali-

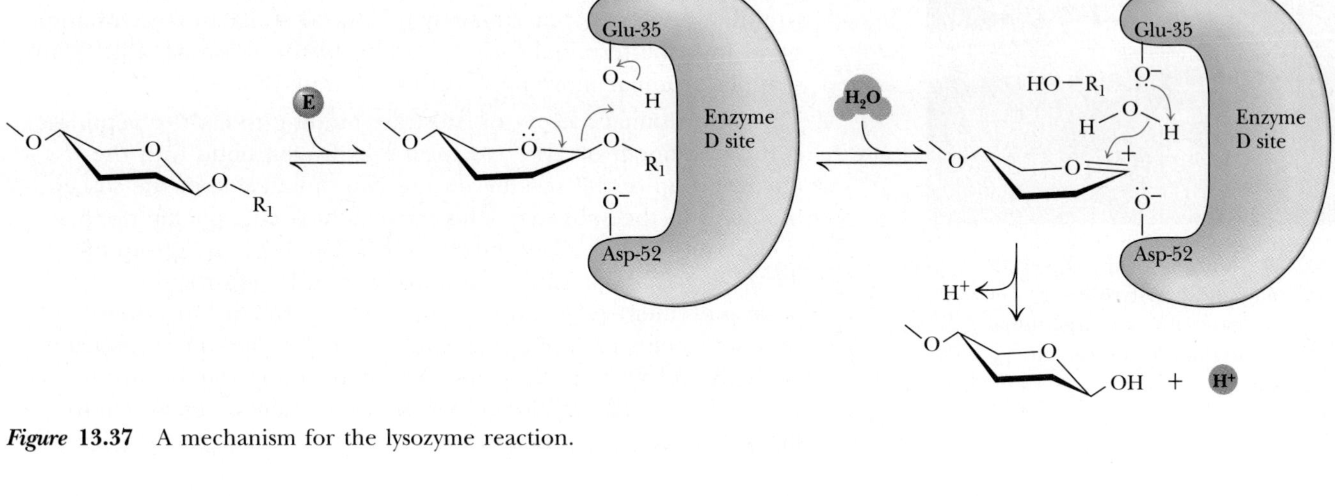

***Figure* 13.37** A mechanism for the lysozyme reaction.

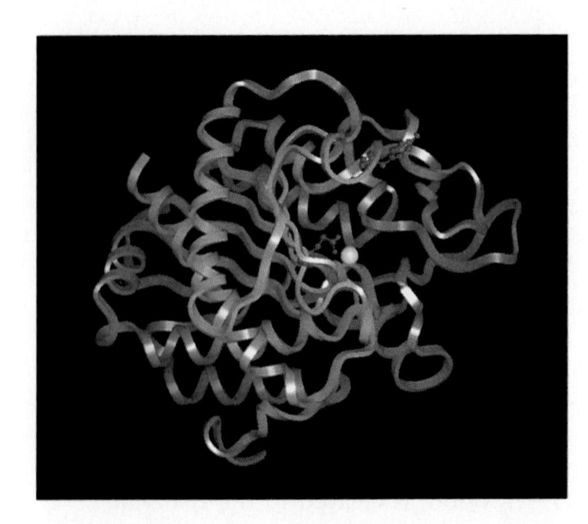

***Figure* 13.38** The structure of carboxypeptidase A. Residues Tyr^{248} (upper right) and Glu^{270} (center) are shown in red. The Zn atom lies to the right of Glu^{270}.

phatic residues, whereas the related carboxypeptidase B functions best with basic residues. Like the serine proteases, these carboxypeptidases are secreted as inactive precursors and then activated by proteolysis. Bovine carboxypeptidase A initially consists of a complex of three subunits, which is proteolyzed by trypsin to yield the active form of the enzyme.

Carboxypeptidase A is a good example of a *zinc metalloenzyme,* and is a notable case of an enzyme that undergoes a large conformational change upon substrate binding. The three-dimensional structure, as determined by Lipscomb and others, is shown in Figure 13.38. The enzyme is approximately ellipsoidal in shape, with dimensions of $5 \times 4.2 \times 3.8$ nm, and contains about 38% α-helix and 17% β-sheet. The active site is located in a depression or groove in the surface of the enzyme where the essential zinc atom is coordinated to His^{69}, to His^{196}, and to both carboxylate oxygens of Glu^{72}. The zinc is pentacoordinate in the absence of substrate, and the fifth ligand is a water molecule. When substrate binds, the zinc becomes hexacoordinate, making an additional bond with the carbonyl oxygen of the peptide bond to be cleaved. Residues important for substrate binding include Arg^{71}, Arg^{127}, Asn^{144}, Arg^{145}, Tyr^{248}, and Glu^{270} (Figure 13.38).

As with the serine proteases and lysozyme, much of our information about the mechanism of carboxypeptidase A has come from studies using inhibitors or weakly reactive substrates. The enzyme is specific for peptides of L-amino acids, and the presence of a second peptide bond is important. Dipeptides thus make rather poor substrates, but dipeptides with an *N*-terminal acetyl group are hydrolyzed at normal rates. Like the serine proteases, carboxypeptidase exhibits esterase activity, particularly if the C-terminal side chain is aromatic and the C-terminal residue is of the L-configuration. Carboxypeptidase is a rather slow enzyme. The most rapidly cleaved peptide substrates are hydrolyzed with a k_{cat} of about 10^2/sec. The maximal rates with esters are approximately 10^3/sec.

X-ray crystallographic studies of the carboxypeptidase A complex with the very slowly hydrolyzed dipeptide glycyl-L-tyrosine provided many of the early insights on the interactions of substrates with carboxypeptidase A. As shown in Figure 13.39, this dipeptide interacts with the enzyme in a number of important ways:

1. The tyrosine side chain of the dipeptide fits into a hydrophobic pocket in the enzyme. This pocket accommodates either aromatic or bulky aliphatic groups, explaining the enzyme's preference for substrates with large, hydrophobic side chains.
2. The active-site zinc coordinates directly to the carbonyl oxygen of the peptide bond to be cleaved. The substrate effectively displaces the water molecule that previously occupied this site.
3. $Arginine^{145}$ moves approximately 0.2 nm to make an electrostatic bond with the terminal carboxyl group of the substrate. Glu^{270} undergoes a similar movement as the substrate binds.
4. The most dramatic event of substrate binding to carboxypeptidase is the movement of Tyr^{248} to form a hydrogen bond with the amino group of the residue on the N-terminal side of the susceptible bond in the substrate. This movement occurs mainly by rotation about the C_α—C_β bond of Tyr^{248}. The hydroxyl group of Tyr^{248} moves about 1.2 nm to make this hydrogen bond—a very large movement relative to the size of the protein. Together with the movements of Arg^{145} and Glu^{270}, this amounts to an **induced fit** of the substrate at the active site, confirming Daniel Koshland's view (Chapter 12) of enzyme-substrate interaction. These conformational changes also function to close the active-site cavity, displacing

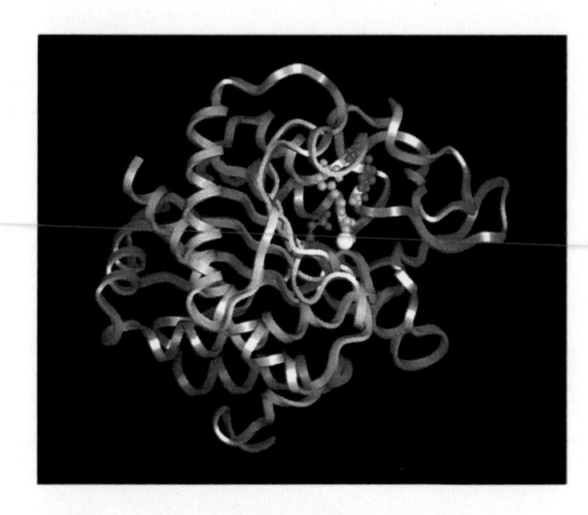

***Figure* 13.39** Binding of the dipeptide substrate analog glycyl-L-tyrosine (green) causes several dramatic conformational changes near the active site of carboxypeptidase A. Note the substantial reorientation of Tyr^{248} relative to its position in Figure 13.38.

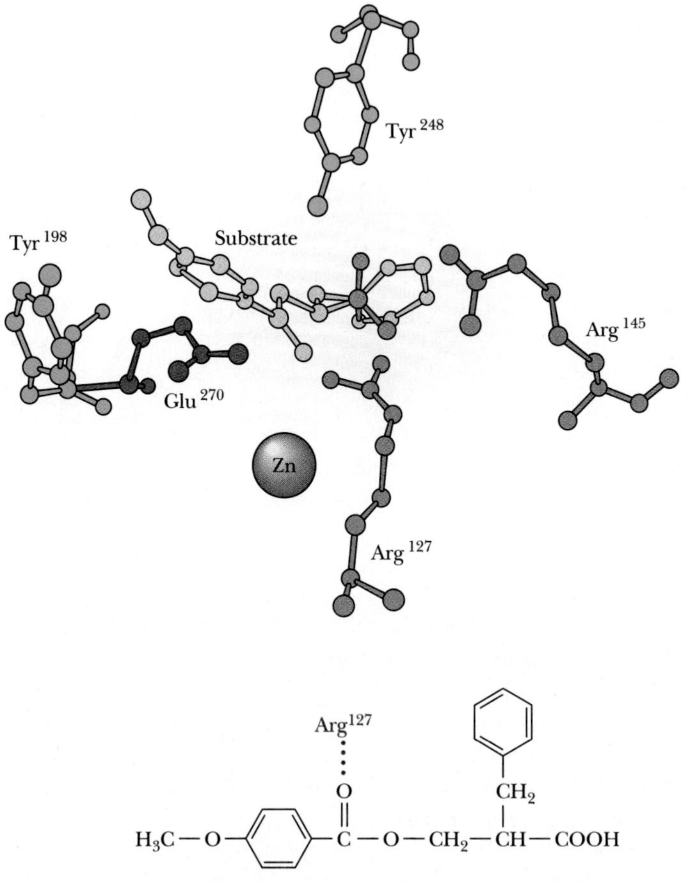

Figure **13.40** The CPA inhibitor (-)-2 benzyl-3-(p-methoxybenzoyl)propanoic acid binds to CPA with its carbonyl oxygen hydrogen-bonded to Arg^{127}.

(Adapted from Christianson, D., and Lipscomb, W., 1989. Accounts of Chemical Research 22:62.)

Tyr 248

Tyr 198

Substrate

Arg 145

Glu 270

Zn

Arg 127

Arg^{127}

$H_3C-O-C_6H_4-C(=O)-O-CH_2-CH(CH_2C_6H_5)-COOH$

(–)–2–benzyl–3–(p–methoxybenzoyl)propanoic acid

at least four water molecules from the active-site cleft (in addition to the one water displaced from the zinc by substrate).

Despite its early usefulness, glycyl-L-tyrosine is not an ideal substrate analog for carboxypeptidase A. A better representation of CPA-substrate interactions has been obtained in X-ray diffraction studies using the substrate analog *(-)-2-benzyl-3-(p-methoxybenzoyl)propanoic acid* (Figure 13.40). In this complex, the carbonyl group does not coordinate the Zn^{2+}ion, but is hydrogen-bonded to the guanidinium group of Arg^{127}.

A mechanism consistent with this structure and other kinetic and structural data is shown in Figure 13.41. This is called a **promoted-water pathway,** since it involves an attack on the substrate carbonyl by water *promoted* by Zn^{2+}. In the ES complex, Glu^{270} acts as a general base, abstracting a proton from the Zn^{2+}-coordinated water. The *gem*-diol intermediate is stabilized by coordination to Zn^{2+} (which is hexacoordinate in this complex, as noted). Breakdown to products occurs as Glu^{270} acts as a general acid to donate a proton to the nitrogen of the departing amino acid.

If the mechanism of carboxypeptidase A resembles Figure 13.41, then molecules that mimic the *gem*-diol transition state should behave as transition-state analogs for carboxypeptidase. Since *gem*-diols are relatively unstable, we must resort to some chemical "tricks" to continue this line of reasoning. Two interesting examples are shown in Figure 13.42. The **phosphonamidate** is an effective inhibitor of carboxypeptidase. The tetrahedral phosphorus with a negative charge distributed over the two oxygen atoms is similar to the pre-

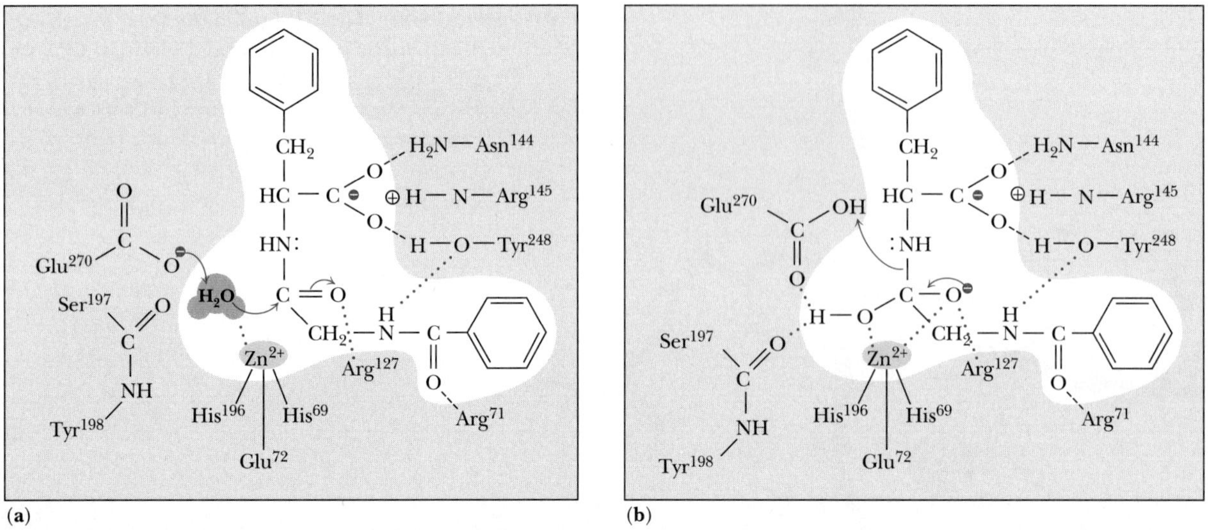

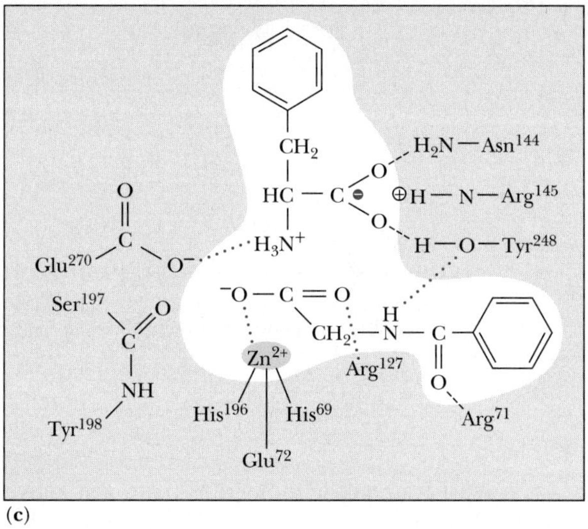

(a)

(b)

(c)

***Figure* 13.41** A mechanism for carboxypeptidase A cleavage of the model substrate benzoylglycylphenylalanine.

(Adapted from Christianson, D., and Lipscomb, W., 1989. Accounts of Chemical Research 22:62.)

sumed *gem*-diol transition state. With a K_i of 90 nM, this analog is bound much more tightly than typical carboxypeptidase substrates, consistent with stabilization of a *gem*-diol transition state by the enzyme. Even stronger evidence for stabilization of the *gem*-diol comes from studies of the ketomethylene substrate analog. *X-ray crystallographic studies have shown that this analog binds to carboxypeptidase A as the* ***ketone hydrate*** *(Figure 13.42).* This is quite surprising, since such ketones are rarely hydrated in solution. (For example, acetone exists in aqueous solution as 0.2% hydrate. The ketomethylene analog in Figure 13.42 is even less reactive and less likely to hydrate than acetone.) The preference of carboxypeptidase for binding of the hydrated form of this analog is dramatic evidence of preferential stabilization of the *gem*-diol transition state, relative to the simple ketone that is structurally similar to the carboxypeptidase substrate. *X-ray crystallographic studies of these complexes show that the preferential stabilization of these transition-state-like structures is achieved through coordination of the oxygen atoms (of the phosphonamidate or the ketone hydrate) by the active-site* Zn^{2+}*, as well as by* Glu^{270} *and* Arg^{127}*, just as predicted by the mechanism of Figure 13.41.*

An important site-directed mutagenesis experiment (Chapter 8) has forced enzymologists to reconsider a long-held belief about the role of Tyr^{248}

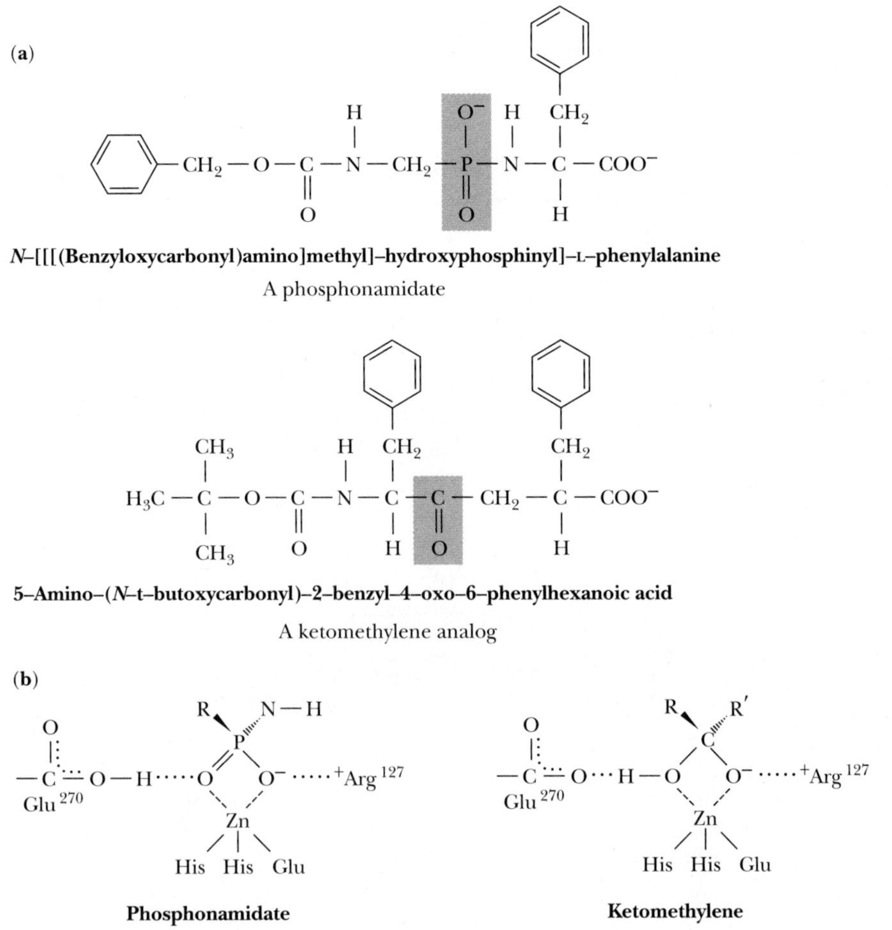

(a)

N-[[[(Benzyloxycarbonyl)amino]methyl]-hydroxyphosphinyl]-L-phenylalanine

A phosphonamidate

5-Amino-(N-t-butoxycarbonyl)-2-benzyl-4-oxo-6-phenylhexanoic acid

A ketomethylene analog

(b)

Phosphonamidate

Ketomethylene

***Figure* 13.42** (a) Transition-state analogs for carboxypeptidase A, including a phosphonamidate and a ketomethylene analog. (b) Structures mimicking the *gem*-diol transition state of carboxypeptidase A. The structure for the ketomethylene complex is that of a ketone hydrate.

in the carboxypeptidase mechanism. Ever since Lipscomb's early X-ray diffraction studies of CPA, Tyr^{248} had been thought to act as a general acid, donating a proton to the departing amine formed from the cleaved peptide bond. However, a genetically engineered mutant CPA containing phenylalanine instead of tyrosine at position 248 displays nearly the same activity as wild-type enzyme. This experiment shows that Tyr^{248} is not absolutely required for catalysis. Interestingly, the Phe^{248} mutant enzyme binds substrates weakly compared to native CPA, indicating that Tyr^{248} is important for substrate binding, probably by hydrogen bonding through the tyrosine hydroxyl group (Figure 13.41).

13.15 Liver Alcohol Dehydrogenase

The various alcohol dehydrogenases are zinc metalloenzymes with broad specificity, able to oxidize a wide range of alcohols to the corresponding ketones and aldehydes. By far, the most intensively studied and best-understood alcohol dehydrogenases are those from liver and yeast. Liver alcohol dehydrogenase (LAD) is a dimer consisting of 374 amino acids. The sequence is known (Figure 13.43) and the crystal structures of the native enzyme with and without metal and substrates have been determined (refer to Figure 5.39). Each subunit of the enzyme is divided into two domains—the NAD^+ coenzyme-binding domain and the catalytic domain. These domains are sepa-

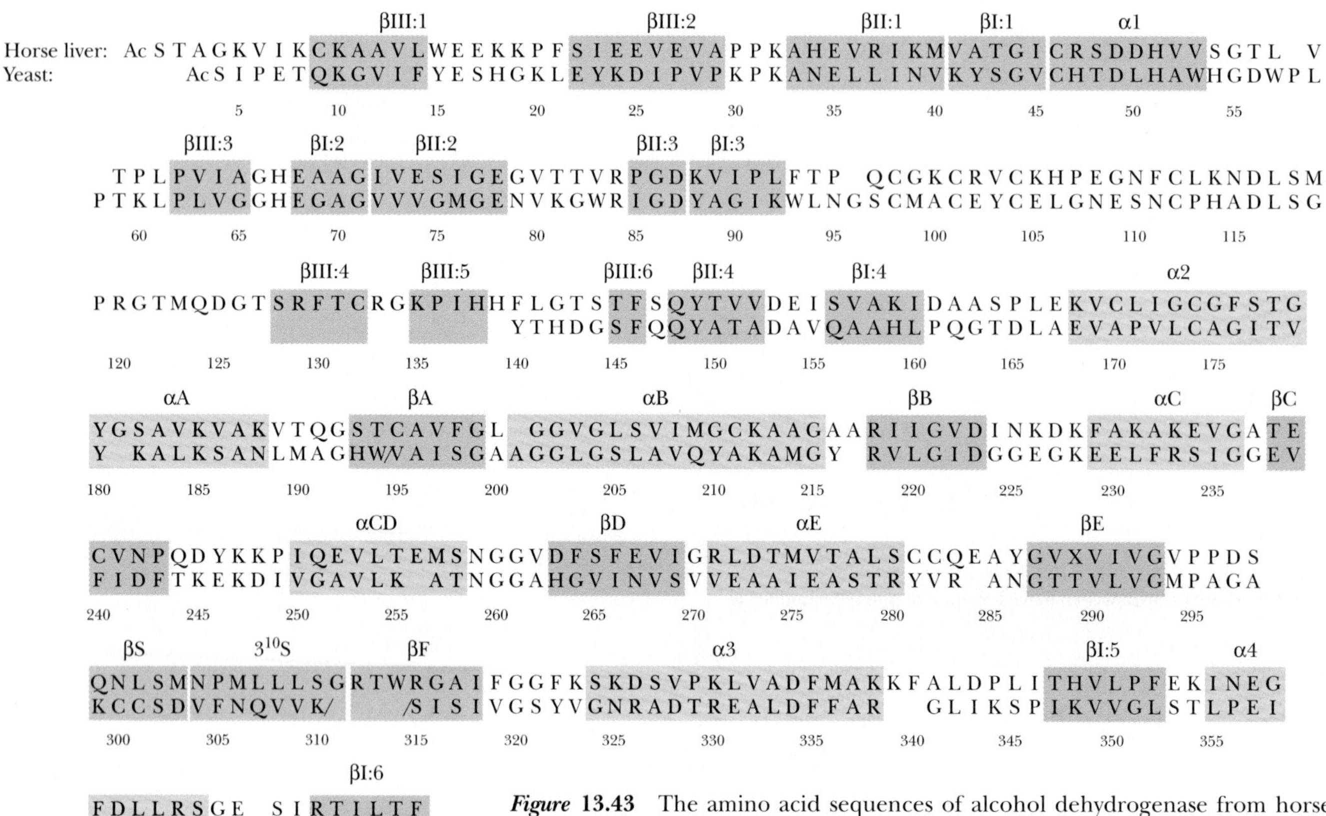

***Figure* 13.43** The amino acid sequences of alcohol dehydrogenase from horse liver (upper line in each row) and yeast (lower line). Green-shaded segments form α-helices; purple-shaded segments form β-sheets.

rated by a cleft containing a deep pocket. The substrate and the nicotinamide moiety of the NAD^+ coenzyme fit into this pocket. The catalytic domain, which is the larger of the two, consists of residues 1 through 175 and 319 through 374, and it determines not only the catalytic events but also the substrate specificity. The NAD^+-binding domain consists of residues 176 through 318. NAD^+ binds to the coenzyme-binding domain with the nicotinamide ring extended toward the catalytic domain. The coenzyme-binding domains of the two subunits are joined together in a central, compact core (Figure 5.39). This domain is a typical parallel α/β array, with a central, pleated sheet of six parallel strands, with α-helices on both sides. This structure (and the way the coenzyme binds to this structure) is more or less common to all the NAD^+-dependent dehydrogenases. The catalytic domain is an intricate network of antiparallel pleated-sheet regions and a few short helices.

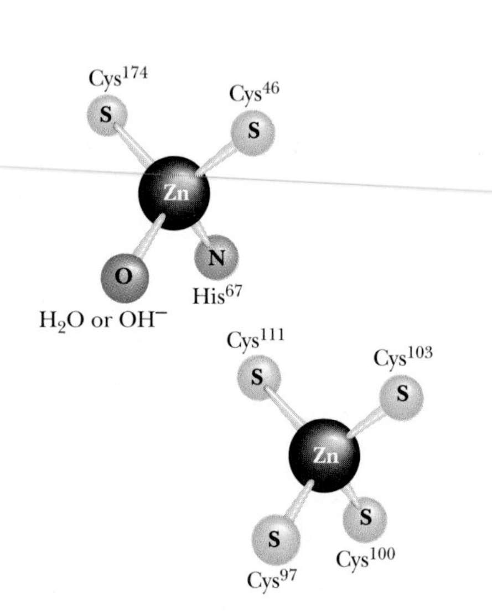

***Figure* 13.44** The zinc ligands in LAD. *Above,* catalytic Zn^{2+}; *below,* structural Zn^{2+}.

The LAD Active Site Includes Structural and Catalytic Zinc Ions

There are two zinc atoms bound per monomer. One of these zinc atoms presumably plays only a structural role, but the other, referred to as the active-site zinc, is essential for catalysis. It is located at the bottom of the active-site cleft, 2.5 nm from the surface, coordinated to Cys^{46}, His^{67}, Cys^{174}, and a water molecule or hydroxide ion. This H_2O or OH^- is hydrogen bonded to the hydroxyl group of Ser^{48}, which itself is hydrogen bonded to His^{51}. The ligand geometry around this catalytic zinc atom is a distorted tetrahedron (Figure 13.44). This active-site zinc is bound near the nicotinamide ring of NAD^+. The structural or noncatalytic zinc is near the surface of the molecule, about 2.5 nm from the catalytic site, and is coordinated to the four thiol groups of

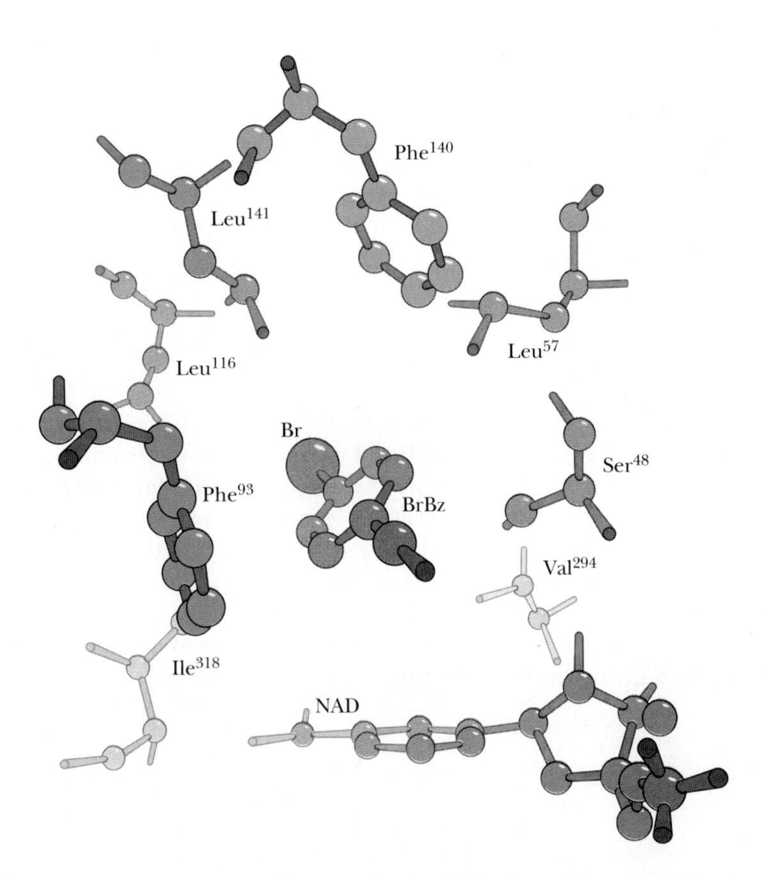

***Figure* 13.45** The substrate-binding pocket in LAD, viewed from the zinc atom. Bromobenzyl alcohol is shown bound in the pocket. Fainter, smaller residues lie farther away.

(Adapted from Eklund, H., et al., 1982. Journal of Biological Chemistry 257:14349.)

Cys^{97}, Cys^{100}, Cys^{103}, and Cys^{111}. The substrate-binding portion of the peptide consists of a large hydrophobic pocket, 0.5 to 1 nm wide and about 2 nm long from the protein surface to the zinc atom, and the water/hydroxide liganding position of the catalytic zinc points out into this pocket. The only polar atoms in the pocket are the zinc ligands, the nicotinamide moiety of the NAD$^+$, and the side chain of Ser^{48}. The remainder of the wall of the pocket is lined with the hydrophobic side chains of Leu^{57}, Phe^{93}, Phe^{110}, Leu^{116}, Phe^{140}, Leu^{141}, Val^{294}, Pro^{295}, Pro^{296}, and Ile^{318} from the same subunit as the ligands to zinc, and also Met^{306} and Leu^{309} from the other subunit (Figure 13.45). This large substrate-binding pocket may explain why the enzyme shows a broad specificity and acts on a variety of aliphatic and cyclic ketones, aldehydes, and alcohols.

Studies of the pH dependence of the binding of NAD$^+$ and measurements of proton release occurring on binding of NAD$^+$ indicate that the **apoenzyme** (the enzyme without NAD$^+$) has a functional group with a pK_a of 9.6, which changes to 7.6 in the **holoenzyme** (the complex of the enzyme and NAD$^+$). This is the pK_a for the ionization of the water bound to the active-site zinc. The oxygen atom of the alcohol or aldehyde (or ketone) substrate binds directly to the zinc ion with the hydrophobic side chain of the substrate extending down into the hydrophobic pocket of the active-site cleft.

Binding of the coenzyme causes a conformational change in the enzyme, consisting of a rotation of each catalytic domain by about 10° with respect to the core of the enzyme dimer (a hingelike motion). This movement closes the cleft or crevice between the catalytic and coenzyme-binding domains and brings the catalytic zinc atoms about 0.1 nm closer to the coenzyme-binding domain and brings the substrate-binding site closer to the nicotinamide ring.

When the NAD^+ is bound in this manner and a substrate is bound as well, the zinc atom, its ligands, and the nicotinamide ring are no longer in contact with the solvent that surrounds the enzyme. All the water molecules that can occupy the active site and the cleft in the open form are pushed out in these ternary complexes, making the active site of the enzyme completely water-free in the closed form.

The LAD Mechanism—Proton Abstraction Followed by Hydride Transfer

The mechanism of alcohol oxidation to aldehyde by LAD involves proton abstraction from the alcohol —OH and hydride transfer from the alcohol carbon to NAD^+ to form NADH (Figure 13.46). The complete mechanism for these steps is not known with certainty, but evidence has accumulated to support several possibilities. The binding of alcohol to the active-site zinc substantially lowers the pK_a of the alcohol group from >14 to approximately 6.4. Proton release from bound ethanol occurs in the closed form of the enzyme, with no access to solvent at all. Thus, the proton abstracted must be picked up by a group at the active site. It can either be released directly into

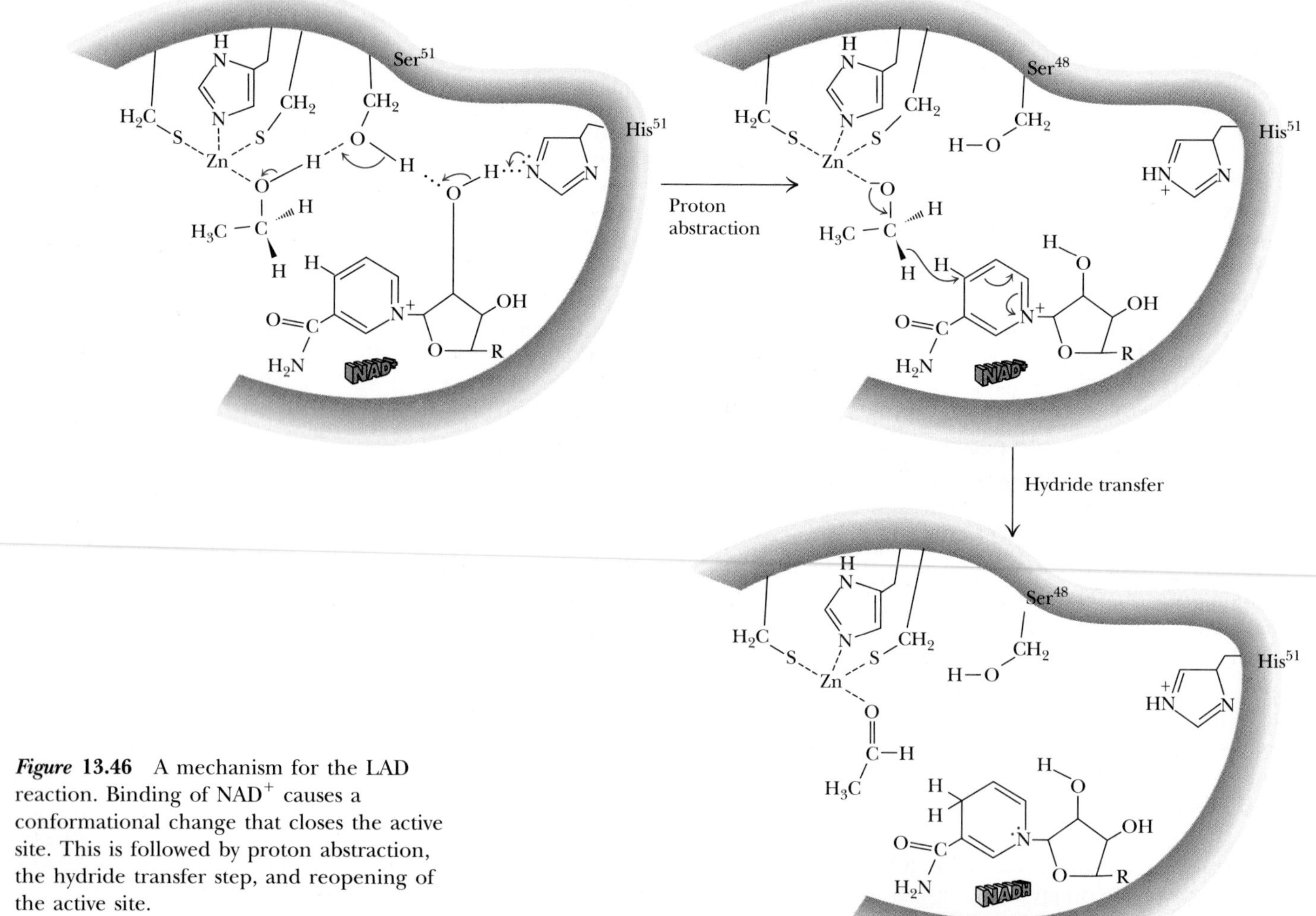

***Figure* 13.46** A mechanism for the LAD reaction. Binding of NAD^+ causes a conformational change that closes the active site. This is followed by proton abstraction, the hydride transfer step, and reopening of the active site.

(Adapted from Eklund, H., et al., 1982. Journal of Biological Chemistry 257:14349.)

solution when the cleft opens again or be transferred from group to group on the enzyme until it reaches the surrounding solution. There is, in fact, some evidence for the latter process, through a network of hydrogen bonds on the enzyme surface. The groups that might be involved in such a transfer include the Ser[48] hydroxyl, the amide oxygen atom of the nicotinamide, and the imidazole of His[51].

The second necessary step in the alcohol dehydrogenase mechanism is the transfer of a hydride anion from the alcohol carbon to the nicotinamide. The positive charges on the catalytic zinc atom and the NAD$^+$ (Figure 13.46) facilitate this hydride transfer by stabilizing the alcoholate anion on the substrate. Just as for the proton abstraction, the hydride transfer step also occurs in a completely water-free environment. The stabilization of the alcoholate anion by the catalytic zinc atom is enhanced in the absence of water. The absence of solvent also prevents possible side reactions between water and the hydride to be removed. A complete mechanism showing the closing of the active-site cleft on NAD$^+$ binding, the chemical steps of proton abstraction and hydride transfer, and the reopening of the active-site cleft is shown in Figure 13.46.

Problems

1. Tosyl-L-phenylalanine chloromethyl ketone (TPCK) specifically inhibits chymotrypsin by covalently labeling His[57].

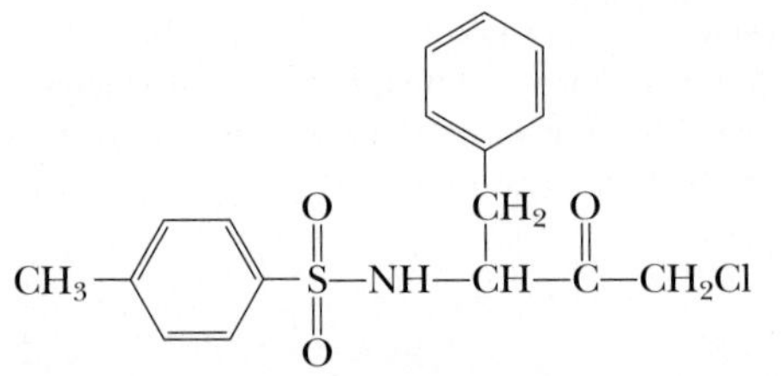

   CH_3–C$_6$H$_4$–S(=O)$_2$–NH–CH(CH$_2$–C$_6$H$_5$)–C(=O)–CH_2Cl

   Tosyl-L-phenylalanine chloromethyl ketone (TPCK)

   a. Propose a mechanism for the inactivation reaction, indicating the structure of the product(s).
   b. State why this inhibitor is specific for chymotrypsin.
   c. Propose a reagent based on the structure of TPCK that might be an effective inhibitor of trypsin.

2. In this chapter, the experiment in which Craik and Rutter replaced Asp[102] with Asn in trypsin (reducing activity 10,000-fold) was discussed.
   a. On the basis of your knowledge of the catalytic triad structure in trypsin, suggest a structure for the "uncatalytic triad" of Asn-His-Ser in this mutant enzyme.
   b. Explain why the structure you have proposed explains the reduced activity of the mutant trypsin.
   c. See the original journal articles (Sprang, et al., 1987. *Science* **237:**905–909 and Craik, et al., 1987. *Science* **237:**909–913) to see what Craik and Rutter's answer to this question was.

3. As noted previously, abstraction of the hydroxyl proton of ethanol is a vital step in the liver alcohol dehydrogenase mechanism. Ehrig et al. (1991. *Biochemistry* **30:**1062–1068) produced a mutant of human liver alcohol dehydrogenase in which His[51], a candidate for the general base in this proton abstraction event, was changed to Gln. The H51Q (in which His[51] is replaced by glutamine) mutant had an ethanol oxidation activity about 1/6 that of the normal enzyme, but full activity could be restored in the presence of high concentrations of glycylglycine, glycine, and phosphate buffers. What conclusions may be drawn about the mechanism of human LAD based on these observations?

4. Pepstatin (below) is an extremely potent inhibitor of the monomeric aspartic proteases, with K_I values of less than 1 n*M*.
   a. On the basis of the structure of pepstatin, suggest an explanation for the strongly inhibitory properties of this peptide.
   b. Would pepstatin be expected to also inhibit the HIV-1 protease? Explain your answer.

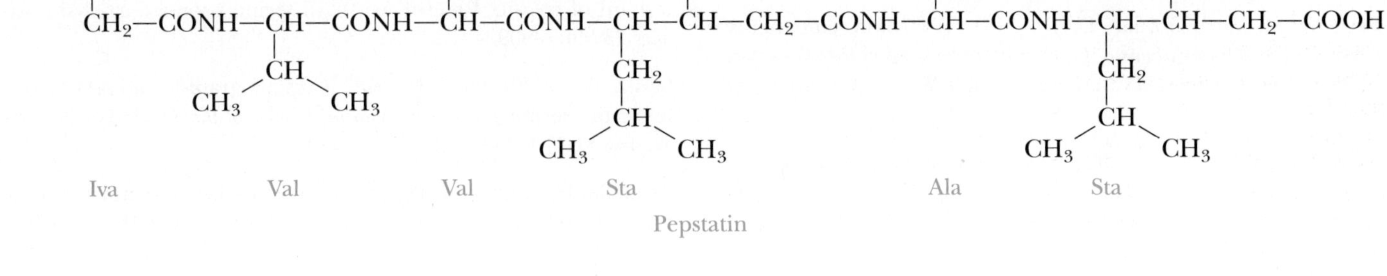

$(CH_3)_2CH$–CH_2–CONH–CH[CH(CH$_3$)$_2$]–CONH–CH[CH(CH$_3$)$_2$]–CONH–CH[CH$_2$CH(CH$_3$)$_2$]–CH(OH)–CH$_2$–CONH–CH(CH$_3$)–CONH–CH[CH$_2$CH(CH$_3$)$_2$]–CH(OH)–CH$_2$–COOH

Iva Val Val Sta Ala Sta

Pepstatin

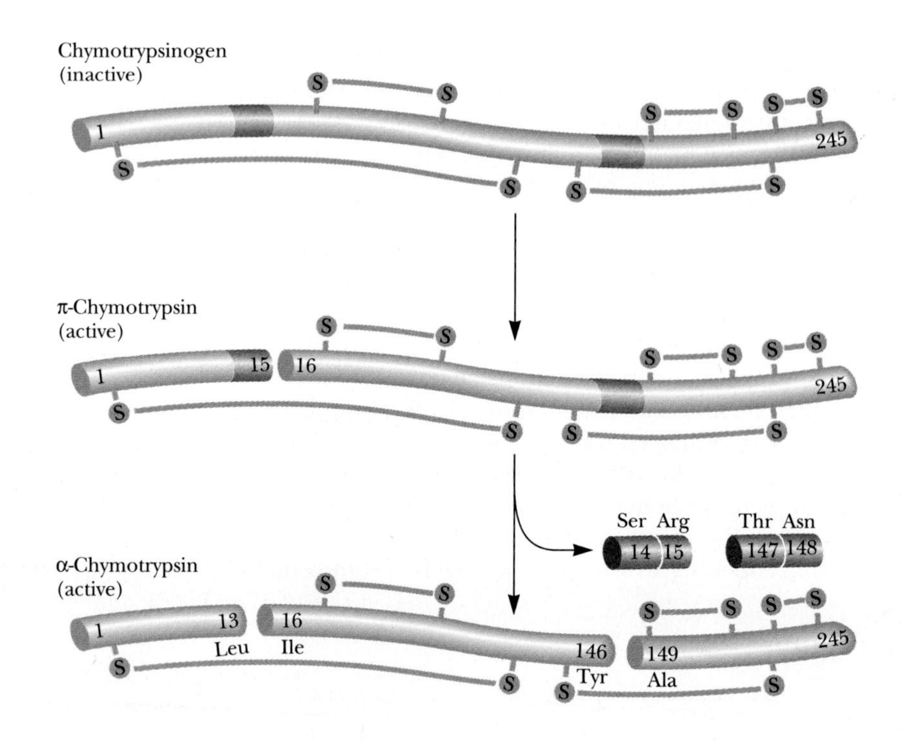

5. The k_{cat} for alkaline phosphatase-catalyzed hydrolysis of methylphosphate is approximately 14/sec at pH 8 and 25°C. The rate constant for the uncatalyzed hydrolysis of methylphosphate under the same conditions is approximately 1×10^{-15}/sec. What is the difference in the free energies of activation of these two reactions?

6. Active α-chymotrypsin is produced from chymotrypsinogen, an inactive precursor, as shown in the figure above.

   The first intermediate—π-chymotrypsin—displays chymotrypsin activity. Suggest proteolytic enzymes that might carry out these cleavage reactions effectively.

Further Reading

General

Eigen, M., 1964. Proton transfer, acid–base catalysis, and enzymatic hydrolysis. *Angewandte Chemie, Int. Ed.* **3:**1–72.

Fersht, A., 1985. *Enzyme Structure and Mechanism.* 2nd ed. New York: W. H. Freeman and Company.

Jencks, W. P., 1969. *Catalysis in Chemistry and Enzymology.* New York: McGraw-Hill.

Page, M. I., and Williams, A., eds., 1987. *Enzyme Mechanisms.* London, England: Royal Society of London.

Walsh, C., 1979. *Enzymatic Reaction Mechanisms.* San Francisco: W. H. Freeman and Company.

Transition-State Stabilization and Transition-State Analogs

Kraut, J., 1988. How do enzymes work? *Science* **242:**533–540.

Kreevoy, M., and Truhlar, D. G., 1986. Transition-state theory. Chapter 1 in *Investigations of Rates and Mechanisms of Reactions,* Vol. 6, Part 1, edited by C. F. Bernasconi. New York: John Wiley and Sons.

Lolis, E., and Petsko, G., 1990. Transition-state analogues in protein crystallography: Probes of the structural source of enzyme catalysis. *Annual Review of Biochemistry* **59:**597–630.

Wolfenden, R., 1972. Analogue approaches to the structure of the transition state in enzyme reactions. *Accounts of Chemical Research* **5:**10–18.

Wolfenden, R., and Frick, L., 1987. Transition state affinity and the design of enzyme inhibitors. Chapter 7 in *Enzyme Mechanisms,* edited by M. I. Page and A. Williams. London, England: Royal Society of London.

Wolfenden, R., and Kati, W. M., 1991. Testing the limits of protein-ligand binding discrimination with transition-state analogue inhibitors. *Accounts of Chemical Research* **24:**209–215.

Serine Proteases

Craik, C. S., et al., 1987. The catalytic role of the active site aspartic acid in serine proteases. *Science* **237:**909–919.

Sprang, S., et al., 1987. The three-dimensional structure of Asn^{102} mutant of trypsin: Role of Asp^{102} in serine protease catalysis. *Science* **237:**905–909.

Steitz, T., and Shulman, R., 1982. Crystallographic and NMR studies of the serine proteases. *Annual Review of Biophysics and Bioengineering* **11:**419–444.

Tsukada, H., and Blow, D., 1985. Structure of α-chymotrypsin refined at 1.68 Å resolution. *Journal of Molecular Biology* **184:**703–711.

Aspartic Proteases

Fruton, J., 1976. The mechanism of the catalytic action of pepsin and related acid proteinases. *Advances in Enzymology* **44:**1–36.

Polgar, L., 1987. The mechanism of action of aspartic proteases involves "push-pull" catalysis. *FEBS Letters* **219:**1–4.

HIV-1 Protease

Blundell, T., et al., 1990. The 3-D structure of HIV-1 proteinase and the design of antiviral agents for the treatment of AIDS. *Trends in Biochemical Sciences* **15:**425–430.

Hyland, L., et al., 1991. Human immunodeficiency virus-1 protease 1: Initial velocity studies and kinetic characterization of reaction intermediates by ^{18}O isotope exchange. *Biochemistry* **30:**8441–8453.

Hyland, L., Tomaszek, T., and Meek, T., 1991. Human immunodeficiency virus-1 protease 2: Use of pH rate studies and solvent isotope effects to elucidate details of chemical mechanism. *Biochemistry* **30:**8454–8463.

Carboxypeptidase

Christianson, D., and Lipscomb, W., 1989. Carboxypeptidase A. *Accounts of Chemical Research* **22:**62–69.

Gardell, S., Craik, C., Hilvert D., et al., 1985. Site-directed mutagenesis shows that tyrosine 248 of carboxypeptidase does not play a crucial role in catalysis. *Nature* **317:**551–555.

Hilvert, D., et al., 1986. Evidence against a crucial role for the phenolic hydroxyl of Tyr-248 in peptide and ester hydrolysis catalyzed by carboxypeptidase: Comparative studies of the pH dependencies of the native and Phe-248-mutant forms. *Journal of the American Chemical Society* **108:**5298–5304.

Lysozyme

Chipman, D., and Sharon, N., 1969. Mechanism of lysozyme action. *Science* **165:**454–465.

Ford, L., et al., 1974. Crystal structure of a lysozyme-tetrasaccharide lactone complex. *Journal of Molecular Biology* **88:**349–371.

Kirby, A., 1987. Mechanism and stereoelectronic effects in the lysozyme reaction. *CRC Critical Reviews in Biochemistry* **22:**283–315.

Phillips, D., 1966. The three-dimensional structure of an enzyme molecule. *Scientific American* **215:**75–80.

Alcohol Dehydrogenase

Biellmann, J.-F., 1986. Chemistry and structure of alcohol dehydrogenase: Some general considerations on binding mode variability. *Accounts of Chemical Research* **19:**321–328.

Colonna-Cesari, F., et al., 1986. Interdomain motion in liver alcohol dehydrogenase. *Journal of Biological Chemistry* **261:**15273–15280.

Eklund, H., Samama, J.-P., and Jones, T., 1984. Crystallographic investigations of nicotinamide adenine dinucleotide binding to horse liver alcohol dehydrogenase. *Biochemistry* **23:**5982–5996.

Eklund, H., and Brändén, C., 1983. The role of zinc in alcohol dehydrogenase. Chapter 4 in *Zinc Enzymes,* edited by T. G. Spiro, pages 123–152. New York: Wiley-Interscience.

Eklund, H., Plapp, B., Samama, J.-P., and Brändén, C., 1982. Binding of substrate in a ternary complex of horse liver alcohol dehydrogenase. *Journal of Biological Chemistry* **257:**14349–14358.

"Upon the rich and poor alike its blessings flow,
And its neglect to young and old alike brings woe."
Horace (65 B.C.–8 B.C.)

Chapter 14

Coenzymes and Vitamins

Fresco from the Villa Issogna, Val d'Aosta, Italy, showing the interior of an early pharmacy.

Outline

14.1 Vitamin B_1: Thiamine and Thiamine Pyrophosphate
14.2 Vitamins Containing Adenine Nucleotides
14.3 Nicotinic Acid and the Nicotinamide Coenzymes
14.4 Riboflavin and the Flavin Coenzymes
14.5 Pantothenic Acid and Coenzyme A
14.6 Vitamin B_6: Pyridoxine and Pyridoxal Phosphate
14.7 Vitamin B_{12}: Cyanocobalamin
14.8 Vitamin C: Ascorbic Acid
14.9 Biotin
14.10 Lipoic Acid
14.11 Folic Acid
14.12 The Vitamin A Group
14.13 The Vitamin D Group
14.14 Vitamin E: Tocopherol
14.15 Vitamin K: Naphthoquinone

Vitamins are essential nutrients that are required in the diet, usually in trace amounts, because they cannot be synthesized by the organism itself. The requirement for any given vitamin depends on the organism. Not all "vitamins" are required by all organisms. Vitamins required in the human diet are listed in Table 14.1. These important substances are traditionally distinguished as being either water-soluble or fat-soluble. Except for vitamin C (ascorbic acid), the water-soluble vitamins are all components or precursors of important biological substances known as **coenzymes.** These are low-molecular-weight molecules that bring unique chemical functionality to certain enzyme reactions. Coenzymes may also act as *carriers* of specific functional groups, such as methyl groups and acyl groups.

Table 14.1
Vitamins and Coenzymes

Vitamin	Coenzyme Form
Water-Soluble	
Thiamine (vitamin B_1)	Thiamine pyrophosphate
Niacin (nicotinic acid)	Nicotinamide adenine dinucleotide (NAD^+) Nicotinamide adenine dinucleotide phosphate ($NADP^+$)
Riboflavin (vitamin B_2)	Flavin adenine dinucleotide (FAD) Flavin mononucleotide (FMN)
Pantothenic acid	Coenzyme A
Pyridoxal, pyridoxine, pyridoxamine (vitamin B_6)	Pyridoxal phosphate
Cobalamin (vitamin B_{12})	5′-Deoxyadenosylcobalamin Methylcobalamin
Biotin	Biotin-lysine complexes (biocytin)
Lipoic acid	Lipoyl-lysine complexes (lipoamide)
Folic acid	Tetrahydrofolate
Fat-Soluble	
Retinol (vitamin A)	
Ergocalciferol (vitamin D_2)	
Cholecalciferol (vitamin D_3)	
α-Tocopherol (vitamin E)	
Vitamin K	

The side chains of the common amino acids provide only a limited range of chemical reactivities and carrier properties. Coenzymes, acting in concert with appropriate enzymes, provide a broader range of catalytic properties for the reactions of metabolism. Coenzymes are typically modified by these reactions, and are then converted back to their original forms by other enzymes, so that small amounts of these substances can be used repeatedly. The coenzymes derived from the water-soluble vitamins are listed in Table 14.1. Each of these will be discussed in this chapter. The fat-soluble vitamins are not directly related to coenzymes, but they play essential roles in a variety of critical biological processes, including vision, maintenance of bone structure, and blood coagulation. The mechanisms of action of fat-soluble vitamins are not as well understood as their water-soluble counterparts, but modern research efforts are gradually closing this gap.

14.1 Vitamin B_1: Thiamine and Thiamine Pyrophosphate

As shown in Figure 14.1, thiamine is composed of a substituted *thiazole* ring joined to a substituted pyrimidine by a methylene bridge. It is the precursor of **thiamine pyrophosphate (TPP),** a coenzyme involved in reactions of carbohydrate metabolism in which bonds to carbonyl carbons (aldehydes or ketones) are synthesized or cleaved. In particular, the *decarboxylations of α-keto acids and*

Thiamine (vitamin B_1)

Thiamine pyrophosphate (TPP)

***Figure* 14.1** Thiamine pyrophosphate (TPP), the active form of vitamin B_1, is formed by the action of TPP-synthetase.

the formation and cleavage of α-hydroxyketones depend on thiamine pyrophosphate. The first of these is illustrated in Figure 14.2 by (a) the decarboxylation of pyruvate by **yeast pyruvate decarboxylase** to yield carbon dioxide and acetaldehyde. Three examples of the formation and cleavage of α-hydroxyketones are presented in Figure 14.2: (b) the condensation of two molecules of pyruvate in the **acetolactate synthase** reaction, (c) a reaction from the pentose phosphate pathway (Chapter 21) called the **transketolase** reaction, and (d) a reaction used by anaerobic bacteria, the **phosphoketolase** reaction. The latter two are referred to as **α-ketol transfers** for obvious reasons. All of these reactions depend on accumulation of negative charge on the carbonyl carbon at which cleavage occurs (Figure 14.3). Thiamine pyrophosphate facilitates these reactions by *stabilizing this negative charge.*

The key to these reactions is the *quaternary nitrogen* of the *thiazolium* group of thiamine pyrophosphate. As shown in the yeast pyruvate decarboxylase mechanism (Figure 14.4), this cationic *imine* nitrogen plays two distinct and important roles in TPP-catalyzed reactions:

1. It provides electrostatic stabilization of the carbanion formed upon removal of the C-2 proton. (The sp^2 hybridization and the availability of vacant *d* orbitals on the adjacent sulfur probably also facilitate proton removal at C-2.) The carbanion thus formed can react with

A Deeper Look

Thiamine and Beriberi

Thiamine, whose structure is shown in Figure 14.1, is known as vitamin B_1, and is essential for the prevention of **beriberi,** a nervous system disease that has occurred in the Far East for centuries and has resulted in considerable sickness and death in these countries. (As recently as 1958, it was the fourth leading cause of death in the Philippine Islands.) It was shown in 1882 by the director-general of the medical department of the Japanese navy that beriberi could be prevented by dietary modifications. Ten years later, Christiaan Eijkman, a Dutch medical scientist working in Java, began research that eventually showed that thiamine was the "anti-beriberi" substance. He found that chickens fed polished rice exhibited paralysis and head retractions and that these symptoms could be reversed if the rice polishings (the outer layers and embryo of the rice kernel) were fed to the birds. In 1911, Casimir Funk prepared a crystalline material from rice bran that cured beriberi in birds. He named it **beriberi vitamine,** since he viewed it as a "vital amine," and thus he is credited with coining the word *vitamin.* The American biochemist R. R. Williams and his research group were the first to establish the structure of thiamine (in 1935) and a route for its synthesis.

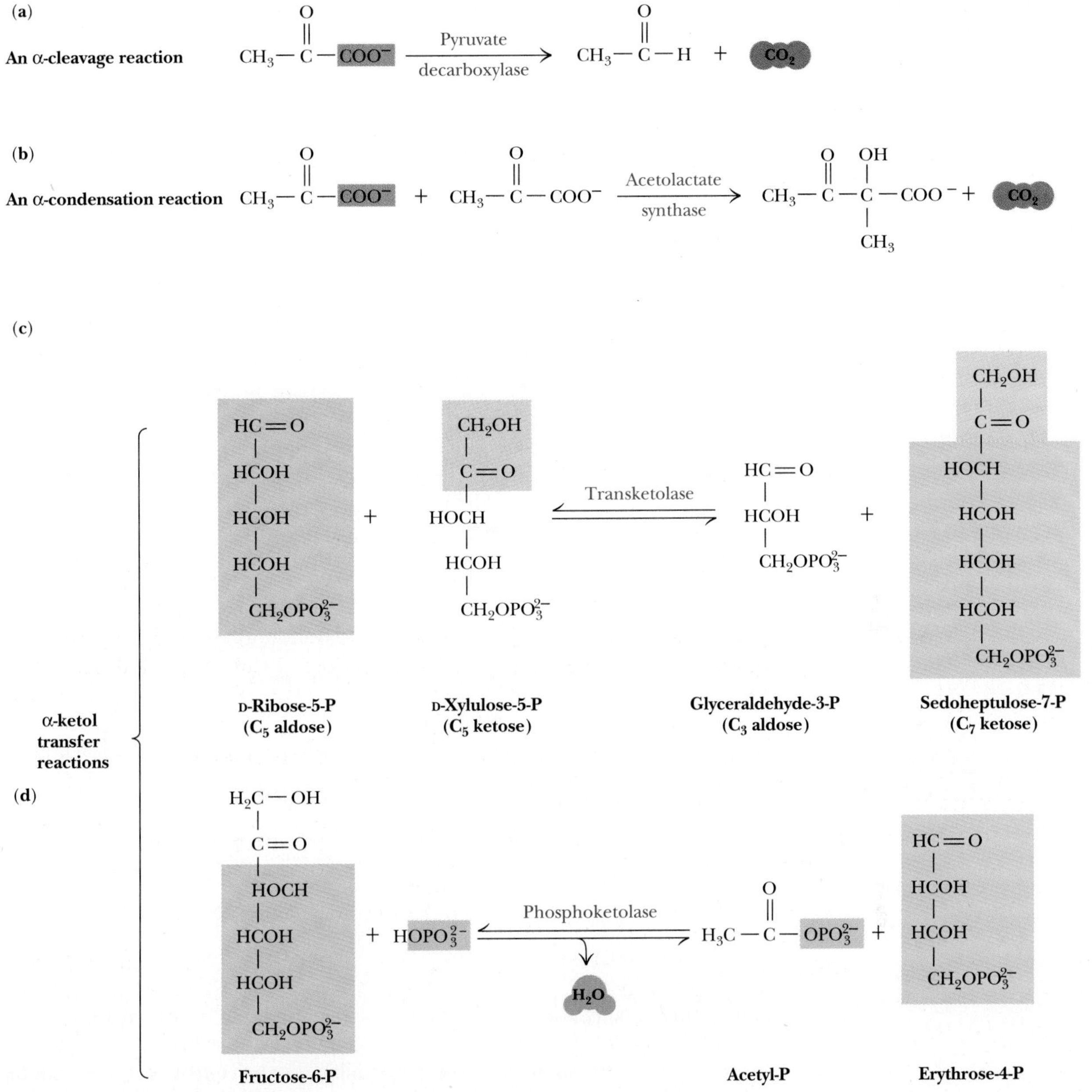

(a)

An α-cleavage reaction $CH_3-C(=O)-COO^-$ —Pyruvate decarboxylase→ $CH_3-C(=O)-H$ + CO_2

(b)

An α-condensation reaction $CH_3-C(=O)-COO^-$ + $CH_3-C(=O)-COO^-$ —Acetolactate synthase→ $CH_3-C(=O)-C(OH)(CH_3)-COO^-$ + CO_2

(c)

α-ketol transfer reactions

D-Ribose-5-P (C5 aldose): HC=O, HCOH, HCOH, HCOH, $CH_2OPO_3^{2-}$ + D-Xylulose-5-P (C5 ketose): CH_2OH, C=O, HOCH, HCOH, $CH_2OPO_3^{2-}$ ⇌ (Transketolase) Glyceraldehyde-3-P (C3 aldose): HC=O, HCOH, $CH_2OPO_3^{2-}$ + Sedoheptulose-7-P (C7 ketose): CH_2OH, C=O, HOCH, HCOH, HCOH, HCOH, $CH_2OPO_3^{2-}$

(d)

Fructose-6-P: H_2C-OH, C=O, HOCH, HCOH, HCOH, $CH_2OPO_3^{2-}$ + $HOPO_3^{2-}$ ⇌ (Phosphoketolase; H_2O) Acetyl-P: $H_3C-C(=O)-OPO_3^{2-}$ + Erythrose-4-P: HC=O, HCOH, HCOH, $CH_2OPO_3^{2-}$

Figure **14.2** Thiamine pyrophosphate participates in (a) the decarboxylation of α-keto acids and (b–d) the formation and cleavage of α-hydroxyketones. The transketolase and phosphoketolase reactions are examples of α-ketol transfers.

substrates such as α-keto acids (for example, pyruvate) and α-keto alcohols by addition to the carbonyl carbon (Figure 14.4).

2. Once TPP attack on the substrate has occurred, the cationic imine nitrogen can act as an effective electron sink to stabilize the negative charge that must develop on the carbon that has been attacked (Figure 14.4). This stabilization takes place by resonance interaction through the double bond to the nitrogen atom.

This resonance-stabilized intermediate can be protonated to give **hydroxyethyl-TPP.** This well-characterized intermediate was once thought to be so unstable that it could not be synthesized or isolated. However, its synthesis and isolation are actually routine. (In fact, a substantial amount of the thiamine pyrophosphate in living things exists as the hydroxyethyl form.) Abstrac-

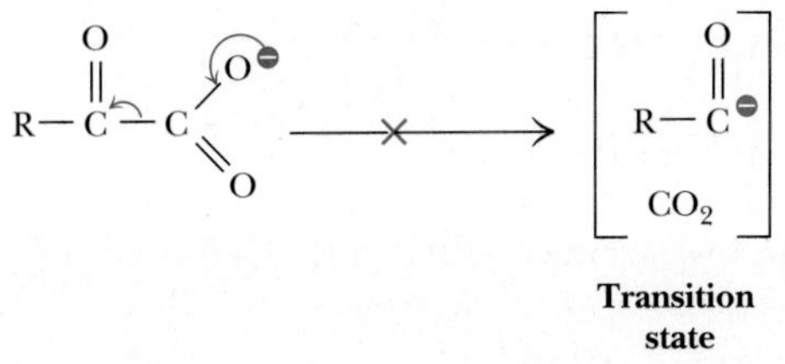

R—C(=O)—CO_2^- —✕→ [R—C(=O)⁻ CO_2] Transition state

Figure **14.3** The unassisted decarboxylation of an α-keto acid leads to an unfavorable accumulation of negative charge on the carbonyl carbon in the transition state. It is the function of thiamine pyrophosphate to stabilize this negative charge.

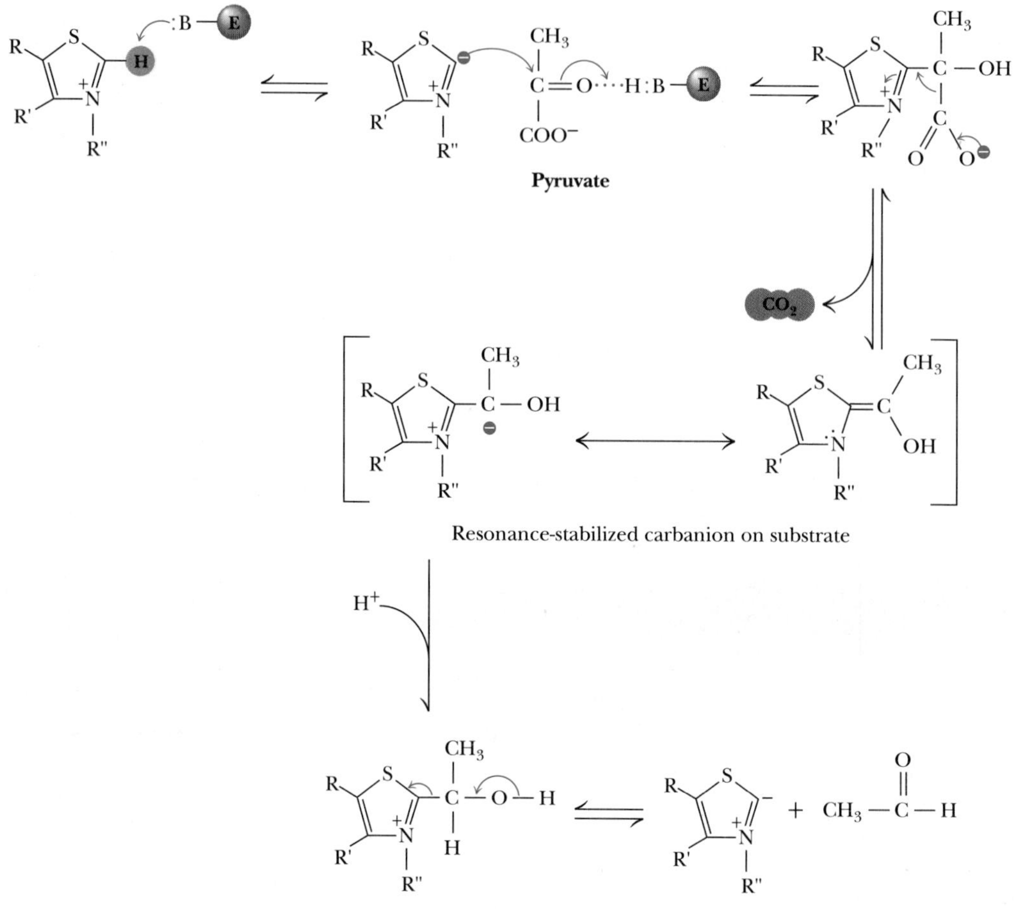

Pyruvate

Resonance-stabilized carbanion on substrate

Hydroxyethyl-TPP

Acetaldehyde

***Figure* 14.4** The mechanism of the yeast pyruvate decarboxylase reaction.

tion of a proton by a base at the enzyme active site leads to dissociation of the product acetaldehyde from the enzyme (Figure 14.4). This is a *nonoxidative decarboxylation.* We shall encounter a similar, but *oxidative,* TPP-dependent decarboxylation in the pyruvate dehydrogenase reaction, which is discussed in Section 14.10.

The acetolactate synthase reaction (Figure 14.5) is an example of formation of an α-hydroxyketone. Here, the carbanion formed by decarboxylation of pyruvate can react with another molecule of pyruvate to form α-acetolactate, an intermediate in the biosynthesis of valine and leucine. The α-ketol transfers in the transketolase and phosphoketolase reactions occur by similar mechanisms (Figure 14.6). Transketolase involves exchange of a two-carbon ketol group between two aldose acceptor molecules. The phosphoketolase reaction is more complicated and involves oxidation of the carbon atom linked to TPP and reduction of the hydroxymethyl group to a methyl group. This is referred to as an **internal oxidation-reduction** reaction.

14.2 Vitamins Containing Adenine Nucleotides

Several classes of vitamins are related to, or are precursors of, coenzymes that contain adenine nucleotides as part of their structure. These coenzymes include the flavin dinucleotides, the pyridine dinucleotides, and coenzyme A.

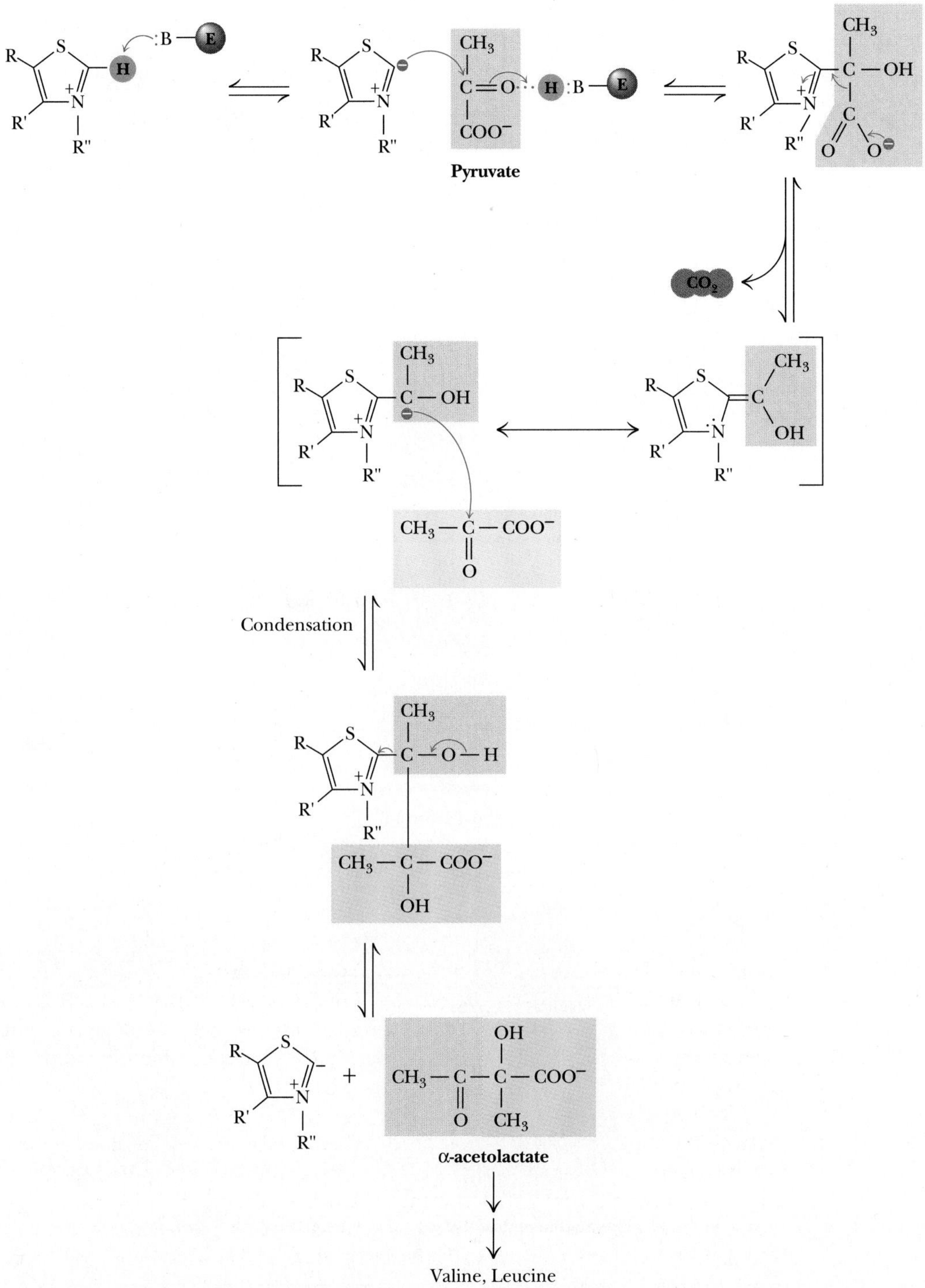

***Figure* 14.5** The mechanism of the acetolactate synthase reaction.

The adenine nucleotide portion of these coenzymes does not participate actively in the reactions of these coenzymes; rather, it enables the proper enzyme to recognize the coenzyme. Specifically, the adenine nucleotide greatly increases both the *affinity* and the *specificity* of the coenzyme for its site on the enzyme, owing to its numerous sites for hydrogen bonding, and also the hydrophobic and ionic bonds it brings to the coenzyme structure.

***Figure* 14.6** The mechanisms of the (a) transketolase and (b) phosphoketolase reactions.

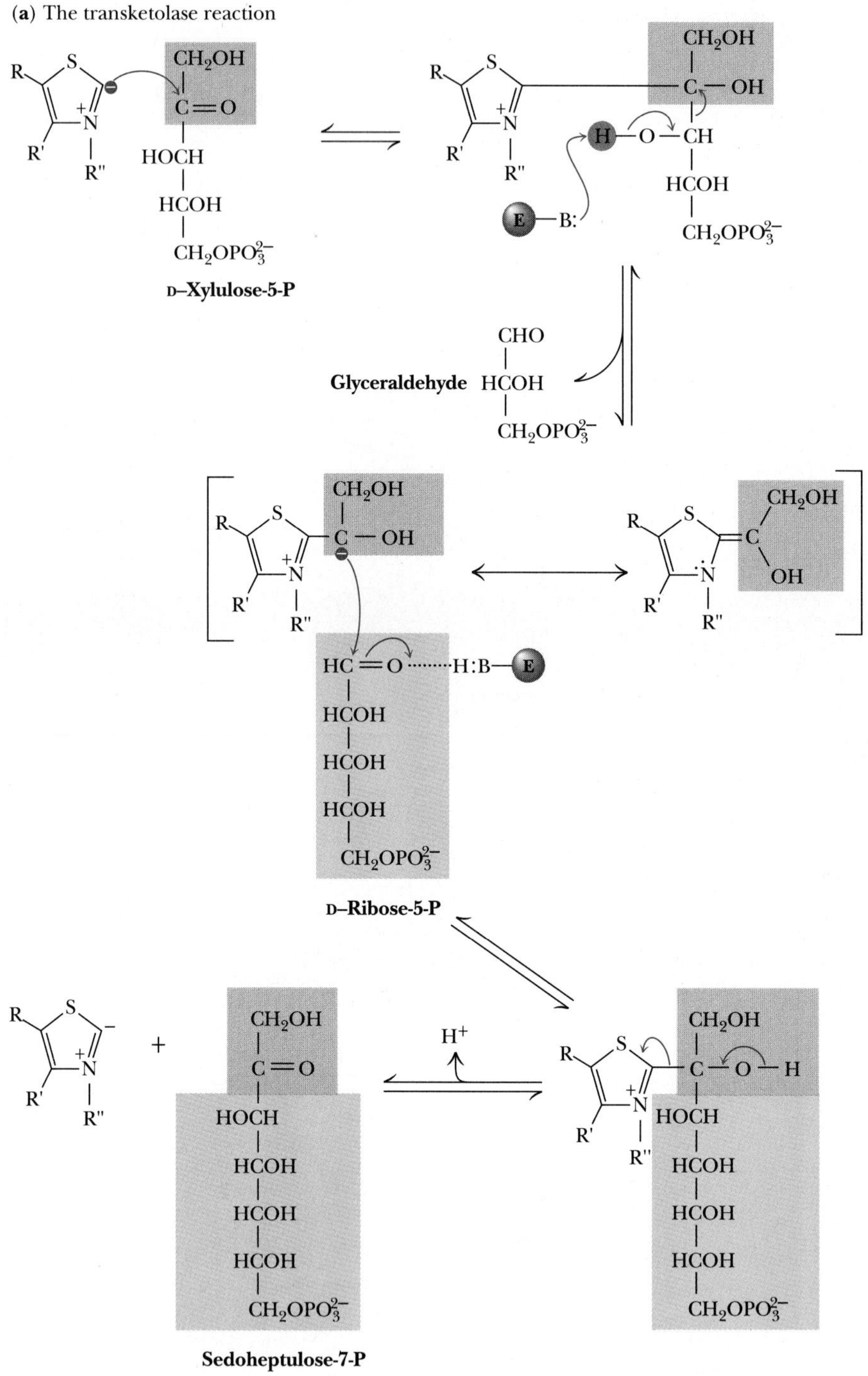

(**a**) The transketolase reaction

D–Xylulose-5-P

Glyceraldehyde

D–Ribose-5-P

Sedoheptulose-7-P

14.3 Nicotinic Acid and the Nicotinamide Coenzymes

Nicotinamide is an essential part of two important coenzymes: **nicotinamide adenine dinucleotide (NAD^+)**, and **nicotinamide adenine dinucleotide phosphate ($NADP^+$)** (Figure 14.7). The reduced forms of these coenzymes are NADH and NADPH. *The nicotinamide coenzymes (also known as pyridine nucleotides) are **electron carriers.*** They play vital roles in a variety of enzyme-catalyzed oxidation-reduction reactions. [NAD^+ is an electron acceptor in oxidative (catabolic) pathways and NADPH is an electron donor in reductive (biosynthetic) pathways.] These reactions involve direct transfer of hydride anion

(b) The phosphoketolase reaction

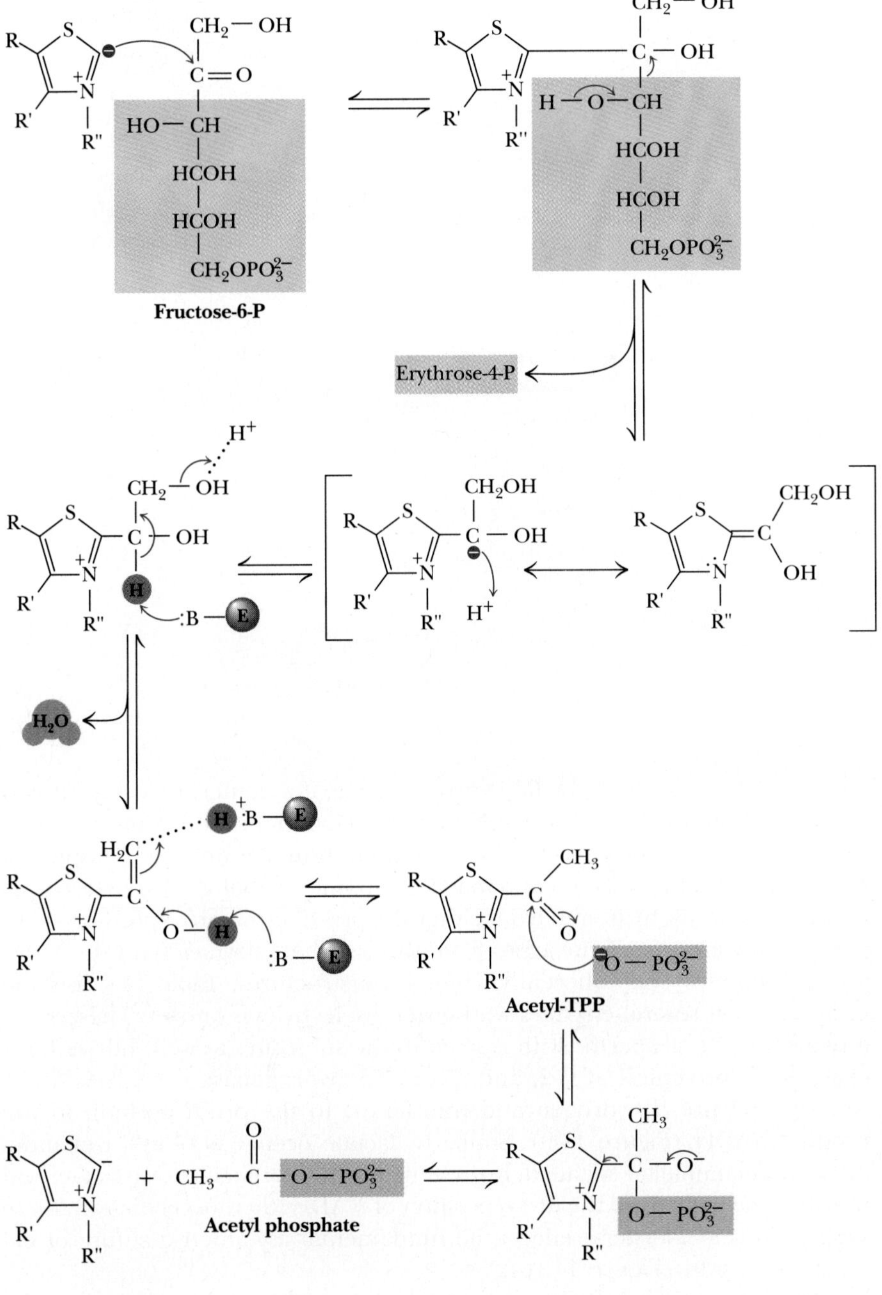

CH2—OH
C=O
HO—CH
HCOH
HCOH
CH2OPO3 2−
Fructose-6-P
Erythrose-4-P
H+
H2O
Acetyl-TPP
Acetyl phosphate

either to NAD$^+$ or from NADH. The enzymes that facilitate such transfers are thus known as **dehydrogenases.** The hydride anion contains two electrons, and thus NAD$^+$ and NADP$^+$ act exclusively as **two-electron carriers.** The C-4 position of the pyridine ring, which can either accept or donate hydride ion, is the reactive center of both NAD and NADP. The quaternary nitrogen of the nicotinamide ring functions as an electron sink to facilitate hydride transfer to NAD$^+$, as shown in Figure 14.8. The adenine portion of the molecule is not directly involved in redox processes.

Examination of the structures of NADH and NADPH reveals that the 4-position of the nicotinamide ring is **pro-chiral,** meaning that while this carbon is not chiral, it *would* be if either of its hydrogens were replaced by something else. As shown in Figure 14.8, the hydrogen "projecting" out of the

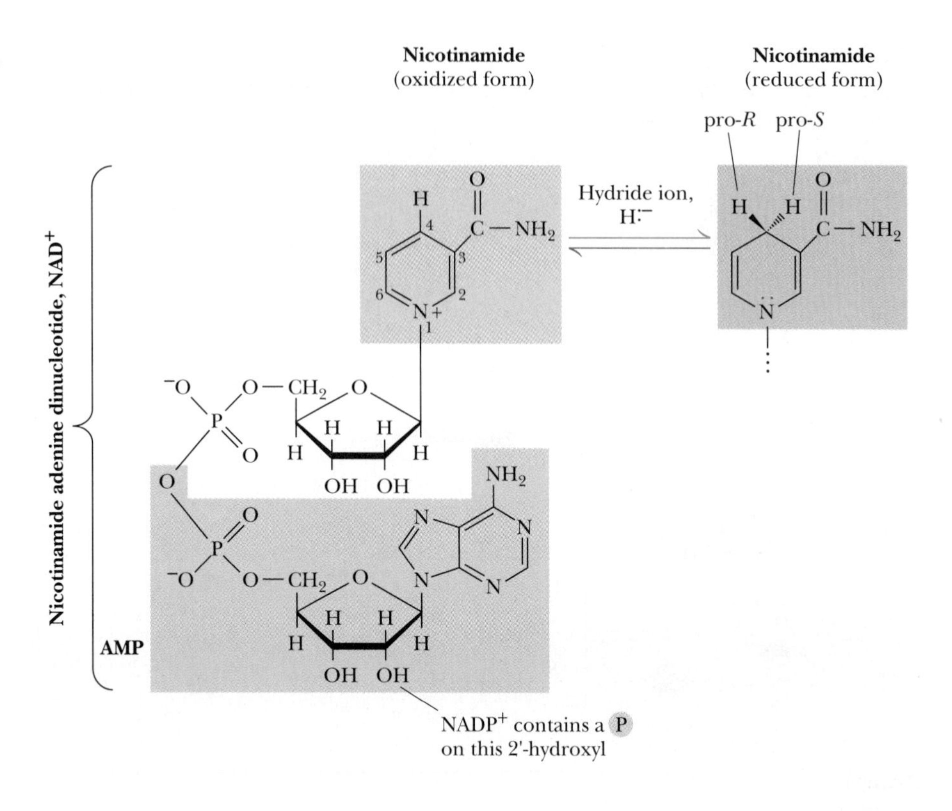

***Figure* 14.7** The structures and redox states of the nicotinamide coenzymes. Hydride ion (H:$^-$, a proton with two electrons) transfers to NAD$^+$ to produce NADH.

page toward you is the "pro-*R*" hydrogen since, if a deuterium is substituted at this position, the molecule would have the *R*-configuration. Substitution of the other hydrogen would yield an *S*-configuration. An interesting aspect of the enzymes that require nicotinamide coenzymes is that they are **stereospecific** and withdraw hydrogen from either the pro-*R* or the pro-*S* position selectively. This stereospecificity arises from the fact that enzymes (and the active sites of enzymes) are inherently asymmetric structures. Table 14.2 lists the preferences of several enzymes whose stereoselectivity is known. These same enzymes are stereospecific with respect to the substrates as well. Ethanol, for example, is pro-chiral at C-2, and alcohol dehydrogenase stereospecifically removes the pro-*R* hydrogen and transfers it to the pro-*R* position in the product NADH (Figure 14.9). Similarly, lactate occurs as D- and L-enantiomers, but mammalian lactate dehydrogenase is stereospecific for L-lactate and transfers the hydride to the pro-*R* position of NADH. Stereospecificity in these simple hydride transfers reflects the fundamental asymmetric nature of enzyme active sites (Figure 14.10).

The NAD- and NADP-dependent dehydrogenases catalyze at least six different types of reactions: simple hydride transfer, deamination of an amino acid to form an α-keto acid, oxidation of β-hydroxy acids followed by decarboxylation of the β-keto acid intermediate, oxidation of aldehydes, reduction of isolated double bonds, and the oxidation of carbon–nitrogen bonds (as with dihydrofolate reductase). These six types of reactions are summarized in Figure 14.11, which lists examples for each type of enzyme reaction.

Some NAD$^+$ and NADP$^+$-dependent enzymes catalyze net transformations of a substrate with no apparent net oxidation or reduction of either the substrate or the nicotinamide coenzyme. A good example is **UDP galactose-4-epimerase,** which catalyzes the reaction shown in Figure 14.12. The enzymatic cycle involves oxidation at the 4-position of the substrate, followed by reduction at the same position, but with opposite stereochemistry. NAD$^+$ also serves as a donor of ADP-ribosyl groups in protein covalent modification reactions catalyzed by cholera toxin and pertussis toxin (Chapter 37).

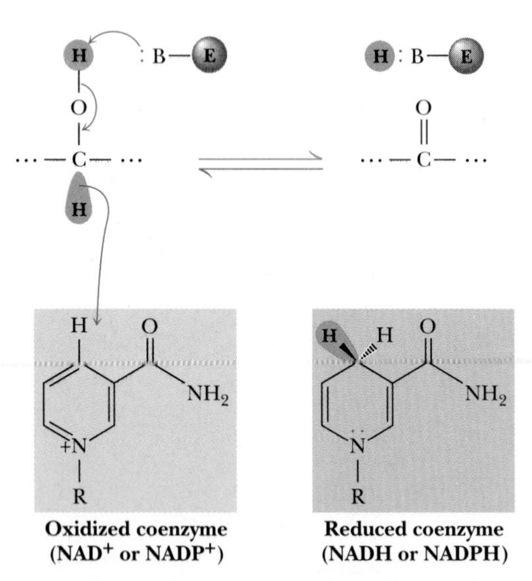

***Figure* 14.8** NAD$^+$ and NADP$^+$ participate exclusively in two-electron transfer reactions. For example, alcohols can be oxidized to ketones or aldehydes via hydride transfer to NAD(P)$^+$.

***Table* 14.2**
Steric Specificity for NAD of Various Pyridine Nucleotide-Linked Enzymes

Dehydrogenase	Source	Steric Specificity
Alcohol (with ethanol)	Yeast, *Pseudomonas,* liver, wheat germ	H_R
Alcohol (with isopropyl alcohol)	Yeast	H_R
Acetaldehyde	Liver	H_R
L-Lactate	Heart muscle, *Lactobacillus*	H_R
L-Malate	Pig heart, wheat germ	H_R
D-Glycerate	Spinach	H_R
Dihydroorotate	*Zymobacterium oroticum*	H_R
α-Glycerophosphate	Muscle	H_S
Glyceraldehyde-3-P	Yeast, muscle	H_S
L-Glutamate	Liver	H_S
D-Glucose	Liver	H_S
β-Hydroxysteroid	*Pseudomonas*	H_S
NADH cytochrome *c* reductase	Rat liver mitochondria, pig heart	H_S
NADPH (transhydrogenase)	*Pseudomonas*	H_S
NADH diaphorase	Pig heart	H_S
L-β-Hydroxybutyryl-CoA	Heart muscle	H_S

Adapted from Kaplan, N. O., 1960. In *The Enzymes,* vol. 3, p. 115, edited by Boyer, Lardy, and Myrbäck. New York: Academic Press.

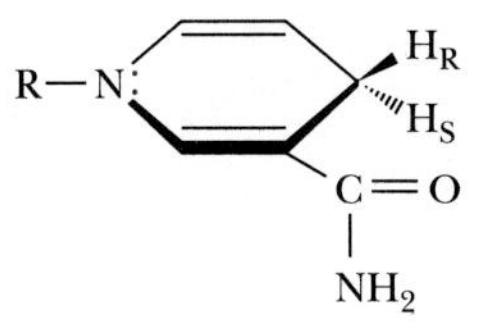

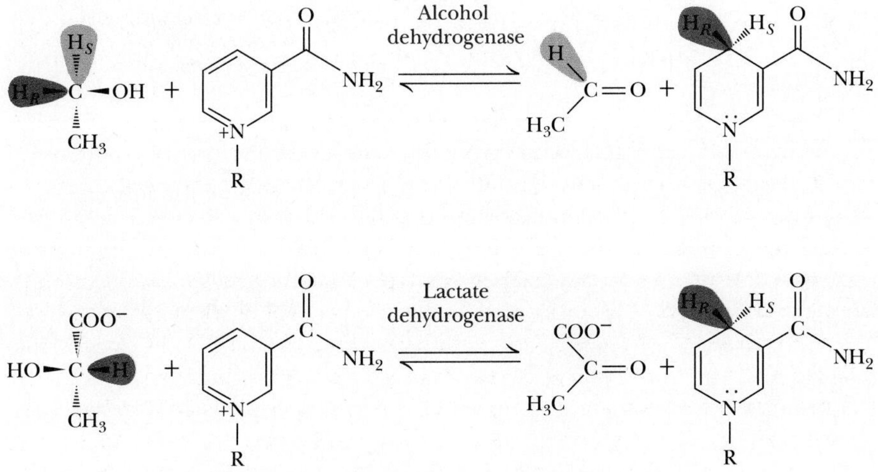

***Figure* 14.9** NAD(P)$^+$-dependent enzymes are stereospecific. Alcohol dehydrogenase, for example, removes hydrogen from the pro-*R* position of ethanol and transfers it to the pro-*R* position of NADH.

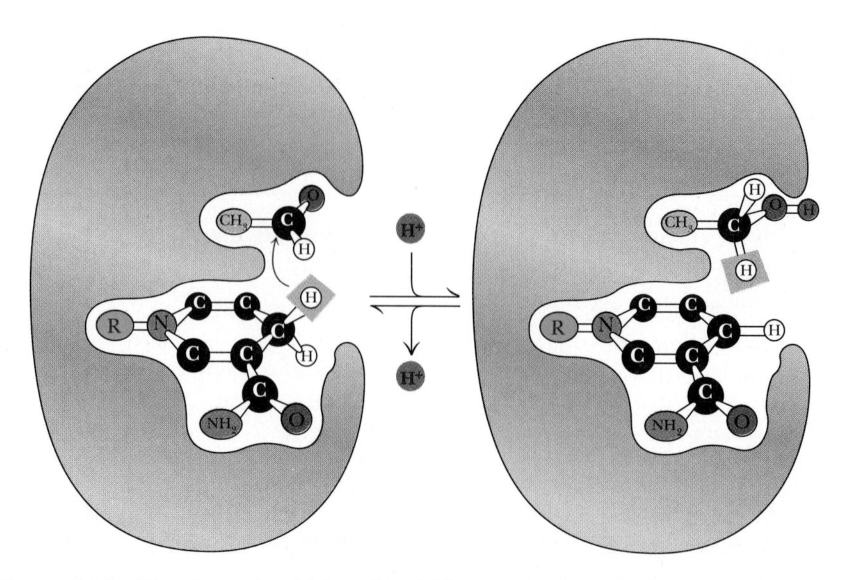

***Figure* 14.10** The stereospecificity of hydride transfer in dehydrogenases is a consequence of the asymmetric nature of the active site.

Category	Reaction	Examples
1	$-\mathrm{CH(OH)}- \rightleftharpoons \;>\mathrm{C}{=}\mathrm{O} + 2\mathrm{H}^+ + 2e^-$	Malate dehydrogenase Lactate dehydrogenase Alcohol dehydrogenase
2	$-\mathrm{CH(NH_3^+)}- \xrightleftharpoons{\mathrm{H_2O}} \;>\mathrm{C}{=}\mathrm{O} + \mathrm{NH_4^+} + \mathrm{H}^+ + 2e^-$	Glutamate dehydrogenase
3	$-\mathrm{C(OH)}-\mathrm{CH(R)}-\mathrm{COO}^- \rightleftharpoons \left[\;>\mathrm{C}(=\mathrm{O})-\mathrm{CH(R)}-\mathrm{COO}^-\right] \rightleftharpoons -\mathrm{C}(=\mathrm{O})-\mathrm{CH_2R} + \mathrm{CO_2}$	Isocitrate dehydrogenase 6-Phosphogluconate dehydrogenase
4	$-\mathrm{CH}{=}\mathrm{O} \xrightleftharpoons{\mathrm{H_2O}} -\mathrm{C}(=\mathrm{O})-\mathrm{O}^- + 3\mathrm{H}^+ + 2e^-$	Aldehyde dehydrogenase
5	$-\mathrm{CH}-\mathrm{CH}- \rightleftharpoons \;>\mathrm{C}{=}\mathrm{C}< + 2\mathrm{H}^+ + 2e^-$	Dihydrosteroid dehydrogenase (Steroid reductase)
6	$-\mathrm{CH}-\mathrm{NH}- \rightleftharpoons \;>\mathrm{C}{=}\mathrm{N}- + 2\mathrm{H}^+ + 2e^-$	Dihydrofolate reductase

***Figure* 14.11** The six classes of $NAD(P)^+$-dependent enzyme reactions. *(Adapted from Walsh, C. T., 1979. Enzymatic Reaction Mechanisms. San Francisco: W. H. Freeman.)*

A Deeper Look

Niacin and Pellegra

Pellegra, a disease characterized by dermatitis, diarrhea, and dementia, has been known for centuries. It was once prevalent in the southern part of the United States and is still a common problem in some parts of Spain, Italy, and Romania. Pellegra was once thought to be an infectious disease, but Joseph Goldberger showed early in this century that it could be cured by dietary actions. Soon thereafter, it was found that brewer's yeast would prevent pellegra in humans. Studies of a similar disease in dogs, called **blacktongue,** eventually led to the identification of **nicotinic acid** as the relevant dietary factor. Elvehjem and his colleagues at the University of Wisconsin in 1937 isolated **nicotinamide** from liver, and showed that it and nicotinic acid could prevent and cure blacktongue in dogs. That same year, nicotinamide and nicotinic acid were both shown to be able to cure pellegra in humans. Interestingly, plants and many animals can synthesize nicotinic acid from tryptophan and other precursors, and nicotinic acid is thus not a true vitamin for these species. However, if dietary intake of tryptophan is low, nicotinic acid is required for optimal health. Nicotinic acid, which is beneficial to humans and animals, is a part of the structure of **nicotine,** a highly toxic tobacco alkaloid. In order to avoid confusion of nicotinic acid and nicotinamide with nicotine itself, **niacin** was adopted as a common name for nicotinic acid. Cowgill, at Yale University, suggested the name from the letters of three words—*ni*cotinic, *ac*id, and vitam*in*.

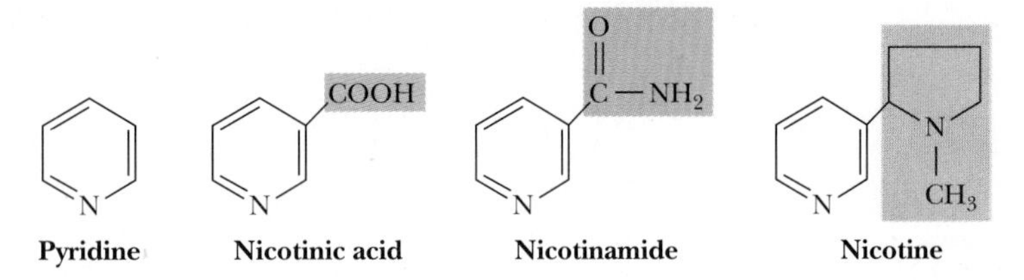

The structures of pyridine, nicotinic acid, nicotinamide, and nicotine.

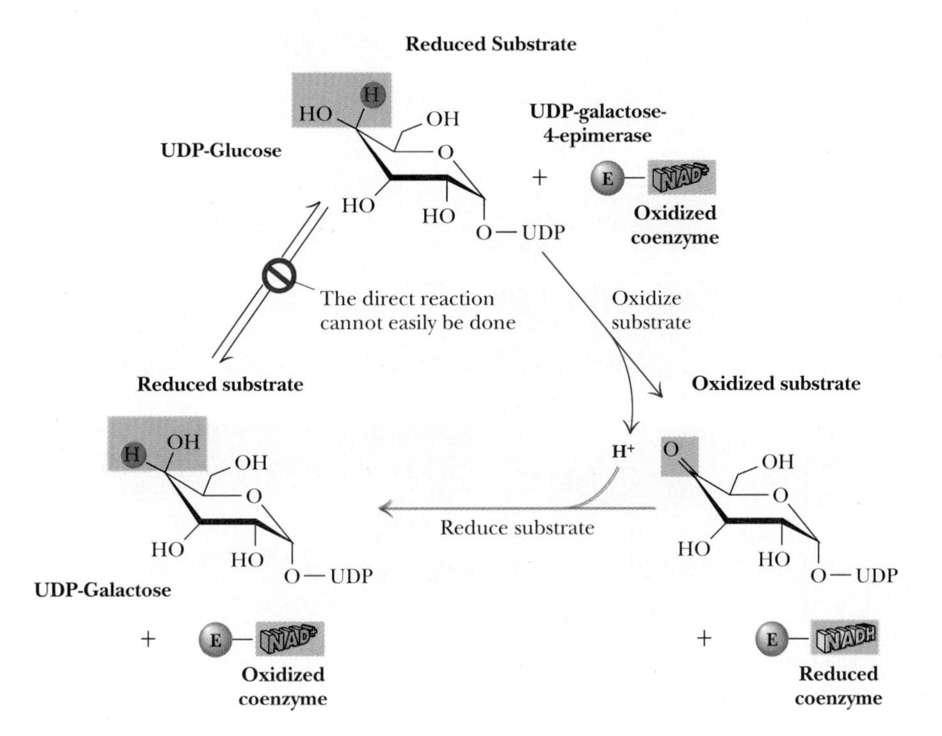

Figure **14.12** The UDP galactose-4-epimerase reaction mechanism.

14.4 Riboflavin and the Flavin Coenzymes

Riboflavin, or **vitamin B_2,** is a constituent and precursor of both **riboflavin 5′-phosphate,** also known as **flavin mononucleotide (FMN),** and **flavin adenine dinucleotide (FAD).** The name *riboflavin* is a synthesis of the names for the molecule's component parts, **ribitol** and **flavin.** The structures of riboflavin, FMN, and FAD are shown in Figure 14.13. The *isoalloxazine ring* is the core structure of the various flavins. Since the ribityl group is not a true pentose sugar (it is a sugar alcohol) and is not joined to riboflavin in a glycosidic bond, the molecule is not truly a "nucleotide" and the terms *flavin mononucleotide* and *dinucleotide* are incorrect. Nonetheless, these designations are so deeply ingrained in common biochemical usage that the erroneous nomenclature persists. The flavins have a characteristic bright yellow color and take their name from the Latin *flavius* for "yellow." As shown in Figure 14.14, the oxidized form of the isoalloxazine structure absorbs light around 450 nm (in the visible region) and also at 350 to 380 nm. The color is lost, however, when the ring is reduced or "bleached." Similarly, the enzymes that bind flavins, known as **flavoenzymes,** can be yellow, red, or green in their oxidized states. Nevertheless, these enzymes also lose their color on reduction of the bound flavin group.

Flavin coenzymes bind tightly to the enzymes that use them, with typical dissociation constants in the range of 10^{-8} to 10^{-11} *M,* so that very low levels of free flavin coenzymes are found in most cells. Even in organisms that rely

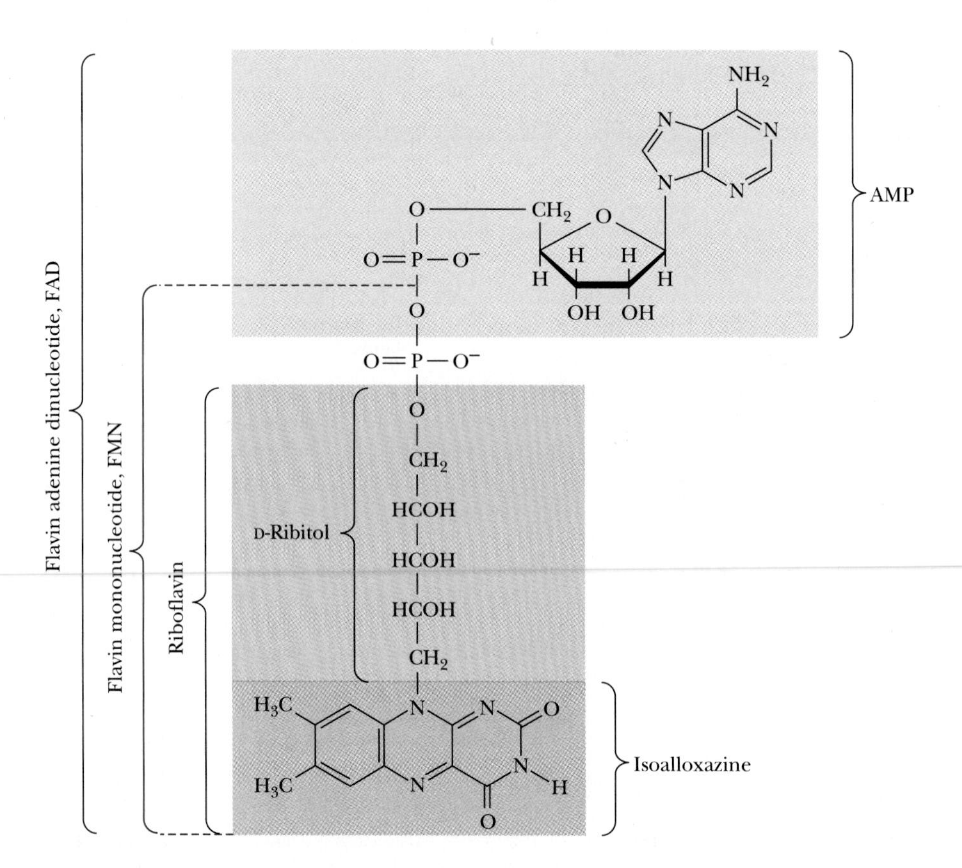

***Figure* 14.13** The structures of riboflavin, flavin mononucleotide (FMN), and flavin adenine dinucleotide (FAD).

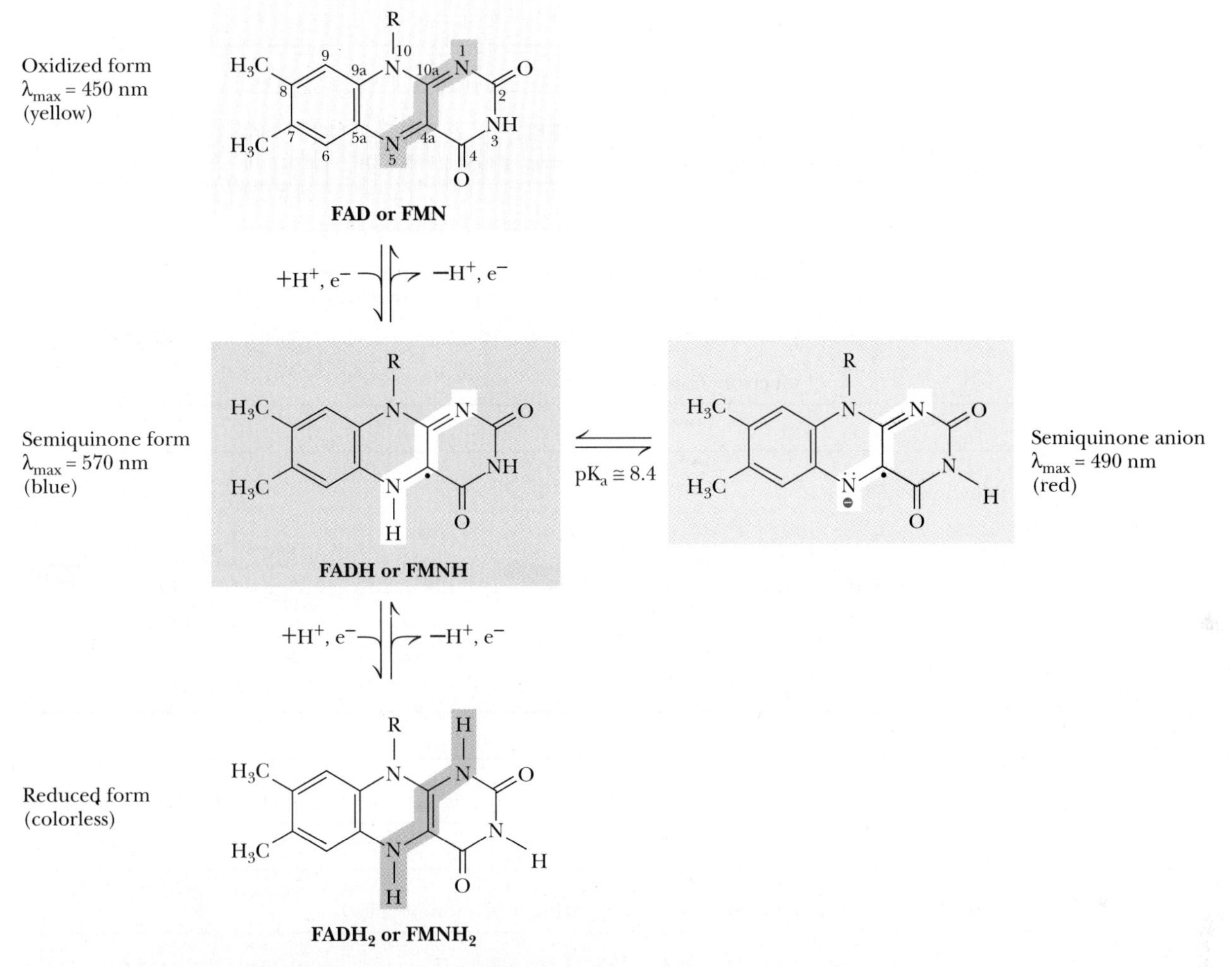

***Figure* 14.14** The redox states of FAD and FMN. The boxes correspond to the colors of each of these forms. The atoms primarily involved in electron transfer are indicated by red shading in the oxidized form, white in the semiquinone form, and blue in the reduced form.

on the nicotinamide coenzymes (NADH and NADPH) for many of their oxidation-reduction cycles, the flavin coenzymes fill essential roles. Flavins are stronger oxidizing agents than NAD^+ and $NADP^+$. They can be reduced by both one-electron and two-electron pathways and can be reoxidized easily by molecular oxygen. Enzymes that use flavins to carry out their reactions—*flavoenzymes*—are involved in many kinds of oxidation-reduction reactions, as shown in Table 14.3.

Flavin coenzymes can exist in any of three different redox states. Fully oxidized flavin is converted to a **semiquinone** by a one-electron transfer, as shown in Figure 14.14. At physiological pH, the semiquinone is a neutral radical, blue in color, with a λ_{max} of 570 nm. The semiquinone possesses a pK_a of about 8.4. When it loses a proton at higher pH values, it becomes a radical anion, displaying a red color with a λ_{max} of 490 nm. The semiquinone radical is particularly stable, owing to extensive delocalization of the unpaired electron across the π-electron system of the isoalloxazine. A second one-electron transfer converts the semiquinone to the completely reduced dihydroflavin as shown in Figure 14.14.

Table 14.3
Flavoprotein-Catalyzed Reactions

Flavoprotein Class	Reaction Catalyzed
Dehydrogenase	
Acyl-CoA dehydrogenase	$RCH_2CH_2CO—SCoA + [FAD] \rightleftharpoons RCH{=}CHCO—SCoA + [FADH_2]$
Glutathione reductase	$GSSG + NADPH + H^+ \rightleftharpoons 2\ GSH + NADP^+$
Dioxygenase	
2-Nitropropane dioxygenase	$2\ (CH_3)_2CHNO_2 + O_2 \rightleftharpoons 2\ (CH_3)_2C{=}O + 2\ NO_2^-$
Flavodoxin	
Clostridium flavodoxin	1-electron transfer
Metalloflavoenzyme	
Clostridium oroticum dihydroorotate dehydrogenase	**Orotate (oxidized)** + NADH + H^+ $\rightleftharpoons$ **Dihydroorotate (reduced)** + NAD^+
Monooxygenase	
Lactate oxidase	$H_3C—CHOH—COO^- + O_2 \rightleftharpoons H_3C—COO^- + CO_2 + H_2O$
Oxidase	
Glucose oxidase	D-Glucose + $O_2 \rightleftharpoons$ D-gluconolactone + H_2O_2

A Deeper Look

Riboflavin and Old Yellow Enzyme

Riboflavin was first isolated from whey in 1879 by Blyth, and the structure was determined by Kuhn and co-workers in 1933. For the structure determination, this group isolated 30 mg of pure riboflavin from the whites of about 10,000 eggs. The discovery of the actions of riboflavin in biological systems arose from the work of Otto Warburg in Germany and Hugo Theorell in Sweden, both of whom identified yellow substances bound to a yeast enzyme involved in the oxidation of pyridine nucleotides. Theorell showed that riboflavin 5′-phosphate was the source of the yellow color in this *old yellow enzyme.* By 1938, Warburg had identified FAD, the second common form of riboflavin, as the coenzyme in D-amino acid oxidase, another yellow protein. Riboflavin deficiencies are not at all common. Humans require only about 2 mg per day, and the vitamin is prevalent in many foods. This vitamin is extremely light sensitive, and it is degraded in foods (milk, for example) left in the sun.

The milling and refining of wheat, rice, and other grains causes a loss of riboflavin and other vitamins. In order to correct and prevent dietary deficiencies, the Committee on Food and Nutrition of the National Research Council began in the 1940s to recommend enrichment of cereal grains sold in the United States. Thiamine, riboflavin, niacin, and iron were the first nutrients originally recommended for enrichment by this group. As a result of these actions, generations of American children have become accustomed to reading (on their cereal boxes and bread wrappers) that their foods contain certain percentages of the "U.S. Recommended Daily Allowance" of various vitamins and nutrients.

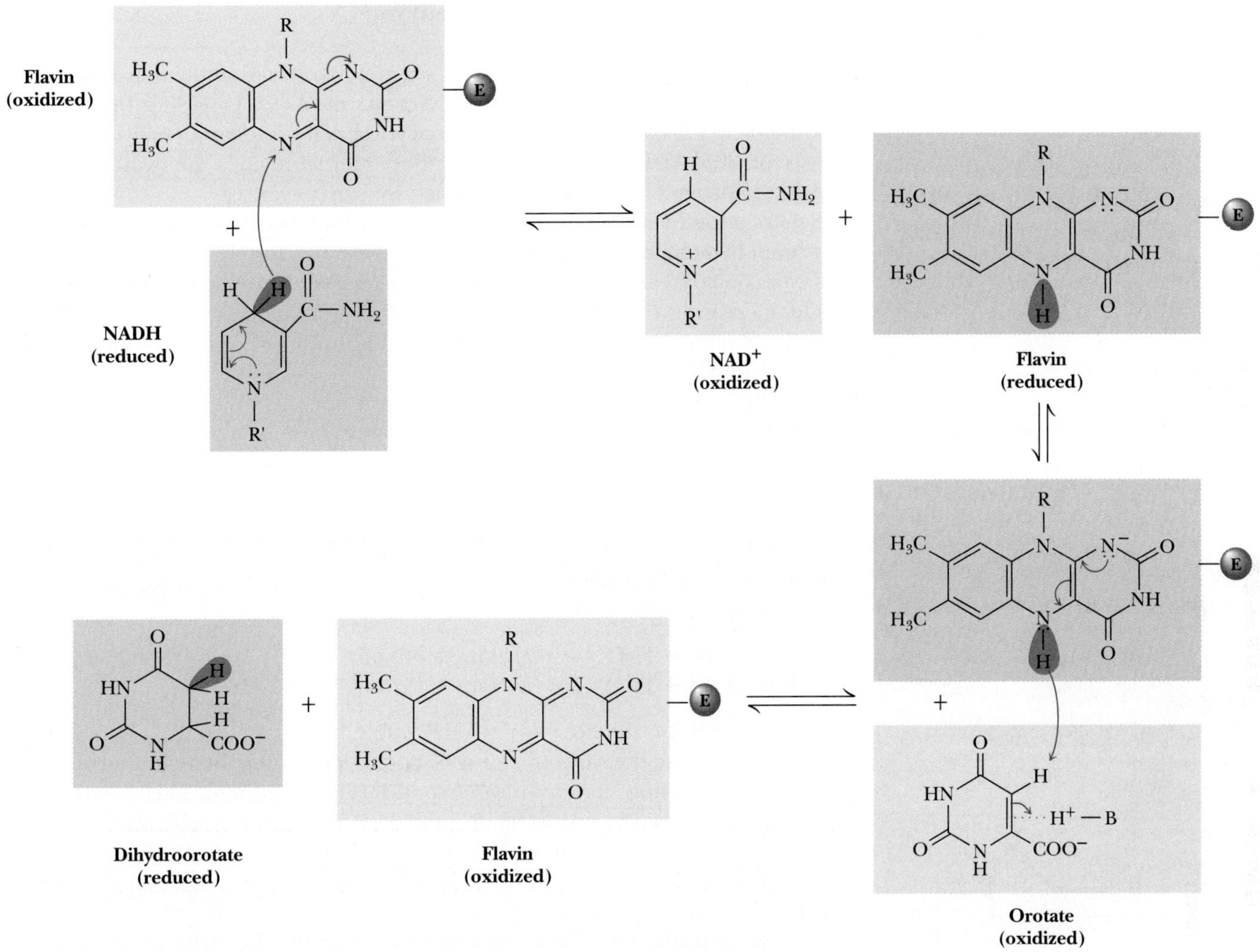

***Figure* 14.15** The bacterial dihydroorotate dehydrogenase reaction mechanism—an example of a flavin-catalyzed, two-electron transfer reaction. Oxidized and reduced forms of each species are indicated by red and blue shading, respectively.

Access to three different redox states allows flavin coenzymes to participate in *one-electron transfer* and *two-electron transfer reactions.* Partly because of this, flavoproteins catalyze many different reactions in biological systems and work together with many different electron acceptors and donors. These include two-electron acceptor/donors, such as NAD^+ and $NADP^+$, one- or two-electron carriers, such as quinones, and a variety of one-electron acceptor/donors, such as cytochrome proteins. Many of the components of the respiratory electron transport chain are one-electron acceptor/donors. The stability of the flavin semiquinone state allows flavoproteins to function as effective electron carriers in respiration processes (Chapter 20).

Bacterial **dihydroorotate dehydrogenase** provides a good example of a flavin-catalyzed reaction involving a two-electron transfer (Figure 14.15). Dihydroorotate is oxidized to orotate in the biosynthesis of pyrimidines (Chapter 27). In addition, certain bacteria can use orotate as a carbon source and reduce orotate in a two-electron transfer to form dihydroorotate. The latter mechanism involves hydride transfer from NADH to FAD to form NAD^+ and $FADH_2$. In the second half of the mechanism in Figure 14.15, hydride transfer to orotate reoxidizes $FADH_2$. (*Note:* Some NAD^+-dependent dihydroorotate dehydrogenases in bacteria contain both FMN *and* FAD.)

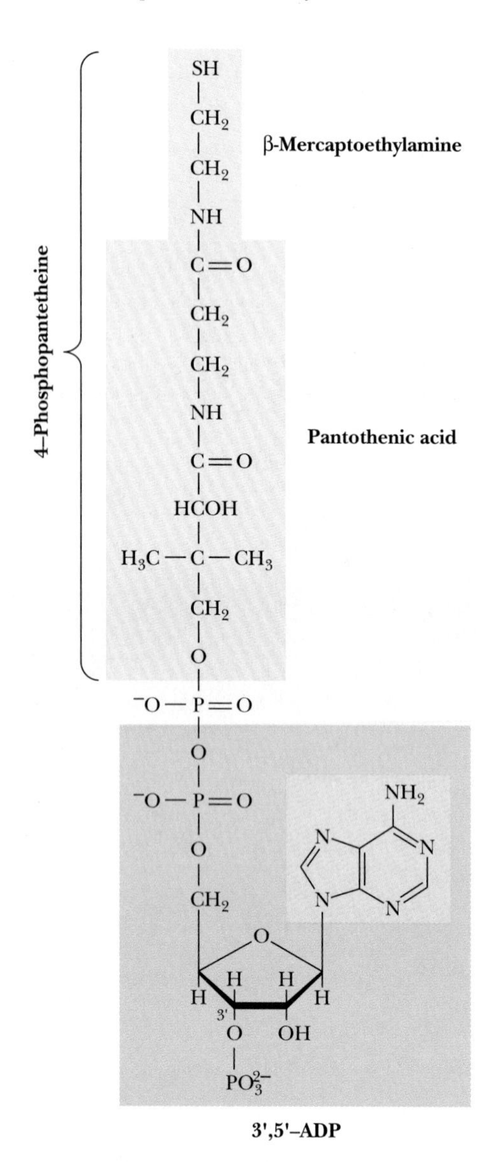

Figure **14.16** The structure of coenzyme A. Acyl groups form thioester linkages with the —SH group of the β-mercaptoethylamine moiety.

14.5 Pantothenic Acid and Coenzyme A

Pantothenic acid, sometimes called vitamin B_3, is a vitamin that makes up one part of a complex coenzyme called **coenzyme A (CoA)** (Figure 14.16). Pantothenic acid is also a constituent of **acyl carrier proteins.** Coenzyme A consists of 3′,5′-ADP joined to 4-phosphopantetheine in a phosphoric anhydride linkage. Phosphopantetheine in turn consists of three parts: β-mercaptoethylamine linked to β-alanine, which makes an amide bond with a branched-chain dihydroxy acid. As was the case for the nicotinamide and flavin coenzymes, the adenine nucleotide moiety of CoA acts as a recognition site, increasing the affinity and specificity of CoA binding to its enzymes. In acyl carrier proteins (Chapter 23), the 4-phosphopantetheine is covalently linked to a serine hydroxyl group.

The two main functions of coenzyme A are

(a) *activation of acyl groups for transfer by nucleophilic attack* and

(b) *activation of the α-hydrogen of the acyl group for abstraction as a proton.*

Both of these functions are mediated by the reactive sulfhydryl group on CoA, which forms **thioester** linkages with acyl groups.

The activation of acyl groups for transfer by CoA can be appreciated by comparing the hydrolysis of the thioester bond of acetyl-CoA with hydrolysis of a simple oxygen ester:

$$\text{Ethyl acetate} + H_2O \longrightarrow \text{acetate} + \text{ethanol} + H^+ \qquad \Delta G^{\circ\prime} = -20.0 \text{ kJ/mol}$$

$$\text{Acetyl-CoA} + H_2O \longrightarrow \text{acetate} + \text{CoA} + H^+ \qquad \Delta G^{\circ\prime} = -31.5 \text{ kJ/mol}$$

Hydrolysis of the thioester is more favorable than that of oxygen esters, presumably because the carbon–sulfur bond has less double bond character than the corresponding carbon–oxygen bond (Figure 14.17). This means that transfer of the acetyl group from acetyl-CoA to a given nucleophile (Figure 14.18) will be more spontaneous than transfer of an acetyl group from an oxygen ester. For this reason, acetyl-CoA is said to have a high group-transfer potential.

The activation of the α-hydrogen of thioesters for abstraction (Figure 14.19) can also be understood from the resonance forms of Figure 14.17. The small contribution of resonance form (2) leaves the carbonyl group of a thioester relatively unperturbed (compared to that of an oxygen ester). As a result, the α-carbanion formed by proton abstraction (Figure 14.19) is itself more readily resonance stabilized. Such **enolate anions** are effective nucleophiles for condensation reactions.

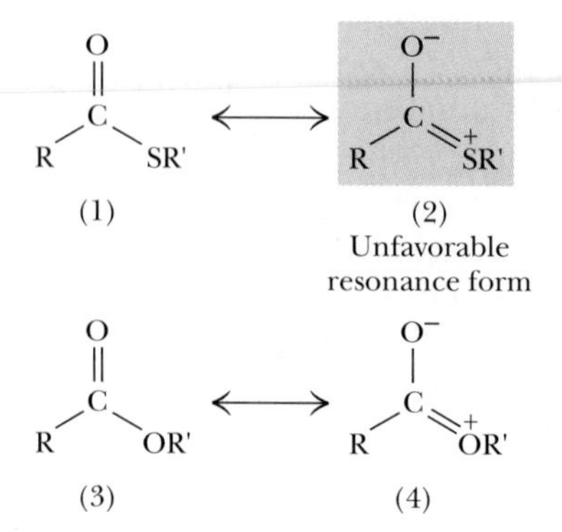

Figure **14.17** Resonance structures for thioesters and oxygen esters. The carbon–sulfur bond has less double bond character than the corresponding carbon–oxygen bond.

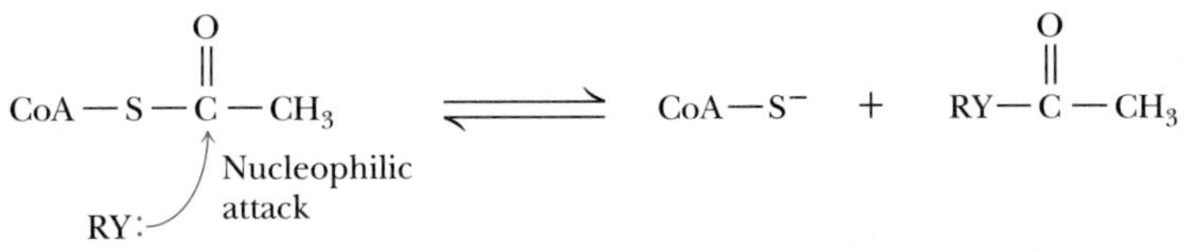

***Figure* 14.18** Acyl transfer from acyl-CoA to a nucleophile is more favorable than transfer of an acyl group from an oxygen ester.

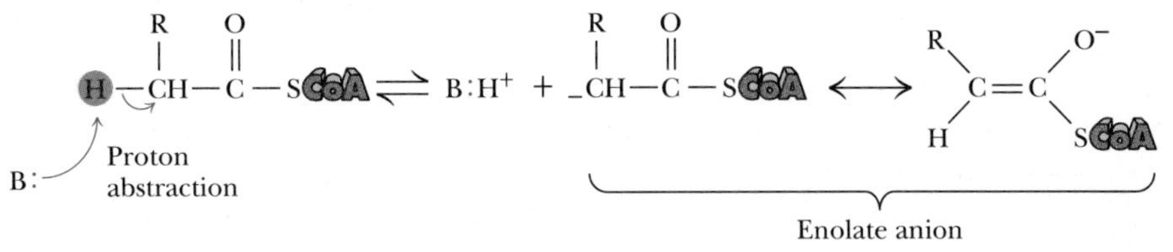

***Figure* 14.19** The α-hydrogens of a thioester are acidic and readily abstracted owing to resonance stabilization of the resulting carbanion. Enolate anions formed in this way are good nucleophiles for condensation reactions.

Both of these special properties of acyl-CoA are evident in the **β-ketothiolase reaction**—a condensation of two molecules of acetyl-CoA to form acetoacetyl-CoA (Figure 14.20). Proton abstraction from one acetyl-CoA yields a reactive enolate anion that readily attacks the carbonyl carbon of the second acetyl-CoA, forming the tetrahedral intermediate. Departure of the CoA thiolate anion (CoA-S^-), a good leaving group, from the tetrahedral intermediate yields the desired product.

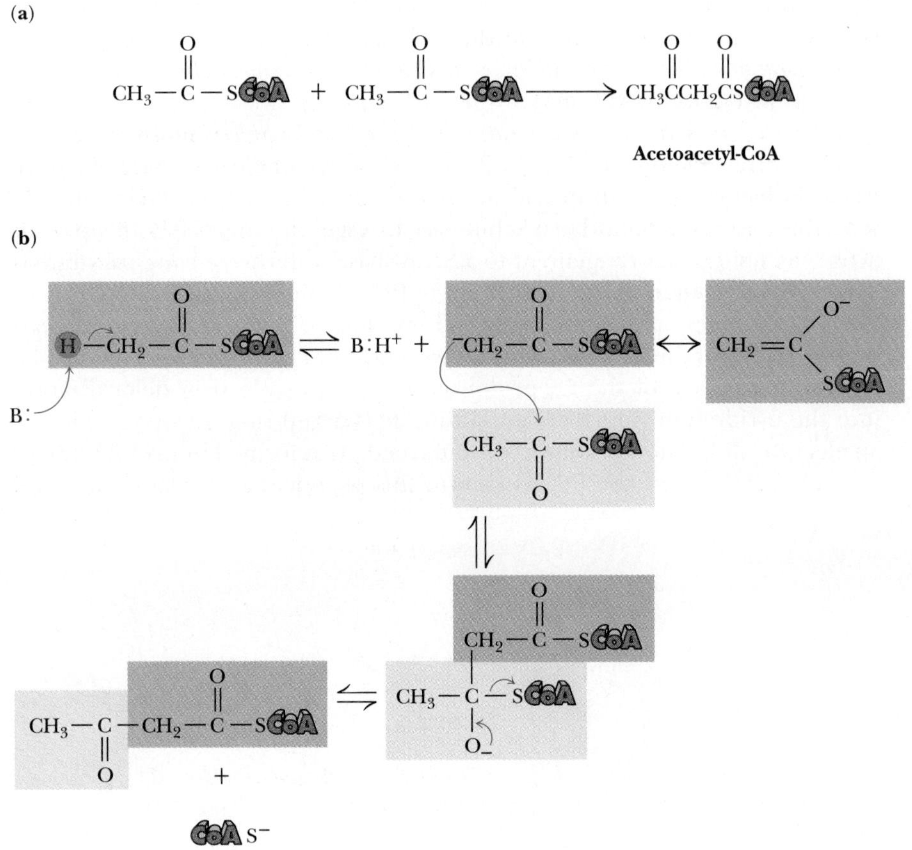

***Figure* 14.20** (a) The β-ketothiolase reaction demonstrates two special properties of acyl-CoA: activation of the acyl group for transfer by nucleophilic attack and the acidic nature of the α-hydrogen.
(b) Removal of the α-hydrogen of one acetyl-CoA yields the nucleophilic carbanion, which attacks the carbonyl carbon of a second acetyl-CoA.

A Deeper Look

Fritz Lipmann and Coenzyme A

Pantothenic acid is found in extracts from nearly all plants, bacteria, and animals, and the name derives from the Greek *pantos,* meaning "everywhere." It is required in the diet of all vertebrates, but some microorganisms produce it in the rumens of animals such as cattle and sheep. This vitamin is widely distributed in foods common to the human diet, and deficiencies are only observed in cases of severe malnutrition. The eminent American biochemist Fritz Lipmann was the first to show that a coenzyme was required to facilitate biological acetylation reactions. (The "A" in coenzyme A in fact stands for *acetylation.*) In studies of acetylation of sulfanilic acid (chosen because of a favorable colorimetric assay) by liver extracts, Lipmann found that a heat-stable cofactor was required. Eventually Lipmann isolated and purified the required cofactor—coenzyme A—from both liver and yeast. For his pioneering work in elucidating the role of this important coenzyme, Fritz Lipmann received the Nobel Prize in physiology or medicine in 1953.

14.6 Vitamin B_6: Pyridoxine and Pyridoxal Phosphate

The biologically active form of vitamin B_6 is **pyridoxal-5-phosphate (PLP),** a coenzyme that exists under physiological conditions in two tautomeric forms (Figure 14.21). PLP participates in the catalysis of a wide variety of reactions involving amino acids, including transaminations, α- and β-decarboxylations, β- and γ-eliminations, racemizations, and aldol reactions (Figure 14.22). Note that these reactions include cleavage of any of the bonds to the amino acid alpha carbon, as well as several bonds in the side chain. The remarkably versatile chemistry of PLP is due to its ability to

(a) *form stable Schiff base (aldimine) adducts with α-amino groups of amino acids,* as in Figure 3.9(e), and

(b) *act as an effective electron sink to stabilize reaction intermediates.*

The Schiff base formed by PLP and its role as an electron sink are illustrated in Figure 14.23. In nearly all PLP-dependent enzymes, PLP in the absence of substrate is bound in a Schiff base linkage with the ϵ-NH_2 group of an active site lysine. Rearrangement to a Schiff base with the arriving substrate is a **transaldiminization** reaction. One key to PLP chemistry is the protonation of the Schiff base, which is stabilized by H bonding to the ring oxygen, increasing the acidity of the C_α proton [as shown in (3) of Figure 14.23]. The carbanion formed by loss of the C_α proton is stabilized by electron delocalization into the pyridinium ring, with the positively charged ring nitrogen acting as an electron sink. Another important intermediate is formed by protonation of the aldehyde carbon of PLP. As shown, this produces a new substrate–PLP

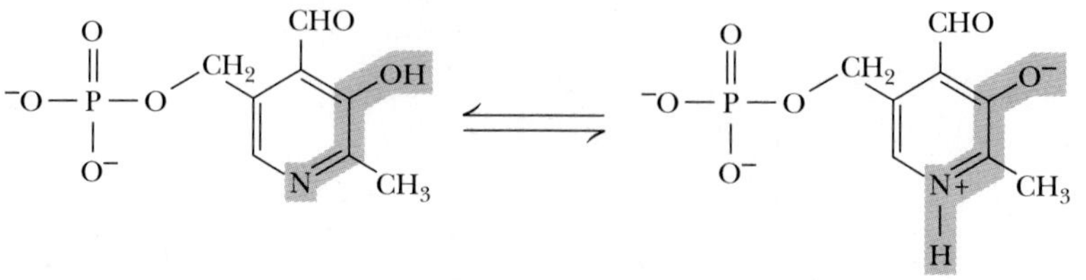

***Figure* 14.21** The tautomeric forms of pyridoxal-5-phosphate (PLP).

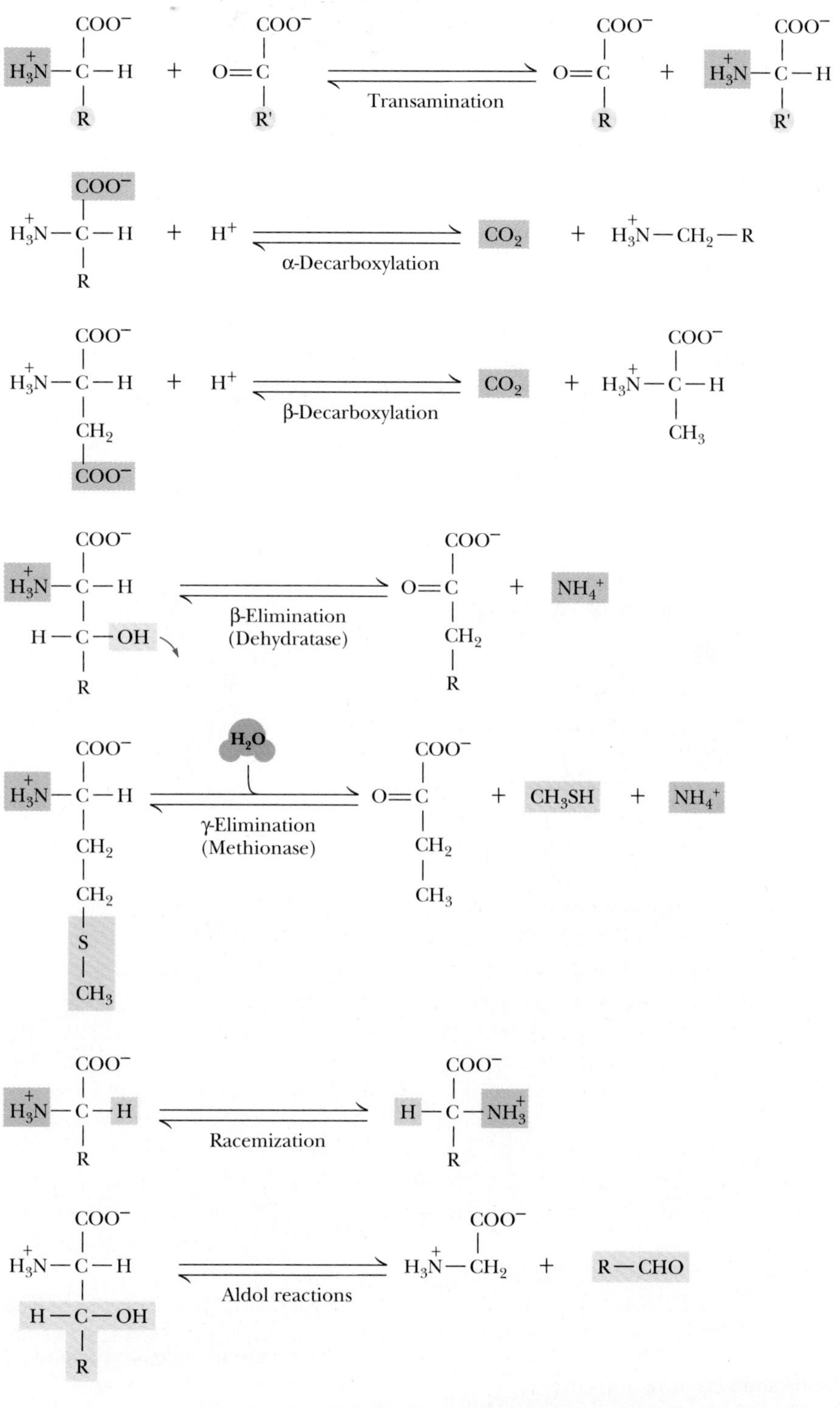

$$\overset{+}{H_3N}-CH(R)-COO^- + O{=}C(R')-COO^- \xrightleftharpoons[\text{Transamination}]{} O{=}C(R)-COO^- + \overset{+}{H_3N}-CH(R')-COO^-$$

$$\overset{+}{H_3N}-CH(R)-COO^- + H^+ \xrightleftharpoons[\alpha\text{-Decarboxylation}]{} CO_2 + \overset{+}{H_3N}-CH_2-R$$

$$\overset{+}{H_3N}-CH(CH_2-COO^-)-COO^- + H^+ \xrightleftharpoons[\beta\text{-Decarboxylation}]{} CO_2 + \overset{+}{H_3N}-CH(CH_3)-COO^-$$

$$\overset{+}{H_3N}-CH(COO^-)-CH(OH)-R \xrightleftharpoons[\substack{\beta\text{-Elimination} \\ \text{(Dehydratase)}}]{} O{=}C(COO^-)-CH_2-R + NH_4^+$$

$$\overset{+}{H_3N}-CH(COO^-)-CH_2-CH_2-S-CH_3 \xrightleftharpoons[\substack{\gamma\text{-Elimination} \\ \text{(Methionase)}}]{H_2O} O{=}C(COO^-)-CH_2-CH_3 + CH_3SH + NH_4^+$$

$$\overset{+}{H_3N}-C(H)(COO^-)-R \xrightleftharpoons[\text{Racemization}]{} H-C(COO^-)(\overset{+}{NH_3})-R$$

$$\overset{+}{H_3N}-CH(COO^-)-CH(OH)-R \xrightleftharpoons[\text{Aldol reactions}]{} \overset{+}{H_3N}-CH_2-COO^- + R-CHO$$

***Figure* 14.22** The seven classes of reactions catalyzed by pyridoxal-5-phosphate.

Schiff base, which plays a role in transamination reactions and increases the acidity of the proton at C_β, a feature important in γ-elimination reactions.

Mechanisms for a variety of PLP-catalyzed reactions are presented in Figures 14.24 through 14.26 and also in the problems at the end of the chapter. Each of these mechanisms employs one or more of the intermediates shown in Figure 14.23. For example, transamination is facilitated by loss of the C_α

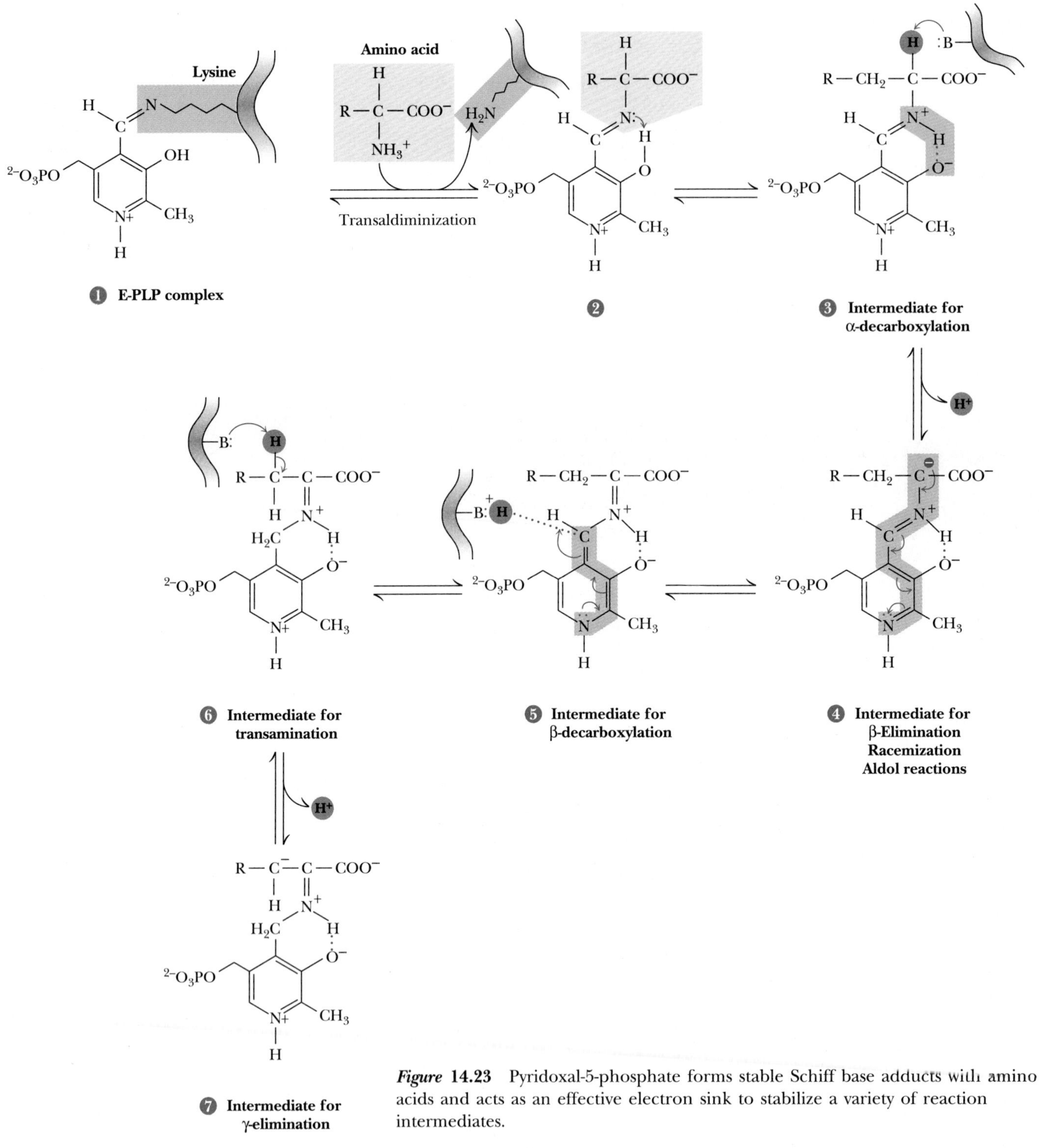

***Figure* 14.23** Pyridoxal-5-phosphate forms stable Schiff base adducts with amino acids and acts as an effective electron sink to stabilize a variety of reaction intermediates.

proton, followed by an *aldimine-ketimine tautomerization* (Figure 14.24). Hydrolysis of the ketimine Schiff base yields the product α-keto acid and pyridoxamine phosphate. Schiff base formation with the second substrate, another α-keto acid, followed by reversal of the process gives the product amino acid.

Most of the common amino acids undergo α-decarboxylation reactions. In the brain, for example, the important neurotransmitters γ-aminobutyric acid, dopamine, and histamine are formed by α-decarboxylations of glutamic

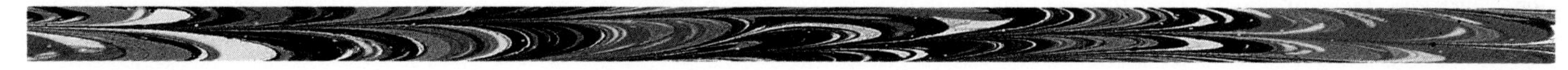

A Deeper Look

Vitamin B_6

Goldberger and Lillie in 1926 found that rats fed certain nutritionally deficient diets developed **dermatitis acrodynia,** a skin disorder characterized by edema and lesions of the ears, paws, nose, and tail. Szent-Györgyi later found that a factor he had isolated prevented these skin lesions in the rat. He proposed the name **vitamin B_6** for his factor. **Pyridoxine,** a form of this vitamin found in plants (and the form of B_6 sold commercially), was isolated in 1938 by three research groups working independently. **Pyridoxal** and **pyridoxamine,** the forms that predominate in animals, were identified in 1945. A metabolic role for pyridoxal was postulated by Esmond Snell, who had shown that when pyridoxal was heated with glutamate (in the absence of any enzymes), the amino group of glutamate was transferred to pyridoxal, forming pyridoxamine. Snell postulated (correctly) that pyridoxal might be a component of a coenzyme needed for transamination reactions in which the α-amino group of an amino acid is transferred to the α-carbon of an α-keto acid.

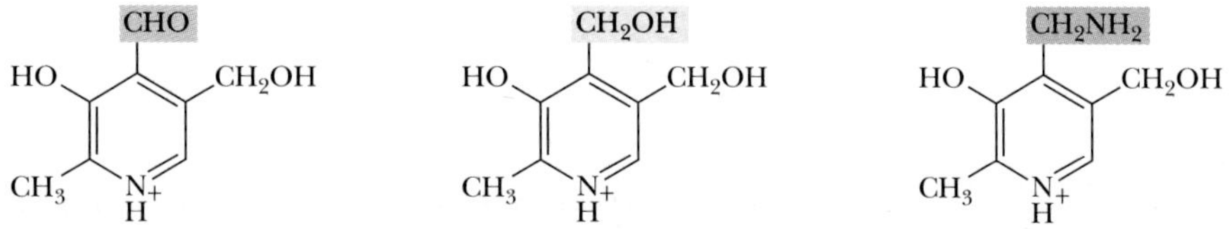

Pyridoxal **Pyridoxine or pyridoxol** **Pyridoxamine**

The structures of pyridoxal, pyridoxine, and pyridoxamine.

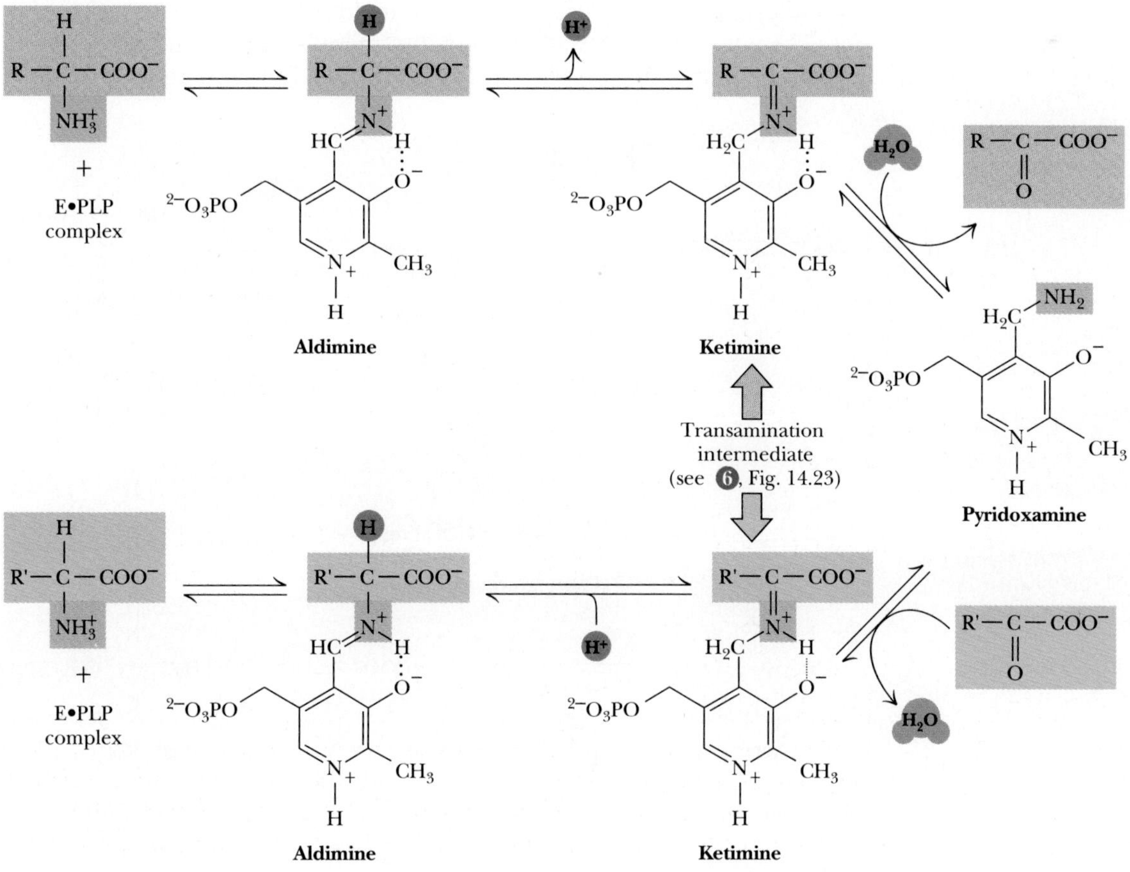

E•PLP complex — Aldimine — Ketimine — Pyridoxamine

Transamination intermediate (see 6, Fig. 14.23)

E•PLP complex — Aldimine — Ketimine

***Figure* 14.24** The mechanism of PLP-catalyzed transamination reactions.

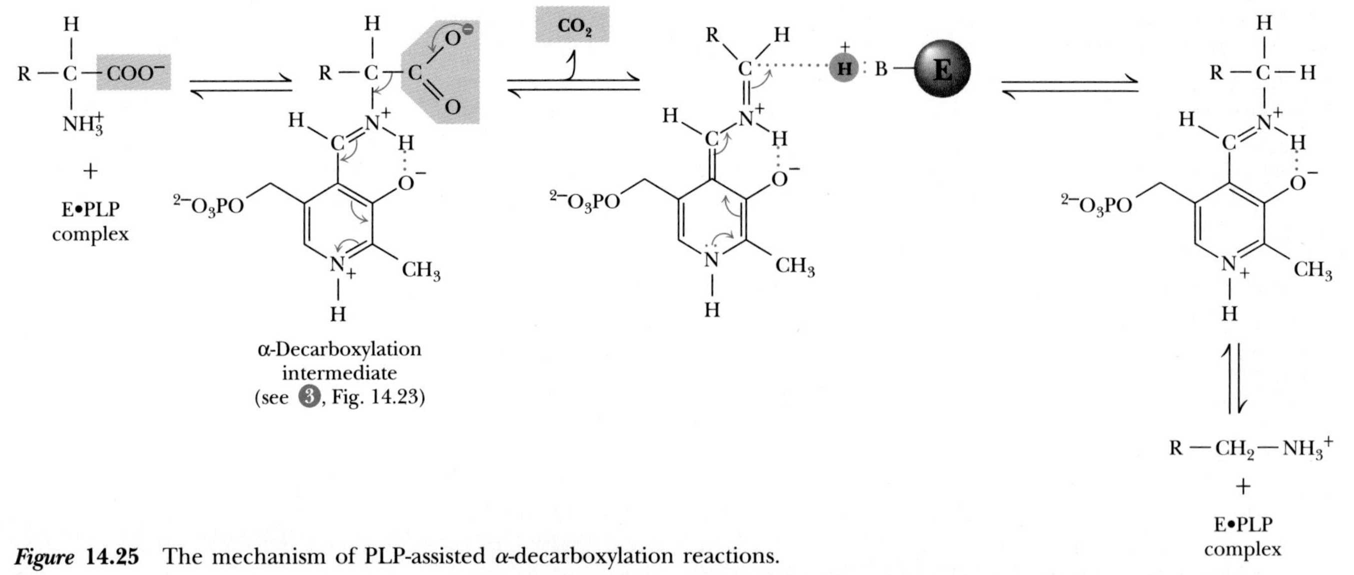

Figure **14.25** The mechanism of PLP-assisted α-decarboxylation reactions.

acid, 3,4-dihydroxyphenylalanine (DOPA), and histidine, respectively. α-Decarboxylation (Figure 14.25) is facilitated by electron delocalization into the pyridinium ring. The C_α hydrogen of the original amino acid is preserved in α-decarboxylations, in contrast to transaminations, where this proton is lost.

PLP-catalyzed β-elimination reactions are illustrated in Figure 14.26 with the serine dehydratase reaction. β-Eliminations mediated by PLP yield products that have undergone a two-electron oxidation at C_α. Serine is thus oxidized to pyruvate, with release of ammonium ion (Figure 14.26). At first, this looks like a transaminase half-reaction, but there is an important difference. In each transaminase half-reaction (Figure 14.24), PLP undergoes a net two-

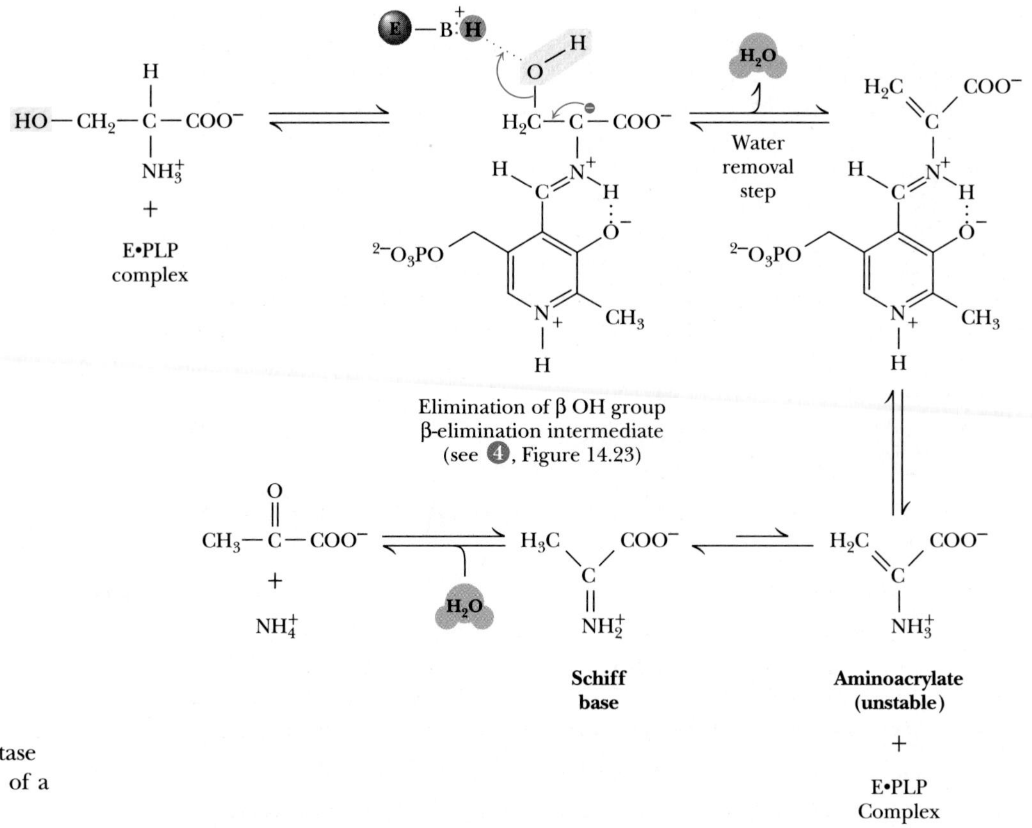

Figure **14.26** The serine dehydratase reaction mechanism—an example of a β-elimination reaction.

electron reduction or oxidation (depending on the direction), whereas β-eliminations occur with no net oxidation or reduction of PLP. Note too that the aminoacrylate released from PLP is unstable in aqueous solution. It rapidly tautomerizes to the preferred imine form, which is spontaneously hydrolyzed to yield the α-keto acid product.

The versatile chemistry of pyridoxal phosphate offers a rich learning experience for the student of mechanistic chemistry. William Jencks, in his classic text, *Catalysis in Chemistry and Enzymology,* writes:

> It has been said that God created an organism especially adapted to help the biologist find an answer to every question about the physiology of living systems; if this is so it must be concluded that pyridoxal phosphate was created to provide satisfaction and enlightenment to those enzymologists and chemists who enjoy pushing electrons, for no other coenzyme is involved in such a wide variety of reactions, in both enzyme and model systems, which can be reasonably interpreted in terms of the chemical properties of the coenzyme. Most of these reactions are made possible by a common structural feature. That is, electron withdrawal toward the cationic nitrogen atom of the imine and into the electron sink of the pyridoxal ring from the α carbon atom of the attached amino acid activates all three of the substituents on this carbon atom for reactions which require electron withdrawal from this atom.[1]

14.7 Vitamin B_{12}: Cyanocobalamin

Vitamin B_{12}, or **cyanocobalamin,** is converted in the body into two coenzymes. The predominant coenzyme form is **5′-deoxyadenosylcobalamin** (Figure 14.27), but smaller amounts of **methylcobalamin** also exist in liver, for example. The crystal structure of 5′-deoxyadenosylcobalamin was determined by X-ray diffraction in 1961 by Dorothy Hodgkin and co-workers in England. The structure consists of a *corrin ring* with a *cobalt ion* in the center. The corrin ring, with four *pyrrole groups,* is similar to the heme porphyrin ring, except that two of the pyrrole rings are linked directly. Methylene bridges form the other pyrrole–pyrrole linkages, as for porphyrin. The cobalt is coordinated to the four (planar) pyrrole nitrogens. One of the axial cobalt ligands is a nitrogen of the dimethylbenzimidazole group. The other axial cobalt ligand may be —CN, —CH_3, —OH, or the 5′-carbon of a 5′-deoxyadenosyl group, depending on the form of the coenzyme. The most striking feature of Hodgkin's structure of 5′-deoxyadenosylcobalamin is the cobalt–carbon bond distance of 0.205 nm. *This bond is predominantly covalent* and the structure is actually an **alkyl cobalt.** Such alkyl cobalts were thought to be highly unstable until Hodgkin's pioneering X-ray study. The Co–carbon–carbon bond angle of 130° indicates partial ionic character.

The B_{12} coenzymes participate in three types of reactions (Figure 14.28):

1. *Intramolecular rearrangements,*
2. *Reductions of ribonucleotides to deoxyribonucleotides* (in certain bacteria), and
3. *Methyl group transfers.*

The first two of these are mediated by 5′-deoxyadenosylcobalamin, whereas methyl transfers are effected by methylcobalamin. The mechanism of ribonucleotide reductase is discussed in Chapter 27. Methyl group transfers that employ *tetrahydrofolate* as a coenzyme are described later in this chapter.

[1]Jencks, William P., 1969. *Catalysis in Chemistry and Enzymology.* New York: McGraw-Hill.

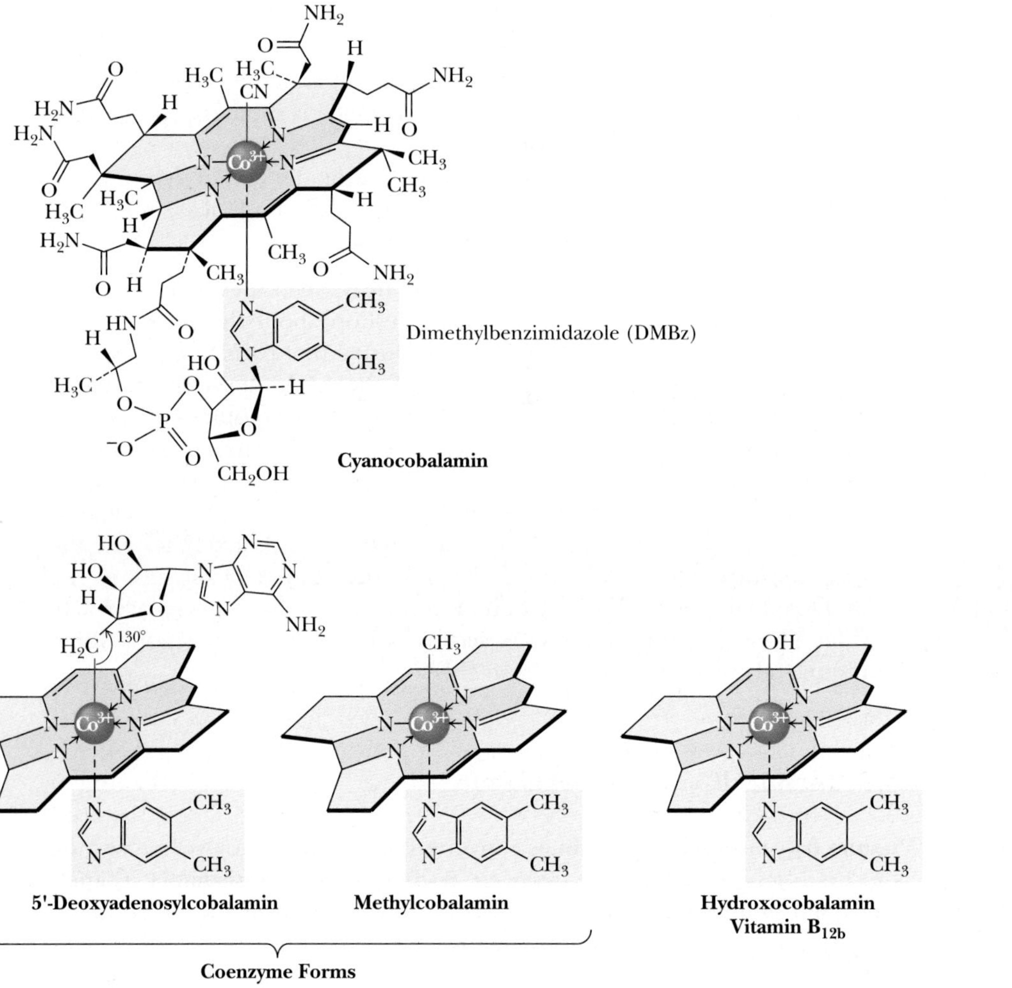

***Figure* 14.27** The structure of cyanocobalamin (top) and simplified structures showing several coenzyme forms of vitamin B_{12}. The Co—C bond of 5′-deoxyadenosylcobalamin is predominantly covalent (note the short bond length of 0.205 nm) but with some ionic character. Note that the convention of writing the cobalt atom as Co^{3+} attributes the electrons of the Co—C and Co—N bonds to carbon and nitrogen, respectively.

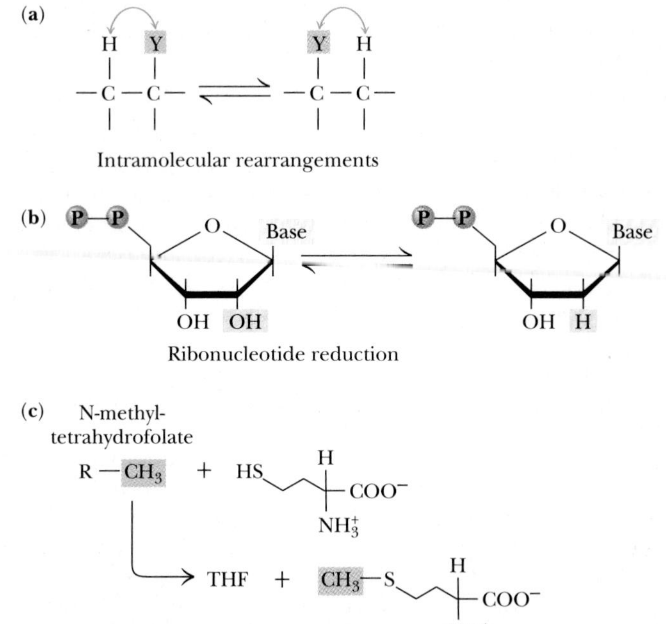

***Figure* 14.28** Vitamin B_{12} functions as a coenzyme in intramolecular rearrangements, reduction of ribonucleotides, and methyl group transfers.

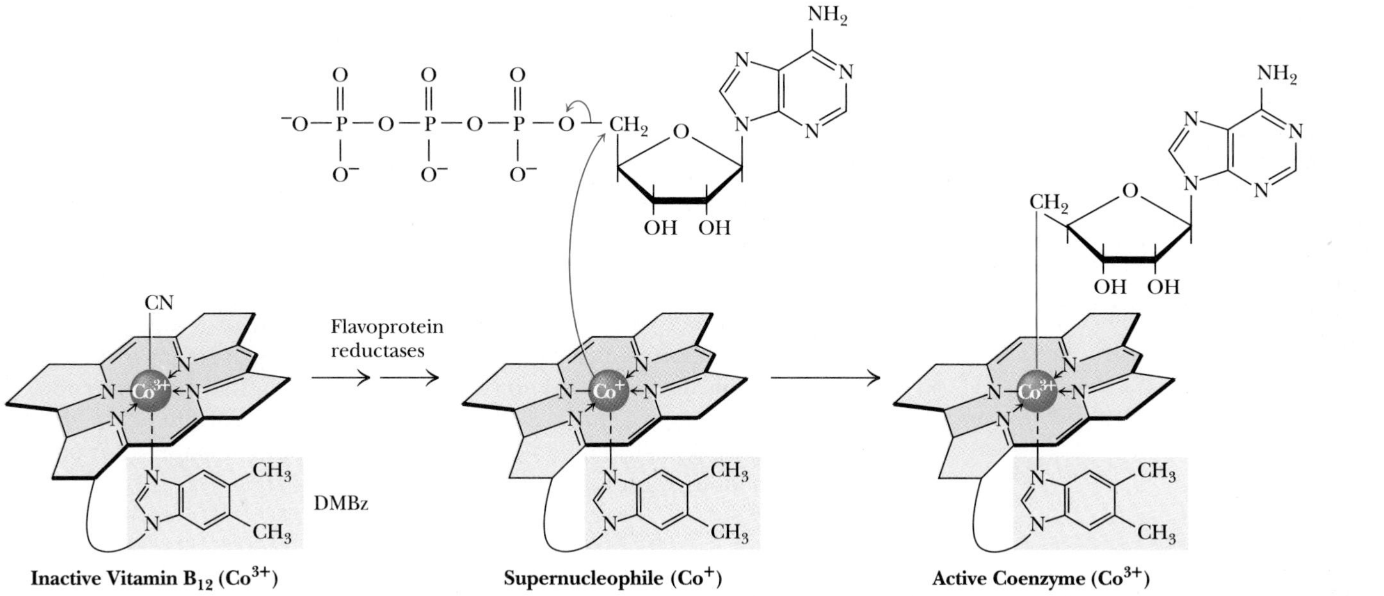

Flavoprotein reductases

DMBz

Inactive Vitamin B_{12} (Co^{3+}) **Supernucleophile (Co^{+})** **Active Coenzyme (Co^{3+})**

***Figure* 14.29** Formation of the active coenzyme 5′-deoxyadenosylcobalamin from inactive vitamin B_{12} is initiated by the action of flavoprotein reductases. The resulting Co^{+} species, dubbed a supernucleophile, attacks the 5′-carbon of ATP in an unusual adenosyl transfer.

Conversion of inactive *vitamin B_{12}* to active *5′-deoxyadenosylcobalamin* is thought to involve three steps (Figure 14.29). Two *flavoprotein reductases* sequentially convert Co^{3+} in cyanocobalamin to the Co^{2+} state and then to the Co^{+} state. The Co^{+} is an extremely powerful nucleophile. It attacks the C-5′ carbon of ATP as shown, expelling the triphosphate anion to form 5′-deoxyadenosylcobalamin. Since the two electrons from Co^{+} are both donated to the Co–carbon bond, the oxidation state of cobalt reverts to Co^{3+} in the active coenzyme. This is one of only two known adenosyl transfers (that is, nucleophilic attack on the ribose 5′-carbon of ATP) in biological systems. (The other is the formation of S-adenosylmethionine—see Chapter 26.)

A Deeper Look

Vitamin B_{12} and Pernicious Anemia

The most potent known vitamin (that is, the one needed in the smallest amounts) was the last to be discovered. Vitamin B_{12} is best known as the vitamin that prevents **pernicious anemia.** Minot and Murphy in 1926 demonstrated that such anemia could be prevented by eating large quantities of liver, but the active agent was not identified for many years. In 1948, Rickes and co-workers (in the U.S.) and Smith (in England) both reported the first successful isolation of vitamin B_{12}. West showed that injections of the vitamin induced dramatic beneficial responses in pernicious anemia patients. Eventually, two different crystalline preparations of the vitamin were distinguished. The first appeared to be true cyanocobalamin. The second showed the same biological activity as cyanocobalamin, but had a different spectrum and was named **vitamin B_{12b}** and also **hydroxocobalamin.** It was eventually found that the cyanide group in cyanocobalamin originated from the charcoal used in the purification process!

Vitamin B_{12} is not synthesized by animals or by plants. Only a few species of bacteria synthesize this complex substance. Carnivorous animals easily acquire sufficient amounts of B_{12} from meat in their diet, but herbivorous creatures typically depend on intestinal bacteria to synthesize B_{12} for them. This is sometimes not sufficient, and certain animals, including rabbits, occasionally eat their feces in order to accumulate the necessary quantities of B_{12}.

The nutritional requirement for vitamin B_{12} is low. Adult humans require only about 3 micrograms per day, an amount easily acquired with normal eating habits. However, since plants do not synthesize vitamin B_{12}, pernicious anemia symptoms are sometimes observed in strict vegetarians.

B_{12}-Catalyzed Intramolecular Rearrangements

Intramolecular rearrangements catalyzed by B_{12} involve the interchange of hydrogen and another substituent on adjacent carbons. The mechanism for such rearrangements (Figure 14.30) involves **homolytic cleavage** of the cobalt–carbon bond of the coenzyme as shown. This reduces cobalt to the Co^{2+} state and produces a $—CH_2\cdot$ radical, which abstracts a hydrogen atom from the substrate, forming 5′-deoxyadenosine and leaving a radical (unpaired electron) on the substrate. Rearrangement of this intermediate, with Y· moving from one carbon to the other, is followed by hydrogen atom transfer from the methyl group of 5′-deoxyadenosine and regeneration of 5′-deoxyadenosylcobalamin. Several examples of such rearrangements are shown in Figure 14.31. The only one of these known to occur in mammals is the conversion of methylmalonyl-CoA to succinyl-CoA. This reaction is important in the degradation of fatty acids having an odd number of carbons (Chapter 23) and in the degradation of isoleucine and valine (Chapter 26).

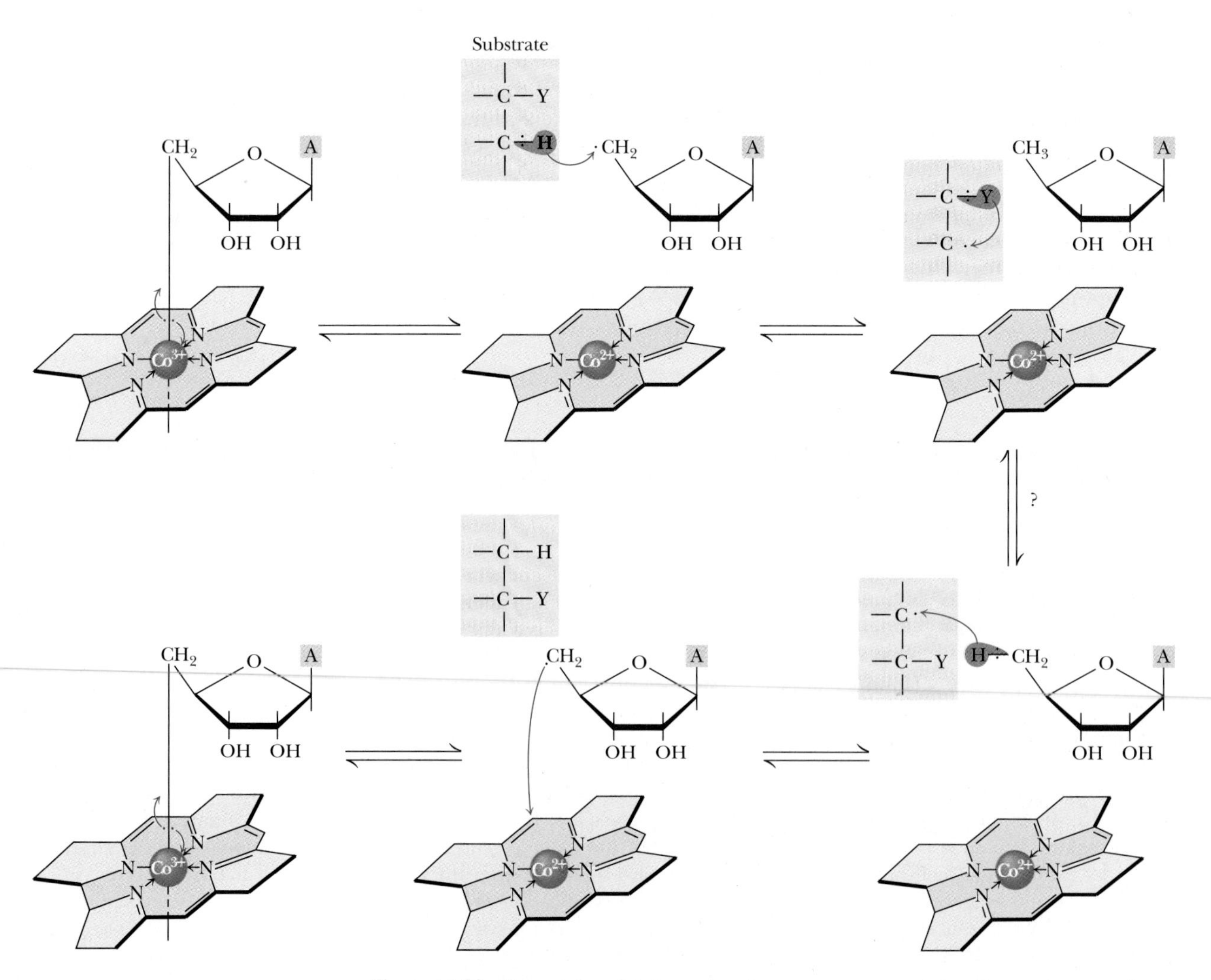

Figure **14.30** B_{12}-catalyzed intramolecular rearrangements involve exchange of hydrogen and another substituent on adjacent carbon atoms.

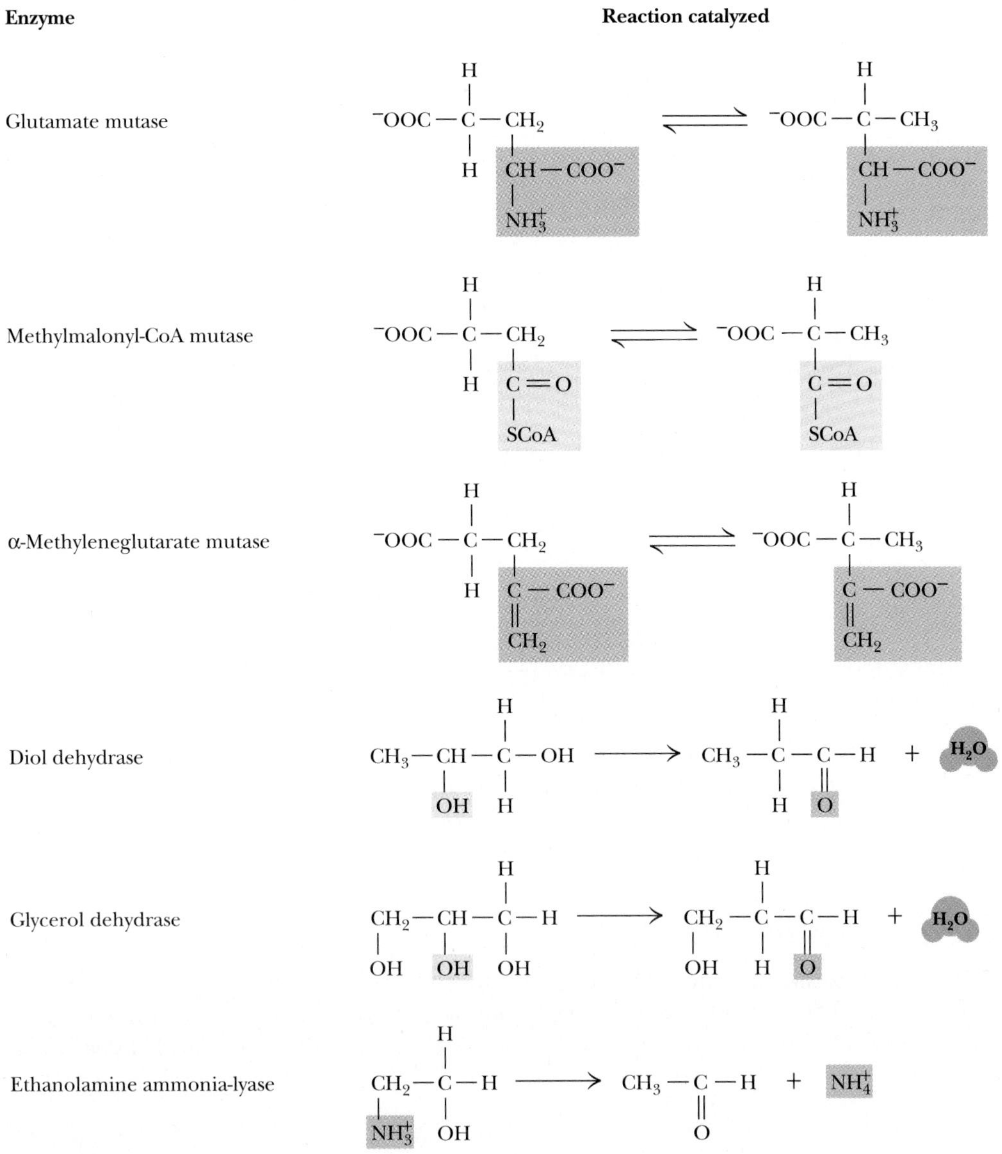

Enzyme	Reaction catalyzed
Glutamate mutase	$^-OOC-CH_2-CH_2(CH(NH_3^+)-COO^-) \rightleftharpoons {}^-OOC-CH(CH_3)-CH(NH_3^+)-COO^-$
Methylmalonyl-CoA mutase	$^-OOC-CH_2-CH_2-C(=O)-SCoA \rightleftharpoons {}^-OOC-CH(CH_3)-C(=O)-SCoA$
α-Methyleneglutarate mutase	$^-OOC-CH_2-CH_2-C(=CH_2)-COO^- \rightleftharpoons {}^-OOC-CH(CH_3)-C(=CH_2)-COO^-$
Diol dehydrase	$CH_3-CH(OH)-CH_2-OH \longrightarrow CH_3-CH_2-C(=O)-H + H_2O$
Glycerol dehydrase	$CH_2(OH)-CH(OH)-CH_2(OH) \longrightarrow CH_2(OH)-CH_2-C(=O)-H + H_2O$
Ethanolamine ammonia-lyase	$CH_2(NH_3^+)-CH(OH)-H \longrightarrow CH_3-C(=O)-H + NH_4^+$

***Figure* 14.31** Typical rearrangement reactions catalyzed by vitamin B_{12}.

14.8 Vitamin C: Ascorbic Acid

L-Ascorbic acid, better known as **vitamin C,** has the simplest chemical structure of all the vitamins (Figure 14.32). It is widely distributed in the animal and plant kingdoms, and only a few vertebrates—man and other primates, guinea pigs, fruit-eating bats, certain birds, and some fish (rainbow trout, carp, and Coho salmon, for example)—are unable to synthesize it. In all these organisms, the inability to synthesize ascorbic acid stems from a lack of a liver enzyme, L-gulono-γ-lactone oxidase. As shown in Figure 14.33, ascorbic acid is normally synthesized from D-glucuronic acid, via gulonic acid and L-gulonolactone.

Ascorbic acid is a reasonably strong reducing agent. The biochemical and physiological functions of ascorbic acid most likely derive from its reducing properties—it functions as an electron carrier. Loss of one electron due to interactions with oxygen or metal ions leads to **semidehydro-L-ascorbate,** a

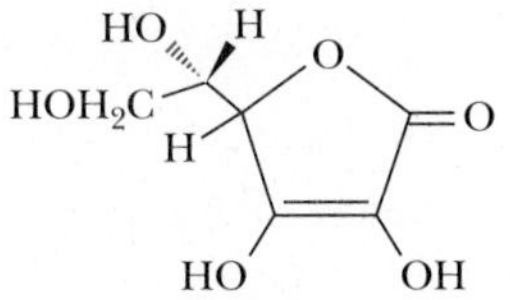

HO, H, HOH_2C, H, O, O, HO, OH

Ascorbic acid (vitamin C)

***Figure* 14.32** The structure of ascorbic acid (vitamin C).

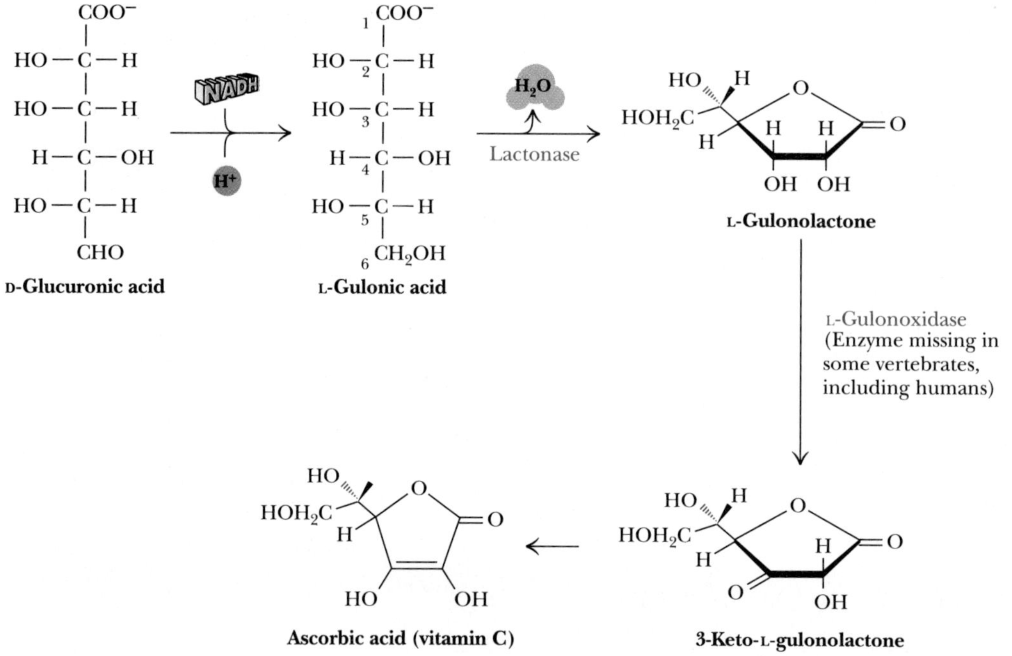

Figure **14.33** The synthesis of ascorbic acid from D-glucuronic acid. Animals that produce the liver enzyme L-gulono-γ-lactone oxidase can synthesize ascorbic acid. Humans lack this enzyme.

D-Glucuronic acid → (NADH, H^+) → L-Gulonic acid → (H_2O, Lactonase) → L-Gulonolactone → (L-Gulonoxidase (Enzyme missing in some vertebrates, including humans)) → 3-Keto-L-gulonolactone → Ascorbic acid (vitamin C)

reactive free radical (Figure 14.34) that can be reduced back to L-ascorbic acid by various enzymes in animals and plants. A characteristic reaction of ascorbic acid is its oxidation to *dehydro-L-ascorbic acid.* Ascorbic acid and dehydroascorbic acid form an effective redox system.

Ascorbic Acid in the Brain

Ascorbic acid may play several important roles in the brain and in central nervous system tissue. The metabolism of L-tyrosine in the brain involves two different mixed-function oxidases that are ascorbic acid–dependent.

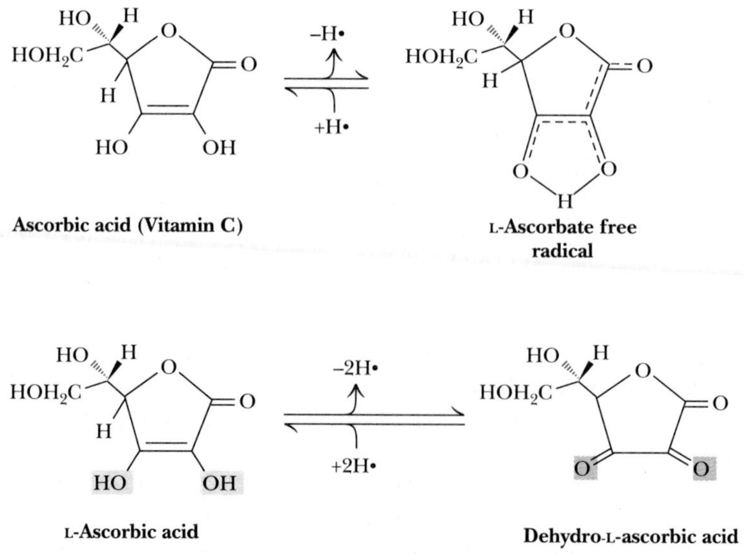

Ascorbic acid (Vitamin C) ⇌ (−H•, +H•) L-Ascorbate free radical

L-Ascorbic acid ⇌ (−2H•, +2H•) Dehydro-L-ascorbic acid

Figure **14.34** The physiological effects of vitamin C are the result of its action as a reducing agent. A two-electron oxidation of ascorbic acid yields dehydro-L-ascorbic acid.

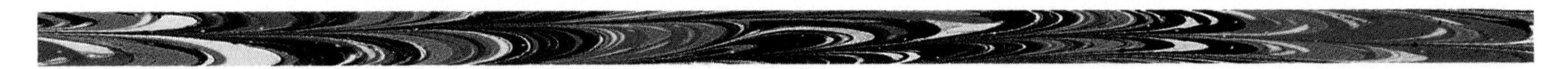

A Deeper Look

Ascorbic Acid and Scurvy

Ascorbic acid is effective in the treatment and prevention of **scurvy,** a potentially fatal disorder characterized by anemia, alteration of protein metabolism, and weakening of collagenous structures in bone, cartilage, teeth, and connective tissues (see Chapter 5). Western world diets are now routinely so rich in vitamin C that it is easy to forget that scurvy affected many people in ancient Egypt, Greece, and Rome, and that, in the Middle Ages, it was endemic in northern Europe in winter when fresh fruits and vegetables were scarce. Ascorbic acid is a vitamin that has routinely altered the course of history, ending ocean voyages and military campaigns when food supplies became depleted of vitamin C and fatal outbreaks of scurvy occurred.

The isolation of ascorbic acid was first reported by Albert Szent-Györgyi (who called it *hexuronic acid*) in 1928. The structure was determined by Hirst and Haworth in 1933, and, simultaneously, Reichstein reported its synthesis. Haworth and Szent-Györgyi, who together suggested that the name be changed to L-ascorbic acid to describe its **antiscorbutic** (antiscurvy) activity, were awarded the Nobel Prize in 1937 for their studies of vitamin C.

p-Hydroxyphenylpyruvate dioxygenase, which acts in converting tyrosine to homogentisic acid, requires ascorbate for the oxidation and decarboxylation of the intermediate, *p*-hydroxyphenylpyruvic acid (see Figure 26.45). Individuals with ascorbate deficiencies excrete abnormally high levels of *p*-hydroxyphenylpyruvic acid. Tyrosine is also converted to catecholamines in ascorbate-dependent processes. A series of hydroxylation and decarboxylation reactions form in succession dopamine, norepinephrine, and epinephrine. The dopamine–norepinephrine conversion is catalyzed by dopamine-β-hydroxylase, which is ascorbate-dependent.

Ascorbic Acid Mobilizes Iron and Prevents Anemia

Ascorbic acid plays a role in the prevention of anemia, and also serves to prevent the toxic effects of some transition metal ions. Iron mobilization from the spleen (but not the liver) is an ascorbate-dependent process.

Ascorbic Acid Ameliorates Allergic Responses

Another important role for ascorbate appears to involve histamine metabolism and the allergic response. In the presence of copper ions, ascorbic acid prevents the accumulation of histamine and assists in its degradation and elimination. Evidence also exists that ascorbic acid modulates prostaglandin synthesis so as to mediate histamine sensitivity and effect relaxation.

Ascorbic Acid Can Stimulate the Immune System

Ascorbic acid is important for the prevention and treatment of infections because of its effects in the stimulation of the immune system. Mononuclear leukocytes, which are important to the immune system, exhibit the highest concentrations of ascorbate of any of the cells in the blood. Ascorbate inhibits the oxidative destruction of white blood cells and increases their mobility. Serum levels of immunoglobulins are increased in the presence of ascorbate. The stimulation of the immune system by ascorbic acid has led Linus Pauling to suggest that vitamin C may be effective in preventing the common cold. Subsequent studies of this hypothesis have been inconclusive.

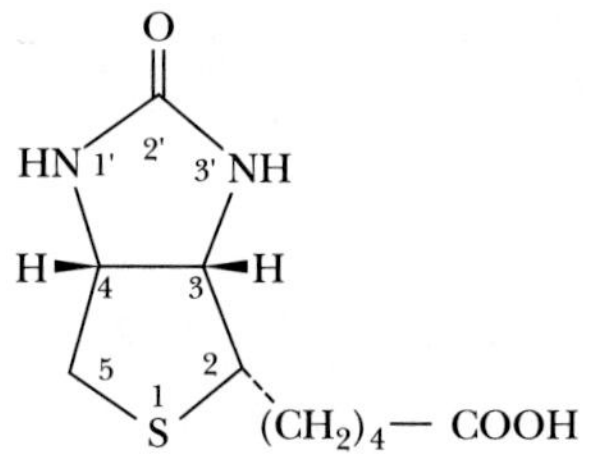

14.9 Biotin

Biotin (Figure 14.35) acts as a **mobile carboxyl group carrier** in a variety of enzymatic carboxylation reactions. In each of these, biotin is bound covalently to the enzyme as a prosthetic group via the ϵ-amino group of a lysine residue on the protein (Figure 14.36). The biotin-lysine function is referred to as a **biocytin residue.** The result is that *the biotin ring system is tethered to the protein by a long, flexible chain.* The 10 atoms in this chain separate the biotin ring and the lysine α-carbon by approximately 1.5 nm. This chain allows biotin to acquire carboxyl groups at one subsite of the enzyme active site and deliver them to a substrate acceptor at another subsite.

***Figure* 14.35** The structure of biotin.

***Figure* 14.36** Biotin is covalently linked to a protein via the ϵ-amino group of a lysine residue. The biotin ring is thus tethered to the protein by a 10-atom chain. It functions by carrying carboxyl groups between distant sites on biotin-dependent enzymes.

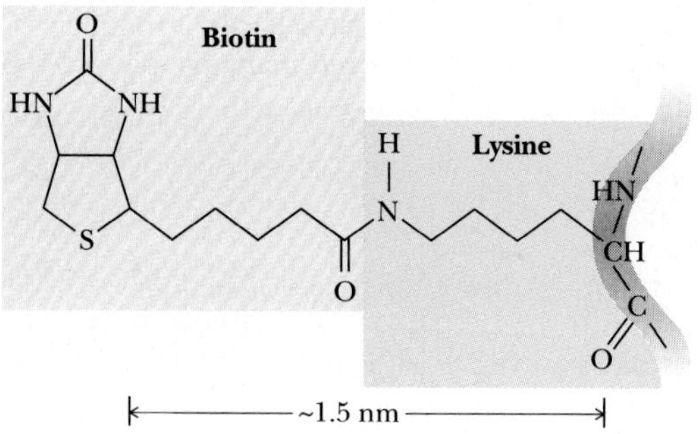

The biotin-lysine (biocytin) complex

***Figure* 14.37** Bicarbonate is activated for carboxylation reactions by formation of *N*-carboxybiotin. ATP drives the reaction forward, with transient formation of a carbonyl-phosphate intermediate (Step 1). In a typical biotin-dependent reaction, nucleophilic attack on the carboxyl carbon of *N*-carboxybiotin—a transcarboxylation—yields the carboxylated product (Step 2).

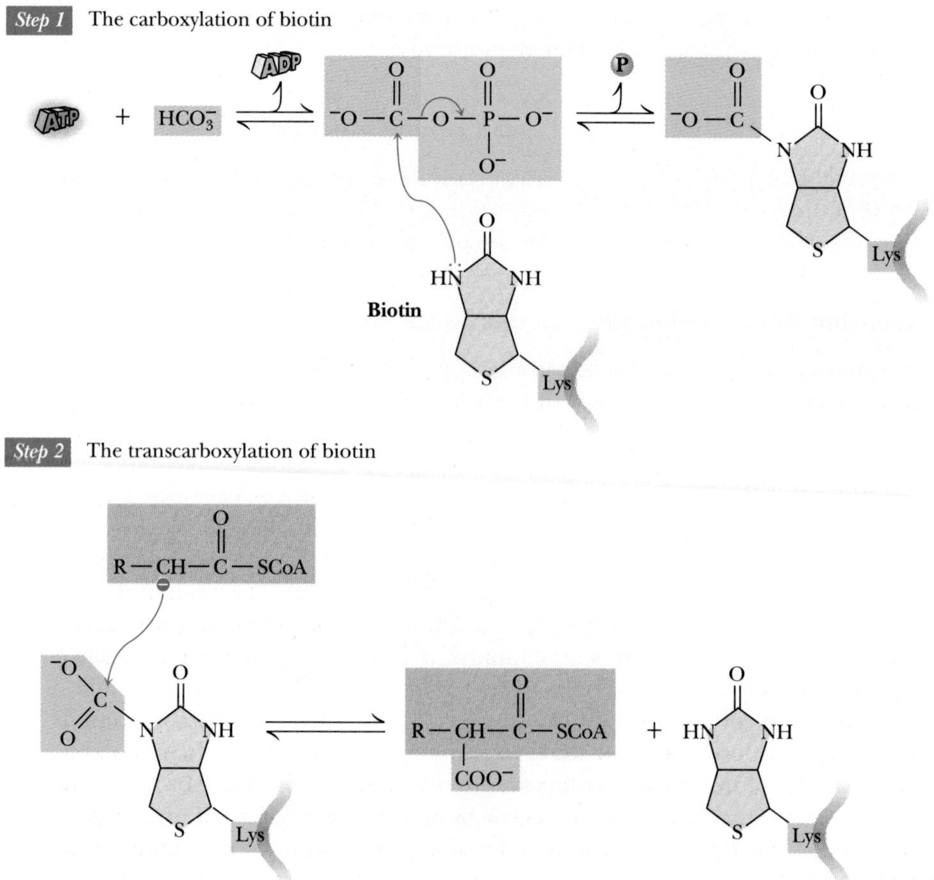

Step 1 The carboxylation of biotin

Step 2 The transcarboxylation of biotin

Table **14.4**
Biotin-Dependent Carboxylations

ATP-dependent

$$ATP + HCO_3^- + H_3C{-}C(=O){-}COO^- \longrightarrow {}^-OOC{-}CH_2{-}C(=O){-}COO^- + ADP + P$$

Pyruvate → **Oxaloacetate**

$$ATP + HCO_3^- + H_3C{-}C(=O){-}SCoA \longrightarrow {}^-OOC{-}CH_2{-}C(=O){-}SCoA + ADP + P$$

Acetyl-CoA → **Malonyl-CoA**

$$ATP + HCO_3^- + H_3C{-}CH_2{-}C(=O){-}SCoA \longrightarrow H_3C{-}CH(COO^-){-}C(=O){-}SCoA + ADP + P$$

Propionyl-CoA → **Methylmalonyl-CoA**

$$ATP + HCO_3^- + (H_3C)_2C{=}CH{-}C(=O){-}SCoA \longrightarrow {}^-OOC{-}CH_2{-}C(CH_3){=}CH{-}C(=O){-}SCoA + ADP + P$$

β-Methylcrotonyl-CoA → **β-Methylglutaconyl-CoA**

ATP + HCO_3^- + Geranyl-CoA ⟶ γ-Carboxygeranyl-CoA + ADP + P

Geranyl-CoA → **γ-Carboxygeranyl-CoA**

$$ATP + HCO_3^- + H_2N{-}C(=O){-}NH_2 \longrightarrow {}^-OOC{-}NH{-}C(=O){-}NH_2 + ADP + P$$

Urea → ***N*-Carboxyurea**

Transcarboxylase

$$H_3C{-}CH(COO^-){-}C(=O){-}SCoA + H_3C{-}C(=O){-}COO^- \rightleftharpoons H_3C{-}CH_2{-}C(=O){-}SCoA + {}^-OOC{-}CH_2{-}C(=O){-}COO^-$$

Methylmalonyl-CoA + **Pyruvate** ⇌ **Propionyl-CoA** + **Oxaloacetate**

Adapted from Walsh, C. T., 1979. *Enzymatic Reaction Mechanisms.* San Francisco: W. H. Freeman.

Most biotin-dependent carboxylations (Table 14.4) use *bicarbonate* as the carboxylating agent and transfer the carboxyl group to a *substrate carbanion.* Bicarbonate is plentiful in biological fluids, but it is a poor electrophile at carbon and must be "activated" for attack by the substrate carbanion. The activation is driven by ATP (Figure 14.37) and involves formation of a **carbonyl-phosphate intermediate**—a mixed anhydride of carbonic and phosphoric acids.

All biotin-dependent enzymes are multimeric and display ping-pong kinetics. The formation of *N*-carboxybiotin and its breakdown occur on different subsites of the active site. The site of attachment of biotin to the enzyme represents a third subsite in the active site. Some biotin-dependent enzymes carry all three of these subsites on a single polypeptide, but many biotin enzymes consist of aggregations of unifunctional subunits, with the carboxylation subsite, the carboxyl carrier subsite, and the carboxyl transferase subsite each existing on separate protein subunits. *Escherichia coli* acetyl-CoA carboxylase, for example, is composed of three kinds of subunits. Of these, the biotin carboxylase is a dimer of 100 kD subunits, the biotin carboxyl carrier protein (BCCP) is a 22-kD monomer that contains covalently bound biotin, and the transcarboxylase function is found on a dimer of 90 kD subunits. Since the biotin cofactor is attached to the protein via a flexible chain 1.5 nm long, it is able to carry activated carboxyl groups from the carboxylase subsite to the transcarboxylase subsite, as diagrammed in Figure 14.38.

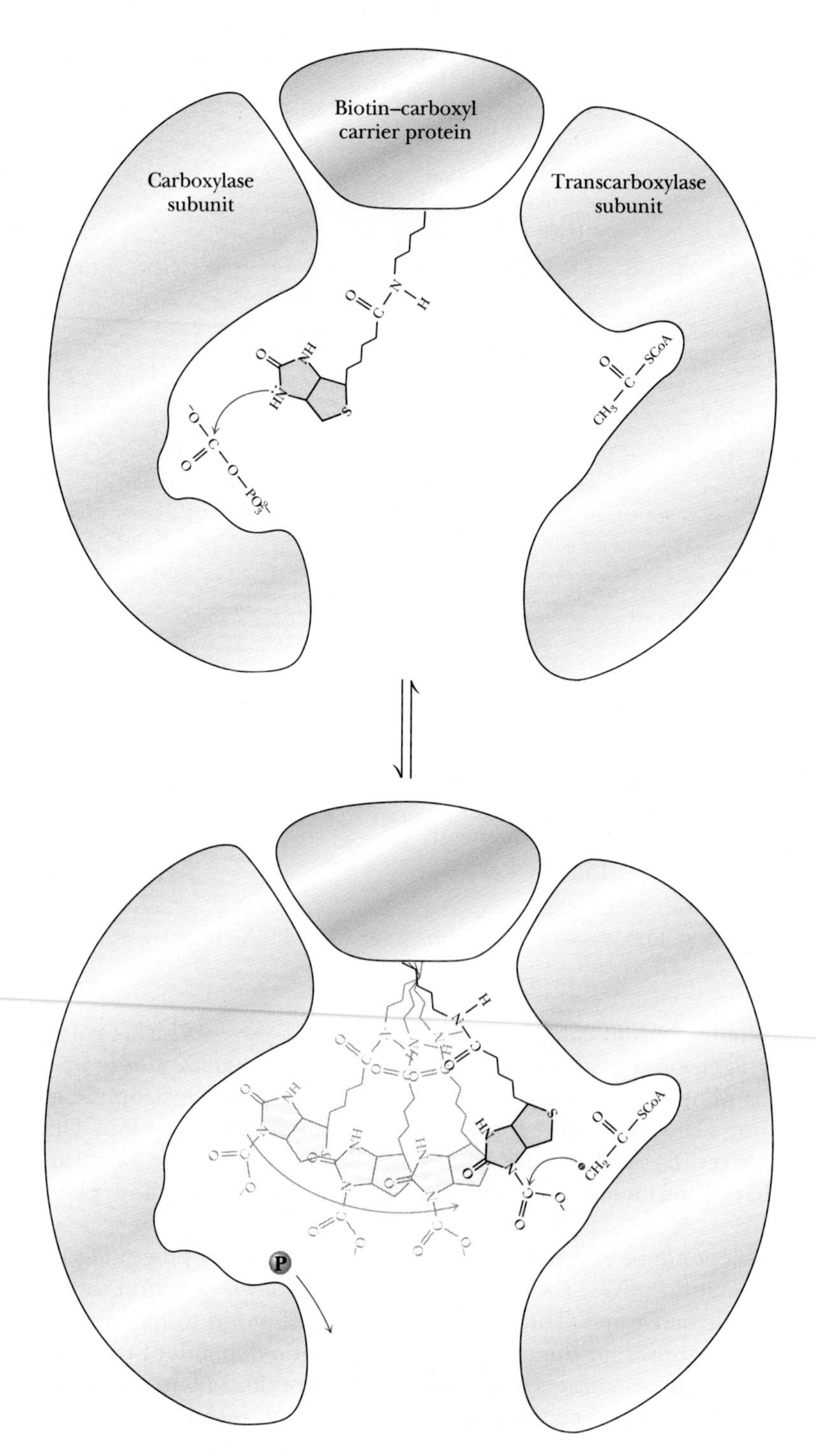

***Figure* 14.38** In the acetyl-CoA carboxylase reaction, the biotin ring, on its flexible tether, acquires carboxyl groups from carbonyl phosphate on the carboxylase subunit and transfers them to acyl-CoA molecules on the transcarboxylase subunit.

A Deeper Look

Biotin

Early in the 1900s, it was observed that certain strains of yeast required a material called **bios** for growth. Bios was eventually found to contain four different substances: myoinositol, β-alanine, pantothenic acid, and a compound later shown to be *biotin.* Kögl and Tönnis first isolated biotin from egg yolk in 1936. Boas, in 1927, and György, in 1931, found substances in liver that were capable of curing and preventing the dermatitis, loss of hair, and paralysis that occurred in rats fed large amounts of raw egg whites (a condition known as *egg white injury*). Boas called the factor "protective factor X" and György named the substance *vitamin H* (from the German *haut,* meaning "skin"), but both were soon shown to be identical to biotin. It is now known that egg white contains a basic protein called **avidin,** which has an extremely high affinity for biotin ($K_D = 10^{-15}\ M$). The sequestering of biotin by avidin is the cause of the egg white injury condition.

The structure of biotin was determined in the early 1940s by Kögl in Europe and by du Vigneaud and coworkers in the U.S. Interestingly, the biotin molecule contains three asymmetric carbon atoms, and biotin could thus exist as eight different stereoisomers. Only one of these shows biological activity.

14.10 Lipoic Acid

Lipoic acid exists as a mixture of two structures: a closed-ring disulfide form and an open chain reduced form (Figure 14.39). Oxidation-reduction cycles interconvert these two species. As is the case for biotin, lipoic acid does not often occur free in nature, but rather is covalently attached in amide linkage with lysine residues on enzymes. The enzyme that catalyzes the formation of

Lipoic acid, oxidized form

S—S, H_2C, $CHCH_2CH_2CH_2CH_2C$, CH_2, O, O^-

Reduced form

HS, HS, H_2C, $CHCH_2CH_2CH_2CH_2C$, CH_2, O, O^-

Lipoamide complex

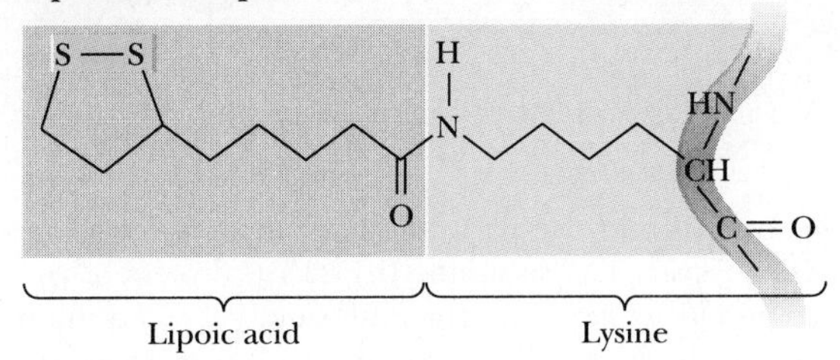

***Figure* 14.39** The oxidized and reduced forms of lipoic acid and the structure of the lipoic acid–lysine conjugate.

A Deeper Look

Lipoic Acid

Lipoic acid (6,8-dithiooctanoic acid) was isolated and characterized in 1951 in studies that showed that it was required for the growth of certain bacteria and protozoa. This accomplishment was one of the most impressive feats of isolation in the early history of biochemistry. Eli Lilly and Co., in cooperation with Lester J. Reed at the University of Texas and I. C. Gunsalus at the University of Illinois, isolated just 30 mg of lipoic acid from approximately 10 tons of liver! No evidence exists of a dietary lipoic acid requirement by humans, and strictly speaking, it is not considered a vitamin. Nevertheless, it is an essential component of several enzymes of intermediary metabolism and is present in body tissues in small amounts.

the *lipoamide* linkage requires ATP and produces lipoamide-enzyme conjugates, AMP, and pyrophosphate as products of the reaction.

Lipoic acid is an **acyl group carrier.** It is found in *pyruvate dehydrogenase* and *α-ketoglutarate dehydrogenase,* two multienzyme complexes involved in carbohydrate metabolism. *Lipoic acid functions to couple acyl-group transfer and electron transfer during oxidation and decarboxylation of α-keto acids.*

The special properties of lipoic acid arise from the ring strain experienced by oxidized lipoic acid. The closed ring form is approximately 20 kJ higher in energy than the open-chain form, and this results in a strong negative reduction potential of about −0.30 V. The oxidized form readily oxidizes cyanides to isothiocyanates and sulfhydryl groups to mixed disulfides.

E. coli pyruvate dehydrogenase, studied extensively by Lester Reed, is a multienzyme complex of 60 subunits, including 24 *pyruvate dehydrogenase* (E_1) subunits (existing as 12 dimers of 192 kD each), which utilize thiamine pyrophosphate as a coenzyme; 24 *dihydrolipoyl transacetylase* (E_2) subunits (70 kD), each of which contains a lysine-linked lipoamide; and 12 *dihydrolipoyl dehydro-*

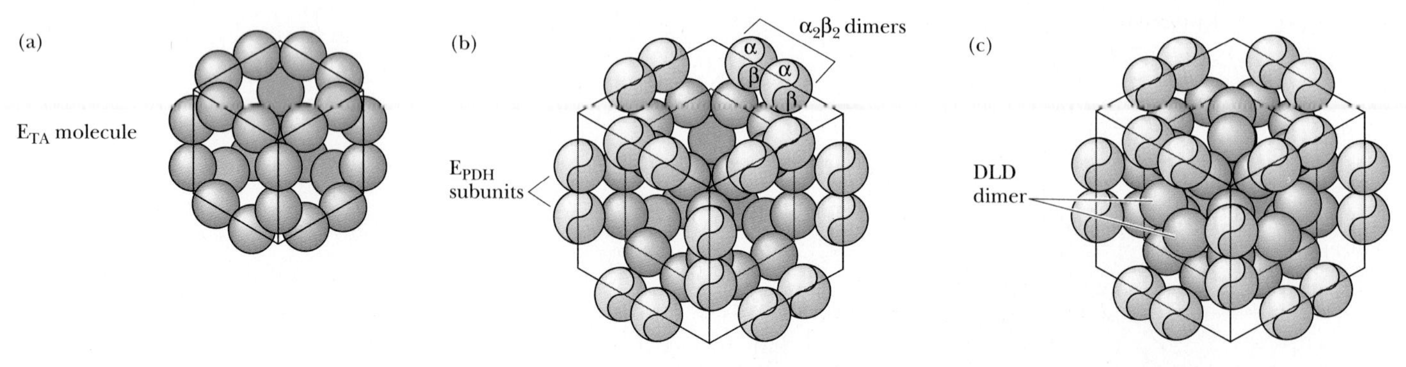

***Figure* 14.40** The structure of the pyruvate dehydrogenase complex. (a) 24 dihydrolipoyl transacetylase (TA) subunits. (b) 24 $\alpha\beta$ dimers of pyruvate dehydrogenase are added to the cube (two per edge). (c) Addition of 12 dihydrolipoyl dehydrogenase subunits (two per face) completes the complex.

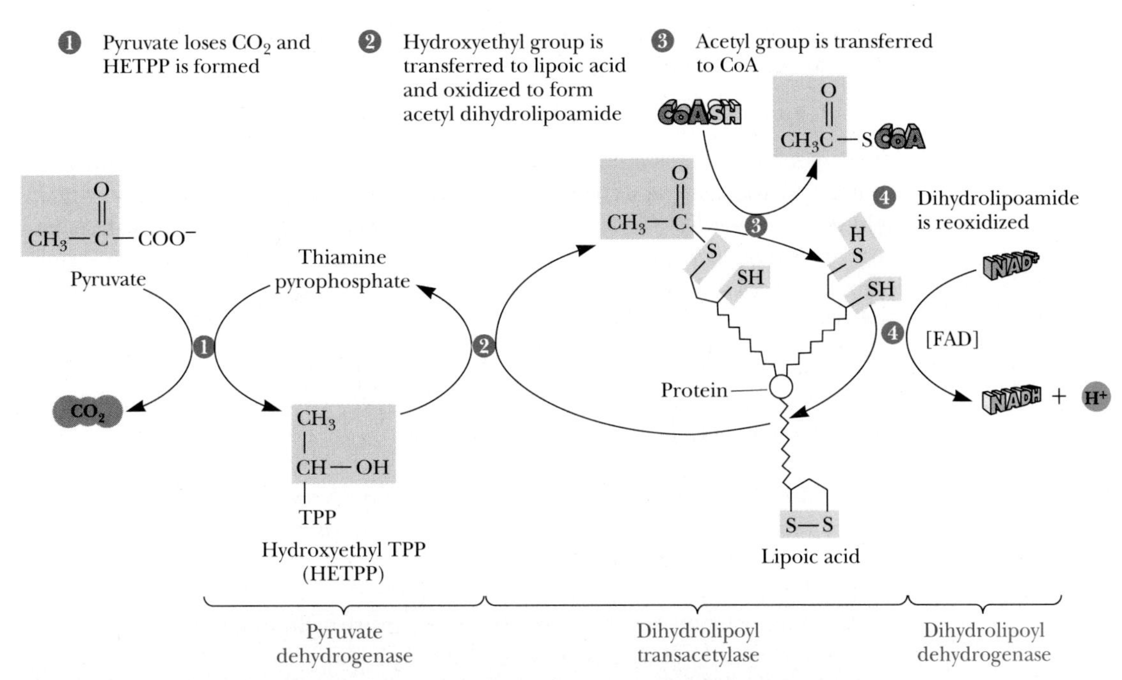

***Figure* 14.41** The mechanism of the pyruvate dehydrogenase reaction. Decarboxylation of pyruvate occurs with formation of hydroxyethyl-TPP (Step 1). Transfer of the two-carbon unit to lipoic acid in Step 2 is followed by formation of acetyl-CoA in Step 3. Lipoic acid is reoxidized in Step 4 of the reaction.

genase (E_3) subunits (existing as 6 dimers of 112 kD), each of which contains a flavin cofactor. As shown in Figure 14.40, the transacetylase subunits form a cubic core, around which are arranged the dehydrogenase subunits. This multienzyme complex forms acetyl-CoA via an oxidative decarboxylation of pyruvate:

$$\text{Pyruvate} + \text{CoA} + \text{NAD}^+ \longrightarrow \text{acetyl-CoA} + \text{CO}_2 + \text{NADH} + \text{H}^+$$

The reaction mechanism (Figure 14.41) involves four distinct steps. In the *first step,* pyruvate is decarboxylated with formation of hydroxyethyl-thiamine pyrophosphate (HETPP). This intermediate transfers its two-carbon unit to a lipoamide group on one of the transacetylase subunits. In the *second step,* the hydroxyethyl group is oxidized to form an *acetyldihydrolipoamide* intermediate and TPP is released. This step involves nucleophilic attack of the resonance-stabilized carbanion of HETPP on the lipoic acid disulfide, followed by oxidation of the attacking carbon atom to give an acetyl-lipoic acid. The *third step* is the transfer of the acetyl group from lipoamide to CoA to form acetyl-CoA, leaving a reduced dihydrolipoamide. In the *fourth step,* dihydrolipoamide is reoxidized (returning to the disulfide form) by dihydrolipoyl dehydrogenase. The reaction involves enzyme-bound FAD (denoted as [FAD]) and a reactive disulfide on the E_3 subunit and results in the net reduction of NAD^+ to NADH. The overall reaction is strongly exergonic, with a $\Delta G°'$ of about -33.5 kJ/mol.

The role of the lipoamide group in this complex process is to carry acetyl groups from TPP on the E_1 subunits to CoA. As was the case with biotin, the extended side chain of the lipoamide group provides the mobility and flexibility to carry the acetyl group between subsites on this multienzyme complex.

A similar multienzyme complex in *E. coli* carries out the oxidative decarboxylation of α-ketoglutarate:

$$\alpha\text{-Ketoglutarate} + \text{CoA} + \text{NAD}^+ \longrightarrow \text{succinyl-CoA} + \text{CO}_2 + \text{NADH} + \text{H}^+$$

The mechanism is similar to that described above for pyruvate dehydrogenase. α-Ketoglutarate dehydrogenase consists of three different subunits, including (a) *α-ketoglutarate dehydrogenase,* (b) *dihydrolipoyl transsuccinylase,* and (c) *dihydrolipoyl dehydrogenase.* The first two are unique to this complex, but the dihydrolipoyl dehydrogenase subunit is identical to that found in the pyruvate dehydrogenase complex.

14.11 Folic Acid

Folates are important acceptors and donors of one-carbon units for all oxidation levels of carbon except that of CO_2 (where biotin is the relevant carrier). The active coenzyme form of folic acid, **tetrahydrofolate (THF),** is formed in two successive reductions of folate by *dihydrofolate reductase* (Figure 14.42). Tetrahydrofolates typically contain from one to seven (or even more) glutamates, all of these in γ-carboxyl amide linkages. One-carbon units in three different oxidation states may be bound to tetrahydrofolate at the N^5 and/or N^{10} nitrogens. As shown in Table 14.5, one-carbon units carried by THF may exist at the oxidation levels of methanol, formaldehyde, or formate (carbon atom oxidation states of −2, 0, and 2, respectively).

An elaborate ensemble of enzymatic reactions serves to introduce one-carbon units into THF and to interconvert the various oxidation states (Figure 14.43). N^5-methyltetrahydrofolate can be oxidized directly to N^5,N^{10}-methylenetetrahydrofolate, which can be further oxidized to N^5,N^{10}-methenyltetrahydrofolate. The N^5-formimino-, N^5-formyl-, and N^{10}-formyltetrahydrofolates can be formed from N^5,N^{10}-methenyltetrahydrofolate (all of these being at the same oxidation level), or they can be formed by one-carbon

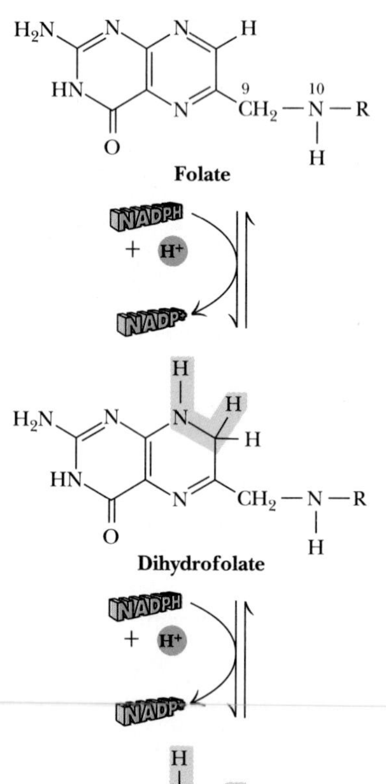

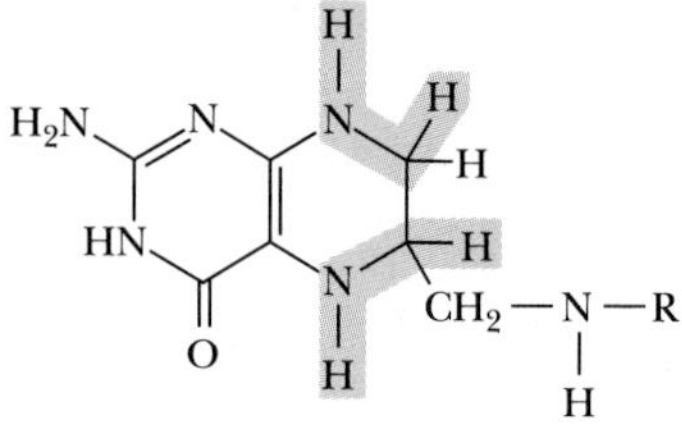

Figure 14.42 The dihydrofolate reductase reaction.

***Table* 14.5**

Oxidation States of Carbon in 1-Carbon Units Carried by Tetrahydrofolate

Oxidation Number*	Oxidation Level	One-Carbon Form	Tetrahydrofolate Form
−2	Methanol (most reduced)	$-CH_3$	N^5-Methyl-THF
0	Formaldehyde	$-CH_2-$	N^5,N^{10}-Methylene-THF
2	Formate (most oxidized)	$-CH{=}O$	N^5-Formyl-THF
		$-CH{=}O$	N^{10}-Formyl-THF
		$-CH{=}NH$	N^5-Formimino-THF
		$-CH{=}$	N^5,N^{10}-Methenyl-THF

*Calculated by assigning valence bond electrons to the more electronegative atom and then counting the charge on the quasi ion. A carbon assigned four valence electrons would have an oxidation number of 0. The carbon in N^5-methyl-THF is assigned six electrons from the three C—H bonds and thus has an oxidation number of −2.

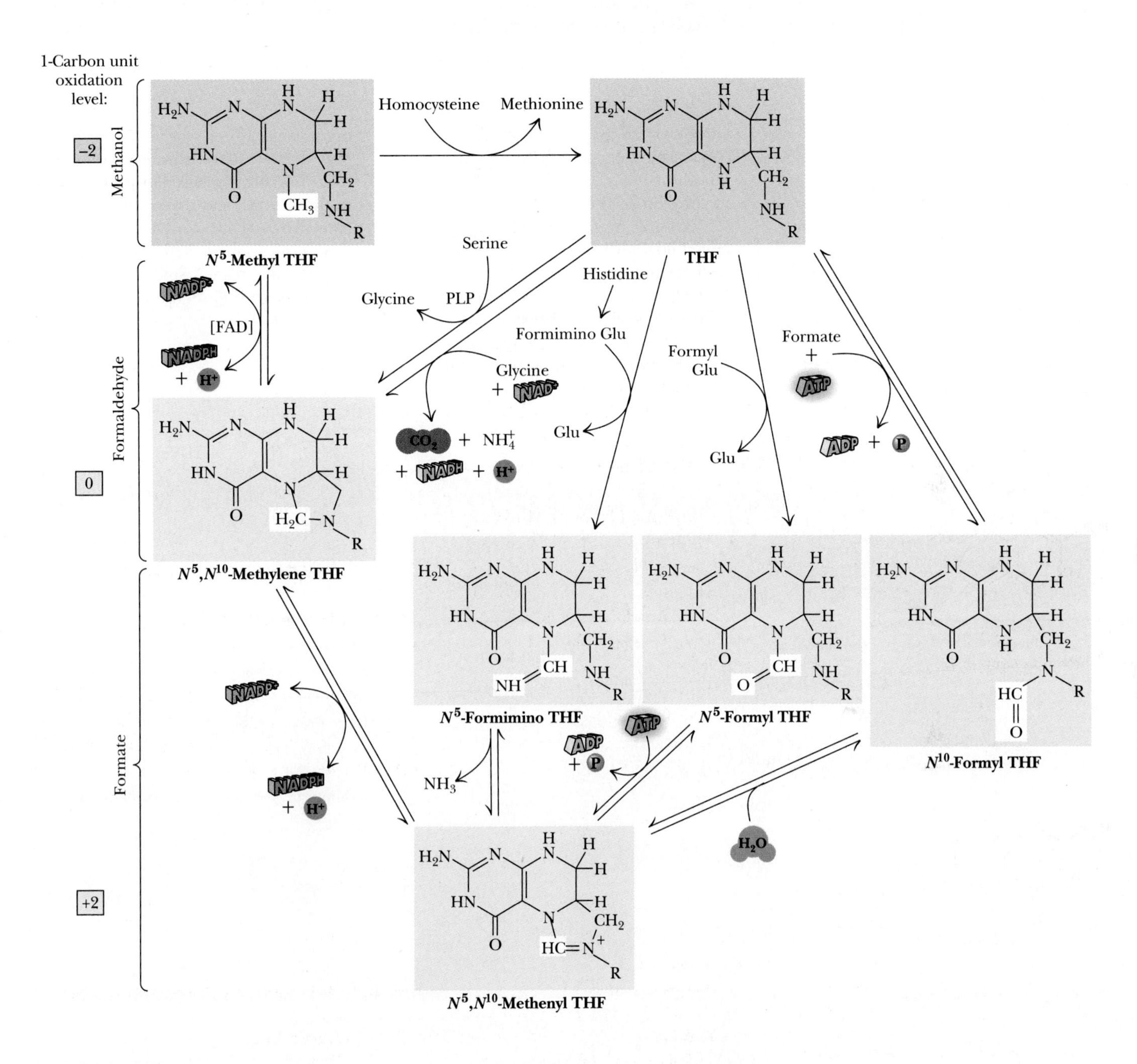

Figure **14.43** The reactions that introduce one-carbon units into tetrahydrofolate (THF) link seven different folate intermediates that carry one-carbon units in three different oxidation states (−2, 0, and +2).

(Adapted from Brody, T., et al., in Machlin, L. J., 1984. Handbook of Vitamins. New York: Marcel Dekker.)

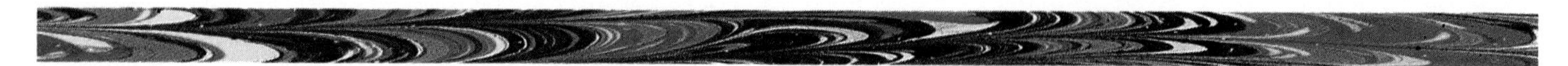

A Deeper Look

Folic Acid, Pterins, and Insect Wings

A number of animal growth factors, antianemia factors, and lactic acid bacteria growth factors were identified in the 1930s from yeast, liver, and spinach, as well as other sources. Eventually, it was found that these factors had common structures based on **folic acid,** first synthesized in 1946. Folic acid is a derivative of the *pteridine* and *pterin* structures, which occur widely in nature, lending color to many insect wings and to the skin of fish and amphibians.

Two of these pterins are familiar to any child who has seen (and probably chased) the ubiquitous yellow sulfur butterfly and its white counterpart, the cabbage butterfly. *Xanthopterin* and *leucopterin* are the respective pigments in these butterflies' wings. Mammalian organisms cannot synthesize pterins; they derive folates from their diet or from microorganisms active in the intestines. Folic acid derives its name from *folium,* Latin for "leaf." Pterin compounds are named from the Greek πτέρυξ, for "wing," since these substances were first identified in insect wings.

Folic acid

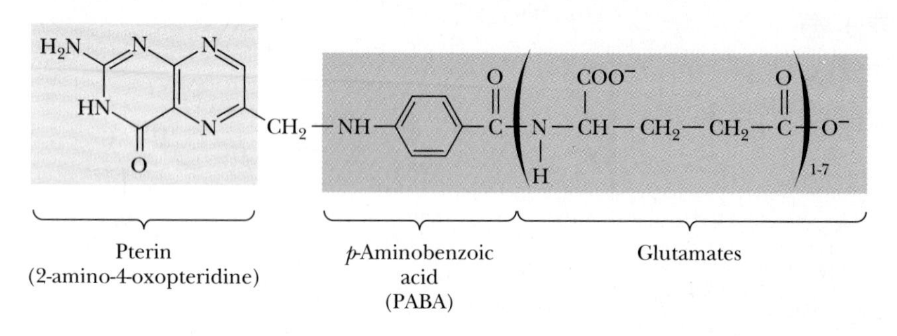

Pteridine

Pterin: 2-amino-4-oxopteridine

Xanthopterin (yellow)

Leucopterin (white)

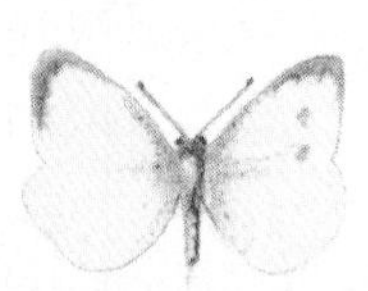

addition reactions from tetrahydrofolate itself. The principal pathway for incorporation of one-carbon units into tetrahydrofolate is the **serine hydroxymethyltransferase** reaction, which converts serine to glycine and forms N^5,N^{10}-methylenetetrahydrofolate. Glycine, histidine, and formate are also sources of one-carbon units, as shown in Figure 14.43.

The biosynthetic pathways for methionine, purines, and the pyrimidine thymine rely on the incorporation of one-carbon units from tetrahydrofolate derivatives (Figure 14.44). The methyl group of methionine is derived from N^5-methyltetrahydrofolate in a vitamin B_{12}-dependent reaction (see above). A methyl group is introduced at the C-5 position of deoxyuridine monophos-

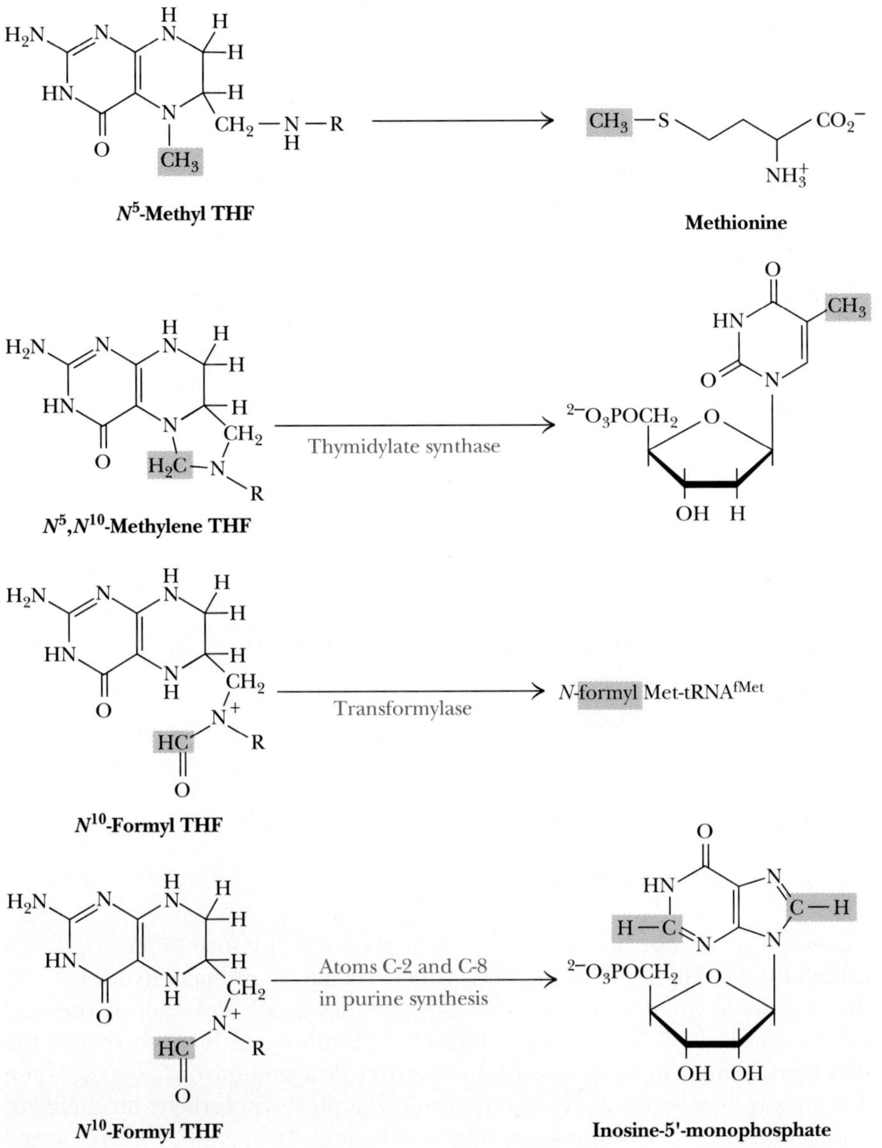

N^5-Methyl THF → **Methionine**

N^5,N^{10}-Methylene THF → Thymidylate synthase

N^{10}-Formyl THF → Transformylase → *N*-formyl Met-tRNA^{fMet}

N^{10}-Formyl THF → Atoms C-2 and C-8 in purine synthesis → **Inosine-5'-monophosphate**

***Figure* 14.44** Some typical one-carbon transfer reactions catalyzed by forms of THF.

phate (dUMP) to form dTMP in the *thymidylate synthase* reaction. N^{10}-formyltetrahydrofolate and N^5,N^{10}-methenyltetrahydrofolate are formyl-group donors for a variety of *transformylase* reactions. The two indicated carbon atoms of purine rings (Figure 14.44) are derived from N^{10}-formyltetrahydrofolate.

Dihydrofolate reductase, which interconverts folate, dihydrofolate, and tetrahydrofolate as noted previously, has received a great deal of attention in recent years because it is the apparent site of action of several anticancer agents, including **amethopterin** (or **methotrexate**) and **aminopterin** (Figure 14.45). These molecules are potent inhibitors of dihydrofolate reductase. Since growing cells require one-carbon THF compounds to synthesize purines and thymine, methotrexate and its analogs are effective blockers of tumor growth. These agents are also toxic to normal cells, and methotrexate can only be used for short-term therapy.

***Figure* 14.45** Inhibitors of dihydrofolate reductase are effective inhibitors of tumor growth.

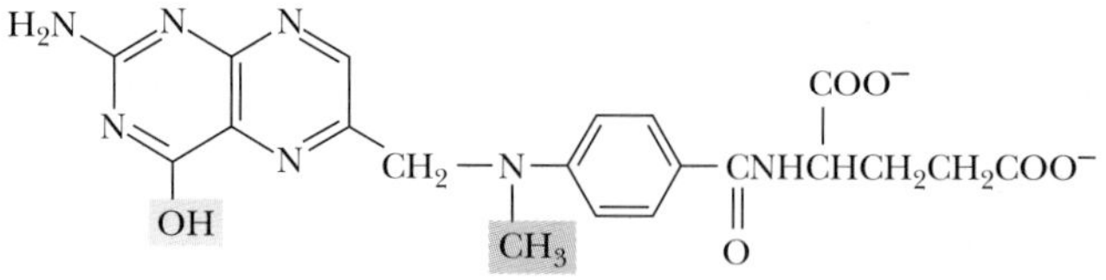

Methopterin

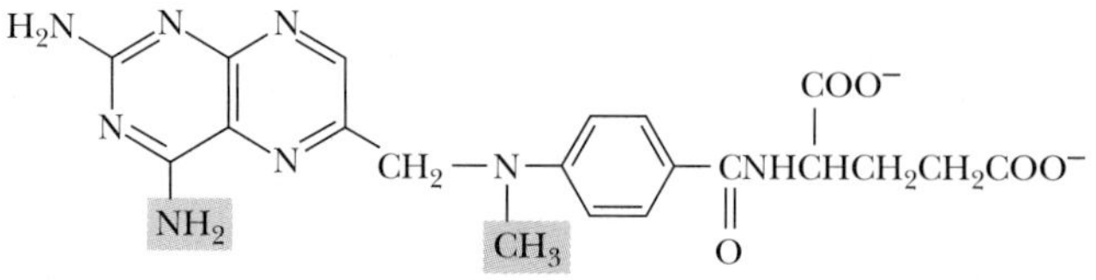

Methotrexate

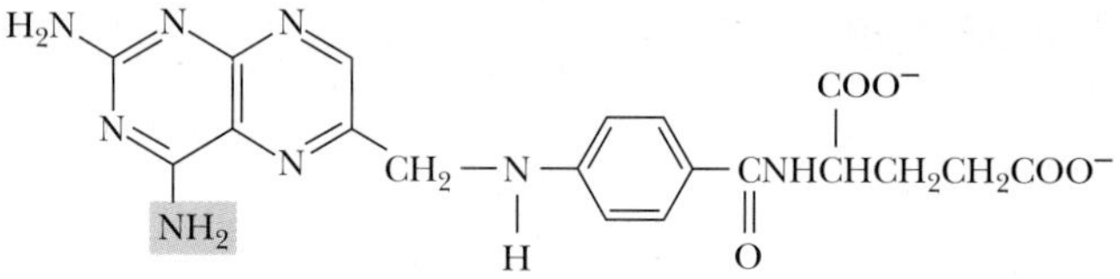

Aminopterin

14.12 The Vitamin A Group

Vitamin A or **retinol** (Figure 14.46) often occurs in the form of esters, called **retinyl esters.** The aldehyde form is called **retinal** or **retinaldehyde.** Like all the fat-soluble vitamins, retinol is an *isoprenoid* molecule and is biosynthesized from isoprene building blocks (Chapter 9). Retinol can be absorbed in the diet from animal sources or synthesized from β-carotene from plant sources. The absorption by the body of fat-soluble vitamins proceeds by mechanisms different from those of the water-soluble vitamins. Once ingested, preformed vitamin A or β-carotene and its analogs are released from proteins by the action of proteolytic enzymes in the stomach and small intestine. The free

A Deeper Look

Mechanism of Methotrexate Inhibition

In recent years, a large number of studies have provided important insights into the mechanism of inhibition of dihydrofolate reductase by methotrexate. Data from X-ray crystallographic studies demonstrate the existence of a salt bridge in the enzyme-methotrexate complex between Asp^{27} of the enzyme and the *N*-4 amino group of methotrexate. This observation has been confirmed by subsequent UV, fluorescence, and NMR studies. The binding of methotrexate to the reductase involves an initial encounter complex followed by isomerization to a tighter complex. Site-specific mutagenesis studies by Stephen Benkovic and his co-workers have shown that hydrophobic interactions between methotrexate and Phe^{31} are at least as important as the (Asp^{27})–(N-4-amino) salt bridge in determining the tight binding of methotrexate to dihydrofolate reductase.

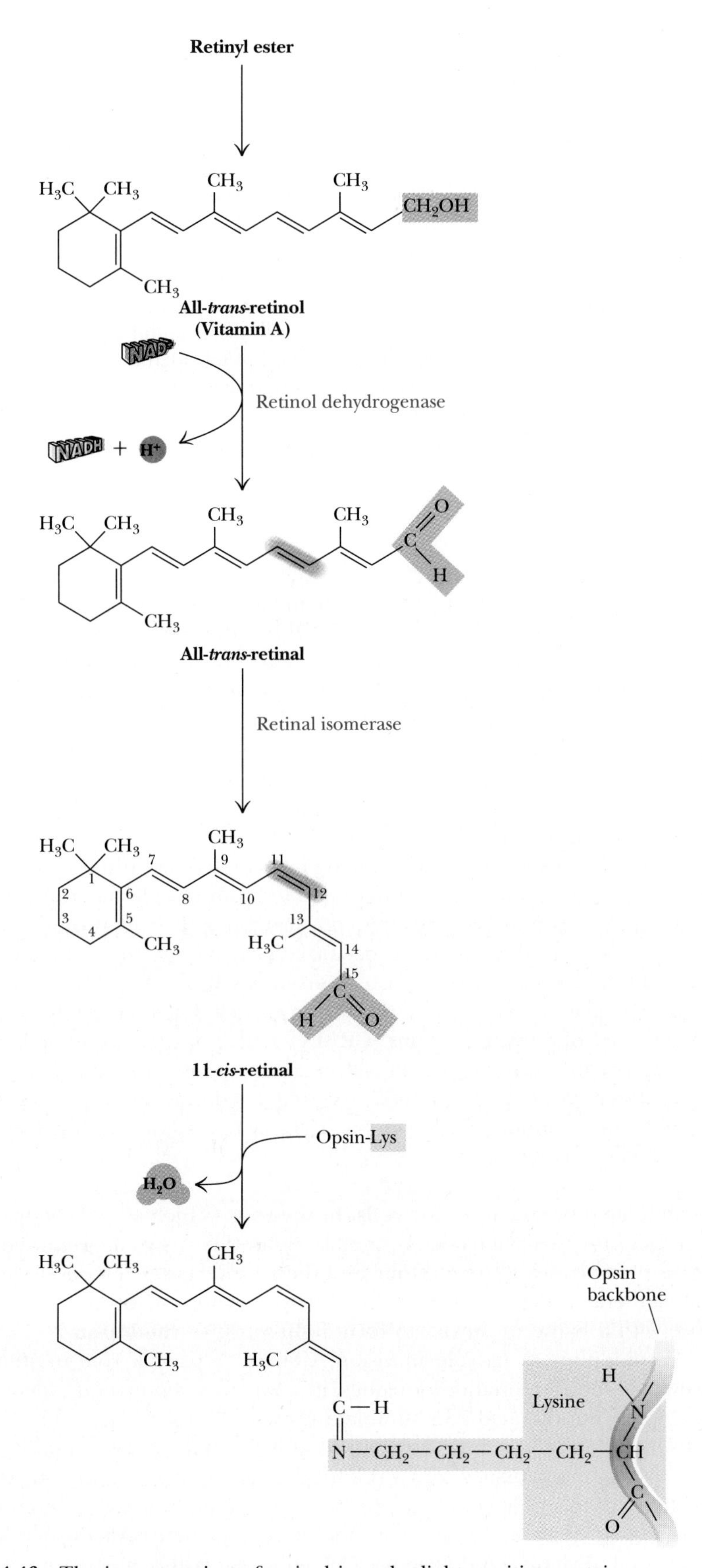

Figure **14.46** The incorporation of retinal into the light-sensitive protein rhodopsin involves several steps. All-*trans*-retinol is oxidized by retinol dehydrogenase and then isomerized to 11-*cis*-retinal, which forms a Schiff base linkage with opsin to form light-sensitive rhodopsin.

A Deeper Look

β-Carotene and Vision

Night blindness was probably the first disorder to be ascribed to a nutritional deficiency. The ancient Egyptians left records as early as 1500 B.C. of recommendations that the juice squeezed from cooked liver could cure night blindness if applied topically, and the method may have been known much earlier. Frederick Gowland Hopkins, working in England in the early 1900s, found that alcoholic extracts of milk contained a growth-stimulating factor. Marguerite Davis and Elmer McCollum at Wisconsin showed that egg yolk and butter contain a similar growth-stimulating lipid, which, in 1915, they called "fat soluble A." Moore in England showed that **β-carotene,** the plant pigment, could be converted to the colorless form of the liver-derived vitamin. In 1935, George Wald of Harvard showed that *retinene* found in visual pigments of the eye was identical with *retinaldehyde,* a derivative of vitamin A.

carotenoids and retinyl esters aggregate in fatty globules that enter the duodenum. The detergent actions of bile salts break these globules down into small aggregates that can be digested by pancreatic lipase, cholesteryl ester hydrolase, retinyl ester hydrolase, and similar enzymes. The product compounds form *mixed micelles* (see Chapter 9) containing the retinol, carotenoids, and other lipids, which are absorbed into mucosal cells in the upper half of the intestinal tract. Retinol is esterified (usually with palmitic acid) and transported to the liver in a lipoprotein complex.

The mobilization of retinol from storage in the liver is a carefully regulated process. A specific **retinol-binding protein, RBP,** synthesized in the liver, binds all-*trans*-retinol, and it is then secreted into the plasma, to be delivered to peripheral tissues. Target tissues appear to recognize the protein portion of this complex, not the retinol, at specific cell surface receptors. The K_D for the RBP–retinol complex is very small, approximately $5 \times 10^{-12} M$, but at the cell surface, the retinol is released into the target cell. Once inside the target cell, **cellular retinol-binding proteins (CRBPs)** protect retinol from oxidation and deliver it to intracellular sites of action. Further transformations to derivative molecules must act on these retinol-aggregate complexes and not on the free retinol. Since retinol is hydrophobic, it tends to associate with proteins, lipid complexes, or membranes.

The retinol that is delivered to the retinas of the eyes in this manner is accumulated by **rod** and **cone cells.** In the rods (which are the better characterized of the two cell types), retinol is oxidized by a specific **retinol dehydrogenase** to become all-*trans* retinal and then converted to 11-*cis* retinal by **retinal isomerase** (Figure 14.46). The aldehyde group of retinal forms a Schiff base with a lysine on **opsin,** to form light-sensitive **rhodopsin.**

In addition to its role in vision, vitamin A plays a role in stimulating growth and differentiation of tissues, in a way that is not yet understood. For example, retinoic acid can stimulate growth in test animals, but does not substitute for retinal in the visual process. Vitamin A also stimulates RNA synthesis in many tissues, and a retinol derivative functions as a sugar carrier in the synthesis of specific glycoproteins. Another process of differentiation affected by vitamin A is the immune response, many aspects of which are depressed in vitamin A deficiency. Cell adhesion is also affected by vitamin A. When some types of cells are grown in culture in vitamin A-free media, addition of vitamin A restores contact inhibition of growth and enhances adhesion between cells.

14.13 The Vitamin D Group

The two most prominent members of the **vitamin D** family are **ergocalciferol** (known as vitamin D_2) and **cholecalciferol** (vitamin D_3). Cholecalciferol is produced in the skin of animals by the action of ultraviolet light (sunlight, for example) on its precursor molecule, 7-dehydrocholesterol (Figure 14.47). The absorption of light energy induces a photoisomerization via an excited

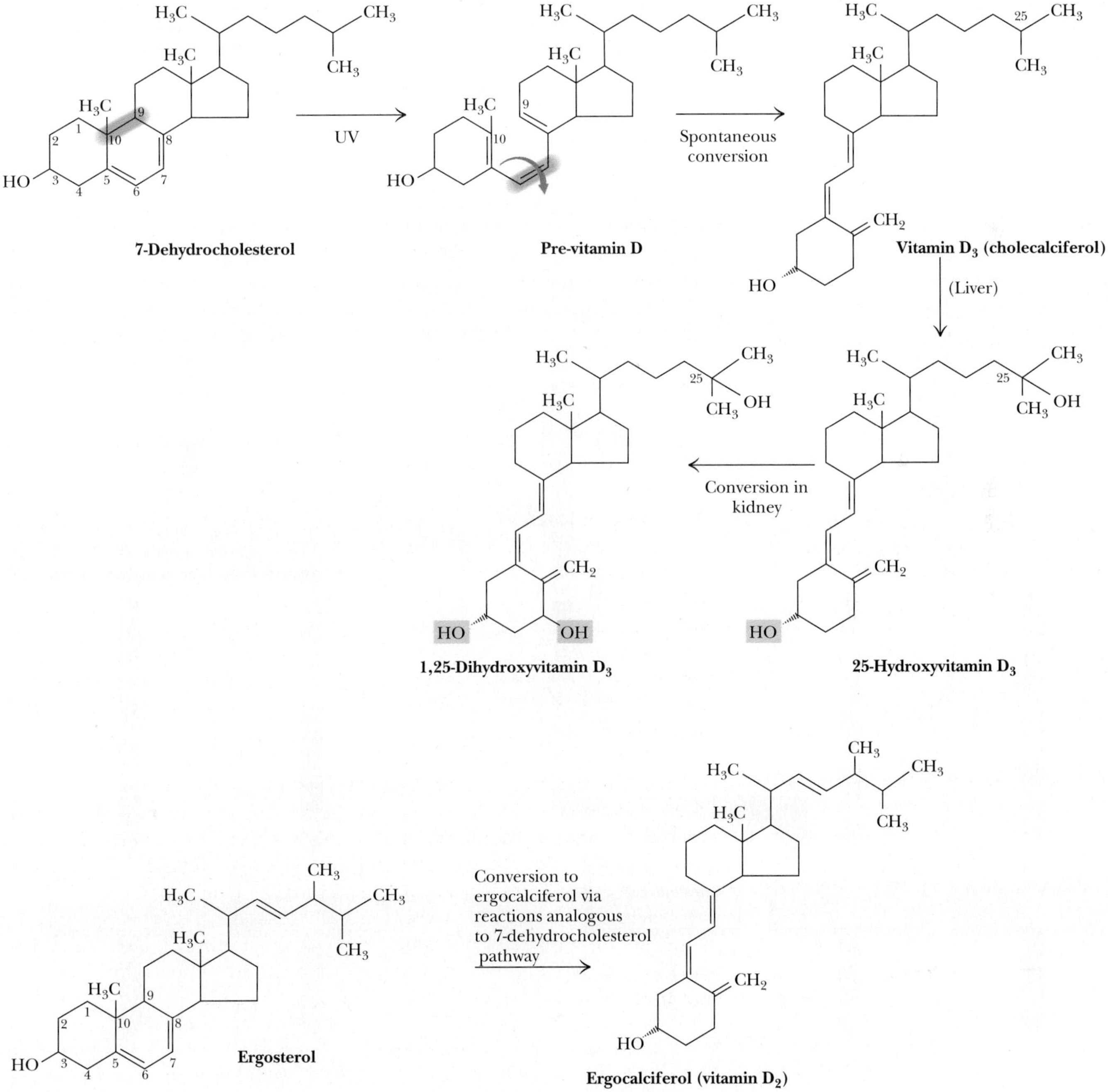

***Figure* 14.47** Vitamin D_3 (cholecalciferol) is produced in the skin by the action of sunlight on 7-dehydrocholesterol. The successive action of mixed-function oxidases in the liver and kidney produces 1,25-dihydroxyvitamin D_3, the active form of vitamin D. Ergocalciferol is produced in analogous fashion from ergosterol.

singlet state, which results in breakage of the 9,10 carbon bond and formation of **previtamin D_3.** The next step is a spontaneous isomerization to yield vitamin D_3, cholecalciferol. Ergocalciferol, which differs from cholecalciferol only in the side-chain structure, is similarly produced by the action of sunlight on the plant sterol **ergosterol.** (Ergosterol is so named because it was first isolated from ergot, a rye fungus.) Since humans can produce vitamin D_3 by the action of sunlight on the skin, "vitamin D" is not strictly speaking a vitamin at all.

On the basis of its mechanism of action in the body, cholecalciferol should be called a **prohormone,** a hormone precursor. Dietary forms of vitamin D are absorbed through the aid of bile salts in the small intestine. Whether absorbed in the intestine or photosynthesized in the skin, cholecalciferol is then transported to the liver by a specific **vitamin D-binding protein**

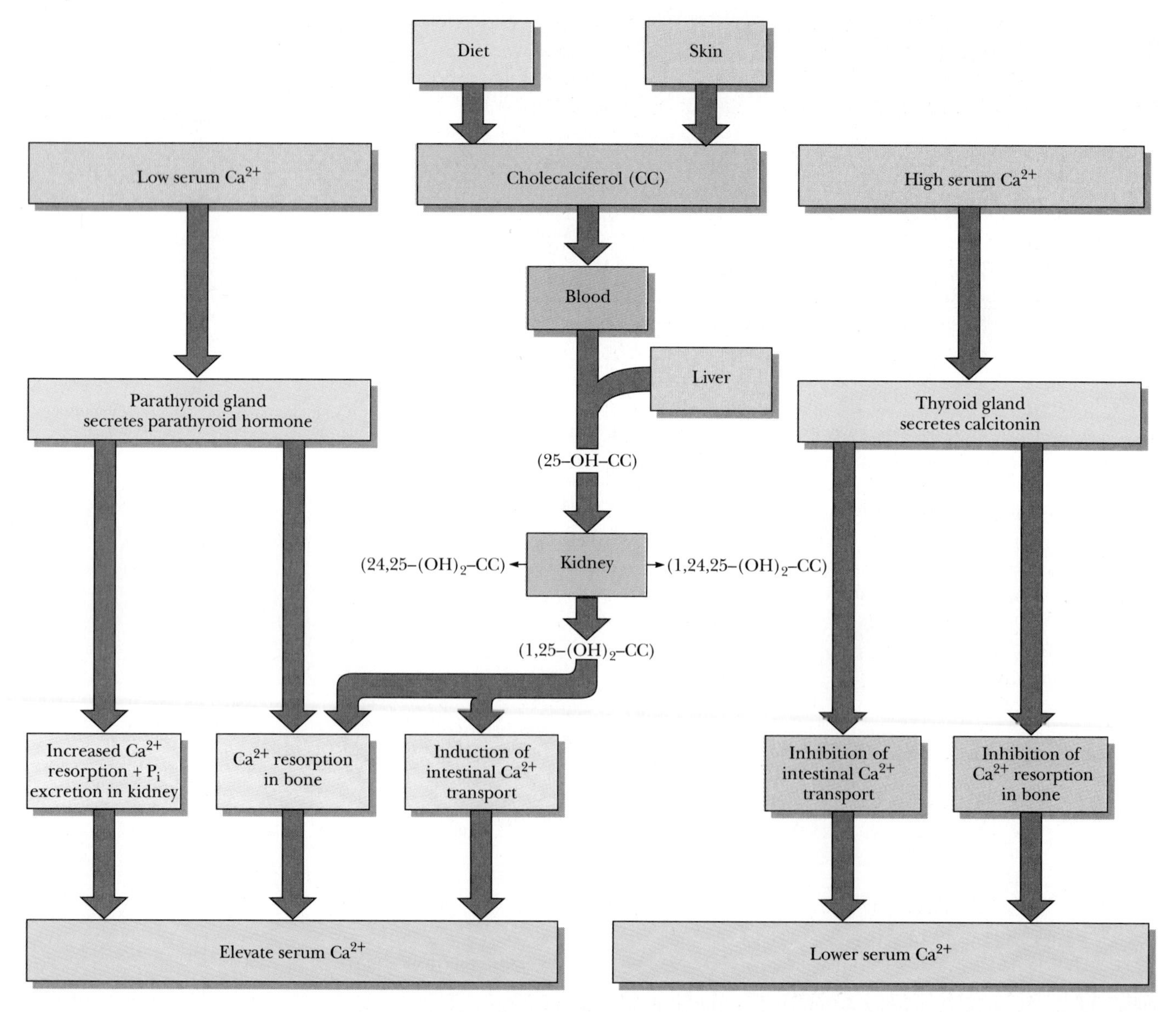

***Figure* 14.48** Calcium homeostasis involves coordination of calcium absorption in the intestine, deposition of calcium in the bones, and excretion of calcium by the kidneys.

A Deeper Look

Vitamin D and Rickets

Vitamin D is a family of closely related molecules that prevent **rickets,** a childhood disease characterized by inadequate intestinal absorption and kidney reabsorption of calcium and phosphate. These inadequacies eventually lead to the demineralization of bones. The symptoms of rickets include bowlegs, knock-knees, curvature of the spine, and pelvic and thoracic deformities, the results of normal mechanical stresses on demineralized bones. Vitamin D deficiency in adults leads to a weakening of bones and cartilage known as **osteomalacia.**

(DBP), also known as **transcalciferin.** In the liver, cholecalciferol is hydroxylated at the C-25 position by a mixed-function oxidase to form *25-hydroxyvitamin D* (that is, *25-hydroxycholecalciferol*). Although this is the major circulating form of vitamin D in the body, 25-hydroxyvitamin D possesses far less biological activity than the final active form. To form this latter species, 25-hydroxyvitamin D is returned to the circulatory system and transported to the kidneys. There it is hydroxylated at the C-1 position by a mitochondrial mixed-function oxidase to form *1,25-dihydroxyvitamin* D_3 (that is, *1,25-dihydroxycholecalciferol*), the active form of vitamin D. 1,25-Dihydroxycholecalciferol is then transported to target tissues where it acts like a hormone to regulate calcium and phosphate metabolism.

1,25-Dihydroxyvitamin D_3, together with two peptide hormones, *calcitonin* and *parathyroid hormone (PTH),* functions to regulate calcium homeostasis and plays a role in phosphorus homeostasis. As described elsewhere in this text, calcium is important for many processes, including muscle contraction, nerve impulse transmission, blood clotting, and membrane structure. Phosphorus, of course, is of critical importance to DNA, RNA, lipids, and many metabolic processes. Phosphorylation of proteins is an important regulatory signal for many biological processes. Phosphorus and calcium are also critically important for the formation of bones. Any disturbance of normal serum phosphorus and calcium levels will result in alterations of bone structure, as in rickets. The mechanism of calcium homeostasis involves precise coordination of calcium (a) absorption in the intestine, (b) deposition in the bones, and (c) excretion by the kidneys (Figure 14.48). If a decrease in serum calcium occurs, vitamin D is converted to its active form, which acts in the intestine to increase calcium absorption. PTH and vitamin D act on bones to enhance absorption of calcium, and PTH acts on the kidney to cause increased calcium reabsorption. If serum calcium levels get too high, calcitonin induces calcium excretion from the kidneys and inhibits calcium mobilization from bone, while inhibiting vitamin D metabolism and PTH secretion.

14.14 Vitamin E: Tocopherol

The structure of **vitamin E** in its most active form, **α-tocopherol,** is shown in Figure 14.49. α-Tocopherol is a potent antioxidant, and its function in animals and humans is often ascribed to this property. On the other hand, the molecular details of its function are almost entirely unknown. One possible role for vitamin E may relate to the protection of unsaturated fatty acids in

Vitamin E (α-tocopherol)

***Figure* 14.49** The structure of vitamin E (α-tocopherol).

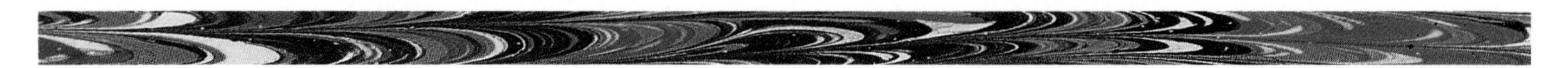

A Deeper Look

Vitamin E

In a study of the effect of nutrition on reproduction in the rat in the 1920s, Herbert Evans and Katherine Bishop found that rats failed to reproduce on a diet of rancid lard, unless lettuce or whole wheat was added to the diet. The essential factor was traced to a vitamin in the wheat germ oil. Named *vitamin E* by Evans (using the next available letter following on the discovery of vitamin D), the factor was purified by Emerson, who named it *tocopherol,* from the Greek *tokos,* for "childbirth," and *pherein,* for "to bring forth." Vitamin E is now recognized as a generic term for a family of substances, all of them similar in structure to the most active form, *α-tocopherol.*

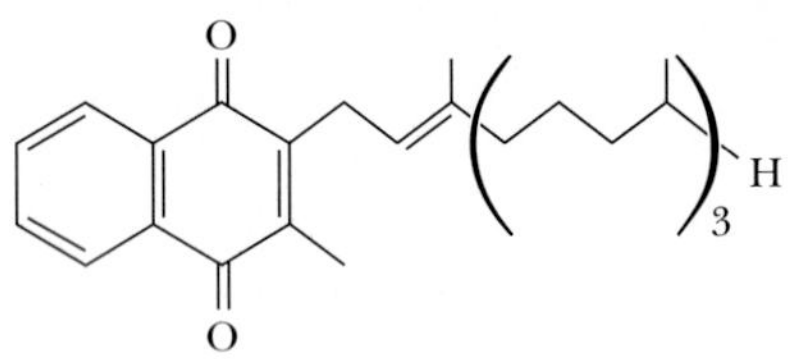

membranes, since these fatty acids are particularly susceptible to oxidation. When human plasma levels of α-tocopherol are low, red blood cells are increasingly subject to oxidative hemolysis. Infants, especially premature infants, are deficient in vitamin E. When low-birth-weight infants are exposed to high oxygen levels, for the purpose of alleviating respiratory distress, the risk of oxygen-induced retina damage can be reduced with vitamin E administration. The mechanism(s) of action of vitamin E remain obscure.

Vitamin K_1
(phylloquinone)

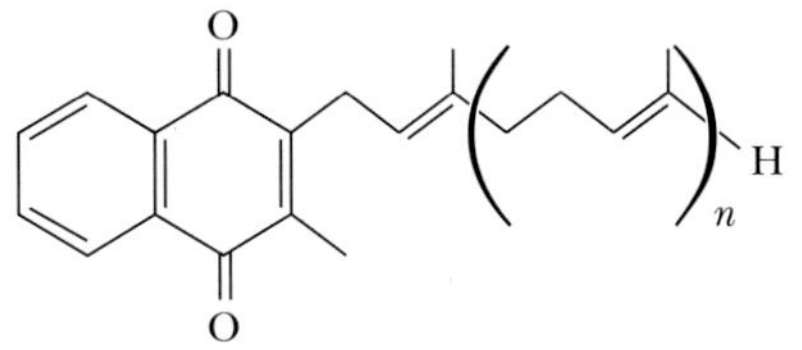

Vitamin K_2
(menaquinone series)

Figure **14.50** The structures of the K vitamins.

14.15 Vitamin K: Naphthoquinone

The function of **vitamin K** (Figure 14.50) in the activation of blood clotting was not elucidated until the early 1970s, when it was found that animals and humans treated with coumarin-type anticoagulants contained an inactive form of **prothrombin** (an essential protein in the coagulation cascade). It was soon shown that a post-translational modification of prothrombin is essential to its structure and function. In this modification, 10 glutamic acid residues

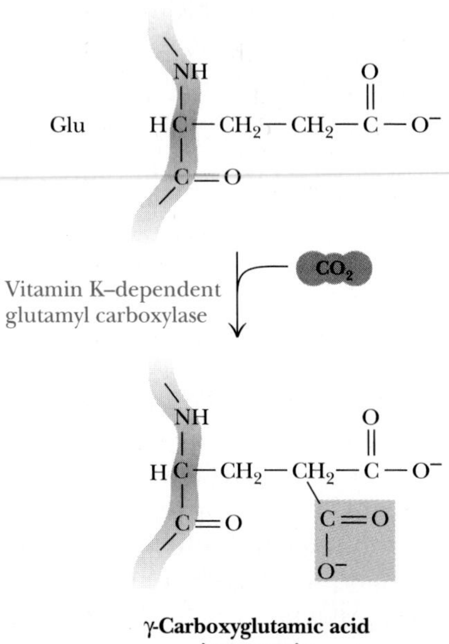

Glu

Vitamin K–dependent glutamyl carboxylase

CO_2

γ-Carboxyglutamic acid in a protein

Figure **14.51** The glutamyl carboxylase reaction is vitamin K-dependent. This enzyme activity is essential for the formation of γ-carboxyglutamyl residues in several proteins of the blood-clotting cascade (Figure 12.5), accounting for the vitamin K dependence of coagulation.

A Deeper Look

Vitamin K and Blood Clotting

In studies in Denmark in the 1920s, Henrik Dam noticed that chicks fed a diet extracted with nonpolar solvents developed hemorrhages. Moreover, blood taken from such animals clotted slowly. Further studies by Dam led him to conclude in 1935 that the antihemorrhagic factor was a new fat-soluble vitamin, which he called *vitamin K* (from *koagulering,* the Danish word for "coagulation"). Dam, along with Karrar of Zurich, isolated the pure vitamin from alfalfa as a yellow oil. Another form, which was crystalline at room temperature, was soon isolated from fish meal. These two compounds were named *vitamins* K_1 and K_2. Vitamin K_2 can actually occur as a family of structures with different chain lengths at the C-3 position.

on the amino terminal end of prothrombin are carboxylated to form γ-carboxyglutamyl residues. These residues are effective in the coordination of calcium, which is required for the coagulation process. The enzyme responsible for this modification, a liver microsomal *glutamyl carboxylase,* requires vitamin K for its activity (Figure 14.51). Not only prothrombin (called "factor II" in the clotting pathway) but also clotting factors VII, IX, and X and several plasma proteins—proteins C, M, S, and Z—contain γ-carboxyglutamyl residues in a manner similar to prothrombin. Other examples of γ-carboxyglutamyl residues in proteins are known.

Problems

1. The enzyme methionase catalyzes the following reaction:

$$H_3C{-}S{-}CH_2CH_2\underset{\displaystyle NH_3^+}{\underset{|}{CH}}{-}COO^- + H_2O \longrightarrow$$

$$H_3C{-}SH + H_3C{-}CH_2{-}\underset{\displaystyle O}{\underset{\|}{C}}{-}COO^- + NH_4^+$$

Suggest a suitable coenzyme for this reaction and write a reasonable mechanism.

2. Design a set of experiments with radioisotopically labeled substrates to demonstrate that the transamination reactions catalyzed by PLP-dependent enzymes occur via a two-step mechanism.

3. Briefly describe the catalytic functions of each of the several coenzymes involved in the pyruvate dehydrogenase reaction.

4. Glyoxylate carboligase catalyzes the reaction shown:

$$H{-}\underset{\displaystyle O}{\underset{\|}{C}}{-}COO^- + H{-}\underset{\displaystyle O}{\underset{\|}{C}}{-}COO^- \longrightarrow$$

$$H{-}\underset{\displaystyle O}{\underset{\|}{C}}{-}\overset{\displaystyle H}{\overset{|}{\underset{\displaystyle OH}{\underset{|}{C}}}}{-}COO^- + CO_2$$

Choose an appropriate coenzyme and write a mechanism for this reaction.

5. Choose a suitable coenzyme and write a valid mechanism for the glutamate mutase reaction:

$$^-OOC{-}CH_2{-}CH_2{-}\underset{\displaystyle NH_3^+}{\underset{|}{CH}}{-}COO^- \longrightarrow$$

$$^-OOC{-}\overset{\displaystyle CH_3}{\overset{|}{\underset{\displaystyle H}{\underset{|}{C}}}}{-}\underset{\displaystyle NH_3^+}{\underset{|}{CH}}{-}COO^-$$

6. The glutamate dehydrogenase reaction is dependent upon the presence of either NAD^+ or $NADP^+$ and is stimulated by the presence of ADP. Suggest roles for $NAD(P)^+$ and ADP and write an appropriate mechanism.

7. Maple syrup urine disease (MSUD) is an autosomal recessive genetic disease characterized by progressive neurological dysfunction and a sweet, burnt-sugar or maple-syrup smell in the urine. Affected individuals carry high levels of branched-chain amino acids (leucine, isoleucine, and valine) and their respective branched-chain α-keto acids in cells and body fluids. The genetic defect has been traced to the mitochondrial branched-chain α-keto acid dehydrogenase (BCKD). Affected individuals exhibit mutations in their BCKD, but these mutant enzymes exhibit normal levels of activity. Nonetheless, treatment of MSUD patients with substantial doses of thiamine can alleviate the symptoms of the disease. Suggest an explanation for the symptoms described and for the role of thiamine in ameliorating the symptoms of MSUD.

8. Choose an appropriate coenzyme and write a suitable mechanism for aspartate-β-decarboxylase, which converts L-aspartate to L-alanine.

9. Alanine racemase converts L-alanine to a racemic mixture of D- and L-alanine. Suggest an appropriate coenzyme for this reaction and write a suitable mechanism.

10. The serine hydroxymethylase reaction converts serine to formaldehyde and glycine:

$$HO\text{—}CH_2\text{—}\underset{NH_3^+}{\underset{|}{C}H}\text{—}COO^- \longrightarrow \underset{H\quad H}{\overset{O}{\overset{\|}{C}}} + H_2\underset{NH_3^+}{\underset{|}{C}}\text{—}COO^-$$

Suggest a suitable coenzyme for this reaction and write an appropriate mechanism. An interesting twist on this problem is the fact that the cleavage of serine does not occur at significant rates unless tetrahydrofolate is present. The THF may act simply as a trap for formaldehyde in this reaction. Several other possible roles for THF have been ruled out by labeling experiments (see Walsh, C. T., 1979. *Enzymatic Reaction Mechanisms.* San Francisco: W. H. Freeman).

Further Reading

Boyer, P. D., 1970. *The Enzymes,* 3rd ed. New York: Academic Press.

——1970. *The Enzymes,* Vol. 6. New York: Academic Press. See discussion of carboxylation and decarboxylation involving TPP, PLP, lipoic acid, and biotin; B_{12}-dependent mutases.

——1972. *The Enzymes,* Vol. 7. New York: Academic Press. See especially eliminations involving PLP.

——1974. *The Enzymes,* Vol. 10. New York: Academic Press. See discussion of pyridine nucleotide-dependent enzymes.

——1976. *The Enzymes,* Vol. 13. New York: Academic Press. See discussion of flavin-dependent enzymes.

DeLuca, H., and Schnoes, H., 1983. Vitamin D: Recent advances. *Annual Review of Biochemistry* **52:**411–439.

Jencks, W. P., 1969. *Catalysis in Chemistry and Enzymology.* New York: McGraw-Hill.

Knowles, J. R., 1989. The mechanism of biotin-dependent enzymes. *Annual Review of Biochemistry* **58:**195–221.

Page, M. I., and Williams, A., eds., 1987. *Enzyme Mechanisms.* London: Royal Society of London.

Reed, L., 1974. Multienzyme complexes. *Accounts of Chemical Research* **7:**40–46.

Walsh, C. T., 1979. *Enzymatic Reaction Mechanisms.* San Francisco: W. H. Freeman.

Wood, H., and Barden, R., 1977. Biotin enzymes. *Annual Review of Biochemistry* **46:**385–414.

Chapter 15

Thermodynamics of Biological Systems

"A theory is the more impressive the greater is the simplicity of its premises, the more different are the kinds of things it relates and the more extended is its range of applicability. Therefore, the deep impression which classical thermodynamics made upon me. It is the only physical theory of universal content which I am convinced, that within the framework of applicability of its basic concepts, will never be overthrown."

Albert Einstein

Sun emblem of Louis XIV on a gate at Versailles. The sun is the prime source of energy for life, and thermodynamics is the gateway to understanding metabolism.

Outline

15.1 Basic Thermodynamic Concepts

15.2 The First Law: Heat, Work, and Other Forms of Energy

15.3 Enthalpy: A More Useful Function for Biological Systems

15.4 The Second Law and Entropy: An Orderly Way of Thinking About Disorder

15.5 The Third Law: Why Is "Absolute Zero" So Important?

15.6 Free Energy: A Hypothetical but Useful Device

15.7 The Physical Significance of Thermodynamic Properties

15.8 The Effect of pH on Standard-State Free Energies

15.9 The Important Effect of Concentration on Net Free Energy Changes

15.10 Irreversible Thermodynamics—Life in the Nonequilibrium Lane

15.11 The Importance of Coupled Processes in Living Things

In previous chapters, the structure and chemistry of biological molecules were discussed. We are now poised and ready to explore metabolism and bioenergetics. To do this requires familiarity with **thermodynamics,** a collection of laws and principles describing the flows and interchanges of heat, energy, and matter in systems of interest. Thermodynamics also allows us to determine whether or not chemical processes and reactions will occur spontaneously. The student should appreciate the power and practical value of thermodynamic reasoning and realize that this is well worth the effort needed to understand it.

Even the most complicated aspects of thermodynamics are based ultimately on three rather simple and straightforward laws. These laws and their extensions sometimes run counter to our intuition. However, once truly understood, the basic principles of thermodynamics become powerful devices for sorting out complicated chemical and biochemical problems. At this milestone in our scientific development, thermodynamic thinking becomes an enjoyable and satisfying activity.

Several basic thermodynamic principles are presented in this chapter, including the analysis of heat flow, entropy production, and free energy functions and the relationship between entropy and information. In addition, some ancillary concepts are considered, including the concept of standard states, the effect of pH on standard-state free energies, the effect of concentration on the net free energy change of a reaction, and the importance of coupled processes in living things.

15.1 Basic Thermodynamic Concepts

In any consideration of thermodynamics, a distinction must be made between the system and the surroundings. The **system** is that portion of the *universe* with which we are concerned, while the **surroundings** include everything else in the universe (Figure 15.1). The nature of the system must also be specified. There are three basic systems: closed, isolated, and open. An **isolated system** cannot exchange matter or energy with its surroundings. A **closed system** may exchange energy, but not matter, with the surroundings. An **open system** may exchange matter, energy, or both with the surroundings. Living things are typically open systems that exchange matter (nutrients and waste products) and heat (from metabolism, for example) with their surroundings.

Isolated system

No exchange of matter or energy

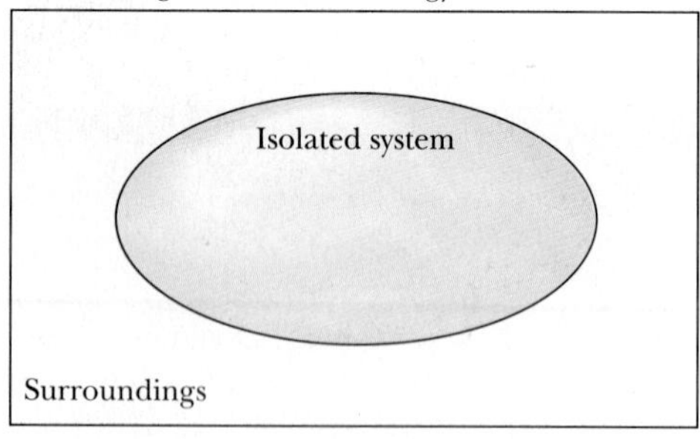

Closed system

Energy exchange may occur

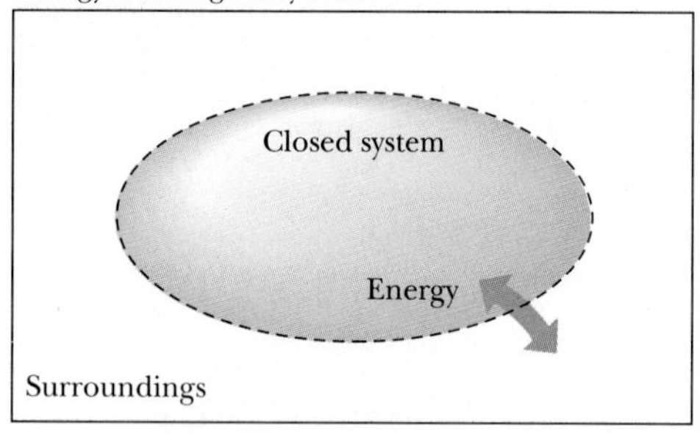

Open system

Energy exchange and/or matter exchange may occur

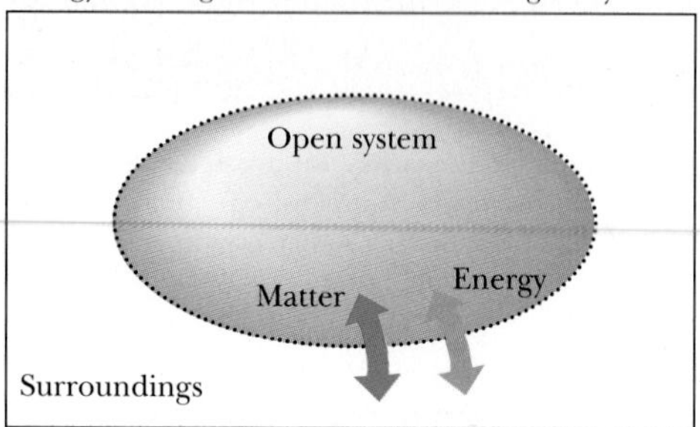

***Figure* 15.1** The characteristics of closed, isolated, and open systems. Closed systems exchange neither matter nor energy with their surroundings. Isolated systems may exchange energy, but not matter, with their surroundings. Open systems may exchange either matter or energy with the surroundings.

15.2 The First Law: Heat, Work, and Other Forms of Energy

It was first realized early in the development of thermodynamics that heat could be converted into other forms of energy, and moreover that all forms of energy could ultimately be converted to some other form. The **first law of thermodynamics** states that *the total energy of an isolated system is conserved.* Thermodynamicists have formulated a mathematical function for keeping track of heat transfers and work expenditures in thermodynamic systems. This function is called the **internal energy,** commonly designated as ***E*** or ***U.*** The internal energy is dependent only on the present state of a system and hence is referred to as a **state function.** The internal energy does not depend on how the system got there and is thus **independent of path.** An extension of this thinking is that we can manipulate the system through any possible pathway of changes, and as long as the system returns to the original state, the internal energy, E, will not have been changed by these manipulations.

The internal energy, E, of any system can change only if energy flows in or out of the system in the form of heat or work. For any process that converts one state (state 1) into another (state 2), the change in internal energy, ΔE, is given as

$$\Delta E = E_2 - E_1 = q + w \tag{15.1}$$

where the quantity q is the *heat absorbed by the system from the surroundings,* and w is the *work done on the system by the surroundings.* **Mechanical work** is defined as *movement through some distance caused by the application of a force.* Both of these must occur for work to have occurred. For example, if a person strains to lift a heavy weight but fails to move the weight at all, then, in the thermodynamic sense, no work has been done. (The energy expended in the muscles of the would-be weight lifter is given off in the form of heat.) In chemical and biochemical systems, work is often concerned with the pressure and volume of the system under study. The mechanical work done on the system is defined as $w = -P\Delta V$, where P is the pressure and ΔV is the volume change and is equal to $V_2 - V_1$. When work is defined in this way, the sign on the right side of Equation (15.1) is positive. (Sometimes w is defined as work done *by* the system; in this case, the equation is $\Delta E = q - w$.) Work may occur in many forms, such as mechanical, electrical, magnetic, and chemical. ΔE, q, and w must all have the same units. The **calorie,** abbreviated **cal,** and **kilocalorie (kcal),** have been traditional choices of chemists and biochemists, but the SI unit, the **joule,** is now recommended.

15.3 Enthalpy: A More Useful Function for Biological Systems

If the definition of work is limited to mechanical work, an interesting simplification is possible. In this case, ΔE is merely the *heat exchanged at constant volume.* This is so because if the volume is constant, no mechanical work can be done on or by the system. Then $\Delta E = q$. Thus ΔE is a very useful quantity in constant volume processes. However, chemical and especially biochemical processes and reactions are much more likely to be carried out at constant pressure. In constant pressure processes, ΔE is not necessarily equal to the heat transferred. For this reason, chemists and biochemists have defined a function that is especially suitable for constant pressure processes. It is called the **enthalpy,** ***H,*** and it is defined as

$$H = E + PV \tag{15.2}$$

The clever nature of this definition is not immediately apparent. However, if the pressure is constant, then we have

$$\Delta H = \Delta E + P\Delta V = q + w + P\Delta V = q - P\Delta V + P\Delta V = q \tag{15.3}$$

Clearly, ΔH is equal to the heat transferred in a constant pressure process. Often, because biochemical reactions normally occur in liquids or solids, rather than in gases, volume changes are small and *enthalpy and internal energy are often essentially equal.*

In order to compare the thermodynamic parameters of different reactions, it is convenient to define a *standard state.* For solutes in a solution, the standard state is normally unit activity (often simplified to 1 *M* concentration). Enthalpy, internal energy, and other thermodynamic quantities are often given or determined for standard-state conditions and are then denoted by a superscript degree sign ("°"), as in $\Delta H°$, $\Delta E°$, and so on.

Enthalpy changes for biochemical processes can be determined experimentally by measuring the heat absorbed (or given off) by the process in a *calorimeter* (Figure 15.2). Alternatively, for any process $A \rightleftharpoons B$ at equilibrium, the standard state enthalpy change for the process can be determined from the temperature dependence of the equilibrium constant:

$$\Delta H° = -R\frac{d(\ln K_{eq})}{d(1/T)} \tag{15.4}$$

Here R is the *gas constant,* defined as $R = 8.314\ \text{J/mol}\cdot\text{K}$. A plot of $R(\ln K_{eq})$ versus $1/T$ is called a **van't Hoff plot.**

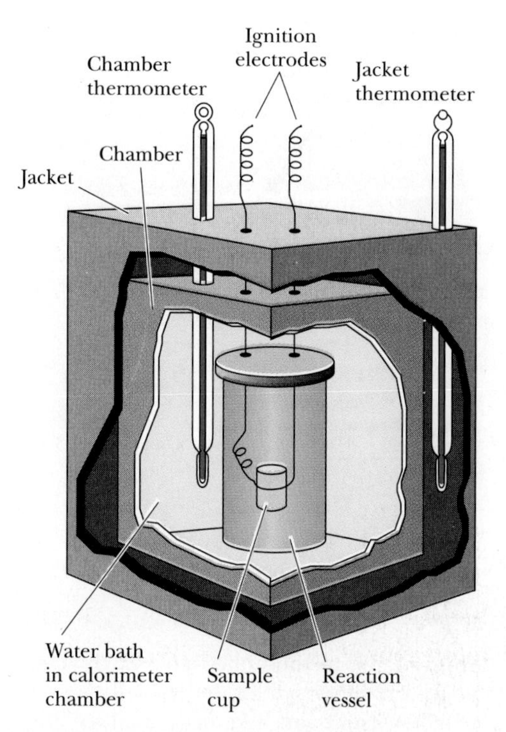

***Figure* 15.2** Diagram of a calorimeter. The reaction vessel is completely submerged in a water bath. The heat evolved by a reaction is determined by measuring the rise in temperature of the water bath.

Example 15.1

In a study[1] of the temperature-induced reversible denaturation of chymotrypsinogen,

$$\text{Native state (N)} \rightleftharpoons \text{denatured state (D)}$$
$$K_{eq} = [D]/[N]$$

[1]Brandts, J. F., 1964. The thermodynamics of protein denaturation. I. The denaturation of chymotrypsinogen. *Journal of the American Chemical Society* **86:**4291–4301.

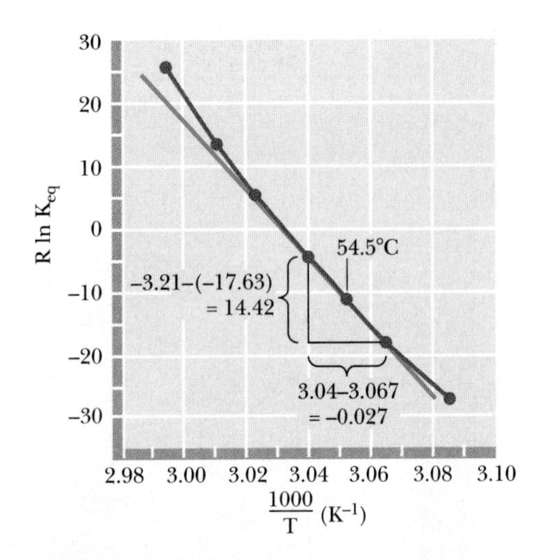

Figure 15.3 The enthalpy change, $\Delta H°$, for a reaction can be determined from the slope of a plot of $R \ln K_{eq}$ versus $\frac{1}{T}$. To illustrate the method, the values of the data points on either side of the 327.5 K (54.5°C) data point have been used to calculate $\Delta H°$ at 54.5°C. Regression analysis would normally be preferable.

(Adapted from Brandts, J. F., 1964. The thermodynamics of protein denaturation. I. The denaturation of chymotrypsinogen. Journal of the American Chemical Society 86:4291–4301.)

***Table* 15.1**

Thermodynamic Parameters for Protein Denaturation

Protein (and conditions)	$\Delta H°$ kJ/mol	$\Delta S°$ kJ/mol · K	$\Delta G°$ kJ/mol	ΔC_p kJ/mol · K
Chymotrypsinogen (pH 3, 25°C)	164	0.440	31	10.9
β-Lactoglobulin (5 *M* urea, pH 3, 25°C)	−88	−0.300	2.5	9.0
Myoglobin (pH 9, 25°C)	180	0.400	57	5.9
Ribonuclease (pH 2.5, 30°C)	240	0.780	3.8	8.4

Adapted from Cantor, C., and Schimmel, P., 1980. *Biophysical Chemistry*. San Francisco: W. H. Freeman, and Tanford, C., 1968. Protein denaturation. *Advances in Protein Chemistry* **23:**121–282.

John F. Brandts measured the equilibrium constants for the denaturation over a range of pH and temperatures. The data for pH 3:

T(K):	324.4	326.1	327.5	329.0	330.7	332.0	333.8
K_{eq}:	0.041	0.12	0.27	0.68	1.9	5.0	21

A plot of $R(\ln K_{eq})$ versus $1/T$ (a van't Hoff plot) is shown in Figure 15.3. $\Delta H°$ for the denaturation process at any temperature is the negative of the slope of the plot at that temperature. As shown, $\Delta H°$ at 54.5°C (327.5 K) is

$$\Delta H° = -[-3.2 - (-17.6)]/[(3.04 - 3.067) \times 10^{-3}] = +533 \text{ kJ/mol}$$

What does this value of $\Delta H°$ mean for the unfolding of the protein? Positive values of $\Delta H°$ would be expected for the breaking of hydrogen bonds as well as for the exposure of hydrophobic side chains from the interior of the native, folded protein during the unfolding process. Such events would raise the energy of the protein–water solution. The magnitude of this enthalpy change (533 kJ/mol) at 54.5°C is large, compared to similar values of $\Delta H°$ for other proteins and for this same protein at 25°C (Table 15.1). If we consider only this positive enthalpy change for the unfolding process, the native, folded state is strongly favored. As we shall see, however, other parameters must be taken into account.

15.4 The Second Law and Entropy: An Orderly Way of Thinking About Disorder

The **second law of thermodynamics** has been described and expressed in many different ways, including the following.

1. Systems tend to proceed from *ordered* (*low entropy* or low probability) states to *disordered* (*high entropy* or *high probability*) states.
2. The *entropy* of the system plus surroundings is unchanged by *reversible processes;* the entropy of the system plus surroundings increases for *irreversible processes.*
3. All naturally occurring processes proceed toward **equilibrium,** that is, to a state of minimum potential energy.

A Deeper Look

Entropy, Information, and the Importance of "Negentropy"

When a thermodynamic system undergoes an increase in entropy, it becomes more disordered. On the other hand, a decrease in entropy reflects an increase in order. A more ordered system is more highly organized and possesses a greater information content. To appreciate the implications of decreasing the entropy of a system, consider the random collection of letters in the figure. This disorganized array of letters possesses no inherent information content, and nothing can be learned by its perusal. On the other hand, this particular array of letters can be systematically arranged to construct the first sentence of the Einstein quotation that opened this chapter: "A theory is the more impressive the greater is the simplicity of its premises, the more different are the kinds of things it relates and the more extended is its range of applicability."

Arranged in this way, this same collection of 151 letters possesses enormous information content—the profound words of a great scientist. Just as it would have required significant effort to rearrange these 151 letters in this way, so large amounts of energy are required to construct and maintain living organisms. Energy input is required to produce information-rich, organized structures such as proteins and nucleic acids. Information content can be thought of as *negative entropy*. In 1945 Erwin Schrödinger took time out from his studies of quantum mechanics to publish a delightful book entitled WHAT IS LIFE? In it, Schrödinger coined the term *negentropy* to describe the negative entropy changes that confer organization and information content to living organisms. Schrödinger pointed out that organisms must "acquire negentropy" to sustain life.

Several of these statements of the second law invoke the concept of **entropy,** which is a measure of disorder and randomness in the system (or the surroundings). An organized or ordered state is a low entropy state, whereas a disordered state is a high entropy state. All else being equal, reactions involving large, positive entropy changes, ΔS, are more likely to occur than reactions with small, positive changes or negative changes in entropy.

Entropy can be defined in several quantitative ways. If W is the number of ways to arrange the components of a system without changing the internal energy or enthalpy (that is, the number of microscopic states at a given temperature, pressure, and amount of material), then the entropy is given by

$$S = k \ln W \tag{15.5}$$

where k is Boltzmann's constant ($k = 1.38 \times 10^{-23}$ J/K). This definition is useful for statistical calculations (it is in fact a foundation of *statistical thermodynamics*), but a more common form relates entropy to the heat transferred in a process:

$$dS_{\text{reversible}} = \frac{dq}{T} \tag{15.6}$$

where $dS_{\text{reversible}}$ is the entropy change of the system in a reversible process, q is the heat transferred, and T is the temperature at which the heat transfer occurs.

15.5 The Third Law: Why Is "Absolute Zero" So Important?

The **third law of thermodynamics** states that the entropy of any crystalline, perfectly ordered substance must approach zero as the temperature approaches 0 K, and at $T = 0$ K *entropy is exactly zero.* Based on this, it is possible to establish a quantitative, absolute entropy scale for any substance as

$$S = \int_0^T C_P d \ln T \qquad (15.7)$$

where C_P is the *heat capacity* at constant pressure. The heat capacity of any substance is the amount of heat one mole of it can store as the temperature of that substance is raised by one degree. For a constant pressure process, this is described mathematically as

$$C_P = \frac{dH}{dT} \qquad (15.8)$$

If the heat capacity can be evaluated at all temperatures between 0 K and the temperature of interest, an absolute entropy can be calculated. For biological processes, *entropy changes* are more useful than absolute entropies. The entropy change for a process can be calculated if the enthalpy change and *free energy change* are known.

15.6 Free Energy: A Hypothetical but Useful Device

An important question for chemists, and particularly for biochemists, is, "Will the reaction proceed in the direction written?" J. Willard Gibbs, one of the founders of thermodynamics, realized that the answer to this question lay in a comparison of the enthalpy change and the entropy change for a reaction at a given temperature. The **Gibbs free energy, *G*,** is defined as

$$G = H - TS \qquad (15.9)$$

For any process $A \rightleftharpoons B$ at constant pressure and temperature, the *free energy change* is given by

$$\Delta G = \Delta H - T\Delta S \qquad (15.10)$$

If ΔG is equal to 0, the process is at *equilibrium,* and there will be no net flow either in the forward or reverse directions. When $\Delta G = 0$, $\Delta S = \Delta H/T$, and the enthalpic and entropic changes are exactly balanced. Any process with a nonzero ΔG will proceed spontaneously to a final state of lower free energy. If ΔG is negative, the process will proceed spontaneously in the direction written. If ΔG is positive, the reaction or process will proceed spontaneously in the reverse direction. (The sign and value of ΔG do not allow us to determine *how fast* the process will go.) If the process has a negative ΔG, it is said to be **exergonic,** while processes with positive ΔG values are **endergonic.**

The Standard-State Free Energy Change

The free energy change, ΔG, for any reaction depends upon the nature of the reactants and products, but it is also affected by the conditions of the reaction, including temperature, pressure, pH, and the concentrations of the reactants and products. As explained above, it is useful to define a standard state for such processes. If the free energy change for a reaction is sensitive to solution conditions, then what is the particular significance of the standard-state free energy change? To answer this question, consider a reaction between two reactants A and B to produce the products C and D.

$$A + B \rightleftharpoons C + D \qquad (15.11)$$

The free energy change for non-standard-state concentrations is given by

$$\Delta G = \Delta G^\circ + RT\ln\frac{[C][D]}{[A][B]} \tag{15.12}$$

At equilibrium, $\Delta G = 0$ and $[C][D]/[A][B] = K_{eq}$. We then have

$$\Delta G^\circ = -RT\ln K_{eq} \tag{15.13}$$

or, in base 10 logarithms,

$$\Delta G^\circ = -2.3RT\log_{10} K_{eq} \tag{15.14}$$

This can be rearranged to

$$K_{eq} = 10^{-\Delta G^\circ/2.3RT} \tag{15.15}$$

In any of these forms, this relationship allows the standard-state free energy change for any process to be determined if the equilibrium constant is known. More importantly, it states that *the equilibrium established for a reaction in solution is a function of the standard-state free energy change for the process.* That is, ΔG° is another way of writing an equilibrium constant.

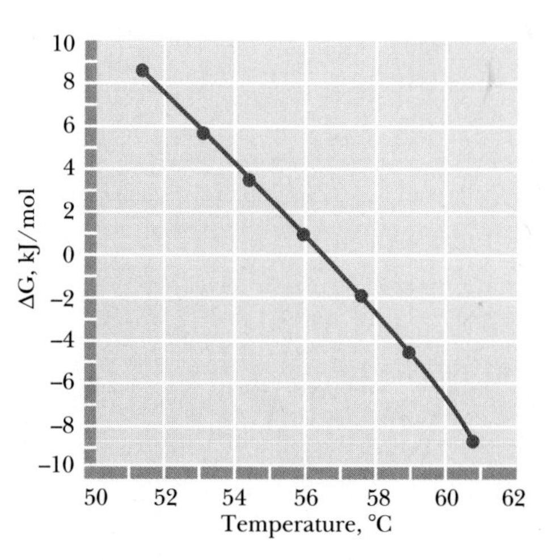

***Figure* 15.4** The dependence of ΔG° on temperature for the denaturation of chymotrypsinogen.

(Adapted from Brandts, J. F., 1964. The thermodynamics of protein denaturation. I. The denaturation of chymotrypsinogen. Journal of the American Chemical Society 86:4291–4301.)

Example 15.2

The equilibrium constants determined by Brandts at several temperatures for the denaturation of chymotrypsinogen (see Example 15.1) can be used to calculate the free energy changes for the denaturation process. For example, the equilibrium constant at 54.5°C is 0.27, so

$$\Delta G^\circ = -(8.314\,\text{J/mol}\cdot\text{K})(327.5\,\text{K})\ln(0.27)$$
$$\Delta G^\circ = -(2.72\,\text{kJ/mol})\ln(0.27)$$
$$\Delta G^\circ = 3.56\,\text{kJ/mol}$$

The positive sign of ΔG° means that the unfolding process is unfavorable, that is, the stable form of the protein at 54.5°C is the folded form. On the other hand, the relatively small magnitude of ΔG° means that the folded form is only slightly favored. Figure 15.4 shows the dependence of ΔG° on temperature for the denaturation data at pH 3 (from the data given in Example 15.1).

Having calculated both ΔH° and ΔG° for the denaturation of chymotrypsinogen, we can also calculate ΔS°, using Equation (15.10):

$$\Delta S^\circ = -\frac{(\Delta G^\circ - \Delta H^\circ)}{T} \tag{15.16}$$

At 54.5°C (327.5 K),

$$\Delta S^\circ = -(3560 - 533{,}000\,\text{J/mol})/327.5\,\text{K}$$
$$\Delta S^\circ = 1{,}620\,\text{J/mol}\cdot\text{K}$$

Figure 15.5 presents the dependence of ΔS° on temperature for chymotrypsinogen denaturation at pH 3. A positive ΔS° indicates that the protein solution has become more disordered as the protein unfolds. Comparison of the value of 1.62 kJ/mol · K with the values of ΔS° in Table 15.1 shows that the present value (for chymotrypsinogen at 54.5°C) is quite large. The physical significance of the thermodynamic parameters for the unfolding of chymotrypsinogen will become clear in the next section.

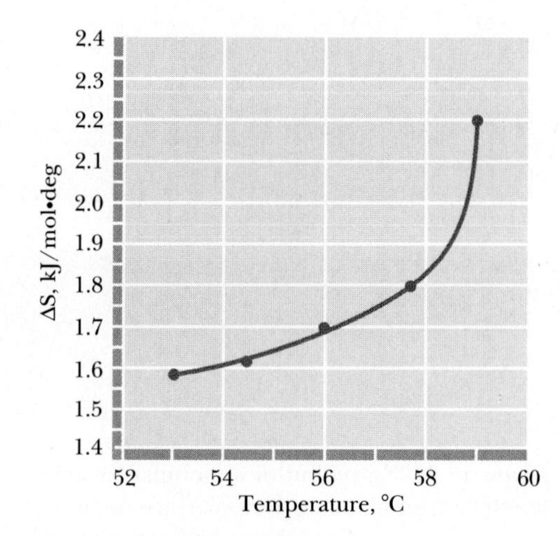

***Figure* 15.5** The dependence of ΔS° on temperature for the denaturation of chymotrypsinogen.

(Adapted from Brandts, J. F., 1964. The thermodynamics of protein denaturation. I. The denaturation of chymotrypsinogen. Journal of the American Chemical Society 86:4291–4301.)

15.7 The Physical Significance of Thermodynamic Properties

What can thermodynamic parameters tell us about biochemical events? The best answer to this question is that a single parameter (ΔH or ΔS, for example) is not very meaningful. A positive $\Delta H°$ for the unfolding of a protein might reflect *either* the breaking of hydrogen bonds within the protein or the exposure of hydrophobic side chains to water (Figure 15.6). However, *comparison of several thermodynamic parameters can provide meaningful insights about a process.* For example, the transfer of Na^+ and Cl^- ions from the gas phase to aqueous solution involves a very large negative $\Delta H°$ (thus a very favorable stabilization of the ions) and a comparatively small $\Delta S°$ (Table 15.2). The negative entropy term reflects the ordering of water molecules in the hydration shells of the Na^+ and Cl^- ions. This unfavorable effect is more than offset by the large heat of hydration, which makes the hydration of ions a very favorable process overall. The negative entropy change for the dissociation of acetic acid in water also reflects the ordering of water molecules in the ion hydration shells. In this case, however, the enthalpy change is much smaller in magnitude. As a result, $\Delta G°$ for dissociation of acetic acid in water is positive, and acetic acid is thus a weak (largely undissociated) acid.

The transfer of a nonpolar hydrocarbon molecule from its pure liquid to water is an appropriate model for the exposure of nonpolar protein side chains to solvent when a globular protein unfolds. The transfer of toluene from liquid toluene to water involves a negative $\Delta S°$, a positive $\Delta G°$, and a $\Delta H°$ that is small compared to $\Delta G°$ (a pattern similar to that observed for the dissociation of acetic acid). *What distinguishes these two very different processes is the change in heat capacity* (Table 15.2). A positive heat capacity change for a process indicates that the molecules have acquired new ways to move (and thus to store heat energy). A negative ΔC_p means that the process has resulted in less freedom of motion for the molecules involved. ΔC_p is negative for the dissociation of acetic acid and positive for the transfer of toluene to water. The explanation is that polar and nonpolar molecules *both* induce organization of nearby water molecules, *but in different ways.* The water molecules near a nonpolar solute are *organized but labile.* Hydrogen bonds formed by water molecules near nonpolar solutes rearrange more rapidly than the hydrogen bonds of pure water. On the other hand, the hydrogen bonds formed between water molecules near an ion are less labile (rearrange more slowly) than they would be in pure water. This means that ΔC_p should be negative for the dissociation of ions in solution, as observed for acetic acid (Table 15.2).

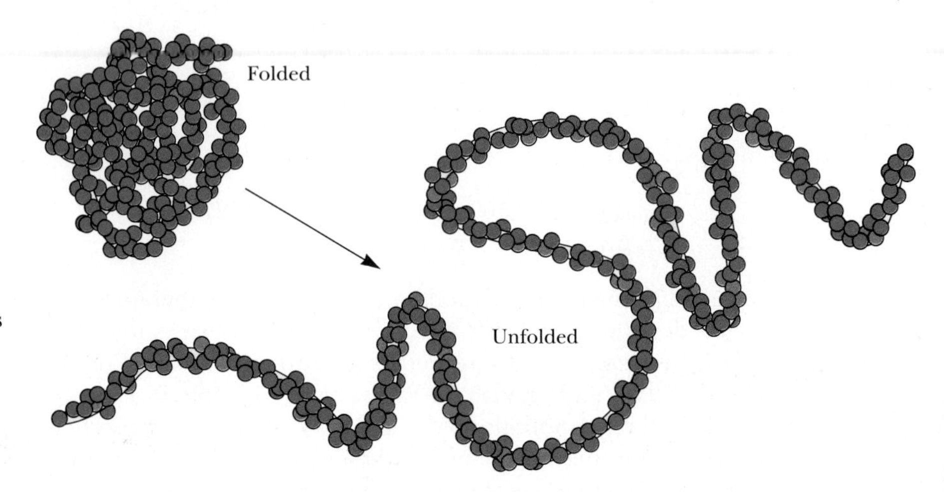

***Figure* 15.6** Unfolding of a globular protein exposes significant numbers of nonpolar groups to water, forcing order on the solvent and resulting in a negative $\Delta S°$ for the unfolding process. Yellow spheres represent nonpolar side chains; blue spheres are polar and/or charged side chains.

Table **15.2**

Thermodynamic Parameters for Several Simple Processes*

Process	$\Delta H°$ kJ/mol	$\Delta S°$ kJ/mol · K	$\Delta G°$ kJ/mol	ΔC_p kJ/mol · K
Hydration of ions†				
$Na^+(g) + Cl^-(g) \longrightarrow Na^+(aq) + Cl^-(aq)$	−760.0	−0.185	−705.0	
Dissociation of ions in solution‡				
$H_2O + CH_3COOH \longrightarrow H_3O^+ + CH_3COO^-$	−10.3	−0.126	27.26	−0.143
Transfer of hydrocarbon from pure liquid to water‡				
Toluene (in pure toluene) $\longrightarrow$ toluene (aqueous)	1.72	−0.071	22.7	0.265

*All data collected for 25°C.

†Berry, R. S., Rice, S. A., and Ross, J., 1980. *Physical Chemistry.* New York: John Wiley.

‡Tanford, C., 1980. *The Hydrophobic Effect.* New York: John Wiley.

15.8 The Effect of pH on Standard-State Free Energies

For biochemical reactions in which hydrogen ions (H^+) are consumed or produced, the usual definition of the standard state is awkward. Standard state for the H^+ ion is 1 *M*, which corresponds to pH 0. At this pH, nearly all enzymes would be denatured, and biological reactions could not occur. It makes more sense to use free energies and equilibrium constants determined at pH 7. Biochemists have thus adopted a modified standard state, designated with prime (′) symbols, as in $\Delta G°'$, K'_{eq}, $\Delta H°'$, and so on. For values determined in this way, a standard state of 10^{-7} *M* H^+ and unit activity (1 *M* for solutions, 1 atm for gases and pure solids defined as unit activity) for all other components (in the ionic forms that exist at pH 7) is assumed. The two standard states can be related easily. For a reaction in which H^+ is produced,

$$A \longrightarrow B^- + H^+ \tag{15.17}$$

the relation of the equilibrium constants for the two standard states is

$$K'_{eq} = K_{eq}[H^+] \tag{15.18}$$

and $\Delta G°'$ is given by

$$\Delta G°' = \Delta G° + RT \ln [H^+] \tag{15.19}$$

For a reaction in which H^+ is consumed,

$$A^- + H^+ \longrightarrow B \tag{15.20}$$

the equilibrium constants are related by

$$K'_{eq} = \frac{K_{eq}}{[H^+]} \tag{15.21}$$

and $\Delta G°'$ is given by

$$\Delta G°' = \Delta G° + RT \ln \left(\frac{1}{[H^+]}\right) = \Delta G° - RT \ln [H^+] \tag{15.22}$$

15.9 The Important Effect of Concentration on Net Free Energy Changes

Equation (15.12) shows that the free energy change for a reaction can be very different from the standard-state value if the concentrations of reactants and products differ significantly from unit activity (1 *M* for solutions). The effects can often be dramatic. Consider the hydrolysis of phosphocreatine:

$$\text{Phosphocreatine} + H_2O \longrightarrow \text{creatine} + P_i \qquad (15.23)$$

This reaction is strongly exergonic and $\Delta G°$ at 37°C is −42.8 kJ/mol. Physiological concentrations of phosphocreatine, creatine, and inorganic phosphate are normally between 1 m*M* and 10 m*M*. Assuming 1 m*M* concentrations, and using Equation (15.12), the ΔG for the hydrolysis of phosphocreatine is

$$\Delta G = -42.8 \text{ kJ/mol} + (8.314 \text{ J/mol} \cdot \text{K})(310 \text{ K}) \ln\left(\frac{[0.001][0.001]}{[0.001]}\right) \qquad (15.24)$$

$$\Delta G = -60.5 \text{ kJ/mol} \qquad (15.25)$$

At 37°C, the difference between standard-state and 1 m*M* concentrations for such a reaction is thus approximately −17.7 kJ/mol.

15.10 Irreversible Thermodynamics—Life in the Nonequilibrium Lane

Classical studies of thermodynamics and the treatments of thermodynamics presented in most undergraduate physical chemistry courses focus primarily on *equilibrium processes*—those with a ΔG of zero. Such approaches are insufficient for the study of living systems, which, by definition, are systems that are *not* at equilibrium. In fact, a suitable thermodynamic description of a living thing would be "a collection of molecules that *maintains* itself in a state far from equilibrium." When such systems reach equilibrium, they are dead. To develop a means of dealing with such systems, we begin by considering Equation (15.10). Any chemical process can be pictured as having available to it the energy embodied in the ΔH term. *This is the free energy that can be obtained from the process if the entropy change is zero.* On the other hand, if the $T\Delta S$ term is positive, some of the available enthalpy will be *dissipated* by the process in the form of entropy change. This entropy change may be viewed as a sum of reversible and irreversible[2] terms:

$$\Delta G = \Delta H - T\Delta S_{\text{reversible}} - T\Delta S_{\text{irreversible}} \qquad (15.26)$$

The last term is zero if the process occurs reversibly, but all reactions occurring in real systems will have an irreversible component. For these reactions, the $T\Delta S_{\text{irreversible}}$ term is always positive. The entropy increase due to process irreversibility represents free energy that is irretrievably lost. The processes by which an organism gives off heat, discharges waste products, and fashions new biomolecules out of old ones reduce the entropy of the organism, so that it can maintain itself in an ordered, highly organized state far from equilibrium.

[2]In the thermodynamic sense, a process is said to be reversible if it can be carried out along a path that keeps the system so close to equilibrium that a microscopic nudge in the opposite direction would "reverse" the process. Reversible processes are maximally efficient, meaning that maximal work is obtained from them as they move along the path from one state to another. Irreversible processes are less efficient, in that less than the maximum work is obtained.

15.11 The Importance of Coupled Processes in Living Things

Many of the reactions necessary to keep cells and organisms alive must run against their **thermodynamic potential,** that is, in the direction of positive ΔG. Among these are the synthesis of adenosine triphosphate and other high-energy molecules (discussed in the next chapter), and the creation of ion gradients in all mammalian cells. These processes are driven in the thermodynamically unfavorable direction via *coupling* with highly favorable processes. Many such *coupled processes* will be discussed later in this text. They are crucially important in intermediary metabolism, oxidative phosphorylation, and membrane transport, as we shall see.

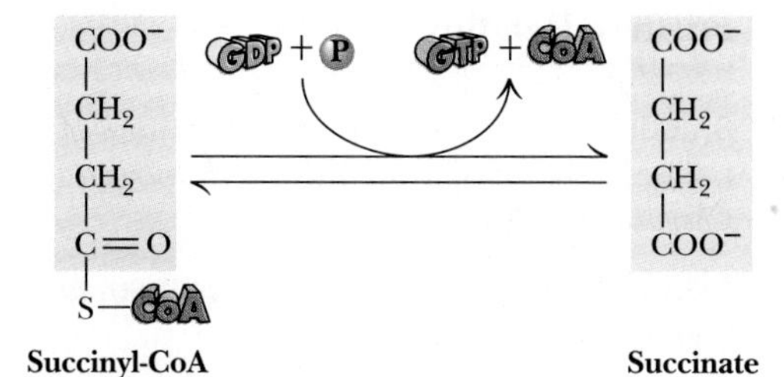

***Figure* 15.7** The succinyl-CoA synthetase reaction.

We can predict whether pairs of coupled reactions will proceed spontaneously by simply summing the free energy changes for each reaction. For example, consider the citric acid cycle reaction (to be discussed in Chapter 19) involving the conversion of succinyl-CoA to succinate (Figure 15.7). The hydrolysis of the thioester bond of succinyl-CoA is energetically very favorable and it is used to drive the synthesis of GTP (or ATP), which is energetically unfavorable.

$$\text{Succinyl-CoA} + H_2O \longrightarrow \text{succinate} + \text{CoA} \quad \Delta G^{\circ\prime} = -37.4 \text{ kJ/mol} \tag{15.27}$$

$$\text{GDP} + P_i \longrightarrow \text{GTP} + H_2O \quad \Delta G^{\circ\prime} = +34 \text{ kJ/mol} \tag{15.28}$$

$$\text{Succinyl-CoA} + \text{GDP} + P_i \longrightarrow \text{succinate} + \text{CoA} + \text{GTP} \quad \text{Total } \Delta G^{\circ\prime} = -3.4 \text{ kJ/mol} \tag{15.29}$$

The net reaction catalyzed by this enzyme depends upon coupling between the two reactions shown in Equations (15.27) and (15.28) to produce the net reaction shown in Equation (15.29) with a net negative $\Delta G^{\circ\prime}$. Many other examples of coupled reactions will be considered in our discussions of intermediary metabolism (Part III). In addition, many of the complex biochemical systems discussed in the later chapters of this text involve reactions and processes with positive $\Delta G^{\circ\prime}$ values that are driven forward by coupling to reactions with a negative $\Delta G^{\circ\prime}$.

Problems

1. An enzymatic hydrolysis of fructose-1-P,

$$\text{Fructose-1-P} + H_2O \rightleftharpoons \text{fructose} + P_i$$

was allowed to proceed to equilibrium at 25°C. The original concentration of fructose-1-P was 0.2 *M*, but when the system had reached equilibrium the concentration of fructose-1-P was only 6.52×10^{-5} *M*. Calculate the equilibrium constant for this reaction and the free energy of hydrolysis of fructose-1-P.

2. The equilibrium constant for some process $A \rightleftharpoons B$ is 0.5 at 20°C and 10 at 30°C. Assuming that ΔH° is independent of temperature, calculate ΔH° for this reaction. Determine ΔG° and ΔS° at 20° and at 30°C. Why is it important in this problem to assume that ΔH° is independent of temperature?

3. The standard-state free energy of hydrolysis for acetyl phosphate is $\Delta G^\circ = -42.3$ kJ/mol.

$$\text{Acetyl-P} + H_2O \longrightarrow \text{acetate} + P_i$$

Calculate the free energy change for acetyl phosphate hydrolysis in a solution of 2 m*M* acetate, 2 m*M* phosphate, and 3 n*M* acetyl phosphate.

4. Define a state function. Name three thermodynamic quantities that are state functions and three that are not.

5. ATP hydrolysis at pH 7.5 is accompanied by release of a hydrogen ion to the medium

$$ATP^{4-} + H_2O \rightleftharpoons ADP^{3-} + HPO_4^{\ 2-} + H^+$$

If the $\Delta G^{\circ\prime}$ for this reaction is -30.5 kJ/mol, what is ΔG° (that is, the free energy change for the same reaction with all components, including H^+, at a standard state of 1 *M*)?

6. For the process $A \rightleftharpoons B$, $K_{eq}(AB)$ is 0.01 at 37°C. For the process $B \rightleftharpoons C$, $K_{eq}(BC) = 1000$ at 37°C.
   a. Determine $K_{eq}(AC)$, the equilibrium constant for the overall process $A \rightleftharpoons C$, from $K_{eq}(AB)$ and $K_{eq}(BC)$.
   b. Determine standard-state free energy changes for all three processes, and use $\Delta G^\circ(AC)$ to determine $K_{eq}(AC)$. Make sure that this value agrees with that determined in part a, above.

Further Reading

Brandts, J. F., 1964. The thermodynamics of protein denaturation. I. The denaturation of chymotrypsinogen. *Journal of the American Chemical Society* **86**:4291–4301.

Cantor, C. R., and Schimmel, P. R., 1980. *Biophysical Chemistry.* San Francisco: W. H. Freeman.

Dickerson, R. E., 1969. *Molecular Thermodynamics.* New York: Benjamin Co.

Edsall, John T., and Gutfreund, H., 1983. *Biothermodynamics: The Study of Biochemical Processes at Equilibrium.* New York: John Wiley.

Edsall, John T., and Wyman, Jeffries, 1958. *Biophysical Chemistry.* New York: Academic Press.

Klotz, I. M., 1967. *Energy Changes in Biochemical Reactions.* New York: Academic Press.

Morris, J. G., 1968. *A Biologist's Physical Chemistry.* Reading, MA: Addison-Wesley.

Patton, A. R., 1965. *Biochemical Energetics and Kinetics.* Philadelphia: Saunders.

Segel, I. H., 1976. *Biochemical Calculations,* 2nd ed. New York: John Wiley.

Schrödinger, E., 1945. *What Is Life?* New York: Macmillan.

Tanford, C., 1980. *The Hydrophobic Effect,* 2nd ed. New York: John Wiley.

"The cell is a machine driven by energy. It can thus be approached by studying matter, or by studying energy."

Albert Szent-Györgyi
Bioelectronics—A Study in Cellular Regulations, Defense, and Cancer (1968)

Chapter 16

ATP and Energy-Rich Compounds

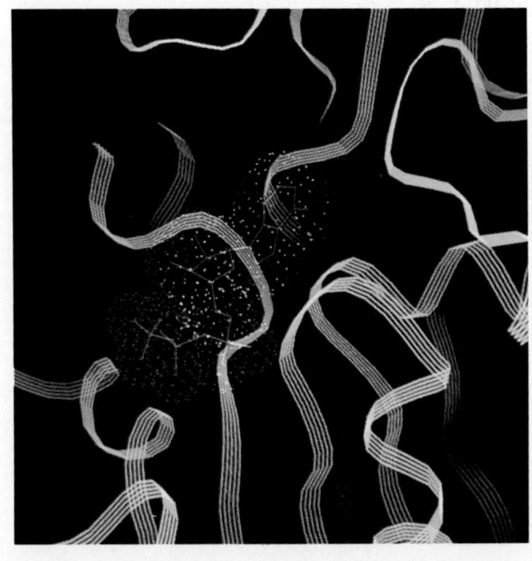

Computer graphic of the ATP-binding site of phosphoglycerate kinase (green ribbon). ATP is depicted as a stick structure with superimposed van der Waals radii (nitrogens in blue and oxygens in red).

Outline

16.1 The High-Energy Biomolecules
16.2 Classes of High-Energy Compounds
16.3 Complex Equilibria Involved in ATP Hydrolysis
16.4 The Effect of Concentration on the Free Energy of Hydrolysis of ATP
16.5 The Daily Human Requirement for ATP

irtually all life on earth depends on energy from the sun. Among life forms, there is a hierarchy of energetics: Certain organisms capture solar energy directly, whereas others derive their energy from this group in subsequent processes. Organisms that absorb light energy directly are called **phototrophic organisms.** These organisms store solar energy in the form of various organic molecules. Organisms that feed on these latter molecules, releasing the stored energy in a series of oxidative reactions, are called **chemotrophic organisms.** Despite these differences, both types of organisms share common mechanisms for generating a useful form of chemical energy. Once captured in chemical form, energy can be released in controlled exergonic reactions to drive a variety of life processes (which require energy). A small family of universal biomolecules mediates the flow of energy from exergonic reactions to the energy-requiring processes of life. These molecules are the *reduced coenzymes* (see Chapter 14) and the *high-energy phosphate compounds.* Phosphate compounds are considered high-energy if they exhibit large negative free energies of hydrolysis (that is, if $\Delta G^{\circ\prime}$ is more negative than -25 kJ/mol).

16.1 The High-Energy Biomolecules

Table 16.1 lists the most important members of the high-energy phosphate compounds. Such molecules include *phosphoric anhydrides* (ATP, ADP), an *enol phosphate* (PEP), an *acyl phosphate* (acetyl phosphate), and a *guanidino phosphate* (creatine phosphate). Also included are thioesters, such as acetyl-CoA, which do not contain phosphorus, but which have a high free energy of hydrolysis. As we saw in Chapter 15, the exact amount of chemical free energy available from the hydrolysis of such compounds will vary depending on concentration, pH, temperature, and so on, but the $\Delta G^{\circ\prime}$ values for hydrolysis of these substances are substantially more negative than for most other metabolic species. Two important points: First, high-energy phosphate compounds are not long-term energy storage substances. They are transient forms of stored energy, meant to carry energy from point to point, from one enzyme system to another, in the minute-to-minute existence of the cell. (As we shall see in subsequent chapters, other molecules bear the responsibility for long-term storage of energy supplies.) Second, the term "high-energy compound" should not be construed to imply that these molecules are unstable and will hydrolyze or decompose unpredictably. ATP, for example, is quite a stable molecule. A substantial activation energy must be delivered to ATP to hydrolyze the terminal, or γ, phosphate group. In fact, as shown in Figure 16.1, the *activation energy* that must be absorbed by the molecule to break the O—P_γ bond is normally 200 to 400 kJ/mol, which is substantially larger than the net 30.5 kJ/mol released in the hydrolysis reaction. Biochemists are much more concerned with the *net release* of 30.5 kJ/mol than with the activation energy for the reaction (since suitable enzymes cope with the latter). The net release of large quantities of free energy distinguishes the high-energy phosphoric anhydrides from their "low-energy" ester cousins, such as glycerol-3-phosphate (Table 16.1). The next section will provide a quantitative framework for understanding these comparisons.

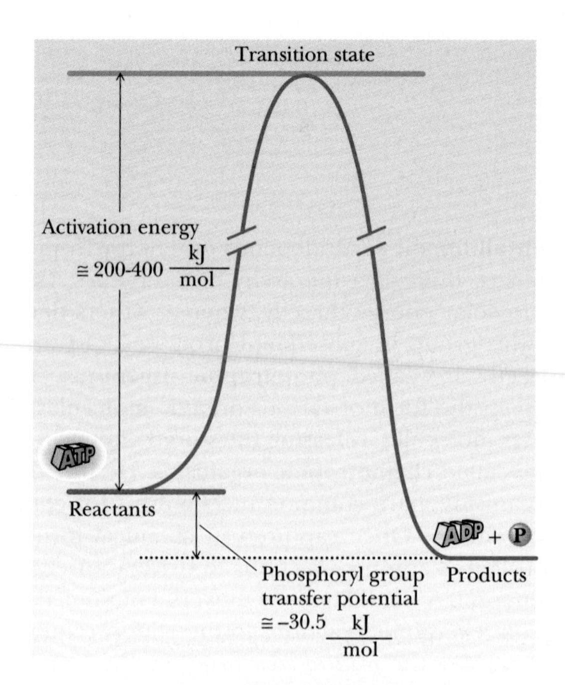

***Figure* 16.1** The activation energies for phosphoryl group-transfer reactions (200 to 400 kJ/mol) are substantially larger than the free energy of hydrolysis of ATP (−30.5 kJ/mol).

***Table* 16.1**
Free Energies of Hydrolysis of Some High-Energy Compounds*

Compound (and Hydrolysis Product)	$\Delta G^{\circ\prime}$ (kJ/mol)	Structure
Phosphoenolpyruvate (pyruvate + P_i)	−62.2	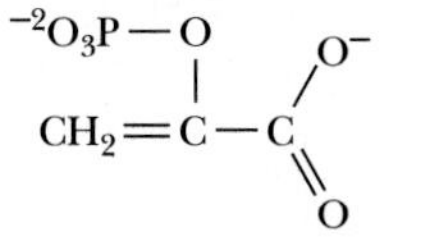 $^{-2}O_3P{-}O$; $CH_2{=}C{-}C(O^-){=}O$
3′,5′-Cyclic adenosine monophosphate (5′-AMP)	−50.4	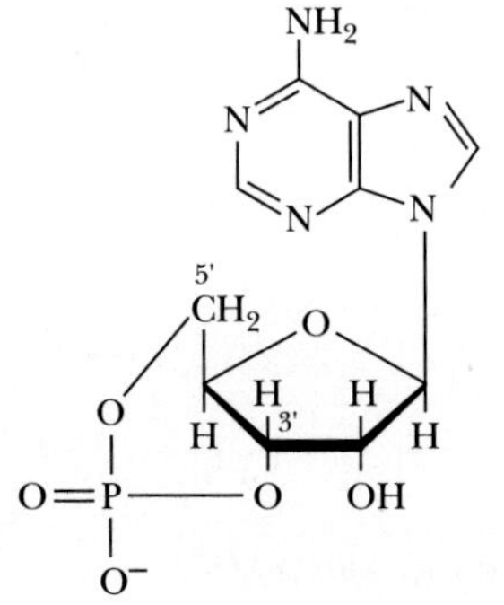
1,3-Bisphosphoglycerate (3-phosphoglycerate + P_i)	−49.6	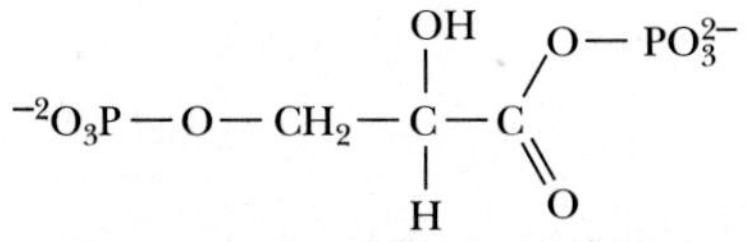 $^{-2}O_3P{-}O{-}CH_2{-}CH(OH){-}C(O{-}PO_3^{2-}){=}O$
Creatine phosphate (creatine + P_i)	−43.3	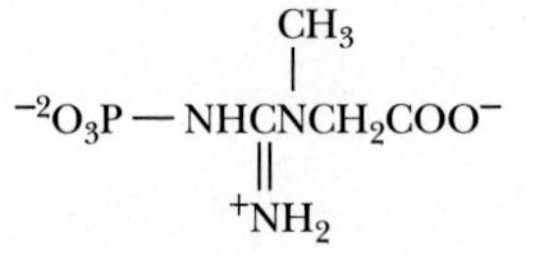 $^{-2}O_3P{-}NHC({=}^{+}NH_2)N(CH_3)CH_2COO^-$
Acetyl phosphate (acetate + P_i)	−43.3	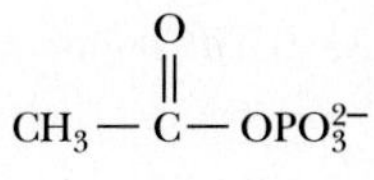 $CH_3{-}C({=}O){-}OPO_3^{2-}$
Adenosine-5′-triphosphate (ADP + P_i)	−35.7†	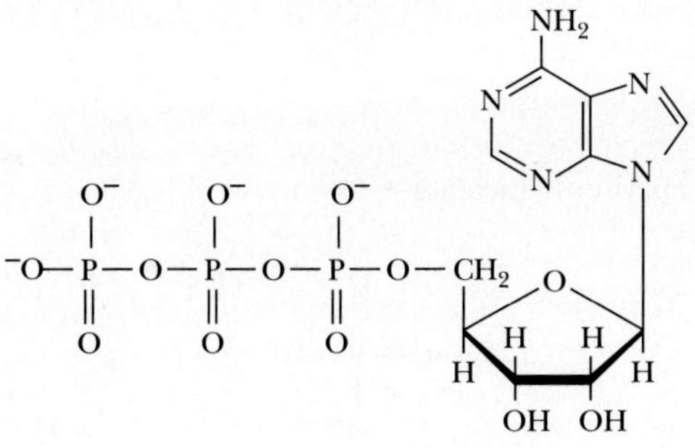
Adenosine-5′-triphosphate (ADP + P_i), excess Mg^{2+}	**−30.5**	

(continued)

***Table* 16.1**
continued

Compound (and Hydrolysis Product)	$\Delta G°'$ (kJ/mol)	Structure
Adenosine-5′-diphosphate (AMP + P_i)	−35.7	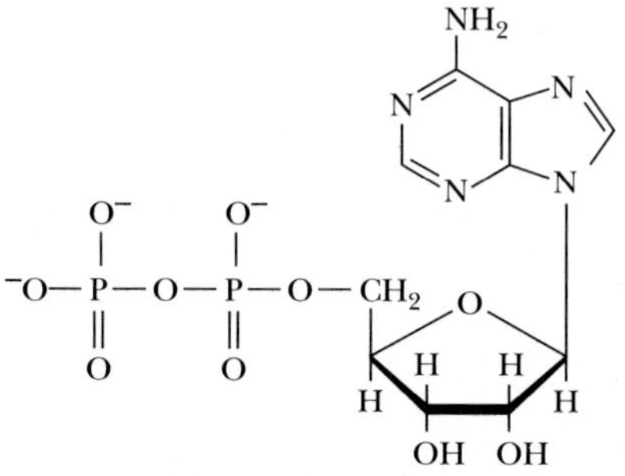
Pyrophosphate (P_i + P_i) in 5 mM Mg^{2+}	−33.6	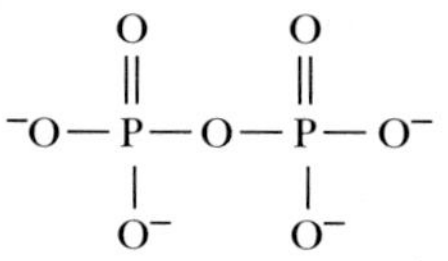
Adenosine-5′-triphosphate (AMP + PP_i), excess Mg^{2+}	−32.3	(See ATP structure on previous page)
Uridine diphosphoglucose (UDP + glucose)	−31.9	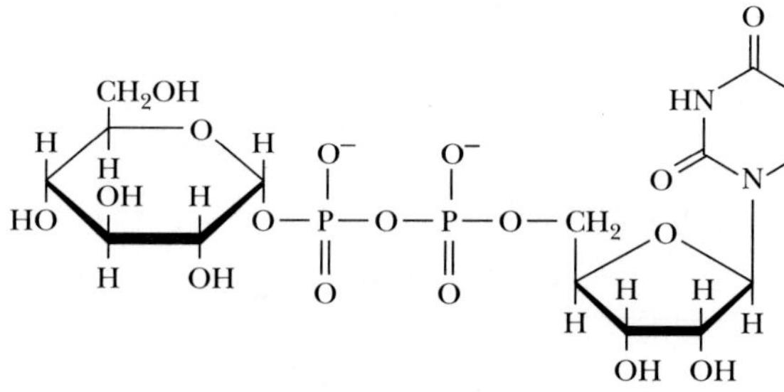
Acetyl-coenzyme A (acetate + CoA)	−31.5	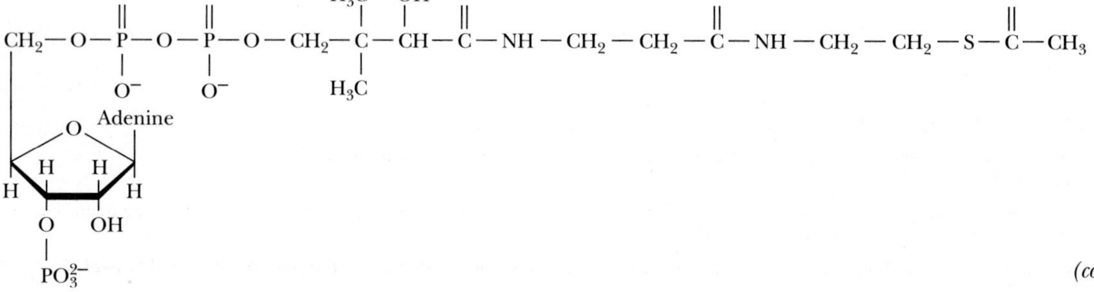

(continued)

***Table* 16.1**
continued

Compound (and Hydrolysis Product)	$\Delta G°'$ (kJ/mol)	Structure
S-adenosyl methionine (methionine + adenosine)	−25.6‡	$^-OOCCH(NH_3^+)CH_2CH_2-\overset{+}{S}(CH_3)-CH_2$-adenosine
Lower-Energy Phosphate Compounds		
Glucose-1-P (glucose + P_i)	−21.0	glucose ring with $O-P(=O)(O^-)_2$ at C-1
Fructose-1-P (fructose + P_i)	−16.0	fructose ring with $CH_2-O-PO_3^{2-}$
Glucose-6-P (glucose + P_i)	−13.9	$^{-2}O_3P-O-CH_2$-glucose ring
sn-Glycerol-3-P (glycerol + P_i)	−9.2	$^{-2}O_3P-O-CH_2-CH(OH)-CH_2OH$
Adenosine-5′-monophosphate (adenosine + P_i)	−9.2	$^-O-P(=O)(O^-)-O-CH_2$-adenosine

*Adapted primarily from *Handbook of Biochemistry and Molecular Biology*, 1976, 3rd ed. In *Physical and Chemical Data*, G. Fasman, ed., Vol. 1, pp. 296–304. Boca Raton, FL: CRC Press.

†From Gwynn, R. W., and Veech, R. L., 1973. The equilibrium constants of the adenosine triphosphate hydrolysis and the adenosine triphosphate-citrate lyase reactions. *Journal of Biological Chemistry* **248**:6966–6972.

‡From Mudd, H., and Mann, J., 1963. Activation of methionine for transmethylation. *Journal of Biological Chemistry* **238**:2164–2170.

ATP Is an Intermediate Energy-Shuttle Molecule

One last point about Table 16.1 deserves mention. Given the central importance of ATP as a high-energy phosphate in biology, students are sometimes surprised to find that ATP holds an intermediate place in the rank of high-energy phosphates. PEP, cyclic-AMP, 1,3-BPG, phosphocreatine, acetyl phosphate, and pyrophosphate all exhibit higher values of $\Delta G°'$. This is not a biological anomaly. ATP is uniquely situated between the very high energy phosphates synthesized in the breakdown of fuel molecules and the numerous lower-energy acceptor molecules that are phosphorylated in the course of further metabolic reactions. ADP can accept both phosphates and energy from the higher-energy phosphates, and the ATP thus formed can donate both phosphates and energy to the lower-energy molecules of metabolism. The ATP/ADP pair is an intermediately placed acceptor/donor system among high-energy phosphates. In this context, ATP functions as a very versatile but intermediate energy shuttle device that interacts with many different energy-coupling enzymes of metabolism.

Group Transfer Potential

Many reactions in biochemistry involve the transfer of a functional group from a donor molecule to a specific receptor molecule or to water. The concept of **group transfer potential** explains the tendency for such reactions to occur. Biochemists define the group transfer potential as the free energy change that occurs upon hydrolysis, that is, upon transfer of the particular group to water. This concept and its terminology are preferable to the more qualitative notion of *high-energy bonds.*

The concept of group transfer potential is not particularly novel. Other kinds of transfer (of hydrogen ions and electrons, for example) are commonly characterized in terms of appropriate measures of transfer potential (pK_a and reduction potential, $\mathscr{E}_o$, respectively). As shown in Table 16.2, the notion of group transfer is fully analogous to those of ionization potential and reduction potential. The similarity is anything but coincidental, since all of these are really specific instances of free energy changes. If we write

$$AH \longrightarrow A^- + H^+ \qquad (16.1a)$$

we really don't mean that a proton has literally been removed from the acid AH. In the gas phase at least, this would require the input of approximately

Table 16.2
Types of Transfer Potential

	Proton Transfer Potential (Acidity)	**Standard Reduction Potential (Electron Transfer Potential)**	**Group Transfer Potential (High-Energy Bond)**
Simple equation	$AH \rightleftharpoons A^- + H^+$	$A \rightleftharpoons A^+ + e^-$	$A \sim P \rightleftharpoons A + P_i$
Equation including acceptor	$AH + H_2O \rightleftharpoons A^- + H_3O^+$	$A + H^+ \rightleftharpoons A^+ + \frac{1}{2}H_2$	$A \sim PO_4^{2-} + H_2O \rightleftharpoons A{-}OH + HPO_4^{2-}$
Measure of transfer potential	$pK_a = \frac{\Delta G°}{2.303\,RT}$	$\Delta\mathscr{E}_o = \frac{-\Delta G°}{n\mathscr{F}}$	$\ln K_{eq} = \frac{-\Delta G°}{RT}$
Free energy change of transfer is given by:	$\Delta G°$ per mole of H^+ transferred	$\Delta G°$ per mole of e^- transferred	$\Delta G°$ per mole of phosphate transferred

Adapted from: Klotz, I. M., 1986. *Introduction to Biomolecular Energetics.* New York: Academic Press.

1200 kJ/mol! What we really mean is that the proton has been *transferred* to a suitable acceptor molecule, usually water:

$$AH + H_2O \longrightarrow A^- + H_3O^+ \quad (16.1b)$$

The appropriate free energy relationship is of course

$$pK_a = \frac{\Delta G°}{2.303RT} \quad (16.2)$$

Similarly, in the case of an oxidation-reduction reaction

$$A \longrightarrow A^+ + e^- \quad (16.3a)$$

we don't really mean that A oxidizes independently. What we really mean (and what is much more likely in biochemical systems) is that the electron is transferred to a suitable acceptor:

$$A + H^+ \longrightarrow A^+ + \tfrac{1}{2}H_2 \quad (16.3b)$$

and the relevant free energy relationship is

$$\mathscr{E}_o = \frac{\Delta G°}{n\mathscr{F}} \quad (16.4)$$

where n is the number of equivalents of electrons transferred, and $\mathscr{F}$ is **Faraday's constant.**

Similarly, the release of free energy that occurs upon the hydrolysis of ATP and other "high-energy phosphates" can be treated quantitatively in terms of *group transfer*. It is common to write for the hydrolysis of ATP

$$ATP + H_2O \longrightarrow ADP + P_i \quad (16.5)$$

The free energy change, which we henceforth call the *group transfer potential* is given by

$$\Delta G° = -RT \ln K_{eq} \quad (16.6)$$

where K_{eq} is the equilibrium constant for the group transfer, which is normally written as

$$K_{eq} = \frac{[ADP][P_i]}{[ATP][H_2O]} \quad (16.7)$$

Even this set of equations represents an approximation, since ATP, ADP, and P_i all exist in solutions as a mixture of ionic species. This problem will be discussed in a later section. For now, it is enough to note that the free energy changes listed in Table 16.1 are the group transfer potentials observed for transfers to water.

16.2 Classes of High-Energy Compounds

Phosphoric Acid Anhydrides

ATP contains two *pyrophosphoryl* or *phosphoric acid anhydride* linkages, as shown in Figure 16.2. Other common biomolecules possessing phosphoric acid anhydride linkages include ADP, GTP, GDP and the other nucleoside triphosphates, sugar nucleotides such as UDP-glucose, and inorganic pyrophosphate itself. All exhibit large negative free energies of hydrolysis, as shown in Table 16.1. The chemical reasons for the large negative $\Delta G°'$ values for the hydrolysis reactions include destabilization of the reactant due to bond strain caused

Figure **16.2** The triphosphate chain of ATP contains two pyrophosphate linkages, both of which release large amounts of energy upon hydrolysis.

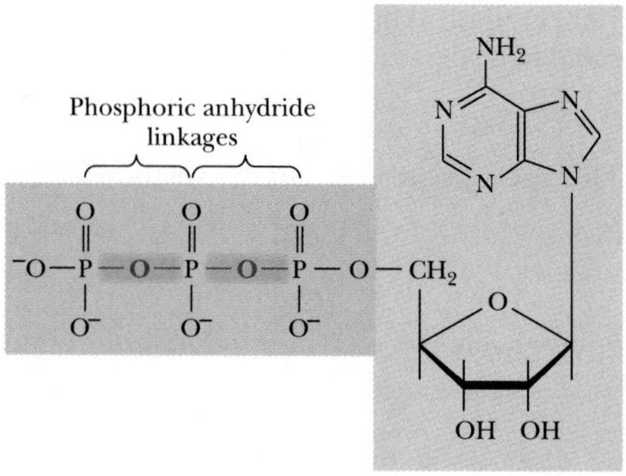

Phosphoric anhydride linkages

ATP
(adenosine-5'-triphosphate)

by electrostatic repulsion, stabilization of the products by ionization and resonance, and entropy factors due to hydrolysis and subsequent ionization.

Destabilization Due to Electrostatic Repulsion

Electrostatic repulsion in the reactants is best understood by comparing these phosphoric anhydrides with other reactive anhydrides, such as acetic anhydride. As shown in Figure 16.3a, the electronegative carbonyl oxygen atoms withdraw electrons from the C=O bonds, producing partial negative charges on the oxygens and partial positive charges on the carbonyl carbons. Each of these electrophilic carbonyl carbons is further destabilized by the other acetyl group, which is also electron-withdrawing in nature. As a result, acetic anhydride is unstable with respect to the products of hydrolysis.

The situation with phosphoric anhydrides is similar. The phosphorus atoms of the pyrophosphate anion are electron-withdrawing and destabilize PP_i with respect to its hydrolysis products. Furthermore, the reverse reaction, re-formation of the anhydride bond from the two anionic products, requires that the electrostatic repulsion between these anions be overcome (see below).

Stabilization of Hydrolysis Products by Ionization and Resonance

The pyrophosphate moiety possesses three negative charges at pH values above 7.5 or so (note the pK_a values, Figure 16.3a). The hydrolysis products, two molecules of inorganic phosphate, both carry about two negative charges each at pH values above 7.2. The increased ionization of the hydrolysis products helps to stabilize the electrophilic phosphorus nuclei.

Resonance stabilization in the products is best illustrated by the reactant anhydrides (Figure 16.3b). The unpaired electrons of the bridging oxygen atoms in acetic anhydride (and phosphoric anhydride) cannot participate in resonance structures with both electrophilic centers at once. This **competing resonance** situation is relieved in the product acetate or phosphate molecules.

Entropy Factors Arising from Hydrolysis and Ionization

For the phosphoric anhydrides, and for most of the high-energy compounds discussed here, there is an additional "entropic" contribution to the free

(a)

Acetic anhydride:

Phosphoric anhydrides:

Pyrophosphate:

Most likely form between pH 6.7 and 9.4

$pK_1 = 0.8$
$pK_2 = 2.0$
$pK_3 = 6.7$
$pK_4 = 9.4$

(b)

Competing resonance in acetic anhydride

These can only occur alternately

Simultaneous resonance in the hydrolysis products

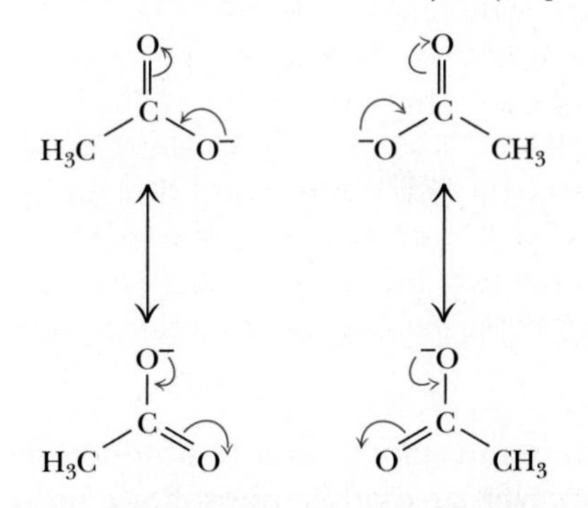

These resonances can occur simultaneously

***Figure* 16.3** (a) Electrostatic repulsion between adjacent partial positive charges (on carbon and phosphorus, respectively) is relieved upon hydrolysis of the anhydride bonds of acetic anhydride and phosphoric anhydrides. The predominant form of pyrophosphate at pH values between 6.7 and 9.4 is shown. (b) The competing resonances of acetic anhydride and the simultaneous resonance forms of the hydrolysis product, acetate.

energy of hydrolysis. Most of the hydrolysis reactions of Table 16.1 result in an increase in the number of molecules in solution. As shown in Figure 16.4, the hydrolysis of ATP (at pH values above 7) creates three species—ADP, inorganic phosphate (P_i), and a hydrogen ion—from only two reactants (ATP and H_2O). The entropy of the solution increases because the more particles, the more disordered the system.[1] (This effect is ionization-dependent, since, at low pH, the hydrogen ion created in many of these reactions will simply protonate one of the phosphate oxygens, and one fewer "particle" will have resulted from the hydrolysis.)

[1] Imagine the "disorder" created by hitting a crystal with a hammer and breaking it into many small pieces.

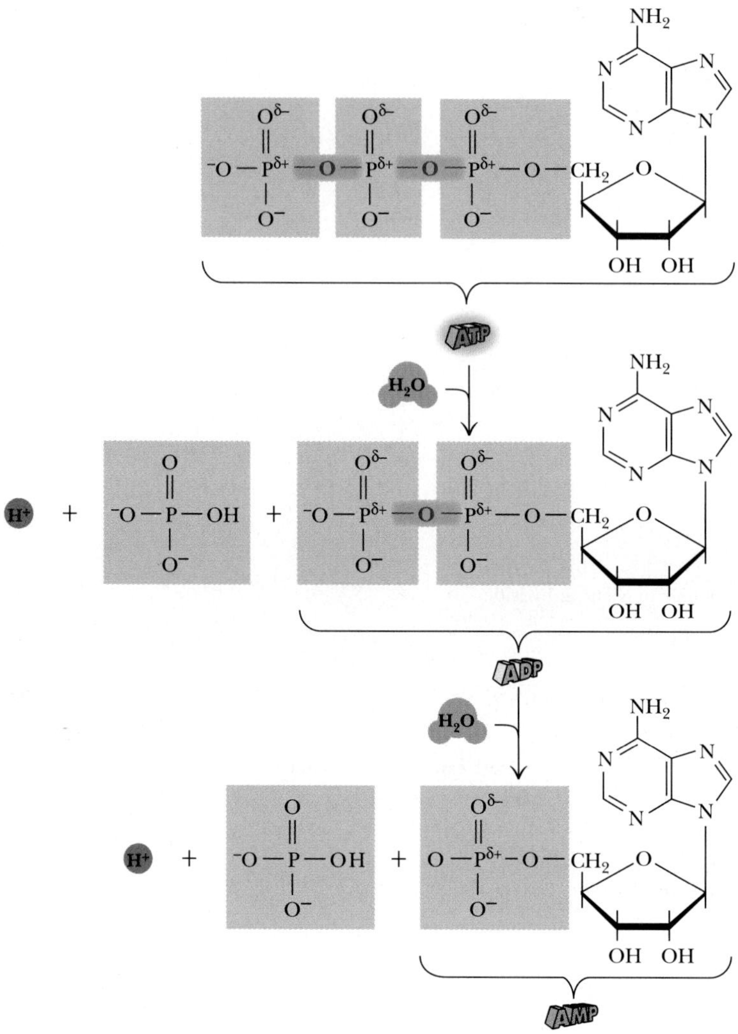

***Figure* 16.4** Hydrolysis of ATP to ADP (and/or of ADP to AMP) leads to relief of electrostatic repulsion.

A Comparison of the Free Energy of Hydrolysis of ATP, ADP, and AMP

The concepts of destabilization of reactants and stabilization of products described for pyrophosphate also apply for ATP and other phosphoric anhydrides (Figure 16.4). ATP and ADP are destabilized relative to the hydrolysis products by electrostatic repulsion, competing resonance, and entropy. AMP, on the other hand, is a phosphate ester (not an anhydride), possessing only a single phosphoryl group, and is not markedly different from the product inorganic phosphate in terms of electrostatic repulsion and resonance stabilization. Thus, the $\Delta G^{\circ\prime}$ for hydrolysis of AMP is much smaller than the corresponding values for ATP and ADP.

Phosphoric-Carboxylic Anhydrides

The mixed anhydrides of phosphoric and carboxylic acids, frequently called acyl phosphates, are also energy-rich. Two biologically important acyl phosphates are acetyl phosphate and 1,3-bisphosphoglycerate. Hydrolysis of these species yields acetate and 3-phosphoglycerate, respectively, in addition to inorganic phosphate (Figure 16.5). Once again, the large $\Delta G^{\circ\prime}$ values indicate that the reactants are destabilized relative to products. This arises from bond strain, which can be traced to the partial positive charges on the carbonyl

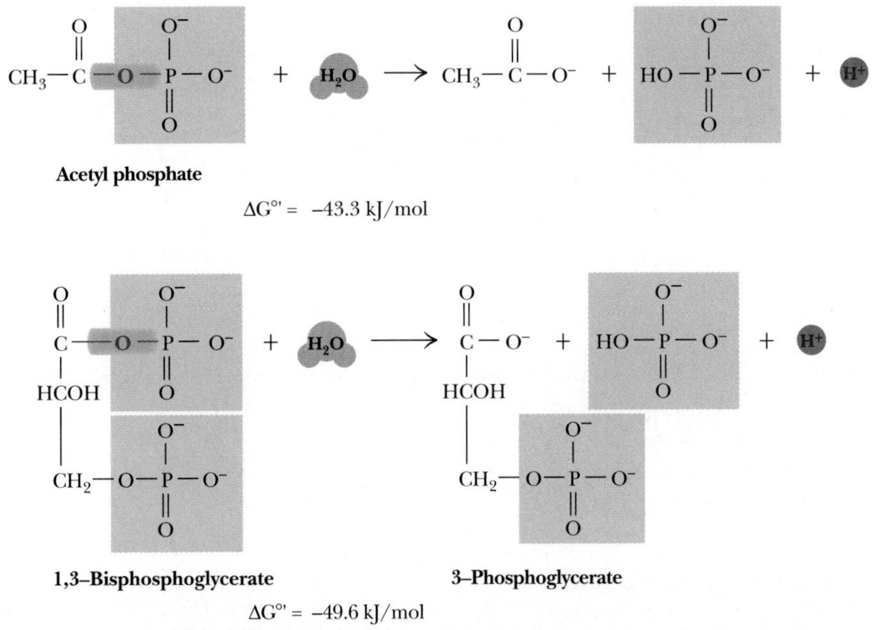

Acetyl phosphate

$\Delta G^{\circ\prime} = -43.3$ kJ/mol

1,3–Bisphosphoglycerate

3–Phosphoglycerate

$\Delta G^{\circ\prime} = -49.6$ kJ/mol

***Figure* 16.5** The hydrolysis reactions of acetyl phosphate and 1,3-bisphosphoglycerate.

carbon and phosphorus atoms of these structures. The energy stored in the mixed anhydride bond (which is required to overcome the charge–charge repulsion) is released upon hydrolysis. Increased resonance possibilities in the products relative to the reactants also contribute to the large negative $\Delta G^{\circ\prime}$ values. The value of $\Delta G^{\circ\prime}$ is dependent on the pK_a values of the starting anhydride and the product phosphoric and carboxylic acids, and of course also on the pH of the medium.

Enol Phosphates

The largest value of $\Delta G^{\circ\prime}$ in Table 16.1 belongs to *phosphoenolpyruvate* or *PEP*, an example of an enolic phosphate. This molecule is an important intermediate in carbohydrate metabolism and, due to its large negative $\Delta G^{\circ\prime}$, it is a potent phosphorylating agent. PEP is formed via dehydration of 2-phosphoglycerate by enolase during fermentation and glycolysis. PEP is subsequently transformed into pyruvate upon transfer of its phosphate to ADP by pyruvate kinase (Figure 16.6). The very large negative value of $\Delta G^{\circ\prime}$ for the latter

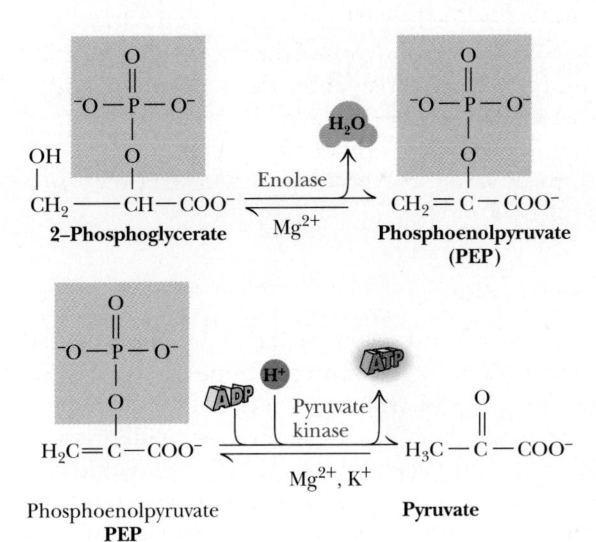

***Figure* 16.6** Phosphoenolpyruvate (PEP) is produced by the enolase reaction (in glycolysis; see Chapter 18) and in turn drives the phosphorylation of ADP to form ATP in the pyruvate kinase reaction.

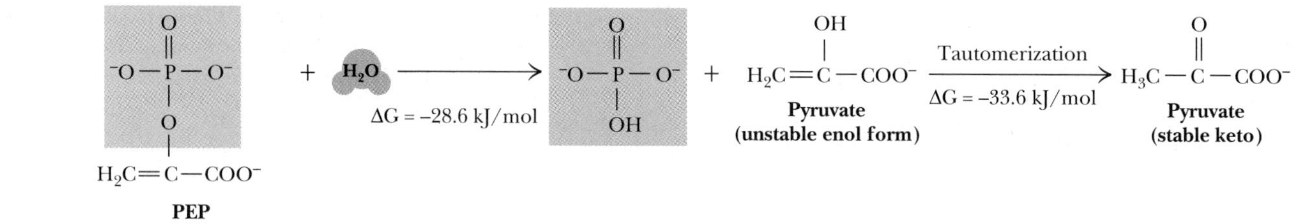

PEP + H_2O → ($\Delta G = -28.6$ kJ/mol) HPO_4^{2-} + Pyruvate (unstable enol form) → Tautomerization ($\Delta G = -33.6$ kJ/mol) → Pyruvate (stable keto)

***Figure* 16.7** Hydrolysis and the subsequent tautomerization account for the very large $\Delta G^{\circ\prime}$ of PEP.

reaction is to a large extent the result of a secondary reaction of the *enol* form of pyruvate. Upon hydrolysis, the unstable enolic form of pyruvate immediately converts to the keto form with a resulting large negative $\Delta G^{\circ\prime}$ (Figure 16.7). Together, the hydrolysis and subsequent *tautomerization* result in an overall $\Delta G^{\circ\prime}$ of -62.2 kJ/mol.

Guanidinium Phosphates

Another important class of high-energy phosphates is the guanidinium phosphates, the most important members being *creatine phosphate* and *arginine phosphate* (Figure 16.8). The high-energy nature of the guanidinium phosphates is a consequence of effective resonance stabilization of the hydrolysis product guanidinium and only limited resonance stabilization of the guanidinium phosphates themselves (Figure 16.9). The guanidinium cation and alkyl guanidinium structures are highly symmetrical species, deriving stability from the several stable resonance structures available to such ions. The formation of guanidinium phosphates from guanidine lowers the symmetry of these species, and fewer favorable resonance structures exist. Most of the possible resonance forms are unfavorable, as in the resonance structure with a positively charged nitrogen adjacent to the partially positive phosphorus atom (Figure 16.9). Guanidinium phosphates thus do not benefit as much from resonance stabilization as guanidinium itself.

Creatine phosphate possesses a more negative free energy of hydrolysis than ATP. In muscle tissue, under conditions of high ATP, creatine phosphate is synthesized in the creatine kinase reaction (Figure 16.10). When ATP levels are low, the reaction runs in the thermodynamically favored reverse direction to produce new ATP.

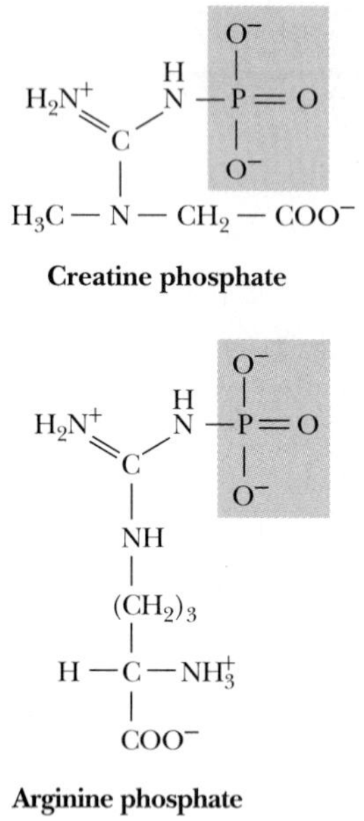

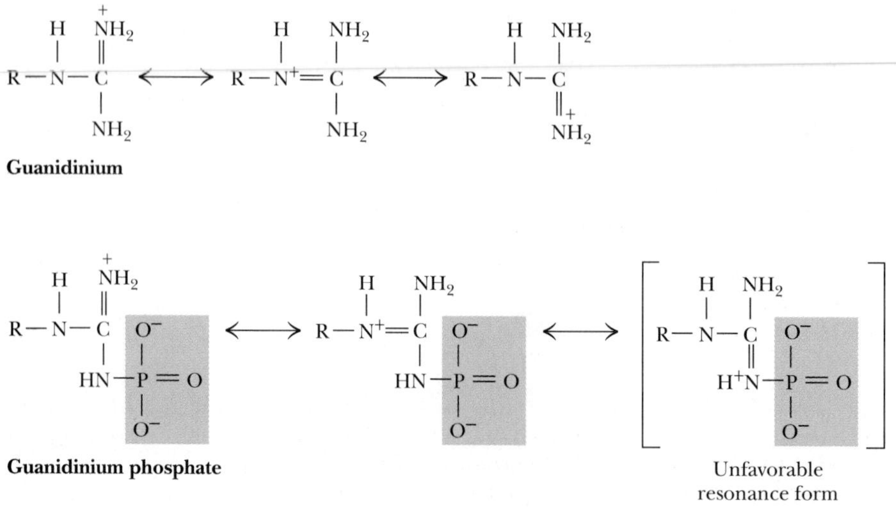

Creatine phosphate

Arginine phosphate

***Figure* 16.8**

Guanidinium

Guanidinium phosphate

Unfavorable resonance form

***Figure* 16.9** A larger number of favorable resonances are available to the symmetric guanidinium cation than to the guanidinium phosphates.

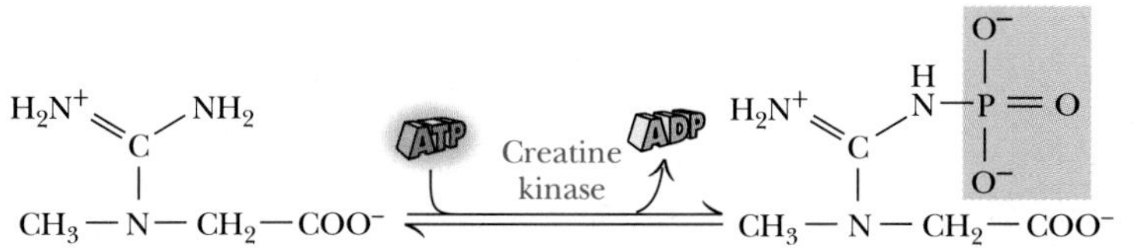

$$H_2N^+{=}C(NH_2){-}N(CH_3){-}CH_2{-}COO^- + ATP \underset{\text{kinase}}{\overset{\text{Creatine}}{\rightleftharpoons}} H_2N^+{=}C(NH{-}PO_3^{2-}){-}N(CH_3){-}CH_2{-}COO^- + ADP$$

***Figure* 16.10** The creatine kinase reaction. Muscle tissue typically contains large amounts of creatine phosphate, and its conversion to creatine produces the large amounts of ATP needed during periods of vigorous exercise.

Cyclic Nucleotides

As will be seen in later chapters, the **cyclic nucleotides,** including 3′,5′-cyclic AMP and 3′,5′-cyclic GMP (Figure 16.11), are important molecules involved in transmission of metabolic signals in most biological systems. These molecules are synthesized from ATP and GTP by appropriate *cyclase* enzymes (adenylyl cyclase and guanylyl cyclase). The hydrolyses of the cyclic nucleotides to yield 5′-nucleoside monophosphates are characterized by very high standard free energies of hydrolysis. Since electrostatic repulsion can play no particular part in the energetics of these hydrolyses, the large values of $\Delta G^{\circ\prime}$ must be attributed to the strained nature of the phosphodiester ring. Cyclic nucleotides play critical roles in modulating the activity of regulatory proteins in a wide variety of biochemical pathways. However, the high-energy nature of the cyclic nucleotides is not critical to the regulatory properties these compounds display.

Amino Acid Esters of tRNA Molecules

In order for amino acids to be incorporated into proteins, they must first be covalently attached to tRNA molecules (Chapter 31). This reaction proceeds

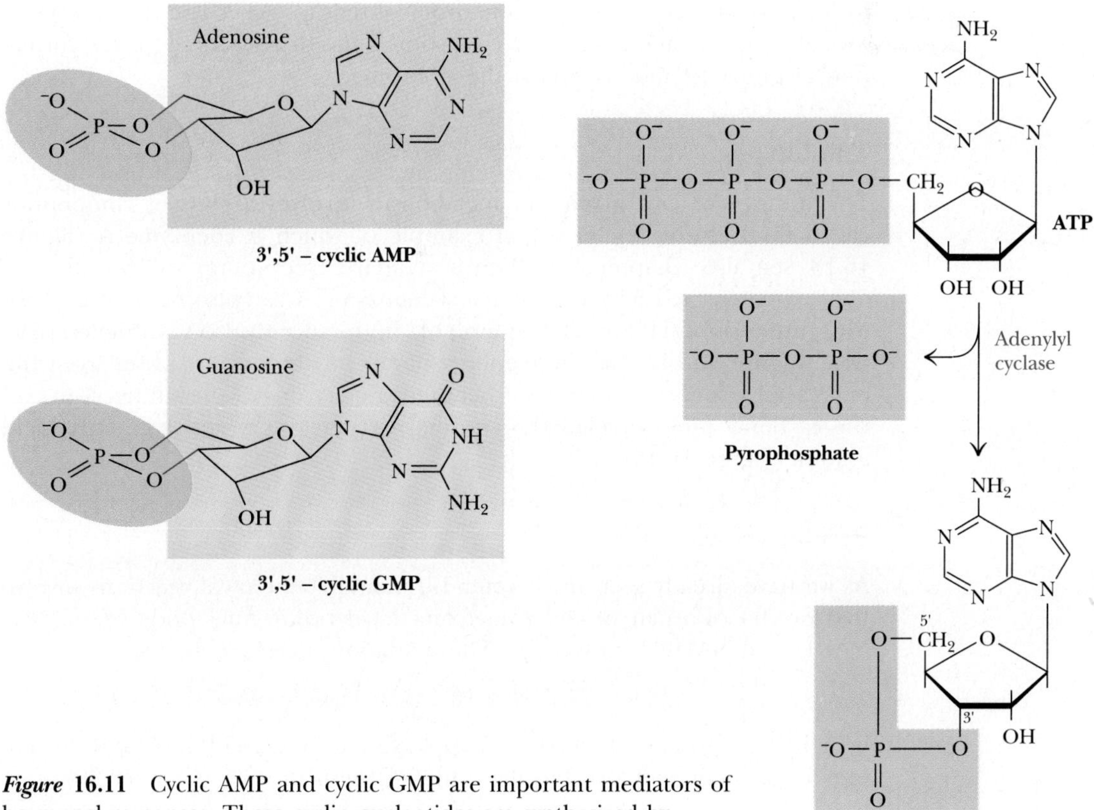

***Figure* 16.11** Cyclic AMP and cyclic GMP are important mediators of hormonal responses. These cyclic nucleotides are synthesized by adenylyl cyclase and guanylyl cyclase, respectively.

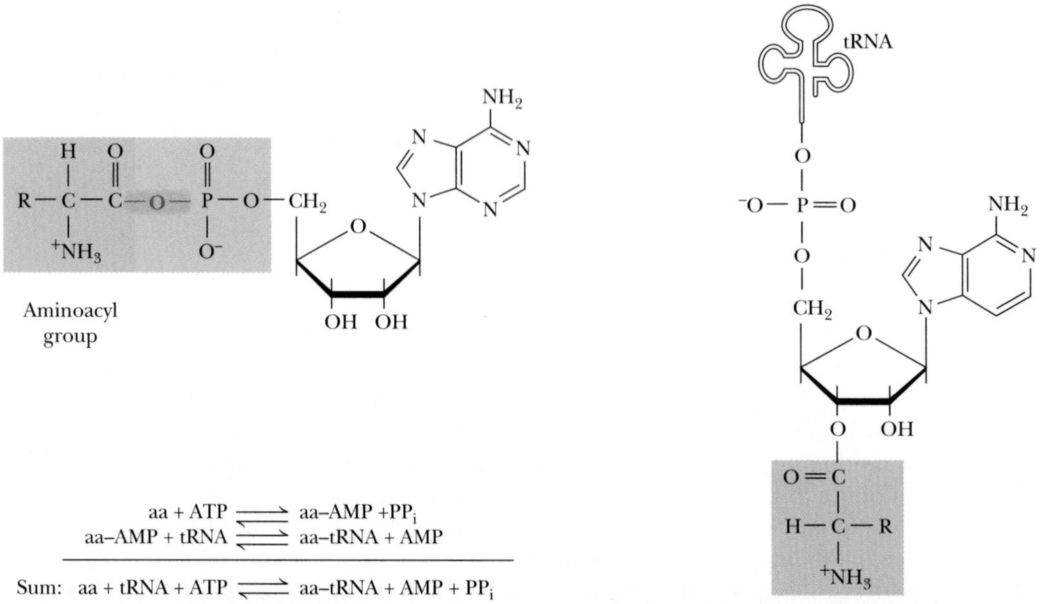

Aminoacyl group

$$\text{aa} + \text{ATP} \rightleftharpoons \text{aa–AMP} + \text{PP}_i$$
$$\text{aa–AMP} + \text{tRNA} \rightleftharpoons \text{aa–tRNA} + \text{AMP}$$

$$\text{Sum: } \text{aa} + \text{tRNA} + \text{ATP} \rightleftharpoons \text{aa–tRNA} + \text{AMP} + \text{PP}_i$$

***Figure* 16.12** The structures of the aminoacyl-adenylate intermediate formed in aminoacyl-tRNA synthesis and the aminoacyl-tRNA ester.

via an aminoacyl-adenylate intermediate (Figure 16.12). The aminoacyl adenylate is another example of a carboxylic-phosphoric anhydride; the aminoacyl-tRNA is an **amino acid ester.** Both exhibit high standard free energies of hydrolysis, with $\Delta G°'$ values on the order of that for ATP. The free energy of hydrolysis of the aminoacyl-tRNA constitutes the driving force for the formation of a new peptide bond on the ribosome.

Thiol Esters

The transfer of acyl groups in metabolism involves a class of compounds called **thiol esters,** the principal example of which is **coenzyme A** (Figure 16.13; see also Chapter 14). With a standard free energy of hydrolysis of approximately -31.5 kJ/mol, succinyl-coenzyme A can also be used to drive the synthesis of GTP (an ATP equivalent) in the succinyl-CoA synthetase reaction (Figure 16.14). The high-energy nature of thiol esters arises from the decreased resonance interaction between π electrons of the sulfur atom and the carbonyl moiety, relative to normal oxygen esters and carboxylic acid anions (Figure 16.15).

Pyridine Nucleotides

As we have already seen in Chapter 14, many biochemical reactions involve the transfer of a pair of electrons from the pyridine nucleotide coenzymes, NADH and NADPH, to oxygen. The oxidation of NADH by O_2

$$\tfrac{1}{2}\,O_2 + \text{NADH} + H^+ \rightleftharpoons H_2O + \text{NAD}^+$$

is highly exergonic, with a $\Delta G°'$ of approximately -220 kJ/mol, and this energy is normally employed to drive the synthesis of ATP (Chapter 20).

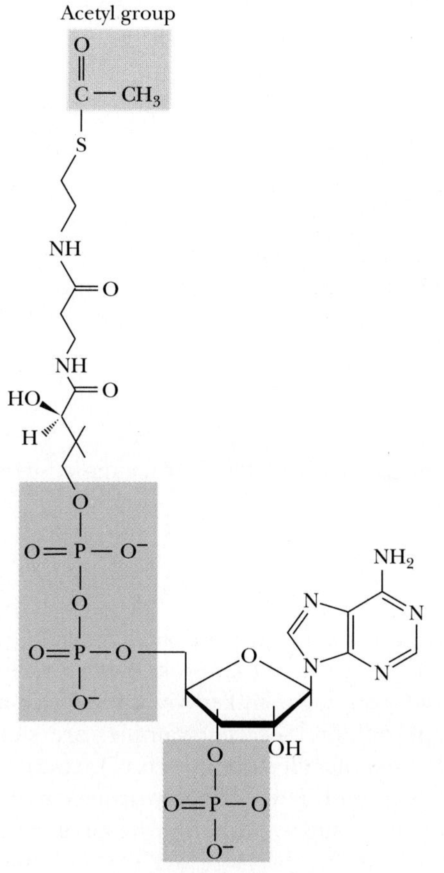

Acetyl group

Acetyl-coenzyme A

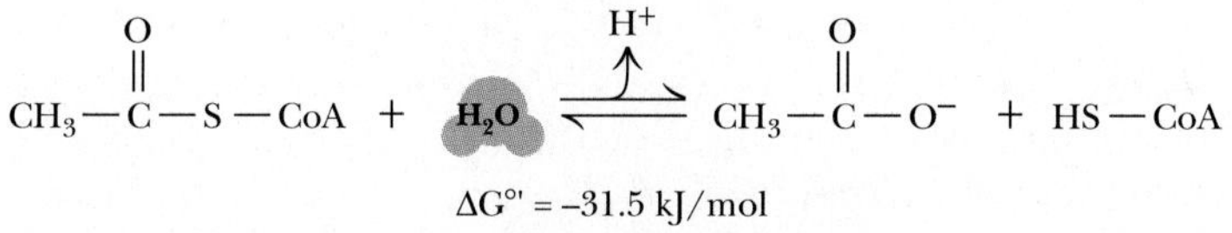

$$CH_3-\overset{O}{\overset{\|}{C}}-S-CoA + H_2O \underset{}{\overset{H^+}{\rightleftharpoons}} CH_3-\overset{O}{\overset{\|}{C}}-O^- + HS-CoA$$

$\Delta G°' = -31.5$ kJ/mol

Figure **16.13** The hydrolysis of acetyl-CoA produces acetate and coenzyme A, with a $\Delta G°'$ of -31.5 kJ/mol.

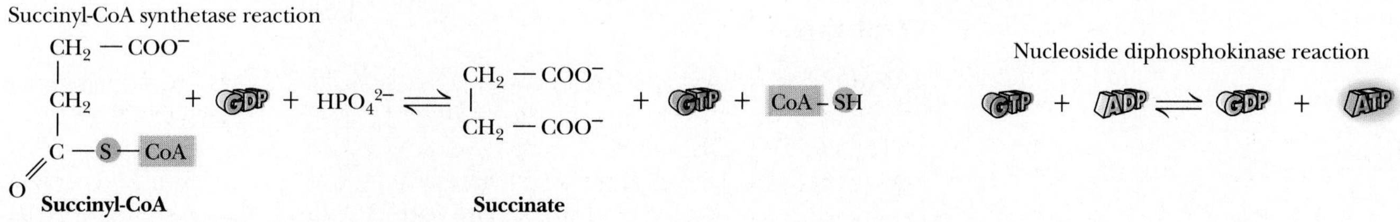

Succinyl-CoA synthetase reaction

Succinyl-CoA + GDP + HPO_4^{2-} ⇌ Succinate + GTP + CoA-SH

Nucleoside diphosphokinase reaction

GTP + ADP ⇌ GDP + ATP

Figure **16.14**

Oxygen esters:

$$R-\overset{O}{\overset{\|}{C}}-O-R' \longleftrightarrow R-\overset{O^-}{\overset{|}{C}}=O^+-R'$$

Thiol esters:

$$\left[R-\overset{:\ddot{O}:^-}{\overset{|}{C}}=\ddot{S}^+-R'\right] \longleftrightarrow R-\overset{\ddot{O}}{\overset{\|}{C}}-S-R' \longleftrightarrow \left[R-\overset{\ddot{O}^+}{\overset{|}{C}}=\ddot{S}^--R'\right]$$

Unfavorable resonance form

Unfavorable resonance form

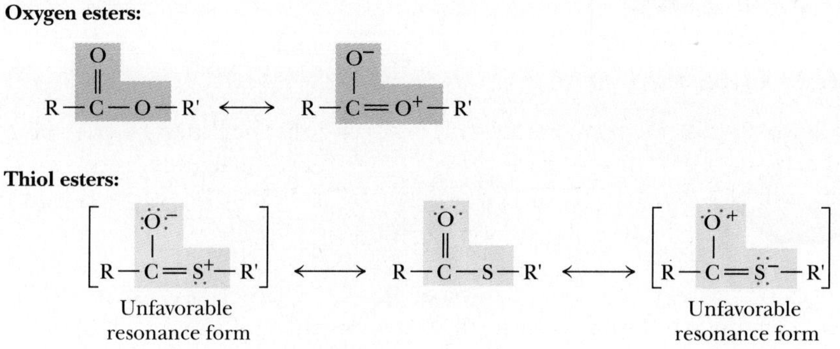

Figure **16.15** Resonance structures available to oxygen and thiol esters. It is difficult for the electrons of the relatively large sulfur atom to overlap favorably with the carbon atom to form a C═S double bond. As a result, thiol esters (also referred to as thioesters) are not stabilized by resonance as effectively as oxygen esters.

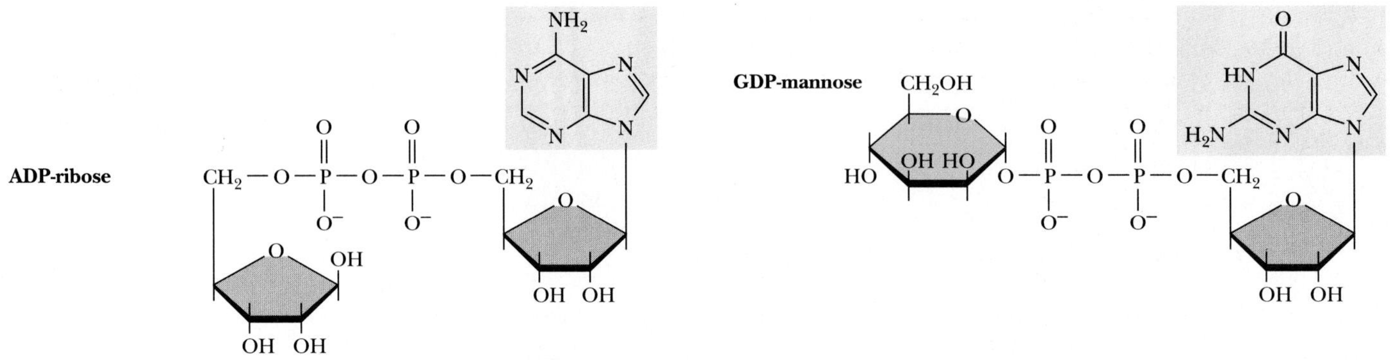

ADP-ribose

GDP-mannose

***Figure* 16.16** The structures of ADP-ribose and GDP-mannose.

Other Compounds

Many other biological molecules could be justifiably considered high-energy molecules. The so-called **sugar nucleotides,** also known as **nucleoside diphosphate sugars,** are composed of a sugar linked via a C—O—P bond arrangement to a nucleoside diphosphate. Several examples are shown in Figure 16.16. The hydrolysis of these molecules (Figure 16.17) typically occurs with a $\Delta G°'$ of approximately −30 kJ/mol. The normal hydrolytic pathway generates a nucleoside diphosphate and a sugar. As before, resonance stabilization in the products is a primary reason for the large $\Delta G°'$ values observed. As mentioned in Chapter 10, the sugar nucleotides deliver sugar units for the synthesis of polysaccharides and other complex carbohydrates.

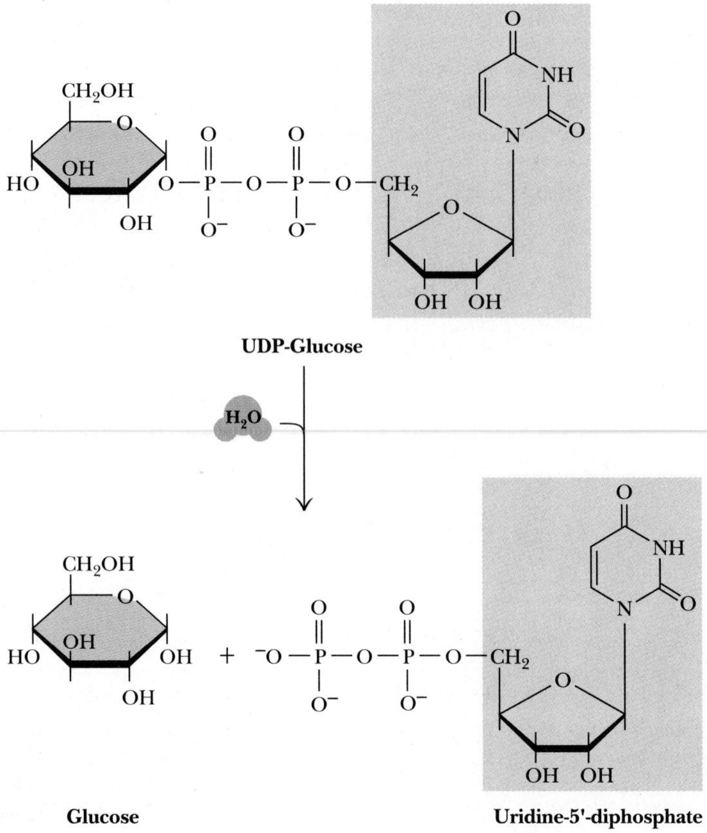

UDP-Glucose

H_2O

Glucose

Uridine-5'-diphosphate

***Figure* 16.17** The hydrolysis of UDP-glucose.

S-adenosyl methionine (Figure 16.18) is another important high-energy molecule, with a $\Delta G°'$ for its hydrolysis of approximately -25.6 kJ/mol. It is involved in the transfer of methyl groups in many metabolic processes. S-adenosyl methionine and methyl group transfer will be discussed in greater detail in Chapters 24 and 26.

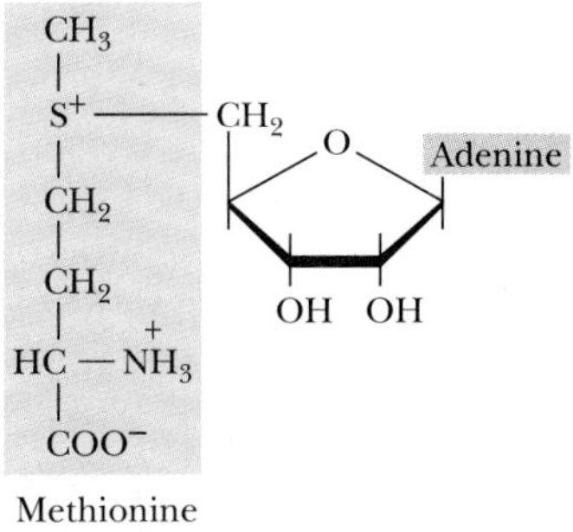

Methionine

Adenine

S-adenosyl methionine

***Figure* 16.18**

16.3 Complex Equilibria Involved in ATP Hydrolysis

So far, as in Equation (16.5), the hydrolyses of ATP and other high-energy phosphates have been portrayed as simple processes. The situation in a real biological system is far more complex, owing to the operation of several ionic equilibria. First of all, ATP, ADP, and the other species in Table 16.1 can exist in several different ionization states that must be accounted for in any quantitative analysis. Second, phosphate compounds bind a variety of divalent and monovalent cations with substantial affinity, and the various metal complexes must also be considered in such analyses. Consideration of these special cases makes the quantitative analysis far more realistic. The importance of these multiple equilibria in group transfer reactions will be illustrated for the hydrolysis of ATP, but the principles and methods presented are general and can be applied to any similar hydrolysis reaction.

The Multiple Ionization States of ATP and the pH Dependence of $\Delta G°'$

ATP has five dissociable protons, as indicated in Figure 16.19. Three of the protons on the triphosphate chain dissociate at very low pH. The adenine ring amino group exhibits a pK_a of 4.06, whereas the last proton to dissociate from the triphosphate chain possesses a pK_a of 6.95. At higher pH values, ATP is completely deprotonated. ADP and phosphoric acid also undergo multiple ionizations. These multiple ionizations make the equilibrium constant for ATP hydrolysis more complicated than the simple expression in Equation (16.7). Multiple ionizations must also be taken into account when the pH dependence of $\Delta G°$ is considered. The calculations are beyond the scope of this text, but Figure 16.20 shows the variation of $\Delta G°$ as a function of pH. The free energy of hydrolysis is nearly constant from pH 4 to pH 6. At higher values of pH, $\Delta G°$ varies linearly with pH, becoming more negative by 5.7 kJ/mol for every pH unit of increase at 37°C. Since the pH of most biological tissues and fluids is near neutrality, the effect on $\Delta G°$ is relatively small, but it must be taken into account in certain situations.

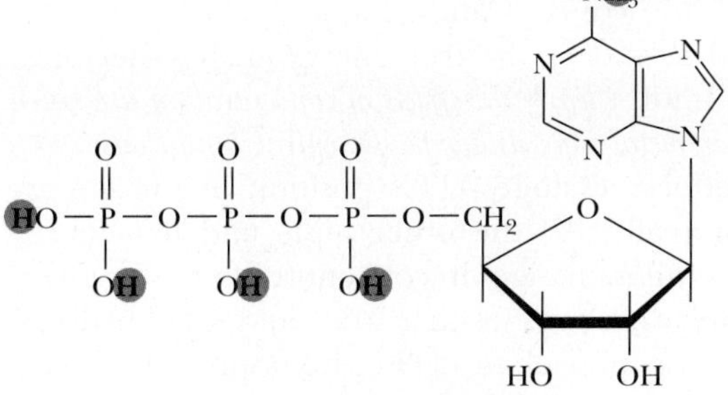

Color indicates the locations of the five dissociable protons of ATP.

***Figure* 16.19**

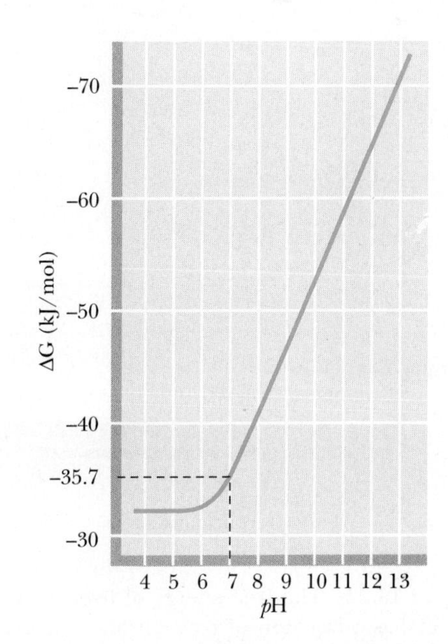

***Figure* 16.20** The pH dependence of the free energy of hydrolysis of ATP. Since pH varies only slightly in biological environments, the effect on ΔG is usually small.

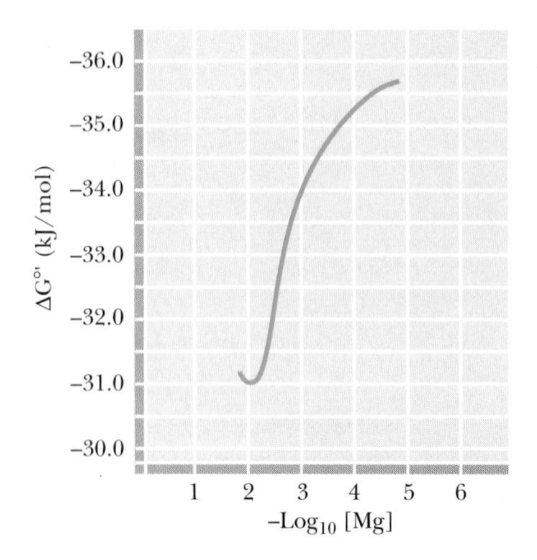

***Figure* 16.21** The free energy of hydrolysis of ATP as a function of total Mg^{2+} ion concentration at 38°C and pH 7.0.

(Adapted from Gwynn, R. W., and Veech, R. L., 1973. The equilibrium constants of the adenosine triphosphate hydrolysis and the adenosine triphosphate-citrate lyase reactions. Journal of Biological Chemistry 248:6966–6972.)

The Effect of Metal Ions on the Free Energy of Hydrolysis of ATP

Most biological environments contain substantial amounts of divalent and monovalent metal ions, including Mg^{2+}, Ca^{2+}, Na^+, K^+, and so on. What effect do metal ions have on the equilibrium constant for ATP hydrolysis and the associated free energy change? Figure 16.21 shows the change in $\Delta G°'$ with pMg (that is, $-\log_{10}[Mg^{2+}]$) at pH 7.0 and 38°C. The free energy of hydrolysis of ATP at zero Mg^{2+} is −35.7 kJ/mol, and at 5 m*M* free Mg^{2+} (the minimum in the plot) the ΔG°_{obs} is approximately −31 kJ/mol. Thus, in most real biological environments (with pH near 7 and Mg^{2+} concentrations of 5 m*M* or more) the free energy of hydrolysis of ATP is altered more by metal ions than by protons. A widely used "consensus value" for $\Delta G°'$ of ATP in biological systems is **−30.5 kJ/mol** (Table 16.1). This value, cited in the 1976 *Handbook of Biochemistry and Molecular Biology* (3rd ed., *Physical and Chemical Data,* Vol. 1, pp. 296–304, Boca Raton, FL: CRC Press), was determined in the presence of "excess Mg^{2+}". *This is the value we shall use for metabolic calculations in the balance of this text.*

16.4 The Effect of Concentration on the Free Energy of Hydrolysis of ATP

Through all these calculations of the effect of pH and metal ions on the ATP hydrolysis equilibrium, we have assumed "standard conditions" with respect to concentrations of all species except for protons. The levels of ATP, ADP, and other high-energy metabolites never even begin to approach the standard state of 1 *M*. In most cells, the concentrations of these species are more typically 1 to 5 m*M* or even less. In Chapter 15, we described the effect of concentration on equilibrium constants and free energies in the form of Equation (15.12). For the present case, we can rewrite this as

$$\Delta G = \Delta G° + RT \ln \frac{[\Sigma\text{ADP}][\Sigma\text{P}_i]}{[\Sigma\text{ATP}]} \qquad (16.8)$$

where the terms in brackets represent the sum (Σ) of the concentrations of all the ionic forms of ATP, ADP, and P_i.

It is clear that changes in the concentrations of these species can have large effects on ΔG. The concentrations of ATP, ADP, and P_i may, of course, vary rather independently in real biological environments, but if, for the sake of some model calculations, we assume that all three concentrations are equal, then the effect of concentration on ΔG is as shown in Figure 16.22. The free energy of hydrolysis of ATP, which is −35.7 kJ/mol at 1 *M*, becomes −49.4 kJ/mol at 5 m*M* (that is, the concentration for which $pC = -2.3$ in Figure 16.22). At 1 m*M* ATP, ADP, and P_i, the free energy change becomes even more negative at −53.6 kJ/mol. *Clearly the effects of concentration are much greater than the effects of protons or metal ions under physiological conditions.*

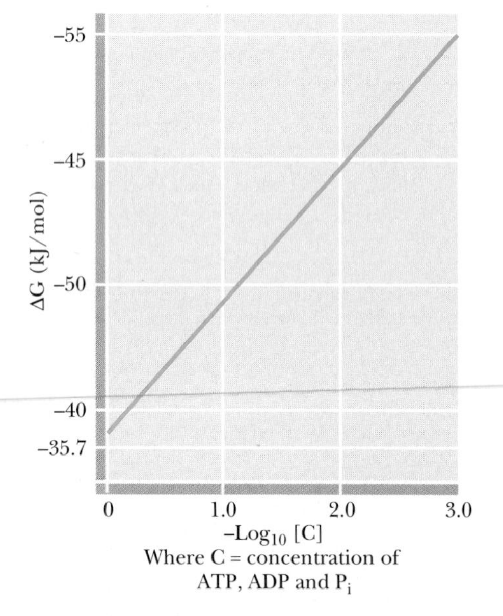

***Figure* 16.22** The free energy of hydrolysis of ATP as a function of concentration at 38°C, pH 7.0. The plot follows the relationship described in Equation (16.8), with the concentrations [C] of ATP, ADP, and P_i assumed to be equal.

Does the "concentration effect" change ATP's position in the energy hierarchy (in Table 16.1)? Not really. All the other high- and low-energy phosphates experience roughly similar changes in concentration under physiological conditions and thus similar changes in their free energies of hydrolysis. The roles of the very high energy phosphates (PEP, 1,3-bisphosphoglycerate, and creatine phosphate) in the synthesis and maintenance of ATP in the cell will be considered in our discussions of metabolic pathways. In the meantime, several of the problems at the end of this chapter address some of the more interesting cases.

16.5 The Daily Human Requirement for ATP

We can end this discussion of ATP and the other important high-energy compounds in biology by discussing the daily metabolic consumption of ATP by humans. An approximate calculation gives a somewhat surprising and impressive result. Assume that the average adult human consumes approximately 11,700 kJ (2800 kcal, that is, 2,800 Calories) per day. Assume also that the tricarboxylic acid cycle and the other metabolic pathways operate at a thermodynamic efficiency of approximately 50%. Thus, of the 11,700 kJ a person consumes as food, about 5,860 kJ end up in the form of synthesized ATP. As indicated in Section 16.4, the hydrolysis of 1 mole of ATP yields approximately 50 kJ of free energy under cellular conditions. This means that the body cycles through 5860/50 = 117 moles of ATP each day. The disodium salt of ATP has a molecular weight of 551 g/mol, so that an average person hydrolyzes about

$$(117 \text{ moles})\frac{551 \text{ g}}{\text{mole}} = 64{,}467 \text{ g of ATP per day}$$

The average adult human, with a typical weight of 70 kg or so, thus consumes approximately 65 kilograms of ATP per day, an amount nearly equal to his/her own body weight! Fortunately, we have a highly efficient recycling system for ATP/ADP utilization. The energy released from food is stored transiently in the form of ATP. Once ATP energy is used and ADP and phosphate are released, our bodies recycle it to ATP through intermediary metabolism, so that it may be reused. The typical 70-kg body contains only about 50 grams of ATP/ADP total. Therefore, each ATP molecule in our bodies must be recycled nearly 1300 times each day! Were it not for this fact, at current commercial prices of about $10 per gram, our ATP "habit" would cost approximately $650,000 per day! In these terms, the ability of biochemistry to sustain the marvelous activity and vigor of organisms gains our respect and fascination.

Problems

1. Draw all possible resonance structures for creatine phosphate and arginine phosphate (other than those shown in Figure 16.9) and discuss their possible effects on resonance stabilization of the molecule.

2. Write the equilibrium constant, K_{eq}, for the hydrolysis of creatine phosphate and calculate a value for K_{eq} at 25°C from the value of $\Delta G°'$ in Table 16.1.

3. Imagine that creatine phosphate, rather than ATP, is the universal energy carrier molecule in the human body. Repeat the calculation presented in Section 16.5, calculating the weight of creatine phosphate that would need to be consumed each day by a typical adult human if creatine phosphate could not be recycled. If recycling of creatine phosphate were possible, and if the typical adult human body contained 20 grams of creatine phosphate, how many times would each creatine phosphate molecule need to be turned over or recycled each day? Repeat the calculation assuming that glycerol-3-phosphate is the universal energy carrier, and that the body contains 20 grams of glycerol-3-phosphate.

4. Calculate the free energy of hydrolysis of ATP in a rat liver cell in which the ATP, ADP, and P_i concentrations are 3.4, 1.3, and 4.8 mM, respectively.

5. Hexokinase catalyzes the phosphorylation of glucose from ATP, yielding glucose-6-P and ADP. Using the values of Table 16.1, calculate the standard-state free energy change and equilibrium constant for the hexokinase reaction.

6. Would you expect the free energy of hydrolysis of acetoacetyl-coenzyme A (see diagram) to be greater than, equal to, or less than that of acetyl-coenzyme A? Provide a chemical rationale for your answer.

$$CH_3-\overset{\overset{O}{\|}}{C}-CH_2-\overset{\overset{O}{\|}}{C}-S-CoA$$

7. Consider carbamoyl phosphate, a precursor in the biosynthesis of pyrimidines:

$$H_3\overset{+}{N}-\overset{\overset{O}{\|}}{C}-O-PO_3^{2-}$$

Based on the discussion of high-energy phosphates in this chapter, would you expect carbamoyl phosphate to possess a high free energy of hydrolysis? Provide a chemical rationale for your answer.

Further Reading

Alberty, R. A., 1968. Effect of pH and metal ion concentration on the equilibrium hydrolysis of adenosine triphosphate to adenosine diphosphate. *Journal of Biological Chemistry* **243**:1337–1343.

Alberty, R. A., 1969. Standard Gibbs free energy, enthalpy, and entropy changes as a function of pH and pMg for reactions involving adenosine phosphates. *Journal of Biological Chemistry* **244**:3290–3302.

Cantor, C. R., and Schimmel, P. R., 1980. *Biophysical Chemistry.* San Francisco: W. H. Freeman.

Dickerson, R. E., 1969. *Molecular Thermodynamics.* New York: Benjamin Co.

Gwynn, R. W., and Veech, R. L., 1973. The equilibrium constants of the adenosine triphosphate hydrolysis and the adenosine triphosphate-citrate lyase reactions. *Journal of Biological Chemistry* **248**:6966–6972.

Klotz, I. M., 1967. *Energy Changes in Biochemical Reactions.* New York: Academic Press.

Lehninger, A. L., 1972. *Bioenergetics,* 2nd ed. New York: Benjamin Co.

Patton, A. R., 1965. *Biochemical Energetics and Kinetics.* Philadelphia: Saunders.

Segel, I. H., 1976. *Biochemical Calculations,* 2nd ed. New York: John Wiley.

Part III

Metabolism and Its Regulation

"All is flux, nothing stays still. Nothing endures but change."

Heraclitus (c. 540–c. 480 B.C.)

Chapter 17

Metabolism—An Overview

Anise swallowtail butterfly (*Papilio zelican*) with its pupal case. Metamorphosis of butterflies is a dramatic example of metabolic change.

Outline

17.1 Virtually All Organisms Have the Same Basic Set of Metabolic Pathways

17.2 Metabolism Consists of Catabolism (Degradative Pathways) and Anabolism (Biosynthetic Pathways)

17.3 Intermediary Metabolism Is a Tightly Regulated, Integrated Process

17.4 Experimental Methods To Reveal Metabolic Pathways

The word *metabolism* derives from the Greek word for "change." **Metabolism** represents the sum of the chemical changes that convert **nutrients,** the "raw materials" necessary to nourish living organisms, into energy and the chemically complex finished products of cells. Metabolism consists of literally hundreds of enzymatic reactions organized into discrete pathways. These pathways proceed in a stepwise fashion, transforming substrates into end products through many specific chemical **intermediates.** Metabolism is sometimes referred to as **intermediary metabolism** to reflect this aspect of the process. Metabolic maps (Figure 17.1) portray virtually all of the principal reactions of the intermediary metabolism of carbohydrates, lipids, amino acids, nucleotides, and their derivatives. These maps are very complex at first glance and would seem to be virtually impossible to learn easily. Despite their appearance, these maps become easy to follow once the major metabolic routes are known and their functions are understood. The underlying order of metabolism and the important interrelationships between the various pathways then appear as simple patterns against the seemingly complicated background.

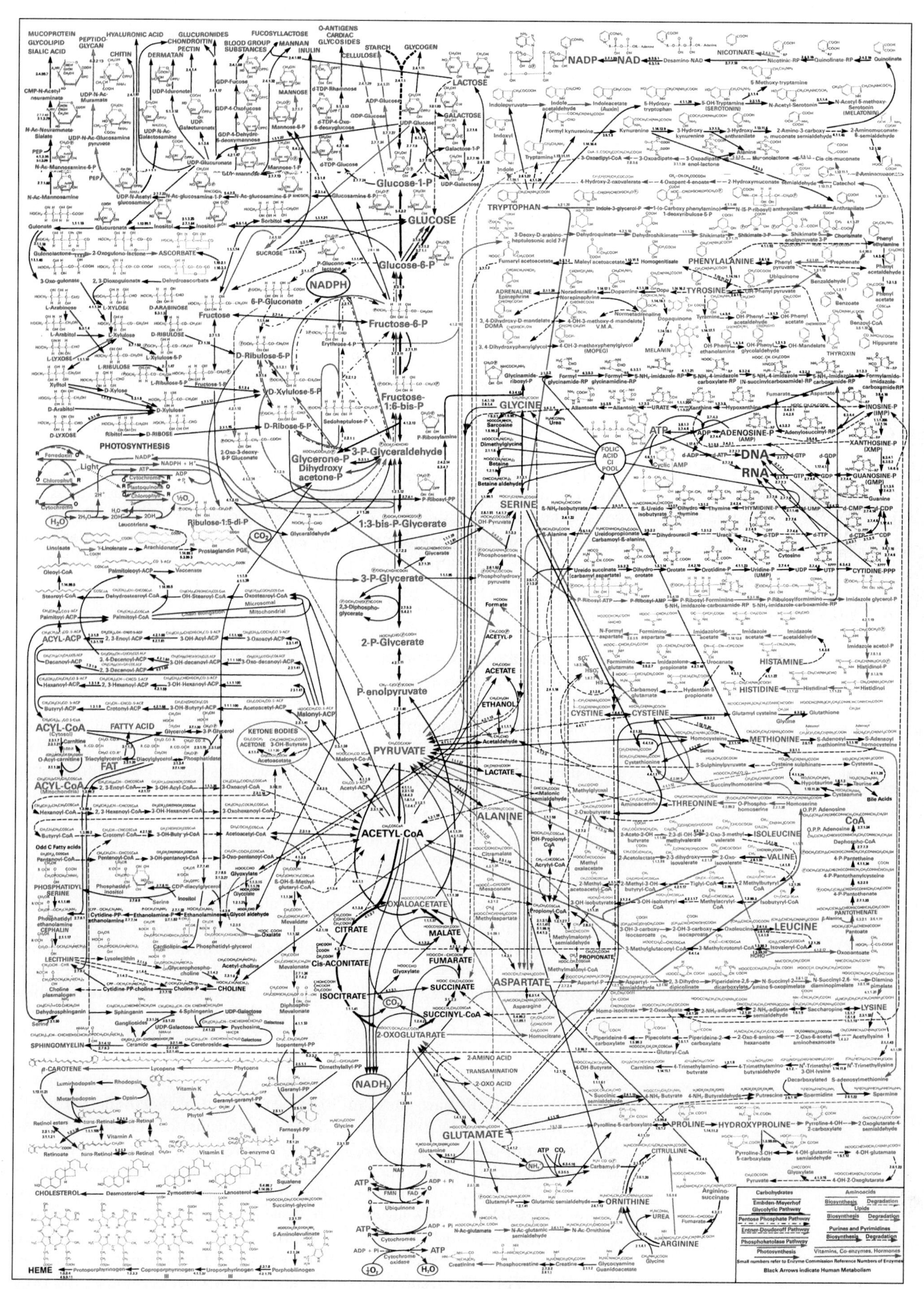

Figure 17.1 A metabolic map, indicating the reactions of intermediary metabolism and the enzymes that catalyze them. Over 500 different chemical intermediates, or **metabolites**, and a greater number of enzymes are represented here.

(Courtesy of D. E. Nicholson, University of Leeds, U.K., and the Sigma Chemical Co.)

Table **17.1**

Number of Dots (Intermediates) in the Metabolic Map of Figure 17.2, and the Number of Lines Associated with Them

Lines	Dots
1 or 2	410
3	71
4	20
5	11
6 or more	8

The Metabolic Map as a Set of Dots and Lines

One interesting transformation of the intermediary metabolism map is to represent each intermediate as a black dot and each enzyme as a line (Figure 17.2). Then, the more than 1000 different enzymes and substrates are represented by just two symbols. This chart has about 520 dots (intermediates). Table 17.1 lists the numbers of dots that have one or two or more lines (enzymes) associated with them. Thus, this table classifies intermediates by the number of enzymes that act upon them. A dot connected to just a single line must be either a nutrient, a storage form, an end product, or an excretory product of metabolism. Also, since many pathways tend to proceed in only one direction (that is, they are essentially irreversible under physiological conditions), a dot connected to just two lines is probably an intermediate in only one pathway and has only one fate in metabolism. If three lines are connected to a dot, that intermediate has at least two possible metabolic fates; four lines, three fates; and so on. Note that about 80% of the intermediates connect to only one or two lines and thus have only a limited purpose in the cell. However, many intermediates are subject to a variety of fates. In such instances, the pathway followed is an important regulatory choice. Indeed, whether any substrate is routed down a particular metabolic pathway is the consequence of a regulatory decision, made in response to the cell's (or organism's) momentary requirements for energy or nutrition. The regulation of metabolism is an interesting and important subject to which we will return often.

17.1 Virtually All Organisms Have the Same Basic Set of Metabolic Pathways

One of the great unifying principles of modern biology is that organisms show marked similarity in their major pathways of metabolism. Given the almost unlimited possibilities within organic chemistry, this generality would appear most unlikely. Yet it's true, and it provides strong evidence that all life has descended from a common ancestral form. All forms of nutrition and almost all metabolic pathways evolved in early prokaryotes prior to the appearance of eukaryotes one billion years ago. For example, **glycolysis,** the metabolic pathway by which energy is released from glucose and captured in the form of ATP under anaerobic conditions, is common to almost every cell. It is believed to be the most ancient of metabolic pathways, having arisen prior to the appearance of oxygen in abundance in the atmosphere. All organisms, even those that can synthesize their own glucose, are capable of glucose degradation and ATP synthesis via glycolysis. Other prominent pathways are also virtually ubiquitous among organisms.

Metabolic Diversity

Although most cells have the same basic set of central metabolic pathways, different cells (and, by extension, different organisms) are characterized by the alternative pathways they might express. These pathways offer a wide diversity of metabolic possibilities. For instance, organisms are often classified according to the major metabolic pathways they exploit to obtain carbon or energy. Classification based on carbon requirements defines two major groups, autotrophs and heterotrophs. **Autotrophs** are organisms that can use just carbon dioxide as their sole source of carbon. **Heterotrophs** require an organic form of carbon, such as glucose, in order to synthesize other essential carbon compounds.

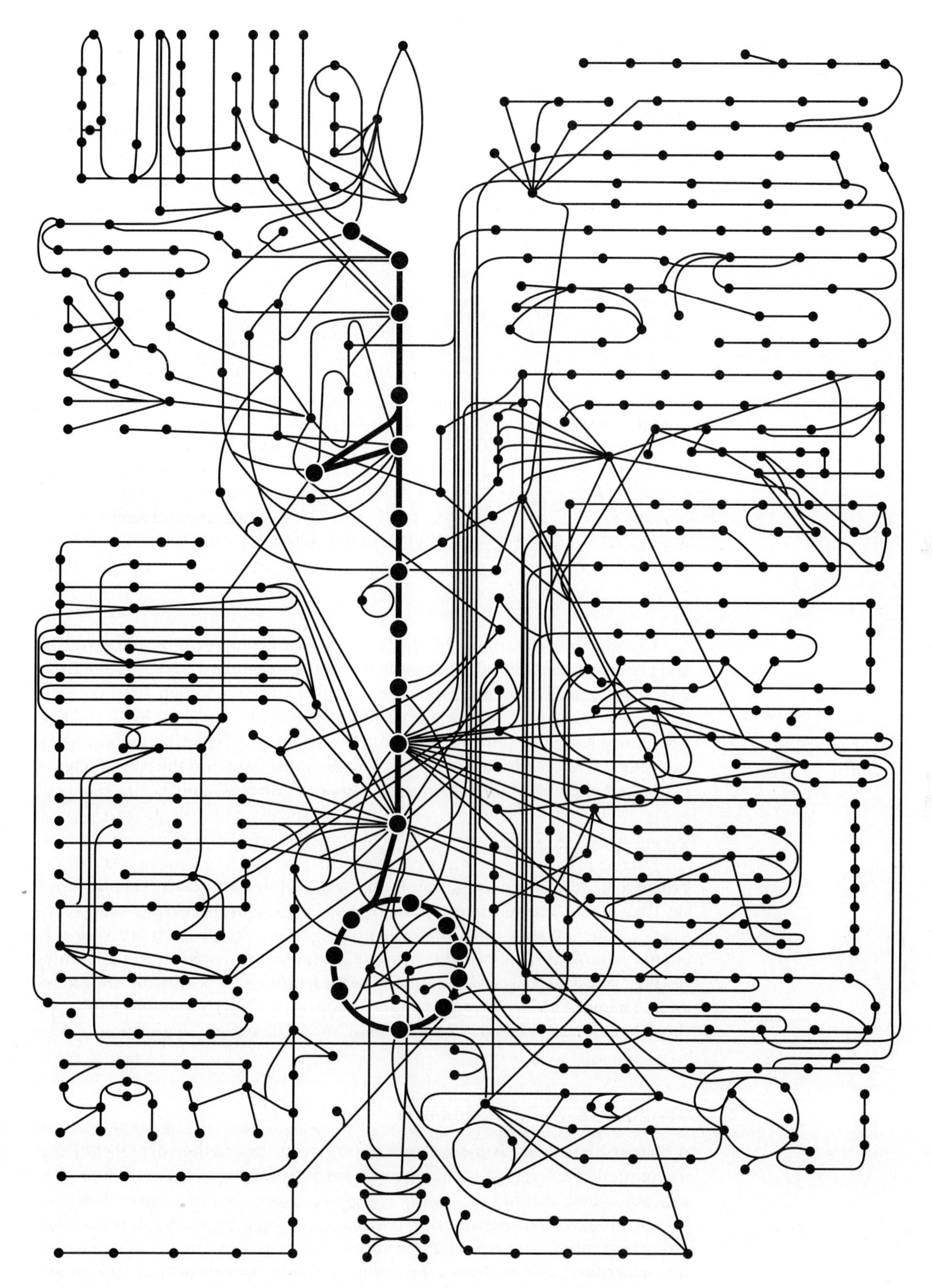

***Figure* 17.2** The metabolic map as a set of dots and lines. The heavy dots and lines trace the central energy-releasing pathways known as glycolysis and the citric acid cycle.

(Adapted from Alberts, B., et al., 1989. Molecular Biology of the Cell, 2nd ed. New York: Garland Publishing Co.)

Table **17.2**
Metabolic Classification of Organisms According to Their Carbon and Energy Requirements

Classification	Carbon Source	Energy Source	Electron Donors	Examples
Photoautotrophs	CO_2	Light	H_2O, H_2S, S, other inorganic compounds	Green plants, algae, cyanobacteria, photosynthetic bacteria
Photoheterotrophs	Organic compounds	Light	Organic compounds	Nonsulfur purple bacteria
Chemoautotrophs	CO_2	Oxidation–reduction reactions	Inorganic compounds: H_2, H_2S, NH_4^+, NO_2^-, Fe^{2+}, Mn^{2+}	Nitrifying bacteria; hydrogen, sulfur, and iron bacteria
Chemoheterotrophs	Organic compounds	Oxidation–reduction reactions	Organic compounds, e.g., glucose	All animals, most microorganisms, nonphotosynthetic plant tissue such as roots, photosynthetic cells in the dark

Classification based on energy sources also gives two groups: phototrophs and chemotrophs. **Phototrophs** are *photosynthetic organisms,* which use light as a source of energy. **Chemotrophs** use organic compounds such as glucose or, in some instances, oxidizable inorganic substances such as Fe^{2+}, NO_2^-, NH_4^+, or elemental sulfur as sole sources of energy. Typically, the energy is extracted through oxidation–reduction reactions. Based on these characteristics, every organism falls into one of four categories (Table 17.2).

Metabolic Diversity Among the Five Kingdoms

Prokaryotes (the kingdom Monera—bacteria) show a greater metabolic diversity than all the four eukaryotic kingdoms (Protoctista [previously called Protozoa], Fungi, Plants, and Animals) put together. Prokaryotes are variously chemoheterotrophic, photoautotrophic, photoheterotrophic, or chemoautotrophic. No protoctista are chemoautotrophs; fungi and animals are exclusively chemoheterotrophs; plants are characteristically photoautotrophs, although some are heterotrophic in their mode of carbon acquisition.

The Role of O_2 in Metabolism

A further metabolic distinction among organisms is whether or not they can use oxygen as an electron acceptor in energy-producing pathways. Those that can are called **aerobes** or *aerobic organisms;* others, termed **anaerobes,** can subsist without O_2. Organisms for which O_2 is obligatory for life are called **obligate aerobes;** humans are an example. Some species, the so-called **facultative anaerobes,** can adapt to anaerobic conditions by substituting other electron acceptors for O_2 in their energy-producing pathways; *Escherichia coli* is an example. Yet others cannot use oxygen at all and are even poisoned by it; these are the **obligate anaerobes.** *Clostridium botulinum,* the bacterium that produces botulin toxin, is representative.

The Flow of Energy in the Biosphere and the Carbon and Oxygen Cycles Are Intimately Related

The primary source of energy for life is the sun. Photoautotrophs utilize light energy to drive the synthesis of organic molecules, such as carbohydrates, from atmospheric carbon dioxide and water (Figure 17.3). Heterotrophic cells then use these organic products of photosynthetic cells both as fuels and as building blocks, or precursors, for the biosynthesis of their own unique complement of biomolecules. Ultimately, carbon dioxide is the end product of heterotrophic carbon metabolism, and CO_2 is returned to the atmosphere for reuse by the photoautotrophs. In effect, solar energy is converted to the chemical energy of organic molecules by photoautotrophs, and heterotrophs recover this energy by metabolizing the organic substances. The flow of energy in the biosphere is thus conveyed within the carbon cycle, and the impetus driving the cycle is light energy.

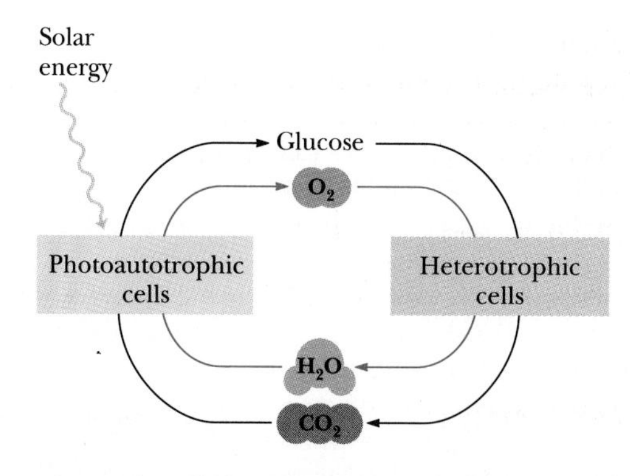

***Figure* 17.3** The flow of energy in the biosphere is coupled primarily to the carbon and oxygen cycles.

Oxygen Cycles Among Water, Carbon Dioxide, and O_2

An exchange of oxygen among water, carbon dioxide, and atmospheric O_2 accompanies the carbon cycle. Most photosynthetic organisms use H_2O as the reductant for carbon dioxide fixation, and O_2 is liberated in the process:

$$CO_2 + H_2O \longrightarrow (CH_2O) + O_2$$

where (CH_2O) represents a carbohydrate unit. Solar energy drives this endergonic reaction. Oxygen comprises about 21% of the atmosphere; essentially all of it came from photosynthesis.

Heterotrophic cells metabolize carbohydrate in a process that is the reverse of the preceding reaction. Oxygen is an obligatory participant in the total combustion and maximal energy release in these cells:

$$(CH_2O)_n + nO_2 \longrightarrow nCO_2 + nH_2O$$

Thus, atmospheric oxygen is produced by photoautotrophs and consumed by chemoheterotrophs, and an oxygen cycle is part of the global carbon and energy cycle (Figure 17.3). As an index of the magnitude of these processes, more than 3×10^{14} kg of carbon dioxide are cycled through the biosphere annually.

A Deeper Look

Calcium Carbonate—A Biological Sink for CO_2

A major biological sink for carbon dioxide that is often overlooked is the calcium carbonate shells of corals, molluscs, and crustacea. These invertebrate animals deposit $CaCO_3$ in the form of protective exoskeletons. In some invertebrates, such as the *scleractinians* (hard corals) of tropical seas, photosynthetic dinoflagellates (kingdom Protoctista) known as *zooxanthellae* live within the animal cells as **endosymbionts.** These phototrophic cells use light to drive the resynthesis of organic molecules from CO_2 released (as bicarbonate ion) by the animal's metabolic activity. In the presence of Ca^{2+}, the photosynthetic carbon dioxide fixation "pulls" the deposition of $CaCO_3$, as summarized in the following coupled reactions:

$$Ca^{2+} + 2\,HCO_3^- \rightleftharpoons CaCO_{3(s)}\downarrow + H_2CO_3$$

$$H_2CO_3 \rightleftharpoons H_2O + CO_2$$

$$H_2O + CO_2 \xrightarrow{\text{Light}} \text{carbohydrate} + O_2$$

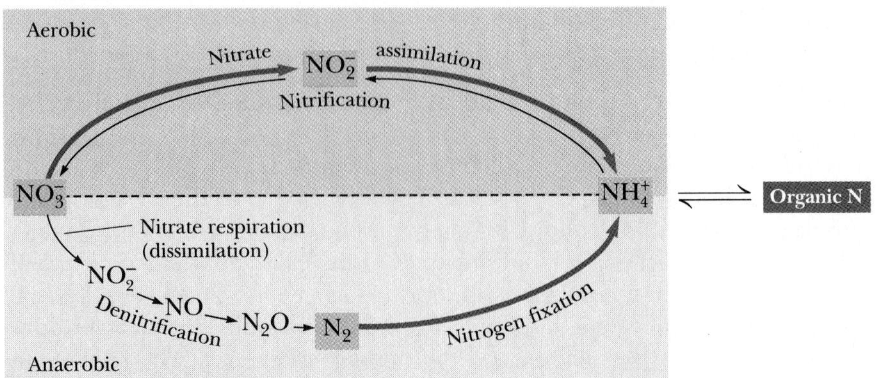

***Figure* 17.4** The nitrogen cycle. Organic nitrogenous compounds are formed by the incorporation of NH_4^+ into carbon skeletons. Ammonium can be formed from oxidized inorganic nitrogen precursors by reductive reactions: **nitrogen fixation** reduces N_2 to NH_4^+; **nitrate assimilation** reduces NO_3^- to NH_4^+. Nitrifying bacteria can oxidize NH_4^+ back to NO_3^- and obtain energy for growth in the process of **nitrification. Denitrification** is a form of bacterial respiration whereby nitrogen oxides serve as electron acceptors in place of O_2 under anaerobic conditions.

The Nitrogen Cycle

Nitrogen is critical to organisms because it is an integral component of proteins and nucleic acids, the two preeminent classes of biological macromolecules. Like carbon and oxygen, nitrogen cycles through the biosphere. Nitrogen exists predominantly in an oxidized state in the inanimate environment, occurring principally as N_2 in the atmosphere or as nitrate ion (NO_3^-) in the soils and oceans. Its acquisition by biological systems is accompanied by its reduction to ammonium ion (NH_4^+) and the incorporation of NH_4^+ into organic linkage as amino or amido groups (Figure 17.4). The formation of NH_4^+ from N_2 gas is termed **nitrogen fixation.** N_2 fixation is an exclusively prokaryotic process, although bacteria in symbiotic association with certain green plants also carry out nitrogen fixation in nodules on the plants' roots. The reduction of NO_3^- to NH_4^+ occurs in green plants, various fungi, and certain bacteria in a two-step metabolic pathway known as **nitrate assimilation.** No animals are capable of either nitrogen fixation or nitrate assimilation, so they are totally dependent on plants and microorganisms for the synthesis of organic nitrogenous compounds, such as amino acids and proteins, to satisfy their requirements for this essential element.

Animals release excess nitrogen in a reduced form, either as NH_4^+ or as organic nitrogenous compounds such as urea. The release of N occurs both during life and as a consequence of microbial decomposition following death. Various bacteria return the reduced forms of nitrogen back to the environment by oxidizing them. The oxidation of NH_4^+ to NO_3^- by **nitrifying bacteria,** a group of chemoautotrophs, provides the sole source of chemical energy for the life of these microbes. Nitrate nitrogen also returns to the atmosphere as N_2 as a result of the metabolic activity of **denitrifying bacteria.** These bacteria are capable of using NO_3^- and similar oxidized inorganic forms of nitrogen as electron acceptors in place of O_2 in energy-producing pathways. The NO_3^- is reduced ultimately to *dinitrogen* (N_2). These bacteria thus deplete the levels of *combined nitrogen,*[1] important as a natural fertilizer, that might otherwise be available. However, such bacterial activity is being exploited in water treatment plants to reduce the load of combined nitrogen that might otherwise enter our lakes, streams, and bays.

Like the carbon and oxygen cycles, the nitrogen cycle involves the participation of diverse kinds of organisms, each ultimately dependent on the others for the smooth turning of the cycle and the restoration of essential resources. Implicit in these cycles is the interdependence of life forms—the metabolic

[1] N joined with other elements in chemical compounds.

activity of one group provides nutritional sustenance for another. In this manner, virtually all organisms are united into mutually dependent communities, or **ecosystems.** Modern humanity shuns an appreciation for these interrelationships at its peril.

17.2 Metabolism Consists of Catabolism (Degradative Pathways) and Anabolism (Biosynthetic Pathways)

Metabolism serves two fundamentally different purposes: the generation of energy to drive vital functions, and the synthesis of biological molecules. To achieve these ends, metabolism consists largely of two contrasting processes, catabolism and anabolism. *Catabolic pathways are characteristically energy-yielding, whereas anabolic pathways are energy-requiring.* **Catabolism** involves the oxidative degradation of complex nutrient molecules (carbohydrates, lipids, and proteins) obtained either from the environment or from cellular reserves. The breakdown of these molecules by catabolism leads to the formation of simpler molecules such as lactic acid, ethanol, carbon dioxide, urea, or ammonia. Catabolic reactions are usually exergonic, and often the chemical energy released is captured in the form of ATP (Chapter 16). Since catabolism is oxidative for the most part, part of the chemical energy may be conserved as energy-rich electrons transferred to the coenzymes NAD^+ and $NADP^+$ (Chapter 14). These two reduced coenzymes have very different metabolic roles: *NAD^+ reduction is part of catabolism; NADPH oxidation is an important aspect of anabolism.* The energy released upon oxidation of NADH is coupled to the phosphorylation of ADP in aerobic cells, and so NADH oxidation back to NAD^+ serves to generate more ATP; in contrast, NADPH is the source of the reducing power needed to drive reductive biosynthetic reactions.

Thermodynamic considerations demand that the energy necessary for biosynthesis of any substance exceed the energy available from its catabolism. Otherwise, organisms could achieve the status of perpetual motion machines: A few molecules of substrate whose catabolism yielded more ATP than required for its resynthesis would allow the cell to cycle this substance and harvest an endless supply of energy.

Anabolism Is Biosynthesis

Anabolism is a synthetic process in which the varied and complex biomolecules (proteins, nucleic acids, polysaccharides, and lipids) are assembled from simpler precursors. Such biosynthesis involves the formation of new covalent bonds, and an input of chemical energy is necessary to drive such endergonic processes. The ATP generated by catabolism provides this energy. Furthermore, NADPH is an excellent donor of high-energy electrons for the reductive reactions of anabolism. Despite their divergent roles, anabolism and catabolism are interrelated in that the products of one provide the substrates of the other (Figure 17.5). Many metabolic intermediates are shared between the two processes, and the precursors needed by anabolic pathways are found among the products of catabolism.

Anabolism and Catabolism Are Not Mutually Exclusive

Interestingly, anabolism and catabolism occur simultaneously in the cell. The conflicting demands of concomitant catabolism and anabolism are managed by cells in two ways. First, the cell maintains tight and separate regulation of both catabolism and anabolism, so that metabolic needs are served in an

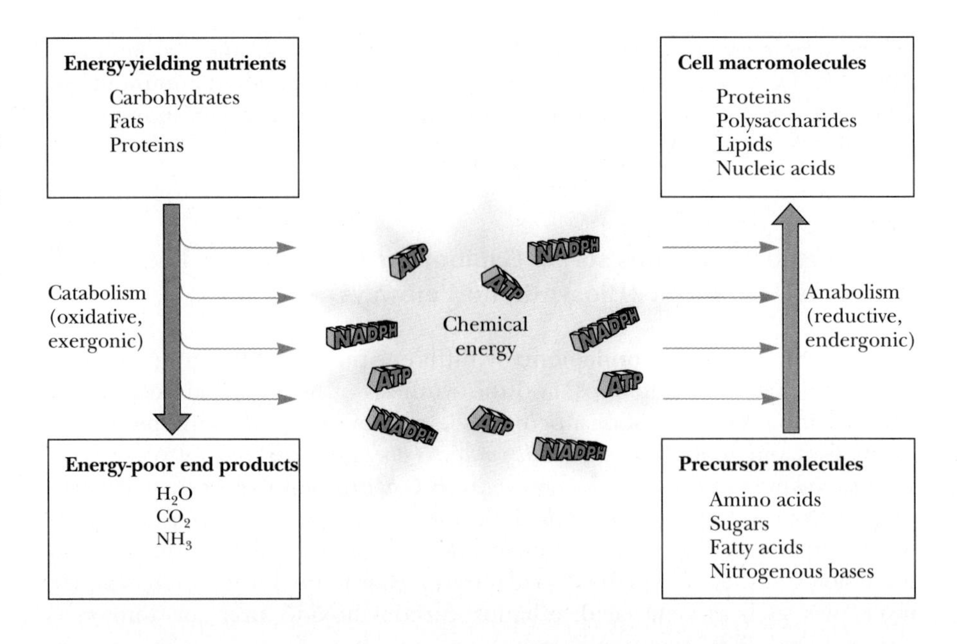

***Figure* 17.5** Energy relationships between the pathways of catabolism and anabolism. Oxidative, exergonic pathways of catabolism release free energy and reducing power that are captured in the form of ATP and NADPH, respectively. Anabolic processes are endergonic, consuming chemical energy in the form of ATP and using NADPH as a source of high-energy electrons for reductive purposes.

immediate and orderly fashion. Second, competing metabolic pathways are often localized within different cellular compartments. Isolating opposing activities within distinct compartments, such as separate organelles, avoids interference between them. For example, the enzymes responsible for catabolism of fatty acids, the *fatty acid oxidation pathway,* are localized within mitochondria. In contrast, *fatty acid biosynthesis* takes place in the cytosol. In subsequent chapters, we shall see that the particular molecular interactions responsible for the regulation of metabolism become important to an understanding and appreciation of metabolic biochemistry.

Modes of Enzyme Organization in Metabolic Pathways

The individual metabolic pathways of anabolism and catabolism consist of sequential enzymatic steps (Figure 17.6). Several types of organization are possible. The enzymes of some multienzyme systems may exist as physically separate, soluble entities, with diffusing intermediates (Figure 17.6a). In other instances, the enzymes of a pathway are collected to form a discrete *multienzyme complex,* and the substrate is sequentially modified as it is passed along from enzyme to enzyme (Figure 17.6b). This type of organization has the advantage that intermediates are not lost or diluted by diffusion. In a third pattern of organization, the enzymes common to a pathway reside together as a *membrane-bound system* (Figure 17.6c). In this case, the enzyme participants (and perhaps the substrates as well) must diffuse in just the two dimensions of the membrane to interact with their neighbors.

As research reveals the ultrastructural organization of the cell in ever greater detail, more and more of the so-called soluble enzyme systems are found to be physically united into functional complexes. Thus, in many (perhaps all) metabolic pathways, the consecutively acting enzymes are associated into stable multienzyme complexes that are sometimes referred to as **metabolons,** a word meaning "units of metabolism."

Why Metabolic Pathways Have So Many Steps

It is interesting to ask why metabolism is organized as it is. For instance, why are there so many individual steps in metabolic pathways? (For example, more

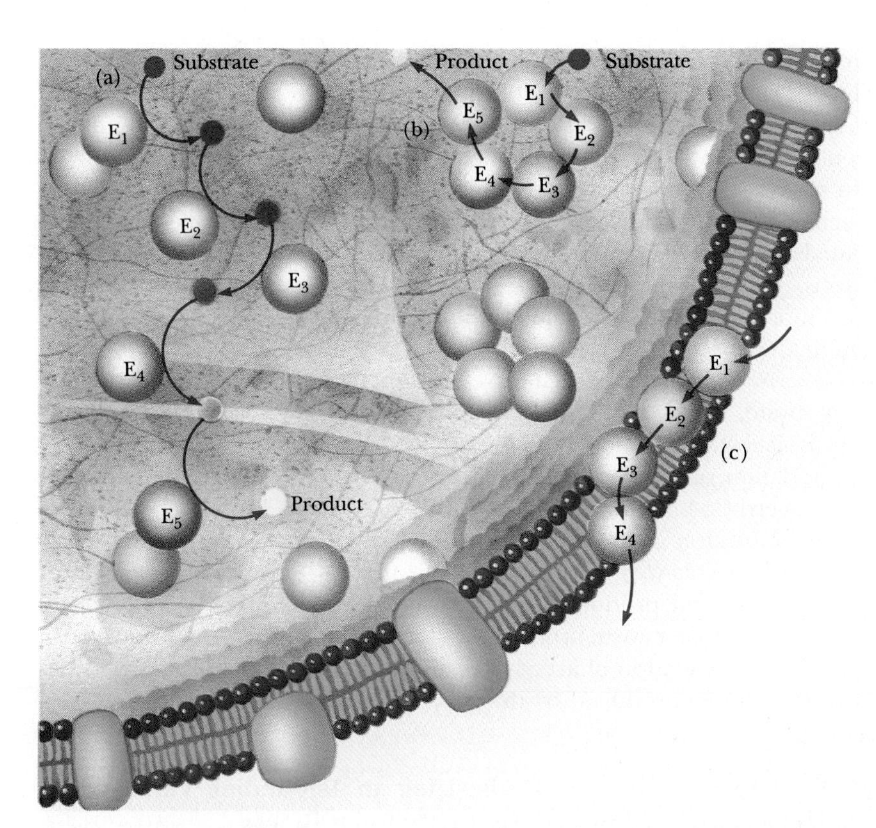

***Figure* 17.6** Schematic representation of types of multienzyme systems carrying out a metabolic pathway: (a) Physically separate, soluble enzymes with diffusing intermediates. (b) A multienzyme complex. Substrate enters the complex, becomes covalently bound and then sequentially modified by enzymes E_1 to E_5 before product is released. No intermediates are free to diffuse away. (c) A membrane-bound multienzyme system.

than 20 different enzymes act in sequence to accomplish the oxidation of glucose to carbon dioxide.) Chemical considerations do not dictate so many steps. Couldn't fewer enzymes serve? Wouldn't fewer enzymes be more efficient and thus favored by evolution? One answer is that a multiplicity of sequential steps provides more chemical possibilities, adding much more versatility and flexibility to metabolism.

A better answer is found in an examination of the energetics of metabolism. Free energy from catabolism is captured in the form of ATP. The "packet" of energy necessary to cause the phosphorylation of ADP by P_i is a finite quantity, about 40 to 50 kJ/mol under cellular conditions. If the potential energy in a molecule (say, glucose) undergoing catabolism is released stepwise over many steps, the number of ATP molecules that can be formed in the process is potentially greater. Similarly, the amount of energy that can be delivered to an anabolic reaction by ATP is limited by the free energy available from its hydrolysis. That is, metabolism consists of many reactions in order to tailor the energy transaction at any step to the amount of free energy that accompanies ATP synthesis or hydrolysis.

Another reason for many steps is the need for metabolic control. Cells disassemble or assemble molecules one chemical step at a time so that the overall process can be carefully regulated.

The Pathways of Catabolism Converge to a Few End Products

If we survey the catabolism of the principal energy-yielding nutrients (carbohydrates, lipids, and proteins) in a typical heterotrophic cell, we see that the degradation of these substances involves a succession of enzymatic reactions. In the presence of oxygen (*aerobic catabolism*), these molecules are degraded ultimately to carbon dioxide, water, and ammonia. Aerobic catabolism consists of three distinct stages. In **stage 1,** the nutrient macromolecules are broken down into their respective building blocks. Given the great diversity of macromolecules, these building blocks represent a rather limited number of products. Proteins yield up their 20 component amino acids, polysaccharides give rise to carbohydrate units that are convertible to glucose, and lipids are broken down into glycerol and fatty acids (Figure 17.7).

In **stage 2,** the collection of product building blocks generated in stage 1 are further degraded to yield an even more limited set of simpler metabolic intermediates. The deamination of amino acids leaves α-keto acid carbon skeletons. Several of these α-keto acids are citric acid cycle intermediates and are fed directly into stage 3 catabolism via this cycle. Others are converted either to the three-carbon α-keto acid *pyruvate* or to the acetyl groups of *acetyl-Coenzyme A* (acetyl-CoA). Glucose and the glycerol from lipids also generate pyruvate, while the fatty acids are broken into two-carbon units that appear as *acetyl-CoA.* Since pyruvate also gives rise to acetyl-CoA, we see that the degradation of macromolecular nutrients converges to a common end product, acetyl-CoA (Figure 17.7).

The combustion of the acetyl groups of acetyl-CoA by the *citric acid cycle* and *oxidative phosphorylation* to produce CO_2 and H_2O represents **stage 3** of catabolism. The end products of the citric acid cycle, carbon dioxide and water, are the ultimate waste products of aerobic catabolism. As we shall see in Chapter 19, the oxidation of acetyl-CoA during stage 3 metabolism generates most of the energy produced by the cell.

Anabolic Pathways Diverge, Synthesizing an Astounding Variety of Biomolecules from a Limited Set of Building Blocks

A rather limited collection of simple precursor molecules is sufficient to provide for the biosynthesis of virtually any cellular constituent, be it protein, nucleic acid, lipid, or polysaccharide. All of these substances are constructed from appropriate building blocks via the pathways of anabolism. In turn, the building blocks (amino acids, nucleotides, sugars, and fatty acids) can be generated from metabolites in the cell. For example, amino acids can be formed by amination of the corresponding α-keto acid carbon skeletons, and pyruvate can be converted to hexoses for polysaccharide biosynthesis.

Amphibolic Intermediates

amphi: from the Greek for "on both sides"

Certain of the central pathways of intermediary metabolism, such as the citric acid cycle, and many of the metabolites of other pathways have dual purposes—they serve in both catabolism and anabolism. This dual nature is reflected in the designation of such pathways as **amphibolic** rather than solely catabolic or anabolic. In any event, in contrast to catabolism—which converges to the common intermediate, acetyl-CoA—the pathways of anabolism diverge from a small group of simple metabolic intermediates to yield a spectacular variety of cellular constituents.

Corresponding Pathways of Catabolism and Anabolism Differ in Important Ways

The anabolic pathway for synthesis of a given end product usually does not precisely match the pathway used for catabolism of the same substance. Some of the intermediates may be common to steps in both pathways, but different enzymatic reactions and unique metabolites characterize other steps. A good example of these differences is found in a comparison of the catabolism of glucose to pyruvic acid by the pathway of glycolysis and the biosynthesis of

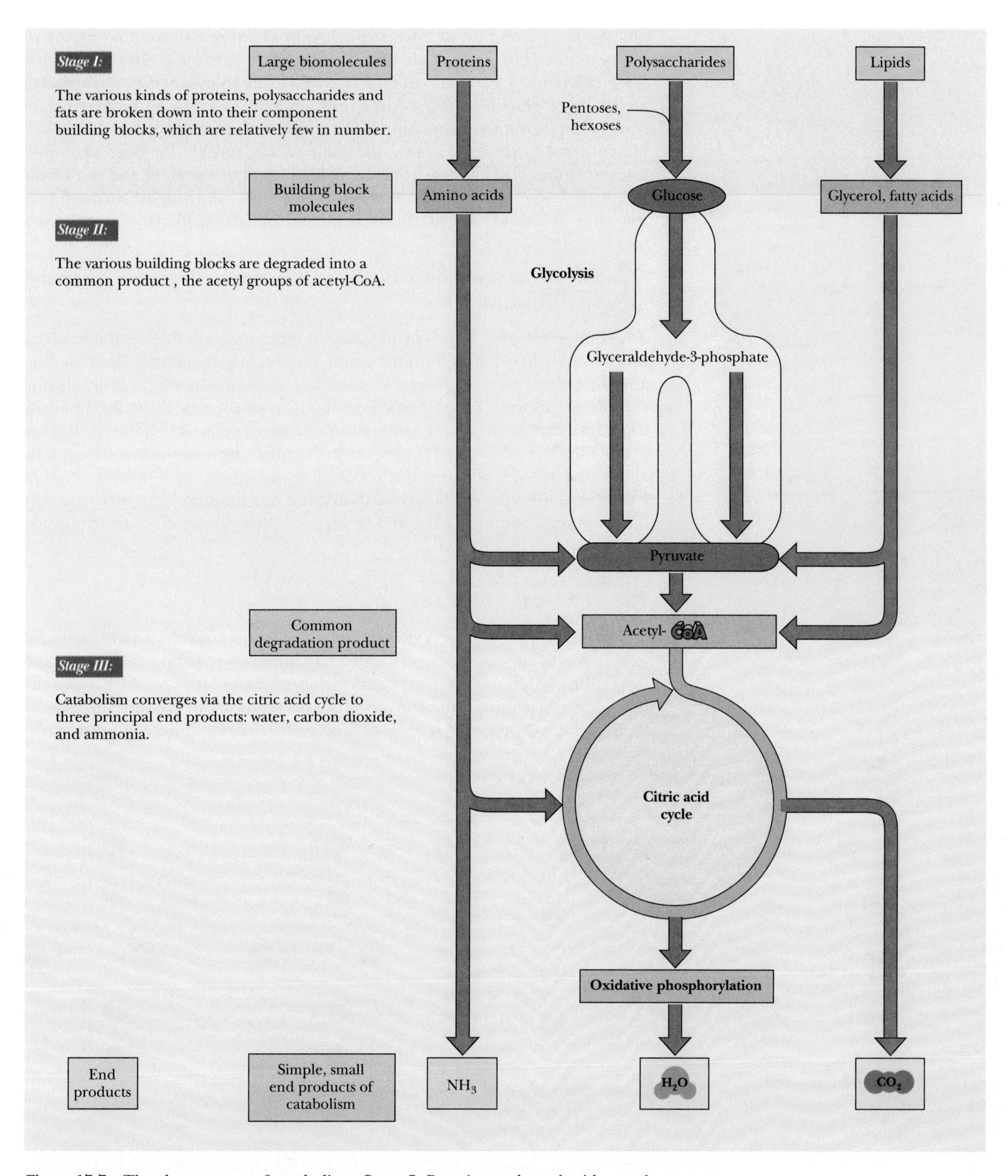

***Figure* 17.7** The three stages of catabolism. **Stage I:** Proteins, polysaccharides, and lipids are broken down into their component building blocks, which are relatively few in number. **Stage II:** The various building blocks are degraded into the common product, the acetyl groups of acetyl-CoA. **Stage III:** Catabolism converges to three principal end products: water, carbon dioxide, and ammonia.

glucose from pyruvate by the pathway called *gluconeogenesis.* The glycolytic pathway from glucose to pyruvate consists of 10 enzymes. Although it may seem efficient for glucose synthesis from pyruvate to proceed by a reversal of all 10 steps, gluconeogenesis uses only seven of the glycolytic enzymes in reverse, replacing the remaining three with four enzymes specific to glucose biosynthesis. In similar fashion, the pathway responsible for degrading proteins to amino acids differs from the protein synthesis system, and the oxidative degradation of fatty acids to two-carbon acetyl-CoA groups does not follow the same reaction path as the biosynthesis of fatty acids from acetyl-CoA.

Metabolic Regulation Favors Different Pathways for Oppositely Directed Metabolic Sequences

A second reason for different pathways serving in opposite metabolic directions is that such pathways must be independently regulated. If catabolism and anabolism passed along the same set of metabolic tracks, equilibrium considerations would dictate that slowing the traffic in one direction by inhibiting a particular enzymatic reaction would necessarily slow traffic in the opposite direction. Independent regulation of anabolism and catabolism can be accomplished only if these two contrasting processes move along different routes *or,* in the case of shared pathways, the rate-limiting steps serving as the points of regulation are catalyzed by enzymes that are unique to each opposing sequence (Figure 17.8).

The ATP Cycle

We saw in Chapter 16 that ATP is the energy currency of cells. In phototrophs, ATP is one of the two energy-rich primary products resulting from the transformation of light energy into chemical energy. (The other is NADPH; see the following discussion.) In heterotrophs, the pathways of catabolism have as their major purpose the release of free energy that can be captured in the

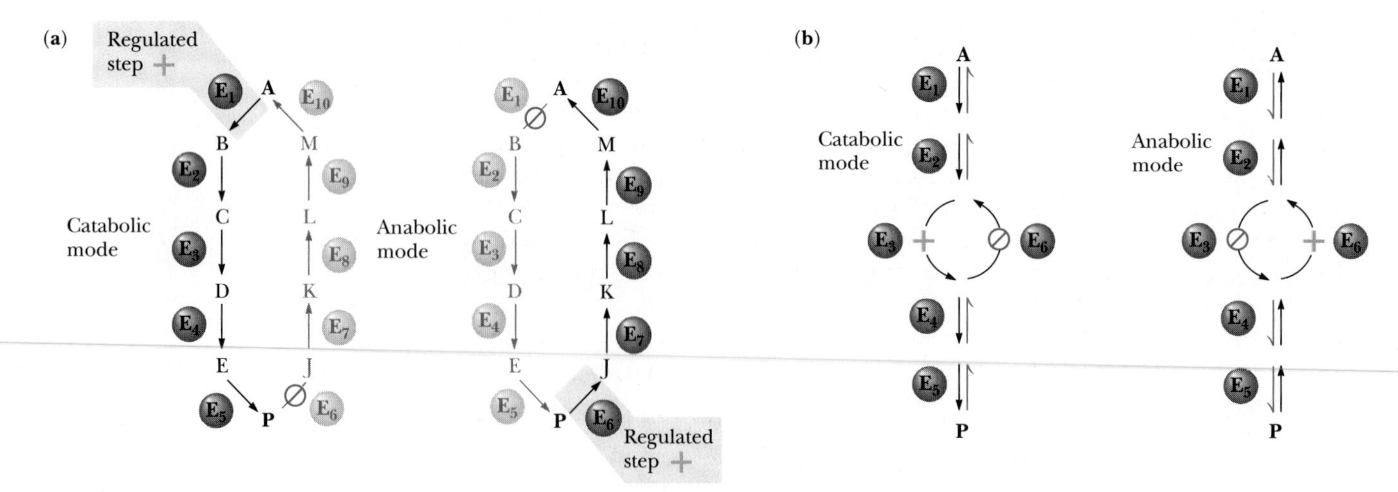

***Figure* 17.8** Parallel pathways of catabolism and anabolism must differ in at least one metabolic step in order that they can be regulated independently. Shown here are two possible arrangements of opposing catabolic and anabolic sequences between A and P. In (a), the parallel sequences proceed via independent routes. In (b), only one reaction has two different enzymes, a catabolic one (E_3) and its anabolic counterpart (E_6). These provide sites for regulation.

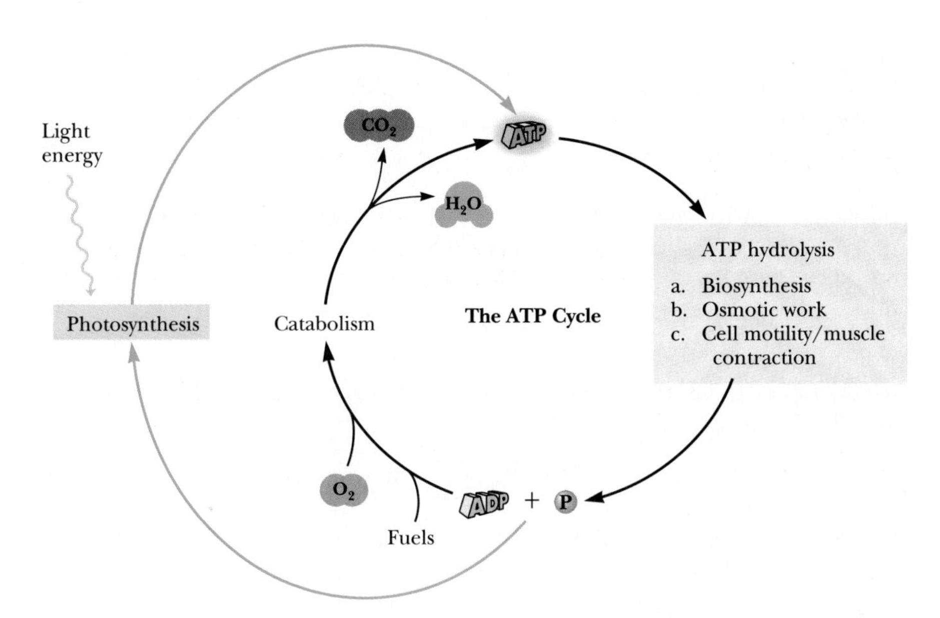

***Figure* 17.9** The ATP cycle in cells. ATP is formed via photosynthesis in phototrophic cells or catabolism in heterotrophic cells. Energy-requiring cellular activities are powered by ATP hydrolysis, liberating ADP and P_i.

form of energy-rich phosphoric anhydride bonds in ATP. In turn, ATP provides the energy that drives the manifold activities of all living cells—the synthesis of complex biomolecules, the osmotic work involved in transporting substances into cells, the work of cell motility, the work of muscle contraction. These diverse activities are all powered by energy released in the hydrolysis of ATP to ADP and P_i. Thus, there is an energy cycle in cells where ATP serves as the vessel carrying energy from photosynthesis or catabolism to the energy-requiring processes unique to living cells (Figure 17.9).

NAD^+ Collects Electrons Released in Catabolism

The substrates of catabolism—proteins, carbohydrates, and lipids—are good sources of chemical energy because the carbon atoms in these molecules are in a relatively reduced state (Figure 17.10). In the oxidative reactions of catabolism, reducing equivalents are released from these substrates, often in the form of **hydride ions** (a proton coupled with two electrons, $H:^-$). These hydride ions are transferred in enzymatic **dehydrogenase** reactions from the substrates to NAD^+ molecules, reducing them to NADH. A second proton accompanies these reactions, appearing in the overall equation as H^+ (Figure 17.11). In turn, NADH is oxidized back to NAD^+ when it transfers its reducing equivalents to electron acceptor systems that are part of the metabolic apparatus of the mitochondria. The ultimate oxidizing agent (e^- acceptor) is O_2, becoming reduced to H_2O.

Oxidation reactions are exergonic, and the energy released is coupled with the formation of ATP in a process called **oxidative phosphorylation.** The

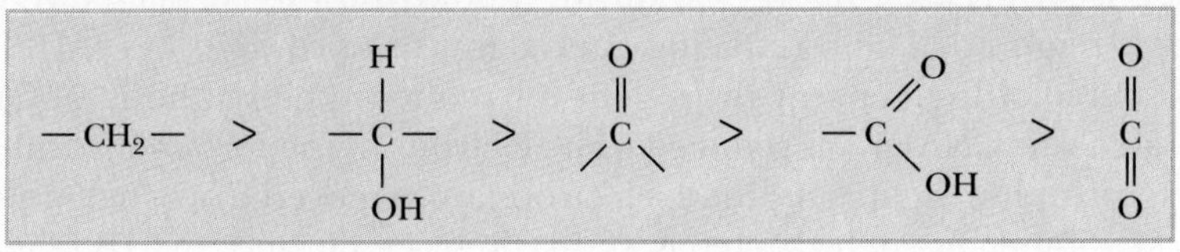

$$-CH_2- \;>\; -\underset{OH}{\overset{H}{C}}- \;>\; \overset{O}{\overset{\|}{C}} \;>\; -C\langle^{O}_{OH} \;>\; O{=}C{=}O$$

***Figure* 17.10** Comparison of the state of reduction of carbon atoms in biomolecules: $—CH_2—$ (fats) > —CHOH— (carbohydrates) > —C=O (carbonyls) > —COOH (carboxyls) > CO_2 (carbon dioxide, the final product of catabolism).

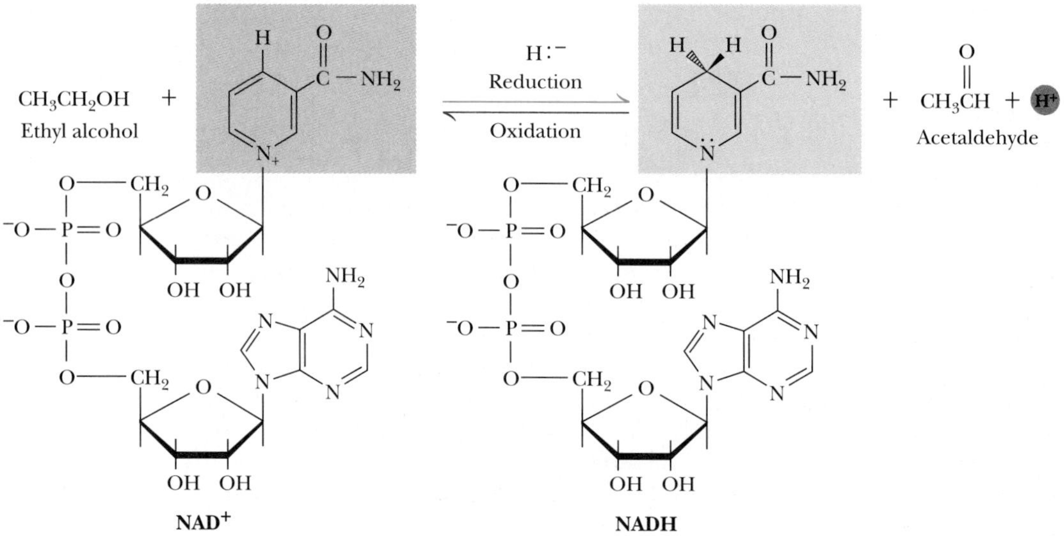

CH_3CH_2OH + NAD⁺ $\underset{\text{Oxidation}}{\overset{H:^- \text{ Reduction}}{\rightleftharpoons}}$ NADH + CH_3CHO + H^+

Ethyl alcohol — NAD$^+$ — NADH — Acetaldehyde

***Figure* 17.11** Hydrogen and electrons released in the course of oxidative catabolism are transferred as hydride ions to the pyridine nucleotide, NAD^+, to form NADH + H^+ in dehydrogenase reactions of the type

$$AH_2 + NAD^+ \rightarrow A + NADH + H^+$$

The reaction shown is catalyzed by alcohol dehydrogenase.

NAD^+–NADH system can be viewed as a *shuttle* that carries the electrons released from catabolic substrates to the mitochondria, where they are transferred to O_2, the ultimate electron acceptor in catabolism. In the process, the free energy released is trapped in ATP. The NADH cycle is an important player in the transformation of the chemical energy of carbon compounds into the chemical energy of phosphoric anhydride bonds. Such transformations of energy from one form to another are referred to as **energy transduction.** Oxidative phosphorylation is one cellular mechanism for energy transduction. Chapter 20 is devoted to electron transport reactions and oxidative phosphorylation.

NADPH Provides the Reducing Power for Anabolic Processes

Whereas catabolism is fundamentally an oxidative process, anabolism is, by its contrasting nature, reductive. The biosynthesis of the complex constituents of the cell begins at the level of intermediates derived from the degradative pathways of catabolism; or, less commonly, biosynthesis begins with oxidized substances available in the inanimate environment, such as carbon dioxide. When the hydrocarbon chains of fatty acids are assembled from acetyl-CoA units, activated hydrogens are needed to reduce the carbonyl (C=O) carbon of acetyl-CoA into a —CH_2— at every other position along the chain. When glucose is synthesized from CO_2 during photosynthesis in plants, reducing power is required. These reducing equivalents are provided by NADPH, the usual source of high-energy hydrogens for reductive biosynthesis. NADPH is generated when $NADP^+$ is reduced with electrons in the form of hydride ions. In heterotrophic organisms, these electrons are removed from fuel molecules by $NADP^+$-specific dehydrogenases. In these organisms, NADPH can be viewed as the carrier of electrons from catabolic reactions to anabolic reactions (Figure 17.12). In photosynthetic organisms, the energy of light is used to pull electrons from water and transfer them to $NADP^+$; O_2 is a by-product of this process.

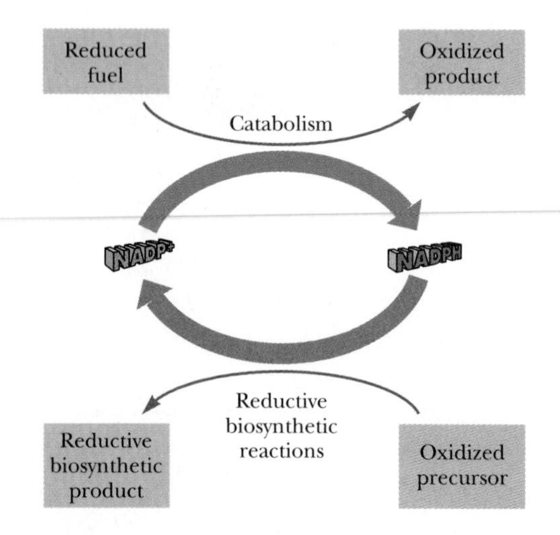

***Figure* 17.12** Transfer of reducing equivalents from catabolism to anabolism via the NADPH cycle.

17.3 Intermediary Metabolism Is a Tightly Regulated, Integrated Process

Cellular metabolism is a tightly regulated, integrated process designed by evolution to operate at maximum economy. Fuel molecules are degraded in catabolic pathways only at the rate necessary to satisfy the cell's need for reducing power (NADPH) and **phosphorylation potential** (defined as the $[ATP]/([ADP][P_i])$ ratio). In similar fashion, biosynthetic activity is matched to cellular demands for essential biomolecules; any excess intermediates are stored as fat or polysaccharide. The anabolic assembly of each cell's unique complement of informational macromolecules proceeds only at rates appropriate to both the maintenance of vital activities and the need for growth. Each of the amino acids is formed at a pace adjusted to the need for it in the synthesis of new proteins. Ribonucleotides are synthesized at rates that satisfy the requirements for RNA synthesis, and just before the time of DNA replication prior to cell division, deoxyribonucleotides are produced from ribonucleotide precursors.

Organisms Have Carbon Reserves but Not Nitrogen Reserves

Most organisms store carbon- and energy-supplying nutrients, such as carbohydrates and fats, so that these fuels can sustain cellular activities in times of need. However, organisms generally do not store proteins or nucleic acids or their simple building blocks, amino acids and nucleotides. Animal eggs and plant seeds are an exception to this rule in that they typically contain storage proteins. The embryos that develop from them require a ready source of amino acids for protein synthesis, and these amino acids come from storage protein breakdown.

Energy Metabolism Quickly Adjusts to Demands

The response of catabolic pathways to changes in the energy requirements of the organism is often dramatic. When a sprinter dashes down the track, her momentary need for ATP surpasses the production capacity of aerobic metabolism. To meet the demand, the rate of glycolysis may be increased more than 2000-fold. Lactic acid is a by-product of this glycolytic activity, and its accumulation soon leads to fatigue. In contrast, the long-distance runner uses ATP at a rate that can be sustained by aerobic metabolism, with an unavoidable sacrifice in running speed. Or consider a housefly. Its rate of ATP synthesis must increase 100-fold in less than a second in order to power its wing muscles as it leaps from the table into full flight. Obviously, regulatory mechanisms can control metabolic rates quickly and flexibly so that responses to different demands for energy are appropriate to the situation.

Metabolic Regulation Is Achieved by Regulating Enzyme Activity

The regulation of metabolism is accomplished by controlling the activities of enzymes. Three levels of control are prevalent: *allosteric regulation, covalent modification,* and *enzyme synthesis* and *degradation* (Chapter 12).

Allosteric Regulation

Allosteric regulation is a characteristic of enzymes situated at crucial junctures in metabolic pathways—for example, the first enzyme in a pathway leading to an essential end product, such as an amino acid (Figure 17.13). *Enzyme 1* is

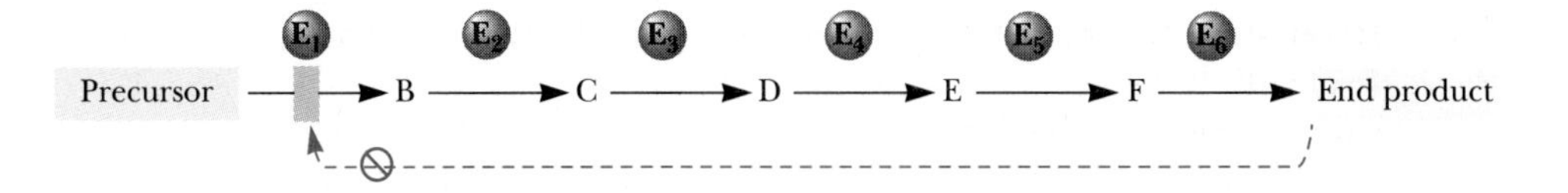

***Figure* 17.13** Allosteric regulation of a metabolic pathway by feedback or *end product* inhibition. The first enzyme in the pathway, E_1, is an allosteric enzyme; it is inhibited by the end product of the pathway.

inhibited by the end product of the pathway, a mechanism that is elegantly efficient, since if the essential end product is present, it is unnecessary to metabolize additional precursor down the pathway. Allosteric enzymes are also positioned at branch points in metabolism so that the flux of metabolites toward various fates can be precisely regulated by appropriate allosteric inhibitors and activators. In addition, allosteric enzymes may respond to inhibition or activation by intermediates that are participants in other pathways. In this manner, the rates of different pathways can be integrated with each other. In catabolic pathways, AMP, ADP, and ATP often act as important allosteric effectors. Usually, ATP signals cellular energy sufficiency and inhibits further catabolism of energy-yielding nutrients; AMP or ADP exerts an opposing effect. AMP is sometimes referred to as the "universal hunger signal" because its accumulation indicates an exhaustion of the phosphorylation potential in ATP and a need to generate ATP energy by catabolism of fuel molecules. AMP allosterically activates the catabolism of major fuel reserves, such as glycogen and fat, in the cell.

Enzyme Regulation by Covalent Modification

Covalent modification is a mechanism for "locking" an enzyme into a stable state of activity or inactivity (see Figure 12.2 in Chapter 12). A classic example is the covalent modification of *glycogen phosphorylase,* the enzyme that initiates glycogen catabolism. When a particular serine residue of glycogen phosphorylase is phosphorylated by ATP, the enzyme is locked into a more active state and glycogen degradation proceeds at a great rate (Figure 17.14). A specific modifying enzyme known as *glycogen phosphorylase kinase* (usually called simply *phosphorylase kinase*) catalyzes this covalent modification. Removal of this phosphoryl group by a different enzyme, a specific *phosphoprotein phosphatase,* returns glycogen phosphorylase to a less active state. Hormonal regulation of metabolism is often achieved through such covalent modification mecha-

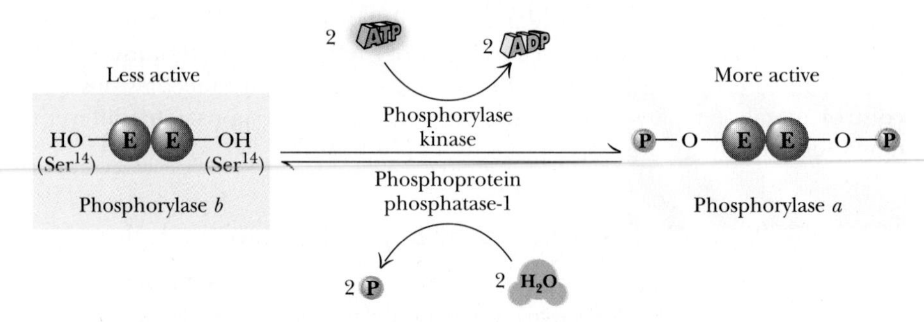

***Figure* 17.14** Covalent modification of glycogen phosphorylase. *Glycogen phosphorylase* is an interconvertible enzyme; it can exist in two forms, *a* and *b*. The phosphorylase *a* form is more active than the phosphorylase *b* form. Phosphorylase is a dimer of identical 90-kD subunits. Each polypeptide has a Ser residue at position 14. In phosphorylase *b*, this Ser is not phosphorylated. In phosphorylase *a*, each Ser^{14} residue is phosphorylated. The converter enzyme, *phosphorylase kinase,* catalyzes the ATP-dependent phosphorylation of Ser^{14} residues in phosphorylase *b*, converting it to the more active phosphorylase *a* form. *Phosphoprotein phosphatase-1* hydrolyzes the Ser-P bonds, restoring phosphorylase to the less active *b* form.

nisms. In the instance discussed here, the hormone *adrenaline* (also known as *epinephrine*) activates glycogen degradation in muscle and liver cells by initiating a chain of reactions that culminates in activation of phosphorylase kinase and phosphorylation of glycogen phosphorylase. The regulatory circuits involving hormone activation and covalent modification reactions will be presented in detail in subsequent chapters, as we examine particular aspects of metabolism.

Enzyme Regulation Through Synthesis and Degradation

The third level of metabolic regulation consists of **enzyme synthesis** and **degradation,** which act to control the concentration of the enzyme in the cell. The amount of an enzyme at any given moment is the result of a balance between its rate of synthesis and its rate of degradation; changing either rate alters the cellular concentration of the enzyme. Usually (but not always) this level of regulation occurs through controls acting on enzyme synthesis at the level of gene expression. Two general mechanisms, *induction* and *repression,* regulate gene expression.

Induction of Enzyme Synthesis

In **induction,** enzyme synthesis is increased by activating transcription of the gene encoding the enzyme. Translation of the mRNA transcript leads to synthesis of enzyme protein and increased enzyme activity (Figure 17.15). (In some cases, induction occurs via increased translation of a preexisting stable mRNA.) Induction is typically initiated indirectly by the availability of nutrients. To take advantage of these nutrients, the appropriate metabolic machin-

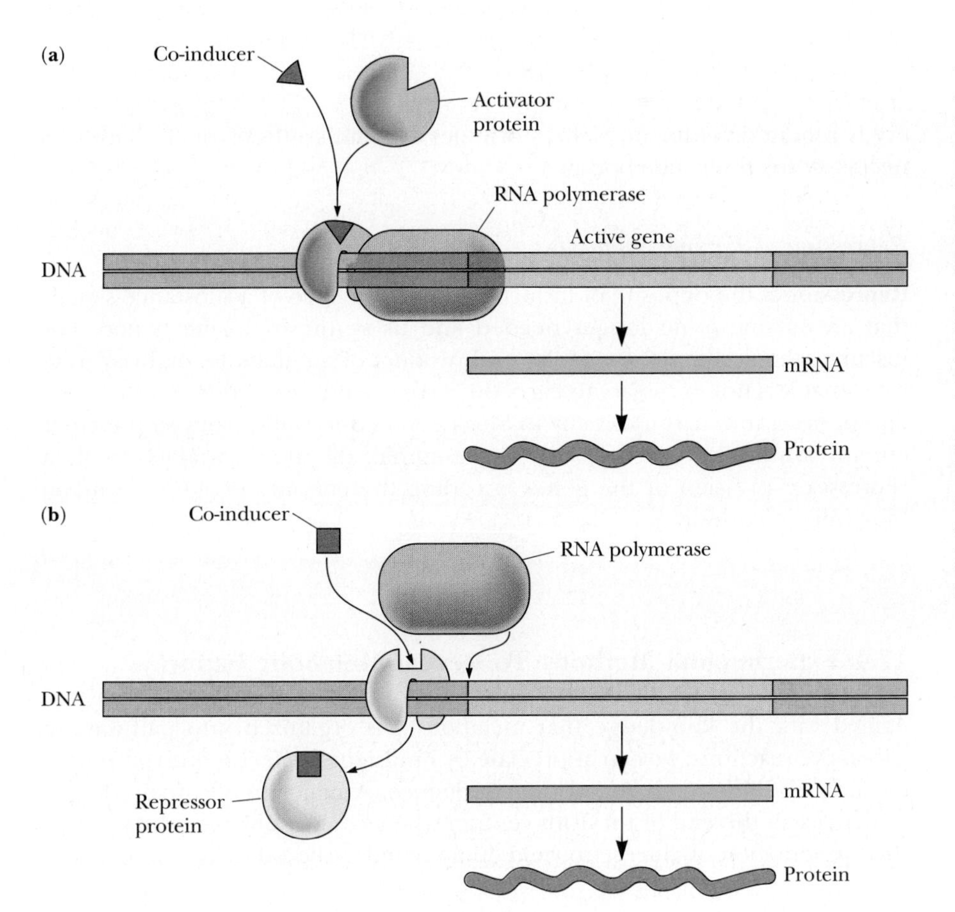

***Figure* 17.15** Model of enzyme induction. Activation of transcription results from co-inducer accumulation, which leads to transcription of the gene, translation of the mRNA formed, and an increase in enzyme protein; increased enzyme activity is the consequence. The co-inducer is typically a small metabolite, usually a substrate of the pathway dependent on the enzymes induced. (a) **Positive control:** Induction can result from binding of the co-inducer to a protein that activates gene transcription. (b) **Negative control:** Binding of the co-inducer to an apo-repressor protein that is blocking transcription allows the gene to be expressed. In this case, the co-inducer:repressor protein complex has a lower DNA-binding affinity than the free apo-repressor.

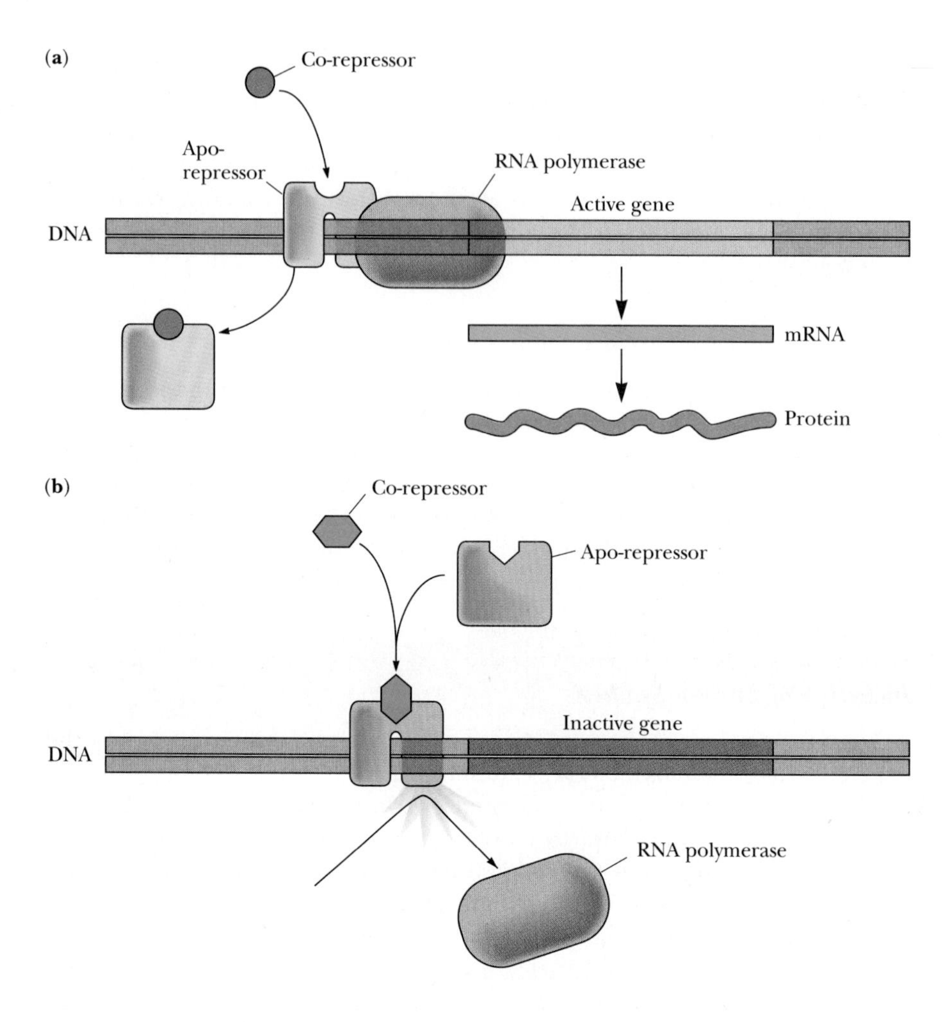

***Figure* 17.16** Model of gene repression. In repression, accumulation of the co-repressor causes cessation of gene expression and decay in the enzyme levels in the cell. Co-repressors are typically end products of pathways; their accumulation signals the end of any need for the enzymes comprising the pathway. (a) **Positive control:** The apo-repressor is an activator of gene transcription. When the co-repressor binds, it causes dissociation of the activator from the gene and a halt in transcription. (b) **Negative control:** The repressor protein blocks transcription but has no affinity for its DNA-binding site unless it has bound the co-repressor.

ery is required. Thus, nutrient molecules activate synthesis of the enzymes necessary for their metabolism.

Repression of Enzyme Synthesis

Repression is the opposite of induction—the presence of a substance signals that an enzyme is no longer needed and its synthesis can be halted. For instance, the accumulation of the end product of an anabolic pathway indicates that it is not necessary to carry on further synthesis of this product. The end product thus acts indirectly to block gene transcription and stop further enzyme synthesis (Figure 17.16). For example, the amino acid tryptophan represses expression of the genes encoding the enzymes of the tryptophan biosynthetic pathway.

17.4 Experimental Methods To Reveal Metabolic Pathways

Armed with the knowledge that metabolism is organized into pathways of successive reactions, we can appreciate by hindsight the techniques employed by early biochemists to reveal their sequence. A major intellectual advance took place at the end of the 19th century when Eduard Buchner showed that the fermentation of glucose to yield ethanol and carbon dioxide can occur in

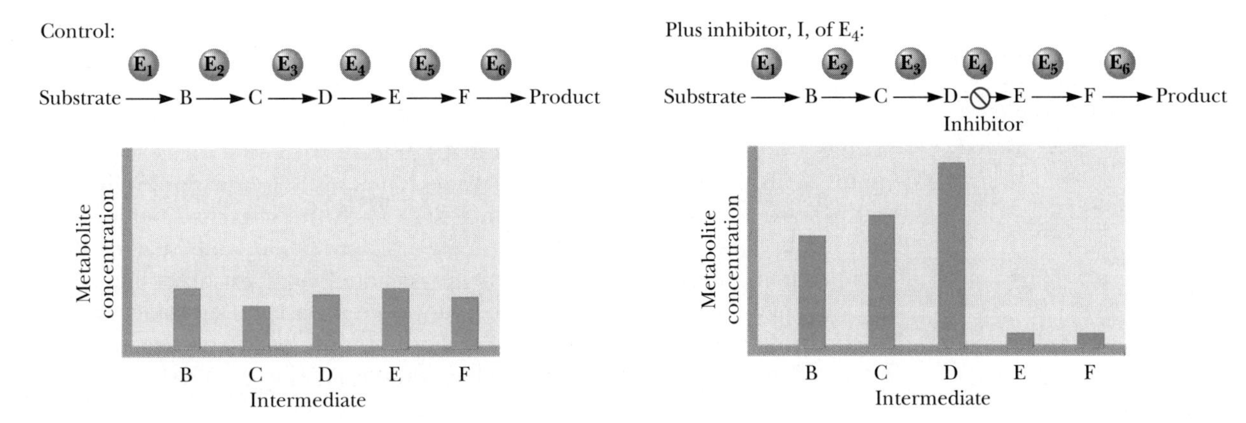

***Figure* 17.17** The use of inhibitors to reveal the sequence of reactions in a metabolic pathway. (a) **Control:** Under normal conditions, the steady-state concentrations of a series of intermediates will be determined by the relative activities of the enzymes in the pathway. (b) **Plus inhibitor:** In the presence of an inhibitor (in this case, an inhibitor of *enzyme 4*), intermediates upstream of the metabolic block (B, C, and D) accumulate, revealing themselves as intermediates in the pathway. The concentration of intermediates lying downstream (E and F) will fall.

extracts of broken yeast cells. Until this discovery, many thought that metabolism was a vital property, unique to intact cells; even the eminent microbiologist Louis Pasteur, who contributed so much to our understanding of fermentation, was a *vitalist,* one of those who believed that the processes of living substance transcend the laws of chemistry and physics. After Buchner's revelation, biochemists searched for intermediates in the transformation of glucose and soon learned that inorganic phosphate was essential to glucose breakdown. This observation gradually led to the discovery of a variety of phosphorylated organic compounds that serve as intermediates along the fermentative pathway.

An important tool for elucidating the steps in the pathway was the use of *metabolic inhibitors.* Adding an enzyme inhibitor to a cell-free extract caused an accumulation of intermediates in the pathway prior to the point of inhibition (Figure 17.17). Each inhibitor was specific for a particular site in the sequence of metabolic events. As the arsenal of inhibitors was expanded, the individual steps in metabolism were revealed.

Mutations Create Specific Metabolic Blocks

Genetics provides an approach to the identification of intermediate steps in metabolism that is somewhat analogous to inhibition. Mutation in a gene encoding an enzyme often results in an inability to synthesize the enzyme in an active form. Such a defect leads to a block in the metabolic pathway at the point where the enzyme acts, and the enzyme's substrate accumulates. Such genetic disorders are lethal if the end product of the pathway is essential or if the accumulated intermediates have toxic effects. In microorganisms, however, it is often possible to manipulate the growth medium so that essential end products are provided. Then the biochemical consequences of the mutation can be investigated. Studies on mutations in genes of the filamentous fungus *Neurospora crassa* led G. W. Beadle and E. L. Tatum to hypothesize in 1941 that genes are units of heredity that encode enzymes (the "one gene–one enzyme" hypothesis; see Chapter 28).

Isotopic Tracers as Metabolic Probes

Another widely used approach to the elucidation of metabolic sequences is to "feed" cells a substrate or metabolic intermediate labeled with a particular isotopic form of an element that can be traced. Two sorts of isotopes are useful in this regard: radioactive isotopes, such as ^{14}C, and stable "heavy" isotopes, such as ^{18}O or ^{15}N (Table 17.3). Since the chemical behavior of isotopically labeled compounds is rarely distinguishable from that of their unlabeled counterparts, isotopes provide reliable "tags" for observing metabolic changes. The metabolic fate of a radioactively labeled substance can be traced by determining the presence and position of the radioactive atoms in intermediates derived from the labeled compound (Figure 17.18).

Heavy Isotopes

Heavy isotopes endow the compounds in which they appear with slightly greater masses than their unlabeled counterparts. These compounds can be separated and quantitated by mass spectrometry (or density gradient centrifugation, if they are macromolecules). For example, ^{18}O was used in separate experiments as a tracer of the fate of the oxygen atoms in water and carbon dioxide to determine whether the atmospheric oxygen produced in photosynthesis arose from H_2O, CO_2, or both:

$$CO_2 + H_2O \longrightarrow (CH_2O) + O_2$$

If ^{18}O-labeled CO_2 was presented to a green plant carrying out photosynthesis, none of the ^{18}O was found in O_2. Curiously, it was recovered as $H_2{}^{18}O$. In contrast, when plants fixing CO_2 were equilibrated with $H_2{}^{18}O$, $^{18}O_2$ was evolved. These latter labeling experiments established that photosynthesis is best described by the equation

$$C^{16}O_2 + 2\,H_2{}^{18}O \longrightarrow (CH_2{}^{16}O) + {}^{18}O_2 + H_2{}^{16}O$$

***Table* 17.3**

Properties of Radioactive and Stable "Heavy" Isotopes Used as Tracers in Metabolic Studies

Isotope	Type	Radiation Type	Half-Life	Relative Abundance*
^{2}H	Stable			0.0154%
^{3}H	Radioactive	β^-	12.1 yr	
^{13}C	Stable			1.1%
^{14}C	Radioactive	β^-	5700 yr	
^{15}N	Stable			0.365%
^{18}O	Stable			0.204%
^{24}Na	Radioactive	β^-, γ	15 hr	
^{32}P	Radioactive	β^-	14.3 days	
^{35}S	Radioactive	β^-	87.1 days	
^{36}Cl	Radioactive	β^-	310,000 yr	
^{42}K	Radioactive	β^-	12.5 hr	
^{45}Ca	Radioactive	β^-	152 days	
^{59}Fe	Radioactive	β^-, γ	45 days	
^{131}I	Radioactive	β^-, γ	8 days	

*The relative natural abundance of a stable isotope is important since, in tracer studies, the amount of stable isotope is typically expressed in terms of atoms percent excess over the natural abundance of the isotope.

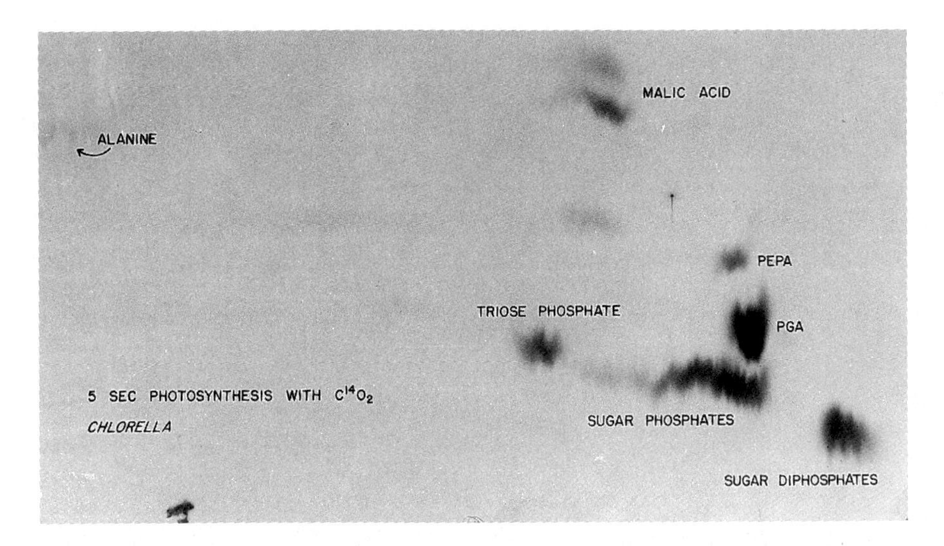

***Figure* 17.18** One of the earliest experiments using a radioactive isotope as a metabolic tracer. Cells of *Chlorella* (a green alga) synthesizing carbohydrate from carbon dioxide were exposed briefly (5 sec) to ^{14}C-labeled CO_2. The products of CO_2 incorporation were then quickly isolated from the cells, separated by two-dimensional paper chromatography, and observed via autoradiographic exposure of the chromatogram. Such experiments identified radioactive 3-phosphoglycerate (PGA) as the primary product of CO_2 fixation. The 3-phosphoglycerate was labeled in the 1-position (in its carboxyl group). Radioactive compounds arising from the conversion of 3-phosphoglycerate to other metabolic intermediates included phosphoenolpyruvate (PEP), malic acid, triose phosphate, alanine, and sugar phosphates and diphosphates.

(Photograph courtesy of Professor Melvin Calvin, Lawrence Berkeley Laboratory, University of California, Berkeley.)

That is, in the process of photosynthesis, the two oxygen atoms in O_2 come from two H_2O molecules. One O is lost from CO_2 and appears in H_2O, and the other O of CO_2 is retained in the carbohydrate product. Two of the four H atoms are accounted for in (CH_2O), and two reduce the O lost from CO_2 to H_2O.

NMR as a Metabolic Probe

A technology analogous to isotopic tracers is provided by **nuclear magnetic resonance (NMR) spectroscopy.** The atomic nuclei of certain isotopes, such as the naturally occurring isotope of phosphorus, ^{31}P, have *magnetic moments.* The resonance frequency of a magnetic moment is influenced by the local chemical environment. That is, the NMR signal of the nucleus is influenced in an identifiable way by the chemical nature of its neighboring atoms in the compound. In many ways, these nuclei are ideal tracers since their signals contain a great deal of structural information about the environment around the atom, and thus the nature of the compound containing the atom. Transformations of substrates and metabolic intermediates labeled with magnetic nuclei can be traced by following changes in NMR spectra. Furthermore, NMR spectroscopy is a noninvasive procedure. Whole-body NMR spectrometers are being used today in hospitals to directly observe the metabolism (and clinical condition) of living subjects (Figure 17.19). NMR promises to be a revolutionary tool for clinical diagnosis and for the investigation of metabolism *in situ* (literally "in site," meaning, in this case, "where and as it happens").

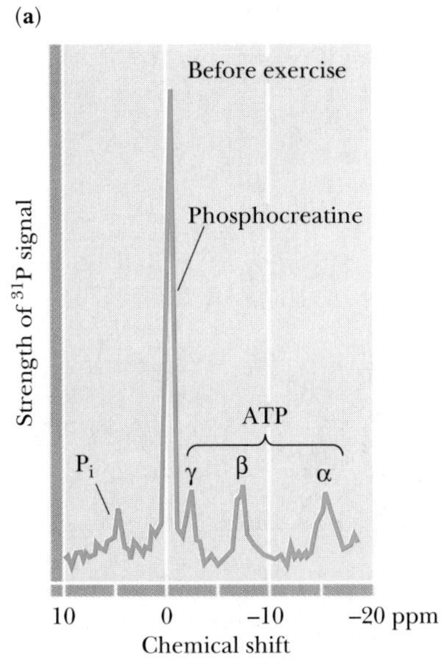

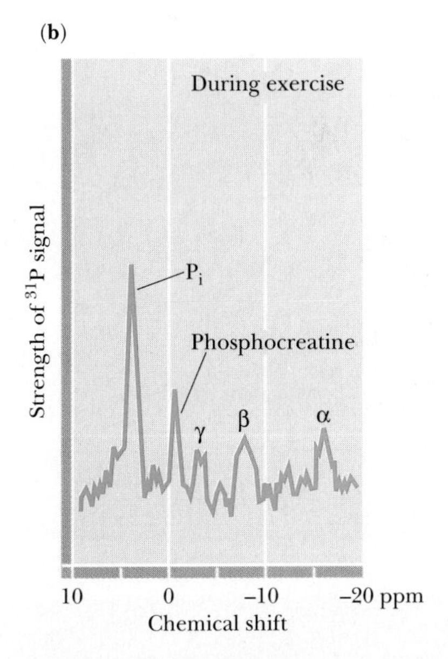

***Figure* 17.19** With NMR spectroscopy one can observe the metabolism of a living subject in real time. These NMR spectra show the changes in ATP, creatine-P (phosphocreatine), and P_i levels in the forearm muscle of a human subjected to 19 minutes of exercise. Note that the three P atoms of ATP (α, β, and γ) have different chemical shifts, reflecting their different chemical environments.

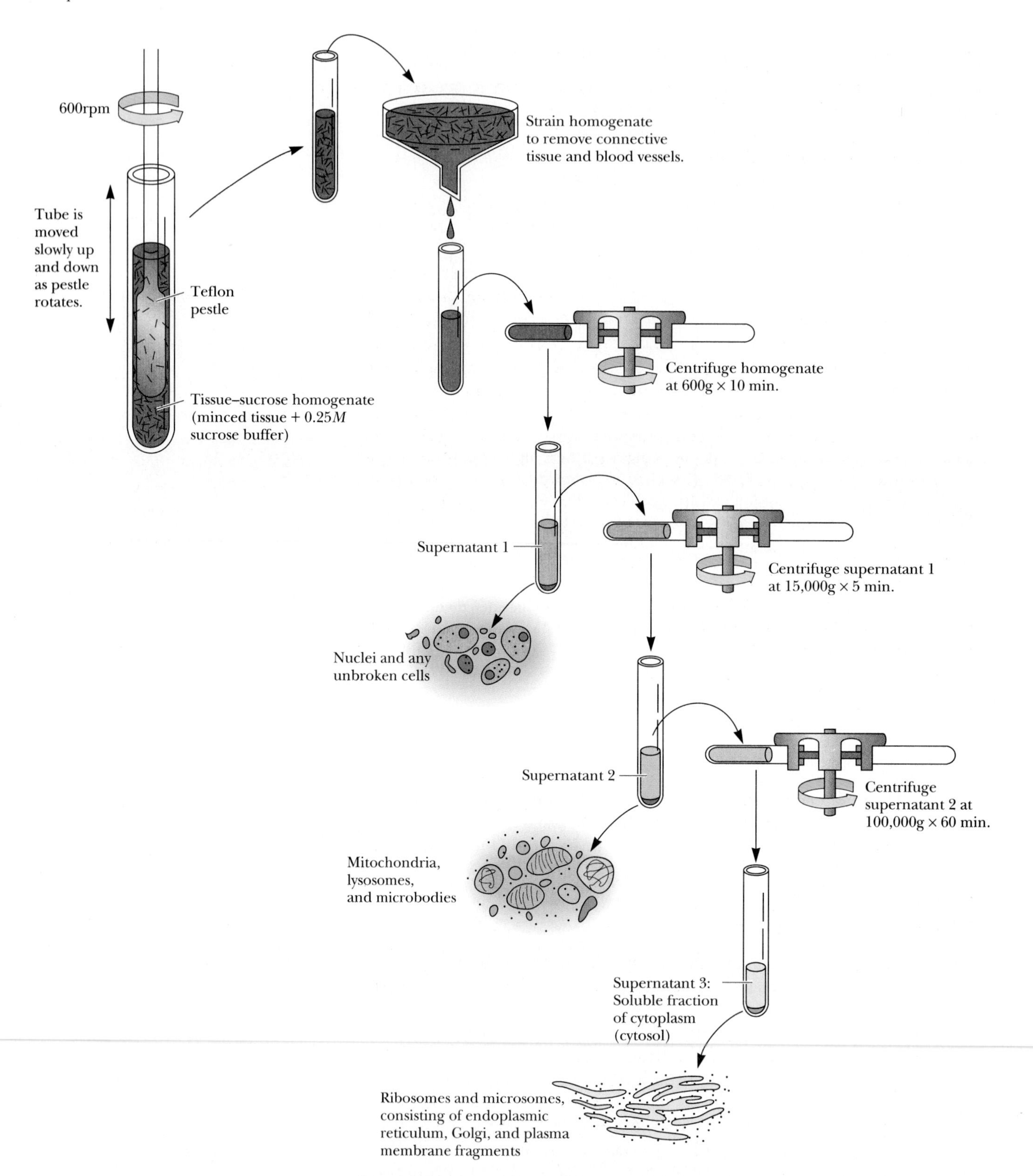
600rpm
Tube is moved slowly up and down as pestle rotates.
Teflon pestle
Tissue–sucrose homogenate (minced tissue + 0.25M sucrose buffer)
Strain homogenate to remove connective tissue and blood vessels.
Centrifuge homogenate at 600g × 10 min.
Supernatant 1
Nuclei and any unbroken cells
Centrifuge supernatant 1 at 15,000g × 5 min.
Supernatant 2
Mitochondria, lysosomes, and microbodies
Centrifuge supernatant 2 at 100,000g × 60 min.
Supernatant 3: Soluble fraction of cytoplasm (cytosol)
Ribosomes and microsomes, consisting of endoplasmic reticulum, Golgi, and plasma membrane fragments

***Figure* 17.20** Fractionation of a cell extract by differential centrifugation. It is possible to separate organelles and subcellular particles in a centrifuge because their inherent size and density differences give them different rates of sedimentation in an applied centrifugal field. Nuclei are pelleted in relatively low centrifugal fields, mitochondria in somewhat stronger fields, whereas strong centrifugal fields are necessary to pellet ribosomes and fragments of the endomembrane system.

Metabolic Pathways Are Compartmentalized Within Cells

Although the interior of a prokaryotic cell is not subdivided into compartments by internal membranes, the cell still shows some segregation of metabolism. For example, certain metabolic pathways, such as phospholipid synthesis and oxidative phosphorylation, are localized in the plasma membrane. Also, protein biosynthesis is carried out on ribosomes.

In contrast, eukaryotic cells are extensively compartmentalized by an endomembrane system. Each of these cells has a true nucleus bounded by a double membrane called the *nuclear envelope.* The nuclear envelope is continuous with the endomembrane system, which is composed of differentiated regions: the endoplasmic reticulum; the Golgi complex; various membrane-bounded vesicles such as lysosomes, vacuoles, and microbodies; and, ultimately, the plasma membrane itself. Eukaryotic cells also possess mitochondria and, if they are photosynthetic, chloroplasts. Disruption of the cell membrane and fractionation of the cell contents into the component organelles have allowed an analysis of their respective functions (Figure 17.20). Each compartment is dedicated to specialized metabolic functions, and the enzymes appropriate to these specialized functions are confined together within the organelle. In many instances, the enzymes of a metabolic sequence occur together within the organellar membrane. Thus, *the flow of metabolic intermediates in the cell is spatially as well as chemically segregated.* For example, the 10 enzymes of glycolysis are found in the cytosol, but pyruvate, the product of glycolysis, is fed into the mitochondria. These organelles contain the citric acid cycle enzymes, which oxidize pyruvate to CO_2. The great amount of energy released in the process is captured by the oxidative phosphorylation system of mitochondrial membranes and used to drive the formation of ATP (Figure 17.21).

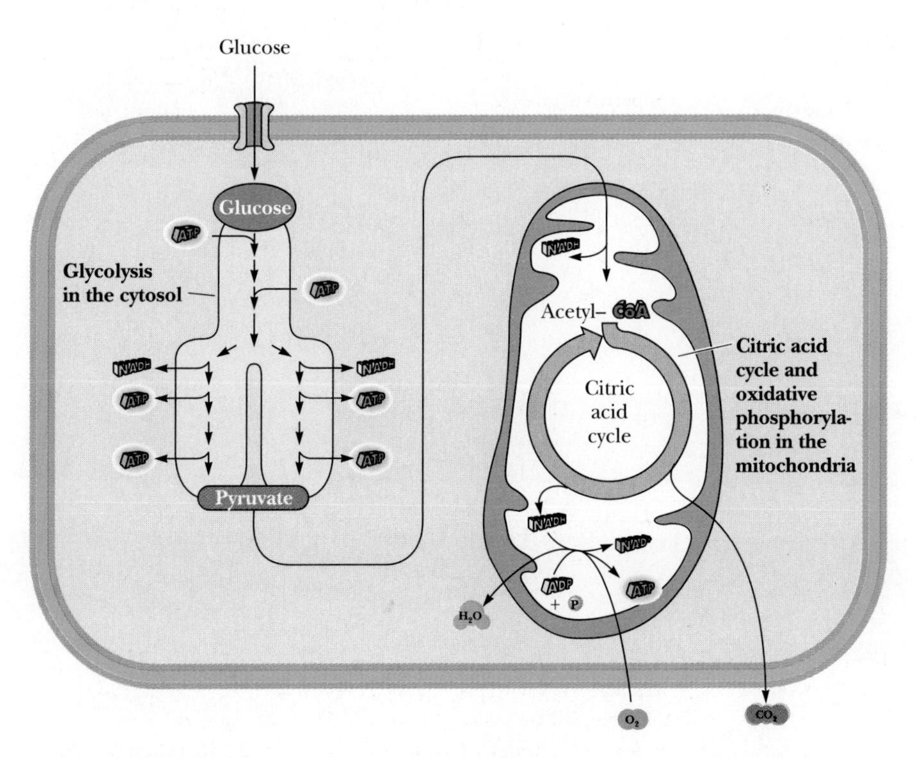

***Figure* 17.21** Compartmentalization of glycolysis, the citric acid cycle, and oxidative phosphorylation.

Problems

1. If 3×10^{14} kg of CO_2 are cycled through the biosphere annually, how many human equivalents (70-kg persons composed of 18% carbon by weight) could be produced each year from this amount of CO_2?

2. Define the differences in carbon and energy metabolism between *photoautotrophs* and *photoheterotrophs,* and between *chemoautotrophs* and *chemoheterotrophs.*

3. What are the three principal inorganic sources of oxygen atoms that are commonly available in the inanimate environment and readily accessible to the biosphere?

4. Keeping in mind the earth's predominantly oxidative environment, what are the two most common forms of inorganic nitrogen available to organisms?

5. What are the features that generally distinguish pathways of catabolism from pathways of anabolism?

6. Name the three principal modes of enzyme organization in metabolic pathways.

7. Why do metabolic pathways have so many different steps?

8. Why is the pathway for the biosynthesis of a biomolecule at least partially different from the pathway for its catabolism? Why is the pathway for the biosynthesis of a biomolecule inherently more complex than the pathway for its degradation?

9. What are the metabolic roles of ATP, NAD^+, and NADPH?

10. Metabolic regulation is achieved via regulating enzyme activity in three prominent ways: allosteric regulation, covalent modification, and enzyme synthesis and degradation. Which of these three modes of regulation is likely to be the quickest; which the slowest? For each of these general enzyme regulatory mechanisms, cite conditions in which cells might employ that mode in preference to either of the other two.

11. What are the advantages of compartmentalizing particular metabolic pathways within specific organelles?

12. If *phosphorylation potential,* Γ (the potential for ATP to serve as a phosphoryl donor), is defined as the ratio of [ATP] to the product of the concentrations of its hydrolysis products, [ADP] and [P_i], what are the units of Γ?

    What is the value of Γ in a cell containing 1 mM P_i and an [ATP]/[ADP] ratio of 10? What is the ΔG of ATP hydrolysis under these conditions?

Further Reading

Atkinson, Daniel E., 1977. *Cellular Energy Metabolism and Its Regulation.* New York: Academic Press. A monograph on energy metabolism that is filled with novel insights regarding the ability of cells to generate energy in a carefully regulated fashion while contending with the thermodynamic realities of life.

Cooper, Terrance G., 1977. *The Tools of Biochemistry.* New York: Wiley-Interscience. Chapter 3, "Radiochemistry," discusses techniques for using radioisotopes in biochemistry.

Srere, Paul A., 1987. Complexes of sequential metabolic enzymes. *Annual Review of Biochemistry* **56:**89–124. A review of how enzymes in some metabolic pathways are organized into complexes.

Chapter 18

Glycolysis

"Living organisms, like machines, conform to the law of conservation of energy, and must pay for all their activities in the currency of catabolism."

Ernest Baldwin, *Dynamic Aspects of Biochemistry* (1952)

Outline

18.1 Overview of Glycolysis
18.2 The Importance of Coupled Reactions in Glycolysis
18.3 The First Phase of Glycolysis
18.4 The Second Phase of Glycolysis
18.5 The Metabolic Fates of NADH and Pyruvate—The Products of Glycolysis
18.6 Anaerobic Pathways for Pyruvate
18.7 The Energetic Elegance of Glycolysis
18.8 Utilization of Other Substrates in Glycolysis

Louis Pasteur in his laboratory. Pasteur's scientific investigations into fermentation of sugar were sponsored by the French wine industry.

Nearly every living cell carries out a catabolic process known as **glycolysis**—the stepwise degradation of glucose (and other simple sugars). Glycolysis is a paradigm of metabolic pathways. Carried out in the cytosol of cells, it is basically an anaerobic process; its principal steps occur with no requirement for oxygen. Living things first appeared in an environment lacking O_2, and glycolysis was an early and important pathway for extracting energy from nutrient molecules. It played a central role in anaerobic metabolic processes during the first 2 billion years of biological evolution on earth. Modern organisms still employ glycolysis to provide precursor molecules for aerobic catabolic pathways (such as the tricarboxylic acid cycle) and as a short-term energy source when oxygen is limiting.

glycolysis: from the Greek *glyk-*, sweet, and *lysis*, splitting

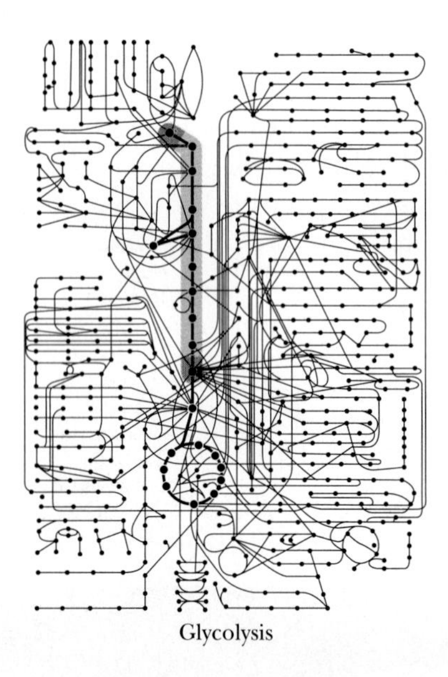
Glycolysis

18.1 Overview of Glycolysis

An overview of the glycolytic pathway is presented in Figure 18.1, page 571. Most of the details of this pathway (the first metabolic pathway to be elucidated) were worked out in the first half of the 20th century by the German biochemists Otto Warburg, G. Embden, and O. Meyerhof. In fact, the sequence of reactions in Figure 18.1 is often referred to as the **Embden–Meyerhof pathway.**

Glycolysis consists of two phases. In the first, a series of five reactions, glucose is broken down to two molecules of glyceraldehyde-3-phosphate. In the second phase, five subsequent reactions convert these two molecules of glyceraldehyde-3-phosphate into two molecules of pyruvate. Phase 1 consumes two molecules of ATP (Figure 18.2). The later stages of glycolysis result in the production of four molecules of ATP. The net is $4 - 2 = 2$ molecules of ATP produced per molecule of glucose.

Rates and Regulation of Glycolytic Reactions Vary Among Species

Microorganisms, plants, and animals (including humans) carry out the 10 reactions of glycolysis in more or less similar fashion, although the rates of the individual reactions and the means by which they are regulated differ from species to species. The most significant difference among species, however, is the way in which the product pyruvate is utilized. The three possible paths for pyruvate are shown in Figure 18.1. In aerobic organisms, including humans, pyruvate is oxidized (with loss of the carboxyl group as CO_2), and the remaining two-carbon unit becomes the acetyl group of acetyl-coenzyme A. This acetyl group is metabolized by the tricarboxylic acid cycle (and fully oxidized) to yield CO_2. The electrons removed in this oxidation process are subsequently passed through the mitochondrial electron transport system and used to generate molecules of ATP by oxidative phosphorylation, thus capturing most of the metabolic energy available in the original glucose molecule.

18.2 The Importance of Coupled Reactions in Glycolysis

The process of glycolysis converts some, but not all, of the metabolic energy of the glucose molecule into ATP. The free energy change for the conversion of glucose to two molecules of lactate (the anaerobic route shown in Figure 18.1) is -183.6 kJ/mol:

$$C_6H_{12}O_6 \longrightarrow 2\ H_3C\text{—CHOH—COO}^- + 2H^+ \qquad \Delta G°' = -183.6\ \text{kJ/mol} \tag{18.1}$$

This process occurs with no net oxidation or reduction. Though several individual steps in the pathway involve oxidation or reduction, these steps compensate each other exactly. Thus, the conversion of a molecule of glucose to two molecules of lactate involves simply a rearrangement of bonds, with no net loss or gain of electrons. The energy made available through this rearrangement into a more stable (lower energy) form is a relatively small part of the total energy obtainable from glucose.

The production of two molecules of ATP in glycolysis is an energy-requiring process:

$$2\ \text{ADP} + 2\ P_i \longrightarrow 2\ \text{ATP} + 2\ H_2O \qquad \Delta G°' = 2 \times 30.5\ \text{kJ/mol} = 61.0\ \text{kJ/mol} \tag{18.2}$$

Glycolysis couples these two reactions:

$$\text{Glucose} + 2\ \text{ADP} + 2\ P_i \longrightarrow 2\ \text{lactate} + 2\ \text{ATP} + 2\ H^+ + 2\ H_2O \qquad \Delta G°' = -183.6 + 61 = -122.6\ \text{kJ/mol} \tag{18.3}$$

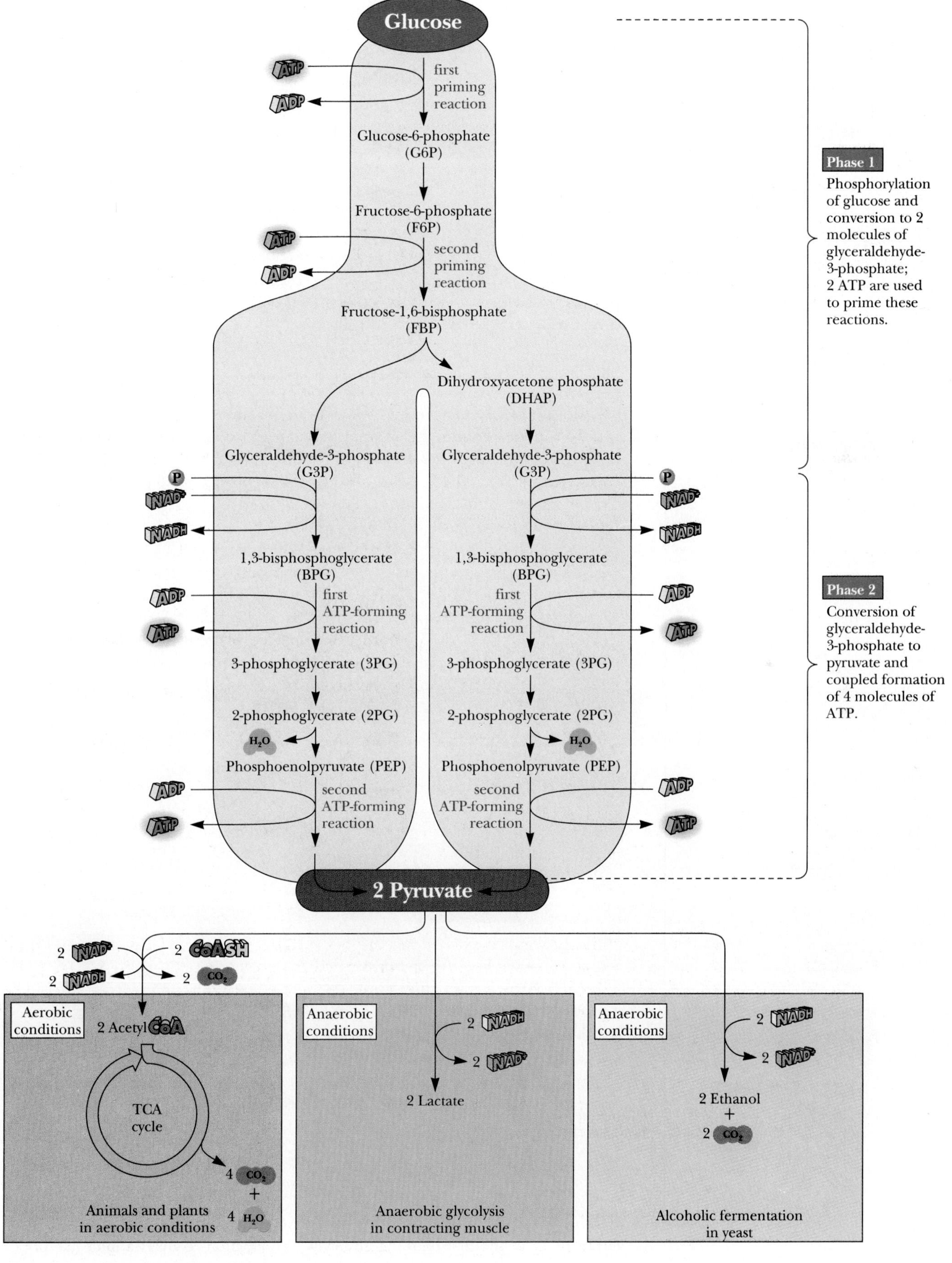

***Figure* 18.1** The glycolytic pathway.

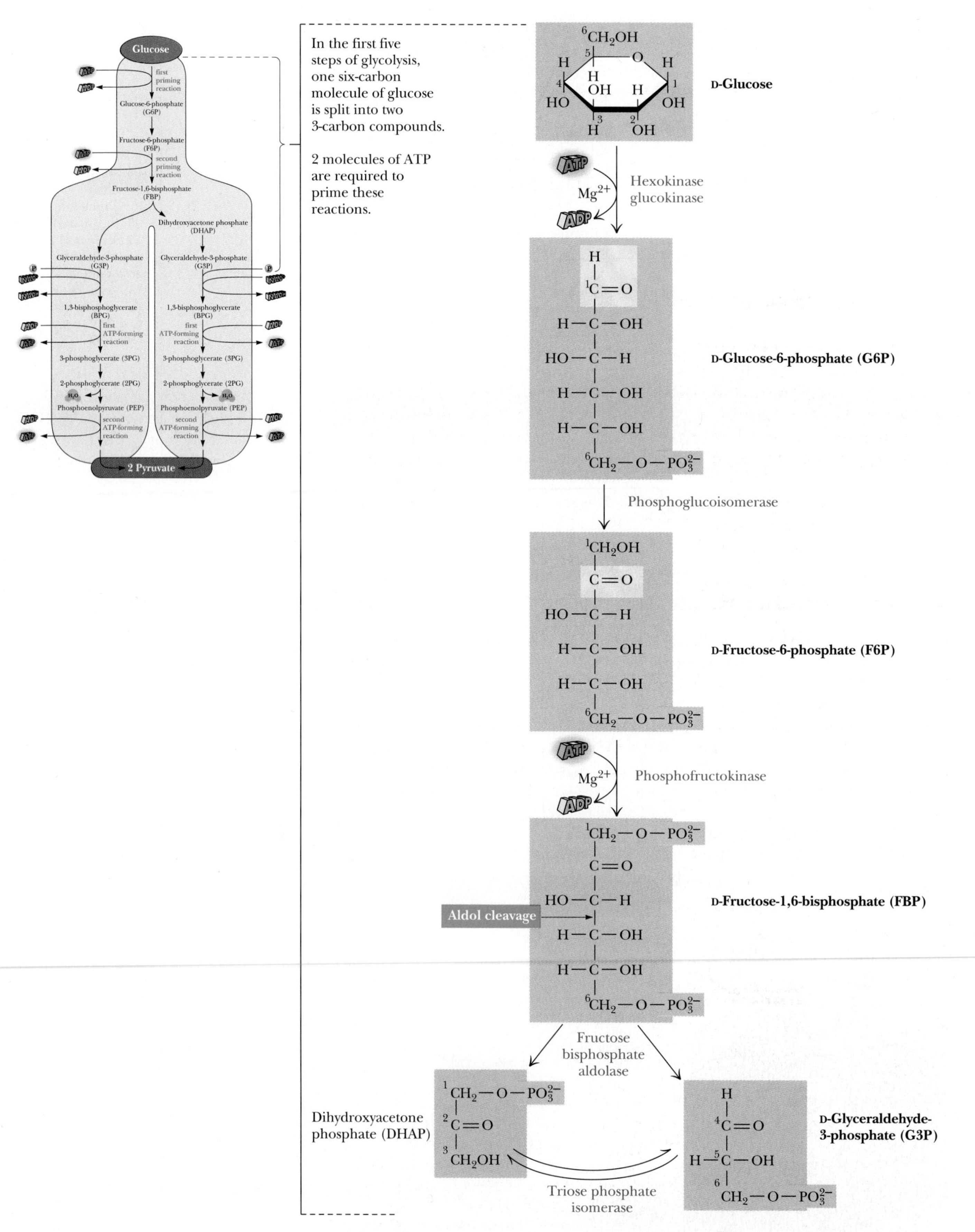

Figure **18.2** In the first phase of glycolysis, five reactions convert a molecule of glucose to two molecules of glyceraldehyde-3-phosphate.

Thus, under standard-state conditions, $(61/183.6) \times 100\%$, or 33%, of the free energy released is preserved in the form of ATP in these reactions. However, as we discussed in Chapter 16, the various solution conditions, such as pH, concentration, ionic strength, and presence of metal ions, can substantially alter the free energy change for such reactions. Under actual cellular conditions, the free energy change for the synthesis of ATP (Equation 18.2) is much larger, and approximately 50% of the available free energy is converted into ATP. Clearly, then, more than enough free energy is available in the conversion of glucose into lactate to drive the synthesis of two molecules of ATP.

18.3 The First Phase of Glycolysis

One way to synthesize ATP using the metabolic free energy contained in the glucose molecule would be to convert glucose into one (or more) of the high-energy phosphates in Table 16.1, which have standard hydrolysis free energies greater than that of ATP. Those molecules in Table 16.1 that can be synthesized easily from glucose are phosphoenolpyruvate, 1,3-bisphosphoglycerate, and acetyl phosphate. In fact, in the first stage of glycolysis, glucose is converted into two molecules of glyceraldehyde-3-phosphate. Energy released from this high-energy molecule in the second phase of glycolysis is then used to synthesize ATP.

Reaction 1: Phosphorylation of Glucose by Hexokinase or Glucokinase—The First Priming Reaction

The initial reaction of the glycolysis pathway involves phosphorylation of glucose at carbon atom 6 by either hexokinase or glucokinase. The formation of such a phosphoester is thermodynamically unfavorable and requires energy input to operate in the forward direction (Chapter 16). The energy comes from ATP, a requirement that at first seems counterproductive. Glycolysis is designed to *make* ATP, not consume it. However, the hexokinase, glucokinase reaction (Figure 18.2) is one of two **priming reactions** in the cycle. Just as old-fashioned, hand-operated water pumps (Figure 18.3) have to be primed

***Figure* 18.3** Just as a water pump must be "primed" with water to get more water out, the glycolytic pathway is primed with ATP in steps 1 and 3, in order to achieve net production of ATP in the second phase of the pathway.

Table **18.1**
Reactions and Thermodynamics of Glycolysis

Reaction	Enzyme
α-D-Glucose + ATP^{4-} ⇌ glucose-6-phosphate + ADP^{3-} + H^+	Hexokinase
	Hexokinase
	Glucokinase
Glucose-6-phosphate ⇌ fructose-6-phosphate	Phosphoglucoisomerase
Fructose-6-phosphate + ATP^{4-} ⇌ fructose-1,6-bisphosphate + ADP^{3-} + H^+	Phosphofructokinase
Fructose-1,6-bisphosphate ⇌ dihydroxyacetone-P + glyceraldehyde-3-P	Fructose bisphosphate aldolase
Dihydroxyacetone-P ⇌ glyceraldehyde-3-P	Triose phosphate isomerase
Glyceraldehyde-3-P + P_i + NAD^+ ⇌ 1,3-bisphosphoglycerate + NADH + H^+	Glyceraldehyde-3-P dehydrogenase
1,3-Bisphosphoglycerate + ADP^{3-} ⇌ 3-P-glycerate + ATP^{4-}	Phosphoglycerate kinase
3-Phosphoglycerate ⇌ 2-phosphoglycerate	Phosphoglycerate mutase
2-Phosphoglycerate ⇌ phosphoenolpyruvate + H_2O	Enolase
Phosphoenolpyruvate + ADP^{3-} + H^+ ⇌ pyruvate + ATP^{4-}	Pyruvate kinase
Pyruvate + NADH + H^+ ⇌ lactate + NAD^+	Lactate dehydrogenase

continued

with a small amount of water to deliver more water to the thirsty pumper, the glycolysis pathway requires two priming ATP molecules to start the sequence of reactions and delivers four molecules of ATP in the end.

The complete reaction for the first step in glycolysis is

$$\alpha\text{-D-Glucose} + ATP^{4-} \longrightarrow \alpha\text{-D-glucose-6-phosphate}^{2-} + ADP^{3-} + H^+ \quad (18.4)$$
$$\Delta G^{\circ\prime} = -16.7 \text{ kJ/mol}$$

The hydrolysis of ATP makes 30.5 kJ/mol available in this reaction, and the phosphorylation of glucose "costs" 13.8 kJ/mol (see Table 16.1). Thus, the reaction liberates 16.7 kJ/mol under standard-state conditions (1 *M* concentrations), and the equilibrium of the reaction lies far to the right (K_{eq} = 850 at 25°C; see Table 18.1).

Under cellular conditions, this first reaction of glycolysis is even more favorable than at standard state. As pointed out in Chapters 15 and 16, the free energy change for any reaction depends on the concentrations of reactants and products. Equation 15.12 in Chapter 15 and the data in Table 18.2 can be used to calculate a value for ΔG for the hexokinase, glucokinase reaction in erythrocytes:

$$\Delta G = \Delta G^{\circ\prime} + RT \ln\left(\frac{[\text{G-6-P}][\text{ADP}]}{[\text{Glu}][\text{ATP}]}\right) \quad (18.5)$$

$$\Delta G = -16.7 \text{ kJ/mol} + (8.314 \text{ J/mol} \cdot \text{K})(310 \text{ K}) \ln\left(\frac{[0.083][0.14]}{[5.0][1.85]}\right)$$

$$\Delta G = -33.9 \text{ kJ/mol}$$

Thus, ΔG is even more favorable under cellular conditions than at standard state. As we will see later in this chapter, the hexokinase, glucokinase reaction is one of several that drive glycolysis forward.

Table **18.2**
Steady-State Concentrations of Glycolytic Metabolites in Erythrocytes

Metabolite	m*M*
Glucose	5.0
Glucose-6-phosphate	0.083
Fructose-6-phosphate	0.014
Fructose-1,6-bisphosphate	0.031
Dihydroxyacetone phosphate	0.14
Glyceraldehyde-3-phosphate	0.019
1,3-Bisphosphoglycerate	0.001
2,3-Bisphosphoglycerate	4.0
3-Phosphoglycerate	0.12
2-Phosphoglycerate	0.030
Phosphoenolpyruvate	0.023
Pyruvate	0.051
Lactate	2.9
ATP	1.85
ADP	0.14
P_i	1.0

Adapted from Minakami, S., and H. Yoshikawa. 1965. *Biochem. Biophys. Res. Comm.* **18:**345.

Table **18.1**
continued

Source	Subunit Molecular Weight (M_r)	Oligomeric Composition	$\Delta G°'$ (kJ/mol)	K_{eq} at 25°C	ΔG (kJ/mol)
Mammals	100,000	Monomer	−16.7	850	−33.9*
Yeast	55,000	Dimer			
Mammalian liver	50,000	Monomer			
Human	65,000	Dimer	+1.67	0.51	−2.92
Rabbit muscle	78,000	Tetramer	−14.2	310	−18.8
Rabbit muscle	40,000	Tetramer	+23.9	6.43×10^{-5}	−0.23
Chicken muscle	27,000	Dimer	+7.56	0.0472	+2.41
Rabbit muscle	37,000	Tetramer	+6.30	0.0786	−1.29
Rabbit muscle	64,000	Monomer	−18.9	2060	+0.1
Rabbit muscle	27,000	Dimer	+4.4	0.169	+0.83
Rabbit muscle	41,000	Dimer	+1.8	0.483	+1.1
Rabbit muscle	57,000	Tetramer	−31.7	3.63×10^{5}	−23.0
Rabbit muscle	35,000	Tetramer	−25.2	2.63×10^{4}	−14.8

*ΔG values calculated for 310 K (37°C) using the data in Table 18.2 for metabolite concentrations in erythrocytes. $\Delta G°'$ values are assumed to be constant between 25°C and 37°C.

The Cellular Advantages of Phosphorylating Glucose

The incorporation of a phosphate into glucose in this energetically favorable reaction is important for several reasons. First, phosphorylation keeps the substrate in the cell. Glucose is a neutral molecule and could diffuse across the cell membrane, but phosphorylation confers a negative charge on glucose, and the plasma membrane is essentially impermeable to glucose-6-phosphate (Figure 18.4). Moreover, rapid conversion of glucose to glucose-6-phosphate keeps the *intracellular* concentration of glucose low, favoring diffusion of glucose *into* the cell. In addition, since regulatory control can be imposed only on reactions not at equilibrium, the favorable thermodynamics of this first reaction makes it an important site for regulation.

Hexokinase

In most animal, plant, and microbial cells, the enzyme that phosphorylates glucose is **hexokinase.** Magnesium ion (Mg^{2+}) is required for this reaction, as

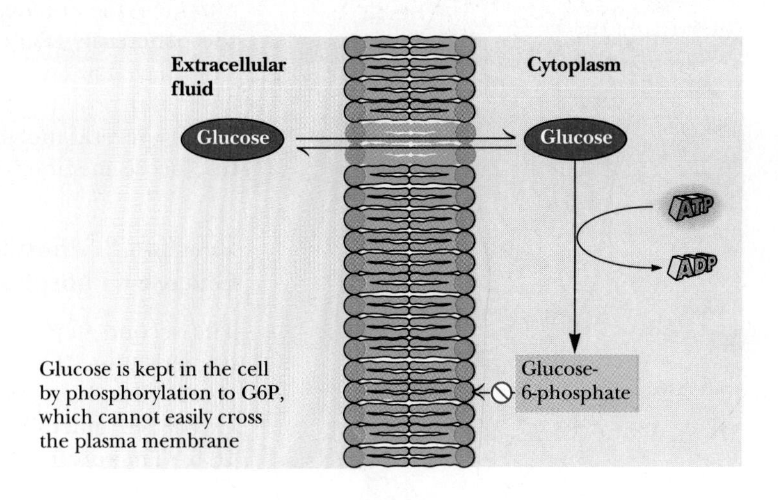

***Figure* 18.4** Phosphorylation of glucose to glucose-6-phosphate by ATP creates a charged molecule that cannot easily cross the plasma membrane.

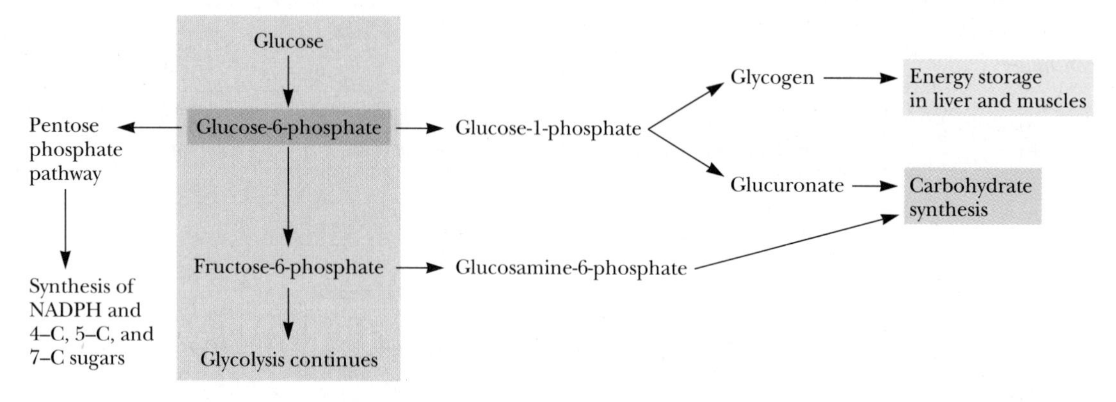

***Figure* 18.5** Glucose-6-phosphate is the branch point for several metabolic pathways.

for the other kinase enzymes in the glycolytic pathway. The true substrate for the hexokinase reaction is $MgATP^{2-}$. The apparent K_m for glucose of the animal skeletal muscle enzyme is approximately 0.1 mM, and the enzyme thus operates efficiently at normal blood glucose levels of 4 mM or so. Different body tissues possess different isozymes of hexokinase, each exhibiting somewhat different kinetic properties. The animal enzyme is allosterically inhibited by the product, glucose-6-phosphate. High levels of glucose-6-phosphate inhibit hexokinase activity until consumption by glycolysis lowers its concentration. The hexokinase reaction is one of three points in the glycolysis pathway that are *regulated.* As the generic name implies, hexokinase can phosphorylate a variety of hexose sugars, including glucose, mannose, and fructose.

Glucokinase

Liver contains an enzyme called **glucokinase,** which also carries out the reaction in Figure 18.4 but is highly specific for D-glucose, has a much higher K_m for glucose (approximately 10.0 mM), and is not product-inhibited. With such a high K_m for glucose, glucokinase becomes important metabolically only when liver glucose levels are high (for example, when the individual has consumed large amounts of sugar). When glucose levels are low, hexokinase is primarily responsible for phosphorylating glucose. However, when glucose levels are high, glucose is converted by glucokinase to glucose-6-phosphate and is eventually stored in the liver as glycogen. Glucokinase is an *inducible* enzyme—the amount present in the liver is controlled by *insulin* (secreted by the pancreas). (Patients with **diabetes mellitus** produce insufficient insulin. They have low levels of glucokinase, cannot tolerate high levels of blood glucose, and produce little liver glycogen.) Since glucose-6-phosphate is common to several metabolic pathways (Figure 18.5), it occupies a branch point in glucose metabolism.

Reaction 2: Phosphoglucoisomerase Catalyzes the Isomerization of Glucose-6-Phosphate

The second step in glycolysis is a common type of metabolic reaction: the isomerization of a sugar. In this particular case, the carbonyl oxygen of glucose-6-phosphate is shifted from C-1 to C-2. This amounts to isomerization of an aldose (glucose-6-phosphate) to a ketose—fructose-6-phosphate (Figure 18.6). The reaction is necessary for two reasons. First, the next step in glycoly-

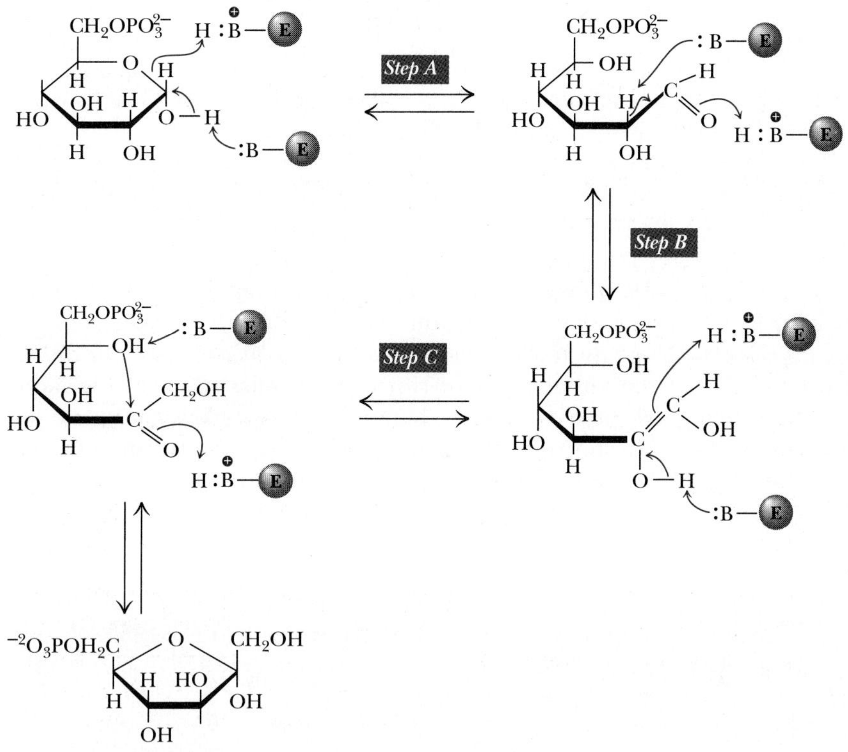

***Figure* 18.6** The phosphoglucoisomerase mechanism involves opening of the pyranose ring (Step A), proton abstraction leading to enediol formation (Step B), and proton addition to the double bond, followed by ring closure (Step C).

sis is phosphorylation at C-1, and the hemiacetal —OH of glucose would be more difficult to phosphorylate than a simple primary hydroxyl. Second, the isomerization to fructose (with a carbonyl group at position 2 in the linear form) activates carbon C-3 for cleavage in the fourth step of glycolysis. The enzyme responsible for this isomerization is **phosphoglucoisomerase,** also known as **glucose phosphate isomerase.** In humans, the enzyme requires Mg^{2+} for activity and is highly specific for glucose-6-phosphate. The $\Delta G°'$ is 1.67 kJ/mol, and the value of ΔG under cellular conditions (Table 18.1) is −2.92 kJ/mol. This small value means that the reaction operates near equilibrium in the cell and is readily reversible. Phosphoglucoisomerase proceeds through an *enediol* intermediate, as shown in Figure 18.6. Although the predominant forms of glucose-6-phosphate and fructose-6-phosphate in solution are the ring forms (Figure 18.6), the isomerase interconverts the open-chain form of G-6-P with the open-chain form of F-6-P. The first reaction catalyzed by the isomerase is the opening of the pyranose ring (Figure 18.6, Step A). In the next step, the C-2 proton is removed from the substrate by a basic residue on the enzyme, facilitating formation of the enediol intermediate (Figure 18.6, Step B). This process then operates somewhat in reverse (Figure 18.6, Step C), creating a carbonyl group at C-2 to complete the formation of fructose-6-phosphate. The furanose form of the product is formed in the usual manner by attack of the C-5 hydroxyl on the carbonyl group, as shown.

Reaction 3: Phosphofructokinase—The Second Priming Reaction

The action of phosphoglucoisomerase, "moving" the carbonyl group from C-1 to C-2, creates a new primary alcohol function at C-1 (see Figure 18.5). The next step in the glycolytic pathway is the phosphorylation of this group by **phosphofructokinase.** Once again, the substrate that provides the phosphoryl group is ATP. Like the hexokinase, glucokinase reaction, the phosphorylation of fructose-6-phosphate is a priming reaction, and is endergonic:

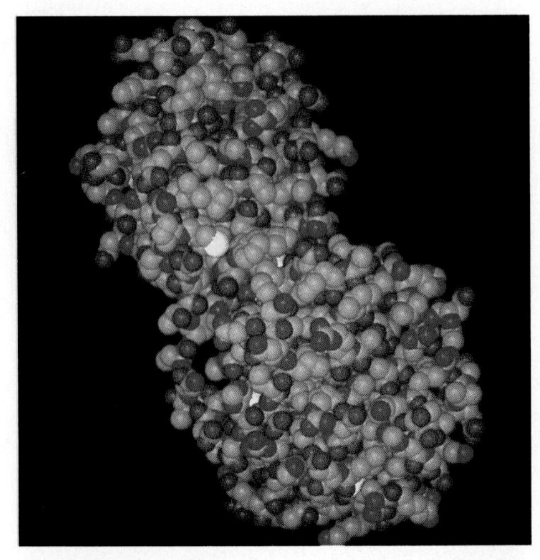

Phosphofructokinase

$$\text{Fructose-6-P} + P_i \longrightarrow \text{fructose-1,6-bisphosphate} \qquad (18.6)$$
$$\Delta G^{\circ\prime} = 16.3 \text{ kJ/mol}$$

When coupled (by phosphofructokinase) with the hydrolysis of ATP, the overall reaction (Figure 18.7) is strongly exergonic:

$$\text{Fructose-6-P} + \text{ATP} \longrightarrow \text{fructose-1,6-bisphosphate} + \text{ADP} \qquad (18.7)$$
$$\Delta G^{\circ\prime} = -14.2 \text{ kJ/mol}$$
$$\Delta G \text{ (in erythrocytes)} = -18.8 \text{ kJ/mol}$$

At pH 7 and 37°C, the phosphofructokinase reaction equilibrium lies far to the right. Just as the hexokinase reaction commits the cell to taking up glucose, *the phosphofructokinase reaction commits the cell to metabolizing glucose* rather than converting it to another sugar or storing it. Similarly, just as the large free energy change of the hexokinase reaction makes it a likely candidate for regulation, so the phosphofructokinase reaction is an important site of regulation—indeed, the most important site in the glycolytic pathway.

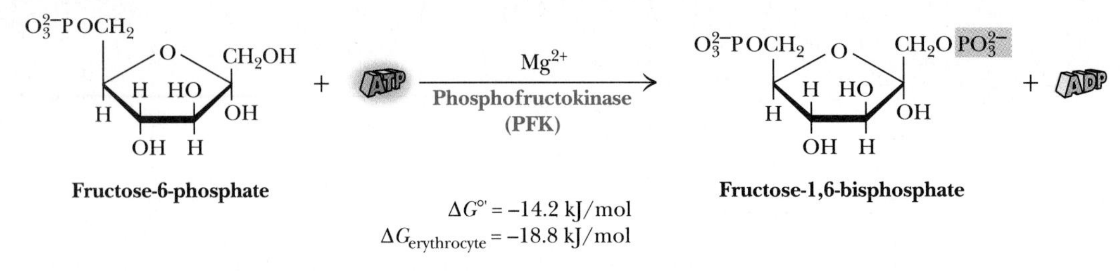

Figure **18.7** The phosphofructokinase reaction.

Regulation of Phosphofructokinase

Phosphofructokinase is the "valve" controlling the rate of glycolysis. ATP is an allosteric inhibitor of this enzyme. In the presence of high ATP concentrations, phosphofructokinase behaves cooperatively, plots of enzyme activity versus fructose-6-phosphate are sigmoid, and the K_m for fructose-6-phosphate is increased (Figure 18.8). Thus when ATP levels are sufficiently high in the cytosol, glycolysis "turns off." Under most cellular conditions, however, the ATP concentration does not vary over a large range. The ATP concentration in muscle during vigorous exercise, for example, is only about 10% lower than that during the resting state. The rate of glycolysis, however, varies much more. A large range of glycolytic rates cannot be directly accounted for by only a 10% change in ATP levels.

AMP reverses the inhibition due to ATP, and AMP levels in cells *can* rise dramatically when ATP levels decrease, due to the action of the enzyme *adenylate kinase,* which catalyzes the reaction

$$\text{ADP} + \text{ADP} \rightleftharpoons \text{ATP} + \text{AMP}$$

with the equilibrium constant:

$$K_{eq} = \frac{[\text{ATP}][\text{AMP}]}{[\text{ADP}]^2} = 0.44 \qquad (18.8)$$

Adenylate kinase rapidly interconverts ADP, ATP, and AMP to maintain this equilibrium. ADP levels in cells are typically 10% of ATP levels, and AMP levels are often less than 1% of the ATP concentration. Under such condi-

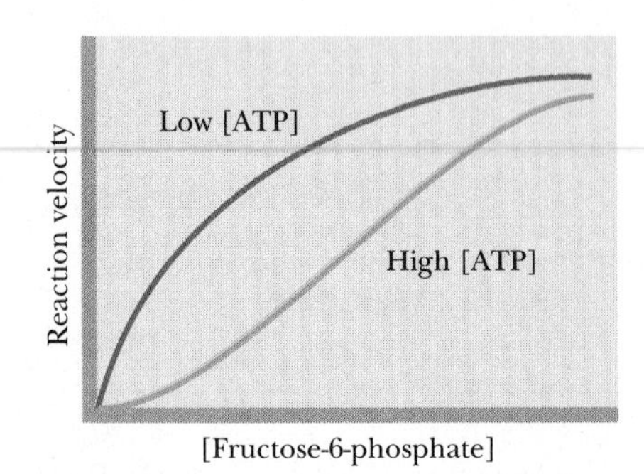

Figure **18.8** At high [ATP], phosphofructokinase behaves cooperatively, and the plot of enzyme activity versus [fructose-6-phosphate] is sigmoid. High [ATP] thus inhibits PFK, decreasing the enzyme's affinity for fructose-6-phosphate.

Example

Calculate the change in concentration in AMP that would occur if 8% of the ATP in an erythrocyte (red blood cell) were suddenly hydrolyzed to ADP. In erythrocytes (Table 18.2), the concentration of ATP is typically 1850 μM, the concentration of ADP is 145 μM, and the concentration of AMP is 5 μM. The total adenine nucleotide concentration is 2000 μM.

Solution

The problem can be solved using the equilibrium expression for the adenylate kinase reaction:

$$K_{eq} = 0.44 = \frac{[\text{ATP}][\text{AMP}]}{[\text{ADP}]^2}$$

If 8% of the ATP is hydrolyzed to ADP, then [ATP] becomes 1850(0.92) = 1702 μM, and [AMP] + [ADP] becomes 2000 − 1702 = 298 μM, and [AMP] may be calculated from the adenylate kinase equilibrium:

$$0.44 = \frac{[1702\ \mu M][\text{AMP}]}{[\text{ADP}]^2}$$

Since [AMP] = 298 μM − [ADP],

$$0.44 = \frac{1702(298 - [\text{ADP}])}{[\text{ADP}]^2}$$

$$[\text{ADP}] = 278\ \mu M$$
$$[\text{AMP}] = 20\ \mu M$$

Thus, an 8% decrease in [ATP] results in a 20/5 or *fourfold increase in the concentration of AMP.*

tions, a small net change in ATP concentration due to ATP hydrolysis results in a much larger relative increase in the AMP levels because of adenylate kinase activity.

Clearly, the activity of phosphofructokinase depends both on ATP and AMP levels and is a function of the cellular energy status. Phosphofructokinase activity is increased when the energy status falls and is decreased when the energy status is high. The rate of glycolysis activity thus decreases when ATP is plentiful and increases when more ATP is needed.

Glycolysis and the citric acid cycle (to be discussed in Chapter 19) are coupled via phosphofructokinase, because *citrate,* an intermediate in the citric acid cycle, is an allosteric inhibitor of phosphofructokinase. When the citric acid cycle reaches saturation, glycolysis (which "feeds" the citric acid cycle under aerobic conditions) slows down. The citric acid cycle directs electrons into the electron transport chain (for the purpose of ATP synthesis in oxidative phosphorylation) and also provides precursor molecules for biosynthetic pathways. Inhibition of glycolysis by citrate ensures that glucose will not be committed to these activities if the citric acid cycle is already saturated.

Phosphofructokinase is also regulated by **β-D-fructose-2,6-bisphosphate,** a potent allosteric activator that increases the affinity of phosphofructokinase for the substrate fructose-6-phosphate (Figure 18.9). Stimulation of phosphofructokinase is also achieved by decreasing the inhibitory effects of ATP (Figure 18.10). Fructose-2,6-bisphosphate increases the net flow of glucose through glycolysis by stimulating phosphofructokinase and, as we shall see in Chapter 21, by inhibiting fructose-1,6-bisphosphatase, the enzyme that catalyzes this reaction in the opposite direction.

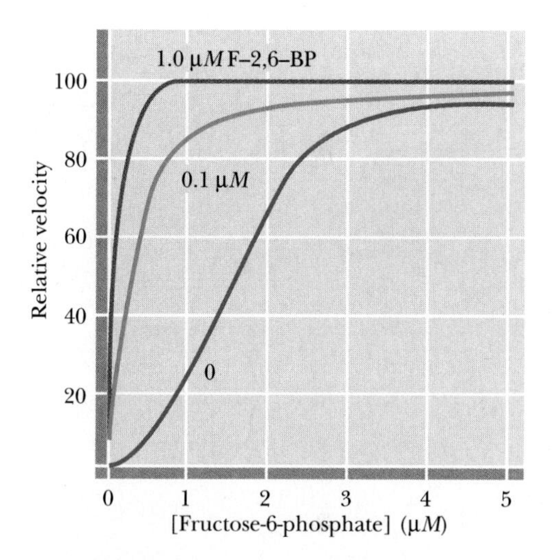

***Figure* 18.9** Fructose-2,6-bisphosphate activates PFK, increasing the affinity of the enzyme for fructose-6-phosphate and restoring the hyperbolic dependence of enzyme activity on substrate.

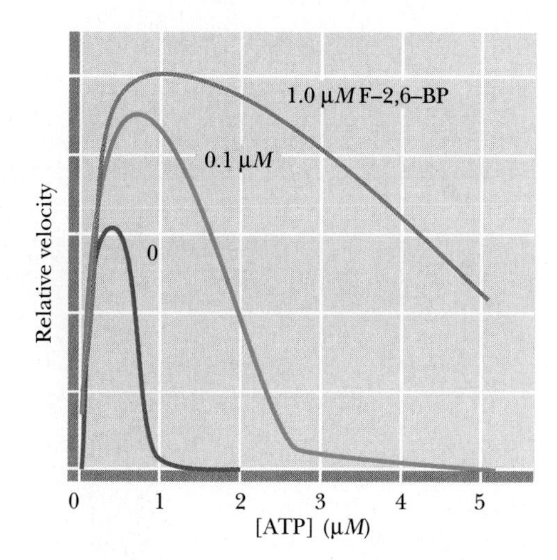

***Figure* 18.10** Fructose-2,6-bisphosphate decreases the inhibition of PFK due to ATP.

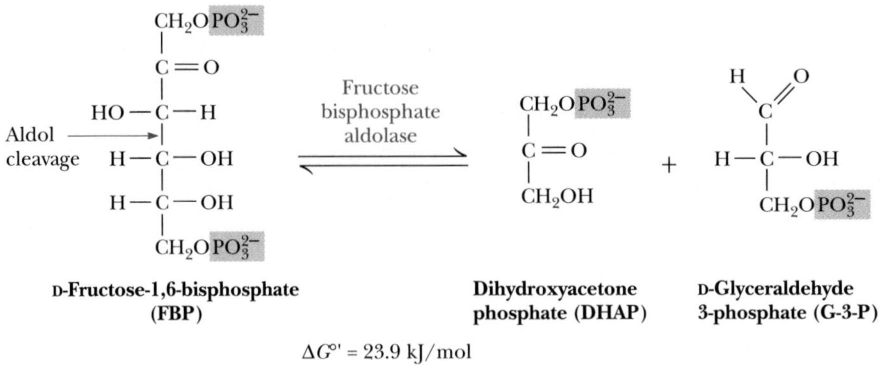

$$
\begin{array}{c}
CH_2OPO_3^{2-}\\
|\\
C{=}O\\
|\\
HO{-}C{-}H\\
|\\
H{-}C{-}OH\\
|\\
H{-}C{-}OH\\
|\\
CH_2OPO_3^{2-}
\end{array}
\xrightleftharpoons[]{\text{Fructose bisphosphate aldolase}}
\begin{array}{c}
CH_2OPO_3^{2-}\\
|\\
C{=}O\\
|\\
CH_2OH
\end{array}
+
\begin{array}{c}
H\diagdown\ \diagup O\\
C\\
|\\
H{-}C{-}OH\\
|\\
CH_2OPO_3^{2-}
\end{array}
$$

Aldol cleavage (between C-3 and C-4)

D-Fructose-1,6-bisphosphate (FBP) | Dihydroxyacetone phosphate (DHAP) | D-Glyceraldehyde 3-phosphate (G-3-P)

$\Delta G°' = 23.9$ kJ/mol

Figure **18.11** The fructose-1,6-bisphosphate aldolase reaction.

Reaction 4: Cleavage of Fructose-1,6-bisP by Fructose Bisphosphate Aldolase

Fructose bisphosphate aldolase cleaves fructose-1,6-bisphosphate between the C-3 and C-4 carbons to yield two triose phosphates. The products are dihydroxyacetone phosphate (DHAP) and glyceraldehyde-3-phosphate. The reaction (Figure 18.11) has an equilibrium constant of approximately 10^{-4} *M*, and a corresponding $\Delta G°'$ of +23.9 kJ/mol. These values might imply that the reaction does not proceed effectively from left to right as written. However, the reaction makes two molecules (glyceraldehyde-3-P and dihydroxyacetone-P) from one molecule (fructose-1,6-bisphosphate), and the equilibrium is thus greatly influenced by concentration. The value of ΔG in erythrocytes is actually −0.23 kJ/mol (see Table 18.1). At physiological concentrations, the reaction is essentially at equilibrium.

Two classes of aldolase enzymes are found in nature. Animal tissues produce a Class I aldolase, characterized by the formation of a covalent Schiff base intermediate between an active-site lysine and the carbonyl group of the substrate. Class I aldolases do not require a divalent metal ion (and thus are not inhibited by EDTA) but are inhibited by sodium borohydride, $NaBH_4$, in the presence of substrate (see A Deeper Look, page 582). Class II aldolases are produced mainly in bacteria and fungi and are not inhibited by borohydride, but do contain an active-site metal (normally zinc, Zn^{2+}) and are inhibited by EDTA. Cyanobacteria and some other simple organisms possess both classes of aldolase.

The aldolase reaction is merely the reverse of the **aldol condensation** well known to organic chemists. The latter reaction involves an attack by a nucleophilic enolate anion of an aldehyde or ketone on the carbonyl group of an aldehyde (Figure 18.12). The opposite reaction, aldol cleavage, begins with removal of a proton from the β-hydroxyl group, which is followed by the elimination of the enolate anion. A mechanism for the aldol cleavage reaction of fructose-1,6-bisphosphate in the Class I–type aldolases is shown in Figure 18.13a. In Class II aldolases, an active-site metal such as Zn^{2+} behaves as an electrophile, polarizing the carbonyl group of the substrate and stabilizing the enolate intermediate (Figure 18.13b).

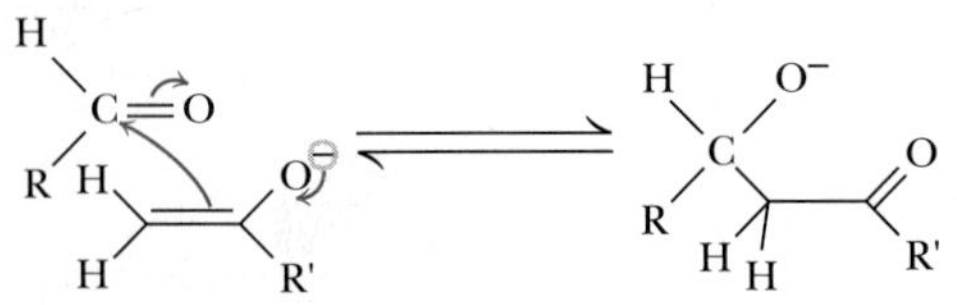

R' = H (aldehyde)
R' = alkyl, etc. (ketone)

Figure **18.12** An aldol condensation reaction.

Reaction 5: Triose Phosphate Isomerase

Of the two products of the aldolase reaction, only glyceraldehyde-3-phosphate goes directly into the second phase of glycolysis. The other triose phosphate, dihydroxyacetone phosphate, must be converted to glyceraldehyde-3-phosphate by the enzyme **triose phosphate isomerase** (Figure 18.14). This reaction

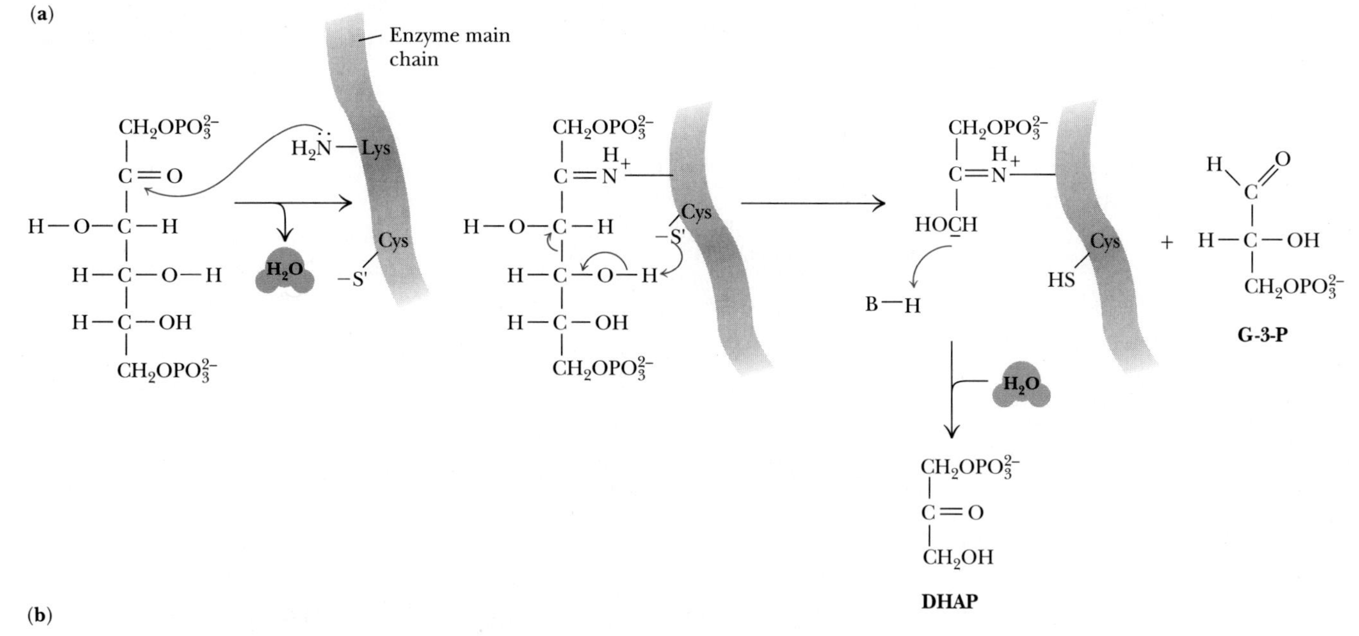

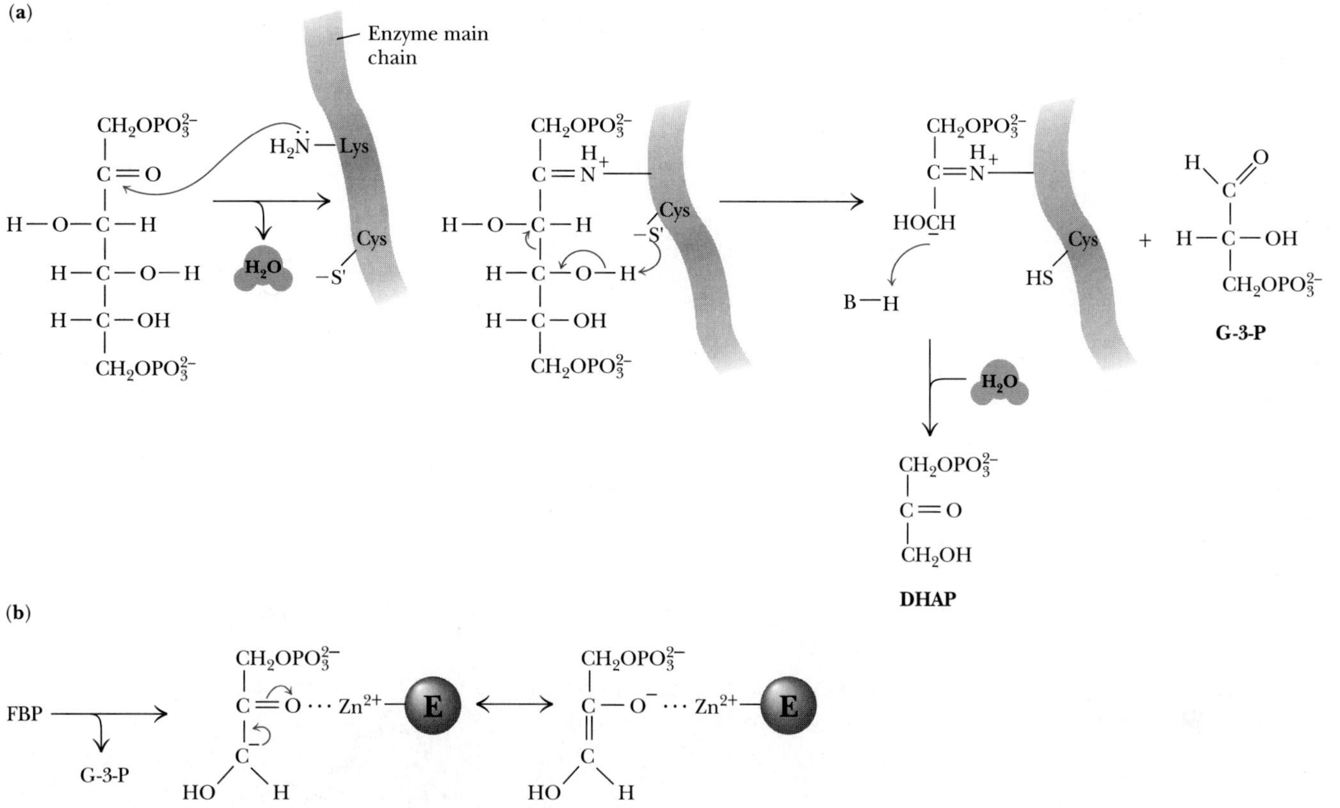

(b)

FBP → G-3-P; C=O ··· Zn^{2+}—E ⟷ C—O^- ··· Zn^{2+}—E

Figure **18.13** (a) A mechanism for the fructose-1,6-bisphosphate aldolase reaction. The Schiff base formed between the substrate carbonyl and an active-site lysine acts as an electron sink, increasing the acidity of the β-hydroxyl group and facilitating cleavage as shown. (b) In class II aldolases, an active-site Zn^{2+} stabilizes the enolate intermediate, leading to polarization of the substrate carbonyl group.

thus permits both products of the aldolase reaction to continue in the glycolytic pathway, and in essence makes the C-1, C-2, and C-3 carbons of the starting glucose molecule equivalent to the C-6, C-5, and C-4 carbons, respectively. The reaction mechanism involves an enediol intermediate that can donate either of its hydroxyl protons to a basic residue on the enzyme and thereby become either dihydroxyacetone phosphate or glyceraldehyde-3-phosphate (Figure 18.15). Triose phosphate isomerase is one of the enzymes that have evolved to a state of "catalytic perfection," with a turnover number near the diffusion limit (Chapter 11, Table 11.5).

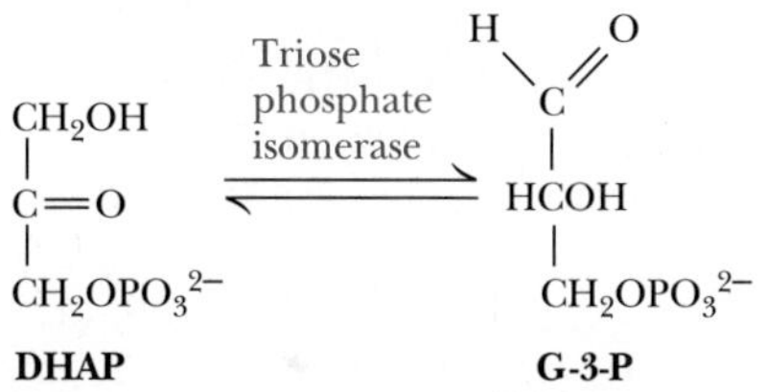

DHAP ⇌ (Triose phosphate isomerase) G-3-P

$\Delta G = +7.56$ kJ/mol

Figure **18.14** The triose phosphate isomerase reaction.

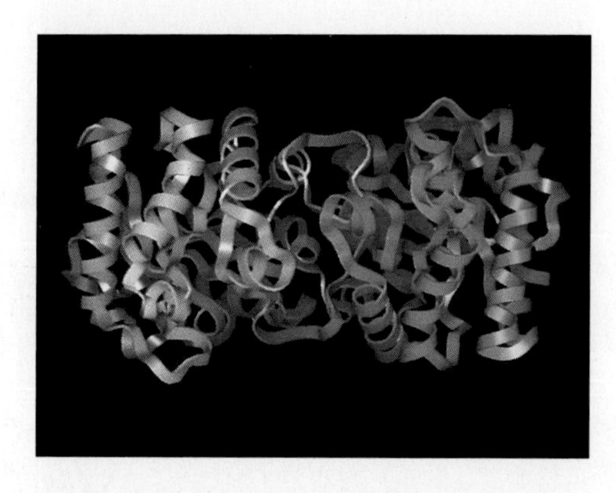

Ribbon structure of triose phosphate isomerase.

DHAP ⇌ Enediol intermediate ⇌ Glyceraldehyde-3-P

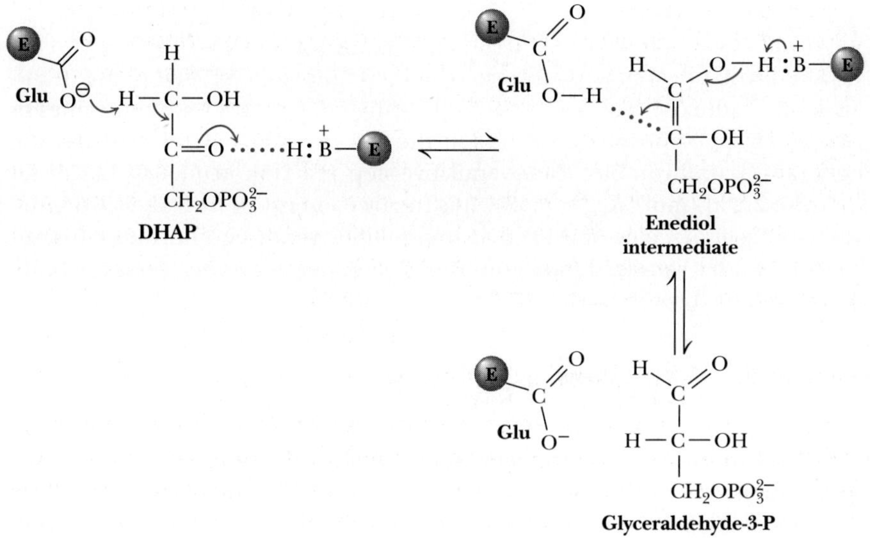

Figure **18.15** A reaction mechanism for triose phosphate isomerase.

A Deeper Look

The Chemical Evidence for the Schiff Basé Intermediate in Class I Aldolases

Fructose bisphosphate aldolase of animal muscle is a Class I aldolase, which forms a Schiff base or *imine,* intermediate between the substrate (fructose-1,6-bisP or dihydroxyacetone-P) and a lysine amino group at the enzyme active site. The chemical evidence for this intermediate comes from studies with the aldolase and the reducing agent sodium borohydride, $NaBH_4$. Incubation of fructose bisphosphate aldolase with dihydroxyacetone-P and $NaBH_4$ inactivates the enzyme. Interestingly, no inactivation is observed if $NaBH_4$ is added to the enzyme in the absence of substrate.

These observations are explained by the mechanism shown in the figure. $NaBH_4$ inactivates Class I aldolases by transfer of a hydride ion ($H:^-$) to the imine carbon atom of the enzyme–substrate adduct. The resulting secondary amine is stable to hydrolysis, and the active-site lysine is thus permanently modified and inactivated. $NaBH_4$ inactivates Class I aldolases in the presence of either dihydroxyacetone-P or fructose-1,6-bisP, but inhibition doesn't occur in the presence of glyceraldehyde-3-P.

Definitive identification of lysine as the modified active-site residue has come from radioisotope-labeling studies. $NaBH_4$ reduction of the aldolase Schiff base intermediate formed from ^{14}C-labeled dihydroxyacetone-P yields an enzyme covalently labeled with ^{14}C. Acid hydrolysis of the inactivated enzyme liberates a novel ^{14}C-labeled amino acid, *N^6-dihydroxypropyl-L-lysine.* This is the product anticipated from reduction of the Schiff base formed between a lysine residue and the ^{14}C-labeled dihydroxyacetone-P. (The phosphate group is lost during acid hydrolysis of the inactivated enzyme.) The use of ^{14}C labeling in a case such as this facilitates the separation and identification of the telltale amino acid.

Formation of Schiff base intermediate

Borohydride reduction of Schiff base intermediate

Stable (trapped) E–S derivative

Degradation of enzyme (acid hydrolysis)

N^6-dihydroxypropyl-L-lysine

The triose phosphate isomerase reaction completes the first phase of glycolysis, each glucose that passes through being converted to two molecules of glyceraldehyde-3-phosphate. Although the last two steps of the pathway are energetically unfavorable, the overall five-step reaction sequence has a net $\Delta G°'$ of $+2.2$ kJ/mol ($K_{eq} \approx 0.43$). It is the free energy of hydrolysis from the two priming molecules of ATP that brings the overall equilibrium constant close to 1 under standard-state conditions. The net ΔG under cellular conditions is -53.4 kJ/mol.

18.4 The Second Phase of Glycolysis

The second half of the glycolytic pathway involves the reactions that convert the metabolic energy in the glucose molecule into ATP. Altogether, four new ATP molecules are produced. If two are considered to offset the two ATPs

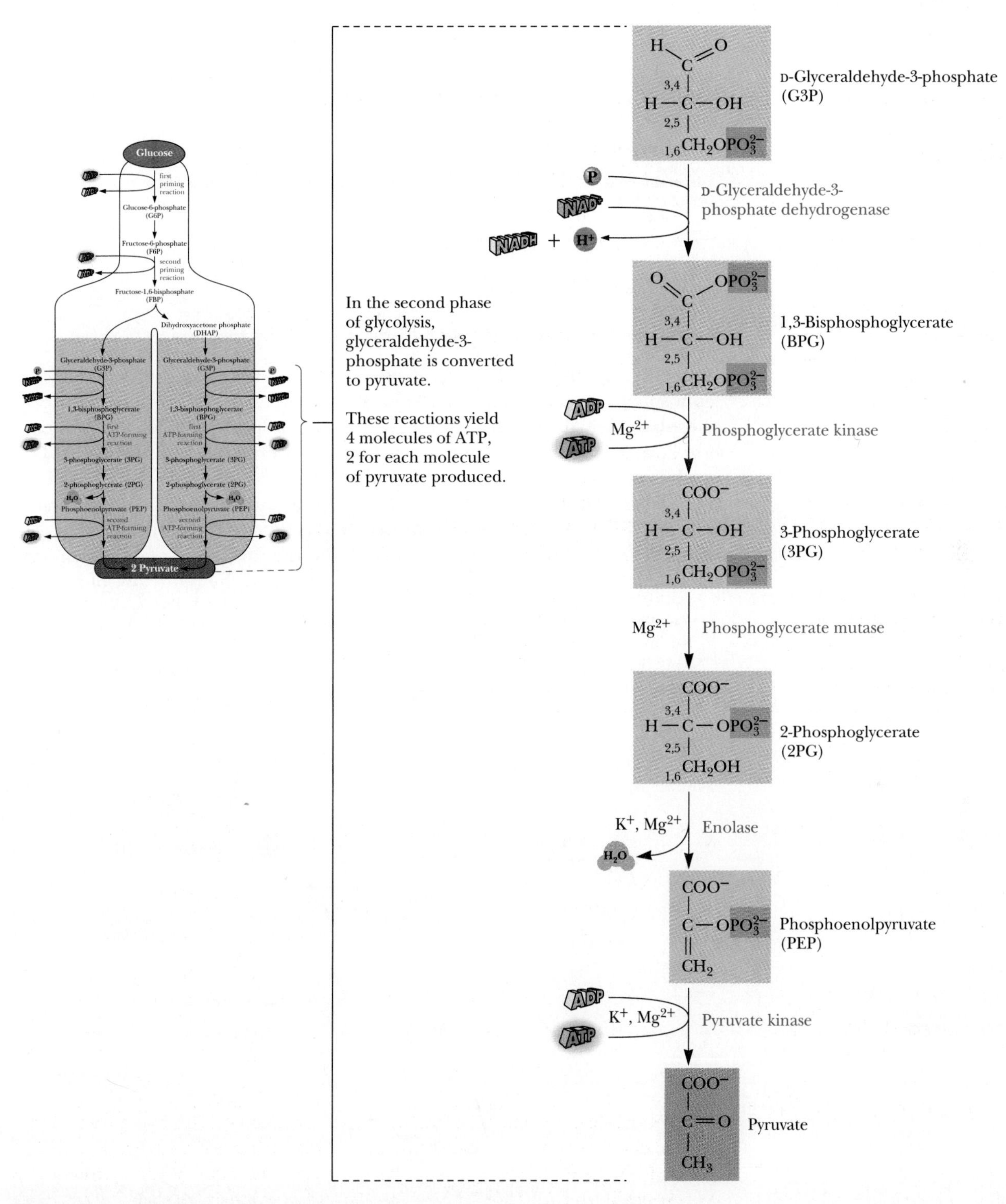

***Figure* 18.16** The second phase of glycolysis. Carbon atoms are numbered to show their original positions in glucose.

consumed in phase 1, a net yield of 2 ATPs per glucose is realized. Phase II starts with the oxidation of glyceraldehyde-3-phosphate, a reaction with a large enough energy "kick" to produce a high-energy phosphate, namely, 1,3-bisphosphoglycerate (Figure 18.16). Phosphoryl transfer from 1,3-BPG to ADP to make ATP is highly favorable. The product, 3-phosphoglycerate, is converted via several steps to phosphoenolpyruvate (PEP), another high-energy phosphate. PEP readily transfers its phosphoryl group to ADP in the pyruvate kinase reaction to make another ATP.

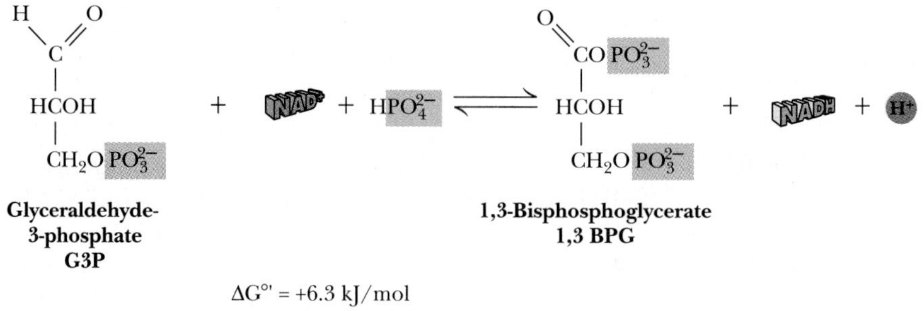

Figure **18.17** The glyceraldehyde-3-phosphate dehydrogenase reaction.

Reaction 6: Glyceraldehyde-3-Phosphate Dehydrogenase

In the first glycolytic reaction to involve oxidation–reduction, glyceraldehyde-3-phosphate is oxidized to 1,3-bisphosphoglycerate by **glyceraldehyde-3-phosphate dehydrogenase.** Although the oxidation of an aldehyde to a carboxylic acid is a highly exergonic reaction, the overall reaction (Figure 18.17) involves both formation of a carboxylic-phosphoric anhydride and the reduction of NAD^+ to NADH and is therefore slightly endergonic at standard state, with a $\Delta G°'$ of +6.30 kJ/mol. The free energy that might otherwise be released as heat in this reaction is directed into the formation of a high-energy phosphate compound, 1,3-bisphosphoglycerate, and the reduction of NAD^+.

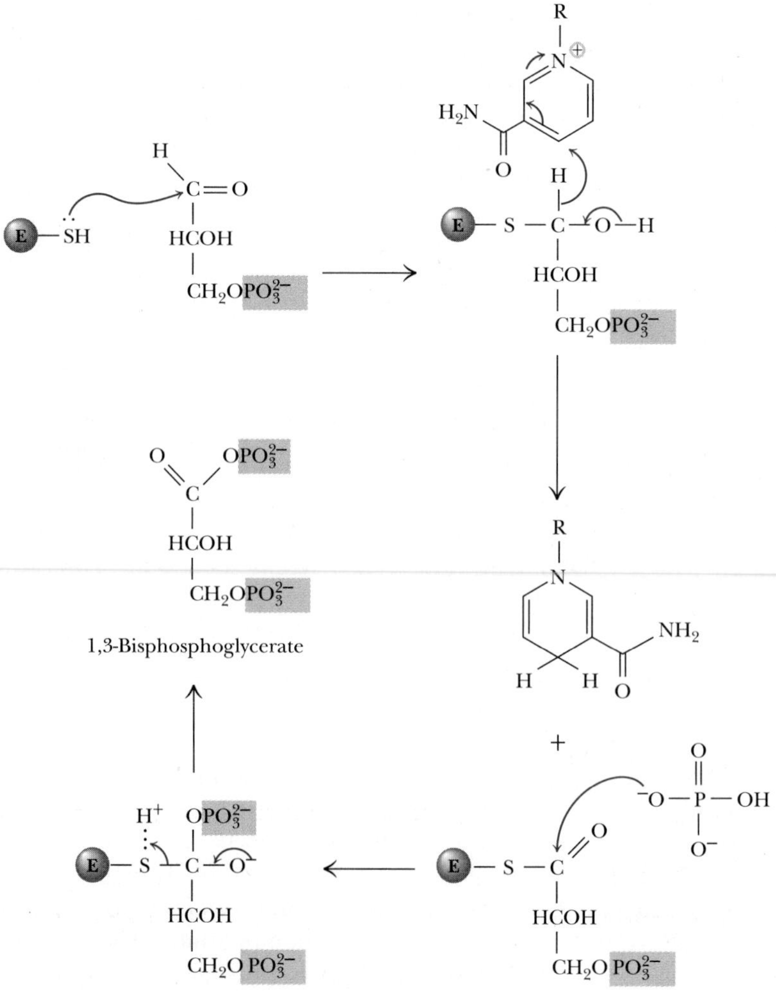

Figure **18.18** A mechanism for the glyceraldehyde-3-phosphate dehydrogenase reaction. Reaction of the enzyme sulfhydryl with the carbonyl carbon of glyceraldehyde-3-P forms a hemithioacetal, which loses a hydride to NAD^+ to become a thioester. Phosphorolysis of this thioester releases 1,3-bisphosphoglycerate.

The reaction mechanism involves nucleophilic attack by a cysteine —SH group on the carbonyl carbon of glyceraldehyde-3-phosphate to form a hemithioacetal (Figure 18.18). The hemithioacetal intermediate decomposes by hydride ($H:^-$) transfer to NAD^+ to form a high-energy thioester. Nucleophilic attack by phosphate displaces the product, 1,3-bisphosphoglycerate, from the enzyme. The enzyme can be inactivated by reaction with iodoacetate, which reacts with and blocks the essential cysteine sulfhydryl.

The glyceraldehyde-3-phosphate dehydrogenase reaction is the site of action of *arsenate* (AsO_4^{3-}), an anion analogous to phosphate. Arsenate is an effective substrate in this reaction, forming *1-arseno-3-phosphoglycerate* (Figure 18.19), but acyl arsenates are quite unstable and are rapidly hydrolyzed. 1-Arseno-3-phosphoglycerate breaks down to yield *3-phosphoglycerate,* the product of the seventh reaction of glycolysis. The result is that glycolysis continues in the presence of arsenate, but the molecule of ATP formed in reaction 7 (phosphoglycerate kinase) is not made because this step has been bypassed. The lability of 1-arseno-3-phosphoglycerate effectively uncouples the oxidation and phosphorylation events, which are normally tightly coupled in the glyceraldehyde-3-phosphate dehydrogenase reaction.

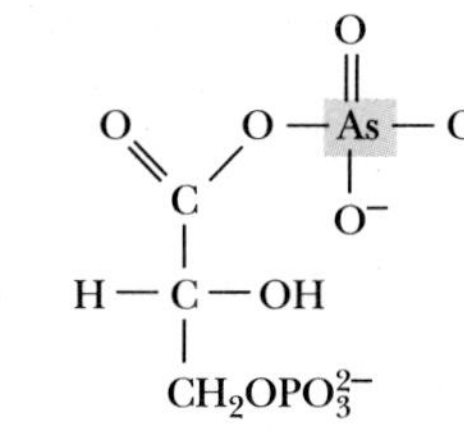

Figure 18.19

Reaction 7: Phosphoglycerate Kinase

The glycolytic pathway breaks even in terms of ATPs consumed and produced with this reaction. The enzyme **phosphoglycerate kinase** transfers a phosphoryl group from 1,3-bisphosphoglycerate to ADP to form an ATP (Figure 18.20). Since each glucose molecule sends two molecules of glyceraldehyde-3-phosphate into the second phase of glycolysis, and since two ATPs were consumed per glucose in the first-phase reactions, the phosphoglycerate kinase reaction "pays off" the ATP debt created by the priming reactions. As might be expected for a phosphoryl transfer enzyme, Mg^{2+} ion is required for activity, and the true nucleotide substrate for the reaction is $MgADP^-$. It is appropriate to view the sixth and seventh reactions of glycolysis as a coupled pair, with 1,3-bisphosphoglycerate as an intermediate. The phosphoglycerate kinase reaction is sufficiently exergonic at standard state to pull the G-3-P dehydrogenase reaction along. (In fact, the aldolase and triose phosphate isomerase are also pulled forward by phosphoglycerate kinase.) The net result of these coupled reactions is

$$\text{Glyceraldehyde-3-phosphate} + \text{ADP} + P_i + NAD^+ \longrightarrow \text{3-phosphoglycerate} + \text{ATP} + \text{NADH} + H^+ \quad \Delta G°' = -12.6 \text{ kJ/mol} \quad (18.9)$$

Another reflection of the coupling between these reactions lies in their values of ΔG under cellular conditions (Table 18.1). In spite of its strongly negative $\Delta G°'$, the phosphoglycerate kinase reaction operates at equilibrium in the erythrocyte ($\Delta G = 0.1$ kJ/mol). In essence, the free energy available in the phosphoglycerate kinase reaction is used to bring the three previous reac-

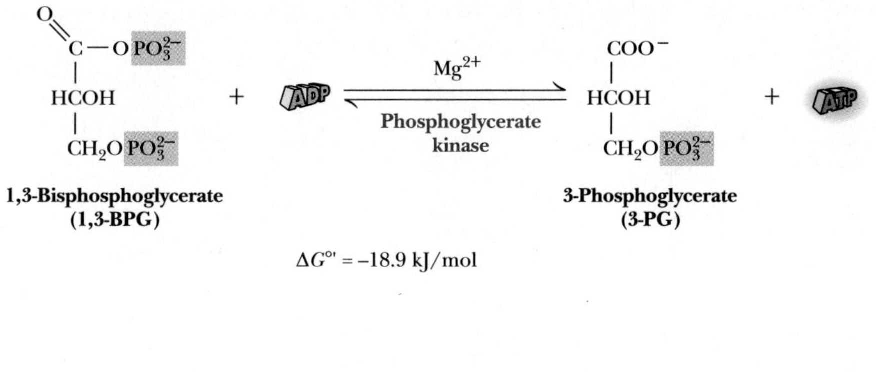

Figure 18.20 The phosphoglycerate kinase reaction.

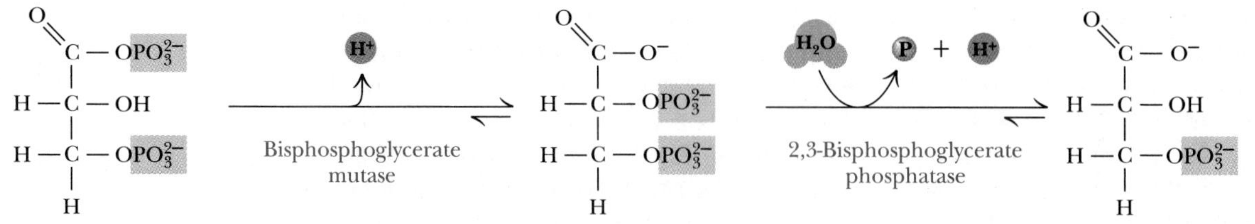

1,3-Bisphosphoglycerate (1,3-BPG) → (Bisphosphoglycerate mutase; H^+) → 2,3-Bisphosphoglycerate (2,3-BPG) → (2,3-Bisphosphoglycerate phosphatase; H_2O → P + H^+) → 3-Phosphoglycerate

Figure **18.21** Formation and decomposition of 2,3-bisphosphoglycerate.

tions closer to equilibrium. Viewed in this context, it is clear that ADP has been phosphorylated to form ATP at the expense of a substrate, namely, glyceraldehyde-3-phosphate. This is an example of **substrate-level phosphorylation,** a concept that will be encountered again. (The other kind of phosphorylation, *oxidative phosphorylation,* is driven energetically by the transport of electrons from appropriate coenzymes and substrates to oxygen. Oxidative phosphorylation will be covered in detail in Chapter 20.) Even though the coupled reactions exhibit a very favorable $\Delta G°'$, there are conditions (i.e., high ATP and 3-phosphoglycerate levels) under which Equation (18.9) can be reversed, so that 3-phosphoglycerate is phosphorylated from ATP.

An important regulatory molecule, *2,3-bisphosphoglycerate,* is synthesized and metabolized by a pair of reactions that make a detour around the phosphoglycerate kinase reaction. 2,3-BPG, which stabilizes the deoxy form of hemoglobin and which is primarily responsible for the cooperative nature of oxygen binding by hemoglobin (see Chapter 12), is formed from 1,3-bisphosphoglycerate by **bisphosphoglycerate mutase** (Figure 18.21). Interestingly, 3-phosphoglycerate is required for this reaction, which involves phosphoryl transfer from the C-1 position of 1,3-bisphosphoglycerate to the C-2 position of 3-phosphoglycerate (Figure 18.22). Hydrolysis of 2,3-BPG is carried out by *2,3-bisphosphoglycerate phosphatase.* Though other cells contain only a trace of 2,3-BPG, erythrocytes typically contain 4 to 5 m*M* 2,3-BPG.

Reaction 8: Phosphoglycerate Mutase

The remaining steps in the glycolytic pathway prepare for synthesis of the second ATP equivalent. This begins with the **phosphoglycerate mutase** reaction (Figure 18.23), in which the phosphoryl group of 3-phosphoglycerate is moved from C-3 to C-2. (The term *mutase* is applied to enzymes that catalyze migration of a functional group within a substrate molecule.) The free energy change for this reaction is very small under cellular conditions, with ΔG = 0.83 kJ/mol. Phosphoglycerate mutase enzymes isolated from different sources exhibit different reaction mechanisms. As shown in Figure 18.24, the enzymes isolated from yeast and from rabbit muscle form *phosphoenzyme* intermediates, use *2,3-bisphosphoglycerate* as a cofactor, and undergo *inter*molecular

Figure **18.22** The mutase that forms 2,3-BPG from 1,3-BPG requires 3-phosphoglycerate. The reaction is actually an intermolecular phosphoryl transfer from C-1 of 1,3-BPG to C-2 of 3-PG.

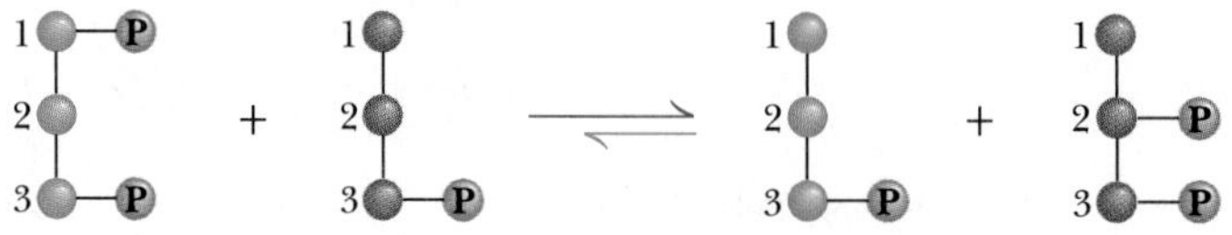

phosphoryl group transfers (in which the phosphate of the product 2-phosphoglycerate is not that from the 3-phosphoglycerate substrate). The prevalent form of phosphoglycerate mutase is a *phosphoenzyme,* with a phosphoryl group covalently bound to a histidine residue at the active site. This phosphoryl group is transferred to the C-2 position of the substrate to form a transient, enzyme-bound 2,3-bisphosphoglycerate, which then decomposes by a second phosphoryl transfer from the C-3 position of the intermediate to the

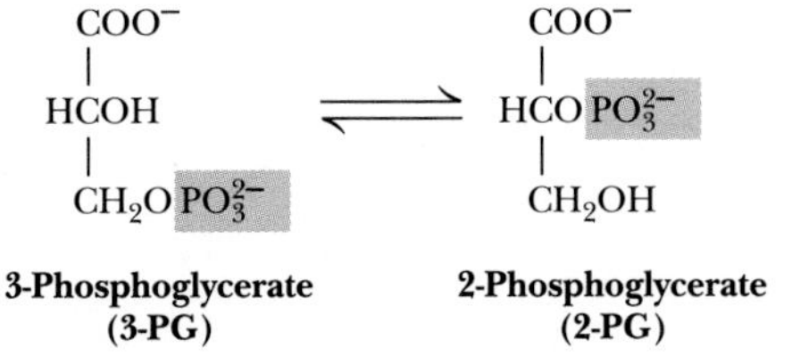

3-Phosphoglycerate (3-PG) ⇌ **2-Phosphoglycerate (2-PG)**

$\Delta G^{\circ\prime} = +4.4$ kJ/mol

***Figure* 18.23** The phosphoglycerate mutase reaction.

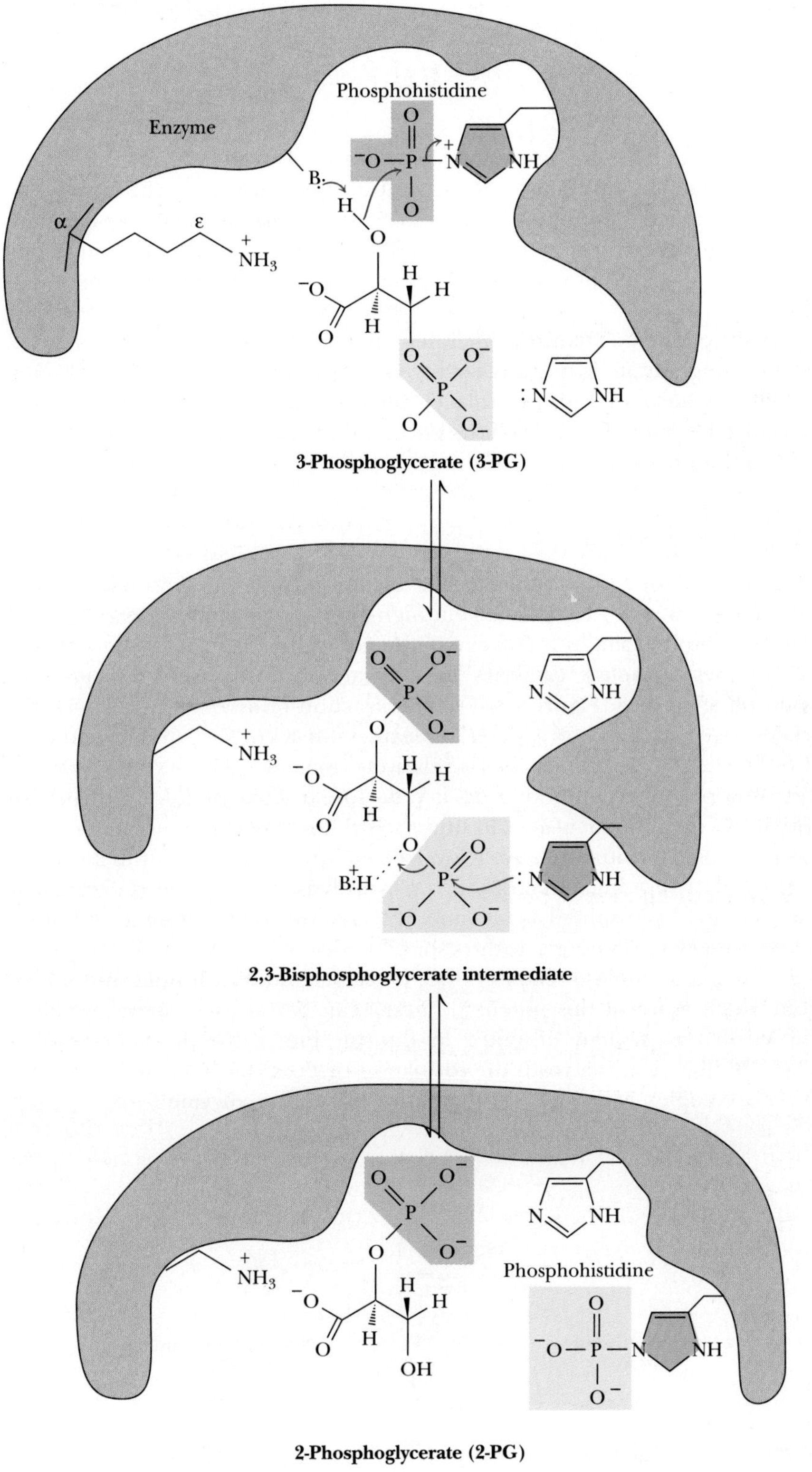

***Figure* 18.24** A mechanism for the phosphoglycerate mutase reaction in rabbit muscle and in yeast.

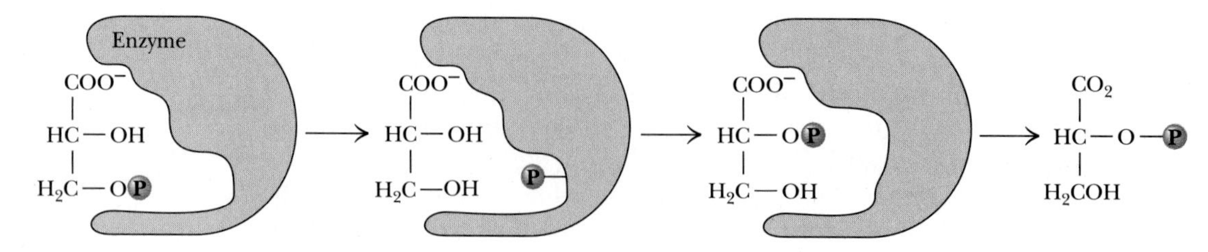

Figure **18.25** The phosphoglycerate mutase of wheat germ catalyzes an intramolecular phosphoryl transfer.

histidine residue on the enzyme. About once in every 100 enzyme turnovers, the intermediate, 2,3-bisphosphoglycerate, dissociates from the active site, leaving an inactive, unphosphorylated enzyme. The unphosphorylated enzyme can be reactivated by binding 2,3-BPG. For this reason, maximal activity of phosphoglycerate mutase requires the presence of small amounts of 2,3-BPG.

A different mechanism operates in the wheat germ enzyme. 2,3-Bisphosphoglycerate is not a cofactor. Instead, the enzyme carries out *intra*molecular phosphoryl group transfer (Figure 18.25). The C-3 phosphate is transferred to an active-site residue and then to the C-2 position of the original substrate molecule to form the product, 2-phosphoglycerate.

Reaction 9: Enolase

Recall that, prior to synthesizing ATP in the phosphoglycerate kinase reaction, it was necessary to first make a high-energy phosphate substrate. Reaction 9 of glycolysis similarly makes a high-energy phosphate in preparation for ATP synthesis. **Enolase** catalyzes the formation of *phosphoenolpyruvate* from 2-phosphoglycerate (Figure 18.26). The reaction in essence involves a dehydration—the removal of a water molecule—to form the enol structure of PEP. The $\Delta G°'$ for this reaction is relatively small at 1.8 kJ/mol ($K_{eq} = 0.5$); and, under cellular conditions, ΔG is very close to zero. In light of this condition, it may be difficult at first to understand how the enolase reaction transforms a substrate with a relatively low free energy of hydrolysis into a product (PEP) with a very high free energy of hydrolysis. This puzzle is clarified by realizing that 2-phosphoglycerate and PEP contain about the same amount of *potential* metabolic energy, with respect to decomposition to P_i, CO_2, and H_2O. What the enolase reaction does is rearrange the substrate into a form from which more of this potential energy can be released upon hydrolysis. The enzyme is strongly inhibited by fluoride ion in the presence of phosphate. Inhibition arises from the formation of *fluorophosphate* (FPO_3^{2-}), which forms a complex with Mg^{2+} at the active site of the enzyme.

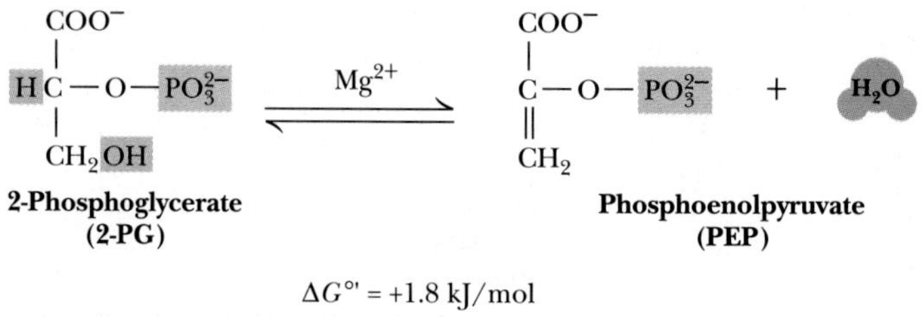

COO^- / $HC—O—PO_3^{2-}$ / CH_2OH $\xrightarrow{Mg^{2+}}$ (⇌) COO^- / $C—O—PO_3^{2-}$ / $\|$ CH_2 + H_2O

2-Phosphoglycerate (2-PG) — **Phosphoenolpyruvate (PEP)**

$\Delta G°' = +1.8$ kJ/mol

Figure **18.26** The enolase reaction.

$$\text{PEP} + H^{+} + ADP^{3-} \underset{K^{+}}{\overset{Mg^{2+}}{\rightleftharpoons}} \text{Pyruvate} + ATP^{4-}$$

$$\Delta G^{\circ\prime} = -31.7 \text{ kJ/mol}$$

***Figure* 18.27** The pyruvate kinase reaction.

Reaction 10: Pyruvate Kinase

The second ATP-synthesizing reaction of glycolysis is catalyzed by **pyruvate kinase,** which brings the pathway at last to its pyruvate branch point. Pyruvate kinase mediates the transfer of a phosphoryl group from phosphoenolpyruvate to ADP to make ATP and pyruvate (Figure 18.27). The reaction requires Mg^{2+} ion and is stimulated by K^{+} and certain other monovalent cations.

Reaction	$\Delta G^{\circ\prime}$ (kJ/mol)
$\text{Phosphoenolpyruvate}^{2-} + H_2O \longrightarrow \text{pyruvate}^{-} + HPO_4^{\ 2-}$	−62.2
$ADP^{3-} + HPO_4^{\ 2-} \longrightarrow ATP^{4-} + H_2O$	+30.5
$\text{Phosphoenolpyruvate}^{2-} + ADP^{3-} \longrightarrow \text{pyruvate}^{-} + ATP^{4-}$	−31.7

The corresponding K_{eq} at 25°C is 3.63×10^5, and it is clear that the pyruvate kinase reaction equilibrium lies very far to the right. Concentration effects reduce the magnitude of the free energy change somewhat in the cellular environment, but the ΔG in erythrocytes is still quite favorable at −23.0 kJ/mol. The high free energy change for the conversion of PEP to pyruvate is due largely to the highly favorable and spontaneous conversion of the enol tautomer of pyruvate to the more stable keto form (Figure 18.28) following the phosphoryl group transfer step.

The large negative ΔG of this reaction makes pyruvate kinase a suitable target site for regulation of glycolysis. For each glucose molecule in the glycolysis pathway, two ATPs are made at the pyruvate kinase stage (because two triose molecules were produced per glucose in the aldolase reaction). Since the pathway broke even in terms of ATP at the phosphoglycerate kinase reaction (two ATPs consumed and two ATPs produced), the two ATPs produced by pyruvate kinase represent the "payoff" of glycolysis—a net yield of two ATP molecules.

Pyruvate kinase possesses allosteric sites for numerous effectors. It is activated by AMP and fructose-1,6-bisphosphate and inhibited by ATP, acetyl-CoA, and alanine. (Note that alanine is the α-amino acid counterpart of the α-keto acid, pyruvate.) Furthermore, liver pyruvate kinase is regulated by covalent modification. Hormones such as *glucagon* activate a cAMP-dependent protein kinase, which transfers a phosphoryl group from ATP to the enzyme. The phosphorylated form of pyruvate kinase is more strongly inhibited by

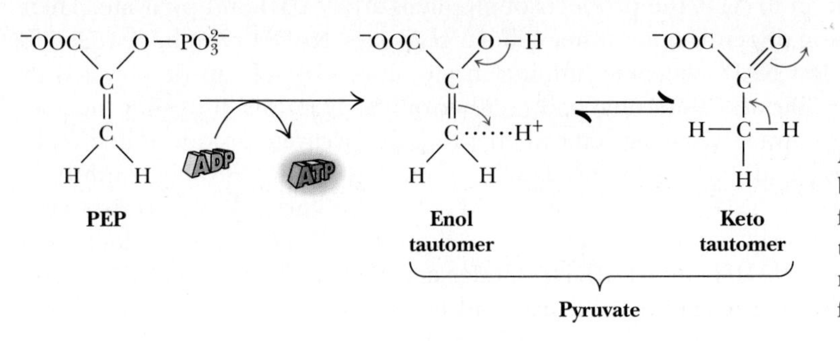

***Figure* 18.28** The conversion of PEP to pyruvate may be viewed as involving two steps: phosphoryl transfer followed by an enol-keto tautomerization. The tautomerization is spontaneous ($\Delta G^{\circ\prime} \sim -35$–$40$ kJ/mol) and accounts for much of the free energy change for PEP hydrolysis.

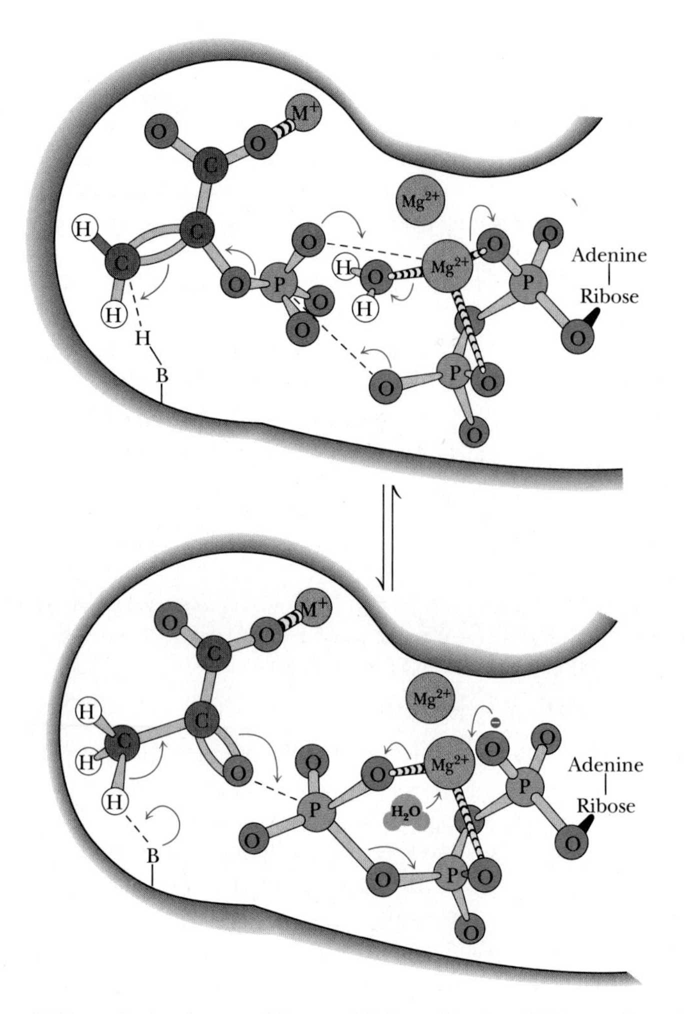

***Figure* 18.29** A mechanism for the pyruvate kinase reaction, based on NMR and EPR studies by Mildvan and colleagues. Phosphoryl transfer from PEP to ADP occurs in four steps: (a) a water on the Mg^{2+} ion coordinated to ADP is replaced by the phosphoryl group of PEP; (b) Mg^{2+} dissociates from the α-P of ADP; (c) the phosphoryl group is transferred; and (d) the enolate of pyruvate is protonated. *(Adapted from Mildvan, A., 1979, Advances in Enzymology 49:103–126.)*

ATP and alanine and has a higher K_m for PEP, so that, in the presence of physiological levels of PEP, the enzyme is inactive. Then PEP is used as a substrate for glucose synthesis in the *gluconeogenesis* pathway (to be described in Chapter 21), instead of going on through glycolysis and the citric acid cycle (or fermentation routes). A suggested active-site geometry for pyruvate kinase, based on NMR and EPR studies by Mildvan and colleagues, is presented in Figure 18.29. The carbonyl oxygen of pyruvate and the γ-phosphorus of ATP lie within 0.3 nm of each other at the active site, consistent with direct transfer of the phosphoryl group without formation of a phosphoenzyme intermediate.

18.5 The Metabolic Fates of NADH and Pyruvate—The Products of Glycolysis

In addition to ATP, the products of glycolysis are NADH and pyruvate. Their processing depends upon other cellular pathways. NADH must be recycled to NAD^+, lest NAD^+ become limiting in glycolysis. NADH can be recycled by both aerobic and anaerobic paths, either of which results in further metabolism of pyruvate. What a given cell does with the pyruvate produced in glycolysis depends in part on the availability of oxygen. Under aerobic conditions, pyruvate can be sent into the citric acid cycle (also known as the tricarboxylic acid cycle; see Chapter 19), where it is oxidized to CO_2 with the production of additional NADH (and $FADH_2$). Under aerobic conditions, the NADH produced in glycolysis and the citric acid cycle is reoxidized to NAD^+ in the mitochondrial electron transport chain (Chapter 20).

enzymes by allosteric effectors brings glycolysis to a halt. When we consider **gluconeogenesis**—the biosynthesis of glucose—in Chapter 21, we will see that different enzymes are used to carry out reactions 1, 3, and 10 in reverse, effecting the net synthesis of glucose. The maintenance of reactions 2 and 4 through 9 at or near equilibrium permits these reactions (and their respective enzymes!) to operate effectively in *either* the forward or reverse direction.

18.8 Utilization of Other Substrates in Glycolysis

The glycolytic pathway described in this chapter begins with the breakdown of glucose, but other sugars, both simple and complex, can enter the cycle if they can be converted by appropriate enzymes to one of the intermediates of glycolysis. Figure 18.32 shows the mechanisms by which several simple metabolites can enter the glycolytic pathway. **Fructose,** for example, which is produced by breakdown of sucrose, may participate in glycolysis by at least two different routes. In the liver, fructose is phosphorylated at C-1 by the enzyme *fructokinase:*

$$\text{D-Fructose} + \text{ATP}^{4-} \longrightarrow \text{D-fructose-1-phosphate}^{2-} + \text{ADP}^{3-} + \text{H}^{+} \quad (18.10)$$

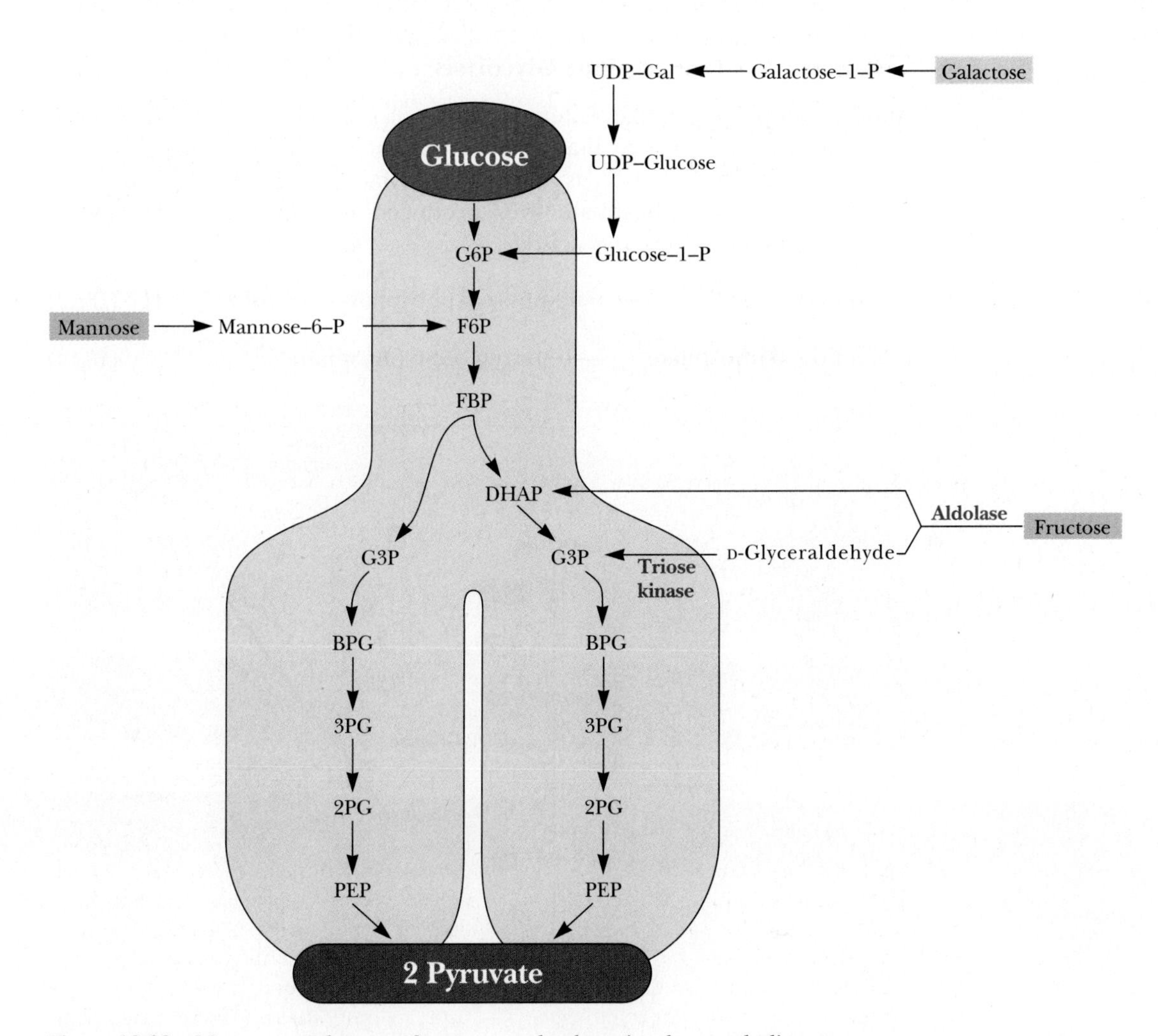

***Figure* 18.32** Mannose, galactose, fructose, and other simple metabolites can enter the glycolytic pathway.

Subsequent action by **fructose-1-phosphate aldolase,**

$$\text{D-Fructose-1-P}^{2-} \longrightarrow \text{D-glyceraldehyde} + \text{dihydroxyacetone phosphate}^{2-} \quad (18.11)$$

cleaves fructose-1-P in a manner like the fructose bisphosphate aldolase reaction to produce dihydroxyacetone phosphate and D-glyceraldehyde. Dihydroxyacetone phosphate is of course an intermediate in glycolysis. D-Glyceraldehyde can be phosphorylated by *triose kinase* in the presence of ATP to form D-glyceraldehyde-3-phosphate, another glycolytic intermediate.

In the kidney and in muscle tissues, fructose is readily phosphorylated by hexokinase, which, as pointed out above, can utilize several different hexose substrates. The free energy of hydrolysis of ATP drives the reaction forward:

$$\text{D-Fructose} + \text{ATP}^{4-} \longrightarrow \text{D-fructose-6-phosphate}^{2-} + \text{ADP}^{3-} + \text{H}^{+} \quad (18.12)$$

Fructose-6-phosphate generated in this way enters the glycolytic pathway directly in step 3, the second priming reaction. This is the principal means for channeling fructose into glycolysis in adipose tissue, which contains high levels of fructose. On the other hand, fructose-6-phosphate is not readily formed in the liver, since glucose levels are high in the liver, and glucokinase binds fructose much more weakly than it binds glucose.

The Entry of Mannose into Glycolysis

Another simple sugar that enters glycolysis at the same point as fructose is **mannose,** which occurs in many glycoproteins, glycolipids, and polysaccharides (Chapter 10). Mannose is also phosphorylated from ATP by hexokinase, and the mannose-6-phosphate thus produced is converted to fructose-6-phosphate by *phosphomannoisomerase.*

$$\text{D-Mannose} + \text{ATP}^{4-} \longrightarrow \text{D-mannose-6-phosphate}^{2-} + \text{ADP}^{3-} + \text{H} \quad (18.13)$$

$$\text{D-Mannose-6-phosphate}^{2-} \longrightarrow \text{D-fructose-6-phosphate}^{2-} \quad (18.14)$$

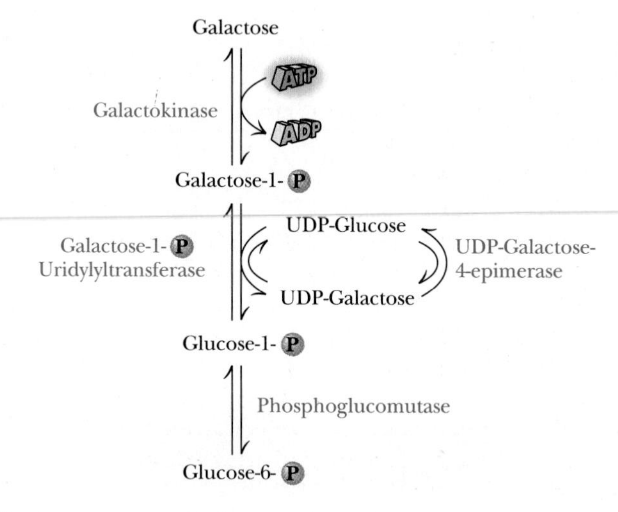

***Figure* 18.33** Galactose metabolism via the Leloir pathway.

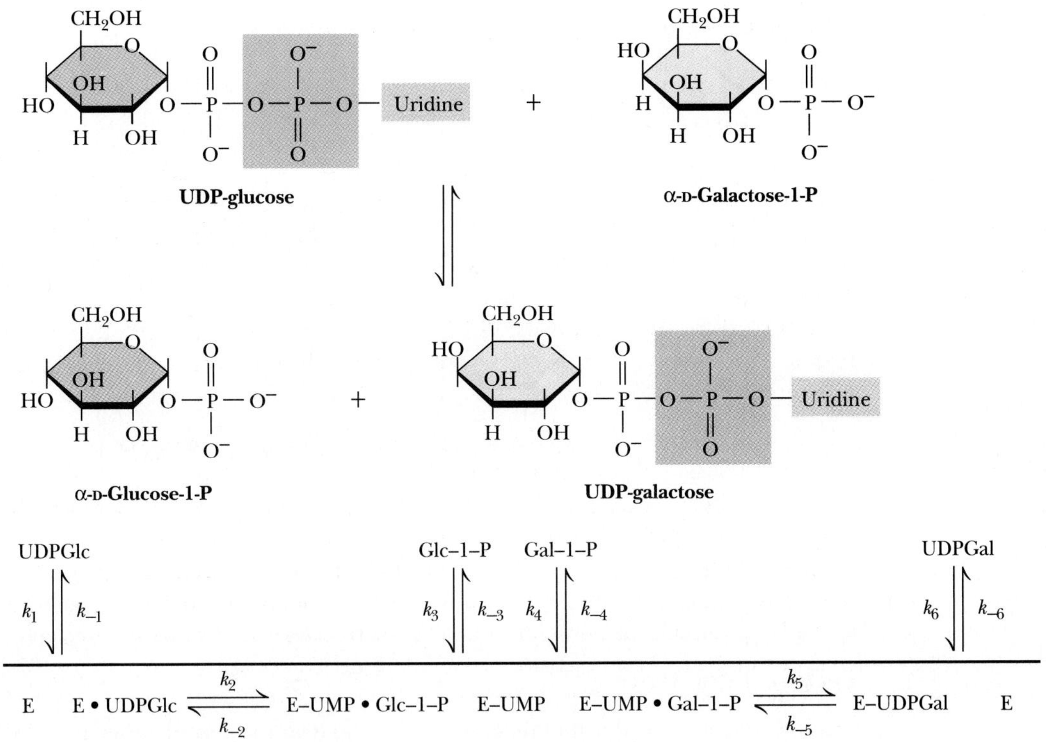

***Figure* 18.34** The galactose-1-phosphate uridylyltransferase reaction involves a "ping-pong" kinetic mechanism.

The Special Case of Galactose

A somewhat more complicated route into glycolysis is followed by **galactose,** another simple hexose sugar. The process, called the **Leloir pathway** after Luis Leloir, its discoverer, begins with phosphorylation from ATP at the C-1 position by *galactokinase:*

$$\text{D-Galactose} + \text{ATP}^{4-} \longrightarrow \text{D-galactose-1-phosphate}^{2-} + \text{ADP}^{3-} + \text{H}^{+} \quad (18.15)$$

Galactose-1-phosphate is then converted into *UDP-galactose* (a sugar nucleotide) by *galactose-1-phosphate uridylyltransferase* (Figure 18.33), with concurrent production of glucose-1-phosphate and consumption of a molecule of UDP-glucose. The uridylyltransferase reaction proceeds via a "ping-pong" mechanism (Figure 18.34) with a covalent enzyme-UMP intermediate. The glucose-1-phosphate produced by the transferase reaction is a substrate for the *phosphoglucomutase* reaction (Figure 18.33), which produces glucose-6-phosphate, a glycolytic substrate. The other transferase product, UDP-galactose, is converted to UDP-glucose by *UDP-glucose-4-epimerase.* The combined action of the uridylyltransferase and epimerase thus produces glucose-1-P from galactose-1-P, with regeneration of UDP-glucose.

A rare hereditary condition known as **galactosemia** involves defects in galactose-1-P uridylyltransferase that render the enzyme inactive. Toxic levels of galactose accumulate in afflicted individuals, causing cataracts and permanent neurological disorders. These problems can be prevented by removing galactose and lactose from the diet. In adults, the toxicity of galactose appears to be less severe, due in part to the metabolism of galactose-1-P by *UDP-glucose pyrophosphorylase,* which apparently can accept galactose-1-P in place of glu-

α-D-Galactose-1-P + UTP ⇌ Pyrophosphate + UDP-galactose (UDPGal)

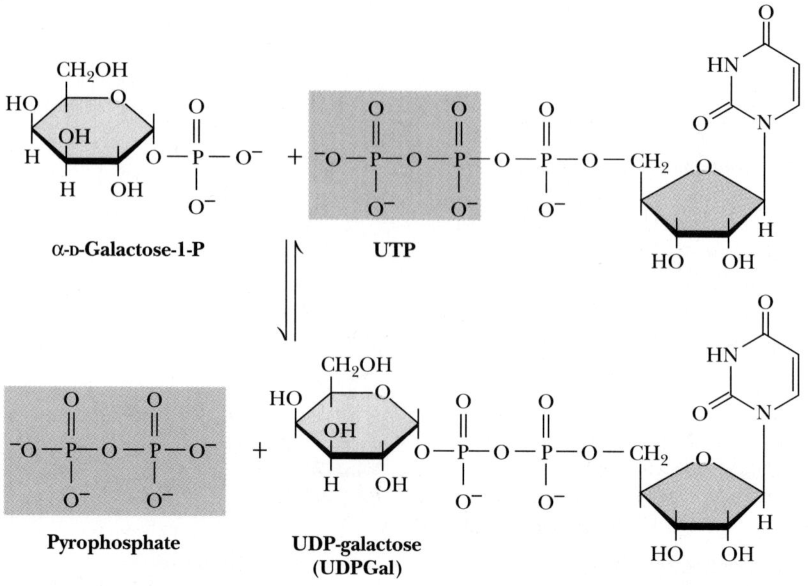

Figure **18.35** The UDP-glucose pyrophosphorylase reaction.

cose-1-P (Figure 18.35). The levels of this enzyme may increase in galactosemic individuals, in order to accommodate the metabolism of galactose.

Lactose Intolerance

A much more common metabolic disorder, **lactose intolerance,** occurs commonly in most parts of the world (notable exceptions being some parts of Africa and northern Europe). Lactose intolerance is an inability to digest lactose because of the absence of the enzyme *lactase* in the intestines of adults. The symptoms of this disorder, which include diarrhea and general discomfort, can be relieved by eliminating milk from the diet.

Glycerol Can Also Enter Glycolysis

Glycerol is the last important simple substance whose ability to enter the glycolytic pathway must be considered. This metabolite, which is produced in substantial amounts by the decomposition of *triacylglycerols* (see Chapter 23) can be converted to glycerol-3-phosphate by the action of *glycerol kinase* and then oxidized to dihydroxyacetone phosphate by the action of *glycerol phosphate dehydrogenase,* with NAD^+ as the required cofactor (Figure 18.36). The dihydroxyacetone phosphate thereby produced enters the glycolytic pathway as a substrate for triose phosphate isomerase.

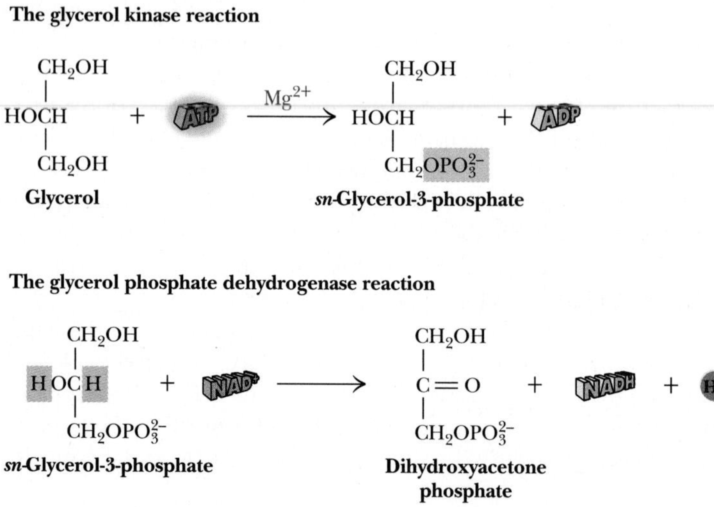

The glycerol kinase reaction

Glycerol + ATP $\xrightarrow{Mg^{2+}}$ *sn*-Glycerol-3-phosphate + ADP

The glycerol phosphate dehydrogenase reaction

sn-Glycerol-3-phosphate + NAD⁺ ⟶ Dihydroxyacetone phosphate + NADH + H⁺

Figure **18.36**

Problems

1. List the reactions of glycolysis that
   a. are energy-consuming (under standard-state conditions).
   b. are energy-yielding (under standard-state conditions).
   c. consume ATP.
   d. yield ATP.
   e. are strongly influenced by changes in concentration of substrate and product because of their molecularity.
   f. are at or near equilibrium in the erythrocyte (see Table 18.2).
2. Determine the anticipated location in pyruvate of labeled carbons if glucose molecules labeled (in separate experiments) with ^{14}C at each position of the carbon skeleton proceed through the glycolytic pathway.
3. In an erythrocyte undergoing glycolysis, what would be the effect of a sudden increase in the concentration of (a) ATP? (b) AMP? (c) fructose-1,6-bisphosphate? (d) fructose-2,6-bisphosphate? (e) citrate? (f) glucose-6-phosphate?
4. Discuss the cycling of NADH and NAD^+ in glycolysis and the related fermentation reactions.
5. For each of the following reactions, name the enzyme that carries out this reaction in glycolysis and write a suitable mechanism for the reaction.

$$\begin{array}{c} CH_2OPO_3^{2-} \\ | \\ C{=}O \\ | \\ HOCH \\ | \\ HCOH \\ | \\ HCOH \\ | \\ CH_2OPO_3^{2-} \end{array} \rightleftharpoons \begin{array}{c} CH_2OPO_3^{2-} \\ | \\ C{=}O \\ | \\ CH_2OH \end{array} + \begin{array}{c} CHO \\ | \\ HCOH \\ | \\ CH_2OPO_3^{2-} \end{array}$$

$$\begin{array}{c} CHO \\ | \\ HCOH \\ | \\ CH_2OPO_3^{2-} \end{array} \rightleftharpoons \begin{array}{c} O{=}C{-}OPO_3^{2-} \\ | \\ HCOH \\ | \\ CH_2OPO_3^{2-} \end{array}$$

6. Write the reactions that permit galactose to be utilized in glycolysis. Write a suitable mechanism for one of these reactions.
7. How might iodoacetic acid affect the glyceraldehyde-3-phosphate dehydrogenase reaction in glycolysis? Justify your answer.
8. If ^{32}P-labeled inorganic phosphate were introduced to erythrocytes undergoing glycolysis, would you expect to detect ^{32}P in glycolytic intermediates? If so, describe the relevant reactions and the ^{32}P incorporation you would observe.
9. Sucrose can enter glycolysis by either of two routes:

   Sucrose phosphorylase:
   Sucrose + P_i $\rightleftharpoons$ fructose + glucose-1-phosphate
   Invertase:
   Sucrose + H_2O $\rightleftharpoons$ fructose + glucose

   Would either of these reactions offer an advantage over the other in the preparation of sucrose for entry into glycolysis?
10. What would be the consequences of a Mg^{2+} ion deficiency for the reactions of glycolysis?

Further Reading

Beitner, R., 1985. *Regulation of Carbohydrate Metabolism.* Boca Raton, FL: CRC Press.

Bioteux, A., and Hess, A., 1981. Design of glycolysis. *Philosophical Transactions, Royal Society of London B* **293:**5–22.

Bodner, G. M., 1986. Metabolism: Part I, glycolysis. *Journal of Chemical Education* **63:**566–570.

Bosca, L., and Corredor, C., 1984. Is phosphofructokinase the rate-limiting step of glycolysis? *Trends in Biochemical Sciences* **9:**372–373.

Boyer, P. D., 1972. *The Enzymes,* 3rd ed., vols. 5–9. New York: Academic Press.

Fothergill-Gilmore, L., 1986. The evolution of the glycolytic pathway. *Trends in Biochemical Sciences* **11:**47–51.

Knowles, J., and Albery, W., 1977. Perfection in enzyme catalysis: The energetics of triose phosphate isomerase. *Accounts of Chemical Research* **10:**105–111.

Newsholme, E., Challiss, R., and Crabtree, B., 1984. Substrate cycles: Their role in improving sensitivity in metabolic control. *Trends in Biochemical Sciences* **9:**277–280.

Pilkus, S., and El-Maghrabi, M., 1988. Hormonal regulation of hepatic gluconeogenesis and glycolysis. *Annual Review of Biochemistry* **57:**755–783.

Saier, M., Jr., 1987. *Enzymes in Metabolic Pathways.* New York: Harper and Row.

Walsh, C. T., 1979. *Enzymatic Reaction Mechanisms.* San Francisco: W. H. Freeman.

Thus times do shift, each thing his turn does hold;
New things succeed, as former things grow old.

Robert Herrick (*Hesperides* [1648], "Ceremonies for Christmas Eve")

Chapter 19

The Tricarboxylic Acid Cycle

A time-lapse photograph of a ferris wheel at night. Aerobic cells use a metabolic wheel—the tricarboxylic acid cycle—to generate energy by acetyl-CoA oxidation.

Outline

19.1 Hans Krebs and the Discovery of the TCA Cycle
19.2 The TCA Cycle—A Brief Summary
19.3 The Bridging Step: Oxidative Decarboxylation of Pyruvate
19.4 Entry into the Cycle: The Citrate Synthase Reaction
19.5 The Isomerization of Citrate by Aconitase
19.6 Isocitrate Dehydrogenase—The First Oxidation in the Cycle
19.7 α-Ketoglutarate Dehydrogenase—A Second Decarboxylation
19.8 Succinyl-CoA Synthetase—A Substrate-Level Phosphorylation
19.9 Succinate Dehydrogenase—An Oxidation Involving FAD
19.10 Fumarase Catalyzes *Trans*-Hydration of Fumarate
19.11 Malate Dehydrogenase—Completing the Cycle
19.12 A Summary of the Cycle
19.13 The TCA Cycle Provides Intermediates for Biosynthetic Pathways
19.14 The Anaplerotic, or "Filling Up," Reactions
19.15 Regulation of the TCA Cycle
19.16 The Glyoxylate Cycle of Plants and Bacteria

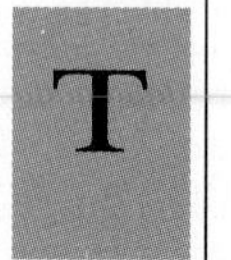

he glycolytic pathway converts glucose to pyruvate and produces two molecules of ATP per glucose—only a small fraction of the potential energy yield from glucose. Under anaerobic conditions, pyruvate is reduced to lactate in animals and to ethanol in yeast, and much of the potential energy of the glucose molecule remains untapped. *In the presence of oxygen,* however, a much more interesting and thermodynamically complete story unfolds. Under aerobic conditions, NADH is oxidized in the electron transport chain, rather than becoming oxidized through reduction of pyruvate to lactate or acetaldehyde to ethanol, for example. Further, pyruvate is converted to *acetyl-coenzyme A* and oxidized to CO_2 in the **tricarboxylic acid (TCA) cycle** (also called the **citric acid cycle**). The electrons liberated by this oxidative process are then passed through an elaborate, membrane-associated **electron transport pathway** to O_2, the final electron acceptor. Elec-

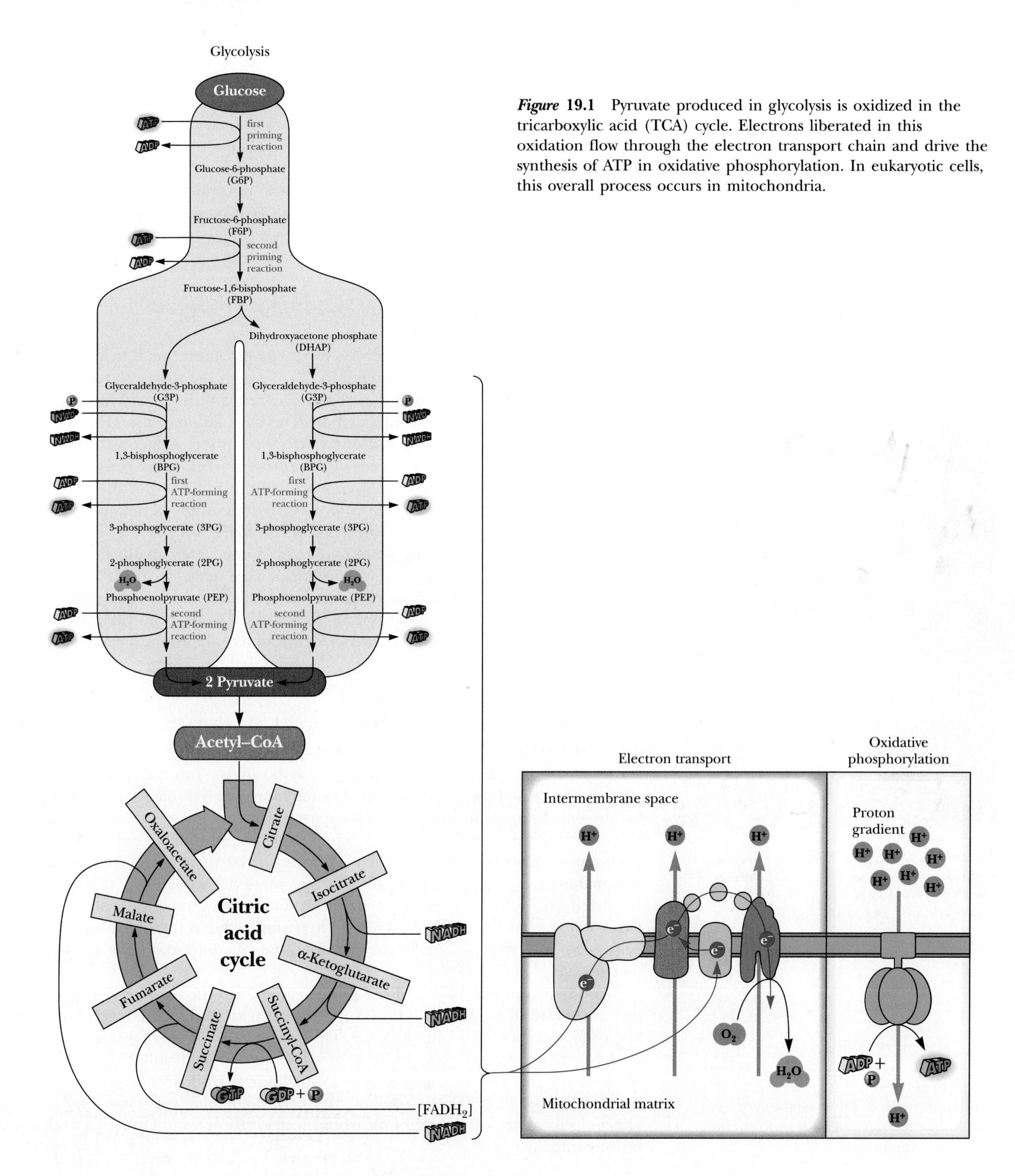

Figure **19.1** Pyruvate produced in glycolysis is oxidized in the tricarboxylic acid (TCA) cycle. Electrons liberated in this oxidation flow through the electron transport chain and drive the synthesis of ATP in oxidative phosphorylation. In eukaryotic cells, this overall process occurs in mitochondria.

tron transfer is coupled to creation of a proton gradient across the membrane. Such a gradient represents an energized state, and the energy stored in this gradient is used to drive the synthesis of many equivalents of ATP.

ATP synthesis as a consequence of electron transport is termed **oxidative phosphorylation;** the complete process is diagrammed in Figure 19.1. *Aerobic pathways* permit the production of 30 to 38 molecules of ATP per glucose oxidized. Although two molecules of ATP come from glycolysis and two more

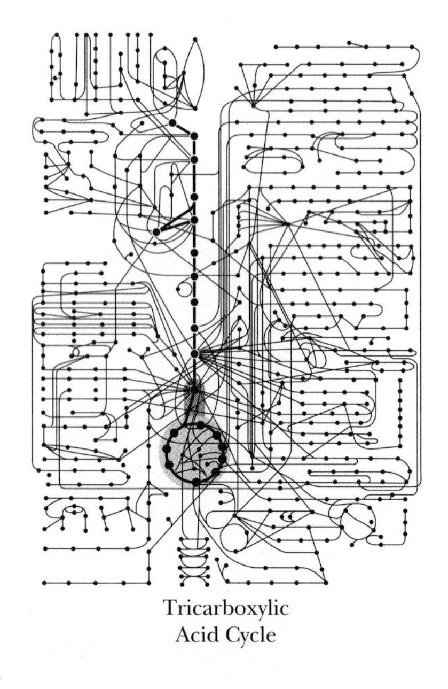

Tricarboxylic Acid Cycle

directly out of the TCA cycle, most of the ATP arises from oxidative phosphorylation. Specifically, reducing equivalents released in the oxidative reactions of glycolysis, pyruvate decarboxylation, and the TCA cycle are captured in the form of NADH and enzyme-bound $FADH_2$, and these reduced coenzymes fuel the electron transport pathway and oxidative phosphorylation. The path to oxidative phosphorylation winds through the TCA cycle, and we will examine this cycle in detail in this chapter.

19.1 Hans Krebs and the Discovery of the TCA Cycle

Within the orderly and logical confines of a textbook, it is difficult to appreciate the tortuous path of the research scientist through the labyrinth of scientific discovery, the patient sifting and comparing of hypotheses, and the often plodding progress toward new information. The elucidation of the TCA cycle in this century is a typical case, and one worth recounting. Armed with accumulated small contributions—pieces of the puzzle—from many researchers over many years, Hans Krebs, in a single, seminal inspiration, put the pieces together and finally deciphered the cyclic nature of pyruvate oxidation. In his honor, the TCA cycle is often referred to as the **Krebs cycle.**

In 1932 Krebs was studying the rates of oxidation of small organic acids by kidney and liver tissue. Only a few substances were active in these experiments—notably succinate, fumarate, acetate, malate, and citrate (Figure 19.2). Later it was found that oxaloacetate could be made from pyruvate in such tissues, and that it could be further oxidized like the other dicarboxylic acids.

In 1935 in Hungary, a crucial discovery was made by Albert Szent-Györgyi, who was studying the oxidation of similar organic substrates by pigeon breast muscle, an active flight muscle with very high rates of oxidation and metabolism. Carefully measuring the amount of oxygen consumed, he observed that addition of any of three four-carbon dicarboxylic acids—fumarate, succinate, or malate—caused the consumption of much more oxygen than was required for the oxidation of the added substance itself. He concluded that these substances were limiting in the cell and, when provided, stimulated oxidation of endogenous glucose and other carbohydrates in the tissues. He also found that **malonate,** a competitive inhibitor of succinate dehydrogenase (Chapter 11), inhibited these oxidative processes; this finding suggested that succinate oxidation is a crucial step. Szent-Györgyi hypothesized that these dicarboxylic acids were linked by an enzymatic pathway that was important for aerobic metabolism.

Another important piece of the puzzle came from the work of Carl Martius and Franz Knoop, who showed that citric acid could be converted to isocitrate and then to α-ketoglutarate. This finding was significant because it was already known that α-ketoglutarate could be enzymatically oxidized to succinate. At this juncture, the pathway from citrate to oxaloacetate seemed to be as shown in Figure 19.3. Whereas the pathway made sense, the *catalytic* effect of succinate and the other dicarboxylic acids from Szent-Györgyi's studies remained a puzzle.

Succinate: H_2C-COO^- / H_2C-COO^-

Fumarate: $H(COO^-)C{=}C(H)\,{}^-OOC$

Acetate: CH_3COO^-

Malate: $HO-CH(COO^-)-H_2C-COO^-$

Citrate: $HO-C(COO^-)(CH_2COO^-)_2$

Oxaloacetate: $O{=}C(COO^-)-H_2C-COO^-$

Figure **19.2** The organic acids observed by Krebs to be oxidized in suspensions of liver and kidney tissue. These substances were pieces in the TCA puzzle that Krebs and others eventually solved.

In 1937 Krebs found that citrate could be formed in muscle suspensions if oxaloacetate and either pyruvate or acetate were added. He saw that he now had a cycle, not a simple pathway, and that addition of any of the intermediates could generate all of the others. The existence of a cycle, together with the entry of pyruvate into the cycle in the synthesis of citrate, provided a clear explanation for the accelerating properties of succinate, fumarate, and malate. If all these intermediates led to oxaloacetate, which combined with pyruvate from glycolysis, they could stimulate the oxidation of many substances besides themselves. (Krebs's conceptual leap to a cycle was not his first. Together with medical student Kurt Henseleit, he had already elucidated the details of the *urea cycle* in 1932.) The complete tricarboxylic acid (Krebs) cycle, as it is now understood, is shown in Figure 19.4.

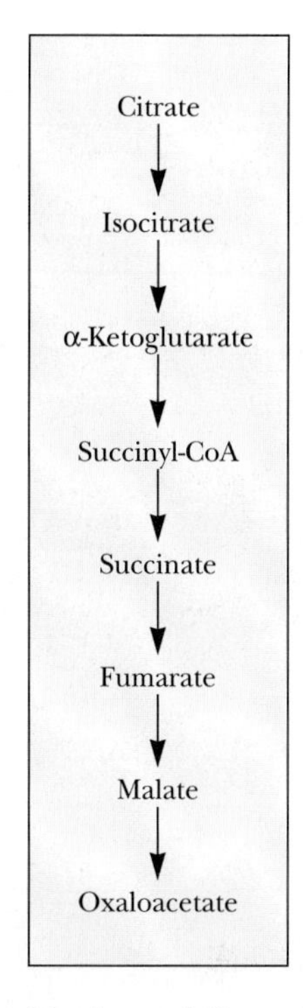

***Figure* 19.3** Martius and Knoop's observation that citrate could be converted to isocitrate and then α-ketoglutarate provided a complete pathway from citrate to oxaloacetate.

19.2 The TCA Cycle—A Brief Summary

The entry of new carbon units into the cycle is through acetyl-CoA. This entry metabolite can be formed either from pyruvate (from glycolysis) or from oxidation of fatty acids (to be discussed in Chapter 23). Transfer of the two-carbon acetyl group from acetyl-CoA to the four-carbon oxaloacetate to yield six-carbon citrate is catalyzed by *citrate synthase.* A dehydration–rehydration rearrangement of citrate yields isocitrate. Two successive decarboxylations produce α-ketoglutarate and then succinyl-CoA, a CoA conjugate of a four-carbon unit. Several steps later, oxaloacetate is regenerated and can combine with another two-carbon unit of acetyl-CoA. Thus, carbon enters the cycle as acetyl-CoA and exits as CO_2. In the process, metabolic energy is captured in the form of ATP, NADH, and enzyme-bound $FADH_2$ (symbolized as $[FADH_2]$).

The Chemical Logic of the TCA Cycle

The cycle shown in Figure 19.4 at first appears to be a complicated way to oxidize acetate units to CO_2, but there is a chemical basis for the apparent complexity. Oxidation of an acetyl group to a pair of CO_2 molecules requires C—C cleavage:

$$CH_3COO^- \longrightarrow CO_2 + CO_2$$

In many instances, C—C cleavage reactions in biological systems occur between carbon atoms α- and β- to a carbonyl group:

$$-\overset{\overset{\displaystyle O}{\|}}{C}-C_{\alpha}\underset{\underset{\text{Cleavage}}{\uparrow}}{-}C_{\beta}-$$

A good example of such a cleavage is the fructose bisphosphate aldolase reaction (see Chapter 18, Figure 18.14a).

Another common type of C—C cleavage is α-cleavage of an α-hydroxyketone:

$$-\overset{\overset{\displaystyle O}{\|}}{C}\underset{\underset{\text{Cleavage}}{\uparrow}}{-}\overset{\overset{\displaystyle OH}{|}}{C_{\alpha}}-$$

(We will see this type of cleavage in the *transketolase* reaction described in Chapter 21.)

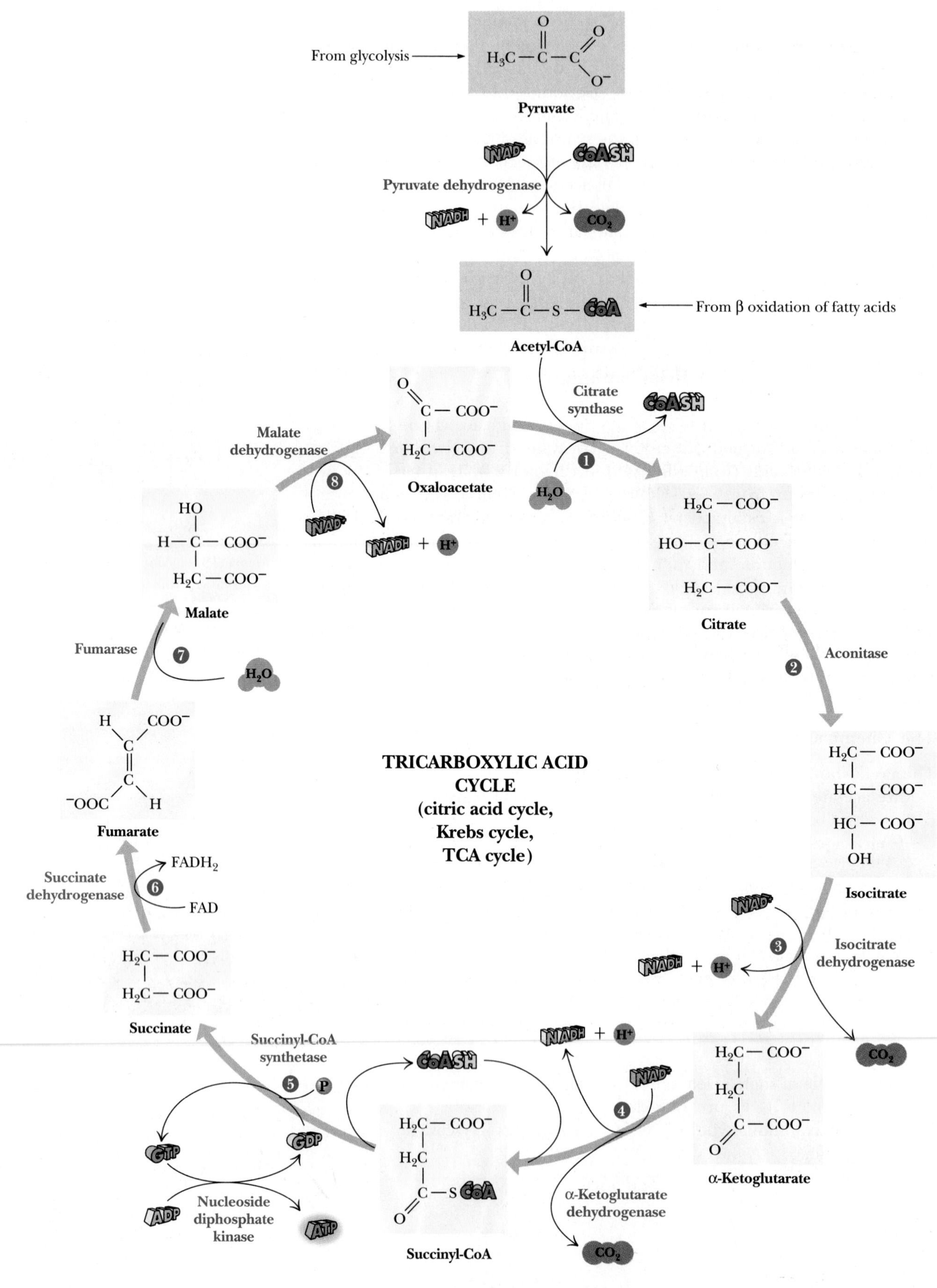

Figure **19.4** The tricarboxylic acid (TCA) cycle.

Neither of these cleavage strategies is suitable for acetate. It has no β-carbon, and the second method would require hydroxylation—not a favorable reaction for acetate. Instead, living things have evolved the clever chemistry of condensing acetate with oxaloacetate and *then* carrying out a β-cleavage. *The TCA cycle combines this β-cleavage reaction with oxidation to form CO_2, regenerate oxaloacetate, and capture the liberated metabolic energy in NADH and ATP.*

19.3 The Bridging Step: Oxidative Decarboxylation of Pyruvate

Pyruvate produced by glycolysis is a significant source of acetyl-CoA for the TCA cycle. Since, in eukaryotic cells, glycolysis occurs in the cytoplasm, whereas the TCA cycle reactions and all subsequent steps of aerobic metabolism take place in the mitochondria, pyruvate must first enter the mitochondria to enter the TCA cycle. The oxidative decarboxylation of pyruvate to acetyl-CoA,

$$\text{Pyruvate} + \text{CoA} + \text{NAD}^+ \longrightarrow \text{acetyl-CoA} + \text{CO}_2 + \text{NADH} + \text{H}^+$$

is the connecting link between glycolysis and the TCA cycle. The reaction is catalyzed by pyruvate dehydrogenase, a multienzyme complex.

The **pyruvate dehydrogenase complex (PDC)** is a noncovalent assembly of three different enzymes operating in concert to catalyze successive steps in the conversion of pyruvate to acetyl-CoA. The active sites of all three enzymes are not far removed from one another, and the product of the first enzyme is passed directly to the second enzyme and so on, without diffusion of substrates and products through the solution. The overall reaction (which was discussed in detail in Chapter 14; see Figure 14.41) involves a total of five coenzymes: thiamine pyrophosphate, coenzyme A, lipoic acid, NAD, and FAD.

19.4 Entry into the Cycle: The Citrate Synthase Reaction

The first reaction within the TCA cycle, the one by which carbon atoms are introduced, is the **citrate synthase reaction** (Figure 19.5). Here acetyl-CoA reacts with oxaloacetate in a **Perkin condensation** (a carbon–carbon condensation between a ketone or aldehyde and an ester). In Chapter 14 it was noted that the acyl group is activated in two ways in an acyl-CoA molecule: the carbonyl carbon is activated for attack by nucleophiles, and the C_α carbon is more acidic and can be deprotonated to form a carbanion. The citrate synthase reaction depends upon the latter mode of activation. As shown in Figure 19.5, a general base on the enzyme accepts a proton from the methyl group of acetyl-CoA, producing a stabilized α-carbanion of acetyl-CoA. This strong nu-

***Figure* 19.5** Citrate is formed in the citrate synthase reaction from oxaloacetate and acetyl-CoA. The mechanism involves nucleophilic attack by the carbanion of acetyl-CoA on the carbonyl carbon of oxaloacetate, followed by thioester hydrolysis.

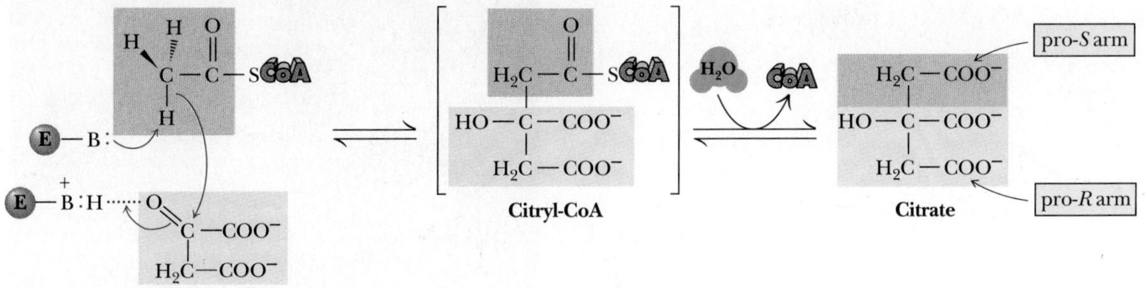

Table **19.1**
The Enzymes and Reactions of the TCA Cycle

Reaction	Enzyme
1. Acetyl-CoA + oxaloacetate + $H_2O \rightleftharpoons$ CoASH + citrate	Citrate synthase
2. Citrate $\rightleftharpoons$ isocitrate	Aconitase
3. Isocitrate + $NAD^+ \rightleftharpoons$ α-ketoglutarate + NADH + CO_2 + H^+	Isocitrate dehydrogenase
4. α-Ketoglutarate + CoASH + $NAD^+ \rightleftharpoons$ succinyl-CoA + NADH + CO_2 + H^+	α-Ketoglutarate dehydrogenase complex
5. Succinyl-CoA + GDP + $P_i \rightleftharpoons$ succinate + GTP + CoASH	Succinyl-CoA synthetase
6. Succinate + [FAD] $\rightleftharpoons$ fumarate + [$FADH_2$]	Succinate dehydrogenase
7. Fumarate + $H_2O \rightleftharpoons$ L-malate	Fumarase
8. L-Malate + $NAD^+ \rightleftharpoons$ oxaloacetate + NADH + H^+	Malate dehydrogenase
Net for reactions 1–8:	
Acetyl-CoA + 3 NAD^+ + [FAD] + GDP + P_i + 2 $H_2O \rightleftharpoons$ CoASH + 3 NADH + [$FADH_2$] + GTP + 2 CO_2 + 3 H^+	
Simple combustion of acetate: Acetate + 2 O_2 + $H^+ \rightleftharpoons$ 2 CO_2 + 2 H_2O	

cleophile attacks the α-carbonyl of oxaloacetate, yielding citryl-CoA. This part of the reaction has an equilibrium constant near 1, but the overall reaction is driven to completion by the subsequent hydrolysis of the high-energy thioester to citrate and free CoA. The overall $\Delta G^{\circ\prime}$ is −31.4 kJ/mol, and under standard conditions the reaction is essentially irreversible. Although the mitochondrial concentration of oxaloacetate is very low (much less than 1 μM—see example in Section 19.11), the strong, negative $\Delta G^{\circ\prime}$ drives the reaction forward.

The Structure of Citrate Synthase

Citrate synthase in mammals is a dimer of 49-kD subunits (Table 19.1). On each subunit, oxaloacetate and acetyl-CoA bind to the active site, which lies in a cleft between two domains and is surrounded mainly by α-helical segments (Figure 19.6). Binding of oxaloacetate induces a conformational change that

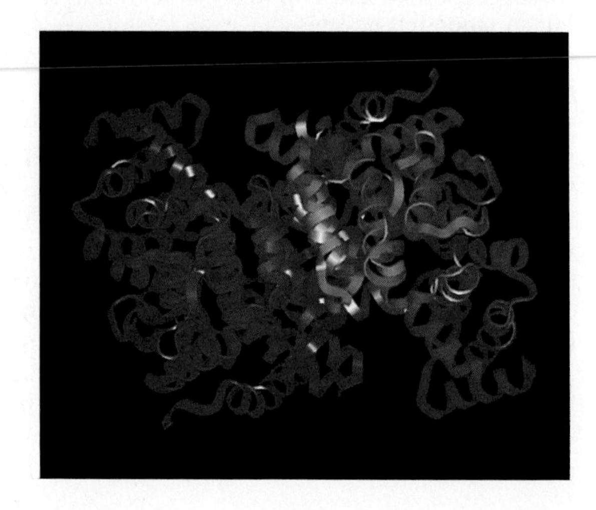

***Figure* 19.6** Citrate synthase is a dimer. The large active-site clefts shown here close upon binding of oxaloacetate.

Table **19.1**
continued

Subunit M_r	Oligomeric Composition	$\Delta G^{\circ\prime}$ (kJ/mol)	K_{eq} at 25°C	ΔG (kJ/mol)
49,000*	Dimer	−31.4	3.2×10^5	−53.9
44,500	Dimer	+6.7	0.067	+0.8
	$\alpha_2\beta\gamma$	−8.4	29.7	−17.5
E_1 96,000	Dimer			
E_2 70,000	24-mer	−30	1.8×10^5	−43.9
E_3 56,000	Dimer			
α 34,500	$\alpha\beta$	−3.3	3.8	≈0
β 42,500				
α 70,000	$\alpha\beta$	+0.4	0.85	≠0
β 27,000				
48,500	Tetramer	−3.8	4.6	≈0
35,000	Dimer	+29.7	6.2×10^{-6}	≈0
		−40		≈(−115)
		−849		

*CS in mammals, A in pig heart, αKDC in *E. coli,* S-CoA S in pig heart, SD in bovine heart, F in pig heart, MD in pig heart. ΔG values from Newsholme, E. A. and Leech, A. R., 1983. *Biochemistry for the Medical Sciences,* New York: Wiley.

facilitates the binding of acetyl-CoA and closes the active site, so that the reactive carbanion of acetyl-CoA is protected from protonation by water.

Regulation of Citrate Synthase

Citrate synthase is the first step in this metabolic pathway, and as stated the reaction has a large negative $\Delta G^{\circ\prime}$. As might be expected, it is a highly regulated enzyme. NADH, a product of the TCA cycle, is an allosteric inhibitor of citrate synthase, as is succinyl-CoA, the product of the fifth step in the cycle (and an acetyl-CoA analog).

19.5 The Isomerization of Citrate by Aconitase

Citrate itself poses a problem: it is a poor candidate for further oxidation because it contains a tertiary alcohol, which could be oxidized only by breaking a carbon–carbon bond. An obvious solution to this problem is to isomerize the tertiary alcohol to a secondary alcohol, which the cycle proceeds to do in the next step.

Citrate is isomerized to isocitrate by **aconitase** in a two-step process involving aconitate as an intermediate (Figure 19.7). In this reaction, the elements of water are first abstracted from citrate to yield aconitate, which is then rehydrated with H— and HO— adding back in opposite positions to produce isocitrate. The net effect is the conversion of a tertiary alcohol (citrate) to a secondary alcohol (isocitrate). Oxidation of the secondary alcohol of isocitrate involves breakage of a C—H bond, a simpler matter than the C—C cleavage required for the direct oxidation of citrate.

Inspection of the citrate structure shows a total of four chemically equivalent hydrogens, but only one of these—the pro-*R* H atom of the pro-*R* arm of citrate—is abstracted by aconitase, which is quite stereospecific. Formation of the double bond of aconitate following proton abstraction requires departure

(a)

Citrate: H_2C-COO^- / $HO-C-COO^-$ / H_R-C-H_S / COO^- ; E—B: ; pro-*S* arm ; pro-*R* arm

H_2O ⇌ H_2O

cis-Aconitate: H_2C-COO^- / $C-COO^-$ / $HC-COO^-$

H_2O ⇌ H_2O

Isocitrate: H_2C-COO^- (*S*) / $H-C-COO^-$ / $HC-COO^-$ (*R*) / OH

Citrate **cis-Aconitate** **Isocitrate**

Aconitase removes the pro-*R* H of the pro-*R* arm of citrate

(b)

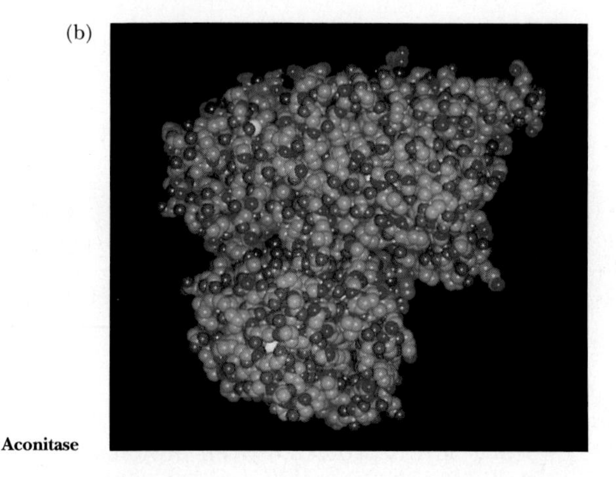

Aconitase

***Figure* 19.7** (a) The aconitase reaction converts citrate to *cis*-aconitate and then to isocitrate. Aconitase is stereospecific and removes the pro-*R* hydrogen from the pro-*R* arm of citrate. (b) A space-filling model of aconitase.

of hydroxide ion from the C-3 position. Hydroxide is a relatively poor leaving group, and its departure is facilitated in the aconitase reaction by coordination with an iron atom in an iron–sulfur cluster.

Aconitase Utilizes an Iron–Sulfur Cluster

Aconitase contains an **iron–sulfur cluster** consisting of three iron atoms and four sulfur atoms in a near-cubic arrangement (Figure 19.8). This cluster is bound to the enzyme via three cysteine groups from the protein. One corner of the cube is vacant and binds Fe^{2+}, which activates aconitase. The iron atom in this position can coordinate the C-3 carboxyl and hydroxyl groups of citrate. This iron atom thus acts as a Lewis acid, accepting an unshared pair of electrons from the hydroxyl group and making it a better leaving group. The equilibrium for the reaction favors citrate, and an equilibrium mixture typically contains about 90% citrate, 4% *cis*-aconitate, and 6% isocitrate. The $\Delta G°'$ is +6.7 kJ/mol.

Fluoroacetate Blocks the TCA Cycle

Fluoroacetate is an extremely poisonous agent that blocks the TCA cycle *in vivo, although it has no apparent effect on any of the isolated enzymes.* Its LD_{50}, the *l*ethal *d*ose for 50% of animals consuming it, is 0.2 mg per kilogram of body weight; it has been used as a rodent poison. The action of fluoracetate has been traced to aconitase, which is inhibited *in vivo* by *fluorocitrate,* which is

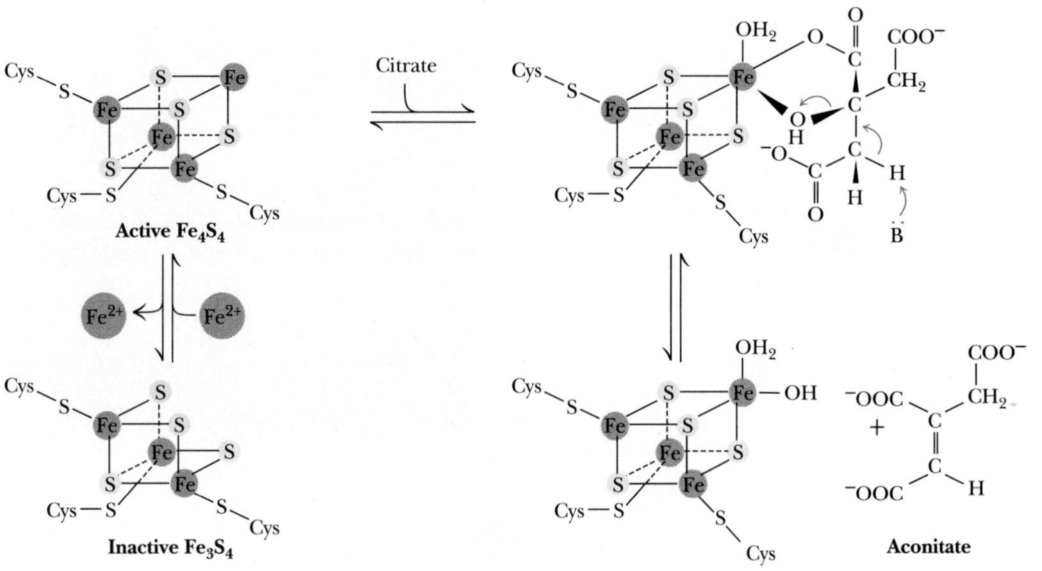

Figure **19.8** The iron–sulfur cluster of aconitase. Binding of Fe^{2+} to the vacant position of the cluster activates aconitase. The added iron atom coordinates the C-3 carboxyl and hydroxyl groups of citrate and acts as a Lewis acid, accepting an electron pair from the hydroxyl group and making it a better leaving group.

formed from fluoroacetate in two steps (Figure 19.9). Fluoroacetate readily crosses both the cellular and mitochondrial membranes, and in mitochondria it is converted to fluoroacetyl-CoA by *acetyl-CoA synthetase.* Fluoroacetyl-CoA is a substrate for citrate synthase, which condenses it with oxaloacetate to form fluorocitrate. Fluoroacetate may thus be viewed as a **trojan horse inhibitor.** Analogous to the giant Trojan Horse of legend—which the soldiers of Troy took into their city, not knowing that Greek soldiers were hidden inside it and waiting to attack—fluoroacetate enters the TCA cycle innocently enough, in the citrate synthase reaction. Citrate synthase converts fluoroacetate to inhibitory fluorocitrate for its TCA cycle partner, aconitase, blocking the cycle.

19.6 Isocitrate Dehydrogenase—The First Oxidation in the Cycle

In the next step of the TCA cycle, isocitrate is oxidatively decarboxylated to yield α-ketoglutarate, with concomitant reduction of NAD^+ to NADH in the isocitrate dehydrogenase reaction (Figure 19.10). The reaction has a net $\Delta G°'$ of −8.4 kJ/mol, and it is sufficiently exergonic to pull the aconitase reaction forward. This two-step reaction involves (1) oxidation of the C-2 alcohol of

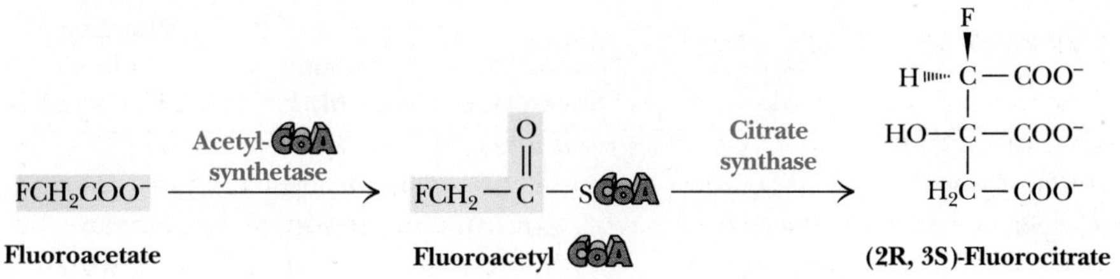

Figure **19.9** The conversion of fluoroacetate to fluorocitrate.

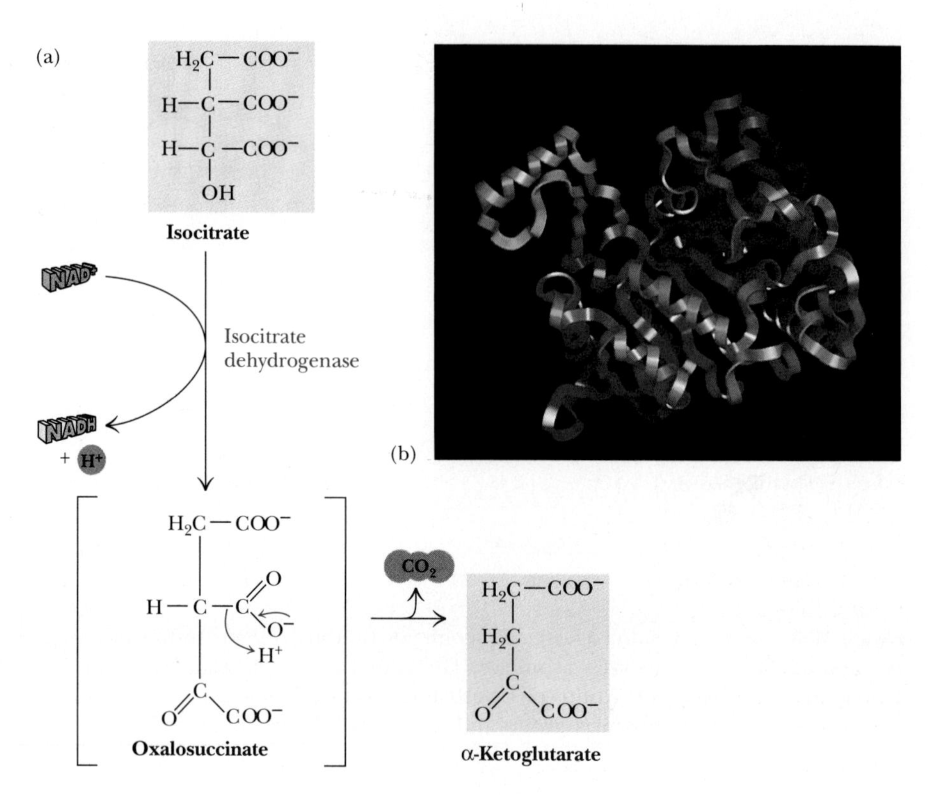

Figure **19.10** (a) The isocitrate dehydrogenase reaction. (b) A ribbon diagram of isocitrate dehydrogenase.

isocitrate to form oxalosuccinate, followed by (2) a β-decarboxylation reaction that expels the central carboxyl group as CO_2, leaving the product α-ketoglutarate. Oxalosuccinate, the β-keto acid produced by the initial dehydrogenation reaction, is unstable and thus is readily decarboxylated.

Isocitrate Dehydrogenase Links the TCA Cycle and Electron Transport

Isocitrate dehydrogenase provides the first connection between the TCA cycle and the electron transport pathway and oxidative phosphorylation, via its production of NADH. As a connecting point between two metabolic pathways, isocitrate dehydrogenase is a regulated reaction. NADH and ATP are allosteric inhibitors, whereas ADP acts as an allosteric activator, lowering the K_m for isocitrate by a factor of 10. The enzyme is virtually inactive in the absence of ADP. Also, the product, α-ketoglutarate, is a crucial α-keto acid for aminotransferase reactions (see Chapters 11 and 14), connecting the TCA cycle (that is, carbon metabolism) with nitrogen metabolism.

Table **19.2**
Composition of α-Ketoglutarate Dehydrogenase Complex from *E. coli*

Enzyme	Coenzyme	Enzyme M_r	Number of Subunits	Subunit M_r	Number of Subunits per Complex
α-Ketoglutarate dehydrogenase	Thiamine pyrophosphate	192,000	2	96,000	24
Dihydrolipoyl transsuccinylase	Lipoic acid, CoASH	1,700,000	24	70,000	24
Dihydrolipoyl dehydrogenase	FAD, NAD^+	112,000	2	56,000	12

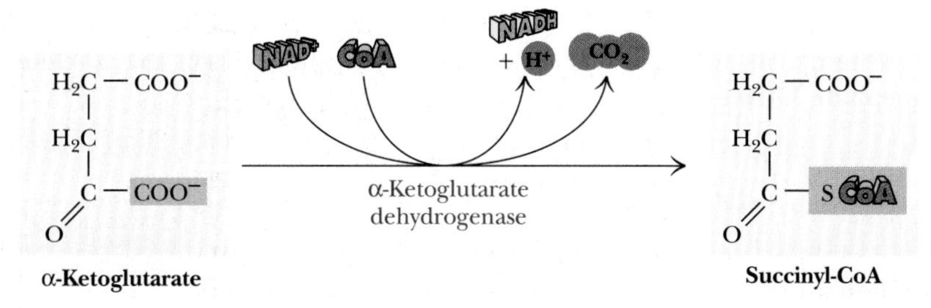

***Figure* 19.11** The α-ketoglutarate dehydrogenase reaction.

19.7 α-Ketoglutarate Dehydrogenase—A Second Decarboxylation

A second oxidative decarboxylation occurs in the **α-ketoglutarate dehydrogenase reaction** (Figure 19.11). Like the pyruvate dehydrogenase complex, α-ketoglutarate dehydrogenase is a multienzyme complex—consisting of *α-ketoglutarate dehydrogenase, dihydrolipoyl transsuccinylase,* and *dihydrolipoyl dehydrogenase*—that employs five different coenzymes (Table 19.2). The dihydrolipoyl dehydrogenase in this reaction is identical to that in the pyruvate dehydrogenase reaction. The mechanism is analogous to that of pyruvate dehydrogenase, and the free energy changes for these reactions are −29 to −33.5 kJ/mol. As with the pyruvate dehydrogenase reaction, this reaction produces NADH and a thioester product—in this case, succinyl-CoA. Succinyl-CoA and NADH products are energy-rich species that are important sources of metabolic energy in subsequent cellular processes.

19.8 Succinyl-CoA Synthetase—A Substrate-Level Phosphorylation

The NADH produced in the foregoing steps can be routed through the electron transport pathway to make high-energy phosphates via oxidative phosphorylation. However, succinyl-CoA is itself a high-energy intermediate and is utilized in the next step of the TCA cycle to drive the phosphorylation of GDP to GTP (in mammals) or ADP to ATP (in plants and bacteria). The reaction (Figure 19.12) is catalyzed by **succinyl-CoA synthetase,** sometimes called **succinate thiokinase.** The free energies of hydrolysis of succinyl-CoA and GTP or ATP are similar, and the net reaction has a $\Delta G°'$ of −3.3 kJ/mol. Succinyl-CoA synthetase provides another example of a **substrate-level phosphorylation** (Chapter 18), in which a substrate, rather than an electron transport chain or proton gradient, provides the energy for phosphorylation. It is the only such

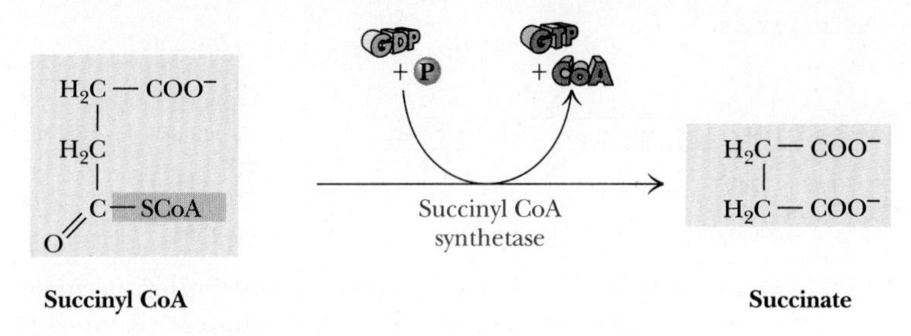

***Figure* 19.12** The succinyl-CoA synthetase reaction.

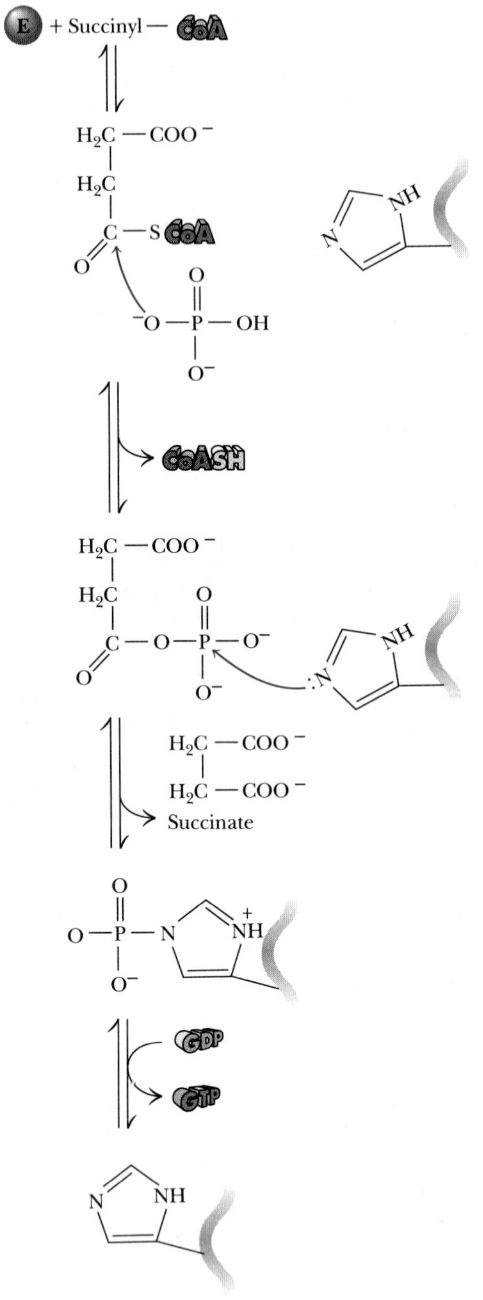

reaction in the TCA cycle. The GTP produced by mammals in this reaction can exchange its terminal phosphoryl group with ADP via the **nucleoside diphosphate kinase reaction:**

$$\text{GTP} + \text{ADP} \xrightleftharpoons[\text{kinase}]{\text{Nucleoside diphosphate}} \text{ATP} + \text{GDP}$$

The Mechanism of Succinyl-CoA Synthetase

The mechanism of succinyl-CoA synthetase is postulated to involve displacement of CoA by phosphate, forming succinyl phosphate at the active site, followed by transfer of the phosphoryl group to an active-site histidine (making a *phosphohistidine* intermediate) and release of succinate. The phosphoryl moiety is then transferred to GDP to form GTP (Figure 19.13). This sequence of steps "preserves" the energy of the thioester bond of succinyl-CoA in a series of high-energy intermediates that lead to a molecule of ATP:

$$\text{Thioester} \longrightarrow [\text{succinyl-P}] \longrightarrow [\text{phosphohistidine}] \longrightarrow \text{GTP} \longrightarrow \text{ATP}$$

Figure 19.13 The mechanism of the succinyl-CoA synthetase reaction.

The First Five Steps of the TCA Cycle Produce NADH, CO_2, GTP (ATP), and Succinate

This is a good point to pause in our trip through the TCA cycle and see what has happened. A two-carbon acetyl group has been introduced as acetyl-CoA and linked to oxaloacetate, and two CO_2 molecules have been liberated. The cycle has produced two molecules of NADH and one of GTP or ATP, and has left a molecule of succinate.

The TCA cycle can now be completed by converting succinate to oxaloacetate. This latter process represents a net oxidation. The TCA cycle breaks it down into (consecutively) an oxidation step, a hydration reaction, and a second oxidation step. The oxidation steps are accompanied by the reduction of an [FAD] and an NAD^+. The reduced coenzymes, [$FADH_2$] and NADH, subsequently provide reducing power in the electron transport chain. (We will see in Chapter 23 that virtually the same chemical strategy is used in β oxidation of fatty acids.)

19.9 Succinate Dehydrogenase—An Oxidation Involving FAD

The oxidation of succinate to fumarate (Figure 19.14) is carried out by **succinate dehydrogenase,** a membrane-bound enzyme that is actually part of the electron transport chain. As will be seen in Chapter 20, succinate dehydrogenase is part of the succinate–coenzyme Q reductase of the electron transport chain. In contrast with all of the other enzymes of the TCA cycle, which are soluble proteins found in the mitochondrial matrix, succinate dehydrogenase

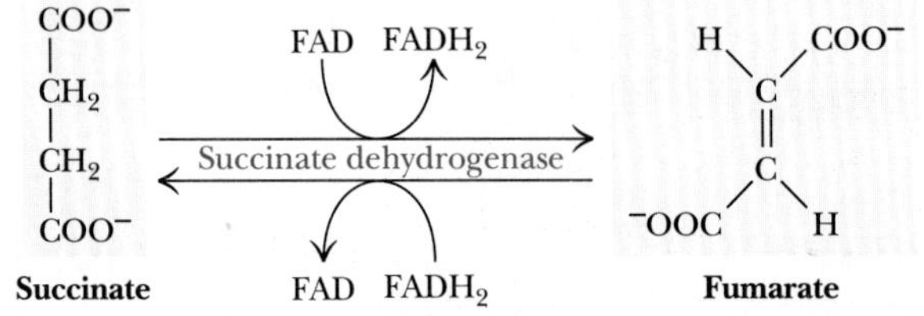

Figure 19.14 The succinate dehydrogenase reaction. Oxidation of succinate occurs with reduction of [FAD]. Reoxidation of $FADH_2$ transfers electrons to coenzyme Q.

is an integral membrane protein tightly associated with the inner mitochondrial membrane. Succinate oxidation involves removal of H atoms across a C—C bond, rather than a C—O or C—N bond, and produces the *trans*-unsaturated fumarate. This reaction (the oxidation of an alkane to an alkene) is not sufficiently exergonic to reduce NAD^+, but it does yield enough energy to reduce [FAD]. (By contrast, oxidations of alcohols to ketones or aldehydes are more energetically favorable and provide sufficient energy to reduce NAD^+.) This important point will be illustrated and clarified in an example in Chapter 20.

Succinate dehydrogenase is a dimeric protein, with subunits of molecular masses 70 kD and 27 kD (see Table 19.1). FAD is covalently bound to the larger subunit; the bond involves a methylene group at C-8a of FAD and N-3 of a histidine on the protein (Figure 19.15). Succinate dehydrogenase also contains three different iron–sulfur clusters (Figure 19.16). Viewed from either end of the succinate molecule, the reaction involves dehydrogenation α,β to a carbonyl (actually, a carboxyl) group. The dehydrogenation is stereospecific (Figure 19.14), with the pro-*S* hydrogen removed from one carbon atom and the pro-*R* hydrogen removed from the other. The electrons captured by [FAD] in this reaction are passed directly into the iron–sulfur clusters of the enzyme and on to *coenzyme Q(UQ)*. The covalently bound FAD is first reduced to [$FADH_2$] and then reoxidized to form [FAD] and the reduced form of coenzyme Q, *UQH_2*. Electrons captured by UQH_2 then flow through the rest of the electron transport chain in a series of events that will be discussed in detail in Chapter 20.

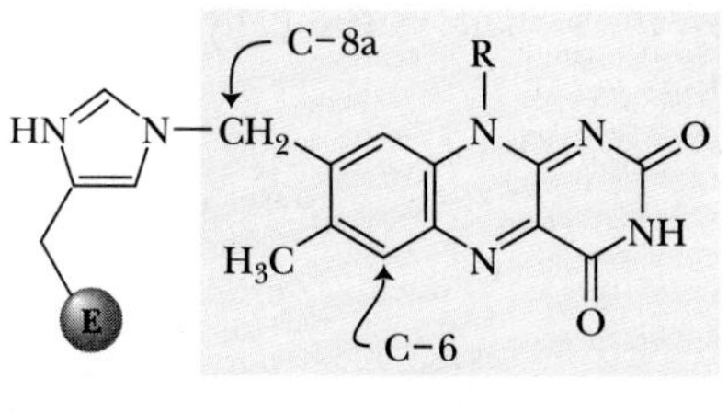

***Figure* 19.15** The covalent bond between FAD and succinate dehydrogenase involves the C-8a methylene group of FAD and the N-3 of a histidine residue on the enzyme.

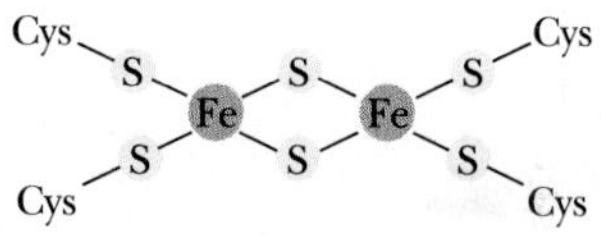

***Figure* 19.16** The Fe_2S_2 cluster of succinate dehydrogenase.

19.10 Fumarase Catalyzes *Trans*-Hydration of Fumarate

Fumarate is hydrated in a stereospecific reaction by fumarase to give L-malate (Figure 19.17). The reaction involves *trans*-addition of the elements of water across the double bond. Recall that aconitase carries out a similar reaction, and that *trans*-addition of —H and —OH occurs across the double bond of *cis*-aconitate. Though the exact mechanism is uncertain, it may involve protonation of the double bond to form an intermediate carbonium ion (Figure 19.18) or possibly attack by water or OH^- anion to produce a carbanion, followed by protonation.

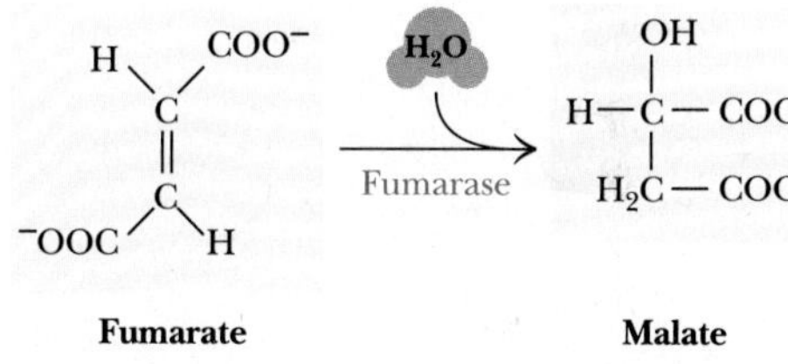

***Figure* 19.17** The fumarase reaction.

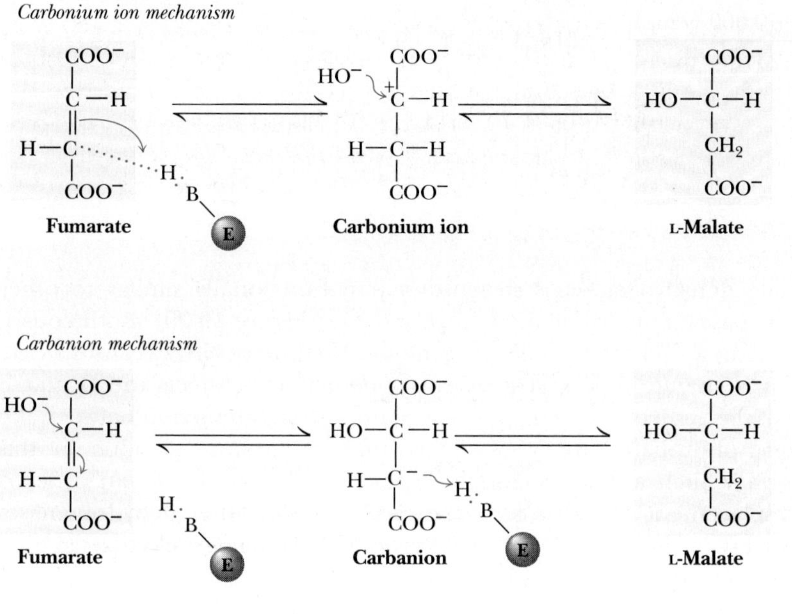

***Figure* 19.18** Two possible mechanisms for the fumarase reaction.

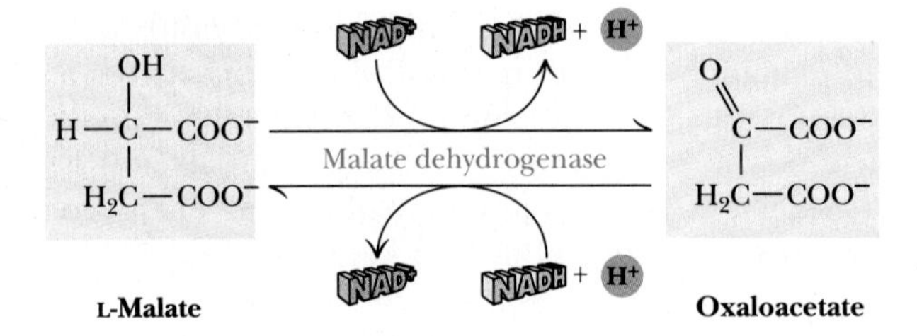

***Figure* 19.19** The malate dehydrogenase reaction.

19.11 Malate Dehydrogenase—Completing the Cycle

In the last step of the TCA cycle, L-malate is oxidized to oxaloacetate by **malate dehydrogenase** (Figure 19.19). This reaction is very endergonic, with a $\Delta G^{\circ\prime}$ of +30 kJ/mol. Consequently, the concentration of oxaloacetate in the mitochondrial matrix is usually quite low (see the following example). The reaction, however, is pulled forward by the favorable citrate synthase reaction. Oxidation of malate is coupled to reduction of yet another molecule of NAD^+, the third one of the cycle. Counting the [FAD] reduced by succinate dehydrogenase, this makes the fourth coenzyme reduced through oxidation of a single acetate unit.

Example

A typical intramitochondrial concentration of malate is 0.22 m*M*. If the $[NAD^+]/[NADH]$ ratio in mitochondria is 20 and if the malate dehydrogenase reaction is at equilibrium, calculate the intramitochondrial concentration of oxaloacetate at 25°C.

Solution

For the malate dehydrogenase reaction,

$$\text{Malate} + NAD^+ \rightleftharpoons \text{oxaloacetate} + NADH + H^+$$

with the value of $\Delta G^{\circ\prime}$ being +30 kJ/mol. Then

$$\Delta G^{\circ\prime} = -RT \ln K_{eq}$$

$$= -(8.314\,\text{J/mol-K})(298)\ln\left(\frac{[1]x}{[20][2.2\times10^{-4}]}\right)$$

$$\frac{-30{,}000\,\text{J/mol}}{2478\,\text{J/mol}} = \ln\,(x/4.4\times10^{-3})$$

$$-12.1 = \ln\,(x/4.4\times10^{-3})$$

$$x = (5.6\times10^{-6})(4.4\times10^{-3})$$

$$x = [\text{oxaloacetate}] = 0.024\ \mu M$$

Malate dehydrogenase is structurally and functionally similar to other dehydrogenases, notably lactate dehydrogenase (Figure 19.20). Both consist of alternating β-sheet and α-helical segments. Binding of NAD^+ causes a conformational change in the 20-residue segment that connects the D and E strands of the β-sheet. The change is triggered by an interaction between the adenosine phosphate moiety of NAD^+ and an arginine residue in this loop region. Such a conformational change is consistent with an ordered single-displacement mechanism for NAD^+-dependent dehydrogenases (Chapter 11).

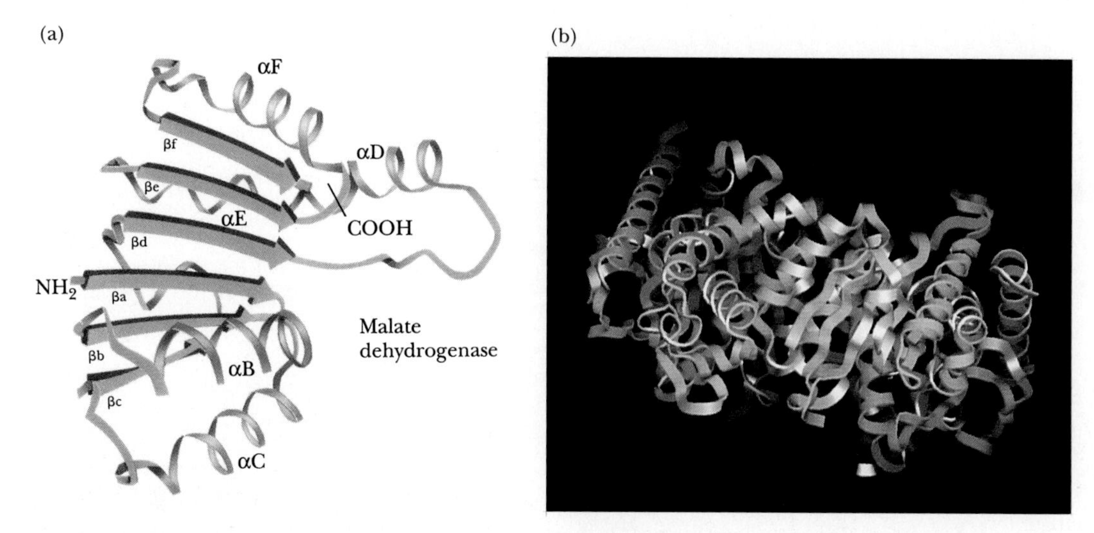

Figure **19.20** (a) The structure of malate dehydrogenase (monomer). (b) A ribbon structure of the dimer of malate dehydrogenase.

19.12 A Summary of the Cycle

The net reaction accomplished by the TCA cycle, as follows, shows two molecules of CO_2, one ATP, and four reduced coenzymes produced per acetate group oxidized. The cycle is exergonic, with a net $\Delta G°'$ for one pass around the cycle of approximately −40 kJ/mol. Table 19.1 compares the $\Delta G°'$ values for the individual reactions with the overall $\Delta G°'$ for the net reaction.

$$\text{Acetyl-CoA} + 3\,\text{NAD}^+ + [\text{FAD}] + \text{ADP} + \text{P}_i + 2\,\text{H}_2\text{O} \rightleftharpoons$$
$$2\,\text{CO}_2 + 3\,\text{NADH} + 3\,\text{H}^+ + [\text{FADH}_2] + \text{ATP} + \text{CoASH}$$
$$\Delta G°' = -40\ \text{kJ/mol}$$

Glucose metabolized via glycolysis produces two molecules of pyruvate and thus two molecules of acetyl-CoA, which can enter the TCA cycle. Combining glycolysis and the TCA cycle gives the net reaction shown:

$$\text{Glucose} + 6\,\text{H}_2\text{O} + 10\,\text{NAD}^+ + 2\,[\text{FAD}] + 4\,\text{ADP} + 4\,\text{P}_i \rightleftharpoons$$
$$6\,\text{CO}_2 + 10\,\text{NADH} + 10\,\text{H}^+ + 2\,[\text{FADH}_2] + 4\,\text{ATP}$$

All six carbons of glucose are liberated as CO_2, and a total of four molecules of ATP are formed thus far in substrate-level phosphorylations. The 12 reduced coenzymes produced up to this point can eventually produce a maximum of 34 molecules of ATP in the electron transport and oxidative phosphorylation pathways. A stoichiometric relationship for these subsequent processes is

$$\text{NADH} + \text{H}^+ + \tfrac{1}{2}\,\text{O}_2 + 3\,\text{ADP} + 3\,\text{P}_i \rightleftharpoons \text{NAD}^+ + 3\,\text{ATP} + 4\,\text{H}_2\text{O}$$
$$[\text{FADH}_2] + \tfrac{1}{2}\,\text{O}_2 + 2\,\text{ADP} + 2\,\text{P}_i \rightleftharpoons [\text{FAD}] + 2\,\text{ATP} + 3\,\text{H}_2\text{O}$$

Thus, a total of 3 ATP per NADH and 2 ATP per $FADH_2$ may be produced through the processes of electron transport and oxidative phosphorylation.

The Fate of the Carbon Atoms of Acetyl-CoA in the TCA Cycle

It is instructive to consider how the carbon atoms of a given acetate group are routed through several turns of the TCA cycle. As shown in Figure 19.21, neither of the carbon atoms of a labeled acetate unit is lost as CO_2 in the first turn of the cycle. The CO_2 evolved in any turn of the cycle derives from the carboxyl groups of the oxaloacetate acceptor (from the previous turn), not from incoming acetyl-CoA. On the other hand, succinate labeled on one end

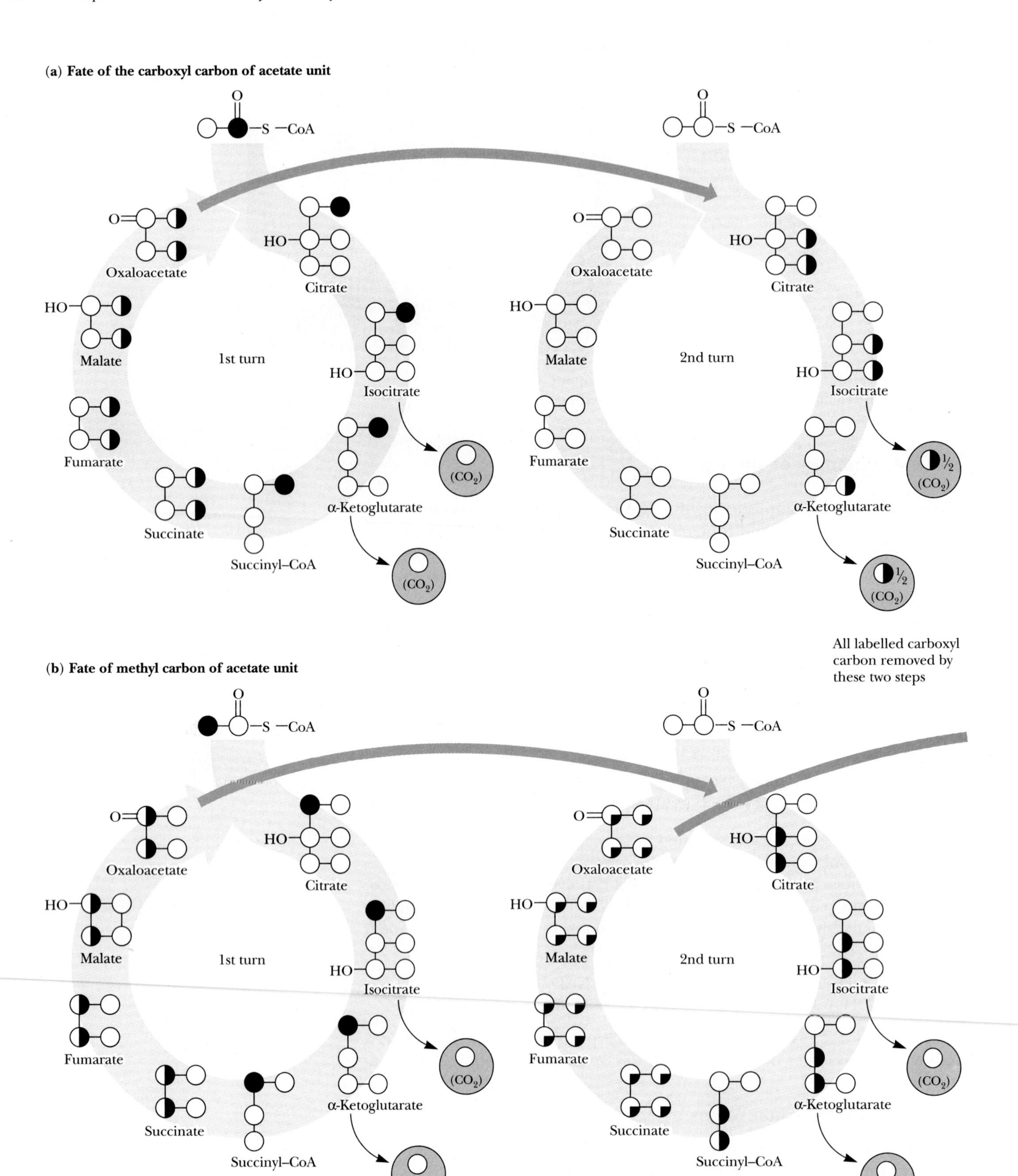
(a) Fate of the carboxyl carbon of acetate unit
S —CoA
Oxaloacetate
Citrate
Malate
1st turn
Isocitrate
Fumarate
Succinate
Succinyl–CoA
α-Ketoglutarate
(CO_2)
2nd turn
½
All labelled carboxyl carbon removed by these two steps
(b) Fate of methyl carbon of acetate unit

from the original labeled acetate forms two different labeled oxaloacetates. The carbonyl carbon of acetyl-CoA is evenly distributed between the two carboxyl carbons of oxaloacetate, and the labeled methyl carbon of incoming acetyl-CoA ends up evenly distributed between the methylene and carbonyl carbons of oxaloacetate.

When these labeled oxaloacetates enter a second turn of the cycle, both of the carboxyl carbons are lost as CO_2, but the methylene and carbonyl carbons survive through the second turn. Thus, the methyl carbon of a labeled acetyl-CoA survives two full turns of the cycle. In the third turn of the cycle, one-half of the carbon from the original methyl group of acetyl-CoA has become one of the carboxyl carbons of oxaloacetate and is thus lost as CO_2. In the fourth turn of the cycle, further ''scrambling'' results in loss of half of the remaining labeled carbon (one-fourth of the original methyl carbon label of acetyl-CoA), and so on.

It can be seen that the carbonyl and methyl carbons of labeled acetyl-CoA have very different fates in the TCA cycle. The carbonyl carbon survives the first turn intact but is completely lost in the second turn. The methyl carbon survives two full turns, then undergoes a 50% loss through each succeeding turn of the cycle.

Figure **19.21** The fate of the carbon atoms of acetate in successive TCA cycles. (a) The carbonyl carbon of acetyl-CoA is fully retained through one turn of the cycle but is lost completely in a second turn of the cycle. (b) The methyl carbon of a labeled acetyl-CoA survives two full turns of the cycle but becomes equally distributed among the four carbons of oxaloacetate by the end of the second turn. In each subsequent turn of the cycle, one half of this carbon (the original labeled methyl group) is lost.

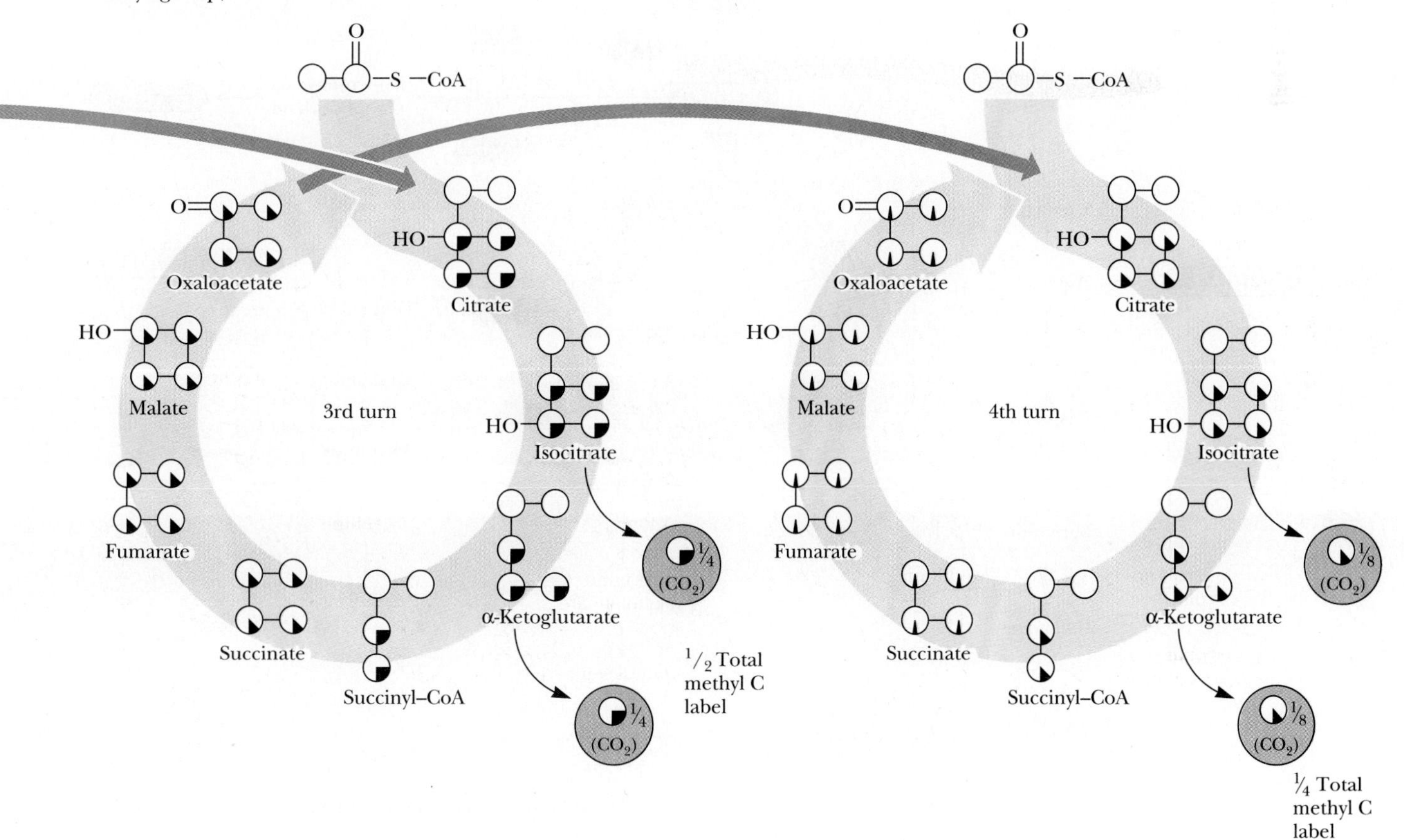

It is worth noting that the carbon–carbon bond cleaved in the TCA pathway entered as an acetate unit in the previous turn of the cycle. Thus, the isocitrate dehydrogenase reaction that cleaves this bond is just a cleverly disguised acetate C—C cleavage and oxidation.

19.13 The TCA Cycle Provides Intermediates for Biosynthetic Pathways

Until now we have viewed the TCA cycle as a catabolic process, since it oxidizes acetate units to CO_2 and converts the liberated energy to ATP and reduced coenzymes. The TCA cycle is, after all, the end point for breakdown of food materials, at least in terms of carbon turnover. However, as shown in Figure 19.22, four-, five-, and six-carbon species produced in the TCA cycle also fuel a variety of **biosynthetic processes.** α-Ketoglutarate, succinyl-CoA, fumarate, and oxaloacetate are all precursors of important cellular species. (In order to participate in eukaryotic biosynthetic processes, however, they must first be transported out of the mitochondria.) A transamination reaction

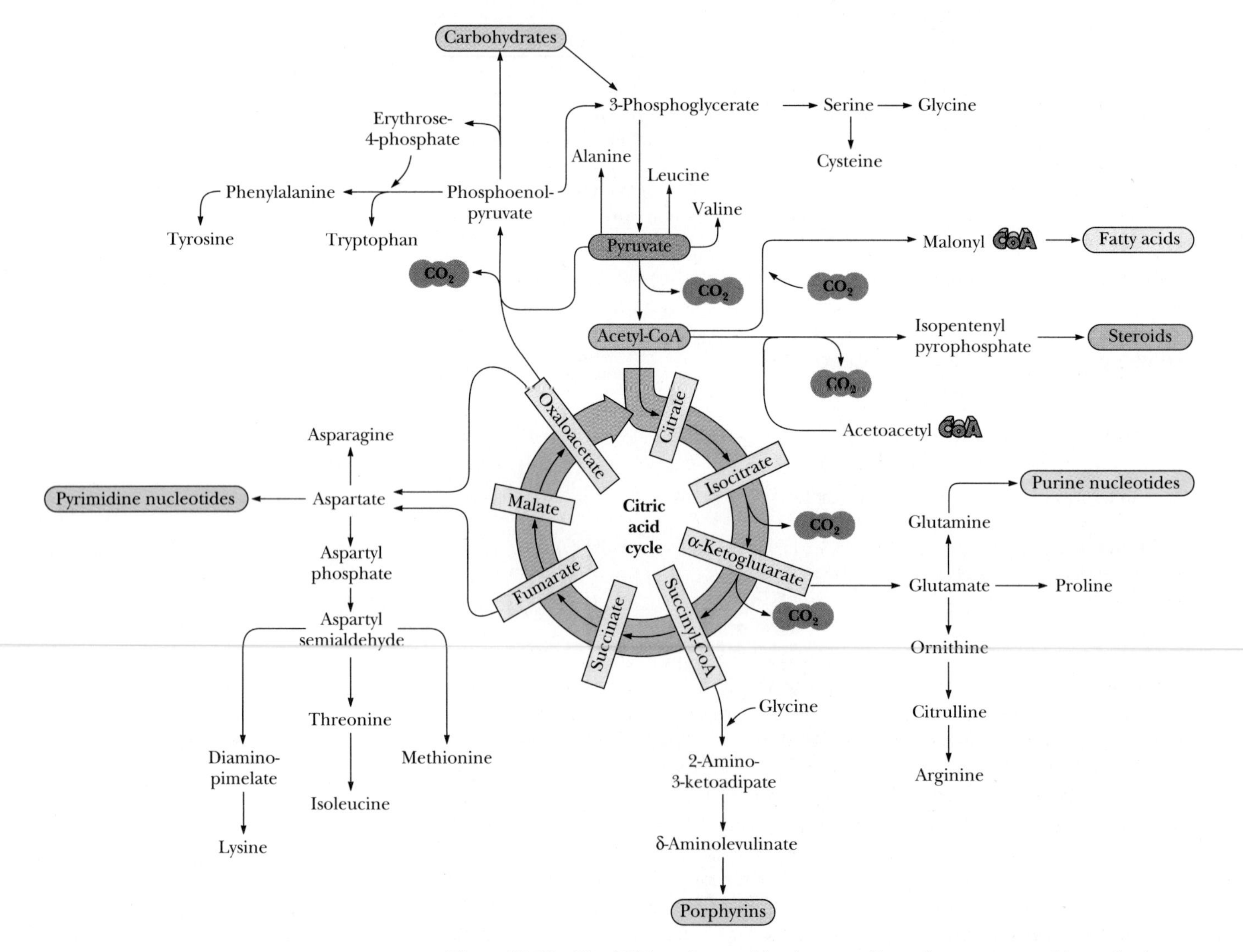

***Figure* 19.22** The TCA cycle provides intermediates for numerous biosynthetic processes in the cell.

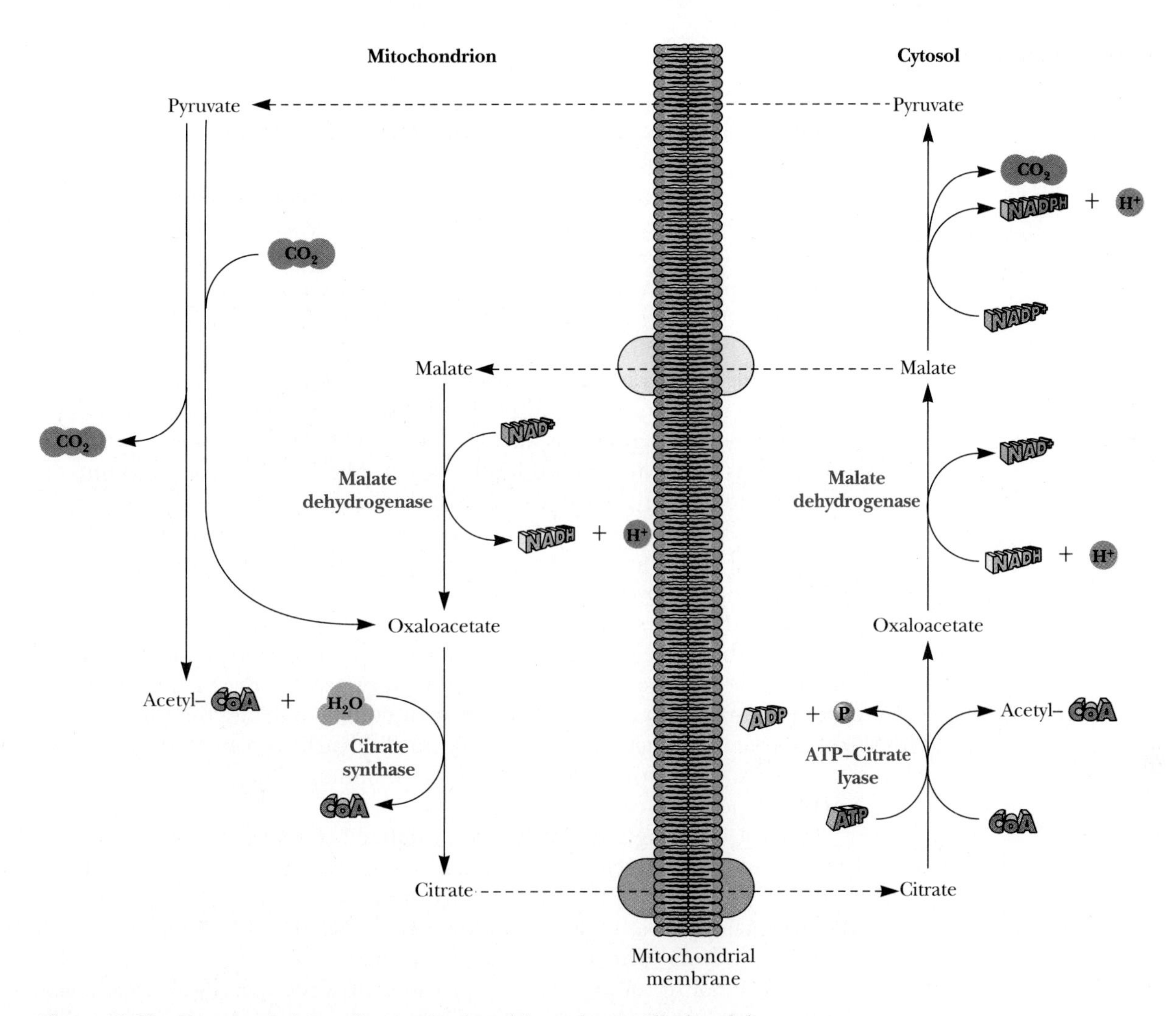

Figure **19.23** Export of citrate from mitochondria and cytosolic breakdown produces oxaloacetate and acetyl-CoA. Oxaloacetate is recycled to malate or pyruvate, which reenters the mitochondria. This cycle provides acetyl-CoA for fatty acid synthesis in the cytoplasm.

converts α-ketoglutarate directly to glutamate, which can then serve as a versatile precursor for proline, arginine, and glutamine (as described in Chapter 26). Succinyl-CoA provides most of the carbon atoms of the porphyrins. Oxaloacetate can be transaminated to produce aspartate. Aspartic acid itself is a precursor of the pyrimidine nucleotides and, in addition, is a key precursor for the synthesis of asparagine, methionine, lysine, threonine, and isoleucine. Oxaloacetate can also be decarboxylated to yield PEP, which is a key element of several pathways, namely, (1) synthesis (in plants and microorganisms) of the aromatic amino acids phenylalanine, tyrosine, and tryptophan; (2) formation of 3-phosphoglycerate and conversion to the amino acids serine, glycine, and cysteine; and (3) *gluconeogenesis,* which, as we will see in Chapter 21, is the pathway that synthesizes new glucose and many other carbohydrates.

Finally, citrate can be exported from the mitochondria and then broken down by *ATP-citrate lyase* to yield oxaloacetate and acetyl-CoA, a precursor of fatty acids (Figure 19.23). Oxaloacetate produced in this reaction is rapidly reduced to malate, which can then be processed in either of two ways: it may be transported into mitochondria, where it is reoxidized to oxaloacetate, or it may be oxidatively decarboxylated to pyruvate by *malic enzyme,* with subsequent mitochondrial uptake of pyruvate. This cycle permits citrate to provide acetyl-CoA for biosynthetic processes, with return of the malate and pyruvate by-products to the mitochondria.

19.14 The Anaplerotic, or "Filling Up," Reactions

In a sort of reciprocal arrangement, the cell also feeds many intermediates back into the TCA cycle from other reactions. Since such reactions replenish the TCA cycle intermediates, H. L. Kornberg proposed that they be called **anaplerotic reactions** (literally, the "filling up" reactions). Thus, *PEP carboxylase* and *pyruvate carboxylase* synthesize oxaloacetate from their respective substrates, and malic enzyme produces malate from pyruvate (Figure 19.24).

Pyruvate carboxylase is the most important of the anaplerotic reactions. It exists in the mitochondria of animal cells but not in plants, and it provides a direct link between glycolysis and the TCA cycle. The enzyme is tetrameric and contains covalently bound biotin and an Mg^{2+} site on each subunit. (It will be examined in greater detail in our discussion of gluconeogenesis in Chapter 21.) Pyruvate carboxylase has an absolute allosteric requirement for acetyl-CoA. Thus, when acetyl-CoA levels exceed the oxaloacetate supply, allosteric activation of pyruvate carboxylase by acetyl-CoA raises oxaloacetate levels, so that the excess acetyl-CoA can enter the TCA cycle.

PEP carboxylase occurs in yeast, bacteria, and higher plants, but not in animals. The enzyme is specifically inhibited by aspartate, which is produced by transamination of oxaloacetate. Thus, organisms utilizing this enzyme control aspartate production by regulation of PEP carboxylase. Malic enzyme is found in the cytosol or mitochondria of many animal and plant cells and is an NADPH-dependent enzyme.

It is worth noting that the reaction catalyzed by *PEP carboxykinase* (Figure 19.25) could also function as an anaplerotic reaction, were it not for the particular properties of the enzyme. CO_2 binds weakly to PEP carboxykinase, whereas oxaloacetate binds very tightly ($K_D = 2 \times 10^{-6}$ *M*), and, as a result, the enzyme favors formation of PEP from oxaloacetate.

The catabolism of amino acids provides pyruvate, acetyl-CoA, oxaloacetate, fumarate, α-ketoglutarate, and succinate, all of which may be oxidized by the TCA cycle. In this way, proteins may serve as excellent sources of nutrient energy, as will be seen in Chapter 26.

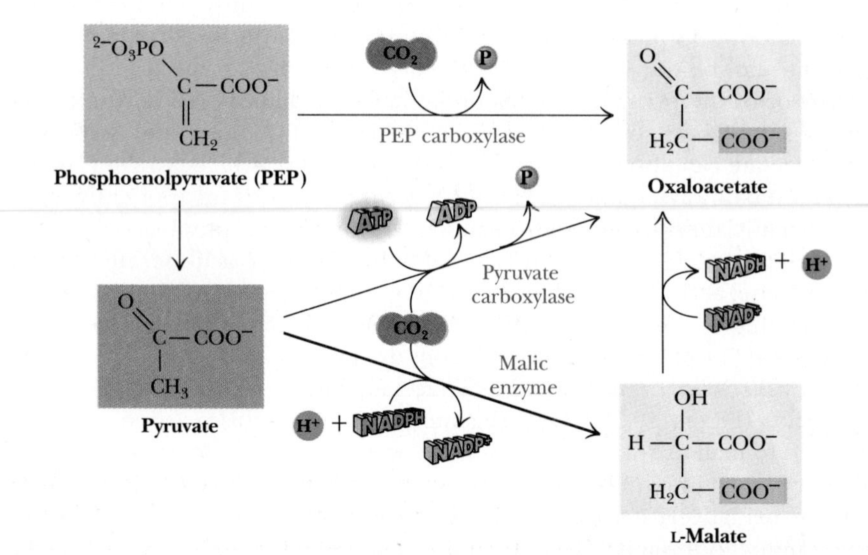

***Figure* 19.24** PEP carboxylase, pyruvate carboxylase, and malic enzyme catalyze anaplerotic reactions, replenishing TCA cycle intermediates.

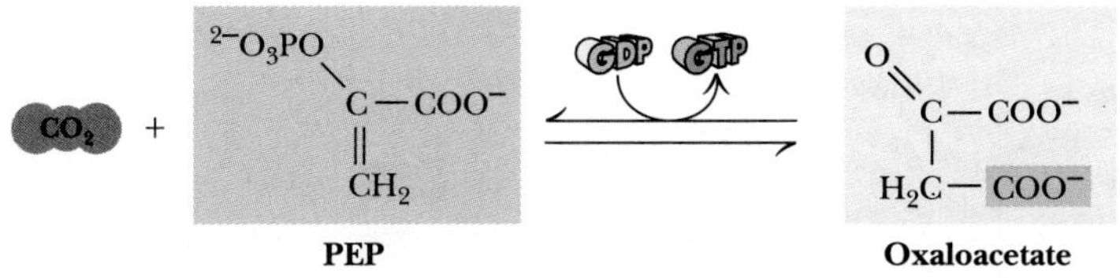

CO_2 + $^{2-}O_3PO-C(=CH_2)-COO^-$ (PEP) $\rightleftharpoons$ (GDP → GTP) Oxaloacetate ($O=C(COO^-)-H_2C-COO^-$)

Figure **19.25** The PEP carboxykinase reaction.

19.15 Regulation of the TCA Cycle

Situated as it is between glycolysis and the electron transport chain, the TCA cycle must be carefully controlled by the cell. If the cycle were permitted to run unchecked, large amounts of metabolic energy could be wasted in overproduction of reduced coenzymes and ATP; conversely, if it ran too slowly, ATP would not be produced rapidly enough to satisfy the needs of the cell. Also, as just seen, the TCA cycle is an important source of precursors for biosynthetic processes and must be able to provide them as needed.

What are the sites of regulation in the TCA cycle? Based upon our experience with glycolysis (Figure 18.31), we might anticipate that some of the reactions of the TCA cycle would operate near equilibrium under cellular conditions (with $\Delta G \approx 0$), whereas others—the sites of regulation—would be characterized by large, negative ΔG values. Estimates have been made for the values of ΔG in mitochondria, based on estimates of mitochondrial concentrations of metabolites, and these values are summarized in Table 19.1. Three reactions of the cycle—citrate synthase, isocitrate dehydrogenase, and α-ketoglutarate dehydrogenase—operate with large, negative ΔG values under mitochondrial conditions and are thus the primary sites of regulation in the cycle.

The regulatory actions that control the TCA cycle are shown in Figure 19.26. As one might expect, the principal regulatory "signals" are the concentrations of acetyl-CoA, ATP, NAD^+, and NADH, with additional effects provided by several other metabolites. The main sites of regulation are pyruvate dehydrogenase, citrate synthase, isocitrate dehydrogenase, and α-ketoglutarate dehydrogenase. All of these enzymes are inhibited by NADH, so that when the cell has produced all the NADH that can conveniently be turned into ATP, the cycle shuts down. For similar reasons, ATP is an inhibitor of pyruvate dehydrogenase and isocitrate dehydrogenase. The TCA cycle is turned on, however, when either the ADP/ATP or NAD^+/NADH ratio is high, an indication that the cell has run low on ATP or NADH. Regulation of the TCA cycle by NADH, NAD^+, ATP, and ADP thus reflects the energy status of the cell. On the other hand, succinyl-CoA is an *intracycle regulator,* inhibiting citrate synthase and α-ketoglutarate dehydrogenase. Acetyl-CoA acts as a signal to the TCA cycle that glycolysis or fatty acid breakdown is producing two-carbon units. Acetyl-CoA activates *pyruvate carboxylase,* the anaplerotic reaction that provides oxaloacetate, the acceptor for increased flux of acetyl-CoA into the TCA cycle.

Regulation of Pyruvate Dehydrogenase

As we shall see in Chapter 21, most organisms can synthesize sugars such as glucose from pyruvate. However, animals cannot synthesize glucose from acetyl-CoA. For this reason, the pyruvate dehydrogenase complex, which converts pyruvate to acetyl-CoA, plays a pivotal role in metabolism. Conversion to acetyl-CoA commits nutrient carbon atoms either to oxidation in the TCA

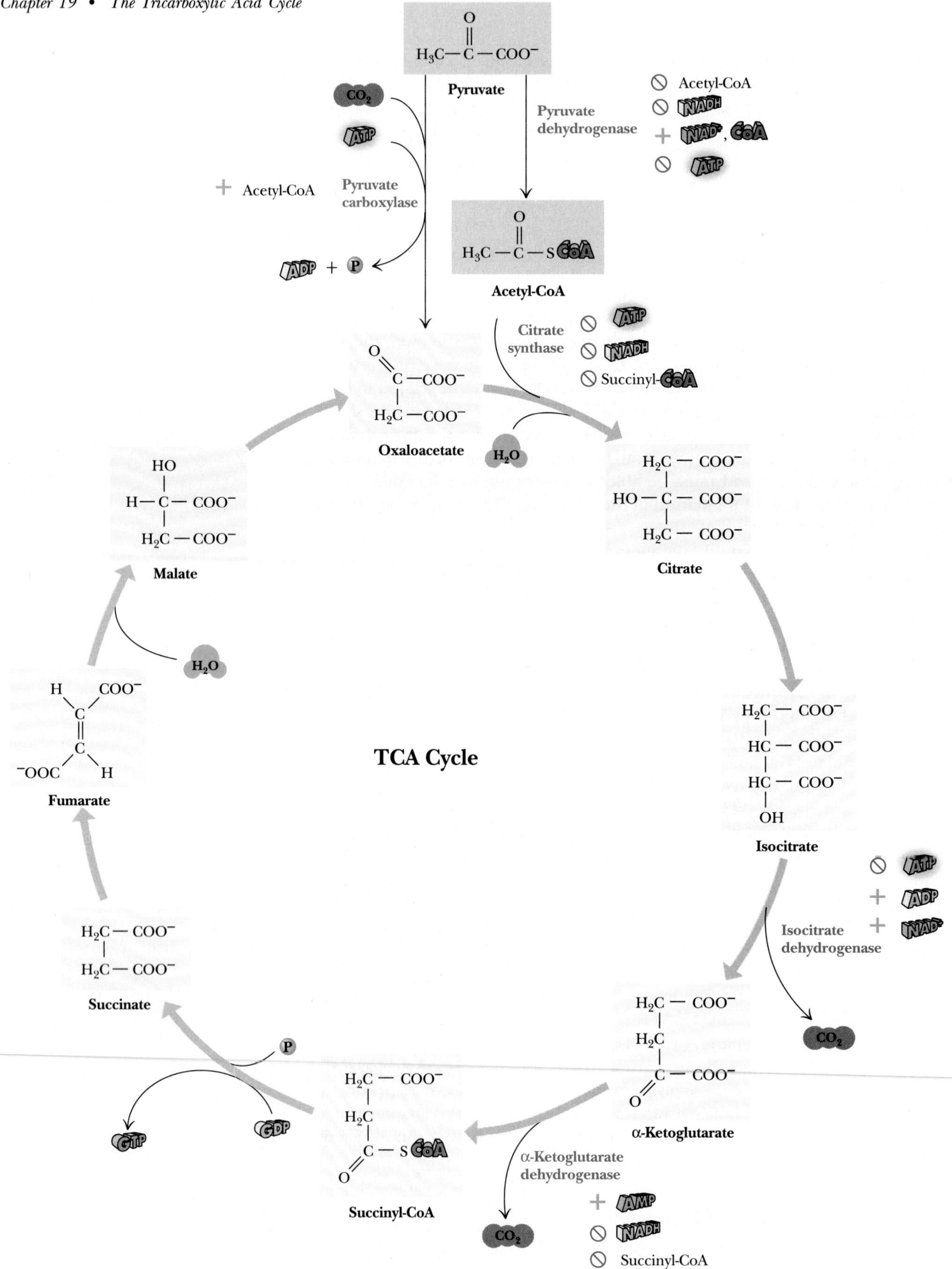

Figure **19.26** Regulation of the TCA cycle.

cycle or to fatty acid synthesis (see Chapter 24). Because this choice is so crucial to the organism, pyruvate dehydrogenase is a carefully regulated enzyme. It is subject to product inhibition and is further regulated by nucleotides. Finally, activity of pyruvate dehydrogenase is regulated by phosphorylation and dephosphorylation of the enzyme complex itself.

High levels of either product, acetyl-CoA or NADH, allosterically inhibit the pyruvate dehydrogenase complex. Acetyl-CoA specifically blocks dihydrolipoyl transacetylase, and NADH acts on dihydrolipoyl dehydrogenase. The mammalian pyruvate dehydrogenase is also regulated by covalent modifications. As shown in Figure 19.27, a Mg^{2+}-dependent *pyruvate dehydrogenase kinase* is associated with the enzyme in mammals. This kinase is allosterically activated by NADH and acetyl-CoA, and when levels of these metabolites rise in the mitochondrion, they stimulate phosphorylation of a serine residue on the pyruvate dehydrogenase subunit, blocking the first step of the pyruvate dehydrogenase reaction, the decarboxylation of pyruvate. Inhibition of the dehydrogenase in this manner eventually lowers the levels of NADH and acetyl-CoA in the matrix of the mitochondrion. Reactivation of the enzyme is carried out by *pyruvate dehydrogenase phosphatase,* a Ca^{2+}-activated enzyme that

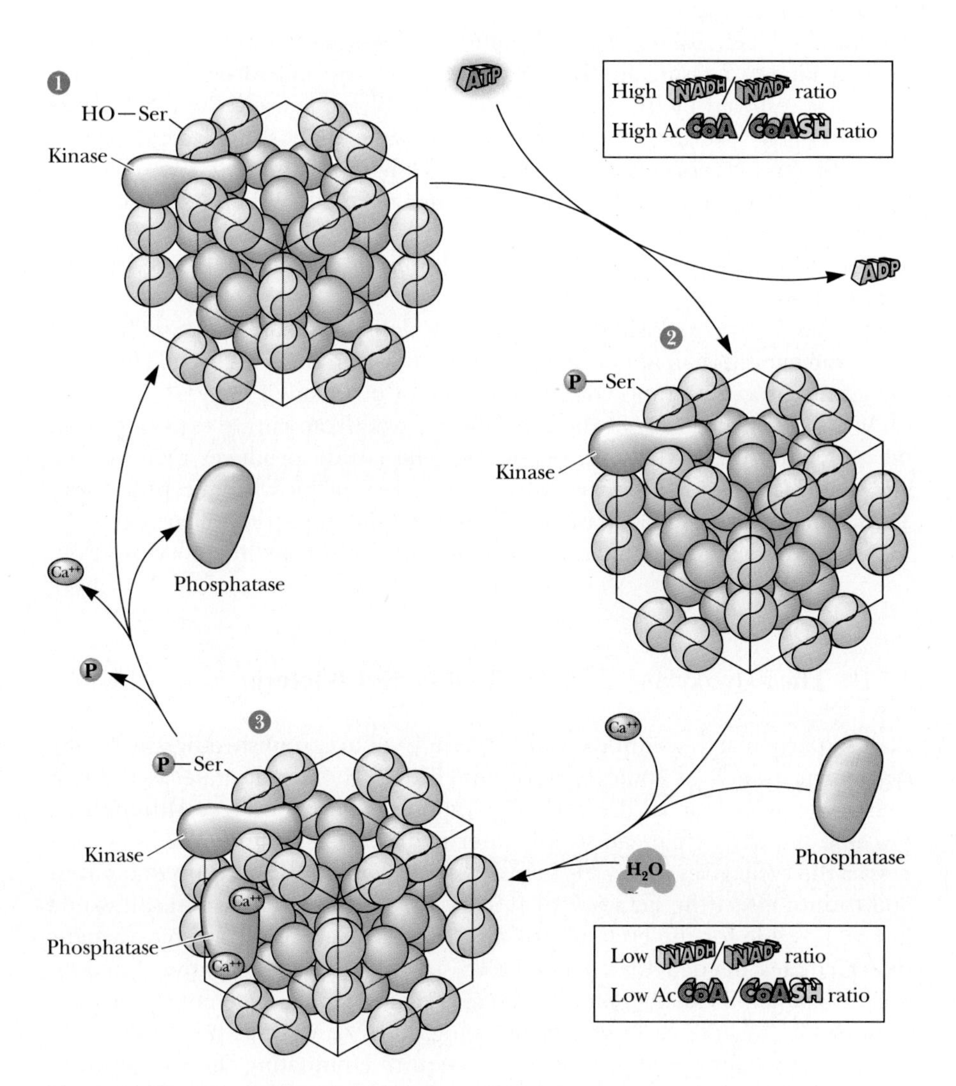

***Figure* 19.27** Regulation of the pyruvate dehydrogenase reaction.

binds to the dehydrogenase complex and hydrolyzes the phosphoserine moiety on the dehydrogenase subunit. At low ratios of NADH to NAD^+ and low acetyl-CoA levels, the phosphatase maintains the dehydrogenase in an activated state, but a high level of acetyl-CoA or NADH once again activates the kinase and leads to the inhibition of the dehydrogenase. Insulin and Ca^{2+} ions activate dephosphorylation, and pyruvate inhibits the phosphorylation reaction.

Pyruvate dehydrogenase is also sensitive to the energy status of the cell. AMP activates pyruvate dehydrogenase, whereas GTP inhibits it. High levels of AMP are a sign that the cell is energy-poor. Activation of pyruvate dehydrogenase under such conditions commits pyruvate to energy production.

Regulation of Isocitrate Dehydrogenase

The mechanism of regulation of isocitrate dehydrogenase is in some respects the reverse of pyruvate dehydrogenase. The mammalian isocitrate dehydrogenase is subject only to allosteric activation by ADP and NAD^+ and to inhibition by ATP and NADH. Thus, high NAD^+/NADH and ADP/ATP ratios stimulate isocitrate dehydrogenase and TCA cycle activity. The *Escherichia coli* enzyme, on the other hand, is regulated by covalent modification. Serine residues on each subunit of the dimeric enzyme are phosphorylated by a protein kinase, causing inhibition of the isocitrate dehydrogenase activity. Activity is restored by the action of a specific phosphatase. When TCA cycle and glycolytic intermediates—such as isocitrate, 3-phosphoglycerate, pyruvate, PEP, and oxaloacetate—are high, the kinase is inhibited, the phosphatase is activated, and the TCA cycle operates normally. When levels of these intermediates fall, the kinase is activated, isocitrate dehydrogenase is inhibited, and isocitrate is diverted to the *glyoxylate pathway,* as explained in the next section.

It may seem surprising that isocitrate dehydrogenase is strongly regulated, since it is not an apparent branch point within the TCA cycle. However, the citrate/isocitrate ratio controls the rate of production of cytosolic acetyl-CoA, because acetyl-CoA in the cytosol is derived from citrate exported from the mitochondrion. (Breakdown of cytosolic citrate produces oxaloacetate and acetyl-CoA, which can be used in a variety of biosynthetic processes.) Thus, isocitrate dehydrogenase activity in the mitochondrion favors catabolic TCA cycle activity over anabolic utilization of acetyl-CoA in the cytosol.

19.16 The Glyoxylate Cycle of Plants and Bacteria

Plants (particularly seedlings, which cannot yet accomplish efficient photosynthesis), as well as some bacteria and algae, can use acetate as the *only* source of carbon for all the carbon compounds they produce. Although we saw that the TCA cycle can supply intermediates for some biosynthetic processes, the cycle gives off 2 CO_2 for every two-carbon acetate group that enters and cannot effect the *net synthesis* of TCA cycle intermediates. Thus, it would not be possible for the cycle to produce the massive amounts of biosynthetic intermediates needed for acetate-based growth unless alternative reactions were available. In essence, the TCA cycle is geared primarily to energy production, and it "wastes" carbon units by giving off CO_2. Modification of the cycle to support acetate-based growth would require eliminating the CO_2-producing reactions and enhancing the net production of four-carbon units (i.e., oxaloacetate). Plants and bacteria employ a modification of the TCA cycle called the **glyoxylate cycle** to produce four-carbon dicarboxylic acids (and

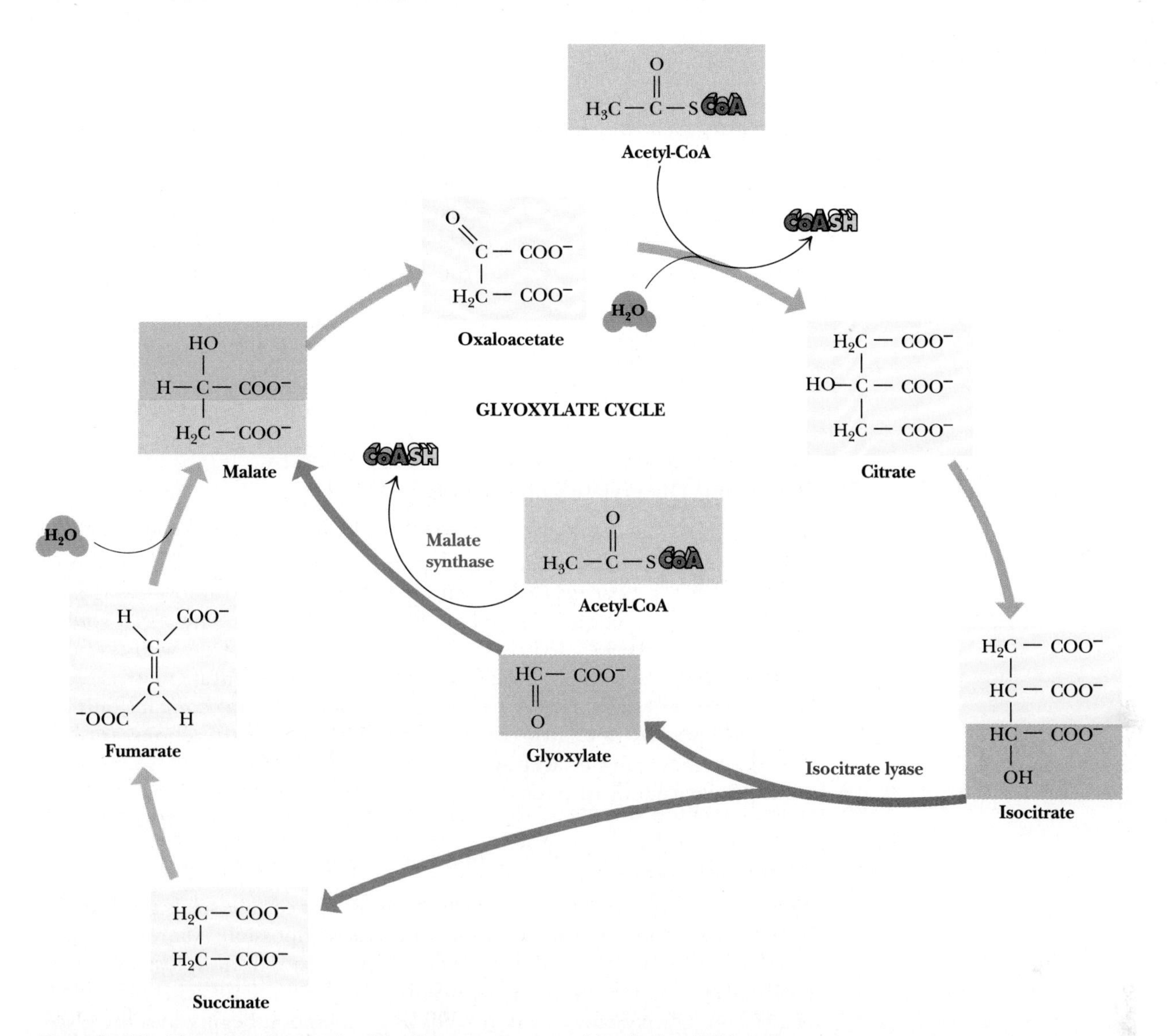

Figure **19.28** The glyoxylate cycle. The first two steps are identical to TCA cycle reactions. The third step bypasses the CO_2-evolving steps of the TCA cycle to produce succinate and glyoxylate. The malate synthase reaction forms malate from glyoxylate and another acetyl-CoA. The result is that one turn of the cycle consumes one oxaloacetate and two acetyl-CoA molecules but produces two molecules of oxaloacetate. The net for this cycle is one oxaloacetate from two acetyl-CoA molecules.

eventually even sugars) from two-carbon acetate units. The glyoxylate cycle bypasses the two oxidative decarboxylations of the TCA cycle, and instead routes isocitrate through the *isocitrate lyase* and *malate synthase* reactions (Figure 19.28). *Glyoxylate* produced by isocitrate lyase reacts with a second molecule of acetyl-CoA to form L-malate. The net effect is to conserve carbon units, using two acetyl-CoA molecules per cycle to generate oxaloacetate. Some of this is converted to PEP and then to glucose by pathways discussed in Chapter 21.

The Glyoxylate Cycle Operates in Specialized Organelles

The enzymes of the glyoxylate cycle in plants are contained in **glyoxysomes,** organelles devoted to this cycle. Yeast and algae carry out the glyoxylate cycle in the cytoplasm. The enzymes common to both the TCA and glyoxylate pathways exist as isozymes, with spatially and functionally distinct enzymes operating independently in the two cycles.

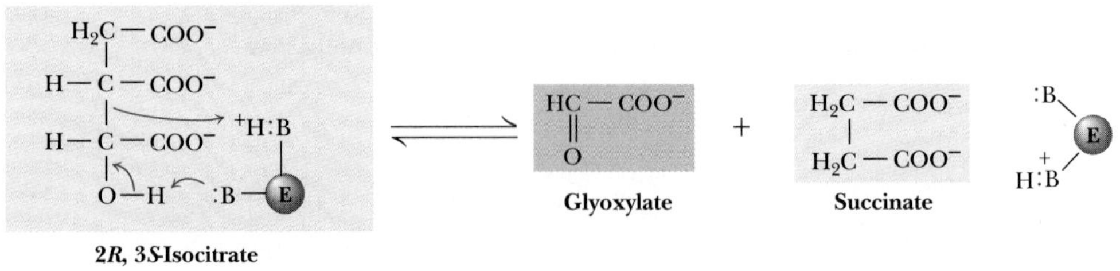

Glyoxylate

Succinate

2*R*, 3*S*-Isocitrate

***Figure* 19.29** The isocitrate lyase reaction.

Isocitrate Lyase Short-Circuits the TCA Cycle by Producing Glyoxylate and Succinate

The *isocitrate lyase reaction* (Figure 19.29) produces succinate, a four-carbon product of the cycle, as well as glyoxylate, which can then combine with a second molecule of acetyl-CoA. Isocitrate lyase catalyzes an aldol cleavage and is similar to the reaction mediated by aldolase in glycolysis. The *malate synthase reaction* (Figure 19.30), a Claisen condensation of acetyl-CoA with the aldehyde of glyoxylate to yield malate, is quite similar to the citrate synthase reaction. Compared with the TCA cycle, the glyoxylate cycle (a) contains only five steps (as opposed to eight), (b) lacks the CO_2-liberating reactions, (c) consumes two molecules of acetyl-CoA per cycle, and (d) produces four-carbon units (oxaloacetate) as opposed to one-carbon units.

The Glyoxylate Cycle Helps Plants Grow in the Dark

The existence of the glyoxylate cycle explains how certain seeds grow underground (or in the dark), where photosynthesis is impossible. Many seeds (peanuts, soybeans, and castor beans, for example) are rich in lipids; and, as we shall see in Chapter 23, most organisms degrade the fatty acids of lipids to acetyl-CoA. Glyoxysomes form in seeds as germination begins, and the glyoxylate cycle uses the acetyl-CoA produced in fatty acid oxidation to provide large amounts of oxaloacetate and other intermediates for carbohydrate synthesis. Once the growing plant begins photosynthesis and can fix CO_2 to produce carbohydrates (see Chapter 22), the glyoxysomes disappear.

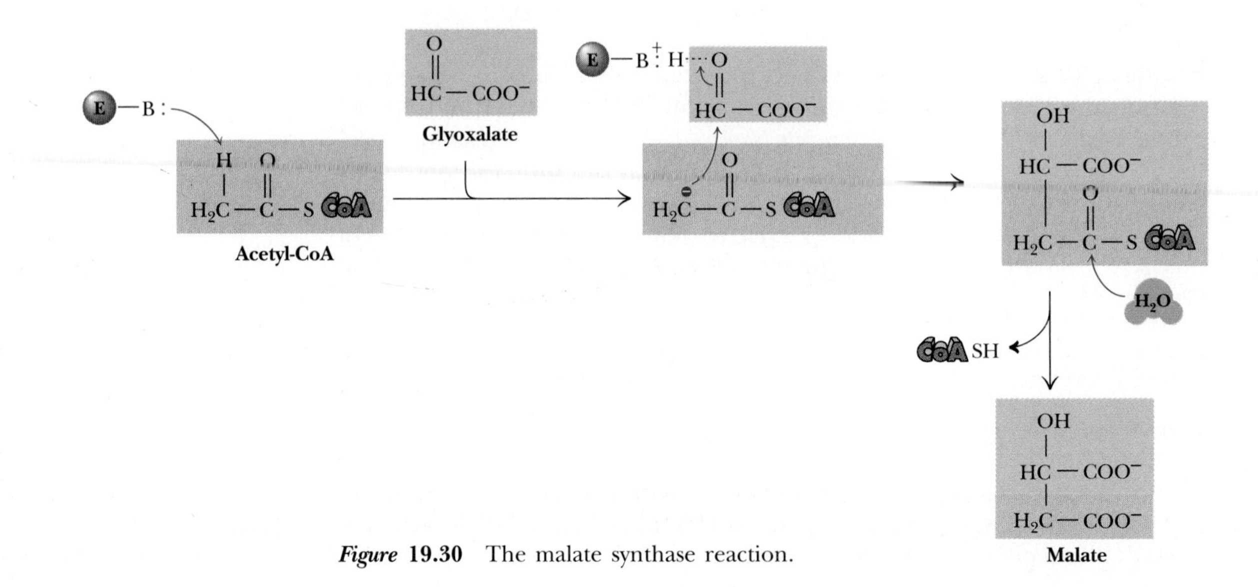

***Figure* 19.30** The malate synthase reaction.

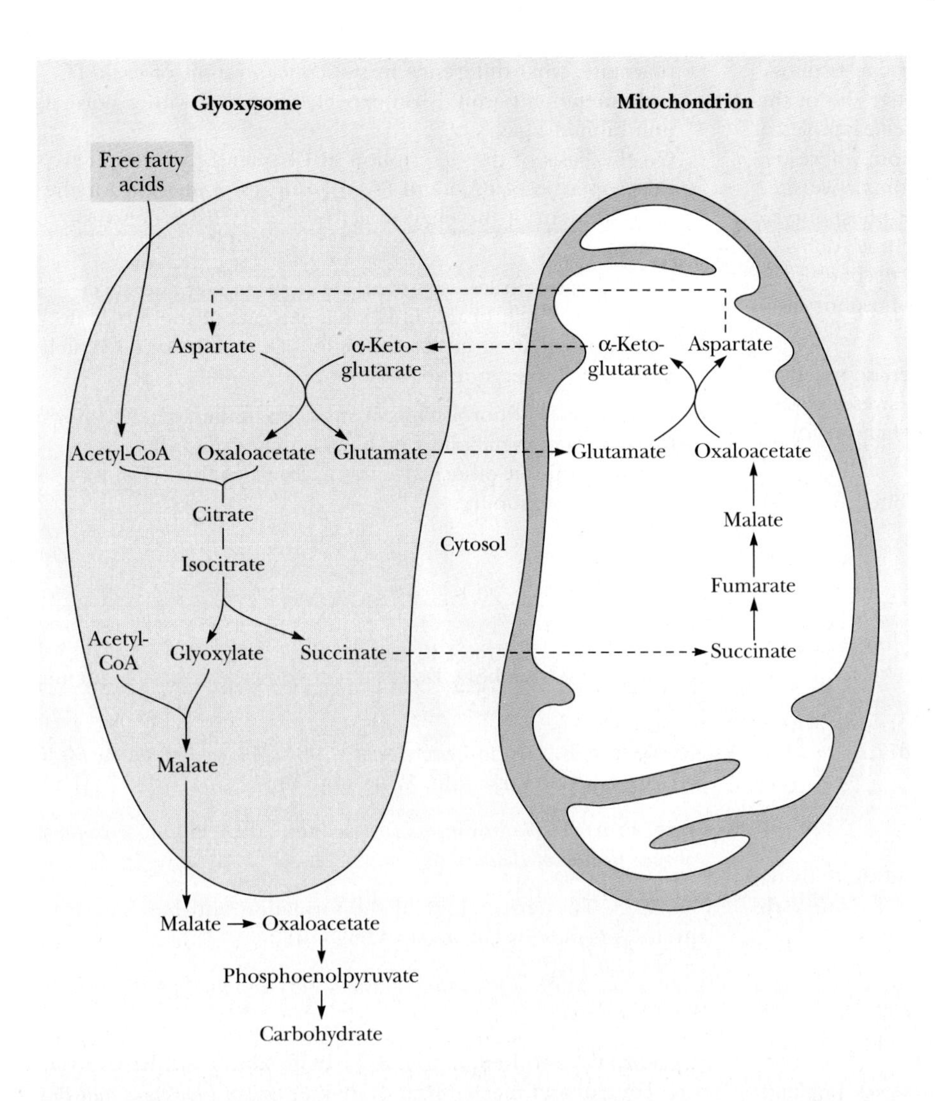

***Figure* 19.31** Glyoxysomes lack three of the enzymes needed to run the glyoxylate cycle. Succinate dehydrogenase, fumarase, and malate dehydrogenase are all "borrowed" from the mitochondria in a shuttle in which succinate is passed to the mitochondria and glutamate, α-ketoglutarate, and aspartate are passed to the glyoxysome.

Glyoxysomes Must Borrow Three Reactions from Mitochondria

Glyoxysomes do not contain all the enzymes needed to run the glyoxylate cycle; succinate dehydrogenase, fumarase, and malate dehydrogenase are absent. Consequently, glyoxysomes must cooperate with mitochondria to run their cycle (Figure 19.31). Succinate travels from the glyoxysomes to the mitochondria, where it is converted to oxaloacetate. Transamination to aspartate follows, because oxaloacetate cannot be transported out of the mitochondria. Aspartate formed in this way then moves from the mitochondria back to the glyoxysomes, where a reverse transamination with α-ketoglutarate forms oxaloacetate, completing the shuttle. Finally, to balance the transaminations, glutamate shuttles from glyoxysomes to mitochondria.

Problems

1. Describe the labeling pattern that would result from the introduction into the TCA cycle of glutamate labeled at the C_γ with ^{14}C.
2. Describe the effect on the TCA cycle of (a) increasing the concentration of NAD^+, (b) reducing the concentration of ATP, and (c) increasing the concentration of isocitrate.

3. The serine residue of isocitrate dehydrogenase that is phosphorylated by protein kinase lies within the active site of the enzyme. This situation contrasts with most other examples of covalent modification by protein phosphorylation, where the phosphorylation occurs at a site remote from the active site. What direct effect do you think such active-site phosphorylation might have on the catalytic activity of isocitrate dehydrogenase? (See Barford, D., 1991. Molecular mechanisms for the control of enzymic activity by protein phosphorylation. *Biochimica et Biophysica Acta* 1133:55–62.)
4. The first step of the α-ketoglutarate dehydrogenase reaction involves decarboxylation of the substrate and leaves a covalent TPP intermediate. Write a reasonable mechanism for this reaction.
5. In a tissue where the TCA cycle has been inhibited by fluoroacetate, what difference in the concentration of each TCA cycle metabolite would you expect, compared with a normal, uninhibited tissue?
6. On the basis of the description in Chapter 14 of the physical properties of FAD and $FADH_2$, suggest a method for the measurement of the enzyme activity of succinate dehydrogenase.
7. Starting with citrate, isocitrate, α-ketoglutarate, and succinate, state which of the individual carbons of the molecule undergo oxidation in the next step of the TCA cycle. Which molecules undergo a net oxidation?
8. In addition to fluoroacetate, consider whether other analogs of TCA cycle metabolites or intermediates might be introduced to inhibit other, specific reactions of the cycle. Explain your reasoning.

Further Reading

Atkinson, D. E., 1977. *Cellular Energy Metabolism and Its Regulation.* New York: Academic Press.

Bodner, G. M., 1986. The tricarboxylic acid (TCA), citric acid or Krebs cycle. *Journal of Chemical Education* **63:**673–677.

Gibble, G. W., 1973. Fluoroacetate toxicity. *Journal of Chemical Education* **50:**460–462.

Hansford, R. G., 1980. Control of mitochondrial substrate oxidation. In *Current Topics in Bioenergetics,* vol. 10, 217–78. New York: Academic Press.

Hawkins, R. A., and Mans, A. M., 1983. Intermediary metabolism of carbohydrates and other fuels. In *Handbook of Neurochemistry,* 2nd ed., edited by Lajtha, A., 259–294. New York: Plenum Press.

Krebs, H. A., 1981. *Reminiscences and Reflections.* Oxford, England: Oxford University Press.

——1970. The history of the tricarboxylic acid cycle. *Perspectives in Biology and Medicine* **14:**154–170.

Lowenstein, J. M., ed., 1969. *Citric Acid Cycle: Control and Compartmentation.* New York: Marcel Dekker.

——1967. The tricarboxylic acid cycle. In *Metabolic Pathways,* 3rd ed., edited by Greenberg, D., vol. 1, 146–270. New York: Academic Press.

Newsholme, E. A., and Leech, A. R., 1983. *Biochemistry for the Medical Sciences.* New York: John Wiley and Sons.

Srere, P. A., 1987. Complexes of sequential metabolic enzymes. *Annual Review of Biochemistry* **56:**89–124.

——1975. The enzymology of the formation and breakdown of citrate. *Advances in Enzymology* **43:**57–101.

Walsh, C., 1979. *Enzymatic Reaction Mechanisms.* San Francisco: W. H. Freeman.

Wiegand, G., and Remington, S. J., 1986. Citrate synthase: Structure, control and mechanism. *Annual Review of Biophysics and Biophysical Chemistry* **15:**97–117.

Williamson, J. R., 1980. Mitochondrial metabolism and cell regulation. In *Mitochondria: Bioenergetics, Biogenesis and Membrane Structure,* edited by Packer, L., and Gomez-Puyou, A. New York: Academic Press.

"In all things of nature there is something of the marvelous."
Aristotle (384–322 B.C.)

Chapter 20

Electron Transport and Oxidative Phosphorylation

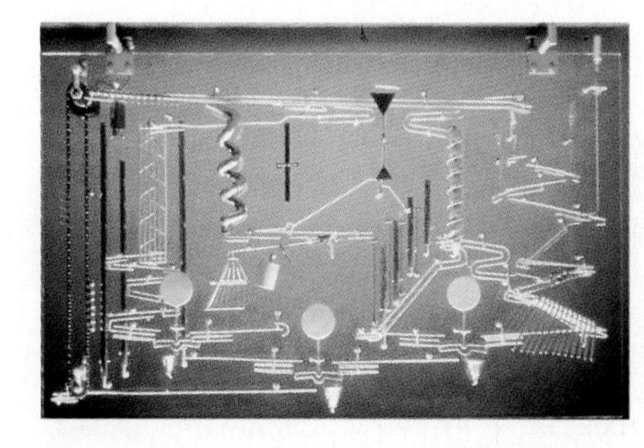

Wall Piece #IV (1985), a kinetic sculpture by George Rhoads. This complex mechanical art form can be viewed as a metaphor for the molecular apparatus underlying electron transport and ATP synthesis by oxidative phosphorylation.

Outline

20.1 Electron Transport and Oxidative Phosphorylation Are Membrane-Associated Processes
20.2 Reduction Potentials—An Accounting Device for Free Energy Changes in Redox Reactions
20.3 The Electron Transport Chain—An Overview
20.4 Complex I: NADH–Coenzyme Q Reductase
20.5 Complex II: Succinate–Coenzyme Q Reductase
20.6 Complex III: Coenzyme Q–Cytochrome *c* Reductase
20.7 Complex IV: Cytochrome *c* Oxidase
20.8 The Thermodynamic View of Chemiosmotic Coupling
20.9 ATP Synthase
20.10 Inhibitors of Oxidative Phosphorylation
20.11 Uncouplers Disrupt the Coupling of Electron Transport and ATP Synthase
20.12 Shuttle Systems Feed the Electrons of Cytosolic NADH into Electron Transport
20.13 ATP Exits the Mitochondria via an ATP–ADP Translocase
20.14 What Is the P/O Ratio for Electron Transport and Oxidative Phosphorylation?

Living cells save up metabolic energy predominantly in the form of fats and carbohydrates, and they "spend" this energy for biosynthesis, membrane transport, and movement. In both directions, energy is exchanged and transferred in the form of ATP. In Chapters 18 and 19 we saw that glycolysis and the TCA cycle convert some of the energy available from stored and dietary sugars directly to ATP. However, most of the metabolic energy that is obtainable from substrates entering gly-

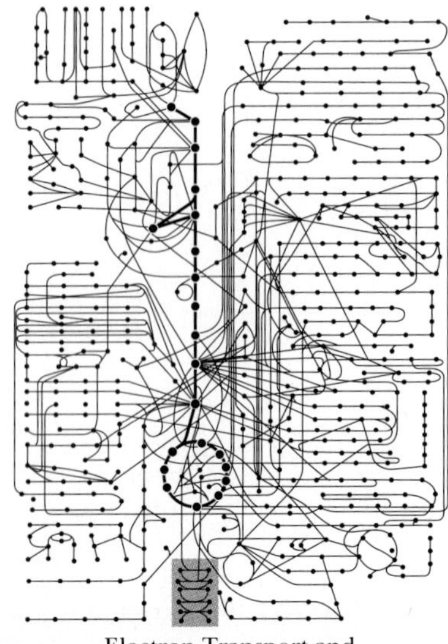
Electron Transport and Oxidative Phosphorylation

colysis and the TCA cycle is funneled via oxidation–reduction reactions into NADH and reduced flavoproteins, symbolized by [$FADH_2$]. We now embark on the discovery of how cells convert the stored metabolic energy of NADH and [$FADH_2$] into ATP.

Whereas ATP made in glycolysis and the TCA cycle is the result of substrate-level phosphorylation, NADH-dependent ATP synthesis is the result of **oxidative phosphorylation.** Electrons stored in the form of the reduced coenzymes, NADH or [$FADH_2$], are passed through an elaborate and highly organized chain of proteins and coenzymes, the so-called **electron transport chain,** finally reaching O_2 (molecular oxygen), the terminal electron acceptor. Each component of the chain can exist in (at least) two oxidation states, and each component is successively reduced and reoxidized as electrons move through it from NADH (or [$FADH_2$]) to O_2. In the course of electron transport, released energy is stored in the form of a proton gradient across the inner mitochondrial membrane. It is this proton gradient that provides the energy to drive ATP synthesis.

20.1 Electron Transport and Oxidative Phosphorylation Are Membrane-Associated Processes

The processes of electron transport and oxidative phosphorylation are **membrane-associated.** Bacteria are the simplest life form, and bacterial cells typically consist of a single cellular compartment surrounded by a plasma membrane and a more rigid cell wall. In such a system, the conversion of energy from NADH and [$FADH_2$] to the energy of ATP via electron transport and oxidative phosphorylation is carried out at (and across) the plasma membrane. In eukaryotic cells, electron transport and oxidative phosphorylation are localized in mitochondria, which are also the sites of citric acid cycle activity and (as we shall see in Chapter 23) fatty acid oxidation. Mammalian cells contain from 800 to 2500 mitochondria; other types of cells may have as few as one or two or as many as half a million mitochondria. Human erythrocytes, whose purpose is simply to transport oxygen to tissues, contain *no* mitochondria at all. The typical mitochondrion is about 0.5 ± 0.3 microns in diameter and from 0.5 micron to several microns long; its overall shape is sensitive to metabolic conditions in the cell.

Mitochondria are surrounded by a simple **outer membrane** and a more complex **inner membrane** (Figure 20.1). The space between the inner and outer membranes is referred to as the **intermembrane space.** Several enzymes that utilize ATP (such as creatine kinase and adenylate kinase) are found in the intermembrane space. The smooth outer membrane is about 30 to 40% lipid and 60 to 70% protein, and has a relatively high concentration of phosphatidylinositol. The outer membrane contains significant amounts of **porin**—a transmembrane protein, rich in β-sheets, that forms large channels across the membrane, permitting free diffusion of molecules with molecular weights of about 10,000 or less. Apparently, the outer membrane functions mainly to maintain the shape of the mitochondrion. The inner membrane is richly packed with proteins, which account for nearly 80% of its weight; thus, its density is higher than that of the outer membrane. The fatty acids of inner-membrane lipids are highly unsaturated. Cardiolipin and diphosphatidylglycerol (Chapter 9) are abundant. By contrast, the inner membrane lacks cholesterol and is quite impermeable to molecules and ions. Species that must cross the mitochondrial inner membrane—ions, substrates, fatty acids for oxidation, and so on—are carried by specific transport proteins in the membrane. Notably, the inner membrane is extensively folded (Figure 20.1).

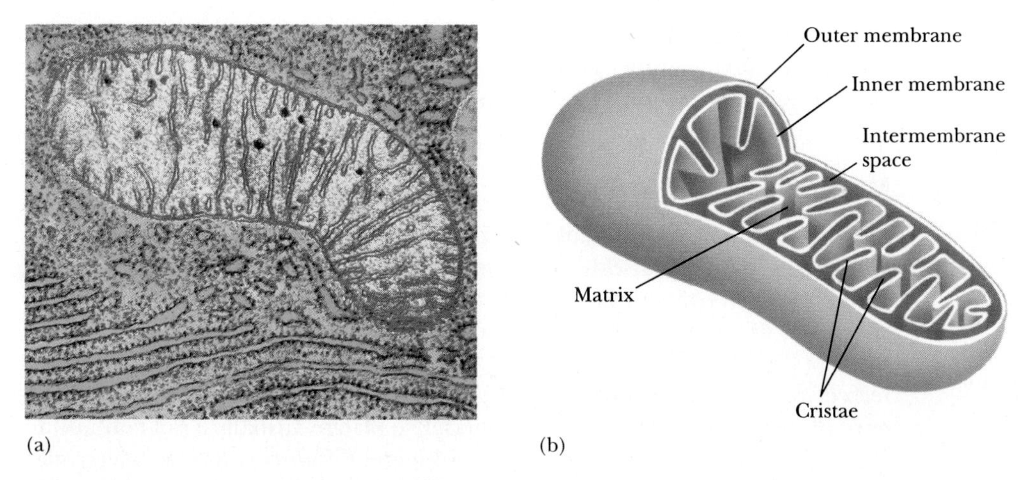

***Figure* 20.1** (a) An electron micrograph of a mitochondrion. (b) A drawing of a mitochondrion with components labeled.

The folds, known as **cristae,** provide the inner membrane with a large surface area in a small volume. During periods of active respiration, the inner membrane appears to shrink significantly, leaving a comparatively large intermembrane space.

The Mitochondrial Matrix Contains the Enzymes of the TCA Cycle

The space inside the inner mitochondrial membrane is called the **matrix,** and it contains most of the enzymes of the TCA cycle and fatty acid oxidation. (An important exception, succinate dehydrogenase of the TCA cycle, is located in the inner membrane itself.) In addition, mitochondria contain circular DNA molecules, along with ribosomes and the enzymes required to synthesize proteins coded within the mitochondrial genome. Although some of the mitochondrial proteins are made this way, most are encoded by nuclear DNA and synthesized by cytoplasmic ribosomes.

20.2 Reduction Potentials—An Accounting Device for Free Energy Changes in Redox Reactions

On numerous occasions in earlier chapters, we have stressed that NADH and reduced flavoproteins ([$FADH_2$]) are forms of metabolic energy. These reduced coenzymes have a strong tendency to be oxidized—that is, to transfer electrons to other species. The electron transport chain converts the energy of electron transfer into the energy of phosphoryl transfer stored in the phosphoric anhydride bonds of ATP. Just as the *group transfer potential* was used in Chapter 16 to quantitate the energy of phosphoryl transfer, the **standard reduction potential,** denoted by $\mathscr{E}_o'$, quantitates the tendency of chemical species to be reduced or oxidized. The standard reduction potential describing electron transfer between two species,

$$\text{Reduced donor} \rightarrow \text{Oxidized donor} \quad ne^- \quad \text{Oxidized acceptor} \rightarrow \text{Reduced acceptor} \qquad (20.1)$$

is related to the free energy change for the process by

$$\Delta G^{\circ\prime} = -n\mathscr{F}\Delta\mathscr{E}_o' \qquad (20.2)$$

where n is the number of electrons transferred; $\mathscr{F}$ is Faraday's constant,

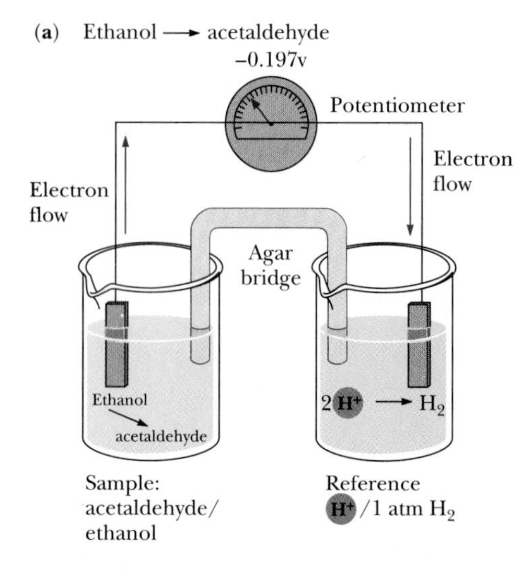

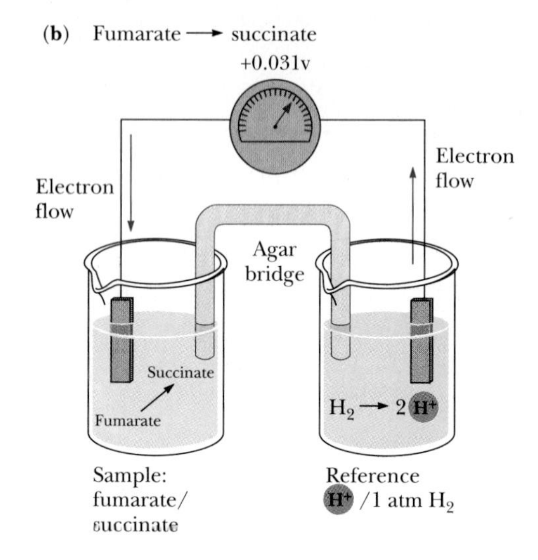

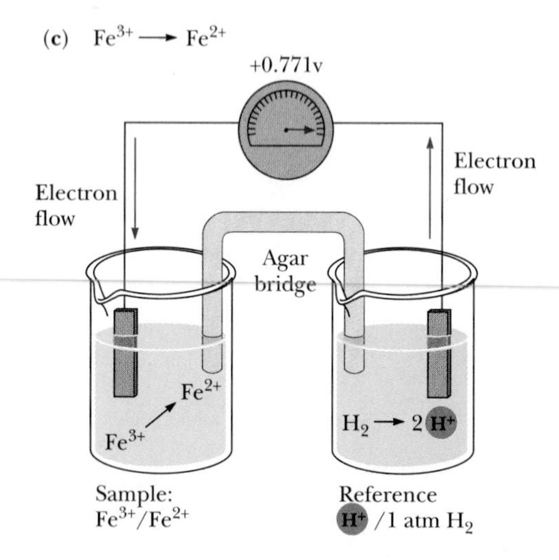

***Figure* 20.2** Experimental apparatus used to measure the standard reduction potential of the indicated redox couples. (a) The acetaldehyde/ethanol couple, (b) the fumarate/succinate couple, (c) the Fe^{3+}/Fe^{2+} couple.

96,485 J/V · mol; and $\Delta\mathscr{E}_o'$ is the difference in reduction potentials between the donor and acceptor. This relationship is straightforward, but it depends on a *standard* of reference by which reduction potentials are defined.

Measurement of Standard Reduction Potentials

Standard reduction potentials are determined by measuring the voltages generated in **reaction half-cells** (Figure 20.2). A half-cell consists of a solution containing 1 *M* concentrations of both the oxidized and reduced forms of the substance whose reduction potential is being measured, and a simple electrode. (Together, the oxidized and reduced forms of the substance are referred to as a **redox couple.**) Such a **sample half-cell** is connected to a **reference half-cell** and electrode via a conductive bridge (usually a salt-containing agar gel). A sensitive potentiometer (voltmeter) connects the two electrodes so that the electrical potential (voltage) between them can be measured. The reference half-cell normally contains 1 *M* H^+ in equilibrium with H_2 gas at a pressure of 1 atm. The H^+/H_2 reference half-cell is arbitrarily assigned a standard reduction potential of 0.0 V. The standard reduction potentials of all other redox couples are defined relative to the H^+/H_2 reference half-cell on the basis of the sign and magnitude of the voltage (electromotive force, emf) registered on the potentiometer (Figure 20.2).

If electron flow between the electrodes is toward the sample half-cell, reduction occurs spontaneously in the sample half-cell, and the reduction potential is said to be positive. If electron flow between the electrodes is away from the sample half-cell and toward the reference cell, the reduction potential is said to be negative, since electron loss (oxidation) is occurring in the sample half-cell. Strictly speaking, the standard reduction potential, $\mathscr{E}_o'$, is the electromotive force generated at 25°C and pH 7.0 by a sample half-cell (containing 1 *M* concentrations of the oxidized and reduced species) with respect to a reference half-cell. (Note that the reduction potential of the hydrogen half-cell is pH-dependent. The standard reduction potential, 0.0 V, assumes 1 *M* H^+. The hydrogen half-cell measured at pH 7.0 has an $\mathscr{E}_o'$ of −0.421 V.)

Several Examples

Figure 20.2a shows a sample/reference half-cell pair for measurement of the standard reduction potential of the acetaldehyde/ethanol couple. Since electrons flow toward the reference half-cell and away from the sample half-cell, the standard reduction potential is negative, specifically −0.197 V. In contrast, the fumarate/succinate couple and the Fe^{3+}/Fe^{2+} couple both cause electrons to flow from the reference half-cell to the sample half-cell; that is, reduction occurs spontaneously in each system, and the reduction potentials of both are thus positive. The standard reduction potential for the Fe^{3+}/Fe^{2+} half-cell is much larger than that for the fumarate/succinate half-cell, with values of +0.771 V and +0.031 V, respectively. For each half-cell, a **half-cell reaction** describes the reaction taking place. For the fumarate/succinate half-cell, since the reaction occurring in this half-cell is indeed a reduction of fumarate,

$$\text{Fumarate} + 2\,H^+ + 2e^- \longrightarrow \text{succinate} \qquad \mathscr{E}_o' = +0.031\text{ V} \quad (20.3)$$

Similarly, for the Fe^{3+}/Fe^{2+} half-cell,

$$Fe^{3+} + e^- \longrightarrow Fe^{2+} \qquad \mathscr{E}_o' = +0.771\text{ V} \quad (20.4)$$

However, the reaction occurring in the acetaldehyde/ethanol half-cell is the oxidation of ethanol:

$$\text{Ethanol} \longrightarrow \text{acetaldehyde} + 2\,H^+ + 2e^- \qquad \mathscr{E}_o' = -0.197\text{ V} \quad (20.5)$$

The Significance of $\mathscr{E}_o'$

Some typical half-cell reactions and their respective standard reduction potentials are listed in Table 20.1. Whenever reactions of this type are tabulated, they are uniformly written as *reduction* reactions, regardless of what occurs in the given half-cell. The sign of the standard reduction potential indicates which reaction really occurs when the given half-cell is combined with the reference hydrogen half-cell. Redox couples that have large positive reduction potentials have a strong tendency to accept electrons, and the oxidized form of such a couple (O_2, for example) is a strong oxidizing agent. Redox

Table **20.1**

Standard Reduction Potentials for Several Biological Reduction Half-Reactions

Reduction Half-Reaction	$\mathscr{E}_o'$ **(V)**
$\frac{1}{2}O_2 + 2H^+ + 2e^- \longrightarrow H_2O$	0.816
$Fe^{3+} + e^- \longrightarrow Fe^{2+}$	0.771
Photosystem P700	0.430
$NO_3^- + 2H^+ + 2e^- \longrightarrow NO_2^- + H_2O$	0.421
Cytochrome $f(Fe^{3+}) + e^- \longrightarrow$ cytochrome $f(Fe^{2+})$	0.365
Cytochrome $a_3(Fe^{3+}) + e^- \longrightarrow$ cytochrome $a_3(Fe^{2+})$	0.350
Cytochrome $a(Fe^{3+}) + e^- \longrightarrow$ cytochrome $a(Fe^{2+})$	0.290
Rieske Fe-S$(Fe^{3+}) + e^- \longrightarrow$ Rieske Fe-S(Fe^{2+})	0.280
Cytochrome $c(Fe^{3+}) + e^- \longrightarrow$ cytochrome $c(Fe^{2+})$	0.254
Cytochrome $c_1(Fe^{3+}) + e^- \longrightarrow$ cytochrome $c_1(Fe^{2+})$	0.220
$UQH\cdot + H^+ + e^- \longrightarrow UQH_2$ (UQ = coenzyme Q)	0.190
$UQ + 2H^+ + 2e^- \longrightarrow UQH_2$	0.060
Cytochrome $b_H(Fe^{3+}) + e^- \longrightarrow$ cytochrome $b_H(Fe^{2+})$	0.050
Fumarate $+ 2H^+ + 2e^- \longrightarrow$ succinate	0.031
$UQ + H^+ + e^- \longrightarrow UQH\cdot$	0.030
Cytochrome $b_5(Fe^{3+}) + e^- \longrightarrow$ cytochrome $b_5(Fe^{2+})$	0.020
$[FAD] + 2H^+ + 2e^- \longrightarrow [FADH_2]$	0.003–0.091*
Cytochrome $b_L(Fe^{3+}) + e^- \longrightarrow$ cytochrome $b_L(Fe^{2+})$	−0.100
Oxaloacetate $+ 2H^+ + 2e^- \longrightarrow$ malate	−0.166
Pyruvate $+ 2H^+ + 2e^- \longrightarrow$ lactate	−0.185
Acetaldehyde $+ 2H^+ + 2e^- \longrightarrow$ ethanol	−0.197
$FMN + 2H^+ + 2e^- \longrightarrow FMNH_2$	−0.219
$FAD + 2H^+ + 2e^- \longrightarrow FADH_2$	−0.219
Glutathione (oxidized) $+ 2H^+ + 2e^- \longrightarrow$ 2 glutathione (reduced)	−0.230
Lipoic acid $+ 2H^+ + 2e^- \longrightarrow$ dihydrolipoic acid	−0.290
1,3-Bisphosphoglycerate $+ 2H^+ + 2e^- \longrightarrow$ glyceraldehyde-3-phosphate $+ P_i$	−0.290
$NAD^+ + 2H^+ + 2e^- \longrightarrow NADH + H^+$	−0.320
$NADP^+ + 2H^+ + 2e^- \longrightarrow NADPH + H^+$	−0.320
Lipoyl dehydrogenase $[FAD] + 2H^+ + 2e^- \longrightarrow$ lipoyl dehydrogenase $[FADH_2]$	−0.340
α-Ketoglutarate $+ CO_2 + 2H^+ + 2e^- \longrightarrow$ isocitrate	−0.380
$2H^+ + 2e^- \longrightarrow H_2$	−0.421
Ferredoxin (spinach) $(Fe^{3+}) + e^- \longrightarrow$ ferredoxin (spinach) (Fe^{2+})	−0.430
Succinate $+ CO_2 + 2H^+ + 2e^- \longrightarrow \alpha$-ketoglutarate $+ H_2O$	−0.670

*Typical values for reduction of bound FAD in flavoproteins such as succinate dehydrogenase (see Bonomi, F., Pagani, S., Cerletti, P., and Giori, C., 1983. *European Journal of Biochemistry* **134:**439–445).

couples with large negative reduction potentials have a strong tendency to undergo oxidation (that is, donate electrons), and the reduced form of such a couple (NADPH, for example) is a strong reducing agent.

Coupled Redox Reactions

The half-reactions and reduction potentials in Table 20.1 can be used to analyze energy changes in redox reactions. The oxidation of NADH to NAD^+ can be coupled with the reduction of α-ketoglutarate to isocitrate:

$$\text{NAD}^+ + \text{isocitrate} \longrightarrow \text{NADH} + \text{H}^+ + \alpha\text{-ketoglutarate} + \text{CO}_2 \quad (20.6)$$

This is the isocitrate dehydrogenase reaction of the TCA cycle. Writing the two half-cell reactions, we have

$$\text{NAD}^+ + 2\,\text{H}^+ + 2e^- \longrightarrow \text{NADH} + \text{H}^+ \qquad \mathcal{E}_o{}' = -0.32\text{ V} \quad (20.7)$$

$$\alpha\text{-Ketoglutarate} + \text{CO}_2 + 2\,\text{H}^+ + 2e^- \longrightarrow \text{isocitrate} \qquad \mathcal{E}_o{}' = -0.38\text{ V} \quad (20.8)$$

In a spontaneous reaction, electrons are donated by (flow away from) the reaction with the more negative reduction potential and are accepted by (flow toward) the reaction with the more positive reduction potential. Thus, in the present case, isocitrate will donate electrons and NAD^+ will accept electrons. The convention defines $\Delta\mathcal{E}_o{}'$ as

$$\Delta\mathcal{E}_o{}' = \mathcal{E}_o{}'(\text{acceptor}) - \mathcal{E}_o{}'(\text{donor}) \quad (20.9)$$

In the present case, isocitrate is the donor and NAD^+ the acceptor, so we write

$$\Delta\mathcal{E}_o{}' = -0.32\text{ V} - (-0.38\text{ V}) = +0.06\text{ V} \quad (20.10)$$

From Equation 20.2, we can now calculate $\Delta G^{\circ\prime}$ as

$$\Delta G^{\circ\prime} = -(2)(96.485\text{ kJ/V}\cdot\text{mol})(0.06\text{ V}) \quad (20.11)$$
$$\Delta G^{\circ\prime} = -11.58\text{ kJ/mol}$$

Note that a reaction with a net positive $\Delta\mathcal{E}_o{}'$ yields a negative $\Delta G^{\circ\prime}$, indicating a spontaneous reaction.

The Dependence of the Reduction Potential on Concentration

We have already noted that the standard free energy change for a reaction, $\Delta G^{\circ\prime}$, does not reflect the actual conditions in a cell, where reactants and products are not at standard-state concentrations (1 *M*). Equation (15.12) was introduced to permit calculations of actual free energy changes under non–standard-state conditions. Similarly, standard reduction potentials for redox couples must be modified to account for the actual concentrations of the oxidized and reduced species. For any redox couple,

$$\text{ox} + ne^- \rightleftharpoons \text{red} \quad (20.12)$$

the actual reduction potential is given by

$$\mathcal{E} = \mathcal{E}_o{}' + (RT/n\mathcal{F})\ln\frac{[\text{ox}]}{[\text{red}]} \quad (20.13)$$

Reduction potentials can also be quite sensitive to molecular environment. The influence of environment is especially important for flavins, such as $FAD/FADH_2$ and $FMN/FMNH_2$. These species are normally bound to their respective flavoproteins; the reduction potential of bound FAD, for example, can be very different from the value shown in Table 20.1 for the free FAD–$FADH_2$ couple of −0.219 V. A problem at the end of the chapter addresses this case.

20.3 The Electron Transport Chain—An Overview

As we have seen, the metabolic energy from oxidation of food materials—sugars, fats, and amino acids—is funneled into formation of reduced coenzymes (NADH) and reduced flavoproteins ([$FADH_2$]). The electron transport chain reoxidizes the coenzymes, and channels the free energy obtained from these reactions into the synthesis of ATP. This reoxidation process involves the removal of both protons and electrons from the coenzymes. Electrons move from NADH and [$FADH_2$] to molecular oxygen, O_2, which is the terminal acceptor of electrons in the chain. This means a drop in energy of more than 200 kJ/mol. The reoxidation of NADH,

$$\text{NADH(reductant)} + H^+ + \tfrac{1}{2}\,O_2\text{(oxidant)} \longrightarrow NAD^+ + H_2O \qquad (20.14)$$

involves the following half-reactions:

$$NAD^+ + 2\,H^+ + 2e^- \longrightarrow NADH + H^+ \qquad \mathscr{E}_o{}' = -0.32\text{ V} \qquad (20.15)$$

$$\tfrac{1}{2}\,O_2 + 2\,H^+ + 2e^- \longrightarrow H_2O \qquad \mathscr{E}_o{}' = +0.816\text{ V} \qquad (20.16)$$

Here, half-reaction (20.16) is the electron acceptor and half-reaction (20.15) is the electron donor. Then

$$\Delta\mathscr{E}_o{}' = 0.816 - (-0.32) = 1.136\text{ V}$$

and, according to Equation (20.2), the standard-state free energy change, $\Delta G^{\circ\prime}$, is -219 kJ/mol. Molecules along the electron transport chain have reduction potentials between the values for the NADH–NAD^+ couple and the oxygen–H_2O couple, so that electrons move down the energy scale toward progressively more positive reduction potentials (Figure 20.3).

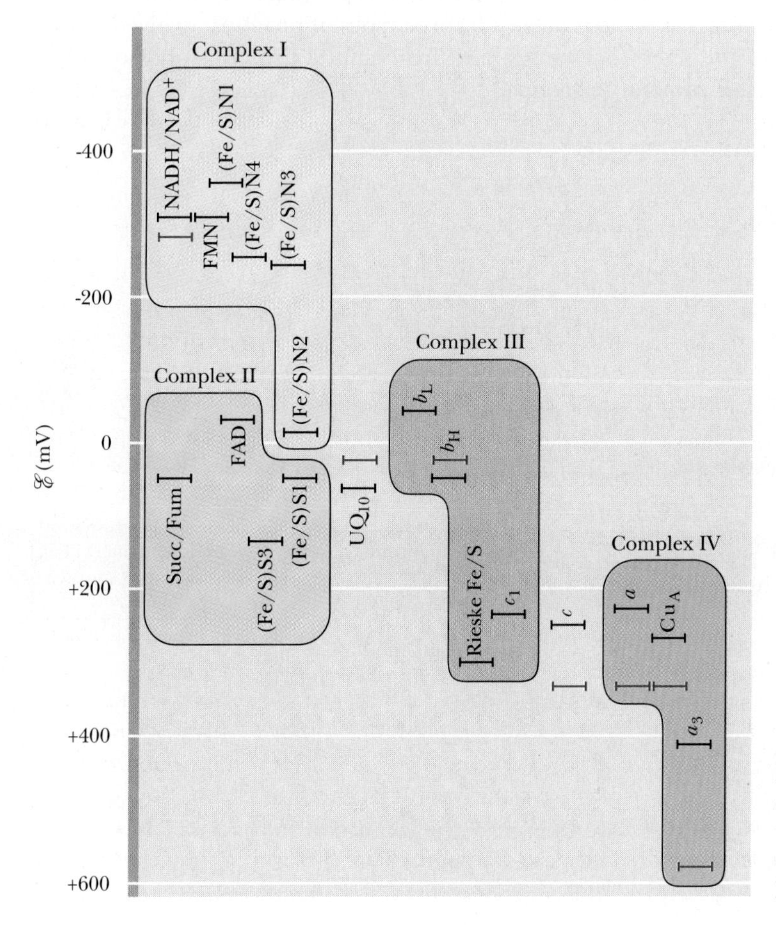

***Figure* 20.3** $\mathscr{E}_o{}'$ and $\mathscr{E}$ values for the components of the mitochondrial electron transport chain. Values indicated are consensus values for animal mitochondria. Black bars represent $\mathscr{E}_o{}'$; red bars, $\mathscr{E}$.

Although electrons move from more negative to more positive reduction potentials in the electron transport chain, it should be emphasized that the electron carriers do not operate in a simple linear sequence. This will become evident when the individual components of the electron transport chain are discussed in the following paragraphs.

The Electron Transport Chain Can Be Isolated in Four Complexes

The electron transport chain involves several different molecular species, including:

(a) **Flavoproteins,** which contain tightly bound FMN or FAD as prosthetic groups, and which (as noted in Chapter 14) may participate in one- or two-electron transfer events

(b) **Coenzyme Q,** also called **ubiquinone** (and abbreviated **CoQ** or **UQ**) (Figure 9.18), which can function in either one- or two-electron transfer reactions

(c) Several **cytochromes** (proteins containing heme prosthetic groups [see Chapter 4], which function by carrying or transferring electrons), including cytochromes *b*, *c*, c_1, *a*, and a_3. Cytochromes are one-electron transfer agents, in which the heme iron is converted from Fe^{2+} to Fe^{3+} and back

(d) A number of **iron–sulfur proteins,** which participate in one-electron transfers involving the Fe^{2+} and Fe^{3+} states

(e) Protein-bound **copper,** a one-electron transfer site, which converts between Cu^{+} and Cu^{2+}

All these intermediates except for cytochrome *c* are membrane-associated (either in the mitochondrial inner membrane of eukaryotes or in the cytoplasmic membrane of prokaryotes). All three types of proteins involved in this chain—flavoproteins, cytochromes, and iron–sulfur proteins—possess electron-transferring **prosthetic groups.**

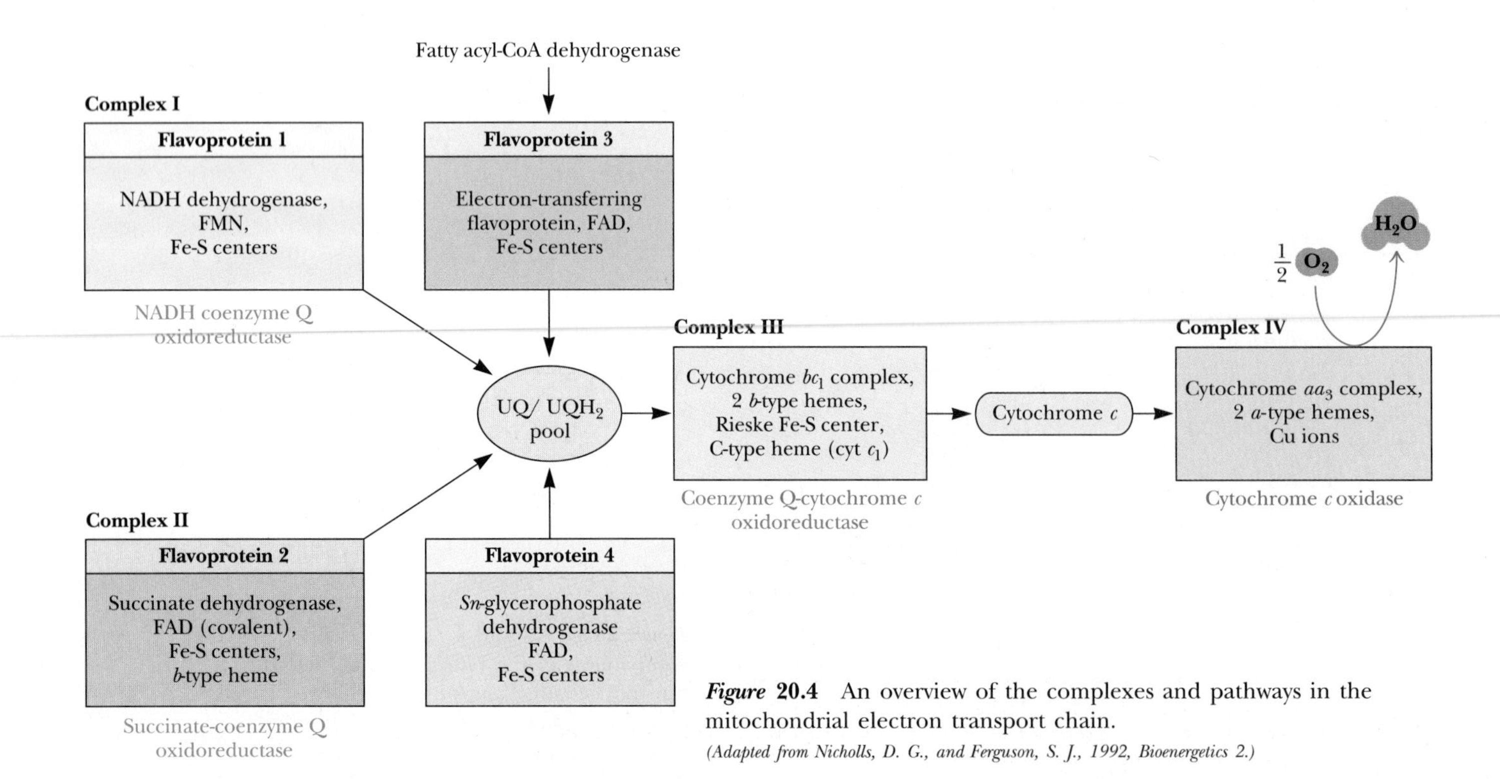

***Figure* 20.4** An overview of the complexes and pathways in the mitochondrial electron transport chain.
(Adapted from Nicholls, D. G., and Ferguson, S. J., 1992, Bioenergetics 2.)

The components of the electron transport chain can be purified from the mitochondrial inner membrane. Solubilization of the membranes containing the electron transport chain results in the isolation of four distinct protein complexes, and the complete chain can thus be considered as composed of four parts: (I) *NADH–coenzyme Q reductase,* (II) *succinate–coenzyme Q reductase,* (III) *coenzyme Q–cytochrome c reductase,* and (IV) *cytochrome c oxidase* (Figure 20.4). Complex I accepts electrons from NADH, serving as a link between glycolysis, the TCA cycle, fatty acid oxidation, and the electron transport chain. Complex II includes succinate dehydrogenase and thus forms a direct link between the TCA cycle and electron transport. Complexes I and II produce a common product, reduced coenzyme Q (UQH_2), which is the substrate for coenzyme Q–cytochrome *c* reductase (Complex III). As shown in Figure 20.4, there are two other ways to feed electrons to UQ: the **electron-transferring flavoprotein,** which transfers electrons from the flavoprotein-linked step of fatty acyl-CoA dehydrogenase, and ***sn*-glycerophosphate dehydrogenase.** Complex III oxidizes UQH_2 while reducing cytochrome *c*, which in turn is the substrate for Complex IV, cytochrome *c* oxidase. Complex IV is responsible for reducing molecular oxygen. Each of the complexes shown in Figure 20.4 is a large multisubunit complex embedded within the inner mitochondrial membrane.

20.4 Complex I: NADH–Coenzyme Q Reductase

As its name implies, this complex transfers a pair of electrons from NADH to coenzyme Q, a small, hydrophobic yellow compound. Another common name for this enzyme complex is *NADH dehydrogenase.* The complex (with an estimated mass of 850 kD) involves more than 30 polypeptide chains, one molecule of flavin mononucleotide (FMN), and as many as seven Fe-S clusters, together containing a total of 20 to 26 iron atoms (Table 20.2). By virtue of its dependence on FMN, NADH–UQ reductase is a *flavoprotein.*

***Table* 20.2**
Protein Complexes of the Mitochondrial Electron-Transport Chain

Complex	Mass (kD)	Subunits	Prosthetic Group	Binding Site for:
NADH-UQ reductase	850	>30	FMN Fe-S	NADH (matrix side) UQ (lipid core)
Succinate-UQ reductase	140	4	FAD Fe-S	Succinate (matrix side) UQ (lipid core)
UQ-Cyt *c* reductase	250	9–10	Heme b_L Heme b_H Heme c_1 Fe-S	Cyt *c* (intermembrane space side)
Cytochrome *c*	13	1	Heme *c*	Cyt c_1 Cyt *a*
Cytochrome *c* oxidase	162	>10	Heme *a* Heme a_3 Cu_A Cu_B	Cyt *c* (intermembrane space side)

Adapted from: Hatefi, Y., 1985. The mitochondrial electron transport chain and oxidative phosphorylation system. *Annual Review of Biochemistry* **54:**1015–1069; and DePierre, J., and Ernster, L., 1977. Enzyme topology of intracellular membranes. *Annual Review of Biochemistry* **46:**201–262.

(a)

Coenzyme Q, oxidized form (Q, ubiquinone) ⇌ (e^- + H^+) Semiquinone intermediate (QH•) ⇌ (e^- + H^+) Coenzyme Q, reduced form (QH_2, ubiquinol)

(b)

Figure **20.5** (a) The three oxidation states of coenzyme Q. (b) A space-filling model of coenzyme Q.

Although the precise mechanism of the NADH–UQ reductase is not known, the first step involves binding of NADH to the enzyme on the *matrix* side of the inner mitochondrial membrane, and transfer of electrons from NADH to tightly bound FMN:

$$\text{NADH} + [\text{FMN}] + \text{H}^+ \longrightarrow [\text{FMNH}_2] + \text{NAD}^+ \qquad (20.17)$$

The second step involves the transfer of electrons from the reduced $[FMNH_2]$ to a series of Fe-S proteins, including both 2Fe-2S and 4Fe-4S clusters (see Figures 19.8 and 19.16). The unique redox properties of the flavin group of FMN are probably important here. NADH is a two-electron donor, whereas the Fe-S proteins are one-electron transfer agents. (Iron in these clusters shuttles between Fe^{3+} (oxidized) and Fe^{2+}(reduced) states.) The flavin of FMN has three redox states—the oxidized, semiquinone, and reduced states. It can act as *either* a one-electron *or* a two-electron transfer agent, and may serve as a critical link between NADH and the Fe-S proteins.

The final step of the reaction involves the transfer of two electrons from iron–sulfur clusters to coenzyme Q. Coenzyme Q is a **mobile electron carrier.** Its isoprenoid tail makes it highly hydrophobic, and it diffuses freely in the hydrophobic core of the inner mitochondrial membrane. As a result, it shuttles electrons from Complexes I and II to Complex III. The redox cycle of UQ is shown in Figure 20.5, and the overall scheme is shown schematically in Figure 20.6.

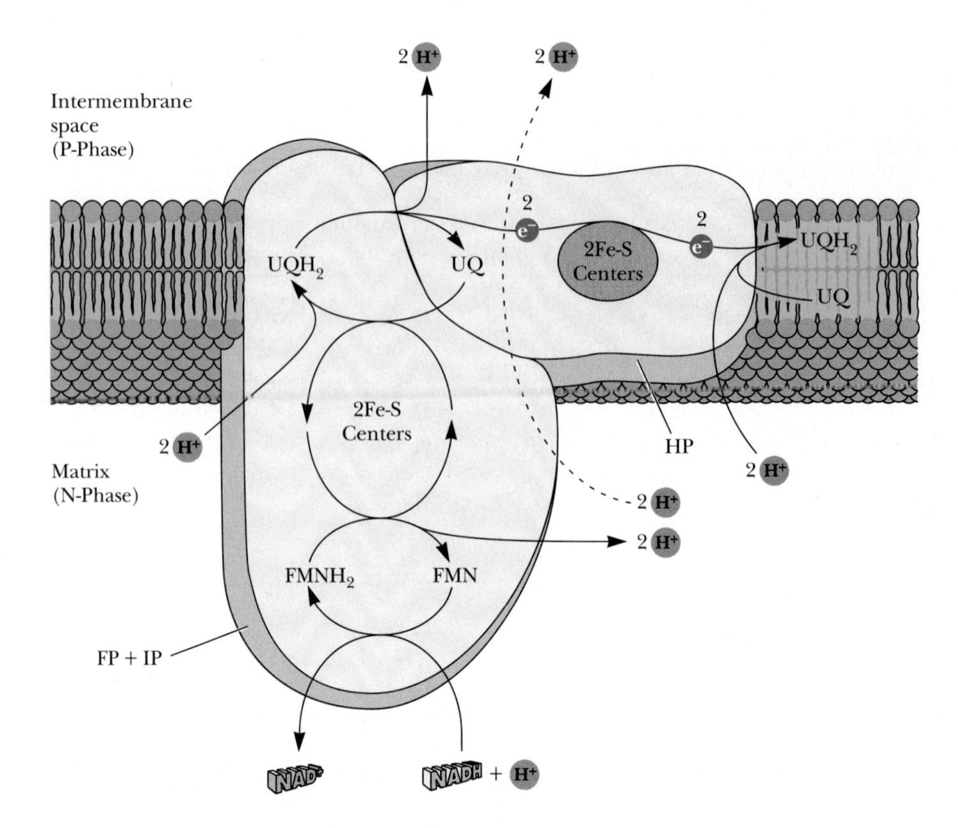

Figure **20.6** Proposed structure and electron transfer pathway for Complex I. Three protein complexes have been isolated, including the **flavoprotein (FP), iron–sulfur protein (IP),** and **hydrophobic protein (HP).** FP contains three peptides (of mass 51, 24, and 10 kD) and bound FMN and has two Fe-S centers (a 2Fe-2S center and a 4Fe-4S center). IP contains six peptides and at least three Fe-S centers. HP contains at least seven peptides and one Fe-S center.

Complex I Transports Protons from the Matrix to the Cytosol

The oxidation of one NADH and the reduction of one UQ by NADH–UQ reductase results in the net transport of protons from the matrix side to the cytosolic side of the inner membrane. The cytosolic side, where H^+ accumulates, is referred to as the **P** (for *positive*) face; thus the matrix side is the **N** (for *negative*) face. Some of the energy liberated by the flow of electrons through this complex is used in a *coupled process* to drive the transport of protons across the membrane. (This is an example of *active transport,* a phenomenon examined in detail in Chapter 35.) Available experimental evidence suggests a stoichiometry of four H^+ transported per two electrons passed from NADH to UQ.

20.5 Complex II: Succinate–Coenzyme Q Reductase

Complex II is perhaps better known by its other name—**succinate dehydrogenase,** the only TCA cycle enzyme that is an integral membrane protein in the inner mitochondrial membrane. This enzyme has a mass of approximately 100 to 140 kD and is composed of four subunits: two Fe-S proteins of masses 70 kD and 27 kD, and two other peptides of masses 15 kD and 13 kD. Also known as *flavoprotein 2 (FP_2),* it contains an FAD covalently bound to a histidine residue (see Figure 19.15), and three Fe-S centers: a 4Fe-4S cluster, a 3Fe-4S cluster, and a 2Fe-2S cluster. When succinate is converted to fumarate in the TCA cycle, concomitant reduction of bound FAD to $FADH_2$ occurs in succinate dehydrogenase. This $FADH_2$ transfers its electrons immediately to Fe-S centers, which pass them on to UQ. Electron flow from succinate to UQ,

$$\text{Succinate} \longrightarrow \text{fumarate} + 2\,H^+ + 2e^- \qquad (20.18)$$

$$UQ + 2\,H^+ + 2e^- \longrightarrow UQH_2 \qquad (20.19)$$

$$\text{Net rxn:}\quad \text{Succinate} + UQ \longrightarrow \text{fumarate} + UQH_2$$

$$\Delta\mathscr{E}_o{}' = 0.029\text{ V} \qquad (20.20)$$

$$\Delta G^{\circ\prime} = -5.6\text{ kJ/mol}$$

yields a net reduction potential of 0.029 V. (Note that the first half-reaction is written in the direction of the e^- flow. As always, $\Delta\mathscr{E}_o{}'$ is calculated according to Equation 20.9.) The small free energy change of this reaction is not sufficient to drive the transport of protons across the inner mitochondrial membrane. This is a crucial point, since (as we will see) proton transport is coupled with ATP synthesis. Oxidation of one $FADH_2$ in the electron transport chain results in synthesis of approximately two molecules of ATP, compared with the approximately three ATPs produced by the oxidation of one NADH. Other enzymes can also supply electrons to UQ, including mitochondrial *sn*-glycerophosphate dehydrogenase, an inner membrane–bound shuttle enzyme, and the fatty acyl–CoA dehydrogenases, three soluble matrix enzymes involved in fatty acid oxidation (Figure 20.7; also see Chapter 23). The path of electrons from succinate to UQ is shown in Figure 20.8.

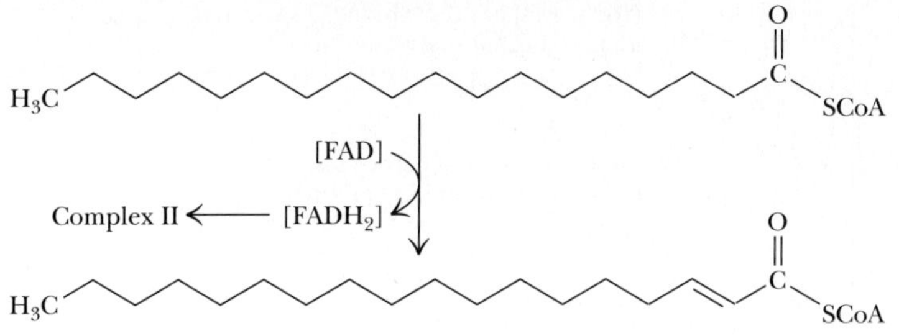

H_3C … SCoA
[FAD]
Complex II ← $[FADH_2]$
H_3C … SCoA

***Figure* 20.7** The fatty acyl-CoA dehydrogenase reaction, emphasizing that the reaction involves reduction of enzyme-bound FAD (indicated by brackets).

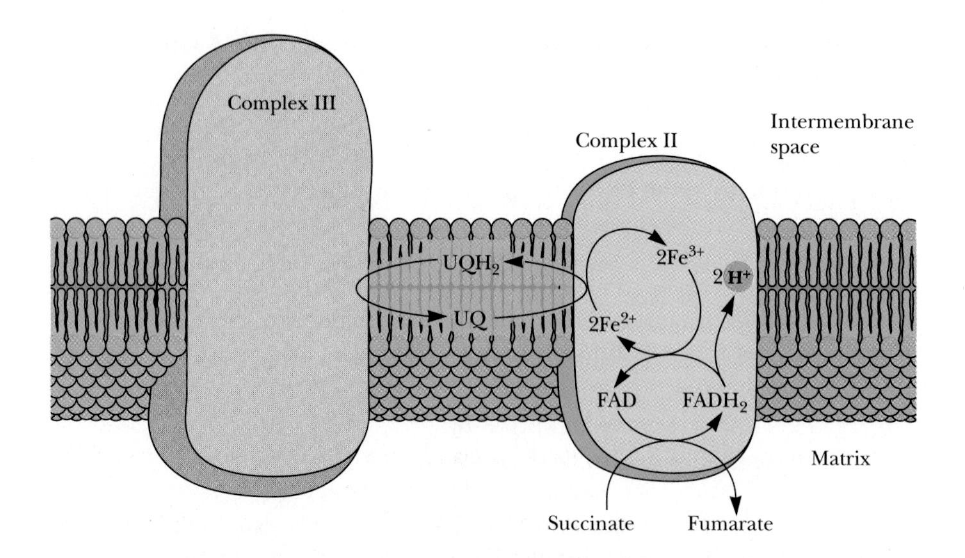

***Figure* 20.8** A probable scheme for electron flow in Complex II. Oxidation of succinate occurs with reduction of [FAD]. Electrons are then passed to Fe-S centers and then to coenzyme Q (UQ). Proton transport does not occur in this complex.

20.6 Complex III: Coenzyme Q–Cytochrome *c* Reductase

In the third complex of the electron transport chain, reduced coenzyme Q (UQH_2) passes its electrons to cytochrome *c* via a unique redox pathway known as the **Q cycle.** *UQ–cytochrome c reductase* (UQ–cyt *c* reductase), as this complex is known, involves three different cytochromes and an Fe-S protein. In the cytochromes of these and similar complexes, the iron atom at the center of the porphyrin ring cycles between the reduced Fe^{2+} (ferrous) and oxidized Fe^{3+} (ferric) states.

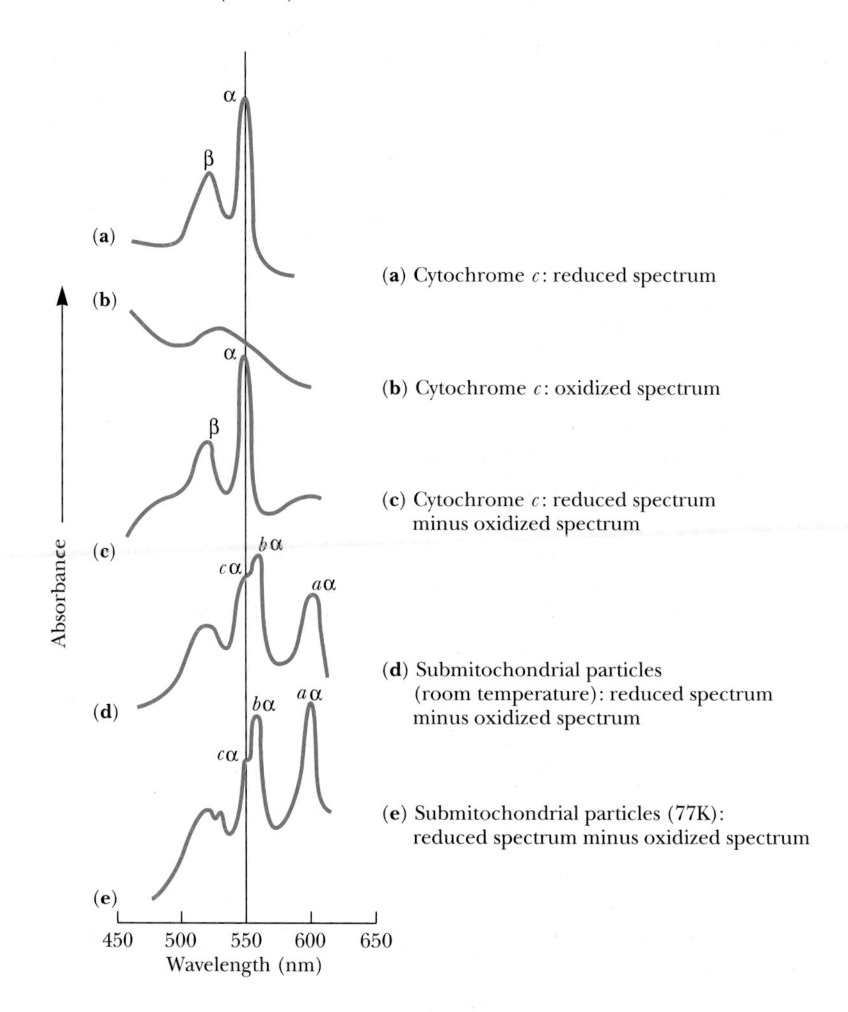

***Figure* 20.9** Typical visible absorption spectra of cytochromes. (a) Cytochrome *c*, reduced spectrum; (b) cytochrome *c*, oxidized spectrum; (c) the difference spectrum: (a) minus (b); (d) beef heart submitochondrial particles: room temperature difference (reduced minus oxidized) spectrum; (e) beef heart submitochondrial particles: same as (d) but at 77 K. α and β bands are labeled, and in (d) and (e) the bands for cytochromes *a*, *b*, and *c* are indicated.

Cytochromes were first named and classified on the basis of their absorption spectra (Figure 20.9), which depend upon the structure and environment of their heme groups. The **b cytochromes** contain *iron–protoporphyrin IX* (Figure 20.10), the same heme found in hemoglobin and myoglobin. The **c cytochromes** contain *heme c*, derived from iron–protoporphyrin IX by the covalent attachment of cysteine residues from the associated protein. UQ–cyt *c* reductase contains a unique *b*-type cytochrome, of 30 to 40 kD, with two different heme sites (Figure 20.11) and one *c*-type cytochrome. (One other variation, heme *a*, contains a 15-carbon isoprenoid chain on a modified vinyl group, and a formyl group in place of one of the methyls [see Figure 20.10]. Cytochrome *a* is found in two forms in Complex IV of the electron transport chain, as we shall see.) The two hemes on the *b* cyctochrome polypeptide in UQ–cyt *c* reductase are distinguished by their reduction potentials and the wavelength (λ_{max}) of the so-called **α band** (see Figure 20.9). One of these hemes, known as b_L or b_{566}, has a standard reduction potential, $\mathscr{E}_o'$, of −0.100 V and a wavelength of maximal absorbance (λ_{max}) of 566 nm. The other, known as b_H or b_{562}, has a standard reduction potential of +0.050 V and a λ_{max} of 562 nm. (*H* and *L* here refer to *high* and *low* reduction potential.)

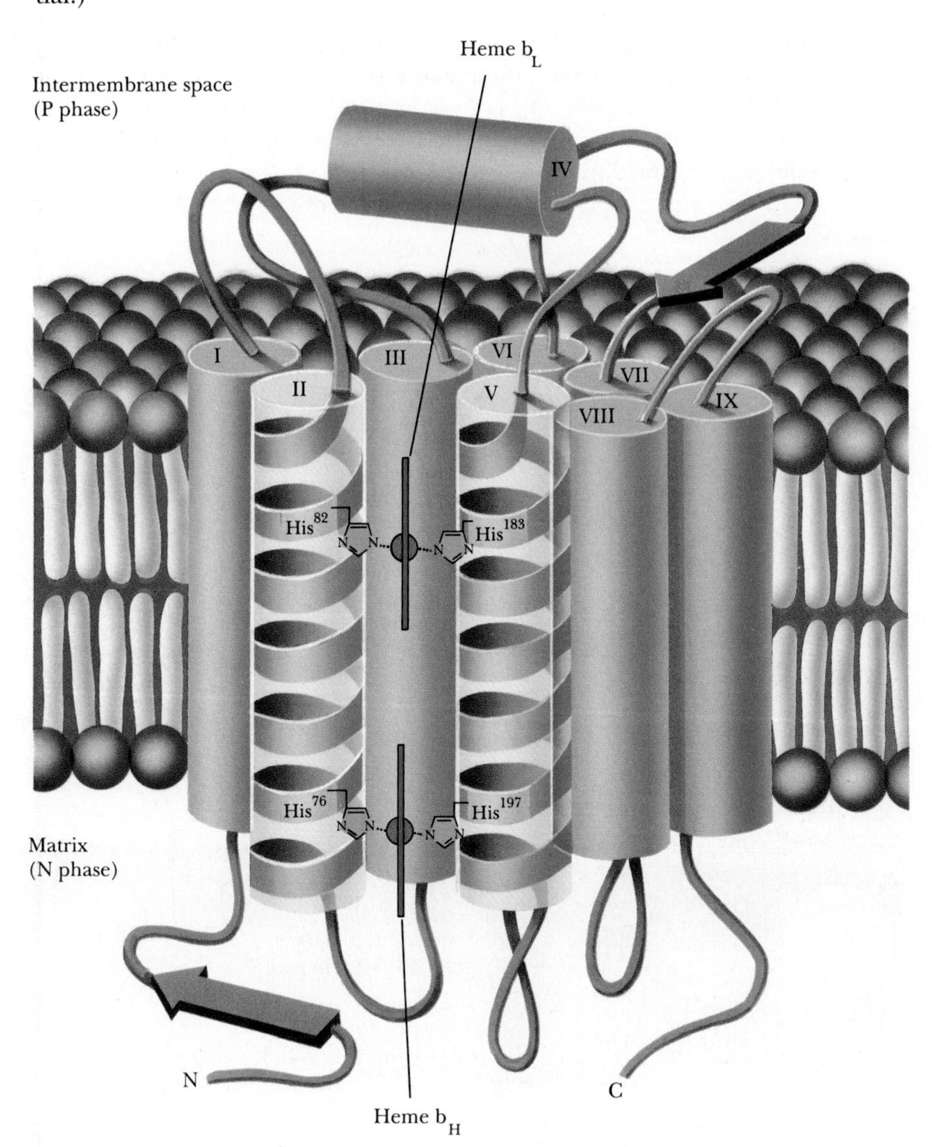

Figure **20.11** The α-helices flanking the heme groups of cytochrome *b*, showing the histidine residues that coordinate the heme iron.

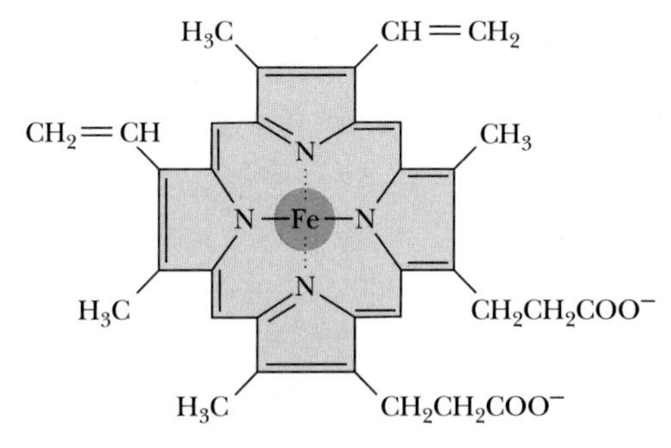

Iron protoporphyrin IX
(found in cytochrome *b*, myoglobin, and hemoglobin)

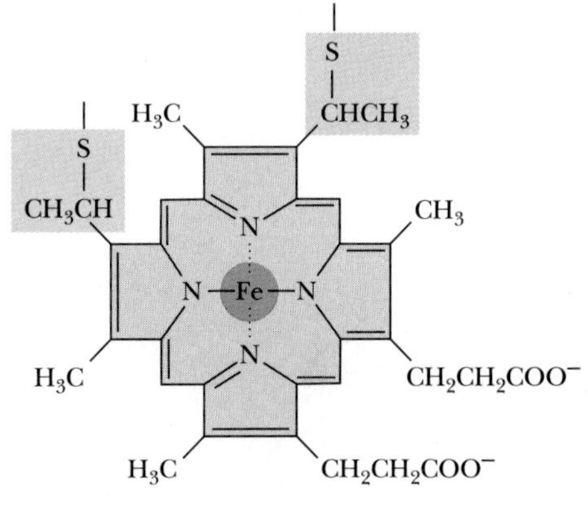

Heme C
(found in cytochrome *c*)

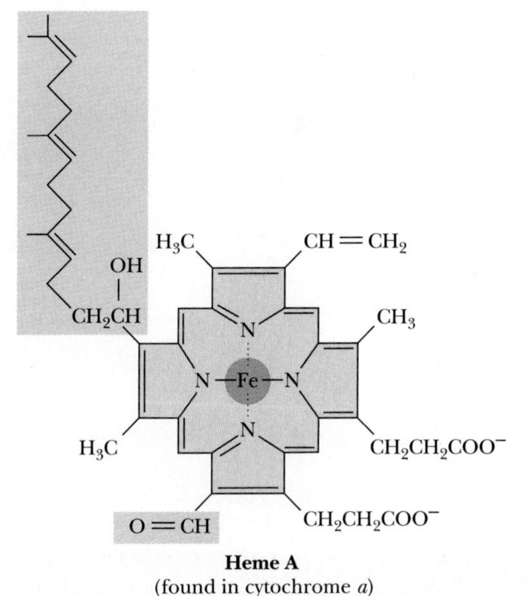

Heme A
(found in cytochrome *a*)

Figure **20.10** The structures of iron protoporphyrin IX, heme *c*, and heme *a*.

Complex III Drives Proton Transport

As with Complex I, passage of electrons through the Q cycle of Complex III is accompanied by proton transport across the inner mitochondrial membrane. The postulated pathway for electrons in this system is shown in Figure 20.12. A large pool of UQ and UQH_2 exists in the inner mitochondrial membrane. The Q cycle is initiated when a molecule of UQH_2 from this pool diffuses to a site (called $\mathbf{Q_p}$) on Complex III near the cytosolic face of the membrane.

Oxidation of this UQH_2 occurs in two steps. First, an electron from UQH_2 is transferred to the *Rieske protein* (an Fe-S protein named for its discoverer) and then to cytochrome c_1. This releases two H^+ to the cytosol and leaves $\mathbf{UQ\cdot^-}$, a semiquinone anion form of UQ, at the Q_p site. The second electron is then transferred to the b_L heme, converting $UQ\cdot^-$ to UQ. The Rieske protein and cytochrome c_1 are similar in structure; each has a globular domain and is anchored to the inner membrane by a hydrophobic segment. However, the hydrophobic segment is N-terminal in the Rieske protein and C-terminal in cytochrome c_1.

The electron on the b_L heme facing the cytosolic side of the membrane is now passed to the b_H heme on the matrix side of the membrane. This electron transfer occurs against a membrane potential of 0.15 V and is driven by the loss of redox potential as the electron moves from b_L ($\mathscr{E}_o{}' = -0.100$ V) to b_H ($\mathscr{E}_o{}' = +0.050$ V). The electron is then passed from b_H to a molecule of UQ at a second quinone-binding site, $\mathbf{Q_n}$, converting this UQ to $UQ\cdot^-$. The resulting $UQ\cdot^-$ remains firmly bound to the Q_n site. This completes the first half of the Q cycle (Figure 20.12a).

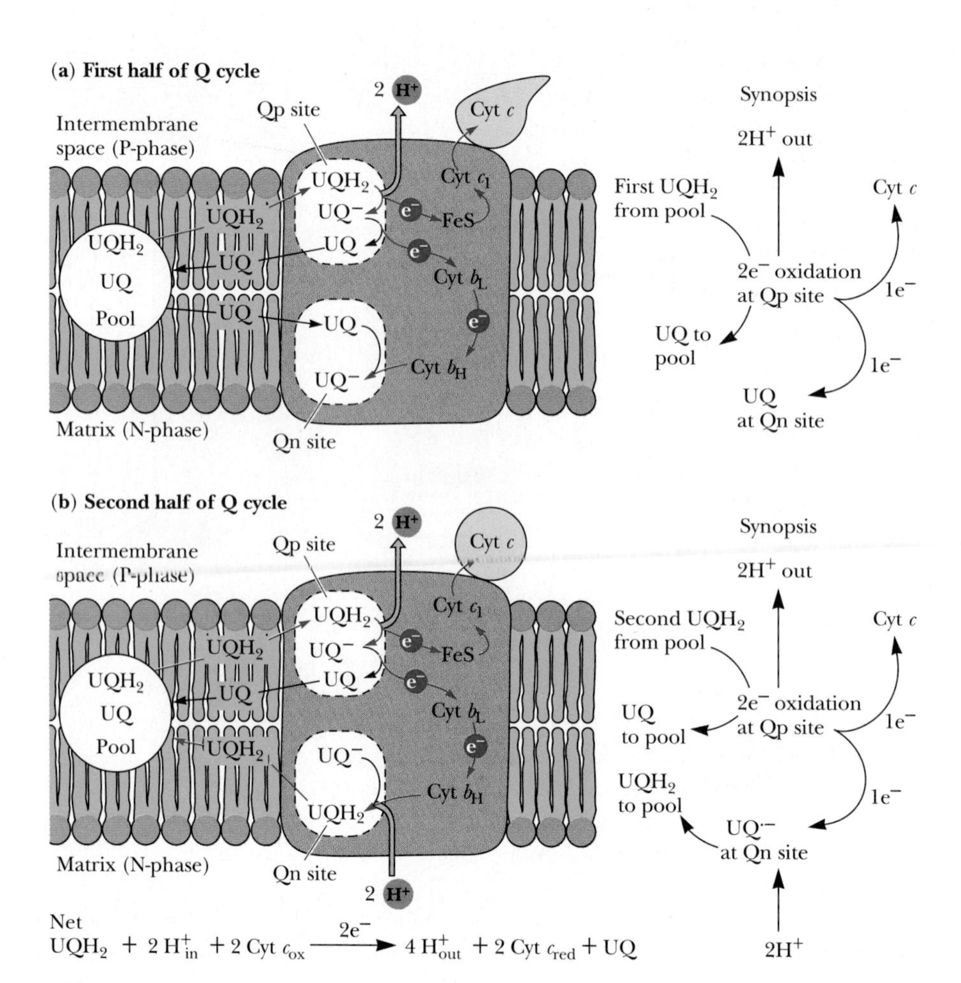

Figure **20.12** The Q cycle in mitochondria. (a) The electron transfer pathway following oxidation of the first UQH_2 at the Q_p site near the cytosolic face of the membrane. (b) The pathway following oxidation of a second UQH_2.

The second half of the cycle (Figure 20.12b) is similar to the first half, with a second molecule of UQH_2 oxidized at the Q_p site, one electron being passed to cytochrome c_1 and the other transferred to heme b_L and then to heme b_H. In this latter half of the Q cycle, however, the b_H electron is transferred to the semiquinone anion, $UQ\cdot^-$, at the Q_n site. With the addition of two H^+ from the mitochondrial matrix, this produces a molecule of UQH_2, which is released from the Q_n site and returns to the coenzyme Q pool, completing the Q cycle.

The Q Cycle Is an Unbalanced Proton Pump

Why has nature chosen this rather convoluted path for electrons in Complex III? First of all, Complex III takes up two protons on the matrix side of the inner membrane and releases four protons on the cytoplasmic side for each pair of electrons that passes through the Q cycle. The apparent imbalance of *two protons in* for *four protons out* will be resolved in the following discussion of Complex IV, the cytochrome oxidase complex. The other significant feature of this mechanism is that it offers a convenient way for a two-electron carrier, UQH_2, to interact with the b_L and b_H hemes, the Rieske protein Fe-S cluster, and cytochrome c_1, all of which are one-electron carriers.

Cytochrome *c* Is a Mobile Electron Carrier

Electrons traversing Complex III are passed through cytochrome c_1 to cytochrome *c*. Cytochrome *c* is the only one of the cytochromes that is water-soluble. Its structure, determined by X-ray crystallography (Figure 20.13), is globular; the planar heme group lies near the center of the protein, surrounded predominantly by hydrophobic protein residues. The iron in the porphyrin ring is coordinated both to a histidine nitrogen and to the sulfur atom of a methionine residue. Coordination with ligands in this manner on both sides of the porphyrin plane precludes the binding of oxygen and other ligands, a feature that distinguishes the cytochromes from hemoglobin (Chapter 12).

Cytochrome *c*, like UQ, is a mobile electron carrier. It associates loosely with the inner mitochondrial membrane (in the *intermembrane space* on the cytoplasmic side of the inner membrane) to acquire electrons from the Fe-S–cyt c_1 aggregate of Complex III, and then it migrates along the membrane surface in the reduced state, carrying electrons to *cytochrome c oxidase,* the fourth complex of the electron transport chain.

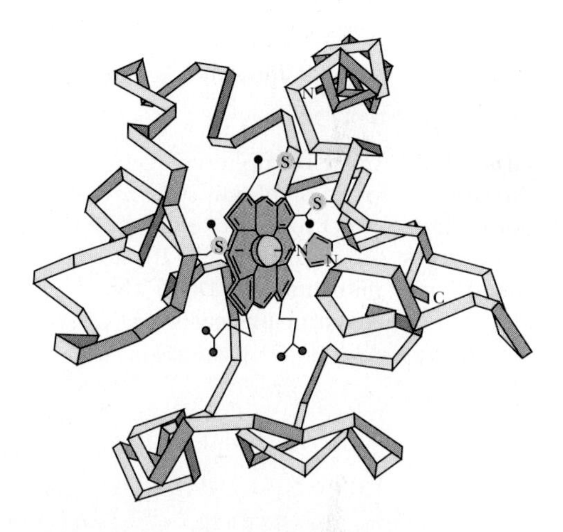

***Figure* 20.13** The structure of mitochondrial cytochrome *c*. The heme is shown at the center of the structure, covalently linked to the protein via its two sulfur atoms (yellow). A third sulfur from a methionine residue coordinates the iron.

20.7 Complex IV: Cytochrome *c* Oxidase

Complex IV is called cytochrome *c* oxidase because it accepts electrons from cytochrome *c* and directs them to the four-electron reduction of O_2 to form H_2O:

$$4\ \text{cyt}\ c\ (Fe^{2+}) + 4\ H^+ + O_2 \longrightarrow 4\ \text{cyt}\ c\ (Fe^{3+}) + 2\ H_2O \quad (20.21)$$

Thus, O_2 and cytochrome *c* oxidase are the final destination for the electrons derived from the oxidation of food materials. In concert with this process, cytochrome *c* oxidase also drives transport of protons across the inner mitochondrial membrane. These important functions are carried out by a transmembrane protein complex consisting of more than ten subunits (Table 20.2).

An electrophoresis gel of the bovine heart complex is shown in Figure 20.14. The mass of the complex, composed of more than 10 subunits, is

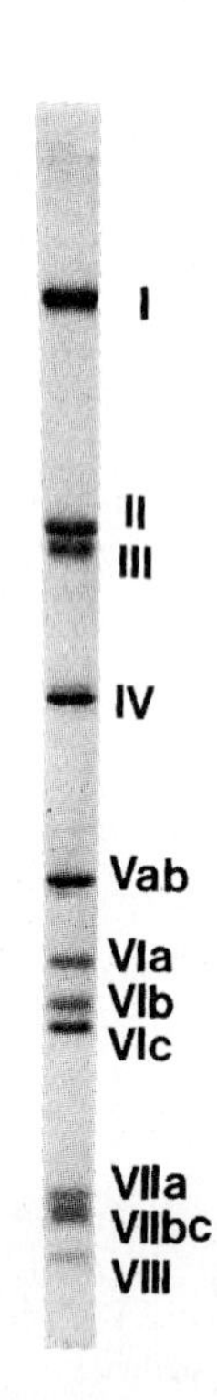

Figure **20.14** An electrophoresis gel showing the complex subunit structure of bovine heart cytochrome *c* oxidase. The three largest subunits, I, II, and III, are coded for by mitochondrial DNA. The others are encoded by nuclear DNA.
(Photo kindly provided by Professor Roderick Capaldi)

approximately 162 kD. Subunits I through III, the largest ones, are encoded by mitochondrial DNA, synthesized in the mitochondrion, and inserted into the inner membrane from the matrix side. The smaller subunits are coded by nuclear DNA and synthesized in the cytosol. Reconstituted preparations of cytochrome *c* oxidase form regular two-dimensional arrays in synthetic vesicle membranes. This phenomenon has special significance for structural studies, because cytochrome *c* oxidase, like most membrane enzymes, has resisted attempts to form true three-dimensional crystals for X-ray diffraction studies. Electron microscopy and image reconstruction techniques (Figure 20.15) applied to these two-dimensional crystals have shown that the enzyme in the membrane is Y-shaped (Figure 20.16) and spans the membrane. Reduced cytochrome *c* is oxidized on the cytoplasmic surface of the inner membrane, but O_2 reduction takes place on the matrix side of the membrane, so the electrons provided by cytochrome *c* must traverse the membrane.

Electron Transfer in Complex IV Involves Two Hemes and Two Copper Sites

Cytochrome *c* oxidase contains two heme centers (cytochromes *a* and a_3) as well as two copper atoms (Figure 20.17). The copper sites, Cu_A and Cu_B, are associated with cytochromes *a* and a_3, respectively. The copper sites participate in electron transfer by cycling between the reduced *(cuprous)* Cu^+ state and the oxidized *(cupric)* Cu^{2+} state. (Remember, the cytochromes and copper sites are one-electron transfer agents.) Reduction of one oxygen molecule requires passage of four electrons through these carriers—one at a time (Figure 20.17).

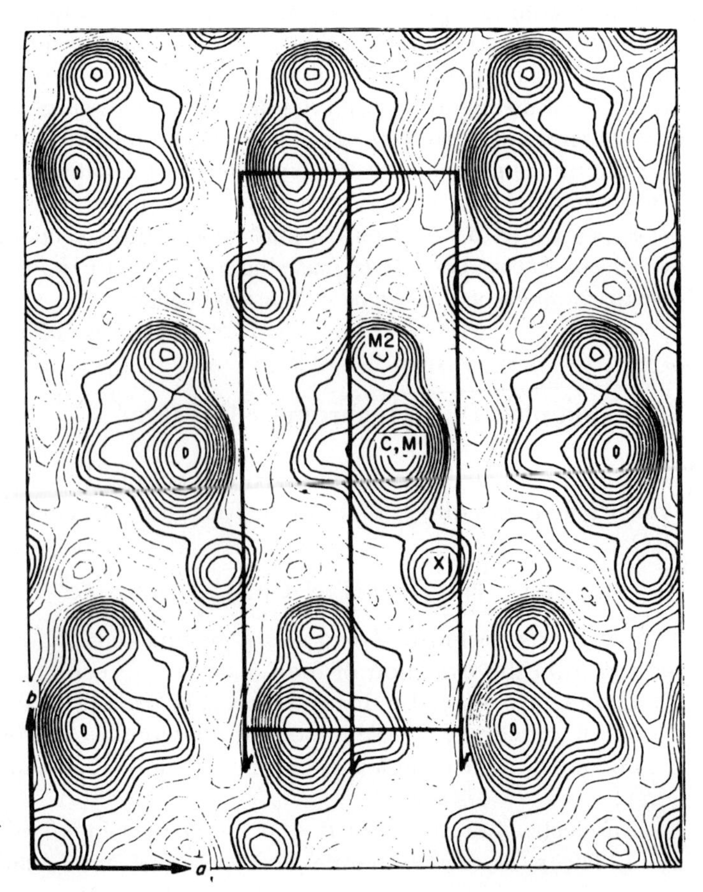

Figure **20.15** Electron density map of a crystalline array of cytochrome oxidase.

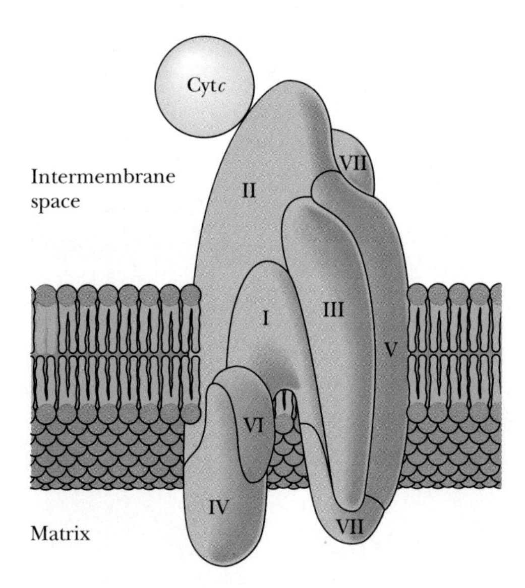

Figure **20.16** Models of cytochrome oxidase structure obtained from electron density map shown in Figure 20.15.

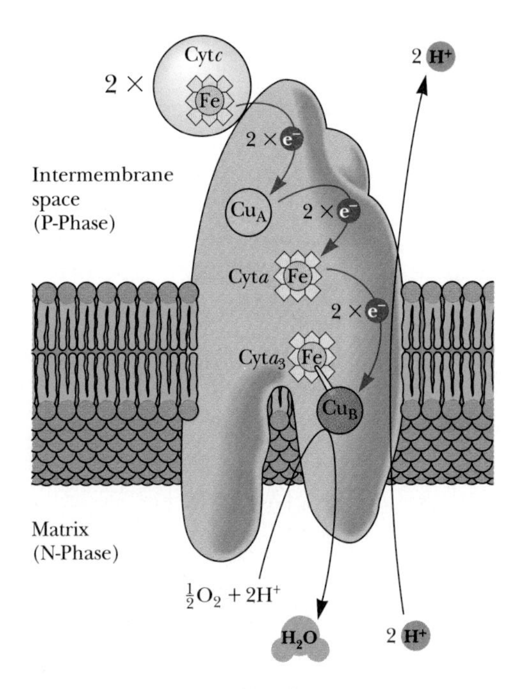

Figure **20.17** The electron transfer pathway for cytochrome oxidase. Cytochrome *c* binds on the cytosolic side, transferring electrons through the copper and heme centers to reduce O_2 on the matrix side of the membrane.

Electrons from cytochrome *c* are transferred to Cu_A sites and then passed to the heme iron of cytochrome *a*. Cu_A is liganded by two cysteines and two histidines (Figure 20.18). The heme of cytochrome *a* is approximately normal to the plane of the membrane (Figure 20.17) and is liganded by imidazole rings of histidine residues (Figure 20.18). The Cu_A and the Fe of cytochrome *a* are within 1.5 nm of each other.

Cu_B and the iron atom of cytochrome a_3 are also situated close to each other and are thought to share a ligand, which may be a cysteine sulfur (Figure 20.19). This closely associated pair of metal ions is referred to as a **binuclear center.** The heme of cytochrome a_3 is approximately normal to the plane of the membrane. Its other axial ligand is probably an imidazole nitrogen.

As shown in Figure 20.20, the electron pathway through Complex IV continues as Cu_B accepts a single electron from cytochrome *a* (O → H). A second electron then reduces the iron center to Fe^{2+} (H → R), leading to the binding of O_2 (R → A) and the formation of a peroxy bridge between heme a_3 and Cu_B (A → P). *This amounts to the transfer of two electrons from the binuclear center to the bound* O_2. The next step involves uptake of two H^+ and a third electron (P → F), which leads to cleavage of the O—O bond and generation

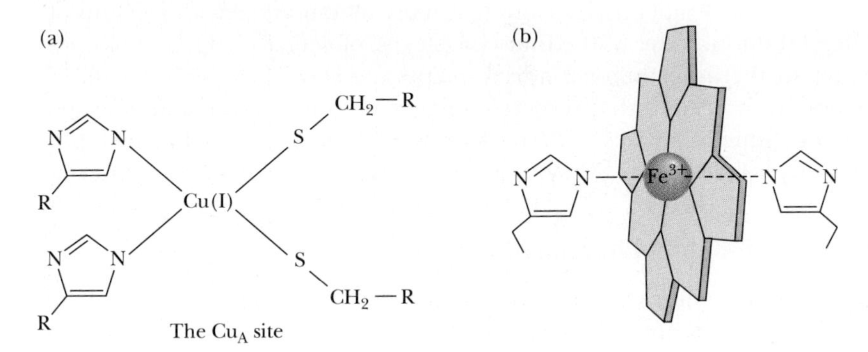

Figure **20.18** (a) The Cu_A site of cytochrome oxidase. Copper ligands include two histidine imidazole groups and two cysteine side chains from the protein. (b) The coordination of histidine imidazole ligands to the iron atom in the heme *a* center of cytochrome oxidase.

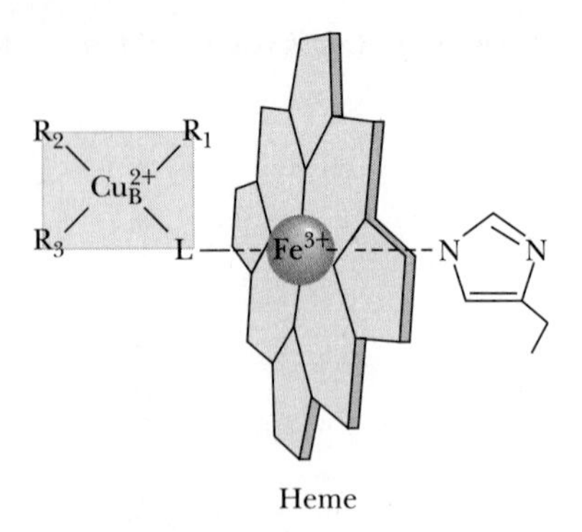

Figure **20.19** The binuclear center of cytochrome oxidase. A ligand, L (probably a cysteine *S*) is shown bridging the Cu_B and Fe_{a3} metal sites.

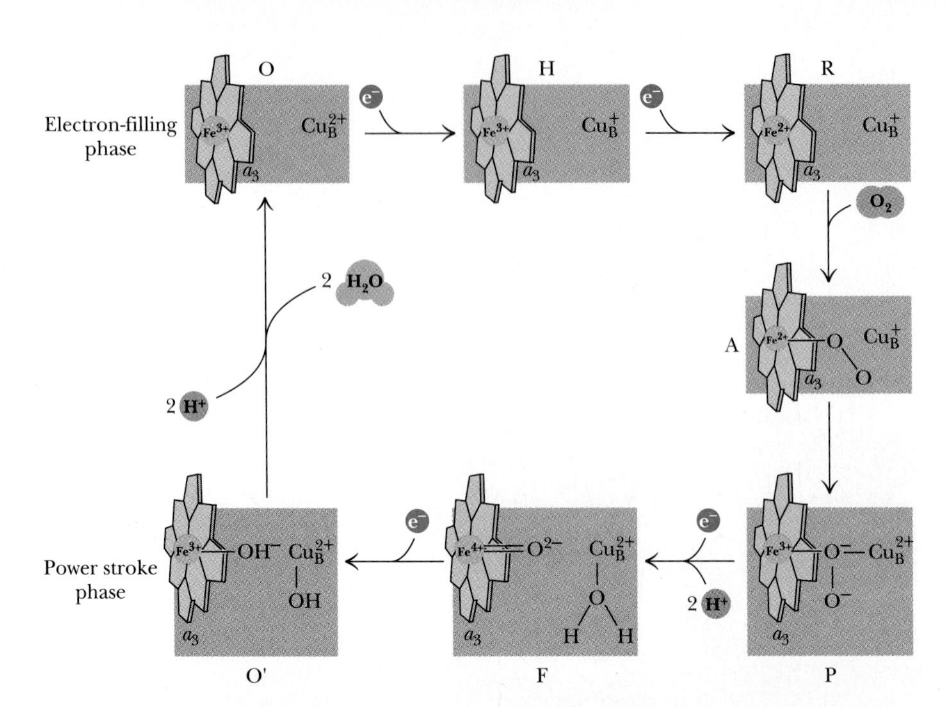

Figure **20.20** A model of the mechanism of O_2 reduction by cytochrome oxidase. *(Adapted from Nicholls, D. G., and Ferguson, S. J., 1992. Bioenergetics 2; and Babcock, G. T., and Wikström, M., 1992. Nature* ***356****:301–309.)*

of Fe^{4+} at the heme. Uptake of a fourth e^- facilitates formation of ferric hydroxide at the heme center ($F \rightarrow O'$). In the final step of the cycle $O' \rightarrow O$), protons from the mitochondrial matrix are accepted by the coordinated hydroxyl groups, and the resulting water molecules dissociate from the binuclear center.

Complex IV Also Transports Protons Across the Inner Mitochondrial Membrane

The reduction of oxygen in Complex IV is accompanied by transport of protons across the inner mitochondrial membrane. Transfer of four electrons through this complex drives the transport of approximately four protons. The mechanism of proton transport is unknown but is thought to involve the steps from state P to state O (Figure 20.20). Four protons are taken up on the matrix side for every two protons transported to the cytoplasm (see Figure 20.17).

Independence of the Four Carrier Complexes

It should be emphasized here that the four major complexes of the electron transport chain operate quite independently in the inner mitochondrial membrane. Each is a multiprotein aggregate maintained by numerous strong associations between peptides of the complex, but there is no evidence that the complexes associate with one another in the membrane. Measurements of the lateral diffusion rates of the four complexes, of coenzyme Q, and of cytochrome *c* in the inner mitochondrial membrane show that the rates differ considerably, indicating that these complexes do not move together in the membrane. Kinetic studies with reconstituted systems show that electron transport does not operate by means of connected sets of the four complexes.

A Dynamic Model of Electron Transport

The model that emerges for electron transport is shown in Figure 20.21. The four complexes migrate independently in the membrane. Coenzyme Q collects electrons from NADH–UQ reductase and succinate–UQ reductase and

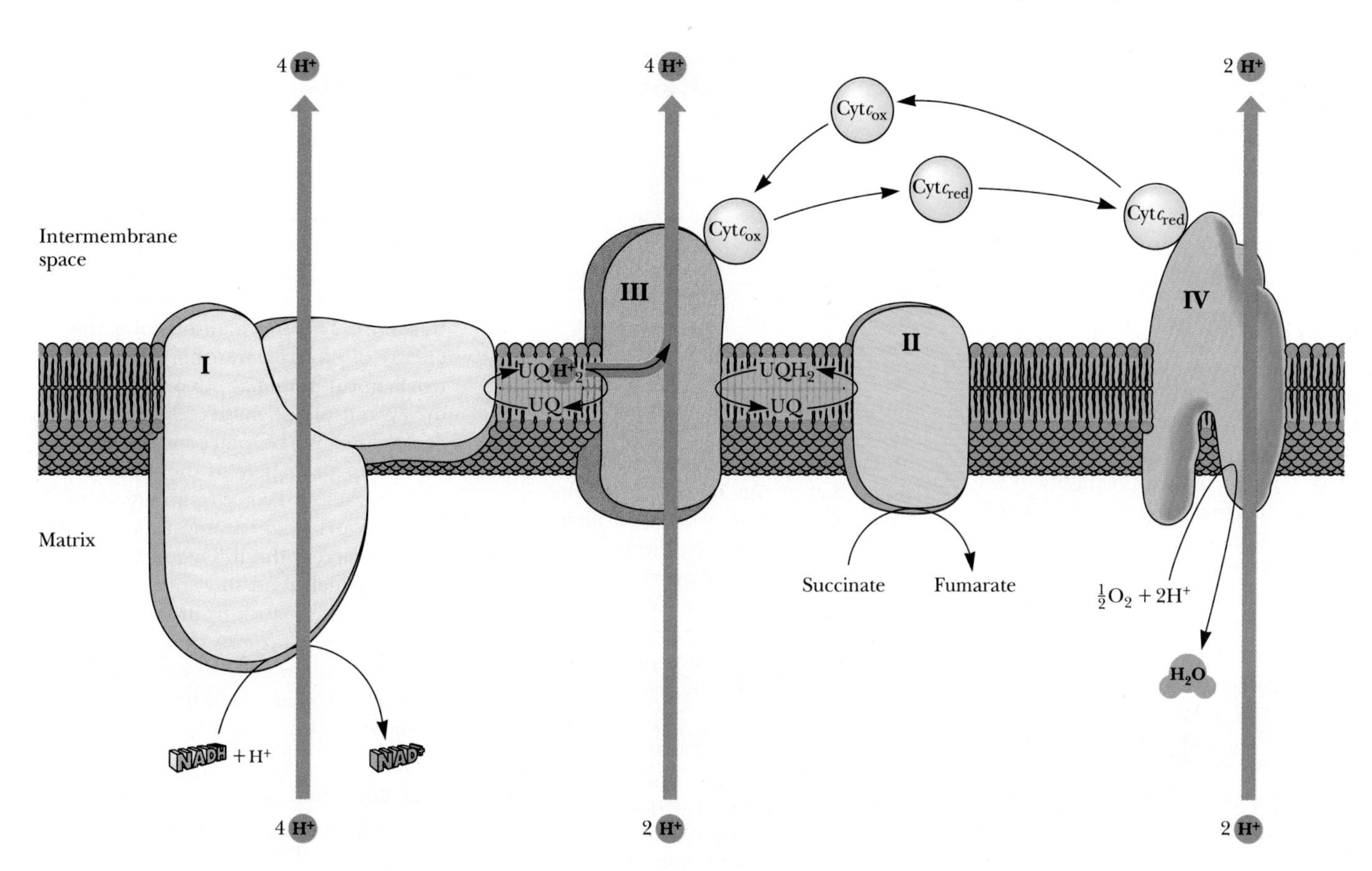

***Figure* 20.21** A model for the electron transport pathway in the mitochondrial inner membrane. UQ/UQH_2 and cytochrome *c* are mobile electron carriers and function by transferring electrons between the complexes. The proton transport driven by Complexes I, III, and IV is indicated.

delivers them (by diffusion through the membrane core) to UQ–cyt *c* reductase. Cytochrome *c* is water-soluble and moves freely, carrying electrons from UQ-cyt *c* reductase to cytochrome *c* oxidase. In the process of these electron transfers, protons are driven across the inner membrane (from the matrix side to the intermembrane space). The proton gradient generated by electron transport represents an enormous source of potential energy. As will be seen in the next section, this potential energy is used to synthesize ATP as protons flow back into the matrix.

The $H^+/2e^-$ Ratio for Electron Transport Is Uncertain

In 1961, Peter Mitchell, a British biochemist, proposed that the energy stored in a proton gradient across the inner mitochondrial membrane by electron transport drives the synthesis of ATP in cells. This proposal became known as **Mitchell's chemiosmotic hypothesis.** The ratio of protons transported per pair of electrons passed through the chain—the so-called **$H^+/2e^-$ ratio**—has been an object of great interest for many years. Nevertheless, the ratio has remained extremely difficult to determine. The consensus estimate for the electron transport pathway from succinate to O_2 is 6 $H^+/2e^-$. The ratio for Complex I by itself remains uncertain, but recent best estimates place it as high as 4 $H^+/2e^-$. On the basis of this value, the stoichiometry of transport for the pathway from NADH to O_2 is 10 $H^+/2e^-$. Although this is the value assumed in Figure 20.21, it is important to realize that this represents a consensus drawn from many experiments; the "true value" may be nonintegral. There is, after all, no reason for this ratio to be an integer.

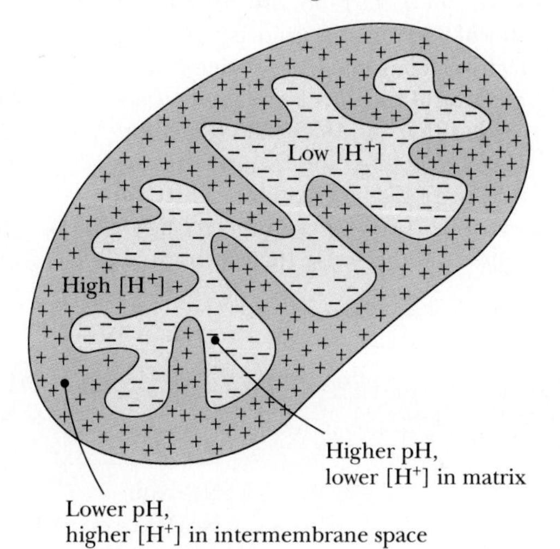

***Figure* 20.22** The proton and electrochemical gradients existing across the inner mitochondrial membrane. The electrochemical gradient is generated by the transport of protons across the membrane.

Critical Developments in Biochemistry

Oxidative Phosphorylation—The Clash of Ideas and Energetic Personalities

For many years, the means by which electron transport and ATP synthesis are coupled was unknown. It is no exaggeration to say that the search for the coupling mechanism was one of the largest, longest, most bitter fights in the history of biochemical research. Since 1777, when the French chemist Lavoisier determined that foods undergo oxidative combustion in the body, chemists and biochemists have wondered how energy from food is captured by living things. A piece of the puzzle fell into place in 1929, when Fiske and Subbarow first discovered and studied adenosine 5′-triphosphate in muscle extracts. Soon it was understood that ATP hydrolysis provides the energy for muscle contraction and other processes.

Engelhardt's experiments in 1930 led to the notion that ATP is synthesized as the result of electron transport, and, by 1940, Severo Ochoa had carried out a measurement of the **P/O ratio,** the number of molecules of ATP generated per atom of oxygen consumed in the electron transport chain. Since two electrons are transferred down the chain per oxygen atom reduced, the P/O ratio also reflects the ratio of ATPs synthesized per pair of electrons consumed. After many tedious and careful measurements, scientists decided that the P/O ratio was 3 for NADH oxidation and 2 for succinate (that is, [$FADH_2$]) oxidation. Electron flow and ATP synthesis are very tightly coupled in the sense that, in normal mitochondria, neither occurs without the other.

A High-Energy Chemical Intermediate Coupling Oxidation and Phosphorylation Proved Elusive

Many models were proposed to account for the coupling of electron transport and ATP synthesis. A persuasive model, advanced by E. C. Slater in 1953, proposed that energy derived from electron transport was stored in a **high-energy intermediate** (symbolized as X~P). This chemical species—in essence an activated form of phosphate—functioned according to certain relations according to Equations (20.22)–(20.25) (see below) to drive ATP synthesis.

This hypothesis was based on the model of substrate-level phosphorylation in which a high-energy substrate intermediate is a precursor to ATP. A good example is the 3-phosphoglycerate kinase reaction of glycolysis, where 1,3-bisphosphoglycerate serves as a high-energy intermediate leading to ATP. Literally hundreds of attempts were made to isolate the high-energy intermediate, X~P. Among the scientists involved in the research, rumors that one group or another had isolated X~P circulated frequently, but none was substantiated. Eventually it became clear that the intermediate could not be isolated, because it did not exist.

Peter Mitchell's Chemiosmotic Hypothesis

In 1961, Peter Mitchell proposed a novel coupling mechanism involving a proton gradient across the inner mitochondrial membrane. In Mitchell's **chemiosmotic hypothesis,** protons are driven across the membrane from the matrix to the intermembrane space and cytoplasm by the events of electron transport. This mechanism stores the energy of electron transport in an **electrochemical potential.** As protons are driven out of the matrix, the pH rises and the matrix becomes negatively charged with respect to the cytoplasm (Figure 20.22). Proton pumping thus creates a pH gradient and an electrical gradient across the inner membrane, both of which tend to attract protons back into the matrix from the cytoplasm. Flow of protons down this electrochemical gradient, an energetically favorable process, then drives the synthesis of ATP.

Paul Boyer and the Conformational Coupling Model

Another popular model invoked what became known as **conformational coupling.** If the energy of electron transport was not stored in some high-energy intermediate, perhaps it was stored in a **high-energy protein conformation.** Proposed by Paul Boyer, this model suggested that reversible conformation changes transferred energy from proteins of the electron transport chain to the enzymes involved in ATP synthesis. This model was consistent with some of the observations made by others, and it eventually evolved into the **binding change mechanism** (the basis for the model in Figure 20.28). Boyer's model is supported by a variety of binding experiments and is essentially consistent with Mitchell's chemiosmotic hypothesis.

$$\mathrm{NADH + H^+ + FMN + X \longrightarrow NAD^+{-}X + FMNH_2} \quad (20.22)$$
$$\mathrm{NAD^+{-}X + P_i \longrightarrow NAD^+ + X{\sim}P} \quad (20.23)$$
$$\mathrm{X{\sim}P + ADP \longrightarrow X + ATP + H_2O} \quad (20.24)$$

Net rxn:

$$\mathrm{NADH + H^+ + FMN + ADP + P_i \longrightarrow NAD^+ + FMNH_2 + ATP + H_2O} \quad (20.25)$$

20.8 The Thermodynamic View of Chemiosmotic Coupling

Peter Mitchell's chemiosmotic hypothesis revolutionized our thinking about the energy coupling that drives the synthesis of ATP by means of an electrochemical gradient. How much energy is stored in this electrochemical gradient? For the transmembrane flow of protons across the inner membrane (from inside [matrix] to outside), we could write

$$H^+_{in} \longrightarrow H^+_{out} \qquad (20.26)$$

The free energy difference for protons across the inner mitochondrial membrane includes a term for the concentration difference and a term for the electrical potential. This is expressed as

$$\Delta G = RT \ln \frac{[c_2]}{[c_1]} + Z\mathcal{F}\Delta\Psi \qquad (20.27)$$

where c_1 and c_2 are the proton concentrations on the two sides of the membrane, Z is the charge on a proton, $\mathcal{F}$ is Faraday's constant, and $\Delta\Psi$ is the potential difference across the membrane. For the case at hand, this equation becomes

$$\Delta G = RT \ln \frac{[H^+_{out}]}{[H^+_{in}]} + \mathcal{F}\Delta\Psi \qquad (20.28)$$

In terms of the matrix and cytoplasm pH values, the free energy difference is

$$\Delta G = -2.303RT(\text{pH}_{out} - \text{pH}_{in}) + \mathcal{F}\Delta\Psi \qquad (20.29)$$

Reported values for $\Delta\Psi$ and ΔpH vary, but the membrane potential is always found to be positive outside and negative inside, and the pH is always more acidic outside and more basic inside. Taking typical values of $\Delta\Psi = 0.18$ V and ΔpH = 1 unit, the free energy change associated with the movement of one mole of protons from inside to outside is

$$\Delta G = 2.3RT + \mathcal{F}(0.18\text{ V}) \qquad (20.30)$$

With $\mathcal{F} = 96.485$ kJ/V · mol, the value of ΔG at 37°C is

$$\Delta G = 5.9\text{ kJ} + 17.4\text{ kJ} = 23.3\text{ kJ} \qquad (20.31)$$

which is the free energy change for movement of a mole of protons across a typical inner membrane. Note that the free energy terms for *both* the pH difference and the potential difference are unfavorable for the outward transport of protons, with the latter term making the greater contribution. On the other hand, the ΔG for *inward* flow of protons is −23.3 kJ/mol. It is this energy that drives the synthesis of ATP, in accord with Mitchell's model.

20.9 ATP Synthase

The mitochondrial complex that carries out ATP synthesis is called **ATP synthase** or sometimes **F_1F_0–ATPase** (for the reverse reaction it catalyzes). ATP synthase was observed in early electron micrographs of submitochondrial particles (prepared by sonication of inner membrane preparations) as round, 8.5-nm-diameter projections or particles on the inner membrane (Figure 20.23). In micrographs of native mitochondria, the projections appear on the matrix-facing surface of the inner membrane. Mild agitation removes the particles from isolated membrane preparations, and the isolated spherical particles catalyze ATP hydrolysis, the reverse reaction of the ATP synthase. Stripped of these particles, the membranes can still carry out electron transfer but cannot synthesize ATP. In one of the first *reconstitution* experiments with membrane proteins, Efraim Racker showed that adding the particles back to stripped membranes restored electron transfer–dependent ATP synthesis.

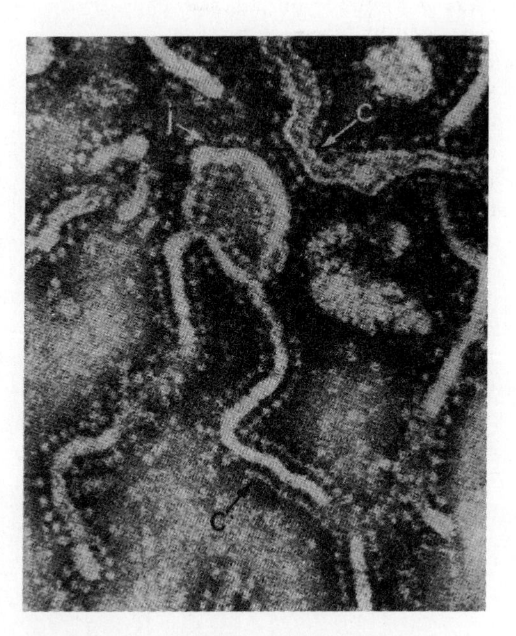

***Figure* 20.23** Electron micrograph of submitochondrial particles showing the 8.5-nm projections or particles on the inner membrane, eventually shown to be F_1 ATP synthase.

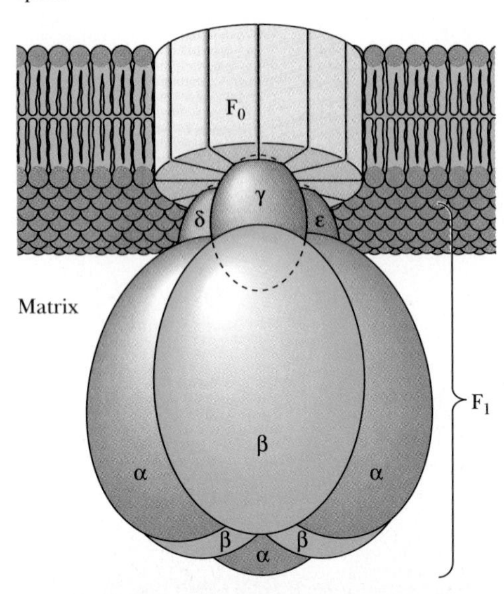

Figure **20.24** A model for the ATP synthase showing the individual component peptides and the relationship between F_0 and F_1.

Table **20.3**

***Escherichia coli* F_1F_0 ATP Synthase Subunit Organization**

Complex	Protein Subunit	Mass (kD)	Stoichiometry
F_1	α	55	3
	β	50	3
	γ	31	1
	δ	19	1
	ϵ	14	1
F_0	a	30	1
	b	17	2
	c	8	12

ATP Synthase Consists of Two Complexes—F_1 and F_0

ATP synthase actually consists of two principal complexes. The spheres observed in electron micrographs make up the **F_1 unit,** which catalyzes ATP synthesis. These F_1 spheres are attached to an integral membrane protein aggregate called the **F_0 unit.** F_1 consists of five polypeptide chains named α, β, γ, δ, and ϵ, with a subunit stoichiometry $\alpha_3\beta_3\gamma\delta\epsilon$ (Table 20.3). The subunits range in mass from 8 kD to 55 kD, and the F_1 unit has a total mass of approximately 360 kD to 380 kD. Minimally, F_0 consists of three hydrophobic subunits denoted by *a*, *b*, and *c* (Table 20.3), with an apparent stoichiometry of $a_1b_2c_{10-12}$. The α and β subunits of F_1 form the catalytic site for ATP synthesis; subunits δ and ϵ control the interaction of F_1 with F_0, and the γ subunit serves as a gate controlling proton access to F_1. F_0 forms the transmembrane pore or channel, through which protons move to F_1 to drive ATP synthesis (Figure 20.24). Polypeptide *c*, with a mass of about 8 kD, has primary responsibility for formation of the channel. The amino acid sequence of polypeptide *c* suggests that the protein possesses two transmembrane helices with a short hairpin loop (a β-turn—see Chapter 5) between them. Acidic amino acid residues in membrane-spanning α-helices of *c* aid in the creation of a proton pathway. For example, if Asp^{61} of polypeptide *c* is replaced by an Asn or Gly through mutation, a proton-impermeable F_0 results.

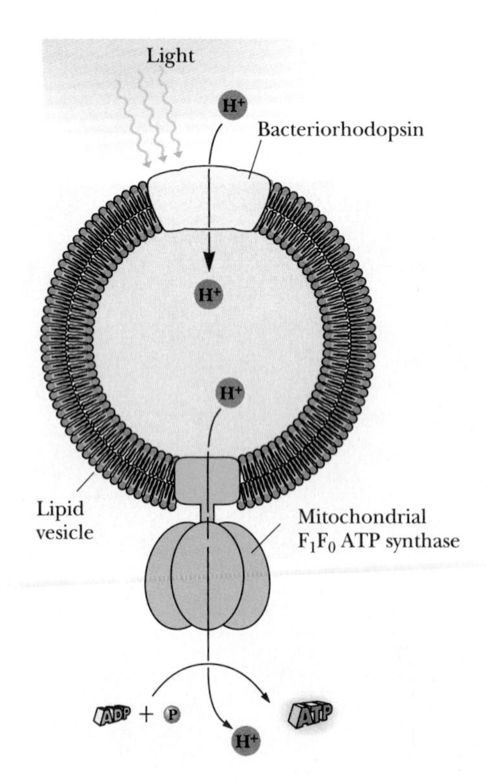

Figure **20.25** The reconstituted vesicles containing ATP synthase and bacteriorhodopsin used by Stoeckenius and Racker to confirm the Mitchell chemiosmotic hypothesis.

Racker and Stoeckenius Confirmed the Mitchell Model in a Reconstitution Experiment

When Mitchell first described his chemiosmotic hypothesis in 1961, little evidence existed to support it, and it was met with considerable skepticism by the scientific community. Eventually, however, considerable evidence accumulated to support this model. It it now clear that the electron transport chain generates a proton gradient, and careful measurements have shown that ATP is synthesized when a pH gradient is applied to mitochondria that cannot carry out electron transport. Even more relevant is a simple but crucial experiment reported in 1974 by Efraim Racker and Walther Stoeckenius, which provided specific confirmation of the Mitchell hypothesis. In this experiment, the bovine mitochondrial *ATP synthase* was reconstituted in simple lipid vesicles with **bacteriorhodopsin,** a light-driven proton pump from *Halobacterium halobium.* As shown in Figure 20.25, upon illumination, bacteriorhodopsin pumped protons into these vesicles, and the resulting proton gradient was sufficient to drive ATP synthesis by the ATP synthase. Since the only two kinds of proteins present were one that produced a proton gradient and one that used such a gradient to make ATP, this experiment essentially verified Mitchell's chemiosmotic hypothesis.

ATP Hydrolysis by ATP Synthase Is a One-Step S_N2 Displacement

The ultimate questions for the ATP synthase are "How does the enzyme synthesize ATP?" and "How is the energy of electron transport used to drive this synthesis?" A partial answer was provided by Webb and Trentham, who showed that hydrolysis of ATP by ATP synthase results in inversion of configuration at the γ-P of ATP. As shown in Figure 20.26, this is consistent with a simple in-line displacement reaction, essentially an S_N2 reaction. In this respect, mitochondrial ATP synthase is quite different from the transport ATPases in other membranes, which form phosphoenzyme intermediates and which carry out the hydrolysis of ATP in two steps.

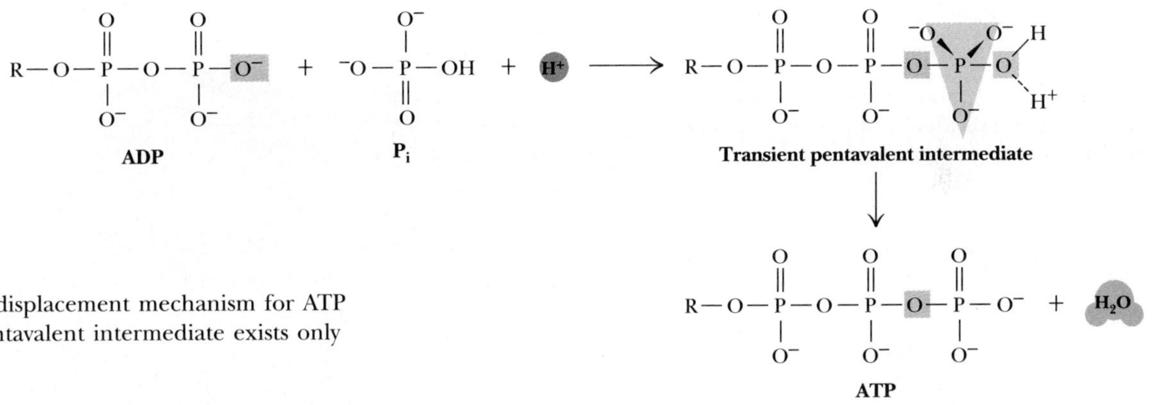

***Figure* 20.26** An S_N2 displacement mechanism for ATP synthesis. Here the pentavalent intermediate exists only transiently ($\sim 10^{-13}$ s).

Boyer's ^{18}O Exchange Experiment Identified the Energy-Requiring Step

The elegant studies by Paul Boyer of ^{18}O exchange in ATP synthase have provided other important insights into the mechanism of the enzyme. Boyer and his colleagues studied the ability of the synthase to incorporate labeled oxygen from $H_2{}^{18}O$ into P_i. This reaction (Figure 20.27) occurs in synthesis of ATP from ADP and P_i, followed by hydrolysis with incorporation of oxygen from the solvent. Although *net production of ATP* requires coupling with a proton gradient, Boyer observed that this *exchange reaction* occurs readily, even in the absence of a proton gradient. His finding indicated that the formation of *enzyme-bound ATP* does not require energy. Indeed, movement of protons through the F_0 channel causes the *release of newly synthesized ATP* from the enzyme. Thus, the energy provided by electron transport appears to drive

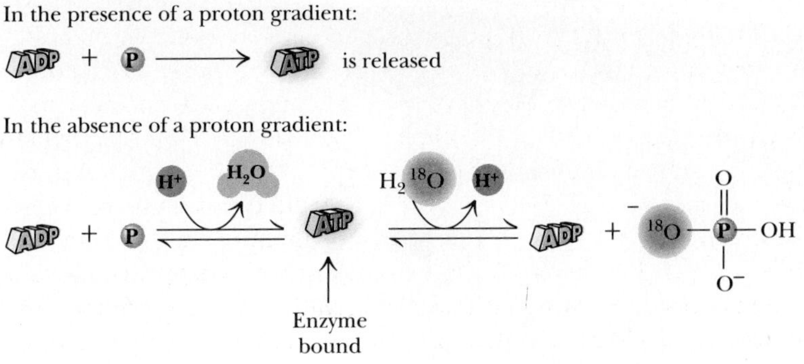

***Figure* 20.27** ATP production in the presence of a proton gradient and ATP/ADP exchange in the absence of a proton gradient. Exchange leads to incorporation of ^{18}O in phosphate, as shown.

enzyme conformational changes that regulate the binding of substrates on ATP synthase, ATP synthesis, and the release of products. The mechanism involves catalytic cooperativity between three interacting sites (Figure 20.28).

20.10 Inhibitors of Oxidative Phosphorylation

The unique properties and actions of an inhibitory substance can often help to identify aspects of an enzyme mechanism. Many details of electron trans-

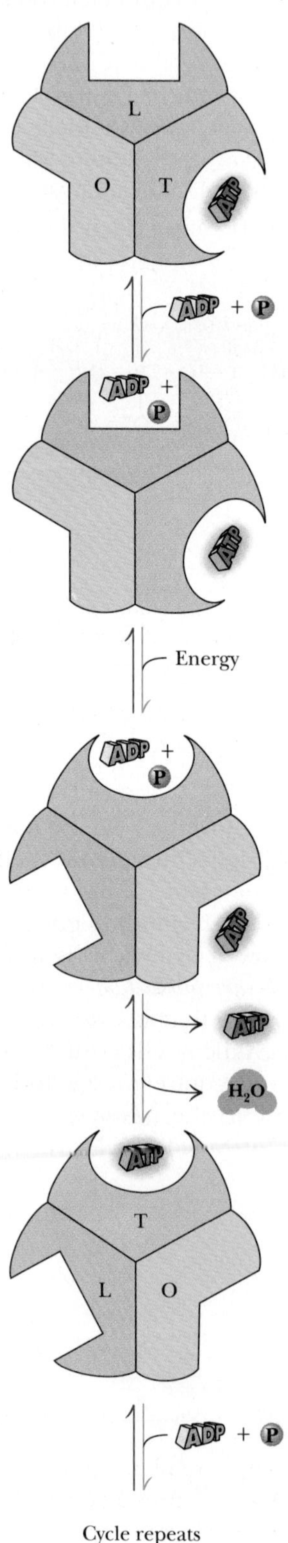

***Figure* 20.28** The binding change mechanism for ATP synthesis by ATP synthase. This model assumes that F_1 has three interacting and conformationally distinct active sites. The open (O) conformation is inactive and has a low affinity for ligands; the L conformation (with "loose" affinity for ligands) is also inactive; the tight (T) conformation is active and has a high affinity for ligands. Synthesis of ATP is initiated (step 1) by binding of ADP and P_i to an L site. In the second step, an energy-driven conformational change converts the L site to the T conformation and also converts T to O and O to L. In the third step, ATP is synthesized at the T site and released from the O site. Two additional passes through this cycle produce two more ATPs and return the enzyme to its original state.

port and oxidative phosphorylation mechanisms have been gained from studying the effects of particular inhibitors. Figure 20.29 presents the structures of some electron transport and oxidative phosphorylation inhibitors. The sites of inhibition by these agents are indicated in Figure 20.30.

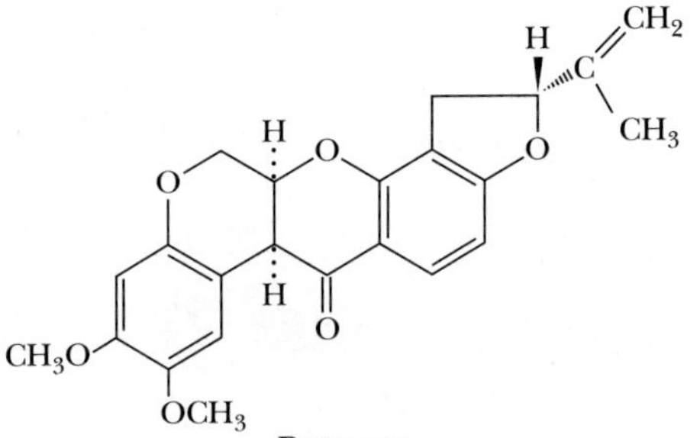

Rotenone 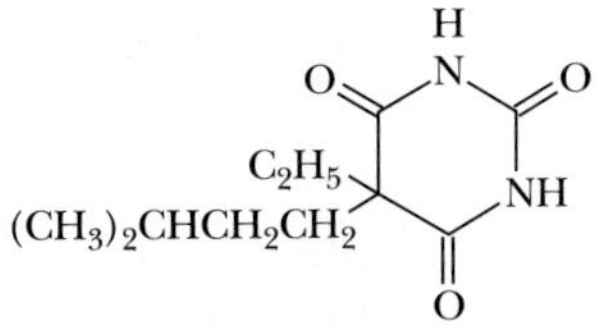 **Amytal (amobarbital)** 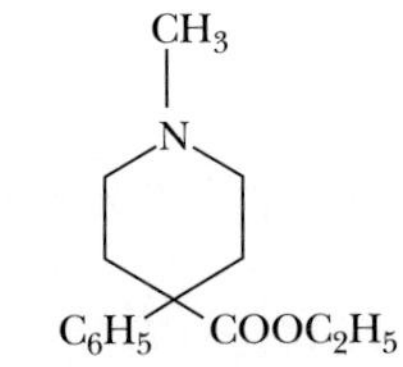 **Demerol (meperdine)** 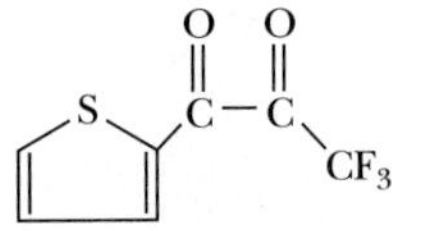 **2-Thenoyltrifluoroacetone** 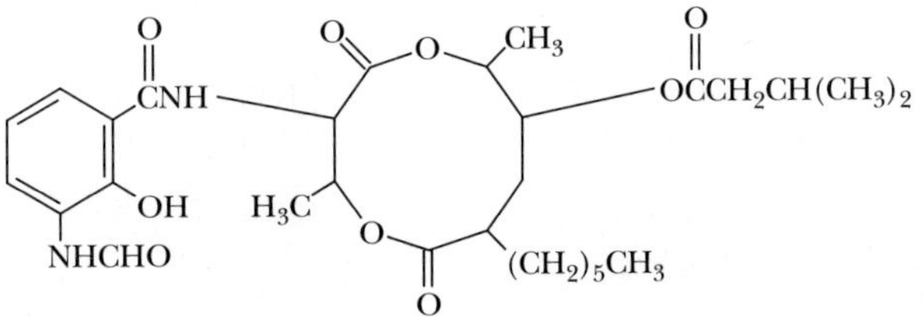 **Antimycin A_1** 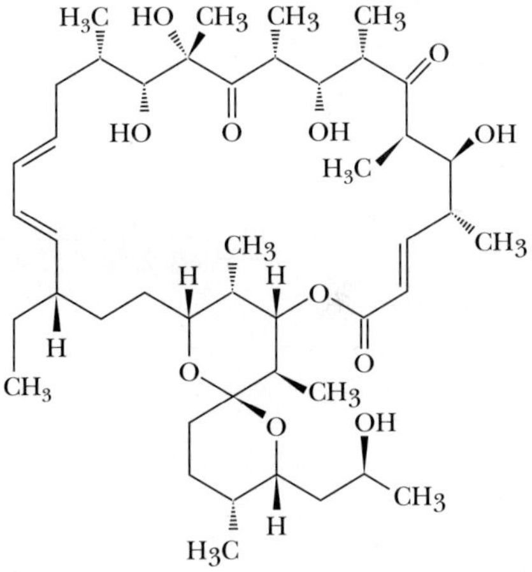 **Oligomycin A** 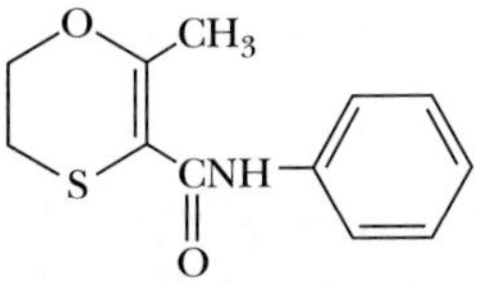 **Carboxin** 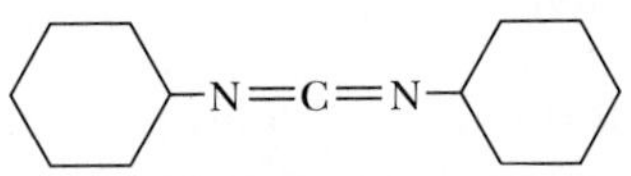 **Dicyclohexylcarbodiimide (DCCD)**

***Figure* 20.29** The structures of several inhibitors of electron transport and oxidative phosphorylation.

Inhibitors of Complexes I, II, and III Block Electron Transport

Rotenone is a common insecticide that strongly inhibits the NADH–UQ reductase. Rotenone is obtained from the roots of several species of plants. Tribes in certain parts of the world have made a practice of beating the roots of trees along riverbanks to release rotenone into the water, where it paralyzes fish and makes them easy prey. Ptericidin, Amytal and other barbiturates, mercurial agents, and the widely prescribed painkiller Demerol also exert their inhibitory actions on this enzyme complex. All these substances appear to inhibit reduction of coenzyme Q and the oxidation of the Fe-S clusters of NADH–UQ reductase.

2-Thenoyltrifluoroacetone and carboxin and its derivatives specifically block Complex II, the succinate–UQ reductase. Antimycin, an antibiotic produced by *Streptomyces griseus,* inhibits the UQ–cytochrome *c* reductase by blocking electron transfer between b_H and coenzyme Q. Myxothiazol inhibits the same complex by acting at the Q_p site.

Cyanide, Azide, and Carbon Monoxide Inhibit Complex IV

Complex IV, the cytochrome *c* oxidase, is specifically inhibited by cyanide (CN^-), azide (N_3^-), and carbon monoxide (CO). Cyanide and azide bind tightly to the ferric form of cytochrome a_3, whereas carbon monoxide binds

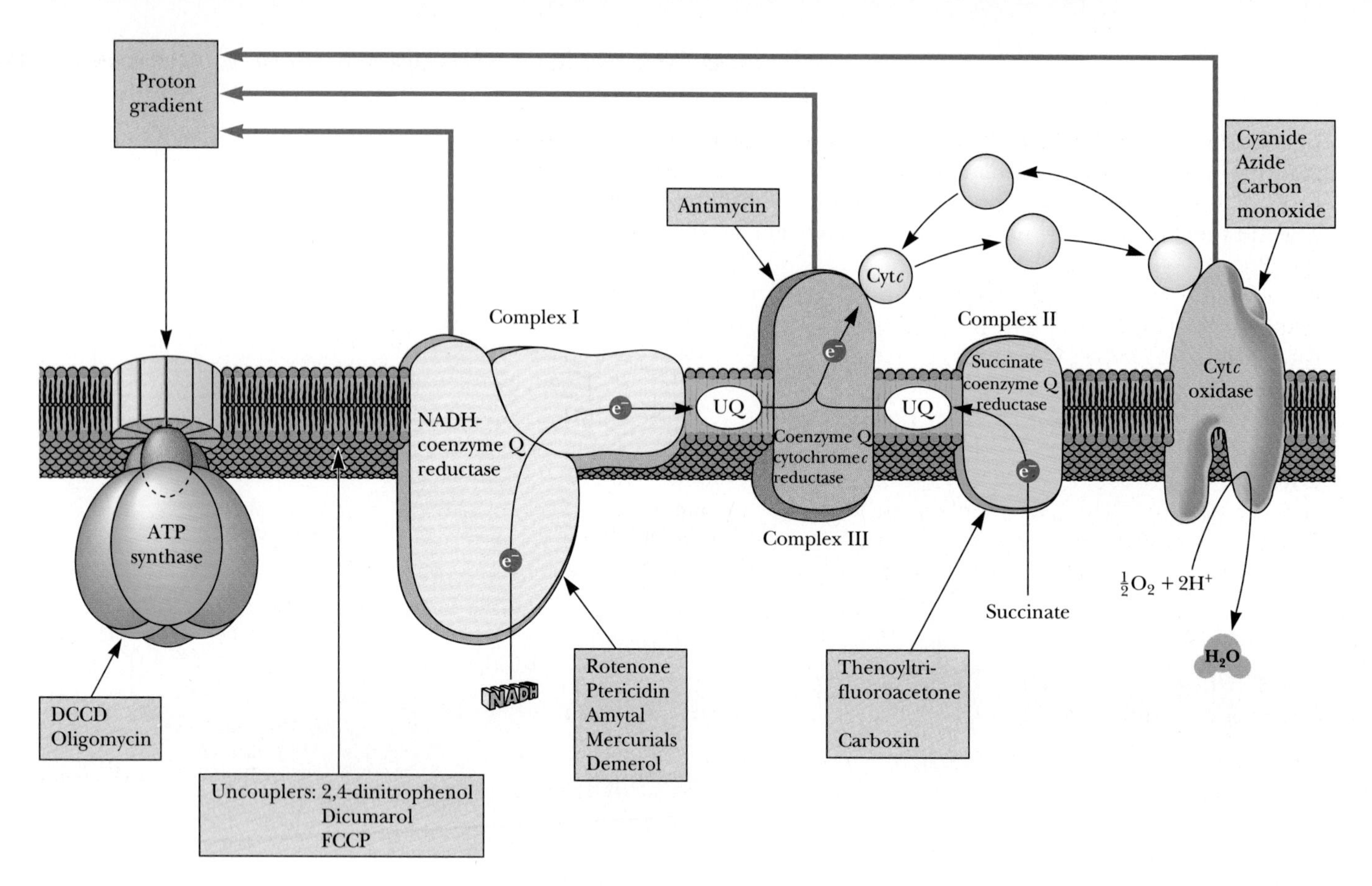

***Figure* 20.30** The sites of action of several inhibitors of electron transport and/or oxidative phosphorylation.

only to the ferrous form. The inhibitory actions of cyanide and azide at this site are very potent, whereas the principal toxicity of carbon monoxide arises from its affinity for the iron of hemoglobin. Herein lies an important distinction between the poisonous effects of cyanide and carbon monoxide. Since animals (including humans) carry many, many hemoglobin molecules, they must inhale a large quantity of carbon monoxide to die from it. These same organisms, however, possess comparatively few molecules of cytochrome a_3. Consequently, a limited exposure to cyanide can be lethal. The sudden action of cyanide attests to the organism's constant and immediate need for the energy supplied by electron transport.

Oligomycin and DCCD Are ATP Synthase Inhibitors

Inhibitors of ATP synthase include dicyclohexylcarbodiimide (DCCD) and oligomycin (Figure 20.29). DCCD bonds covalently to carboxyl groups in hydrophobic domains of proteins in general, and to a glutamic acid residue of the *c* subunit of F_0, the proteolipid forming the proton channel of the ATP synthase, in particular. If the *c* subunit is labeled with DCCD, proton flow through F_0 is blocked and ATP synthase activity is inhibited. Likewise, oligomycin acts directly on the ATP synthase. By binding to a subunit of F_0, oligomycin also blocks the movement of protons through F_0.

20.11 Uncouplers Disrupt the Coupling of Electron Transport and ATP Synthase

Another important class of reagents affects ATP synthesis, but in a manner that does not involve direct binding to any of the proteins of the electron

The Net Yield of ATP from Glucose Oxidation Depends on the Shuttle Used

The complete route for the conversion of the metabolic energy of glucose to ATP has now been described, in Chapters 18 through 20. Assuming appropriate P/O ratios, the number of ATP molecules produced by the complete oxidation of a molecule of glucose can be estimated. Keeping in mind that P/O ratios must be viewed as approximate, for all the reasons previously cited, we will assume the values of 2.5 and 1.5 for the mitochondrial oxidation of NADH and succinate, respectively. In eukaryotic cells, the combined pathways of glycolysis, the TCA cycle, electron transport, and oxidative phosphorylation then yield a net of approximately 30 to 32 molecules of ATP per molecule of glucose oxidized, depending on the shuttle route employed (Table 20.4).

Table **20.4**
Yield of ATP from Glucose Oxidation

Pathway	**ATP Yield per Glucose**	
	Glycerol-Phosphate Shuttle	Malate-Aspartate Shuttle
Glycolysis: glucose to pyruvate (cytosol)		
Phosphorylation of glucose	−1	−1
Phosphorylation of fructose-6-phosphate	−1	−1
Dephosphorylation of 2 molecules of 1,3-BPG	+2	+2
Dephosphorylation of 2 molecules of PEP	+2	+2
Oxidation of 2 molecules of glyceraldehyde-3-phosphate yields 2 NADH		
Pyruvate conversion to acetyl-CoA (mitochondria)		
2 NADH		
Citric acid cycle (mitochondria)		
2 molecules of GTP from 2 molecules of succinyl-CoA	+2	+2
Oxidation of 2 molecules each of isocitrate, α-ketoglutarate, and malate yields 6 NADH		
Oxidation of 2 molecules of succinate yields 2 [$FADH_2$]		
Oxidative phosphorylation (mitochondria)		
2 NADH from glycolysis yield 1.5 ATP each if NADH is oxidized by glycerol-phosphate shuttle; 2.5 ATP by malate-aspartate shuttle	+3	+5
Oxidative decarboxylation of 2 pyruvate to 2 acetyl-CoA: 2 NADH produce 2.5 ATP each	+5	+5
2 [$FADH_2$] from each citric acid cycle produce 1.5 ATP each	+3	+3
6 NADH from citric acid cycle produce 2.5 ATP each	+15	+15
Net Yield	30	32

Note: These P/O ratios of 2.5 and 1.5 for mitochondrial oxidation of NADH and [$FADH_2$] are "consensus values." Since they may not reflect actual values and since these ratios may change depending on metabolic conditions, these estimates of ATP yield from glucose oxidation are approximate.

The net stoichiometric equation for the oxidation of glucose, using the glycerol phosphate shuttle, is

$$\text{Glucose} + 6\ O_2 + \sim 30\ \text{ADP} + \sim 30\ P_i \longrightarrow 6\ CO_2 + \sim 30\ \text{ATP} + \sim 36\ H_2O \quad (20.32)$$

Since the 2 NADH formed in glycolysis are "transported" by the glycerol phosphate shuttle in this case, they each yield only 1.5 ATP, as already described. On the other hand, if these 2 NADH take part in the malate–aspartate shuttle, each yields 2.5 ATP, giving a total (in this case) of 32 ATP formed per glucose oxidized. Most of the ATP—26 out of 30 or 28 out of 32—is produced by oxidative phosphorylation; only 4 ATP molecules result from direct synthesis during glycolysis and the TCA cycle.

The situation in bacteria is somewhat different. Prokaryotic cells need not carry out ATP/ADP exchange. Thus, bacteria have the potential to produce approximately 38 ATP per glucose.

3.5 Billion Years of Evolution Have Resulted in a System That Is 54% Efficient

Hypothetically speaking, how much energy does a eukaryotic cell extract from the glucose molecule? Taking a value of 50 kJ/mol for the hydrolysis of ATP under cellular conditions (Chapter 16), the production of 32 ATP per glucose oxidized yields 1600 kJ/mol of glucose. The cellular oxidation (combustion) of glucose yields $\Delta G = -2937$ kJ/mol. We can calculate an efficiency for the pathways of glycolysis, the TCA cycle, electron transport, and oxidative phosphorylation of

$$\frac{1600}{2937} \times 100\% = 54\%$$

This is the result of approximately 3.5 billion years of evolution.

Problems

1. For the following reaction,

$$[\text{FAD}] + 2\ \text{cyt}\ c\ (Fe^{2+}) + 2\ H^+ \rightleftharpoons [\text{FADH}_2] + 2\ \text{cyt}\ c\ (Fe^{3+})$$

determine which of the redox couples is the electron acceptor and which is the electron donor under standard-state conditions, calculate the value of $\Delta\mathscr{E}_o'$, and determine the free energy change for the reaction.

2. Calculate the value of $\Delta\mathscr{E}_o'$ for the glyceraldehyde-3-phosphate dehydrogenase reaction, and calculate the free energy change for the reaction.

3. For the following redox reaction,

$$\text{NAD}^+ + 2\ H^+ + 2e^- \longrightarrow \text{NADH} + H^+$$

suggest an equation (analogous to Equation 20.13) that predicts the pH dependence of this reaction, and calculate the reduction potential for this reaction at pH 8.

4. Sodium nitrite ($NaNO_2$) is used by emergency medical personnel as an antidote for cyanide poisoning (for this purpose, it must be administered immediately). Based on the discussion of cyanide poisoning in Section 20.10, suggest a mechanism for the life-saving effect of sodium nitrite.

5. A wealthy investor has come to you for advice. She has been approached by a biochemist who seeks financial backing for a company that would market dinitrophenol and dicumarol as weight-loss medications. The biochemist has explained to her that these agents are uncouplers and that they would dissipate metabolic energy as heat. The investor wants to know if you think she should invest in the biochemist's company. How do you respond?

6. Assuming that 3 H^+ are transported per ATP synthesized in the mitochondrial matrix, the membrane potential difference is 0.18 V (negative inside), and the pH difference is 1 unit (acid outside, basic inside), calculate the largest ratio of [ATP]/[ADP][P_i] under which synthesis of ATP can occur.

7. Of the dehydrogenase reactions in glycolysis and the TCA cycle, all but one use NAD^+ as the electron acceptor. The lone exception is the succinate dehydrogenase reaction, which uses covalently bound FAD of a flavoprotein as the electron acceptor. The standard reduction potential for this bound FAD is in the range of 0.003 to 0.091 V (Table 20.1). Compared to the other dehydrogenase reactions of glycolysis and the TCA cycle, what is unique about succinate dehydrogenase? Why is bound FAD a more suitable electron acceptor in this case?

Further Reading

Babcock, G. T., and Wikström, M., 1992. Oxygen activation and the conservation of energy in cell respiration. *Nature* **356:**301–309.

Bonomi, F., Pagani, S., Cerletti, P., and Giori, C., 1983. Modification of the thermodynamic properties of the electron-transferring groups in mitochondrial succinate dehydrogenase upon binding of succinate. *European Journal of Biochemistry* **134:**439–445.

Boyer, Paul D., 1989. A perspective of the binding change mechanism for ATP synthesis. *The FASEB Journal* **3:**2164–2178.

Boyer, P., et al., 1977. Oxidative phosphorylation and photophosphorylation. *Annual Review of Biochemistry* **46:**955–966.

Dickerson, R. E., 1980. Cytochrome *c* and the evolution of energy metabolism. *Scientific American* **242**(3):137–153.

Ernster, L., ed., 1980. *Bioenergetics.* New York: Elsevier Press.

Fillingame, R. H., 1980. The proton-translocating pump of oxidative phosphorylation. *Annual Review of Biochemistry* **49:**1079–1113.

Futai, M., Noumi, T., and Maeda, M., 1989. ATP synthase: results by combined biochemical and molecular biological approaches. *Annual Review of Biochemistry* **58:**111–136.

Harold, F. M., 1986. *The Vital Force: A Study of Bioenergetics.* New York: W. H. Freeman and Company.

Mitchell, P., 1979. Keilin's respiratory chain concept and its chemiosmotic consequences. *Science* **206:**1148–1159.

Mitchell, P., and Moyle, J., 1965. Stoichiometry of proton translocation through the respiratory chain and adenosine triphosphatase systems of rat mitochondria. *Nature* **208:**147–151.

Moser, C. C., et al., 1992. Nature of biological electron transfer. *Nature* **355:**796–802.

Naqui, A., Chance, B., and Cadenas, E., 1986. Reactive oxygen intermediates in biochemistry. *Annual Review of Biochemistry* **55:**137.

Nicholls, David G., and Ferguson, Stuart J., 1992. *Bioenergetics 2.* London: Academic Press.

Nicholls, D. G., and Rial, E., 1984. Brown fat mitochondria. *Trends in Biochemical Sciences* **9:**489–491.

Pedersen, P., and Carafoli, E., 1987. Ion-motive ATPases. I. Ubiquity, properties and significance to cell function. *Trends in Biochemical Sciences* **12:**146–150.

Slater, E. C., 1983. The Q cycle: An ubiquitous mechanism of electron transfer. *Trends in Biochemical Sciences* **8:**239–242.

Trumpower, B. L., 1990. Cytochrome *bc1* complexes of microorganisms. *Microbiological Reviews* **54:**101–129.

Trumpower, B. L., 1990. The protonmotive Q cycle—Energy transduction by coupling of proton translocation to electron transfer by the cytochrome bc_1 complex. *Journal of Biological Chemistry* **265:**11409–11412.

Vignais, P. V., and Lunardi, J., 1985. Chemical probes of mitochondrial ATP synthesis and translocation. *Annual Review of Biochemistry* **54:**977–1014.

von Jagow, G., 1980. *b*-Type cytochromes. *Annual Review of Biochemistry* **49:**281–314.

Walker, J. E., 1992. The NADH:ubiquinone oxidoreductase (Complex I) of respiratory chains. *Quarterly Reviews of Biophysics* **25:**253–324.

Weiss, H., Friedrich, T., Hofhaus, G., and Preis, D., 1991. The respiratory-chain NADH dehydrogenase (complex I) of mitochondria. *European Journal of Biochemistry* **197:**563–576.

"Con pan y vino se anda el camino."
(With bread and wine you can walk your road.)
Spanish proverb

Chapter 21

Gluconeogenesis, Glycogen Metabolism, and the Pentose Phosphate Pathway

Bread and pastries on a rack at a French bakery, Paris. Carbohydrates such as these provide a significant portion of human caloric intake.

Outline

21.1 Gluconeogenesis
21.2 Regulation of Gluconeogenesis
21.3 Glycogen Catabolism
21.4 Glycogen Synthesis
21.5 Control of Glycogen Metabolism
21.6 The Pentose Phosphate Pathway

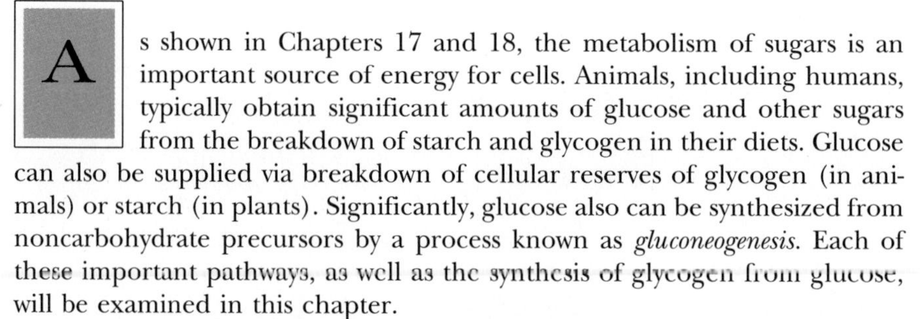

As shown in Chapters 17 and 18, the metabolism of sugars is an important source of energy for cells. Animals, including humans, typically obtain significant amounts of glucose and other sugars from the breakdown of starch and glycogen in their diets. Glucose can also be supplied via breakdown of cellular reserves of glycogen (in animals) or starch (in plants). Significantly, glucose also can be synthesized from noncarbohydrate precursors by a process known as *gluconeogenesis.* Each of these important pathways, as well as the synthesis of glycogen from glucose, will be examined in this chapter.

Another pathway of glucose catabolism, the *pentose phosphate pathway,* is the primary source of NADPH, the reduced coenzyme essential to most reductive biosynthetic processes. For example, NADPH is crucial to the biosynthesis of fatty acids (Chapter 24) and amino acids (Chapter 26). The pentose phosphate pathway also results in the production of ribose-5-phosphate, an important component of ATP, NAD^+, FAD, coenzyme A, and particularly DNA and RNA. This important pathway will also be considered in this chapter.

21.1 Gluconeogenesis

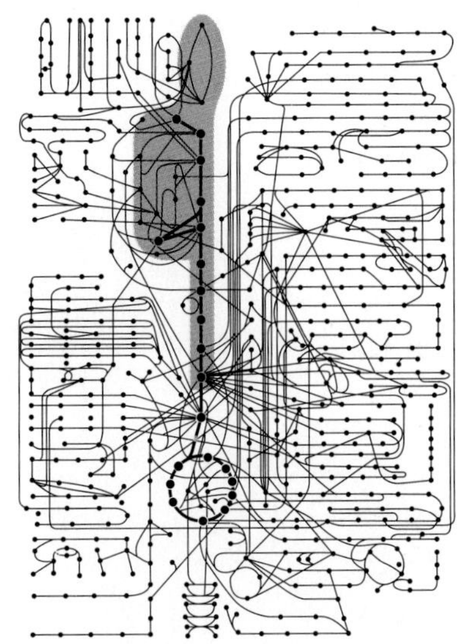

Gluconeogenesis, Glycogen Metabolism, and the Pentose Phosphate Pathway

The ability to synthesize glucose from common metabolites is very important to most organisms. Human metabolism, for example, consumes about 160 ± 20 grams of glucose per day, about 75% of this in the brain. Body fluids carry only about 20 grams of free glucose, and glycogen stores normally can provide only about 180 to 200 grams of free glucose. Thus, the body carries only a little more than a one-day supply of glucose. If glucose is not obtained in the diet, the body must produce new glucose from noncarbohydrate precursors. The term for this activity is **gluconeogenesis,** which means the generation (*genesis*) of new (*neo*) glucose.

Further, muscles consume large amounts of glucose via glycolysis, producing large amounts of pyruvate. In vigorous exercise, muscle cells become anaerobic and pyruvate is converted to lactate. Gluconeogenesis salvages this pyruvate and lactate and reconverts it to glucose.

The Substrates of Gluconeogenesis

In addition to pyruvate and lactate, other noncarbohydrate precursors can be used as substrates for gluconeogenesis in animals. These include most of the amino acids, as well as glycerol and all the TCA cycle intermediates. On the other hand, fatty acids are not substrates for gluconeogenesis in animals, because most fatty acids yield only acetyl-CoA upon degradation, and animals cannot carry out net synthesis of sugars from acetyl-CoA. Lysine and leucine are the only amino acids that are not substrates for gluconeogenesis. These amino acids produce only acetyl-CoA upon degradation.

Nearly All Gluconeogenesis Occurs in the Liver and Kidneys in Animals

Interestingly, the mammalian organs that consume the most glucose, namely, brain and muscle, carry out very little glucose synthesis. The major sites of gluconeogenesis are the liver and kidneys, which account for about 90% and 10% of the body's gluconeogenic activity, respectively. Glucose produced by gluconeogenesis in the liver and kidney is released into the blood and is subsequently absorbed by brain, heart, muscle, and red blood cells to meet their metabolic needs. In turn, pyruvate and lactate produced in these tissues are returned to the liver and kidney to be used as gluconeogenic substrates.

Gluconeogenesis Is Not Merely the Reverse of Glycolysis

In some ways, gluconeogenesis is the reverse, or antithesis, of glycolysis. Glucose is synthesized, not catabolized; ATP is consumed, not produced; and NADH is oxidized to NAD^+, rather than the other way around. However, gluconeogenesis cannot be *merely* the reversal of glycolysis, for two reasons. First, glycolysis is exergonic, with a $\Delta G^{\circ\prime}$ of approximately −74 kJ/mol. If gluconeogenesis were merely the reverse, it would be a strongly endergonic process and could not occur spontaneously. Somehow the energetics of the process must be augmented so that gluconeogenesis can proceed spontaneously. Second, the processes of glycolysis and gluconeogenesis must be regulated in a reciprocal fashion so that when glycolysis is active, gluconeogenesis is inhibited, and when gluconeogenesis is proceeding, glycolysis is turned off. Both of these limitations are overcome by having unique reactions within the routes of glycolysis and gluconeogenesis, rather than a completely shared pathway.

Gluconeogenesis—Something Borrowed, Something New

The complete route of gluconeogenesis is shown in Figure 21.1, side by side with the glycolytic pathway. Gluconeogenesis employs three different reactions, catalyzed by three different enzymes, for the three steps of glycolysis that are highly exergonic (and highly regulated). In essence, seven of the ten steps of glycolysis are merely reversed in gluconeogenesis. The six reactions between fructose-1,6-bisphosphate and PEP are shared by the two pathways, as is the isomerization of glucose-6-P to fructose-6-P. The three exergonic regulated reactions—the hexokinase (glucokinase), phosphofructokinase, and pyruvate kinase reactions—are replaced by alternative reactions in the gluconeogenic pathway.

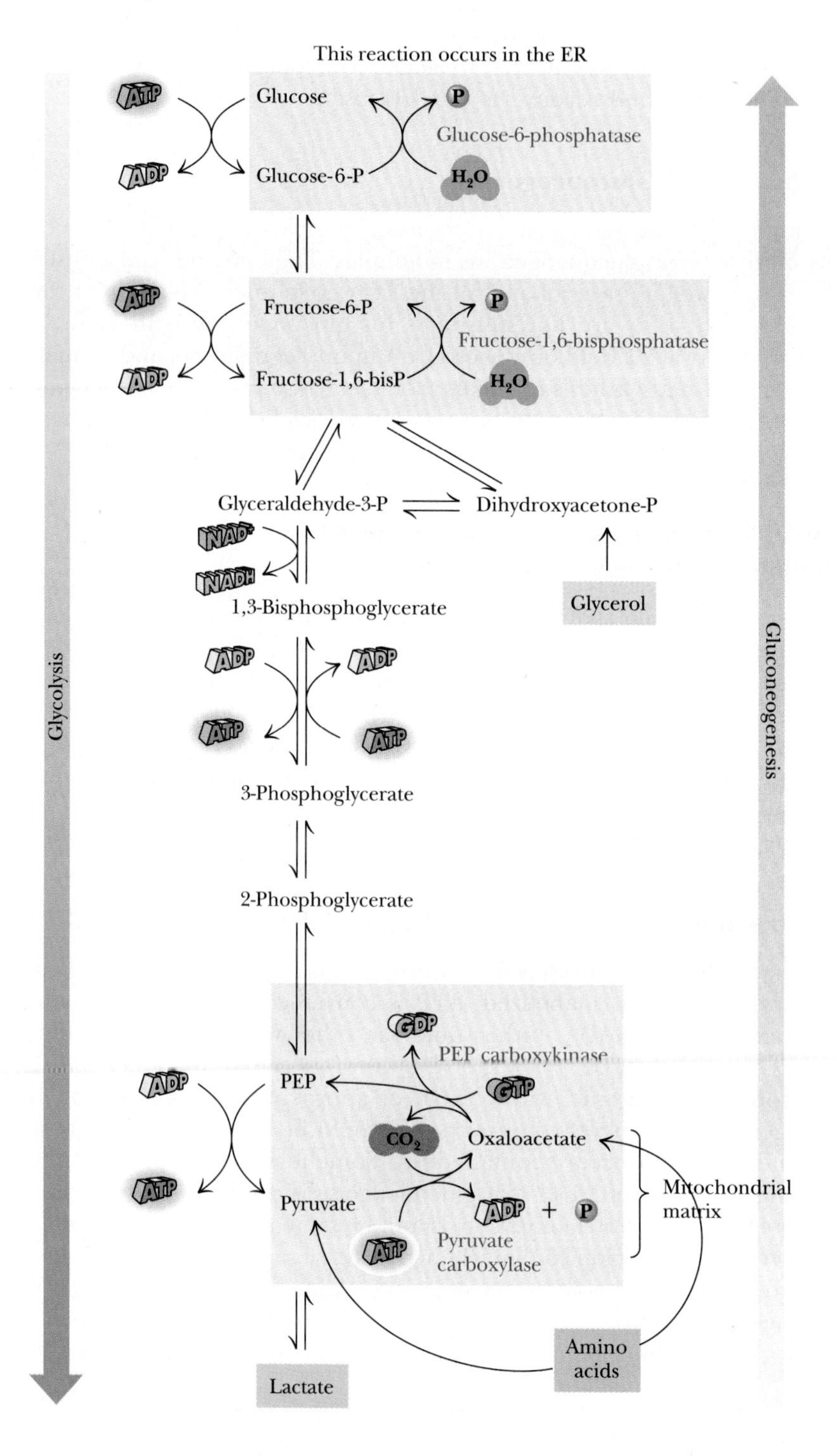

***Figure* 21.1** The pathways of gluconeogenesis and glycolysis. Species in blue, green, and peach-colored shaded boxes indicate other entry points for gluconeogenesis (in addition to pyruvate).

$$CH_3C(=O)-COO^- + HCO_3^- + ATP \longrightarrow {}^-OOC\,CH_2C(=O)-COO^- + ADP + P$$

Pyruvate **Bicarbonate** **Oxaloacetate**

***Figure* 21.2** The pyruvate carboxylase reaction.

The conversion of pyruvate to PEP that initiates gluconeogenesis is accomplished by two unique reactions. *Pyruvate carboxylase* catalyzes the first, converting pyruvate to oxaloacetate. Then, *PEP carboxykinase* catalyzes the conversion of oxaloacetate to PEP. Conversion of fructose-1,6-bisphosphate to fructose-6-phosphate is catalyzed by a specific phosphatase, *fructose-1,6-bisphosphatase.* The final step to produce glucose, hydrolysis of glucose-6-phosphate, is mediated by *glucose-6-phosphatase.* Each of these steps is considered in detail in the following paragraphs. The overall conversion of pyruvate to PEP by pyruvate carboxylase and PEP carboxykinase has a $\Delta G^{\circ\prime}$ close to zero but is pulled along by subsequent reactions. The conversion of fructose-1,6-bisphosphate to glucose in the last three steps of gluconeogenesis is strongly exergonic with a $\Delta G^{\circ\prime}$ of about -30.5 kJ/mol. This sequence of two phosphatase reactions separated by an isomerization accounts for most of the free energy release that makes the gluconeogenesis pathway spontaneous.

The Unique Reactions of Gluconeogenesis

(1) Pyruvate Carboxylase—A Biotin-Dependent Enzyme

Initiation of gluconeogenesis occurs in the **pyruvate carboxylase reaction**—the conversion of pyruvate to oxaloacetate (Figure 21.2). The reaction takes place in two discrete steps, involves ATP and bicarbonate as substrates, and utilizes biotin as a coenzyme and acetyl-coenzyme A as an allosteric activator. Pyruvate carboxylase is a tetrameric enzyme (with a molecular mass of about 500 kD). Each monomer possesses a biotin covalently linked to the ϵ-amino group of a lysine residue at the active site (Figure 21.3). The first step of the reaction involves nucleophilic attack of a bicarbonate oxygen at the γ-P of ATP to form **carbonylphosphate,** an activated form of CO_2, and ADP (Figure

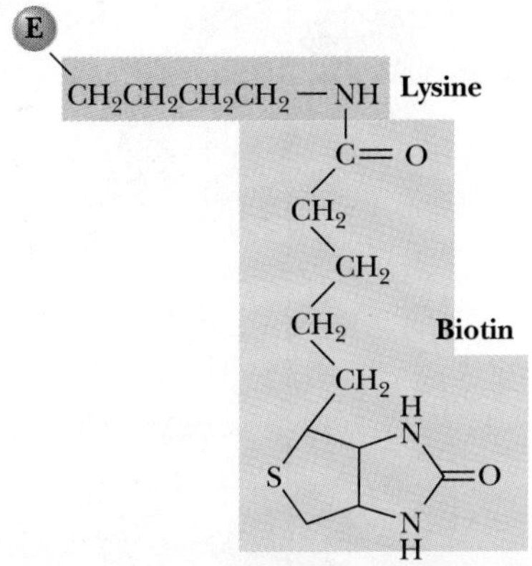

E
$CH_2CH_2CH_2CH_2-NH$ **Lysine**
C=O
CH_2
CH_2
CH_2 **Biotin**
CH_2
S H N =O N H

***Figure* 21.3** Covalent linkage of biotin to an active-site lysine in pyruvate carboxylase.

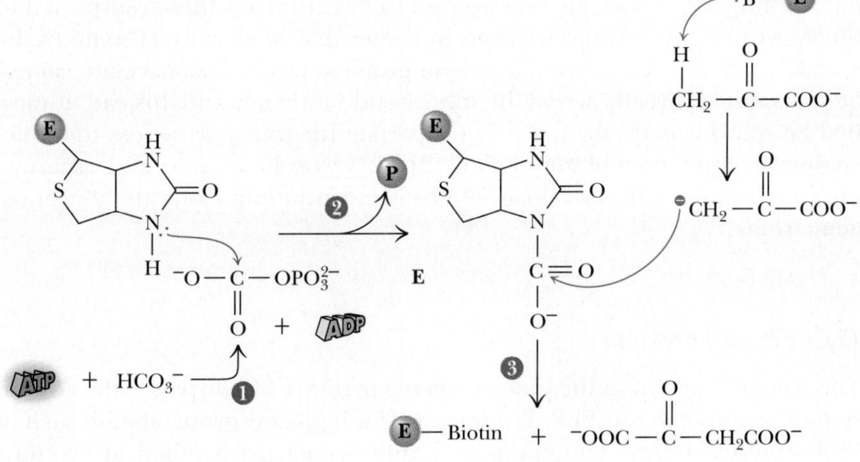

E S H N =O N H
$^-O-C-OPO_3^{2-}$
O + ADP
ATP + HCO_3^- ❶
❷ P
E
E S H N =O N C=O O$^-$
:B—E
H O
$CH_2-C-COO^-$
O
$CH_2-C-COO^-$
❸
E—Biotin + $^-OOC-C(=O)-CH_2COO^-$

***Figure* 21.4** A mechanism for the pyruvate carboxylase reaction.

21.4). Reaction of carboxyphosphate with biotin occurs rapidly to form N-carboxybiotin, liberating inorganic phosphate. The third step involves abstraction of a proton from the C-3 of pyruvate, forming a carbanion, which can attack the carbon of N-carboxybiotin to form oxaloacetate.

Pyruvate Carboxylase Is Allosterically Activated by Acyl-Coenzyme A Two particularly interesting aspects of the pyruvate carboxylase reaction are (a) allosteric activation of the enzyme by acyl-coenzyme A derivatives and (b) compartmentation of the reaction in the mitochondrial matrix. The carboxylation of biotin requires the presence (at an allosteric site) of acetyl-coenzyme A or other acylated coenzyme A derivatives. The second half of the carboxylase reaction—the attack by pyruvate to form oxaloacetate—is not affected by CoA derivatives.

Activation of pyruvate carboxylase by acetyl-CoA provides an important physiological regulation. Acetyl-CoA is the primary substrate for the TCA cycle, and oxaloacetate (formed by pyruvate carboxylase) is an important intermediate in both the TCA cycle and the gluconeogenesis pathway. If levels of ATP and/or acetyl-CoA (or other acyl-CoAs) are low, pyruvate is directed primarily into the TCA cycle, which eventually promotes the synthesis of ATP. If ATP and acetyl-CoA levels are high, pyruvate is converted to oxaloacetate and consumed in gluconeogenesis. Clearly, high levels of ATP and CoA derivatives are signs that energy is abundant and that metabolites will be converted to glucose (and perhaps even glycogen). If the energy status of the cell is low (in terms of ATP and CoA derivatives), pyruvate is consumed in the TCA cycle. Also, as noted in Chapter 19, pyruvate carboxylase is an important anaplerotic enzyme. Its activation by acetyl-CoA leads to oxaloacetate formation, replenishing the level of TCA cycle intermediates.

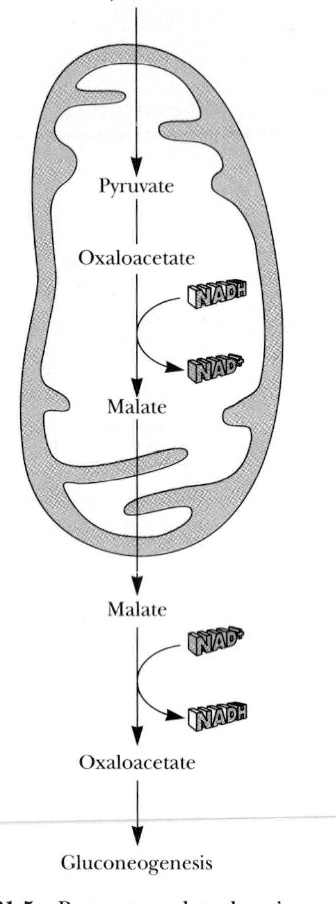

***Figure* 21.5** Pyruvate carboxylase is a compartmentalized reaction. Pyruvate is converted to oxaloacetate in the mitochondria. Since oxaloacetate cannot be transported across the mitochondrial membrane, it must be reduced to malate, transported to the cytosol, and then oxidized back to oxaloacetate before gluconeogenesis can continue.

Compartmentalized Pyruvate Carboxylase Depends on Metabolite Conversion and Transport The second interesting feature of pyruvate carboxylase is that it is found only in the *matrix* of the mitochondria. By contrast, the next enzyme in the gluconeogenic pathway, PEP carboxykinase, may be localized in the cytosol or in the mitochondria or both. For example, rabbit liver PEP carboxykinase is predominantly mitochondrial, whereas the rat liver enzyme is strictly cytosolic. In human liver, PEP carboxykinase is found both in the cytosol and in the mitochondria. Pyruvate is transported into the mitochondrial matrix, where it can be converted to acetyl-CoA (for use in the TCA cycle) or to citrate (for use in fatty acid synthesis; see Figure 21.5). Alternatively, it may be converted to oxaloacetate by pyruvate carboxylase and used in gluconeogenesis. In tissues where PEP carboxykinase is found only in the mitochondria, oxaloacetate is converted to PEP, which is then transported to the cytosol (Figure 21.5). However, in tissues that must convert some oxaloacetate to PEP in the cytosol, a unique problem arises. Oxaloacetate cannot be transported directly across the mitochondrial membrane. Instead, it must first be transformed into malate or aspartate for transport across the mitochondrial inner membrane (Figure 21.5). Cytosolic malate and aspartate must be reconverted to oxaloacetate before continuing along the gluconeogenic route.

(2) PEP Carboxykinase

The second reaction in the gluconeogenic pyruvate-PEP bypass is the conversion of oxaloacetate to PEP. Production of a high-energy metabolite such as PEP requires energy. The energetic requirements are handled in two ways here. First, the CO_2 added to pyruvate in the pyruvate carboxylase step is

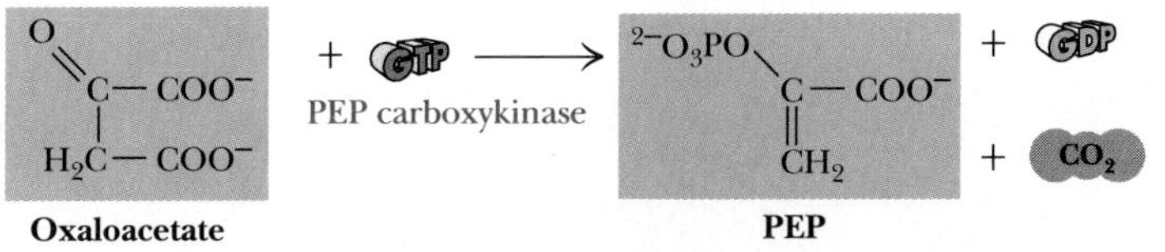

Figure **21.6** The PEP carboxykinase reaction. GTP formed in this reaction can be converted to ATP by nucleoside diphosphate kinase, though liver cells in some species may not contain this enzyme.

removed in the PEP carboxykinase reaction. Decarboxylation is a favorable process and helps to drive the formation of the very high energy enol phosphate in PEP. This decarboxylation drives a reaction that would otherwise be highly endergonic. Note the inherent metabolic logic in this pair of reactions: pyruvate carboxylase consumed an ATP to drive a carboxylation, so that the PEP carboxykinase could use the decarboxylation to facilitate formation of PEP. Second, as shown in Figure 21.6, another high-energy phosphate is consumed by the carboxykinase. Mammals and several other species use GTP in this reaction, rather than ATP. The use of GTP here is equivalent to the consumption of an ATP, due to the activity of the *nucleoside diphosphate kinase* (see Figure 19.4). The substantial free energy of hydrolysis of GTP is crucial to the synthesis of PEP in this step. The overall ΔG for the pyruvate carboxylase and PEP carboxykinase reactions under physiological conditions in the liver is -22.6 kJ/mol). Once PEP is formed in this way, phosphoglycerate mutase, phosphoglycerate kinase, glyceraldehyde-3-P dehydrogenase, aldolase, and triose phosphate isomerase reactions act to eventually form fructose-1,6-bisphosphate, as in Figure 21.1.

(3) Fructose-1,6-Bisphosphatase

The hydrolysis of fructose-1,6-bisphosphate to fructose-6-phosphate (Figure 21.7), like all phosphate ester hydrolyses, is a thermodynamically favorable (exergonic) reaction under standard-state conditions ($\Delta G°' = -16.7$ kJ/mol). Under physiological conditions in the liver, the reaction is also exergonic ($\Delta G = -8.6$ kJ/mol). Fructose-1,6-bisphosphatase is an allosterically regulated enzyme. Citrate stimulates bisphosphatase activity, but *fructose-2,6-bisphosphate* is a potent allosteric inhibitor. AMP also inhibits the bisphosphatase; the inhibition by AMP is enhanced by fructose-2,6-bisphosphate.

(4) Glucose-6-Phosphatase

The final step in the gluconeogenesis pathway is the conversion of glucose-6-phosphate to glucose by the action of *glucose-6-phosphatase.* This enzyme is present in the membranes of the endoplasmic reticulum of liver and kidney cells, but is absent in muscle and brain. For this reason, gluconeogenesis is not carried out in muscle and brain. Its membrane association is important to its function, since (Figure 21.8) the substrate is hydrolyzed as it passes into the

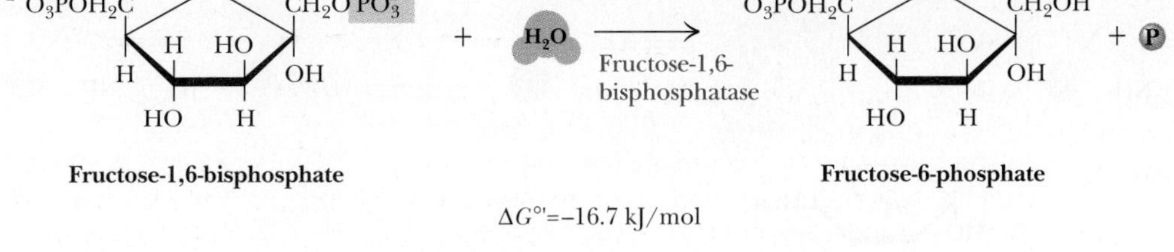

Figure **21.7** The fructose-1,6-bisphosphatase reaction.

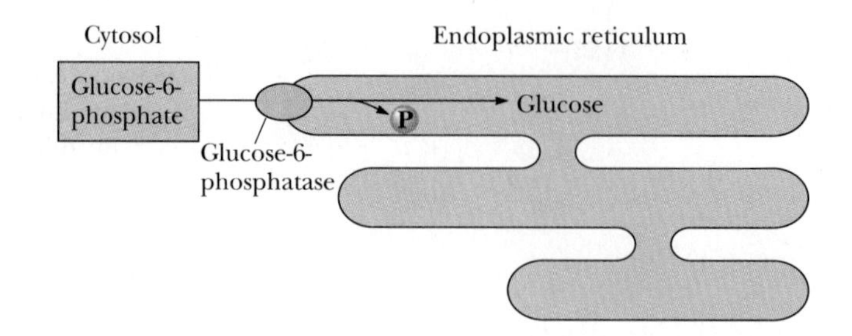

***Figure* 21.8** Glucose-6-phosphatase is localized in the endoplasmic reticulum membrane. Conversion of glucose-6-phosphate to glucose occurs during transport into the ER.

endoplasmic reticulum itself. Vesicles form from the endoplasmic reticulum membrane and diffuse to the plasma membrane and fuse with it, releasing their glucose contents into the bloodstream. The glucose-6-phosphatase reaction involves a phosphorylated enzyme intermediate, which may be a phosphohistidine (Figure 21.9). The ΔG in liver is -5.1 kJ/mol.

Coupling with Hydrolysis of ATP and GTP Drives Gluconeogenesis The net reaction for the conversion of pyruvate to glucose in gluconeogenesis is

$$2\ \text{Pyruvate} + 4\ \text{ATP} + 2\ \text{GTP} + 2\ \text{NADH} + 2\ \text{H}^+ + 6\ \text{H}_2\text{O}$$
$$\downarrow$$
$$\text{glucose} + 4\ \text{ADP} + 2\ \text{GDP} + 6\ \text{P}_\text{i} + 2\ \text{NAD}^+$$

The net free energy change, $\Delta G°'$, for this conversion is -37.7 kJ/mol. The consumption of a total of six nucleoside triphosphates drives this process forward. If glycolysis were merely reversed to achieve the net synthesis of glucose from pyruvate, the net reaction would be

$$2\ \text{Pyruvate} + 2\ \text{ATP} + 2\ \text{NADH} + 2\ \text{H}^+ + 2\ \text{H}_2\text{O}$$
$$\downarrow$$
$$\text{glucose} + 2\ \text{ADP} + 2\ \text{P}_\text{i} + 2\ \text{NAD}^+$$

and the overall $\Delta G°'$ would be about $+74$ kJ/mol. Such a process would be highly endergonic, and therefore thermodynamically unfeasible. Hydrolysis of four additional high-energy phosphate bonds makes gluconeogenesis thermodynamically favorable. Under physiological conditions, however, gluconeogenesis is somewhat less favorable than at standard state, with an overall ΔG of -15.6 kJ/mol for the conversion of pyruvate to glucose.

Lactate Formed in Muscles Is Recycled to Glucose in the Liver A final point on the redistribution of lactate and glucose in the body serves to emphasize the metabolic interactions between organs. Vigorous exercise can lead to oxygen shortage (anaerobic conditions), and energy requirements must be met by increased levels of glycolysis. Under such conditions, glycolysis converts NAD^+ to NADH, yet O_2 is unavailable for regeneration of NAD^+ via cellular respiration. Instead, large amounts of NADH are reoxidized by the reduction of pyruvate to lactate. The lactate thus produced can be transported from

***Figure* 21.9** The glucose-6-phosphatase reaction involves formation of a phosphohistidine intermediate.

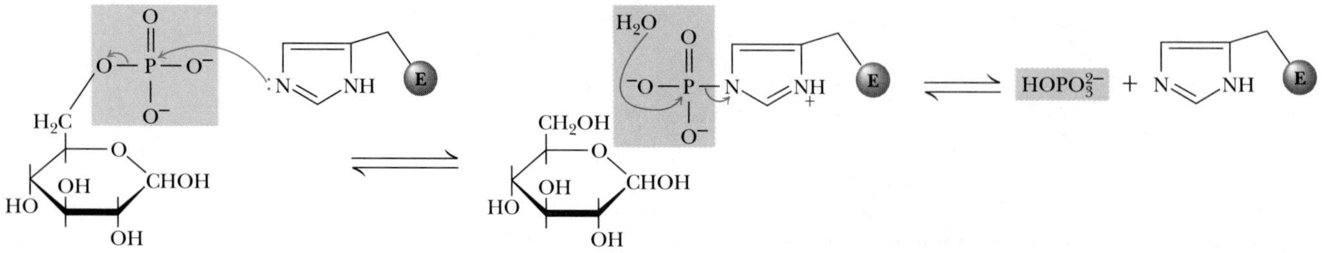

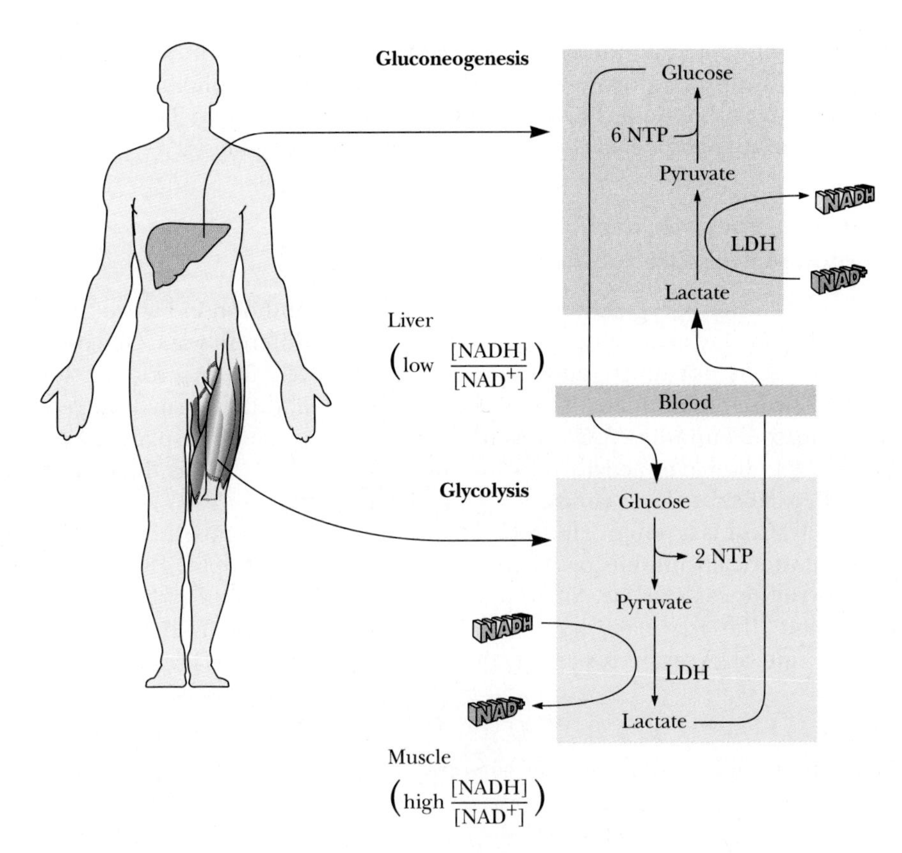

***Figure* 21.10** The Cori cycle.

muscle to the liver, where it is reoxidized by liver lactate dehydrogenase to yield pyruvate, which is converted eventually to glucose. In this way, the liver shares in the metabolic stress created by vigorous exercise. It exports glucose to muscle, which produces lactate, which can be processed by the liver into new glucose. This is referred to as the **Cori cycle** (Figure 21.10). Liver, with a typically high NAD^+/NADH ratio (about 700), readily produces more glucose than it can use. Muscle that is vigorously exercising will enter anaerobiosis and show a decreasing NAD^+/NADH ratio, which favors reduction of pyruvate to lactate.

21.2 Regulation of Gluconeogenesis

Nearly all of the reactions of glycolysis and gluconeogenesis take place in the cytosol. If metabolic control were not exerted over these reactions, glycolytic degradation of glucose and gluconeogenic synthesis of glucose could operate simultaneously, with no net benefit to the cell and with considerable consumption of ATP. This is prevented by a sophisticated system of **reciprocal control,** so that glycolysis is inhibited when gluconeogenesis is active, and vice versa. Reciprocal regulation of these two pathways depends largely on the energy status of the cell. When the energy status of the cell is low, glucose is rapidly degraded to produce needed energy. When the energy status is high, pyruvate and other metabolites are utilized for synthesis (and storage) of glucose.

In glycolysis, the three regulated enzymes are those catalyzing the strongly exergonic reactions: hexokinase (glucokinase), phosphofructokinase, and pyruvate kinase. As noted, the gluconeogenic pathway replaces these three reactions with corresponding reactions which are exergonic in the direction of glucose synthesis: glucose-6-phosphatase, fructose-1,6-bisphosphatase, and the pyruvate carboxylase–PEP carboxykinase pair, respectively. These are the three most appropriate sites of regulation in gluconeogenesis.

Gluconeogenesis Is Regulated by Allosteric and Substrate-Level Control Mechanisms

The mechanisms of regulation of gluconeogenesis are shown in Figure 21.11. Control is exerted at all of the predicted sites, but in different ways. Glucose-6-phosphatase is not under allosteric control. However, the K_m for the substrate, glucose-6-phosphate, is considerably higher than the normal range of substrate concentrations. As a result, glucose-6-phosphatase displays a near-linear dependence of activity on substrate concentrations and is thus said to be under **substrate-level control** by glucose-6-phosphate.

Acetyl-CoA is a potent allosteric effector of glycolysis and gluconeogenesis. It allosterically inhibits pyruvate kinase (as noted in Chapter 18) and activates pyruvate carboxylase. Since it also allosterically inhibits pyruvate dehydrogenase (the enzymatic link between glycolysis and the TCA cycle), the cellular fate of pyruvate is strongly dependent on acetyl-CoA levels. A rise in

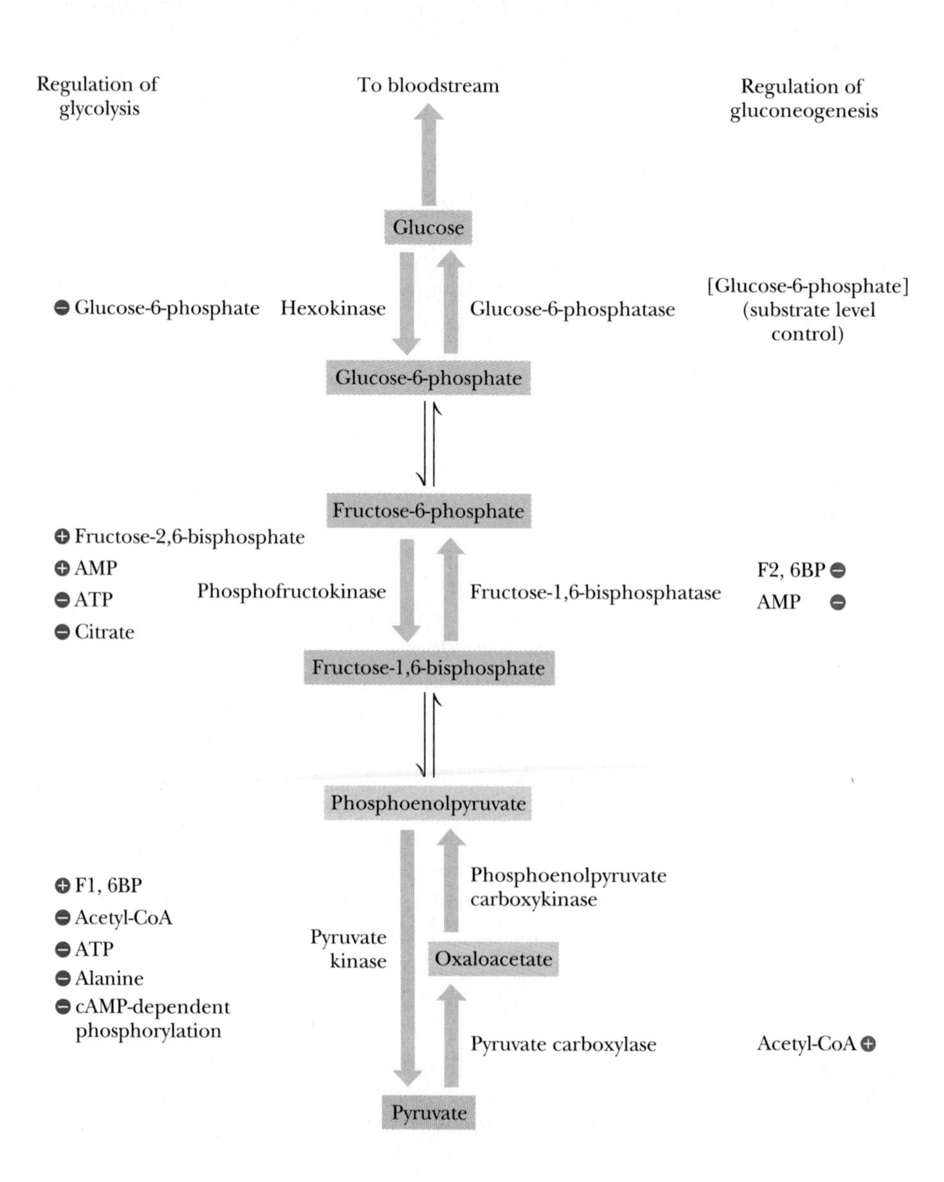

Figure **21.11** The principal regulatory mechanisms in glycolysis and gluconeogenesis. Activators are indicated by plus signs and inhibitors by minus signs.

[acetyl-CoA] indicates that cellular energy levels are high and that carbon metabolites can be directed to glucose synthesis and storage. When acetyl-CoA levels drop, the activities of pyruvate kinase and pyruvate dehydrogenase increase and flux through the TCA cycle increases, providing needed energy for the cell.

Fructose-1,6-bisphosphatase is another important site of gluconeogenic regulation. This enzyme is inhibited by AMP and activated by citrate. These effects by AMP and citrate are the opposites of those exerted on phosphofructokinase in glycolysis, providing another example of reciprocal regulatory effects. When AMP levels increase, gluconeogenic activity is diminished and glycolysis is stimulated. An increase in citrate concentration signals that TCA cycle activity can be curtailed and that pyruvate should be directed to sugar synthesis instead.

Fructose-2,6-Bisphosphate—Allosteric Regulator of Gluconeogenesis

$^{2-}O_3POCH_2$ O OPO_3^{2-} H HO H CH_2OH OH H

Fructose-2,6-bisphosphate

As described in Chapter 18, Emile Van Schaftingen and Henri-Géry Hers demonstrated in 1980 that fructose-2,6-bisphosphate is a potent stimulator of phosphofructokinase. Cognizant of the reciprocal nature of regulation in glycolysis and gluconeogenesis, Van Schaftingen and Hers also considered the possibility of an opposite effect—inhibition—for fructose-1,6-bisphosphatase. In 1981 they reported that fructose-2,6-bisphosphate was indeed a powerful inhibitor of fructose-1,6-bisphosphatase (Figure 21.12). Inhibition occurs in either the presence or absence of AMP, and the effects of AMP and fructose-2,6-bisphosphate are synergistic.

Cellular levels of fructose-2,6-bisphosphate are controlled by *phosphofructokinase-2* (PFK-2), an enzyme distinct from the phosphofructokinase of the glycolytic pathway, and by *fructose-2,6-bisphosphatase* (F-2,6-BPase). Remarkably, these two enzymatic activities are both found in the same protein molecule, which is an example of a **bifunctional,** or **tandem, enzyme** (Figure 21.13). The opposing activities of this bifunctional enzyme are themselves

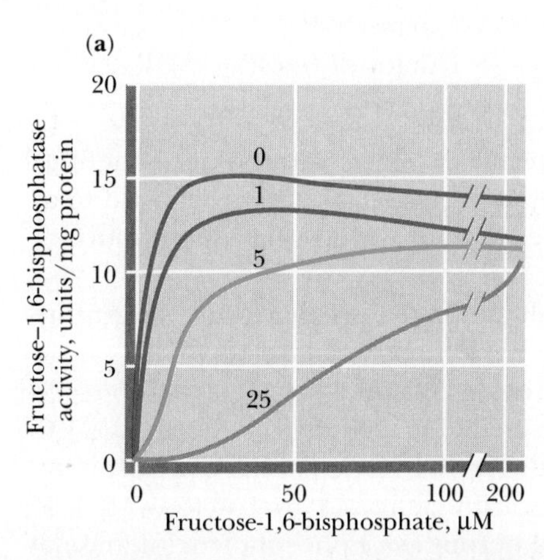

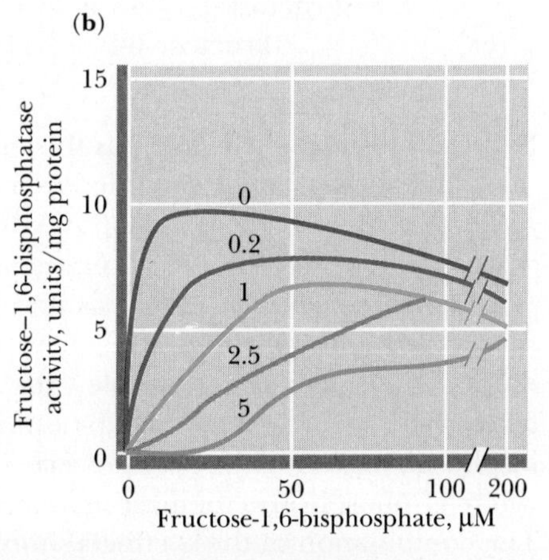

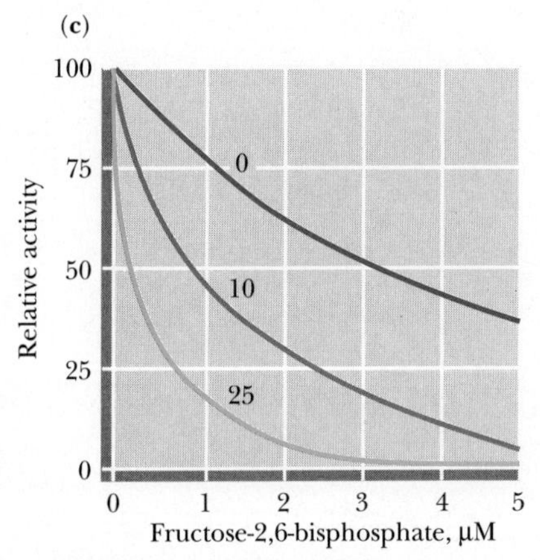

***Figure* 21.12** Inhibition of fructose-1,6-bisphosphatase by fructose-2,6-bisphosphate in the (a) absence and (b) presence of 25 μM AMP. In (a) and (b), enzyme activity is plotted against substrate (fructose-1,6-bisphosphate) concentration. Concentrations of fructose-2,6-bisphosphate (in μM) are indicated above each curve. (c) The effect of AMP (0, 10, and 25 μM) on the inhibition of fructose-1,6-bisphosphatase by fructose-2,6-bisphosphate. Activity was measured in the presence of 10 μM fructose-1,6-bisphosphate.

(Adapted from Van Schaftingen, E., and Hers, H.-G., 1981. Inhibition of fructose-1,6-bisphosphatase by fructose-2,6-bisphosphate. Proceedings of the National Academy of Science, USA 78:2861–2863.)

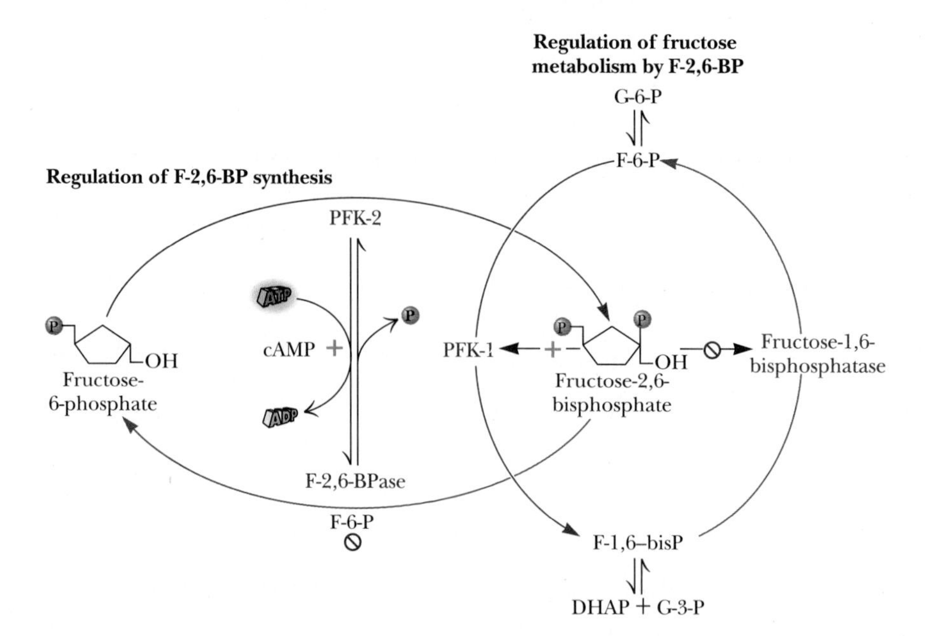

Figure **21.13** Synthesis and degradation of fructose-2,6-bisphosphate are catalyzed by the same bifunctional enzyme.

regulated in two ways. First, fructose-6-phosphate, the substrate of phosphofructokinase and the product of fructose-1,6-bisphosphatase, allosterically activates PFK-2 and inhibits F-2,6-BPase. Second, the phosphorylation by *cAMP-dependent protein kinase* of a single Ser residue on the 49-kD subunit of this dimeric enzyme exerts reciprocal control of the PFK-2 and F-2,6-BPase activities. Phosphorylation inhibits PFK-2 activity (by increasing the K_m for fructose-6-phosphate) and stimulates F-2,6-BPase activity.

Substrate Cycles Provide Metabolic Control Mechanisms

If fructose-1,6-bisphosphatase and phosphofructokinase acted simultaneously, they would constitute a **substrate cycle** in which fructose-1,6-bisphosphate and fructose-6-phosphate became interconverted with net consumption of ATP:

$$\begin{array}{lrcl} & \text{Fructose-1,6-bisP} + H_2O & \longrightarrow & \text{fructose-6-P} + P_i \\ & \text{Fructose-6-P} + \text{ATP} & \longrightarrow & \text{fructose-1,6-bisP} + \text{ADP} \\ \hline \text{Net:} & \text{ATP} + H_2O & \longrightarrow & \text{ADP} + P_i \end{array}$$

Because substrate cycles such as this appear to operate with no net benefit to the cell, they were once regarded as metabolic quirks and were referred to as *futile cycles.* More recently, substrate cycles have been recognized as important devices for controlling metabolite concentrations.

The three steps in glycolysis and gluconeogenesis that differ constitute three such substrate cycles, each with its own particular metabolic raison d'être. Consider, for example, the regulation of the fructose-1,6-bisP–fructose-6-P cycle by fructose-2,6-bisphosphate. As already noted, fructose-1,6-bisphosphatase is subject to allosteric inhibition by fructose-2,6-bisphosphate, whereas phosphofructokinase is allosterically activated by fructose-2,6-bisP. The combination of these effects should permit *either* phosphofructokinase *or* fructose-1,6-bisphosphatase (but not both) to operate at any one time and should thus prevent futile cycling. For instance, in the **fasting state,** when food (i.e., glucose) intake is zero, phosphofructokinase (and therefore glycolysis) is inactive due to the low concentration of fructose-2,6-bisphosphate. In the liver, gluconeogenesis operates to provide glucose for the brain. However, in the fed state, up to 30% of fructose-1,6-bisphosphate formed from phosphofructokinase is recycled back to fructose-6-P (and then to glucose). Because the dependence of fructose-1,6-bisphosphatase activity on fructose-1,6-bisphosphate is sigmoidal in the presence of fructose-2,6-bisphosphate (Figure 21.12), substrate cycling occurs only at relatively high levels of fructose-1,6-bisphosphate. Substrate cycling in this case prevents the accumulation of excessively high levels of fructose-1,6-bisphosphate.

21.3 Glycogen Catabolism

Dietary Glycogen and Starch Breakdown

As noted earlier, well-fed adult human beings normally metabolize about 160 g of carbohydrates each day. A balanced diet easily provides this amount, mostly in the form of starch, with smaller amounts of glycogen. If too little carbohydrate is supplied by the diet, glycogen reserves in liver and muscle tissue can also be mobilized. The reactions by which ingested starch and glycogen are digested are shown in Figure 21.14. The enzyme known as **α-amylase** is an important component of saliva and pancreatic juice. (**β-Amylase** is found in plants. The α- and β- designations for these enzymes serve only to distinguish the two, and do not refer to glycosidic linkage nomenclature.) α-Amylase is an *endoglycosidase* that hydrolyzes α-(1 → 4) linkages of amylopectin and glycogen at random positions, eventually producing a mixture of maltose, maltotriose [with three α-(1 → 4)-linked glucose residues], and other small oligosaccharides. α-Amylase can cleave on either side of a glycogen or amylopectin branch point, but activity is reduced in highly branched regions of the polysaccharide and stops four residues from any branch point.

The highly branched polysaccharides that are left after extensive exposure to α-amylase are called **limit dextrins.** These structures can be further degraded by the action of a **debranching enzyme,** which carries out two distinct reactions. The first of these, known as **oligo(α1,4 → α1,4) glucantransferase** activity, removes a trisaccharide unit and transfers this group to the end of another, nearby branch (Figure 21.15). This leaves a

α–Amylase
β–Amylase
α-(1 → 6)-glucosidase

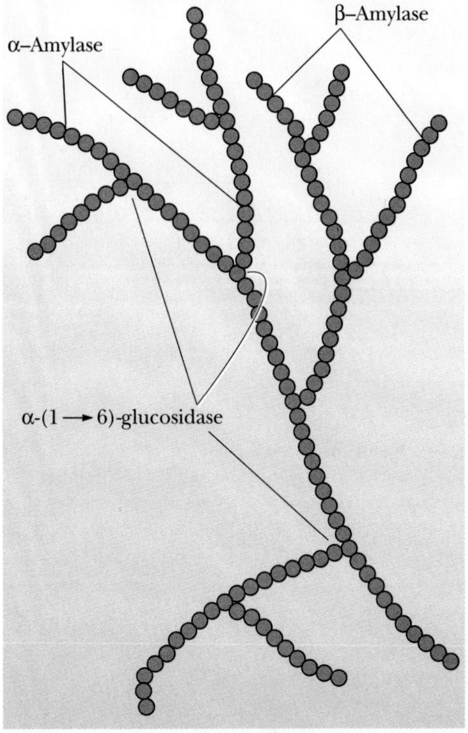

Figure **21.14** Hydrolysis of glycogen and starch by α-amylase and β-amylase.

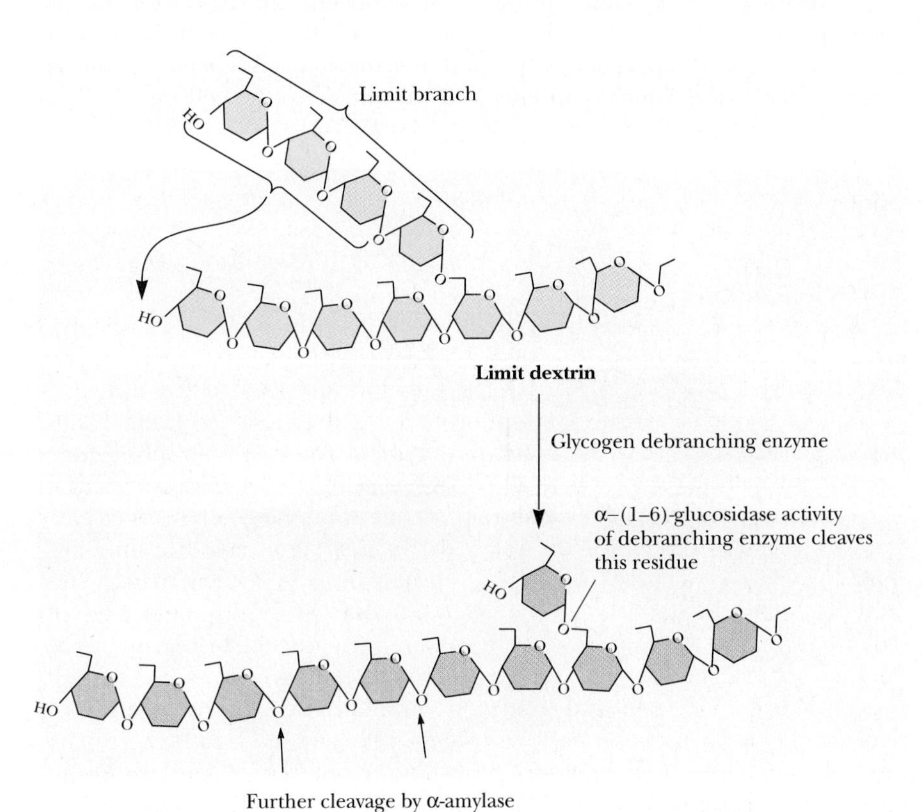

Figure **21.15** The reactions of glycogen debranching enzyme. Transfer of a group of three α-(1 → 4)-linked glucose residues from a limit branch to another branch is followed by cleavage of the α-(1 → 6) bond of the residue that remains at the branch point.

single glucose residue in α-(1 → 6) linkage to the main chain. The **α-(1 → 6) glucosidase** activity of the debranching enzyme then cleaves this residue from the chain, leaving a polysaccharide chain with one branch fewer. Repetition of this sequence of events leads to complete degradation of the polysaccharide.

β-Amylase is an *exoglycosidase* that cleaves maltose units from the free, nonreducing ends of amylopectin branches, as in Figure 21.14. Like α-amylase, however, β-amylase does not cleave either the α-(1 → 6) bonds at the branch points or the α-(1 → 4) linkages near the branch points.

Metabolism of Tissue Glycogen

Digestion itself is a highly efficient process in which almost 100% of ingested food is absorbed and metabolized. Digestive breakdown of starch and glycogen is an unregulated process. On the other hand, tissue glycogen represents an important reservoir of potential energy, and it should be no surprise that the reactions involved in its degradation and synthesis are carefully controlled and regulated. Glycogen reserves in liver and muscle tissue are stored in the cytosol as granules exhibiting a molecular weight range from 6×10^6 to 1600×10^6 (Figure 21.16). These granular aggregates contain the enzymes required to synthesize and catabolize the glycogen as well as all the enzymes of glycolysis.

The Glycogen Phosphorylase Reaction

The cleavage of glucose units from the nonreducing ends of glycogen molecules is catalyzed by *glycogen phosphorylase.* The enzymatic reaction involves phosphorolysis of the bond between C-1 of the departing glucose unit and the glycosidic oxygen, to yield *glucose-1-phosphate* and a glycogen molecule that is shortened by one residue (Figure 21.17). (Because the reaction involves attack by phosphate instead of H_2O, it is referred to as a **phosphorolysis** rather than a hydrolysis.) The standard-state free-energy change ($\Delta G°'$) for the glycogen phosphorylase reaction is +3.1 kJ/mol, but the intracellular ratio of [P_i] to [glucose-1-P] approaches 100, and thus the actual ΔG *in vivo* is approximately −6 kJ/mol. There is an energetic advantage to the cell in this phos-

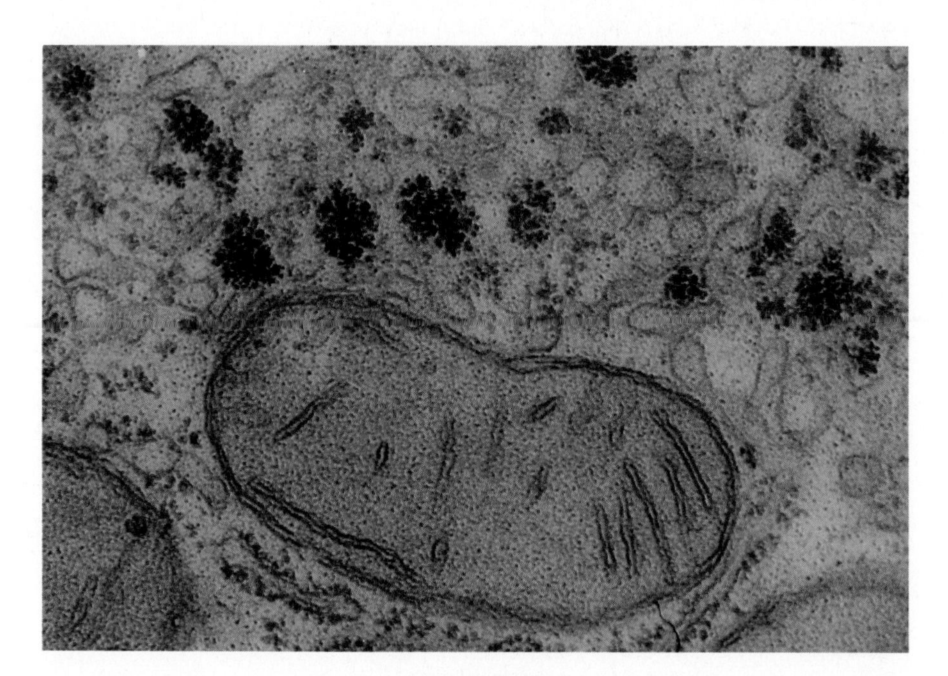

***Figure* 21.16** Electron micrograph of a liver cell. The dark particles grouped in clusters are glycogen granules.

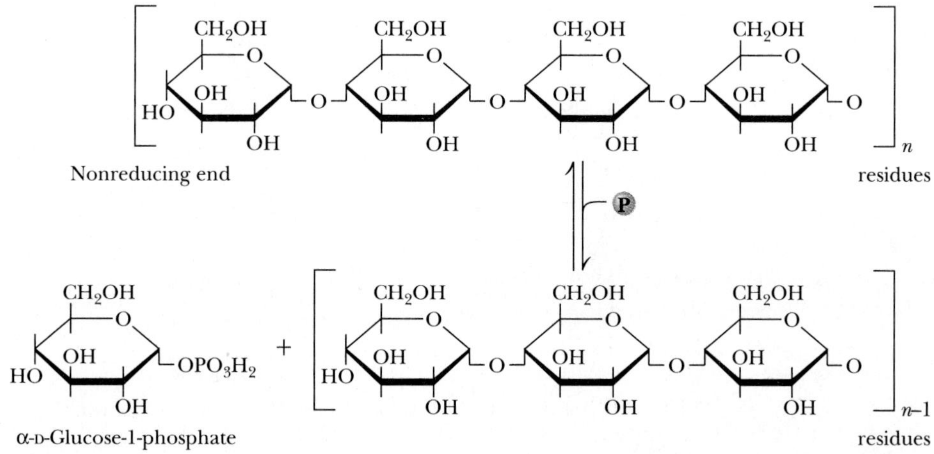

Nonreducing end

n residues

α-D-Glucose-1-phosphate

$n-1$ residues

Figure **21.17** The glycogen phosphorylase reaction.

phorolysis reaction. If glycogen breakdown were hydrolytic and yielded glucose as a product, it would be necessary to phosphorylate the product glucose (with the expenditure of a molecule of ATP) to initiate its glycolytic degradation. Instead, phosphorolysis produces a phosphorylated sugar product, glucose-1-P, which is converted to the glycolytic substrate, glucose-6-P, by *phosphoglucomutase* (Figure 21.18). In muscle, glucose-6-P proceeds into glycolysis, providing needed energy for muscle contraction. In liver, hydrolysis of glucose-6-P yields glucose, which is exported to other tissues via the circulatory system.

The glycogen phosphorylase reaction degrades glycogen to produce limit dextrins, which are further degraded by debranching enzyme, as already described.

The Structure of Glycogen Phosphorylase

Glycogen phosphorylase is a dimer of two identical subunits (842 residues, 97.44 kD). Each subunit contains a pyridoxal phosphate cofactor, covalently linked as a Schiff base to Lys^{680}. Each subunit contains an active site (at the center of the subunit) and an allosteric effector site near the subunit interface (Figure 21.19). A regulatory phosphorylation site is located at Ser^{14} on each subunit. There is a glycogen-binding site on each subunit, which facilitates prior association of glycogen phosphorylase with its substrate and also exerts regulatory control on the enzymatic reaction.

Each subunit contributes a **tower helix** (residues 262 to 278) to the subunit–subunit contact interface in glycogen phosphorylase. In the phosphorylase dimer, the tower helices extend from their respective subunits and pack against each other in an antiparallel manner.

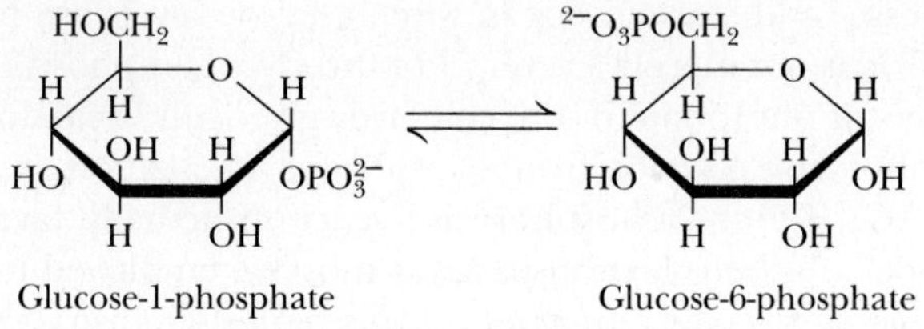

Glucose-1-phosphate

Glucose-6-phosphate

Figure **21.18** The phosphoglucomutase reaction.

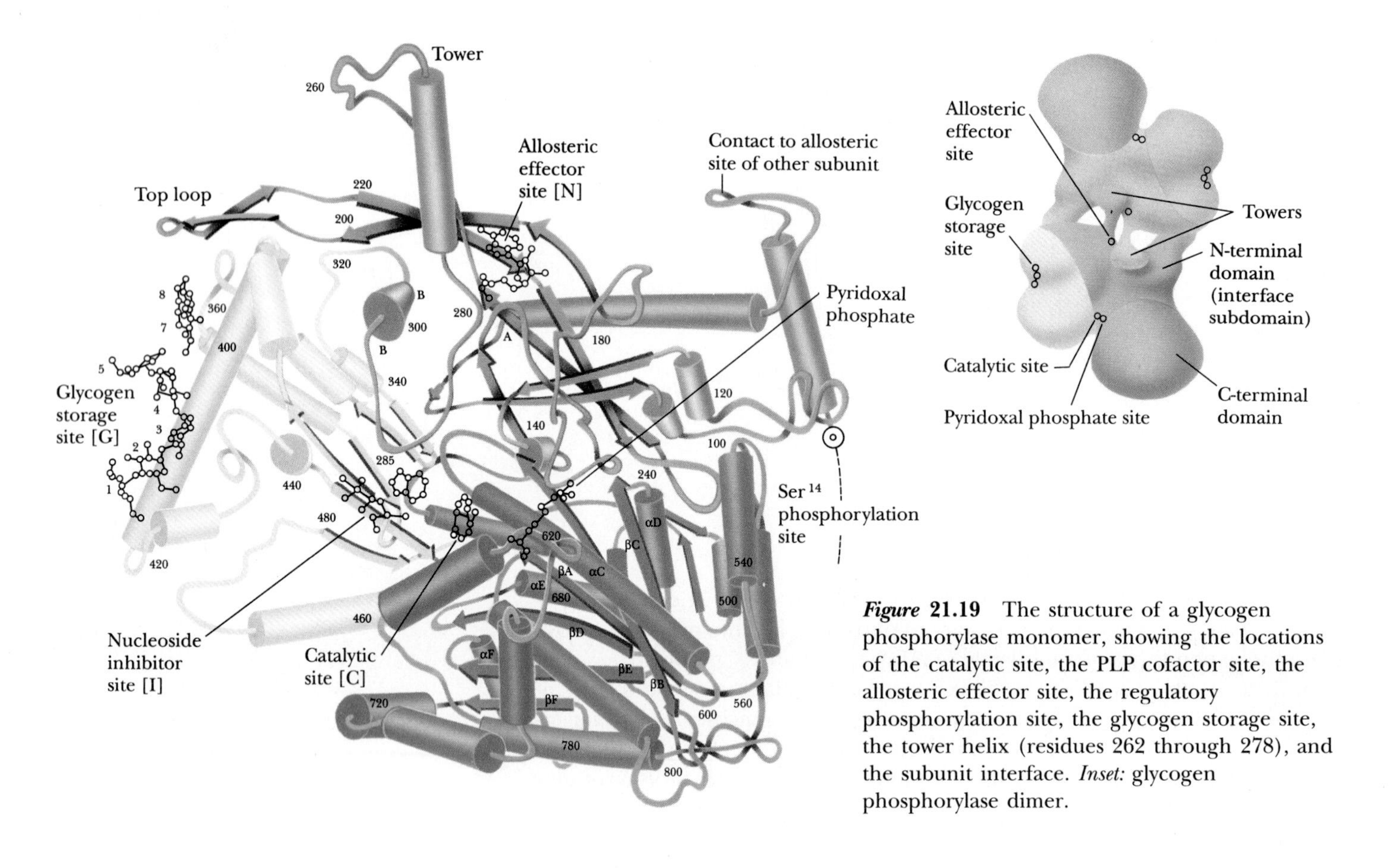

***Figure* 21.19** The structure of a glycogen phosphorylase monomer, showing the locations of the catalytic site, the PLP cofactor site, the allosteric effector site, the regulatory phosphorylation site, the glycogen storage site, the tower helix (residues 262 through 278), and the subunit interface. *Inset:* glycogen phosphorylase dimer.

The Phosphorylase Reaction Mechanism

An intriguing aspect of the phosphorylase mechanism is the role of the pyridoxal phosphate cofactor. In contrast with conventional PLP-dependent enzymes, the Lys^{680} Schiff base of glycogen phosphorylase can be reduced (with sodium borohydride, for example) with no loss of enzymatic activity. From the crystal structure of the enzyme, it is clear that the phosphate substrate is hydrogen-bonded to the phosphate moiety of enzyme-bound PLP (Figure 21.19). In the reaction mechanism (Figure 21.20), the substrate phosphate acquires a proton from the PLP phosphate and immediately donates it to the glycosidic O atom of the glycogen substrate, forming an **oxonium ion intermediate** at the active site. *This step constitutes general acid catalysis by the substrate phosphate.* Subsequent nucleophilic attack by the substrate phosphate forms the product α-D-glucose-1-phosphate. The reaction occurs with retention of configuration at C-1 of the glucose unit removed from glycogen.

21.4 Glycogen Synthesis

Animals synthesize and store glycogen when glucose levels are high, but the synthetic pathway is not merely a reversal of the glycogen phosphorylase reaction. High levels of phosphate in the cell favor glycogen breakdown and prevent the phosphorylase reaction from synthesizing glycogen *in vivo,* in spite of the fact that $\Delta G°'$ for the phosphorylase reaction actually favors glycogen synthesis. Hence, another reaction pathway must be employed in the cell for the net synthesis of glycogen. In essence, this pathway must activate glucose units for transfer to glycogen chains.

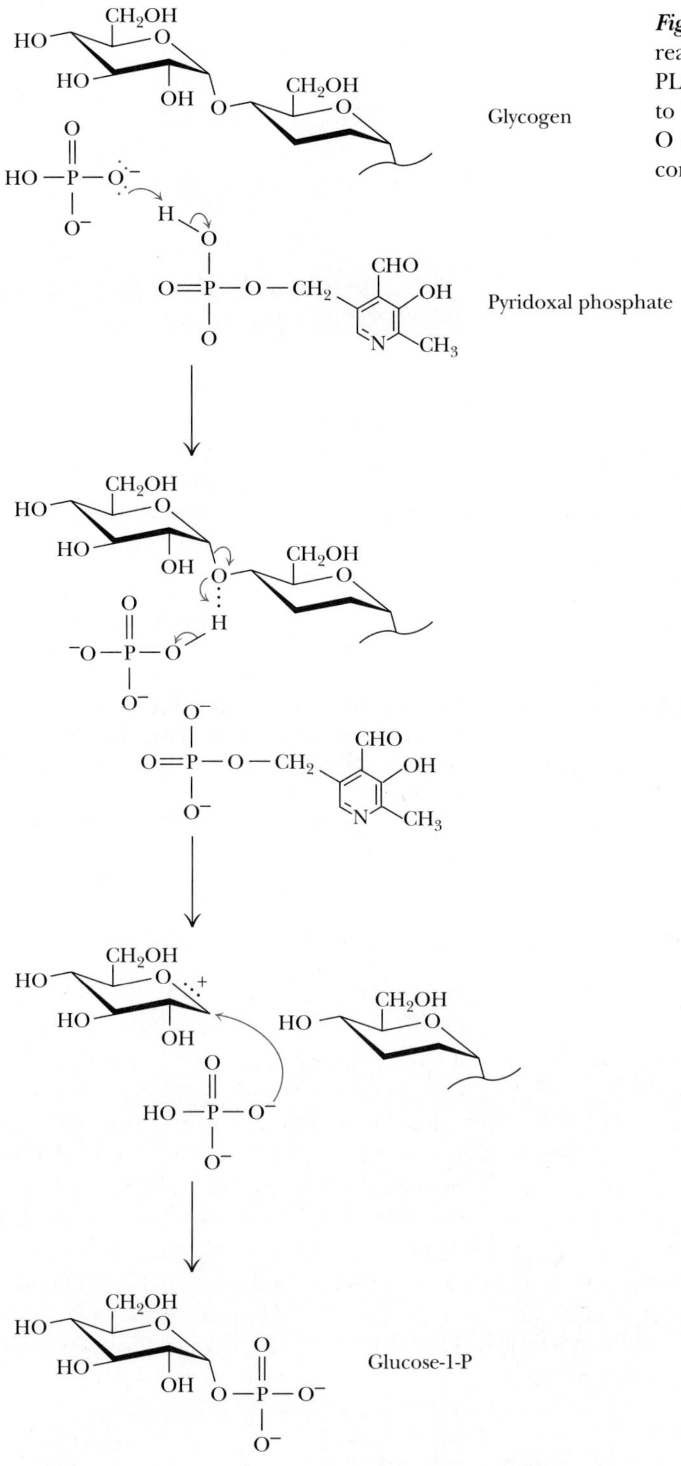

Glycogen

Pyridoxal phosphate

Glucose-1-P

***Figure* 21.20** A mechanism for the glycogen phosphorylase reaction. The phosphate substrate makes a hydrogen bond with PLP at the active site. Proton transfer from the PLP phosphate to the substrate phosphate, and then to the O atom of the bond to be cleaved, probably occurs in a concerted fashion.

Glucose Units Are Activated for Transfer by Formation of Sugar Nucleotides

We are familiar with several examples of chemical activation as a strategy for group transfer reactions. Acetyl-CoA is an activated form of acetate, biotin and tetrahydrofolate activate one-carbon groups for transfer, and ATP is an activated form of phosphate. Luis Leloir, a biochemist in Argentina, showed in the 1950s that glycogen synthesis depended upon **sugar nucleotides,** which

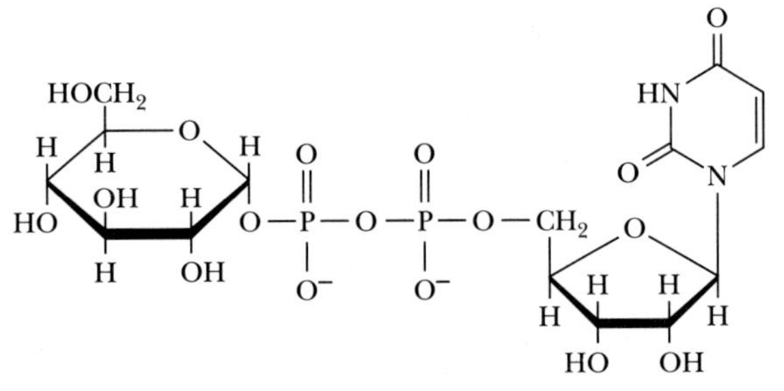

Uridine diphosphate glucose (UDPG)

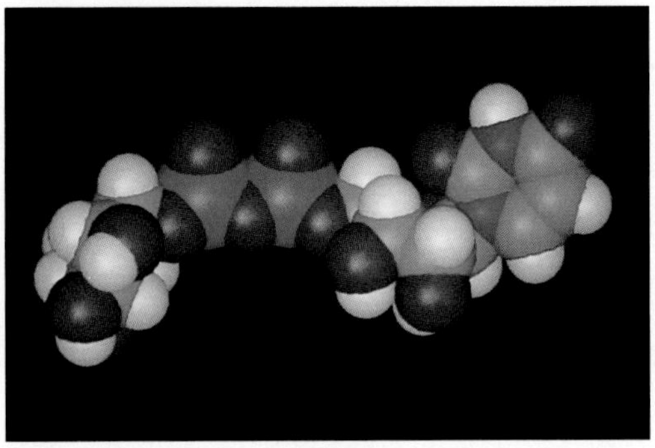

Figure **21.21** The structure of UDP-glucose, a sugar nucleotide.

may be thought of as activated forms of sugar units (Figure 21.21). For example, formation of an ester linkage between the C-1 hydroxyl group and the β-phosphate of UDP activates the glucose moiety of *UDP-glucose.*

UDP-Glucose Synthesis Is Driven by Pyrophosphate Hydrolysis

Sugar nucleotides are formed from sugar-1-phosphates and nucleoside triphosphates by specific *pyrophosphorylase* enzymes (Figure 21.22). For example, *UDP-glucose pyrophosphorylase* catalyzes the formation of UDP-glucose from glucose-1-phosphate and uridine 5′-triphosphate:

$$\text{Glucose-1-P} + \text{UTP} \longrightarrow \text{UDP-glucose} + \text{pyrophosphate}$$

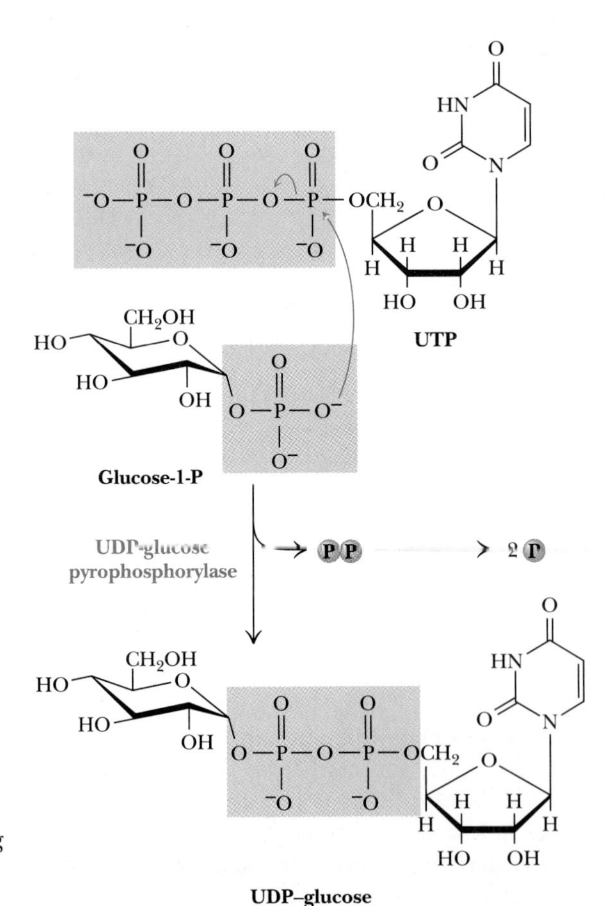

Figure **21.22** The UDP-glucose pyrophosphorylase reaction is a phosphoanhydride exchange, with the phosphoryl oxygen of glucose-1-P attacking the α-phosphorus of UTP to form UDP-glucose and pyrophosphate.

The reaction proceeds via attack by the phosphate oxygen of glucose-1-phosphate on the α-phosphorus of UTP, with departure of the pyrophosphate anion. The reaction is a reversible one, but—as is the case for many biosynthetic reactions—it is driven forward by subsequent hydrolysis of pyrophosphate:

$$\text{Pyrophosphate} + H_2O \longrightarrow 2\ P_i$$

The net reaction for sugar nucleotide formation (combining the preceding two equations) is thus

$$\text{Glucose-1-P} + \text{UTP} + H_2O \longrightarrow \text{UDP-glucose} + 2\ P_i$$

Sugar nucleotides of this type act as donors of sugar units in the biosynthesis of oligo- and polysaccharides.

Glycogen Synthase Catalyzes Formation of α-(1 → 4) Glycosidic Bonds in Glycogen

The very large glycogen polymer is built around a tiny protein core. The first glucose residue is covalently joined to the protein **glycogenin** via an acetal linkage to a tyrosine–OH group on the protein. Sugar units are added to the glycogen polymer by the action of *glycogen synthase.* The reaction involves transfer of a glucosyl unit from UDP-glucose to the C-4 hydroxyl group at a nonreducing end of a glycogen strand. The mechanism proceeds by cleavage of the C—O bond between the glucose moiety and the β-phosphate of UDP-glucose, leaving an oxonium ion intermediate, which is rapidly attacked by the C-4 hydroxyl oxygen of a terminal glucose unit on glycogen (Figure 21.23). The reaction is exergonic and has a $\Delta G°'$ of -13.3 kJ/mol.

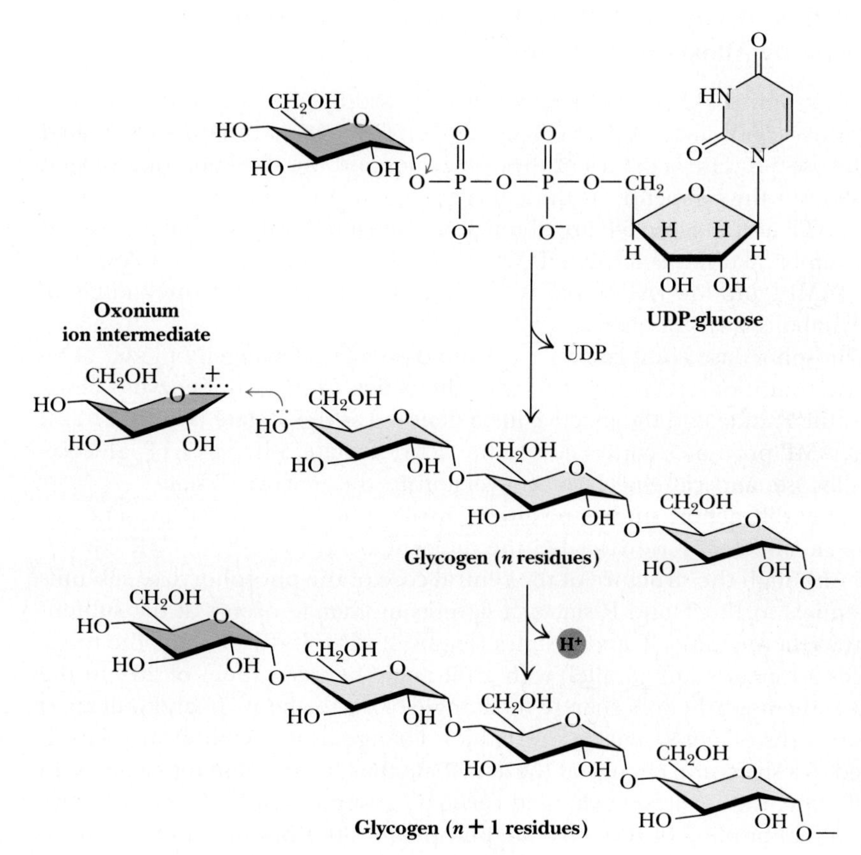

***Figure* 21.23** The glycogen synthase reaction. Cleavage of the C—O bond of UDP-glucose yields an oxonium intermediate (similar to that formed in the glycogen phosphorylase mechanism). Attack by the hydroxyl oxygen of the terminal residue of a glycogen molecule completes the reaction.

Glycogen Branching Occurs by Transfer of Terminal Chain Segments

Glycogen is a branched polymer of glucose units. The branches arise from α-(1 → 6) linkages which occur every 8 to 12 residues. As noted in Chapter 10, the branches provide multiple sites for rapid degradation or elongation of the polymer, and also increase its solubility. Glycogen branches are formed by *amylo-(1,4→ 1,6)-transglycosylase,* also known as *branching enzyme.* The reaction involves the transfer of a six- or seven-residue segment from the nonreducing end of a linear chain at least eleven residues in length to the C-6 hydroxyl of a glucose residue of the same chain or another chain (Figure 21.24). For each branching reaction, the resulting polymer has gained a new terminus at which growth can occur.

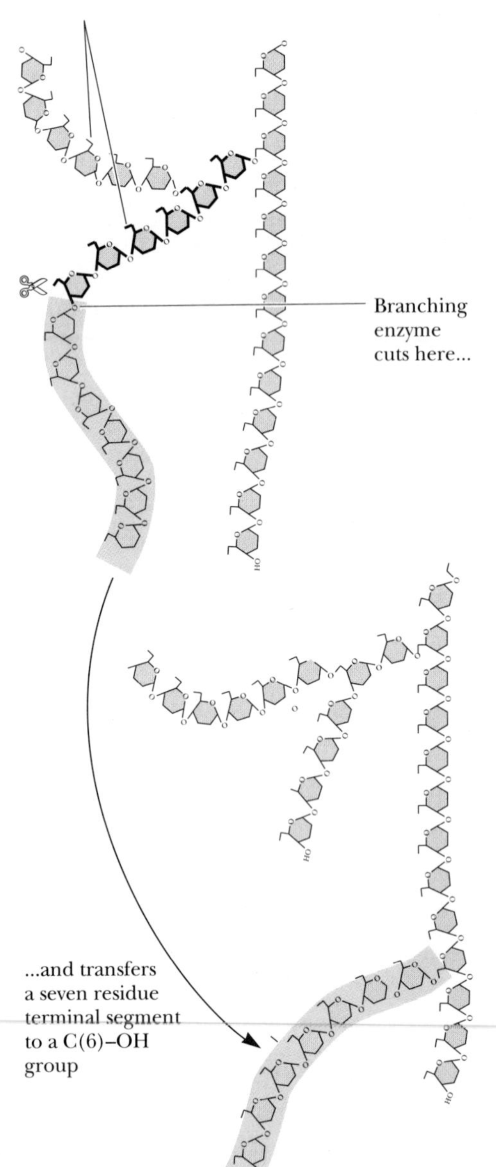
(1 ⟶ 4)-terminal chains of glycogen
Branching enzyme cuts here...
...and transfers a seven residue terminal segment to a C(6)–OH group

Figure **21.24** Formation of glycogen branches by the branching enzyme. Six- or seven-residue segments of a growing glycogen chain are transferred to the C-6 hydroxyl group of a glucose residue on the same or a nearby chain.

21.5 Control of Glycogen Metabolism

Glycogen Metabolism Is Highly Regulated

Synthesis and degradation of glycogen must be carefully controlled so that this important energy reservoir can properly serve the metabolic needs of the organism. Glucose is the principal metabolic fuel for the brain, and the concentration of glucose in circulating blood must be maintained at about 5 m*M* for this purpose. Glucose derived from glycogen breakdown is also a primary energy source for muscle contraction. Control of glycogen metabolism is effected via reciprocal regulation of glycogen phosphorylase and glycogen synthase. Thus, activation of glycogen phosphorylase is tightly linked to inhibition of glycogen synthase, and vice versa. Regulation involves both allosteric control and covalent modification, with the latter being under hormonal control.

Regulation of Glycogen Phosphorylase and Glycogen Synthase by Allosteric Effectors

Glycogen phosphorylase is allosterically activated by AMP and is inhibited by ATP, glucose-6-P, and caffeine (Figure 21.25). Glycogen synthase is stimulated by glucose-6-P. The activities of these enzymes and the direction of glycogen metabolism thus depend on the energy status of the (muscle or liver) cell. When ATP and glucose-6-P are abundant, glycogen synthesis is activated and glycogen breakdown is inhibited. When cellular energy reserves are low (i.e., high [AMP] and low [ATP] and [G-6-P]), glycogen synthesis stops and glycogen catabolism is stimulated.

Phosphorylase conforms to the Monod–Wyman–Changeux model of allosteric transitions (Chapter 12), with the active form of the enzyme designated the **R state** and the inactive form denoted as the **T state** (Figure 21.25). Thus, AMP promotes conversion to the active R state, whereas ATP, glucose-6-P, glucose, and caffeine favor conversion to the inactive T state.

X-ray diffraction studies of glycogen phosphorylase in the presence of allosteric effectors have revealed the molecular basis for the T ⇌ R conversion. Although the structure of the central core of the phosphorylase subunits is identical in the T and R states, a significant change occurs at the subunit interface between the T and R states (Figure 21.26). In the T state, the tower helices are nearly antiparallel, with a tilt angle between them of 20°. In the R state, the tower helices change their angle of tilt to about 70° and pull apart by two turns of helix. This conformation change at the subunit interface is linked to a structural change at the active site that is important for catalysis. In the T state, the negatively charged carboxyl group of Asp^{283} faces the active site, so that binding of the substrate phosphate is unfavorable. In the conver-

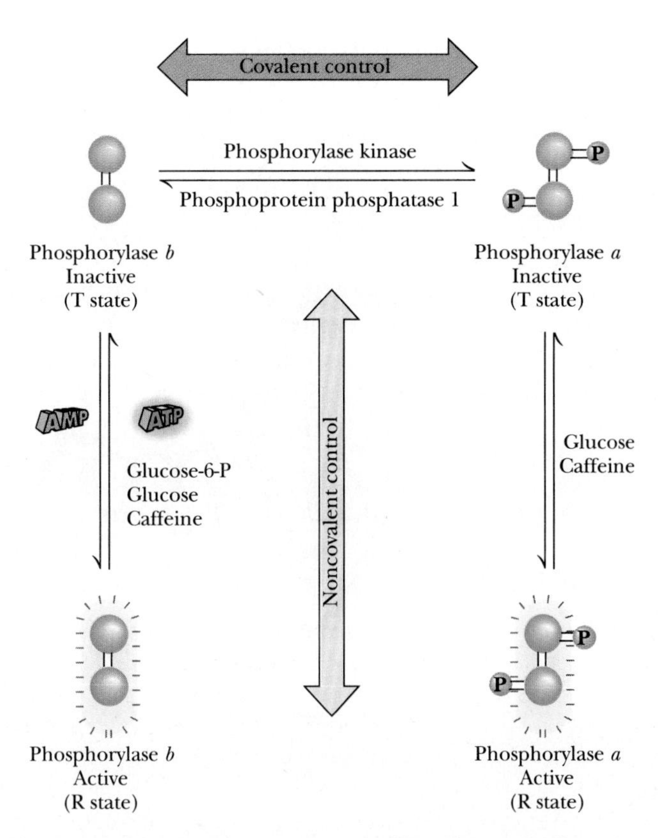

***Figure* 21.25** The mechanism for covalent modification and allosteric regulation of glycogen phosphorylase. The T states are blue and the R states blue-green.

sion to the R state, Asp^{283} is displaced from the active site and replaced by Arg^{569}. The exchange of negatively charged aspartate for positively charged arginine at the active site provides a favorable binding site for phosphate. These allosteric controls provide a mechanism for adjusting the activity of glycogen phosphorylase to meet normal metabolic demands. These controls can be overridden, however, by a covalent modification of glycogen phosphorylase that converts the enzyme from a less active, allosterically regulated form (the *b* form) to a more active, allosterically unresponsive form (the *a* form).

Regulation of Glycogen Phosphorylase and Glycogen Synthase by Covalent Modification

As early as 1938, it was known that glycogen phosphorylase existed in two forms: the less active **phosphorylase *b*** and the more active **phosphorylase *a*.** In 1956, Edwin Krebs and Edmond Fischer reported that a "converting enzyme" could convert phosphorylase *b* to phosphorylase *a*. Three years later, Krebs and Fischer demonstrated that the conversion of phosphorylase *b* to phosphorylase *a* involved covalent phosphorylation, as in Figure 21.25. For their groundbreaking work in elucidating this control mechanism, Krebs and Fischer were awarded the 1992 Nobel Prize in physiology or medicine.

Phosphorylation of Ser^{14} causes a dramatic conformation change in phosphorylase. Upon phosphorylation, the amino-terminal end of the protein (including residues 10 through 22) swings through an arc of 120°, moving into the subunit interface (Figure 21.27). This conformation change moves Ser^{14} by more than 3.6 nm.

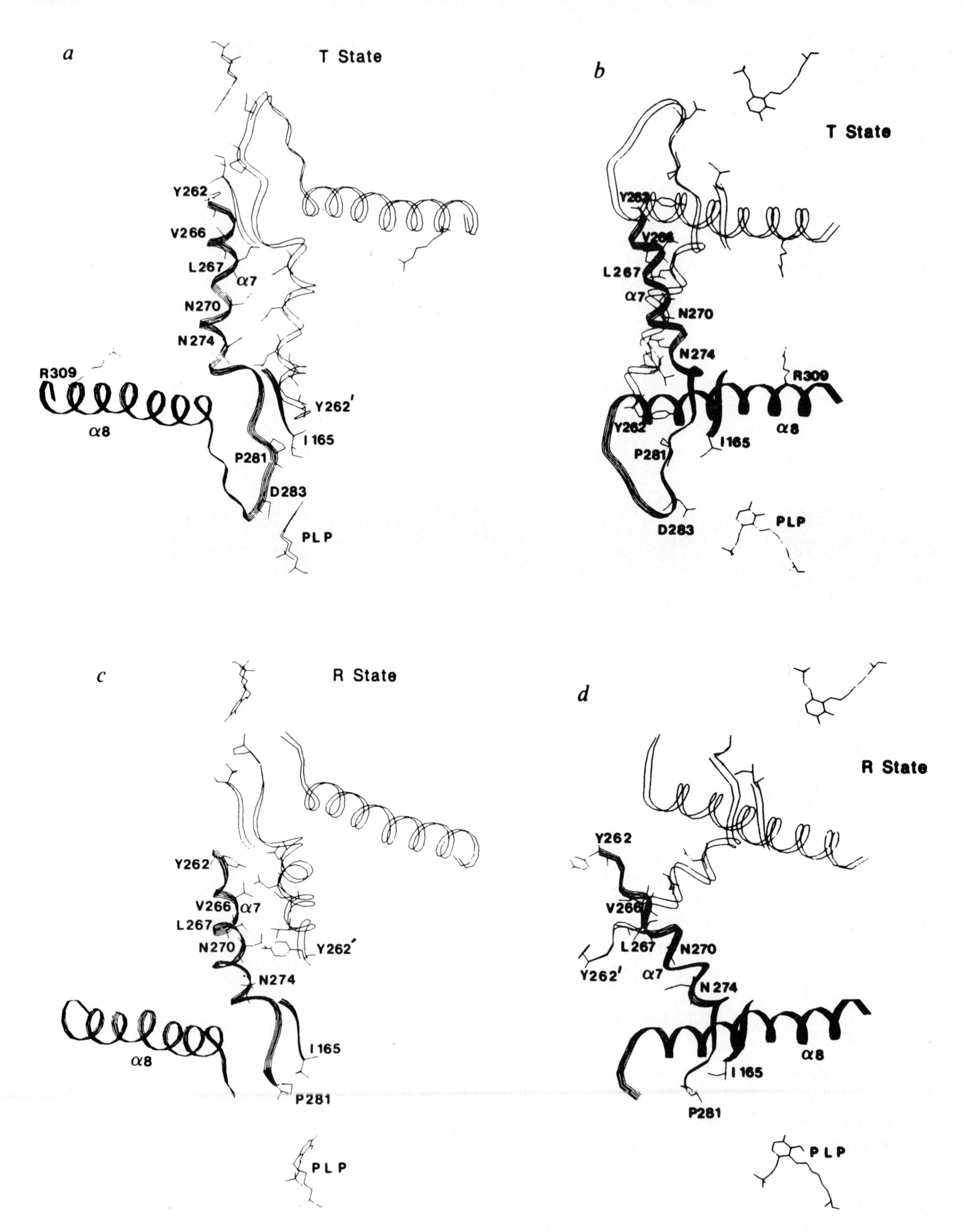

Figure **21.26** The rearrangement of the tower helices in the T-to-R conversion of glycogen phosphorylase involves a change in tilt angle and a sliding that moves the helices about two turns farther apart. The view in (a) and (c) is along the twofold symmetry axis of the T-state dimer. The view in (b) and (d) is perpendicular to the twofold axis.

(Figure courtesy of Louise Johnson, Oxford University)

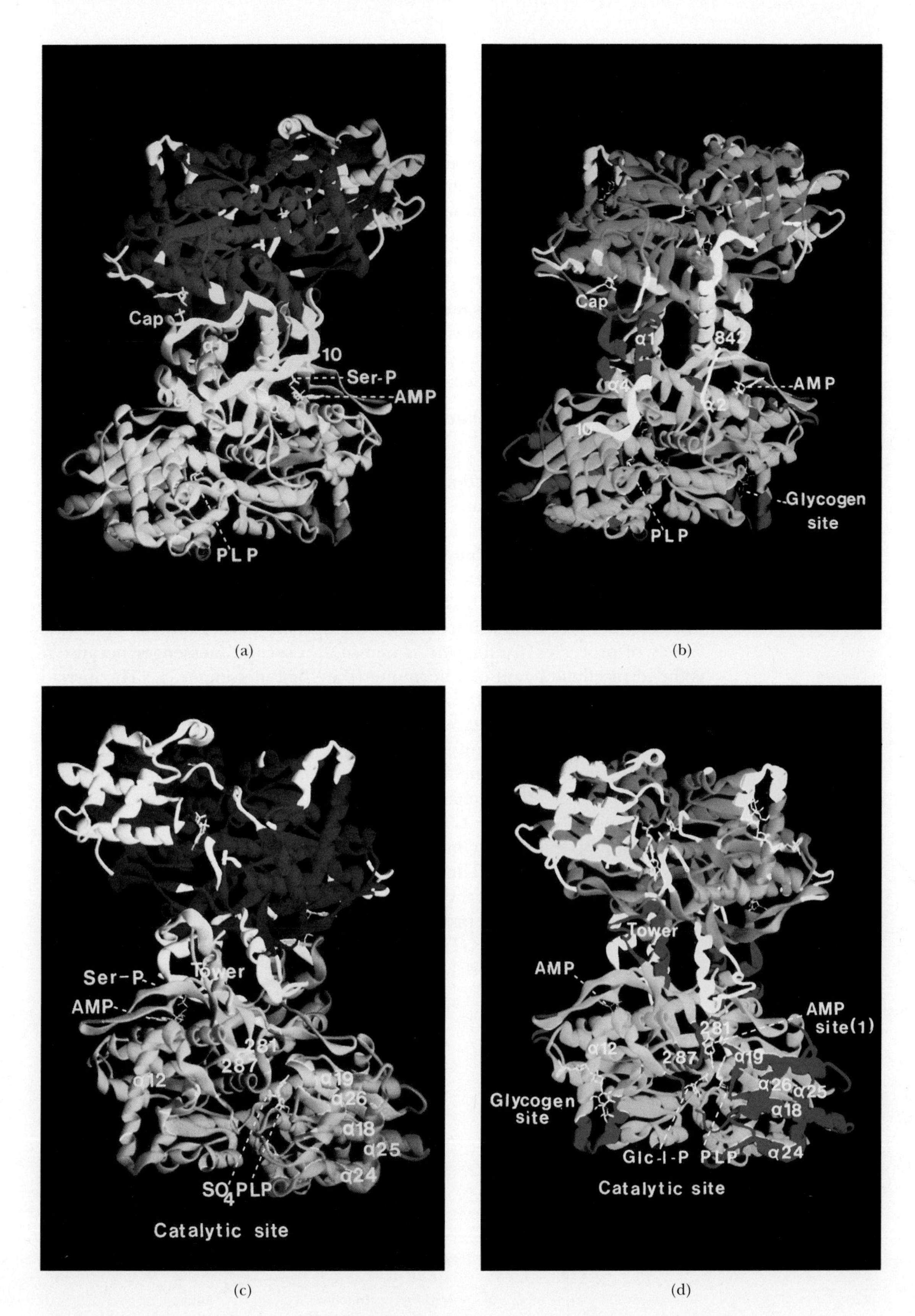

***Figure* 21.27** Dimers of glycogen phosphorylase in (a, c) the phosphorylated R state GP*a* and (b, d) the unphosphorylated T state GP*b*. View is down the twofold axis with (a, b) the allosteric and Ser-P sites toward the viewer. In panels (c) and (d), the catalytic site and tower helices are toward the viewer. (a, c) Green and dark blue and (b, d) light blue and rose indicate domains that are unchanged in the R to T conversion. (a, c) Orange and pink and (b, d) red and yellow indicate domains that differ by more than 1 Å in C_α positions. *Cap* indicates a peptide segment that caps the AMP activation site. SO_4^{2-}, an analog of P_i, binds within the catalytic site.

(Photo courtesy of Louise Johnson, Oxford University)

Glycogen synthase also exists in two distinct forms which can be interconverted by the action of specific enzymes: active, dephosphorylated **glycogen synthase I** (glucose-6-P-independent) and less active phosphorylated **glycogen synthase D** (glucose-6-P-dependent). The nature of phosphorylation is more complex with glycogen synthase. As many as nine serine residues on the enzyme appear to be subject to phosphorylation, each site's phosphorylation having some effect on enzyme activity.

Dephosphorylation of both glycogen phosphorylase and glycogen synthase is carried out by **phosphoprotein phosphatase 1.** The action of phosphoprotein phosphatase 1 inactivates glycogen phosphorylase and activates glycogen synthase.

Enzyme Cascades Regulate Glycogen Phosphorylase and Glycogen Synthase

The phosphorylation reactions which activate glycogen phosphorylase and inactivate glycogen synthase are mediated by an **enzyme cascade** (Figure 21.28). The first part of the cascade leads to hormonal stimulation (described in the next section) of **adenylyl cyclase,** a membrane-bound enzyme which converts ATP to *3′,5′-cyclic adenosine monophosphate,* denoted as *cyclic AMP* or simply *cAMP* (Figure 21.29). This regulatory molecule is found in all eukaryotic cells and acts as an intracellular messenger molecule, controlling a wide variety of processes. Cyclic AMP is known as a **second messenger** because it is the intracellular agent of a hormone (the "first messenger"). (The myriad cellular roles of cyclic AMP are described in detail in Chapter 37.)

The hormonal stimulation of adenylyl cyclase is effected by a transmembrane signaling pathway consisting of three components, all membrane-associated. Binding of hormone to the external surface of a hormone receptor causes a conformational change in the transmembrane receptor protein, which in turn stimulates a **GTP-binding protein** (abbreviated **G protein**). G proteins are heterotrimeric proteins consisting of α- (45–47 kD), β- (35 kD), and γ- (7–9 kD) subunits. The α-subunit binds GDP or GTP and has an intrinsic, slow GTPase activity. In the inactive state, the $G_{\alpha\beta\gamma}$ complex has GDP at the nucleotide site. When a G protein is stimulated by a hormone-receptor complex, GDP dissociates and GTP binds to G_α, causing it to dissociate from $G_{\beta\gamma}$ and to associate with adenylyl cyclase. *Binding of G_α(GTP) activates adenylyl*

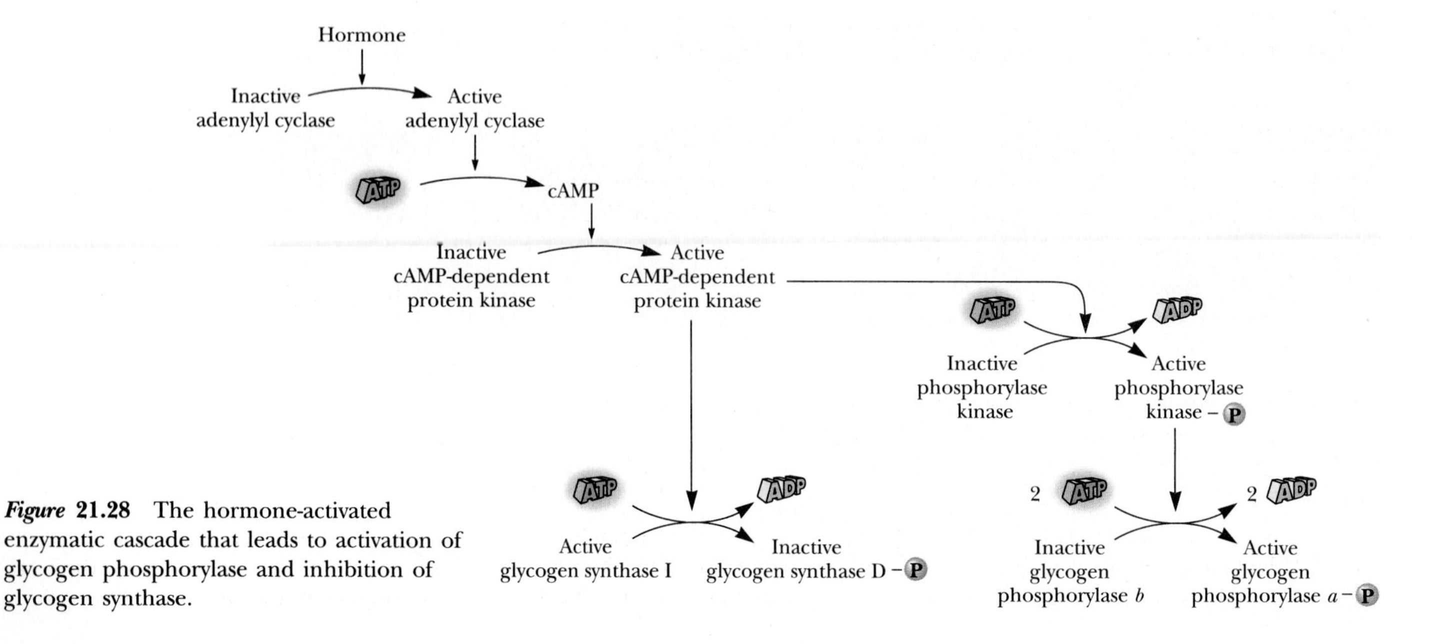

***Figure* 21.28** The hormone-activated enzymatic cascade that leads to activation of glycogen phosphorylase and inhibition of glycogen synthase.

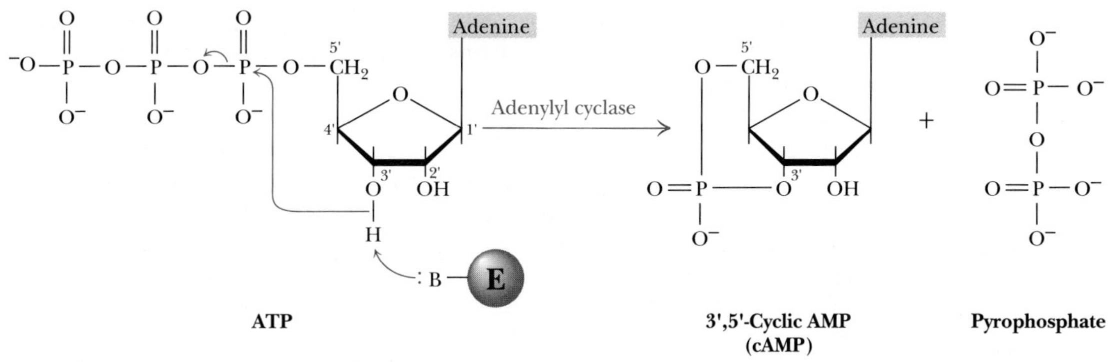

***Figure* 21.29** The adenylyl cyclase reaction yields 3′,5′-cyclic AMP and pyrophosphate. The reaction is driven forward by subsequent hydrolysis of pyrophosphate by the enzyme inorganic pyrophosphatase.

cyclase to form cAMP from ATP. However, the intrinsic GTPase activity of G_α eventually hydrolyzes GTP to GDP, leading to dissociation of G_α(GDP) from adenylyl cyclase and reassociation with $G_{\beta\gamma}$ to form the inactive $G_{\alpha\beta\gamma}$ complex. This cascade amplifies the hormonal signal, since a single hormone-receptor complex can activate many G proteins before the hormone dissociates from the receptor, and since the G_α-activated cyclase can synthesize many cAMP molecules before bound GTP is hydrolyzed by G_α. More than 100 different G protein–coupled receptors and at least 21 distinct G_α proteins are known. These include a class of **inhibitory G proteins** that inactivate adenylyl cyclase, and G proteins that stimulate a variety of transmembrane ion channels for Ca^{2+}, Na^+, and K^+.

Cyclic AMP is an essential activator of *cAMP-dependent protein kinase (cAPK).* This enzyme is normally inactive, because its two catalytic subunits (C) are strongly associated with a pair of regulatory subunits (R), which serve to block activity. Binding of cyclic AMP to the regulatory subunits induces a conformation change that causes the dissociation of the C monomers from the R dimer. The free C subunits are active and can phosphorylate other proteins. Two of the many proteins phosphorylated by cAPK are *phosphorylase kinase* and glycogen synthase (Figure 21.28). As noted earlier, glycogen synthase is inactivated by phosphorylation. In contrast, phosphorylase kinase is inactive in the unphosphorylated state and active in the phosphorylated form. As its name implies, phosphorylase kinase functions to phosphorylate (and activate) glycogen phosphorylase. Thus, stimulation of adenylyl cyclase leads to activation of glycogen breakdown and inhibition of glycogen synthesis.

Hormones Regulate Glycogen Synthesis and Degradation

Storage and utilization of tissue glycogen, maintenance of blood glucose concentration, and other aspects of carbohydrate metabolism are meticulously regulated by hormones, including *insulin, glucagon, epinephrine,* and the *glucocorticoids.*

Insulin Is a Response to Increased Blood Glucose

The primary hormone responsible for conversion of glucose to glycogen is **insulin** (Figure 5.34). Insulin is secreted by special cells in the pancreas called the **islets of Langerhans.** *Secretion of insulin is a response to increased glucose in the blood.* When blood glucose levels rise (after a meal, for example), insulin is secreted from the pancreas into the *pancreatic vein,* which empties into the **portal vein system** (Figure 21.30), so that insulin traverses the liver before it enters the systemic blood supply. Insulin acts to rapidly lower blood glucose

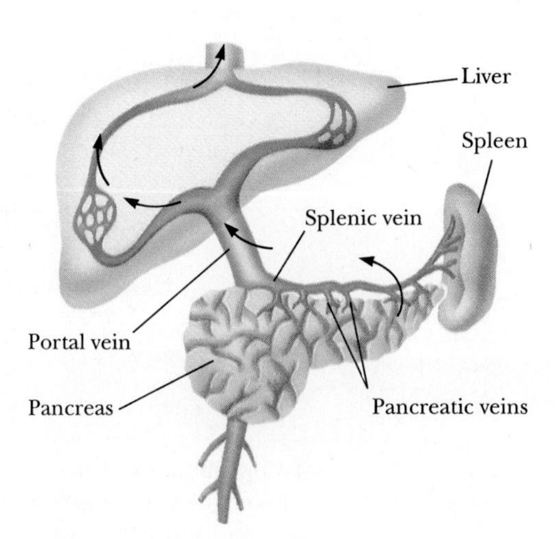

***Figure* 21.30** The portal vein system carries pancreatic secretions such as insulin and glucagon to the liver and then into the rest of the circulatory system.

A Deeper Look

Carbohydrate Utilization in Exercise

Animals have a remarkable ability to "shift gears" metabolically during periods of strenuous exercise or activity. Metabolic adaptations allow the body to draw on different sources of energy (all of which produce ATP) for different types of activity. During periods of short-term, high-intensity exercise (e.g., a 100-m dash), most of the required energy is supplied directly by existing stores of ATP and creatine phosphate (Figure, part a). Long-term, low-intensity exercise (a 10-km run or a 42.2-km marathon) is fueled almost entirely by aerobic metabolism. Between these extremes is a variety of activities (an 800-m run, for example) that rely on anaerobic glycolysis—conversion of glucose to lactate in the muscles and utilization of the Cori cycle.

For all these activities, breakdown of muscle glycogen provides much of the needed glucose. The rate of glycogen consumption depends upon the intensity of the exercise (Figure, part b). By contrast, glucose derived from gluconeogenesis makes only small contributions to total glucose consumed during exercise. During prolonged mild exercise, gluconeogenesis accounts for only about 8% of the total glucose consumed. During heavy exercise, this percentage becomes even lower.

Choice of diet has a dramatic effect on glycogen recovery following exhaustive exercise. A diet consisting mainly of protein and fat results in very little recovery of muscle glycogen, even after five days (Figure, part c). On the other hand, a high-carbohydrate diet provides faster restoration of muscle glycogen. Even in this case, however, complete recovery of glycogen stores takes about two days.

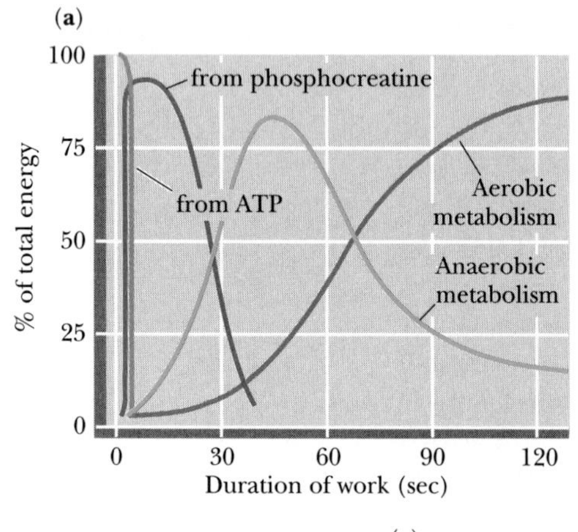

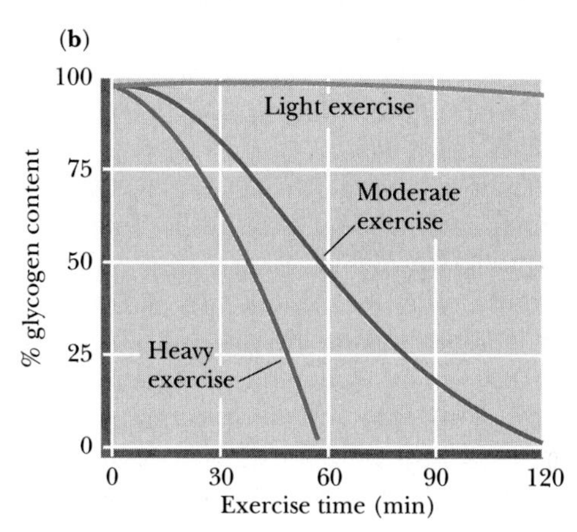

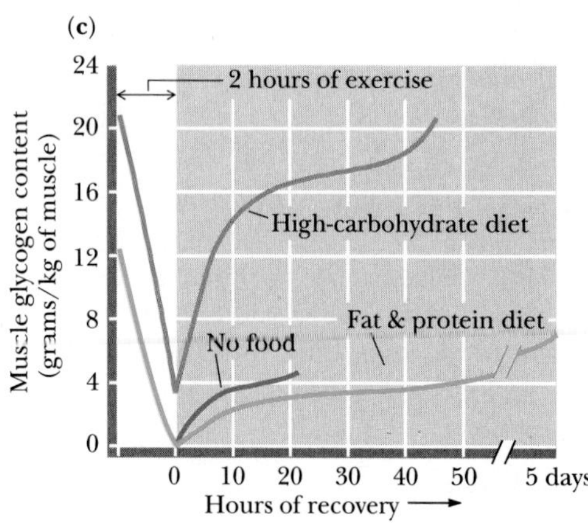

(a) Contributions of the various energy sources to muscle activity during mild exercise. (b) Consumption of glycogen stores in fast-twitch muscles during light, moderate, and heavy exercise. (c) Rate of glycogen replenishment following exhaustive exercise.

(a and c adapted from Rhodes and Pflanzer, 1992. Human Physiology. Philadelphia: Saunders College Publishing; b adapted from Horton and Terjung, 1988. Exercise, Nutrition and Energy Metabolism. New York: Macmillan.)

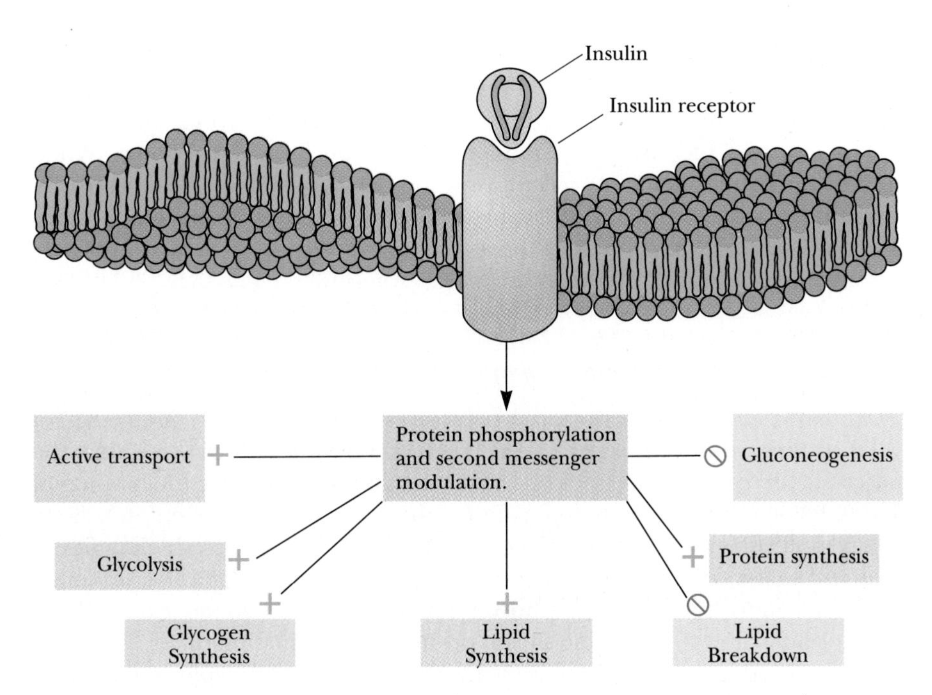

***Figure* 21.31** The metabolic effects of insulin. As described in Chapter 37, binding of insulin to membrane receptors stimulates the protein kinase activity of the receptor. Subsequent phosphorylation of target proteins modulates the effects indicated.

concentration in several ways. Insulin stimulates glycogen synthesis and inhibits glycogen breakdown in liver and muscle.

Several other physiological effects of insulin also serve to lower blood and tissue glucose levels (Figure 21.31). Insulin stimulates the active transport of glucose (and amino acids) across the plasma membranes of muscle and adipose tissue. Insulin also increases cellular utilization of glucose by inducing the synthesis of several important glycolytic enzymes, namely, glucokinase, phosphofructokinase, and pyruvate kinase. In addition, insulin acts to inhibit several enzymes of gluconeogenesis. These various actions enable the organism to respond quickly to increases in blood glucose levels.

Glucagon and Epinephrine Stimulate Glycogen Breakdown

Catabolism of tissue glycogen is triggered by the actions of the hormones **epinephrine** and **glucagon** (Figure 21.32). *In response to decreased blood glucose,* glucagon is released from the α cells in pancreatic islets of Langerhans. This peptide hormone travels through the blood to specific receptors on liver cell

H_3^+N—His—Ser—Glu—Gly—Thr—Phe—Thr—Ser—Asp—Tyr—Ser—Lys—Tyr—Leu—Asp—Ser—Arg—Arg—Ala—Gln—Asp—Phe—Val—Gln—Trp—Leu—Met—Asn—Thr—COO^-

***Figure* 21.32** The amino acid sequence of glucagon.

membranes. (Glucagon is active in liver and adipose tissue, but not in other tissues.) Similarly, signals from the central nervous system cause release of *epinephrine* (Figure 21.33)—also known as adrenaline—from the adrenal glands into the bloodstream. Epinephrine acts on liver and muscles. When either hormone binds to its receptor on the outside surface of the cell membrane, a cascade is initiated that activates glycogen phosphorylase and inhibits glycogen synthase. The result of these actions is *tightly coordinated stimulation of glycogen breakdown and inhibition of glycogen synthesis.*

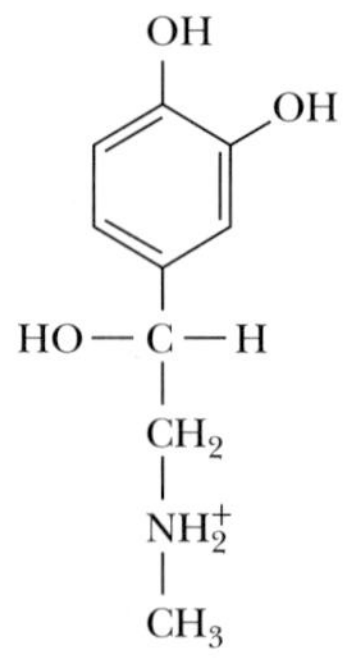

***Figure* 21.33** Epinephrine.

The Phosphorylase Cascade Amplifies the Hormonal Signal

Stimulation of glycogen breakdown involves consumption of molecules of ATP at three different steps in the cascade (Figure 21.28). Note that the cascade mechanism is a means of chemical amplification, since the binding of just a few molecules of epinephrine or glucagon results in the synthesis of many molecules of cyclic AMP, which, through the action of cAMP-dependent protein kinase, can activate many more molecules of phosphorylase kinase and even more molecules of phosphorylase. For example, an extracellular level of 10^{-10} to 10^{-8} *M* epinephrine prompts the formation of 10^{-6} *M* cyclic AMP, and for each protein kinase activated by cyclic AMP, approximately 30 phosphorylase kinase molecules are activated; these in turn activate some 800 molecules of phosphorylase. Each of these catalyzes the formation of many molecules of glucose-1-P.

The Difference Between Epinephrine and Glucagon

Although both epinephrine and glucagon exert glycogenolytic effects, they do so for quite different reasons. Epinephrine is secreted as a response to anger or fear and may be viewed as an alarm or danger signal for the organism. Called the "fight or flight" hormone, it prepares the organism for mobilization of large amounts of energy. Among the many physiological changes elicited by epinephrine, one is the initiation of the enzyme cascade, as in Figure 21.28, which leads to rapid breakdown of glycogen, inhibition of glycogen synthesis, stimulation of glycolysis, and production of energy. The burst of energy produced is the result of a 2000-fold amplification of the rate of glycolysis. Because a fear or anger response must include generation of energy (in the form of glucose)—both immediately in localized sites (the muscles) and eventually throughout the organism (as supplied by the liver)—epinephrine must be able to activate glycogenolysis in both liver and muscles.

Glucagon is involved in the long-term maintenance of steady-state levels of glucose in the blood and other tissues. It performs this function by stimulating the liver to release glucose from glycogen stores into the bloodstream. To further elevate glucose levels, glucagon also activates liver gluconeogenesis. It is important to note, however, that stabilization of blood glucose levels is managed almost entirely by the liver. Glucagon does not activate the phosphorylase cascade in muscle (muscle membranes do not contain glucagon receptors). Muscle glycogen breakdown occurs only in response to epinephrine release, and muscle tissue does not participate in maintenance of steady-state glucose levels in the blood.

Cortisol and Glucocorticoid Effects on Glycogen Metabolism

Glucocorticoids are a class of steroid hormones that exert distinct effects on liver, skeletal muscle, and adipose tissue. The effects of cortisol, a typical glucocorticoid, are best described as *catabolic,* since cortisol promotes protein

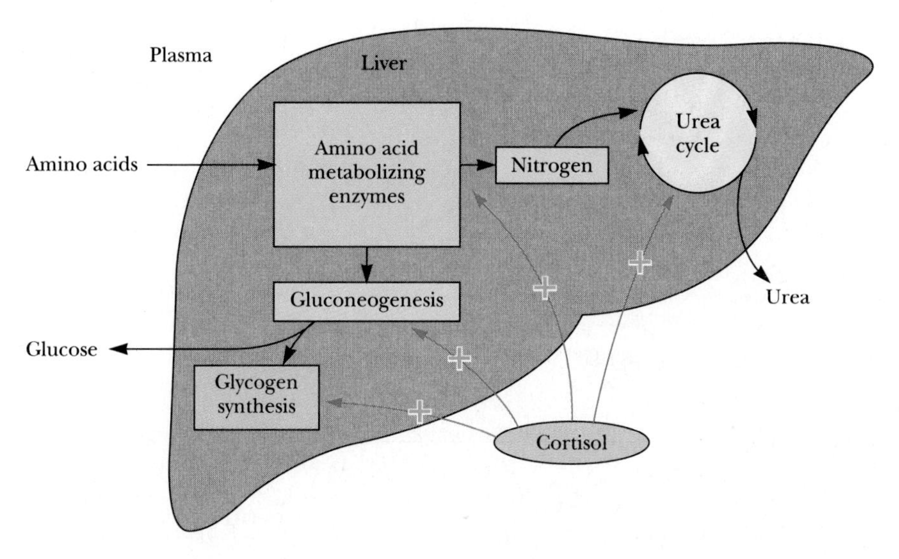

***Figure* 21.34** The effects of cortisol on carbohydrate and protein metabolism in liver.

breakdown and decreases protein synthesis in skeletal muscle. In the liver, however, it stimulates gluconeogenesis and increases glycogen synthesis. Cortisol-induced gluconeogenesis results primarily from increased conversion of amino acids into glucose (Figure 21.34). Specific effects of cortisol in the liver include increased gene expression of several of the enzymes of the gluconeogenic pathway, activation of enzymes involved in amino acid metabolism, and stimulation of the urea cycle, which disposes of nitrogen liberated during amino acid catabolism (Chapter 26).

21.6 The Pentose Phosphate Pathway

Cells require a constant supply of NADPH for reductive reactions vital to biosynthetic purposes. Much of this requirement is met by a glucose-based metabolic sequence variously called the **pentose phosphate pathway,** the **hexose monophosphate shunt,** or the **phosphogluconate pathway.** In addition to providing NADPH for biosynthetic processes, this pathway produces *ribose-5-phosphate,* which is essential for nucleic acid synthesis. Several metabolites of the pentose phosphate pathway can also be shuttled into glycolysis.

An Overview of the Pathway

The pentose phosphate pathway begins with glucose-6-phosphate, a six-carbon sugar, and produces three-, four-, five-, six-, and seven-carbon sugars (Figure 21.35). As we will see, two successive oxidations lead to the reduction of $NADP^+$ to NADPH and the release of CO_2. Five subsequent nonoxidative steps produce a variety of carbohydrates, some of which may enter the glycolytic pathway. The enzymes of the pentose phosphate pathway are particularly abundant in the cytoplasm of liver and adipose cells. These enzymes are largely absent in muscle, where glucose-6-phosphate is utilized primarily for energy production via glycolysis and the TCA cycle. These pentose phosphate pathway enzymes are located in the cytosol, which is the site of fatty acid synthesis, a pathway heavily dependent on NADPH for reductive reactions.

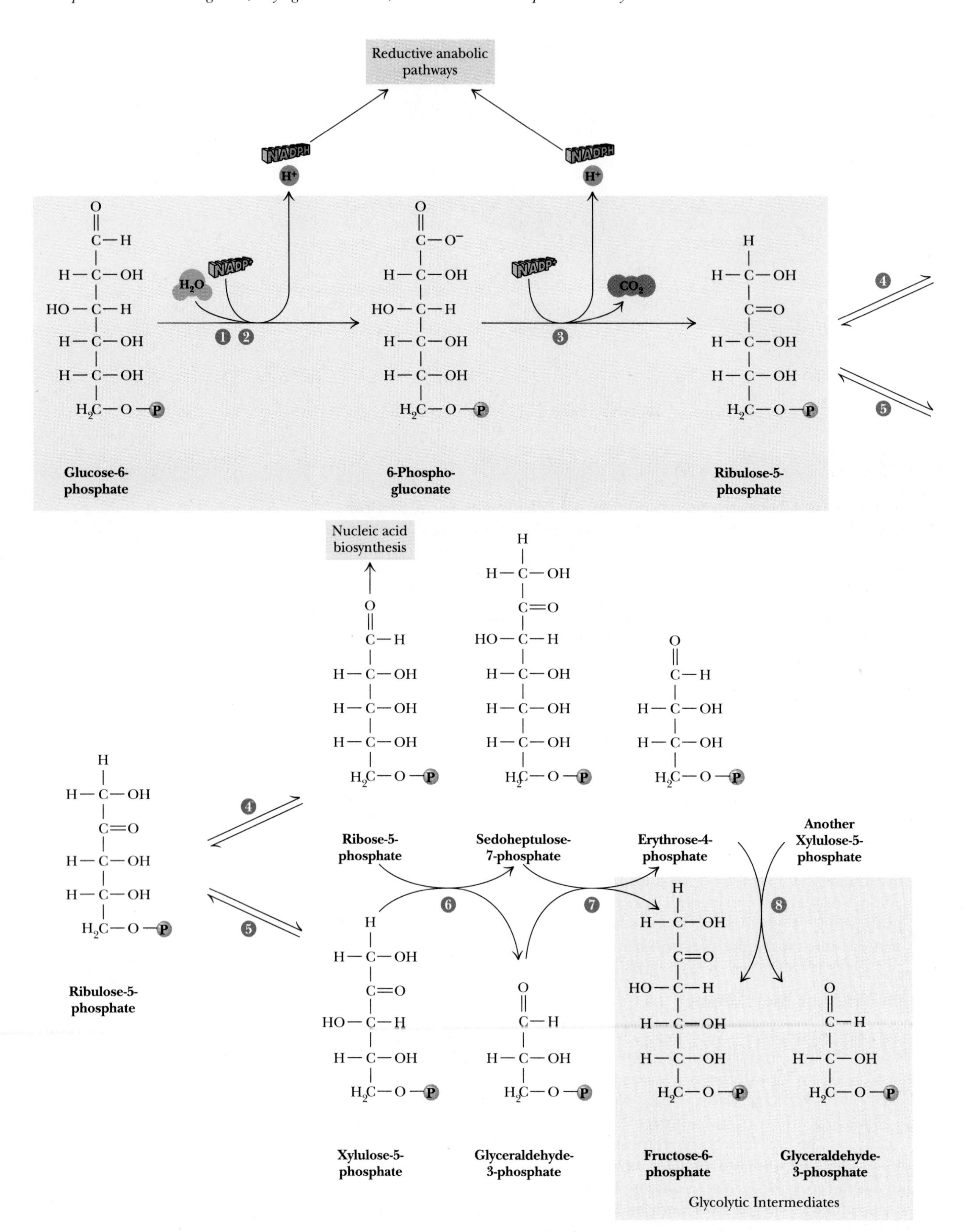

***Figure* 21.35** The pentose phosphate pathway. The numerals in the blue circles indicate the steps discussed in the text.

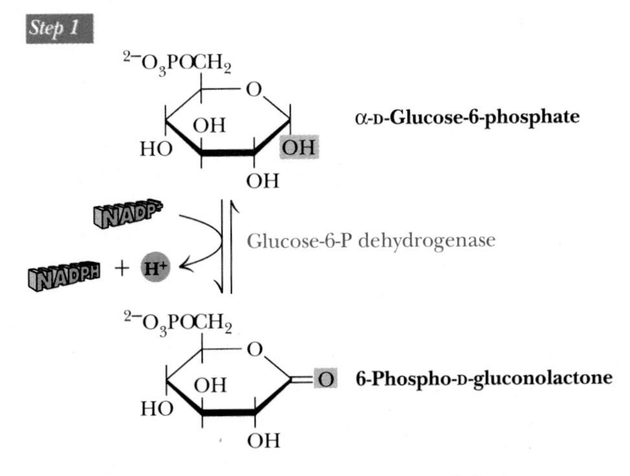

Figure **21.36** The glucose-6-phosphate dehydrogenase reaction is the committed step in the pentose phosphate pathway.

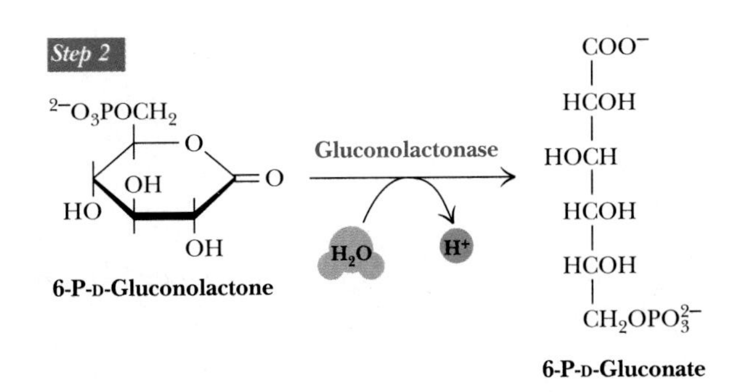

Figure **21.37** The gluconolactonase reaction.

The Oxidative Steps of the Pentose Phosphate Pathway

(1) Glucose-6-Phosphate Dehydrogenase

The pentose phosphate pathway begins with the oxidation of glucose-6-phosphate. The products of the reaction are a cyclic ester, the lactone of phosphogluconic acid, and NADPH (Figure 21.36). *Glucose-6-phosphate dehydrogenase,* which catalyzes this reaction, is highly specific for $NADP^+$. As the first step of a major pathway, the reaction is irreversible and highly regulated. Glucose-6-phosphate dehydrogenase is strongly inhibited by the product coenzyme, NADPH, and also by fatty acid esters of coenzyme A (which are intermediates of fatty acid biosynthesis). Inhibition due to NADPH depends upon the cytosolic $NADP^+$/NADPH ratio, which in the liver is about 0.015 (compared to about 725 for the NAD^+/NADH ratio in the cytosol).

(2) Gluconolactonase

The gluconolactone produced in step 1 is hydrolytically unstable and readily undergoes a spontaneous ring-opening hydrolysis, although an enzyme, *gluconolactonase,* accelerates this reaction (Figure 21.37). The linear product, the sugar acid 6-phospho-D-gluconate, is further oxidized in step 3.

(3) 6-Phosphogluconate Dehydrogenase

The oxidative decarboxylation of 6-phosphogluconate by *6-phosphogluconate dehydrogenase* yields D-ribulose-5-phosphate and another equivalent of NADPH. There are two distinct steps in this reaction (Figure 21.38): the ini-

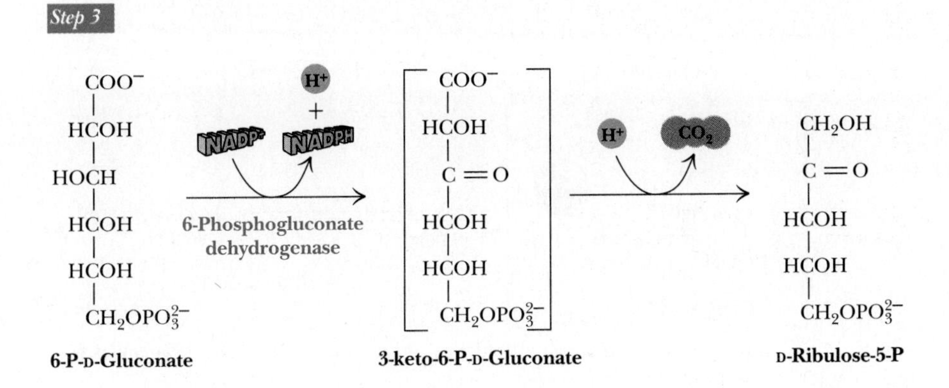

Figure **21.38** The 6-phosphogluconate dehydrogenase reaction.

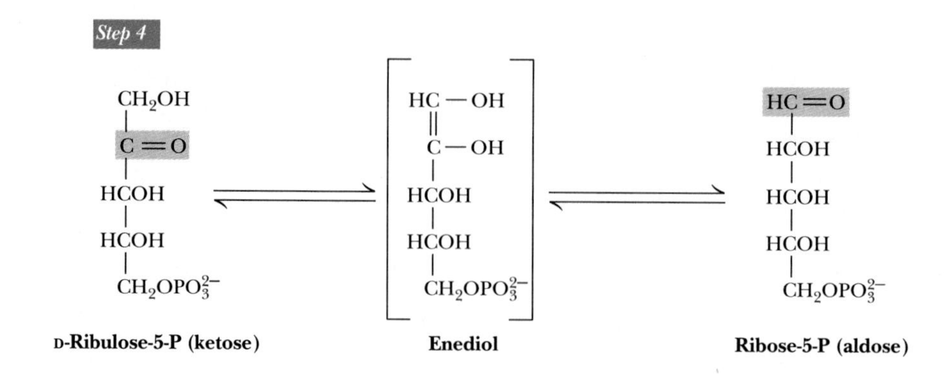

***Figure* 21.39** The phosphopentose isomerase reaction involves an enediol intermediate.

tial $NADP^+$-dependent dehydrogenation yields a β-keto acid, 3-keto-6-phosphogluconate, which is very susceptible to decarboxylation (the second step). The resulting product, D-ribulose-5-P, is the substrate for the nonoxidative reactions comprising the rest of this pathway.

The Nonoxidative Steps of the Pentose Phosphate Pathway

This portion of the pathway begins with an isomerization and an epimerization, and it leads to the formation of either D-ribose-5-phosphate or D-xylulose-5-phosphate. These intermediates can then be converted into glycolytic intermediates or directed to biosynthetic processes.

(4) Phosphopentose Isomerase

This enzyme interconverts ribulose-5-P and ribose-5-P via an enediol intermediate (Figure 21.39). The reaction (and mechanism) is quite similar to the phosphoglucoisomerase reaction of glycolysis, which interconverts glucose-6-P and fructose-6-P. The ribose-5-P produced in this reaction is utilized in the biosynthesis of coenzymes (including NADH, NADPH, FAD, and B_{12}), nucleotides, and nucleic acids (DNA and RNA). The net reaction for the first four steps of the pentose phosphate pathway is

$$\text{Glucose-6-P} + 2\ NADP^+ \longrightarrow \text{ribose-5-P} + 2\ NADPH + 2\ H^+ + CO_2$$

(5) Phosphopentose Epimerase

This reaction converts ribulose-5-P to another ketose, namely, xylulose-5-P. This reaction also proceeds by an enediol intermediate, but involves an inversion at C-3 (Figure 21.40). In the reaction, an acidic proton located α- to a carbonyl carbon is removed to generate the enediolate, but the proton is

***Figure* 21.40** The phosphopentose epimerase reaction interconverts ribulose-5-P and xylulose-5-phosphate. The mechanism involves an enediol intermediate and occurs with inversion at C-3.

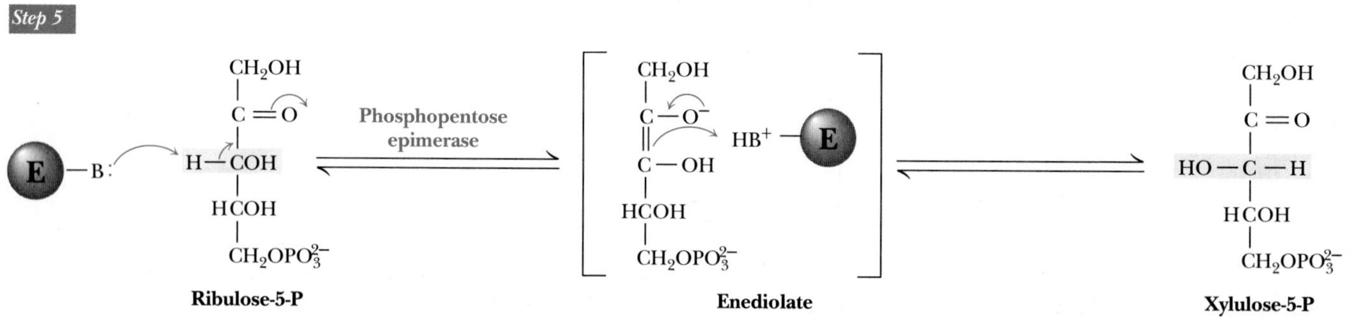

Step 5

Phosphopentose epimerase

Ribulose-5-P

Enediolate

Xylulose-5-P

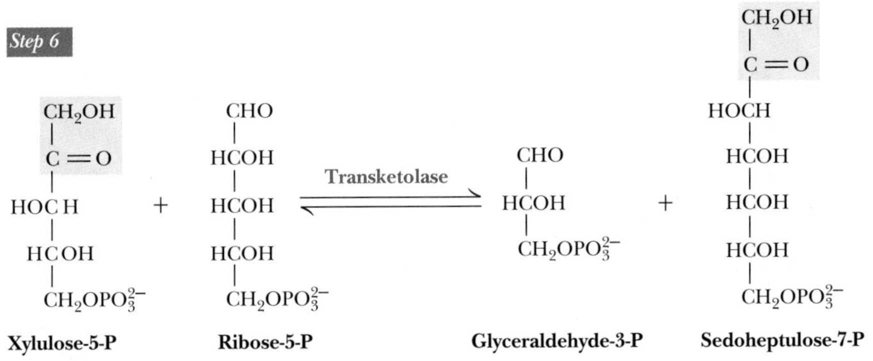

Step 6

Xylulose-5-P + Ribose-5-P ⇌ (Transketolase) Glyceraldehyde-3-P + Sedoheptulose-7-P

Figure **21.41** The transketolase reaction of step 6 in the pentose phosphate pathway.

added back to the same carbon from the opposite side. Note the distinction in nomenclature here. Interchange of groups on a single carbon is an epimerization, and interchange of groups between carbons is referred to as an isomerization.

To this point, the pathway has generated a pool of pentose phosphates. The $\Delta G^{\circ\prime}$ for each of the last two reactions is small, and the three pentose-5-phosphates coexist at equilibrium. The pathway has also produced two molecules of NADPH for each glucose-6-P converted to pentose-5-phosphate. The next three steps rearrange the five-carbon skeletons of the pentoses to produce three-, four-, six-, and seven-carbon units, which can be used for various metabolic purposes. Why should the cell do this? Very often, the cellular need for NADPH is considerably greater than the need for ribose-5-phosphate. The last three steps thus return some of the five-carbon units to glyceraldehyde-3-phosphate and fructose-6-phosphate, which can enter the glycolytic pathway. The advantage of this is that the cell has met its needs for NADPH and ribose-5-phosphate in a single pathway, yet at the same time it can return the excess carbon metabolites to glycolysis.

(6) and (8) Transketolase

The transketolase enzyme acts at both steps 6 and 8 of the pentose phosphate pathway. In both cases, the enzyme catalyzes the transfer of two-carbon units. In these reactions (and also in step 7, the transaldolase reaction, which transfers three-carbon units), the donor molecule is a ketose and the recipient is an aldose. In step 6, xylulose-5-phosphate transfers a two-carbon unit to ribose-5-phosphate to form glyceraldehyde-3-phosphate and sedoheptulose-7-phosphate (Figure 21.41). Step 8 involves a two-carbon transfer from xylulose-5-phosphate to erythrose 4-phosphate to produce another glyceraldehyde-3-phosphate and a fructose-6-phosphate (Figure 21.42). Three of these products enter directly into the glycolytic pathway. (The sedoheptulose-7-phos-

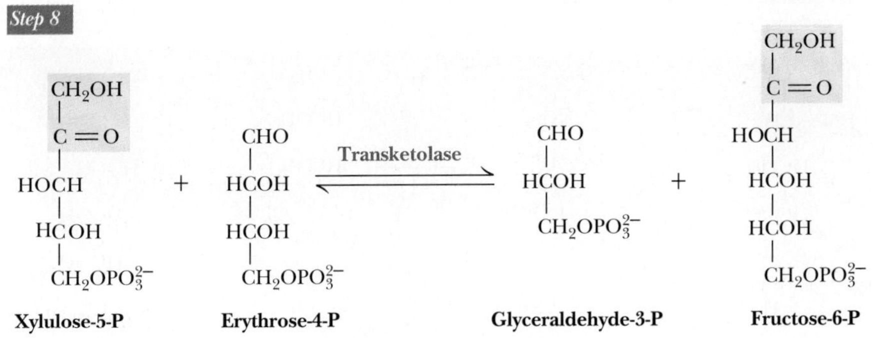

Step 8

Xylulose-5-P + Erythrose-4-P ⇌ (Transketolase) Glyceraldehyde-3-P + Fructose-6-P

Figure **21.42** The transketolase reaction of step 8 in the pentose phosphate pathway.

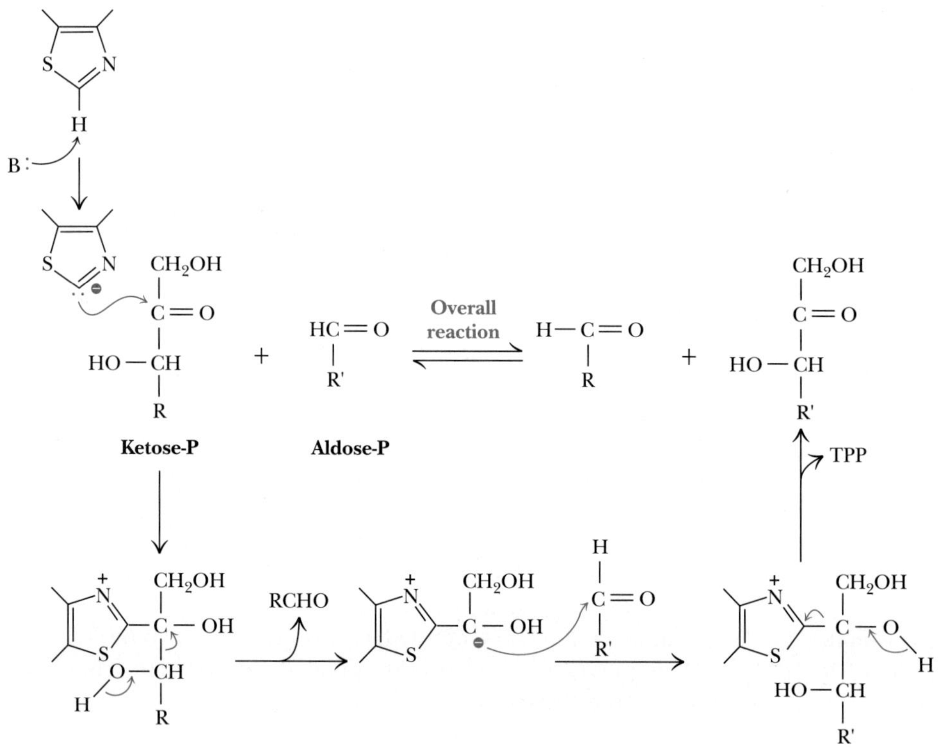

***Figure* 21.43** The mechanism of the TPP-dependent transketolase reaction. (Ironically, the group transferred in the transketolase reaction might best be described as an aldol, whereas the transferred group in the transaldolase reaction is actually a ketol. Despite the irony, these names persist for historical reasons.)

phate is taken care of in step 7, as we shall see.) Transketolase is a thiamine pyrophosphate–dependent enzyme, and the mechanism (Figure 21.43; see also Chapter 14) involves abstraction of the acidic thiazole proton of TPP, attack by the resulting carbanion at the carbonyl carbon of the ketose phosphate substrate, expulsion of the glyceraldehyde-3-phosphate product, and transfer of the two-carbon unit. Transketolase can process a variety of 2-keto sugar phosphates in a similar manner. It is specific for ketose substrates with the configuration shown, but can accept a variety of aldose phosphate substrates.

(7) Transaldolase

The transaldolase functions primarily to make a useful glycolytic substrate from the sedoheptulose-7-phosphate produced by the first transketolase reaction. This reaction (Figure 21.44) is quite similar to the aldolase reaction of

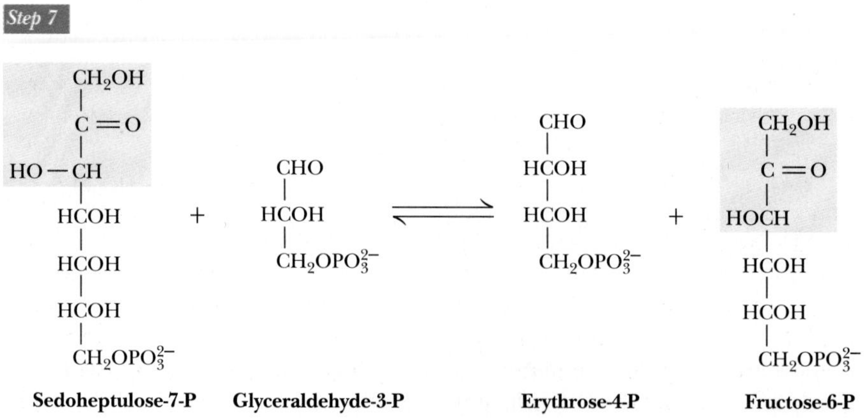

Step 7

***Figure* 21.44** The transaldolase reaction.

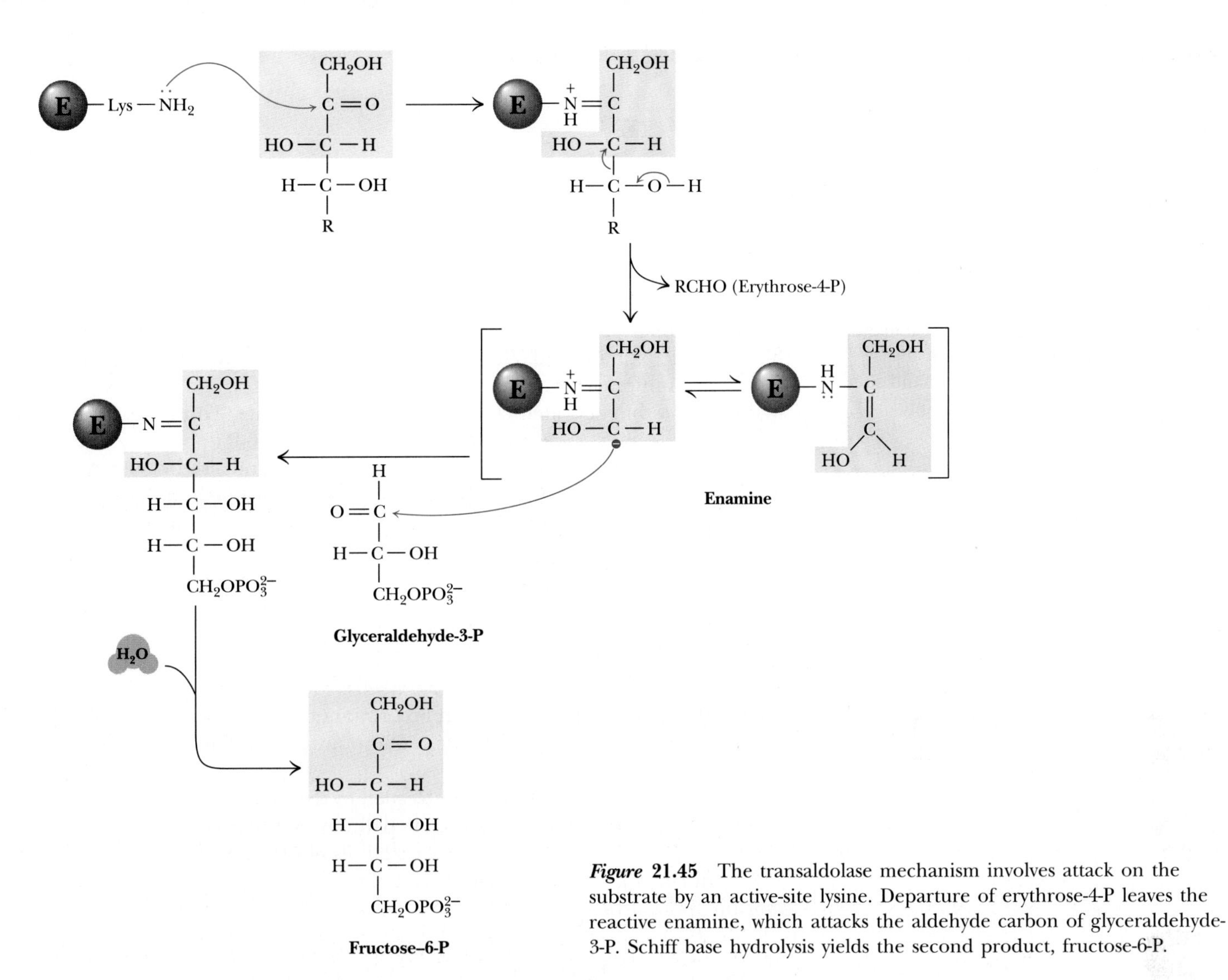

Figure **21.45** The transaldolase mechanism involves attack on the substrate by an active-site lysine. Departure of erythrose-4-P leaves the reactive enamine, which attacks the aldehyde carbon of glyceraldehyde-3-P. Schiff base hydrolysis yields the second product, fructose-6-P.

glycolysis, involving formation of a Schiff base intermediate between the sedoheptulose-7-phosphate and an active-site lysine residue (Figure 21.45). Elimination of the erythrose-4-phosphate product leaves an enamine of dihydroxyacetone, which remains stable at the active site (without imine hydrolysis) until the other substrate comes into position. Attack of the enamine carbanion at the carbonyl carbon of glyceraldehyde-3-phosphate is followed by hydrolysis of the Schiff base (imine) to yield the product fructose-6-phosphate.

Utilization of Glucose-6-P Depends on the Cell's Need for ATP, NADPH, and Ribose-5-P

It is clear that glucose-6-phosphate can be used as a substrate either for glycolysis or for the pentose phosphate pathway. The cell makes this choice on the basis of its relative needs for biosynthesis and for energy from metabolism. ATP can be produced in abundance if glucose-6-phosphate is channeled into glycolysis. On the other hand, if NADPH or ribose-5-phosphate is needed, glucose-6-phosphate can be directed to the pentose phosphate pathway. The molecular basis for this regulatory decision depends on the enzymes which metabolize glucose-6-phosphate in glycolysis and the pentose phosphate pathway. In glycolysis, phosphoglucoisomerase converts glucose-6-phosphate to fructose-6-phosphate, which is utilized by phosphofructokinase (a highly regulated enzyme) to produce fructose-1-6-bisphosphate. In the pentose phosphate pathway, glucose-6-phosphate dehydrogenase (also highly regulated) produces gluconolactone from glucose-6-phosphate. Thus, the fate of glu-

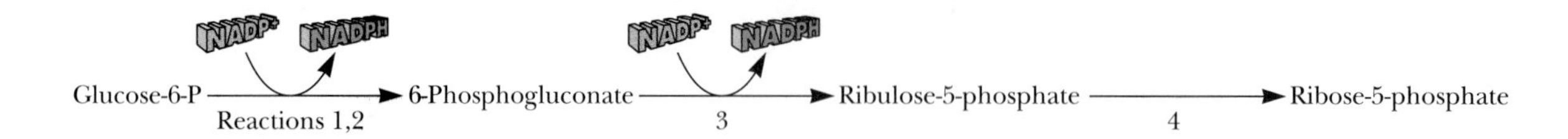

***Figure* 21.46** When biosynthetic demands dictate, the first four reactions of the pentose phosphate predominate, and the principal products are ribose-5-P and NADPH.

cose-6-phosphate is determined to a large extent by the relative activities of phosphofructokinase and glucose-6-P dehydrogenase. Recall (Chapter 18) that PFK is inhibited when the ATP/AMP ratio increases, and that it is inhibited by citrate but activated by fructose-2,6-bisphosphate. Thus, when the energy charge is high, glycolytic flux decreases. Glucose-6-P dehydrogenase, on the other hand, is inhibited by high levels of NADPH and also by the intermediates of fatty acid biosynthesis. Both of these are indicators that biosynthetic demands have been satisfied. If that is the case, glucose-6-phosphate dehydrogenase and the pentose phosphate pathway are inhibited. If NADPH levels drop, the pentose phosphate pathway turns on, and NADPH and ribose-5-phosphate are made for biosynthetic purposes.

Even when the latter choice has been made, however, the cell must still be "cognizant" of the relative needs for ribose-5-phosphate and NADPH (as well as ATP). Depending on these relative needs, the reactions of glycolysis and the pentose phosphate pathway can be combined in novel ways to emphasize the synthesis of needed metabolites. There are four principal possibilities.

(1) Both Ribose-5-P and NADPH Are Needed by the Cell In this case, the first four reactions of the pentose phosphate pathway predominate (Figure 21.46). NADPH is produced by the oxidative reactions of the pathway, and ribose-5-P is the principal product of carbon metabolism. As stated earlier, the net reaction for these processes is

$$\text{Glucose-6-P} + 2\ \text{NADP}^+ + \text{H}_2\text{O} \longrightarrow$$
$$\text{ribose-5-P} + \text{CO}_2 + 2\ \text{NADPH} + 2\ \text{H}^+$$

(2) More Ribose-5-P Than NADPH Is Needed by the Cell Synthesis of ribose-5-P can be accomplished without production of NADPH if the oxidative steps of the pentose phosphate pathway are bypassed. The key to this route is the extraction of fructose-6-P and glyceraldehyde-3-P, but not glucose-6-P, from glycolysis (Figure 21.47). The action of transketolase and transaldolase on fructose-6-P and glyceraldehyde-3-P produces three molecules of ribose-5-P from two molecules of fructose-6-P and one of glyceraldehyde-3-P. In this route, as in case 1, no carbon metabolites are returned to glycolysis. The net reaction for this route is

$$5\ \text{Glucose-6-P} + \text{ATP} \longrightarrow 6\ \text{ribose-5-P} + \text{ADP} + \text{H}^+$$

(3) More NADPH Than Ribose-5-P Is Needed by the Cell Large amounts of NADPH can be supplied for biosynthesis without concomitant production of ribose-5-P, if ribose-5-P produced in the pentose phosphate pathway is recycled to produce glycolytic intermediates. As shown in Figure 21.48, this alternative involves a complex interplay between the transketolase and transaldolase reactions to convert ribose-5-P into fructose-6-P and glyceraldehyde-3-P, which can be recycled to glucose-6-P via gluconeogenesis. The net reaction for this process is

$$6\ \text{Glucose-6-P} + 12\ \text{NADP}^+ + 6\ \text{H}_2\text{O} \longrightarrow$$
$$6\ \text{ribose-5-P} + 6\ \text{CO}_2 + 12\ \text{NADPH} + 12\ \text{H}^+$$
$$6\ \text{Ribose-5-P} \longrightarrow 5\ \text{glucose-6-P} + \text{P}_i$$

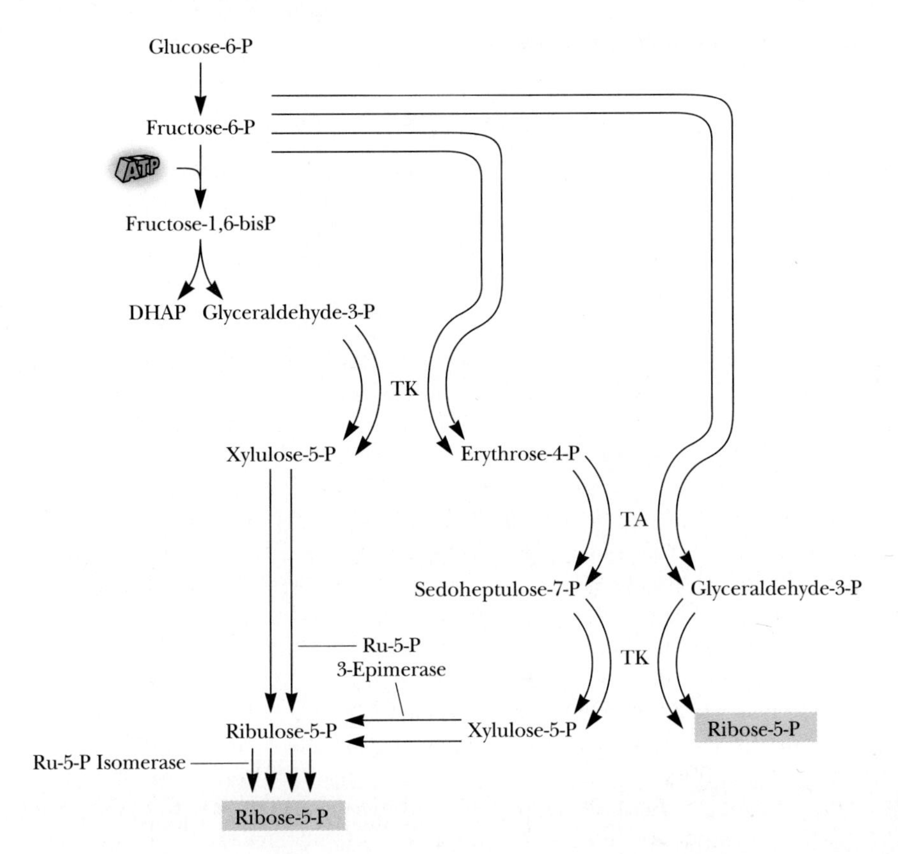

Figure **21.47** The oxidative steps of the pentose phosphate pathway can be bypassed if the primary need is for ribose-5-P.

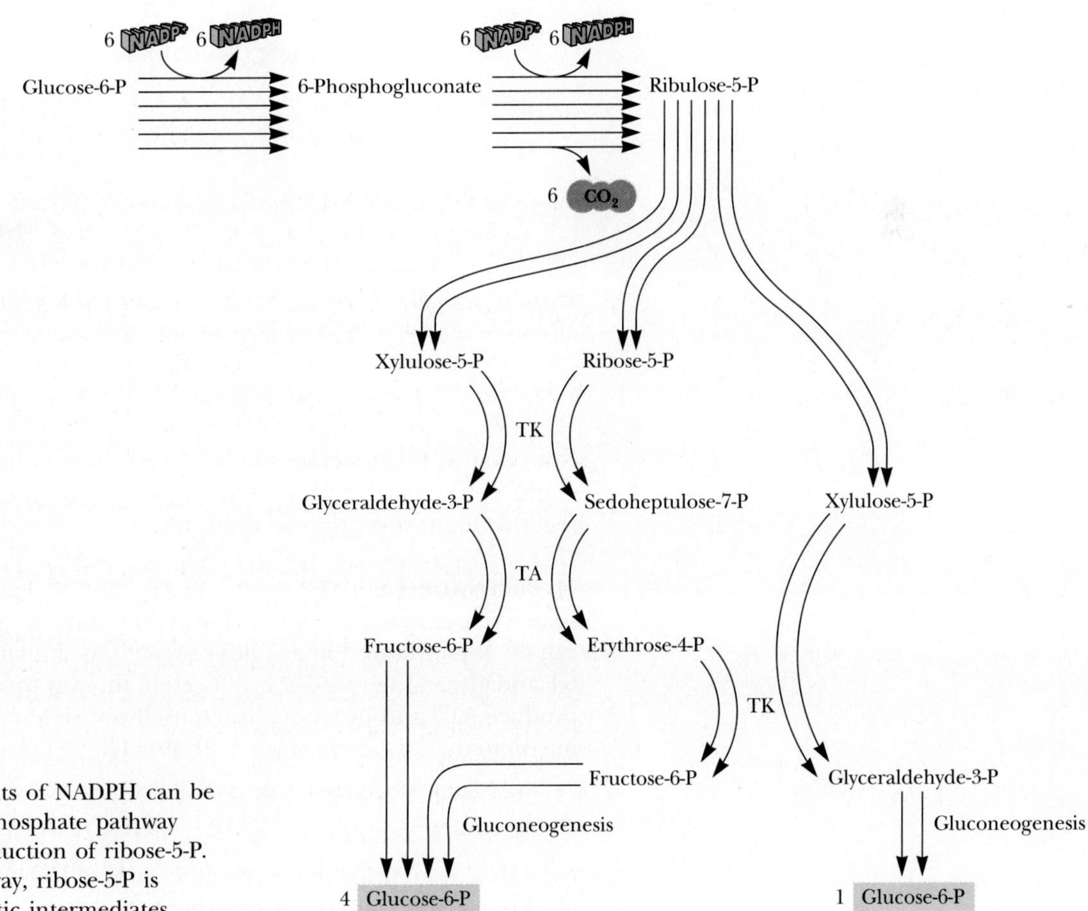

Figure **21.48** Large amounts of NADPH can be produced by the pentose phosphate pathway without significant net production of ribose-5-P. In this version of the pathway, ribose-5-P is recycled to produce glycolytic intermediates.

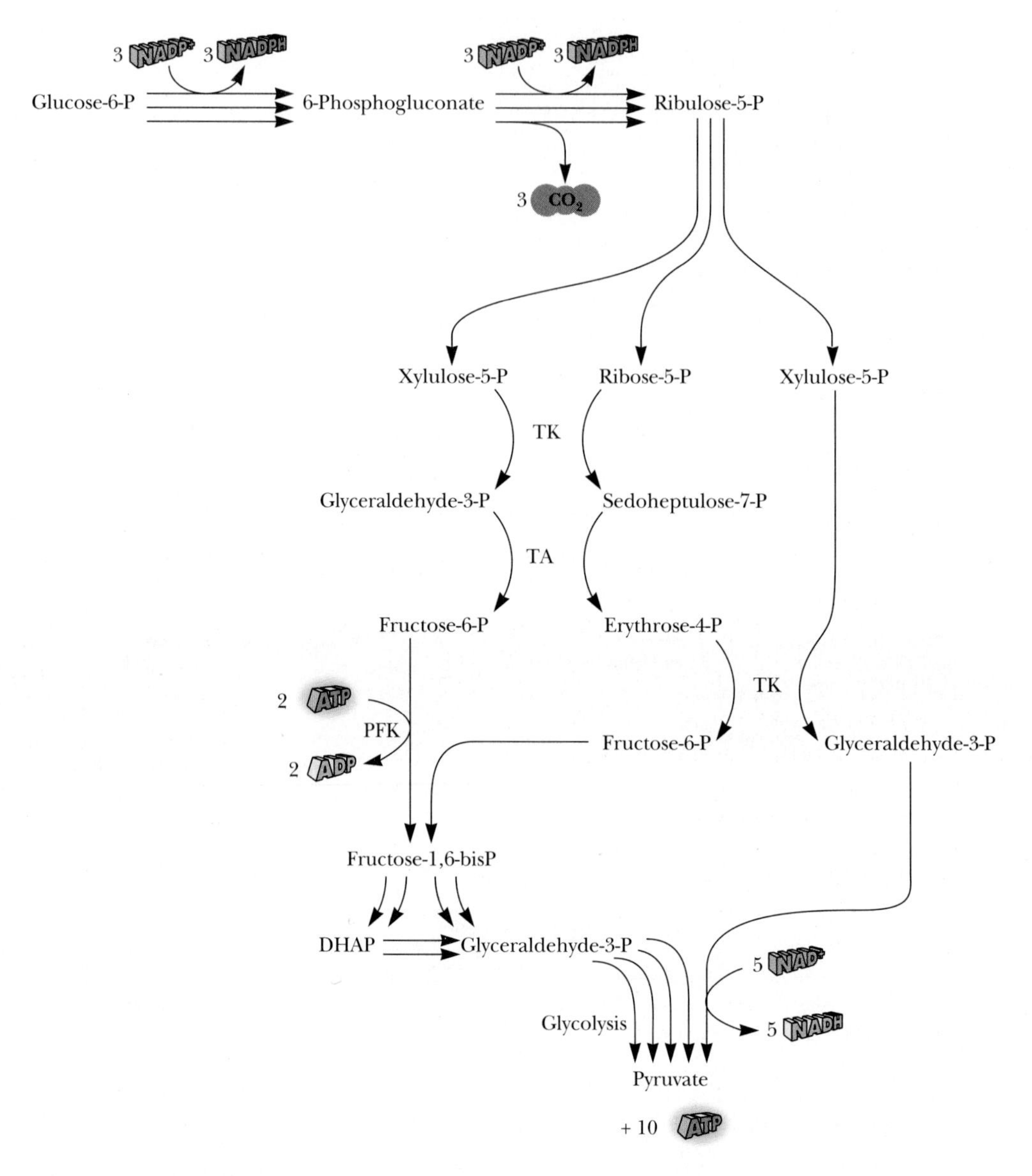

***Figure* 21.49** Both ATP and NADPH (as well as NADH) can be produced by this version of the pentose phosphate and glycolytic pathways.

Note that in this scheme, the six hexose sugars have been converted to six pentose sugars with release of six molecules of CO_2, and the six pentoses are reconverted to five glucose molecules.

(4) Both NADPH and ATP Are Needed by the Cell, but Ribose-5-P Is Not Under some conditions, both NADPH and ATP must be provided in the cell. This can be accomplished in a series of reactions similar to case 3, if the fructose-6-P and glyceraldehyde-3-P produced in this way proceed through glycolysis to produce ATP and pyruvate, which itself can yield even more ATP by continuing on to the TCA cycle (Figure 21.49). The net reaction for this alternative is

$$\text{3 Glucose-6-P} + 5\ \text{NAD}^+ + 6\ \text{NADP}^+ + 8\ \text{ADP} + 5\ \text{P}_\text{i} \longrightarrow$$
$$\text{5 pyruvate} + 3\ \text{CO}_2 + 5\ \text{NADH} + 6\ \text{NADPH} + 8\ \text{ATP} + 2\ \text{H}_2\text{O} + 8\ \text{H}^+$$

Note that, except for the three molecules of CO_2, all the other carbon from glucose-6-P is recovered in pyruvate.

Problems

1. Consider the balanced equation for gluconeogenesis in Section 21.1. Account for each of the components of this equation and the indicated stoichiometry.

2. Calculate $\Delta G^{\circ\prime}$ and ΔG for gluconeogenesis in the erythrocyte (using data in Table 18.2) and see how closely your values match those in Section 21.1.

3. Use the data of Figure 21.12 to calculate the percent inhibition of fructose-1,6-bisphosphatase by 25 μM fructose-2,6-bisphosphate when fructose-1,6-bisphosphate is (a) 25 μM and (b) 100 μM.

4. Discuss the relative advantages of α- and β-amylases for breakdown of glycogen.

5. Suggest an explanation for the exergonic nature of the glycogen synthase reaction ($\Delta G^{\circ\prime} = -13.3$ kJ/mol). Consult Chapter 16 to review the energetics of high-energy phosphate compounds if necessary.

6. Using the values in Table 23.1 for body glycogen content and the data in part b of the illustration for A Deeper Look (page 684), calculate the rate of energy consumption by muscles in heavy exercise (in J/sec). Use the data for fast-twitch muscle.

7. What would be the distribution of carbon from positions 1, 3, and 6 of glucose after one pass through the pentose phosphate pathway if the primary need of the organism is for ribose-5-P and the oxidative steps are bypassed (Figure 21.47)?

8. What is the fate of carbon from positions 2 and 4 of glucose-6-P after one pass through the scheme shown in Figure 21.49?

9. Which reactions of the pentose phosphate pathway would be inhibited by $NaBH_4$? Why?

10. Imagine a glycogen molecule with 8000 glucose residues. If branches occur every eight residues, how many reducing ends does the molecule have? If branches occur every 12 residues, how many reducing ends does it have? How many nonreducing ends does it have in each of these cases?

11. Explain the effects of each of the following on the rates of gluconeogenesis and glycogen metabolism:
    a. Increasing the concentration of tissue fructose-1,6-bisphosphate
    b. Increasing the concentration of blood glucose
    c. Increasing the concentration of blood insulin
    d. Increasing the amount of blood glucagon
    e. Decreasing levels of tissue ATP
    f. Increasing the concentration of tissue AMP
    g. Decreasing the concentration of fructose-6-phosphate

12. The free-energy change of the glycogen phosphorylase reaction is $\Delta G^{\circ\prime} = +3.1$ kJ/mol. If $[P_i] = 1$ mM, what is the concentration of glucose-1-P when this reaction is at equilibrium?

Further Reading

Browner, M. F., and Fletterick, R. J., 1992. Phosphorylase: A biological transducer. *Trends in Biochemical Sciences* **17:**66–71.

Fox, E. L., 1984. *Sports Physiology,* 2nd ed. Philadelphia: Saunders College Publishing.

Hers, H. G., and Hue, L., 1983. Gluconeogenesis and related aspects of glycolysis. *Annual Review of Biochemistry* **52:**617–653.

Horton, E. S., and Terjung, R. L., eds. 1988. *Exercise, Nutrition and Energy Metabolism.* New York: Macmillan.

Johnson, L. N., 1992. Glycogen phosphorylase: Control by phosphorylation and allosteric effectors. *FASEB Journal* **6:**2274–2282.

Larner, J., 1990. Insulin and the stimulation of glycogen synthesis: The road from glycogen structure to glycogen synthase to cyclic AMP-dependent protein kinase to insulin mediators. *Advances in Enzymology* **63:**173–231.

Newsholme, E. A., Chaliss, R. A. J., and Crabtree, B., 1984. Substrate cycles: Their role in improving sensitivity in metabolic control. *Trends in Biochemical Sciences* **9:**277–280.

Newsholme, E. A., and Leech, A. R., 1983. *Biochemistry for the Medical Sciences.* New York: Wiley.

Pilkis, S. J., El-Maghrabi, M. R., and Claus, T. H., 1988. Hormonal regulation of hepatic gluconeogenesis and glycolysis. *Annual Review of Biochemistry* **57:**755–783.

Rhoades, R., and Pflanzer, R., 1992. *Human Physiology.* Philadelphia: Saunders College Publishing.

Sies, H., ed., 1982. *Metabolic Compartmentation.* London: Academic Press.

Sukalski, K. A., and Nordlie, R. C., 1989. Glucose-6-phosphatase: Two concepts of membrane-function relationship. *Advances in Enzymology* **62:**93–117.

Taylor, S. S., et al., 1993. A template for the protein kinase family. *Trends in Biochemical Sciences* **18:**84–89.

Van Schaftingen, E., and Hers, H.-G., 1981. Inhibition of fructose-1,6-bisphosphatase by fructose-2,6-bisphosphate. *Proceedings of the National Academy of Sciences,* USA **78:**2861–2863.

Williamson, D. H., Lund, P., and Krebs, H. A., 1967. The redox state of free nicotinamide–adenine dinucleotide in the cytoplasm and mitochondria of rat liver. *Biochemical Journal* **103:**514–527.

Woodget, J. R., 1991. A common denominator linking glycogen metabolism, nuclear oncogenes, and development. *Trends in Biochemical Sciences* **16:**177–181.

"Is it so small a thing
To have enjoyed the sun,
To have lived light in the spring,
To have loved, to have thought, to have done . . . ?"

Matthew Arnold, *Empedocles on Etna* (1852)

Chapter 22

Photosynthesis

"Irises," Vincent van Gogh

Outline

22.1 General Aspects of Photosynthesis
22.2 Photosynthesis Depends on the Photoreactivity of Chlorophyll
22.3 Eukaryotic Phototrophs Possess Two Distinct Photosystems
22.4 The *Z* Scheme of Photosynthetic Electron Transfer
22.5 The Molecular Architecture of Photosynthetic Reaction Centers
22.6 The Quantum Yield of Photosynthesis
22.7 Light-Driven ATP Synthesis—Photophosphorylation
22.8 Carbon Dioxide Fixation
22.9 The Ribulose Bisphosphate Oxygenase Reaction: Photorespiration
22.10 The Calvin–Benson Cycle
22.11 Regulation of Carbon Dioxide Fixation
22.12 The C-4 Pathway of CO_2 Fixation
22.13 Crassulacean Acid Metabolism

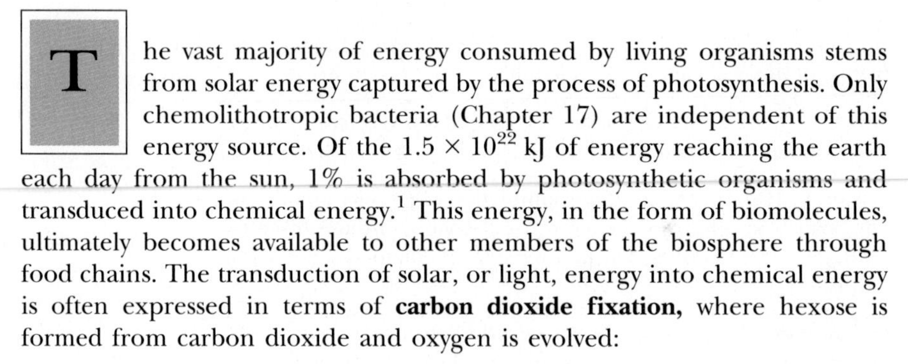

The vast majority of energy consumed by living organisms stems from solar energy captured by the process of photosynthesis. Only chemolithotropic bacteria (Chapter 17) are independent of this energy source. Of the 1.5×10^{22} kJ of energy reaching the earth each day from the sun, 1% is absorbed by photosynthetic organisms and transduced into chemical energy.[1] This energy, in the form of biomolecules, ultimately becomes available to other members of the biosphere through food chains. The transduction of solar, or light, energy into chemical energy is often expressed in terms of **carbon dioxide fixation,** where hexose is formed from carbon dioxide and oxygen is evolved:

$$6\ CO_2 + 6\ H_2O \xrightarrow{\text{Light}} C_6H_{12}O_6 + 6\ O_2 \qquad (22.1)$$

[1]Of the remaining 99%, two-thirds is absorbed by the earth and oceans, thereby heating the planet; the remaining one-third is lost as light reflected back into space.

Estimates indicate that 10^{11} tons of carbon dioxide are fixed globally per year, of which one-third is fixed in the oceans, primarily by photosynthetic marine microorganisms.

Although photosynthesis is traditionally equated with CO_2 fixation, light energy (or rather the chemical energy derived from it) can be used to drive virtually any cellular process. The assimilation of inorganic forms of nitrogen and sulfur into organic molecules (Chapter 25) represents two other metabolic conversions driven by light energy in green plants. Our previous considerations of aerobic metabolism (Chapters 18 through 20) treated cellular respiration (precisely the reverse of Equation 22.1) as the central energy-releasing process in life. It necessarily follows that the formation of hexose from carbon dioxide and water, the products of cellular respiration, must be endergonic. The requisite energy is derived from light. Note that in the carbon dioxide fixation reaction described, light is used to drive a chemical reaction against its thermodynamic potential.

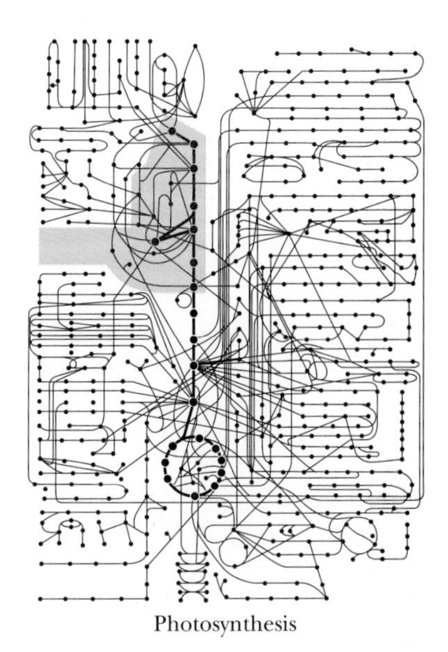

Photosynthesis

22.1 General Aspects of Photosynthesis

Photosynthesis Occurs in Membranes

Organisms capable of photosynthesis are very diverse, ranging from simple prokaryotic forms to the largest organisms of all, *Sequoia gigantea,* the giant redwood trees of California. Despite this diversity, we can, as comparative biochemists, anticipate certain generalities regarding photosynthesis. An important one is that *photosynthesis occurs in membranes.* In photosynthetic prokaryotes, the photosynthetic membranes fill up the cell interior; in photosynthetic eukaryotes, the photosynthetic membranes are localized in **chloroplasts** (Figures 22.1 and 22.2). Chloroplasts are one member in a family of related plant-specific organelles known as **plastids.** Chloroplasts themselves show a range of diversity, from the single, spiral chloroplast that gives *Spirogyra* its name to the multitude of ellipsoidal plastids typical of higher plant cells (Figure 22.3).

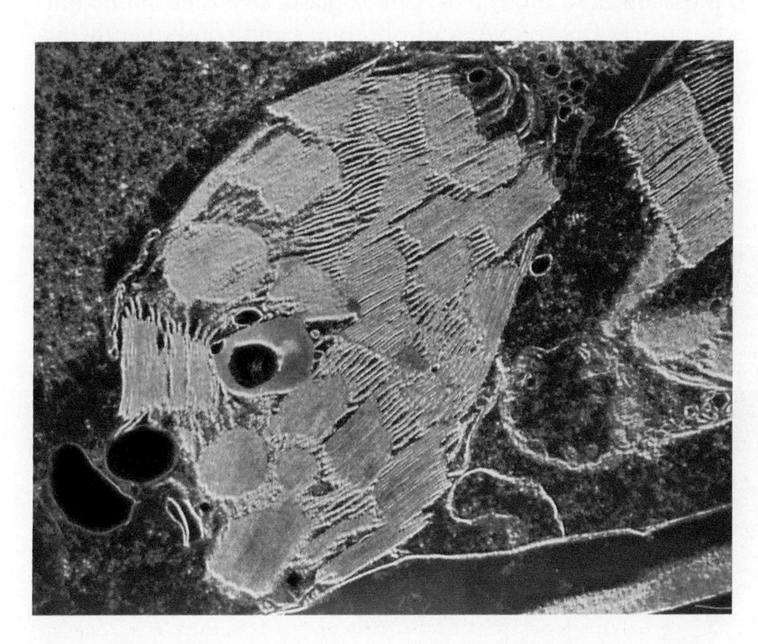

***Figure* 22.1** Electron micrograph of a representative chloroplast.

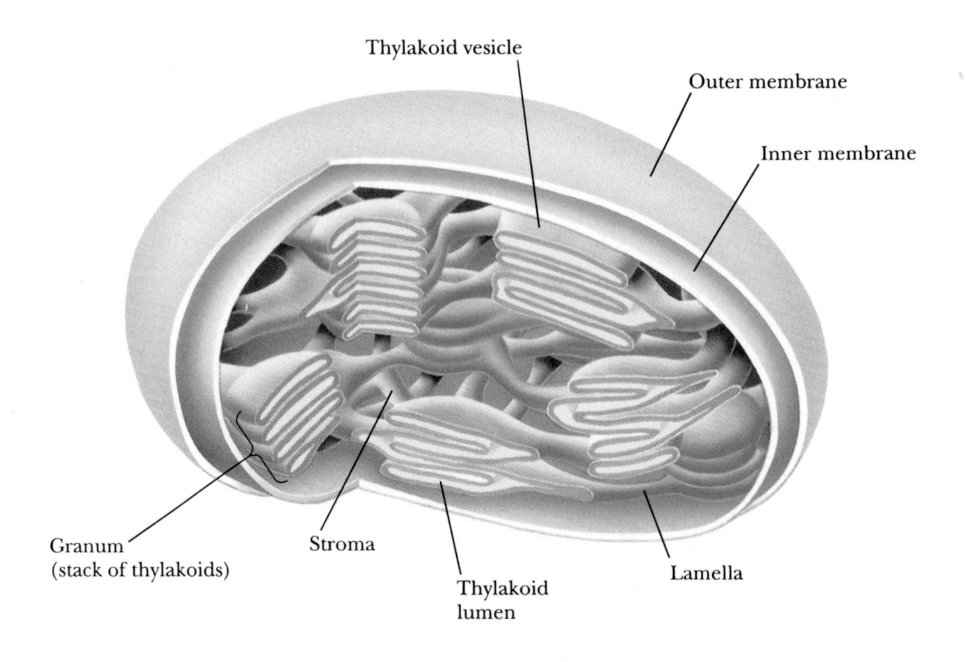

***Figure* 22.2** Schematic diagram of an idealized chloroplast.

Characteristic of all chloroplasts, however, is the organization of the inner membrane system, the so-called **thylakoid membrane.** The thylakoid membrane is organized into paired folds that extend throughout the organelle, as in Figure 22.1. These paired folds, or **lamellae,** give rise to flattened sacs or disks, **thylakoid vesicles** (from the Greek *thylakos,* meaning "sack"), which occur in stacks called **grana.** A single stack, or **granum,** may contain dozens of thylakoid vesicles, and different grana are joined by lamellae running through the soluble portion, or **stroma,** of the organelle. Chloroplasts thus possess three membrane-bound compartments: the intermembrane space, the stroma, and the interior of the thylakoid vesicles, the so-called **thylakoid space** (also known as the **thylakoid lumen**). As we shall see, this third compartment serves an important function in the transduction of light energy into ATP formation. The thylakoid membrane that surrounds it has a highly characteristic lipid composition and, like the inner membrane of the mitochondrion, is impermeable to most ions and molecules. Chloroplasts, like their mitochondrial counterparts, possess DNA, RNA, and ribosomes and consequently display a considerable amount of autonomy. However, many critical chloroplast components are encoded by nuclear genes, so autonomy is far from absolute.

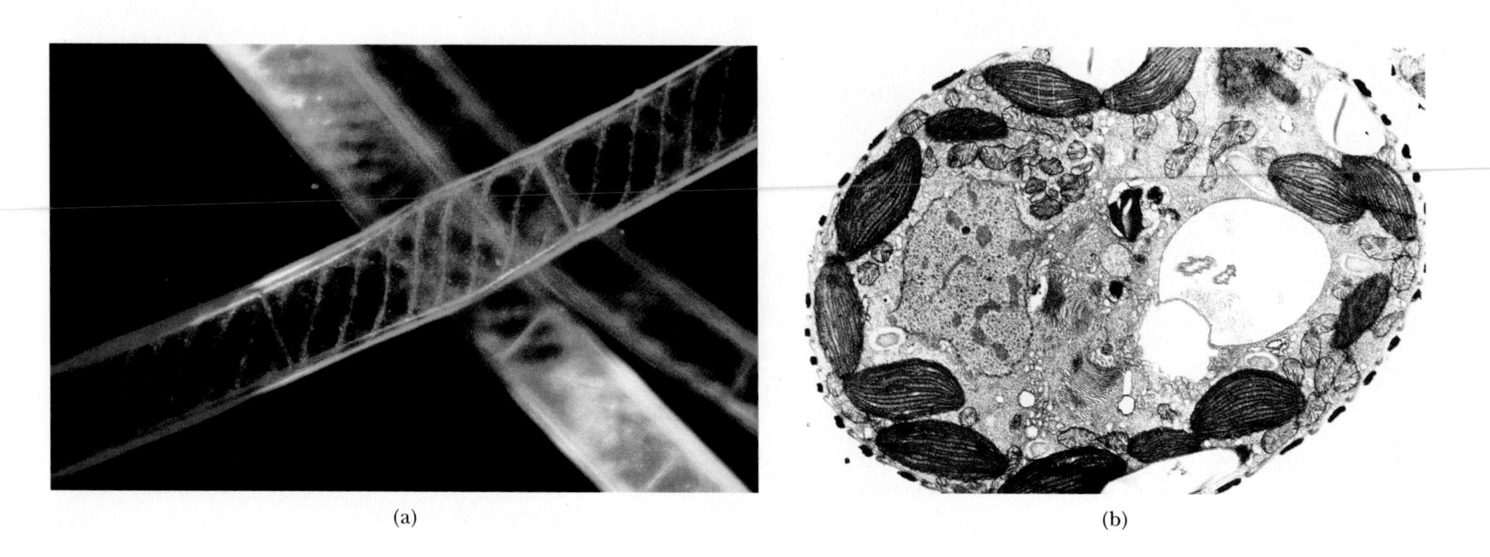

***Figure* 22.3** (a) *Spirogyra*—a freshwater green alga. (b) A higher plant cell.

Photosynthesis Consists of Both Light Reactions and Dark Reactions

If a chloroplast suspension is illuminated in the absence of carbon dioxide, oxygen is evolved. Furthermore, if the illuminated chloroplasts are now placed in the dark and supplied with CO_2, net hexose synthesis can be observed (Figure 22.4). Thus, the evolution of oxygen can be temporally separated from carbon dioxide fixation and also has a light dependency that CO_2 fixation lacks. The **light reactions** of photosynthesis, of which O_2 evolution is only one part, are associated with the thylakoid membranes. In contrast, the **dark reactions,** notably carbon dioxide fixation, are located in the stroma. A concise summary of the photosynthetic process is that radiant electromagnetic energy (light) is transformed by a specific photochemical system located in the thylakoids to yield chemical energy in the form of reducing potential (NADPH) and high-energy phosphate (ATP). NADPH and ATP can then be used to drive the endergonic process of hexose formation from CO_2 by a series of enzymatic reactions found in the stroma (see Equation 22.3, which follows).

temporally: with regard to time

Water Is the Ultimate e^- Donor for Photosynthetic $NADP^+$ Reduction

In green plants, water serves as the ultimate electron donor for the photosynthetic generation of reducing equivalents. The reaction sequence

$$2\,H_2O + 2\,NADP^+ + x\,ADP + x\,P_i \xrightarrow{nh\nu} O_2 + 2\,NADPH + 2\,H^+ + x\,ATP + x\,H_2O \quad (22.2)$$

describes the process, where $nh\nu$ symbolizes *light energy* (n is some number of photons of energy $h\nu$, where h is Planck's constant and ν is the frequency of the light). The stoichiometry of ATP formation depends on the pattern of photophosphorylation operating in the cell at the time and on the ATP yield in terms of the ATP/H^+ ratio, as we will see later. Nevertheless, the stoichiometry of the metabolic pathway of CO_2 fixation is certain:

$$12\,NADPH + 12\,H^+ + 18\,ATP + 6\,CO_2 + 12\,H_2O \longrightarrow C_6H_{12}O_6 + 12\,NADP^+ + 18\,ADP + 18\,P_i \quad (22.3)$$

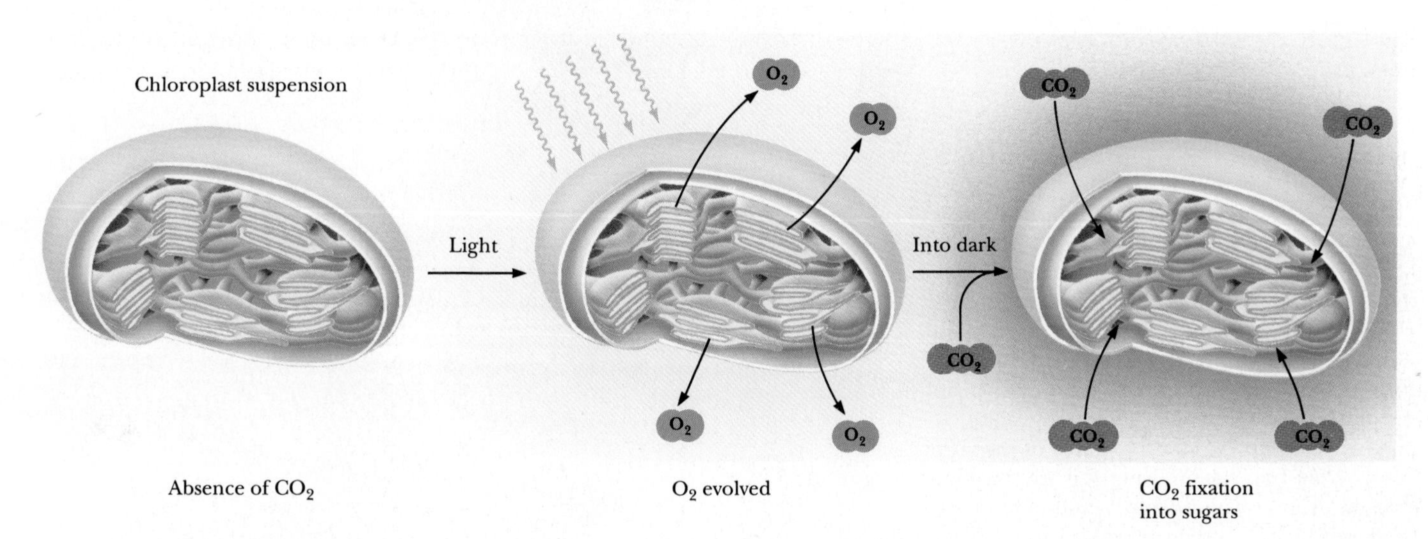

Figure **22.4** The light and dark reactions of photosynthesis. Light reactions are associated with the thylakoid membranes, and dark reactions are associated with the stroma.

Generalized Equation for the Photochemical Reaction

In 1931, comparative study of photosynthesis in bacteria led van Niel to a more general formulation of the photochemical reaction:

$$\underset{\text{Hydrogen acceptor}}{CO_2} + \underset{\text{Hydrogen donor}}{2\,H_2A} \xrightarrow{\text{Light}} \underset{\text{Reduced acceptor}}{(CH_2O)} + \underset{\text{Oxidized donor}}{2\,A} + H_2O \qquad (22.4)$$

In photosynthetic bacteria, H_2A is variously H_2S (photosynthetic green and purple sulfur bacteria), isopropanol, or some similar oxidizable substrate. [(CH_2O) symbolizes a carbohydrate unit.]

$$CO_2 + 2\,H_2S \longrightarrow (CH_2O) + H_2O + 2\,S$$

$$CO_2 + 2\,CH_3\text{—}CHOH\text{—}CH_3 \longrightarrow (CH_2O) + H_2O + 2\,CH_3\text{—}\overset{O}{\overset{\|}{C}}\text{—}CH_3$$

In cyanobacteria and the eukaryotic photosynthetic cells of algae and higher plants, H_2A is H_2O, as implied earlier, and 2 A is O_2. The accumulation of O_2 to constitute 20% of the earth's atmosphere is the direct result of eons of global oxygenic photosynthesis.

22.2 Photosynthesis Depends on the Photoreactivity of Chlorophyll

Chlorophylls are magnesium-containing substituted tetrapyrroles whose basic structure is reminiscent of heme, the iron-containing porphyrin (Chapters 4 and 20). Chlorophylls differ from heme in a number of properties: magnesium instead of iron is coordinated in the center of the planar conjugated

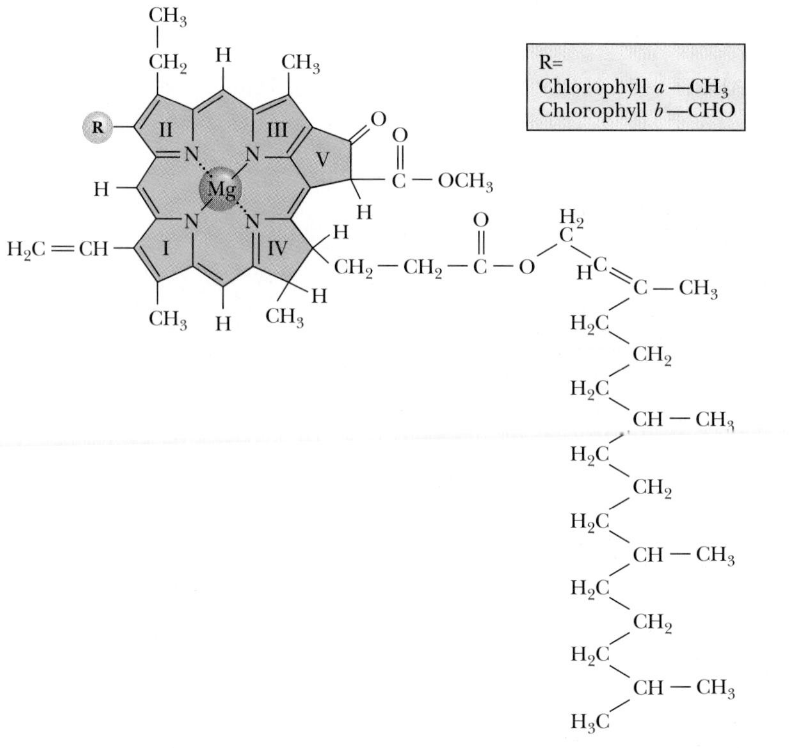

R=
Chlorophyll *a* —CH_3
Chlorophyll *b* —CHO

Hydrophobic phytyl side chain

***Figure* 22.5** Structures of chlorophyll *a* and *b*. Chlorophylls are structurally related to hemes, except Mg^{2+} replaces Fe^{2+} and ring II is more reduced than the corresponding ring of the porphyrins. This chlorophyll tetrapyrrole ring system is a chlorin. R = CH_3 in chlorophyll *a*; R = CHO in chlorophyll *b*. Note that the aldehyde C=O bond of chlorophyll *b* introduces an additional double bond into conjugation with the double bonds of the tetrapyrrole ring system. Ring V is the additional ring created by interaction of the substituent of the methine bridge between pyrroles III and IV with the side chain of ring III. The phytyl side chain of ring IV provides a hydrophobic tail to anchor the chlorophyll in membrane protein complexes.

ring structure; a long-chain alcohol, **phytol,** is esterified to a pyrrole ring substituent; and the methine bridge linking pyrroles III and IV is substituted and cross-linked to ring III, leading to the formation of a fifth five-membered ring. The structures of chlorophyll *a* and *b* are shown in Figure 22.5.

Chlorophylls are excellent light absorbers because of their aromaticity. That is, they possess delocalized π electrons above and below the planar ring structure. The energy differences between electronic states in these π orbitals correspond to the energies of visible light photons. Light energy is absorbed, promoting electrons to higher orbitals and enhancing the potential for transfer of these electrons to suitable acceptors. Loss of such a photo-excited electron to an acceptor is an oxidation–reduction reaction. The essence of the process is the transduction of light energy into the chemical energy of a redox reaction. Light absorption leads to electron excitation, activating photoreceptive molecules for oxidation–reduction reactions. George Wald has made an insightful generalization regarding life in the universe. He noted that extraterrestrial life can be expected to share our dependency on light because the energies of that portion of the electromagnetic radiation spectrum defined as light are equivalent to those energies that result in chemical reactions.

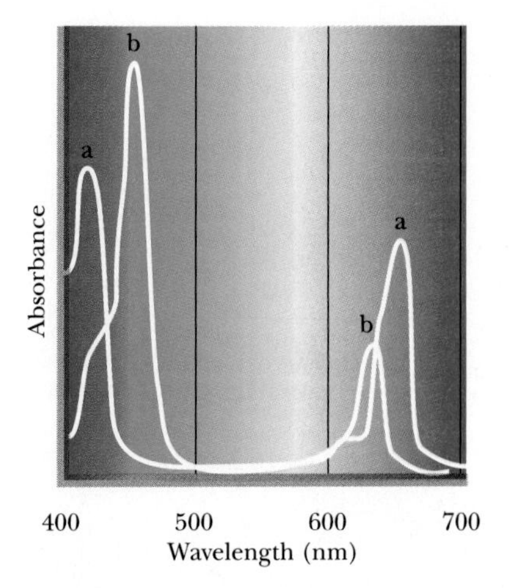

***Figure* 22.6** Absorption spectra of chlorophylls *a* and *b.*

Chlorophylls and Accessory Light-Harvesting Pigments

The absorption spectra of chlorophylls *a* and *b* (Figure 22.6) differ somewhat. Plants that possess both chlorophylls can harvest a wider spectrum of incident energy. Other pigments in photosynthetic organisms, so-called **accessory light-harvesting pigments** (Figure 22.7), increase the possibility for absorption of incident light of wavelengths not absorbed by the chlorophylls. These accessory pigments, such as *carotenoids* and *phycobilins,* are also responsible for the magnificent colors of autumn. They persist longer after leaf death than the more labile green chlorophylls, finally imparting their particular hues to the plant. These pigments, like chlorophyll, possess many conjugated double bonds and thus absorb electromagnetic radiation of wavelengths characteristic of visible light.

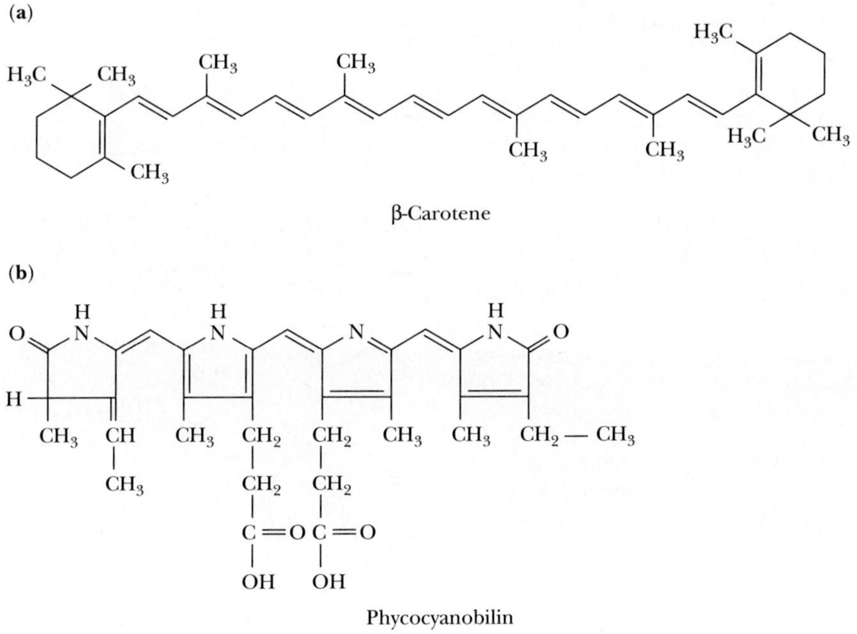

(a)

β-Carotene

(b)

Phycocyanobilin

***Figure* 22.7** Structures of representative accessory light-harvesting pigments in photosynthetic cells. (a) *β*-Carotene, an accessory light-harvesting pigment in leaves. Note the many conjugated double bonds. (b) Phycocyanobilin, a blue pigment found in cyanobacteria. It is a linear or open pyrrole.

The Fate of Light Energy Absorbed by Photosynthetic Pigments

The quantum of light energy absorbed by the photosynthetic pigments has four possible fates (Figure 22.8):

A. **Loss as heat.** The energy can be dissipated as heat through redistribution of the energy among the vibrational modes of the pigment molecule. In terms of photochemical potentiality, this fate represents energy lost from the system.

B. **Loss as light.** Energy of excitation reappears as **fluorescence** (light emission); a photon of fluorescence is emitted as the e^- returns to a lower orbital. This fate is common only in saturating light intensities. For thermodynamic reasons, the photon of fluorescence is necessarily of longer wavelength and hence lower energy than the quantum of excitation.

C. **Resonance energy transfer.** The excitation energy can be transferred by resonance energy transfer, a radiationless process, to a neighboring molecule if their energy level difference corresponds to the quantum of excitation energy. In this process, the quantum, or so-called **exciton,** is transferred, raising an electron in the receptor molecule to a higher energy state as the photo-excited e^- in the original absorbing molecule returns to ground state. This so-called *Förster resonance energy transfer* is the mechanism whereby quanta of light falling anywhere within an array of pigment molecules can be transferred ultimately to specific photochemically reactive sites.

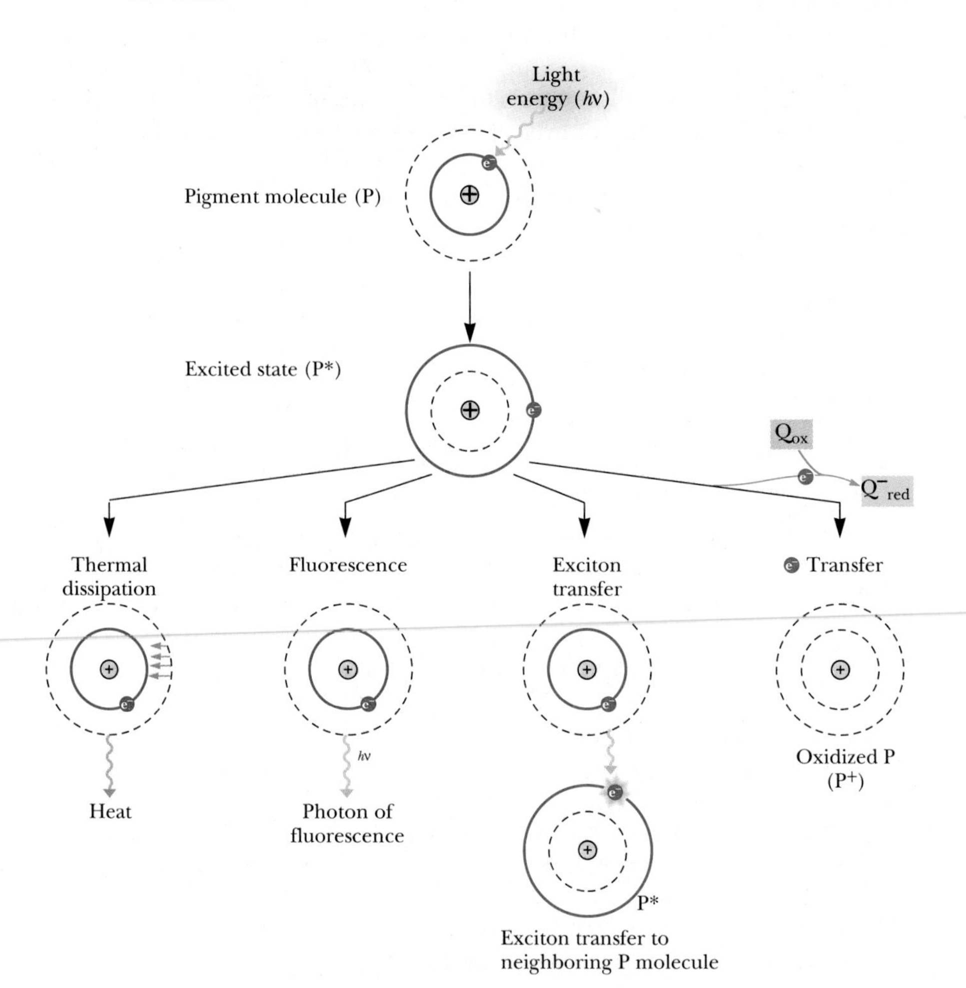

***Figure* 22.8** Possible fates of the quantum of light energy absorbed by photosynthetic pigments.

D. **Energy transduction.** The energy of excitation, in raising an electron to a higher energy orbital, dramatically changes the standard reduction potential, $\mathscr{E}_o'$, of the photoreceptor such that it becomes a much more effective electron donor. That is, the excited-state species, by virtue of having an electron at a higher energy level through light absorption, has become a potent electron donor. Reaction of this excited-state electron donor with an electron acceptor appropriately situated in its vicinity leads to the transduction of light energy (photons) to chemical energy (reducing power, the potential for electron-transfer reactions). *Transduction into chemical energy is the photochemical event that is the essence of photosynthesis.*

Chlorophyll serves two roles in photosynthesis. It is involved in light harvesting and the transfer of light energy to photoreactive sites by exciton transfer, and it participates directly in the photochemical events whereby light energy becomes chemical energy. Oxidation of chlorophyll leaves a **cationic free radical,** $Chl\cdot^+$, whose properties as an electron acceptor have important consequences for photosynthesis. Note that the Mg^{2+} ion does not change in valence during these redox reactions.

Photosynthetic Units Consist of Many Chlorophyll Molecules but Only a Single Reaction Center

In the early 1930s, Emerson and Arnold investigated the relationship between the amount of incident light energy, the amount of chlorophyll present, and the amount of oxygen evolved by illuminated algal cells (this relationship is the so-called **quantum yield of photosynthesis**). Their studies gave an unexpected result: When algae were illuminated with very brief light flashes that could excite every chlorophyll molecule at least once, only one molecule of O_2 was evolved per 2400 chlorophyll molecules. This result implied that not all chlorophyll molecules are photochemically reactive, and it led to the concept that photosynthesis occurs in functionally discrete units. A **photosynthetic unit** can be envisioned as an antenna of several hundred light-harvesting chlorophyll molecules plus a special pair of photochemically reactive chlorophyll *a* molecules within the unit called the **reaction center.** The purpose of the vast majority of chlorophyll is to harvest light incident within the unit and funnel it, via resonance energy transfer, to special reaction center chlorophyll molecules that are photochemically active. The great preponderance of chlorophyll thus acts as a large light-collecting antenna, and it is at the reaction centers that the photochemical event occurs (Figure 22.9).

22.3 Eukaryotic Phototrophs Possess Two Distinct Photosystems

The identity of the specialized Chl *a* molecules and the existence of two separate but interacting photosystems in photosynthetic eukaryotes were demonstrated through analysis of the photochemical action spectrum of photosynthesis. That is, the oxygen-evolving capacity as a function of light wavelength was determined, revealing a curious phenomenon known as the *red drop* in photosynthesis (Figure 22.10).

Although chlorophyll *a* has some capacity to absorb 700-nm light, light of this wavelength is relatively inefficient in driving photosynthesis. However, if light of shorter wavelength (less than 680 nm) is used to supplement 700-nm light, an enhancement effect, the so-called *Emerson enhancement effect,* is observed. In other words, these two wavelengths are synergistic: When given

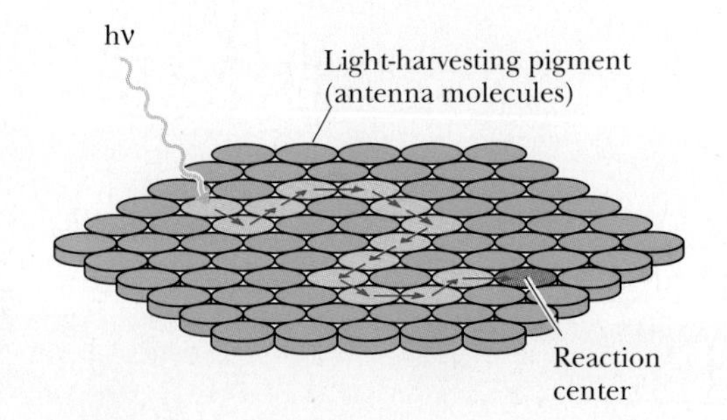

***Figure* 22.9** Schematic diagram of a photosynthetic unit. The light-harvesting pigments, or antenna molecules (green) absorb and transfer light energy to the specialized chlorophyll dimer that constitutes the reaction center (orange).

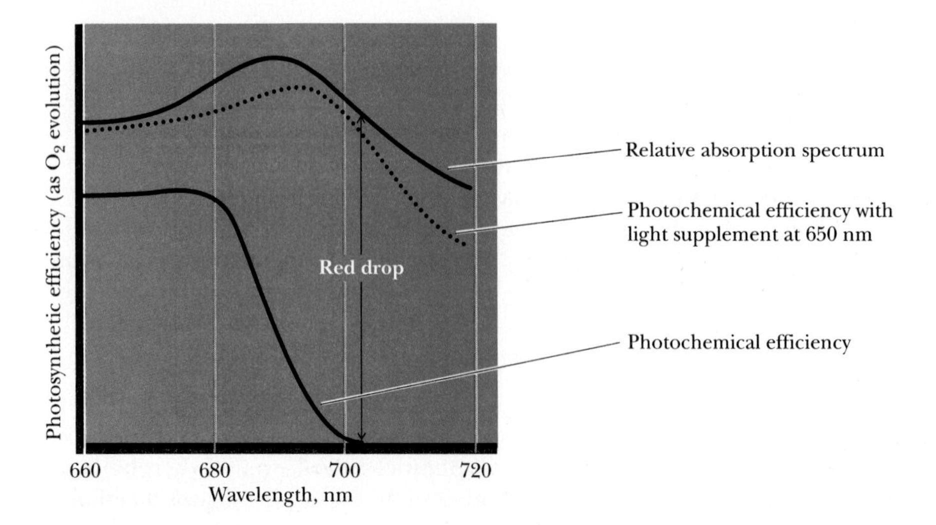

***Figure* 22.10** The photochemical action spectrum of photosynthesis. The quantum yield of photosynthesis as a function of wavelength of incident light shows an abrupt decrease above 700 nm, the so-called *red drop.*

together, these wavelengths elicit more O_2 evolution than expected from the sum of the amounts when each wavelength of light is given alone. One interpretation is that two light reactions participate in oxygen-evolving photosynthetic cells, one utilizing light of 700 nm and the other needing light of wavelength 680 nm or less. Duysens used this interpretation to advance a hypothesis suggesting two photosytems, I and II. **Photosystem I** (PSI) is defined as dependent on forms of chlorophyll absorbing light of longer wavelength; it is not involved in O_2 evolution. **Photosystem II** (PSII) functions in O_2 evolution and uses chlorophyll species that absorb shorter wavelengths.

All photosynthetic cells contain some form of photosystem. Photosynthetic bacteria, unlike cyanobacteria and eukaryotic phototrophs, have only one photosystem. Interestingly, bacterial photosystems resemble eukaryotic PSII more than PSI, even though photosynthetic bacteria lack O_2-evolving capacity.

P700 and P680 Are the Reaction Centers of PSI and PSII, Respectively

B. Kok provided direct experimental evidence for a unique photoreactive species in O_2-evolving photosynthetic cells. He used precise spectrophotometric measurements to show that a small amount of pigment absorbing 700-nm light **(P700)** is bleached when light of this wavelength is used to illuminate cell suspensions. Since this bleaching, or disappearance, of the 700-nm absorbance can be mimicked by adding an electron acceptor such as ferricyanide, bleaching is correlated with electron loss from P700. The concentration of P700 is small, only 0.25% of the total amount of chlorophyll in plants. However, this low concentration is consistent with the notion of specific photoreactive sites, or **reaction centers.** P700 is the reaction center of photosystem I. Similar studies using shorter-wavelength light identified an analogous pigment, **P680,** which constitutes the reaction center of photosystem II. Both P700 and P680 are chlorophyll *a* dimers situated within specialized protein complexes.

Chlorophyll Exists in Plant Membranes in Association with Proteins

Detergent treatment of chloroplast membranes results in their dissolution, and complexes containing both chlorophyll and protein can be isolated. Their composition varies with the method of preparation. Nevertheless, several distinctive chlorophyll–protein complexes have been identified and ana-

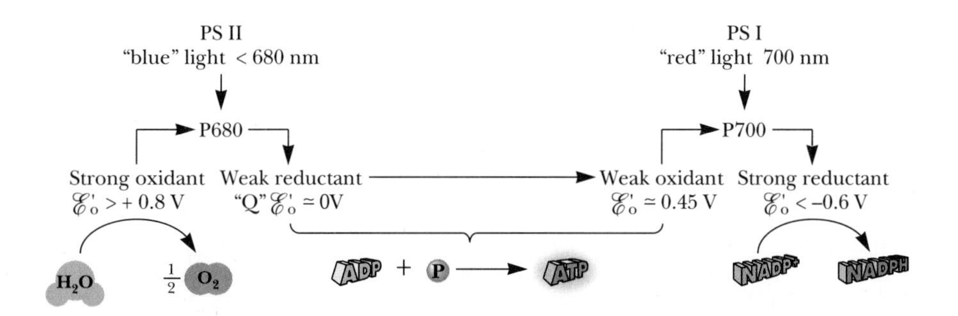

***Figure* 22.11** Roles of the two photosystems, PSI and PSII.

lyzed, and conclusions have been drawn regarding their functions. These chlorophyll–protein complexes represent integral components of the thylakoid membrane, and their organization reflects their roles as either **light-harvesting complexes** (LHC), **PSI complexes,** or **PSII complexes.** All chlorophyll is apparently localized within these three macromolecular assemblies. Their relative constitutions of chlorophyll *a* and *b*, the presence of accessory pigments such as β-carotene, as well as the presence or absence of P700 or P680, and their unique complement of polypeptides all serve to identify them.

The Roles of PSI and PSII

What are the roles of the two photosystems, and what is their relationship to each other? As seen in Equation (22.2), photosynthesis can be represented by the reduction of $NADP^+$ at the expense of electrons derived from water and activated by light, $h\nu$. Some ATP is also generated in the process. The standard reduction potential for the $NADP^+$/NADPH couple is -0.32 V. Thus, a strong reductant with $\mathscr{E}_o'$ more negative than -0.32 V is required to reduce $NADP^+$ under standard conditions. By similar reasoning, a very strong oxidant will be required to oxidize water to oxygen since the $\mathscr{E}_o'(\frac{1}{2}\ O_2/H_2O)$ is $+0.82$ V. The separation of the oxidizing and reducing aspects of Equation (22.2) is accomplished in nature by devoting PSI to $NADP^+$ reduction and PSII to water oxidation. PSI and PSII are linked via an electron transport chain wherein the weak reductant generated by PSII provides the electron to reduce $P700^+$, a weak oxidant, restoring it for another cycle of photochemical activity (Figure 22.11). Thus, photosystems I and II are linked in series, and *electrons flow from H_2O to $NADP^+$*, driven by light energy absorbed at the reaction centers. Oxygen is a by-product of the **photolysis,** literally "light splitting," of water. Phosphorylation of ADP results because a proton gradient is established across the thylakoid membrane as a consequence of the light-induced electron transfer reactions. This light-driven phosphorylation is termed **photophosphorylation.**

22.4 The Z Scheme of Photosynthetic Electron Transfer

Photosystems I and II contain unique complements of electron carriers, and these carriers mediate the stepwise transfer of electrons from water to $NADP^+$. When the individual redox components of PSI and PSII are arranged as an e^- transport chain according to their standard reduction potentials, the zigzag result is akin to the letter *Z* (Figure 22.12). The various electron carriers are indicated as follows: "Mn complex" symbolizes the manganese-containing oxygen-evolving complex; D is its e^- acceptor and the immediate e^- donor to $P680^+$; Q_A and Q_B represent special plastoquinone molecules

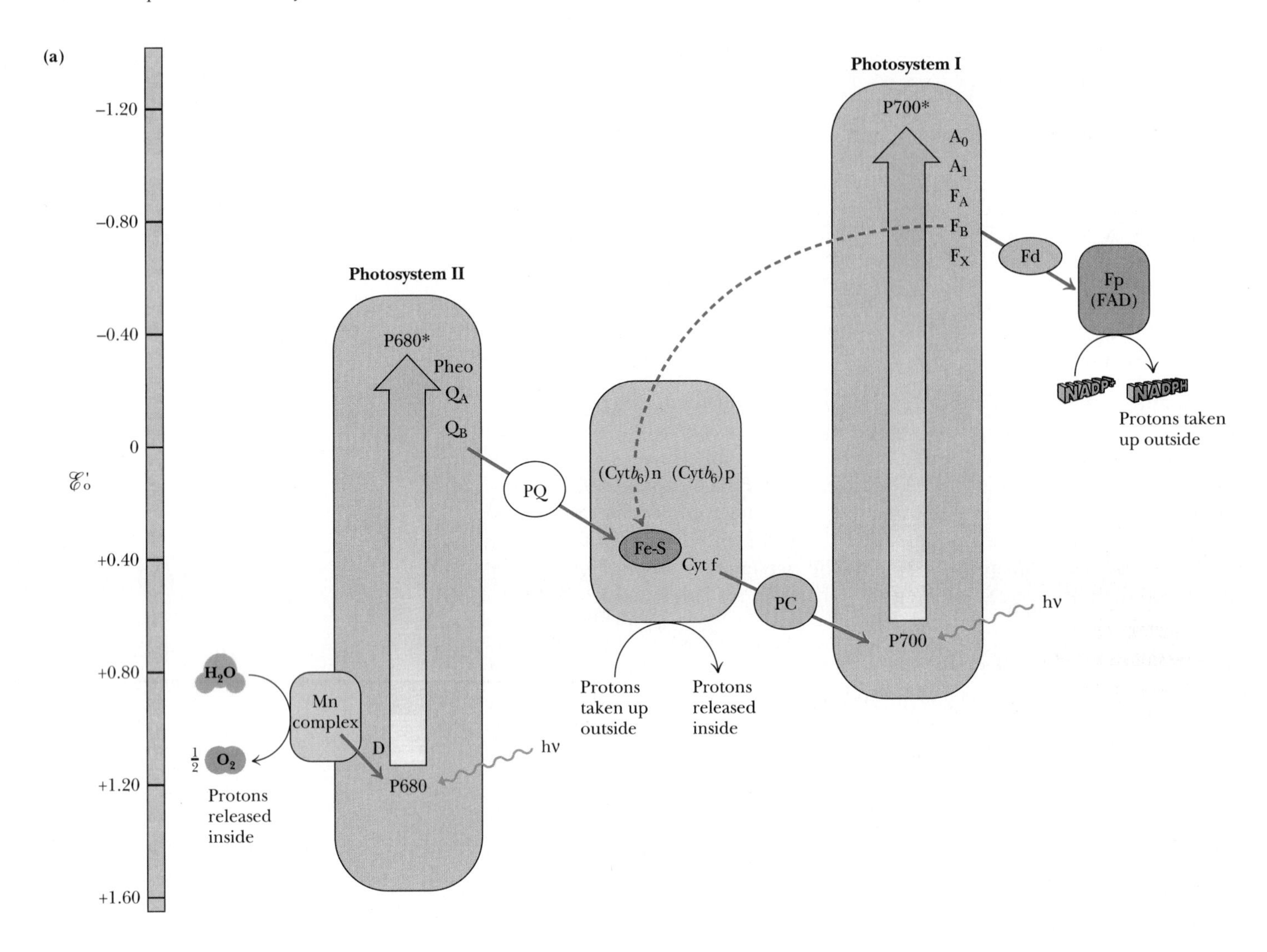
(a)
Photosystem I
Photosystem II
P700*
A_0
A_1
F_A
F_B
F_X
Fd
Fp (FAD)
NADP+
NADPH
Protons taken up outside
P680*
Pheo
Q_A
Q_B
PQ
$(Cytb_6)n$ $(Cytb_6)p$
Fe-S
Cyt f
PC
P700
hν
Protons taken up outside
Protons released inside
H_2O
Mn complex
D
$\frac{1}{2}$ O_2
P680
Protons released inside
-1.20
-0.80
-0.40
0
+0.40
+0.80
+1.20
+1.60
$\mathscr{E}'_o$

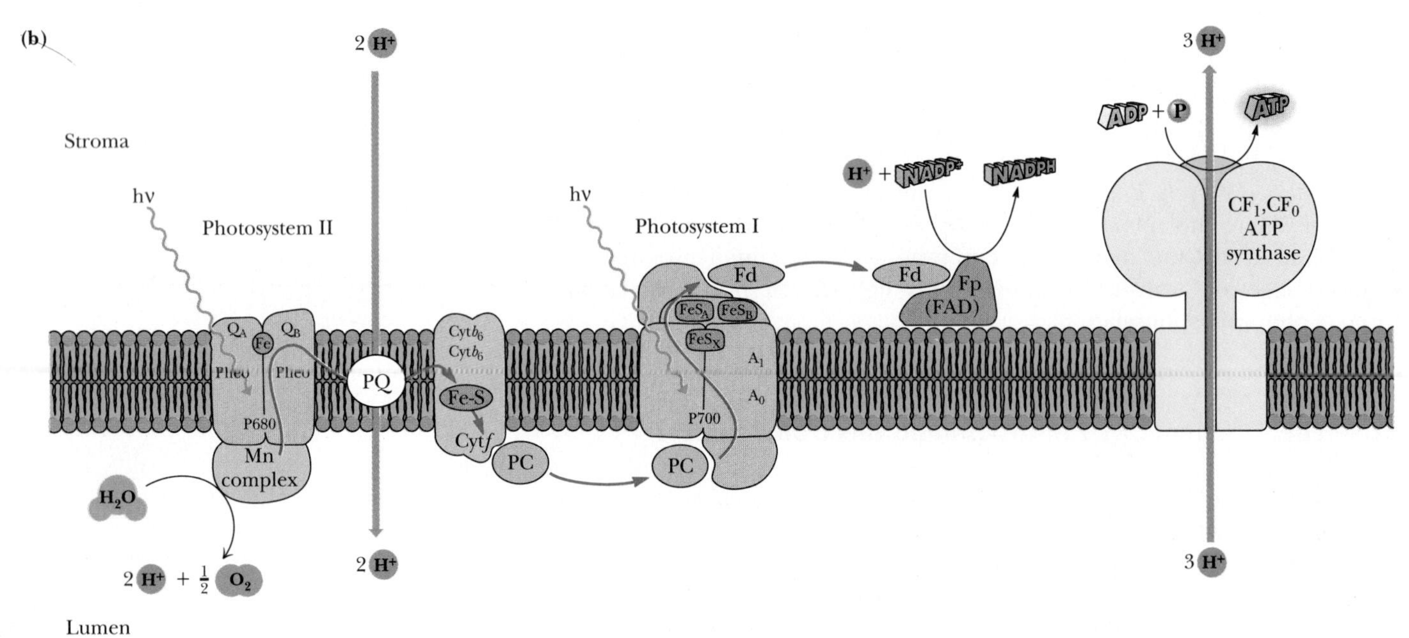
(b)
2 H^+
3 H^+
Stroma
ADP + P
ATP
H^+ + NADP+
NADPH
hν
Photosystem II
Photosystem I
CF_1,CF_0 ATP synthase
Fd
Fd
Fp (FAD)
Q_A
Fe
Q_B
Pheo
Pheo
P680
PQ
$Cytb_6$
$Cytb_6$
Fe-S
Cyt f
FeS_A
FeS_B
FeS_X
A_1
A_0
P700
PC
PC
Mn complex
H_2O
2 H^+ + $\frac{1}{2}$ O_2
2 H^+
3 H^+
Lumen

***Figure* 22.12** The *Z* scheme of photosynthesis. (a) The *Z* scheme is a diagrammatic representation of photosynthetic electron flow from H_2O to $NADP^+$. The energy relationships can be derived from the $\mathcal{E}_o'$ scale beside the *Z* diagram, with lower standard potentials and hence greater energy as you go from bottom to top. Energy input as light is indicated by two broad arrows, one photon appearing in P680, the other in P700. P680* and P700* represent photoexcited states. Electron loss from P680* and P700* creates $P680^+$ and $P700^+$. The representative components of the three supramolecular complexes (*PSI, PSII,* and the *cytochrome* b_6*–cytochrome f complex*) are in shaded boxes enclosed by solid black lines. Proton translocations which establish the proton-motive force driving ATP synthesis are illustrated as well. (b) Inset showing the functional relationships among PSII, the cytochrome *b*–cytochrome *f* complex, PSI, and the photosynthetic CF_1CF_0 ATP synthase within the thylakoid membrane. Note that e^- acceptors Q_A (for PSII) and A_1 (for PSI) are at the stromal side of the thylakoid membrane, whereas the e^- donors to $P680^+$ and $P700^+$ are situated at the lumenal side of the membrane. The consequence is charge separation ($-_{out}$, $+_{in}$) across the membrane. Also note that protons are translocated into the thylakoid lumen, giving rise to a chemiosmotic gradient that is the driving force for ATP synthesis by CF_1CF_0 ATP synthase.

(see Figure 22.15) and PQ the plastoquinone pool; Fe-S stands for the Rieske iron–sulfur center, and *cyt f,* cytochrome *f.* PC is the abbreviation for plastocyanin, the immediate e^- donor to $P700^+$; and F_A, F_B, and F_X represent the membrane-associated ferredoxins downstream from A_0 (a specialized Chl *a*) and A_1 (a specialized PSI quinone). Fd is the soluble ferredoxin pool that serves as the e^- donor to the flavoprotein (Fp), called *ferredoxin–NADP*$^+$ *reductase,* which catalyzes reduction of $NADP^+$ to NADPH. $Cyt(b_6)_n$, $(b_6)_p$ symbolizes the cytochrome b_6 moiety functioning to transfer e^- from F_A/F_B back to $P700^+$ during cyclic photophosphorylation (the pathway symbolized by dashed arrows).

Overall photosynthetic electron transfer is accomplished by three **membrane-spanning supramolecular complexes,** composed of intrinsic and extrinsic polypeptides (shown as shaded boxes bounded by solid black lines in Figure 22.12). These complexes are the PSII complex, the cytochrome b_6/cytochrome *f* complex, and the PSI complex. The PSII complex is aptly described as a light-driven *water:plastoquinone oxidoreductase;* it is the enzyme system responsible for photolysis of water, and as such, it is also referred to as the **oxygen-evolving complex,** or OEC. Within this complex, a manganese-containing protein is intimately involved in the evolution of oxygen, perhaps through formation of a bimetallic center consisting of 2 Mn^{2+} coordinating two equivalents of water. Both protons and electrons are abstracted from these water molecules, and O_2 is released as P680 undergoes four cycles of oxidation (Figure 22.13).

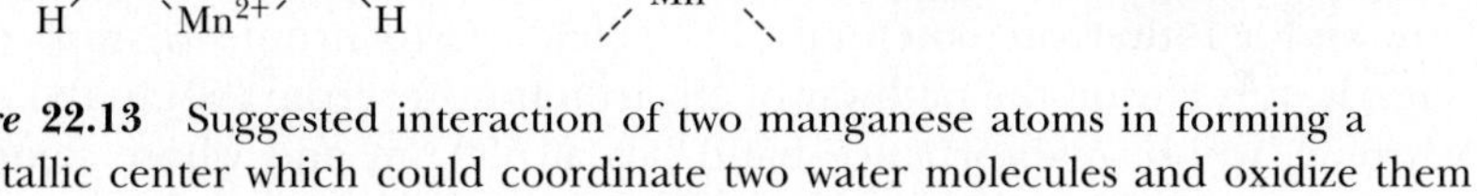

$$2\,H_2O\cdot(Mn^{2+})_2 \xrightarrow{4\,h\nu} (Mn^{2+})_2 + O_2 + 4\,H^+ + 4\,e^-$$

***Figure* 22.13** Suggested interaction of two manganese atoms in forming a bimetallic center which could coordinate two water molecules and oxidize them to yield a molecule of O_2. This photo-oxidation, or photolysis, of water would proceed as P680 undergoes four cycles of light-induced oxidation. The four oxidizing equivalents accumulate in the manganese-containing active site of the O_2-evolving complex.

Oxygen Evolution Requires the Accumulation of Four Oxidizing Equivalents in PSII

When isolated chloroplasts which have been held in the dark are illuminated with very brief flashes of light, O_2 evolution reaches a peak on the third flash and every fourth flash thereafter (Figure 22.14a). The oscillation in O_2 evolved dampens over repeated flashes and converges to an average value. These data are interpreted to mean that the P680 reaction center complex cycles through five different oxidation states, numbered $\mathbf{S_0}$ to $\mathbf{S_4}$. One electron is removed photochemically in each step. When $\mathbf{S_4}$ is attained, an O_2 molecule is released (Figure 22.14b) and PSII returns to oxidation state $\mathbf{S_0}$. The reason the first pulse of O_2 release occurred on the third flash (Figure 22.14a) is that the PSII reaction centers in the isolated chloroplasts were already poised at $\mathbf{S_1}$ reduction level.

(a)

O_2 evolved/flash

4 8 12 16 20 24

Flash number

(b)

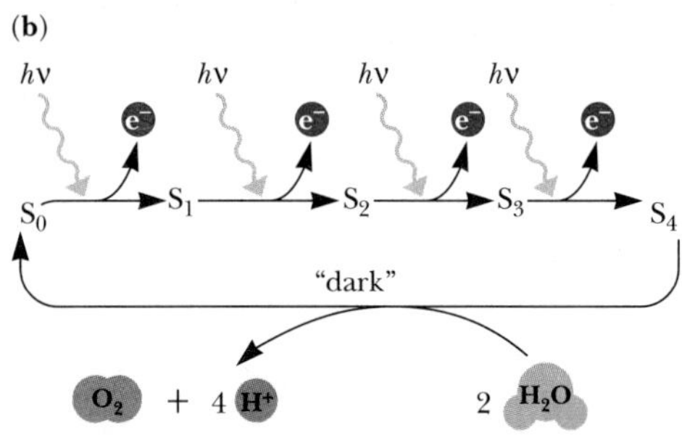

Figure **22.14** Oxygen evolution requires the accumulation of four oxidizing equivalents in PSII. (a) Dark-adapted chloroplasts show little O_2 evolution after two brief light flashes. Oxygen evolution then shows a peak on the third flash and every fourth flash thereafter. The oscillation in O_2 evolution is dampened by repeated flashes and converges to an average value after 20 or so flashes. (b) The oscillation in O_2 evolution per light flash is due to the cycling of the PSII reaction center through five different oxidation states, $\mathbf{S_0}$ to $\mathbf{S_4}$. When $\mathbf{S_4}$ is reached, O_2 is released. One e^- is removed photochemically at each light flash, moving the reaction center successively through $\mathbf{S_1}$, $\mathbf{S_2}$, $\mathbf{S_3}$, and $\mathbf{S_4}$. $\mathbf{S_4}$ decays spontaneously to $\mathbf{S_0}$ by oxidizing 2 H_2O to O_2. The peak of O_2 evolution at flash 3 in part (a) is due to the fact that the isolated chloroplast suspension is already at the $\mathbf{S_1}$ stage.

Light-Driven Electron Flow from H_2O Through PSII

The events intervening between H_2O and P680 involve *D*, the name assigned to a specific protein **tyrosine residue** that mediates e^- transfer from H_2O via the Mn complex to $P680^+$ (Figure 22.12). The oxidized form of *D* is a tyrosyl free radical species, $D\cdot^+$. $P680^+$ is generated when an exciton of energy excites P680 to P680*, rendering it capable of reducing a special molecule of **pheophytin,** symbolized by "Pheo" in Figure 22.12. Pheophytin is like chlorophyll *a*, except 2 H^+ replace the centrally coordinated Mg^{2+} ion. This special pheophytin is thought to be the direct electron acceptor from P680*. Electrons flow from Pheo via specialized molecules of **plastoquinone,** represented by "Q" in Figure 22.12, to a pool of plastoquinone within the membrane. Because of its lipid nature, plastoquinone is mobile within the membrane and hence serves to shuttle electrons from the PSII supramolecular complex to the cytochrome b_6/cytochrome *f* complex. Alternate oxidation–reduction of plastoquinone to its hydroquinone form involves the uptake of protons (Figure 22.15). The asymmetry of the thylakoid membrane is designed to exploit this proton uptake and release so that protons (H^+) accumulate within the thylakoid vesicle, establishing an electrochemical gradient. Note that plastoquinone is an analog of coenzyme Q, the mitochondrial electron carrier (Chapter 20).

Electron Transfer Within the Cytochrome b_6/Cytochrome *f* Complex

The cytochrome b_6/cytochrome *f* complex, or *plastoquinol:plastocyanin oxidoreductase,* includes the two heme-containing electron transfer proteins for which it is named as well as *iron–sulfur clusters* (Chapter 20), which also participate in electron transport. The purpose of this complex is to mediate the transfer of electrons from PSII to PSI. Cytochrome *f* (*f* from the Latin *folium,* meaning "foliage") is a *c*-type cytochrome, with an α-absorbance band at 553 nm and a reduction potential of +0.365 V. Cytochrome b_6 apparently does not lie directly on the pathway of electron transfer from PSII to PSI. This cytochrome, whose α-absorbance band lies at 559 nm and whose standard reduction potential is −0.06 V, is thought to participate in an alternative **cyclic e^- transfer pathway.** Under certain conditions, electrons derived from P700* are not passed on to $NADP^+$ but instead cycle down an alternative path via ferredoxins in the PSI complex to cytochrome b_6 and ultimately back to $P700^+$. This cyclic flow yields no O_2 evolution or $NADP^+$ reduction but can lead to ATP synthesis via so-called *cyclic photophosphorylation,* discussed later.

Electron Transfer from the Cytochrome b_6/Cytochrome f Complex to PSI

Plastocyanin ("PC" in Figure 22.12) is an electron carrier capable of rapid diffusion and migration in and out of the membrane. As such, this molecule is aptly suited to its role in shuttling electrons between the cytochrome b_6/cytochrome f complex and PSI. Plastocyanin is a low-molecular-weight (10.4 kD) protein containing a single copper atom. PC functions as a single-electron carrier ($\mathscr{E}_o{}' = +0.32$ V) as its copper atom undergoes alternate oxidation–reduction between the cuprous (Cu^{+}) and cupric (Cu^{2+}) states. PSI is a light-driven *plastocyanin:ferredoxin oxidoreductase.* When P700, the specialized chlorophyll *a* dimer of PSI, is excited by light and oxidized by transferring its e^- to an adjacent chlorophyll *a* molecule, which serves as its immediate e^- acceptor, $P700^+$ is formed. (The standard reduction potential for the $P700^+/P700$ couple lies near +0.45 V.) $P700^+$ readily gains an electron from plastocyanin.

The chemical identity of the immediate electron acceptor for P700* (and the one for P680*, for that matter) eluded investigators for a long time. Evidence now supports a special molecule of chlorophyll *a* as this agent. This unique Chl *a* (A_0) rapidly passes the electron to a specialized quinone (A_1), which in turn passes the e^- to the first in a series of *membrane-bound ferredoxins* (Fd, Chapter 20). This Fd series ends with a soluble form of ferredoxin, Fd_s, which serves as the immediate electron donor to the flavoprotein (Fp) that catalyzes $NADP^+$ reduction, namely, *ferredoxin:NADP⁺ reductase.*

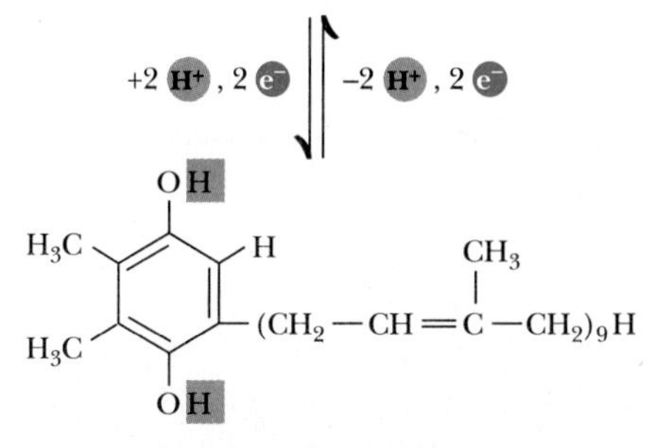

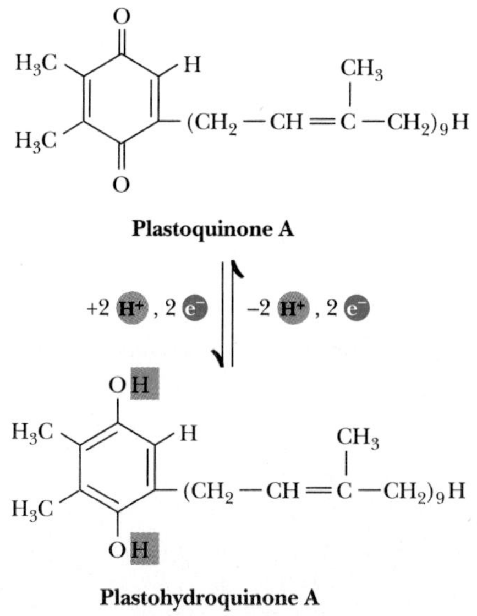

Figure **22.15** The structures of plastoquinone and its reduced form, plastohydroquinone (or plastoquinol). The oxidation of the hydroquinone releases 2 H^+ as well as 2 e^-. The form shown (plastoquinone A) has nine isoprene units and is the most abundant plastoquinone in plants and algae. Other plastoquinones have different numbers of isoprene units and may vary in the substitutions on the quinone ring.

The Initial Events in Photosynthesis Are Very Rapid Electron-Transfer Reactions

Electron transfer from P680 to Q and from P700 to Fd occurs on a picosecond-to-microsecond time scale. The necessity for such rapid reaction becomes obvious when one realizes that light-induced Chl excitation followed by electron transfer leads to separation of opposite charges in close proximity, that is, $P700^+ : A_0{}^-$. Accordingly, subsequent electron transfer reactions occur rapidly in order to shuttle the electron away quickly, before the wasteful back reaction of charge recombination (and dissipation of excitation energy) can happen. The time scale of events following P700 excitation illustrates this point (Figure 22.16).

22.5 The Molecular Architecture of Photosynthetic Reaction Centers

What molecular architecture couples the absorption of light energy to rapid electron-transfer events, in turn coupling these e^- transfers to proton translocations so that ATP synthesis is possible? Part of the answer to this question

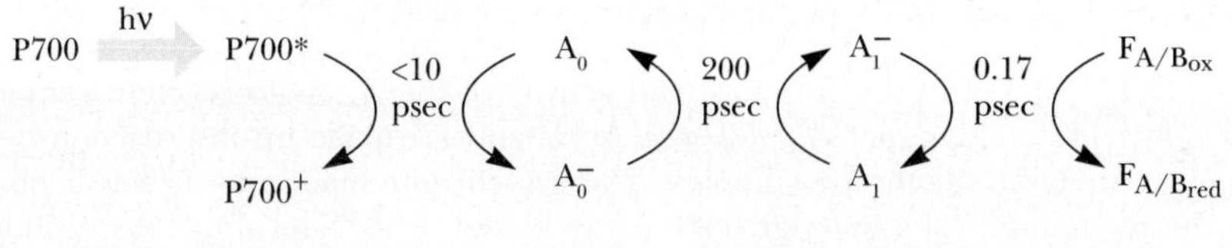

Figure **22.16** Electron transfer from P700* to membrane-bound Fe-S centers (F_A/F_B) occurs on a picosecond time scale. These electron transfers must be very rapid to prevent recombination of the opposite charges ($P700^+$:$A_0{}^-$) created when P700* transfers its e^- to A_0. A_1 is a PSI-associated quinone mediating e^- transfer between A_0 and F_A/F_B.

lies in the membrane-associated nature of the photosystems. Unfortunately, it has been very difficult to study the properties of membrane proteins due to their insolubility in the usual aqueous solvents employed in protein biochemistry. A major breakthrough occurred in 1984 when Johann Deisenhofer, Hartmut Michel, and Robert Huber reported the first X-ray crystallographic analysis of a membrane protein. To the great benefit of photosynthesis research, this protein was the reaction center from the photosynthetic purple bacterium *Rhodopseudomonas viridis.* This research earned these three scientists the 1984 Nobel Prize in chemistry.

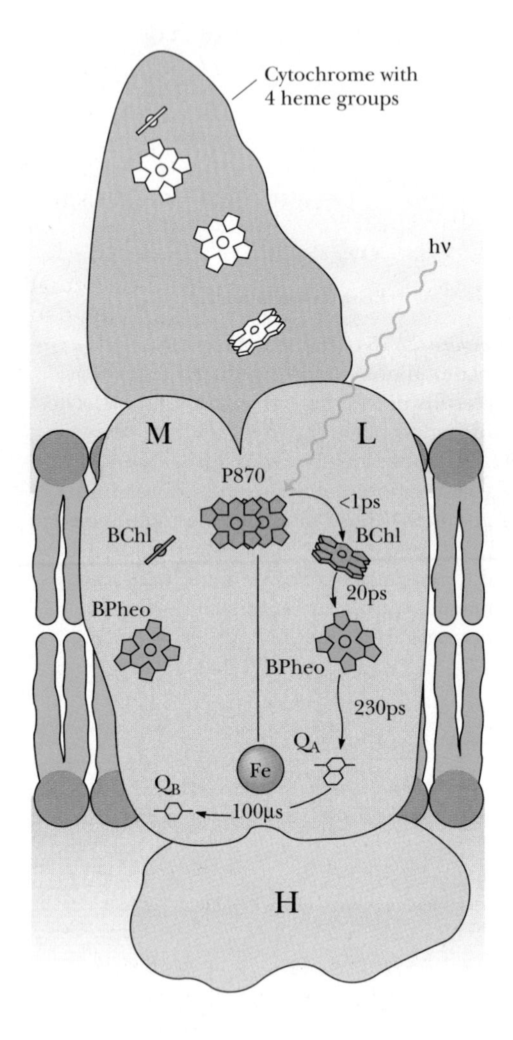

Note: The cytochrome subunit is membrane-associated via a diacylglycerol moiety on its N-terminal Cys residue:

$$H_3\overset{+}{N}-\underset{\underset{\displaystyle S-CH_2-CH(O-C{=}O)-CH_2(O-C{=}O)}{\displaystyle CH_2}}{\overset{H}{C}}-\overset{O}{\overset{\|}{C}}-\cdots$$

Membrane anchor

Structure of the *R. viridis* Photosynthetic Reaction Center

The *R. viridis* reaction center (145 kD) is localized in the plasma membrane of these photosynthetic bacteria and is composed of four different polypeptides, designated *L* (273 amino acid residues), *M* (323 residues), *H* (258 residues), and *cytochrome* (333 amino acid residues). *L* and *M* each consist of five membrane-spanning α-helical segments; *H* has one such helix, the majority of the protein forming a globular domain in the cytoplasm (Figure 22.17). The cytochrome subunit contains four heme groups; the N-terminal amino acid of this protein is cysteine. This cytochrome is anchored to the periplasmic face of the membrane via the hydrophobic chains of two fatty acid groups that are esterified to a glyceryl moiety joined via a thioether bond to the Cys (Figure 22.17). *L* and *M* each bear two *bacteriochlorophyll* molecules (the bacterial version of Chl) and one *bacteriopheophytin. L* also has a bound quinone molecule, Q_A. Together, *L* and *M* coordinate an Fe atom. The photochemically active species of the *R. viridis* reaction center, **P870,** is composed of two bacteriochlorophylls, one contributed by *L* and the other by *M.*

Photosynthetic Electron Transfer in the *R. viridis* Reaction Center

The prosthetic groups of the *R. viridis* reaction center (P870, BChl, BPheo, and the bound quinones) enjoy a special spatial relationship that is conducive to photosynthetic e^- transfer (Figure 22.17). Photoexcitation of P870 (creation of P870*) leads to e^- loss ($P870^+$) via electron transfer to the bacteriochlorophyll (BChl) that is *not* part of P870. The e^- is then transferred via the *L* bacteriopheophytin (BPheo) to Q_A, which is also an *L* prosthetic group. The corresponding site on *M* is occupied by a loosely bound quinone, Q_B, and electron transfer from Q_A to Q_B takes place. An interesting aspect of the system is that *no* electron transfer occurs through *M,* even though it has components apparently symmetrical to and identical with the *L* e^- transfer pathway.

The reduced quinone formed at the Q_B site is free to diffuse to a neighboring *cytochrome b/cytochrome* c_1 membrane complex, where its oxidation is coupled to H^+ translocation (and, hence, ultimately to ATP synthesis) (Fig-

***Figure* 22.17** Model of the structure and activity of the *R. viridis* reaction center. Four polypeptides (designated *cytochrome, M, L,* and *H*) make up the reaction center, an integral membrane complex. The cytochrome maintains its association with the membrane via a diacylglyceryl group linked to its N-terminal Cys residue by a thioether bond. *M* and *L* both consist of five membrane-spanning α-helices; *H* has a single membrane-spanning α-helix. The prosthetic groups are spatially situated so that rapid e^- transfer from P870* to Q_B is facilitated. Photoexcitation of P870 leads in less than 1 picosecond (psec) to reduction of the *L*-branch BChl only. $P870^+$ is re-reduced via an e^- provided through the heme groups of the cytochrome.

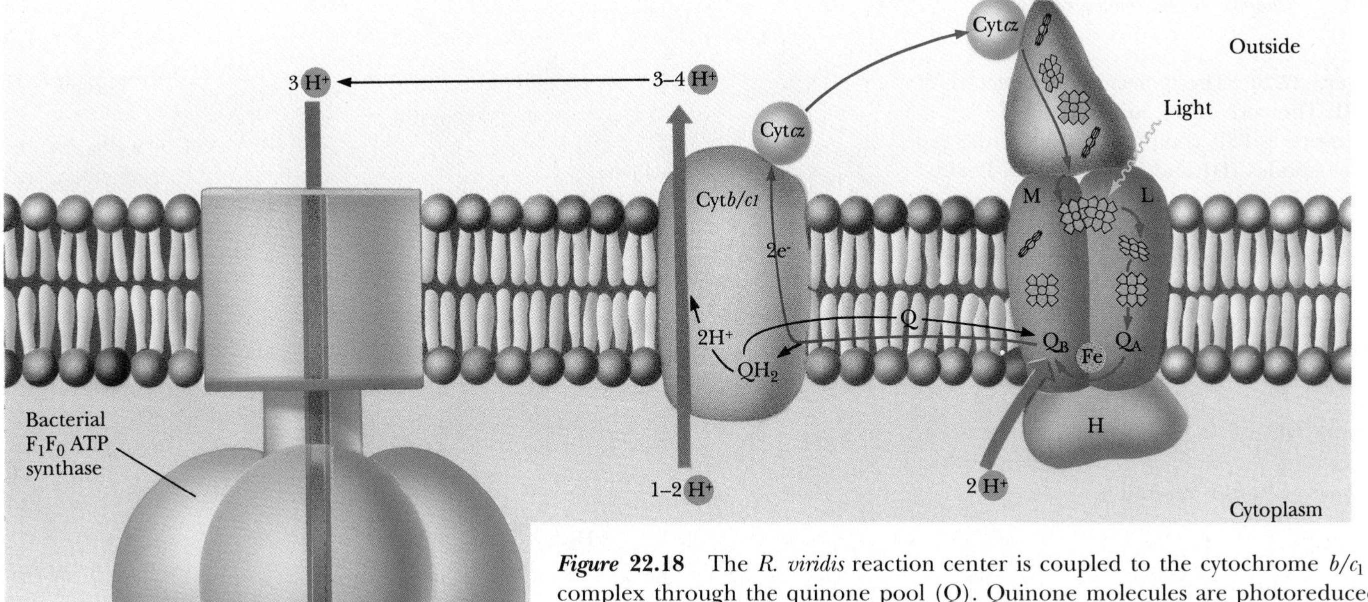

***Figure* 22.18** The *R. viridis* reaction center is coupled to the cytochrome b/c_1 complex through the quinone pool (Q). Quinone molecules are photoreduced at the reaction center Q_B site ($2\ e^-$ [$2\ h\nu$] per Q reduced) and then diffuse to the cytochrome b/c_1 complex, where they are reoxidized. Note that e^- flow from cytochrome b/c_1 back to the reaction center occurs via the periplasmic protein cytochrome c_2. Note also that 3 to 4 H^+ are translocated into the periplasmic space for each Q molecule oxidized at cytochrome b/c_1. The resultant proton-motive force drives ATP synthesis by the bacterial F_1F_0 ATP synthase.

(Adapted from Deisenhofer, J., and Michel, H., 1989. The photosynthetic reaction center from the purple bacterium Rhodopseudomonas viridis. *Science 245:1463.)*

ure 22.18). A periplasmic **cytochrome c_2** serves to cycle electrons back to $P870^+$ via the four hemes of the reaction center cytochrome subunit. A specific tyrosine residue of *L* (Tyr^{162}) is situated between P870 and the closest cytochrome heme. This Tyr is the immediate e^- donor to $P870^+$ and completes the light-driven electron transfer cycle. The structure of the *R. viridis* reaction center (derived from X-ray crystallographic data) is modeled in Figure 22.19.

Eukaryotic Reaction Centers: The Molecular Architecture of PSII

The structure of the *R. viridis* reaction center turns out to be a fairly good model for PSII. P680 and its immediate e^- acceptor, Pheo, are located on a pair of integral membrane proteins designated **D1** (32 kD) and **D2** (34 kD)

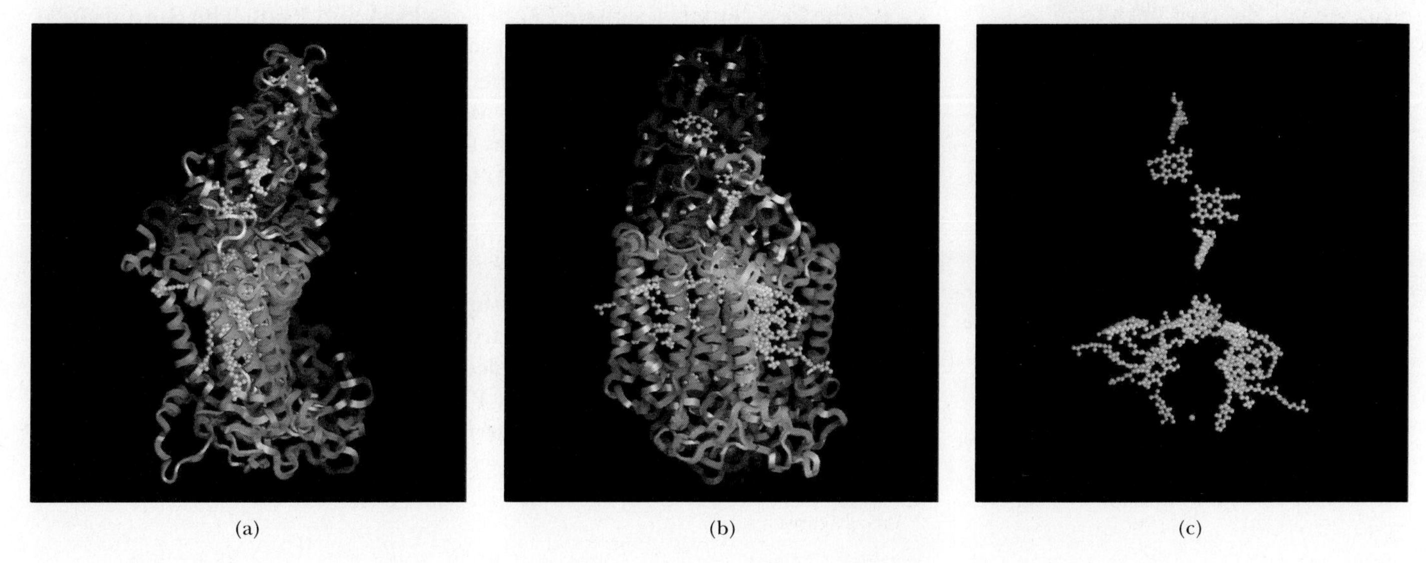

***Figure* 22.19** Model of the *R. viridis* reaction center. (a, b) Two views of the ribbon diagram of the reaction center. *M* and *L* subunits appear in purple and blue, respectively. Cytochrome subunit is brown; H subunit is green. These proteins provide a scaffold upon which the prosthetic groups of the reaction center are situated for effective photosynthetic electron transfer. Panel (c) shows the spatial relationship between the various prosthetic groups (4 hemes, P870, 2 BChl, 2 BPheo, 2 quinones, and the Fe atom) in the same view as in (b), but with protein chains deleted.

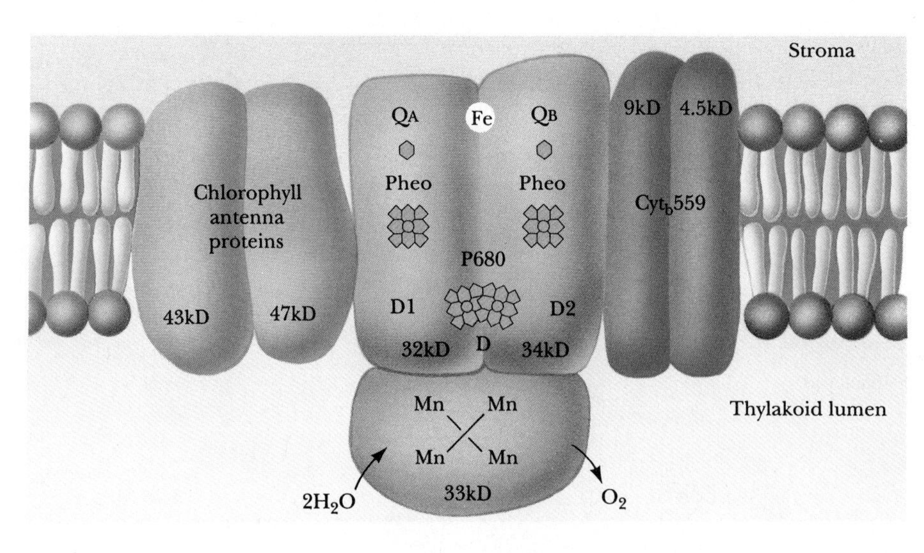

***Figure* 22.20** The molecular architecture of PSII. The core of the minimal O_2-evolving PSII complex consists of the two polypeptides (D1 and D2) that bind P680, pheophytin (Pheo), and the e^--transferring quinones, Q_A and Q_B, and constitute the reaction center. Additional components of this complex include cytochrome b_{559} (a heterodimer of 9-kD and 4.5-kD polypeptides), two additional intrinsic proteins (47 and 43 kD) that serve an accessory light-harvesting function, and a 33-kD extrinsic protein involved in stabilizing the Mn^{2+} prosthetic complex that is essential to O_2 evolution.

(Figure 22.20). The tyrosine species D is Tyr^{161} in the **D1** amino acid sequence. Also complexed to **D1** and **D2** are two special quinone molecules, Q_A and Q_B, which function in e^- transfer from Pheo to the plastoquinone pool. A cytochrome species, **cytochrome b_{559}**, composed of two polypeptides (4.5 and 9 kD), co-purifies with PSII; its function is as yet unclear. Three additional polypeptides (47, 43, and 33 kD) are ubiquitous components of all O_2-evolving PSII preparations. The 47- and 43-kD polypeptides are intrinsic proteins serving as Chl a–binding antenna complexes. The 33-kD polypeptide is an extrinsic protein implicated with Mn^{2+} in O_2 evolution.

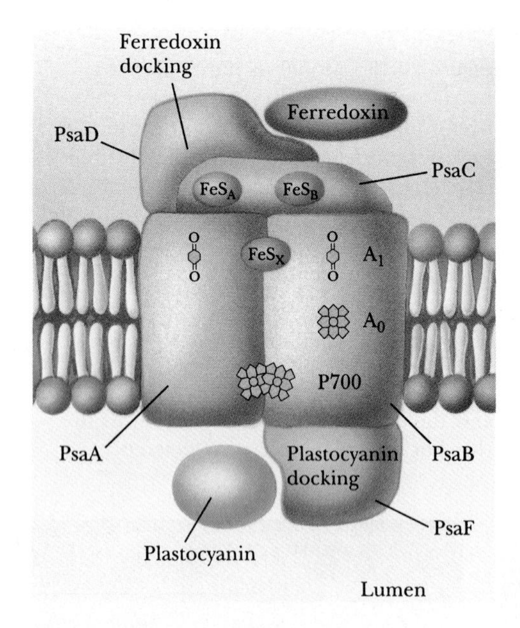

***Figure* 22.21** The molecular architecture of PSI. PsaA and PsaB constitute the reaction center dimer, an integral membrane complex; P700 is located at the lumenal side of this dimer. PsaC, which bears Fe-S centers F_A and F_B, and PsaD, the interaction site for ferredoxin, are on the stromal side of the thylakoid membrane. PsaF, which provides the plastocyanin interaction site, is on the lumen side.

(Adapted from Golbeck, J. H., 1992, Annual Review of Plant Physiology and Plant Molecular Biology 43:293–324.)

The Molecular Architecture of PSI

All the pigments and electron-transferring prosthetic groups essential to PSI function are localized to three polypeptides. Two of these, the **PsaA** (83 kD) and **PsaB** (82 kD) proteins, comprise the reaction center heterodimer, a structural pattern that thus seems universal in photosynthetic energy-transducing protein systems. Together **PsaA** and **PsaB** contain 100 or so chlorophyll molecules, including the pair comprising P700 and the one serving as immediate e^- acceptor for P700, namely, A_0 (Figure 22.21). Quinones are also found on the **PsaA/PsaB** heterodimer, including the intermediate e^- carrier, A_1. The Fe-S center designated F_x bridges **PsaA** and **PsaB**. The third protein, **PsaC** (8.9 kD), bears the two Fe-S clusters designated F_A and F_B, and occupies a position on the stromal face of the reaction center complex. Several additional polypeptides are necessary to assemble the complete PSI holocomplex, including one **(PsaF)** which provides the interaction site for plastocyanin (on the lumen side) and another **(PsaD)** serving a similar role for ferredoxin (on the stromal side).

Although many structural and functional details remain to be resolved, the mode of action envisioned for PSI begins with exciton absorption at P700, almost instantaneous electron transfer and charge separation ($P700^+:A_0^-$), followed by transfer of the electron from A_0 through A_1 and F_x to F_A/F_B where it is used to reduce a ferredoxin molecule at the stromal side of the thylakoid membrane. The positive charge at $P700^+$ and the e^- at F_A/F_B represent a charge separation across the membrane, an energized condition created by light.

PSI and PSII Show a Nonuniform Distribution in the Thylakoid Membrane

Recent studies on the concentrations of PSI and PSII and their organization within the thylakoid membrane indicate that the relative amounts of PSI and PSII vary as much as eightfold in different photosynthetic membranes. Further, PSI appears localized in stroma-exposed membrane regions, whereas PSII and LHC (the major light-harvesting complex, which contains about 50% of the chlorophyll found in chloroplasts) are preferentially localized in the appressed membrane regions (i.e., the grana of thylakoids) (Figure 22.22). Note the continuity of the thylakoid lumen in this diagram. The coupling factor, CF_1CF_0 ATP synthase, is discussed in subsequent paragraphs.

22.6 The Quantum Yield of Photosynthesis

The **quantum yield** of photosynthesis, the amount of product formed per equivalent of light input, has traditionally been expressed as the ratio of CO_2 fixed or O_2 evolved per quantum absorbed. At each reaction center, one photon or quantum yields one electron. Interestingly, an overall stoichiometry of one H^+ translocated into the thylakoid vesicle for each photon has also been observed. Two photons per center would allow a pair of electrons to flow from H_2O to $NADP^+$ (Figure 22.12), resulting in the formation of 1 NADPH and $\frac{1}{2}$ O_2. If one ATP were formed for every 3 H^+ translocated during photosynthetic electron transport, $1\frac{1}{3}$ ATP would be synthesized. More appropriately, 4 $h\nu$ per center (8 quanta total) would drive the evolution of 1 O_2, the reduction of 2 $NADP^+$, and the phosphorylation of $2\frac{2}{3}$ ATP.

The energy of a photon depends on its wavelength, according to the equation $E = h\nu = hc/\lambda$, where E is energy, c is the speed of light, and λ is its wavelength. Expressed in molar terms, an *Einstein* is the amount of energy in

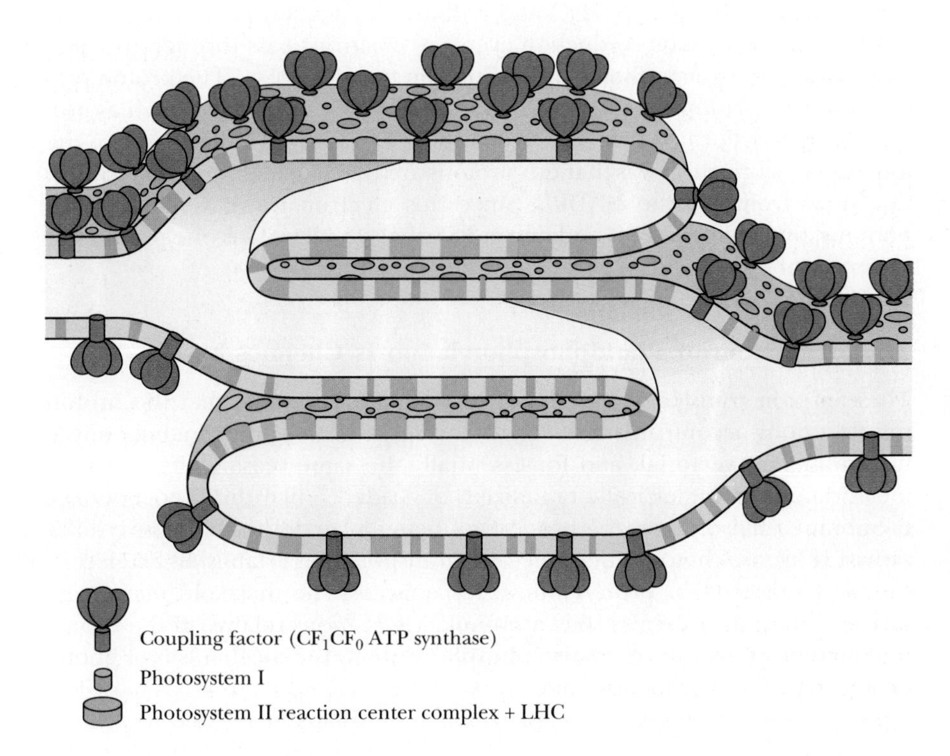

***Figure* 22.22** Organization of the photosystems within the thylakoid membrane. PSI is localized in the stroma-exposed regions, whereas PSII and LHC are localized in the appressed membrane regions, i.e., the grana. Note the continuity of the thylakoid lumen.

Avogadro's number of photons: $E = Nhc/\lambda$. Light of 700-nm wavelength is the longest-wavelength and the lowest-energy light acting in the eukaryotic photosystems discussed here. An Einstein of 700-nm light is equivalent in energy to approximately 170 kJ. Eight Einsteins of this light, 1360 kJ, theoretically generate 2 moles of NADPH, $2\frac{2}{3}$ moles of ATP, and 1 mole of O_2.

Photosynthetic Energy Requirements for Hexose Synthesis

The fixation of carbon dioxide to form hexose, *the dark reactions of photosynthesis,* requires considerable energy. The overall stoichiometry of this process (Eq. 22.3) involves 12 NADPH and 18 ATP. To generate 12 equivalents of NADPH necessitates the consumption of 48 Einsteins of light, minimally 170 kJ each. However, if the preceding ratio of $1\frac{1}{3}$ ATP per NADPH were correct, insufficient ATP for CO_2 fixation would be produced. Six additional Einsteins would provide the necessary two additional ATP. From 54 Einsteins, or 9180 kJ, one hexose would be synthesized. The standard free energy change, $\Delta G°'$, for hexose formation from carbon dioxide and water (the exact reverse of cellular respiration) is +2870 kJ/mol.

22.7 Light-Driven ATP Synthesis—Photophosphorylation

Light-driven ATP synthesis, termed **photophosphorylation,** is a fundamental part of the photosynthetic process. The conversion of light energy to chemical energy results in electron-transfer reactions leading to the generation of reducing power (NADPH). Coupled with these electron transfers are proton translocations across the thylakoid membrane. These proton translocations occur in a manner analogous to the proton translocations accompanying mitochondrial electron transport that provide the driving force for oxidative phosphorylation (Chapter 20). Figure 22.12 indicates that proton translocations can occur at a number of sites. For example, protons may be translocated by reactions between H_2O and PSII as a consequence of the photolysis of water. The oxidation–reduction events as electrons pass through the plastoquinone pool are another source of proton translocations. The proton production that accompanies the reduction of $NADP^+$ can be envisioned as protons being translocated from the stromal side of the thylakoid vesicle into the lumen. The current view is that two protons are translocated for each electron that flows from H_2O to $NADP^+$. Since this electron transfer requires two photons, one falling at PSII and one at PSI, the overall yield is one proton per quantum of light.

The Mechanism of Photophosphorylation Is Chemiosmotic

These proton translocations are vectorial in the same way that the proton translocations accompanying electron transport in the inner mitochondrial membrane are vectorial, and for essentially the same reason: the thylakoid membrane is asymmetrically organized, or "sided," like the mitochondrial membrane. It also shares the property of being a barrier to the passive diffusion of H^+ ions. Photosynthetic electron transport thus establishes an electrochemical gradient, or proton-motive force, across the thylakoid membrane with the interior, or lumen, side accumulating H^+ ions relative to the stroma of the chloroplast. Like oxidative phosphorylation, the mechanism of photophosphorylation is chemiosmotic.

A proton-motive force of approximately −250 mV is needed to achieve ATP synthesis. This proton-motive force, Δp, is comprised of a membrane

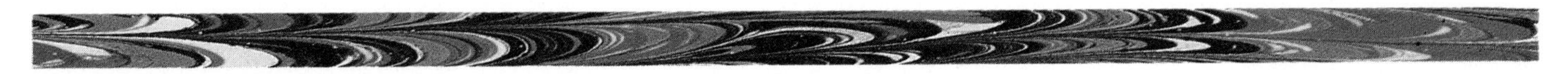

Critical Developments in Biochemistry

Experiments with Isolated Chloroplasts Provided the First Direct Evidence for the Chemiosmotic Hypothesis

Experimental proof that the mechanism of photophosphorylation is chemiosmotic was provided by an elegant experiment by Andre Jagendorf and Ernest Uribe in 1966 (see figure). Jagendorf and Uribe reasoned that if photophosphorylation were indeed driven by an electrochemical gradient established by photosynthetic electron transfer reactions, they might artificially generate such a gradient by first incubating chloroplasts in an acid bath in the dark and then quickly raising the pH of the external medium. The resulting inequality in hydrogen ion electrochemical activity across the membrane should mimic the conditions normally found upon illumination of chloroplasts and should provide the energized condition necessary to drive ATP formation. To test this interpretation, Jagendorf and Uribe bathed isolated chloroplasts in a weakly acidic (pH 4) medium for 60 seconds, allowing the pH inside the chloroplasts to equilibrate with the external medium. The pH of the solution was then quickly raised to slightly alkaline pH (pH 8), artificially creating a pH gradient across the thylakoid membranes. When ADP and P_i were added, ATP synthesis was observed as the pH gradient collapsed. This classic experiment was the first real proof of Mitchell's chemiosmotic hypothesis and directed the scientific community to a greater acceptance of Mitchell's interpretations. Mitchell's chemiosmotic hypothesis for ATP synthesis now occupies the position of dogma as the weight of evidence has accumulated in its favor. Photophosphorylation then can be concisely summarized by noting that thylakoid vesicles accumulate H^+ upon illumination and that the consequent electrochemical gradient, which represents an energized state, can be tapped to drive ATP synthesis. Collapse of the gradient—that is, equilibration of the ion concentration difference across the membrane—is the energy-transducing mechanism: the chemical potential of a concentration difference is transduced into synthesis of ATP.

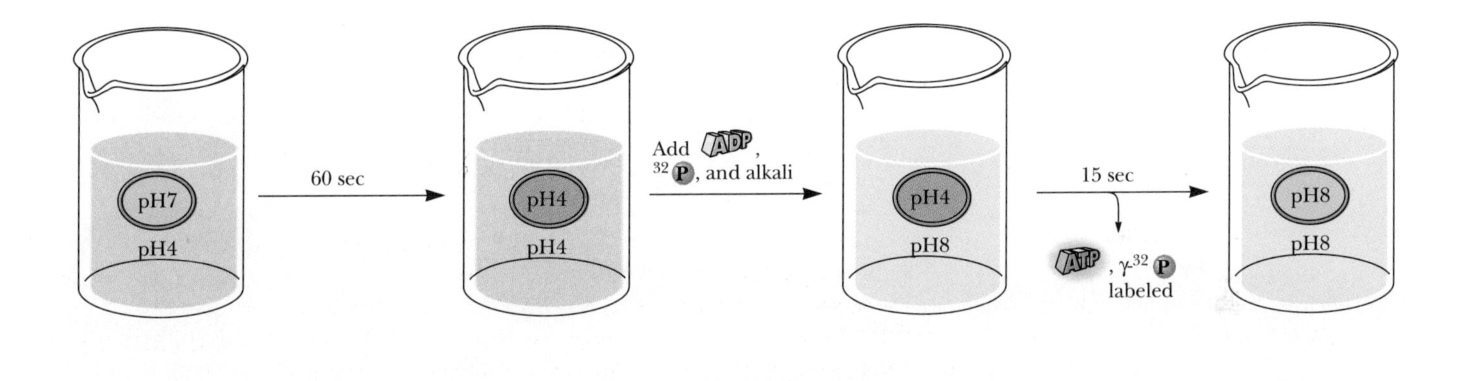

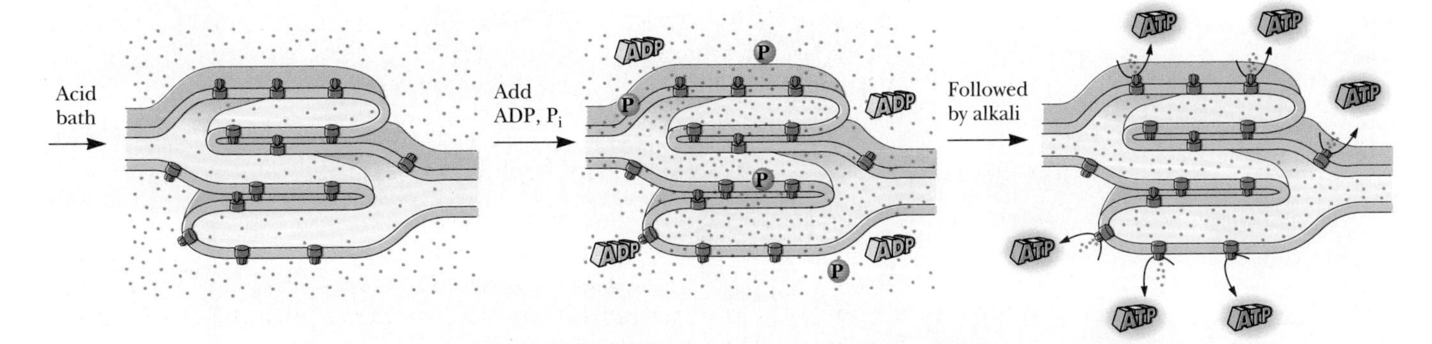

The mechanism of photophosphorylation is chemiosmotic. In 1966, Jagendorf and Uribe experimentally demonstrated for the first time that establishment of an electrochemical gradient across the membrane of an energy-transducing organelle could lead to ATP synthesis. They equilibrated isolated chloroplasts for 60 seconds in a pH 4 bath, adjusted the pH to 8 in the presence of ADP and P_i, and allowed phosphorylation to proceed for 15 seconds. The entire experiment was carried out in the dark.

potential, $\Delta\Psi$, and a pH gradient, ΔpH (Chapter 20), as defined in the relationship

$$\Delta p = \Delta\Psi - (2.3RT/\mathcal{F})\Delta\text{pH} \qquad (22.5)$$

In chloroplasts, the value of $\Delta\Psi$ is typically -50 to -100 mV, and the pH gradient is equivalent to about 3 pH units, so that $-(2.3\,RT/\mathcal{F})\Delta\text{pH} = -200$ mV. This situation contrasts with the mitochondrial proton-motive force, where the membrane potential contributes relatively more to Δp than does the pH gradient.

CF_1CF_0 ATP Synthase Is the Chloroplast Equivalent of the Mitochondrial F_1F_0 ATP Synthase

The transduction of the electrochemical gradient into the chemical energy represented by ATP is carried out by the chloroplast ATP synthase, which is highly analogous to the mitochondrial F_1F_0 ATP synthase. The chloroplast enzyme complex is called **CF_1CF_0 ATP synthase,** "C" symbolizing chloroplast. Like the mitochondrial complex, CF_1CF_0 ATP synthase is a heteromultimer of α, β, γ, δ, and ϵ subunits (Chapter 20), consisting of a knoblike structure some 9 nm in diameter (CF_1) attached to a stalked base (CF_0) embedded in the thylakoid membrane. The mechanism of action of CF_1CF_0 ATP synthase in coupling ATP synthesis to the collapse of the pH gradient is similar to that of the mitochondrial ATP synthase described in Chapter 20. The mechanism of photophosphorylation is summarized schematically in Figure 22.23.

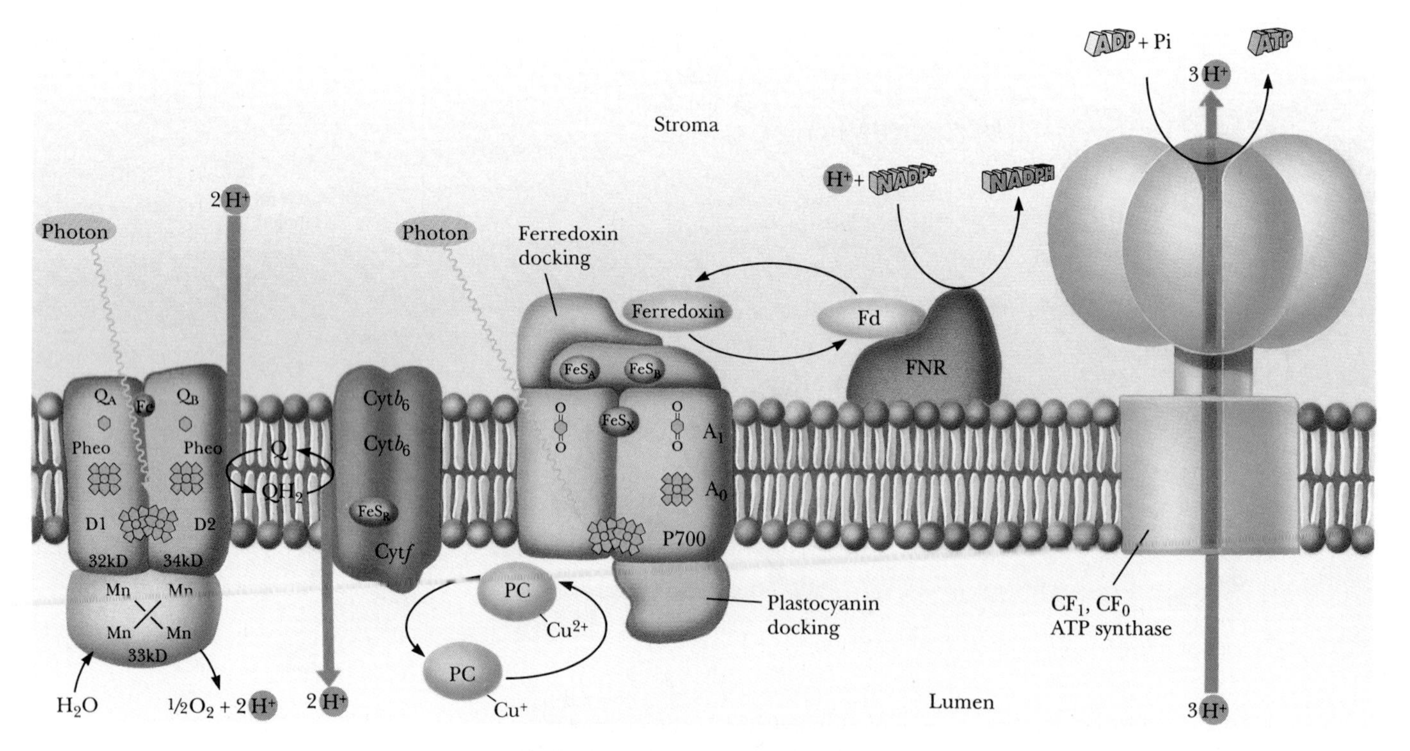

Figure **22.23** The mechanism of photophosphorylation. Photosynthetic electron transport establishes a proton gradient that is tapped by the CF_1CF_0 ATP synthase to drive ATP synthesis. Critical to this mechanism is the fact that the membrane-bound components of light-induced electron transport and ATP synthesis are asymmetrical with respect to the thylakoid membrane so that vectorial discharge and uptake of H^+ ensue, generating the proton-motive force.

Cyclic and Noncyclic Photophosphorylation

Photosynthetic electron transport, which pumps H^+ into the thylakoid lumen, can occur in two modes, both of which lead to the establishment of a transmembrane proton-motive force. Thus, both modes are coupled to ATP synthesis and are considered alternative mechanisms of photophosphorylation even though they are distinguished by differences in their electron transfer pathways. The two modes are cyclic and noncyclic photophosphorylation. **Noncyclic photophosphorylation** has been the focus of our discussion and is represented by the scheme in Figure 22.12, where electrons activated by quanta at PSII and PSI flow from H_2O to $NADP^+$, with concomitant establishment of the proton-motive force driving ATP synthesis. Note that in noncyclic photophosphorylation, O_2 is evolved and $NADP^+$ is reduced.

Cyclic Photophosphorylation

In **cyclic photophosphorylation,** the "electron hole" in $P700^+$ created by electron loss from P700 is filled *not* by an electron derived from H_2O via PSII but by a cyclic pathway in which the photoexcited electron returns ultimately to $P700^+$. This pathway is schematically represented in Figure 22.12 by the dashed line connecting F_B and cytochrome b_6. Thus, the function of cytochrome b_6 (b_{563}) is to couple the bound ferredoxin carriers of the PSI complex with the cytochrome b_6/cytochrome f complex. This pathway diverts the activated e^- from $NADP^+$ reduction back through plastocyanin to re-reduce $P700^+$ (Figure 22.24).

Proton translocations accompany these cyclic electron transfer events, so ATP synthesis can be achieved. In cyclic photophosphorylation, ATP is the

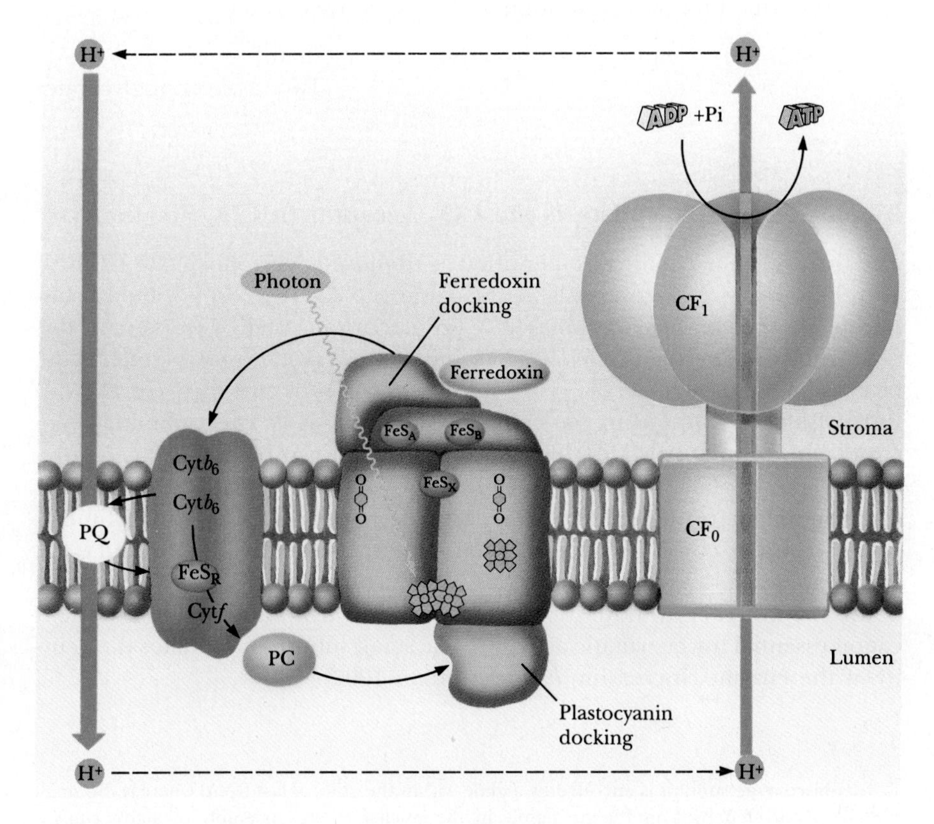

***Figure* 22.24** The pathway of cyclic photophosphorylation by PSI.
(Adapted from Arnon, D. I., 1984, Trends in Biochemical Sciences 9:258.)

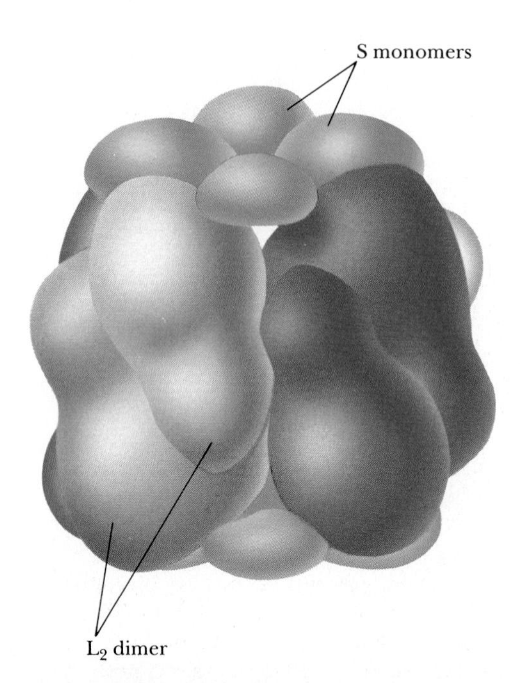

***Figure* 22.25** Schematic diagram of the subunit organization of ribulose bisphosphate carboxylase as revealed by X-ray crystallography. The enzyme consists of eight equivalents each of two types of subunits, large *L* (55 kD) and small *S* (15 kD). Clusters of four small subunits are located at each end of the symmetrical octamer formed by four L_2 dimers.

(From Knight, Andersson, and Branden, 1990, Journal of Molecular Biology 215:113–160.)

sole product of energy conversion. No NADPH is generated, and, since PSII is not involved, no oxygen is evolved. Cyclic photophosphorylation depends only on PSI and serves as a useful model for the photosynthetic reactions found in green and purple photosynthetic bacteria, which lack a second photosystem and the oxygen-evolving complex.

22.8 Carbon Dioxide Fixation

As we began this chapter, we saw that photosynthesis traditionally is equated with the process of carbon dioxide fixation, that is, the net synthesis of carbohydrate from CO_2. Indeed, the capacity to perform net accumulation of carbohydrate from CO_2 distinguishes the phototrophic (and autotrophic) organisms from heterotrophs. While animals possess enzymes capable of linking CO_2 to organic acceptors, they cannot achieve a net accumulation of organic material by these reactions. For example, fatty acid biosynthesis is primed by covalent attachment of CO_2 to acetyl-CoA to form malonyl-CoA (Chapter 24). Nevertheless, this "fixed CO_2" is liberated in the very next reaction, so no net CO_2 incorporation occurs.

Elucidation of the pathway of CO_2 fixation represents one of the earliest applications of radioisotope tracers to the study of biology. In 1945, Melvin Calvin and his colleagues at the University of California, Berkeley, were investigating photosynthetic CO_2 fixation in *Chlorella.* Using $^{14}CO_2$, they traced the incorporation of radioactive ^{14}C into organic products and found that the earliest labeled product was **3-phosphoglycerate** (see Figure 17.18). Though this result suggested that the CO_2 acceptor was a two-carbon compound, further investigation revealed that, in reality, two equivalents of 3-phosphoglycerate were formed following addition of CO_2 to a five-carbon (pentose) sugar:

$$CO_2 + \text{5-carbon acceptor} \longrightarrow [\text{6-carbon intermediate}] \longrightarrow \text{Two 3-phosphoglycerates}$$

Ribulose-1,5-Bisphosphate Is the CO_2 Acceptor in CO_2 Fixation

The five-carbon acceptor was identified as **ribulose-1,5-bisphosphate** (RuBP), and the enzyme catalyzing this key reaction of CO_2 fixation is **ribulose bisphosphate carboxylase/oxygenase,** or, in the jargon used by workers in this field, **rubisco.** The name *ribulose bisphosphate carboxylase/oxygenase* reflects the fact that rubisco catalyzes the reaction of either CO_2 or, alternatively, O_2 with RuBP. Rubisco is found in the chloroplast stroma. It is a very abundant enzyme, constituting over 15% of the total chloroplast protein. Given the preponderance of plant material in the biosphere, rubisco is probably the world's most abundant protein. Rubisco is large: it is a 550-kD heteromultimeric ($\alpha_8\beta_8$) complex consisting of eight identical large subunits (55 kD) and eight small subunits (15 kD) (Figure 22.25). The large subunit is the catalytic unit of the enzyme. It binds both substrates (CO_2 and RuBP) and Mg^{2+} (a divalent cation essential for enzymatic activity). The small subunit modulates the activity of the enzyme, increasing k_{cat} more than 100-fold.[2]

[2] The rubisco large subunit is encoded by a gene within the chloroplast DNA, whereas the small subunit is encoded by a multigene family in the nuclear DNA. Assembly of active rubisco heteromultimers occurs within chloroplasts following transit of the small subunit polypeptide across the chloroplast membrane.

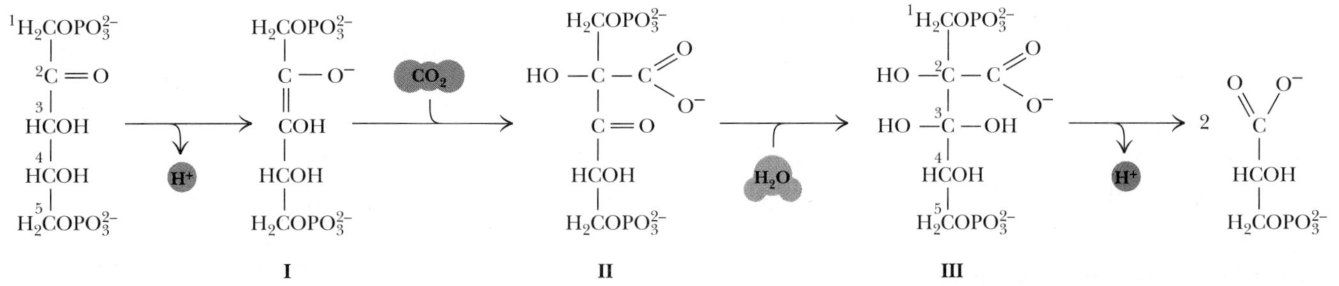

Figure **22.26** The ribulose bisphosphate carboxylase reaction. Enzymic abstraction of the C-3 proton of RuBP yields a 2,3-enediol intermediate (I), which is stereospecifically carboxylated at C-2 to create the six-carbon β-keto acid intermediate (II) known as 3-keto-arabinitol. Intermediate II is rapidly hydrated to give the gem-diol form (III). Deprotonation of the C-3 hydroxyl and cleavage yield two 3-phosphoglycerates. Mg^{2+} at the active site aids in stabilizing the 2,3-enediol transition state for CO_2 addition and in facilitating the carbon–carbon bond cleavage that leads to product formation. Note that CO_2, not HCO_3^- (its hydrated form), is the true substrate.

The Ribulose-1,5-Bisphosphate Carboxylase Reaction

The addition of CO_2 to ribulose-1,5-bisphosphate results in the formation of an enzyme-bound intermediate, **2-carboxy, 3-keto ribitol** (Figure 22.26). This intermediate arises when CO_2 adds to the enediol intermediate generated when Mg^{2+} forms a complex with ribulose-1,5-bisphosphate. Hydrolysis of the C_2—C_3 bond of the intermediate generates two molecules of 3-phosphoglycerate. The CO_2 ends up as the carboxyl group of one of the two molecules.

Regulation of Ribulose-1,5-Bisphosphate Carboxylase Activity

Rubisco exists in three forms: an inactive form designated *E;* a carbamylated, but inactive, form designated *EC;* and an active form, *ECM,* which is carbamylated and has Mg^{2+} at its active sites as well. Carbamylation of rubisco takes place by addition of CO_2 to its Lys^{201} ϵ-NH_2 groups (to give ϵ—NH—COO^- derivatives). The CO_2 molecules used to carbamylate Lys residues do not become substrates. The carbamylation reaction is promoted by slightly alkaline pH (pH 8). Carbamylation of rubisco completes the formation of a binding site for the Mg^{2+} that participates in the catalytic reaction. Once Mg^{2+} binds to *EC,* rubisco achieves its active *ECM* form. Activated rubisco displays a K_m for CO_2 of 10 to 20 μM.[3]

Substrate RuBP binds much more tightly to the inactive, *E* form of rubisco (K_D = 20 nM) than to the active, *ECM* form (K_m for RuBP = 20 μM). Thus, RuBP is also a potent inhibitor of rubisco activity. Release of RuBP from the active site of rubisco is mediated by **rubisco activase.** Rubisco activase is a regulatory protein; it binds to *E*-form rubisco and, in an ATP-dependent reaction, promotes the release of RuBP. Rubisco then becomes activated by carbamylation and Mg^{2+} binding. Rubisco activase itself is activated in an indirect manner by light. Thus, light is the ultimate activator of rubisco.

22.9 The Ribulose Bisphosphate Oxygenase Reaction: Photorespiration

As indicated, ribulose bisphosphate carboxylase/oxygenase catalyzes an alternative reaction in which O_2 replaces CO_2 as the substrate added to RuBP (Figure 22.27a). The *ribulose-1,5-bisphosphate oxygenase* reaction has profound significance for plant productivity because it leads to the wasteful loss of

[3]The relative abundance of CO_2 in the atmosphere is low, about 0.03%. The concentration of CO_2 dissolved in aqueous solutions equilibrated with air is about 10 μM.

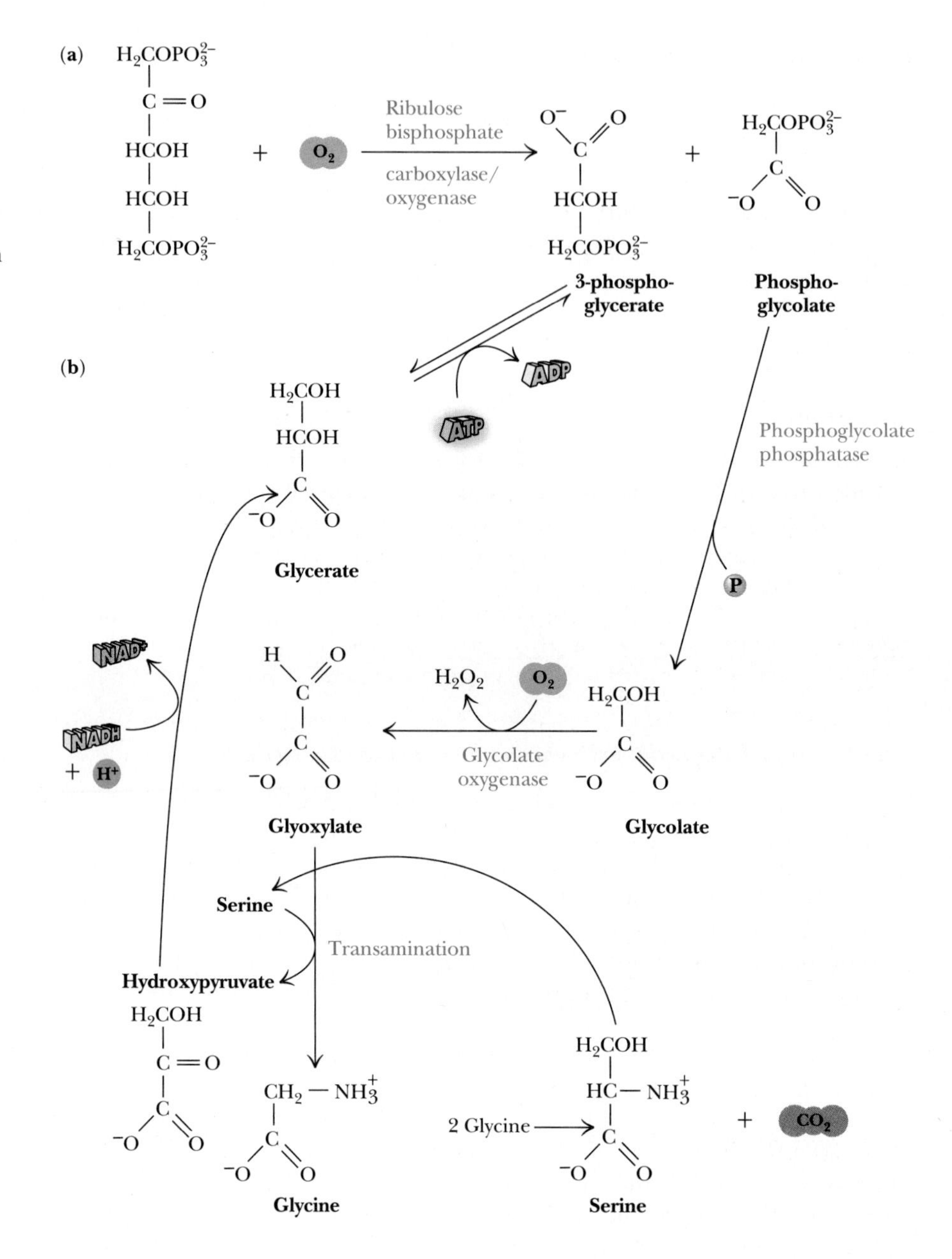

***Figure* 22.27** The oxygenase reaction of rubisco. (a) The reaction of ribulose bisphosphate carboxylase with O_2 in the presence of ribulose bisphosphate leads to wasteful cleavage of RuBP to yield 3-phosphoglycerate and phosphoglycolate. (b) Conversion of phosphoglycolate to glycine. In mitochondria, two glycines from photorespiration are converted into one serine plus CO_2. This step is the source of the CO_2 evolved in photorespiration. Transamination of glyoxylate to glycine by the product serine yields hydroxypyruvate; reduction of hydroxypyruvate yields glycerate, which can be phosphorylated to 3-phosphoglycerate to fuel resynthesis of ribulose bisphosphate by the Calvin cycle.

RuBP, the essential CO_2 acceptor. The K_m for O_2 in this oxygenase reaction is about 200 μM. Given the relative abundance of CO_2 and O_2 in the atmosphere and their relative K_m values in these rubisco-mediated reactions, the ratio of carboxylase to oxygenase activity *in vivo* is about 3 or 4 to 1.

The products of ribulose bisphosphate oxygenase activity are *3-phosphoglycerate* and *phosphoglycolate.* Dephosphorylation and oxidation convert phosphoglycolate to **glyoxylate,** the α-keto acid of glycine (Figure 22.27b). Transamination yields glycine. Other fates of phosphoglycolate are also possible, including oxidation to CO_2, with the released energy being dissipated as heat. Obviously, agricultural productivity is dramatically lowered by this phenomenon, which, because it is a light-related uptake of O_2 *and* release of CO_2, is termed **photorespiration.** As we shall see, certain plants, particularly tropical grasses, have evolved means to circumvent photorespiration. These plants are more efficient users of light for carbohydrate synthesis.

22.10 The Calvin–Benson Cycle

The immediate product of CO_2 fixation, 3-phosphoglycerate, must undergo a series of transformations before the net synthesis of carbohydrate is realized. Among carbohydrates, hexoses (particularly glucose) occupy center stage. Glucose is the building block for both cellulose and starch synthesis. These plant polymers constitute the most abundant organic material in the living world, and thus, the central focus on glucose as the ultimate end product of CO_2 fixation is amply justified. Also, sucrose (α-D-glucopyranosyl-($1 \rightarrow 2$)-β-D-fructofuranoside) is the major carbon form translocated out of leaves to other plant tissues.

The set of reactions that transform 3-phosphoglycerate into hexose is named the **Calvin–Benson cycle** (often referred to simply as the *Calvin cycle*) for its discoverers. The reaction series is indeed cyclic since not only must carbohydrate appear as an end product, but the 5-carbon acceptor, RuBP, must be regenerated to provide for continual CO_2 fixation. Balanced equations that schematically represent this situation are

$$\begin{array}{rcl} 6(1) + 6(5) & \longrightarrow & 12(3) \\ 12(3) & \longrightarrow & 1(6) + 6(5) \\ \hline \textit{Net:}\ 6(1) & \longrightarrow & 1(6) \end{array}$$

Each number in parentheses represents the number of carbon atoms in a compound, and the number preceding the parentheses indicates the stoichiometry of the reaction. Thus, 6(1), or 6 CO_2, condense with 6(5) or 6 RuBP to give 12 3-phosphoglycerates. These 12(3)s are then rearranged in the Calvin cycle to form one hexose, 1(6), and regenerate the six 5-carbon (RuBP) acceptors.

The Enzymes of the Calvin Cycle

The Calvin cycle enzymes serve three important ends:

1. They constitute the only net CO_2 fixation pathway in nature.
2. They accomplish the reduction of 3-phosphoglycerate to glyceraldehyde-3-phosphate so that carbohydrate synthesis becomes feasible.
3. They catalyze reactions that transform 3-carbon compounds into 4-, 5-, 6-, and 7-carbon compounds.

With two exceptions, the enzymes mediating the reactions of the Calvin cycle are enzymes familiar to us from glycolysis (Chapter 18) and the pentose phosphate pathway (Chapter 21). The aim of the Calvin scheme is to account for hexose formation from 3-phosphoglycerate. In the course of this metabolic sequence, we shall see that NADPH and ATP are consumed, as indicated earlier in Equation (22.3).

The Calvin cycle of reactions starts with *ribulose bisphosphate carboxylase* and concludes with **ribulose-5-phosphate kinase** (also called *phosphoribulose kinase*), which forms RuBP (Figure 22.28 and Table 22.1). The carbon balance is given at the right side of the table. Several features of the reactions in Table 22.1 merit discussion. Note that the 18 equivalents of ATP consumed in hexose formation are expended in reactions 2 and 15: 12 to form 12 equivalents of 1,3-bisphosphoglycerate from 3-phosphoglycerate by a reversal of the normal glycolytic reaction catalyzed by *3-phosphoglycerate kinase,* and six to phosphorylate Ru-5-P to regenerate 6 RuBP. All 12 NADPH equivalents are used in reaction 3. Plants possess an *NADPH-specific glyceraldehyde-3-phosphate dehydro-*

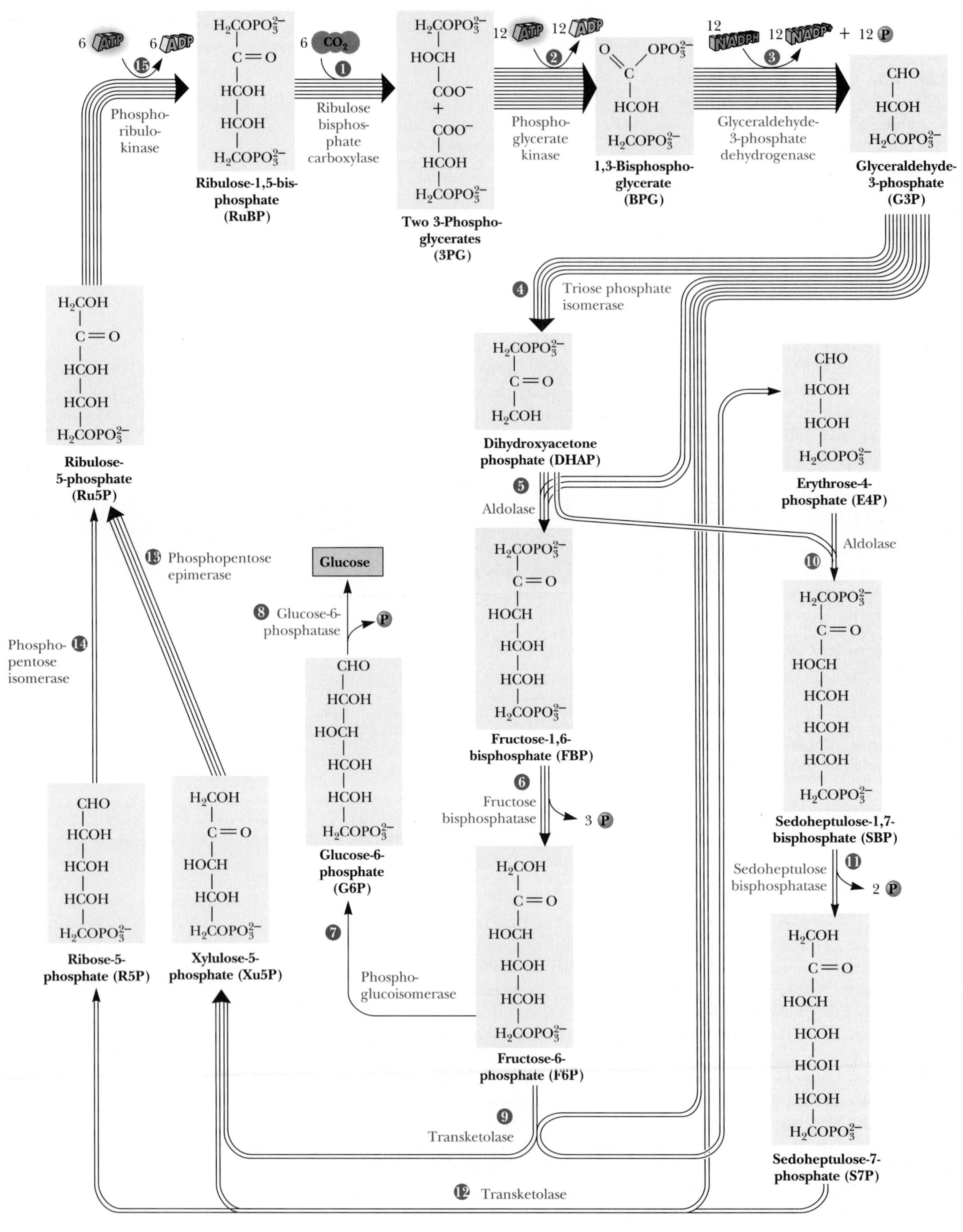

***Figure* 22.28** The Calvin–Benson cycle of reactions. The number of arrows at each step indicates the number of molecules reacting in a turn of the cycle that produces one molecule of glucose. Reactions are numbered as in Table 22.1.

Table **22.1**

The Calvin Cycle Series of Reactions

Reactions 1 through 15 constitute the cycle that culminates in the formation of one equivalent of glucose. The enzyme catalyzing each step, a concise reaction, and the overall carbon balance is given. Numbers in parentheses denote the numbers of carbon atoms in the substrate and product molecules. Prefix numbers indicate in a stoichiometric fashion how many times each step is carried out in order to provide a balanced net reaction.

Reaction	Carbon balance
1. Ribulose bisphosphate carboxylase: 6 CO_2 + 6 H_2O + 6 RuBP $\longrightarrow$ 12 3-PG	6(1) + 6(5) $\longrightarrow$ 12(3)
2. 3-Phosphoglycerate kinase: 12 3-PG + 12 ATP $\longrightarrow$ 12 1,3-BPG + 12 ADP	12(3) $\longrightarrow$ 12(3)
3. $NADP^+$-glyceraldehyde-3-P dehydrogenase: 12 1,3-BPG + 12 NADPH $\longrightarrow$ 12 $NADP^+$ + 12 G3P + 12 P_i	12(3) $\longrightarrow$ 12(3)
4. Triose-P isomerase: 5 G3P $\longrightarrow$ 5 DHAP	5(3) $\longrightarrow$ 5(3)
5. Aldolase: 3 G3P + 3 DHAP $\longrightarrow$ 3 FBP	3(3) + 3(3) $\longrightarrow$ 3(6)
6. Fructose bisphosphatase: 3 FBP + 3 H_2O $\longrightarrow$ 3 F6P + 3 P_1	3(6) $\longrightarrow$ 3(6)
7. Phosphoglucoisomerase: 1 F6P $\longrightarrow$ 1 G6P	1(6) $\longrightarrow$ 1(6)
8. Glucose phosphatase: 1 G6P + 1 H_2O $\longrightarrow$ 1 GLUCOSE + 1 P_i	1(6) $\longrightarrow$ 1(6)
The remainder of the pathway involves regenerating six RuBP acceptors (= 30C) from the leftover two F6P (12C), four G3P (12C), and two DHAP (6C).	
9. Transketolase: 2 F6P + 2 G3P $\longrightarrow$ 2 Xu5P + 2 E4P	2(6) + 2(3) $\longrightarrow$ 2(5) + 2(4)
10. Aldolase: 2 E4P + 2 DHAP $\longrightarrow$ 2 sedoheptulose-1,7-bisphosphate (SBP)	2(4) + 2(3) $\longrightarrow$ 2(7)
11. Sedoheptulose bisphosphatase: 2 SBP + 2 H_2O $\longrightarrow$ 2 S7P + 2 P_i	2(7) $\longrightarrow$ 2(7)
12. Transketolase: 2 S7P + 2 G3P $\longrightarrow$ 2 Xu5P + 2 R5P	2(7) + 2(3) $\longrightarrow$ 4(5)
13. Phosphopentose epimerase: 4 Xu5P $\longrightarrow$ 4 Ru5P	4(5) $\longrightarrow$ 4(5)
14. Phosphopentose isomerase: 2 R5P $\longrightarrow$ 2 Ru5P	2(5) $\longrightarrow$ 2(5)
15. Phosphoribulose kinase: 6 Ru5P + 6 ATP $\longrightarrow$ 6 RuBP + 6 ADP	6(5) $\longrightarrow$ 6(5)
Net: 6 CO_2 + 18 ATP + 12 NADPH + 12 H^+ + 12 H_2O $\longrightarrow$ glucose + 18 ADP + 18 P_i + 12 $NADP^+$	6(1) $\longrightarrow$ 1(6)

genase, which contrasts with its glycolytic counterpart in its specificity for NADP over NAD and in the direction in which the reaction normally proceeds.

Balancing the Calvin Cycle Reactions To Account for Net Hexose Synthesis

When carbon rearrangements are balanced to account for net hexose synthesis, five of the glyceraldehyde-3-phosphate molecules are converted to dihydroxyacetone phosphate (DHAP). Three of these DHAPs then condense with three glyceraldehyde-3-P via the aldolase reaction to yield hexose in the form of fructose bisphosphate. (Recall that the $\Delta G°'$ for the aldolase reaction in the glycolytic direction is +23.9 kJ/mol. Thus, the aldolase reaction running "in reverse" in the Calvin cycle would be thermodynamically favored under standard-state conditions.) Taking one FBP to glucose, the desired product of this scheme, leaves 30 carbons, distributed as two fructose-6-phosphates, four glyceraldehyde-3-phosphates, and 2 DHAP. These 30 Cs are reorganized into 6 RuBP by reactions 9 through 15. Step 9 and steps 12 through 14 involve pentose phosphate pathway enzymes. Reaction 11 is mediated by **sedoheptulose-1,7-bisphosphatase.** This phosphatase is unique to plants; it serves to generate the sedoheptulose-7-P form of the seven-carbon sugar that acts as the transketolase substrate. Likewise, **phosphoribulose kinase** carries out the unique plant function of providing RuBP from Ru-5-P (reaction 15). The net conversion accounts for the fixation of six equivalents of carbon dioxide into one hexose at the expense of 18 ATP and 12 NADPH.

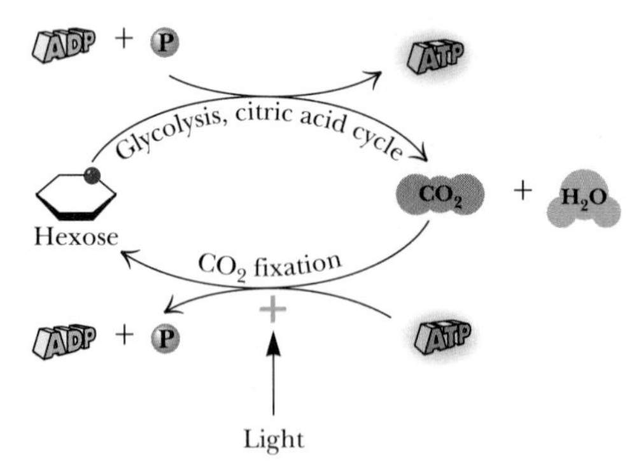

***Figure* 22.29** Light regulation of CO_2 fixation prevents a substrate cycle between cellular respiration and hexose synthesis by carbon dioxide fixation. Since plants possess mitochondria and are capable of deriving energy from hexose catabolism (glycolysis and the citric acid cycle), regulation of photosynthetic CO_2 fixation by light activation controls the net flux of carbon between these opposing routes.

22.11 Regulation of Carbon Dioxide Fixation

Plant cells contain mitochondria and can carry out cellular respiration (glycolysis, the citric acid cycle, and oxidative phosphorylation) to provide energy in the dark. Futile cycling of carbohydrate to CO_2 by glycolysis and the citric acid cycle in one direction, and CO_2 to carbohydrate by the CO_2 fixation pathway in the opposite direction, is thwarted through regulation of the Calvin cycle (Figure 22.29). In this regulation, the activities of key Calvin cycle enzymes are coordinated with the output of photosynthesis. In effect, these enzymes respond indirectly to *light activation.* Thus, when light energy is available to generate ATP and NADPH for CO_2 fixation, the Calvin cycle proceeds. In the dark, when ATP and NADPH are no longer photosynthetically produced, fixation of CO_2 ceases. The light-induced effects in the chloroplast which serve to modulate the activities of key Calvin cycle enzymes include (1) *pH changes,* (2) *generation of reducing power,* and (3) *Mg^{2+} efflux from the thylakoid vesicles.*

Light-Induced pH Changes in Chloroplast Compartments

Illumination of chloroplasts leads to light-driven proton pumping and consequent pH changes in the chloroplast compartments, specifically the stroma and the thylakoid space (Figure 22.30). Since the enzymes of the Calvin cycle reside in the stroma, CO_2 fixation is activated as the stromal pH rises in response to proton pumping into the thylakoid vesicles. *Ribulose bisphosphate carboxylase* (rubisco) is activated at pH 8, and the rise in pH is implicated in *rubisco activase* activation as well. *Fructose-1,6-bisphosphatase, ribulose-5-phosphate kinase,* and *glyceraldehyde-3-phosphate dehydrogenase* all have alkaline pH optima. Thus, their activities increase as a result of the light-induced pH increase in the stroma.

Light-Induced Generation of Reducing Power

Illumination of chloroplasts initiates photosynthetic electron transport, which generates reducing power in the form of reduced ferredoxin and NADPH. Several enzymes of CO_2 fixation, notably *fructose-1,6-bisphosphatase, sedoheptulose-1,7-bisphosphatase,* and *ribulose-5-phosphate kinase,* are activated upon reduction. Reductive activation also can be achieved *in vitro* in the presence of reduced ferredoxin. Reduced sulfhydryl reagents such as *dithiothreitol* also accomplish this activation. *In vivo,* the reduced form of *thioredoxin* mediates this reaction. Thioredoxin is a small (12 kD) protein possessing in its reduced state a pair of sulfhydryls (—SH HS—), which upon oxidation form a disulfide bridge (—S—S—). Thioredoxin serves as the hydrogen carrier between NADPH or Fd_{red} and enzymes regulated by light (Figure 22.31).

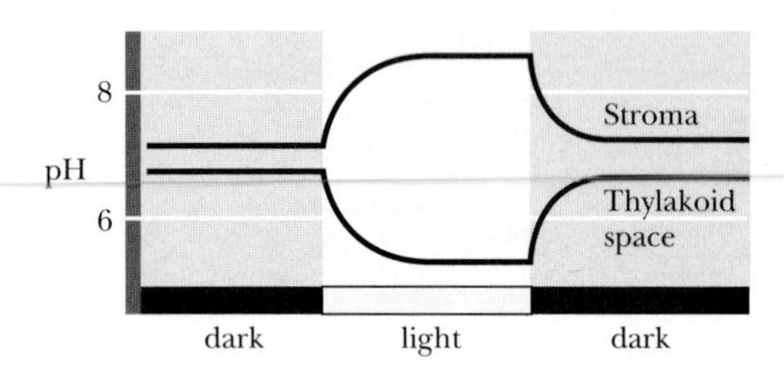

***Figure* 22.30** Light-induced pH changes in chloroplast compartments. Illumination of chloroplasts leads to proton pumping and pH changes in the chloroplast, such that the pH within the thylakoid space falls and the pH of the stroma rises. These pH changes modulate the activity of key Calvin cycle enzymes.

Light-Induced Mg^{2+} Efflux from Thylakoid Vesicles

When light-driven proton pumping across the thylakoid membrane occurs, a concomitant efflux of Mg^{2+} ions from vesicles into the stroma is observed. This efflux of Mg^{2+} somewhat counteracts the charge accumulation due to H^+ influx and is one reason why the membrane potential change in response to proton pumping is less in chloroplasts than in mitochondria (Eq. 22.5). Both *ribulose bisphosphate carboxylase* and *fructose-1,6-bisphosphatase* are Mg^{2+}-activated enzymes, and Mg^{2+} flux into the stroma as a result of light-driven proton pumping stimulates the CO_2 fixation pathway at these key steps. Activity measurements have indicated that fructose bisphosphatase may be the

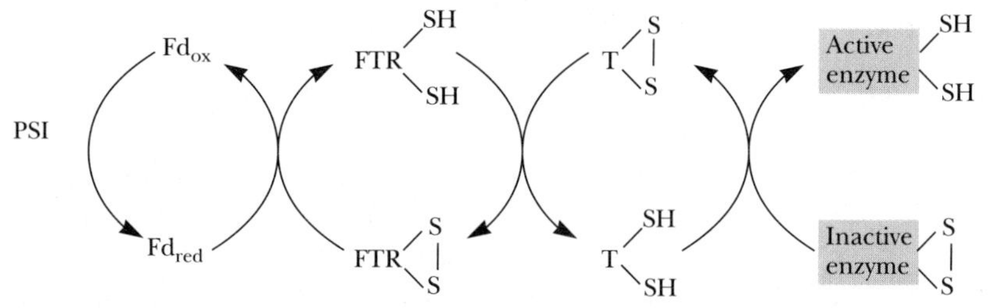

PSI
Fd_{ox}
Fd_{red}
FTR(SH)(SH)
FTR(S–S)
T(S–S)
T(SH)(SH)
Active enzyme (SH)(SH)
Inactive enzyme (S–S)

Figure **22.31** The pathway for light regulation of Calvin cycle enzymes. Light-generated reducing power (Fd_{red} = reduced ferredoxin) provides e^- for reduction of thioredoxin (T) by FTR (ferredoxin–thioredoxin reductase). Several Calvin cycle enzymes have pairs of Cys residues that are involved in the disulfide-sulfhydryl transition between an inactive (—S—S—) form and an active (—SH HS—) form, as shown here. These enzymes include *fructose-1,6-bisphosphatase* (residues Cys^{174} and Cys^{179}), *$NADP^+$-malate dehydrogenase* (residues Cys^{10} and Cys^{15}), and *ribulose-5-P kinase* (residues Cys^{16} and Cys^{55}).

rate-limiting step in the Calvin cycle. The recurring theme of fructose bisphosphatase as the target of the light-induced changes in the chloroplasts implicates this enzyme as a key point of control in the Calvin cycle.

22.12 The C-4 Pathway of CO_2 Fixation

Tropical grasses are less susceptible to the deleterious effects of photorespiration, as noted earlier. Studies employing $^{14}CO_2$ as a tracer indicated that the first organic intermediate labeled in these plants was not a three-carbon compound but a four-carbon compound. Hatch and Slack, two Australian biochemists, first discovered this C-4 product of CO_2 fixation, and, appropriately, the C-4 pathway of CO_2 incorporation is named the *Hatch–Slack pathway* after them. The C-4 pathway is not an alternative to the Calvin cycle series of reactions or even a net CO_2 fixation scheme. Instead, it functions as a *CO_2 delivery system,* carrying carbon dioxide from the relatively oxygen-rich surface of the leaf to interior cells where oxygen is lower in concentration and hence less effective in competing with CO_2 in the rubisco reaction. Thus, the C-4 pathway is a means of avoiding photorespiration by sheltering the rubisco reaction in a cellular compartment away from high $[O_2]$. The C-4 compounds serving as CO_2 transporters are *oxaloacetate* (OAA) and *malate.*

The compartmentation of these reactions to thwart photorespiration involves the interaction of two cell types, *mesophyll cells* and *bundle sheath cells.* The mesophyll cells take up CO_2 at the leaf surface, where O_2 is abundant, and use it to carboxylate phosphoenolpyruvate to yield OAA in a reaction catalyzed by *PEP carboxylase* (Figure 22.32). This four-carbon dicarboxylic acid is then either reduced to malate by an *NADPH-specific malate dehydrogenase* or transaminated to give *aspartate* in the mesophyll cells.[4] The 4-C CO_2 carrier (OAA or malate) then is transported to the bundle sheath cells, where it is decarboxylated to yield CO_2 and a 3-C product. The CO_2 is then fixed into organic carbon by the Calvin cycle localized within the bundle sheath cells, and the 3-C product is returned to the mesophyll cells, where it is reconverted to PEP in preparation to accept another CO_2 (Figure 22.32).

Intercellular Transport of Each CO_2 via a C-4 Intermediate Costs 2 ATP

The transport of each CO_2 requires the expenditure of two high-energy phosphate bonds. The energy of these bonds is expended in the phosphorylation of pyruvate to PEP (phosphoenolpyruvate) by the plant enzyme *pyruvate-P_i dikinase;* the products are PEP, AMP, and pyrophosphate (PP_i). This repre-

[4]A number of different biochemical subtypes of C-4 plants are known. They differ in whether OAA or malate is the CO_2 carrier to the bundle sheath cell and in the nature of the reaction by which the CO_2 carrier is decarboxylated to regenerate a 3-C product. In all cases, the 3-C product is returned to the mesophyll cell and reconverted to PEP.

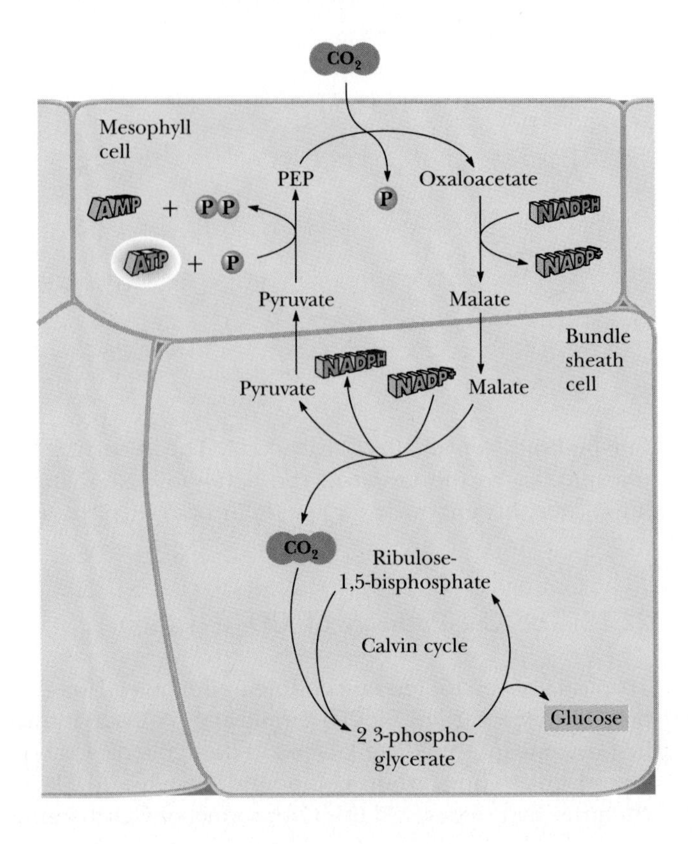

Figure **22.32** Essential features of the compartmentation and biochemistry of the Hatch–Slack pathway of carbon dioxide uptake in C-4 plants. Carbon dioxide is fixed into organic linkage by PEP carboxylase of mesophyll cells, forming OAA. Either malate (the reduced form of OAA) or aspartate (the aminated form) serves as the carrier transporting CO_2 to the bundle sheath cells. Within the bundle sheath cells, CO_2 is liberated by decarboxylation of malate or Asp; the C-3 product is returned to the mesophyll cell. Formation of PEP by pyruvate:P_i dikinase reinitiates the cycle. The CO_2 liberated in the bundle sheath cell is used to synthesize hexose by the conventional rubisco–Calvin cycle series of reactions.

sents a unique phosphotransferase reaction in that both the β- and γ-phosphates of a single ATP are used to phosphorylate the two substrates, pyruvate and P_i. The reaction mechanism involves an enzyme phosphohistidine intermediate. The γ-phosphate of ATP is transferred to P_i, whereas formation of E-His-P occurs by addition of the β-phosphate from ATP:

$$\begin{aligned} \text{E—His} + \text{AMP}_\alpha\text{—P}_\beta\text{—P}_\gamma + \text{P}_i &\longrightarrow \text{E—His—P}_\beta + \text{AMP}_\alpha + \text{P}_\gamma\text{P}_i \\ \text{E—His—P}_\beta + \text{pyruvate} &\longrightarrow \text{PEP} + \text{E—His} \\ \hline \textit{Net:}\ \text{ATP} + \text{pyruvate} + \text{P}_i &\longrightarrow \text{AMP} + \text{PEP} + \text{PP}_i \end{aligned}$$

Pyruvate-P_i dikinase is regulated by reversible phosphorylation of a threonine residue, the nonphosphorylated form being active. Interestingly, ADP is the phosphate donor in this interconvertible regulation. Despite the added metabolic expense of two phosphodiester bonds for each equivalent of carbon dioxide taken up, CO_2 fixation is still considerably more efficient in plants which possess the C-4 pathway. Tropical grasses which are C-4 plants include sugarcane, maize, and crabgrass. In terms of photosynthetic efficiency, cultivated fields of sugarcane represent the pinnacle of light-harvesting efficiency. Approximately 8% of the incident light energy on a sugarcane field appears as chemical energy in the form of CO_2 fixed into carbohydrate. This efficiency compares dramatically with the estimated photosynthetic efficiency of 0.2% for uncultivated plant areas. Research on photorespiration is actively pursued in hopes of enhancing the efficiency of agriculture by controlling this wasteful process.

22.13 Crassulacean Acid Metabolism

In contrast to C-4 plants, which have evolved spatial separation of CO_2 uptake and fixation into distinct cells in order to thwart photorespiration, succulent plants native to semiarid and tropical environments exploit a temporal separation. Carbon dioxide (as well as O_2) enters the leaf through microscopic pores known as **stomata,** and water vapor escapes from plants via these same openings. In nonsucculent plants, the stomata are open during the day, when light can drive photosynthetic CO_2 fixation, and closed at night. Succulent plants, such as the *Cactaceae* (cacti) and *Crassulaceae,* cannot open their stomata during the heat of day because any loss of precious H_2O in their arid habitats would doom them. Instead, these plants open their stomata to take up CO_2 only at night, when temperatures are lower and water loss is less likely. This carbon dioxide is immediately incorporated into PEP to form OAA by PEP carboxylase; OAA is then reduced to malate by malate dehydrogenase and stored within vacuoles until morning. During the day, the malate is released from the vacuoles and decarboxylated to yield CO_2 and a 3-C product. The CO_2 is then fixed into organic carbon by rubisco and the reactions of the Calvin cycle. Because this process involves the accumulation of organic acids (OAA, malate) and is common to succulents of the *Crassulaceae* family, it is referred to as *crassulacean acid metabolism,* and plants capable of it are called *CAM plants.*

Problems

1. In photosystem I, P700 in its ground state has an $\mathcal{E}_o' = +0.4$ V. Excitation of P700 by a photon of 700-nm light alters the $\mathcal{E}_o'$ of P700* to -0.6 V. What is the efficiency of energy capture in this light reaction of P700?
2. What is the $\mathcal{E}_o'$ for the light-generated primary oxidant of photosystem II if the light-induced oxidation of water (which leads to O_2 evolution) proceeds with a $\Delta G°'$ of -25 kJ/mol?
3. Assuming that the concentrations of ATP, ADP, and P_i in chloroplasts are 3 mM, 0.1 mM, and 10 mM, respectively, what is the ΔG for ATP synthesis under these conditions? Photosynthetic electron transport establishes the proton-motive force driving photophosphorylation. What redox potential difference is necessary to achieve ATP synthesis under the foregoing conditions, assuming an electron pair is transferred per molecule of ATP generated?
4. ^{14}C-labeled carbon dioxide is administered to a green plant, and shortly thereafter the following compounds are isolated from the plant: 3-phosphoglycerate, glucose, erythrose-4-phosphate, sedoheptulose-1,7-bisphosphate, ribose-5-phosphate. In which carbon atoms will radioactivity be found?
5. A worker in a radioisotope laboratory is accidently exposed to high levels of ^{14}C carbon dioxide. In which carbon atoms of the person's blood glucose do you initially expect to find radioactivity?
6. If noncyclic photosynthetic electron transport leads to the translocation of 3 H^+/e^- and cyclic photosynthetic electron transport leads to the translocation of 2 H^+/e^-, what is the relative photosynthetic efficiency of ATP synthesis (expressed as the number of photons absorbed per ATP synthesized) for noncyclic *versus* cyclic photophosphorylation? (Assume that the CF_1CF_0 ATP synthase yields 1 ATP/3 H^+).
7. The photosynthetic CO_2 fixation pathway is regulated in response to specific effects induced in chloroplasts by light. What is the nature of these effects, and how do they regulate this metabolic pathway?
8. The overall equation for photosynthetic CO_2 fixation is

$$6\,CO_2 + 6\,H_2O \longrightarrow C_6H_{12}O_6 + 6\,O_2$$

All the O atoms evolved as O_2 come from water; *none* come from carbon dioxide. But 12 O atoms are evolved as 6 O_2, and only 6 O atoms appear as 6 H_2O in the equation. Also, 6 CO_2 have 12 O atoms, yet there are only 6 O atoms in $C_6H_{12}O_6$. How can you account for these discrepancies? (*Hint:* Consider the partial reactions of photosynthesis: ATP synthesis, $NADP^+$ reduction, photolysis of water, and the overall reaction for hexose synthesis in the Calvin–Benson cycle.)

Further Reading

Anderson, J. M., 1986. Photoregulation of the composition, function and structure of the thylakoid membrane. *Annual Review of Plant Physiology* **37:**93–136.

Anderson, J. M., and Andersson, B., 1988. The dynamic photosynthetic membrane and regulation of solar energy conversion. *Trends in Biochemical Sciences* **13:**351–355.

Arnon, D. I., 1984. The discovery of photosynthetic phosphorylation. *Trends in Biochemical Sciences* **9:**258–262. A historical account of photophosphorylation by its discoverer.

Barber, J., 1987. Photosynthetic reaction centres: A common link. *Trends in Biochemical Sciences* **12:**321–326.

Barber, J., 1987. Rethinking the structure of the photosystem II reaction centre. *Trends in Biochemical Sciences* **12:**123–124.

Blankenship, R. E., and Parson, W. W., 1978. The photochemical electron transfer reactions of photosynthetic bacteria and plants. *Annual Review of Biochemistry* **47:**635–653.

Burnell, J. N., and Hatch, M. D., 1985. Light–dark modulation of leaf pyruvate, P_i dikinase. *Trends in Biochemical Sciences* **10:**288–291. Regulation of a key enzyme in C-4 CO_2 fixation.

Chapman, M. S., et al., 1988. Tertiary structure of plant rubisco: Domains and their contacts. *Science* **241:**71–74. Structural details of rubisco.

Clayton, R. K., 1980. Photosynthesis: Physical mechanisms and chemical patterns. *IUPAB Biophysics Series* **4,** Cambridge University Press.

Cramer, W. A., Widger, W. R., Hermann, R. G., and Trebst, A., 1985. Topography and function of thylakoid membrane proteins. *Trends in Biochemical Sciences* **10:**125–129.

Cramer, W. A., and Knaff, D. B., 1990. *Energy Transduction in Biological Membranes: A Textbook of Bioenergetics.* New York: Springer-Verlag. A textbook on bioenergetics by two prominent workers in photosynthesis.

Deisenhofer, J., and Michel, H., 1989. The photosynthetic reaction center from the purple bacterium *Rhodopseudomonas viridis. Science* **245:**1463–1473. Published version of the Nobel laureate address by two of the researchers who first elucidated the molecular structure of a photosynthetic reaction center.

Deisenhofer, J., Michel, H., and Huber, R., 1985. The structural basis of light reactions in bacteria. *Trends in Biochemical Sciences* **10:**243–248.

Deisenhofer, J., et al., 1985. Structure of the protein subunits in the photosynthetic reaction center of *Rhodopseudomonas viridis* at 3 Å resolution. *Nature* **318:**618–624.
Also 1984, *Journal of Molecular Biology* **180:**385–398. These papers are the original reports of the crystal structure of a photosynthetic reaction center.

Ghanotakis, D. F., and Yocum, C. F., 1990. Photosystem II and the oxygen-evolving complex. *Annual Review of Plant Physiology and Plant Molecular Biology* **41:**255–276.

Glazer, A. N., 1983. Comparative biochemistry of photosynthetic light-harvesting pigments. *Annual Review of Biochemistry* **52:**125–157.

Glazer, A. N., and Melis, A., 1987. Photochemical reaction centers: Structure, organization and function. *Annual Review of Plant Physiology* **38:**11–45.

Golbeck, J. H., 1992. Structure and function of photosystem I. *Annual Review of Plant Physiology and Plant Molecular Biology* **43:**292–324.

Haehnel, W., 1984. Photosynthetic electron transport in higher plants. *Annual Review of Plant Physiology* **35:**659–693.

Harold, F. M., 1986. *The Vital Force: A Study of Bioenergetics,* Chap. 8: Harvesting the Light. New York: Freeman & Co.

Hatch, M. D., 1987. C_4 photosynthesis: A unique blend of modified biochemistry, anatomy and ultrastructure. *Biochimica et Biophysica Acta* **895:**81–106. A review of the biochemistry of the C-4 pathway by its discoverer.

Jagendorf, A. T., and Uribe, E., 1966. ATP formation caused by acid–base transition of spinach chloroplasts. *Proceedings of the National Academy of Sciences, U.S.A.* **55:**170–177. The classic paper providing the first experimental verification of Mitchell's chemiosmotic hypothesis.

Knaff, D. B., 1991. Regulatory phosphorylation of chloroplast antenna proteins. *Trends in Biochemical Sciences* **16:**82–83. Additional discussion of the structure of light-harvesting antenna complexes associated with photosynthetic reaction centers can be found in *Trends in Biochemical Sciences* **11:**414 (1986), **14:**72 (1989), and **16:**181 (1991).

Knaff, D. B., 1989. The regulatory role of thioredoxin in chloroplasts. *Trends in Biochemical Sciences* **14:**433–434.

Knight, S., Andersson, I., and Branden, C. I., 1990. Crystallographic analysis of ribulose 1,5-bisphosphate carboxylase from spinach at 2.4 Å resolution: Subunit interactions and active site. *Journal of Molecular Biology* **215:**113–160.

Krauss, N., et al., 1993. Three-dimensional structure of system I of photosynthesis at 6 Å resolution. *Nature* **361:**326–331.

Miziorko, H. M., and Lorimer, G. H., 1983. Ribulose-1,5-bisphosphate carboxylase/oxygenase. *Annual Review of Biochemistry* **52:**507–535. A thorough review of the enzymological properties of rubisco.

Murata, N., and Miyao, M., 1985. Extrinsic membrane proteins in the photosynthetic oxygen-evolving complex. *Trends in Biochemical Sciences* **10:**122–124.

Nitschke, W., and Rutherford, A. W., 1991. Photosynthetic reaction centers: Variations on a common structural theme. *Trends in Biochemical Sciences* **16:**241–245.

Ogren, W. L., 1984. Photorespiration: Pathways, regulation and modification. *Annual Review of Plant Physiology* **35:**415–442.

Portis, A. R., Jr., 1992. Regulation of ribulose 1,5-bisphosphate carboxylase/oxygenase activity. *Annual Review of Plant Physiology and Plant Molecular Biology* **43:**415–437.

Sharit, N., 1980. Energy transduction in chloroplasts: Structure and function of the ATPase complex. *Annual Review of Biochemistry* **49:**111–138.

Ting, I. P., 1985. Crassulacean acid metabolism. *Annual Review of Plant Physiology* **36:**595–622.

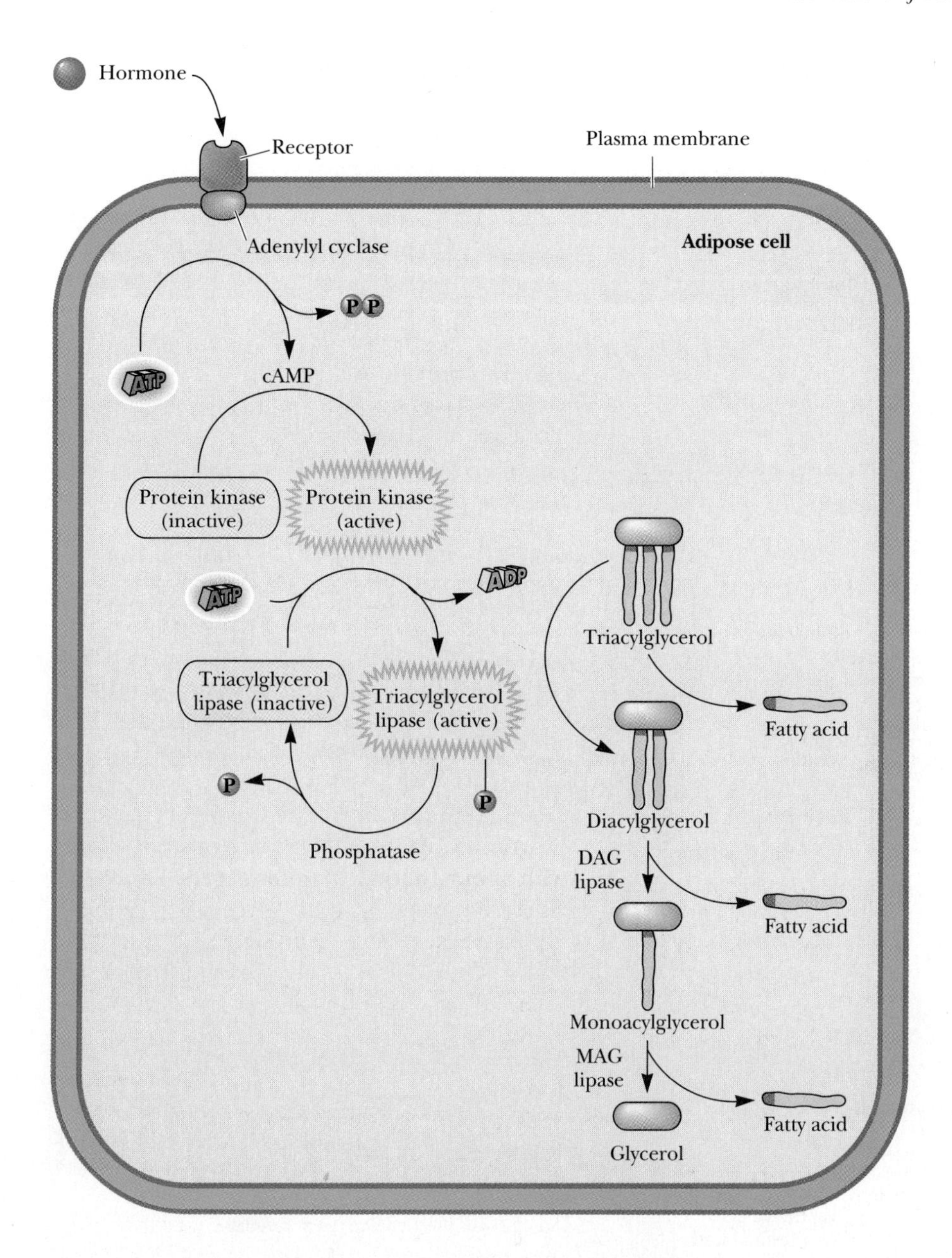

***Figure* 23.2** Liberation of fatty acids from triacylglycerols in adipose tissue is hormone-dependent.

nal molecules bind to receptors on the plasma membrane of adipose cells and lead to the activation of adenylyl cyclase, which forms cyclic AMP from ATP. (Second messengers and hormonal signaling are discussed in Chapter 37.) In adipose cells, cAMP activates a protein kinase, which phosphorylates and activates a **triacylglycerol lipase** (also termed **hormone-sensitive lipase**) that hydrolyzes a fatty acid from C-1 or C-3 of triacylglycerols. Subsequent actions of **diacylglycerol lipase** and **monoacylglycerol lipase** yield fatty acids and glycerol. The cell then releases the fatty acids into the blood, where they are carried (in complexes with *serum albumin*) to sites of utilization.

Degradation of Dietary Fatty Acids Occurs Primarily in the Duodenum

Dietary triacylglycerols are degraded to a small extent (via fatty acid release) by lipases in the low-pH environment of the stomach, but mostly pass untouched into the duodenum. Alkaline pancreatic juice secreted into the duo-

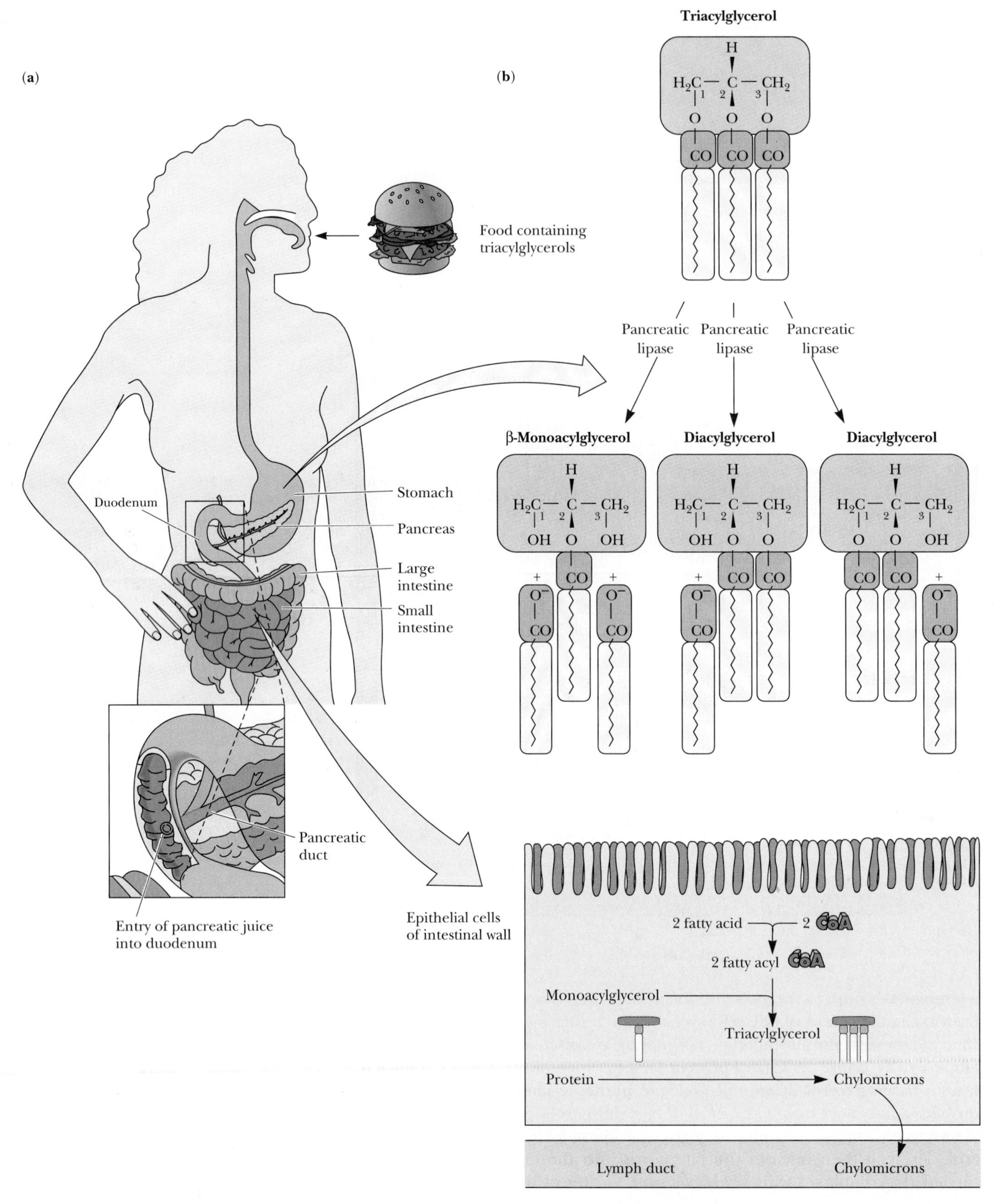

***Figure* 23.3** (a) A duct at the junction of the pancreas and duodenum secretes pancreatic juice into the duodenum, the first portion of the small intestine. (b) Hydrolysis of triacylglycerols by pancreatic and intestinal lipases. Pancreatic lipases cleave fatty acids at the C-1 and C-3 positions. Resulting monoacylglycerols with fatty acids at C-2 are hydrolyzed by intestinal lipases. Fatty acids and monoacylglycerols are absorbed through the intestinal wall and assembled into lipoprotein aggregates termed *chylomicrons* (discussed in Chapter 24).

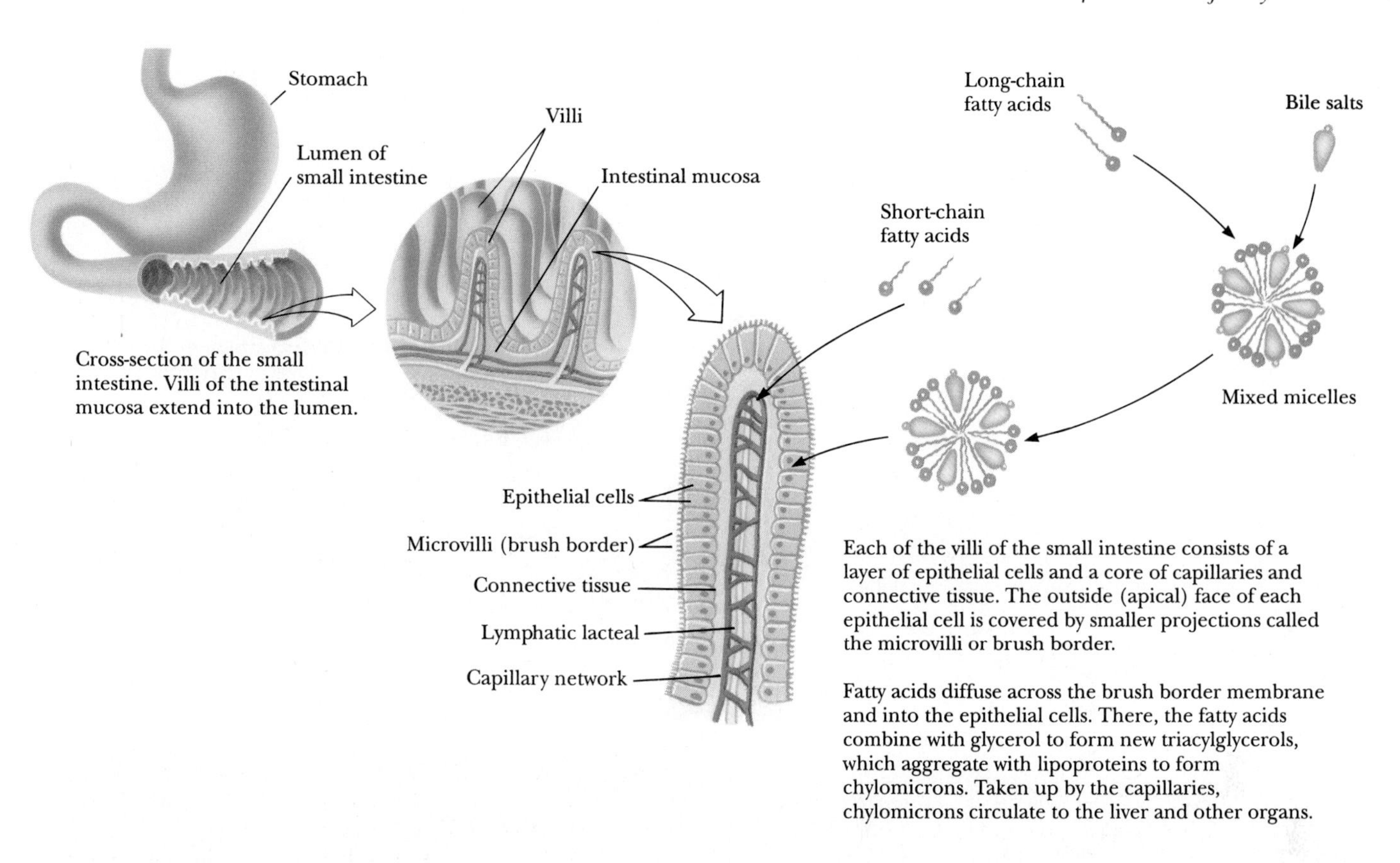
Stomach
Lumen of small intestine
Villi
Intestinal mucosa
Long-chain fatty acids
Bile salts
Short-chain fatty acids
Cross-section of the small intestine. Villi of the intestinal mucosa extend into the lumen.
Mixed micelles
Epithelial cells
Microvilli (brush border)
Connective tissue
Lymphatic lacteal
Capillary network
Each of the villi of the small intestine consists of a layer of epithelial cells and a core of capillaries and connective tissue. The outside (apical) face of each epithelial cell is covered by smaller projections called the microvilli or brush border.
Fatty acids diffuse across the brush border membrane and into the epithelial cells. There, the fatty acids combine with glycerol to form new triacylglycerols, which aggregate with lipoproteins to form chylomicrons. Taken up by the capillaries, chylomicrons circulate to the liver and other organs.

***Figure* 23.4** In the small intestine, fatty acids combine with bile salts in mixed micelles, which deliver fatty acids to epithelial cells that cover the intestinal villi. Triacylglycerols are formed within the epithelial cells.

denum (Figure 23.3a) raises the pH of the digestive mixture, allowing hydrolysis of the triacylglycerols by pancreatic lipase and by nonspecific esterases, which hydrolyze the fatty acid ester linkages. Pancreatic lipase cleaves fatty acids from the C-1 and C-3 positions of triacylglycerols, and other lipases and esterases attack the C-2 position (Figure 23.3b). These processes depend upon the presence of **bile salts,** a family of carboxylic acid salts with steroid backbones (see also Chapter 24). These agents act as detergents to emulsify the triacylglycerols and facilitate the hydrolytic activity of the lipases and esterases. Short-chain fatty acids (10 carbons or less) released in this way are absorbed directly into the villi of the intestinal mucosa, whereas long-chain fatty acids, which are less soluble, form mixed micelles with bile salts, and are carried in this fashion to the surfaces of the epithelial cells that cover the villi (Figure 23.4). The fatty acids pass into the epithelial cells, where they are condensed with glycerol to form new triacylglycerols. These triacylglycerols aggregate with lipoproteins to form particles called **chylomicrons,** which are then transported into the lymphatic system and on to the bloodstream, where they circulate to the liver, lungs, heart, muscles, and other organs (Chapter 24). At these sites, the triacylglycerols are hydrolyzed to release fatty acids, which can then be oxidized in a highly exergonic metabolic pathway known as *β-oxidation.*

23.2 β-Oxidation of Fatty Acids

Franz Knoop and the Discovery of β-Oxidation

The earliest clue to the secret of fatty acid oxidation and breakdown came in the early 1900s, when Franz Knoop carried out experiments in which he fed dogs fatty acids in which the terminal methyl group had been replaced with a phenyl ring (Figure 23.5). Knoop discovered that fatty acids containing an

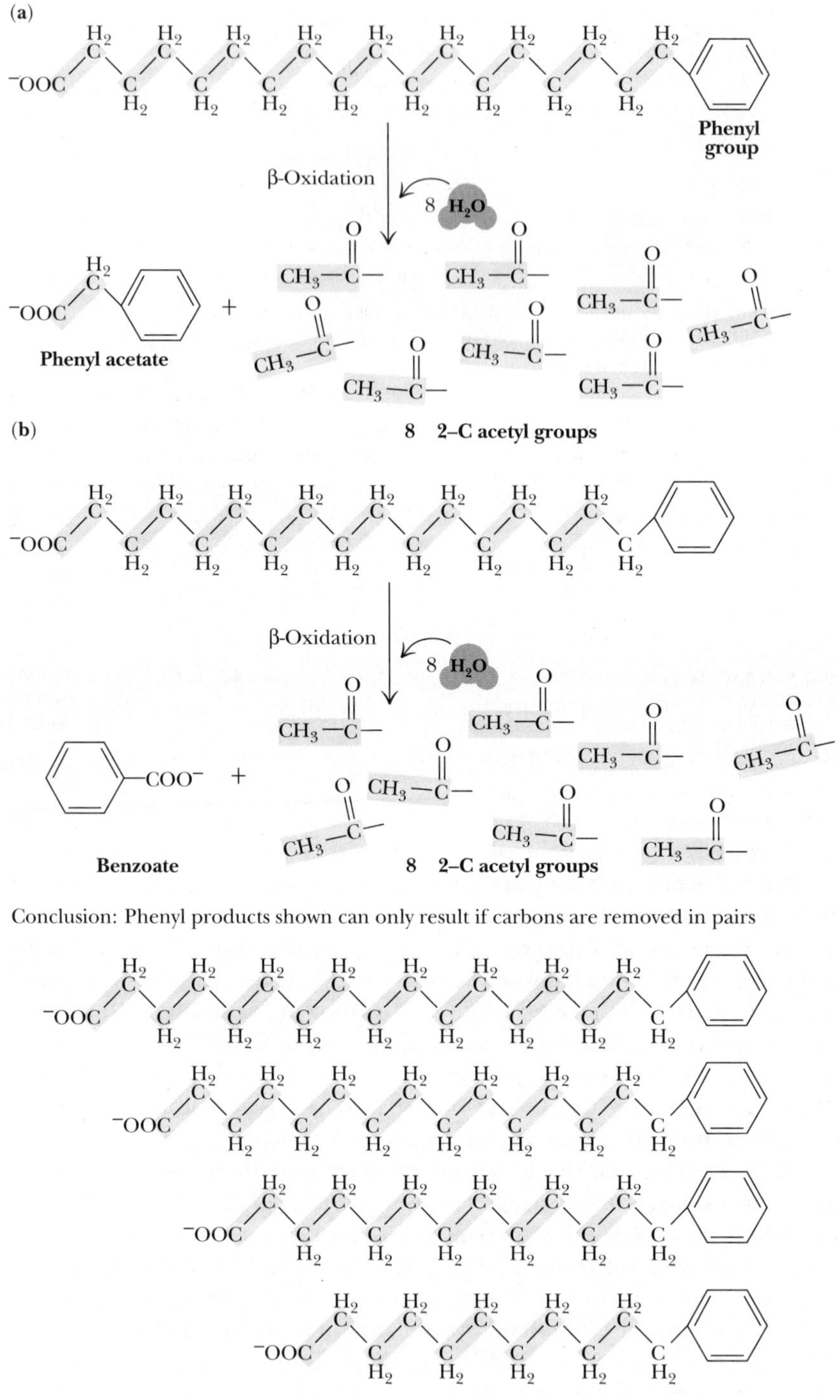

Figure 23.5 The oxidative breakdown of phenyl fatty acids observed by Franz Knoop. He observed that fatty acid analogs with even numbers of carbon atoms yielded phenyl acetate, whereas compounds with odd numbers of carbon atoms produced only benzoate.

(a)

Phenyl group

β-Oxidation

8 H_2O

Phenyl acetate

8 2–C acetyl groups

(b)

β-Oxidation

8 H_2O

Benzoate

8 2–C acetyl groups

Conclusion: Phenyl products shown can only result if carbons are removed in pairs

even number of carbon atoms were broken down to yield phenyl acetate as the final product, whereas fatty acids with an odd number of carbon atoms yielded benzoate as the final product (Figure 23.5). From these experiments, Knoop concluded that the fatty acids must be degraded by *oxidation at the β-carbon* (Figure 23.6), followed by cleavage of the C_α—C_β bond. Repetition of this process yielded 2-carbon units, which Knoop assumed must be acetate. Much later, Albert Lehninger showed that this degradative process took place in the mitochondria, and F. Lynen and E. Reichart showed that the 2-carbon unit released is *acetyl-CoA,* not free acetate. Since the entire process begins with oxidation of the carbon that is "β" to the carboxyl carbon, the process has come to be known as ***β*-oxidation.**

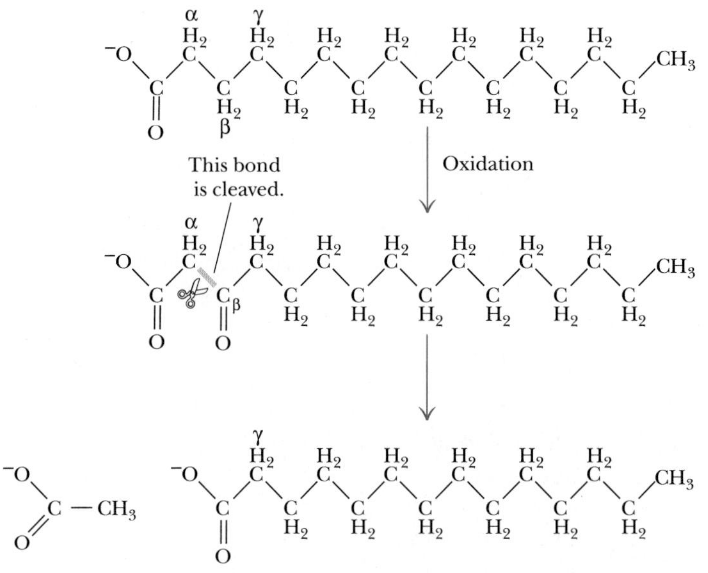

This bond is cleaved.

Oxidation

***Figure* 23.6** Fatty acids are degraded by repeated cycles of oxidation at the *β*-carbon and cleavage of the C_α—C_β bond to yield acetate units.

Coenzyme A Activates Fatty Acids for Degradation

The process of *β*-oxidation begins with the formation of a thiol ester bond between the fatty acid and the thiol group of coenzyme A. This reaction, shown in Figure 23.7, is catalyzed by *acyl-CoA synthetase,* which is also called *acyl-CoA ligase* or *fatty acid thiokinase.* This condensation with CoA activates the fatty acid for reaction in the *β*-oxidation pathway. For long-chain fatty acids, this reaction normally occurs at the outer mitochondrial membrane, prior to entry of the fatty acid into the mitochondrion, but it may also occur at the surface of the endoplasmic reticulum. Short- and medium-length fatty acids undergo this activating reaction in the mitochondria. In all cases, the reaction is accompanied by the hydrolysis of ATP to form AMP and pyrophosphate. As shown in Figure 23.7, the two combined reactions have a net $\Delta G°'$ of about -0.8 kJ/mol, so that the reaction is favorable but easily reversible. However, there is more to the story. As we have seen in several similar cases, the pyrophosphate produced in this reaction is rapidly hydrolyzed by inorganic pyrophosphatase to two molecules of phosphate, with a net $\Delta G°'$ of about -33.6 kJ/mol. Thus, pyrophosphate is maintained at a low concentration in the cell (usually less than 1 m*M*) and the synthetase reaction is strongly promoted. The mechanism of the acyl-CoA synthetase reaction is shown in Figure 23.8 and involves attack of the fatty acid carboxylate on ATP to form an *acyl-adenylate intermediate,* which is subsequently attacked by CoA, forming a fatty acyl-CoA thioester.

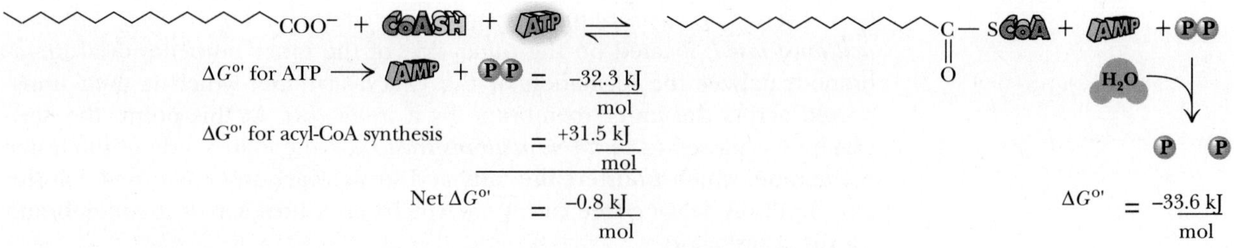

$\Delta G°'$ for ATP ⟶ AMP + PP $= -32.3$ kJ/mol

$\Delta G°'$ for acyl-CoA synthesis $= +31.5$ kJ/mol

Net $\Delta G°' = -0.8$ kJ/mol

$\Delta G°' = -33.6$ kJ/mol

***Figure* 23.7** The acyl-CoA synthetase reaction activates fatty acids for *β*-oxidation. The reaction is driven by hydrolysis of ATP to AMP and pyrophosphate and by the subsequent hydrolysis of pyrophosphate.

Figure **23.8** The mechanism of the acyl-CoA synthetase reaction involves fatty acid carboxylate attack on ATP to form an acyl-adenylate intermediate. The fatty acyl-CoA thioester product is formed by CoA attack on this intermediate.

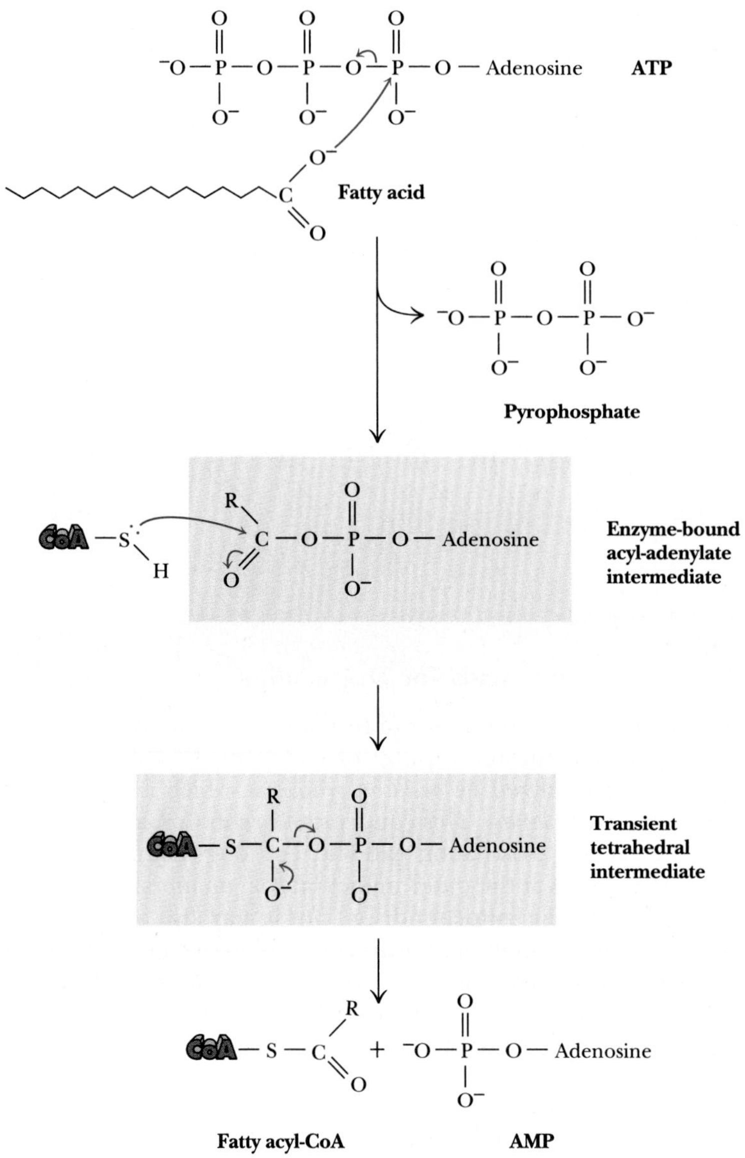

ATP

Fatty acid

Pyrophosphate

Enzyme-bound acyl-adenylate intermediate

Transient tetrahedral intermediate

Fatty acyl-CoA

AMP

Carnitine Carries Fatty Acyl Groups Across the Inner Mitochondrial Membrane

All of the other enzymes that catalyze the reactions of the β-oxidation pathway are located in the mitochondrial matrix. Short-chain fatty acids, as already mentioned, are transported into the matrix as free acids and form the acyl-CoA derivatives there. However, long-chain fatty acyl-CoA derivatives cannot be transported into the matrix directly. These long-chain derivatives must first be converted to *acylcarnitine* derivatives, as shown in Figure 23.9. *Carnitine acyltransferase I,* located on the outer side of the inner mitochondrial membrane, catalyzes the formation of the O-acylcarnitine, which is then transported across the inner membrane by a *translocase.* At this point, the acyl-carnitine is passed to *carnitine acyltransferase II* on the matrix side of the inner membrane, which transfers the fatty acyl group back to CoA to re-form the fatty acyl-CoA, leaving free carnitine, which can return across the membrane via the translocase.

Several additional points should be made. First, although oxygen esters usually have lower group-transfer potentials than thiol esters, the O—acyl

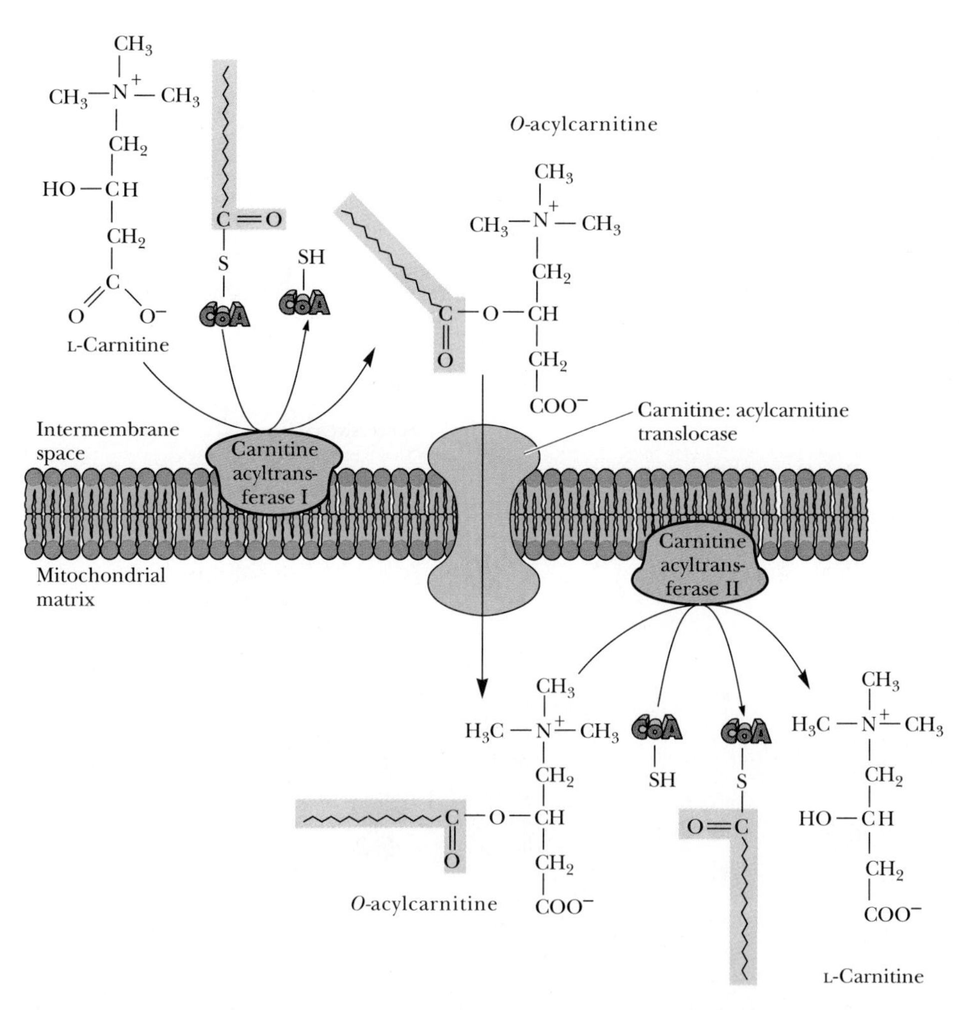

***Figure* 23.9** The formation of acylcarnitines and their transport across the inner mitochondrial membrane. The process involves the coordinated actions of carnitine acyltransferases on both sides of the membrane, and of a translocase, which shuttles O-acylcarnitines across the membrane.

bonds in acylcarnitines have high group-transfer potentials, and the trans-esterification reactions mediated by the acyltransferases have equilibrium constants close to 1. Second, note that eukaryotic cells maintain separate pools of CoA in the mitochondria and in the cytosol. The cytosolic pool is utilized principally in fatty acid biosynthesis (Chapter 24), and the mitochondrial pool is important in the oxidation of fatty acids and pyruvate, as well as some amino acids.

β-Oxidation Involves a Repeated Sequence of Four Reactions

For saturated fatty acids, the process of β-oxidation involves a recurring cycle of four steps, as shown in Figure 23.10. The overall strategy in the first three steps is to create a carbonyl group on the β-carbon by oxidizing the C_α—C_β bond to form an olefin, with subsequent hydration and oxidation. In essence, this cycle is directly analogous to the sequence of reactions converting succinate to oxaloacetate in the TCA cycle. The fourth reaction of the cycle cleaves the β-keto ester in a reverse Claisen condensation, producing an acetate unit and leaving a fatty acid chain that is two carbons shorter than it began. (Recall from Chapter 19 that Claisen condensations involve attack by a nucleophilic agent on a carbonyl carbon to yield a β-keto acid.)

Figure **23.10** The β-oxidation of saturated fatty acids involves a cycle of four enzyme-catalyzed reactions. Each cycle produces single molecules of $FADH_2$, NADH, and acetyl-CoA and yields a fatty acid shortened by two carbons. (The delta [Δ] symbol connotes a double bond, and its superscript indicates the lowest-numbered carbon involved.)

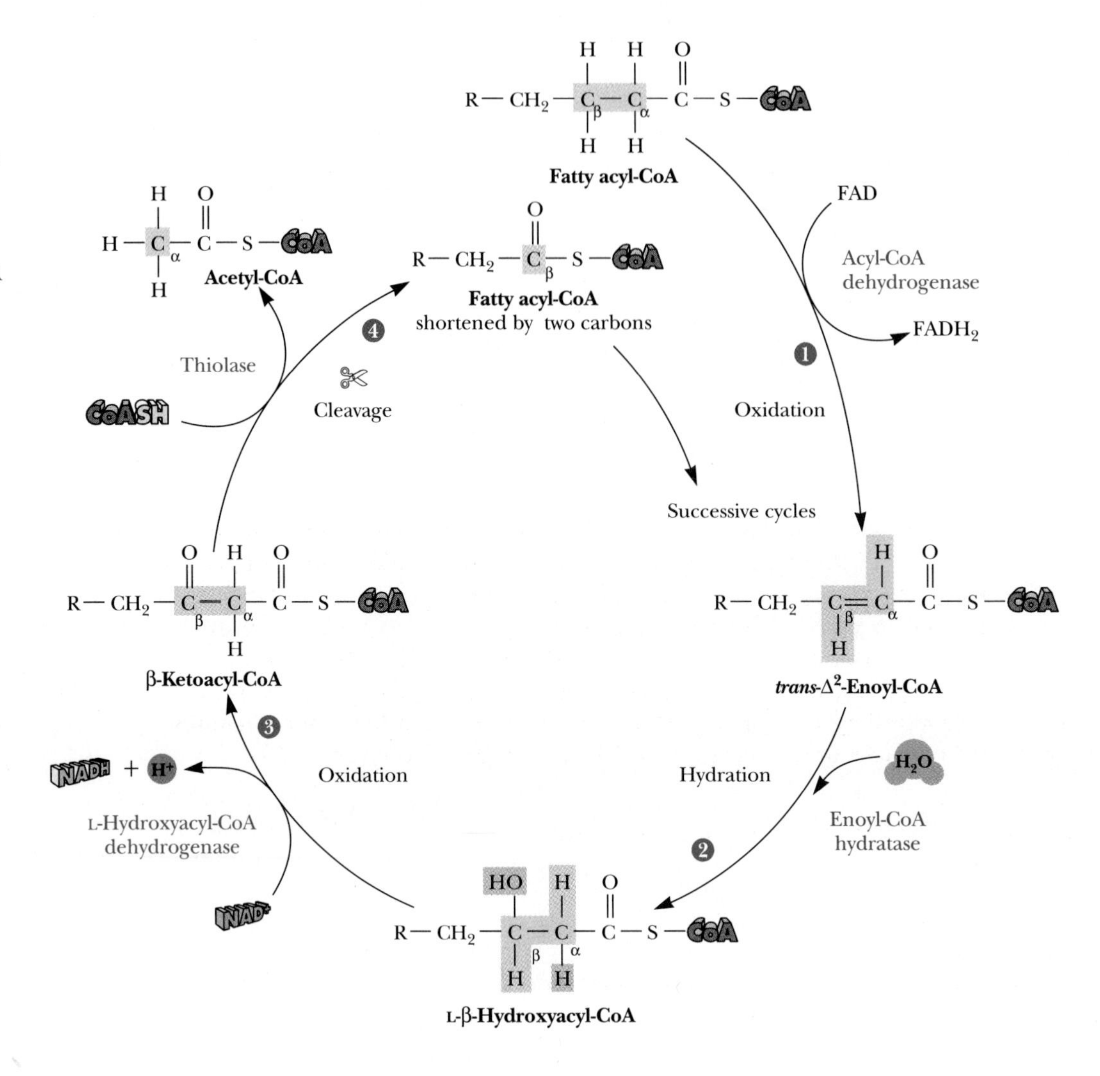

Acyl-CoA Dehydrogenase—The First Reaction of β-Oxidation

The first reaction, the oxidation of the C_α—C_β bond, is catalyzed by **acyl-CoA dehydrogenases,** a family of three soluble matrix enzymes (with molecular weights of 170 to 180 kD), which differ in their specificity for either long-, medium-, or short-chain acyl-CoAs. They carry noncovalently (but tightly) bound FAD, which is reduced during the oxidation of the fatty acid. As shown in Figure 23.11, $FADH_2$ transfers its electrons to an *electron transfer flavoprotein* (ETF). Reduced ETF is reoxidized by a specific oxidoreductase (an iron–

Figure **23.11** The acyl-CoA dehydrogenase reaction. The two electrons removed in this oxidation reaction are delivered to the electron transport chain in the form of reduced coenzyme Q (UQH_2).

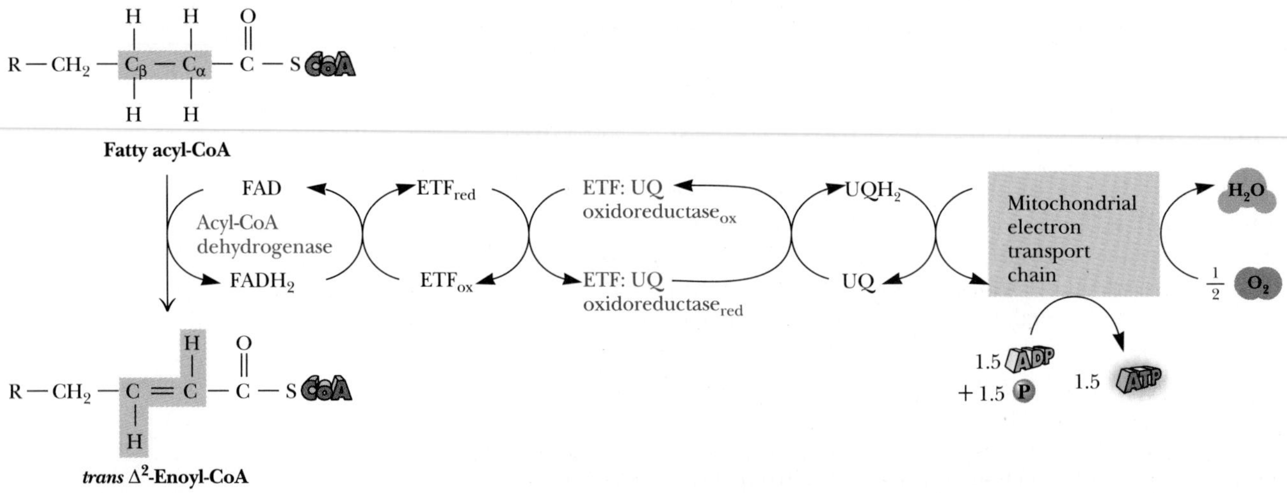

Fatty acyl-CoA
FAD
Acyl-CoA dehydrogenase
$FADH_2$
ETF_{red}
ETF_{ox}
ETF: UQ oxidoreductase$_{ox}$
ETF: UQ oxidoreductase$_{red}$
UQH_2
UQ
Mitochondrial electron transport chain
H_2O
$\frac{1}{2}$ O_2
1.5 ADP + 1.5 P
1.5 ATP
trans Δ^2-Enoyl-CoA

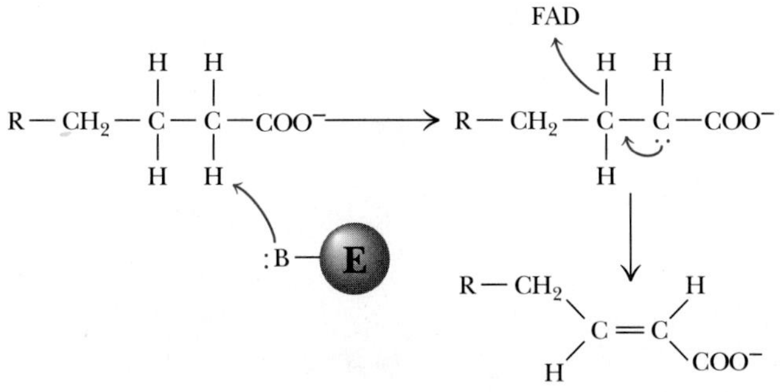

$$R-CH_2-CH_2-CH_2-COO^- \xrightarrow{:B-E} R-CH_2-CH_2-\overset{..}{C}H-COO^- \xrightarrow{FAD} R-CH_2-CH=CH-COO^-$$

Figure **23.12** The mechanism of acyl-CoA dehydrogenase. Removal of a proton from the α-C is followed by hydride transfer from the β-carbon to FAD.

sulfur protein), which in turn sends the electrons on to the electron transport chain at the level of coenzyme Q. As always, mitochondrial oxidation of FAD in this way eventually results in the net formation of about 1.5 ATP. The mechanism of the acyl-CoA dehydrogenase (Figure 23.12) involves deprotonation of the fatty acid chain at the α-carbon, followed by hydride transfer from the β-carbon to FAD. The structure of the medium-chain dehydrogenase from pig liver places an FAD molecule in an extended conformation between a bundle of α-helices and a distorted β-barrel (Figure 23.13).

A Metabolite of Hypoglycin from Akee Fruit Inhibits Acyl-CoA Dehydrogenase

The unripened fruit of the **akee tree** contains **hypoglycin,** a rare amino acid (Figure 23.14). Metabolism of hypoglycin yields *methylenecyclopropylacetyl-CoA* (MCPA-CoA). Acyl-CoA dehydrogenase will accept MCPA-CoA as a substrate,

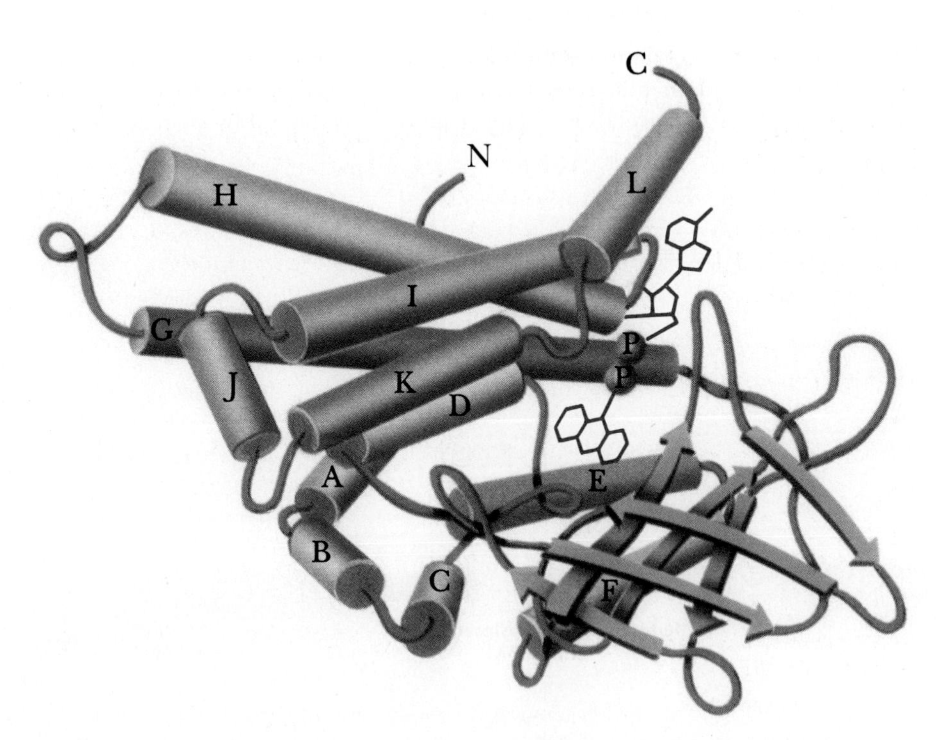

Figure **23.13** The structure of medium-chain acyl-CoA dehydrogenase from pig liver mitochondria. Note the location of the bound FAD (red).

(Adapted from Kim, J.-T., and Wu, J., 1988. Structure of the medium-chain acyl-CoA dehydrogenase from pig liver mitochondria at 3-Å resolution. Proceedings of the National Academy of Sciences, USA **85:**6671–6681.)

A Deeper Look

The Akee Tree

The akee (also spelled *ackee*) tree is native to West Africa and was brought to the Caribbean by African slaves. It was introduced to science by William Bligh, captain of the infamous sailing ship the *Bounty,* and its botanical name is (appropriately) *Blighia sapida* (the latter name from the Latin *sapidus* meaning "tasty"). A popular dish in the Caribbean consists of akee and salt fish.

> "*Akee, rice, salt fish are nice,*
> *And the rum is fine any time of year.*"
> —*From the song* Jamaica Farewell

removing a proton from the α-carbon to yield an intermediate that irreversibly inactivates acyl-CoA dehydrogenase by reacting covalently with FAD on the enzyme. For this reason, consumption of unripened akee fruit can lead to vomiting and, in severe cases, convulsions, coma, and death. The condition is most severe in individuals with low levels of acyl-CoA dehydrogenase.

Enoyl-CoA Hydratase Adds Water Across the Double Bond

The next step in β-oxidation is the addition of the elements of H_2O across the new double bond in a stereospecific manner, yielding the corresponding hydroxyacyl-CoA (Figure 23.15). The reaction is catalyzed by *enoyl-CoA hydra-*

$H_2C{=}C{-}CH{-}CH_2{-}CH(NH_3^+){-}COO^-$ (with CH_2 bridging the ring)

Hypoglycin A

↓ CoASH

$H_2C{=}C{-}CH{-}CH_2{-}C(=O){-}S\text{CoA}$ (with CH_2 bridging the ring)

Methylenecyclopropylacetyl-CoA (MCPA-CoA)

↓ H^+

$(H_2C{=})(H_2C{=})C{-}CH{=}CH{-}C(=O){-}S\text{CoA}$ ⇌ $H_2C{=}C{-}CH{-}\bar{C}H{-}C(=O){-}S\text{CoA}$ (with CH_2 bridging the ring)

Reactive intermediate

Figure **23.14** The conversion of hypoglycin from akee fruit to a form that inhibits acyl-CoA dehydrogenase.

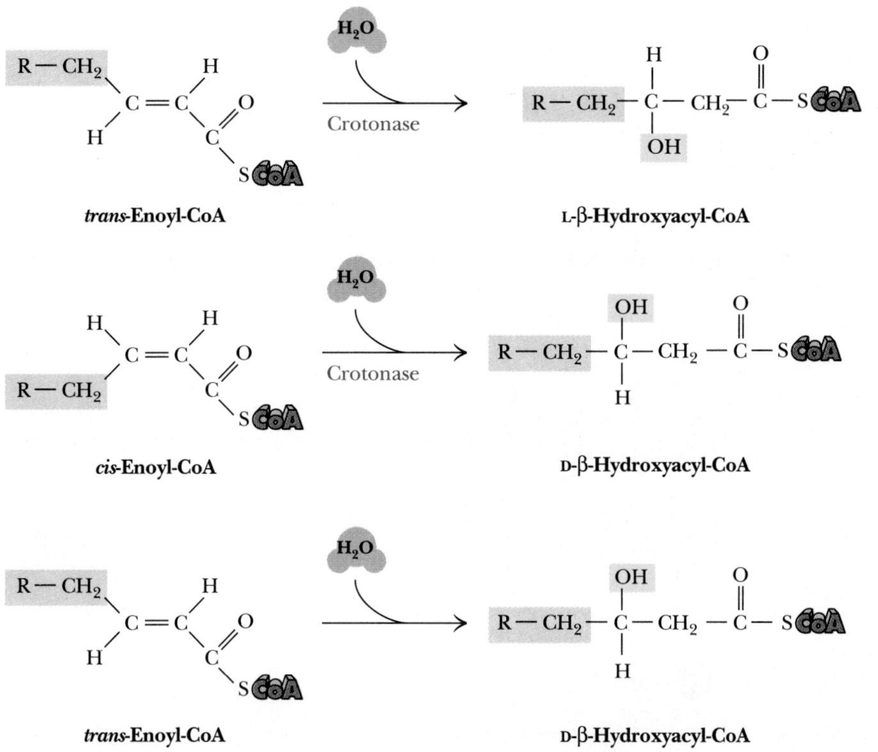

H_2O — Crotonase

***trans*-Enoyl-CoA** → **L-β-Hydroxyacyl-CoA**

H_2O — Crotonase

***cis*-Enoyl-CoA** → **D-β-Hydroxyacyl-CoA**

H_2O

***trans*-Enoyl-CoA** → **D-β-Hydroxyacyl-CoA**

***Figure* 23.15** (a) The conversion of *trans-* and *cis*-enoyl-CoA derivatives to L- and D-β-hydroxyacyl-CoA, respectively. These reactions are catalyzed by enoyl-CoA hydratases (also called crotonases), enzymes that vary in their acyl-chain-length specificity. (b) A recently discovered enzyme converts *trans*-enoyl-CoA directly to D-β-hydroxyacyl-CoA.

tase. At least three different enoyl-CoA hydratase activities have been detected in various tissues. Also called **crotonases,** these enzymes specifically convert *trans*-enoyl-CoA derivatives to L-β-hydroxyacyl-CoA. As shown in Figure 23.15, these enzymes will also metabolize *cis*-enoyl-CoA (at slower rates) to give specifically D-β-hydroxyacyl-CoA. Recently, a novel enoyl-CoA hydratase was discovered, which converts *trans*-enoyl-CoA to D-β-hydroxyacyl-CoA, as shown in Figure 23.15.

L-Hydroxyacyl-CoA Dehydrogenase Oxidizes the β-Hydroxyl Group

The third reaction of this cycle is the oxidation of the hydroxyl group at the β-position to produce a β-ketoacyl-CoA derivative. This second oxidation reaction is catalyzed by *L-hydroxyacyl-CoA dehydrogenase,* an enzyme that requires NAD^+ as a coenzyme. NADH produced in this reaction represents metabolic energy. Each NADH produced in mitochondria by this reaction drives the synthesis of 2.5 molecules of ATP in the electron transport pathway. L-Hydroxyacyl-CoA dehydrogenase shows absolute specificity for the L-hydroxyacyl isomer of the substrate (Figure 23.16). (D-Hydroxyacyl isomers, which arise mainly from oxidation of unsaturated fatty acids, are handled differently, as we will see.)

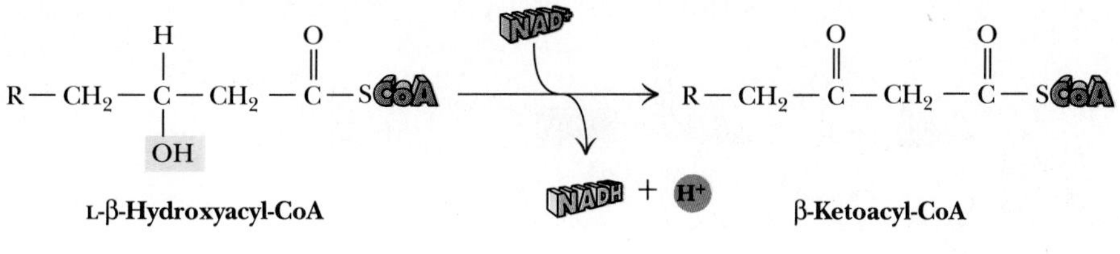

***Figure* 23.16** The L-β-hydroxyacyl-CoA dehydrogenase reaction.

NAD^+ → NADH + H^+

L-β-Hydroxyacyl-CoA → **β-Ketoacyl-CoA**

β-Ketoacyl-CoA Intermediates Are Cleaved in the Thiolase Reaction

The final step in the β-oxidation cycle is the cleavage of the β-ketoacyl-CoA. This reaction, catalyzed by *thiolase* (also known as *β-ketothiolase*), involves the attack of a cysteine thiolate from the enzyme on the β-carbonyl group, followed by cleavage to give the enolate of acetyl-CoA and an enzyme-thioester intermediate (Figure 23.17). Subsequent attack by the thiol group of a second CoA and departure of the cysteine thiolate yields a new (shorter) acyl-CoA. If the reaction in Figure 23.17 is read in reverse, it is easy to see that it is a

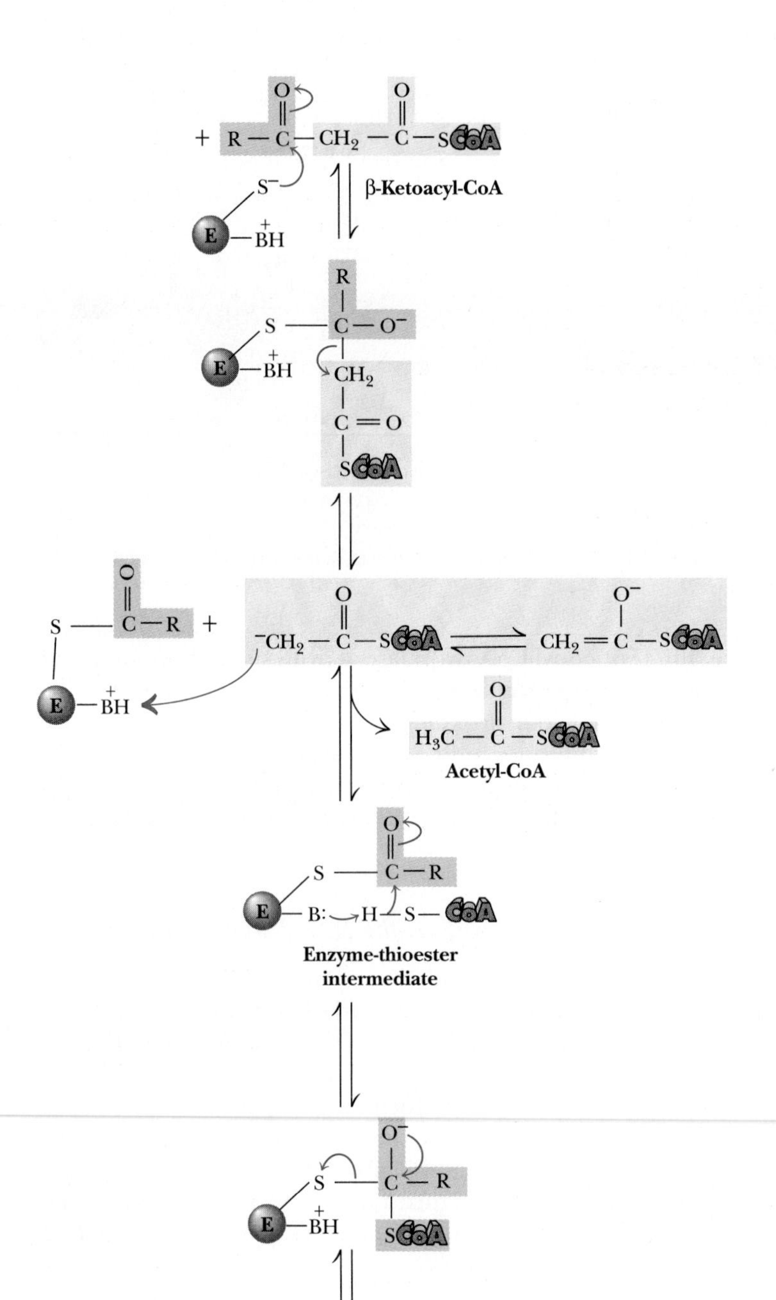

***Figure* 23.17** The mechanism of the thiolase reaction. Attack by an enzyme cysteine thiolate group at the β-carbonyl carbon produces a tetrahedral intermediate, which decomposes with departure of acetyl-CoA, leaving an enzyme thioester intermediate. Attack by the thiol group of a second CoA yields a new (shortened) acyl-CoA.

Claisen condensation—an attack of the enolate anion of acetyl-CoA on a thioester. Despite the formation of a second thioester, this reaction has a very favorable K_{eq}, and it drives the three previous reactions of β-oxidation.

Repetition of the β-Oxidation Cycle Yields a Succession of Acetate Units

In essence, this series of four reactions has yielded a fatty acid (as a CoA ester) that has been shortened by two carbons, and one molecule of acetyl-CoA. The shortened fatty acyl-CoA can now go through another β-oxidation cycle, as shown in Figure 23.10. Repetition of this cycle with a fatty acid with an even number of carbons eventually yields two molecules of acetyl-CoA in the final step. As noted in the first reaction in Table 23.2, complete β-oxidation of palmitic acid yields eight molecules of acetyl-CoA as well as seven molecules of $FADH_2$ and seven molecules of NADH. The acetyl-CoA can be further metabolized in the TCA cycle (as we have already seen). Alternatively, acetyl-CoA can also be used as a substrate in amino acid biosynthesis (Chapter 26). As noted in Chapter 21, however, acetyl-CoA cannot be used as a substrate for gluconeogenesis.

Complete β-Oxidation of One Palmitic Acid Yields 106 Molecules of ATP

If the acetyl-CoA is directed entirely to the TCA cycle in mitochondria, it can eventually generate approximately 10 high-energy phosphate bonds—that is, 10 molecules of ATP synthesized from ADP (Table 23.2). Including the ATP formed from $FADH_2$ and NADH, complete β-oxidation of a molecule of palmitoyl-CoA in mitochondria yields 108 molecules of ATP. Subtracting the two high-energy bonds needed to form palmitoyl-CoA, the substrate for β-oxidation, one concludes that β-oxidation of a molecule of palmitic acid yields 106 molecules of ATP. The $\Delta G°'$ for complete combustion of palmitate to CO_2 is −9790 kJ/mol. The hydrolytic energy embodied in 106 ATPs is 106 × 30.5 kJ/mol = 3233 kJ/mol, so the overall efficiency of β-oxidation under standard-state conditions is approximately 33%. The large energy yield from fatty acid oxidation is a reflection of the highly reduced state of the

Table **23.2**

Equations for the Complete Oxidation of Palmitoyl-CoA to CO_2 and H_2O

Equation	ATP Yield	Free Energy (kJ/mol) Yield
$CH_3(CH_2)_{14}CO—CoA + 7\ [FAD] + 7\ H_2O + 7\ NAD^+ + 7\ CoA \longrightarrow$		
$8\ CH_3CO\text{-}CoA + 7\ [FADH_2] + 7\ NADH + 7\ H^+$		
$7\ [FADH_2] + 10.5\ P_i + 10.5\ ADP + 3.5\ O_2 \longrightarrow$		
$7\ [FAD] + 17.5\ H_2O + 10.5\ ATP$	10.5	320
$7\ NADH + 7\ H^+ + 17.5\ P_i + 17.5\ ADP + 3.5\ O_2 \longrightarrow$		
$7\ NAD^+ + 24.5\ H_2O + 17.5\ ATP$	17.5	534
$8\text{-Acetyl-CoA} + 16\ O_2 + 80\ ADP + 80\ P_i \longrightarrow$		
$8\ CoA + 88\ H_2O + 16\ CO_2 + 80\ ATP$	80	2440
$CH_3(CH_2)_{14}CO—CoA + 108\ P_i + 108\ ADP + 23\ O_2 \longrightarrow$		
$108\ ATP + 16\ CO_2 + 130\ H_2O + CoA$	108	3294
Energetic "cost" of forming palmitoyl-CoA	−2	−61
	106	3233

(a) Gerbil

(b) Ruby-throated hummingbird

(c) Golden plover

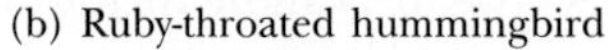

***Figure* 23.18** Animals whose existence is strongly dependent on fatty acid oxidation: (a) gerbil, (b) ruby-throated hummingbird, (c) golden plover, (d) orca (killer whale), and (e) camels.

(d) Orca

(e) Camels

carbon in fatty acids. Sugars, in which the carbon is already partially oxidized, produce much less energy, carbon for carbon, than do fatty acids. The breakdown of fatty acids is regulated by a variety of metabolites and hormones. The details of this regulation will be described in Chapter 24, following a discussion of fatty acid synthesis.

Migratory Birds Travel Long Distances on Energy from Fatty Acid Oxidation

Because they represent the most highly concentrated form of stored biological energy, fatty acids are the metabolic fuel of choice for sustaining the incredibly long flights of many migratory birds. Though some birds migrate over land masses and dine frequently, other species fly long distances without stopping to eat. The American golden plover flies directly from Alaska to Hawaii, a 3300-kilometer flight requiring 35 hours (at an average speed of nearly 60 miles/hr) and more than 250,000 wing beats! The ruby-throated hummingbird, which winters in Central America and nests in southern Canada, often flies nonstop across the Gulf of Mexico. These and similar birds accomplish these prodigious feats by storing large amounts of fatty acids (as triacylglycerols) in the days before their migratory flights. The percentage of dry-weight body fat in these birds may be as high as 70% when migration begins (compared with values of 30% and less for nonmigratory birds).

Fatty Acid Oxidation Is an Important Source of Metabolic Water for Some Animals

Large amounts of metabolic water are generated by β-oxidation (130 H_2O per palmitoyl-CoA). For certain animals—including desert animals, such as gerbils, and killer whales (which do not drink seawater)—the oxidation of fatty acids can be a significant source of dietary water. A striking example is the camel (Figure 23.18), whose hump is essentially a large deposit of fat. Metabolism of fatty acids from this store provides needed water (as well as metabolic energy) during periods when drinking water is not available. It might well be said that "the ship of the desert" sails on its own metabolic water!

23.3 β-Oxidation of Odd-Carbon Fatty Acids

β-Oxidation of Odd-Carbon Fatty Acids Yields Propionyl-CoA

Fatty acids with odd numbers of carbon atoms are rare in mammals, but fairly common in plants and marine organisms. Humans and animals whose diets include these food sources metabolize odd-carbon fatty acids via the β-oxidation pathway. The final product of β-oxidation in this case is the 3-carbon propionyl-CoA instead of acetyl-CoA. Three specialized enzymes then carry out the reactions that convert propionyl-CoA to succinyl-CoA, a TCA cycle intermediate. (Because propionyl-CoA is a degradation product of methionine, valine, and isoleucine, this sequence of reactions is also important in amino acid catabolism, as we shall see in Chapter 26.) The pathway involves an initial carboxylation at the α-carbon of propionyl-CoA to produce D-methylmalonyl-CoA (Figure 23.19). The reaction is catalyzed by a biotin-

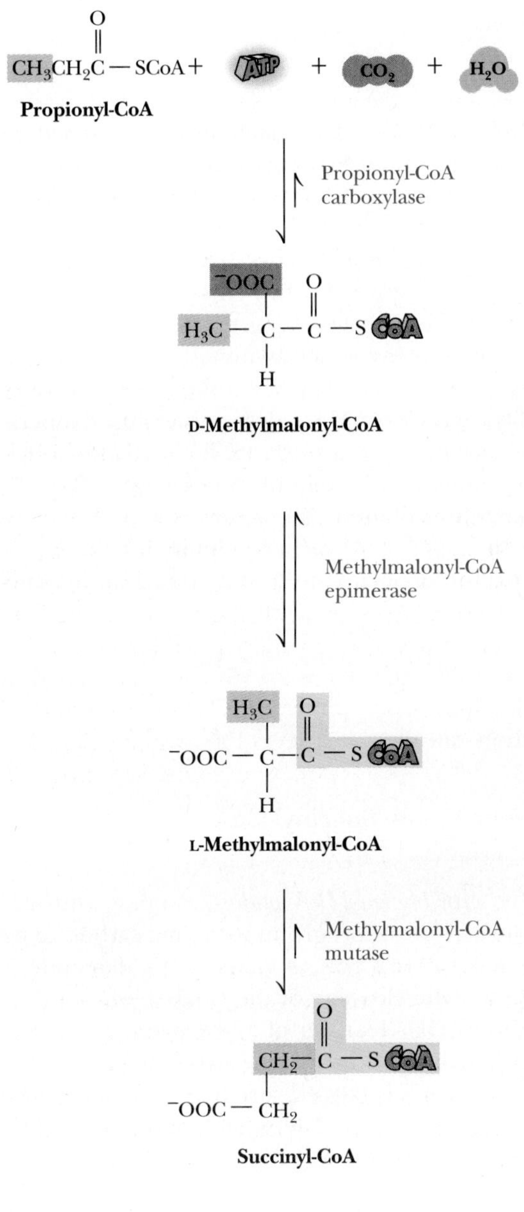

$CH_3CH_2C(=O)—SCoA$ + ATP + CO_2 + H_2O

Propionyl-CoA

Propionyl-CoA carboxylase

D-Methylmalonyl-CoA

Methylmalonyl-CoA epimerase

L-Methylmalonyl-CoA

Methylmalonyl-CoA mutase

Succinyl-CoA

Figure **23.19** The conversion of propionyl-CoA (formed from β-oxidation of odd-carbon fatty acids) to succinyl-CoA is carried out by a trio of enzymes as shown. Succinyl-CoA can enter the TCA cycle or be converted to acetyl-CoA.

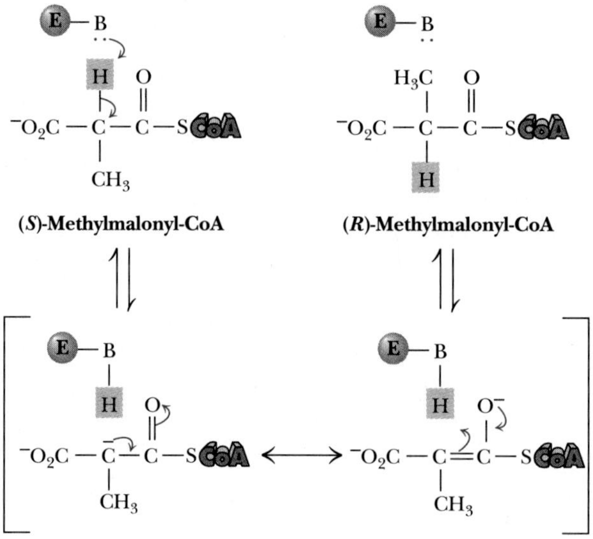

(S)-Methylmalonyl-CoA **(R)-Methylmalonyl-CoA**

Resonance-stabilized carbanion intermediate

***Figure* 23.20** The methylmalonyl-CoA epimerase mechanism involves a resonance-stabilized carbanion at the α-position.

dependent enzyme, *propionyl-CoA carboxylase.* The mechanism (discussed in detail in Chapter 14) involves ATP-driven carboxylation of biotin at N_1, followed by nucleophilic attack by the α-carbanion of propionyl-CoA in a stereospecific manner.

D-Methylmalonyl-CoA, the product of this reaction, is converted to the L-isomer by *methylmalonyl-CoA epimerase* (Figure 23.19). (This enzyme has often and incorrectly been called "methylmalonyl-CoA racemase." It is not a racemase, since the CoA moiety contains five other asymmetric centers.) The epimerase reaction also appears to involve a carbanion at the α-position (Figure 23.20). The reaction is readily reversible and involves a reversible dissociation of the acidic α-proton. The L-isomer is the substrate for methylmalonyl-CoA mutase. Methylmalonyl-CoA epimerase is an impressive catalyst. The pK_a for the proton that must dissociate to initiate this reaction is approximately 21! If binding of a proton to the α-anion is diffusion-limited, with $k_{on} = 10^9\ M^{-1}\ sec^{-1}$, then the initial proton dissociation must be rate-limiting, and the rate constant must be

$$k_{off} = K_a \cdot k_{on} = (10^{-21}\ M) \cdot (10^9\ M^{-1}\ sec^{-1}) = 10^{-12}\ sec^{-1}$$

The turnover number of methylmalonyl-CoA epimerase is 100 sec^{-1}, and thus the enzyme enhances the reaction rate by a factor of 10^{14}.

A B_{12}-Catalyzed Rearrangement Yields Succinyl-CoA From L-Methylmalonyl-CoA

The third reaction, catalyzed by *methylmalonyl-CoA mutase,* is quite unusual, since it involves a migration of the carbonyl-CoA group from one carbon to its neighbor (Figure 23.21). The mutase reaction is vitamin B_{12}–dependent (Chapter 14) and begins with homolytic cleavage of the Co(III)—C bond in cobalamin, reducing the cobalt to Co(II). Transfer of a hydrogen atom from the substrate to the deoxyadenosyl group produces a methylmalonyl-CoA radical, which then can undergo a classic B_{12}-catalyzed rearrangement to yield a succinyl-CoA radical. Hydrogen transfer from the deoxyadenosyl group yields succinyl-CoA and regenerates the B_{12} coenzyme.

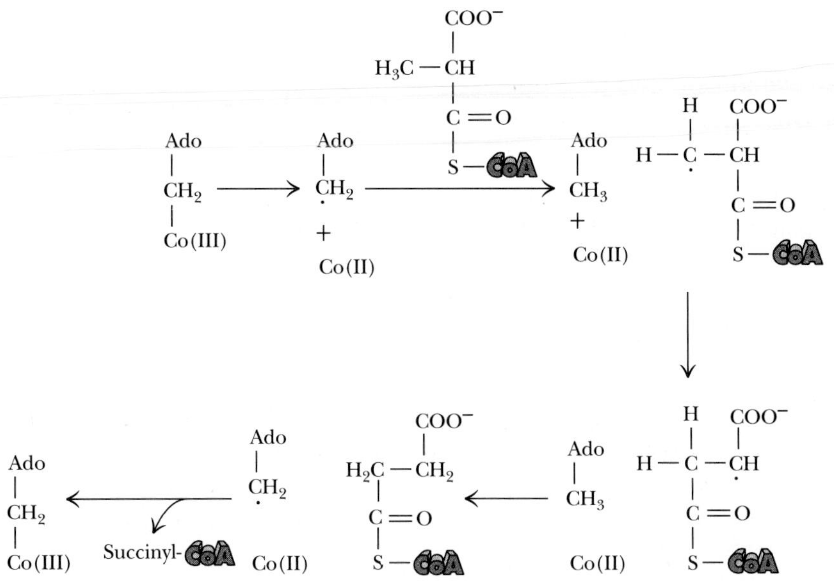

***Figure* 23.21** A mechanism for the methylmalonyl-CoA mutase reaction. In the first step, Co(III) is reduced to Co(II) due to homolytic cleavage of the Co(III)—C bond in cobalamin. Hydrogen atom transfer from methylmalonyl-CoA yields a methylmalonyl-CoA radical that can undergo rearrangement to form a succinyl-CoA radical. Transfer of an H atom regenerates the coenzyme and yields succinyl-CoA.

Net Oxidation of Succinyl-CoA Requires Conversion to Acetyl-CoA

Succinyl-CoA derived from propionyl-CoA can enter the TCA cycle. Oxidation of succinate to oxaloacetate provides a substrate for glucose synthesis. Thus, though the acetate units produced in β-oxidation cannot be utilized in gluconeogenesis, the occasional propionate produced from oxidation of odd-carbon fatty acids *can* be used for sugar synthesis. Alternatively, succinate introduced to the TCA cycle from odd-carbon fatty acid oxidation may be oxidized to CO_2. However, all of the 4-carbon intermediates in the TCA cycle are regenerated in the cycle and thus should be viewed as catalytic species. Net consumption of succinyl-CoA thus does not occur directly in the TCA cycle. Rather, the succinyl-CoA generated from β-oxidation of odd-carbon fatty acids must be converted to pyruvate and then to acetyl-CoA (which is completely oxidized in the TCA cycle). To follow this latter route, succinyl-CoA entering the TCA cycle must be first converted to malate in the usual way, and then transported from the mitochondrial matrix to the cytosol, where it is oxidatively decarboxylated to pyruvate and CO_2 by *malic enzyme,* as shown in Figure 23.22. Pyruvate can then be transported back to the mitochondrial matrix, where it enters the TCA cycle via pyruvate dehydrogenase.

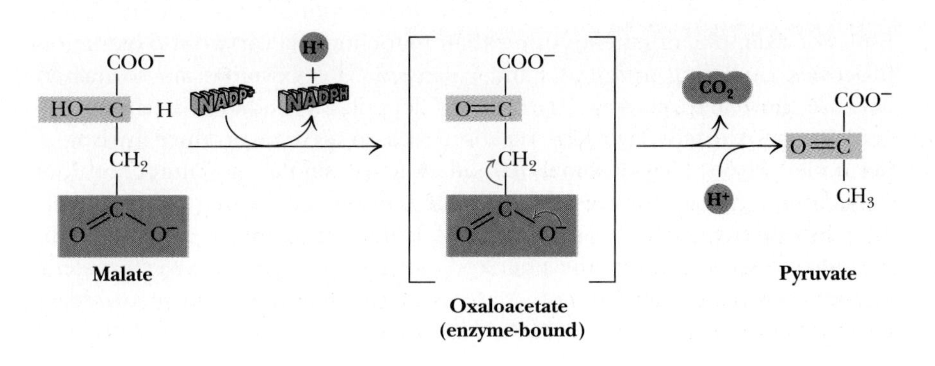

***Figure* 23.22** The malic enzyme reaction proceeds by oxidation of malate to oxaloacetate, followed by decarboxylation to yield pyruvate.

23.4 β-Oxidation of Unsaturated Fatty Acids

An Isomerase and a Reductase Facilitate the β-Oxidation of Unsaturated Fatty Acids

Unsaturated fatty acids are also catabolized by β-oxidation, but two additional mitochondrial enzymes—an isomerase and a novel reductase—are required to handle the *cis-* double bonds of naturally occurring fatty acids. As an example, consider the breakdown of oleic acid, an 18-carbon chain with a double bond at the 9,10-position. The reactions of β-oxidation proceed normally through three cycles, producing three molecules of acetyl-CoA and leaving the degradation product *cis*-Δ^3-dodecenoyl-CoA, shown in Figure 23.23. This intermediate is not a substrate for acyl-CoA dehydrogenase. With a double bond at the 3,4-position, it is not possible to form another double bond at the 2,3- (or β-) position. As shown in Figure 23.23, this problem is solved by *enoyl-CoA isomerase,* an enzyme that rearranges this *cis*-Δ^3 double bond to a *trans*-Δ^2 double bond. This latter species can proceed through the normal route of β-oxidation.

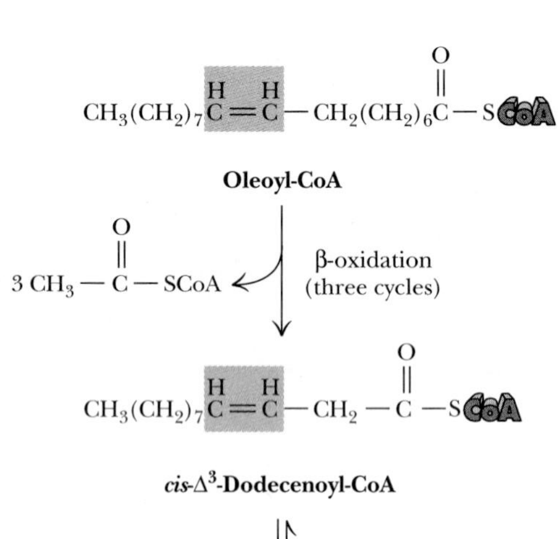

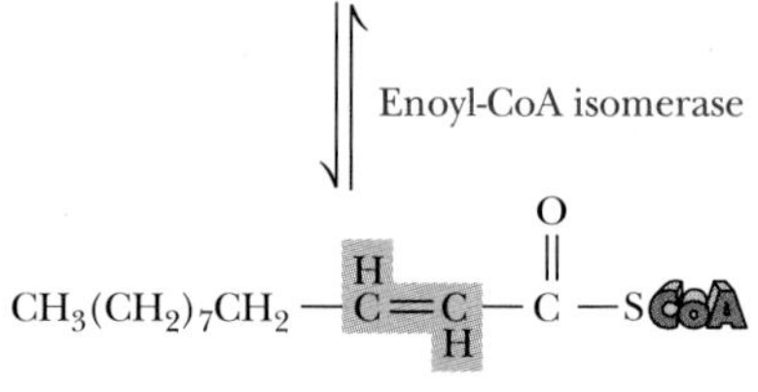

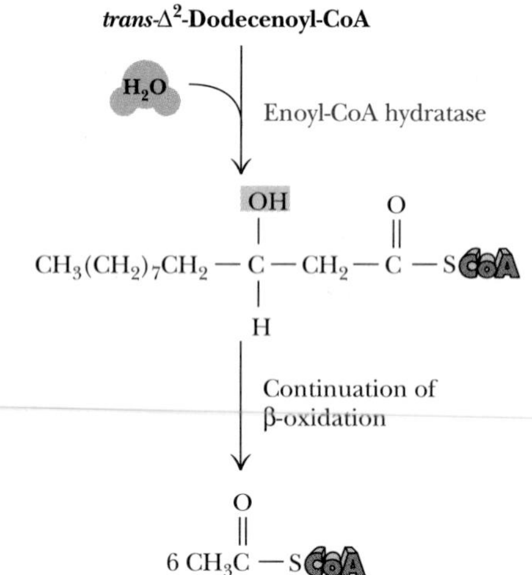

***Figure* 23.23** β-Oxidation of unsaturated fatty acids. In the case of oleoyl-CoA, three β-oxidation cycles produce three molecules of acetyl-CoA and leave *cis*-Δ^3-dodecenoyl-CoA. Rearrangement by enoyl-CoA isomerase gives the *trans*-Δ^2 species, which then proceeds normally through the β-oxidation pathway.

Degradation of Polyunsaturated Fatty Acids Requires 2,4-Dienoyl-CoA Reductase

Polyunsaturated fatty acids pose a slightly more complicated situation for the cell. Consider, for example, the case of linoleic acid shown in Figure 23.24. As with oleic acid, β-oxidation proceeds through three cycles, and enoyl-CoA isomerase converts the *cis*-Δ^3 double bond to a *trans*-Δ^2 double bond to permit one more round of β-oxidation. What results this time, however, is a *cis*-Δ^4 enoyl-CoA, which is converted normally by acyl-CoA dehydrogenase to a *trans*-Δ^2, *cis*-Δ^4 species. This, however, is a poor substrate for the enoyl-CoA hydratase. This problem is solved by *2,4-dienoyl-CoA reductase,* the product of which depends on the organism. The mammalian form of this enzyme produces a *trans*-Δ^3 enoyl product, as shown in Figure 23.24, which can be converted by an enoyl-CoA isomerase to the *trans*-Δ^2 enoyl-CoA, which can then proceed normally through the β-oxidation pathway. *Escherichia coli* possesses a 2,4-enoyl-CoA reductase that reduces the double bond at the 4,5-position to yield the *trans*-Δ^2 enoyl-CoA product in a single step.

23.5 Other Aspects of Fatty Acid Oxidation

Peroxisomal β-Oxidation Requires FAD-Dependent Acyl-CoA Oxidase

Although β-oxidation in mitochondria is the principal pathway of fatty acid catabolism, several other minor pathways play important roles in fat catabolism. For example, organelles other than mitochondria carry out β-oxidation processes, including *peroxisomes* and *glyoxysomes.* **Peroxisomes** are so named because they carry out a variety of flavin-dependent oxidation reactions, regenerating oxidized flavins by reaction with oxygen to produce hydrogen peroxide, H_2O_2. Peroxisomal β-oxidation is similar to mitochondrial β-oxidation, except that the initial double bond formation is catalyzed by an FAD-dependent *acyl-CoA oxidase* (Figure 23.25). The action of this enzyme in the peroxisomes transfers the liberated electrons directly to oxygen instead of the electron transport chain. As a result, each 2-carbon unit oxidized in peroxisomes produces fewer ATPs. The enzymes responsible for fatty acid

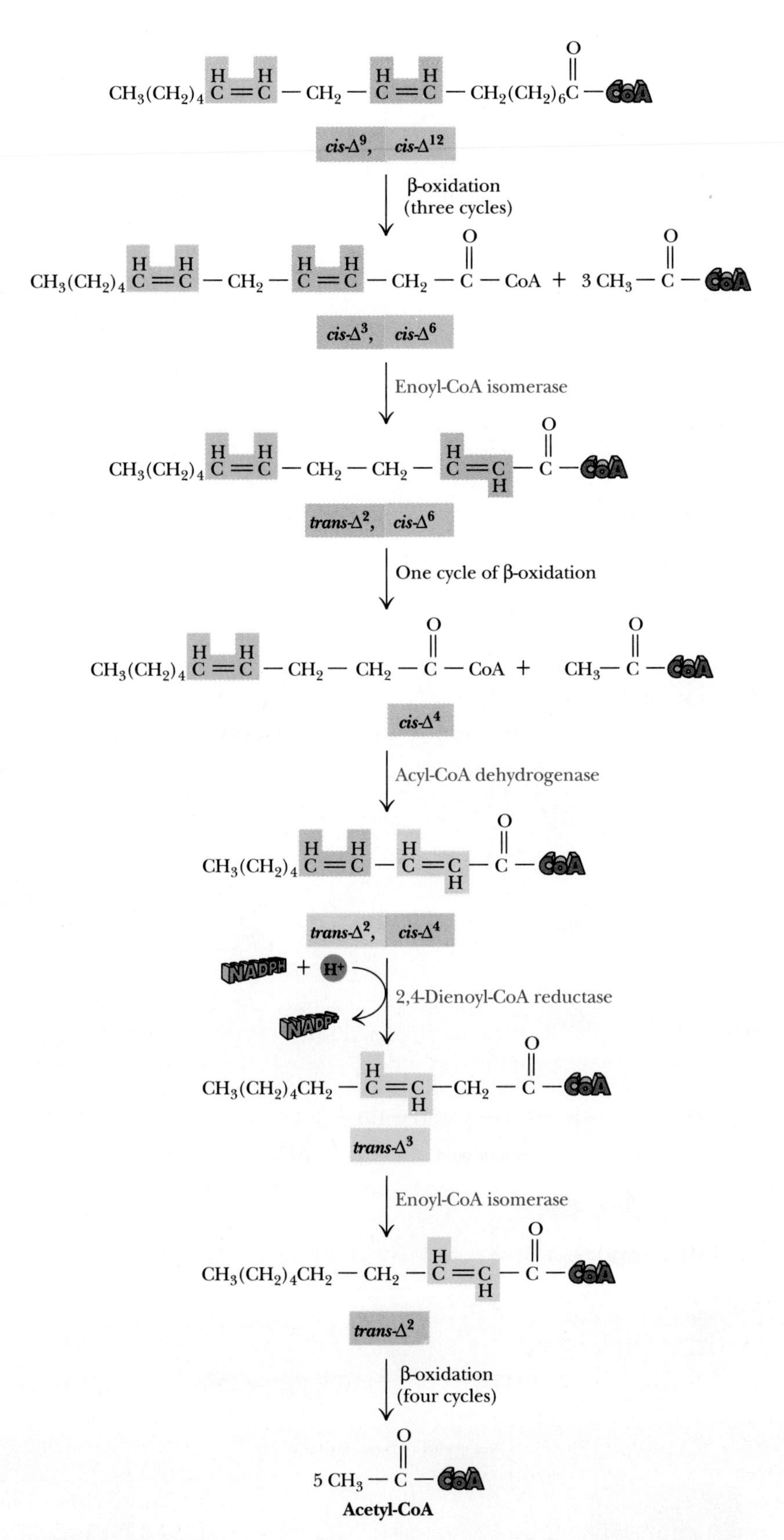

***Figure* 23.24** The oxidation pathway for polyunsaturated fatty acids, illustrated for linoleic acid. Three cycles of β-oxidation on linoleoyl-CoA yield the *cis*-Δ^3, *cis*-Δ^6 intermediate, which is converted to a *trans*-Δ^2, *cis*-Δ^6 intermediate. An additional round of β-oxidation gives *cis*-Δ^4 enoyl-CoA, which is oxidized to the *trans*-Δ^2, *cis*-Δ^4 species by acyl-CoA dehydrogenase. The subsequent action of 2,4-dienoyl-CoA reductase yields the *trans*-Δ^3 product, which is converted by enoyl-CoA isomerase to the *trans*-Δ^2 form. Normal β-oxidation then produces five molecules of acetyl-CoA.

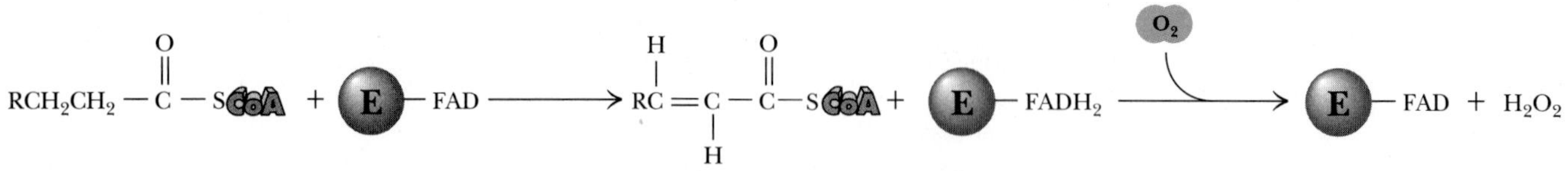

$$RCH_2CH_2-\overset{O}{\overset{\|}{C}}-S\text{CoA} + \text{E}-FAD \longrightarrow RC\overset{H}{\overset{|}{}}=\underset{H}{\underset{|}{C}}-\overset{O}{\overset{\|}{C}}-S\text{CoA} + \text{E}-FADH_2 \xrightarrow{O_2} \text{E}-FAD + H_2O_2$$

Figure **23.25** The acyl-CoA oxidase reaction in peroxisomes.

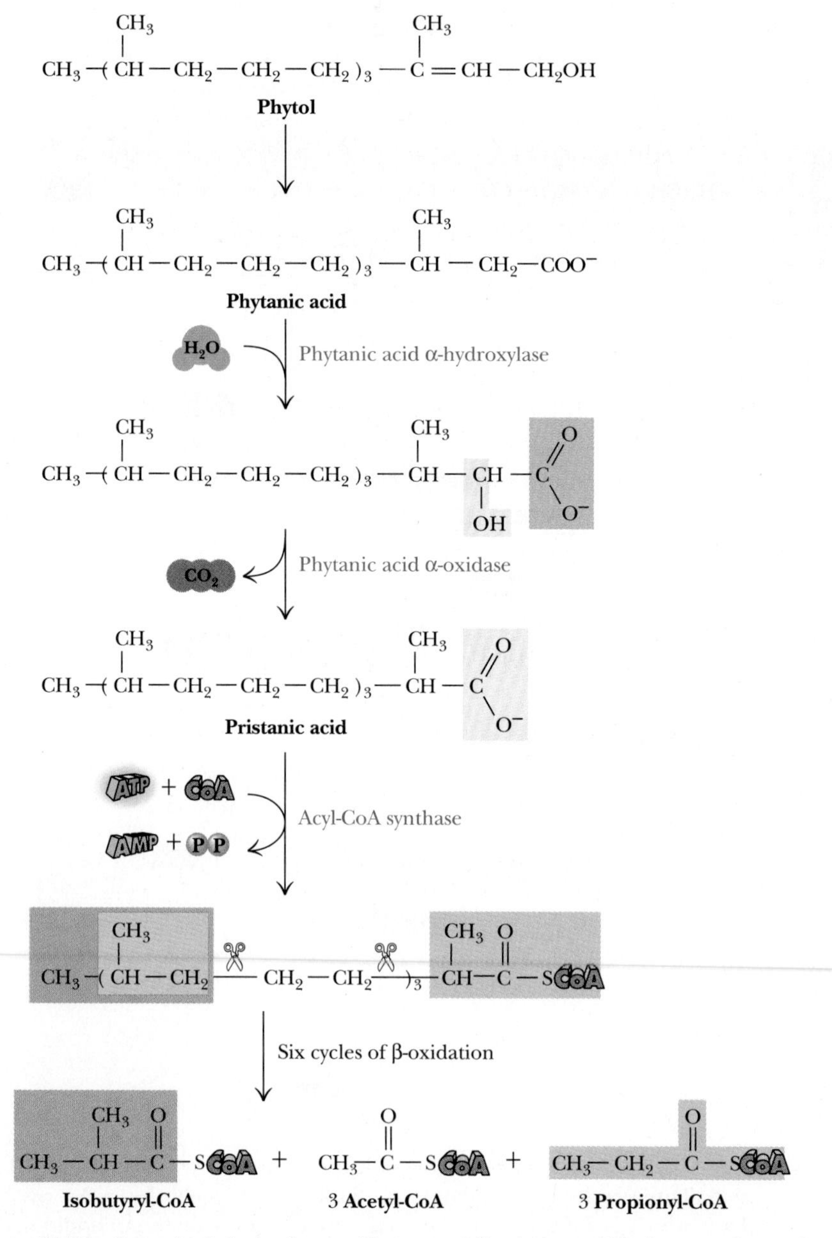

Figure **23.26** Branched-chain fatty acids are oxidized by α-oxidation, as shown for phytanic acid. The product of the phytanic acid oxidase, pristanic acid, is a suitable substrate for normal β-oxidation. Isobutyryl-CoA and propionyl-CoA can both be converted to succinyl-CoA, which can enter the TCA cycle.

oxidation in peroxisomes are inactive with carbon chains of six to eight or fewer. Such short-chain products must be transferred to the mitochondria for further breakdown. Similar β-oxidation enzymes are also found in **glyoxysomes**—peroxisomes in plants which also carry out the reactions of the glyoxylate pathway.

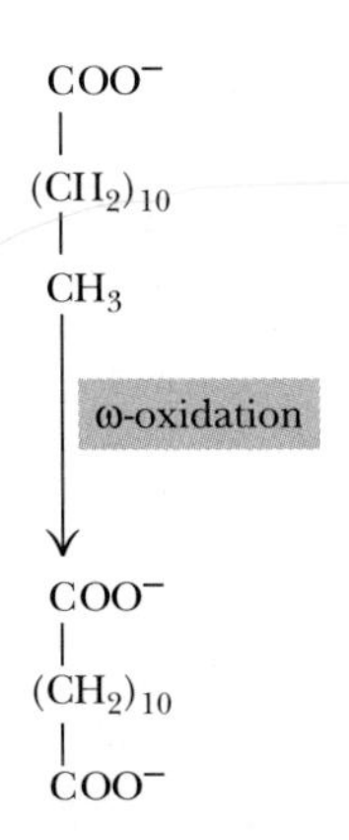

Figure **23.27** Dicarboxylic acids can be formed by oxidation of the methyl group of fatty acids in a cytochrome P-450–dependent reaction.

Branched-Chain Fatty Acids and α-Oxidation

Although β-oxidation is universally important, there are some instances in which it cannot operate effectively. For example, branched-chain fatty acids with alkyl branches at odd-numbered carbons are not effective substrates for β-oxidation. For such species, **α-oxidation** is a useful alternative. Consider **phytol,** a breakdown product of chlorophyll that occurs in the fat of ruminant animals such as sheep and cows and also in dairy products. The methyl group at C-3 will block β-oxidation, but, as shown in Figure 23.26, *phytanic acid α-hydroxylase* places an —OH group at the α-carbon, and *phytanic acid α-oxidase* decarboxylates to yield *pristanic acid.* The CoA ester of this metabolite can undergo β-oxidation in the normal manner. The terminal product, isobutyryl-CoA, can be sent into the TCA cycle by conversion to succinyl-CoA.

Refsum's Disease Is a Result of Defects in α-Oxidation

The α-oxidation pathway is defective in **Refsum's disease,** an inherited metabolic disorder that results in defective night vision, tremors, and other neurologic abnormalities. These symptoms are caused by accumulation of phytanic acid in the body. Treatment of Refsum's disease requires a diet free of chlorophyll, the precursor of phytanic acid. This regimen is difficult to implement, since all green vegetables and even meat from plant-eating animals, such as cows, pigs, and poultry, must be excluded from the diet.

ω-Oxidation of Fatty Acids Yields Small Amounts of Dicarboxylic Acids

In the endoplasmic reticulum of eukaryotic cells, the oxidation of the terminal carbon of a normal fatty acid—a process termed ω-oxidation—can lead to the synthesis of small amounts of dicarboxylic acids (Figure 23.27). **Cytochrome P-450,** a monooxygenase enzyme that requires NADPH as a coenzyme and uses O_2 as a substrate, places a hydroxyl group at the terminal carbon. Subsequent oxidation to a carboxyl group produces a dicarboxylic acid. Either end can form an ester linkage to CoA and be subjected to β-oxidation, producing a variety of smaller dicarboxylic acids. (Cytochrome P-450–dependent monooxygenases also play an important role as agents of **detoxication,** the degradation and metabolism of toxic hydrocarbon agents.)

23.6 Ketone Bodies

Ketone Bodies Are a Significant Source of Fuel and Energy for Certain Tissues

Most of the acetyl-CoA produced by the oxidation of fatty acids in liver mitochondria undergoes further oxidation in the TCA cycle, as stated earlier. However, some of this acetyl-CoA is converted to three important metabolites: acetone, acetoacetate, and β-hydroxybutyrate. The process is known as **ketogenesis,** and these three metabolites are traditionally known as **ketone bodies,**

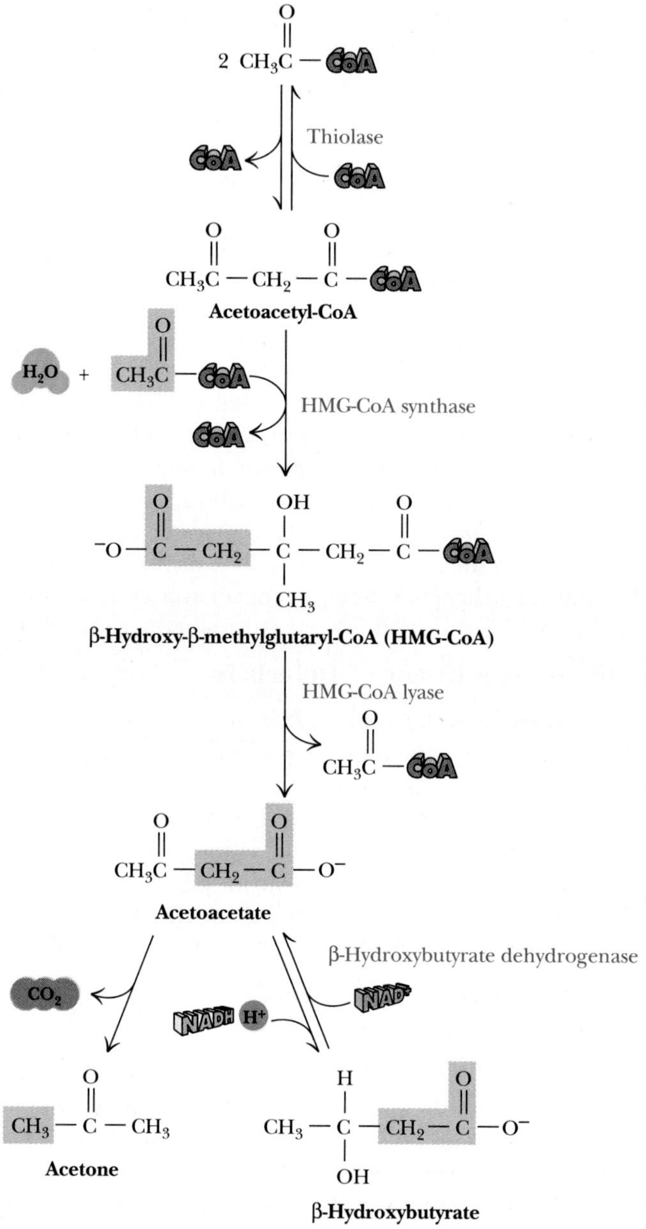

***Figure* 23.28** The formation of ketone bodies, synthesized primarily in liver mitochondria.

in spite of the fact that *β*-hydroxybutyrate does not contain a ketone function. These three metabolites are synthesized primarily in the liver but are important sources of fuel and energy for many peripheral tissues, including brain, heart, and skeletal muscle. The brain, for example, normally uses glucose as its source of metabolic energy. However, during periods of starvation, ketone bodies may be the major energy source for the brain. Acetoacetate and 3-hydroxybutyrate are the preferred and normal substrates for kidney cortex and for heart muscle.

Ketone body synthesis occurs only in the mitochondrial matrix. The reactions responsible for the formation of ketone bodies are shown in Figure 23.28. The first reaction—the condensation of two molecules of acetyl-CoA to form acetoacetyl-CoA—is catalyzed by *thiolase,* which is also known as *acetoacetyl-CoA thiolase* or *acetyl-CoA acetyltransferase.* This is the same enzyme that carries out the thiolase reaction in β-oxidation, but here it runs in reverse. The second reaction adds another molecule of acetyl-CoA to give *β-hydroxy-β-methylglutaryl-CoA,* commonly abbreviated HMG-CoA. These two mitochondrial matrix reactions are analogous to the first two steps in cholesterol biosynthesis, a cytosolic process, as we shall see in Chapter 24. HMG-CoA is converted to acetoacetate and acetyl-CoA by the action of *HMG-CoA lyase* in a mixed aldol-Claisen ester cleavage reaction. This reaction is mechanistically similar to the reverse of the citrate synthase reaction in the TCA cycle. A membrane-bound enzyme, *β-hydroxybutyrate dehydrogenase,* then can reduce acetoacetate to β-hydroxybutyrate.

Acetoacetate and β-hydroxybutyrate are transported through the blood from liver to target organs and tissues, where they are converted to acetyl-CoA (Figure 23.29). *Ketone bodies are easily transportable forms of fatty acids that move through the circulatory system without the need for complexation with serum albumin and other fatty acid–binding proteins.*

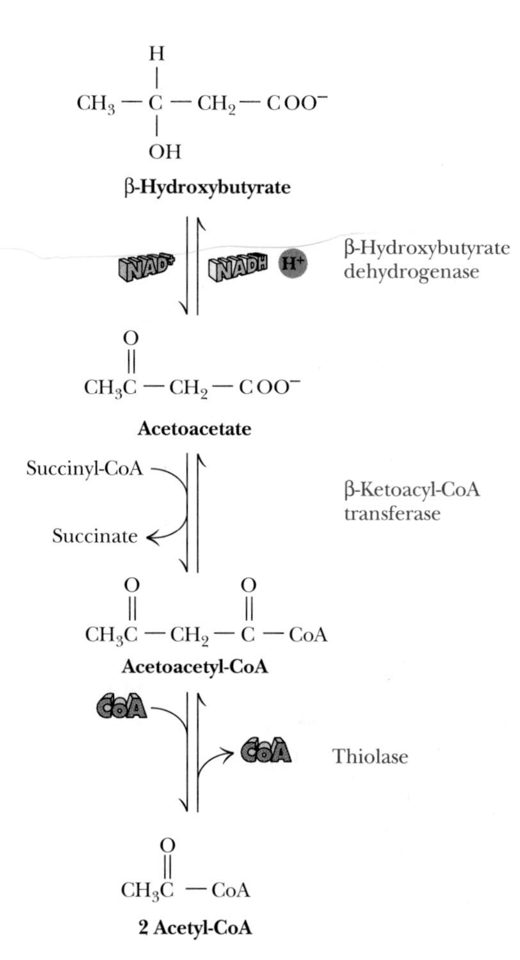

Figure **23.29** Reconversion of ketone bodies to acetyl-CoA in the mitochondria of many tissues (other than liver) provides significant metabolic energy.

Ketone Bodies and Diabetes Mellitus

Diabetes mellitus is the most common endocrine disease and the third leading cause of death in the United States, with approximately 6 million diagnosed cases and an estimated 4 million more borderline but undiagnosed cases. Diabetes is characterized by an abnormally high level of glucose in the blood. In **type I diabetes** (representing 10% or fewer of all cases), elevated blood glucose results from inadequate secretion of insulin by the islets of Langerhans in the pancreas. **Type II diabetes** (at least 90% of all cases) results from an insensitivity to insulin. Type II diabetics produce normal or even elevated levels of insulin, but owing to a shortage of insulin receptors (Chapter 37), their cells are not responsive to insulin. In both cases, transport of glucose into muscle, liver, and adipose tissue is significantly reduced, and, despite abundant glucose in the blood, the cells are metabolically starved. They respond by turning to increased gluconeogenesis and catabolism of fat and protein. In type I diabetes, increased gluconeogenesis consumes most of the available oxaloacetate, but breakdown of fat (and, to a lesser extent, protein) produces large amounts of acetyl-CoA. This increased acetyl-CoA would normally be directed into the TCA cycle, but, with oxaloacetate in short supply, it is used instead for production of unusually large amounts of ketone bodies. Acetone can often be detected on the breath of type I diabetics, an indication of high plasma levels of ketone bodies.

Problems

1. Calculate the volume of metabolic water available to a camel through fatty acid oxidation if it carries 30 lb of triacylglycerol in its hump.
2. Calculate the approximate number of ATP molecules that can be obtained from the oxidation of *cis*-11-heptadecenoic acid to CO_2 and water.
3. Phytanic acid, the product of chlorophyll that causes problems for individuals with Refsum's disease, is 3,7,11,15-tetramethyl hexadecanoic acid. Suggest a route for its oxidation that is consistent with what you have learned in this chapter. (*Hint:* The methyl group at C-3 effectively blocks hydroxylation and normal β-oxidation. You may wish to initiate breakdown in some other way.)
4. Even though acetate units, such as those obtained from fatty acid oxidation, cannot be used for *net* synthesis of carbohydrate in animals, labeled carbon from ^{14}C-labeled acetate *can* be found in newly synthesized glucose (for example, in liver glycogen) in animal tracer studies. Explain how this can be. Which carbons of glucose would you expect to be the first to be labeled by ^{14}C-labeled acetate?
5. What would you expect to be the systemic metabolic effects of consuming unripened akee fruit?
6. Overweight individuals who diet to lose weight often view fat in negative ways, since adipose tissue is the repository of excess caloric intake. However, the "weighty" consequences might be even worse if excess calories were stored in other

forms. Consider a person who is 10 lb "overweight," and estimate how much more he or she would weigh if excess energy were stored in the form of carbohydrate instead of fat.

7. What would be the consequences of a deficiency in vitamin B_{12} for fatty acid oxidation? What metabolic intermediates might accumulate?

8. Write properly balanced chemical equations for the oxidation to CO_2 and water of (a) myristic acid, (b) stearic acid, (c) α-linolenic acid, and (d) arachidonic acid.

9. How many tritium atoms are incorporated in acetate if a molecule of palmitic acid is oxidized in 100% tritiated water?

10. What would be the consequences of a carnitine deficiency for fatty acid oxidation?

Further Reading

Bieber, L. L., 1988. Carnitine. *Annual Review of Biochemistry* **88:**261–283.

Boyer, P. D., ed., 1983. *The Enzymes,* 3rd ed., vol. 16. New York: Academic Press.

Halpern, J., 1985. Mechanisms of coenzyme B_{12}–dependent rearrangements. *Science* **227:**869–875.

Hiltunen, J. K., Palosaari, P., and Kunau, W.-H., 1989. Epimerization of 3-hydroxyacyl-CoA esters in rat liver. *Journal of Biological Chemistry* **264:**13535–13540.

McGarry, J. D., and Foster, D. W., 1980. Regulation of hepatic fatty acid oxidation and ketone body production. *Annual Review of Biochemistry* **49:**395–420.

Newsholme, E. A., and Leech, A. R., 1983. *Biochemistry for the Medical Sciences.* New York: Wiley.

Schulz, H., 1987. Inhibitors of fatty acid oxidation. *Life Sciences* **40:**1443–1449.

Schulz, H., and Kunau, W.-H., 1987. β-Oxidation of unsaturated fatty acids: A revised pathway. *Trends in Biochemical Sciences* **12:**403–406.

Stanbury, J. B., Wyngaarden, J. B., Fredrickson, D. S., Goldstein, J. L., and Brown, M. S., 1989. *The Metabolic Basis for Inherited Disease,* 6th ed. New York: McGraw-Hill.

Tolbert, N. E., 1981. Metabolic pathways in peroxisomes and glyoxysomes. *Annual Review of Biochemistry* **50:**133–157.

Vance, D. E., and Vance, J. E., eds., 1985. *Biochemistry of Lipids and Membranes.* Menlo Park, CA: Benjamin/Cummings.

Chapter 24

"To every thing there is a season, and a time for every purpose under heaven . . . A time to break down, a time to build up."

Ecclesiastes 3:1–3

Lipid Biosynthesis

Outline

24.1 The Fatty Acid Biosynthesis and Degradation Pathways Are Different

24.2 Biosynthesis of Complex Lipids

24.3 Eicosanoid Biosynthesis and Function

24.4 Cholesterol Biosynthesis

24.5 Transport of Many Lipids Occurs via Lipoprotein Complexes

24.6 Biosynthesis of Bile Acids

24.7 Synthesis and Metabolism of Steroid Hormones

Southern elephant seal

(Mirounga leonina)

We turn now to the biosynthesis of lipid structures. We will begin with a discussion of the biosynthesis of fatty acids, stressing the basic pathways, additional means of elongation, mechanisms for the introduction of double bonds, and regulation of fatty acid synthesis. Sections then follow on the biosynthesis of glycerophospholipids, sphingolipids, eicosanoids, and cholesterol. The transport of lipids through the body in lipoprotein complexes is described, and the chapter closes with discussions of the biosynthesis of bile salts and steroid hormones.

24.1 The Fatty Acid Biosynthesis and Degradation Pathways Are Different

We have already seen several cases in which the *synthesis* of a class of biomolecules is conducted differently from degradation (glycolysis versus gluconeogenesis and glycogen or starch breakdown versus polysaccharide synthesis, for example). Likewise, the synthesis of fatty acids and other lipid components is different from their degradation. Fatty acid synthesis involves a set of reac-

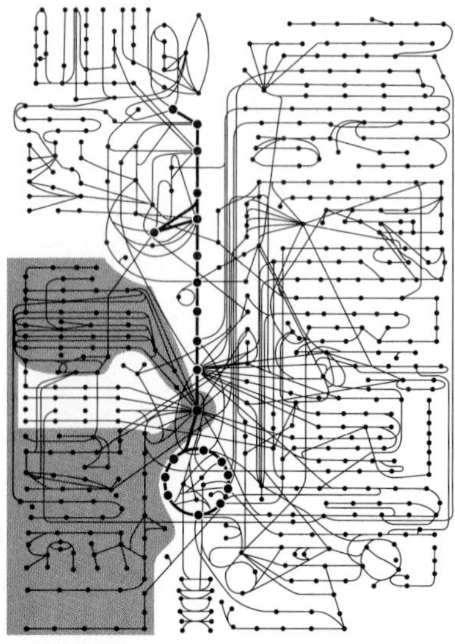
Lipid Biosynthesis

tions that follow a synthetic strategy different in several ways from the corresponding degradative process:

1. Intermediates in fatty acid synthesis are linked covalently to the sulfhydryl groups of special proteins—called **acyl carrier proteins.** In contrast, fatty acid breakdown intermediates are bound to the —SH group of coenzyme A.
2. Fatty acid synthesis occurs in the cytosol, whereas fatty acid degradation takes place in mitochondria.
3. In animals, the enzymes of fatty acid synthesis are components of one long polypeptide chain, the **fatty acid synthase,** whereas no similar association exists for the degradative enzymes. (Plants and bacteria employ separate enzymes to carry out the biosynthetic reactions.)
4. The coenzyme for the oxidation–reduction reactions of fatty acid synthesis is $NADPH/NADP^+$, whereas degradation involves the $NADH/NAD^+$ couple.

Formation of Malonyl-CoA Activates Acetate Units for Fatty Acid Synthesis

The design strategy for fatty acid synthesis is this:

a. Fatty acid chains are constructed by the addition of two-carbon units derived from *acetyl-CoA.*
b. The acetate units are activated by formation of *malonyl-CoA* (at the expense of ATP).
c. The addition of two-carbon units to the growing chain is driven by decarboxylation of malonyl-CoA.
d. The elongation reactions are repeated until the growing chain reaches 16 carbons in length (palmitic acid).
e. Other enzymes then add double bonds and additional carbon units to the chain.

Fatty Acid Biosynthesis Depends on the Reductive Power of NADPH

The net reaction for the formation of palmitate from acetyl-CoA is

$$\text{Acetyl-CoA} + 7\ \text{malonyl-CoA}^- + 14\ \text{NADPH} + 14\ \text{H}^+ \longrightarrow \text{palmitoyl-CoA} + 7\ \text{HCO}_3^- + 7\ \text{CoASH} + 14\ \text{NADP}^+ \quad (24.1)$$

(Levels of free fatty acids are very low in the typical cell. The palmitate made in this process is rapidly converted to CoA esters in preparation for the formation of triacylglycerols and phospholipids.)

Providing Cytosolic Acetyl-CoA and Reducing Power for Fatty Acid Synthesis

Eukaryotic cells face a dilemma in providing suitable amounts of substrate for fatty acid synthesis. Sufficient quantities of acetyl-CoA, malonyl-CoA, and NADPH must be generated *in the cytosol* for fatty acid synthesis. Malonyl-CoA is made by carboxylation of acetyl-CoA, so the problem reduces to generating sufficient acetyl-CoA and NADPH.

There are three principal sources of acetyl-CoA (Figure 24.1):

1. Amino acid degradation produces cytosolic acetyl-CoA.
2. Fatty acid oxidation produces mitochondrial acetyl-CoA.
3. Glycolysis yields cytosolic pyruvate, which (after transport into the mitochondria) is converted to acetyl-CoA by pyruvate dehydrogenase.

The acetyl-CoA derived from amino acid degradation is normally insufficient for fatty acid biosynthesis, and the acetyl-CoA produced by pyruvate dehydrogenase and by fatty acid oxidation cannot cross the mitochondrial membrane to participate directly in fatty acid synthesis. Instead, acetyl-CoA is linked with oxaloacetate to form citrate, which is transported from the mitochondrial matrix to the cytosol (Figure 24.1). Here it can be converted back into acetyl-

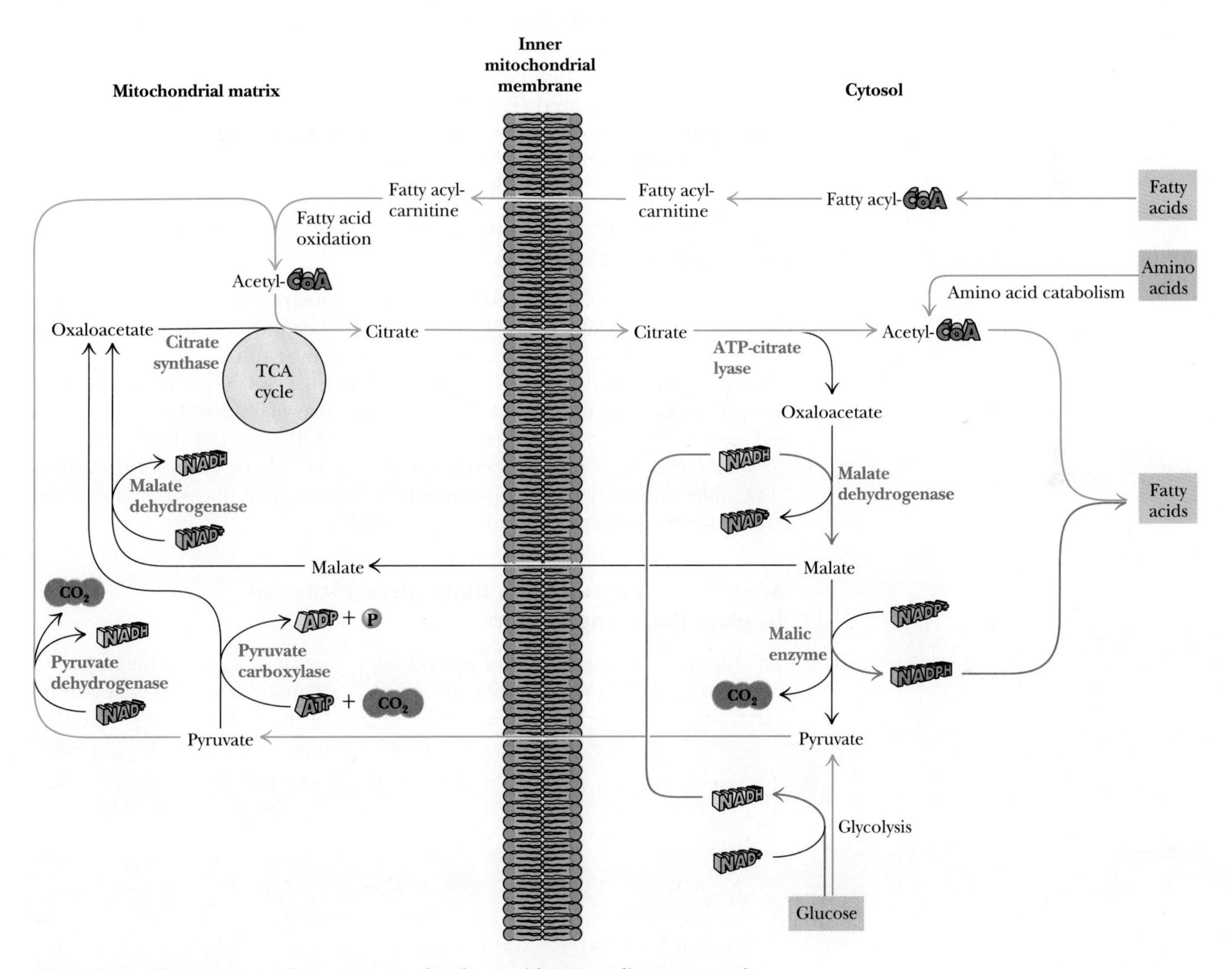

***Figure* 24.1** The citrate–malate–pyruvate shuttle provides cytosolic acetate units and reducing equivalents (electrons) for fatty acid synthesis. The shuttle collects carbon substrates, primarily from glycolysis but also from fatty acid oxidation and amino acid catabolism. Most of the reducing equivalents are glycolytic in origin. Pathways that provide carbon for fatty acid synthesis are shown in blue; pathways that supply electrons for fatty acid synthesis are shown in red.

CoA and oxaloacetate by *ATP-citrate lyase.* In this manner, mitochondrial acetyl-CoA becomes the substrate for cytosolic fatty acid synthesis. (Oxaloacetate returns to the mitochondria in the form of either pyruvate or malate, which is then reconverted to acetyl-CoA and oxaloacetate, respectively.)

NADPH can be produced in the pentose phosphate pathway as well as by malic enzyme (Figure 24.1). Reducing equivalents (electrons) derived from glycolysis in the form of NADH can be transformed into NADPH by the combined action of malate dehydrogenase and malic enzyme:

$$\text{Oxaloacetate} + \text{NADH} \longrightarrow \text{malate} + \text{NAD}^+$$
$$\text{Malate} + \text{NADP}^+ \longrightarrow \text{pyruvate} + \text{CO}_2 + \text{NADPH}$$

How many of the 14 NADPH needed to form one palmitate (Eq. 24.1) can be made in this way? The answer depends on the status of malate. Every citrate entering the cytosol produces one acetyl-CoA and one malate (Figure 24.1). Every malate oxidized by malic enzyme produces one NADPH, at the expense of a decarboxylation to pyruvate. Thus, when malate is oxidized, one NADPH is produced for every acetyl-CoA. Conversion of 8 acetyl-CoA units to one palmitate would then be accompanied by production of 8 NADPH. (The other 6 NADPH required (Eq. 24.1) would be provided by the pentose phosphate pathway.) On the other hand, for every malate returned to the mitochondria, one NADPH fewer is produced.

Acetate Units Are Committed to Fatty Acid Synthesis by Formation of Malonyl-CoA

Rittenberg and Bloch showed in the late 1940s that acetate units are the building blocks of fatty acids. Their work, together with the discovery by Salih Wakil that bicarbonate is required for fatty acid biosynthesis, eventually made clear that this pathway involves synthesis of *malonyl-CoA*. The carboxylation of acetyl-CoA to form malonyl-CoA is essentially irreversible and is the **committed step** in the synthesis of fatty acids (Figure 24.2). The reaction is catalyzed by *acetyl-CoA carboxylase,* which contains a biotin prosthetic group. This carboxylase is the only enzyme of fatty acid synthesis in animals that is not part of the multienzyme complex called fatty acid synthase.

Acetyl-CoA Carboxylase Is Biotin-Dependent and Displays Ping-Pong Kinetics

The biotin prosthetic group of acetyl-CoA carboxylase is covalently linked to the ϵ-amino group of an active-site lysine, in a manner similar to pyruvate carboxylase (Figure 21.3). The reaction mechanism is also analogous to that

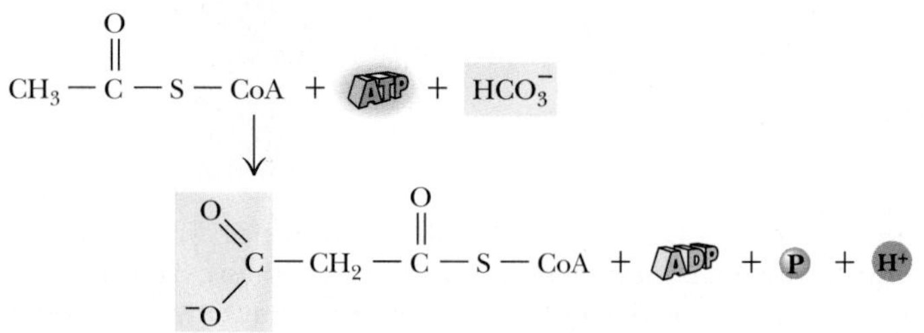

$$\text{CH}_3-\overset{\overset{\text{O}}{\|}}{\text{C}}-\text{S}-\text{CoA} + \text{ATP} + \text{HCO}_3^-$$
$$\downarrow$$
$$^-\text{O}(\text{O}=)\text{C}-\text{CH}_2-\overset{\overset{\text{O}}{\|}}{\text{C}}-\text{S}-\text{CoA} + \text{ADP} + \text{P} + \text{H}^+$$

***Figure* 24.2** The acetyl-CoA carboxylase reaction produces malonyl-CoA for fatty acid synthesis.

of pyruvate carboxylase (Figure 21.4): ATP-driven carboxylation of biotin is followed by transfer of the activated CO_2 to acetyl-CoA to form malonyl-CoA. As noted in Chapter 14, *all* biotin-dependent enzymes display ping-pong kinetics, and acetyl-CoA carboxylase is no exception. ADP and P_i are released before acetyl-CoA is bound (Figure 24.3). The enzyme from *Escherichia coli* has three subunits: (1) a **biotin carboxyl carrier protein** (a dimer of 22.5-kD subunits); (2) **biotin carboxylase** (a dimer of 51-kD subunits), which adds CO_2 to the prosthetic group; and (3) **transcarboxylase** (an $\alpha_2\beta_2$ tetramer with 30-kD and 35-kD subunits), which transfers the activated CO_2 unit to acetyl-CoA. The long, flexible biotin–lysine chain (biocytin) enables the activated carboxyl group to be carried between the biotin carboxylase and the transcarboxylase.

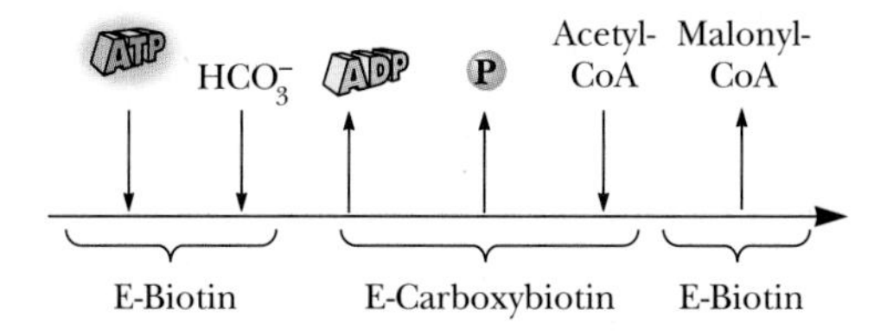

***Figure* 24.3** Acetyl-CoA carboxylase is biotin-dependent and follows a ping-pong mechanism. Binding of ATP and bicarbonate leads to the formation of N-carboxybiotin at the active site, followed by the dissociation of ADP and P_i. Binding of acetyl-CoA is followed by formation and release of malonyl-CoA.

Acetyl-CoA Carboxylase in Animals Is a Multifunctional Protein

In animals, acetyl-CoA carboxylase (ACC) is a filamentous polymer (4 to 8 × 10^6 D) composed of 230-kD protomers. Each of these subunits contains the biotin carboxyl carrier moiety, biotin carboxylase, and transcarboxylase activities, as well as allosteric regulatory sites. Animal ACC is thus a multifunctional protein. The polymeric form is active, but the 230-kD protomers are inactive. The activity of ACC is thus dependent upon the position of the equilibrium between these two forms:

$$\text{Inactive protomers} \rightleftharpoons \text{active polymer}$$

Since this enzyme catalyzes the committed step in fatty acid biosynthesis, it is carefully regulated. *Palmitoyl-CoA,* the final product of fatty acid biosynthesis, shifts the equilibrium toward the inactive protomers, whereas *citrate,* an important allosteric activator of this enzyme, shifts the equilibrium toward the active polymeric form of the enzyme. Acetyl-CoA carboxylase shows the kinetic behavior of a Monod–Wyman–Changeux V-system allosteric enzyme (Chapter 12).

Phosphorylation of ACC Modulates Activation by Citrate and Inhibition by Palmitoyl-CoA

The regulatory effects of citrate and palmitoyl-CoA are dependent on the phosphorylation state of acetyl-CoA carboxylase. The animal enzyme is phosphorylated at 8 to 10 sites on each enzyme subunit (Figure 24.4). Some of these sites are regulatory, whereas others are "silent" and have no effect on enzyme activity. Unphosphorylated acetyl-CoA carboxylase binds citrate with high affinity and thus is active at very low citrate concentrations (Figure 24.5). Phosphorylation of the regulatory sites decreases the affinity of the enzyme for citrate, and in this case high levels of citrate are required to activate the carboxylase. The inhibition by fatty acyl-CoAs operates in a similar but opposite manner. Thus, low levels of fatty acyl-CoA inhibit the phosphorylated carboxylase, but the dephosphoenzyme is inhibited only by high levels of fatty acyl-CoA. Specific phosphatases act to dephosphorylate ACC, thereby increasing the sensitivity to citrate.

Acyl Carrier Proteins Carry the Intermediates in Fatty Acid Synthesis

The basic building blocks of fatty acid synthesis are acetyl and malonyl groups, but they are not transferred directly from CoA to the growing fatty acid chain. Rather, they are first passed to **acyl carrier protein** (or simply ACP), discov-

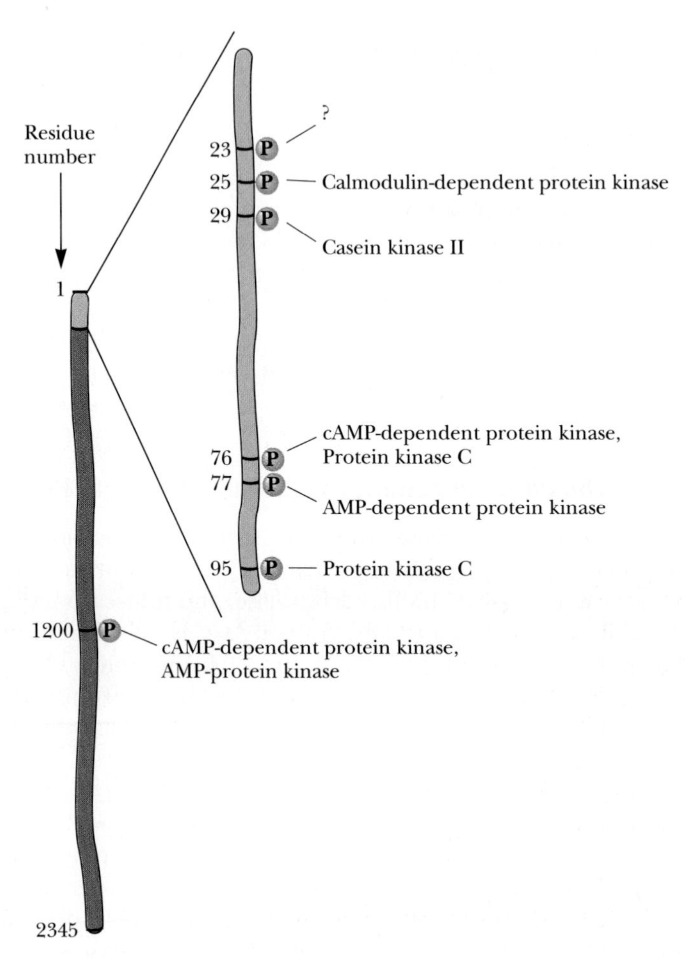

***Figure* 24.4** Models of the acetyl-CoA carboxylase polypeptide, with phosphorylation sites indicated, along with the protein kinases responsible. Phosphorylation at Ser^{1200} is primarily responsible for decreasing the affinity for citrate.

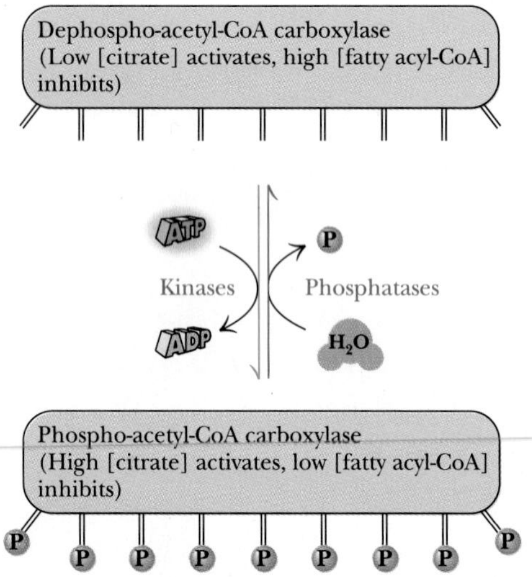

Dephospho-acetyl-CoA carboxylase
(Low [citrate] activates, high [fatty acyl-CoA] inhibits)

ATP
Kinases
ADP
P
Phosphatases
H_2O

Phospho-acetyl-CoA carboxylase
(High [citrate] activates, low [fatty acyl-CoA] inhibits)

***Figure* 24.5** The activity of acetyl-CoA carboxylase is modulated by phosphorylation and dephosphorylation. The dephospho form of the enzyme is activated by low [citrate] and inhibited only by high levels of fatty acyl-CoA. In contrast, the phosphorylated form of the enzyme is activated only by high levels of citrate, but is very sensitive to inhibition by fatty acyl-CoA.

ered by P. Roy Vagelos. This protein consists (in *Escherichia coli*) of a single polypeptide chain of 77 residues to which is attached (on a serine residue) a **phosphopantetheine group,** the same group that forms the ''business end'' of coenzyme A. Thus, acyl carrier protein is a somewhat larger version of coenzyme A, specialized for use in fatty acid biosynthesis (Figure 24.6).

The enzymes that catalyze formation of acetyl-ACP and malonyl-ACP and the subsequent reactions of fatty acid synthesis are organized quite differently in different organisms. We will first discuss fatty acid biosynthesis in bacteria and plants, where the various reactions are catalyzed by separate, independent proteins. Then we will discuss the animal version of fatty acid biosynthesis, which involves a single multienzyme complex called *fatty acid synthase.*

Fatty Acid Synthesis in Bacteria and Plants

The individual steps in the elongation of the fatty acid chain are quite similar in bacteria, fungi, plants, and animals. The ease of purification of the separate enzymes from bacteria and plants made it possible in the beginning to sort out each step in the pathway, and then by extension to see the pattern of biosynthesis in animals. The reactions are summarized in Figure 24.7. The elongation reactions begin with the formation of acetyl-ACP and malonyl-ACP, which are formed by **acetyl transacylase (acetyl transferase)** and **malonyl**

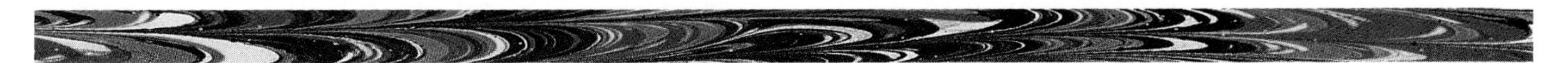

A Deeper Look

Choosing the Best Organism for the Experiment

The selection of a suitable and relevant organism is an important part of any biochemical investigation. The studies that revealed the secrets of fatty acid synthesis are a good case in point.

The paradigm for fatty acid synthesis in plants has been the avocado, which has one of the highest fatty acid contents in the plant kingdom. Early animal studies centered primarily on pigeons, which are easily bred and handled and which possess high levels of fats in their tissues. Other animals, richer in fatty tissues, might be even more attractive but more challenging to maintain. Grizzly bears, for example, carry very large fat reserves but are difficult to work with in the lab!

transacylase (malonyl transferase), respectively. The acetyl transacylase enzyme is not highly specific—it can transfer other acyl groups, such as the propionyl group, but at much lower rates. (Fatty acids with odd numbers of carbons are made beginning with a propionyl group transfer by this enzyme.) Malonyl transacylase, on the other hand, is highly specific.

Decarboxylation Drives the Condensation of Acetyl-CoA and Malonyl-CoA

Another transacylase reaction transfers the acetyl group from ACP to **β-ketoacyl-ACP synthase (KSase),** also known as **acyl-malonyl-ACP condensing enzyme.** The first actual elongation reaction involves the condensation of acetyl-ACP and malonyl-ACP by the β-ketoacyl-ACP synthase to form acetoacetyl-ACP (Figure 24.7). *One might ask at this point: Why is the three-carbon malonyl group used here as a two-carbon donor?* The answer is that this is yet another example of a decarboxylation driving a desired but otherwise thermodynamically unfavorable reaction. The decarboxylation that accompanies the reaction with malonyl-ACP drives the synthesis of acetoacetyl-ACP. Note that hydrolysis of ATP drove the carboxylation of acetyl-CoA to form malonyl-

$HS-CH_2-CH_2-NH-C(=O)-CH_2-CH_2-NH-C(=O)-CH(OH)-C(CH_3)_2-CH_2-O-P(=O)(O^-)-O-P(=O)(O^-)-O-CH_2$–(ribose: O, Adenine, H, H, H, H, $^{2-}O_3PO$, OH)

Phosphopantetheine group of coenzyme A

$HS-CH_2-CH_2-NH-C(=O)-CH_2-CH_2-NH-C(=O)-CH(OH)-C(CH_3)_2-CH_2-O-P(=O)(O^-)-O-CH_2-Ser-$Acyl carrier protein

Phosphopantetheine prosthetic group of ACP

***Figure* 24.6** Fatty acids are conjugated both to coenzyme A and to acyl carrier protein through the sulfhydryl of phosphopantetheine prosthetic groups.

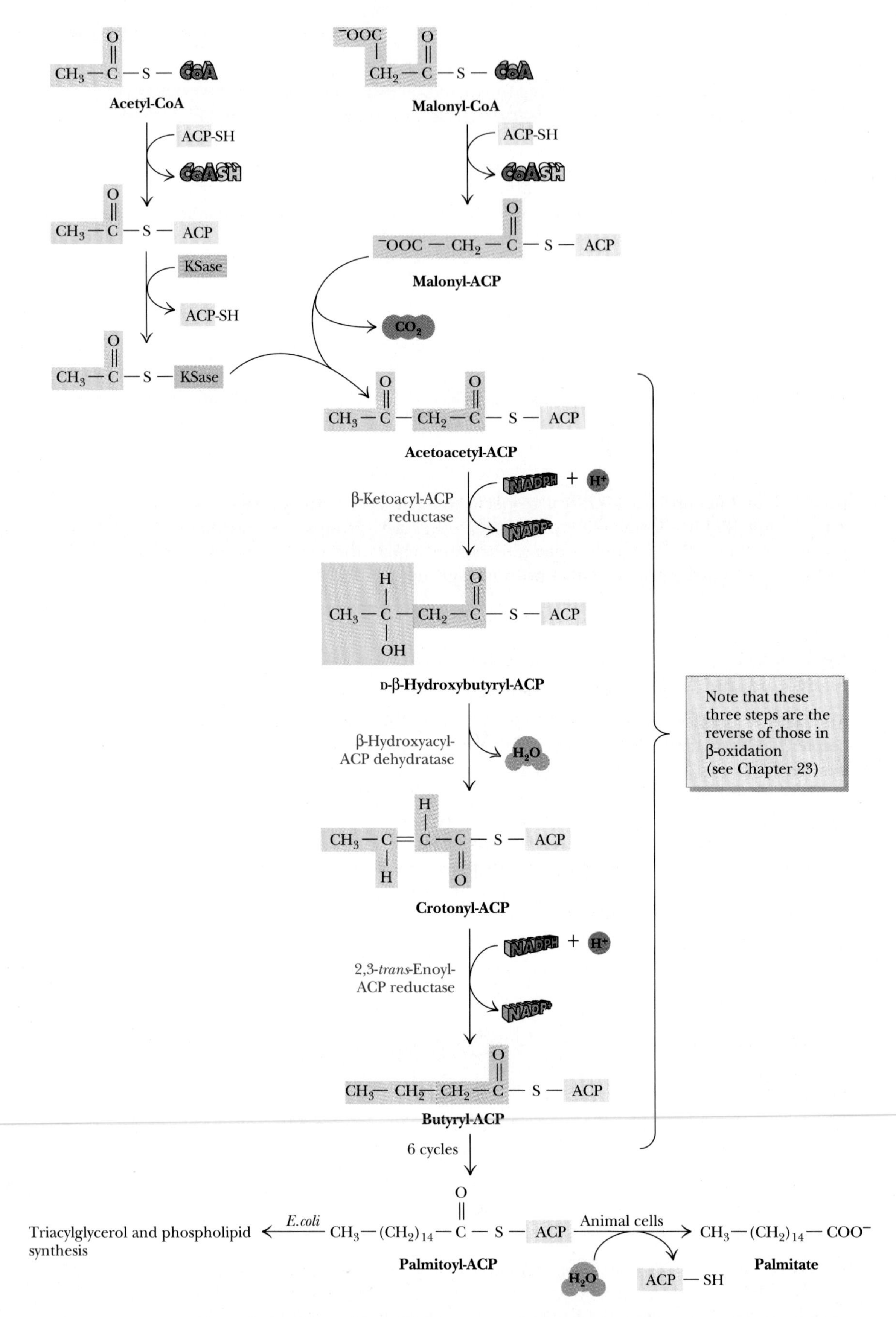

***Figure* 24.7** The pathway of palmitate synthesis from acetyl-CoA and malonyl-CoA. Acetyl and malonyl building blocks are introduced as acyl carrier protein conjugates. Decarboxylation drives the β-ketoacyl-ACP synthase and results in the addition of two-carbon units to the growing chain. Concentrations of free fatty acids are extremely low in most cells, and newly synthesized fatty acids exist primarily as acyl-CoA esters.

ACP, so, indirectly, ATP is responsible for the condensation reaction to form acetoacetyl-ACP. Malonyl-CoA can be viewed as a form of stored energy for driving fatty acid synthesis.

It is also worth noting that the carbon of the carboxyl group that was added to drive this reaction is the one removed by the condensing enzyme. Thus, all the carbons of acetoacetyl-ACP (*and* of the fatty acids to be made) are derived from acetate units of acetyl-CoA.

Reduction of the *β*-Carbonyl Group Follows a Now-Familiar Route

The next three steps—reduction of the *β*-carbonyl group to form a *β*-alcohol, followed by dehydration and reduction to saturate the chain (Figure 24.7)—will look very similar to the fatty acid degradation pathway in reverse. However, there are two crucial differences between fatty acid biosynthesis and fatty acid oxidation (besides the fact that different enzymes are involved): First, the alcohol formed in the first step has the D configuration rather than the L form seen in catabolism, and, second, the reducing coenzyme is NADPH, though NAD^+ and FAD are the oxidants in the catabolic pathway.

The net result of this biosynthetic cycle is the synthesis of a four-carbon unit, a butyryl group, from two smaller building blocks. In the next cycle of the process, this butyryl-ACP condenses with another malonyl-ACP to make a six-carbon *β*-ketoacyl-ACP and CO_2. Subsequent reduction to a *β*-alcohol, dehydration, and another reduction yield a six-carbon saturated acyl-ACP. This cycle continues with the net addition of a two-carbon unit in each turn until the chain is 16 carbons long (Figure 24.7). The *β*-ketoacyl-ACP synthase cannot accommodate larger substrates, so the reaction cycle ends with a 16-carbon chain. Hydrolysis of the C_{16}-acyl-ACP yields a palmitic acid and the free ACP.

In the end, seven malonyl-CoA molecules and one acetyl-CoA yield a palmitate (shown here as palmitoyl-CoA):

$$\text{Acetyl-CoA} + 7\ \text{malonyl-CoA}^- + 14\ \text{NADPH} + 14\ \text{H}^+ \longrightarrow \text{palmitoyl-CoA} + 7\ \text{HCO}_3^- + 14\ \text{NADP}^+ + 7\ \text{CoASH}$$

The formation of seven malonyl-CoA molecules requires

$$7\ \text{Acetyl-CoA} + 7\ \text{HCO}_3^- + 7\ \text{ATP}^{4-} \longrightarrow 7\ \text{malonyl-CoA}^- + 7\ \text{ADP}^{3-} + 7\ \text{P}_i^{2-} + 7\ \text{H}^+$$

Thus, the overall reaction of acetyl-CoA to yield palmitic acid is

$$8\ \text{Acetyl-CoA} + 7\ \text{ATP}^{4-} + 14\ \text{NADPH} + 7\ \text{H}^+ \longrightarrow \text{palmitoyl-CoA} + 14\ \text{NADP}^+ + 7\ \text{CoASH} + 7\ \text{ADP}^{3-} + 7\ \text{P}_i^{2-}$$

Note: These equations are stoichiometric and are charge balanced. See Problem 1 at the end of the chapter for practice in balancing these equations.

Fatty Acid Synthesis in Eukaryotes Occurs on a Multienzyme Complex

In contrast with bacterial and plant systems, the reactions of fatty acid synthesis beyond the acetyl-CoA carboxylase in animal systems are carried out by a special multienzyme complex called **fatty acid synthase (FAS).** In yeast, this 2.4×10^6 D complex contains two different peptide chains, an *α* subunit of 213 kD and a *β* subunit of 203 kD, arranged in an $\alpha_6\beta_6$ dodecamer. The separate enzyme activities associated with each chain are shown in Figure 24.8. In animal systems, FAS is a dimer of identical 250-kD *multifunctional polypeptides*. Studies of the action of proteolytic enzymes on this polypeptide

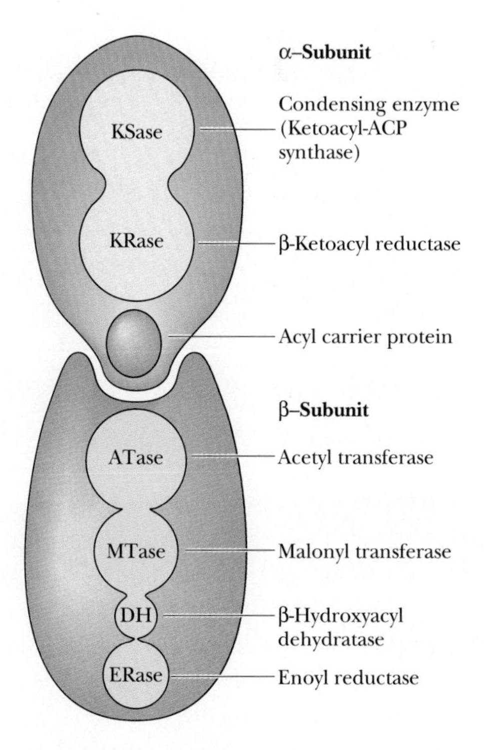

***Figure* 24.8** In yeast, the functional groups and enzyme activities required for fatty acid synthesis are distributed between *α* and *β* subunits.

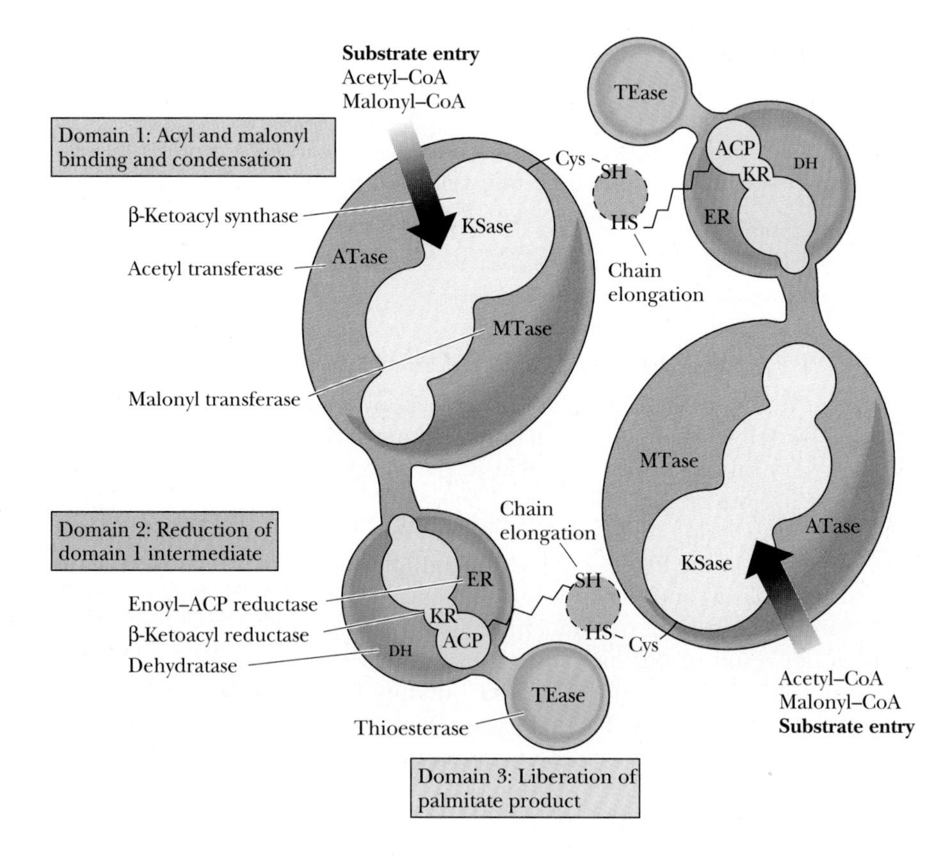

***Figure* 24.9** Fatty acid synthase in animals contains all the functional groups and enzyme activities on a single multifunctional subunit. The active enzyme is a head-to-tail dimer of identical subunits.

(Adapted from Wakil, S. J., Stoops, J. K., and Joshi, V. C., 1983, Annual Review of Biochemistry 52:556.)

have led to a model involving three separate domains joined by flexible connecting sequences (Figure 24.9). The first domain is responsible for the binding of acetyl and malonyl building blocks and for the condensation of these units. This domain includes the acetyl transferase, the malonyl transferase, and the acyl-malonyl ACP condensing enzyme (also known as the β-ketoacyl synthase). The second domain is primarily responsible for the reduction of the intermediate synthesized in domain 1, and contains the acyl carrier protein, the β-ketoacyl reductase, the dehydratase, and the enoyl-ACP reductase. The third domain contains the thioesterase that liberates the product palmitate when the growing acyl chain reaches its limit length of 16 carbons. The close association of activities in this complex permits efficient exposure of intermediates to one active site and then the next. The presence of all these activities on a single polypeptide ensures that the cell will simultaneously synthesize all the enzymes needed for fatty acid synthesis.

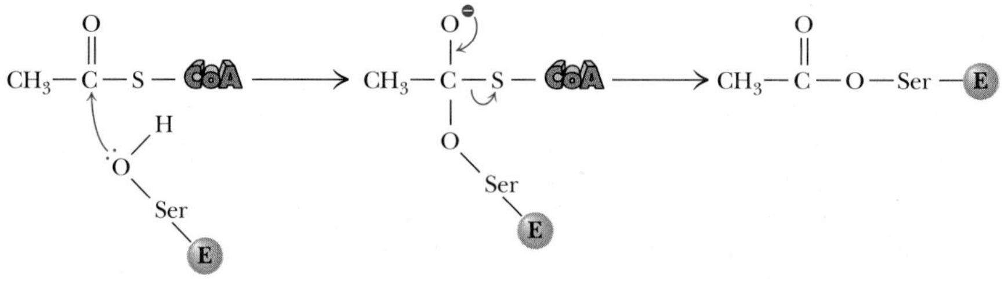

$$CH_3-\overset{O}{\overset{\|}{C}}-S-CoA \;+\; E-Ser-OH \longrightarrow CH_3-\overset{O^-}{\overset{|}{C}}(-O-Ser-E)-S-CoA \longrightarrow CH_3-\overset{O}{\overset{\|}{C}}-O-Ser-E$$

***Figure* 24.10** Acetyl units are covalently linked to a serine residue at the active site of the acetyl transferase in eukaryotes. A similar reaction links malonyl units to the malonyl transferase.

The Mechanism of Fatty Acid Synthase

The first domain of one subunit of the fatty acid synthase interacts with the second and third domains of the other subunit; that is, the subunits are arranged in a head-to-tail fashion (Figure 24.9). The first step in the fatty acid synthase reaction is the formation of an acetyl-O-enzyme intermediate between the acetyl group of an acetyl-CoA and an active-site serine of the acetyl transferase (Figure 24.10). In a similar manner, a malonyl-O-enzyme intermediate is formed between malonyl-CoA and a serine residue of the malonyl transferase. The acetyl group on the acetyl transferase is then transferred to the —SH group of the acyl carrier protein, as shown in Figure 24.11. The next step is the transfer of the acetyl group to the β-ketoacyl-ACP synthase, or condensing enzyme. This frees the acyl carrier protein to acquire the malonyl group from the malonyl transferase. The next step is the condensation reaction, in which decarboxylation facilitates the concerted attack of the remaining two-carbon unit of the acyl carrier protein at the carbonyl carbon of the acetate group on the condensing enzyme. Note that decarboxylation forms a transient, highly nucleophilic carbanion which can attack the acetate group.

The next three steps—reduction of the carbonyl to an alcohol, dehydration to yield a *trans*-α,β double bond, and reduction to yield a saturated chain—

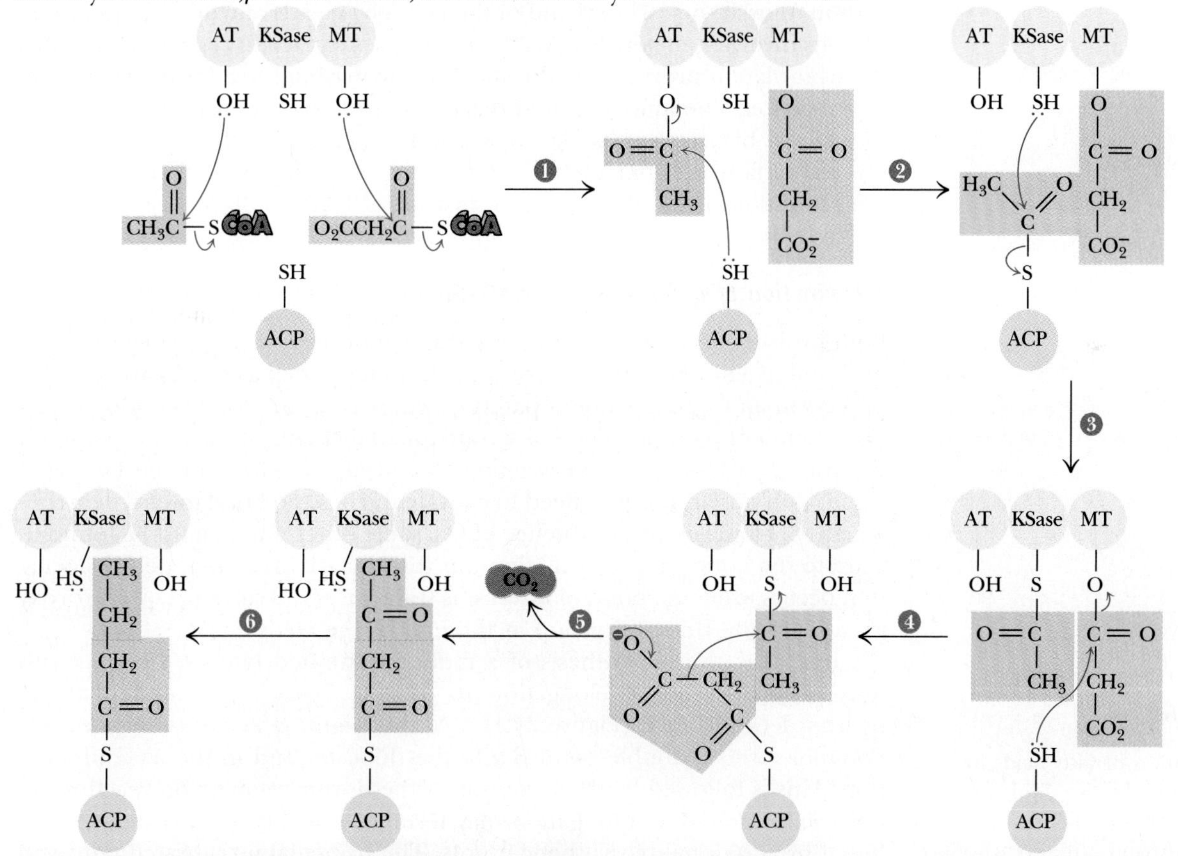

***Figure* 24.11** The mechanism of the fatty acyl synthase reaction in eukaryotes. (1) Acetyl and malonyl groups are loaded onto acetyltransferase and malonyl transferase, respectively. (2) The acetate unit that forms the base of the nascent chain is transferred first to the acyl carrier protein domain and (3) then to the β-ketoacyl synthase. (4) Attack by ACP on the carbonyl carbon of a malonyl unit on malonyl transferase forms malonyl-ACP. (5) Decarboxylation leaves a reactive, transient carbanion that can attack the carbonyl carbon of the acetyl group on the β-ketoacyl synthase. (6) Reduction of the keto group, dehydration, and saturation of the resulting double bond follow, leaving an acyl group on ACP, and steps 3 through 6 repeat to lengthen the nascent chain.

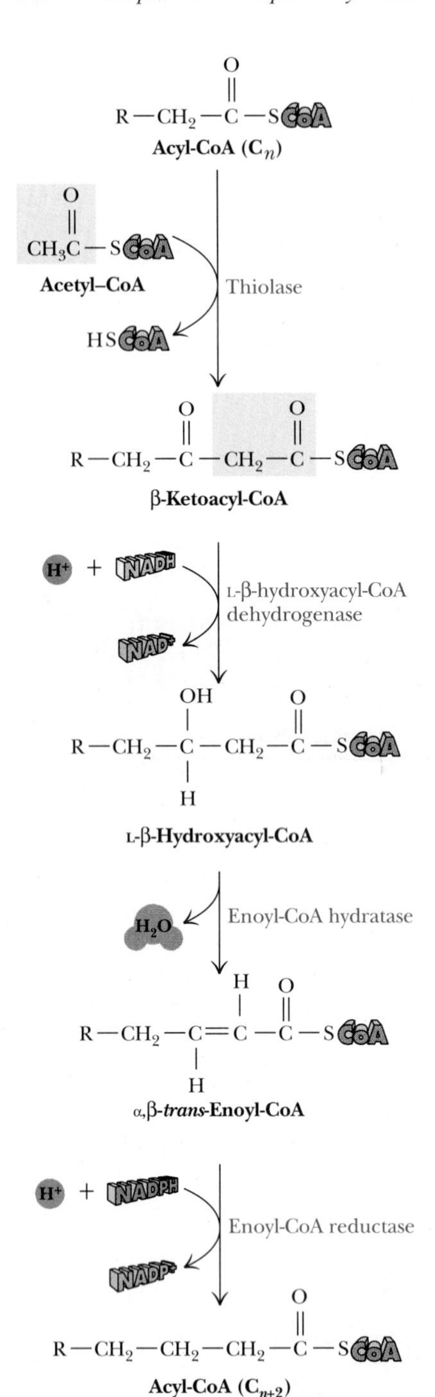

***Figure* 24.12** Elongation of fatty acids in mitochondria is initiated by the thiolase reaction. The β-ketoacyl intermediate thus formed undergoes the same three reactions (in reverse order) that are the basis of β-oxidation of fatty acids. Reduction of the β-keto group is followed by dehydration to form a double bond. Reduction of the double bond yields a fatty acyl-CoA which is elongated by two carbons. Note that the reducing coenzyme for the second step is NADH, whereas the reductant for the fourth step is NADPH.

are identical to those occurring in bacteria and plants (Figure 24.7) and resemble the reverse of the reactions of fatty acid oxidation (and the conversion of succinate to oxaloacetate in the TCA cycle). This synthetic cycle now repeats until the growing chain is 16 carbons long. It is then released by the thioesterase domain on the synthase. The amino acid sequence of the thioesterase domain is homologous with serine proteases; the enzyme has an active-site serine that carries out nucleophilic attack on the carbonyl carbon of the fatty acyl thioester to be cleaved.

Further Processing of C_{16} Fatty Acids

Additional Elongation

As seen already, palmitate is the primary product of the fatty acid synthase. Cells synthesize many other fatty acids. Shorter chains are easily made if the chain is released before reaching 16 carbons in length. Longer chains are made through special elongation reactions, which occur both in the mitochondria and at the surface of the endoplasmic reticulum. The ER reactions are actually quite similar to those we have just discussed: addition of two-carbon units at the carboxyl end of the chain by means of oxidative decarboxylations involving malonyl-CoA. As was the case for the fatty acid synthase, this decarboxylation provides the thermodynamic driving force for the condensation reaction. The mitochondrial reactions involve addition (and subsequent reduction) of acetyl units. These reactions (Figure 24.12) are essentially a reversal of fatty acid oxidation, with the exception that NADPH is utilized in the saturation of the double bond, instead of $FADH_2$.

Introduction of a Single cis *Double Bond*

Both prokaryotes and eukaryotes are capable of introducing a single *cis* double bond in a newly synthesized fatty acid. Bacteria such as *E. coli* carry out this process in an O_2-independent pathway, whereas eukaryotes have adopted an O_2-dependent pathway. There is a fundamental chemical difference between the two. The O_2-dependent reaction can occur anywhere in the fatty acid chain, with no (additional) need to activate the desired bond toward dehydrogenation. However, in the absence of O_2, some other means must be found to activate the bond in question. Thus, in the bacterial reaction, dehydrogenation occurs while the bond of interest is still near the β-carbonyl or β-hydroxy group and the thioester group at the end of the chain.

In *E. coli,* the biosynthesis of a monounsaturated fatty acid begins with four normal cycles of elongation to form a 10-carbon intermediate, β-hydroxydecanoyl-ACP (Figure 24.13). At this point, *β-hydroxydecanoyl thioester dehydrase* forms a double bond β,γ to the thioester and in the *cis* configuration. This is followed by three rounds of the normal elongation reactions to form *palmitoleoyl-ACP.* Elongation may terminate at this point or may be followed by additional biosynthetic events. The principal unsaturated fatty acid in *E. coli, cis-vaccenic acid,* is formed by an additional elongation step, using palmitoleoyl-ACP as a substrate.

Unsaturation Reactions Occur in Eukaryotes in the Middle of an Aliphatic Chain

The addition of double bonds to fatty acids in eukaryotes does not occur until the fatty acyl chain has reached its full length (usually 16 to 18 carbons). Dehydrogenation of stearoyl-CoA occurs in the middle of the chain despite the absence of any useful functional group on the chain to facilitate activation:

$$CH_3{-}(CH_2)_{16}CO{-}SCoA \longrightarrow CH_3{-}(CH_2)_7CH{=}CH(CH_2)_7CO{-}SCoA$$

This impressive reaction is catalyzed by *stearoyl-CoA desaturase,* a 53-kD enzyme containing a nonheme iron center. NADH and oxygen (O_2) are required, as are two other proteins: *cytochrome b_5 reductase* (a 43-kD flavoprotein) and *cytochrome b_5* (16.7 kD). All three proteins are associated with the endoplasmic reticulum membrane. Cytochrome b_5 reductase transfers a pair of electrons from NADH through FAD to cytochrome b_5 (Figure 24.14). Oxidation of reduced cytochrome b_5 is coupled to reduction of Fe^{3+} to Fe^{2+} in the desaturase. The Fe^{3+} accepts a pair of electrons (one at a time in a cycle) from cytochrome b_5 and creates a *cis* double bond at the 9,10-position of the stearoyl-CoA substrate. O_2 is the terminal electron acceptor in this fatty acyl desaturation cycle. Note that two water molecules are made, which means that four electrons are transferred overall. Two of these come through the reaction sequence from NADH, and two come from the fatty acyl substrate that is being dehydrogenated.

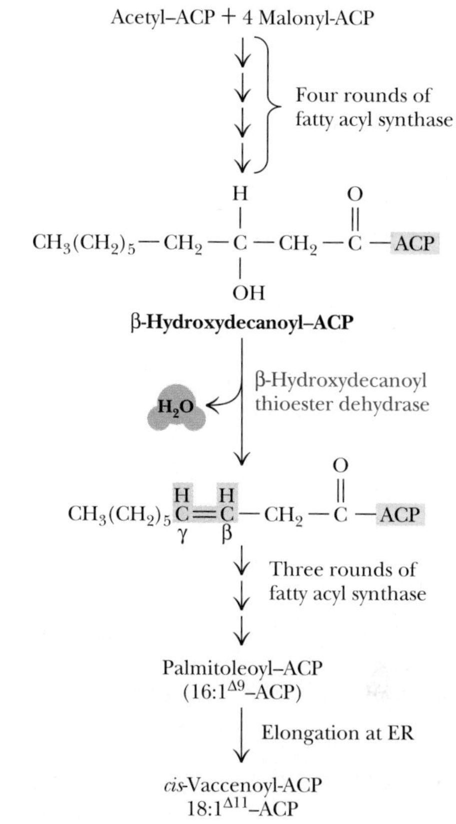

Acetyl–ACP + 4 Malonyl-ACP

↓ Four rounds of fatty acyl synthase

$CH_3(CH_2)_5{-}CH_2{-}CH(OH){-}CH_2{-}C(=O){-}ACP$

β-Hydroxydecanoyl–ACP

↓ β-Hydroxydecanoyl thioester dehydrase (H_2O)

$CH_3(CH_2)_5\overset{H}{C}{=}\overset{H}{C}{-}CH_2{-}C(=O){-}ACP$ (γ, β)

↓ Three rounds of fatty acyl synthase

Palmitoleoyl–ACP
($16{:}1^{\Delta 9}$–ACP)

↓ Elongation at ER

cis-Vaccenoyl-ACP
$18{:}1^{\Delta 11}$–ACP

***Figure* 24.13** Double bonds are introduced into the growing fatty acid chain in *E. coli* by specific dehydrases. Palmitoleoyl-ACP is synthesized by a sequence of reactions involving four rounds of chain elongation, followed by double bond insertion by *β*-hydroxydecanoyl thioester dehydrase and three additional elongation steps. Another elongation cycle produces *cis*-vaccenic acid.

The Unsaturation Reaction May Be Followed by Chain Elongation

Additional chain elongation can occur following this single desaturation reaction. The oleoyl-CoA produced can be elongated by two carbons to form a 20:1 *cis*-Δ^{11} fatty acyl-CoA. If the starting fatty acid is palmitate, reactions similar to the preceding scheme yield palmitoleoyl-CoA (16:1 *cis*-Δ^9), which subsequently can be elongated to yield *cis*-vaccenic acid (18:1 *cis*-Δ^{11}). Similarly, C_{16} and C_{18} fatty acids can be elongated to yield C_{22} and C_{24} fatty acids, such as are often found in sphingolipids.

Biosynthesis of Polyunsaturated Fatty Acids

Organisms differ with respect to formation, processing, and utilization of polyunsaturated fatty acids. *Escherichia coli,* for example, does not have any polyunsaturated fatty acids. Eukaryotes *do* synthesize a variety of polyunsaturated fatty acids, certain organisms more than others. For example, plants manufacture double bonds between the Δ^9 and the methyl end of the chain, but mammals cannot. Plants readily desaturate oleic acid at the 12-position (to give linoleic acid) or at both the 12- and 15-positions (producing linolenic

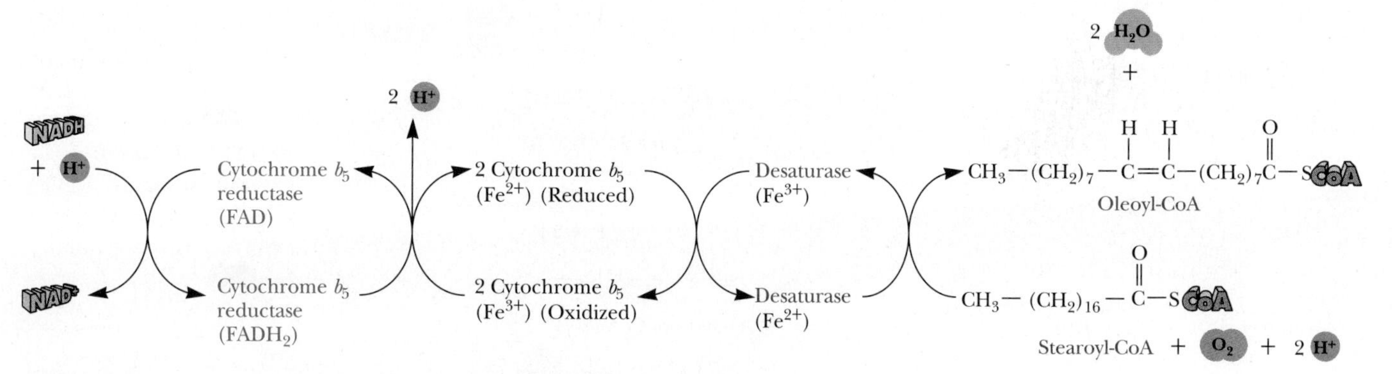

***Figure* 24.14** The conversion of stearoyl-CoA to oleoyl-CoA in eukaryotes is catalyzed by stearoyl-CoA desaturase in a reaction sequence that also involves cytochrome b_5 and cytochrome b_5 reductase. Two electrons are passed from NADH through the chain of reactions as shown, and two electrons are also derived from the fatty acyl substrate.

acid). Mammals require polyunsaturated fatty acids, but must acquire them in their diet. As such, they are referred to as **essential fatty acids.** On the other hand, mammals *can* introduce double bonds between the double bond at the 8- or 9-position and the carboxyl group. Enzyme complexes in the endoplasmic reticulum desaturate the 5-position, provided a double bond exists at the 8-position, and form a double bond at the 6-position if one already exists at the 9-position. Thus, oleate can be unsaturated at the 6,7-position to give an 18:2 *cis*-Δ^6,Δ^9 fatty acid.

Arachidonic Acid Is Synthesized from Linoleic Acid by Mammals

Mammals can add additional double bonds to unsaturated fatty acids in their diets. Their ability to make arachidonic acid from linoleic acid is one example (Figure 24.15). This fatty acid is the precursor for prostaglandins and other biologically active derivatives such as leukotrienes. Synthesis involves formation of a linoleoyl ester of CoA from dietary linoleic acid, followed by introduction of a double bond at the 6-position. The triply unsaturated product is then elongated (by malonyl-CoA with a decarboxylation step) to yield a 20-carbon fatty acid with double bonds at the 8-, 11-, and 14-positions. A second desaturation reaction at the 5-position followed by an *acyl-CoA synthetase* reaction (Chapter 23) liberates the product, a 20-carbon fatty acid with double bonds at the 5-, 8-, 11-, and 14-positions.

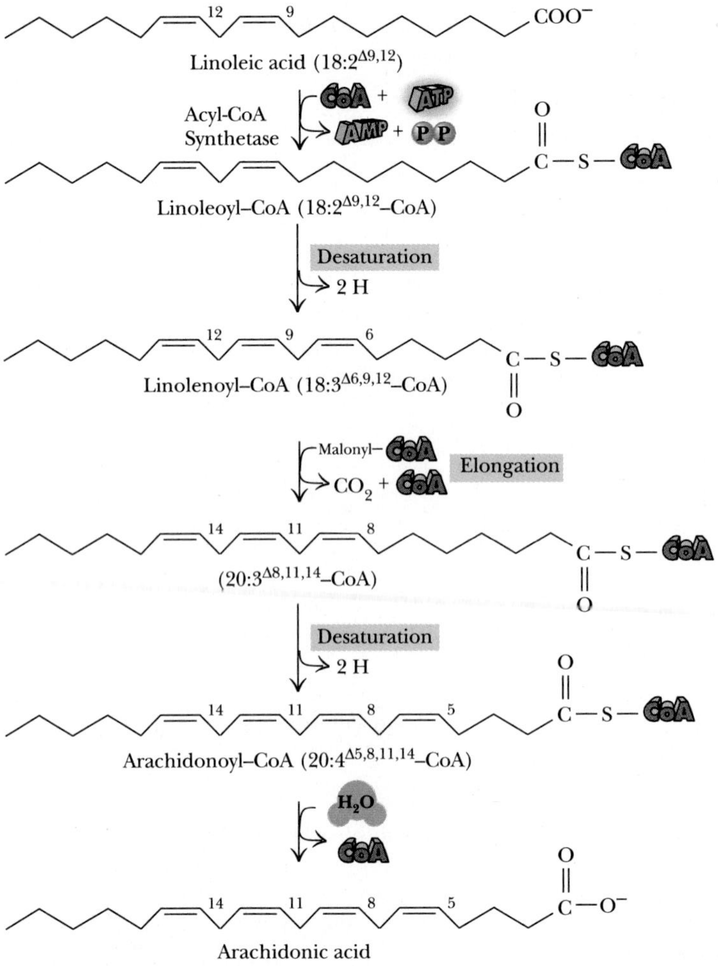

***Figure* 24.15** Arachidonic acid is synthesized from linoleic acid in eukaryotes. This is the only means by which animals can synthesize fatty acids with double bonds at positions beyond C-9.

Regulatory Control of Fatty Acid Metabolism—An Interplay of Allosteric Modifiers and Phosphorylation–Dephosphorylation Cycles

The control and regulation of fatty acid synthesis is intimately related to regulation of fatty acid breakdown, glycolysis, and the TCA cycle. Acetyl-CoA is an important intermediate metabolite in all these processes. In these terms, it is easy to appreciate the interlocking relationships in Figure 24.16. Malonyl-CoA can act to prevent fatty acyl-CoA derivatives from entering the mitochondria by inhibiting the carnitine acyltransferase that is responsible for this transport. In this way, when fatty acid synthesis is turned on (as signaled by higher levels of malonyl-CoA), β-oxidation is inhibited. As we pointed out earlier, citrate is an important allosteric activator of acetyl-CoA carboxylase, and fatty acyl-CoAs are inhibitors. The degree of inhibition is proportional to the chain length of the fatty acyl-CoA; longer chains show a higher affinity for the allosteric inhibition site on acetyl-CoA carboxylase. Palmitoyl-CoA, stearoyl-CoA, and arachidyl-CoA are the most potent inhibitors of the carboxylase.

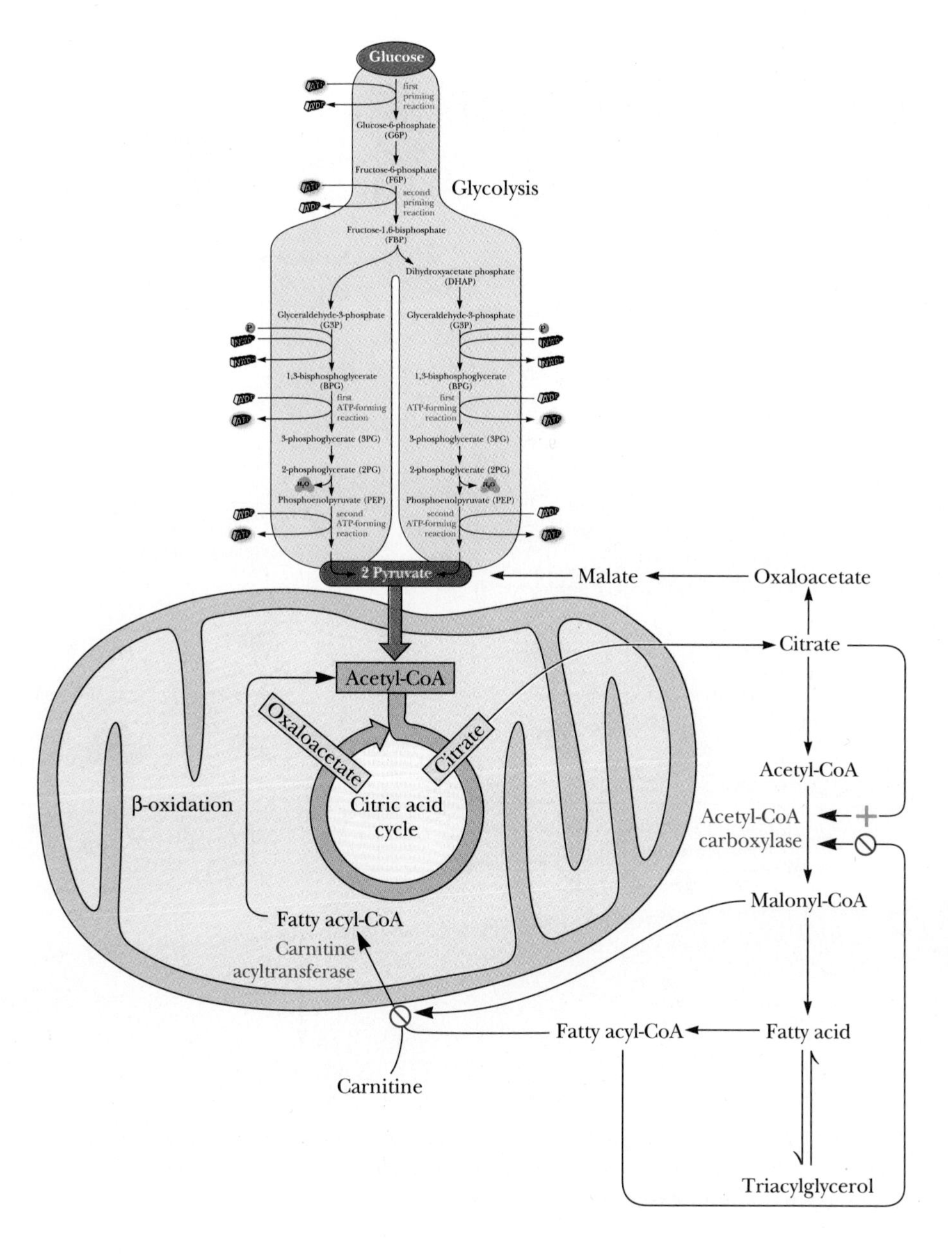

Figure **24.16** Regulation of fatty acid synthesis and regulation of fatty acid oxidation are coupled as shown. Malonyl-CoA, produced during fatty acid synthesis, inhibits the uptake of fatty acylcarnitine (and thus fatty acid oxidation) by mitochondria. When fatty acyl-CoA levels rise, fatty acid synthesis is inhibited and fatty acid oxidation activity increases. Rising citrate levels (which reflect an abundance of acetyl-CoA) similarly signal the initiation of fatty acid synthesis.

Hormonal Signals Regulate ACC and Fatty Acid Biosynthesis

As described earlier, citrate activation and palmitoyl-CoA inhibition of acetyl-CoA carboxylase are strongly dependent on the phosphorylation state of the enzyme. This provides a crucial connection to hormonal regulation. Many of the enzymes that act to phosphorylate acetyl-CoA carboxylase (Figure 24.4) are controlled by hormonal signals. Glucagon is a good example (Figure 24.17). As noted in Chapter 21, glucagon binding to membrane receptors activates an intracellular cascade involving activation of adenylyl cyclase. Cyclic AMP produced by the cyclase activates a protein kinase, which then phosphorylates acetyl-CoA carboxylase. Unless citrate levels are high, phosphorylation causes inhibition of fatty acid biosynthesis. The carboxylase (and fatty acid synthesis) can be reactivated by a specific phosphatase, which dephosphorylates the carboxylase. Also indicated in Figure 24.17 is the simultaneous activation by glucagon of triacylglycerol lipases, which hydrolyze triacylglycer-

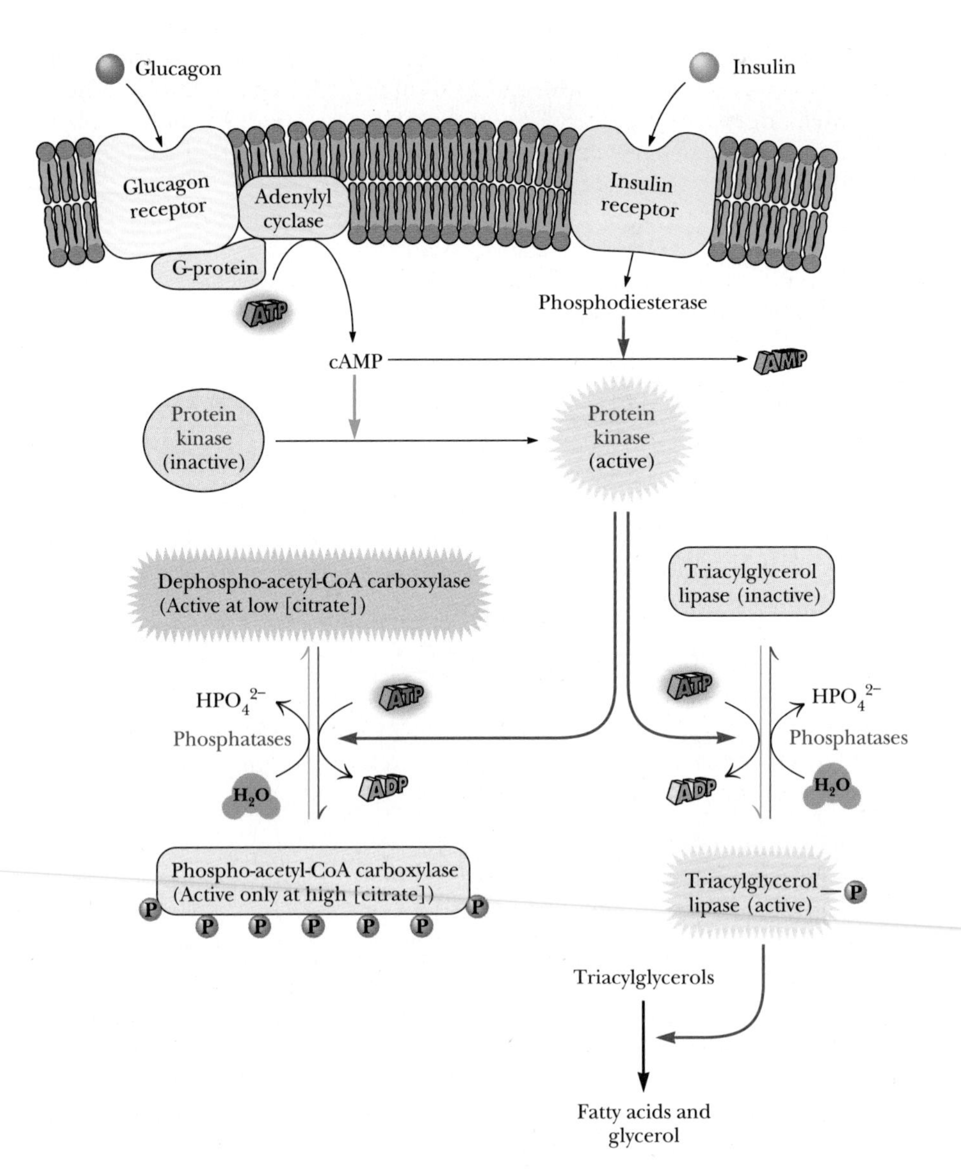

***Figure* 24.17** Hormonal signals regulate fatty acid synthesis, primarily through actions on acetyl-CoA carboxylase. Availability of fatty acids also depends upon hormonal activation of triacylglycerol lipase.

ols, releasing fatty acids for β-oxidation. Both the inactivation of acetyl-CoA carboxylase and the activation of triacylglycerol lipase are counteracted by insulin, whose receptor acts to stimulate a phosphodiesterase that converts cAMP to AMP.

24.2 Biosynthesis of Complex Lipids

Complex lipids consist of backbone structures to which fatty acids are covalently bound. Principal classes include the **glycerolipids,** for which glycerol is the backbone, and **sphingolipids,** which are built on a sphingosine backbone. The two major classes of glycerolipids are **glycerophospholipids** and **triacylglycerols.** The **phospholipids,** which include both glycerophospholipids and sphingomyelin, are crucial components of membrane structure. They are also precursors of hormones such as the *eicosanoids* (e.g., *prostaglandins*) and signal molecules, such as the breakdown products of *phosphatidylinositol.*

Different organisms possess greatly different complements of lipids and therefore invoke somewhat different lipid biosynthetic pathways. For example, sphingolipids and triacylglycerols are produced only in eukaryotes. In contrast, bacteria usually have rather simple lipid compositions. Phosphatidylethanolamine accounts for at least 75% of the phospholipids in *Escherichia coli,* with phosphatidylglycerol and cardiolipin accounting for most of the rest. *E. coli* membranes possess no phosphatidylcholine, phosphatidylinositol, sphingolipids, or cholesterol. On the other hand, some bacteria (such as *Pseudomonas*) can synthesize phosphatidylcholine, for example. In this section, we will consider some of the pathways for the synthesis of glycerolipids, sphingolipids, and the eicosanoids, which are derived from phospholipids.

Glycerolipid Biosynthesis

A common pathway operates in nearly all organisms for the synthesis of **phosphatidic acid,** the precursor to other glycerolipids. *Glycerokinase* catalyzes the phosphorylation of glycerol to form glycerol-3-phosphate, which is then acylated at both the 1- and 2-positions to yield phosphatidic acid (Figure 24.18). The first acylation, at position 1, is catalyzed by *glycerol-3-phosphate acyltransferase,* an enzyme that in most organisms is specific for saturated fatty acyl groups. Eukaryotic systems can also utilize *dihydroxyacetone phosphate* as a starting point for synthesis of phosphatidic acid (Figure 24.18). Again a specific acyltransferase adds the first acyl chain, followed by reduction of the backbone keto group by *acyldihydroxyacetone phosphate reductase,* using NADPH as the reductant. Alternatively, dihydroxyacetone phosphate can be reduced to glycerol-3-phosphate by *glycerol-3-phosphate dehydrogenase.*

Eukaryotes Synthesize Glycerolipids from CDP-Diacylglycerol or Diacylglycerol

In eukaryotes, phosphatidic acid is converted directly either to diacylglycerol or to *cytidine diphosphodiacylglycerol* (or simply *CDP-diacylglycerol;* Figure 24.19). From these two precursors, all other glycerophospholipids in eukaryotes are derived. Diacylglycerol is a precursor for synthesis of triacylglycerol, phosphatidylethanolamine, and phosphatidylcholine. Triacylglycerol is synthesized mainly in adipose tissue, liver, and intestines and serves as the principal energy storage molecule in eukaryotes. Triacylglycerol biosynthesis in liver and adipose tissue occurs via *diacylglycerol acyltransferase,* an enzyme bound to the cytoplasmic face of the endoplasmic reticulum. A different route is used, however, in intestines. Recall (Figure 23.3) that triacylglycerols from the diet

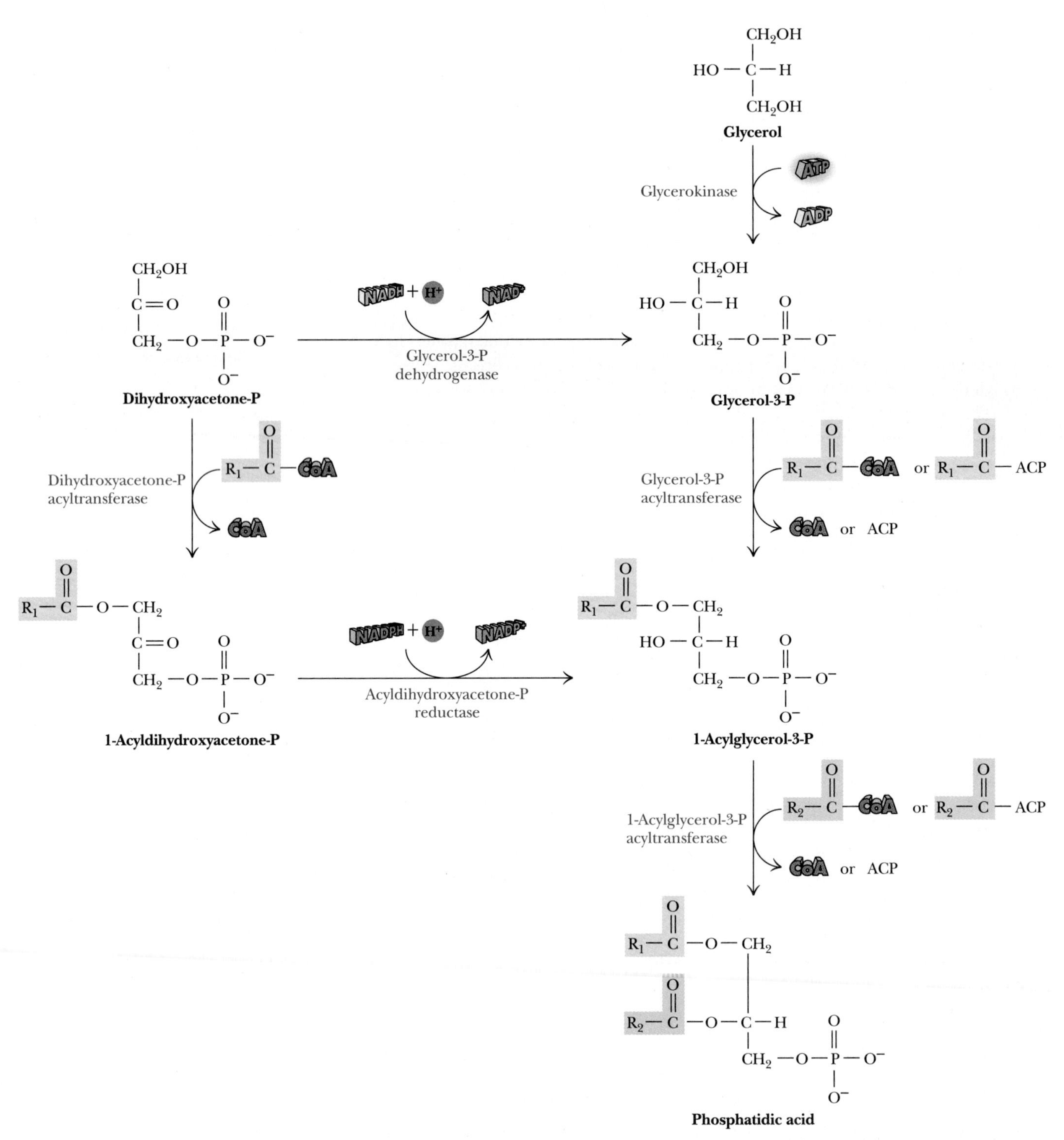

***Figure* 24.18** Synthesis of glycerolipids in eukaryotes begins with the formation of phosphatidic acid, which may be formed from dihydroxyacetone phosphate or glycerol, as shown.

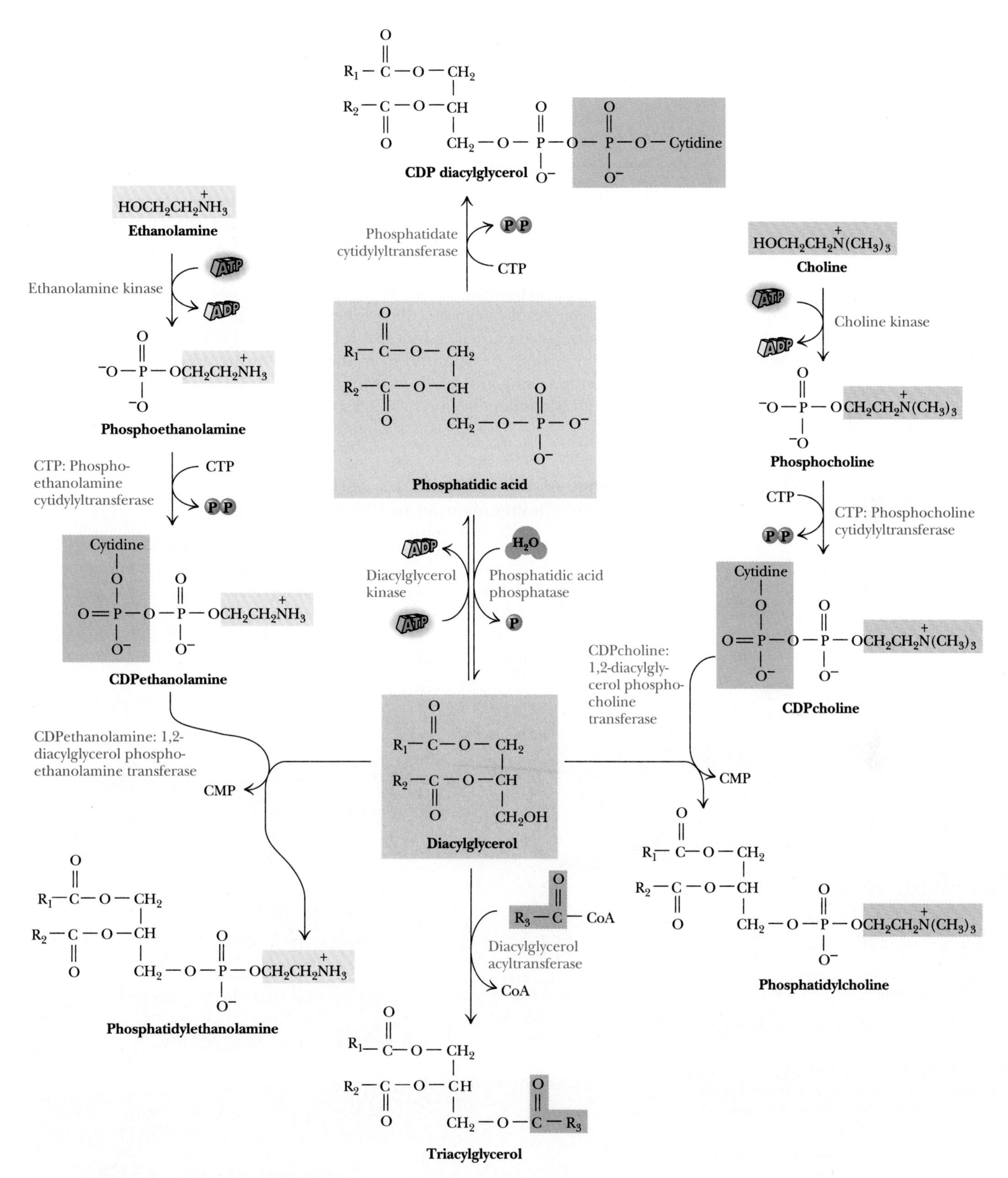

***Figure* 24.19** Diacylglycerol and CDP-diacylglycerol are the principal precursors of glycerolipids in eukaryotes. Phosphatidylethanolamine and phosphatidylcholine are formed by reaction of diacylglycerol with CDP-ethanolamine and with CDP-choline, respectively.

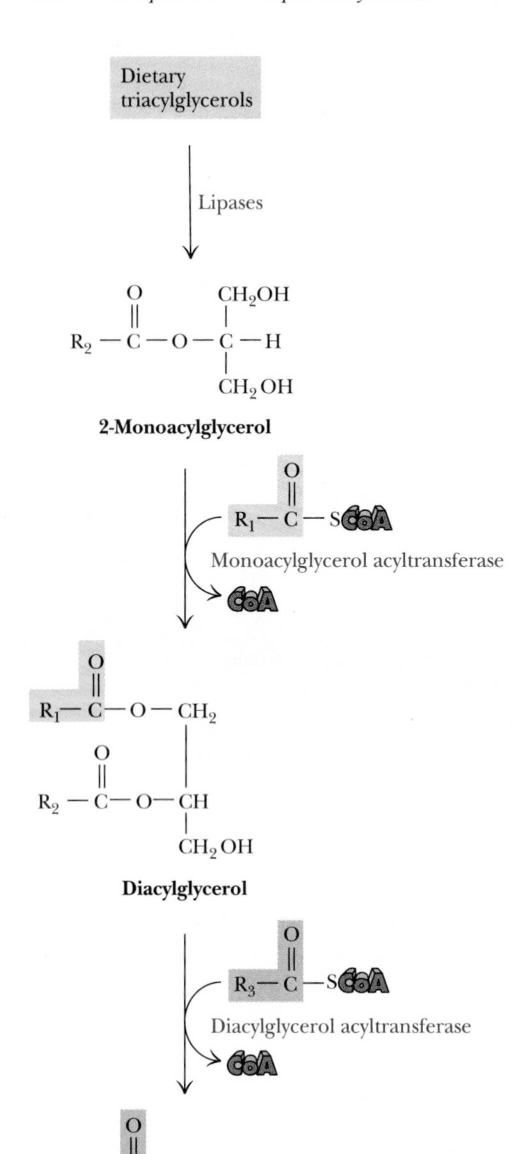

Figure **24.20** Triacylglycerols are formed primarily by the action of acyltransferases on mono- and diacylglycerol. Acyltransferase in *E. coli* is an integral membrane protein (83 kD) and can utilize either fatty acyl-CoAs or acylated acyl carrier proteins as substrates. It shows a particular preference for palmitoyl groups. Eukaryotic acyltransferases use only fatty acyl-CoA molecules as substrates.

are broken down to 2-monoacylglycerols by specific lipases. Acyltransferases then acylate 2-monoacylglycerol to produce new triacylglycerols (Figure 24.20).

Phosphatidylethanolamine Is Synthesized from Diacylglycerol and CDP-Ethanolamine

Phosphatidylethanolamine synthesis begins with phosphorylation of ethanolamine to form phosphoethanolamine (Figure 24.19). The next reaction involves transfer of a cytidylyl group from CTP to form CDP-ethanolamine and pyrophosphate. As always, PP_i hydrolysis drives this reaction forward. A specific phosphoethanolamine transferase then links phosphoethanolamine to the diacylglycerol backbone. Biosynthesis of phosphatidylcholine is entirely analogous, since animals synthesize it directly. All of the choline utilized in this pathway must be acquired from the diet. On the other hand, yeast, certain bacteria, and animal livers can convert phosphatidylethanolamine to phosphatidylcholine by methylation reactions involving S-adenosylmethionine.

Exchange of Ethanolamine for Serine Converts Phosphatidylethanolamine to Phosphatidylserine

Mammals synthesize phosphatidylserine (PS) in a calcium ion–dependent reaction involving aminoalcohol exchange (Figure 24.21). The enzyme catalyzing this reaction is associated with the endoplasmic reticulum and will accept phosphatidylethanolamine (PE) and other phospholipid substrates. A mitochondrial PS decarboxylase can subsequently convert PS to PE. No other pathway converting serine to ethanolamine has been found.

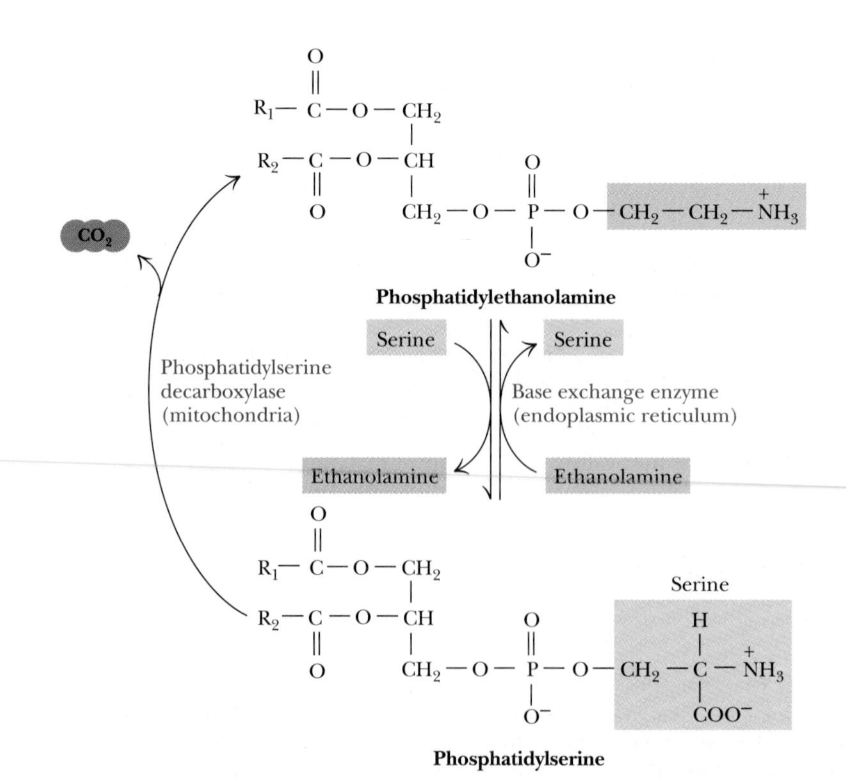

Figure **24.21** The interconversion of phosphatidylethanolamine and phosphatidylserine in mammals.

Eukaryotes Synthesize Other Phospholipids via CDP-Diacylglycerol

Eukaryotes also use CDP-diacylglycerol, derived from phosphatidic acid, as a precursor for several other important phospholipids, including phosphatidylinositol (PI), phosphatidylglycerol (PG), and cardiolipin (Figure 24.22). PI accounts for only about 2 to 8% of the lipids in most animal membranes, but breakdown products of PI, including inositol-1,4,5-trisphosphate and diacylglycerol, are second messengers in a vast array of cellular signaling processes.

Dihydroxyacetone Phosphate Is a Precursor to the Plasmalogens

Certain glycerophospholipids possess alkyl or alkenyl ether groups at the 1-position in place of an acyl ester group. These glyceroether phospholipids are synthesized from dihydroxyacetone phosphate (Figure 24.23). Acylation of dihydroxyacetone phosphate (DHAP) is followed by an exchange reaction, in which the acyl group is removed as a carboxylic acid and a long-chain alcohol adds to the 1-position. This long-chain alcohol is derived from the corresponding acyl-CoA by means of an acyl-CoA reductase reaction involving oxidation of two molecules of NADH. The *2-keto* group of the DHAP backbone is then reduced to an alcohol, followed by acylation. The resulting 1-alkyl-2-acylglycero-3-phosphate can react in a manner similar to phosphatidic acid to produce ether analogs of phosphatidylcholine, phosphatidylethanolamine, and so forth (Figure 24.23). In addition, specific *desaturase* enzymes associated with the endoplasmic reticulum can desaturate the alkyl ether chains of these lipids as shown. The products, which contain α,β-unsaturated ether-linked chains at the C-1 position, are **plasmalogens;** they are abundant in cardiac tissue and in the central nervous system. The desaturases catalyzing these reactions are distinct from but similar to those which introduce unsaturations in fatty acyl-CoAs. These enzymes use cytochrome b_5 as a cofactor, NADH as a reductant, and O_2 as a terminal electron acceptor.

Platelet Activating Factor

A particularly interesting ether phospholipid with unusual physiological properties has recently been characterized. As shown in Figure 24.24, **1-alkyl-2-acetylglycerophosphocholine,** also known as **platelet activating factor,** possesses an alkyl ether at C-1 and an acetyl group at C-2. The very short chain at C-2 makes this molecule much more water-soluble than typical glycerolipids. Platelet activating factor displays a dramatic ability to dilate blood vessels (and thus reduce blood pressure in hypertensive animals) and to aggregate platelets.

Sphingolipid Biosynthesis

Sphingolipids, ubiquitous components of eukaryotic cell membranes, are present at high levels in neural tissues. The myelin sheath that insulates nerve axons is particularly rich in sphingomyelin and other related lipids. Prokaryotic organisms normally do not contain sphingolipids. Sphingolipids are built upon sphingosine backbones rather than glycerol. The initial reaction, which involves condensation of serine and palmitoyl-CoA with release of bicarbonate, is catalyzed by *3-ketosphinganine synthase,* a PLP-dependent enzyme (Figure 24.25). Reduction of the ketone product to form *sphinganine* is catalyzed by

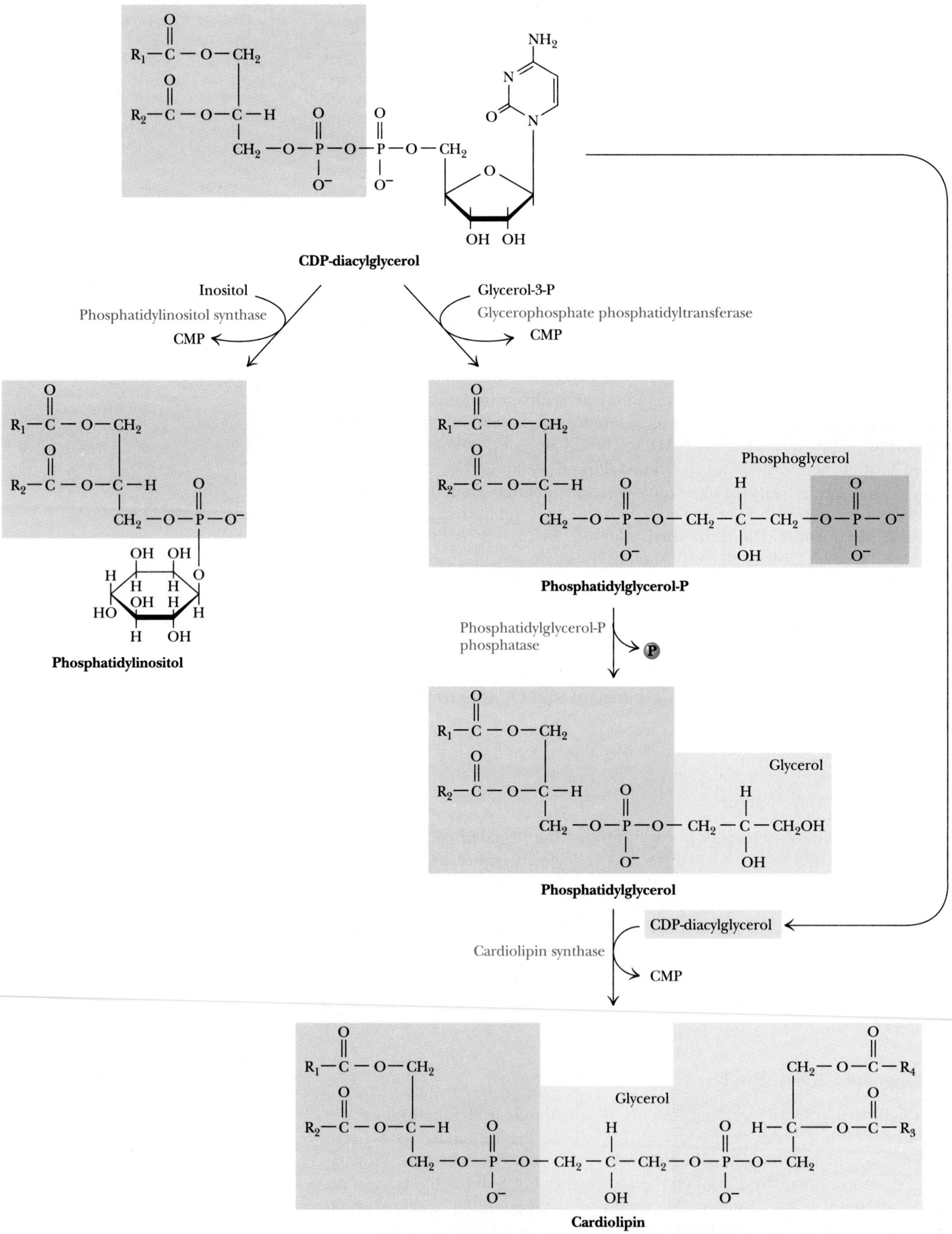

CDP-diacylglycerol

Inositol
Phosphatidylinositol synthase
CMP

Glycerol-3-P
Glycerophosphate phosphatidyltransferase
CMP

Phosphatidylinositol

Phosphoglycerol

Phosphatidylglycerol-P

Phosphatidylglycerol-P phosphatase
P

Glycerol

Phosphatidylglycerol

CDP-diacylglycerol
Cardiolipin synthase
CMP

Glycerol

Cardiolipin

***Figure* 24.22** CDP-diacylglycerol is a precursor of phosphatidylinositol, phosphatidylglycerol, and cardiolipin in eukaryotes.

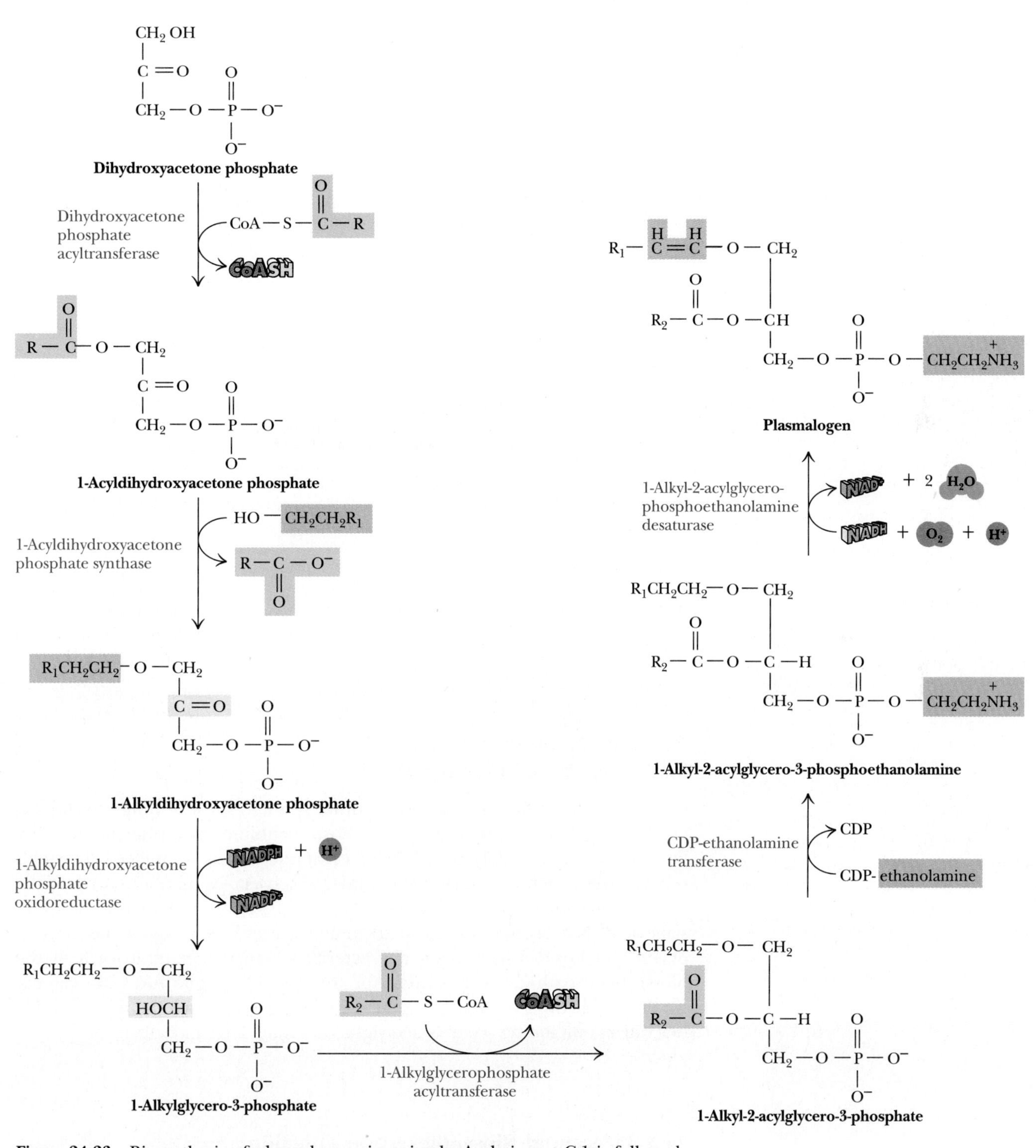

Figure **24.23** Biosynthesis of plasmalogens in animals. Acylation at C-1 is followed by exchange of the acyl group for a long-chain alcohol. Reduction of the keto group at C-2 is followed by transferase reactions, which add an acyl group at C-2 and a polar head-group moiety, and a desaturase reaction that forms a double bond in the alkyl chain. The first two enzymes are of cytoplasmic origin, and the last transferase is located at the endoplasmic reticulum.

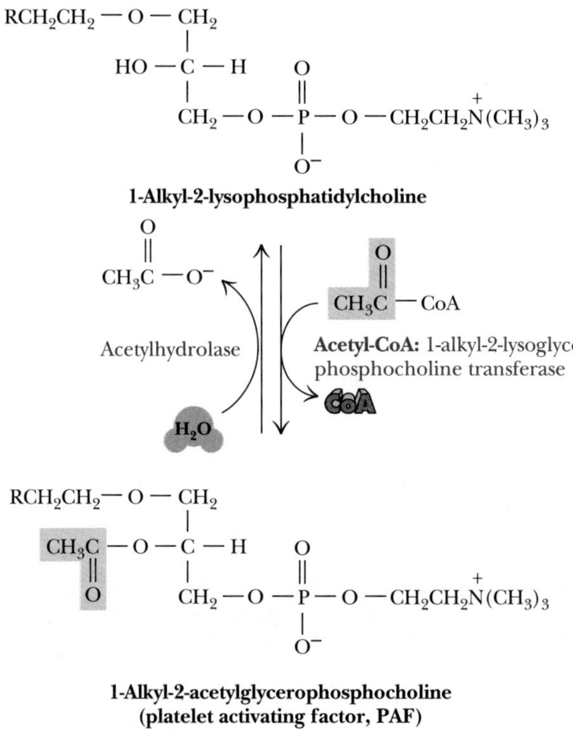

$$RCH_2CH_2-O-CH_2-CH(OH)-CH_2-O-P(=O)(O^-)-O-CH_2CH_2\overset{+}{N}(CH_3)_3$$

1-Alkyl-2-lysophosphatidylcholine

$CH_3C(=O)-O^-$ Acetylhydrolase H_2O

$CH_3C(=O)-CoA$ **Acetyl-CoA:** 1-alkyl-2-lysoglycero-phosphocholine transferase CoA

$$RCH_2CH_2-O-CH_2-CH(O-C(=O)CH_3)-CH_2-O-P(=O)(O^-)-O-CH_2CH_2\overset{+}{N}(CH_3)_3$$

1-Alkyl-2-acetylglycerophosphocholine (platelet activating factor, PAF)

***Figure* 24.24** Platelet activating factor, formed from 1-alkyl-2-lysophosphatidylcholine by acetylation at C-2, is degraded by the action of acetylhydrolase.

3-ketosphinganine reductase, with NADPH as a reactant. In the next step, sphinganine is acylated to form N-acyl sphinganine, which is *then* desaturated to form ceramide. Sphingosine itself does not appear to be an intermediate in this pathway in mammals.

Ceramide Is the Precursor for Other Sphingolipids and Cerebrosides

Ceramide is the building block for all other sphingolipids. Sphingomyelin, for example, is produced by transfer of phosphocholine from phosphatidylcholine (Figure 24.26). Glycosylation of ceramide by sugar nucleotides yields **cerebrosides,** such as galactosylceramide, which makes up about 15% of the lipids of myelin sheath structures. Cerebrosides that contain one or more sialic acid (N-acetylneuraminic acid) moieties are called **gangliosides.** Several dozen gangliosides have been characterized, and the general form of the biosynthetic pathway is illustrated for the case of ganglioside GM_2 (Figure 24.26). Sugar units are added to the developing ganglioside from nucleotide derivatives, including UDP-N-acetylglucosamine, UDP-galactose, and UDP-glucose.

24.3 Eicosanoid Biosynthesis and Function

Eicosanoids, so named because they are all derived from 20-carbon fatty acids, are ubiquitous breakdown products of phospholipids. In response to appropriate stimuli, cells activate the breakdown of selected phospholipids (Figure 24.27). Phospholipase A_2 (Chapter 9) selectively cleaves fatty acids from the

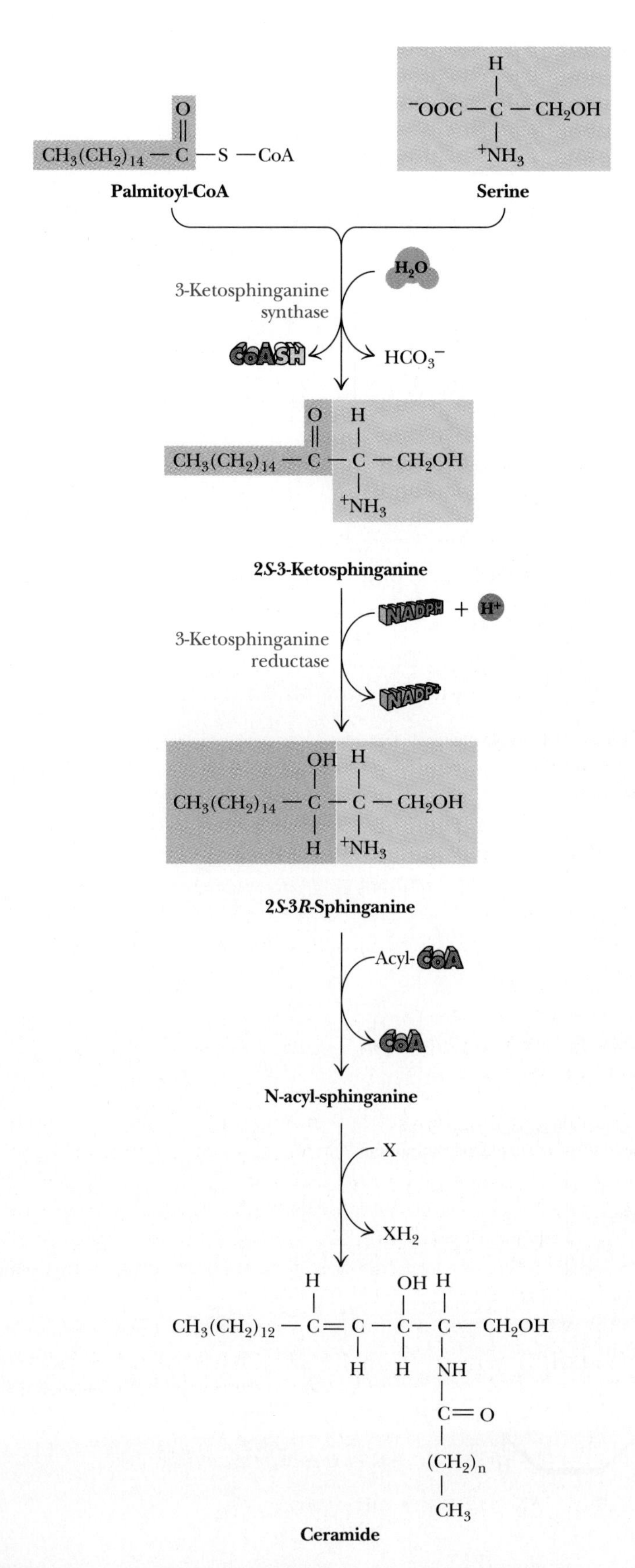

***Figure* 24.25** Biosynthesis of sphingolipids in animals begins with the 3-ketosphinganine synthase reaction, a PLP-dependent condensation of palmitoyl-CoA and serine. Subsequent reduction of the keto group, acylation, and desaturation (via reduction of an electron acceptor, X) form ceramide, the precursor of other sphingolipids.

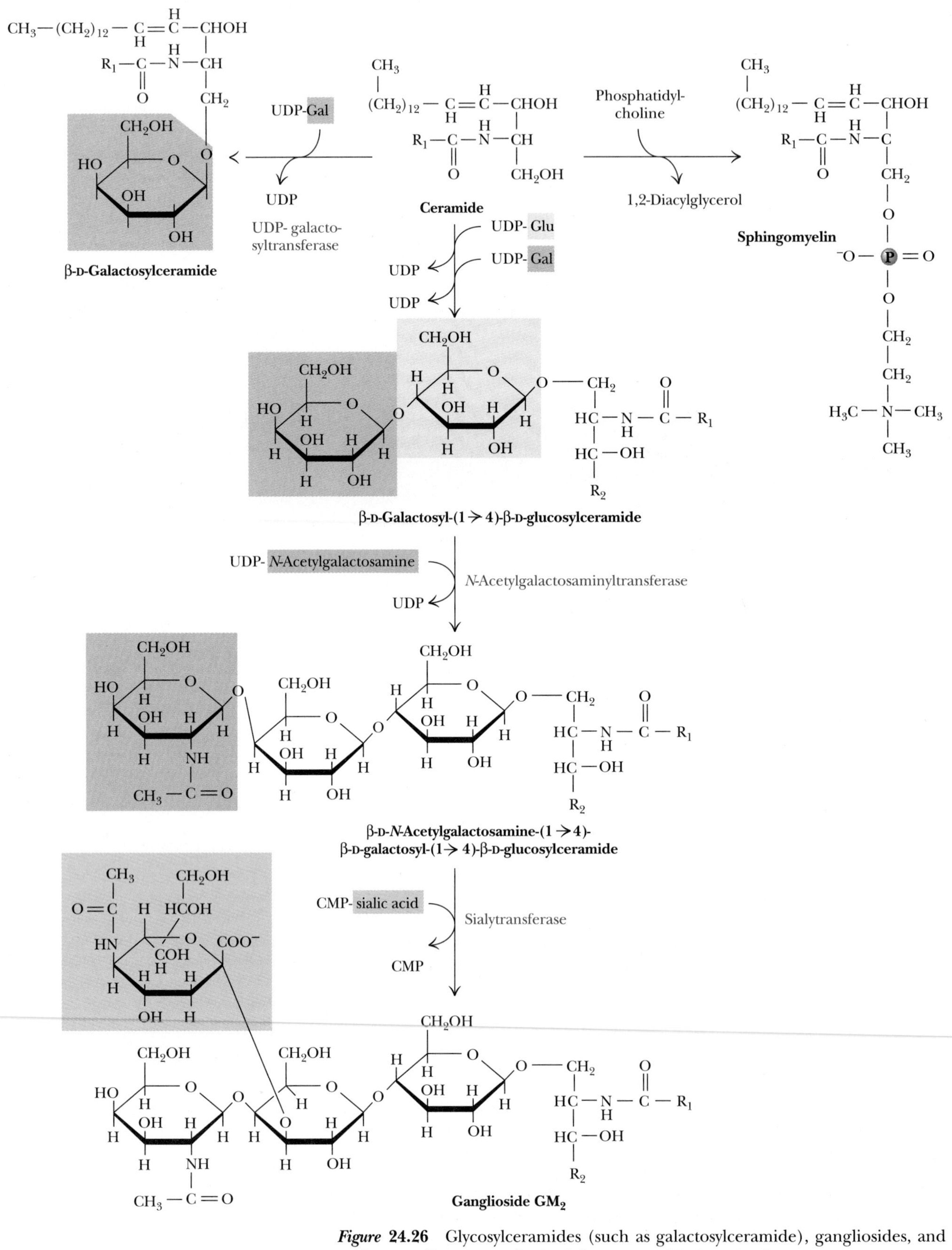

UDP-Gal

UDP

UDP-galactosyltransferase

β-D-Galactosylceramide

Ceramide

Phosphatidylcholine

1,2-Diacylglycerol

Sphingomyelin

UDP-Glu

UDP-Gal

UDP

UDP

β-D-Galactosyl-(1→4)-β-D-glucosylceramide

UDP-N-Acetylgalactosamine

N-Acetylgalactosaminyltransferase

UDP

β-D-N-Acetylgalactosamine-(1→4)-β-D-galactosyl-(1→4)-β-D-glucosylceramide

CMP-sialic acid

Sialytransferase

CMP

Ganglioside GM_2

Figure **24.26** Glycosylceramides (such as galactosylceramide), gangliosides, and sphingomyelins are synthesized from ceramide in animals.

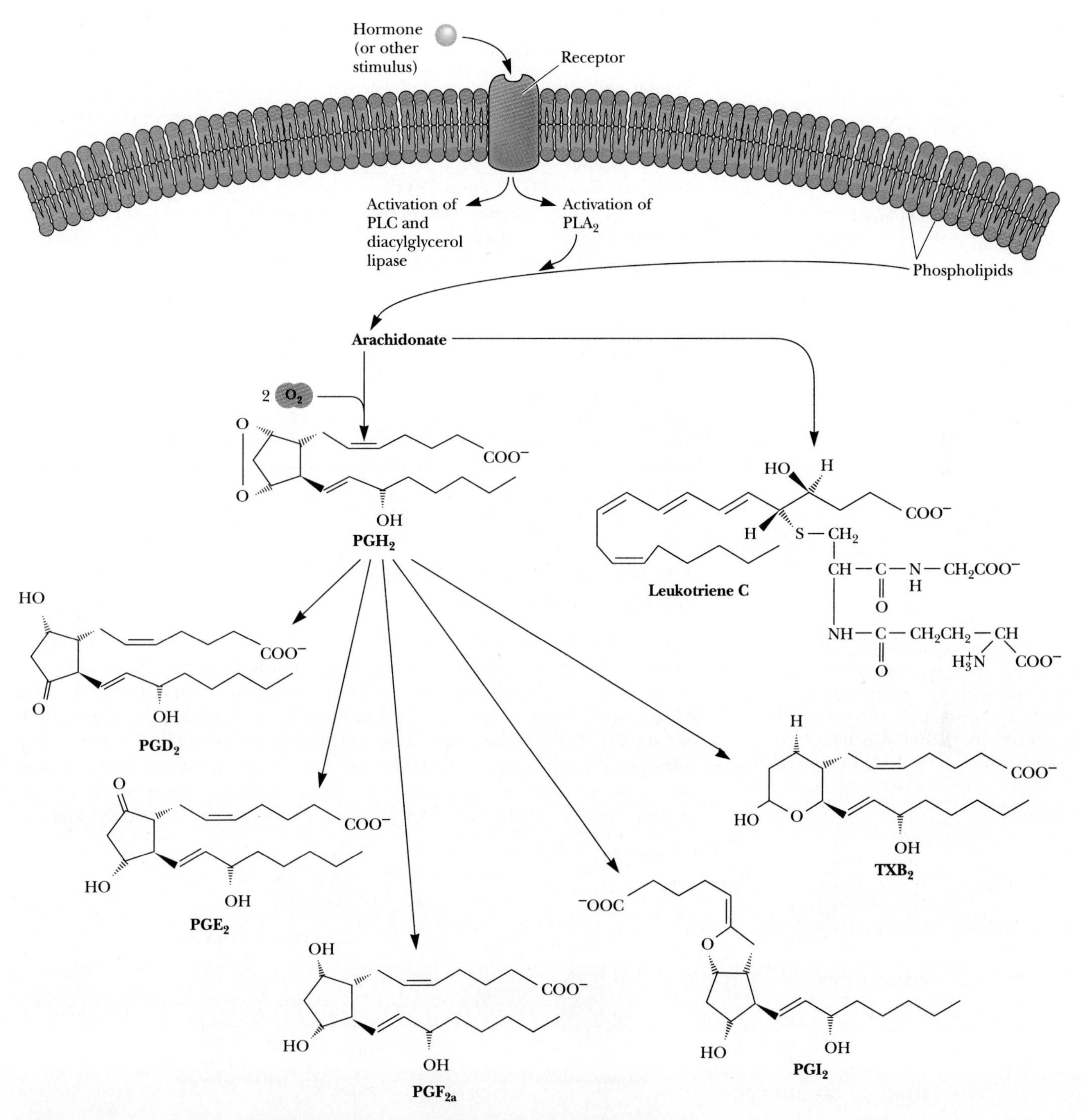

***Figure* 24.27** Arachidonic acid, derived from breakdown of phospholipids (PL), is the precursor of prostaglandins, thromboxanes, and leukotrienes. The letters used to name the prostaglandins are assigned on the basis of similarities in structure and physical properties. The class denoted PGE, for example, consists of β-hydroxyketones that are soluble in ether, whereas PGF denotes 1,3-diols that are soluble in phosphate buffer; PGA denotes prostaglandins possessing α,β-unsaturated ketones. The number following the letters refers to the number of carbon–carbon double bonds. Thus, PGE_2 contains two double bonds.

C-2 position of phospholipids. Often these are unsaturated fatty acids, among which is arachidonic acid. Arachidonic acid may also be released from phospholipids by the combined actions of phospholipase C (which yields diacylglycerols) and diacylglycerol lipase (which releases fatty acids).

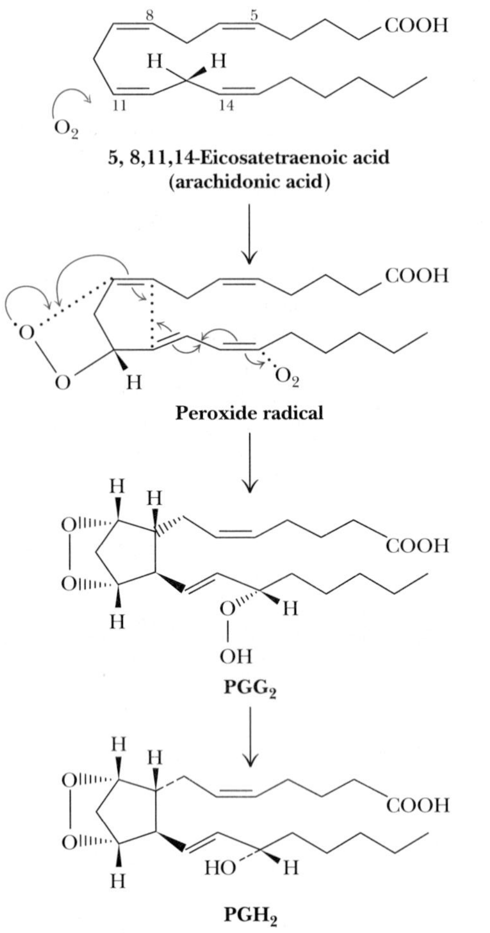

***Figure* 24.28** Endoperoxide synthase, the enzyme that converts arachidonic acid to prostaglandin PGH_2, possesses two distinct activities: cyclooxygenase (steps 1 and 2) and glutathione (GSSG)-dependent hydroperoxidase (step 3). Cyclooxygenase is the site of action of aspirin and many other analgesic agents.

Eicosanoids Are Local Hormones

Animal cells can modify arachidonic acid and other polyunsaturated fatty acids, in processes often involving cyclization and oxygenation, to produce so-called local hormones that (1) exert their effects at very low concentrations and (2) usually act near their sites of synthesis. These substances include the **prostaglandins** (PG) (Figure 24.27) as well as **thromboxanes** (Tx), **leukotrienes,** and other **hydroxyeicosanoic acids.** Thromboxanes, discovered in blood platelets (*thrombo*cytes), are cyclic ethers (TxB_2 is actually a hemiacetal; see Figure 24.27) with a hydroxyl group at C-15.

Prostaglandins Are Formed from Arachidonate by Oxidation and Cyclization

All prostaglandins are cyclopentanoic acids derived from arachidonic acid. The biosynthesis of prostaglandins is initiated by an enzyme associated with the endoplasmic reticulum, called **prostaglandin endoperoxide synthase.** The enzyme catalyzes simultaneous oxidation and cyclization of arachidonic acid. The enzyme is viewed as having two distinct activities, cyclooxygenase and peroxidase, as shown in Figure 24.28.

A Variety of Stimuli Trigger Arachidonate Release and Eicosanoid Synthesis

The release of arachidonate and the synthesis or interconversion of eicosanoids can be initiated by a variety of stimuli, including histamine, hormones such as epinephrine and bradykinin, proteases such as thrombin, and even serum albumin. An important mechanism of arachidonate release and eicosanoid synthesis involves tissue injury and inflammation. When tissue damage or injury occurs, special inflammatory cells, **monocytes** and **neutrophils,** invade the injured tissue and interact with the resident cells (e.g., smooth muscle cells and fibroblasts). *This interaction typically leads to*

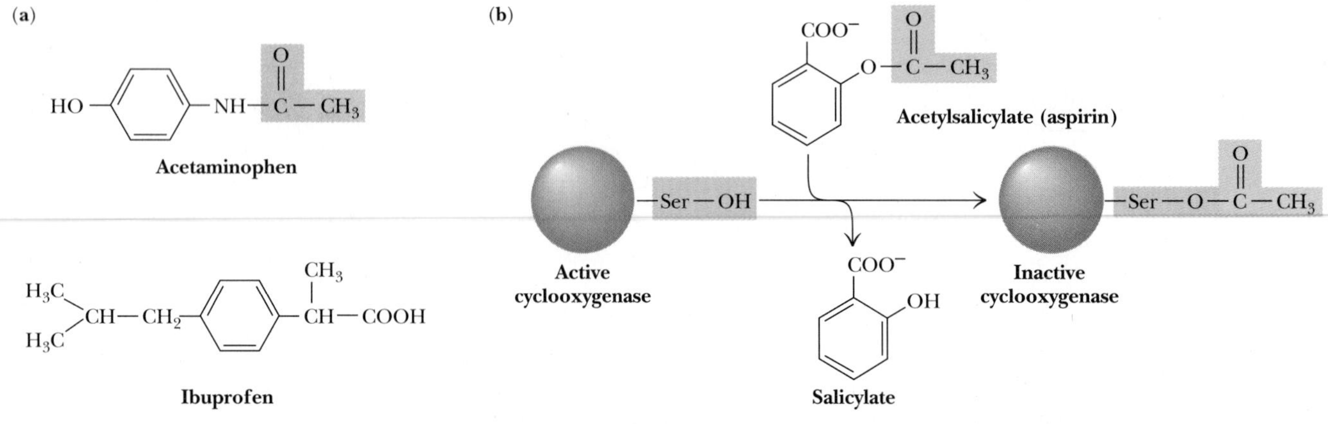

***Figure* 24.29** (a) The structures of several common analgesic agents. Acetaminophen is marketed under the trade name Tylenol®. Ibuprofen is sold as Motrin®, Nuprin®, and Advil®. (b) Acetylsalicylate (aspirin) inhibits the cyclooxygenase activity of endoperoxide synthase via acetylation (covalent modification) of Ser^{530}.

A Deeper Look

The Discovery of Prostaglandins

The name *prostaglandin* was given to this class of compounds by Ulf von Euler, their discoverer, in Sweden in the 1930s. He extracted fluids containing these components from human semen. Since he thought they originated in the prostate gland, he named them prostaglandins. Actually, they were synthesized in the seminal vesicles, and it is now known that similar substances are synthesized in most animal tissues (both male and female). Von Euler observed that injection of these substances into animals caused smooth muscle contraction and dramatic lowering of blood pressure.

Von Euler (and others) soon found that it is difficult to analyze and characterize these obviously interesting compounds since they are present at extremely low levels. Prostaglandin $E_{2\alpha}$, or $PGE_{2\alpha}$, is present in human serum at a level of less than 10^{-14} *M*! In addition, they often have half-lives of only 30 seconds to a few minutes, not lasting long enough to be easily identified. Moreover, most animal tissues upon dissection and homogenization rapidly synthesize and degrade a variety of these substances, so the amounts obtained in isolation procedures are extremely sensitive to the methods used and highly variable even when procedures are carefully controlled.

Sune Bergstrom and his colleagues described the first structural determinations of prostaglandins in the late 1950s. In the early 1960s, dramatic advances in laboratory techniques such as NMR spectroscopy and mass spectrometry made further characterization possible.

arachidonate release and eicosanoid synthesis. Examples of tissue injury where eicosanoid synthesis has been characterized include heart attack (myocardial infarction), rheumatoid arthritis, and ulcerative colitis.

"Take Two Aspirin and . . . " Inhibit Your Prostaglandin Synthesis

In 1971, biochemist John Vane was the first to show that **aspirin** (acetylsalicylate; Figure 24.29) exerts most of its effects by inhibiting the biosynthesis of prostaglandins. Its site of action is prostaglandin endoperoxide synthase. Cyclooxygenase activity is destroyed when aspirin O-acetylates Ser^{530} on the enzyme. From this you may begin to infer something about how prostaglandins (and aspirin) function. Prostaglandins are known to enhance inflammation in animal tissues. Aspirin exerts its powerful antiinflammatory effect by inhibiting this first step in their synthesis. Aspirin does not have any measurable effect on the peroxidase activity of the synthase. Other nonsteroid antiinflammatory agents, such as ibuprofen (Figure 24.29) and phenylbutazone, inhibit the cyclooxygenase by competing at the active site with arachidonate or with the peroxyacid intermediate (PGG_2, Figure 24.28).

24.4 Cholesterol Biosynthesis

The most prevalent steroid in animal cells is **cholesterol** (Figure 24.30). Plants contain no cholesterol, but they *do* contain other steroids very similar to cholesterol in structure (Figure 9.19). Cholesterol serves as a crucial component

Figure **24.30** The structure of cholesterol, drawn (a) in the traditional planar motif and (b) in a form that more accurately describes the conformation of the ring system.

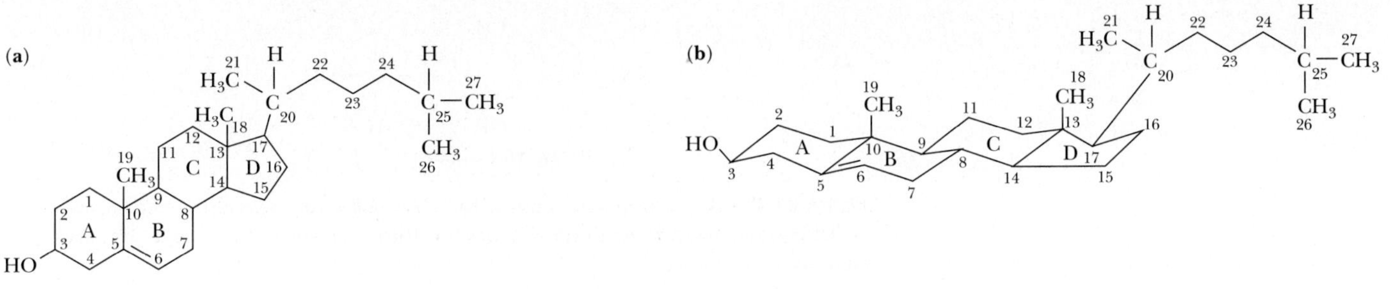

of cell membranes and as a precursor to bile acids (e.g., cholate, glycocholate, taurocholate) and steroid hormones (e.g., testosterone, estradiol, progesterone). Also, vitamin D_3 is derived from *7-dehydrocholesterol,* the immediate precursor of cholesterol. Liver is the primary site of cholesterol biosynthesis.

Mevalonate Is Synthesized from Acetyl-CoA via HMG-CoA Synthase

The cholesterol biosynthetic pathway begins in the cytosol with the synthesis of mevalonate from acetyl-CoA (Figure 24.31). The first step is the *β-ketothiolase*-catalyzed Claisen condensation of two molecules of acetyl-CoA to form acetoacetyl-CoA. In the next reaction, acetyl-CoA and acetoacetyl-CoA join to form *3-hydroxy-3-methylglutaryl-CoA,* which is abbreviated *HMG-CoA.* The reaction—a second Claisen condensation—is catalyzed by *HMG-CoA synthase.* The third step in the pathway is the rate-limiting step in cholesterol biosynthesis. Here, HMG-CoA undergoes two NADPH-dependent reductions to produce *3R-mevalonate* (Figure 24.32). The reaction is catalyzed by *HMG-CoA*

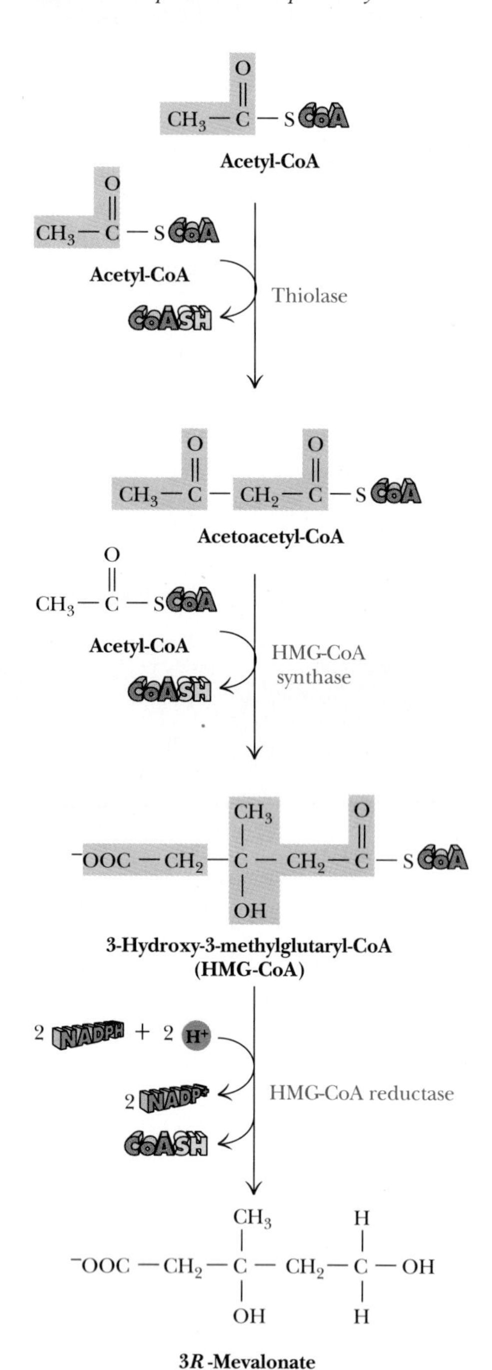

Figure **24.31** The biosynthesis of 3*R*-mevalonate from acetyl-CoA.

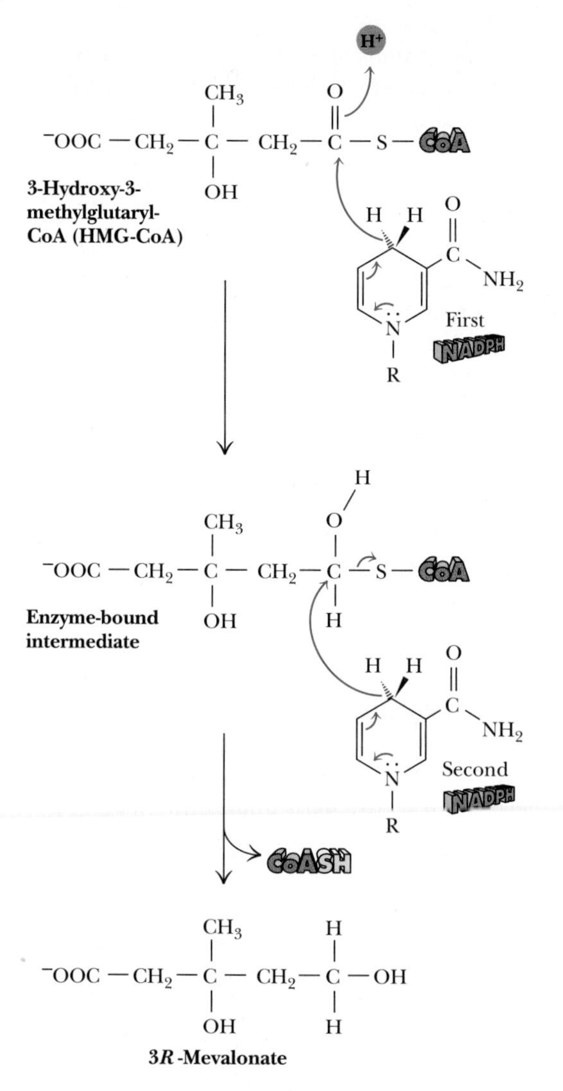

Figure **24.32** A reaction mechanism for HMG-CoA reductase. Two successive NADPH-dependent reductions convert the thioester, HMG-CoA, to a primary alcohol.

reductase, a 97-kD glycoprotein that traverses the endoplasmic reticulum membrane with its active site facing the cytosol. As the rate-limiting step, HMG-CoA reductase is the principal site of regulation in cholesterol synthesis.

Three different regulatory mechanisms are involved:

1. Phosphorylation by cAMP-dependent protein kinases inactivates the reductase. This inactivation can be reversed by two specific phosphatases (Figure 24.33).
2. Degradation of HMG-CoA reductase. This enzyme has a half-life of only three hours, and the half-life itself depends on cholesterol levels: high [cholesterol] means a short half-life for HMG-CoA reductase.
3. Gene expression—cholesterol levels control the amount of mRNA. If [cholesterol] is high, levels of mRNA coding for the reductase are reduced. If [cholesterol] is low, more mRNA is made. (Regulation of gene expression will be discussed in Chapter 30.)

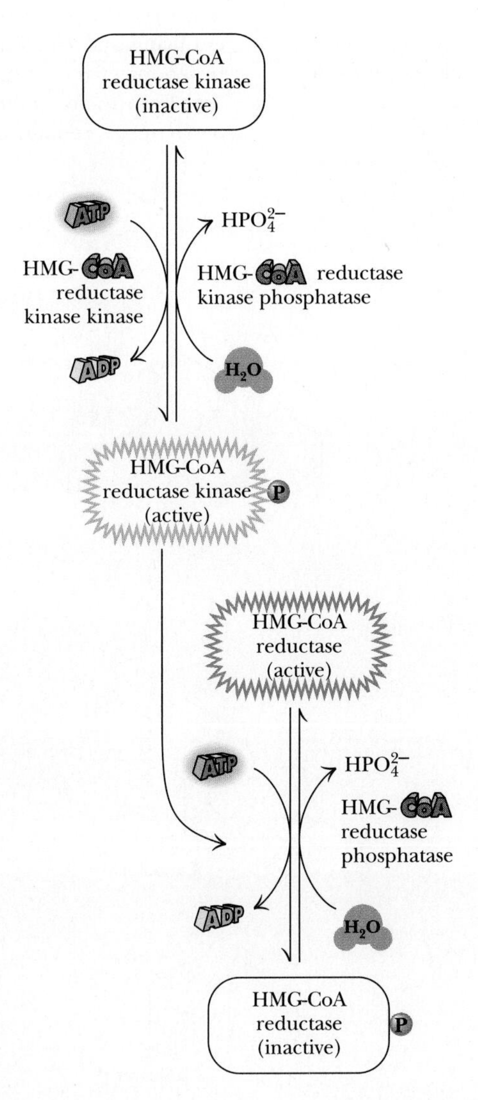

***Figure* 24.33** HMG-CoA reductase activity is modulated by a cycle of phosphorylation and dephosphorylation.

Critical Developments in Biochemistry

The Long Search for the Route of Cholesterol Biosynthesis

Heilbron, Kamm, and Owens suggested as early as 1926 that squalene is a precursor of cholesterol. That same year, H. J. Channon demonstrated that animals fed squalene from shark oil produced more cholesterol in their tissues. Bloch and Rittenberg showed in the 1940s that a significant amount of the carbon in the tetracyclic moiety and in the aliphatic side chain of cholesterol was derived from acetate. In 1934, Sir Robert Robinson suggested a scheme for the cyclization of squalene to form cholesterol before the biosynthetic link between acetate and squalene was understood. Squalene is actually a polymer of isoprene units, and Bonner and Arreguin suggested in 1949 that three acetate units could join to form 5-carbon *isoprene* units (see figure part a).

In 1952, Konrad Bloch and Robert Langdon showed conclusively that labeled squalene is synthesized rapidly from labeled acetate and also that cholesterol is derived from squalene. Langdon, a graduate student of Bloch's, performed the critical experiments in Bloch's laboratory at the University of Chicago, while Bloch spent the summer in Bermuda attempting to demonstrate that radioactively labeled squalene would be converted to cholesterol in shark livers. As Bloch himself admitted, "All I was able to learn was that sharks of manageable length are very difficult to catch and their oily livers impossible to slice."[a]

In 1953, Bloch, together with the eminent organic chemist R. B. Woodward, proposed a new scheme (see figure part b) for the cyclization of squalene. (Together with Fyodor Lynen, Bloch received the Nobel Prize in medicine or physiology in 1964 for his work.) The picture was nearly complete, but one crucial question remained: How could isoprene be the intermediate in the transformation of acetate into squalene? In 1956, Karl Folkers and his colleagues at Merck, Sharpe and Dohme isolated mevalonic acid and also showed that mevalonate was the precursor of isoprene units. The search for the remaining details (described in the text) made the biosynthesis of cholesterol one of the most enduring and challenging bioorganic problems of the forties, fifties, and sixties. Even today, several of the enzyme mechanisms remain poorly understood.

(a)

$$CH_2{=}C(CH_3){-}CH{=}CH_2$$

Isoprene

(b)

Squalene

↓

OH

Lanosterol

(Many steps)

Cholesterol

(a) An isoprene unit and a scheme for head-to-tail linking of isoprene units. (b) The cyclization of squalene to form lanosterol, as proposed by Bloch and Woodward.

[a]Bloch, K., 1987. Summing up, *Annual Review of Biochemistry* 56:1–19.

A Thiolase Brainteaser

If acetate units can be condensed by the thiolase reaction to yield acetoacetate in the first step of cholesterol synthesis, why couldn't this same reaction also be used in fatty acid synthesis, avoiding all the complexity of the fatty acyl synthase? The answer is that the thiolase reaction is more or less reversible but slightly favors the cleavage reaction. In the cholesterol synthesis pathway, subsequent reactions, including HMG-CoA reductase and the following kinase reactions, pull the thiolase-catalyzed condensation forward. However, in the case of fatty acid synthesis, a succession of eight thiolase condensations would be distinctly unfavorable from an energetic perspective. Given the necessity of repeated reactions in fatty acid synthesis, it makes better energetic sense to use a reaction that is favorable and spontaneous in the desired direction.

Squalene Is Synthesized from Mevalonate

The biosynthesis of squalene involves conversion of mevalonate to two key 5-carbon intermediates, isopentenyl pyrophosphate and dimethylallyl pyrophosphate, which join to yield farnesyl pyrophosphate and then squalene. A series of four reactions converts mevalonate to isopentenyl pyrophosphate and then to dimethylallyl pyrophosphate (Figure 24.34). The first three steps each consume an ATP, two for the purpose of forming a pyrophosphate at the 5-position and the third to drive the decarboxylation and double bond formation in the third step. Pyrophosphomevalonate decarboxylase phosphorylates the 3-hydroxyl group, and this is followed by *trans* elimination of the phosphate and carboxyl groups to form the double bond in isopentenyl pyrophosphate. Isomerization of the double bond yields the dimethylallyl pyrophosphate. Condensation of these two 5-carbon intermediates produces geranyl pyrophosphate; addition of another 5-carbon isopentenyl group gives farnesyl pyrophosphate. Both steps in the production of farnesyl pyrophosphate occur with release of pyrophosphate, hydrolysis of which drives these reactions forward. Note too that the linkage of isoprene units to form farnesyl pyrophosphate occurs in a head-to-tail fashion. This is the general rule in biosynthesis of molecules involving isoprene linkages. The next step—the joining of two farnesyl pyrophosphates to produce squalene—is a "tail-to-tail" condensation and represents an important exception to the general rule.

Squalene monooxygenase, an enzyme bound to the endoplasmic reticulum, converts squalene to *squalene-2,3-epoxide* (Figure 24.35). This reaction employs FAD and NADPH as coenzymes and requires O_2 as well as a cytosolic protein called *soluble protein activator.* A second ER membrane enzyme, *2,3-oxidosqualene lanosterol cyclase,* catalyzes the second reaction, which involves a succession of 1,2 shifts of hydride ions and methyl groups.

Conversion of Lanosterol to Cholesterol Requires 20 Additional Steps

Although lanosterol may appear similar to cholesterol in structure, another *20 steps* are required to convert lanosterol to cholesterol (Figure 24.35). The enzymes responsible for this are all associated with the endoplasmic reticulum. The primary pathway involves *7-dehydrocholesterol* as the penultimate intermediate. An alternative pathway, also composed of many steps, produces the intermediate *desmosterol.* Reduction of the double bond at C-24 yields cholesterol. Cholesterol esters—a principal form of circulating cholesterol—are synthesized by *acyl-CoA:cholesterol acyltransferases* (ACAT) on the cytoplasmic face of the endoplasmic reticulum.

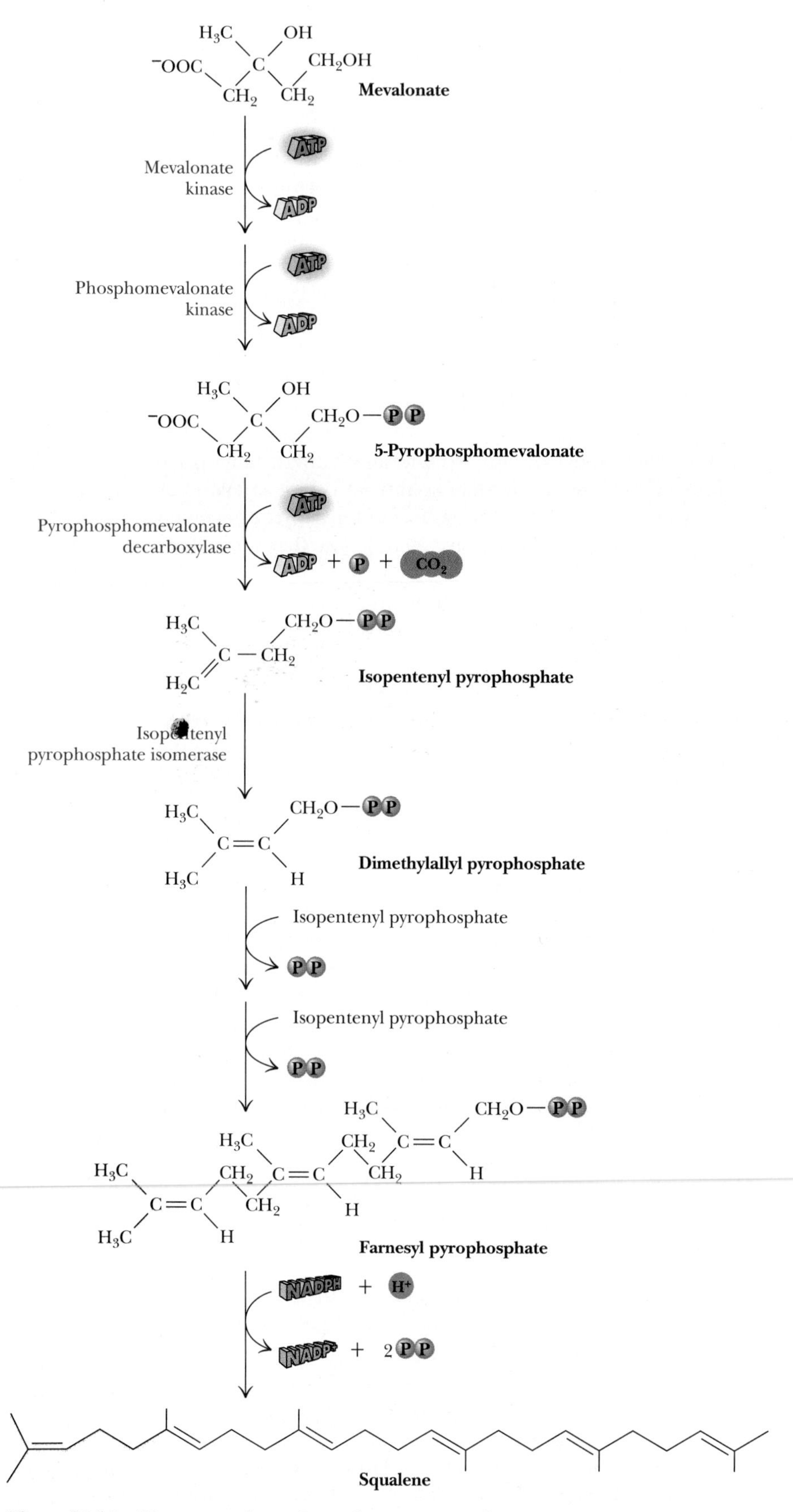

***Figure* 24.34** The conversion of mevalonate to squalene.

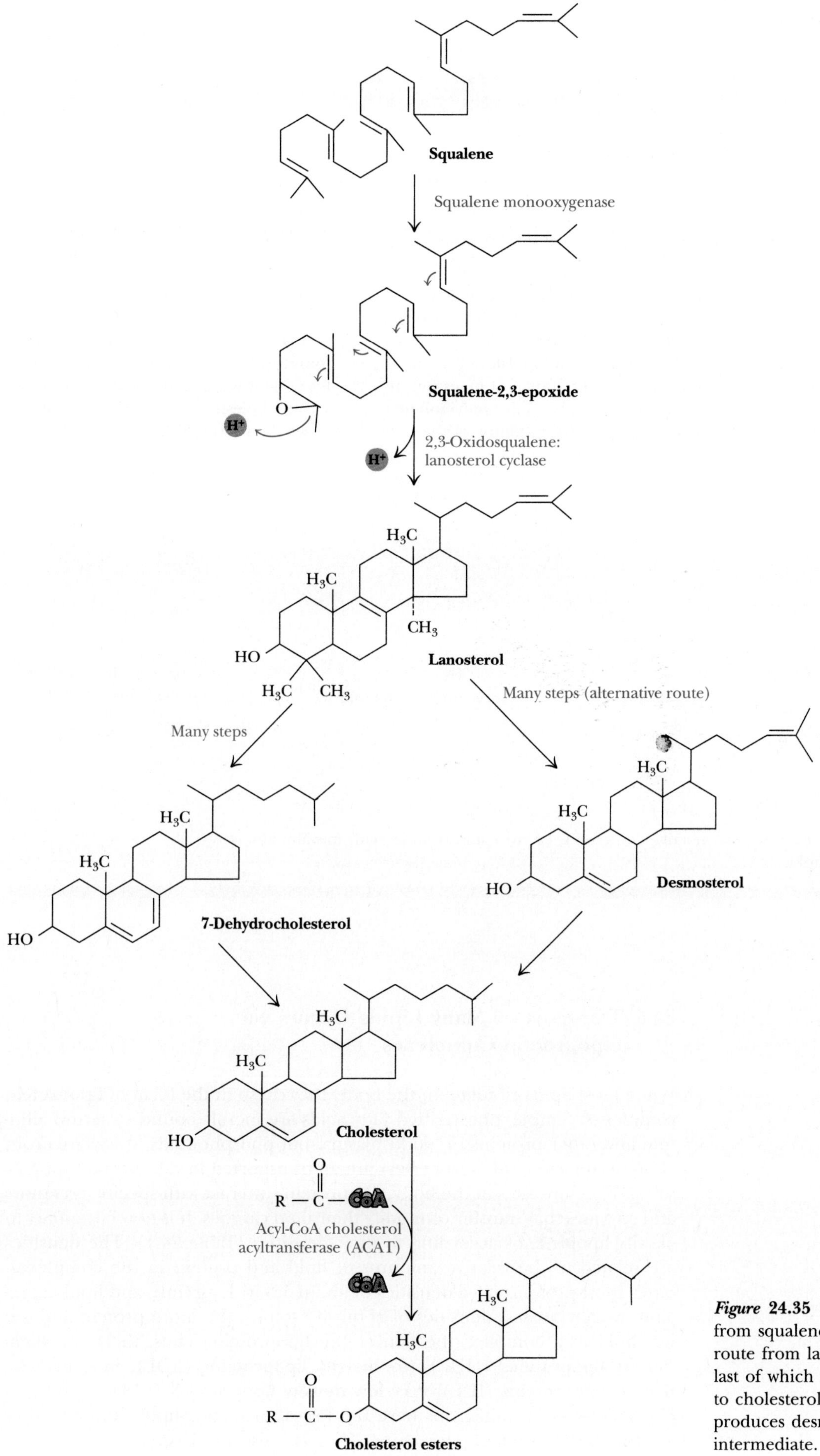

Squalene

Squalene monooxygenase

Squalene-2,3-epoxide

H^+

2,3-Oxidosqualene:
lanosterol cyclase

H^+

H_3C

H_3C

CH_3

HO

H_3C CH_3

Lanosterol

Many steps

Many steps (alternative route)

H_3C

H_3C

HO

7-Dehydrocholesterol

H_3C

H_3C

HO

Desmosterol

H_3C

H_3C

HO

Cholesterol

$R-\overset{O}{\overset{\|}{C}}-$CoA

Acyl-CoA cholesterol
acyltransferase (ACAT)

CoA

H_3C

H_3C

$R-\overset{O}{\overset{\|}{C}}-O$

Cholesterol esters

***Figure* 24.35** Cholesterol is synthesized from squalene via lanosterol. The primary route from lanosterol involves 20 steps, the last of which converts 7-dehydrocholesterol to cholesterol. An alternative route produces desmosterol as the penultimate intermediate.

Critical Developments in Biochemistry

Lovastatin Lowers Serum Cholesterol Levels

Chemists and biochemists have long sought a means of reducing serum cholesterol levels to reduce the risk of heart attack and cardiovascular disease. Since HMG-CoA reductase is the rate-limiting step in cholesterol biosynthesis, this enzyme is a likely drug target. **Mevinolin,** also known as **lovastatin** (see figure), was isolated from a strain of *Aspergillus terreus* and developed at Merck, Sharpe and Dohme for this purpose. It is now a widely prescribed cholesterol-lowering drug. Dramatic reductions of serum cholesterol are observed at doses of 20 to 80 mg per day.

Lovastatin is administered as an inactive lactone. After oral ingestion, it is hydrolyzed to the active **mevinolinic acid,** a competitive inhibitor of the reductase with a K_I of 0.6 n*M*. Mevinolinic acid is thought to behave as a transition-state analog (Chapter 13) of the tetrahedral intermediate formed in the HMG-CoA reductase reaction (see figure).

Derivatives of lovastatin have been found to be even more potent in cholesterol-lowering trials. **Synvinolin** lowers serum cholesterol levels at much lower doses than lovastatin.

1 R=H Mevinolin (Lovastatin, MEVACOR®)
2 R=CH_3 Synvinolin (Simnastatin, ZOCOR®)

Mevinolinic acid

Mevalonate

Tetrahedral intermediate in HMG-CoA reductase mechanism

The structures of (inactive) lovastatin, (active) mevinolinic acid, mevalonate, and synvinolin.

24.5 Transport of Many Lipids Occurs via Lipoprotein Complexes

When most lipids circulate in the body, they do so in the form of **lipoprotein complexes.** Simple, unesterified fatty acids are merely bound to serum albumin and other proteins in blood plasma, but phospholipids, triacylglycerols, cholesterol, and cholesterol esters are all transported in the form of lipoproteins. At various sites in the body, lipoproteins interact with specific receptors and enzymes that transfer or modify their lipid cargoes. It is now customary to classify lipoproteins according to their densities (Table 24.1). The densities are related to the relative amounts of lipid and protein in the complexes. Since most proteins have densities of about 1.3 to 1.4 g/mL, and lipid aggregates usually possess densities of about 0.8 g/mL, the more protein and the less lipid in a complex, the denser the lipoprotein. Thus, there are **high-density lipoproteins** (HDL), **low-density lipoproteins** (LDL), **intermediate-density lipoproteins** (IDL), **very low density lipoproteins** (VLDL), and also **chylomicrons.** Chylomicrons have the lowest protein-to-lipid ratio and thus are the lowest-density lipoproteins. They are also the largest.

Table **24.1**
Composition and Properties of Human Lipoproteins

Lipoprotein Class	Density (g/mL)	Diameter (nm)	Composition (% dry weight)			
			Protein	Cholesterol	Phospholipid	Triacylglycerol
HDL	1.063–1.21	5–15	33	30	29	8
LDL	1.019–1.063	18–28	25	50	21	4
IDL	1.006–1.019	25–50	18	29	22	31
VLDL	0.95–1.006	30–80	10	22	18	50
Chylomicrons	<0.95	100–500	1–2	8	7	84

Adapted from Brown, M., and Goldstein, J., 1987. *In* Braunwald, E., et al., eds., *Harrison's Principles of Internal Medicine,* 11th ed. New York: McGraw-Hill; and Vance, D., and Vance, J., eds., 1985. *Biochemistry of Lipids and Membranes.* Menlo Park, CA: Benjamin/Cummings.

The Structure and Synthesis of the Lipoproteins

HDL and VLDL are assembled primarily in the endoplasmic reticulum of the liver (with smaller amounts produced in the intestine), whereas chylomicrons form in the intestine. LDL is not synthesized directly, but is made from VLDL. LDL appears to be the major circulatory complex for cholesterol and cholesterol esters. The primary task of chylomicrons is to transport triacylglycerides. Despite all this, it is extremely important to note that each of these lipoprotein classes contains some of each type of lipid. The relative amounts of HDL and LDL are important in the disposition of cholesterol in the body and in the development of arterial plaques (Figure 24.36). The structures of the various lipoproteins are approximately similar, and they consist of a core of mobile triacylglycerols or cholesterol esters surrounded by a single layer of phospholipid, into which is inserted a mixture of cholesterol and proteins (Figure 24.37). Note that the phospholipids are oriented with their polar head groups facing outward to interact with solvent water, and that the phospholipids thus shield the hydrophobic lipids inside from the solvent water outside. The proteins also function as recognition sites for the various lipo-

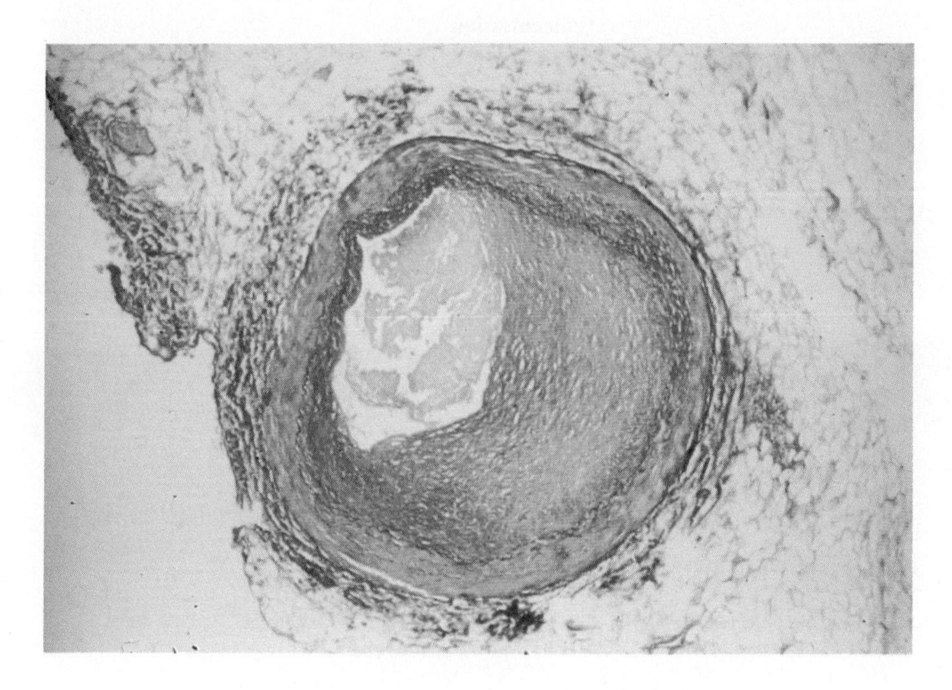

Figure **24.36** Photograph of an arterial plaque.

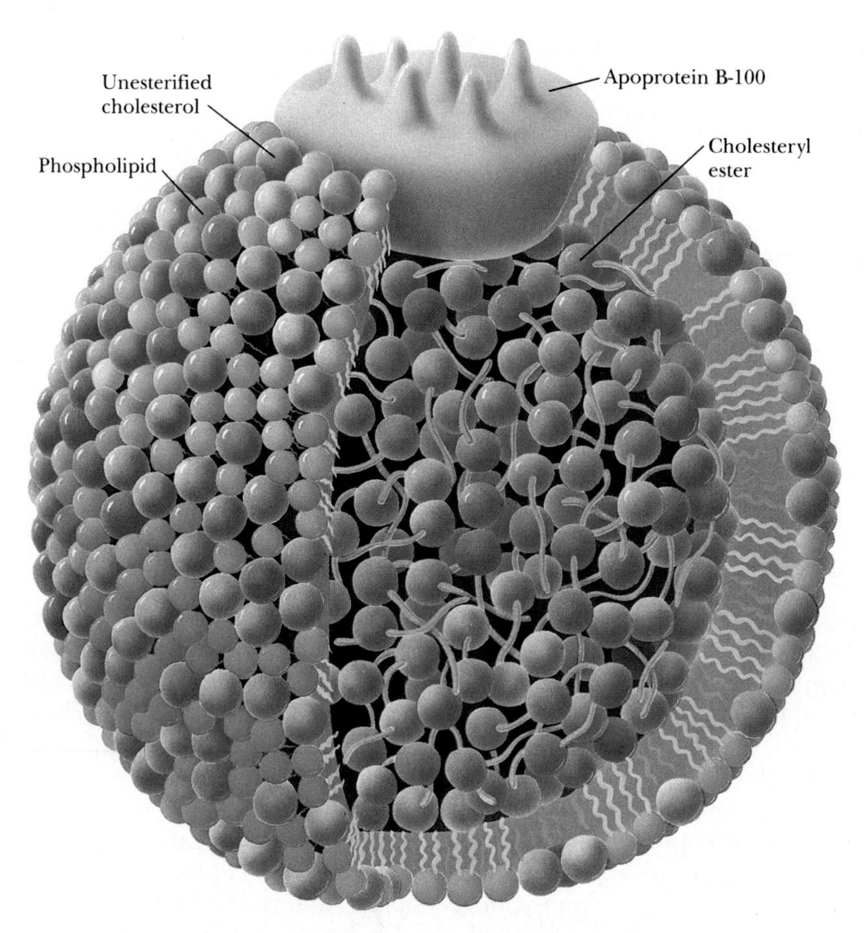

Figure **24.37** Structure of a typical lipoprotein. A core of cholesterol and cholesteryl esters is surrounded by a phospholipid (monolayer) membrane in which apolipoproteins are embedded.

Table **24.2**
Apoproteins of Human Lipoproteins

Apoprotein	M_r	Concentration in Plasma (mg/100 mL)	Distribution
A-1	28,300	90–120	Principal protein in HDL
A-2	8,700	30–50	Occurs as dimer mainly in HDL
B-48	240,000	<5	Found only in chylomicrons
B-100	500,000	80–100	Principal protein in LDL
C-1	7,000	4–7	Found in chylomicrons, VLDL, HDL
C-2	8,800	3–8	Found in chylomicrons, VLDL, HDL
C-3	8,800	8–15	Found in chylomicrons, VLDL, IDL, HDL
D	32,500	8–10	Found in HDL
E	34,100	3–6	Found in chylomicrons, VLDL, IDL, HDL

Adapted from Brown, M., and Goldstein, J., 1987. *In* Braunwald, E., et al., eds., *Harrison's Principles of Internal Medicine,* 11th ed. New York: McGraw-Hill; and Vance, D., and Vance, J., eds., 1985. *Biochemistry of Lipids and Membranes.* Menlo Park, CA: Benjamin/Cummings.

protein receptors throughout the body. A number of different apoproteins have been identified in lipoproteins (Table 24.2), and others may exist as well. The apoproteins are abundant in hydrophobic amino acid residues, as is appropriate for interactions with lipids. A *cholesterol ester transfer protein* also associates with lipoproteins.

Lipoproteins in Circulation Are Progressively Degraded by Lipoprotein Lipase

The livers and intestines of animals are the primary sources of circulating lipids. Chylomicrons carry triacylglycerol and cholesterol esters from the intestines to other tissues, and VLDLs carry lipid from liver, as shown in Figure 24.38. At various target sites, particularly in the capillaries of muscle and adipose cells, these particles are degraded by *lipoprotein lipase,* which hydrolyzes triacylglycerols. Lipase action causes progressive loss of triacylglycerol (and apoprotein) and makes the lipoproteins smaller. This process gradually con-

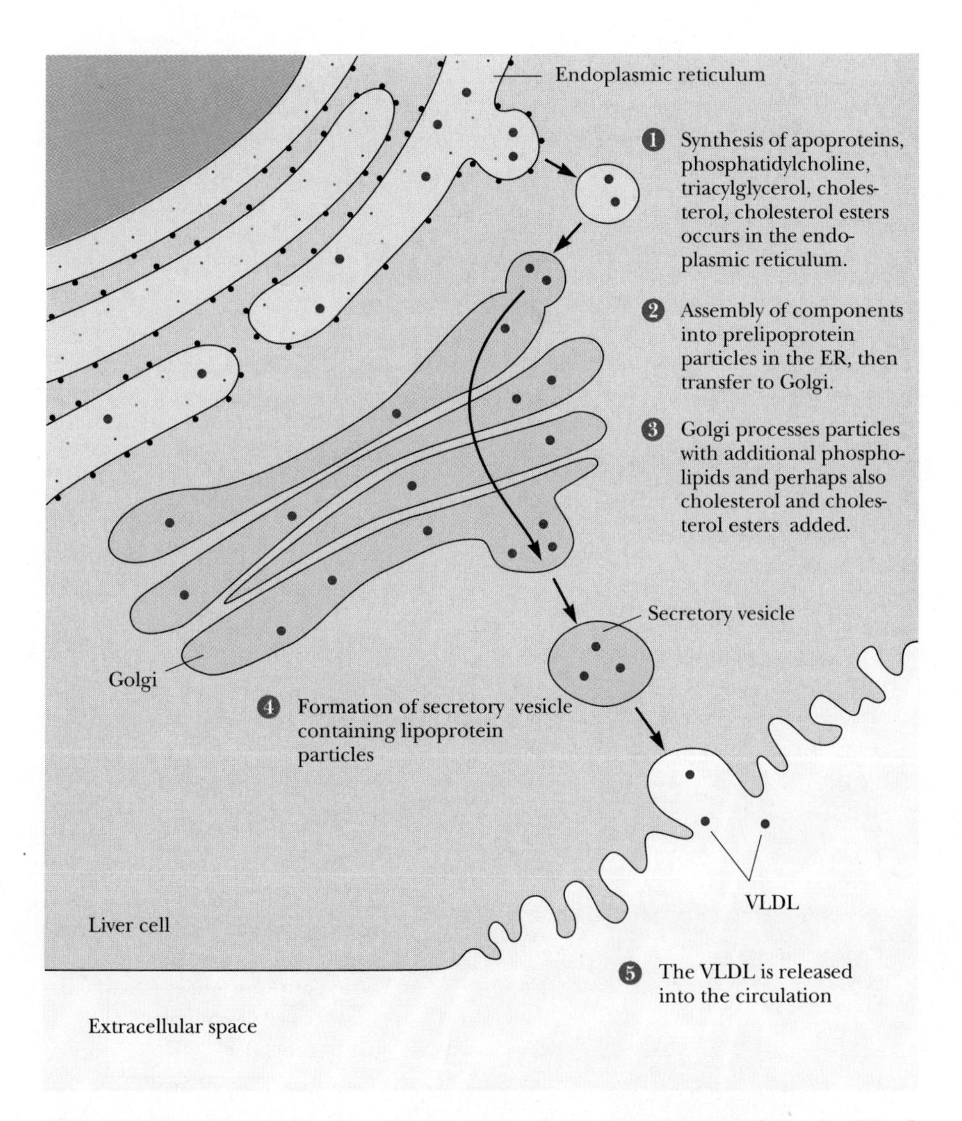
Endoplasmic reticulum
1 Synthesis of apoproteins, phosphatidylcholine, triacylglycerol, cholesterol, cholesterol esters occurs in the endoplasmic reticulum.
2 Assembly of components into prelipoprotein particles in the ER, then transfer to Golgi.
3 Golgi processes particles with additional phospholipids and perhaps also cholesterol and cholesterol esters added.
Secretory vesicle
Golgi
4 Formation of secretory vesicle containing lipoprotein particles
VLDL
Liver cell
5 The VLDL is released into the circulation
Extracellular space

***Figure* 24.38** Lipoprotein components are synthesized predominantly in the ER of liver cells. Following assembly of lipoprotein particles (red dots) in the ER and processing in the Golgi, lipoproteins are packaged in secretory vesicles for export from the cell (via exocytosis) and release into the circulatory system.

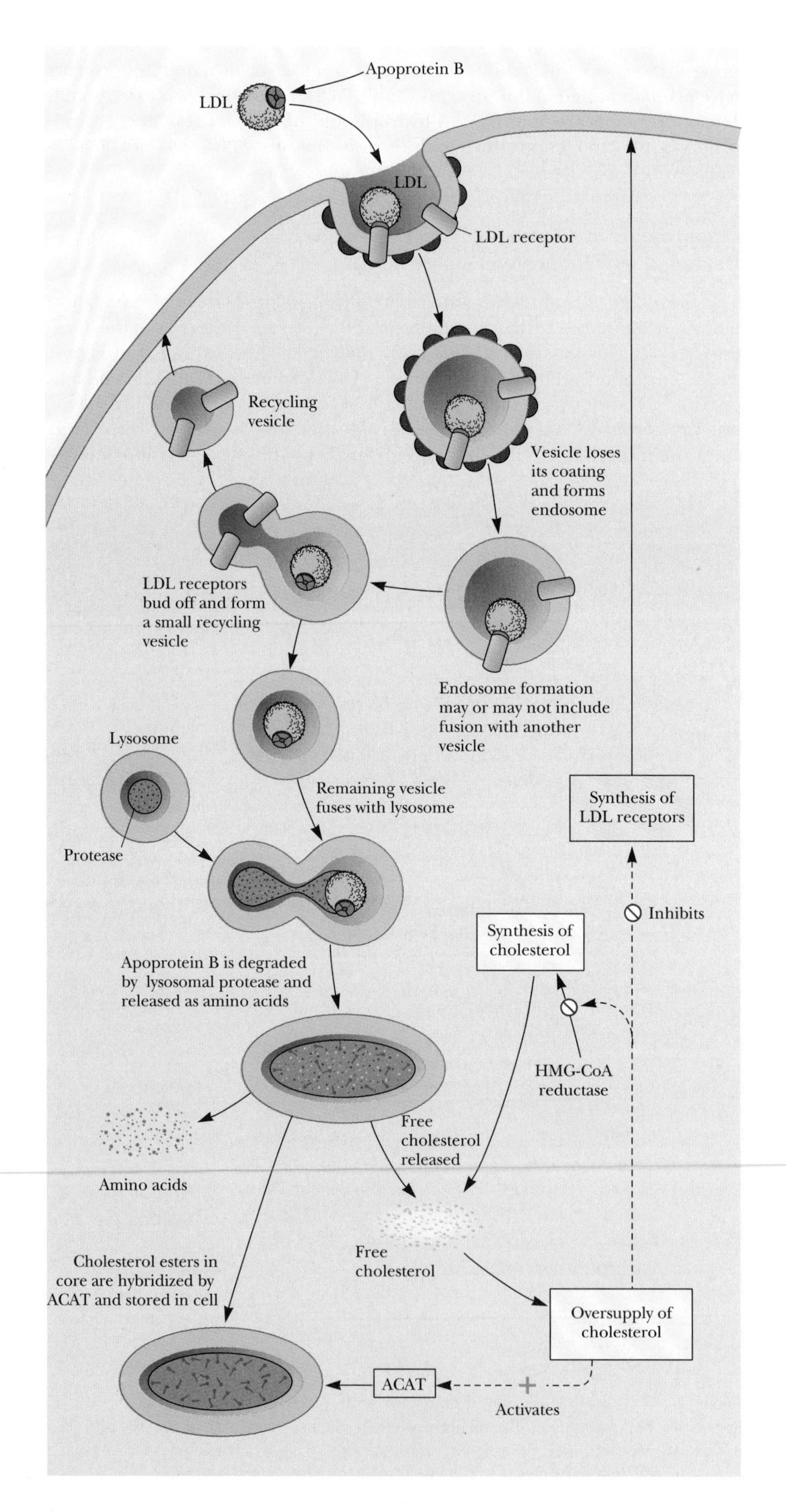

***Figure* 24.39** Endocytosis and degradation of lipoprotein particles. (ACAT is acyl-CoA cholesterol acyltransferase.)

verts VLDL particles to IDL and then LDL particles, which are either returned to the liver for reprocessing or redirected to adipose tissues and adrenal glands. Every 24 hours, nearly half of all circulating LDL is removed from circulation in this way. The LDL binds to specific LDL receptors, which cluster in domains of the plasma membrane known as **coated pits** (discussed in subsequent paragraphs). These domains eventually invaginate to form **coated vesicles** (Figure 24.39). Within the cell, these vesicles fuse with lysosomes, and the LDLs are degraded by *lysosomal acid lipases.*

High-density lipoproteins (HDL) have much longer life spans in the body (5 to 6 days) than other lipoproteins. Newly formed HDL contains virtually no cholesterol ester. However, over time, cholesterol esters are accumulated through the action of *lecithin:cholesterol acyltransferase* (LCAT), a 59-kD glycoprotein associated with HDLs. Another associated protein, *cholesterol ester transfer protein,* transfers some of these esters to VLDL and LDL. Alternatively, HDLs function to return cholesterol and cholesterol esters to the liver. This latter process apparently explains the correlation between high HDL levels and reduced risk of cardiovascular disease. (High LDL levels, on the other hand, are correlated with an *increased* risk of coronary artery and cardiovascular disease.)

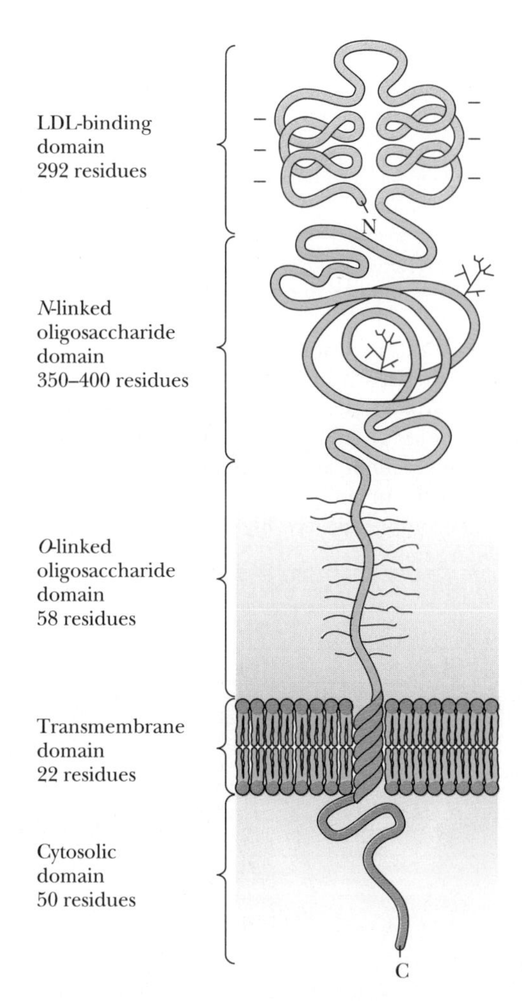

***Figure* 24.40** The structure of the LDL receptor. The amino-terminal binding domain is responsible for recognition and binding of LDL apoprotein. The O-linked oligosaccharide-rich domain may act as a molecular spacer, raising the binding domain above the glycocalyx. The cytosolic domain is required for aggregation of LDL receptors during endocytosis.

Structure of the LDL Receptor

The LDL receptor in plasma membranes (Figure 24.40) consists of 839 amino acid residues and is composed of five domains. These domains include an LDL-binding domain of 292 residues, a segment of about 350 to 400 residues containing N-linked oligosaccharides, a 58-residue segment of O-linked oligosaccharides, a 22-residue membrane-spanning segment, and a 50-residue segment extending into the cytosol. The clustering of receptors prior to the formation of coated vesicles requires the presence of this cytosolic segment.

Defects in Lipoprotein Metabolism Can Lead to Elevated Serum Cholesterol

The mechanism of LDL metabolism and the various defects which can occur therein have been studied extensively by Michael Brown and Joseph Goldstein, who received the Nobel Prize in medicine or physiology in 1985. **Familial hypercholesterolemia** is the term given to a variety of inherited metabolic defects that lead to greatly elevated levels of serum cholesterol—much of it in the form of LDL particles. The general genetic defect responsible for familial hypercholesterolemia is the absence or dysfunction of LDL receptors in the body. Only about half the normal level of LDL receptors is found in heterozygous individuals (persons carrying one normal gene and one defective gene). Homozygotes (with two copies of the defective gene) have few if any functional LDL receptors. In such cases, LDLs (and cholesterol) cannot be absorbed, and plasma levels of LDL (and cholesterol) are very high. Typical heterozygotes display serum cholesterol levels of 300 to 400 mg/dL, but homozygotes carry serum cholesterol levels of 600 to 800 mg/dL or even higher. There are two possible causes of an absence of LDL receptors. Either receptor synthesis does not occur at all, or the newly synthesized protein does not successfully reach the plasma membrane, due to faulty processing in the Golgi or faulty transport to the plasma membrane. Even when LDL receptors are made and reach the plasma membrane, they may fail to function for two reasons. They may be unable to form clusters competent in coated pit formation because of folding or sequence anomalies in the carboxy-terminal domain, or they may be unable to bind LDL because of sequence or folding anomalies in the LDL-binding domain.

24.6 Biosynthesis of Bile Acids

Bile acids, which exist mainly as **bile salts,** are polar carboxylic acid derivatives of cholesterol that are important in the digestion of food, especially the solubilization of ingested fats. The Na^+ and K^+ salts of *glycocholic acid* and *taurocholic acid* are the principal bile salts (Figure 24.41). Glycocholate and taurocholate are conjugates of *cholic acid* with glycine and taurine, respec-

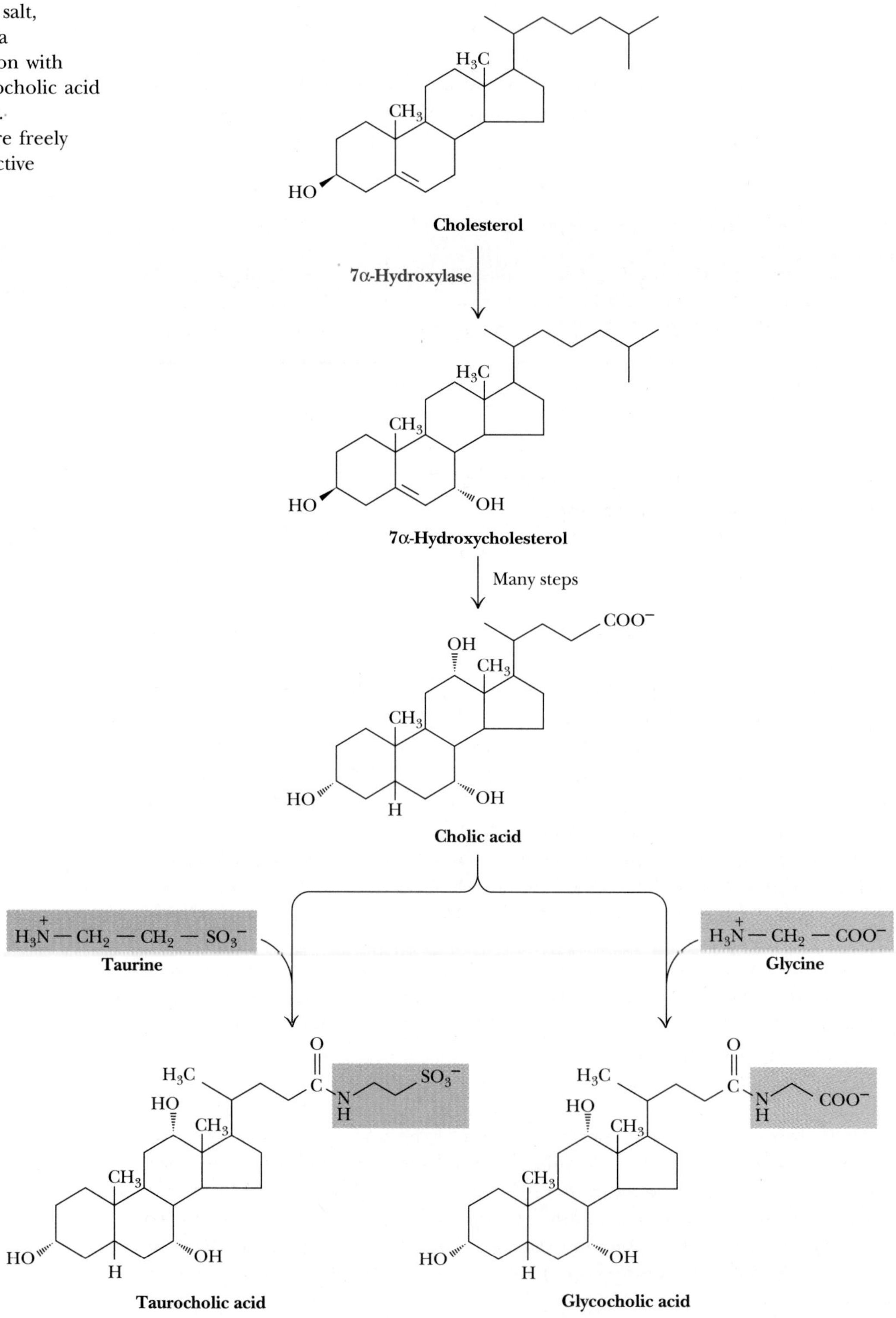

Figure **24.41** Cholic acid, a bile salt, is synthesized from cholesterol via 7α-hydroxycholesterol. Conjugation with taurine or glycine produces taurocholic acid and glycocholic acid, respectively. Taurocholate and glycocholate are freely water-soluble and are highly effective detergents.

tively. Since they contain both nonpolar and polar domains, these bile salt conjugates are highly effective as detergents. These substances are made in the liver, stored in the gallbladder, and secreted as needed into the intestines.

The formation of bile salts represents the major pathway for cholesterol degradation. The first step involves hydroxylation at C-7 (Figure 24.41). *7α-Hydroxylase,* which catalyzes the reaction, is a mixed-function oxidase involving *cytochrome P450.* **Mixed-function oxidases** use O_2 as substrate. One oxygen atom goes to hydroxylate the substrate, while the other is reduced to water (Figure 24.42). The function of cytochrome P450 is to activate O_2 for the hydroxylation reaction. Such hydroxylations are quite common in the synthetic routes for cholesterol, bile acids, and steroid hormones and also in detoxification pathways for aromatic compounds. Several of these will be considered in the next section. 7α-Hydroxycholesterol is the precursor for cholic acid.

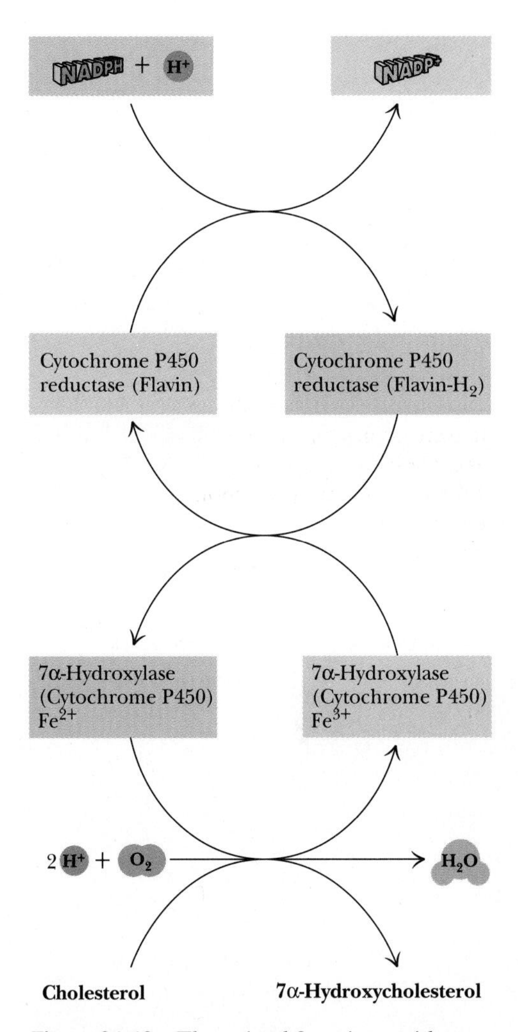

***Figure* 24.42** The mixed-function oxidase activity of 7α-hydroxylase.

24.7 Synthesis and Metabolism of Steroid Hormones

Steroid hormones are crucial signal molecules in mammals. (The details of their physiological effects will be described in Chapter 37.) Their biosynthesis begins with the **desmolase reaction,** which converts cholesterol to pregnenolone (Figure 24.43). Desmolase is found in the mitochondria of tissues that synthesize steroids (mainly the adrenal glands and gonads). Desmolase activity includes two hydroxylases and utilizes cytochrome P450.

Pregnenolone and Progesterone Are the Precursors of All Other Steroid Hormones

Pregnenolone is transported from the mitochondria to the ER, where a hydroxyl oxidation and migration of the double bond yield progesterone. Pregnenolone synthesis in the adrenal cortex is activated by **adrenocorticotropic hormone** (ACTH), a peptide of 39 amino acid residues secreted by the anterior pituitary gland.

Progesterone is secreted from the corpus luteum during the latter half of the menstrual cycle and prepares the lining of the uterus for attachment of a fertilized ovum. If an ovum attaches, progesterone secretion continues to ensure the successful maintenance of a pregnancy. Progesterone is also the precursor for synthesis of the **sex hormone steroids** and the **corticosteroids.** Male sex hormone steroids are called **androgens,** and female hormones, **estrogens.** Testosterone is an androgen synthesized in males primarily in the testes (and in much smaller amounts in the adrenal cortex). Androgens are necessary for sperm maturation. Even nonreproductive tissue (liver, brain, and skeletal muscle) is susceptible to the effects of androgens.

Testosterone is also produced primarily in the ovaries (and in much smaller amounts in the adrenal glands) of females as a precursor for the estrogens. *β*-Estradiol is the most important estrogen (Figure 24.43).

Steroid Hormones Modulate Transcription in the Nucleus

Steroid hormones act in a different manner from most hormones we have considered. They do not bind to plasma membrane receptors, but rather pass easily across the plasma membrane. Steroids may bind directly to receptors in the nucleus or may bind to cytosolic steroid hormone receptors, which then enter the nucleus. In the nucleus, the hormone-receptor complex binds directly to specific nucleotide sequences in DNA, increasing transcription of DNA to RNA (Chapters 30 and 37).

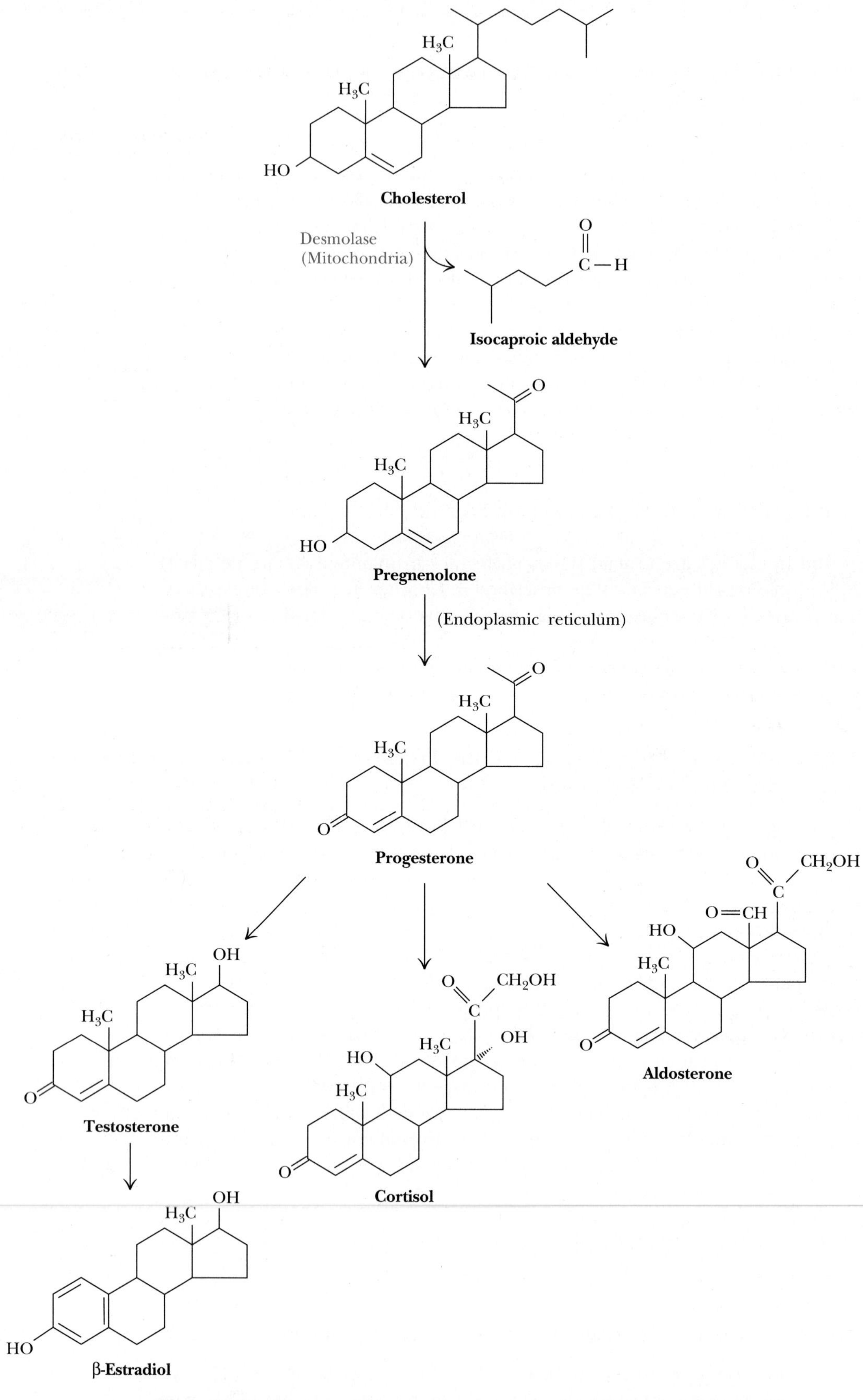

Cholesterol

Desmolase
(Mitochondria)

Isocaproic aldehyde

Pregnenolone

(Endoplasmic reticulum)

Progesterone

Testosterone

Cortisol

Aldosterone

β-Estradiol

Figure **24.43** The steroid hormones are synthesized from cholesterol, with intermediate formation of pregnenolone and progesterone. Testosterone, the principal male sex hormone steroid, is a precursor to β-estradiol. Cortisol, a glucocorticoid, and aldosterone, a mineralocorticoid, are also derived from progesterone.

Cortisol and Other Corticosteroids Regulate a Variety of Body Processes

Corticosteroids, including the *glucocorticoids* and *mineralocorticoids,* are made by the cortex of the adrenal glands on top of the kidneys. **Cortisol** (Figure 24.43) is representative of the **glucocorticoids,** a class of compounds that (1) stimulate gluconeogenesis and glycogen synthesis in liver (by signaling the synthesis of PEP carboxykinase, fructose-1,6-bisphosphatase, glucose-6-phosphatase, and glycogen synthase); (2) inhibit protein synthesis and stimulate protein degradation in peripheral tissues such as muscle; (3) inhibit allergic and inflammatory responses; (4) exert an immunosuppressive effect, inhibiting DNA replication and mitosis and repressing the formation of antibodies and lymphocytes; and (5) inhibit formation of fibroblasts involved in healing wounds and slow the healing of broken bones.

Aldosterone, the most potent of the **mineralocorticoids** (Figure 24.43), is involved in the regulation of sodium and potassium balances in tissues. Aldosterone increases the kidney's capacity to absorb Na^+, Cl^-, and H_2O from the glomerular filtrate in the kidney tubules.

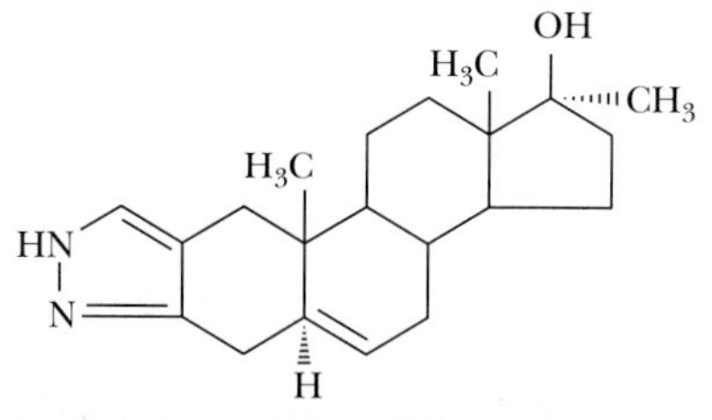

Stanozolol

***Figure* 24.44** The structure of stanozolol, an anabolic steroid.

Anabolic Steroids Have Been Used Illegally To Enhance Athletic Performance

The dramatic effects of androgens on protein biosynthesis have led many athletes to the use of *synthetic androgens,* which go by the blanket term **anabolic steroids.** Despite numerous warnings from the medical community about side effects, which include kidney and liver disorders, sterility, and heart disease, abuse of such substances is epidemic. **Stanozolol** (Figure 24.44) was one of the agents found in the blood and urine of Ben Johnson following his record-setting performance in the 100-meter dash in the 1988 Olympic Games. Because use of such substances is disallowed, Johnson lost his gold medal, and Carl Lewis was declared the official winner.

Problems

1. Carefully count and account for each of the atoms and charges in the equations for the synthesis of palmitoyl-CoA, the synthesis of malonyl-CoA, and the overall reaction for the synthesis of palmitoyl-CoA from acetyl-CoA.

2. Use the relationships shown in Figure 24.1 to determine which carbons of glucose will be incorporated into palmitic acid. Consider the cases of both citrate that is immediately exported to the cytosol following its synthesis and citrate that enters the TCA cycle.

3. Based on the information presented in the text and in Figures 24.4 and 24.5, suggest a model for the regulation of acetyl-CoA carboxylase. Consider the possible roles of subunit interactions, phosphorylation, and conformation changes in your model.

4. Consider the role of the pantothenic acid groups in animal fatty acyl synthase and the size of the pantothenic acid group itself, and estimate a maximal separation between the malonyl transferase and the ketoacyl-ACP synthase active sites.

5. Carefully study the reaction mechanism for the stearoyl-CoA desaturase in Figure 24.14, and account for all of the electrons flowing through the reactions shown. Also account for all of the hydrogen and oxygen atoms involved in this reaction, and convince yourself that the stoichiometry is correct as shown.

6. Write a balanced, stoichiometric reaction for the synthesis of phosphatidylethanolamine from glycerol, fatty acyl-CoA, and ethanolamine. Make an estimate of the $\Delta G°'$ for the overall process.

7. Write a balanced, stoichiometric reaction for the synthesis of cholesterol from acetyl-CoA.

8. Trace each of the carbon atoms of mevalonate through the synthesis of cholesterol, and determine the source (i.e., the position in the mevalonate structure) of each carbon in the final structure.

9. Suggest a structural or functional role for the O-linked saccharide domain in the LDL receptor (Figure 24.40).

10. Identify the lipid synthetic pathways that would be affected by abnormally low levels of CTP.

11. Determine the number of ATP equivalents needed to form palmitic acid from acetyl-CoA. (Assume for this calculation that each NADPH is worth 3.5 ATP.)

Further Reading

Bloch, K., 1987. Summing up. *Annual Review of Biochemistry* **56:**1–19.

Bloch, K., 1965. The biological synthesis of cholesterol. *Science* **150:**19–28.

Boyer, P. D., ed., 1983. *The Enzymes,* 3rd ed., Vol. 16. New York: Academic Press.

Carman, G. M., and Henry, S. A., 1989. Phospholipid biosynthesis in yeast. *Annual Review of Biochemistry* **58:**635–669.

Chang, S. I., and Hammes, G. G., 1990. Structure and mechanism of action of a multifunctional enzyme: Fatty acid synthase. *Accounts of Chemical Research* **23:**363–369.

Hansen, H. S., 1985. The essential nature of linoleic acid in mammals. *Trends in Biochemical Sciences* **11:**263–265.

Jeffcoat, R., 1979. The biosynthesis of unsaturated fatty acids and its control in mammalian liver. *Essays in Biochemistry* **15:**1–36.

Kim, K-H., et al., 1989. Role of reversible phosphorylation of acetyl-CoA carboxylase in long-chain fatty acid synthesis. *The FASEB Journal* **3:**2250–2256.

Lardy, H., and Shrago, E., 1990. Biochemical aspects of obesity. *Annual Review of Biochemistry* **59:**689–710.

McCarthy, A. D., and Hardie, D. G., 1984. Fatty acid synthase—An example of protein evolution by gene fusion. *Trends in Biochemical Sciences* **9:**60–63.

Needleman, P., et al., 1986. Arachidonic acid metabolism. *Annual Review of Biochemistry* **55:**69–102.

Samuelsson, B., et al., 1987. Leukotrienes and lipoxins: Structures, biosynthesis and biological effects. *Science* **237:**1171–1176.

Schewe, T., and Kuhn, H., 1991. Do 15-lipoxygenases have a common biological role? *Trends in Biochemical Sciences* **16:**369–373.

Vance, D. E., and Vance, J. E., eds., 1985. *Biochemistry of Lipids and Membranes.* Menlo Park, CA: Benjamin/Cummings.

Wakil, S., 1989. Fatty acid synthase, a proficient multifunctional enzyme. *Biochemistry* **28:**4523–4530.

Wakil, S., Stoops, J. K., and Joshi, V. C., 1983. Fatty acid synthesis and its regulation. *Annual Review of Biochemistry* **52:**537–579.

"Study of an enzyme, a reaction, or a sequence can be biologically relevant only if its position in the hierarchy of function is kept in mind."

Daniel E. Atkinson (1921–) *Cellular Energy Metabolism and Its Regulation* (1977)

Chapter 25

Metabolic Integration and the Unidirectionality of Pathways

Outline

25.1 A Systems Analysis of Metabolism

25.2 Metabolic Stoichiometry and ATP Coupling

25.3 Unidirectionality

25.4 Metabolism in a Multicellular Organism

Ballet class, University of San Francisco. Metabolic integration is achieved through the highly regulated choreography of thousands of enzymatic reactions.

In the preceding chapters in this section (Part III: Metabolism and Its Regulation), we have explored the major metabolic pathways—glycolysis, the citric acid cycle, electron transport and oxidative phosphorylation, gluconeogenesis, photosynthesis, fatty acid oxidation, and lipid biosynthesis. Several of these pathways are catabolic and serve to generate chemical energy that is useful to the cell; others are anabolic and use this energy to drive the synthesis of essential biomolecules. Despite their opposing purposes, these reactions typically occur at the same time, so that food molecules are broken down to provide the building blocks and energy for ongoing biosynthesis. Cells maintain a dynamic steady state through processes that involve considerable metabolic flux. We can gain a broader understanding of metabolism and biological processes in general if we step back and consider intermediary metabolism at a systems level of organization. Our goal is to integrate metabolic pathways into a regulated, orderly, responsive whole that is in accord with the vitality and stability of cells.[1]

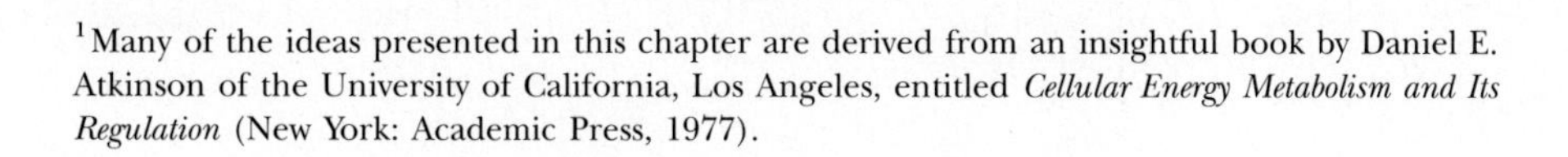

[1]Many of the ideas presented in this chapter are derived from an insightful book by Daniel E. Atkinson of the University of California, Los Angeles, entitled *Cellular Energy Metabolism and Its Regulation* (New York: Academic Press, 1977).

25.1 A Systems Analysis of Metabolism

The metabolism of a typical aerobic heterotrophic cell can be portrayed by a schematic diagram consisting of just three interconnected functional blocks: (1) catabolism, (2) anabolism, and (3) macromolecular synthesis and growth (Figure 25.1).

1. Catabolism. Foods are oxidized to CO_2 and H_2O in catabolism, and most of the electrons liberated are passed to oxygen via an electron transport pathway coupled to oxidative phosphorylation, so that ATP is formed. Some electrons go to reduce $NADP^+$ to NADPH, the source of reducing power for anabolism. Glycolysis, the citric acid cycle, electron transport and oxidative phosphorylation, and the pentose phosphate pathway are the principal pathways within this block. The metabolic intermediates in these pathways also serve as substrates for processes within the anabolic block.

2. Anabolism. The biosynthetic reactions that form the great variety of cellular molecules are included in anabolism. For thermodynamic reasons, the chemistry of anabolism is more complex than that of catabolism (i.e., it takes

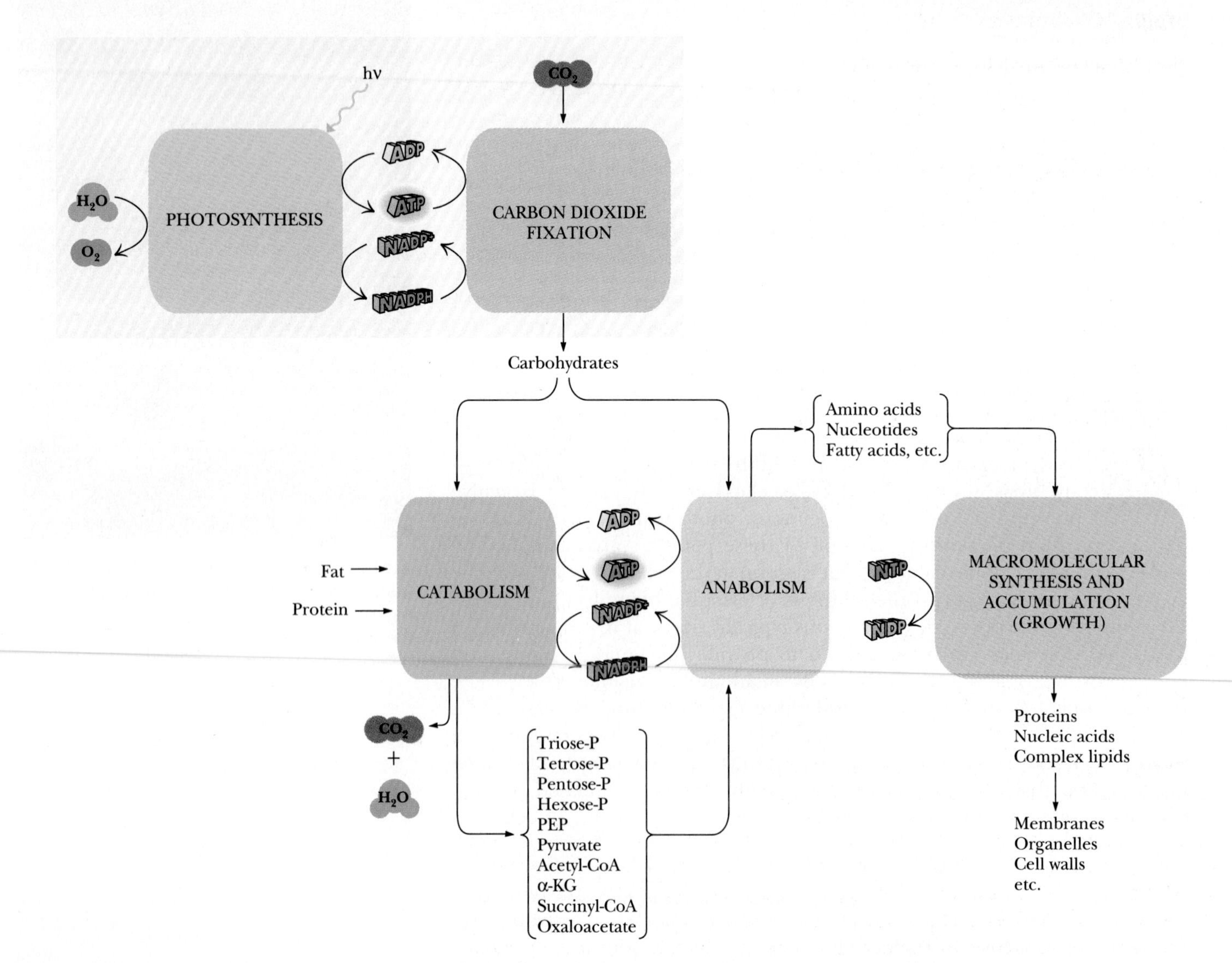

***Figure* 25.1** Block diagram of intermediary metabolism.

more energy [and often more steps] to synthesize a molecule than can be produced from its degradation). Metabolic intermediates derived from glycolysis and the citric acid cycle are the precursors for this synthesis, with NADPH supplying the reducing power and ATP the coupling energy.

3. Macromolecular Synthesis and Growth. The organic molecules produced in anabolism are the building blocks for creation of macromolecules. Like anabolism, macromolecular synthesis is driven by ATP energy, though indirectly in some cases: GTP is the principal energy source for protein synthesis, CTP for phospholipid synthesis, and UTP for polysaccharide synthesis. However, keep in mind that ATP is the ultimate phosphorylating agent for formation of GTP, CTP, and UTP from GDP, CDP, and UDP, respectively. Macromolecules are the agents of biological function and information—proteins, nucleic acids, lipids that self-assemble into membranes, and so on. Growth can be represented as cellular accumulation of macromolecules and the partitioning of these materials of function and information into daughter cells in the process of cell division.

Only a Few Intermediates Interconnect the Major Metabolic Systems

Despite the complexity of processes taking place within each block, the connections between blocks involve only a limited number of substances. Just 10 or so kinds of catabolic intermediates from glycolysis, the pentose phosphate pathway, and the citric acid cycle serve as the raw material for most of anabolism: four kinds of sugar phosphates (triose-P, tetrose-P, pentose-P, hexose-P), three α-keto acids (pyruvate, oxaloacetate, and α-ketoglutarate), two coenzyme A derivatives (acetyl-CoA and succinyl-CoA), and PEP (phosphoenolpyruvate).

ATP and NADPH Couple Anabolism and Catabolism

Metabolic intermediates are consumed by anabolic reactions and must be continuously replaced by catabolic processes. In contrast, the energy-rich compounds ATP and NADPH are recycled rather than replaced. When these substances are used in biosynthesis, the products are ADP and $NADP^+$, and ATP and NADPH are regenerated from them by oxidative reactions that occur in catabolism. ATP and NADPH are unique in that they are the only compounds whose purpose is to couple the energy-yielding processes of catabolism to the energy-consuming reactions of anabolism. Certainly, other coupling agents serve essential roles in metabolism. For example, NADH and [$FADH_2$] participate in the transfer of electrons from substrates to O_2 during oxidative phosphorylation. However, these reactions are solely catabolic, and the functions of NADH and [$FADH_2$] are fulfilled within the block called catabolism.

Phototrophs Have an Additional Metabolic System—The Photochemical Apparatus

The systems in Figure 25.1 reviewed thus far are representative only of metabolism as it exists in aerobic heterotrophs. The photosynthetic production of ATP and NADPH in photoautotrophic organisms entails a fourth block, the photochemical system (Figure 25.1). This block consumes H_2O and releases O_2. When this fourth block operates, energy production within the catabolic block can be largely eliminated. Yet another block, one to account for the fixation of carbon dioxide into carbohydrates, is also required for photoautotrophs. The inputs to this fifth block are the products of the photochemical system—ATP and NADPH—and CO_2 derived from the environment. The carbohydrate products of this block may enter catabolism, but not primarily for energy production. In photoautotrophs, carbohydrates are fed into catabolism to generate the metabolic intermediates needed to supply the block of anabolism. Although these diagrams are oversimplifications of the total metabolic repertoire of heterotrophic or phototrophic cells, they are useful illustrations of functional relationships between the major metabolic subdivisions. This general pattern provides an overall perspective on metabolism, making its purpose easier to understand.

25.2 Metabolic Stoichiometry and ATP Coupling

Stoichiometry: measurement of the amounts of chemical elements and molecules involved in chemical reactions (from the Greek *stoicheion,* element, and *metria,* measure).

Virtually every metabolic pathway either consumes or produces ATP. Furthermore, the physiological role of a pathway and its relationship to other pathways is determined in large part by whether it consumes or produces ATP. The amount of ATP involved—that is, the stoichiometry of ATP synthesis or hydrolysis—lies at the heart of metabolic relationships. It is the ATP stoichiometry that determines the overall thermodynamics of metabolic sequences. By this we mean that the overall reaction mediated by any metabolic pathway is energetically favorable because of its particular ATP stoichiometry. A significant part of the energy released in the highly exergonic reactions of catabolism is captured in ATP synthesis. In turn, energy released upon ATP hydrolysis drives the thermodynamically unfavorable reactions of anabolism. Ultimately, then, all of metabolism is thermodynamically efficient because of the involvement of ATP.

To illustrate this principle, we must first consider the three types of stoichiometries. The first two are fixed by the laws of chemistry, but the third is unique to living systems and reveals a fundamental difference between the world of chemistry and physics and the world of functional design—that is, the world of living organisms. The fundamental difference is the stoichiometry of ATP coupling.

1. Reaction Stoichiometry

This is simple chemical stoichiometry—the number of each kind of atom in any chemical reaction remains the same, and thus equal numbers must be present on both sides of the equation. This requirement holds even for a process as complex as cellular respiration:

$$C_6H_{12}O_6 + 6\ O_2 \longrightarrow 6\ CO_2 + 6\ H_2O$$

The six carbons in glucose appear as 6 CO_2, the 12 H of glucose appear as the 12 H in six molecules of water, and the 18 oxygens are distributed between CO_2 and H_2O.

2. Obligate Coupling Stoichiometry

Cellular respiration is an oxidation–reduction process, and the oxidation of glucose is coupled to the reduction of NAD^+ and [FAD]. (Brackets here denote that the relevant FAD is covalently linked to succinate dehydrogenase [Chapter 20].) The NADH and [$FADH_2$] thus formed are oxidized in the electron transport pathway:

(a) $C_6H_{12}O_6 + 10\ NAD^+ + 2\ [FAD] + 6\ H_2O \longrightarrow 6\ CO_2 + 10\ NADH + 10\ H^+ + 2\ [FADH_2]$

(b) $10\ NADH + 10\ H^+ + 2\ [FADH_2] + 6\ O_2 \longrightarrow 12\ H_2O + 10\ NAD^+ + 2\ [FAD]$

Sequence (a) accounts for the oxidation of glucose via glycolysis and the citric acid cycle. Sequence (b) is the overall equation for electron transport per glucose. The stoichiometry of coupling by the biological e^- carriers, NAD^+ and FAD, is fixed by the chemistry of electron transfer; each of the coenzymes serves as an e^- pair acceptor. Reduction of each O atom takes an e^- pair. Metabolism must obey these facts of chemistry: biological oxidation of glucose releases 12 e^- pairs, creating an obligate requirement for 12 equivalents of e^- pair acceptors, which transfer the electrons to 12 O atoms.

3. Evolved Coupling Stoichiometries

The participation of ATP is fundamentally different from the role played by pyridine nucleotides and flavins. The stoichiometry of adenine nucleotides in metabolic sequences is not fixed by chemical necessity. Instead, the "stoichiometries" we observe are the consequences of evolutionary design. The overall equation for cellular respiration, including the coupled formation of ATP by oxidative phosphorylation, is[2]

$$C_6H_{12}O_6 + 6\ O_2 + 38\ ADP + 38\ P_i \longrightarrow 6\ CO_2 + 38\ ATP + 44\ H_2O$$

The "stoichiometry" of ATP formation, $38\ ADP + 38\ P_i \rightarrow 38\ ATP + 38\ H_2O$, cannot be predicted from any chemical considerations. The value of 38 ATP is an end result of biological adaptation. It is a phenotypic character of organisms—that is, a trait acquired through interaction of heredity and environment over the course of evolution. Like any evolved phenotypic character, this ATP stoichiometry is the result of compromise. As we shall see, more than 38 ATP equivalents are available from the free energy of glucose oxidation, and certainly the biological system could also yield less. Nevertheless, the final trait is one particularly suited to the fitness of the organism.

The Significance of 38 ATP/Glucose in Cellular Respiration

The standard free energy of ATP hydrolysis, −30.5 kJ/mol, is a large negative number. Under physiological conditions of prevailing cellular ATP, ADP, and P_i concentrations, the actual free energy of ATP hydrolysis is probably closer to −50 kJ/mol. The important point is that a change in the ATP "stoichiometry" dramatically affects the K_{eq} in a coupled reaction. The K_{eq} of interest here can be calculated from the standard free energy change for glucose oxidation, −2870 kJ/mol. Under cellular conditions where the concentrations of glucose, O_2, and CO_2 may be taken as 10 m*M*, 0.13 atm, and 0.05 atm, respectively,[3] the free energy change, ΔG, is −2867 kJ/mol (virtually identical to $\Delta G°'$). The K_{eq} for the coupling of glucose oxidation is

$$K_{eq} = [CO_2]^6[ATP]^n/[glucose][ADP]^n[P_i]^n[O_2]^6$$

[2]This overall equation for cellular respiration is for the reaction within an uncompartmentalized (bacterial) cell. In eukaryotes, where much of the cellular respiration is compartmentalized within mitochondria, mitochondrial ADP/ATP exchange imposes a metabolic cost on the proton gradient of 1 H^+ per ATP, so that the overall yield of ATP per glucose is 32, not 38.

[3]The concentrations for O_2 and CO_2 taken as approximating cellular conditions are derived from the partial pressure for these gases in arterial blood, namely, 0.13 atm (100 mm Hg) for pO_2 and 0.05 atm (40 mm Hg) for pCO_2.

The superscript n for ATP, ADP, and P_i denotes that the stoichiometry for ATP is not fixed and could vary. If n = 38 and each ATP "costs" 50 kJ/mol, then the formation of 38 ATP coupled to glucose oxidation will have an overall ΔG of -967 kJ/mol (-2867 kJ/mol + 1900 kJ/mol). Since $\Delta G = -RT\ln K_{eq}$, this free energy change corresponds to a K_{eq} of 10^{170}, an extremely large number! Almost 58 ATP (58 × 50 = 2900 kJ/mol) could be formed from the oxidation of a mole of glucose, if the coupled reactions had an overall $\Sigma\Delta G$ of about 0 ($K_{eq} = 1$). At the physiological value of n = 38, the astronomically large K_{eq} for cellular respiration means that under such conditions the combustion of glucose is highly favorable from a thermodynamic point of view. That is, the reaction is emphatically spontaneous and will go essentially to completion. In the other case, where n = 58 and $K_{eq} = 1$, the reaction will come to equilibrium before much glucose has been oxidized. Clearly, this would not be advantageous to an organism using glucose for energy, since at equilibrium, no more energy could be obtained even though some glucose was still available for metabolism. The limitation in not being able to utilize all the glucose offsets any advantage resulting from an apparently greater yield of ATP per mole of glucose (58 versus 38).

The number 38 is not magical. Recall that in eukaryotes, the net yield of ATP per glucose is 30–32, not 38 (Table 20.4). Also, the value 38 was established a long time ago in evolution, when the prevailing atmospheric conditions and the competitive situation were undoubtedly very different from those today. The significance of this number is that it provides a high yield of ATP for every glucose, yet the yield is still low enough that glucose is metabolized essentially to completion.

The Significance of Large K_{eq}

In a nutshell, if ATP yields per glucose were greater, the K_{eq} for the overall process would be reduced and the minimal level of glucose that could be utilized effectively would be higher. If [glucose] becomes limiting, the organism is at a disadvantage. A very large overall K_{eq} is desirable since it means that [glucose] will not become limiting unless it falls to a very low concentration. Another important consequence of a large K_{eq} is that the reaction will be far from equilibrium at physiological concentrations of reactants and products. Metabolic control then becomes feasible, since regulation can be imposed only on processes displaced from equilibrium.

The ATP Equivalent

Since ATP is coupled with virtually every metabolic process, it is of interest to know the evolved stoichiometric relationships that accompany reactions and pathways. We can define the **coupling coefficient** for a process as the moles of ATP produced or consumed per mole of substrate converted (or product formed). Cellular respiration of glucose thus has a coupling coefficient ranging from 30 to 38, depending on cell type. A reaction such as pyruvate kinase has a coupling coefficient of 1 in the physiological direction, whereas the phosphofructokinase and hexokinase reactions, each of which consumes 1 ATP, have coupling coefficients of -1. These coefficients allow us to put a **metabolic price** on transactions in metabolism. A glucose, then, is maximally "worth" 38 ATP, and it "costs" 1 ATP to make glucose-6-phosphate from glucose via the hexokinase reaction. These points justify the oft-repeated statement that "ATP is the energy currency of the cell."

The metabolic unit of exchange is the **ATP equivalent,** defined as the conversion of ATP to ADP (or ADP to ATP). In some metabolic reactions, such as the activation of fatty acids, ATP is converted to AMP and PP_i. Since two phosphorylation events are required to re-form ATP from AMP, the ATP equivalent for such reactions is -2.

The ATP Value of NADH and $FADH_2$

Because of metabolic interrelationships, it is possible to express all metabolic conversions in terms of ATP equivalents and to assign values, or "prices," to metabolic intermediates. Such expressions are particularly useful for describing the values of common coupling agents such as NADH, NADPH, and $FADH_2$ in metabolic transactions. The value of NADH is 3 ATP (2.5 in mitochondria)—the number of ATPs formed as a consequence of NADH oxidation in the process of oxidative phosphorylation. In like fashion, $FADH_2$ oxidation yields 2 ATP (1.5 in mitochondria), so $FADH_2$ has an ATP value of 2.

The ATP Value of NADPH

The metabolic value of NADPH is less obvious. NADPH is not oxidized via electron transport to generate ATP. Also, NAD and NADP do not serve the same purposes in metabolism, despite their great chemical similarity. Most dehydrogenases show a marked preference for one or the other of these two pyridine nucleotides. In particular, the dehydrogenases of catabolism are typically NAD-specific, whereas anabolic dehydrogenases are characteristically NADP-dependent. Catabolism is an oxidative process, and substrate oxidation is coupled to NAD^+ reduction to give NADH. NADH is subsequently reconverted to NAD^+ by oxidative phosphorylation. In contrast, anabolism is inherently reductive in nature and NADPH is the usual reductant. $NADP^+$ is reduced to NADPH via catabolic sequences dedicated to this purpose, as in the glucose-6-phosphate dehydrogenase and 6-phosphogluconate dehydrogenase of the pentose phosphate pathway. In effect, NADPH acts as a primary coupling agent that carries reducing power released in certain catabolic sequences to reductive biosynthetic processes.

Several lines of reasoning suggest that the metabolic value of NADPH is greater than that of NADH, even though the standard reduction potentials for the $NAD^+/NADH$ and $NADP^+/NADPH$ couples are the same, -320 mV (Table 20.1). First, in the cell, the concentration of NAD^+ is significantly greater than that of NADH, whereas for $NADP^+$ and NADPH the situation is reversed (i.e., $[NAD^+] > [NADH]$ and $[NADP^+] < [NADPH]$; see Chapter 21). Because of these inequalities, the cellular reduction potentials of these two pyridine nucleotides are not equivalent, and under prevailing cellular conditions, the $NADP^+/NADPH$ couple is a much better electron donating system than the $NAD^+/NADH$ couple.

Second, mitochondria express a *transhydrogenase activity* whereby reducing equivalents are transferred from NADH in order to reduce $NADP^+$, with the expenditure of 1 ATP:

$$NADH + NADP^+ + ATP \rightleftharpoons NAD^+ + NADPH + ADP + P_i$$

Since each NADH is worth 2.5–3 ATP, we can surmise from this reaction that each NADPH has a value of 3.5–4 ATP.

The Nature and Magnitude of the ATP Equivalent

The fundamental biological purpose of ATP as an energy coupling agent is to drive thermodynamically unfavorable reactions. (As a corollary, metabolic sequences composed of thermodynamically favorable reactions are exploited to drive the phosphorylation of ADP to make ATP.) Nature has devised enzymatic mechanisms that couple unfavorable reactions with ATP hydrolysis. We have cited many such reactions in preceding chapters. These mechanisms provide new routes for the transformation of substances, routes that are thermodynamically favorable. In effect, the energy release accompanying ATP hydrolysis is transmitted to the unfavorable reaction so that the overall free

A Deeper Look

ATP Changes the K_{eq} for a Process by a Factor of 10^8

Consider a process, $A \rightleftharpoons B$. It could be a biochemical reaction, or the transport of an ion against a concentration gradient, or even a mechanical process (such as muscle contraction). Assume that it is a thermodynamically unfavorable reaction. Let's say, for purposes of illustration, that $\Delta G^{\circ\prime} = +13.8$ kJ/mol. From the equation,

$$\Delta G^{\circ\prime} = -RT \ln K_{eq}$$

we have

$$+13{,}800 = -(8.31 \text{ kJ/K} \cdot \text{mol})(298 \text{ K}) \ln K_{eq}$$

which yields

$$\ln K_{eq} = -5.57$$

Therefore,

$$K_{eq} = 0.0038 = [B_{eq}]/[A_{eq}]$$

This reaction is clearly unfavorable (as we could have foreseen from its positive $\Delta G^{\circ\prime}$). At equilibrium, there is one molecule of product B for every 263 molecules of reactant A. Not much A was transformed to B.

Now suppose the reaction $A \rightleftharpoons B$ is coupled to ATP hydrolysis, as is often the case in metabolism:

$$A + ATP \rightleftharpoons B + ADP + P_i$$

The thermodynamic properties of this coupled reaction are the same as the sum of the thermodynamic properties of the partial reactions:

$$\begin{array}{ll} A \rightleftharpoons B & \Delta G^{\circ\prime} = +13.8 \text{ kJ/mol} \\ ATP + H_2O \rightleftharpoons ADP + P_i & \Delta G^{\circ\prime} = -30.5 \text{ kJ/mol} \\ \hline A + ATP + H_2O \rightleftharpoons B + ADP + P_i & \Delta G^{\circ\prime} = -16.7 \text{ kJ/mol} \end{array}$$

That is,

$$\Delta G^{\circ\prime}_{overall} = -16.7 \text{ kJ/mol}$$

So

$$\begin{aligned} -16{,}700 = RT \ln K_{eq} &= -(8.31)(298) \ln K_{eq} \\ \ln K_{eq} &= -16{,}700/-2476 = 6.75 \\ K_{eq} &= 850 \end{aligned}$$

Using this equilibrium constant, let us now consider the cellular situation in which the concentrations of A and B are brought to equilibrium in the presence of typical prevailing concentrations of ATP, ADP, and P_i.[4]

$$K_{eq} = \frac{[B_{eq}][ADP][P_i]}{[A_{eq}][ATP]}$$

$$850 = \frac{[B_{eq}][8 \times 10^{-3}][10^{-3}]}{[A_{eq}][8 \times 10^{-3}]}$$

$$[B_{eq}]/[A_{eq}] = 850{,}000$$

Comparison of the $[B_{eq}]/[A_{eq}]$ ratio for the simple $A \rightleftharpoons B$ reaction with the coupling of this reaction to ATP hydrolysis gives

$$\frac{850{,}000}{0.0038} = 2.2 \times 10^8$$

The equilibrium ratio of B to A is more than 10^8 greater when the reaction is coupled to ATP hydrolysis. A reaction that was clearly unfavorable ($K_{eq} = 0.0038$) has become emphatically spontaneous!

The involvement of ATP has raised the equilibrium ratio of B/A by more than 200 million–fold. It is informative to realize that this multiplication factor does not depend on the nature of the reaction. Recall that we defined $A \rightleftharpoons B$ in the most general terms. Also, the value of this equilibrium constant ratio, some 2.2×10^8, is not at all dependent on the particular reaction chosen or its standard free energy change, $\Delta G^{\circ\prime}$. You can satisfy yourself on this point by choosing some value for $\Delta G^{\circ\prime}$ other than +13.8 kJ/mol and repeating these calculations (keeping the concentrations of ATP, ADP, and P_i at 8, 8, and 1 mM, as before).

Phosphoric anhydride linkages

ATP
(adenosine-5'-triphosphate)

[4]The concentrations of ATP, ADP, and P_i in a normal, healthy bacterial cell growing at 25°C are maintained at roughly 8 mM, 8 mM, and 1 mM, respectively. Therefore, the ratio $[ADP][P_i]/[ATP]$ is about 10^{-3}. Under these conditions, ΔG for ATP hydrolysis is approximately −47.6 kJ/mol.

energy change for the coupled process is negative, i.e., favorable. The involvement of ATP serves to alter the free energy change for a reaction; or, to put it another way, the role of ATP is to change the equilibrium ratio of [reactants] to [products] for a reaction.

Another way of viewing these relationships is to note that, at equilibrium, the concentrations of ADP and P_i will be vastly greater than that of ATP because $\Delta G°'$ for ATP hydrolysis is a large negative number.[5] However, the cell where this reaction is at equilibrium is a dead cell. The living cell metabolizes food molecules to generate ATP. These catabolic reactions proceed with a very large overall decrease in free energy. Kinetic controls over the rates of the catabolic pathways are designed to ensure that the $[ATP]/([ADP][P_i])$ ratio is maintained very high. *The cell, by employing kinetic controls over the rates of metabolic pathways, maintains a very high* $[ATP]/([ADP][P_i])$ *ratio so that ATP hydrolysis can serve as the driving force for virtually all biochemical events.*

ATP and the Solvent Capacity of the Cell

The fact that ATP hydrolysis renders metabolic reactions thermodynamically favorable also has important consequences for the solvent capacity of the cell. By **solvent capacity** we mean the capacity of the cell to keep all of its essential metabolites and macromolecules in an appropriate state of solvation. Any cell contains several thousands of solutes—metabolites, proteins, nucleic acids—and their concentrations must be controlled for at least two reasons. First, for so many compounds to exist together in the same solution, their individual concentrations must be very low. Second, many of these compounds are rather reactive substances, and the likelihood of undesirable side reactions between them is directly proportional to their concentrations. Hence, the cell is challenged to maintain each metabolite at a low effective concentration in order to sustain its solvent capacity and metabolic order.

The role of ATP in the avoidance of high metabolite concentrations and the protection of the solvent capacity of the cell can be illustrated by the following example. Consider the activation of glucose by phosphorylation to glucose-6-phosphate. This reaction initiates the catabolism of glucose via cellular respiration by feeding glucose into glycolysis. Suppose that this activation of glucose could occur in two different ways: (1) the direct phosphorylation of glucose by P_i or (2) the transfer of a phosphoryl group from ATP.

$$\text{Glucose} + P_i \rightleftharpoons \text{glucose-6-phosphate} + H_2O \qquad (1)$$

The free energy change for direct phosphorylation of glucose is +13.9 kJ/mol, and $K_{eq} = 0.0037$. At equilibrium,

$$[P_i] = \frac{[\text{glucose-6-phosphate}]}{0.0037[\text{glucose}]}$$

Therefore, to maintain an effective ratio of [G-6-P]/[glucose]—say, 10:1—the concentration of P_i would have to be almost 2700 *M*! Even if [G-6-P] equaled [glucose], $[P_i]$ would have to equal 270 *M*, a physical impossibility since there is not enough room in 1 liter for the more than 200 kg of phosphate required. Note that this analysis has nothing to do with the nature of the enzyme that might catalyze the phosphorylation; the argument is strictly thermodynamic. As a thermodynamically stable intermediate, inorganic phosphate would have to be present in enormous amounts, with deleterious consequences to the solvent capacity of the cell.

[5] Since $\Delta G°' = -30.5$ kJ/mol, $\ln K_{eq} = 12.3$. So $K_{eq} = 2.2 \times 10^5$. Choosing starting conditions of [ATP] = 8 m*M*, [ADP] = 8 m*M*, and $[P_i]$ = 1 m*M*, we can assume that, at equilibrium, [ATP] has fallen to some insignificant value *x*, [ADP] = approximately 16 m*M*, and $[P_i]$ = approximately 9 m*M*. The concentration of ATP at equilibrium, *x*, then calculates to be about 1 n*M*.

$$\text{ATP} + \text{glucose} \rightleftharpoons \text{glucose-6-phosphate} + \text{ADP} \qquad (2)$$

For the ATP-dependent phosphorylation of glucose, $\Delta G°' = -16.7$ kJ/mol. This is the reaction catalyzed by hexokinase, the first enzyme of glycolysis; $K_{eq} = 850$.

At equilibrium, 850 = [G-6-P][ADP]/[glucose][ATP].

If ATP and ADP are approximately equal under physiological conditions, a ratio of 850 G-6-P to1 glucose can be maintained. So if a cell were provided with 10 m*M* glucose, and the hexokinase reaction were allowed to come to equilibrium (with no removal of G-6-P down the glycolytic pathway—an unlikely scenario), the concentration of G-6-P would be 9.988 m*M*, whereas the concentration of glucose would be 0.012 m*M*.

The essential feature here is that, in case (1), virtually all of the glucose and phosphate remains, since P_i is thermodynamically stable. The formation of G-6-P is not favored. In case (2), ATP can be viewed as an activated form of phosphate, and its phosphoryl group is readily transferred to glucose. The thermodynamic favorability of phosphoryl transfer by ATP makes it possible for the cell to carry out reactions with efficiency, even though the substances are present in low concentrations. Maintenance of low metabolite concentrations is one of the most important benefits that cells receive from the participation of ATP in metabolism. A variety of metabolic activating agents has evolved, including ATP, NADPH, and coenzyme A. These activating agents have in common the ability to provide metabolic routes that avoid accumulation of thermodynamically stable intermediates.

Substrate Cycles Revisited

If the ATP coupling coefficient for a metabolic sequence in one direction differs from the ATP coupling coefficient for the same sequence in the opposite direction, then it is conceivable that the two reactions could constitute a *substrate cycle* (Chapter 21), where ATP would be hydrolyzed with no net conversion of substrate to product in either direction.

Reconsider, for example, the ATP-dependent formation of fructose-1,6-bisphosphate by phosphofructokinase (PFK) in glycolysis versus the conversion of fructose-1,6-bisphosphate to fructose-6-phosphate by fructose bisphosphatase (FBPase) during gluconeogenesis:

$$\begin{array}{lrcl} & \text{Fructose-6-phosphate} + \text{ATP} & \xrightarrow{\text{PFK}} & \text{fructose-1,6-bisphosphate} + \text{ADP} \\ & \text{Fructose-1,6-bisphosphate} + \text{H}_2\text{O} & \xrightarrow{\text{FBPase}} & \text{fructose-6-phosphate} + \text{P}_i \\ \hline \textit{Net:} & \text{ATP} + \text{H}_2\text{O} & \longrightarrow & \text{ADP} + \text{P}_i \end{array}$$

The ATP coupling coefficient for the PFK reaction is −1; for FBPase it is 0, since ATP is neither produced nor consumed. Now, in living cells, such substrate cycles are usually prevented by kinetic controls that govern the activity of the enzymes catalyzing the reactions in such putative cycles (Figure 25.2). That is, allosteric effectors reciprocally regulate the two enzymes so that only one is significantly active, and ATP energy is not dissipated fruitlessly (Chapter 21).

The $\Delta G°'$ for the FBPase reaction is −16.7 kJ/mol, so K_{eq} is about 850. Therefore, if [P_i] were 1 m*M*, the equilibrium ratio of [F-6-P]/[FBP] would be 850,000! Under virtually any likely cellular condition, the FBPase reaction is thermodynamically favorable in the direction of F-6-P formation. (If [FBP] were as low as 10^{-6} *M* (an unlikely situation), the FBPase reaction would remain favorable until [F-6-P] > 850 m*M*. For the PFK reaction, $\Delta G°'$ is

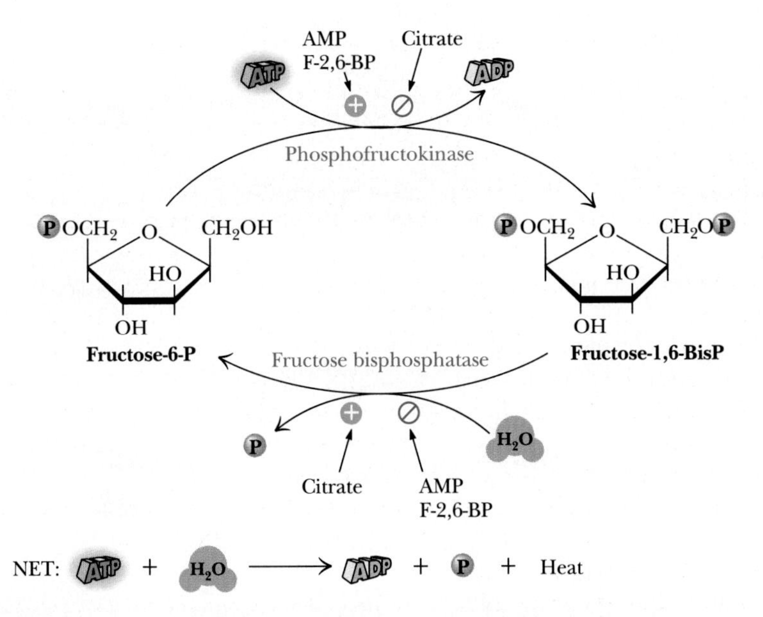

***Figure* 25.2** A substrate cycle, in which the interconversion of fructose-6-phosphate and fructose-1,6-bisphosphate by phosphofructokinase (PFK) and fructose bisphosphatase (FBPase) is accompanied by a net hydrolysis of ATP. Such substrate cycles, or *futile cycles,* as they are also called, are generally prevented by reciprocal regulatory controls: AMP and fructose-2,6-bisphosphate stimulate PFK and inhibit FBPase, whereas citrate inhibits PFK and stimulates FBPase.

In some circumstances, substrate cycles may be advantageous by generating heat through ATP hydrolysis. For example, a bumblebee must maintain a temperature of 30°C in its thorax in order to fly. If the ambient temperature is only 10°C, simultaneous high activity of both PFK and FBPase releases heat by ATP hydrolysis, permitting flight. In bumblebees, FBPase is not inhibited by AMP—an adaptation that favors heat generation by this substrate cycle.

−14.2 kJ/mol, and K_{eq} = 310. If the concentrations of ATP and ADP are roughly equal, the equilibrium ratio of [FBP]/[F-6-P] = 310. So the PFK reaction is favorable in the direction of FBP synthesis until [FBP] > 310[F-6-P]. Consequently, *both* the PFK and the FBPase reactions are favorable in their respective physiological directions as long as [FBP]/[F-6-P] is between 0.0000012 and 310.[6] Since it is certain that the ratio of [FBP] to [F-6-P] will always be within this range, the reaction that actually occurs depends on levels of allosteric effectors that exert kinetic control over these two regulatory enzymes. Or, to put it another way, either reaction is favorable; the one that occurs is determined by the metabolic needs of the cell for either glycolysis or gluconeogenesis.[7]

A very important feature of all pairs of oppositely directed metabolic sequences is illustrated by such substrate cycles. *The ATP coefficients for the opposing metabolic sequences always differ, and this difference allows both sequences to be thermodynamically favorable at all times.* The choice as to which sequence operates is decided by metabolic needs as signaled by the changing concentrations of allosteric effectors.

[6]Note that the cellular conditions stipulated in this analysis are [P_i] = 1 m*M* and [ATP] = [ADP]. These are reasonable approximations of real conditions. If other conditions prevail, these values will change somewhat. Note also that the ratio between 310 and 0.0000012 is about 2.6×10^8, this factor being the amount by which ATP coupling shifts the equilibrium ratio of [reactants] to [products].

[7]End-of-chapter Problem 1 analyzes another substrate cycle: the pyruvate/PEP cycle.

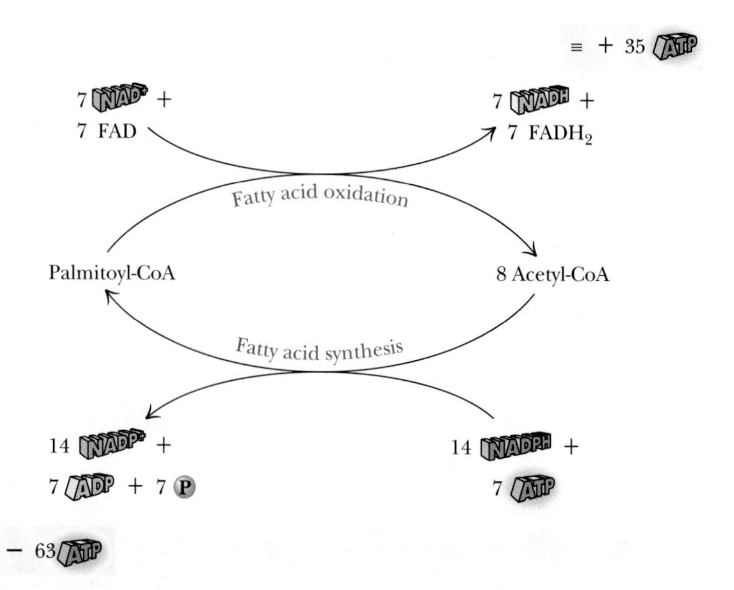

***Figure* 25.3** ATP coupling coefficients for fatty acid oxidation and fatty acid synthesis. *Fatty acid oxidation:* Each NADH is worth 3 ATP, and each $FADH_2$ is worth 2, so the ATP coupling coefficient for palmitoyl-CoA oxidation is +35. *Fatty acid synthesis:* Each NADPH is worth 4 ATP, so the ATP coupling coefficient for palmitoyl-CoA synthesis is −63.

The difference in coupling coefficients for these two sequences is 28 ATP equivalents, which guarantees that each sequence is thermodynamically favorable and that the net conversion between palmitoyl-CoA and acetyl-CoA is not determined by equilibrium considerations, but instead by metabolic need.

25.3 Unidirectionality

Although opposing metabolic sequences may share steps in common (glycolysis and gluconeogenesis provide a good example), in a functional sense, metabolic pathways can be treated as *unidirectional.* That is, they are either catabolic or anabolic, *not* both, and they proceed in only one direction, fulfilling only one physiological purpose—synthesis *or* degradation. The rates of these opposing metabolic sequences are controlled by allosteric regulators that modulate the activity of regulatory enzymes at key positions in each pathway. Thus, the net flux of substrate depends on the relative rates of opposing pathways, rates that are set by allosteric modulators responding to the immediate metabolic needs of the cell.

For this to be possible, both members of any pair of opposing metabolic pathways, such as fatty acid oxidation and fatty acid biosynthesis, must be thermodynamically favorable at essentially the same time under the same conditions. Remember that regulation can be imposed only on reactions displaced far from equilibrium. Indeed, as Daniel Atkinson succinctly states, "Escape from simple equilibrium control, and acquisition of the ability to control the direction of metabolic conversions on the basis of metabolic need, was possible only by the evolution of oppositely directed pairs of sequences for virtually every metabolic conversion."[8] Given an adequate ATP coupling coefficient, essentially any metabolic sequence can be thermodynamically favorable. Here, then, is perhaps the most fundamental, yet most overlooked, role of ATP in metabolism: *The ATP coupling coefficient for any metabolic sequence evolved so that the overall equilibrium for the conversion is highly favorable.* This role of ATP can be termed its **stoichiometric role.** By provision of an appropriate number of ATP equivalents in a metabolic pathway, a very large K_{eq} for the overall pathway can be established. That is, in thermodynamic terms, the pathway is emphatically **unidirectional** (the pathway in the reverse direction is highly unfavorable). Thus, metabolic sequences running in opposite directions, such as fatty acid oxidation and fatty acid synthesis (Figure 25.3), have different ATP coupling coefficients, and so both are thermodynamically favorable and each is unidirectional.

[8]In Atkinson, Daniel E., 1977. *Cellular Energy Metabolism and Its Regulation.* New York: Academic Press.

(**a**) *Cellular respiration:*

$$\text{Glucose} + 6\,O_2 + 38\,ADP + 38\,P \longrightarrow 6\,CO_2 + 38\,ATP + 44\,H_2O \quad K_{eq} = 10^{170}$$

(**b**) *Hypothetical ATP coupling coefficient where glucose oxidation occurs with $K_{eq} = 1$:*

$$\text{Glucose} + 6\,O_2 + 58\,ADP + 58\,P \longrightarrow 6\,CO_2 + 58\,ATP + 64\,H_2O \quad K_{eq} = 1$$

(**c**) *Glucose oxidation with an ATP coupling coefficient of 66 is unfavorable:*

$$\text{Glucose} + 6\,O_2 + 66\,ADP + 66\,P \longrightarrow 6\,CO_2 + 66\,ATP + 72\,H_2O \quad K_{eq} = 10^{-76}$$

(**d**) *The reverse of* (**c**) *describes the photosynthetic fixation of carbon dioxide to form glucose:*

$$6\,CO_2 + 66\,ATP + 72\,H_2O \longrightarrow \text{Glucose} + 6\,O_2 + 66\,ADP + 66\,P \quad K_{eq} = 10^{76}$$

***Figure* 25.4** ATP coupling coefficients between the aerobic respiration of glucose to carbon dioxide and the photosynthetic fixation of CO_2 to glucose.

Cellular Respiration Versus CO_2 Fixation—A Vivid Illustration of the Role of ATP Stoichiometry

The most vivid illustration of the stoichiometric role of ATP is provided by the most important pair of oppositely directed metabolic pathways in the biosphere, namely, photosynthetic carbon dioxide fixation and cellular respiration. As pointed out earlier, the ATP coupling coefficient for cellular respiration is 38, and an ATP coupling coefficient of 58 for this metabolic sequence would yield a K_{eq} close to 1 (Figure 25.4). Note that the conversion of glucose to CO_2 becomes thermodynamically unfavorable if the ATP coupling coefficient is greater than 58, which of necessity means the conversion of CO_2 to glucose becomes favorable. The biological fixation of 6 CO_2 to form 1 glucose requires 12 NADPH and 18 ATP, for an ATP coupling coefficient of −66. Given this coupling coefficient, the carbon dioxide fixation pathway has an overall K_{eq} of 10^{76} (Figure 25.4).[9] Therefore, both cellular respiration and CO_2 fixation have very large overall equilibrium constants, and thus, in thermodynamic terms, both are emphatically favorable because of their respective ATP coupling coefficients. For the same reason, both of these pathways are unidirectional.

ATP Has Two Metabolic Roles

The role of ATP in metabolism is twofold:

1. It serves in a stoichiometric role to establish large equilibrium constants for metabolic conversions and give metabolic sequences a unidirectional character. This is the role referred to when we call ATP the *energy currency* of the cell.
2. ATP also serves as an important allosteric effector in the kinetic regulation of metabolism. Its concentration (relative to those of

[9] This calculation presumes that the cellular conditions during CO_2 fixation are the same as those assumed during cellular respiration: [glucose] = 10 m*M*, pO_2 = 0.13 atm, pCO_2 = 0.05 atm, and $\Delta G_{\text{ATP hydrolysis}} = -50$ kJ/mol.

ADP and AMP) is an index of the energy status of the cell and determines the rates of regulatory enzymes situated at key points in metabolism, such as PFK in glycolysis and FBPase in gluconeogenesis.

Energy Storage in the Adenylate System

Energy transduction and energy storage in the adenylate system—ATP, ADP, and AMP—lie at the very heart of metabolism. The amount of ATP used per minute by a cell is roughly equivalent to the steady-state amount of ATP it contains. Thus, the metabolic lifetime of an ATP molecule is brief. ATP, ADP, and AMP are all important effectors in exerting kinetic control on regulatory enzymes situated at key points in metabolism, so uncontrolled changes in their concentrations could have drastic consequences. The regulation of metabolism by adenylates in turn requires close control of the relative concentrations of ATP, ADP, and AMP. Some ATP-consuming reactions produce ADP; PFK and hexokinase are examples. Others lead to the formation of AMP, as in fatty acid activation by acetyl-CoA synthetases:

$$\text{Fatty acid} + \text{ATP} + \text{coenzyme A} \longrightarrow \text{AMP} + \text{PP}_i + \text{fatty acyl-CoA}$$

Adenylate Kinase Interconverts ATP, ADP, and AMP

Adenylate kinase (Chapter 18), by catalyzing the reversible phosphorylation of AMP by ATP, provides a direct connection among all three members of the adenylate pool:

$$\text{ATP} + \text{AMP} \rightleftharpoons 2\ \text{ADP}$$

The free energy of hydrolysis of a phosphoanhydride bond is essentially the same in ADP and ATP (Chapter 16), and the standard free energy change for this reaction is close to zero ($K_{eq} = 0.44$). Since 2 ADP can lead to 1 ATP by this reaction, an ADP is worth $\frac{1}{2}$ ATP.

Energy Charge

The role of the adenylate system is to provide phosphoryl groups at high group-transfer potential in order to drive thermodynamically unfavorable reactions. The capacity of the adenylate system to fulfill this role depends on

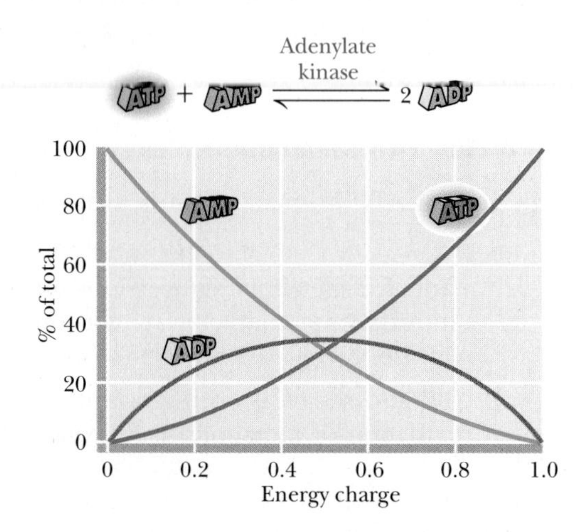

***Figure* 25.5** Relative concentrations of AMP, ADP, and ATP as a function of energy charge. (This graph was constructed assuming that the adenylate kinase reaction is at equilibrium and that $\Delta G°'$ for the reaction is -473 J/mol; $K_{eq} = 1.2$.)

how fully charged it is with phosphoric anhydrides. **Energy charge** is an index of this capacity:

$$\text{Energy charge} = \frac{1}{2}\left(\frac{2\,[\text{ATP}] + [\text{ADP}]}{[\text{ATP}] + [\text{ADP}] + [\text{AMP}]}\right)$$

The denominator represents the total adenylate pool ([ATP] + [ADP] + [AMP]); the numerator is the number of phosphoric anhydride bonds in the pool, two for each ATP and one for each ADP. The factor $\frac{1}{2}$ normalizes the equation so that energy charge, or **E.C.,** has the range 0 to 1.0. If all the adenylate is in the form of ATP, E.C. = 1.0, and the potential for phosphoryl transfer is maximal. At the other extreme, if AMP is the only adenylate form present, E.C. = 0. It is reasonable to assume that the adenylate kinase reaction is never far from equilibrium in the cell. Then the relative amounts of the three adenine nucleotides are fixed by the energy charge. Figure 25.5 shows the relative changes in the concentrations of the adenylates as energy charge varies from 0 to 1.0.

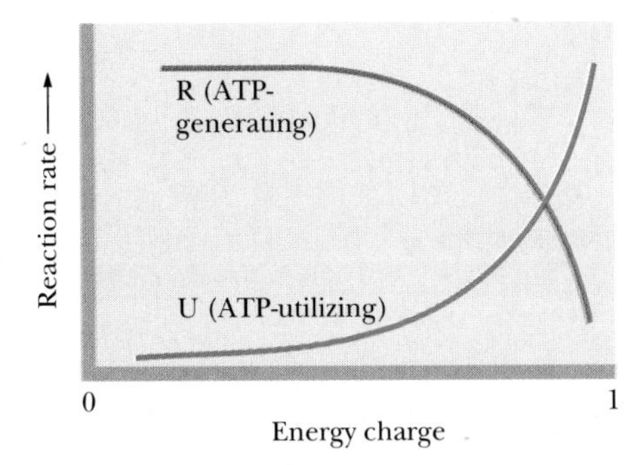

***Figure* 25.6** Responses of regulatory enzymes to variation in energy charge. Enzymes in catabolic pathways have as their ultimate metabolic purpose the regeneration of ATP from ADP. Such enzymes show an **R** pattern of response to energy charge. Enzymes in biosynthetic pathways utilize ATP to drive anabolic reactions; these enzymes follow the **U** curve in response to energy charge.

The Response of Enzymes to Energy Charge

Regulatory enzymes typically respond in reciprocal fashion to adenine nucleotides. For example, PFK is stimulated by AMP and inhibited by ATP. Clearly, the degree of activity expressed by such enzymes *in vivo* is determined by the relative amounts of these substances. Energy charge is a useful parameter to illustrate these responses. If the activities of various regulatory enzymes are examined *in vitro* as a function of energy charge, an interesting relationship appears. Regulatory enzymes in energy-producing catabolic pathways show greater activity at low energy charge, but the activity falls off abruptly as E.C. approaches 1.0. In contrast, regulatory enzymes of anabolic sequences are not very active at low energy charge, but their activities increase exponentially as E.C. nears 1.0 (Figure 25.6). These contrasting responses are termed **R**, for ATP-regenerating, and **U**, for ATP-utilizing. Regulatory enzymes such as PFK and pyruvate kinase in glycolysis follow the **R** response curve as E.C. is varied. Note that PFK itself is an ATP-utilizing enzyme, using ATP to phosphorylate fructose-6-phosphate to yield fructose-1,6-bisphosphate. Nevertheless, since PFK acts physiologically as the valve controlling the flux of carbohydrate down the catabolic pathways of cellular respiration that lead to ATP regeneration, it responds as an "**R**" enzyme to energy charge. Regulatory enzymes in anabolic pathways, such as acetyl-CoA carboxylase, which initiates fatty acid biosynthesis, respond as "**U**" enzymes.

The overall purposes of the **R** and **U** pathways are diametrically opposite in terms of ATP involvement. Note in Figure 25.6 that the **R** and **U** curves intersect at a rather high E.C. value. As E.C. increases past this point, **R** activities decline precipitously and **U** activities rise. That is, when E.C. is very high, biosynthesis is accelerated while catabolism diminishes. The consequence of these effects is that ATP is used up faster than it is regenerated, and so E.C. begins to fall. As E.C. drops below the point of intersection, **R** processes are favored over **U**. Then, ATP is generated faster than it is consumed, and E.C. rises again. The net result is that the value of energy charge oscillates about a point of **steady state** (Figure 25.7). The experimental results obtained from careful measurement of the relative amounts of AMP, ADP, and ATP in living cells reveals that normal cells have an energy charge in the neighborhood of 0.85 to 0.88. Maintenance of this steady-state value is one criterion of cell health and normalcy.

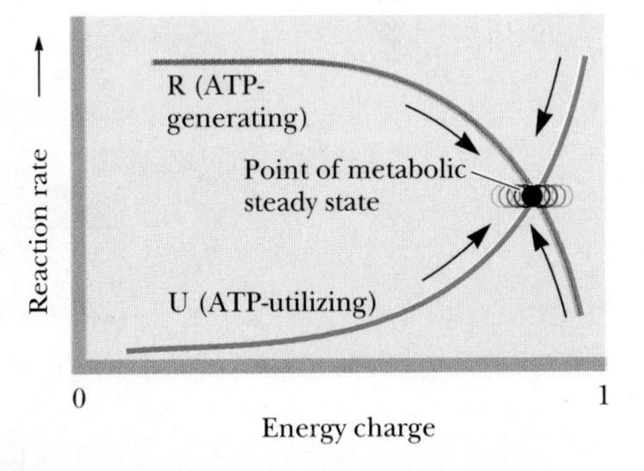

***Figure* 25.7** The oscillation of energy charge (E.C.) about a steady-state value as a consequence of the offsetting influences of **R** and **U** processes on the production and consumption of ATP. As E.C. increases, the rates of **R** reactions decline, but **U** reactions go faster. ATP is consumed, and E.C. drops. Below the point of intersection, **R** processes are more active and **U** processes are slower, so E.C. recovers. Energy charge oscillates about a steady-state value determined by the intersection point of the **R** and **U** curves.

Phosphorylation Potential

Because energy charge is maintained at a relatively constant value in normal cells, it is not an informative index of cellular capacity to carry out phosphorylation reactions. Since such reactions provide the thermodynamic drive to metabolism, a more illustrative index of phosphorylation potential would be useful. The relative concentrations of ATP, ADP, and P_i provide such a relationship, and a function called **phosphorylation potential** has been defined in terms of these concentrations:

$$ADP + P_i \rightleftharpoons ATP + H_2O$$

Phosphorylation potential, Γ, is equal to $[ATP]/([ADP][P_i])$.

Note that this expression includes a term for the concentration of inorganic phosphate. $[P_i]$ does not enter into the energy charge equation, yet its value has substantial influence on ATP-dependent reactions. In contrast with energy charge, phosphorylation potential varies over a significant range

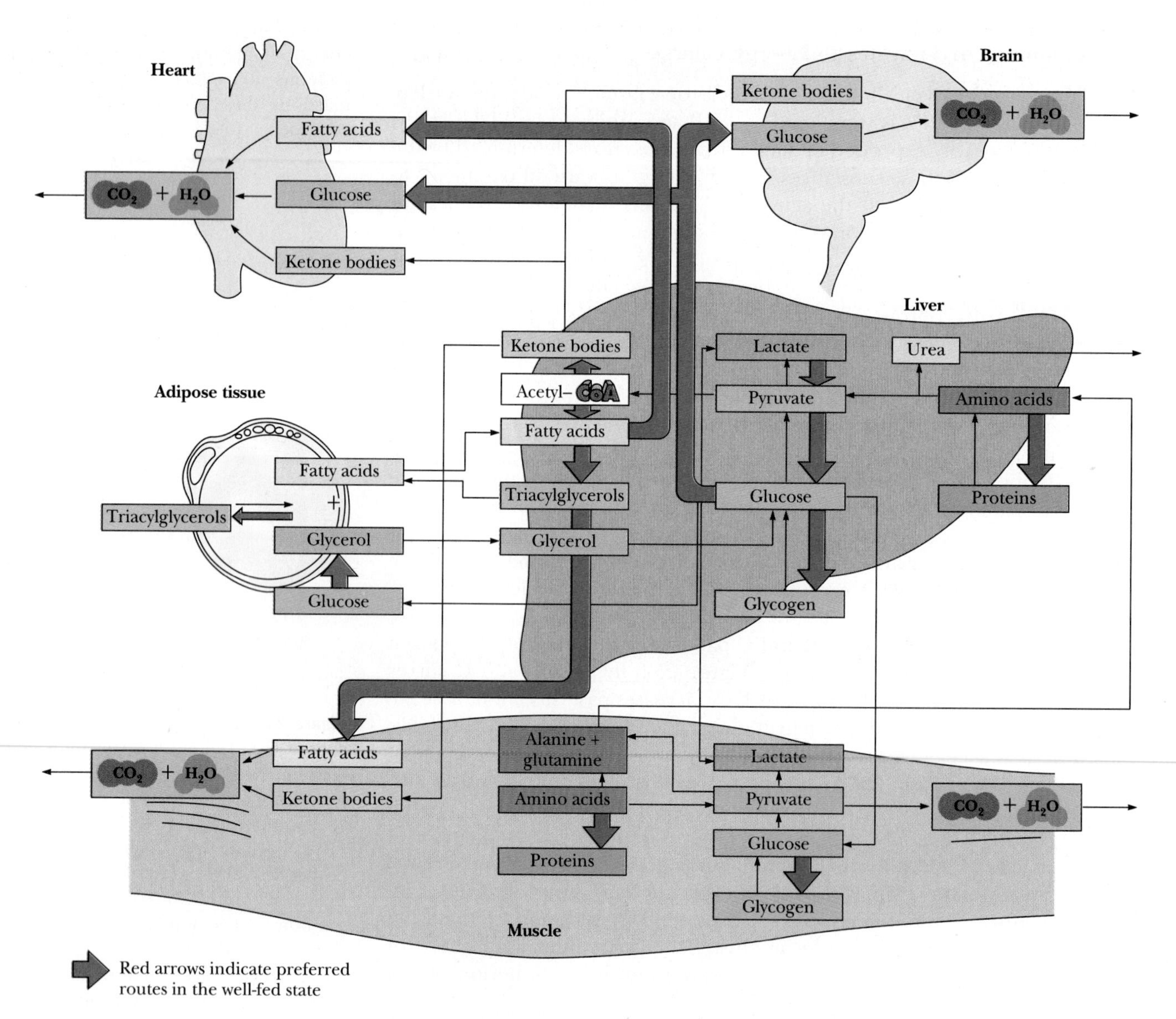

***Figure* 25.8** Metabolic relationships among the major human organs: brain, muscle, heart, adipose tissue, and liver.

under physiological conditions, and thus acts as a more sensitive index of the energy status of the cell. The actual proportions of ATP, ADP, and P_i in cells vary in response to metabolic state, so that Γ ranges from 200 to 800 M^{-1}, higher levels signifying more ATP and correspondingly greater phosphorylation potential.

25.4 Metabolism in a Multicellular Organism

In complex multicellular organisms, organ systems have arisen to carry out specific physiological functions. Each organ expresses a repertoire of metabolic pathways that is consistent with its physiological purpose. Such specialization depends on coordination of metabolic responsibilities among organs so that the organism as a whole may thrive. Essentially all cells in animals have the set of enzymes common to the central pathways of intermediary metabolism, especially the enzymes involved in the formation of ATP and the synthesis of glycogen and lipid reserves. Nevertheless, organs differ in the metabolic fuels they prefer as substrates for energy production. Important differences also occur in the ways ATP is used to fulfill the organs' specialized metabolic functions. To illustrate these relationships, we will consider the metabolic interactions among the major organ systems found in humans: brain, skeletal muscle, heart, adipose tissue, and liver. In particular, the focus will be on energy metabolism in these organs (Figure 25.8). The major fuel depots in animals are *glycogen* in liver and muscle, *triacylglycerols* (fats) stored in adipose tissue, and *protein,* most of which is in skeletal muscle. In general, the order of preference for the use of these fuels is the order given: glycogen>triacylglycerol > protein. Nevertheless, the tissues of the body work together to maintain **caloric homeostasis,** defined as *a constant availability of fuels in the blood.*

Organ Specializations

Table 25.1 summarizes the important functions of the major human organs in energy metabolism.

***Table* 25.1**
Energy Metabolism in Major Vertebrate Organs

Organ	Energy Reservoir	Preferred Substrate	Energy Sources Exported
Brain	None	Glucose (ketone bodies during starvation)	None
Skeletal muscle (resting)	Glycogen	Fatty acids	None
Skeletal muscle (prolonged exercise)	None	Glucose	Lactate
Heart muscle	Glycogen	Fatty acids	None
Adipose tissue	Triacylglycerol	Fatty acids	Fatty acids, glycerol
Liver	Glycogen, triacylglycerol	Amino acids, glucose, fatty acids	Fatty acids, glucose, ketone bodies

Brain. The brain has two remarkable metabolic features. First, it has a very high respiratory metabolism. In resting adult humans, 20% of the oxygen consumed is used by the brain, even though it constitutes only 2% or so of body mass. Interestingly, this level of oxygen consumption is independent of mental activity, continuing even during sleep. Second, the brain is an organ with no significant fuel reserves—no glycogen, usable protein, or fat (even in "fatheads"!). Normally, the brain uses only glucose as a fuel and is totally dependent on the blood for a continuous incoming supply. Interruption of glucose supply for even brief periods of time (as in a stroke) can lead to irreversible losses in brain function. Brain uses glucose to carry out ATP synthesis via cellular respiration. High rates of ATP production are necessary to power the plasma membrane Na^+/K^+-ATPase so that the membrane potential essential for transmission of nerve impulses is maintained.

During prolonged fasting or starvation, the body's glycogen reserves are depleted. Under such conditions, the brain adapts to use β-hydroxybutyrate (Figure 25.9) as a source of fuel, converting it to acetyl-CoA for energy production via the citric acid cycle. β-Hydroxybutyrate (Chapter 23) is formed from fatty acids in the liver. Although the brain cannot use free fatty acids or lipids directly from the blood as fuel, the conversion of these substances to β-hydroxybutyrate in the liver allows the brain to use body fat as a source of energy. The brain's other potential source of fuel during starvation is glucose obtained from gluconeogenesis in the liver (Chapter 21), using the carbon skeletons of amino acids derived from muscle protein breakdown. The adaptation of the brain to use β-hydroxybutyrate from fat spares protein from degradation until more drastic times.

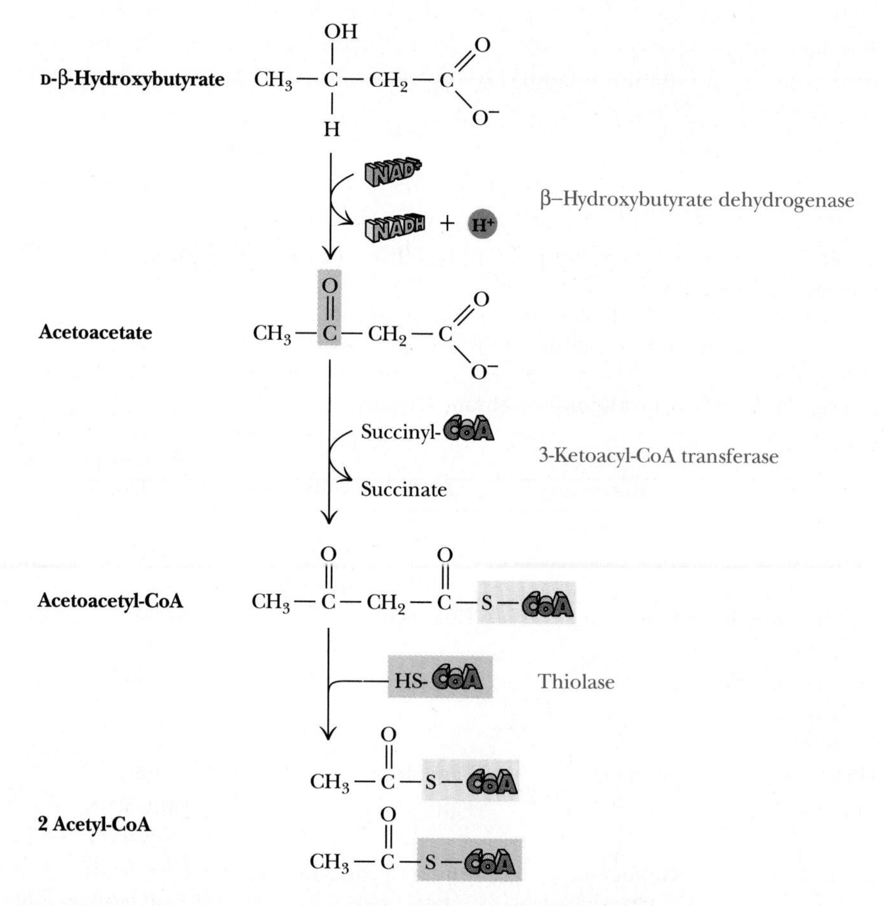

***Figure* 25.9** The structure of β-hydroxybutyrate and its conversion to acetyl-CoA for combustion in the citric acid cycle.

Muscle. Skeletal muscle is responsible for about 30% of the O_2 consumed by the human body at rest. During periods of maximal exertion, skeletal muscle can account for over 90% of the total metabolism. Muscle metabolism is primarily dedicated to the production of ATP as the source of energy for contraction and relaxation. Muscle contraction occurs when a motor nerve impulse causes Ca^{2+} release from specialized endomembrane compartments (the transverse tubules and sarcoplasmic reticulum). Ca^{2+} floods the *sarcoplasm* (the term denoting the cytosolic compartment of muscle cells), where it binds to **troponin C,** a regulatory protein, initiating a series of events that culminate in the sliding of *actin* thin filaments along *myosin* thick filaments. This mechanical movement is driven by energy released upon hydrolysis of ATP (Chapter 36). The net result is that the muscle shortens. Relaxation occurs when the Ca^{2+} ions are pumped back into the sarcoplasmic reticulum by the action of a Ca^{2+}-transporting membrane ATPase. Two Ca^{2+} are translocated per ATP hydrolyzed. The amount of ATP used during relaxation is almost as much as that consumed during contraction.

Since muscle contraction is an intermittent process that occurs upon demand, muscle metabolism is designed for a demand response. Muscle at rest uses free fatty acids, glucose, or ketone bodies as fuel and produces ATP via oxidative phosphorylation. Resting muscle also contains about 2% glycogen by weight and an amount of phosphocreatine (Figure 25.10) capable of providing enough ATP to power about 4 seconds of exertion. During strenuous exertion, such as a 100-meter sprint, once the phosphocreatine is depleted, muscle relies solely on its glycogen reserves, making the ATP for contraction via glycolysis. In contrast with the citric acid cycle and oxidative phosphorylation pathways, glycolysis is capable of explosive bursts of activity, and the flux of glucose-6-phosphate through this pathway can increase 2000-fold almost instantaneously. The triggers for this activation are Ca^{2+} and the "fight or flight" hormone *epinephrine* (Chapters 18 and 21). Little interorgan cooperation occurs during strenuous (anaerobic) exercise.

Muscle fatigue is the inability of a muscle to maintain power output. During maximum exertion, the onset of fatigue takes only 20 seconds or so. Fatigue is not the result of exhaustion of the glycogen reserves, nor is it a consequence of lactate accumulation in the muscle. Instead, it is caused by a decline in intramuscular pH as protons are generated during glycolysis. (The overall conversion of glucose to two lactate in glycolysis is accompanied by the release of two H^+). The pH may fall as low as 6.4. It is likely that the decline in PFK activity at low pH leads to a lowered flux of hexose through glycolysis and inadequate ATP levels, causing a feeling of fatigue. One benefit of PFK inhibition is that the ATP remaining is not consumed in the PFK reaction, and the cell is spared the more serious consequences of losing all of its ATP.

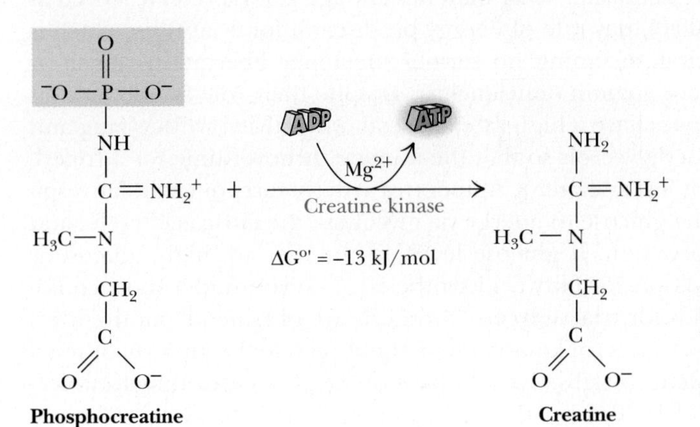

Figure **25.10** Phosphocreatine serves as a reservoir of ATP-synthesizing potential. When ADP accumulates as a consequence of ATP hydrolysis, creatine kinase catalyzes the formation of ATP at the expense of phosphocreatine. During periods of rest, when ATP levels are restored by oxidative phosphorylation, creatine kinase acts in reverse to restore the phosphocreatine supply.

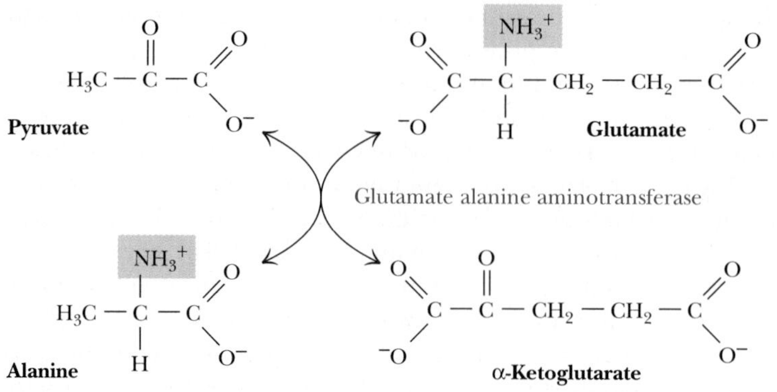

Figure **25.11** The transamination of pyruvate to alanine by glutamate:alanine aminotransferase.

Pyruvate + Glutamate → (Glutamate alanine aminotransferase) → Alanine + α-Ketoglutarate

During fasting or excessive activity, skeletal muscle protein is degraded to amino acids so that their carbon skeletons can be used as fuel. Many of the skeletons are converted to pyruvate, which can be transaminated back into alanine for export via the circulation (Figure 25.11). Alanine is carried to the liver, which in turn transaminates it back into pyruvate so that it can serve as a substrate for gluconeogenesis. Although muscle protein can be mobilized as an energy source, it is not efficient for an organism to consume its muscle and lower its overall fitness for survival. Muscle protein represents a fuel of last resort.

Heart. In contrast with the intermittent work of skeletal muscle, the activity of heart muscle is constant and rhythmic. The range of activity in heart is also much less than that in muscle. Consequently, the heart can function as a completely aerobic organ and, as such, is very rich in mitochondria. Roughly half the cytoplasmic volume of heart muscle cells is occupied by mitochondria. Under normal working conditions, the heart prefers fatty acids as fuel, oxidizing acetyl-CoA units via the citric acid cycle and producing ATP for contraction via oxidative phosphorylation. Heart tissue has minimal energy reserves: a small amount of phosphocreatine and limited quantities of glycogen. As a result, the heart must be continually nourished with oxygen and free fatty acids, glucose, or ketone bodies as fuel.

Adipose Tissue. Adipose tissue is an amorphous tissue that is widely distributed about the body—around blood vessels, in the abdominal cavity and mammary glands, and, most prevalently, as deposits under the skin. It consists principally of cells known as **adipocytes** that no longer replicate. However, adipocytes can increase in number as adipocyte precursor cells divide, and obese individuals tend to have more of them. As much as 65% of the weight of adipose tissue is triacylglycerol that is stored in adipocytes, essentially as oil droplets. A normal 70-kg man has enough caloric reserve stored as fat to sustain a 6000-kJ/day rate of energy production for 3 months, which is adequate for survival, assuming no serious metabolic aberrations (such as nitrogen, mineral, or vitamin deficiencies). Despite their role as energy storage depots, adipocytes have a high rate of metabolic activity, synthesizing and breaking down triacylglycerols so that the average turnover time for a triacylglycerol molecule is just a few days. Adipocytes actively carry out cellular respiration, transforming glucose to energy via glycolysis, the citric acid cycle, and oxidative phosphorylation. If glucose levels in the diet are high, glucose is converted to acetyl-CoA for fatty acid synthesis. However, under most conditions, free fatty acids for triacylglycerol synthesis are obtained from the liver. Since adipocytes lack glycerol kinase, they cannot recycle the glycerol of triacylglycerol, but depend on glycolytic conversion of glucose to dihydroxyacetone-3-phosphate (DHAP) and the reduction of DHAP to glycerol-3-phosphate for triacylglycerol biosynthesis. Adipocytes also require glucose to feed the pentose phosphate pathway for NADPH production.

Glucose plays a pivotal role for adipocytes. If glucose levels are adequate, glycerol-3-phosphate is formed in glycolysis, and the free fatty acids liberated in triacylglycerol breakdown are reesterified to glycerol to re-form triacylglycerols. However, if glucose levels are low, [glycerol-3-phosphate] falls, and free fatty acids are released to the bloodstream (Chapter 23).

"Brown Fat." A specialized type of adipose tissue, so-called **brown fat,** is found in newborns and hibernating animals. The abundance of mitochondria, with their rich complement of cytochromes, is responsible for the brown color of this fat. As usual, these mitochondria are very active in electron transport–driven proton translocation, but these particular mitochondria contain in their inner membranes a protein, **thermogenin** (Chapter 20), that creates a passive proton channel, permitting the H^+ ions to reenter the mitochondrial matrix without generating ATP. Instead, the energy of oxidation is dissipated as heat. Indeed, brown fat is specialized to oxidize fatty acids for heat production rather than ATP synthesis.

Liver. The liver serves as the major metabolic processing center in vertebrates. Except for dietary triacylglycerols, which are metabolized principally by adipose tissue, most of the incoming nutrients that pass through the intestinal tract are routed via the portal vein to the liver for processing and distribution. Much of the liver's activity centers around conversions involving glucose-6-phosphate (Figure 25.12). Glucose-6-phosphate can be converted to glycogen, released as blood glucose, used to generate NADPH and pentoses via the pentose phosphate cycle, or catabolized to acetyl-CoA for fatty acid synthesis or for energy production via oxidative phosphorylation. Most of the liver glucose-6-phosphate arises from dietary carbohydrate, from degradation of glycogen reserves, or from muscle lactate that enters the gluconeogenic pathway.

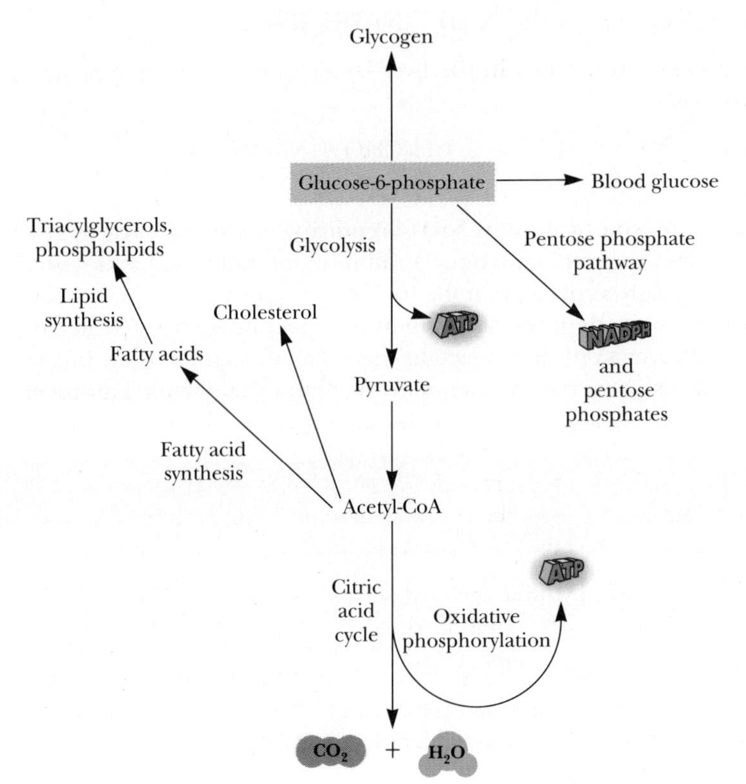

Figure **25.12** Metabolic conversions of glucose-6-phosphate in the liver.

The liver plays an important regulatory role in metabolism by buffering the level of blood glucose. Liver has two enzymes for glucose phosphorylation, hexokinase and glucokinase. Unlike hexokinase, glucokinase has a low affinity for glucose. Its K_m for glucose is high, on the order of 10 mM. When blood glucose levels are high, glucokinase activity augments hexokinase in phosphorylating glucose as an initial step leading to its storage in glycogen. The major metabolic hormones—epinephrine, glucagon, and insulin—all influence glucose metabolism in the liver to keep blood glucose levels relatively constant (Chapter 37).

The liver is a major center for fatty acid turnover. When the demand for metabolic energy is high, triacylglycerols are broken down and fatty acids are degraded in the liver to acetyl-CoA to form ketone bodies, which are exported to the heart, brain, and other tissues. If energy demands are low, fatty acids are incorporated into triacylglycerols that are carried to adipose tissue for deposition as fat. Cholesterol is also synthesized in the liver from two-carbon units derived from acetyl-CoA.

In addition to these central functions in carbohydrate and fat-based energy metabolism, the liver serves other purposes. For example, the liver can use amino acids as metabolic fuels. Amino acids are first converted to their corresponding α-keto acids by aminotransferases. The amino group is excreted after incorporation into urea in the urea cycle. The carbon skeletons of gluconeogenic amino acids can be used for glucose synthesis, whereas those of ketogenic amino acids appear in ketone bodies (Chapter 26). The liver is also the principal detoxification organ in the body. The endoplasmic reticulum of liver cells is rich in enzymes that convert biologically active substances such as hormones, poisons, and drugs into less harmful by-products.

Liver disease leads to serious metabolic derangements, particularly in amino acid metabolism. In cirrhosis, the liver becomes defective in converting NH_4^+ to urea for excretion, and blood levels of NH_4^+ rise. Ammonia is toxic to the central nervous system, and coma ensues.

Ethanol Metabolism Alters the NAD^+/NADH Ratio

Ethanol is metabolized to acetate in the liver by alcohol dehydrogenase and aldehyde dehydrogenase:

$$CH_3CH_2OH + NAD^+ \longrightarrow CH_3CHO + NADH + H^+$$
$$CH_3CHO + NAD^+ \longrightarrow CH_3COO^- + NADH + H^+$$

The excess NADH produced inhibits NAD^+-requiring reactions, such as gluconeogenesis and fatty acid oxidation. Inhibition of fatty acid oxidation causes elevated triacylglycerol levels in the liver. Over time, these triacylglycerols accumulate as fatty deposits. Inhibition of gluconeogenesis leads to buildup of this pathway's substrate, lactate. Lactic acid accumulation in the blood causes acidosis. A further consequence is that acetaldehyde can form adducts with protein $-NH_2$ groups, which may impair protein function.

Problems

1. The **pyruvate/PEP substrate cycle.** Another substrate cycle involving reactions of glycolysis and gluconeogenesis (in addition to the F-6-P/FBP cycle) is the conversion between pyruvate and phosphoenolpyruvate (PEP). The direction of net conversion is determined by the relative concentrations of allosteric regulators that exert kinetic control over pyruvate kinase, pyruvate carboxylase, and PEP carboxykinase. Recall that the last step in glycolysis is catalyzed by pyruvate kinase:

$$PEP + ADP \longrightarrow pyruvate + ATP$$

The ATP coupling coefficient for this reaction is +1; the standard free energy change is −31.7 kJ/mol.

a. Calculate the equilibrium constant for this reaction.
b. For this reaction to proceed in the reverse direction, by what factor must [pyruvate] exceed [PEP]?

The reversal of this reaction in eukaryotic cells is essential to gluconeogenesis and proceeds in two steps, each requiring an equivalent of nucleoside triphosphate energy:

$$\text{Pyruvate} + CO_2 + \text{ATP} \xrightarrow{\textit{Pyruvate carboxylase}} \text{oxaloacetate} + \text{ADP} + P_i$$

$$\text{Oxaloacetate} + \text{GTP} \xrightarrow{\textit{PEP carboxykinase}} \text{PEP} + CO_2 + \text{GDP}$$

$$\textit{Net: } \text{Pyruvate} + \text{ATP} + \text{GTP} \longrightarrow \text{PEP} + \text{ADP} + \text{GDP} + P_i$$

c. What is the coupling coefficient for this sequence?
d. Do the two halves of the substrate cycle differ in ATP coupling coefficient?
e. The $\Delta G°'$ for the overall reaction is +0.8 kJ/mol. What is the value of K_{eq}?
f. Assuming [ATP] = [ADP], [GTP] = [GDP], and P_i = 1 mM when this reaction reaches equilibrium, what is the ratio of [PEP]/[pyruvate]?
g. Are both directions in the substrate cycle likely to be strongly favored under physiological conditions?

2. Assume the following intracellular concentrations in muscle tissue: ATP = 8 mM, ADP = 0.9 mM, AMP = 0.04 mM, P_i = 8 mM. What is the *energy charge* in muscle? What is the *phosphorylation potential*?

3. Strenuous muscle exertion (as in the 100-meter dash) rapidly depletes ATP levels. How long will 8 mM ATP last if 1 gram of muscle consumes 300 μmol of ATP per minute? (Assume muscle is 70% water.) Muscle contains phosphocreatine as a reserve of phosphorylation potential. Assuming [phosphocreatine] = 40 mM, [creatine] = 4 mM, and $\Delta G°'$ (phosphocreatine + H_2O → creatine + P_i) = −43.3 kJ/mol, how low must [ATP] become before it can be replenished by the reaction: phosphocreatine + ADP → ATP + creatine? [Remember, $\Delta G°'$ (ATP hydrolysis) = −30.5 kJ/mol.]

4. The standard reduction potentials for the (NAD^+/NADH) and ($NADP^+$/NADPH) couples are identical, namely, −320 mV. Assuming the *in vivo* concentration ratios NAD^+/NADH = 20 and $NADP^+$/NADPH = 0.1, what is ΔG for the following reaction?

$$\text{NADPH} + \text{NAD}^+ \longrightarrow \text{NADP}^+ + \text{NADH}$$

Calculate how many ATP equivalents can be formed from ADP + P_i by the energy released in this reaction.

5. Assume the total intracellular pool of adenylates (ATP + ADP + AMP) = 8 mM, 90% of which is ATP. What are [ADP] and [AMP] if the adenylate kinase reaction is at equilibrium? Suppose [ATP] drops suddenly by 10%. What are the concentrations now for ADP and AMP, assuming the adenylate kinase reaction is at equilibrium? By what factor has the AMP concentration changed?

6. The **PFK/FBPase substrate cycle.** PFK is AMP-activated; FBPase is AMP-inhibited. In muscle, the maximal activity of PFK (μmol of substrate transformed per minute) is 10 times times greater than FBPase activity. If the increase in [AMP] described in Problem 5 raised PFK activity from 10% to 90% of its maximal value but lowered FBPase activity from 90% to 10% of its maximal value, by what factor is the flux of fructose-6-P through the glycolytic pathway changed? (*Hint:* Let PFK maximal activity = 10, FBPase maximal activity = 1; calculate the relative activities of the two enzymes at low [AMP] and at high [AMP]; let J, the flux of F-6-P through the substrate cycle under any condition, equal the velocity of the PFK reaction *minus* the velocity of the FBPase reaction.)

Further Reading

Atkinson, D. E., 1977. *Cellular Energy Metabolism and Its Regulation.* New York: Academic Press. A very readable book on the design and purpose of cellular energy metabolism. Its emphasis is the evolutionary design of metabolism within the constraints of chemical thermodynamics. The book is filled with novel insights regarding why metabolism is organized as it is and why ATP occupies a central position in biological energy transformations.

Harris, R., and Crabb, D. W., 1992. Metabolic Interrelationships. In *Textbook of Biochemistry with Clinical Correlations,* 3rd ed., edited by T. M. Devlin. New York: Wiley-Liss. A synopsis of the interdependence of metabolic processes in the major tissues of the human body—brain, liver, muscle, kidney, gut, and adipose tissue. Metabolic aberrations that occur in certain disease states are also discussed.

Newsholme, E. A., Challiss, R. A. J., and Crabtree, B., 1984. Substrate cycles: Their role in improving sensitivity in metabolic control. *Trends in Biochemical Sciences* **9**:277–280. A review suggesting that substrate cycles provide a mechanism for greater responsiveness to regulatory signals. (See end-of-chapter problems 1 and 6.)

Sugden, M. C., Holness, M. J., and Palmer, T. N., 1989. Fuel selection and carbon flux during the starved-to-fed transition. *Biochemical Journal* **263**:313–323. Changes in lipid and carbohydrate metabolism of rats upon carbohydrate feeding after prolonged starvation.

"One generation passes away, and another generation comes: but the earth abides forever."
Ecclesiastes 1:4

Chapter 26

Nitrogen Acquisition and Amino Acid Metabolism

Wheat harvest. Only plants and certain microorganisms are able to transform the oxidized, inorganic forms of nitrogen available in the inanimate environment into reduced, biologically useful forms.

Outline

26.1 The Two Major Pathways of Biological N Acquisition
26.2 The Fate of Ammonium
26.3 *Escherichia coli* Glutamine Synthetase: A Case Study in Enzyme Regulation
26.4 Amino Acid Biosynthesis
26.5 Metabolic Degradation of Amino Acids

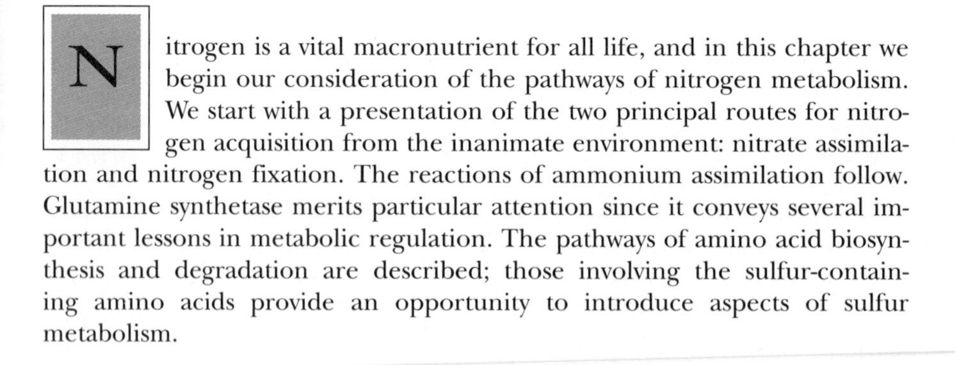

Nitrogen is a vital macronutrient for all life, and in this chapter we begin our consideration of the pathways of nitrogen metabolism. We start with a presentation of the two principal routes for nitrogen acquisition from the inanimate environment: nitrate assimilation and nitrogen fixation. The reactions of ammonium assimilation follow. Glutamine synthetase merits particular attention since it conveys several important lessons in metabolic regulation. The pathways of amino acid biosynthesis and degradation are described; those involving the sulfur-containing amino acids provide an opportunity to introduce aspects of sulfur metabolism.

26.1 The Two Major Pathways of Biological N Acquisition

The biological nitrogen cycle was introduced when we first surveyed metabolism (Chapter 17, Figure 17.4). Two general observations are particularly relevant to the biochemistry of this cycle:

1. Virtually all biologically important N-compounds contain nitrogen in a reduced form.
2. The principal inorganic forms of N in the environment are in an oxidized state. Thus, nitrogen acquisition by the biosphere (i.e., the entry of N into organisms) depends on the reduction of the oxidized inorganic forms (N_2 and NO_3^-) to NH_4^+. For the most part,

the reactions involving inorganic N-compounds occur only in microorganisms and green plants. Animals acquire their N from the catabolism of organic N-compounds, mainly proteins, obtained in the diet.

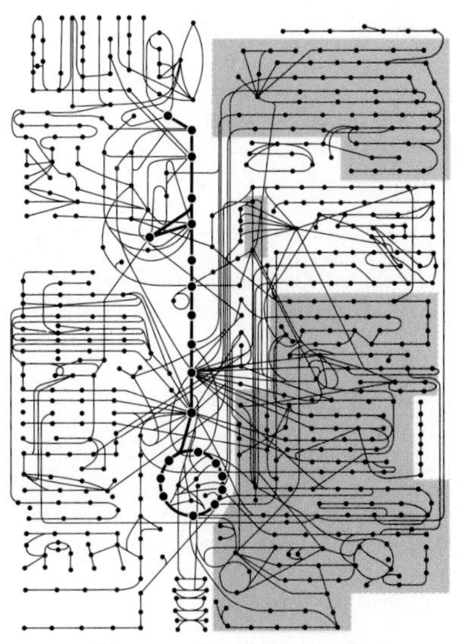

Nitrogen Acquisition and Amino Acid Metabolism

An Overview of the Biochemistry of N Acquisition

Ammonium Formation from NO_3^- and N_2

The two pathways for N acquisition in the biosphere are *nitrate assimilation* and *nitrogen fixation;* both lead to the formation of ammonium that is subsequently incorporated into organic compounds. **Nitrate assimilation** occurs in two steps: the two-electron reduction of nitrate to nitrite, followed by the six-electron reduction of nitrite to ammonium.

$$(1)\quad NO_3^- + 2\,H^+ + 2\,e^- \longrightarrow NO_2^- + H_2O \qquad (26.1)$$
$$(2)\quad NO_2^- + 8\,H^+ + 6\,e^- \longrightarrow NH_4^+ + 2\,H_2O \qquad (26.2)$$

The first reaction is catalyzed by *nitrate reductase,* and the second, by *nitrite reductase.* Nitrate assimilation is the preponderant means by which many microorganisms, algae, and green plants acquire nitrogen. *In toto,* it accounts for over 99% of the N entering the biosphere.

Nitrogen fixation involves the reduction of atmospheric nitrogen gas via an enzyme system that is exclusively prokaryotic. The heart of this enzyme system is a protein called **nitrogenase,** which catalyzes the reaction

$$N_2 + 8\,H^+ + 8\,e^- \longrightarrow 2\,NH_3 + H_2 \qquad (26.3)$$

Note that an obligatory reduction of two protons to hydrogen gas accompanies the biological reduction of N_2 to ammonia. Less than 1% of the N acquired by the biosphere can be attributed to nitrogen fixation.

The Enzymology of Nitrate Assimilation

The pathway of nitrate assimilation is found in green plants, as well as in some fungi and bacteria, and consists of two soluble electron-transfer enzymes, nitrate reductase and nitrite reductase.

A. Nitrate Reductase.

$$NADH \rightarrow [\text{—SH} \rightarrow FAD \rightarrow \text{cytochrome } b_{557} \rightarrow MoCo] \longrightarrow NO_3^-$$

A pair of electrons is transferred from NADH via enzyme-associated sulfhydryl groups, FAD, cytochrome *b,* and **MoCo** (an essential molybdenum-containing cofactor) to nitrate, reducing it to nitrite. The brackets [ ] denote the protein-bound prosthetic groups that constitute an e^- transport chain between NADH and nitrate. Nitrate reductases are typically α_2-type homodimers of 210 to 270 kD molecular weight. The structure of the molybdenum cofactor (MoCo) is shown in Figure 26.1a. Molybdenum cofactor is necessary for both nitrate reductase activity and the assembly of nitrate reductase subunits into the active holoenzyme dimer form. Molybdenum cofactor is also an essential cofactor for a variety of enzymes that catalyze hydroxylase-type reactions, including xanthine dehydrogenase, aldehyde oxidase, and sulfite oxidase.

B. Nitrite Reductase in Green Plants.

$$\text{Light} \longrightarrow 6\,Fd_{red} \longrightarrow [(4Fe\text{-}4S) \longrightarrow \text{siroheme}] \longrightarrow NO_2^-$$

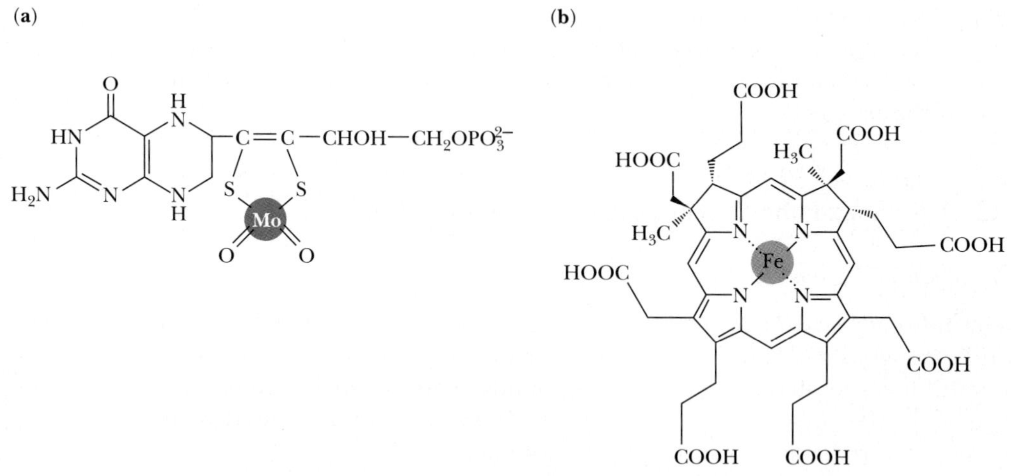

***Figure* 26.1** The novel prosthetic groups of nitrate reductase and nitrite reductase. Both nitrate reductase and nitrite reductase are metal-containing hemoproteins (proteins possessing heme as well as other metal-containing prosthetic groups). (a) The molybdenum cofactor of nitrate reductase. The molybdenum-free version of this compound is a pterin derivative called **molybdopterin.** (b) Siroheme, a uroporphyrin derivative, is a member of the **isobacteriochlorin class** of hemes, a group of porphyrins in which adjacent pyrrole rings are reduced. Siroheme is novel in having eight carboxylate-containing side chains; it is quite polar.

Six electrons are required to reduce NO_2^- to NH_4^+, and in photosynthetic organisms capable of assimilating nitrate, these electrons are provided by six equivalents of photosynthetically reduced ferredoxin (Fd_{red}). Photosynthetic nitrite reductases are 63-kD monomeric proteins having a tetranuclear iron-sulfur cluster and a novel heme, termed **siroheme,** as prosthetic groups. The [4Fe-4S] cluster and the siroheme act as a coupled e^- transfer center. Nitrite binds directly to siroheme, providing the sixth ligand, much as O_2 binds to the heme of hemoglobin. Nitrite is reduced to ammonium while liganded to siroheme. The structure of siroheme is shown in Figure 26.1b.

In higher plants, nitrite reductase is found in chloroplasts, where it has ready access to its primary reductant, photosynthetically reduced ferredoxin. Nitrate reductase is located in the cytosol, perhaps in association with the cytosolic faces of chloroplasts. It is interesting to note that plants use NADH instead of NADPH as the electron donor for this first reaction in their principal (often only) pathway of N acquisition. In fungi and bacteria, NADPH serves as the primary e^- donor for both the nitrate reductase and nitrite reductase reactions. Microbial nitrite reductases are larger and more complex than plant nitrite reductases. Indeed, microbial nitrite reductases closely resemble nitrate reductases in having essential —SH groups and FAD prosthetic groups to couple enzyme-mediated NADPH oxidation to nitrogen oxide reduction.

The Enzymology of Nitrogen Fixation

Nitrogen fixation is a process found only in certain prokaryotic cells, either free-living or acting as symbionts with higher plants (Table 26.1). A group of particular interest is the *Rhizobia*. ***Rhizobia*** are bacteria that fix nitrogen in symbiotic association with leguminous plants. Since nitrogen in a metabolically useful form is often the limiting nutrient for plant growth, the *Rhizobia*:legume symbiosis has great agricultural significance. *Rhizobia* infect the roots of host plants and proliferate in the root tissue, forming nodules

Table **26.1**
Representative Nitrogen-Fixing Bacteria

Free-Living	Symbiotic (Symbiont)
Methanogenic archaebacteria	Cyanobacteria (blue-green algae):
Anacystis (a cyanobacterium)	*Anabena* (*Azolla** symbiont)
Azospirillum spp.†	*Franckia*‡ (*Casuarina*§)
A few *Bacillus* spp.	Rhizobia:‖ 2 genera
Azotobacter spp.	*Rhizobium* (clover)
Clostridium pasteurianum#	*Bradyrhizobium* (soybeans)
Enterobacteria (several genera):	
Klebsiella pneumonia	
Escherichia spp.**	
Citrobacter spp.††	

* *Azolla* is commonly known as duckweed; it grows in rice paddies.

† *Azospirillum* is found in the rhizosphere (the plant root environment).

‡ *Franckia* is an actinomycete (actinomycetes are fungilike bacteria).

§ *Franckia* forms N_2-fixing nodules on the roots of *Casuarina* trees. (These trees are commonly known as Australian pines; they are able to colonize N-deficient coastal areas in the tropics and subtropics because of the nitrogen derived from their symbiotic association with *Franckia*.)

‖ Rhizobia form nodules on the roots of various legumes. These nodules are the sites of nitrogen fixation. *Rhizobia* are separated into two genera: *Rhizobium*, "fast-growing," and *Bradyrhizobium*, "slow-growing."

Clostridium pasteurianum is an obligate anaerobe of the soil.

** *Escherichia coli* is not one of the N_2-fixing *Escherichia* spp.

†† *Citrobacter* can fix N_2 as a free-living bacterium. However, these bacteria also thrive in the guts of termites and shipworms, enabling their animal hosts to subsist on a woody diet that is very poor in nitrogen content.

(Figure 26.2). These **nodules** are the sites of nitrogen fixation. In the symbiosis, fixed nitrogen is transferred from the bacteria to the host plant, augmenting the plant's nitrogen nutrition. In turn, the plant provides the bacteria with carbon substrates for growth.

All nitrogen-fixing enzyme systems are nearly identical, despite the wide diversity of bacteria in which nitrogen fixation takes place. Apparently, nitrogen fixation is an idea that has occurred only once in nature's rich imagination. All N_2-fixing systems have five basic requirements:

1. *Nitrogenase,* the protein that catalyzes the reduction of N_2 to ammonium
2. A powerful reductant, such as reduced *ferredoxin* (or flavodoxin)
3. ATP
4. Oxygen-free conditions
5. Regulatory controls

Figure **26.2** Nitrogen-fixing nodules on the roots of a legume.

To a first approximation, two regulatory controls are paramount (Figure 26.3):

a. ADP inhibits the activity of nitrogenase; thus, as the ATP/ADP ratio drops, nitrogen fixation is blocked.

b. NH_4^+ represses the expression of the *nif* genes, the genes that encode the proteins of the nitrogen-fixing system. To date, as many as 20 *nif* genes have been identified with the nitrogen fixation process. Repression of *nif* gene expression by ammonium, the primary product of nitrogen fixation, is an efficient and effective way of shutting down N_2 fixation when its end product is not needed.

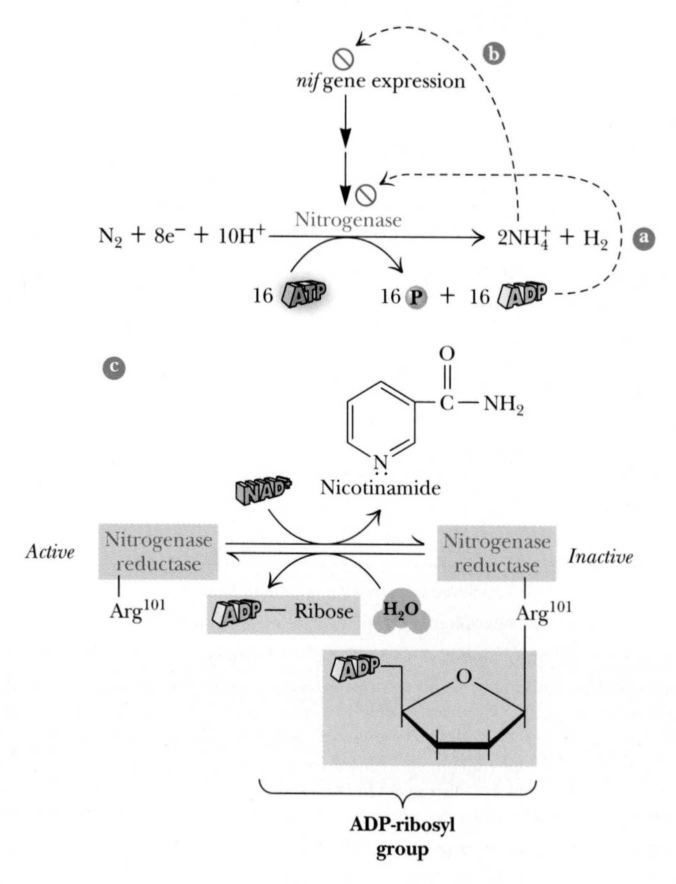

***Figure* 26.3** Regulation of nitrogen fixation. (a) ADP inhibits nitrogenase activity. (b) NH_4^+ represses *nif* gene expression. (c) In some organisms, the nitrogenase complex is regulated by covalent modification. ADP-ribosylation of the nitrogenase reductase component (see Figure 26.6) leads to its inactivation.

The Nitrogenase Complex. Two proteins constitute the nitrogenase complex, nitrogenase reductase and nitrogenase proper. **Nitrogenase reductase** is a 60-kD homodimer possessing a single [4Fe-4S] cluster as a prosthetic group. Nitrogenase reductase is extremely O_2-sensitive, and careful precautions must be taken to ensure anaerobic conditions to prevent it from being inactivated

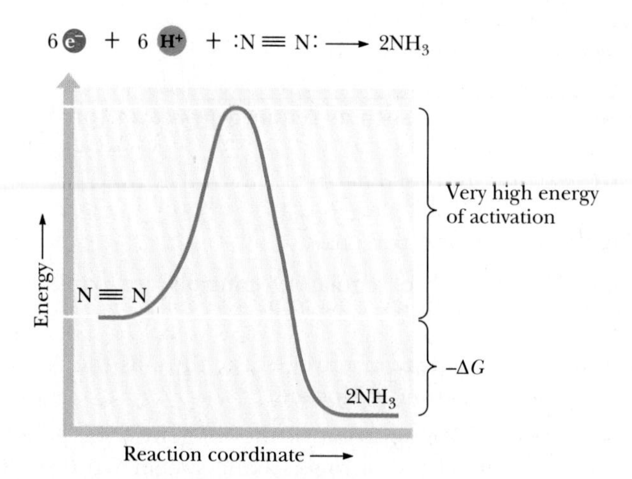

***Figure* 26.4** The triple bond in N_2 must be broken during nitrogen fixation. A substantial energy input is needed to overcome this thermodynamic barrier, even though the overall free energy change, $\Delta G^{\circ\prime}$, is negative.

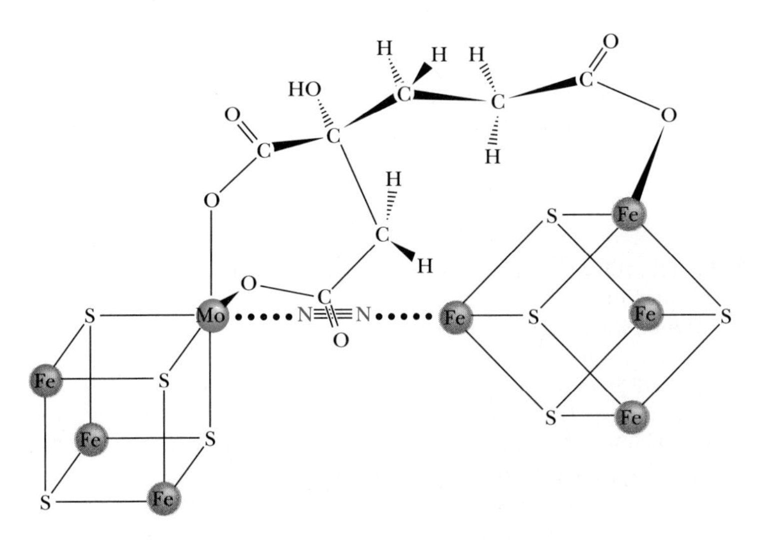

***Figure* 26.5** A model of the structure of the iron-molybdenum cofactor of nitrogenase. Note the homocitrate bridge spanning the two iron–sulfur clusters.

in vitro during biochemical analysis. Nitrogenase reductase also binds MgATP. There is a requirement for four ATP equivalents per pair of electrons transferred during nitrogen fixation. Since reduction of N_2 to 2 NH_3 + H_2 requires four pairs of electrons, 16 ATP equivalents are consumed per N_2 reduced.

This ATP requirement seems paradoxical since the reduction of N_2 to NH_4^+ is thermodynamically favorable (i.e., the free energy change for the reaction is negative). The solution to the paradox is found in the very strong bonding between the two N atoms in N_2 (Figure 26.4). Substantial energy input is needed to overcome this large activation energy and break the N—N triple bond. In this biological system, the energy is provided by ATP.

Nitrogenase is a 220-kD $\alpha_2\beta_2$-type heterotetramer. Each molecule of this protein contains 2 atoms of molybdenum (Mo), 32 atoms of iron (Fe), and about 30 equivalents of acid-labile sulfide (S^{2-}). These atoms are combined to form four [4Fe-4S] clusters and two unique **molybdenum-iron cofactors** (abbreviated FeMoco). The chemical identity of FeMoco remains elusive, but it is believed to be an iron-sulfur type cluster that also contains *homocitrate* (Figure 26.5). Nitrogenase under unusual circumstances may contain an **iron:vanadium cofactor** instead of the molybdenum-containing one. Like nitrogenase reductase, nitrogenase is very oxygen-labile.

The Nitrogenase Reaction. In the nitrogenase reaction (Figure 26.6), electrons from reduced ferredoxin pass to nitrogenase reductase, which serves as the electron donor for nitrogenase, the enzyme that actually catalyzes N_2 fixation. Nitrogenase is a rather slow enzyme: its optimal rate of e^- transfer is about 12 e^- pairs per second per enzyme molecule; that is, it reduces only three molecules of nitrogen gas per second. Because its activity is so weak, nitrogen-fixing cells maintain large amounts of nitrogenase so that their requirements for reduced N can be met. As much as 5% of the cellular protein may be nitrogenase. As indicated earlier, nitrogenase catalyzes the concomitant reduction of protons to hydrogen gas, its so-called **hydrogenase activity.** This activity accompanies N_2 reduction *in vivo* and is energy-depleting. One equivalent of H_2 is formed for every 2 NH_3 (the obligate nature of this reaction leads to the stoichiometry given in Equation 26.3). Proton reduction results in an unavoidable loss in reducing power and a corresponding decrease in nitrogen-fixing potential.

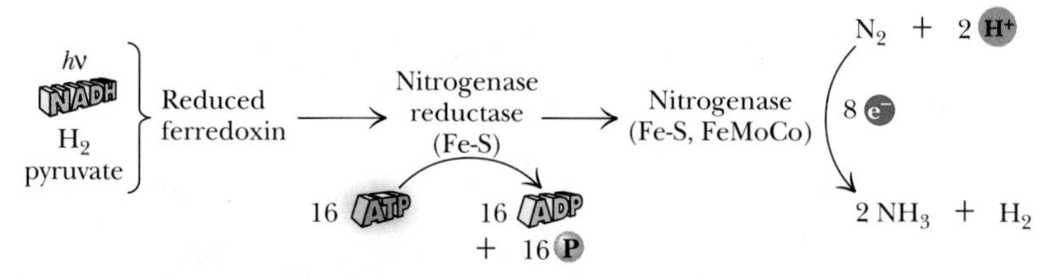

***Figure* 26.6** The nitrogenase reaction. Depending on the bacterium, electrons for N_2 reduction may derive from light, NADH, hydrogen gas, or pyruvate. The primary e^- donor for the nitrogenase system is reduced ferredoxin. (Under Fe-limited conditions, some bacteria synthesize a flavoprotein very similar to ferredoxin, called **flavodoxin.**) Reduced ferredoxin passes electrons directly to nitrogenase reductase. The redox potential ($\mathscr{E}_o'$) of the Fe-S cluster in nitrogenase reductase is -0.29 V in the absence of ATP; in the presence of MgATP, this $\mathscr{E}_o'$ is -0.40 V. A total of six electrons are required to reduce N_2 to $2\ NH_4^+$, and another two electrons are consumed in the obligatory reduction of $2\ H^+$ to H_2. Nitrogenase reductase transfers e^- to nitrogenase one electron at a time. N_2 is bound at the critical FeMoCo prosthetic group of nitrogenase until all electrons and protons are added; no free intermediates such as HN=NH or H_2N—NH_2 are detectable.

26.2 The Fate of Ammonium

Ammonium enters organic linkage via three major reactions that are found in all cells. The enzymes mediating these reactions are (1) *carbamoyl-phosphate synthetase I,* (2) *glutamate dehydrogenase,* and (3) *glutamine synthetase.*

Carbamoyl-phosphate synthetase I catalyzes one of the steps in the *urea cycle.* Two equivalents of ATP are expended, one in the activation of HCO_3^- for reaction with ammonium, and the other in the phosphorylation of the carbamate formed:

$$NH_4^+ + HCO_3^- + 2\ ATP \longrightarrow H_2N\text{—}\overset{\overset{\displaystyle O}{\|}}{C}\text{—}O\text{—}PO_3^{2-} + 2\ ADP + P_i + 2H^+$$

N-acetylglutamate is an essential allosteric activator for this enzyme.

Glutamate dehydrogenase (GDH) catalyzes the reductive amination of α-ketoglutarate to yield glutamate. Reduced pyridine nucleotides (NADH or NADPH) provide the reducing power:

$$NH_4^+ + \alpha\text{-ketoglutarate} + NADPH + H^+ \longrightarrow \text{glutamate} + NADP^+ + H_2O$$

This reaction provides an important interface between nitrogen metabolism and cellular pathways of carbon and energy metabolism because α-ketoglutarate is a citric acid cycle intermediate. As a result, GDH represents

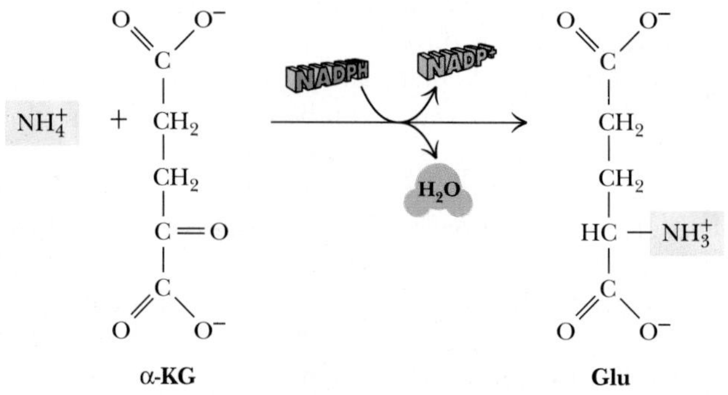

***Figure* 26.7** The glutamate dehydrogenase reaction.

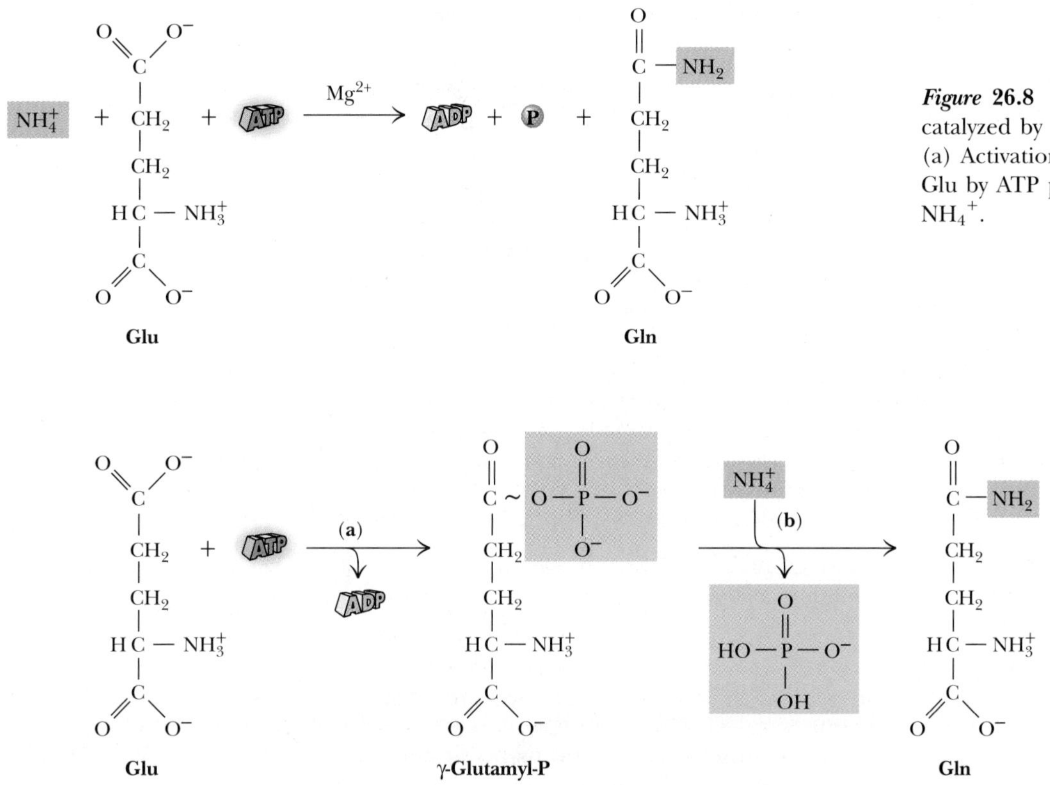

Figure **26.8** The enzymatic reaction catalyzed by glutamine synthetase. (a) Activation of the γ-carboxyl group of Glu by ATP precedes (b) amidation by NH_4^+.

an ideal site for metabolic regulation. Liver GDH is an α_6-type multimer that uses NADPH as electron donor when operating in the biosynthetic direction (the direction of glutamate synthesis) (Figure 26.7). In contrast, when GDH acts in the catabolic direction to generate α-ketoglutarate from glutamate, NAD^+, not $NADP^+$, is usually the electron acceptor. The catabolic activity is allosterically activated by ADP and inhibited by GTP. Some organisms (the fungus *Neurospora crassa* is one example) have two GDH isozymes, an $NADP^+$-specific cytosolic enzyme that functions in the direction of glutamate synthesis and an NAD^+-specific mitochondrial enzyme acting in the catabolic direction to convert excess glutamate into α-ketoglutarate for energy metabolism.

Glutamine synthetase (GS) catalyzes the ATP-dependent amidation of the γ-carboxyl group of glutamate to form glutamine. The reaction proceeds via a γ-glutamyl-phosphate intermediate, and GS activity depends on the presence of divalent cations such as Mg^{2+} (Figure 26.8). **Glutamine** is a major N donor in the biosynthesis of many organic N compounds such as purines, pyrimidines, and other amino acids, and GS activity is tightly regulated, as we shall soon see. The amide-N of glutamine provides the nitrogen atom in these biosyntheses. In quantitative terms, GDH and GS are responsible for most of the ammonium assimilated into organic compounds.

Pathways of Ammonium Assimilation

In organisms that enjoy environments rich in nitrogen, GDH and GS acting in sequence furnish the principal route of NH_4^+ incorporation (Figure 26.9).

(**a**) $NH_4^+ + \alpha\text{-Ketoglutarate} + NADPH \xrightarrow{GDH} \text{Glutamate} + NADP^+ + H_2O$

(**b**) $\text{Glutamate} + NH_4^+ + ATP \xrightarrow{GS} \text{Glutamine} + ADP + P_i$

SUM: $2\ NH_4^+ + \alpha\text{-Ketoglutarate} + NADPH + ATP \longrightarrow \text{Glutamine} + NADP^+ + ADP + P_i + H_2O$

Figure **26.9** The GDH/GS pathway of ammonium assimilation. The sum of these reactions is the conversion of 1 α-ketoglutarate to 1 glutamine at the expense of 2 NH_4^+, 1 ATP, and 1 NADPH.

$$\left.\begin{array}{l} H^+ + NADH \text{ (yeast, } N.\ crassa) \\ H^+ + NADPH \text{ (}E.\ coli\text{) or} \\ 2H^+ + 2 \text{ reduced ferredoxin (plants)} \end{array}\right\} + \alpha\text{-KG} + \text{Gln} \longrightarrow 2\ \text{Glu} + \begin{array}{l} NAD^+ \\ NADP^+ \text{ or} \\ 2 \text{ oxidized ferredoxin} \end{array}$$

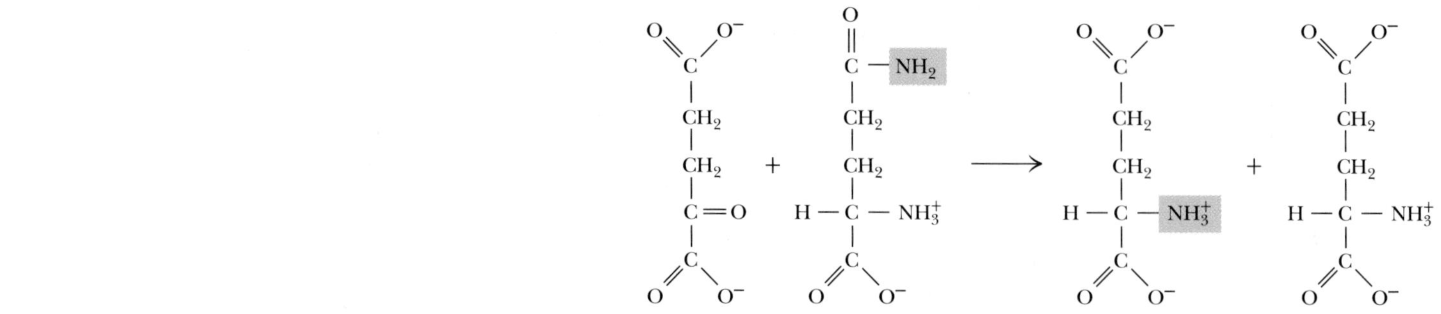

α-KG Gln Glu Glu

Figure **26.10** The glutamate synthase reaction, showing the reductants exploited by different organisms in this reductive amination reaction.

However, GDH has a significantly higher K_m for NH_4^+ than does GS. Consequently, in organisms confronting N limitation, GDH is not effective and GS is the only NH_4^+-assimilative reaction. Such a situation creates the need for an alternative mode of glutamate synthesis to replenish the glutamate consumed by the GS reaction. This need is filled by *glutamate synthase* (also known as GOGAT, the acronym for the other name of this enzyme—*g*lutamate: *o*xo-*g*lutarate *a*mino-*t*ransferase). Glutamate synthase mediates the reductive amination of α-ketoglutarate where the amide-N of glutamine serves as the N donor:

$$\text{Reductant} + \alpha\text{-KG} + \text{Gln} \longrightarrow 2\ \text{Glu} + \text{oxidized reductant}$$

Two equivalents of glutamate are formed—one from amination of α-ketoglutarate and the other from deamidation of Gln (Figure 26.10). These glutamates can now serve as ammonium acceptors for glutamine synthesis by GS. Organisms variously use NADH (yeast, *Neurospora crassa*), NADPH (*E. coli*), or reduced ferredoxin (plants) as reductant. Glutamate synthases are typically large, complex proteins; in *E. coli,* GOGAT is an 800-kD flavoprotein containing both FMN and FAD as well as [4Fe-4S] clusters.

Together, GS and GOGAT constitute a second pathway of ammonium assimilation, in which GS is the only NH_4^+-fixing step; the role of GOGAT is to regenerate glutamate (Figure 26.11). Note that this pathway consumes two equivalents of ATP and 1 NADPH (or similar reductant) per pair of N atoms introduced into Gln, in contrast with the GDH/GS pathway, in which only 1 ATP and 1 NADPH are used up per pair of NH_4^+ fixed. Clearly, coping within a nitrogen-limited environment has its cost.

$$\text{(a) } 2\ NH_4^+ + 2\ ATP + 2\ \text{Glutamate} \xrightarrow{GS} 2\ \text{Glutamine} + 2\ ADP + 2\ P_i$$

$$\text{(b) } NADPH + \alpha\text{-Ketoglutarate} + \text{Glutamine} \xrightarrow{GOGAT} 2\ \text{Glutamate} + NADP^+$$

$$\text{SUM: } 2\ NH_4^+ + \alpha\text{-Ketoglutarate} + NADPH + 2\ ATP \longrightarrow \text{Glutamine} + NADP^+ + 2\ ADP + 2\ P_i$$

Figure **26.11** The GS/GOGAT pathway of ammonium assimilation. The sum of these reactions results in the conversion of 1 α-ketoglutarate to 1 glutamine at the expense of 2 ATP and 1 NADPH.

26.3 *Escherichia coli* Glutamine Synthetase: A Case Study in Enzyme Regulation

As indicated earlier, glutamine plays a pivotal role in nitrogen metabolism by donating its amide nitrogen to the biosynthesis of many important organic N compounds. Consistent with its metabolic importance, in enteric bacteria such as *E. coli*, GS is regulated at three different levels:

1. Its activity is regulated allosterically by *cumulative feedback inhibition.*
2. GS is interconverted between active and inactive forms by *covalent modification.*
3. Cellular amounts of GS are carefully controlled at the level of *gene expression* and *protein synthesis.*

Eukaryotic versions of glutamine synthetase show none of these regulatory features.

Escherichia coli GS is a 600-kD dodecamer (α_{12}-type subunit organization) of identical 52-kD monomers (each monomer contains 468 amino acid residues). These monomers are arranged as a stack of two hexagons (Figure 26.12). The active sites are located at subunit interfaces within the hexagons; these active sites are recognizable in the X-ray crystallographic structure by the pair of divalent cations that occupy them. Adjacent subunits contribute to each active site, thus accounting for the fact that GS monomers are catalytically inactive.

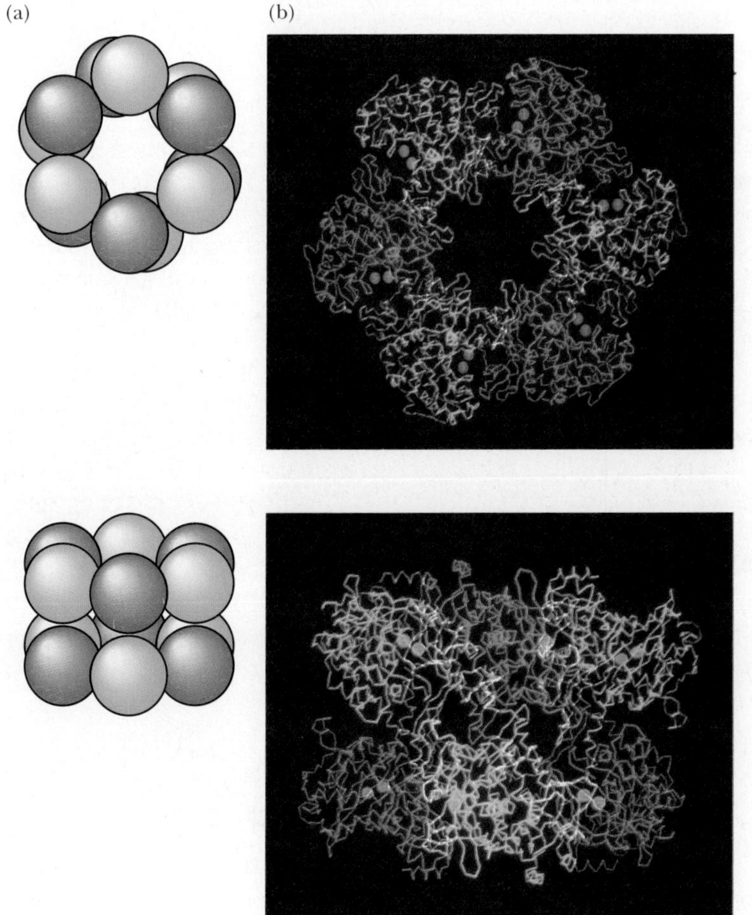

***Figure* 26.12** The subunit organization of bacterial glutamine synthetase. (a) Diagram showing its dodecameric structure as a stack of two hexagons. (b) Molecular structure of glutamine synthetase from *Salmonella typhimurium* (a close relative of *E. coli*), as revealed by X-ray crystallographic analysis.

(From Almassy, R. J., Janson, C. A., Hamlin, R., Xuong, N.-H., and Eisenberg, D., 1986. Nature 323:304. Photos courtesy of S.-H. Liaw and D. Eisenberg.)

Allosteric Regulation of GS. Nine distinct feedback inhibitors (Gly, Ala, Ser, His, Trp, CTP, AMP, carbamoyl-P, and glucosamine-6-P) act on GS. Gly, Ala, and Ser are key indicators of amino acid metabolism in the cell; each of the other six compounds represents an end product of a biosynthetic pathway dependent on Gln (Figure 26.13). These feedback inhibitors show a *cumulative pattern* of inhibition. Each alone exerts only 11% inhibition of GS. In the presence of saturating amounts of more than one feedback inhibitor, the inhibition is additive. For example, in the presence of any three inhibitors, GS is 67% active; in the presence of any six, GS is only 33% active. All nine effect complete inhibition of GS activity. Such an inhibitory pattern implies a distinct binding site on each subunit for each inhibitor. Thus, each GS polypeptide must provide binding sites for a variety of allosteric effectors, three substrates (NH_4^+, ATP, and Glu), and a pair of divalent cations.

Covalent Modification of GS. Each GS subunit can be adenylylated at a specific tyrosine residue (Tyr^{397}) in an ATP-dependent reaction (Figure 26.14). Adenylylation inactivates GS. If we define *n* as the average number of adenylyl groups per GS molecule, GS activity is inversely proportional to *n*. The number *n* varies from 0 (no adenylyl groups) to 12 (every subunit in each GS molecule is adenylylated). Adenylylation of GS is catalyzed by the *converter*

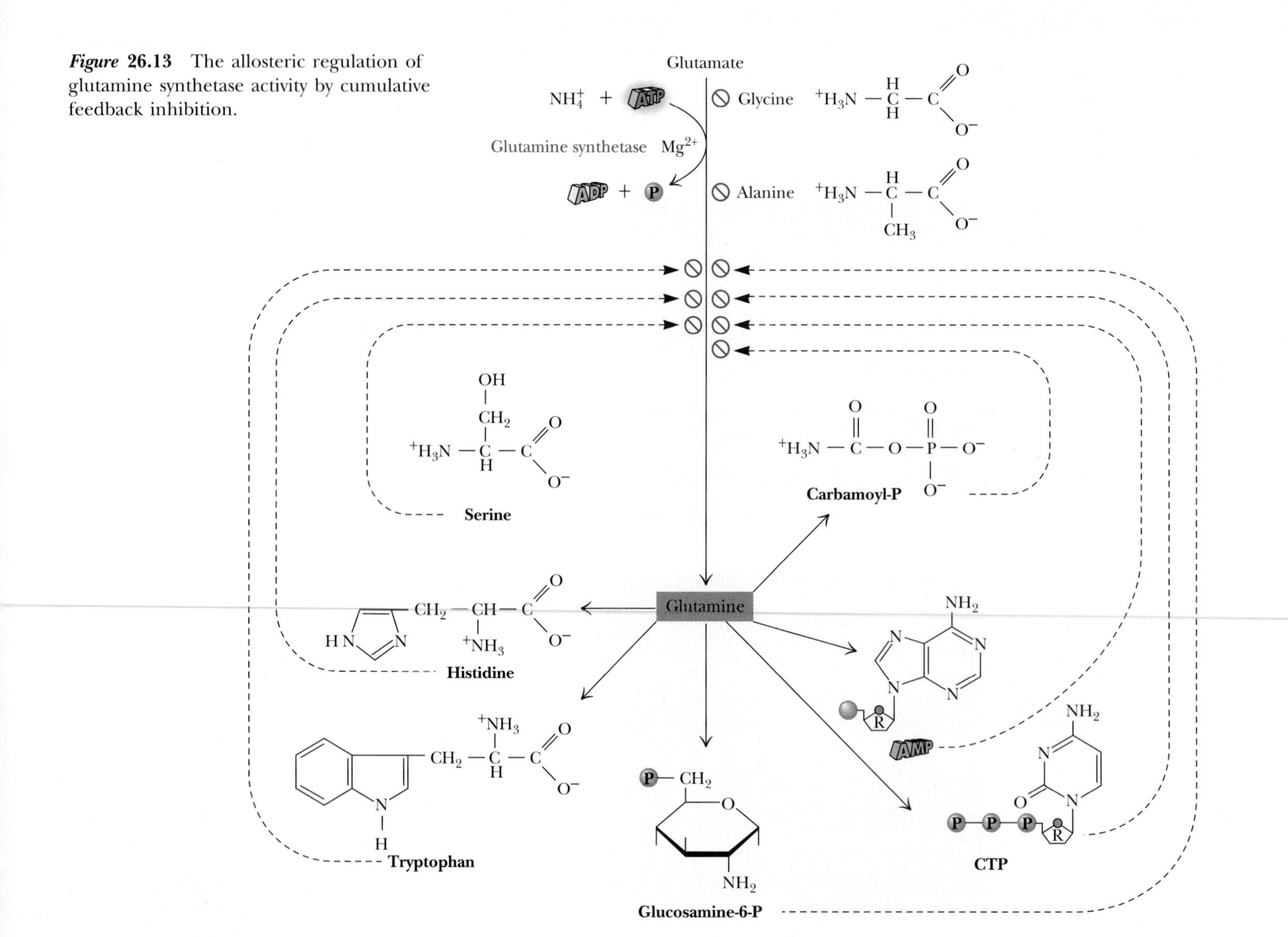

***Figure* 26.13** The allosteric regulation of glutamine synthetase activity by cumulative feedback inhibition.

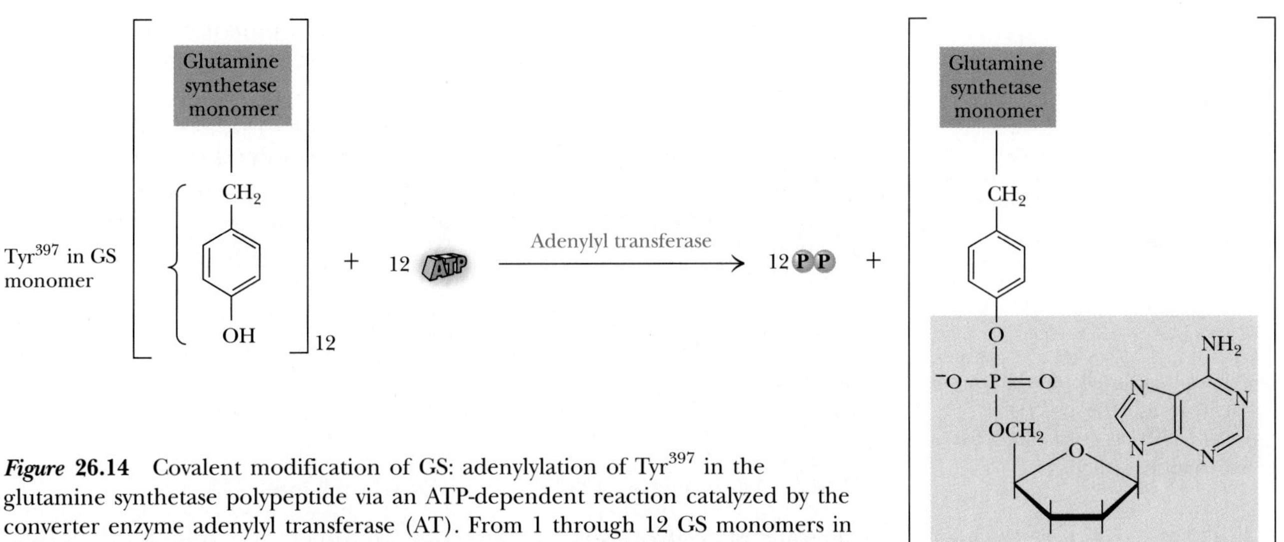

***Figure* 26.14** Covalent modification of GS: adenylylation of Tyr^{397} in the glutamine synthetase polypeptide via an ATP-dependent reaction catalyzed by the converter enzyme adenylyl transferase (AT). From 1 through 12 GS monomers in the GS holoenzyme can be modified, with progressive inactivation as the ratio of [modified]/[unmodified] GS subunits increases.

enzyme ATP:GS:adenylyl transferase. However, whether or not this covalent modification occurs is determined by a highly regulated **cyclic cascade system** (Figure 26.15) composed of two regulatory cycles, one catalyzed by the *adenylyl transferase* (AT), a 110-kD protein, and the other catalyzed by a second converter enzyme, *uridylyl transferase* (UT), a 95-kD protein. A third regulatory protein, P_{II}, is also intimately involved.

The Bicyclic Cascade–Covalent Modification System Regulating GS Activity. AT not only catalyzes adenylylation of GS, it also catalyzes **deadenylylation**—the

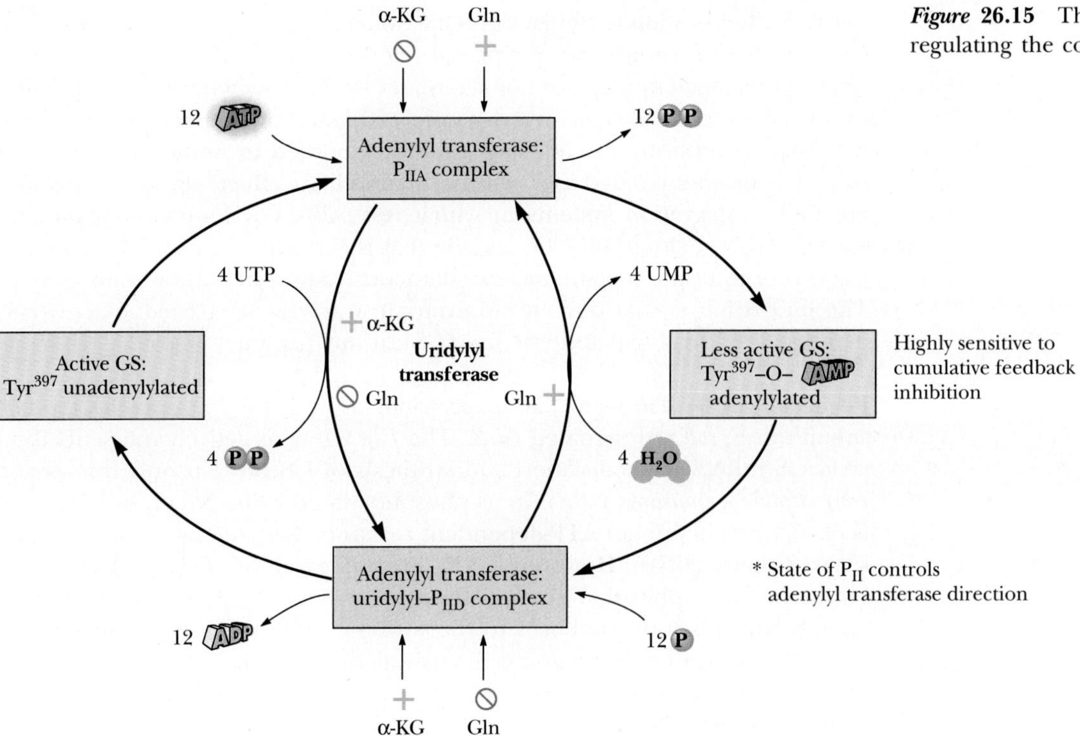

***Figure* 26.15** The cyclic cascade system regulating the covalent modification of GS.

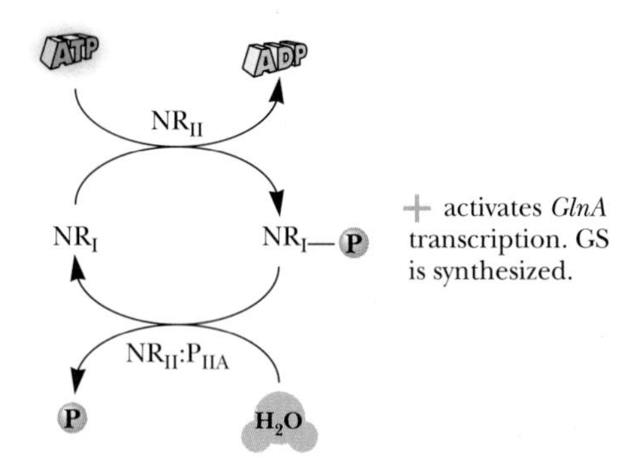

Figure **26.16** Transcriptional regulation of *GlnA* expression through the reversible phosphorylation of NR_I, as controlled by NR_{II} and its association with P_{IIA}.

phosphorolytic removal of the Tyr-linked adenylyl groups as ADP. The direction in which AT operates depends on the nature of the regulatory protein P_{II} associated with it. $\mathbf{P_{II}}$ is a 44-kD protein (tetramer of 11-kD subunits), and each P_{II} subunit is itself a target of covalent modification (Figure 26.15). P_{II} is the substrate of the converter enzyme UT, which catalyzes the *UTP-dependent uridylylation* of P_{II} subunits at a specific tyrosine residue. This reaction is allosterically activated by α-ketoglutarate (α-KG) and inhibited by glutamine. UT also catalyzes the hydrolytic removal of UMP from these same Tyr residues (**deuridylylation**). UT activity in this hydrolytic direction is strongly activated by Gln. Thus, the interconversion of P_{II} between the uridylylated and deuridylylated states is ultimately a function of the cellular [Gln]/[α-KG] ratio.

AT:P_{IIA} Adenylylates GS; AT:P_{IID} Deadenylylates GS

The state of P_{II} controls the direction in which AT acts. If P_{II} is not uridylylated—that is, if it is in its so-called P_{IIA} form—the AT:P_{IIA} complex acts to adenylylate GS. When P_{II} is in its uridylylated state—the so-called P_{IID} form—the AT:P_{IID} complex catalyzes the deadenylylation of GS. The active sites of AT:P_{IIA} and AT:P_{IID} are different, consistent with the difference in their catalytic roles. In addition, the AT:P_{IIA} and AT:P_{IID} complexes are allosterically regulated in a reciprocal fashion by the effectors α-KG and Gln. Gln activates AT:P_{IIA} activity and inhibits AT:P_{IID} activity; the effect of α-KG on the activities of these two complexes is diametrically opposite (Figure 26.15).

Clearly, the determining factor regarding the degree of adenylylation, n, and hence the relative activity of GS, is the [Gln]/[α-KG] ratio. A high [Gln] level signals cellular nitrogen sufficiency, and GS becomes adenylylated and inactivated. In contrast, a high [α-KG] level is an indication of nitrogen limitation and a need for ammonium fixation by GS. Note that Gln affects all four reactions of the cyclic cascade system, whereas α-KG affects only three. Thus, Gln is ultimately the dominant effector in this system.

Covalent Modification Is Not an ON/OFF Switch. It is important to stress that covalent modification is not reciprocal regulation in an all-or-none manner. That is, this mechanism does not act as an ON/OFF switch. Covalent modification and cyclic cascades are dynamic processes such that, for any given metabolic condition, a steady-state ratio of modified to unmodified enzyme (e.g., GS to adenylylated GS) will be attained. In effect, these systems are **metabolic integration systems** in which reversible covalent modification is modulated by a closed bicyclic cascade that is capable of sensing the concentrations of multiple metabolites simultaneously and integrating their effects. The end result is that only an appropriate fraction of enzymatic activity is expressed by the regulated enzyme (GS) at any instant.

Regulation of GS Through Gene Expression. The gene that encodes the GS subunit in *E. coli* is designated *GlnA*. The *GlnA* gene is actively transcribed to yield GS mRNA for translation and synthesis of GS protein only if a *specific transcriptional enhancer, NR_I,* is in its phosphorylated form, NR_I-P. In turn, NR_I is phosphorylated in an ATP-dependent reaction catalyzed by NR_{II}, a protein kinase (Figure 26.16). However, if NR_{II} is complexed with P_{IIA}, it acts not as a kinase but as a phosphatase, and the transcriptionally active form of NR_I, namely NR_I-P, is converted back to NR_I with the result that *GlnA* transcription halts. Recall from the foregoing discussion that a high [Gln]/[α-KG] ratio favors P_{IIA} at the expense of P_{IID}. Under such conditions, GS gene expression is not necessary.

26.4 Amino Acid Biosynthesis

Organisms show substantial differences in their capacity to synthesize the 20 amino acids common to proteins. Typically, plants and microorganisms can form all of their nitrogenous metabolites, including all of the amino acids, from inorganic forms of N such as NH_4^+ and NO_3^-. In these organisms, the α-amino group for all amino acids is derived from glutamate, usually via transamination of the corresponding α-keto acid analog of the amino acid (Figure 26.17). In many cases, amino acid biosynthesis is thus a matter of synthesizing the appropriate α-keto acid carbon skeleton, followed by transamination with Glu. The amino acids can be classified according to the source of intermediates for the α-keto acid biosynthesis (Table 26.2). For example, the amino acids Glu, Gln, Pro, and Arg (and, in some instances, Lys) are all members of the α-ketoglutarate family since they are all derived from the citric acid cycle intermediate, α-ketoglutarate. We will return to this classification scheme later when we discuss the individual biosynthetic pathways.

Mammals can synthesize only 10 of the 20 common amino acids (Table 26.3); the others must be obtained in the diet. Those that can be synthesized are classified as **nonessential,** meaning it is not essential that these amino acids be part of the diet. In effect, mammals can synthesize the α-keto acid analogs of nonessential amino acids and form the amino acids by transamination. In contrast, mammals are incapable of constructing the carbon skeletons of **essential** amino acids, and so must rely on dietary sources for these essential metabolites. As a matter of fact, excess dietary amino acids cannot be stored for future use, nor are they excreted unused. Instead, they are converted to

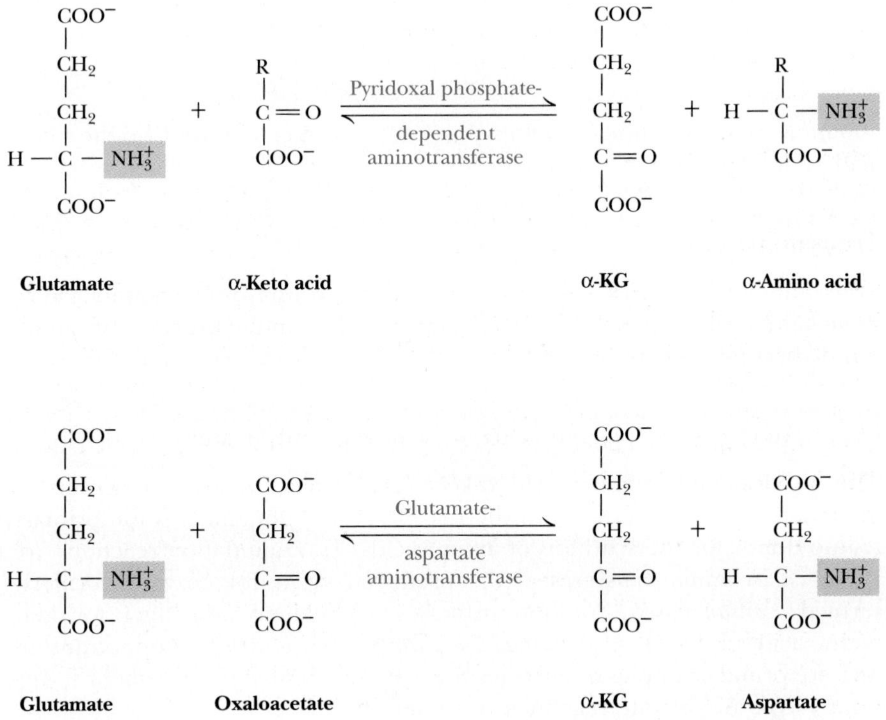

$$\underset{\textbf{Glutamate}}{{}^{-}OOC{-}CH_2{-}CH_2{-}CH(NH_3^+){-}COO^-} + \underset{\alpha\textbf{-Keto acid}}{R{-}C(=O){-}COO^-} \xrightleftharpoons[\text{aminotransferase}]{\text{Pyridoxal phosphate-dependent}} \underset{\alpha\textbf{-KG}}{{}^{-}OOC{-}CH_2{-}CH_2{-}C(=O){-}COO^-} + \underset{\alpha\textbf{-Amino acid}}{R{-}CH(NH_3^+){-}COO^-}$$

$$\underset{\textbf{Glutamate}}{{}^{-}OOC{-}CH_2{-}CH_2{-}CH(NH_3^+){-}COO^-} + \underset{\textbf{Oxaloacetate}}{{}^{-}OOC{-}CH_2{-}C(=O){-}COO^-} \xrightleftharpoons[\text{aminotransferase}]{\text{Glutamate-aspartate}} \underset{\alpha\textbf{-KG}}{{}^{-}OOC{-}CH_2{-}CH_2{-}C(=O){-}COO^-} + \underset{\textbf{Aspartate}}{{}^{-}OOC{-}CH_2{-}CH(NH_3^+){-}COO^-}$$

***Figure* 26.17** Glutamate-dependent transamination of α-keto acid carbon skeletons is a primary mechanism for amino acid synthesis. The generic transamination–aminotransferase reaction involves the transfer of the α-amino group of glutamate to an α-keto acid acceptor (see Figure 14.24). The transamination of oxaloacetate by glutamate to yield aspartate and α-ketoglutarate is a prime example.

Table 26.2

The Grouping of Amino Acids into Families According to the Metabolic Intermediates That Serve as Their Progenitors

α-Ketoglutarate Family	Aspartate Family
Glutamate	Aspartate
Glutamine	Asparagine
Proline	Methionine
Arginine	Threonine
Lysine*	Isoleucine
	Lysine*
Pyruvate Family	**3-Phosphoglycerate Family**
Alanine	Serine
Valine	Glycine
Leucine	Cysteine
Phosphoenolpyruvate and Erythrose-4-P Family	
The aromatic amino acids	
Phenylalanine	
Tyrosine	
Tryptophan	
The remaining amino acid, *histidine,* is derived from PRPP (5-phosphoribosyl-1-pyrophosphate) and ATP.	

*Different organisms use different precursors to synthesize lysine.

Table 26.3

Essential and Nonessential Amino Acids in Mammals

Essential	Nonessential
Arginine*	Alanine
Histidine*	Asparagine
Isoleucine	Aspartate
Leucine	Cysteine
Lysine	Glutamate
Methionine	Glutamine
Phenylalanine	Glycine
Threonine	Proline
Tryptophan	Serine
Valine	Tyrosine†

*Arginine and histidine are essential in the diets of juveniles, not adults.

†Tyrosine is classified as nonessential only because it is readily formed from essential phenylalanine.

common metabolic intermediates that can be either oxidized by the citric acid cycle or used to form glucose (see Section 26.5).

Transamination

Transamination involves transfer of an α-amino group from an amino acid to the α-keto position of an α-keto acid (Figure 26.17). In the process, the amino donor becomes an α-keto acid while the α-keto acid acceptor becomes an α-amino acid:

$$\text{Amino acid}_1 + \alpha\text{-keto acid}_2 \longrightarrow \alpha\text{-keto acid}_1 + \text{amino acid}_2$$

The predominant amino acid/α-keto acid pair in these reactions is glutamate/α-ketoglutarate, with the net effect that glutamate is the primary amino donor for the synthesis of amino acids. Transamination reactions are catalyzed by **aminotransferases** (the preferred name for enzymes formerly termed *transaminases*). Aminotransferases are named according to their amino acid substrates, as in *glutamate-aspartate aminotransferase.* Aminotransferases are prime examples of enzymes that catalyze double displacement (ping-pong)–type bisubstrate reactions (Chapter 11).

The Pathways of Amino Acid Biosynthesis

As indicated in Table 26.2, the amino acids can be grouped into families on the basis of the metabolic intermediates that serve as their precursors.

The α-Ketoglutarate Family of Amino Acids

Amino acids derived from α-ketoglutarate include glutamate (Glu), glutamine (Gln), proline (Pro), arginine (Arg), and, in fungi and protoctists such as *Euglena,* lysine (Lys). We discussed the routes for Glu and Gln synthesis when we considered pathways of ammonium assimilation.

Proline is derived from glutamate via a series of four reactions involving activation, then reduction, of the γ-carboxyl group to an aldehyde (*glutamate-5-semialdehyde*), which spontaneously cyclizes to yield the internal Schiff base, *Δ^1-pyrroline-5-carboxylate* (Figure 26.18). NADPH-dependent reduction of the pyrroline double bond gives proline.

Arginine biosynthesis involves enzymatic steps that are also part of the **urea cycle,** a metabolic pathway that functions in N excretion in certain animals. Net synthesis of arginine depends on the formation of **ornithine.** Interestingly, ornithine is derived from glutamate via a reaction pathway reminiscent of the proline biosynthetic pathway (Figure 26.19). Glutamate is first N-acetylated in an acetyl-CoA–dependent reaction to yield *N-acetylglutamate* (Figure 26.19). An ATP-dependent phosphorylation of *N*-acetylglutamate to give *N-acetylglutamate-5-phosphate* primes this substrate for a reduced pyridine nucleotide–dependent reduction to the semialdehyde. *N-acetylglutamate-5-semialdehyde* then is aminated by a glutamate-dependent aminotransferase, giving *N-acetylornithine,* which is deacylated to ornithine.

Ornithine has three metabolic roles: (1) to serve as a precursor to arginine, (2) to function as an intermediate in the urea cycle, and (3) to act as an intermediate in Arg degradation. In either case, the δ-NH_3^+ of ornithine is carbamoylated in a reaction catalyzed by *ornithine transcarbamoylase.* The carbamoyl group is derived from carbamoyl-P synthesized by *carbamoyl phosphate synthetase I (CPS-I).* **CPS-I** is the mitochondrial CPS isozyme; it uses two ATP equivalents and catalyzes the formation of carbamoyl-P from NH_3 and HCO_3^- (Figure 26.20). CPS-I represents the committed step in the urea cycle, and CPS-I is allosterically activated by *N*-acetylglutamate. Since *N*-acetylglutamate is both a precursor to ornithine synthesis and essential to the operation of the urea cycle, it serves to coordinate these related pathways.

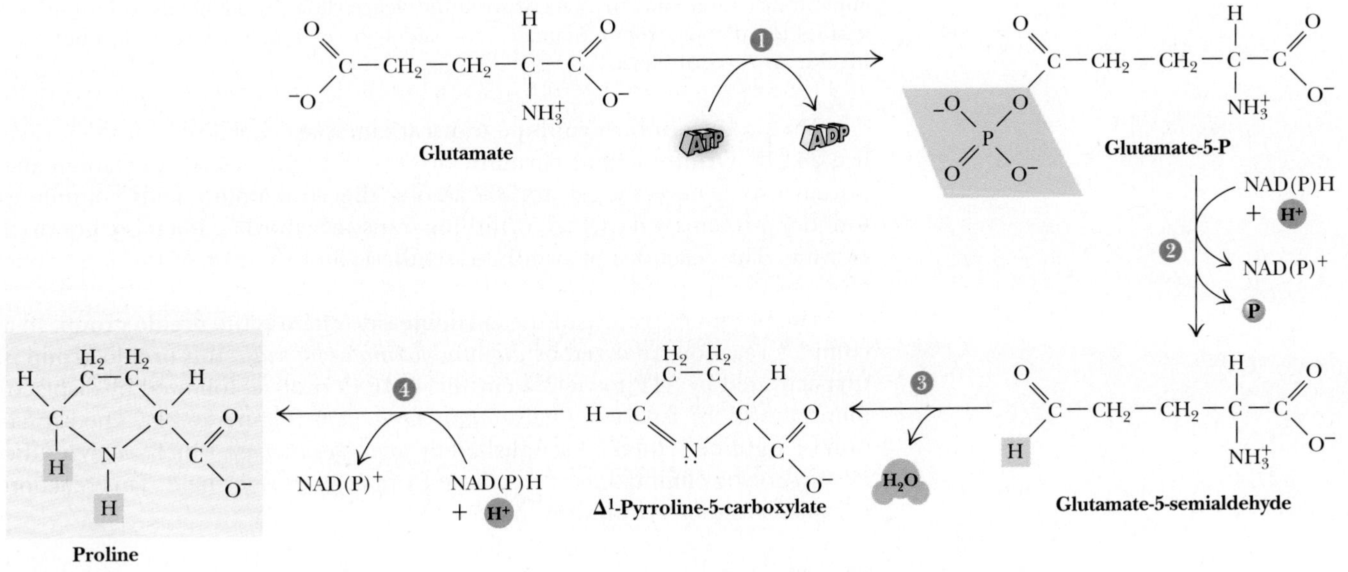

Figure **26.18** The pathway of proline biosynthesis from glutamate. The enzymes are (1) γ-glutamyl kinase, (2) glutamate-5-semialdehyde dehydrogenase, and (4) Δ^1-pyrroline-5-carboxylate reductase; reaction (3) occurs nonenzymatically.

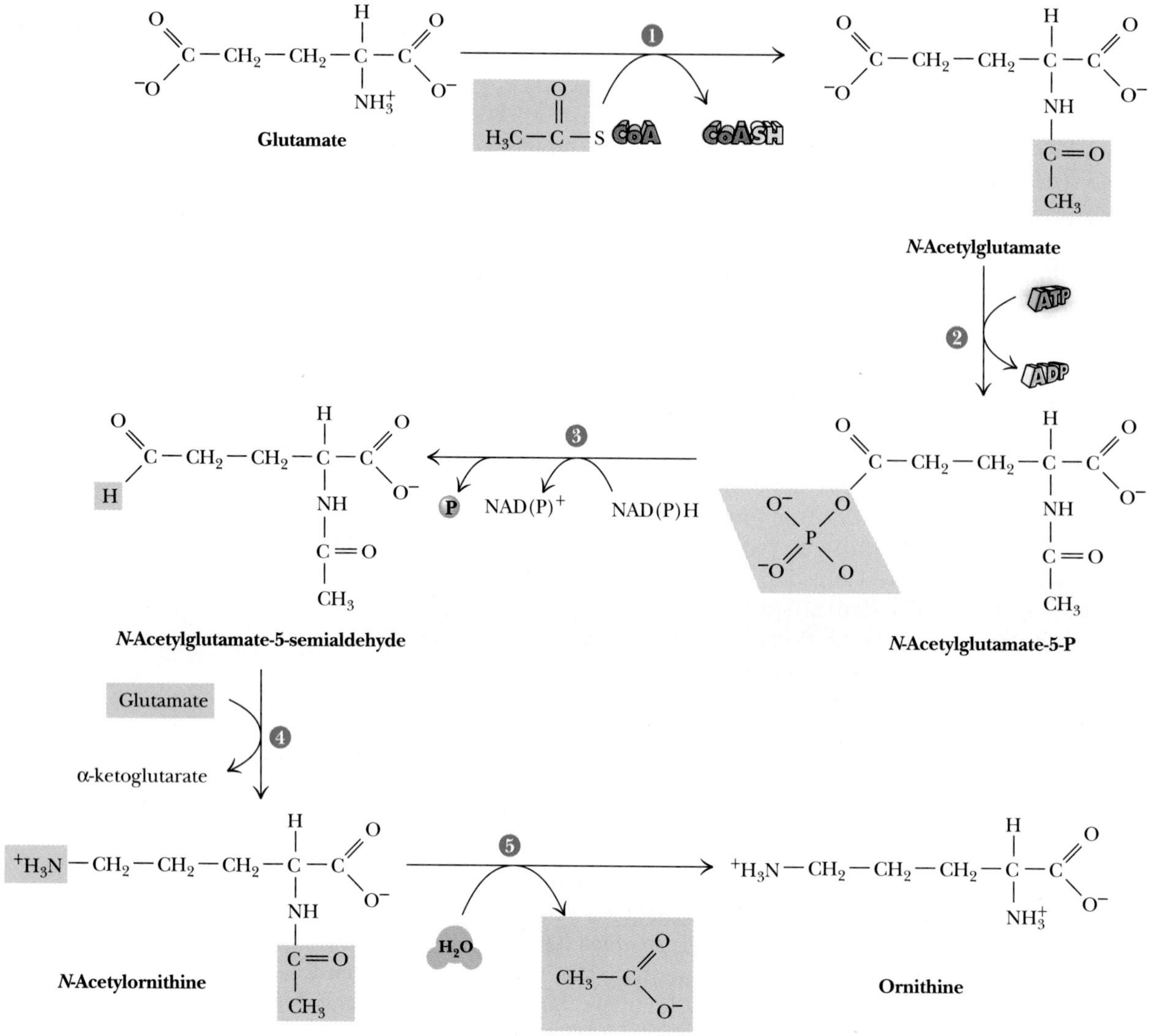

Figure **26.19** The bacterial pathway of ornithine biosynthesis from glutamate. The enzymes are (1) *N*-acetylglutamate synthase, (2) *N*-acetylglutamate kinase, (3) *N*-acetylglutamate-5-semialdehyde dehydrogenase, (4) *N*-acetylornithine δ-aminotransferase, and (5) *N*-acetylornithine deacetylase. In mammals, ornithine is synthesized directly from glutamate-5-semialdehyde by a pathway that does not involve an *N*-acetyl block.

The product of the ornithine transcarbamoylase reaction is *citrulline* (Figure 26.21). Ornithine and citrulline are two α-amino acids of metabolic importance that nevertheless are *not* among the 20 α-amino acids commonly found in proteins. Like CPS-I, ornithine transcarbamoylase is a mitochondrial enzyme. The reactions of ornithine synthesis and the rest of the urea cycle enzymes occur in the cytosol.

The pertinent feature of the citrulline side chain is the **ureido group.** In a complex reaction catalyzed by *argininosuccinate synthetase,* this ureido group is first activated by ATP to yield a citrullyl-AMP derivative, followed by displacement of AMP by aspartate to give *argininosuccinate* (Figure 26.21). The formation of arginine is then accomplished by *argininosuccinase,* which catalyzes the nonhydrolytic elimination of fumarate from argininosuccinate. This reaction completes the biosynthesis of Arg.

The Urea Cycle

The carbon skeleton of arginine is derived principally from α-ketoglutarate, but the N and C atoms composing the **guanidino group** (Figure 26.21) of the Arg side chain come from NH_4^+, HCO_3^- (as carbamoyl-P), and the α-NH_2

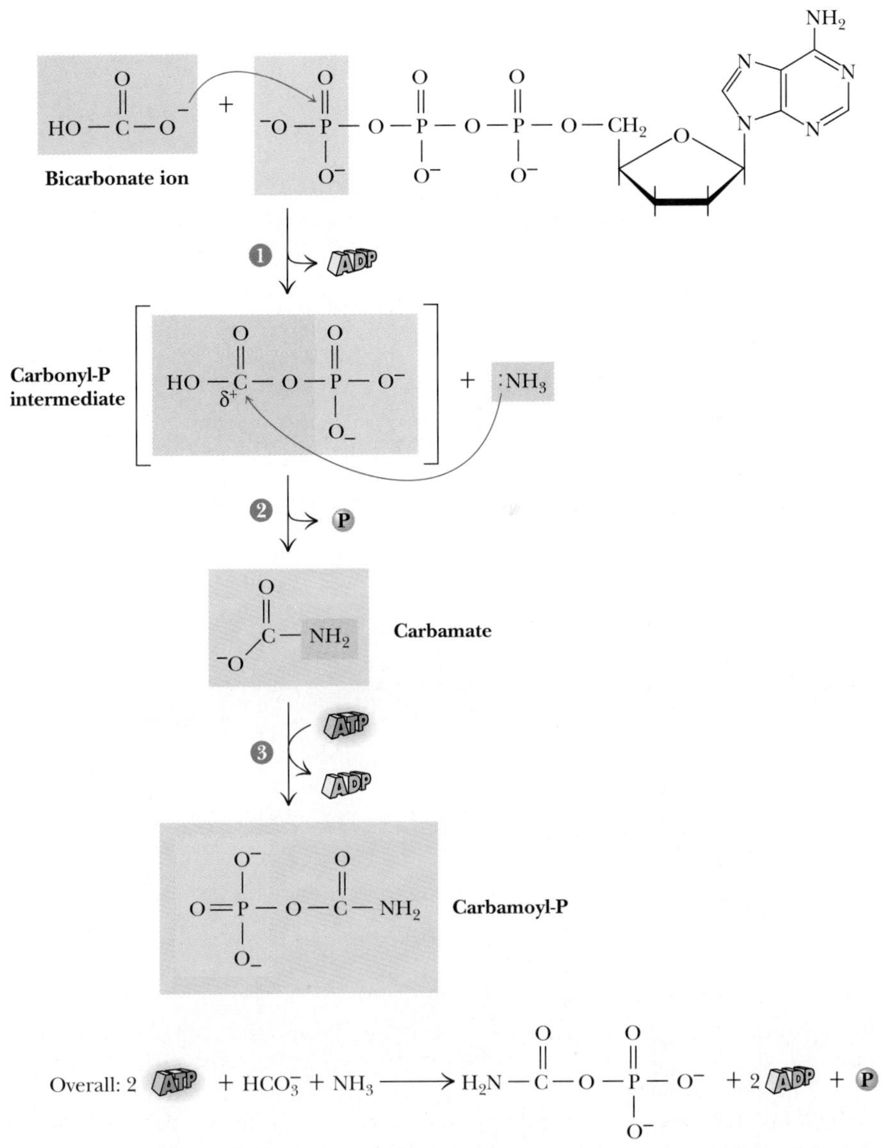

***Figure* 26.20** The mechanism of action of CPS-I, the NH_3-dependent mitochondrial CPS isozyme. (1) HCO_3^- is activated via an ATP-dependent phosphorylation. (2) Ammonia attacks the carbonyl carbon of carbonyl-P, displacing P_i to form carbamate. (3) Carbamate is phosphorylated via a second ATP to give carbamoyl-P.

groups of glutamate and aspartate. The circle of the urea cycle is closed when ornithine is regenerated from Arg by the *arginase*-catalyzed hydrolysis of arginine. Urea is the other product of this reaction and lends its name to the cycle. In terrestrial vertebrates, urea synthesis is required to excrete excess nitrogen generated by increased amino acid catabolism—for example, following dietary consumption of more than adequate amounts of protein. Urea formation is basically confined to the liver. Increases in amino acid catabolism lead to elevated glutamate levels and a rise in N-acetylglutamate, the allosteric activator of CPS-I. Stimulation of CPS-I raises overall urea cycle activity, since activities of the remaining enzymes of the cycle simply respond to increased substrate availability. Removal of potentially toxic NH_4^+ by CPS-I is an important aspect of this regulation.

The urea cycle is linked to the citric acid cycle through *fumarate,* a by-product of the action of *argininosuccinase* (Figure 26.21, reaction 3). In the citric acid cycle, fumarate is hydrated to form malate, and in turn malate is oxidized to oxaloacetate. In addition to its role in the citric acid cycle, oxaloacetate has a number of alternative fates, including (1) amination to form aspartate (Figure 26.17) and (2) conversion to PEP and entry into gluconeogenesis (Chapter 21).

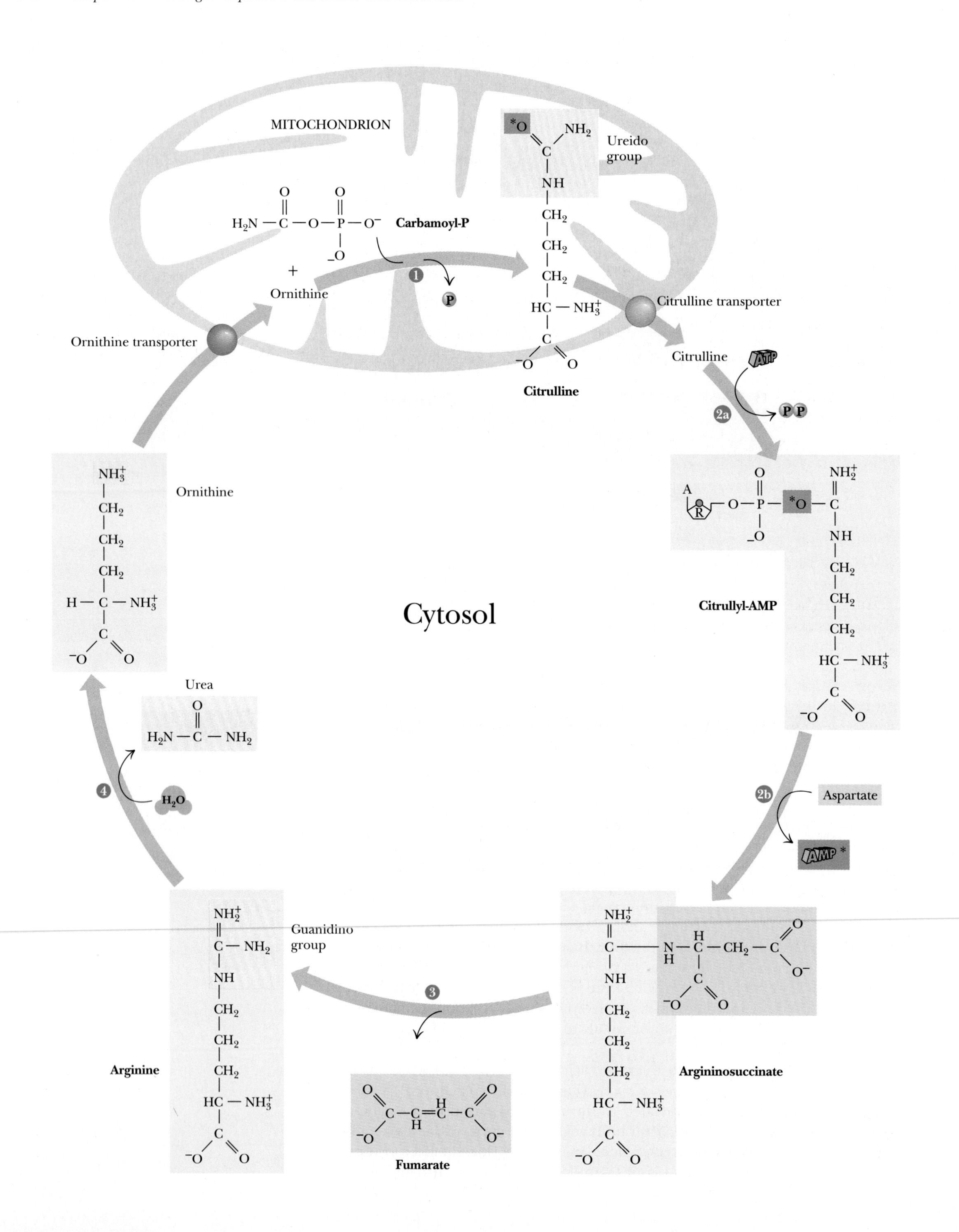
MITOCHONDRION
Carbamoyl-P
+
Ornithine
Ureido group
Citrulline
Ornithine transporter
Citrulline transporter
Citrulline
ATP
PP
2a
Citrullyl-AMP
Ornithine
Cytosol
Urea
4
H2O
2b
Aspartate
AMP *
Guanidino group
3
Arginine
Fumarate
Argininosuccinate

A Deeper Look

The Urea Cycle as Both an Ammonium and a Bicarbonate Disposal Mechanism

Excretion of excess NH_4^+ in the innocuous form of urea has traditionally been viewed as the physiological role of the urea cycle. However, the urea cycle also provides a mechanism for the excretion of excess HCO_3^- arising principally from α-carboxyl groups generated during the catabolism of α-amino acids. The following equations illustrate this property:

(1) $HCO_3^- + 2\ NH_4^+ \longrightarrow H_2NCONH_2 + 2\ H_2O + H^+$
(2) $HCO_3^- + H^+ \longrightarrow H_2O + CO_2$
Sum: $2\ HCO_3^- + 2\ NH_4^+ \longrightarrow H_2NCONH_2 + CO_2 + 3\ H_2O$

That is, *two* moles of HCO_3^- are eliminated in the synthesis of each mole of urea: one is incorporated into the product, urea (reaction 1), and the second is simply protonated to form CO_2 (reaction 2), which is easily excreted.

One interpretation of the above is that these coupled reactions allow a weak acid (NH_4^+) to protonate the conjugate base of a stronger acid (HCO_3^-). At first glance, this protonation would appear thermodynamically unfavorable, but recall that in the urea cycle, 4 equivalents of ATP are consumed per equivalent of urea synthesized: 2 ATP in the synthesis of carbamoyl-P, and 2 more as 1 ATP is converted to AMP + PP_i in the synthesis of argininosuccinate from citrulline (Figure 26.21). If this interpretation is correct, *the urea cycle may be considered an ATP-driven proton pump that transfers H^+ ions from NH_4^+ to HCO_3^- against a thermodynamic barrier. In the process, the potentially toxic waste products, ammonium and bicarbonate, are rendered innocuous and excreted.*

Lysine biosynthesis in some fungi and in the protoctist *Euglena* also stems from α-ketoglutarate, making lysine a member of the α-ketoglutarate family of amino acids in these organisms. (As we shall see, the other organisms capable of lysine synthesis—namely, bacteria, other fungi, algae, and green plants—use aspartate as precursor.) To make lysine from α-ketoglutarate requires a lengthening of the carbon skeleton by one CH_2 unit to yield *α-ketoadipate* (Figure 26.22). This addition is accomplished by a series of reactions reminiscent of the initial stages of the citric acid cycle. First, a two-carbon acetyl-CoA unit is added to the α-carbon of α-ketoglutarate to form *homocitrate.* Then, in a reaction sequence like that catalyzed by aconitase, *homoisocitrate* is formed from homocitrate. Oxidative decarboxylation (as in isocitrate dehydrogenase) removes one carbon (the original α-carboxyl group of α-ketoglutarate), leaving *α-ketoadipate.* A glutamate-dependent aminotransferase enzyme then aminates α-ketoadipate to give *α-aminoadipate.* Next, the δ-COO^- group is activated in an ATP-dependent adenylylation reaction, priming this δ-COO^- group for reduction to an aldehyde by NADPH. *α-Aminoadipic-6-semialdehyde* is then reductively aminated by addition of glutamate to its aldehydic carbon in

◀ ***Figure* 26.21** The urea cycle series of reactions: Transfer of the carbamoyl group of carbamoyl-P to ornithine by *ornithine transcarbamoylase* (OTCase, reaction 1) yields citrulline. The citrulline ureido group is then activated by reaction with ATP to give a citrullyl-AMP intermediate (reaction 2a); AMP is then displaced by aspartate, which is linked to the carbon framework of citrulline via its α-NH_2 group (reaction 2b). The course of reaction 2 was verified using ^{18}O-labeled citrulline. The ^{18}O label (indicated by the asterisk, *) was recovered in AMP. Citrulline and AMP are joined via the ureido *O atom. The product of this reaction is argininosuccinate; the enzyme catalyzing the two steps of reaction 2 is *argininosuccinate synthetase.* The next step (reaction 3) is carried out by *argininosuccinase,* which catalyzes the nonhydrolytic removal of fumarate from argininosuccinate to give arginine. Hydrolysis of Arg by *arginase* (reaction 4) yields urea and ornithine, completing the urea cycle.

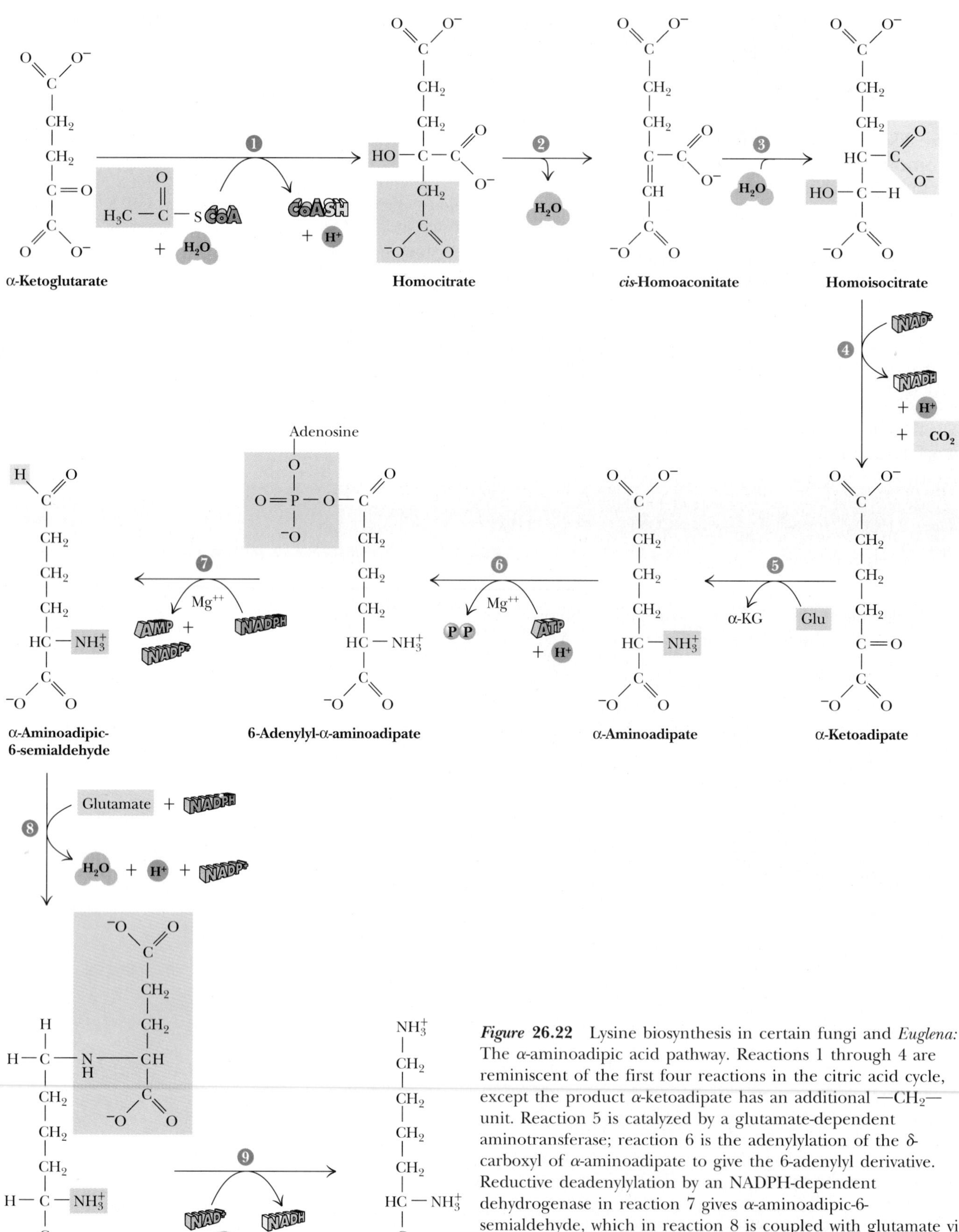

Figure **26.22** Lysine biosynthesis in certain fungi and *Euglena:* The α-aminoadipic acid pathway. Reactions 1 through 4 are reminiscent of the first four reactions in the citric acid cycle, except the product α-ketoadipate has an additional $—CH_2—$ unit. Reaction 5 is catalyzed by a glutamate-dependent aminotransferase; reaction 6 is the adenylylation of the δ-carboxyl of α-aminoadipate to give the 6-adenylyl derivative. Reductive deadenylylation by an NADPH-dependent dehydrogenase in reaction 7 gives α-aminoadipic-6-semialdehyde, which in reaction 8 is coupled with glutamate via its amino group by a second NADPH-dependent dehydrogenase. Oxidative removal of the α-ketoglutarate moiety by NAD^+-dependent saccharopine dehydrogenase in reaction 9 leaves this amino group as the ε-NH_3^+ of lysine.

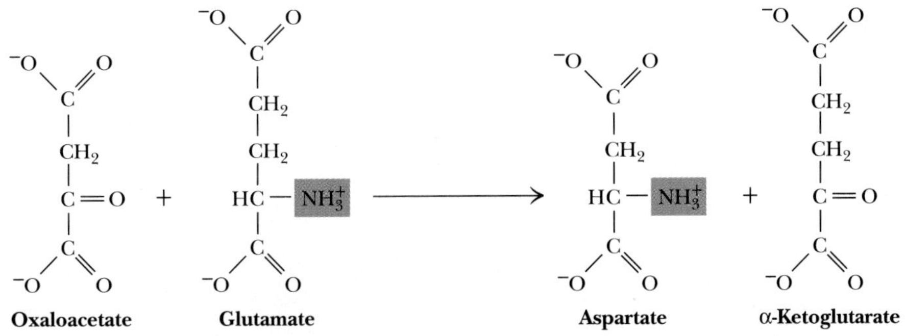

Oxaloacetate + Glutamate → Aspartate + α-Ketoglutarate

Figure **26.23** Aspartate biosynthesis via transamination of oxaloacetate by glutamate. The enzyme responsible is PLP-dependent glutamate:aspartate aminotransferase.

an NADPH-dependent reaction leading to the formation of *saccharopine.* Oxidative cleavage of saccharopine by way of an NAD^+-dependent dehydrogenase activity yields α-ketoglutarate and *lysine.* This pathway is known as the α-aminoadipic acid pathway of lysine biosynthesis. Interestingly, lysine degradation in animals leads to formation of α-aminoadipate by a reverse series of reactions identical to those occurring along the last steps of this biosynthetic pathway.

The Aspartate Family of Amino Acids

The members of the aspartate family of amino acids include aspartate (Asp), asparagine (Asn), lysine (via the diaminopimelic acid pathway), methionine (Met), threonine (Thr), and isoleucine (Ile).

Aspartate is formed from the citric acid cycle intermediate, oxaloacetate, by transfer of an amino group from glutamate via an aminotransferase reaction (Figure 26.23). Like glutamate synthesis from α-ketoglutarate, aspartate synthesis is a drain on the citric acid cycle. As we already saw, the Asp amino group serves as the N donor in the conversion of citrulline to arginine. In Chapter 27, we shall see that this $—NH_2$ is also incorporated as one of the N atoms of the purine ring system during nucleotide biosynthesis, and provides the 6-amino group of the major purine, adenine. In addition, the entire aspartate molecule is used in the biosynthesis of pyrimidine nucleotides.

Asparagine is formed by amidation of the β-carboxyl group of aspartate. In bacteria, in analogy with glutamine synthesis, the nitrogen added in this amidation comes directly from NH_4^+. In other organisms, *asparagine synthetase* catalyzes the ATP-dependent transfer of the amido-N of glutamine to aspartate to yield glutamate and asparagine (Figure 26.24).

Threonine, methionine, and **lysine** biosynthesis in bacteria proceeds from the common precursor, aspartate, which is converted first to *aspartyl-β-phosphate* and then to *β-aspartatyl-semialdehyde.* The first reaction is an ATP-dependent phosphorylation catalyzed by *aspartokinase* (Figure 26.25). In *E. coli,* there are three isozymes of aspartokinase, designated **aspartokinases I, II,** and **III.** Each of these isozymes is uniquely controlled by one of the three end-product amino acids (Table 26.4). Threonine is a specific feedback inhib-

Table **26.4**
Regulation of the Three Aspartokinase Isozymes of *E. coli*

Enzyme	Feedback Inhibitor	Co-repressor*
Aspartokinase I	Threonine	Threonine and isoleucine
Aspartokinase II	None	Methionine
Aspartokinase III	Lysine	Lysine

* *Co-repressor* is the term given to metabolites that can act in repressing expression of specific genes.

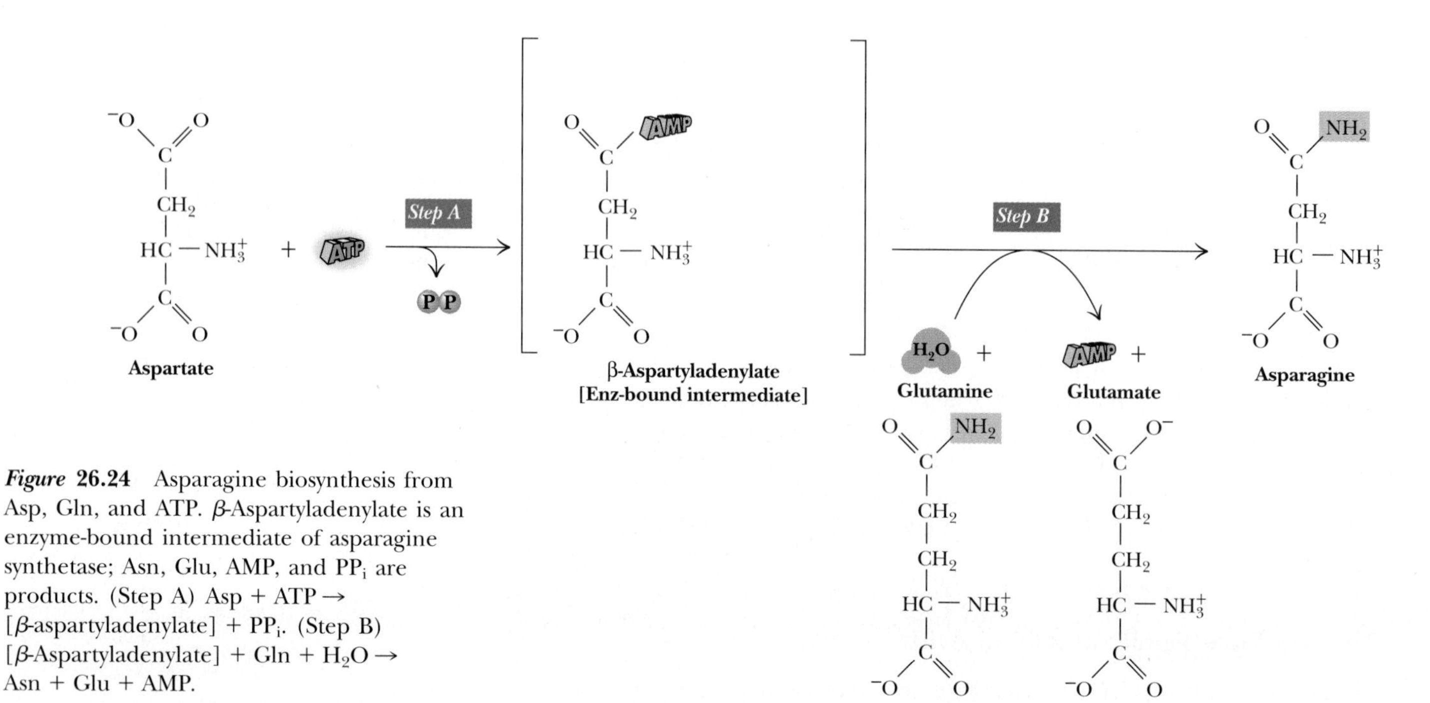

Figure **26.24** Asparagine biosynthesis from Asp, Gln, and ATP. β-Aspartyladenylate is an enzyme-bound intermediate of asparagine synthetase; Asn, Glu, AMP, and PP_i are products. (Step A) Asp + ATP → [β-aspartyladenylate] + PP_i. (Step B) [β-Aspartyladenylate] + Gln + H_2O → Asn + Glu + AMP.

itor of aspartokinase I, and lysine feedback inhibits aspartokinase III. Although aspartokinase II is unaffected by feedback inhibition, methionine acts as a co-repressor of the aspartokinase II structural gene. Both threonine and isoleucine are co-repressors of aspartokinase I gene transcription, and lysine plays the same role for aspartokinase III. Thus, the biosynthesis of each of the three amino acids may be independently regulated through controls exerted on the formation or activity of a particular aspartokinase isozyme.

β-Aspartyl-semialdehyde is formed via NADPH-dependent reduction of aspartyl-β-phosphate in a reaction catalyzed by *β-aspartyl-semialdehyde dehydrogenase* (Figure 26.25). From here, the pathway of lysine synthesis diverges. The methyl carbon of pyruvate is condensed with β-aspartyl-semialdehyde, and H_2O is eliminated to yield the cyclic compound *2,3-dihydropicolinate* (Figure

Figure **26.25** *(opposite)* Biosynthesis of threonine, methionine, and lysine, members of the aspartate family of amino acids. β-Aspartyl-semialdehyde is a common precursor to all three. It is formed by aspartokinase (reaction 1) and β-aspartyl-semialdehyde dehydrogenase (reaction 2). From here, the pathways diverge. Reduction of β-aspartyl-semialdehyde by homoserine dehydrogenase (reaction 3) gives homoserine, a precursor to threonine and methionine but not lysine. The branch designated by reactions 4 and 5 (catalyzed by homoserine kinase and threonine synthase) gives rise to threonine. The other branch from homoserine (reactions 6 through 9) leads to methionine (the enzymes are, in order, homoserine acyltransferase, cystathionine synthase, cystathionine-β-lyase, and homocysteine methyltransferase). The route to lysine from β-aspartyl-semialdehyde is the so-called diaminopimelate pathway (reactions 10 through 16). Pyruvate is condensed with β-aspartyl-semialdehyde to yield 2,3-dihydropicolinate (reaction 10, dihydropicolinate synthase), which is then reduced by Δ^1-piperidine-2,6-dicarboxylate dehydrogenase (reaction 11). Succinylation (reaction 12, *N*-succinyl-2-amino-6-ketopimelate synthase) is accompanied by opening of the ring; amination ensues (reaction 13, succinyl-diaminopimelate aminotransferase), followed by desuccinylation (reaction 14, succinyl-diaminopimelate desuccinylase) to give L,L-α,ε-diaminopimelate. Epimerization to the *meso* form (reaction 15, diaminopimelate epimerase), then decarboxylation (reaction 16, diaminopimelate decarboxylase), yields lysine.

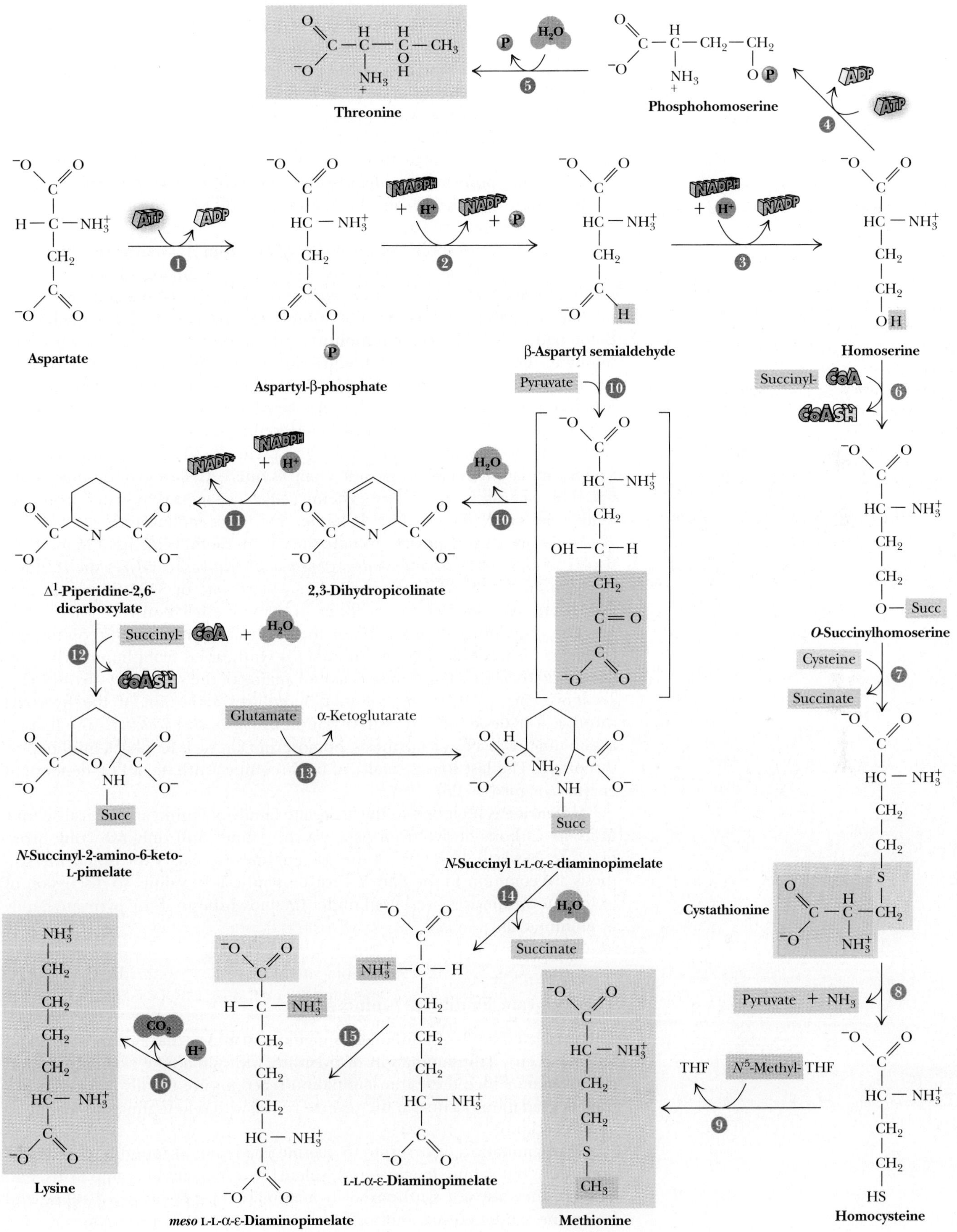

Threonine

Phosphohomoserine

Aspartate

Aspartyl-β-phosphate

β-Aspartyl semialdehyde

Homoserine

1 ATP → ADP

2 NADPH + H^+ → $NADP^+$ + P

3 NADPH + H^+ → $NADP^+$

4 ATP → ADP

5 H_2O → P

6 Succinyl-CoA → CoASH

O-Succinylhomoserine

7 Cysteine → Succinate

Cystathionine

8 → Pyruvate + NH_3

Homocysteine

9 N^5-Methyl-THF → THF

Methionine

10 Pyruvate; H_2O

2,3-Dihydropicolinate

11 NADPH + H^+ → $NADP^+$

Δ^1-Piperidine-2,6-dicarboxylate

12 Succinyl-CoA + H_2O → CoASH

***N*-Succinyl-2-amino-6-keto-L-pimelate**

13 Glutamate → α-Ketoglutarate

***N*-Succinyl L-L-α-ε-diaminopimelate**

14 H_2O → Succinate

L-L-α-ε-Diaminopimelate

15

***meso* L-L-α-ε-Diaminopimelate**

16 H^+ → CO_2

Lysine

26.25). Lysine thus must be considered a member of both the aspartate and the pyruvate families of amino acids. Lysine is a feedback inhibitor of this branch-point enzyme. *Dihydropicolinate* is then reduced in an NADPH-dependent reaction to give *Δ^1-piperidine-2,6-dicarboxylate.* A series of reactions, including a hydrolytic opening of the piperidine ring, a succinylation, a glutamate-dependent amination, and the hydrolytic removal of succinate, results in the formation of the symmetrical *L,L-α,ε-diaminopimelate.* Epimerization of this intermediate to the *meso* form, followed by decarboxylation, yields the end product, *lysine.* Since this pathway proceeds through the symmetrical L,L-α,ε-diaminopimelate, one-half of the CO_2 evolved in the terminal decarboxylase step is derived from the carboxyl group of pyruvate, and one-half from the α-carboxyl of Asp.

The other metabolic branch diverging from β-aspartyl-semialdehyde leads to *threonine and methionine* via *homoserine,* an analog of serine that is formed by the NADPH-dependent reduction of β-aspartyl-semialdehyde (Figure 26.25) catalyzed by *homoserine dehydrogenase.* From homoserine, the biosynthetic pathways leading to methionine and threonine separate. To form **methionine,** the —OH group of homoserine is first succinylated by *homoserine acyltransferase.* Methionine is a feedback inhibitor of this enzyme. The succinyl group of *O-succinylhomoserine* is then displaced by cysteine to yield *cystathionine* (Figure 26.25). The sulfur atom in methionine is contributed by a cysteine sulfhydryl. *Cystathionine* is then split to give pyruvate, NH_4^+, and *homocysteine,* a nonprotein amino acid whose side chain is one $—CH_2—$ group longer than Cys. Methylation of the homocysteine —SH via methyl transfer from the methyl donor, N^5-methyl-THF (Chapter 14), gives methionine.

In passing, it is important to note the role of methionine itself in methylation reactions. The enzyme *S-adenosylmethionine synthase* catalyzes the reaction of methionine with ATP to form *S-adenosylmethionine,* or SAM (Figure 26.26). SAM serves as a methyl group donor in many methylation reactions.

The remaining amino acids of the aspartate family are threonine and isoleucine. **Threonine,** like methionine, is synthesized from homoserine. Indeed, homoserine is the primary alcohol analog of the secondary alcohol Thr. To move this —OH from C-4 to C-3 requires activation of the hydroxyl through ATP-dependent phosphorylation by *homoserine kinase.* As the first reaction unique to Thr biosynthesis, homoserine kinase is feedback-inhibited by threonine. The last step is catalyzed by threonine synthase, a PLP-dependent enzyme (Figure 26.25).

Isoleucine is included in the aspartate family of amino acids because four of its six carbons derive from Asp (via threonine) and only two come from pyruvate. Nevertheless, four of the five enzymes necessary for isoleucine synthesis are common to the pathway for biosynthesis of valine, so discussion of isoleucine synthesis is presented under the biosynthesis of the pyruvate family of amino acids.

The Pyruvate Family of Amino Acids

The pyruvate family of amino acids includes alanine (Ala), valine (Val), and leucine (Leu). Transamination of pyruvate, with glutamate as amino donor, gives **alanine.** Since these transamination reactions are readily reversible, alanine degradation occurs via the reverse route, with α-ketoglutarate serving as amino acceptor.

Transamination of pyruvate to alanine is a reaction found in virtually all organisms, but valine, leucine, and isoleucine are essential amino acids, and as such, they are not synthesized in mammals. The pathways of **valine** and **isoleucine** can be considered together since one set of four enzymes is com-

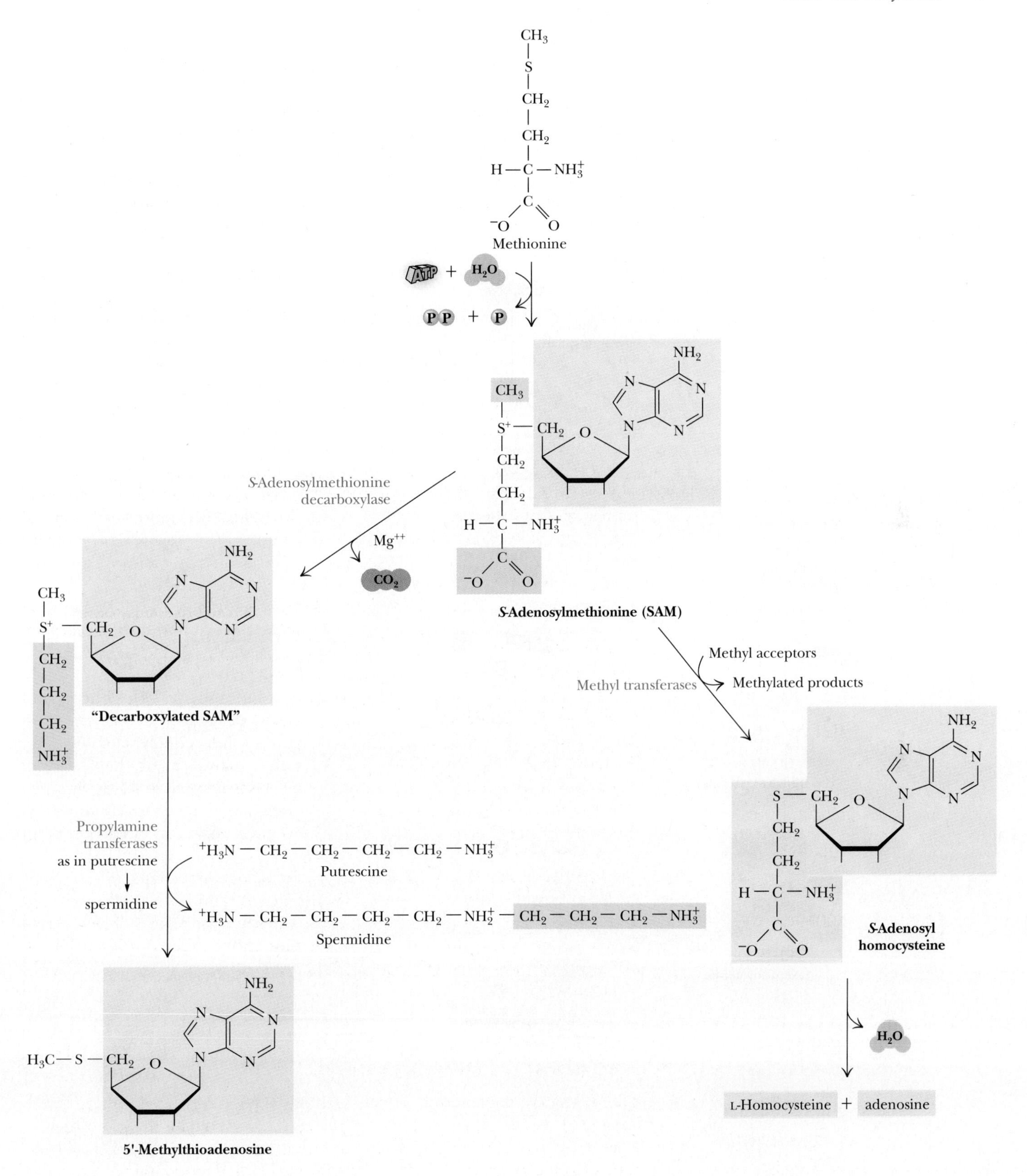

***Figure* 26.26** The synthesis of *S*-adenosylmethionine (SAM) from methionine plus ATP, and the role of SAM as a substrate of methyltransferases in methyl donor reactions and in propylamine transfer reactions, as in the synthesis of polyamines.

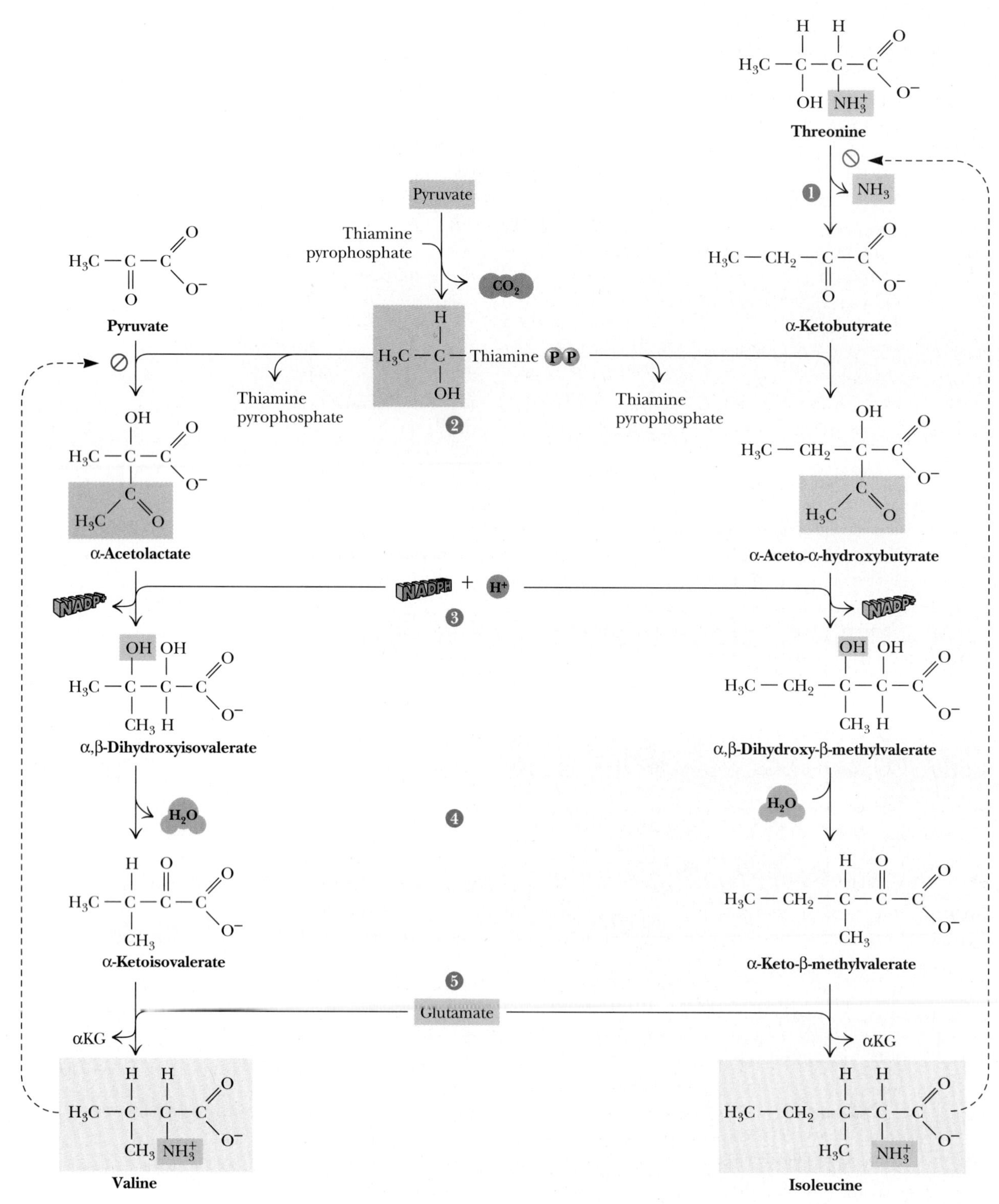

***Figure* 26.27** Biosynthesis of valine and isoleucine. The enzymes are (1) threonine deaminase, (2) acetohydroxy acid synthase, (3) acetohydroxy acid isomeroreductase, (4) dihydroxy acid dehydratase, and (5) glutamate-dependent aminotransferase. Feedback inhibition regulates this pathway: Enzyme 1 is isoleucine-sensitive and enzyme 2 is valine-sensitive.

mon to the last four steps of both pathways (Figure 26.27). Both pathways begin with an α-keto acid. Isoleucine can be considered a structural analog of valine that has one extra $—CH_2—$ unit, and its α-keto acid precursor, namely, *α-ketobutyrate,* is one carbon longer than the valine precursor, pyruvate. Interestingly, α-ketobutyrate is formed from threonine by the action of *threonine deaminase,* an enzyme that both deaminates and dehydrates Thr. Threonine deaminase, a PLP-dependent enzyme, is feedback-sensitive to isoleucine (this enzyme is also known as threonine dehydratase and serine dehydratase). So, part of the carbon skeleton for Ile comes from Asp by way of Thr. From here on, the Val and Ile pathways employ the same set of enzymes. The first reaction involves the generation of hydroxyethyl-thiamine pyrophosphate from pyruvate in a reaction analogous to those catalyzed by transketolase and the pyruvate dehydrogenase complex. The two-carbon hydroxyethyl group is transferred from TPP to the respective keto acid acceptor by *acetohydroxy acid synthase* (acetolactate synthase) to give *α-acetolactate* or *α-aceto-α-hydroxybutyrate.* NAD(P)H-dependent reduction of these α-keto hydroxy acids yields the dihydroxy acids *α,β-dihydroxy-isovalerate* and *α,β-dihydroxy-β-methylvalerate.* Dehydration of each of these dihydroxy acids by *dihydroxy acid dehydratase* gives the appropriate α-keto acid carbon skeletons *α-ketoisovalerate* and *α-keto-β-methylvalerate.* Transamination by the *branched-chain amino acid* or *glutamate aminotransferase* yields Val or Ile, respectively (Figure 26.27).

Leucine synthesis depends on these reactions as well, since α-ketoisovalerate is a precursor common to both Val and Leu (Figure 26.28). Although Val and Leu differ by only a single $—CH_2—$ in their respective side chains, the carboxyl group of α-ketoisovalerate first picks up *two* carbons from acetyl-CoA to give *α-isopropylmalate* in a reaction catalyzed by *isopropylmalate synthase;* the enzyme is sensitive to feedback inhibition by Leu (Figure 26.28). *Isopropylmalate dehydratase* converts the α-isomer to the β form, which undergoes an NAD^+-dependent oxidative decarboxylation by *isopropylmalate dehydrogenase,* so that the carboxyl group of α-ketoisovalerate is lost as CO_2. Amination of α-ketoisocaproate by *leucine aminotransferase* gives Leu.

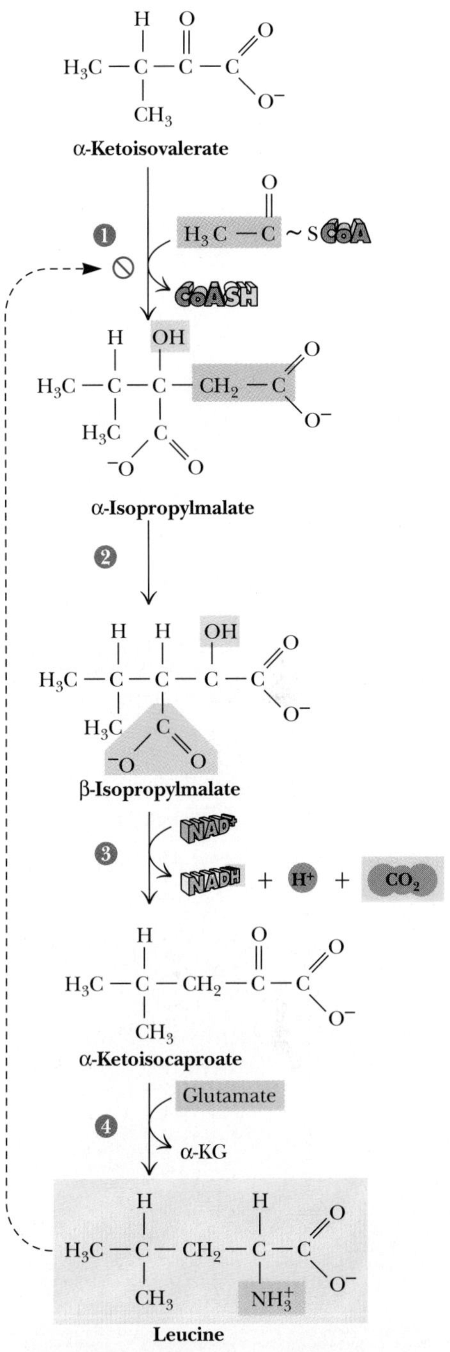

α-Ketoisovalerate

α-Isopropylmalate

β-Isopropylmalate

α-Ketoisocaproate

Leucine

***Figure* 26.28** Biosynthesis of leucine. The enzymes are (1) α-isopropylmalate synthase, (2) α-isopropylmalate dehydratase, (3) isopropylmalate dehydrogenase, and (4) leucine aminotransferase. Enzyme 1 is feedback-inhibited by leucine.

The 3-Phosphoglycerate Family of Amino Acids

Serine, glycine, and cysteine are derived from the glycolytic intermediate 3-phosphoglycerate. The diversion of 3-PG from glycolysis is achieved via *3-phosphoglycerate dehydrogenase* (Figure 26.29). This NAD^+-dependent oxidation of 3-PG yields *3-phosphohydroxypyruvate*—which, as an α-keto acid, is a substrate for transamination by glutamate to give *3-phosphoserine. Serine phosphatase* then generates **serine.** Serine inhibits the first enzyme, 3-PG dehydrogenase, and thereby feedback-regulates its own synthesis.

Glycine is made from serine via two related enzymatic processes. In the first, *serine hydroxymethyltransferase,* a PLP-dependent enzyme, catalyzes the transfer of the serine β-carbon to tetrahydrofolate (THF), the principal agent of one-carbon metabolism (Figure 26.30a). Glycine and N^5,N^{10}-methylene-THF are the products. In addition, glycine can be synthesized by a reversal of the *glycine oxidase* reaction (Figure 26.30b). Here, glycine is formed when N^5,N^{10}-methylene-THF condenses with NH_4^+ and CO_2. Via this route, the β-carbon of serine becomes part of glycine. The conversion of serine to glycine is a prominent means of generating one-carbon derivatives of THF, which are so important for the biosynthesis of purines and the C-5 methyl group of thymine (a pyrimidine), as well as the amino acid methionine. Glycine itself contributes to both purine and heme synthesis.

Cysteine synthesis is accomplished by sulfhydryl transfer to serine. In some bacteria, H_2S condenses directly with serine via a PLP-dependent en-

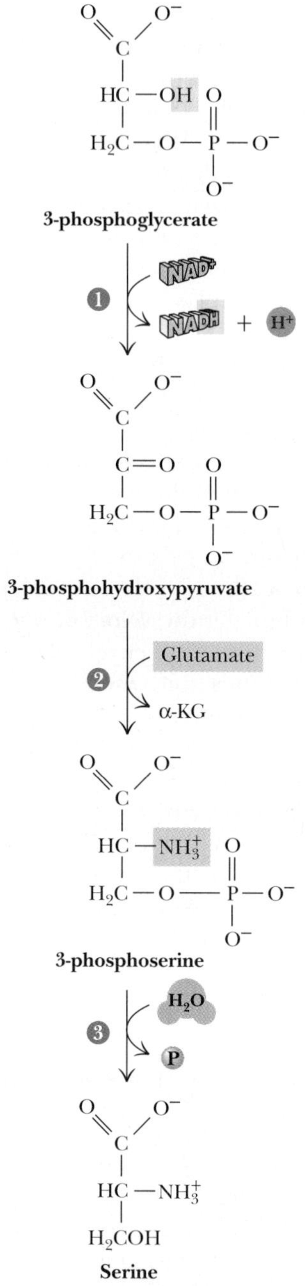

3-phosphoglycerate

1 NAD^+ → $NADH + H^+$

3-phosphohydroxypyruvate

2 Glutamate → α-KG

3-phosphoserine

3 H_2O → P

Serine

***Figure* 26.29** Biosynthesis of serine from 3-phosphoglycerate. The enzymes are (1) 3-phosphoglycerate dehydrogenase, (2) 3-phosphoserine aminotransferase, and (3) phosphoserine phosphatase.

zyme-catalyzed reaction, but in most microorganisms and green plants, the sulfhydrylation reaction requires an activated form of serine, *O-acetylserine* (Figure 26.31). *O*-acetylserine is made by *serine acetyltransferase,* with the transfer of an acetyl group from acetyl-CoA to the —OH of Ser. This enzyme is inhibited by Cys. *O*-Acetylserine then undergoes sulfhydrylation by H_2S with elimination of acetate; the enzyme is *O-acetylserine sulfhydrylase.*

Sulfate Assimilation. Given the prevailing oxidative nature of our environment and the reactivity and toxicity of H_2S, the source of sulfide for Cys synthesis merits discussion. In microorganisms and plants, sulfide is the prod-

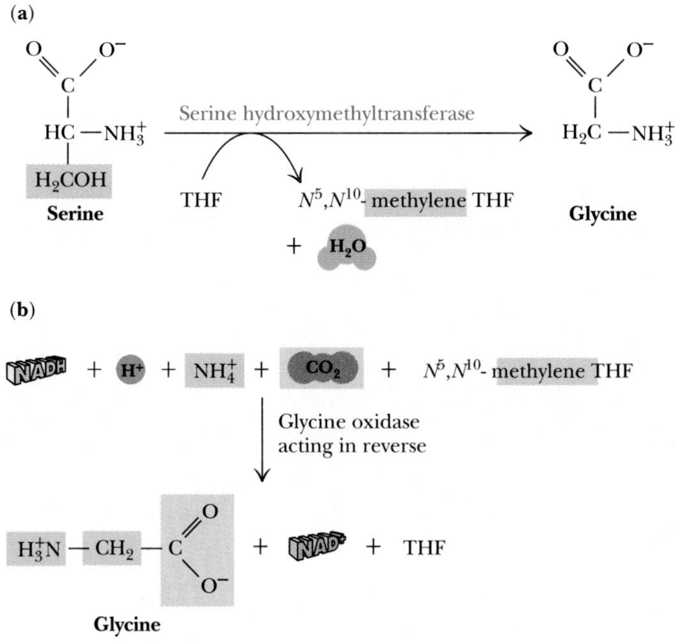

(a) Serine → Glycine (Serine hydroxymethyltransferase; THF → N^5,N^{10}-methylene THF + H_2O)

(b) NADH + H^+ + NH_4^+ + CO_2 + N^5,N^{10}-methylene THF → (Glycine oxidase acting in reverse) → Glycine + NAD^+ + THF

***Figure* 26.30** Biosynthesis of glycine from serine (a) via serine hydroxymethyltransferase and (b) via glycine oxidase.

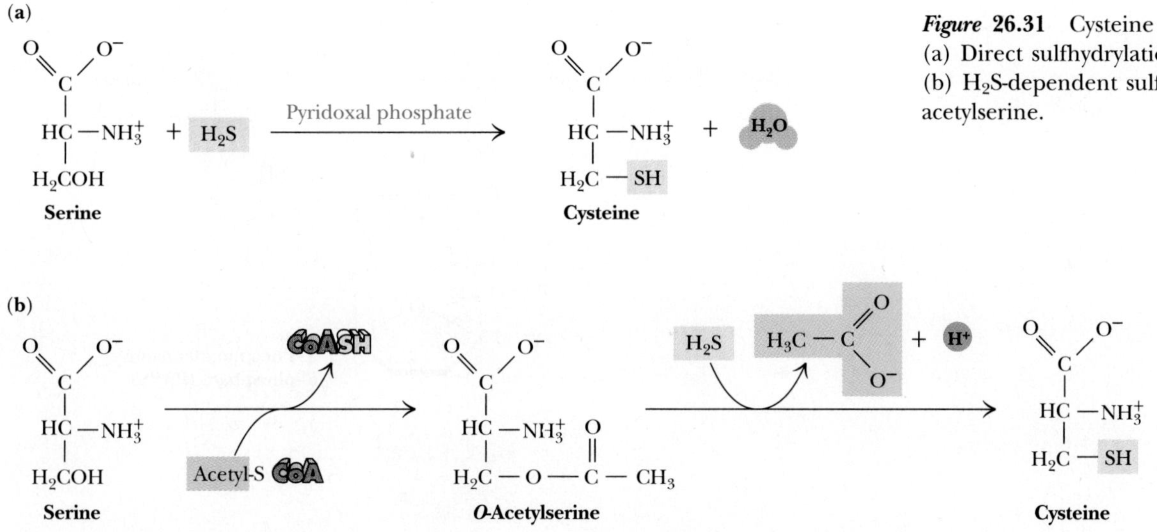

(a) Serine + H_2S → (Pyridoxal phosphate) → Cysteine + H_2O

(b) Serine + Acetyl-S CoA → O-Acetylserine + CoASH; O-Acetylserine + H_2S → Cysteine + acetate + H^+

***Figure* 26.31** Cysteine biosynthesis. (a) Direct sulfhydrylation of serine by H_2S. (b) H_2S-dependent sulfhydrylation of *O*-acetylserine.

uct of sulfate assimilation. Sulfate is the common inorganic form of combined sulfur, and its assimilation involves several interesting ATP derivatives (Figure 26.32). **3′-Phosphoadenosine-5′-phosphosulfate** (PAPS) is not only an intermediate in sulfate assimilation, it also serves as the substrate for synthesis of sulfate esters, such as the sulfated polysaccharides found in the glycocalyx of animal cells. The "activated sulfate" of PAPS is reduced to sulfite (SO_3^{2-}) in a thioredoxin-dependent reaction, and sulfite is then reduced to sulfide (S^{2-}).

Biosynthesis of the Aromatic Amino Acids

The aromatic amino acids, phenylalanine, tyrosine, and tryptophan, are derived from a shared pathway that has **chorismic acid** (Figure 26.33) as a key intermediate. Indeed, chorismate is common to the synthesis of cellular compounds having benzene rings, including these amino acids, the fat-soluble

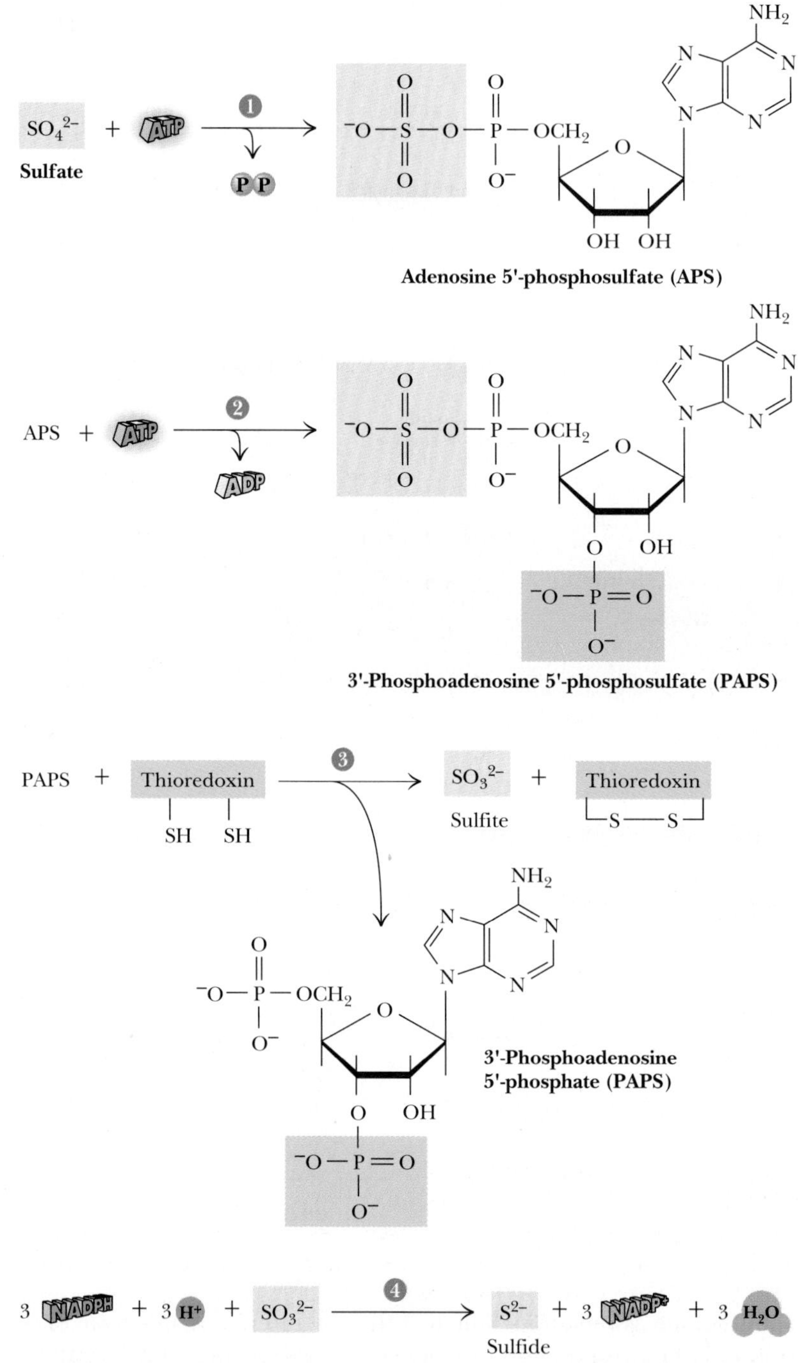

SO_4^{2-} + ATP →(1; PP) Adenosine 5'-phosphosulfate (APS)

Sulfate

APS + ATP →(2; ADP) 3'-Phosphoadenosine 5'-phosphosulfate (PAPS)

PAPS + Thioredoxin (SH SH) →(3) SO_3^{2-} (Sulfite) + Thioredoxin (S—S) + 3'-Phosphoadenosine 5'-phosphate (PAPS)

3 NADPH + 3 H^+ + SO_3^{2-} →(4) S^{2-} (Sulfide) + 3 $NADP^+$ + 3 H_2O

***Figure* 26.32** Sulfate assimilation and the generation of sulfide for synthesis of organic S compounds. In reaction 1, ATP sulfurylase catalyzes the formation of adenosine-5′-phosphosulfate (APS) + PP_i. In reaction 2, adenosine-5′-phosphosulfate 3′-phosphokinase catalyzes the reaction of adenosine 5′-phosphosulfate with a second ATP to form 3′-phosphoadenosine-5′-phosphosulfate (PAPS) + ADP. Both enzymes are Mg^{2+}-dependent. In reaction 3, PAPS is reduced to sulfite (SO_3^{2-}) in a *thioredoxin-dependent* reaction. Thioredoxin is a small (12 kD) protein that functions in a number of biological reductions (see Chapter 27). In reaction 4, *sulfite reductase* catalyzes the six-electron reduction of sulfite to sulfide. NADPH is the electron donor. Sulfite reductase possesses siroheme as a prosthetic group, the same heme found in nitrite reductase (Figure 26.1), which also catalyzes a six-electron transfer reaction.

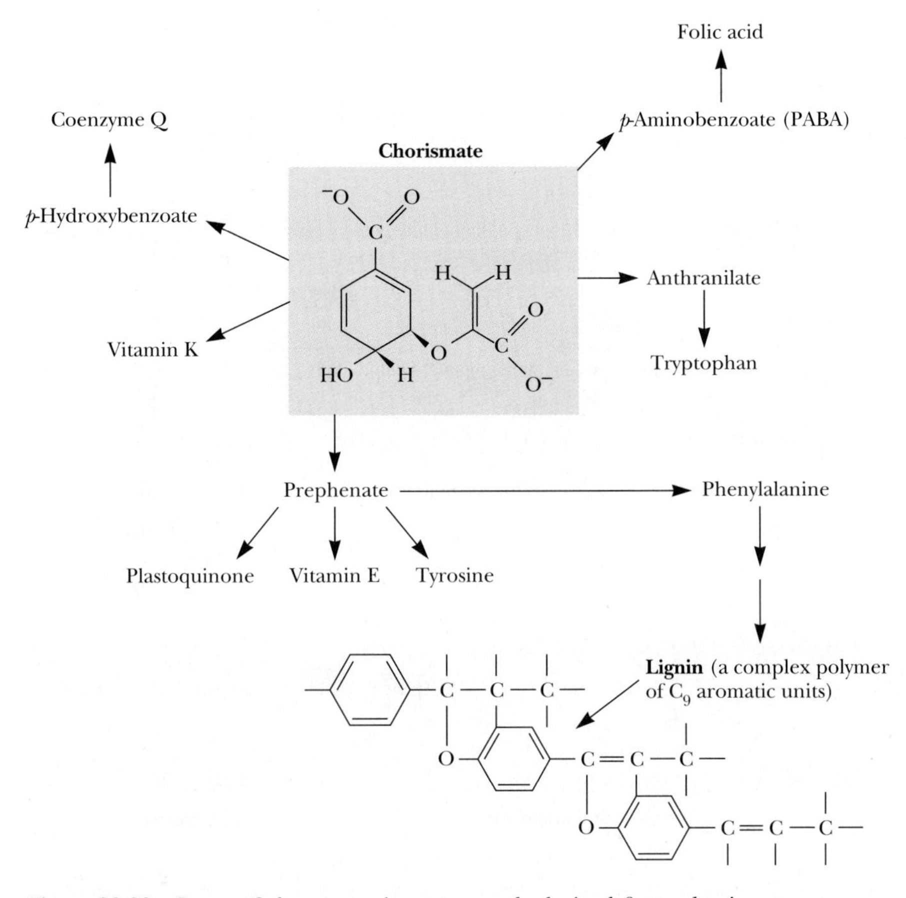

***Figure* 26.33** Some of the aromatic compounds derived from chorismate.

vitamins E and K, folic acid, and coenzyme Q and plastoquinone (the two quinones necessary to electron transport during respiration and photosynthesis, respectively). **Lignin,** a polymer of nine-carbon aromatic units, is also a derivative of chorismate. Lignin and related compounds can account for as much as 35% of the dry weight of higher plants; clearly, enormous amounts of carbon pass through the chorismate biosynthetic pathway.

The Shikimate Pathway. Chorismate biosynthesis occurs via the **shikimate pathway** (Figure 26.34). The precursors for this pathway are the common metabolic intermediates *phosphoenolpyruvate* and *erythrose-4-phosphate.* These intermediates are linked to form *3-deoxy-D-arabino-heptulosonate-7-phosphate* (DAHP) by *DAHP synthase.* Though this reaction is remote from the ultimate aromatic amino acid end products, it is an important point for regulation of aromatic amino acid biosynthesis, as we shall see. In the next step on the way to chorismate, DAHP is cyclized to form a six-membered saturated ring compound, *3-dehydroquinate.* A sequence of reactions ensues that introduces unsaturations into the ring, yielding *shikimate,* then *chorismate.* Note that the side chain of chorismate is derived from a second equivalent of phosphoenolpyruvate.

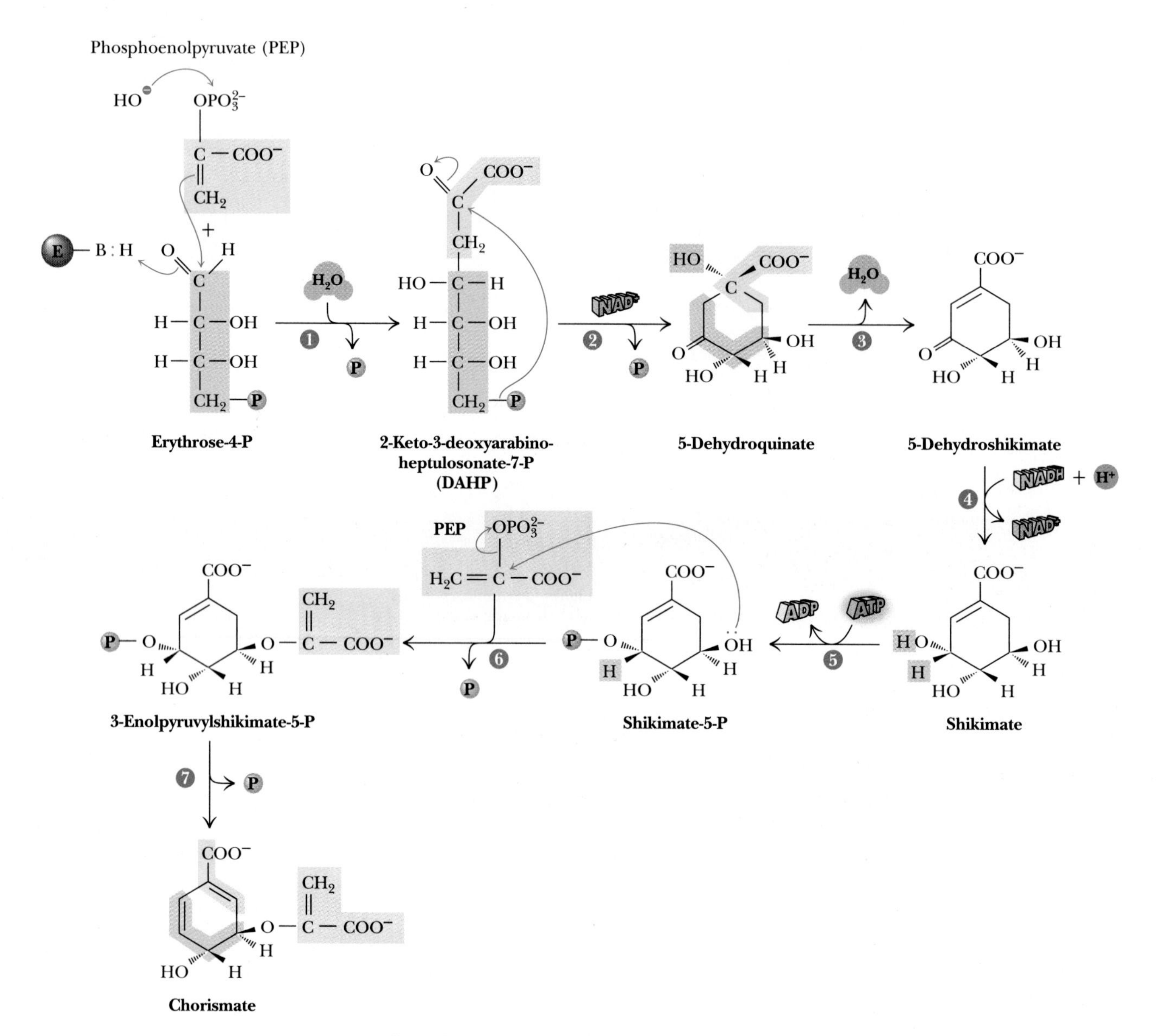

Figure **26.34** The shikimate pathway leading to the synthesis of chorismate. The starting substrates are phosphoenolpyruvate and erythrose-4-phosphate. The enzymes are (1) 2-keto-3-deoxy-D-arabino-heptulosonate-7-P synthase, (2) dehydroquinate synthase (note that the coenzyme NAD^+ is not altered in this reaction), (3) 5-dehydroquinate dehydratase, (4) shikimate dehydrogenase, (5) shikimate kinase, (6) 3-enolpyruvylshikimate-5-phosphate synthase, and (7) chorismate synthase.

Phenylalanine and Tyrosine. At chorismate, the pathway separates into three branches, each leading specifically to one of the aromatic amino acids. The branches leading to phenylalanine and tyrosine both pass through *prephenate* (Figure 26.35). In some organisms, such as *E. coli,* the branches are truly distinct because prephenate does not occur as a free intermediate, but remains bound to the bifunctional enzyme that catalyzes the first two reactions after chorismate. In any case, *chorismate mutase* is the first reaction leading to Phe or Tyr. In the Phe branch, the —OH group *para* to the prephenate carboxyl is removed by a *dehydratase;* in the Tyr branch, this —OH is retained

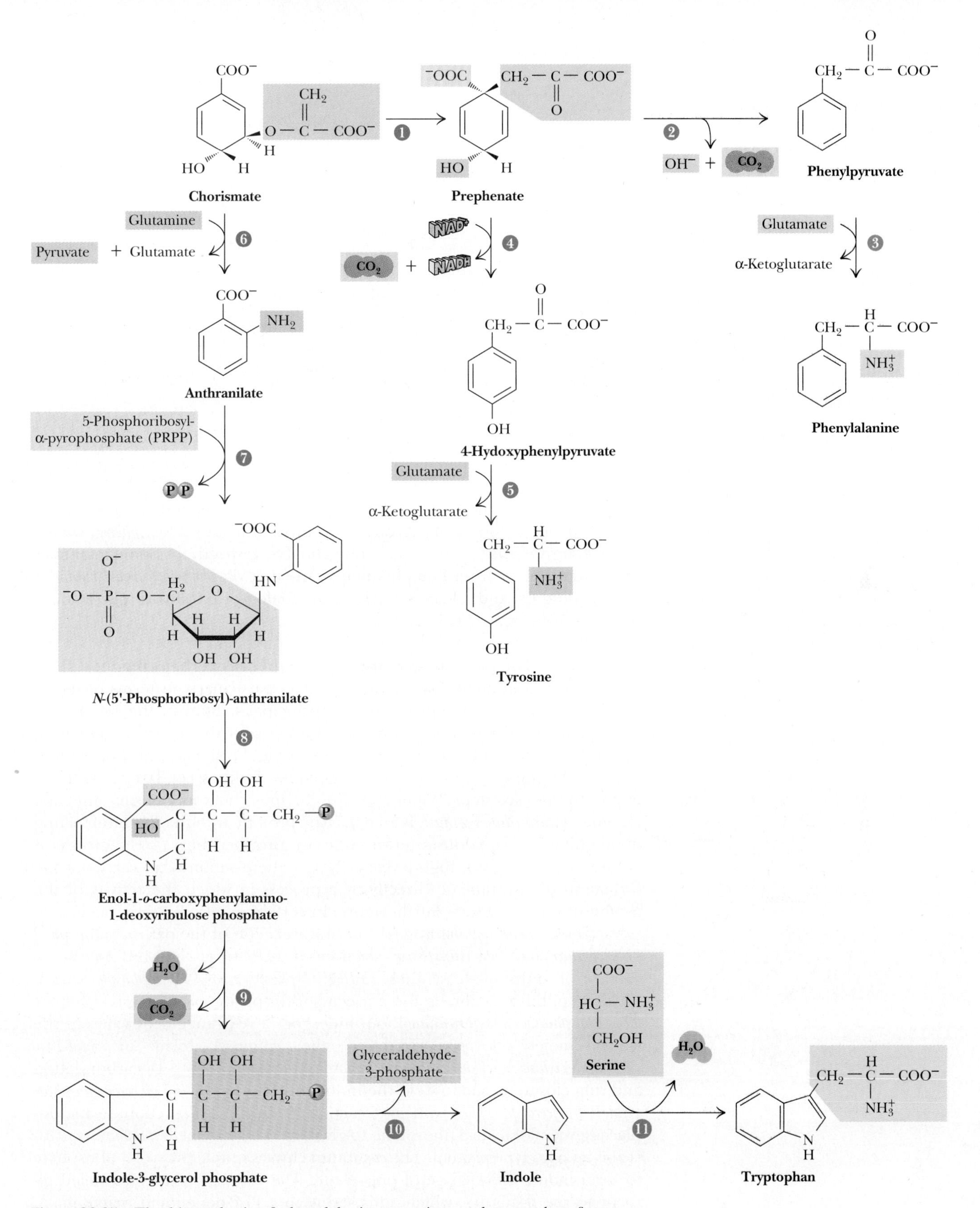

Figure **26.35** The biosynthesis of phenylalanine, tyrosine, and tryptophan from chorismate. The enzymes are (1) chorismate mutase, (2) prephenate dehydratase, (3) phenylalanine aminotransferase, (4) prephenate dehydrogenase, (5) tyrosine aminotransferase, (6) anthranilate synthase, (7) anthranilate-phosphoribosyl transferase, (8) *N*-(5′-phosphoribosyl)-anthranilate isomerase, (9) indole-3-glycerol phosphate synthase, (10) tryptophan synthase (α-subunit), and (11) tryptophan synthase (β-subunit).

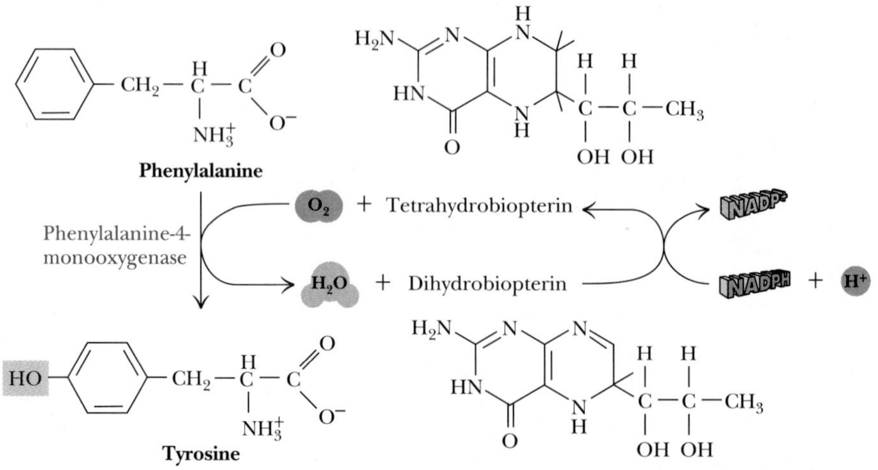

***Figure* 26.36** The formation of tyrosine from phenylalanine. This reaction is normally the first step in phenylalanine degradation in most organisms; in mammals, however, it provides a route for the biosynthesis of Tyr from Phe.

and becomes the phenolic —OH of Tyr. Glutamate-dependent aminotransferases introduce the amino groups into the two α-keto acids, *phenylpyruvate* and *4-hydroxyphenylpyruvate,* to give Phe and Tyr, respectively. Some mammals can synthesize Tyr from Phe obtained in the diet via *phenylalanine-4-monooxygenase,* using O_2 and *tetrahydrobiopterin,* an analog of tetrahydrofolic acid, as co-substrates (Figure 26.36).

Tryptophan. The pathway of tryptophan synthesis is perhaps the most thoroughly studied of any biosynthetic sequence, particularly in terms of its genetic organization and expression. The principal stalwart of this research is Charles Yanofsky of Stanford University, and his contributions in illuminating the details of this pathway represent many original insights regarding themes of general significance in metabolic regulation. Synthesis of Trp from chorismate requires six steps (Figure 26.35). In most microorganisms, the first enzyme, *anthranilate synthase,* is an $\alpha_2\beta_2$-type protein, with the β-subunit acting in a glutamine–amidotransferase role to provide the $—NH_2$ group of anthranilate. Or, given high levels of NH_4^+, the α-subunit can carry out the formation of anthranilate directly by a process in which the activity of the β-subunit is unnecessary. Furthermore, in certain enteric bacteria, such as *E. coli* and *Salmonella typhimurium,* the second reaction of the pathway, the *phosphoribosyl-anthranilate transferase* reaction, is an activity catalyzed by the α-subunit of anthranilate synthase. *PRPP (5-phosphoribosyl-1-pyrophosphate),* the substrate of this reaction, is also a precursor for purine biosynthesis (Chapter 27). *Phosphoribosyl-anthranilate* then undergoes a rearrangement wherein the ribose moiety is isomerized to the ribulosyl form in *enol-1-(o-carboxyphenylamino)-1-deoxyribulose-5-phosphate* (reaction 8). Decarboxylation and ring closure ensue to yield the indole nucleus as *indole-3-glycerol phosphate* (*indole-3-glycerol phosphate synthase,* reaction 9). In the last reaction, serine displaces glyceraldehyde-3-phosphate to give Trp. The enzyme *tryptophan synthase* is also an $\alpha_2\beta_2$-type protein. The α-subunit cleaves indoleglycerol-3-phosphate to form indole and 3-glycerol phosphate. The indole is then channeled directly to the β-subunit, which adds serine in a PLP-dependent reaction.

X-ray crystallographic analysis of the structure of tryptophan synthase from *S. typhimurium* reveals that the active sites of the α- and β-subunits of the enzyme, though separated from each other by 2.5 nm, are connected by a hydrophobic tunnel wide enough to accommodate the bound indole intermediate (Figure 26.37). Thus, indole can be transferred directly from one active site to the other without being lost from the enzyme complex and

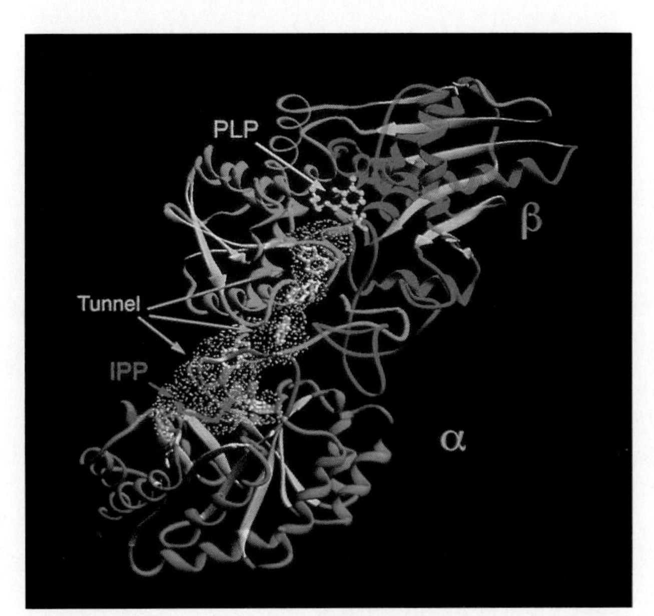

Figure 26.37 Tryptophan synthase is an example of a "channeling" multienzyme complex in which indole, the product of the α-reaction catalyzed by the α-subunit, passes intramolecularly to the β-subunit. In the β-subunit, the hydroxyl of the substrate L-serine is replaced with indole via a complicated pyridoxal phosphate–catalyzed reaction to produce the final product, L-tryptophan. The schematic figure shown here is a ribbon diagram of one α-subunit (blue) and neighboring β-subunit (the N-terminal domain of the β-subunit is in orange, C-terminal domain in red). The tunnel is outlined by the yellow dot surface and is shown with several indole molecules (green) packed in head-to-tail fashion. The labels "IPP" and "PLP" point to the active sites of the α- and the β-subunits, respectively, in which a competitive inhibitor (indole propanol phosphate, IPP) and the coenzyme PLP are bound. Figure courtesy of C. Craig Hyde, National Institutes of Health. The figure was generated using the RIBBONS program by Mike Carson, University of Alabama.

(Adapted from Hyde, C. C., et al. (1985). Journal of Biological Chemistry 260:3716.)

diluted in the surrounding milieu. This phenomenon of direct transfer of enzyme-bound metabolic intermediates, or **channeling,** increases the efficiency of the overall pathway by preventing loss and dilution of the intermediate. Channeling is a widespread mechanism for substrate transfer in metabolism, particularly among the enzymes of higher organisms.

Regulation of Aromatic Amino Acid Biosynthesis. Regulation occurs at several levels. One regulatory pattern impinges at the very beginning where DAHP is synthesized. *Escherichia coli* and related enteric bacteria have three *DAHP synthase isozymes* (encoded by separate DAHP synthase genes), each uniquely inhibited by one of the three aromatic amino acid end products. In addition, synthesis of the Trp-sensitive DAHP synthase is repressed by Trp, the Tyr-specific enzyme is repressed by Tyr, and the Phe-specific enzyme is repressed by Phe. Also, in enteric bacteria, the genes encoding the five enzymes unique to Trp biosynthesis are not expressed if Trp is present. A different mode of regulation is exemplified in *Bacillus subtilis: DAHP synthase* and *shikimate kinase* are inhibited by chorismate and prephenate; chorismate mutase is inhibited by prephenate; and prephenate dehydratase and prephenate dehydrogenase are inhibited by Phe and Tyr, respectively (see Figure 26.35).

Histidine. Like aromatic amino acid biosynthesis, **histidine** biosynthesis shares metabolic intermediates with the pathway of purine nucleotide synthesis. The pathway involves 10 separate steps, the first being an unusual reaction that links ATP and PRPP (Figure 26.38). Five carbon atoms from PRPP and one from ATP end up in histidine. Step 5 involves some novel chemistry, the details of which remain unknown. In this reaction, the substrate, *phosphoribulosylformimino-5-aminoimidazole-4-carboxamide ribonucleotide,* picks up an amino group (from the amide of glutamine) in a reaction accompanied by cleavage and ring closure to yield *two* imidazole compounds: the histidine precursor, *imidazole glycerol phosphate,* and a purine nucleotide precursor, *5-aminoimidazole-4-carboxamide ribonucleotide* (AICAR). Note that AICAR as a purine nucleotide precursor can ultimately replenish the ATP consumed in reaction 1. Nine enzymes act in histidine's 10 synthetic steps. Reactions 9 and 10, the successive NAD^+-dependent oxidations of an alcohol to an aldehyde and then to a carboxylic acid, are catalyzed by the same dehydrogenase.

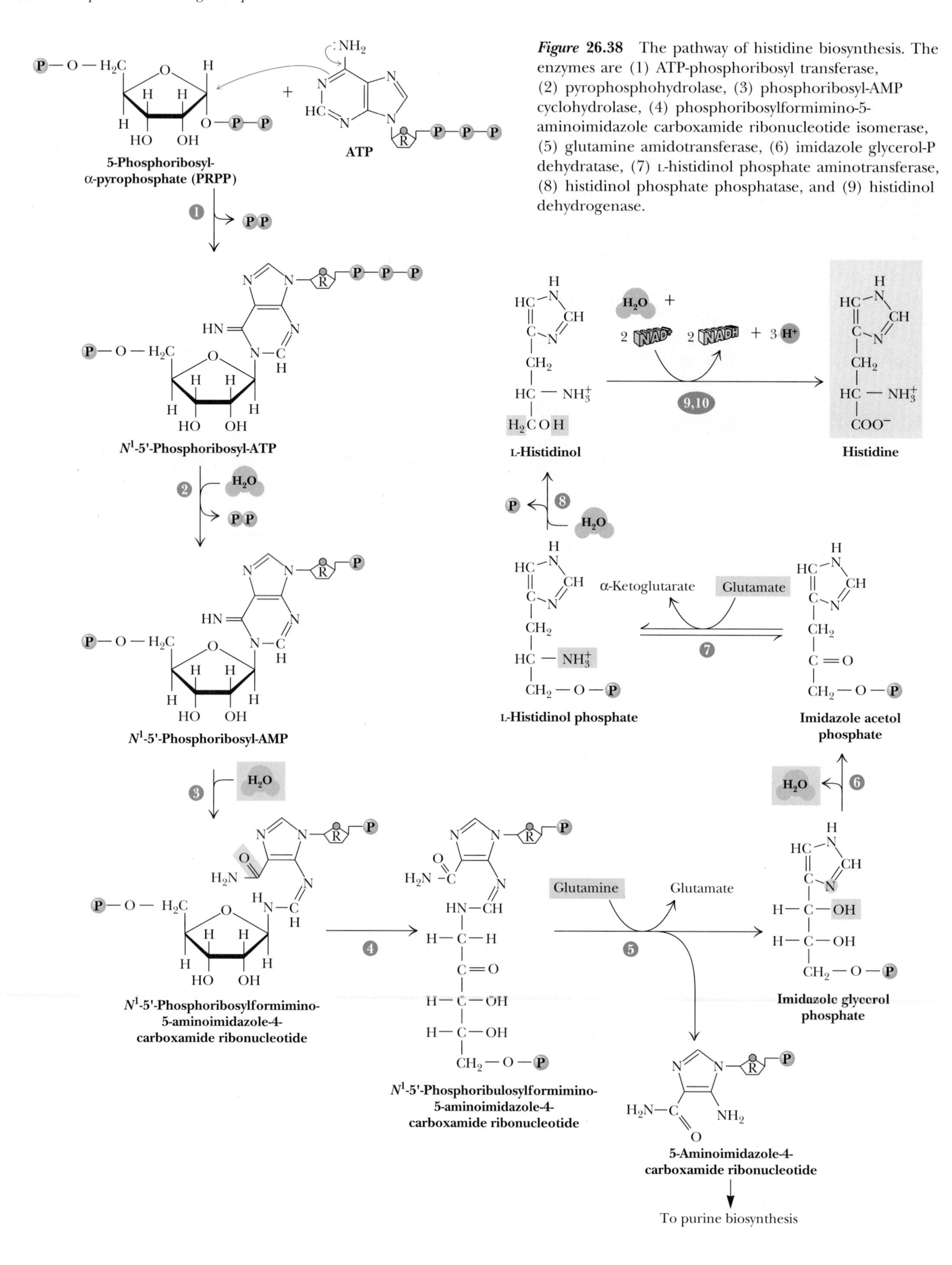

***Figure* 26.38** The pathway of histidine biosynthesis. The enzymes are (1) ATP-phosphoribosyl transferase, (2) pyrophosphohydrolase, (3) phosphoribosyl-AMP cyclohydrolase, (4) phosphoribosylformimino-5-aminoimidazole carboxamide ribonucleotide isomerase, (5) glutamine amidotransferase, (6) imidazole glycerol-P dehydratase, (7) L-histidinol phosphate aminotransferase, (8) histidinol phosphate phosphatase, and (9) histidinol dehydrogenase.

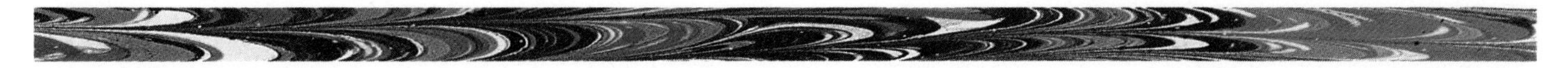

A Deeper Look

Amino Acid Biosynthesis Inhibitors as Herbicides

Unlike animals, plants can synthesize all 20 of the common amino acids. Inhibitors acting specifically on the plant enzymes that are capable of carrying out the biosynthesis of the "essential" amino acids (i.e., enzymes that animals lack) have been developed. These substances appear to be ideal for use as herbicides, since they should show no effect on animals. **Glyphosate,** sold commercially as *Roundup*®, is a PEP analog that acts as a specific inhibitor of 3-enolpyruvylshikimate-5-P synthase (Figure 26.34). **Sulfmeturon methyl,** a sulfonylurea herbicide that inhibits *acetohydroxy acid synthase,* an enzyme common to Val, Leu, and Ile biosynthesis (Figure 26.27), is the active ingredient in *Oust*®. **Aminotriazole,** sold as *Amitrole*®, blocks His biosynthesis by inhibiting *imidazole glycerol-P dehydratase* (Figure 26.38). **PPT (phosphinothricin)** is a potent inhibitor of *glutamine synthetase.* Although Gln is a nonessential amino acid and glutamine synthetase is a ubiquitous enzyme, PPT is relatively safe for animals since it does not cross the blood–brain barrier and is rapidly cleared by the kidneys.

$$^-O-P(=O)(O^-)-CH_2-NH-CH_2-COO^-$$

Glyphosate

$$C_6H_4(C(=O)-O-CH_3)-SO_2-NH-C(=O)-NH-C_4HN_2(CH_3)_2$$

Sulfmeturon methyl

$$C_2HN_3(H)-NH_2$$

Aminotriazole

$$CH_3-P(=O)(O^-)-CH_2-CH_2-CH(NH_2)-COO^-$$

DL-Phosphinothricin (PPT)

In enteric bacteria, the structural genes encoding the enzymes of His biosynthesis are, like those unique to the Trp biosynthetic pathway, arranged in a single contiguous array, a pattern of organization referred to as an **operon.** This arrangement allows expression of all the enzymes of the pathway to be regulated coordinately. Transcription of the *his* operon yields a single large mRNA, and translation of this message gives rise to the nine enzymes of His synthesis. Histidine controls its own synthesis by repressing transcription of the *his* operon. Elucidation of the His biosynthetic pathway is credited to the pioneering efforts of two outstanding biochemical geneticists, Philip Hartman of Johns Hopkins University and Bruce Ames of the University of California, Berkeley.

26.5 Metabolic Degradation of Amino Acids

In normal human adults, close to 90% of the energy requirement is met by oxidation of carbohydrates and fats; the remainder comes from oxidation of the carbon skeletons of amino acids. The primary physiological purpose of amino acids is to serve as the building blocks for protein biosynthesis. The

A Deeper Look

Histidine—A Clue to Understanding Early Evolution?

Histidine residues in the active sites of enzymes often act directly in the enzyme's catalytic mechanism. Catalytic participation by the imidazole group of His and the presence of imidazole as part of the purine ring system support the current speculation that life before the full evolution of protein molecules must have been RNA-based. This notion correlates with recent revelations that RNA molecules can have catalytic activity, an idea captured in the term *ribozyme* (Chapter 11).

dietary amount of free amino acids is trivial under most circumstances. However, if excess protein is consumed in the diet or if the amount of amino acids released during normal turnover of cellular proteins exceeds the requirements for new protein synthesis, the amino acid surplus must be catabolized. Also, if carbohydrate intake is insufficient (as during fasting or starvation) or if carbohydrates cannot be appropriately metabolized due to disease (as in *diabetes mellitus*), body protein becomes an important fuel for metabolic energy.

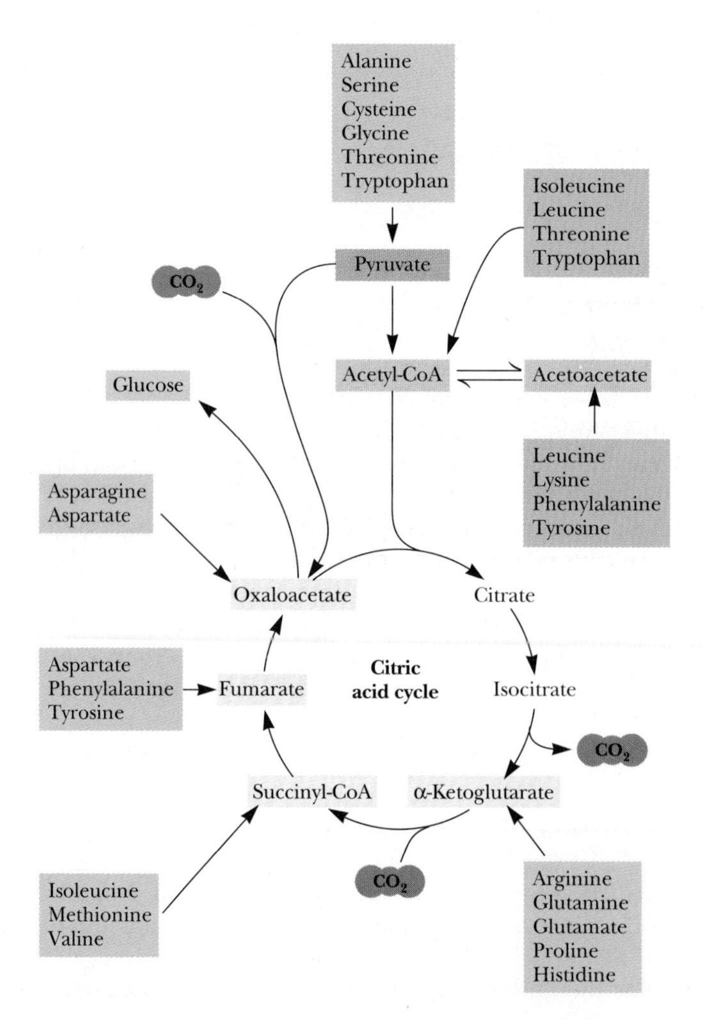

***Figure* 26.39** Metabolic degradation of the common amino acids. The 20 common amino acids can be classified according to their degradation products. Those that give rise to precursors for glucose synthesis, such as α-ketoglutarate, succinyl-CoA, fumarate, oxaloacetate, and pyruvate, are termed *glucogenic* (red). Those degraded to acetyl-CoA or acetoacetate are called *ketogenic* (blue) because they can be converted to fatty acids or ketone bodies. Some amino acids are both glucogenic and ketogenic.

The 20 Common Amino Acids Are Degraded by 20 Different Pathways That Converge to Just 7 Metabolic Intermediates

Since the 20 common amino acids of proteins are distinctive in terms of their carbon skeletons, each amino acid requires its own unique degradative pathway. Because amino acid degradation normally supplies only 10% of the body's energy, then, on average, degradation of any given amino acid will satisfy less than 1% of energy needs. Therefore, we will not discuss these pathways in detail. It so happens, however, that degradation of the carbon skeletons of the 20 common α-amino acids converges to just 7 metabolic intermediates: *acetyl-CoA, succinyl-CoA, pyruvate, α-ketoglutarate, fumarate, oxaloacetate,* and *acetoacetate.* Since succinyl-CoA, pyruvate, α-ketoglutarate, fumarate, and oxaloacetate can serve as precursors for glucose synthesis, amino acids giving rise to these intermediates are termed **glucogenic.** Those degraded to yield acetyl-CoA or acetoacetate are termed **ketogenic,** since these substances can be used to synthesize fatty acids or ketone bodies. Some amino acids are both glucogenic and ketogenic (Figure 26.39).

The C-3 Family of Amino Acids: Alanine, Serine, and Cysteine

The carbon skeletons of alanine, serine, and cysteine all converge to *pyruvate* (Figure 26.40). Transamination of alanine yields pyruvate:

$$\text{Alanine} + \alpha\text{-ketoglutarate} \rightleftharpoons \text{pyruvate} + \text{glutamate}$$

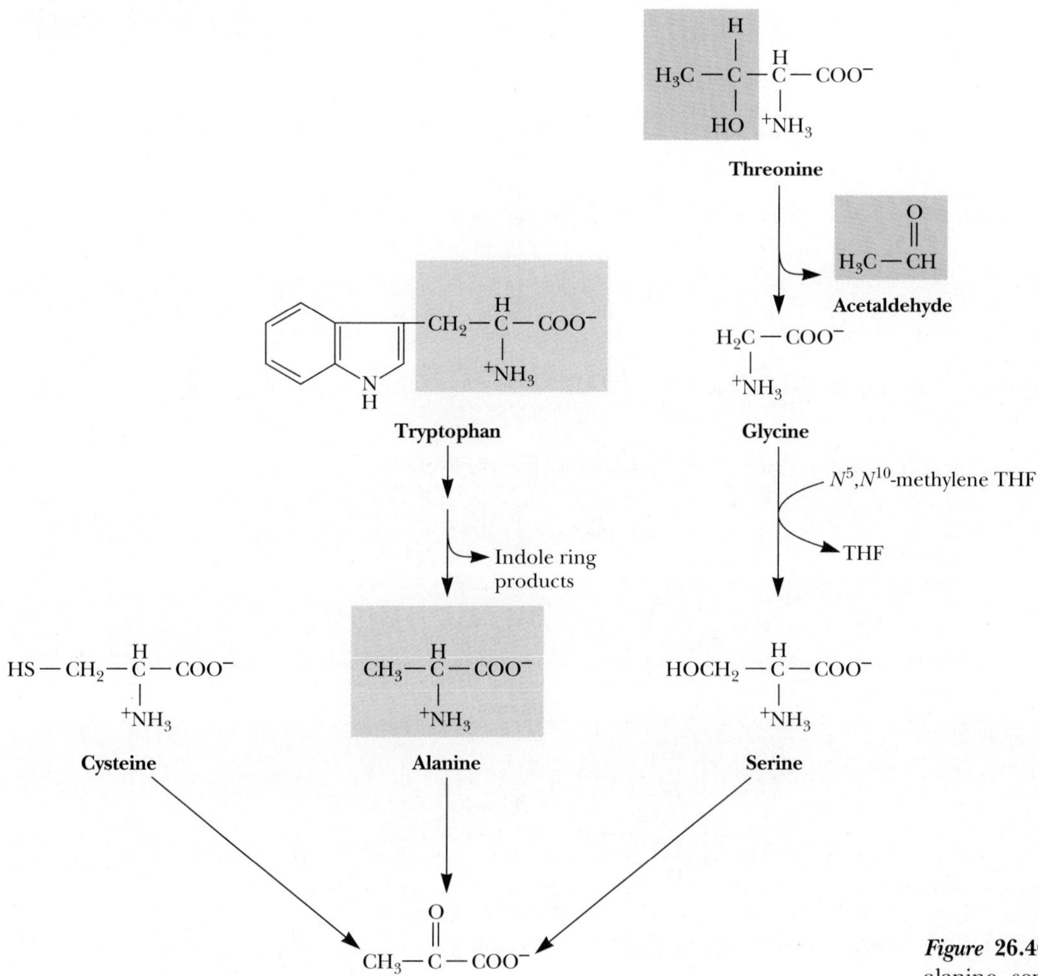

***Figure* 26.40** Formation of pyruvate from alanine, serine, cysteine, glycine, tryptophan, or threonine.

Deamination of serine by *serine dehydratase* also yields pyruvate. Cysteine is converted to pyruvate via a number of paths.

The carbon skeletons of three other amino acids also become pyruvate. *Glycine* is convertible to serine and thus to pyruvate. The three carbon atoms of *tryptophan* that are not part of its indole ring appear as alanine (and, hence, pyruvate) upon Trp degradation. *Threonine* by one of its degradation routes is cleaved to glycine and acetaldehyde. The glycine is then converted to pyruvate via serine; the acetaldehyde is oxidized to acetyl-CoA (Figure 26.40).

The C-4 Family of Amino Acids: Aspartate and Asparagine

Transamination of aspartate gives *oxaloacetate:*

$$\text{Aspartate} + \alpha\text{-ketoglutarate} \rightleftharpoons \text{oxaloacetate} + \text{glutamate}$$

Hydrolysis of asparagine by *asparaginase* yields aspartate and NH_4^+. Alternatively, aspartate degradation via the urea cycle leads to a different citric acid cycle intermediate, namely, *fumarate* (Figure 26.21).

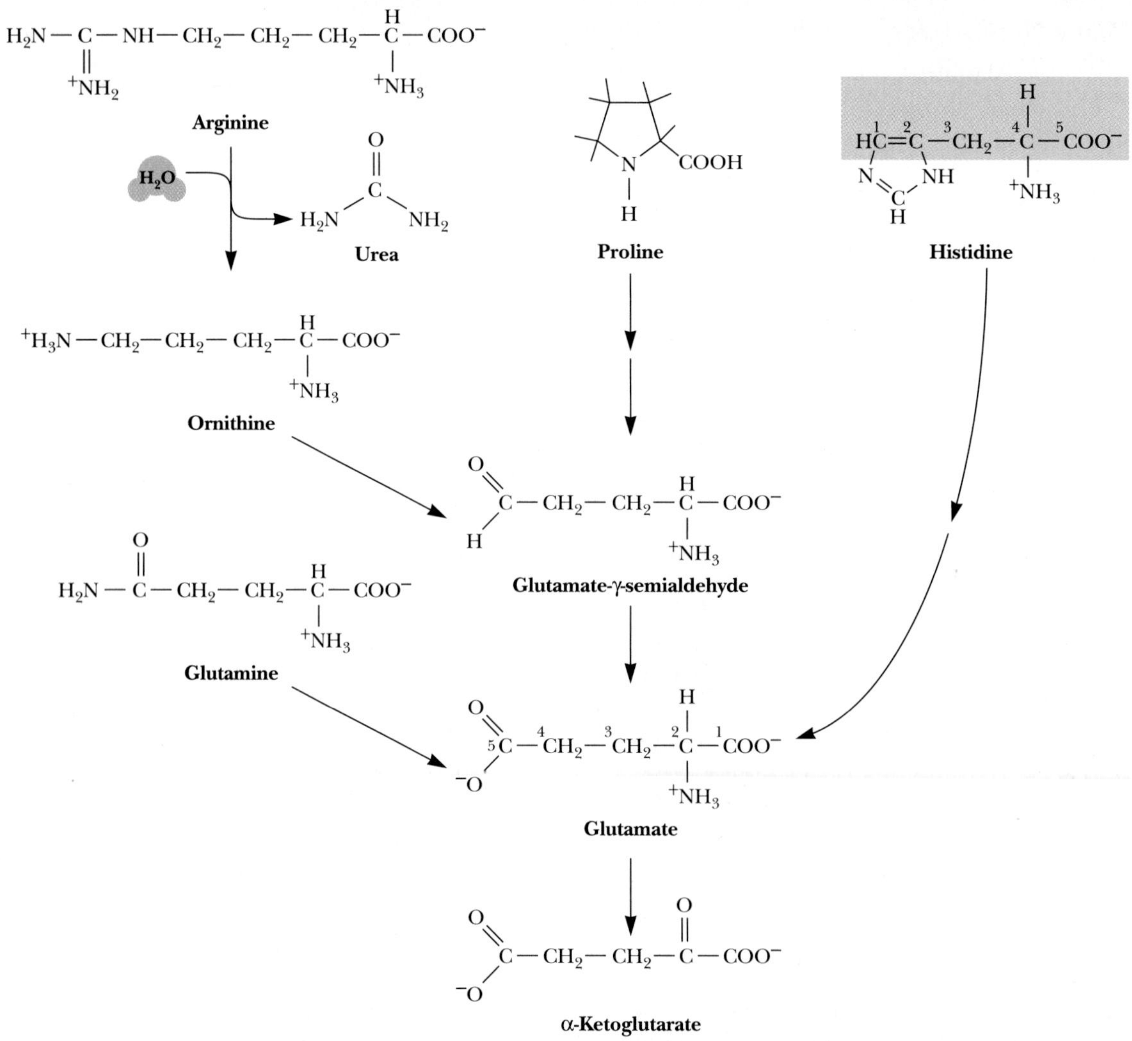

Arginine (H_2O) → **Urea** + **Ornithine**

Proline

Histidine

Glutamate-γ-semialdehyde

Glutamine

Glutamate

α-Ketoglutarate

***Figure* 26.41** The degradation of the C-5 family of amino acids leads to α-ketoglutarate via glutamate. The histidine carbons, numbered 1 through 5, become carbons 1 through 5 of glutamate, as indicated.

The C-5 Family of Amino Acids Are Converted to α-Ketoglutarate via Glutamate

The five-carbon citric acid cycle intermediate α-ketoglutarate is always a product of transamination reactions involving *glutamate.* Thus, glutamate *and* any amino acid convertible to glutamate are classified within the C-5 family. These amino acids include *glutamine, proline, arginine,* and *histidine* (Figure 26.41).

Degradation of Valine, Isoleucine, and Methionine Leads to Succinyl-CoA

The breakdown of valine, isoleucine, and methionine converges at *propionyl-CoA* (Figure 26.42). Propionyl-CoA is subsequently converted to methylmalonyl-CoA and thence to *succinyl-CoA* via the same reactions mediating the oxidation of fatty acids that have odd numbers of carbon atoms (Chapter 23).

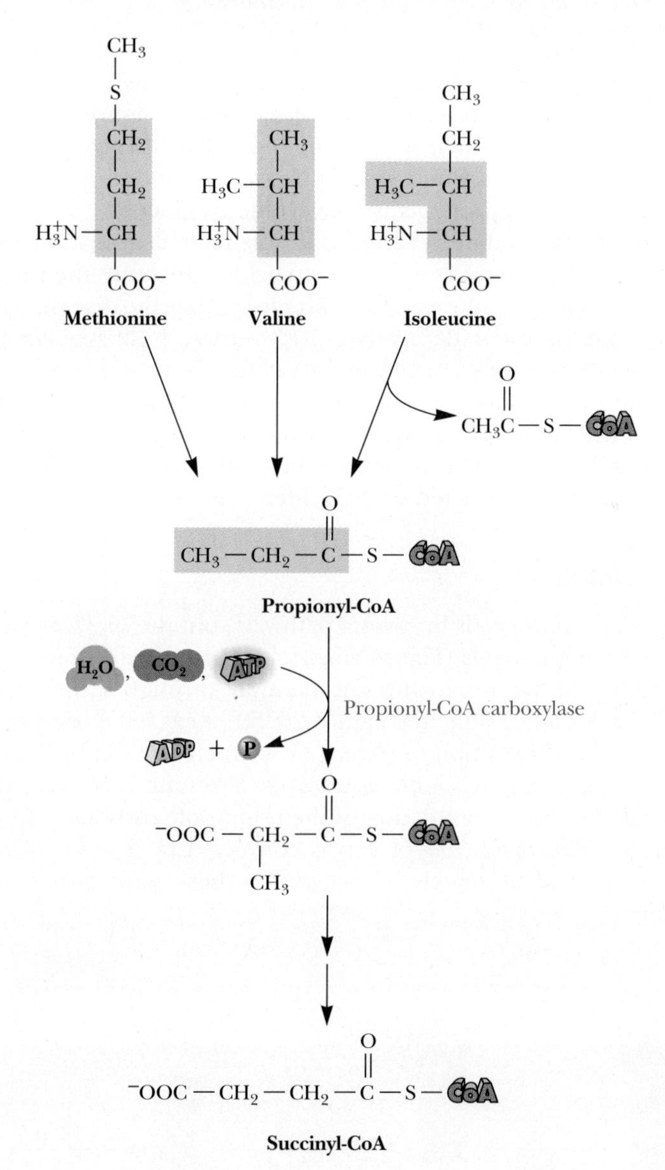

***Figure* 26.42** Valine, isoleucine, and methionine are converted via propionyl-CoA to succinyl-CoA for entry into the citric acid cycle. The shaded carbon atoms of the three amino acids give rise to propionyl-CoA. All three amino acids lose their α-carboxyl group as CO_2. Methionine first becomes *S*-adenosylmethionine, then homocysteine (see Figure 26.26). The terminal two carbons of isoleucine become acetyl-CoA.

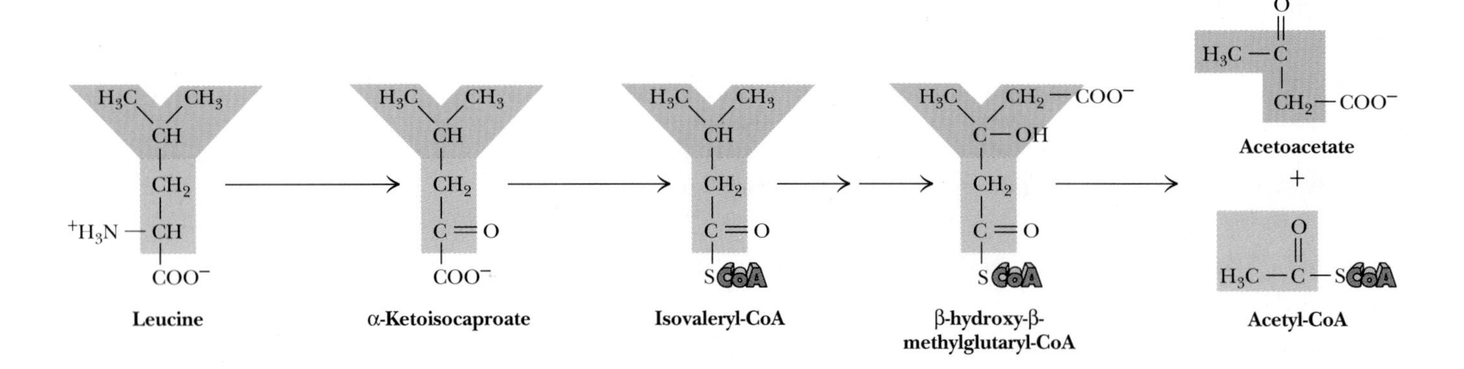

Figure **26.43** Leucine is degraded to acetyl-CoA and acetoacetate via *β*-hydroxy-*β*-methylglutaryl-CoA, which is also the intermediate in ketone body formation from fatty acids (Chapter 23).

Leucine Is Degraded to Acetyl-CoA and Acetoacetate

Leucine is one of only two purely ketogenic amino acids; the other is lysine. Deamination of leucine via a transamination reaction yields *α-ketoisocaproate,* which is oxidatively decarboxylated to *isovaleryl-CoA* (Figure 26.43). Subsequent reactions, one of which is a biotin-dependent carboxylation, give *β-hydroxy-β-methylglutaryl-CoA,* which is then cleaved to yield *acetyl-CoA* and *acetoacetate.* Neither of these products is convertible to glucose.

The initial steps in valine, leucine, and isoleucine degradation are identical. All three are first deaminated to *α*-keto acids, which are then oxidatively decarboxylated to form CoA derivatives. **Maple syrup urine disease** is a hereditary defect in the oxidative decarboxylation of these branched-chain *α*-keto acids. The metabolic block created by this defect leads to elevated levels of valine, leucine, and isoleucine (and their corresponding branched-chain *α*-keto acids) in the blood and urine. The urine of individuals with this disease smells like maple syrup. The defect is fatal unless dietary intake of these amino acids is greatly restricted early in life.

Lysine Degradation

Lysine degradation proceeds by several pathways, but the *saccharopine pathway* found in liver predominates (Figure 26.44). This degradative route proceeds backward along the lysine biosynthetic pathway through saccharopine and *α*-aminoadipate to *α*-ketoadipate (Figure 26.22). Next, *α*-ketoadipate undergoes oxidative decarboxylation to *glutaryl-CoA,* which is then transformed into *acetoacetyl-CoA* and ultimately into *acetoacetate,* a ketone body (Chapter 23).

As indicated earlier, degradation of the nonindole carbons of tryptophan yields pyruvate. The *indole ring* of Trp is converted by a series of reactions to *α*-ketoadipate and ultimately *acetoacetate* by these same reactions of Lys degradation.

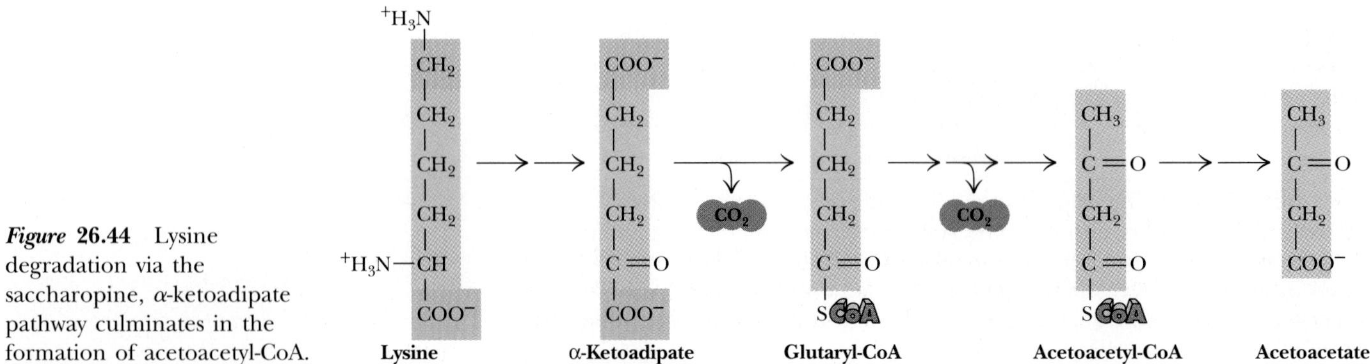

Figure **26.44** Lysine degradation via the saccharopine, *α*-ketoadipate pathway culminates in the formation of acetoacetyl-CoA.

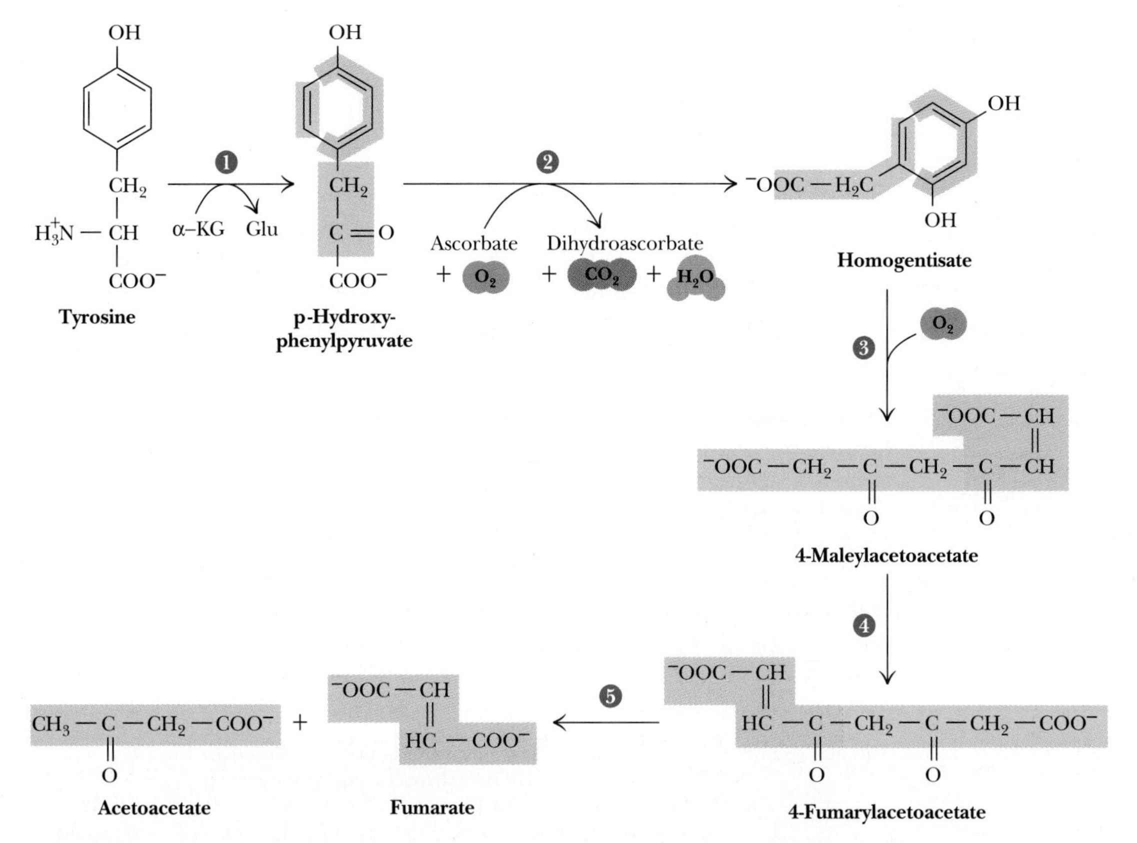

Figure **26.45** Phenylalanine and tyrosine degradation. (1) Transamination of Tyr gives *p*-hydroxyphenylpyruvate, which (2) is oxidized to homogentisate by *p*-hydroxyphenylpyruvate dioxygenase in an ascorbic acid (vitamin C)–dependent reaction. (3) The ring opening of homogentisate by homogentisate dioxygenase gives 4-maleylacetoacetate. (4) 4-Maleylacetoacetate isomerase gives 4-fumarylacetoacetate, which (5) is hydrolyzed by fumarylacetoacetase.

Phenylalanine and Tyrosine Are Degraded to Acetoacetate and Fumarate

The first reaction in phenylalanine degradation is the hydroxylation reaction of *tyrosine* biosynthesis (Figure 26.36). Both these amino acids thus share a common degradative pathway. Transamination of Tyr yields the α-keto acid *p-hydroxyphenylpyruvate* (Figure 26.45). *p-Hydroxyphenylpyruvate dioxygenase,* a vitamin C–dependent enzyme, then carries out a ring hydroxylation–oxidative decarboxylation to yield *homogentisate.* Ring opening and isomerization give *4-fumarylacetoacetate,* which is hydrolyzed to *acetoacetate* and *fumarate.*

Hereditary Defects in Phe Catabolism: Alcaptonuria and Phenylketonuria

Alcaptonuria and phenylketonuria are two human genetic diseases arising from specific enzyme defects in phenylalanine degradation. **Alcaptonuria** is characterized by urinary excretion of large amounts of homogentisate and results from a deficiency in homogentisate dioxygenase. Air oxidation of homogentisate causes urine to turn dark on standing, but the only malady suffered by carriers of this disease is a tendency toward arthritis later in life.

In contrast, **phenylketonurics,** whose urine contains excessive *phenylpyruvate* (Figure 26.46), suffer severe mental retardation if the defect is not recognized immediately after birth and treated by putting the victim on a diet low in phenylalanine. These individuals are deficient in phenylalanine hydroxylase, and the excess Phe that accumulates is transaminated to phenylpyruvate and excreted.

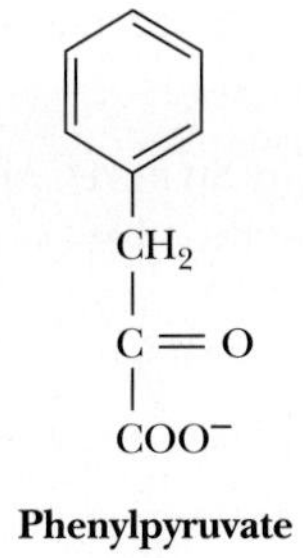

Phenylpyruvate

Figure **26.46** The structure of phenylpyruvate.

Nitrogen Excretion

Animals are the only group of organisms that commonly enjoy a dietary surplus of nitrogen. Excess nitrogen liberated upon metabolic degradation of amino acids is excreted by animals in three different ways, in accord with the availability of water. Aquatic animals simply release free ammonia to the surrounding water; such animals are termed **ammonotelic** (from the Greek *telos,* end). On the other hand, terrestrial and aerial species employ mechanisms that convert ammonium to less toxic waste compounds that require little H_2O for excretion. Many terrestrial vertebrates are **ureotelic,** meaning that they excrete excess N as **urea,** a highly water-soluble nonionic substance. Urea is formed by ureoteles via the urea cycle. The **uricotelic** organisms are those animals using the third means of N excretion, conversion to **uric acid,** a rather insoluble purine analog. Birds and reptiles are uricoteles. Uric acid metabolism is discussed in the next chapter. Some animals can switch from ammonotelic to ureotelic to uricotelic metabolism, depending on water availability.

Problems

1. What is the oxidation number of N in nitrate, nitrite, NO, N_2O, and N_2?
2. How many ATP are consumed per N atom of ammonium formed by (a) the nitrate assimilation pathway? (b) the nitrogen fixation pathway?
3. Suppose at certain specific metabolite concentrations *in vivo* the cyclic cascade regulating *E. coli* glutamine synthetase has reached a dynamic equilibrium where the average state of GS adenylylation is poised at $n = 6$. Predict what change in n will occur if:
   a. [ATP] increases.
   b. [UTP] increases.
   c. [α-KG]/[Gln] increases.
   d. [P_i] decreases.
4. How many ATP equivalents are consumed in the production of one equivalent of urea by the urea cycle?
5. Why are persons on a high-protein diet advised to drink lots of water?
6. How many ATP equivalents are consumed in the biosynthesis of lysine from aspartate by the pathway shown in Figure 26.25?
7. If PEP labeled with ^{14}C in the 2-position serves as the precursor to chorismate synthesis, which C atom in chorismate is radioactive?
8. Write a balanced equation for the synthesis of glucose (by gluconeogenesis) from aspartate.

Further Reading

Atkinson, D. E., and Camien, M. N., 1982. The role of urea synthesis in the removal of metabolic bicarbonate and the regulation of blood pH. *Current Topics in Cellular Regulation* **26:**261–302. Describes the reasoning behind the proposal that the urea cycle eliminates bicarbonate as well as ammonium when urea is formed.

Bender, David A., 1985. *Amino Acid Metabolism.* New York: Wiley. A general review of amino acid metabolism.

Burris, R. H., 1991. Nitrogenases. *The Journal of Biological Chemistry* **266:**9339–9342.

Kishore, G. M., and Shah, D. M., 1988. Amino acid biosynthesis inhibitors as herbicides. *Annual Review of Biochemistry* **57:**627–663.

Rhee, C., and Stadtman, E. R., 1989. Regulation of *E. coli* glutamine synthetase. *Advances in Enzymology* **62:**37–92.

Srere, P. A., 1987. Complexes of sequential metabolic enzymes. *Annual Review of Biochemistry* **56:**89–124. A review of the evidence that enzymes in a pathway sequence are often physically associated with one another *in vivo,* particularly in eukaryotic cells.

Wray, J. L., and Kinghorn, J. R., 1989. *Molecular and Genetic Aspects of Nitrate Assimilation.* New York: Oxford Science Publications.

Chapter 27

The Synthesis and Degradation of Nucleotides

"Guano, a substance found on some coasts frequented by sea birds, is composed chiefly of the birds' partially decomposed excrement. . . . The name for the purine guanine *derives from the abundance of this base in guano.*

The Chincha Islands . . . are . . . off the coast of Peru . . . Myriads of birds, after satisfying their voracious appetites upon the finny tenants of the deep, have for ages made the islands their nightly abode, and the receptacle of their fecal offerings."

J. C. Nesbit. *On Agricultural Chemistry and the Nature and Properties of Peruvian Guano* (1850)

Pigeon drinking at Gaia Fountain, Siena, Italy. The basic features of purine biosynthesis were elucidated initially from metabolic studies of nitrogen metabolism in pigeons. Pigeons excrete excess N as uric acid, a purine analog.

Outline

27.1 Nucleotide Biosynthesis
27.2 The Biosynthesis of Purines
27.3 Purine Salvage
27.4 Purine Degradation
27.5 The Biosynthesis of Pyrimidines
27.6 Pyrimidine Degradation
27.7 Deoxyribonucleotide Biosynthesis
27.8 Synthesis of Thymine Nucleotides

Nucleotides are ubiquitous constituents of life, actively participating in the majority of biochemical reactions. Recall that ATP is the "energy currency" of the cell, that uridine nucleotide derivatives of carbohydrates are common intermediates in cellular transformations of carbohydrates (Chapter 21), and that biosynthesis of phospholipids proceeds via cytidine nucleotide derivatives (Chapter 24). In Chapter 32, we will see that GTP serves as the immediate energy source driving the endergonic reactions of protein synthesis. Many of the coenzymes (such as coenzyme A, NAD, NADP, and FAD) are derivatives of nucleotides. Nucleotides also act in metabolic regulation, as in the response of key enzymes of intermediary metabolism to the relative concentrations of AMP, ADP, and ATP (PFK is a prime example here; see also Chapter 18). Further, cyclic derivatives of purine nucleotides, cAMP and cGMP, have no other role in metabolism than regulation. Last and not least, nucleotides are the monomeric units of nucleic acids. Deoxynucleoside triphosphates (dNTPs) and nucleoside triphosphates (NTPs) serve as the immediate substrates for the biosynthesis of DNA and RNA, respectively (see Part IV, Genetic Information). Without RNA, protein biosynthesis is not possible; in the absence of DNA synthesis, the genetic material is not replicated and cell division cannot occur.

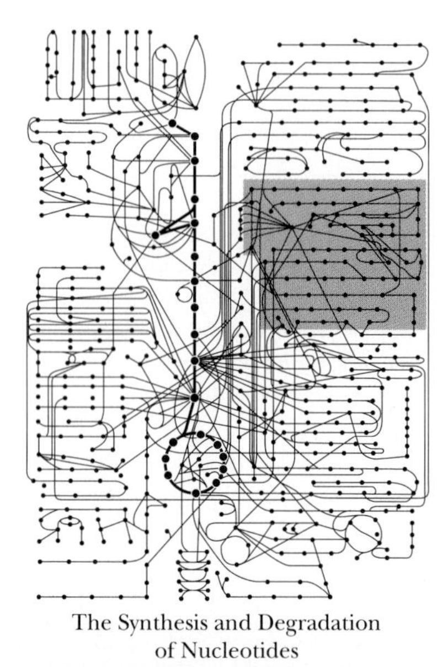

The Synthesis and Degradation of Nucleotides

27.1 Nucleotide Biosynthesis

Nearly all organisms can make the purine and pyrimidine nucleotides via so-called *de novo* biosynthetic pathways. (*De novo* means "anew"; a less literal but more apt translation might be "from scratch" since *de novo* pathways are metabolic sequences that lead to the synthesis of complex end products from rather simple precursors commonly available to the cell.) Many organisms also possess salvage pathways to recover purine and pyrimidine compounds obtained in the diet or released during nucleic acid turnover and degradation. While the ribose of nucleotides can be catabolized to generate energy, the nitrogenous bases do *not* serve as energy sources; their catabolism does not lead to products used by pathways of energy conservation. Compared to slowly dividing cells, rapidly proliferating cells synthesize larger amounts of DNA and RNA per unit time. To meet the increased demand for nucleic acid synthesis, substantially greater quantities of nucleotides must be produced. The pathways of nucleotide biosynthesis thus become attractive targets for the clinical control of rapidly dividing cells such as cancers or infectious bacteria. Many antibiotics and anticancer drugs are inhibitors of purine or pyrimidine nucleotide biosynthesis.

27.2 The Biosynthesis of Purines

Uric acid

***Figure* 27.1** Nitrogen waste is excreted by birds principally as the purine uric acid.

Substantial insight into the *de novo* pathway for purine biosynthesis was provided in 1948 by John Buchanan, who cleverly exploited the fact that birds excrete excess nitrogen principally in the form of uric acid, a water-insoluble purine. Buchanan fed isotopically labeled compounds to pigeons and then examined the distribution of the labeled atoms in *uric acid* (Figure 27.1). By tracing the metabolic source of the various atoms in this end product, it was possible to show that the nine atoms of the purine ring system (Figure 27.2) are contributed by aspartic acid (N-1), glutamine (N-3 and N-9), glycine (C-4, C-5, and N-7), CO_2 (C-6), and THF one-carbon derivatives (C-2 and C-8). The coenzyme THF and its role in one-carbon metabolism were introduced in Chapter 14.

IMP Biosynthesis: Inosinic Acid Is the Immediate Precursor to GMP and AMP

The *de novo* synthesis of purines occurs in an interesting manner: The atoms forming the purine ring are successively added to *ribose-5-phosphate;* thus, purines are directly synthesized as nucleotide derivatives by assembling the atoms that comprise the purine ring system directly on the ribose. In Step 1, ribose-5-phosphate is activated via the direct transfer of a pyrophosphoryl group from ATP to C-1 of the ribose, yielding *5-phosphoribosyl-α-pyrophosphate (PRPP)* (Figure 27.3). The enzyme is *ribose-5-phosphate pyrophosphokinase.* PRPP

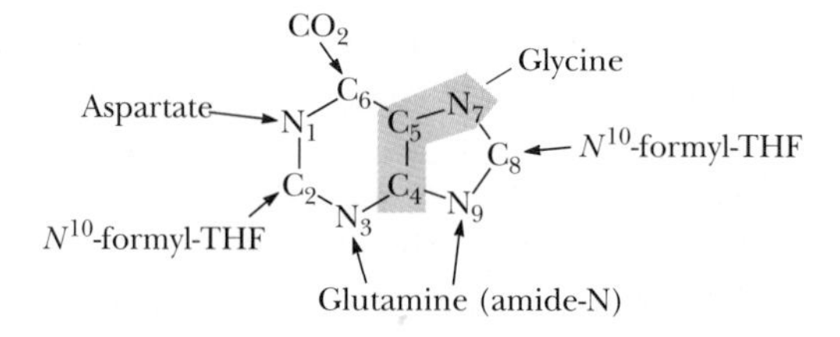

***Figure* 27.2** The metabolic origin of the nine atoms in the purine ring system.

is the limiting substance in purine biosynthesis. The two major purine nucleoside diphosphates, ADP and GDP, are negative effectors of ribose-5-phosphate pyrophosphokinase. However, since PRPP serves additional metabolic needs, the next reaction is actually the committed step in the pathway.

Step 2 (Figure 27.3) is catalyzed by *glutamine phosphoribosyl pyrophosphate amidotransferase.* The anomeric carbon of the substrate PRPP is in the α-configuration; the product is a β-glycoside (recall that all the biologically important nucleotides are β-glycosides). The N atom of this *N*-glycoside will become N-9 of the nine-membered purine ring; it is the first atom added in the construction of this ring. Glutamine phosphoribosyl pyrophosphate amidotransferase is subject to feedback inhibition by GMP, GDP, and GTP as well as AMP, ADP, and ATP. The G series of nucleotides interacts at a guanine-specific allosteric site on the enzyme, whereas the adenine nucleotides act at an A-specific site. The pattern of inhibition by these nucleotides is competitive, thus ensuring that residual enzyme activity is expressed until sufficient amounts of both adenine and guanine nucleotides are synthesized. Glutamine phosphoribosyl pyrophosphate amidotransferase is also sensitive to inhibition by the glutamine analog *azaserine* (Figure 27.4). Azaserine has been employed as an antitumor agent since it causes inactivation of glutamine-dependent enzymes in the purine biosynthetic pathway.

Step 3 is carried out by *glycinamide ribonucleotide synthetase (GAR synthetase)* via its ATP-dependent condensation of the glycine carboxyl group with the amine of *5-phosphoribosyl-β-amine* (Figure 27.3). The reaction proceeds in two stages. First, the glycine carboxyl group is activated via ATP-dependent phosphorylation.

$$H_3{}^{+}N{-}CH_2{-}\overset{\overset{\displaystyle O}{\|}}{C}{-}O^{-} + ATP \longrightarrow H_3{}^{+}N{-}CH_2{-}\overset{\overset{\displaystyle O}{\|}}{C}{-}OPO_3{}^{2-} + ADP$$

Next, an amide bond is formed between the activated carboxyl of glycine and the β-amine. Glycine contributes C-4, C-5, and N-7 of the purine.

Step 4 is the first of two THF-dependent reactions in the purine pathway. *GAR transformylase* transfers the N^{10}-formyl group of N^{10}-formyl-THF to the free amino group of GAR to yield *α-N-formylglycinamide ribonucleotide (FGAR).* Thus, C-8 of the purine is "formyl-ly" introduced. Although all of the atoms of the imidazole portion of the purine ring are now present, the ring is not closed until Reaction 6.

Step 5 is catalyzed by *FGAR amidotransferase* (also known as *FGAM synthetase*). ATP-dependent transfer of the glutamine amido group to the C-4-carbonyl of FGAR yields *formylglycinamidine ribonucleotide (FGAM).* First, ATP is used to form a phosphoryl ester intermediate with the oxygen of the C-4-carbonyl. Then, when the enzyme forms a thiol ester with the γ-carboxyl of glutamine, the amide group of Gln is released as so-called "active NH_3," displacing the phosphoryl group from C-4 to become the C-4 imine-N. As a glutamine-dependent enzyme, FGAR amidotransferase is, like glutamine phosphoribosyl pyrophosphate amidotransferase (Reaction 2), irreversibly inactivated by azaserine. The imino-N will be N-3 of the purine.

Step 6 is an ATP-dependent dehydration that leads to formation of the imidazole ring. ATP is used to phosphorylate the oxygen atom of the formyl group, activating it for the ring closure step that follows. The enzyme, *FGAM cyclase,* is K^+-activated. Since the product is *5-aminoimidazole ribonucleotide,* or *AIR,* this enzyme is also called *AIR synthetase.* Tautomerization of the C-4 imine to an enamine contributes to the aromatic character of the imidazole. In avian liver, the enzymatic activities for Steps 3, 4, and 6 (GAR synthetase, GAR transformylase, and AIR synthetase) reside on a single, 110-kD multifunctional polypeptide.

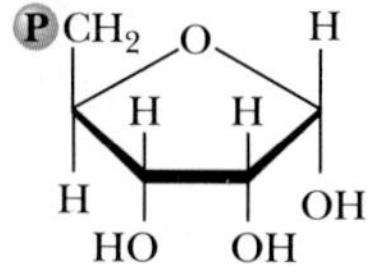

α-D-Ribose-5-phosphate

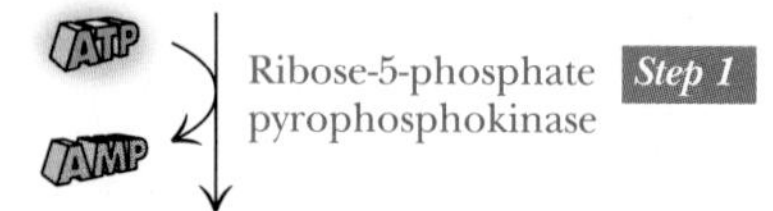

5-Phosphoribosyl-α-pyrophosphate (PRPP)

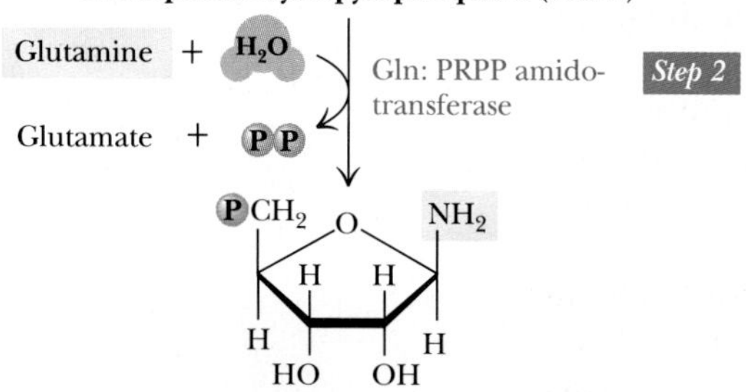

Phosphoribosyl-β-amine

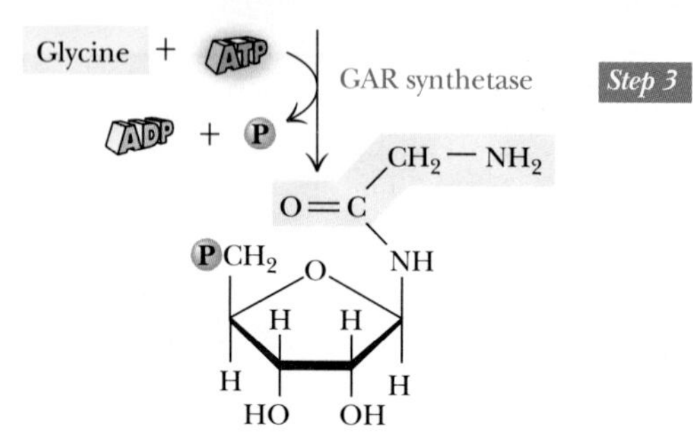

Glycinamide ribonucleotide (GAR)

N^{10}-Formyl-THF

THF

GAR transformylase

Step 4

Formylglycinamide ribonucleotide (FGAR)

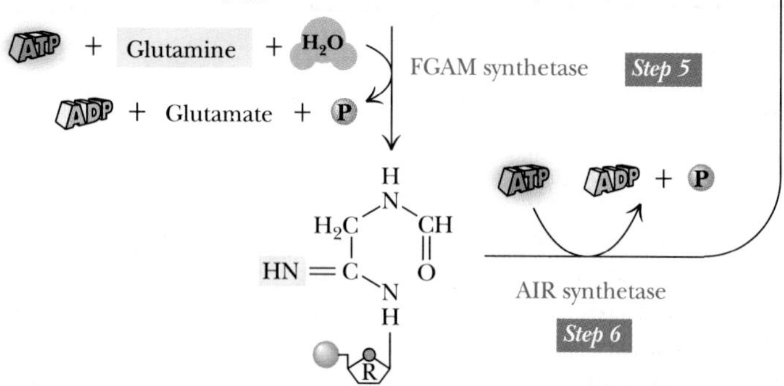

Formylglycinamidine ribonucleotide (FGAM)

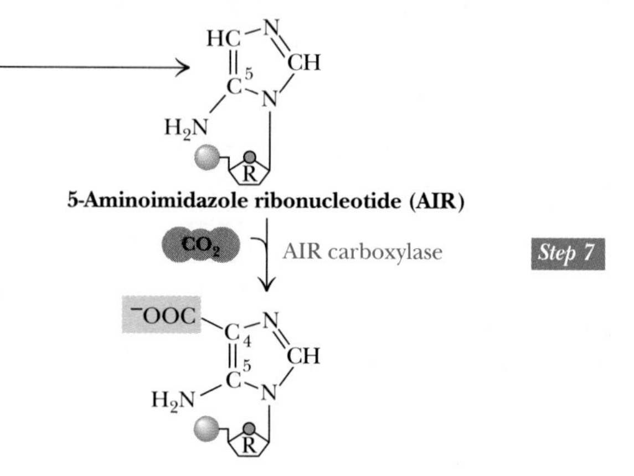

5-Aminoimidazole ribonucleotide (AIR)

Carboxyaminoimidazole ribonucleotide (CAIR)

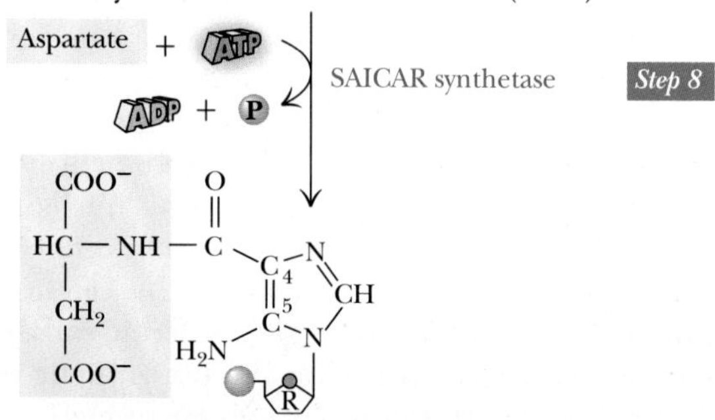

N-succinylo-5-aminoimidazole-4-carboxamide ribonucleotide (SAICAR)

Fumarate

Adenylosuccinate lyase

Step 9

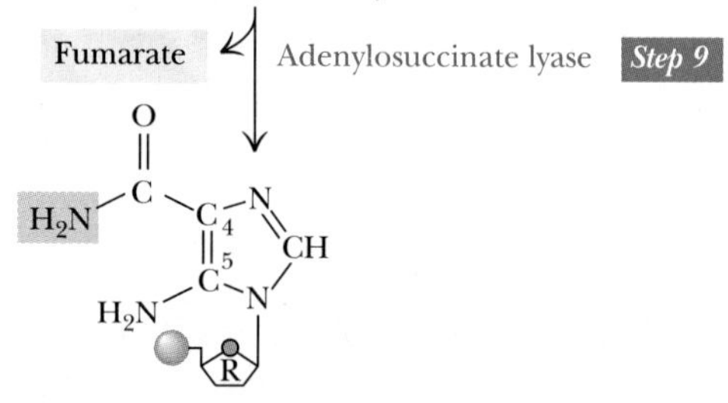

5-Aminoimidazole-4-carboxamide ribonucleotide (AICAR)

N^{10}-Formyl-THF

THF

AICAR transformylase

Step 10

N-formylaminoimidazole-4-carboxamide ribonucleotide (FAICAR)

H_2O

IMP synthase

Step 11

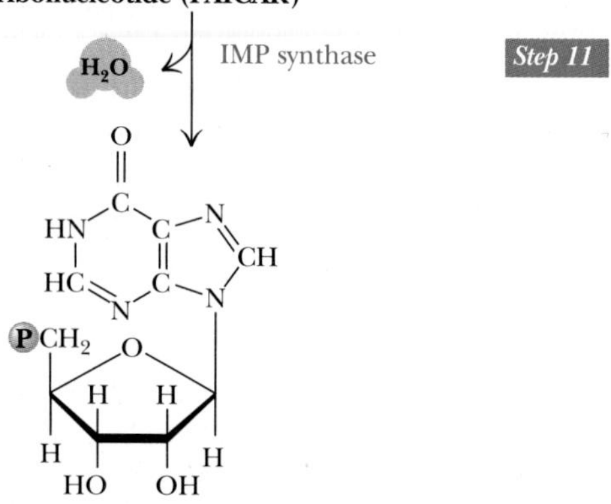

Inosine monophosphate (IMP)

Step 7 accomplishes the acquisition of C-6 from carbon dioxide. This carboxylation reaction is exceptional in that it does not require biotin, nor is ATP energy expended. CO_2 is added to the imidazole C-4 by *AIR carboxylase.* The product is *carboxyaminoimidazole ribonucleotide (CAIR).*

In Step 8, the amino-N of aspartate provides N-1 through linkage to the C-6 carboxyl function of CAIR. ATP hydrolysis drives the condensation of Asp with CAIR. The product is *N-succinylo-5-aminoimidazole-4-carboxamide ribonucleotide (SAICAR). SAICAR synthetase* catalyzes the reaction. The enzymatic activities for Steps 7 and 8 reside on a single, bifunctional polypeptide in avian liver.

Step 9 removes the four carbons of Asp as fumaric acid in a nonhydrolytic cleavage. The product is *5-aminoimidazole-4-carboxamide ribonucleotide (AICAR);* the enzyme is *adenylosuccinase (adenylosuccinate lyase).* Adenylosuccinase acts again in that part of the purine pathway leading from IMP to AMP and derives its name from this latter activity (see following). AICAR is also an intermediate in the histidine biosynthetic pathway (see Chapter 26), but since ATP serves as the precursor to AICAR in that pathway, no net purine synthesis is achieved.

Step 10 adds the formyl carbon of N^{10}-formyl-THF as the ninth and last atom necessary for forming the purine nucleus. The enzyme is called *AICAR transformylase;* the products are THF and *N-formylaminoimidazole-4-carboxamide ribonucleotide,* or *FAICAR.*

Step 11 involves dehydration and ring closure and completes the initial phase of purine biosynthesis. The enzyme is *IMP cyclohydrolase* (also known as *IMP synthase* and *inosinicase*). Unlike Step 6, this ring closure does not require ATP. In avian liver, the enzymatic activities catalyzing Steps 10 and 11 (AICAR transformylase and inosinicase) activities reside on α-67-kD bifunctional polypeptides organized into 135-kD dimers.

Note that five ATPs are required in the purine biosynthetic pathway from ribose-5-phosphate to IMP: one each at Steps 1, 3, 5, 6, and 8. However, 6 high-energy phosphate bonds (equal to 6 ATP equivalents) are consumed since α-PRPP formation in Reaction 1 followed by PP_i release in Reaction 2 represents the loss of 2 ATP equivalents.

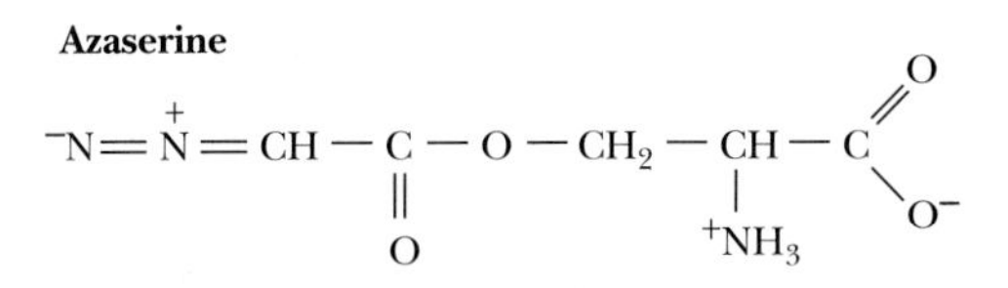

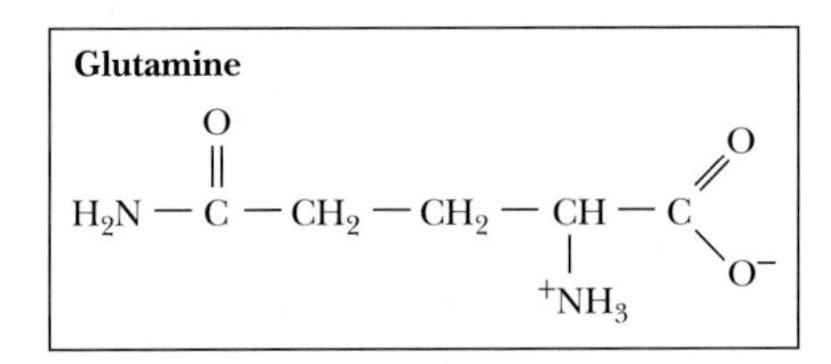

***Figure* 27.4** The structure of azaserine. Azaserine acts as an irreversible inhibitor of glutamine-dependent enzymes by covalently attaching to nucleophilic groups in the glutamine-binding site.

◀ ***Figure* 27.3** The *de novo* pathway for purine synthesis. The first purine product of this pathway, IMP (inosinic acid or inosine monophosphate), serves as a precursor to AMP and GMP. *Step 1:* PRPP synthesis from ribose-5-phosphate and ATP by ribose-5-phosphate pyrophosphokinase. *Step 2:* 5-Phosphoribosyl-β-1-amine synthesis from α-PRPP, glutamine, and H_2O by glutamine phosphoribosyl pyrophosphate amidotransferase. *Step 3:* Glycinamide ribonucleotide (GAR) synthesis from glycine, ATP, and 5-phosphoribosyl-β-amine by glycinamide ribonucleotide synthetase. *Step 4:* Formylglycinamide ribonucleotide synthesis from N^{10}-formyl-THF and GAR by GAR transformylase. *Step 5:* Formylglycinamidine ribonucleotide (FGAM) synthesis from FGAR, ATP, and H_2O by FGAM synthetase (FGAR amidotransferase). The other products are ADP, P_i, and glutamate. *Step 6:* 5-Aminoimidazole ribonucleotide (AIR) synthesis is achieved via the ATP-dependent closure of the imidazole ring, as catalyzed by FGAM cyclase (AIR synthetase). (Note that the ring closure changes the numbering system.) *Step 7:* Carboxyaminoimidazole ribonucleotide (CAIR) synthesis from CO_2 and AIR by AIR carboxylase. *Step 8:* *N*-succinylo-5-aminoimidazole-4-carboxamide ribonucleotide (SAICAR) synthesis from aspartate, CAIR, and ATP by SAICAR synthetase. The products are SAICAR, ADP, and P_i. *Step 9:* 5-Aminoimidazole carboxamide ribonucleotide (AICAR) formation by the nonhydrolytic removal of fumarate from SAICAR. The enzyme is adenylosuccinase (also called adenylosuccinate lyase). *Step 10:* 5-Formylaminoimidazole carboxamide ribonucleotide (FAICAR) formation from AICAR and N^{10}-formyl-THF by AICAR transformylase. *Step 11:* Dehydration/ring closure yields the authentic purine ribonucleotide IMP. The enzyme is IMP synthase.

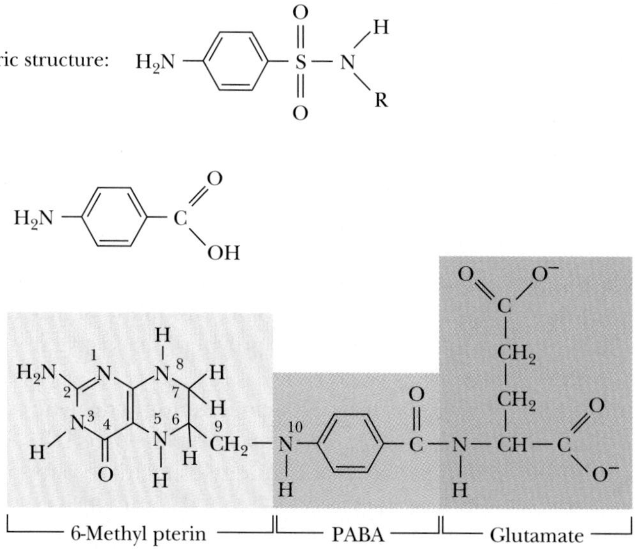

Sulfonamides have the generic structure: $H_2N-C_6H_4-SO_2-NHR$

PABA (*p*-aminobenzoic acid) $H_2N-C_6H_4-COOH$

THF (tetrahydrofolate)

6-Methyl pterin — PABA — Glutamate

Additional γ-glutamyl residues (up to a maximum of seven) may add here

***Figure* 27.5** Sulfa drugs, or sulfonamides, owe their antibiotic properties to their similarity to *p*-aminobenzoate (PABA), an important precursor in folic acid synthesis. Sulfonamides block folic acid formation by competing with PABA.

The dependence of purine biosynthesis on folic acid compounds at Steps 4 and 10 means that antagonists of folic acid metabolism (for example, *methotrexate,* as described in Chapter 14) will indirectly inhibit purine formation and, in turn, nucleic acid synthesis, cell growth, and cell division. Clearly, rapidly dividing cells such as malignancies or infective bacteria are more susceptible to these antagonists than slower-growing normal cells. Also among the folic acid antagonists are *sulfonamides* (Figure 27.5). Folic acid is a vitamin for animals and is obtained in the diet (Chapter 14). In contrast, bacteria synthesize folic acid from precursors, including *p-aminobenzoic acid (PABA)* (Chapter 14), and thus are more susceptible to sulfonamides than are animal cells.

AMP and GMP Are Synthesized from IMP

IMP is the precursor to both AMP and GMP. These major purine nucleotides are formed via distinct two-step metabolic pathways that diverge from IMP. The branch leading to AMP (adenosine 5′-monophosphate) involves the displacement of the 6-O group of inosine with aspartate (Figure 27.6) in a GTP-dependent reaction, followed by the nonhydrolytic removal of the 4-carbon skeleton of Asp as fumarate; the Asp amino group remains as the 6-amino group of AMP. *Adenylosuccinate synthetase* and *adenylosuccinase* are the two enzymes. Recall that adenylosuccinase also acted at Step 9 in the pathway from ribose-5-phosphate to IMP.

The formation of GMP from IMP requires an oxidation at C-2 of the purine ring, followed by a glutamine-dependent amidotransferase reaction that replaces the oxygen on C-2 with an amino group to yield *2-amino,6-oxy purine nucleoside monophosphate,* or as this compound is commonly known, *guanosine monophosphate.* The enzymes in the GMP branch are *IMP dehydrogenase* and *GMP synthetase.* Note that, starting from ribose-5-phosphate, 7 ATP equivalents are consumed in the synthesis of AMP and 8 in the synthesis of GMP.

Regulation of the Purine Biosynthetic Pathway

The regulatory network that controls purine synthesis is schematically represented in Figure 27.7. To recapitulate, the purine biosynthetic pathway from ribose-5-phosphate to IMP is allosterically regulated at the first two steps.

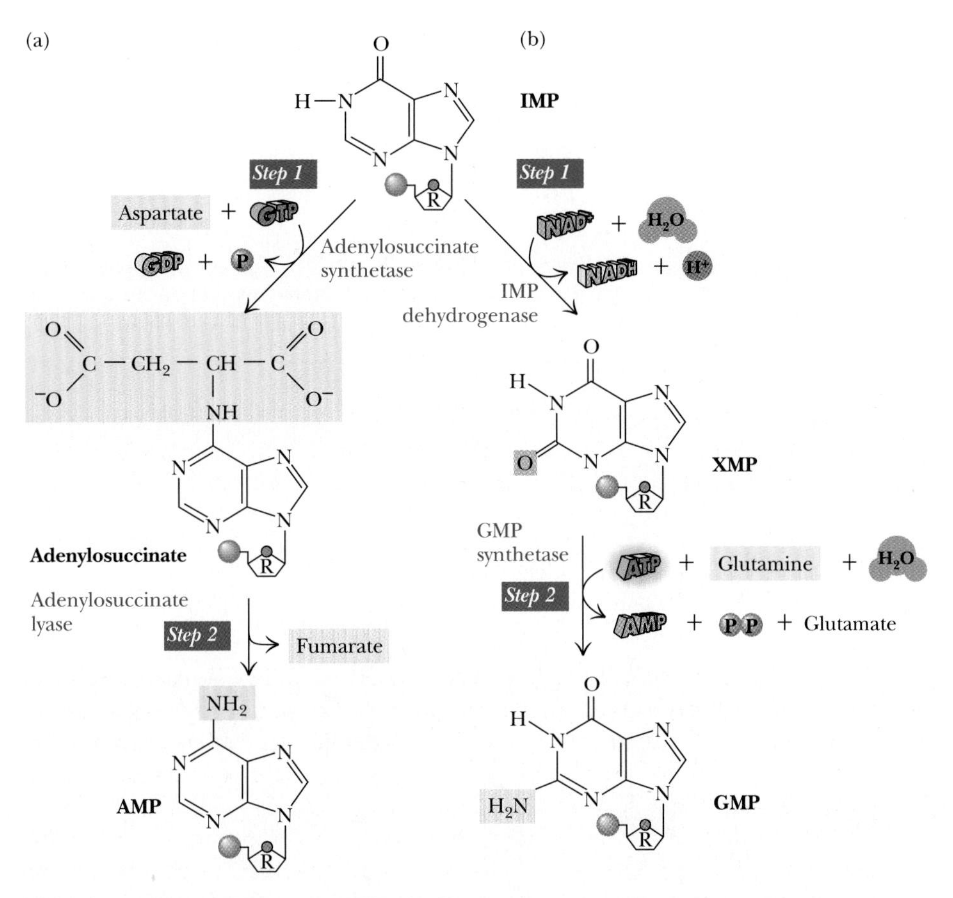

Figure **27.6** The synthesis of AMP and GMP from IMP. (a) AMP synthesis: The two reactions of AMP synthesis mimic Steps 8 and 9 in the purine pathway leading to IMP. In *Step 1,* the 6-O of inosine is displaced by aspartate to yield adenylosuccinate. The energy required to drive this reaction is derived from GTP hydrolysis. The enzyme is adenylosuccinate synthetase. AMP is a competitive inhibitor (with respect to the substrate IMP) of adenylosuccinate synthetase. In *Step 2,* adenylosuccinase (also known as adenylosuccinate lyase, the same enzyme catalyzing Step 9 in the purine pathway) carries out the nonhydrolytic removal of fumarate from adenylosuccinate, leaving AMP. (b) GMP synthesis: The two reactions of GMP synthesis are an NAD^+-dependent oxidation followed by an amidotransferase reaction. In *Step 1,* IMP dehydrogenase employs the substrates NAD^+ and H_2O in catalyzing oxidation of IMP at C-2. The products are xanthylic acid (XMP or xanthosine monophosphate), NADH, and H^+. GMP is a competitive inhibitor (with respect to IMP) of IMP dehydrogenase. In *Step 2,* transfer of the amido-N of glutamine to the C-2 position of XMP yields GMP. This ATP-dependent reaction is catalyzed by GMP synthetase. Besides GMP, the products are glutamate, AMP, and PP_i. Hydrolysis of PP_i to two P_i by ubiquitous pyrophosphatases pulls this reaction to completion.

Ribose-5-phosphate pyrophosphokinase, although not the committed step in purine synthesis, is subject to feedback inhibition by ADP and GDP. The enzyme catalyzing the next step, glutamine phosphoribosyl-pyrophosphate amidotransferase, has two allosteric sites, one where the "A" series of nucleoside phosphates (AMP, ADP, and ATP) binds and feedback-inhibits, and another where the corresponding "G" series binds and inhibits. Further, PRPP is a "feed-forward" activator of this enzyme. Thus, the rate of IMP formation by this pathway is governed by the levels of the final end products, the adenine and guanine nucleotides.

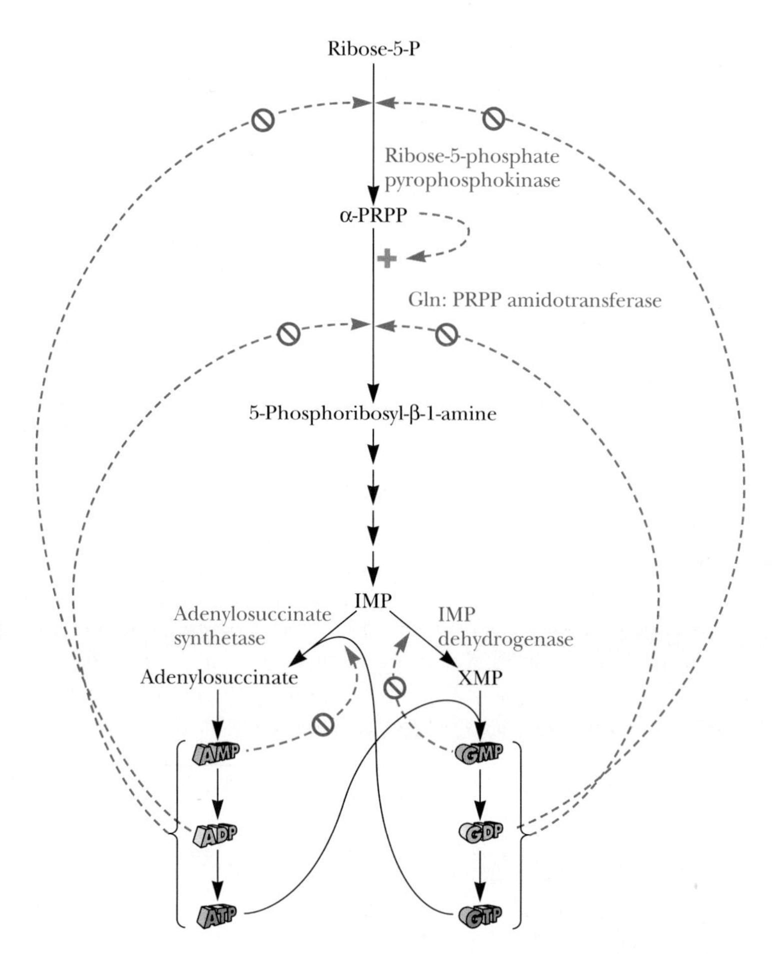

***Figure* 27.7** The regulatory circuit controlling purine biosynthesis. ADP and GDP are feedback inhibitors of ribose-5-phosphate pyrophosphokinase, the first reaction in the pathway. The second enzyme, glutamine phosphoribosyl pyrophosphate amidotransferase, has two distinct feedback inhibition sites, one for A nucleotides and one for G nucleotides. Also, this enzyme is allosterically activated by PRPP. In the branch leading from IMP to AMP, the first enzyme is feedback-inhibited by AMP, while the corresponding enzyme in the branch from IMP to GMP is feedback-inhibited by GMP. Last, ATP is the energy source for GMP synthesis, whereas GTP is the energy source for AMP synthesis.

The purine pathway splits at IMP, with respective branches leading to AMP and GMP. The first enzyme in the AMP branch, adenylosuccinate synthetase, is competitively inhibited by AMP; its counterpart in the GMP branch, IMP dehydrogenase, is inhibited in like fashion by GMP. Thus, the fate of IMP is determined by the relative levels of AMP and GMP, so that any deficiency in the amount of either of the principal purine nucleotides is self-correcting. This reciprocity of regulation is an effective mechanism for balancing the formation of AMP and GMP to satisfy cellular needs. Note also that reciprocity is even manifested at the level of energy input: GTP provides the energy to drive AMP synthesis, while ATP serves this role in GMP synthesis (Figure 27.7).

ATP-Dependent Kinases Form Nucleoside Diphosphates and Triphosphates from the Nucleoside Monophosphates

The products of *de novo* purine biosynthesis are the nucleoside monophosphates AMP and GMP. These nucleotides are converted by successive phosphorylation reactions into their metabolically prominent triphosphate forms,

ATP and GTP. The first phosphorylation, to give the nucleoside diphosphate forms, is carried out by two base-specific, ATP-dependent kinases, *adenylate kinase* and *guanylate kinase.*

Adenylate kinase: AMP + ATP → 2 ADP

Guanylate kinase: GMP + ATP → GDP + ADP

These nucleoside monophosphate kinases also act on deoxynucleoside monophosphates to give dADP or dGDP.

Oxidative phosphorylation (see Chapter 20) is primarily responsible for the conversion of ADP into ATP. ATP then serves as the phosphoryl donor for synthesis of the other nucleoside triphosphates from their corresponding NDPs in a reaction catalyzed by *nucleoside diphosphate kinase,* a nonspecific enzyme. For example,

$$\text{GDP} + \text{ATP} \rightleftharpoons \text{GTP} + \text{ADP}$$

Since this enzymatic reaction is readily reversible and nonspecific with respect to both phosphoryl acceptor and donor, in effect any NDP can be phosphorylated by any NTP, and vice versa. The preponderance of ATP over all other nucleoside triphosphates means that, in quantitative terms, it is the principal nucleoside diphosphate kinase substrate. The enzyme does not discriminate between the ribose moieties of nucleotides and thus functions in phosphoryl transfers involving deoxy-NDPs and deoxy-NTPs as well.

27.3 Purine Salvage

Nucleic acid turnover (synthesis and degradation) is an ongoing metabolic process in most cells. Messenger RNA in particular is actively synthesized and degraded. These degradative processes can lead to the release of free purines in the form of adenine, guanine, and hypoxanthine (the base in IMP). These substances represent a metabolic investment by cells. So-called salvage pathways exist to recover them in useful form. Salvage reactions involve resynthesis of nucleotides from bases via phosphoribosyltransferases.

$$\text{Base} + \text{PRPP} \rightleftharpoons \text{nucleoside-5}'\text{-phosphate} + \text{PP}_i$$

The subsequent hydrolysis of PP_i to inorganic phosphate by pyrophosphatases renders the phosphoribosyltransferase reaction effectively irreversible.

The purine phosphoribosyltransferases are *adenine phosphoribosyltransferase (APRT),* which mediates AMP formation, and *hypoxanthine-guanine phosphoribosyltransferase (HGPRT),* which can act on either hypoxanthine to form IMP or guanine to form GMP (shaded area, Figure 27.8).

Lesch-Nyhan Syndrome: HGPRT Deficiency Leads to Severe Clinical Disorder

The symptoms of **Lesch-Nyhan syndrome** are tragic: a crippling gouty arthritis due to excessive uric acid accumulation (uric acid is a purine degradation product, discussed in the next section) and, worse, severe malfunctions in the nervous system that lead to mental retardation, spasticity, aggressive behavior, and self-mutilation. Lesch-Nyhan syndrome results from a complete deficiency in HGPRT activity. The structural gene for HGPRT is located on the X chromosome, and the disease is a congenital, recessive, sex-linked trait manifested only in males. The severe consequences of HGPRT deficiency argue that purine salvage has greater metabolic importance than simply the energy-saving recovery of bases. Although HGPRT might seem to play a minor role in purine metabolism, its absence has profound consequences: *de novo*

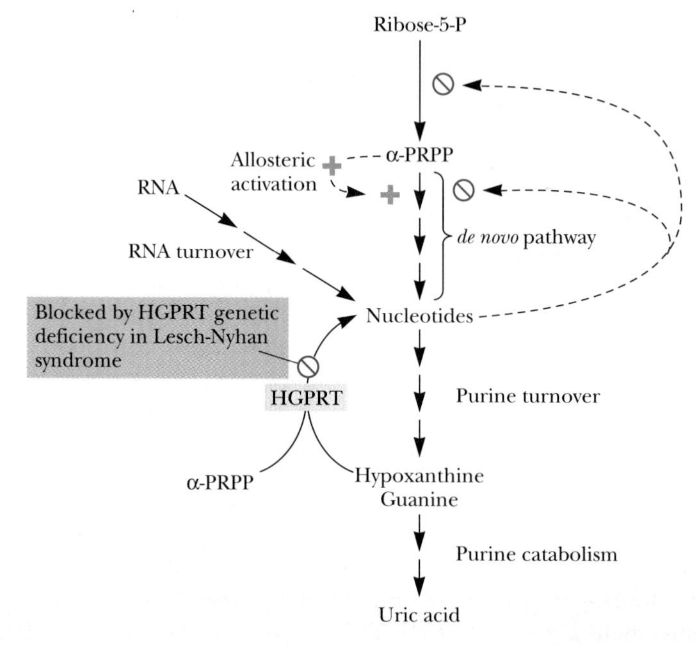

***Figure* 27.8** The metabolic consequences of congenital HGPRT deficiency in Lesch-Nyhan syndrome. Loss of HGPRT leads to elevated PRPP levels and stimulation of *de novo* purine synthesis. One ultimate consequence is increased production of uric acid.

purine biosynthesis is increased 200-fold and uric acid levels in the blood are greatly elevated. Presumably, these changes ensue because lack of consumption of PRPP by HGPRT elevates its availability for glutamine-PRPP amidotransferase, enhancing overall *de novo* purine synthesis and, ultimately, uric acid production (Figure 27.8). Despite these explanations, it remains unclear why deficiency in this single enzyme leads to the particular neurological aberrations characteristic of the syndrome. Fortunately, deficiencies in HGPRT activity in fetal cells can be detected following amniocentesis.

27.4 Purine Degradation

Since nucleic acids are ubiquitous in cellular material, significant amounts are ingested in the diet. Nucleic acids are degraded in the digestive tract to nucleotides by various nucleases and phosphodiesterases. Nucleotides are then converted to nucleosides by base-specific nucleotidases and nonspecific phosphatases.

$$\text{NMP} + H_2O \longrightarrow \text{nucleoside} + P_i$$

Nucleosides are hydrolyzed by nucleosidases or nucleoside phosphorylases to release the purine base:

$$\text{Nucleoside} + H_2O \xrightarrow{\textit{nucleosidase}} \text{base} + \text{ribose}$$

$$\text{Nucleoside} + P_i \xrightarrow{\textit{nucleoside phosphorylase}} \text{base} + \text{ribose-1-P}$$

The pentoses liberated in these reactions provide the only source of metabolic energy available from purine nucleotide degradation.

Feeding experiments using radioactively labeled nucleic acids as metabolic tracers have demonstrated that little of the amount of nucleotide ingested in the diet is incorporated into cellular nucleic acids. These findings confirm the *de novo* pathways of nucleotide biosynthesis as the primary source of nucleic acid precursors. Ingested bases are, for the most part, excreted. Nevertheless, cellular nucleic acids do undergo degradation in the course of the continuous recycling of cellular constituents.

The Major Pathways of Purine Catabolism Lead to Uric Acid

The major pathways of purine catabolism in animals are outlined in Figure 27.9. The various nucleotides are first converted to nucleosides by **intracellular nucleotidases.** These nucleotidases are under strict metabolic regulation so that their substrates, which act as intermediates in many vital processes, are not depleted below critical levels. Nucleosides are then degraded by the enzyme *purine nucleoside phosphorylase (PNP)* to release the purine base and the ribose sugar. Note that neither adenosine nor deoxyadenosine is a substrate for PNP. Instead, these nucleosides are first converted to inosine by *adenosine deaminase.* The PNP products are merged into *xanthine* by *guanine deaminase* and *xanthine oxidase,* and xanthine is then oxidized to uric acid by this latter enzyme.

Severe Combined Immunodeficiency Syndrome: A Lack of Adenosine Deaminase Is One Cause of This Inherited Disease

Severe combined immunodeficiency syndrome, or **SCID,** is a group of related inherited disorders characterized by the lack of an immune response to infectious disease. This immunological insufficiency is attributable to the inability of B and T lymphocytes to proliferate and produce antibodies in reaction to an antigenic challenge. About 30% of SCID patients suffer from a deficiency in the enzyme *adenosine deaminase (ADA).* ADA deficiency is also implicated in a variety of other diseases, including AIDS, anemia, and various lymphomas and leukemias. The first clinical trial of **gene therapy,** *the repair of a genetic deficiency by introduction of a functional recombinant version of the gene,* was conducted on an individual with SCID due to a defective ADA gene. ADA is a Zn^{2+}-dependent enzyme, and Zn^{2+} deficiency is associated with reduced immune function.

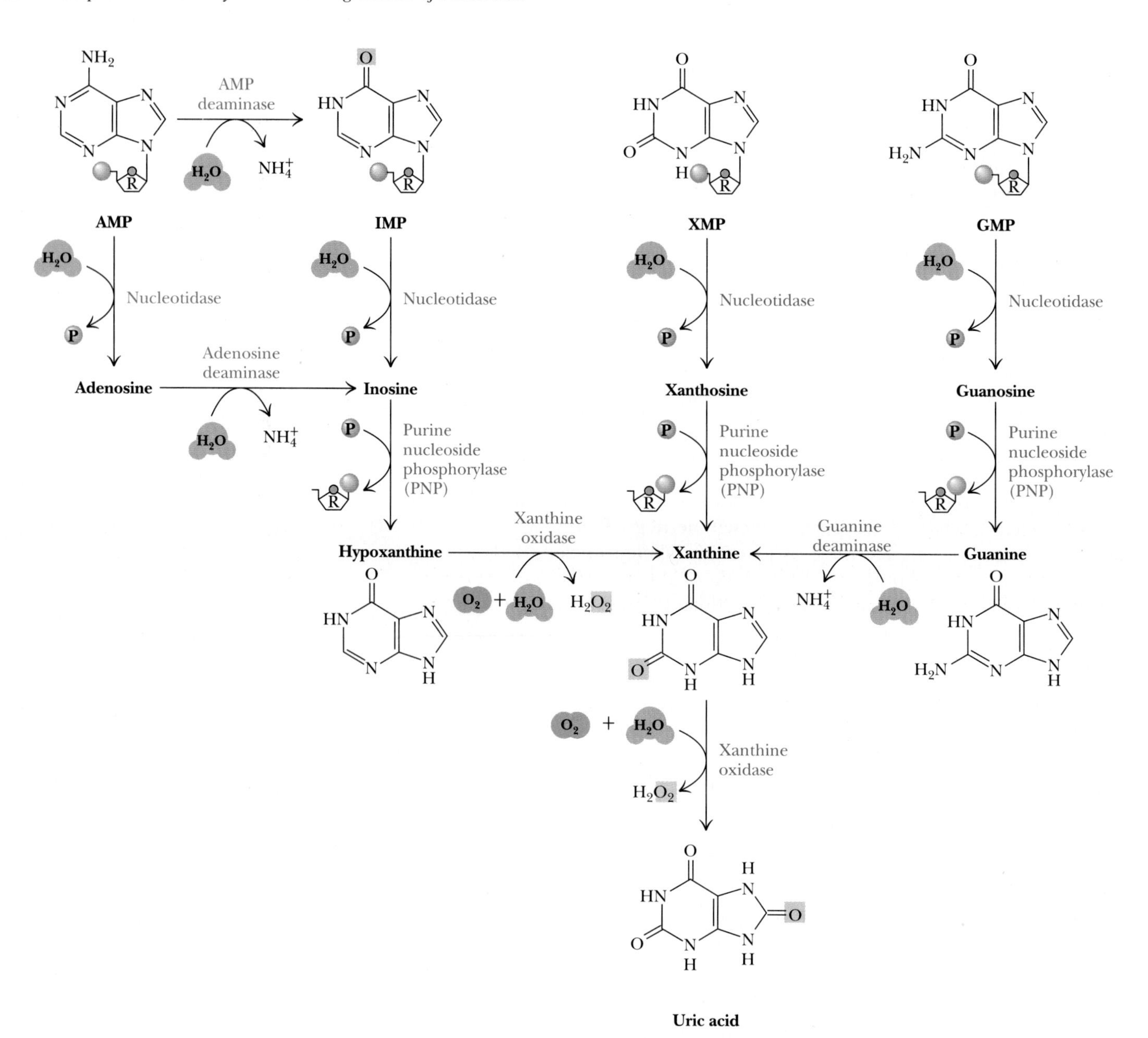

Figure **27.9** The major pathways for purine catabolism in animals. Catabolism of the different purine nucleotides converges in the formation of uric acid.

In the absence of ADA, deoxyadenosine is not degraded but instead is converted into dAMP and then into dATP. dATP is a potent feedback inhibitor of deoxynucleotide biosynthesis (discussed later in this chapter). Without deoxyribonucleotides, DNA cannot be replicated and cells cannot divide (Figure 27.10). Rapidly proliferating cell types such as lymphocytes are particularly susceptible to diminished DNA synthesis.

The Purine Nucleoside Cycle: An Anaplerotic Pathway in Skeletal Muscle

Deamination of AMP to IMP by *AMP deaminase* (Figure 27.9) followed by resynthesis of AMP from IMP by the *de novo* purine pathway enzymes, *adenylosuccinate synthetase* and *adenylosuccinate lyase,* constitutes a purine nucleoside

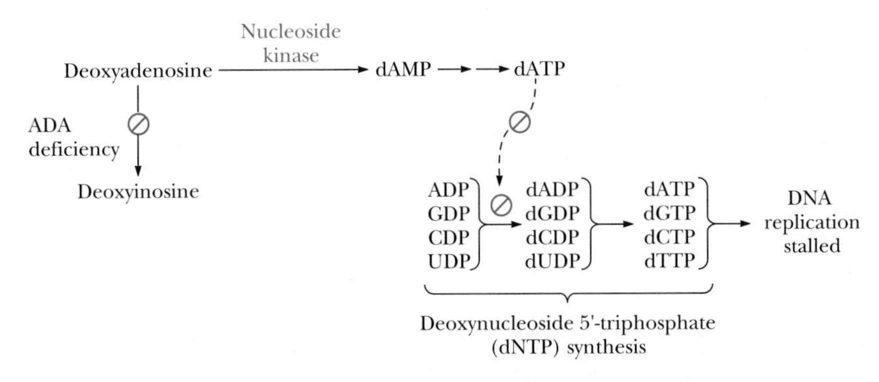

***Figure* 27.10** The effect of elevated levels of deoxyadenosine on purine metabolism. If ADA is deficient or absent, deoxyadenosine is not converted into deoxyinosine as normal (see Figure 27.9). Instead, it is salvaged by a nucleoside kinase, which converts it to dAMP, leading to accumulation of dATP and inhibition of deoxynucleotide synthesis (see Figure 27.27). Thus, DNA replication is stalled.

cycle (Figure 27.11). This cycle has the net effect of converting aspartic acid to fumaric acid plus NH_3. Although this cycle might seem like senseless energy consumption, it plays an important role in energy metabolism in skeletal muscle: The fumarate that it generates replenishes the levels of citric acid cycle intermediates lost in amphibolic side reactions (see Chapter 19). Skeletal muscle lacks the usual complement of anaplerotic enzymes and relies on enhanced levels of AMP deaminase, adenylosuccinate synthetase, and adenylosuccinate lyase to compensate.

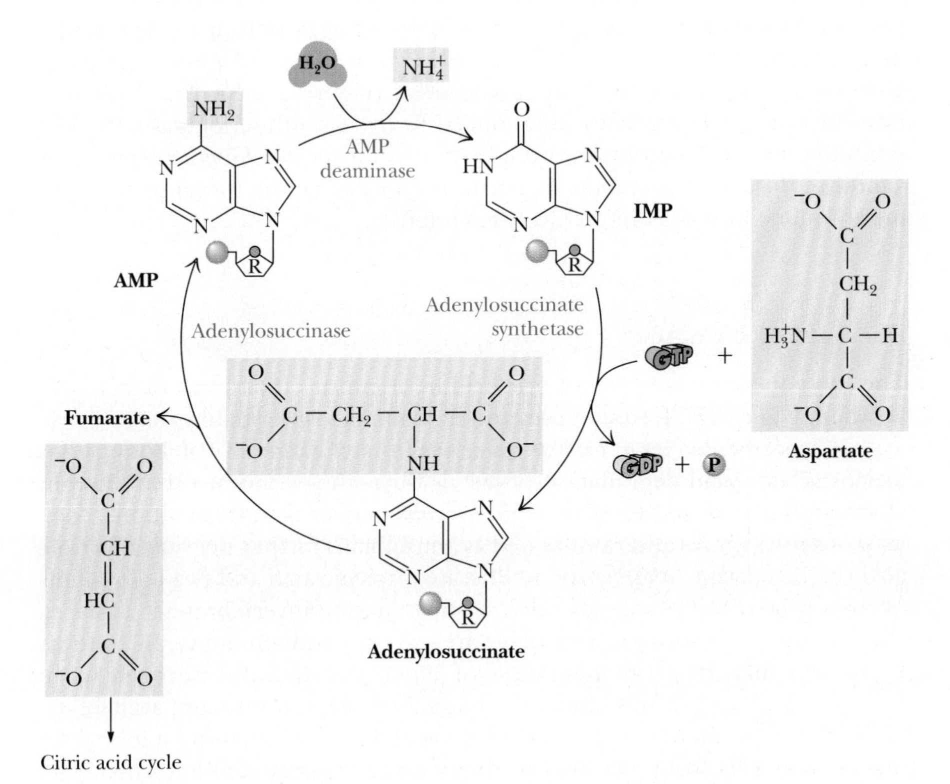

***Figure* 27.11** The purine nucleoside cycle for anaplerotic replenishment of citric acid cycle intermediates in skeletal muscle.

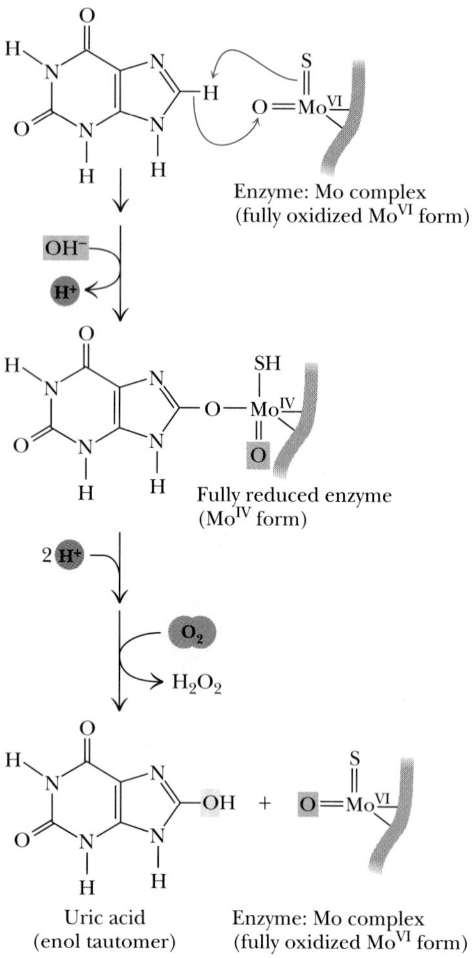

Figure **27.12** The mechanism of action of xanthine oxidase.

Xanthine Oxidase

Xanthine oxidase is present in large amounts in liver, intestinal mucosa, and milk. It oxidizes hypoxanthine to xanthine and xanthine to uric acid. Xanthine oxidase is a rather indiscriminate enzyme, using molecular oxygen to oxidize a wide variety of purines, pteridines, and aldehydes, producing H_2O_2 as a product. Xanthine oxidase possesses FAD, nonheme Fe-S centers, and a *molybdenum cofactor* (see Figure 26.1) as electron-transferring prosthetic groups. Its mechanism of action is diagrammed in Figure 27.12.

In humans and other primates, uric acid is the end product of purine catabolism and is excreted in the urine. Birds, terrestrial reptiles, and many insects also excrete uric acid, but, in these organisms, uric acid represents the major nitrogen excretory compound, since, unlike mammals, they do not also produce urea (Chapter 26). Instead, the catabolism of all nitrogenous compounds, including amino acids, is channeled into uric acid. This route of nitrogen catabolism allows these animals to conserve water by excreting crystals of uric acid in paste-like solid form.

Gout: An Excess of Uric Acid

Gout is the clinical term describing the physiological consequences accompanying excessive uric acid accumulation in body fluids. Uric acid and urate salts are rather insoluble in water and tend to precipitate from solution if produced in excess. The most common symptom of gout is arthritic pain in the joints as a result of urate deposition in cartilaginous tissue. The joint of the big toe is particularly susceptible. Urate crystals may also appear as kidney stones and lead to painful obstruction of the urinary tract. **Hyperuricemia,** chronic elevation of blood uric acid levels, occurs in about 3% of the population as a consequence of impaired excretion of uric acid or overproduction of purines. Purine-rich foods (such as caviar—fish eggs rich in nucleic acids) may exacerbate the condition. The biochemical causes of gout are varied. However, a common treatment is *allopurinol* (Figure 27.13). This hypoxanthine analog binds tightly to the reduced form of xanthine oxidase, thereby inhibiting its activity and preventing uric acid formation. Hypoxanthine and xanthine do not accumulate to harmful concentrations, because they are more soluble and thus more easily excreted.

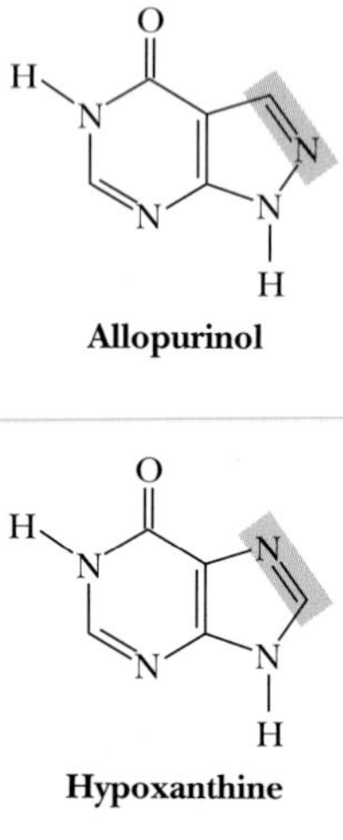

Figure **27.13** Allopurinol, an analog of hypoxanthine, is a potent inhibitor of xanthine oxidase.

The Fate of Uric Acid

The subsequent metabolism of uric acid in organisms that don't excrete it is shown in Figure 27.14. In molluscs and in mammals other than primates, uric acid is oxidized by *urate oxidase* to *allantoin* and excreted. In bony fishes (teleosts), uric acid degradation proceeds through yet another step wherein allantoin is hydrolyzed to *allantoic acid* by *allantoinase* before excretion. Cartilaginous fish (sharks and rays) as well as amphibians further degrade allantoic acid via the enzyme, *allantoicase,* to liberate glyoxylic acid and two equivalents of *urea.* Even simpler animals, such as most marine invertebrates (crustacea and so forth), use *urease* to hydrolyze urea to CO_2 and ammonia. In contrast to animals that must rid themselves of potentially harmful nitrogen waste products, microorganisms often are limited in growth by nitrogen availability. Many possess an identical pathway of uric acid degradation, using it instead to liberate the NH_3 from uric acid so that it can be assimilated into organic-N compounds essential to their survival.

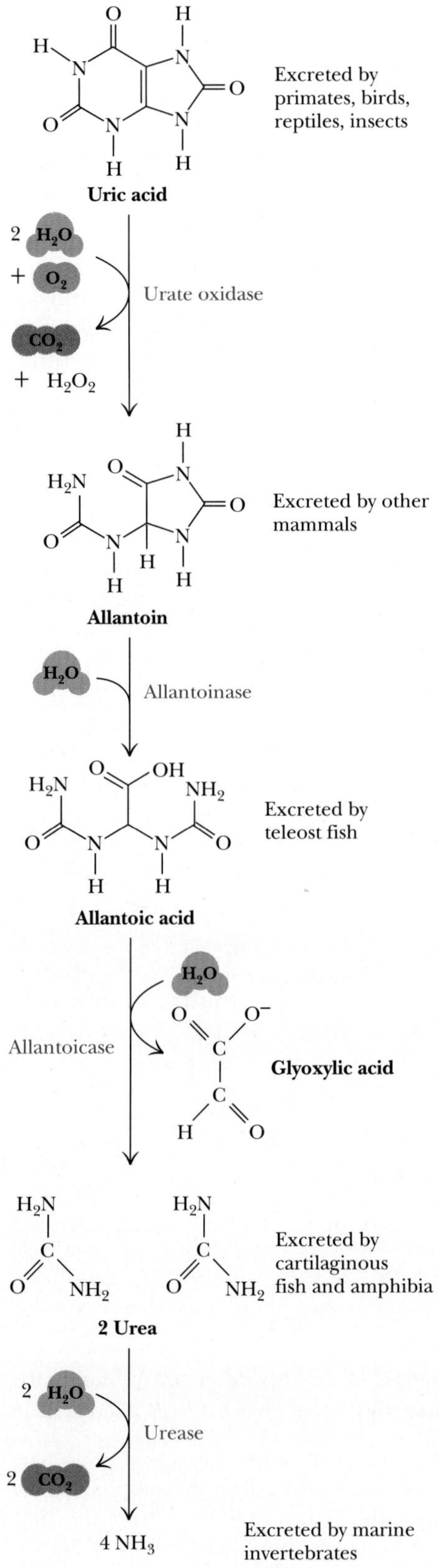

Uric acid

Excreted by primates, birds, reptiles, insects

$2\ H_2O + O_2$ → Urate oxidase → $CO_2 + H_2O_2$

Allantoin

Excreted by other mammals

H_2O → Allantoinase

Allantoic acid

Excreted by teleost fish

H_2O → Allantoicase → Glyoxylic acid

2 Urea

Excreted by cartilaginous fish and amphibia

$2\ H_2O$ → Urease → $2\ CO_2$

$4\ NH_3$

Excreted by marine invertebrates

***Figure* 27.14** The catabolism of uric acid to allantoin, allantoic acid, urea, or ammonia in various animals.

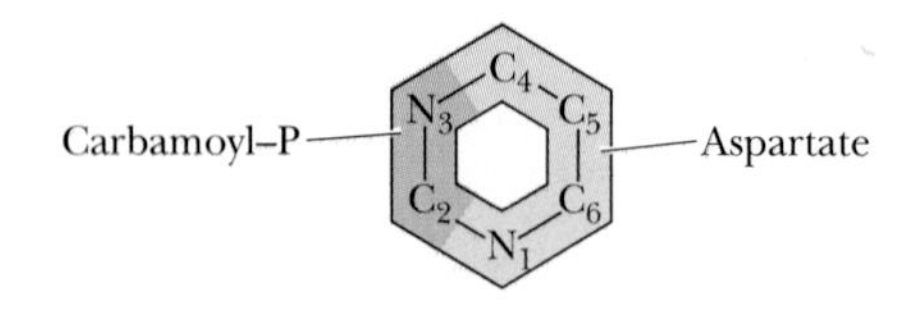

***Figure* 27.15** The metabolic origin of the six atoms of the pyrimidine ring.

27.5 The Biosynthesis of Pyrimidines

In contrast to purines, pyrimidines are not synthesized as nucleotide derivatives. Instead, the pyrimidine ring system is completed before a ribose-5-P moiety is attached. Also, only two precursors, carbamoyl-P and aspartic acid, contribute atoms to the six-membered pyrimidine ring (Figure 27.15), compared to seven precursors for the 9 purine atoms.

Mammals have two enzymes for carbamoyl phosphate synthesis. Carbamoyl phosphate for pyrimidine biosynthesis is formed by *carbamoyl phosphate synthetase II (CPS II)*, a cytosolic enzyme. Recall that carbamoyl phosphate synthetase I is a mitochondrial enzyme dedicated to the urea cycle and arginine biosynthesis (Chapter 26). The substrates of carbamoyl phosphate synthetase II are HCO_3^-, H_2O, glutamine, and 2 ATPs (Figure 27.16). Since carbamoyl phosphate made by CPS II in mammals has no fate other than incorporation into pyrimidines, mammalian CPS II can be viewed as the committed step in the pyrimidine *de novo* pathway. Bacteria have but one CPS, and its carbamoyl

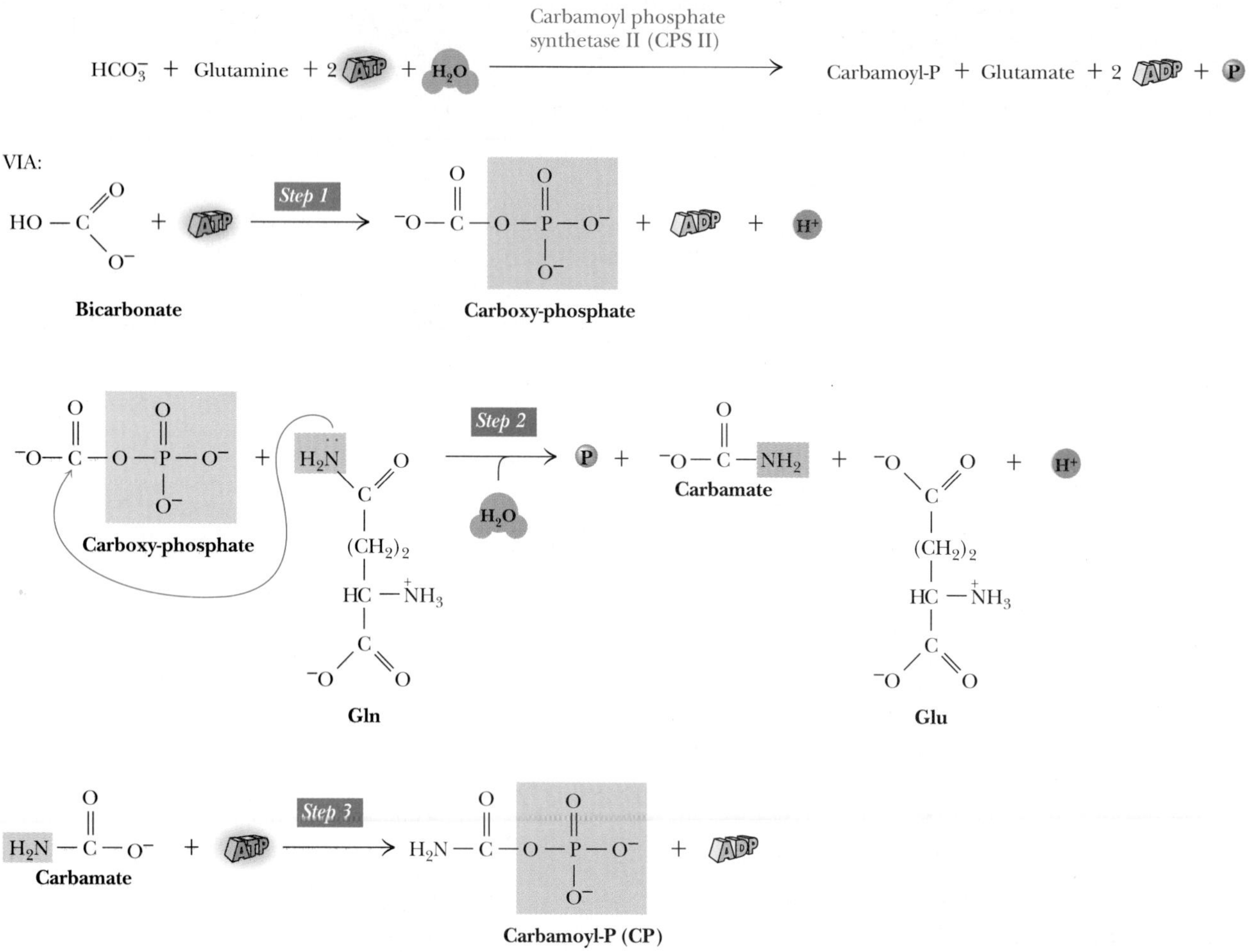

$$HCO_3^- + \text{Glutamine} + 2\,\text{ATP} + H_2O \xrightarrow{\text{Carbamoyl phosphate synthetase II (CPS II)}} \text{Carbamoyl-P} + \text{Glutamate} + 2\,\text{ADP} + \text{P}$$

VIA:

Step 1: Bicarbonate + ATP → Carboxy-phosphate + ADP + H^+

Step 2: Carboxy-phosphate + Gln (H_2O) → P + Carbamate + Glu + H^+

Step 3: Carbamate + ATP → Carbamoyl-P (CP) + ADP

***Figure* 27.16** The reaction catalyzed by carbamoyl phosphate synthetase II (CPS II). Note that, in contrast to carbamoyl phosphate synthetase I, CPS II uses the amide of glutamine, not NH_4^+, to form carbamoyl-P. *Step 1:* The first ATP consumed in carbamoyl phosphate synthesis is used in forming carboxy-phosphate, an activated form of CO_2. *Step 2:* Carboxy-phosphate (also called carbonyl-phosphate) then reacts with the glutamine amide to yield carbamate and glutamate. *Step 3:* Carbamate is phosphorylated by the second ATP to give ADP and carbamoyl phosphate.

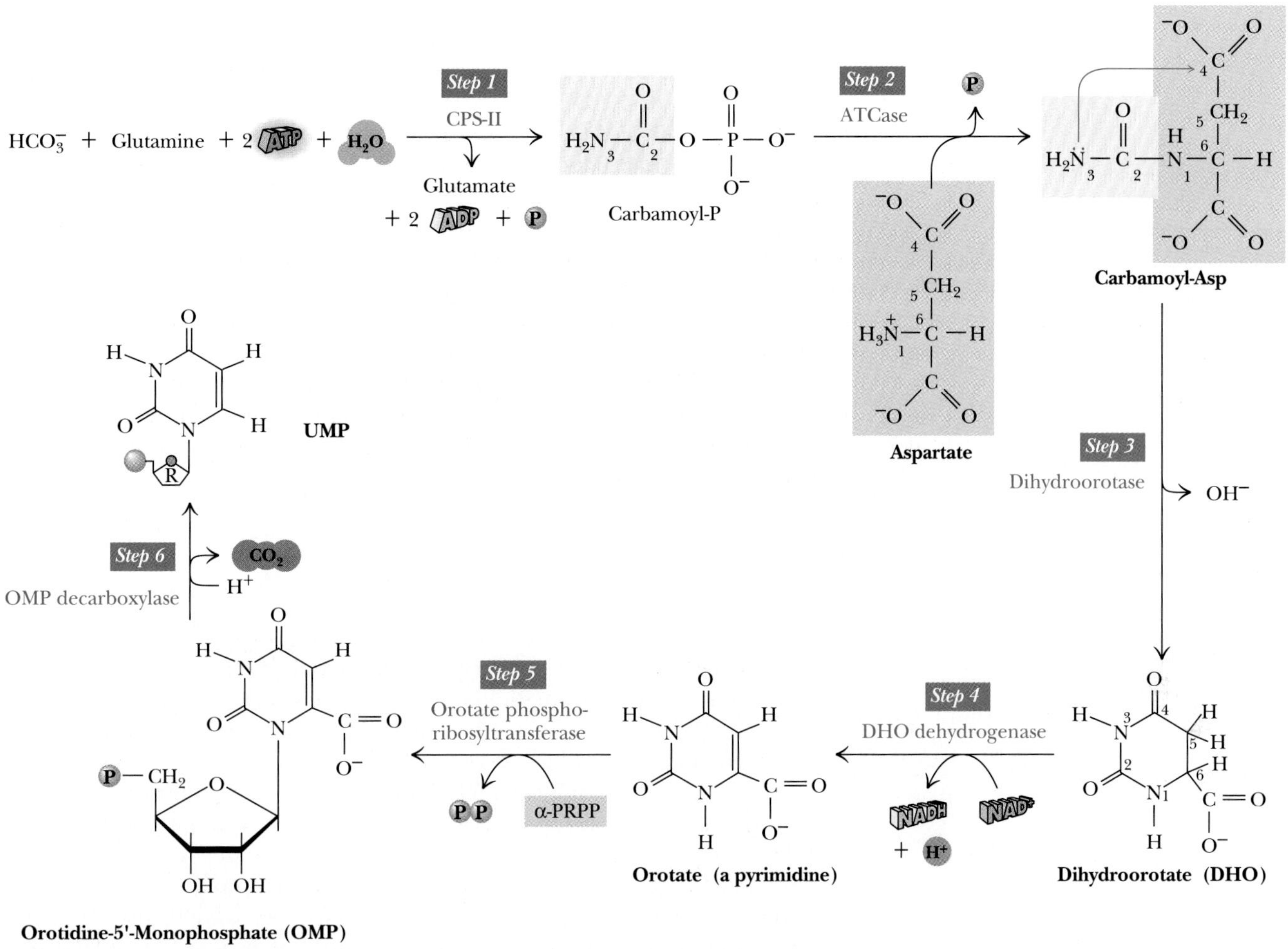

Figure **27.17** The *de novo* pyrimidine biosynthetic pathway. *Step 1:* Carbamoyl-P synthesis. *Step 2:* Condensation of carbamoyl phosphate and aspartate to yield carbamoyl-aspartate is catalyzed by aspartate transcarbamoylase. *Step 3:* An intramolecular condensation catalyzed by dihydroorotase gives the six-membered heterocyclic ring characteristic of pyrimidines. The product is dihydroorotate (DHO). *Step 4:* The oxidation of DHO by dihydroorotate dehydrogenase gives orotate. *Step 5:* PRPP provides the ribose-5-P moiety that transforms orotate into orotidine-5′-monophosphate, a pyrimidine nucleotide. Note that orotate phosphoribosyltransferase joins N-1 of the pyrimidine to the ribosyl group in appropriate β-configuration. PP_i hydrolysis renders this reaction thermodynamically favorable. *Step 6:* Decarboxylation of OMP by OMP decarboxylase yields UMP.

phosphate product is incorporated into arginine as well as pyrimidines. Thus, the committed step in bacterial pyrimidine synthesis is the next reaction, which is mediated by *aspartate transcarbamoylase.* The structure and regulation of *E. coli* aspartate transcarbamoylase, a paradigm for allosteric enzymes, is discussed in Chapter 12.

Aspartate transcarbamoylase catalyzes the condensation of carbamoyl phosphate with aspartate to form carbamoyl-aspartate (Figure 27.17). No ATP input is required at this step because carbamoyl phosphate represents an "activated" carbamoyl group. Ring closure and dehydration via linkage of the —NH_2 group introduced by carbamoyl-P with the former β-COO^- of aspartate is mediated by the enzyme *dihydroorotase.* The product of the reaction is *dihydroorotate,* a six-membered ring compound (Step 3). Dihydroorotate is not a true pyrimidine, but its oxidation yields *orotate,* which is. This oxidation (Step 4) is catalyzed by *dihydroorotate dehydrogenase.* (The dihydroorotate dehydrogenase mechanism is discussed in Chapter 14.) Bacterial dihydroorotate dehydrogenases are NAD^+-linked flavoproteins, which are somewhat unusual in possessing both FAD and FMN; these enzymes also have nonheme Fe-S centers as additional redox prosthetic groups. The eukaryotic version of dihydroorotate dehydrogenase is a protein component of the inner mitochondrial membrane; its immediate e^- acceptor is a quinone and the reducing equivalents drawn from dihydroorotate can be used to drive ATP synthesis via oxidative phosphorylation. At this stage, ribose-5-phosphate is joined to N-1 of orotate, giving the pyrimidine nucleotide *orotidine-5′-monophosphate,* or *OMP* (Step 5, Figure 27.17). The ribose phosphate donor is PRPP; the enzyme is *orotate phosphoribosyltransferase.* The next reaction is catalyzed by *OMP decarboxylase.* Decarboxylation of OMP gives *UMP (uridine-5′-monophosphate,* or *uridylic acid),* one of the two common pyrimidine ribonucleotides.

Pyrimidine Biosynthesis in Mammals Is Another Example of "Metabolic Channeling"

In bacteria, the six enzymes of *de novo* pyrimidine biosynthesis exist as distinct proteins, each independently catalyzing its specific step in the overall pathway. In contrast, in mammals, the six enzymatic activities are distributed among only three proteins, two of which are **multifunctional polypeptides:** single polypeptide chains having two or more enzymic centers. The first three steps of pyrimidine synthesis, CPS-II, aspartate transcarbamoylase, and dihydroorotase, are all localized on a single 210-kD cytosolic polypeptide. This multifunctional enzyme is the product of a solitary gene, yet it is equipped with the active sites for all three enzymatic activities. Step 4 (Figure 27.17) is catalyzed by *DHO dehydrogenase,* a separate enzyme associated with the outer surface of the inner mitochondrial membrane, but the enzymatic activities mediating Steps 5 and 6, namely, *orotate phosphoribosyltransferase* and *OMP decarboxylase* in mammals, are also found on a single cytosolic polypeptide known as *UMP synthase.*

The purine biosynthetic pathway of avian liver also provides examples of metabolic channeling. Recall that Steps 3, 4, and 6 of *de novo* purine synthesis are catalyzed by three enzymatic activities localized on a single multifunctional polypeptide, and Steps 7 and 8 and Steps 10 and 11 by respective bifunctional polypeptides (see Figure 27.3).

Such multifunctional enzymes confer an advantage: The product of one reaction in a pathway is the substrate for the next. In multifunctional enzymes, such products remain bound and are channeled directly to the next active site, rather than dissociated into the surrounding medium for diffusion to the next enzyme. This **metabolic channeling** is more efficient since substrates are not diluted into the milieu and no pools of intermediates accumulate. There are a number of documented instances in eukaryotes where such multifunctional polypeptides achieve metabolic channeling and thus greater metabolic efficiency. Fatty acid biosynthesis is one example that we considered earlier (Chapter 24).

Synthesis of the Prominent Ribonucleotides UTP and CTP

The two prominent pyrimidine ribonucleotide products are derived from UMP via the same unbranched pathway. First, UDP is formed from UMP via an ATP-dependent *nucleoside monophosphate kinase.*

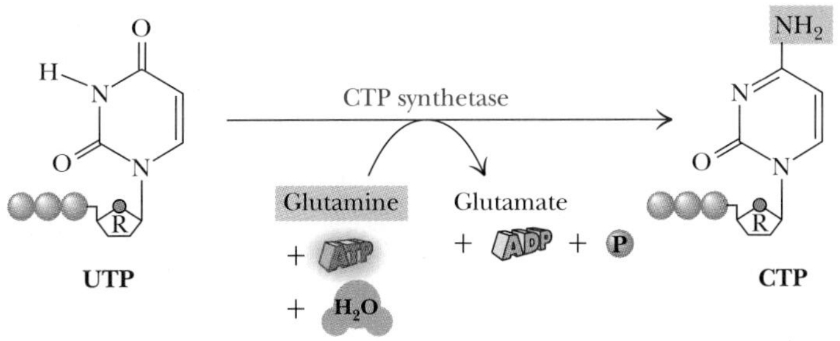

Figure **27.18** CTP synthesis from UTP. CTP synthetase catalyzes amination of the 6-position of the UTP pyrimidine ring, yielding CTP. In eukaryotes, this —NH_2 comes from the amide-N of glutamine; in bacteria, NH_4^+ serves this role.

$$\text{UMP} + \text{ATP} \rightleftharpoons \text{UDP} + \text{ADP}$$

Then, UTP is formed by *nucleoside diphosphate kinase.*

$$\text{UDP} + \text{ATP} \rightleftharpoons \text{UTP} + \text{ADP}$$

Amination of UTP at the 6-position gives CTP. The enzyme, *CTP synthetase,* is a glutamine amidotransferase (Figure 27.18). ATP hydrolysis provides the energy to drive the reaction.

Regulation of Pyrimidine Biosynthesis

Pyrimidine biosynthesis in bacteria is allosterically regulated at aspartate transcarbamoylase (ATCase; see Chapter 12). *Escherichia coli* ATCase is feedback-inhibited by the end product, CTP. ATP, which can be viewed as a signal of both energy availability and purine sufficiency, is an allosteric activator of ATCase. CTP and ATP compete for a common allosteric site on the enzyme. In many bacteria, UTP, not CTP, acts as the ATCase feedback inhibitor.

In animals, CPS-II catalyzes the committed step in pyrimidine synthesis and serves as the focal point for allosteric regulation. UDP and UTP are feedback inhibitors of CPS-II, while PRPP and ATP are allosteric activators. With the exception of ATP, none of these compounds are substrates of CPS-II or of either of the two other enzymic activities residing with it on the trifunctional polypeptide. Figure 27.19 compares the regulatory circuits governing pyrimidine synthesis in bacteria and animals.

27.6 Pyrimidine Degradation

Like purines, free pyrimidines can be salvaged and recycled to form nucleotides via phosphoribosyltransferase reactions similar to those discussed earlier. Pyrimidine catabolism results in degradation of the pyrimidine ring to products reminiscent of the original substrates, aspartate, CO_2, and ammonia (Figure 27.20). β-Alanine can be recycled into the synthesis of coenzyme A (see Chapter 14). Catabolism of the pyrimidine base, thymine (5-methyluracil) yields β-aminoisobutyric acid instead of β-alanine.

Pathways presented thus far in this chapter account for the synthesis of the four principal ribonucleotides: ATP, GTP, UTP, and CTP. These compounds serve important coenzymic functions in metabolism and are the immediate precursors for ribonucleic acid (RNA) synthesis. Roughly 90% of the total nucleic acid in cells is RNA, with the remainder being deoxyribonucleic acid (DNA). DNA differs from RNA in being a polymer of deoxyribonucleotides, one of which is deoxy-thymidylic acid. We now turn to the synthesis of these compounds.

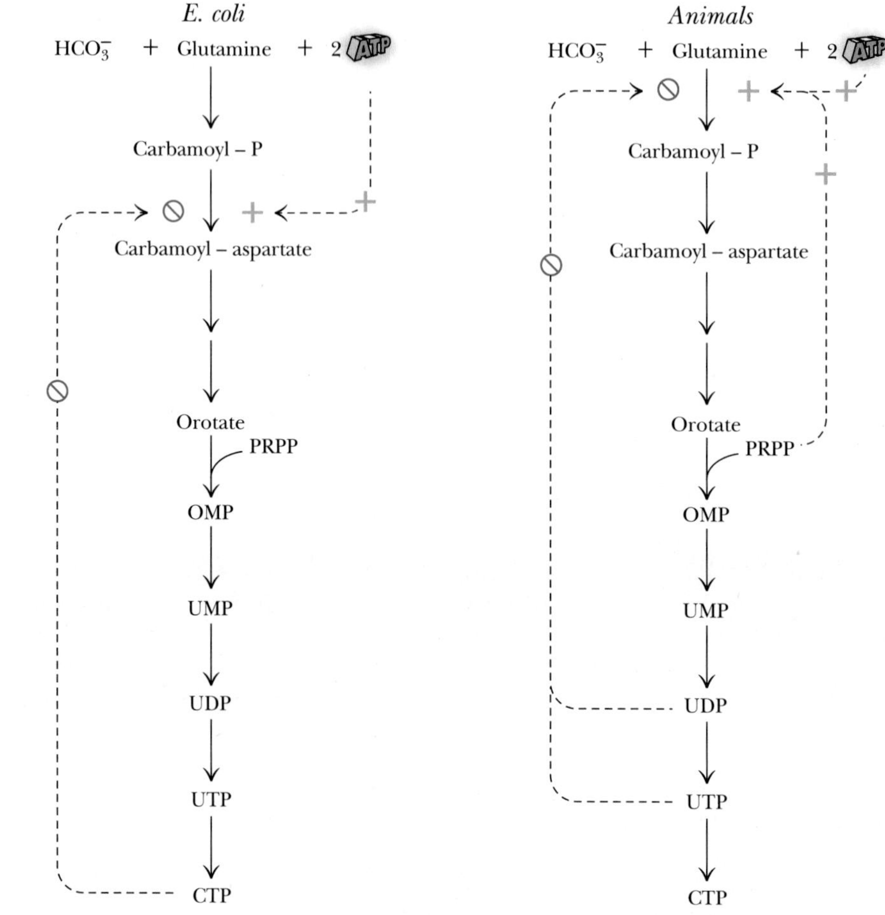

Figure **27.19** A comparison of the regulatory circuits that control pyrimidine synthesis in *E. coli* and animals.

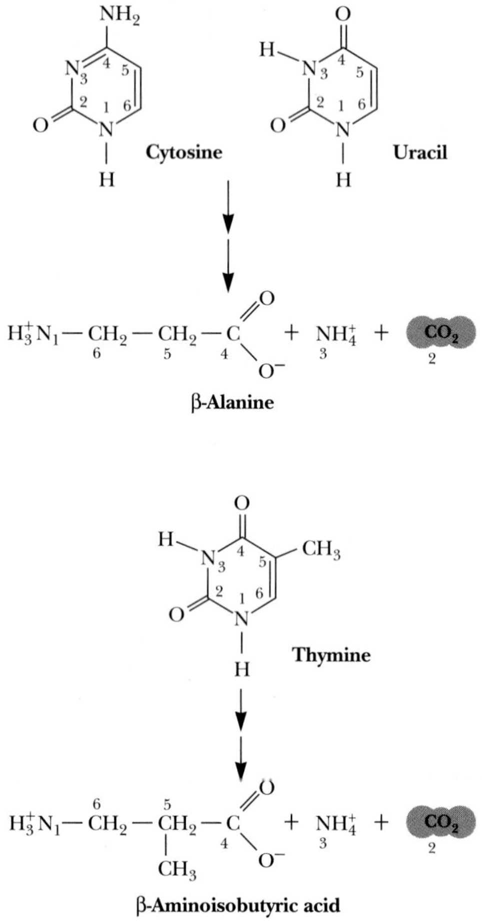

Cytosine **Uracil**

$$H_3^+N_1-CH_2-CH_2-COO^- + NH_4^+ + CO_2$$

β-Alanine

Thymine

$$H_3^+N_1-CH_2-CH(CH_3)-COO^- + NH_4^+ + CO_2$$

β-Aminoisobutyric acid

Figure **27.20** Pyrimidine degradation. Carbons 4, 5, and 6 plus N-1 are released as β-alanine, N-3 as NH_4^+, and C-2 as CO_2. (The pyrimidine thymine yields β-aminoisobutyric acid.) Recall that aspartate was the source of N-1 and C-4, 5, and 6, while C-2 came from CO_2 and N-3 from NH_4^+ via glutamine.

27.7 Deoxyribonucleotide Biosynthesis

The deoxyribonucleotides have only one metabolic purpose: to serve as precursors for DNA synthesis. In most organisms, ribonucleoside diphosphates (NDPs) are the substrates for deoxyribonucleotide formation. Reduction at the 2′-position of the ribose ring in NDPs produces 2′-deoxy forms of these nucleotides (Figure 27.21). This reaction involves replacement of the 2′-OH by a hydride ion ($H:^-$) and is catalyzed by an enzyme known as *ribonucleotide reductase.* Three types of ribonucleotide reductases have been described: (a) Fe-dependent, (b) Mn-dependent, and (c) 5′-deoxyadenosylcobalamin-dependent (see Chapter 14). The Fe-dependent enzyme is found in *E. coli* and in virtually all eukaryotes. This enzyme has been exceptionally well characterized by Peter Reichard and his colleagues and represents the highest extent of our understanding of deoxynucleotide biosynthesis.

E. coli Ribonucleotide Reductase

The enzyme system for dNDP formation consists of four proteins, two of which constitute the ribonucleotide reductase proper, an enzyme of the $\alpha_2\beta_2$ type. The other two proteins, *thioredoxin* and *thioredoxin reductase,* function in the delivery of reducing equivalents, as we shall see shortly. The two proteins

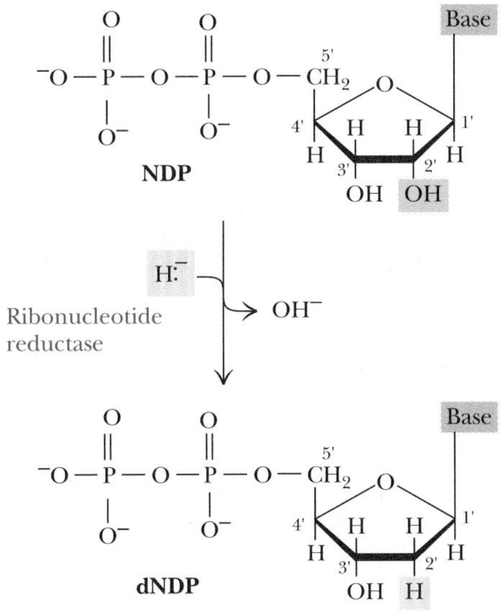

NDP

Ribonucleotide reductase

dNDP

***Figure* 27.21** Deoxyribonucleotide synthesis involves reduction at the 2′-position of the ribose ring of nucleoside diphosphates.

of ribonucleotide reductase are designated *B1* (85 kD) and *B2* (45 kD) and each is a homodimer in the holoenzyme (Figure 27.22). Each B1 possesses a pair of reactive sulfhydryl groups and two types of regulatory sites. Both of these sites bind effectors. One is termed the **substrate specificity site:** Occupation of this site by various effectors determines the substrate specificity of the enzyme. The other effector site is the **overall activity site:** Depending on the effector occupying this site, the enzyme is either active or inactive. The actual substrate binding site lies at the interface between the B1 and B2 subunits (Figure 27.22). Each B2 has an Fe^{3+}. Replacing Tyr^{122} of the B2 protein with Phe by genetic engineering results in an inactive protein. This observation suggests that the phenolic-OH of Tyr^{122} is crucial to the enzyme's activity.

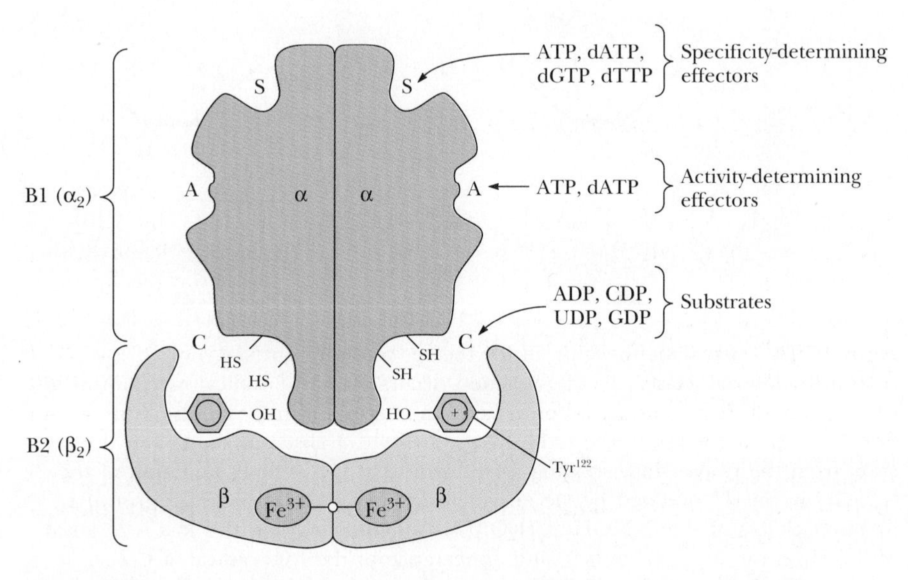

***Figure* 27.22** *E. coli* ribonucleotide reductase: its binding sites and subunit organization. Two proteins, B1 and B2 (each a dimer of identical subunits), combine to form the holoenzyme. The holoenzyme has three classes of nucleotide binding sites: S, the specificity-determining sites; A, the activity-determining sites; and C, the catalytic or active sites. These various sites bind different nucleotide ligands. Note that the holoenzyme apparently possesses only one active site formed by interaction between Fe^{3+} atoms in each B2 subunit.

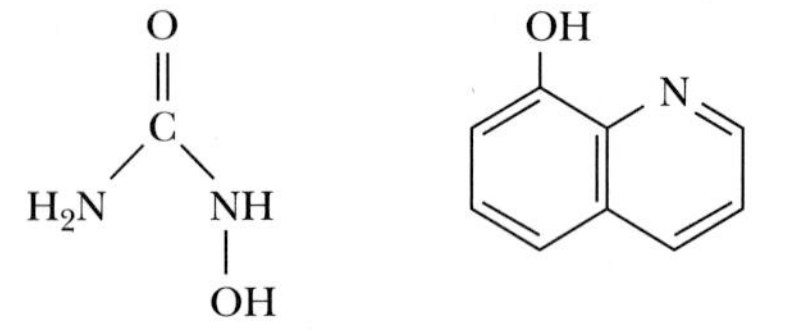

Hydroxyurea **8-Hydroxyquinoline**

***Figure* 27.23** Hydroxyurea and 8-hydroxyquinoline, inhibitors of ribonucleotide reductase.

Hydroxyurea, a free radical quenching agent, and **8-hydroxyquinoline,** an Fe^{3+}-specific chelator, are inhibitors of ribonucleotide reductase (Figure 27.23). The catalytic mechanism for ribonucleotide reductase proposed by JoAnn Stubbe of MIT is summarized in Figure 27.24.

The Reducing Power for Ribonucleotide Reductase

NADPH is the ultimate source of reducing equivalents for ribonucleotide reduction, but the immediate source is reduced **thioredoxin,** a small (12 kD) protein with reactive Cys-sulfhydryl groups situated next to one another in the sequence Cys—Gly—Pro—Cys. These Cys residues are able to undergo revers-

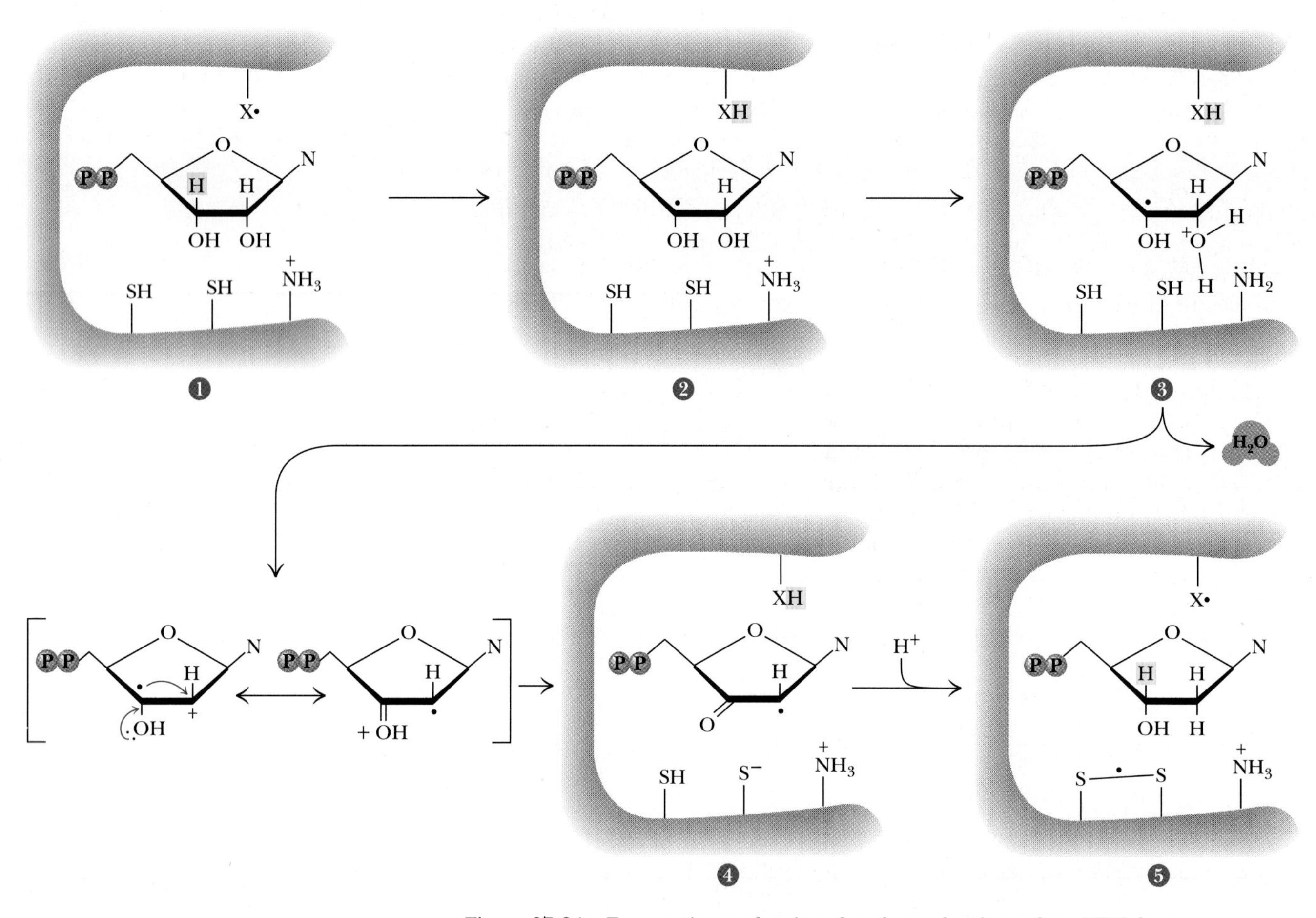

***Figure* 27.24** Enzymatic mechanism for the reduction of an NDP by ribonucleotide reductase. (1) The reaction occurs via a free radical mechanism in which the reducing equivalents come from an enzyme sulfhydryl-to-disulfide transition. (2) An enzyme free radical X · (at Tyr^{122}?) abstracts the C-3′ hydrogen atom from the ribose ring, creating a free radical at C-3′. (3) Protonation of the C-2′-OH by an H^+ donated by the enzyme (shown here as an $—NH_3^+$ group) is followed by loss of the C-2′-OH as H_2O. (4) Concomitant with this loss is removal of the H on the hydroxyl at C-3′ and appearance of the free radical at C-2′. (5) Reduction at the C-2′ and C-3′ positions by enzyme thiol groups and return of the H transferred to X back to C-3′ leads to formation of the deoxynucleoside diphosphate.

To restore the enzyme to an active form, the reactive thiol groups must be regenerated. This restoration is achieved by reduction of the disulfide through intervention of the other proteins of the ribonucleotide reductase system, thioredoxin and thioredoxin reductase.

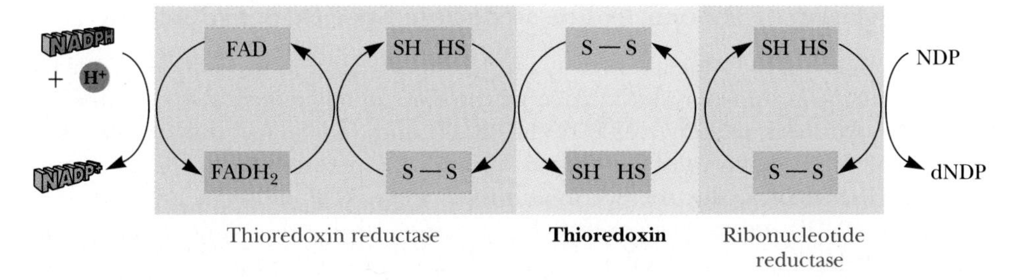

***Figure* 27.25** The (—S—S—)/(—SH HS—) oxidation–reduction cycle involving ribonucleotide reductase, thioredoxin, thioredoxin reductase, and NADPH.

ible oxidation–reduction between (—S—S—) and (—SH HS—) and, in their reduced form, serve as primary electron donors to regenerate the reactive —SH pair of the ribonucleotide reductase active site (Figure 27.24). In turn, the sulfhydryls of thioredoxin must be restored to the (—SH HS—) state for another catalytic cycle. **Thioredoxin reductase,** an α_2-type enzyme composed of 58-kD flavoprotein subunits, mediates the NADPH-dependent reduction of thioredoxin (Figure 27.25). Thioredoxin functions in a number of metabolic roles besides deoxyribonucleotide synthesis, the common denominator of which is reversible sulfide : sulfhydryl transitions. Another sulfhydryl protein similar to thioredoxin, called *glutaredoxin,* can also function in ribonucleotide reduction. Oxidized glutaredoxin is re-reduced by two equivalents of *glutathione* (γ-glutamylcysteinylglycine; Figure 27.26), which in turn is re-reduced by glutathione reductase, another NADPH-dependent flavoenzyme.

The substrates for ribonucleotide reductase are CDP, UDP, GDP, and ADP, and the corresponding products are dCDP, dUDP, dGDP, and dADP. Since CDP is not an intermediate in pyrimidine nucleotide synthesis, it must arise by dephosphorylation of CTP, for instance, via nucleoside diphosphate kinase action. Although uridine nucleotides do not occur in DNA, UDP is a substrate. The formation of dUDP is justified since it is a precursor to dTTP, a necessary substrate for DNA synthesis (see following discussion).

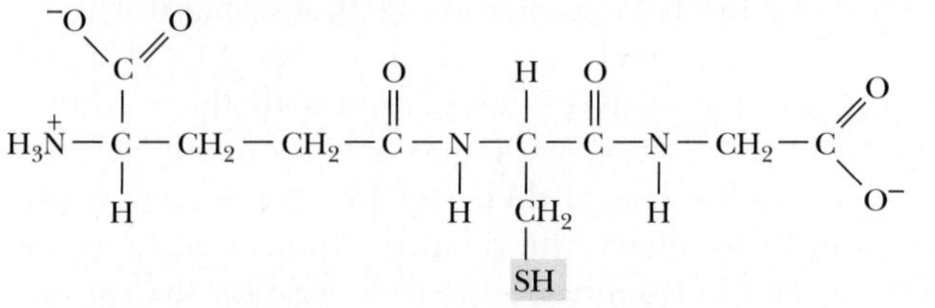

Reduced glutathione (γ-glutamylcysteinylglycine, GSH)

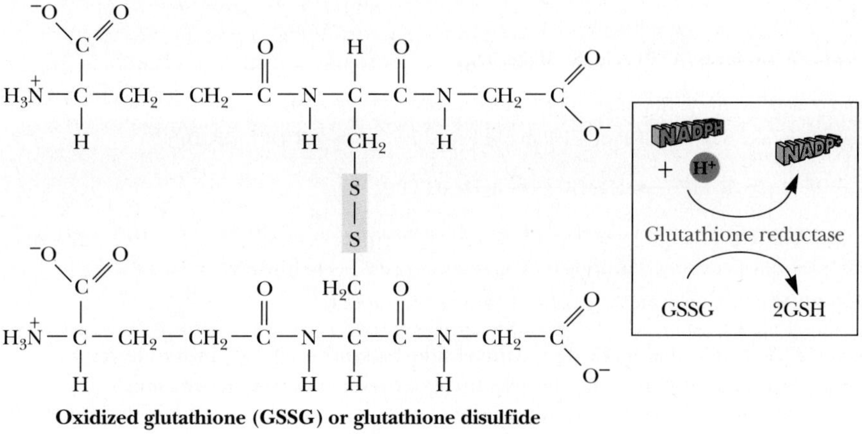

Oxidized glutathione (GSSG) or glutathione disulfide

NADPH + H+ → NADP+
Glutathione reductase
GSSG → 2GSH

***Figure* 27.26** The structure of glutathione.

Regulation of Ribonucleotide Reductase Specificity and Activity

Ribonucleotide reductase activity must be modulated in two ways in order to maintain an appropriate balance of the four deoxynucleotides essential to DNA synthesis, namely, dATP, dGTP, dCTP, and dTTP. First, the overall activity of the enzyme must be turned on and off in response to the need for dNTPs. Second, the relative amounts of each NDP substrate transformed into dNDP must be controlled in order that the right balance of dATP:dGTP:dCTP:dTTP is produced. The two different sets of effector binding sites on ribonucleotide reductase, *discrete from the substrate-binding active site,* are designed to serve these purposes. These two regulatory sites are designated the *overall activity site* and the *substrate specificity site.* Only two nucleotides, ATP and dATP, are able to bind at the overall activity site. If ATP is bound, the enzyme is active, while if its deoxy counterpart, dATP, occupies this site, the enzyme is inactive. That is, ATP is a positive effector and dATP is a negative effector with respect to enzyme activity, and they compete for the same site. Binding of dATP also promotes aggregation of ribonucleotide reductase.

The second effector site, the *substrate specificity site,* can bind either ATP, dTTP, dGTP, or dATP, and the substrate specificity of the enzyme is determined by which of these nucleotides occupies this site. If ATP is in the substrate specificity site, ribonucleotide reductase preferentially binds pyrimidine nucleotides (UDP or CDP) at its active site and reduces them to dUDP and dCDP. With dTTP in the specificity-determining site, GDP is the preferred substrate. When dGTP binds to this specificity site, ADP becomes the favored substrate for reduction. The rationale for these varying affinities is as follows (Figure 27.27): High [ATP] signals healthy energy status for the cell, a condition consistent with cell growth and division and, consequently, the need for DNA synthesis. Thus, ATP binds in the activity-determining site of ribonucleotide reductase, turning it on and promoting production of dNTPs for DNA synthesis. Under these conditions, ATP is also likely to occupy the substrate specificity site, so that UDP and CDP are reduced to dUDP and dCDP. As we shall soon see, both of these pyrimidine deoxynucleoside diphosphates are precursors to dTTP. Thus, elevation of dUDP and dCDP levels leads to an increase in [dTTP]. High dTTP levels increase the likelihood that it will occupy the substrate specificity site, in which case GDP becomes the preferred substrate, and dGTP levels rise. Upon dGTP association with the substrate specificity site, ADP is the favored substrate, leading to ADP reduction and the eventual accumulation of dATP. Binding of dATP to the overall activity site then shuts the enzyme down. In summary, the relative affinities of the three classes of nucleotide binding sites in ribonucleotide reductase for the various substrates, activators, and inhibitors are such that the formation of dNDPs

Energy status of cell is robust; [ATP] is high. Make DNA

1. ATP occupies activity site A: ribonucleotide reductase *ON*
2. ATP in specificity site S favors CDP or UDP in catalytic site C, ⟶ [dCDP], [dUDP]↑
3. dCDP, dUDP ⟶ ⟶ dUMP ⟶ dTMP ⟶ ⟶ dTTP
4. dTTP occupies specificity site S, favoring GDP or ADP in catalytic site C. GDP ⟶ dGDP ⟶ dGTP
5. dGTP occupies specificity site S, favoring ADP in catalytic site C ⟶ [dADP]↑
6. dATP replaces ATP in activity site A: ribonucleotide reductase *OFF*

***Figure* 27.27** Regulation of deoxynucleotide biosynthesis: The rationale for the various affinities displayed by the two nucleotide-binding regulatory sites on ribonucleotide reductase.

From dUDP: dUDP ⟶ dUTP ⟶ dUMP ⟶ dTMP

From dCDP: dCDP ⟶ dCMP ⟶ dUMP ⟶ dTMP

Figure **27.28** Pathways of dTMP synthesis. dTMP production is dependent on dUMP formation from dCDP and dUDP synthesis. If the dCDP pathway is traced from the common pyrimidine precursor, UMP, it will proceed as follows:

UMP ⟶ UDP ⟶ UTP ⟶ CTP ⟶ CDP ⟶ dCDP ⟶ dCMP ⟶ dTMP

proceeds in an orderly and balanced fashion. As these dNDPs are formed in amounts consistent with cellular needs, their phosphorylation by nucleoside diphosphate kinases produces dNTPs, the actual substrates of DNA synthesis.

27.8 Synthesis of Thymine Nucleotides

The synthesis of thymine nucleotides proceeds from other pyrimidine deoxyribonucleotides. Cells have no requirement for free thymine ribonucleotides and do not synthesize them. Small amounts of thymine ribonucleotides do occur in tRNA (an RNA species harboring a number of unusual nucleotides) but these Ts arise via methylation of U residues already incorporated into the tRNA. Both dUDP and dCDP can lead to dUMP formation, the immediate precursor for dTMP synthesis (Figure 27.28). Interestingly, formation of dUMP from dUDP passes through dUTP, which is then cleaved by *dUTPase,* a pyrophosphatase that removes PP_i from dUTP. The action of dUTPase prevents dUTP from serving as a substrate in DNA synthesis. An alternative route to dUMP formation starts with dCDP, which is dephosphorylated to dCMP and then deaminated by *dCMP deaminase* (Figure 27.29), leaving dUMP. dCMP deaminase provides a second point for allosteric regulation of dNTP synthesis; it is allosterically activated by dCTP and feedback-inhibited by dTTP. Note that, of the four dNTPs, only dCTP does not interact with either of the regulatory sites on ribonucleotide reductase (Figure 27.27). Instead it acts upon dCMP deaminase.

Synthesis of dTMP from dUMP is catalyzed by *thymidylate synthase* (Figure 27.30). This enzyme methylates dUMP at the 5-position to create dTMP; the methyl donor is the one-carbon folic acid derivative N^5,N^{10}-methylene-THF. The reaction is actually a reductive methylation in which the one-carbon unit is transferred at the methylene level of reduction and then reduced to the methyl level. The THF cofactor is oxidized at the expense of methylene reduction to yield dihydrofolate, or DHF. Dihydrofolate reductase then reduces

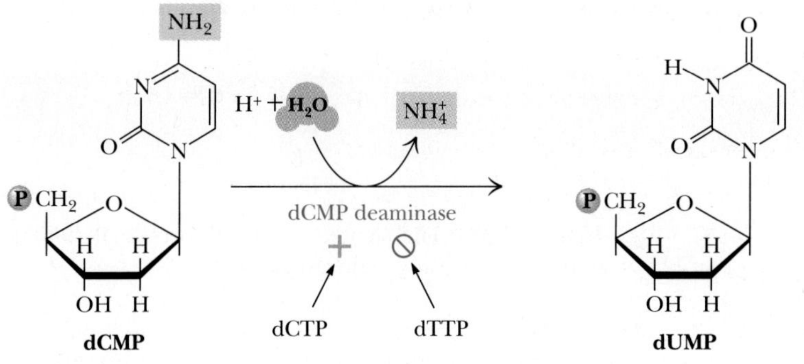

Figure **27.29** The dCMP deaminase reaction.

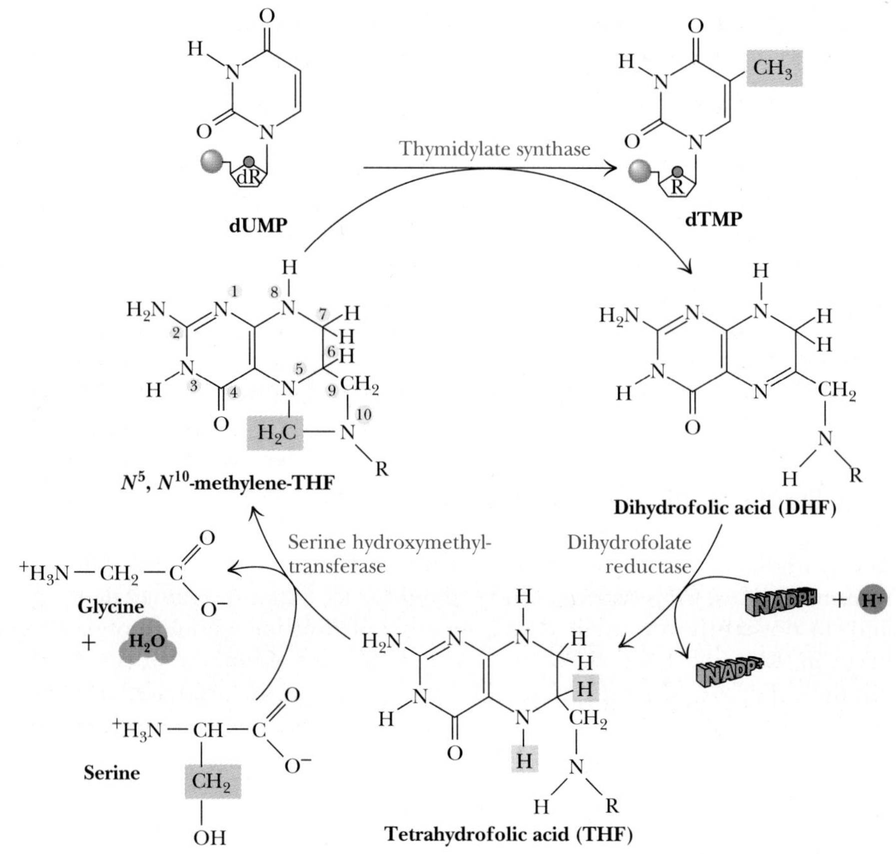

Figure 27.30 The thymidylate synthase reaction. The 5-CH_3 group is ultimately derived from the β-carbon of serine.

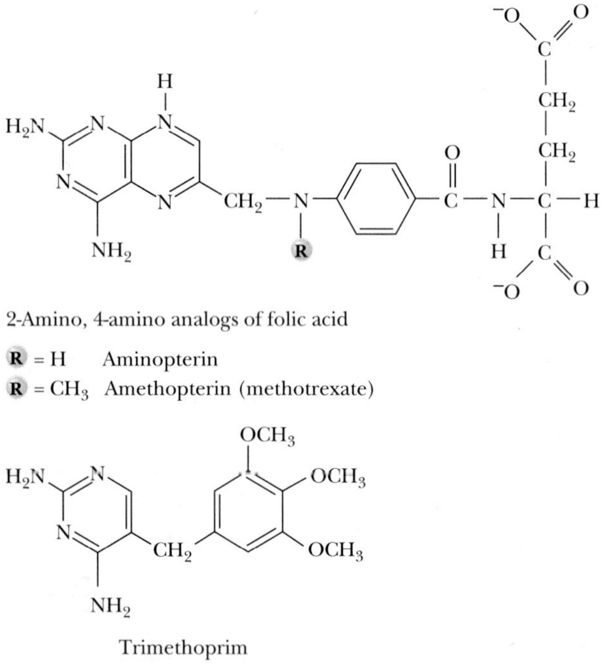

dUMP
Thymidylate synthase
dTMP
N^5, N^{10}-methylene-THF
Dihydrofolic acid (DHF)
Serine hydroxymethyl-transferase
Dihydrofolate reductase
NADPH + H^+
$NADP^+$
Glycine + H_2O
Serine
Tetrahydrofolic acid (THF)

2-Amino, 4-amino analogs of folic acid

R = H Aminopterin

R = CH_3 Amethopterin (methotrexate)

Trimethoprim

Figure 27.31 Precursors and analogs of folic acid employed as antimetabolites: sulfonamides (see Figure 27.5), methotrexate, aminopterin, and trimethoprim. The latter three compounds bind to dihydrofolate reductase with about one thousandfold greater affinity than DHF and thus act as virtually irreversible inhibitors.

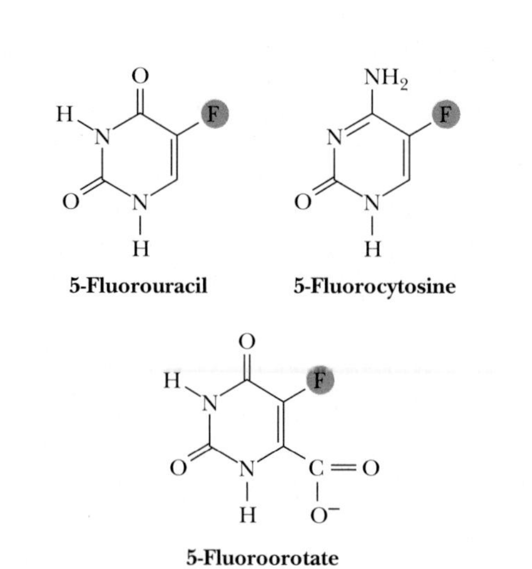

Figure 27.32 The structures of 5-fluorouracil (5-FU), 5-fluorocytosine, and 5-fluroorotate.

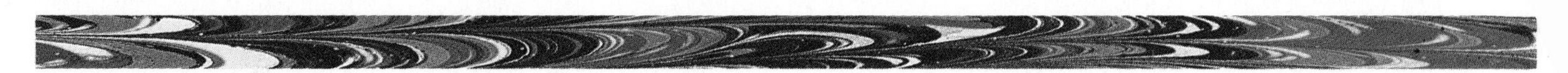

Critical Developments in Biochemistry

Enzyme Inhibition by Fluoro Compounds

Carbon–fluorine bonds are exceedingly rare in nature and fluorine is an uncommon constituent of biological molecules. F has three properties attractive to drug designers: (a) It is the smallest replacement for an H atom in organic synthesis, (b) fluorine is the most electronegative element, and (c) the F-C bond is relatively unreactive. This steric compactness and potential for strong inductive effects through its electronegativity renders F a useful substituent in the construction of inhibitory analogs of enzyme substrates. One interesting strategy is to devise fluorinated precursors that are taken up and processed by normal metabolic pathways to generate a potent antimetabolite. A classic example is *fluoroacetate* (see Chapter 19). FCH_2COO^- is exceptionally toxic because it is readily converted to fluorocitrate by citrate synthase of the citric acid cycle. In turn, fluorocitrate is a powerful inhibitor of aconitase. The metabolic transformation of an otherwise innocuous compound into a poisonous derivative is termed **lethal synthesis.** *5-Fluorouracil* and *5-fluorocytosine* are also examples of this strategy (see text).

Unlike hydrogen, which is often abstracted from substrates as H^+, electronegative fluorine cannot be readily eliminated as the corresponding F^+. Thus, enzyme inhibitors can be fashioned in which F replaces H at positions where catalysis involves H removal as H^+. Thymidylate synthase catalyzes removal of H from dUMP as H^+, through a covalent catalysis mechanism. A thiol group on this enzyme normally attacks the 6-position of the uracil moiety of 2′-deoxyuridylic acid so that C-5 can act as a carbanion in attack on the methylene carbon of N^5,N^{10}-methylene-THF (see figure). Regeneration of free enzyme then occurs through loss of the C-5 H atom as H^+ and dissociation of product dTMP. If F replaces H at C-5 as in 2′-deoxy-5-fluorouridylate (dFUMP), the enzyme is immobilized in a very stable ternary [enzyme: dFUMP:methylene-THF] complex and effectively inactivated. Enzyme inhibitors like dFUMP whose adverse properties are only elicited through direct participation in the catalytic cycle are variously called **mechanism-based inhibitors, suicide substrates,** or **Trojan horse substrates.**

E — Cys — SH

N^5, N^{10}-methylene- THF

Ternary complex

The effect of the 5-fluoro substitution on the mechanism of action of thymidylate synthase. An enzyme thiol group (from a Cys side chain) ordinarily attacks the 6-position of dUMP so that C-5 can react as a carbanion with N^5,N^{10}-methylene-THF. Normally, free enzyme is regenerated following release of the hydrogen at C-5 as a proton. Since release of fluorine as F^+ cannot occur, the ternary (three-part) complex of [enzyme:fluorouridylate:methylene-THF] is stable and persists, preventing enzyme turnover. (The N^5,N^{10}-methylene-THF structure is given in abbreviated form.)

DHF back to THF for service again as a one-carbon vehicle (see Chapter 14). Thymidylate synthase sits at a junction connecting dNTP synthesis with folate metabolism. It has become a preferred target for inhibitors designed to disrupt DNA synthesis. An indirect approach is to employ folic acid precursors or analogs as antimetabolites of dTMP synthesis (Figure 27.31). *5-Fluorouracil (5-FU;* Figure 27.32) is a thymine analog. It is converted to *5′-fluorouridylate in vivo* by a PRPP-dependent phosphoribosyltransferase, and passes through the reactions of dNTP synthesis, culminating ultimately as *2′-deoxy-5-fluorouridylic acid,* a potent inhibitor of dTMP synthase. 5-FU is used as a chemotherapeutic agent in the treatment of cancer. Similarly, *5-fluorocytosine* is used as an antifungal drug, since, unlike mammals, fungi can convert it to 2′-deoxy-5-fluorouridylate. Further, malarial parasites can use exogenous orotate to make pyrimidines for nucleic acid synthesis whereas mammals cannot. Thus, *5-fluoroorotate* is an effective antimalarial drug, since it is selectively toxic to these parasites.

Problems

1. Draw the purine and pyrimidine ring structures, indicating the metabolic source of each atom in the rings.
2. Starting from glutamine, aspartate, glycine, CO_2 and N^{10}-formyl-THF, how many ATP equivalents are expended in the synthesis of (a) ATP, (b) GTP, (c) UTP, and (d) CTP?
3. Illustrate the key points of regulation in (a) the biosynthesis of IMP, AMP, and GMP; (b) *E. coli* pyrimidine biosynthesis; and (c) mammalian pyrimidine biosynthesis.
4. Indicate which reactions of purine or pyrimidine metabolism are affected by the inhibitors (a) azaserine, (b) methotrexate, (c) sulfonamides, (d) allopurinol, and (e) 5-fluorouracil.
5. Since dUTP is not a normal component of DNA, why do you suppose ribonucleotide reductase has the capacity to convert UDP to dUDP?
6. Describe the underlying rationale for the regulatory effects exerted on ribonucleotide reductase by ATP, dATP, dTTP, and dGDP.
7. By what pathway(s) does the ribose released upon nucleotide degradation enter intermediary metabolism and become converted to cellular energy? How many ATP equivalents can be recovered from one equivalent of ribose?
8. At which steps does the purine biosynthetic pathway resemble the pathway for biosynthesis of the amino acid histidine?

Further Reading

Abeles, R. H., and Alston, T. A., 1990. Enzyme inhibition by fluoro compounds. *The Journal of Biological Chemistry* **265:**16705–16708. A brief review of the usefulness of fluoro derivatives in probing reaction mechanisms.

Benkovic, S. J., 1984. The transformylase enzymes in *de novo* purine biosynthesis. *Trends in Biochemical Sciences* **9:**320–322. These enzymes provide an instance of metabolic channeling in one-carbon metabolism.

Henikoff, S., 1987. Multifunctional polypeptides for purine *de novo* synthesis. *BioEssays* **6:**8–13.

Holmgren, A., 1989. Thioredoxin and glutaredoxin systems. *The Journal of Biological Chemistry* **264:**13963–13966.

Jones, M. E., 1980. Pyrimidine nucleotide biosynthesis in animals: Genes, enzymes and regulation of UMP biosynthesis. *Annual Review of Biochemistry* **49:**253–279.

Reichard, P., 1988. Interactions between deoxyribonucleotide and DNA synthesis. *Annual Review of Biochemistry* **57:**349–374. A review of the regulation of ribonucleotide reductase by the scientist who discovered these phenomena.

Scriver, C. R., Beaudet, A. L., Sly, W. S., and Valle, D., eds., 1989. *The Metabolic Basis of Inherited Disease,* vols. 1 and 2. New York: McGraw-Hill. Part 6 deals with inherited disorders of purine and pyrimidine metabolism, including gout, Lesch-Nyhan syndrome, and immunodeficiency disease caused by adenosine deaminase deficiency or purine nucleoside phosphorylase deficiency.

Srere, P. A., 1987. Complexes of sequential metabolic enzymes. *Annual Review of Biochemistry* **56:**89–124. A discussion of how the enzymes acting sequentially in a metabolic pathway are often organized into multienzyme complexes, or even synthesized as multifunctional proteins, especially in eukaryotes.

Stubbe, JoAnn, 1990. Ribonucleotide reductases: Amazing and confusing. *The Journal of Biological Chemistry* **265:**5329–5332. An analysis of the complex catalytic mechanism that underlies the overtly simple reaction mediated by ribonucleotide reductase.

Watts, R. W. E., 1983. Some regulatory and integrative aspects of purine nucleotide synthesis and its control: An overview. *Advances in Enzyme Regulation* **21:**33–51.

Wilson, D. K., Rudolph, F. B., and Quiocho, F. A., 1991. Atomic structure of adenosine deaminase complexed with a transition-state analog: Understanding catalysis and immunodeficient mutations. *Science* **252:**1279–1284.

Wyngaarden, J. B., 1977. Regulation of purine biosynthesis and turnover. *Advances in Enzyme Regulation* **14:**25–42.

Part IV

Genetic Information

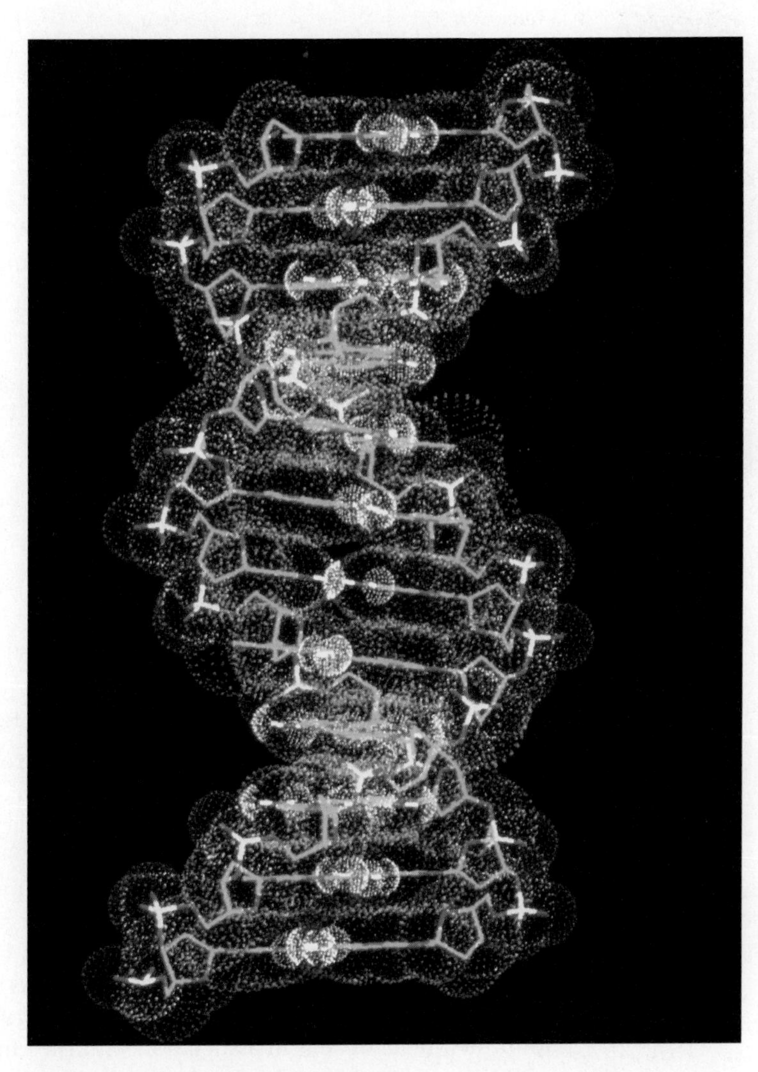

β-DNA, van der Waals model, side view.

ls now are ended. These our

told you, were all spirits and
ted into air, into thin air:
ike the baseless fabric of this vision,
cloud-capp'd towers, the gorgeous palaces,
The solemn temples, the great globe itself,
Yea, all which it inherit, shall dissolve
And, like this insubstantial pageant faded,
Leave not a rack behind. We are such stuff
As dreams are made on, and our little life
Is rounded with a sleep.

William Shakespeare (1564–1616) *The Tempest* (1611–1612)

Chapter 28

DNA: Genetic Information, Recombination, and Mutation

"Ahu Akivi," Easter Island, South Pacific

Outline

28.1 Genetic Information: The One-Gene, One-Enzyme Hypothesis

28.2 The Discovery That DNA Carries Genetic Information

28.3 Genetic Information in Bacteria: Its Organization, Transfer, and Rearrangement

28.4 The Molecular Mechanism of Recombination

28.5 The Immunoglobulin Genes: Generating Protein Diversity Using Genetic Recombination

28.6 The Molecular Nature of Mutation

28.7 RNA as Genetic Material

28.8 Transgenic Animals

The fact that DNA is the material of heredity is common knowledge today, even though no one could have successfully defended such a proposition as recently as 50 years ago. **Heredity,** which we can define generally as *the tendency of an organism to possess the characteristics of its parent(s),* was clearly evident throughout nature and since the dawn of history had served to justify the classification of organisms according to shared similarities. The molecular basis of heredity, however, was not obvious. Early geneticists demonstrated that **genes,** the elements or units carrying and transferring inherited characteristics from parent to offspring, are contained within the nuclei of cells in association with the chromosomes. Yet the chemical identity of genes remained unknown, and genetics was an abstract science. Even the realization that chromosomes are composed of proteins and nucleic acids did little to define the molecular nature of the gene, because at the time no one understood either of these substances.

Preliminary attempts to learn what genes are and what they do proved most fruitful in studies of the inheritance of metabolic disorders. The English physician A. E. Garrod suggested in 1909 that such diseases are "inborn errors in metabolism." These errors in metabolism arise because the enzyme responsible for a specific metabolic conversion is congenitally absent or defective. For example, in **phenylketonuria,** a hereditary disease in humans, afflicted individuals are unable to convert phenylalanine to tyrosine and, as a consequence, excrete excess amounts of phenylpyruvic acid in their urine (see Chapter 26). Most victims of phenylketonuria lack phenylalanine hydroxylase activity due to a defect in the gene for this enzyme. The consequence of phenylketonuria is severe mental retardation.

28.1 Genetic Information: The One-Gene, One-Enzyme Hypothesis

A direct relationship between genes and enzymes was established in 1941 by the work of George Beadle and Edward Tatum in their studies on the metabolism of the orange bread mold *Neurospora crassa.* This fungus normally grows very well on sucrose, inorganic salts (including a simple nitrogen source such as NH_4NO_3), and the vitamin biotin. Because this basic set of nutrients is all the fungus needs, it is called **minimal growth medium.** Beadle and Tatum irradiated spores of this fungus with X-rays to cause mutations (Figure 28.1a1) and then tested the growth requirements of progeny derived from the irradiated spores. Progeny were first grown on **complete growth medium,** consisting of minimal growth medium supplemented with all 20 amino acids and various other nutrients (Figure 28.1a2), and then tested for growth on minimal medium. Some failed to grow (Figure 28.1a3). Beadle and Tatum postulated that *N. crassa* strains unable to grow on minimal medium were defective in enzymes necessary for the biosynthesis of essential metabolites. By adding various amino acids to minimal medium and testing for growth (Figure 28.1a4), they identified mutant strains able to grow normally if provided with a specific nutritional supplement. They concluded that such mutant strains were no longer capable of synthesizing certain essential substances. Presumably, X-irradiation had caused a mutation in the gene responsible for an enzyme in the biosynthetic pathway for the needed supplement, since the nutritional requirement of a given strain was inherited by its progeny. Since mutants defective in specific steps of a metabolic pathway could be identified (Figure 28.1b), Beadle and Tatum were able to demonstrate a one-to-one correspondence between a mutation and the absence of a specific enzyme. Publication of this research marked the birth of biochemical genetics, but the molecular identity of genes still remained uncertain.

28.2 The Discovery That DNA Carries Genetic Information

The material of heredity should have certain properties:

1. It must be very stable so that genetic information can be stored in it and transmitted countless times to subsequent generations.
2. It must be capable of precise copying or replication so that its information is not lost or altered.

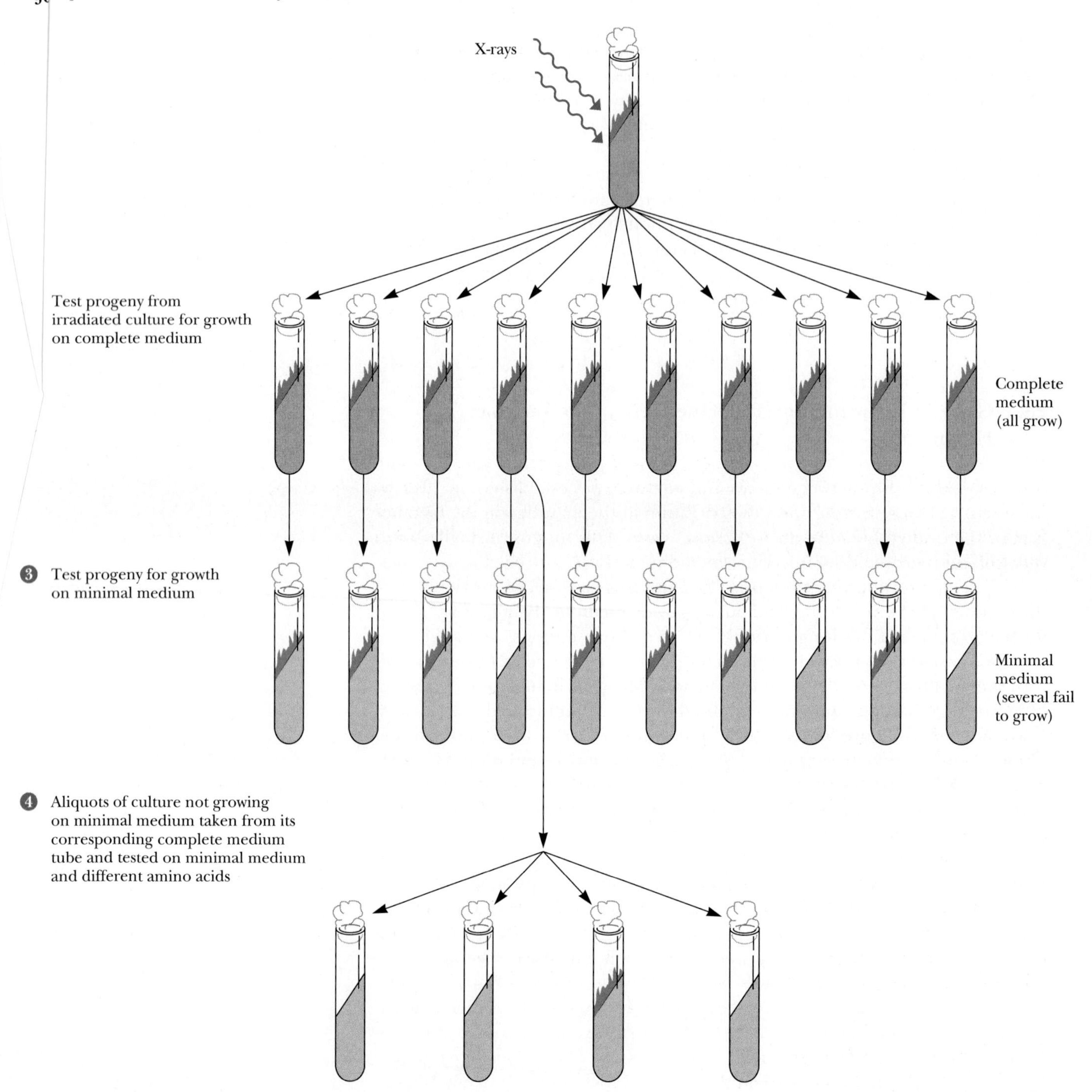
X-rays
Test progeny from irradiated culture for growth on complete medium
Complete medium (all grow)
3 Test progeny for growth on minimal medium
Minimal medium (several fail to grow)
4 Aliquots of culture not growing on minimal medium taken from its corresponding complete medium tube and tested on minimal medium and different amino acids

(b) Analysis of various *Arg* auxotrophs isolated by the above procedure:

	Mutants		
Growth on:	I	II	III
Minimal medium	NO	NO	NO
Minimal medium + ornithine	YES	NO	NO
Minimal medium + citrulline	YES	YES	NO
Minimal medium + arginine	YES	YES	YES

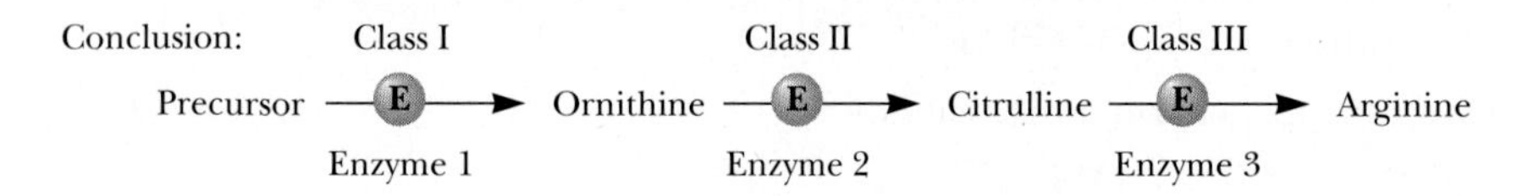
Conclusion:
Class I
Class II
Class III
Precursor
E
Ornithine
E
Citrulline
E
Arginine
Enzyme 1
Enzyme 2
Enzyme 3

3. Although stable, it must also be subject to change in order to account, in the short term, for the appearance of mutant forms and, in the long term, for evolution.

The first evidence that **deoxyribonucleic acid,** or **DNA,** might be the material of heredity came from investigations on *Streptococcus pneumoniae,* one of the types of bacteria that cause pneumonia. In 1928, Frederick Griffith, an English microbiologist, was comparing the properties of two strains of pneumococcus bacteria. One strain, **Type S** (S for smooth colonial morphology), is virulent because it is enclosed within a slippery polysaccharide coat, or capsule, that protects it from the immune system of its host. The other strain, **Type R** (R for rough-looking colonies), lacks an enzyme for the biosynthesis of the polysaccharide coat and is not virulent because it cannot resist attack by the host's immune system. When Griffith injected Type S bacteria into mice, the blood became filled with S bacteria and the mice died. Heat-killed Type S bacteria had no effect on the mice, but if mice were injected with nonvirulent Type R bacteria that had been mixed with heat-killed Type S bacteria, the mice died and virulent Type S bacteria could be recovered from their blood. Somehow, the heat-killed Type S bacteria had transformed the nonvirulent R Type into the virulent S Type (Figure 28.2). In 1931, M. H. Dawson and R. H. P. Sia showed that extracts of heat-killed Type S cells could transform nonpathogenic R cells into genetically stable, pathogenic S cells.

The "Transforming Principle" Is DNA

In 1944, Oswald T. Avery and his associates Colin M. MacLeod and Maclyn McCarty at the Rockefeller Institute made the discovery that the substance active in transforming Type R bacteria to virulence was, in fact, DNA. This finding was surprising and not immediately accepted, since most scientists at the time thought that proteins, substances chemically more complex and diverse than nucleic acids, were the genetic material. Avery, MacLeod, and McCarty showed that highly purified preparations of "transforming principle" contained no detectable protein and were unaffected by trypsin or chymotrypsin (two proteolytic enzymes) or by pancreatic RNase (which hydrolyzes RNA). However, the transforming substance was readily inactivated by treatment with pancreatic DNase, an enzyme that specifically degrades DNA. Thus, DNA must have been the agent carrying the information that transforms R bacteria to virulence. Since transformation was stably inherited, DNA merited strong consideration as the actual material of heredity.

◀ ***Figure* 28.1** Beadle and Tatum's one-gene, one-enzyme concept. (a) Mutant strains unable to grow on minimal medium unless supplemented with an essential metabolite, such as an amino acid, were identified. Such mutations are said to be **auxotrophic.** A mutant is auxotrophic for arginine, for example, if it requires Arg for growth. (Strains requiring no nutritional supplements are called **prototrophic,** meaning they are able to grow on minimal medium.) (b) A number of mutants unable to grow unless provided with Arg in the medium could be further subdivided into different classes depending on their ability to grow on citrulline and ornithine, compounds known to be intermediates in the Arg biosynthetic pathway (see Chapter 26). Assuming that each mutant was defective in a single gene, Beadle and Tatum postulated that the different mutant classes each lacked a different enzyme of Arg synthesis: Class I mutants lacked an enzyme responsible for catalyzing the formation of ornithine, "enzyme 1"; class II mutants lacked the enzyme mediating the formation of citrulline from ornithine, "enzyme 2"; and class III mutants could not convert citrulline to arginine, owing to a defect in "enzyme 3." Thus was born the "one-gene, one-enzyme" hypothesis.

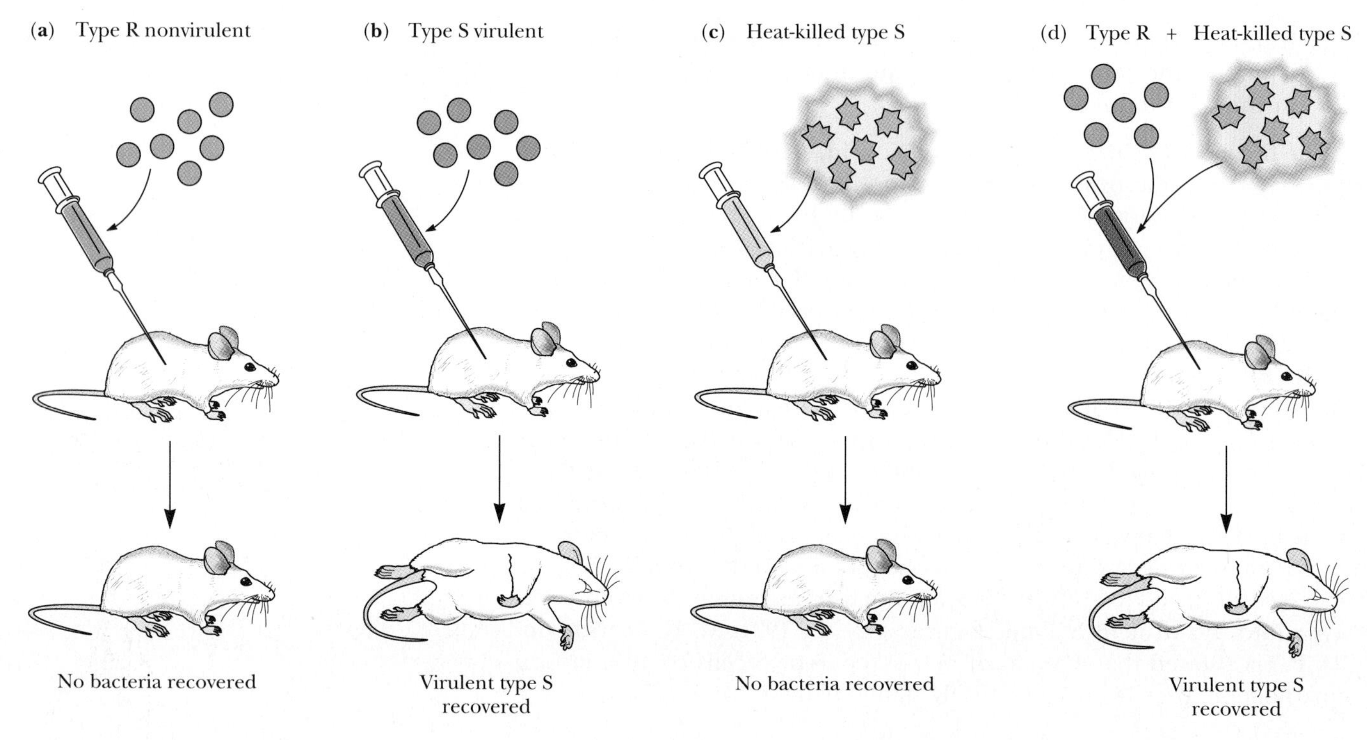

***Figure* 28.2** Griffith experiment on pneumococcal transformation. (a) Mice are resistant to Type R *Streptococcus pneumoniae bacteria* but (b) are killed by injection with virulent Type S *S. pneumoniae* bacteria. (c) Injection with heat-killed virulent bacteria does not kill mice, but (d) if heat-killed Type S bacteria are mixed with nonvirulent Type R bacteria, they have the capacity to *transform* nonvirulent Type R bacteria into the virulent Type S form.

DNA Is the Hereditary Molecule of Bacteriophage

Further proof that DNA is the material of heredity came from the study of bacteriophage. In 1952, Alfred Hershey and Martha Chase devised an elegant experiment to trace the fates of the two major components of bacteriophage—coat protein and DNA—following infection. They took advantage of the fact that nucleic acids lack sulfur and proteins lack phosphorus to uniquely label bacteriophage DNA with ^{32}P and bacteriophage protein with ^{35}S. Bacteriophage labeled with either isotope were obtained from cultures of *bacteriophage T2* grown on *Escherichia coli* in medium containing radioactive ^{32}P-labeled inorganic phosphate or radioactive ^{35}S-labeled methionine.

Phage infection of bacteria involves attachment of the bacteriophage to the bacterial cell at specific attachment sites. The phage DNA enters the bacterial cell, leaving its protein coat behind on the surface of the bacterium (Figure 28.3). Hershey and Chase mixed labeled bacteriophage T2 with unlabeled *E. coli* cells, permitting sufficient time for the phage to attach. Then they vigorously agitated the culture in a blender to shear the phage coats from the bacterial surface. Following centrifugation of the culture, infected bacteria could be recovered in the pellet, whereas the phage coats containing most of the ^{35}S label remained suspended in the supernatant. In contrast, when *E. coli* cells were infected with ^{32}P-labeled T2 phage, the bacterial pellet contained most of the ^{32}P. Furthermore, upon lysis, 30% of the original ^{32}P but only 1% of the ^{35}S was recovered in the bacteriophage progeny produced by the infection (Figure 28.4). Hershey and Chase surmised that the bacteriophage DNA was sufficient for bacteriophage reproduction. That is, DNA must be the material of heredity.

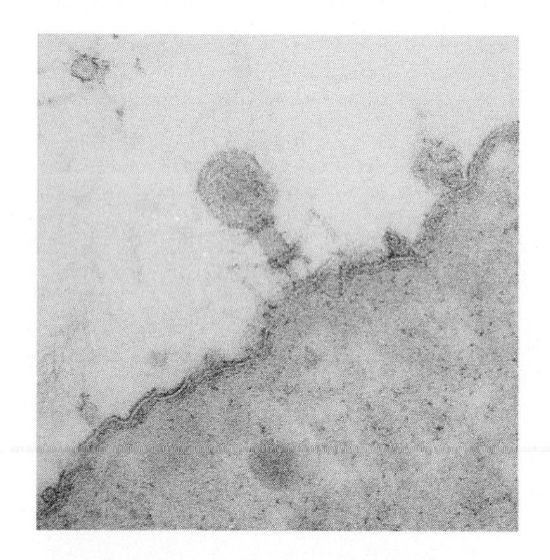

***Figure* 28.3** Electron micrograph of bacteriophage particle attached to a bacterial cell. A single T2 bacteriophage weighs 5×10^{-13} g and consists of 60% DNA and 40% protein. Its volume is about 1/1000 the volume of an *E. coli* cell. T2 phage heads are 100 nm × 65 nm icosahedra attached to tails 100 nm long by 25 nm in diameter.

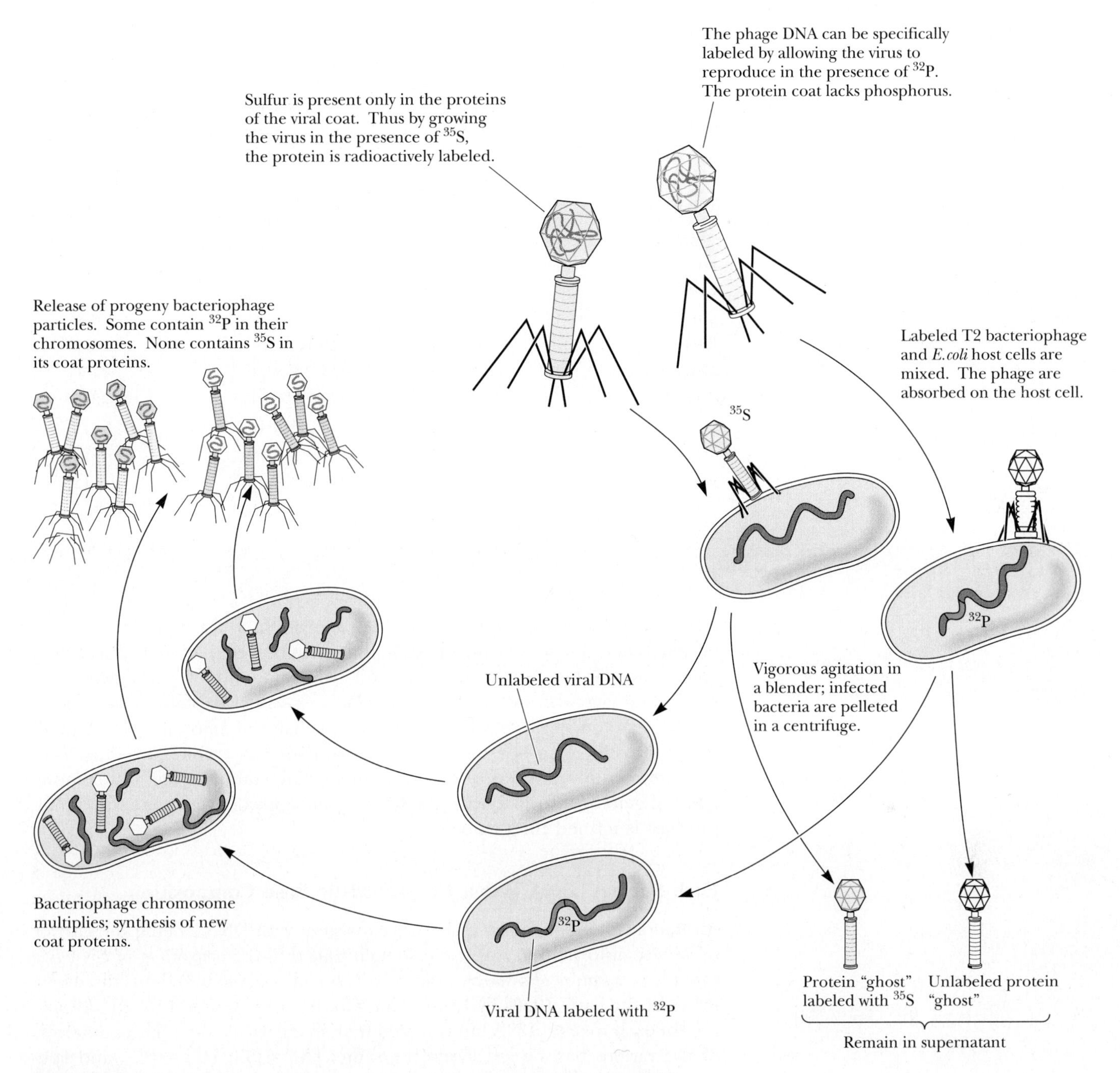
The phage DNA can be specifically labeled by allowing the virus to reproduce in the presence of ^{32}P. The protein coat lacks phosphorus.
Sulfur is present only in the proteins of the viral coat. Thus by growing the virus in the presence of ^{35}S, the protein is radioactively labeled.
Release of progeny bacteriophage particles. Some contain ^{32}P in their chromosomes. None contains ^{35}S in its coat proteins.
Labeled T2 bacteriophage and *E. coli* host cells are mixed. The phage are absorbed on the host cell.
^{35}S
^{32}P
Unlabeled viral DNA
Vigorous agitation in a blender; infected bacteria are pelleted in a centrifuge.
Bacteriophage chromosome multiplies; synthesis of new coat proteins.
^{32}P
Viral DNA labeled with ^{32}P
Protein "ghost" labeled with ^{35}S
Unlabeled protein "ghost"
Remain in supernatant

***Figure* 28.4** The Hershey and Chase experiment demonstrated that the DNA component of bacteriophage T2 carried the requisite genetic information for bacteriophage reproduction.

The Quantity and Composition of DNA Per Cell Is Constant

Other evidence consistent with the role of DNA as the repository of genetic information accumulated. First, the amount of DNA found in the cell(s) of any species of organism is the same from cell to cell, changing only to double in amount prior to cell division, whereupon half is allocated to each daughter cell. In contrast to other biomolecules such as metabolites, proteins, or even RNA, whose amounts often vary depending on the metabolic or nutritional state of the cell, DNA content seems to be fixed, as might be expected for the material of heredity. Also, as a general rule, the amount of DNA in the cells of

Table **28.1**

The DNA Content of Various Cells and Viruses

Species	Base Pairs in Haploid Genome	pg DNA Per Cell (or Virion)	Approximate Number of Genes
Flowering plants	$0.3–100 \times 10^9$	0.6–24	<50,000
Mammals	$2–4 \times 10^9$	4–9	<25,000
Amphibia	$0.9–80 \times 10^9$	2–175	
Fish	$0.3–3 \times 10^9$	0.6–6	
Crustacea	$0.1–5 \times 10^9$	0.2–5.5	
Fungi	$1–3 \times 10^7$	$1.1–3.3 \times 10^{-2}$	~4,000
Bacteria	$2–9 \times 10^6$	$2.2–10 \times 10^{-3}$	~2,000
Mycoplasma*	$0.6–2 \times 10^6$	$0.7–2.2 \times 10^{-3}$	~750
dsDNA viruses	$5–200 \times 10^3$	$0.6–22 \times 10^{-5}$	6–300
ssDNA viruses	$\sim 5 \times 10^3$	$\sim 6 \times 10^{-6}$	5–12
dsRNA viruses	$\sim 20 \times 10^3$	$\sim 2 \times 10^{-5}$	~20
ssRNA viruses	$1–20 \times 10^3$	$0.1–2.2 \times 10^{-5}$	1–12

*Mycoplasma are the smallest bacteria.

different species tends to be somewhat proportional to the complexity of the organism and the genetic information that must be necessary to support such complexity (Table 28.1). Moreover, the DNA content of **germ cells** (the sperm and eggs) of multicellular organisms, so-called **haploid cells** because they contain only one set of chromosomes, is half the amount of DNA in all the other cells of an organism (so-called **somatic cells,** which are **diploid** in that they contain two sets of chromosomes). The complete haploid complement of genetic information, representing one copy of every gene the organism has, is termed the **genome.**

Each Species' DNA Has a Characteristic Base Composition

An important biochemical discovery consistent with DNA's being the carrier of genetic information was the evidence that *the base composition of DNA is a characteristic feature of a species* (Table 28.2; see also Table 6.3). This discovery, made in the early 1950s by Erwin Chargaff, not only pointed to the uniqueness of each species' DNA but revealed that, in any DNA, the relative amounts of the various bases were always the same: [A] = [T], [G] = [C], and [purines] = [pyrimidines]. These concentration relationships provided a crucial clue for Watson and Crick in their elucidation of the double helical nature of DNA (see Chapter 6).

28.3 Genetic Information in Bacteria: Its Organization, Transfer, and Rearrangement

Bacteria are very useful organisms for genetic analysis: Under optimal conditions of growth and reproduction, some bacteria (such as *E. coli*) divide every 20 minutes, the progeny of each division being a new generation. A genetic experiment can be completed with bacteria in hours, while an analogous experiment with a multicellular organism would take months or years because the generation times of such organisms are months or even years in duration.

Table **28.2**

Base Composition of DNA from Various Organisms

	Base Composition (Mole %)					
Organism	**A**	**G**	**C**	**5-Methyl C**	**T**	$\frac{A + T}{G + C + 5mC}$ **Ratio**
Animals						
Human	29.3	20.7	19.9	0.7	30.0	1.43
Cow	27.3	22.5	21.2	1.3	27.7	1.22
Rat	28.6	21.4	20.5	1.1	28.4	1.33
Chicken	28.0	22.0	21.6	—	28.4	1.29
Frog	26.3	23.5	21.8	2.0	26.8	1.12
Salmon	28.0	22.0	20.0	1.8	27.8	1.27
Sea urchin	28.4	19.5	19.3	—	32.8	1.58
Fruit fly	30.7	19.6	20.2	—	29.5	1.51
Plants						
Corn	26.8	22.8	17.0	6.2	27.2	1.17
Carrot	26.7	23.1	17.3	5.9	26.9	1.16
Clover	29.9	21.0	15.6	4.8	28.6	1.41
Fungi						
Neurospora crassa	23.0	27.1	26.6	—	23.3	0.86
Bacteria						
Escherichia coli	23.8	26.8	26.3	—	23.1	0.88
Staphylococcus aureus	30.8	21.0	19.0	—	29.2	1.50
Bacteriophage						
T7	26.0	23.8	23.6	—	26.6	1.11
T4	32.3	17.6	—	16.7*	33.4	1.92
λ	21.3	28.6	27.2	—	22.9	0.79

*In bacteriophage T4, all of the C is in the hydroxymethyl form.

Further, a single milliliter of bacterial culture can contain enormous numbers of bacteria—as many as 10^{10}—all derived from a single parental bacterium:

> A single bacterium growing with a generation time of 20 min can give rise to 10^{10} progeny in less than 11 hr. N, the number of cells after n number of generations, is given by $N = 2^n$. For $N = 10^{10} = 2^n$, $n = 33.22$ ($2^{33.22} = 10^{10}$). At 0.33 hr per generation, 33.22 generations (the time to accumulate 10^{10} cells from a single bacterium) occur in about 11 hours.

Because of these vast numbers, very rare genetic events can be observed. That is, a one-in-a-million occurrence could be present in thousands of bacteria in a culture. In addition, because bacteria are haploid organisms, each cell contains but one set of genetic instructions. Consequently, any mutation in a gene is not masked or corrected by a second, normal copy of the gene, as it usually is in diploid organisms. In haploid organisms like bacteria, the **phenotype,** or perceptible characteristics of the organism, reflects its **genotype,** or genetic composition. In contrast, diploid organisms may exhibit a **wild-type,** or **normal, phenotype** for any trait, even though their genotype might contain one mutant copy and one wild-type copy of the gene responsible for the trait.

Identifying Bacterial Mutants

Beadle and Tatum's discovery of nutritional mutants in *Neurospora* paved the way in the early 1940s for analogous studies with bacteria such as *E. coli*. Numerous techniques for identifying and recovering such mutants were developed, and virtually every aspect of bacterial metabolism and function became accessible to genetic analysis. Mutants dependent on exogenous supplementation with particular amino acids, nitrogenous bases, or vitamins were soon obtained. One interesting strategy led to the selection of mutants that grew normally under usual conditions but were unable to grow if these conditions were modified. A prime example of these **conditional lethal mutants** are **temperature-sensitive mutants,** which can grow at one temperature (the **permissive temperature**) but are unable to grow at another temperature (the **restrictive temperature**). Typically, these temperature-sensitive mutants can grow at 30°C but not at 42°C because some essential protein is rendered slightly unstable as a consequence of the mutation and denatures at somewhat lower temperature than normal. Another powerful strategy was to devise conditions where only mutant cells would grow. For example, nontoxic substrate analogs may be converted to toxic products by enzyme action. Cells mutant or otherwise defective in the responsible enzyme would be **analog resistant.** For example, chlorate (ClO_3^-), an analog of nitrate (NO_3^-), is reduced by the enzyme nitrate reductase (see Chapter 26) to chlorite (ClO_2^-), a very toxic substance. Selection for *E. coli* cells able to grow in the presence of chlorate has led to the identification of seven mutant classes defective in the expression of nitrate reductase activity (Figure 28.5).

Mapping the Structure of Bacterial Chromosomes

In 1946, Joshua Lederberg and Edward Tatum discovered that genetic information could be transferred between bacteria. They used two strains of *E. coli* that differed in their growth requirements due to mutations each carried

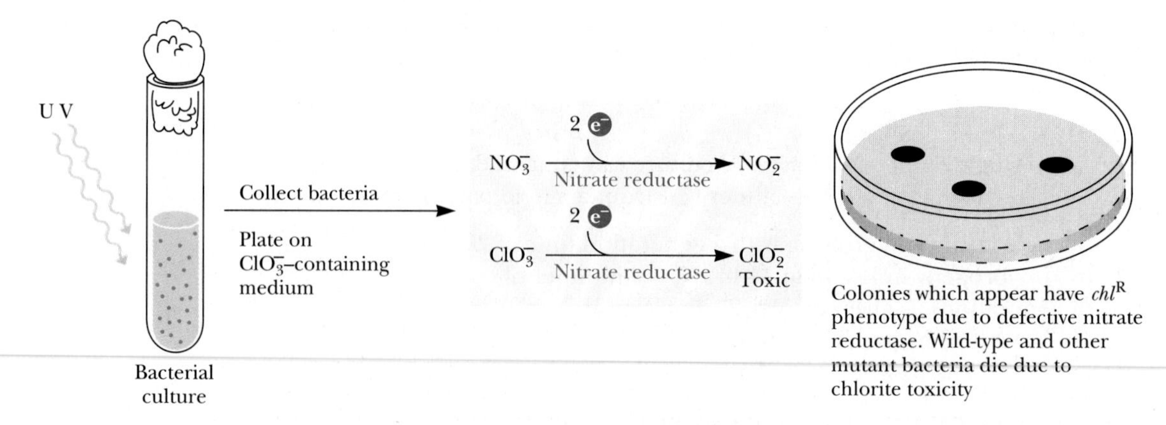

***Figure* 28.5** Chlorate as an example of a positive mutant selection scheme for the isolation of mutants defective in nitrate reductase activity. Bacteria previously exposed to an agent causing mutations are plated onto medium containing chlorate (ClO_3^-), an analog of nitrate (NO_3^-). Chlorate is an alternative substrate for nitrate reductase, which reduces it to chlorite (ClO_2^-), a toxic agent that kills any cells capable of expressing nitrate reductase activity. Only mutant cells defective in the formation of active nitrate reductase survive and grow on the plate. This **chlorate resistance** (chl^R) selection scheme has led to the identification of at least seven different genes in *E. coli* that affect nitrate reductase activity, the so-called *chl* genes.

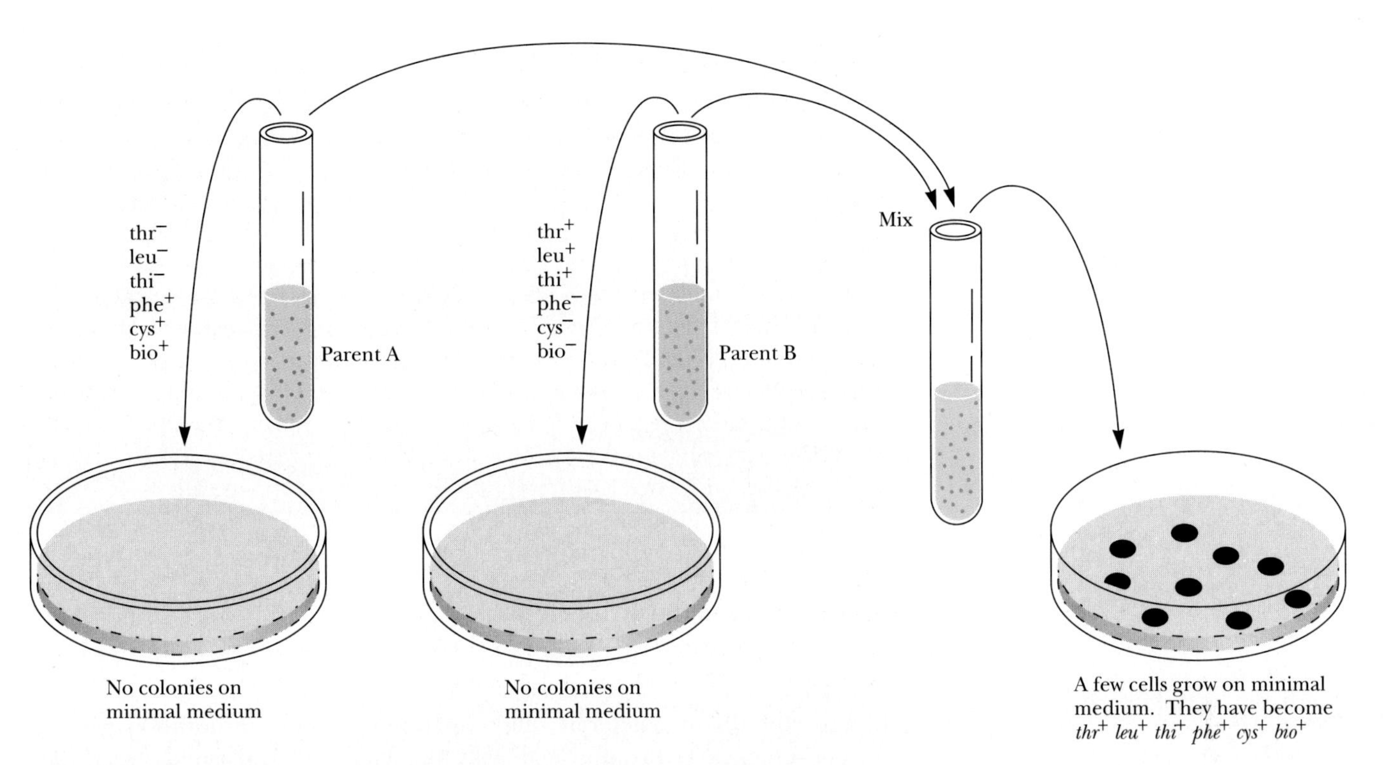

***Figure* 28.6** The use of nutritional mutants to demonstrate sexuality in bacteria. The genetic markers are thr^-, leu^-, thi^- (inability to grow in the absence of threonine, leucine, and thiamine, respectively) on one chromosome (Parent A), and phe^-, cys^-, bio^- (defects in genes for phenylalanine, cystine, and biotin synthesis, respectively) on the other chromosome (Parent B). A very few of the bacteria from the mixed culture grew; these bacteria have become thr^+ leu^+ thi^+ phe^+ cys^+ bio^+ as a result of **genetic recombination:** the formation of chromosomes with combinations of gene types different from those found in the parental chromosomes. That is, genetic recombination is the process of forming new combinations of genes. The production of offspring through sex is one mechanism for genetic recombination.

(Figure 28.6). One strain (thr^-, leu^-, thi^-) required threonine, leucine, and thiamine to grow; the other (phe^-, cys^-, bio^-) required phenylalanine, cystine, and biotin. These two strains were mixed together and spread on the surface of a petri plate of minimal medium lacking any of the required supplements. After a day, a very small number of bacterial colonies were observed to be growing. Somehow, these growing bacteria had acquired functional (wild-type) copies of each of the mutant genes. This remarkable result suggested strongly that the chromosomes of the two different cell types were brought together in a process akin to sexual exchange. In order for the progeny cells (which contain but one chromosome) to have acquired genetic information from the parental strains, **genetic recombination** must have occurred. This represents, in the words of Lederberg and Tatum, "the assortment of genes in new combinations." Apparently, at some point in time, parental DNA molecules must have aligned along regions of homology (sequence similarity), and segments from one of these molecules must have been interchanged with similar segments from the other parent so that some DNA molecules (chromosomes) now carried wild-type thr^+ leu^+ thi^+ phe^+ cys^+ bio^+ genes (Figure 28.6). Lederberg and Tatum speculated that, in order for the various genes to have had the opportunity to recombine, the cells of one strain must have interacted with the cells of the other.

***Figure* 28.7** Electron micrograph of two *E. coli* cells, one F^+, the other F^-, joined in sexual conjugation.

Sexual Conjugation in Bacteria

The transfer of DNA between bacteria takes place via a process known as **sexual conjugation,** a phenomenon unsuspected prior to the Lederberg-Tatum experiment. Bacterial cells sometimes contain, in addition to their chromosome, extrachromosomal DNA molecules called **plasmids** (see Chapter 8). Plasmids represent "extra" or auxiliary genetic information. Bacterial cells are capable of conjugation if they possess a particular plasmid called the **F factor** (F for fertility). Such *F^+*, or *donor, cells* have thin, hollow tubes projecting from their surface known as **sex pili** or **F pili** (singular = *pilus*). One or more pili can bind to specific receptors on the surface of cells that lack an F factor (*F^-*, or *recipient, cells;* Figure 28.7). The pilus provides a connection between the two cells. Upon conjugation, a single strand of the F factor is passed to the F^- cell, where its complementary strand is synthesized (Figure 28.8). The recipient F^- cell thus becomes F^+ by virtue of now having a double-stranded F factor plasmid. The F factor plasmid consists of about 94,000 base pairs; about one-third of this DNA is devoted to about 25 genes that function specifically in the transfer of genetic material from F^+ to F^- cells. Among these genes are those necessary for the formation of pili. In reality, the F factor is an infectious agent.

High Frequency of Recombination

In rare instances, the F factor will integrate into the bacterial host chromosome. (Plasmids capable of chromosomal integration are termed **episomes.**) Cells harboring F factor integrated into the chromosome show a much higher frequency of recombination of chromosomal genes upon conjugation, or "mating," with F^- cells and so are referred to as ***Hfr* cells,** for "high frequency of recombination." In *Hfr* cells, the conjugal process determined by the F factor operates as it does when the F factor is acting autonomously (Figure 28.9). That is, a single strand is passed to the recipient F^- cell, where its complementary strand is synthesized. However, because of its integrated position within the *Hfr* chromosome, the F factor carries along genes adjacent to it on the chromosome. If conjugation continues long enough, a single-stranded copy of the entire host chromosome is passed to the F^- cell. However, conjugation rarely persists the 100 minutes or more required for complete transfer, so usually only part of the *Hfr* chromosome is transferred.

Genes from the *Hfr* chromosomes are transferred into the F^- cell in a fixed order. Therefore, the order of the genes along the *Hfr* chromosome

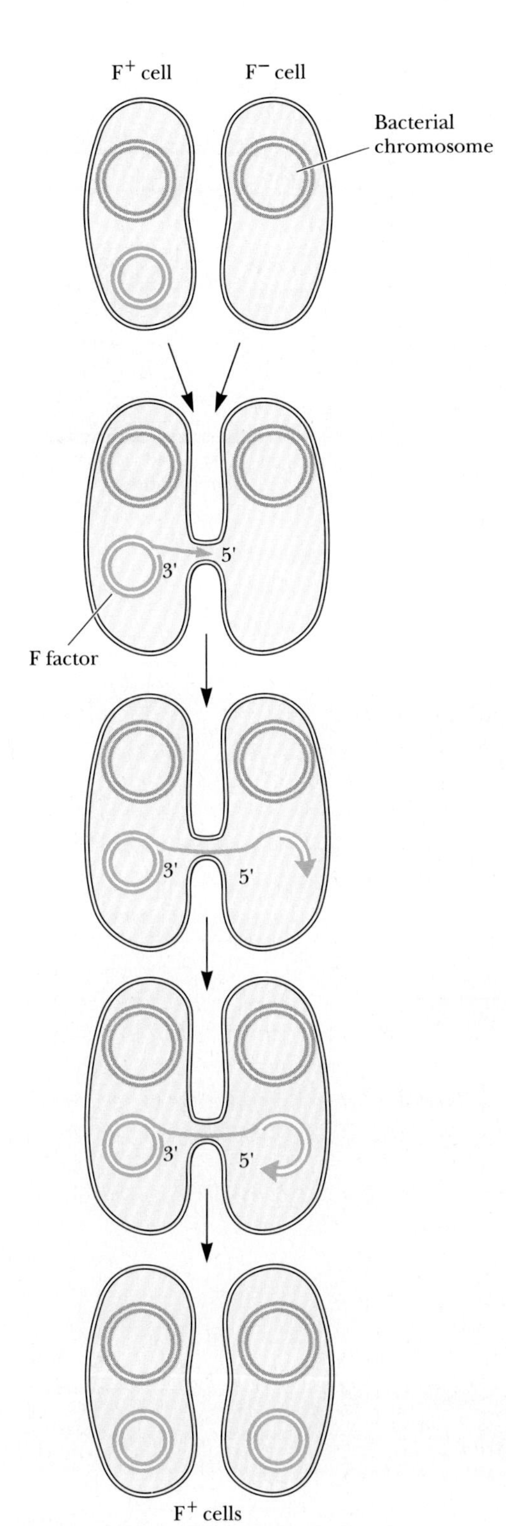

***Figure* 28.8** Diagram showing the transfer of F factor from an F$^+$ to an F$^-$ cell. A single strand of the F factor is nicked and transferred into the recipient F$^-$ cell. The complementary strand is then synthesized within the recipient F$^-$ cell to create a new double-stranded F factor, transforming the F$^-$ cell into an F$^+$ one.

must be fixed, and this order can be mapped by the technique of **interrupted mating** (Figure 28.9). Further, since the F factor is integrated at different sites in different *Hfr* strains of *E. coli,* genes difficult to map because they were transferred very late (and hence rarely) in one *Hfr* strain are readily mapped in another. The genetic map obtained by the interrupted mating method reveals a circular arrangement of genes, consistent with the circular organization of the *E. coli* chromosome (Figure 28.10). Other bacterial chromosomes show a similar circular organization.

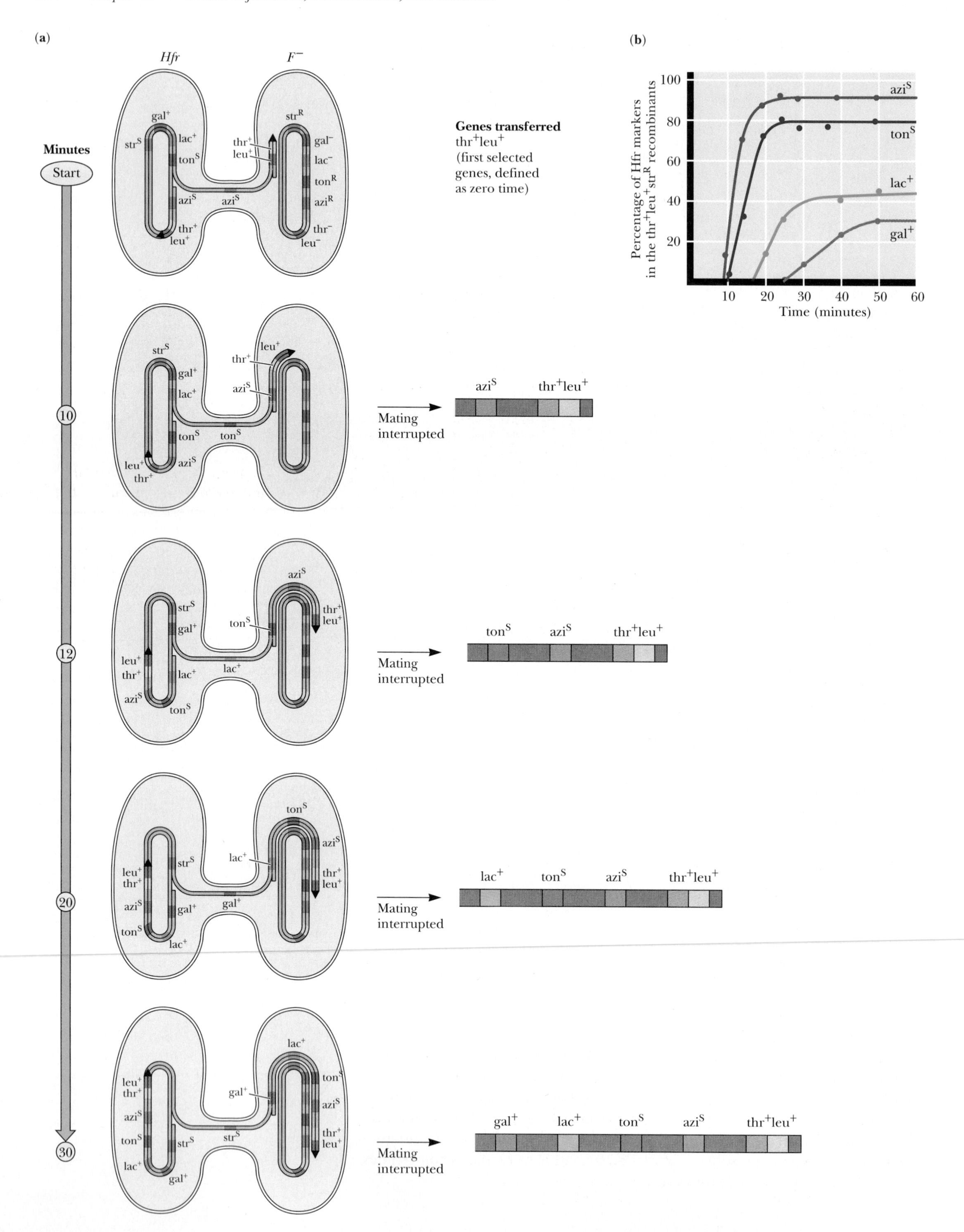
(a)
Hfr
F⁻
Minutes
Start
Genes transferred
thr⁺leu⁺
(first selected genes, defined as zero time)
10
Mating interrupted
aziˢ thr⁺leu⁺
12
Mating interrupted
tonˢ aziˢ thr⁺leu⁺
20
Mating interrupted
lac⁺ tonˢ aziˢ thr⁺leu⁺
30
Mating interrupted
gal⁺ lac⁺ tonˢ aziˢ thr⁺leu⁺
(b)
Percentage of Hfr markers in the thr⁺leu⁺strᴿ recombinants
Time (minutes)
aziˢ
tonˢ
lac⁺
gal⁺

◀ ***Figure* 28.9** Transfer of segments of the bacterial chromosome from the donor *Hfr* to the recipient F$^-$ cell. Since complete transfer of the *Hfr* chromosome rarely happens and since replication begins at a site *within* the F factor, transfer of the entire F factor is seldom achieved. Thus, the recipient cell usually remains F$^-$. The *E. coli* chromosome can be mapped by interrupted mating of *Hfr* strains with F$^-$ strains: The genetic markers here are *thr*$^-$, *leu*$^-$ (requirement for threonine, or leucine, in order to grow), *gal*$^-$, *lac*$^-$ (inability to grow on galactose, or lactose, as sole carbon source), *azi*^{R} (resistance to azide), *ton*^{R} (resistance to bacteriophage T1), and *str*^{R} (resistance to streptomycin). Superscripts "$^+$, ^{R}, ^{S}" denote wild-type, resistance, and sensitivity, respectively. (a) Ordered transfer of the *Hfr* chromosome during mating. Mating is interrupted by the shearing of the joined cells in a blender at chosen intervals; this process separates cells at various stages in the transfer of the *Hfr* chromosome. The cells are then plated onto selective medium and scored for their sensitivity to bacteriophage T1 and azide, and their ability to grow on galactose or lactose as sole carbon source. (b) The frequencies of genetic markers *azi*^{S}, *ton*^{S}, *lac*$^+$, and *gal*$^+$ among the recombinants as a function of mating time. Extrapolation to zero gives an indication of when the various markers enter the recipient cell.

(Adapted from Jacob, F., and Wollman, E., 1961. Sexuality and the Genetics of Bacteria, New York: Academic Press, p. 135.)

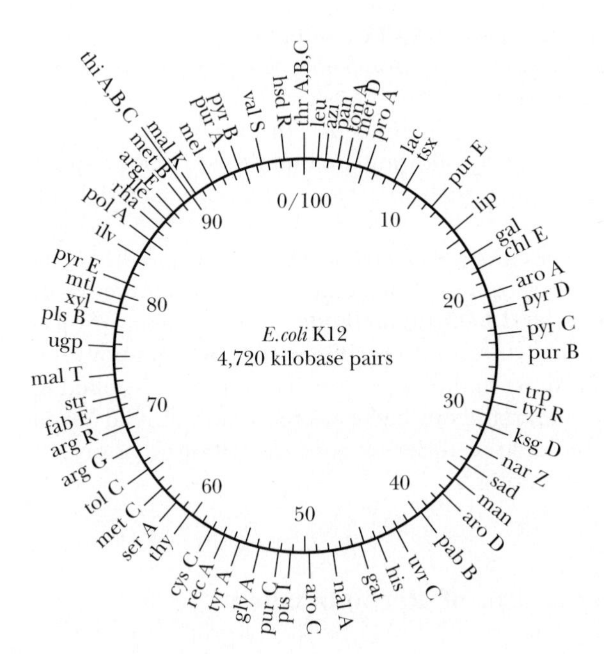

***Figure* 28.10** The genetic map of the *E. coli* chromosome. This circular map is divided into 100 minutes. The 100 minutes arose historically as the time period necessary for complete gene transfer in interrupted mating experiments. The marker *thrA* is arbitrarily chosen as minute 0. The most recent linkage map for *E. coli* (see B. J. Bachmann, 1990. *Microbiological Reviews* **54**:130–197) lists the positions of 1403 separate genetic loci on the circular *E. coli* chromosome; it has been estimated that *E. coli* has 2800 to 3000 genes.

Transduction

In the course of bacterial infection by certain bacterial viruses (bacteriophages), segments of the bacterial chromosome may become accidentally packaged into bacteriophage particles. Mistakes of this type are uncommon, usually accounting for only 0.01% of the bacteriophage progeny. Although they are able to attach to host cells and to inject the DNA they carry, such bacteriophage particles are incapable of reproduction because they lack part or all of the viral chromosome. Nevertheless, these defective phage particles provide a route for the passage of bacterial genes to new host cells in which they can recombine with the host chromosome. This mode of genetic transmission is known as **transduction.** For example, bacteriophage P1 particles obtained following infection of an *E. coli* host capable of metabolizing lactose (*lac*$^+$ cells) will have among their number a few phage particles carrying the genes for lactose utilization (Figure 28.11). If the DNA within such a phage is

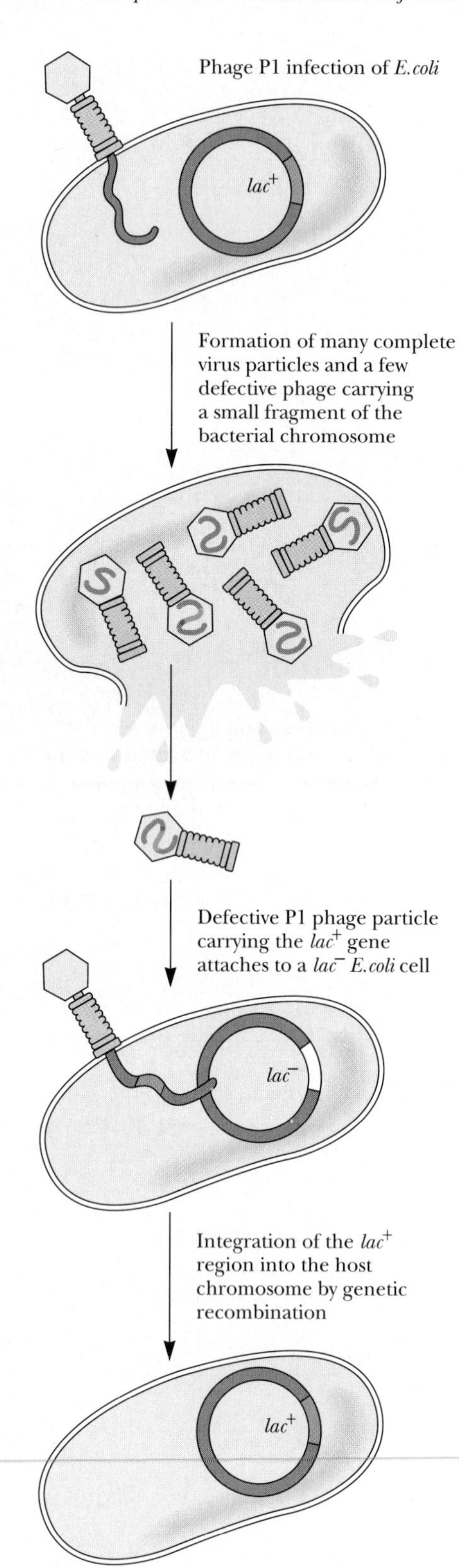

Phage P1 infection of E.coli
lac+
Formation of many complete virus particles and a few defective phage carrying a small fragment of the bacterial chromosome
Defective P1 phage particle carrying the lac+ gene attaches to a lac− E.coli cell
lac−
Integration of the lac+ region into the host chromosome by genetic recombination
lac+

***Figure* 28.11** Transduction, the passive delivery of bacterial genes from one bacterium to another by means of defective phage particles.

injected into an *E. coli* cell unable to grow on lactose as sole carbon source (lac^-), the recipient cell may be transduced to lac^+ by genetic recombination.

The capsids of bacteriophage P1 particles typically hold no more than 63 kb of DNA. The *E. coli* chromosome is 4720 kbp, so one transducing phage particle could maximally carry only 1% or so of the total genetic content of *E. coli* (or about 1.3 minutes of the *E. coli* genetic map). Phage transduction has proven very useful in mapping the fine structure (the intimate details) of gene order in chromosomes, since only genes closely linked to one another will be transduced together. Much of the fine structure of the genetic map for *E. coli* shown in Figure 28.10 was elucidated through transduction experiments using phage P1.

Transformation and Transfection

Transformation, defined as *the uptake, integration, and expression of naked DNA by competent cells,* occurs naturally in a few genera of bacteria (such as *Streptococcus* and *Bacillus* spp.). Other genera, including *Escherichia,* are not naturally competent to take up exogenous DNA. To circumvent this limitation, artificial means for incorporating DNA into recipient cells have been devised and adapted to a large number of different experimental organisms. **Transfection,** defined as the uptake of *viral or plasmid DNA* by competent cells, is similar to transformation. Unfortunately, the terms *transfection* and *transformation* are often used interchangeably. These *in vitro* methods entail the introduction of naked DNA (devoid of protein) into recipient cells. For example, *E. coli* cells become competent for DNA uptake following treatment with $CaCl_2$ and a brief (42°C) heat shock. Apparently, such treatment alters the bacterial cell wall such that it no longer presents an impenetrable barrier to DNA passage. Other transformation procedures include **electroporation,** in which the membranes of cells exposed to pulses of high voltage are rendered momentarily permeable to DNA molecules, and **ballistic methods,** in which microprojectiles coated with DNA are actually fired into recipient cells by a gas-powered gun. Transformation (transfection) in the laboratory has now been applied to the introduction of genetic material from bacteria, yeast, and animal (even human) cells into cells of the same or different species. Transformation is a powerful tool in genetic engineering.

28.4 The Molecular Mechanism of Recombination

Genetic recombination is the natural process by which genetic information is rearranged to form new associations. At the molecular level, *genetic recombination is the exchange (or incorporation) of one DNA sequence with (or into) another.* For example, homologous recombination involves an exchange of DNA sequences between homologous chromosomes, resulting in the arrangement of genes into new combinations. The process underlying homologous recombination is termed **general recombination** because the enzymatic machinery that mediates the exchange can use essentially any pair of homologous DNA sequences as substrates. Homologous recombination occurs during the production of gametes (*meiosis*) in diploid organisms. In higher animals, that is, those with immune systems, recombination also occurs in the DNA of somatic cells responsible for expressing proteins of the immune response, such as the immunoglobulins. This **somatic recombination** rearranges the immunoglobulin genes, dramatically increasing the potential diversity of immunoglobulins available from a fixed amount of genetic information (see Section 28.5). Homologous recombination also occurs in bacteria during conjugation, transformation, and transduction. Indeed, even viral chromosomes undergo recombination. For example, if two mutant viral particles simultaneously infect a host cell, a recombination event between the two viral genomes can lead to the formation of a recombinant virus chromosome that is wild-type.

Bacteriophage genomes can insert into bacterial chromosomes by a form of recombination, but since the integration of the bacteriophage DNA into the host DNA occurs only at a unique site on the host chromosome and involves specific DNA sequences on both phage DNA and bacterial DNA, the process is called **site-specific recombination.** Only a short length of homology (often less than 15 bp) is necessary for site-specific recombination events, and the enzymes involved act only on these sequences. In **transposition,** yet another type of recombination, particular DNA sequences known as *transposons* (see discussion later in this chapter) are inserted somewhat independently of any sequence homology on the DNA into which the insertion occurs. However, transposons themselves carry a specific sequence essential to insertion. Transposition serves as a mechanism by which genetic material may be moved from one chromosomal location to another. A fourth, rare form of recombination, **illegitimate recombination,** occurs between nonhomologous DNA independently of any unique sequence element.

General Recombination

Recombination occurs by the breakage and reunion of DNA strands, so that a physical exchange of parts takes place. Matthew Meselson and J. J. Weigle demonstrated in 1961 that this happens by coinfecting *E. coli* with two genetically distinct bacteriophage λ strains, one of which had been density-labeled by growth in ^{13}C and ^{15}N-containing media (Figure 28.12). The phage progeny were recovered and separated by CsCl density centrifugation. Phage particles that displayed recombinant genotypes were distributed throughout the gradient while parental (nonrecombinant) genotypes were found within discrete "heavy" and "light" bands in the density gradient. The results showed that recombinant phage contained DNA derived in varying proportions from both parents. The obvious explanation is that these recombinant DNAs arose via the breakage and rejoining of DNA molecules.

A second important observation made during this type of experiment was that some of the plaques formed by the phage progeny contained phage of two different genotypes, even though each plaque was caused by a single phage infecting one bacterium. Therefore, some infecting phage chromosomes must have contained a region of **heteroduplex DNA,** duplex DNA in which a part of each strand is contributed by a different parent (Figure 28.13).

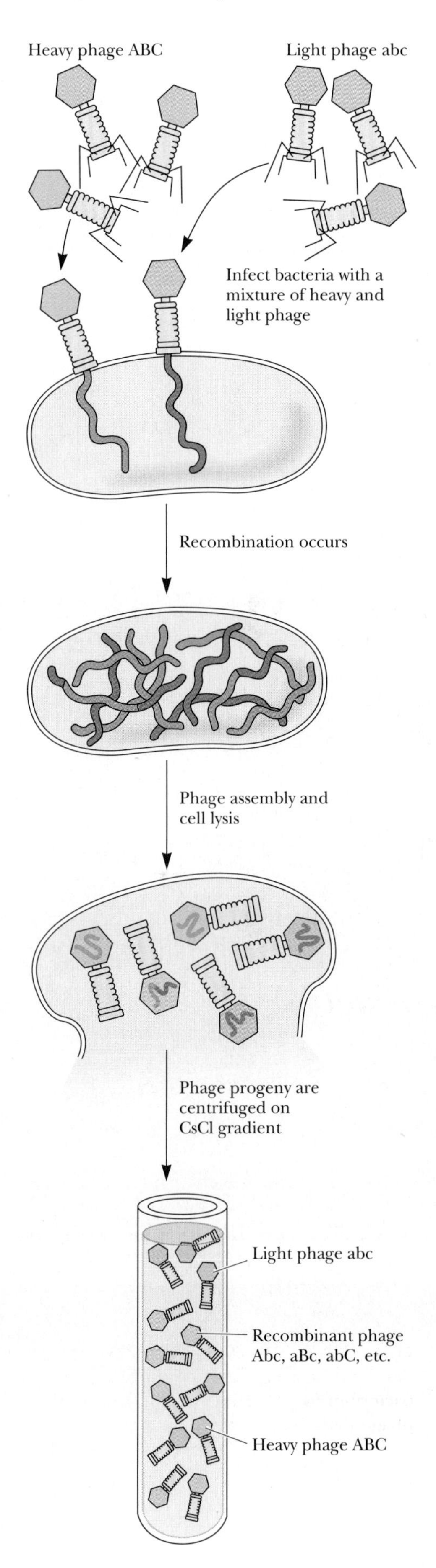

***Figure* 28.12** Meselson and Weigle's experiment demonstrated that a physical exchange of chromosome parts actually occurs during recombination. Density-labeled, "heavy" phage, symbolized as ABC phage in the diagram, were used to coinfect bacteria along with "light" phage, the abc phage. The progeny from the infection were collected and subjected to CsCl density gradient centrifugation. Parental-type ABC and abc phage were well separated in the gradient, but recombinant phage (ABc, Abc, aBc, aBC, and so on) were distributed diffusely between the two parental bands because they contained chromosomes constituted from fragments of both "heavy" and "light" DNA. These recombinant chromosomes were formed by breakage and reunion of parental "heavy" and "light" chromosomes.

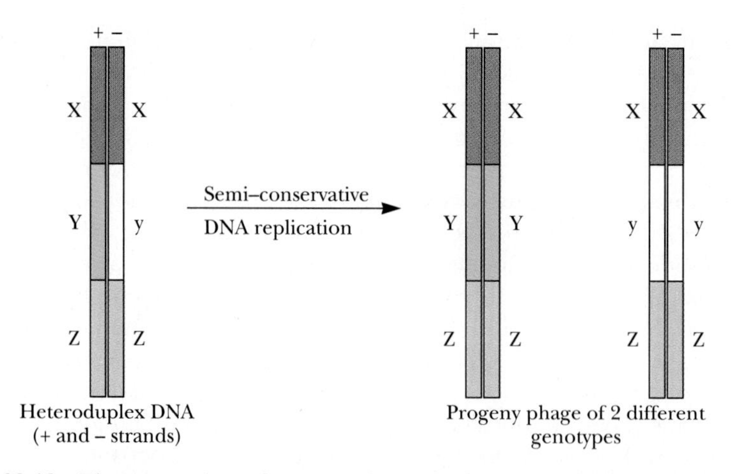

Figure **28.13** The generation of progeny bacteriophage of two different genotypes from a single phage particle carrying a heteroduplex DNA region within its chromosome. The heteroduplex DNA is composed of one strand that is genotypically XYZ (the + strand), and another strand that is genotypically XyZ (the − strand). That is, the genotype of the two parental strands for gene Y is different (one is Y, the other, y).

The Holliday Model

In 1964, Robin Holliday proposed a model for homologous recombination that has proven influential (Figure 28.14). The two homologous DNA duplexes are first juxtaposed so that their sequences are aligned. This process of **chromosome pairing** is called **synapsis** (Figure 28.14a). Holliday suggested that recombination begins by introduction of single-stranded nicks at homologous sites on the two paired chromosomes (Figure 28.14b). The two duplexes partially unwind, and the free, single-stranded end of one duplex begins to base-pair with its nearly complementary, single-stranded region along the intact strand in the other duplex, and vice versa (Figure 28.14c). This **strand invasion** is followed by ligation of the free ends from different duplexes to create a cross-stranded intermediate known as a **Holliday junction** (Figure 28.14d). The cross-stranded junction can now migrate in either direction (**branch migration**) by unwinding and rewinding of the two duplexes (Figure 28.14e). Branch migration results in **strand exchange;** heteroduplex regions of varying length are possible. In order for the joint molecule formed by strand exchange to be resolved into two DNA duplex molecules, another pair of nicks must be introduced. Resolution can be represented best if the duplexes are drawn with the chromosome arms bent "up" or "down" to give a planar representation (Figure 28.14f). Nicks then take place, either in the − strands that were originally nicked (see Figure 28.14b) or in the + strands (the strands not previously nicked). Duplex resolution is most easily kept straight by remembering that + strands are complementary to − strands and any resultant duplex must have one of each. Nicks made in the strands originally nicked lead to DNA duplexes in which one strand of each remains intact. Although these duplexes contain heteroduplex regions, they are not recombinant for the markers (*AZ, az*) that flank the heteroduplex region; such heteroduplexes are called **patch recombinants** (Figure 28.14g). Nicks introduced into the two strands not previously nicked yield DNA molecules that are both heteroduplex and recombinant for the markers *A*/*a* and *Z*/*z*; these heteroduplexes are termed **splice recombinants** (Figure 28.14h). Although this Holliday model explains the outcome of recombination, it provides no mechanistic explanation for the strand exchange reactions and other molecular details of the process.

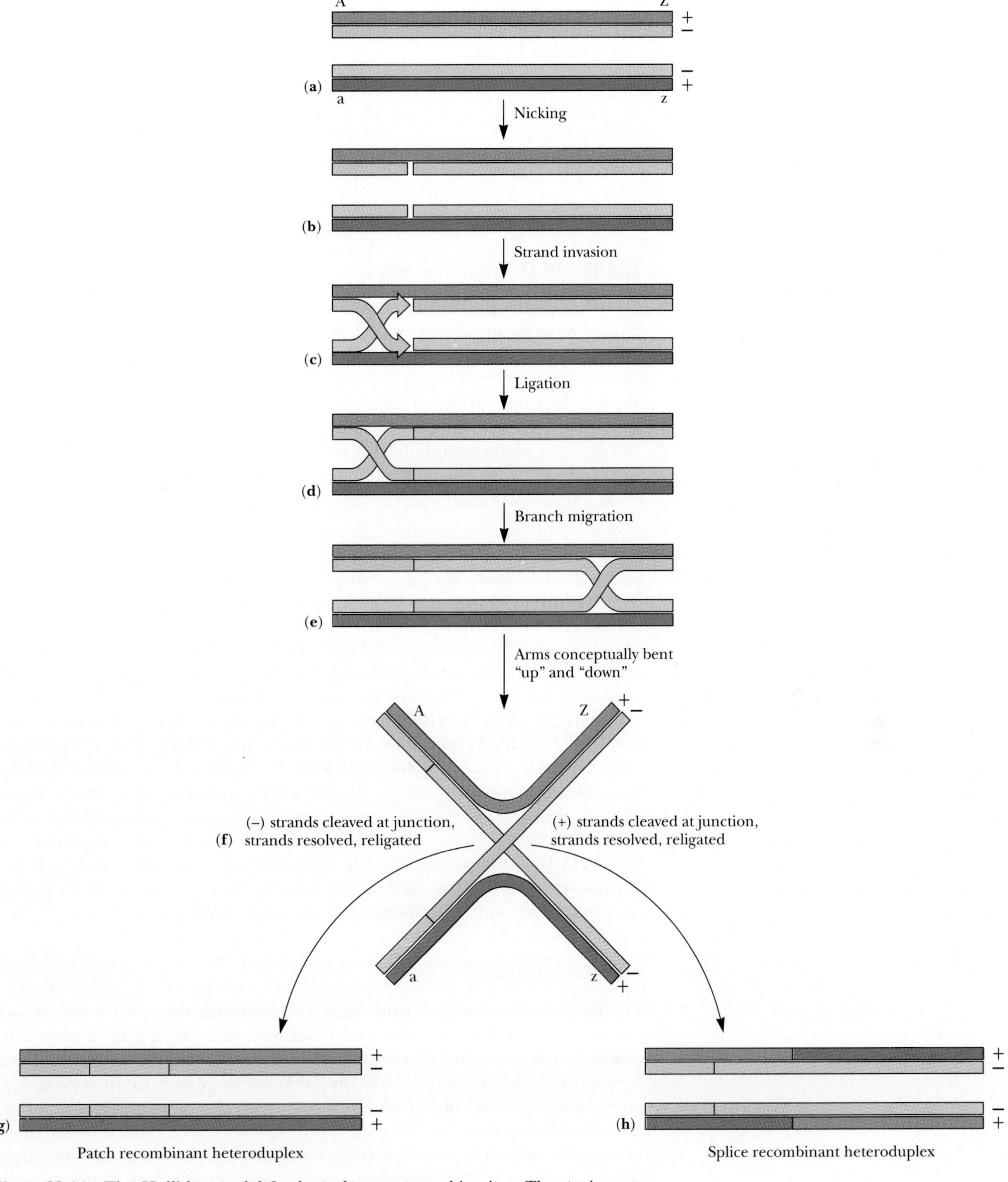

***Figure* 28.14** The Holliday model for homologous recombination. The + signs and − signs label strands of like polarity. For example, assume the two strands running 5′ → 3′ as read left to right are labeled as +; and the two strands running 3′ → 5′ as read left to right are labeled −. Only strands of like polarity exchange DNA during recombination. (See text for detailed description.)

The Enzymology of General Recombination

To illustrate recombination mechanisms, we focus on general recombination as it occurs in *E. coli*. Analysis of mutant strains of *E. coli* defective in general recombination, so-called ***rec*** mutants, led to the discovery of various proteins responsible for mediating recombination. The basic framework of an enzymatic pathway for genetic recombination believed to be generally applicable to other organisms has emerged. Although many details regarding the protein machinery of general recombination remain to be elucidated, the principal players in the process are the **RecA** protein, the **RecBCD** enzyme complex, and **SSB,** the name given *s*ingle-*s*tranded, DNA-*b*inding protein. Proteins analogous to RecA have been identified in yeast and humans.

The RecBCD Enzyme Complex

The proteins **RecB** (140 kD), **RecC** (130 kD), and **RecD** (58 kD) form a multifunctional enzyme complex. The RecBCD complex initiates recombination by attaching to the end of a DNA duplex and unwinding it in an ATP-dependent reaction (Figure 28.15a). As RecBCD progresses along, unwinding the duplex, the single-stranded tails left behind re-anneal into duplex DNA (Figure 28.15b). Since rewinding is somewhat slower than unwinding, a growing "bubble" of single-stranded DNA is created in the DNA duplex. This "bubble" consists of two single-stranded loops, or "rabbit ears," that propagate along the DNA behind the advancing RecBCD (Figure 28.15c).

RecBCD Enzyme Cleaves ssDNA at *Chi* Sites

Another important feature of the RecBCD protein complex is its site-specific endonuclease activity: RecBCD enzyme cleaves ssDNA adjacent to so-called *Chi* sites characterized by the sequence *5′-GCTGGTGG-3′*. When the advancing RecBCD enzyme encounters a *Chi* site within the bubble (Figure 28.15d), it cleaves only the strand with the 5′-GCTGGTGG-3′ sequence, 5 or 6 nucleotides to the 3′-side. *Chi* sites are recombinational "hot spots" that occur naturally in the *E. coli* chromosome, about once every 5 kb of DNA. Endonucleolytic cleavage at a *Chi* site followed by continued RecBCD-mediated unwinding/rewinding of the duplex generates a 3′-terminated single strand of DNA; the 5′ tail from this strand is rewound into dsDNA, resulting in the disappearance of one of the "rabbit ears" (Figure 28.15e). Continuation of the RecBCD complex progressively lengthens the 3′-terminated ssDNA, perhaps to several kilobases. This ssDNA becomes the substrate of RecA protein for assimilation into an adjacent DNA duplex (Figure 28.15f).

The RecA Protein

The **RecA recombinase,** or **protein,** is a multifunctional, 352-residue (38 kD) enzyme that acts in general recombination to catalyze the **DNA strand exchange reaction** (Figure 28.15f through i). RecA is believed to have two DNA-binding sites, one for ssDNA and the other for homologous duplex DNA. In effect, RecA protein promotes conformational changes in the structure of ssDNA and duplex DNA, allowing these DNAs to hybridize to each other. RecA binds to single-stranded DNA with a stoichiometry of one RecA per three nucleotides. SSB, binding to ssDNA as a tetramer of 19-kD subunits, promotes the interaction between RecA and ssDNA, in part by diminishing secondary structure in the ssDNA strand. This binding of RecA protein to the 3′-terminal ssDNA creates a **nucleoprotein filament** that has affinity for double-stranded DNA molecules. RecA protein also forms nucleoprotein fila-

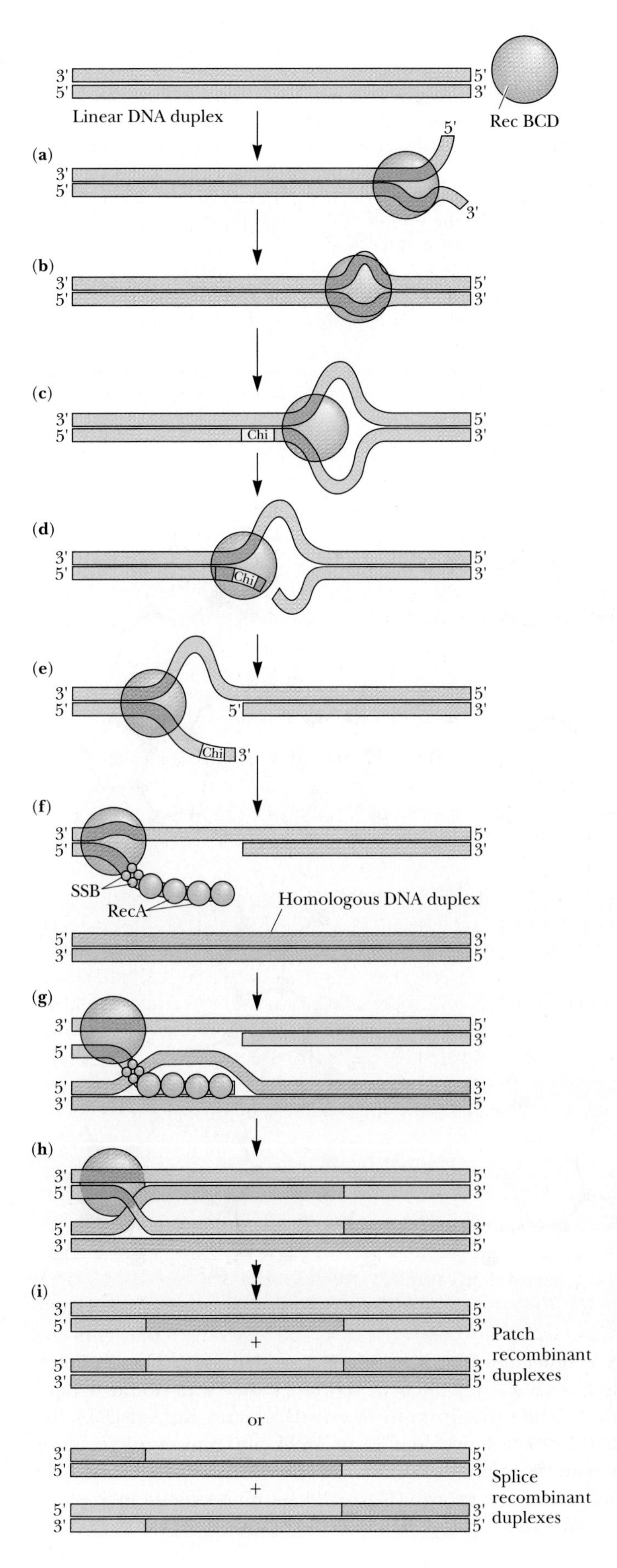

***Figure* 28.15** Model for homologous recombination as promoted by RecBCD enzyme, the *Chi* site, and RecA enzyme. Steps (a) through (c) show the RecBCD-mediated unwinding/rewinding of DNA. At (d), RecBCD enzyme cleaves the *Chi*-site strand to the 3′-side of *Chi*. Continued unwinding of the duplex by RecBCD enzyme elongates the single-stranded 3′-tail; (e) the 5′-end is rewound into the duplex. (f) RecA protein and SSB aid strand invasion of the 3′-ssDNA into a homologous DNA duplex, (g) forming a D-loop. (h) The D-loop strand that has been displaced by strand invasion pairs with its complementary strand in the original duplex to form a Holliday junction as strand invasion continues. (i) Resolution of the Holliday junction via strand cleavage yields patch or splice recombinant heteroduplex molecules, depending on which two strands of like polarity are cleaved.

(Adapted from Smith, G., 1988. Microbiological Reviews 52:1–28.)

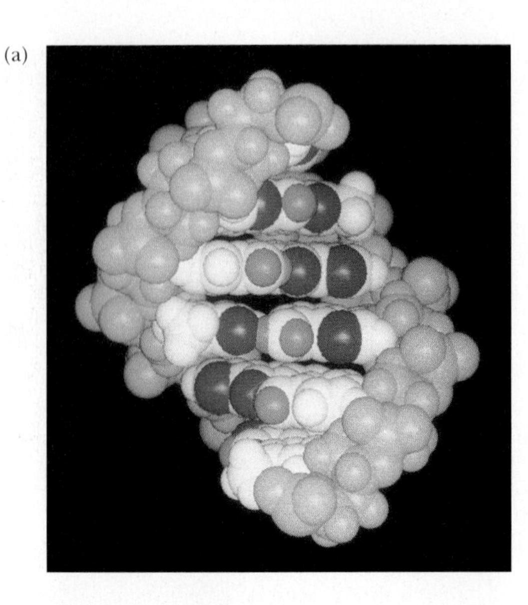

***Figure* 28.16** (a) A view of the major groove of B-DNA, revealing the matrix of H-bond donors (green) and acceptors (red) available for base-pairing with a third DNA strand.

(Adapted from Camerini-Otero and Hsieh, 1993. Cell 73:217; photo courtesy of Peggy Hsieh and R. Daniel Camerini-Otero of the National Institutes of Health.)

(b and c) Pairing scheme proposed for triple helix formation by recombinase-type proteins. (b) Pairing of the third strand within the major groove of a homologous duplex DNA. The strands of duplex DNA are labeled W and C; the invading strand is shown in the middle. (The sugar-phosphate backbones of each of the three strands are indicated by shading.) (c) Non-Watson–Crick base pairings involving T, C, A, or G residues in the invading ssDNA and the homologous DNA duplex. Note that C residues in the ssDNA must be protonated (C^+) in order to present an appropriate H-bonding pattern to the duplex.

(Adapted from Hsieh, Camerini-Otero, and Camerini-Otero, 1993. Genes & Development 4:1951.) Alternative triplex base-pairing schemes have been proposed, notably by Rao, B. J., Chiu, S. K., and Radding, C. M., 1993. *Journal of Molecular Biology* 229:328–343; see end-of-chapter problem 10.

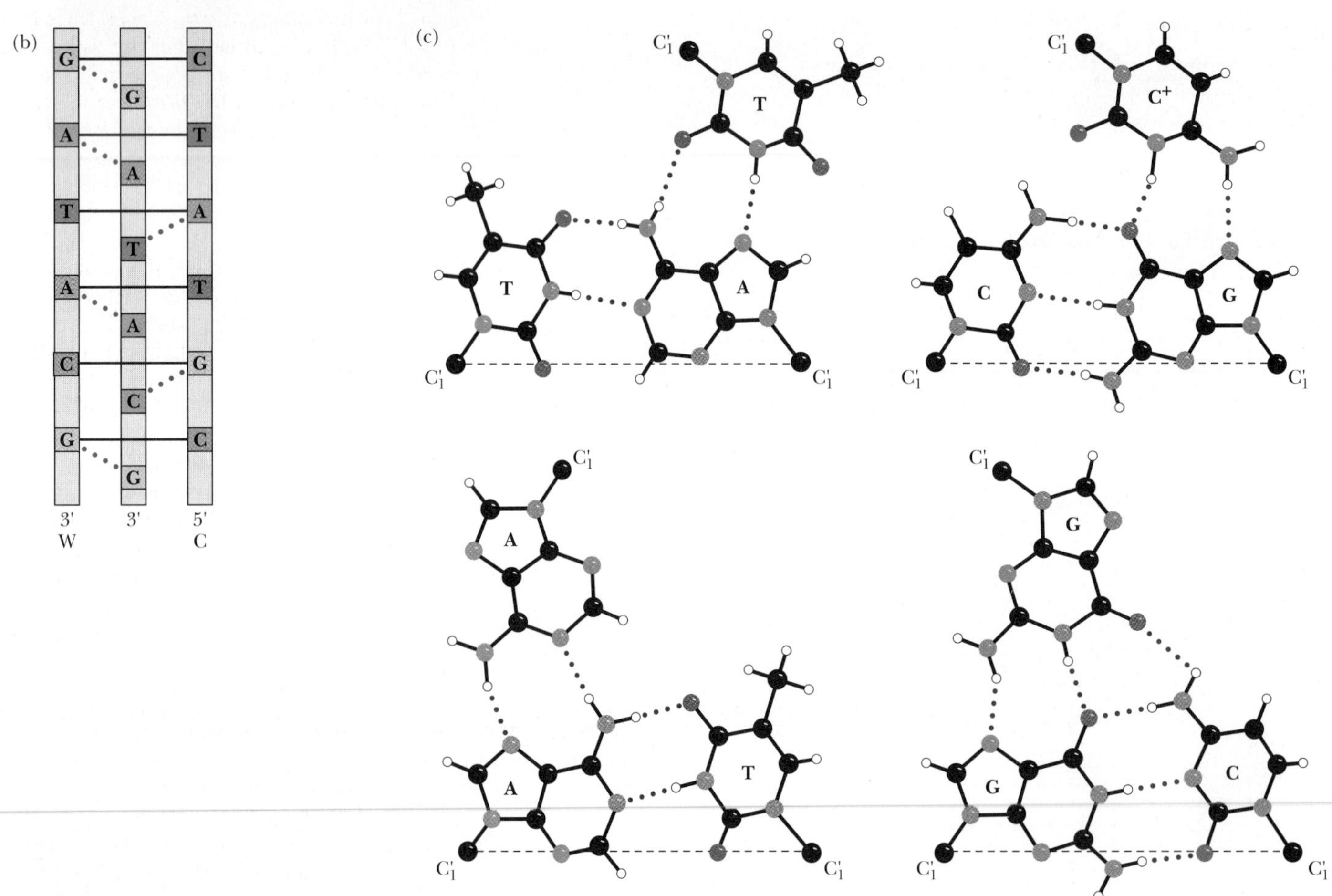

ments with dsDNA molecules that have single-stranded gaps. These nucleoprotein filaments have a helical pitch of 8.5 to 10 nm, with about 6 RecA monomers per turn. The DNA in both RecA:ssDNA and RecA:dsDNA filaments is extended 150% relative to B-form DNA and unwound to about 18.6 bp per turn. The unwinding of dsDNA exposes a matrix of H-bond donors and acceptors in its major groove (Figure 28.16a), making them available for pairing with a third DNA strand. These observations have led to the suggestion that a three-stranded DNA molecule, or **DNA triplex** (Figure 28.17), is the intermediate in recombination. The specificity of pairing in this DNA triplex (Figure 28.16) would ensure that recombination is homologous, right down to the level of single base pairs.

In forming the recombination intermediate, the 3′-end of the RecA:ssDNA complex invades a homologous dsDNA, creating a **D-loop** (see Figure 28.15g). RecA protein partially unwinds the DNA duplex and scans it, searching for a region of homology with its complexed ssDNA. This search for homology culminates in *synapsis* between homologous nucleotide sequences. Along the region of homology, RecA further unwinds the dsDNA, allowing base pairing of its ssDNA to the complementary sequence within the DNA duplex. The procession of this base-unpairing and re-pairing along the DNA duplex represents **branch migration** (see Figure 28.15h). Branch migration drives the displacement of the homologous DNA strand from the DNA duplex and its replacement with the ssDNA strand, a process known as **single-stranded assimilation** (or **single-stranded uptake**). RecA-driven displacement of the homologous strand by the invading ssDNA requires ATP. Strand assimilation will not occur if there is no homology between the ssDNA end and the invaded DNA duplex. The DNA strand displaced by the invading 3′-terminal ssDNA is free to anneal with the single-stranded gap left in the original DNA; this step is also mediated by RecA protein and SSB (see Figure 28.15h). The result is a Holliday junction, the classic intermediate in genetic recombination. Depending on how the strands in the Holliday junction are cleaved and resolved, patch or splice recombinant duplexes result (see Figure 28.15i).

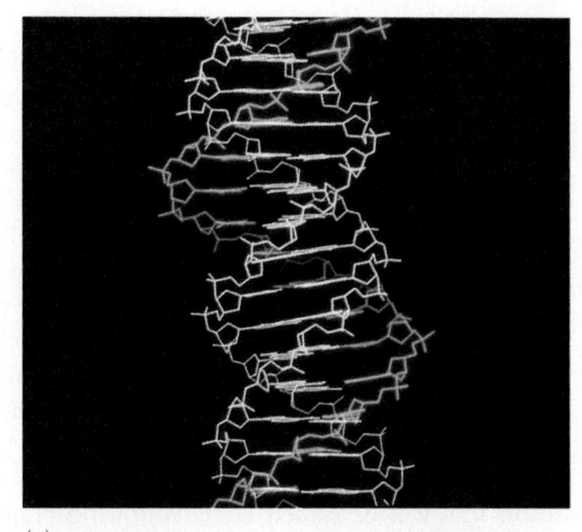

(a)

***Figure* 28.17** (a) A model for triple-stranded DNA (a DNA triplex), based on chemical and enzymatic analysis and computer modeling. B-form duplex DNA is shown in yellow, with its major groove occupied by the third strand, colored blue. Note how the bases of the third strand are aligned with the base pairs of the duplex DNA. The third strand divides the major groove into two unequal parts, creating from it a new "major" and a new "minor" groove.

(Photo courtesy of Charles M. Radding of Yale University.)

(b) Model structure of a DNA triplex molecule suggested by 2-D NMR spectroscopy. The 31-base oligonucleotide 5′-dAGAGAGAACCCCTTCTCTCTTTTTCTCT-CTT-3′ folds to form a stable intramolecular triple helix. Seven base triplets and one Watson–Crick base pair form the core. Residues 1–11 in red, 12–21 in blue, and 22–31 in yellow. The red and blue strands are hydrogen bonded via formation of Watson–Crick base pairs. The bases of the yellow strand lie within the major groove formed by the red:blue duplex, interacting via H bonds with the duplex *WC* base pairs.

(Adapted from Macaya, R., Wang, E., Schultze, P., Sklenar, V., and Feigon, J. 1992. Journal of Molecular Biology 225: 755–760.) Image courtesy of Peter Schultze and Juli Feigon of UCLA.

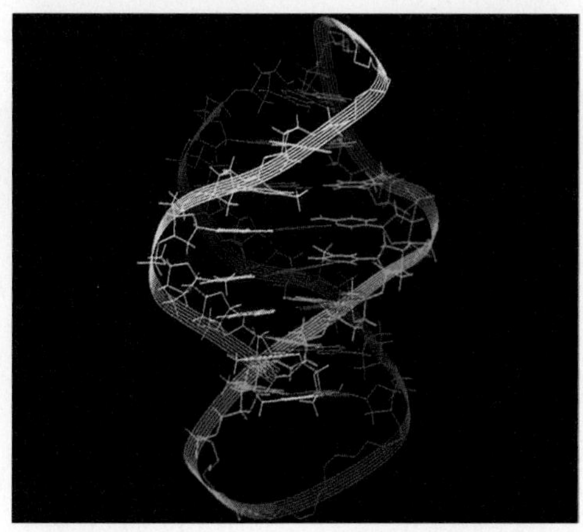

(b)

Resolution of the Holliday Junction Yields "Patch" or "Splice" Heteroduplex DNA Molecules

Holliday junctions are cleaved by distinctive endonucleases, known as the **Holliday junction resolvases,** to yield the two DNA heteroduplexes. In *E. coli,* at least one Holliday junction resolvase is the *ruvC* gene product. The Holliday junction is actually two homologous DNA molecules in alignment (Figure 28.18). We can designate these two duplexes *WC* and *wc,* for Watson and Crick, of course. (Let *W* and *w* represent the $5' \rightarrow 3'$ strands of each duplex, and *C* and *c* their complementary $3' \rightarrow 5'$ strands.) *Strand exchange* occurs if RecA protein mediates base rotation so that *W* now pairs with *c,* whereas *w* pairs with *C* (Figure 28.18). The result is two heteroduplex regions (*Wc* and *wC*). Cleavage of strands with **like polarity**—that is, both *Watson strands or* both *Crick strands*—at the Holliday junction by *resolvase,* followed by branch migration and a second cleavage of strands, yields either patch *or* splice recombinant molecules. If the same pairs of like strands (*Ww* or *Cc*) are cleaved both times, patch heteroduplexes are formed. If different pairs of like strands are cleaved (*Ww* one time, *Cc* the other), splice heteroduplexes result (see Figure 28.14).

Transposons

In 1950, Barbara McClintock reported the results of her studies on an **activator gene** in maize (*Zea mays,* or as it's usually called, *corn*) that was recognizable principally by its ability to cause mutations in a second gene. Activator genes were thus an internal source of mutation. A most puzzling property was their ability to move relatively freely about the genome. As we have seen, scientists had labored to establish that chromosomes consisted of genes arrayed in a fixed order, so most geneticists viewed as incredible this idea of genes moving around. The recognition that McClintock so richly deserved for her explanation of this novel phenomenon had to await verification by molecular biologists. In 1983, Barbara McClintock was finally awarded the Nobel

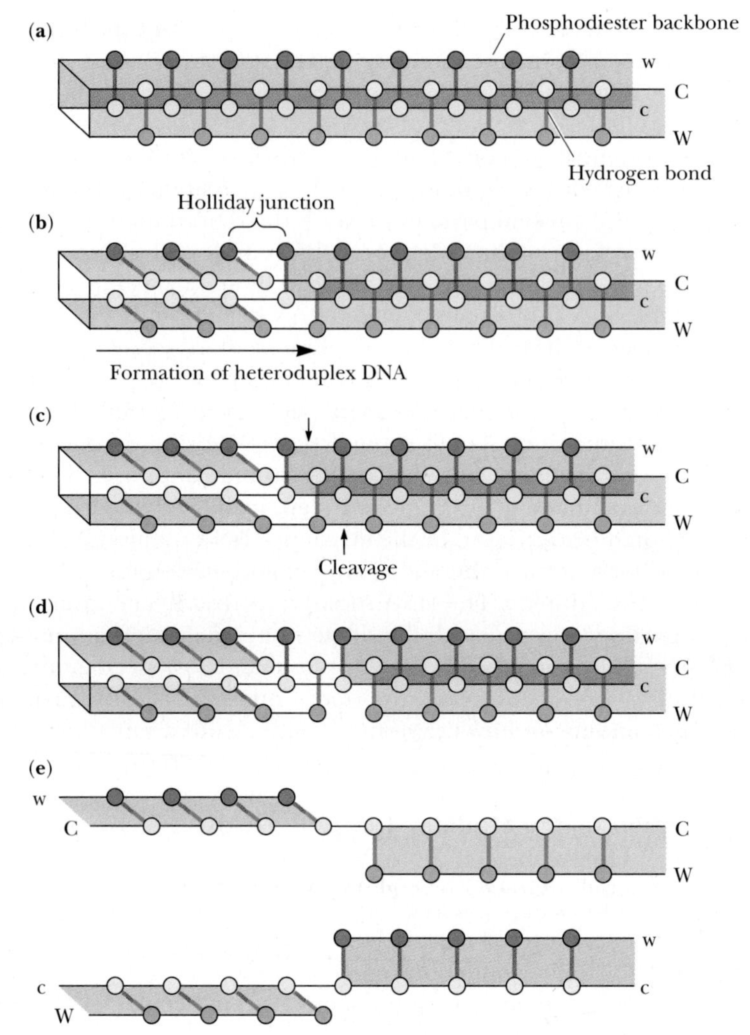

***Figure* 28.18** Resolution of a Holliday junction by enzymatic cleavage. (a) Two homologous DNA duplexes in alignment (designated *WC* and *wc*) form (b) a heteroduplex region via RecA protein-mediated base rotation so that a region of *W* now pairs with *c*, and the corresponding region of *w* pairs with *C*. Note that the Holliday junction (the point of base rotation) does not involve any crossed strands or unpaired bases. Recognition of the heteroduplex junction by *ruvC* endonuclease (*resolvase*) and cleavage of strands (c, d) with like polarity (either *W* and *w or C* and *c*) creates heteroduplex molecules. (Here, cleavage of *W* and *w* is arbitrarily chosen.) Cleavage is arbitrarily shown to occur one base from the junction. (e) The two DNA heteroduplexes then separate, each having a nick in one strand that will be sealed later. Following branch migration (beyond the boundary of this figure) and a second cleavage of like strands (not shown), the two heteroduplexes will be resolved to become patch or splice recombinants. In the case shown here, if the *W* and *w* strands are cleaved again, a pair of patch recombinant dsDNA molecules is formed; if the *C* and *c* strands are now cleaved, a pair of splice recombinant dsDNA molecules results. (See also Figure 28.14.)

(Adapted from Müller, B. C., et al., 1990. Enzymatic formation and resolution of Holliday junctions in vitro. Cell 60:329–336.)

Prize in physiology or medicine. By this time, it was appreciated that many organisms, from bacteria to humans, possessed similar "*jumping genes*" able to move from one site to another in the genome. This mobility led to their designation as **mobile elements, transposable elements,** or, simply, **transposons.**

Transposons are segments of DNA that are moved enzymatically from place to place in the genome. That is, their location within the DNA is unstable. Transposons range in size from several hundred bp to more than 8 kbp. Transposons contain a gene encoding an enzyme necessary for insertion into a chromosome and for the remobilization of the transposon to different locations. These movements are termed **transposition events.** The smallest transposons are called **insertion sequences,** or **IS**s, signifying their ability to insert apparently at random in the genome. Insertion into a new site causes mutation because it disrupts the DNA sequence at that site. Insertion occurs at sites that show little homology to the insertion sequence or transposon. Although certain transposons (such as *E. coli* transposon Tn 7) may undergo transposition once per cell generation, most transposition events are infrequent, taking place only once every 10^4 to 10^7 generations. Larger and more complex transposons also carry genes that are not involved in the enzymology of insertion and excision of the transposon, such as genes conferring resistance to antibiotics. Episomes, the class of plasmids that reversibly integrate into the genome, contain transposons.

28.5 The Immunoglobulin Genes: Generating Protein Diversity Using Genetic Recombination

The immunoglobulin genes are a highly evolved system for maximizing protein diversity from a finite amount of genetic information. This diversity is essential for gaining immunity to the great variety of infectious organisms and foreign substances that cause disease.

The Immune Response

Only vertebrates show an immune response. If a foreign substance, called an **antigen,** gains entry to the bloodstream of a vertebrate, the animal responds via a protective system called the immune response. The immune response involves production of proteins capable of recognizing and destroying the antigen. This response is mounted by certain white blood cells—the **B** and **T cell lymphocytes** and the **macrophages.** B cells are so named because they mature in the bone marrow; T cells mature in the thymus gland. Each of these cell types is capable of gene rearrangement as a mechanism for producing proteins essential to the immune response. **Antibodies,** which can recognize and bind antigens, are immunoglobulin proteins secreted from B cells. Since antigens can be almost anything, the immune response must have an incredible repertoire of structural recognition. Thus, vertebrates must have the potential to produce immunoglobulins of great diversity in order to recognize virtually any antigen.

The Immunoglobulin G Molecule

Immunoglobulin G (**IgG** or **γ-globulin**) is the major class of antibody molecules found circulating in the bloodstream. IgG is a very abundant protein, amounting to 12 mg per mL of serum. It is a 150-kD $\alpha_2\beta_2$-type tetramer. The α or H (for *heavy*) chain is 50 kD; the β or L (for *light*) chain is 25 kD. A preparation of IgG from serum is heterogeneous in terms of the amino acid sequences represented in its L and H chains. However, the IgG L and H chains produced from any given B lymphocyte are homogeneous in amino acid sequence. L chains consist of 214 amino acid residues and are organized into two roughly equal segments, the $\mathbf{V_L}$ and $\mathbf{C_L}$ regions. The V_L designation reflects the fact that L chains isolated from serum IgG show variations in amino acid sequence over the first 108 residues, V_L symbolizing this "variable" region of the L polypeptide. The amino acid sequence for residues 109 to 214 of the L polypeptide is constant, as represented by its designation as the "constant light," or C_L, region. The heavy, or H, chains consist of 446 amino acid residues. Like L chains, the amino acid sequence for the first 108 residues of H polypeptides is variable, ergo its designation as the V_H region, while residues 109 to 446 are constant in amino acid sequence. This "constant heavy" region consists of three quite equivalent domains of homology designated C_H1, C_H2, and C_H3. Each L chain has two intrachain disulfide bonds, one in the V_L region and the other in the C_L region. The C-terminal amino acid in L chains is cysteine, and it forms an interchain disulfide bond to a neighboring H chain. Each H chain has four intrachain disulfide bonds, one in each of the four regions. Figure 28.19 presents a diagram of IgG organization. Within the variable regions of the L and H chains, certain positions are **hypervariable** with regard to amino acid composition. These hypervariable residues occur at positions 24 to 34, 50 to 55, and 89 to 96 in the L chains and at positions 31 to 35, 50 to 65, 81 to 85, and 91 to 102 in the H chains. The hypervariable regions are also called **complementarity-determining regions,** or **CDRs,** since it is these regions that form the structural site that is complementary to some

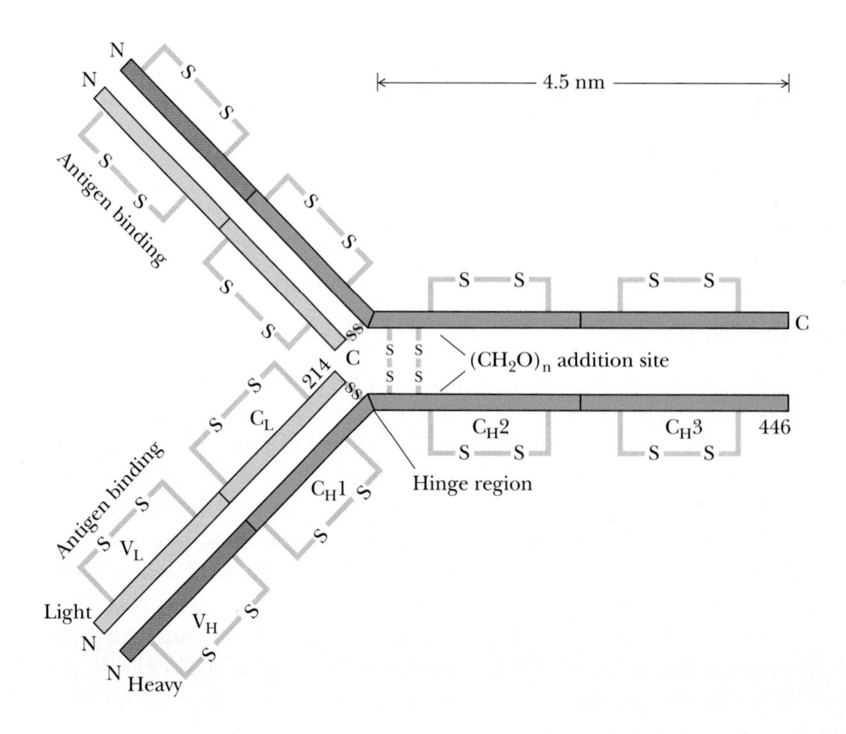

Figure **28.19** Diagram of the organization of the IgG molecule. Two identical L chains are joined with two identical H chains. Each L chain is held to an H chain via an interchain disulfide bond. The variable regions of the four polypeptides lie at the ends of the arms of the Y-shaped molecule. These regions are responsible for the antigen recognition function of antibody molecules. The actual antigen-binding site is constituted from hypervariable residues within the V_L and V_H regions. For purposes of illustration, some features are shown on only one or the other L chain or H chain, but all features are common to both chains.

part of an antigen's structure, providing the basis for antibody:antigen recognition.

In the immunoglobulin genes, the arrangement of exons correlates with protein structure. In terms of its tertiary structure, the immunoglobulin G molecule is composed of 12 discrete *collapsed β-barrel domains,* each domain having a *Greek key* motif (see Figure 5.32). The characteristic structure of this domain is referred to as the **immunoglobulin fold** (Figure 28.20). Each of IgG's two heavy chains contributes four of these domains and each of its light chains contributes two. The four *variable-region* domains (one on each chain) are encoded by multiple exons, but the eight *constant-region* domains are each the product of a single exon. All of these constant-region exons are derived from a single ancestral exon encoding an immunoglobulin fold. The major variable-region exon probably derives from this ancestral exon also. Contemporary immunoglobulin genes are a consequence of multiple duplications of the ancestral exon.

The discovery of variability in amino acid sequence in otherwise identical polypeptide chains was surprising and almost heretical to protein chemists. For geneticists, it presented a genuine enigma. They noted that mammals, which can make millions of different antibodies, don't have millions of different antibody genes. How can the mammalian genome encode the diversity seen in L and H chains?

The Organization of Immunoglobulin Genes

The answer to the enigma of immunoglobulin sequence diversity is found in the organization of the immunoglobulin genes. The genetic information for an immunoglobulin polypeptide chain is scattered among multiple gene segments along a chromosome in germline cells (sperm and eggs). During vertebrate development and the formation of B lymphocytes, these segments are brought together and assembled by **DNA rearrangement** (that is, genetic recombination) into complete genes. DNA rearrangement, or **gene reorganization,** provides a mechanism for generating a variety of protein isoforms from a limited number of genes. DNA rearrangement occurs in only a few genes,

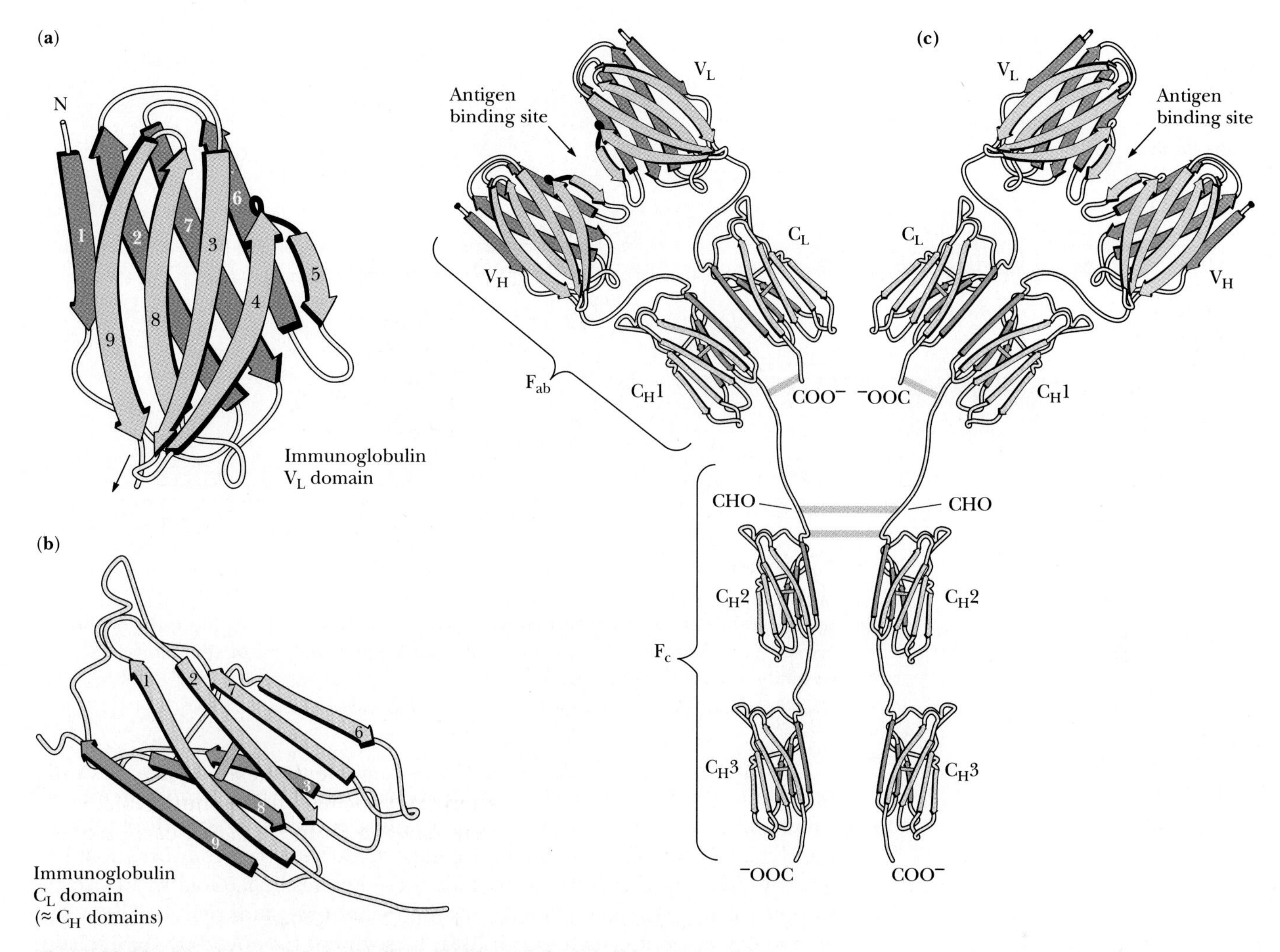

Figure **28.20** The characteristic "collapsed β-barrel domain" known as the *immunoglobulin fold.* The β-barrel structures for both (a) *variable* and (b) *constant regions* are shown. (c) This diagram shows a schematic diagram of the 12 collapsed β-barrel domains that make up an IgG molecule. CHO indicates the carbohydrate addition site; F_{ab} denotes one of the two antigen-binding fragments of IgG, and F_c, the proteolytic fragment consisting of the pairs of C_H2 and C_H3 domains.

namely those encoding the antigen-binding proteins of the immune response—the immunoglobulins and the T cell receptors. The gene segments encoding the amino-terminal portion of immunoglobulin polypeptides are also unusually susceptible to mutation events. The result is a population of B cells whose antibody-encoding genes collectively show great sequence diversity even though a given cell can make only a limited set of immunoglobulin chains. Hence, at least one cell among the B cell population will likely be capable of producing an antibody that will specifically recognize a particular antigen.

DNA Rearrangements Assemble an L-Chain Gene by Combining Three Separate Genes

The organization of various immunoglobulin gene segments in the mouse genome is shown in Figure 28.21. L-chain variable-region genes are assembled from two kinds of **germline genes,** $\mathbf{V_L}$ and $\mathbf{J_L}$ (*J* stands for *joining*). In mammals, there are two different families of **L-chain genes,** the $\boldsymbol{\kappa}$, or **kappa, gene**

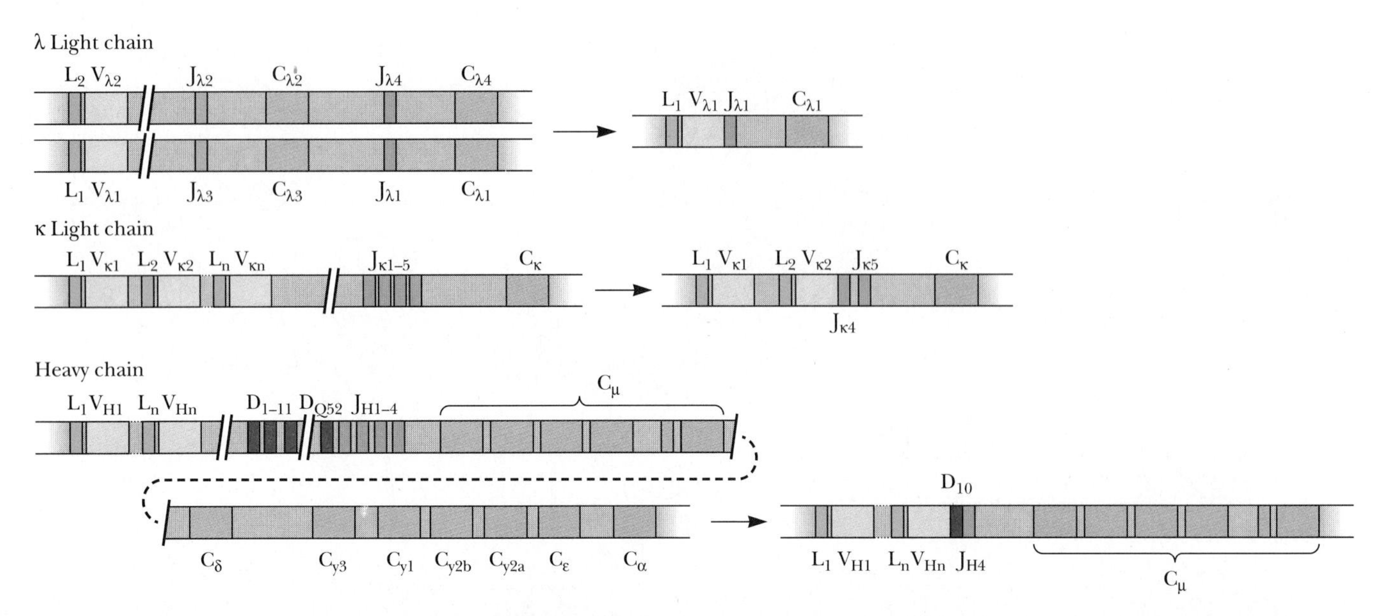

Figure **28.21** The organization of mouse immunoglobulin gene segments. The organization in germline cells is shown on the left, and the rearranged organization characteristic of mature B lymphocytes is shown to the right of the arrows. The rearranged states shown are but single examples of the many possibilities for each gene family.

(Adapted from Tonegawa, S., 1983. Somatic generation of antibody diversity. Nature 302:575.)

family and the **λ,** or **lambda, gene family;** each family has V and J members. These families are on different chromosomes. In mice, 90% of the L chains are κ chains; λ L chains are a minor component. Mice have four functional J_κ genes (and a fifth nonfunctional one); these J genes lie 2.5 to 4 kb upstream from the single $\mathbf{C_\kappa}$ **gene** that encodes the L-chain constant region. There are at least 200 V_κ genes, each with its own L_κ segment for encoding the L-chain leader peptide that targets the L chain to the endoplasmic reticulum for IgG assembly and secretion. (This leader peptide is cleaved once the L chain reaches the ER lumen.) The λ family of L-chain genes is organized a little differently, with only two V_λ genes, each of which is followed downstream by a pair of J_λ-C_λ units (Figure 28.21). In different mature B lymphocyte cells, V_κ and J_κ genes have joined in different combinations, and along with the C_κ gene, form complete L_κ chains with a variety of V_κ regions. However, any given B lymphocyte expresses only one V_κ-J_κ combination. Construction of the mature B lymphocyte L-chain gene has occurred by DNA rearrangements that combine three genes (L-$V_{\kappa,\lambda}$, $J_{\kappa,\lambda}$, $C_{\kappa,\lambda}$) to make one polypeptide!

DNA Rearrangements Assemble an H-Chain Gene by Combining Four Separate Genes

The first 98 amino acids of the 108-residue, H-chain variable region are encoded by a $\mathbf{V_H}$ **gene.** Each V_H gene has an accompanying L_H gene that encodes its essential leader peptide. It is estimated that there are from 200 to 1000 V_H genes and they can be subdivided into eight distinct families based on nucleotide sequence homology. The members of a particular V_H family are grouped together on the chromosome, separated from one another by 10 to 20 bp. In assembling a mature H-chain gene, a V_H gene is joined to a **D gene** (*D* for *diversity*), which encodes amino acids 99 to 113 of the H chain. These amino acids comprise the core of the third CDR in the variable region of H chains. The V_H-D gene assemblage is linked in turn to a $\mathbf{J_H}$ **gene,** which

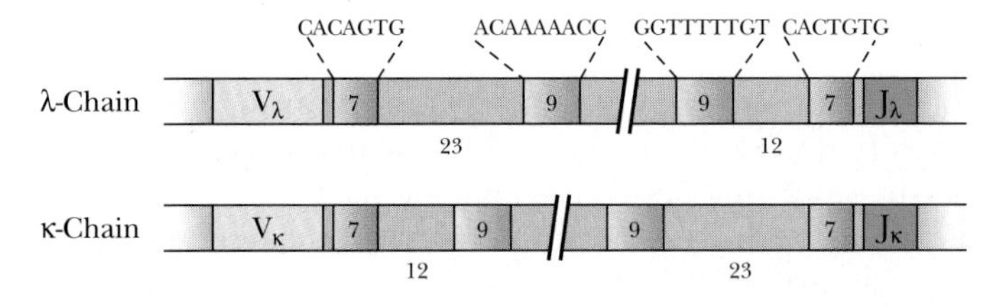

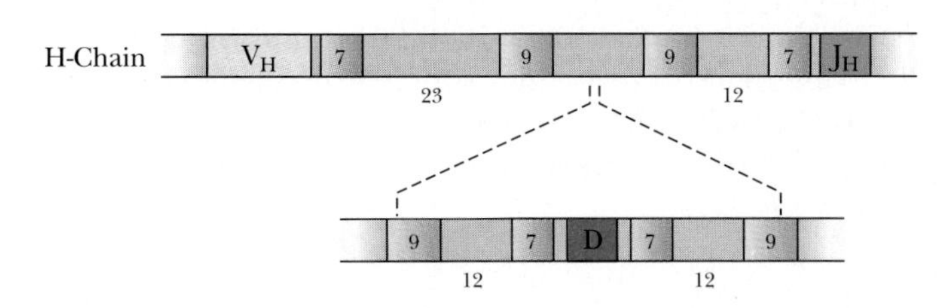

***Figure* 28.22** Consensus elements are located above and below germline variable-region genes that recombine to form genes encoding immunoglobulin chains. These consensus elements are complementary and are arranged in a heptamer-nonamer, 12-bp to 23-bp spacer pattern.

(Adapted from Tonegawa, S., 1983. Somatic generation of antibody diversity. Nature 302:575.)

encodes the remaining part of the variable region of the H chain. The V_H, D, and J_H genes are grouped in three separate clusters on the same chromosome. The four J_H genes lie 7 kb upstream of the eight C genes, the closest of which is C_μ. Any of four **C genes** may encode the constant region of IgG H chains: $C_{\gamma 1}$, $C_{\gamma 2a}$, $C_{\gamma 2b}$, and $C_{\gamma 3}$. Each C gene is composed of multiple exons (as shown in Figure 28.21 for C_μ, but not the other C genes). Ten to twenty D genes are found 1 to 80 kb farther upstream. The V_H genes lie even farther upstream. In B lymphocytes, the variable region of a heavy-chain gene is composed of one each of the L_H-V_H genes, a D gene, and a J_H gene joined head to tail. Because the H-chain variable region is encoded in three genes and the joinings can occur in various combinations, the heavy chains have a greater potential for diversity than the light-chain variable regions that are assembled from just two genes (for example, L_κ-V_κ and J_κ). In making heavy-chain genes, four genes have been brought together and reorganized by DNA rearrangement to produce a single polypeptide!

The Mechanism of V-J and V-D-J Joining in Light- and Heavy-Chain Gene Assembly

Specific nucleotide sequences adjacent to the various variable-region genes suggest a mechanism in which these sequences act as joining signals. All germline V and D genes are followed by a consensus CACAGTG heptamer separated by a short, unconserved 23-bp spacer from a consensus ACAAAAACC nonamer. Likewise, all germline D and J genes are immediately preceded by a consensus GGTTTTTGT nonamer separated by a short, unconserved 12-bp spacer from a consensus CACTGTG heptamer (Figure 28.22). Note that the consensus elements downstream of a gene are complementary to those upstream from the gene with which it recombines. Indeed, it is these complementary consensus sequences that govern the recombination of the variable-region genes. Functionally meaningful recombination only happens between 2 genes where one has a 12-bp spacer and the other has a 23-bp spacer (Figure 28.22). Then, a **stem-loop structure** can form via base pairing between complementary heptamer units and nonamer units with looping out of the respective 12- and 23-bp spacers (Figure 28.23). Apparently, the **recombinase** that catalyzes the breakage and reunion events of immunoglobulin gene rearrangement contains two DNA-binding proteins, one specific for the heptamer-nonamer with a 12-bp spacer, the other specific for the heptamer-nonamer with a 23-bp spacer.

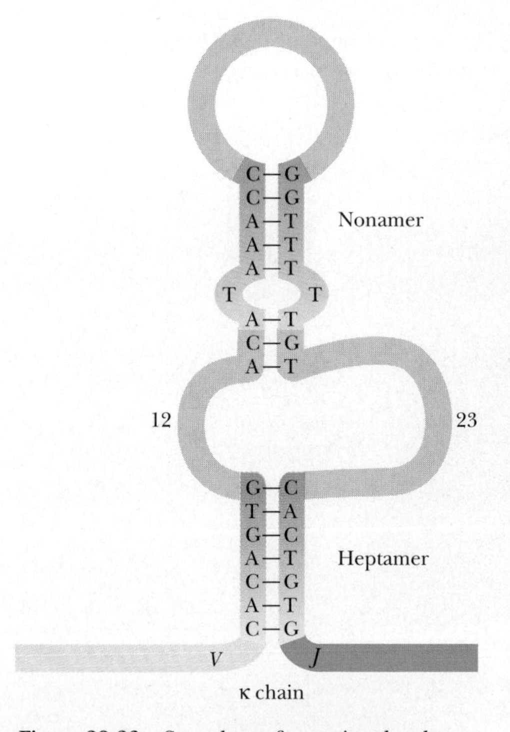

***Figure* 28.23** Stem-loop formation by the heptamer-nonamer, 12-bp to 23-bp rule as it occurs in a germline κ gene family. This stem-loop type of structure is believed to mediate the recombination event joining a V_κ and a J_κ gene in constructing an immunoglobulin L-chain gene. Analogous structures form in H-chain gene assembly.

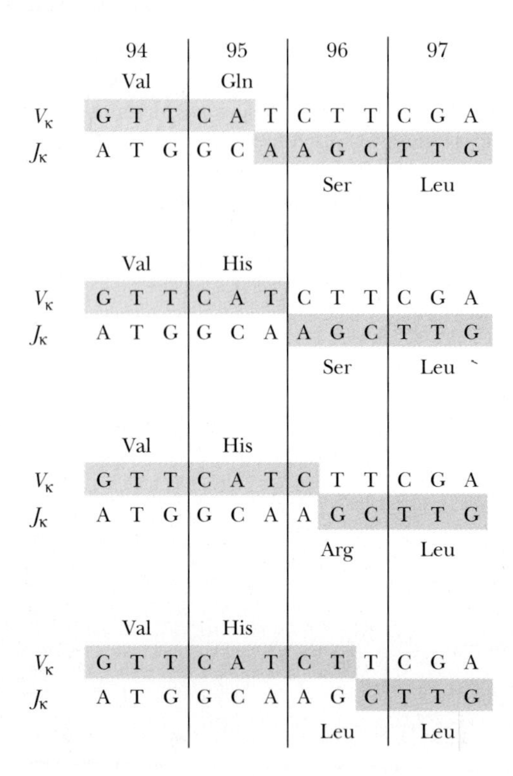

***Figure* 28.24** Recombination between the V_κ and J_κ genes can vary by several nucleotides, giving rise to variations in amino acid sequence and hence diversity in immunoglobulin L chains.

Imprecise Joining

Joining of the ends of the immunoglobulin-coding regions during gene reorganization is somewhat imprecise. This imprecision actually leads to even greater antibody diversity since new coding arrangements result. Position 96 in κ chains is typically encoded by the first triplet in the J_κ element. Most κ chains have one of four amino acids here, depending on which J_κ gene was recruited in gene assembly. However, occasionally only the second and third bases or just the third base of the codon for position 96 is contributed by the J_κ gene, with the other one or two nucleotides supplied by the V_κ segment (Figure 28.24). So the precise point where recombination occurs during gene reorganization can vary over several nucleotides, creating even more diversity.

Antibody Diversity

Taking as an example the mouse with perhaps 300 V_κ genes, 4 J_κ genes, 200 V_H genes, 12 D genes, and 4 J_H genes, the number of possible combinations is given by $300 \times 4 \times 200 \times 12 \times 4$. Thus, greater than 10^7 different antibody molecules can be created from roughly 500 or so different mouse variable-region genes. Including the possibility for V_κ-J_κ joinings occurring within codons adds to this diversity, as does the high rate of somatic mutation associated with the variable-region genes. (Somatic mutations are mutations that arise in diploid cells and are transmitted to the progeny of these cells within the organism, but not to the offspring of the organism.) Clearly, gene rearrangement is a powerful mechanism for dramatically enhancing the protein-coding potential of genetic information.

28.6 The Molecular Nature of Mutation

Genes are normally transmitted unchanged from generation to generation, owing to the great precision and fidelity with which genes are copied during chromosome duplication. However, on rare occasions, genetically heritable changes (**mutations**) occur that result in altered forms. Most mutated genes function less effectively than the unaltered, wild-type allele, but occasionally mutations arise that give the organism a selective advantage. When this occurs, they are propagated to many offspring. Together with recombination, mutation provides for genetic variability within species and, ultimately, the evolution of new species.

Mutations change the sequence of bases in DNA, either by the substitution of one base pair for another (so-called **point mutations**), or by the insertion or deletion of one or more base pairs (**insertions** and **deletions**).

Point Mutations

Point mutations are the class of mutations in which one base pair is substituted for another. The two possible kinds of point mutations are **transitions,** where one purine (or pyrimidine) is replaced by another, as in A → G (or T → C), and **transversions,** where a purine is substituted for a pyrmidine or *vice versa.*

Point mutations arise by the pairing of bases with inappropriate partners, by the introduction of base analogs into DNA, or by chemical mutagens. Bases may rarely mispair, because of their tautomeric properties (see Chapter 6). For example, an amino group ($-NH_2$), usually an H-bond donor, can tautomerize to an imino form ($=NH$) and become an H-bond acceptor. Or a keto group ($C=O$), normally an H-bond acceptor, can tautomerize to an

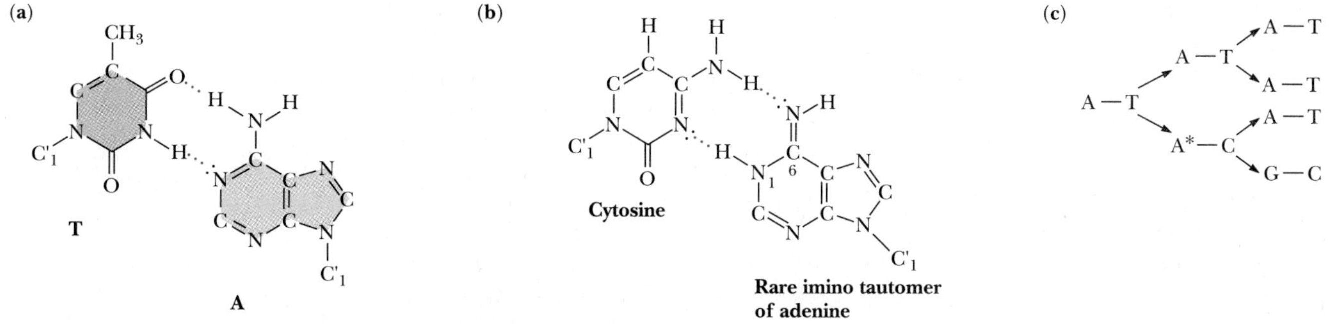

Figure **28.25** The rare imino tautomer of adenine base-pairs with cytosine rather than thymine. (a) The normal A-T base pair. (b) The A*-C base pair is possible due to the adenine tautomer that results from the transfer of a proton from the 6-NH_2 of adenine to N-1. (c) The pairing of C with the rare tautomer of adenine (A*) leads to a transition mutation (A-T to G-C) appearing in the next generation.

enol C—OH, an H-bond donor. About 0.01% of bases are in these rare tautomeric forms at any given moment. Figure 28.25 depicts the imino form of adenine, which base-pairs with cytosine instead of thymine, its usual partner. This mispairing can cause the change of an A-T base pair to a G-C pair in the next generation. Although adenine exists as the imino tautomer at a frequency of 1 in 10^4, proofreading mechanisms operating during DNA replication catch most mispairings. The frequency of spontaneous mutation in both *E. coli* and fruit flies (*Drosophila melanogaster*) is about 10^{-10} per base pair per replication.

Mutations Induced by Base Analogs

Base analogs that become incorporated into DNA can induce mutations through changes in base-pairing possibilities. Two examples are **5-bromouracil (5-BU)** and **2-aminopurine (2-AP).** 5-Bromouracil is a thymine analog and becomes inserted into DNA at sites normally occupied by T; its 5-Br group sterically resembles thymine's 5-methyl group. However, since 5-BU frequently assumes the enol tautomeric form and pairs with G instead of A, a point mutation of the transition type may be induced (Figure 28.26). Less often, 5-BU is inserted into DNA at cytosine sites, not T sites. Then, if it base-pairs in its keto form, mimicking T, a C-G to T-A transition ensues. The adenine analog, 2-aminopurine (recall adenine is 6-aminopurine) normally behaves like A and base-pairs with T. However, 2-AP can form a single H bond of sufficient stability with cytosine (Figure 28.27) that occasionally C replaces T in DNA replicating in the presence of 2-AP. Hypoxanthine (Figure 28.28) is an adenine analog that arises *in situ* in DNA through oxidative deamination of A. Hypoxanthine base-pairs with cytosine, creating an A-T to G-C transition.

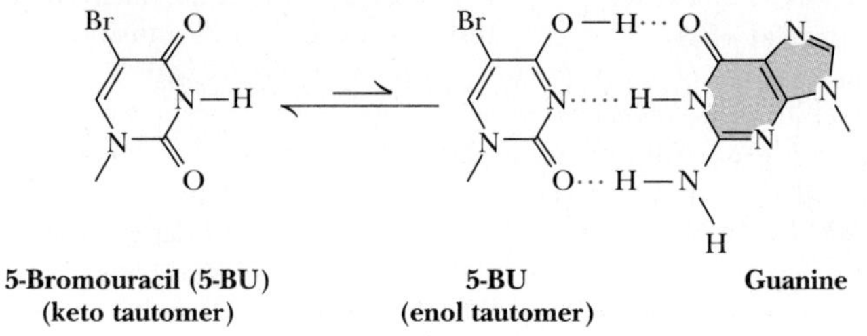

Figure **28.26** 5-Bromouracil usually favors the keto tautomer that mimics the base-pairing properties of thymine, but it frequently shifts to the enol form, whereupon it can base-pair with guanine, causing a T-A to C-G transition.

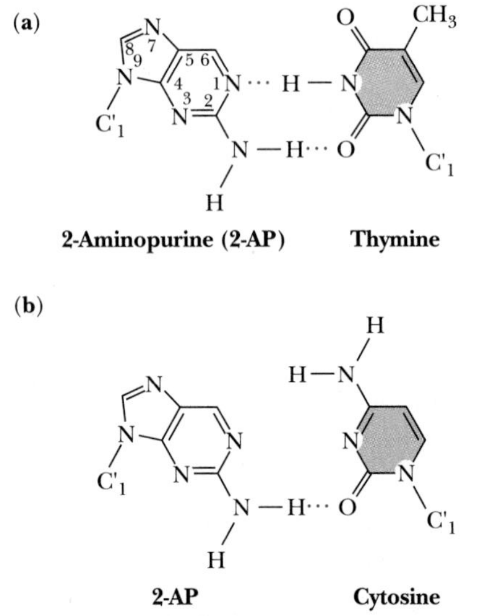

(a) 2-Aminopurine (2-AP) Thymine

(b) 2-AP Cytosine

Figure **28.27** (a) 2-Aminopurine (2-AP) normally base-pairs with T, but (b) may also pair with cytosine through a single hydrogen bond.

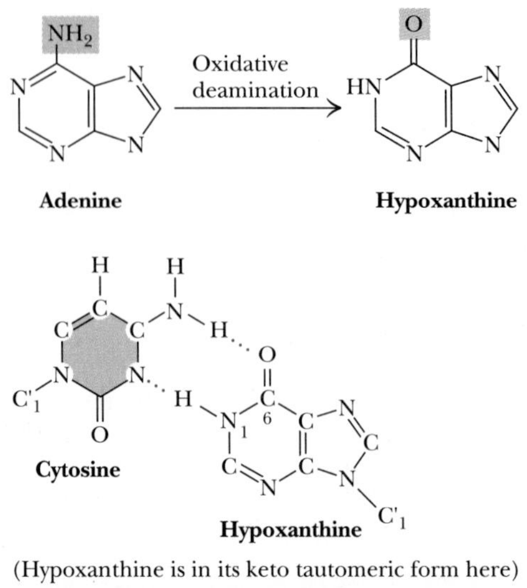

Adenine → Oxidative deamination → Hypoxanthine

Cytosine Hypoxanthine

(Hypoxanthine is in its keto tautomeric form here)

Figure **28.28** Oxidative deamination of adenine in DNA yields hypoxanthine, which base-pairs with cytosine, resulting in an A-T to G-C transition.

Chemical Mutagens

Chemical mutagens are agents that chemically modify bases so that their base-pairing characteristics are altered. For instance, *nitrous acid* (HNO_2) causes the oxidative deamination of primary amines such as adenine and cytosine. Oxidative deamination of cytosine yields uracil, which base-pairs the way T does and gives a C-G to T-A transition (Figure 28.29a). *Hydroxylamine* specifically causes C-G to T-A transitions because it reacts specifically with cytosine, converting it to a derivative that base-pairs with adenine instead of guanine (Figure 28.29c). **Alkylating agents** are also chemical mutagens. Alkylation of reactive sites on the bases with methyl or ethyl groups alters their H-bonding and hence base pairing. For example, methylation of O^6 on guanine (giving O^6-methylguanine) causes this G to mispair with thymine, resulting in a G-C to A-T transition (Figure 28.29d). Alkylating agents can also induce point mutations of the transversion type. Alkylation of N^7 of guanine labilizes its *N*-glycosidic bond, which leads to elimination of the purine ring, creating a gap in the base sequence. An enzyme, **apurinic acid endonuclease,** then cleaves the sugar-phosphate backbone of the DNA on the 5′-side, and the gap can be repaired by enzymatic removal of the 5′-sugar phosphate and insertion of a new nucleotide. A transversion results if a pyrimidine nucleotide is inserted in place of the purine during enzymatic repair of this gap. A number of alkylating agents are shown in Figure 28.29e.

Figure **28.29** Chemical mutagens. (a) HNO_2 (nitrous acid) converts cytosine to uracil and adenine to hypoxanthine. (b) Nitrosoamines, organic compounds that react to form nitrous acid, also lead to the oxidative deamination of A and C. (c) Hydroxylamine (NH_2OH) reacts with cytosine, converting it to a derivative that base-pairs with adenine instead of guanine. The result is a C-G to T-A transition. (d) Alkylation of G residues gives O^6-methylguanine, which base-pairs with T. (e) Alkylating agents include *nitrosoamines, nitrosoguanidines, nitrosoureas, alkyl sulfates,* and *nitrogen mustards.* Note that nitrosoamines are mutagenic in two ways: They can react to yield HNO_2 or they can act as alkylating agents. The nitrosoguanidine *N-methyl-N′-nitro-N-nitrosoguanidine* is a very potent mutagen used in laboratories to induce mutations in experimental organisms such as *Drosophila melanogaster. Ethylmethane sulfonate (EMS)* and *dimethyl sulfate* are also favorite mutagens among geneticists.

(a)

Cytosine $\xrightarrow{HNO_2}$ Uracil ··· Adenine

Adenine $\xrightarrow{HNO_2}$ Hypoxanthine ··· Cytosine

(b) Generic structure of nitrosoamines

$R_1(R_2)N-N=O$

(c)

Cytosine $\xrightarrow{NH_2OH}$ ··· Adenine

(d)

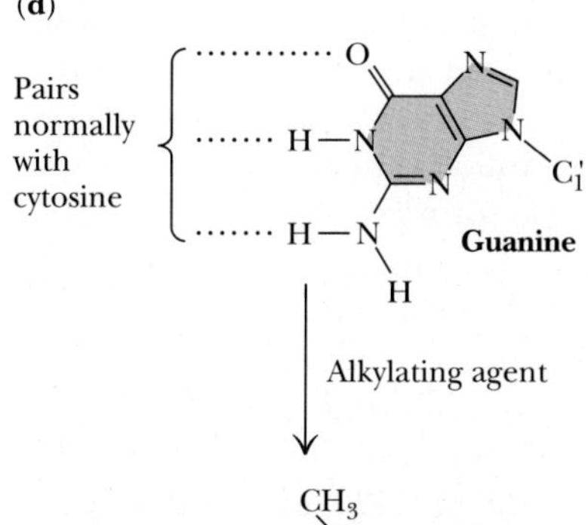

(e)

Nitrosoamines:

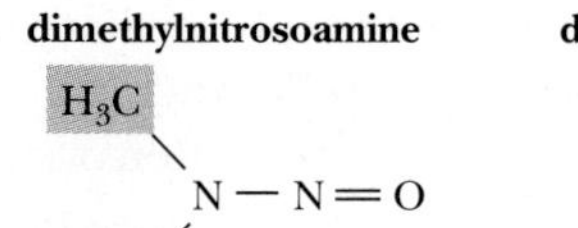

dimethylnitrosoamine $(H_3C)_2N-N=O$

diethylnitrosoamine $(CH_3CH_2)_2N-N=O$

Nitrosoguanidine:

N-methyl-N'-nitro-N-nitrosoguanidine $O=N-N(CH_3)-C(=NH)-NH-NO_2$

Nitrosourea:

Ethyl nitrosourea $H_2N-C(=O)-N(CH_2CH_3)-N=O$

Alkyl Sulfates:

Ethylmethane sulfonate $H_3C-S(=O)_2-O-CH_2CH_3$

Dimethyl sulfate $H_3C-O-S(=O)_2-O-CH_3$

Nitrogen Mustard:

$H_3C-N(CH_2-CH_2-Cl)(NH_2-CH_2-Cl)$

Insertions and Deletions

The addition or removal of one or more base pairs leads to *insertion* or *deletion* mutations, respectively. Such mutations can arise when flat aromatic molecules such as *acridine orange* (see Figure 7.19) insert themselves between successive bases in one or both strands of the double helix. This insertion, or more aptly, **intercalation,** doubles the distance between the bases as measured along the helix axis. This distortion of the DNA (see Figure 7.19) results in bases being inappropriately inserted or deleted when the DNA is replicated. Disruptions that arise from the insertion of a transposon within a gene also fall in this category of mutation.

28.7 RNA as Genetic Material

Whereas the genetic material of cells is double-stranded DNA, virtually all plant viruses, several bacteriophage, and many animal viruses have genomes consisting of RNA. In most cases, this RNA is single-stranded. Viruses with single-stranded genomes use the single strand as template for synthesis of a complementary strand, which can then serve as template in replicating the original strand. **Retroviruses** are an interesting group of eukaryotic viruses having single-stranded RNA genomes that replicate through a double-stranded DNA intermediate. Further, the life cycle of retroviruses includes an obligatory step in which the dsDNA is inserted into the host cell genome in a transposition event. Retroviruses are responsible for many diseases, including tumors and other disorders. **HIV-1,** the **human immunodeficiency virus** that causes **AIDS,** is a retrovirus. (Retroviruses will be considered in detail in subsequent chapters.) **Tobacco mosaic virus (TMV),** an RNA virus infecting plants, was instrumental in establishing that nucleic acids are the substance of heredity. TMV has a molecular mass of 40×10^3 kD and consists of an RNA genome (3×10^3 kD) packaged in a protein coat made of 2130 identical protein chains of 18 kD each (see Figures 1.24, 34.9, and 34.11). In 1956, Gierer and Schramm demonstrated that the RNA itself was able to produce viral lesions on the surfaces of tobacco leaves, if the leaf surface was lightly scratched so

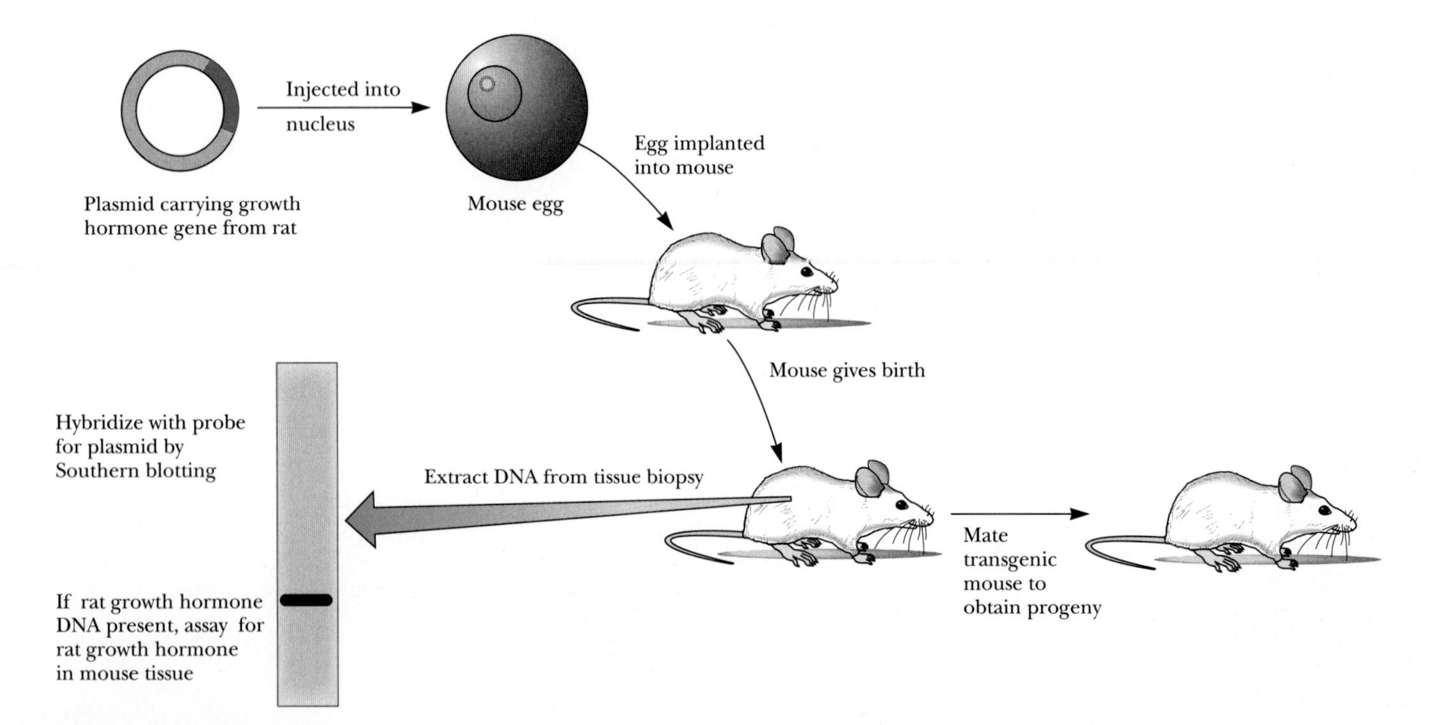

***Figure* 28.30** Transfection can introduce new genes into animals. The rat growth hormone gene carried on a plasmid is injected into a mouse oocyte or fertilized egg that is then implanted in a receptive female mouse. Integration of the plasmid into the mouse genome can be ascertained by Southern analysis of DNA from the newborn mouse. Expression of the foreign gene can be determined by assaying for the gene product, in this case, rat growth hormone.

the RNA could gain access to the cells. In 1957, Fraenkel-Conrat and Singer used two different strains of the virus, HR and TMV, and reconstructed virus particles *in vitro* by mixing isolated proteins and RNAs in the 4 possible combinations:

TMV protein + HR RNA
TMV protein + TMV RNA
HR protein + HR RNA
HR protein + TMV RNA

These reconstituted virus particles were infective, and, when the virus progeny obtained after their infection of host plants were examined, it was found that the protein coat borne by the progeny virus particles was determined by the source of RNA in the virus infecting the plant: TMV RNA always yielded TMV protein coats in the progeny; HR RNA yielded HR protein coats. This experiment was early proof that nucleic acids, not proteins, are the repository of genetic information.

28.8 Transgenic Animals

An exciting new advance in gene transfer techniques is the ability to introduce genes into animals by transfection. Animals that have acquired new genetic information as a consequence of the introduction of foreign genes are termed **transgenic.** The methodology involves the injection of plasmids carrying the gene of interest into the nucleus of an oocyte or fertilized egg, followed by implantation of the egg into a receptive female. The technique has been perfected for mice (Figure 28.30). In a small number of cases, 10% or so, the mice that develop from the injected eggs carry the transfected gene integrated into a single chromosomal site. The gene is subsequently inherited by the progeny of the transfected animal as if it were a normal gene. Expression of the donor gene in the transgenic animals is variable, because the gene is randomly integrated into the host genome and gene expression is often influenced by chromosomal location. Nevertheless, transfection of animals has produced some startling results, as in the case of the transfection of mice with the gene encoding the **rat growth hormone (rGH).** The transgenic mice grew to nearly twice the normal size (Figure 28.31). Growth hormone levels in

***Figure* 28.31** Photograph showing a transgenic mouse with an active rat growth hormone gene (left). This transgenic mouse is twice the size of a normal mouse (right). *(Photo courtesy of Ralph L. Brinster, School of Veterinary Medicine, University of Pennsylvania.)*

these animals were several hundred times greater than normal. Similar results were obtained in transgenic mice transfected with the **human growth hormone (hGH)** gene.

The biotechnology of transfection has been extended to farm animals, and transgenic chickens, cows, pigs, rabbits, sheep, and even fish have been produced. These transgenic animals presently experience negative side effects. However, such genetic engineering is anticipated eventually to have a major impact on agriculture. The human genes encoding the α- and β-globin chains of hemoglobin have recently been microinjected into fertilized mouse eggs, and the transgenic mice that developed contained authentic human hemoglobin. Human Hb isolated from transgenic mouse erythrocytes had an oxygen-binding curve identical to that of human HbA, demonstrating that functional human hemoglobin can be synthesized in mice. Transgenic pigs producing human Hb are touted as a source of "human blood substitute" potentially useful during surgical procedures. Such transfection technology also holds promise as a mechanism for "gene therapy" by replacing defective genes in animals with functional genes (Chapter 8). Problems concerning integration and regulation of the transfected gene, including its appropriate expression in the right cells at the proper time during development and growth of the organism, must be brought under control before such therapy becomes commonplace.

Problems

1. Design an experiment to select for spontaneous arginine$^-$, ornithine$^+$ mutations in the arginine biosynthetic pathway of a microorganism.

2. F factors can integrate into the genome in either orientation. Assuming an F factor has integrated into the *E. coli* chromosome near the *polA* marker (minute 85), what are the two possibilities in terms of the order of gene transfer in an interrupted mating experiment of this *Hfr* strain with F$^-$ strain?

3. From the information in Figure 28.14, diagram the recombinational event leading to the formation of a heteroduplex DNA region within a bacteriophage chromosome.

4. From the base-pairing "rules" for triplex DNA formation outlined in Figure 28.16b, diagram how an invading ssDNA strand of sequence ACGCTCAGA will interact with its homologous region on a dsDNA molecule.

5. If RecA protein unwinds duplex DNA so that there are about 18.6 bp per turn, what is the *change* in $\Delta\phi$, the helical twist of DNA, compared to its value in B-DNA?

6. Using Figure 28.18 as a guide, diagram a Holliday junction between two duplex DNA molecules and show how the action of resolvase might give rise to either patch or splice recombinant DNA molecules.

7. From the data in Figure 28.24, how many different amino acid sequences are possible across the V_κ-J_κ light-chain region from residue 94 to 97?

8. Show the nucleotide sequence changes that might arise in a dsDNA (coding strand segment GCTA) upon mutagenesis with (a) HNO_2 (b) bromouracil, and (c) 2-aminopurine.

9. Transposons are mutagenic agents. Why?

10. Figure 28.16(c) shows a postulated set of hydrogen-bonding patterns between the bases in triplex DNA. Alternative hydrogen-bonding patterns between the bases in triple-stranded DNA have been proposed. One widely accepted alternative suggests that the N7 atoms of purines in Watson–Crick base pairs are *not* involved in H-bonding with the corresponding third-strand base, but the pyrimidine of each Watson–Crick base pair *is*. Further, in this alternative scheme, the C residues are *not* protonated but are in their usual tautomeric form (as in Figure 6.3). Draw a hydrogen-bonded set of the four respective pairing possibilities (A-T with A; A-T with T; G-C with G; and G-C with C) that demonstrates H-bond interactions in which the Watson–Crick pyrimidines participate but the N7 of purines does not.
(*Hint:* Make a photocopy of Figure 28.16(c), separate the Watson–Crick base pairs and their triplex partners by cutting between them with scissors; then see if you can realign them into alternative H-bonding interactions. [For the third-strand A with the A-T base pair, flip the A over, hold it up to the light, and trace its purine ring system on the blank side. Use this "flipped" A in constructing your new triplex set.])

Further Reading

Alt, F. W., Blackwell, T. K., and Yancopoulos, G. D., 1987. Development of the primary antibody repertoire. *Science* **238:**1079–1087.

Behringer, R. R., et al., 1989. Synthesis of functional human hemoglobin in transgenic mice. *Science* **245:**971–979.

Camerini-Otero, R. D., and Hsieh, P., 1993. Parallel DNA triplexes, homologous recombination, and other recombination-dependent DNA interactions. *Cell* **73:**217–223.

Chiu, S. K., Rao, B. J., Story, R. M., and Radding, C. M., 1993. Interactions of three strands in joints made by RecA protein. *Biochemistry* **32:**13146–13155.

Connolly, B., et al., 1991. Resolution of Holliday junctions *in vitro* requires the *Escherichia coli ruvC* gene product. *Proceedings of the National Academy of Science, USA* **88:**6063–6067.

Cox, M. M., and Lehman, I. R., 1987. Enzymes of general recombination. *Annual Review of Biochemistry* **56:**229–262. An excellent review of the enzymology of homologous recombination.

Dixon, D. A., and Kowalczykowski, S. C., 1991. Homologous pairing *in vitro* stimulated by the recombination hotspot *Chi. Cell* **66:**361–371. Experimental evidence for the role of *Chi* sequences in homologous recombination.

Hayes, W., 1968. *The Genetics of Bacteria and Their Viruses,* 2nd ed. New York: John Wiley & Sons. An early, advanced text covering many of the historical developments in bacterial genetics that led to the science we now know as molecular biology.

Holliday, R., 1964. A mechanism for gene conversion in fungi. *Genetic Research* **5:**282–304. The classic model for the mechanism of DNA strand exchange during homologous recombination.

Hsieh, P., Camerini-Otero, C. S., and Camerini-Otero, R. D., 1993. Pairing of homologous DNA sequences by proteins: Evidence for three-stranded DNA. *Genes & Development* **4:**1951–1963.

Htun, H., and Dahlberg, J. E., 1989. Topology and formation of triple-stranded H-DNA. *Science* **243:**1571–1576. A model for the structure of triple-stranded DNA is proposed.

Lambowitz, A. M., and Belfort, M., 1993. Introns as mobile genetic elements. *Annual Review of Biochemistry* **62:**587–622.

Lewin, B., 1994. *Genes V.* New York: Oxford University Press, and Cambridge, MA: Cell Press. A contemporary genetics text that seeks to explain heredity in terms of molecular structures.

Meselson, M., and Weigle, J. J., 1961. Chromosome breakage accompanying genetic recombination in bacteriophage. *Proceedings of the National Academy of Sciences, U.S.A.* **47:**857–869. The experiments demonstrating that physical exchange of DNA occurs during recombination.

Morgan, R. A., and Anderson, W. F., 1993. Human gene therapy. *Annual Review of Biochemistry* **62**:192–217.

Müller, B., et al., 1990. Enzymatic formation and resolution of Holliday junctions *in vitro. Cell* **60:**329–336. A reevaluation of the Holliday model in light of recent advances in the enzymology of homologous recombination.

Neihardt, F. C., et al., eds., 1987. *Escherichia coli and Salmonella typhimurium: Cellular and Molecular Biology,* vols. 1 and 2. Washington, DC: American Society for Microbiology Press. Definitive treatise on the two bacteria most widely used in molecular biological research.

Palmiter, R. D., et al., 1982. Dramatic growth of mice that develop from eggs microinjected with metallothionein-growth hormone fusion genes. *Nature* **300:**611–615.

Palmiter, R. D., et al., 1983. Metallothionein-human GH fusion genes stimulate growth of mice. *Science* **222:**809–814.

Pursel, V. G., et al., 1989. Genetic engineering of livestock. *Science* **244:**1281–1288.

Radding, C. M., 1991. Helical interactions in homologous pairing and strand exchange driven by *recA* protein. *The Journal of Biological Chemistry* **266:**5355–5358.

Rao, B. J., Chiu, S. K., and Radding, C. M., 1993. Homologous recognition and triplex formation promoted by RecA protein between duplex oligonucleotides and single-stranded DNA. *Journal of Molecular Biology* **229:**328–343.

Smith G. R., 1988. Homologous recombination in prokaryotes. *Microbiological Reviews* **52:**1–28. An authoritative review on homologous recombination by one of its leading investigators.

Story, R. M., et al., 1993. Structural relationship of the bacterial *RecA* proteins to recombination proteins from bacteriophage T4 and yeast. *Science* **259:**1892–1899.

Story, R. M., Weber, I. T., and Steitz, T. A., 1992. The structure of the *E. coli RecA* protein monomer and polymer. *Nature* **355:**318–325.

Tonegawa, S., 1983. Somatic generation of antibody diversity. *Nature* **302:**575–581.

Watson, J. D., et al., 1987. *Molecular Biology of the Gene,* 4th ed., vol. 1. Menlo Park, CA: Benjamin/Cummings. The leading textbook of molecular biology.

West, S. C., 1992. Enzymes and molecular mechanisms of genetic recombination. *Annual Review of Biochemistry* **61:**603–640.

Yancopoulos, G. D., and Alt, F. W., 1986. Assembly and expression of variable-region genes. *Annual Review of Immunology* **4:**339–368.

"When you come to a fork in the road, take it."

Lawrence Peter (Yogi) Berra
Baseball Hall-of-Famer (1925–)

Chapter 29

DNA Replication and Repair

Noah's Ark, by Andre Normil

Outline

29.1 DNA Replication Is Semiconservative
29.2 The Enzymology of DNA Replication
29.3 General Features of DNA Replication
29.4 The Mechanism of DNA Replication in *E. coli*
29.5 Eukaryotic DNA Replication
29.6 Reverse Transcriptase: An RNA-Directed DNA Polymerase
29.7 DNA Repair

The publication of Watson and Crick's famous paper titled *Molecular Structure of Nucleic Acids: A Structure for Deoxyribose Nucleic Acid* (Figure 29.1), marked the dawn of a new scientific epoch, the age of molecular biology. As these authors drew to a close their brief but far-reaching description of the DNA double helix, they pointedly commented, "It has not escaped our notice that the specific [base] pairing we have postulated immediately suggests a possible copying mechanism for the genetic material." The mechanism for DNA replication that Watson and Crick viewed as intuitively obvious is *strand separation followed by the copying of each strand.* In the process, each separated strand acts as a **template** for the synthesis of a new complementary strand whose nucleotide sequence is fixed by the base-pairing rules Watson and Crick proposed. Strand separation is achieved by untwisting the double helix (Figure 29.2). Base pairing then dictates an accurate replication of the original DNA double helix.

template: something whose edge is shaped in a particular way so that it can serve as a guide in making a similar object with a corresponding contour

No. 4356 **April 25, 1953** NATURE 737

MOLECULAR STRUCTURE OF NUCLEIC ACIDS

A Structure for Deoxyribose Nucleic Acid

WE wish to suggest a structure for the salt of deoxyribose nucleic acid (D.N.A.). This structure has novel features which are of considerable biological interest.

A structure for nucleic acid has already been proposed by Pauling and Corey[1]. They kindly made their manuscript available to us in advance of publication. Their model consists of three intertwined chains, with the phosphates near the fibre axis, and the bases on the outside. In our opinion, this structure is unsatisfactory for two reasons: (1) We believe that the material which gives the X-ray diagrams is the salt, not the free acid. Without the acidic hydrogen atoms it is not clear what forces would hold the structure together, especially as the negatively charged phosphates near the axis will repel each other. (2) Some of the van der Waals distances appear to be too small.

Another three-chain structure has also been suggested by Fraser (in the press). In his model the phosphates are on the outside and the bases on the inside, linked together by hydrogen bonds. This structure as described is rather ill-defined, and for his reason we shall not comment on it.

We wish to put forward a radically different structure for the salt of deoxyribose nucleic acid. This structure has two helical chains each coiled round the same axis (see diagram). We have made the usual chemical assumptions, namely, that each chain consists of phosphate diester groups joining β-D-deoxyribofuranose residues with 3′,5′ linkages. The two chains (but not their bases) are related by a dyad perpendicular to the fibre axis. Both chains follow right-handed helices, but owing to the dyad the sequences of the atoms in the two chains run in opposite directions. Each chain loosely resembles Furberg's[2] model No. 1; that is, the bases are on the inside of the helix and the phosphates on the outside. The configuration of the sugar and the atoms near it is close to Furberg's 'standard configuration', the sugar being roughly perpendicular to the attached base. There is a residue on each chain every 3·4 A. in the z-direction. We have assumed an angle of 36° between adjacent residues in the same chain, so that the structure repeats after 10 residues on each chain, that is, after 34 A. The distance of a phosphorus atom from the fibre axis is 10 A. As the phosphates are on the outside, cations have easy access to them.

Tl diagrammatic. The two ribbons symbolize the two phosphate—sugar chains, and the horizontal rods the pairs of bases holding the chains together. The vertical line marks the fibre axis

The structure is an open one, and its water content is rather high. At lower water contents we would expect the bases to tilt so that the structure could become more compact.

The novel feature of the structure is the manner in which the two chains are held together by the purine and pyrimidine bases. The planes of the bases are perpendicular to the fibre axis. They are joined together in pairs, a single base from one chain being hydrogen-bonded to a single base from the other chain, so that the two lie side by side with identical z-co-ordinates. One of the pair must be a purine and the other a pyrimidine for bonding to occur. The hydrogen bonds are made as follows: purine position 1 to pyrimidine position 1; purine position 6 to pyrimidine position 6.

If it is assumed that the bases only occur in the structure in the most plausible tautomeric forms (that is, with the keto rather than the enol configurations) it is found that only specific pairs of bases can bond together. These pairs are: adenine (purine) with thymine (pyrimidine), and guanine (purine) with cytosine (pyrimidine).

In other words, if an adenine forms one member of a pair, on either chain, then on these assumptions the other member must be thymine; similarly for guanine and cytosine. The sequence of bases on a single chain does not appear to be restricted in any way. However, if only specific pairs of bases can be formed, it follows that if the sequence of bases on one chain is given, then the sequence on the other chain is automatically determined.

It has been found experimentally[3,4] that the ratio of the amounts of adenine to thymine, and the ratio of guanine to cytosine, are always very close to unity for deoxyribose nucleic acid.

It is probably impossible to build this structure with a ribose sugar in place of the deoxyribose, as the extra oxygen atom would make too close a van der Waals contact.

The previously published X-ray data[5,6] on deoxyribose nucleic acid are insufficient for a rigorous test of our structure. So far as we can tell, it is roughly compatible with the experimental data, but it must be regarded as unproved until it has been checked against more exact results. Some of these are given in the following communications. We were not aware of the details of the results presented there when we devised our structure, which rests mainly though not entirely on published experimental data and stereochemical arguments.

It has not escaped our notice that the specific pairing we have postulated immediately suggests a possible copying mechanism for the genetic material.

Full details of the structure, including the conditions assumed in building it, together with a set of co-ordinates for the atoms, will be published elsewhere.

We are much indebted to Dr. Jerry Donohue for constant advice and criticism, especially on interatomic distances. We have also been stimulated by a knowledge of the general nature of the unpublished experimental results and ideas of Dr. M. H. F. Wilkins, Dr. R. E. Franklin and their co-workers at King's College, London. One of us (J. D. W.) has been aided by a fellowship from the National Foundation for Infantile Paralysis.

J. D. WATSON
F. H. C. CRICK

Medical Research Council Unit for the Study of the Molecular Structure of Biological Systems,
Cavendish Laboratory, Cambridge.
April 2.

[1] Pauling, L., and Corey, R. B., *Nature*, **171**, 346 (1953); *Proc. U.S. Nat. Acad. Sci.*, **39**, 84 (1953).
[2] Furberg, S., *Acta Chem. Scand.*, **6**, 634 (1952).
[3] Chargaff, E., for references see Zamenhof, S., Brawerman, G., and Chargaff, E., *Biochim. et Biophys. Acta*, **9**, 402 (1952).
[4] Wyatt, G. R., *J. Gen. Physiol.*, **36**, 201 (1952).
[5] Astbury, W. T., Symp. Soc. Exp. Biol. 1, Nucleic Acid, 66 (Camb. Univ. Press, 1947).
[6] Wilkins, M. H. F., and Randall, J. T., *Biochim. et Biophys. Acta*, **10**, 192 (1953).

***Figure* 29.1** Watson and Crick's famous paper, in its entirety.
(Watson, J. D., and Crick, F. H. C., 1953. Nature 171:737–738.)

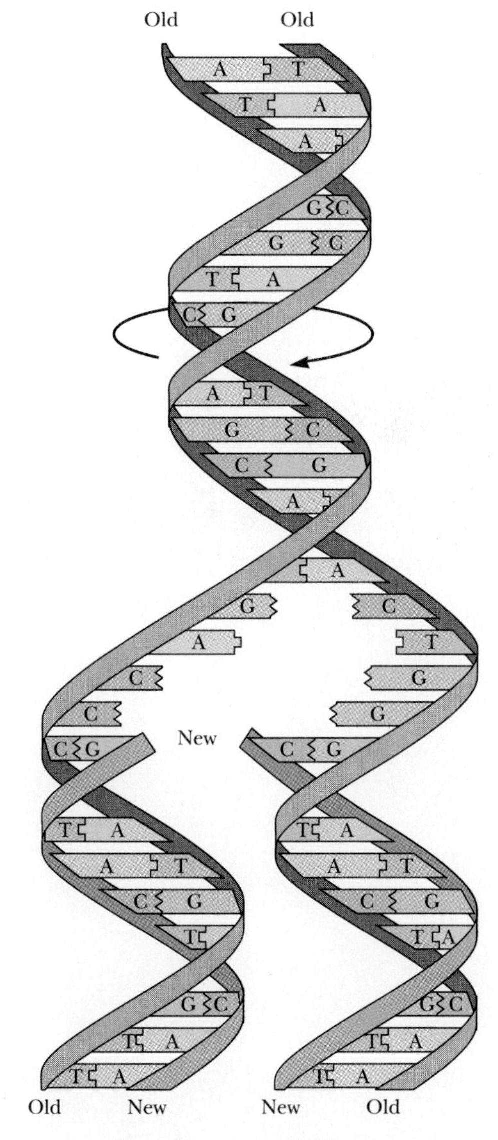

***Figure* 29.2** Untwisting of DNA strands exposes their bases for hydrogen bonding. Base pairing ensures that appropriate nucleotides are inserted in the correct positions as the new complementary strands are synthesized. By this mechanism, the nucleotide sequence of one strand dictates a complementary sequence in its daughter strand. The original strands untwist by rotating about the axis of the unreplicated DNA double helix.

29.1 DNA Replication Is Semiconservative

Actually, three basic models for DNA replication are consistent with the requirement that the nucleotide sequence in one strand dictate, through Watson and Crick's base-pairing rules, the sequence of nucleotides in the other. These three models—conservative, semiconservative, and dispersive—are diagrammed in Figure 29.3. In 1958, Matthew Meselson and Franklin Stahl provided the experimental proof for the **semiconservative model** of DNA replication. *Escherichia coli* cells were grown for many generations in medium containing $^{15}NH_4Cl$ as the sole nitrogen source. Thus, the nitrogen atoms in the purine and pyrimidine bases of the DNA in these cells were mostly ^{15}N, the stable heavy isotope of nitrogen. Then, a tenfold excess of ordinary $^{14}NH_4Cl$ was added to the growing culture, and, at appropriate intervals, cells were collected from the culture and lysed. The DNA they contained was analyzed by CsCl density gradient ultracentrifugation (Chapter 7, Appendix). This technique can resolve macromolecules differing in density by less than 0.01 g/mL.

DNA isolated from cells grown on $^{15}NH_4^+$ (the "0" generation cells) banded in the ultracentrifuge at a density corresponding to 1.724 g/mL,

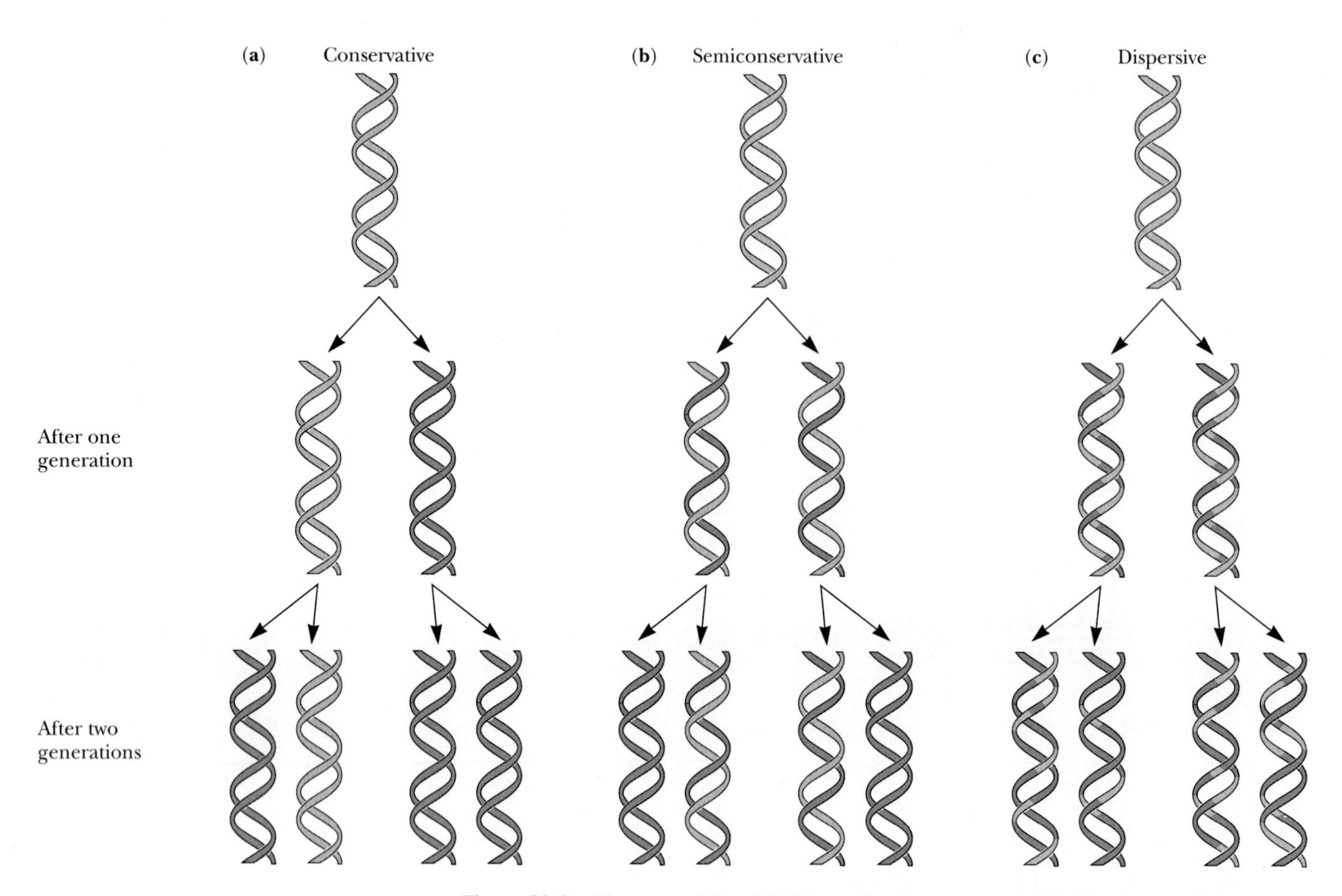

***Figure* 29.3** Three models of DNA replication prompted by Watson and Crick's double helix structure for DNA. (a) *Conservative:* Each strand of the DNA duplex is replicated and the two newly synthesized strands join to form one DNA double helix while the two parental strands remain associated with each other. The products are one completely new DNA duplex and the original DNA duplex. (b) *Semiconservative:* The two strands separate and each strand is copied to generate a complementary strand. Each parental strand remains associated with its newly synthesized complement, so that each DNA duplex contains one parental strand and one new strand. (c) *Dispersive:* This model predicts that each of the four strands in the two daughter DNA duplexes contains both newly synthesized segments and segments derived from the parental strands.

whereas DNA from cells grown for 4.1 generations on ^{14}N had a density of 1.710 g/mL (Figure 29.4). These bands represent "heavy" and "light" DNA, respectively. Significantly, DNA isolated from cells grown for just one generation on ^{14}N yielded just a single band corresponding to a density of 1.717 g/mL, halfway between heavy and light DNA, indicating that *each* DNA duplex molecule contained equal amounts of ^{15}N and ^{14}N. This result is consistent with the semiconservative model for DNA replication, at the same time ruling out the conservative model. After approximately two generations on ^{14}N, cells yielded DNA that gave two essentially equal bands upon ultracentrifugation, one at the intermediate density of 1.717 g/mL and the other at the light position, 1.710 g/mL, also in accord with semiconservative replication.

It remained conceivable that DNA replication might follow a dispersive model, so that both strands of DNA in the chromosomes of cells after one generation on ^{14}N might be intermediate in density. Meselson and Stahl eliminated this possibility by the following experiment: DNA isolated from ^{15}N-labeled cells kept one generation on ^{14}N was heated at 100°C so that the DNA duplexes were denatured into their component single strands. When this

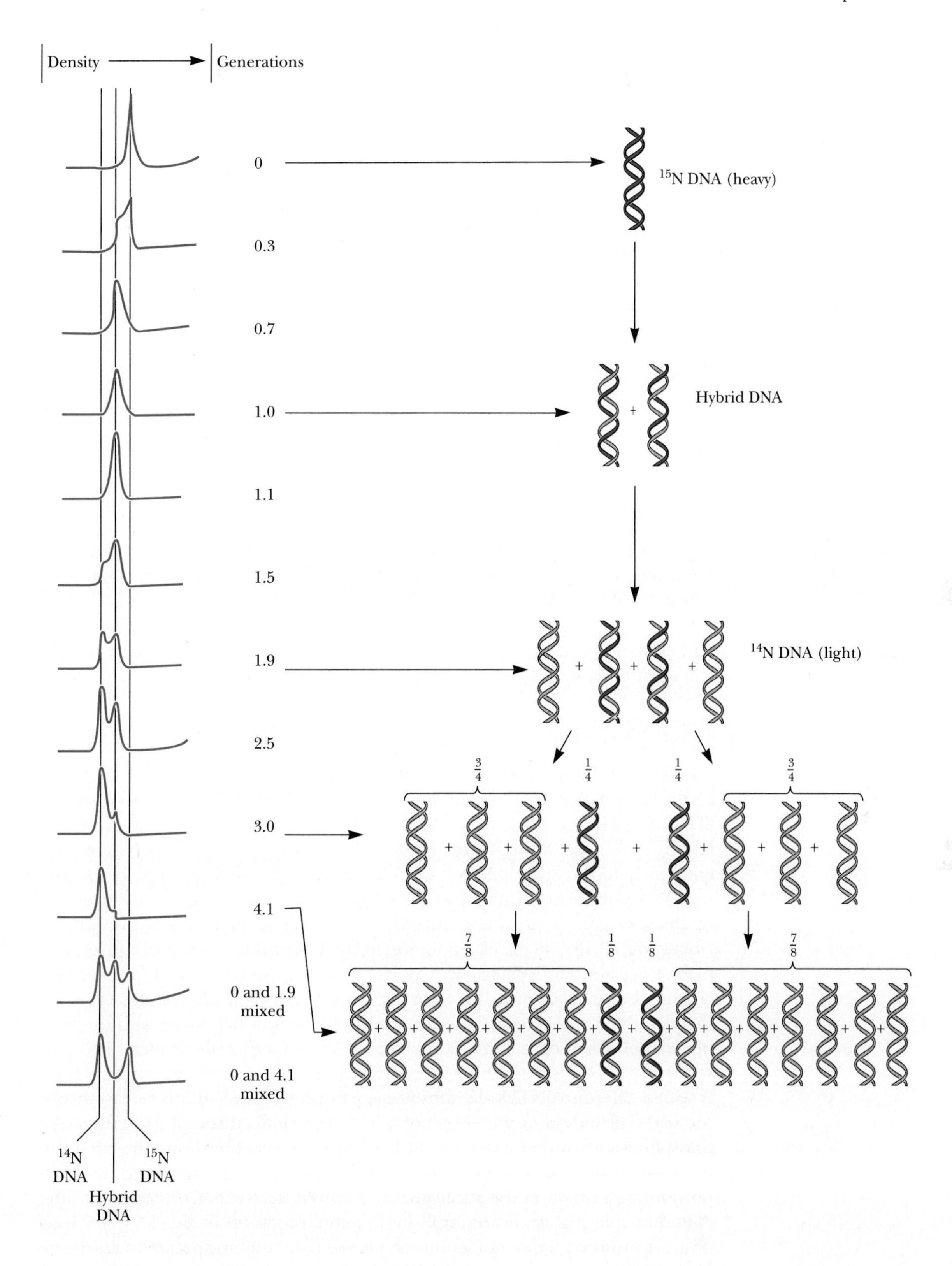

***Figure* 29.4** The Meselson and Stahl experiment demonstrating that DNA replication is semiconservative. On the left are shown densitometric traces made of UV absorption photographs taken of the ultracentrifugation cells containing DNA isolated from *E. coli* grown for various generation times after ^{15}N-labeling. The photographs were taken once the migration of the DNA in the density gradient had reached equilibrium. Density increases from left to right. The peaks reveal the positions of the banded DNA with respect to the density of the solution. The number of generations that the *E. coli* cells were grown (following 14 generations of ^{15}N density-labeling) is shown down the middle of the figure. A schematic representation interpreting the pattern expected of semiconservative replication is shown on the right side of this figure.

(Adapted from Meselson, M. and Stahl, F. W., 1958. Proceedings of the National Academy of Sciences, USA 44:671–682.)

heat-denatured DNA was analyzed by CsCl density gradient ultracentrifugation, two distinct bands were observed, showing that one strand was ^{15}N-labeled and the other strand was ^{14}N-labeled. A dispersive mode of replication would have yielded a single band of DNA of intermediate density. An interesting feature of density gradient ultracentrifugation is that the width of the band occupied by a macromolecular species in the gradient is inversely proportional to the molecular weight of the macromolecule. The band widths of the two bands seen with heat-denatured DNA both correspond to molecular masses one-half that of native DNA. This experiment established that DNA replication proceeded by a semiconservative mechanism and also verified that DNA was indeed composed of two polynucleotide strands of equal size.

29.2 The Enzymology of DNA Replication

Watson and Crick's model for the structure of DNA spurred a search for the enzyme capable of mediating its synthesis. Presumably, such an enzyme would require a DNA strand to serve as a template and would make a complementary strand by polymerizing deoxynucleotides in the order specified by their base-pairing with bases in the template. All DNA polymerases discovered thus far, whether from prokaryotic or eukaryotic sources, share the following properties: (a) The incoming base is selected as directed by Watson–Crick base pairing with the template, (b) chain growth is in the $5' \rightarrow 3'$ direction and is antiparallel to the template strand, and (c) DNA polymerases cannot initiate DNA synthesis *de novo*—all require a primer to build upon.

E. coli DNA Polymerase I

In 1957, Arthur Kornberg and his colleagues observed that an *E. coli* extract could catalyze the incorporation of ^{14}C-labeled thymidine into acid-precipitable material in the presence of ATP. This acid-precipitable material was DNA, and this assay was subsequently exploited in purifying and characterizing the enzyme system responsible for DNA synthesis. The enzyme they discovered was **DNA polymerase I.** Purified DNA polymerase I will catalyze the synthesis of DNA *in vitro* if provided with all four deoxynucleoside-5′-triphosphates (dATP, dTTP, dCTP, dGTP), a template DNA strand to copy, *and* a **primer.** A primer is essential because DNA polymerases can only elongate pre-existing chains; they cannot join two deoxyribonucleoside-5′-triphosphates together to make the initial phosphodiester bond. The primer base-pairs with the template DNA, forming a short, double-stranded region. This primer must possess a free 3′-OH end to which an incoming deoxynucleoside monophosphate is added. All four dNTPs are substrates, pyrophosphate (PP_i) is released, and the dNMP is linked to the 3′-OH of the primer chain through formation of a phosphoester bond (Figure 29.5). The deoxynucleoside monophosphate to be incorporated is chosen through base pairing with the template. As DNA polymerase I catalyzes the successive addition of deoxynucleotide units to the 3′-end of the primer, it accomplishes chain elongation in the $5' \rightarrow 3'$ direction, synthesizing a polynucleotide sequence that runs antiparallel to the template but complementary to it. DNA polymerase I can proceed along the template strand, synthesizing a complementary strand of about 20 bases before it "falls off" (dissociates from) the template. The degree to which the enzyme remains associated with the template through successive cycles of nucleotide addition is referred to as its **processivity.** As DNA polymerases go, DNA polymerase I is a moderately processive enzyme. Arthur Kornberg was awarded the Nobel Prize in physiology or medicine in 1959 for his discovery of DNA polymerase.

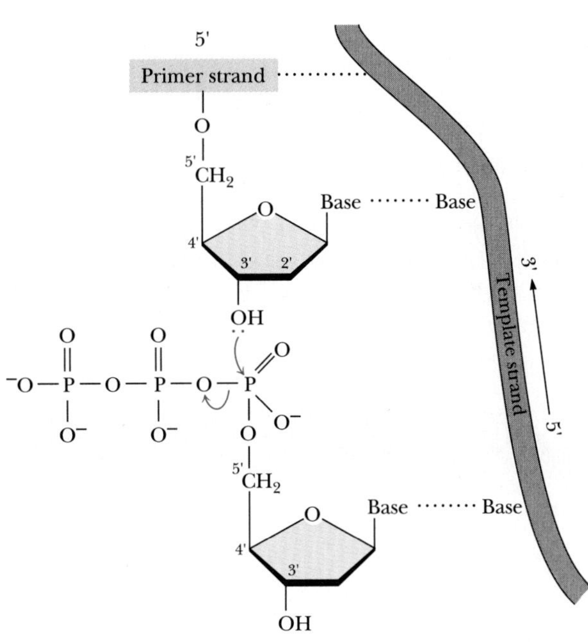

Figure **29.5** The chain elongation reaction catalyzed by DNA polymerase. DNA polymerase I joins deoxynucleoside monophosphate units to the 3′-OH end of the primer, employing dNTPs as substrates. The 3′-OH carries out a nucleophilic attack on the α-phosphoryl group of the incoming dNTP to form a phosphoester bond, and PP_i is released. The subsequent hydrolysis of PP_i by inorganic pyrophosphatase renders the reaction effectively irreversible.

E. coli DNA Polymerase I Is Its Own Proofreader and Editor

E. coli DNA polymerase I is a 109-kD protein consisting of a single polypeptide of 928 amino acid residues. In addition to its 5′ → 3′ polymerase activity, DNA polymerase I has two other catalytic functions, a *3′ → 5′ exonuclease (3′-exonuclease)* activity and a *5′ → 3′ exonuclease (5′-exonuclease)* activity. The 3′-exonuclease activity removes nucleotides from the 3′-end of the growing chain (Figure 29.6), an action that apparently negates the effects of the polymerase activity. Its purpose, however, is to remove incorrect (mismatched) bases. Although the 3′-exonuclease works slowly when compared to the polymerase, the polymerase cannot elongate an improperly base-paired primer terminus. Thus, the relatively slow 3′-exonuclease has time to act and remove the mispaired nucleotide. Therefore, the polymerase active site is a proofreader and the 3′-exonuclease activity is an editor. This check on the accuracy of base pairing during DNA replication enhances the overall precision of the process, improving the fidelity of replication.

The 5′-exonuclease of DNA polymerase I acts upon duplex DNA, degrading it from the 5′-end by releasing mono- and oligonucleotides. It can remove distorted (mispaired) segments lying in the path of the advancing polymerase. Its biological roles depend on the ability of DNA polymerase I to bind at

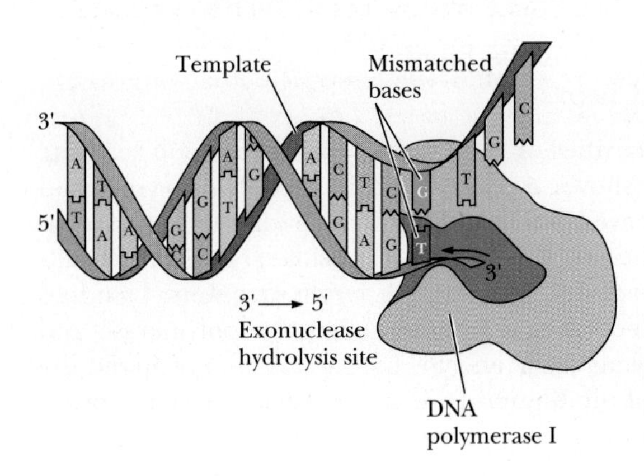

Figure **29.6** The 3′ → 5′ exonuclease activity of DNA polymerase I removes nucleotides from the 3′-end of the growing DNA chain.

Figure **29.7** (a) The $5' \rightarrow 3'$ exonuclease activity of DNA polymerase I can remove up to 10 nucleotides in the 5′-direction downstream from a 3′-OH single-strand nick. (b) If the $5' \rightarrow 3'$ polymerase activity fills in the gap, the net effect is nick translation by DNA polymerase.

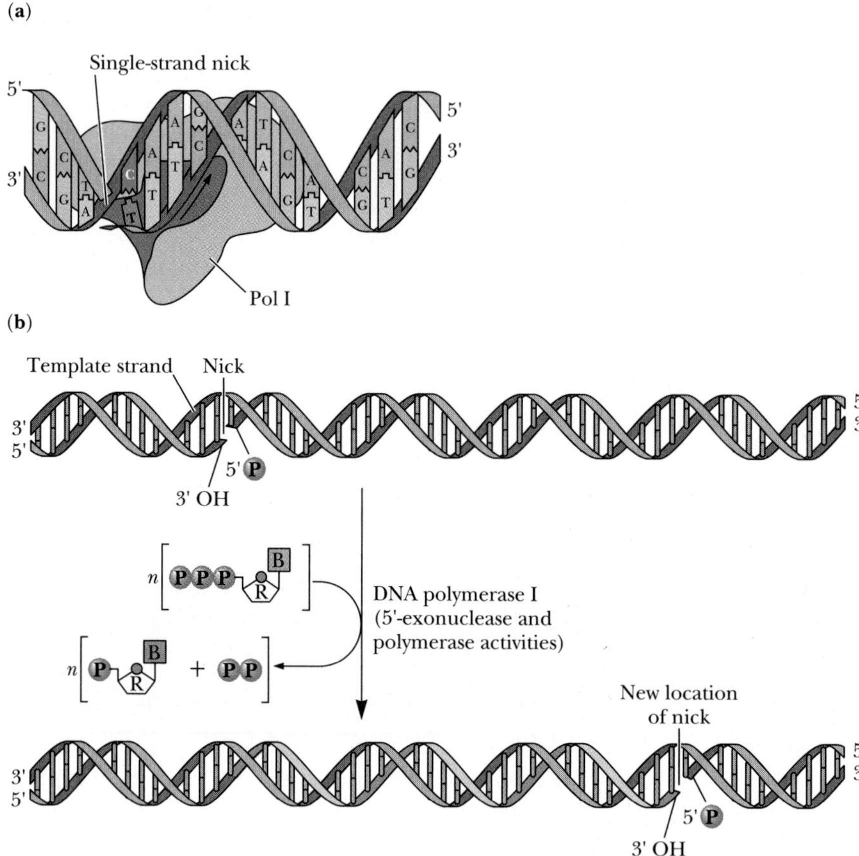

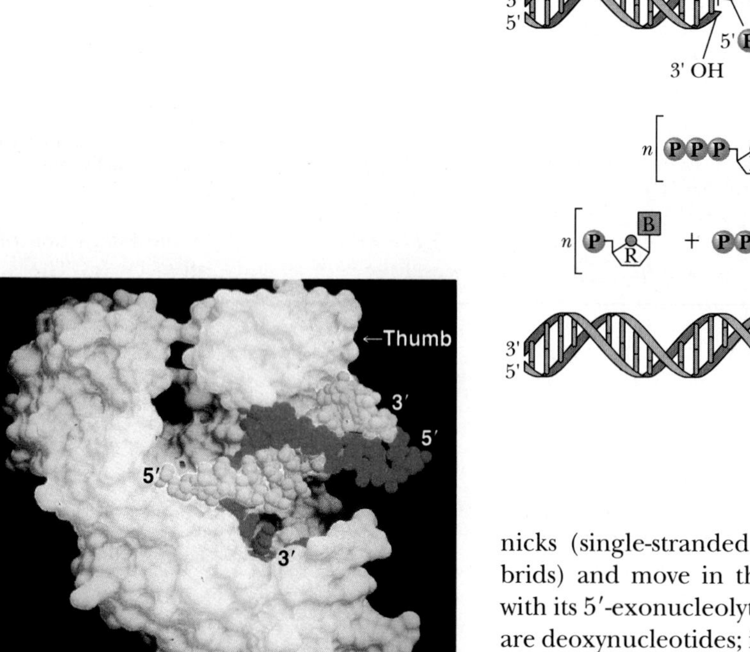

Figure **29.8** A representation of the solvent-accessible surface of the DNA polymerase I Klenow fragment with bound DNA. The "template" strand (12 nucleotides) is blue; the "primer" strand (14 nucleotides) is red. The thumblike region of the protein that changes conformation upon DNA binding is indicated. Note that the vertical cleft is an extension of the horizontal cleft in the protein which DNA occupies. In this picture, the 3′-end of the primer strand lies within the 3′-exonuclease catalytic site of the enzyme. The polymerase site of the enzyme lies almost directly above the 3′-exonuclease site, in the vertical cleft (see Figure 29.9).

(Adapted from Beese, L. S., Derbyshire, V., and Steitz, T. A., 1993. Science 260:352–355. Photograph courtesy of Thomas A. Steitz of Yale University.)

nicks (single-stranded breaks) in double-stranded DNA (or RNA:DNA hybrids) and move in the $5' \rightarrow 3'$ direction, removing successive nucleotides with its 5′-exonucleolytic activity. (If the substrate is dsDNA, these nucleotides are deoxynucleotides; if the substrate is an RNA:DNA hybrid, the products are ribonucleotides.) In an error-correcting role, the $5' \rightarrow 3'$ polymerase activity fills in the sequence behind the 5′-exonuclease activity, so that the enzyme carries out a **nick translation** (*translation* meaning movement) along the DNA molecule (Figure 29.7). No *net* synthesis of DNA results.

The nick translation reaction is very useful for radioactively labeling DNA molecules *in vitro* by incubating nicked dsDNA with DNA polymerase I and ^{32}P-labeled dNTPs. *In vivo,* this action of DNA polymerase I can "edit out" sections of damaged DNA, if a nick with a free 3′-OH is present. This excision is coordinated with $5' \rightarrow 3'$ polymerase-catalyzed replacement of the damaged nucleotides so that DNA of the right sequence is restored.

DNA Polymerase I Has Three Active Sites on Its Single Polypeptide Chain

The three distinct catalytic activities of DNA polymerase I reside in separate active sites in the enzyme. As shown by Hans Klenow, the DNA polymerase I polypeptide chain can be cleaved into two fragments by limited proteolysis with subtilisin or trypsin. The smaller fragment (residues 1 through 323) contains the 5′-exonuclease activity, whereas the larger fragment (residues 324 through 928, the so-called **Klenow fragment**) has the polymerase and 3′-exonuclease activities. Thomas Steitz and his colleagues have analyzed the three-dimensional structure of the Klenow fragment via X-ray crystallography (Figure 29.8). Upon DNA binding, a conformational change takes place in a thumblike region of the protein that makes direct contacts with duplex portions of DNA. The polymerase and 3′-exonuclease active sites of the Klenow fragment lie within a common cleft that forms virtually a right-angle bend within the protein, but these two catalytic sites are surprisingly far apart—about 3.5 nm—and lie in separate arms of the bend. The DNA enters the cleft from the end closest to the 3′-exonuclease site. Apparently, the 3′-terminus of the DNA primer is shuttled between each catalytic site—the polymerase active site and the 3′ → 5′ exonuclease site (Figure 29.9), and the polymerization and editing functions of the enzyme can be carried out without dissociation of the DNA from the protein.

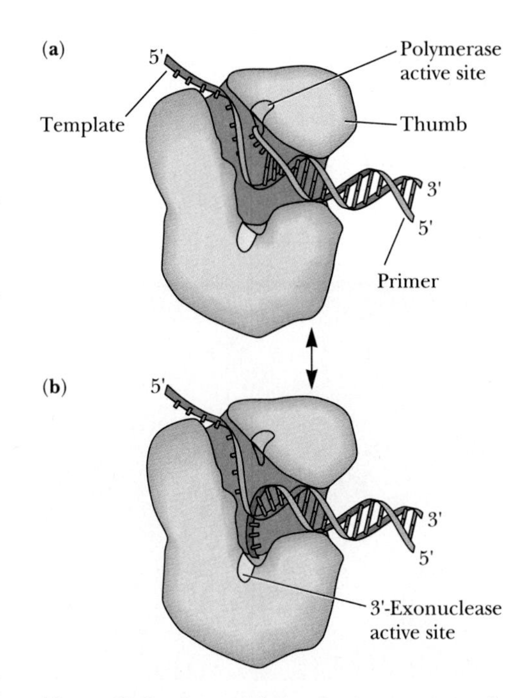

Figure **29.9** A model for the interaction of the Klenow fragment with DNA. (a) The 3′-end of the primer/growing chain resides in the polymerase active site. (b) The 3′-end is situated in the 3′-exonuclease active site. The protein must slide along the DNA in shifting the 3′-end of the growing chain from one of its active sites to the other. Thus, polymerization and the 3′ editing functions may occur without dissociation of the DNA from the Klenow fragment.

(Adapted from Beese, L. S., Derbyshire, V., and Steitz, T. A., 1993. Science 260:352–355.)

E. coli DNA Polymerase III

In 1969, Paula DeLucia and John Cairns isolated an *E. coli* mutant (named *polA1*) having only 1% of wild-type DNA polymerase I activity. Nevertheless, they found that the *polA1* mutant cells grew and divided at essentially normal rates! The *polA1* cells were sensitive to UV light. Cairns suggested that DNA replication proceeded normally in *polA1* mutants, but that DNA repair was defective, implying that the physiological role of DNA polymerase I was DNA repair. Two more *E. coli* DNA polymerases were then discovered: DNA polymerases II and III. Table 29.1 compares the properties of the various *E. coli* DNA polymerases. Like DNA polymerase I, DNA polymerase II apparently functions in DNA repair. **DNA polymerase III** is the chief DNA-replicating enzyme of *E. coli*. Only 10 or so molecules are present per cell.

"Core" DNA Polymerase III

The simplest form of DNA polymerase III showing any DNA-synthesizing activity *in vitro* is 165 kD in size and consists of three polypeptides: α (130 kD), ϵ (27.5 kD), and θ (10 kD). This *"core" DNA polymerase III* binds at short, single-stranded regions (<100 nucleotides) created by nuclease treatment of dsDNA and fills in the gaps to reestablish the duplex structure. It cannot initiate synthesis on intact duplex DNA. The α subunit carries out the catalytic polymerase function. The ϵ subunit performs the 3′ → 5′ exonuclease activity and contributes proofreading ability to the "core" polymerase. The role of subunit θ is unknown.

Table **29.1**
Properties of the DNA Polymerases of *E. coli*

Property	Pol I	Pol II	Pol III (core)*
Mass (kD)	103	90	130, 27.5, 10
Molecules/cell	400	?	10–20
Turnover number†	600	30	9000
Structural gene	*polA*	*polB*	*dnaE* (α subunit) *dnaQ* (ϵ subunit) ? (θ subunit)
Polymerization 5′ → 3′	Yes	Yes	Yes
Exonuclease 3′ → 5′	Yes	Yes	Yes
Exonuclease 5′ → 3′	Yes	No	No

*α, ϵ, and θ subunits.

†Nucleotides polymerized @ 37°C/minute/molecule of enzyme.

Source: Adapted from Kornberg, A., and Baker, T. A., 1991. *DNA Replication,* 2nd ed.: New York: W. H. Freeman and Co.

Table **29.2**
The Subunits of *E. coli* DNA Polymerase III Holoenzyme

Subunit	Mass (kD)	Structural Gene	Function
α	130	*polC (dnaE)*	Polymerase
ϵ	27.5	*dnaQ*	3′-exonuclease
θ	10	?	α, ϵ assembly?
τ	71	*dnaX*	Assembly of holoenzyme on DNA
β	41	*dnaN*	Sliding clamp, processivity
γ	47.5	*dnaX(Z)*	Part of the γ complex*
δ	34	?	Part of the γ complex*
δ'	32	?	Part of the γ complex*
χ	13	?	Part of the γ complex*
ψ	15	?	Part of the γ complex*

*Subunits γ, δ, δ', χ, and ψ form the so-called γ-complex responsible for adding β-subunits (the sliding clamp) to DNA.

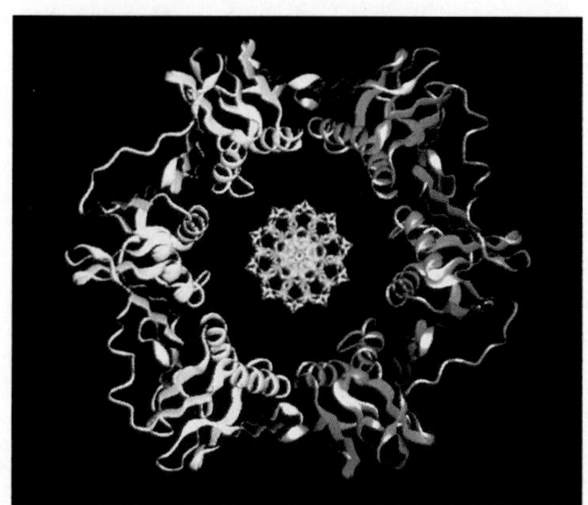

(a)

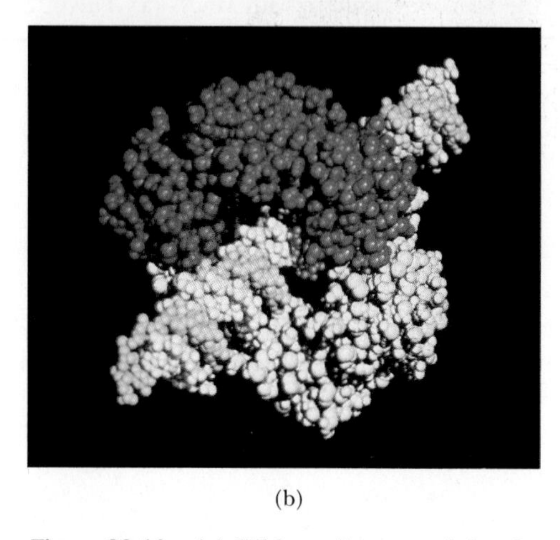

(b)

Figure **29.10** (a) Ribbon diagram of the β subunit dimer of the DNA polymerase III holoenzyme on B-DNA, viewed down the axis of the DNA. One monomer of the β subunit dimer is colored red and the other yellow. The centrally located DNA is mostly blue. (b) Space-filling model of the β subunit dimer of the DNA polymerase III holoenzyme on B-DNA. One monomer is shown in red, the other in yellow. The B-DNA has one strand colored white and the other blue. The hole formed by the β subunits (diameter $\simeq$ 3.5 nm) is large enough to easily accommodate either A-form or B-form DNA (diameter $\simeq$ 2.5 nm) with no steric repulsion. The rest of pol III holoenzyme ("core" polymerase + γ complex) associates with this sliding clamp to form the replicative polymerase (not shown).

(Adapted from Kong, X.-P., et al., 1992. Cell 69:425–437; photos courtesy of John Kuriyan of the Rockefeller University.)

DNA Polymerase III Holoenzyme

In vivo, "core" DNA polymerase III functions as part of a multisubunit complex, the **DNA polymerase III holoenzyme,** consisting of at least 10 different subunits (Table 29.2). The various auxiliary subunits increase both the polymerase activity of the "core" enzyme and its processivity. Reconstitution of DNA polymerase III holoenzyme *in vitro* from purified subunits suggests that the assembly pathway proceeds in two stages. In the first stage, the γ complex, consisting of 5 proteins (γ, δ, δ', χ, and ψ), catalyzes the ATP-dependent transfer of a pair of β subunits to the DNA template. These β subunits form a closed ring around the template, acting as a tight clamp that can slide along the DNA (Figure 29.10). In the second stage of assembly, the "core" polymerase III binds to the β dimer to yield *DNA polymerase III holoenzyme.* The *sliding clamp* created by the β subunits tethers the "core" polymerase to the template, accounting for the great processivity of DNA polymerase III holoenzyme. This complex can replicate an entire strand of the *E. coli* genome (nearly 5 megabases) without dissociating. Compare this to the processivity of DNA polymerase I, which is only 20!

29.3 General Features of DNA Replication

Replication Is Bidirectional

Replication of DNA molecules begins at one or more unique sites called **origin(s) of replication** (discussed in section 29.4), and, excepting certain bacteriophage chromosomes and plasmids, proceeds in both directions from this origin (Figure 29.11). For example, replication of *E. coli* DNA begins at *oriC,* a unique 245-bp site. From this site, replication advances in both directions around the circular chromosome. That is, bidirectional replication involves two **replication forks,** which move in opposite directions.

Unwinding the DNA Helix

Semiconservative replication depends on unwinding the DNA double helix to expose single-stranded templates to polymerase action. For a double helix to unwind, it must either rotate about its axis (while the end of its strands are

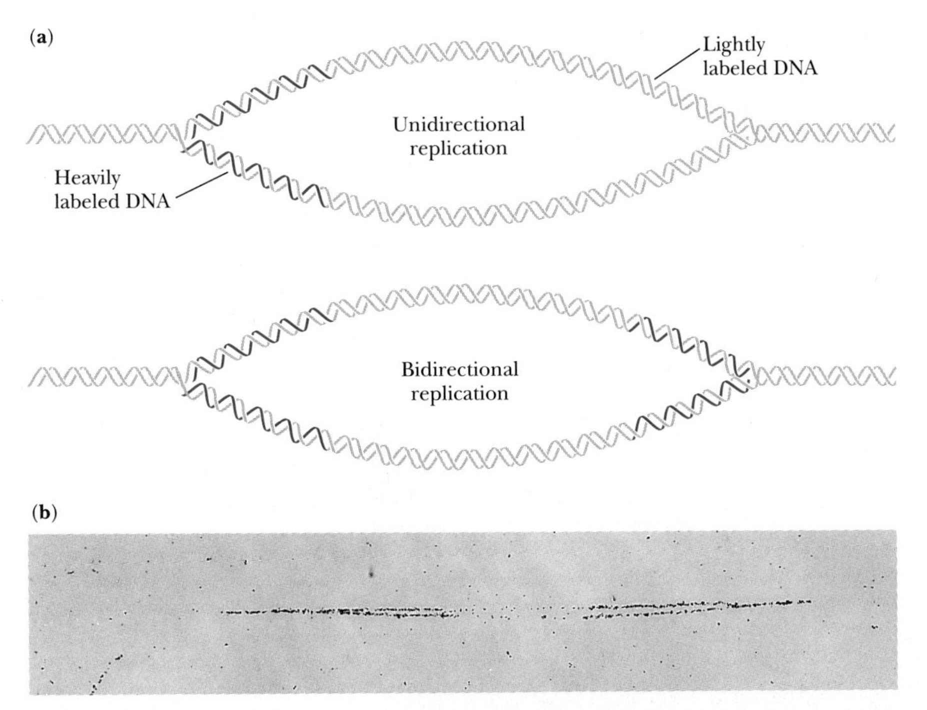

Figure **29.11** Bidirectional replication of the *E. coli* chromosome. (a) Unidirectional versus bidirectional replication can be resolved if cells grown several generations in the presence of low amounts of radioactive ^{3}H-thymidine to lightly label the chromosome are then exposed briefly to high levels of ^{3}H-thymidine to cause heavy labeling in newly synthesized regions of DNA. If replication is unidirectional, only one advancing replication fork is present and only the DNA adjacent to it should be heavily labeled. If replication is bidirectional, autoradiograms of replicating chromosomes should show two replication forks heavily labeled with radioactive thymidine. (b) An autoradiogram of the chromosome from a dividing *E. coli* cell shows bidirectional replication. Unidirectional replication has been observed only in bacteriophages and some plasmids.
(Photo courtesy of David M. Prescott, University of Colorado.)

held fixed), or positive supercoils must be introduced, one for each turn of the helix unwound (Chapter 7). If the chromosome is circular, as in *E. coli*, only the latter alternative is available. Since DNA replication in *E. coli* proceeds at a rate approaching 1000 nucleotides per second, and there are about 10 bp per helical turn, the chromosome would accumulate 100 positive supercoils per second! In effect, the DNA would become too tightly supercoiled to allow unwinding of the strands.

DNA gyrase, by introducing negative supercoils at the expense of ATP hydrolysis, acts to overcome the torsional stress imposed by unwinding. Recall that negative supercoils have some of the characteristics of local single-strandedness; that is, they create regions topologically equivalent to single-stranded DNA. DNA gyrase is a Type II topoisomerase (Chapter 7). The unwinding reaction is driven by **helicases,** a class of proteins that catalyze the ATP-dependent unwinding of DNA double helices. Unlike topoisomerases that alter the linking number of dsDNA through phosphodiester bond breakage and reunion (Chapter 7), helicases simply disrupt the hydrogen bonds that hold the two strands of duplex DNA together. A helicase molecule requires a single-stranded region for binding. It then moves along the DNA strand, its translocation coupled to ATP hydrolysis and to strand unwinding. SSB (ssDNA-binding protein, Chapter 28) binds to the unwound strands, preventing their re-annealing. At least 10 distinct DNA helicases involved in different aspects of DNA and RNA metabolism have been found in *E. coli* alone.

Replication Is Semidiscontinuous

The autoradiographic results (Figure 29.11) indicate that the two strands of duplex DNA are both replicated at each advancing replication fork. Recall that DNA polymerases only synthesize DNA in a $5' \rightarrow 3'$ direction, reading the antiparallel template strand in a $3' \rightarrow 5'$ sense. A dilemma arises: How does DNA polymerase copy the parent strand that runs in the $5' \rightarrow 3'$ direction at the replication fork? It turns out that the two daughter strands are synthesized in different ways so that *replication is semidiscontinuous* (Figure 29.12). As the DNA helix is unwound during its replication, the $3' \rightarrow 5'$

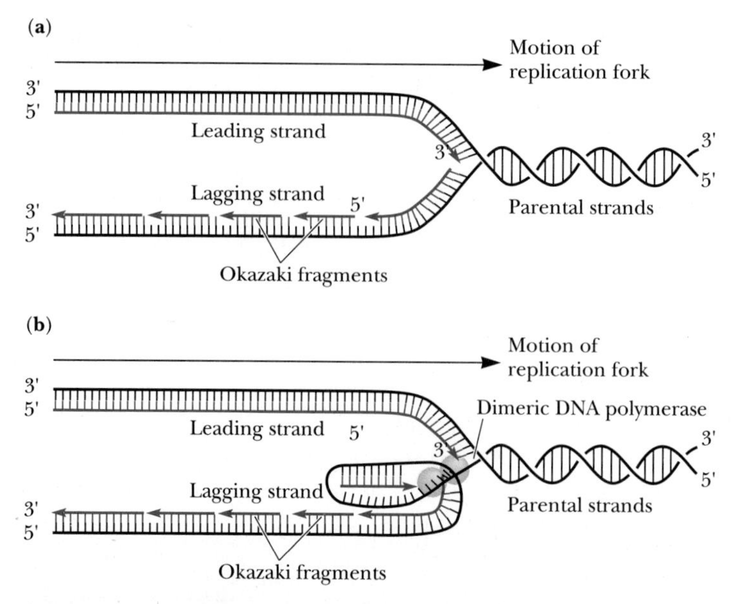

***Figure* 29.12** The semidiscontinuous model for DNA replication. Newly synthesized DNA is shown as red. Because DNA polymerases only polymerize nucleotides $5' \rightarrow 3'$, both strands must be synthesized in the $5' \rightarrow 3'$ direction. Thus, the copy of the parental $3' \rightarrow 5'$ strand is synthesized continuously; this newly made strand is designated the **leading strand.** (a) As the helix unwinds, the other parental strand (the $5' \rightarrow 3'$ strand) is copied in a discontinuous fashion through synthesis of a series of fragments 1000 to 2000 nucleotides in length, called the **Okazaki fragments;** the strand constructed from the Okazaki fragments is called the **lagging strand.** (b) Since both strands are synthesized in concert by a dimeric DNA polymerase situated at the replication fork, the $5' \rightarrow 3'$ parental strand must wrap around in *trombone fashion* so that the unit of the dimeric DNA polymerase replicating it can move along it in the $3' \rightarrow 5'$ direction. This parental strand is copied in a discontinuous fashion because the DNA polymerase must occasionally dissociate from this strand and rejoin it further along. The Okazaki fragments are then covalently joined by DNA ligase to form an uninterrupted DNA strand.

strand (as defined by the direction that the replication fork is moving) can be copied continuously by DNA polymerase proceeding in the $5' \rightarrow 3'$ direction behind the replication fork. The other parental strand is copied only when a sufficient stretch of its sequence has been exposed for DNA polymerase to move along it in the $5' \rightarrow 3'$ mode. Thus, one parental strand is copied continuously to give a newly synthesized copy, the **leading strand;** the other parental strand is copied in an intermittent, or discontinuous, mode to yield a set of fragments. These fragments are then joined to form an intact **lagging strand.**

The Lagging Strand Is Formed from Okazaki Fragments

In 1968, Tuneko and Reiji Okazaki provided biochemical verification of the semidiscontinuous pattern of DNA replication just described. The Okazakis exposed a rapidly dividing *E. coli* culture to ^{3}H-labeled thymidine for 30 seconds, quickly collected the cells, and found that half of the label incorporated into nucleic acid appeared in short ssDNA chains just 1000 to 2000 nucleotides in length. (The other half of the radioactivity was recovered in very large DNA molecules.) Subsequent experiments by others used a *pulse-chase* protocol to trace the fate of the short ssDNA chains (the so-called **Okazaki fragments**). A culture of growing bacteria was given a brief (30 sec) exposure to

^{3}H-thymidine, and then the cells were exposed to excess unlabeled thymidine to dilute the radiolabeled compound and to minimize the likelihood of further ^{3}H-thymidine incorporation into DNA. These experiments demonstrated that, with time, the newly synthesized Okazaki fragments became covalently joined to form longer polynucleotide chains, in accord with a semidiscontinuous mode of replication. The generality of this mode of replication has been corroborated in electron micrographs of DNA undergoing replication in eukaryotic cells (Figure 29.13).

The Chemistry of DNA Synthesis Favors Semidiscontinuous Replication

A consideration of the chemistry of chain elongation suggests why semidiscontinuous DNA replication and the inherently complex mechanism of lagging strand synthesis evolved. Note that $5' \rightarrow 3'$ synthesis takes an activated intermediate, the deoxynucleoside-5′-triphosphate, and forms a phosphoester bond by joining its 5′-α-phosphate to a 3′-OH group with elimination of PP_i:

$$dN^{*}TP + pNpNpNpN\text{-}3'OH \longrightarrow PP_i + pNpNpNpNpN^{*}\text{-}3'OH$$

The 3′-OH of this newly added deoxynucleotide provides the acceptor for the next incoming deoxynucleotide. The alternative, linking of deoxynucleoside-5′-triphosphates in the $3' \rightarrow 5'$ direction, would require the growing chain to present a reactive phosphoric anhydride at its 5′-end to activate condensation with the 3′-OH of an incoming nucleotide:

$$dN^{*}TP + pppNpNpNpN\text{-}3'OH \longrightarrow pppN^{*}pNpNpNpN\text{-}3'OH + PP_i$$

This mechanism is energetically feasible, but error correction by proofreading would create an energetically unfavorable situation: The proofreading activity would be a $5' \rightarrow 3'$ exonuclease

$$pppN^{*}pNpNpNpN\text{-}3'OH \longrightarrow pppN^{*}\text{-}3'(p \text{ or } OH) + (OH \text{ or } p)NpNpNpN\text{-}3'OH$$

and its action would leave a 5′-OH or a 5′-phosphate ester group at the growing end of the chain, neither of which will form a phosphodiester linkage unless chemically activated. Without activation, chain extension would be aborted.

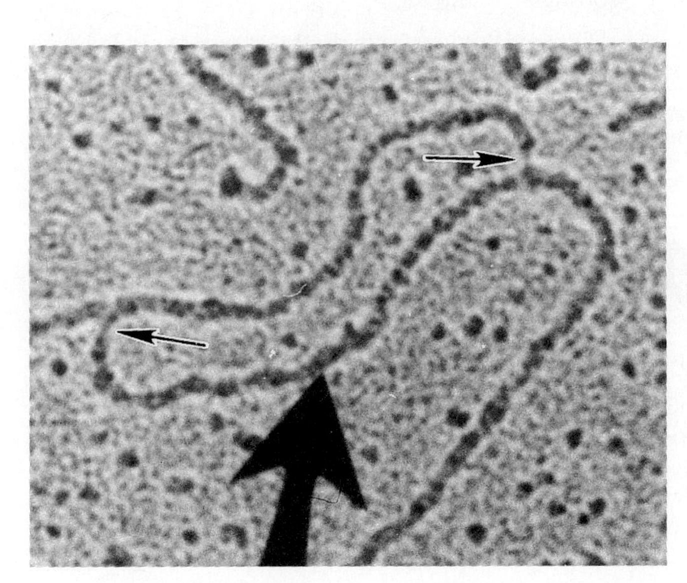

***Figure* 29.13** An electron micrograph of DNA replication in cultured Chinese hamster ovary (CHO) cells. A replication bubble is indicated by the large arrow. Note the two replication forks that define the bubble; each fork is characterized by a thinner, single-stranded region (indicated by the small arrows) on one of its two branches. Also note that the two single-stranded regions are *trans* to each other, as predicted by the semidiscontinuous model for DNA replication. *(Photo courtesy of Joyce L. Hamlin, University of Virginia.)*

Synthesis of DNA Is Primed by RNA

Since all DNA polymerases require a primer with a 3′-OH end to build upon, the question arises: Where does the primer come from? An important clue was the discovery that the 5′-ends of Okazaki fragments contain short RNA segments (1 to 60 nucleotides in length, depending on species), and these segments are complementary in sequence to the template DNA. It was also known that RNA synthesis is essential for the initiation of DNA synthesis. In *E. coli,* **primase,** a specific RNA polymerase (60 kD) encoded by the *dnaG* gene, catalyzes the synthesis of these RNA primers; primers in *E. coli* are RNA chains about 5 nucleotides long.

E. coli DNA Polymerase I Excises the RNA Primer

No RNA is found in replicated DNA. An important role of *E. coli* DNA polymerase I 5′-exonuclease activity is the removal of primer RNA base-paired to DNA. Note that the place where two Okazaki fragments meet resembles a nick (Figure 29.14). DNA polymerase I binds here, and the ribonucleotides are excised by the 5′-exonuclease activity and replaced with deoxynucleotides by the 5′ → 3′ polymerase function (Figure 29.14). **RNase H,** an RNase specific for the RNA strands of RNA:DNA hybrid duplexes, also participates in removal of RNA primers, but it cannot hydrolyze a phosphodiester bond linking a ribonucleotide to a deoxyribonucleotide. After the ribonucleotides are removed, the nick that remains is sealed by DNA ligase.

Several biochemical features provide a plausible explanation for why RNA primers initiate DNA synthesis. Unlike DNA polymerases, RNA polymerases can initiate polynucleotide synthesis *de novo.* Also, RNA polymerases do not check the pairing accuracy of the preceding base pair; RNA polymerases are low-fidelity enzymes in comparison to DNA polymerases. Further, RNA, as the priming oligonucleotide, provides a chemical signal that the primer is temporary and will be replaced; it doesn't matter if its synthesis is error-prone. Our emerging view of life indicates that RNA served as the repository of genetic information before the appearance of DNA in early evolution. Thus, RNA was prepared for its role in DNA replication.

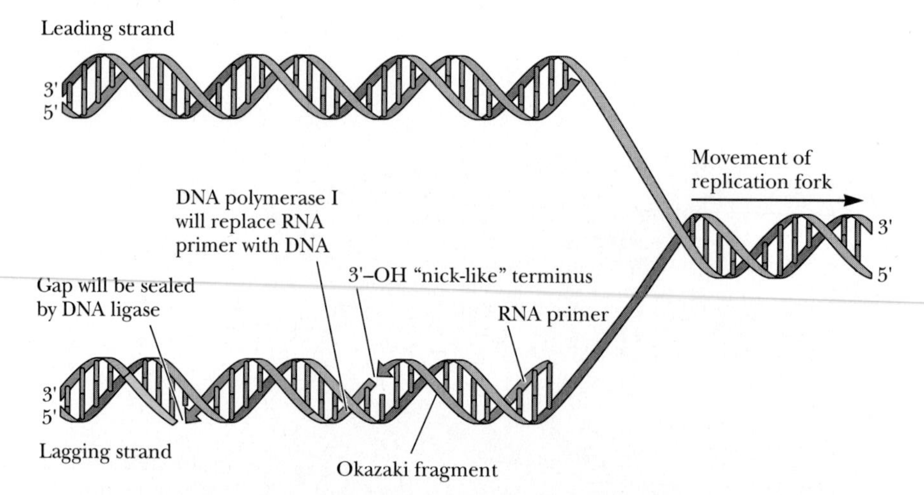

***Figure* 29.14** The junction of two adjacent Okazaki fragments resembles a 3′-OH nick. DNA polymerase I binds here, its 5′ → 3′ exonuclease activity removes the RNA bases that primed synthesis of the fragment, and its 5′ → 3′ polymerase activity fills in with deoxynucleotides to generate authentic DNA. DNA ligase then covalently closes the chain.

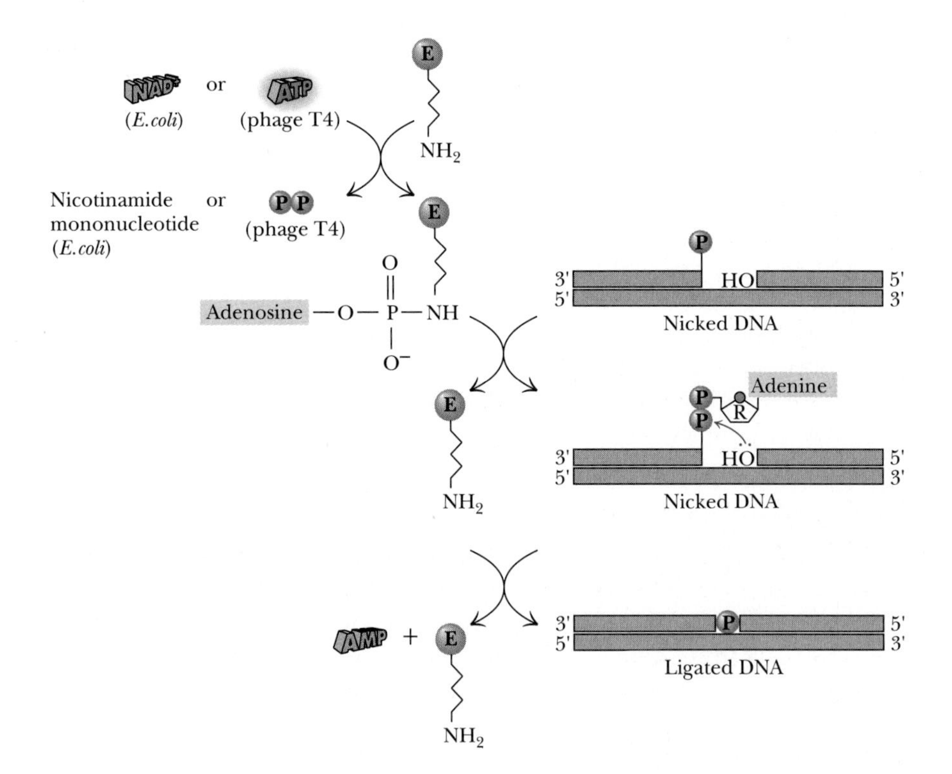

Figure **29.15** The mechanism of action of DNA ligases. Both *E. coli* and bacteriophage T4 DNA ligase proceed via a ping-pong kinetic mechanism involving an enzyme-linked covalent intermediate (E′) in which the side chain ϵ-NH_2 of a Lys residue is adenylylated. The two enzymes differ in that NAD^+ is the AMP donor for the *E. coli* enzyme, whereas ATP serves this role for the phage DNA ligase.

DNA Ligase

DNA ligase (see Chapter 8) seals nicks in double-stranded DNA where a 3′-OH and a 5′-phosphate are juxtaposed. This enzyme is responsible for joining Okazaki fragments together to make the lagging strand a covalently contiguous polynucleotide chain. DNA ligase from eukaryotes and bacteriophage T4 is ATP-dependent; the *E. coli* enzyme requires NAD^+. Both types of DNA ligase act via an adenylylated ϵ-amino group of a Lys residue (Figure 29.15). Adenylylation of the 5′-phosphoryl group activates it for formation of a phosphoester bond with the 3′-OH, covalently sealing the sugar-phosphate backbone of DNA.

General Features of a Replication Fork

We now can present a snapshot of the enzymatic apparatus assembled at a replication fork (Figure 29.16 and Table 29.3). DNA gyrase (topoisomerase) and helicase unwind the DNA double helix, and the unwound, single-

Figure **29.16** General features of a replication fork. The DNA duplex is unwound by the action of DNA gyrase and helicase, and the single strands are coated with SSB (ssDNA-binding protein). Primase periodically primes synthesis on the lagging strand. Each half of the dimeric replicative polymerase is a holoenzyme bound to its template strand by a β-subunit sliding clamp. DNA polymerase I and DNA ligase act downstream on the lagging strand to remove RNA primers, replace them with DNA, and ligate the Okazaki fragments.

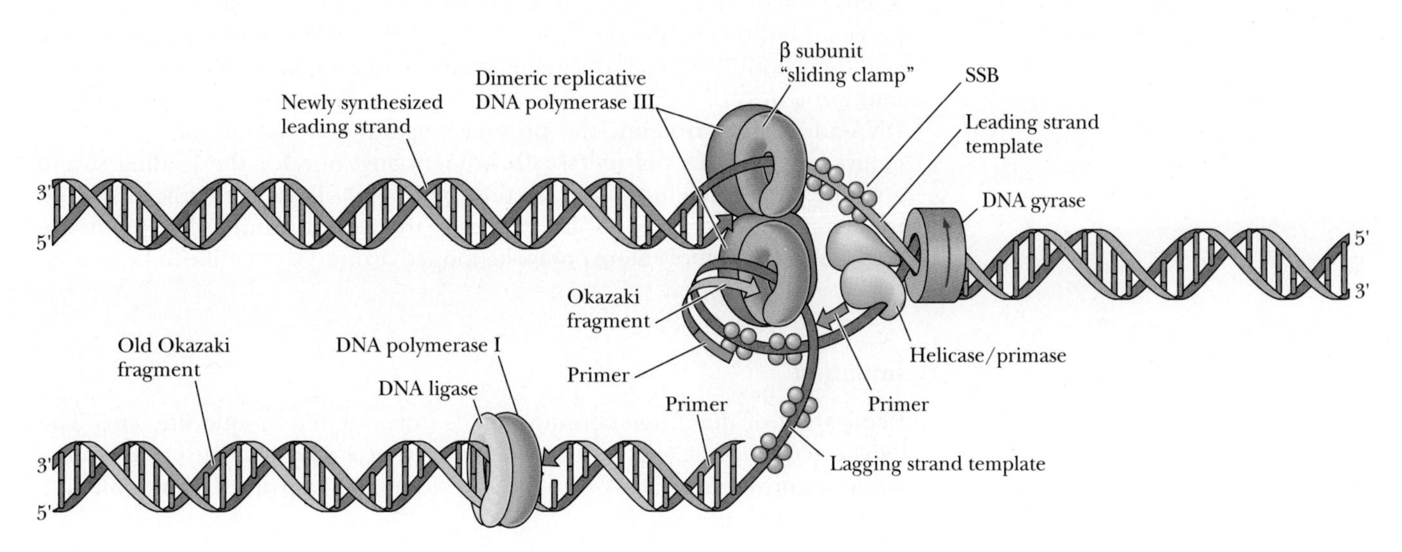

Table **29.3**
Proteins Necessary for DNA Replication in *E. coli*

Protein	Function
DNA gyrase	Unwinding DNA
SSB	Single-stranded DNA binding
DnaA	Initiation factor
HU	Histone-like (DNA binding)
PriA	Primosome assembly, $3' \rightarrow 5'$ helicase
PriB	Primosome assembly
PriC	Primosome assembly
DnaB	$5' \rightarrow 3'$ helicase (DNA unwinding)
DnaC	DnaB chaperone
DnaT	Assists DnaC in delivery of DnaB
Primase	Synthesis of RNA primer
DNA polymerase III holoenzyme	Elongation (DNA synthesis)
DNA polymerase I	Excises RNA primer, fills in with DNA
DNA ligase	Covalently links Okazaki fragments
Ter	Termination

stranded regions of DNA are maintained through interaction with SSB. Primase synthesizes an RNA primer on the lagging strand; the leading strand, which needs priming only once, was primed when replication was initiated. The lagging strand template is looped around, and the replicative DNA polymerase (DNA polymerase III holoenzyme in *E. coli*), as a dimer, moves $5' \rightarrow 3'$ relative to each strand, copying template and synthesizing new DNA strands. This replicative polymerase is tethered to the DNA by its β-subunit sliding clamps. The DNA pol III γ complex periodically unclamps and then reclamps the β subunits on the lagging strand as the primer for each new Okazaki fragment is encountered. Downstream on the lagging strand, DNA polymerase I excises the RNA primer and replaces it with DNA, and DNA ligase seals the remaining nick.

29.4 The Mechanism of DNA Replication in *E. coli*

Replication of the *E. coli* chromosome begins at a single replication origin and proceeds bidirectionally until the two replication forks meet. At each replication fork, both leading and lagging strand syntheses are catalyzed by a single multiprotein replication machine, the so-called **replisome,** which consists of DNA-unwinding proteins; the priming apparatus, or **primosome;** and two equivalents of DNA polymerase III holoenzyme, one for the leading strand and one for the lagging strand. As this replisome follows the replication fork, the template for lagging strand synthesis (the strand running $5' \rightarrow 3'$ in the direction of fork movement) must be looped around so that it can be read in the $3' \rightarrow 5'$ direction.

Initiation

Replication of the *E. coli* chromosome is initiated at a unique site, *oriC*. The 245-bp sequence of *oriC* contains elements that are highly conserved among Gram-negative bacteria. Within *oriC* are four 9-bp repeats. **DnaA protein,** a

52-kD polypeptide, is the *initiation factor,* which recognizes and binds to these repeats. DnaA protein binding is cooperative; once the four 9-bp repeats are occupied, 20 to 40 additional DnaA monomers bind so that the entire *oriC* region is complexed with DnaA protein. This assembly process is facilitated by ***HU,*** a histone-like protein (Chapter 7). The resulting complex resembles a nucleosome, with negatively supercoiled *oriC* DNA wrapped around a DnaA core. The DnaA protein then mediates the separation of the strands of the DNA duplex by acting on three AT-rich tandem repeats (consensus sequence = 5′-GATCTNTNTTNTT, where N = any nucleotide) located at the 5′-end of the sequence defining *oriC.* Formation of this 45-bp "open complex" by DnaA protein is ATP-dependent. Several other proteins (the PriA, PriB, PriC proteins) also participate in the prepriming process. Next, **DnaB** protein (a hexamer of 50-kD subunits) binds to the "open" *oriC.* DnaB protein is delivered to *oriC* by **DnaC** protein (29 kD) in the form of a hexameric $(DnaB{:}DnaC{:}ATP)_6$ complex, but DnaC protein does not enter the protein assemblage at *oriC.* Delivery of DnaB protein by DnaC protein is assisted by DnaT protein. The addition of DnaB protein completes assembly of the **pre-priming complex.** ATP hydrolysis drives the formation of this complex (Figure 29.17). DnaB protein has helicase activity and it further unwinds the DNA in the pre-priming complex in both directions, assisted by DNA gyrase. SSB tetramers coat single-stranded regions as they arise. Unwinding exposes the base sequence of the strands so that RNA primers can be synthesized by primase, and the strands can be read as templates by the replicative polymerase.

The addition of primase to the pre-priming complex completes the formation of the primosome, a protein machine of about 700 kD that synthesizes the RNA primer essential to DNA synthesis (Figure 29.18a–e). Interestingly, the AT-rich tandem repeats of *oriC* resemble sequences with which RNA polymerase interacts (see Chapter 30). Note that *two* primosomes form at *oriC,* one for each replication fork. Once the primosome has primed leading strand synthesis, it remains associated with the advancing replication fork and periodically primes Okazaki fragment synthesis, as required.

Elongation

The double helix must be unwound ahead of the advancing replication fork. Unwinding is driven by ATP hydrolysis; DnaB protein has the helicase activity associated with the primosome. DnaB moves in the 5′ → 3′ direction along the leading strand template, hydrolyzing 2 ATPs for each base pair that it separates. PriA protein is, like DnaB, a helicase associated with the primosome, but PriA moves 3′ → 5′ along the lagging strand. The binding of SSB protein to the single-stranded regions created behind these proteins prevents re-annealing.

Figure 29.19 depicts a model for the replication of *E. coli* DNA as it occurs at one of the replication forks. Two DNA polymerase III holoenzymes are present. One is synthesizing the leading strand, using the parental 3′ → 5′ strand as template. The other synthesizes the lagging strand using the 5′ → 3′ parental strand as template. The lagging strand template must be looped out so that it can be copied in the 3′ → 5′ direction. The *pol III holoenzyme* on this lagging strand has completed the synthesis of an RNA-primed Okazaki fragment when it encounters the 5′-end of the previous fragment. Synthesis of the next RNA primer by the primosome triggers *pol III holoenzyme* to release the template, reassociate with it at the 3′-OH end of the RNA primer, and begin a new round of lagging strand synthesis.

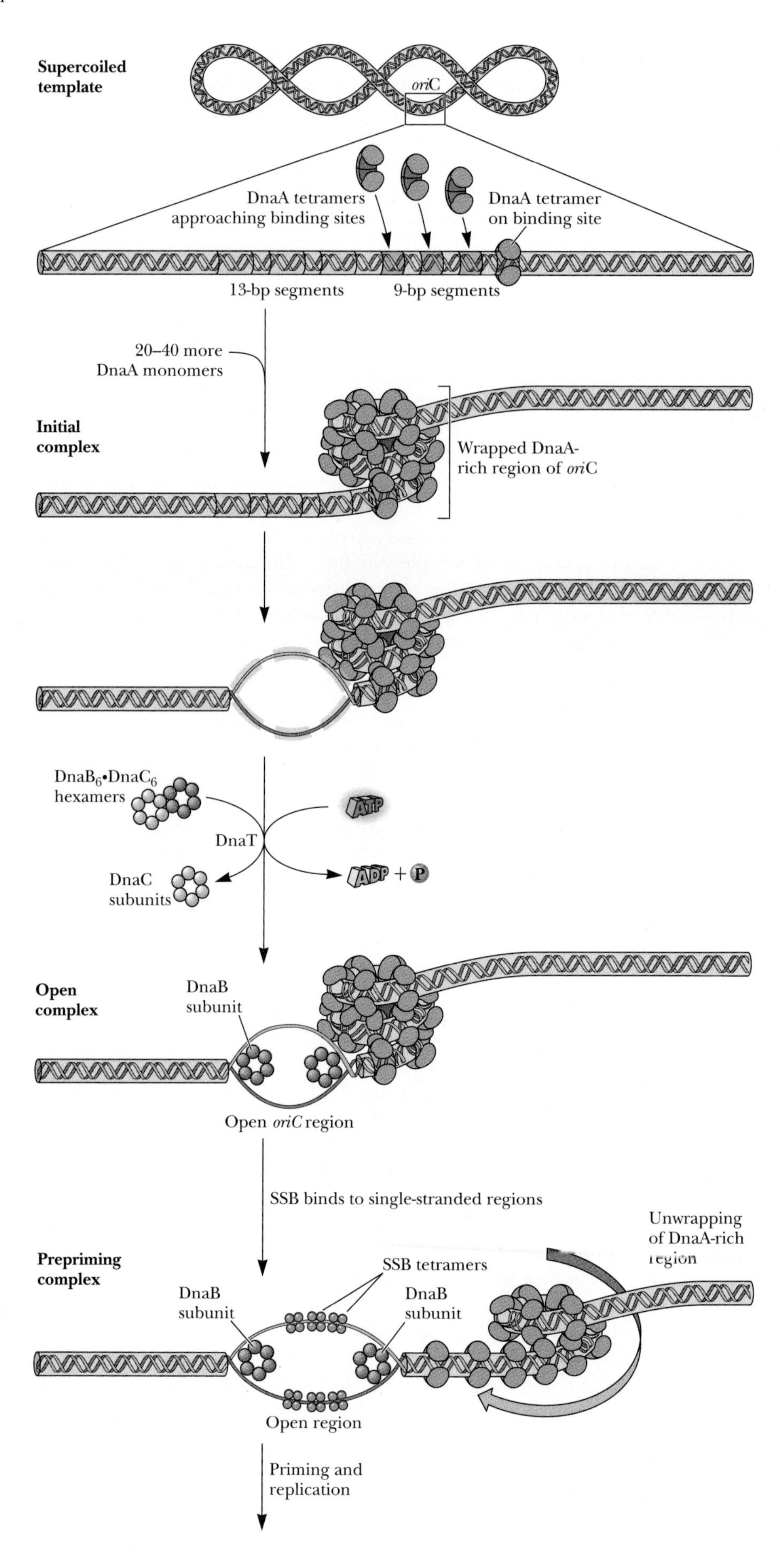

***Figure* 29.17** The probable course of events during initiation of *E. coli* replication at *oriC.* Tetramers of DnaA protein bind at each of the four 9-bp repeats in *oriC,* then additional DnaA protein binds to give a nucleosome-like structure with DNA on the outside. The three 13-bp A:T-rich regions are then "melted" to yield an open complex, and the DnaB protein delivered from the DnaB:DnaC protein complex enters. The helicase activity of DnaB unwinds the duplex, displacing DnaA protein. SSB protein binds to single-stranded DNA as it is generated, preventing its re-annealing.

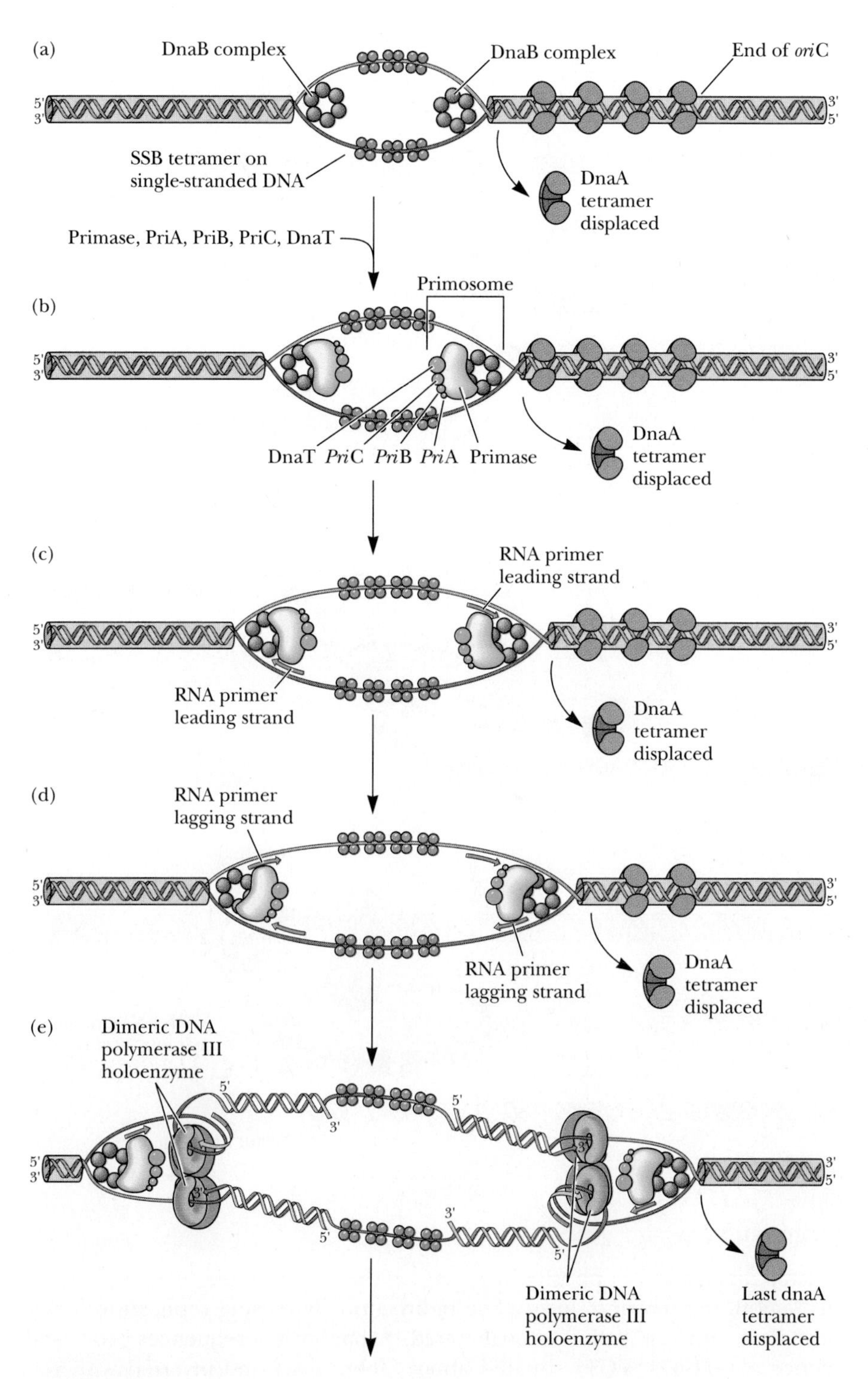

***Figure* 29.18** A representation of replisome assembly at each replication fork. (a) The "pre-priming complex" of Figure 29.17 is the starting point. (b) Auxiliary proteins PriA, PriB, PriC, and DnaT (subunits of the primosome) join, followed by primase. This completes assembly of the primosome. An RNA primer is synthesized at each replication fork by primase, (c) one for the leading strand and (d) one for the lagging strand. Two DNA polymerase III holoenzyme complexes (e) at each replication fork carry out DNA elongation, one synthesizing the leading strand and the other, the lagging strand.

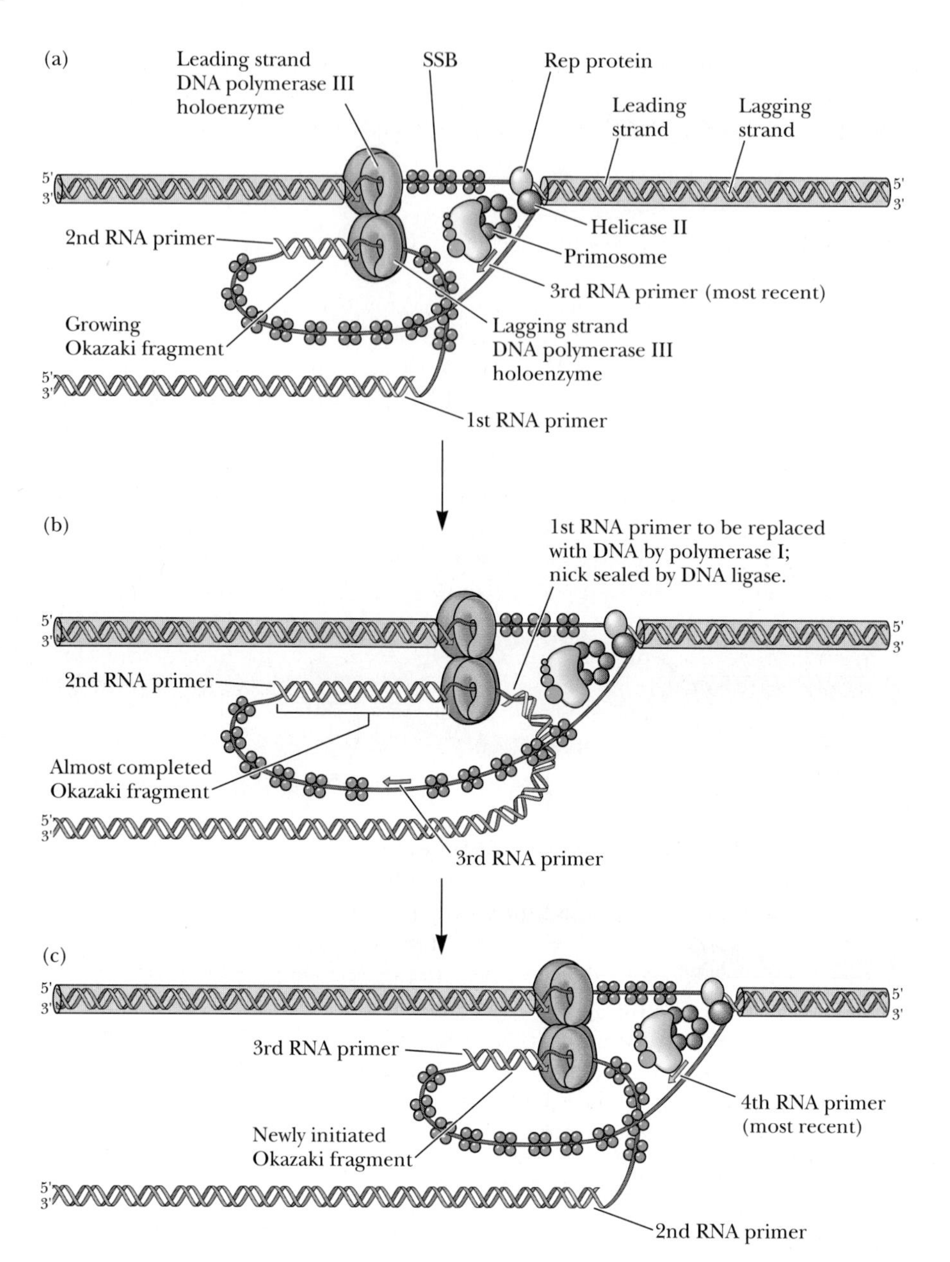

***Figure* 29.19** The asymmetric dimer model for *E. coli* DNA replication. Two helicases (Rep protein and helicase II) act in concert to unwind the DNA. The replisome contains two DNA polymerase III holoenzyme complexes that remain physically associated even though their tasks are different: One performs leading strand synthesis while the other synthesizes Okazaki fragments, the precursors to the lagging strand. Note that (a) the lagging strand template must loop around so that the pol III holoenzyme replicating it can move along it in the $3' \rightarrow 5'$ direction. As seen in (b), the lagging strand pol III holoenzyme must release the template when it comes up against the end of the Okazaki fragment it previously synthesized. In (c), the lagging strand pol III holoenzyme then shifts back to a position farther along the template in the 3′ direction to resume a new round of Okazaki fragment synthesis at the next RNA primer. Looping of the lagging strand template, the so-called **trombone model,** allows these events to happen. One of the two pol III holoenzymes is believed to have biochemical properties suited to continuous DNA synthesis, while the other one has properties consistent with its role in discontinuous DNA synthesis.

Termination

Located diametrically opposite from *oriC* on the *E. coli* circular map is a terminus region, the **ter,** or τ, locus. The bidirectionally moving replication forks meet here and replication is terminated. A pair of ter sequences (core sequence 5′-GTGTGTTGT) situated about 20 bp apart and inverted with respect to each other constitute the termination site. One ter orientation blocks the clockwise-moving replication fork; its inverted counterpart blocks the counterclockwise-moving fork. A ter sequence impedes replication fork progress *only if oriented in the proper direction* with respect to the approaching fork and then only if a specific 36-kD replication termination protein, **Ter protein,** is bound to it. Ter protein is a **contrahelicase.** That is, the Ter protein prevents DNA duplex from unwinding by blocking the ATP-dependent DnaB helicase activity; progress of the replication fork is thereby arrested. Final synthesis of both duplexes is completed. Replication usually leaves the circular progeny chromosomes intertwined by 20 to 30 coils about each other, a so-called **catenated** state. In order to disengage the individual duplexes from each

catenated: connected in a series of links, as in a chain other prior to their distribution to daughter cells, double-stranded cuts must be made so that the double helices can pass through one another. Topoisomerase II (DNA gyrase) can catalyze this process.

29.5 Eukaryotic DNA Replication

DNA replication in eukaryotic cells shows strong parallels with prokaryotic DNA replication. Although the mechanism of replication in *E. coli* provides a useful paradigm, the situation in eukaryotes is vastly more complex. For example, in a growing human cell, some 6 million kilobase pairs of DNA must be duplicated with high fidelity once each cell cycle. The events associated with cell growth and cell division in eukaryotic cells fall into a general sequence having four distinct phases, M, G_1, S, and G_2 (Figure 29.20). Eukaryotic cells have solved the problem of replicating their enormous genomes in the few hours allotted to the S phase by initiating DNA synthesis at multiple origins of replication. There is one origin of replication for anywhere from 3 to 300 kbp, depending on the organism and cell type (for example, at least 200 or so on an average human chromosome). Thus, DNA replication is occurring concomitantly in many units of replication, so-called **replicons,** distributed throughout the genome.

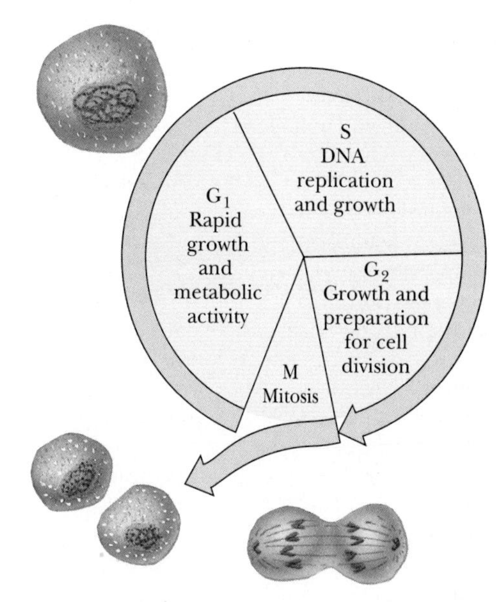

***Figure* 29.20** The eukaryotic cell cycle. The stages of mitosis and cell division define the M phase ("M" for *mitosis*). G_1 ("G" for *gap,* not *growth*) is typically the longest part of the cell cycle; G_1 is characterized by rapid growth and metabolic activity. Cells that are quiescent, that is, not growing and dividing (such as neurons), are said to be in G_0. The S phase is the time of DNA replication. S is followed by G_2, a relatively short period of growth when the cell prepares for cell division. Cell cycle times vary from less than 24 hours (rapidly dividing cells such as the epithelial cells lining the mouth and gut) to hundreds of days.

Eukaryotic DNA Polymerases

A number of different DNA polymerases have been described in animal cells and named in their order of discovery (Table 29.4). **DNA polymerase α,** a complex multimeric protein, is found only in the cell nucleus and functions in chromosome replication. It requires a template and a primer and carries out chain extension in the $5' \rightarrow 3'$ direction. It consists of four subunits: a 180-kD polymerase, two associated primase subunits (60 kD and 50 kD, respectively), and an additional 70-kD polypeptide whose function is uncertain. Interestingly, eukaryotic DNA polymerase α from most sources lacks $3' \rightarrow 5'$ exonuclease activity. **DNA polymerase δ** is similar to DNA polymerase α in

***Table* 29.4**
The Biochemical Properties of Eukaryotic DNA Polymerases

	α	δ	ϵ	β	γ
Mass (kD)					
Native	>250	170	256	36–38	160–300
Catalytic core	165–180	125	215	36–38	125
Other subunits	70, 50, 60	48	55	None	35, 47
Location	Nucleus	Nucleus	Nucleus	Nucleus	Mitochondria
Associated functions					
$3' \rightarrow 5'$ exonuclease	No	Yes	Yes	No	Yes
Primase	Yes	No	No	No	No
Properties					
Processivity	Low	High	High	Low	High
Fidelity	High	High	High	Low	High
Replication	Yes	Yes	Yes	No	Yes
Repair	No	?	Yes	Yes	No

Source: Adapted from Kornberg, A., and Baker, T. A., 1992. *DNA Replication,* 2nd ed. New York: W. H. Freeman and Co.

many respects and may be a companion enzyme in replication. DNA polymerase δ has $3' \rightarrow 5'$ exonuclease activity, as well as an associated 37-kD subunit, the **PCNA** protein (for *proliferating cell nuclear antigen*), which endows it with apparently unlimited processivity. PCNA shows homology to the prokaryotic DNA polymerase III β subunit. (The processivity of DNA polymerase α is a modest 200 nucleotides or so.) These two polymerases represent the eukaryotic counterparts of the prokaryotic dimer, one for the leading strand and one for the lagging strand. DNA polymerase α catalyzes the extension of a polynucleotide chain at the rate of 50 nucleotides per second, about one twentieth the rate of *E. coli* DNA polymerase III holoenzyme, further emphasizing the need for multiple origins of replication in eukaryotic cells. **DNA polymerase ϵ** is highly processive, despite lacking a PCNA counterpart; its precise function remains uncertain, although it has a major role in yeast and mammalian chromosome replication. The small **DNA polymerase β** of eukaryotes functions in DNA repair. **DNA polymerase γ** is the DNA-replicating enzyme of mitochondria.

SV40 DNA Replication

Studies on animal viruses such as SV40 (SV for simian virus) have provided a useful model for eukaryotic DNA replication, since the SV40 genome can be considered to represent a single replicon. This viral genome is a closed circular DNA duplex consisting of about 5000 bp with a single origin of replication. The virus makes extensive use of the host cell's replication machinery; only a single replication protein, **T antigen,** is viral DNA-encoded. Viral DNA replication is bidirectional and two nascent chains are synthesized at each replication fork: a leading strand and a lagging strand. Both are initiated by RNA primers. The results invite the hypothesis that the highly processive DNA polymerase δ catalyzes leading strand synthesis and the less processive DNA polymerase α mediates lagging strand synthesis in an "asymmetric dimer" replicase. Thus, SV40 DNA replication is very similar to replication in *E. coli.*

nascent: newly synthesized. From the Latin *nascor,* to be born

T Antigen and the Initiation of SV40 Replication

SV40 T antigen is a 95-kD protein playing an essential role in the initiation of viral DNA replication. The SV40 *ori* (origin of replication) is a 64-bp sequence consisting of at least three functionally distinct domains: (a) a central set of four copies of the pentameric sequence GAGGC organized as an inverted repeat. This sequence element is the T antigen binding site. On one side of the T antigen binding domain is (b) a 17-bp region composed exclusively of A and T residues; on the other side is (c) a 15-bp imperfect inverted repeat whose function is as yet unknown. One T antigen molecule binds to each of the four pentameric GAGGC repeats. T antigen is an ATP-dependent helicase, and once it forms a nucleoprotein complex with the four pentamers, it enters the duplex at the adjacent AT region and unwinds the helix. The AT region, with only two H bonds per base pair, is susceptible to strand separation. In association with **replication factor A (RF A),** a host cell protein that is the eukaryotic counterpart of SSB, T antigen carries out a set of changes at the replication origin not unlike those mediated by DnaA and DnaB at the prokaryotic *ori.* T antigen also binds DNA polymerase α and recruits host auxiliary replication proteins to the replication origin. In brief, the T antigen-mediated reaction is a critical step in the overall process—it establishes two replication forks and generates the substrate for assembly of the primase and DNA polymerase complexes that accomplish SV40 replication.

T Antigen Function Is Regulated by Protein Phosphorylation

Phosphorylation at a single residue (Thr^{124}) activates T antigen to bind at the SV40 ori and initiate replication. This phosphorylation is catalyzed by a *CDC2-type protein kinase.* Such kinases are implicated in triggering eukaryotic cells to enter mitosis. Dephosphorylation at different amino acid phosphorylation sites on T antigen also activates it. **Replication protein C (RP-C),** a protein essential to the early stages of SV40 replication, is a *protein phosphatase subunit.* The combined actions of CDC2-PK and RP-C generate an appropriately phosphorylated form of T antigen competent in replication initiation. These various phosphorylations/dephosphorylations represent paradigms for the central role of protein phosphorylation in regulating eukaryotic replication and cell division.

29.6 Reverse Transcriptase: An RNA-Directed DNA Polymerase

The information encoded in DNA can be replicated into new DNA chains by DNA polymerases (this chapter) or transcribed into RNA chains by DNA-dependent RNA polymerases (Chapter 30). Nature has also found a role for the synthesis of DNA chains from an RNA template. In 1964, Howard Temin noted that inhibitors of DNA synthesis prevented infection of cells in culture by RNA tumor viruses such as avian sarcoma virus. On the basis of this observation, Temin made the bold proposal that DNA is an intermediate in the replication of such viruses; that is, *an RNA tumor virus can use viral RNA as the template for DNA synthesis.*

RNA viral chromosome → DNA intermediate → RNA viral chromosome

In 1970, Temin and David Baltimore independently discovered a viral enzyme capable of mediating such a process, namely an **RNA-directed DNA polymerase,** or as it is usually called, **reverse transcriptase.** All RNA tumor viruses contain such an enzyme within their virions (viral particles), so they are now classified as **retroviruses.** Note that not all retroviruses are tumor viruses; HIV (human immunodeficiency virus) is a retrovirus that causes AIDS.

Like other DNA and RNA polymerases, reverse transcriptase synthesizes polynucleotides in the $5' \rightarrow 3'$ direction, and like all DNA polymerases, reverse transcriptase requires a primer. Interestingly, the primer is a specific tRNA molecule captured by the virion from the host cell in which it was produced. The 3′-end of the tRNA is base-paired with the viral RNA template at the site where DNA synthesis initiates and its free 3′-OH accepts the initial deoxynucleotide once transcription commences. Reverse transcriptase then transcribes the RNA template into a complementary DNA (cDNA) strand to form a double-stranded DNA:RNA hybrid.

The Enzymatic Activities of Reverse Transcriptases

Reverse transcriptases possess three enzymatic activities, all of which are essential to viral replication:

1. *RNA-directed DNA polymerase activity,* for which the enzyme is named (see Figure 8.15, page 265).
2. *RNase H activity.* Recall that RNase H is an exonuclease activity that specifically degrades RNA chains in DNA:RNA hybrids (Figure 8.15). The RNase H function of reverse transcriptase degrades the template genomic RNA and also removes the priming tRNA after DNA synthesis is completed.

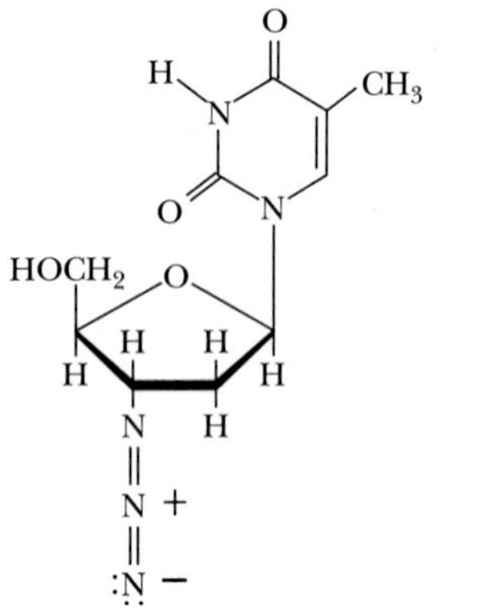

3. *DNA-directed DNA polymerase activity.* This activity replicates the ssDNA remaining after RNase H degradation of the viral genome, yielding a DNA duplex. This DNA duplex directs the remainder of the viral infection process or it becomes integrated into the host chromosome, where it can lie dormant for many years as a **provirus.** Activation of the provirus restores the infectious state.

HIV reverse transcriptase is of great clinical interest as an enzyme required for replication of the AIDS virus. It is a heterodimer of 66- and 51-kD subunits. The two polypeptides have identical N-terminal sequences, indicating that the 51-kD subunit is derived from the 66-kD polypeptide by proteolytic cleavage. The 66-kD polypeptide consists of two domains: an N-terminal polymerase domain and a C-terminal RNase H domain. The N-terminal domain shows significant amino acid sequence homology to bacterial and viral DNA polymerases, whereas the C-terminal half is homologous to bacterial RNase H. The active sites for the polymerase and RNase H are physically separate and functionally distinct. DNA synthesis by HIV reverse transcriptase is blocked by AZT (Figure 29.21). HIV reverse transcriptase is error-prone: It incorporates the wrong base at a frequency of 1 per 2000 to 4000 nucleotides polymerized. This high error rate during replication of the HIV genome means that the virus is ever changing, a feature that confounds attempts to devise a vaccine.

Figure **29.21** The structure of AZT (3′-azido-2′,3′-dideoxythymidine). This nucleoside is a major drug in the treatment of AIDS. AZT is phosphorylated *in vivo* to give AZTTP (AZT 5′-triphosphate), a substrate analog that binds to HIV reverse transcriptase. HIV reverse transcriptase incorporates AZTTP into growing DNA chains in place of dTTP. Incorporated AZTMP blocks further chain elongation because its 3′-azido group cannot form a phosphodiester bond with an incoming nucleotide. Host cell DNA polymerases have little affinity for AZTTP.

29.7 DNA Repair

Biological macromolecules are susceptible to chemical alterations that arise from environmental damage or errors during synthesis. For RNAs, proteins, or other cellular molecules, most consequences of such damage are circumvented by replacement of these molecules through normal turnover (synthesis and degradation). However, the integrity of DNA is vital to cell survival and reproduction. Its information content must be protected over the life span of the cell and preserved from generation to generation. Safeguards include (a) high-fidelity replication systems and (b) repair systems that correct DNA damage that might alter its information content. DNA is the only molecule that, if damaged, is repaired by the cell. Such repair is possible because the information content of duplex DNA is inherently redundant. The most common forms of damage are (a) a missing, altered, or incorrect base; (b) bulges

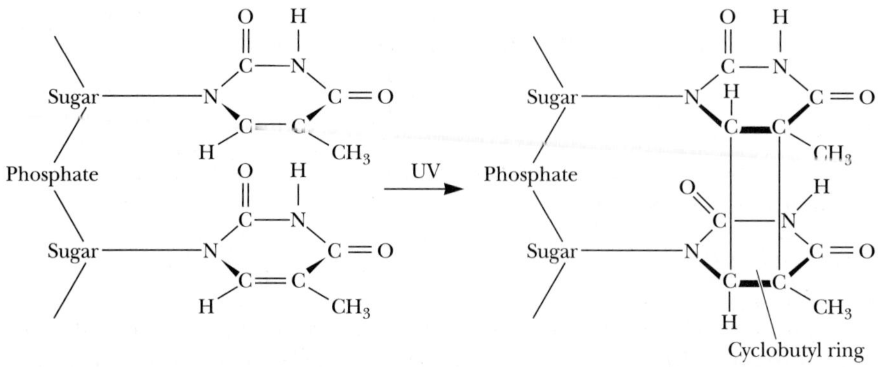

Figure **29.22** UV irradiation causes dimerization of adjacent thymine bases. A cyclobutyl ring is formed between carbons 5 and 6 of the pyrimidine rings. Normal base pairing is disrupted by the presence of such dimers.

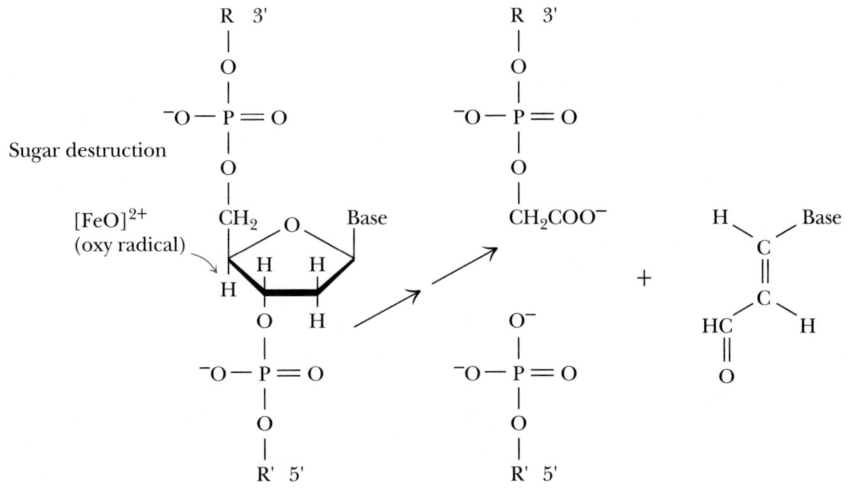

***Figure* 29.23** Oxygen radicals, in the presence of metal ions such as Fe^{2+}, can destroy sugar rings in DNA, breaking the strand.

due to deletions or insertions; (c) UV-induced pyrimidine dimers (Figure 29.22); (d) strand breaks at phosphodiester bonds or within deoxyribose rings (Figure 29.23); and (e) covalent cross-linking of strands. Cells have extraordinarily diverse and effective DNA repair systems to deal with these problems. When repair fails, the genome may still be preserved if an ''error-prone'' mode of replication allows the lesion to be bypassed.

Usually, the complementary structure of duplex DNA ensures that information lost through damage to one strand can be recovered from the other. However, even errors involving information on both strands can be corrected. For example, deletions or insertions can be repaired by replacing the region through recombination (see Chapter 28). Double-stranded breaks, potentially the most serious lesions, can be repaired by DNA ligases or recombination events.

Molecular Mechanisms of DNA Repair

Two fundamental types of molecular mechanisms for DNA repair can be distinguished: (a) mechanisms that excise and replace damaged regions by replication, recombination, or **mismatch repair,** and (b) mechanisms that reverse damaging chemical changes in DNA.

Mismatch Repair

Mismatch repair systems scan duplex DNA for mispaired bases, excise the mispaired region, and replace it by DNA polymerase-mediated local replication. The key in such replacement is to know which base of the mismatched pair is incorrect. One mismatch repair system is the **methyl-directed pathway** of *E. coli.* The nascent DNA strand in a duplex can be distinguished from the parental strand because it lacks methyl groups on its bases. DNA methylation, often an identifying and characteristic feature of a prokaryote's DNA, occurs subsequent to DNA replication. When the methyl-directed mismatch repair system encounters a mismatched base pair, it searches along the DNA, through thousands of base pairs if necessary, until it finds a methylated base. The system then identifies the strand bearing the methylated base as parental, assumes its sequence is the correct one, and replaces the entire stretch of nucleotides within the nascent strand from this recognition point to and including the mismatched base.

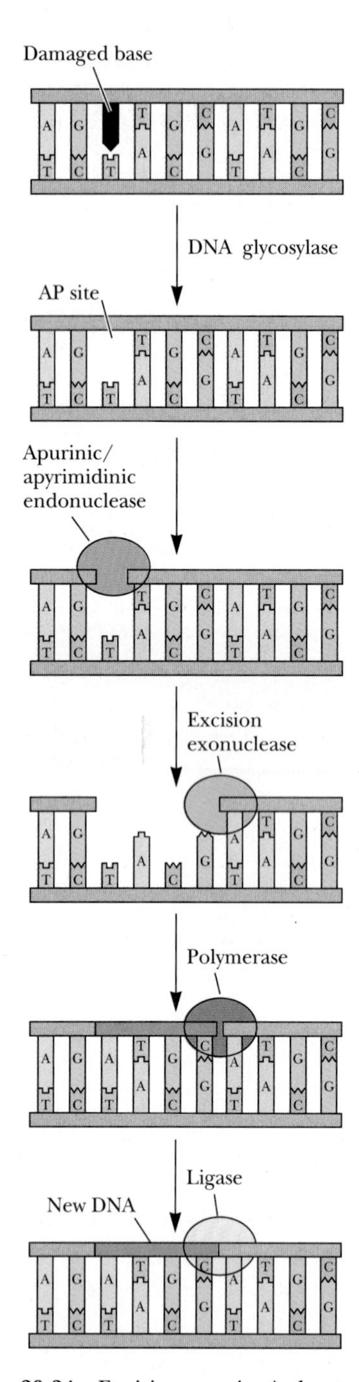

***Figure* 29.24** Excision repair. A damaged base (■) is excised from the sugar-phosphate backbone by DNA glycosylase, creating an AP site. Then, an apurinic/apyrimidinic endonuclease severs the DNA strand, and an excision nuclease removes the AP site and several nucleotides. DNA polymerase I and DNA ligase then repair the gap.

Reversing Chemical Damage

Photoreactivation of Pyrimidine Dimers. UV irradiation promotes the formation of covalent bonds between adjacent thymine residues in a DNA strand, creating a cyclobutyl ring (Figure 29.22). Since the C—C bonds in this ring are shorter than the normal 0.34-nm base stacking in B-DNA, the DNA is distorted at this spot and is no longer a proper template for either replication or transcription. **Photolyase** (also called **photoreactivating enzyme**), a flavin- and pterin-dependent enzyme, binds at the dimer and uses the energy of visible light to break the cyclobutyl ring, restoring the pyrimidines to their original form.

Excision Repair. Replacement of many damaged or modified bases occurs via **excision repair systems.** These systems include DNA glycosylases, which excise a damaged base by cleaving the glycosidic bond, creating an AP site where the sugar-phosphate backbone of a DNA strand is intact but a purine (*apurinic site*) or pyrimidine (*apyrimidinic site*) is missing. An AP endonuclease then cleaves the backbone, an exonuclease removes the deoxyribose-P and several additional residues, and the gap is repaired by DNA polymerase and ligase (Figure 29.24).

In mammalian cells, excision repair is the main pathway for removal of carcinogenic (cancer-causing) lesions caused by sunlight or other mutagenic agents. Such lesions are recognized by **XP-A** protein, named for *xeroderma pigmentosum,* an inherited human syndrome whose victims suffer serious skin lesions if exposed to sunlight. At sites recognized by XP-A, a multiprotein endonuclease is assembled and the damaged strand is cleaved. After the damaged oligonucleotide is displaced, DNA synthesis takes place to form a short patch, using the undamaged strand as template. Since PCNA (Section 29.5) is required for mammalian excision repair, DNA polymerase δ (or ϵ) is believed responsible for such repair synthesis. DNA ligase then seals the nick, completing the repair.

The SOS Response

E. coli DNA lesions that block replication activate the **SOS response,** a system that converts the lesion to an error-prone site and restores replication. Such lesions include pyrmidine dimers, cross-linked strands, and quinolone antibiotic-induced breaks in DNA. Quinolone antibiotics such as **nalidixic acid** (Figure 29.25) inhibit the nick-ligating activity of bacterial DNA gyrases. The SOS response is activated when the replication fork stalls and RecA protein (Chapter 28) binds to exposed ssDNA (or UV-damaged dsDNA). The RecA protein:DNA complex binds a protein, LexA. The LexA protein has proteolytic activity, and its binding induces a conformational change that causes LexA protein to cleave itself. Because LexA protein blocks expression of many genes encoding a set of proteins mediating error-prone replication, LexA protein's self-destruction leads to synthesis of these proteins. These proteins then assemble at the lesion and form a **mutasome,** an error-prone replication apparatus that allows DNA polymerase to replicate past the lesion.

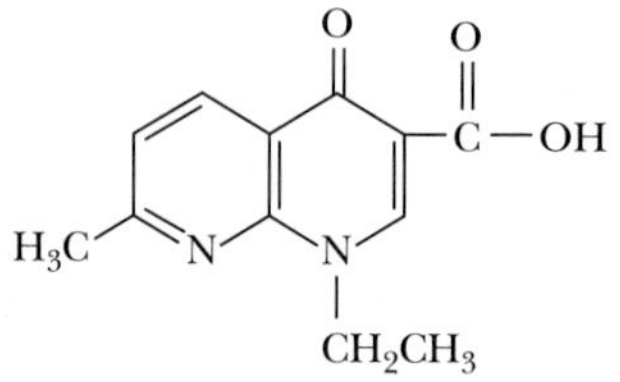

***Figure* 29.25** Nalidixic acid. Quinolone antibiotics such as nalidixic acid are DNA topoisomerase inhibitors. In the presence of one of these compounds, the DNA gyrase A subunits become covalently linked to the 5′-end of the DNA strands.

A Deeper Look

DNA Repair and Aging

Aging is a virtually universal disease in multicellular organisms. Certainly many factors, from environmental insults to genetically programmed cellular events, underlie the aging process. Since accumulation of DNA lesions must result in loss of cellular functions and hence viability, inadequate DNA repair is a prime culprit for those seeking biochemical causes for aging. An intriguing correlation has been noted between animal life span and competency in repairing UV-induced DNA lesions. Set against this observation are many findings that DNA repair capacity is not different in other animals with markedly different life spans, nor does DNA repair ability necessarily decline with age. While overall DNA repair may be a poor indicator of the aging process, subtle changes in the repair of specific genes may be important. The biochemistry of aging is a fascinating scientific problem.

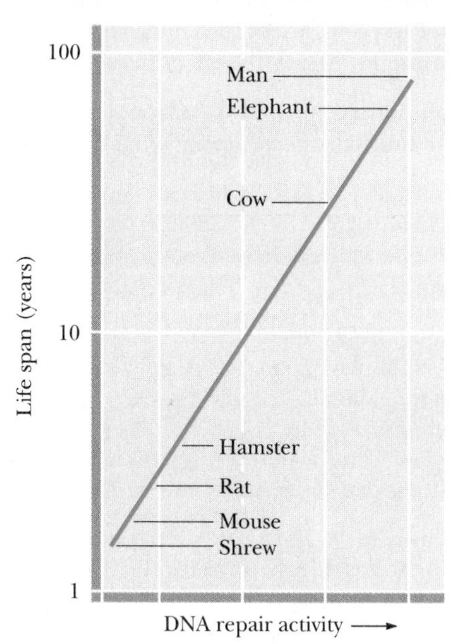

Correlation between DNA repair activity in young fibroblast cells from various mammalian species and the species' life span.

Problems

1. If ^{15}N-labeled *E. coli* DNA has a density of 1.724 g/mL, ^{14}N-labeled DNA has a density of 1.710 g/mL, and *E. coli* cells grown for many generations on $^{14}NH_4^+$ as nitrogen source are transferred to media containing $^{15}NH_4^+$ as sole N-source, what will be the density of the DNA after one generation, since replication is semiconservative? Supposing the mode of replication were dispersive, what would be the density of DNA after one generation? Design an experiment to distinguish between semiconservative and dispersive modes of replication.
2. What are the respective roles of the 5′-exonuclease and 3′-exonuclease activities of DNA polymerase I? What would be the phenotype of an *E. coli* strain that lacked DNA polymerase I 3′-exonuclease activity?
3. Assuming DNA replication proceeds at a rate of 750 base pairs per second, calculate how long it will take to replicate the entire *E. coli* genome. Under optimal conditions, *E. coli* cells divide every 20 minutes. What is the minimal number of replication forks per *E. coli* chromosome in order to sustain such a rate of cell division?
4. It is estimated that there are 10 molecules of DNA polymerase III per *E. coli* cell. Is it likely that *E. coli* growth rate is limited by DNA polymerase III levels?
5. Approximately how many Okazaki fragments are synthesized in the course of replicating an *E. coli* chromosome?
6. How do DNA gyrases and helicases differ in their respective functions and modes of action?
7. None of the mechanisms of DNA replication described in this chapter can account for the *complete* synthesis of a *linear* duplex DNA molecule. Why?
8. Homologous recombination in *E. coli* leads to the formation of regions of heteroduplex DNA. By definition, such regions contain mismatched bases. Why doesn't the mismatch repair system of *E. coli* eliminate these mismatches?

Further Reading

Beese, L. S., Derbyshire, V., and Steitz, T. A., 1993. Structure of DNA polymerase I Klenow fragment bound to duplex DNA. *Science* **260:**352–355.

Campbell, J. L., 1993. Yeast DNA replication. *The Journal of Biological Chemistry* **268**:25261–25264.

DePamphilis, M. E., 1993. Eukaryotic DNA replication: anatomy of an origin. *Annual Review of Biochemistry* **62:**29–63.

Johnson, K. A., 1993. Conformational coupling in DNA polymerase fidelity. *Annual Review of Biochemistry* **62:**685–713.

Kelly, T. J., 1988. SV40 DNA replication. *The Journal of Biological Chemistry* **263:**17889–17892. A review of this viral DNA replication system, which is viewed as a paradigm of eukaryotic DNA replication.

Kong, X.-P., et al., 1992. Three-dimensional structure of the β-subunit of *E. coli* DNA polymerase III holoenzyme: A sliding clamp model. *Cell* **69:**425–437. The crystal structure of the β-subunit of DNA pol III holoenzyme reveals that this processivity factor forms a closed ring around DNA, tightly clamping the pol III enzyme to its substrate.

Kornberg, A., 1988. DNA replication. *The Journal of Biological Chemistry* **263:**1–4. A brief update on the enzymology of DNA replication by its foremost investigator.

Kornberg, A., and Baker, T. A., 1992. *DNA Replication,* 2nd ed. New York: W. H. Freeman and Co. A comprehensive detailed account of the enzymology of DNA metabolism, including replication, recombination, repair, and more.

Lee, H. E., et al., 1989. *Escherichia coli* replication termination protein impedes the action of helicases. *The Journal of Biological Chemistry* **86:**9104–9108. How Ter protein achieves termination of replication.

Lehman, I. R., and Kaguni, L. S., 1989., DNA polymerase α. *The Journal of Biological Chemistry* **264:**4265–4268. A short review of the properties of the major eukaryotic DNA polymerase.

Lohman, T. M., 1993. Helicase-catalyzed DNA unwinding. *The Journal of Biological Chemistry* **268**:2269–2272.

Matson, S. W., and Kaiser-Rogers, K. A., 1990. DNA helicases. *Annual Review of Biochemistry* **59:**289–329. Biochemistry of the enzymes that unwind DNA.

McHenry, C. S., 1988. DNA polymerase III holoenzyme of *Escherichia coli. Annual Review of Biochemistry* **57:**519–550. A comprehensive review of the biochemistry of the DNA-replicating enzyme of *E. coli.*

McHenry, C. S., 1991. DNA polymerase III holoenzyme. *The Journal of Biological Chemistry* **266:**19127–19130.

Meselson, M., and Stahl, F. W., 1958. The replication of DNA in *Escherichia coli. Proceedings of the National Academy of Sciences, USA* **44:**671–682. The classic paper showing that DNA replication is semiconservative.

Ogawa, T., and Okazaki, T., 1980. Dicontinuous DNA replication. *Annual Review of Biochemistry* **49:**421–457. Okazaki fragments and their implications for the mechanism of DNA replication.

Siata, P. R., Hutchinson, C. A., III, and Bastia, D., 1991. DNA-protein interaction at the replication termini of plasmid R6K. *Genes and Development* **5:**74–82. Studies indicating the mechanism by which DNA replication is terminated.

Wickner, R. B., 1993. Double-stranded RNA virus replication and packaging. *The Journal of Biological Chemistry* **268**:3797–3800.

Chapter 30

"A day will come when some laborious monk
Will bring to light my zealous, nameless toil,
Kindle . . . his lamp, and from the parchment
Shaking the dust of ages, will transcribe
My chronicles."

Alexander Pushkin from *Boris Godunov* (1825)

Transcription and the Regulation of Gene Expression

"Monk Transcribing Manuscript," ca. 1470.

Outline

30.1 Transcription in Prokaryotes

30.2 Transcription in Eukaryotes

30.3 The Regulation of Transcription in Prokaryotes

30.4 Transcription Regulation in Eukaryotes

30.5 Structural Motifs in DNA-Binding Regulatory Proteins

30.6 Post-Transcriptional Processing of mRNA in Eukaryotic Cells

In 1958, Francis Crick enunciated the "central dogma of molecular biology" (Figure 30.1). This scheme outlined the residue-by-residue transfer of biological information as encoded in the primary structure of the informational biopolymers, nucleic acids and proteins. The predominant path of information transfer, DNA → RNA → protein, postulated that RNA was an information carrier between DNA and proteins, the agents of biological function. In 1961, François Jacob and Jacques Monod extended this hypothesis to predict that the RNA intermediate, which they dubbed **messenger RNA,** or **mRNA,** would have the following properties:

1. Its base composition would reflect the base composition of DNA (a property consistent with genes as protein-encoding units).
2. It would be very heterogeneous with respect to molecular mass, yet the average molecular mass would be several hundred kD. (A 200-kD RNA contains roughly 750 nucleotides, which could encode a protein of about 250 amino acids—approximately 30 kD—a reasonable estimate for the average size of polypeptides.)
3. It would be able to associate with ribosomes, since ribosomes are the site of protein synthesis.

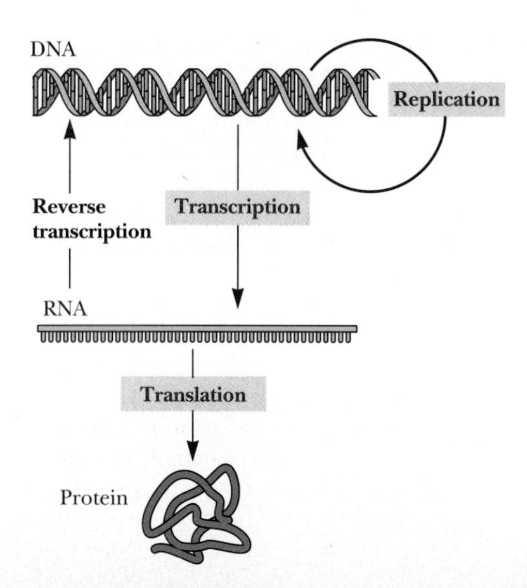

***Figure* 30.1** Crick's 1958 view of the "central dogma of molecular biology": Directional flow of detailed sequence information includes DNA→DNA (replication), DNA→RNA (transcription), RNA→protein (translation), RNA→DNA (reverse transcription). Note that no pathway exists for the flow of information from proteins to nucleic acids, that is, protein→RNA or DNA. A possible path from DNA to protein has since been discounted. Interestingly, in 1958, mRNA had not yet been discovered.

4. It would have a high rate of turnover. (That is, mRNA would rapidly degrade. Lability of mRNA would allow the rate of mRNA synthesis to control the rate of protein synthesis.)

Since Jacob and Monod's 1961 hypothesis, it has been realized that cells contain three major classes of RNA—mRNA, ribosomal RNA (rRNA), and transfer RNA (tRNA)—all of which participate in protein synthesis (Chapter 6). All of these RNAs are synthesized from DNA templates by **DNA-dependent RNA polymerases** in the process known as **transcription.** However, only mRNAs direct the synthesis of proteins. Thus, not all genes encode proteins; some encode rRNAs or tRNAs. Protein synthesis occurs via the process of **translation,** wherein the instructions encoded in the sequence of bases in mRNA are translated into a specific amino acid sequence by ribosomes, the "workbenches" of polypeptide synthesis (Chapter 32).

Transcription is tightly regulated in all cells. In prokaryotes, only about 3% of the genes are undergoing transcription at any given time. The metabolic conditions and the growth status of the cell dictate which gene products are needed at any moment. In a differentiated eukaryotic cell, the figure is around 0.01%. Such differentiated cells express only the information characteristic of the biological functions of their cell type, not the full genetic potential encoded in their chromosomes.

30.1 Transcription in Prokaryotes

In prokaryotes, virtually all RNA is synthesized by a single species of *DNA-dependent RNA polymerase.* (The only exception is the short RNA primers formed by primase during DNA replication.) Like DNA polymerases, RNA polymerase links ribonucleoside 5′-triphosphates (ATP, GTP, CTP, and UTP, represented generically as *NTPs*) in an order specified by base pairing with a DNA template:

$$n \text{ NTP} \longrightarrow (\text{NMP})_n + n \text{ PP}_i$$

The enzyme moves along a DNA strand in the 3′→5′ direction, joining the 5′-phosphate of an incoming ribonucleotide to the 3′-OH of the previous residue. Thus, the RNA chain grows 5′→3′ during transcription, just as DNA chains do during replication. The reaction is driven by subsequent hydrolysis of PP_i to inorganic phosphate by ubiquitous pyrophosphatase activity.

The Structure and Function of *E. coli* RNA Polymerase

The RNA polymerase of *E. coli,* so-called **RNA polymerase holoenzyme,** is a complex multimeric protein (465 kD) large enough to be visible in the electron microscope. Its subunit composition is $\alpha_2\beta\beta'\sigma$. The largest subunit, β' (160 kD), functions in DNA binding; β (150 kD) binds the nucleoside triphosphate substrates and interacts with σ (82 kD). Any of a number of related proteins, **the sigma (σ) factors,** can serve as the σ subunit. Sigma subunits function in recognizing specific sequences on DNA called **promoters** that identify the location of *transcription start sites,* where transcription begins. Both β and β' contribute to formation of the catalytic site. The two α subunits (36.5 kD each) are essential for assembly of the enzyme and activation by some regulatory proteins. Dissociation of the σ subunit from the holoenzyme leaves the so-called **core polymerase** ($\alpha_2\beta\beta'$), which is catalytically competent but unable to recognize promoters.

A Deeper Look

Conventions Employed in Expressing the Sequences of Nucleic Acids and Proteins

Certain conventions are useful in tracing the course of information transfer from DNA to protein. The strand of duplex DNA that is used as a template by RNA polymerase is termed the **sense strand.** Thus, the strand that is not copied is the **antisense strand.** Since the sense strand is read by the RNA polymerase moving 3′→5′ along it, the RNA product, the so-called **transcript,** grows in the 5′→3′ direction (see figure). Note that the antisense strand has a nucleotide sequence and direction identical to the RNA transcript, except that the transcript has U residues in place of T. The RNA transcript will eventually be translated into the amino acid sequence of a protein (Chapters 31, 32) by a process in which successive triplets of bases (termed **codons**), read 5′→3′, specify a particular amino acid. Polypeptide chains are synthesized in the N→C direction, and the 5′-end of mRNA encodes the N-terminus of the protein.

By convention, when the order of nucleotides in DNA is specified, it is the 5′→3′ sequence of nucleotides in the antisense strand that is presented. Consequently, if convention is followed, DNA sequences are rendered in terms that correspond directly to mRNA sequences, which correspond in turn to the amino acid sequences of proteins as read beginning with the N-terminus.

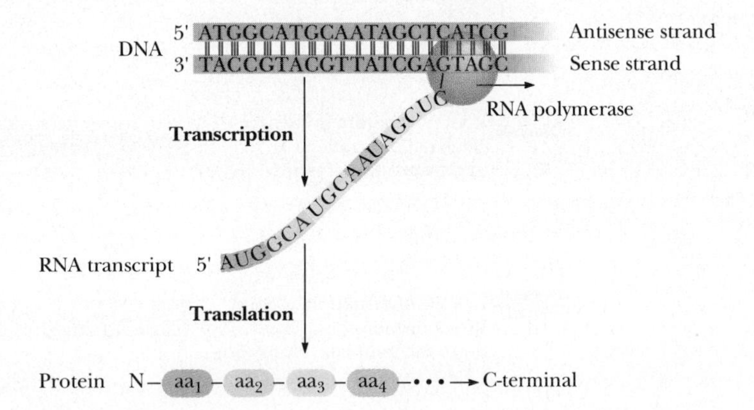

The DNA strand copied by RNA polymerase is the **sense strand;** the other strand is the **antisense strand.** Note that the nucleotide sequence of the antisense strand as read in the 5′→3′ direction is identical to that of the RNA transcript from the "sense" strand, except that the RNA contains U residues where Ts occur in the DNA. The nucleotide sequence of the transcript, or mRNA, as read in the 5′→3′ direction specifies the amino acid sequence of the protein that it encodes, as read in the N→C direction.

The Steps of Transcription in Prokaryotes

Transcription can be divided into four stages: (a) binding of RNA polymerase holoenzyme at promoter sites, (b) initiation of polymerization, (c) chain elongation, and (d) chain termination. A discussion of these stages follows.

Binding of RNA Polymerase to Template DNA

RNA polymerase binds nonspecifically to DNA with low affinity and then migrates along it, searching for a promoter (Figure 30.2, Step 1). The process of transcription begins when the σ subunit of RNA polymerase recognizes a promoter sequence, and RNA polymerase holoenzyme and the promoter form a so-called **closed promoter complex** (Figure 30.2, Step 2). Dissociation constants for RNA polymerase holoenzyme:closed promoter complexes range

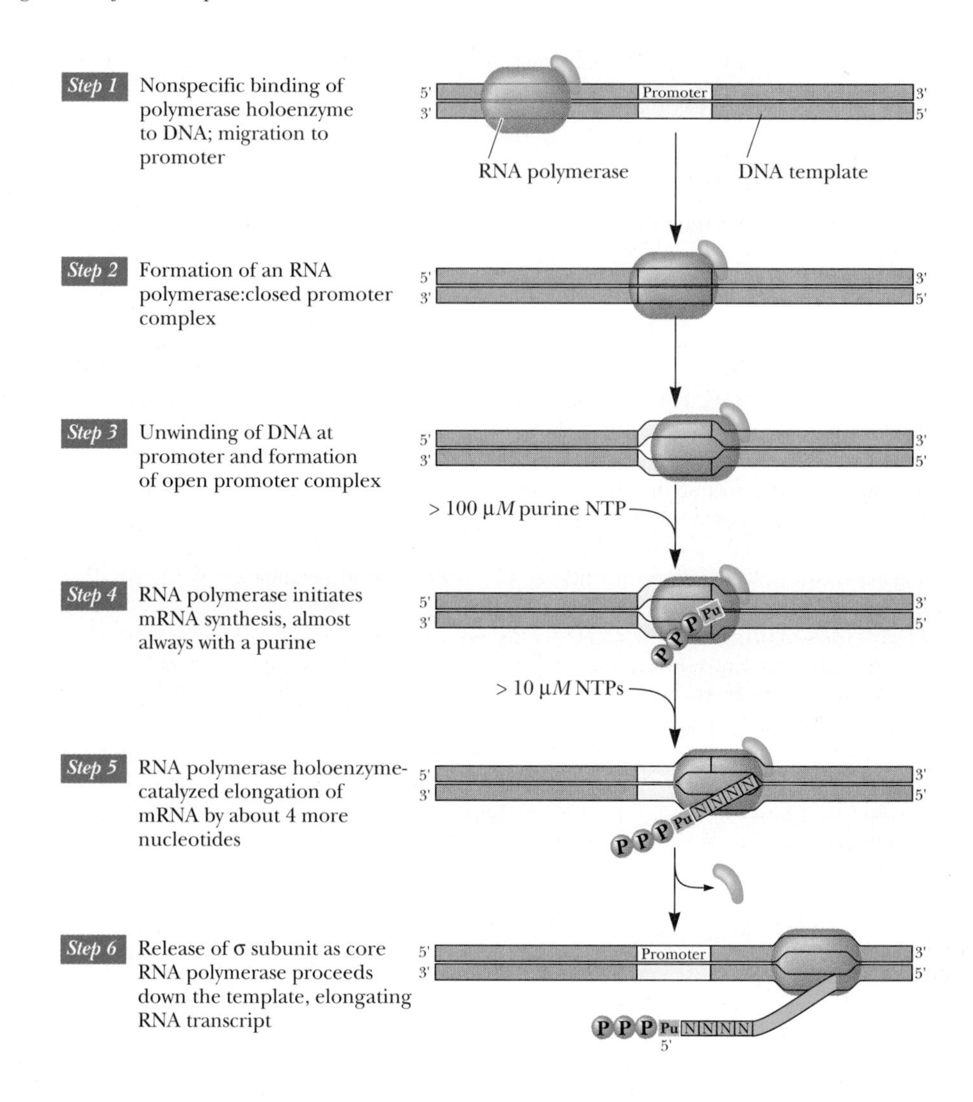
Step 1
Nonspecific binding of polymerase holoenzyme to DNA; migration to promoter
5'
3'
Promoter
3'
5'
RNA polymerase
DNA template
Step 2
Formation of an RNA polymerase:closed promoter complex
Step 3
Unwinding of DNA at promoter and formation of open promoter complex
> 100 μM purine NTP
Step 4
RNA polymerase initiates mRNA synthesis, almost always with a purine
P P P Pu
> 10 μM NTPs
Step 5
RNA polymerase holoenzyme-catalyzed elongation of mRNA by about 4 more nucleotides
P P P Pu N N N N
Step 6
Release of σ subunit as core RNA polymerase proceeds down the template, elongating RNA transcript
P P P Pu N N N N

***Figure* 30.2** Sequence of events in the initiation and elongation phases of transcription as it occurs in prokaryotes. Nucleotides in this region are numbered with reference to the base at the transcription start site, which is designated +1.

from 10^{-6} to 10^{-9} *M*. Although the DNA strands must be unwound so that the RNA polymerase can read and transcribe the DNA sense strand into a complementary RNA sequence, this stage in RNA polymerase:DNA interaction is referred to as the *closed* promoter complex because the DNA is not yet unwound.

Once the closed promoter complex is established, the RNA polymerase holoenzyme unwinds about 12 base pairs of DNA (from −9 to +3), forming the very stable **open promoter complex** (Figure 30.2, Step 3). In this complex, the RNA polymerase holoenzyme is bound very tightly to the DNA ($K_D \simeq 10^{-14}$ *M*). Indeed, promoter sequences can be identified *in vitro* by binding RNA polymerase holoenzyme to a putative promoter sequence in a DNA duplex and then treating the DNA:protein complex with DNase I. DNase I degrades any DNA not protected by bound protein, so the DNA fragment left after exhaustive DNase digestion defines the RNA polymerase holoenzyme binding site, which by one definition is the promoter[1]. RNA polymerase binding typically protects a nucleotide sequence spanning the region from −40 to +20. Limited DNase digestion protects a region extending up to −60, suggesting RNA polymerase interacts with this region too, although less securely.

[1]Another definition is mutational: nucleotide changes in this region may block gene expression because they inactivate the promoter.

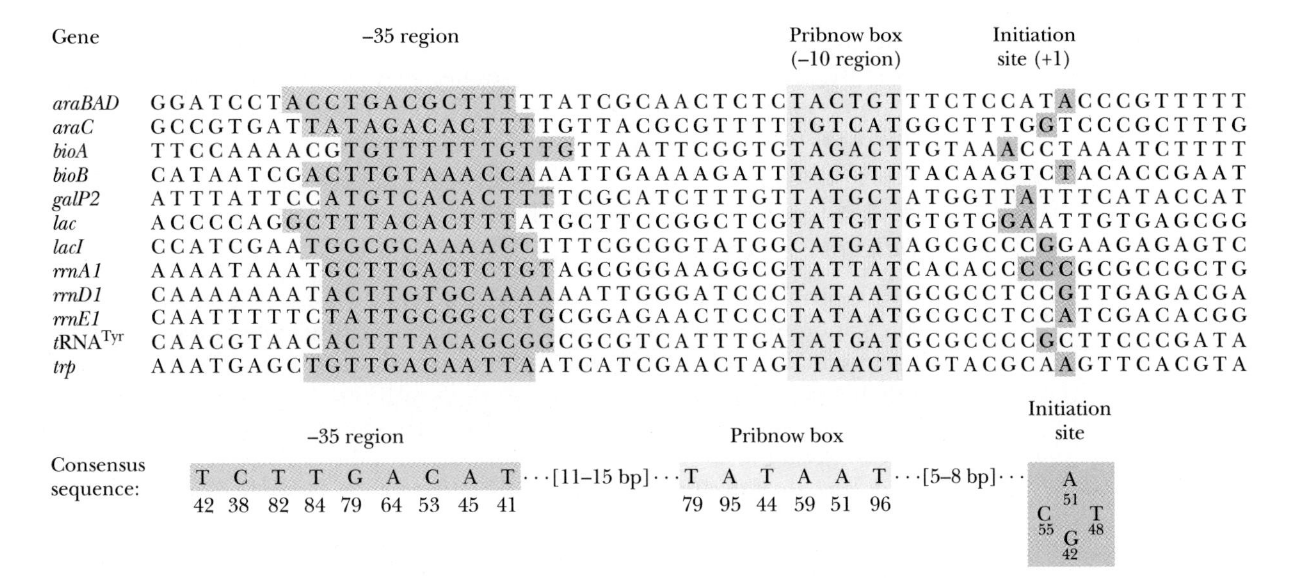

Gene
−35 region
Pribnow box (−10 region)
Initiation site (+1)
araBAD GGATCCTACCTGACGCTTTTTATCGCAACTCTCTACTGTTTCTCCATACCCGTTTTT
araC GCCGTGATTATAGACACTTTTGTTACGCGTTTTTGTCATGGCTTTGGTCCCGCTTTG
bioA TTCCAAAACGTGTTTTTTGTTGTTAATTCGGTGTAGACTTGTAAACCTAAATCTTTT
bioB CATAATCGACTTGTAAACCAAATTGAAAAGATTTAGGTTTACAAGTCTACACCGAAT
galP2 ATTTATTCCATGTCACACTTTTCGCATCTTTGTTATGCTATGGTTATTTCATACCAT
lac ACCCCAGGCTTTACACTTTATGCTTCCGGCTCGTATGTTGTGTGGAATTGTGAGCGG
lacI CCATCGAATGGCGCAAAACCTTTCGCGGTATGGCATGATAGCGCCCGGAAGAGAGTC
rrnA1 AAAATAAATGCTTGACTCTGTAGCGGGAAGGCGTATTATCACACCCCGCGCCGCTG
rrnD1 CAAAAAAATACTTGTGCAAAAAATTGGGATCCCTATAATGCGCCTCCGTTGAGACGA
rrnE1 CAATTTTTCTATTGCGGCCTGCGGAGAACTCCCTATAATGCGCCTCCATCGACACGG
tRNATyr CAACGTAACACTTTACAGCGGCGCGTCATTTGATATGATGCGCCCCGCTTCCCGATA
trp AAATGAGCTGTTGACAATTAATCATCGAACTAGTTAACTAGTACGCAAGTTCACGTA
−35 region
Pribnow box
Initiation site
Consensus sequence:
T C T T G A C A T ···[11–15 bp]··· T A T A A T ···[5–8 bp]··· A
42 38 82 84 79 64 53 45 41
79 95 44 59 51 96
A 51 C 55 T 48 G 42

***Figure* 30.3** The nucleotide sequences of representative *E. coli* promoters. (In accordance with convention, these sequences are those of the antisense strand where RNA polymerase binds.) Consensus sequences for the −35 region, the Pribnow box, and the initiation site are shown at the bottom. The numbers represent the percent occurrence of the indicated base. (*Note:* The −35 region is only roughly 35 nucleotides from the transcription start site; the Pribnow box [the −10 region] likewise is located at approximately position −10.) In this figure, sequences are aligned relative to the Pribnow box.

Properties of Prokaryotic Promoters. Prokaryotic promoters vary in size from 20 to 200 bp, but typically consist of a 40-bp region located on the 5′-side of the **transcription start site.** The transcription start site is the base in DNA, almost always a purine, specifying the first base of the transcript. The base at the transcription start site is designated +1 and bases on the 3′-side of it (that is, reading 5′→3′ from +1) are numbered +2, +3, and so on, in order. The bases on the 5′-side of the transcription start site are numbered consecutively in the 3′→5′ direction, −1, −2, and so on. (Note that there is no zero.) The "minus" direction is considered to lie "upstream" of the transcription start site.

Within the promoter are two **consensus sequence elements.** (A consensus sequence can be defined as *the bases that appear with highest frequency at each position when a series of sequences believed to have a common function are compared.*) These two elements are the **Pribnow box**[2] near −10, whose consensus sequence is the hexameric TATAAT, and a sequence in the **−35 region** containing the hexameric consensus TTGACA (Figure 30.3). This sequence is believed to be the site of initial binding by the σ subunit. The more closely the −35 region sequence corresponds to this consensus, the greater is the efficiency of transcription of the gene. Thus, variations from these consensus sequences affect the stability of RNA polymerase binding, and in turn, transcriptional efficiency. Such variation provides one way that gene expression is modulated.

Once bound at the −35 region, RNA polymerase holoenzyme moves into contact with the Pribnow box and associates with it to form the closed promoter complex. In order for transcription to begin, the DNA duplex must be "opened" so that RNA polymerase has access to single-stranded template.

[2]Named for David Pribnow, who, along with David Hogness, first recognized the importance of this sequence element in transcription.

The efficiency of initiation is inversely proportional to the melting temperature, T_m, in the Pribnow box, suggesting that the A:T-rich nature of this region is aptly suited for facile "melting" of the DNA duplex and creation of the open promoter complex (Figure 30.2). Negative supercoiling facilitates transcription initiation by favoring DNA unwinding.

Initiation of Polymerization

RNA polymerase has two binding sites for NTPs—the initiation site and the elongation site. The **initiation site** binds the purine nucleotides ATP and GTP preferentially; most RNAs begin with a purine at the 5′-end. The first nucleotide binds at the initiation site, H-bonding with the +1 base exposed within the *open promoter complex* (Figure 30.2, Step 4). The second incoming nucleotide binds at the elongation site, H-bonding with the +2 base. The ribonucleotides are then united when the 3′-OH of the first nucleotide makes a nucleophilic attack on the α-phosphorus atom of the second nucleotide. A phosphoester bond is formed, and PP_i is eliminated. Note that the 5′-end of the transcript starts out with a triphosphate attached to it. Movement of RNA polymerase along the sense strand (*translocation*) to the next base prepares the RNA polymerase to add the next nucleotide (Figure 30.2, Step 5). Once an oligonucleotide 6 to 10 residues long has been formed, the σ subunit dissociates from RNA polymerase, signaling the completion of initiation (Figure 30.2, Step 6). The core RNA polymerase goes on to synthesize the remainder of the mRNA. As the core RNA polymerase progresses, advancing the 3′-end of the RNA chain, the DNA duplex is unwound just ahead of it. About 12 base pairs of the growing RNA remain base-paired to the DNA template at any time, with the RNA strand becoming displaced as the DNA duplex rewinds behind the advancing RNA polymerase.

Rifamycin B and its analog, **rifampicin,** are inhibitors of initiation. Despite their structural similarity (Figure 30.4), they act in different ways. Rifamycin binds to the β subunit of RNA polymerase and blocks binding of incoming NTP at the initiation site. Rifampicin allows the first phosphodiester bond to be formed, but it prevents the translocation of RNA polymerase along the DNA template. However, once the second phosphodiester bond is formed, creating an RNA trinucleotide, rifampicin is without effect. Thus, like rifamycin, rifampicin inhibits initiation but not elongation.

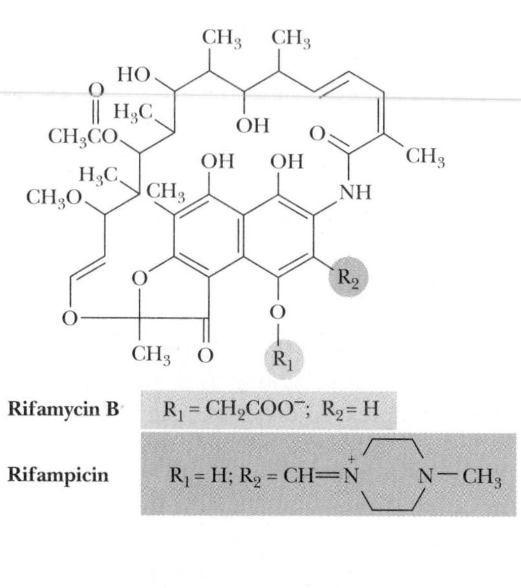

Rifamycin B $R_1 = CH_2COO^-$; $R_2 = H$

Rifampicin $R_1 = H$; $R_2 = CH{=}N^+$(piperazine ring)$N{-}CH_3$

Figure **30.4** The structures of rifamycin B and rifampicin, specific inhibitors of prokaryotic RNA polymerases. Since these compounds do not inhibit eukaryotic RNA polymerases, they have proven useful in the treatment of tuberculosis and infections caused by Gram-positive bacteria.

Chain Elongation

Elongation of the RNA transcript is catalyzed by the *core polymerase,* since once a short oligonucleotide chain has been synthesized, the σ subunit dissociates. **Cordycepin** (Figure 30.5) is an inhibitor of chain elongation in prokaryotes. This nucleoside can be phosphorylated *in vivo* to give 3′-deoxyadenosine 5′-triphosphate, which can bind to the core polymerase and add to the growing RNA. However, since cordycepin lacks a 3′-OH, it aborts further elongation. The accuracy of transcription is high; about once every 10^4 nucleotides, an error is made and the wrong base is inserted. Since many transcripts are made per gene, this error rate is acceptable. Also, the nature of the genetic code is such that errors are often innocuous (Chapter 31).

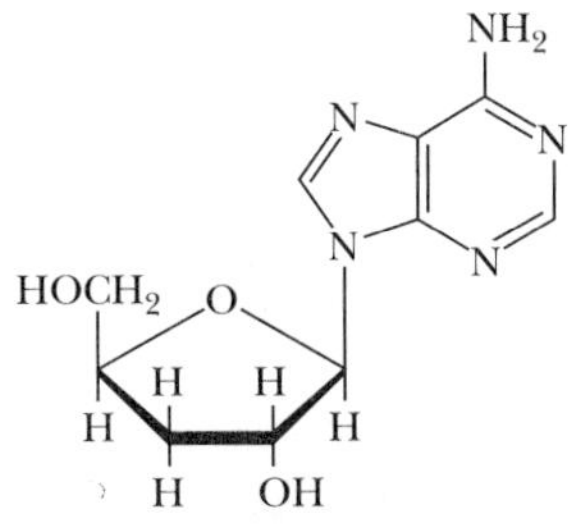

Cordycepin

Figure **30.5** *Cordycepin* is the name given 3′-deoxyadenosine.

Chain elongation does not proceed at a constant rate, but varies between 20 to 50 nucleotides per second. The RNA polymerase slows down and even pauses in G:C-rich regions due to the greater difficulty in unwinding G:C base pairs. As the RNA polymerase moves along the template, the DNA double helix is unwound ahead of it and recloses after the polymerase has passed by. Only a short stretch of RNA:DNA hybrid duplex exists at any time. Two possibilities can be envisioned: one in which RNA polymerase follows the template strand around the axis of the DNA duplex, although this seems unlikely due to its potential for tangling the nucleic acid strands (Figure 30.6a). The other possibility involves supercoiling of the DNA, so that positive supercoils are created ahead of the transcription bubble and negative supercoils are created behind it (Figure 30.6b). To prevent torsional stress from inhibiting transcription, topoisomerases act to remove these supercoils from the DNA segment undergoing transcription (Figure 30.6b).

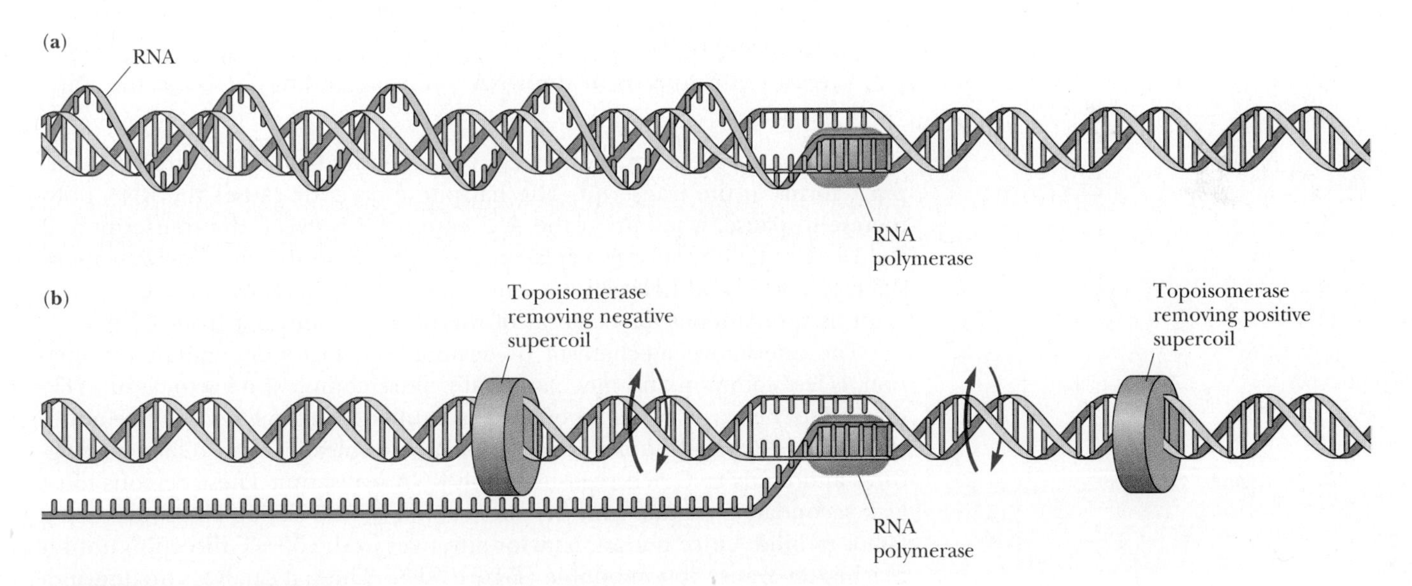

Figure **30.6** Supercoiling versus transcription. (a) If the RNA polymerase follows the template strand around the axis of the DNA duplex, no supercoiling of the DNA would occur, but the RNA chain would be wrapped around the double helix once every 10 bp. This possibility seems unlikely since it would be difficult to disentangle the transcript from the DNA duplex. (b) Alternatively, topoisomerases could remove the supercoils. A topoisomerase capable of relaxing positive supercoils situated ahead of the advancing transcription bubble would "relax" the DNA. A second topoisomerase behind the bubble would remove the negative supercoils.

(Adapted from Futcher, B., 1988. Supercoiling and transcription, or vice versa? Trends in Genetics 4:271–272.)

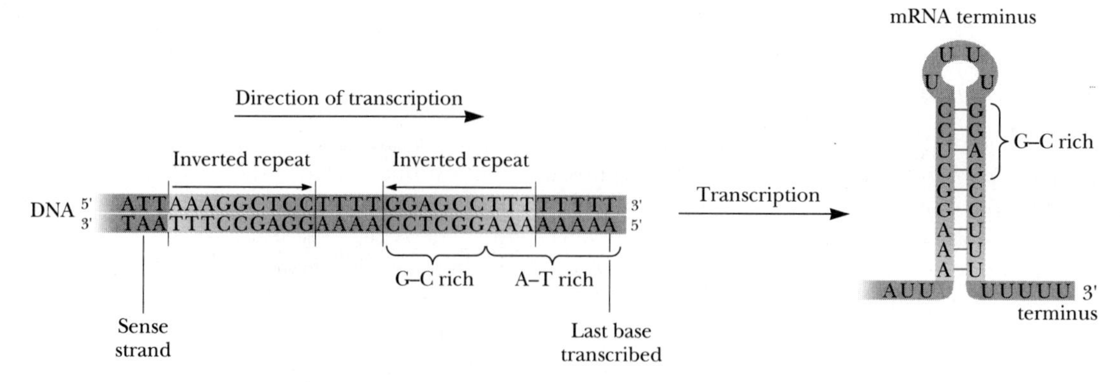

Figure **30.7** The termination site for the *E. coli trp* operon, from which is synthesized a transcript encoding the enzymes of tryptophan biosynthesis. The inverted repeats give rise to a stem-loop or "hairpin" structure ending in a series of U residues.

Chain Termination

Two types of transcription termination mechanisms operate in bacteria: one that is dependent on a specific protein **termination factor** called ***ρ*** (pronounced "rho") and another that is not dependent on this protein. In the latter, termination of transcription is determined by specific sequences in the DNA called **termination sites.** These sites are not characterized by a unique base where transcription halts. Instead, these sites consist of three structural features whose base-pairing possibilities ordain termination:

1. An inverted repeat sequence in which a nonrepeating segment punctuates the inverted repeats.
2. Inverted repeats, which are typically G:C-rich, so a stable **stem-loop structure** can form in the transcript via intrachain hydrogen bonding (Figure 30.7).
3. A run of 6 to 8 As in the DNA template, coding for Us in the transcript.

Termination then occurs as follows: A G:C-rich, stem-loop structure, or "hairpin," forms in the transcript. The hairpin apparently causes the RNA polymerase to pause, whereupon the A:U base pairs between the transcript and the DNA sense strand are displaced through formation of somewhat more stable A:T base pairs between the sense and antisense strands of the DNA. The result is spontaneous dissociation of the nascent transcript from DNA.

The alternative mechanism of termination, factor-dependent termination, is less common and mechanistically more complex. ***ρ*** **Factor** is an ATP-dependent helicase (hexamer of 50-kD subunits) that catalyzes the unwinding of RNA:DNA hybrid duplexes (or RNA:RNA duplexes). The ρ factor recognizes and binds to C-rich regions in the RNA transcript. These regions must lack secondary structure and be unoccupied by translating ribosomes for ρ factor to bind. Once bound, ρ factor advances in the $5' \rightarrow 3'$ direction until it reaches the transcription bubble (Figure 30.8). There it catalyzes the unwinding of the transcript and template, releasing the nascent RNA chain. It is likely that the RNA polymerase must stall at a G:C-rich termination region, allowing ρ factor to overtake it.

30.2 Transcription in Eukaryotes

Eukaryotic cells have three classes of RNA polymerase, each of which synthesizes a different class of RNA. All three enzymes are found in the nucleus. **RNA polymerase I** is localized to the nucleolus and transcribes the major

(a)

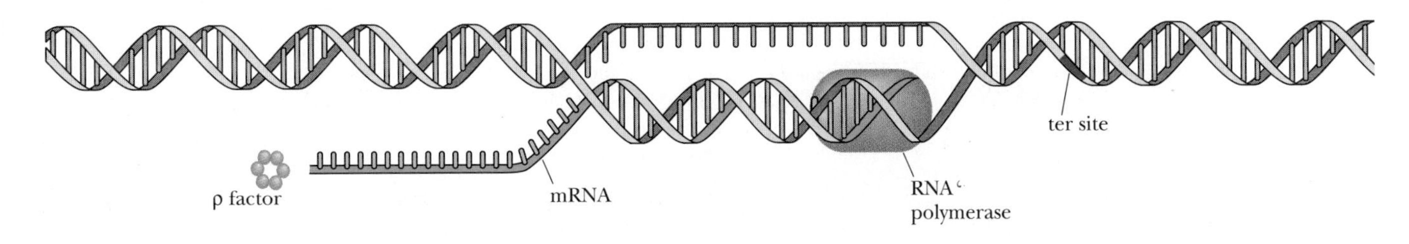

(b)

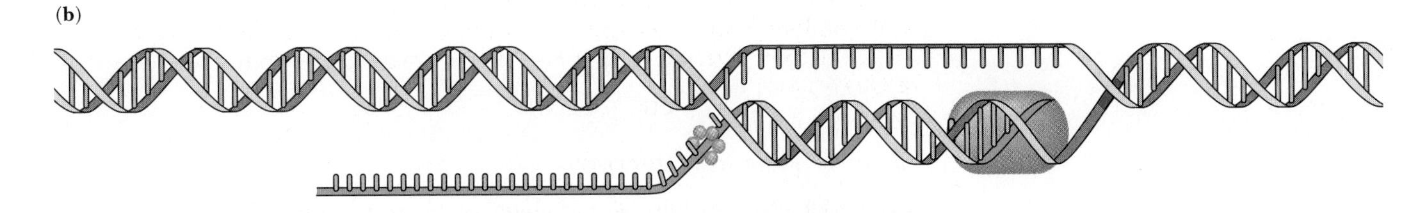

(c)

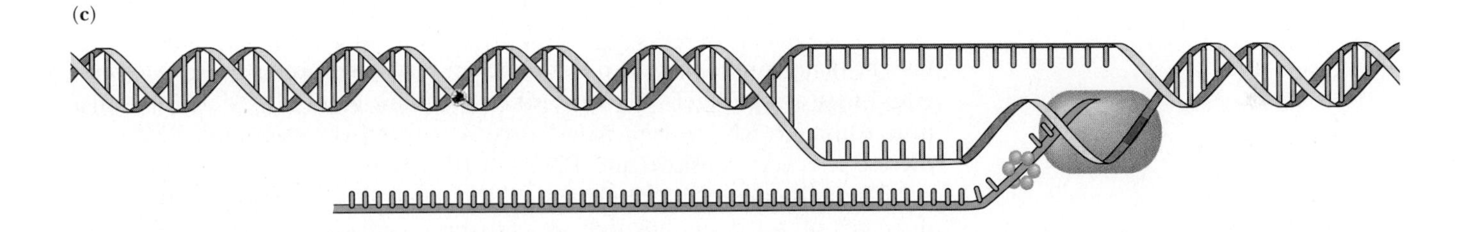

(d)

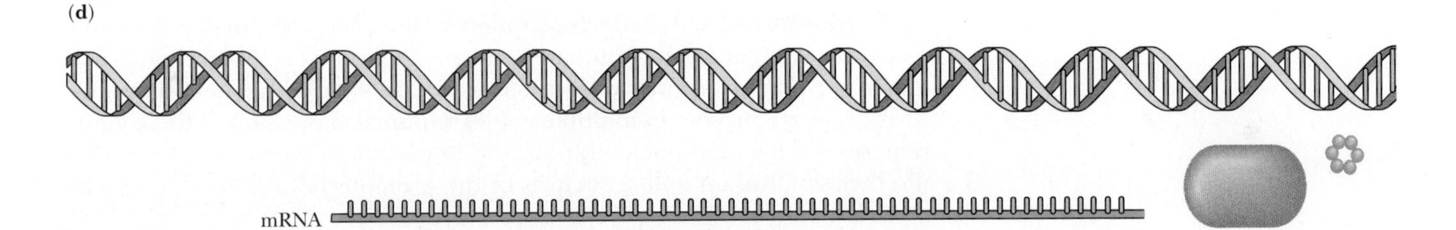

***Figure* 30.8** The ρ factor mechanism of transcription termination. Rho (ρ) factor (a) attaches to a recognition site on mRNA and (b) moves along it behind RNA polymerase. (c) When RNA polymerase pauses at the termination site, ρ factor unwinds the DNA:RNA hybrid in the transcription bubble, (d) releasing the nascent mRNA.

ribosomal RNA genes. **RNA polymerase II** transcribes protein-encoding genes, and thus it is responsible for the synthesis of mRNA. **RNA polymerase III** transcribes tRNA genes, the ribosomal RNA genes encoding 5S rRNA, and a variety of other small RNAs, including several involved in mRNA processing and protein transport.

All three RNA polymerase types are large, complex multimeric proteins (500 to 700 kD), consisting of 10 or more types of subunits. Although the three differ in overall subunit composition, they have several smaller subunits in common. Further, all possess two large subunits (each 140 kD or greater) having sequence similarity to the large β and β' subunits of *E. coli* RNA polymerase, indicating that the fundamental catalytic site of RNA polymerase is conserved among its various forms.

In addition to their different functions, the three classes of RNA polymerase can be distinguished by their sensitivity to **α-amanitin** (Figure 30.9), a

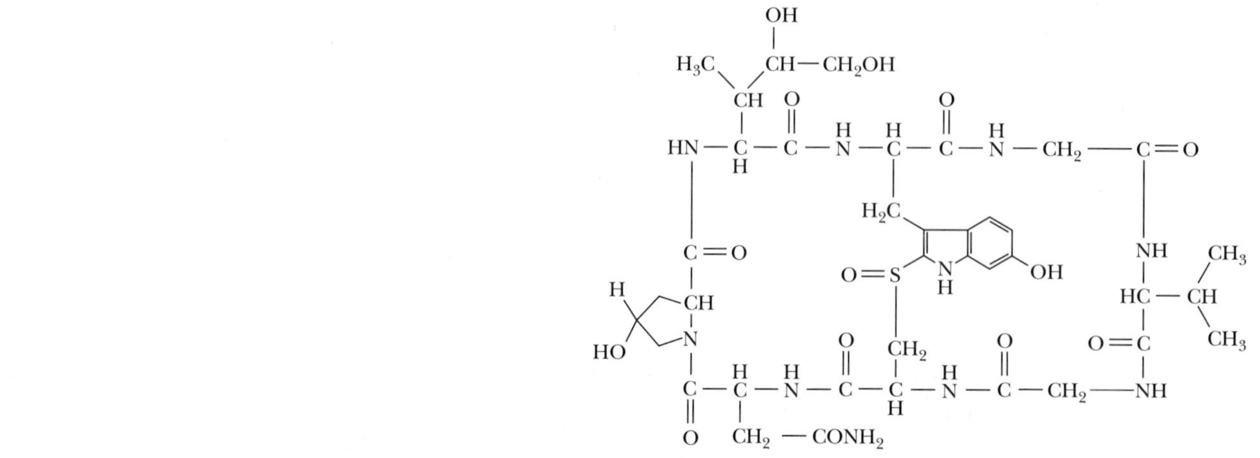

α–Amanitin

***Figure* 30.9** The structure of α-amanitin, one of a series of toxic compounds known as amatoxins that are found in the mushroom *Amanita phalloides.*

bicyclic octapeptide produced by the poisonous mushroom *Amanita phalloides* (the "destroying angel" mushroom). α-Amanitin blocks RNA chain elongation. Although RNA polymerase I is resistant to this compound, RNA polymerase II is very sensitive and RNA pol III is less sensitive.

The existence of three classes of RNA polymerases acting on three distinct sets of genes implies that at least three categories of promoters must exist in order to maintain this specificity. All three polymerases interact with their promoters via so-called **transcription factors,** DNA-binding proteins that recognize and accurately initiate transcription at specific promoter sequences. For RNA polymerase I, its templates are the rRNA genes. Ribosomal RNA genes are present in multiple copies. Optimal expression of these genes requires the first 150 nucleotides in the immediate 5′-upstream region, but the precise locations and sequences of the promoter(s) are not known with certainty.

RNA polymerase III interacts with transcription factors **TFIIIA, TFIIIB,** and **TFIIIC.** Interestingly, TFIIIA and/or TFIIIC bind to specific recognition sequences that in some instances are located *within* the coding regions of the genes, not in the 5′-untranscribed region upstream from the transcription start site. TFIIIB associates with TFIIIA or TFIIIC already bound to the DNA and in turn facilitates the association of RNA pol III to establish an initiation complex.

The Structure and Function of RNA Polymerase II

As the enzyme responsible for the regulated synthesis of mRNA, RNA polymerase II has aroused greater interest than RNA pol I and pol III. RNA pol II must be capable of transcribing a great diversity of genes, yet it must carry out its function at any moment only on those genes whose products are appropriate to the needs of the cell in its everchanging metabolism and growth. The RNA pol II from yeast *(Saccharomyces cerevisiae)* has been extensively characterized. Yeast is viewed by molecular biologists as an excellent eukaryotic prototype. The yeast RNA pol II consists of 10 different polypeptides, designated RPB1 through RPB10, ranging in size from 220 to 10 kD (Table 30.1).[3] RPB1

[3] *RPB* stands for *RNA polymerase B*; RNA pol I, II, and III are sometimes called RNA pol A, B, and C.

Table **30.1**

Yeast RNA Polymerase II Subunits

Subunit	Size (kD)*	Features	Prokaryotic Homolog
RPB1	220	PTSPSYS CTD	β'
RPB2	150	NTP binding	β
RPB3	45	Core assembly	α
RPB4	32	Promoter recognition	σ
RPB5	27	In pol I, II, and III	
RPB6	23	In pol I, II, and III	
RPB7	17	Unique to pol II	
RPB8	14	In pol I, II, and III	
RPB9	13		
RPB10	10	In pol I, II, and III	

*Protein sizes estimated from protein mobilities in SDS-polyacrylamide gel electrophoresis. Actual protein molecular weights deviate somewhat from these values.

Source: Adapted from Woychik, N. A., and Young, R. A., 1990. RNA polymerase II: Subunit structure and function. *Trends in Biochemical Sciences* **15:**347–351.

and RPB2 functions are homologous to those of the prokaryotic RNA polymerase β and β' subunits: RPB1 has a DNA-binding site, RPB2 binds nucleotide substrates, and both contribute to the catalytic site. RPB3 is the functional homolog of the prokaryotic α; there are two RPB3 subunits per enzyme and RPB3 is essential for assembly of the polymerase. RPB4 resembles σ subunit in amino acid sequence. RPB 3, 4, and 7 are unique to RNA pol II, whereas RPB 5, 6, 8, and 10 are common to all three eukaryotic RNA polymerases. RPB 4 and 7 readily dissociate from RNA pol II.

The RPB1 subunit has an unusual structural feature not found in prokaryotes: Its *C-terminal domain (CTD)* contains 27 repeats of the amino acid sequence PTSPSYS. (The analogous subunit in RNA pol II enzymes of other eukaryotes also has this heptapeptide tandemly repeated as many as 52 times.) Note that the side chains of 5 of the 7 residues in this repeat have —OH groups, endowing the CTD with considerable hydrophilicity *and* multiple sites for phosphorylation. This domain may project more than 50 nm from the globular enzyme. The CTD is essential to RNA pol II function. Only RNA pol II whose CTD is *not* phosphorylated can initiate transcription. However, transcription elongation proceeds only after protein phosphorylation within the CTD, suggesting that phosphorylation triggers the conversion of an initiation complex into an elongation complex.

Transcription Initiation by RNA Polymerase II

Promoters

Two DNA elements are common for many promoters of protein-encoding genes—the **TATA** motif and the **initiator element (Inr).** The TATA box is a TATAAA consensus promoter element (Figure 30.10) usually located at about position −25. One important role of the TATA box is to indicate the initiation site. The initiator element *Inr* encompasses the transcription start site. The sequence of *Inr* is not highly conserved between genes; a consensus *Inr* for one family of genes is $_{-3}$YYCAYYYYY$_{+6}$ (where Y represents any pyrimidine).

Adenovirus late GGGGCTATAAAAGGGGGTGGGGGCGCGTTCGTCCTCACTC

Chicken ovalbumin GAGGCTATATATTCCCCAGGGCTCAGCCAGTGTCTGTACA

Mouse β globin major GAGCATATAAGGTGAGGTAGGATCAGTTGCTCCTCACATTT

Rabbit β globin TTGGGCATAAAAGGCAGAGCAGGGCAGCTGCTGCTGCTAACACT

↑ +1 Transcription start site

T	A	T	A	A	A	A
82	97	93	85	63	83	50
				T		T
				37		37

***Figure* 30.10** The TATA box in selected eukaryotic genes. The consensus sequence of a number of such promoters is presented in the lower part of the figure, the numbers giving the percent occurrence of various bases at the positions indicated.

General Transcription Factors

General transcription factors are proteins required for transcription by RNA pol II. Seven general transcription factors have been defined thus far (Table 30.2). Six are required for transcription: **TFIIB, TFIID, TFIIE, TFIIF, TFIIH,** and **TFIIJ.** One **(TFIIA)** shows stimulatory activity. TFIID is the *TATA box-binding factor.* Once TFIID binds to a TATA box, TFIIB joins it, followed by RNA pol IIA in association with TFIIF. Then other factors join (Figure 30.11), establishing a competent transcription *preinitiation complex.* An *open complex* then forms and transcription begins.

30.3 The Regulation of Transcription in Prokaryotes

In bacteria, genes encoding the enzymes of a particular metabolic pathway are often grouped adjacent to one another in a cluster on the chromosome. Such clusters, together with the regulatory sequences that control their transcription, are called **operons.** This pattern of organization allows all of the

(a)

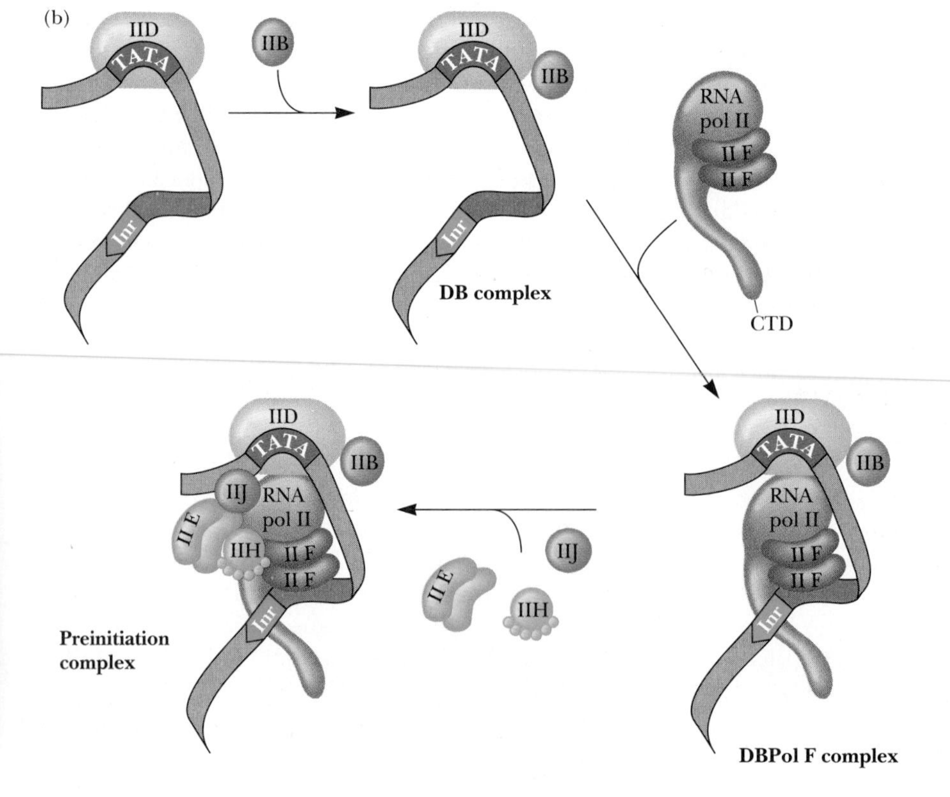

***Figure* 30.11** Transcription initiation. (a) Model of the yeast TATA-binding protein (TBP) in complex with a yeast DNA TATA sequence. The sugar-phosphate backbone of the TATA box is shown in yellow; the TATA base pairs are in red; adjacent DNA segments are in blue. The saddle-shaped TBP (green) is unusual in that it binds in the minor groove of DNA, sitting on the DNA like a saddle on a horse. TBP-binding pries open the minor groove, creating a 100° bend in the DNA axis and unwinding the DNA within the TATA sequence. The other components of the TFIID heteromer (Table 30.2) sit on TBP, like a "cowboy on a saddle." All known eukaryotic genes (those lacking a TATA box as well as those transcribed by RNA polymerase I or III) rely on TBP. (*Photo courtesy of Paul B. Sigler of Yale University.*) (b) Formation of a preinitiation complex at a TATA-containing promoter. Binding of TFIID, the multisubunit protein (>100 kD) consisting of the TATA-binding protein (TBP) and other polypeptides, is stimulated by TFIIA. TFIID bound to the TATA motif recruits TFIIB, forming a DB complex. In association with TFIIF, RNA pol IIA (the nonphosphorylated form of RNA pol II) joins the DB complex to give the DBpol F complex. TFIIE, TFIIH, and TFIIJ then associate to yield the preinitiation complex. Melting of the DNA duplex around *Inr* generates the *open complex* and transcription ensues.
(Adapted from Weiss, L., and Rheinberg, D., 1992. FASEB Journal 6:3300, Figure 1.)

Table **30.2**
General Transcription Factors Interacting with RNA Polymerase II

Factor	Size	Function
TFIIA	67-kD heteromer	Stimulatory
TFIIB	33-kD monomer	TFIID, pol II recognition
TFIID	>100-kD heteromer	TATA motif recognition
TFIIE	180-kD $\alpha_2\beta_2$ tetramer	Preinitiation complex assembly
TFIIF	208-kD $\alpha_2\beta_2$ tetramer	Pol II, DB complex recognition
TFIIH	>200-kD heteromer	Preinitiation complex assembly
TFIIJ	?	Preinitiation complex assembly

Source: Adapted from Weiss, L., and Rheinberg, D., 1992. *FASEB Journal* **6:**3300–3309.

genes in the group to be expressed in a coordinated fashion through transcription into a **single polycistronic mRNA** encoding all the enzymes of the metabolic pathway[4]. A regulatory sequence lying adjacent to this unit of transcription determines whether it is transcribed. This sequence is termed the **operator** (Figure 30.12). The operator is located next to the promoter. Interaction of a **regulatory protein** with the operator controls transcription of the operon by governing the accessibility of RNA polymerase to the promoter. Although this is the paradigm for prokaryotic gene regulation, it must be emphasized that many regulated prokaryotic genes do not contain operators and are regulated in ways that do not involve protein:operator interactions.

Transcription of Operons Is Controlled by Induction and Repression

In prokaryotes, regulation is ultimately responsive to small molecules serving as signals of the nutritional or environmental conditions confronting the cell. Increased synthesis of enzymes in response to the presence of a particular substrate is termed **induction.** For example, lactose (Figure 30.13) can serve as both carbon and energy source for *Escherichia coli.* Metabolism of this substrate depends on hydrolysis into its component sugars, glucose and galactose, by the enzyme **β-galactosidase.** In the absence of lactose, *E. coli* cells contain very little β-galactosidase (less than 5 molecules per cell). However, lactose availability *induces* the synthesis of β-galactosidase by activating transcription of the *lac operon.* One of the genes in the *lac* operon, *lacZ,* is the structural gene for β-galactosidase. When its synthesis is fully induced, β-galactosidase

[4]A **polycistronic mRNA** is a single RNA transcript that encodes more than one polypeptide. "Cistron" is a genetic term for a DNA region representing a protein; "cistron" and "gene" are essentially equivalent terms.

Figure **30.12** The general organization of operons. Operons consist of transcriptional control regions and a set of related structural genes, all organized in a contiguous linear array along the chromosome. The transcriptional control regions are the *promoter* and the *operator,* which lie next to, or overlap, each other, upstream from the structural genes they control. Operators may lie at various positions relative to the promoter, either upstream or downstream. Expression of the operon is determined by access of RNA polymerase to the promoter, and occupancy of the operator by regulatory proteins influences this access. Induction activates transcription from the promoter; repression prevents it.

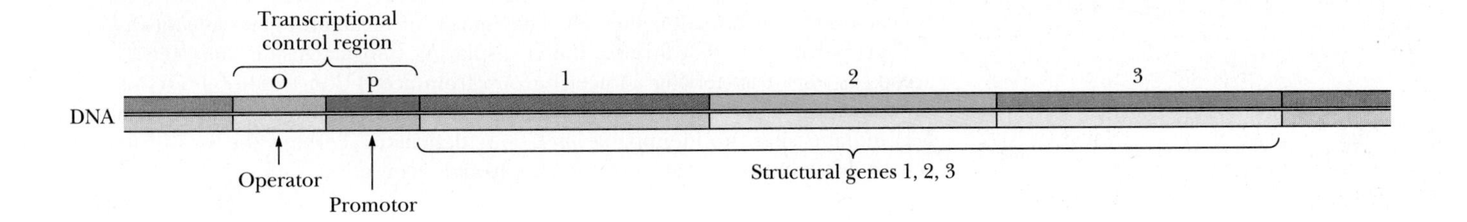

Lactose
(O-β-D-galactopyranosyl (1 → 4) β-D-glucopyranose)

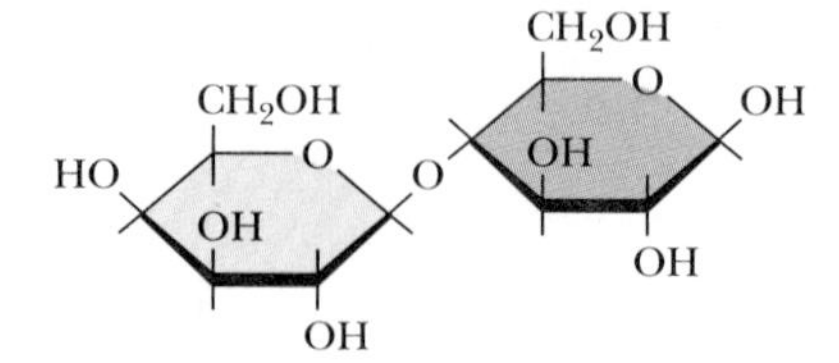

Figure **30.13** The structure of lactose, a β-galactoside.

Isopropyl β-thiogalactoside (IPTG)

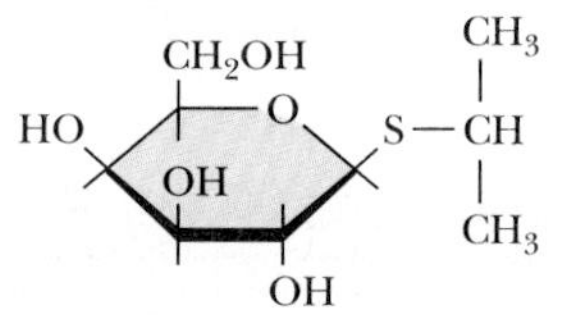

Figure **30.14** The structure of IPTG (isopropyl β-thiogalactoside).

can amount to almost 10% of the total soluble protein in *E. coli*. When lactose is removed from the culture, synthesis of β-galactosidase halts.

The alternative to induction, namely *decreased* synthesis of enzymes in response to a specific metabolite, is termed **repression.** For example, the enzymes of tryptophan biosynthesis in *E. coli* are encoded in the *trp* operon. If sufficient Trp is available to the growing bacterial culture, the *trp* operon is not transcribed, so the Trp biosynthetic enzymes are not made; that is, their synthesis is *repressed.* Repression of the *trp* operon in the presence of Trp is an eminently logical control mechanism: If the end product of the pathway is present, why waste cellular resources making unneeded enzymes?

Induction and repression are two faces of the same phenomenon. In induction, a substrate activates enzyme synthesis. Substrates capable of activating synthesis of the enzymes that metabolize them are called **inducers.** Some substrate analogs can induce enzyme synthesis even though the enzymes are incapable of metabolizing them. These analogs are called **gratuitous inducers.** A number of thiogalactosides, such as **IPTG** (isopropylthiogalactoside, Figure 30.14), are excellent gratuitous inducers of β-galactosidase activity in *E. coli.* In repression, a metabolite, typically an end product, depresses synthesis of its own biosynthetic enzymes. Such metabolites are called **co-repressors.**

lac: The Paradigm of Operons

In 1961, François Jacob and Jacques Monod proposed the **operon hypothesis** to account for the coordinate regulation of related metabolic enzymes. The operon was considered to be the unit of gene expression, consisting of two classes of genes: the structural genes for the enzymes, and regulatory elements or genes that controlled expression of the structural genes. The two kinds of genes could be distinguished by mutation. Mutations in a structural gene would abolish one particular enzymatic activity, but mutations in a regulatory

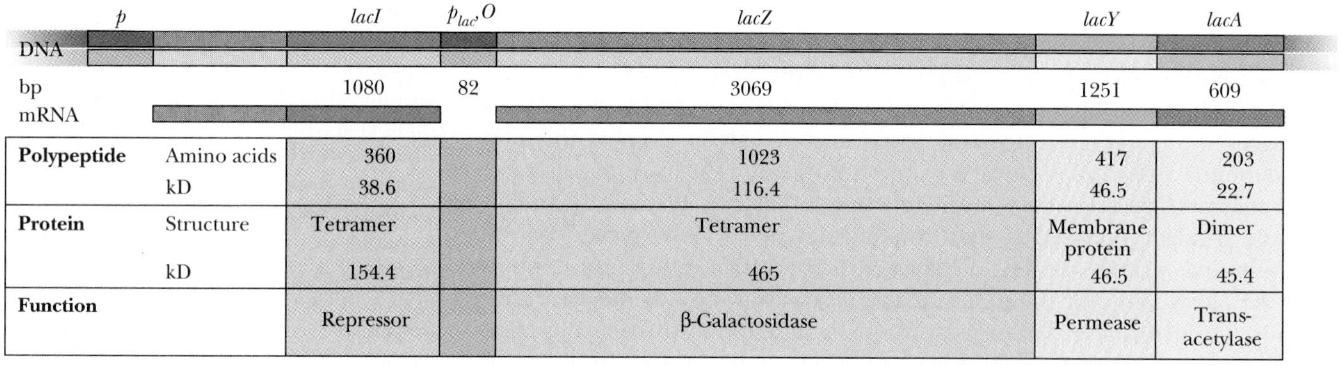

DNA	*p*	*lacI*	p_{lac}, *O*	*lacZ*	*lacY*	*lacA*
bp		1080	82	3069	1251	609
Polypeptide Amino acids		360		1023	417	203
kD		38.6		116.4	46.5	22.7
Protein Structure		Tetramer		Tetramer	Membrane protein	Dimer
kD		154.4		465	46.5	45.4
Function		Repressor		β-Galactosidase	Permease	Trans-acetylase

Figure **30.15** The *lac* operon. The operon consists of two transcription units. In one unit, there are three structural genes, *lacZ, lacY,* and *lacA*, under control of the promoter, p_{lac}, and the operator, *O.* In the other unit, there is a regulator gene, *lacI,* with its own promoter, *p lacI. lacI* encodes a 360-residue, 38.6-kD polypeptide that forms a tetrameric *lac* repressor protein. *lacZ* encodes β-galactosidase, a tetrameric enzyme of 116-kD subunits. *lacY* is the β-galactoside permease structural gene, a 46.5-kD integral membrane protein active in β-galactoside transport into the cell. The remaining structural gene encodes a 22.7-kD polypeptide that forms a dimer displaying thiogalactoside transacetylase activity *in vitro,* transferring an acetyl group from acetyl-CoA to the C-6 OH of thiogalactosides, but the metabolic role of this protein *in vivo* remains uncertain. *lacA* mutants show no identifiable metabolic deficiency. Perhaps the *lacA* protein acts to detoxify toxic analogs of lactose through acetylation.

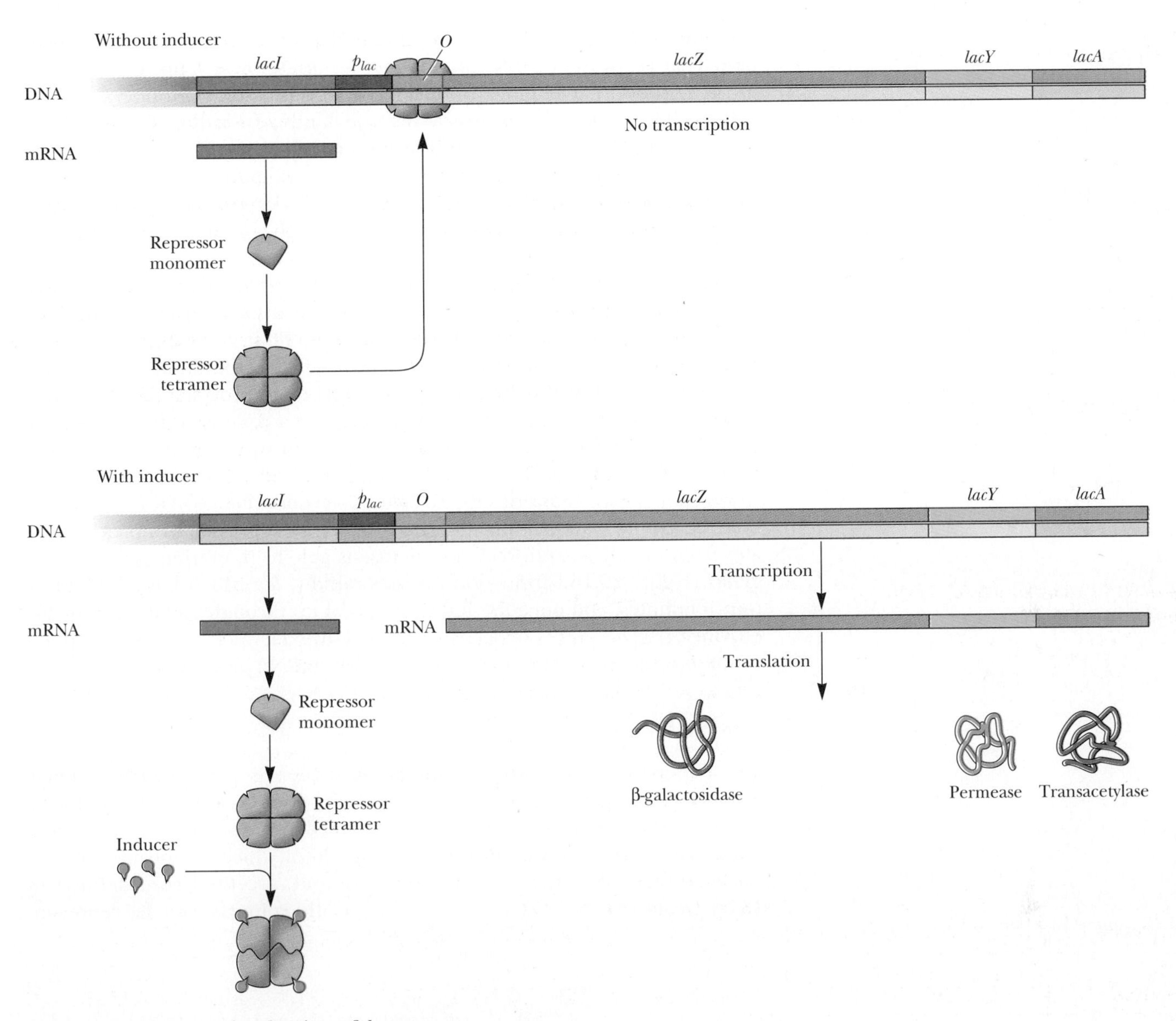

***Figure* 30.16** The mode of action of *lac* repressor.

gene would affect all of the different enzymes under its control. Mutations of both kinds were known in *E. coli* for lactose metabolism. Bacteria with mutations in either the *lacZ* gene or the *lacY* gene (Figure 30.15) could no longer metabolize lactose—the *lacZ* mutants ($lacZ^-$ strains) because β-galactosidase activity was absent, the *lacY* mutants because lactose was no longer transported into the cell. Lactose transport could still be induced in *lacZ* mutants, and *lacY* mutants displayed lactose-inducible, β-galactosidase activity. Other mutations defined another gene, the *lacI* gene. *lacI* mutants were different because they both expressed β-galactosidase activity and immediately transported lactose, *without prior exposure to an inducer.* That is, a single mutation led to the expression of lactose metabolic functions independently of inducer. Expression of genes independently of regulation is termed **constitutive expression.** Thus, *lacI* had the properties of a regulatory gene. The *lac* operon includes the regulatory gene *lacI*, its promoter *p*, and three structural genes, *lacZ, lacY,* and *lacA,* with their own promoter p_{lac} and operator *O* (Figure 30.15).

The structural genes of the *lac* operon are controlled by **negative regulation.** That is, they are transcribed to give an mRNA unless turned off by the *lacI* gene product. This gene product is the ***lac* repressor,** a tetrameric protein (Figure 30.16). The *lac* repressor has two kinds of binding sites—one for inducer and another for DNA. In the absence of inducer, *lac* repressor blocks *lac* gene expression. It accomplishes repression by binding to the *operator* DNA site upstream from the *lac* structural genes, blocking transcription by RNA polymerase from the p_{lac} promoter (Figure 30.16). In *lacI* mutants, the *lac* repressor is absent or defective in binding to operator DNA, *lac* gene transcription is not blocked, and the *lac* operon is constitutively expressed in these mutants. Note that *lacI* is normally expressed constitutively from its promoter, so that *lac* repressor protein is always available to fill its regulatory role. About 10 molecules of *lac* repressor are present in an *E. coli* cell.

Derepression of the *lac* operon occurs when appropriate β-galactosides occupy the inducer site on *lac* repressor, causing a conformational change in the protein that lowers the repressor's affinity for operator DNA. As a tetramer, *lac* repressor has 4 inducer binding sites and its response to inducer shows cooperative allosteric effects. Thus, as a consequence of the "inducer"-induced conformational change, the inducer:*lac* repressor complex dissociates from the DNA, and RNA polymerase is able to transcribe the structural genes (Figure 30.16). Induction reverses rapidly. *lac* mRNA has a half-life of only 3 minutes, and once the inducer is used up through metabolism by the enzymes, free *lac* repressor re-associates with the operator DNA, transcription of the operon is halted, and the residual *lac* mRNA decays.

The *lac* Operator

The *lac* operator is a palindromic DNA sequence (Figure 30.17). **Palindromes,** or "inverted repeats" (Chapter 7), provide a twofold, or *dyad*, symmetry, a structural feature common at sites in DNA where proteins specifically bind. While the operator consists of 35 bp, 26 of which are protected from nuclease digestion when *lac* repressor is bound, a central core defined by 13 bp (from +5 to +17) is involved in specific contacts with *lac* repressor.

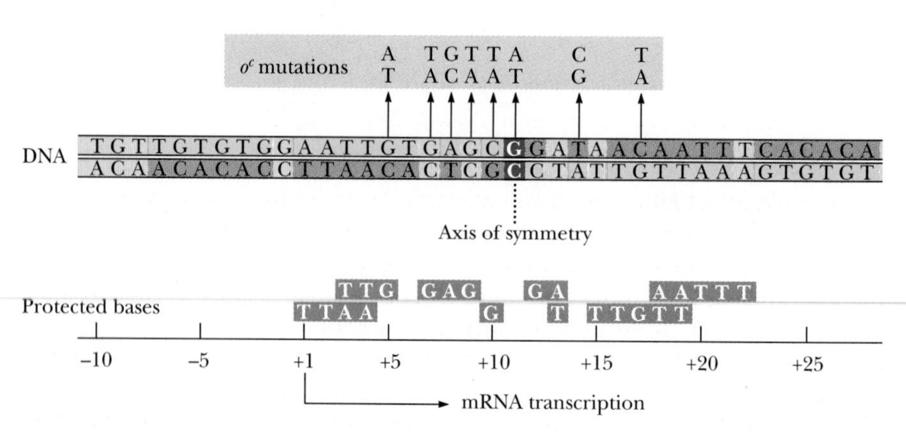

***Figure* 30.17** The nucleotide sequence of the *lac* operator. This sequence comprises 35 bp showing nearly palindromic symmetry. The inverted repeats that constitute this approximate twofold symmetry are shaded in rose. The bases are numbered relative to the +1 start site for transcription. The G:C base pair at position +11 represents the axis of symmetry. *In vitro* studies show that bound *lac* repressor protects a 26-bp region from −5 to +21 against nuclease digestion. O^c mutants are shown above the operator. Bases that are protected against methylation by dimethyl sulfate or that undergo UV-induced cross-linking to bound *lac* repressor are indicated below the operator. Note the symmetry of protection at +1 through +4 TTAA to +18 through +21 AATT.

A Deeper Look

Quantitative Evaluation of *lac* Repressor:DNA Interactions

The affinity of lac *repressor for random DNA ensures that effectively all repressor is DNA bound.* Assume that *E. coli* DNA has a single specific *lac* operator site for repressor binding and 4.72×10^6 nonspecific sites. (Since the *E. coli* genome consists of 4.72×10^6 base pairs, and any nucleotide sequence even one base out of phase with the operator constitutes a nonspecific binding site, there are 4.72×10^6 nonspecific sites for repressor binding.)

The binding of repressor to DNA is given by the association constant, K_A:

$$K_A = \frac{[\text{repressor:DNA}]}{[\text{repressor}][\text{DNA}]}$$

where [repressor:DNA] is the concentration of the repressor:DNA complex, [repressor] is the concentration of free repressor, and [DNA] is the concentration of nonspecific binding sites.

Rearranging gives the following:

$$\frac{[\text{repressor}]}{[\text{repressor:DNA}]} = \frac{1}{K_A[\text{DNA}]}$$

If the number of nonspecific binding sites is 4.72×10^6, there are $(4.72 \times 10^6)/(6.023 \times 10^{23}) = 0.78 \times 10^{-17}$ "moles" of binding sites contained in the volume of a bacterial cell (roughly 10^{-15} liters). Therefore, [DNA] = $(0.78 \times 10^{-17})/(10^{-15}) = 0.78 \times 10^{-2}$ *M*. Since $K_A = 2 \times 10^6\ M^{-1}$ (Table 30.3),

$$\frac{[\text{repressor}]}{[\text{repressor:DNA}]} = \frac{1}{(2 \times 10^6)(0.78 \times 10^{-2})} = \frac{1}{(1.56 \times 10^4)}$$

So, the ratio of free repressor to DNA-bound repressor is 6.4×10^{-5}. *Less than 0.01% of repressor is not bound to DNA!* (See the appendix to this chapter for a general quantitative evaluation of the specificity of DNA:protein interactions.)

Mutations at eight sites in this restricted region lead to constitutive expression of the *lac* operon because repressor can no longer bind (Figure 30.17). These mutants are so-called ***O^c*, or operator-constitutive, mutants.** Note that the distribution of O^c mutants is not symmetrical about the axis of symmetry. Further, certain O^c mutations, as in G:C → A:T changes at positions 7 or 9, actually render the palindrome more perfect. The distribution of O^c mutants indicates that repressor contacts with the left half of the palindrome may be more crucial than those with the right half. The operator and promoter (p_{lac}) sites overlap: *lac* repressor protects a region roughly covering nucleotides −5 to +21 from nuclease digestion, whereas RNA polymerase binding and nuclease protection defines p_{lac} as falling within the −45 to +18 region.

Interactions of *lac* Repressor with DNA

Limited digestion of *lac* repressor with trypsin removes an N-terminal, 59-residue fragment from each subunit, leaving a "core" tetramer that is no longer capable of binding to operator DNA. IPTG binding by the "core" tetramer is unaffected. The N-terminal, 59-residue fragment retains DNA-binding ability. Thus, the protein is comprised of an N-terminal, DNA-binding domain, with the rest of the protein functioning in inducer binding and tetramer formation. In the absence of inducer, intact *lac* repressor nonspecifically binds to duplex DNA with an association constant, K_A, of $2 \times 10^6\ M^{-1}$ (Table 30.3), and to the *lac* operator DNA sequence with much higher affinity, $K_A = 2 \times 10^{13}\ M^{-1}$. Thus, *lac* repressor binds 10^7 times better to *lac* operator DNA than to any random DNA sequence. IPTG binds to *lac* repressor with an association constant of about 10^6 *M*. The IPTG:*lac* repressor complex binds to operator DNA with an association constant, $K_A = 2 \times 10^{10}\ M^{-1}$. Although this affinity is high, it is 3 orders of magnitude *less* than the affinity of inducer-free repressor for *lac* operator. There is no difference in the affinity of free *lac* repressor and *lac* repressor with IPTG bound for nonoperator DNA. The *lac* repressor apparently acts by binding to DNA and sliding along it, testing sequences in a one-dimensional search until it finds the *lac* operator.

***Table* 30.3**
The Affinity of *lac* Repressor for DNA*

DNA	Repressor	Repressor + Inducer
lac operator	$2 \times 10^{13}\ M^{-1}$	$2 \times 10^{10}\ M^{-1}$
All other DNA	$2 \times 10^6\ M^{-1}$	$2 \times 10^6\ M^{-1}$
Specificity†	10^7	10^4

*Values for repressor:DNA binding are given as association constants, K_A, for the formation of DNA:repressor complex from DNA and repressor.

†Specificity is defined as the ratio (K_A for repressor binding to operator DNA)/(K_A for repressor binding to random DNA).

Positive Control of the *lac* Operon by CAP

The previous discussion assumed that RNA polymerase initiates transcription whenever it gains access to the promoter. However, RNA polymerase cannot initiate transcription at some promoters unless assisted by an accessory protein that acts as a *positive regulator.* One such protein is **CAP,** or **catabolite activator protein.** Its name derives from the phenomenon of catabolite repression in *E. coli.* Catabolite repression is a global control that coordinates gene expression with the total physiological state of the cell: As long as glucose is available, *E. coli* catabolizes it in preference to any other energy source, such as lactose or galactose. Catabolite repression ensures that the operons necessary for metabolism of these alternative energy sources, that is, the *lac* and *gal* operons, remain repressed until the supply of glucose is exhausted. Catabolite repression overrides the influence of any inducers that might be present.

Catabolite repression is mediated by cAMP levels, which in turn are regulated by glucose. Transport of glucose into the cell is accompanied by deactivation of ***E. coli* adenylyl cyclase,** leading to lower cAMP levels. The action of CAP as a positive regulator is cAMP-dependent. CAP, also referred to as **CRP** (for **cAMP receptor protein**), is a dimer of identical 210-residue (22.5-kD) polypeptides. The N-terminal domains bind cAMP; the C-terminal domains constitute the DNA-binding site. Two molecules of cAMP are bound per dimer. The CAP-$(cAMP)_2$ complex binds to specific target sites near the promoters of operons (Figure 30.18). Its presence assists closed promoter complex formation by RNA polymerase holoenzyme. For example, CAP binding at the −72 to −52 region of *lac* DNA promotes formation of an RNA polymer-

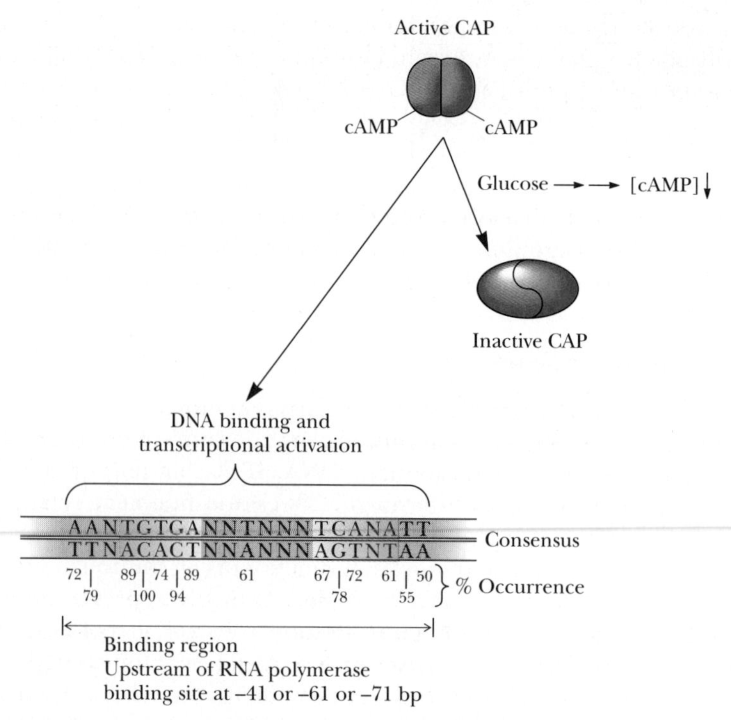

***Figure* 30.18** The mechanism of catabolite repression and CAP action. Glucose instigates catabolite repression by lowering cAMP levels. cAMP is necessary for CAP binding near promoters of operons whose gene products are involved in the metabolism of alternative energy sources such as lactose, galactose, and arabinose. The binding sites for the CAP-$(cAMP)_2$ complex are consensus DNA sequences containing the conserved pentamer TGTGA and a less well conserved inverted repeat, TCANA (where N is any nucleotide).

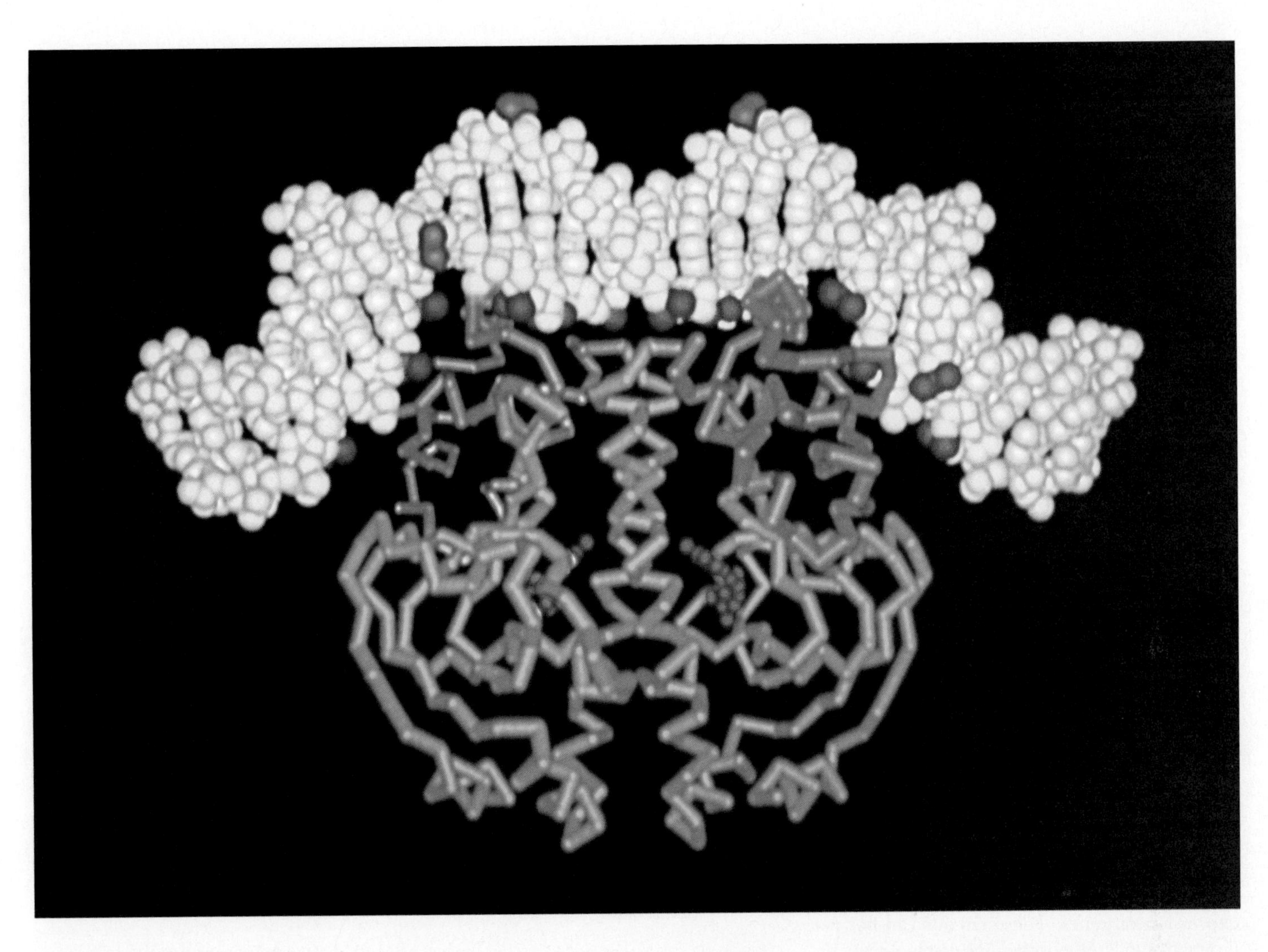

ase holoenzyme:p_{lac} DNA closed promoter complex. Analysis of the structure of the CAP:DNA complex reveals that the DNA is bent more than 90° about the center of dyad symmetry (Figure 30.19). This bend may be related to the ability of CAP to assist in transcription initiation.

Figure* 30.19** Binding of CAP-$(cAMP)_2$ induces a severe bend in DNA about the center of dyad symmetry at the CAP-binding site. The CAP dimer with 2 molecules of cAMP bound interacts with 27 to 30 base pairs of duplex DNA. The cAMP-binding domain of CAP protein is shown in blue and the DNA-binding domain in purple. The 2 cAMP molecules bound by the CAP dimer are indicated in red. For DNA, the bases are shown in white and the sugar-phosphate backbone in yellow. DNA phosphates which interact with CAP are highlighted in red. Binding of CAP-$(cAMP)_2$ to its specific DNA site involves H-bonding and ionic interactions between protein functional groups and DNA phosphates, as well as H-bonding interactions in the DNA major groove between amino acid sidechains of CAP and DNA base pairs. *(Adapted from Schultz, S. C., Shields, G. C., and Steitz, T. A. (1991). Crystal Structure of a CAP-DNA Complex: The DNA Is Bent By 90°. Science* ***253*:1001–1007. Photograph courtesy of Professor Thomas A. Steitz of Yale University.)*

Positive Versus Negative Control

Negative- and positive-control systems are fundamentally different. Genes under negative control are transcribed *unless* they are turned off by the presence of a repressor protein. Often, transcription activation is essentially *anti-inhibition;* that is, the reversal of negative control. In contrast, genes under positive control are expressed *only if* an active regulator protein is present. The *lac* operon illustrates these differences. The action of *lac* repressor is negative. It binds to operator DNA and blocks transcription; expression of the operon can only be attained if this negative control is lifted through release of the repressor. In contrast, regulation of the *lac* operon by CAP is positive: Transcription of the operon by RNA polymerase is stimulated by CAP's action as a positive regulator.

Operons can also be classified as **inducible** or **repressible,** or both, depending on how they respond to the small molecules that mediate their expression. Repressible operons are expressed only in the absence of their *co-repressors.* Inducible operons are transcribed only in the presence of small-molecule *co-inducers* (Figure 30.20).

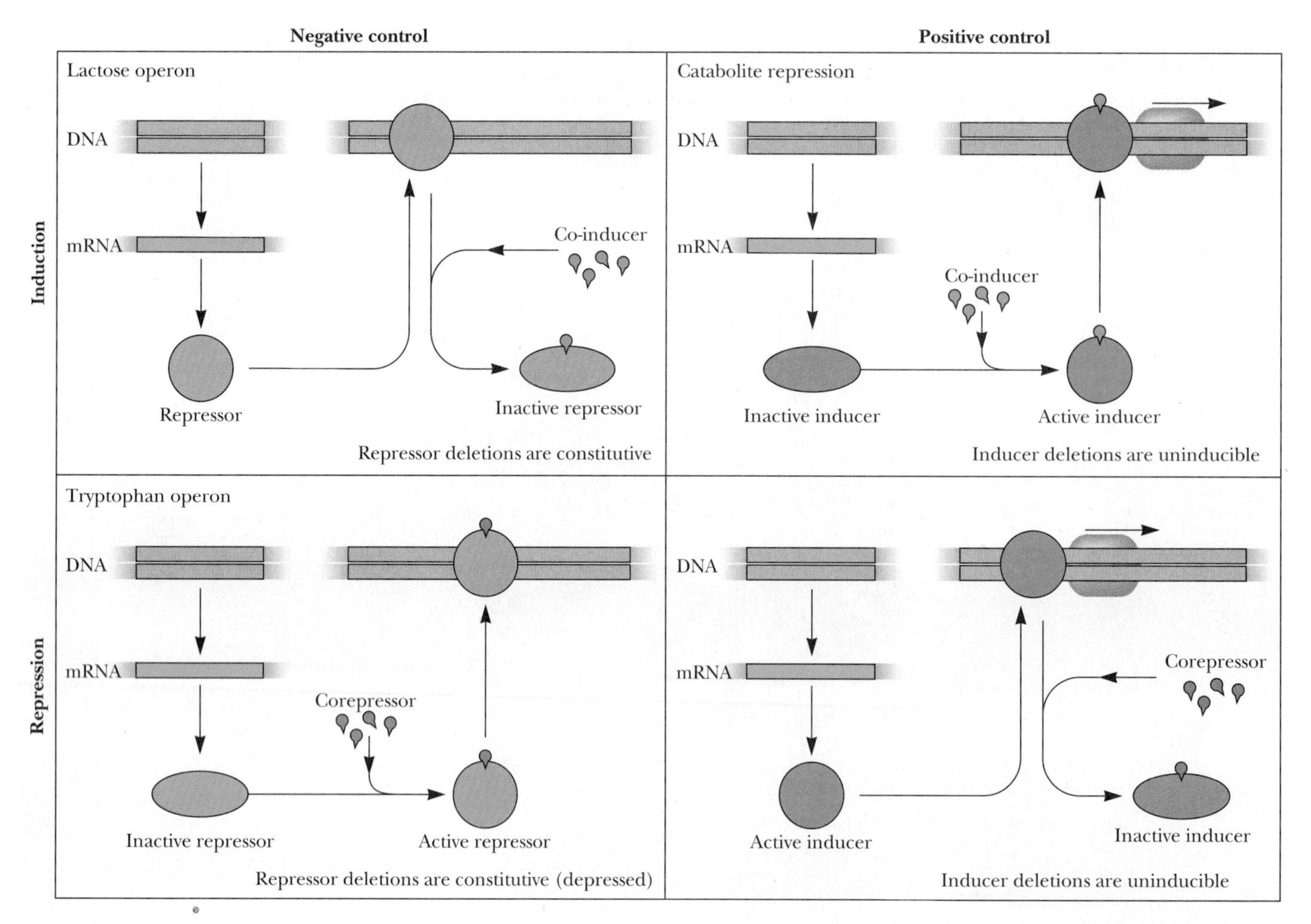

***Figure* 30.20** Control circuits governing the expression of genes. These circuits can be either negative or positive, inducible or repressible.

The *araBAD* Operon: Positive and Negative Control by *AraC*

E. coli can use the plant pentose L-arabinose as sole source of carbon and energy. Arabinose is metabolized via conversion to D-xylulose-5-P (a pentose-phosphate pathway intermediate and transketolase substrate [Chapter 21]) by three enzymes encoded in the ***araBAD*** **operon.** Transcription of this operon is regulated by both catabolite repression and arabinose-mediated induction. CAP functions in catabolite repression; arabinose induction is achieved via the product of the *araC* gene, which lies next to the *araBAD* operon on the *E. coli* chromosome. The *araC* gene product, the protein **AraC**[5], is a DNA-binding protein. Regulation of *araBAD* by AraC is novel in that it acts both negatively and positively. The *ara* operon has three binding sites for AraC: *ara*O_1, located at nucleotides −106 to −144 relative to the *araBAD* transcription start site; *ara*O_2 (spanning positions −265 to −294); and *araI*, the *araBAD* promoter. The *araI* site consists of two "half-sites": *ara*I_1 (nucleotides −56 to −78) and *ara*I_2 (−35 to −51). (The *ara*O_1 site contributes minimally to *ara* operon regulation.)

The details of *araBAD* regulation are as follows: When AraC protein levels are low, the *araC* gene is transcribed from its promoter p_c (adjacent to *ara*O_1) by RNA polymerase (Figure 30.21). *araC* is transcribed in the direction away from *araBAD*. When cAMP levels are low and arabinose is absent, an AraC

[5]Proteins are often named for the gene encoding them. By convention, the name of the protein is capitalized but not italicized.

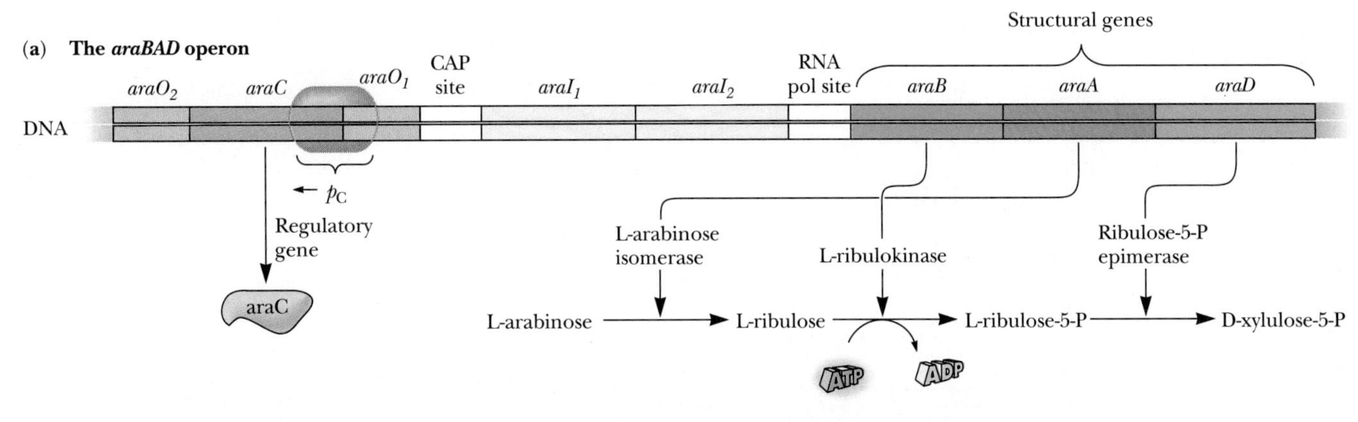

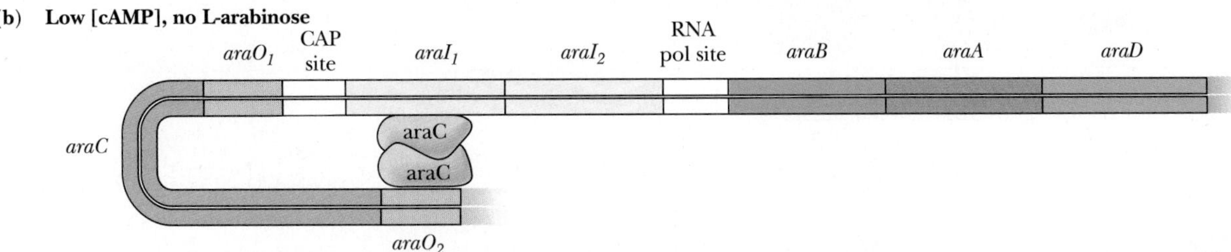

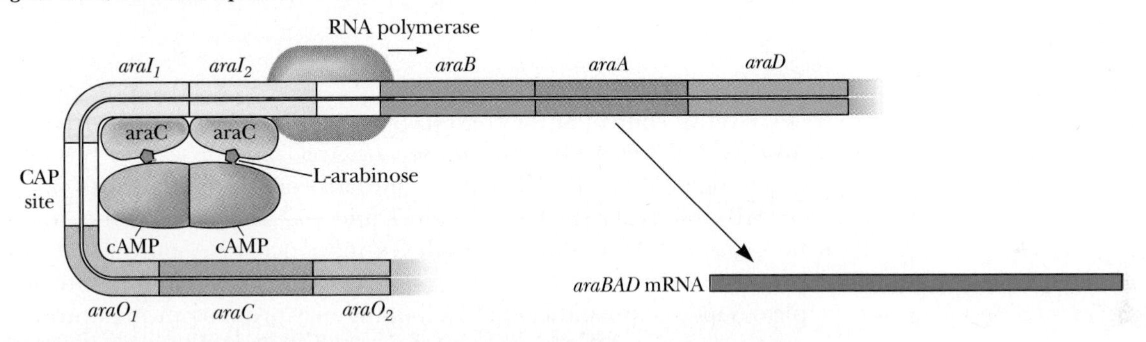

***Figure* 30.21** Regulation of the *araBAD* operon by the combined action of CAP and AraC protein.

protein dimer binds to two sites, *araO*$_2$ and the *araI*$_1$ half-site, forming a DNA loop between them and restricting transcription of *araBAD* (Figure 30.21). In the presence of L-arabinose, the monomer of AraC bound to the *araO*$_2$ site is released from that site; it then associates with the unoccupied *araI* half-site, *araI*$_2$. L-arabinose thus behaves as an allosteric effector that alters the conformation of AraC. In the arabinose-liganded conformation, the AraC dimer interacts with CAP-(cAMP)$_2$ to activate transcription by RNA polymerase. Thus, AraC protein is both a repressor *and* an activator.

Deletion studies reveal that both *araO*$_2$ and *araI* must be present on the chromosome in order for AraC protein to repress *araBAD*. The DNA loop created when AraC binds both *araO*$_2$ and *araI*$_1$ consists of some 210 bp. If 5 bp of DNA (one-half a helical turn) are added or deleted in this intervening region, the two AraC-binding sites are rotated away from each other, so that interaction of AraC with both sites is not possible, and repression is no longer observed (Figure 30.22). The creation of DNA loops by sequence-specific, DNA-binding proteins is a mechanism common to many regulatory phenomena involving DNA. (DNA looping is considered in greater detail later in this chapter.)

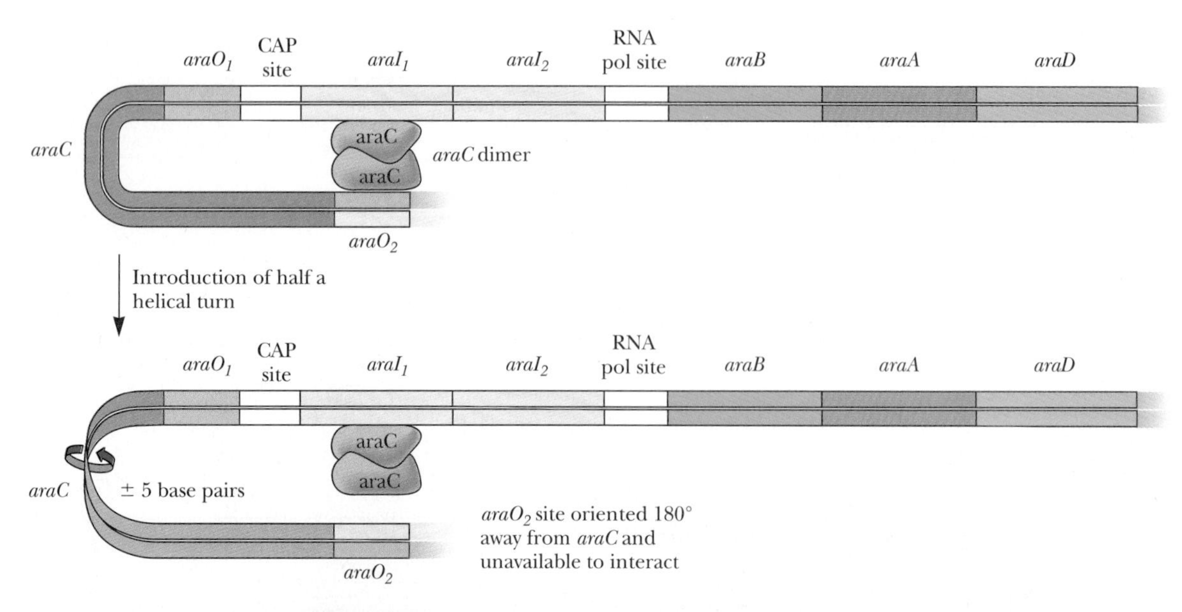

***Figure* 30.22** Introduction or removal of half a helical turn in the DNA between *araO*$_2$ and *araI* prevents AraC protein from interacting with both sites and achieving *araBAD* repression.

(Adapted from Schleif, R., 1987. The L-Arabinose Operon, in Escherichia coli and Salmonella typhimurium, vol. 2. Edited by Neidhardt, F. C., et al. Washington, DC: American Society for Microbiology.)

Positive control of the *araBAD* operon occurs in the presence of L-arabinose *and* cAMP. Arabinose binding by AraC protein causes the release of *araO*$_2$, opening of the DNA loop, and association of AraC with *araI*$_2$. CAP-(cAMP)$_2$ binds at a site between *araO*$_1$ and *araI,* and together the AraC-(arabinose)$_2$ and CAP-(cAMP)$_2$ complexes influence RNA polymerase, through protein:protein interactions, to create an active transcription initiation complex. Supercoiling-induced DNA looping may promote protein:protein interactions between DNA-binding proteins by bringing them into juxtaposition.

The *trp* Operon: Attenuation as a Mechanism to Regulate Gene Expression

The *trp* operon of *E. coli* (and *S. typhimurium*) encodes a leader peptide sequence *(trpL)* and five polypeptides, *trpE* through *trpA* (Figure 30.23). The five polypeptides comprise three enzymes that catalyze the formation of tryptophan from chorismate (Chapter 26). Expression of the *trp* operon is under the control of **Trp repressor,** a dimer of 108-residue polypeptide chains. When tryptophan is plentiful, Trp repressor binds two Trp equivalents and associates with the *trp* operator that is located within the *trp* promoter. Trp repressor binding excludes RNA polymerase from the promoter, preventing transcription of the *trp* operon. When Trp becomes limiting, repression is lifted because Trp repressor lacking bound Trp (Trp apo-repressor) has a lowered affinity for the *trp* promoter. Thus, the behavior of Trp repressor corresponds to a co-repressor-mediated, negative control circuit (Figure 30.20). The Trp repressor regulates two other operons: *trpR* and *aroH* (Figure 30.24). Trp repressor is itself encoded by the *trpR* operon, and its regulation of this operon serves as an example of **autogenous regulation (autoregulation),** which is regulation of gene expression by the product of the gene. The *aroH* operon encodes the Trp-sensitive DAHP synthase isozyme of aromatic amino acid biosynthesis (Chapter 26).

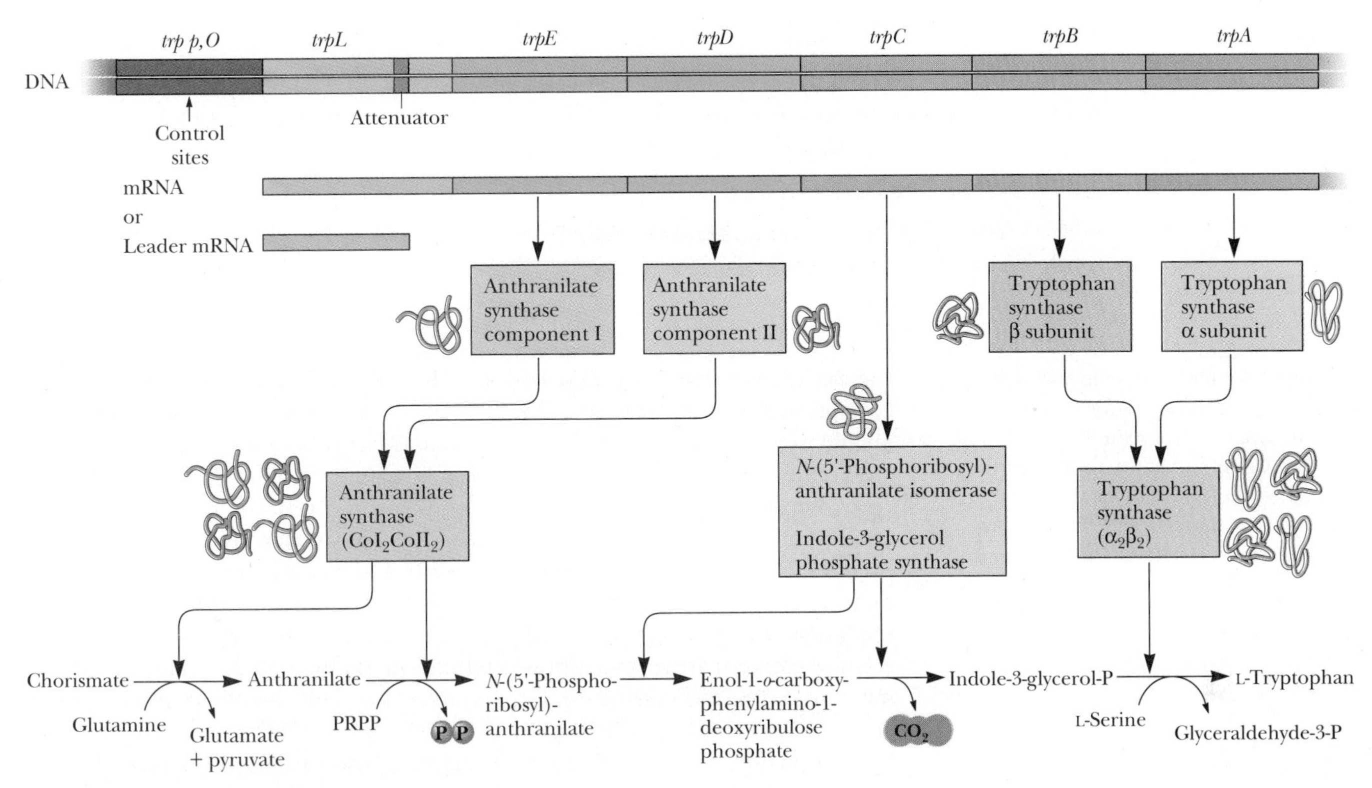

***Figure* 30.23** The *trp* operon of *E. coli*.

Attenuation

In addition to repression, the *trp* operon is controlled by **transcription attenuation.** Unlike the mechanisms discussed thus far, attenuation regulates transcription *after* it has begun. Charles Yanofsky, the discoverer of this phenomenon, has defined attenuation as *any regulatory mechanism that manipulates transcription termination or transcription pausing to regulate gene transcription downstream.* In prokaryotes, transcription and translation (Chapters 6 and 32) are coupled, and the translating ribosome is affected by the formation and persistence of pause and termination structures in the mRNA. Attenuation occurs under normal conditions but is blocked when levels of specific **charged tRNAs** (aminoacyl-tRNAs) are lowered on account of amino acid limitation. In many operons encoding enzymes of amino acid biosynthesis, a transcribed 150- to 300-bp leader region is positioned between the promoter and the first major structural gene. These regions encode a short leader peptide containing **multiple codons** for the pertinent amino acid. For example, the leader peptide of

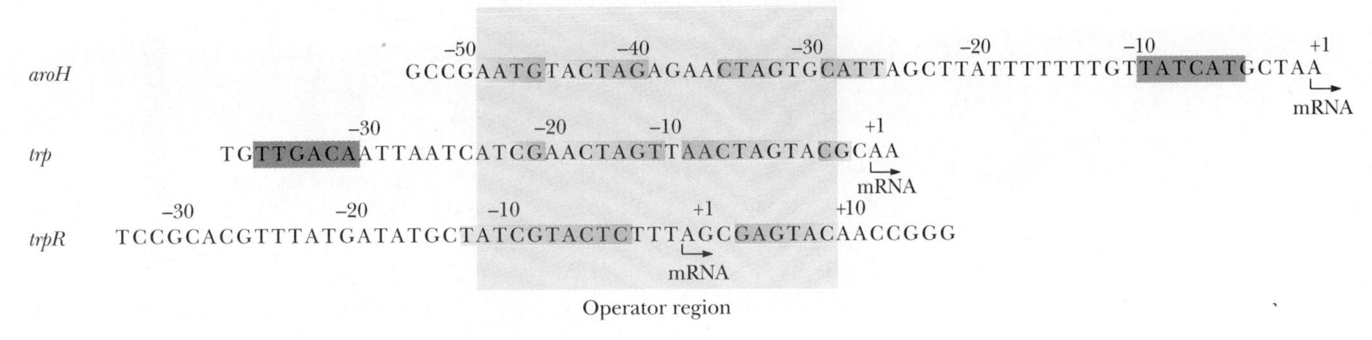

***Figure* 30.24** The three operators recognized by Trp repressor.

Operon	Amino acid Sequence
his	Met–Thr–Arg–Val–Gln–Phe–Lys–His–His–His–His–His–His–His–Pro–Asp
ilv	Met–Thr–Ala–Leu–Leu–Arg–Val–Ile–Ser–Leu–Val–Val–Ile–Ser–Val–Val–Val–Ile–Ile–Ile–Pro–Pro–Cys–Gly–Ala–Ala–Leu–Gly–Arg–Gly–Lys–Ala
leu	Met–Ser–His–Ile–Val–Arg–Phe–Thr–Gly–Leu–Leu–Leu–Leu–Asn–Ala–Phe–Ile–Val–Arg–Gly–Arg–Pro–Val–Gly–Gly–Ile–Gln–His
pheA	Met–Lys–His–Ile–Pro–Phe–Phe–Phe–Ala–Phe–Phe–Phe–Thr–Phe–Pro
thr	Met–Lys–Arg–Ile–Ser–Thr–Thr–Ile–Thr–Thr–Thr–Ile–Thr–Ile–Thr–Thr–Gln–Asn–Gly–Ala–Gly
trp	Met–Lys–Ala–Ile–Phe–Val–Leu–Lys–Gly–Trp–Trp–Arg–Thr–Ser

***Figure* 30.25** Amino acid sequences of leader peptides in various amino acid biosynthetic operons regulated by attenuation. Color indicates amino acids synthesized in the pathway catalyzed by the operon's gene products (The *ilv* operon encodes enzymes of isoleucine, leucine, and valine biosynthesis.).

the *leu* operon has four Leu codons, the *trp* operon has two tandem Trp codons, and so forth (Figure 30.25). Translation of these codons depends on an adequate supply of the relevant aminoacyl-tRNA, which in turn rests on the availability of the amino acid. When Trp is scarce, the entire *trp* operon from *trpL* to *trpA* is transcribed to give a polycistronic mRNA. But, as [Trp] increases, more and more of the *trp* transcripts consist of only a 140-nucleotide fragment corresponding to the 5′-end of *trpL.* Trp availability is causing premature termination of *trp* transcription, that is, transcription attenuation. The secondary structure of the 160-bp leader region transcript is the principal control element in transcription attenuation (Figure 30.26). This RNA segment includes the coding region for the 14-residue leader peptide. Three critical base-paired hairpins can form in this RNA: the **1:2 pause** structure, the **3:4 terminator,** and the **2:3 antiterminator.** Obviously, the 1:2 pause, 3:4 terminator, and the 2:3 antiterminator represent mutually exclusive alternatives. A significant feature of this coding region is the tandem UGG Trp codons.

Transcription by RNA polymerase begins and progresses until position 92 is reached, whereupon the 1:2 hairpin is formed, causing RNA polymerase to pause in its elongation cycle. While RNA polymerase is paused, a ribosome begins to translate the leader region of the transcript. Translation by the ribosome releases the paused RNA polymerase and transcription continues,

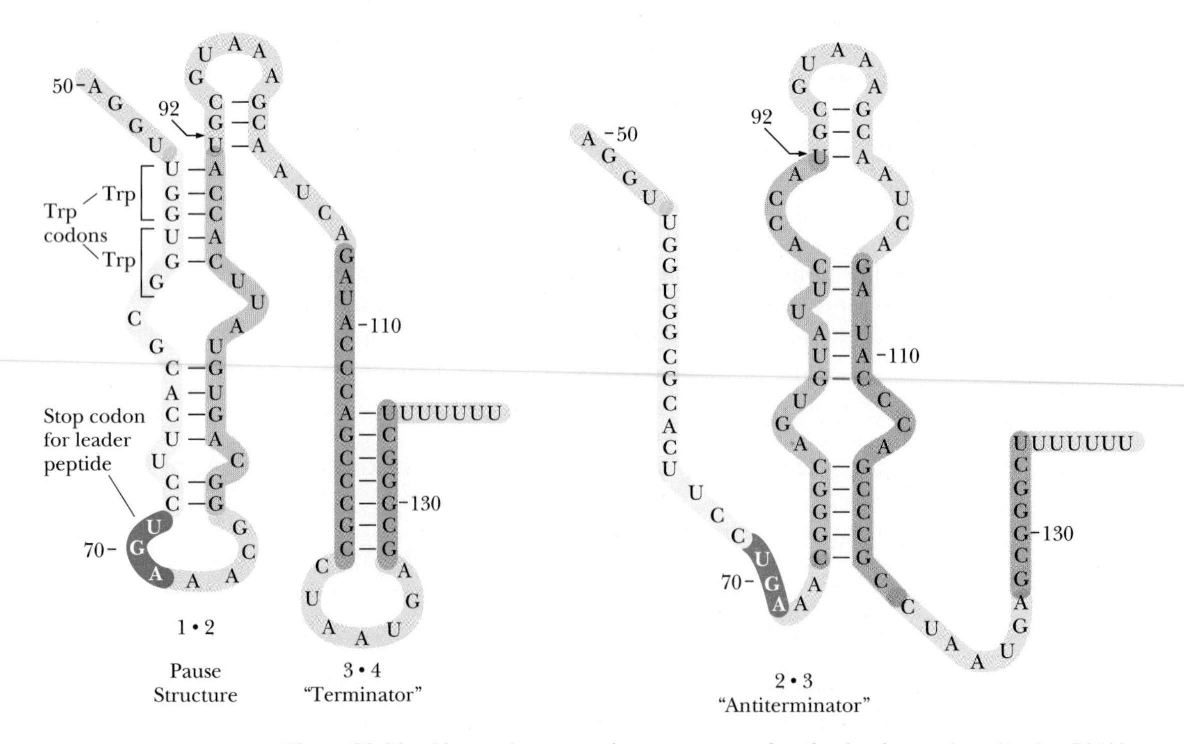

***Figure* 30.26** Alternative secondary structures for the leader region (*trpL* mRNA) of the *trp* operon transcript.

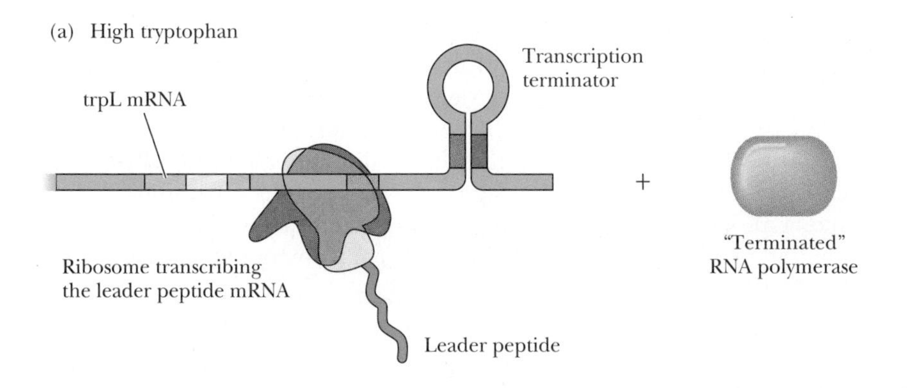

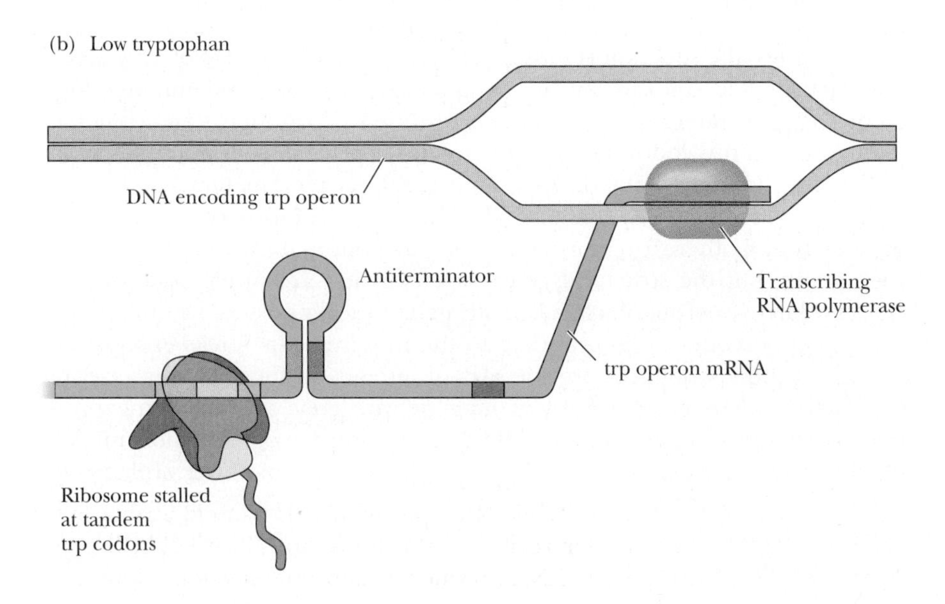

***Figure* 30.27** The mechanism of attenuation in the *trp* operon.

with RNA polymerase and the ribosome moving in unison. As long as Trp is plentiful enough that Trp-tRNA^{Trp} is not limiting, the ribosome is not delayed at the two Trp codons and follows closely behind RNA polymerase, translating the message soon after it is transcribed. The presence of the ribosome atop segment 2 blocks formation of the 2:3 antiterminator hairpin, allowing the alternative 3:4 terminator hairpin to form (Figure 30.27). Stable hairpin structures followed by a run of Us are features typical of *rho*-independent transcription termination signals, so the RNA polymerase perceives this hairpin as a transcription stop signal and transcription is terminated at this point. On the other hand, a paucity of Trp and hence Trp-tRNA^{Trp} causes the ribosome to stall on segment 1. This leaves segment 2 free to pair with segment 3 and to form the 2:3 antiterminator hairpin in the transcript. Since this hairpin precludes formation of the 3:4 terminator, termination is prevented and the entire operon is transcribed. Thus, transcription attenuation is determined by the availability of charged tRNA^{Trp} and its transitory influence over the formation of alternative secondary structures in the mRNA.

Transcription Is Regulated by a Diversity of Mechanisms

A surprising variety of control mechanisms operate in transcriptional regulation, as we have just seen. Several organizing principles materialize. First, DNA:protein interactions are a central feature in transcriptional control, and the DNA sites where regulatory proteins bind commonly display at least partial dyad symmetry or inverted repeats. Further, DNA-binding proteins themselves are generally even-numbered oligomers (for example, dimers, tetramers) that have an innate twofold rotational symmetry. Second, protein:protein interactions are an essential component of transcriptional activation. We see this latter feature in the activation of RNA polymerase by CAP-$(cAMP)_2$ or AraC-$(arabinose)_2$, to select just two examples. Third, the regulator proteins receive cues that signal the status of the environment (for example, Trp, lactose, cAMP) and act to communicate this information to the genome, typically via the medium of conformational changes and DNA:protein interactions.

30.4 Transcription Regulation in Eukaryotes

In eukaryotes, the situation is substantially more complicated. Not only metabolic activity and cell division but complex patterns of embryonic development and cell differentiation must be coordinated through transcriptional regulation. All this coordinated regulation takes place in cells where the relative quantity (and diversity) of DNA is very great: A typical mammalian cell has 1500 times as much DNA as an *Escherichia coli* cell. Eukaryotic genes have promoters and other regulatory elements analogous to those found in prokaryotic genes, but the structural genes of eukaryotes are rarely organized in clusters akin to operons. Each eukaryotic gene typically possesses a discrete set of regulatory sequences appropriate to the requirements for its expression. Certain of these sequences provide sites of interaction for general transcription factors, whereas others endow the gene with great specificity in expression by providing targets for specific transcription factors. Further, mRNA stability plays a greater role in eukaryotic gene expression; unlike prokaryotic mRNAs, eukaryotic mRNAs show a wide range in relative half-lives. The longer-lived an mRNA is, the greater the potential for its genetic information to be persistently expressed.

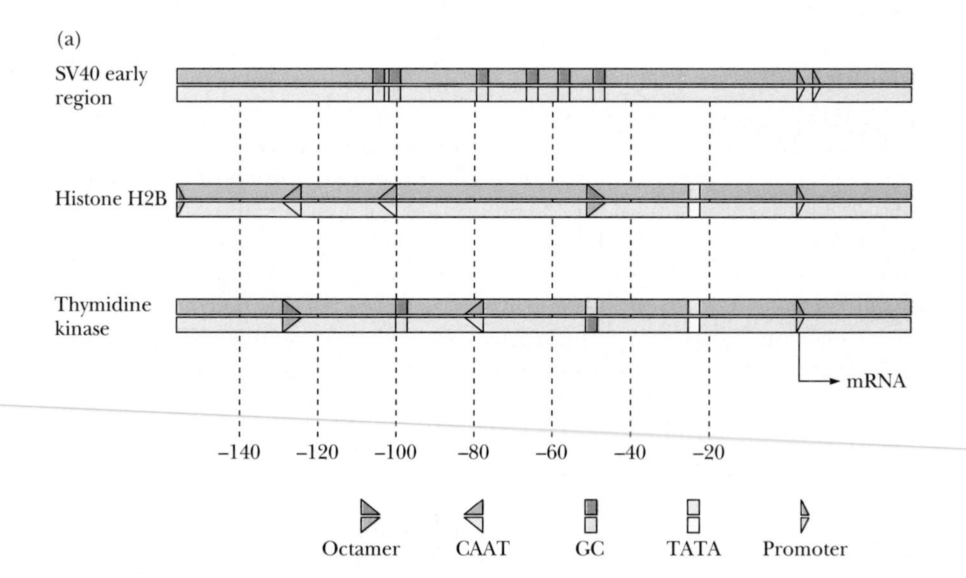

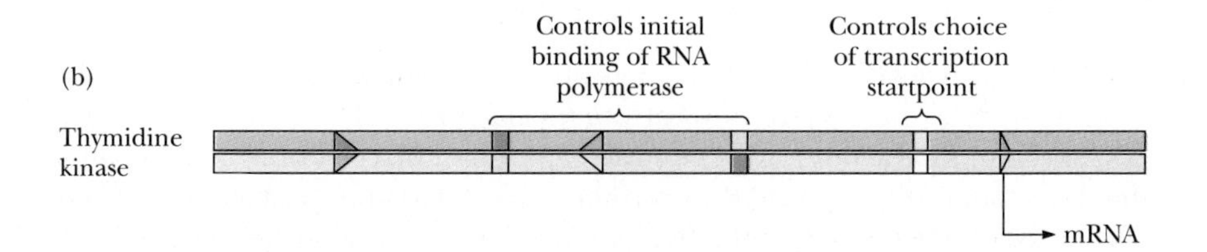

***Figure* 30.28** Promoter regions of several representative eukaryotic genes. (a) The SV40 early genes, the histone H2B gene, and the thymidine kinase gene. Note that these promoters contain different combinations of the various modules. In (b), the function of the modules within the thymidine kinase gene is shown.

Eukaryotic Promoters, Enhancers, and Response Elements

Promoters

The promoters of eukaryotic genes encoding proteins are defined by **modules** of short conserved sequences, such as the **TATA box,** the **CAAT box,** and the **GC box.** For example, one or more copies of the sequence GGGCGG or its complement (referred to as the GC box) have been found upstream from the transcription start sites of so-called ''housekeeping genes.'' Housekeeping genes encode proteins commonly present in all cells and essential to normal function; such genes are typically transcribed at more or less steady levels. Sets of the various sequence modules are embedded in the upstream region of such genes and loosely define the promoter. Figure 30.28a–b depicts the promoter regions of several representative eukaryotic genes. Table 30.4 lists transcription factors that bind to respective modules. These transcription factors typically behave as positive regulatory proteins essential to transcriptional activation by RNA polymerase II at these promoters.

Enhancers

In addition to these promoter elements, eukaryotic genes are characterized by additional regulatory sequences known as **enhancers.** Enhancers (also called **upstream activation sequences,** or **UASs**) assist initiation. Enhancers differ from promoters in two fundamental ways. First, the location of enhancers relative to the transcription start site is not fixed. Enhancers may be several thousand nucleotides away from the promoter, and they act to enhance transcription initiation even if positioned *downstream* from the gene. Second, enhancer sequences are *bidirectional* in that they function in either orientation. That is, enhancers can be removed and then reinserted in the reverse sequence orientation without any diminution in their function. Like promoters, enhancers represent modules of consensus sequence. Enhancers are promiscuous, since they will stimulate transcription from any promoter that happens to be in their vicinity. Nevertheless, *enhancer function is dependent on recognition by a specific transcription factor.* A specific transcription factor bound at an enhancer element interacts with RNA pol II at a nearby promoter via a looping mechanism (Figure 30.29).

***Table* 30.4**

The Consensus Sequences That Define Various RNA Polymerase II Promoter Modules and the Transcription Factors That Bind to Them

Sequence Module	Consensus Sequence	DNA Bound	Factor	Size (kD)	Abundance (molecules/cell)
TATA box	TATAAAA	~10 bp	TBP	27	?
CAAT box	GGCCAATCT	~22 bp	CTF/NF1	60	300,000
GC box	GGGCGG	~20 bp	SP1	105	60,000
Octamer	ATTTGCAT	~20 bp	Oct-1	76	?
"	"	23 bp	Oct-2	52	?
κB	GGGACTTTCC	~10 bp	NFκB	44	?
"	"	~10 bp	H2-TF1	?	?
ATF	GTGACGT	~20 bp	ATF	?	?

Source: Adapted from Lewin, B., 1994. *Genes V.* Cambridge, MA: Cell Press.

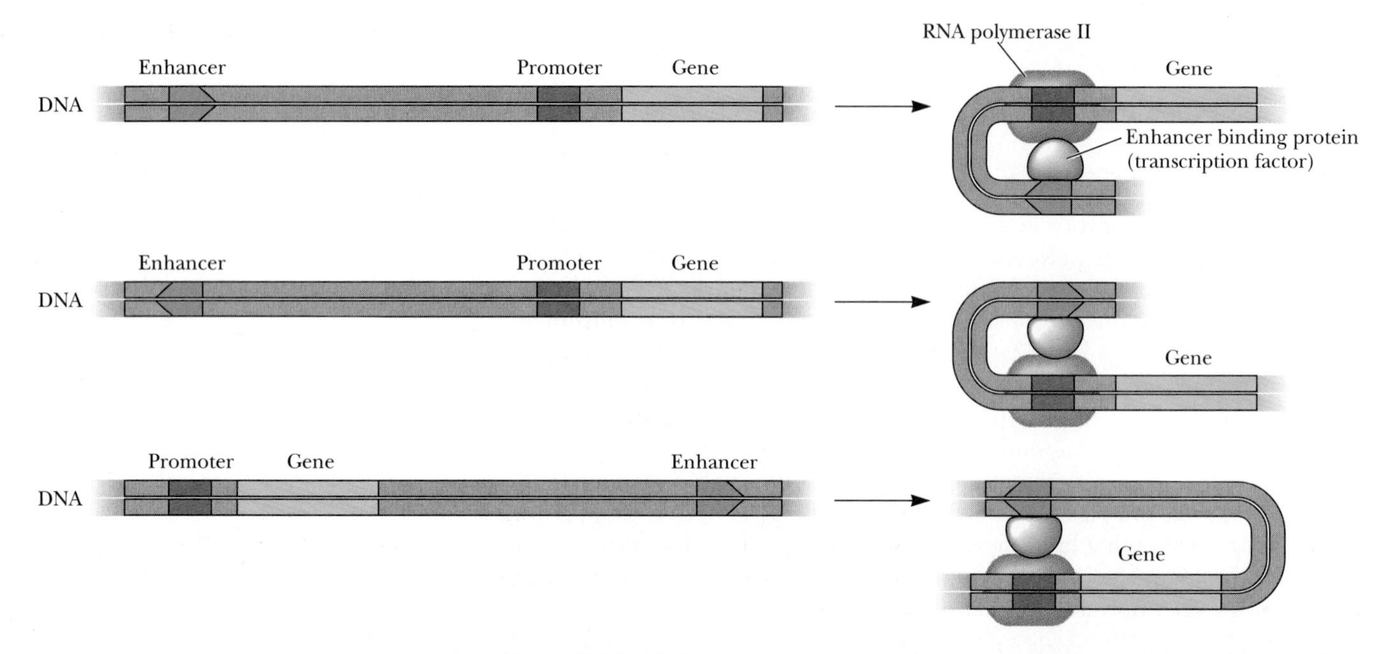

***Figure* 30.29** Enhancers are sequence elements located at varying positions and orientation relative to the promoter that act to enhance transcription initiation. Transcription factors (proteins) bind to enhancers and stimulate RNA polymerase II binding at a nearby promoter.

Response Elements

Promoter modules in genes responsive to common regulation are termed **response elements.** Examples include the **heat shock element (HSE),** the **glucocorticoid response element (GRE),** and the **metal response element (MRE).** These various elements are found in the promoter regions of genes whose transcription is activated in response to a sudden increase in temperature (heat shock), glucocorticoid hormones, or toxic heavy metals, respectively (Table 30.5). HSE sequences are recognized by a specific transcription factor, **HSTF** (for **heat shock transcription factor**). HSEs are located about 15 bp upstream from the transcription start site of a variety of genes whose expression is dramatically enhanced in response to elevated temperature. Similarly, the response to steroid hormones depends on the presence of a GRE positioned 250 bp upstream of the transcription start point. Binding of a specific transcription factor, the **steroid receptor,** at a GRE occurs when certain steroids bind to the steroid receptor.

Many genes are subject to a multiplicity of regulatory influences. Regulation of such genes is achieved through the presence of an array of different regulatory elements. The **metallothionein** gene is a good example (Figure

***Table* 30.5**

Response Elements That Identify Genes Coordinately Regulated in Response to Particular Physiological Challenges

Physiological Challenge	Response Element	Consensus Sequence	DNA Bound	Factor	Size (kD)
Heat shock	HSE	CNNGAANNTCCNNG	27 bp	HSTF	93
Glucocorticoid	GRE	TGGTACAAATGTTCT	20 bp	Receptor	94
Cadmium	MRE	CGNCCCGGNCNC	?	?	?
Phorbol ester	TRE	TGACTCA	22 bp	AP1	39
Serum	SRE	CCATATTAGG	20 bp	SRF	52

Source: Adapted from Lewin, B., 1994. *Genes V.* Cambridge, MA: Cell Press.

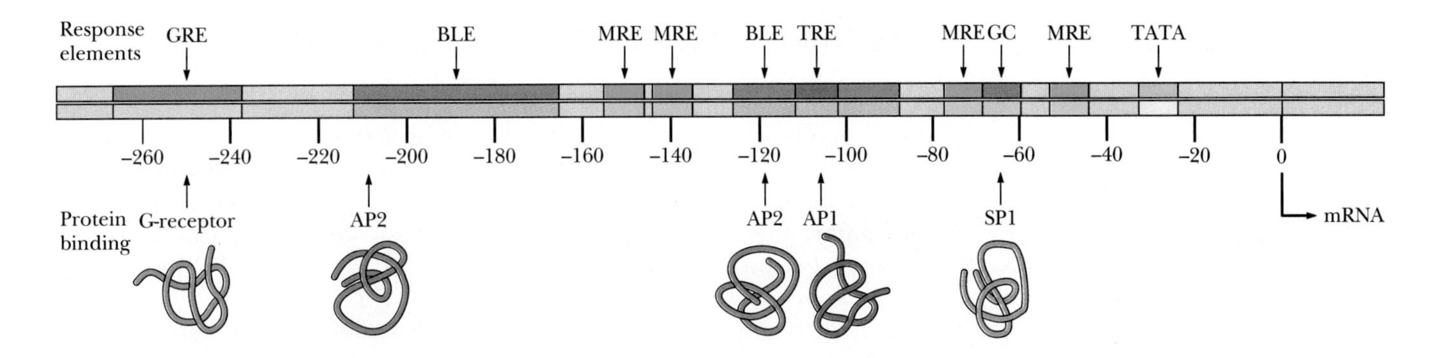

30.30). Metallothionein is a metal-binding protein that protects cells against metal toxicity by binding excess amounts of heavy metals and removing them from the cell. This protein is always present at low levels, but its concentration increases in response to heavy metal ions such as cadmium or in response to glucocorticoid hormones. The metallothionein gene promoter consists of two general promoter elements, namely, a TATA box and a GC box; two basal-level enhancers; four MREs; and one GRE. These elements function independently of one another; any one is able to activate transcription of the gene.

***Figure* 30.30** The metallothionein gene possesses several constitutive elements in its promoter (the TATA and GC boxes) as well as specific response elements such as MREs and a GRE. The BLEs are elements involved in basal level expression (constitutive expression). TRE is a tumor response element activated in the presence of tumor-promoting phorbol esters such as TPA (tetradecanoyl phorbol acetate).

DNA Looping

Since transcription must respond to a variety of regulatory signals, multiple proteins are essential for appropriate regulation of gene expression. These regulatory proteins are the **sensors** of cellular circumstances, and they communicate this information to the genome by binding at specific nucleotide sequences. However, DNA is virtually a one-dimensional polymer, and there is little space for a lot of proteins to bind at (or even near) a transcription initiation site. DNA looping permits additional proteins to convene at the initiation site and to exert their influence on creating and activating an RNA pol II initiation complex (Figure 30.31). The repertoire of transcriptional regulation is greatly expanded by DNA looping. Further, DNA looping is greatly influenced by negative supercoiling.

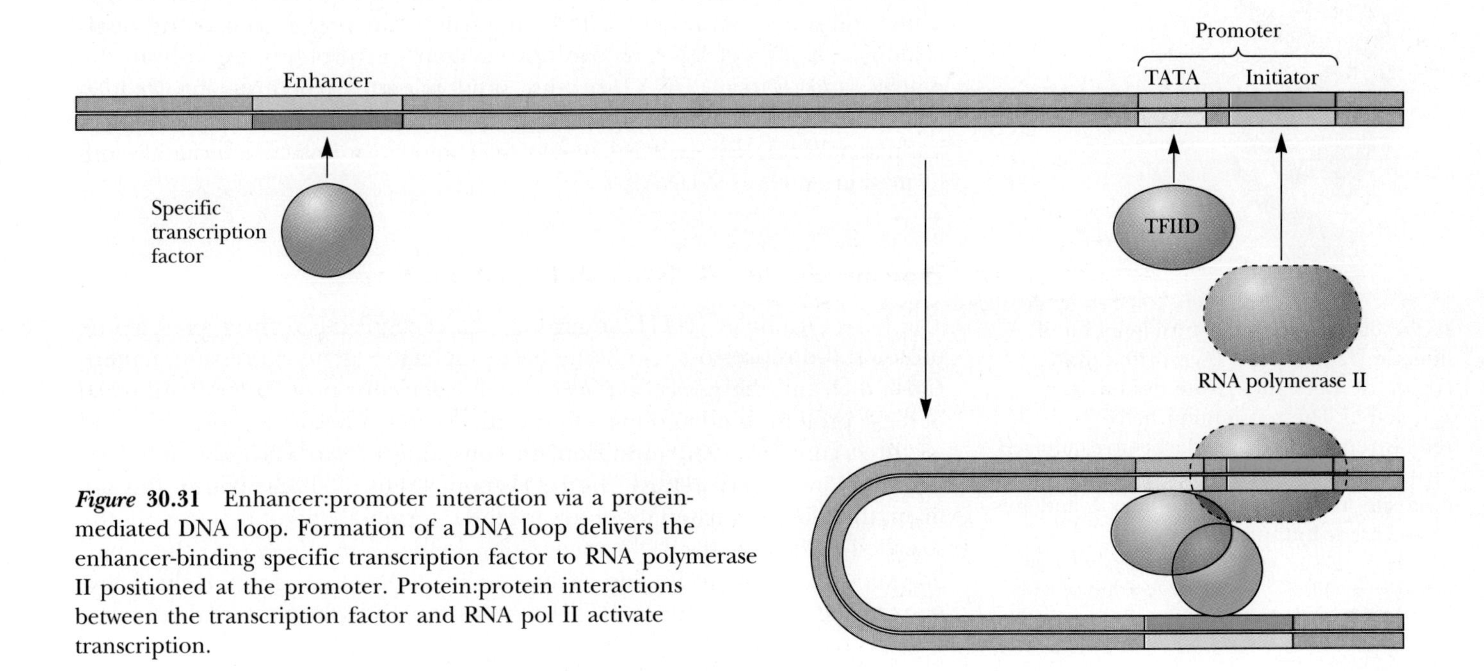

***Figure* 30.31** Enhancer:promoter interaction via a protein-mediated DNA loop. Formation of a DNA loop delivers the enhancer-binding specific transcription factor to RNA polymerase II positioned at the promoter. Protein:protein interactions between the transcription factor and RNA pol II activate transcription.

30.5 Structural Motifs in DNA-Binding Regulatory Proteins

Proteins that recognize nucleic acids do so by the basic rule of macromolecular recognition. That is, such proteins present a three-dimensional shape or contour that is structurally and chemically complementary to the surface of a DNA sequence. When the two molecules come into close contact, the numerous atomic interactions that underlie recognition and binding can take place between the two. Nucleotide sequence-specific recognition by the protein involves a set of atomic contacts with the bases and the sugar-phosphate backbone. Hydrogen bonding is critical for recognition, with amino acid side chains providing most of the critical contacts with DNA. Most protein contacts with the bases of DNA occur within the major groove; protein contacts with the DNA backbone involve both H bonds and salt bridges with oxygen atoms of the phosphodiester bridges. Rapidly advancing studies on regulatory proteins that bind to specific DNA sequences have revealed that roughly 80% of such proteins can be assigned to one of just three classes based on their possession of one of three small, distinctive structural motifs: the **helix-turn-helix** (or **HTH**), the **zinc finger** (or **Zn-finger**), and the **leucine zipper** (or **bZIP**).

In addition to their DNA-binding domains, these proteins commonly possess other structural domains that function in protein:protein recognitions essential to oligomerization (for example, dimer formation), DNA looping, transcriptional activation, and signal reception (for example, effector binding). Substantial progress has been made toward revealing how these proteins bind to DNA, as we shall soon see. However, little is known regarding the molecular mechanisms by which these proteins promote transcription. These mechanisms are based on protein:protein interactions involving RNA polymerase and the various DNA-binding proteins, but the details remain a mystery.

α-Helices Fit Snugly into the Major Groove of B-DNA

A recurring structural feature in DNA-binding proteins is the presence of α-helical segments that fit directly into the major groove of B-form DNA. The diameter of an α-helix (including its side chains) is about 1.2 nm. The dimensions of the major groove in B-DNA are 1.2 nm wide by 0.6 to 0.8 nm deep. Thus, one side of an α-helix can fit snugly into the major groove. Although examples of β-sheet DNA recognition elements in proteins are known, the α-helix and B-form DNA are the predominant structures involved in protein:DNA interactions. Significantly, proteins can recognize specific sites in "normal" B-DNA; the DNA need not assume any unusual, alternative conformation (such as Z-DNA).

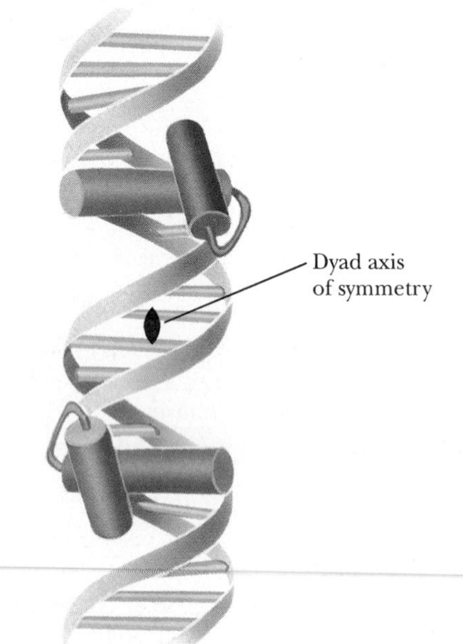

***Figure* 30.32** The helix-turn-helix motif. A dimeric HTH protein bound to a dyad-symmetric DNA site. (Note dyad axis of symmetry.) The recognition helix (helix 3, red) is represented as a dark barrel situated in the major groove of each half of the dyad site. Helix 2 sits above helix 3 and aids in locking it into place.

(Adapted from Johnson, P. F., and McKnight, S. L., 1989. Eukaryotic transcriptional regulatory proteins. Annual Review of Biochemistry 58:799, Figure 1.)

Proteins with the Helix-Turn-Helix Motif

The helix-turn-helix (HTH) motif was first recognized in three prokaryotic proteins: two repressors regulating bacteriophage λ gene expression, namely ***Cro*** and ***cI,*** and the *Escherichia coli* catabolite activator protein, CAP. All three of these proteins bind as dimers to specific dyad-symmetric sites on DNA, and all three contain a structural domain consisting of two successive α-helices separated by a sharp β-turn, the HTH motif (Figure 30.32). Within this domain, the α-helix situated more toward the C-terminal end of the protein, the so-called **helix 3,** is the DNA recognition helix; it fits nicely into the major groove, with several of its side chains touching DNA base pairs. **Helix 2,** the helix at the beginning of the HTH motif, through hydrophobic interactions with helix 3, creates a stable structural domain that locks helix 3 into its DNA interface. The two helix 3 cylinders are antiparallel to each other, such that their N→C orientations match the inverted relationship of nucleotide sequence in the dyad-symmetric DNA-binding site.

The HTH Motif

The "orthodox" HTH motif is an amino acid sequence approximately 20 residues in length found in a number of DNA-binding proteins, both prokaryotic and eukaryotic (Table 30.6). Residues 1 through 7 form the first helix (helix 2), and 12 through 20 the second, or "recognition," helix (helix 3). Residue 9 is usually the β-turn maker, Gly. Figure 30.33a–h depicts a gallery of HTH-containing domains. Note the strong structural similarity shared by the DNA-binding motif in this otherwise dissimilar family of proteins.

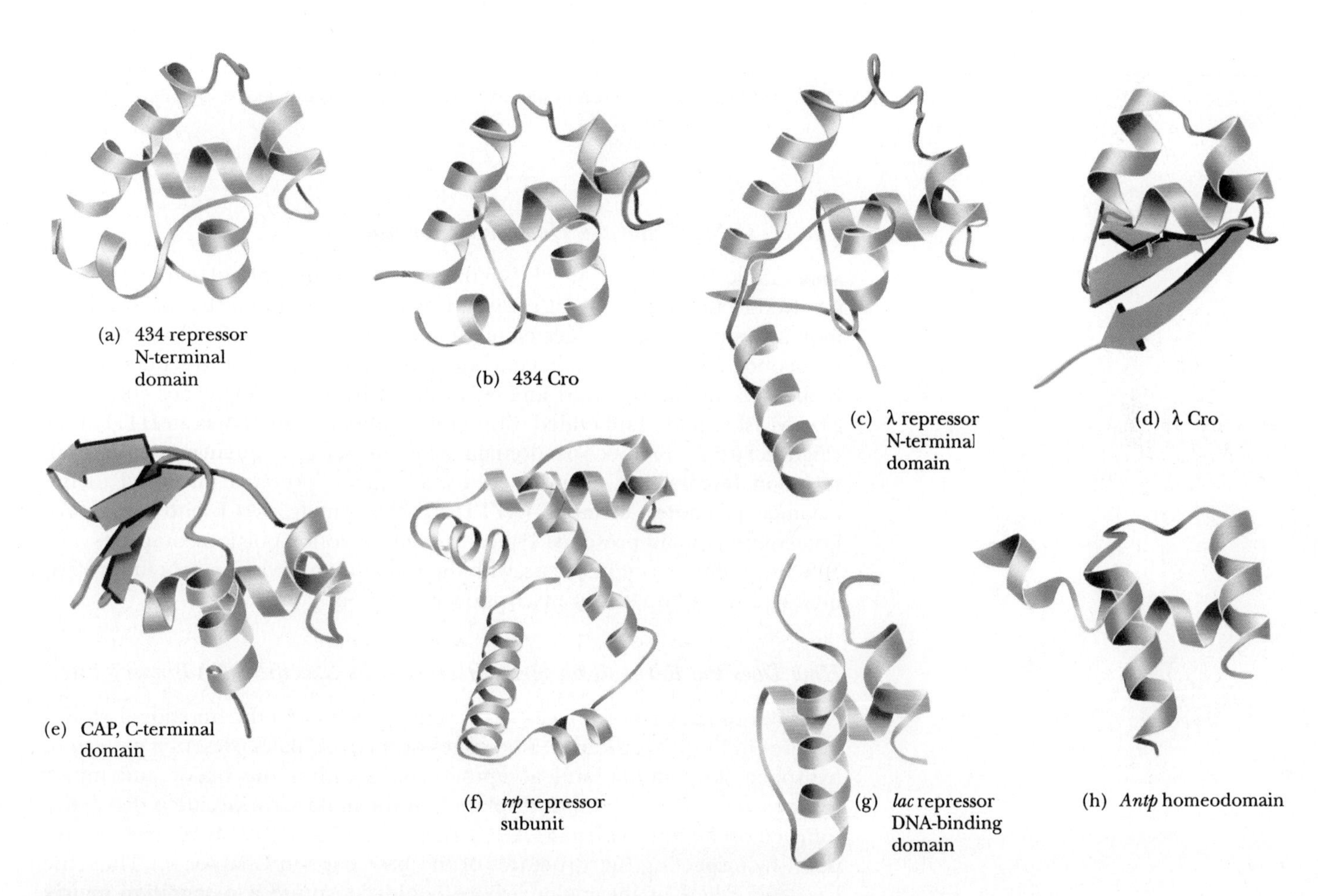

***Figure* 30.33** HTH domains in various sequence-specific, DNA-binding proteins. All HTH domains are oriented the same; the view is from the "rear" (the side opposite the DNA-binding face). The first HTH helix runs vertically down from the upper right; the turn is right center; and the "recognition" helix runs across the center from right to left "behind" the domain in this perspective. (a) 434 repressor N-terminal domain; (b) 434 Cro; (c) λ repressor N-terminal domain; (d) λ Cro; (e) CAP, C-terminal domain; (f) *trp* repressor subunit; (g) *lac* repressor DNA-binding domain; (h) *Antp* homeodomain.

(Adapted from Harrison, S. C., and Aggarwal, A. K., 1990. Annual Review of Biochemistry 59:933, Figure 2.)

***Table* 30.6**

Amino Acid Sequences in the HTH Regions of Selected Transcription Regulatory Proteins

434 *Rep* and *Cro* are bacteriophage 434 proteins; *Lam Rep* and *Cro* are bacteriophage λ proteins; CAP, *Trp Rep,* and *Lac Rep* are catabolite activator protein, *trp* repressor, and *lac* repressor of *E. coli,* respectively. *Antp* is the homeodomain protein of the *Antennapedia* gene of the fruit fly *Drosophila melanogaster.* The numbers in each sequence indicate the location of the HTH within the amino acid sequences of the various polypeptides.

		Helix							Turn				Helix									
		1	2	3	4	5	6	7	8	9	10	11	12	13	14	15	16	17	18	19	20	
434 Rep	17 -	Gln	Ala	Glu	**Leu**	**Ala**	Gln	Lys	**Val**	**Gly**	Thr	Thr	Gln	Gln	Ser	**Ile**	Glu	Gln	**Leu**	Glu	Asn	- 36
434 Cro	17 -	Gln	Thr	Glu	**Leu**	**Ala**	Thr	Lys	**Ala**	**Gly**	Val	Lys	Gln	Gln	Ser	**Ile**	Gln	Leu	**Ile**	Glu	Ala	- 36
Lam Rep	33 -	Gln	Glu	Ser	**Val**	**Ala**	Asp	Lys	**Met**	**Gly**	Met	Gly	Gln	Ser	Gly	**Val**	Gly	Ala	**Leu**	Phe	Asn	- 52
Lam Cro	16 -	Gln	Thr	Lys	**Thr**	**Ala**	Lys	Asp	**Leu**	**Gly**	Val	Tyr	Gln	Ser	Ala	**Ile**	Asn	Lys	**Ala**	Ile	His	- 35
CAP	169 -	Arg	Gln	Glu	**Ile**	**Gly**	Glu	Ile	**Val**	**Gly**	Cys	Ser	Arg	Glu	Thr	**Val**	Gly	Arg	**Ile**	Leu	Lys	- 188
Trp Rep	68 -	Gln	Arg	Glu	**Leu**	**Lys**	Asn	Glu	**Leu**	**Gly**	Ala	Gly	Ile	Ala	Thr	**Ile**	Thr	Arg	**Gly**	Ser	Asn	- 87
Lac Rep	6 -	Leu	Tyr	Asp	**Val**	**Ala**	Arg	Leu	**Ala**	**Gly**	Val	Ser	Tyr	Gln	Thr	**Val**	Ser	Arg	**Val**	Val	Asn	- 25
Antp	31 -	Arg	Ile	Glu	**Ile**	**Ala**	His	Ala	**Leu**	**Cys**	Leu	Thr	Glu	Arg	Gln	**Ile**	Lys	Ile	**Trp**	Phe	Gln	- 50

Source: Adapted from Harrison, S. C., and Aggarwal, A. K., 1990. DNA recognition by proteins with the helix-turn-helix motif. *Annual Review of Biochemistry* **59:**933–969.

The HTH Motif and the Homeobox Domain

Antp (Table 30.6, Figure 30.33) is a member of a family of eukaryotic proteins involved in the regulation of early embryonic development that have in common an amino acid sequence element known as the **homeobox[6] domain.** The homeobox is a DNA motif that encodes a related 60-amino acid sequence (the homeobox domain) found among proteins from virtually every eukaryote, from yeast to man. Embedded within the homeobox domain is an HTH motif (Figure 30.34). Homeobox domain proteins act as **sequence-specific transcription factors.** For example, two mammalian proteins that bind to the **octamer promoter element** (ATTTGCAT), namely, *Oct*-1 and *Oct*-2, are homeobox domain proteins. Typically, the homeobox portion comprises only 10% or so of the protein's mass, with the remainder serving in protein:protein interactions essential to transcription regulation.

How Does the Recognition Helix Recognize Its Specific DNA-Binding Site?

Consider the possible contacts between the protein and the functional groups of bases in the DNA grooves. The edges of the base pairs present a pattern of hydrogen-bond donor and acceptor groups within the major and minor grooves, but only the pattern displayed on the major-groove side is distinctive for each of the four base pairs A:T, T:A, C:G, and G:C. (Satisfy yourself on this point by inspecting the structures of the base pairs in Chapter 6.) Thus, the base-pair edges in the major groove could constitute a **recognition matrix** identifiable through H bonding with a specific protein, without requiring melting of the base pairs to reveal the base sequence. While formation of such H bonds is very important in DNA:protein recognition, other interactions also play a significant role. For example, the C-5-methyl groups unique to thymine residues are nonpolar "knobs" projecting into the major groove.

[6] *Homeo* derives from *homeotic genes,* a set of genes originally discovered in the fruit fly *Drosophila melanogaster* through their involvement in the specification of body parts during development.

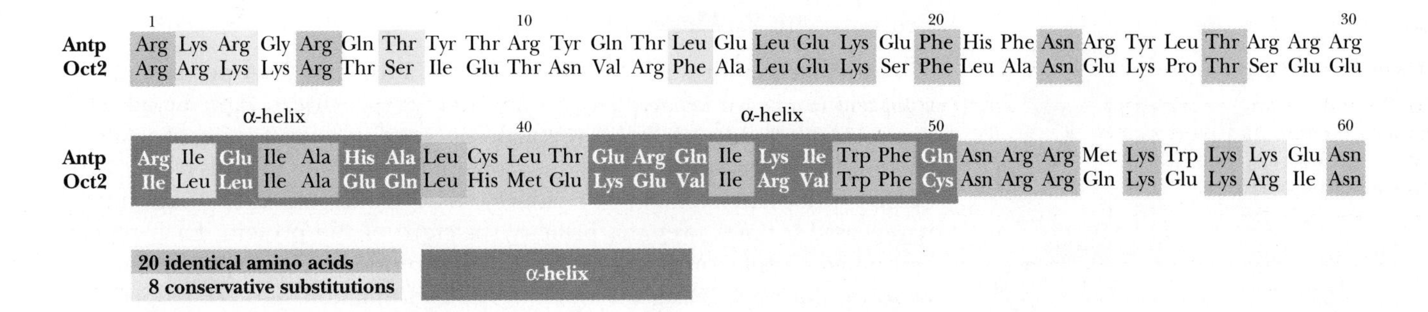

Is there a side chain:DNA base-pair recognition code? Certain amino acid side chains do form unique H-bonding patterns with particular bases. For example, binding of bacteriophage λ *Cro* repressor to its operator involves a bidentate ("two-toothed") H-bonding pattern between a glutamine side chain and an adenine (Figure 30.35). An Arg residue and a guanine show a similar interaction. However, an examination of known DNA:protein structures does not reveal any simple universal recognition code of this type. For one thing, the periodicity of functional groups in an α-helix shows no general spatial relationship to the periodicity of bases in B-form DNA.

***Figure* 30.34** The homeo domains of *Antp* and *Oct-2*. *Antp* is an insect protein while *Oct-2* is a mammalian protein. Trp^{48}, Phe^{49}, Asn^{51}, and Arg^{53} are universally conserved in all known homeodomains. The apparent HTH motif within these sequences is indicated.

Do Proteins Recognize DNA via "Indirect Readout"?

Indirect readout is the term for the ability of a protein to indirectly recognize a particular nucleotide sequence by recognizing local conformational variations resulting from the effects that base sequence has on DNA structure. Superficially, the B-form structure of DNA appears to be a uniform cylinder. Nevertheless, the conformation of DNA over local regions is markedly influenced by the local base sequence. In particular, base pairs in A:T-rich regions can show high propeller twist (Chapter 7), leading to diminution in the planar stacking of the base pairs perpendicular to the double helix axis and a corresponding change in the conformation of the DNA's sugar-phosphate backbone. These changes generate unique contours that proteins can recognize. In *E. coli* Trp repressor:*trp* operator DNA cocrystals, the Trp repressor engages in 30 specific hydrogen bonds to the DNA—28 involve phosphate groups in the backbone; only two are to bases.

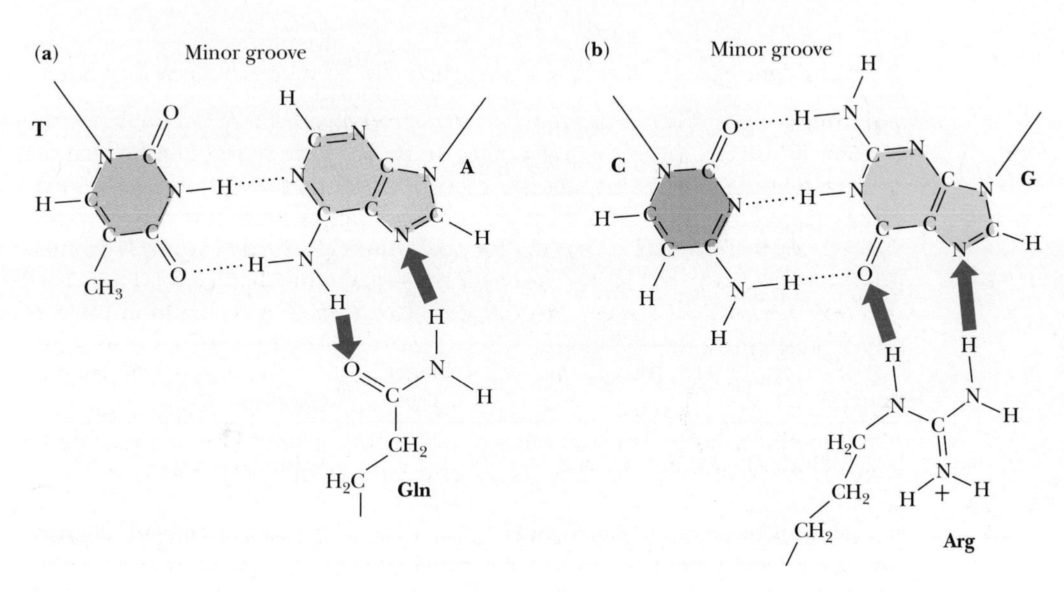

***Figure* 30.35** The bidentate H-bonding interactions between (a) the glutamine amide group and N6 and N7 of adenine, and (b) the arginine guanidine group and O6 and N7 of guanine.

(Source: Pabo, C. O. and Sauer, R. T., 1984. Annual Review of Biochemistry 53:293.)

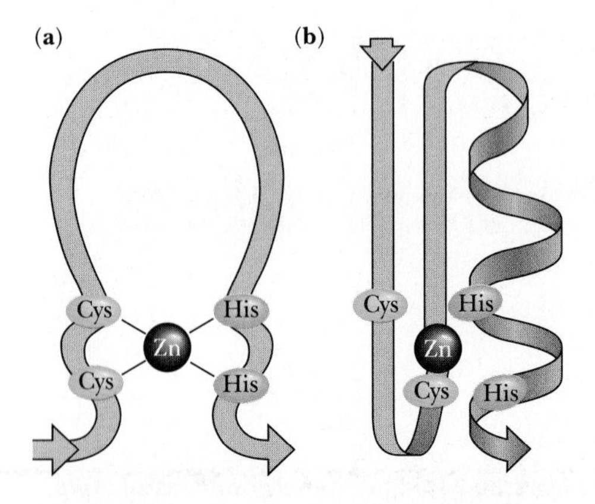

***Figure* 30.36** The Zn-finger motif of the C_2H_2 type showing (a) the coordination of Cys and His residues to Zn and (b) the secondary structure.

(Adapted from Evans, R. M., and Hollenberg, S. M., 1988. Cell 52:1, Figure 1.)

Proteins with Zn-Finger Motifs

The Zn-finger motif was first discovered in TFIIIA, an RNA polymerase III transcription factor isolated from *Xenopus laevis,* the African clawed toad. The amino acid sequence of TFIIIA reveals a repetitive motif consisting of a pair of Cys residues intervened by 2 residues separated by 12 amino acids from a pair of His residues intervened by 3 residues (Cys-x_2-Cys-x_{12}-His-x_3-His). This motif is repeated 9 times over the primary structure of the protein. Each repeat coordinates a zinc ion via its 2 Cys and 2 His residues (Figure 30.36). The 12 or so residues separating the Cys and His coordination sites are looped out and form a distinct DNA interaction module, the so-called Zn-finger. When TFIIIA associates with DNA, each Zn-finger binds in the major groove and interacts with about five nucleotides, adjacent fingers interacting with contiguous stretches of DNA. Other DNA-binding proteins with this motif have been identified, including transcription factor **SP1,** a protein that gives 10- to 20-fold stimulation of genes with GC boxes. In all cases, the finger motif is repeated at least two times, with at least a 7 to 8 amino acid linker between Cys/Cys and His/His sites (Table 30.7). Proteins endowed with this general pattern are assigned to the **C_2H_2 class** of Zn-finger proteins to distinguish them from proteins bearing another kind of Zn-finger, the **C_x type.**

***Table* 30.7**

Zinc Finger Motifs of the C_2H_2 and C_x Classes

The consensus C_2H_2 finger sequence is Cys-X_{2-4}-Cys-X_3-Phe-X_5-Leu-X_2-His-X_{3-4}-His.

Protein	Number of C_2H_2 Repeats	Organism
TFIIIA	9	*Xenopus*
ADR1	2	Yeast
SP1	3	Human
NGF1-A	3	Rat
Krüppel h	2(+)	*Drosophila*
Krüppel	4	*Drosophila*
Hunchback	4 + 2	*Drosophila*
Serendipity β	5	*Drosophila*
Serendipity δ	6 + 1	*Drosophila*
Snail	4	*Drosophila*
MKR1	7(+)	Mouse
MKR2	9(+)	Mouse
TDF	13(+)	Human
Xfin	6 + 6 + 8 + 7 + 3 + 5	*Xenopus*

The C_x sequences are:

C_4: C-X_2-C-X_{13}-C-X_2-C
C_5: C-X_5-C-X_9-C-X_2-C-X_4-C
C_6: C-X_2-C-X_6-C-X_6-C-X_2-C-X_6-C

Protein	Finger Type	Organism
GAL4/PPRI/ARGRII/LAC9/*qa*-1F	C_6	Yeast, *Neurospora*
E1A	C_4	Adenovirus
Steroid hormone receptor superfamily	C_4 + C_5	Human/rat/mouse/chicken

Source: Evans, R. M., and Hollenberg, S. M., 1988. *Cell* **52:**1–3.

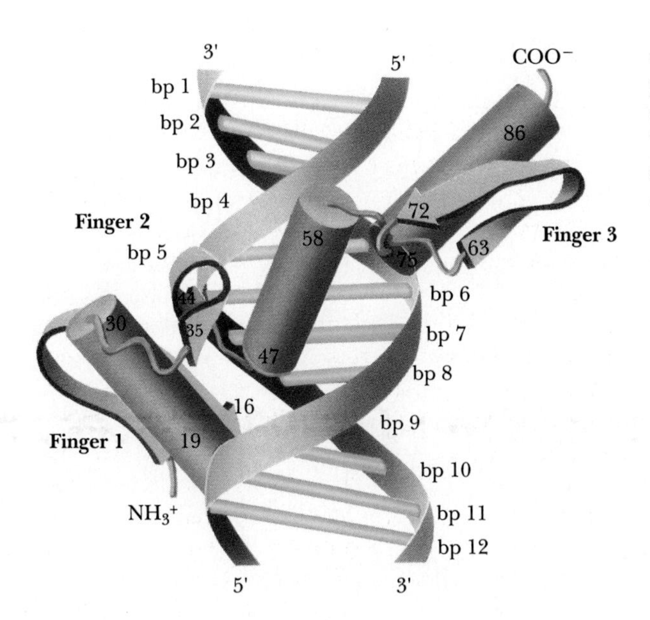

***Figure* 30.37** X-ray crystallographic analysis of a C_2H_2 Zn-finger protein:DNA complex reveals that three zinc fingers are arranged in a semicircular structure that follows the major groove around the DNA. The α-helix of each finger fits directly into the major groove and amino acid residues contact base pairs in the major groove.

(Adapted from Pavletich, N., and Pabo, C. O., 1991. Science 252:809, Figure 2.)

X-ray crystallographic analysis of a C_2H_2 Zn-finger:DNA complex reveals that the zinc fingers bind in the major groove of the DNA and wrap partway around the double helix (Figure 30.37). Each zinc finger makes its primary DNA contacts with a particular subsite consisting of 3 base pairs, with amino acid residues from the N-terminal portion of the α-helix contacting the guanine-rich strand in its recognition site. Arginine side chains are prominent in making these contacts with G (Figure 30.38), and these Arg:G contacts are believed to be a common feature in Zn-finger:DNA recognition.

The C_x proteins have a variable number of Cys residues available for Zn **chelation.** For example, GAL4 protein, the yeast transcription factor regulating the enzymes of galactose metabolism, has a cluster of 6 Cys residues (Table 30.7), whereas the vertebrate steroid receptors have two sets of Cys residues, one with four conserved cysteines (C_4) and the other with five (C_5). The GAL4 polypeptide has only a single Zn-finger. The C_x Zn-finger motif

chelation: from the Greek *chele,* claw; binding of a metal ion to two or more nonmetallic atoms of the same molecule

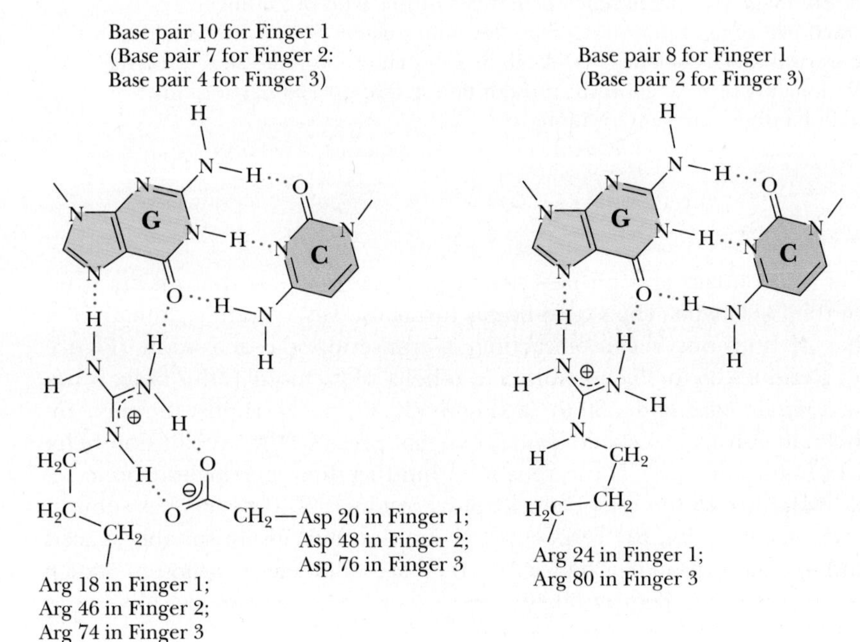

***Figure* 30.38** The important protein:DNA contacts in C_2H_2 Zn-finger:DNA binding include hydrogen bonds formed between arginine side chains and G residues localized to one strand of the DNA. Arg residues occur immediately preceding the Zn-finger α-helix and at the second, third, and sixth positions in the helix.

(Adapted from Pavletich, N., and Pabo, C. O., 1991. Science 252:809, Figure 3.)

(a)

$C_4 + C_5$ (estrogen receptor)

Motif 1 K E T R Y C A V C N D Y A S G Y H Y G V W S C E G C K A F F R R S I Q

Motif 2 G H N D Y M C P A T N Q C T I D K N R R K S C Q A C R L R K C Y E V G M M

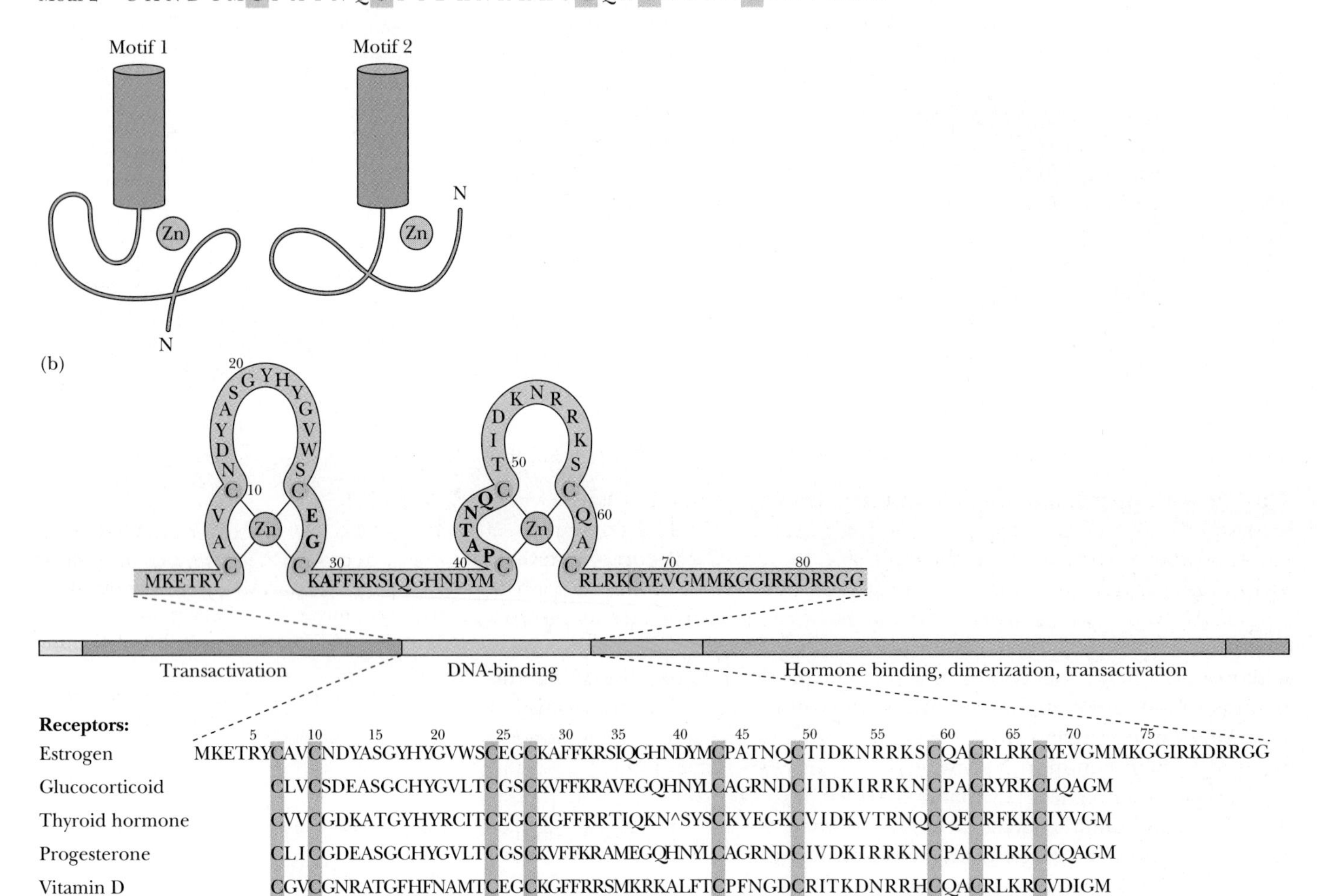

***Figure* 30.39** Features of the C_x family of Zn-finger proteins. (a) DNA-binding motifs in the estrogen hormone receptor. Note the secondary structure of the C_x type of Zn-finger and the absence of β-sheet in this type of Zn-finger.
(b) Functional organization of C_x-type hormone receptor proteins. In addition to the conserved metal-chelating, DNA-binding domain, note that these proteins possess domains that function in transcription activation (*trans*activation), hormone binding, and dimerization.

(Adapted from Schwabe, J. W. R., and Rhodes, D., 1991. Trends in Biochemical Sciences 16:291, Figures 1 and 2.)

displays a secondary and tertiary structure (Figure 30.39) that differs from that of the C_2H_2 type. The C_x Zn-finger forms the DNA-binding domain of a number of hormone receptors acting in transcriptional activation (Figure 30.40). Residues 25 to 35 constitute the α-helix of C_x motif 1; this helix is the *DNA recognition helix.* Helix 60 to 72 (Figure 30.40) packs against the recognition helix to construct a stable framework that presents the recognition helix to its DNA-binding site. These receptors bind as dimers to palindromic response elements. In the estrogen receptor, residues 44 to 48 serve as subunit contacts, ensuring that the respective recognition helices are suitably spaced to bind the palindromic "half-sites" that are situated in adjacent major grooves of the DNA (Figure 30.40).

Proteins with the Leucine Zipper Motif

The leucine zipper is a structural motif characterizing the third major class of sequence-specific, DNA-binding proteins. This motif was first recognized by Steve McKnight in C/EBP, a heat-stable, DNA-binding protein isolated from rat liver nuclei that binds to both CCAAT promoter elements and certain enhancer core elements.[7] The DNA-binding domain of C/EBP was localized to the C-terminal region of the protein. This region shows a notable absence of Pro residues, suggesting it might be arrayed in an α-helix. When a 28-residue sequence of this region is displayed end-to-end down the axis of a hypothetical α-helix, beginning at Leu^{315}, an amphipathic cylinder is generated (Figure 30.41). One side of this amphipathic helix consists principally of hydrophobic residues (particularly leucines), whereas the other side has an array of negatively and positively charged side chains (Asp, Glu, Arg, and Lys), as well as many uncharged polar side chains (glutamines, threonines, and serines).

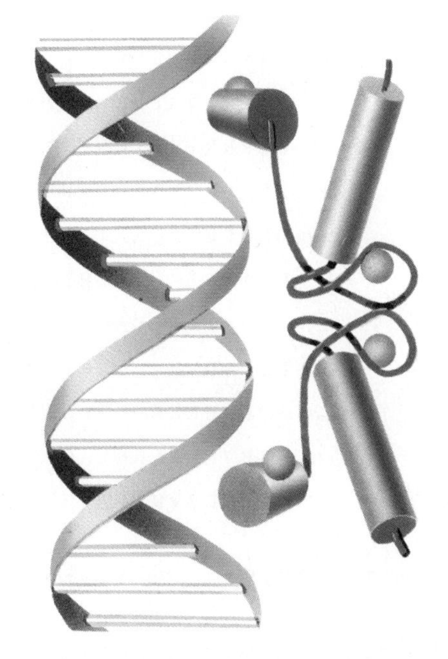

***Figure* 30.40** A model for the binding of steroid receptor dimers to their palindromic recognition sequences in DNA. Note that the DNA recognition helices are antiparallel and oriented to bind symmetrically in the major groove.

(Adapted from Schwabe, J. W. R., and Rhodes, D., 1991. Trends in Biochemical Sciences 16:291, Figure 5.)

The Zipper Motif: Intersubunit Interaction of Leucine Side Chains

The leucine zipper motif arises from the periodic repetition of leucine residues within this helical region. The periodicity causes the Leu side chains to protrude from the same side of the helical cylinder, where they can enter into hydrophobic interactions with a similar set of Leu side chains extending from a matching helix in a second polypeptide. These hydrophobic interactions establish a stable noncovalent linkage, fostering dimerization of the two polypeptides (as you'll note in Figure 30.43). Note that, unlike the HTH and Zn-finger motifs, the leucine zipper is not a DNA-binding domain. Instead, it functions in protein dimerization. Similar leucine zippers have been found in other mammalian transcriptional regulatory proteins, including *Myc, Fos,* and *Jun.*

The actual DNA contact surface of these proteins is contributed by a 16-residue segment that ends exactly 7 residues before the first Leu residue of the Leu zipper. This DNA contact region is rich in basic residues and hence is referred to as the **basic region;** its consensus sequence is BB-BN--AA-B-R-BB, where B is a basic residue (Arg or Lys), N is Asn, AA represents a pair of invariant alanines, and R is an invariant Arg. The alignment between the

[7]The acronym *C/EBP* designates this protein as a "*C*CAAT and *E*nhancer *B*inding *P*rotein."

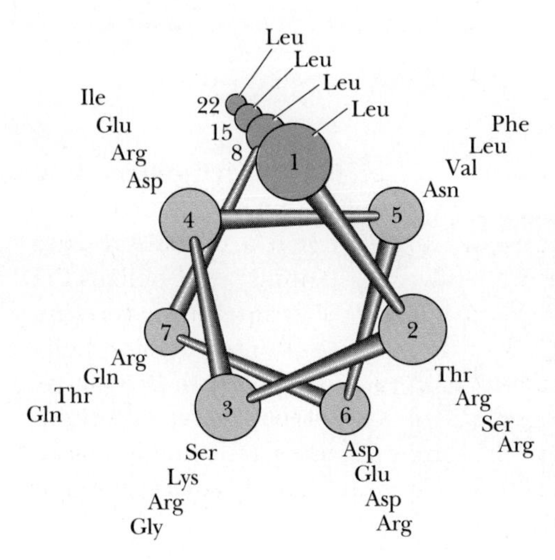

***Figure* 30.41** Helical wheel analysis of the C-terminal region of C/EBP. The amino acid sequence from Leu^{315} to Gln^{342} is displayed as a hypothetical α-helix viewed down its axis. Leu^{315} is labeled Leu^1, Thr^{316} is Thr^2, and so on. Seven successive residues complete about two turns of the helix. (Recall that α-helices have 3.6 amino acid residues per turn.) Note that the amino acid sequence in this region generates an amphipathic α-helix. In particular, Leu residues are spaced every seven residues (positions 1, 8, 15, and 22) and line up on one side of this helix to form a hydrophobic "spine."

(Adapted from Landschulz, W. H., Johnson, P. F., and McKnight, S. L., 1988. Science 240:1759–1764, Figure 1.)

Protein	Basic region A / Basic region B / Leucine zipper
C/EBP	278–DKNSNEYRVRRERNNIAVRKSHDKAKQRNVETQQKVLELTSDNDRLRKRVEQLSRELDTLRG–341
Jun	257–SQERIKAERKRMRNRIAASKCHKRKLERIARLEEKVKTLKAQNSELASTANMLTEQVAQLKO–320
Fos	233–EERRRIRRIRRERNKMAAAKCRNRRRELTDTLQAETDQLEDKKSALQTEIANLLKEKEKLEF–296
GCN4	221–PESSDPAALKRARNTEAARRSRARKLQRMKOLEDKVEELLSKNYHLENEVARLKKLVGER–COOH
YAP1	60–DLDPETKQKRTAQNRAAQRAFHERKERKMKELEKKVQSLESIQQQNEVEATFLRDQLITLVN–123
CREB	279–EEAARKREVRLMKNREAARECRRKKKEYVKCLENRVAVLENQNKTLIEELKALKDLYCHKSD–342
Cys-3	95–ASRLAAEEDKRKRNTAASARFRIKKKQREQALEKSAKEMSEKVTQLEGRIQALETENKYLKG–148
CPC1	211–EDPSDVVAMKRARNTLAARKSBERKAQRLEELEAKIEELIAERDRYKNLALAHGASTE–COOH
HBP1	176–WDERELKKQKRLSNRESARRSRLRKQAECEELGQRAEALKSENSSLRIELDRIKKEYEELLS–239
TGA1	68–SKPVEKVLRRLAQRNEAARKSRLRKKAYVQQLENSKLKLIQLEQELERARKQGMCVGGGVDA–131
Opaque2	223–MPTEERVRKRKESNRESARRSRYRKAAHLKELEDQVAQLKAENSCLLRRIAALNQKYNDANV–286
Consensus	---------BB-BN--AA-B-R-BB-------L(E/Q)-----L------L------L------L--

***Figure* 30.42** Amino acid sequence comparisons in the basic region and leucine zipper region of 11 sequence-specific, DNA-binding proteins. These proteins represent both animal, plant, and fungal organisms. *(Adapted from Vinson, C. R., Sigler, P. B., and McKnight, S. L., 1989. Science 246:912, Figure 1.)*

basic region and the Leu zipper is precisely the same in different Leu zipper proteins (Figure 30.42). The juxtaposition of this basic region next to the Leu zipper has led to the designation of these proteins as *bZIP* proteins. One model for the secondary structure of *bZIP* proteins suggests that the basic region can be divided into 2 α-helical parts, BR-A and BR-B. In this model, the invariant Asn residue serves to terminate the BR-A helix, allowing the protein to bend between the BR-A and BR-B segments.

The *bZIP* proteins recognize specific binding sites on DNA having dyad symmetry. Two *bZIP* polypeptides join via a Leu zipper to form a Y-shaped molecule in which the stem of the Y corresponds to a coiled pair of α-helices held by the leucine zipper. The arms of the Y are the respective basic regions of each polypeptide; they act as a linked set of DNA contact surfaces (Figure 30.43). The dimer interacts with a DNA target site by situating the bifurcation point of the Y at the center of the dyad-symmetric DNA sequence. The two arms of the Y can then track along the major groove of the DNA in opposite directions, reading the specific dyad-symmetric recognition sequence (Figure 30.44). The rationale behind the proposal for two helices in the basic region

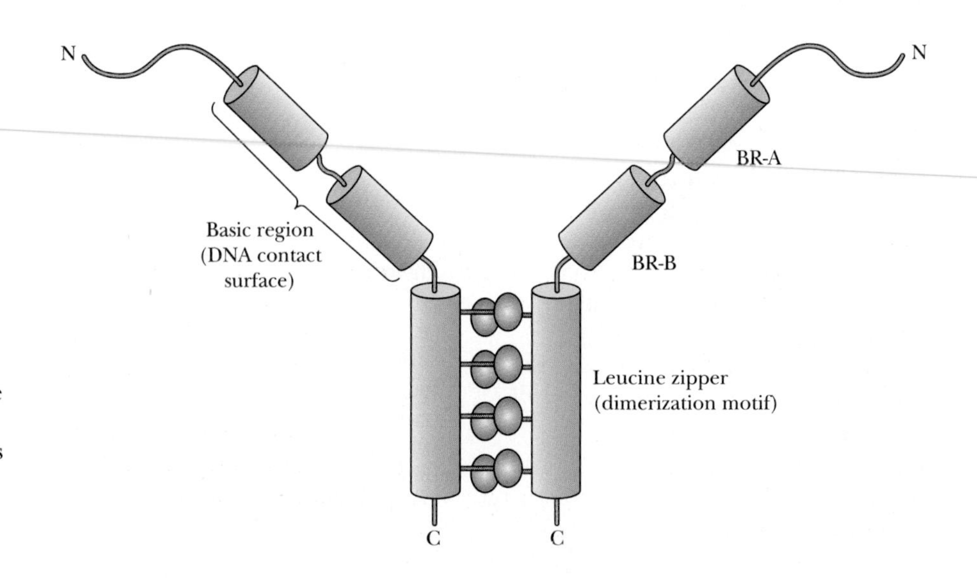

***Figure* 30.43** Model for a dimeric *bZIP* protein. Two *bZIP* polypeptides dimerize to form a Y-shaped molecule. The stem of the Y is the Leu zipper and it holds the two polypeptides together. Each arm of the Y is the basic region from one polypeptide. Each arm is composed of two α-helical segments: BR-A and BR-B.

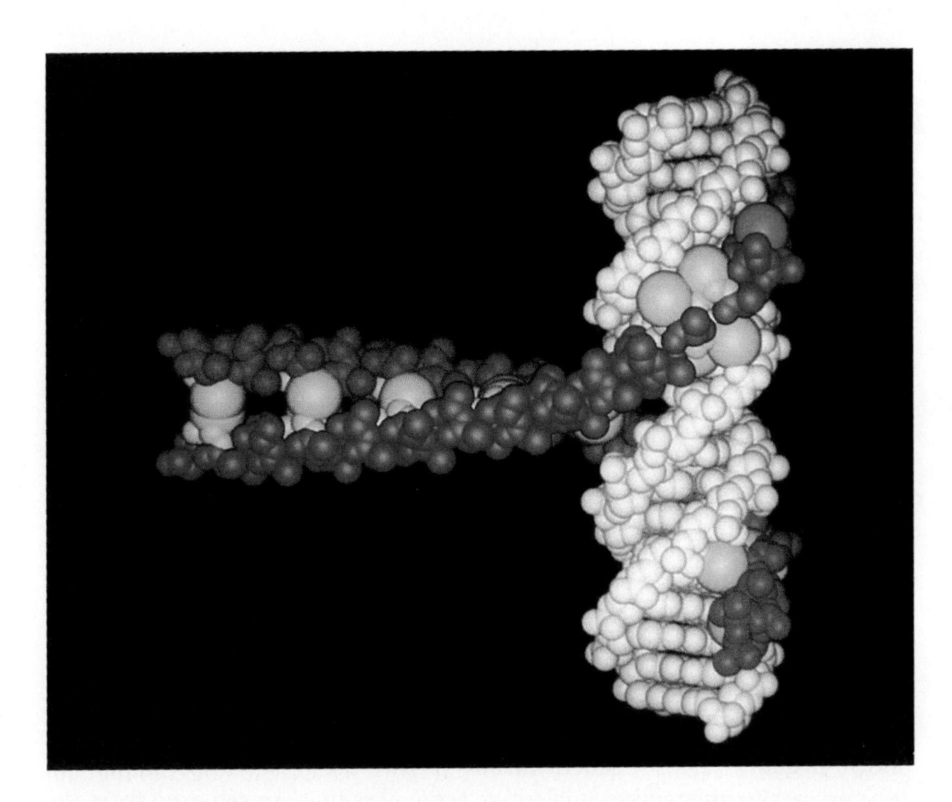

***Figure* 30.44** Model for the dimeric *bZIP* protein C/EBP interacting with one of its dyad-symmetric DNA target sequences (5′-TGCAGATTGC•GCAATCTGCA-3′). One monomer is displayed in red and the other in blue. Helices of the zipper are in a coiled coil conformation. The leucines of the zipper and the consensus basic residues are shown as large yellow balls; the invariant Arg is green.

(Adapted from Vinson, C. R., Sigler, P. B., and McKnight, S. L., 1989. Science 246:911, Figure 2; photo courtesy of Charles M. Vinson of the National Institutes of Health.).

The authors examined a large number of related proteins containing the same DNA-binding motif and from them deduced the conserved sequence elements essential to the structure of the motif. From conclusions based on such comparative molecular anatomy, they proposed the model shown here for binding of bZIP proteins to DNA. Subsequent crystallographic studies have substantiated the principal results of this modeling strategy.

now becomes clear. The Asn-induced break after BR-A allows the polypeptide to bend in order to follow the major groove around the DNA double helix. Helix BR-B then falls nicely into the continuing major groove. The positively charged side chains of conserved basic amino acids in BR-A and BR-B are directed toward the negatively charged sugar-phosphate backbone of the DNA.

An interesting aspect of *bZIP* proteins is that the two polypeptides need not be identical. Heterodimers can form provided both polypeptides possess a leucine zipper region. An important consequence of heterodimer formation is that the DNA target site no longer has to be dyad symmetric. The respective BR-A and BR-B regions of the two different *bZIP* polypeptides (for example, *Fos* and *Jun*) can track along the major groove reading two different (that is, asymmetric) half-sites. Heterodimer formation expands enormously the DNA recognition and regulatory possibilities of this set of proteins.

30.6 Post-Transcriptional Processing of mRNA in Eukaryotic Cells

Transcription and translation are concomitant processes in prokaryotes, but in eukaryotes, the two processes are spatially and temporally disconnected (Chapter 6). *Transcription occurs on DNA in the nucleus and translation occurs on ribosomes in the cytoplasm.* Consequently, transcripts must move from the nucleus to the cytosol to be translated. On the way, these transcripts undergo **processing:** alterations that convert the newly synthesized RNAs, or *primary transcripts,* into mature messenger RNAs. Also, unlike prokaryotes, in which many mRNAs encode more than one polypeptide (that is, they are polycistronic), eukaryotic mRNAs encode only one polypeptide (that is, they are exclusively monocistronic).

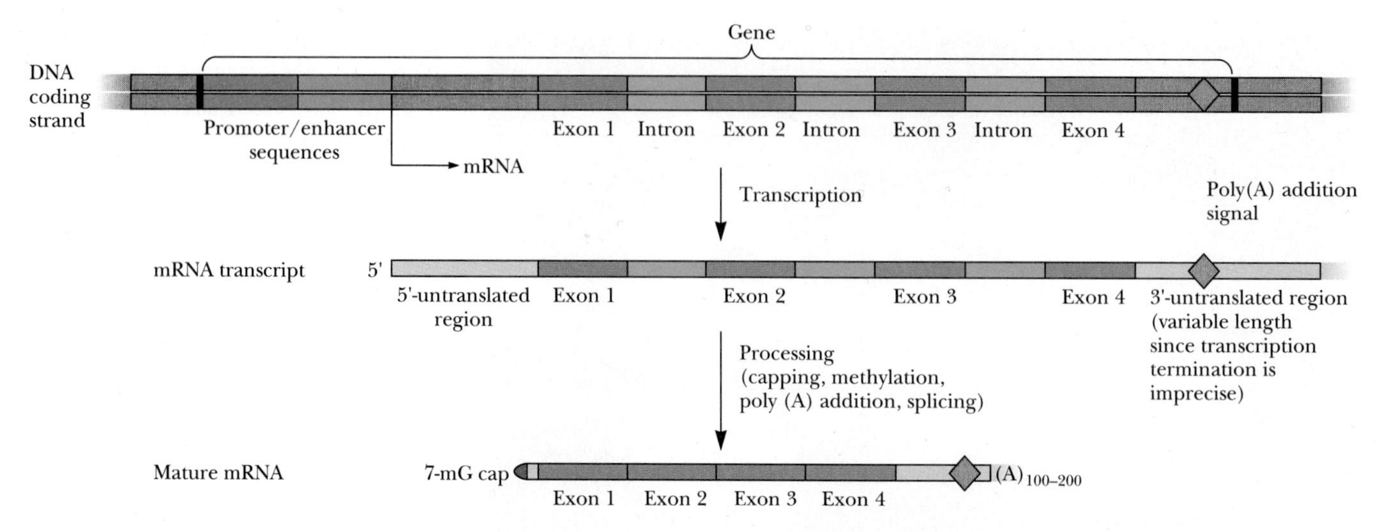

***Figure* 30.45** The organization of split eukaryotic genes.

Eukaryotic Genes Are Split Genes

Most genes in higher eukaryotes are split into coding regions, called **exons,**[8] and noncoding regions, called **introns** (Figure 30.45; see also Chapter 6). Introns are the intervening nucleotide sequences that are removed from the primary transcript when it is processed into a mature RNA. Gene expression in eukaryotes entails not only transcription but also the *processing of primary transcripts* to yield the mature RNA molecules we classify as mRNAs, tRNAs, rRNAs, and so forth.

The Organization of Split Genes

Split genes occur in an incredible variety of interruptions and sizes. The yeast **actin gene** is a simple example, having only a single 309-bp intron that separates the first 3 amino acids from the remaining 350 or so amino acids in the protein. The chicken **ovalbumin gene** is composed of 8 exons and 7 introns. The two **vitellogenin genes** of the African clawed toad *Xenopus laevis* are both spread over more than 21 kbp of DNA; their primary transcripts consist of just 6 kb of message that is punctuated by 33 introns. The chicken **pro α-2 collagen gene** has a length of about 40 kbp; the coding regions constitute only 5 kb distributed over 51 exons within the primary transcript. The exons are quite small, ranging from 45 to 249 bases in size.

Clearly, the mechanism by which introns are removed and multiple exons are spliced together to generate a continuous, translatable mRNA must be both precise and complex. If one base too many or too few is excised during splicing, the coding sequence in the mRNA will be disrupted. The mammalian **DHFR (dihydrofolate reductase) gene** is split into 6 exons spread over more than 31 kbp of DNA. The 6 exons are spliced together to give a 6-kb mRNA (Figure 30.46). Note that, in three different mammalian species, the size and position of the exons are essentially the same but that the lengths of the corresponding introns vary considerably. Indeed, the lengths of introns in

[8]Although the term *exon* is commonly used to refer to the protein-coding regions of an interrupted or split gene, a more precise definition would specify exons as *sequences that are represented in mature RNA molecules.* This definition encompasses not only protein-coding genes but also the genes for various RNAs (such as tRNAs or rRNAs) from which intervening sequences must be excised in order to generate the mature gene product.

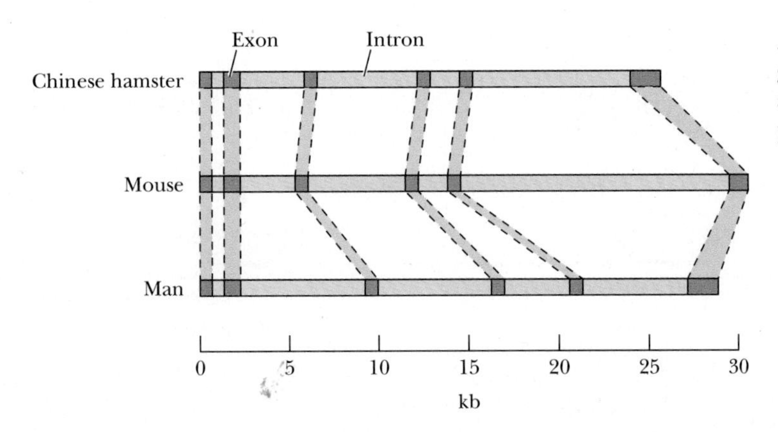

***Figure* 30.46** The organization of the mammalian DHFR gene in three representative species. Note that the exons are much shorter than the introns. Note also that the exon pattern is more highly conserved than the intron pattern.

vertebrate genes range from a minimum of about 60 bases to more than 10,000 bp.

Despite the remarkable variations on the "split gene theme," interrupted genes do have a number of features in common. First, the order of the exons in the mature RNA is the same as in the primary transcript and the DNA. So, whereas the coding regions of interrupted genes are intervened by noncoding regions, their order is fixed, rather than randomly arranged along the gene. Second, each gene has precisely the same pattern and size of exons and introns in all tissues and cells of the organism, and, other than genes of the immune response and the major histocompatibility complex, no cell-specific rearrangements are known. Third, many introns have nonsense codons in all three reading frames, so nuclear introns are nontranslatable. Introns are found in the genes of mitochondria and chloroplasts as well as in nuclear genes. Although introns have been observed in archaebacteria and even bacteriophage T4, none are known in the genomes of eubacteria.

Post-Transcriptional Processing of Messenger RNA Precursors

Capping and Methylation of Eukaryotic mRNAs

The protein-coding genes of eukaryotes are transcribed by RNA polymerase II to form primary transcripts or **pre-mRNAs** that serve as precursors to mRNA. As a population, these RNA molecules are very large and their nucleotide sequences are very heterogeneous since they represent the transcripts of many different genes, hence the designation **heterogeneous nuclear RNA,** or **hnRNA.** Shortly after transcription of hnRNA is initiated, the 5′-end of the growing transcript is capped by addition of a guanylyl residue. This reaction is catalyzed by the nuclear enzyme **guanylyl transferase** using GTP as substrate (Figure 30.47). The **cap structure** is methylated at the 7-position of the G residue. Additional methylations may occur at the 2′-O positions of the two nucleosides following the 7-methyl-G cap and at the 6-amino group of a first base adenine (Figure 30.48).

3′-Polyadenylation of Eukaryotic mRNAs

Transcription by RNA polymerase II typically continues past the 3′-end of the mature messenger RNA. Little is known regarding the signals that regulate transcription termination in eukaryotes, and primary transcripts show heterogeneity in sequence at their 3′-ends, indicating that the precise point where termination occurs is nonspecific. However, termination does not normally occur until RNA polymerase II has transcribed past a consensus AAUAAA sequence known as the **poly(A)$^+$ addition site.**

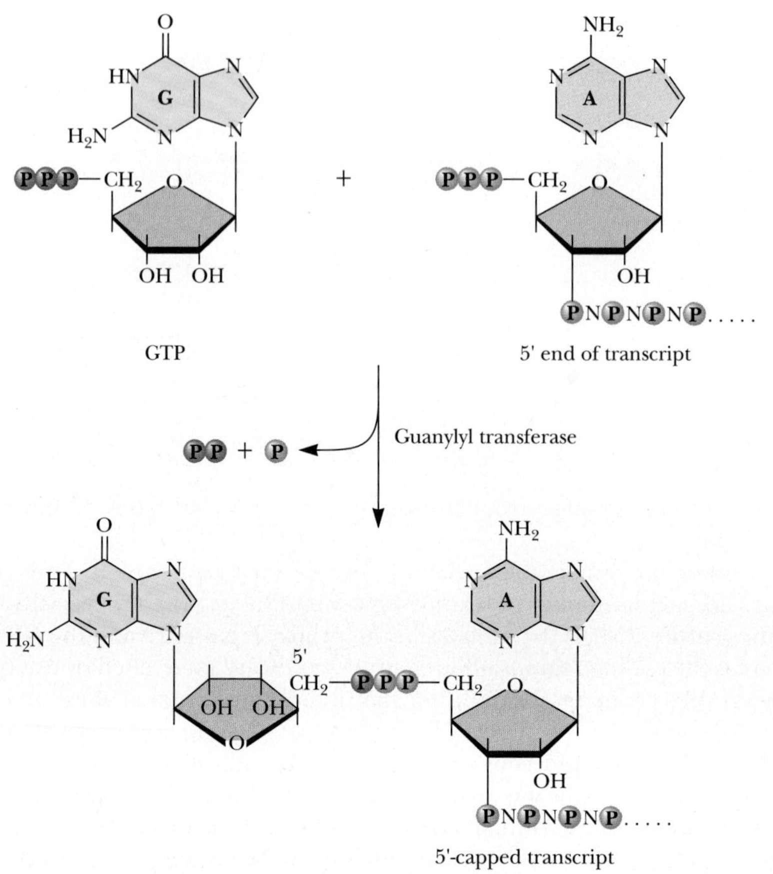

GTP

5' end of transcript

Guanylyl transferase

5'-capped transcript

***Figure* 30.47** The capping of eukaryotic pre-mRNAs. Guanylyl transferase catalyzes the addition of a guanylyl residue (*Gp*) derived from GTP to the 5′-end of the growing transcript, which has a 5′-triphosphate group already there. In the process, pyrophosphate (*pp*) is liberated from GTP and the terminal phosphate (p) is removed from the transcript.

$$Gppp + pppApNpNpNp \ldots \longrightarrow GppppApNpNpNp \ldots + pp + p$$

(A is often the initial nucleotide in the primary transcript.)

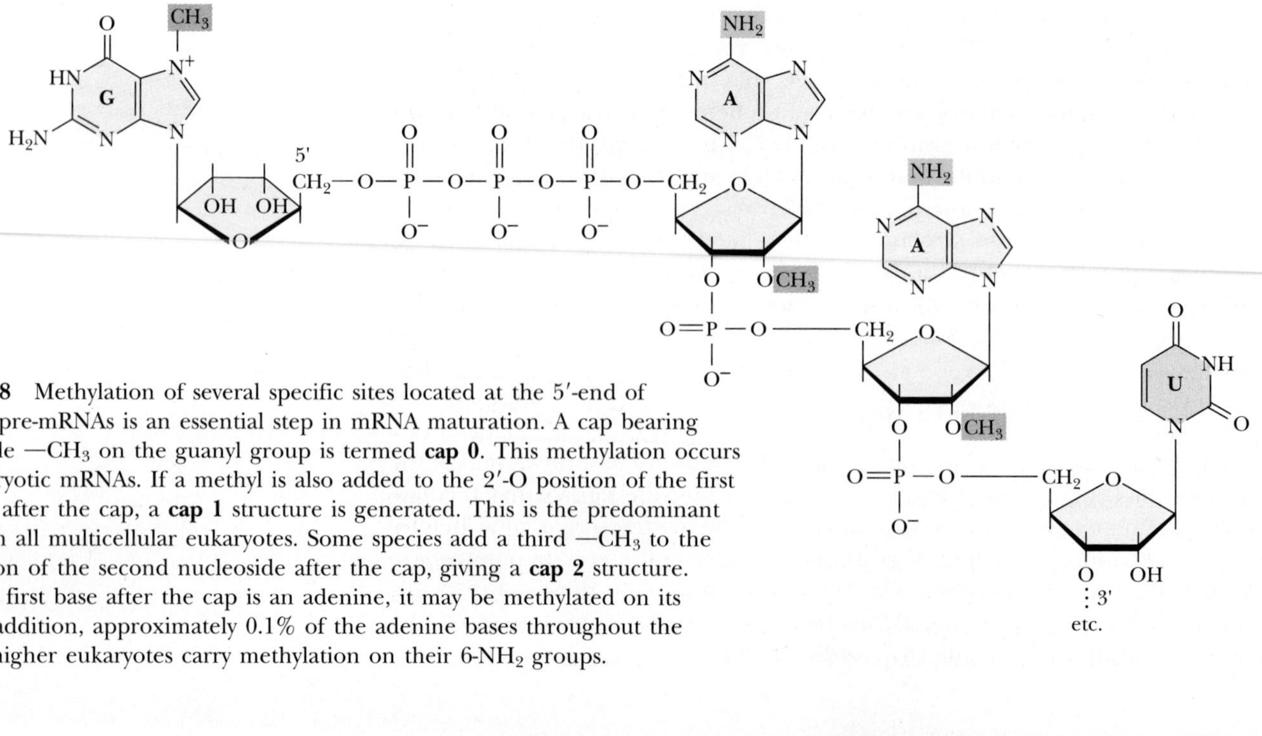

***Figure* 30.48** Methylation of several specific sites located at the 5′-end of eukaryotic pre-mRNAs is an essential step in mRNA maturation. A cap bearing only a single —CH_3 on the guanyl group is termed **cap 0**. This methylation occurs in all eukaryotic mRNAs. If a methyl is also added to the 2′-O position of the first nucleoside after the cap, a **cap 1** structure is generated. This is the predominant cap form in all multicellular eukaryotes. Some species add a third —CH_3 to the 2′-O position of the second nucleoside after the cap, giving a **cap 2** structure. Also, if the first base after the cap is an adenine, it may be methylated on its 6-NH_2. In addition, approximately 0.1% of the adenine bases throughout the mRNA of higher eukaryotes carry methylation on their 6-NH_2 groups.

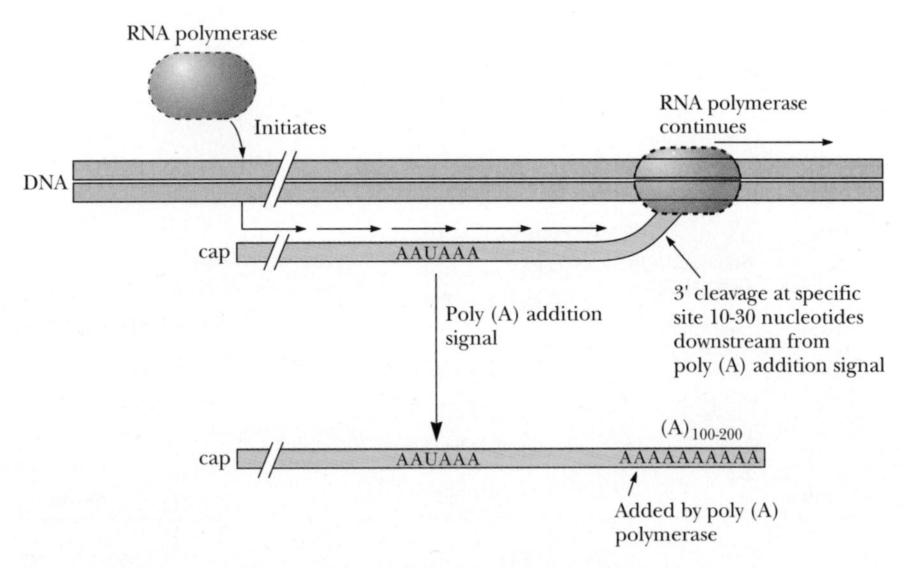

***Figure* 30.49** Poly(A) addition to the 3′-ends of transcripts occurs 10 to 30 nucleotides downstream from a consensus AAUAAA sequence, defined as the poly(A) addition site. A new 3′-OH end is generated by endonucleolytic cleavage of the nascent transcript, and poly(A) polymerase successively adds up to 200 or so adenylyl residues to the 3′-end.

Most eukaryotic mRNAs have 100 to 200 adenine residues attached at their 3′-end, the **poly(A)$^+$ tail.** (Histone mRNAs are the only common mRNAs that lack poly(A)$^+$ tails.) These A residues are not encoded in the DNA but are added post-transcriptionally by the enzyme **poly(A) polymerase,** using ATP as substrate. The consensus AAUAAA is not itself the poly(A)$^+$ addition site; instead it defines the position where poly(A)$^+$ addition occurs (Figure 30.49). The consensus AAUAAA is found 10 to 30 nucleotides upstream from where the nascent primary transcript is cleaved by an endonuclease to generate a new 3′-OH end. This end is where the poly(A) tail is added. The processing events of mRNA capping and poly(A) addition take place before splicing of the primary transcript creates the mature mRNA.

Nuclear Pre-mRNA Splicing

Within the nucleus, hnRNA forms **ribonucleoprotein particles (RNPs)** through association with a characteristic set of nuclear proteins. These proteins interact with the nascent RNA chain as it is synthesized, maintaining the hnRNA in an untangled, accessible conformation. The substrate for splicing, that is, intron excision and exon ligation, is the capped, polyadenylated RNA in the form of an RNP complex. Splicing occurs exclusively in the nucleus. The mature mRNA that results is then exported to the cytoplasm to be translated. Splicing requires precise cleavage at the 5′- and 3′-ends of introns and the accurate joining of the two ends. Consensus sequences define the exon/intron junctions in eukaryotic mRNA precursors, as indicated from an analysis of the splice sites in vertebrate genes (Table 30.8). Note that the sequences GU and AG are found at the 5′- and 3′-ends, respectively, of introns in pre-mRNAs from higher eukaryotes. In addition to the splice junctions, a conserved sequence within the intron, the **branch site,** is also essential to pre-mRNA splicing. The site lies 18 to 40 nucleotides upstream from the 3′-splice site and is represented in higher eukaryotes by the consensus $Y_{80}NY_{87}R_{75}A_{100}Y_{95}$, where Y is any pyrimidine, R is any purine, and N is any nucleotide. The subscripts give the percent frequency of occurrence of each particular base in the consensus sequence. The A is invariant.

Table **30.8**

Consensus Sequences at the Splice Sites in Vertebrate Genes*

5′-Splice Site

	Exon		Intron					
%G	13	77	100	0	32	12	84	18
%A	62	8	0	0	60	74	9	15
%U	13	8	0	100	5	7	3	50
%C	12	8	0	0	3	7	4	17
Consensus	A	G	G	U	A	A	G	U

3′-Splice Site

	Intron												Exon	
%G	10	7	7	10	9	5	5	5	24	0	0	100	55	27
%A	7	4	9	8	10	8	4	9	26	2	100	0	20	21
%U	56	59	43	49	41	46	42	46	23	19	0	0	8	32
%C	27	30	42	33	40	40	49	41	27	78	0	0	17	20
Consensus	Py	Py	Py	Py	Py	Py	Py	Py	–	C	A	G	G	–

*Vertebrate genes have introns beginning with GU and ending with AG. (Rarely, introns begin with a GC dinucleotide; these rarities are excluded from the table.) Note the 5′ consensus AG:GU(A)AGU where the colon indicates the exon:intron junction/cleavage site. Note also the 3′ consensus (Py series) NCAG:G, the colon designating the 3′ intron:exon junction/cleavage site.

Source: Adapted from Padgett, R. A., et al., 1986. *Annual Review of Biochemistry* **55:**1119–1150.

The Splicing Reaction: Lariat Formation

A universal mechanism is proposed for the splicing of all nuclear mRNA precursors (Figure 30.50). A covalently closed loop of RNA, the **lariat,** is formed by attachment of the 5′-phosphate group of the intron's invariant 5′-G to the 2′-OH at the invariant branch site A to form a 2′-5′ phosphodiester bond. Note that lariat formation creates an unusual branched nucleic acid. The lariat structure is excised when the 3′-OH of the consensus G at the 3′-end of the 5′ exon (Exon 1, Figure 30.50) covalently joins with the 5′-phosphate at the 5′-end of the 3′ exon (Exon 2). The reactions that occur are transesterification reactions where an OH group reacts with a phosphodiester bond, displacing an —OH to form a new phosphodiester link (Figure 30.51). Since the reactions lead to no net change in the number of phosphodiester linkages, no energy input, for example, as ATP, is needed. The lariat product is unstable; the 2′-5′ phosphodiester branch is quickly cleaved to give a linear excised intron that is rapidly degraded in the nucleus.

Splicing Depends on snRNPs

The hnRNA (pre-mRNA) substrate is not the only RNP complex involved in the splicing process. Splicing also depends on a unique set of small nuclear ribonucleoprotein particles, so-called **snRNPs** (pronounced "snurps"). In higher eukaryotes, each snRNP consists of a small RNA molecule 100 to 200 nucleotides long and a set of about 10 different proteins. (The snRNAs in yeast are much bigger, more than 1000 nucleotides in length.) Some of the different proteins form a "core" set common to all snRNPs, while others are unique to a specific snRNP. The major snRNP species are very abundant, present at greater than 100,000 copies per nucleus. The RNAs of snRNPs are

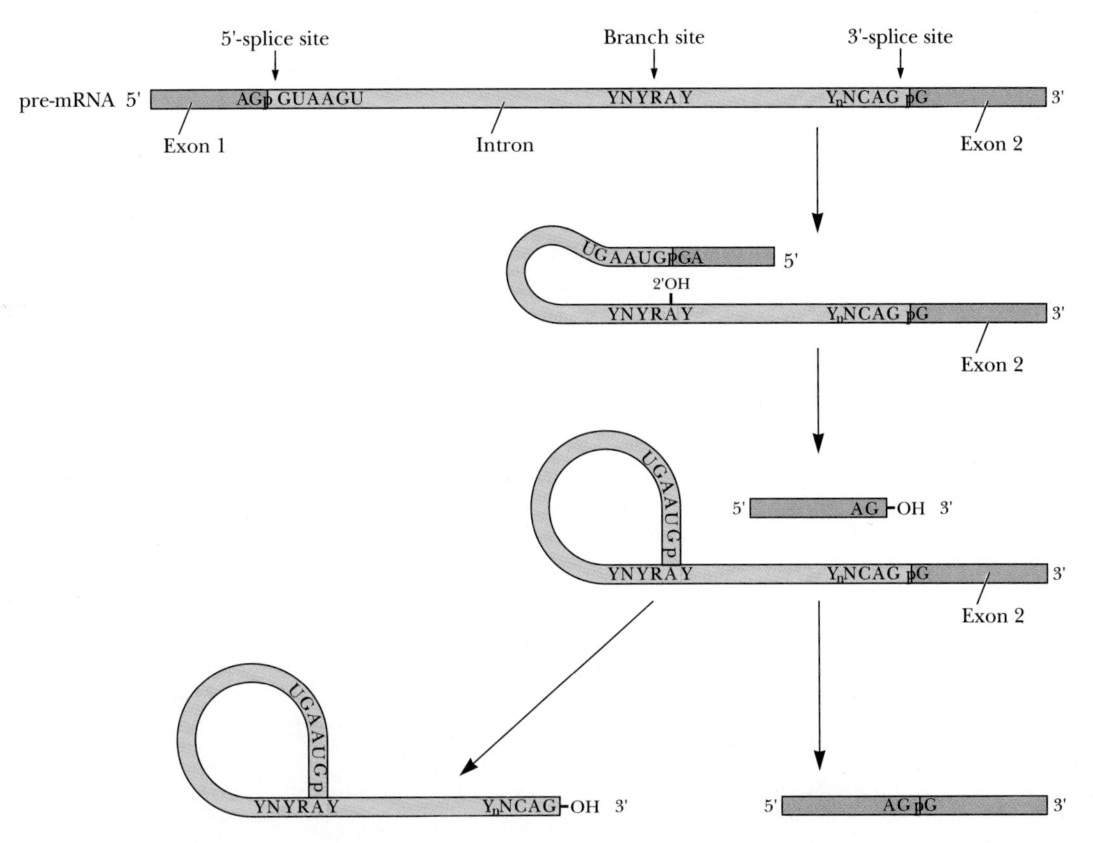

***Figure* 30.50** Splicing of mRNA precursors. A representative precursor mRNA is depicted. Exon 1 and Exon 2 indicate two exons separated by an intervening sequence (or intron) with consensus 5′, 3′, and branch sites. The fate of the phosphates at the 5′ and 3′ splice sites can be followed by tracing the fate of the respective *p*s. The products of the splicing reaction, the lariat form of the excised intron and the united exons, are shown at the bottom of the figure. The lariat intermediate is generated when the invariant G at the 5′-end of the intron attaches via its 5′-phosphate to the 2′-OH of the invariant A within the branch site. The consensus guanosine residue at the 3′ end of Exon 1 (the 5′-splice site) then reacts with the 5′-phosphate at the 3′-splice site (the 5′-end of Exon 2), ligating the two exons and releasing the lariat structure. Although the reaction is shown here in a stepwise fashion, 5′ cleavage, lariat formation, and exon ligation/lariat excision are believed to occur in a concerted fashion.
(Adapted from Sharp, P. A., 1987. Science 235:766, Figure 1.)

typically rich in uridine, hence the classification of particular snRNPs as U1, U2, and so on. snRNP RNAs share several distinctive properties: a trimethylguanosine cap on their 5′-end and an internal AUUUUUG consensus sequence. The properties of some of the most abundant snRNPs are given in Table 30.9. U1 snRNA can be folded into a secondary structure that leaves the 11 nucleotides at its 5′-end single-stranded. The 5′-end of U1 snRNA is complementary to the consensus sequence at the 5′-splice junction of the pre-mRNA (Figure 30.52). U2 snRNA is complementary to the consensus branch-site sequence.

***Table* 30.9**
The Properties of Higher Eukaryotic snRNPs

snRNP	Length (nt)	Splicing Target
U1	165	5′ splice
U2	189	Branch
U4	145	?
U5	115	3′ splice
U6	106	?

snRNPs Form the Spliceosome

Splicing occurs when the various snRNPs come together with the pre-mRNA to form a multicomponent complex called the **spliceosome.** The spliceosome is a large complex, roughly equivalent to a ribosome in size, and its assembly

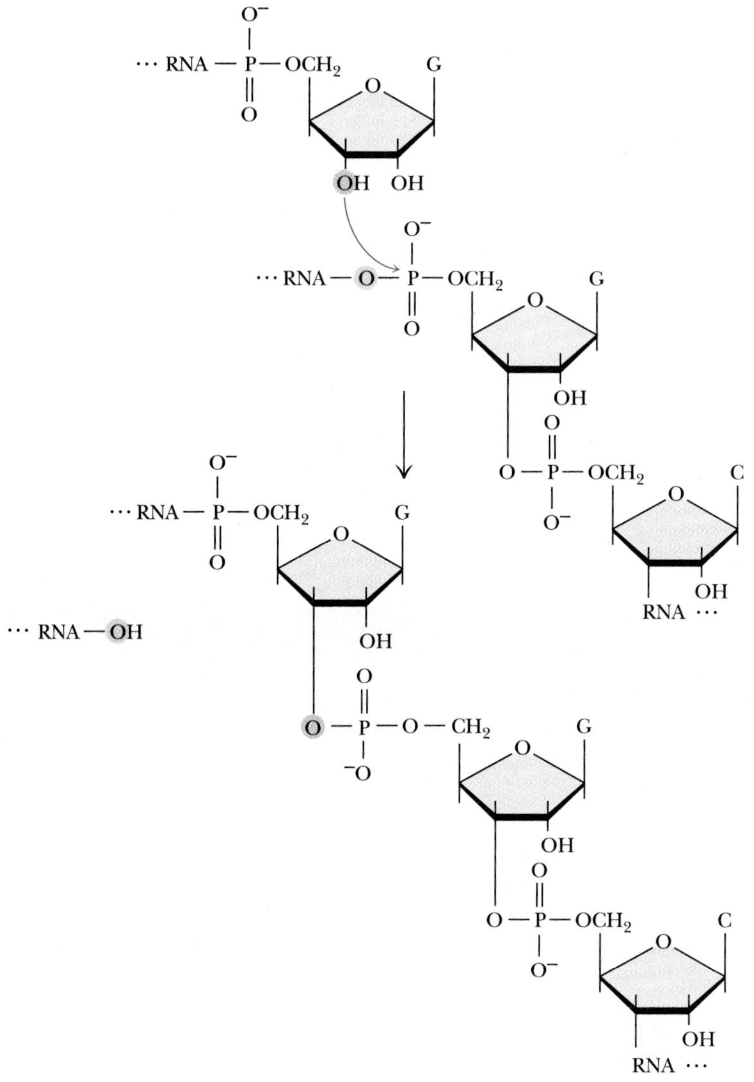

***Figure* 30.51** Transesterification. An incoming —OH group reacts with an existing phosphodiester bridge, displacing a hydroxyl and forming a new phosphodiester bond.

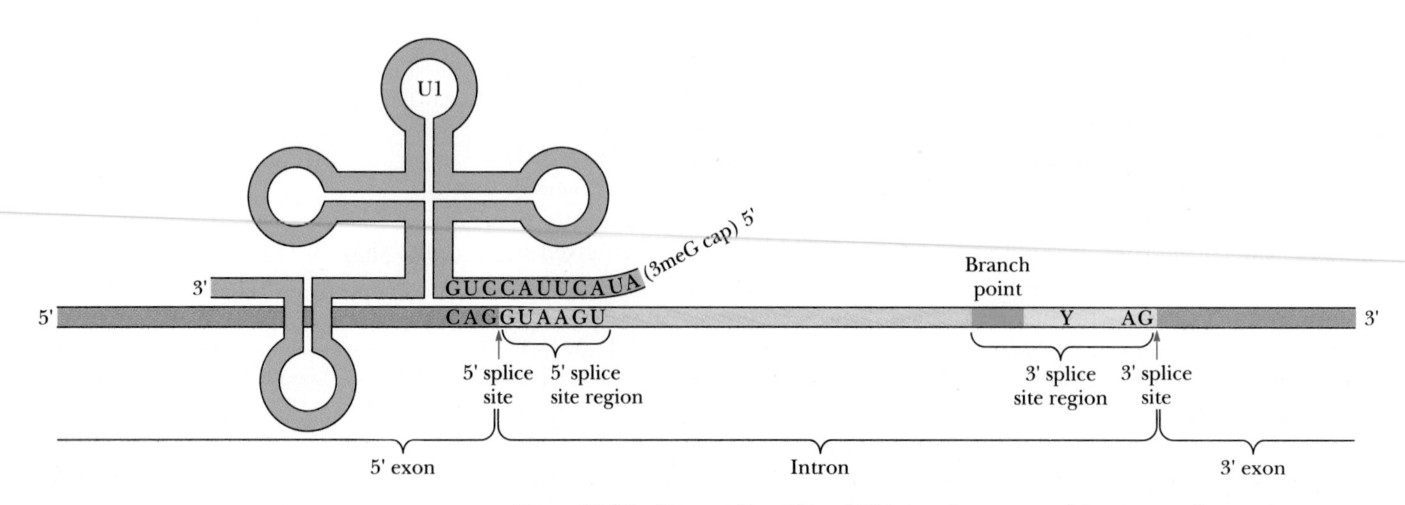

***Figure* 30.52** Mammalian U1 snRNA can be arranged in a secondary structure where its 5′-end is single-stranded and can base-pair with the consensus 5′-splice site of the intron.

(Adapted from Rosbash, M., and Seraphin, B., 1991. Trends in Biochemical Sciences 16:187, Figure 1.)

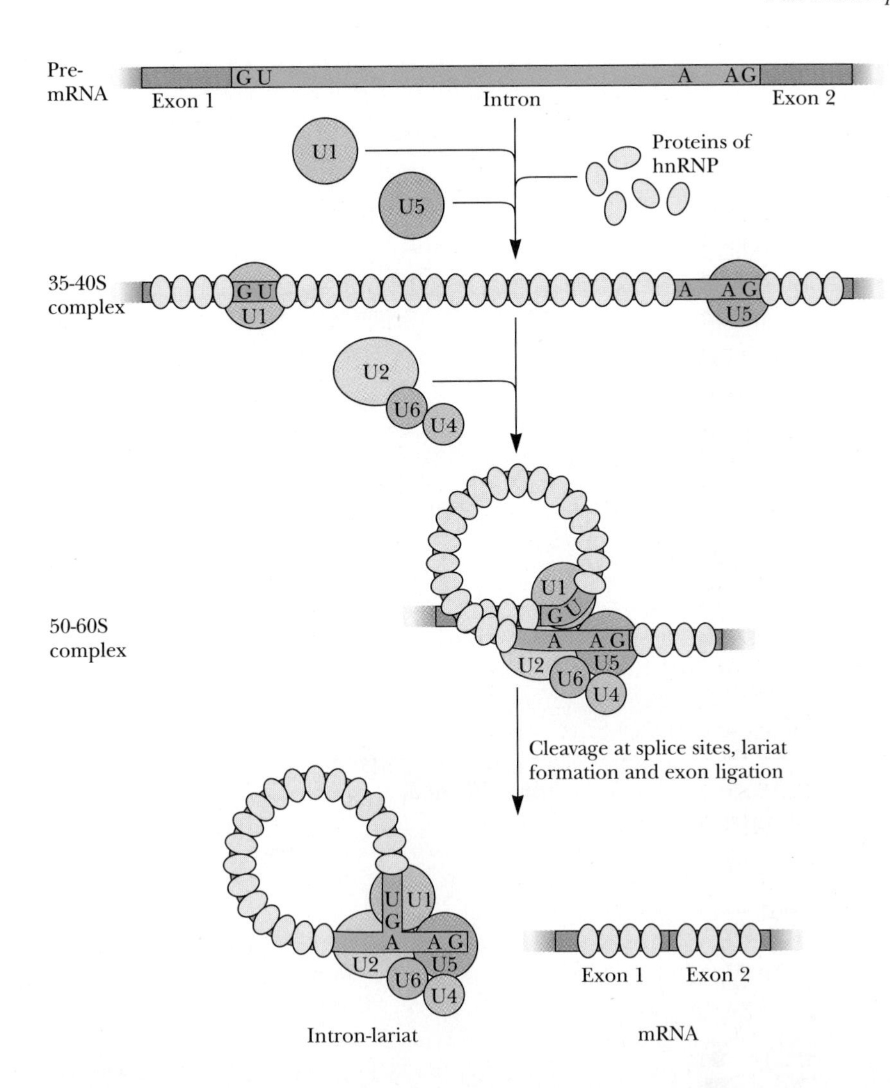

***Figure* 30.53** A model for spliceosome assembly. U1 and U5 snRNPs bind at the 5′- and 3′-splice sites, respectively, followed by U2 snRNP binding at the branch site. Proteins characteristic of hnRNP are also present. U2 snRNP binding is dependent on the presence of U1 snRNP. The 5′- and 3′-splice sites are aligned via collaboration between U5 snRNP and U6 snRNP, and U2 snRNP delivers the branch site to the 5′-splice site. U6 is believed to be the catalytic core of the spliceosome. The complete spliceosome, consisting of snRNPs U1, U2, U4, U5, and U6 bound to a pre-mRNA within a hnRNP complex, can be isolated as a 50-60S complex.

(Adapted from Maniatis, T., and Reed, R., 1987. Nature 325:673, Figure 2, and Wise, J. A., 1994. Science 262:1978–1979.)

requires ATP. A model for spliceosome assembly has been proposed (Figure 30.53), but details remain to be resolved. U1 and U5 snRNPs bind at the 5′- and 3′-splice sites, while U2 snRNP binds at the branch site. The U4 and U6 snRNP-binding sites have not been identified yet. Interaction between the snRNPs in assembly of the spliceosome brings the 5′- and 3′-splice sites into proximity so that lariat formation and exon ligation can take place. The spliceosome is thus a dynamic structure that uses the pre-mRNA as a template for assembly, carries out its transesterification reactions, and then disassembles when the splicing reaction is over. Thomas Cech's discovery of catalytic RNA (see ribozymes, Chapter 12) raises the possibility that the transesterification reactions are catalyzed not by snRNP proteins but by the snRNAs themselves.

Alternative RNA Splicing

In one mode of splicing, every intron is removed and every exon is incorporated into the mature RNA without exception. This type of splicing, termed **constitutive splicing,** results in a single form of mature mRNA from the primary transcript. However, many eukaryotic genes can give rise to multiple forms of mature RNA transcripts. The mechanisms for production of multiple transcripts from a single gene include utilization of different promoters, selection of different polyadenylation sites, **alternative splicing** of the primary transcript, or even a combination of the three.

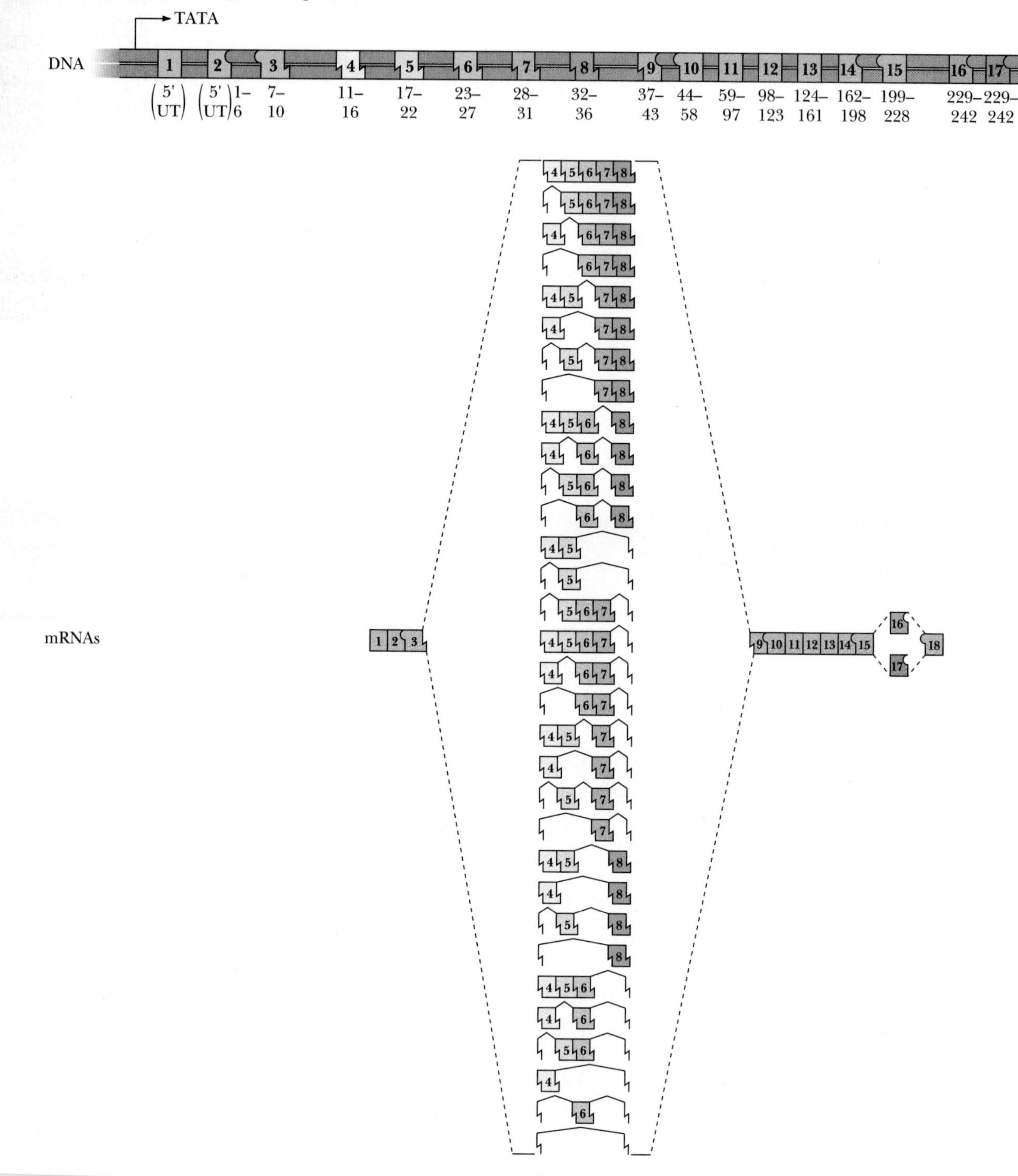

Figure **30.54** Organization of the fast skeletal muscle troponin T gene and the 64 possible mRNAs that can be generated from it. Exons are *constitutive* (yellow), *combinatorial* (green), or *mutually exclusive* (blue or orange). Exon 1 is composed of 5′-untranslated (UT) sequences and exon 18 includes the polyadenylation site (AATAAA) and 3′-UT sequences. The TATA box indicates the transcription start site. The amino acid residues encoded by each exon are indicated below. Many exon:intron junctions fall between codons. The "sawtooth" exon boundaries indicate that the splice site falls between the first and second nucleotides of a codon, the "concave/convex" exon boundaries indicate that the splice site falls between the second and third nucleotides of a codon, and flush boundaries between codons signify that the splice site falls between intact codons. Each mRNA includes all constitutive exons, one of the 32 possible combinations of exons 4 to 8, and either exon 16 or 17.

(Adapted from Breitbart, R. E., Andreadis, A., and Nadal-Ginard, B., 1987. Annual Review of Biochemistry 56:467, Figure 2.)

Different transcripts from a single gene make possible a set of related polypeptides, or **protein isoforms,** each with slightly altered functional capability. Such variation serves as a useful mechanism for increasing the apparent coding capacity of the genome. Further, alternative splicing offers another level at which regulation of gene expression can operate. For example, mRNAs unique to particular cells, tissues, or developmental stages could be formed from a single gene by choosing different 5′- or 3′-splice sites or by omitting entire exons. Translation of these mature mRNAs produces cell-specific protein isoforms that display properties tailored to the needs of the particular cell. Such regulated expression of distinct protein isoforms is a fundamental characteristic of eukaryotic cell differentiation and development. The factors that determine splice-site selection in alternative splicing remain unknown but likely include regulatory proteins that recognize specific splice sites as well as alternative secondary structures that form within the primary transcript.

The Fast Skeletal Muscle Troponin T Gene—An Example of Alternative Splicing

In addition to many other instances, alternative splicing is a prevalent mechanism for generating protein isoforms from the genes encoding muscle proteins (Chapter 36), allowing distinctive isoforms aptly suited to the function of each muscle. An impressive manifestation of alternative splicing is seen in the expression possibilities for the rat **fast skeletal muscle troponin T gene** (Figure 30.54). This gene consists of 18 exons, 11 of which are found in all mature mRNAs (exons 1 through 3, 9 through 15, and 18) and thus are **constitutive.** Five exons, those numbered 4 through 8, are **combinatorial** in that they may be individually included or excluded, in any combination, in the mature mRNA. Two exons, 16 and 17, are *mutually exclusive:* one or the other is always present, but never both. Sixty-four different mature mRNAs can be generated from the primary transcript of this gene by alternative splicing. Since each exon represents a *cassette* of genetic information encoding a subsegment of protein, alternative splicing is a versatile way to introduce functional variation within a common protein theme.

Problems

1. The 5′-end of an mRNA has the sequence

. . . AGAUCCGUAUGGCGAUCUCGACGAAGACUCCUAGGGAAUCC . . .

   What is the nucleotide sequence of the DNA "sense" strand from which it was transcribed? If this mRNA is translated beginning with the first AUG codon in its sequence, what is the N-terminal amino acid sequence of the protein it encodes? (See Table 31.3 for the genetic code.)

2. Describe the sequence of events involved in the initiation of transcription by *E. coli* RNA polymerase. Include in your description those features a gene must have for proper recognition and transcription by RNA polymerase.

3. RNA polymerase has two binding sites for ribonucleoside triphosphates, the *initiation site* and the *elongation site.* The initiation site has a greater K_m for NTPs than the elongation site. Suggest what possible significance this fact might have for the control of transcription in cells.

4. How does transcription in eukaryotes differ from transcription in prokaryotes?

5. Use the information in the box "A Deeper Look: Quantitative Evaluation of *lac* Repressor:DNA Interactions," to determine what will be the nonspecific association constant for the binding of *lac* repressor to DNA if the ratio of free to DNA-bound repressor is 0.10.

6. Is transcription attenuation likely to be an important mechanism of transcriptional regulation in eukaryotic cells? Explain your answer.

7. DNA-binding proteins may recognize specific DNA regions by either reading the base sequence or by "indirect readout." How do these two modes of protein:DNA recognition differ?

8. Why do you suppose that DNA looping in transcriptional regulation is a mechanism common to both prokaryotes and eukaryotes?

9. Describe why the ability of *bZIP* proteins to form heterodimers increases the repertoire of genes whose transcription might be responsive to regulation by these proteins.

10. Suppose exon 17 were deleted from the fast skeletal muscle troponin T gene (Figure 30.54). How many different mRNAs could now be generated by alternative splicing? Suppose that exon 7 in a wild-type troponin T gene were duplicated. How many different mRNAs might be generated from a transcript of this new gene by alternative splicing?

Further Reading

Berg, O. G., and von Hippel, P. H., 1988. Selection of DNA binding sites by regulatory proteins. *Trends in Biochemical Sciences* **13:**207–211. A discussion of the quantitative binding aspects of DNA:protein interactions.

Breitbart, R. E., Andreadis, A., and Nadal-Ginard, B., 1987. Alternative splicing: A ubiquitous mechanism for the generation of multiple protein isoforms from single genes. *Annual Review of Biochemistry* **56:**467–495.

Conaway, R. C., and Conaway, J. W., 1993. General initiation factors for RNA polymerase II. *Annual Review of Biochemistry* **62:**161–190.

Corden, J. L., 1990. Tails of RNA polymerase II. *Trends in Biochemical Sciences* **15:**383–387. Possible roles for the repetitive YSPTSPS tail domain of RNA polymerase II.

Das, A., 1993. Control of transcription termination by DNA-binding proteins. *Annual Review of Biochemistry* **62:**893–930.

de Crombrugghe, B., Busby, S., and Buc, H., 1985. Cyclic AMP receptor protein: Role in transcription activation, *Science* **224:**831–838. The role of CAP in activating expression of prokaryotic genes.

Dreyfuss, G., et al., 1993. hnRNP proteins and the biogenesis of mRNA. *Annual Review of Biochemistry* **62:**289–321.

Edelman, G. M., and Jones, F. S., 1993. Outside and downstream of the homeobox. *Journal of Biological Chemistry* **268**:20683–20686.

Edmondson, D. G., and Olson, E. N., 1993. Helix-loop-helix proteins as regulators of muscle-specific transcription. *Journal of Biological Chemistry* **268**:755–758.

Evans, R. M., and Hollenberg, S. M., 1988. Zinc fingers: Gilt by association. *Cell* **52:**1–3.

Harrison, S. C., 1991. A structural taxonomy of DNA-binding proteins. *Nature* **353:**715–719. The various structural motifs found in DNA-binding proteins are described.

Harrison, S. C., and Aggarwal, A. K., 1990. DNA recognition by proteins with the helix-turn-helix motif. *Annual Review of Biochemistry* **59:**933–969.

Jacob, F., and Monod, J., 1961. Genetic regulatory mechanisms in the synthesis of proteins. *The Journal of Molecular Biology* **3:**318–356. The classic paper presenting the operon hypothesis.

Johnson, P. F., and McKnight, S. L., 1989. Eukaryotic transcriptional regulatory proteins. *Annual Review of Biochemistry* **58:**799–839. A review of the structure and function of eukaryotic DNA-binding proteins that activate transcription.

Kim, J. L., Nikolov, D. B., and Burley, S. K., 1993. Co-crystal structure of TBP recognizing the groove of a TATA element. *Nature* **365:**520–527.

Kim, Y., Geiger, J. H., Hahn, S., and Sigler, P. B., 1993. Crystal structure of a yeast TBP/TATA-box complex. *Nature* 365:512–519.

Klevit, R. E., 1991. Recognition of DNA by Cys_2,His_2 zinc fingers. *Science* **253:**1367, 1393.

Kolb, A., et al., 1993. Transcriptional regulation by cAMP and its receptor protein. *Annual Review of Biochemistry* **62:**749–795.

Kornberg, T. B., 1993. Understanding the homeodomain. *Journal of Biological Chemistry* **268**:26813–26816.

Landschulz, W. H., Johnson, P. F., and McKnight, S. L., 1988. The leucine zipper: A hypothetical structure common to a new class of DNA-binding proteins. *Science* **240:**1759–1764. The initial structural description of the leucine zipper dimerization motif found in certain eukaryotic DNA-binding proteins.

Leff, S. E., Rosenfeld, M. G., and Evans, R. M., 1986. Complex transcriptional units: Diversity in gene expression by alternative RNA processing. *Annual Review of Biochemistry* **55:**1091–1117.

Levine, M., and Hoey, T., 1988. Homeobox proteins as sequence-specific transcription factors. *Cell* **55:**537–540.

Lin, S.-Y., and Riggs, A. D., 1975. The general affinity of *lac* repressor for *E. coli* DNA: Implications for gene regulation in procaryotes and eucaryotes. *Cell* **4:**107–111. A quantitative analysis of the binding of *lac* repressor to DNA.

Lobell, R. B., and Schleif, R. F., 1990. DNA looping and unlooping by AraC protein. *Science* **250:**528–532. Regulation of *araBAD* expression involves DNA looping mediated by AraC protein.

Lucas, P. C., and Granner, D. K., 1992. Hormone response domains in gene transcription. *Annual Review of Biochemistry* **61:**1131–1173.

Maniatis, T., and Reed, R., 1987., The role of small nuclear ribonucleoprotein particles in pre-mRNA splicing. *Nature* **325:**673–678.

Matthews, K. S., 1992. DNA looping. *Microbiological Reviews* **56:**123–136. A review of DNA looping as a general mechanism in the regulation of gene expression.

Mermelstein, F. H., Flores, O., and Rheinberg, D., 1989. Initiation of transcription by RNA polymerase II. *Biochimica et Biophysica Acta* **1009:**1–10. The interplay of RNA polymerase and transcription factors (TFs) in the initiation of eukaryotic transcription.

Mitchell, P. J., and Tjian, R., 1989. Transcriptional regulation in mammalian cells by sequence-specific DNA binding proteins. *Science* **245:**371–378.

Neidhardt, F. C., et al., eds., 1987. *Escherichia coli* and *Salmonella typhimurium: Cellular and Molecular Biology,* vol. 2, section C. Washington, D.C.: American Society for Microbiology. Compendium of the molecular biology and biochemistry of the two enteric bacteria most commonly used in research.

Pabo, C. O., and Sauer, R. T., 1992. Transcription factors: Structural families and principles of DNA recognition. *Annual Review of Biochemistry* **61:**1053–1095.

Padgett, R. A., et al., 1986. Splicing of messenger RNA precursors. *Annual Review of Biochemistry* **55:**1119–1150.

Palmer, J. M., and Folk, W. F., 1990. Unraveling the complexities of transcription by RNA polymerase III. *Trends in Biochemical Sciences* **15:**300–304.

Pavletich, N., and Pabo, C. O., 1991. Zinc finger-DNA recognition: Crystal structure of a Zif268-DNA complex at 2.1 Å. *Science* **252:**809–817.

Ptashne, M., 1988. How eukaryotic transcriptional activators work. *Nature* **335:**683–689.

Rosbash, M., and Seraphin, B., 1991. Who's on first? The U1 snRNP-5′ splice site interaction and splicing. *Trends in Biochemical Sciences* **16:**187–190.

Sachs, A., and Wahle, E., 1993. Poly (A) tail metabolism and function in eucaryotes. *Journal of Biological Chemistry* **268:**22955–22958.

Saltzman, A. G., and Weinmann, R., 1989. Promoter specificity and modulation of RNA polymerase II transcription. *FASEB Journal* **3:**1723–1733.

Sawadogo, M., and Sentenac, A., 1990. RNA polymerase B (II) and general transcription factors. *Annual Review of Biochemistry* **59:**711–754.

Schleif, R., 1992. DNA looping. *Annual Review of Biochemistry* **61:**199–223. An excellent review of DNA looping as a regulatory mechanism in gene expression.

Schultz, S. C., Shields, G. C., and Steitz, T. A., 1991. Crystal structure of CAP-DNA complex: The DNA is bent by 90°. *Science* **253:**1001–1007.

Schwabe, J. W. R., and Rhodes, D., 1991. Beyond zinc fingers: Steroid receptors have a novel structural motif for DNA recognition. *Trends in Biochemical Sciences* **16:**291–296.

Shadel, G. S., and Clayton, D. A., 1993. Mitochondrial transcription initiation. Variation and conservation. *Journal of Biological Chemistry* **268:**16083–16086.

Sharp, P. A., 1987. Splicing of messenger RNA precursors. *Science* **235:**766–771.

Steitz, T. A., 1990. Structural studies of protein-nucleic acid interaction: The sources of sequence-specific binding. *Quarterly Review of Biophysics* **23:**205–280. A comprehensive account of the structural properties of DNA-binding proteins and their interaction with DNA.

Struhl, K., 1989. Molecular mechanisms of transcriptional regulation in yeast. *Annual Review of Biochemistry* **58:**1051–1077.

Touchette, N., 1993. Molecular saddle offers new view of transcription. *The Journal of NIH Research* **5:**54–58.

Vinson, C. R., Sigler, P. B., and McKnight, S. L., 1989. Scissors-grip model for DNA recognition by a family of leucine zipper proteins. *Science* **246:**911–916. A model for the interaction of a *bZIP* protein with its DNA site.

Weiss, L., and Rheinberg, D., 1992. Transcription by RNA polymerase II: Initiator-directed formation of transcription-competent complexes. *FASEB Journal* **6:**3300–3309.

Wiener, A. M., 1993. mRNA splicing and autocatalytic introns: Distant cousins or the products of chemical determinism? *Cell* **72:**161–164. This article considers the possibility that small nuclear RNAs (such as U2 and U6) are the catalysts mediating mRNA splicing.

Woychik, N. A., and Young, R. A., 1990. RNA polymerase II: Subunit structure and function. *Trends in Biochemical Sciences* **15:**347–351.

Yanofsky, C., 1988. Transcription attenuation. *Journal of Biological Chemistry* **263:**609–612. A mini-review of the mechanism of transcription attenuation.

Appendix to Chapter 30

DNA:Protein Interactions

Peter von Hippel has provided the following conceptual framework, which interprets the specificity of DNA:protein interactions in terms of (a) the relative affinity of a particular protein for unique versus random DNA sequences, and (b) the relative number of unique versus random (nonspecific) DNA sites.

Consider a DNA-binding protein (such as a repressor). The total amount of this protein will be distributed between free (unbound) protein, protein bound to unique target sites, and protein nonspecifically bound to DNA:

$$[\text{total protein}] = [\text{free protein}] + [\text{protein:site}] + [\text{protein:DNA}]$$

The total number of unique target sites on the DNA is distributed between those that are occupied by protein and those that are not:

$$[\text{total site}] = [\text{free site}] + [\text{protein:site}]$$

The specific binding of a protein to its unique target site is described by the equilibrium

$$K_{sp} = \frac{[\text{protein:site}]}{[\text{free protein}][\text{free site}]}$$

The competing nonspecific binding of protein to random DNA sequences is described by a similar equation:

$$K_{ran} = \frac{[\text{protein:DNA}]}{[\text{free protein}][\text{DNA}]}$$

The ratio of K_{sp} to K_{ran} is an index of the specificity of the protein for its target site versus all other DNA sequences:

$$\text{Specificity} = \frac{K_{sp}}{K_{ran}} = \frac{[\text{protein:site}]}{[\text{free site}]} \times \frac{[\text{DNA}]}{[\text{protein:DNA}]}$$

Substituting [total protein] − [protein:site] − [free protein] in the denominator for [protein:DNA] gives

$$\text{Specificity} = \frac{[\text{protein:site}]}{[\text{free site}]} \times \frac{[\text{DNA}]}{[\text{total protein}] - [\text{protein:site}] - [\text{free protein}]}$$

Simplifying Assumptions

Assuming the affinity of the protein for random DNA sequences is sufficiently high so that all protein is bound by DNA (as was the case with *lac* repressor), then [free protein] is so small as to be negligible relative to [total protein]. [Protein:site] is also small relative to [total protein] and may be replaced by [total site] under conditions of saturation. Then we have the following:

$$\text{Specificity} = \frac{[\text{protein:site}]}{[\text{free site}]} \times \frac{[\text{DNA}]}{[\text{total protein}] - [\text{total site}]}$$

The distribution of free versus bound sites is as follows:

$$\frac{[\text{free site}]}{[\text{protein:site}]} = \frac{[\text{DNA}]}{\text{specificity}([\text{total protein}] - [\text{total site}])}$$

Taking as our units for calculation purposes the number of sites per cell, [DNA] = 4.72×10^6 (the size of the *E. coli* genome), specificity = 10^7, total molecules of protein = 10, and total number of unique target sites = 1 gives a [free site]/[protein:site] ratio of $\frac{(4.72 \times 10^6)}{10^7(10 - 1)} = 0.524 \times 10^{-1} = 0.05$. So, for every 5 free sites, there are 100 sites occupied by protein. In other words, 100 out of every 105 times, the single site is occupied by protein. If the site is an operator and the protein is a repressor, then the operon is repressed 95% of the time.

Note that only 10 molecules of repressor are needed to effectively saturate the single operator target among 4.72×10^6 competing nonspecific DNA sequences, *provided* specificity = 10^7.

Induction

Now suppose that in the presence of **inducer,** the specificity of the protein for its unique target drops to 10^4; [total protein] is still 10 molecules. The [protein:site] value, which we replaced with [total site] in the above considerations, is likely to be less than 1, and cannot exceed 1 if the target site is truly unique in the genome. Let us take 1 as a "worst case" scenario for induction. (1 really means *no induction* because repressor occupies the target site.) Using these values, $\frac{[\text{free site}]}{[\text{protein site}]} = \frac{(4.72 \times 10^6)}{10^4(10 - 1)} = 0.524 \times 10^2$. The ratio of [free sites] to [protein sites] is 52.4. That is, 98% of the sites (52.4 out of 53.4) are not occupied by repressor and the operon is transcribed.

Induction causes the release of repressor from its target sequence (that is, the operator). However, the repressor is *not* free in solution; virtually all of it is bound randomly at nonspecific DNA sequences. In effect, repressor is stored on DNA and can rapidly "diffuse" along the one-dimensional DNA strand back to its target site.

Parameters Influencing the Specificity of DNA:Protein Interactions

The parameters that determine whether a regulatory protein saturates its unique target sequence are revealed in the binding equilibria presented previously and the influences listed below.

1. *The size of the genome.* DNA-binding proteins bind with low affinity to random DNA sequences and with high affinity to their unique target sites. As genome size increases, the ability of a protein to bind uniquely to its target site is diluted by the increasing number of nonspecific binding sites.
2. *The specificity of the regulatory protein* offsets a vast preponderance of nonspecific sites in the DNA. The much greater affinity of the protein for its particular target ensures that the target sequence will be located and occupied, despite the overwhelming number of competing nonspecific sequences. The absolute values of the association constants of the protein for specific versus nonspecific DNA sequences is *not* important. It is the ratio, *the specificity,* of these values that is important.
3. *The amount of regulatory protein* required for effective gene regulation increases with increasing genome size and decreases with increasing specificity.
4. *The number of molecules of regulatory protein must be in moderate excess over the number of target sites in the DNA.* Although regulatory proteins are typically in low abundance in cells, their numbers must be sufficient to ensure saturation of their unique target(s). (Note that we assumed 10 molecules of regulatory protein per cell and one unique target per genome for this protein.) If the DNA-binding protein associates with multiple regulatory sites in the genome, it must occur in greater quantity compared to regulators with only one target sequence.

Chapter 31

The Genetic Code

"It is impossible to dissociate language from science or science from language, because every natural science always involves three things: the sequence of phenomena on which the science is based; the abstract concepts which call these phenomena to mind; and the words in which the concepts are expressed. To call forth a concept, a word is needed; to portray a phenomenon, a concept is needed. All three mirror one and the same reality."

Antoine-Laurent Lavoisier 1743–1794
Traité Elémentaire de Chimie (1789)

The Maya encoded their history in hieroglyphs carved on stelae and temples like these in Guatemala.

Outline

31.1 The Collinearity of Gene Structure and Protein Structure

31.2 Elucidating the Genetic Code

31.3 The Nature of the Genetic Code

31.4 Amino Acid Activation for Protein Synthesis: Aminoacyl-tRNA Synthetases

31.5 Codon–Anticodon Pairing, Third-Base Degeneracy, and the Wobble Hypothesis

31.6 Codon Usage

31.7 Nonsense Suppression

We turn now to the problem of how the sequence of nucleotides in an mRNA molecule is translated into the specific amino acid sequence of a protein. The problem raises both informational and mechanical questions. First, what is the genetic code that allows the information specified in a sequence of bases to be translated into the amino acid sequence of a polypeptide? That is, how is the 4-letter language of nucleic acids translated into the 20-letter language of proteins? Implicit in this question is a mechanistic dilemma for structural biologists: It is easy to see how base pairing established a one-to-one correspondence that allowed the template-directed synthesis of polynucleotide chains in the processes of replication and transcription. However, there is no obvious chemical affinity between the purine and pyrimidine bases and the 20 different amino acids. Nor is there any structural complementarity or stereochemical connection between polynucleotides and amino acids that might guide the translation of information.

Francis Crick reasoned that **adapter molecules** must bridge this informational gap. These adapter molecules must interact specifically with both nucleic acids (mRNAs) and amino acids. At least 20 different adapter molecules would be needed, at least one for each amino acid. The various adapter molecules would be able to read the genetic code in an mRNA template and align the amino acids according to the template's directions so that they could be

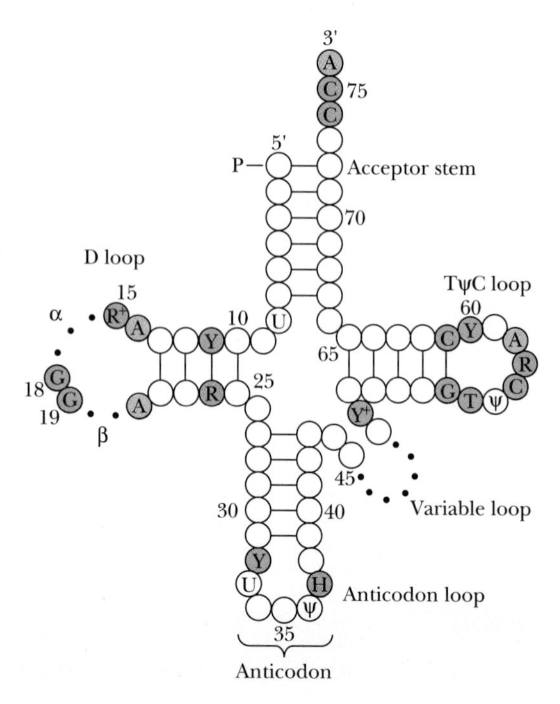

***Figure* 31.1** The general structure of tRNA molecules. Circles represent nucleotides in the tRNA sequence. The numbers given indicate the standardized numbering system for tRNAs (which differ in total number of nucleotides). Dots indicate places where the number of nucleotides may vary in different tRNA species.

polymerized into a unique polypeptide. Transfer RNAs (tRNAs; Figure 31.1) are the adapter molecules (Chapters 6 and 7). Amino acids are attached to the 3′-OH at the 3′-CCA end of tRNAs as aminoacyl esters. The formation of these aminoacyl-tRNAs, so-called "charged tRNAs," is catalyzed by specific **aminoacyl-tRNA synthetases.** There is one of these enzymes for each of the 20 amino acids and each aminoacyl-tRNA synthetase loads its amino acid only onto tRNAs designed to carry it. In turn, these tRNAs specifically recognize unique sequences of bases in the mRNA through complementary base pairing.

31.1 The Collinearity of Gene Structure and Protein Structure

The possibility that the linear sequence of nucleotides in a gene specifies the linear sequence of amino acids in a protein seemed very likely following Watson and Crick's double-helical structure for DNA and Frederick Sanger's revelation in the same year (1953) from work on insulin that the amino acid sequence of any given protein is very specific. In 1964, Charles Yanofsky at Stanford University provided experimental proof to validate the assumption that genes and proteins are collinear. Yanofsky and his associates isolated a number of *Escherichia coli* mutants in the gene encoding the α-subunit of tryptophan synthase. They then compared the location of these mutations along the *trpA* genetic map with the position of the amino acid substitutions in the α-subunit amino acid sequence and noted that a collinear relationship could be established between the genetic map and the primary structure of the polypeptide (Figure 31.2). Furthermore, the relative distances between mutations (that is, base substitutions) along the genetic map corresponded to the distances between amino acid substitutions in the protein, so that the maps are not only collinear but also proportional, at least in prokaryotes. In recent decades, recombinant DNA technology and nucleic acid sequencing have provided innumerable examples that verify collinearity.

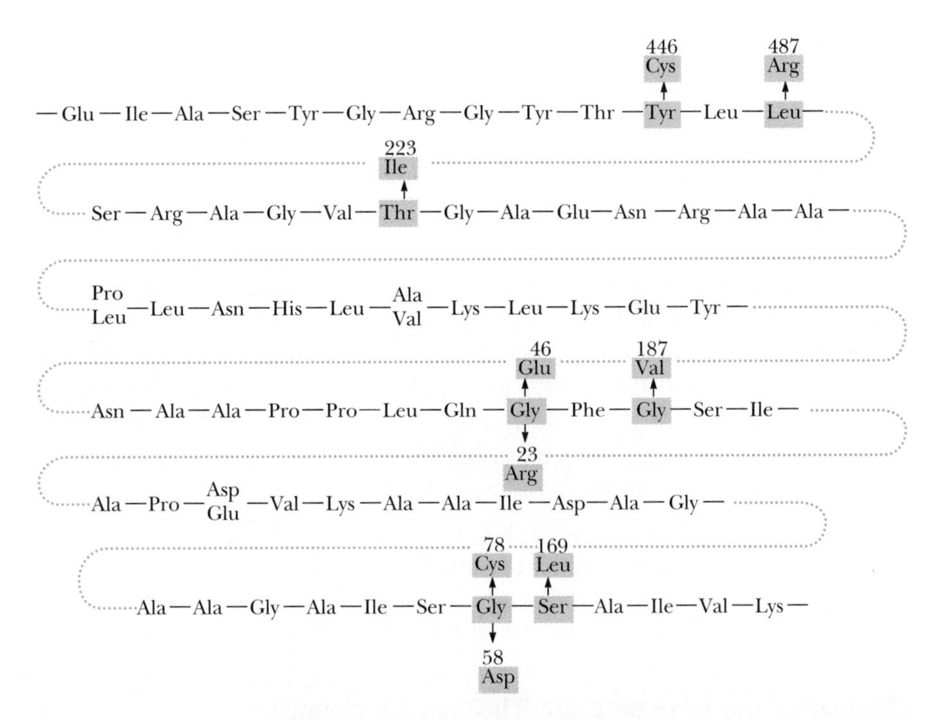

***Figure* 31.2** The collinear relationship between the order of mutations in the *E. coli trpA* gene and the position of amino acid substitutions in its gene product, the α-subunit of tryptophan synthase. Genetic analysis ordered the various mutants in the sequence (the numbers were assigned to these mutants in the order that they were isolated)

A446-A487-A223-(A23, A46)-A187-(A58, A78)-A169

The peptide and partial amino acid sequence of an α-subunit protein segment encompasses these mutations. The positions of the amino acid substitutions caused by the various mutations are indicated.

(Adapted from Yanofsky, C., et al., 1964. On the colinearity of gene structure and protein structure. Proceedings of the National Academy of Science, USA 51:266–272.)

31.2 Elucidating the Genetic Code

Once it was realized that the sequence of bases in a gene specified the sequence of amino acids in a protein, various possibilities for the genetic code came under consideration. How many bases were necessary to specify each amino acid? Is the code overlapping or nonoverlapping (Figure 31.3)? Is the code punctuated or continuous? Mathematical considerations favored a triplet of bases as the minimal code word, or *codon,* for each amino acid: A doublet code based on pairs of the four possible bases, A, C, G, and U, has $4^2 = 16$ unique arrangements, an insufficient number to encode the 20 amino acids. A triplet code of four bases has $4^3 = 64$ possible code words, more than enough for the task. Genetic results gave early answers to several of the other questions. For example, nitrous acid-induced mutations in the coat protein of TMV (tobacco mosaic virus, Chapters 28 and 34) caused single amino acid substitutions, discounting the possibility that the code was overlapping. A single base change in an overlapping code should cause multiple amino acid changes in the protein. For example, three changes would occur if the code is a triplet one (Figure 31.3).

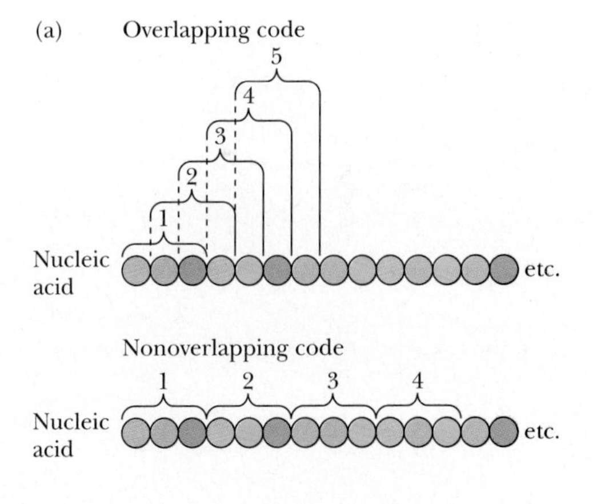

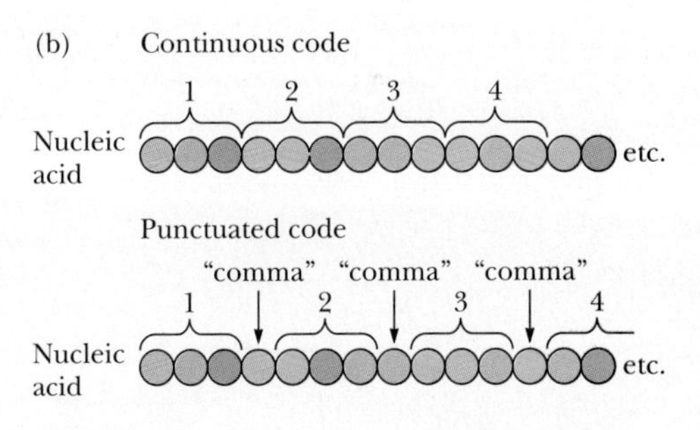

***Figure* 31.3** (a) An overlapping versus a nonoverlapping code. (b) A continuous versus a punctuated code.

The General Nature of the Genetic Code

The genetic code is a triplet code read continuously from a fixed starting point in each mRNA. Specifically, it is defined by the following:

1. A group of three bases codes for one amino acid.
2. The code is not overlapping.
3. The base sequence is read from a fixed starting point without punctuation. That is, the mRNA sequences contain no "commas" signifying appropriate groupings of triplets. If the reading frame is displaced by one base, it remains shifted throughout the subsequent message; no "commas" are present to restore the "correct" frame.
4. The code is **degenerate,** meaning that, in most cases, each amino acid can be coded by any of several triplets.

Regarding this latter point, recall that a triplet code yields 64 codons for 20 amino acids. If only 20 of these are used, then the majority of codons would be *nonsense* in that they would not code for any amino acid. A consequence of degeneracy is that most codons (61 of 64) code for some amino acid.

Elucidating the Genetic Code Through Biochemistry

The actual assignment of codons to the respective amino acids came from *in vitro* studies using synthetic oligo- and polyribonucleotides as messenger RNAs. Marshall Nirenberg and Heinrich Matthaei discovered that a cell-free system from *Escherichia coli* catalyzed the synthesis of polyphenylalanine (poly[Phe]) in the presence of polyuridylic acid (poly[U]). This cell-free system contained, among other things, ribosomes, tRNAs, and the soluble enzymes necessary to activate amino acids for protein synthesis. Even though the other 19 amino acids were present in the reaction mixture, only phenylalanine was incorporated into protein when poly[U] served as mRNA. The first codon had been deciphered: *UUU codes for Phe.* Similar experiments with polyadenylic acid (poly[A]) and polycytidylic acid (poly[C]) yielded polylysine and polyproline, respectively, showing that AAA codes for Lys and CCC codes for Pro.[1]

Synthetic mRNAs from Nucleotide Copolymers

The strategy of using synthetic polyribonucleotides as messengers in the cell-free, protein-synthesizing system was extended to mixed copolymers formed by the enzyme **polynucleotide phosphorylase,** which catalyzes the random polymerization of ribonucleotides:

$$x\,\text{NDP} \longrightarrow (\text{NMP})_x + x\,\text{P}_\text{i}$$

This enzyme synthesizes polynucleotide chains *de novo,* without a template, using ribonucleoside diphosphates as its substrates. The base composition of the RNA product is determined by the mix of NDPs present during the reaction. For example, a random 5:1 AC copolymer could be synthesized if the reaction mixture contained 5 parts of ADP and 1 part of CDP. Poly (A,C) has 8 possible codons—AAA, AAC, ACA, CAA, ACC, CAC, CCA, and CCC—and their relative frequency will depend on the [A]/[C] ratio during copolymer synthesis. When poly (AC) was used as template in cell-free protein synthesis,

[1] Because polyguanylic acid (poly[G]) has a very strong tendency to form multistranded helices, it was a poor template for protein synthesis. The fact that GGG codes for Gly was not learned until later.

Table **31.1**

Amino Acid Incorporation into Proteins Synthesized from Random (AC) Copolymer Synthetic mRNAs

The relative frequencies of the various codons are a function of the probability that a nucleotide will occur in a codon. For example, when the A/C ratio is 5:1, the ratio of AAA/AAC is $(5 \times 5 \times 5)/(5 \times 5 \times 1) = 5/1$. If the 3A codon frequency is assigned the value of *100,* then the 2A1C codon has a frequency of 20. Correlation of relative frequencies of amino acid incorporation with calculated codon frequencies allows tentative codon assignments to particular amino acids.

Poly AC (5:1)

Amino Acid	Calculated Triplet Frequency 3A	2A1C	1A2C	3C	Sum of Calculated Triplet Frequencies	Amino Acid Incorporation
Asparagine	—	20	—	—	20	24.2
Glutamine	—	20	—	—	20	23.7
Histidine	—	—	4.0	—	4	6.5
Lysine	100	—	—	—	100	*100*
Proline	—	—	4.0	0.8	4.8	7.2
Threonine	—	20	4.0	—	24	26.5

Poly AC (1:5)

Amino Acid	Calculated Triplet Frequency 3A	2A1C	1A2C	3C	Sum of Calculated Triplet Frequencies	Amino Acid Incorporation
Asparagine	—	3.3	—	—	3.3	5.3
Glutamine	—	3.3	—	—	3.3	5.2
Histidine	—	—	16.7	—	16.7	23.4
Lysine	0.7	—	—	—	0.7	1.0
Proline	—	—	16.7	83.3	100	*100*
Threonine	—	3.3	16.7	—	20	20.8

Source: Adapted from Speyer, J. F., et al., 1963. Synthetic polynucleotides and the amino acid code. *Cold Spring Harbor Symposium on Quantitative Biology* **28:**559–567.

the amino acids asparagine, glutamine, threonine, and histidine as well as lysine and proline were incorporated into protein (Table 31.1). The relative abundances of these amino acids in proteins made from AC copolymers of different composition suggested that the codons containing 2 A and 1 C encoded Asn, Gln, and Thr and those containing 2 C and 1 A encoded His, Thr, and Pro. Since the actual base sequence in these random copolymers was unknown, unique assignments of AAC, ACA, or CAA to Asn, Thr, and Gln could not be decided.

Alternating Nucleotide Copolymers as mRNAs

Further advances followed the synthesis of polyribonucleotides of defined sequence in the laboratory of H. Gobind Khorana. The first, a copolymer of alternating U and C residues, when used as template for cell-free protein synthesis gave a protein product that was a copolymer of alternating Ser and Leu residues. The possible triplet codons presented by poly (UC) are UCU and CUC. These results alone do not reveal whether UCU codes for Ser and CUC for Leu, or vice versa. Further experiments showed that poly (AG) di-

Table **31.2**

Protein Products from Alternating Polyribonucleotide Copolymers

Alternating RNA Copolymer	Possible Codons	Protein Product
UC	UCU and CUC	$(\text{Ser-Leu})_n$
AG	AGA and GAG	$(\text{Arg-Glu})_n$
UG	UGU and GUG	$(\text{Cys-Val})_n$
AC	ACA and CAC	$(\text{Thr-His})_n$

Source: Adapted from Nishimura, S., Jones, D. S., and Khorana, H. G., 1965. Studies on polynucleotides, XLVIII, The *in vitro* synthesis of a co-polypeptide containing two amino acids in alternating sequence dependent upon a DNA-like polymer containing two nucleotides in alternating sequence. *Journal of Molecular Biology* **13:**302–324.

rected the synthesis of an Arg-Glu alternating co-polypeptide, poly (UG) gave repeating Cys-Val copolymer, and poly (AC) produced a Thr-His copolymer (Table 31.2). Note that these results provide conclusive evidence that the genetic code must consist of an odd number of letters: An even-numbered code would have yielded homopolypeptides from the alternating dinucleotide copolymers.

Repeating Nucleotide Copolymers as mRNAs

Khorana's group next achieved the synthesis of polyribonucleotides composed of repeating trinucleotide and tetranucleotide sequences. The repeating polymer poly (AAG) directed the formation of polylysine, polyglutamate, and polyarginine, depending on which reading frame was adopted during the protein synthesis process (Figure 31.4). A repeating tetranucleotide polymer would be expected to produce a polypeptide that also had a repeating tetrameric sequence if the code was indeed a triplet code (and the message contained no nonsense codons). Poly (UAUC) directed the synthesis of the repeating polypeptide $(\text{Tyr-Leu-Ser-Ile})_n$. Chymotrypsin cleavage of this tetrameric polypeptide followed by N-terminal amino acid analysis of the products revealed that the carboxyl groups of Tyr residues were linked to the amino groups of Leu residues, establishing that, as the messenger RNA was read in the 5′→3′ direction, amino acids were specified in the N→C direction (Figure 31.5).

Trinucleotides Bound to Ribosomes Promote the Binding of Specific Aminoacyl-tRNAs

In 1964, Marshall Nirenberg and Philip Leder reported that trinucleotides bound to ribosomes directed the binding of specific aminoacyl-tRNAs. That is, ternary ribosome:trinucleotide:aminoacyl-tRNA complexes could be

AAGAAGAAGAAGAAGAAGAAGAAG . . . etc. Synthetic mRNA

Lys Lys Lys Lys Lys Lys Lys Lys . . . Polylysine (AAG = Lys)

Arg Arg Arg Arg Arg Arg Arg Arg . . . Polyarginine (AGA = Arg)

Glu Glu Glu Glu Glu Glu Glu Glu . . . Polyglutamate (GAA = Glu)

***Figure* 31.4** The synthetic polyribonucleotide mRNAs might be read in either of the three possible reading frames, each of which yields a different homopolypeptide.

(a)
UAU CUA UCU AUC UAU CUA UCU AUC UAU CUA UCU AUC

–Tyr – Leu – Ser – Ile – Tyr – Leu – Ser – Ile – Tyr – Leu – Ser – Ile–

(b)
Leu–Ser–Ile–Tyr   Leu–Ser–Ile–Tyr

***Figure* 31.5** (a) Cell-free protein synthesis using the repeating tetranucleotide polymer poly (UAUC) as template yielded a polypeptide whose amino acid sequence consisted of repeating tetrameric units. (b) Chymotrypsin acts preferentially to cleave polypeptide chains on the C-side of aromatic amino acid residues. When the polypeptide was cleaved with chymotrypsin, the products had Leu, *not* Ile, as the N-terminal amino acid residue. Therefore, as the mRNA was read 5′→3′, the protein was synthesized in the N→C direction, *not* in the C→N direction.

formed, *provided* the right trinucleotide and aminoacyl-tRNA combination was present. Aminoacyl-tRNAs were prepared by adding all 20 amino acids to a purified tRNA mixture in the presence of a soluble *E. coli* fraction containing the necessary aminoacyl-tRNA synthetases. Only one of the amino acids was ^{14}C-labeled in any one binding assay. Trinucleotides are the equivalent of codons, so if a specific trinucleotide promoted the binding of a particular ^{14}C-labeled aminoacyl-tRNA, the base sequence of the trinucleotide must be the code word for that amino acid. Binding was detected because the ribosomes were retained on a nitrocellulose filter while free aminoacyl-tRNAs passed through; only aminoacyl-tRNAs bound by ribosomes were retained (Figure 31.6).

This system was quickly exploited to elucidate the genetic code. Elucidation of the genetic code was probably the greatest scientific achievement of the 1960s. For their roles in it, Marshall Nirenberg and H. Gobind Khorana shared in the 1968 Nobel Prize for physiology or medicine.

31.3 The Nature of the Genetic Code

The complete translation of the genetic code is presented in Table 31.3. Codons, like other nucleotide sequences, are read 5′→3′. Codons represent triplets of bases in mRNA or, replacing U with T, triplets along the nontranscribed (antisense) strand of DNA. Several noteworthy features characterize the genetic code:

1. *All the codons have meaning.* Sixty-one of the 64 codons specify particular amino acids. The remaining 3—UAA, UAG, and UGA—specify no amino acid and thus they are **nonsense codons.** Nonsense codons serve as **termination codons**—they are "stop" signals indicating that the end of the protein has been reached.
2. *The genetic code is unambiguous.* Each of the 61 "sense" codons encodes only one amino acid.
3. *The genetic code is degenerate.* With the exception of Met and Trp, every amino acid is coded by more than one codon. Several—Arg, Leu, and Ser—are represented by 6 different codons. Codons coding for the same amino acid are called **synonymous codons.**
4. *Codons representing the same amino acid or chemically similar amino acids tend to be similar in sequence.* Often the third base in a codon is irrelevant, so that, for example, all 4 codons in the GGX family specify

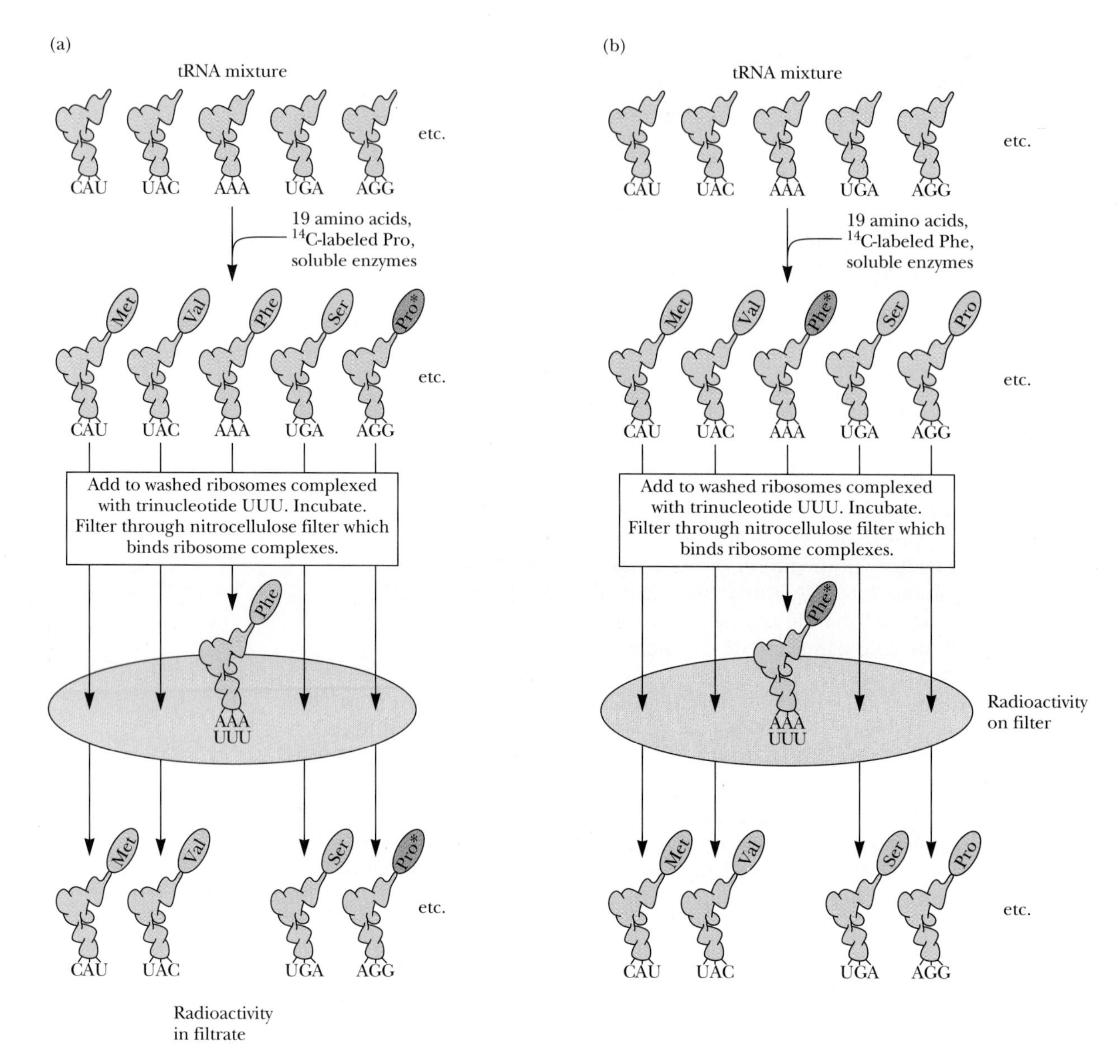

***Figure* 31.6** The filter-binding assay for elucidation of the genetic code. Reaction mixture includes washed ribosomes, Mg^{2+}, a particular trinucleotide (pUpUpU in this example), and all 20 aminoacyl-tRNAs, one of which is radioactively (^{14}C) labeled. (a) ^{14}C-labeled prolyl-tRNA. (b) ^{14}C-labeled phenylalanyl-tRNA. Only the aminoacyl-tRNA whose binding is directed by the trinucleotide codon will become bound to the ribosomes and retained on the nitrocellulose filter. The amount of radioactivity retained by the filter is a measure of trinucleotide-directed binding of a particular labeled aminoacyl-tRNA by ribosomes. Use of this binding assay to test the 64 possible codon trinucleotides against the 20 different amino acids quickly enabled researchers to assign triplet code words to the individual amino acids. The genetic code was broken.

(Adapted from Nirenberg, M. W., and Leder, P., 1964. RNA codewords and protein synthesis. Science 145:1399–1407.)

Gly, and the UCX family specifies Ser (Table 31.3). This feature, known as **third-base degeneracy,** is one indication that codon assignments to various amino acids are not random, as is readily apparent from the nonrandom arrangement in the table. Note also that codons with a pyrimidine as second base likely encode amino acids with hydrophobic side chains, while codons with a purine in

Table 31.3
The Genetic Code

First Position (5′-end)	Second Position U	C	A	G	Third Position (3′-end)
U	UUU Phe	UCU Ser	UAU Tyr	UGU Cys	U
	UUC Phe	UCC Ser	UAC Tyr	UGC Cys	C
	UUA Leu	UCA Ser	UAA Stop	UGA Stop	A
	UUG Leu	UCG Ser	UAG Stop	UGG Trp	G
C	CUU Leu	CCU Pro	CAU His	CGU Arg	U
	CUC Leu	CCC Pro	CAC His	CGC Arg	C
	CUA Leu	CCA Pro	CAA Gln	CGA Arg	A
	CUG Leu	CCG Pro	CAG Gln	CGG Arg	G
A	AUU Ile	ACU Thr	AAU Asn	AGU Ser	U
	AUC Ile	ACC Thr	AAC Asn	AGC Ser	C
	AUA Ile	ACA Thr	AAA Lys	AGA Arg	A
	AUG Met*	ACG Thr	AAG Lys	AGG Arg	G
G	GUU Val	GCU Ala	GAU Asp	GGU Gly	U
	GUC Val	GCC Ala	GAC Asp	GGC Gly	C
	GUA Val	GCA Ala	GAA Glu	GGA Gly	A
	GUG Val	GCG Ala	GAG Glu	GGG Gly	G

*AUG signals translation initiation as well as coding for Met residues.

Third-Base Degeneracy Is Color-Coded

Third-Base Relationship	Third Bases with Same Meaning	Number of Codons
Third base irrelevant	U, C, A, G	32 (8 families)
Purines	A or G	12 (6 pairs)
Pyrimidines	U or C	14 (7 pairs)
Three out of four	U, C, A	3 (AUX = Ile)
Unique definitions	G only	2 (AUG = Met) (UUG = Trp)
Unique definition	A only	1 (UGA = Stop)

the second-base position typically specify polar or charged amino acids. The two negatively charged amino acids, Asp and Glu, are encoded by GAX codons; GA-pyrimidine gives Asp and GA-purine specifies Glu. The consequence of these similarities is that the mutations are less likely to be deleterious since single base changes in a codon will result either in no change or in a substitution with an

***Table* 31.4**

Natural Variation in the Standard Genetic Code

Included are mitochondrial variants as well as genomic variants. *Mycoplasma:* (prokaryote) UGA = Trp; ciliated protozoa (*Tetrahymena, Paramecium*): UAA, UGA = Gln instead of "Stop"; and in *E. coli:* UGA = selenocysteine in a small number of proteins.

Known Variant Mitochondrial Codon Assignments

	Codons				
Organisms	**UGA (Term)***	**AUA (Ile)**	**AGA/G (Arg)**	**CUN (Leu)**	**CGG (Arg)**
Animals					
Vertebrates	Trp	Met	Term	+†	+
Drosophila	Trp	Met	Ser	+	+
Yeasts					
Saccharomyces cerevisiae	Trp	Met	+	Thr	+
Torulopsis glabrata	Trp	Met	+	Thr	?‡
Schizosaccharomyces pombe	Trp	+	+	+	+
Filamentous fungi	Trp	+	+	+	+
Trypanosomes	Trp	+	+	+	+
Higher plants	+	+	+	+	Trp
Chlamydomonas reinhardtii	?	+	+	+	?

Genomic Codon Variants

Prokaryotes
Mycoplasma: UGA = Tryptophan
E. coli: UGA = Selenocysteine§
Ciliated protozoa (*Tetrahymena, Paramecium*): UAA and UGA = Glutamine

*Standard codon assignments indicated in parentheses.
†Indicates standard assignment.
‡Indicates that the codon has not been observed to occur.
§Selenocysteine has a selenium atom in place of the sulfur atom of cysteine:

$$\begin{array}{c} H \\ | \\ Se \\ | \\ CH_2 \\ | \\ H_3{}^+N\text{—}C\text{—}H \\ | \\ C \\ {}^-O \diagup \;\; \diagdown\!\!\diagdown O \end{array}$$

Source: Adapted from Fox, T. D., 1987. Natural variation in the genetic code. *Annual Review of Genetics* **21**:67–91.

amino acid similar to the original amino acid. The degeneracy of the code is evolution's buffer against mutational disruption.

5. *The genetic code is "universal."* While certain minor exceptions in codon usage occur in the genomes of mitochondria, some prokaryotes, and certain lower eukaryotes (Table 31.4), the more striking feature of the code is its universality: Codon assignments are virtually the same throughout microorganisms, plants, and animals. This conformity provides strong evidence that all extant organisms evolved from a common primordial ancestor.

31.4 Amino Acid Activation for Protein Synthesis: Aminoacyl-tRNA Synthetases

Codon recognition is achieved by aminoacyl-tRNAs. In order for accurate translation to occur, the appropriate aminoacyl-tRNA must ''read'' the codon through base pairing via its **anticodon loop** (Chapter 7). Once an aminoacyl-tRNA has been synthesized, the amino acid part makes no contribution to accurate translation of the mRNA. Von Ehrenstein proved this point by loading ^{14}C-cysteine onto its particular tRNA, tRNA^{Cys}. The product cysteinyl-tRNA^{Cys} was chemically reduced so that its —SH group was removed to yield Ala-tRNA^{Cys} (Figure 31.7). The Ala-tRNA^{Cys} was then added to an *in vitro* hemoglobin-synthesizing system and the product Hb was analyzed. Alanine was found at positions in the Hb amino acid sequence normally occupied by Cys. The protein-synthesizing machinery was unable to recognize Ala-tRNA^{Cys} as ''foreign'' or inappropriate. Thus, the amino acid attached to a tRNA vehicle is passively chauffeured and becomes inserted into a growing peptide chain as dictated through codon–anticodon recognition.

The Aminoacyl-tRNA Synthetase Reaction

Formation of aminoacyl-tRNAs serves two purposes: It activates the amino acid so that it will readily react to form a peptide bond and it bridges the information gap between amino acids and codons. Clearly, the proper amino acids must be loaded onto the various tRNAs so that the mRNA is translated with fidelity. **Aminoacyl-tRNA synthetases** are responsible for this task. Cells have 20 different aminoacyl-tRNA synthetases, one for each amino acid. Each enzyme catalyzes the ATP-dependent esterification of its specific amino acid to the 3′-end of its **cognate tRNA molecules** (Figure 31.8). The immediate hydrolysis of product PP$_i$ by ubiquitous pyrophosphatases renders amino acid activation thermodynamically favorable and essentially irreversible.

cognate: kindred; in this sense, *cognate* refers to those tRNAs having anticodons that can read one or more of the codons that specify one particular amino acid.

Because activation is a two-step process, two levels of specificity exist—one in formation of the aminoacyl adenylate and a second in transfer of the aminoacyl group to a specific tRNA. The specificity at the first level is not absolute, as shown by the ability of **isoleucyl-tRNA synthetase** to catalyze an ATP $\rightleftharpoons$ PP$_i$ exchange reaction with either isoleucine or valine, that is, reaction (i) of Figure 31.8. Although valyl adenylate is synthesized, no valyl-tRNA^{Ile} is

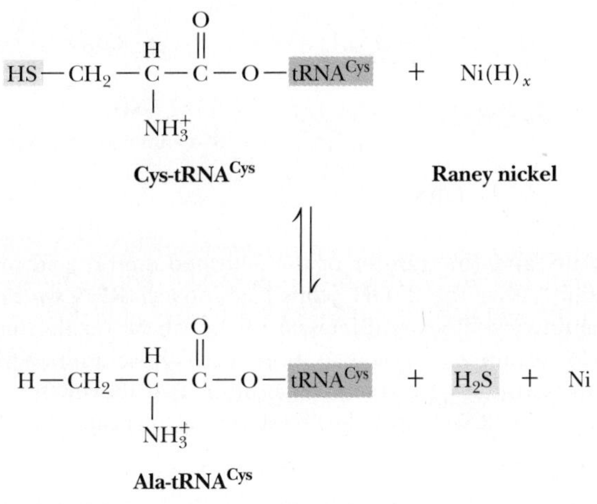

$$\underset{\textbf{Cys-tRNA}^{\textbf{Cys}}}{HS{-}CH_2{-}\underset{NH_3^+}{\overset{H}{C}}{-}\overset{O}{\overset{\|}{C}}{-}O{-}tRNA^{Cys}} + \underset{\textbf{Raney nickel}}{Ni(H)_x}$$

$$\rightleftharpoons$$

$$\underset{\textbf{Ala-tRNA}^{\textbf{Cys}}}{H{-}CH_2{-}\underset{NH_3^+}{\overset{H}{C}}{-}\overset{O}{\overset{\|}{C}}{-}O{-}tRNA^{Cys}} + H_2S + Ni$$

***Figure* 31.7** Reduction of cysteinyl-tRNA^{Cys} with Raney nickel converts the cysteine —CH_2SH R group to —CH_3. That is, Cys is transformed into Ala.

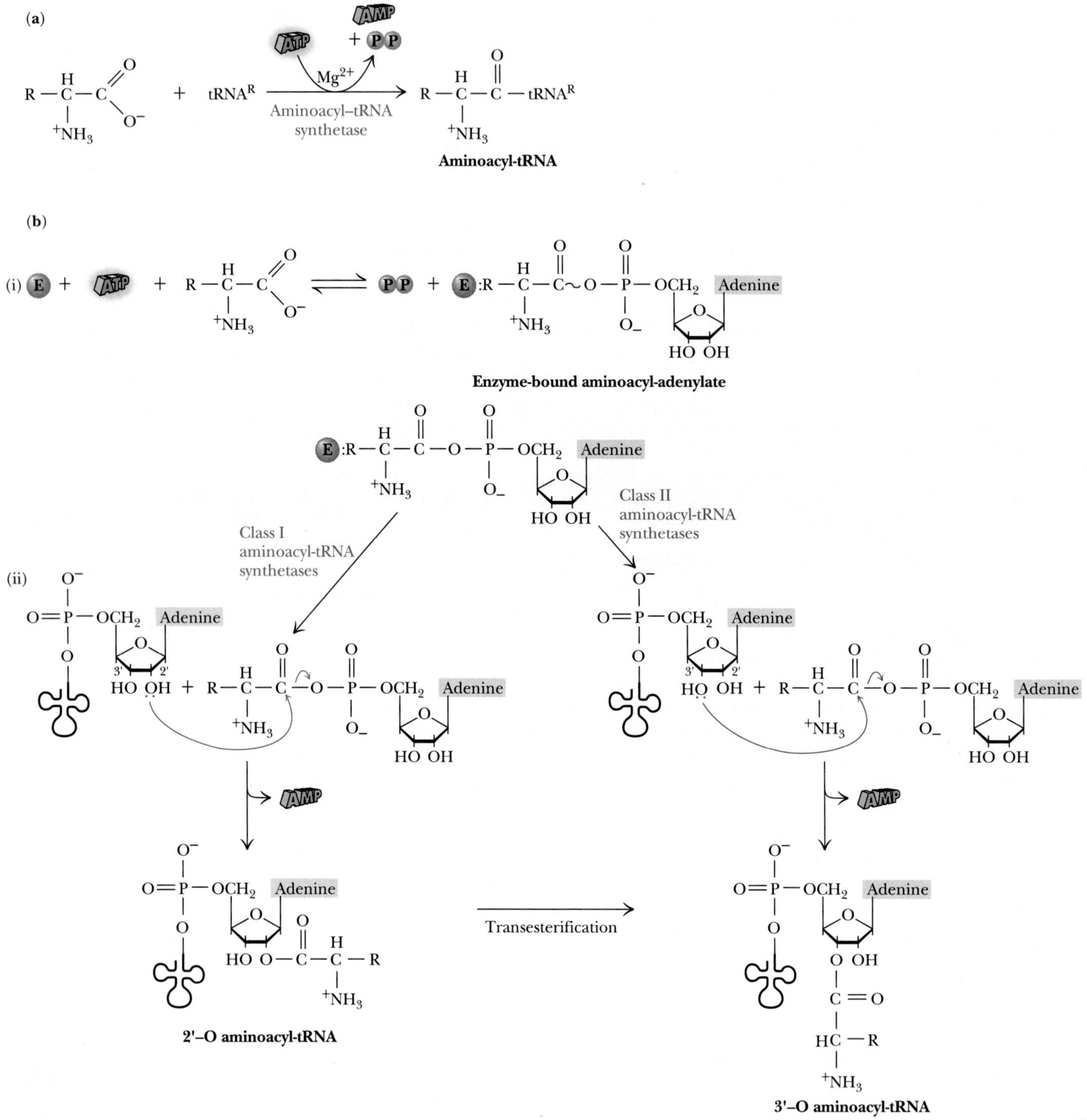

***Figure* 31.8** The aminoacyl-tRNA synthetase reaction. (a) The overall reaction. (b) The overall reaction commonly proceeds in two steps: (i) formation of an aminoacyl-adenylate, and (ii) transfer of the activated amino acid moiety of the mixed anhydride to either the 2′-OH (class I aminoacyl-tRNA synthetases) or 3′-OH (class II aminoacyl-tRNA synthetases) of the ribose on the terminal adenylic acid at the 3′-CCA terminus common to all tRNAs. Those aminoacyl-tRNAs formed as 2′-aminoacyl esters undergo a transesterification that moves the aminoacyl function to the 3′-O of tRNA. Only the 3′-esters are substrates for protein synthesis.

released. That is, the overall specificity of the reaction is virtually absolute. The enzyme has an editing function that establishes this specificity: Synthesis of misacylated valyl-$tRNA^{Ile}$ triggers an editing deacylase site in the enzyme that hydrolyzes the misacylated aminoacyl-tRNA.

Despite their common enzymatic function, aminoacyl-tRNA synthetases are a diverse group of proteins with four different patterns of quaternary organization—α, α_2, α_4, and $\alpha_2\beta_2$—and subunits ranging from 334 to more than 1000 amino acid residues. Aminoacyl-tRNA synthetases can be divided into two classes (I and II of Table 31.5) on the basis of shared amino acid sequence motifs, oligomeric state, and acylation function. Class I enzymes are chiefly monomeric, whereas class II aminoacyl-tRNA synthetases are always oligomeric. Furthermore, class I aminoacyl-tRNA synthetases add the amino acid to the 2′-OH of the terminal adenylate residue of tRNA before shifting it to the 3′-OH, whereas class II enzymes add it directly to the 3′-OH (Figure 31.8). Only the 3′-aminoacyl-tRNA esters are substrates for protein synthesis. The results suggest that the catalytic domains of these enzymes evolved from two different ancestral predecessors. Apparently, different forms of these enzymes were present very early in evolution, since aminoacyl-tRNA synthetases are ranked among the oldest proteins. In higher eukaryotes, at least some aminoacyl-tRNA synthetases assemble into large multiprotein complexes.

***Table* 31.5**
The Two Classes of Aminoacyl-tRNA Synthetases

Class I	Class II
Arg	Ala
Cys	Asn
Gln	Asp
Glu	Gly
Ile	His
Leu	Lys
Met	Phe
Trp	Pro
Tyr	Ser
Val	Thr

Selective tRNA Recognition by Aminoacyl-tRNA Synthetases

Aside from the need to uniquely recognize their cognate amino acids, aminoacyl-tRNA synthetases must be able to discriminate between the various tRNAs. Most synthetases are believed to bind tRNAs through contacts along and around the inside of the L-shaped tRNA (Figure 31.9). Not only the anticodon, but the acceptor stem, the anticodon stem, and the D loop of the tRNA all contribute contact regions to the tRNA:enzyme interface. Despite the fact that the anticodon is the crucial tRNA feature in codon recognition, the anticodon is not always the determinant used by aminoacyl-tRNA synthetases to selectively identify their cognate tRNAs. Its importance varies from enzyme to enzyme: Some aminoacyl-tRNA synthetases seem to rely only on the anticodon in selecting tRNAs for aminoacylation, others rely on it partially, and still others do not rely on it at all. Figure 31.10 highlights the major identity elements in four different tRNAs recognized by their respective aminoacyl-tRNA synthetases. A discussion follows on several of these tRNAs. Apparently, a common set of rules does not govern tRNA recognition by these enzymes.

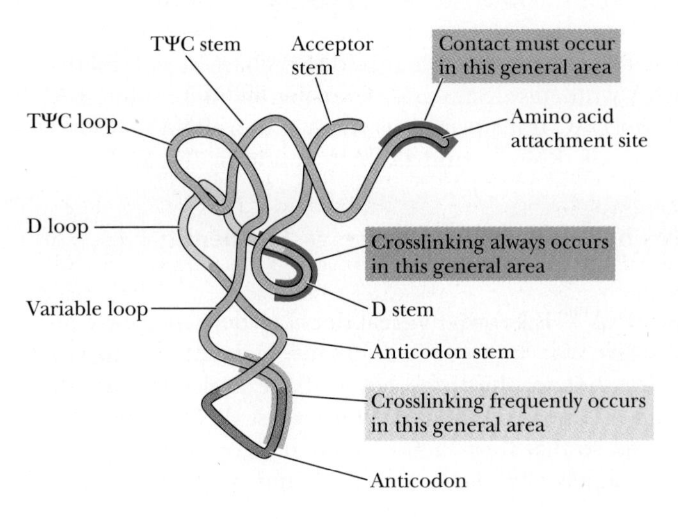

***Figure* 31.9** Aminoacyl-tRNA synthetase: tRNA contacts important in tRNA binding. Chemical studies creating covalent crosslinks between tRNAs and their cognate aminoacyl-tRNA synthetases have revealed some of these contacts.

(Adapted from Schweizer, M. P., et al., 1984. ^{13}C NMR studies of dynamics and synthetase interaction of [4-^{13}C]-uracil-labeled Escherichia coli tRNAs. Federation Proceedings 43:2984–2986.)

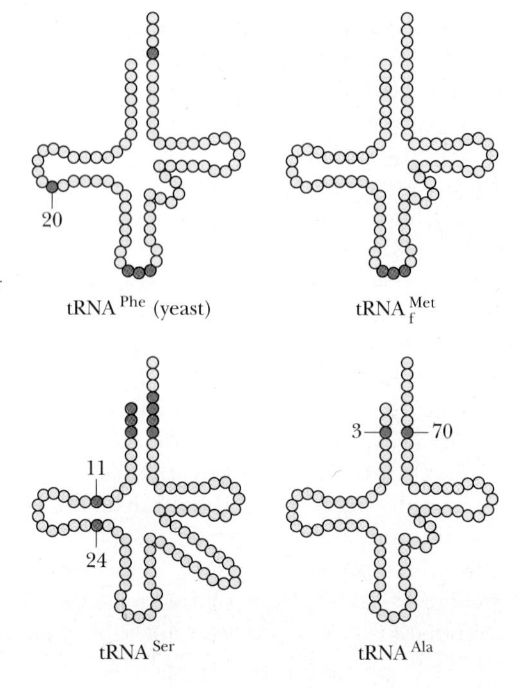

***Figure* 31.10** Major identity elements in four tRNA species. Each base in the tRNA is represented by a circle. Numbered filled circles indicate positions of identity elements within the tRNA that are recognized by its specific aminoacyl-tRNA synthetase.

(Adapted from Schulman, L. H., and Abelson, J., 1988. Recent excitement in understanding transfer RNA identity. Science 240:1591–1592.)

tRNA Recognition Sites in *E. coli* Glutaminyl-tRNA^{Gln} Synthetase

E. coli glutaminyl-tRNA^{Gln} synthetase is a 63.4-kD (553-residue) class I monomeric enzyme. The crystal structure of glutaminyl-tRNA^{Gln} synthetase complexed with tRNA^{Gln} reveals that the enzyme shares a continuous interaction with its cognate tRNA that extends from the anticodon to the acceptor stem along the entire inside of the L-shaped tRNA (Figure 31.11), much as suggested in Figure 31.9. Specific recognition elements include enzyme contacts with the 2-NH_2 groups of acceptor-stem guanine residues G2 and G3 lying within the minor groove, as well as interactions between the enzyme and anticodon, particularly the central U in the CUG anticodon. The carboxylate group of Asp^{235} makes sequence-specific H bonds in the tRNA minor groove with base pair G3:C70 of the acceptor stem. A mutant glutaminyl-tRNA^{Gln} synthetase with Asn substituted for Asp at position 235 shows relaxed specificity; that is, it will now incorrectly acylate a noncognate tRNA with Gln.

The Identity Elements for Some tRNAs Reside in the Anticodon

Alteration of the anticodons of either tRNA^{Trp} or tRNA^{Val} to CAU, the anticodon for the methionine codon AUG, transforms each of these tRNAs into tRNA^{Met}. That is, these tRNAs are now recognized and charged with methionine by methionyl-tRNA synthetase. Similarly, reversing the methionine CAU anticodon of tRNA^{Met} to UAC transforms this tRNA into a tRNA^{Val}. Clearly, crucial identity elements of tRNA^{Met} and tRNA^{Val} reside in the anticodon.

Five Different Bases in Yeast tRNA^{Phe} Serve as Its Identity Elements

The structure of yeast tRNA^{Phe} is known in great detail (Figure 31.10; see also Figures 7.37 and 7.38). Five of its bases serve as identity elements for the yeast phenylalanyl-tRNA^{Phe} synthetase: the three bases of the anticodon, residue G20 in the D loop, and A73 near the 3′-end. When yeast tRNA^{Arg}, tRNA^{Met}, and tRNA^{Tyr} were altered so that they each contained a complete set of the five identity elements (namely G20, G34, A35, A36, and A73), they became

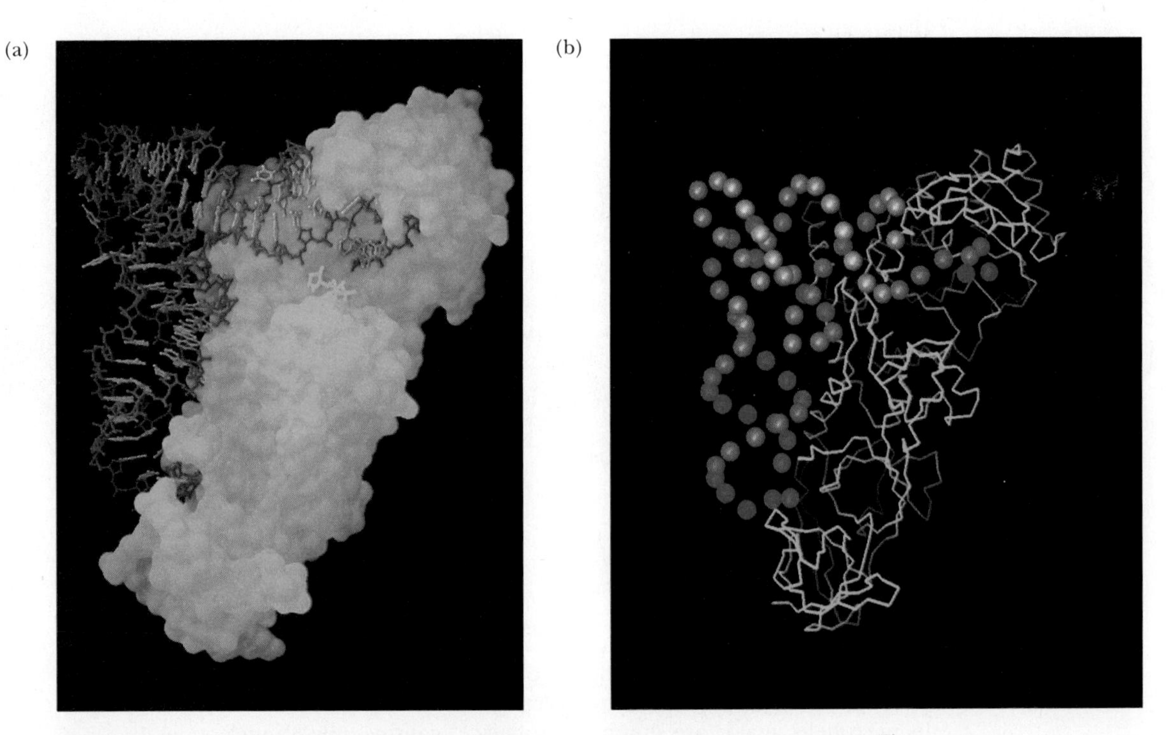

Figure **31.11** (a) A solvent-accessible representation of *E. coli* glutaminyl-$tRNA^{Gln}$ synthetase complexed with $tRNA^{Gln}$ and ATP, derived from analysis of the crystal structure of the complex. The protein is colored blue. The sugar-phosphate backbone of the tRNA is red; its bases are yellow. The protein:tRNA contact region extends along one side of the entire length of this extended protein. The acceptor stem of the tRNA and the ATP (green) fit into a cleft at the top of the protein in this view. The enzyme also interacts extensively with the anticodon (lower tip of $tRNA^{Gln}$). (b) Diagram showing the structure of $tRNA^{Gln}$, as represented by its phosphorus atoms (purple spheres), in complex with *E. coli* glutaminyl-$tRNA^{Gln}$ synthetase, as represented in the terms of its Cα atoms (blue).

(Adapted from Rould, et al., 1989. Structure of E. coli *glutaminyl-tRNA synthetase complexed with $tRNA^{Gln}$ and ATP at 2.8 Å resolution. Science 246:1135; photograph courtesy of Thomas A. Steitz of Yale University.)*

excellent substrates of yeast phenylalanyl-$tRNA^{Phe}$ synthetase. The G20 of $tRNA^{Phe}$ may be an important discriminatory nucleotide since a G has not been found at this position in any other yeast tRNA.

Twelve Nucleotides in Common Define the $tRNA^{Ser}$ Family

Six codons specify serine. Six distinct isoacceptor tRNAs can be aminoacylated by the *E. coli* seryl-$tRNA^{Ser}$ synthetase. Five of these tRNAs are the product of *E. coli* genes; the sixth is encoded by the phage-T4 genome. These 6 tRNAs have anticodons that include UGA, CGA, GGA, and GCU; thus, variations occur at all 3 anticodon base positions. The nucleotide sequences of these 6 tRNAs have been compared and only 12 positions are held in common. These nucleotides include G1, G2, A3 (or U3) and U70 (or A70), C71, C72, and G73 in the acceptor stem, and C11 and G24 in the dihydrouridine (D) stem. All of these nucleotides except G73 are involved in intrachain H bonds (Figure 31.10). When a leucine-specific tRNA was modified so that it shared all 12 $tRNA^{Ser}$ identities, it was transformed into a $tRNA^{Ser}$.

A Single G:U Base Pair Defines $tRNA^{Ala}$s

In contrast, a single, noncanonical base pair, G3:U70, is the principal element by which alanyl-$tRNA^{Ala}$ synthetase recognizes tRNAs as its substrates. All cytoplasmic $tRNA^{Ala}$ representatives that have been sequenced thus far, from ar-

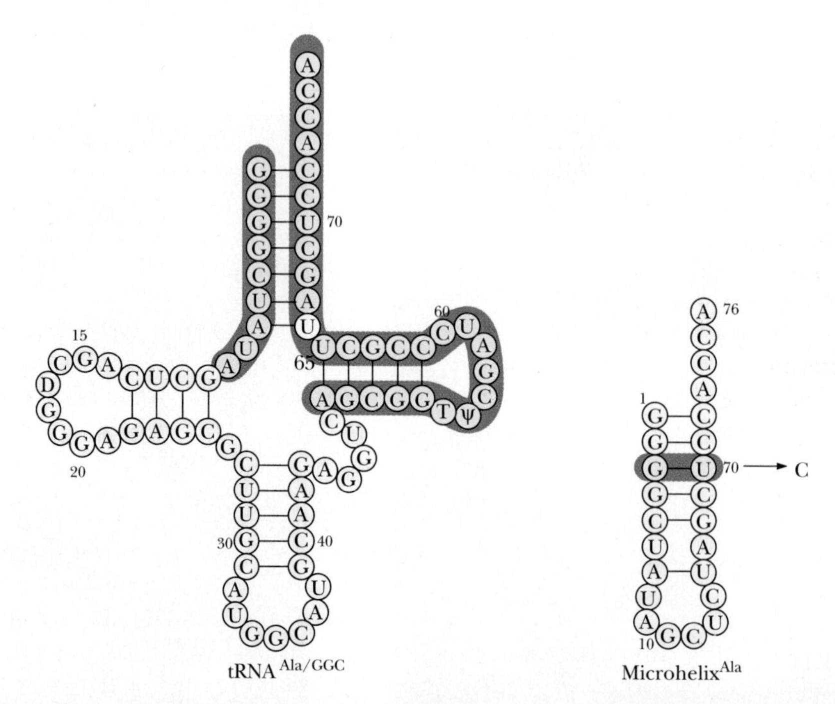

Figure **31.12** A microhelix analog of tRNA^{Ala} is aminoacylated by alanyl-tRNA^{Ala} synthetase provided it retains the G3:U70 base pair. Substituting C for U at position 70 abolishes its ability to accept Ala. The sequences of tRNA$^{Ala/GGC}$ and its microhelix analog are shown. Microhelix^{Ala} consists only of nucleotides 1 through 13 directly connected to 66 through 76 to re-create the tRNA^{Ala} 7-bp acceptor stem.

(Adapted from Schimmel, P., 1989. Parameters for the molecular recognition of transfer RNAs. Biochemistry 28:2747–2759.)

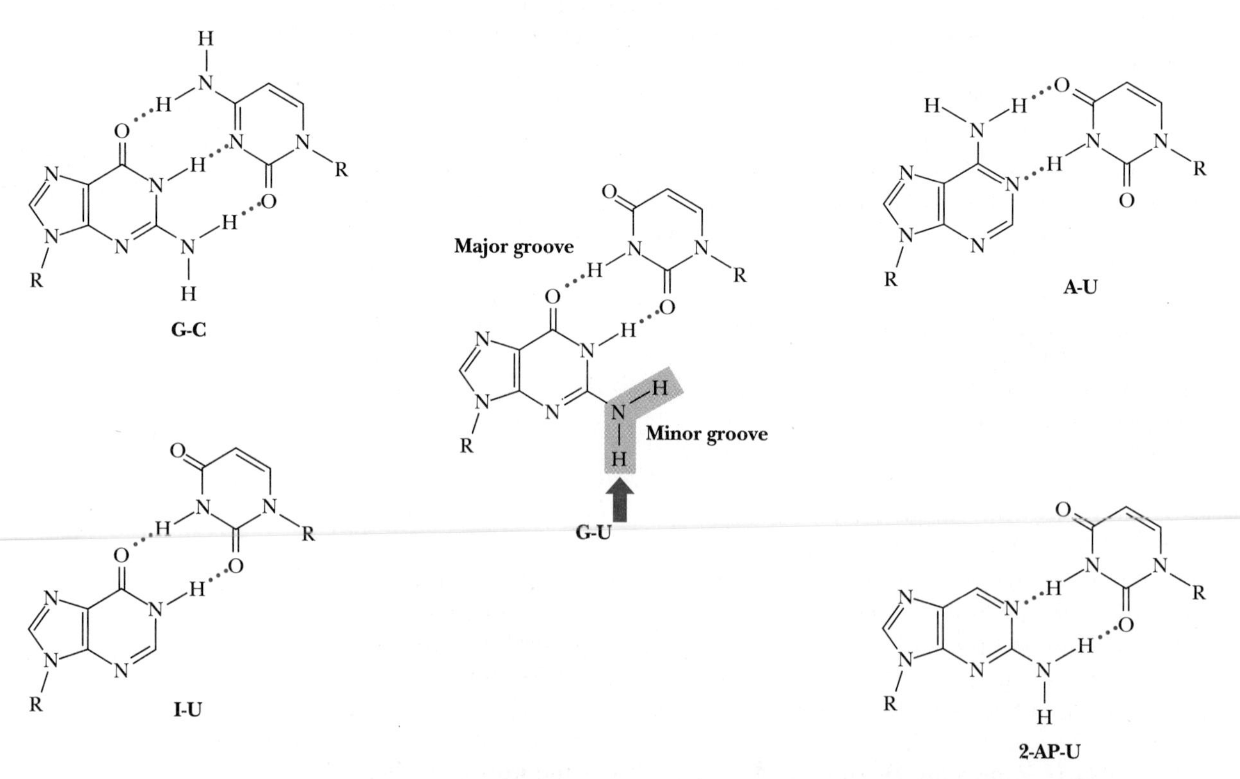

Figure **31.13** Structures of various base pairs in relationship to the 3:70 position in tRNA molecules. The double-helical regions of transfer RNAs adopt the A-form double-helical conformation of nucleic acids (Chapter 7), which has a deep, narrow major groove on one side and a shallow, wide minor groove on the other.

chaebacteria to eukaryotes, possess this G3:U70. Altering the G3:C70 base pairs found in Lys-specific, Cys-specific, and Phe-specific tRNAs to G3:U70 confers alanine acceptability on these tRNAs. Altering the unusual G3:U70 base pair of $tRNA^{Ala}$ to G:C, A:U, or even U:G abolishes its ability to be aminoacylated with Ala. On the other hand, provided the G3:U70 base pair is present, alanyl-$tRNA^{Ala}$ synthetase will aminoacylate a 24-nucleotide "microhelix" analog of $tRNA^{Ala}$ (Figure 31.12).

Paul Schimmel has deduced that the key structural feature of the G3:U70 determinant is the 2-NH_2 group of G3. A series of analogs was prepared in which other base pairs replaced the "G3:U70" base pair in an analog of $tRNA^{Ala}$ (Figure 31.13). Only the original G3:U70 analog was aminoacylated with Ala by Ala-tRNA synthetase. Note that in RNA A-form double-helical structures such as tRNAs, the G3 2-amino group is exposed in the minor groove of the helix (Figure 31.13). If G3 base-pairs with U at position 70, this —NH_2 is not H-bonded. In the G:C and 2-AP:U analogs, this —NH_2 is not free, but is hydrogen-bonded with C or U; the I:U and A:U analogs lack a 2-NH_2. The inverse U3:G70 analog (not shown) places the 2-NH_2 in the major groove. Paul Schimmel and his colleagues thus concluded that an unpaired guanine 2-amino group at the proper location within the minor groove earmarks a tRNA for aminoacylation by Ala-tRNA synthetase. Since this structural feature is common to all cytoplasmic $tRNA^{Ala}$s, the various tRNA recognition elements must have been decided very early in evolution.

31.5 Codon–Anticodon Pairing, Third-Base Degeneracy, and the Wobble Hypothesis

Protein synthesis depends on the codon-directed binding of the proper aminoacyl-tRNAs so that the right amino acids are sequentially aligned according to the specifications of the mRNA undergoing translation. This alignment is achieved via codon–anticodon pairing in antiparallel orientation (Figure 31.14). However, considerable degeneracy exists in the genetic code at the third position. Conceivably, this degeneracy could be handled in either of two ways: (a) codon–anticodon recognition could be highly specific so that a complementary anticodon is required for each codon, or (b) fewer than 61 anticodons could be used for the "sense" codons if certain allowances were made in the base-pairing rules. Then, some anticodons could recognize more than one codon. Nirenberg demonstrated as early as 1965 that poly (U) bound *all* the Phe-$tRNA^{Phe}$ even though UUC is also a Phe codon. This result suggested that the phenylalanine-specific tRNAs could recognize both UUU and UUC. The yeast $tRNA^{Ala}$ isolated by Robert Holley in 1965 (Chapter 7) bound to three codons: GCU, GCC, and GCA.

Anticodon
3'—C—G—G—5'
5'—G—C—C—3'
Codon

***Figure* 31.14** Codon–anticodon pairing. Complementary trinucleotide sequence elements align in antiparallel fashion.

The Wobble Hypothesis

Francis Crick considered these results and tested alternative base-pairing possibilities by model building. He hypothesized that the first two bases of the codon and the last two bases of the anticodon form canonical Watson–Crick A:U or G:C base pairs, but pairing between the third base of the codon and the first base of the anticodon follows less stringent rules. That is, a certain amount of play, or **wobble,** might occur in base pairing at this position. The first base of the anticodon is sometimes referred to as the **wobble position.** Crick examined the steric consequences of various noncanonical base pairs. The purine inosine was included since it was known to be a component of tRNAs. In some pairs, the bases were rather close together, as revealed by the

(a)

The guanine-adenine base pair

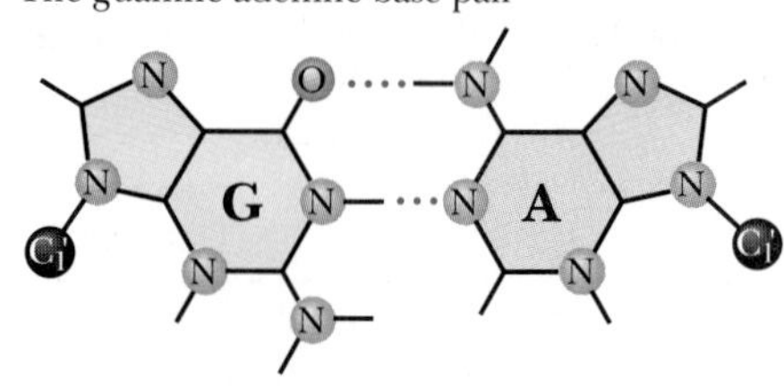

The two possible uracil-uracil base pairs

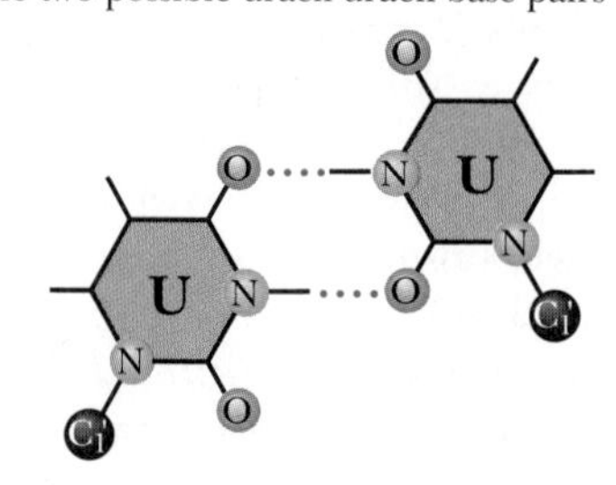

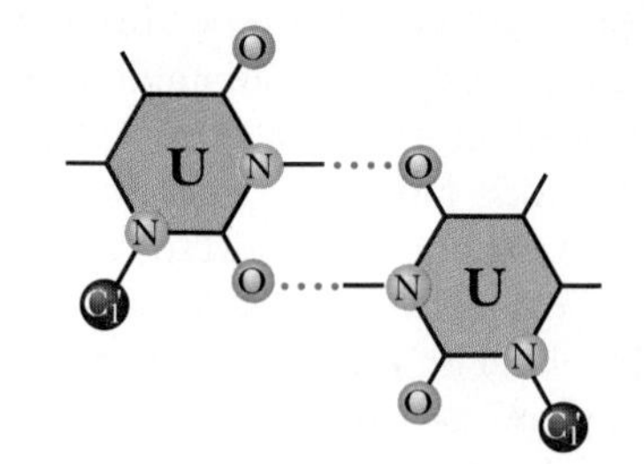

The uracil-cytosine base pair

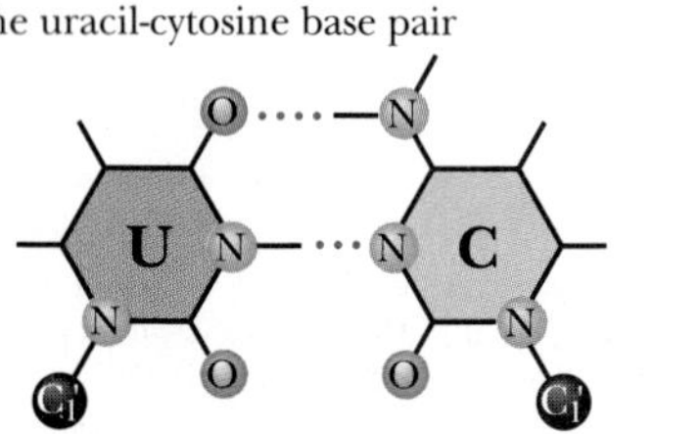

The guanine-uracil and inosine-uracil base pairs are similar

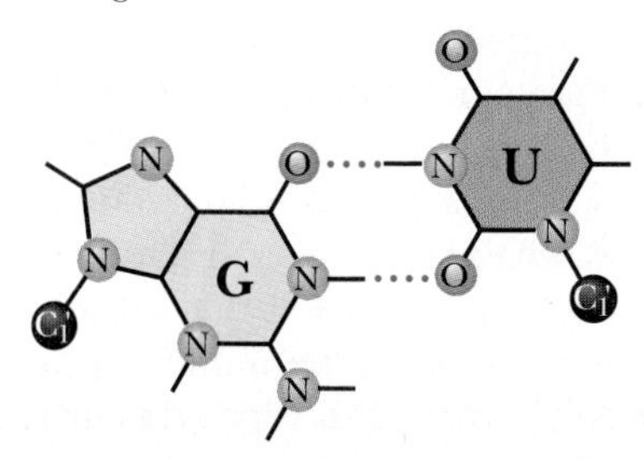

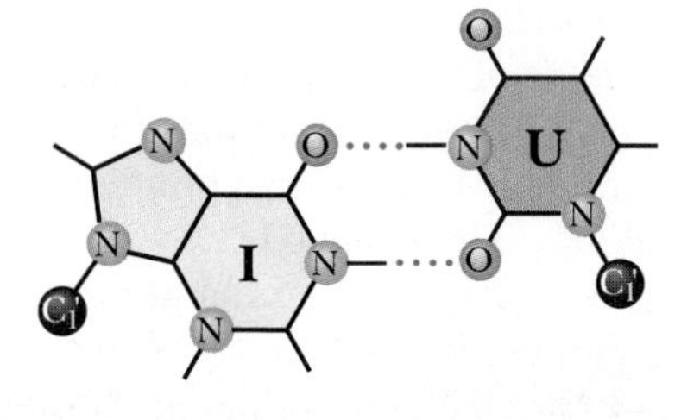

The inosine-adenine base pair

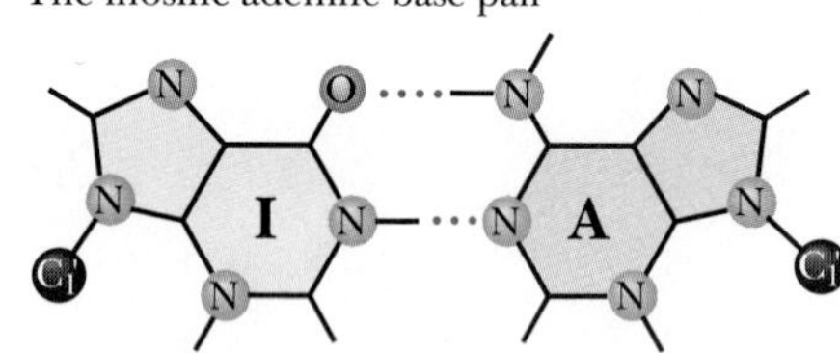

(b)

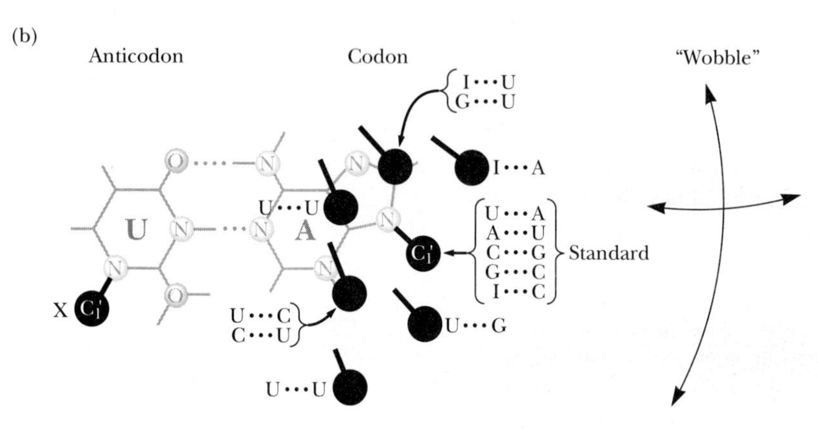

***Figure* 31.15** Various base-pairing alternatives. (a) G:A is unlikely because the 2-NH_2 of G cannot form one of its H bonds; even water is sterically excluded. U:C may be possible even though the two C=O are juxtaposed. Two U:U arrangements are feasible. G:U and I:U are both possible and somewhat similar. The purine pair I:A is also possible. (b) The relative positions of the glycosidic C_1' atoms in various base-pairing alternatives. The positional variation seen for the codon C_1' carbon atom is a measure of wobble. The U:C, C:U, and either of the two possible U:U base pairs bring the respective glycosidic C_1' atoms closer than the standard position; C_1' atoms in I:U, G:U, and U:G pairs are spaced similar to the standard; the I:A pair moves them farther apart.

(Adapted from Crick, F. H. C., 1966. Codon–anticodon pairing: The wobble hypothesis. Journal of Molecular Biology 19:548–555.)

relative positions of the glycosidic C_1' atoms (Figure 31.15). In Figure 31.15, the C_1' of the first nucleotide of the anticodon is taken as fixed and the relative position of the corresponding codon third-nucleotide C_1' is shown. Since the genetic code must distinguish between U or C versus A or G in the third position (as in the codons for Phe versus Leu or His versus Gln), the three close positions must be intolerable; otherwise, anticodon U might read any base—U:U, U:C, U:G, and U:A pairs all could form, not just A and G.

This constraint leads to a set of rules for pairing between the third base of the codon and the first base of the anticodon (Table 31.6). The wobble rules indicate that a first-base anticodon U could recognize either an A or G in the codon third-base position; first-base anticodon G might recognize either U or C in the third-base position of the codon; and first-base anticodon I might interact with U, C, or A in the codon third position.[2]

Note that inosine is a versatile base in establishing degeneracy. (Inosine arises in tRNAs from specific A residues that undergo deamination.) Yeast $tRNA^{Ala}$ (Figure 7.36) has I in the wobble position. The wobble rules also predict that four-codon families (like Pro or Thr) where any of the four bases may be in the third position require at least two different tRNAs. Such four-codon families could be read by two tRNAs whose recognition patterns are either UC and AG or are UCA and G.

The only codons for a given amino acid that differ in either of the first two bases are the 6-codon families for Leu, Ser, and Arg; these amino acids require at least three different tRNAs. Altogether a minimum of 31 tRNAs are necessary to interpret the 61 sense codons. However, most cells have more than 32 different tRNA species. We saw that *E. coli* has 5 distinct tRNAs for the 6 Ser codons. Some tRNAs have the same anticodon but differ in their nucleotide sequences. For example, there are two distinct $tRNA^{Tyr}$ species in *E. coli,* both having a GUA anticodon capable of reading the UAU and UAC codons for Tyr. All members of the set of tRNAs specific for a particular amino acid—termed **isoacceptor tRNAs**—are served by one aminoacyl-tRNA synthetase.

Table **31.6**

Base-Pairing Possibilities at the Third Position of the Codon

Base on the Anticodon	Bases Recognized on the Codon
U	A, G
C	G
A	U
G	U, C
I	U, C, A

Source: Adapted from Crick, F. H. C., 1966. Codon–anticodon pairing: The wobble hypothesis. *Journal of Molecular Biology* **19:**548–555.

The Purpose of Wobble

The first two bases of the codon confer most of the codon:anticodon specificity. The wobble position also contributes to codon recognition and specificity, but hydrogen bonds between noncanonical base pairs are weaker, and thus the pairing here is "looser." Wobbling is possible because the 5′-side of the anticodon is situated in a conformationally flexible part of the tRNA anticodon loop. There is a kinetic advantage to wobble: If all three base pairs of the codon–anticodon complex were of the strong Watson–Crick type, codon–anticodon associations would be more stable and the tRNAs would dissociate less readily from the mRNA, slowing the rate of protein synthesis. However, since the wobble position makes only a marginal contribution to codon–anticodon interaction, wobble tends to accelerate the process of translation.

31.6 Codon Usage

Since more than one codon exists for most amino acids, the possibility for variation in codon usage arises. Indeed, variation in codon usage accommodates the fact that the DNA of different organisms varies in relative A:T/G:C content. However, even in organisms of average base composition, codon usage may be biased. Table 31.7 summarizes data from *E. coli* and humans reflecting the nonrandom usage of codons. Of over 109,000 Leu codons tabulated in human genes, CUG was used over 48,000 times, CUC over 23,000 times, but UUA just 6,000 times.

The occurrence of codons in *E. coli* mRNAs correlates well with the relative abundance of the tRNAs that read them. *Preferred codons are represented by the most abundant isoacceptor tRNAs.* Further, mRNAs for proteins that are syn-

[2]Thus, the first base of the anticodon indicates whether the tRNA can read one, two, or three different codons: Anticodons beginning with A or C read only one codon, those beginning with G or U read two, while anticodons beginning with I can read three codons.

***Table* 31.7**

Codon Usage in *E. coli* and Human Genes

The results for *E. coli* are tabulated from 524,410 codons recorded in 1562 genes and are expressed as frequency of occurrence per 1000 codons tabulated. (For example, the Arg codon AGG was present at a frequency of 1.4 per 1000 codons, or 7341 times out of 524,410 codons tabulated.) The results for human genes are tabulated from

Amino Acid	Codon	*E. coli* Gene Frequency/1000	Human Gene Frequency/1000
Arg	CGA	3.1	5.6
	CGC	22.1	11.2
	CGG	4.6	10.7
	CGU	24.1	4.6
	AGA	2.1	9.8
	AGG	1.4	10.8
Leu	CUA	3.2	6.1
	CUC	9.9	20.1
	CUG	54.6	42.1
	CUU	10.2	10.8
	UUA	10.9	5.4
	UUG	11.5	11.1
Ser	UCA	6.8	9.7
	UCC	9.4	17.8
	UCG	8.0	4.1
	UCU	10.4	13.3
	AGC	15.2	18.7
	AGU	7.2	9.9
Thr	ACA	6.5	14.3
	ACC	24.3	22.6
	ACG	12.7	6.6
	ACU	10.2	12.6
Pro	CCA	8.2	15.4
	CCC	4.3	20.6
	CCG	23.8	6.8
	CCU	6.6	16.1
Ala	GCA	15.6	14.4
	GCC	34.4	29.7
	GCG	32.9	7.2
	GCU	13.4	18.9
Gly	GGA	7.0	17.4
	GGC	30.2	25.3
	GGG	9.2	17.5
	GGU	27.6	11.5
Val	GUA	11.6	6.1
	GUC	14.2	16.2
	GUG	25.3	30.7
	GUU	20.1	10.2

Table 31.7

(*continued*)

1,145,022 codons recorded in 2681 genes and are expressed as frequency of occurrence of a particular codon per 1000 codons tabulated. (For example, the Arg codon CGA was present at a frequency of 5.6 per 1000 codons, or 6412 times out of 1,145,022 codons tabulated.)

Amino Acid	Codon	*E. coli* Gene Frequency/1000	Human Gene Frequency/1000
Lys	AAA	36.5	21.9
	AAG	12.0	35.2
Asn	AAC	23.9	22.3
	AAU	16.3	16.5
Gln	CAA	13.2	10.8
	CAG	30.1	33.8
His	CAC	10.7	14.7
	CAU	11.6	9.3
Glu	GAA	43.4	26.4
	GAG	19.2	41.6
Asp	GAC	21.8	28.9
	GAU	32.3	21.5
Tyr	UAC	13.4	18.0
	UAU	15.4	12.3
Cys	UGC	6.1	13.8
	UGU	4.7	9.9
Phe	UUC	18.2	22.1
	UUU	19.2	15.8
Ile	AUA	4.1	6.1
	AUC	26.5	24.3
	AUU	27.2	15.0
Met	AUG	26.5	22.3
Trp	UGG	12.8	13.7
stop	UAA	2.0	0.7
	UAG	0.2	0.5
	UGA	0.8	0.1

Source: Adapted from Wada, K., et al., 1992. Codon usage tabulated from the GenBank genetic sequence data. *Nucleic Acids Research* **20:**2111–2118.

thesized in abundance tend to employ preferred codons. Rare tRNAs correspond to rarely used codons, and messages containing such codons might experience delays in translation.

31.7 Nonsense Suppression

Mutations that alter a sense codon to one of the three nonsense codons—UAA, UAG, or UGA—result in premature termination of protein synthesis and the release of truncated (incomplete) polypeptides. Geneticists found that second mutations elsewhere in the genome were able to *suppress* the effects of nonsense mutations so that the organism survived. The molecular basis for such intergenic suppression was a mystery until it was realized that **suppressors** were mutations in tRNA genes that altered the anticodon so that the mutant tRNA could now read a particular ''stop'' codon and insert an amino acid. For example, alteration of the anticodon of a $tRNA^{Tyr}$ from GUA to CUA allows this tRNA to read the so-called *amber* stop codon, UAG, and insert Tyr. (The nonsense codons are whimsically named *amber* [UAG], *ochre* [UAA], and *opal* [UGA]). **Suppressor tRNAs** are typically generated from minor tRNA species within a set of isoacceptor tRNAs, so their recruitment to a new role via mutation is not particularly deleterious to the organism. Several different suppressor tRNAs for each of the stop codons have been characterized in *E. coli.*

An interesting amber suppressor mutation results when the anticodon of a $tRNA^{Trp}$ is altered from CAA to CUA. Surprisingly, this suppressor tRNA inserts glutamine, *not* tryptophan. Thus, suppressor tRNAs don't necessarily carry the same amino acid as the wild-type tRNA. (Cells carrying this suppressor must also have a wild-type copy of $tRNA^{Trp}$ to survive.) This suppressor tRNA is no longer a good substrate for tryptophanyl $tRNA^{Trp}$ synthetase, which evidently selects its tRNAs via anticodon recognition. Instead, glutamine tRNA synthetase charges it with Gln. Apparently, glutamine tRNA synthetase has a relaxed specificity for this tRNA substrate. A single base change has influenced both codon–anticodon recognition and the interaction of the tRNA with aminoacyl-tRNA synthetases.

Problems

1. The following sequence represents part of the nucleotide sequence of a cloned cDNA:

   . . . CAATACGAAGCAATCCCGCGACTAGACCTTAAC . . .

   Can you reach an unambiguous conclusion from these data about the partial amino acid sequence of the protein encoded by this cDNA?

2. A random (AG) copolymer was synthesized with polynucleotide phosphorylase using a mixture of 5 parts of ADP, 1 part GDP as substrate. If this random copolymer is used as an mRNA in a cell-free protein synthesis system, which amino acids will be incorporated into the polypeptide product? What will be the relative abundances of these amino acids in the product?

3. List the reagents and cellular components you would need to carry out an elucidation of the genetic code, using the aminoacyl-tRNA:trinucleotide:ribosome-binding assay of Marshall Nirenberg and Philip Leder. Describe the experimental protocol to be followed.

4. Francis Crick reasoned that tRNAs are adapter molecules and, as such, they play no decision-making role in the translation of mRNA into protein. What experimental evidence proves that this interpretation is correct?

5. Review the evidence establishing that aminoacyl-tRNA synthetases bridge the information gap between amino acids and codons. Indicate the various levels of specificity possessed by aminoacyl-tRNA synthetases that are essential for high-fidelity translation of messenger RNA molecules.

6. Draw base-pair structures for (a) a G:C base pair, (b) a C:G base pair, (c) a G:U base pair, and (d) a U:G base pair. Note how these various base pairs differ in the potential hydrogen-bonding patterns they present within the major groove and minor groove of a double-helical nucleic acid.

7. Point out why Crick's wobble hypothesis would allow fewer than 61 anticodons to be used to translate the 61 sense codons. How does "wobble" tend to accelerate the rate of translation?

8. How many codons can mutate to become nonsense codons through a single base change? Which amino acids do they encode?

9. Nonsense suppression occurs when a suppressor mutant arises that reads a nonsense codon and inserts an amino acid, as if the nonsense codon were actually a sense codon. Which amino acids do you think are most likely to be incorporated by nonsense suppressor mutants?

Further Reading

Burbaum, J. J., and Schimmel, P., 1991. Structural relationships and the classification of aminoacyl-tRNA synthetases. *Journal of Biological Chemistry* **266:**16965–16968.

Carter, C. W., Jr., 1993. Cognition, mechanism, and evolutionary relationships in aminoacyl-tRNA synthetases. *Annual Review of Biochemistry* **62:**715–748.

Crick, F. H. C., 1966. Codon–anticodon pairing: The wobble hypothesis. *Journal of Molecular Biology* **19:**548–555. Crick's original paper on wobble interactions between tRNAs and mRNA.

Crick, F. H. C., et al., 1961. General nature of the genetic code for proteins. *Nature* **192:**1227–1232. An insightful paper on insertion/deletion mutants providing convincing genetic arguments that the genetic code was a triplet code, read continuously from a fixed starting point. This genetic study foresaw the nature of the genetic code, as later substantiated by biochemical results.

Eriani, G., et al., 1990. Partition of tRNA synthetases into two classes based on mutually exclusive sets of sequence motifs. *Nature* **347:**203–206.

Francklyn, C., Shi, J.-P., and Schimmel, P., 1992. Overlapping nucleotide determinants for specific aminoacylation of RNA microhelices. *Science* **255:**1121–1125. One of a series of papers from Schimmel's laboratory elucidating tRNA features that serve as aminoacyl-tRNA synthetase recognition elements.

Khorana, H. G., et al., 1966. Polynucleotide synthesis and the genetic code. Cold Spring Harbor Symposium on Quantitative Biology **31:**39–49. The use of synthetic polyribonucleotides in elucidating the genetic code.

Musier-Forsyth, K., et al., 1991. Specificity for aminoacylation of an RNA helix: An unpaired, exocyclic amino group in the minor groove. *Science* **253:**784–786. An unpaired amino group at a particular location in the minor groove of $tRNA^{Ala}$ determines its amino acid acceptor specificity.

Nirenberg, M. W., and Leder, P., 1964. RNA codewords and protein synthesis. *Science* **145:**1399–1407. The use of simple trinucleotides in the ribosome-binding assay to decipher the genetic code.

Nirenberg, M. W., and Matthaei, J. H., 1961. The dependence of cell-free protein synthesis in *E. coli* upon naturally occurring or synthetic polyribonucleotides. *Proceedings of the National Academy of Sciences, USA* **47:**1588–1602.

Nishimura, S., Jones, D. S., and Khorana, H. G., 1965. Studies on polynucleotides, XLVIII, The *in vitro* synthesis of a co-polypeptide containing two amino acids in alternating sequence dependent upon a DNA-like polymer containing two nucleotides in alternating sequence. *The Journal of Molecular Biology* **13:**302–324.

Normanly, J., and Abelson, J., 1989. tRNA identity. *Annual Review of Biochemistry* **58:**1029–1049. Review of the structural features of tRNA that are recognized by aminoacyl-tRNA synthetases.

Rould, M. A., et al., 1989. Structure of *E. coli* glutaminyl-tRNA synthetase complexed with $tRNA^{Gln}$ and ATP at 2.8 Å resolution. *Science* **246:**1135–1142. One of the first high-resolution, three-dimensional structures of an aminoacyl-tRNA synthetase complexed with its cognate tRNA provides insights into the features employed by these enzymes in recognizing unique tRNAs and translating the genetic code.

Schimmel, P., 1989. Parameters for the molecular recognition of transfer RNAs. *Biochemistry* **28:**2747–2759.

Schimmel, P., 1987. Aminoacyl-tRNA synthetases: General scheme of structure-function relationships in the polypeptides and recognition of transfer RNAs. *Annual Review of Biochemistry* **56:**125–158.

Schulman, L. H., and Abelson, J., 1988. Recent excitement in understanding transfer RNA identity. *Science* **240:**1591–1592.

Speyer, J. F., et al., 1963. Synthetic polynucleotides and the amino acid code. Cold Spring Harbor Symposium on Quantitative Biology **28:**559–567.

Yanofsky, C., et al., 1964. On the colinearity of gene structure and protein structure. *Proceedings of the National Academy of Sciences, USA* **51:**266–272. Yanofsky's classic work establishing the collinearity of genes and proteins.

"... *we are a spectacular, splendid manifestation of life. We have language and can build metaphors as skillfully and precisely as ribosomes make proteins. We have affection. We have genes for usefulness, and usefulness is about as close to a 'common goal' of nature as I can guess at. And finally, and perhaps best of all, we have music.*"

Lewis Thomas (1913–1994)
"The Youngest and Brightest Thing Around" in THE MEDUSA AND THE SNAIL (1979)

Chapter 32

Protein Synthesis and Degradation

Learning sign language. Sign language is a means of translating information from a verbal to a visual form.

Outline

32.1 Ribosome Structure and Assembly
32.2 The Mechanics of Protein Synthesis
32.3 Protein Synthesis in Eukaryotic Cells
32.4 Inhibitors of Protein Synthesis
32.5 Protein Folding
32.6 Post-Translational Processing of Proteins
32.7 Protein Degradation

Protein biosynthesis is achieved by the process of **translation.** Translation converts the language of genetic information embodied in the base sequence of a messenger RNA molecule into the amino acid sequence of a polypeptide chain. During translation, proteins are synthesized on ribosomes by linking amino acids together in the specific linear order stipulated by the sequence of codons in an mRNA. *Ribosomes* are the agents of protein synthesis.

32.1 Ribosome Structure and Assembly

Ribosomes are compact ribonucleoprotein particles found in the cytosol of all cells, as well as in the matrix of mitochondria and the stroma of chloroplasts. The general structure of ribosomes was described in Chapter 6; here we consider their structure in light of their function in synthesizing proteins. Ribosomes are mechano-chemical systems that move along mRNA templates, orchestrating the interactions between successive codons and the corresponding anticodons presented by aminoacyl-tRNAs. As they align successive amino acids via codon:anticodon recognition, ribosomes also catalyze the formation of peptide bonds between adjacent amino acid residues.

The Composition of Prokaryotic Ribosomes

E. coli ribosomes are representative of the structural organization of the prokaryotic versions of these supramolecular protein-synthesizing machines (Table 32.1, see also Figure 6.25). The *E. coli* ribosome is a roughly globular particle with a diameter of 25 nm, a sedimentation coefficient of 70S, and a mass of about 2520 kD. It consists of two unequal subunits that dissociate from each other at Mg^{2+} concentrations below 1 m*M*. The smaller, or **30S,** subunit has a mass of 930 kD and is composed of 21 different proteins and a so-called **16S ribosomal RNA (rRNA)** molecule 1542 nucleotides long. The 21 proteins are designated **S1** to **S21** in decreasing order of size and collectively contribute 370 kD to the mass of the 30S subunit. The larger **50S** subunit has a mass of 1590 kD and consists of 31 different polypeptides (**L1** to **L34,** see below) and two rRNAs, a 2904-nucleotide **23S rRNA** and a 120-nucleotide **5S rRNA.** Together, the L proteins contribute 487 kD to the 50S subunit mass. Thus, ribosomes are roughly two-thirds RNA and one-third protein by mass. An *E. coli* cell contains around 20,000 ribosomes, constituting about 20% of the dry cell mass.

Ribosomal Proteins

There is one copy of each ribosomal protein per 70S ribosome, excepting protein L7/L12 (L7 and L12 have identical amino acid sequences and differ only in the degree of N-terminal acetylation). Four copies of L7/L12 are present per ribosome; together with one copy of L10 they form a complex originally designated L8, accounting in part for the historical misassignment of L1 to L34 given to the L set of 31 proteins. Only one protein is common to both the small and large subunit: S20 = L26. The amino acid sequences of all 52 *E. coli* ribosomal proteins are known. The largest is S1 (557 residues, 61.2 kD); the smallest is L34 (46 residues, 5.4 kD). The sequences of ribosomal proteins share little similarity. These proteins are typically rich in the cationic amino acids Lys and Arg and have few aromatic amino acid residues, properties appropriate to proteins intended to interact strongly with polyanionic RNAs. Little is known regarding the three-dimensional structure of ribosomal proteins. Ribosomal proteins tend to be insoluble in aqueous solutions and are thus difficult to study.

***Table* 32.1**
Structural Organization of *E. coli* Ribosomes

	Ribosome	Small Subunit	Large Subunit
Sedimentation coefficient	70S	30S	50S
Mass (kD)	2520	930	1590
Major RNAs		16S = 1542 bases	23S = 2904 bases
Minor RNAs			5S = 120 bases
RNA mass (kD)	1664	560	1104
RNA proportion	66%	60%	70%
Protein number		21 polypeptides	31 polypeptides
Protein mass (kD)	857	370	487
Protein proportion	34%	40%	30%

rRNAs

The three *E. coli* rRNA molecules—23S, 16S, and 5S—are derived from a single 30S rRNA precursor transcript that also includes several tRNAs (Figure 32.1). Ribosomal RNAs show extensive potential for intrachain hydrogen bonding and assume secondary structures reminiscent of tRNAs, although substantially more complex (Figures 7.39 and 7.40). Almost half the bases in 16S rRNA are base paired. Relatively small double-helical regions are punctuated by short, single-stranded stretches, generating hairpin configurations that dominate the molecule; four distinct domains (I through IV) can be discerned in the secondary structure. The three-dimensional structure of rRNAs remains unsolved. Electron micrographs of 16S rRNA show an intriguing resemblance to pictures of complete 30S subunits.

Self-Assembly of Ribosomes

Ribosomal subunit self-assembly is one of the paradigms for the spontaneous formation of supramolecular complexes from their macromolecular components (Chapter 34). If the individual proteins and rRNAs comprising ribosomal subunits are mixed together under appropriate conditions of pH and ionic strength, spontaneous self-assembly into functionally competent subunits takes place without the intervention of any additional factors or chaperones. Apparently, the rRNA acts as a scaffold upon which the various ribosomal proteins convene. 30S subunit assembly is diagramed in Figure 32.2; the 50S subunit assembles in similar fashion. This pattern has been traced by observing the consequences of stepwise omission of selected proteins from the reconstitution mixture. The results show that ribosomal proteins bind in a specified order. The first proteins to bind to 16S rRNA, namely S4, S8, S15, as well as S13 and S7, show cooperative interactions. The assembly passes through intermediate stages where prior binding of certain proteins is essential to formation of an appropriate scaffold for subsequent protein binding, so the overall process is sequential in nature.

***Figure* 32.1** The seven ribosomal RNA operons in *E. coli.* These operons, or gene clusters, are transcribed to give a precursor RNA that is subsequently cleaved by RNase III and other nucleases, at the sites indicated, to generate 23S, 16S, and 5S rRNA molecules, as well as several tRNAs that are unique to each operon. Numerals to the right of the brackets indicate the number of species of tRNA encoded by each transcript.

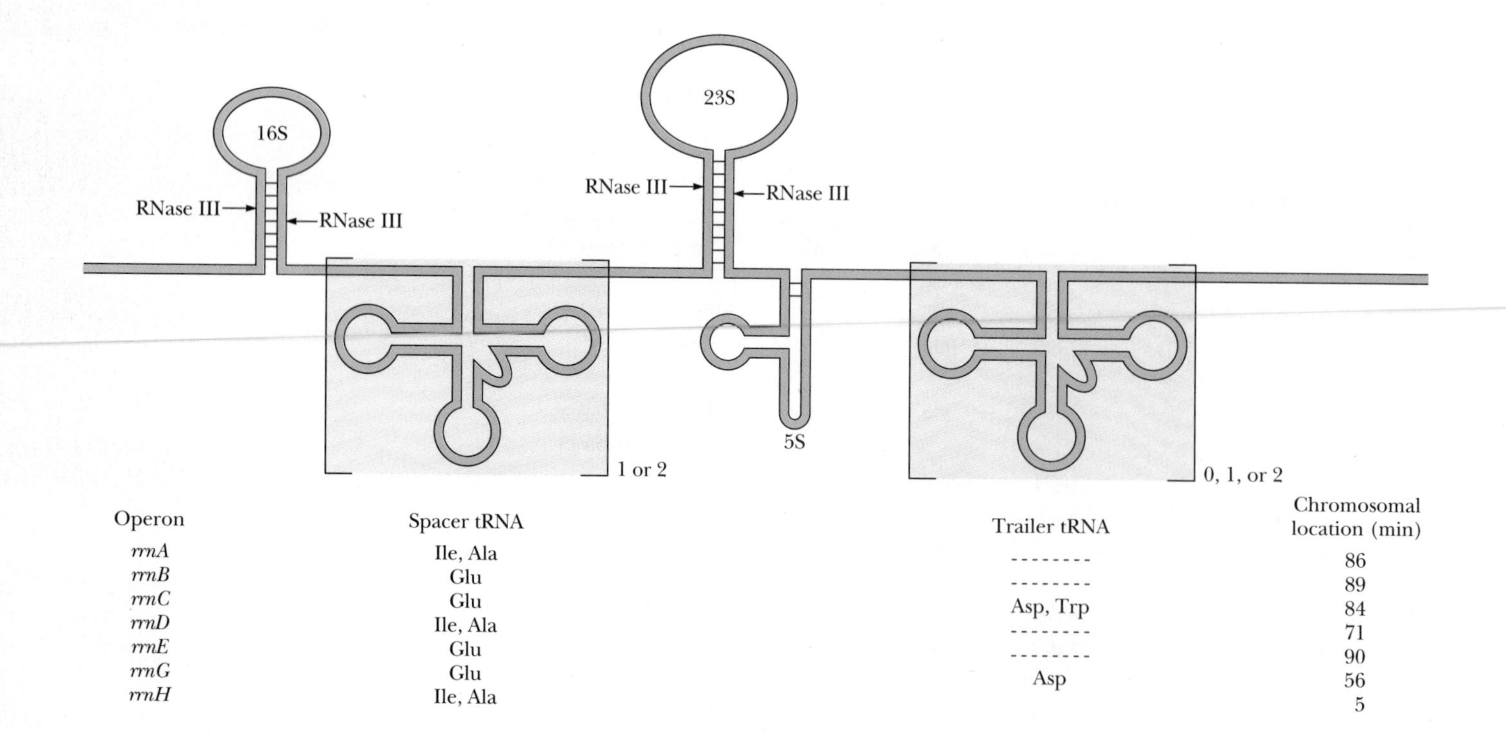

16S
RNase III
RNase III
23S
RNase III
RNase III
5S
1 or 2
0, 1, or 2
Operon
Spacer tRNA
Trailer tRNA
Chromosomal location (min)
rrnA Ile, Ala -------- 86
rrnB Glu -------- 89
rrnC Glu Asp, Trp 84
rrnD Ile, Ala -------- 71
rrnE Glu -------- 90
rrnG Glu Asp 56
rrnH Ile, Ala 5

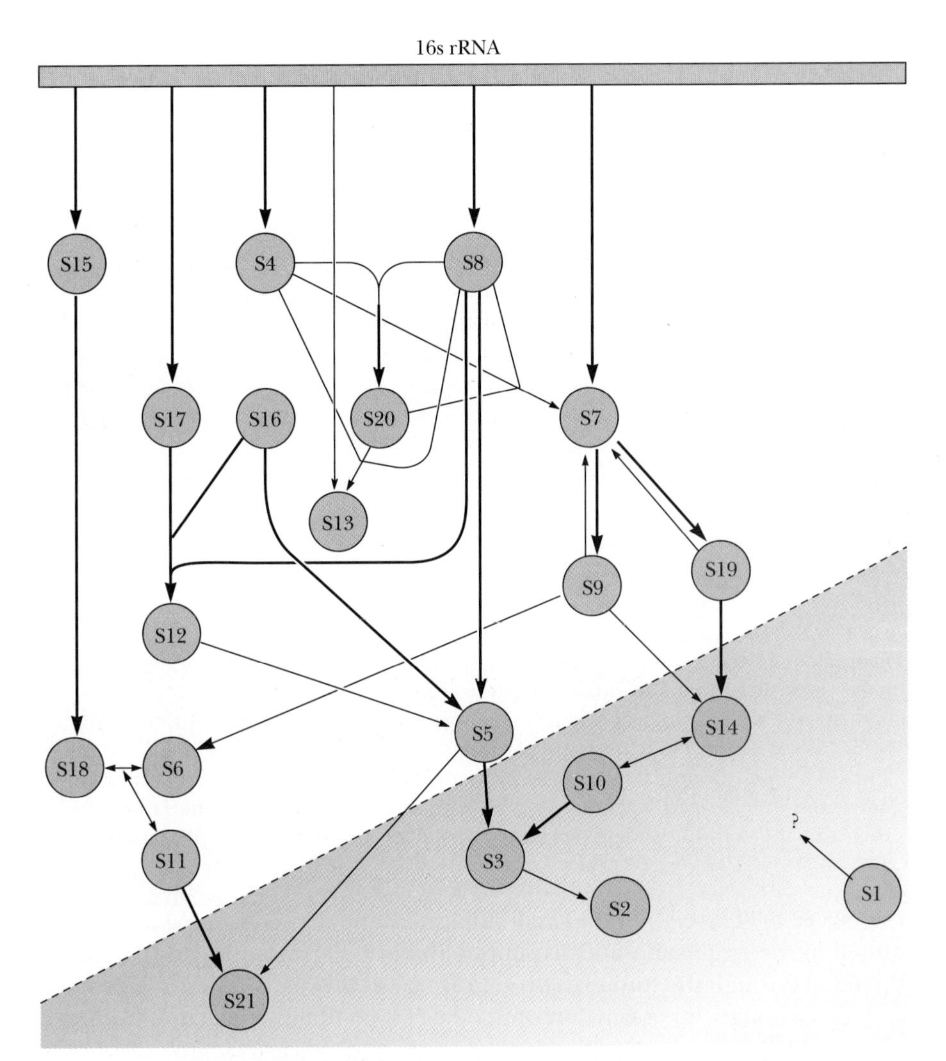

***Figure* 32.2** Assembly of the *E. coli* 30S ribosomal subunit. Note the essential role of 16S rRNA as scaffold. Thick arrows connect the RNA with proteins that bind directly, whereas thin arrows connect rRNA and/or proteins whose binding requires the participation of additional proteins. For example, S7 binds cooperatively with S4, S8, S9, S19, and S20 to 16S rRNA. Double arrows signify mutual effects. The dashed line separates a precursor subassembly from proteins that are the last to bind.

Proteins that depend on the presence of another protein on the scaffold end up adjacent to those proteins in the final assembly, as expected if physical interactions between such proteins are a requirement for binding. Nevertheless, conformational changes accompany protein binding and are probably necessary to expose new sites on the ribonucleoprotein complex for additional proteins to occupy. Assembly of 30S subunits begins even as the rRNA precursor is being transcribed. Interestingly, the first part of the 16S rRNA to be transcribed, the 5′-region, possesses a cluster of the strongest protein-binding sites.

Ribosomal Architecture

Ribosomal subunits have a characteristic three-dimensional architecture that has been revealed by image reconstructions from electron micrographs. Such an analysis of the small ribosomal subunit leads to the model depicted in Figure 32.3. The small subunit features a "head" and a "base" from which a "platform" projects. A "cleft" is defined by the spatial relationship between the head, base, and platform. The large subunit is a globular structure with three distinctive projections: a "central protuberance," the "stalk," and a winglike ridge. The two subunits associate with each other so that the side of the small subunit nestles into the cleft of the large subunit, with the platform of the small subunit oriented toward the "wing" of the large subunit

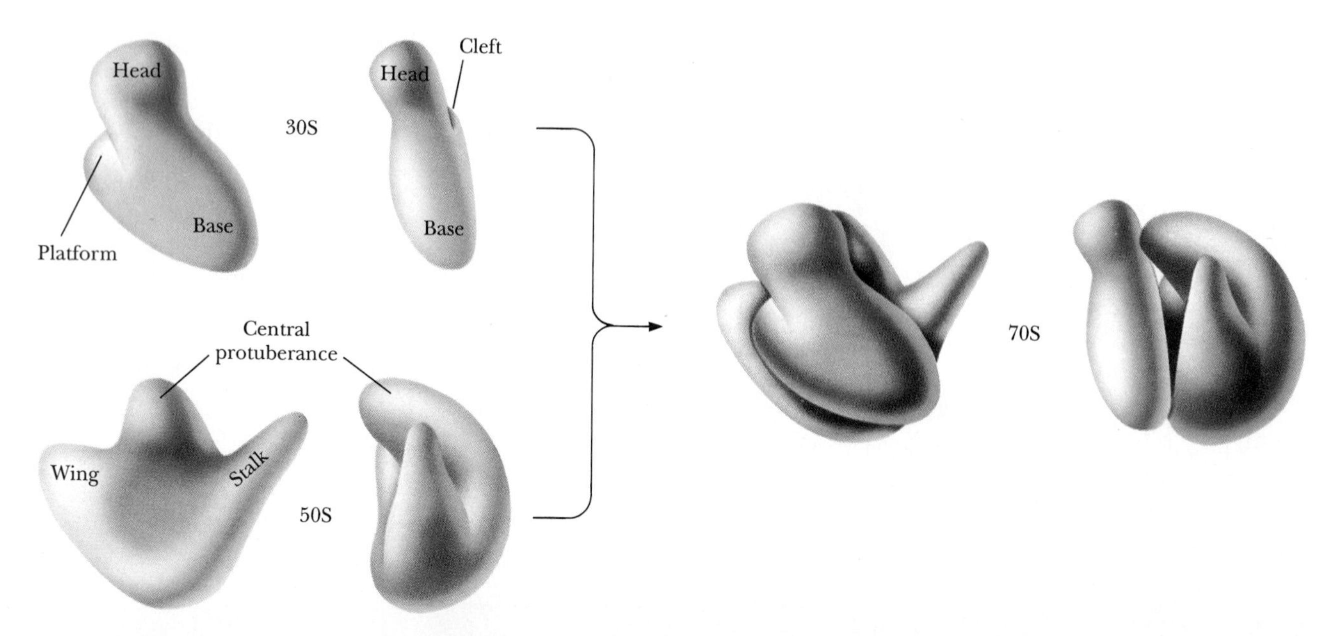

***Figure* 32.3** A three-dimensional model for the *E. coli* ribosome as deduced by image reconstruction. The small and large ribosomal subunits each have characteristic morphological features, such as protuberances and clefts. The 30S subunit is somewhat elongated and asymmetric and has the dimensions 5.5 × 22 × 22 nm. The 50S subunit is predominantly spheroidal but has three projections; its dimensions are 15 × 20 × 20 nm. A 2.5-nm tunnel passes through the large subunit in the central "valley" region between the subunit's three protrusions.

(Figure 32.3). The cleft of the small subunit is thus aligned somewhat with the tunnel in the large subunit. It is supposed that the growing peptidyl chain is threaded through the tunnel as protein synthesis proceeds. The 70S ribosome is large enough to easily accommodate two tRNA molecules at once (Figure 32.4).

Eukaryotic Ribosomes

Eukaryotic cells have ribosomes in their mitochondria (and chloroplasts) as well as in the cytosol. The mitochondrial and chloroplastic ribosomes resemble prokaryotic ribosomes in size, overall organization, structure, and function, a fact reflecting the prokaryotic origins of these organelles. Whereas eukaryotic cytosolic ribosomes retain many of the structural and functional properties of their prokaryotic counterparts, they are larger and considerably more complex. Further, higher eukaryotes have more complex ribosomes than lower eukaryotes. For example, the yeast cytosolic ribosomes have major rRNAs of 3392 (large subunit) and 1799 nucleotides (small subunit); the major rRNAs of mammalian cytosolic ribosomes are 4718 and 1874 nucleotides, respectively. Table 32.2 lists the properties of cytosolic ribosomes in a representative mammal, the rat. Their mass is almost 1.7 times the mass of *E. coli* ribosomes, and proteins contribute a relatively greater proportion of this mass. Small (40S) subunits have 33 different proteins and large (60S) subunits have 49. Large subunits have three characteristic rRNAs: 28S, 5.8S, and 5S. The sequence of the 5.8S rRNA shows homology to the 5′-end of prokaryotic 23S rRNA, suggesting it may be an evolutionary derivative of it. This 5.8S rRNA forms a secondary structure with 28S rRNA through complementary base pairing. Comparison of base sequences and secondary structures of rRNAs from different organisms suggests that evolution has worked to conserve the secondary structure of these molecules, although not necessarily

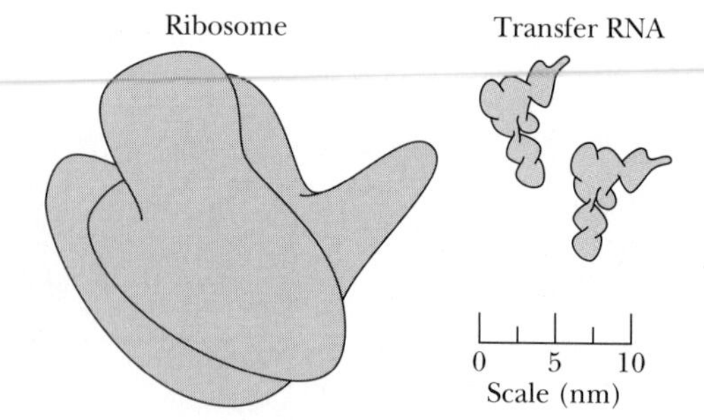

***Figure* 32.4** A comparison of their relative sizes indicates that ribosomes are large enough to accommodate two tRNAs simultaneously.

Table 32.2
Structural Organization of Mammalian (Rat Liver) Cytosolic Ribosomes

	Ribosome	Small Subunit	Large Subunit
Sedimentation coefficient	80S	40S	60S
Mass (kD)	4220	1400	2820
Major RNAs		18S = 1874 bases	28S = 4718 bases
Minor RNAs			5.8S = 160 bases
			5S = 120 bases
RNA mass (kD)	2520	700	1820
RNA proportion	60%	50%	65%
Protein number		33 polypeptides	49 polypeptides
Protein mass (kD)	1700	700	1000
Protein proportion	40%	50%	35%

the nucleotide sequences creating such structure. That is, the retention of a base pair at a particular location seems more important than whether the base pair is G:C or A:U. The morphology of eukaryotic cytosolic ribosomes resembles that of prokaryotic ribosomes.

Conserved Bases in rRNA Are Clustered in Single-Stranded Regions

Although retention of secondary structure apparently plays a leading role in rRNA evolution, conservation of rRNA primary structure at particular places is also a significant feature of these molecules. Comparison of the base sequences of 16S-like (rRNAs) (the small subunit rRNA) from diverse phylogenetic sources reveals that about one-third of the nucleotides are universally conserved. Interestingly, many of these nucleotides are clustered in a few single-stranded regions of the molecule. Several such sequences on the order of 10 to 20 nucleotides long are essentially invariant in all organisms examined. Such conservation strongly suggests that these unpaired stretches play some functional role during protein synthesis.

32.2 The Mechanics of Protein Synthesis

Like chemical polymerization processes, protein biosynthesis in all cells is characterized by three distinct phases: initiation, elongation, and termination. At each stage, the energy driving the assembly process is provided by GTP hydrolysis, and specific soluble protein factors participate in the events.

Initiation involves binding of mRNA by the small ribosomal subunit, followed by association of a particular **initiator aminoacyl-tRNA** that recognizes the first codon. This codon often lies within the first 30 nucleotides or so of mRNA spanned by the small subunit. The large ribosomal subunit then joins the initiation complex, preparing it for the elongation stage.

Elongation includes the synthesis of all peptide bonds from the first to the last. The ribosome remains associated with the mRNA throughout elongation, moving along it and translating its message into an amino acid sequence. This is accomplished via a repetitive cycle of events in which successive aminoacyl-tRNAs add to the ribosome:mRNA complex as directed by codon binding, and the polypeptide chain grows by one amino acid at a time.

(a) Each tRNA binding site extends over both subunits and matches a triplet codon in the mRNA

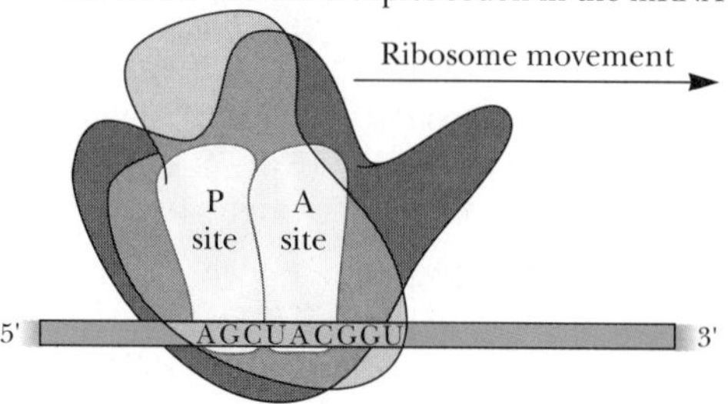

(b) Prior to peptide bond formation, an aminoacyl-tRNA is present in the A site and polypeptidyl-tRNA is situated in the P site

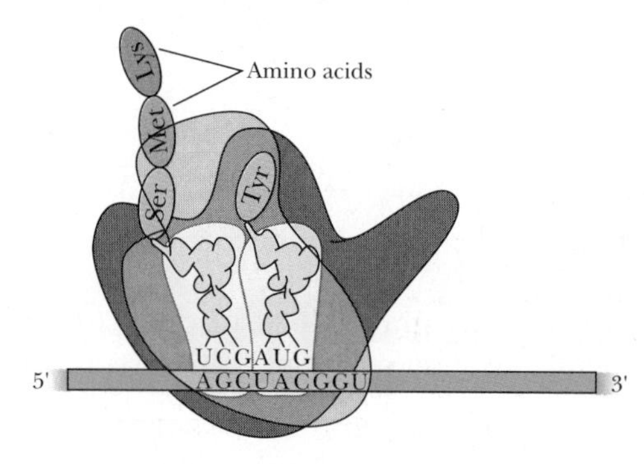

(c) Peptide bond formation involves transfer of the polypeptide to the amino group of the amino acid carried by the tRNA in the A site

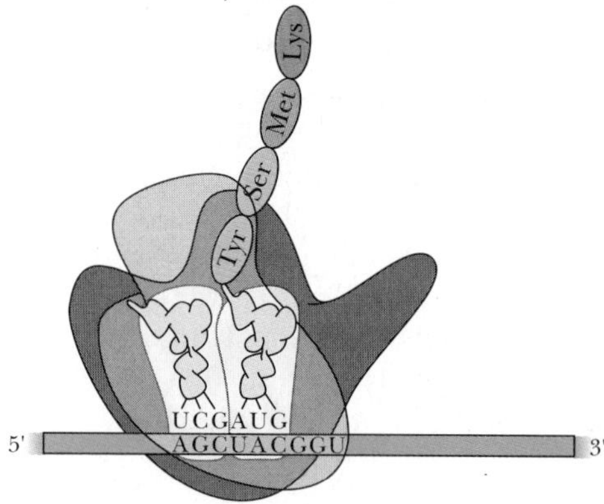

(d) The ribosome then translocates one codon further along the mRNA and the uncharged tRNA is expelled. Translocation places the polypeptidyl-tRNA in the P site and aligns a new codon within the A site, ready to accept the next incoming aminoacyl tRNA

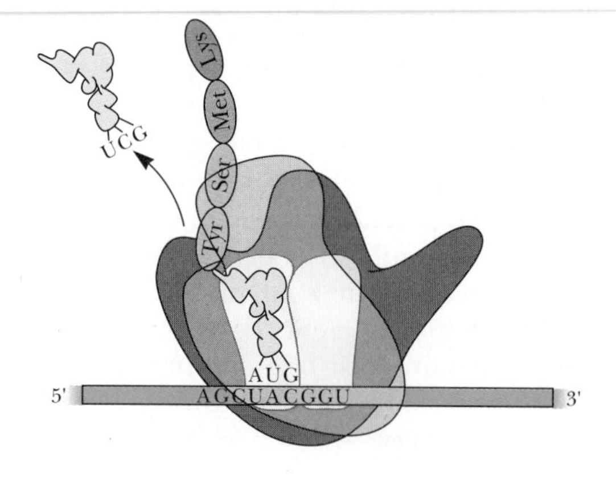

***Figure* 32.5** The basic steps in protein synthesis. Note that the ribosome has two principal sites for binding tRNA, the A, or acceptor, site and the P, or peptidyl, site.

Only two tRNA molecules are part of the ribosome:mRNA complex at any moment. Each lies in a distinct site (Figure 32.5). The **A,** or **acceptor, site** is the attachment site for an incoming aminoacyl-tRNA. The **P,** or **peptidyl, site** is occupied by peptidyl-tRNA, the tRNA carrying the growing polypeptide chain. The elongation reaction transfers the peptide chain from the peptidyl-tRNA in the P site to the aminoacyl-tRNA in the A site. This transfer occurs through covalent attachment of the peptidyl α-carboxyl to the α-amino of the aminoacyl-tRNA, forming a new peptide bond. The new, longer peptidyl-tRNA now moves from the A site into the P site as the ribosome moves one codon further along the mRNA. The A site, left vacant by this translocation, can accept the next incoming aminoacyl-tRNA. These events are summarized in Figure 32.5. A third tRNA-binding site on the ribosome, the **E,** or **exit, site,** is transiently occupied by deacylated tRNAs as they exit the P site, having lost their peptidyl chains. Figure 32.6 depicts plausible locations for the tRNA-binding sites on ribosomes. Elongation is the most rapid phase of protein synthesis.

Termination is triggered when the ribosome reaches a "stop" codon on the mRNA. At this point, the polypeptide chain is released, and the ribosomal subunits dissociate from the mRNA.

Protein synthesis proceeds rapidly. In vigorously growing bacteria, about 15 amino acid residues are added to a growing polypeptide chain each second. So an average protein molecule of about 300 amino acid residues is synthesized in only 20 seconds. Eukaryotic protein synthesis is only about 10% as fast. We will focus first on protein synthesis in *E. coli,* the most intensively studied system.

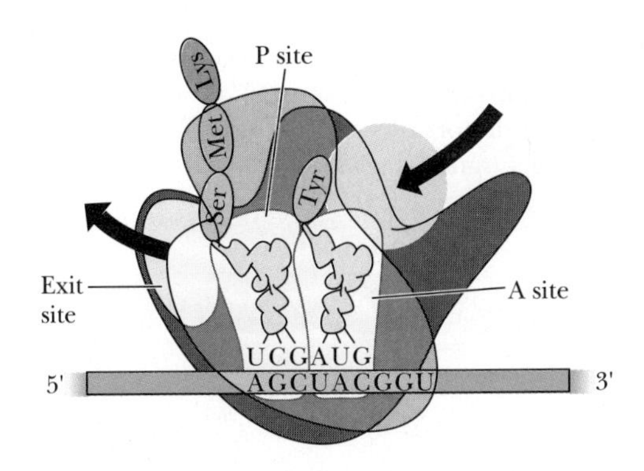

***Figure* 32.6** Plausible locations for tRNA-binding sites within the ribosome. The A, or acceptor site, binds the incoming aa-tRNA, as directed by the mRNA codon underlying it, the pink-shaded area representing the approximate site of interaction of EF-Tu. The P, or peptidyl, site is occupied by the peptidyl-tRNA in association with the mRNA codon previously occupying the A site. The likely path of a tRNA as it transits the events of protein synthesis is shown by the arrows. Juxtaposition of the acceptor stems of the tRNAs in the A and P sites will facilitate peptidyl transfer. Although not obvious in this figure, tRNA binding is considered to be a 50S subunit role, with participation by the 30S subunit being secondary.

(Adapted from Noller, H. F., 1991. Ribosomal RNA and translation. Annual Review of Biochemistry 60:191–227.)

Peptide Chain Initiation in Prokaryotes

The components required for peptide chain initiation include (a) mRNA, (b) 30S and 50S ribosomal subunits, (c) a set of proteins known as **initiation factors,** (d) GTP, and (e) a specific charged tRNA, **f-Met-tRNA$_f^{Met}$.** A discussion of the properties of these components and their interaction follows.

Initiator tRNA

tRNA$_f^{Met}$ is a particular tRNA for reading an AUG (or GUG, or even UUG) codon that signals the start site, or N-terminus, of a polypeptide chain. This tRNA$_f^{Met}$ does not read internal AUG codons, so it does not participate in chain elongation. Instead, that role is filled by another methionine-specific tRNA, referred to as tRNA$_m^{Met}$, which cannot replace tRNA$_f^{Met}$ in peptide chain initiation. (However, both of these tRNAs are loaded with Met by the same methionyl-tRNA synthetase.) The synthesis of all *E. coli* polypeptides begins with the incorporation of a modified methionine residue, *N*-formyl-Met, as *N*-terminal amino acid. However, in about half of the *E. coli* proteins, this Met residue is cleaved off once the growing polypeptide is 10 or so residues long; as a consequence, many mature proteins in *E. coli* lack *N*-terminal Met.

The methionine contributed in peptide chain initiation by tRNA$_f^{Met}$ is unique in that its amino group has been formylated. This reaction is catalyzed by a specific enzyme, **methionyl-tRNA$_f^{Met}$ formyl transferase** (Figure 32.7). Note that the addition of the formyl group to the Met-NH_2 creates an N-terminal block resembling a peptidyl grouping. That is, the initiating Met is transformed into a minimal analog of a peptidyl chain.

The structure of *E. coli* tRNA$_f^{Met}$ has several distinguishing features (Figure 32.8). Unlike the case with all other tRNAs, the 5′-terminal base is not matched with a complementary base in the tRNA$_f^{Met}$ acceptor stem, and thus

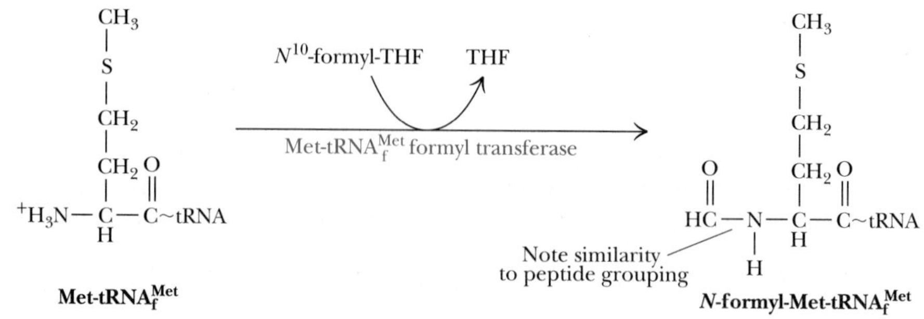

N^{10}-formyl-THF THF

Met-$tRNA_f^{Met}$ formyl transferase

Met-$tRNA_f^{Met}$

Note similarity to peptide grouping

***N*-formyl-Met-$tRNA_f^{Met}$**

***Figure* 32.7** Methionyl-$tRNA_f^{Met}$ formyl transferase catalyzes the transformylation of methionyl-$tRNA_f^{Met}$ using N^{10}-formyl-THF as formyl donor. $tRNA_m^{Met}$ is not a substrate for this transformylase.

no base pair forms here. $tRNA_f^{Met}$ also has a unique CCU sequence in its D loop and an exclusive set of three G:C base pairs in its anticodon stem. Collectively, these features identify this tRNA as essential to initiation and inappropriate for chain elongation.

mRNA Recognition and Alignment

In order for the mRNA to be translated accurately, its sequence of codons must be brought into proper register with the translational apparatus. Recognition of translation initiation sequences on mRNAs involves the 16S rRNA component of the 30S ribosomal subunit. Base pairing between a pyrimidine-rich sequence at the 3′-end of 16S rRNA and complementary purine-rich tracts at the 5′-end of prokaryotic mRNAs brings the 30S ribosomal subunit into proper alignment with respect to translation "start" sites on the mRNA. The purine-rich mRNA sequence, the **ribosome-binding site,** is often called the **Shine-Dalgarno sequence** in honor of its discoverers. Figure 32.9 shows various Shine-Dalgarno sequences found in prokaryotic mRNAs, along with the complementary 3′-tract on *E. coli* 16S rRNA. The 3′-end of 16S rRNA resides in the "head" region of the 30S small subunit.

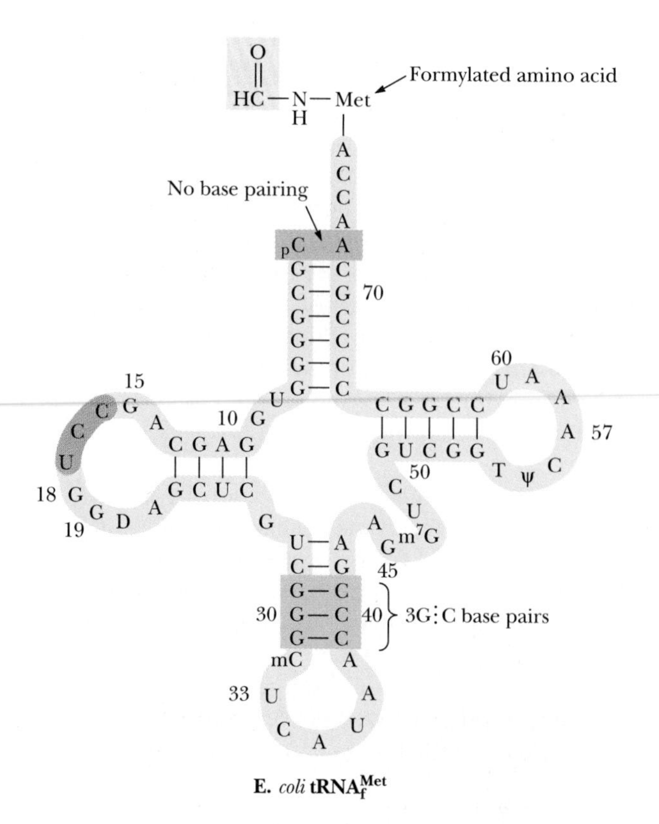

***Figure* 32.8** The structure of *E. coli* N-formyl-methionyl-$tRNA_f^{Met}$. The features distinguishing it from noninitiator tRNAs are highlighted.

	Initiation codon
araB	– U U U G G A U G G A G U G A A A C G A U G G C G A U U –
galE	– A G C C U A A U G G A G C G A A U U A U G A G A G U U –
lacI	– C A A U U C A G G G U G G U G A A U U G U G A A A C C A –
lacZ	– U U C A C A C A G G A A A C A G C U A U G A C C A U G –
Q β phage replicase	– U A A C U A A G G A U G A A A U G C A U G U C U A A G –
φX174 phage A protein	– A A U C U U G G A G G C U U U U U U A U G G U U C G U –
R17 phage coat protein	– U C A A C C G G G G U U U G A A G C A U G G C U U C U –
ribosomal protein S12	– A A A A C C A G G A G C U A U U U A A U G G C A A C A –
ribosomal protein L10	– C U A C C A G G A G C A A A G C U A A U G G C U U U A –
trpE	– C A A A A U U A G A G A A U A A C A A U G C A A A C A –
trpL leader	– G U A A A A A G G G U A U C G A C A A U G A A A G C A –
3′-end of 16S rRNA	3′ $_{HO}$A U U C C U C C A C U A G – 5′

***Figure* 32.9** Various Shine-Dalgarno sequences recognized by *E. coli* ribosomes. These sequences lie about 10 nucleotides upstream from their respective AUG initiation codon and are complementary to the UCCU core sequence element of *E. coli* 16S rRNA. G:U as well as canonical G:C and A:U base pairs are involved here.

These recognition events are verified by studies of prokaryotic ribosomes treated with the bacteriocidal protein **colicin E3.** This protein is a phosphodiesterase that specifically cleaves the bond after position 1493 in 16S rRNA, removing the 3′-terminal 49 nucleotides of 16S rRNA that include the pyrimidine-rich Shine-Dalgarno-binding sequence (Figure 7.39). Colicin E3–treated ribosomes are competent in the elongation of polypeptide chains whose synthesis has already been initiated, but these ribosomes can no longer initiate mRNA translation.

Initiation Factors

Initiation involves interaction of the **initiation factors (IFs)** with GTP, *N*-formyl-Met-tRNA$_f^{Met}$, mRNA, and the 30S subunit to give a **30S initiation complex** to which the 50S subunit then adds to form a **70S initiation complex.** The initiation factors are soluble proteins required for assembly of proper initiation complexes. Their properties are summarized in Table 32.3. The requirement for these proteins was discovered when it was found that 30S ribosomal subunits "washed" with 1 *M* NH_4Cl were inactive in initiating protein synthesis, unless loosely associated proteins that had been removed in the "wash" were added back.

Events in Initiation

Initiation begins when a 30S subunit:(IF-3:IF-1) complex binds mRNA and a complex of IF-2, GTP, and f-Met-tRNA$_f^{Met}$. The sequence of events is summarized in Figure 32.10. Although IF-3 is absolutely essential for mRNA binding by the 30S subunit, it is not involved in locating the proper translation initia-

***Table* 32.3**

Properties of *E. coli* Initiation Factors

Factor	Mass (kD)	Molecules/ Ribosome	Function
IF-1	9	0.15	Assists IF-3 function
IF-2	97		Binds initiator tRNA and GTP
IF-3	23	0.25	Binds to 30S subunits and directs mRNA binding

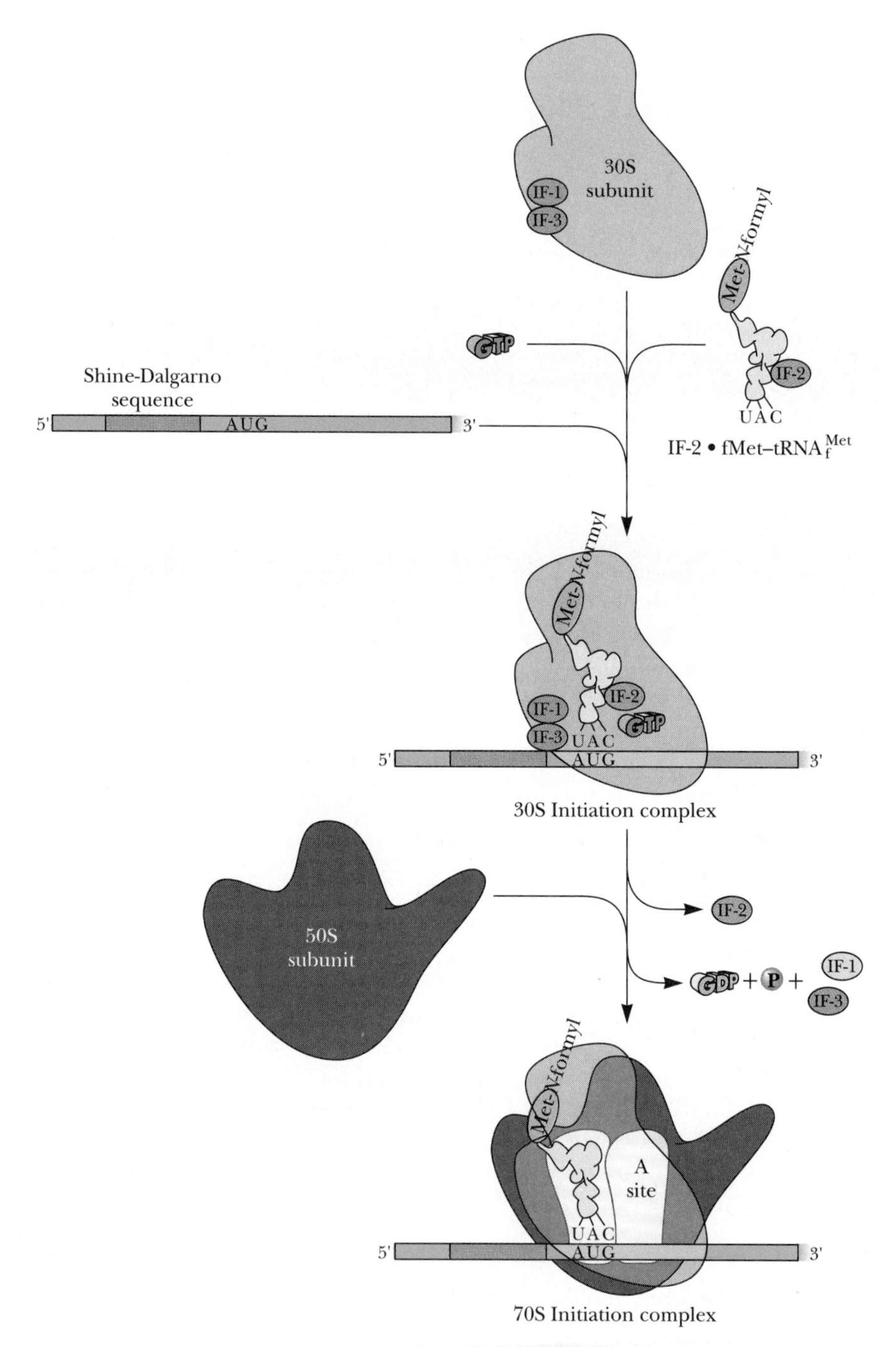

***Figure* 32.10** The sequence of events in peptide chain initiation.

tion site on the message. The presence of IF-3 on 30S subunits also prevents them from reassociating with 50S subunits. IF-3 must dissociate before the 50S subunit will associate with the mRNA:30S subunit complex.

IF-2 delivers the initiator f-Met-tRNA$_f^{Met}$ in a GTP-dependent process. Apparently, the 30S subunit is aligned with the mRNA such that the initiation codon is situated within the "30S part" of the P site. Upon binding, f-Met-tRNA$_f^{Met}$ enters this 30S portion of the P site. The GTP analog, GMPPCP (Figure 32.11), can replace GTP in promoting IF-2-mediated binding of initiator tRNA to mRNA and the 30S subunit. However, GTP hydrolysis is necessary to form an active 70S ribosome. GTP hydrolysis is accompanied by IF-1 and IF-2 release and probably occurs when the 50S subunit joins. It is likely that

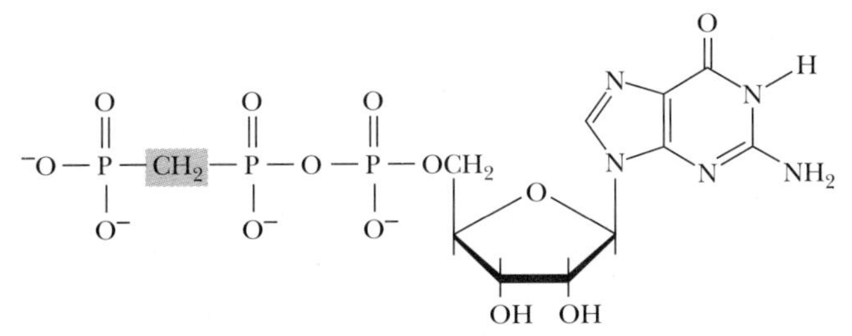

GMPPCP

Figure **32.11** The structure of GMPPCP, a nonhydrolyzable analog of GTP. The advantage of GMPPCP is that it allows the events promoted by GTP binding to be separated from those promoted by GTP hydrolysis.

GTP hydrolysis is catalyzed by a ribosomal protein, not IF-2. In any event, GTP hydrolysis is believed to drive a conformational alteration that renders the 70S ribosome competent in chain elongation. The A site of the *70S initiation complex* is poised to accept an incoming aminoacyl-tRNA.

Peptide Chain Elongation

The requirements for peptide chain elongation are (a) an mRNA:70S ribosome:peptidyl-tRNA complex (peptidyl-tRNA in the P site), (b) aminoacyl-tRNAs, (c) a set of proteins known as **elongation factors,** and (d) GTP. Chain elongation can be divided into three principal steps:

1. Codon-directed binding of the incoming aminoacyl-tRNA at the A site.
2. Peptide bond formation: transfer of the peptidyl chain from the tRNA bearing it to the $-NH_2$ group of the new amino acid.
3. Translocation of the "one-residue-longer" peptidyl-tRNA to the P site to make room for the next aminoacyl-tRNA at the A site. These shifts are coupled with movement of the ribosome one codon further along the mRNA.

The Elongation Cycle

The properties of the soluble proteins essential to peptide chain elongation are summarized in Table 32.4. These proteins are present in large quantities, reflecting the great importance of protein synthesis to cell vitality. For example, **elongation factor Tu (EF-Tu)** is the most abundant protein in *E. coli,* accounting for 5% of total cellular protein.

Table **32.4**
Properties of *E. coli* Elongation Factors

Factor	Mass (kD)	Molecules/ Cell	Function
EF-Tu	43	70,000	Binds aminoacyl-tRNA in presence of GTP
EF-Ts	74	10,000	Displaces GDP from EF-Tu
G	77	20,000	Binds GTP, promotes translocation of ribosome along mRNA

Aminoacyl-tRNA Binding

EF-Tu binds aminoacyl-tRNA and GTP. There is only one EF-Tu species serving all the different aminoacyl-tRNAs, and aminoacyl-tRNAs will bind to the A site of active 70S ribosomes only in the form of aminoacyl-tRNA:EF-Tu:GTP complexes, which likely consist of two EF-Tu:GTP per aminoacyl-tRNA. Once the aminoacyl-tRNA is situated in the A site, the GTP is hydrolyzed to GDP and P_i, and the EF-Tu molecules are released as EF-Tu:GDP complexes (Figure 32.12). (The nonhydrolyzable GTP analog GMPPCP permits aminoacyl-tRNA:EF-Tu binding, but no EF-Tu release occurs, and elongation is arrested.) EF-Tu will not interact with f-Met-tRNA$_f^{Met}$.

Elongation factor Ts (EF-Ts) promotes the recycling of EF-Tu by mediating the displacement of GDP from EF-Tu and its replacement by GTP. EF-Ts accomplishes its job through entry into a transient complex with EF-Tu. GTP then displaces EF-Ts from EF-Tu (Figure 32.12).

Peptidyl Transfer

Peptidyl transfer, or **transpeptidation,** is the central reaction of protein synthesis, the actual peptide bond-forming step. No energy input (for example, in the form of ATP) is needed; the ester bond linking the peptidyl moiety to tRNA is intrinsically reactive (Figure 32.13). Peptidyl transferase, the activity catalyzing peptide bond formation, is associated with the 50S ribosomal subunit.

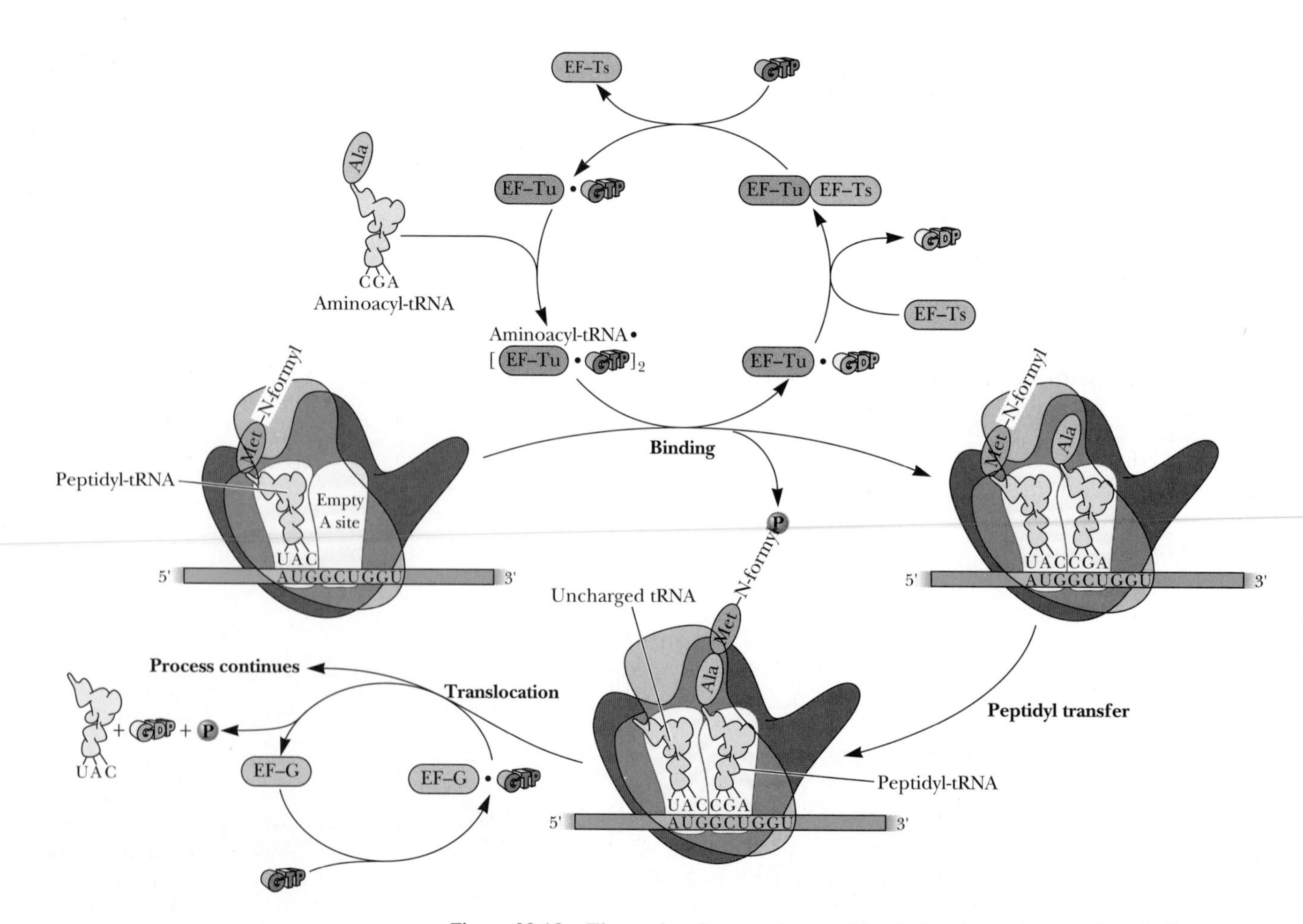

Figure **32.12** The cycle of events in peptide chain elongation on *E. coli* ribosomes.

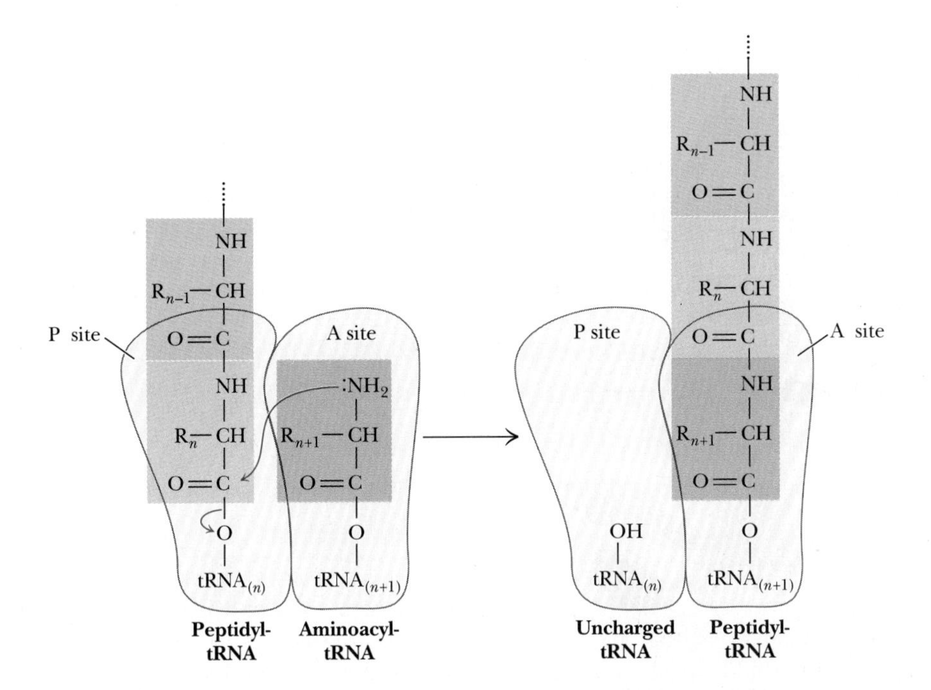

***Figure* 32.13** Reaction of the tRNA-linked peptidyl chain with the α-amino group of an adjacent aminoacyl-tRNA requires no external energy of activation (as, for example, from ATP).

23S rRNA Is the Peptidyl Transferase Enzyme

Remarkably, *E. coli* 50S ribosomal subunits from which virtually all ribosomal proteins have been removed retain significant peptidyl transferase activity (as assayed by a simplified model reaction which mimics the usual peptidyl transferase assay, Figure 11.24). These experiments, carried out by Harry Noller and his colleagues, imply that *the peptidyltransferase enzyme is the 23S rRNA.* The putative *peptidyl transferase center* of 23S rRNA is depicted in Figure 32.14. Nucleotide sequences in this region of 23S rRNA are among the most highly conserved in all biology.

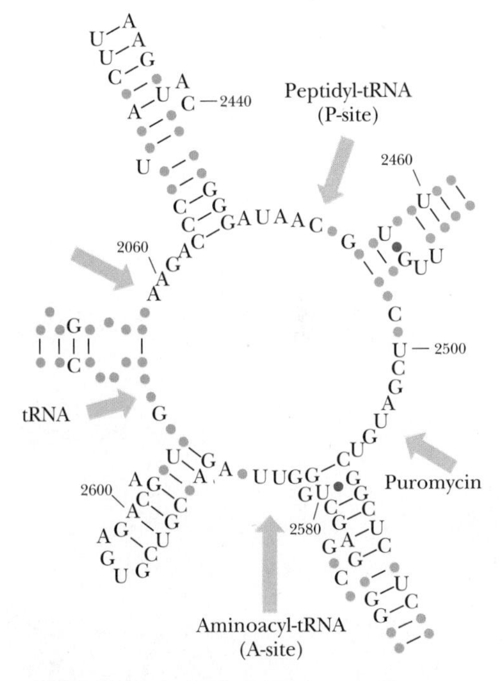

***Figure* 32.14** The peptidyl transferase center of 23S rRNA. 23S rRNA has a highly conserved secondary structure reminiscent of that of 16S rRNA (see Figure 7.39). Only the sequence implicated in the peptidyl transferase function is presented here. Numbers indicate base positions in the 23S rRNA base sequence. Bases shown are conserved in over 90% of all species. Green dots symbolize bases that are not conserved. Arrows point out 23S rRNA sites involved in interactions with various substrates (or inhibitors) of protein synthesis.

(Adapted from Pace, N. R., 1992. New horizons for RNA catalysis. Science 256:1402–1403.)

Translocation

Three things remain to be accomplished in order to return the active 70S ribosome:mRNA complex to the starting point in the elongation cycle:

1. The deacylated tRNA must be removed from the P site.
2. The peptidyl-tRNA must be moved (translocated) from the A site to the P site.
3. The ribosome must move one codon down the mRNA so that the next codon is positioned in the A site.

The precise events in translocation are only poorly understood, but the process appears to involve two distinct steps. In the first step, the **acceptor ends** of the A- and P-site tRNAs move with respect to the 50S subunit concomitantly with peptidyl transfer (Figure 32.15c–d). That is, peptide bond formation brings the acceptor end of the A-site tRNA into the P site as it picks up the peptidyl chain, and the acceptor end of the P-site tRNA is shunted into the E site. This results in two hybrid states of tRNA binding: the *E/P state* and the *P/A state* (Figure 32.15). These shifts allow the nascent polypeptide chain to remain in a fixed position during peptidyl transfer. In the second step, the tRNAs and mRNA move together with respect to the 30S subunit (Figure 32.15e). This movement requires translocation protein **elongation factor G**

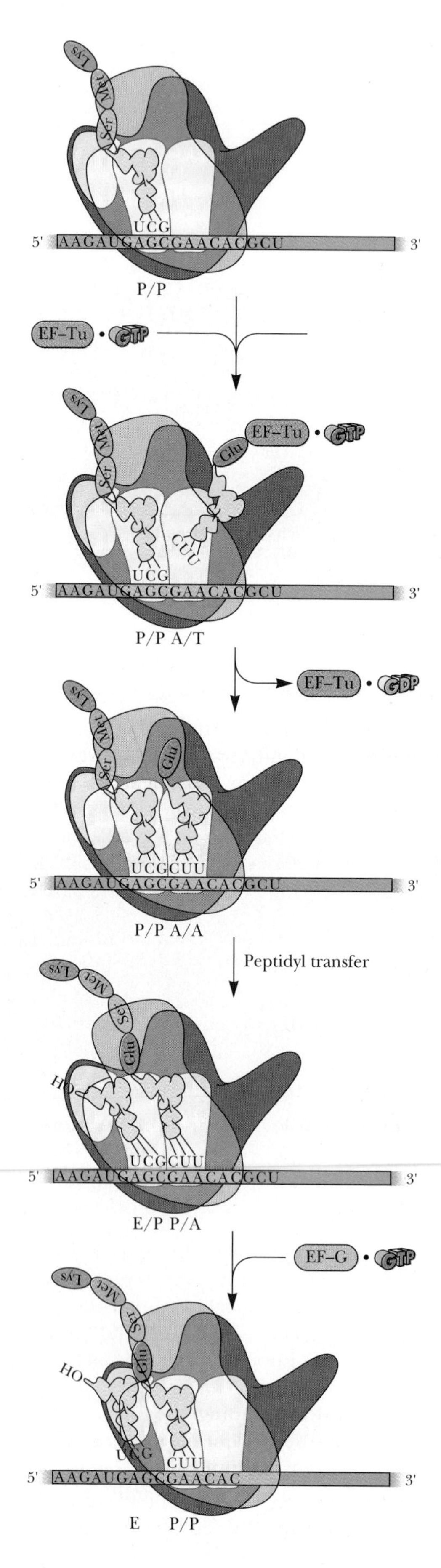

***Figure* 32.15** Model of the binding states, (a) through (e), for the movement of tRNAs during translation. In this model, the hybrid states are A/T, E/P, and P/A. The numerator represents the hypothetical 50S condition and the denominator is the corresponding 30S condition.

(Adapted from Noller, H. F., 1991. Ribosomal RNA and translation. Annual Review of Biochemistry 60:191–227.)

(EF-G), and the energy released by GTP hydrolysis (see Figures 32.12 and 32.15). Translocation of the small subunit relative to the mRNA delivers the next codon to the A site. EF-G binds to the ribosome together with GTP. GTP hydrolysis is essential not only for translocation but also for subsequent EF-G dissociation. Also, since ribosome binding of EF-G and EF-Tu are mutually exclusive, EF-G release is a prerequisite for return of the 70S ribosome:mRNA to the beginning point in the elongation cycle.

The two-step model for translocation (Figure 32.15) identifies six different states of tRNA binding: **P/P** (peptidyl-tRNA in the P site), **A/T** (an aminoacyl-tRNA entering the A site), **A/A** (an aminoacyl-tRNA in the A site), **P/A** (the peptide chain has been transferred to the aminoacyl-tRNA in the A site), **E/P** (the deacylated tRNA exiting the P site), and **E** (the deacylated tRNA in the "exit" site). The A/T state is the first state in tRNA selection at the A site and involves interaction of the ribosome with an aminoacyl-tRNA:EF-Tu:GTP complex. Codon:anticodon recognition and consequent translational fidelity must be determined at this stage. The two-step model proposes that the small and large subunits move relative to each other, as opposed to moving as a unit. *The requirement for relative movement provides a convincing explanation for why ribosomes are universally organized into a two-subunit architecture.*

GTP Hydrolysis Fuels the Conformational Changes That Drive Ribosomal Functions

Note that three GTPs are hydrolyzed for each amino acid residue incorporated into peptide during chain elongation, two upon EF-Tu-mediated binding of aa-tRNA and one more in translocation. The role of GTP (with EF-Tu as well as EF-G) is believed to be mechanical, in analogy with the role of ATP in driving muscle contraction (Chapter 36). GTP binding induces conformational changes in ribosomal components that actively engage these components in the mechanics of protein synthesis; subsequent GTP hydrolysis and GDP and P_i release relax the system back to the initial conformational state so another turn in the cycle can take place. The energy expenditure for protein synthesis is at least five high-energy phosphoric anhydride bonds per amino acid. In addition to the three provided by GTP, two are expended in amino acid activation via aminoacyl-tRNA synthesis (Figure 31.8).

Peptide Chain Termination

The elongation cycle of polypeptide synthesis continues until the 70S ribosome encounters a "stop" codon. At this point, polypeptidyl-tRNA occupies the P site and the arrival of a "stop" or nonsense codon in the A site signals that the end of the polypeptide chain has been reached (Figure 32.16). These nonsense codons are not "read" by any "terminator tRNAs" but instead are recognized by specific proteins known as **release factors,** so named because they promote polypeptide release from the ribosome. The release factors bind at the A site. **RF-1** (36 kD) recognizes UAA and UAG, while **RF-2** (41 kD) recognizes UAA and UGA. There is about one molecule each of RF-1 and RF-2 per 50 ribosomes. Ribosomal binding of RF-1 or RF-2 is competitive with EF-G. The binding of RF-1 or RF-2 is promoted by a third release factor, **RF-3** (46 kD). RF-3 function requires GTP.

The presence of release factors with a nonsense codon in the A site creates a **70S ribosome:RF-1** (or **RF-2**)**:RF-3-GTP:termination signal** complex that transforms the ribosomal peptidyl transferase into a hydrolase. That is, instead of catalyzing the transfer of the polypeptidyl chain from a polypeptidyl-tRNA to an acceptor aminoacyl-tRNA, the peptidyl transferase hydrolyzes the ester bond linking the polypeptidyl chain to its tRNA carrier. In actuality,

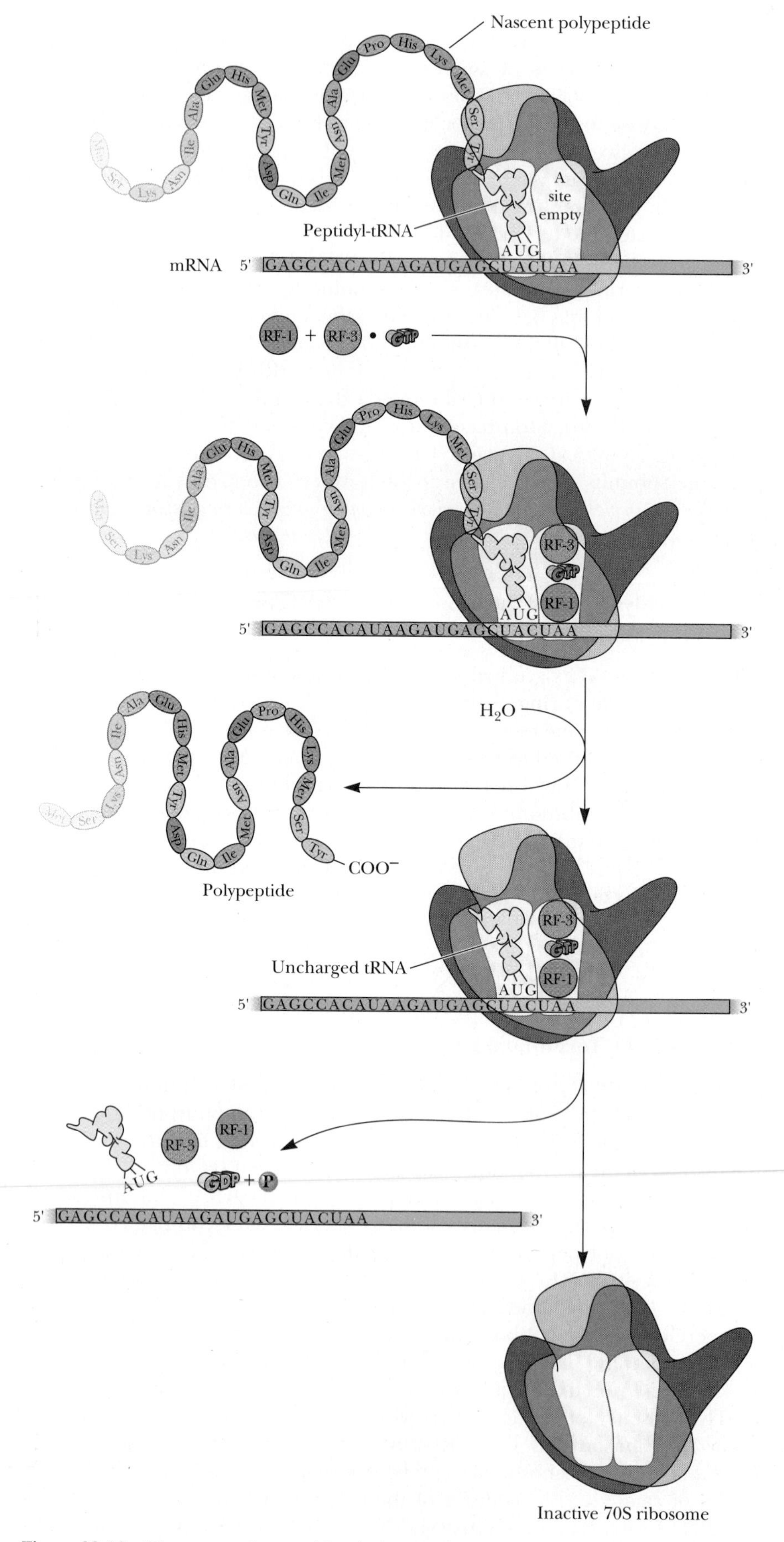

***Figure* 32.16** The events in peptide chain termination.

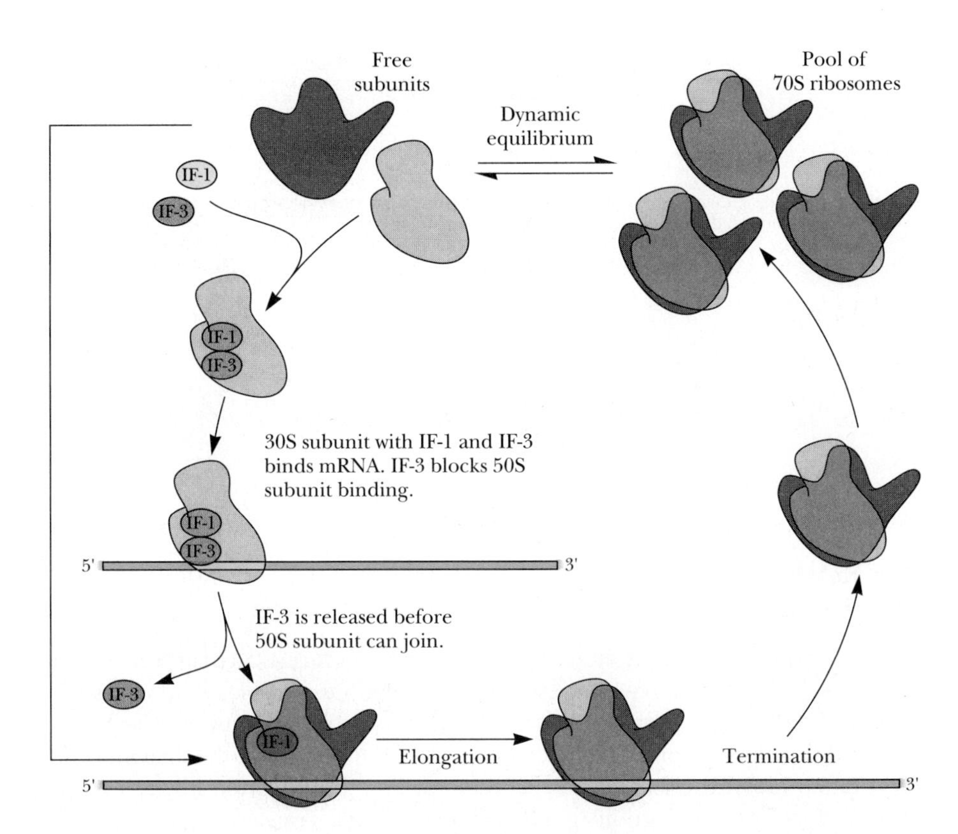

***Figure* 32.17** The ribosome life cycle. Note that IF3 is released prior to 50S addition.

peptidyl transferase transfers the polypeptidyl chain to a water molecule instead of an aminoacyl-tRNA. GTP hydrolysis now drives conformational events leading to the dissociation of the uncharged tRNA and expulsion of the release factors from the ribosome (Figure 32.16).

The Ribosome Life Cycle

Ribosomal subunits cycle rapidly through protein synthesis. In actively growing bacteria, 80% of the ribosomes are engaged in protein synthesis at any instant. Once a polypeptide chain is synthesized and the nascent polypeptide chain is released, the 70S ribosome dissociates from the mRNA and separates into free 30S and 50S subunits (Figure 32.17). Intact 70S ribosomes are inactive in protein synthesis, because only free 30S subunits can interact with the initiation factors. Binding of initiation factor IF-3 by 30S subunits and interaction of 30S subunits with 50S subunits are mutually exclusive. 30S subunits with bound initiation factors associate with mRNA, but 50S subunit addition requires IF-3 release from the 30S subunit.

Polyribosomes Are the Active Structures of Protein Synthesis

Active protein-synthesizing units consist of an mRNA with several ribosomes attached to it. Such structures are **polyribosomes,** or, simply, **polysomes** (Figure 32.18). All protein synthesis occurs on polysomes. In the polysome, each ribosome is traversing the mRNA and independently translating it into polypeptide. The further a ribosome has moved along the mRNA, the greater the length of its associated polypeptide product. In prokaryotes, as many as 10 ribosomes may be found in a polysome. Ultimately, as many as 300 ribosomes may translate an mRNA, so as many as 300 enzyme molecules may be produced from a single transcript. Eukaryotic polysomes typically contain fewer than 10 ribosomes.

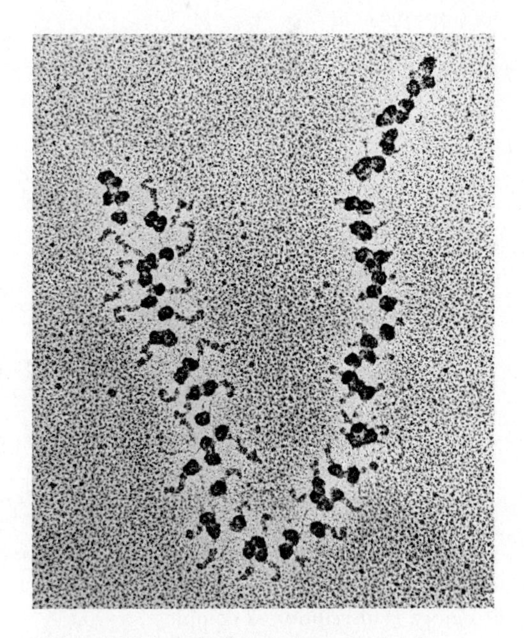

***Figure* 32.18** Electron micrograph of polysomes: multiple ribosomes translating the same mRNA.

(From Franke, C., et al., 1982. Electron microscopic visualization of a discrete class of giant translation units in salivary gland cells of Chironomus tentans. The EMBO Journal 1:59–62. Photo courtesy of Oscar L. Miller, University of Virginia.)

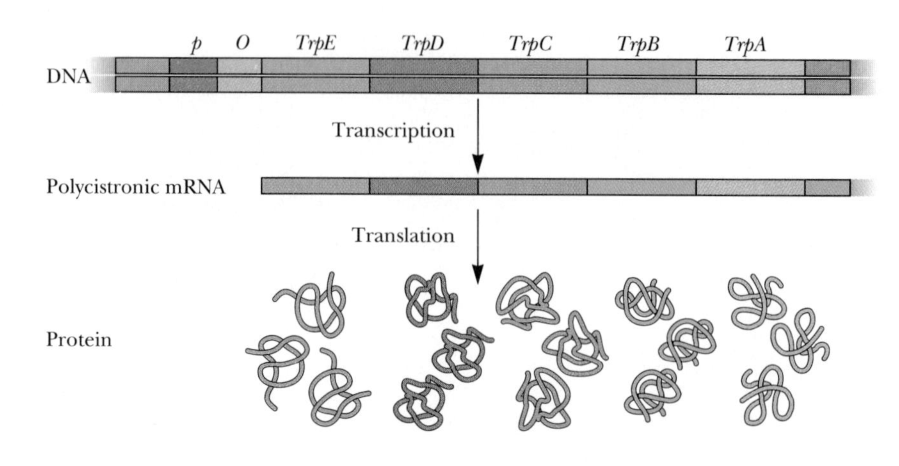

***Figure* 32.19** The *E. coli trp* operon. This operon consists of a promoter, *p*, an operator region, *O*, and five genes encoding the various enzymes necessary for tryptophan biosynthesis, arranged in the order E, D, C, B, and A. All these genes are transcribed under the control of *O* and *p* to give a single polycistronic mRNA.

The Relationship Between Transcription and Translation in Prokaryotes

In prokaryotes, ribosomes attach to mRNA even before transcription of the mRNA is completed, and as a consequence, polysomes can be found in association with DNA. Biochemical evidence for this relationship between transcription and translation comes from a study on the expression of the enzymes for tryptophan biosynthesis in *E. coli.* These enzymes are encoded in the *trp* operon, a set of five contiguous genes (Figure 32.19). Transcription of the *trp* operon only occurs in the absence of tryptophan, and transcription of the entire operon takes more than 5 minutes. The first two genes, E and D, encode the subunits of the enzyme, anthranilate synthase. Within 2½ minutes of the start of *trp* operon transcription, anthranilate synthase activity is detectable, demonstrating that translation has begun before transcription is completed. That is, the mRNA is translated while the operon is still being transcribed. Oscar Miller, Barbara Hamkalo, and Charles Thomas provided visual evidence for concomitant transcription and translation, as shown in Figure 32.20. The leading ribosome may actually make physical contact with RNA polymerase.

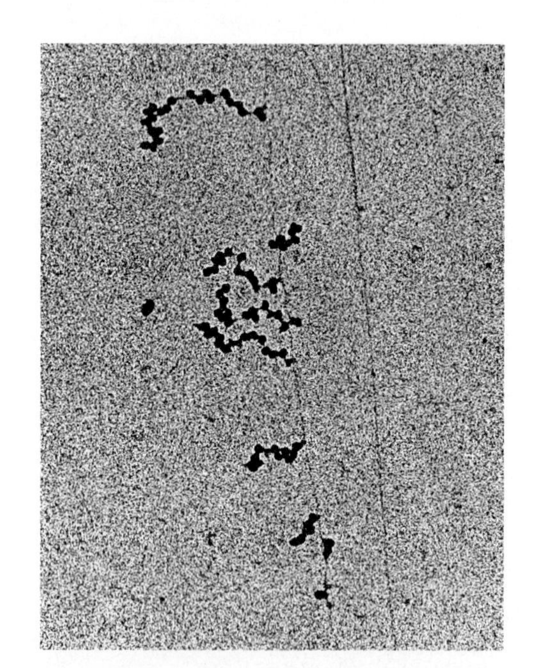

***Figure* 32.20** Electron micrograph demonstrating that translation occurs concomitantly with transcription. RNA polymerase initiates transcription (bottom) and moves along the DNA, transcribing it into a complementary RNA strand. Successive ribosomes (dark bodies) attach to the 5′-end of the RNA and begin translating it before transcription is complete, as evidenced by the progressively longer RNA chains with more and more associated ribosomes (top), as RNA polymerase progresses along the DNA strand from the initiation site.

(Adapted from Miller, O. L., Jr., Hamkalo, B., and Thomas, C., 1970. Visualization of bacterial genes in action. Science 169:392–395. Photo courtesy of Oscar L. Miller, University of Virginia.)

32.3 Protein Synthesis in Eukaryotic Cells

Eukaryotic mRNAs are characterized by two post-transcriptional modifications: the **5′-^{7}methyl-GTP cap** and the **poly(A) tail** (Figure 32.21). The ^{7}methyl-GTP cap is essential for ribosomal binding of mRNAs in eukaryotes and also enhances the stability of these mRNAs by preventing their degradation by 5′-exonucleases. The poly(A) tail is believed to enhance both the stability and translational efficiency of eukaryotic mRNAs. The Shine-Dalgarno sequences found at the 5′-end of prokaryotic mRNAs are absent in eukaryotic mRNAs.

Peptide Chain Initiation in Eukaryotes

The events in eukaryotic peptide chain initiation are summarized in Figure 32.22, and the properties of **eukaryotic initiation factors,** symbolized as **eIFs,** are presented in Table 32.5. The eukaryotic initiator tRNA is a unique tRNA functioning only in initiation. Like the prokaryotic initiator tRNA, the eukaryotic version carries only Met. However, unlike prokaryotic f-Met-$tRNA_f^{Met}$, the Met on this tRNA is not formylated. Hence, the eukaryotic initiator tRNA is usually designated $\mathbf{tRNA_i^{Met}}$, with the "i" indicating "initiation."

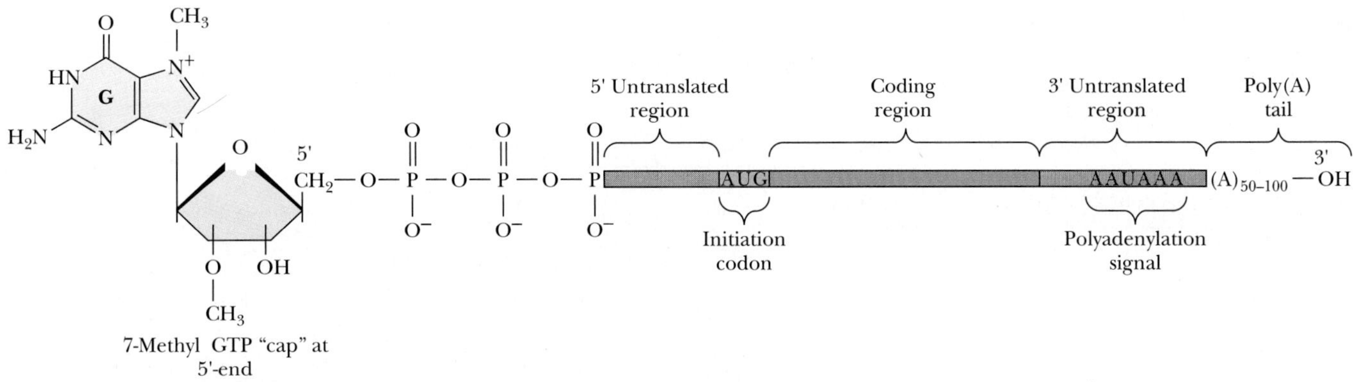

***Figure* 32.21** The characteristic structure of eukaryotic mRNAs. Untranslated regions ranging between 40 and 150 bases in length occur at both the 5′- and 3′-ends of the mature mRNA. An initiation codon at the 5′-end, invariably AUG, signals the translation start site.

Initiation begins with the formation of a ternary complex containing eIF-2, GTP, and Met-tRNA$_i^{Met}$. This ternary complex then associates with a 40S ribosomal subunit:(eIF-3:eIF-4C) complex to yield a **40S preinitiation complex** (Figure 32.22). Thus, eIF-2 is analogous to prokaryotic IF-2 in function. Unlike in prokaryotes, binding of Met-tRNA$_i$ by eukaryotic ribosomes occurs in the absence of mRNA, so Met-tRNA$_i^{Met}$ binding is not codon-directed. The **40S initiation complex** results when mRNA, in association with a multimeric protein complex consisting of eIF-4A, eIF-4B, **cap binding protein (CBP-I),** and perhaps other proteins, binds to the 40S preinitiation complex. This multimeric protein complex, which delivers mRNA to the 40S subunit, is known collectively as **eIF-4F** (or **CBP-II**). ATP is required in eukaryotic initiation, and the energy released upon its hydrolysis drives the unwinding of secondary structure at the 5′-end of the mRNA, which might prevent proper initiation.

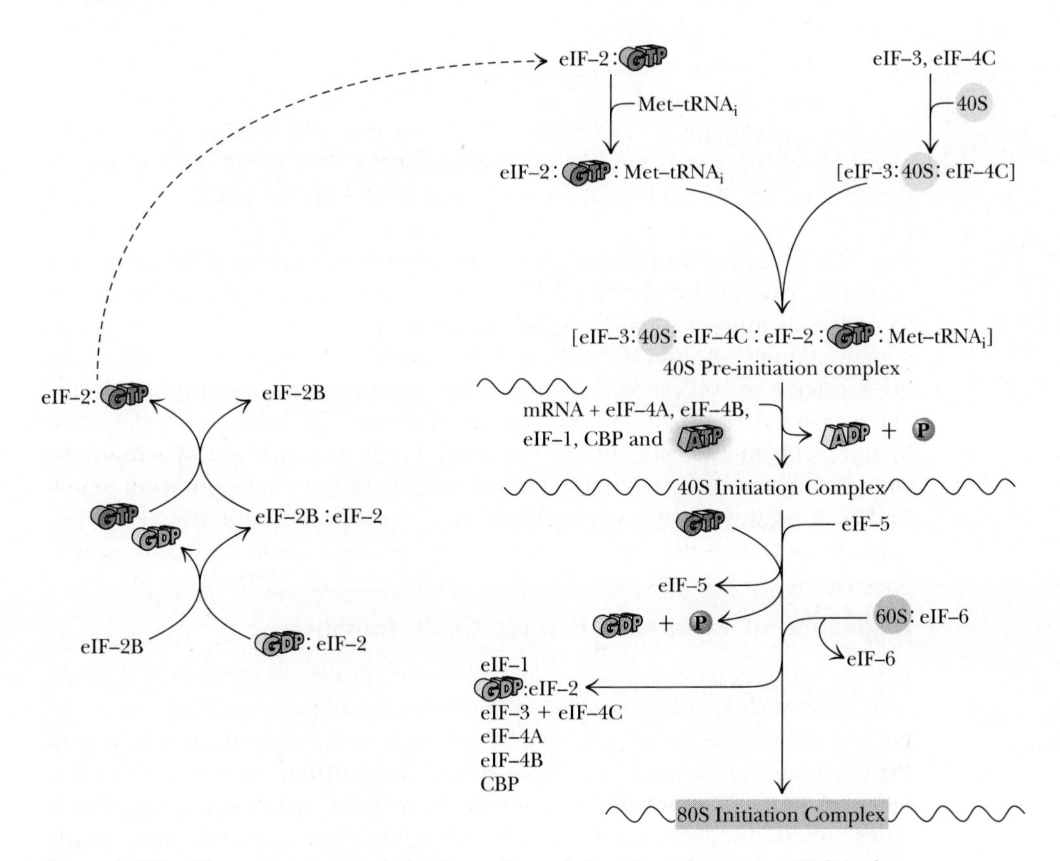

***Figure* 32.22** The events in the initiation of translation in eukaryotes. eIF-2 is the eukaryotic analog of IF-2. Since eIF-2 binds GDP very tightly, eIF-2B is necessary to mediate its replacement by GTP.

Table **32.5**
Properties of Eukaryotic Initiation Factors

Factor	Mass (kD)	Composition	Function
eIF-1	15	Monomer	Assists mRNA binding
eIF-2	130	Trimer	Binds GTP and Met-tRNA$_i^{Met}$
		α (36 kD)	Binds GTP
		β (38 kD)	?
		γ (55 kD)	Binds Met-tRNA$_i^{Met}$ and mRNA
eIF-2B	272	5 polypeptides	Recycles eIF-2:GDP to eIF-2:GTP
eIF-3	550	7–10 polypeptides (35–170 kD each)	Binds 40S subunit, prevents 60S association
eIF-4C	20	Monomer	Part of the eIF-3 complex
eIF-4F	270	Multimer	Binding 5′-end of mRNA and 2° structure unwinding
eIF-4A	44	Monomer	Binds mRNA and ATP
eIF-4B	80	Monomer	mRNA binding and 2° structure unwinding
CBP-I	72	Trimer	Binds ^{7}methyl-GTP cap of mRNA
eIF-5	150	Monomer	GTP hydrolysis and release of eIF-2 and eIF-3 complex
eIF-6	24	Monomer	Binds 60S subunit, prevents 40S association

Since eukaryotic mRNAs lack a Shine-Dalgarno sequence that aligns the ribosomal small subunit with the AUG codon where translation begins, a "scanning mechanism" is believed to direct the 40S subunit to the AUG translational start site. That is, the 40S:Met-tRNA$_i^{Met}$:eIF-2/3/4C complex is directed to the 5′-end of mRNA via interactions with the mRNA:eIF-1/4A,B protein complex, and then it scans along the mRNA until it encounters the first AUG codon. (Most, although not all, eukaryotic proteins start at the first AUG.) Pairing of the Met-tRNA$_i^{Met}$ anticodon with this AUG completes 40S initiation complex formation. eIF-5 now acts to mediate GTP hydrolysis and concomitant eIF-2 and eIF-3 release. GTP hydrolysis is necessary for eIF-2 and eIF-3 release as well as 60S addition. Presumably, essential conformational changes are driven by the energy released upon GTP hydrolysis. eIF-6 must dissociate from the 60S subunit in order for it to enter the **80S initiation complex.** The mRNA poly(A) tail may act in recruiting the 60S subunit to join the 40S initiation complex. The Met-tRNA$_i^{Met}$ resides in the P site of this 80S complex, which is now poised to enter the peptide chain elongation cycle.

Regulation of Eukaryotic Peptide Chain Initiation

Regulation of gene expression can be exerted post-transcriptionally through control of mRNA translation. Phosphorylation/dephosphorylation of translational components is a dominant mechanism for control of protein synthesis. Peptide chain initiation, the initial phase of the synthetic process, is the optimal place for such control. Phosphorylation of 40S ribosomal protein S6 facilitates initiation of protein synthesis, resulting in a shift of the ribosomal population from inactive ribosomes to actively translating polysomes. S6 phosphorylation is stimulated by serum growth factors (Chapter 37). The

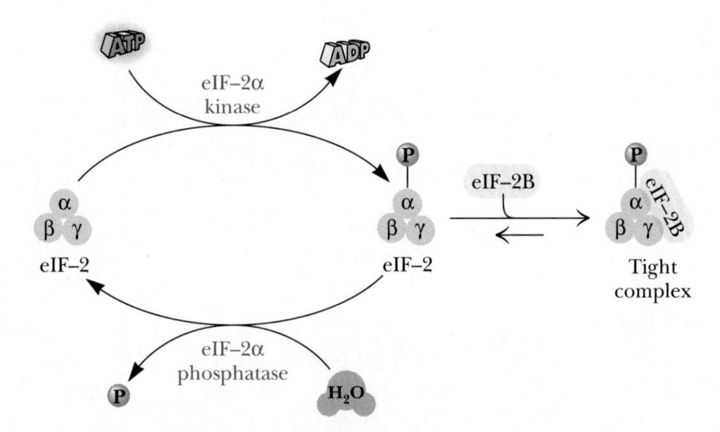

***Figure* 32.23** Control of eIF-2 functions through reversible phosphorylation of a Ser residue on its α-subunit. The phosphorylated form of eIF-2 (eIF-2-P) enters a tight complex with eIF-2B and is unavailable for initiation.

action of eIF-4F, the mRNA cap-binding complex, is promoted by phosphorylation. On the other hand, phosphorylation of other translational components inhibits protein synthesis. For example, the α-subunit of eIF-2 can be reversibly phosphorylated at a specific Ser residue by an eIF-2α kinase/phosphatase system (Figure 32.23). Phosphorylation of eIF-2α ultimately inhibits peptide chain initiation, not because eIF-2α-P is ineffectual, but because phosphorylated eIF-2 binds eIF-2B much more tightly than does eIF-2. All of the eIF-2B, which is present at only 20% to 30% of eIF-2 levels, becomes sequestered in eIF-2:eIF-2B complexes, and eIF-2 cannot be regenerated from eIF-2:GDP for further cycles of initiation. Reversible phosphorylation of eIF-2 is an important control governing globin synthesis in reticulocytes. If heme for hemoglobin synthesis becomes limiting in these cells, eIF-2α is phosphorylated so globin mRNA is not translated and chains are not synthesized. Availability of heme reverses the inhibition through phosphatase-mediated removal of the phosphate group from the Ser residue.

Peptide Chain Elongation in Eukaryotes

Eukaryotic peptide elongation occurs in very similar fashion to the process in prokaryotes. An incoming aminoacyl-tRNA enters the ribosomal A site while peptidyl-tRNA occupies the P site. Peptidyl transfer then occurs, followed by translocation of the ribosome one codon further along the mRNA. Two elongation factors, EF-1 and EF-2, mediate the elongation steps. EF-1 consists of two components: EF-1α, a 50-kD protein, and EF-1βγ, a complex of 26-kD (β) and 46-kD(γ) protein subunits. EF-1α is the eukaryotic counterpart of EF-Tu; it serves as the aminoacyl-tRNA binding factor and requires GTP. EF-1βγ is the eukaryotic equivalent of prokaryotic EF-Ts; it catalyzes the exchange of bound GDP on EF-1:GDP for GTP so active EF-1:GTP can be regenerated. EF-2, a 110-kD polypeptide, is the eukaryotic translocation factor. Like its prokaryotic kin, EF-G, EF-2 binds GTP, and GTP hydrolysis accompanies translocation.

Eukaryotic Peptide Chain Termination

Whereas prokaryotic termination involves three different release factors (RFs), just one RF is sufficient for eukaryotic termination. Eukaryotic RF (110 kD) is an α_2 dimer of 55-kD subunits. Eukaryotic RF binding to the ribosomal A site is GTP-dependent, and RF:GTP binds at this site when it is occupied by a termination codon. Then, hydrolysis of the peptidyl-tRNA ester bond, hydrolysis of GTP, release of nascent polypeptide and deacylated tRNA, and ribosome dissociation from mRNA ensue.

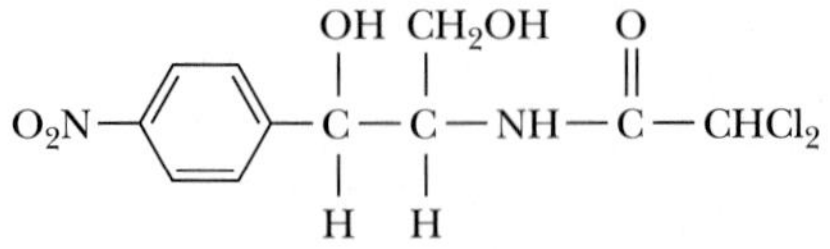

Chloramphenicol

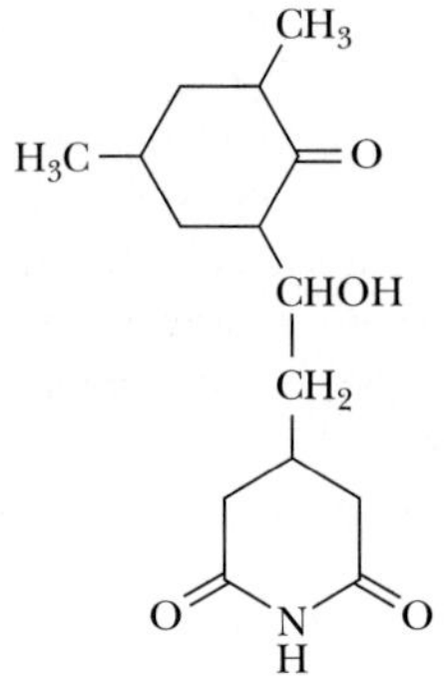

Cycloheximide

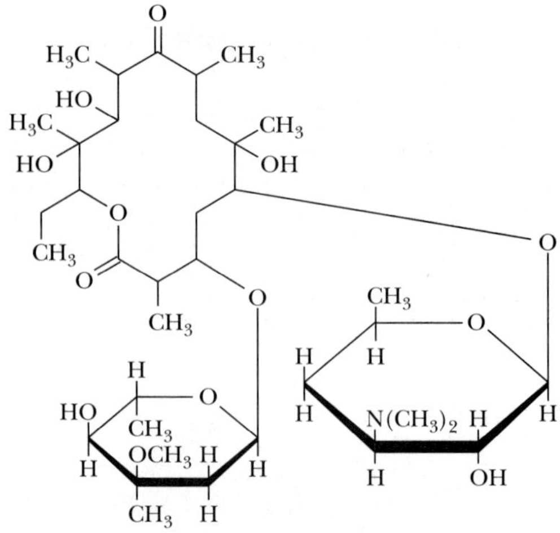

Erythromycin

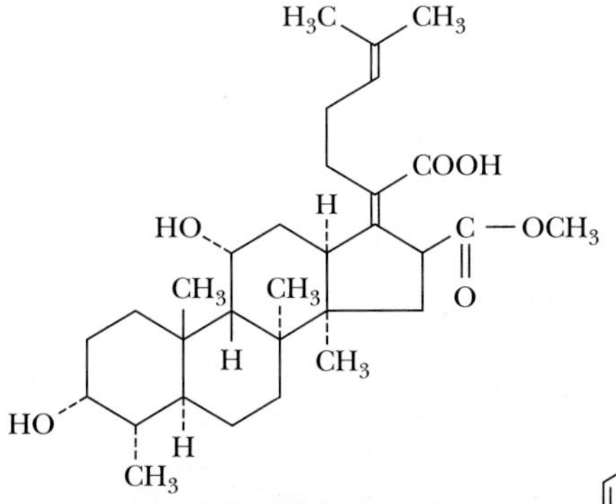

Fusidic acid

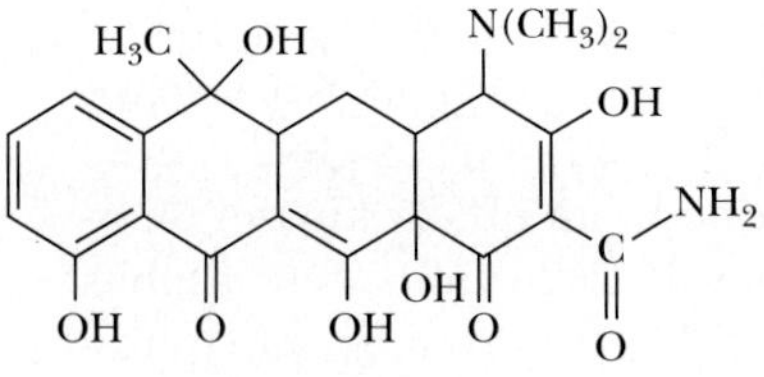

Tetracycline

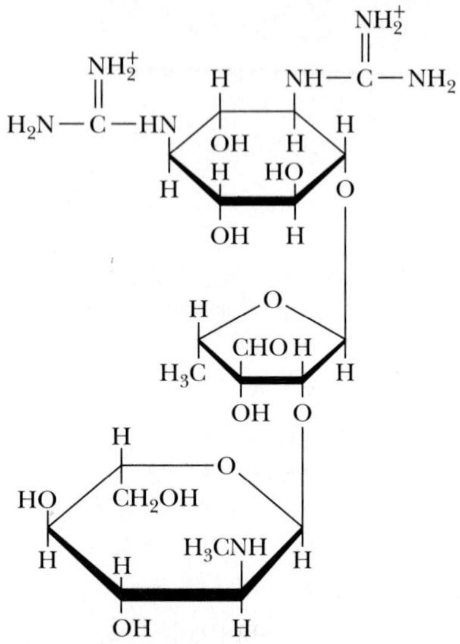

Streptomycin

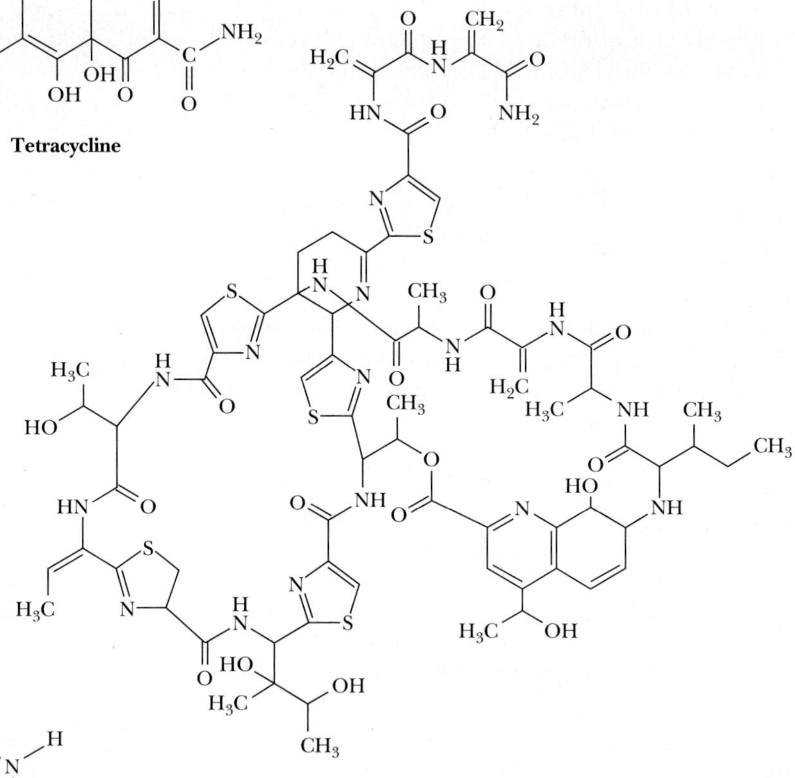

Thiostrepton

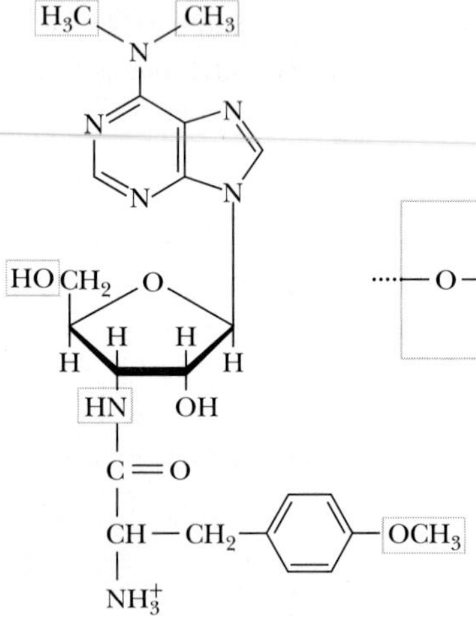

Puromycin

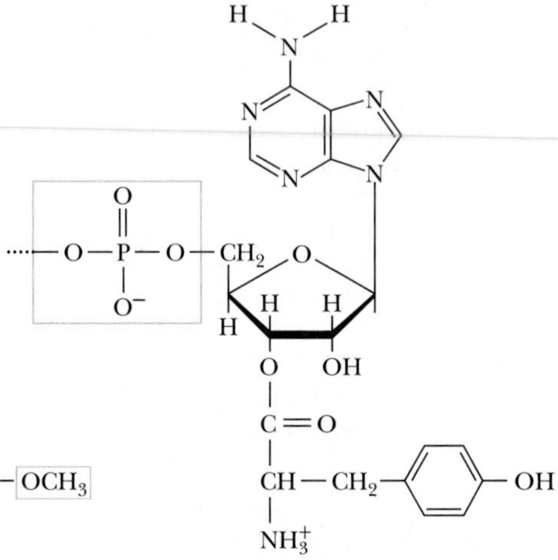

Tyrosyl-tRNA

***Figure* 32.24** The structures of various antibiotics that act as protein synthesis inhibitors. Puromycin mimics the structure of aminoacyl-tRNA, in that it resembles the 3′-terminus of a Tyr-tRNA.

32.4 Inhibitors of Protein Synthesis

Protein synthesis inhibitors have served two major, and perhaps complementary, purposes. First, they have been very useful scientifically in elucidating the biochemical mechanisms of protein synthesis. Second, some of these inhibitors affect prokaryotic but not eukaryotic protein synthesis and thus are medically important antibiotics. Table 32.6 is a partial list of these inhibitors and their mode of action. The structures of some of these compounds are given in Figure 32.24.

Streptomycin

Streptomycin is an aminoglycoside antibiotic that affects the function of the prokaryotic 30S subunit, as demonstrated by the fact that streptomycin-resistant mutations map either to the gene encoding 30S protein S12 or to position 912 in the 16S rRNA sequence. Low concentrations of streptomycin induce mRNA misreading, so that improper amino acids are incorporated into the polypeptide. Codons with pyrimidines in the first and second positions are

Table **32.6**
Some Protein Synthesis Inhibitors

Inhibitor	System Inhibited	Mode of Action
Initiation		
Aurintricarboxylic acid	Prokaryotic	Prevents IF binding to 30S subunit
Kasugamycin	Prokaryotic	Inhibits F-Met-$tRNA_f^{Met}$ binding
Streptomycin	Prokaryotic	Prevents formation of initiation complexes
Elongation: Aminoacyl-tRNA Binding		
Tetracycline	Prokaryotic	Inhibits aminoacyl-tRNA binding at A site
Streptomycin	Prokaryotic	Codon misreading, insertion of improper amino acid
Elongation: Peptide Bond Formation		
Sparsomycin	Prokaryotic	Peptidyl transferase inhibitor
Chloramphenicol	Prokaryotic	Binds to 50S subunit, blocks peptidyl transferase activity
Erythromycin	Prokaryotic	Binds to 50S subunit, blocks peptidyl transferase activity
Cycloheximide	Eukaryotic	Inhibits translocation of peptidyl-tRNA
Elongation: Translocation		
Fusidic acid	Both	Inhibits EF-G:GDP dissociation from ribosome
Thiostrepton	Prokaryotic	Inhibits ribosome-dependent EF-Tu and EF-G GTPase
Diphtheria toxin	Eukaryotic	Inactivates eEF-2 through ADP-ribosylation
Premature Termination		
Puromycin	Both	Aminoacyl-tRNA analog, which acts as a peptidyl acceptor, aborting further peptide elongation
Ribosome Inactivation		
Ricin	Eukaryotic	Catalytic inactivation of 28S rRNA

particularly susceptible to streptomycin-induced misreading. These reading errors are *not* frameshift mistakes, so totally aberrant proteins are not made at low streptomycin levels. Thus, susceptible cells are not killed, but their growth rate is severely depressed. When streptomycin is present at high concentrations, 70S nonproductive ribosome:mRNA complexes accumulate, preventing the formation of active initiation complexes with new mRNA.

Puromycin

Puromycin is a structural analog of the aminoacyl-adenylyl grouping characteristic of the 3′-end of aminoacyl-tRNAs (Figure 32.24). Puromycin binds at the A site of both prokaryotic and eukaryotic ribosomes. Puromycin binding is not dependent on EF-Tu (or EF-1). Puromycin serves as an acceptor of the peptidyl chain from peptidyl-tRNA in the P site, in a reaction in which peptidyl transferase catalyzes the attachment of the peptidyl chain to the free NH_3^+ group of puromycin. Peptidyl-puromycin is a dead-end product because the peptidyl chain is now linked via an amide bond to the 3′-NH of the modified adenosine moiety, *not* via the usual ester bond to the 3′-OH terminus found in all tRNAs. Puromycin aborts protein synthesis through premature termination, leading to the release of nonfunctional, truncated polypeptides.

Diphtheria Toxin

Diphtheria arises from infection by *Corynebacterium diphtheriae* bacteria carrying bacteriophage *corynephage* β. Diphtheria toxin is a phage-encoded enzyme secreted by these bacteria that is capable of inactivating a number of GTP-dependent enzymes through covalent attachment of an ADP-ribosyl moiety derived from NAD^+. That is, diphtheria toxin is an NAD^+-dependent ADP-ribosylase. One target of diphtheria toxin is the eukaryotic translocation factor, EF-2. This protein has a modified His residue known as **diphthamide.** Diphthamide is generated post-translationally on EF-2; its biological function is unknown. (EF-G of prokaryotes lacks this unusual modification and is not susceptible to diphtheria toxin.) Diphtheria toxin specifically ADP-ribosylates an imidazole-N within the diphthamide moiety of EF-2 (Figure 32.25). ADP-ribosylated EF-2 retains the ability to bind GTP, but is unable to function in protein synthesis. Since diphtheria toxin is an enzyme and can act catalytically to modify many molecules of its target proteins, just a few micrograms suffice to cause death.

Ricin

Ricin is an extremely toxic glycoprotein produced by the plant *Ricinus communis* (castor bean). The protein is a disulfide-linked, $\alpha\beta$ heterodimer of roughly equal 30-kD subunits. The A subunit (32 kD) is an enzyme and serves as the toxic subunit; it gains entry to cells because the B subunit (33 kD) is a **lectin.** (Lectins form a class of proteins that bind to specific carbohydrate moieties commonly displayed by glycoproteins and glycolipids on cell surfaces.) Endocytosis of bound ricin followed by disulfide reduction releases the A chain, which gains access to the cytosol and there catalytically inactivates eukaryotic large ribosomal subunits. A single molecule of ricin A chain in the cytosol can inactivate 50,000 ribosomes and kill a eukaryotic cell! Ricin A chain specifically attacks a single, highly conserved adenosine (an A at or near position 4324) in the eukaryotic 28S rRNA, through an N-glycosidase activity that removes the adenine base, leaving the rRNA sugar-phosphate backbone intact. Removal of this single base is sufficient to inactivate a 60S large subunit. The adenine in this highly conserved region of the 28S rRNA sequence is believed to be crucial to functions of the 60S subunit that involve EF-1 and EF-2.

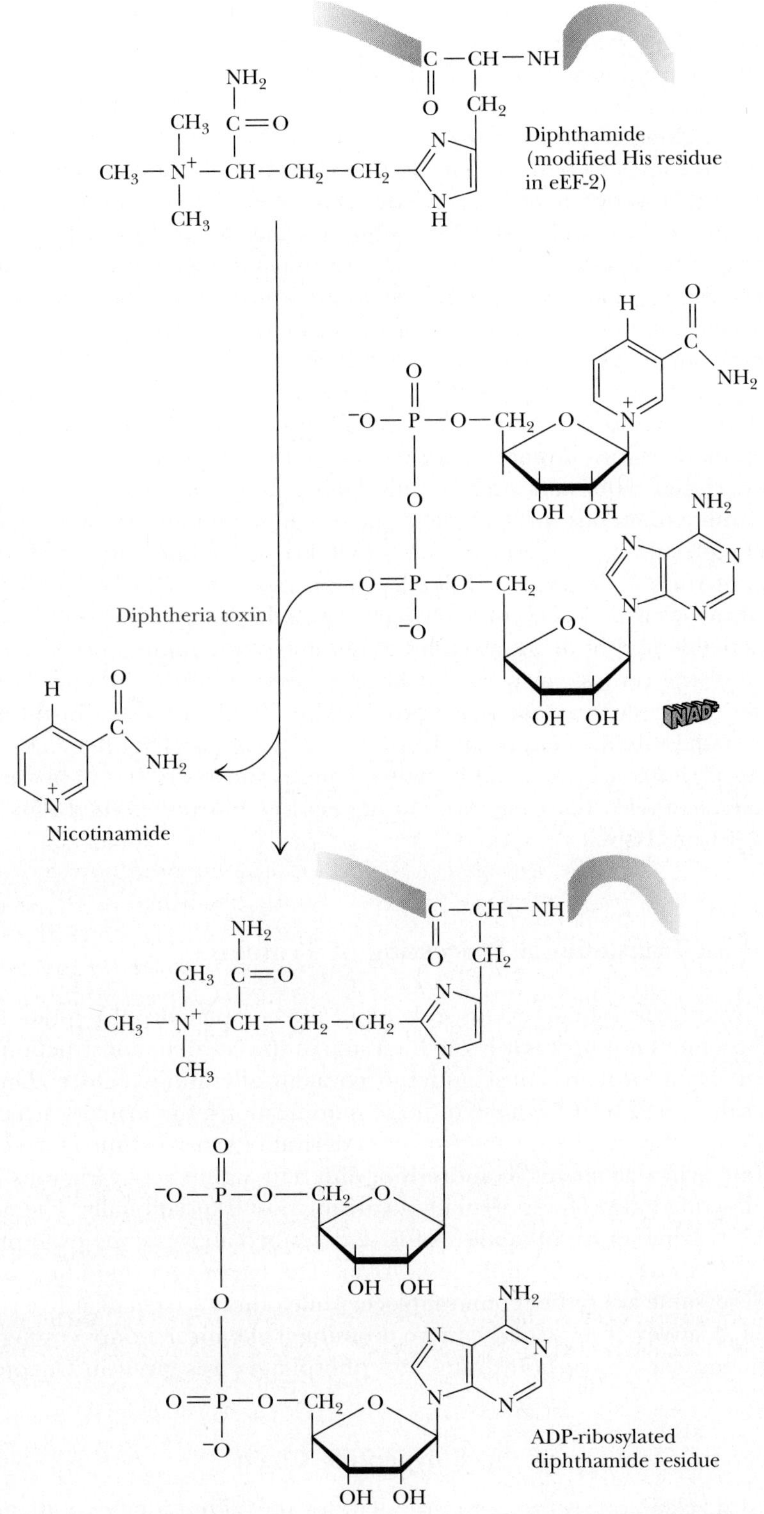

Diphthamide (modified His residue in eEF-2)

Diphtheria toxin

Nicotinamide

ADP-ribosylated diphthamide residue

***Figure* 32.25** Diphtheria toxin catalyzes the NAD^+-dependent ADP-ribosylation of selected proteins. ADP-ribosylation of the diphthamide moiety of eukaryotic EF-2. (Diphthamide = 2-[3-carboxamido-3-(trimethylammonio)propyl]histidine.)

32.5 Protein Folding

The information for folding of each protein into its unique three-dimensional architecture resides within its amino acid sequence (primary structure). Proteins begin to fold even as they are being synthesized on ribosomes. Nevertheless, proteins are generally assisted in folding by a family of helper proteins, or **molecular chaperones** (Chapter 5). Chaperones also serve to shepherd proteins to their ultimate cellular destinations.

The Hsp70 and Hsp60 classes of chaperones are believed to interact sequentially along a normal pathway of protein folding (Figure 32.26). The Hsp70 class binds to nascent polypeptide chains while they are still on ribosomes. Hsp70 recognizes exposed, extended regions of polypeptides that are rich in hydrophobic residues. By binding to these regions, Hsp70 prevents nonproductive associations and keeps the polypeptide in an unfolded (or partially folded) state until productive folding interactions can occur. Completion of folding requires release of the protein from Hsp70; release is energy-dependent and is driven by ATP hydrolysis.

Upon release from Hsp70, the partially folded protein would be susceptible again to inappropriate associations or aggregation through its exposed hydrophobic regions. Completion of folding involves the Hsp60 class of chaperones. Hsp60 sequesters the partially folded molecules from one another (and from extraneous interactions), but still allows folding to proceed. The Hsp60 molecules are organized as α_{14} (840 kD) assemblies, arranged as two 7-subunit rings. The partially folded protein molecule is bound within the central cavity of the 14-subunit complex, where its folding is facilitated in an ATP-dependent fashion. Many cycles of binding of the folding protein to the surface of the central cavity, ATP hydrolysis, release of the protein, and rebinding to the surface take place, with folding steps occurring in the brief intervals when the protein is free from the cavity surface (Figure 32.26). The folding of rhodanese, a 33-kD protein, requires the hydrolysis of about 130 equivalents of ATP. Once the protein has achieved its fully folded state, it is released from Hsp60.

32.6 Post-Translational Processing of Proteins

Aside from these folding events, release of the completed polypeptide from the ribosome is not necessarily the final step in the covalent construction of a protein. Many proteins must undergo covalent alterations before they become functional. In the course of these modifications, the primary structure of a protein may be altered and/or novel derivations may be introduced into its amino acid side chains. Hundreds of different amino acid variations have been described in proteins, virtually all arising post-translationally. The modified His residue, or diphthamide, of EF-2 described above is one example, as is its ADP-ribosylation by diphtheria toxin. The list of such modifications is very large; some are rather commonplace, while others are peculiar to a single protein. A survey of some of the more prominent chemical groups conjugated to proteins, such as carbohydrates and phosphates, was given in Chapter 4.

Proteolytic Cleavage of the Polypeptide Chain

Proteolytic cleavage, as the most prevalent form of protein post-translational modification, merits special attention. The very occurrence of proteolysis as a processing mechanism seems strange: Why join a number of amino acids in sequence and then eliminate some of them? Three reasons can be cited. First,

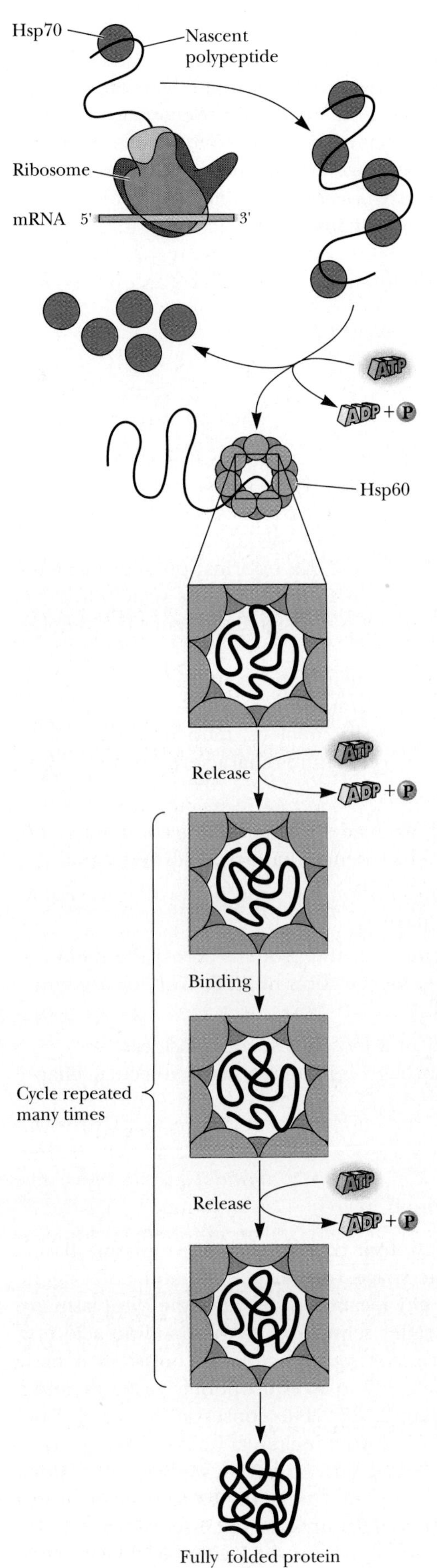

***Figure* 32.26** Model for Hsp70 and Hsp60 functions in the protein folding pathway. *Hsp* is an acronym for heat shock protein; 60 and 70 indicate the mass (kD) of the two major Hsp species. Hsp70 binds to nascent polypeptides on ribosomes, preventing inappropriate liaisons. The partially folded protein is then released from Hsp70 in an ATP-dependent step and bound by the 14-subunit Hsp60 complex. Successive cycles of protein binding within the Hsp60 central cavity, ATP-dependent release, and progressive folding of the protein ends in release of the protein from Hsp60 in its fully folded state.

diversity can be introduced where none exists. For example, a simple form of proteolysis, enzymatic removal of N-terminal Met residues, occurs in many proteins. Met-aminopeptidase, by removing the invariant Met initiating all polypeptide chains, introduces diversity at N-termini. Second, proteolysis serves as an activation mechanism so that expression of the biological activity of a protein can be delayed until appropriate. A number of metabolically active proteins, including digestive enzymes and hormones, are synthesized as larger inactive precursors termed **pro-proteins** that are activated through proteolysis (see **zymogens,** Chapter 12). Third, proteolysis is involved in the targeting of proteins to their proper destinations in the cell, a process known as **protein translocation.**

Protein Translocation

Proteins targeted for service in membranous organelles or for export from the cell are synthesized in precursor form carrying an N-terminal stretch of amino acid residues, or **leader peptide,** that serves as a **signal sequence.** In effect, signal sequences serve as "zip codes" for sorting and dispatching proteins to their proper compartments. Thus, the information specifying the correct cellular localization of a protein is found within its structural gene. Once the protein is routed to its destination, the signal sequence is proteolytically clipped from the protein.

In addition to the bacterial plasma membrane, a number of eukaryotic membranes are competent in protein translocation, including the membranes of the endoplasmic reticulum (ER), nucleus, mitochondria, chloroplasts, and peroxisomes. Several common features characterize protein translocation systems:

1. Proteins to be translocated are made as preproteins containing contiguous blocks of amino acid sequence that act as sorting signals.
2. Membranes involved in translocation have specific protein receptors exposed on their cytosolic faces.
3. **Translocases** catalyze movement of the proteins across the membrane, and metabolic energy in the form of ATP, GTP, or a membrane potential is essential.
4. Preproteins are maintained in a loosely folded, translocation-competent conformation through interaction with molecular chaperones.

Prokaryotic Protein Translocation

Gram-negative bacteria typically have four compartments: cytoplasm, plasma (or inner) membrane, periplasmic space (or periplasm), and outer membrane. Most proteins destined for any location other than the cytoplasm are synthesized with amino-terminal leader sequences 16 to 26 amino acid residues long. These leader sequences, or *signal sequences,* consist of a basic N-terminal region, a central domain of 7 to 13 hydrophobic residues, and a nonhelical C-terminal region (Figure 32.27). The conserved features of the last part of the leader, the C-terminal region, include a helix-breaking Gly or Pro residue and amino acids with small side chains located one and three residues before the proteolytic cleavage site. Unlike the basic N-terminal and nonpolar central regions, the C-terminal features are not essential for translocation but instead serve as recognition signals for the **leader peptidase,** which

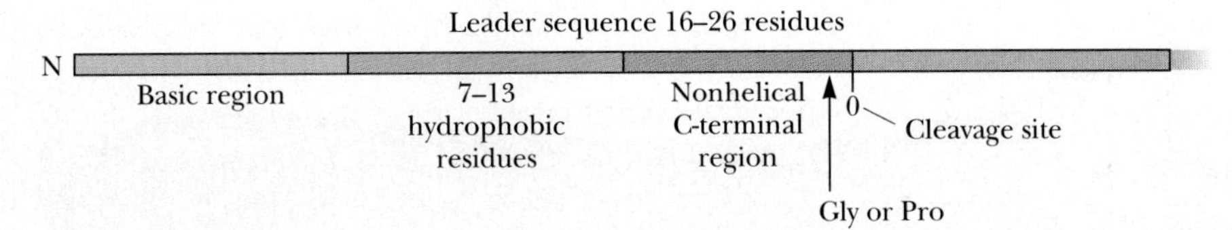

Figure **32.27** General features of the N-terminal signal sequences on *E. coli* proteins destined for translocation: a basic N-terminal region, a central apolar domain, and a nonhelical C-terminal region.

removes the leader sequence. The exact amino acid sequence of the leader peptide is unimportant. Nonpolar residues in the center and a few Lys residues at the amino terminus are sufficient for successful translocation. The functions of leader peptides are to retard the folding of the preprotein so that molecular chaperones have a chance to interact with it and to provide recognition signals for the translocation machinery and leader peptidase.

Eukaryotic Protein Sorting and Translocation

Eukaryotic cells are characterized by many membrane-bounded compartments. In general, signal sequences targeting proteins to their appropriate compartments are located at the N-terminus as *cleavable presequences,* although many proteins have internal, noncleaved targeting sequences. Proteolytic removal of the leader sequences is also catalyzed by specialized proteases, but removal is not essential to translocation. No sequence similarity is found among the targeting signals for each compartment. Thus, the targeting information resides in more generalized features of the leader sequences such as charge distribution, relative polarity, and secondary structure.

The Synthesis of Secretory Proteins and Many Membrane Proteins Is Coupled to Translocation Across the ER Membrane

The signals recognized by the endoplasmic reticulum translocation system are virtually indistinguishable from bacterial signal sequences; indeed the two are interchangeable *in vitro.* In higher eukaryotes, translation and translocation of many proteins destined for processing via the ER are tightly coupled. That is, translocation across the ER occurs as the protein is being translated on the ribosome. As the N-terminal signal sequence of a preprotein undergoing synthesis emerges from the ribosome, it is detected by a so-called **signal recognition particle (SRP;** Figure 32.28). SRP is a 325-kD nucleoprotein assembly that contains six polypeptides and a 300-nucleotide *7S RNA.* SRP binding of the signal sequence halts further protein synthesis on the ribosome. This prevents release of the growing protein into the cytosol before it reaches the ER and its destined translocation. The SRP-ribosome complex then diffuses to the cytosolic face of the ER, where it binds to the **docking protein** (or **SRP receptor**), a heterodimeric transmembrane protein of 69- and 30-kD subunits. Dissociation of SRP then occurs in a GTP-dependent process. Docking of the ribosome on the ER allows protein synthesis to resume, and the growing polypeptide chain passes through the ER membrane into the lumen (Figure 32.28). Translocation is ATP-driven.

Soon after it enters the ER lumen, the signal peptide is clipped off by membrane-bound **signal peptidase.** Other modifying enzymes within the lumen introduce additional post-translational alterations into the polypeptide, such as glycosylation with specific carbohydrate residues. ER-processed proteins destined for secretion from the cell or inclusion in vesicles such as lysosomes end up contained within the soluble phase of the ER lumen, but

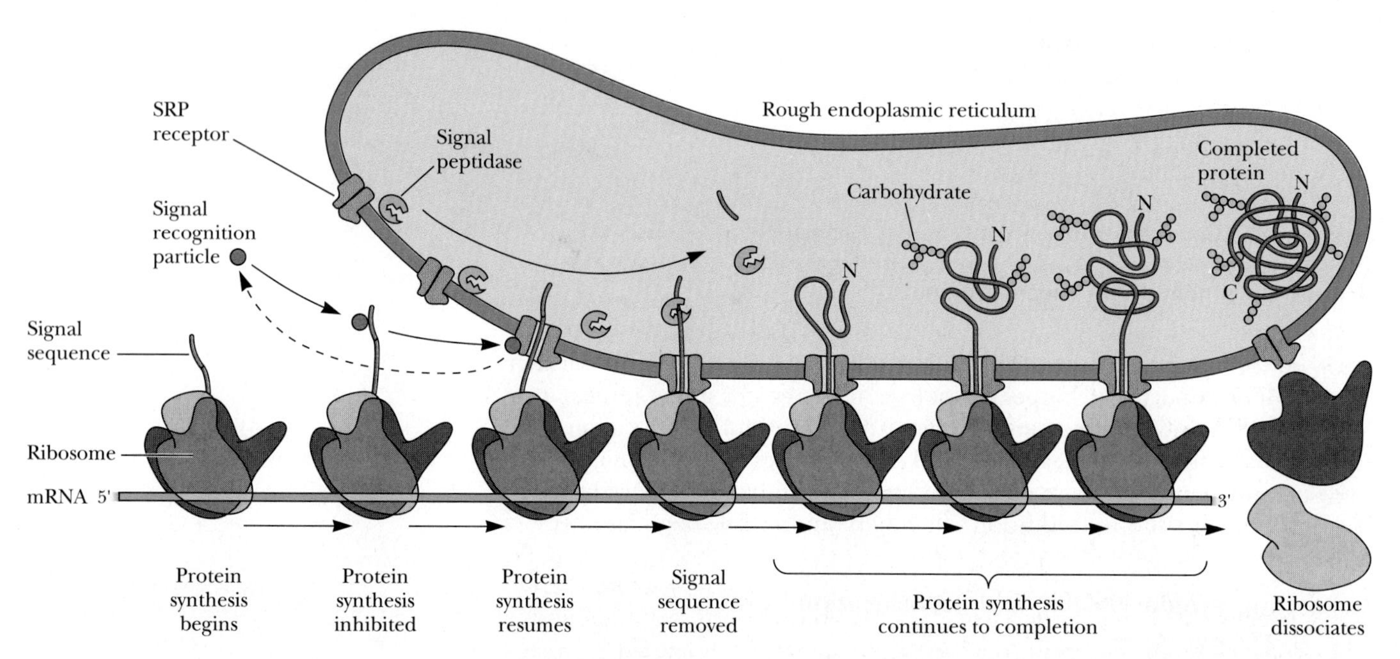

***Figure* 32.28** Synthesis of a eukaryotic integral membrane protein and its translocation via the endoplasmic reticulum.

polypeptides destined to become membrane proteins carry 20-residue hydrophobic **stop-transfer** sequences within their mature domains that arrest their passage across the ER membrane. These proteins remain embedded in the ER membrane with their C-termini on the cytosolic face of the ER. Such membrane proteins arrive at their intended destinations via subsequent processing of the ER.

Mitochondrial Protein Import

Most mitochondrial proteins are encoded by the nuclear genome and synthesized on cytosolic ribosomes. Mitochondria consist of three subcompartments: the outer membrane, the inner membrane, and the matrix. Thus, mitochondrial proteins must not only find mitochondria, they must gain access to the proper subcompartment, and once there they must attain a functionally active conformation. In principle, comparable considerations apply to protein import to chloroplasts, organelles with four subcompartments (outer membrane, inner/thylakoid membrane, stroma, and thylakoid lumen; Chapter 22).

Signal sequences on nuclear-encoded proteins destined for the mitochondria are N-terminal cleavable presequences 10 to 70 residues long. These mitochondrial presequences lack contiguous hydrophobic regions. Instead, they have positively charged and hydroxyl-amino acid residues spread along their entire length. These sequences form **amphiphilic α-helices** (Figure 32.29) with basic residues on one side of the helix and uncharged and hydrophobic residues on the other; that is, mitochondrial presequences are positively charged amphiphiles. In general, mitochondrial targeting sequences share no sequence homology. Once synthesized, mitochondrial preproteins are retained in an unfolded state with their target sequences exposed, through association with heat shock proteins of the Hsp70 family (Chapter 5). Import involves binding of a preprotein to a receptor protein on the mitochondrial outer membrane and subsequent uptake via the mitochondrial protein translocation apparatus.

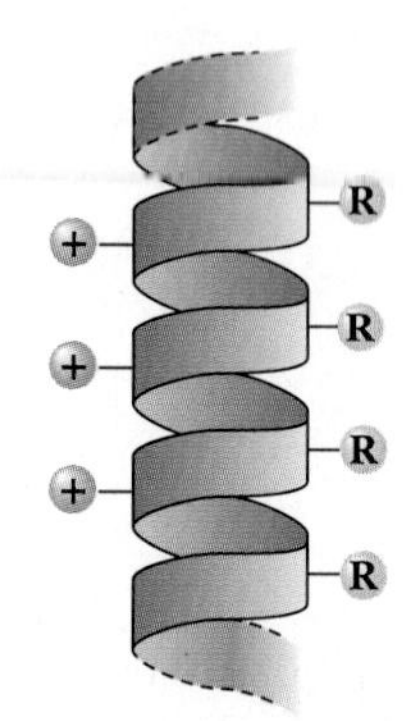

***Figure* 32.29** Structure of an amphiphilic α-helix having basic (+) residues on one side and uncharged and hydrophobic (R) residues on the other.

32.7 Protein Degradation

Cellular proteins are in a dynamic state of turnover, with the relative rates of protein synthesis and protein degradation ultimately determining the amount of protein present at any point in time. In many instances, transcriptional regulation determines the concentrations of specific proteins expressed within cells, with protein degradation playing a minor role. In other instances, protein synthesis is constitutive, and the amounts of key enzymes and regulatory proteins are controlled via selective protein degradation. In addition, abnormal proteins arising from biosynthetic errors or postsynthetic damage must be destroyed to prevent the deleterious consequences of their buildup. The elimination of proteins typically follows first-order kinetics, with half-lives ($t_{1/2}$) of different proteins ranging from several minutes to many days.

A number of mechanisms for protein degradation exist. Some, such as lysosome-mediated protein breakdown, are largely nonselective. A single, random proteolytic break introduced into the polypeptide backbone of a protein is believed sufficient to cause its rapid disappearance, since no partially degraded proteins are normally observed in cells. On the other hand, there is also a selective, ATP-dependent mechanism for protein degradation found in the cytosol of eukaryotic cells, the ubiquitin-mediated pathway.

The Ubiquitin Pathway for Protein Degradation in Eukaryotes

Ubiquitin is a highly conserved, 76-residue (8.5 kD) polypeptide widespread in eukaryotes. The amino acid sequences of yeast and human ubiquitin are 53% identical. Proteins are committed to degradation through ligation to ubiquitin. Three proteins in addition to ubiquitin are involved in the ligation process: $\mathbf{E_1}$, $\mathbf{E_2}$, and $\mathbf{E_3}$ (Figure 32.30). E_1 is the **ubiquitin-activating enzyme** (105-kD dimer). It becomes attached via a thioester bond to the C-terminal Gly residue of ubiquitin through ATP-driven formation of an activated ubiq-

E1 : Ubiquitin-activating enzyme

1. Ubiquitin—COO⁻ (C-term. Gly) + ATP —(E1)→ PP + Ubiquitin—C(=O)~AMP (Ubiquitinyl acyladenylate)
2. Ubiquitin—C(=O)~AMP + E1—SH → AMP + E1—S~C(=O)—Ubiquitin (Thioester)

E2 : Ubiquitin-carrier protein

E1—S~C(=O)—Ubiquitin + E2—SH → E1—SH + E2—S~C(=O)—Ubiquitin

E3 : Ligase

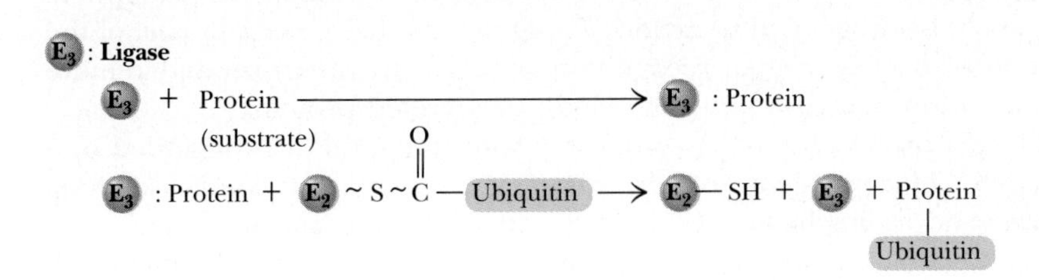

***Figure* 32.30** Enzymatic reactions in the ligation of ubiquitin to proteins. Ubiquitin is attached to selected proteins via isopeptide bonds formed between the ubiquitin carboxy-terminus and free amino groups (α-NH_2 terminus, Lys ϵ-NH_2 side chains) on the protein.

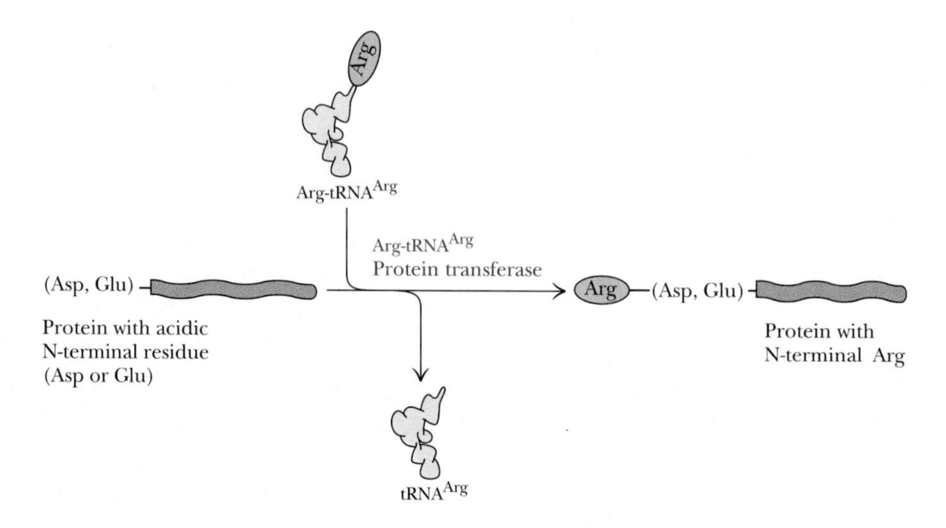

***Figure* 32.31** Proteins with acidic N-termini show a tRNA requirement for degradation. Arginyl-tRNA^{Arg}:protein transferase catalyzes the transfer of Arg to the free α-NH_2 of proteins with Asp or Glu N-terminal residues. Arg-tRNA^{Arg}:protein transferase serves as part of the protein degradation recognition system.

uitin-adenylate intermediate. Ubiquitin is then transferred from E_1 to an SH group on E_2, the **ubiquitin-carrier protein.** (E_2 is actually a family of at least seven different small proteins, several of which are heat shock proteins.) E_2-S ~ ubiquitin has two possible fates: It can directly transfer its ubiquitinyl moiety to free amino groups (N-terminus or Lys side chains) of certain proteins, such as histones, although this path is probably not important in protein degradation. Alternatively, E_2-S ~ ubiquitin transfers ubiquitin to free amino groups on proteins selected by E_3 (180 kD), the **ubiquitin-protein ligase.** Upon binding a protein substrate, E_3 catalyzes the transfer of ubiquitin from E_2-S ~ ubiquitin to free amino groups (usually Lys ϵ-NH_2) on the protein. More than one ubiquitin may be attached to a protein substrate, and tandemly linked chains of ubiquitin also occur via *isopeptide bonds* between the C-terminus of one ubiquitin and the ϵ-amino of Lys^{48} of another.

E_3 plays a central role in recognizing and selecting proteins for degradation. E_3 selects proteins by the nature of the N-terminal amino acid. Proteins must have a free α-NH_2 to be susceptible. Proteins with either Met, Ser, Ala, Thr, Val, Gly, or Cys at the amino terminus are resistant to the ubiquitin-mediated degradation pathway. However, proteins having Arg, Lys, His, Phe, Tyr, Trp, Leu, Asn, Gln, Asp, or Glu N-termini have half-lives of only 2 to 30 minutes. E_3 has three distinct protein substrate-binding sites: Type I for proteins with a basic N-terminus as in Arg, Lys, or His; Type II for proteins with bulky hydrophobic N-terminal residues (Phe, Tyr, Trp, Leu); and Type III for proteins whose N-termini are neither basic nor hydrophobic.

Interestingly, proteins with acidic N-termini (Asp or Glu) show a tRNA requirement for degradation (Figure 32.31). Transfer of Arg from Arg-tRNA to the N-terminus of these proteins alters their N-terminus from acidic to basic, rendering the protein susceptible to Type I E_3. It is also interesting that Met is less likely to be cleaved from the N-terminus if the next amino acid in the chain is one particularly susceptible to ubiquitin-mediated degradation.

Most proteins with susceptible N-terminal residues are *not* normal intracellular proteins but tend to be secreted proteins in which the susceptible residue has been exposed by action of a signal peptidase. Perhaps part of the function of the N-terminal recognition system is to recognize and remove from the cytosol any invading "foreign" or secreted proteins.

Once a protein has been conjugated with ubiquitin, it is degraded by a large **26S ATP-dependent protease complex.** This complex consists of three **conjugate-degrading factors:** CF-1, CF-2, and a 20S multicatalytic protease designated CF-3. The action of this 26S complex is not yet completely elucidated.

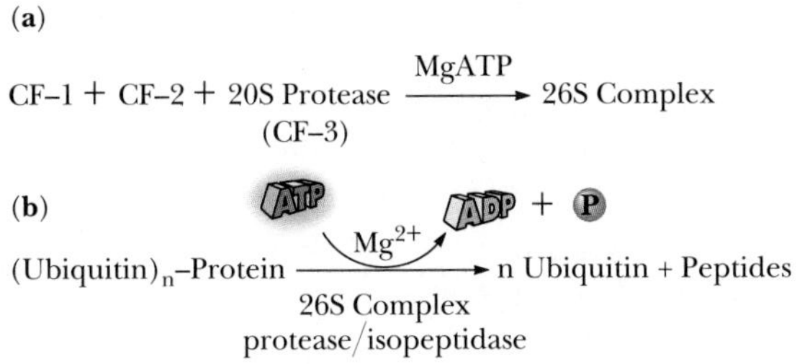

(a)

$$\text{CF–1} + \text{CF–2} + \underset{(\text{CF–3})}{\text{20S Protease}} \xrightarrow{\text{MgATP}} \text{26S Complex}$$

(b)

$$(\text{Ubiquitin})_n\text{–Protein} \xrightarrow[\text{26S Complex protease/isopeptidase}]{\text{ATP} \searrow \text{Mg}^{2+} \nearrow \text{ADP} + \text{P}} \text{n Ubiquitin} + \text{Peptides}$$

***Figure* 32.32** Breakdown of protein:ubiquitin conjugates by the 26S protease complex. (a) Formation of active 26S complex. (b) Degradation of protein ligated to ubiquitin.

It liberates peptide products from the ubiquitin-protein conjugates and cleaves the isopeptide bonds joining the conjugates, allowing ubiquitin to be recycled (Figure 32.32).

Problems

1. Why do you suppose eukaryotic protein synthesis is only 10% as fast as prokaryotic protein synthesis?
2. Eukaryotic ribosomes are larger and more complex than prokaryotic ribosomes. What advantages and disadvantages might this greater ribosomal complexity bring to a eukaryotic cell?
3. What ideas can you suggest to explain why ribosomes invariably exist as two-subunit structures, instead of a larger, single-subunit entity?
4. How do prokaryotic cells determine whether a particular methionyl-tRNA^{Met} is intended to initiate protein synthesis or to deliver a Met residue for internal incorporation into a polypeptide chain? How do the Met codons for these two different purposes differ? How do eukaryotic cells handle these problems?
5. What is the Shine-Dalgarno sequence? What does it do? The efficiency of protein synthesis initiation may vary by as much as 100-fold for different mRNAs. How might the Shine-Dalgarno sequence be responsible for this difference?
6. In the protein synthesis elongation events described under the section entitled *Translocation,* which of the following seems the most apt account of the peptidyl transfer reaction: (a) The peptidyl-tRNA delivers its peptide chain to the newly arrived aminoacyl-tRNA situated in the A site, *or* (b) the aminoacyl end of the aminoacyl-tRNA moves toward the P site to accept the peptidyl chain? Which of these two scenarios makes most sense to you? Why?
7. Why might you suspect that the elongation factors EF-Tu and EF-Ts are evolutionarily related to the G proteins of membrane signal transduction pathways described in Chapter 21?
8. How do the polypeptide products produced in the presence of (a) streptomycin and (b) puromycin differ from polypeptides synthesized in the absence of these inhibitors?
9. Which of the protein synthesis inhibitors listed in Table 32.6 might have clinical value as antibiotics?
10. Human rhodanese (33 kD) consists of 296 amino acid residues. Approximately how many ATP equivalents are consumed in the synthesis of the rhodanese polypeptide chain from its constituent amino acids and the folding of this chain into an active tertiary structure?
11. A single proteolytic break in a polypeptide chain of a native protein is often sufficient to initiate its total degradation. What does this fact suggest to you regarding the structural consequences of proteolytic nicks in proteins?

Further Reading

Agard, D. A., 1993. To fold or not to fold . . . . *Science* **260:**1903–1904.

Caskey, C. T., 1973. Inhibitors of protein synthesis. In Hochster, R. M., Kates, M., and Quastel, J. H., eds. *Metabolic Inhibitors,* vol. 4. New York: Academic Press.

Craig, E. A., 1993. Chaperones: Helpers along the pathways to protein folding. *Science* **260:**1902–1903.

Creighton, T. E., 1984. *Proteins: Structures and Molecular Properties.* New York: Freeman and Co.

Endo, Y., et al., 1987. The mechanism of action of ricin and related toxic lectins on eukaryotic ribosomes. The site and the characteristics of the modification in 28S ribosomal RNA caused by the toxins. *Journal of Biological Chemistry* **262:**5908–5912.

Giri, L., Hill, W. E., and Wittman, H. G., 1984. Ribosomal proteins: Their structure and spatial arrangement in prokaryotic ribosomes. *Advances in Protein Chemistry* **36:**1–78. The properties of *E. coli* ribosomal proteins and their associations within ribosomes is given in detail.

Guttell, R. R., et al., 1985. Comparative anatomy of 16S-like ribosomal RNA. *Progress in Nucleic Acid Research and Molecular Biology* **32:**155–216.

Hartl, F.-U., and Neupert, W., 1990. Protein sorting to mitochondria: Evolutionary conservations of folding and assembly. *Science* **247:**930–938.

Hendrick, J. P., and Hartl, F.-U., 1993. Molecular chaperone functions of heat-shock proteins. *Annual Review of Biochemistry* **62:**349–384.

Hershey, J. W. B., 1991. Translational control in mammalian cells. *Annual Review of Biochemistry* **60:**717–755.

Hershko, A., 1991. The ubiquitin pathway for protein degradation. *Trends in Biochemical Sciences* **16:**265–268.

Hershko, A., 1988. Ubiquitin-mediated protein degradation. *Journal of Biological Chemistry* **263:**15237–15240. Minireview of the ubiquitin-mediated pathway for selective protein degradation in eukaryotic cells.

Hill, W. E., et al., eds., 1990. *The Ribosome: Structure, Function and Evolution.* Washington DC: American Society for Microbiology Press. The properties of ribosomes, as reported at a research conference on the structure, function, and evolution of ribosomes held in 1989.

Moldave, K., 1985. Eukaryotic protein synthesis. *Annual Review of Biochemistry* **54:**1109–1149. The detailed features of protein synthesis as it occurs in eukaryotic cells.

Noller, H. F., 1991. Ribosomal RNA and translation. *Annual Review of Biochemistry* **60:**191–227. The intimate role of ribosomal RNA in the protein biosynthetic process.

Noller, H. F., Hoffarth, V., and Zimniak, L., 1992. Unusual resistance of peptidyl transferase to protein extraction procedures. *Science* **256:**1416–1419. Research paper presenting evidence that the peptide bond-forming step in protein synthesis—the peptidyl transferase reaction—is catalyzed by 23S rRNA.

Rapoport, T. A., 1991. Protein transport across the endoplasmic reticulum membrane: Facts, models, mysteries. *The FASEB Journal* **5:**2792–2798.

Rechsteiner, M., Hoffman, L., and Dubiel, W., 1993. The multicatalytic and 26S proteases. *Journal of Biological Chemistry* **268:**6065–6068.

Rhoads, R. E., 1993. Regulation of eukaryotic protein synthesis by initiation factors. *Journal of Biological Chemistry* **268:**3017–3020.

Riis, B., et al., 1990. Eukaryotic protein elongation factors. *Trends in Biochemical Sciences* **15:**420–424. Properties of the elongation factors in eukaryotic cells.

Samuel, C. E., 1993. The eIF-2α protein kinases, regulators of translation in eukaryotes from yeast to humans. *Journal of Biological Chemistry* **268:**7603–7606.

Weijland, A., and Parmeggiani, A., 1993. Toward a model for the interaction between elongation factor Tu and the ribosome. *Science* **259:**1311–1314.

Wickner, W., Driessen, A. J. M., and Hartl, F.-U., 1991. The enzymology of protein translocation across the *Escherichia coli* plasma membrane. *Annual Review of Biochemistry* **60:**101–124.

Chapter 33

Molecular Evolution

"It is interesting to contemplate an entangled bank, clothed with many plants of many kinds, with birds singing on the bushes, with various insects flitting about, and with worms crawling through the damp earth, and to reflect that these elaborately constructed forms, so different from each other, and dependent on each other in so complex a manner, have all been produced by laws acting around us."

Charles Darwin (1809–1882), *On the Origin of Species* (1859)

"Swamp Animals and Birds on the River Gambia" (ca. 1912), H. H. Johnston

Outline

33.1 An Overview of Evolution
33.2 Evolutionary Change in Nucleotide Sequences
33.3 Molecular Clocks
33.4 Nonrandom Codon Usage
33.5 Evolution by Gene Duplication and Exon Shuffling
33.6 Gene Sharing
33.7 Proton-Translocating ATPases: Clues to Early Evolution

DNA molecules are not only the repositories of genetic information, they are also the chronicles of evolutionary history. The DNA of every organism provides a historical record of its origin and evolution. Such records are now accessible because the methods of molecular biology can reveal the nucleotide sequences of genes and genomes. These sequences are the epics of evolution, but the epics are fragmentary. Mutations and rearrangements that have accumulated in the DNA over time often veil, disguise, or even obliterate the record so that its historical roots are obscured and difficult to trace. The science of **molecular evolution** seeks to decipher these records in order to understand the evolution of macromolecules and to reconstruct the *phylogeny* of organisms from molecular information. This latter aspect of molecular evolution is sometimes called *molecular phylogeny.*

phylogeny: the origin and evolution of the many types and species of organisms; from th Greek *phyle,* tribe, and *gen,* to produce

The primary work of molecular evolutionists is the reading, comparison, and interpretation of nucleotide and amino acid sequences in an evolutionary context. Genes evolve through sequence changes such as **nucleotide substitutions** as well as through processes occurring on a larger scale, such as exon rearrangements, or **shuffling; gene duplication; transposition** (movement of the gene from one chromosomal location to another); and **gene conversion** (a recombination process whereby two different sequences in a genome interact in such a way that one is converted to the other). Viewed over time, *genomes are both constant and changing:* the quantity and quality of an organism's DNA are replicated with great fidelity over countless generations (the "constancy" part), yet alterations sporadically arise that have profound evolutionary consequences.

33.1 An Overview of Evolution

Evolution can be defined as the transformations that have altered the form of life from its earliest origins to its contemporary complexity and diversity. In the final analysis, evolution is the consequence of genetic changes that occur within populations of organisms. When a mutant gene arises, its course of evolution is determined by the probability that it will persist in the population or even ultimately replace an old gene. The chromosomal location of a gene is termed its **locus.** Alternative forms of a gene at a given locus are called **alleles.** Within a population, a particular locus may be represented by more than one allele—i.e., more than one version of a gene—and their relative proportions in the population are defined as allele *frequencies.* Together, the alleles constitute the **gene pool.** A locus represented by two or more alleles in the population is said to be **polymorphic,** and the population of organisms is said to carry a *genetic polymorphism* at that particular locus.

The differential reproduction of genetically distinct individuals within a population is called **natural selection.** Natural selection will occur if an allele affects the **fitness** of the organism, that is, its ability to survive and reproduce. The view that natural selection determines the frequency of alleles in a population and, ultimately, the genetic makeup of populations is known as the **neo-Darwinian theory of evolution.** On the other hand, if the alleles at a locus are selectively neutral—that is, their effects on fitness are equivalent—changes in allele frequency may still occur in the population over time simply by chance. Such random change in allelic frequency is called **random genetic drift.** Solely by random genetic drift, a particular allele may, over time, become permanently established, or **fixed,** in the population; or it may become lost. With the accumulation of protein sequence data during the last several decades, it has been postulated that the majority of molecular changes in evolution are due to the random fixation of neutral mutations within the population. This hypothesis has become known as the **neutral theory of molecular evolution.** In the neutral theory, most genetic polymorphism is considered transitory—the alleles at a polymorphic locus are on their way to either fixation or extinction. Molecular evolution and genetic polymorphism are two sides of the same coin. Evolution, then, is the consequence of two different processes at work in populations: natural selection and random genetic drift.

33.2 Evolutionary Change in Nucleotide Sequences

Evolution proceeds in part by the appearance of nucleotide changes in DNA sequences over time. These changes give rise to new alleles, which may become fixed in the population. Since the rate of nucleotide substitution is usually very slow and the time required for fixation is typically long, genes appear stable in the time frame of a human observer. On the other hand, if genes are viewed in a geological time scale, their nucleotide sequences change continuously. Given these constraints, the easiest way to document evolutionary changes in nucleotide sequences is to compare two different sequences that share a common ancestry.

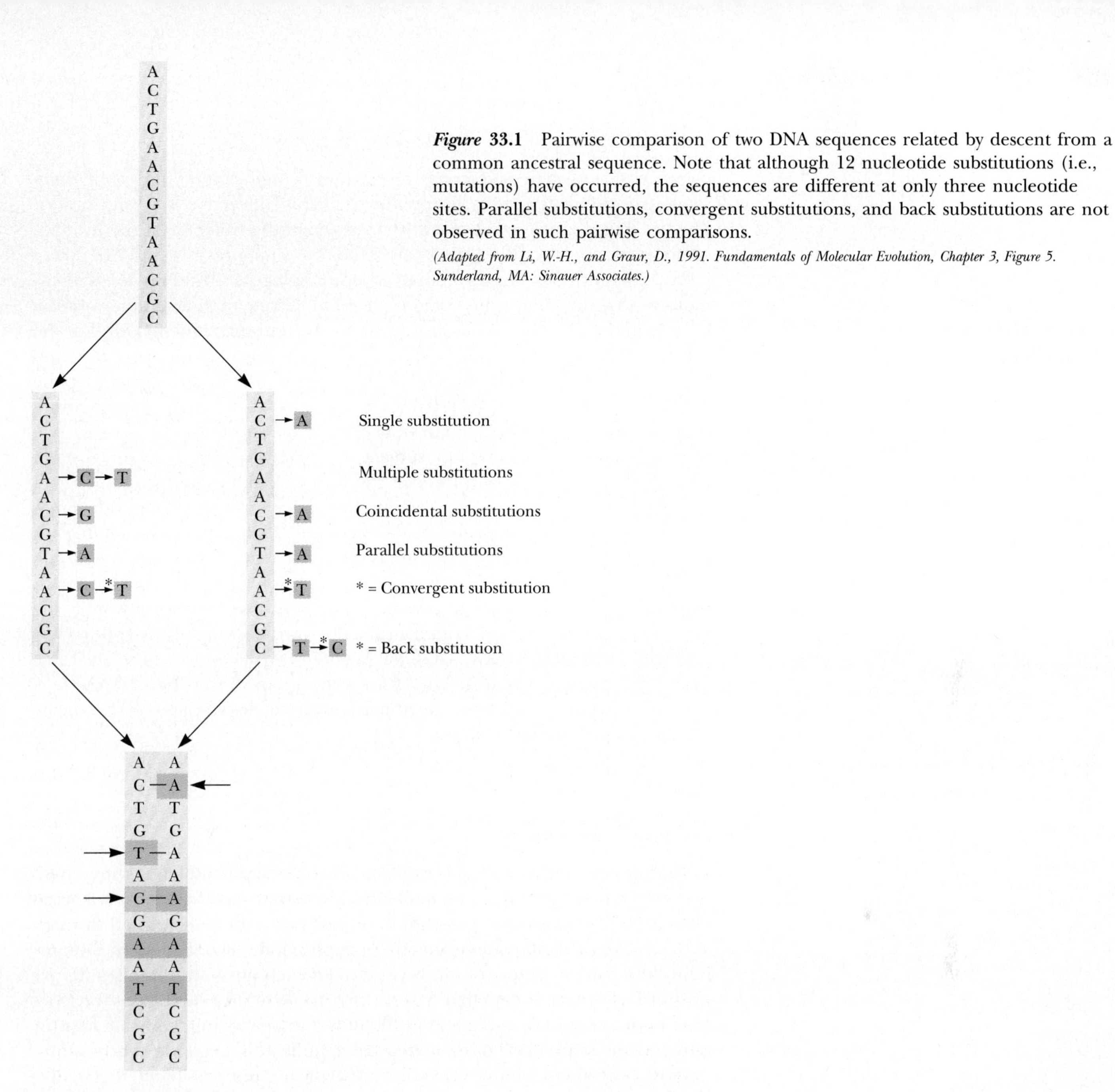

Figure **33.1** Pairwise comparison of two DNA sequences related by descent from a common ancestral sequence. Note that although 12 nucleotide substitutions (i.e., mutations) have occurred, the sequences are different at only three nucleotide sites. Parallel substitutions, convergent substitutions, and back substitutions are not observed in such pairwise comparisons.

(Adapted from Li, W.-H., and Graur, D., 1991. Fundamentals of Molecular Evolution, Chapter 3, Figure 5. Sunderland, MA: Sinauer Associates.)

Nucleotide substitutions can be either transitions or transversions. **Transitions** occur when a purine is replaced by a different purine or a pyrimidine by a different pyrimidine—for example, A → G or C → T. **Transversions** involve replacement of a purine by a pyrimidine or vice versa (Chapter 28). Because transversions are less common than transitions, the substitution of one nucleotide for another is not random.

Number of Nucleotide Substitutions Between Two Sequences

Nucleotide substitutions are inferred from a pairwise comparison of two sequences that share a common evolutionary origin. The number of substitutions that have arisen since two sequences diverged from each other is the most commonly used parameter in molecular evolution. If the sequences are closely related, then the number of observed differences probably represents the actual number of substitutions. On the other hand, if the sequences have diverged substantially, multiple substitutions may have occurred at the same site, so that the number of differences is less than the actual number of substitutions that have taken place (Figure 33.1).

Coding Regions Versus Noncoding Regions

Comparisons of gene sequences reveal that protein-coding and noncoding sequences usually evolve at different rates. This difference is a reflection of the fact that there is selective pressure to retain the function of the protein, whereas most noncoding regions are under no such pressure and can evolve freely. Furthermore, within coding regions, some substitutions are **synonymous** in that the same amino acid is encoded despite the nucleotide substitution, as in the glutamine codons CAA and CAG. Substitution of G for A at the third position is a synonymous substitution in terms of protein coding, and the codons CAA and CAG are said to be synonymous. On the other hand, the codons CAA (Gln) and CAC (His) are **nonsynonymous.** Synonymous substitutions are much more likely than nonsynonymous substitutions. Because synonymous substitutions are not subject to selective pressure at the protein level, they are unlikely to influence the fitness of the organism.

Nucleotide sites in coding regions can also be classified as nondegenerate, twofold degenerate, or fourfold degenerate. If all the possible changes at the site are nonsynonymous, the site is said to be a **nondegenerate** site; if one of the three possible substitutions at a site is synonymous, the site is termed **twofold degenerate;** and if all possible changes at a site are synonymous, the site is **fourfold degenerate.** For example, the third position in the codon CCC is fourfold degenerate: CCC, CCT, CCA, and CCG all code for Pro (see Table 31.3, "The Genetic Code"). The first two positions of the codon AAA (Lys) are nondegenerate, whereas the third is twofold degenerate (AAX can be either a Lys or an Asn codon).

Sequence Alignments

In addition to nucleotide substitutions, insertions and deletions may have occurred in two homologous nucleotide or amino acid sequences derived from a common ancestor. Insertions and deletions can be identified through *sequence alignment.* **Alignments** involve juxtaposing the two sequences and analyzing the pairs of bases (or amino acids). At each position, three results are possible: (1) pairs of matched bases, (2) pairs of mismatched bases, or (3) a base in one sequence and a gap in the other. Matched pairs indicate identity between the sequences, mismatched bases indicate nucleotide substitutions, and gaps signal insertions or deletions. Usually it is not possible to distinguish whether an insertion has occurred in one sequence or a deletion in the other.

Several methods are available for sequence alignment analyses, including the dot-matrix method and the sequence-distance method. The **dot-matrix method** is straightforward. The two sequences to be aligned are written out, one as column headings in a matrix, the other as row headings, and a dot is placed in the matrix where the nucleotides are identical (Figure 33.2). If the sequences are identical at every position, the dots are aligned along the diagonal elements of the matrix (Figure 33.2a). If the sequences are not identical but contain no gaps, dots are aligned in most of the diagonal elements (Figure 33.2b). If an insertion or deletion has occurred, the presence of a gap is revealed by a vertical or horizontal shift in the alignment diagonal (Figure 33.2c). If the two sequences differ by both gaps and substitutions (Figure 33.2d), it may be difficult to decide between several alternative alignments.

Sequence-distance methods attempt to reach the best possible alignment by minimizing the number of gaps and mismatches through application of certain criteria. However, reducing the number of gaps usually increases the number of mismatches, and vice versa. Consider the sequences CTAGTCATTG ($n = 11$) and CTGGTGCTG ($n = 9$).

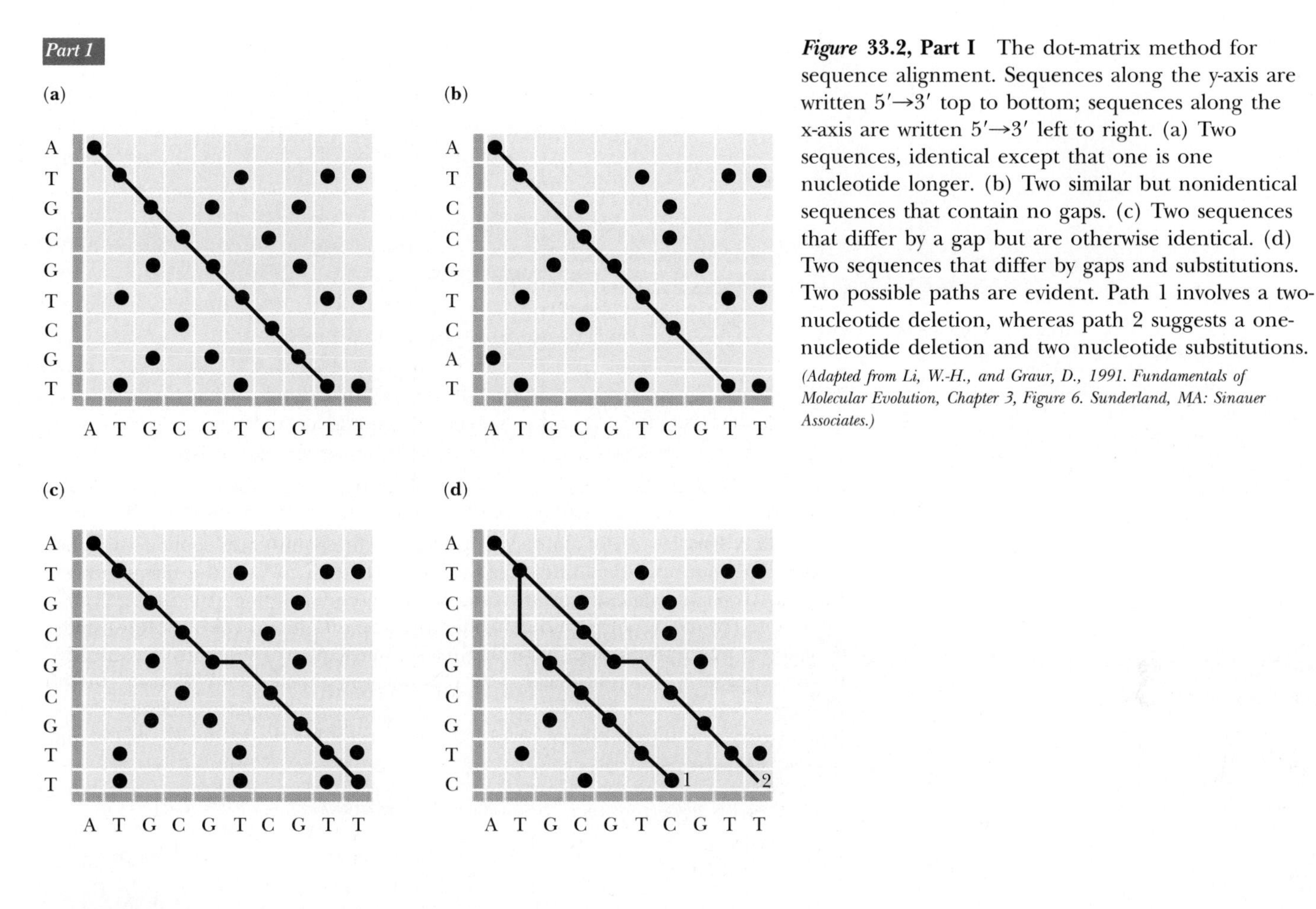

Figure **33.2, Part I** The dot-matrix method for sequence alignment. Sequences along the y-axis are written 5′→3′ top to bottom; sequences along the x-axis are written 5′→3′ left to right. (a) Two sequences, identical except that one is one nucleotide longer. (b) Two similar but nonidentical sequences that contain no gaps. (c) Two sequences that differ by a gap but are otherwise identical. (d) Two sequences that differ by gaps and substitutions. Two possible paths are evident. Path 1 involves a two-nucleotide deletion, whereas path 2 suggests a one-nucleotide deletion and two nucleotide substitutions. *(Adapted from Li, W.-H., and Graur, D., 1991. Fundamentals of Molecular Evolution, Chapter 3, Figure 6. Sunderland, MA: Sinauer Associates.)*

The alignment I

CTAG–TCG–ATTG
CT–GGT–GC–T–G

reduces the mismatches to zero but creates six gaps. Alternatively, the alignment II

CTAGTCGATTG
CTGGTGCTG—

has only a single terminal gap but five mismatches (the underlined bases). A possible third alignment (III) has two mismatches and four gaps:

CTAG–TCGATTG
CT–GGT–GCTG–

How can mismatches and gaps be compared? One criterion is to assume *a priori* that gaps are less likely than substitutions. This assumption imposes a penalty (the **gap penalty**) on sequences with gaps. One method uses the equation

$$\mathbf{D} = \mathbf{y} + \mathbf{wz}$$

where **D** is a measure of the evolutionary distance between two sequences, obtained by adding **y** (defined as the number of mismatches) and **wz** (where **w** is the gap penalty and **z** is the total length [in nucleotides] of the gap). If **w** = 1, so that a gap is assumed to be equally as likely as a nucleotide substitution, then **D** for alignment I is $0 + (1 \times 6) = 6$; **D** for alignment II is $5 + (1 \times 2) = 7$; and **D** for alignment III is $2 + (1 \times 4) = 6$. Thus, alignments I and III

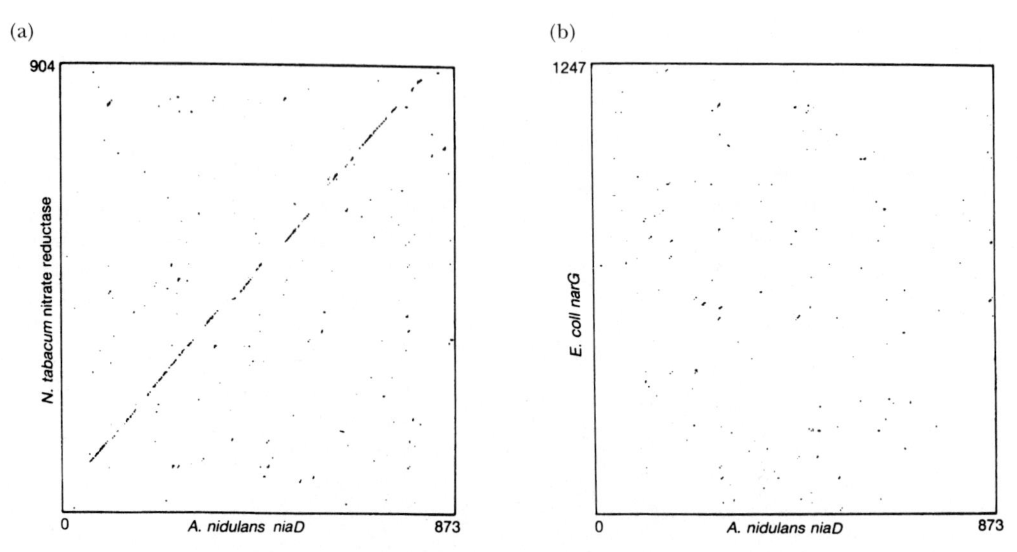

Figure **33.2, Part II** A dot-matrix comparison of the amino acid sequences of (a) tobacco *(Nicotiana tabacum)* nitrate reductase (904 residues) and the filamentous fungus *Aspergillus nidulans* nitrate reductase (873 residues); and (b) the *Escherichia coli* nitrate reductase polypeptide encoded by the *narG* gene (1247 residues) and *A. nidulans* nitrate reductase. Note that the strong sequence similarity evidenced by the diagonal in (a) is absent in (b). (No diagonal elements are seen in part b.) Thus, whereas the nitrate reductases of tobacco and the fungus are highly homologous, the fungal nitrate reductase shows no similarity to the *E. coli* enzyme. *(Adapted from Kinghorn, J. R., and Campbell, E. I., 1989. Amino Acid Sequence Relationships Between Bacterial, Fungal and Plant Nitrate Reductase and Nitrite Reductase Proteins, pp. 385–403 in Molecular and Genetic Aspects of Nitrate Assimilation, edited by Wray, J. L., and Kinghorn, J. R. Oxford: Oxford Science Publications.)*

are equally preferable, and alignment II is less preferable. However, if **w** = 2, **D** for alignment I is $0 + (2 \times 6) = 12$; **D** for alignment II is $5 + (2 \times 2) = 9$, and **D** for alignment III is $2 + (2 \times 4) = 10$. Then alignment II gets the nod.

Clearly, the task of deciding between alternative sequence alignments is complicated. A variety of computer *algorithms* have been devised for assessing gap penalties and for aligning sequences.

algorithm: a set of steps to be followed in a prescribed order to solve a particular problem

Rates of Nucleotide Substitutions

The **rate of nucleotide substitution** is defined as the number of substitutions per site per year. Typical rates of substitution range from 0.1 to 10 nucleotide substitutions per site per 1 billion years ($= 0.1$ to 10×10^{-9} per year). Rates of nonsynonymous substitution are much less than rates of synonymous substitution (Table 33.1). Furthermore, different regions of genes differ in their average rates of nucleotide substitution (Figure 33.3). The results demonstrate that sequences that are not subject to functional constraints, such as introns and pseudogenes, evolve at higher rates. In contrast, coding regions as well as flanking regions that contain sequences involved in regulation of gene expression evolve less rapidly. The contrast between nonsynonymous and synonymous rates in a gene reflects an important principle of molecular evolution: *The greater the functional constraints on a macromolecule, the slower its rate of evolution.*

If we consider the variation in rates of nonsynonymous substitutions among genes, two explanations for these differences offer themselves: (1) The rate of mutation for genes in different locations may be different;

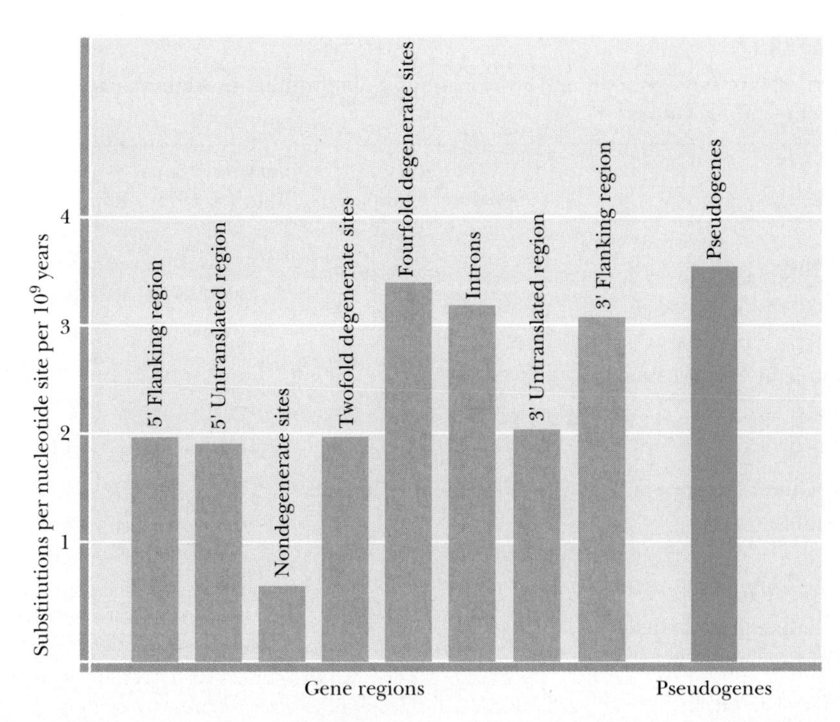

***Figure* 33.3** Average rates of substitution in different regions of genes and in pseudogenes. (A pseudogene is an inactive copy of a gene derived by duplication and mutation of an ancestral functional gene.)

(Adapted from Li, W.-H., and Graur, D., 1991. Fundamentals of Molecular Evolution. Chapter 4, Figure 2. Sunderland, MA: Sinauer Associates.)

(2) the selection pressure on gene products determines the rate of substitution. Estimates indicate that different parts of the mammalian nuclear genome may vary in their rates of mutation by a factor of 2. But rates of nonsynonymous substitution vary by orders of magnitude from effectively zero for the histone 3 and 4 genes to close to 3×10^{-9} per year for interferon γ (Table 33.1). Thus, selection constraints are the principal factor determining the rate of mutation that can be tolerated by a gene.

Histone 3 is a case in point. Most amino acids in histone 3 interact directly with other core histones or with DNA in the formation of a nucleosome (see Figure 7.30), so the histone 3 molecule cannot tolerate the structural alterations that would be the consequence of amino acid substitutions. In contrast, the **apolipoproteins** show a nonsynonymous rate of substitution of about 1×10^{-9}/year. These proteins serve as lipid carriers in the bloodstreams of vertebrates. The lipid-binding domains of these proteins are constructed from hydrophobic residues, and replacement of one hydrophobic amino acid by another usually has minimal effect on the lipid transport properties of the protein. Thus, apolipoproteins can tolerate a higher rate of nonsynonymous substitution than can histone 3.

33.3 Molecular Clocks

In Chapter 4, using cytochrome *c* as our example, we asserted that homologous proteins from different organisms have homologous amino acid sequences. Using myoglobin and hemoglobin as examples, we argued that related proteins share a common evolutionary origin. The rates of amino acid substitution in these proteins are about the same in different mammalian lineages. Based on these similarities in rate, Emile Zuckerkandl and Linus

Table **33.1**

Rates of Nonsynonymous and Synonymous Substitutions in Mammalian Protein-Coding Genes*

Gene	Number of Codons Compared	Nonsynonymous Rate ($\times 10^9$)	Synonymous Rate ($\times 10^9$)
Histones			
Histone 3	135	0.00	6.4
Histone 4	101	0.00	6.1
Contractile System Proteins			
Actin α	376	0.01	3.7
Actin β	349	0.03	3.1
Hormones, Neuropeptides, and Other Active Peptides			
Somatostatin-28	28	0.00	4.0
Insulin	51	0.13	4.0
Thyrotropin	118	0.33	4.7
Insulinlike growth factor II	179	0.52	2.3
Erythropoietin	191	0.72	4.3
Insulin C-peptide	35	0.91	6.8
Parathyroid hormone	90	0.94	4.2
Luteinizing hormone	141	1.02	3.3
Growth hormone	189	1.23	5.0
Urokinase-plasminogen activator	435	1.28	3.9
Interleukin I	265	1.42	4.6
Relaxin	54	2.51	7.5
Globins			
Myoglobin	153	0.56	4.4
Hemoglobin α-chain	141	0.55	5.1
Hemoglobin β-chain	144	0.80	3.1

Adapted from Li, W.-H., and Graur, D., 1991. *Fundamentals of Molecular Evolution,* Chapter 4, Table 1. Sunderland, MA: Sinauer Associates.

*All rates are based on comparisons between human and rodent genes, and the time of divergence was set at 80 million years ago. Rates are in units of substitutions per site per 10^9 years.

Pauling suggested in 1965 that, for any given protein, the rate of amino acid substitution is fairly constant over time in all lineages. Therefore, these rates constitute a **molecular clock** for measuring the pace of evolution. Although this idea was well received, many evolutionists argued strongly that the rate of evolution was erratic—at times fast, at other times slow. For example, rates of evolution are known to increase dramatically following gene duplication, such as the duplication leading to the separate α and β hemoglobin genes (Chapter 4). This issue was investigated by comparing rates of synonymous substitutions in the genes of related mammals. Synonymous substitutions were compared because they are functionally neutral. DNA sequence data readily reveal synonymous substitutions. Comparisons of related species—mice versus rats and humans versus monkeys—revealed that rats and mice show nearly equal rates (about 8×10^{-9} synonymous substitutions per year), whereas human genes (1.3×10^{-9} synonymous substitutions per year) are apparently

Table 33.1
(*continued*)

Gene	Number of Codons Compared	Nonsynonymous Rate ($\times 10^9$)	Synonymous Rate ($\times 10^9$)
Apolipoproteins			
E	283	1.0	4.0
A-I	243	1.6	4.5
A-IV	371	1.6	4.2
Immunoglobulins			
Ig V_H	100	1.1	5.7
Ig $\gamma 1$	321	1.5	5.1
Ig κ	106	1.9	5.9
Interferons			
$\alpha 1$	166	1.4	3.5
$\beta 1$	159	2.2	5.9
γ	136	2.8	8.6
Enzymes and Other Proteins			
Aldolase A	363	0.07	3.6
Hypoxanthine-guanine phosphoribosyltransferase	217	0.13	2.1
Creatine kinase M	380	0.15	3.1
Glyceraldehyde-3-phosphate dehydrogenase	331	0.20	2.8
Lactate dehydrogenase A	331	0.20	5.0
Acetylcholine receptor γ subunit	540	0.29	3.2
Fibrinogen γ	411	0.55	5.8
Albumin	590	0.91	6.6
Average†		0.85	4.6

†Averages are arithmetic means.

evolving more slowly than those of monkeys (2.2×10^{-9} synonymous substitutions per year). Furthermore, the average rate of substitution in rodents is as much as six times higher than that of primates.

An obvious difference in these organisms is the time it takes them to complete one generation (i.e., their *generation time*). Most mutations (nucleotide substitutions) arise during germ-line DNA replications in the process of gamete formation. Let us assume a correlation between generation time and germ-line replication cycles; rodents, with a much shorter generation time than that of primates, would be expected to accumulate more mutations. Rodents may also have a less effective DNA repair system. In any case, the molecular clock concept works reasonably well for comparison of nucleotide substitution rates in organisms with similar generation times (such as mice and rats). Thus, although the concept of a universal molecular clock is dubious, closely related organisms do keep time with one another.

Small Subunit rRNA

Ribosomes are found in all organisms, and thus their origins must be ancient. The rRNA molecule associated with the small ribosomal subunit (e.g., the 30S ribosomal subunit of *Escherichia coli*) has become a molecule of choice as a molecular chronometer for evolutionary comparisons and the elucidation of phylogenetic trees. Its structure is highly conserved (Chapter 7), and therefore it is slow to evolve, meaning it can provide more reliable information about early events in evolution. Sequence comparisons of small subunit rRNAs have led to hypotheses regarding evolutionary relationships between the major forms of life (see Chapter 7 and later discussions in this chapter).

33.4 Nonrandom Codon Usage

The degeneracy of the genetic code means that most amino acids are represented by more than one codon (Chapter 31). Since synonymous codons encode the same amino acid, they should be used with essentially equal frequency. However, DNA sequence data reveal a decided bias in codon usage in both prokaryotes and eukaryotes (see Table 31.7). As an example, the *E. coli* outer membrane protein *OmpA* has 23 Leu residues, 21 of which are encoded by the codon CUG even though there are five other Leu codons.

Interestingly, a given organism tends to use the same nonrandom set of codons in encoding its proteins; that is, there is a species-specific bias in codon usage. Correlated with this bias is a correspondence between the relative frequency of usage of a particular codon and the relative abundance of its cognate tRNA: Codons that are used frequently are represented by greater concentrations of their cognate tRNA. And, in a given organism, highly expressed genes are much more likely to show a strong bias in codon usage. These findings are illustrated in Figure 33.4 for leucine codon usage, relative tRNA abundance, and level of gene expression in *E. coli* and yeast. Apparently, highly expressed genes, whose protein products are needed in large amounts, rely on a bias in codon usage to exploit the differential amounts of cognate tRNAs present in the cell. This strategy enhances their expression through optimization of translational efficiency. Given these biases, the assumption that synonymous substitutions are selectively neutral in evolution may not be completely accurate.

33.5 Evolution by Gene Duplication and Exon Shuffling

Thus far, we have restricted our attention to the role of nucleotide substitutions in evolution. Larger changes in gene structure, namely *gene duplication* and *exon shuffling*, also influence the course of evolution. **Gene duplication** leads to a second copy of an existing gene. Since the gene's function is fulfilled by a single copy, the duplicate is redundant and consequently is free of functional constraints on its evolution. Hence, it may diverge to eventually become a new gene, giving rise to a new function. Gene duplication is by far the predominant mechanism for the appearance of new genes. We will be concerned only with gene and partial gene duplications here; complete and partial chromosome duplication, and even entire genome duplication, are important factors in evolution but beyond the scope of this molecular discussion.

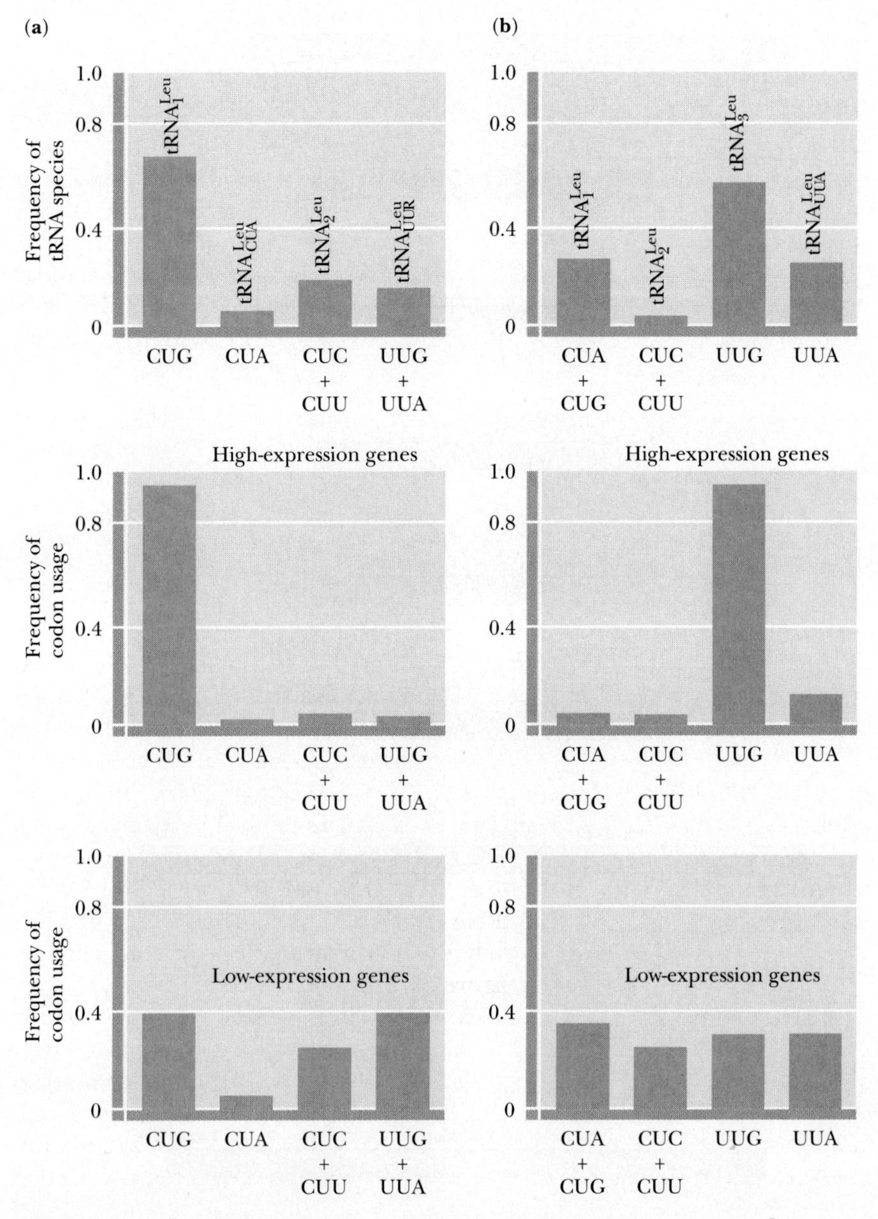

***Figure* 33.4** The relationships between relative amounts of various tRNA^{Leu} species and relative frequency of Leu codon usage in genes that are expressed at high levels or low levels in (a) *Escherichia coli* and (b) yeast.

(Adapted from Li, W.-H., and Graur, D., 1991. Fundamentals of Molecular Evolution. Chapter 4, Figure 8. Sunderland, MA: Sinauer Associates.)

Exons and the Evolution of Proteins

In reflecting on the nature of split genes, Gilbert proposed in 1978 that the organization of eukaryotic genes into introns and exons allows for the exchange of exons between genes **(exon shuffling)** via recombination events taking place in introns. If exons correspond to modules of protein function, such recombination could arrange these modules to create proteins with novel functions. Such proteins would be **mosaics** of exon-encoded structural units, and the collective activities of the various modules would endow the protein with its characteristic function. An alternative suggestion is that exons represent not functional units but integrally folded units of protein structure that could be recruited and assembled into novel structures by exon shuffling.

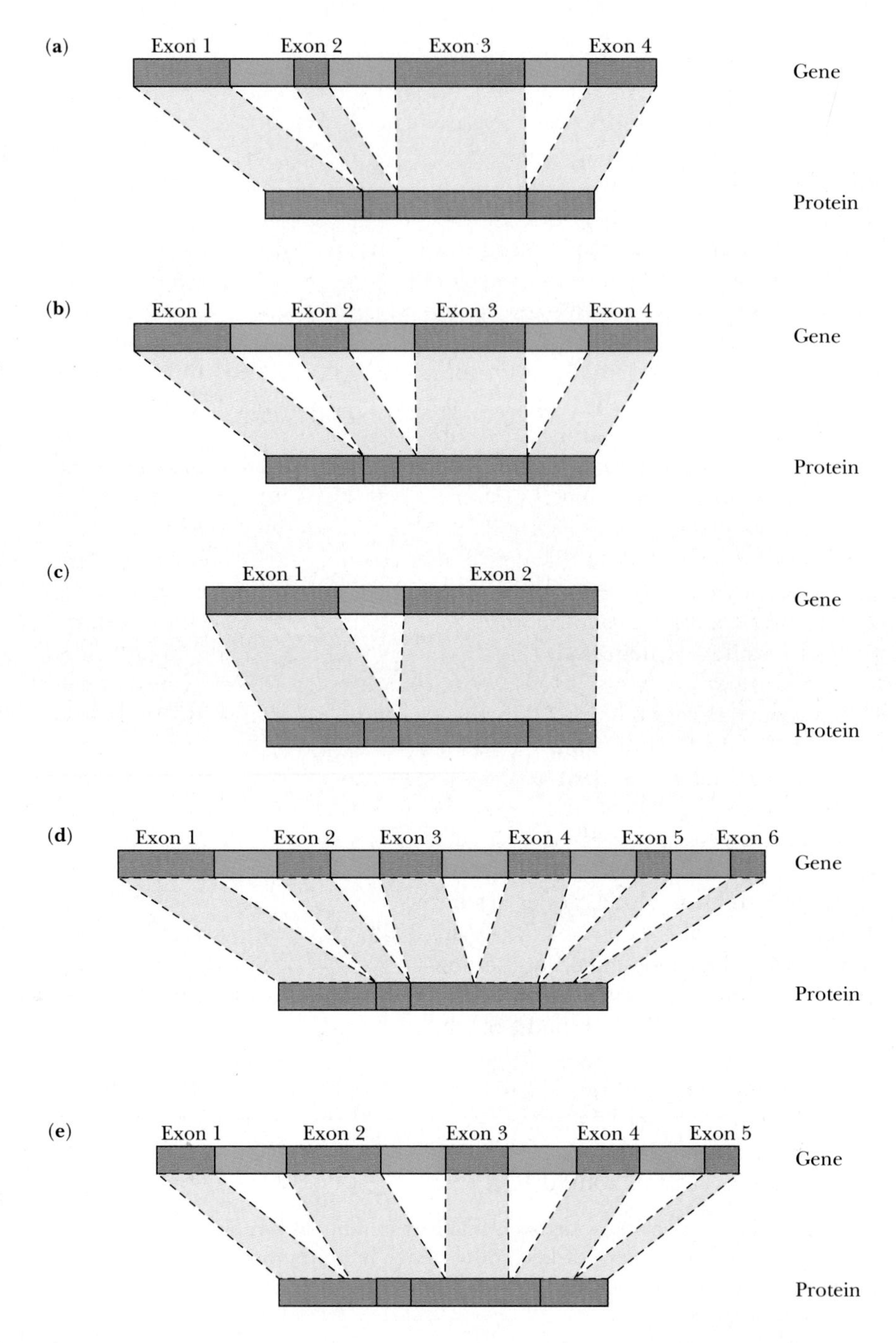

***Figure* 33.5** Possible relationships between exon arrangements and domain organization in proteins. In (a), each exon corresponds exactly to a structural domain; in (b), the correspondence is only approximate. In (c), each exon encodes more than one domain. The possibility in (d) suggests that structural domains may be encoded by more than one exon. Finally, in (e), there is no real correspondence between exons and structural domains.

(Adapted from Li, W.-H., and Graur, D., 1991. Fundamentals of Molecular Evolution. Chapter 6, Figure 1. Sunderland, MA: Sinauer Associates.)

Figure 33.5 illustrates possible patterns of exon organization in a gene with respect to the structural domains in its protein product. The question became: Is it possible to relate structural domains in proteins to the exons of genes?

Correlations of Exons with Protein Structure

Collagen. Because it is a fibrous protein, **collagen** lacks discrete functional and structural domains. Collagen is a triple helix; each of its three polypeptide chains consists of 1014 amino acid residues. One-third of the residues in each chain are glycine. The collagen polypeptide chains actually consist of the triplet (Gly—x—y) repeated 338 times. The chicken *α2-procollagen gene* comprises 52 exons (Figure 33.6). Most of these exons are just 54 bp long and encode 18 amino acids or 6 (Gly—x—y) triplets. Others are 108 bp (exactly twice the size of the majority class), and a few are 45 bp and 99 bp in length—that is, exactly 9 bp, or 1 (Gly—x—y) unit, shorter than the 54-bp and 108-bp class. This organization suggests a model for collagen evolution in which a 54-bp exon was amplified by successive duplications to give multiple copies separated by intervening sequences (introns) (Figure 33.7). Later, if an intron was deleted, a 108-bp exon was created. Subsequent deletion of 9-bp units in certain exons accounts for the formation of 45-bp and 99-bp exons.

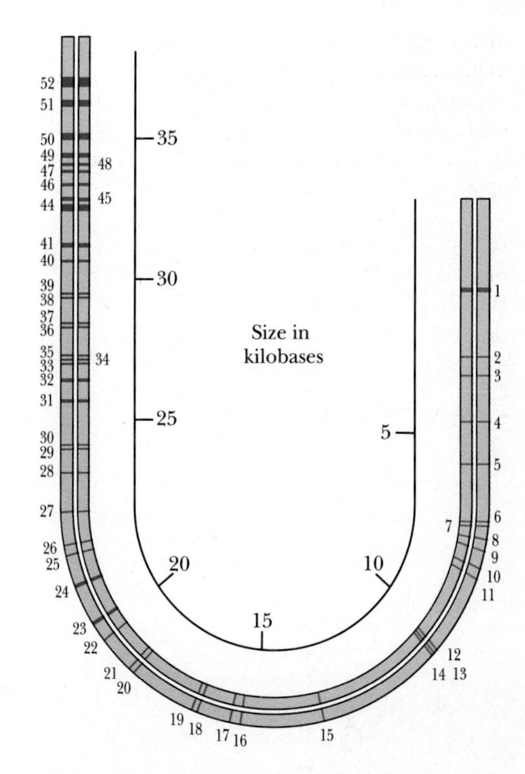

***Figure* 33.6** The α-procollagen gene of the chicken consists of 52 short exons distributed over almost 40 kbp of DNA.
(Adapted from Blake, C. C. F., 1985. International Review of Cytology 93:149–185, Figure 9.)

Hemoglobin. A globular $\alpha_2\beta_2$-type protein molecule, **hemoglobin** has four very similar globular globin chains that show no obvious organization reflecting the presence of any domainlike substructure. Every known vertebrate globin gene contains three exons. Furthermore, the exon boundaries in all these genes coincide in terms of globin tertiary structure, corresponding to the coding pattern shown in Figure 33.8. Although any relationship between exon organization and globin structure or function may be arbitrary, analysis of interresidue distances in the globin tertiary structure reveals a suborganization into four compact structural units, with exon 2 encoding the two central units (represented by helices C–D–E–F) and exons 1 and 3 encoding helices A–B and G–H, respectively. The dividing line between the two central units falls between residues 66 and 71 in the β-globin chain; indeed, the results suggest that exon 2 arose from the merger of two ancestral exons.

Sequence analysis of plant genes for **leghemoglobin,** the plant counterpart of hemoglobin, reveals the existence of an intron between residues 68 and 69, in the region predicted by analysis of the vertebrate β-globin gene. The results lead to a model for globin gene evolution postulating that contemporary vertebrate globin genes are derived from an ancestral globin line in which the central intron was lost (Figure 33.9). On the other hand, the two introns within the genes encoding hemoglobin in nematodes (*Caenorhabditis elegans* and *Pseudoterranova decipiens*) show no relationship to the vertebrate globin gene introns, casting doubt on this model.

For the most part, it appears that the presence of duplicate (or highly similar) domains in a protein can be attributed to an exon duplication in the gene. Exon duplication then becomes an important consideration in protein evolution. Table 33.2 lists a number of proteins having internal domain duplications. Sometimes such a protein shows a repetitive structure extending over

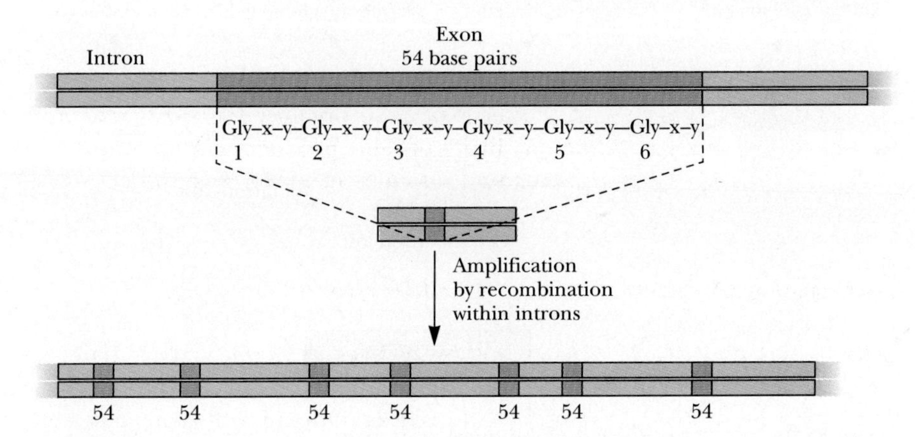

***Figure* 33.7** A model for the evolution of the collagen gene.
(Adapted from Blake, C. C. F., 1985. International Review of Cytology 93:149–185, Figure 10.)

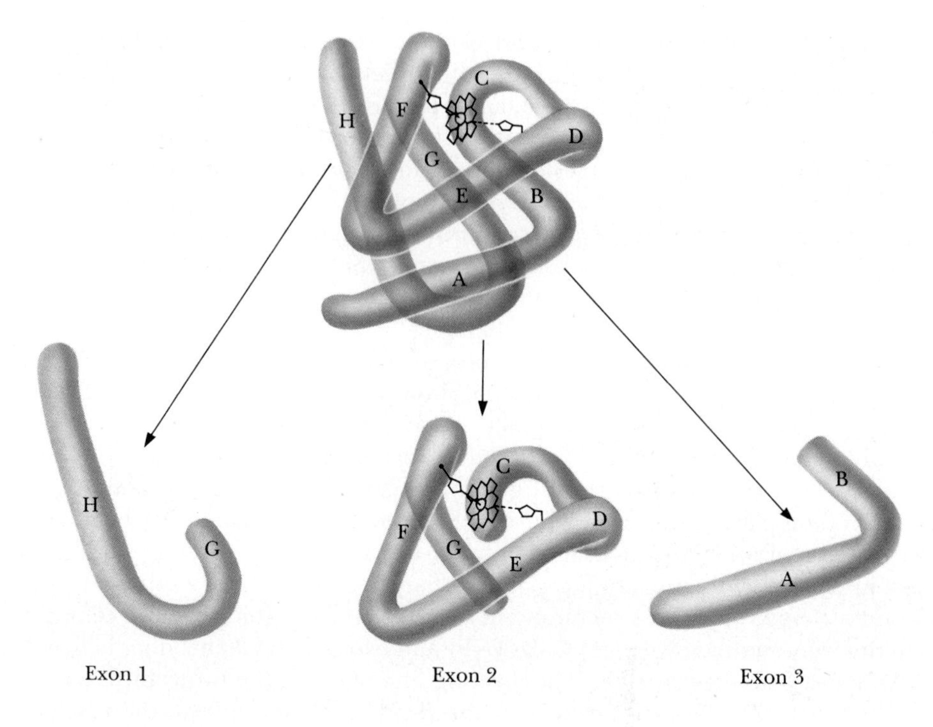

***Figure* 33.8** Drawings showing the relationship between the tertiary structure of globin chains and their three exon-encoded regions. Recall that globin chains typically consist of eight successive α-helical segments, labeled A through H.

(Adapted from Blake, C. C. F., 1985. International Review of Cytology 93:149–185, Figure 14.)

its entire length. Indeed, internal duplications are probably more common than the nucleotide or amino acid sequence data might indicate, since evolution often operates on function at the level of the protein's secondary or tertiary structure as opposed to its primary structure.

Mosaic Proteins

Exon shuffling, the exchange of exons between genes, occurs in two ways: exon duplication (just discussed) or exon insertion. **Exon insertion** is the process by which exons encoding particular functional or structural domains of proteins from one gene are inserted into another. This process creates a new gene whose protein product is a mosaic of the structural and functional domains, or **modules,** that it carries. A good place to look for proteins of mosaic design might be in organisms of recent evolutionary origin. Vertebrate animals as a group are relatively recent arrivals in the evolutionary scheme of

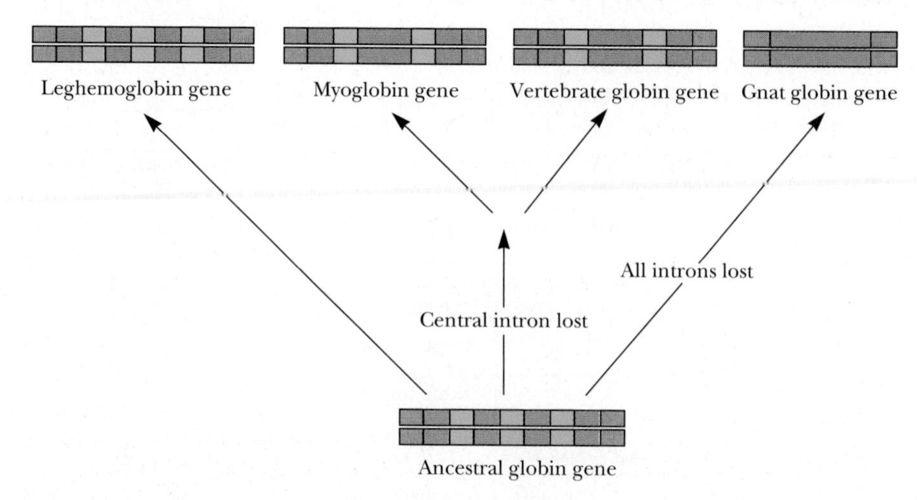

***Figure* 33.9** Intron loss during globin gene evolution.

(Adapted from Li, W.-H., and Graur, D., 1991. Fundamentals of Molecular Evolution. Chapter 6, Figure 2. Sunderland, MA: Sinauer Associates.)

Table 33.2
Proteins Having Internal Domain Duplications

Sequence	**Length of Protein***	**Length of Repeat**	**Number of Repeats**	**Percent Repetition**†
Immunoglobulin ϵ-chain C region (humans)	423	108	4	100
Immunoglobulin γ-chain C region (humans)	329	108	3	98
Serum albumin (humans)	584	195	3	100
Parvalbumin (humans)	108	39	2	72
Protease inhibitor, Bowman–Birk type (soybean)	71	28	2	79
Protease inhibitor, submandibular-gland type (rodents)	115	54	2	94
Ferredoxin (*Clostridium pasteurianum*)	55	28	2	100
Plasminogen (humans)	790	79	5	50
Calcium-dependent regulator protein (humans)	148	74	2	100
Tropomyosin α-chain (humans)	284	42	7	100

Adapted from Li, W.-H., and Graur, D., 1991. *Fundamentals of Molecular Evolution,* Chapter 6, Table 1. Sunderland, MA: Sinauer Associates.
*Number of amino acid residues.
†Percentage of the total length of the protein occupied by repeated sequences.

things, first appearing about 500 million years ago. Uniquely vertebrate proteins, such as those involved in blood coagulation, provide several examples of mosaic proteins (see also Chapter 5).

Tissue plasminogen activator (TPA) is a classic example. TPA is an important component in the proteolytic pathway for the removal of blood clots. TPA is a serine protease that proteolytically activates plasminogen to form **plasmin,** another serine protease that then acts to cleave **fibrin,** the major structural protein of clots. TPA binds to both plasminogen and fibrin, liberating plasmin directly on its target. Many heart attacks are caused by clots restricting blood flow in coronary arteries. Intravenous administration of TPA within 90 minutes of the first symptoms of a heart attack markedly enhances likelihood of survival; TPA for treatment of heart attacks was one of the first clinical products of recombinant DNA technology.

TPA has four structural modules in addition to its serine protease domain: a **fingerlike module** of structure that is responsible for fibrin affinity, a **growth-factor module** that is homologous to epidermal growth factor (EGF), and two **kringle** domains (Figure 33.10). The serine protease domain of TPA is itself homologous to the trypsin family of serine proteases. These various modules are small, compact, cysteine-rich structural units typically stabilized by several disulfide bonds. As such, modules are unlikely to serve any mechanistic or catalytic function; instead, they provide recognition or binding structures that account for the biological specificity of the protease. The disulfide pattern of the finger module yields a structure resembling a finger; the three internal disulfides of the kringle module create a structure that looks like the Danish pastry of this name. The conclusion is that during its evolution the TPA serine protease gene acquired at least four DNA sequence elements from other genes: kringles from the plasminogen gene, a finger from the fibronectin gene, and a growth factor module from the EGF gene. The exon junctions for these modules in the TPA gene correspond precisely to exon–intron borders in the donor genes, strongly suggesting that these exons have been transferred from one gene to another, as envisioned in the exon shuffling concept.

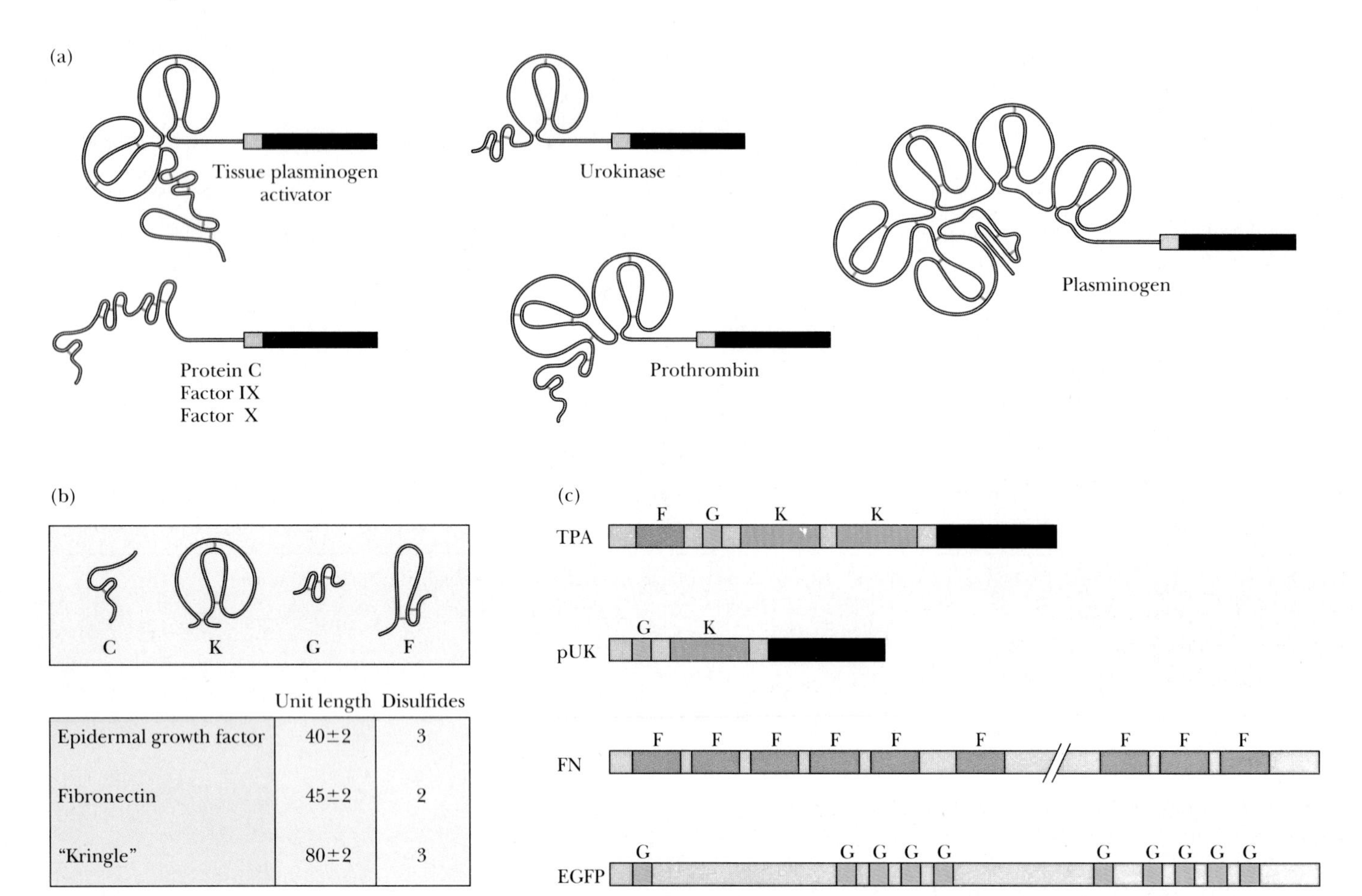

	Unit length	Disulfides
Epidermal growth factor	40±2	3
Fibronectin	45±2	2
"Kringle"	80±2	3

***Figure* 33.10** (a) Structural modules found in tissue plasminogen activator (TPA) and other proteins of blood coagulation and fibrinolysis. The inset (b) shows the structures, unit lengths, and numbers of disulfides of the modules. F = finger, G = epidermal growth factor, K = kringle, C = a vitamin K–dependent calcium-binding module. Black bars denote the serine protease regions homologous to trypsin. The proteins represented are TPA (tissue plasminogen activator), pUK (pro-urokinase, the precursor to urokinase, a plasminogen activator similar to TPA that lacks fibrin-binding ability), FN (fibronectin, a cell-surface protein that allows cells to bind to the extracellular matrix; its finger modules enable cells to associate with fibrin in clots to facilitate repair of wounds), and EGFP (the epidermal growth factor precursor). Note that plasminogen itself has five kringle modules, whereas prothrombin, another coagulation protease precursor, has two. (c) Diagram of exon organization in TPA, pUK, FN, and EGFP, showing correspondence of modules to exons (the serine protease regions of TPA and pUK are represented simply as black bars).

(Adapted from Li, W.-H., and Graur, D., 1991. Fundamentals of Molecular Evolution, Chapter 6, Figure 9, as well as Doolittle, R. F., 1985. Trends in Biochemical Sciences 10:233–237, Table III, and Patthy, L., 1985. Cell 41:657–663, Figure 1.)

Is Exon Shuffling a Major Mechanism in Eukaryotic Evolution?

Wen-Hsiung Li has pointed out that many complex genes in contemporary organisms likely evolved by internal duplication and subsequent modification of primordial genes that presumably contained only a few exons, each encoding a simple biological function. Exons, as separate, transferable blocks of coding sequence, provide a major stimulus to the evolution of proteins. Nevertheless, *exons generally do* not *exactly correspond to protein domains*. Instead, domains are typically encoded on a number of exons, and the idea that exons

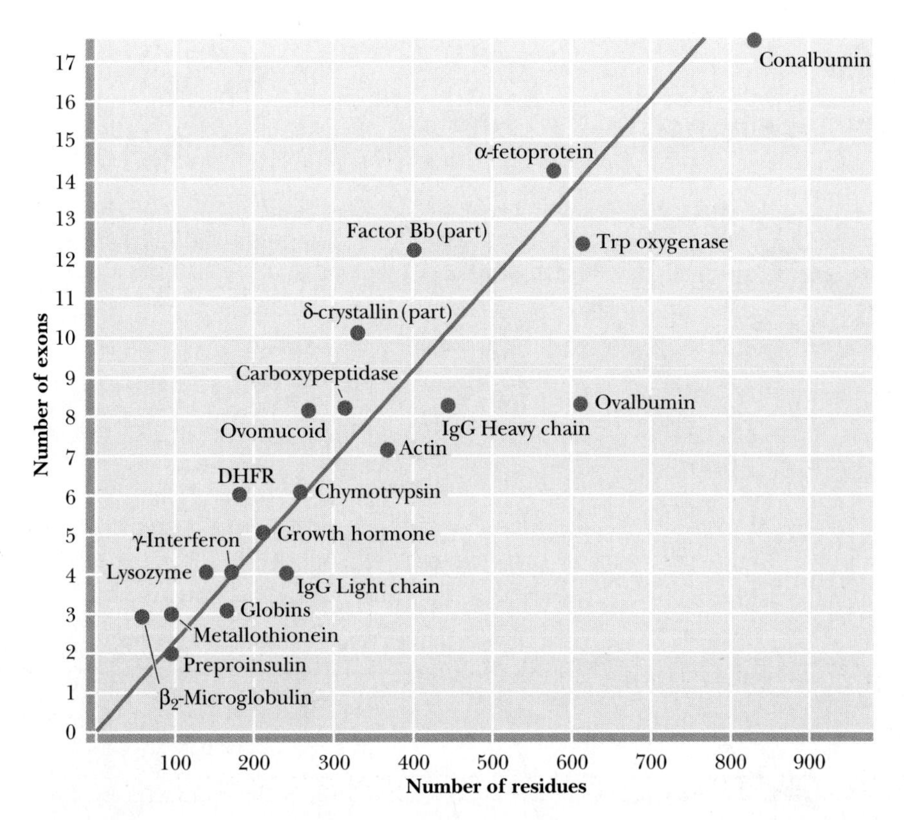

***Figure* 33.11** A plot of the number of exons in a gene versus the length of the polypeptide chain (in amino acid residues) that it encodes, for a number of protein genes.

(Adapted from Blake, C. C. F., 1985. International Review of Cytology, 93:149–185, Figure 19.)

delineate compact units of protein structure is, despite certain exceptions, not universally valid. Most structural domains in proteins comprise 100 or more amino acid residues, whereas the sizes of most exons average about 140 bp (45 or so codons). (Only a small fraction of exons are smaller than 30 bp or larger than 300.) A plot of amino acid residues versus number of exons in a protein (Figure 33.11) has a slope of about 1 exon per 47 amino acids.

Perhaps exons code for a range of small units of supersecondary structure. Figure 33.12 shows the overall tertiary structures of two proteins, carboxypeptidase (a) and chymotrypsin (b), as compared to their exon-encoded parts. In most instances, these parts appear to be units of supersecondary structure. That is, the breaks between organized secondary structures, such as α-helices and β-pleated units, usually occur at exon junctions, so that an exon corresponds to a "unit of supersecondary structure." Another interesting feature that emerges from this deconstruction is that exon junctions (splice junctions) tend to map to the protein surface.

33.6 Gene Sharing

Instead of using parts of genes to construct mosaic proteins, occasionally the entire gene product, without any change in amino acid sequence, is utilized for two entirely different functions. This phenomenon has been termed *gene sharing.* **Gene sharing** means that a single gene, without being duplicated, may

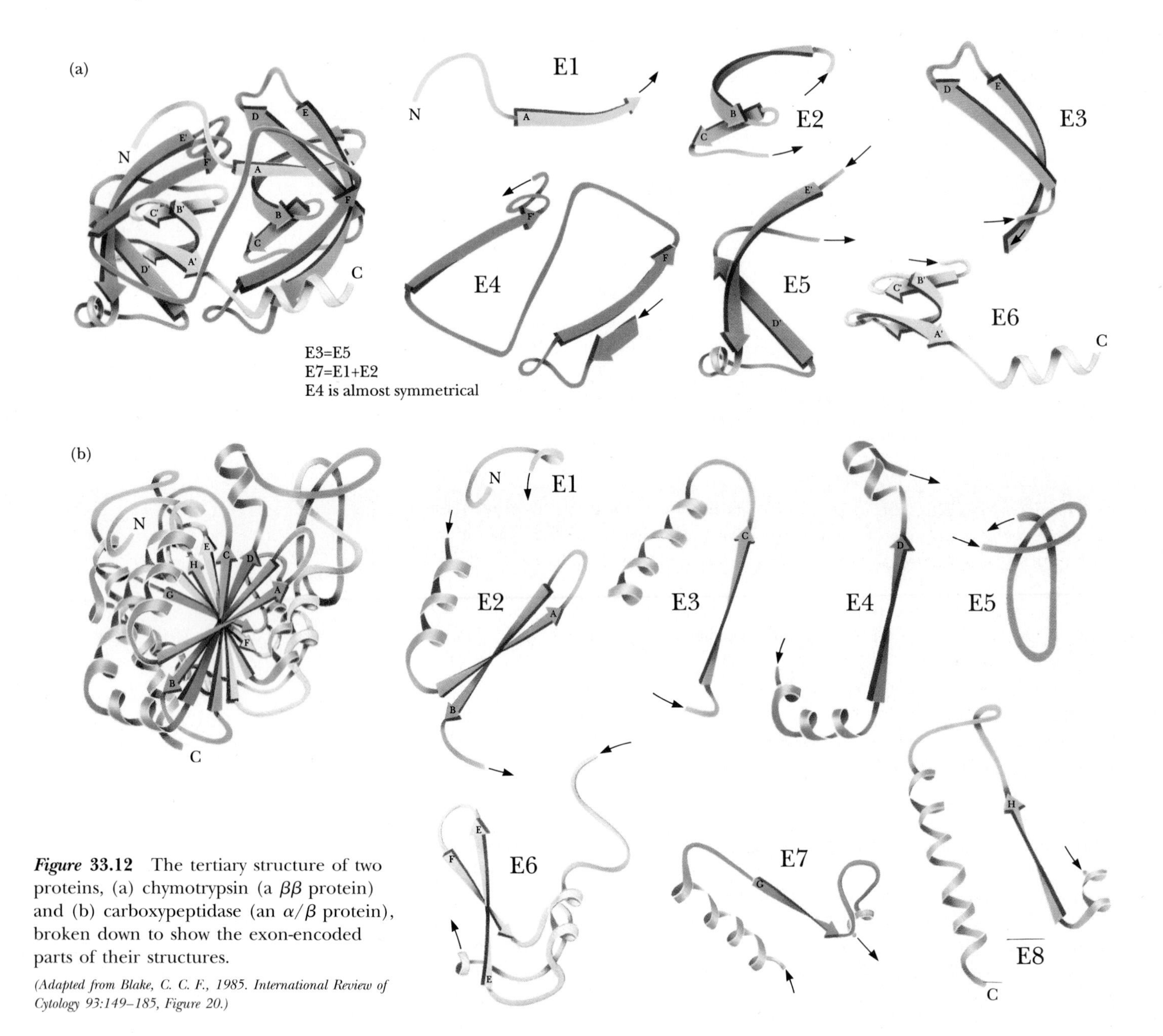

***Figure* 33.12** The tertiary structure of two proteins, (a) chymotrypsin (a $\beta\beta$ protein) and (b) carboxypeptidase (an α/β protein), broken down to show the exon-encoded parts of their structures.

(Adapted from Blake, C. C. F., 1985. International Review of Cytology 93:149–185, Figure 20.)

gain a second function without losing its primary function. Gene sharing was first observed in the proteins making up the lenses of vertebrate eyes. The optical properties of the lens—transparency and proper light diffraction—depend on high concentrations of soluble structural proteins known as **crystallins.** Two categories of crystallins are recognized: the **ubiquitous crystallins** (α, β, and γ) found in all vertebrate lenses, and the various **species-specific crystallins** present only in certain species. For example, ϵ-crystallin is found only in some birds and their close relatives, the crocodiles. In 1987, the amino acid sequence of duck ϵ-crystallin was observed to be essentially identical to lactate dehydrogenase B4 (LDH B4). LDH is an ancient, highly conserved glycolytic enzyme. Further, ϵ-crystallin and LDH B4 display identical LDH enzyme activity. In fact, the proteins are one and the same, and they are encoded by the same gene. Another crystallin, τ-crystallin, which is found in lampreys, bony fishes, birds, and reptiles, is identical to and encoded by the same gene as enolase, another glycolytic enzyme. The amino acid sequence of

Thus, sequence comparisons indicate that energy-transducing H^+-ATPases form two major families, the **F**-ATPases and the **V**-ATPases, that are distinct but related. A conclusion of great evolutionary significance can be drawn from these comparisons: The evolutionary split between **F**- and **V**-ATPases was an ancient event, predating the appearance of eukaryotic cells. Figure 33.16 depicts a scheme for these evolutionary relationships. This scheme suggests a previously unrecognized relationship: that eukaryotes and

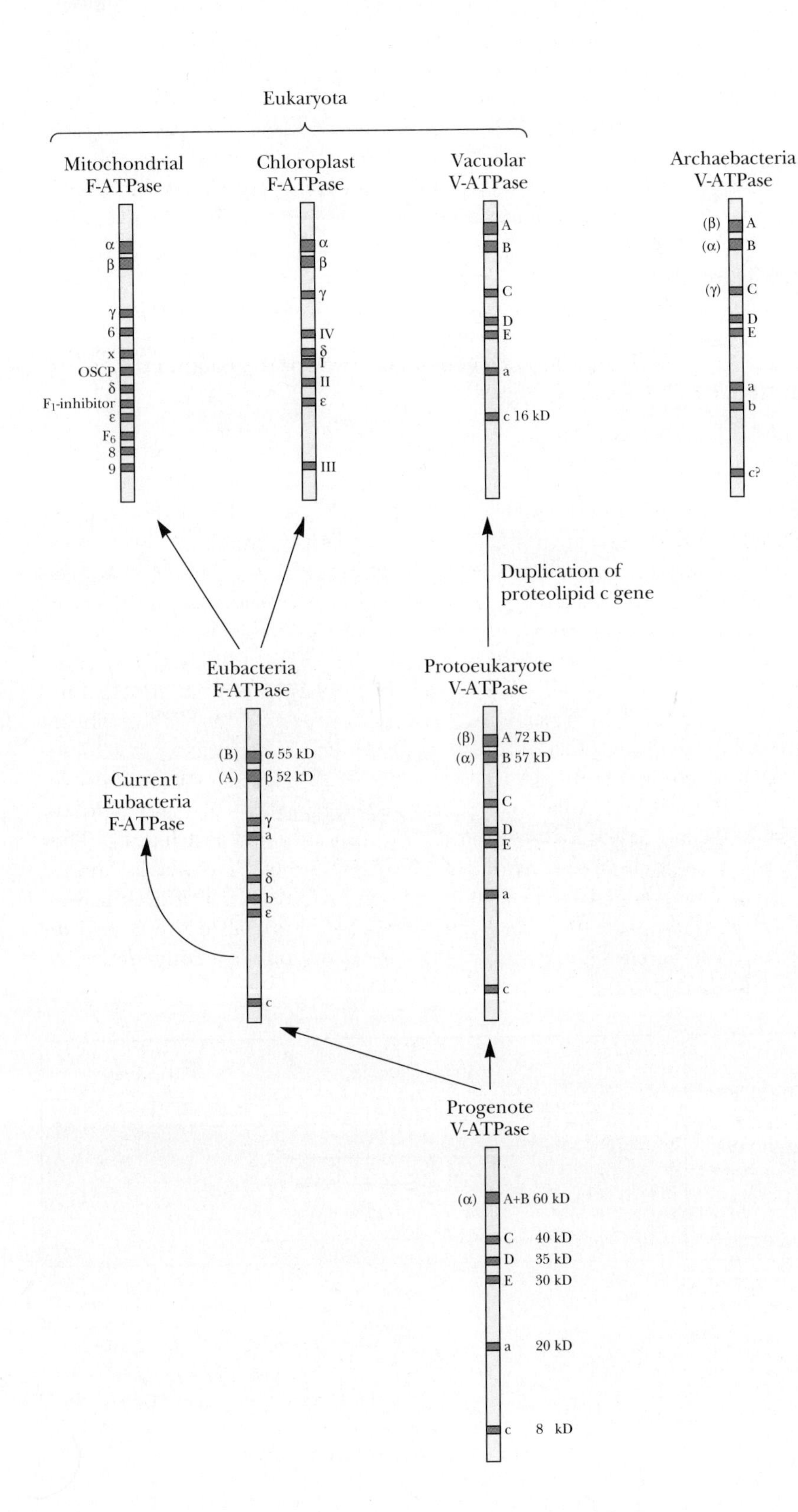

Figure **33.16** Proposed scheme for the evolution of **F**-ATPases and **V**-ATPases. The proteins in the various ATPases are illustrated as bands in descending order of molecular weight, much as they would appear when separated in an SDS-polyacrylamide gel electrophoresis experiment. The primordial ATPase precursor ("progenote ATPase") is shown at the bottom, and all contemporary **F**- and **V**-ATPases are arrayed at the top. The catalytic subunit and regulatory subunit genes arose via duplication of a single gene in the progenote, shown as (A + B). The transition from protoeukaryote to vacuolar **V**-ATPase apparently was accompanied by duplication of the proteolipid-encoding *c* gene.

(Adapted from Nelson, N., and Taiz, L., 1989. Trends in Biochemical Sciences 14:113–116, Figure 2.)

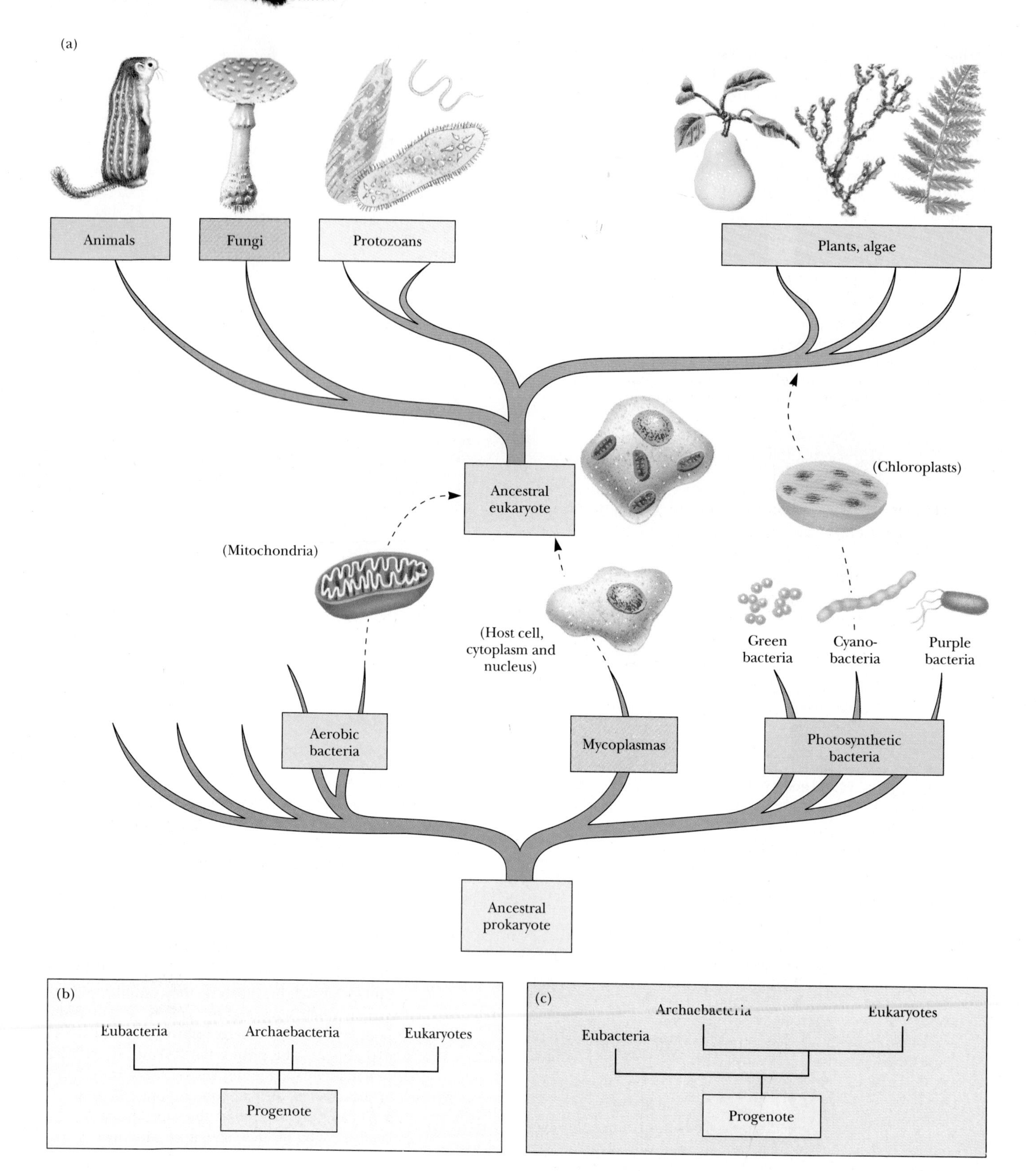

(a)
Animals
Fungi
Protozoans
Plants, algae
Ancestral eukaryote
(Chloroplasts)
(Mitochondria)
(Host cell, cytoplasm and nucleus)
Green bacteria
Cyano-bacteria
Purple bacteria
Aerobic bacteria
Mycoplasmas
Photosynthetic bacteria
Ancestral prokaryote
(b)
Eubacteria
Archaebacteria
Eukaryotes
Progenote
(c)
Archaebacteria
Eukaryotes
Eubacteria
Progenote

◀ *Figure* **33.17** The evolution of cells. (a) The view of Margulis. This scheme, which was the prevalent view until 1977, is based primarily on the origins of mitochondria and chloroplasts. In this view, the cell wall–less mycoplasmas gave rise to the eukaryotic host cell cytoplasm and nucleus; mitochondria were derived from aerobic bacteria, and chloroplasts from the cyanobacteria line. (b) The view of Woese and Fox. This scheme is based on 16S rRNA sequence analyses. All cells have ribosomes, and ribosomes are organized similarly in all cells. Hence, analysis of homologies between rRNA sequences can provide glimpses into the evolutionary relationships uniting all cells. A precursor cell type, the "progenote," diverged into three equidistant lines of descent: eubacteria, archaebacteria, and "urkaryotes," which served as the forerunners of eukaryotes by providing the cytoplasm hosting the symbiotic invasion of mitochondria and chloroplasts. (c) A recent scheme proposed by Iwabe et al. (see Further Reading), based on comparisons of genes whose duplication must have preceded divergence of archaebacteria, eubacteria, and eukaryotes from one another. The compared genes encode *EF-Tu* and *EF-G,* elongation factors in protein synthesis, and the α- *(A-)* and β- *(B-)* subunits of **F**- and **V**-ATPases, protein pairs found in all extant cells. Thus, the genes encoding these pairs must have arisen from a common ancestral gene via a gene duplication event that preceded divergence of the primary kingdoms.

archaebacteria share a common ancestor. Earlier models suggested that eukaryotes arose either from eubacterial precursors or from a common ancestor, or **progenote,** that diverged into three equidistant lines of descent (Figure 33.17). Other recent studies postulate that archaebacteria are more closely related to eukaryotes than to eubacteria. Although our appreciation and understanding of these relationships remain controversial and are themselves evolving and changing with new evidence, it is clear that nucleotide and amino acid sequences offer unique windows on the evolutionary past.

Problems

1. Align the following pairs of nucleotide sequences by the dot-matrix method.
   a. CAGCTGCGTA and CAGCGCGTA
   b. CAGCTGCGTA and CATTGCGTC

2. Indicate various alignments for the following pair of sequences, using the sequence-distance method. Calculate **D,** the evolutionary distance between them, using the equation **D** = **y** + **wz.**

   GTTACTGAAG and GTCGGGAG

3. Which of the following amino acid substitutions in a protein is more likely to represent random genetic drift as opposed to a change in fitness: (a) the substitution of a Val residue for a Leu residue, (b) the substitution of an Asp residue for an Arg residue, (c) the substitution of a Lys residue for a Phe residue?

4. How many single-base changes are possible within any given codon? For the codon GCA, how many of these single nucleotide substitutions are synonymous; how many are nonsynonymous? Which of the nonsynonymous changes lead to conservative amino acid substitutions (replacement by an amino acid whose R group displays similar polarity)? Answer these same questions for the codon AAG.

5. The rate of nonsynonymous substitution in human hemoglobin β-chain genes (146 codons) is 0.8×10^{-9} nonsynonymous substitutions per site per year (Table 33.1). If the human per capita birth rate is 30 per 1000 persons per year, and the human population is 5.8×10^{9}, how many individuals with nonsynonymous substitutions in their hemoglobin β-chain genes are likely to be born each year? Assume for the purposes of this calculation that nonsynonymous substitutions in the β-globin gene are not lethal at or before birth (since expression of the β-globin gene does not begin until a few weeks prior to birth). Assume also that such substitutions may occur anywhere in the coding region of the gene.

6. It is often assumed that synonymous substitutions are selectively neutral in evolution. Given the correlation between relative amounts of various tRNA species and the relative frequency of codon usage (as in Figure 33.4), do you think this assumption is valid or not?

7. Why is exon duplication likely to be of greater consequence to molecular evolution than duplication of a random segment of the genome?

8. Ribosomal RNAs and H^{+}-translocating ATPases represent two macromolecules with early evolutionary origins. From your knowledge of biochemistry, list five proteins whose functions suggest to you that their origins date back to the earliest days of cellular evolution.

Further Reading

Blake, C. C. F., 1985. Exons and the evolution of proteins. *International Review of Cytology* **93:**149–185.

Dixon, B., and Pohajdak, B., 1992. Did the ancestral globin gene of plants and animals contain only two introns? *Trends in Biochemical Sciences* **17:**486–488. Globin gene sequences obtained from two nematode worms raise questions regarding the nature of the ancestral globin gene and the hypothesis that globin exons define modules of structure.

Doolittle, R. F., 1985. The genealogy of some recently evolved vertebrate proteins. *Trends in Biochemical Sciences* **10:**233–237.

Doolittle, R. F., and Bork, P., 1993. Evolutionarily mobile modules in proteins. *Scientific American* **269:**50–56.

Fox, G. E., et al., 1980. The phylogeny of prokaryotes. *Science* **209:**457–463. The nucleotide sequence of small-subunit (16S) rRNA is used to determine phylogenetic relationships among the major classes of prokaryotes.

Gilbert, W., 1978. Why genes in pieces? *Nature* **271:**501.

Gouy, M., and Li, W.-H., 1990. "Evolutionary relationships among primary lineages of life inferred from rRNA sequences." Chap. 50 in *The Ribosome: Structure, Function and Evolution,* edited by Hill, W. E., et al. Washington, DC: American Society for Microbiology Press.

Iwabe, N., et al., 1989. Evolutionary relationship of archaebacteria, eubacteria and eukaryotes inferred from phylogenetic trees of duplicated genes. *Proceedings of the National Academy of Science, USA* **86:**9355–9359.

Lake, J., 1991. Tracing origins with molecular sequences: Metazoan and eukaryotic beginnings. *Trends in Biochemical Sciences* **16:**46–50.

Li, W.-H., and Graur, D., 1991. *Fundamentals of Molecular Evolution.* Sunderland, MA: Sinauer Associates.

Nelson, N., and Taiz, L. 1989. The evolution of H^+-ATPases. *Trends in Biochemical Sciences* **14:**113–116.

Patthy, L., 1985. Evolution of the proteases of blood coagulation and fibrinolysis by assembly from modules. *Cell* **41:**657–663.

Petsko, G. A., et al., 1993. On the origin of enzymatic species. *Trends in Biochemical Sciences* **18:**372–376.

Piatigorsky, J., and Wistow, G. J., 1989. Enzyme/crystallins: Gene sharing as an evolutionary strategy. *Cell* **57:**197–199.

Zimmer, E. A., et al., eds., 1993. Molecular evolution: Producing the biochemical data. *Methods in Enzymology* **224.**

Abbreviated Answers to Problems

For detailed answers to the end-of-chapter problems as well as additional problems to solve, see the *Student Study Guide* by David Jemiolo that accompanies this textbook.

Chapter 1

1. Since bacteria (compared with humans) have simple nutritional requirements, their cells obviously contain enzyme systems that allow them to convert rudimentary precursors (even inorganic substances such as NH_4^+, NO_3^-, N_2, and CO_2) into the complex biomolecules—proteins, nucleic acids, polysaccharides, and complex lipids. On the other hand, animals have an assortment of different cell types designed for specific physiological functions; these cells possess a correspondingly greater repertoire of complex biomolecules to accomplish their intricate physiology.
2. Consult Figures 1.20, 1.21, and 1.22 to confirm your answer.
3. a. Laid end to end, 250 *E. coli* cells would span the head of a pin.
   b. The volume of an *E. coli* cell is about 10^{-15} L.
   c. The surface area of an *E. coli* cell is about 6.3×10^{-12} m^2. Its surface-to-volume ratio is 6.3×10^6 m^{-1}.
   d. 600,000 molecules.
   e. 1.7 n*M*.
   f. Since we can calculate the volume of one ribosome to be 4.2×10^{-24} m^3 (or 4.2×10^{-21} L), 15,000 ribosomes would occupy 6.3×10^{-17} L, or 6.3% of the total cell volume.
   g. Since the *E. coli* chromosome contains 4700 kilobase pairs (4.7×10^3 bp) of DNA, its total length would be 1.6 mm—approximately 800 times the length of an *E. coli* cell. This DNA could encode 4350 different proteins, each 360 amino acids long.
4. a. The volume of a single mitochondrion is about 4.2×10^{-16} L (about 40% the volume calculated for an *E. coli* cell in problem 3).
   b. A mitochondrion would contain on average less than 8 molecules of oxaloacetate.
5. a. Laid end to end, 25 liver cells would span the head of a pin.
   b. The volume of a liver cell is about 8×10^{-12} liters (8000 times the volume of an *E. coli* cell).
   c. The surface area of a liver cell is 2.4×10^{-9} m^2; its surface-to-volume ratio is 3×10^5 m^{-1}, or about 0.05 (1/20) that of an *E. coli* cell. Cells with lower surface-to-volume ratios are limited in their exchange of materials with the environment.
   d. The number of base pairs in the DNA of a liver cell is 6.1×10^9 bp, which would amount to a total DNA length of 1.82 m (or 6 feet of DNA!) contained within a cell that is only 20 μm on a side. Maximal information content of liver-cell DNA = 3.05×10^9, which, expressed in proteins 400 amino acids in length, could encode 2.5×10^6 proteins.
6. The amino acid side chains of proteins provide a range of shapes, polarity, and chemical features that allow a protein to be tailored to fit almost any possible molecular surface in a complementary way.
7. Biopolymers may be informational molecules because they are constructed of different monomeric units ("letters") joined head to tail in a particular order ("words, sentences"). Polysaccharides are often linear polymers composed of only one (or two repeating) monosaccharide unit(s), and thus display little information content. Polysaccharides with a variety of monosaccharide units may convey information, through specific recognition by other biomolecules. Also, most monosaccharide units are typically capable of forming branched polysaccharide structures that are potentially very rich in information content (as in cell surface molecules that act as the unique labels displayed by different cell types in multicellular organisms).
8. Molecular recognition is based on structural complementarity. If complementary interactions involved covalent bonds (strong forces), stable structures would be formed that would be less responsive to the continually changing dynamic interactions that characterize living processes.
9. Slight changes in temperature, pH, ionic concentrations, and so forth may be sufficient to disrupt weak forces (H bonds, ionic bonds, van der Waals interactions, hydrophobic interactions).
10. Living systems are maintained by a continuous flow of matter and energy through them. Despite the ongoing transformations of matter and energy by these highly organized, dynamic systems, no overt changes seem to occur in them: They are in a *steady state*.

Chapter 2

1. a. 3.3; b. 9.85; c. 5.7; d. 12.5; e. 4.4; f. 6.97.
2. a. 1.26 m*M*; b. 0.25 μ*M*; c. 4×10^{-12} *M*; d. 2×10^{-4} *M*; e. 3.16×10^{-10} *M*; f. 1.26×10^{-7} *M* (0.126 μ*M*).

3. a. $[H^+] = 2.51 \times 10^{-5}$ *M*; b. $K_a = 3.13 \times 10^{-8}$; $pK_a = 7.5$.

4. a. pH = 2.38; b. pH = 4.23.

5. Combine 187 mL of 0.1 *M* acetic acid with 813 mL of 0.1 *M* sodium acetate.

6. $[HPO_4^{2-}]/[H_2PO_4^-] = 0.398$.

7. Combine 555.7 mL of 0.1 *M* Na_3PO_4 with 444.3 mL of 0.1 *M* H_3PO_4. Final concentrations of ions will be $[H_2PO_4^-] =$ 0.0333 *M*; $[HPO_4^{2-}] = 0.0667$ *M*; $[Na^+] = 0.1667$ *M*; $[H^+] =$ 3.16×10^{-8} *M*.

8. Add 432 mL of 0.1 *N* HCl to 1 L 0.1 *M* BICINE. [BICINE] = 0.05 *M*/1.432 L = 0.0349 *M*.

9. a. Fraction of H_3PO_4: @pH 0 = 0.993; @pH 2 = 0.58; @pH 4 = 0.01; negligible @pH 6.
   b. Fraction of $H_2PO_4^-$: @pH 0 = 0.007; @pH 2 = 0.41; @pH 6 = 0.94; @pH 8 = 0.14; negligible @pH 10.
   c. Fraction of HPO_4^{2-}: negligible @pH 0, 2, and 4; @pH 6 = 0.06; @pH 8 = 0.86; @pH 10 ≅ 1.0; @pH 12 = 0.72.
   d. Fraction of PO_4^{3-}: negligible at any pH < 10; @pH 12 = 0.28.

10. At pH 5.2, $[H_3A] = 4.33 \times 10^{-5}$ *M*; $[H_2A^-] = 0.0051$ *M*; $[HA^{2-}] = 0.014$ *M*; $[A^{3-}] = 0.0009$ *M*.

11. a. pH = 7.04; $[H_2PO_4^-] = 0.0197$ *M*; $[HPO_4^{2-}] = 0.0137$ *M*.
    b. pH = 7.36; $[H_2PO_4^-] = 0.0137$ *M*; $[HPO_4^{2-}] = 0.0197$ *M*.

Chapter 3

1. Structures for glycine, aspartate, leucine, isoleucine, methionine, and threonine are presented in Figure 3.3.

2. Asparagine = Asn = N.
   Arginine = Arg = R.
   Cysteine = Cys = C.
   Lysine = Lys = K.
   Proline = Pro = P.
   Tyrosine = Tyr = Y.
   Tryptophan = Trp = W.

3.

Alanine dissociation:

$$H_3N^+{-}C(COOH)(CH_3){-}H \rightleftharpoons H_3N^+{-}C(COO^-)(CH_3){-}H \rightleftharpoons H_2N{-}C(COO^-)(CH_3){-}H$$

Glutamate dissociation:

$$H_3N^+{-}C(COOH)(CH_2CH_2COOH){-}H \rightleftharpoons H_3N^+{-}C(COO^-)(CH_2CH_2COOH){-}H \rightleftharpoons H_3N^+{-}C(COO^-)(CH_2CH_2COO^-){-}H \rightleftharpoons H_2N{-}C(COO^-)(CH_2CH_2COO^-){-}H$$

Histidine dissociation:

$$H_3N^+{-}C(COOH)(CH_2{-}\text{imidazolium: HC=CH, }H^+N, NH, CH){-}H \rightleftharpoons H_3N^+{-}C(COO^-)(CH_2{-}\text{imidazolium}){-}H \rightleftharpoons H_3N^+{-}C(COO^-)(CH_2{-}\text{imidazole: HC=CH, N, NH, CH}){-}H \rightleftharpoons H_2N{-}C(COO^-)(CH_2{-}\text{imidazole}){-}H$$

Lysine dissociation:

$$H_3N^+{-}C(COOH)((CH_2)_4NH_3^+){-}H \rightleftharpoons H_3N^+{-}C(COO^-)((CH_2)_4NH_3^+){-}H \rightleftharpoons H_2N{-}C(COO^-)((CH_2)_4NH_3^+){-}H \rightleftharpoons H_2N{-}C(COO^-)((CH_2)_4NH_2){-}H$$

Phenylalanine dissociation:

$$H_3N^+{-}C(COOH)(CH_2{-}C_6H_5){-}H \rightleftharpoons H_3N^+{-}C(COO^-)(CH_2{-}C_6H_5){-}H \rightleftharpoons H_2N{-}C(COO^-)(CH_2{-}C_6H_5){-}H$$

4. The proximity of the α-carboxyl group lowers the pK_a of the α-amino group.

5.

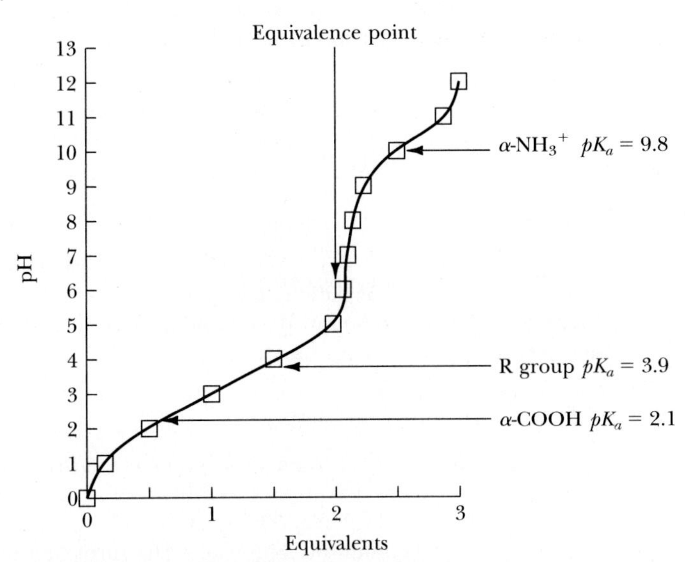

6. Denoting the four histidine species as His^{2+}, His^+, His^0, and His^-, the concentrations are:
pH 2: $[His^{2+}] = 0.097\ M$, $[His^+] = 0.153\ M$, $[His^0] = 1.53 \times 10^{-5}\ M$, $[His^-] = 9.6 \times 10^{-13}\ M$.
pH 6.4: $[His^{2+}] = 1.78 \times 10^{-4}\ M$, $[His^+] = 0.071\ M$, $[His^0] = 0.179\ M$, $[His^-] = 2.8 \times 10^{-4}\ M$.
pH 9.3: $[His^{2+}] = 1.75 \times 10^{-12}\ M$, $[His^+] = 5.5 \times 10^{-5}\ M$, $[His^0] = 0.111\ M$, $[His^-] = 0.139\ M$.

7. $pH = pK_a + \log(2/1) = 4.3 + 0.3 = 4.6$.
The γ-carboxyl group of glutamic acid is 2/3 dissociated at pH = 4.6.

8. $pH = pK_a + \log(1/4) = 10.5 + (-0.6) = 9.9$.

9. a. The pH of a 0.3 *M* leucine hydrochloride solution is approximately 1.46.
b. The pH of a 0.3 *M* sodium leucinate solution is approximately 11.5.
c. The pH of a 0.3 *M* solution of isoelectric leucine is approximately 6.65.

10. $[Arg] = (35\ \text{degrees})/([\alpha]_D^{25} \times 1\ \text{dm})$.
Using $[\alpha]_D^{25}$ for L-arginine = 12.5 (from Table 3.2), then $[Arg] = 2.8$ g/mL.

11. The sequence of reactions shown would demonstrate that L(−)-serine is related stereochemically to L(−)-glyceraldehyde:

COOH / H_2N—C—H / CH_2OH (L-(−)-Serine) → COOH / H_2N—C—H / CH_3 (L-(−)-Alanine) ← (looped) COOH / H—C—Br / CH_3 (2-Bromopropanoic acid)

2-Bromopropanoic acid → (looped) COOH / HO—C—H / CH_3 (2-Hydroxypropanoic acid)

CHO / HO—C—H / CH_2OH (L-(−)-Glyceraldehyde) → COOH / HO—C—H / CH_2OH (L-Glyceric acid) → COOH / HO—C—H / CH_3 (2-Hydroxypropanoic acid)

Straight arrows indicate reactions that occur with retention of configuration.
Looped arrows indicate inversion of configuration.
(From Kopple, K. D., 1966. Peptides and amino acids. New York: Benjamin Co.)

12. Cystine (disulfide-linked cysteine) has two chiral carbons, the two α-carbons of the cysteine moieties. Since each chiral center can exist in two forms, there are four stereoisomers of cystine. However, it is not possible to distinguish the difference between L-cysteine/D-cysteine and D-cysteine/L-cysteine dimers. So three distinct isomers are formed:

L-Cystine: COO^- / $H_3\overset{+}{N}$—C—H, COO^- / $H_3\overset{+}{N}$—C—H; CH_2—S—S—CH_2

Mirror plane

D-Cystine: COO^- / H—C—$\overset{+}{N}H_3$, COO^- / H—C—$\overset{+}{N}H_3$; CH_2—S—S—CH_2

meso-Cystine: COO^- / $H_3\overset{+}{N}$—C—H | COO^- / H—C—$\overset{+}{N}H_3$; CH_2—S—S—CH_2

Mirror plane

L-Cystine D-Cystine *meso*-Cystine

13.

$$\mathrm{I{-}CH_2{-}\overset{O}{\overset{\|}{C}}{-}NH_2} \quad + \quad \mathrm{^{-}S{-}CH_2{-}\overset{+}{H_3N}{-}CH(COO^-)} \longrightarrow \mathrm{H_2N{-}C({=}O){-}CH_2{-}S{-}CH_2{-}\overset{+}{H_3N}{-}CH(COO^-)} + \mathrm{I^-}$$

14. The order of elution should be:
    a. basic, positively charged amino acids,
    b. neutral, uncharged amino acids,
    c. acidic, negatively charged amino acids.
    Therefore, the order of elution should be: arginine, histidine, isoleucine, valine, aspartate. (The pK_a values for valine are slightly lower than those for isoleucine—thus the elution time of valine should be somewhat slower than that of isoleucine.)

15. L-threonine is (2*S*, 3*R*)-threonine.
    D-threonine is (2*R*, 3*S*)-threonine.
    L-allothreonine is (2*S*, 3*S*)-threonine.
    D-allothreonine is (2*R*, 3*R*)-threonine.

Chapter 4

1. At Se = 0.34% by weight, the minimal M_r for glutathione peroxidase is 23,224. Assuming M_r = 80,000 and 1 Se/polypeptide chain, the actual Se content would be 0.30% if glutathione peroxidase were a trimer or 0.39% if it were a tetramer. These analytical results do not allow distinction between these possibilities.
2. Phe-Asp-Tyr-Met-Leu-Met-Lys.
3. Tyr-Asn-Trp-Met-(Glu-Leu)-Lys. Parentheses indicate that the relative positions of Glu and Leu cannot be assigned from the information provided.
4. Ser-Glu-Tyr-Arg-Lys-Lys-Phe-Met-Asn-Pro.
5. Ala-Arg-Met-Tyr-Asn-Ala-Val-Tyr or Asn-Ala-Val-Tyr-Ala-Arg-Met-Tyr sequences both fit the results. (That is, in one-letter code, either *ARMYNAVY* or *NAVYARMY*.)
6. Gly-Arg-Lys-Trp-Met-Tyr-Arg-Phe.
7. Actually, there are four possible sequences: NIGIRVIA, GINIRVIA, VIRNIGIA, and, of course, VIRGINIA.
8. Gly-Trp-Arg-Met-Tyr-Lys-Gly-Pro.
9. Leu-Met-Cys-Val-Tyr-Arg-Cys-Gly-Pro.

10. Ser-Tyr-Arg-Met-Lys-Thr-Trp-Glu.
11. Alanine, attached to a solid-phase matrix via its α-carboxyl group, is reacted with dicyclohexylcarbodiimide-activated lysine. Both the α-amino and ϵ-amino groups of the lysine must be blocked with *t*-butyloxycarbonyl groups. To add leucine to Lys-Ala to form a linear tripeptide, precautions must be taken to prevent the incoming Leu α-carboxyl group from reacting inappropriately with the Lys ϵ-amino group instead of the Lys α-amino group.

Chapter 5

1. The central rod domain of keratin is composed of distorted α-helices, with 3.6 residues per turn, but a pitch of 0.51 nm, compared with 0.54 nm for a true α-helix.

   (0.51 nm/turn) (312 residues)/(3.6 residues/turn) = 44.2 nm = 442 Å.

   For an α-helix, the length would be:

   (0.54 nm/turn) (312 residues)/(3.6 residues/turn) = 46.8 nm = 468 Å.

   The distance between residues is 0.347 nm for antiparallel β-sheets and 0.325 nm for parallel β-sheets. So 312 residues of antiparallel β-sheet amount to 1083 Å and 312 residues of parallel β-sheet amount to 1014 Å.

2. The collagen helix has 3.3 residues per turn and 0.29 nm per residue, or 0.96 nm/turn. Then:

   (4 in/year) (2.54 cm/in) (10^7 nm/cm)/(0.96 nm/turn) = 1.06×10^8 turns/year.
   (1.06×10^8 turns/year) (1 year/365 days) (1 day/24 hours) (1 hour/60 minutes) = 201 turns/minute.

3. **Asp:** The ionizable carboxyl can participate in ionic and hydrogen bonds. Hydrophobic and van der Waals interactions are negligible.
   Leu: The leucine side chain does not participate in hydrogen bonds or ionic bonds, but it will participate in hydrophobic and van der Waals interactions.
   Tyr: The phenolic hydroxyl of tyrosine, with a relatively high pK_a, will participate in ionic bonds only at high pH, but can both donate and accept hydrogen bonds. Protonated tyrosine is capable of hydrophobic interactions. The relatively large size of the tyrosine side chain will permit substantial van der Waals interactions.
   His: The imidazole side chain of histidine can act both as acceptor and donor of hydrogen bonds and, when protonated, can participate in ionic bonds. van der Waals interactions are expected, but hydrophobic interactions are less likely in most cases.

4. As an imido acid, proline has a secondary nitrogen with only one hydrogen. In a peptide bond, this nitrogen possesses no hydrogens and thus cannot function as a hydrogen bond donor in α-helices. On the other hand, proline stabilizes the *cis*-configuration of a peptide bond and is thus well suited to β-sheets, which require the *cis*-configuration.

5. For a right-handed crossover, moving in the N-terminal to C-terminal direction, the crossover moves in a clockwise direction when viewed from the C-terminal side toward the N-terminal side. The reverse is true for a left-handed crossover; that is, movement from N-terminus to C-terminus is accompanied by counterclockwise rotation.

6. The Ramachandran plot reveals allowable values of ϕ and ψ for α-helix and β-sheet formation. The plots consider steric hindrance and will be somewhat specific for individual amino acids. For example, peptide bonds containing glycine can adopt a much wider range of ϕ and ψ angles than can peptide bonds containing tryptophan.

7. The protein appears to be a tetramer of four 60-kD subunits. Each of the 60-kD subunits in turn is a heterodimer of two

peptides, one of 34 kD and one of 26 kD, joined by at least one disulfide bond.

8. Hydrophobic interactions frequently play a major role in subunit–subunit interactions. The surfaces that participate in subunit–subunit interactions in the B_4 tetramer are likely to possess larger numbers of hydrophobic residues than the corresponding surfaces of protein A.

9. The length is given by: (53 residues) × (0.15 nm run/residue) = 7.95 nm. The number of turns in the helix is given by: (53 residues)/(3.6 residues/turn) = 14.7 turns. There are 49 hydrogen bonds in this helix.

Chapter 6

1. See Figure 6.17.
2. $f_A = 0.304$; $f_G = 0.195$; $f_C = 0.182$; $f_T = 0.318$.
3. 5′-TAGTGACAGTTGCGAT-3′.
4. 5′-ATCGCAACTGTCACTA-3′.
5. 5′-TACGGTCTAAGCTGA-3′.
6. pGACCAUCGC.
7. There are two possibilities, a and b. (E = *EcoR1* site; B = *BamH1* site.)

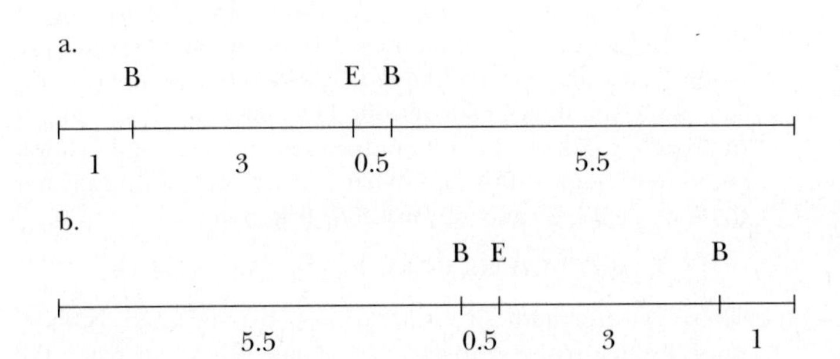

Chapter 7

1. a. Sanger's dideoxy method:

A	*G*	*C*	*T*

b. Maxam–Gilbert chemical cleavage method:
(i) 3′-end-labeled DNA.

G	*A* + *G*	*C*	*C* + *T*

(ii) 5′-end-labeled DNA.

G	*A* + *G*	*C*	*C* + *T*

2. Original nucleotide: 5′-dAGACTTGACGCT
DNA 3′-end labeled and subjected to Maxam–Gilbert sequencing:

G	*A* + *G*	*C*	*C* + *T*

3. $\Delta Z = 0.32$ nm; $P = 3.36$ nm/turn; 10.5 base pairs per turn; $\Delta\phi = 34.3°$; c (true repeat) = 6.72 nm.

4. 27.3 nm; 122 base pairs.

5. 4325 nm (4.33 μm).

6. $L_0 = 160$. If $W = -12$, $L = T + W = 160 + (-12) = 148$. $\sigma = \Delta L/L_0 = -12/160 = -0.075$.

7. For 1 turn of B-DNA (10 base pairs): $L_B = 1.0 + W_B$.

   For Z-DNA, 10 base pairs can only form 10/12 turn (0.833 turn), and $L_Z = 0.833 + W_Z$.

   For the transition B-DNA to Z-DNA, strands are not broken, so $L_B = L_Z$; that is, $1.0 + W_B = 0.833 + W_Z$, or $W_Z - W_B = +0.167$.

   (In going from B-DNA to Z-DNA, the change in *W*, the number of supercoils, is positive.) This result means that, if B-DNA contains negative supercoils, their number will be reduced in Z-DNA. Thus, all else being equal, negative supercoils favor the B → Z transition.

8. 6×10^9 bp/200 bp = 3×10^7 nucleosomes. The length of B-DNA 6×10^9 bp long = (0.34 nm) $(6 \times 10^9) = 2.04 \times 10^9$ nm (over 2 meters!). The length of 3×10^7 nucleosomes = (6 nm)$(3 \times 10^7) = 18 \times 10^7$ nm (0.18 meter).

9. From Figure 7.34: Similarly shaded regions indicate complementary sequences joined via intrastrand hydrogen bonds:

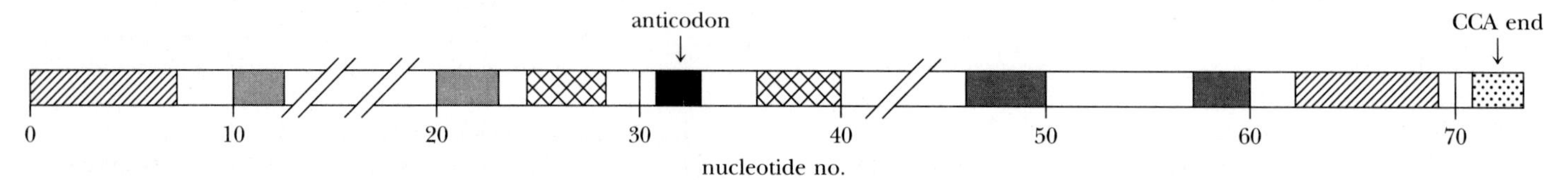

10. Increasing order of T_m: yeast < human < salmon < wheat < *E. coli*.

11. In 0.2 *M* Na^+, T_m(°C) = 69.3 + 0.41(% G + C):
    Rats (%G + C) = 40%, T_m = 69.3 + 0.41(40) = 85.7°C
    Mice (%G + C) = 44%, T_m = 69.3 + 0.41(44) = 87.3°C
    Since mouse DNA differs in GC content from rat DNA, they could be separated by isopycnic centrifugation in a CsCl gradient.

12. GC content = 0.714 (from Table 6.3 and equations used in problem 2, Chapter 6). $\rho = 1.660 + 0.098(GC) = 1.730$ g/mL.

Chapter 8

1. Linear and circular DNA molecules consisting of one or more copies of just the genomic DNA fragment; linear and circular DNA molecules consisting of one or more copies of just the vector DNA; linear and circular DNA molecules containing one or more copies of both the genomic DNA fragment and plasmid DNA.

2. -GAATTCCCGGGGATCCTCTAGAGTCGACCTGCAGGCATGC-

```
GAATTC        GGATCC          GTCGAC         GCATGC
EcoRI         BamHI           SalI           SphI
      CCCGGG          TCTAGA         CTGCAG
      SmaI            XbaI           PstI
```

3. a. AAGCTTGAGCTCGAGATCTAGATCGAT

```
   HindIII   XhoI    XbaI
        SacI     BglII    ClaI
```

   b.

```
Vector:    HindIII: 5'-A.........-gap-.........CGAT-3': ClaI
                    3'-TTCGA....-gap-...........TA-5'
Fragment:  HindIII: 5'-AGCTT(NNNN-etc-NNNN)AT-3'
                      3'-A(NNNN-etc-NNNN)TAGC-5'
```

4. N = 3838.

5. N = 10.4 million.

6. 5′-ATGCCGTAGTCGATCAT and 5′-ATGCTATCTGTCCT-ATG.

7. Thr-Met-Ile-Thr-Asn-Ser-Pro-*Asp-Pro-Phe-Ile-His-Arg-Arg-Ala-Gly-Ile-Pro-Lys-Arg-Arg-Pro . . .*

   The junction between β-galactosidase and the insert amino acid sequence is between Pro and Asp, so the first amino acid encoded by the insert is Asp. (The polylinker itself codes for Asp just at the *BamHI* site, but, in constructing the fusion, this Asp and all of this downstream section of polylinker DNA is displaced to a position after the end of the insert.)

8. 5′-(G)AATTCNGGNATGCAYCCNGGNAAR$_C$^{T}T$_N$^{Y}GCN<u>AGY</u>T-GGTTYGTNGGGAATTCN-

   (Note: The underlined triplet AGY represents the middle Ser residue. Ser codons are either AGY or TCN (where Y = pyrimidine and N = any base); AGY was selected here so that the mutagenesis of this codon to a Cys codon (TGY) would involve only an A → T change in the nucleotide sequence.)

   Since the middle Ser residue lies nearer to the 3′-end of this *EcoRI* fragment, the mutant primer for PCR amplification should encompass this end. That is, it should be the primer for the 3′ → 5′ strand of the *EcoRI* fragment:

   5′-NNNGAATTCCCN$_c$ACR$_c$AACCA<u>R$_c$CA</u>N$_c$GC-3′

   where the mutated Ser → Cys triplet is underlined, NNN = several extra bases at the 5′-end of the primer to place the *EcoRI* site internal, N_c = the nucleotide complementary to the nucleotide at this position in the 5′ → 3′ strand, and R_c = the pyrimidine complementary to the purine at this position in the 5′ → 3′ strand.

Chapter 9

1. See *Student Study Guide* for structures.

2. PL/protein molar ratio in purple patches of *H. halobium* is 10.8.

3. See *Student Study Guide* for plots of sucrose solution density vs. percent by weight and by volume. The plot in terms of percent by weight exhibits a greater curve, because less water is required to form a solution that is, for example, 10% by weight (10 g sucrose/100 g total) than to form a solution that is 10% by volume (10 g sucrose/100 mL solution).

4. $r = (4Dt)^{1/2}$
   According to this equation, a phospholipid with D = 1×10^{-8} cm^2 will move approximately 200 nm in 10 milliseconds.

5. Fibronectin: For $t = 10$ msec, $r = 1.67 \times 10^{-7}$ cm = 1.67 nm. Rhodopsin: $r = 110$ nm.
   All else being equal, the value of D is roughly proportional to $(M_r)^{-1/3}$. Molecular weights of rhodopsin and fibronectin are 40,000 and 460,000, respectively. The ratio of diffusion coefficients is thus expected to be:

$$(40{,}000)^{-1/3}/(460{,}000)^{-1/3} = 2.3$$

   On the other hand, the values given for rhodopsin and fibronectin give an actual ratio of 4286. The explanation is that fibronectin is anchored in the membrane via interactions with cytoskeletal proteins, and its diffusion is severely restricted compared with that of rhodopsin.

6. a. Divalent cations increase T_m.
   b. Cholesterol broadens the phase transition without significantly changing T_m.
   c. Distearoylphosphatidylserine should increase T_m, due to increased chain length and also to the favorable interactions between the more negative PS headgroup and the more positive PC headgroups.
   d. Dioleoylphosphatidylcholine, with unsaturated fatty acid chains, will increase T_m.
   e. Integral proteins will broaden the transition and could raise or lower T_m, depending on the nature of the protein.

Chapter 10

1. See structures in *Student Study Guide.*

2. The systematic name for stachyose (Figure 10.19) is β-D-Fructofuranosyl-O-α-D-galactopyranosyl-(1 → 6)-O-α-D-galactopyranosyl-(1 → 6)-α-D-glucopyranoside.

3. The systematic name for trehalose is α-D-glucopyranosyl-(1 → 1)-α-D-glucopyranoside.
   Trehalose is not a reducing sugar. Both anomeric carbons are occupied in the disaccharide linkage.

4. See structures in *Student Study Guide.*

5. A sample that is 0.69 g α-D-glucose/mL and 0.31 g β-D-glucose/mL will produce a specific rotation of 83°.

6. A 0.2 g sample of amylopectin corresponds to 0.2 g/162 g/mole or 1.23×10^{-3} mole glucose residues. 50 μmole is 0.04 of the total sample or 4% of the residues. Methylation of such a sample should yield 1,2,3,6-tetramethylglucose for the glucose residues on the reducing ends of the sample. Unless these are measured, it is not possible to determine how many reducing ends the amylopectin sample contains.

Chapter 11

1. $v/V_{max} = 0.8$.

2. $v = 91$ μmol/s.

3. $K_s = 1.43 \times 10^{-5}$ *M*; $K_m = 3 \times 10^{-4}$ *M*. Since k_2 is 20 times greater than k_{-1}, the system behaves like a steady-state system.

4. From double-reciprocal Lineweaver–Burk plots: $V_{max} = 51$ μmol/s and $K_m = 3.2$ m*M*. Inhibitor (2) shows competitive inhibition with a $K_I = 2.13$ m*M*. Inhibitor (3) shows noncompetitive inhibition with a $K_I = 4$ m*M*.

5. a. The slope is given by $K_m{}^B/V_{max}(K_S{}^A/[A] + 1)$.
   b. *y*-intercept = $(K_m{}^A/[A] + 1)1/V_{max}$.
   c. The horizontal and vertical coordinates of the point of intersection are: $1/[B] = -K_m{}^A/K_S{}^A K_m{}^B$ and $1/v = 1/V_{max}(1 - K_m{}^A/K_S{}^A)$.

6. Top left: (1) Competitive inhibition (*I* competes with *S* for binding to *E*). (2) *I* binds to and forms a complex with *S*.
   Top right: (1) Pure noncompetitive inhibition. (2) Random, single-displacement bisubstrate reaction, where *A* doesn't affect *B* binding and vice versa. (Other possibilities include [3] Irreversible inhibition of *E* by *I*; [4] $1/v$ versus $1/[S]$ plot at two different concentrations of enzyme, *E*.)
   Bottom: (1) Mixed, noncompetitive inhibition. (2) Ordered, single-displacement bisubstrate reaction.

7. Clancy must drink 694 mL of wine, or about one 750-mL bottle.

8. Base pairing between the 5′-CAACCA of the aminoacylated oligonucleotide substrate and the GUUGGG-5′-end of the *Tetrahymena* ribozyme is the key to ribozyme:substrate recognition.

Chapter 12

1. a. As [*P*] rises, the rate of *P* formation shows an apparent decline, as enzyme-catalyzed conversion of *P* → *S* becomes more likely.
   b. Availability of substrates and cofactors.
   c. Changes in [enzyme] due to enzyme synthesis and degradation.
   d. Covalent modification.
   e. Allosteric regulation.
   f. Specialized controls, such as zymogen activation, isozyme variability, and modulator protein influences.

2. Proteolytic enzymes have the potential to degrade the proteins of the cell in which they are synthesized. Synthesis of these enzymes as zymogens is a way of delaying expression of their activity to the appropriate time and place.

3. Monod, Wyman, Changeux allosteric *K* system:

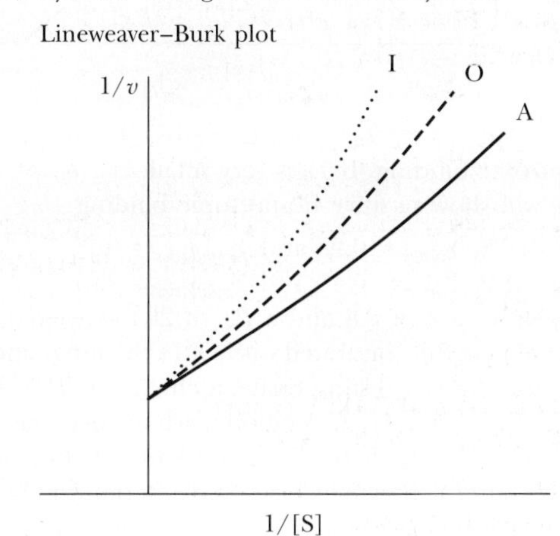

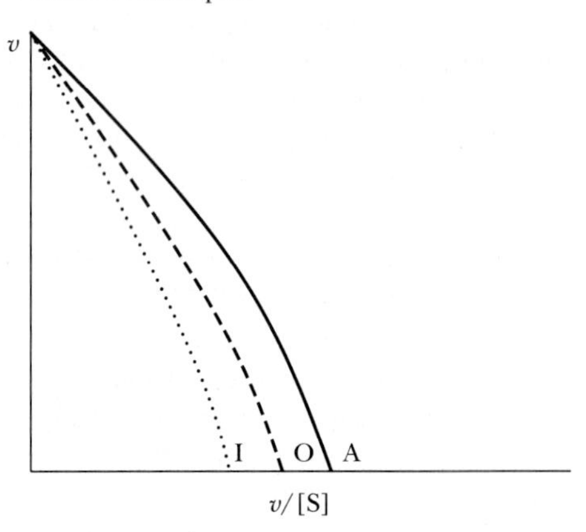

Monod, Wyman, Changeux allosteric *V* system:

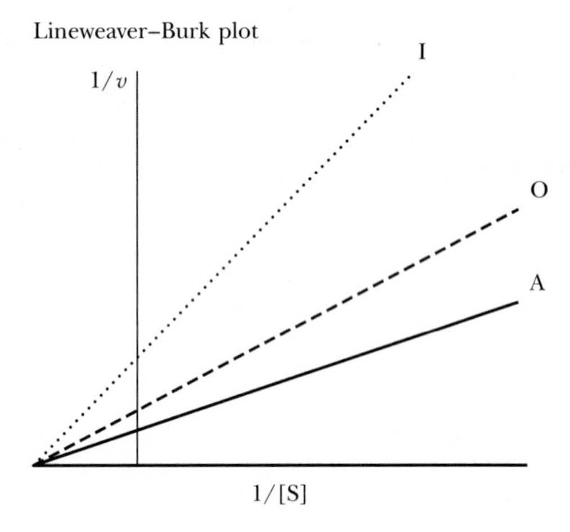

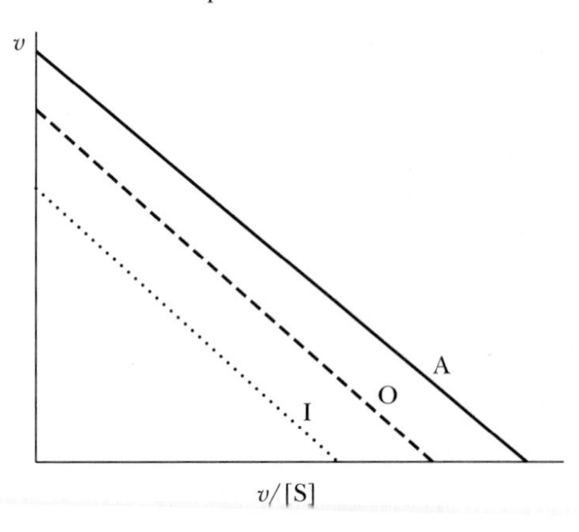

4. If L_0 is large, i.e., $T_0 \gg R_0$, activator, *A*, will show positive homotropic binding. If L_0 is very small, i.e., $R_0 \gg T_0$, inhibitor, *I*, will show positive homotropic binding.

5. For $n = 2.8$, $Y_{lungs} = 0.98$ and $Y_{capillaries} = 0.77$.
For $n = 1.0$, $Y_{lungs} = 0.79$ and $Y_{capillaries} = 0.61$.
Thus, with an n of 2.8 and a P_{50} of 26 torr, hemoglobin becomes almost fully saturated with O_2 in the lungs and drops to 77% saturation in resting tissue, a change of 21%. If n of 1.0 and a P_{50} of 26 torr, hemoglobin would become only 79% saturated with O_2 in the lungs and would drop to 61% in resting tissue, a change of 18%. The difference in hemoglobin O_2 saturation conditions between the values for $n = 2.8$ and 1.0 (21% − 18% or 3% saturation) seems small, but note that the potential for O_2 delivery (98% saturation vs. 79% saturation) is large and becomes crucial when pO_2 in actively metabolizing tissue falls below 40 torr.

6. At altitudes below 10,000 feet, [BPG] increases so that the affinity of Hb for O_2 decreases, allowing a larger fraction of O_2 to be discharged to the tissues.

7. Over time, stored erythrocytes will metabolize 2,3-BPG via the pathway of glycolysis. If [BPG] drops, hemoglobin may bind O_2 with such great affinity that it will not be released to the tissues (see Figure 12.21). The patient receiving a transfusion of [BPG]-depleted blood may actually suffocate.

8. a. By definition, when [Asp] = $K_{0.5}$, $v = 0.5\ V_{max}$.
   b. In the presence of ATP, at [Asp] = $K_{0.5}$, $v \approx 0.85\ V_{max}$ (from Figure 12.30).
   c. In the presence of CTP, at [Asp] = $K_{0.5}$, $v \approx 0.12\ V_{max}$ (from Figure 12.29).

Chapter 13

1. a. Nucleophilic attack by an imidazole nitrogen of His^{57} on the —CH_2— carbon of the chloromethyl group of TPCK covalently inactivates chymotrypsin. (See *Student Study Guide* for structures.)
   b. TPCK is specific for chymotrypsin because the phenyl ring of the phenylalanine residue interacts effectively with the binding pocket of the chymotrypsin active site. This positions the chloromethyl group to react with His^{57}.
   c. Replacement of the phenylalanine residue of TPCK with arginine or lysine produces reagents that are specific for trypsin.

2. a. The structure proposed by Craik, et al., 1987, (*Science* **237**:905–907) is shown below. (If you look up this reference, note that the letters A and B of the figure legend for Figure 3 of this article actually refer to parts B and A of the figure, respectively. Reverse either the letters in the figure or the letters in the figure legend and it will make sense.)

A

57, 56, Asp^{102}, His^{57}, Ser^{214}, Ser^{195}

B

57, 56, Asn^{102}, His^{57}, Ser^{214}, Ser^{195}

b. In the proposed model (shown above), Asn[102] of the mutant enzyme can serve only as a hydrogen-bond donor to His[57]. It is unable to act as a hydrogen-bond acceptor, as aspartate does in native trypsin. As a result, His[57] is unable to act as a general base in transferring a proton from Ser[195]. This presumably accounts for the diminished activity of the mutant trypsin.

3. The results of this paper suggest that His[51] acts as a general base catalyst in the oxidation of alcohol in the native enzyme and that this histidine can be functionally replaced in the mutant enzyme by general base catalysts (such as glycine, glycylglycine, and phosphate) in the solvent.

4. a. The usual explanation for the inhibitory properties of pepstatin is that the central amino acid, statine, mimics the tetrahedral amide hydrate transition state of a good pepsin substrate with its unique hydroxyl group.

   b. Pepsin and other aspartic proteases prefer to cleave peptide chains between a pair of hydrophobic residues, whereas HIV-1 protease preferentially cleaves a Tyr–Pro amide bond. Since pepstatin more closely fits the profile of a pepsin substrate, we would surmise that it is a better inhibitor of pepsin than of HIV-1 protease. In fact, pepstatin is a potent inhibitor of pepsin ($K_I < 1$ nM) but only a moderately good inhibitor of HIV-1 protease, with a K_I of about 1 μM.

5. Using Equation (13.13), we can show that the difference in activation energies for the uncatalyzed and catalyzed hydrolysis reactions ($\Delta G_u - \Delta G_c$) is 92 kJ/mol.

6. Trypsin catalyzes the conversion of chymotrypsinogen to π-chymotrypsin, and chymotrypsin itself catalyzes the conversion of π-chymotrypsin to α-chymotrypsin.

Chapter 14

1. The methionase reaction is a classic PLP-dependent γ-elimination:

The reaction: $H_3C-S-CH_2-CH_2-CH(NH_3^+)-COO^- \longrightarrow H_3C-SH + H_3C-CH_2-C(=O)-COO^- + NH_4^+$

The mechanism:

E-PLP

H_3C-SH

$CH_3-C{=}C-COO^-$ (H, NH_3^+) + E-PLP

$CH_3-CH_2-C(=NH_2^+)-COO^-$

$CH_3-CH_2-C(=O)-COO^-$ + NH_4^+

H_2O

2. If the reactant amino acid and the reactant α-keto acid are each isotopically labeled (with ^{14}C, for example), then one should observe the appearance of the product α-keto acid before the appearance of the product amino acid. Other labeling strategies could be devised (see *Student Study Guide*).

3. TPP assists in the decarboxylation of pyruvate and carries the two-carbon hydroxyethyl group. The covalently bound lipoic acid group of transacetylase couples acyl-group transfer and oxidation of the two-carbon unit. It also transfers these two-carbon units (as acetyl groups) to CoA to form acetyl-CoA. NAD^+ and FAD participate in the reoxidation of the dihydrolipoamide; NAD^+ undergoes net reduction in this process.

4. The glyoxylate carboligase reaction is analogous to the acetolactate synthetase reaction and requires thiamine pyrophosphate as a coenzyme:

$$\mathrm{H{-}C({=}O){-}COO^-} + \mathrm{H{-}C({=}O){-}COO^-} \longrightarrow \mathrm{H{-}C({=}O){-}CH(OH){-}COO^-}$$

H, C—OH, C=O, O⁻, CO_2, H, C, OH, H, C, OH, H—C—COO⁻, O, H, C—OH, B, HC—COO⁻, OH, O H, + H—C—C—COO⁻, OH

5. Glutamate mutase is a B_{12}-dependent enzyme:

Glutamate mutase

$$\mathrm{Co^{3+}{-}CH_2{-}Ad} \rightleftharpoons \mathrm{Co\cdot^{2+}} + \mathrm{Ad{-}CH_2\cdot}$$

Ad—CH_2· + H—$^-OOC{-}CH{-}CH_2{-}CH(NH_3^+){-}COO^-$ ⇌ Ad—CH_3 + $^-OOC{-}\dot{C}H{-}CH_2{-}CH(NH_3^+){-}COO^-$ ⇌ Ad—CH_3 + $^-OOC{-}CH{-}\dot{C}H_2$ / $CH{-}COO^-$ / NH_3^+ ⇌ Ad—CH_2· + $^-OOC{-}CH{-}CH_3$ / $CH{-}COO^-$ / NH_3^+

6. The mechanism of glutamate dehydrogenase is similar to many $NAD(P)^+$-dependent reactions. In the first step, glutamate is oxidized to the imine of α-ketoglutarate [with reduction of $NAD(P)^+$ to NAD(P)H]. In a second step, the imine is hydrolyzed to yield α-ketoglutarate and ammonium ion. ADP is an allosteric activator of glutamate dehydrogenase. When ADP is abundant, enhanced glutamate dehydrogenase activity produces α-ketoglutarate, which in turn stimulates the tricarboxylic acid cycle (Chapter 19).

7. A wide variety of mutations of mitochondrial branched-chain α-keto acid dehydrogenases (BCKD) have been identified in patients with MSUD. Some of these mutations produce BCKD with low enzyme activity, but others appear to shorten the biological half-life of the enzyme. Some mutant BCKD are degraded much more rapidly in the cell than are normal BCKD. The result is that, although such mutant BCKD molecules show normal activity, each cell contains fewer enzyme molecules and displays a lower total enzyme activity. Studies by Louis Elsas at Emory University have led to a proposal that thiamine pyrophosphate stabilizes a conformation of BCKD that is more resistant to cellular degradation. For this reason, treatment of some MSUD patients with thiamine (100–500 mg/day) increases the biological half-life of the BCKD enzyme molecules and thus increases the total mitochondrial BCKD activity.

8. Aspartate β-decarboxylase is a PLP-dependent enzyme:

The aspartate β-decarboxylase mechanism

$^{-}OOC{-}CH_2{-}CH{-}COO^-$
NH_3^+
+
E-PLP
$\longrightarrow$

: B—Enzyme
H
$OOC{-}CH_2{-}C{-}COO^-$
N^+H
CH
$^{2-}O_3PO$
O^-
N
CH_3
H

$H^+{-}B{-}$Enzyme
H_2C
COO^-
C
: NH
CO_2
CH
$^{2-}O_3PO$
O^-
N
CH_3
H

O^-
CH_2
COO^-
C
O
C
N^+H
CH
$^{2-}O_3PO$
O
N
CH_3
H

H_3C
COO^-
C
$-B^+{-}H$
N^+H
CH
$^{2-}O_3PO$
O^-
N
CH_3
H
$\longrightarrow$

H_3C
COO^-
CH
N^+H
CH
$^{2-}O_3PO$
O^-
N
CH_3
H

H_2O

CH_3
COO^-
CH
NH_3^+
+
E-PLP

9. Alanine racemase is also a PLP-dependent enzyme:

Alanine Racemase

The reaction: $H_3C{-}C(H)(NH_3^+){-}COO^- \longrightarrow H{-}C(CH_3)(NH_3^+){-}COO^-$

The mechanism:

E-PLP $\longrightarrow$

B :
H
$H_3C{-}C{-}COO^-$
N^+H
HC
$^{2-}O_3PO$
O^-
N
H
$\longrightarrow$

B^+H
$H_3C{-}C{-}COO^-$
N^+H
HC

CH_3
E-PLP + $H{-}C{-}COO^-$
NH_3^+

10. Serine hydroxymethylase is another PLP-dependent enzyme. Though tetrahydrofolate is required for optimal enzyme activity, it is not necessary to include THF in a reasonable mechanism for this enzyme:

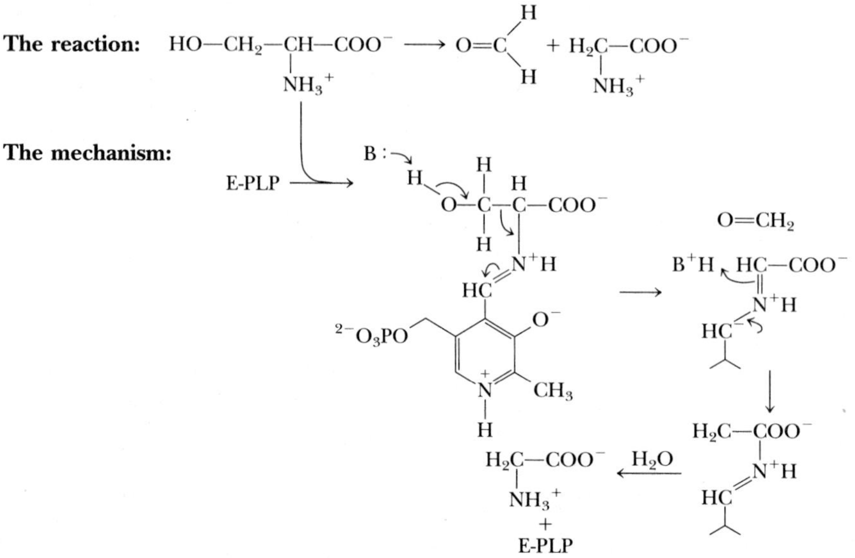

The reaction: $HO{-}CH_2{-}CH(NH_3^+){-}COO^- \longrightarrow O{=}CH_2 + H_2C(NH_3^+){-}COO^-$

The mechanism:

E-PLP → $B{:}$ $H{-}O{-}CH_2{-}C(H)({-}COO^-){-}N^+H{=}CH{-}$(pyridoxal phosphate ring: $^{2-}O_3PO$, O^-, CH_3, $N^+{-}H$) → $O{=}CH_2$ + B^+H, $HC(COO^-){=}N^+H{-}HC$ → $H_2C{-}COO^-$, $N^+H{=}HC$ → (H_2O) → $H_2C(NH_3^+){-}COO^-$ + E-PLP

Chapter 15

1. $K_{eq} = 613\ M$; $\Delta G° = -15.9$ kJ/mol.

2. $\Delta G° = 1.69$ kJ/mol at 20°C; $\Delta G° = -5.80$ kJ/mol at 30°C. $\Delta S° = 0.75$ kJ/mol · K.

3. $\Delta G = -24.8$ kJ/mol.

4. State functions are quantities that depend on the state of the system and not on the path or process taken to reach that state. Volume, pressure, and temperature are state functions. Heat and all forms of work, such as mechanical work and electrical work, are not state functions.

5. $\Delta G°' = \Delta G° - 39.5\ n$ (in kJ/mol), where n is the number of H^+ produced in any process. So $\Delta G° = \Delta G°' + 39.5\ n = -30.5$ kJ/mol + 39.5(1) kJ/mol.
$\Delta G° = 9.0$ kJ/mol at 1 M [H^+].

6. $K_{eq}(AC) = (0.01 \times 1000) = 10$.
$\Delta G°(AB) = 11.8$ kJ/mol.
$\Delta G°(BC) = -17.8$ kJ/mol.
$\Delta G°(AC) = -6.0$ kJ/mol.

Chapter 16

1. See *Student Study Guide* for resonance structures.

2. $K_{eq} = [Cr][P_i]/[CrP][H_2O]$.
$K_{eq} = 3.89 \times 10^7$.

3. CrP in the amount of 135.3 moles would be required per day to provide 5860 kJ energy. This corresponds to 17,730 g of CrP per day. With a body content of 20 g CrP, each molecule would recycle 886 times per day.
Similarly, 637 moles of glycerol-3-P, or 108,300 g of glycerol-3-P, would be required. Each molecule would recycle 5410 times/day.

4. Using 37°C (310 K), the normal temperature of a rat liver cell, and employing Equation (15.12), we obtain $\Delta G = -30.5$ kJ/mol + (-16.2 kJ/mol) = -46.7 kJ/mol.

5. The hexokinase reaction is a sum of the reactions for hydrolysis of ATP and phosphorylation of glucose:

$$ATP + H_2O \rightleftharpoons ADP + P_i$$
$$Glucose + P_i \rightleftharpoons G6P + H_2O$$
$$\overline{Glucose + ATP \rightleftharpoons G6P + ADP}$$

The free energy change for the hexokinase reaction can thus be obtained by summing the free energy changes for the first two reactions above.

$\Delta G°'$ for hexokinase = -30.5 kJ/mol + 13.9 kJ/mol = -16.6 kJ/mol

6. Comparing the acetyl group of acetoacetyl-CoA and the methyl group of acetyl-CoA, it is reasonable to suggest that the acetyl group is more electron-withdrawing in nature. For this reason it tends to destabilize the thiol ester of acetoacetyl-CoA, and the free energy of hydrolysis of acetoacetyl-CoA should be somewhat larger than that of acetyl-CoA. In fact, $\Delta G°' = -43.9$ kJ/mol, compared with -31.5 kJ/mol for acetyl-CoA.

7. Carbamoyl phosphate should have a somewhat larger free energy of hydrolysis than acetyl phosphate, at least in part because of greater opportunities for resonance stabilization in the products. In fact, the free energy of hydrolysis of carbamoyl phosphate is -51.5 kJ/mol, compared with -43.3 kJ/mol for acetyl phosphate.

Chapter 17

1. 6.5×10^{12} (6.5 trillion) people.

2. Consult Table 17.2.

3. O_2, H_2O, and CO_2.
4. N_2 (dinitrogen) and NO_3^- (nitrate ion).
5. See Section 17.2.
6. Consult Figure 17.6.
7. See *Why Metabolic Pathways Have So Many Steps,* p. 552.
8. Consult *Corresponding Pathways of Catabolism and Anabolism Differ in Important Ways,* p. 554. See also Figure 17.8.
9. See *The ATP Cycle,* p. 556; *NAD⁺ Collects Electrons Released in Catabolism,* p. 557; and *NADPH Provides the Reducing Power for Anabolic Processes,* p. 558.
10. See *Metabolic Regulation Is Achieved by Regulating Enzyme Activity,* p. 559. In terms of quickness of response, the order is allosteric regulation > covalent modification > enzyme synthesis and degradation. See the *Student Study Guide* accompanying this textbook for further discussions.
11. See *Metabolic Pathways Are Compartmentalized Within Cells,* p. 567, and discussions in the *Student Study Guide.*
12. When [ATP]/[ADP] = 10 and P_i = 1 m*M*, phosphorylation potential (Γ) = 10,000 M^{-1} and ΔG of ATP hydrolysis = −53.3 kJ/mol.

Chapter 18

1. a. Phosphoglucoisomerase, fructose bisphosphate aldolase, triose phosphate isomerase, glyceraldehyde-3-P dehydrogenase, phosphoglycerate mutase, enolase.
   b. Hexokinase/glucokinase, phosphofructokinase, phosphoglycerate kinase, pyruvate kinase, lactate dehydrogenase.
   c. Hexokinase and phosphofructokinase.
   d. Phosphoglycerate kinase and pyruvate kinase.
   e. According to Equation (15.12), reactions in which the number of reactant molecules differs from the number of product molecules exhibit a strong dependence on concentration. That is, such reactions are extremely sensitive to changes in concentration. Using this criterion, we can predict from Table 18.1 that the free energy changes of the fructose bisphosphate aldolase and glyceraldehyde-3-P dehydrogenase reactions will be strongly influenced by changes in concentration.
   f. Reactions that occur with ΔG near zero operate at or near equilibrium. See Table 18.1 and Figure 18.31.
2. The carboxyl carbon of pyruvate derives from carbons 3 and 4 of glucose. The keto carbon of pyruvate derives from carbons 2 and 5 of glucose. The methyl carbon of pyruvate is obtained from carbons 1 and 6 of glucose.
3. Increased [ATP] or [citrate] inhibits glycolysis. Increased [AMP], [fructose-1,6-bisphosphate], [fructose-2,6-bisphosphate], or [glucose-6-P] stimulates glycolysis.
4. See *Student Study Guide* for discussion.
5. The mechanisms for fructose bisphosphate aldolase and glyceraldehyde-3-P dehydrogenase are shown in Figures 18.13 and 18.18, respectively.
6. The relevant reactions of galactose metabolism are shown in Figure 18.33. See *Student Study Guide* for mechanisms.
7. Iodoacetic acid would be expected to alkylate the reactive active-site cysteine that is vital to the glyceraldehyde-3-P dehydrogenase reaction. This alkylation would irreversibly inactivate the enzyme.
8. Ignoring the possibility that ^{32}P might be incorporated into ATP, the only reaction in glycolysis that utilizes P_i is glyceraldehyde-3-P dehydrogenase, which converts glyceraldehyde-3-P to 1,3-bisphosphoglycerate. $^{32}P_i$ would label the phosphate at carbon 1 of 1,3-bisphosphoglycerate. The label will be lost in the next reaction, and no other glycolytic intermediates will be directly labeled. (Once the label is incorporated into ATP, it will also show up in glucose-6-P, fructose-6-P, and fructose-1, 6-bisphosphate.)
9. The sucrose phosphorylase reaction produces glucose-6-P directly, bypassing the hexokinase reaction. Direct production offers the obvious advantage of saving a molecule of ATP.
10. All of the kinases involved in glycolysis, as well as enolase, are activated by Mg^{2+} ion. A Mg^{2+} deficiency could lead to reduced activity for some or all of these enzymes. However, other systemic effects of a Mg^{2+} deficiency might cause even more serious problems.

Chapter 19

1. Glutamate enters the TCA cycle via a transamination to form α-ketoglutarate. The γ-carbon of glutamate entering the cycle is equivalent to the methyl carbon of an entering acetate. Thus no radioactivity from such a label would be lost in the first or second cycle, but in each subsequent cycle, 50% of the total label would be lost (see Figure 19.21).
2. NAD^+ activates pyruvate dehydrogenase and isocitrate dehydrogenase and thus would increase TCA cycle activity. If NAD^+ increases at the expense of NADH, the resulting decrease in NADH would likewise activate the cycle by stimulating citrate synthase and α-ketoglutarate dehydrogenase. ATP inhibits pyruvate dehydrogenase, citrate synthase, and isocitrate dehydrogenase; reducing the ATP concentration would thus activate the cycle. Isocitrate is not a regulator of the cycle, but increasing its concentration would mimic an increase in acetate flux through the cycle and increase overall cycle activity.
3. For most enzymes that are regulated by phosphorylation, the covalent binding of a phosphate group at a distant site induces a conformation change at the active site that either activates or inhibits the enzyme activity. On the other hand, X-ray crystallographic studies reveal that the phosphorylated and unphosphorylated forms of isocitrate dehydrogenase share identical structures with only small (and probably insignificant) conformational changes at Ser^{113}, the locus of phosphorylation. What phosphorylation *does* do is block isocitrate binding (with no effect on the binding affinity of $NADP^+$). As shown in Figure 2 of the paper cited in the problem (Barford, D., 1991. Molecular mechanisms for the control of enzymic activity by protein phosphorylation. *Biochimica et Biophysica Acta* **1133,** 55–62), the γ-carboxyl group of bound isocitrate forms a hydrogen bond with the hydroxyl group of Ser^{113}. Phosphorylation apparently prevents isocitrate binding by a combination of a loss of the crucial H-bond between substrate and enzyme, and by repulsive electrostatic and steric effects.

4\.

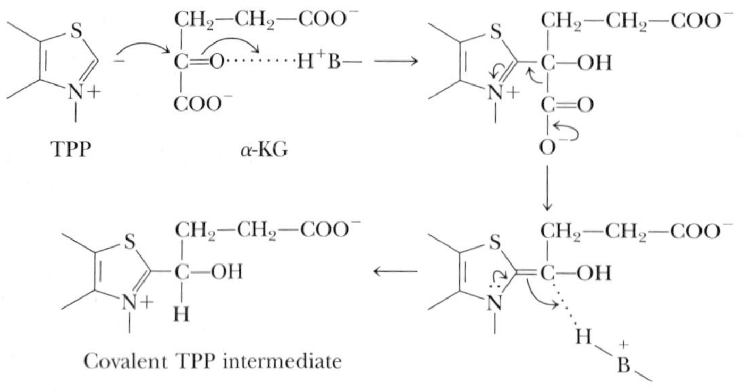

TPP α-KG

Covalent TPP intermediate

5. Aconitase is inhibited by fluorocitrate, the product of citrate synthase action on fluoroacetate. In a tissue where inhibition has occurred, all TCA cycle metabolites should be reduced in concentration. Fluorocitrate would replace citrate, and the concentrations of isocitrate and all subsequent metabolites would be reduced because of aconitase inhibition.
6. $FADH_2$ is colorless, but FAD is yellow, with a maximal absorbance at 450 nm. Succinate dehydrogenase could be conveniently assayed by measuring the decrease in absorbance in a solution of the flavoenzyme and succinate.
7. The central (C-3) carbon of citrate is reduced, and an adjacent carbon is oxidized in the aconitase reaction. The carbon bearing the hydroxyl group is obviously oxidized by isocitrate dehydrogenase. In the α-ketoglutarate dehydrogenase reaction, the departing carbon atom and the carbon adjacent to it are both oxidized. Both of the $—CH_2—$ carbons of succinate are oxidized in the succinate dehydrogenase reaction. Of these four molecules, all but citrate undergo a net oxidation in the TCA cycle.
8. Several TCA metabolite analogs are known, including malonate, an analog of succinate, and 3-nitro-2-S-hydroxypropionate, the anion of which is a transition-state analog for the fumarase reaction.

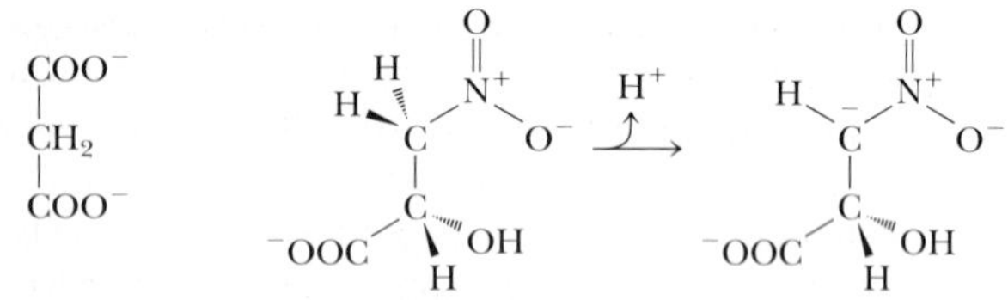

Malonate (succinate analog) 3-Nitro-2-S-hydroxypropionate Transition-state analog for fumarase

Chapter 20

1. The cytochrome couple is the acceptor, and the donor is the (bound) $FAD/FADH_2$ couple, because the cytochrome couple has a higher (more positive) reduction potential.

$$\Delta\mathscr{E}_o' = 0.254\text{V} - 0.02\text{V}^*$$

*This is a typical value for enzyme-bound [FAD].

$$\Delta G - n\mathscr{F}\Delta\mathscr{E}_o' = -45.2 \text{ kJ/mol.}$$

2. $\Delta\mathscr{E}_o' = -0.03$ V; ΔG +5,790 J/mol
3. This situation is analogous to that described in Section 15.8. The net result of the reduction of NAD^+ is that a proton is consumed. The effect on the calculation of free energy change is similar to that described in Equation (15.22). Adding an appropriate term to Equation (20.13) yields:

$$\mathscr{E} = \mathscr{E}_o' - RT \ln [H^+] + (RT/n\mathscr{F}) \ln ([ox]/[red])$$

4. Cyanide acts primarily via binding to cytochrome a_3, and the amount of cytochrome a_3 in the body is much lower than the amount of hemoglobin. Nitrite anion is an effective antidote for cyanide poisoning because of its unique ability to oxidize ferrohemoglobin to ferrihemoglobin, a form of hemoglobin that competes very effectively with cytochrome a_3 for cyanide. The amount of ferrohemoglobin needed to neutralize an otherwise lethal dose of cyanide is small compared with the total amount of hemoglobin in the body. Even though a small amount of hemoglobin is sacrificed by sequestering the cyanide in this manner, a "lethal dose" of cyanide can be neutralized in this manner without adversely affecting oxygen transport.
5. You should advise the wealthy investor that she should decline this request for financial backing. Uncouplers can indeed produce dramatic weight loss, but can also cause death. Dinitrophenol was actually marketed for weight loss at one time, but sales were discontinued when the potentially fatal nature of such "therapy" was fully appreciated. Weight loss is best achieved by simply making sure that the number of calories of food eaten is less than the number of calories metabolized 17,000 kJ of energy (4,000 Cal). A person who metabolizes about 2,000 kJ (about 500 Cal) more than he or she consumes every day will lose about a pound in about eight days.)
6. The calculation in Section 20.8 assumes the same conditions as in this problem (1 pH unit gradient across the membrane and an electrical potential difference of 0.18 V). The transport of 1 mole of H^+ across such a membrane generates 23.3 kJ of energy. For three moles of H^+, the energy yield is 69.9 kJ. Then, from Equation (15.12), we have:

$$69{,}900 \text{ J/mol} = 30{,}500 \text{ J/mol} + RT \ln([ATP]/[ADP][P_i])$$
$$39{,}400 \text{ J/mol} = RT \ln([ATP]/[ADP][P_i])$$
$$[ATP]/[ADP][P_i] = 4.36 \times 10^6 \text{ at } 37°C.$$

In the absence of the proton gradient, this same ratio is 7.25×10^{-6} at equilibrium!

7. The succinate/fumarate redox couple has the highest (i.e., most positive) reduction potential of any of the couples in glycolysis and the TCA cycle. Thus oxidation of succinate by NAD^+ would be very unfavorable in the thermodynamic sense.

Oxidation of succinate by NAD^+:

$$\Delta\mathscr{E}_o' = -0.35\text{V};\ \Delta G^{\circ\prime} = 67{,}500\ \text{J/mol}.$$

On the other hand, oxidation of succinate is quite feasible using enzyme-bound FAD.
Oxidation of succinate by [FAD]:

$\Delta\mathscr{E}_o' \approx 0$, $\Delta G^{\circ\prime} \approx 0$, depending on the exact reduction potential for bound FAD.

Chapter 21

1. The reactions that contribute to the equation on page 666 are:

$$\begin{aligned} 2\ \text{Pyruvate} + 2H^+ + 2H_2O &\longrightarrow \text{glucose} + O_2 \\ 2\text{NADH} + 2H^+ + O_2 &\longrightarrow 2NAD^+ + 2H_2O \\ 4\text{ATP} + 4H_2O &\longrightarrow 4\text{ADP} + 4P_i \\ 2\text{GTP} + 2H_2O &\longrightarrow 2\text{GDP} + 2P_i \end{aligned}$$

Summing these 4 reactions produces the equation on page 666.

2. This problem essentially involves consideration of the three unique steps of gluconeogenesis. The conversion of PEP to pyruvate was shown in Chapter 18 to have a $\Delta G^{\circ\prime}$ of -31.7 kJ/mol. For the conversion of pyruvate to PEP, we need only add the conversion of a GTP to GDP (equivalent to ATP to ADP) to the reverse reaction. Thus:
Pyruvate $\longrightarrow$ PEP: $\Delta G^{\circ\prime} = +31.7$ kJ/mol $-$ 30.5 kJ/mol $=$ 1.2 kJ/mol
Then, using Equation (15.12):

$\Delta G = 1.2$ kJ/mol + RT ln ([PEP][ADP]^2[P_i][Py][ATP]^2)
ΔG (in erythrocytes) $= -14.1$ kJ/mol

In the case of the fructose-1,6-bisphosphatase reaction, $\Delta G^{\circ\prime} = -16.3$ kJ/mol (see Equation [18.6]).

$\Delta G = -16.3$ kJ/mol + RT ln ([F6P]/[F1,6P])
ΔG (in erythrocytes) $= -18.3$ kJ/mol

For the glucose-6-phosphate reaction, $\Delta G^{\circ\prime} = -13.8$ kJ/mol (see Table 16.1). $\Delta G = -13.8$ kJ/mol + RT ln ([Glu]/[G6P]) $= -3.2$ kJ/mol. From these ΔG values and those in Table 18.1, ΔG for gluconeogenesis $= -35.6$ kJ/mol.

3. Inhibition by 25 μM fructose-2,6-bisphosphate is approximately 94% at 25 μM fructose-1,6-bisphosphate and approximately 44% at 100 μM fructose-1,6-bisphosphate.

4. α-Amylase, an endoglycosidase, can produce cleavages throughout a complex glycogen molecule, but it would show low activity near branch points. β-Amylase (a plant exoglycosidase) could presumably cleave maltose units from any free end of a glycogen molecule, but would be unable to proceed beyond branch points.

5. The hydrolysis of UDP-glucose to UDP and glucose is characterized by a $\Delta G^{\circ\prime}$ of -31.9 kJ/mol. This is more than sufficient to overcome the energetic cost of synthesizing a new glycosidic bond in a glycogen molecule. The net $\Delta G^{\circ\prime}$ for the glycogen synthase reaction is -13.3 kJ/mol.

6. According to Table 23.1, a 70-kg person possesses 1,920 kJ of muscle glycogen. Without knowing how much of this is in fast-twitch muscle, we can simply use the fast-twitch data from *A Deeper Look* (page 684) to calculate a rate of energy consumption. The plot on page 684 shows that glycogen supplies are exhausted after 60 minutes of heavy exercise. Ignoring the curvature of the plot, 1,920 kJ of energy consumed in 60 minutes corresponds to an energy consumption rate of 533 J/sec.

7. C-6 of glucose becomes C-5 of ribose for all five glucose molecules passing through the pentose phosphate pathway in Figure 21.47. C-3 of glucose becomes C-3 of ribose in one molecule, C-2 in two molecules of ribose and C-1 in two molecules of ribose produced by one pass of the pathway in Figure 21.47. C-1 of glucose becomes C-5 of one molecule of ribose and C-1 of four molecules of ribose through one pass of the pathway.

8. Three molecules of glucose are required for one pass through the scheme of Figure 21.49. The C-2 of one glucose becomes the C-3 of pyruvate, and the C-4 of the same molecule of glucose becomes C-1 of a different molecule of pyruvate. For another glucose through the pathway, C-2 becomes C-3 of pyruvate and C-4 becomes C-1 of another pyruvate. For the third glucose through the pathway, both C-2 and C-4 carbons become C-1 of pyruvate molecules.

9. Though other inhibitory processes might also occur, enzymes with mechanisms involving formation of Schiff base intermediates with active-site lysine residues are likely to be inhibited by sodium borohydride (see *A Deeper Look*, page 582). The transaldolase reaction of the pentose phosphate pathway involves this type of active-site intermediate and would be expected to be inhibited by sodium borohydride.

10. Glycogen molecules do not have any free reducing ends, regardless of the size of the molecule. If branching occurs every 8 residues and each arm of the branch has 8 residues (or 16 per branch point), a glycogen molecule with 8000 residues would have about 500 ends. If branching occurs every 12 residues, a glycogen molecule with 8000 residues would have about 334 ends.

11. (a) Increased fructose-1,6-bisphosphate would activate pyruvate kinase, stimulating glycolysis. (b) Increased blood glucose would decrease gluconeogenesis and increase glycogen synthesis. (c) Increased blood insulin inhibits gluconeogenesis and stimulates glycogen synthesis. (d) Increased blood glucagon inhibits glycogen synthesis and stimulates glycogen breakdown. (e) Since ATP inhibits both phosphofructokinase and pyruvate kinase, and since its level reflects the energy status of the cell, a decrease in tissue ATP would have the effect of stimulating glycolysis. (f) Increasing AMP would have the same effect as decreasing ATP—stimulation of glycolysis and inhibition of gluconeogenesis. (g) Fructose-6-phosphate is not a regulatory molecule and decreases in its concentration would not markedly affect either glycolysis or gluconeogenesis (ignoring any effects due to decreased [GCP] as a consequence of decreased [FGP]).

12. At 298 K, assuming roughly equal concentrations of glycogen molecules of different lengths, the glucose-1-P concentration would be about 3.5 m*M*.

Chapter 22

1. Efficiency $= \Delta G^{\circ\prime}/(\text{light energy input}) = n\mathscr{F}\Delta\mathscr{E}_o'/(Nhc/\lambda) = 0.564$.

2. $\Delta G^{\circ\prime} = n\mathscr{F}\Delta\mathscr{E}_o'$, so $\Delta G^{\circ\prime}/n\mathscr{F} = \Delta\mathscr{E}_o'$. $n = 4$ for $2H_2O \longrightarrow 4e^- + 4H^+ + O_2$. $\Delta\mathscr{E}_o' = 0.0648\text{V} = (\mathscr{E}_o'$ (primary oxidant) $- \mathscr{E}_o'(\tfrac{1}{2}O_2/H_2O))$. $\mathscr{E}_o'$ (primary oxidant) $= +0.88$V.

3. ΔG for ATP synthesis $= 50{,}336$ J/mol. $\Delta\mathscr{E}_o' = 0.26$V.

4. Radioactivity will be found in C-1 of 3-phosphoglycerate; C-3 and C-4 of glucose; C-1 and C-2 of erythrose-4-P; C-3, C-4, and C-5 of sedoheptulose-1,7-bisP; and C-1, C-2, and C-3 of ribose-5-P.

5. Since animals cannot achieve net synthesis of glucose from CO_2, the person's blood glucose will contain no radioactive C.

6. Efficiency of noncyclic photophosphorylation = 2 $h\nu$/ATP; efficiency of cyclic photophosphorylation = 1.5 $h\nu$/ATP.

7. Light induces three effects in chloroplasts: (1) pH increase in the stroma; (2) generation of reducing power (as ferredoxin); (3) Mg^{2+} efflux from the thylakoid lumen. Key enzymes in the Calvin-Benson CO_2 fixation pathway are activated by one or more of these effects. In addition, rubisco activase is activated indirectly by light, and, in turn, activates rubisco.

8. Considering the reactions involving water separately:
   1. ATP synthesis:

$$18\ ADP + 18\ P_i \rightarrow 18\ ATP + 18\ H_2O$$

   2. $NADP^+$ reduction and the photolysis of water:

$$12\ H_2O + 12\ NADP^+ \rightarrow 12\ NADPH + 12\ H^+ + 6\ O_2$$

   3. Overall reaction for hexose synthesis:

$$6CO_2 + 12NADPH + 12H^+ + 18ATP + 12H_2O \rightarrow \text{glucose} + 12NADP^+ + 6O_2 + 18\ ADP + 18P_i$$

$$\textit{Net: } 6\ CO_2 + 6\ H_2O \rightarrow \text{glucose} + 6\ O_2$$

Of the 12 waters consumed in O_2 production in reaction 2 and the 12 waters consumed in the reactions of the Calvin-Benson cycle (reaction 3), 18 are restored by H_2O release in phosphoric anhydride bond formation in reaction 1.

Chapter 23

1. Assuming that all fatty acid chains in the triacylglycerol are palmitic acid, the fatty acid content of the triacylglycerol is 95% of the total weight. On the basis of this assumption, one can calculate that 30 lb of triacylglycerol will yield 118.7 L of water.

2. 11-cis-Heptadecenoic acid is metabolized by means of seven cycles of β-oxidation, leaving a propionyl-CoA as the final product. However, the fifth cycle bypasses the acyl-CoA dehydrogenase reaction, because a cis-double bond is already present at the proper position. Thus, β-oxidation produces 7 NADH (=17.5 ATP), 6 $FADH_2$ (= 9 ATP), and 7 acetyl-CoA (= 70 ATP), for a total of 96.5 ATP. Propionyl-CoA is converted to succinyl-CoA (with expenditure of 1 ATP), which can be converted to oxaloacetate in the TCA cycle (with production of 1 GTP, 1 $FADH_2$ and 1 NADH). Oxaloacetate can be converted to pyruvate (with no net ATP formed or consumed), and pyruvate can be metabolized in the TCA cycle (producing 1 GTP, 1 $FADH_2$, and 4 NADH). The net for these conversions of propionate is 16.5 ATP. Together with the results of β-oxidation, the total ATP yield for the oxidation of one molecule of 11-cis-heptadecenoic acid is 113 ATP.

3. Instead of invoking hydroxylation and β-oxidation, the best strategy for oxidation of phytanic acid is α-hydroxylation, which places a hydroxyl group at C-2. This facilitates oxidative α-decarboxylation, and the resulting acid can react with CoA to form a CoA ester. This product then undergoes six cycles of β-oxidation. In addition to CO_2, the products of this pathway are three molecules of acetyl-CoA, three molecules of propionyl-CoA and one molecule of 2-propionyl-CoA.

4. Although acetate units cannot be used for net carbohydrate synthesis, oxaloacetate can enter the gluconeogenesis pathway in the PEP carboxykinase reaction. (For this purpose, it must be converted to malate for transport to the cytosol.) Acetate labeled at the carboxyl carbon will first label (equally) the C-3 and C-4 positions of newly formed glucose. Acetate labeled at the methyl carbon will label (equally) the C-1, C-2, C-5, and C-6 positions of newly formed glucose.

5. Hypoglycin in unripened akee fruit irreversibly inactivates acyl-CoA dehydrogenase, thereby blocking β-oxidation of fatty acids, a pathway that is a major source of serum glucose. Thus victims of poisoning by unripened akee fruit are often found to be severely hypoglycemic.

6. Fat is capable of storing more energy (37 kJ/g) than carbohydrate (16 kJ/g). Ten lb of fat contain $10 \times 454 \times 37 =$ 167,980 kJ of energy. This same amount of energy would require 167,980/16 = 10,499 g or 23 lb of stored carbohydrate.

7. The enzyme methylmalonyl-CoA mutase, which catalyzes the third step in the conversion of propionyl-CoA to succinyl-CoA, is B_{12}-dependent. If a deficiency in this vitamin occurs, and if large amounts of odd-carbon fatty acids were ingested in the diet, L-methymalonyl-CoA could accumulate.

8. a. Myristic acid:

$$CH_3(CH_2)_{12}CO\text{-}CoA + 94\ P_i + 94\ ADP + 20\ O_2 \rightarrow 94\ ATP + 14\ CO_2 + 113\ H_2O + CoA$$

   b. Stearic acid:

$$CH_3(CH_2)_{16}CO\text{-}CoA + 122\ P_i + 122\ ADP + 26\ O_2 \rightarrow 122\ ATP + 18\ CO_2 + 147\ H_2O + CoA$$

   c. α-Linolenic acid:

$$C_{17}H_{29}CO\text{-}CoA + 116.5\ P_i + 116.5\ ADP + 24.5\ O_2 \rightarrow 116.5\ ATP + 18\ CO_2 + 138.5\ H_2O + CoA$$

   d. Arachidonic acid:

$$C_{19}H_{31}CO\text{-}CoA + 128\ P_i + 128\ ADP + 27\ O_2 \rightarrow 128\ ATP + 20\ CO_2 + 152\ H_2O + CoA$$

9. During the hydration step, the elements of water are added across the double bond. Also, the proton transferred to the acetyl-CoA carbanion in the thiolase reaction is derived from the solvent, so each acetyl-CoA released by the enzyme would probably contain two tritiums. Seven tritiated acetyl-CoAs would thus derive from each molecule of palmitoyl-CoA metabolized, each with two tritiums at C-2.

10. A carnitine deficiency would presumably result in defective or limited transport of fatty acids into the mitochondrial matrix and reduced rates of fatty acid oxidation.

Chapter 24

1. The equations needed for this problem are found on page 765. See the *Student Study Guide* for details.

2. Carbons C-1 and C-6 of glucose become the methyl carbons of acetyl-CoA that is the substrate for fatty acid synthesis. Carbons C-2 and C-5 of glucose become the carboxyl carbon of acetyl-CoA for fatty acid synthesis. Only citrate that is immediately exported to the cytosol provides glucose carbons for fatty acid synthesis. Citrate that enters the TCA cycle does not immediately provide carbon for fatty acid synthesis.

3. A suitable model, based on the evidence presented in this chapter, would be that the fundamental regulatory mechanism in ACC is a polymerization-dependent conformation change in the protein. All other effectors—palmitoyl-CoA, citrate and phosphorylation-dephosphorylation—may function primarily by shifting the inactive protomer-active polymer equilibrium. Polymerization may bring domains of the protomer (i.e., bicarbonate-, acetyl-CoA-, and biotin-binding domains) closer together, or may bring these domains on separate protomers close to each other. See the *Student Study Guide* for further details.

4. The pantothenic acid group may function, at least to some extent, as a flexible "arm" to carry acyl groups between the malonyl transferase and ketoacyl-ACP synthase active sites. The pantothenic acid moiety is approximately 1.9 nm in length, setting an absolute upper-limit distance between these active sites of 3.8 nm. However, on the basis of modeling considerations, it seems likely that the distance between these sites is smaller than this upper-limit value.

5. Two electrons pass through the chain from NADH to FAD to the two cytochromes of the cytochrome b_5 reductase and then to the desaturase. Together with two electrons from the fatty acyl substrate, these electrons reduce an O_2 to two molecules of water. The hydrogen for the waters thus formed comes from the substrate (2H) and from two protons from solution.

6. Ethanolamine + glycerol + 2 fatty acyl-CoA + 2 ATP + CTP + $H_2O \rightarrow$
Phosphatidylethanolamine + 2 ADP + 2 CoA + CMP + PP_i + P_i

7. The conversion of acetyl-CoA to lanosterol can be written as:

$$\text{18 Acetyl-CoA} + \text{13 NADPH} + \text{13 H}^+ + \text{18 ATP} + 0.5\ O_2 \rightarrow \text{Lanosterol} + \text{18 CoA} + \text{13 NADP}^+ + \text{18 ADP} + 6\ P_i + 6\ PP_i + CO_2$$

The conversion of lanosterol to cholesterol is complicated, but, in terms of carbon counting, three carbons are lost in the conversion to cholesterol. This might be viewed as 1.5 acetate groups for the purpose of completing the balanced equation.

8. The numbers 1–4 in the cholesterol structure indicate the carbon positions of mevalonate as shown:

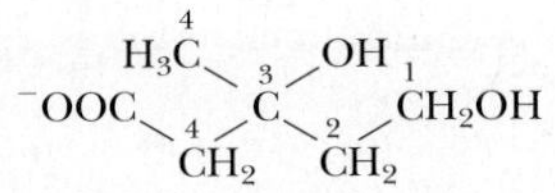

Mevalonate

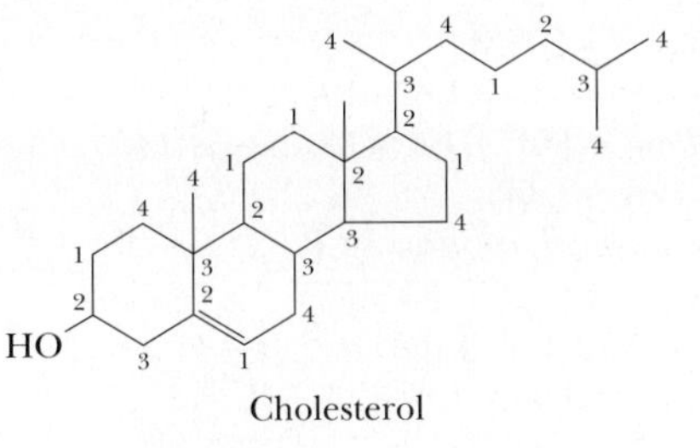

Cholesterol

9. The *O*-linked saccharide domain of the LDL receptor probably functions to extend the receptor domain away from the cell surface and above the glycocalyx coat, so that the receptor can recognize circulating lipoproteins.

10. As shown in Figures 24.19 and 24.22, the syntheses (in eukaryotes) of phosphatidylcholine, phosphatidylethanolamine, phosphatidylinositol and phosphatidylglycerol are dependent upon CTP. A CTP deficiency would be likely to affect all these synthetic pathways.

11. To form a malonyl-CoA "costs" one ATP. Each cycle of the fatty acyl synthase consumes two NADPH molecules, each worth 3.5 ATP. Thus, each of the seven cycles required to form a palmitic acid consumes 8 ATP. A total of 56 ATP are consumed to synthesize one molecule of palmitic acid.

Chapter 25

1. a. $K_{eq} = 360{,}333$.
   b. Assuming [ATP] ≈ [ADP], [pyruvate] must exceed 360,333[PEP] for the reaction to proceed in reverse.
   c. Coupling coefficient = −2.
   d. Yes. The coupling coefficient for the cycle is −1.
   e. $K_{eq} = 0.724$.
   f. [PEP]/[pyruvate] = 724.
   g. Yes. Both reactions will be favorable as long as [PEP]/[pyruvate] falls in the range between 0.0000028 and 724.
2. Energy charge = 0.945; Phosphorylation potential = 1111 M^{-1}.
3. Eight m*M* ATP will last 1.12 sec. Since the equilibrium constant for creatine-P + ADP → Cr + ATP is 175, [ATP] must be less than 1750 [ADP] for the reaction creatine-P + ADP → Cr + ATP to proceed to the right when [Cr-P] = 40 m*M* and [Cr] = 4 m*M*.
4. From Equation (20:13), $\mathscr{E}(\text{NADP}^+/\text{NADPH}) = -0.350$V; $\mathscr{E}(\text{NAD}^+/\text{NADH}) = -0.281$V; Thus, $\Delta\mathscr{E} = 0.069$V, and ΔG = 13,316 J/mol. If a ATP "costs" 50 kJ/mol, this reaction can produce about 0.27 ATP equivalents at these concentrations of NAD^+, NADH, $NADP^+$, and NADPH.
5. Assume that the K_{eq} for ATP + AMP → 2 ADP = 1.2 (legend, Figure 25.5). When [ATP] = 7.2 m*M*, [ADP] = 0.737 m*M*, and [AMP] = 0.063 m*M*. When [ATP] decreases by 10% to 6.48 m*M*, [ADP] + [AMP] = 1.52 m*M*, and thus [ADP] = 1.30 m*M* and [AMP] = 0.22 m*M*. A 10% decrease in [ATP] has resulted in a 0.22/0.063 = 3.5-fold increase in [AMP].
6. *J*, the flux of F-6-P through the substrate cycle, at low [AMP] = 0.1; *J* at high [AMP] = 8.9. Therefore, the flux of F-6-P through the cycle has increased 89-fold.

Chapter 26

1. The oxidation number of N in nitrate is +5; in nitrite, +3; in NO, +2; in N_2O, +1; and in N_2, 0.

2. a. Nitrate assimilation requires 4 NADPH equivalents per NO_3^- reduced to NH_4^+; 4 NADPH have a metabolic value of 16 ATP.
   b. Nitrogen fixation requires 8 e^- (Equation [26.3]) and 16 ATP (Figure 26.6) per N_2 reduced. If 4 NADH provide the requisite 8 e^-, each NADH having a metabolic value of 3 ATP, then 28 ATP equivalents are consumed per N_2 reduced in biological nitrogen fixation (or 14 ATP equivalents per NH_4^+ formed).

3. a. [ATP] increase will favor adenylylation; the value of n will be greater than 6 ($n > 6$).
   b. [UTP] increase will favor deadenylylation; the value of n will be less than 6 ($n < 6$).
   c. An increase in the [αKG]/[Gln] ratio will favor deadenylylation; $n < 6$.
   d. $[P_i]$ decrease will favor adenylylation; $n > 6$.

4. Two ATP are consumed in the carbamoyl-P synthetase-I reaction, and 2 phosphoric anhydride bonds (equal to 2 ATP equivalents) are expended in the argininosuccinate synthetase reaction. Thus, 4 ATP equivalents are consumed in the urea cycle, as 1 urea and 1 fumarate are formed from 1 CO_2, 1 NH_3, and 1 aspartate.

5. Protein catabolism to generate carbon skeletons for energy production releases the amino groups of amino acids as excess nitrogen, which is excreted in the urine, principally as urea.

6. One ATP in reaction 1, one NADPH in reaction 2, one NADPH in reaction 11, and one succinyl-CoA in reaction 12 add up to 10 ATP equivalents. (The succinyl-CoA synthetase reaction of the citric acid cycle (Figure 19.12) fixes the metabolic value of succinyl-CoA versus succinate at 1 GTP [=1 ATP].)

7. From Figure 26.34: ^{14}C-labeled carbon atoms derived from ^{14}C-2 of PEP are shaded.

   COO^-, CH_2, O—C—COO^-, H, HO, H

8. 1. 2 aspartate → *transamination* → 2 oxaloacetate.
   2. 2 oxaloacetate + 2 GTP → *PEP carboxykinase* → 2 PEP + 2 CO_2 + 2 GDP.
   3. 2 PEP + 2 H_2O → *enolase* → 2 2-PG.
   4. 2 2-PG → *phosphoglyceromutase* → 2 3-PG.
   5. 2 3-PG + 2 ATP → *3-P glycerate kinase* → 2 1,3 bisPG + 2 ADP.
   6. 2 1,3 bisPG + 2 NADH + 2 H^+ → *G3P dehydrogenase* → 2 G3P + 2 NAD^+ + 2 P_i.
   7. 1 G3P → *triose-P isomerase* → 1 DHAP.
   8. G3P + DHAP → *aldolase* → Fructose-1,6-bisP.
   9. F-1,6-bisP + H_2O → *FBPase* → F6P + P_i.
   10. F6P → *phosphoglucoisomerase* → G6P.
   11. G6P + H_2O → *glucose phosphatase* → glucose P_i.

   Net: 2 aspartate + 2 GTP + 2 ATP + 2 NADH + 2 H^+ + 4 H_2O → glucose + 2 CO_2 + 2 GDP + 2 ADP + 4 P_i + 2 NAD^+

   (As a consequence of reaction [1], 2 α-keto acids [e.g., α-ketoglutarate] will receive amino groups to become 2 α-amino acids [e.g., glutamate].)

Chapter 27

1. See Figure 27.2 (purines) and Figure 27.15 (pyrimidines).

2. Assume ribose-5-P is available.
   Purine synthesis: Two ATP equivalents in the ribose-5-P pyrophosphokinase reaction, 1 in the GAR synthetase reaction, 1 in the FGAM synthetase reaction, 1 in the CAIR synthetase reaction, and 1 in the SACAIR synthetase reaction yields IMP, the precursor common to ATP and GTP. *Net:* 6 ATP equivalents.
   a. ATP: 1 GTP (an ATP equivalent) is consumed in converting IMP to AMP; 2 more ATP equivalents are needed to convert AMP to ATP. Overall, ATP synthesis from ribose-5-P onward requires 9 ATP equivalents.
   b. *GTP:* 2 high-energy phosphoric anhydride bonds from ATP, but 1 NADH is produced in converting IMP to GMP; 2 more ATP equivalents are needed to convert GMP to GTP. Overall, GTP synthesis from ribose-5-P onward requires 7 ATP equivalents. *Pyrimidine synthesis:* Starting from HCO_3^- and Gln, 2 ATP equivalents are consumed by CPS-II, and an NADH equivalent is produced in forming orotate. OMP synthesis from ribose-5-P plus orotate requires conversion of ribose-5-P to PRPP at a cost of 2 ATP equivalents. Thus, the net ATP investment in UMP synthesis is just 1 ATP.
   c. Formation of UTP from UMP requires 2 ATP equivalents. Net ATP equivalents in UTP biosynthesis = 3.
   d. CTP biosynthesis from UTP by CTP synthetase consumes 1 ATP equivalent. Overall ATP investment in CTP synthesis = 4 ATP equivalents.

3. a. See Figure 27.7.
   b. See Figure 27.19.
   c. See Figure 27.19.

4. a. Azaserine inhibits glutamine-dependent enzymes, as in steps 2 and 5 of IMP synthesis (glutamine:PRPP amidotransferase and FGAM synthetase), as well as GMP synthetase (step 2, Figure 27.6), and CTP synthetase (Figure 27.18).
   b. Methotrexate, an analog of folic acid, antagonizes THF-dependent processes, such as steps 4 and 10 (GAR transformylase and AICAR transformylase) in purine biosynthesis (Figure 27.3), and the thymidylate synthase reaction (Figure 27.30) of pyrimidine metabolism.
   c. Sulfonamides are analogs of *p*-aminobenzoic acid (PABA). Like methotrexate, sulfonamides antagonize THF formation. Thus, sulfonamides affect nucleotide biosynthesis at the same sites as methotrexate, but only in organisms such as prokaryotes that synthesize their THF from simple precursors such as PABA.
   d. Allopurinol is an inhibitor of xanthine oxidase (Figure 27.13).
   e. 5-Fluorouracil inhibits the thymidylate synthase reaction (Figure 27.30).

5. UDP, via conversion to dUDP (Figure 27.38), is ultimately a precursor to dTTP, which is essential to DNA synthesis.

6. See Figure 27.27

7. Ribose, as ribose-5-P, is released during nucleotide catabolism (as in Figure 27.9). Ribose-5-P is catabolized via the pentose phosphate pathway and glycolysis to form pyruvate, which enters the citric acid cycle. From Figure 21.49, note that 3 ribose-

5-P (rearranged to give 1 ribose-5-P and 2 xylulose-5-P) give a net consumption of 2 ATP and a net production of 10 ATP and 5 NADH (=15 ATP), when converted to 5 pyruvate. If each pyruvate is worth 15 ATP equivalents (as in a prokaryotic cell), the overall yield of ATP from 3 ribose-5-P is 75 ATP + 25 ATP − 2 ATP = 98 ATP. Net yield per ribose-5-P is thus 32.7 ATP equivalents.

8. Comparing Figures 27.3 and 26.38, note that AICAR (5-aminoimidazole-4-carboxamide ribonucleotide) is a common intermediate in both pathways. It is a product of step 5 of histidine biosynthesis (Figure 26.38) and step 9 of purine biosynthesis (Figure 27.3). Thus, formation of AICAR as a by-product of histidine biosynthesis from PRPP and ATP bypasses the first 9 steps in purine synthesis. However, cells require greater quantities of purine than of histidine, and these 9 reactions of purine synthesis are essential in satisfying cellular needs for purines.

Chapter 28

1. Cells of the microorganism are plated on complete medium (+ Arg), and then replica-plated (see Figure 8.7) on complete medium in which ornithine replaces the arginine supplement. Comparison of the complete medium (+ Arg) petri plates with those of the complete medium (+ ornithine) plates will reveal colonies which grow on Arg, but not Orn. The colonies can be picked from the "+ Arg" plates and characterized.

2. Consult Figure 28.10. The two possibilities for order of gene transfer are (a): *ilv, pyr E, mtl, xyl, pls B,* etc.; and (b) *rha, ile, arg E, met B, thi A,B,C,* etc.

3. For purposes of illustration, consider how the heteroduplex bacteriophage chromosome in Figure 28.13 might have arisen:

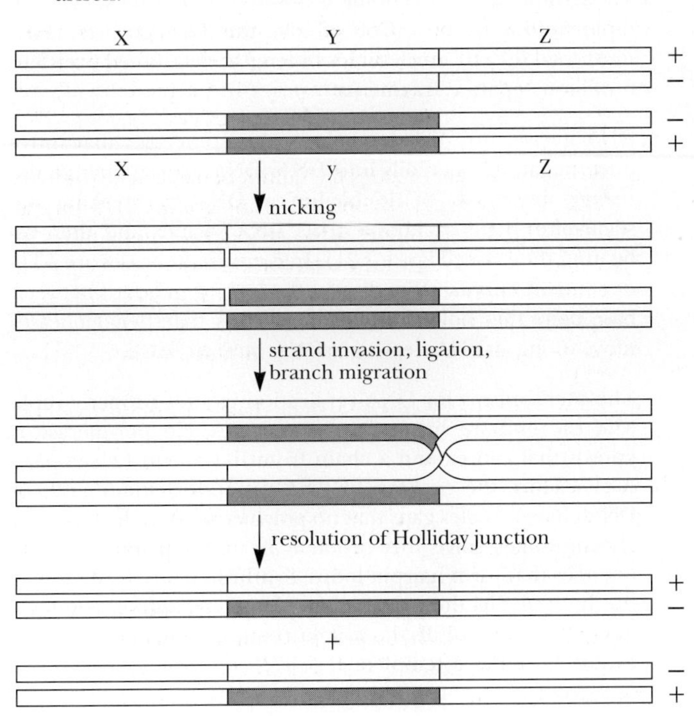

4. The invading ssDNA is shown between the "Watson" and "Crick" strands of homologous duplex DNA:

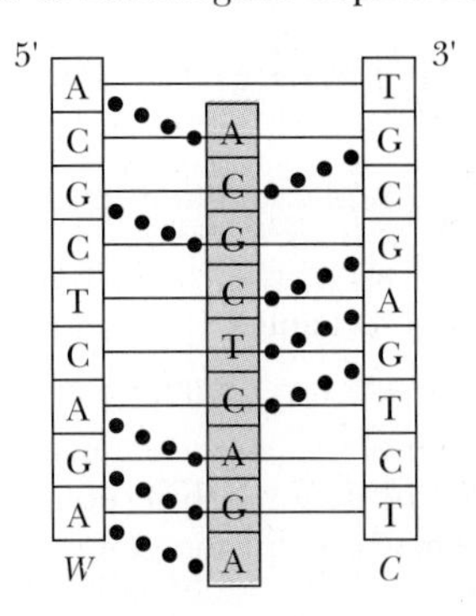

5. B-DNA normally has a helical twist of about 10 bp/turn, so that the rotation per residue (base pair), $\Delta\phi$, is 36°. If the DNA is unwound to 18.6 bp per turn, $\Delta\phi$ becomes 19.3°. Thus, the change in $\Delta\phi$ is 16.7°.

6.

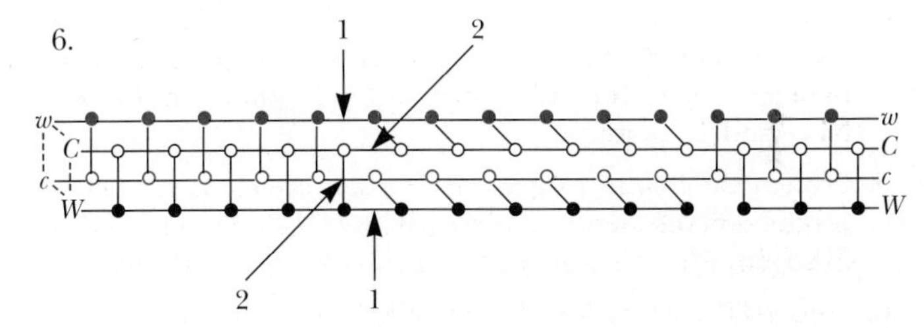

The gap at the right denotes the initial cleavage, in this case of the *W* and *w* strands. Cleavage at the arrows labeled **1** and ligation of like strands (*W* with *w* and *C* with *c*) will yield patch recombinants; cleavage at the arrows labeled **2** and ligation of like strands gives splice recombinants.

7. Assuming only one recombination event that joins V_κ and J_κ *within* codons 94 to 97, we have:

GTT.CAT.CTT.CGG = Val.His.Leu.Arg
GTT.CAT.CTT.CTG = Val.His.Leu.Leu
GTT.CAT.CTT.TTG = Val.His.Leu.Leu
GTT.CAT.CTC.TTG = Val.His.Leu.Leu
GTT.CAT.CGC.TTG = Val.His.Arg.Leu
GTT.CAT.AGC.TTG = Val.His.Ser.Leu
GTT.CAA.AGC.TTG = Val.Gln.Ser.Leu
GTT.CCA.AGC.TTG = Val.Pro.Ser.Leu
GTT.GCA.AGC.TTG = Val.Ala.Ser.Leu

Note that the nine different codon sequences yield only seven different amino acid sequences.

8. The DNA is: GCTA
CGAT

a. HNO_2 causes deamination of C and A. Deamination of C yields U, which pairs the way T does, giving:

GTTA and ACTA and (rarely) ATTA
CAAT TGAT TAAT

Deamination of A yields I, which pairs as G does, giving:

GCTG and GCCA and (rarely) GCCG
CGAC CGGT CGGC

b. Bromouracil usually replaces T, and pairs the way C does:

GCCA and GCTG and (rarely) GCCG
CGGT CGAC CGGC

Less often, bromouracil replaces C, and mimics T in its base-pairing:

GTTA and ACAT and (rarely) ATTA
CAAT TGTA TAAT

c. 2-Aminopurine replaces A and normally base-pairs with T, but may also pair with C:

GCTG and CGGT and (rarely) GCCG
CGAC GCCA CGGC

9. Transposons are mobile genetic elements that can move from place to place in the genome. Insertion of a transposon within or near a gene may disrupt the gene or inactivate its expression.

10.

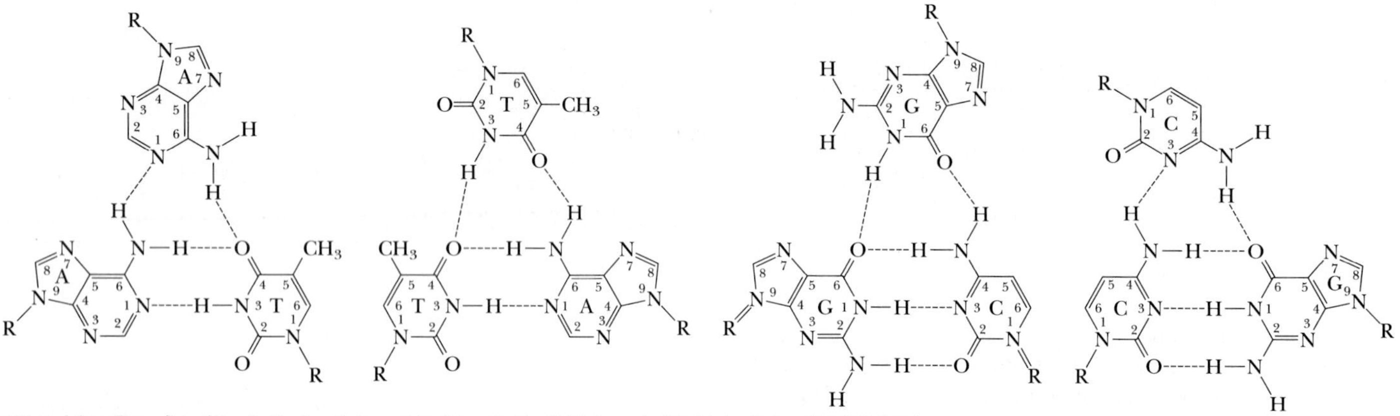

(Adapted from Figure 7 in Chiu, S. K., Rao, J. B., and Radding, C. M., 1993. Journal of Molecular Biology 229:328–343.)

Chapter 29

1. Since DNA replication is semi-conservative, the density of the DNA will be (1.724 + 1.710)/2 = 1.717 g/mL. If replication is dispersive, the density of the DNA will also be 1.717 g/mL. To distinguish between these possibilities, heat the dsDNA obtained after one generation in ^{15}N to 100°C to separate the strands and then examine the strands by density gradient ultracentrifugation. If replication is indeed semi-conservative, two bands of ssDNA will be observed, the "heavy" one containing ^{15}N atoms, the "light" one containing ^{14}N atoms. If replication is dispersive, only a single band of intermediate density would be observed.

2. The 5′-exonuclease activity of DNA polymerase I removes mispaired segments of DNA sequence that lie in the path of the advancing polymerase. Its biological role is to remove mispaired bases during DNA repair. The 3′-exonuclease activity acts as a proofreader to see whether the base just added by the polymerase activity is properly base-paired with the template. If not (that is, if it is an improper base with respect to the template), the 3′-exonuclease removes it, and the polymerase activity can try once more to insert the proper base. An *E. coli* strain lacking DNA polymerase I 3′-exonuclease activity would show a high rate of spontaneous mutation.

3. Assume that the polymerization rate achieved by each half of the DNA polymerase III homodimer is 750 nucleotides per sec. The entire *E. coli* genome consists of 4.72 × 10^6 bp. At a rate of 750 bp/sec per DNA pol III homodimer (one at each replication fork), DNA replication would take almost 3146.7 sec (52.4 min; 0.87 hr). When *E. coli* is dividing at a rate of once every 20 min, *E. coli* cells must be replicating DNA at the rate of 4.72 × 10^6 bp per 20 min (2.36 × 10^5 bp/min or 3933 bp/sec). To achieve this rate of replication would require initiation of DNA replication once every 20 min at *ori*, and a minimum of 3933/1500 = 2.62 replication bubbles per *E. coli* chromosome, or 5.24 replication forks.

4. If there are only 10 molecules of DNA polymerase III (as DNA pol III homodimers) per *E. coli* cell, and *E. coli* growing at its maximum rate has about 5 replication forks per chromosome, it seems likely that DNA polymerase III availability is sufficient to sustain growth at this rate.

5. Okazaki fragments are 1000–2000 nucleotides in length. Since DNA replication in *E. coli* generates Okazaki fragments whose total length must be 4.72 × 10^6 nucleotides, a total of 2300–4700 Okazaki fragments must be synthesized. Consider in comparison a human cell carrying out DNA replication. The haploid human genome is 3 × 10^9 bp, but most cells are diploid (6 × 10^9 bp). Collectively, the Okazaki fragments must total 6 × 10^9 nucleotides in length, distributed over 3 to 6 million separate fragments.

6. DNA gyrases are ATP-dependent topoisomerases that introduce negative supercoils into DNA. DNA gyrases change the *linking number, L,* of double-helical DNA by breaking the sugar-phosphate backbone of its DNA strands and then religating them (see Figure 7.21). In contrast, helicases are ATP-dependent enzymes that disrupt the hydrogen bonds between base pairs that hold double-helical DNA together. Helicases move along dsDNA, leaving ssDNA in their wake.

7. The mechanisms for DNA replication cannot properly replicate the ends of linear DNA molecules. No polymerase is known that can extend a chain from the 5′-end (a; see page A-21). Thus, the excision of RNA primers at the 5′-ends of DNA chains creates gaps that no polymerase can fill. (b) Some chromosomes solve this problem by using palindromic sequences that form hairpin loops. Synthesis from the 3′-end of the hairpin fills the gap; endonuclease cleavage generates a new 3′-end that fulfills the primer requirement, and the palindrome becomes a template.

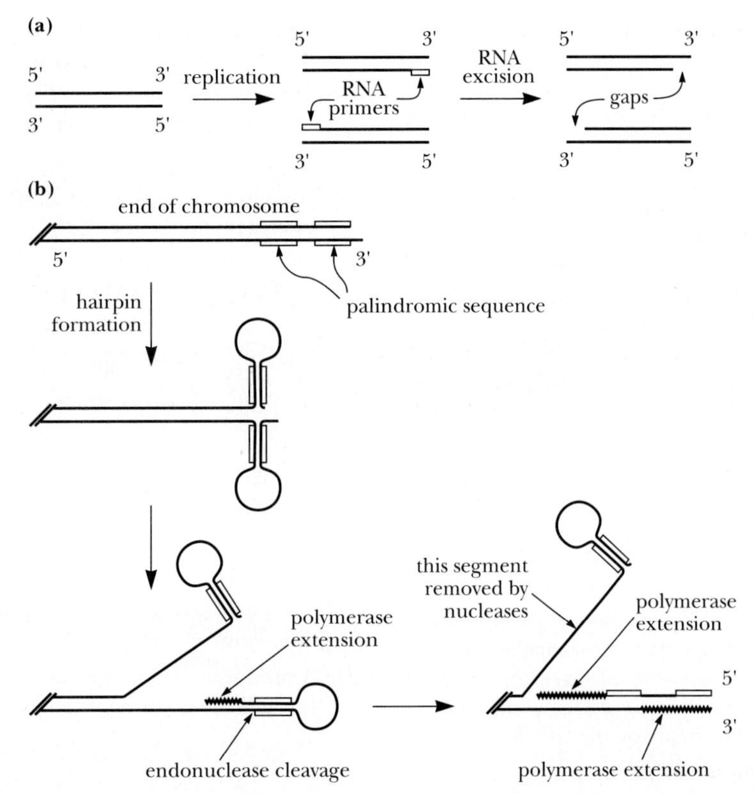

8. The mismatch repair system of *E. coli* relies on DNA methylation to distinguish which DNA strand is "correct" and which is "mismatched"; the unmethylated strand (the newly synthesized one that has not had sufficient time to become methylated is, by definition, the mismatched one). Homologous recombination involves DNA duplexes at similar stages of methylation, neither of which would be interpreted as "mismatched," and thus the mismatch repair system does not act.

Chapter 30

1. mRNA:
5′-AGAUCCGUAUGGCGAUCUCGACGAAGACUCCUAGGAAUCC . . .
3′-TCTAGGCATACCGCTAGAGCTGCTTCTGAGGATCCCTTAGG . . . 5′ = DNA sense strand

   The first AUG codon is underlined:

   5′-AGAUCCGU<u>AUG</u>GCGAUCUC-
   GACGAAGACUCCUAGGGAAUCC. . .

   Reading from the AUG codon, the amino acid sequence of the protein is:

   (N-term)Met-Ala-Ile-Ser-Thr-Lys-Thr-Pro-Arg-Glu-Ser-. . .

2. See Figure 30.2 for a summary of the events in transcription initiation by *E. coli* RNA polymerase. A gene must have a promoter region for proper recognition and transcription by RNA polymerase. The promoter region (where RNA polymerase binds) typically consists of 40 bp to the 5′-side of the transcription start site. The promoter is characterized by two consensus sequence elements: a hexameric TTGACA consensus element in the −35 region and a Pribnow box (consensus sequence TATAAT) in the −10 region (see Figure 30.3).

3. The fact that the initiator site on RNA polymerase for nucleotide binding has a higher K_m for NTP than the elongation site ensures that initiation of mRNA synthesis will not begin unless the available concentration of NTP is sufficient for complete synthesis of an mRNA.

4. In contrast to prokaryotes (which have only a single kind of RNA polymerase), eukaryotes possess three distinct RNA polymerases—I, II, and III—acting on three distinct sets of genes (see Section 30.2). The subunit organization of these eukaryotic RNA polymerases is more complex than that of their prokaryotic counterparts. The three sets of genes are recognized by their respective RNA polymerases because they possess distinctive categories of promoters. In turn, all three polymerases interact with their promoters via particular transcription factors that recognize promoter elements within their respective genes. Protein-coding eukaryotic genes, in analogy with prokaryotic genes, have two consensus sequence elements within their promoters, a TATA box (consensus sequence TATAAA) located in the −25 region, and *Inr*, an initiator element that encompasses the transcription start site. In addition, eukaryotic promoters often contain additional short, conserved sequence modules such as enhancers and response elements for appropriate regulation of transcription. Transcription termination differs as well in prokaryotes and eukaryotes. In prokaryotes, two transcription basic termination mechanisms occur: *rho* protein-dependent termination and DNA-encoded termination sites (see p. 970). Transcription in eukaryotes tends to be imprecise, occurring downstream from consensus AAUAAA sequences known as poly$(A)^+$ addition sites. An important aspect of eukaryotic transcription is post-transcriptional processing of mRNA (Section 30.6).

5. Given [repressor]/[repressor:DNA] = $1/K_A$[DNA] and [DNA] = 0.78×10^{-2} *M*, if [repressor]/[repressor:DNA] = 0.10, then $K_A = 1282\ M^{-1}$.

6. Because transcription occurs in the nucleus in eukaryotic cells and translation by ribosomes occurs in the cytosol, transcription and translation are not coupled in eukaryotes as they are in prokaryotes. Thus, transcription attenuation is not a feasible mechanism for regulation of transcription in eukaryotes.

7. When DNA-binding proteins "read" specific base sequences in DNA, they do so by recognizing a specific matrix of H-bond donors and acceptors displayed within the major groove of B-DNA by the edges of the bases. When DNA-binding proteins recognize a specific DNA region by "indirect readout," the protein is discerning local conformational variations in the cylindrical surface of the DNA double helix. These conformational variations are a consequence of unique sequence information contributed by the base pairs that make up this region of the DNA.

8. DNA looping as a mechanism of transcriptional regulation is universal in both prokaryotic and eukaryotic cells because of the fundamental nature of DNA. The information inherent in the nucleotide sequence of DNA is presented in essentially a one-dimensional arrangement, so the information content of a short length of DNA is limited. To convene additional information (such as a regulatory signal) at a particular site, the DNA must loop. Looping increases the information conveyed by the one-dimensional polymer by bringing additional sequences within a three-dimensional volume of space.

9. If cells contain a variety of proteins that share a leucine zipper dimerization motif but differ in their DNA recognition helices, heterodimeric *bZIP* proteins can form that will contain two different DNA contact *basic regions* (consider this possibility in light of Figure 30.43). These heterodimers are no longer restricted to binding sites on DNA that are dyad-symmetric; the repertoire of genes with which they can interact will thus be dramatically increased.

10. Exon 17 of the fast skeletal muscle troponin T gene is one of the *mutually exclusive* exons (see Figure 30.54). Deletion of this exon would cut by half the alternative splicing possibilities, so that only 32 different mature mRNAs would be available from this gene. If exon 7, a *combinatorial* exon, were duplicated, it would likely occur as a tandem duplication. The different mature mRNAs could contain 0, 1, or 2 copies of exon 7. The 32 combinatorial possibilities shown in the center of Figure 30.54 represent those with 0 or 1 copy of exon 7. Those with 2 copies are 16 in number: *456778; 56778; 46778; 6778; 45778; 4778; 5778; 778; 4577; 577; 5677; 45677; 4677; 477; 677;* and *77.* Given the *mutually exclusive* exons 17 and 18, 96 different mature mRNAs would be possible.

Chapter 31

1. The cDNA sequence as presented:

   CAATACGAAGCAATCCCGCGACTAGACCTTAAC. . .

   represents 6 potential reading frames, the three inherent in the sequence as written and the three implicit in the complementary DNA strand, which written $5' \rightarrow 3'$ is:

   . . .GTTAAGGTCTAGTCGCGGGATTGCTTCGTATTG

   Two of the three reading frames of the cDNA sequence given contain stop codons. The third reading frame is a so-called open reading frame (a stretch of coding sequence devoid of stop codons):

   CAATACGAAGCAATCCCGCGACTAGACCTTAAC. . .
   (alternate codons underlined)

   The amino acid sequence it encodes is:

   Gln-Tyr-Glu-Ala-Ile-Pro-Arg-Leu-Asp-Leu-Asn. . .

   Two of the three reading frames of the complementary DNA sequence also contain stop codons. The third may be an open reading frame:

   . . .GTTAAGGTCTAGTCGCGGGATTGCTTCGTATTG
   (alternate codons underlined)

   An unambiguous conclusion about the partial amino acid sequence of this cDNA cannot be reached.

2. A random (AG) copolymer would contain varying amounts of the following codons

   AAA AAG AGA GAA AGG GAG GGA GGG, codons for Lys Lys Arg Glu Arg Glu Gly Gly, respectively. Therefore, the random (AG) copolymer would direct the synthesis of a polypeptide consisting of Lys, Arg, Glu, and Gly. Using the analysis demonstrated in Table 31.1, the relative abundances of these amino acids in the polypeptide would be: Lys = 120; Arg = 24; Glu = 24; Gly = 4.4
   (Normalized to Lys = 100: Arg = 20; Glu = 20; Gly = 3.67)

3. See Figure 31.6.

4. The fact that tRNA molecules and amino acids display no obvious chemical affinity nor any stereochemical similarity to one another led Francis Crick to suggest the adaptor role for tRNA in the translation of genetic information. The experiment by von Ehrenstein in which Cys-$tRNA^{Cys}$ was treated with Raney nickel, reducing the cysteinyl residue on the $tRNA^{Cys}$ to alanine, and the demonstration that Ala was incorporated into positions in hemoglobin normally occupied by Cys when the Ala-$tRNA^{Cys}$ was used in an *in vitro* hemoglobin-synthesizing system confirmed Crick's hypothesis (Figure 31.7).

5. Review the information on pp. 1027–1039 and Figure 31.8, noting in particular the two levels of specificity exhibited by aminoacyl-tRNA synthetases (1: at the level of $ATP \rightleftharpoons PP_i$ exchange / aminoacyl adenylate synthesis in the presence of amino acid and the absence of tRNA; and 2: at the level of loading the aminoacyl group on an acceptor tRNA).

6. Base pairs are drawn such that the B-DNA major groove is at the top (see figure below).

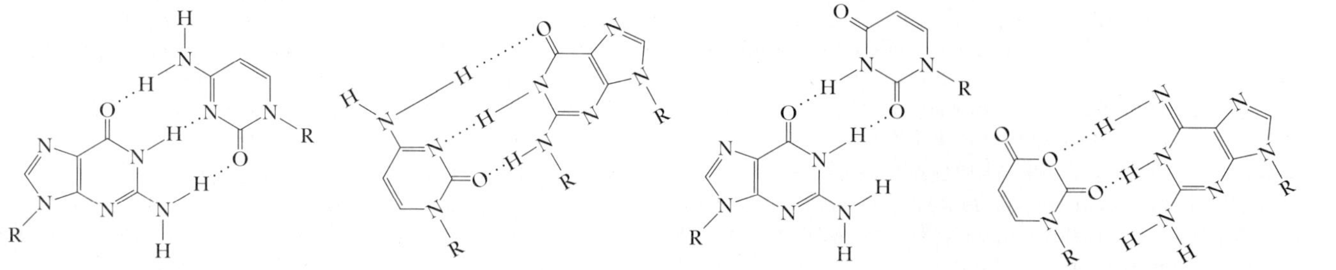

G:C C:G G:U U:G

7. The wobble rules state that a first-base anticodon U can recognize either an A or a G in the codon third-base position; first-base anticodon G can recognize either U or C in the codon third-base position; and first-base anticodon I can recognize either U, C, or A in the codon third-base position. Thus, codons with third-base A or G, which are degenerate for a particular amino acid (Table 31.3 reveals that all codons with third-base purines are degenerate, except those for Met and Trp) could be served by single tRNA species with first-base anticodon U. More emphatically, codons with third-base pyrimidines (C or U) are always degenerate and could be served by single tRNA species with first-base anticodon G. Wobble involving first-base anticodon I further minimizes the number of tRNAs needed to translate the 61 sense codons. Wobble tends to accelerate the rate of translation because the non-canonical base pairs formed between bases in the third position of codons and bases occupying the first-base wobble position of anticodons are less stable. As a consequence, the codon: anticodon interaction is more transient.

8. The stop codons are UAA, UAG, and UGA.
Sense codons that are a single-base change from UAA include: CAA (Pro), AAA (Lys), GAA (Glu), UUA (Leu), UCA (Ser), UAU (Tyr), and UAC (Tyr).
Sense codons that are a single-base change from UAG include: CAG (Gln), AAG (Lys), GAG (Glu), UCG (Ser), UUG (Leu), UGG (Trp), UAU (Tyr), and UAC (Tyr).
Sense codons that are a single-base change from UGA include: CGA (Arg), AGA (Arg), GGA (Gly), UUA (Leu), UCA (Ser), UGU (Cys), UGC (Cys), and UGG (Trp).
That is, 19 of the 61 sense codons are just a single base change from a nonsense codon.

9. The list of amino acids in problem 8 is a good place to start in considering the answer to this question. Those amino acid codons in which the codon base (the wobble position) is but a single base change from a nonsense codon are the more likely among this list, because pairing is less stringent at this position. These include UAU (Tyr) and UAC (Tyr) for nonsense codons UAA and UAG; and UGU (Cys), UGC (Cys), and UGG (Trp) for nonsense codon UGA.

Chapter 32

1. The more obvious answer to this question is that eukaryotic ribosomes are larger, more complex, and hence slower than prokaryotic ribosomes. In addition, initiation of translation requires a greater number of initiation factors in eukaryotes than in prokaryotes. It is also worth noting that eukaryotic cells, in contrast to prokaryotic cells, are typically under less selective pressure to multiply rapidly.

2. Larger, more complex ribosomes offer greater advantages in terms of their potential to respond to the input of regulatory influences and to enter into interactions with subcellular structures. Their larger size may slow the rate of translation, which in some instances may be a disadvantage.

3. The universal organization of ribosomes as two-subunit structures in all cells—archae, eubacteria, and eukaryotes—suggests that such an organization is fundamental to ribosome function. One possibility is that translocation along mRNA, aminoacyl-tRNA binding, peptidyl transfer, and/or tRNA release are processes that require repetitious uncoupling of physical interactions between the large and small ribosomal subunits.

4. Prokaryotic cells rely on *N*-formyl-Met-$tRNA_f^{Met}$ to initiate protein synthesis. The $tRNA_f^{Met}$ molecule has a number of distinctive features, not found in noninitiator tRNAs, that earmark it for its role in translation initiation (see Figure 32.8). Further, *N*-formyl-Met-$tRNA_f^{Met}$ interacts only with initiator codons (AUG or, less commonly, GUG). A second $tRNA^{Met}$, designated $tRNA_m^{Met}$, serves to deliver methionyl residues as directed by internal AUG codons. Both $tRNA_f^{Met}$ and $tRNA_m^{Met}$ are loaded with methionine by the same methionyl-tRNA synthetase, and UAG is the Met codon, both in initiation and elongation. AUG initiation codons are distinctive in that they are situated about 10 nucleotides downstream from the Shine-Dalgarno sequence at the 5′-end of mRNAs; this sequence determines the translation start site (see Figure 32.9). Eukaryotic cells also have two $tRNA^{Met}$ species, one of which is a unique $tRNA_i^{Met}$ that functions only in translation initiation. Eukaryotic mRNAs lack a counterpart to the prokaryotic Shine-Dalgarno sequence; apparently the eukaryotic small ribosomal subunit binds to the 5′-end of a eukaryotic mRNA and scans along it until it encounters an AUG codon. This first AUG codon defines the eukaryotic translation start site.

5. The Shine-Dalgarno sequence is a purine-rich sequence element near the 5′-end of prokaryotic mRNAs (Figure 32.9). It base-pairs with a complementary pyrimidine-rich region near the 3′-end of 16S rRNA, the rRNA component of the prokaryotic 30S ribosomal subunit. Base pairing between the Shine-Dalgarno sequence and 16S rRNA brings the translation start site of the mRNA into the A site on the prokaryotic ribosome. Since the nucleotide sequence of the Shine-Dalgarno element varies somewhat from mRNA to mRNA, while the pyrimidine-rich Shine-Dalgarno-binding sequence of 16S rRNA is invariant, different mRNAs vary in their affinity for binding to 30S ribosomal subunits. Those that bind with highest affinity are more likely to be translated.

6. The most apt account is (b). Polypeptide chains typically contain hundreds of amino acid residues. Such chains, attached as a peptidyl group to a tRNA in the P site, would show significant inertia to movement, compared with an aminoacyl-tRNA in the A site.

7. Elongation factors EF-Tu and EF-Ts interact in a manner analogous to the GTP-binding G proteins of signal transduction pathways (see p. 682). EF-Tu binds GTP and in the GTP-bound form delivers an aminoacyl-tRNA to the ribosome, whereupon the GTP is hydrolyzed to yield EF-Tu:GDP and P_i. EF-Ts mediates an exchange of the bound GDP on EF-Tu:GDP with free GTP, regenerating EF-Tu:GTP for another cycle of aminoacyl-tRNA delivery. The α subunit of the heterotrimeric GTP-binding G proteins also binds GTP. G_α has an intrinsic GTPase activity and the GTP is eventually hydrolyzed to form G_α:GDP and P_i. The β and γ subunits of G proteins facilitate the exchange of bound GDP on G_α for GTP, in analogy with EF-Ts for EF-Tu. The amino acid sequences of EF-Tu and G proteins reveal that they share a common ancestry.

8. a. Streptomycin causes misreading of codons in mRNA, so that the wrong amino acids are inserted in polypeptide chains. As a result, the polypeptides made in the presence of streptomycin, though of the proper size, are nonfunctional.

b. Puromycin is a structural analog of aminoacyl-tRNA. It can bind at the A site of ribosomes and accept growing peptidyl chains from peptidyl-tRNAs in the P site. Peptidyl transfer to puromycin aborts further elongation and the products are a series of truncated polypeptides having puromycin groups at their C-termini.

9. Kasugamycin, streptomycin, tetracycline, sparsomycin, chloramphenicol, erythromycin, and thiostrepton all have potential as antibiotics because they specifically inhibit prokaryotic protein synthesis. However, several of these substances have side effects that render them undesirable for clinical applications. Erythromycin, tetracycline, and, less commonly, chloramphenicol, are prescribed for the treatment of diseases caused by bacterial infection.

10. Human rhodanese has 296 amino acid residues. Its synthesis would require the involvement of 5 ATP equivalents per residue (p. 1055) or 1480 ATP equivalents. Folding of rhodanese by the Hsp60 α_{14} complex consumes another 130 ATP equivalents (see p. 1066). The total number of ATP equivalents expended in the synthesis and folding of rhodanese is approximately 1610.

11. Since a single proteolytic nick in a protein can doom it to total degradation, nicked proteins are clearly not tolerated by cells and are quickly degraded. Cells are virtually devoid of partially degraded protein fragments, which would be the obvious intermediates in protein degradation. The absence of such intermediates and the rapid disappearance of nicked proteins from cells indicates that protein degradation is a rigorously selective, efficient cellular process.

Chapter 33

1.

(a) C A G C G C G T A (vertical axis); C A G C T G C G T A (horizontal axis)

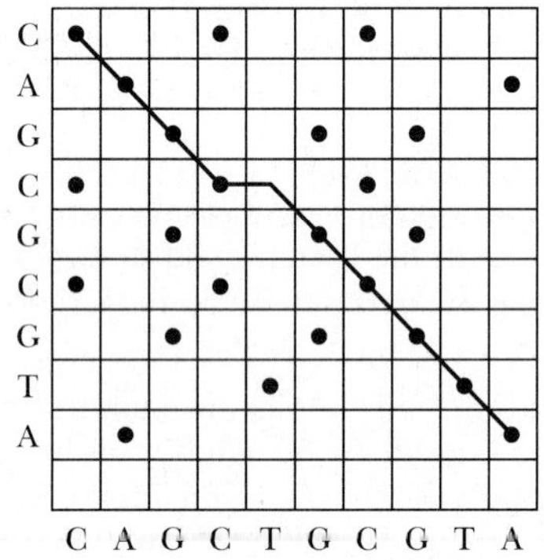

(b) C A T T G C G T C (vertical axis); C A G C T G C G T A (horizontal axis)

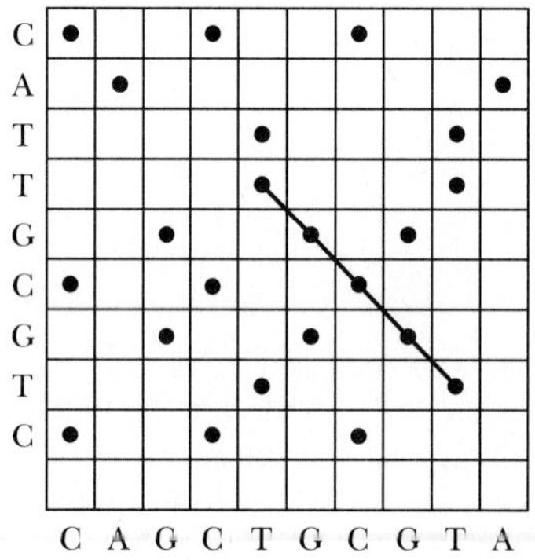

2. Three possible alignments might be:

a. GTTACTGAAG b. GTTACTGAAG c. GTTACTG–AAG
GT– –CGGGAG GTCG–GG–AG GT– –CGGG–AG

Applying **D = y + wz** for w = 1   for w = 2

	for w = 1	for w = 2
a.	**D** = 2 + 1(2) = 4	**D** = 2 + 2(2) = 6
b.	**D** = 3 + 1(2) = 5	**D** = 3 + 2(2) = 7
c.	**D** = 1 + 1(4) = 5	**D** = 1 + 2(4) = 9

For a gap penalty, **w**, of 1 or 2, sequence alignment *a* gives the lowest values for **D**.

3. a. Substitution of Val for Leu is a conservative substitution and is rather unlikely to affect fitness.

b. Asp for Arg replaces a positively charged side chain with a negatively charged one and is likely to have some effect on fitness.

c. Replacing Lys with Phe substitutes a positively charged side chain with a large hydrophobic side chain. Such a substitution is likely to affect fitness.

4. For any given codon, the base at each position in it may be replaced by any of the other three bases. $3 \times 3 = 9$.

GCA = Ala

ACA = Thr	GUA = Val	GCG = Ala
UCA = Ser	GAA = Glu	GCU = Ala
CCA = Pro	GGA = Gly	GCC = Ala

Three single-base substitutions in GCA are synonymous: GCG, GCU, and GCC. Six are nonsynonymous, one of which (GUA = Val) is conservative.

AAG = Lys

CAG = Gln	AGG = Arg	AAA = Lys
GAG = Glu	ACG = Thr	AAC = Asn
UAG = stop	AUG = Met	AAU = Asn

One single-base substitution in AAG is synonymous; AAA = Lys. Of the eight other possibilities, one is a nonsense codon, and 7 are nonsynonymous substitutions, one of which is conservative (AGG = Arg).

5. The gene encoding the β-globin chain (146 residues) experiences a rate of nonsynonymous substitutions per site per year equal to 0.8×10^{-9}. The rate of nonsynonymous substitutions in the β-globin chain gene is thus $146(0.8 \times 10^{-9}) = 116.8 \times 10^{-9}$ nonsynonymous substitutions per β-globin chain gene per year. The human population (5.8×10^{9} people and growing) has a birth rate of 30 per 1000 persons per year. Thus, $(30/1000)(5.8 \times 10^{9}) = 174$ million births per year. $(174 \times 10^{6})(116.8 \times 10^{-9}) = 20$ individuals born per year with nonsynonymous substitutions in their β-globin chains.

6. If a synonymous substitution occurs such that a frequently used codon (for example, UUG in yeast, Figure 33.4) is replaced by an infrequently used codon (UUA in yeast), the level of protein expression due to this codon change may be affected, and synonymous substitution may not be selectively neutral. On the other hand, "wobble" (Table 31.6) might balance some of the differences due to codon usage.

7. Exon duplication gives rise to a second protein-coding segment that is free to accumulate genetic differences and evolve, since the original copy of the exon is still present to satisfy the biological role of the exon. Duplication of a random segment of the genome implies that only by chance will the segment encode a gene product or even be expressed. Such random duplication is of lesser consequence to evolution than exon duplication.

8. DNA polymerase, RNA polymerase, aminoacyl-tRNA synthetases, elongation factors EF-Tu and EF-G (see legend, Figure 33.17), and the enzymes of glycolysis, to name just a few of the more obvious possibilities.

Index

B = box; **D** = definition; **F** = figure; **G** = molecular graphic; **S** = structure; **T** = table

Abeles, Robert, 430
ABO blood group, 303
Absolute zero, temperature, 516
Aburamycin C, 325**F**
Abzyme 381**D**, 385**F**
ACAT, 789, 791**F**
Acceptor stem, tRNA, 238, 238**F**, 239**F**
Acceptor (A) site (aminoacyl-tRNA binding), 1047**F**
Accessory light-harvesting pigments, 703
Acetals, 321, 322**F**
Acetaminophen, 784**F**
Acetate units, fatty acid synthesis and, 760
Acetic acid, titration of, 47**F**
Acetic anhydride, 115**F**
Acetoacetate, degradation and, 868, 868**F**, 869, 869**F**
Acetoacetyl-CoA, 541
  in lysine degradation, 868, 868**F**
Acetoacetyl-CoA thiolase, 754
Acetohydroxy acid isomeroreductase
  feedback inhibition of, 852**F**
  in Val and Ile synthesis, 852**F**
α-Aceto-α-hydroxybutyrate, 852**F**
α-Acetolactate, 852**F**
Acetolactate synthase, 464, 465**F**, 467**F**
Acetyl phosphate, hydrolysis of, 521, 525**T**, 533**F**
Acetyl transacylase, 762
Acetyl transferase, 762, 766**F**
Acetyl-CoA (acetyl-Coenzyme A), 554, 602**F**, 604**T**
  biotin-dependent carboxylation of, 493**T**
  hydrolysis of, 526**T**, 537**F**
  sources of, for fatty acid synthesis, 759
Acetyl-CoA acetyltransferase, 754
Acetyl-CoA:1-alkyl-2-lysoglycerophosphocholine transferase, 780**F**
Acetyl-CoA carboxylase, 356**T**, 494**F**, 760**F**, 761, 762**F**
  regulation of by hormones, 772**F**
Acetyl-CoA synthetase, 607**F**
Acetyl-d-neuraminic acid, 321
Acetylchloramphenicol, 270**F**
Acetylcholinesterase, 365**T**, 366**T**, 436
Acetyldihydrolipoamide, 497
N-Acetylgalactosaminotransferase, 782**F**
N-Acetylglucosamine, 448**F**
N-Acetylglutamate, ornithine biosynthesis and, 841, 842**S**
N-Acetylglutamate kinase, 841, 842**F**
N-Acetylglutamate-5-semialdehyde dehydrogenase, 842**F**
Acetylhydrolase, 780**F**
Acetylimidazole, 69**S**
Acetyllysine (N-), 60, 61**S**
N-Acetylneuraminic acid, 287, 448**F**
N-Acetylornithine, ornithine biosynthesis and, 841, 842**F**
N-Acetylornithine deacylase, 842**F**
N-Acetylornithine δ-aminotransferase, 842**F**
Acetylpepstatin, 445, 445**F**
Acetylphenylalanine methyl ester, 438**S**
Acetylsalicylate, 784**F**, 785
O-Acetylserine sufhydrylase, in cysteine synthesis, 854, 855**F**
O-Acetylserine, in cysteine biosynthesis, 854, 855**F**
Aconitase, 602**F**, 604**T**, 605, 606**F**, 606**G**, 607**F**
ACP, 761
Acridine orange, 226**S**, 932
Acrylonitrile, 68**S**, 107**F**
ACTH, 732, 799
Actin 27**T**, 95**T**, 97
Actin gene, yeast, intron of, 1002
Actin thin filaments, in muscle contraction, 821
Actinomycin D, 226**S**
Activation, free energy of, ($\Delta G^{\ddagger}$), 359**D**
Activator gene, in maize, 921
Activator protein, 561**F**
Activity, specific 104
Acyl group activation, by CoA, 478
Acyl carrier proteins, 758, 761
Acyl adenylate intermediate, in acyl-CoA synthetase reaction, 737
Acyl phosphate, 524
Acyl carrier protein, 478
Acyl-CoA dehydrogenase, 476**T**, 740**F**, 741**F**, 741**S**, 751**F**
Acyl-CoA ligase, 737
Acyl-CoA oxidase, 750, 752**F**
Acyl-CoA synthetase, 770, 737**F**, 752**F**
  mechanism of, 738
Acyl-enzyme intermediate, in chymotrypsin, 438, 439**F**
Acyl-group carrier, lipoic acid as, 496
Acyl-group transfer, 496
Acyl-malonyl-ACP condensing enzyme, 763
Acyl-CoA:cholesterol acyltransferase, 789, 791**F**
Acylcarnitine, transport across inner mitochondrial membrane, 738
1-Acyldihydroxyacetone phosphate, 779**F**
Acyldihydroxyacetone phosphate reductase, 773, 774**F**
1-Acyldihydroxyacetone phosphate synthase, 779**F**
N-Acylsphinganine, 781**F**
Acyltransferases, 355**T**
ADA (adenosine deaminase), deficiency, 881, 882, 883**F**
dATP accumulation and, 883**F**
Adapter molecules, tRNAs as, 1017
Adenine, 182**S**
  imino analog of, 929**S**
Adenine phosphoribosyltransferase, in purine salvage, 879, 880**F**
Adenosine, 184**S**, 187**B**
Adenosine deaminase, 430, 431
  in purine catabolism, 881, 882**F**, 883**F**
Adenosine 5′-diphosphate (see also ADP), 186**S**
  hydrolysis of, 526**T**
Adenosine 5′-monophosphate (*see also* AMP), 185**S**, 186
  hydrolysis of, 527**T**
Adenosine 5′-phosphosulfate, in sulfate assimilation, 856**F**
Adenosine 5′-phosphosulfate-3′-phosphokinase, in SO_4^{2-} assimilation, 856**F**
Adenosine 5′-triphosphate (see also ATP), 186**S**, 320**F**
  hydrolysis of, 525**T**, 526**T**
Adenovirus, 27**F**, 100**T**
Adenovirus late promoter, 974**F**
Adenylate kinase, 816, 816**F**
  in ADP synthesis, 879
Adenylate system, energy storage and, 816
Adenylic acid (see also AMP), 185**S**, 186
Adenylosuccinase
  in AMP biosynthesis, 876, 877**F**
  in IMP biosynthesis, 874**F**, 875
  in the purine nucleoside cycle, 882, 883**F**
Adenylosuccinate lyase, in IMP biosynthesis, 874**F**, 875
Adenylosuccinate synthetase, 876, 877
  in the purine nucleoside cycle, 883**F**

Adenylyl cyclase, 682, 683**F**
E. coli and catabolite repression and, 980
Adenylyl transferase, glutamine synthetase and, 837, 837**F**
6-Adenylyl-α-aminoadipatate, in lysine biosynthesis, 846**F**
Adenylylation of *E. coli* glutamine synthetase, 837, 837**F**
Adhesion molecules, 308
Adipocyte(s), 732**F**
adipose tissue metabolism and, 822
Adipose tissue, energy metabolism in, 819**T,** 822
ADP, 186**S,** 574**T**
as a substrate of ribonucleotide reductase, 891**F,** 894, 894**F**
synthesis from AMP, 878, 879
ADP-ribose, 538**S**
ADP-ribosylation, 470
of nitrogenase reductase, 830**F**
ADR1, (zinc finger protein), 996**T**
Adrenaline, 61, 62**S,** 561, 732
Adrenocorticotropic hormone, 732, 799
chemical synthesis of, 125
Advil®, 784**F**
Aerobes, 548**D**
Aerobic catabolism, 553
Affinity purification, 133, 135**F**
Agar, 334
Agaropectin, 334
Agarose, 132
double helix of, 334**F**
Agglutinin, soybean, oligosaccharide of, 344**F**
AICAR (5-aminoimidazole-4-carboxamide ribonucleotide), 861, 862
AICAR transformylase, in purine biosynthesis, 874**F,** 875
AIDS (acquired immunodeficiency syndrome), 307, 932
AIDS virus, 100**T**
replication of, 958
AIR (5-aminoimidazole ribonucleotide), 873, 874**F**
AIR carboxylase, 874**F,** 875
AIR synthetase, in purine biosynthesis, 873, 874**F**
Akabori reaction, 109, 110**F**
Akee tree, 742**F**
Alanine, 58**S, G**
degradation to pyruvate, 865**F**
NMR spectrum of, 74
R and S isomers of, 73
β-Alanine, 61, 62**S**
in pyrimidine degradation, 889, 890**F**
Alanine biosynthesis, 850
Alanine racemase, 510
Albumin, serum, 85**T,** 95**T**
Alcaptonuria, hereditary defect in phenylalanine catabolism, 869
Alcohol dehydrogenase, 95**T,** 96**F,** 100**T,** 356**T,** 427**T,** 472**T,** 591**F**
amino acid composition of, 88**T**
bisubstrate kinetic mechanism of, 378
liver and, 173**S,** 351**S,** 455
mechanism of, 458**F**
sequence of, 456**F**
steric specificity of, 471**F**
Alcohol dehydrogenase *(Continued)*
ζ-crystallin, gene sharing, 1092**T**
Alcoholic fermentation, 591**F**
Aldaric acids, 319
Aldehyde oxidation, by xanthine oxidase, 884
Aldehyde dehydrogenase, 472**T**
Aldimine, of PLP, 480
Alditols, 319
Aldohexose, 311
Aldol condensation, 580
in collagen crosslinking, 155
Aldol reactions, PLP-assisted, 481**F**
Aldolase(s), 430, 431, 433**T**
class I and class II, 582
in CO_2 fixation, 724**F,** 725**T**
Aldonic acid, 319
Aldopentose, 311
Aldose, 311
Aldosterone Na^+ and K^+ balance and, 801
synthesis of, 800**F**
Aldotetrose, 311
Alginates, 332
Alkaline phosphatase, 427**T,** 433**T,** 460
Alkyl sulfates, as chemical mutagens, 930, 931**F**
1-Alkyl-2-acetylglycerophosphocholine, 777, 780**F**
1-Alkyl-2-acylglycero-3-phosphate, 779**F**
1-Alkyl-2-acylglycero-3-phosphoethanolamine, 779**F**
1-Alkyl-2-acylglycerophosphoethanolamine desaturase, 779**F**
1-Alkyl-2-lysophosphatidylcholine, 780**F**
Alkylating agents, as chemical mutagens, 930, 931**F**
1-Alkyldihydroxyacetone phosphate, 779**F**
1-Alkyldihydroxyacetone phosphate oxidoreductase, 779**F**
1-Alkylglycero-3-phosphate, 779**F**
1-Alkylglycerophosphate acyltransferase, 779**F**
Allantoins, in uric acid metabolism, 884, 885**F**
Allele, 1076**D**
Allergic response, amelioration by ascorbic acid, 491
Alloleucine, stereoisomers of, 72**S**
Allopurinol
as inhibitor of xanthine oxidase, 884, 884**S**
in the treatment of gout, 884, 884**S**
Allose, 312
Allosteric enzymes, 368
Allosteric model, concerted. *See* Allosteric model, Monod-Wyman-Changeux.
Allosteric model, KNF, 414, 414**F**
Allosteric model, Monod-Wyman-Changeux, 410*ff,* 411**F,** 412**F,** 413**F**
Allosteric model, sequential. *See* Allosteric model, KNF.
Allosteric regulation 390**D,** 559, 560**F**
of enzyme activity, 394
Allysine, 155
Alpha helix, structure of, 141, 142, 143**S**
Alternating copolymers as synthetic mRNAs, 1021**T**
Alternative RNA splicing, 1009, 1010**F**
protein isoforms and, 1011
Altman, Sidney, 382
Altrose, 312
Amanita phalloides, 972
α-Amanitin, 971, 972**S**
Amber stop codon, 1038
Amber suppressor mutation, 1038
Ames, Bruce, 863
Amethopterin, 501
DHF and dTMP synthesis and, 896**S**
Amide bond, 82
Amide dihydrate, in aspartic proteases, 444
Amide plane, 83, 84**F**
Amide-linked glycosyl phosphatidylinositol anchors, 306, 307, 308**F**
Amide-linked myristoyl anchors, 306**F**
Amino acid(s), 56**S,** 56**F,** 63**T,** 87
acidic, 58**S,** 58**G,** 60
basic, 59**G,** 59**S,** 60
biosynthesis of, 839, 840**T,** 840*ff,* 841
chromatography of, 74
degradation of, 865, 865**F**
essential, 839, 840**T**
frequency of occurrence in proteins, 118**T**
as metabolic fuel, 824
mRNA-directed incorporation of, 1021**T**
nonessential, 839, 840**T**
NMR spectra of, 73
nonpolar, 57, 58, 59**S,** 58**G**
1-letter code, 57
polar uncharged, 57, 58, 59**S,** 59**G**
protein synthesis of, 1027
reactions of, 66–70
residues, 1019**F**
sequence(s)
databases for, 117
determination, 104, 109, 114, 115**F**
disulfide bridge cleavage and, 106
invariant residues and, 119
number of, 90**B**
side chain ionization of, 65
specific rotation of, 71**T**
spectroscopic properties of, 72
3-letter code, 57
ultraviolet spectra of, 72
2-Amino, 6-oxypurine nucleoside monophosphate (GMP), 876
2-Amino-4-oxopteridine, 500
2-Amino-6-oxypurine (guanine), 182
Amino sugars, 320
Aminoacyl adenylate, intermediate in aminoacyl-tRNA synthesis, 1028**F**
Aminoacyl-tRNA(s), 191, 195, 1018
binding of, 1052
Aminoacyl-tRNA synthetases, 1018, 1027, 1028**F**
class I and class II, 1028**F,** 1029**T**
α-Aminoadipate, in lysine biosynthesis, 845, 846**F**
Aminoadipic acid, 60, 61**S**
α-Aminoadipic-6-semialdehyde, in lysine biosynthesis, 845, 846**F**
p-Aminobenzoate, from chorismate, 857**F**
p-Aminobenzoic acid, 500**F,** 876
γ-Aminobutyric acid, 61, 62**S,** 482
5-Aminoimidazole-4-carboxamide ribonucleotide, 861, 862**F**

5-Aminoimidazole ribonucleotide, 873, 874**F**, 875
β-Aminoisobutyric acid, in thymine degradation, 889, 890**F**
p-Aminophenyl-β-thiogalactoside, 269**T**
β-Aminopropionitrile, 156**S**
Aminopterin, 501
  DHF and dTMP synthesis and, 896**S**
2-Aminopurine, as a mutagenic base analog, 930**S**
6-Aminopurine (adenine), 182
Aminotransferase(s)
  bisubstrate kinetic mechanism of, 379, 380**F**
  α-keto acid carbon skeletons and, 839**F**, 840
Aminotriazole, 863**B**
Amitrole®, 863**B**
Ammonia, excretion of in ammonotelic animals, 870
Ammonium, 9
  assimilation, pathways of, 833, 833**F**, 834, 834**F**
  fate of, 832
  formation, from N_2 or NO_3^-, 827
  ion of as nitrogen source, 549
Ammonium oxidase, 100**T**
Ammonotelic animals, 870
Amniocentesis, detection of Lesch-Nyhan syndrome by, 880
Amobarbital, 652**S**
AMP, 183, 185**S**
  absorption spectrum, 183**F**
  synthesis from IMP, 876, 877**F**
  as feedback inhibitor of AMP synthesis, 877**F**, 878
  as universal hunger signal, 560
AMP deaminase
  in purine catabolism, 882**F**
  in the purine nucleoside cycle, 882, 883**F**
Amphibia
  DNA content of, 906**T**
  uric acid metabolism of, 884, 885**F**
Amphibolic intermediates, 554
Amphipathic, 38, 276**D**
Amphiphilic, 38
  α-helices, presequences on mitochondrial proteins, 1070**F**
Ampicillin, 248, 254**F**
Amplification of DNA, 247
Amygdalin, 324**F**
α-Amylase, 328
  action of, 671**F**
β-Amylase, 328
  action of, 671**F**
Amylo-(1,4 → 1,6)-transglycosylase, 678**F**
Amylopectin, 326**F**, 327
Amyloplasts, 327
Amylose, 326**F**, 327**F**, 330
Amytal, 652**S**, 653
Anabena, nitrogen fixation and, 829**T**
Anabolic steroids, athletic performance and, 801
Anabolism, 551
  energy relationship to catabolism, 552**F**
  systems analysis of, 804, 804**F**
Anaerobes, 548, 548**D**
Analogs, transition state, 430
Anaplerotic reactions
  citric acid cycle and, 882, 883**F**
  in TCA cycle, 618
Androgens, 291, 799
Anemia
  ascorbic acid and, 491
  sickle-cell, 406
Anfinsen, Christian, 167
Angiotensin, 442**T**
Angiotensinogen, 442**T**
Anion exchange media, 76**F**
Anomer(s), 315
Anomeric carbon, 184, 315**D**
Anomeric sugars, 314
Anserine, 50, 61
Anthranilate
  from chorismate, 857**F**
  in tryptophan biosynthesis, 859**F**, 860
Anthranilate-phosphoribosyl transferase, 859**F**, 860
Anthranilate synthase, 859**F**, 860
  as a product of the *trpD* and *trpE* genes, 985**F**
anti, 185**F**
Anti-inhibition, in transcription activation, 981
Antibodies, 923**D**
  catalytic, 384**D**, 385**F**
  cDNA expression library screening and, 266
  diversity of, 928
Anticodon loop, tRNA, 238, 238**F**, 239, 239**F**
Antifreeze proteins, 95**T**, 98, 342
Antigen, 384**D**, 923**D**
  binding region in IgG, 924**F**, 925**F**
Antimycin A_1, 652**S**, 653
Antiparallel α-helix proteins, 161**F**, 162, 163**F**
Antiparallel βsheet proteins, 162, 163**F**
Antisense strand, of DNA, 965**B**
2:3 Antiterminator, in transcription attenuation, 986, 986**F**
Antithrombin III, 334, 347
Antp homeodomain, 993**F**, 994**F**, 995**F**
2-AP. *See* 2-Aminopurine.
Apoenzyme, 356**D**
  LAD, 457
Apolipoprotein(s), 794**T**
  amino acid substitutions in, 1081
Apoprotein B-100, 794**F**, 794**T**
Apo-repressor, 562**F**
  Trp, 984
APRT. *See* Adenine phosphoribosyltransferase.
APS. *See* Adenosine 5′-phosphosulfate.
Apurinic acid, 199
  in nucleic acid sequencing, 213**F**
Apurinic/apyrimidinic acid endonuclease, 930,
  in DNA repair, 960, 960**F**
Aquaspirillum serpens, 292**F**
araB mRNA Shine-Dalgarno sequence, 1049**F**
araBAD
  operon, control by AraC and CAP, 982, 983**F**
  promoter, nucleotide sequence of, 967**T**
Arabinose, 312
Arabinose *(Continued)*
  control of the *araBAD* operon and, 982, 983**F**
  catabolic pathway for, 983**F**
  in nucleic acid sequencing, 210
AraC, control of *araBAD* operon and, 982, 983**F**
araC promoter, nucleotide sequence of, 967**T**
Arachidic acid, 277
Arachidonate, as precursor to eicosanoids, 783**F**
Arachidonic acid, 277, 278**S**, 278**G**
  synthesis from linoleic acid, 770**F**
araI site, and regulation of the *araBAD* operon, 982, 983**F**, 984**F**
araO sites, in the *araBAD* operon, 982, 983**F**, 984**F**
Archae (archaebacteria), 20
Archaebacterial V-ATPase, evolution of, 1097**F**
ARGII, (zinc finger protein), 996**T**
Arginase, 843, 844**F**
Arginine, 59**S**, 59**G**; 69
  auxotrophs and one-gene, one-enzyme hypothesis, 902**F**
  biosynthesis of, and the urea cycle, 841, 844**F**
  degradation to α-ketoglutarate, 866**F**, 867
Arginine phosphate, 534**F**, 541
Argininosuccinase, 842, 843, 844**F**
  δ-crystallin and, 1092**T**
Argininosuccinate, 842, 844**S**
Argininosuccinate synthetase, 842, 844**F**
Arginyl-tRNA synthetase, 364**T**
Arginyl-tRNA^{Arg}:protein transferase in protein degradation, 1072**F**
Arnold, 705
aroH operon, 984
Aromatic amino acid biosynthesis, 855
  regulation of, 861
Arrhenius equation, 359
1-Arseno-3-phosphoglycerate, 585
ARS (autonomously replicating sequence), 257
Ascorbic acid, 489, 490**F**
Asialoglycoprotein receptor, 344
Asparaginase, deamidation of Asn to aspartate, 866
Asparagine, 58**S**, 58**G**
  acid lability of, 87
  biosynthesis, 847, 847**F**, 848**F**
Aspargine synthetase, 847, 848**F**
Aspartate, 407, 408**F**
  in AMP synthesis, 876, 877**F**
  degradation of, 844, 866
  in purine biosynthesis, 874**F**, 875
  in pyrimidine synthesis, 886, 886**F**, 887**F**
  urea cycle and, 844**F**
Aspartate aminotransferase, 356**T**, 364**T**
Aspartate β-decarboxylase, 510
Aspartate transcarbamoylase, *E. coli*, 407, 408**F**
  active site residues, 417
  allosteric mechanism, 415–418, 418**F**
  as allosteric enzyme, 407
  in pyrimidine synthesis, 887, 887**F**

Aspartate transcarbamoylase *(Continued)*
quaternary structure, 416
subunit organization of, 410**F**
Aspartic acid, 58**S**, 58**G**
Aspartic protease, 440
mechanism of, 443**F**
Aspartokinase(s), in Thr, Met, and Lys biosynthesis, 849**F**
Aspartokinase isozymes of *E. coli*, regulation of, 847**T**
β-Aspartyl adenylate, an intermediate in Asn biosynthesis, 848**F**
β-Aspartyl-semialdehyde, in Thr, Met and Lys biosynthesis, 848, 849**F**
β-Aspartyl-semialdehyde dehydrogenase, 847, 848**F**
Aspergillus nidulans nitrate reductase, sequence comparison, 1080**F**
Aspirin, 784**F**
Assimilation, single-stranded, in recombination, 921
Asymmetric molecule, 56
Asymmetric structures in membranes, 296
Asymmetry, phospholipid, 300
ATCase, *See* Aspartate transcarbamoylase.
ATF, 989**T**
Atherosclerosis, 156, 291
Atkinson, Daniel E., 803
Atlas of Protein Sequence and Structure, 117
Atomic fluctuations in proteins, 158
ATP, 3, 186**S**, 574**T**
allosteric activator
of aspartate transcarbamoylase, 408
in pyrimidine synthesis, 889, 890**F**
of ribonucleotide reductase, 891**F**, 894
in anabolism and catabolism, 805
in CO_2 fixation, 725**T**
coupling coefficients and, 808, 813, 814**F**, 815**F**
cycle of, 555, 557**F**
daily requirements for, 541
as energy-shuttle molecule, 528
equivalent of, 808, 809
formation of, 807
$\Delta G^{\circ\prime}$ and, 539**F**, 540**F**
hydrolysis of, 521
ionization of, 539**F**
K_{eq} and, 810**B**
metabolism and, 808, 814, 815, 815**F**
pentose pathway and, 696**F**
in photosynthesis, 701
solvent capacity of cell and, 811
synthesis of, light-driven, 716
value of coenzymes and, 809
yield of, from oxidation, 649**T**, 745**T**
ATP sulfurylase, in sulfate assimilation, 856**F**
ATP synthase, 647, 648
ATP:D-glucose-6-phosphotransferase, 354**D**
ATP:glutamine synthetase adenylyl transferase, 837
ATP:D-hexose-6-phosphotransferase, 354**D**
ATP-ADP translocase, 656, 657**F**
ATP-citrate lyase, 617, 760
ATP-dependent protease complex, 1072, 1073**F**
ATP-phosphoribosyl transferase, in histidine synthesis, 862**F**
Attenuation
alternative secondary structures in mRNA and, 986**F**
mechanism of in the *trp* operon and, 984, 985, 986**F**, 987**F**
regulation of transcription and, 984, 985
Aurintricarboxylic acid (inhibitor of peptide chain initiation), 1063**T**
Autocoid, 187**B**
Autoradiography, 209, 259**F**, 261**F**, 945**F**
Autoregulation of gene expression, 984
Autotrophs, 546**D**
Auxotrophs, 902**F**, 903**D**
Avery, Oswald T., 903
Avian sarcoma virus, 957
Avidin, 495
Axial substituents, 317
Azaserine, 875**S**
as inhibitor of purine biosynthesis, 873
Azide as inhibitor of electron transport, 653
3′-Azido-2′,3′-dideoxythymidine (AZT), 958**F**
Azotobacter spp., nitrogen fixation and, 829**T**
AZT (3′-azido-2′,3′-dideoxythymidine), 958**F**
AZTTP (AZT 5′-triphosphate), 958**F**

B cell lymphocytes, 923
b cytochromes, 639
Bacillus spp., nitrogen fixation and, 829**T**
Bacillus subtilis, 201
Bacteria
base DNA composition of, 906**T**, 907**T**
generation time of, 907
as genetic organisms, 906
nitrifying, 548**T**, 550**F**
photosynthetic, 548**T**
purple nonsulfur, 548**T**
Bacterial F_1F_0 ATP synthase, 713**F**
Bacteriochlorophyll, 712, 712**F**, 713**F**
Bacteriophage, 26
base DNA composition of, 907**T**
DNA as the hereditary material of, 904, 905**F**
filamentous, 255
Bacteriophage λ, 255, 255**F**, 256**F**
Bacteriophage λ *Cro*, an HTH motif DNA-binding protein, 993**F**, 994**F**
Bacteriophage T2, 904**F**, 905**F**
Bacteriophage T4, 27**F**
Bacteriophage φX174 phage A protein mRNA Shine-Dalgarno sequence, 1049**F**
Bacteriophage *SP6* RNA polymerase, 266, 266**F**
Bacteriopheophytin, 712, 712**F**, 713**F**
Bacteriorhodopsin, 298, 304, 305**S**, 648
Bactoprenol, 290**S**
Baculoviruses, 266
BalI restriction endonuclease, 203, 204**T**
Ballistic methods of DNA delivery, 914
Baltimore, David, 957
BamHI restriction endonuclease, 204**T**, 205
Bases
methylated, 195, 197**F**, 241**F**, 931**F**, 1003, 1004**F**
nitrogenous, 181
Base exchange enzyme, in PS-PE conversion, 776**F**
Base pairing, 191. *See also* Wobble hypothesis.
A:T; G:C, 218**S**
in triple helix formation, 920**F**
Base sequence, 190
Basic region, in bZIP DNA-binding proteins, 999
Beadle, George W., 563, 901
Beads on a string, nucleosomes and DNA, 234**F**, 235**F**
Beef, fatty acids in, 279
Behenic acid, 277
Benkovic, Stephen, 502
Benzoyl chloride, 236**F**
Benzoylalanine methyl ester, 438**S**
Benzoylglycylphenylalanine, in carboxypeptidase mechanism, 454
(-)-2-Benzyl-3-(p-methoxybenzoyl)propanoic acid, 453**S**
Bergstrom, Sune, 785
Beriberi, 464
Beta-bend, 147, 148**S**
Beta-bulge, 148, 149**S**
Beta-pleated sheet, 145**D**, 146**S**
antiparallel, 146**D**, 147**S**
parallel, 146**D**, 147**S**
Beta-pleated strand, 91**F**
Beta-turn, 147, 148**S**
Betaine, 61, 62**S**
Bicarbonate buffer, 50, 51**B**
BICINE, 53**F**
Bifunctional enzyme, 669
Bifunctional polypeptides, 875
Bijvoet, J. M., 70
Bilayer(s), 295, 296
lipid, 292, 293
Bile acids, 291
synthesis of, 798**F**
Bile salts, 735
synthesis of, 798**F**
Bimolecular reactions, 358**D**
Binding change mechanism, 646, 651**F**
Binuclear center in cytochrome oxidase, 644**F**
bioA and *bioB* promoters, nucleotide sequence of, 967**T**
Biocytin (biotin), 356**T**, 463**T**, 492**S**
BioGel P®, 131
Bioluminescence, 269, 271**F**
Biomolecule(s), 5
Biomolecular dimensions, 8**T**
Biomolecular recognition, 16
Bios, 495
Biot, Jean-Baptiste, 70
Biotin, 356**T**, 463**T**, 492**S**
Biotin-carboxyl carrier protein, 494**F**
Biotin-lysine complex, 463**T**, 492**S**
Birds, uric acid metabolism of, 884, 885**F**
Bisabolene, 289**S**
Bishop, Katherine, 508
1,3-Bisphosphoglycerate, 574**T**, 583**F**, 584**S**, 586**F**
in CO_2 fixation, 724**F**
hydrolysis of, 525**T**, 533**F**
2,3-Bisphosphoglycerate, 574**T**, 583**F**, 586**F**
as allosteric effector of Hb, 404, 404**S**

2,3-Bisphosphoglycerate *(Continued)*
physiological significance in binding Hb, 405
Bisphosphoglycerate mutase, 586**F**
Bisubstrate enzyme kinetics, 374*ff*
Bisubstrate reaction
double-displacement, 375**D**, 379
ordered, 375**D**, 377, 378
ping-pong, 375**D**, 376**F**, 379
random, 375**D**, 376, 377
single-displacement, 375**D**, 376**F**
Biuret reaction, 102, 102**F**
Blacktongue, 473
Bleomycin A_2, 325**F**
Bloch, Konrad, 788
Blood clotting, pathways of, 392, 393**F**
zymogen activation of, 391, 303**F**
Blood coagulation proteins, structural modules in, 1090**F**
Blood groups, 303
Blotting, DNA, 260**F**
Blue/white selection, 268
Blunt-end ligation, 251, 251**F**
Blyth, 476
Boas, 495
Boat conformation of sugars, 317**F**
Bohr, Christian, 403
Boiling point elevation, 39
Boltzmann's constant, 515
Bovine luteinizing hormone, oligosaccharide of, 344**F**
Bowlegedness, vitamin D deficiency and, 507
Boyer, Herbert, 252
Boyer, Paul, 646, 650
BPG. *See* 2,3-Bisphosphoglycerate.
Bradykinin, chemical synthesis of, 125
Bradyrhizobium, nitrogen fixation and, 829**T**
Brain, energy metabolism in, 819**T**, 820
Branched-chain fatty acids, oxidation of, 752**F**, 753
Branched-chain α-keto acid dehydrogenase, 509
Branching enzyme, 678**F**
Branch migration in homologous recombination, 916, 917**F**, 921
Branch site, RNA splicing and, 1005, 1007**F**
Brandts, John F., 514
Bretscher, Mark, 299
Briggs and Haldane, 361
Bromelain, 113
Bromobenzyl alcohol in LAD active site, 457
5-Bromo-4-chloro-indolyl-β-galactopyranoside (X-gal), 269**S**
N-Bromosuccinimide, 69**S**
5-Bromouracil, 929**S**
Brown adipose tissue, 654, 823
Brown, Michael, 797
5-BU. *See* 5-Bromouracil.
Buchanan, John, 872
Buchner, Eduard, 562
Buffers 48, 49
Bundle sheath cells, C-4 pathway and, 727, 728**F**
Burst kinetics, 438**F**
t-Butyloxycarbonyl chloride 125, 126**F**
t-Butyloxycarbonyl azide 125, 126**F**
bZIP (leucine zipper) protein
as DNA-binding proteins, 992, 999
motifs, 1000**F**, 1001**F**

^{14}C as a metabolic tracer, 565**F**
C2, 170
C-3 family of amino acids, 865, 865**F**
C-4 family of amino acids, 866
C-4 pathway of CO_2 fixation, 727
C-5 family of amino acids 866
C genes of immunoglobulin 926, 926**F**
C regions, in IgG, 923, 924**F**, 925**F**
C_x class of zinc finger proteins 996, 996**T**
features of, 998**T**
$C_{12}E_8$, 293**S**
C/EBP, (bZIP DNA-binding protein), 999**F**
interaction with DNA, 1001**F**
sequence of binding motif, 1000**F**
C-terminal domain, of yeast RNA polymerase II subunit 1, 973
C_2H_2 class of zinc finger proteins, 996, 996**T**, 998**F**
DNA contacts and, 997**F**
Ca^{2+}, in regulating muscle contraction, 821
CAAT box, as eukaryotic promoter module, 989, 989**T**
cab2 gene, 271**F**
Cactaceae, CO_2 fixation and, 729
Cadmium, as physiological challenge, 990**T**
Caenorhabditis elegans globin gene organization, 1087
CAIR. *See* Carboxyaminoimidazole ribonucleotide.
Cairns, John, 943
Calcitonin, 507
Calcium carbonate as sink for CO_2, 551**B**
Calcium homeostasis, 506**F**
Calmodulin, 157
Caloric homeostasis, 819
Calorie, 513
Calorimeter, 513
Calvin, Melvin, 564**F**, 720
Calvin-Benson cycle
enzymes of, 723, 725**T**
of CO_2 fixation, 723, 724**F**, 725**T**, 728**F**
N-CAM, 170**F**
CAM plants. *See* Crassulacean acid metabolism and CO_2 fixation in plants.
Camel, dependence of on fatty acids, 746
cAMP 186**S**, 682
control of the *araBAD* operon, 982, 983**F**
receptor protein (CRP) of *E. coli*, 980
protein kinase and, 306
Camphene, 289**S**
Canonical, 218
CAP (catbolite activator protein), 95**T**, 96, 980, 980**F**, 981**F**
control of the *araBAD* operon and, 982, 983**F**
helix-turn-helix DNA-binding motif and, 992, 993**F**
CAP-binding protein (CBP-I) in eukaryotic protein synthesis, 1059**F**
Cap structures in eukaryotic mRNAs, 1003, 1004**F**
Carbamate
in the carbamoyl-P synthetase reaction, 843**S**, 886**F**
O_2 binding by hemoglobin and, 404
N-Carbamoylaspartate, 408**F**
Carbamoyl aspartate in pyrimidine biosynthesis, 887**F**
Carbamoyl phosphate, 407, 408**F**, 541
as precursor in pyrimidine synthesis, 886
Carbamoyl phosphate synthetase I
allosteric activation of, 841
ammonium assimilation and, 832
in arginine biosynthesis, 841, 843**F**
Carbamoyl phosphate synthetase II, in pyrimidine synthesis, 886, 886**F**
Carbamoyltransferases, 355**T**
Carbohydrate, 310
Carbon
reduction of in organic molecules, 557**F**
reserves of in organisms, 559
Carbon cycle, 549
Carbon dioxide, 51, 51**B**
fixation of, 698, 72**DFF**
Carbon monoxide
binding of to myoglobin, 397, 397**F**
as inhibitor of electron transport, 653
Carbonic acid 50, 51**B**
Carbonic anhydrase, 51**B**, 356**T**, 364**T**, 365**T**, 366**T**, 403, 427**T**
Carbonyl cyanide-*p*-trifluoromethoxyphenyl hydrazone, 653, 654**F**
Carbonyl phosphate, 663
in biotin-dependent reactions, 493**T**
in carbamoyl phosphate synthesis, 843**S**
in the carbamoyl phosphate synthetase reaction, 886**F**
Carboxin, 652**S**
N-Carboxybiotin, 664
formation of, 492**F**
2-Carboxy, 3-keto ribitol, in CO_2 fixation, 721, 721**S**
Carboxyaminoimidazole ribonucleotide synthetase, 874**F**, 875
γ-Carboxygeranyl-CoA, formed by biotin-dependent carboxylation, 493**T**
γ-Carboxyglutamic acid, 60, 61**S**, 508**S**, 509
Carboxyl groups, biotin as carrier of, 492
Carboxymethyl cellulose, 76
Carboxyltransferases, 355**T**
Carboxypeptidase(s), 111 conformation changes in, 452
tertiary structure as exon-encoded units, 1091**F**
Carboxypeptidase A, 111, 451, 452**G**
mechanism of, 454
Carboxypeptidase-glycyl-L-tyrosine complex, 452**G**
Carboxy-phosphate, carbamoyl phosphate synthetase reaction and, 886**F**
N-Carboxyurea, formation by biotin-dependent carboxylation, 493**T**
Cardiac glycoside, 320
Cardiolipin, 282, 283**S**
synthesis from CDP-diacylglycerol, 777, 778**F**
Cardiolipin synthase, 778**F**
Carnauba wax, 288

Carnitine as carrier of fatty acyl groups, 738
Carnitine acyltransferase, 738, 739**F**
Carnitine:acylcarnitine translocase, 739**F**
Carnosine, 61
β-Carotene, 288, 504, 703**S**
Carotenoids, 288, 703
Carrot
  base DNA composition of, 907**T**
  V-ATPase genes of, 1096**F**
Cartilage, 335, 348
Cartilage matrix proteoglycan, 345, 346**F**
Casein, 95**T,** 97, 100**T,** 101
Castor bean (*Ricinus communis*), 1064
Casurina, and nitrogen fixation, 829**T**
CAT gene, 269**T**
  as reporter gene, 269, 270**F**
Catabolism, 550**D**
  compared with anabolism, 552**F,** 554, 555**F**
  stages of, 553, 555**F**
  systems analysis of 804, 804**F**
Catabolite repression
  in *E. coli,* 980
  in regulation of operons, 982**F**
Catabolite activator protein, 95**T,** 96, 980, 980**F,** 981**F**
  bending of DNA by, 981**F**
Catalase, 95**T,** 100**T,** 354, 356**T,** 365**T,** 366**T**
Catalyst, 360**D**
Catalytic efficiency, 365, 366**T**
Catalytic antibody, 384**D,** 385**F**
Catalytic RNA, 381, 382**F**
Catalytic triad of chymotrypsin, 437**F**
Catenated state in DNA replication, 954
Cathepsin D, 442**T**
Cation exchange media, 76**F**
Cavendish Lab, Cambridge, 193
CBP-I (CAP-binding protein), in eukaryotic protein synthesis, 1059**F**
CBP-II (eIF-4F complex), in eukaryotic protein synthesis, 1059**F**
CDC2-type protein kinase, 957
cDNA, 262
  libraries, 262
  synthesis by reverse transcriptase, 265, 265**F**
CDP, 187
  as a substrate of ribonucleotide reductase, 891**F,** 894, 894**F**
CDP-choline, 775**F**
CDP-choline:1,2-diacylglycerol phosphocholine transferase, 775**F**
CDP-diacylglycerol synthesis, 773, 774**F,** 775**F,** 778**F**
CDP-ethanolamine, 775**F**
CDP-ethanolamine transferase, 779**F**
CDP-ethanolamine:1,2-diacylglycerol phosphoethanolamine transferase, 775**F**
CDR. *See* Complementarity-determining region.
Cech, Thomas, 382, 1009
Cell cycle, eukaryotic, 955**F**
Cell extract, fractionation by differential centrifugation, 566**F**
Cellobiose, 323**F**
Cell wall, 28**T,** 335, 337**F,** 338**F**
Cellular respiration *versus* CO_2 fixation and ATP stoichiometry, 815
Cellular retinol-binding protein, 504
Cellulase, 331
Cellulose, 326, 327, 330, 333**F**
Cellulose acetates, 330
Central dogma of molecular biology, 963, 964**F**
Centrifugation, 566**F**
  density gradient of, 245, 246**F**
Ceramide, 286, 286**S,** 286**G**
  as precursor for sphingolipids and cerebrosides, 780
  synthesis of, 781**F**
Cerebroside, 287
  synthesis of, 780
Cesium chloride and density gradient centrifugation, 245, 246**F**
Cetyl palmitate, 288
CF-1, CF-2 (conjugate degrading factors), 1072
CF_1CF_0 ATP synthase
  chloroplase ATP synthase, 718, 718**F**
  distribution in thylakoid, 715**F**
  of photosynthesis 708**F**
cGMP, 186**S**
Chain termination method, nucleic acid sequencing and, 209
Chain elongation in transcription, 969
Chair conformation of sugars, 317**F**
Chang, Annie, 252
Changeux, Jean-Pierre, 410
Chaperones, molecular, 167, 1066
Chaperonins, 167
Chargaff, Erwin, 191, 906
Chargaff's rules, 191, 192**T,** 906
Charged tRNAs
  formation of, 1018
  in transcription attenuation, 985
Charlotte's Web, 152
Chase, Martha, 904
Chelation, 997**D**
Chelex-100, 76
Chemical shift, 74
Chemiosmotic coupling, 647
Chemiosmotic hypothesis, 645, 646
  direct evidence of from photosynthesis, 717**B**
Chemoautotrophs, 548**T**
Chemoheterotrophs, 548**T**
Chemotrophic organisms, 523, 548**D**
Chi sites, in recombination, 918, 919**F**
Chicken
  base DNA composition of, 907**T**
  ovalbumin gene promoter region of, 974**F**
Chiral molecules, 56
Chitin, 320, 327, 331, 333**F**
Chl^+, cationic free radical, 705
Chloramphenicol, 62, 1062**S**
Chloramphenicol acetyltransferase, 269**T,** 270**F**
Chlorate (CLO_3^-), 908
  resistance of, 908**F**
Chlorella, ^{14}C as metabolic tracer and, 565**F,** 720
Chlorophyll, in photosynthesis, 702, 702**S,** 703**F**
Chlorophyll *a,* 702**S**
  dimers of, as photosynthetic reaction centers, 706
  light-absorption spectrum of, 703**F**
Chlorophyll *a/b* binding protein gene (*cab2*), 271**F**
Chlorophyll *b,* 702**S**
  light-absorption spectrum of, 703**F**
Chloroplast, 9, 28**T,** 327, 699, 699**F,** 700**F**
Chloroplast F-ATPase, evolution of, 1097**F**
Cholecalciferol, 463**T,** 505**S,** 506
Cholesterol, 100**T,** 289, 290**S,** 290**G**
  as precursor to bile acids, 798**F**
  biosynthesis of, 785
  esters of, 791**F,** 794**F**
  in lipoproteins, 794**F**
  structure of, 785**F**
  synthesis of steroid hormones from, 800**F**
  synthesis from squalene, 791**F**
Cholesterol ester transfer protein, 797
Cholic acid, 291**S,** 798**F**
Choline, 775**F**
Choline kinase, 775**F**
Choline plasmalogen, 285**S,** 285**G**
Choline sphingomyelin, 286, 286**S,** 286**G**
Chondroitin-4-sulfate, 335**F**
Chondroitin-6-sulfate, 335**F**
Chorismate
  as precursor to aromatic compounds, 857**F**
  synthesis of via shikimate pathway, 857, 858**F**
Chorismate mutase in Phe and Tyr biosynthesis, 858, 859**F**
Chorismate synthase, 858**F**
Chorismic acid, in aromatic amino acid synthesis, 855, 857**F**
Chou, Peter, 171
Chou-Fasman algorithm, 171
Chromatin, 233, 235**F**
  organization of, 234, 235**F**
Chromatography
  affinity, 133, 135**F**
  hydrophobic interaction (HIC), 134
  ion exchange, 75, 77**F,** 78**F**
  molecular sieve, 131, 132**F**
  oligo(dT)-cellulose, 264**F**
  partition, 75
  size exclusion, 102, 131, 132**F**
Chromogenic substrate, 269
Chromosome(s), 24, 190, 193, 233
  of *Escherichia coli,* 190, 193, 194**F**
  organization of, 234, 235**F**
  pairing of, 916
  yeast artificial (YACs), 257
  structure of,
    mapping through genetics, 908
    model of, 235**F**
  walking of, 262, 263**F**
Chylomicrons, 792, 793**T**
  assembly of in intestines, 734, 735
Chymosin, 442**T**
Chymotrypsin 85**T,** 92**F,** 112, 115**T,** 122, 364**T,** 365**T,** 391, 427**T,** 433**T,** 436, 437**G**
  amino acid sequence of, 437**F**
  mechanism of, 441
  substrate binding pocket of, 438**F**

Chymotrypsin *(Continued)*
tertiary structure as exon-encoded units, 1091**F**
α-Chymotrypsin, 460**F**
π-Chymotrypsin, 391, 392**F,** 460**F**
Chymotrypsinogen, 391, 391**T,** 460**F**
denaturation of, 513, 514**T**
proteolytic activation of, 392**F**
cI protein and the helix-turn-helix DNA-binding motif, 992
cis-Vaccenic acid, synthesis of, 768
cis-Vaccenoyl-ACP, 769**F**
Citrate, 602**F,** 604**T**
Citrate synthase, 602**F,** 603**F,** 604**T,** 604**G**
regulation of, 605
Citrate-malate-pyruvate shuttle, 617, 759
Citric acid cycle, 599**F,** 602**F**
regulation of, 619, 620**F**
Citrobacter spp., and nitrogen fixation, 829**T**
Citronellal, 289**S**
Citrulline, 61, 62**S,** 842, 844**S**
transporter of, 844**F**
Citrullyl-AMP, an intermediate in the urea cycle, 844**F**
Claisen condensation in glyoxylate cycle, 624
Clathrate, 37, 37**F**
Cleavable presequences in eukaryotic protein translocation, 1069
α-Cleavage reactions, with TPP, 465
Clocks, molecular, 1081
Cloning, 248
directional, 251, 253**F**
shotgun, 258
site, 249
vector, 249
bacteriophage λ 255
cosmid, 255, 256**F**
Closed promoter complex, in RNA polymerase:DNA interaction, 966**F**
Closed system, thermodynamic, 512**D, F**
Clostrapain, 113, 115**T**
Clostridium botulinum, 548
Clostridium flavodoxin, 476**T**
Clostridium oroticum, 476**T**
Clostridium pasteurianum, nitrogen fixation and, 829**T**
Clover, base DNA composition of, 907**T**
Cloverleaf tRNA, 238, 238**F,** 239**F**
Clr, 170**F**
Cls, 170**F**
CMC, 293**D**
CMP, absorption spectrum, 183**F**
CMV (cytomegalovirus), 266
Co-inducer, 561**F**
Co-inducers, in operon regulation, 981, 982**F**
Co-repressors, 562**F**
in operon regulation, 981, 982**F**
CO_2
in purine biosynthesis, 874**F,** 875
fixation
light activation of, 726
regulation of, 726
Coated pits, in LDL metabolism, 797
Coated vesicles, in LDL metabolism, 797
Cobalamin, 463**T**
Cobra, Indian, 284**F**
Coconut, fatty acids in, 279
Codon(s), 965**B**
multiple, in transcription attenuaton, 985, 986**F**
nonrandom, 1084
nonsense, 1023**D**
synonymous, 1023**D**
termination, 1023**D**
usage of, 1035, 1036**T**
relative frequency of Leu codon usage, 1085**F**
variants, 1026**T**
Codon-anticodon pairing, 1033
Coenzymes, 356**D,** 356**T,** 462**D**
Coenzyme A, 356**T,** 463**T,** 478**S**
activation of fatty acids and, 737
mechanism of, 479**F**
Coenzyme Q, 290**S,** 634, 636**G,** 636**S**
from chorismate, 857**F**
Coenzyme Q-cytochrome *c* reductase, 638
Cofactor, 356**D**
Cognate tRNA molecules in aminoacyl-tRNA synthesis, 1027
Cohen, Stanley, 252
Coil, random, 157**D**
Collagen, 95**T,** 97, 150–156, 154**F**
amino acid composition of, 88**T**
exon organization and protein structure of, 1086, 1087**F**
gene of, model for evolution, 1087**F**
types of, 152
Collapsed β-barrel, in immunoglobulin molecule, 924, 925**F**
Collective motions in proteins, 158
Colligative properties, 39
Collinearity of gene and protein structure, 1018
Colony hybridization experiment, 258, 259**F**
Compartmentalization of metabolic pathways, 567, 567**F**
Competing resonance in high-energy molecules, 530, 531**F**
Competitive inhibition, equations of 371**B**
Complement control protein module, 169**S**
Complement regulatory factor, 348
Complementarity-determining regions, in IgG, 923
Complex I, of electron transport, 635, 636**F**
Complex II, 637, 638**F**
Complex III, 638, 640**F**
Complex IV, 641
Complex lipids, biosynthesis of, 773
Computerized tomography, 294
Concanavalin A, 164**S,** 164**G**
Concatamers, 256**F**
Concentration, effect on net free energy, 520
Concerted step, in chymotrypsin mechanism, 440
α-Condensation reactions, with TPP, 465
Cone cells, 504
Configuration, 94**F**
Conformation, 94**F**
of proteins, 93, 136, 158
Conformational coupling model, 646, 651**F**
Conjugate acid/base, 43
Conjugate-degrading factors (CF-1, CF-2), 1072
Conjugation, sexual, in bacteria, 910, 910**F,** 911**F**
Consensus sequence elements, in promoters, 967
Constant regions, in immunoglobulin G, 924, 924**F**
Constitutive expression of genes, 977**D**
Contact inhibition, 339
Contractile proteins, 97
Contrahelicase, 954
Contrast agent, liposome-encapsulated, 294
Control circuits in gene expression, 982**F**
Control of enzyme synthesis, 561–562
Conventions, in expressing protein, RNA, and DNA sequences, 965**B**
Coomassie blue, 103**B**
Cooperativity, 178, 396
Copper (Cu), 100**T,** 356**T**
in electron transport chain, 634, 642
Copy number, plasmid, 249
Corals, 551**B**
Cordecypin (3′-deoxyadenosine), 969**S**
Co-repressors, 976**D**
Corey, Robert, 142
Cori cycle, 667**F**
Corn
base DNA composition of, 907**T**
fatty acids in, 279
Corticosteroids, 799
body processes and, 801
Cortisol, 291**S,** 686
effects on carbohydrate and protein metabolism, 687
regulation of body processes by, 801
synthesis of, 800**F**
Corynebacterium diphtheriae, 1064
Coryneophage β, 1064
cos (cohesive end) site, 255, 256**F**
Cosmid cloning vectors, 255, 256**F**
c_ot curves, 232**F**
Cotten, 330
Coumarin, 508
Coupled reactions, in glycolysis, 570
Coupling, of favorable and unfavorable processes, 521
Coupling coefficient, ATP, 808
Coupling constant, 74
Covalent modification, 390
enzyme regulation and, 560
Covalent catalysis, 427, 431
Covalent intermediate, formation of, 431
Cow, base DNA composition of, 907**T**
Cowgill, 473
CPC1, amino acid sequence of its bZIP DNA-binding motif, 1000**F**
cpn60, 167
CPS II, in pyrimidine synthesis, 886, 886**F,** 887**F**
Craik, Charles, 440
Crambe abyssinica, 165
Crambin, 165**S,** 165**G**
Crassulaceae, CO_2 fixation and, 729
Crassulacean acid metabolism, CO_2 fixation in plants and, 728
CRBP, 504
Creatine, 377
in muscle contraction 821

Creatine kinase, 427**T**, 535**F**
  bisubstrate kinetic mechanism of, 377
Creatine phosphate, 377, 534**F**, 541
  hydrolysis of, 525**T**
  in muscle contraction, 821
CREB, amino acid sequence of, 1000**F**
Crick, Francis H. C., 180, 191, 936, 937, 963, 1017, 1033
Critical micelle concentration, 293**D**
Cro protein and the helix-turn-helix DNA-binding motif, 992
434 *Cro* protein, an HTH motif DNA-binding protein, 993**F**, 994**F**
Crossovers, protein, 160**F**
Crotalus adamanteus, 284**F**
Crotonase, 366**T**, 743**F**
Cruciforms, 227, 228, 229**F**
Crustacea
  DNA content of, 906**T**
  uric acid metabolism of, 884, 885**F**
Crystallin(s), gene sharing and, 1091, 1092**T**
γ-Crystallin, 164**S**, 164**G**
δ-Crystallin, arginsuccinase and, 1092**T**
ϵ-Crystallin, LDH and, 1091, 1092**T**
λ-Crystallin, hydroxyacyl-CoA dehydrogenase and, 1092**T**
ζ-Crystallin, alcohol dehydrogenase and, 1092**T**
τ-Crystallin, enolase and, 1091, 1092**T**
CsCl, 245
CsCl density gradient ultracentrifugation of DNA, 937, 939**F**
CTD. C-terminal domain. *See.*
CTF, transcription factor for RNA polymerase II, 989**T**
CTP, 187
  as feedback inhibitor of aspartate transcarbamoylase, 408
  as feedback inhibitor in pyrimidine synthesis, 889, 890**F**
  synthesis of from UTP, 889, 889**F**
CTP synthetase, 889, 889**F**
CTP:phosphocholine cytidylyltransferase, 775**F**
CTP:phosphoethanolamine cytidylyltransferase, 775**F**
Cu_A site, in cytochrome oxidase, 643**F**
Cyanide as inhibitor of electron transport, 653
Cyanobacteria, 23, 548**T**
  nitrogen fixation and, 829**T**
Cyanocobalamin, 485, 486**S**
Cyanogen bromide, 113, 114**F**, 115**T**
3′,5′-Cyclic adenosine monophosphate (*See also* cAMP), 186**S**, 682
  hydrolysis of, 525**T**
  -dependent protein kinase, 670, 683
Cyclic cascade system, 837, 837**F**
Cyclic electron transfer in photosynthesis, 710
3′,5′-Cyclic GMP (*See also cGMP*), 186**S**
Cyclic nucleotides, 186
  formation of, 535**F**
Cyclic photophosphorylation, 710, 719, 719**F**
Cycloamylose, 324**F**
Cyclohexane-1,2-dione, 69**S**
Cycloheximide, 1062**S**, 1063**T**
Cyclooxygenase, inactivation by aspirin, 784**F**, 785
Cycloserine, 62**S**
Cys-3, amino acid sequence of its bZIP DNA-binding motif, 1000**F**
Cystathionine, in methionine biosynthesis, 849**F**, 850
Cystathionine synthase, in methionine biosynthesis, 849**F**
Cystathionine-β-lyase, in methionine biosynthesis, 849**F**
Cysteic acid, 107**F**
Cysteine, 59**S**,**G**
  biosynthesis of, 853, 854, 855**F**
  degradation of to pyruvate 865**F**
Cysteinyl-$tRNA^{Cys}$, reduction with Raney nickel, 1027**F**
Cystic fibrosis gene, 273**B**
Cytidine, 184**S**
Cytidine diphosphodiacylglycerol, 773, 774**F**
Cytidine 5′-monophosphate (*see also* CMP), 185**S**, 186
Cytidylic acid (*see also* CMP), 186
Cytochrome, 634
Cytochrome b_5, 769**F**
Cytochrome b_6, 708**F**, 710
Cytochrome b_{559}, in photosystem II, 714
Cytochrome b/cytochrome c_1, in *R. viridis* photosynthesis, 712
Cytochrome *b*-cytochrome *f* complex of photosynthesis, 708**F**, 709
Cytochrome b_5 reductase, 769**F**
Cytochrome *c*, 85**T**, 635**T**
  absorption spectra of, 638
  amino acid sequence of, 119**F**, 120**T**
  evolutionary relationships of, 121**F**
  invariant residues of, 119
Cytochrome *c*′, layer structure of, 160**F**
Cytochrome *c* oxidase, 635**T**, 641, 642**F**, 643**F**, 644**F**
Cytochrome *f*, 708**F**, 710
Cytochrome oxidase, 100**T**, 356**T**
Cytochrome P450, 753, 799**F**
Cytochrome P450 reductase, 799**F**
Cytomegalovirus, 266
Cytosine, 181**S**
  deamination to uracil, 197, 197**F**
  degradation of, 890**F**
  methylation of and Z-DNA, 225
Cytosine 5′-diphosphate, 187. *See also CDP.*
Cytosine 5′-triphosphate (*see also* CTP), 187
Cytoskeleton, 25, 27**T**
Cytosol, 23**T**

D, as primary electron donor to *P*680$^+$ in photosynthesis, 708**F**
D gene, of immunoglobulin heavy chains, 926, 926**F**
D-loop, in homologous recombination, 919**F**, 921
dADP, 187
DAHP, 857, 858**F**. *See also* 3-deoxy-D-arabino-heptulosonate-7-phosphate
DAHP synthase and the synthesis of chorismate, 857, 858**F**
Dam, Henrik, 509
dAMP, 187
Dansyl chloride, 108, 109**F**
Dansyl-amino acid, 109**F**
Dark reactions of photosynthesis, 701, 701**F**, 716
dATP, 187
  as inhibitor of ribonucleotide reductase, 891**F**, 894
Davis, Marguerite, 504
Dawson, M. H. 903
DBP, 507
DCCD, 652**S**
dCDP, 187, 895
dCMP, 187
dCMP deaminase, 895, 895**F**
dCTP, 187, 895
de novo, 872**D**
Debranching enzyme, 671**F**
Decarboxylases, 433**T**
Decarboxylated SAM, 851**S**
Decarboxylation, 466, 497
  PLP and, 481**F**, 484**F**
Decay-accelerating factor, 342**F**
Decorin, 345, 346**F**
Degenerate nature of the genetic code, 1020
Degenerate oligonucleotides, 262, 262**F**
Degenerate sites, 1078
Degradation of amino acids, 863, 864**F**
Dehydratase, PLP-assisted, 481**F**
Dehydro-L-ascorbic acid, 490
7-Dehydrocholesterol, 505**S**, 789
  in cholesterol synthesis, 791**F**
Dehydrogenases, 469**D**
3-Dehydroquinate in chorismate synthesis, 857, 858**F**
5-Dehydroquinate dehydratase in chorismate synthesis, 858**F**
Dehydroquinate synthase in chorismate synthesis, 858**F**
Deisenhofer, Johann, 712
Deletion mutations, 928**D**, 932
DeLucia, Paula, 943
Demerol, 652**S**, 653
Denaturation, 18
Denitrifying bacteria, 550
Density gradient, 245, 246**F**
  ultracentrifugation of DNA, 937, 939**F**
Dental plaque, 329
Deoxy sugars, 319
Deoxyadenosine, effects of on purine metabolism, 883**F**
5′-Deoxyadenosylcobalamin, 356**T**, 463**T**, 485, 486**S**, 487**F**
  in ribonucleotide reductase reaction, 890
3-Deoxy-D-arabino-heptulosonate-7-P and shikimate pathway, 857, 858**F**
Deoxycholic acid, 291**S**
2′-Deoxy-5′-fluorouridylic acid, 898
Deoxyhemoglobin, 400
Deoxymyoglobin, 397
Deoxyribonucleic acid (DNA), 180, 188
Deoxyribonucleoside phosphates, 187
Deoxyribonucleotide biosynthesis, 890, 891**F**
  reduction at NDP 2′-position, 891**F**
  regulation of, 894**F**
2-Deoxyribose, 180, 320**F**

Deoxythymidylic acid (see also dTMP), 187
Depurination of nucleic acids by H^+, 213, 213**F**
Dermatan sulfate, 335**F**
Dermatitis acrodynia, 483
Desensitization, allosteric regulation and, 409
Designer enzyme, 384
Desmolase, 799
  in steroid hormone synthesis, 800**F**
Desmosterol, 789
  in cholesterol synthesis, 791**F**
Desolvation in ES complex, 428
Destabilization of ES complex, 426, 428
"Destroying angel" mushroom, 972
Detergents, 293**F**
Determination, N-terminal analysis, 106
Deuridylylation, P_{II} and *E. coli* glutamine synthetase, 837**F**, 838
Dextran, 329**F**
Dextrantriose, 324**F**
Dextrorotatory, 71**D**
dGDP, 187
dGMP, 187
dGTP, 187
  regulation of ribonucleotide reductase by, 894, 894**F**
DHF, from the thymidylate synthase reaction, 895, 896**F**
DHFR gene of mammals, exons in, 1002, 1003**F**
DHO dehydrogenase in pyrimidine biosynthesis, 887, 887**F**
Diabetes, 178, 319, 576, 755
  amino acid degradation and, 864
Diacylglycerol
  as precursor of glycerolipids in eukaryotes, 775**F**
  synthesis of, 776**F**
  synthesis of glycerolipids from, 773, 774**F**
Diacylglycerol acyltransferase, 775**F**, 776**F**
  synthesis of glycerolipids from, 773, 774**F**
Diacylglycerol kinase, 775**F**
Diacylglycerol lipase, 733**F**
Dialysis, 131, 132**F**
Diaminoaphthtylenesulfonyl chloride, 108, 109**F**
meso-L,L-α,ϵ-Diaminopimelate, in lysine biosynthesis, 849**F**
Diaminopimelate decarboxylase in lysine biosynthesis, 849**F**
Diaminopimelate epimerase in lysine biosynthesis, 849**F**
Diastase, 322
Diastereomer(s), 71**D**
  of sugars, 314
Diazoacetylserine (O-), 62**S**
Dicumarol, 653
Dicyclohexylcarbodiimide, 125, 126**F**, 652**S**
Dicyclohexylurea, 127**F**
Dideoxy method, nucleic acid sequencing, 209
Dideoxynucleotide, 211**S**
Dielectric constants, 36**T**
2,4-Dienoyl-CoA reductase, 750, 751**F**
Dietary lipids, fatty acids in, 279**T**
Diethylaminoethyl cellulose, 76
Diethylnitrosoamine, as chemical mutagen, 931**F**
Differential centrifugation, 566**F**
DIFP, 69, 439, 440**F**
Digestive tract, proteolytic enzymes of, 391
Dihydrobiopterin, 860**F**
Dihydrofolate reductase, 472**T**, 498**F**, 501
  in dTMP synthesis, 896**F**
  gene of, 1002, 1003**F**
Dihydrolipoyl dehydrogenase, 496, 498, 608**T**, 609
Dihydrolipoyl transacetylase, 496
Dihydrolipoyl transsuccinylase, 498, 608**T**, 609
Dihydroorotase, in pyrimidine biosynthesis, 887, 887**F**
Dihydroorotate, 476**T**
Dihydroorotate dehydrogenase, 100**T**, 476**T**, 477**F**
  in pyrimidine biosynthesis, 887**F**, 888
Dihydropicolinate, in lysine biosynthesis, 849**F**, 850
2,3-Dihydropicolinate, in lysine biosynthesis, 847, 849**F**
Dihydropicolinate synthase, in lysine biosynthesis, 849**F**
Dihydrosteroid dehydrogenase, 472**T**
Dihydrouridine, 197**S**
Dihydroxy acid dehydratase, in Val and Ile synthesis, 852**F**
Dihydroxyacetone, 313
Dihydroxyacetone phosphate
  as precursor to plasmalogens, 779**F**
  in CO_2 fixation, 724**F**, 725**T**
  in glycolysis, 572**F**, 574**T**, 580
Dihydroxyacetone phosphate acyltransferase, 779**F**
α,β-Dihydroxyisovalerate (intermediate in Val synthesis), 852**F**
α,β-Dihydroxy-methylvalerate (in isoleucine synthesis), 852**F**
3,4-Dihydroxyphenylalanine, 484
1,25-Dihydroxyvitamin D_3, 505**S**, 507
Diisopropylfluorophosphate, 69, 439, 440**S**
Dimerization motif for leucine zipper proteins, 1000**F**, 1001**F**
Dimethoxytrityl group, 236**S**
Dimethyl sulfate
  as chemical mutagen, 931**F**
  in nucleic acid sequencing, 212**F**
Dimethylallyl pyrophosphate, 789, 790**F**
Dimethylnitrosoamine, as chemical mutagen, 931**F**
Dinitrofluorobenzene, 106, 108**F**
2,4-Dinitrophenol, 653
Dinoflagellates, 551**B**
Diol dehydrase, 489
Dipeptide, 84
Diphosphatidylglycerol, 282, 283**S**
Diphthamide, a modified histidine, 1064, 1065**F**
Diphtheria toxin, 95**T**, 98
  ADP-ribosylation of eukaryotic EF-2 and, 1064, 1065**F**
  as inhibitor of protein synthesis, 1063**T**, 1064
Diploid, 906**D**
Dipole(s), 138**D**
Disaccharides, 322
Dissipation of free energy, 520
Dissociation constants, 46**T**
Dissociation of ions, 519**T**
Distribution coefficient (K_D), equation for, 131
Disulfide cleavage
  diagonal electrophoresis, 117**F**
  performic acid oxidation, 106, 107**F**
  reduction and alkylation, 106
Disulfide-rich proteins, 165**G**, 165**G**
Diterpene, 288
5,5′-Dithiobis (2-nitrobenzoic acid), 68**S**
Dithiothhreitol, 107**F**
D, L system of nomenclature, 71**D**
DNA (deoxyribonucleic acid), 5, 26**T**, 100**T**, 180, 188, 191
  acid lability of, 199
  A-form, 221, 221**T**, 222**S**
  alkali stability of, 197
  antisense strand of, 965**B**
  A:T, G:C content of DNA from various sources, 192**T**
  B-form, 217, 219**F**, 221**T**, 222**S**
  base composition in various organisms, 906, 907**T**
  base pairing of complementary strands of, 192**F**
  breaking and rejoining of, 915
  buoyant density of, 233, 245, 246**F**
  as the carrier of genetic information, 901, 903
  cDNA, 262
  synthesis by reverse transcriptase, 265
  change in conformation from B to Z, 224**F**
  chemical synthesis of, 236, 236**F**
  chromosome structure and, 233
  cloning, 248
  comparison of A, B, and Z double-helical forms, 221**T**, 222**S**
  conformational variations in, 220, 225, 226
  constancy of quantity and composition per cell, 905
  content of various organisms, 906**T**
  c^0t curves, 231, 232**F**
  cruciforms, 227
  CsCl density gradient ultracentrifugation of, 937, 939**F**
  damage to, 958-9
  denaturation, 230
  density as a function of G:C content, 233**F**
  dependence of meting temperature on G:C content, 2230**F**
  double helix, 191,192**F**, 219, 222**S**
    antiparallel nature of, 192**F**
  foreign, 249
  heat denaturation, 230, 230**F**
  helix, unwinding of, 944
  heteroduplex, 915
  hybridization, 232
  hyperchromic shift, 230
  intercalating agents, 226, 226**S**
  libraries of, 257
    cDNA, 262
    genomic, 257

DNA (deoxyribonucleic acid) *(Continued)*
screening, 258, 266
linking number in, 227, 229**F**
looping of, 983, 991, 991**F**
major groove in, 218**F**, 219**F**, 222**S**, 992
matrix of H-bond donors and acceptors, 920**F**
melting temperature (T_m), 230, 230**F**
methylation of, 959
minor groove, 218**F**, 219**F**, 222**S**
molecules, size of, 193
mouse satellite, 232**F**
supercoils in, 227**F**, 229**F**
probe, 233
protein interaction with, 1014, 1016
reannealing, 231, 231**F**
recognition helix of, 998, 999**F**
recombinant, 247, 273
relaxed form of, 226, 227**F**, 229**F**
renaturation of, 231, 231**F**
repair of, 958, 959, 960, 960**F**, 961**F**, 961**B**
replication of. *See* DNA replication.
secondary structure, 216, 217**F**, 221**F**, 222**S**
sense strand of, 965**B**
sequencing of, 210, 211**F**, 216, 217**F**
single-stranded, 200
specific linking difference in supercoiled DNA, 228
strand reactions of, 230, 918
sugar-phosphate backbone of, 189**F**
supercoils of, 227
superhelix density of, 228
synthesizer of, 236
topoisomerases and, 228, 228**F**
as the transforming principle, 903
triple helix formation and, 920**F**, 921**F**
twist, 227, 229**F**
viruses, DNA content of, 906**T**
writhe, 227, 229**F**
Z-form, 221**T**, 222**S**, 223
and gene expression, 225
structural characteristics of, 224
zippering up of, 231**F**
DNA-binding proteins, 992, 992**F**, 993**F**
DNA glycosylase, 960, 960**F**
DNA gyrase, 228, 228**F**
in DNA replication, 945, 949**F**, 950**T**
DNA ligase, 250, 949, 954**F**
in DNA repair, 960, 960**F**
in DNA replication, 948, 948**F**, 949, 949**F**, 950**T**, 954**F**
mechanism of action of, 949**F**
phage T4, 251, 251**F**
DNA polymerase(s), 209, 940, 955, 955**T**, 956, 960
bacteriophage T7, 210
in DNA repair, 960, 960**F**
of *E. coli*, 943**T**
immunoglobulin genes and, 924
from *Thermus aquaticus (Taq)*, 271, 272
DNA polymerase I
chain elongation reaction, 941**F**
E. coli and, 210, 210**F**, 365**T**, 940, 941, 949**F**, 950**T**, 954**F**
excising the RNA primer of, 948, 948**F**, 954**F**
DNA polymerase I, 3′ → 5′ activity of, 941, 941**F**
active sites of, 942
Klenow fragment of, 942, 943**F**
limited proteolysis of, 942
DNA polymerase III
β-subunit dimer of, 944, 944**F**, 949**F**
"core" of, 943, 943**T**
dimer replicative form of, 949**F**
E. coli and, 943, 944, 944**T**, 950**T**, 951, 953**F**
DNA polymerase III
holoenzyme of,
lagging strand of, 954**F**
leading strand, 954**F**
processivity of, 944
DNA replication, 209**F**
bidirectionality of, 944, 945**F**
in CHO cells, 947**F**
conservative model of, 937, 938**F**
daughter strand formation and, 937**F**
dispersive model of, 937, 938**F**
DNA ligase in, 948, 948**F**, 949, 949**F**
DnaT in priming DNA replication, 953**F**
in *E. coli*, 950, 951**F**, 954**F**
elongation and, 951, 953**F**
enzymology of, 940
eukaryotic, 955
helicases and gyrases in, 945, 959**T**
lagging strand in, 946, 946**F**, 949**F**
leading strand in, 946, 946**F**
mechanism of, in *E. coli*, 950, 950**T**
origins of replication, 944
priming of, 948, 948**F**, 951
RNase H in, 948
semiconservative nature of, 193**F**, 937, 938**F**, 939**F**
semidiscontinuous, 945, 946**F**, 947
termination of, 954
unwinding of helix and, 944, 949**F**
viral, 956
DNA:RNA hybrids, A-form of, 221
DnaA, 950, 950**T**, 952**F**, 953**F**
DnaB, 950, 950**T**, 952**F**, 953**F**
DnaC, 950, 950**T**, 952**F**
DNases, 200**T**, 202
DnaT, 950**T**, 953**F**
DNP-amino acid, 107, 108**F**
Docking protein, 1069
Dodecyl octaoxyethylene ether, 293**S**
Dolichol(s), 289
Dolichol phosphate, 290**S**
DOPA, 484
Dopamine, 482
Dot-matrix method of nucleotide sequence alignment, 1078, 1079**F**
Doubly wound parallel β-sheet, 162**S**
Dowex, 76
dsDNA, 200**D**
dTDP, 187
dTMP, 187, 895**F**
DTNB, 68**S**
dTTP, 187
regulation of dCMP deaminase by, 895**F**
regulation of ribonucleotide reductase by, 894, 894**F**
du Vigneaud, V., 495
Duckweed, nitrogen fixation and, 829**T**
dUDP
as precursor to dTMP, 895
as precursor to dTTP, 893
Duodenum, degradation of fatty acids in, 733, 734**F**
dUTPase, in dTMP synthesis, 895
Duysens, 706
Dyad-symmetric DNA target sequence, 1001**F**
Dynamics in proteins, 158
Dynein, 95**T**, 97

E/P state of the ribosome, 1053
E. coli See *Escherichia coli*.
E.C. *See* Energy charge *or* Enzyme Commission
E-F hand, 157, 158**S**
E1A (zinc finger protein), 996**T**
Eadie-Hofstee plot, 366**D**, 367**F**
Early cells, H^+-ATPases and, 1094**F**
Earth, elemental composition of crust, 6**T**
EcoRI restriction endonuclease, 203, 204**T**
Ecosystems, 550**D**
Edidin, M., 296
Edman degradation, 108, 110**F**
Edman sequenators, 108
Edman's reagent 108, 110**F**
EF-1, EF-2 (elongation factors), 1061
EF-G
protein synthesis and, 1051**T**
translocation and, 1053
EF-T, 1051**T**, 1052
Effector(s), heterotropic, 408, 411, 413**F**
Effector molecules, 390**D**
Egg white injury, 495
Ehlers-Danlos syndrome, 156
Eicosanoids, 278
biosynthesis of, 773, 780
as local hormones, 784
eIF (eukaryotic initiation factor), 1058, 1059**F**
eIF proteins properties of, 1060**T**
eIF-2, regulation of, 1061**F**
eIF-4F complex (CBP-II), 1059**F**
Eijkman, Christian, 464
Einstein, (unit of light energy), 715**D**
ELAM-1, 170**F**
Elastase, 113, 122, 433**F**, 436, 437
amino acid sequence of, 437**F**
substrate binding pocket of, 438**F**
Elastin, 95**T**, 97
Electroblotting, 260**F**
Electrolytes, 43
Electron hole, 719
Electron transfer potential, 528**T**
Electron transport, 628**D**
chain of, 633**F**, 634**F**, 645**F**
Electrophilic catalysis, 432
Electrophoresis 132
of DNA molecules, 206**F**, 207
polyacrylamide gel, 209, 261**F**
Electroporation 914**D**
Electrostatic effects, in ES complex, 428, 429
Electrostatic interactions, in proteins, 137
Electrostatic repulsion, in high-energy molecules, 530, 531**F**, 532**F**

Elementary reaction, 357**D**
β-Elimination, PLP-assisted, 481**F**
β-Elimination, PLP-catalyzed mechanism, 484**F**
γ-Elimination, PLP-assisted, 481**F**
Ellman's reagent, 68**S**
Elongation factors, 1051, 1051**T**, 1052**F**, 1054**F**
Elvehjem, 473
Embden, G., 570
Embden-Meyerhof pathway, 570
EMBL (European Molecular Biology Laboratory Data Library), 117
Emerson, 508, 705, 706
Emerson enhancement effect, in photosynthesis, 705
Enantiomer(s), 56**D**, 71**D**
  of sugars, 313
Endergonic process, 516
Endo cleavage, of nucleic acids, 200
Endoamylase, 328
Endocytosis, 344
  of lipoproteins, 796**F**
Endoglycosidase, 671
Endonucleases, 200**T**
Endoperoxide synthase, 784**F**
Endoplasmic reticulum, 25, 27**T**, 28**T**
Endosymbionts, zooxanthellae as, 551**B**
Energetic contributions to protein stability, 166**F**
Energy charge, 816**F**, 817**D**
  metabolic steady state and, 817
  response of enzymes to, 817, 817**F**
Energy flow in biosphere, 549
Energy metabolism
  in adipose tissue, 819**T**
  in brain, 819**T**
  in heart muscle, 819**T**
  in liver, 819**T**
  in skeletal muscle, 819**T**, 821
  in vertebrate organs, 819**T**
Energy storage, 732**T**
  in the adenylate system, 816
Energy transduction, 558**D**, 705
Energy yield from β-oxidation of fatty acids, 745**T**
Enhancer-promoter interactions, via protein-mediated DNA loop, 991**F**
Enhancers
  of eukaryotic transcription, 989**D**
  mechanism of action, 990**T**
Enol phosphates, 524, 533
Enolase, 533, 583**F**, 588**F**
Enolase, τ-crystallin and, 1091, 1092**T**
Enol-1-*o*-carboxyphenylamino-1-deoxyribulose P, Trp synthesis, 859**F**
3-Enolpyruvylshikimate-5-P synthase, in chorismate synthesis, 858**F**
2,3-*trans*-Enoyl-ACP reductase, 764
Enoyl-CoA hydratase, 742, 743**F**, 750**F**, 768
Enoyl-CoA isomerase, 750**F**, 751**F**
Enoyl-CoA reductase, 768
Enterobacteria, nitrogen fixation and, 829**T**
Enthalpy, 513**D**
  of solvation, 429
Entropy, 514, 515**D**, 517
  loss of in ES formation, 426, 428
Enzyme(s), 18, 20**F**, 94, 352**D**
  activity of
    controls over, 389
    feedback regulation of, 394
    organization in metabolic pathways, 552, 553**F**
    pH dependence of, 49**F**
  as agents of metabolic function, 353
  cascade, 682
  catalytic power of, 353
  cofactors, 356**T**
  converter, 390, 390**F**
  inhibition, 369
  interconvertible, 390, 390**F**
  modifiying, 390
  nomenclature, 354, 355**T**
  regulation
    allosteric, 390**D**
    through synthesis and degradation, 561
    by covalent modification, 560
  specificity, 353, 387
  synthesis, 389
  units, 364**D**
  regulation of, 354, 354**D**
Enzyme Commission, 354
Enzyme-substrate complex, 425
Epidermal growth factor, 348
    as precursor to, 1090**F**
Epimerases, 355**T**
Epimers of sugars, 314
Epinephrine, 61, 62**S**, 561, 685, 686**S**
  in muscle contraction, 821
Episome, 910**D**
Epstein-Barr virus, 266
Equatorial substituents, 317
ER *See* Endoplasmic reticulum.
Ergocalciferol, 463**T**, 506**S**
Ergosterol, 505**S**, 506
Ergot, 506
Erythromycin, 1062**S**, 1063**T**
Erythrose, 312
Erythrose-4-P
  in CO_2 fixation, 724**F**, 725**T**
  shikimate pathway and, 857
Erythrulose, 313
Escherichia coli, 22**F**
  base DNA composition of, 907**T**
  bidirectional chromosome replication of, 945**F**
  genetic map of, 913**F**
  sequence comparisons, 1080**F**, 1096**F**
Escherichia spp., and nitrogen fixation, 829**T**
Essential fatty acids, 770
Esterases, 433**T**
Esters, resonance forms of, 478**F**
Estradiol, 291**S**
β-Estradiol, synthesis of, 800**F**
Estrogen receptor, a C_x-type of zinc finger protein, 998**T**
Estrogens, 291, 799
Ethanol metabolism and the NAD^+/NADH ratio, 824
Ethanolamine, 775**F**
Ethanolamine ammonia lyase, 489
Ethanolamine kinase, 775**F**
Ether glycerophospholipids, 284
Ethidium bromide, 226**S**
Ethyl nitrosourea, as chemical mutagen, 931**F**
N-Ethylmaleimide, 68**S**
Ethylmethane sulfonate, as chemical mutagen, 931**F**
Eubacteria, 20
Eubacterial F-ATPase, evolution of, 1097**F**
Eudesmol, 289**S**
Eukaryotes, relation to archaebacteria, 1098**F**
Eukaryotic cells, 9, F24**F**, 25**F**, 26-27**T**, 955**F**
Eukaryotic initiation factors, properties of, 1060**T**
Eukaryotic initiator tRNA ($tRNA_i^{Met}$), 1058
Evans, Herbert, 508
Evolution, 1076
  cellular, 1098**F**
  coding regions *vs.* noncoding regions, 1078
  of F-ATPases and V-ATPases, 1097**F**
  molecular, 1075
Evolved coupling stoichiometries, 807
Exchange reaction, 380**D**
Excision exonuclease, in DNA repair, 960, 960**F**
Exciton transfer, in photosynthesis, 703**F**
Exercise, glycogen utilization in, 684
Exergonic process, 516
Exit (E) site for tRNA release from ribosome, 1047, 1047**F**
Exo cleavage, of nucleic acids, 200
Exoamylase, 328
Exoglycosidase, 672
Exon(s), 194, 195**F**, 1002**D**
  insertion, 1088
  ligation, 1008**F**, 1009**F**
  number *vs.* polypeptide length, 1091**F**
  shuffling, 1075
    evolution by, 1084, 1085, 1086**F**, 1090
Exon-intron junctions, consensus sequence, 1006**T**
  combinatorial, 1010**F**, 1011
  constitutive, 1010**F**, 1011
  correlations with protein structure, 1086, 1086**F**
  mutually exclusive, 1010**F**, 1011
  evolution of proteins and, 1085
Exonucleases, 200**T**, 941, 941**F**, 942**F**, 943**F**
Exopeptidase, 451
Expression library, screening with antibodies, 267
Expression vectors, 265, 266**F**, 267**F**, 268**F**
Extracellular matrix, 26**T**, 339
Extrinsic pathway of blood clotting, 392, 393**F**
Extrinsic protein, 295

Facultative anaerobes, 548
F^+ cells, 910
F^- cells, 910
F factor plasmid, 910
F factor transfer, 911**F**
F_{AB}, of immunoglobulin G, 925**F**
F_C, of immunoglobulin G 925**F**
F-ATPase(s), 1095, 1095**F**, 1095**T**
  V-ATPase evolution and, scheme for, 1097**F**

F_1F_0-ATPase, 647, 648**T**
f-Met-$tRNA^{Met}$ in prokaryotic protein synthesis initiation, 1047
FAB-MS, 115
Factor B, 170**F**
Factor VII, 170**F**
Factor VII_a of blood clotting, 392, 393**F**
Factor IX, 170**F**
Factor IX, structural modules in, 1090**F**
Factor IX_a of blood clotting, 392, 393**F**
Factor X, 170**F**
Factor X_a of blood clotting, 392, 393**F**
Factor X, structural modules in, 1090**F**
Factor XI_a of blood clotting, 392, 393**F**
Factor XII, 170**F**
Factor XII_a of blood clotting, 392, 393**F**
FAD, 100**T,** 463**T,** 473**D,** 474**S,** 475**S**
FADH, semiquinone, 475**S**
$FADH_2$, 475**S**
  ATP value of, 809
FAICAR. *See* *N*-formylaminoimidazole-4-carboxamide ribonucleotide.
Familial hypercholesterolemia, 797
Faraday's constant, 529
Farnesyl pyrophosphate, 789, 790**F**
FAS, 765
Fasman, Gerald, 171
Fast atom bombardment mass spectrometry (FAB-MS), 115
Fasting state, 670
Fat content of modern diets, 732
Fatigue, muscle 821
Fatty acid(s), 277**D,** 277**T**
  metabolism, regulation of, 771**F**
  oxidation
    regulation of, 771**F**
    as a source of metabolic water, 746
  synthesis
    elongation in, 768**F**
    introduction of double bonds in, 768
    regulation of, 771**F,** 772**F**
  triacylglycerol breakdown and, 733**F**
  turnover, in liver, 824
Fatty acid synthase, 758, 762**F,** 765**F**
Fatty acyl synthase
  mechanism of, 767**F**
  structure of, 765**F,** 766
Fatty acid thiokinase, 737
Fatty acyl-CoA dehydrogenase, 637**F**
FCCP, 654**F**
Fe^{2+}, Fe^{3+}, 356**T**
Feedback inhibition of enzyme activity, 394
Feedback regulation, 560**F**
Fehling's solution, 319
FeMoco. *See* Iron-molybedenum cofactor.
Ferredoxin(s), 165**S**
  in the nitrogenase reaction, 832**F**
  in photosynthesis, 708**F,** 711
Ferredoxin:$NADP^+$ reductase, 708**F,** 709, 711
Ferredoxin-thioredoxin reductase, 727**F**
Ferritin, 95**T,** 97, 100**T**
FGAM. *See also* Formylglycinamidine ribonucleotide.
FGAM cyclase, in purine biosynthesis, 873, 874**F**
FGAM synthetase, in purine synthesis, 873, 874**F**
FGAR. *See* α-*N*-formylglycinamide ribonucleotide.
FGAR amidotransferase, in purine synthesis, 873, 874**F**
Fibrin, 1089
  in blood clotting, 392, 393**F**
Fibrinogen, 95**T,** 97
  in blood clotting, 392, 393**F**
Fibrinolysis proteins, structural modules in, 1090**F**
Fibroblast growth factor, 347, 348
Fibroin 95**T,** 97, 150
Fibronectin, 100**T,** 169**S,** 170**F,** 346, 1090**F**
Fibrous proteins, 149**D**
Fight or flight hormone, 686
Filamentous phage, 255
Filaments, intermediate, 27**T**
Filmer, 414
Fingerlike modules, in mosaic proteins, 1089
First law of thermodynamics, 512**D**
First-order reaction, 357**D**
Fischer, Edmund, 390, 679
Fischer, Emil, 70
Fischer convention, 70, 312
Fischer projection, 70**S**
Fish, DNA content of, 906**T**
Flagella, 23
Flavin, 474**S**
Flavin adenine dinucleotide, 356**T,** 463**T,** 474S *See also* FAD.
Flavin mononucleotide, 463**T,** 474**S**
Flavodoxin, 163**S,** 163**G**
  in the nitrogenase reaction, 832**F**
Flavoenzymes, 474
Flavoprotein 2, 100, 100**T,** 634, 637
  -catalyzed reactions, 476**T**
Flavoprotein reductase, 487
Fleming, Alexander, 446
Flippases, 300**F**
Fluid mosaic model, 295**F**
Fluorescein, 296
Fluorescence, in photosynthesis, 704, 704**F**
Fluoro compounds as enzyme inhibitors, 897**B**
Fluoroacetate, 606, 626, 897**B**
Fluoroacetyl-CoA, 607**F**
Fluorocarbonylcyanide phenylhydrazone, 654
Fluorocitrate, 606, 626
5-Fluorocytosine, 896**S,** 897**B,** 898
5-Fluoroorotate, 896**S,** 897**B,** 898
Fluorophosphates, 439
5-Fluorouracil, 896**S,** 897**B,** 898
5′-Fluorouridylate, 897
5-FU. *See* 5-Fluorouracil
FMN, 100**T,** 463**T,** 474**S,** 475**S**
$FMNH_2$, 475**S**
Folate, 498**S**
Folding of proteins, 139, 168**F**
Folic acid, 463**T,** 498, 500**S**
  from chorismate, 857**F**
  purine biosynthesis and, 876
Folin-Ciocalteau reagent, 103**B**
Folkers, Karl, 788
Food chains, 3
N^5-Formimino THF, 499**S,T**
Formylglycinamidine ribonucleotide, in purine synthesis, 873, 874
Formylglycinamidine ribonucleotide cyclase, 873, 874**F**
Formylglycinamidine ribonucleotide synthetase, 873, 874**F**
α-*N*-Formylglycinamide ribonucleotide, in purine synthesis, 873, 874**F**
α-*N*-Formylglycinamide ribonucleotide amidotransferase, 873, 874**F**
N-Formyl Met-$tRNA^{fMet}$, 501
N-Formyl-Met-$tRNA_f^{Met}$ synthesis, 1048**F**
Formylphenylalanine methyl ester, 438**S**
N^5-Formyl THF, 499**S,** 499**T**
N^{10}-Formyl THF, 499**S,** 499**T,** 501**F**
  in protein synthesis initiation, 1048**F**
  in purine biosynthesis, 873, 874**F,** 875
Förster resonance energy transfer, in photosynthesis, 704
Fos (a leucine zipper protein), 999, 1000, 1001
Four-cutter, restriction endonuclease, 205
Fourfold degenerate sites, 1078
Fp, of photosynthesis, 708**F**
Franckia, nitrogen fixation and, 829**T**
Free radical(s),
  in ribonucleotide reductase reaction mechanism, 892**F**
  quenching agents, 892
Free energy, 516**D**
Freeze etch, 297**F,** 299**F**
Freeze fracture, 297**F,** 298**F**
Freezing point depression, 39
Frog, base DNA composition of, 907**T**
Fructofuranose, 315
Fructose, 313
  in glycolysis, 593
Fructose bisphosphatase in CO_2 fixation, 724**F,** 725**T**
Fructose-1,6-bisphosphatase, 320**F,** 665
  activation by reduction, 726, 727**F**
  in CO_2 fixation, 724**F,** 725**T**
  in glycolysis, 572**F,** 574**T,** 580
  inhibition of, 669**F**
  light-induced pH activation of, 726
  substrate cycles and, 812
Fructose bisphosphate aldolase, 572**F,** 574**T,** 580
Fructose-1-P, hydrolysis of, 521, 527**T**
Fructose-2,6-bisphosphate, 665, 669, 669**S,** 670**F**
  glycolysis in, 572**F,** 574, 579
  synthesis and degradation of, 670**F**
Fruit fly, base DNA composition of, 907**T**
Frye, L., 296
FTR. *See* ferredoxin-thioredoxin reductase.
Fucose, 320**F**
Fumarase, 366**T,** 602**T,** 604**T,** 611**F**
Fumarate, 602**F,** 604**T**
  urea cycle and, 844**F**
  from Asp, 883**F**
  from purine biosynthesis, 874**F**
  in tyrosine degradation, 869, 869**F**
4-Fumarylacetoacetate, 869, 869**F**
Fungi, DNA content of, 906**T,** 907**T**
Funk, Casimir, 464
Furan, 315
Furanose, 314, 316**F**
  in nucleic acids, 183**S**

Furanose *(Continued)*
puckered conformations of, 220**F**
Fusidic acid, 1062**S**, 1063**T**
Fusion protein expression, 268, 269**T**
Futile cycle. *See* Substrate cycle.

G proteins, 307, 682
GABA, 61, 62**S**
gag mRNA, 444**F**
gag proteins, 307
gag-pol polyprotein, 444**F**
GAL4 (zinc finger protein), 996**T**
Galactokinase, 594**F**, 595
Galactosamine, 320, 321**F**
Galactose, 312
in glycolysis, 593, 594, 595
Galactose-1-P uridylyltransferase, 594**F**, 595**F**
Galactosemia, 595
β-Galactosidase 364**T**, 975, 976**F**
as gene fusion system, 269**T**
blue/white selection and, 268
β-Galactoside permease, 976**F**
β-D-Galactosylceramide, 782**F**
Galactosylceramide, synthesis of, 780
galE mRNA Shine-Dalgarno sequence, 1049**F**
galP2 promoter, nucleotide sequence of, 967**T**
Ganglioside(s), 287, 287**G**, 287**S**, 782**F**
synthesis of, 780
Gap penalty, in nucleotide sequence alignments, 1079, 1080
GAR. *See* Glycinamide ribonucleotide.
Garrod, A. E., 901
Gauche configuration, lipid chain, 302
GC box, as eukaryotic promoter module, 989, 989**T**
GCN4, amino acid sequence of its bZIP DNA-binding motif, 1000**F**
GDH, see glutamate dehydrogenase
GDP, 187, 878, 879, 891**F**, 894, 894**F**
GDP-mannose, 538**S**
Gel electrophoresis, 2-dimensional, 133, 134**F**
Gel phase, in membranes, 301**F**
GenBank, 117
Gene(s)
assembly of in immunoglobulin heavy-chain genes, 927**F**
chemically synthesized, 237, 238**T**
chimeric, 247
conversion 1075**D**
duplication, 1075, 1084
expression, regulation of, 963ff
negative regulation of, 978
positive *vs.* negative control, 981**F**
fusion systems, 269**T**
jumping, 922
pool, 1076**D**
reorganization, in immunoglobulin genes, 924
replacement therapy, 274
repression, 562**F**
sharing, 1091
therapy 273**B**, 881, 934
General acid/base catalysis, 427, 433
in lysozyme mechanism, 451**F**
Genetic code, 1017, 1025**T**

Genetic code *(Continued)*
elucidation of, 1019, 1020, 1021**T**, 1024**F**
general nature of, 1020, 1023
natural variation in, 1026**T**
overlapping, punctuated, or continuous? 1019**F**
triplet nature of, 1020
universality of, 1026
Genetic drift, 1076**D**
Genetic engineering, 247
Genetic map of *E. coli* K12, 913**F**
Genetic recombination, 909, 909**F**, 914**D**
Genetics in bacteria, 906, 908
Genome, 906**D**
Genomic DNA libraries, 257
Genotype, 907**D**
Geraniol, 288
Geranyl-CoA, biotin-dependent carboxylation of, 493**T**
Gerbil, dependence of on fatty acids, 746
Germline genes, of immunoglobins, 925
Gibberelic acid, 289**S**
Gibbs, J. Willard, 516
Gibbs free energy, 516
Gilbert, Walter, 209, 212
Giraffe, cellulose degradation by, 332
GlnA and *E. coli* glutamine synthetase gene expression, 838
Globin chains, amino acid sequence of, 122**F**
β-Globin gene, promoter region, 974**F**
Globin genes, 120, 122**F**, 123**F**, 1088**F**
Globular proteins, 156**D**
γ-Globulin, 85**T**, 100**T**, 923
Glucagon, 685**S**, 686, 732
regulation of fatty acid synthesis by, 772**F**
Glucans, 326
Glucaric acid, 318**F**
Glucitol, 319**F**
Glucocorticoid(s), 291, 686, 687
receptor (zinc finger protein), 998**T**
regulation of body processes by, 801
response element (GRE), 990, 990**T**
Glucogenic amino acids, 864**F**, 865
Glucokinase, 354
in glycolysis, 572**F**, 573, 574**T**, 576
Glucolactonase, 689
Gluconeogenesis, 661, 662**F**
effect of cortisol on, 687
regulation of, 667, 668
Gluconic acid, 318**F**
Gluconolactone, 318**F**
Glucopyranose, 314
Glucosamine, 320, 321**F**
D-Glucose, 70**S**, 312
Glucose
as product of photosynthetic CO_2 fixation, 723, 724**F**, 725**T**
in glycolysis, 572**F**
Glucose oxidase, 476**T**
Glucose oxidation, reaction profile showing large $\Delta G^{\ddagger}$, 353**F**
Glucose phosphate isomerase, 576
Glucose transporter, 95**T**, 97**F**
Glucose-1-phosphate, 320**F**
hydrolysis of, 527**T**
Glucose-6-P
in CO_2 fixation, 724**F**, 725**T**

Glucose-6-P *(Continued)*
hydrolysis of, 527**T**
metabolic conversions in liver, 823**F**
Glucose-6-P dehydrogenase, 689
Glucose-6-phosphatase, 663, 665
in glycolysis, 572**F**, 574**T**
$\alpha(1 \rightarrow 6)$-Glucosidase, action of, 671**F**, 672
Glucuronic acid, 490**S**
Glue proteins, 95**T**, 98
Glutamine:PRPP amidotransferase, inhibition by A and G nucleotides, 877
Glutamate, degradation to α-ketoglutarate, 866**F**, 867
Glutamate aminotransferase, in Val and Ile synthesis, 852**F**, 853
Glutamate dehydrogenase, 85**T**, 364**T**, 472**T**, 509
ammonium assimilation and, 832, 832**F**
Glutamate mutase, 489, 509
Glutamate oxoglutarate amidotransferase, 834
Glutamate synthase, 834, 834**F**
Glutamate-5-P, and proline biosynthesis, 841**F**
Glutamate-5-semialdehyde, and proline biosynthesis, 841**S**
Glutamate-aspartate aminotransferase, 839**F**, 840
Glutamic acid, 58**S**, 58**G**, 65**F**
Glutamine, 58**S**, 58**G**
acid lability of, 87
in carbamoyl phosphate synthesis, 886, 886**F**
in CTP synthesis, 889
degradation to α-ketoglutarate, 866**F**, 867
in GMP biosynthesis, 876, 877**F**
as major N donor, 833
in purine biosynthesis, 873, 874**F**
Glutamine amidotransferase, 860, 862**F**
Glutamine phosphoribosyl pyrophosphate amidotransferase, 873, 874**F**
Glutamine synthetase, 85**T**, 86
ammonium assimilation and, 833, 833**F**
E. coli and, 835, 835**F**, 836, 836**F**, 837**F**, 838
Glutaminyl-tRNA^{Gln} synthetase, and tRNA^{Gln} recognition, 1030, 1031**F**
Glutamyl carboxylase, vitamin K-dependent, 508**F**
γ-Glutamyl kinase, proline biosynthesis and, 841**F**
Glutaredoxin, in ribonucleotide reduction, 893
Glutaryl-CoA, in lysine degradation, 868, 868**F**
Glutathione, 893**S**
Glutathione peroxidase, 356**T**
Glutathione reductase, 476**T**, 893
domain 3 of, 164**S**, 164**G**
Glutathione-S-transferase, as gene fusion system, 269**T**
Glycans, 326
Glyceraldehyde, 312, 71**S**
R and S isomers of, 73
Glyceraldehyde-3-P
in CO_2 fixation, 724**F**, 725**T**
in glycolysis, 572**F**, 574**T**, 580, 583**F**

Glyceraldehyde-3-P dehydrogenase, 432**F,** 433**T,** 583**F,** 584**F**
kinetic mechanism of, 381
light-induced pH activation, 726
NADPH-specific, 723, 724**F,** 725**T**
Glyceraldehyde-P dehydrogenase, domain 2 of, 164**S**
Glycerokinase, in glycerolipid synthesis, 773, 744**F**
Glycerol, 280**S**
use of, in glycolysis, 596**F**
Glycerol dehydrase, 489
Glycerol kinase, 596**F**
Glycerol-3-phosphate, 282, 596**F**
sn-Glycerol-3-phosphate, hydrolysis of, 527**T**
Glycerol-3-phosphate acyltransferase, 773, 744**F**
Glycerol-3-phosphate dehydrogenase, 773, 774**F**
Glycerol phosphate dehydrogenase, 596**F**
Glycerol phosphatide, 281
Glycerolipids, synthesis of, 773, 774**F**
Glycerophosphate dehydrogenase, 635
Glycerophosphate phosphatidyltransferase, 778**F**
Glycerophosphate shuttle, 654**F,** 655
Glycerophospholipid, 281
degradation of, 284**F**
synthesis of, 773
Glycinamide ribonucleotide synthetase, purine synthesis of, 873, 874**F**
Glycinamide ribonucleotide transformylase, 873, 874**F**
Glycine, 58**S, G;** 62-64; 68
in bile acid synthesis, 798**F**
biosynthesis from serine, 853, 855**F**
degradation of via serine, 865**F,** 866
as a product of photorespiration, 722, 722**F**
in purine biosynthesis, 873, 874**F**
Glycine oxidase, in glycine synthesis from serine, 853, 855**F**
Glycocalyx, 342**F,** 797
Glycocholic acid, synthesis of, 798**F**
Glycoconjugates, 310
Glycogen, 327, 328
as fuel depot, 819
branching of, 678
catabolism of, 671, 672, 685
effect of cortisol on, 687
metabolism, regulation by hormones, 683
replenishment, after exercise, 689**F**
synthesis of, 674
Glycogen phosphorylase, 100**T,** 329, 427**T,** 672, 673, 673**F,** 674**F,** 678, 680**F,** 681**G**
reaction mechanism of, 674, 675**F**
regulation by covalent modification, 560**F**
Glycogen phosphorylase kinase, 560
Glycogen synthase, 677**F**
regulation of, 678, 682
Glycogenin, 677
Glycolate (product of photorespiration), 722, 722**F**
Glycolate oxygenase, 722**F**
Glycolipid, 310
Glycolysis, 569
energetics of, 592**F**
Glycolysis *(Continued)*
regulation of, 668
Glycophorin, 298, 303
Glycoprotein(s) 98, 100**T,** 310, 339, 340
Glycoprotein peptidase, 374**F**
Glycosaminoglycans, 334, 345
Glycosides, 321
Glycosidic bond, 184
Glycosphingolipids, 286
Glycosylceramides, synthesis of, 782**F**
Glycyl-L-tyrosine, 452**G**
Glyoxylate, 623**F,** 722, 722**F**
Glyoxylate carboligase, 509
Glyoxylate cycle, 622, 623**F**
Glyoxysomes, 623, 625**F**
Glyphosate, 863**B**
GMP, 183, 185**S**
absorption spectrum, 183**F**
as feedback inhibitor of GMP synthesis, 877**F,** 878
synthesis from IMP, 876, 877**F**
GMP synthetase, 876, 877**F**
GMPCP (analog of GTP), 1051**F**
GOGAT (Glu oxoglutarate amidotransferase) *See* Glutamate synthase.
Goldberger, Joseph, 473, 483
Goldstein, Joseph, 797
Golgi apparatus, 25, 26**T,** 28**T,** 292**F,** 294**F**
Good's buffers, 52, 53**F**
Gout, 884
GPI anchors, 306, 307, 308**F**
Gram stain, 336
Grana, 700, 700**F**
GRE (glucocorticoid response element), transcription, 990, 990**T**
Greek key topology, in proteins, 164**F**
Griffith, Frederick, 903, 904**F**
GroEL, 167
Ground substance, 97
Group transfer potential, 528**T,** 529**D**
Growth factor modules, 169**S,** 1089
Growth hormone, 274, 933, 933**F,** 934
GS. *See* Glutamine synthetase.
GSH (reduced glutathione), 893**S**
GSSG (oxidized glutathione), 893**S**
GTP, 187
GTP, protein synthesis and, 1047, 1050**F,** 1051, 1052**F**
GTP-binding protein, 682
Guanidinium phosphates, 534**F**
Guanidino group, of arginine, 842, 844**S**
Guanidino phosphate, 524
Guanine, 182**S**
Guanine deaminase, in purine catabolism, 881, 882**F**
Guano, 871
Guanosine, 184**S**
Guanosine 5′-diphosphate (see also GDP), 187
Guanosine 5′-monophosphate (also see GMP), 185**S,** 186
Guanosine 5′-triphosphate (see also GTP), 187
Guanylate kinase, in GDP synthesis, 879
Guanylic acid (see also GMP), 185**S,** 186
Guanylyl transferase, 1003, 1004**F**
Gulonic acid, 490**S**
Gulonolactone, 490**S**
Gulose, 312
Gunsalus, I. C., 496
Györgyi, 495

H chain of immunoglobulin G, 923, 924**F,** 926, 926**F**
H^+-ATPase(s)
from archaebacteria, 1096
evolution of, 1093, 1094**F**
in eukaryotic cells, 1094, 1095**F,** 1095**T**
genes of, sequence comparisons, 1096, 1096**F**
properties of, 1095**T**
hν, as photon, 703
H2-TF1, transcription factor for RNA polymerase II, 989**T**
Hairpin turns, tRNA, 238, 238**F,** 239**F**
Hairpins, in proteins, 159**F**
Half cell, 630**F**
Halobacterium halobium, 298, 304
Halobacterium halobium H^+-ATPase, 1096
Halobacterium volcanii, rRNA, 242, 243**F**
Halophiles, 22
Hamkalo, Barbara, 1058
Hanes-Woolf plot, 367**D,** 367**F**
Haploid cells, 906**D**
Hartman, Philip, 863
Hatch, C-4 pathway of CO_2 fixation and, 727
Haworth, 491
Haworth projection formulas, 314**F,** 315
Hb S, 124. *See* also Sickle-cell anemia.
Hb. *See* Hemoglobin.
Hb F. *See* Hemoglobin, fetal.
HBP1, 1000**F**
HDL. *See* High-density lipoprotein
Heart muscle, energy metabolism in, 819**T,** 822
Heat, thermodynamic, 512
Heat capacity, 516
Heat shock, as physiological challenge, 990**T**
Heat shock element (HSE), in eukaryotic transcription, 990, 990**T**
Heat shock proteins, 167
Helical wheel analysis of C/EBP, 999**F**
Helicase(s), in DNA replication, 945
Helicase II, in DNA replication, 954**F**
α-Helix, 91**F**
amphiphilic, 1070**F**
Helix-breaking amino acids, 144**T**
Helix-forming amino acids, 144**T**
Helix-turn-helix DNA-binding proteins, 992, 993**F**
Helling, Robert, 252
Hematin, 396
Heme, 100, 101**F,** 639
of myoglobin and hemoglobin, 395, 401**B**
regulation of globin synthesis and, 1061
Hemerythrin, 176**S**
Hemiacetal, 314, 322
Hemiketal, 314, 322**F**
Hemoglobin, 84, T85, 93**F,** 95**T,** F97, 398
amino acid sequence of, F122
globin gene exon organization and, 1087, 1088**F**

Hemoglobin *(Continued)*
as allosteric protein, 394ff
β-subunit of, 144**S**
Bohr effect, 403, 403**F**
chains, conformational comparison, 398**F,** 399**F**
conformational changes in, 400**F,** 401**F**
O_2 binding and, 398, 399, 399**B,** 400**F,** 402, 403, 403**F,** 404**F,** 420, 421, 423, 423**F**
evolutionary relationships of, 120, F122, F123
fetal, 405, 406, 406**F**
human, from transgenic mice and pigs, 934
mutant, 124, 124**T**
salt bridges in, 402**F**
sickle-cell (Hb S), 124
Hemoproteins, 100**T,** 101
Henderson, Richard, 305
Henderson-Hasselbalch equation, 44, 64
Heparin, 334, 335**F,** 347
HEPES, 53**F**
Heptulose, 311
Herbicides, amino acid synthesis inhibitors as, 863**B**
Heredity, 900**D**
Hershey, Alfred, 904
Hershey and Chase experiment, 905**F**
Heterocyclic ring, 183
Heterodimer formation, in bZIP proteins, 1001
Heteroduplexes, DNA, 921, 922**F**
Heterogeneous nuclear RNA (*See also* hnRNA), 194, 195**F,** 1003
Heterologous interactions, in proteins, 172, 174**F**
Heteromultimeric, 84
Heteropolysaccharide, 326
Heterotrophs, 546**D**
Hexokinase, 85**T,** 354, 356**T,** 364**T,** 427**T,** 541
in glycolysis, 572**F,** 573, 574**T,** 575
structure of, 389**F**
Hexokinase domain 1, 163**S**
Hexose monophosphate shunt, 687, 688**F**
Hexulose, 311
Hfr, 910
E. coli and, 912**F**
HGPRT (hypoxanthine-guanine phosphoribosyltransferase), 879, 880**F**
in purine salvage, 879, 880**F**
deficiency (Lesch-Nyhan syndrome), 879, 880**F**
High density lipoproprotein(s), 100**T,** 792, 793**T**
High-energy phosphate compounds, 523
High-pressure liquid chromatography (HPLC), 134
High-performance liquid chromatography (HPLC), 134
High-potential iron protein, 165**S**
Hill plot, 421**D,** 421**F,** 423, 423**F**
Hill, Archibald, 421
Hill coefficient, 421**D**
Hirst, 491
his operon, 863
transcription attenuation and, 986**F**
Histamine, 61, 62**S,** 482
Histidine, 50, 59**S,**G, 484
biosynthesis of, 861, 863
degradation of to α-ketoglutarate, 866**F,** 867
evolution of, 864**B**
Histone(s), 193, 233, 234**F,** 234**T**
Histone H3, amino acid composition of 88**T**
Histone H2B gene, promoter, 988**F**
Histone octamer, 194**F,** 234**S**
HIV reverse transcriptase, 958
HIV-1 (human immunodeficiency virus), 100**T,** 307, 932, 957
HIV-1 protease, 442**G,** 442**T,** 444
inhibitor complex of, 445**F**
pH-rate profile of, 447
HMG-CoA, 786**F**
regulation of, 787**F**
HMG-CoA lyase, 754**F,** 755
HMG-CoA reductase, 786**F,** 787**F**
tetrahedral intermediate in mechanism of, 792**F**
HMG-CoA synthase, 754**F,** 786**F**
hnRNA (heterogeneous nuclear RNA), 194, 1003
Hodgkin, Dorothy, 485
Hogness, David, 967
Holliday junction, 916, 917**F**
resolution of, 921, 922**F**
Holliday, Robin, 916
Holoenzyme, 356**D,** 457
Homeobox domain, 994, 995**F**
Homeostasis, caloric, 819
Homocitrate
in lysine biosynthesis, 845, 846**F**
in the iron-molybedenum cofactor of nitrogenase, 831**F**
Homocysteine, 61, 62**S**
in methionine biosynthesis, 849**F,** 850
Homocysteine methyltransferase, 849**F**
Homodisaccharides, 322
Homogentisate in tyrosine degradation, 869, 869**F**
Homogentisate dioxygenase
deficiency of, in alcaptonuria 869**F**
in tyrosine catabolism, 869**F**
Homoglycan, 326
Homoisocitrate, in lysine biosynthesis, 845, 846**F**
Homolytic cleavage, of Co-C bond by B_{12}, 488
Homomultimericity, 84
Homopolysaccharide, 326
Homoserine, 61, 62**S**
in Thr and Met biosynthesis, 849**F,** 850
Homoserine acyltransferase, 849**F,** 850
Homoserine dehydrogenase, 849**F,** 850
Homoserine kinase in threonine biosynthesis, 849**F,** 850
Homoserine lactone, 114**F**
Hopkins, Frederick Gowland, 504
Hormones, 291
Housekeeping genes, 989**D**
HPLC, 134
HSE (heat shock element), in eukaryotic transcription, 990, 990**T**
hsp60, 167, 1066, 1067**F**
hsp70, 167, 1066, 1067**F**
HSTF (heat shock transcription factor), 990, 990**T**
HTH (helix-turn-helix) DNA-binding proteins, 992, 993**F**
HTH domains, 993**F,** 994**F**
HU, histone-like protein in DNA replication, 950**T,** 951
Huber, Robert, 712
Human immunodeficiency virus (HIV), 100**T,** 307, 442**G,** 442**T,** 444, 932, 957, 958
Human(s), base DNA composition of, 907**T**
Human growth hormone (hGH) in transgenic mice, 934
Human Genome Project, 264**B**
Human leukocyte associated proteins, 304
Hummingbird, dependence on fatty acids, 746
Hunchback, zinc finger protein, 996**T**
Hyaluronate, 334, 335**F,** 349**F**
Hyaluronic acid, 327
Hybrid protein, 268
Hydration, 35, 36**F,** 41**F**
of ions, 519**T**
Hydrazine, 109, 110**F**
in nucleic acid sequencing, 213, 214**F**
Hydrazinolysis, 109, 110**F**
Hydride ions and NAD^+ reduction, 557, 558**F**
Hydrocarbon, transfer of, to water, 519**T**
Hydrogen
intrastrand bonding of, 238
Hydrogen bonds 14, 16**F,** 137
ions of, 40
Hydrogenase activity of nitrogenase, 831
Hydrolases, 355**T**
Hydronium ions, 40
Hydrophobic interaction chromatography (HIC), 134
Hydrophobic interactions, 10, 14**T,** 15, 37, 137
Hydroxocobalamin, 486**S,** 487
β-Hydroxyacyl-ACP dehydratase, 764
Hydroxyacyl-CoA dehydrogenase, 743**F,** 768
λ-crystallin gene sharing and, 1092**T**
Hydroxyapatite, 154
p-Hydroxybenzoate, from chorismate 857**F**
β-Hydroxybutyrate and energy metabolism in the brain, 820, 820**S**
β-Hydroxybutyate dehydrogenase, 754**F,** 755
7α-Hydroxycholesterol, 799**F**
in bile acid synthesis, 798**F**
β-Hydroxydecanoyl thioester dehydrase, 768, 769**F**
β-Hydroxydecanoyl-ACP, 769**F**
Hydroxyethyl-TPP, 465, 466**S**
in pyruvate dehydrogenase, 497
Hydroxyl ions, 40
Hydroxylamine, as a chemical mutagen 930, 931**F**
7α-Hydroxylase, 799**F**
Hydroxylysine, 60, 61**S**
5-Hydroxylysine (Hyl), 153

p-Hydroxymercuribenzoate, 68**S**
effect of on aspartate transcarbamoylase 409
3-Hydroxy-3-methylglutaryl-CoA, (β-hydroxy-β-methylglutaryl-CoA), 755, 755, 786**F**
β-Hydroxy-β-methylglutaryl-CoA, in leucine degradation, 868, 868**F**
p-Hydroxyphenylpyruvate, in tyrosine degradation, 869, 869**F**
4-Hydroxyphenylpyruvate, in tyrosine synthesis, 859**F**
p-Hydroxyphenylpyruvate dioxygenase, in tyrosine catabolism, 869**F**
Hydroxyproline, 60, 61**S**
3-Hydroxyproline, 153
4-Hydroxyproline (Hyp), 153
8-Hydroxyquinoline, 892
Hydroxypyridinium, cross link in collagen, 155**S**
Hydroxyurea, 892, 892**S**
25-Hydroxyvitamin D_3, 505**S**, 507
Hypercholesterolemia, familial, 797
Hyperuricemia, 884
Hypervariable residues in IgG, 923, 924**F**
Hypoglycin, 741, 742**F**
Hypoxanthine, 182**S**, 879, 880**S**
from oxidative deamination of adenine, 930**S**
in purine catabolism, 882
oxidation of by xanthine oxidase, 884

Ibuprofen, 784**F**, 785
Ice, structure of, 34, 34**F**
IDLs, 792, 793**T**
Idose, 312
Iduronic acid, 318**F**
IF1, IF2, IF3 (*E. coli* initiation factors 1, 2, 3), 1049, 1050, 1050**F**, 1057**F**
IgG (immunoglobulin G), 923
molecule, organization of, 924**F**
oligosaccharide of, 344**F**
ilv operon, and transcription attenuation, 986**F**
Imidazole acetol phosphate, histidine and, 862**F**
Imidazole glycerol phosphate, histidine and, 861, 862**F**
Imidazole glycerol-P dehydratase, in histidine synthesis, 862**F**
Imidazole ring, 181
Immune response, 923
Immune system, stimulation of by ascorbic acid, 491
Immunoglobulin, 95**T**, 97, 167, 169**S**, 175**F**
fold of, 924, 925**F**
genes of, 923, 924, 927**F**
Immunoglobulin G, 13**F**, 17**F**, 923
molecular organization of, 924**F**
IMP(s), 297, 298**F**
biosynthesis of, 872
in purines and, 874**F**, 875, 882**F**
IMP cyclohydrolase, in purine biosynthesis, 874**F**, 875
IMP dehydrogenase, in GMP biosynthesis, 876, 877**F**
IMP synthase, in purine biosynthesis, 874**F**, 875
Inactivator, mechanism-based, 373
Indigo blue, 269
Indirect readout, recognition of DNA by, 995
Indole, tryptophan and, 859**F**, 860, 861, 861**G**
Indole-3-glycerol phosphate, in Trp synthesis, 859**F**, 860
Indole-3-glycerol-P isomerase, product of the *trpC* gene, 985**F**
Indole-3-glycerol-P synthase, in Trp synthesis, 859**F**, 860
Induced dipole, 138**D**
Induced fit hypothesis, 388
Inducers, 976
Induction
enzyme synthesis and, 389, 561, 561**F**
gene expression and, 975
Influenza virus hemagglutinin, 161**S**
Information transfer, 181**F**
Inhibition, mixed, 369, 370**D**, 370**F**, 370**T**, 371**F**, 374, 394
Inhibitors
metabolic pathways and, 563**F**
as herbicides, 863**B**
of electron transport, 652**F**
Initiation codons (AUG, GUG, or UUG), 1047
Initiation complex, 1049, 1050**F**, 1059**F**
Initiation factors, 1047, 1049**T**, 1049, 1058
Initiator aminoacyl-tRNA, in initiating protein synthesis, 1045
Initiator element, in promoters, 973
Initiator tRNA, 1047
Inosine, 184**S**
Inosine monophosphate, in purine biosynthesis, 874**F**, 875
Inosine 5'-monophosphate, 501
Inosinicase, in purine biosynthesis, 874**F**, 875
Inositol, 319**F**
Inositol hexaphosphate, as ligand of hemoglobins, 405, 405**S**
Inositol pentaphosphate, as ligand of hemoglobins, 405, 405**S**
Insect wings, 500
Insects, uric acid metabolism of, 884, 885**F**
Insert, 249
Insertion
mutations of, 928**D**
sequences (ISs) of, 922
Insulin, 85**T**, 95**T**, 96, 165**S**, 165**G**, 178, 683
amino acid sequence of, 105**F**, 116**F**
chemical synthesis of, 125
in diabetes, 755
metabolic effects of, 685
regulation of fatty acid synthesis by, 772**F**
synthesis as zymogen, 391
Integral protein, 295, 303
Intercalating agents, 226**S**
Intercalation in DNA, 932
Interleukin-2, 170**F**
Intermediary metabolism, 544**D**, 559
Intermediate-density lipoproteins, 792, 793**T**
Intermediates, in metabolism, 544
Internal energy, 512**D**
International Union of Biochemistry, 364
Interrupted mating in bacteria, 912**F**
Intervening sequences, 194, 195**F**
Intestinal calcium-binding protein, 157
Intramembrane particles, 297, 298**F**
Intrinsic binding energy, 427
Intrinsic pathway of blood clotting, 392, 393**F**
Intrinsic protein, 295
Intron, 194, 195**F**, 1002**D**, 1006**T**
Invertase, 597
Inverted repeat (palindrome), in *lac* operator, 978**F**
Iodine, in amylose structure, 327
Iodoacetate, 68**D**, 107**F**
Ionic bonds, 15, 17**F**
IPTG (isopropyl β-thiogalactoside), 266, 976**S**
Iron, 100**T**
mobilization of by ascorbic acid, 491
spin states of, 401**B**
Iron protoporphyrin IX, 639
Iron-molybedenum cofactor, of nitrogenase, 831, 831**S**
Iron-sulfur cluster, 606, 607**S**, 611**F**
Iron-sulfur protein, 634
Irreversible inhibition, 369, 373**D**
Irreversible thermodynamics, 520
Islets of Langerhans, 683**F**
Isoalloxazine, 474
Isobacteriochlorin, 828**F**
Isobutylene, 127**F**
Isobutyryl chloride, 236**F**
Isocaproid aldehyde, in steroid hormone synthesis, 800**F**
Isocitrate, 602**F**, 604**T**
Isocitrate dehydrogenase, 472**T**, 602**F**, 604**T**, 607**F**, 608**F**, 608**G**, 626
regulation of, 622
Isocitrate lyase, 623**F**, 624**F**
Isoelectric focusing, 133, 134**F**
Isoelectric point (*pI*), 133
α-Isoketocaproate, in leucine degradation, 868, 868**F**
Isolated system, thermodynamic, 512**D**, 512**F**
Isoleucine, 59**S**, 59**G**, 850
biosynthesis of from pyruvate, 850, 852**F**
degradation to succinyl-CoA, 867, 867**F**
stereoisomers of, 72**S**
Isoleucyl-tRNA synthetase, 1027
Isologous interactions, in proteins, 172, 174**F**
Isomaltose, 323**F**
Isomerases, 355**T**
Isopentenyl pyrophosphate, 789, 790**F**
Isopentenyl pyrophosphate isomerase, 790**F**
Isopentenyladenosine, 197**S**
Isopeptide bonds, in ubiquitin-mediated protein degradation, 1072
Isoprene, 288
as precursor to cholesterol, 788**F**
Isopropyl β-thiogalactoside (IPTG), 266, 976**S**
α-Isopropylmalate, in leucine biosynthesis, 853, 853**F**
α-Isopropylmalate dehydratase, in leucine biosynthesis, 853, 853**F**

α-Isopropylmalate dehydrogenase, in leucine biosynthesis, 853, 853**F**
α-Isopropylmalate synthase, in leucine biosynthesis, 853, 853**F**
Isopycnic centrifugation, 233, 245, 246**F**
Isoschizomers, restriction endonucleases, 205
Isotopes as metabolic probes, 564
Isotopes, 564, 564**T**
Isovaleryl-CoA, in leucine degradation, 868, 868**F**
Isozymes, 392**D**
ISs (insertion sequences), 922

J_H gene, of immunoglobulin heavy chains, 926, 926**F**
J_L genes, of immunoglobulin light chains, 925, 926**F**
Jacob, Francois, 963, 976
Jagenforf, Andre, 717**B**
Johnson, Ben, 801
Joining of immunoglobulin genes, 925*ff*, 928, 928**F**
Joule, 513
Judson, Horace Freeland, 208
Jun (a leucine zipper protein), 999, 100**F,** 1001

K system, allosteric regulation, 413, 413**F**
κ (kappa) gene family, of immunoglobulins, 925, 926**F**
k_{cat} values for some enzymes, 365**T**
k_{cat}/K_m, 365, 366**T**
K_m (Michaelis constant), 362**D**
K_m, apparent, 371**D**
K_m values for some enzymes, 364**T**
Kallikrein, 392, 393**F**
Karrar, 509
Kasugamycin, inhibition of peptide chain initiation by, 1063**T**
Katal, 364**D**
Kay, Steve A., 271**F**
κβ (kappa B), as eukaryotic promoter module, 989**T**
Kendrew, John, 157
Keratan sulfate, 335**F**
Keratin, 142, 150
α-Keratin, 95**T,** 97, 150
β-Keratin, 151**S**
Ketals, 321, 322**F**
β-Ketoacyl-ACP reductase, 764
β-Ketoacyl-ACP synthase, 763
α-Ketoadipate, in lysine biosynthesis, 845, 846**F**
α-Ketoadipate, in lysine degradation, 868, 868**F**
α-Ketobutyrate, an intermediate in isoleucine biosynthesis, 852**F**
Ketogenesis, 753, 754**F**
Ketogenic amino acids, 864**F,** 865
α-Ketoglutarate, 602**F,** 604**T**
α-Ketoglutarate dehydrogenase, 496, 498, 602**F,** 604**T,** 608**T,** 609**F**
3-Keto-L-gulonolactone, 490**S**
Ketohexose, 311
α-Ketoisovalerate in biosynthesis, 852**F,** 853, 853**F**
α-Ketol transfers, 464
Ketomethylene substrate analog, for carboxypeptidase, 455**S**
α-Keto-β-methylvalerate, in isoleucine biosynthesis, 852**F**
Ketone bodies, 753, 754**F**
Ketopentose, 311
Ketose, 311
3-Ketosphinganine reductase, 780, 781**F**
3-Ketosphinganine synthase, 777, 781**F**
Ketotetrose, 311
β-Ketothiolase, 744**F**
mechanism of, 479**F**
Kevlar, 152
Khorana, H. Gobind, 1021, 1022, 1023
Kilocalorie, 513
Kinases in ADP, GDP, ATP and GTP synthesis, 878, 879
Kinesin, 95**T,** 97
Kinetics, 356, 357
of enzyme-catalyzed reactions, 360
sigmoid or S-shaped, 394, 394**F**
Kininogen, 393**F**
Klebsiella pneumonia, nitrogen fixation and, 829**T**
Klenow fragment, of DNA polymerase I, 942, 943**F**
Klenow, Hans, 942
KNF model of allosteric regulation, 414, 414**F**
Knoop, Franz, 600, 735
Koagulering, 509
Kögl, 495
Kok, B., 706
Kornberg, Arthur, 940
Kornberg, H. L., 618
Koshland, Daniel, 414, 452
Krebs, Edwin, 390, 679
Krebs, Hans, 600
Krebs cycle, 602**F**
Kringle, a structural module in mosaic proteins, 169**S,** 1089
Krüppel h, a C_2H_2-type of zinc finger protein, 996**T**
K_s (enzyme-substrate dissociation constant), 361**D**
KSase, 763
Kuhn, 476
K_w, 41

L chain of immunoglobulin G, 923, 924**F**
L, the allosteric equilibrium constant, 411, 412**F**
L-chain genes, of immunoglobulins, 925, 926**F**
L-Histidinol, histidine and, 862**F**
lac operator, 978
nucleotide sequence of, 978**F**
lac operon, 975, 976**F**
positive control of by CAP, 980, 981
lac promoter, nucleotide sequence of, 967**T**
lac repressor, 95**T,** 96, 976**F,** 993**F,** 994**F**
affinity for DNA, 979**T**
interaction with DNA, 979, 979**B**
lac repressor *(Continued)*
mode of action, 977**F**
LAC9, a C_x-type of zinc finger protein, 996**T**
lacA gene, 976**F**
lacI promoter, nucleotide sequence of, 967**T**
lacI gene, 976**F**
lacI mRNA Shine-Dalgarno sequence, 1049**F**
α-Lactalbumin, human milk, 122, 123**F**
β-Lactamase, 366**T**
Lactam-lactim, 182**S**
Lactate, 574**T**
Lactate dehydrogenase, 365**T,** 472**T,** 591**F**
ε-crystallin and, 1091, 1092**T**
isozymes of, 392, 393**F**
Lactate oxidase, 476**T**
Lactic acid fermentation, 591**F**
Lactobacillic acid, 279**S**
β-Lactoglobulin, 514**T**
Lactose, 323**F,** 976**S**
intolerance of, 596
operon, regulation of, 982**F**
lacY gene, 976**F**
lacZ mRNA Shine-Dalgarno sequence, 1049**F**
lacZ gene, 268**F,** 975, 976**F**
LAD, 173**S,** 455, 456, 457, 458**F**
Laetrile, 324**F**
Lamellae, 23, 700
Langdon, Robert, 788
Lanolin, 288
Lanosterol, 288, 289**S**
in cholesterol biosynthesis, 791**F**
L-Lanthionine, 62**S**
Lariat formation in RNA splicing, 1006, 1007**F,** 1009
Laschtchenko, P., 446
Lateral asymmetry in membranes, 297
Lathyrism, 155
Lathyrus odoratus (sweet pea), 156
Lauric acid, 277
LCAT, 797
LDL *See* Low density lipoproteins.
Leader peptidase, 1068
Leader peptides, in protein translocation, 1068
Leader sequences, prokaryotic N-terminal sequences, 1069**F**
Lecithin:cholesterol acyltransferase, 797
Lectins, 1064**D**
Leder, Philip, 1022
Lederberg, Joshua, 908
Leghemoglobin, 1087, 1088**F**
Lehninger, Albert, 2, 736
Leloir, Luis, 675
Leonardo da Vinci, 247
Lepidopterans, 266, 500
Lesch-Nyhan syndrome, 879, 880**F**
Lethal synthesis, 897**B**
leu operon, and transcription attenuation, 986**F**
Leucine, 58**S,** 58**G,** 868, 868**F**
Leucine aminopeptidase, 109
Leucine biosynthesis, 853, 853**F**
Leucine aminotransferase, 853, 853**F**
Leucine zipper DNA-binding proteins, 992, 1000**F,** 1001**F**
Leukosialin, 342**F**
Leukotrienes, synthesis of, 783**F**

Levorotatory, 71**D**
Lewis, Carl, 801
LexA protein, in the SOS response, 960
Ligand, 16
Ligases, 355**T**
Ligation, blunt-end, 251
Light reactions of photosynthesis, 701, 701**F,** 703, 704, 716, 717**B,** 726, 726**F**
Lignin biosynthesis, from chorismate, 857, 857**F**
Lignoceric acid, 277
Lillie, 483
Limit dextrin(s), 328, 671**F**
Limonene, 289**S**
Lineweaver-Burk double-reciprocal plot, 366**D,** 367**F**
Link protein, in proteoglycan, 349**F,** 350
Linkers, in cloning DNA, 251, 252**F**
Linking number (*L*), in DNA, 227, 229**F**
Linoleic acid, 277, 278**S,** 278**G**
Lipase, hormone-sensitive, 733
Lipid(s), 276
  -anchored proteins, 303
  transfer proteins, 300
  transport of, 792
Lipids, 276
Lipmann, Fritz, 480
Lipoamide, complex, 495**S**
Lipoic acid, 463**T,** 495**S,** 496
Lipopolysaccharide, 337, 339**F**
Lipoprotein(s), 98, 100**T,** 291, 792, 794**F,** 795**F,** 793
  complexes of, 792
  endocytosis and degradation of, 796**F**
Lipoprotein lipase, 795
Liposomes, 294
Lipoyl-lysine complex, 463**T**
Lipscomb, William N., 415
Liquid SIMS, 115
Liquid crystal phase, in membranes, 301**F**
Liquid scattered ion mass spectrometry (liquid SIMS), 115
Lithium aluminum hydride ($LiAlH_4$), 111**F**
Liver alcohol dehydrogenase. *See* LAD.
Liver, 294**F,** 666, 819**T,** 823
L,L-α,ϵ-diaminopimelate, in lysine biosynthesis, 849**F,** 850
Lock and key hypothesis, 388
Locus, genetic, 1076**D**
Lovastatin, 792**F**
Low-density lipoproteins, 98, 100**T,** 792, 793**T**
  receptor, 342, 796**F,** 797**F**
Luciferace, 269, 271**F**
Luciferin, 269, 271**F**
Lumen, 28**T**
Lyases, 355**T**
Lycopene, 288, 289**S**
Lymphocyte homing factor, 348
Lynen, Fyodor, 736, 788
Lysine, 59**S,** 59**G,** 68
  biosynthesis, 845, 846**F,** 847, 848, 849**F**
  degradation of, via the saccharopine pathway, 868, 868**F**
  titration of, 65**F**
Lysogeny, 29
Lysolecithin, 284**S**
Lysosomal acid lipases, 797
Lysosomal enzymes, oligosaccharide of, 344**F**
Lysosome(s), 25, 27**T,** 28**T**
Lysozyme, 85**T,** 122, 123**F,** 156, 157**S,** 157**G,** 364**T,** 365**T,** 446, 448**F,** 448**G,** 450**S,** 451**F**
Lysozyme-$(NAG)_6$ complex, 448**G**
Lysyl oxidase, 155**F**
Lysyl hydroxylase, 153
Lyxose, 312

M13 bacteriophage, as cloning vector, 255
MacLeod, Colin M., 903
Macromolecular synthesis and growth, 804**F,** 805
Macrophages, 923
Magnesium (Mg), 356**T**
  deficiency, in glycolysis, 597
  efflux of, from thylakoid vesicles, 726
Major transplantation antigens, 304
Malate, 602**F,** 604**T**
  -aspartate shuttle, 655**F,** 656
  in the C-4 pathway, 727, 728**F**
Malate dehydrogenase, 472**T,** 602**F,** 604**T,** 612**F,** 613**S,** 613**G**
Malate synthase, 623**F,** 624**F**
4-Maleylacetoacetate, in tyrosine degradation, 869, 869**F**
4-Maleylacetoacetate isomerase, in tyrosine degradation, 869**F**
Malic enzyme, 95**T,** 617, 618**F,** 749**F**
Malonate as competitive inhibitor, 372**F**
Malonyl-CoA, 758
  biotin-dependent carboxylation of, 493**T**
Malonyltransacylase, 762
Malonyl transferase, 762
Maltase, 324
Maltose, 322, 323**F**
Maltose phosphorylase, kinetics of, 380
Maltose-binding protein, as gene fusion system, 269**T**
Mammals
  DNA content of, 906**T**
  uric acid metabolism of, 884, 885**F**
Mandrill, 3**F**
Manganese (Mn), 100**T,** 356**T,** 709**F**
Mannans, 326, 333**F**
Mannitol, 319**F**
Mannose, 312
  utilization, in glycolysis, 593
Maple syrup urine disease, 509, 868
Marfan's syndrome, 156
Margulis view of early evolution, 1098
Marine brown algae, 332
Marker, selectable, 249
Martius, Carl, 600
Mass spectrometry, 115
Matthei, Heinrich, 1020
Mavalonic acid, 226**S**
Maxam, A. M., 209, 212
Maxam-Gilbert nucleic acid sequencing, 212–6, 212**F,** 214**F**
Maximal velocity (V_{max}), 363**D**
Mb. *See* Myoglobin.
MboI restriction endonuclease, 204**T,** 205
McCarty, Maclyn, 903
McClintock, Barbara, 921
McCollum, Elmer, 504
Mechanical work, 512**D**
Mechanism-based inhibitors, fluoro compounds as, 897**B**
Meek, Thomas, 444
Melanocyte-stimulating hormone, chemical synthesis of, 125
Melezitose, 324**F**
Melting temperature of membrane, 301, 302**T**
Membrane(s), 10, 291
  plasma, 22, 12**T,** 26**T,** 28**T**
  protein, structure, 303
  transport proteins of, 97
Menaquinone, 508**S**
Menoidium, EM of, 292
Menthol, 289**S**
Meperidine, 652**S**
Merrifield, Bruce, 125
2-Mercaptoethanol, 107**F**
β-Mercaptoethylamine, 478**S**
Meselson and Weigle experiment, 915**F**
Meselson, Matthew, 915, 937
Meselson and Stahl experiment, 939**F**
Mesophyll cells, C-4 pathway and, 727, 728**F**
Mesosome, 22, 23**T**
Messenger RNA (mRNA), 190, 191**T,** 194, 963**D,** 1003, 1004, 1021**T,** 1022**T,** 1048, 1049**F,** 1058, 1059**F**
  oligo(dT)-cellulose chromatography and, 264**F**
Metabolic channeling, 888
Metabolic diversity, 546, 548
Metabolic energy, stored forms, 732**T**
Metabolic inhibitors, 563
Metabolic integration system, and glutamine synthetase, 837**F,** 838
Metabolic map, 21**F,** 545**F,** 546, 547**F**
Metabolic pathways, similar set in all organisms, 546, 552, 553**F,** 567, 567**F**
Metabolic regulation, 559
Metabolic relationships among organs, 818**F**
Metabolic stoichiometry, ATP coupling and, 806
Metabolism, 18, 19**F,** 21**F,** 544**D**
  biosynthetic pathways of, 551
  degradative pathways of, 550
  intermediary, 544**D**
  in a multicellular organism, 819
  systems analysis of, 804, 804**F**
Metabolon, 552**D**
Metal ion(s) catalysis in, 427
  in enzymes, 356**T,** 434**D**
Metal response element (MRE), in eukaryotic transcription, 990, 990**T**
Metal-rich proteins, 100**T,** 101, 165**S,** 165**G**
Metalloenzymes, 356**T,** 434**D**
Metalloproteins, 100**T,** 101, 165**S,** 165**G**
Metallothionein, 991, 991**F**
Methanogen(s), 22
  nitrogen fixation and, 829**T**
N^5,N^{10}-Methenyl THF, 499**S,** 499**T,** 501**F**
Methionase, 509
  PLP-assisted, 481**F**
Methionine, 59**S,** 59**G**
  biosynthesis of, 847, 848

Methionine *(Continued)*
  degradation of, 867, 867**F**
  THF-dependent formation of, 501
Methionyl-tRNA$_f^{Met}$, 1047, 1048**F,** 1048**S**
Methotrexate, 501, 502, 876
Methyl-directed pathway, in mismatch repair of DNA, 959
Methyl group transfers, B_{12}-catalyzed, 485, 486**F**
Methyl-α-D-glucoside, 322
Methyl-β-D-glucoside, 322
Methyl iodide, 115**F**
Methyl phosphate, 427**T,** 460
Methyl transferases, 851**F**
Methylarginine (N-), 60, 61**S**
Methylation of eukaryotic pre-mRNAs, 1003, 1004**F**
Methylcobalamin, 463**T,** 485, 486**S**
β-Methylcrotonyl-CoA, biotin-dependent carboxylation of, 493**T**
5-Methylcytosine and Z-DNA, 225
Methylenecyclopropylacetyl-CoA, 741, 742**F**
α-Methyleneglutarate mutase, 489
N^5,N^{10}-Methylene-THF, 499**S,** 499**T,** 501**F**
  in the thymidylate synthase reaction 895, 896**F,** 897**B**
β-Methylglutaconyl-CoA, formed by biotin-dependent carboxylation, 493**T**
N-Methylglycine, 62
5′-^{7}Methyl-GTP cap of eukaryotic mRNAs, 1058, 1059**F**
Methylguanosine, 197**S**
Methylhistidine, 60, 61**S**
ε-N-Methyllysine, 60, 61**S**
Methylmalonyl-CoA, formation of, 493**T**
Methylmalonyl-CoA epimerase, 747**F,** 748**F**
Methylmalonyl-CoA mutase, 356**T,** 747**F,** 749**F**
N-Methyl-*N′*-nitro-*N*-nitrosoguanidine, as chemical mutagen, 931**F**
N^5-Methyl THF, 499**S,** 499**T,** 501**F**
5′-Methylthioadenosine, 851**S**
Methyltransferases, 355**T**
5-Methyluracil (thymine), 197
Mevacor®, 792**F**
Mevalonate, 792**F**
  synthesis of, 786**F**
Mevalonate kinase, 790**F**
Mevinolin, 792**F**
Mevinolinic acid, 792**F**
Meyerhof, O., 570
Micelle(s), 38, 39**F,** 292
Michaelis constant (K_m), 362**D**
Michaelis-Menten equation, 361
Michel, Hartmut, 712
Microfilaments, 27**T**
Microhelix analog of tRNA^{Ala}, 1032**F**
Microtubule(s), 27**T,** 177**S**
Milk, fatty acids in, 279
Miller, Oscar L, Jr., 1058
Mineralocorticoids, 291, 801
Miniband, 235**F**
Minot, 487
Mismatch repair, of DNA, 959
Mitchell, Peter, 646
Mitchell's chemiosmotic hypothesis, 645, 646
Mitchell's chemiosmotic hypothesis (*Continued*)
  photosynthesis and, 717**B**
Mitochondria, 9, 25, 26**T,** 28**T,** 628, 629**F**
Mitochondrial F-ATPase, evolution of, 1097**F**
Mitochondrial protein import, 1070
Mitosis, 24
Mixed function oxidase, 799**F**
MKR1 and 2, 996**T**
Mn complex of photosynthesis, 708**F**
MN blood group, 303
Mobile elements, 922
MoCo. *See* Molybdenum cofactor.
Modulator proteins, 392**D,** 393
Modules, protein, 167
Molecular activity of an enzyme, 365
Molecular chaperones, 1066
Molecular clocks, 1081
Molecular evolution, 1075, 1075**D**
Molecularity, 357**D**
Molecular recognition, 387
Molluscs, uric acid metabolism of, 884, 885**F**
Molybdenum (Mo), 100**T,** 356**T**
Molybdenum cofactor, 827, 828**F**
Molybdopterin, 828**F**
Monellin, 95**T,** 98
2-Monoacylglcyerol, synthesis of, 776**F**
Monoacylglycerol acyltransferase, 776**F**
Monoacylglycerol lipase, 733**F**
Monod, Jacques, 410, 963, 976
Monod, Wyman, Changeux model of allosteric regulation, 410, 411**F**
Monolayer, lipid, 292
Monomeric, 84
Monosaccharide, 311
Monoterpene, 288
Moore, Stanford, 77
Moore, 504
MOPS, 53**F**
Mosaic proteins, 1088, 1089
Motion, in proteins, 97, 158, 159**T**
Motrin®, 784**F**
Mouse, transgenic, 274
Mouse satellite DNA, 232**F**
MRE (metal response element), in eukaryotic transcription, 990, 990**T**
mRNA(s) (*See also* Messenger RNA), 190, 191**T,** 194, 963**D,** 1003, 1004, 1020, 1021**T,** 1022**T,** 1048, 1049**F,** 1058, 1059**F**
  copolymers as synthetic mRNAs, 1020, 1021**T**
  eukaryotic, capping and methylation, 1003, 1004**F,** 1058, 1059**F**
  isolation of by oligo(dT)-cellulose chromatography, 264**F**
  poly(A) tails of, 194, 195**F,** 264**F**
  ribosome-binding site of, 1048, 1049**F**
  Shine-Dalgarno sequences and, 1049**F**
  synthetic, 1020, 1021**T,** 1022, 1022**F,** 1022**T**
Mucins, 340
Mucopolysaccharides, 327
20S Multicatalytic protease (a conjugate-degrading factor), 1072
Multi-enzyme complex, 552, 553**F**
Multifunctional polypeptides, in pyrimidine biosynthesis, 888
Multilamellar vesicle, 294
Multimeric, 84
Multiple codons, in transcription attenuation, 985, 986**F**
Multisubstrate reaction, 381
Muramic acid, 321**F**
Murein, 337
Murphy, 487
Muscle
  energy metabolism in, 819**T**
  fatigue, 821
  protein degradation of, 822
Mutagenesis, 273, 415, 930, 931**F**
Mutagens, chemical, 930, 931**F**
Mutants, 908
Mutarotation, 315**D**
Mutasome, in DNA repair, 960
Mutation(s), 124**D,** 928, 928**D,** 929, 932, 1077
  metabolic pathways and, 563
  molecular nature of, 928
  spontaneous, 929
MWC model. *See* Monod, Wyman, Changeux.
Myc (leucine zipper protein), 999
Mycoplasma, DNA content of, 906**T**
Myoglobin, 85**T,** 156, 158, 394, 396**F,** 398**F,** 514**T**
  amino acid composition of, 88**T,** 122**F**
  evolutionary relationships of, 120, 122**F,** 123**F**
  gene exon organization of, 1088**F**
  hemoglobin and, 395, 395**F**
  O_2 binding and, 395**F,** 396, 397**F,** 420, 421, 421**F,** 423**F**
Myohemerythrin, 144**S,** 161**S,** 161**G**
Myosin, 85**T,** 86, 95**T,** 97
  thick filaments in muscle contraction, 821
Myristic acid, 277
Myristoyl-CoA:protein N-myristoyltransferase, 306
N-Myristoylation, 306

n (Hill coefficient), 421**D**
N face, 637
N-linked saccharides, 340, 341**F,** 343
NAD^+, 463**T,** 468, 470**S,** 574**T**
NAD^+-dehydrogenases, bisubstrate kinetic mechanism of, 378
NAD^+-linked dehydrogenases, 471**T,** 472**T,** 472**F**
NAD^+-saccharopine dehydrogenase, in lysine biosynthesis, 846**F**
NADH, 574**T,** 809
NADH-coenzyme Q reductase, 635
NADH dehydrogenase, 635
NADH-UQ reductase, 635**T**
$NADP^+$, 463**T,** 468, 470**S**
$NADP^+$-malate dehydrogenase, activation by reduction, 727**F**
NADPH, 3, 809
  anabolism and catabolism and, 805
  in CO_2 fixation, 725**T**
NADPH
  as electron donor, 558, 828, 892, 893, 893**F**
  fatty acid synthesis and, 758

NADPH *(Continued)*
in photosynthesis, 701
pentose phosphate pathway and, 694, 695**F**
NADPH-specific glyceraldehyde-3-P dehydrogenase, 723, 724**F,** 725**T**
NADPH-specific malate dehydrogenase and C-4 CO_2 fixation, 727
$(NAG)_2$, 448**T**
$(NAG)_3$, 448**F,** 448**T**
(NAG)3, 447, 449**S**
(NAG)4, 448**T**
(NAG)5, 448**T**
(NAG)6, 448**T**
(NAG-NAM)3, 448**T**
Naja naja, 284**F**
Nalidixic acid, 960**S**
Naphthoquinone, 508
Natural selection, 1076**D**
NDP kinases in NTP synthesis, 879
Negative control of enzyme synthesis, 561**F,** 562**F**
Negative heterotropic effector, 408, 411
Negentropy, 515**D**
Nemethey, 414
Neo-Darwinian theory of evolution, 1076**D**
Nervonic acid, 277
Neuraminic acid, 321**F**
Neurospora crassa
base DNA composition of, 907**T**
Beadle and Tatum and, 563
one-gene, one-enzyme hypothesis of, 901, 902**F**
Neutral theory of molecular evolution, 1076**D**
NF*κ*B, transcription factor for RNA polymerase II, 989**T**
NF1, transcription factor for RNA polymerase II, 989**T**
NGF receptor, 170**F**
NGF1-A, a C_2H_2-type of zinc finger protein, 996**T**
Niacin, 463**T,** 473
Nicholson, D. E., 545
Nick translation, 942, 942**F**
Nickel (Ni), 356**T**
Nicolson, G. L., 295
Nicotiana tabacum nitrate reductase, sequence comparison, 1080**F**
Nicotinamide, 473**S**
Nicotinamide adenine dinucleotide (NAD), 356**T,** 463**T,** 468, 470**S**
Nicotine, 473**S**
Nicotinic acid, 463**T,** 473**S**
nif gene expression, and the regulation of nitrogen fixation, 830**F**
Night blindness, 504
Ninhydrin, 67**F,** 67**S**
Nirenberg, Marshall, 1020, 1022, 1023
Nitrate, 9, 549
assimilation of, 549**D,** 550**F,** 827
Nitrate reductase(s), 100**T,** 356**T,** 827
amino acid sequence comparison of, 1080**F**
Nitrifying bacteria, 550
Nitrite reductase, 82
Nitrocellulose, 259**F,** 261**F**
Nitrogen, 9
Nitrogen *(Continued)*
acquisition of, 826, 827
combined, 550
compounds as chemical mutagens, 930, 931**F**
cycle, 549, 550**F,** 826
excretion of, 870
fixation of, 549**D,** 550**F,** 827, 828, 830**F**
Nitrogen gas, 549
reserves of in organisms, 559
Nitrogenase, 827, 829
complex, 830
hydrogenase activity of, 831
reaction of, 831, 832**F**
of regulation of, 830**F**
Nitrogenase reductase, 830, 830**F,** 832**F**
p-Nitrophenylacetate, 438**S**
p-Nitrophenylacetate hydrolysis, 434, 435
2-Nitropropane dioxygenase, 476**T**
NMR spectroscopy, whole-body (*See also* Nuclear magnetic resonance), 565, 565**F**
NMT, 306
Nodules, root and nitrogen fixation, 829, 829**F**
Noller, Harry F., 382
Non-reducing end of disaccharides, 322
Noncompetitive inhibition, 372, 372**F,** 373, 373**F**
Noncyclic photophosphorylation, 719
Nondegenerate sites, 1078
Nonhistone chromosomal proteins, 193, 194, 233
Nonsense condons, 1023**D**
Nonsense suppression, 1038
Nonsynonymous substitutions, 1078
Northern blotting, 260**F**
NR and *E. coli* glutamine synthetase gene expression, 838
Nuclear matrix, 233
Nuclear pre-mRNA splicing, 1005
Nuclear magnetic resonance, 73, 294, 565, 565**F**
Nuclear lamins, 307
Nucleases, 200, 200**T**
Nucleic acids, 180, 188, 209
cleavage of, 199**F**
hybridization of, 232
hydrolysis of, 199**F**
sequencing of, 209, 209**F,** 210, 211**F,** 212, 212**F,** 213**F**
Nucleoid, 23**T**
Nucleophilic catalysis, 432**F**
Nucleoprotein(s), 98, 100**T**
filaments of, 918, 919
Nucleosidases, in purine degradation, 881
Nucleoside(s), 184*ff,* 184**S**
Nucleoside diphosphates (NDPs), 186**S**
Nucleoside diphosphate kinase, 665, 879
Nucleoside diphosphate sugars, 538**F**
Nucleoside phosphates, 185, 186
Nucleoside phosphorylases, in purine degradation, 881
Nucleoside phosphoramidite, 237
Nucleoside triphosphates (NTPs), 186**S**
Nucleosomes, 193, 194**F,** 233
Nucleotidases, in purine catabolism, 881, 882**F**
Nucleotide(s), 180, 185**F**
absorption spectra of, 183**F**
binding sites of, 894
biosynthesis of, 872**F**
proton dissociation constant (pK_a) values of, 182**T**
as recognition units, 187–188
sequence of, 190
alignments, 1078, 1079**F**
evolutionary change in, 1076
pairwise comparisons of, 1077**F**
substitution, 1075, 1077**F,** 1081**F**
rate of, 1080, 1081**F,** 1082**T,** 1083**T**
Nucleotide diphasphate kinase, 602**F,** 604**T**
Nucleus, 9, 24, 26**T,** 28**T**
Nuprin®, 784**F**
Nutrients, 544**D**

O antigens, 338
Obligate anaerobes, 548
O_c mutations in the *lac* operator, 979
Ochre stop codon, 1038
Oct-1 and -2, transcription factors for RNA polymerase II, 989**T,** 994
Oct-2, amino acid sequence of, 995**F**
Octamer, as eukaryotic promoter module, 989**T**
Octyl glucoside, 293**S**
Odd-carbon fatty acids, *β*-oxidation of, 747
OEC. *See* Oxygen-evolving complex.
Okazaki fragments, 946, 946**F,** 949, 954**F**
Okazaki, Tuneko and Reiji, 946
Old yellow enzyme, 476
O-linked saccharides, 340, 341**F,** 342**F**
Oleic acid, 277, 278**S,** 278**G**
Oligo(α1,4 → α1,4) glucantransferase, 671
Oligo(dT)-cellulose chromatography, 264**F**
Oligomers, of protein subunits, 172
Oligomycin A, 652**S,** 653
Oligonucleotides, degenerate, 262, 262**F**
Oligopeptide, 84
Oligosaccharides, 322
Olive oil, fatty acids in, 279
OMP. *See also* Orotidine 5′-monophosphate.
decarboxylation of, 887**F,** 888
OmpA (*E. coli* outer membrane protein) gene, codon usage in, 1084
One-gene, one-enzyme hypothesis, 901
Opal stop codon, 1038
Opaque2, amino acid sequence of, 1000**F**
Open promoter complex, 966**F,** 968
Open system, thermodynamic, 512**D,** 512**F**
Operator region in operon, 975, 975**F**
Operator-constitutive mutations, 979
Operon(s), 974
inducible *vs.* repressible, 981, 982**F**
Jacob and Monod hypothesis, 976
organization of, 975**F**
transcription control of, 975
Opsin, 503, 504
Optical activity, 71**D,** 78
Order of a reaction, 357**D**
Organelles, 9

Ori (Origin of replication), 248, 255, 267**F**
OriC, in *E. coli,* 944, 950, 951, 952**F**
Ornithine, 61, 62**S**
  biosynthesis of, 841, 842**F**
  degradation of to α-ketoglutarate, 866**F**
  metabolic roles of, 841
  transporter of, 844**F**
Ornithine transcarbamoylase, 841, 842
Orotate, 476**T**
  in pyrimidine biosynthesis, 887**F,** 888
Orotate phosphoribosyltransferase, in pyrimidine biosynthesis, 887**F**
Orotidine 5′-monophosphate, in pyrimidine biosynthesis, 887**F,** 888
Osmotic pressure, 39, 40**F**
Osteomalacia, 507
Ouabain, 320**F**
Oust®, 863**B**
Ovalbumin, 95**T,** 97
  gene, exons and introns of, 1002
  oligosaccharide of, 344**F**
Oxaloacetate, 602**F,** 604**T**
  in the C-4 pathway, 727, 728**F**
  formation by biotin-dependent carboxylation, 493**T**
α-Oxidation, 753
β-Oxidation of fatty acids, 735, 737**F,** 740**F**
ω-Oxidation, 753
Oxidation state of THF, 499**F,** 499**T**
Oxidative phosphorylation, 554, 628**D**
Oxidoreductases, 355**T**
2,3-Oxidosqualene lanosterol cyclase, 789
Oxonium ion intermediate, 674, 675**F,** 677**F**
Oxyanion hole, in chymotrypsin, 439**F**
Oxyanion transition state, in chymotrypsin, 439
Oxygen radicals, in DNA damage, 959**F**
Oxygen cycle, 549
Oxygen, role in metabolism, 548
Oxygen-evolving complex (OEC), 709
Oxyhemoglobin, 400
Oxymyoglobin, 397
Oxytocin, chemical synthesis of, 125

P-ATPases 1094, 1095**F,** 1095**T**
P/A state of the ribosome, 1053
P/O ratio, for electron transport/oxidative phosphorylation, 657
P face, of mitochondrial inner membrane, 637
p21[ras] protein, 307
*P*680, as reaction center of photosystem II, 706, 707**F,** 708**F**
*P*700, as reaction center of photosystem I, 706, 707**F,** 708**F**
*P*700 and cyclic photophosphorylation, 719
*P*870, in *Rhodopseudomonas viridis* photosynthesis, 712
PABA, 876
PAF, 284, 285**S,** 285**G,** 780**F**
PALA. See *N*-phosphonacetyl-L-aspartate.
Palindrome, 228, 229**F**
  in *lac* operator, 978**F**
Palindromic "half-sites," 998
Palm oil, fatty acids in, 279
Palmitate, synthesis of, 764, 765
Palmitic acid, 277, 278**S,** 278**G**
Palmitoleic acid, 277
Palmitoleoyl-ACP, 768, 769**F**
Pancreas, role in insulin secretion, 683
Pancreatic acinar cell, EM of, 292
Pancreatic juice, carboxypeptidase in, 451
Pancreatic lipase, 734**F**
Pancreatic trypsin inhibitor, 157**S**
Pantothenic acid, 463**T,** 478**S**
Papain, 113, 433**T**
Papain domain 2, 163**S**
PAPS. *See* 3′-phosphoadenosine 5′-phosphosulfate.
Parallel β-proteins, 162**S**
Parathyroid hormone, 507
Parvalbumin, 157
Pasteur, Louis, 70, 563, 569
Patch recombination, 916, 917**F,** 918**F,** 922**F**
Pauling, Linus, 142, 145**B,** 491, 1081
1:2 Pause structure, in transcription attenuation, 986, 986**F**
PC *See* Plastocyanin.
PCNA (proliferating cell nuclear antigen), 956
PCR (polymerase chain reaction), 271, 272**F,** 273
Pellegra, 473
Penicillamine, 61, 62**S**
Penicillin, 374**S**
Penicillinase, 364**T,** 365**T**
Pentose(s), in nucleic acids, 183**S**
Pentose phosphate pathway, 687, 688**F**
  steps of, 689, 690
Pentulose, 311
PEP carboxylase, 618**F**
  CO_2 fixation and, 728
PEP carboxykinase, 618, 619**F,** 663, 664
Pepsin, 113, 354, 442**G,** 442**T**
  pH-rate profile of, 447
Pepsinogen, 391**T**
Pepstatin, 459**S**
Peptide(s), 84, 101, 102**F**
  bonds of, 56, 57**F,** 82, 140
  chain
    elongation of, 1051, 1061
    initiation of, 1050**F,** 1058, 1060
    termination of, 1055, 1056**F,** 1061
Peptidoglycan, 336**S,** 337**F**
Peptidyl transferase, 382, 1053**F**
Peptidyl (P)
  binding site, 1047**F**
  transfer reaction, 1052**F,** 1053**F,** 1054**F**
Performic acid oxidation of disulfides, 106, 107**F**
Peripheral protein, 295, 303
Perkin condensation, 603**D**
Permanent dipole, 138**D**
Permanent wave, 150
Permease gene, 976**F**
Pernicious anemia, 487
Peroxidase, 356**T**
Peroxisomes, 25, 27**T,** 28**T,** 750
Perutz, Max, 142, 398
PGD_2, synthesis of, 783**F**
PGE_2, synthesis of, 783**F**
$PGE_{2\alpha}$, 785
$PGF_{2\alpha}$, synthesis of, 783**F**
PGG_2, synthesis of, 784**F**
PGH_2, synthesis of, 783**F,** 784**F**
PGI_2, synthesis of, 783**F**
pH, 42, 43**T**
  of blood plasma, 50
  effect of on enzyme activity, 368, 368**F**
  effect of on standard state free energy, 519
  optimum, for various enzymes, 368**F,** 368**T**
  profiles, of enzyme activity, 446**F**
Phase separation, 297
Phase transition, membrane lipid, 301, 302**F**
Phaseolin, 95**T,** 97
pheA operon, and transcription attenuation, 986**F**
Phenotype, 907**D**
Phenylalanine, 59**S,** 59**G**
  biosynthesis of, 858, 859**F**
  from chorismate, 857**F**
  degradation of, 860, 869, 869**F**
Phenylalanine aminotransferase, 859**F,** 860
Phenylalanine hydroxylase deficiency, in phenylketonuria, 869
Phenylalanine-4-monooxygenase, in tyrosine synthesis, 860**F**
Phenylbutazone, 785
Phenylisothiocyanate, 108, 110**F**
Phenylketonuria, 869, 901
Phenylpyruvate, 859**F,** 860, 869, 869**S**
Phenylsepharose®, 134
L-Phenylserine, 62**S**
Phenylthiohydantoin, 108, 110**F**
Pheophytin, 708**F,** 710
Phi (ϕ), angle in peptides, 83, 140**D**
Phillips, David, 447
*p*HMB. *See p*-Hydroxymercuribenzoate.
Phorbol ester, as physiological challenge 990**T**
Phosphatase, 354**D**
Phosphatidate cytidylyltransferase, 775**F**
Phosphatidic acid, 282, 775**F**
  synthesis of, 773, 774**F**
Phosphatidic acid phosphatase, 775**F**
Phosphatidylcholine, 282, 283**S,** 283**G,** 775**F**
Phosphatidylethanolamine, 282, 283**S,** 775**F,** 776, 776**F**
Phosphatidyglycerol, 283**S,** 283**G**
  synthesis of, from CDP-diacylglycerol, 777, 778**F**
Phosphatidylglycerol-P phosphatase, 778**F**
Phosphatidylinositol, 283**S,** 283**G**
  synthesis of from CDP-diacylglycerol, 777, 778**F**
Phosphatidylinositol synthase, 778**F**
Phosphatidylserine, conversion of from PE, 776**F**
Phosphatidylserine decarboxylase, 776**F**
Phosphinothricin, 863**B**
Phosphite group, 236**F**
3′-Phosphoadenosine 5′-phosphosulfate, in sulfate assimilation, 855
Phosphocholine, 775**F**
Phosphocreatine, hydrolysis of, 520
Phosphodiesterase, 200**T,** 201
Phosphodiester bridges, 188–9, 189**F**
Phosphoenolpyruvate, 525**T,** 533, 574**T,** 583**F,** 857

Phosphoethanolamine, 775**F**
Phosphofructokinase, 95**T**, 96**F**
in glycolysis, 572**F**, 574**T**, 577
regulation of, 578
substrate cycles and, 812
Phosphofructokinase-2, 669, 670**F**
Phosphoglucoisomerase
in CO_2 fixation, 724**F**, 725**T**
in glycolysis, 572**F**, 574**T**, 576
Phosphoglucomutase, 433**T**, 594, 673**F**
6-Phosphogluconate dehydrogenase, 472**T**, 689
Phosphogluconate pathway, 687, 688**F**
3-Phosphoglycerate
amino acid family, 574**T**, 583**F**, 586**F**, 587**F**, 853
CO_2 fixation and, 720, 724**F**, 725**T**
in serine biosynthesis, 853, 854**F**
3-Phosphoglycerate dehydrogenase, 853, 854**F**
Phosphoglycerate mutase, 163**S**, 163**G**, 433**T**, 583**F**, 586, 587**F**, 588**F**
Phosphoglycerate kinase, 160**F**, 583**F**, 585
Phosphoglycolate, (product of photorespiration), 722, 722**F**
Phosphoglycolate phosphatase, 722**F**
Phosphoglycolohydroxamate, 430, 431**S**
Phosphohistidine, intermediate in glucose-6-phosphatase, 666
Phosphohomoserine, in threonine biosynthesis, 849**F**
Phosphoketolase, 464, 465**F**, 469**F**
Phospholipase A_2, 165**S**, 165**G**, 780
Phospholipid(s), 100**T**, 281, 794**F**
5-Phosphomevalonate, 790**F**
Phosphomevalonate kinase, 790**F**
Phosphonamidate inhibitor, of carboxypeptidase A, 453, 455**S**
N-Phosphonacetyl-L-aspartate, 415, 415**S**, 417**F**
Phosphopantetheine on acetyl-CoA carboxylase, 762, 763**S**
4-Phosphopantetheine, 478**S**
Phosphopentose epimerase, 690, 724**F**, 725**T**
Phosphopentose isomerase, 690, 724**F**, 725**T**
Phosphoprotein(s) 100**T**, 101
Phosphoprotein phosphatase, 560**F**, 679, 682
3-Phosphopyruvate, in serine biosynthesis, 853, 854**F**
Phosphoramide chemistry, 236**F**, 237
5-Phosphoribosyl-β-amine, in purine biosynthesis, 873, 874**F**
N-(5′-Phosphoribosyl) anthranilate, 859**F**, 860, 985**F**
N^1-Phosphoribosyl-AMP, as precursor to histidine, 862**F**
N^1-Phosphoribosyl-ATP, as precursor to histidine, 862**F**
Phosphoribosyl formimino-5-aminoimidazole-4-carboxamide ribonucleotide, 861
Phosphoribulose kinase, 723, 724**F**, 725**T**
Phosphoric acid, 47
anhydrides of, 529
titration of, 48**F**
Phosphoric anhydride(s), 524
linkage of, 186, 186**S**
Phosphoric-carboxylic anhydrides, 532
Phosphorolysis, 672
Phosphoryl group transfer, 188**F**
Phosphorylase *a*, 560**F**, 679
Phosphorylase *b*, 560**F**, 679
Phosphorylase (domain 2), layer structure of, 160**F**
Phosphorylase reaction, 328
Phosphorylase kinase, 560**F**, 679, 686
Phosphorylation potential (Γ), 818**D**
Phosphodiesterases, 200
3-Phosphoserine, in serine biosynthesis, 853, 854**F**
Photoautotrophs, 548**T**
Photochemical efficiency, 706**F**
Photochemical reaction of photosynthesis, 702
Photoheterotrophs, 548**T**
Photolysis, of water in photosynthesis, 707, 708**F**
Photon ($h\nu$), 703
Photophosphorylation, 707**D**, 708**F**, 716, 719
chemiosmotic nature of, 716, 717**B**, 718**F**
Photorespiration, 721, 722
Photosynthesis, 698**F**, 699
efficiency of, 706**F**
energy requirements of, 716
photochemical action spectrum of, 706**F**
photosystem I of, 706, 707**F**, 708**F**, 714, 714**F**
photosystem II of, 706, 707**F**, 708**F**, 713, 714**F**
quantum yield of, 715
Photosynthetic reaction center(s), 711**F**, 712, 712**F**, 713
Photosynthetic unit, 705, 705**F**
Photosystem I, of photosynthesis, 705, 706, 707**F**, 708**F**, 714, 714**F**, 715**F**
Photosystem II, of photosynthesis, 705, 706, 707**F**, 708**F**, 713, 714**F**, 715**F**
Phototrophs, 523, 548**D**
photochemical apparatus and, 805
Phycobilins, 703
Phycocyanobilin, 703**S**
Phylloquinone, 508**S**
Phylogenetic tree, 119, 121**F**, 123**F**
Phylogeny, 1075**D**
Phytanic acid, 752**S**, 753
Phytanic acid α-oxidase, 752**F**, 753
Phytol, 289**S**, 703, 752**S**, 753
Phytyl side chain, of chlorophyll, 702**S**
P_{II} and *E. coli* glutamine synthetase, 837**F**, 838
Pili (*sing.* pilus), 910
α-Pinene, 289**S**
Ping-pong
mechanisms, 431
reaction, 375**D**, 376**F**, 379
Piperidine, in nucleic acid sequencing, 212**F**
Δ^1-Piperidine-2,6-dicarboxylate dehydrogenase, 849**F**
Δ^1-Piperidine-2,6-dicarboxylate, in lysine biosynthesis, 849**F**, 850
PIR (Protein Resource Identification Protein Sequence Database), 117
Pitch (of DNA helix), 217
pK_a values. *See* Proton dissociation constant values.
p_{lac} promoter for the *lac* operon, 976**F**
Plamids, recombinant, 249**F**
Plants, base DNA composition of, 906**T**, 907**T**
Plant cell, 25**F**
Plaque, arterial, lipoproteins and, 793**F**
Plasmalogens, 285
synthesis of, 777, 779**F**
Plasmid(s), 227, 248, 248**F**
chimeric, 249, 249**F**
as cloning vectors, 249, 249**F**
Plasmid *pBR322*, 248**F**, 254**F**
Plasmin, 436, 1089
Plasminogen, 436, 1089, 1090**F**
Plastids, 23, 699**D**
Plastocyanin, 708**F**, 711
Plastocyanin:ferredoxin oxidoreductase, 711
Plastohydroquinone, 711**F**
Plastoquinol, 711**F**
Plastoquinol:plastocyanin oxidoreductase, 710
Plastoquinone, 708**F**, 709, 710, 711**F**, 857**F**
Platelet activating factor, 284, 285**S**, 285**G**, 777, 780**F**
Platelet-derived growth factor, 170**F**
β-Pleated sheet, 146**D**, 147**S**
β-Pleated strand, 91**F**
Plover, golden, dependence of on fatty acids, 746
PLP, 480**S**, 481**F**, 482**F**
pol mRNA, 444**F**
polA1 mutants, of *E. coli*, lack DNA polymerase I activity, 943
Polar bears, use of triacylglycerols by, 281
Poly(A) tails, 194, 195**F**, 1058, 1059**F**
Poly(A) addition signal, 1002**F**, 1003
Poly(A) polymerase, 1005
$Poly(A)^+$ addition site, 1003, 1005**F**
$Poly(A)^+$ tail, at 3′-end of eukaryotic mRNAs, 1005, 1005**F**
Poly (Gly-Pro-Pro) triple helix, 154**S**
Poly (L-guluronate), 333, 334**F**
Poly (D-mannuronate), 333
Polyacrylamide, 132
Polyacrylamide gel electrophoresis, 133, 133**F**
3′-Polyadenylylation of eukaryotic mRNAs, 1003, 1005**F**
Polyamine biosynthesis, 851**F**
Polyamino acids, 144**D**
Polycistronic RNA, 975, 975**F**
Polyhedrin, 266
Polylinkers, in cloning DNA, 251, 252**F**
Polymerase chain reaction (PCR), 271, 272**F**
Polymorphic, 1076**D**
Polynucleotide(s), 188, 189**F**
Polynucleotide phosphorylase, 1020
Polypeptide, 84
chemical synthesis of, 125
Polyprenols, 289
Polyprotic acids, 187
Polyribosomes, 1057, 1057**F**
Polysaccharides, 326
Polysomes, 1057**F**
Polyunsaturated fatty acids, β-oxidation of, 751**F**
Polyuridylic acid, as synthetic mRNA, 1020
Porin, 628

Porphyrin ring system, in chlorophyll, 702**F**
Positive control of enzyme synthesis, 561**F,** 562**F**
Positive cooperative effect, 396
Positive effectors, 408, 411
Post-transcriptional modifications of eukaryotic mRNAs, 1058
Post-transcriptional processing of RNA, 1001
Post-translational modification, 153**D**
Post-translational processing of proteins, 1066
pp60[src] tyrosine kinase, 306
PPR1, a C_x-type of zinc finger protein, 996**T**
PPT. *See* Phosphinothricin.
PQ. *See* Plastoquinone.
Pre-mRNAs, 1003
Pre-vitamin D, 505**S,** 506
Prealbumin, 174**S**
Predictive algorithms, 170
Pregnenolone, 799, 800**F**
Prehybridization, 260**F**
Preinitiation complex, 974**F,** 1059**F**
Prephenate, 857**F,** 858, 859**F**
Prephenate dehydratase, 858, 859**F,** 860
Pri in DNA replication primosome, 950**T,** 953**F**
Pribnow box, 967, 967**F**
Primary structure, nucleic acids, 209
Primary transcripts, 1001
  processing of, 1002
Primase, an RNA polymerase in DNA replication, 948, 949**F,** 950**T,** 953**F**
Primer(s) in DNA replication, 209, 211**F,** 271, 272**F,** 940**D,** 953**F**
  mutant, 273
Priming reactions in glycolysis, 573
Primosome, in DNA replication, 950, 954**F**
Pristanic acid, 752**S,** 753
Pro α-2 collagen gene, exons and introns in, 1002**F**
Pro-proteins, and post-translational processing, 1068
Probe, DNA 233
  heterologous, 262
  radioactive, 259**F,** 261**F**
Procarboxypeptidase, 391**T**
Processing of RNA, 1001
Processivity in DNA replication, 940**D**
Prochirality, 282**F,** 469**D**
α2-Procollagen gene, exon organization and, 1087**F**
Proelastase, 391**T**
Proenzymes 390**D,** 436
Progenote, 22, 1098**F,** 1099
Progenote V-ATPase, evolution of, 1097**F**
Progesterone, 291**S**
  receptor for, 998**T**
  in steroid hormone synthesis, 799, 800**F**
Progestins, 291
Prohormone, 506**D**
Proinsulin, 391**F**
Prokaryotes, 20
Proline, 58**S,** 58**G**
  biosynthesis of, 841, 841**F**
  degradation of to α-ketoglutarate, 866**F,** 867
Proline racemase, 430
Promoted-water pathway, in carboxypeptidase A, 453
Promoter(s), 251, 266, 267**F,** 964**D,** 967, 967**T,** 989, 989**T**
  region in operon, 975**F,** 988**F**
Propionyl-CoA
  biotin-dependent carboxylation of, 493**T**
  as intermediate in Val, Met, Ile degradation, 867**F**
  as a product of β-oxidation, 747
Propionyl-CoA carboxylase, 356**T,** 747**F**
Propylamine transferases, 851**F**
Prostaglandin(s), 278, 785
  synthesis of, 773, 783**F**
Prostaglandin $E_{2\alpha}$, 785
Prosthetic groups, 98, 356
Protease(s), 354**D**
  serine, of blood clotting, 392, 393**F**
Protective proteins, 97
Protein A, 269**T**
Protein C, structural modules in, 1090**F**
Protein, C-terminal end, 87
  chemical synthesis of, 125
  concentration of, 103**B**
  conformation of, 93
  degradation of, 1071, 1071**F,** 1072, 1073**F**
Protein(s) 81, 84
  amino acid composition of, 86, 87**T,** 88**T,** 1091**F**
  architecture of, 89
  assay of, 103**B**
  binding of to DNA, 997**F,** 1014
  conjugated, 98
  denaturation of, thermodynamic parameters for, 514
  fibrous, 89, 90**F**
  folding of, 1066, 1067**F**
  as fuel depot, 819
  function of, 93**F,** 95**T**
  globular 89, 90**F**
  heteromultimer, 105**D**
  homologous, 119
  with internal domain duplications, 1089**T**
  import of into mitochondria, 1070
  isoforms of, RNA splicing and, 1011
  modulator, 392**D**
  modules of, 167, 169**F**
  mutant, 124
  N-terminal end of, 87
  post-translational processing of, 1066
  purification of, 102, 104, 104**T**
  reactions of 101, 102**F**
  separation methods of, 102
  sequencing strategy of, 105
  shape, 89
  simple, 98
  sorting and translocation of in eukaryotes, 1069
  structure of, 89, 90, 91, 91**F,** 92**B,** 105, 1086, 1086**F,** 1091**F**
  synthesis
    basic steps in, 1046**F**
    degradation and, 1040
    elongation cycle of, 1045, 1051, 1052**F,** 1053, 1061
Protein synthesis *(Continued)*
    in eukaryotic cells, 1058
    GTP hydrolysis of, 1055
    inhibitors of, 1062**F,** 1063, 1063**T**
    initiation of, 1045, 1047, 1049, 1050**F,** 1058, 1060
    initiation factors, 1060**T**
    mechanics of, 1045
    regulation of via eIF-2, 1061**F**
    release factors (RFs) of, 1061
    secretory proteins, 1069, 1070**F**
    sequence of events, 1050**F**
    23SrRNA and, 1053
    termination of, 1047, 1056**F,** 1061
    translocation during, EF-G and, 1053
    translocation of to the ER, 1069, 1070**F**
    termination, 1055, 1056
  translocation of, 1068, 1069, 1070**F**
Protein kinase, 390**F**
  CDC2-type, 957
  cAMP-dependent, 393
Protein phosphatase, 390**F**
Protein:ubiquitin conjugates, breakdown of, 1073**F**
Proteoglycans, 95**T,** 97, 100**T,** 339, 345, 347**T,** 348**F,** 349**F**
Prothrombin, 393**F,** 508
  structural modules in, 1090**F**
Protoeukaryote V-ATPase, evolution of, 1097**F**
Proton(s)
  dissociation constant (pK_a) values of, 46**T,** 182**T**
  jumping of, 41
  in photosynthesis, 716, 718
  transfer potential of, 528**T**
Proton-translocating ATPases, 1093, 1094**F,** 1095**T**
Protoporphyrin IX, 101**F**
Prototrophic growth, 900**D**
Provirus, 958
Proximity, 427
  effects in reactions, 434
Proyl hydroxylase, 153
PRPP, 859**F,** 862**F,** 872, 877, 878**F,** 887**F,** 888, 889, 890**F**
Pyrimidine biosynthesis, 886, 887**F**
Pseudo first-order reaction, 358**D**
Pseudoterranova decipiens, globin gene organization of, 1087
Pseudouridine, 195, 197**S**
  in tRNA, 239
Psi (Ψ), 83
Psi angle in peptides, 140**D**
PS. *See* Photosystem.
Psicose, 313
Pteridine(s), 500
  oxidation, by xanthine oxidase, 884
Pterins, 500
Pulse-chase experiment, in DNA replication, 946
Purine(s)
  biosynthesis of, 872
    regulation of, 876, 878**F**
  degradation of, 881

Purine(s) *(Continued)*
ring system of, 181**S**, 872, 872**F**
salvage of, 879
Purine nucleoside cycle in muscle, 882, 883**F**
Purine nucleoside phsophorylase, 881, 882**F**
Purine riboside, 430, 431**S**
Puromycin, 1062**S**, 0163**T**
in protein synthesis, 1064
in peptidyl transferase assay, 383**F**
Putrescine, 851**S**
Pyran, 314
Pyranose, 314, 316**F**
Pyridine, 473**S**
nucleotides of, 468, 536
Pyridoxal, 463**T**, 483**S**
Pyridoxal phosphate, 356**T**
Pyridoxal-5-phosphate, 463**T**, 480**S**, 481**F**, 482**F**
Pyridoxal-phosphate enzymes, 433**T**
Pyridoxamine, 463**T**, 483**S**
Pyridoxine, 463**T**, 483**S**
Pyridoxol, 483**S**
degradation, 889, 890**F**
regulation of, 889
ring system of, 181**S**, 886**F**
Pyroglutamic acid, 60, 61**S**
Pyrolle, linear or open, 703**S**
Pyrophosphate, hydrolysis of, 526**T**
Pyrophosphohydrolase, in histidine synthesis, 862**F**
Pyrophosphomevalonate decarboxylase, 789, 790**F**
Pyrophosphoryl group transfer, 188**F**
Pyrophosphorylase, 676
Pyrrole-2-carboxylate, 430**S**
Δ^1-Pyrrolidine-5-carboxylase, proline biosynthesis and, 841, 841**F**
Δ^1-Pyrroline-2-carboxylate, 430**S**
Pyruvate, 554, 574**T**, 583**F**, 602**F**
biotin-dependent carboxylation of, 493**T**
Pyruvate carboxylase, 465**F**, 466**F**, 618**F**, 619, 663**F**
allosteric activation of, 664
regulation of, 668
Pyruvate decarboxylase, 464
Pyruvate dehydrogenase, 356**T**, 496, 509, 602**F**, 603
mechanism of, 497
regulation of, 619, 620**F**, 621**F**, 668
Pyruvate dehydrogenase kinase, 621**F**
Pyruvate dehydrogenase phosphatase, 621**F**
Pyruvate family of amino acids, 850
Pyruvate, CO_2 fixation and, 727, 728**F**
Pyruvate kinase, 356**T**, 533, 583**F**, 589**F**, 590**F**
regulation of, 668

Q_{10}, 369**D**
qa-1F, a C_x-type of zinc finger protein, 996**T**
Q cycle, 640**F**, 641
Qβ phage replicase mRNA Shine-Dalgarno sequence, 1049**F**
Quantum yield of photosynthesis, 705, 706**F**, 715
Quaternary structure of proteins, 172**D**, 172**T**, 174, 175**F**
Quinolone antibiotics, 960**F**

R state
of allosteric protein, 411**F**
of glycogen phosphorylase, 678
R. viridis. See Rhodopseudomonas viridis.
R17 phage coat protein, mRNA Shine-Dalgarno sequence of, 1049**F**
Racemases, 355**T**
Racemization, PLP-assisted, 481**F**
Racker, Efraim, 647
Radioactive isotopes, 564**T**
Ramachandran, G. N., 140
Ramachandran diagram, 141**F**
Random coil, 157**D**
Random copolymers as synthetic mRNAs, 1021**T**
Random genetic drift, 1076**D**
ras oncogene, 307
Rat, base DNA composition of, 907**T**
Rat growth hormone (rGH) in transgenic mice, 933, 933**F**
Rate constant, 357**D**
Rate law, 357
Rattlesnake, Eastern diamondback, 284**F**
RBP, 504
proteins of, 973**T**
Reaction center
in photosynthesis, 705, 705**F**, 706
in *Rhodopseudomonas viridis*, 712, 712**F**, 713
Reaction stoichiometry, 806
Rearrangements, B_{12}-catalyzed, 485, 486**F**, 488**F**
RecA protein, 918, 919**F**
in the SOS response, 960
RecA recombinase. *See* RecA protein.
RecBCD complex, 918, 919**F**
Reciprocal control, in regulation of gluconeogenesis, 667
Recognition matrix, of H bonds in DNA major groove, 994
Recombinant DNA, 247, 273
Recombinase, in immunoglobulin gene rearrangement, 927
Recombination
general, 914, 915, 915**F**, 918, 923
genetic, 909, 909**F**, 914**D**
high frequency of (*Hfr*), 910
homologous, 914, 919**F**
Holliday model of, 916, 917**F**
illegitimate, 915
molecular mechanism of, 914
patch, 916, 917**F**, 918**F**, 922**F**
site-specific, 915
somatic, 914
Recommended daily allowance (vitamins), 476
Red drop, in photosynthesis, 706, 706**F**
Redox couple, 630
Redox reaction, 632
Reduced coenzymes, as energy carriers, 523
Reducing end of disaccharides, 322
Reducing power, light-driven generation of, 708**F**, 726, 727**F**
Reducing sugars, 319
Reduction and alkylation of disulfides, 106, 107**F**
Reduction potential, standard, 629**D**, 631**T**

Reed, Lester J., 496
Reference half cell, 630
Refsum's disease, 753
Regulation
allosteric, 390**D**
of CO_2 fixation, 726
metabolic, 19, 20**F**, 559
Reichart, E., 736
Release factors (RFs) in termination of protein synthesis, 1055
Renin, 442**T**
Rennin, 442**T**
Rep protein, in DNA replication, 954**F**
Repeating copolymers as synthetic mRNAs, 1022, 1022**F**
Replica plating, 252**F**
Replication, 181**F**
Replication bubble, in DNA replication, 947**F**
Replication factor A (RF A), 956
Replication fork(s), 944, 945**F**, 947**F**, 949, 949**F**, 950**F**
Replication protein C (RP-C), 957
Replication termination protein, 954
Replicator, plasmid, 249
Replicons, in eukaryotic DNA replication, 955
Replisome, in DNA replication, 950
Reporter gene constructs, 269, 270**F**
Repression
of gene expression, 976**D**
of enzyme synthesis, 389, 562**F**
Repressor, 561**F**
434 Repressor, an HTH motif DNA-binding protein, 993**F**, 994**F**
λ Repressor, an HTH motif DNA-binding protein, 993**F**, 994**F**
Reptiles, uric acid metabolism of, 884, 885**F**
Residue, amino acid, 84
Resilin, 95**T**, 98
Resolvase(s) (*ruvC* endonuclease), 922**F**
Holliday junction, 921
Resonance energy transfer, in photosynthesis, 704**F**
Resonance stabilization energy, in proteins, 140
Response elements, 990**T**
Response of enzymes to energy charge, 817, 817**F**
Restriction digest, of DNA, 206, 206**F**
Restriction endonucleases, 201**D**, 202**F**, 203, 204**T**, 205, 209
Restriction fragment, size, 205
Restriction mapping, 206, 206**F**
Retinal, 289**S**, 502
all-*trans*, 503**F**
11-*cis*, 503**F**
Retinal isomerase, 504
Retinaldehyde, 502, 504
Retinoic acid receptor, 998**T**
Retinol, 463**T**, 502
all-*trans*, 503**F**
binding protein of, 504
Retinol dehydrogenase, 504
Retinyl esters, 502, 503**F**
Retroviruses, 932**D**, 957

Reverse transcriptase, 957
synthesis of cDNA and, 265, 265**F**
Reversible inhibitor, 369
RF-1,2,3 (release factors) in termination of protein synthesis, 1055
rGH (rat growth hormone), in transgenic mice, 933, 933**F**
Rhamnose, 320**F**
Rhizobia, in nitrogen fixation, 828, 829**T**
Rhizobia:legume symbiosis, 828
ρ (Rho) factor, in transcription termination 970, 971**F**
Rhodamine, 296
Rhodanese, folding, 1066
Rhodopseudomonas viridis, 712, 712**F,** 713
Rhodopsin, 504
gene of, 237, 237**F**
incorporation of retinal by, 503**F**
Rubisco activase, pH activation of, 726
Ribitol, 319**F,** 340, 474**S**
Riboflavin, 463**T,** 474
Riboflavin 5′-phosphate, 474**S**
Ribofuranose, 183**S**
Ribonuclease(s), 514**T**
Ribonuclease A, 85**T,** 88**T,** 89**F,** 95**T**
chemical synthesis of, 125, 126
Ribonuclease B, oligosaccharide of, 344**F**
Ribonucleic acid (RNA), 180, 188
Ribonucleoprotein particles, 1005
Ribonucleotide reductase, catalytic site, 891**F,** 894
Ribonucleotide(s), 185
aborption spectra of, 183**F**
reduction of, by B_{12}, 485, 486**F**
Ribonucleotide reductase, 890, 891, 891**F,** 894
reaction mechanism of, 892**F**
regulation of, 894, 894**F**
Ribose, 180, 312
Ribose-5-Phosphate
in CO_2 fixation, 724**F,** 725**T**
in purine biosynthesis, 872
in pentose phosphate pathway, 687, 694, 695**F**
Ribose-5-pyrophosphokinase, in purine biosynthesis, 872, 873, 874**F,** 877
Ribosomal protein(s), 196**F,** 1041, 1041**T,** 1049**F**
eukaryotic, 1045**T**
40S Ribosomal protein, phosphorylation of, 1060
Ribosomal RNA (rRNA), 190**S,** 191**T,** 241, 1041, 1041**T,** 1042**F**
Ribosomal subunit, 382, 383**F,** 1041, 1042**T,** 1043**F,** 1052
Ribosome(s), 23**T,** 27**T,** 100**T,** 190, 195**F,** 196**F,** 1041, 1041**T,** 1044, 1057, 1057**F**
architecture of, 1043, 1044**F**
as binding site, 266, 267**F,** 1046**F,** 1047, 1047**F,** 1054**F**
conformational changes during protein synthesis, 1055
cytosolic, 1044, 1045**T**
E. coli, 1041**T,** 1044**F**
life cycle of, 1057, 1057**F**
structure and assembly of, 1040, 1042
Ribothymidylic acid, 195, 197**S**
Ribozyme, 381**D,** 382**F,** 383
Ribulose, 313
Ribulose bisphosphate carboxylase, 85**T**
Ribulose-1,5-bisphosphate, 720, 721**F,** 723, 724**F**
Ribulose-1,5-bisphosphate carboxylase/oxygenase, 720, 722**F**
Ribulose-1,5-bisphosphate carboxylase, 720, 720**F**
activation of, 726
reaction of, 721, 721**F**
regulation of, 721
Ribulose-1,5-bisphosphate oxygenase reaction, 721
Ribulose-5-P in CO_2 fixation, 724**F,** 725**T**
Ribulose-5-phosphate kinase, 723, 724**F,** 725**T**
activation of, 726, 727**F**
Ribonucleic acid. *See* RNA.
Rich, Alexander, 223
Ricin, 95**T,** 97
as inhibitor of protein synthesis, 1063**T**
mechanism of action of, 1064
Ricinus communis, 1064
Rickes, 487
Rickets, 507
Rieske iron-sulfur center, of photosynthesis, 708**F,** 709
Rieske protein, 640
Rifampicin, 968**S**
Rifamycin B, 968**S**
RNA, (ribonucleic acid), 26**T,** 100**T,** 180, 188, 194
alkaline hydrolysis of, 198**F**
in DNA replication, 948, 949**F,** 951, 953**F**
direction of DNA polymerase by, 957
double-helical A-form of, 221
as genetic material, 932
messenger, 190, 191**T,** 195**F**
RNA core polymerase, 964
RNA polymerase(s), 561**F,** 562**F**
DNA-dependent, 964, 965
E. coli, structure and function, 964
eukaryotic, 970, 971
holoenzyme, 964
inhibitors, 968**F**
initiation site, 968
postranscriptional processing of, 1001
ribosomal, 190, 191**T,** 194, 196**F**
secondary structure of, 238
splicing of, 195**F**
sugar-phosphate backbone of, 189**F**
tertiary structure of, 240
transfer, 190, 191**T**
tumor viruses of, 957
RNA, polymerase I, eukaryotic, 971
RNA polymerase II, eukaryotic, 971
RNA polymerase II, promoter modules, 989**T**
RNA polymerase II, promoters, 973
consensus sequences of, 989**T**
structure and function, 972
transcription initiation by, 973
from yeast, 972
RNase(s), 200**T,** 201
of *Bacillus subtilis,* 200**T,** 201
RNase(s) *(Continued)*
pancreatic, 200**T,** 202
RNase A, 200**T.** *See also* Ribonuclease A.
RNase H, 265, 265**F,** 948, 957
RNPs (ribonucleoprotein particles), 1005
Robinson, Sir Robert, 788
Rod cells, 504
Rosanoff convention, 70
Rosanoff, M. A., 70
Rotational entropy, 428
Rotenone, 652**S,** 653
Roundup®, 863**B**
rrnA of *E. coli,* 1042**F**
rRNA (*see also* ribosomal RNA), 190, 191**T,** 1042
conserved bases in, 1045
of *Halobacterium volcanii,* 242, 243**F**
of *Saccharomyces cerevisiae,* 242, 243**F**
secondary structure, 241, 242, 243**F**
tertiary structure of, 244
4S rRNA, 1041
5S rRNA, eukaryotic, 1044, 1045**T**
16S rRNA, 1041, 1049**F,** 1084
23S rRNA, 1041, 1044, 1045**T**
28S rRNA, eukaryotic, 1044, 1045**T**
rrnA1 promoter, nucleotide sequence of, 967**T**
rrnD1 promoter, nucleotide sequence of, 967**T**
rrnE1 promoter, nucleotide sequence of, 967**T**
Rubisco. *See* Ribulose-1,5-bisphosphate carboxylase.
Rubisco activase, 721
RuBP. *See* Ribulose-1,5-bisphosphate.
Rubredoxin, 163**S,** 163**G**
Ruminants, 331
Rutter, William, 440
ruvC endonuclease resolvase, 922**F**

S (sedimentation coefficient), 135
S-Adenosylhomocysteine, 851**S**
S-Adenosylmethionine, 487, 539**S,** 776, 850, 851**S**
hydrolysis of, 527**T**
S-Adenosylmethionine decarboxylase, 851**S**
S-Adenosylmethionine synthase, 850, 851**F**
S6 (40S ribosomal protein), phosphorylation of, 1060
^{35}S as radioactive tracer in nucleic acid sequencing, 209
Saccharomyces cerevisiae, rRNA of, 242, 243**F**
Saccharopine, in lysine biosynthesis, 846**F,** 847
Safflower oil, fatty acids in, 279
SAICAR (*N*-succinylo-5-aminoimidazole-4-carboxamide ribonucleotide)
synthetase of, in purine biosynthesis, 874**F,** 875
Salmon, base DNA composition of, 907**T**
SAM. *See S*-adenosylmethionine.
Sample half cell, 630

Sanger, Frederick, 105, 209, 1018
Sanger's reagent, 106, 108**F**
Saponification, 280
Sarcosine, 62**S**, 226**S**
Sau3A restriction endonuclease, 204**T**, 205
Schachman, Howard, 415
Schimmel, Paul, 1033
Schrödinger, Erwin, 515
SCID (severe combined immunodeficiency syndrome), 881
Scleractinians, 551**B**
Scurvy, 156, 491
SDS-polyacrylamide gel electrophoresis (SDS-PAGE), 133, 134**F**
Sea urchin, base DNA composition of, 907**T**
Second law of thermodynamics, 514**D**
Second messenger, 682
Second-order reaction, 358**D**
Secondary structure of DNA, 216, 217**F**
  in proteins, 142
Sedimentation coefficient (*S*), 135
Sedoheptulose-1,7-bisphosphatase, in CO_2 fixation, 724**F**, 725**T**, 726, 727**F**
Sedoheptulose-7-phosphate, in CO_2 fixation, 724**F**, 725**T**
Selenium (Se), 356**T**
Selenocysteine, in *E. coli*, 1026**T**
Semidehydro-L-ascorbate, 489
Semiquinone, of FAD, 475**S**
Sense, 12
Sense strand of DNA, 965**B**
Sensors, in regulation of gene expression, 991
Sephacryl®, 131
Sephadex®, 131, 329**F**
Sequenase 2® DNA polymerase, 210
Sequence alignments, 1078
Sequence-specific transcription factors, 994
Sequoia gigantea, 699
Serendipity finger proteins, 996**T**
Serglycin, 345, 346**F**
Serine, 58**S**, 58**G**, 69
  D- and L-isomers of, 71**S**
  degradation of to pyruvate, 865**F**
  as methyl group donor, 896**F**
  as a product of photorespiration, 722, 722**F**
Serine acetyltransferase, in cysteine biosynthesis, 854, 855**F**
Serine dehydratase
  deamination of Ser to give pyruvate, 866
  PLP-catalyzed mechanism, 484**F**
Serine esterase, 436
Serine hydroxymethylase, 510
Serine hydroxylmethyltransferase, 500
  in dTMP synthesis, 896**F**
  in glycine biosynthesis, 853, 855**F**
Serine hydroxymethyltransferase, N^5,N^{10}-methylene-THF synthesis, 896**F**
Serine proteases, 436
  blood clotting and, 392, 393**F**, 1090**F**
  evolutionary relationships of, 120
Serotonin, 61, 62**S**
Serum, as physiological challenge, 990**T**
Sesquiterpene, 288
Sex pili, 910
Sexual conjugation in bacteria, 910, 910**F**, 911**F**
Sharks, uric acid metabolism of, 884, 885**F**
Shikimate, 857, 858**F**
Shine-Dalgarno sequence, 1048, 1049**F**, 1050**F**
Shipworms, 331
Shuttle systems, in electron transport, 654**F**
Sia, R. H. P., 903
Sialic acid(s), 287, 321**F**
Sialyltransferase, 782**F**
Sickle-cell anemia, 406, 407
Sigma (σ) factors, 964, 966**F**
Signal recognition particle (SRP), 1069, 1070**F**
Signal sequences translocation, 1068, 1069, 1070**F**
Simian virus 40, 266
Simnastatin, 792**F**
Singer, S. J., 295
single-stranded DNA-binding protein, 921, 949**F**
  of nitrite reductase, 828, 828**F**
  of sulfite reductase, 856**F**
Site-directed mutagenesis, chymotrypsin, 440, 454, 455, 459
Six-cutter, restriction endonuclease, 205
Slack C-4 pathway of CO_2 fixation and, 727
Slater, E. C., 646
Sliding clamp and the β subunit dimer of DNA polymerase III, 944
Small nuclear ribonucleoprotein particles (*See also* snRNPs), 190, 196, 1006, 1007**T**
Small nuclear RNA (*See also* snRNA), 190, 196, 1007, 1008**F**, 1009**F**
Small subunit rRNA, as a molecular clock, 1084
Smith, 487
sn sytem of nomenclature, 281
Snail zinc finger protein, 996**T**
Snake venom, 200**T**
Snell, Esmond, 483
snRNA (small nuclear RNA), 190, 196, 1007, 1008**F**, 1009**F**
  secondary structure and base pairing with intron, 1008**F**
snRNPs, 190, 196, 1006
  properties of, 1007**T**
snRNP U1 through U6, 1009**F**
Soap, 280
Sodium dodecylsulfate (SDS), 133**S**
Sodium palmitate, 38
Solenoid, 234, 235**F**
Solid phase synthesis
  of DNA, 236**F**
  of polypeptides, 125
Soluble protein activator, 789
Solvent capacity of the cell, 811
Somatic cells, 906**D**
Somatotropin, 95**T**, 96
Sorbitol, 319**F**
Sorbose, 313
Sørenson, 42
SOS reponse, and DNA repair, 960
Southern blotting, 260**F**
Southern, E. M., 260**B**
Southern hybridization, 258, 260**F**
Soybean oil, fatty acids in, 279
Soybean trypsin inhibitor, 163**S**
SP1, transcription factor for RNA polymerase II, 989**T**, 996
SP6 RNA polymerase, 266, 266**F**
Spackman, Darrel, 77
Sparsomycin, 1063**T**
Specific acid/base catalysis, 433
Specific rotation of amino acids, 71**T**
Spermaceti, 288
Spermidine, 851**S**
Sphinganine, 777, 781**F**
Sphingolipids, biosynthesis of, 773, 777, 781**F**
Sphingomyelin, 286, 286**G**, 286**S**
  synthesis of, 782**F**
Sphingosine, 286, 286**G**, 286**S**
Spider dragline silk, 152**S**
Spike proteins, 304
Spirogyra, 699, 700**F**
Spliceosome, 1007, 1009**F**
Splice recombinants, in homologous recombination, 916, 917**F**, 918**F**
Splice recombination, 916, 917**F**, 918**F**, 922**F**
Splice sites,
  consensus sequences of, 1006**T**
Splicing
  lariat formation during, 1006, 1007**F**
  of mRNA precursors, 1005, 1007**F**, 1008**F**
  of RNA, 195**F**, 381, 382**F**, 1009
  snRNPs and, 1006
  transesterification during, 1008**F**
Split genes, organization of, in eukaryotes, 1002, 1002**F**
Spontaneous mutation, frequency of, 929
Squalene, 288, 289**S**
  as precursor to cholesterol, 788**F**
  synthesis of cholesterol from, 791**F**
  synthesis from mevalonate, 789, 790**F**
Squalene monooxygenase, 789
Squalene-2,3-epoxide, 789
  in cholesterol biosynthesis, 791**F**
SRP (signal recognition particle), 1069, 1070**F**
SRP receptor, 1069, 1070**F**
SSB (single-stranded DNA-binding protein), 918, 919**F**, 950**T**, 951
ssDNA, 200**D**
Stachyose, 324**F**
Staggered temini, 203
Stahl, Franklin, 937
Standard state, 513**D**
Stanozolol, 801**S**
Staphylococcus protease, 113, 115**T**
Starch, 327
State function, 512**D**
Statistical thermodynamics, 515
Steady state, 4
Steady-state assumption, 361
Stearic acid, 277, 278**S**, 278**G**
Stearoyl-CoA desaturase, 769
Stein, William, 77
Steitz, Thomas, 942
Stem-loop structure
  in immunoglobulin genes, 927**F**
  in transcription termination, 970**F**

Stereospecific numbering, of glycerol phosphatides, 281
Steric specificity, of NAD^+-linked dehydrogenases, 471**T**
Steroid hormones, 289, 799, 800**F**
Steroid receptor(s), 990, 990**T,** 996**T,** 999**F**
Steroid reductase, 472**T**
Sticky ends, 203, 248, 251
Stoeckenius, Walther, 648
Stoichiometry, 806
  coupling and, 806, 807, 815**F**
Stomata (of plants), 728
Stop-transfer sequence, in protein translocation, 1070
Storage granules, 23**T**
Storage polysaccharides, 327
Storage proteins, 97
Strain, in ES complex, 428
Strand(s)
  recombination and, 916, 917**F**
  scission by β-elimination, 212**F**
Streptavidin, 269**T**
Streptococcus pneumoniae, 903
Streptomycin, 325**F**, 1062**S**, 1063, 1063**T**
Stroma, 28**T**
  of chloroplast, 700
Structural complementarity, 16, 17**F,** 387
Structural polarity, 12
Structural proteins, 97
β-Structure, 91**F**
Stubbe, JoAnn, 892
Subclone, 263**F**
Submitochondrial particles, 647**F**
Substrate
  in a bisubstrate reaction, 375
  cycle of, 670, 812, 813**F**
  of an enzyme, 354
  level
    control of, in gluconeogenesis, 668
    phosphorylation of, 586**D,** 609
  suicide, 373
  Trojan horse, 373
  saturation curve of, 360**F**
Subtilisin, 113, 433**T,** 436
  inhibitor, from *Streptomyces,* 164**S,** 164**G**
Succinate, 602**F,** 604**T**
Succinate dehydrogenase, 100**T,** 356**T,** 602**F,** 604**T,** 610**F,** 637
  inhibition by malonate, 372
Succinate thiokinase, 609
Succinate-coenzyme Q reductase, 635**T,** 637
Succinic anhydride, 68**S**
Succinyl-CoA, 602**F,** 604**T**
Succinyl-CoA synthetase, 433**T,** 537**F,** 602**F,** 604**T,** 609**F,** 610**F**
Succinyl-diaminopimelate amino transferase, 849**F**
Succinyl-diaminopimelate desuccinylase, 849**F**
N-Succinyl-L,L-α,ε-diaminopimelate, in lysine biosynthesis, 849**F**
N-Succinyl-2-amino-6-ketopimelate synthase, 849**F**
O-Succinylhomoserine, in methionine biosynthesis, 849**F,** 850
Sucrase, 324
Sucrose, 323. 323**F**
Sucrose phosphorylase, 597
  kinetics of, 380
Sugar acids, 318
Sugar alcohols, 319
Sugar esters, 320
Sugar nucleotides, 538**F,** 676**G**, 676**S**
  in activation of sugar units, 675
Sugar-phosphate backbone
  degrees of freedom of, 220**F**
  of nucleic acids, 189**F,** 190
Suicide substrate(s), 373, 897**B**
Sulfa drugs, 876**F**
Sulfate assimilation pathway, 854, 856**F**
Sulfatide, 287
Sulfide, as the product of sulfate assimilation, 855, 856**F**
Sulfinyl carbanion, 115**F**
Sulfite, as an intermediate in sulfate assimilation, 855, 856**F**
Sulfite reductase, 100**T**
  in sulfate assimilation, 856**F**
Sulfmeturon methyl, 863**B**
Sulfolobus acidocaldarius H^+-ATPase, 1096, 1096**F**
Sulfonamides, 876**S**
Sulfurmycin B, 325**F**
Sunflower oil, fatty acids in, 279
Supercoil(s), 227**F**
Supercoiling *vs.* transcription, 969**F**
Suppressor tRNAs, 1038
Supramolecular complexes, 9
  of photosynthesis, 708**F,** 709
Surface immunoglobulin receptors, 304
Surroundings, thermodynamic, 512**D**
SV40, 266, 988**F**
  replication of, and T antigen, 956
Symmetry, in proteins, 173
syn configurations, 185**F**
Synapsis 916**D**
Syndecan, 346**F**
Synonymous substitutions, 1078
Synovial fluid, in joints, 335
Synvinolin, 792**F**
System, thermodynamic, 512**D**
Szent-Györgyi, Albert, 483, 491, 600

+T antigen, 956
T antigen, regulation by protein phosphorylation, 957
T cell lymphocytes, 923
T state
  of allosteric protein, 411**F**
  of glycogen phosphorylase, 678
TψC loop, tRNA, 238, 239, 239**F**
Tagatose, 313
Talose, 312
Tandem enzyme, 669
Tandem trp codons, in transcription attenuation, 987, 987**F**
Taq DNA polymerase, 271, 272**F,** 273
TATA-binding protein, 974**F**
TATA box
  as eukaryotic promoter module, 989, 989**T**
  in selected eukaryotic genes, 974**F**
TATA motif, in promoters, 973
Tatum, Edward, 563, 901, 908
Taurine, in bile acid synthesis, 798**F**
Taurocholic acid, synthesis of, 798**F**
Tautomerism, 182
Tautomerization
  of aldimine-ketimine, 482
  of pyruvate, 534
Tay-Sachs disease, 287
TBP (TATA-binding protein), 989**T**
TCA cycle, 602**F**
  regulation of, 619, 620**F**
  role of in biosynthetic pathways, 616
TDF, a C_2H_2-type of zinc finger protein, 996**T**
Teichoic acids, 339, 340**F**
Teleosts, uric acid metabolism of, 884, 885**F**
Temin, Howard, 957
Temperature
  dependence of K_{eq} on, 513
  effect of on enzyme activity, 369, 369**F**
Template, in DNA replication, 209, 211**F,** 936**D,** 940
Tendons, 335
Ter, protein acting in the termination of DNA replication, 950**T,** 954
Ter site, in transcription termination, 971**F**
Teredo navalis, 331
Termination, of DNA replication, 954
Termination codons, 1023**D**
Termination factor, in transcription, 970
3:4 Terminator, in transcription attenuation, 986, 986**F**
Termites, 331
Terpenes, 288
Tertiary structure of proteins, 148**D**
Testosterone, 291**S**
  synthesis of, 800**F**
Tetracycline, 248, 254**F,** 1062**S,** 1063**T**
Tetradecanoyl phorbol acetate (TPA), 991**F**
Tetrahydrobiopterin, in tyrosine synthesis, 860**F**
Tetrahydrofolate (THF), 356**T,** 463**T,** 498**S**
Tetrahymena (ciliated protozoan) catalytic RNA, 381, 382**F**
Tetrahymena ribozyme, 382, 384**F**
Tetrapyrrole, 101**F**
  chlorophyll as Mg^{2+}-containing, 702
Tetraterpene, 288
Tetrazole, 236**F**
Tetrulose, 311
TF proteins, transcription and, 972, 974, 974**F,** 975**T,** 996
TGA1, amino acid sequence of its bZIP DNA-binding motif, 1000**F**
Thales, 32
2-Thenoyltrifluoroacetone, 652**S,** 653
Theorell, Hugo, 476
Thermal cyclers in PCR, 271, 272**F**
Thermoacidophiles, 22
Thermodynamic parameters, for protein denaturation, 514
Thermodynamic potential, 352, 521
Thermodynamics of protein folding, 166**F**
Thermogenin, 654
  in brown fat mitochondria, 823
Thermolysin, 113
Thermus aquaticus, 271

THF, 498**S**, 499**F**
in glycine biosynthesis, 853, 855**F**
in purine biosynthesis, 873, 874**F**, 875
Thiamine, 463, 463**T**, 464**S**
Thiamine pyrophosphate, 356**T**, 463, 463**T**, 464**S**
Thiazolium group, of TPP, 464
Thioester(s), resonance forms of, 478**F**. *See also* Thiolesters.
Thioester-linked fatty acyl anchors, 306**F**
Thioester-linked prenyl anchors, 306, 307**F**
Thiogalactoside transacetylase 976**F**
Thiol esters, as high energy molecules, 536, 537**F**
Thiolase, 744**F**, 754**F**, 768, 786**F**, 789
Thioredoxin,
in deoxyribonucleotide biosynthesis, 890, 891**F**, 892
light activation and, 726, 727**F**
in sulfate assimilation, 856**F**
Thioredoxin reductase, in dNDP synthesis, 890, 893, 893**F**
Thioredoxin reductase, oxidation-reduction cycle, 893**F**
Thiostrepton, 1062**S**, 1063**T**
Thiouridine, 197**S**
Third law of thermodynamics, 516
Third-base degeneracy (in the genetic code), 1024, 1025**F**, 1033
Thomas, Lewis, 1040
Thomas, Charles, 1058
thr operon, and transcription attenuation, 986**F**
Threonine, 59**S**, 59**G**
biosynthesis of, 847
degradation to acetaldehyde and pyruvate, 865**F**, 866
as feedback inhibitor of aspartokinase I, 847**T**, 848
Threonine deaminase, 364**T**
in isoleucine biosynthesis, 852**F**
as precursor for isoleucine synthesis, 852**F**
Threonine synthase, in threonine biosynthesis, 849**F**
Threose, 312
Thrombin, 95**T**, 97, 312, 392, 393**F**, 433**T**, 436
Thromboxanes, synthesis of, 783**F**
Thylakoid, 28**T**
Thylakoid structures, 700, 700**F**
Thymidine, 184
Thymidine kinase gene, promoter, 988**F**
Thymidine nucleotide synthesis, 895
Thymidylate synthase, 356**T**, 501, 895
reaction of, 896**F**
Thymine, 181**S**
degradation, 889, 890**F**
dimers of, 958**F**
Thyroglobulin, 60, 61**S**
oligosaccharide of, 344**F**
Thyroid hormone receptor, a C_x-type of zinc finger protein, 998**T**
Thyrotropin, 95**T**, 96
Thyroxine, 60, 61**S**
Tight turn, 147, 148**S**
Tissue plasminogen activator (TPA), 436, 1089, 1090**F**
Titration curves, 46
TMV. *See* Tobacco mosaic virus.
Tn 7 (*E. coli* transposon 7), 922
Tobacco mosaic virus, 27**F**, 100**T**, 932, 933
coat protein of, 161**S**
α-Tocopherol, 290**S**, 463**T**, 507**S**
Tönnis, 495
Tonoplast, 28**T**
Topoisomerase(s), 228, 945, 949**F**, 969
Tosyl-L-phenylalanine chloromethyl ketone, 459**F**
Tower helix (of glycogen phosphorylase), 673, 674**F**, 680**F**
Toxic shock syndrome, 285
tPA, 170**F**
TPA (tissue plasminogen activator), 1089, 1090**F**
TPA (tetradecanoyl phorbol acetate), 991**F**
TPCK, 459**F**
TPP, 464**S**
TPP synthetase, 464**F**
Tracers as metabolic probes, 564
Trans configuration, lipid chain, 302
Transaldolase, 692**F**
mechanism of, 693**F**
Transamination
of α-keto acid carbon skeletons, 839**F**, 840
in muscle protein degradation, 822
PLP and, 481**F**, 483**F**, 509
Transcalciferin, 507
Transcarboxylase, biotin and, 492**F**, **493T**
Transcript, 965**B**
Transcription, 181**F**, 195**F**, 963**F**, 964, 970
activation of, as anti-inhibition, 981
attenuation of, 985, 986, 986**F**
chain elongation in, 966**F**, 969
factors in, 972, 974, 974**F**, 975**T**
initiation of, 966**F**, 968, 974**F**
modulation of by steroid hormones, 799
regulation, mechanisms of, 987, 988
translation to in prokaryotes, 1058
start sites of, 964, 967
supercoiling and, 969**F**
termination and, 970, 970**F**, 971**F**, 987**F**
Transduction, genetic transmission by, 913, 914**F**
Transesterification, in RNA splicing, 1008**F**
Transfection, 914**D**
of animals, 933, 934
Transfer RNA (tRNA), 190**S**, 191**T**
secondary structure of, 238, 238**F**
Transfer of hydrocarbon to water, 519**T**
Transferase, 354**D**, 355**T**
Transformation, 914**D**
Transformation, DNA-mediated, 252, 252**F**, 254**F**, 903, 904**F**
Transforming growth factor β, 348
Transformylase, 501
Transformylation reaction in *N*-formyl-Met-$tRNA_f^{Met}$ synthesis, 1048**F**
Transgenic animals, 274, 933, 934
Transhydrogenase, 809
Transition(s), 1077
Transition mutations, 928**D**
Transition state, 359**D**, 359**F**, 425**D**, 425**F**, 426, 430**D**
intermediate and induced fit in, 388
stabilization of, 448
Transition temperature, membrane, 301, 302**T**
Transketolase, 464, 465**F**, 468**F**, 691
in CO_2 fixation, 724**F**, 725
mechanism of, 692**F**
Translation (process of protein synthesis), 181**F**, 195**F**, 964, 1040
relationship to transcription, 1058, 1058**F**
start site of, 266
Translational entropy, 428
Translocases, 1068
Translocation, 968, 1069**F**
Transpeptidation, 1052
Transposable elements, 922
Transposition, 915, 1075
Transposition events, 922
Transposons, 921, 922
Transverse asymmetry, in membranes, 298
Transversion mutations, 928**D**
Transversions, 1077
Trehalose, 350
Triacylglycerol(s), 100**T**, 279, 732, 735**F**, 775
as fuel depot, 819
mixed, 280**S**, 280**G**
synthesis of, 776**F**
Triacylglycerol lipase, 733**F**
regulation by hormones, 772**F**
Triantennary oligosaccharides, 344
Tricarboxylic acid cycle, 602**F**
regulation of, 619, 620**F**
TRICINE, 53**F**
Trifluoroacetic acid, 110**F**
Triglycerides, 279
Trimethoprim, DHF and dTMP synthesis, 896**S**
Trimethylglycine (N,N,N-), 62**S**
Trimethyllysine (ε-N,N,N-), 60, 61**S**
Trinucleotides, and aminoacyl-tRNA binding to ribosomes, 1022, 1024**F**
Trioleoylglycerol, 280
Triose phosphate isomerase, 162**S**, 366**T**
in CO_2 fixation, 724**F**, 725**T**
in glycolysis, 572**F**, 574**T**, 580, 581**G**
layer structure of, 160**F**
Triple helix
of collagen, 150, 153, 154**S**
of DNA, models for, 921**F**
formation of, in recombination, 920**F**
Tris, 53
Tristearin, 280**S**, 280**G**
Tristearoylglycerol, 280**S**, 280**G**
Triterpene, 288
Triton X-100, 293**S**
tRNA(s) (*See also* Transfer RNA), 190, 191**T**
amino acid esters of, 535, 536**F**
anticodon loop, 1027
binding on ribosomes and, 1047**F**, 1054**F**
initiator of, 1047
isoacceptor of, 1031, 1035**D**
identity elements of, 1030, 1030**F**
molecular structure of, 1018**S**, 1044**F**
noncanonical base pairing in, 241**F**
protein synthesis and, 1053
recognition by aminoacyl-tRNA synthetases, 1029, 1029**F**
secondary structure, 238, 238**F**
tertiary structure, 240, 240**F**

tRNA(s) *(Continued)*
yeast phenylalanine $tRNA^{Phe}$, 241**S**
$tRNA^{Ala}$, structure, 1031, 1032**F**, 1033
$tRNA^{Gln}$, recognition by glutaminyl-$tRNA^{Gln}$ synthetase, 1030, 1031**F**
$tRNA_i^{Met}$, eukaryotic initiator tRNA, 1058
$tRNA^{Phe}$, bases serving as recognition elements in, 1030, 1030**F**
$tRNA^{Ser}$, nucleotides serving as recognition elements in, 1031, 1030**F**
$tRNA^{Tyr}$ promoter, nucleotide sequence of, 967**T**
Trojan horse inhibitor(s), 607
Trojan horse substrate(s), 373, 897**B**
Trombone fashion, replication, 946**6**, 949**F**, 954**F**
Tropocollagen, 152
Troponin C, 157, 821
Troponin T gene, alternative splicing and, 1010
trp operon, 984, 985**F**
trp promoter, 967**T**, 984
trp repressor(s), 95**T**, 96, 984, 993**F**, 994**F**, 995
trp
mRNA, Shine-Dalgarno sequence and, 1049**F**
trpA gene:protein collinearity and, 1018, 1019**F**
trpR operon, 984
Trypsin, 95**T**, 112, 112**F**, 115**T**, 122, 354, 391, 433**T**, 436
amino acid sequence of, 437**F**
substrate binding pocket of, 438**F**
Trypsinogen, 391, 391**T**
Tryptophan, 59**S**, 59**G**, 69
acid lability of, 86
degradation of, 865**F**, 868
from chorismate, 857**F**
phosphorescence of, 73
operon of, regulation of, 982**F**
Tryptophan synthase, 859**F**, 860, 861**G**, 985**F**
Tuberculostearic acid, 279**S**
Tubulin, 95**T**, 97, 177**F**
Turnover number of an enzyme, 364**D**
Twist, 227, 229**F**
Twitchin, 170**F**
Twofold degenerate sites, 1078
TXB_2, synthesis of, 783**F**
Tylenol®, 784**F**
Tyrosine, 59**S**, 59**G**, 69
biosynthesis of, 858, 860, 860**F**
from chorismate, 857**F**, 858, 859**F**
degradation of, 896, 869**F**
NMR spectrum of, 74
Tyrosine aminotransferase, 859**F**, 860
Tyrosyl-tRNA, 1062**S**

UASs (upstream activation sequences), 989
Ubiquinone. *See* Coenzyme Q.
Ubiquitin, 1071, 1071**F**, 1072
Ubiquitin:protein conjugates, breakdown of, 1073**F**
UDP, 187, 891**F**, 894, 894**F**
UDP-N-acetylglucosamine, 780
UDP-galactose, 780
UDP galactose-4-epimerase, 470, 473**F**, 594**F**, 595
UDP-galactosyltransferase, 782**F**
UDP-glucose (UDPG), 538**F**, 676**S**, 676**G**, 780
UDP-glucose pyrophosphorylase, 676**F**
Ultracentrifugation, 102, 135
Ultrafiltration, 102, 131
UMP, 183, 185**S**
absorption spectrum of, 183**F**
as endproduct of pyrimidine biosynthesis, 887**F**, 888
UMP synthase, in mammalian pyrimidine biosynthesis, 888
Uncouplers, 653
Undecaprenyl alcohol, 290**S**
Unidirectionality, the concept of metabolic unidirectionality, 814
Unilamellar vesicle, 294
Unimolecular reaction, 357**D**
Unsaturated fatty acids, β-oxidation of, 750
3′-Untranslated region of mRNA, 1002**F**
5′-Untranslated region of mRNA, 1002**F**
Unwin, Nigel, 305
Upstream activation sequences, 989**D**
Uptake, single-stranded, in recombination, 921
UQ-cytochrome *c* reductase, 635**T**, 638
Uracil, 181**S**
degradation of, 890**F**
Urate crystals, gout and, 884
Urate oxidase, in uric acid metabolism, 884, 885**F**
Urea
biotin-dependent carboxylation of, 493**T**
cycle of, 842, 844**F**, 845**B**
effect of cortisol on, 687
excretion of, in ureotelic animals, 870
hydrolysis rate of, 353
as product of uric acid metabolism, 884, 885**F**
Urease
jack bean, 20**F**, 353 356**T**, 427**T**
in uric acid metabolism, 884, 885**F**
Ureido group, of citrulline, 842, 844**S**
Ureotelic animals, 870
Uribe, Ernesto, 717**B**
Uric acid, 182**S**, 872**S**
formation of, by xanthine oxidase, 884
as endproduct of purine catabolism, 881, 882**F**
excretion of in uricotelic animals, 870
metabolic fate of, 884, 885**F**
Uricotelic animals, 870
Uridine, 184**S**
Uridine 5′-diphosphate (see also UDP), 187
Uridine 5′-monophosphate (also see UMP), 186
Uridine 5′-triphosphate (see also UTP), 187
Uridine diphosphate glucose, 676**S**, 676**G**
hydrolysis of, 526**T**
Uridylic acid (*See also* UMP), 186
Uridylyl transferase and *E. coli* glutamine synthetase, 837, 837**F**
Urkaryotes, 1099
Urokinase, 1090**F**
Uronic acids, 319
Uroporphyrin, 828**F**
Uteroglobin, 161**S**, 161**G**
UTP, 187
in pyrimidine synthesis, 889, 890**F**
synthesis of, from UMP, 888
UV irradiation and thymine dimer formation in DNA, 958**F**

V system, allosteric regulation of, 413, 413**F**
V-ATPase(s), 1095, 1095**F**, 1095**T**
F-ATPase evolution, 1097**F**
Vacuoles, 25, 28**T**
Vacuum blotting, 260**F**
Vagelos, P. Roy, 762
Valine, 58**S**, 58**G**
biosynthesis of from pyruvate, 850, 852**F**
degradation of to succinyl-CoA, 867, 867**F**
van der Waals contact distance, 139**D**
van der Waals forces, 14, 15**F**
van der Waals interactions, in proteins, 138, 139**F**
van Niel and photosynthesis, 702
van't Hoff plot, 513
Vane, John, 785
Vapor pressure, 39
Variable loop, tRNA, 239
Variable region, 924, 924**F**, 927**F**
Vasopressin, chemical synthesis of, 125
Vector(s), 249
shuttle, 255, 257**F**
expression, 265, 266, 266**F**, 267**F**, 268**F**
Venom, animal, 95**T**, 97
Versican, 345, 346**F**
Very low density lipoproteins, 792, 793**T**
V_H region, in IgG, 923, 924**F**, 925**F**, 926, 926**F**
Viruses, 26, 29**F**
Vitamin A, 463**T**, 502, 503
Vitamin B_2, 463**T**, 474
Vitamin B_6, 463**T**, 480, 483
Vitamin B_{12}, 485
Vitamin B_1, 463**T**, 464**S**
Vitamin C, 156, 489
Vitamin D, 505, 506, 507, 998**T**
Vitamin D_2, 463**T**, 505**S**
Vitamin D_3, 463**T**, 505**S**
Vitamin E, 290**S**, 463**T**, 507**S**, 857**F**
Vitamin K, 463**T**, 857**F**
Vitamin K_1, 290**S**, 508**S**
Vitamin K_2, 290**S**, 508**S**
Vitellogenin gene, *Xenopus laevis*, exons and introns in, 1002**F**
Vitreous humor, 334
V_l region, in IgG, 923, 924**F**, 925**F**
V_L genes, of immunoglobulins, 925, 926**F**
VLDLs, 792, 793**T**
V_{max} (maximal velocity), 363**D**
von Hippel, Peter, 1004
von Euler, Ulf, 785

Wald, George, 504
Warburg, Otto, 476, 570

Water, 33, 33**F,** 52
as electron donor in photosynthesis, 701
ionization of, 40, 40**F,** 41
Watson, James D., 180, 191, 936, 937
Watson-Crick base pairs, 193**S,** 218, 218**F**
Waxes, 287
Weak forces, 14**T**
Weigle, J. J., 915
Western blotting, 260**F**
Whale, killer, dependence of on fatty acids, 746
White, E. B., 152
Williams, R. R., 464
Wobble, 1033, 1034**F,** 1035
Wobble hypothesis, 1033
base-pairing alternatives and, 1034**F,** 1035**T**
Woese and Fox, early evolution and, 1098**F**
Woodward, R. B., 788
Work, 512**D**
Writhe, 227, 229**F**
Wyman, Jeffries, 410

ϕX174 phage A protein mRNA Shine-Dalgarno sequence, 1049**F**
X-gal (5-bromo-4-chloro-indolyl-β-galacto-pyranoside), 269**S**
Xanthine, 182**S**
in purine catabolism, 882**F**
Xanthine dehydrogenase, purification, 104**T**
Xanthine oxidase
mechanism of, 884, 884**F**
in purine catabolism, 881, 882**F**
moybdenum cofactor and, 884
Xenopus laevis, TFIIIA RNA polymerase III transcription factor, 996
Xeroderma pigmentosum, 960
Xfin, (zinc finger protein), 996**T**
XMP, in purine catabolism, 882**F**
XP-A protein, in DNA repair, 960
Xylitol, 319**F**
Xylose, 312
Xylulose, 313
Xylulose-5-P, in CO_2 fixation, 724**F,** 725**T**

YACs (yeast artifical chromosomes), 257
Yanofsky, Charles, 860, 985, 1018
YAP1, amino acid sequence of its bZIP DNA-binding motif, 1000**F**
Yeast mating factors, 307

Z scheme of photosynthesis, 707, 708**F**
Zea mays, 921
Zein 95**T,** 97
Zero-order kinetics, 360**D**
Zinc (Zn), 100**T,** 356**T**
Zinc finger DNA-binding proteins, 992, 996
Zinc metalloenzyme, 452
Zipper motif, interaction of leucine side chains, 999
Zocor®, 792**F**
Zooxanthellae, 551**B**
Zuckerhandl, Emile, 1081
Zwitterion, 56**D**
Zymogens, 391**D,** 436

Photo and Illustration Credits

The following figures were provided by Mr. Irving Geis: 3.3, 3.6, 4.3, 4.4, 4.11, 4.32, 5.3, 5.4, 5.7, 5.11, 5.14, 5.42, 12.9, 12.10, 12.14, 12.15, 12.16, 12.17, 12.19, 12.23, 13.35

The following figures were provided by Ms Jane Richardson: 5.8, 5.12, 5.23, 5.24, 5.27, 5.28, 5.29, 5.30, 5.31, 5.32, 5.33, 5.34, 5.39, 5.41, 19.20

Molecular graphics (unless otherwise indicated) were created by Melinda M. Whaley

Dust jacket photos: (Garrett) Catherine Leigh Touchton and (Grisham) Rosemary A. Grisham

Part I Opener: J/B Woolsey Associates

Chapter 1 opener: "The Indian Hunter" by Alexander-Francois Desportes; Giraudon/Bridgeman Art Library; **1.1a.:** Tony Angermayer/Photo Researchers, Inc.; **1.1b:** Thomas C. Boyden/Marie Selby Botanical Gardens; **1.4a:** Catherine Leigh Touchton; **1.4b:** Randal Harrison Garrett; **1.4c:** Michael Brown; **1.15:** Professor Simon E.V. Phillips; **1.19:** D. E. Nicholson, University of Leeds; **1.20a:** *E. coli* bacteria, © Martin Rotker/Phototake, Inc. **(inset)** *E. coli* showing flagella, © David M. Phillips/ The Population Council/Science Source/ Photo Researchers, Inc.; **1.21** (rough ER): Dwight R. Kuhn; (smooth ER): D. W. Fawcett/Visuals Unlimited; (mitochondrion): Keith Porter/Photo Researchers, Inc.; **1.22** (Golgi body): Dr. Dennis Kunkel/Phototake, NYC, (nucleus): Biophoto Associates; **1.23a:** M. Wurtz/Biozentrum/University of Basel/ SPL/Photo Researchers, Inc.; **1.23b:** Dr. Thomas Broker/Phototake, NYC; **1.23b (inset) :** CNRI/SPL/Photo Researchers, Inc.; **1.23c:** Biology Media, Photo Researchers, Inc.

Chapter 2 opener: © 1989 Bruce Iverson; **2.18:** Sigma Chemical Company, St. Louis, MO

Chapter 3 opener: "The Mirror of Venus" (1898) by Sir Edward Burne-Jones; Museu Calouste Gulbenkian Lisbon/The Bridgeman Art Library

Chapter 4 opener: "Girl Running on a Balcony" (1912) by Giacomo Balla; Galleria d'Arte Moderna, Milan/The Bridgeman Art Library **4.32** © Irving Geis

Chapter 5 opener: "Proteus and All Waters" mosaic, Roman period; photo © Ronald Sheridan; The Ancient Art and Architecture Collection; **5.3** © Irving Geis **5.4** © Irving Geis **5.7** © Irving Geis **5.9** © Irving Geis **5.18:** J. Gross, Biozentrum/ Science Photo Library **5.42** © Irving Geis

Chapter 6 opener: Watson and Crick in front of the DNA model, photographed by A.C. Barrington Brown. From J.D. Watson, 1968. *The Double Helix.* New York: Atheneum Press, courtesy of Cold Spring Harbor Laboratory Archives; **6.22:** Dr. Gopal Murti/CNRI/Phototake NYC

Chapter 7 opener: "Scherzo in D & A" (detail) by David E. Rodale (1955–1985); **7.9:** James D. Colandene, University of Virginia; **7.11** Applied Biosystems, Inc., Foster City, CA; **7.15:** Robert Stodola, Fox Chase Cancer Research Center and Irving Geis; **7.15A,B,Z** © Irving Geis **7.28:** Prof. Oscar L. Miller, Jr., University of Virginia; **7.30:** Prof. Evangelos N. Moudrianakis, Johns Hopkins University

Chapter 8 opener: "The Minotaur" by George Frederick Watts (1817–1904), Tate Gallery, London/The Bridgeman Art Library; **8.8:** Robley C. Williams, University of California/BPS; **8.22:** Dr. Steve A. Kay, University of Virginia

Chapter 9 opener: Humpback whale (*Megaptera novaeangliae)* breaching, Cape Cod, MA; photo © Steven Morello/Peter Arnold, Inc.; **p. 281:** Schafer and Hill/ Peter Arnold, Inc.; **p. 284:** Dr. E. R. Degginger; **9.2a-d:** J. Beveridge/Visuals Unlimited; Biophoto/Photo Researchers, Inc.; S. W. Watson/Visuals Unlimited; **9.21(a)** © T. T. Beveridge/Visuals Unlimited **9.21(b)** © Cabisco/Visuals Unlimited **9.21(c,d)** © D. W. Fawcett/Photo Researchers, Inc. **9.24b:** David Phillips/ Visuals Unlimited; **9.25:** courtesy of Walter Perkins, the Liposome Company, Inc., Princeton, NJ, and Brigham and Women's, Boston, MA; **9.30:** Dr. Dennis Kunkel/Phototake NYC; **9.37:** W. Stoeckenius and R. Henderson

Chapter 10 opener: "The Discovery of Honey" (1462) by Piero di Cosimo; courtesy of the Worcester Art Museum

Part II opener: J/B Woolsey Associates

Chapter 11 opener: "The Queen of Hearts" (detail) by Sir John Tenniel, from *Alice's Adventures in Wonderland*

Chapter 12 opener: "Sea Scape" (1947) by Alexander Calder/Whitney Museum of American Art, New York; photographed by Geoffrey Clements, New York; **12.9(a)** © Irving Geis **12.9(b)** © Irving Geis **12.10** © Irving Geis **12.14** © Irving Geis—all three structures **12.15(a)** © Irving Geis **12.15(b)** © Irving Geis **12.16** © Irving Geis **12.17(a)** © Irving Geis **12.17(b)** © Irving Geis **12.17 center** © Irving Geis **12.19** © Irving Geis **12.40-12.43:** Prof. Evan R. Kantrowitz, Boston College

Chapter 13 opener: Detail of the mechanism of an antique clock, photographed by David Parker/Science Photo Library; **13.19** © Irving Geis **13.34:** John Rupley, University of Arizona

Chapter 14 opener: Fresco from the Villa Issogna, Val d'Aosta, "Interior of a Pharmacy"; Giraudon/Bridgeman Art Library

Chapter 15 opener: Sun emblem of Louis XIV, Versailles; Giraudon/Art Resource, New York

Chapter 16 opener: Oxford Molecular Biophysics Laboratory/Photo Researchers, Inc.

Part III opener: (photo) European champs by Paul J. Sutton/Duomo; (line art) J/B Woolsey Associates

Chapter 17 opener: Anise swallowtail (Papilio zelicaon) adult and pupal case; © 1986 Peter Bryant/Biological Photo Service; **17.1** D. E. Nicholson, University of Leeds; **17.18:** courtesy of Prof. Melvin Calvin, Lawrence Berkeley Lab, University of California, Berkeley

Chapter 18 opener: "Pasteur in His Laboratory," by A. Edelfeldt, Musee d' Orsay, Paris; Giraudon/Art Resource, New York; **18.3:** Michelle Sassi/The Stock Market

Chapter 19 opener: "Ferris Wheel Abstract" © 1991 Xenophoa A. Beake/The Stock Market

Chapter 20 opener: "Wall Piece IV" (1985) by George Rhoads; **20.1a** B. King/BPS; 20.1: © Bill Longacre/Photo Researchers, Inc.; **20.14:** R. Capaldi, F. Malatesta, V. Darley-Usmar; **20.15:** S. Fuller, R. Capaldi, R. Henderson; **20.23:** D. F. Parsons, 1963: Science 140:985

Chapter 21 opener: Breads and pastries at French bakery, Paris; photograph by Steven Rothfield/Tony Stone Wordwide, Ltd.; **21.16:** Don W. Fawcett/Visuals Unlimited; **21.26** and **21.27:** D. Barford and L.N. Johnson, Oxford University

Chapter 22 opener: "Irises" (1889 by Vincent van Gogh/Superstock Intl.; **22.1:** James Dennis/CNRI/Phototake NYC; **22.3a:** Michael Seigel/Phototake NYC; **22.3b:** Biophoto Associates/Science Source

Chapter 23 opener: Bonaparte's Star-Fronted Hummingbird (colored litho) by John Gould (1804–1881); courtesy of the Board of Trustees of the V. & A./The Bridgeman Art Library; **p. 742:** (akee tree) R. R. Head/Earth Scenes; **23.1** Prof. P. Motta, Dept. of Anatomy, University "La Sapienza," Rome/Science Photo Library/Photo Researchers, Inc.; **23.14** (akee tree) R. R. Head/Earth Scenes/Animals, Animals **23.18** (camels) © George Holton/Photo Researchers, Inc.; (orca) © Francois Gohier/Photo Researchers, Inc.; (hummingbird) Tom J. Ulrich/Visuals Unlimited; (golden plover) S. J. Krasemann/Photo Researchers, Inc., (gerbil) Photo Researchers, Inc.

Chapter 24 opener: Gerard Lacz/Peter Arnold, Inc.; **24.36:** Science Photo Library/Photo Researchers, Inc.

Chapter 25 opener: Phil Schermeister/Allstock

Chapter 26 opener: © Dick Durrance II/Woodfin Camp and Associates, Inc.; **26.2:** Adam Hart-Davis/Science Photo Library; **26.12:** Dr. Shwu-Huey Liaw, National Taiwan University and Prof. David Eisenberg, University of California, Los Angeles; **26.37:** Dr. C. Craig Hyde, National Institutes of Health

Chapter 27 opener: Louis Castenada/The Image Bank

Part IV opener: B-DNA double helix © Leonard Lessin/Peter Arnold, Inc.

Chapter 28 opener: Ahu Akivi, Easter Island; Superstock Intl.; **28.3:** J. Broek/Biozentrum, University of Basel Science Photo Library; **28.7:** Fred Marsik/Visuals Unlimited; **28.16a:** Dr. Peggy Hsieh and Dr. R. Daniel Camerini-Otero, National Institutes of Health; **28.17a:** Professor Charles M. Radding, Yale University; **28.17b:** Peter Schultze and Dr. Juli Feigon, University of California, Los Angeles; **28.31:** Prof. Ralph L. Brinster, University of Pennsylvania

Chapter 29 opener: "Noah's Ark" by Andre Normil; Superstock I: **29.8:** Prof. Thomas A. Steitz, Yale University; **29.10:** Prof. John Kuriyan, Rockefeller University; **29.11:** Prof. David M. Prescott, University of Colorado; **29.13:** Prof. Joyce L. Hamlin, University of Virginia

Chapter 30 opener: "Writing" (ca. 1470) Jean Mielot (author of "Miracles de Notre Dame") at his desk with quill and scraping-knife; Bibliotheque National de Paris/ Mary Evans Picture Library; **30.11a:** Prof. Paul B. Sigler, Yale University; **30.44:** Dr. Charles R. Vinson, National Institutes of Health

Chapter 31 opener: Tikal Maya ruins, Guatemala; George Holton/Photo Researchers, Inc.; **31.11a** Prof. Thomas A. Steitz, Yale University

Chapter 32 opener: Learning sign language; © Lynn Johnson/Black Star; **32.18** and **32.20:** Prof. Oscar L. Miller, Jr., University of Virginia

Chapter 33 opener: "Swamp Animals and Birds on the River Gambia" (ca. 1912) by J. H. Johnston; Royal Geographical Society, London/The Bridgeman Art Library

Part V Opener: courtesy of David Barford, Oxford University

Chapter 34 opener: Dallas Skyline at Night; Werner Krutein/Liaison International; **34.1** (influenza): KG Murti/Visuals Unlimited; (rotavirus): CNRI/Photo Researchers, Inc.; (polyoma) Science VA/Visuals Unlimited; (rat sperm) David Phillips/Visuals Unlimited; (connective tissue) Fawcett and Heuser/Photo Researchers Inc.; (ciliate) Eric Grave/Phototake; (nasal ciliate): Veronika Burmeister/Visuals Unlimited; (microtubule) KG Murti/Visuals Unlimited; (soil bacteria) Dr. Tony Brain/Custom Medical Stock; **34.2:** Adapted from a drawing by Gerald Stubbs; **34.8:** Adapted from a drawing by Ronald Vale; **34.4:** M. Schliwa/Visuals Unlimited; **34.9:** Dr. Jeremy Burgess/Photo Researchers, Inc.; **34.19:** J.C. Revy/Phototake

Chapter 35 opener: "Drawbridge at Arles with a Group of Washerwomen" (1888) by Vincent van Gogh; Rikjsmuseum Kroller-Muller/photo by Erich Lessing/Art Resource; **p. 1135:** (viceroy butterfly) Patti Murray/Animals, Animals; (monarch butterfly) E. R. Degginger/Animals/Animals; (foxglove) Zig Leszczynski/Animals, Animals; (lily of the valley) John Bova/Photo Researchers, Inc.; (oleander) Arthur Hill/Visuals Unlimited; (milkweed) L. West/Photo Researchers, Inc.; **35.39:** (cecropia moth) Greg Neise/Visuals Unlimited; (cecropia moth caterpillar) Patti Murray/Animals, Animals

Chapter 36 opener: Michelangelo's "David"; the Firenze Academia/Art Resource; **36.3:** courtesy of S. Fleischer and M. Inui; **36.4** and **36.6** courtesy of Hugh Huxley, Brandeis University; **36.7 (a and b)** courtesy of Linda Rost and David DeRosier, Brandeis University; **(c)** courtesy of George Phillips, Rice University; **36.8:** courtesy of Henry Slayter, Harvard Medical School; **36.9 (ribbon image)** and **36.16 (inset):** courtesy of Ivan Rayment and Hazel M. Holden, University of Wisconsin, Madison; **36.20, 36.21,** and **36.24:** courtesy of S. Fleischer

Chapter 37 opener: Drawing of Embryo in the Uterus by Leonardo da Vinci; Superstock Int.; **37.37:** courtesy of David Barford, Oxford University

Chapter 38 opener: Spiders and Snakes from Albert Seba's "Locupletissimi Rerum Naturalium" (ca 1750) hand-coloured engraving; The Bridgeman Art Library; **p. 1224: (left)** Zig Leszczynski/Animals, Animals; **(right)** Tim Rock/Animals, Animals **38.11:** courtesy of M. Montal; **38.31:** Prof. P. Motta/Dept. of Anatomy/University "La Sapienza," Rome/Science Photo Library; **38.43:** courtesy of Randall Reed, Johns Hopkins University.